second edition

principles of virology

Molecular Biology, Pathogenesis, and Control of Animal Viruses

second edition

principles of virology

Molecular Biology, Pathogenesis, and Control of Animal Viruses

S. J. Flint
Department of Molecular Biology
Princeton University
Princeton, New Jersey

L. W. Enquist
Department of Molecular Biology
Princeton University
Princeton, New Jersey

V. R. Racaniello
Department of Microbiology
College of Physicians & Surgeons
Columbia University
New York, New York

A. M. Skalka
Institute for Cancer Research
Fox Chase Cancer Center
Philadelphia, Pennsylvania

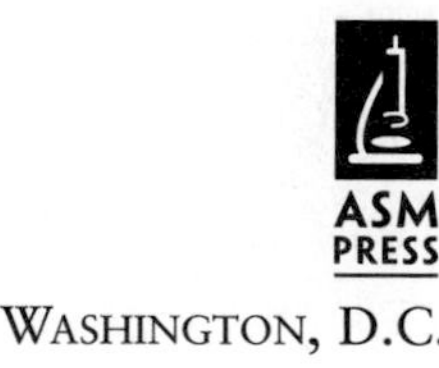

WASHINGTON, D.C.

Front cover illustration: A model of the atomic structure of the poliovirus type 1 Mahoney strain. The model has been highlighted by radial depth cuing so that the portions of the model that are farthest from the center are bright. Prominent surface features include a star-shaped mesa at each of the fivefold axes and a propeller-shaped feature at each of the threefold axes. A deep cleft or canyon surrounds the star-shaped feature. This canyon is the receptor-binding site. Courtesy of Robert Grant, Stéphane Crainic, and James Hogle (Harvard Medical School).

Back cover illustration: Progress in the global eradication of poliomyelitis has been striking, as illustrated by maps showing areas of known or probable circulation of wild-type poliovirus in 1988, 1998, and 2002. Yellow indicates either the presence of virus or insufficient surveillance to exclude the possibility of virus circulation. In 1988, virus was present on all continents except Australia. By 1998, the Americas were free of wild-type poliovirus, and transmission was interrupted in the western Pacific region (including the People's Republic of China) and in the European region (with the exception of southeastern Turkey). By 2002, the number of African countries reporting virus had been dramatically reduced. There are currently two major foci of transmission: southern Asia (Afghanistan, Pakistan, and India) and West Africa (mainly Nigeria).

Address editorial correspondence to ASM Press, 1752 N St. NW, Washington, DC 20036-2904, USA

Send orders to ASM Press, P.O. Box 605, Herndon, VA 20172, USA
Phone: (800) 546-2416 or (703) 661-1593
Fax: (703) 661-1501
E-mail: books@asmusa.org
Online: www.asmpress.org

Library of Congress Cataloging-in-Publication Data

Principles of virology : molecular biology, pathogenesis, and control of animal viruses / S.J. Flint ... [et al.].—2nd ed.
p. ; cm.
Includes bibliographical references and index.
ISBN 1-55581-259-7 (hardcover)
1. Virology. 2. Viruses. 3. Virus diseases.
[DNLM: 1. Viruses. 2. Genetics, Microbial. 3. Molecular Biology.
4. Virology—methods. QW 160 P957 2004] I. Flint, S. Jane.
QR360.P697 2004
579.2—dc21

2003009529

10 9 8 7 6 5 4 3 2 1

Illustrations and illustration concepting: J/B Woolsey Associates
Cover and interior design: Susan Brown Schmidler

We dedicate this book to the students, current and future scientists
and physicians, for whom it was written.
We kept them ever in mind.

We also dedicate it to our families:

Jonn, Gethyn, and Amy Leedham

Kathy and Brian

Doris, Aidan, Devin, and Nadia

Rudy, Jeanne, and Chris

Oh, be wiser thou!
Instructed that true knowledge leads to love.

WILLIAM WORDSWORTH
Lines Left upon a Seat in a Yew-tree
1888

Contents

Preface

The enduring goal of scientific endeavor, as of all human enterprise, I imagine, is to achieve an intelligible view of the universe. One of the great discoveries of modern science is that its goal cannot be achieved piecemeal, certainly not by the accumulation of facts. To understand a phenomenon is to understand a category of phenomena or it is nothing. Understanding is reached through creative acts.

A. D. HERSHEY
Carnegie Institution Yearbook 65

The major goal of both the first and second editions of this book is to define and illustrate the basic principles of animal virus biology. In this information-rich age, the quantity of data describing any given virus can be overwhelming, if not indigestible, for student and expert alike. Furthermore, the urge to write more and more about less and less is the curse of reductionist science and the bane of those who write textbooks meant to be used by students. Consequently, in the second edition, we have continued to distill information to extract essential principles, while retaining some descriptions of how the work is done. We continue to be selective in our choice of topics, viruses, and examples in an effort to make the book readable, rather than comprehensive. Detailed works like *Fields Virology* (2001), in two encyclopedic volumes, have made the best attempt to be all-inclusive, and *Fields* is recommended as a resource for detailed reviews of specific virus families.

What's New

We tested the first edition in our own classes and also received constructive comments and suggestions from other virology instructors and their students. The student comments were particularly useful in finding annoying typographical errors, clarifying confusing or complicated illustrations, and pointing out inconsistencies in text material. We have updated each chapter, adding many new illustrations. We have expanded Appendix A, reorganized it alphabetically according to virus family names, and also added a glossary of important terms. We realized that a cardinal organizing principle based on viral genome structure was not adequately elaborated in the first edition. To remedy this deficiency, we have added a new chapter on viral genomes. The new material provides a more detailed treatment of the Baltimore classification system. In our test lectures, this chapter proved to be an eye-opener for students, who found it easier to understand the complexities of RNA synthesis and translation when these topics were presented in the context of the seven major genome strategies. For those who like to teach "virus by virus," this new chapter provides an opportunity to feature their favorite viruses and to use the subsequent chapters on mechanisms as needed.

Use of the first edition in the classroom also indicated that a more logical and fuller treatment of pathogenesis would be valuable. As a consequence, we have completely reorganized the section of the book that deals with viral pathogenesis. In addition, a new Appendix B, called "Disease, Epidemiology, and Disease Mechanisms of Selected Animal Viruses Discussed in This Book," is included to capture the essential information of agent, epidemiology, and disease for the common viral diseases of humans.

Probably the most time-consuming, yet rewarding, task for the authors was to read every chapter, figure legend, text box, and table out loud with all of the authors present. The rule was that the lead author of the chapter could not read his or her own chapter. As a result, we were able to learn from each other, develop consistency in thought and language use throughout the text, and purge the text of jargon and anthropomorphism. To paraphrase Al Hershey, viruses don't, virologists do.

For purposes of readability, references are again generally omitted from the text, but each chapter ends with an updated and expanded list of relevant books, review articles, and selected research papers for readers who wish to pursue specific topics. If an experiment is featured in a chapter, one or more references are listed to provide more detailed information. In many cases, the figures and their legends also provide important detail that, if included in the text, might interrupt the flow of thought. Color-coded text boxes include general background information (boxes with aqua and pale green backgrounds), definitions of terms or clarification of common sources of confusion in terminology (maroon and pale blue), detailed information relevant to the main topic of discussion (purple and greenish beige), discussions of specific experiments (blue and pale yellow), and cautionary information (red and light mustard yellow).

How To Use This Book

The text follows the general strategy by which all viruses are reproduced within host cells and covers the principles of critical steps in virus replication in the following sections.

Part I: The Science of Virology

Chapter 1 provides a general introduction to the field with historical perspectives and a discussion of how viruses are classified. It summarizes the unifying principles that are the foundations of virology. In chapter 2, we describe general methods by which animal viruses are studied in the laboratory as an entrée into the molecular biology section.

Part II: Molecular Biology

The second section, consisting of chapters 3 to 13, focuses on molecular processes that take place in an infected host cell. Chapter 3 describes the organization of viral genomes and provides an overview of the limited repertoire of viral strategies for genome replication and mRNA synthesis. In chapter 4, the architecture of extracellular virus particles (virions) is discussed in the context of providing both protection and delivery of the viral genome in a single vehicle. Chapters 5 through 13 describe the broad spectrum of molecular processes that characterize the common steps of the reproductive cycle of viruses in a single cell, from decoding genetic information to genome replication and production of progeny virions. The intention is to describe how these common steps are accomplished by diverse but representative viruses, while illuminating principles applicable to all.

Part III: Pathogenesis

This section moves away from events in the single cell and addresses issues related to the interplay between viruses and their host organisms. Because viruses are of paramount importance as disease agents, chapter 14 addresses the basic concepts of pathogenesis and how it is studied. Chapter 15 provides a detailed description of the components of the immune system and viral countermeasures to host defenses, while chapter 16 considers the different types of relationship (some short and others of long duration) that viruses can establish with their hosts. An entire chapter (chapter 17) is devoted to the AIDS virus, not only because it is the causative agent of a serious, worldwide epidemic but also because of its unique and informative interactions with the human immune defenses. Chapter 18 describes the roles of viruses in cell transformation and in oncogenesis, the formation of tumors.

Part IV: Control and Evolution

Chapter 19 addresses the principles involved in developing methods of treatment and control of infection. Chapter 20, the final chapter, is a foray into the future and includes consideration of viral evolution and the ways in which new viruses emerge. This chapter also discusses novel infectious agents such as prions and satellites, and it ends with an analysis of critical public health issues, including those stemming from the potential of viruses as agents of war and terrorism.

Appendices

Appendix A provides a brief description and a general scheme for the reproductive cycles in single cells of viruses that serve as important examples in several chapters of the text. This section is intended to be a reference resource when one reads individual chapters and a convenient visual means by which specific topics may be related to the overall infectious cycles of the selected viruses.

Appendix B summarizes the pathogenesis of common viruses that infect humans in three "slides" (viruses and diseases, epidemiology, and disease mechanisms) for each virus or virus group. This information is intended to provide a simple snapshot of pathogenesis and epidemiology.

Reference

Knipe, D. M., and P. M. Howley (ed. in chief). 2001. *Fields Virology,* 4th ed. Lippincott Williams & Wilkins, Philadelphia, Pa.

Acknowledgments

This book could not have been completed without help and contributions from many individuals. Indeed, this venture would not have begun were it not for the initial encouragement from our colleagues in virology. We are especially indebted to Patrick Fitzgerald (formerly Director, ASM Press) for his enthusiastic support when the project was little more than a rough plan and for his sound advice during its development. The creative energy and ideas of Priscilla Schaffer (Beth Israel Deaconess Medical Center) had a substantial impact on the initial planning. Our sincere thanks also go to colleagues who took considerable time and effort to review the text in its evolving manifestations. Their expert knowledge and advice on issues ranging from teaching virology to organization of individual chapters and style were invaluable, even when orthogonal to our approach, and are inextricably woven into the final form of the book.

We thank the following individuals for their reviews and comments on multiple chapters or the entire book: Nicholas Acheson (McGill University), Karen Beemon and her virology students (Johns Hopkins University), Clifford W. Bond (Montana State University), Martha Brown (University of Toronto Medical School), Teresa Compton (University of Wisconsin), Stephen Dewhurst (University of Rochester Medical Center), Mary K. Estes (Baylor College of Medicine), Ronald Javier (Baylor College of Medicine), Richard Kuhn (Purdue University), Muriel Lederman (Virginia Polytechnic Institute and State University), Richard Moyer (University of Florida College of Medicine), Leonard Norkin (University of Massachusetts), Martin Petric (University of Toronto Medical School), Marie Pizzorno (Bucknell University), Nancy Roseman (Williams College), David Sanders (Purdue University), Dorothea Sawicki (Medical College of Ohio), Bert Semler (University of California, Irvine), and Bill Sugden (University of Wisconsin).

We also are grateful to those who gave so generously of their time to serve as expert reviewers of individual chapters or specific topics: James Alwine (University of Pennsylvania), Edward Arnold (Center for Advanced Biotechnology and Medicine, Rutgers University), Carl Baker (National Institutes of Health), Amiya Banerjee (Cleveland Clinic Foundation), Silvia Barabino (University of Basel), Albert Bendelac (University of Chicago), Susan Berget (Baylor College of Medicine), Kenneth I. Berns (University of Florida), Jim Broach (Princeton University), Michael J. Buchmeier (The Scripps Research Institute), Bruce Chesebro (Rocky Mountain Laboratories, National Institute of Allergy and Infectious Diseases), Marie Chow (University of Arkansas Medical Center), Barclay Clements (University of Glasgow), Don Coen (Harvard Medical School), Richard Condit (University of Florida), Michael Cordingly (Bio-Mega/Boehringer Ingelheim), Ted Cox (Princeton University), Andrew Davison (Institute of Virology, MRC Virology Unit), Ron Desrosiers (Harvard Medical School), Robert Doms (University of Pennsylvania), Emilio Emini (Merck Sharp & Dohme Research Laboratories), Bert Flanagan (University of Florida), Nigel Fraser (University of Pennsylvania Medical School), Charles Grose (Iowa University Hospital), Samuel Gunderson (European Molecular Biology Laboratory), Peter Howley (Harvard Medical School), James Hoxie (University of Pennsylvania), Frederick Hughson (Princeton University), Clinton Jones (University of Nebraska), Walter Keller (University of Basel), Tom Kelly (Memorial Sloan-Kettering Cancer Center), Elliott Kieff (Harvard Medical School), Robert Lamb (Northwestern University), Michael Linden (Mount Sinai School of Medicine), Ihor Lemischka (Princeton University), Arnold Levine (Institute for Advanced Study), Daniel Loeb (University of Wisconsin), Michael Malim (King's College London), James Manley (Columbia University), William Mason (Fox Chase Cancer Center), Malcolm Martin (National Institutes of Health), Edward Mocarski (Stanford University School of Medicine), Bernard Moss (Laboratory of Viral Diseases, National Institutes of Health), Peter O'Hare (Marie Curie Research Institute), Radhakris Padmanabhan (University of Kansas Medical Center), Peter Palese (Mount Sinai School of Medicine), Stuart Peltz (University of Medicine and Dentistry of New Jersey, Robert Wood Johnson Medical School), Roger Pomerantz (Thomas Jefferson University), Glenn Rall (Fox Chase Cancer Center), Charles Rice (Rockefeller University), Jack Rose (Yale University), Barry Rouse (University of Tennessee College of Veterinary Medicine), Rozanne Sandri-Goldin (University of California, Irvine), Nancy Sawtell (Childrens Hospital Medical Center), Priscilla Schaffer (Beth Israel Deaconess Medical Center), Robert Schneider (New York University School of Medicine), Christoph Seeger (Fox Chase Cancer Center), Aaron Shatkin (Center for Advanced Biotechnology and Medicine, Rutgers University), Thomas Shenk (Princeton University), Geoff Smith (Wright-Fleming Institute), Joan Steitz (Yale University), Victor Stollar (University of Medicine and Dentistry of New Jersey), Wesley Sundquist (University of Utah), John M. Taylor (Fox Chase Cancer Center), Alice Telesnitsky (University of Michigan Medical School), Heinz-Jürgen Thiel (Institut für Virologie, Giessen, Germany), Paula Traktman (Medical College of Wisconsin), James van Etten (University of Nebraska—Lincoln), Luis Villarreal (University of California, Irvine), Herbert Virgin (Washington University School of Medicine), Peter Vogt (The Scripps Research Institute), Simon Wain-Hobson (Institut Pasteur), Gerry Waters (Merck Sharp & Dohme Research Laboratories), Robin Weiss (University College London), Sandra Weller (University of Connecticut Health Center), Michael Whitt (University of Tennessee), Lindsay Whitton (The Scripps Research Institute), and Eckard Wimmer (State University of New York at Stony Brook). Their rapid responses to our requests for details and checks on accuracy, as well as their assistance in simplifying complex concepts, were invaluable. All remaining errors or inconsistencies are entirely ours.

Since the inception of this work, our belief has been that the illustrations must complement and enrich the text. Execution of this plan would not have been possible without the support of Jeff Holtmeier (Director, ASM Press) and the technical expertise and craft of our illustrators, J/B Woolsey Associates. The illustrations are an integral part of the exposition of the information and ideas discussed, and credit for their execution goes to the knowledge, insight, and enthusiasm of Patrick Lane of J/B Woolsey Associates. As noted in the figure legends, many of the figures could not have been completed without the help and generosity of our many colleagues who provided original images. Special thanks go to those who crafted figures tailored specifically to our needs or provided multiple pieces: Mark Andrake (Fox Chase Cancer Center), Edward Arnold (Rutgers University), Bruce Banfield (The University of Colorado), Christopher Basler and Peter Palese (Mount Sinai School of Medicine), Amy Brideau (Maxigen), Roger Burnett (Wistar Institute), Rajiv Chopra and Stephen Harrison (Harvard University), Marie Chow (University of Arkansas Medical Center), Richard Compans (Emory University), Friedrich Frischknecht (European Molecular Biology Laboratory), Wade Gibson (Johns Hopkins University School of Medicine), Ramon Gonzalez (Universidad Autónoma del Estado de Morelos), Thomas Leitner (Los Alamos National Laboratory), Paul Masters (New York State Department of Health), Rolf Menzel (Small Molecule Therapeutics), B. V. Venkataram Prasad (Baylor College of Medicine), Alasdair Steven (National Institutes of Health), Wesley Sundquist (University of Utah), Jose Varghese (Commonwealth Scientific and Industrial Research Organization), Robert Webster (St. Jude's Children's Research Hospital), Thomas Wilk (European Molecular Biology Laboratory), and Alexander Wlodawer (National Cancer Institute).

We thank all those who guided and assisted in the preparation and production of the book: Jeff Holtmeier (Director, ASM Press) for steering us through the complexities inherent in a team effort, Ken April (Senior Production Editor, ASM Press) for keeping us on track during production, Greg Payne (Senior Editor, ASM Press) for organizing the reviews of the text, and Susan Schmidler (Susan Schmidler Graphic Design) for her elegant and creative designs for the layout and cover. We are also grateful for the expert secretarial and administrative support from Trisha Barney and Norma Caputo (Princeton University) and Mary Estes and Rose Walsh (Fox Chase Cancer Center) that facilitated preparation of this text.

We owe a very special debt of gratitude to the Rockefeller Foundation for the award of a residency for our team at the Bellagio Study and Conference Center, Bellagio, Italy, in June 1997. The extraordinary beauty of the Center and the gracious hospitality of its staff made that an unforgettable as well as rewarding experience. The unique opportunity for study and extended consultation with each other and for interaction with resident scholars from many countries and disciplines had a major impact on the delineation of the purpose, approach, and scope of this textbook. ASM Press generously provided the financial support for travel to Bellagio, as well as for our many meetings, and a retreat at Woods Hole to focus on the second edition.

This often consuming enterprise was made possible by the emotional, intellectual, and logistical support of our families, to whom the book is dedicated.

Our acknowledgments close with a special tribute to Trisha Barney and Norma Caputo (Princeton University), who carried responsibilities for the first and second editions, respectively, too numerous to list. However, we are especially grateful for their painstaking care, patience, and good humor and particularly for collecting all the permissions necessary for preparation of the figures.

The Science of Virology

1 Foundations
2 Virus Cultivation, Detection, and Genetics

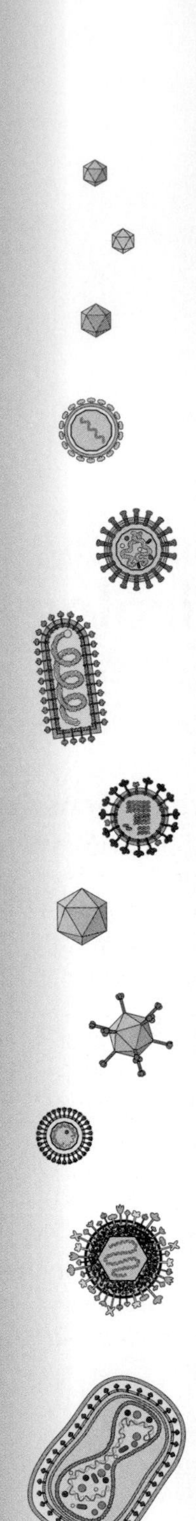

1

Foundations

> *Thus, we cannot reject the assumption that the effect of the filtered lymph is not due to toxicity, but rather to the ability of the agent to replicate.*
>
> F. Loeffler
> 1898

Luria's Credo

Half a century has passed since Salvador Luria wrote the following credo in the introduction to the classic textbook *General Virology:* "There is an intrinsic simplicity of nature and the ultimate contribution of science resides in the discovery of unifying and simplifying generalizations, rather than in the description of isolated situations—in the visualization of simple, overall patterns rather than in the analysis of patchworks."

Despite an explosion of information in biology since Luria wrote these words, his vision of unity in diversity is as relevant now as it was then. That such first principles exist may not be obvious considering the bewildering array of viruses, genes, and proteins recognized in modern virology. Indeed, new viruses are being described regularly (more than 50 since 1988), and viral diseases like acquired immunodeficiency syndrome (AIDS), hepatitis, and influenza continue to defy our efforts to control them. Yet, as discussed below, Luria's credo still stands: all viruses follow a simple three-part strategy to ensure their survival. This insight has been hard won over many years of observation, research, and debate; the history of virology is rich and instructive.

Virus Prehistory

Viruses have been known as distinct biological entities for little more than a century. Consequently, efforts to understand and control these important agents of disease are phenomena of the 20th century. Nevertheless, evidence of viral infection can be found among the earliest recordings of human activity, and methods for combatting viral disease were practiced long before the first virus was recognized.

Viral Infections in Antiquity

Reconstruction of the prehistoric past to provide a plausible account of when or how viruses established themselves in human populations is a challenging task. However, extrapolating from current knowledge, we can deduce that some modern viruses undoubtedly were associated with the earliest precursors of mammals and coevolved with humans. Other viruses entered human populations only recently. It is instructive to consider the last 10,000 years of human history, a time of radical change for humans and our viruses: animals were domesticated, the human population increased dramatically, large population centers appeared, and commerce drove interactions among unprecedented numbers of humans. We can infer from scattered glimpses of ancient history that viruses have long been a part of human experience.

Some viruses that eventually established themselves in human populations were transmitted to early humans from animals, much as still happens today. The initial human contact with viruses from nonhuman sources probably depended on many parameters. For example, early human groups that domesticated and lived with their animals were undoubtedly exposed to different viruses than were nomadic hunter societies. Similarly, as many different viruses are **endemic** in the tropics, human societies in that environment must have been exposed to a greater variety of viruses than societies established in temperate climates. When nomadic groups met others with domesticated animals, or individuals from tropical cultures mingled with those from cooler climates, human-to-human contact could have provided new avenues for virus spread. Even so, it seems unlikely that highly virulent viruses, such as measles and smallpox virus, could have entered a permanent relationship with small groups of early humans. Such virulent viruses either kill their hosts or induce life-long immunity in the survivors. Consequently, they can survive only where large, interacting host populations are available for their continued propagation. Such viruses could not have been established in human populations until large, settled communities appeared. Less virulent viruses that enter into a more benign, long-term relationship with their hosts were therefore more likely to be the first to become adapted to replication in the earliest human populations. These viruses include the modern retroviruses, herpesviruses, and papillomaviruses.

Figure 1.1 References to viral diseases abound in the ancient literature. (A) An image of Hector from an ancient Greek vase. Courtesy of the University of Pennsylvania Museum (object 30-44-4). (B) An Egyptian stele, or stone tablet, from the 18th dynasty (1580–1350 B.C.) depicting a man with a withered leg and the "drop foot" syndrome characteristic of polio. Panel B is reprinted from W. Biddle, *A Field Guide to Germs* (Henry Holt and Co., LLC, New York, N.Y., 1995; © 1995 by Wayne Biddle), with permission from the publisher.

A

B

Here this firebrand, rabid Hector, leads the charge.
HOMER, *The Iliad*,
translated by Robert Fagels
(Viking Penguin)

Evidence of several viral diseases can be found in ancient records (Fig. 1.1). The Greek poet Homer characterizes Hector as rabid in *The Iliad.* Mesopotamian laws describing the responsibilities of the owners of rabid dogs date from before 1,000 B.C. Their existence indicates that the communicable nature of this viral disease was already well known by that time, as it was to other ancient civilizations. Egyptian hieroglyphs illustrating what appear to be the consequences of poliovirus infection (a withered leg typical of poliomyelitis [Fig. 1.1B]) or pustular lesions characteristic of smallpox also date from that period. Smallpox, which was probably endemic in the Ganges river basin by the 5th century B.C. and subsequently spread to other parts of Asia and Europe, has played an important part in human history. Its introduction into the previously unexposed native populations of Central and South America by colonists in the 15th century led to lethal epidemics, which are considered an important factor in the conquests achieved by a small number of European soldiers. Other viral diseases known in ancient times include mumps and, perhaps, influenza. Yellow fever has been described since the discovery of Africa by Europeans, and it has been suggested that this scourge of the tropical trade was the basis for legends about ghost ships, such as the *Flying Dutchman,* in which an entire ship's crew perished mysteriously.

Humans have not only been subject to viral disease throughout much of their history but have also manipulated these agents, albeit unknowingly, for much longer than might be imagined. One classic example is provided by efforts to cultivate marvelously patterned tulips, which were of enormous value in 17th-century Holland. Such cultivation included deliberate spread of a virus (tulip breaking virus or tulip mosaic virus) that we now know causes the striping of tulip petals so highly prized at that time (Fig. 1.2). Efforts to control viral disease have an even more venerable history.

Figure 1.2 ***Three Broken Tulips.*** A painting by Nicolas Robert (1624–1685), now in the collection of the Fitzwilliam Museum, Cambridge, England. Striping patterns (color breaking) in tulips were described in 1576 in western Europe and were caused by a viral infection. This beautiful image depicts the remarkable consequences of infection with the tulip mosaic virus. Courtesy of the Fitzwilliam Museum, University of Cambridge.

The First Vaccines

Measures to control one viral disease have been used with some success for the last millennium. The disease is smallpox (Fig. 1.3), and the practice is called **variolation**, inoculation of healthy individuals with material from a smallpox pustule into a scratch made on the arm. Variolation was widespread in China and India by the 11th century and was based on the recognition that survivors of smallpox infection were protected against subsequent bouts of the disease. Variolation later spread to Asia Minor, where its value was recognized by Lady Mary Wortley Montagu, wife of the British ambassador to the Ottoman Empire. She introduced this practice into England in 1721, where it became quite widespread following successful inoculation of children of the royal family. George Washington is said to have introduced variolation among Continental

Figure 1.3 Characteristic smallpox lesions in a young smallpox victim. Illustrations like these were used to track down individuals infected with smallpox during the World Health Organization campaign to eradicate this disease. Photo courtesy of the Immunization Action Coalition (original source: Centers for Disease Control and Prevention).

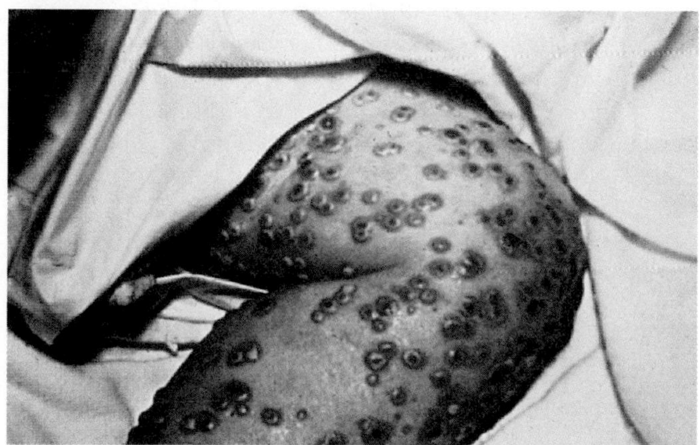

Army soldiers in 1776. However, the consequences of variolation were unpredictable and never pleasant. Serious skin lesions invariably developed at the site of inoculation and might be accompanied by more generalized rash and disease, with a fatality rate of 1 to 2%. From the comfortable viewpoint of a developed country in the 21st century, such a high death rate seems unacceptable. However, in the 18th century, variolation was perceived as a much better alternative than natural smallpox, a disease associated with a fatality rate of 25% in the whole population and 40% in babies and young children.

In the 1790s, Edward Jenner, an English country physician, recognized the principle on which modern methods of viral immunization are based, even though viruses themselves were not to be identified for another 100 years. Jenner himself was variolated as a boy and also practiced this procedure. He was undoubtedly familiar with its effects and risks. Perhaps this experience spurred his great insight, which was that milkmaids who had been exposed to cowpox (a mild disease in humans) were protected against smallpox. Jenner followed up this astute observation with direct experiments. In 1794–1796, he demonstrated that inoculation with extracts from cowpox lesions induced only mild symptoms but protected against the far more dangerous smallpox. It is from these experiments with cowpox that we derive the term **vaccination** (*vacca* = cow in Latin). This term was first given its general meaning by Louis Pasteur in 1881 to honor Jenner's accomplishments.

Initially, the only way to propagate and maintain cowpox vaccine was by serial infection (passage) of human subjects. This method was eventually banned, as it was associated with transmission of other diseases, such as syphilis and hepatitis, at significant frequencies. By 1860, the vaccine had been passaged in cows; later, sheep and water buffaloes were also used. While Jenner's original vaccine was based on cowpox virus, sometime during the human-to-human or cow-to-cow transfers the poxvirus now called vaccinia virus replaced the cowpox virus. Vaccinia virus is the basis for the modern smallpox vaccine, but its origins are not known. It exhibits little genetic similarity to cowpox virus, smallpox virus, or many of the known members of this family. Scientists have recovered the smallpox vaccine used in New York in 1876 and have verified that it is vaccinia virus and not cowpox virus. Speculation about when and how the switch occurred has produced some fascinating scenarios (Box 1.1).

The first deliberately attenuated viral vaccine was made by Pasteur in 1885. He inoculated rabbits with material from rabies-infected cow brain and then used aqueous suspensions of dried spinal cords from these animals to infect other rabbits. After several such passages, the resulting virus preparations caused mild disease (i.e., were **attenuated**) yet produced effective immunity against rabies. Safer and more efficient methods for the production of larger quantities of these first vaccines awaited the recognition of viruses as distinctive biological entities and of their parasitism in the cells of their hosts. Indeed, it took almost 50 years to discover the next antiviral vaccines: a vaccine for yellow fever virus appeared in 1935, and an influenza vaccine was available in 1936. These advances became possible only with radical changes in our knowledge of living organisms and of the causes of disease during the second half of the previous century.

BOX 1.1

Origin of vaccinia virus

Vaccinia virus is now used as a live vaccine to protect against smallpox, but the origin of this virus is obscure. Historical records report that cowpox virus was the original smallpox vaccine formulated by Jenner. Sometime during the years of propagation of the vaccine under ill-defined conditions, a new virus called vaccinia virus replaced cowpox virus. Over the years, at least four hypotheses to explain this curious substitution have been considered:

1. recombination of cowpox with smallpox virus after variolation of humans
2. recombination between cowpox and animal poxviruses during passage in various animals
3. inadvertent infection by a virus maintained in the laboratory, but now extinct
4. genetic drift of cowpox after repeated passage in humans and animals

None of these hypotheses has been proven conclusively.

Microorganisms as Pathogenic Agents

The 19th century was a period of revolution in scientific thought, particularly in ideas about living things and their origins. The publication of Charles Darwin's *The Origin of Species* in 1859 crystallized startling (and to many people, shocking) new ideas about the origin of diversity in plants and animals, until then generally attributed directly to the hand of God. These insights permanently undermined the belief that humans were somehow set apart from all other members of the animal kingdom. From the point of view of the science of virology, the most important changes were in ideas about the causes of disease.

The diversity of macroscopic organisms has been appreciated and cataloged since the dawn of recorded human history. A vast new world of organisms too small to be visible to the naked eye was revealed through the microscopes of Antony van Leeuwenhoek (1632–1723). Among van Leeuwenhoek's vivid and enthusiastic descriptions of living microorganisms as "wee animalcules," seen in such ordinary materials as rain or seawater, are examples of what we now know as protozoa, algae, and bacteria. By the early 19th century, the scientific community had accepted the existence of microorganisms and turned to the question of their origin—a topic of fierce debate. Some believed that microorganisms arose spontaneously, for example, in decomposing matter, where they were especially abundant. Others held the view that all were generated by the reproduction of like microorganisms, as were macroscopic organisms. The death knell of the spontaneous generation hypothesis was sounded with the famous experiments of Pasteur. He demonstrated that boiled (i.e., sterilized) medium remained free of microorganisms as long as it was maintained in special flasks with curved, narrow necks designed to prevent entry of airborne microbes (Fig. 1.4).

Figure 1.4 Four experiments to challenge the spontaneous generation hypothesis. The first step in each experiment was to boil the broth medium very thoroughly to destroy all living organisms. Air was then admitted to the flasks to satisfy the believers in spontaneous generation, who insisted that oxygen must be present for life to originate. However, the air admitted to the broth medium first was freed of living organisms in several ingenious ways. Under these conditions the broths remained perfectly sterile; no microorganisms appeared in them, showing that living things could not be generated spontaneously from the lifeless liquid. (A) Sterilizing air by chemical treatment. The set of bulbs next to the person's face contained alkali, and the other set contained concentrated acid. Air was drawn in through the acid to inactivate microbes before it reached the broth. (B) Sterilizing air by heat. Air could enter the flask of broth only by passing through the coiled glass tube kept hot by the flame. (C) Sterilizing air by physically trapping particles. The aspiration bottle drew air into the flask through the tube containing cotton at the right. The cotton filtered out the microbes in the air, just as the cotton or foam plugs now used in bacteriological culture tubes protect the culture from air contamination. (D) Pasteur's famous Swan-neck flasks provided passive exclusion of microbes from the sterilized broth. Although the flask was freely open to the air, the broth remained sterile so long as the microbe-bearing dust that collected in the neck did not reach the liquid.

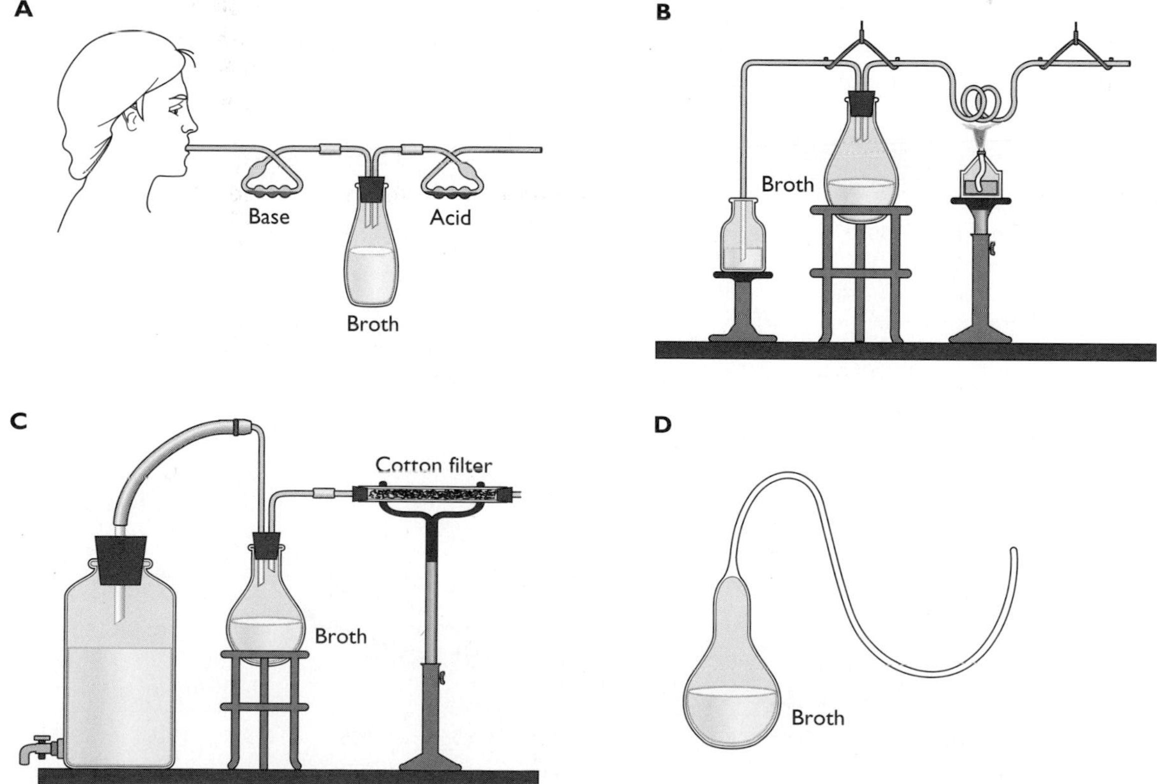

Pasteur also established that particular microorganisms were associated with specific processes, for example, the production of alcohol, lactic acid, or acetic acid (vinegar) by fermentation. This idea was crucial in the development of modern explanations of the causes of disease.

From the earliest times, poisonous air (miasma) was generally invoked to account for **epidemics** of contagious diseases, and there was little recognition of the differences among their causative agents. The association of particular microorganisms, initially bacteria, with specific diseases can be attributed to the ideas of the German physician Robert Koch. He developed and applied a set of criteria for identification of the agent responsible for a specific disease (a **pathogen**). These criteria, **Koch's postulates**, are still applied in the identification of those pathogens that can be propagated in the laboratory and tested in an appropriate animal model. The postulates are as follows.

- The organism must be regularly associated with the disease and its characteristic lesions.
- The organism must be isolated from the diseased host and grown in culture
- The disease must be reproduced when a pure culture of the organism is introduced into a healthy, susceptible host.
- The same organism must be reisolated from the experimentally infected host.

By applying these principles, Koch demonstrated that anthrax, a common disease of cattle, was caused by a specific bacterium (designated *Bacillus anthracis*) and that a second, distinct bacterial species caused tuberculosis in humans. Armed with these principles, as well as methods for the sterile culture and isolation of pure preparations of bacteria developed by Pasteur, Joseph Lister, and Koch, scientists working during the last part of the 19th century identified and classified many pathogenic bacteria (as well as yeasts and fungi) (Fig. 1.5). Understanding of the causes of infectious disease was placed on a secure scientific foundation, the first step toward rational treatment and ultimately control. During the last decade of the 19th century, failures of the paradigm that bacterial or fungal agents are responsible for disease led to the identification of a new class of infectious agents—submicroscopic pathogens that came to be called **viruses.**

Figure 1.5 The pace of early discovery of new infectious agents. Koch's introduction of efficient bacteriological techniques spawned an explosion of new discoveries of bacterial agents in the early 1880s. Similarly, the discovery of filterable agents launched the field of virology in the early 1900s. Despite an early surge of virus discovery, only 19 distinct human viruses had been reported by 1935. TMV, tobacco mosaic virus. Adapted from K. L. Burdon, *Medical Microbiology* (MacMillan Co., New York, N.Y., 1939), with permission.

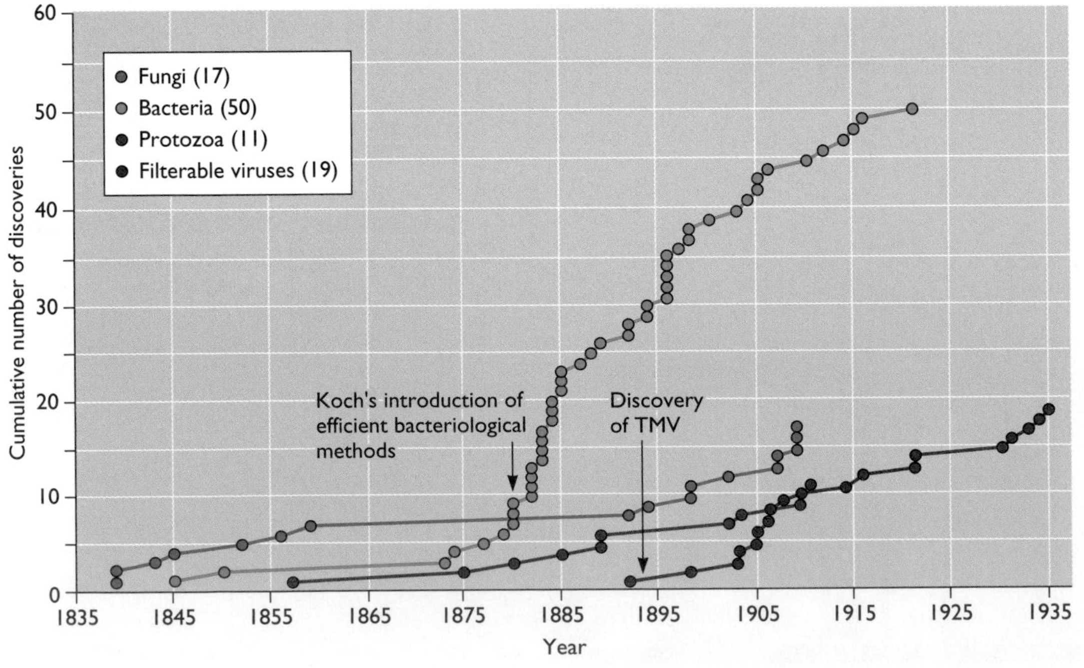

Discovery of Viruses

The First Viruses

The first report of a pathogenic agent smaller than any known bacterium appeared in 1892. The Russian scientist Dimitrii Ivanovsky observed that the causative agent of tobacco mosaic disease was not retained by the unglazed porcelain filters used at that time to remove bacteria from extracts and culture media. Six years later, Martinus Beijerinck independently made the same observation in Holland. More important, Beijerinck made the conceptual leap that the pathogen responsible for tobacco mosaic disease must be a distinctive agent so small that it could pass through filters that trapped all known bacteria (Fig. 1.6).

Figure 1.6 Filter systems used to characterize viruses. Filters for making liquids free of cultivatable organisms were instrumental in the first identification of viruses. Several types of filters were used in the early days of virus research, and four are illustrated here. Berkefeld filters (A) and Chamberland filters (B) are typical of the "candle" style of filter comprising diatomaceous earth pressed into the shape of a hollow candle open at one end. Only the smallest pore sizes retained bacteria and allowed viruses to pass through. For the Chamberland filter (B), the fluid to be filtered was introduced into the open end by means of a funnel. The filter is cut away to illustrate the hollow center. Filters similar to these were probably used by Ivanovsky, Loeffler, and Frosch to isolate the first plant and animal viruses. (C) Example of a collodion membrane "ultrafilter." The thin membrane is the filtering surface and is stretched over a filter plate. (D) A Mudd filtration apparatus illustrating the typical setup for filtering fluids. All the filters were designed to operate under negative pressure by connection with a suction pump or aspirator. Positive pressure was also used. The trap (on the right) was essential to catch back flow of liquid when the suction was released. The manometer (in the middle) provided an accurate means for measuring and regulating pressure. The filtrate was collected in a graduated buret, from which measured samples were collected and tested.

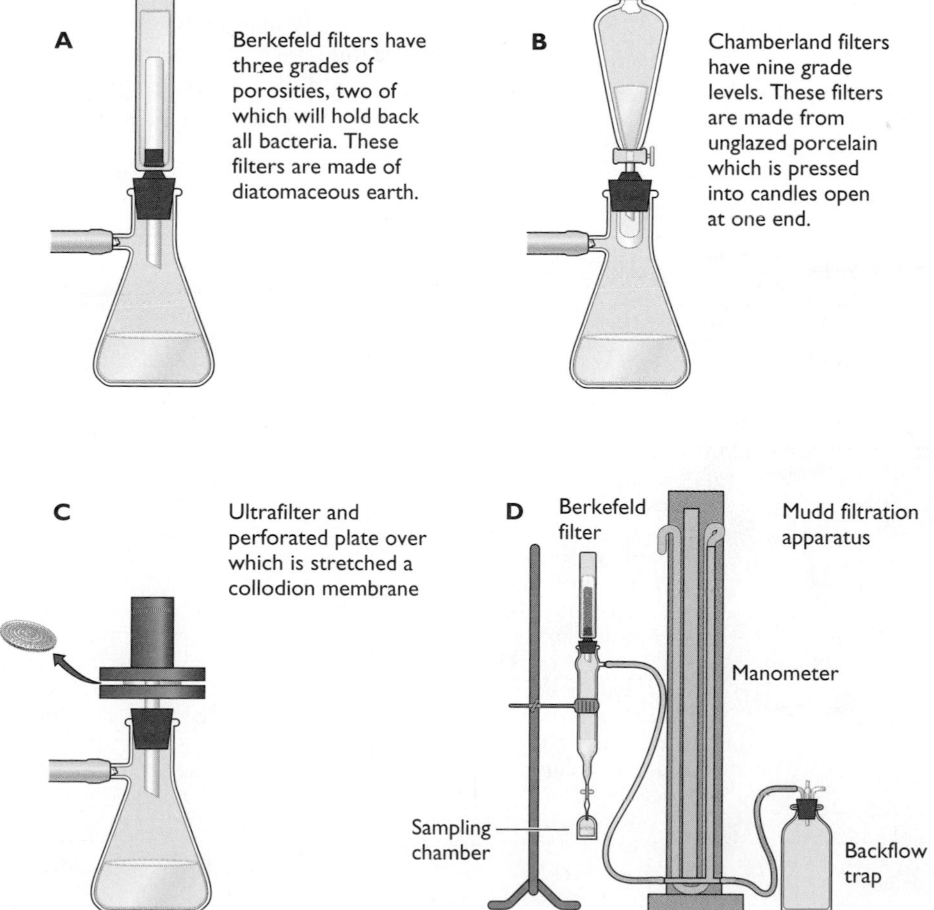

The same year (1898), the German scientists Friedrich Loeffler and Paul Frosch, both former students and assistants of Koch, observed that the causative agent of foot-and-mouth disease, a widespread, devastating infection of cattle and other livestock, was also filterable (Box 1.2). Not only were tobacco mosaic and foot-and-mouth disease pathogens much smaller than any previously recognized microorganism, but they also replicated **only** in their host organisms. For example, extracts of an infected tobacco plant diluted into sterile solution produced no additional infectious agents until introduced into leaves of healthy plants, which subsequently developed tobacco mosaic disease. The serial transmission of infection by diluted extracts established that these diseases were not caused by a bacterial toxin present in the original preparations derived from infected tobacco plants or cattle. The failure of both pathogens to multiply in solutions that readily supported growth of bacteria, and their dependence on host organisms for reproduction, further distinguished these new agents from pathogenic bacteria. Beijerinck termed the submicroscopic agent responsible for tobacco mosaic disease *contagium vivum fluidum* to emphasize its infectious nature and distinctive reproductive and physical properties. Agents passing through filters that retain bacteria came to be called ultrafilterable viruses, appropriating the term **virus** from the Latin for "poison." This term eventually was simplified to viruses.

The discovery of the first virus, tobacco mosaic virus, is often dated to the work of Ivanovsky in 1892. However, Ivanovsky did not identify the tobacco mosaic disease pathogen as a distinctive agent, nor was he convinced that its passage through bacterial filters was not the result of some technical failure. It may be more appropriate to attribute the founding of the field of virology to the astute insights of Beijerinck, Loeffler, and Frosch, who recognized the distinctive nature of the plant and animal pathogens they were studying more than 100 years ago.

The pioneering work on tobacco mosaic virus and foot-and-mouth disease virus was followed by the identification of viruses associated with specific diseases in many other organisms. Important landmarks from this early period include the identification of viruses that cause leukemias or solid tumors in chickens by Vilhelm Ellerman and Olaf Bang in 1908 and by Peyton Rous in 1911, respectively. The study of viruses associated with cancers in chickens, particularly the Rous sarcoma virus, eventually led to an understanding of the molecular basis of cancer.

Bacterial viruses were first described by Frederick Twort in 1915 and Felix d'Hérelle in 1917. d'Hérelle named them **bacteriophages** because of their ability to lyse bacteria on the surface of agar plates (**phage** is derived from the Greek for "eating"). In an interesting twist of serendipity, Twort made his discovery of bacterial viruses while testing the smallpox vaccine virus to see if it would grow on simple

BOX 1.2

The first animal virus discovered remains a scourge today

The first animal virus to be discovered, foot-and-mouth disease virus, infects domestic cattle, pigs, and sheep as well as many species of wild animals. Although mortality is low, morbidity is high and infected domestic animals lose their commercial value. The virus is highly contagious and the most common and effective method of control is by the slaughter of entire herds in affected areas.

Outbreaks of foot-and-mouth disease were widely reported in Europe, Asia, Africa, and South and North America in the 1800s. The largest epidemic ever recorded in the United States occurred in 1914. After gaining entry into the Chicago stockyards the virus spread to more than 3,500 herds in 22 states. This calamity accelerated epidemiologic and disease control programs, eventually leading to the field and laboratory-based systems maintained by the U.S. Department of Agriculture to protect domestic livestock from foreign animal and plant diseases. Similar control systems have been established in other Western countries, but this virus still presents a formidable challenge throughout the world. A 1997 outbreak of foot-and-mouth disease among pigs in Taiwan resulted in economic losses of greater than $10 billion.

In 2001, an epidemic outbreak in the United Kingdom spread to other countries in Europe and led to the slaughter of more than 3 million infected and uninfected farm animals. The associated economic, societal, and political costs threatened to bring down the British government. Images of mass graves and horrific pyres consuming the corpses of dead animals sensitized the public as never before. It is clear that viruses do not have to infect humans to exact a human cost. It is sobering to realize that we have known about this animal virus longer than any other, but have still not learned enough to keep it from exacting its toll.

Murphy, F. A., E. P. J. Gibbs, M. C. Horzinek, and M. J. Studdert. 1999. *Veterinary Virology*, 3rd ed. Academic Press, Inc., San Diego, Calif.

media. He found bacterial contaminants, some of them exhibiting an unusual "glassy transformation" which proved to be the result of lysis by a bacteriophage. Investigation of bacteriophages provided the foundations for the field of molecular biology, as well as important general insights into interactions of viruses with their host cells.

The First Human Viruses

The first human virus to be identified, in 1901, was that responsible for yellow fever. The complicated transmission and life cycle of this virus made this achievement all the more remarkable. Yellow fever is known to have been widespread in tropical countries since the 15th century. It was responsible for devastating epidemics associated with such high rates of mortality (e.g., 28% in the New Orleans epidemic of 1853) that normal life became impossible (Appendix B; see also Box 20.14). This disease was not, however, directly contagious, and an infectious agent could not be demonstrated in yellow fever patients. These puzzling properties defeated efforts to establish the origin of yellow fever until the Cuban physician Carlos Juan Finlay proposed in 1880 that a bloodsucking insect, most likely a mosquito, played a part in the transmission of the disease. A commission to study the etiology of yellow fever was established by the U.S. Army under Colonel Walter Reed in 1899, in part because of the high incidence of the disease among soldiers who were occupying Cuba. Dr. Jesse Lazear, a member of Reed's commission, provided leadership and ultimately gave his life to demonstrate that yellow fever was transmitted by mosquitoes. He was the first experimentally infected person who died from the disease. The results of the Reed Commission's study proved conclusively that mosquitoes are the vector. In 1901, Reed and James Carroll injected diluted, filtered serum from an experimentally infected yellow fever patient into three nonimmune individuals. Two subsequently developed yellow fever. Reed and Carroll concluded that a filterable virus was the cause of the disease. In the same year, Juan Guiteras, a professor of pathology and tropical medicine at the University of Havana, attempted to produce immunity to yellow fever by exposing volunteers to mosquitoes carrying yellow fever virus. Of 19 volunteers, 8 contracted yellow fever and 3 died. One of the dead was Clara Louise Maass, a U.S. Army nurse from New Jersey. Yellow fever had been a constant scourge of Havana for 150 years, but the conclusions of Reed and his colleagues were a revelation. Rapid introduction of mosquito control by the mayor of Havana, William Gorgas, dramatically reduced the incidence of disease within a year. To this day, mosquito control remains an important method for control of yellow fever.

Other human viruses were identified during the early decades of the 20th century (Fig. 1.5). However, the pace of discovery was slow, not least because of the dangers and difficulties associated with experimental manipulation of human viruses so amply illustrated by the experience with yellow fever virus. Consequently, agents of some important human diseases were not identified for many years, and then only with some good luck. A classic case in point is the virus responsible for influenza, a name derived in the mid-1700s from Italian to indicate that the disease resulted from the "influence" of miasma (bad air) and astrological signs. The human disease is now thought to have arisen as a result of the transfer of virus among humans and livestock following human domestication of animals about 6,000 years ago. Worldwide epidemics, called **pandemics**, of influenza have been documented in humans for well over 100 years. These pandemics were typically associated with mortality among the very young and the very old, but the 1918–1919 pandemic following the end of World War I was especially devastating. Over 40 million people died during this pandemic, more than were killed in the preceding war. Despite many efforts, a human influenza virus was not isolated until 1933. This virus was first identified by Wilson Smith, Christopher Andrewes, and Patrick Laidlaw only because they found a host suitable for its propagation. They infected ferrets with human throat washings and isolated the virus now known as influenza A virus. Ferrets may seem to be an unusual animal host, and in fact the success achieved with these animals was serendipitous: Laidlaw was using ferrets in studies of canine distemper virus and they were available in his laboratory. Subsequently, influenza A virus was shown to infect adult mice and chicken embryos. The latter proved to be an especially valuable host system, for vast quantities of the virus are produced in the allantoic sac. Indeed, chicken eggs are still used today to produce influenza vaccines.

The Definitive Properties of Viruses

Throughout the early period of virology, in which many viruses of plants, animals, and bacteria were cataloged, the origin and nature of these distinctive infectious agents were controversial. Arguments centered on whether viruses represented parts of a cell that had somehow acquired a new kind of existence or were built from virus-specific components. Little progress was made toward resolving these issues and establishing the definitive properties of viruses until new techniques that allowed their visualization or propagation in cultured cells were developed.

The Structural Simplicity of Viruses

Dramatic confirmation of the structural simplicity of viruses came in 1935, when crystals of tobacco mosaic virus were obtained by Wendell Stanley. At that time, nothing was known of the structural organization of biologically important macromolecules, such as proteins and DNA. Indeed, the crucial role of DNA as genetic material had not even been recognized. The ability to obtain an infectious agent in crystalline form, a state that is more generally associated with inorganic material, therefore created much wonder and speculation about whether a virus is truly a life form. In retrospect, it is obvious that the relative ease with which tobacco mosaic virus could be crystallized was a direct result of its structural simplicity and the ability of many particles to associate in regular arrays.

The 1930s saw the introduction of the instrument that rapidly revolutionized virology, the electron microscope. The great magnifying power of this instrument (eventually over 100,000-fold) allowed direct visualization of virus particles for the first time. It has always been an exciting experience for investigators to obtain images of viruses, especially as they often prove to be remarkably elegant in appearance (Fig. 1.7). Images of many different viruses confirmed that these agents are very small (Fig. 1.8) and far simpler in structure than any cellular organism. Many appeared as regular helical or spherical particles. The description of the morphology of virus particles made possible by electron microscopy also opened the way for the first rational classification of viruses.

The Intracellular Parasitism of Viruses

Organisms as Hosts

The fundamental characteristic of viruses is their absolute dependence on a living host for reproduction: they are **obligate parasites**. All early studies on viruses relied on the interactions with host organisms.

Transmission of plant viruses such as tobacco mosaic virus can be readily achieved, for example, by applying extracts of an infected plant to a scratch made on the leaf of a healthy plant. Furthermore, as a single infectious particle of many plant viruses induces the characteristic lesion (Fig. 1.9), the concentration of the infectious agent could be readily measured. Plant viruses were therefore the first to be studied in detail. Viruses of animals could also be propagated under experimental conditions, and methods were developed to quantify these agents by determining the lethal dose. Volunteers were initially used in the study of human viruses, but as this practice had potentially lethal consequences, laboratory animals were employed if they could be infected. For example, yellow fever virus was transmitted to mice by Max Theiler in 1930, an achievement that led to the isolation of an attenuated vaccine strain, considered one of the safest and most effective vaccines ever produced.

After specific viruses and host organisms were identified, it became possible to produce sufficient quantities of virus

A

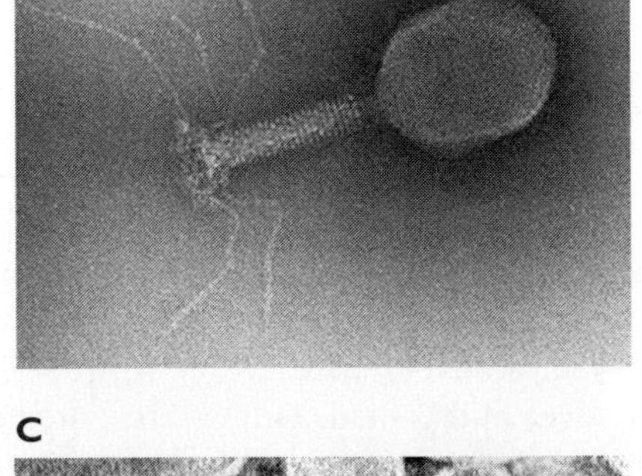

B

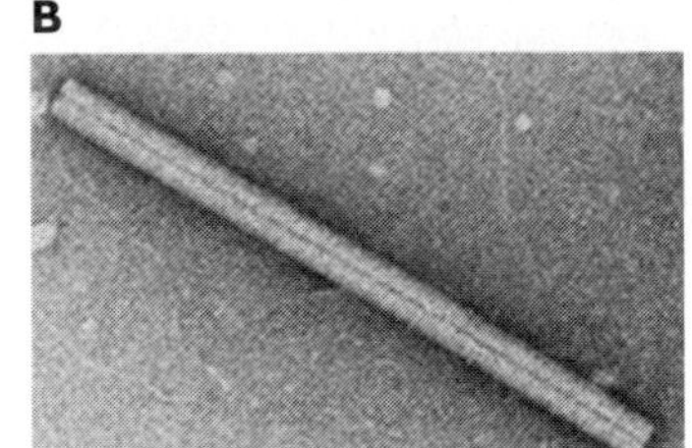

C

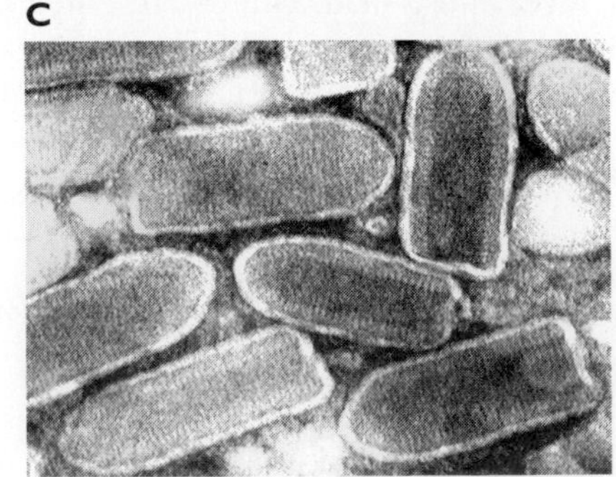

D

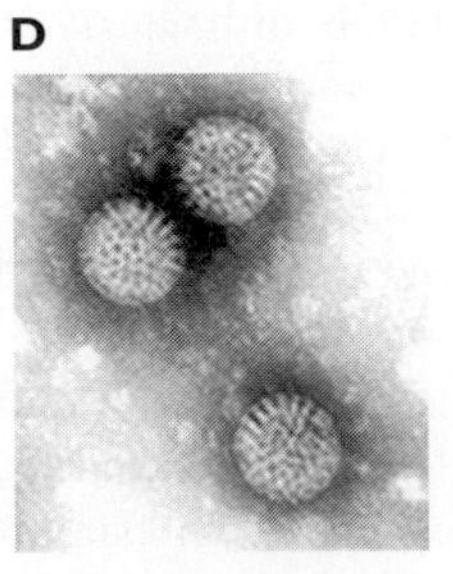

Figure 1.7 Electron micrographs of viruses following negative staining. (A) The complex, nonenveloped virus bacteriophage T4. Courtesy of R. L. Duda, University of Pittsburgh. (B) The helical, nonenveloped particle of tobacco mosaic virus. Reprinted from the Universal Virus Database of the International Committee on Taxonomy of Viruses (http://www.ncbi.nlm.nih.gov/ICTVdb/WIntkey/Images/em_tmv.gif), with permission. (C) Enveloped particles of the rhabdovirus vesicular stomatitis virus. Courtesy of F. P. Williams, University of California, Davis. (D) Nonenveloped, icosahedral human rotavirus particles. Courtesy of F. P. Williams, U.S. Environmental Protection Agency (http://www.epa.gov/nerlcwww/rota.htm).

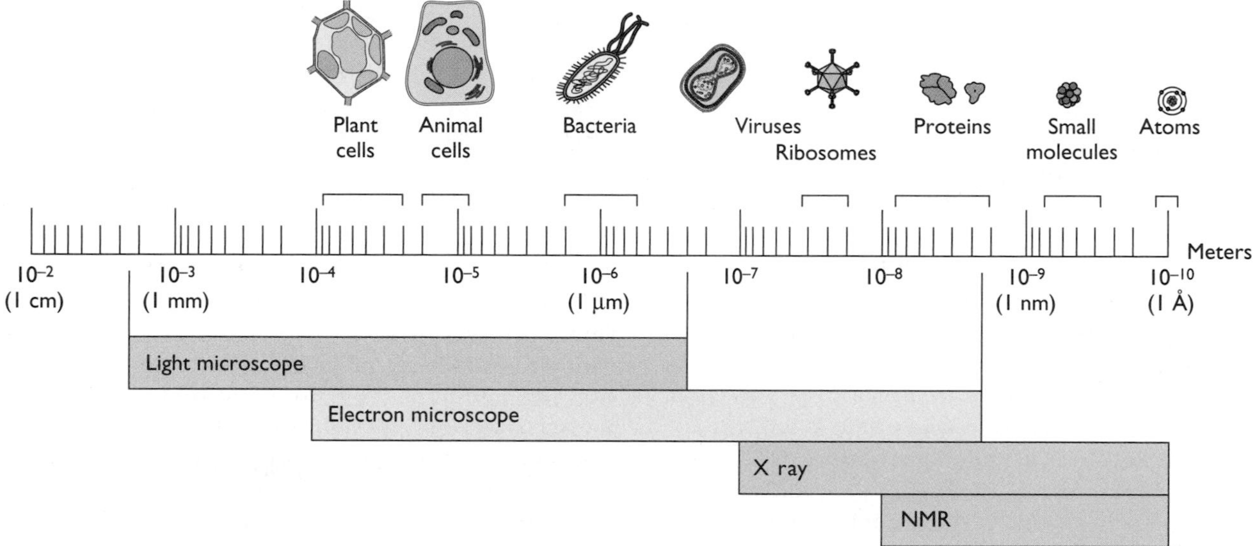

Figure 1.8 The small size of viruses is illustrated with a logarithmic, metric scale. Sizes of animal and plant cells, bacteria, viruses, proteins, molecules, and atoms are indicated. The resolving powers of various techniques used in virology, including light microscopy, electron microscopy, X-ray crystallography, and nuclear magnetic resonance (NMR) spectroscopy, are indicated. Viruses, which are within the resolving power of the electron microscope, are about 2 orders of magnitude smaller than the smallest bacterium. The units commonly used in descriptions of virus particles or their components are the nanometer (nm [10^{-9} m]) and the angstrom (Å [10^{-10} m]). Adapted from A. J. Levine, *Viruses* (Scientific American Library, New York, N.Y., 1991). Used with permission of Henry Holt and Company, LLC.

particles for investigation of their physical and chemical properties. Scientists also were able to determine the consequences of infection for the host. Parameters such as the incubation period, gross symptoms of infection, and effects on specific tissues and organs were investigated. Animal hosts remain an essential tool in investigations of the pathogenesis of viruses that cause disease in humans and other animals. However, real progress toward understanding the mechanisms of virus replication was made only with the development of tissue and cell culture systems. Among the simplest, but of crucial importance to both virology and molecular biology, were cultures of bacterial cells.

Lessons from Bacteriophages

In the late 1930s and early 1940s, bacteriophages, or phages, received increased attention as a result of controversy centering on how they were formed. d'Hérelle, one of the discoverers of bacteriophages, favored the hypothesis that there was only one phage that attacked all bacteria. This hypothesis was disproved by F. Macfarlane Burnet, who showed that there were a great variety of phages with different physical and biological properties. Nevertheless, the precise nature of these agents remained elusive. John Northrup, a biochemist at the Rockefeller Institute in Princeton, N.J., championed the theory that a phage was a metabolic product of a bacterium. Phage formation was said to be analogous to the autocatalytic formation of enzymes from inactive precursors, a direct rejection of the proposal by d'Hérelle that a phage was a living organism. On the other hand, Max Delbrück, in his work with Emory Ellis and later with Luria, regarded phages as autonomous, stable, self-replicating entities characterized by heritable traits. According to this paradigm, phages were seen as

Figure 1.9 Lesions induced by tobacco mosaic virus on an infected tobacco leaf. In 1886, Adolph Mayer first described the characteristic patterns of light and dark green areas on the leaves of tobacco plants infected with tobacco mosaic virus. He demonstrated that the mosaic lesions could be transmitted from an infected plant to a healthy plant by aqueous extracts derived from infected plants. The number of local necrotic lesions that result is directly proportional to the number of infectious particles in the preparation. Photo by J. P. Krausz, from "Lessons in Plant Pathology," The American Phytopathological Society Education Center (http://www.apsnet.org/education/lessonsPlantPath/TMV/text/symptom.htm), reprinted with permission.

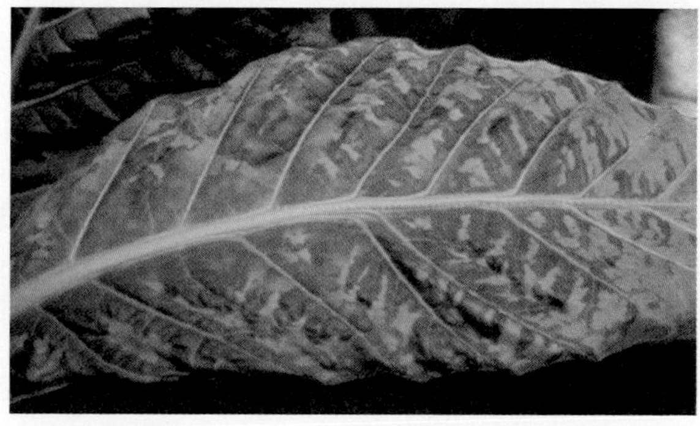

ideal tools with which to understand genes and heredity. Probably the most critical early contribution of Delbrück and Ellis was the perfection of the one-step growth technique for synchronization of the replication of phage. This achievement allowed analysis of a single cycle of phage growth in a population of bacteria. The new technique introduced highly quantitative studies to virology, as well as a rigor of analysis that was unprecedented. The first experiments showed that phages indeed multiplied in the bacterial host and were liberated in a "burst" by lysis of the cell.

Delbrück was a zealot for phage research and recruited talented scientists to pursue the fundamental issues of what is now known as the field of molecular biology. This group of scientists, working together in what came to be called the "phage school," focused their attention on specific phages of the bacterium *Escherichia coli*. The list of Nobel laureates who were trained as phage workers is a testament to Delbrück's leadership. Progress was rapid, primarily because of the simplicity of the phage infectious cycle. Phages replicate in bacterial hosts that can be readily obtained by overnight culture, as opposed to eukaryotic cells, which take much longer to grow to concentrations sufficient for infection. By the mid-1950s, it was evident that viruses from bacteria, animals, and plants share many fundamental properties. However, bacteriophages provided a far more tractable experimental system for analysis. Consequently, their study had a profound impact on the development of virology.

One critical lesson came from definitive experiments establishing that viral nucleic acid carries the genetic information. It was known from studies of the "transforming principle" of *Pneumococcus* by Oswald Avery, Colin MacLeod, and Maclyn McCarty (1944) that nucleic acid was both necessary and sufficient for the transfer of genetic traits. However, in the early 1950s, viral protein was still suspected to be an important component of viral heredity. In a brilliantly simple experiment that included the use of a common kitchen item, a Waring blender, Alfred Hershey and Martha Chase showed that this hypothesis was incorrect. By differentially labeling the nucleic acid and protein components of virus particles with radioactive phosphorus (^{32}P) and radioactive sulfur (^{35}S), respectively, they showed that the protein coat of the infecting virus could be removed soon after infection by agitating the bacteria for a few minutes in the blender. Conversely, ^{32}P-labeled bacteriophage DNA entered and remained associated with the bacterial cells; such cells nonetheless produced a normal burst of new virus particles.

Bacteriophages were originally thought to be lethal agents, killing their hosts after infection. In the early 1920s, a previously unknown interaction, in which the host not only survived the infection but also stably inherited the genetic information of the virus, was described. It was also observed that certain bacterial strains not known to be infected could lyse spontaneously and produce bacteriophages after a period of growth in culture. Such strains were called **lysogenic**, and the phenomenon was called **lysogeny**. Studies of lysogeny identified many previously unrecognized features of virus-host cell interactions (Box 1.3).

BOX 1.3

Studies of lysogeny established several general principles in virology

Some viruses can enter into either destructive (lytic) or relatively benign (lysogenic) relationships with their host cells. Such bacteriophages were called "temperate." In a lysogenic bacterial cell, viral genetic information persists, but viral gene expression is restricted. Such cells are called **lysogens**, and the quiescent viral genome is called a prophage.

For some bacteriophages like lambda and Mu (Mu stands for mutator), prophage DNA is integrated into the host genome of lysogens and passively replicated by the host. Virally encoded enzymes, known as integrase (lambda) and transposase (Mu), mediate the covalent insertion of viral DNA into the chromosome of the host bacterium, establishing it as a prophage. Bacteriophage Mu inserts its genome into many random locations on the host chromosome, causing numerous mutations. This process is called **insertional mutagenesis**. Transposition by Mu establishes an integrated prophage when repressed but also leads to replication of the viral genome during the lytic cycle.

The prophage DNA of other bacteriophages, such as P1, is a plasmid, a self-replicating, autonomous chromosome in the cytoplasm of a lysogen.

Viral gene expression in lysogens is turned off by the action of viral proteins called repressors. Viral gene expression can be turned on when repressors are inactivated (a process called induction).

Viral genomes can pick up cellular genes and deliver them to new cells (a process known as **transduction**). The process can be generalized, with the acquisition by the virus of any gene segment from the host chromosome. Or transduction can be specialized, as is the case for viruses that integrate into specific sites in the host chromosome. For example, occasional mistakes in excision of the lambda prophage after induction result in production of unusual progeny phage that have lost some of their own DNA but have acquired the bacterial DNA adjacent to the prophage.

Recognition of this phenomenon resulted from the work of many scientists, but it began with the elegant experiments of André Lwoff and colleagues at the Institut Pasteur in Paris. Lwoff demonstrated the existence of the viral genome in lysogenic cells as a specific genetic element called the **prophage**. This element determined the ability of lysogenic bacteria to produce infectious bacteriophage. Subsequent studies of *E. coli* bacteriophage lambda established a paradigm for one of the many mechanisms of lysogeny, the integration of a circular phage genome into a specific site on the bacterial chromosome. Other mechanisms were soon found (Box 1.3).

Bacteriophages became inextricably associated with the new field of molecular biology (Table 1.1). Their study also established many fundamental principles. For example, control of the decision to enter a lysogenic or a lytic pathway is encoded in the genome of the virus. The first mechanisms discovered for the control of gene expression, exemplified by the elegant operon theory of Nobel laureates François Jacob and Jacques Monod, were deduced in part from studies of bacteriophage lambda lysogeny. The biology of bacteriophage lambda provided a fertile ground for work on gene regulation, but study of virulent T phages (T1 to T7, where T stands for type) of *E. coli* paved the way for many other important advances (Table 1.1). As we shall see, these systems also provided an extensive preview of mechanisms of animal virus replication (Box 1.4).

Table 1.1 Bacteriophages: landmarks in molecular biology[a]

Year	Discovery
1939	One-step growth of viruses (Ellis and Delbrück)
1946	Mixed phage infection leads to genetic recombination (Delbrück)
1947	Mutation and DNA repair (multiplicity reactivation) (Luria)
1952	Transduction of genetic information (Zinder and Lederberg)
1952	DNA, not protein, found to be the genetic material (Hershey and Chase)
1952	Restriction and modification of DNA (Luria)
1955	Definition of a gene (*cis-trans* test) (Benzer)
1958	Mechanisms of control of gene expression by repressors and activators are established (Pardee, Jacob, and Monod)
1958	Definition of the episome (Jacob and Wollman)
1961	Discovery of mRNA (Brenner, Jacob, and Meselson)
1961	Elucidation of the triplet code by genetic analysis (Crick, Barnett, Brenner, and Watts-Tobin)
1961	Genetic definition of nonsense codons as stop signals for translation (Campbell, Epstein, and Berstein)
1964	Colinearity of the gene with the polypeptide chain (Sarabhai, Stretton, and Brenner)
1966	Pathways of macromolecular assembly (Edgar and Wood)
1974	Vectors for recombinant DNA technology (Murray and Murray, Thomas, Cameron, and Davis)

[a]Sources: T. D. Brock, *The Emergence of Bacterial Genetics* (Cold Spring Harbor Laboratory Press, Cold Spring Harbor, N.Y., 1990); K. Denniston and L. Enquist, *Recombinant DNA* (*Benchmark Papers in Microbiology*, vol. 15; Dowden, Hutchinson and Ross, Inc., Stroudsburg, Pa., 1981); and C. K. Mathews, E. Kutter, G. Mosig, and P. Berget, *Bacteriophage T4* (American Society for Microbiology, Washington, D.C., 1983).

Animal Cells as Hosts

The culture of animal cells in the laboratory was initially more of an art than a science, restricted to cells that grew out of organs or tissues maintained in nutrient solutions under sterile conditions. The finite life span of such **primary cells**, their dependence for growth on natural media such as lymph, plasma, or chicken embryo extracts, and the technical demands of sterile culture prior to the discovery of antibiotics, made reproducible experimentation very difficult. By 1955 the work of many investigators had led to a series of important methodological advances. These included the development of defined media optimal for growth of mammalian cells, incorporation of antibiotics into tissue culture media, and development of immortal cell lines such as the mouse L and human HeLa cells that are still in widespread use. These advances allowed growth of animal cells in culture to become a routine, reproducible exercise.

The availability of well-characterized cell cultures had several important consequences for virology. It allowed the discovery of several new human viruses, such as adenovirus, measles virus, and rubella virus, for which animal hosts were not available. In 1949, John Enders and colleagues used cell cultures to propagate poliovirus, a feat that led to development of polio vaccines a few years later. Cell culture technology revolutionized the ability to investigate the replication of viruses. Viral infectious cycles could be studied under precisely controlled conditions by employing the analog of the one-step growth cycle of bacteriophages and simple methods for quantification of infectious particles. Our current understanding of the molecular basis of parasitism by viruses, the focus of the initial chapters of this book, is based almost entirely on analyses of one-step growth cycles in cultured cells. Such studies established that viruses are **molecular** parasites: their reproduction depends absolutely on their host cell's biosynthetic machinery for synthesis of the components from which they are built. In contrast to cells, viruses do not reproduce by growth and division. Rather, the infecting genome contains information necessary to redirect cellular systems to the production of all the components needed to assemble new virus particles.

Viruses Defined

Advances in knowledge of the structure of virus particles and the mechanisms by which viruses are reproduced in their host cells have been accompanied by increasingly accurate definitions of these unique agents. The earliest,

BOX 1.4

The episome

In 1958, François Jacob and Elie Wollman realized that lambda prophage, the *E. coli* F sex factor, and the colicinogenic factor had many common genetic properties. Their remarkable insight led to the definition of the **episome**.

An episome is an exogenous genetic element not necessary for cell survival. Its defining characteristic is the ability to reproduce in two alternative states, integrating into the host chromosome or by autonomous replication.

Nowadays this term is often applied to genomes that can be maintained in cells by autonomous replication and **never** integrate, for example, certain viral DNA genomes.

Jacob and Wollman immediately understood that the episome had value in understanding larger problems, including cancer, as revealed by this quotation: "... in the no man's land between heredity and infection, between physiology and pathology at the cellular level, episomes provide a new link and a new way of thinking about cellular genetics in bacteria, and perhaps in mice, men and elephants" (F. Jacob and E. Wollman, *Viruses and Genes: Readings from* Scientific American [W. H. Freeman & Co., New York, N.Y., 1961]).

pathogenic agents, distinguished by their small size and dependence on a host organism for reproduction, emphasized the importance of viruses as agents of disease. We can now provide a much more precise definition of viruses, elaborating their relationship with the host cell and the important features of virus particles. The definitive properties of viruses are summarized as follows:

- A virus is an infectious, obligate, intracellular parasite.
- The viral genome comprises either DNA or RNA.
- Within an appropriate host cell, the viral genome is replicated and directs the synthesis, by cellular systems, of other viral components.
- Progeny infectious virus particles, called **virions**, are formed by de novo assembly from newly synthesized components within the host cell.
- A progeny virion assembled during the infectious cycle is the vehicle for transmission of the viral genome to the next host cell or organism, where its disassembly leads to the beginning of the next infectious cycle.

With these properties in mind, we can accurately place viruses within the evolutionary continuum of biological agents. They are far simpler than even the smallest microorganisms and lack the complex energy-generating and biosynthetic systems necessary for independent existence (Box 1.5). On the other hand, viruses are **not** the simplest biologically active agents: even the smallest virus, built from a very limited genome and a single type of protein, is significantly more complex than other pathogens. Some of these minimalist molecular pathogens, **viroids**, which are infectious agents of a variety of economically important plants, comprise a single small molecule of RNA. Others, termed **prions**, are believed to be single protein molecules.

Cataloging Animal Viruses

Many Sizes and Shapes Produced by Evolution

Around 1960, virus classification was a subject of colorful and quite heated controversy. New viruses were being discovered and studied by electron microscopy (Box 1.6). The virus world consisted of a veritable zoo of particles with different sizes, shapes, and composition (see, for example,

BOX 1.5

Viruses don't actually "do" anything!

Many have succumbed to the temptation of ascribing various actions and motives to viruses. While remarkably effective to enliven a lecture or an article, anthropomorphic characterizations are inaccurate and often misleading.

Here are some examples from the anthropomorphic lexicon that should be avoided.

- Viruses **cannot** think, employ, ensure, synthesize, exhibit, display, destroy, deploy, depend, reprogram, avoid, retain, evade, exploit, generate, etc.
- Infected cells and hosts do many things in the presence of viruses, but viruses themselves are passive agents, totally at the mercy of their environments.

It is exceedingly difficult to purge such anthropomorphic terms from virology communications. Indeed, hours were spent doing so in preparation of this textbook. A good exercise to appreciate the difficulty of the problem may be to find all the examples that were missed.

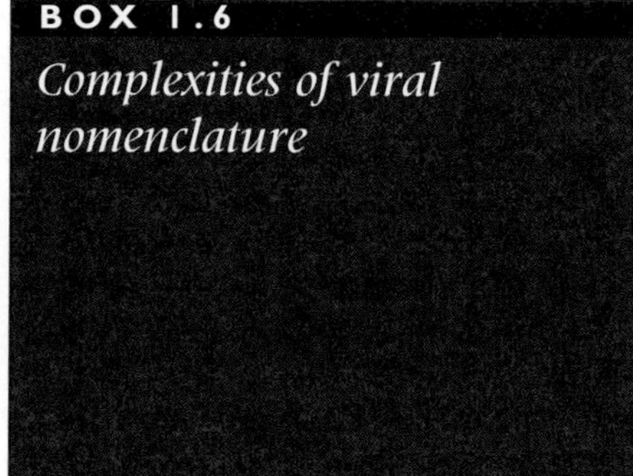

No consistent system for naming viruses has been established by their discoverers. For example, among the vertebrate viruses, some are named for the **associated diseases** (e.g., poliovirus, rabies virus), for the specific **type of disease** they cause (e.g., murine leukemia virus), or for the **sites in the body** that are affected or from which they were first isolated (e.g., rhinovirus, adenovirus). Others are named for the **geographic locations** in which they were first isolated (e.g., Sendai virus [Sendai, Japan] and coxsackievirus [Coxsackie, upstate New York]), or for the **scientists** who first discovered them (e.g., Epstein-Barr virus). In these cases the virus names are capitalized. Some viruses are even named for the way in which people **imagined they were contracted** (e.g., dengue, for "evil spirit," and influenza, for the "influence" of bad air). Finally, **combinations** of the above designations are also used (e.g., Rous sarcoma virus).

Fig. 1.10). Very strong opinions were advanced concerning classification and nomenclature, and opposing camps developed, as in any controversy, involving individuals who tend to focus on differences and those who look for similarities (conventionally known as "splitters" and "lumpers"). Splitters pointed to the inability to infer, from the known properties of viruses, anything about their evolutionary origin or their relationships to one another—the major goal of classical taxonomy. Lumpers maintained that, despite such limitations, there were significant practical advantages in grouping isolates with similar properties. Furthermore, it seemed likely that a good classification might actually stimulate fruitful investigation. A major sticking point, however, was finding agreement on the properties that should be considered most important in constructing a scheme for virus classification.

The Classical System

In 1962, Lwoff, Robert W. Horne, and Paul Tournier advanced a comprehensive scheme for the classification of all viruses (bacterial, plant, and animal) under the classical Linnaean hierarchical system consisting of phylum, class, order, family, genus, and species. Although a subsequently formed international committee on the nomenclature of viruses did not adopt this system in toto, its designation of families, genera, and species was used for the classification of animal viruses.

One of the most important principles embodied in the system advanced by Lwoff and his colleagues was that viruses should be grouped according to **their** shared properties rather than the properties of the cells or organisms they infect. A second principle was a focus on the nucleic acid genome as the primary criterion for classification. The importance of the genome had become clear when it was inferred from the Hershey-Chase experiment that viral nucleic acid alone can be infectious. Four characteristics were to be used in the classification of all viruses:

1. Nature of the nucleic acid in the virion (DNA or RNA)
2. Symmetry of the protein shell (**capsid**)
3. Presence or absence of a lipid membrane (**envelope**)
4. Dimensions of the virion and capsid.

These and a few additional properties are summarized in Fig. 1.10, which lists 23 families of viruses that infect humans and other mammals.

The order in which families are aligned in the figure is arbitrary, as the evolutionary relationships among families are still largely unknown. However, it is clear that the morphological distinctions highlighted in the scheme belie some significant similarities, for example, in the organization of the picornaviral and flaviviral genomes and of the rhabdoviral and paramyxoviral genomes. Genomics, the elucidation of evolutionary relationships by analyses of nucleic acid and protein sequence similarities, increasingly is being used to order members within a virus family, to classify newly isolated viral nucleic acids, and to predict the functions of genes. For example, human herpesvirus 8 was placed in the subfamily *Gammaherpesvirinae* on the basis of sequence analysis of a small segment of its DNA, and hepatitis C virus was classified as a member of the family *Flaviviridae* from the sequence of a cloned DNA copy of its genome. Indeed, as our knowledge of molecular properties of viruses and their replication has increased, it has become apparent that **any** two-dimensional comparison (e.g., virus versus size or virus versus capsid morphology) is somewhat misleading. For example, *Hepadnaviridae*, *Retroviridae*, and some plant viruses are classified as different families on the basis of the nature of their genomes, but reverse transcription is an essential step in all their reproductive cycles. Moreover, the viral polymerases that perform this task exhibit similarities in amino acid sequence.

As of the latest report (2000) of the International Committee on Taxonomy of Viruses (ICTV), approximately 30,000 to 40,000 virus isolates from bacteria, plants, and animals have been assigned to one of 3 orders, 56 families, 9 subfamilies, 233 genera, and 1,550 species, with yet others assigned provisionally. As few new virus families have been identified in recent years, it seems likely that a significant fraction of all existing virus families are now known.

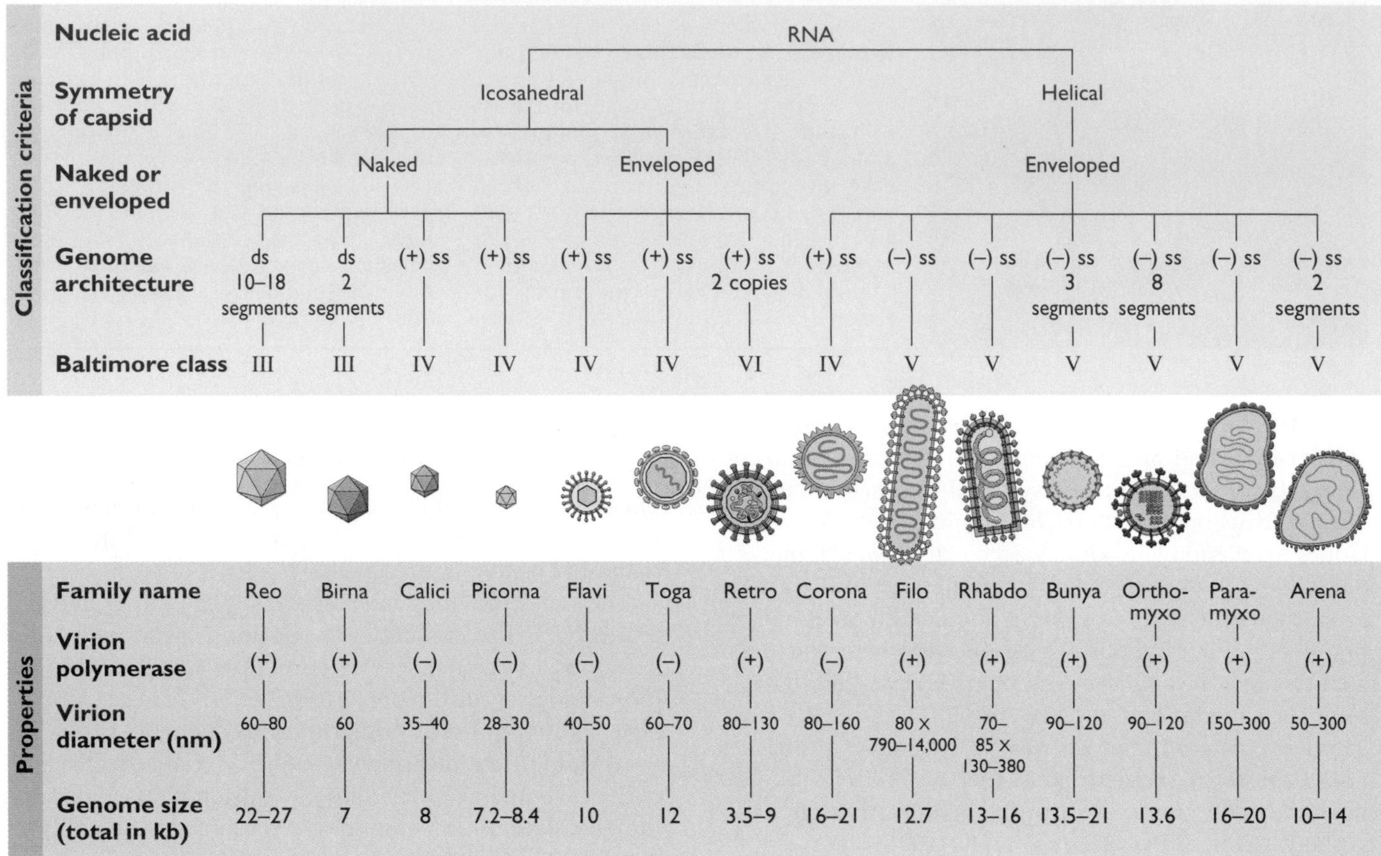

Classification criteria
Nucleic acid
RNA
Symmetry of capsid
Icosahedral
Helical
Naked or enveloped
Naked
Enveloped
Enveloped
Genome architecture
ds 10–18 segments
ds 2 segments
(+) ss
(+) ss
(+) ss
(+) ss
(+) ss 2 copies
(+) ss
(–) ss
(–) ss
(–) ss 3 segments
(–) ss 8 segments
(–) ss
(–) ss 2 segments
Baltimore class
III III IV IV IV IV VI IV V V V V V V
Properties
Family name
Reo Birna Calici Picorna Flavi Toga Retro Corona Filo Rhabdo Bunya Ortho-myxo Para-myxo Arena
Virion polymerase
(+) (+) (–) (–) (–) (–) (+) (–) (+) (+) (+) (+) (+) (+)
Virion diameter (nm)
60–80 60 35–40 28–30 40–50 60–70 80–130 80–160 80 × 790–14,000 70–85 × 130–380 90–120 90–120 150–300 50–300
Genome size (total in kb)
22–27 7 8 7.2–8.4 10 12 3.5–9 16–21 12.7 13–16 13.5–21 13.6 16–20 10–14

However, as we learn more and more about genes, proteins, and reproduction strategies, other relationships will certainly be revealed. Classification refinements can therefore be expected to continue in the future. The ICTV report also includes descriptions of subviral agents (**satellites**, viroids, and prions) and a list of viruses for which information is still insufficient to make assignments. Satellites are composed of nucleic acid molecules that depend for their multiplication on coinfection of a host cell with a helper virus. However, they are not related to this helper. When a satellite encodes the coat protein in which its nucleic acid is encapsidated, it is referred to as a satellite virus (e.g., hepatitis delta virus is a **satellite virus**).

Several years ago, the bacterial virologists who were members of the ICTV agreed to coin similar Latinized family names for the different types of bacteriophages. However, this nomenclature never really took hold, and it has not been widely used by those who do research with bacteriophages. Plant virologists do not classify their viruses into families and genera. Instead, they use group names derived from the prototype virus of each group. For animal viruses, however, the ICTV nomenclature has proved quite useful and, as it has been applied widely in both the scientific and medical literature, we adopt it in this text. In this nomenclature, the Latinized virus family names are recognized as starting with capital letters and ending with *-viridae*, as, for example, in the family name *Parvoviridae*. These names are used interchangeably with their common derivatives as, for example, parvoviruses.

The Baltimore Classification System

The past 3 decades have brought an enormous increase in knowledge of the molecular biology of viruses and cells. Among the most significant advances has been the elucidation of pathways by which information encoded in genomes is expressed. We know that the genes of cells are encoded in their DNA but that cells express this information via messenger RNAs (mRNAs) that are translated in the cytoplasm by ribosomes and associated machinery. This is the so-called central dogma conceptualized by Francis Crick.

$$\text{DNA} \rightarrow \text{RNA} \rightarrow \text{protein}$$

All viruses must direct the synthesis of mRNA to produce proteins, as viral protein synthesis is completely dependent on the cell's translational machinery. Appreciation of the central role of the translational machinery and of the

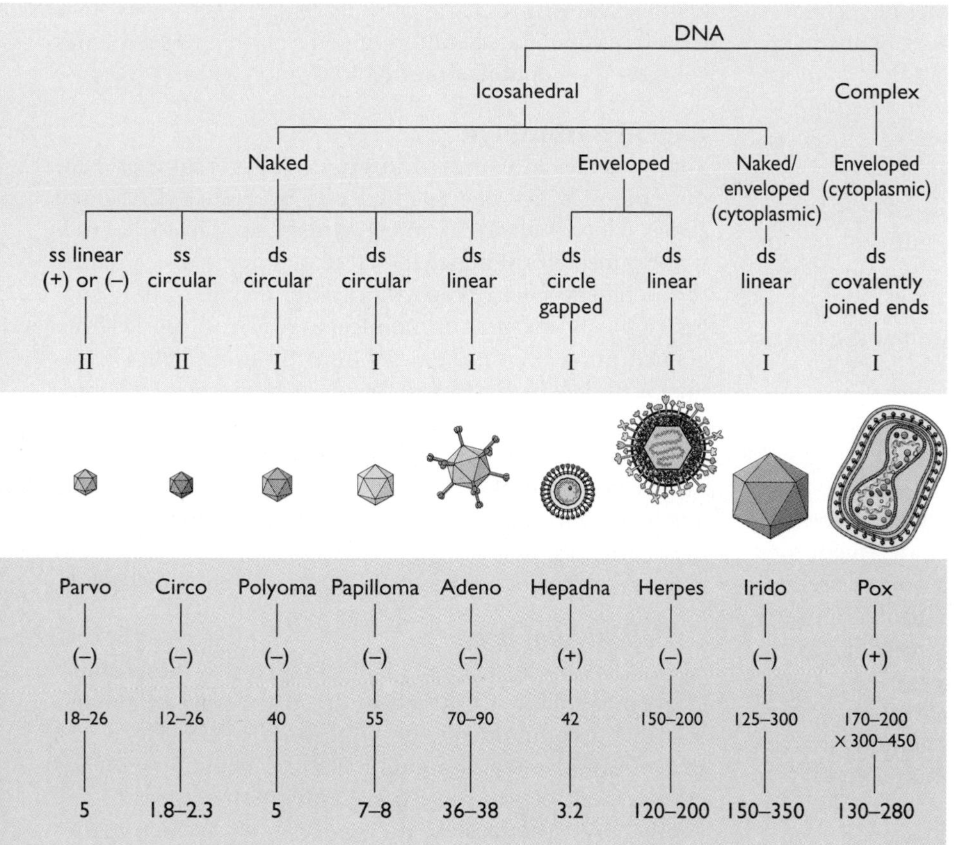

Figure 1.10 Classification schemes for animal viruses. Summary of the major characteristics of 23 representative families of viruses that infect vertebrates. Not all virus families are shown in the figure. Adapted from M. H. V. van Regenmortel et al. (ed.), *Virus Taxonomy: Classification and Nomenclature of Viruses. Seventh Report of the International Committee on Taxonomy of Viruses* (Academic Press, Inc., San Diego, Calif., 2000).

importance of mRNA molecules in the programming of viral protein synthesis inspired an alternative classification scheme devised by David Baltimore (Fig. 1.11). This classification is based on the genetic system of each virus and describes the obligatory relationship between the viral genome and its mRNA.

By the molecular biologist's convention, mRNA is defined as a **positive [(+)] strand** because it contains immediately translatable information. In the Baltimore classification, a strand of DNA that is of equivalent sequence is also designated a (+) strand. The RNA and DNA complements of (+) strands are designated **negative [(–)] strands**. The Baltimore classification is superimposed on the ICTV nomenclature in Fig. 1.10. It is apparent that these two schemes are not mutually exclusive but rather complement one another. In the case of animal viruses, the Baltimore designations can be seen as simply providing additional defining characteristics. The roman numeral designations of the Baltimore classification are rarely used in the animal virus literature. In contrast, the principles embodied in this classification have proved to be extremely valuable, especially for viruses with single-stranded RNA genomes. Designations such as (+) strand

Figure 1.11 The Baltimore classification. All viruses must produce mRNA that can be translated by cellular ribosomes. In this classification system, the unique pathways from various viral genomes to mRNA define specific virus classes on the basis of the nature and polarity of their genomes.

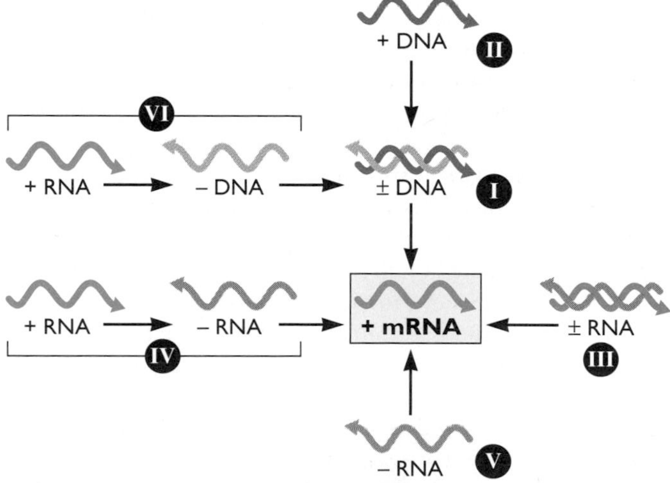

and (–) strand are applied widely to describe genomes (Chapter 3), as will be illustrated in this text. Knowledge of strand polarity provides virologists with immediate insight into the steps that must take place to initiate replication and expression of the viral genome.

The Molecular Biologist's Focus

Because the viral genome carries the entire blueprint for virus propagation, molecular virologists have long considered it the most important characteristic for classification purposes. This feature therefore is placed at the top of our hierarchy of classification (Fig. 1.10).

Genetic Content

DNA Viruses

As shown in Fig. 1.10, the members of all but two families of DNA viruses contain double-stranded genomes (Baltimore's class I). Most of the DNA genomes are not simple linear molecules. For example, polyomavirus genomes are covalently closed, double-stranded circles, and hepadnavirus genomes are duplex circles in which one strand contains a nick and the other contains a gap. The ends of the double-stranded poxvirus genome are covalently joined, whereas herpesvirus genomes contain terminal and internal duplications.

RNA Viruses

In contrast to DNA viruses, most RNA viruses have single-stranded genomes. Those RNA viruses with continuous (+) strand genomes (Baltimore's class IV) include the morphologically diverse picornaviruses and caliciviruses. Retroviruses (the sole members of Baltimore's class VI) also contain (+) strand RNA genomes, but two copies of genomic RNA are included in each virion. This property is unique to the family *Retroviridae*, as all other viruses contain only one copy of their genome.

The genomes of (–) strand RNA viruses (Baltimore's class V) are all packaged within helical **nucleocapsids**. Three members of this class, rhabdoviruses, paramyxoviruses, and filoviruses, are sufficiently related to be organized into the order *Mononegavirales*. As suggested by the name, the genomes of all viruses in this order consist of a single (–) strand RNA molecule. Other RNA viruses in class V have segmented genomes: orthomyxovirus genomes comprise seven or eight separate RNA molecules, and arenavirus and bunyavirus genomes comprise two and three RNA molecules, respectively. The genomes of arenaviruses and some bunyaviruses are **ambisense**, i.e., not only are their complements mRNAs, but the 5′ ends of some of the constituent genomic RNAs also function as mRNAs.

Two families make up the class of viruses with double-stranded RNA genomes (Baltimore's class III). Reovirus genomes consist of 10 to 12 double-stranded linear RNA molecules that encode different genes; birnavirus genomes comprise two double-stranded RNA molecules.

Capsid Symmetry

Viral genomes are encased in protective protein capsids or nucleocapsids. As first pointed out by Francis Crick and James Watson in 1956, most viruses appear to be rod-shaped or spherical under the electron microscope. Because of the limited coding capacity of viral genomes, they suggested that it was most economical to construct capsids from a small number of protein subunits by using helical symmetry (rod-shaped viruses) or the symmetry of Platonic polyhedra (e.g., tetrahedron, octahedron, or icosahedron) (spherical viruses). Indeed, capsids are formed by the assembly of identical structural units in ways that provide maximal contact among them. The proteins that comprise these structures have the ability to associate in highly regular arrays that produce the two symmetries described below.

Helical Symmetry

One can visualize a helix by forming a two-dimensional lattice and rolling it into a cylindrical structure of a diameter that accommodates the viral nucleic acid (Fig. 1.12). Helical capsids may be rigid or flexible, depending on the arrangement of the protein subunits with respect to one another. As summarized in Fig. 1.10, all animal viruses with filamentous nucleocapsids in a helical conformation also possess envelopes.

Icosahedral Symmetry

The most economical way to build a spherical shell of maximal internal volume is to arrange nonsymmetric proteins with icosahedral symmetry. In the simplest case, an icosahedron comprises 20 triangular faces organized with characteristic rotational symmetries (Fig. 1.12B).

Complex Particles

Some viruses fit into neither the helical nor the icosahedral category. They form complex capsids with structures that are still not well understood. Members of the family *Poxviridae* and some members of the *Retroviridae* represent examples among animal viruses.

Enveloped or Naked Particles

Many animal viruses are surrounded by an envelope, which comprises a lipid membrane acquired from cell membranes, and viral proteins required for infectivity. Viruses that are not enveloped are referred to as **naked**.

Virion Enzymes

Many virions contain enzymes that are encoded in the viral genomes (Fig. 1.10). The most important of these are polymerases, which copy all or part of the genomes of

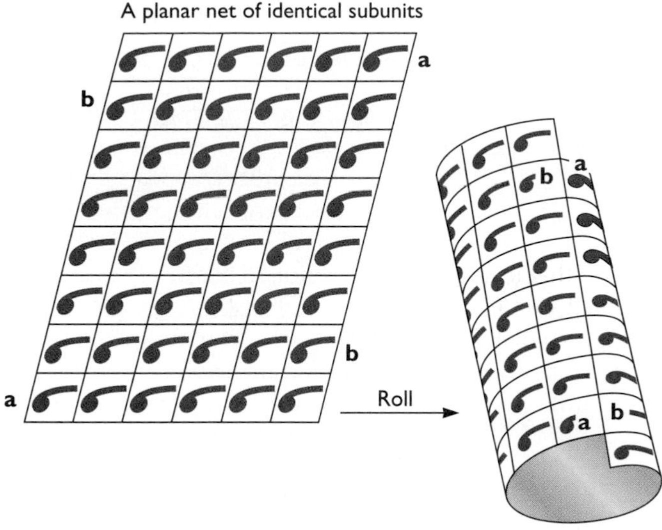

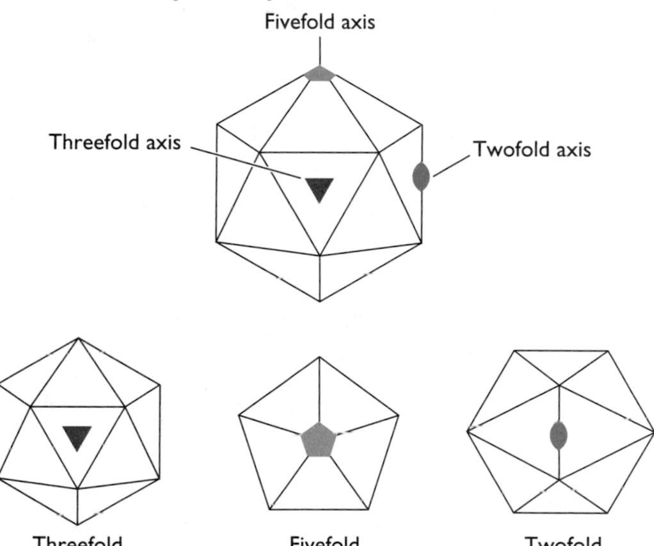

Figure 1.12 Capsid architecture is based on principles of helical and icosahedral symmetry. (A) A planar net of identical subunits and the same planar net rolled to form a helix. Many virions are rod-shaped (e.g., tobacco mosaic virus) with the viral RNA inside bound to the capsid protein subunits arranged on the outside with helical symmetry. (B) An icosahedron comprises 20 equilateral triangular faces with characteristic positions of rotational symmetry. The three views at the bottom highlight these positions.

some viruses, often immediately upon entry into a newly infected cell. These enzymes are essential for viruses with single (–) strand RNA genomes because host cells lack the RNA-dependent RNA polymerase necessary to produce viral mRNAs. The first viral polymerase was discovered in particles of a DNA virus, the poxvirus vaccinia virus. Poxviruses are a special case, though, because unlike most other DNA viruses, they replicate in the cytoplasm and lack access to the cellular transcription machinery for production of their mRNA.

Unifying Principles

A Common Strategy

The basic thesis of this textbook is that **all** viral propagation can be described in the context of **three fundamental properties.**

- All viral genomes are packaged inside particles that mediate their transmission from host to host.
- The viral genome contains the information for initiating and completing an infectious cycle within a susceptible, permissive cell. An infectious cycle includes attachment and entry of the particle, decoding of genome information, translation of viral mRNA by host ribosomes, genome replication, and assembly and release of particles containing the genome.
- All viruses are able to establish themselves in a host population so that virus survival is ensured.

Modern virology is both exciting and challenging because of the many and varied ways this strategy is executed. As viruses are obligate molecular parasites, every tactical solution must of necessity tell us something about the host as well as the virus. The intellectual satisfaction of discovering and understanding new principles is as rewarding as the practical consequences of providing solutions to problems of disease.

Highlights of Virus Strategies Outlined in This Book

Making Viral Proteins

All viral RNAs must be translated by the host's protein-synthesizing machinery; all viruses are parasites of translation. Unless the viral genome can function directly as an mRNA, this requirement means that strategies must exist for synthesizing mRNAs soon after the viral genome enters the cell. Even when the genome is a functional mRNA, successful replication necessarily includes production of many new copies of the genome. Analysis of the various means of producing viral mRNA has taught us much about the cell itself.

Making Viral Genomes

Many viral genomes are copied by the cell's synthetic machinery in cooperation with viral proteins. The cell provides nucleotide substrates, energy, enzymes, and other

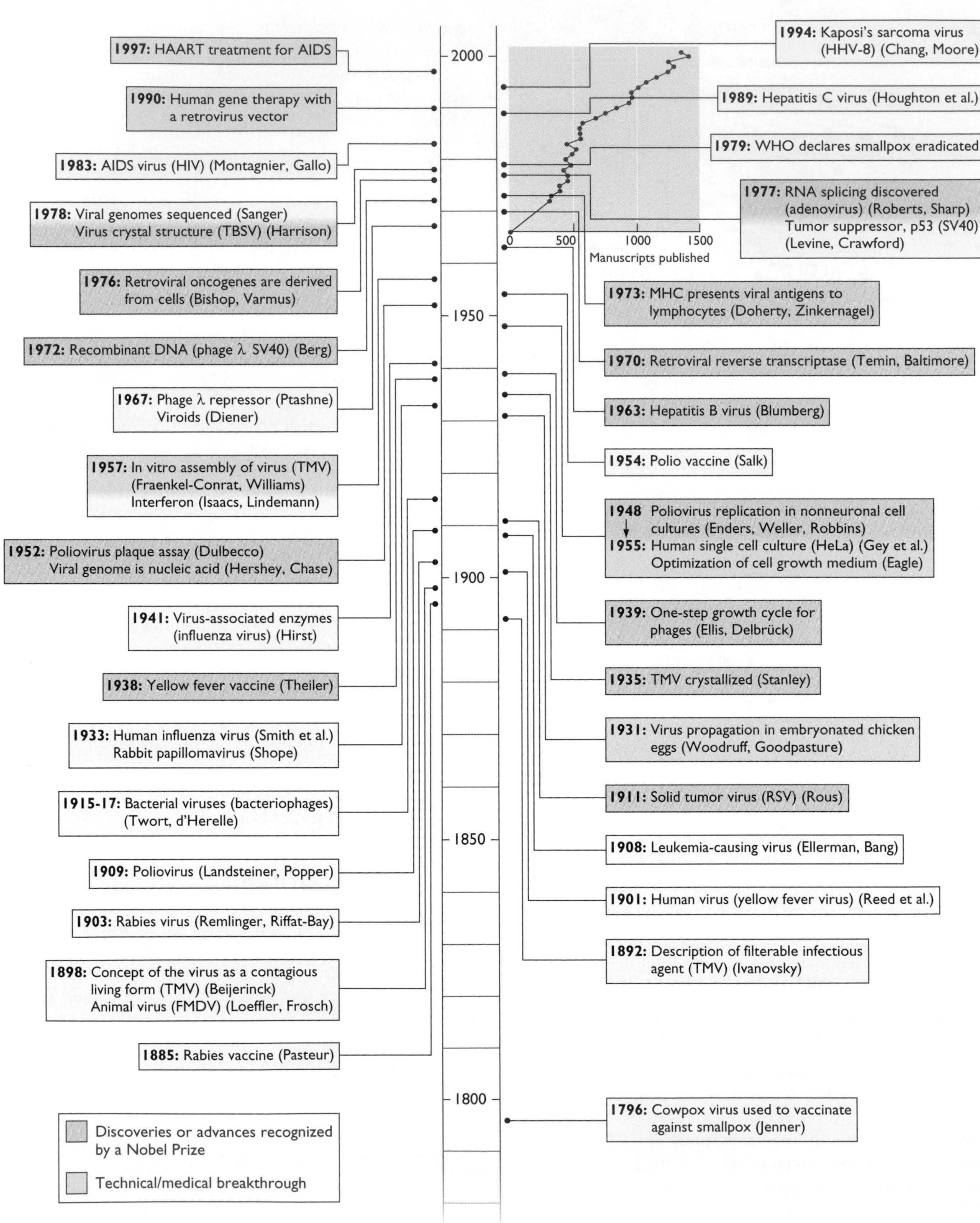

1997: HAART treatment for AIDS
1990: Human gene therapy with a retrovirus vector
1983: AIDS virus (HIV) (Montagnier, Gallo)
1978: Viral genomes sequenced (Sanger) Virus crystal structure (TBSV) (Harrison)
1976: Retroviral oncogenes are derived from cells (Bishop, Varmus)
1972: Recombinant DNA (phage λ SV40) (Berg)
1967: Phage λ repressor (Ptashne) Viroids (Diener)
1957: In vitro assembly of virus (TMV) (Fraenkel-Conrat, Williams) Interferon (Isaacs, Lindemann)
1952: Poliovirus plaque assay (Dulbecco) Viral genome is nucleic acid (Hershey, Chase)
1941: Virus-associated enzymes (influenza virus) (Hirst)
1938: Yellow fever vaccine (Theiler)
1933: Human influenza virus (Smith et al.) Rabbit papillomavirus (Shope)
1915-17: Bacterial viruses (bacteriophages) (Twort, d'Herelle)
1909: Poliovirus (Landsteiner, Popper)
1903: Rabies virus (Remlinger, Riffat-Bay)
1898: Concept of the virus as a contagious living form (TMV) (Beijerinck) Animal virus (FMDV) (Loeffler, Frosch)
1885: Rabies vaccine (Pasteur)
2000
1950
1900
1850
1800
0
500
1000
1500
Manuscripts published
1994: Kaposi's sarcoma virus (HHV-8) (Chang, Moore)
1989: Hepatitis C virus (Houghton et al.)
1979: WHO declares smallpox eradicated
1977: RNA splicing discovered (adenovirus) (Roberts, Sharp) Tumor suppressor, p53 (SV40) (Levine, Crawford)
1973: MHC presents viral antigens to lymphocytes (Doherty, Zinkernagel)
1970: Retroviral reverse transcriptase (Temin, Baltimore)
1963: Hepatitis B virus (Blumberg)
1954: Polio vaccine (Salk)
1948 ↓ 1955: Poliovirus replication in nonneuronal cell cultures (Enders, Weller, Robbins) Human single cell culture (HeLa) (Gey et al.) Optimization of cell growth medium (Eagle)
1939: One-step growth cycle for phages (Ellis, Delbrück)
1935: TMV crystallized (Stanley)
1931: Virus propagation in embryonated chicken eggs (Woodruff, Goodpasture)
1911: Solid tumor virus (RSV) (Rous)
1908: Leukemia-causing virus (Ellerman, Bang)
1901: Human virus (yellow fever virus) (Reed et al.)
1892: Description of filterable infectious agent (TMV) (Ivanovsky)
1796: Cowpox virus used to vaccinate against smallpox (Jenner)
Discoveries or advances recognized by a Nobel Prize
Technical/medical breakthrough

Figure 1.13 Landmarks in the study of animal viruses. Key discoveries and technical advances are listed on either side of the central time line. The graph at the upper right shows the number of publications per year appearing in the *Journal of Virology* (1972 to 2001), as an indication of the continuous increase in virus-related research over the past 3 decades. Pink boxes indicate discoveries recognized by a Nobel Prize; light blue boxes represent technical or medical breakthroughs. Abbreviations: HAART, highly active antiretroviral therapy; HIV, human immunodeficiency virus; TBSV, tomato bushy stunt virus; TMV, tobacco mosaic virus; SV40, simian virus 40; FMDV, foot-and-mouth disease virus; WHO, World Health Organization; MHC, major histocompatibility complex; HHV-8, human herpesvirus 8; RSV, Rous sarcoma virus.

proteins. In all cases, the internal compartmentalization of the cell must be reckoned with, because essential components are found only in the nucleus, are restricted to the cytoplasm, or are present in cellular membranes. Study of the mechanisms of viral genome replication has established fundamental principles of cell biology and nucleic acid synthesis.

Forming Progeny Virions

The various components of a virion, the nucleic acid genome, capsid protein(s), and in some cases envelope proteins, are often synthesized in different cellular compartments. Their trafficking through and among the cell's compartments and organelles requires that they be equipped with the proper homing signals. Virion components must be assembled at some central location, and the information for assembly must be preprogrammed in the component molecules.

Viral Pathogenesis

Viruses command our attention because of their association with animal or plant diseases. The process by which viruses cause disease is called **viral pathogenesis**. To study this process we must investigate not only the relationships of viruses with the specific cells that they infect, but also the consequences of infection to the host organism. The nature of viral disease depends on the effects of viral replication on host cells, the response of the host's defense systems, and the ability of the virus to spread in and among hosts.

Overcoming Host Defenses

Organisms have evolved many physical barriers to protect themselves from dangers in their environment, such as invading parasites. In addition, vertebrates possess an effective immune system to defend against anything recognized as nonself or dangerous. Studies of the interactions among viruses and the immune system are particularly instructive, because of the many viral countermeasures that can defeat this system. Understanding these measures teaches us much about the basis of immunity.

Perspectives

This chapter is about the foundations of virology; some of the important landmarks in animal virology are summarized in Fig. 1.13.

References

Books

Brock, T. D. 1990. *The Emergence of Bacterial Genetics.* Cold Spring Harbor Laboratory Press, Cold Spring Harbor, N.Y.

Brothwell, D., and A. T. Sandison (ed.). 1967. *Diseases in Antiquity.* Charles C Thomas, Publisher, Springfield, Ill.

Cairns, J., G. S. Stent, and J. D. Watson (ed.). 1966. *Phage and the Origins of Molecular Biology.* Cold Spring Harbor Laboratory for Quantitative Biology, Cold Spring Harbor, N.Y.

Creager, A. N. H. 2002. *The Life of a Virus: Tobacco Mosaic Virus as an Experimental Model, 1930–1965.* The University of Chicago Press, Chicago, Ill.

Denniston, K., and L. Enquist. 1981. *Recombinant DNA. Benchmark Papers in Microbiology,* vol. 15. Dowden, Hutchinson and Ross, Inc., Stroudsburg, Pa.

Fiennes, R. 1978. *Zoonoses and the Origins and Ecology of Human Disease.* Academic Press, Inc., New York, N.Y.

Hughes, S. S. 1977. *The Virus: a History of the Concept.* Heinemann Educational Books, London, United Kingdom.

Karlen, A. 1996. *Plague's Progress, a Social History of Man and Disease.* Indigo, Guernsey Press Ltd., Guernsey, Channel Islands.

Knipe, D. M., P. M. Howley, D. E. Griffin, R. A. Lamb, M. A. Martin, B. Roizman, and S. E. Straus (ed.). 2001. *Fields Virology,* 4th ed. Lippincott Williams & Wilkins, Philadelphia, Pa.

Luria, S. E. 1953. *General Virology.* John Wiley & Sons, Inc., New York, N.Y.

Murphy, F. A., C. M. Fauquet, D. H. L. Bishop, S. A. Ghabrial, A. W. Jarvis, and N. Rasmussen. 1997. *Picture Control: the Electron Microscope and the Transformation of Biology in America 1940–1960.* Stanford University Press, Stanford, Calif.

Oldstone, M. B. A. 1998. *Viruses, Plagues and History.* Oxford University Press, New York, N.Y.

Stent, G. S. 1960. *Papers on Bacterial Viruses.* Little, Brown & Co., Boston, Mass.

van Regenmortel, M. H. V., C. M. Fauquet, D. H. L. Bishop, E. B. Carstens, M. K. Estes, S. M. Lemon, J. Maniloff, M. A. Mayo, D. J. McGeoch, C. R. Pringle, and R. B. Wickner (ed.). 2000. *Virus Taxonomy: Classification and Nomenclature of Viruses. Seventh Report of the International Committee on Taxonomy of Viruses.* Academic Press, Inc., San Diego, Calif.

Waterson, A. P., and L. Wilkinson. 1978. *An Introduction to the History of Virology.* Cambridge University Press, London, United Kingdom.

Papers of Special Interest

Baltimore, D. 1971. Expression of animal virus genomes. *Bacteriol. Rev.* **35:**235–241.

Brown, F., J. Atherton, and D. Knudsen. 1989. The classification and nomenclature of viruses: summary of results of meetings of the International Committee on Taxonomy of Viruses. *Intervirology* **30:**181–186.

Burnet, F. M. 1953. Virology as an independent science. *Med. J. Aust.* **40:**842.

Crick, F. H. C., and J. D. Watson. 1956. Structure of small viruses. *Nature* **177:**473–475.

Lustig, A., and A. J. Levine. 1992. One hundred years of virology. *J. Virol.* **66:**4629–4631.

Van Helvoort, T. 1993. *Research Styles in Virus Studies in the Twentieth Century: Controversies and the Formation of Consensus.* Doctoral dissertation, University of Limburg, Maastricht, The Netherlands.

Websites

http://www.tulane.edu/~dmsander/garryfavweb.html *Has links to almost all the virology sites on the Web.*

http://www.ncbi.nlm.gov/ICTVdb/ *ICTV-approved virus names and other information as well as links to virus databases.*

http://life.anu.edu.au/viruses/welcome.html *Useful information on many viruses.*

http://virology.wisc.edu/IMV/ *University of Wisconsin website.*

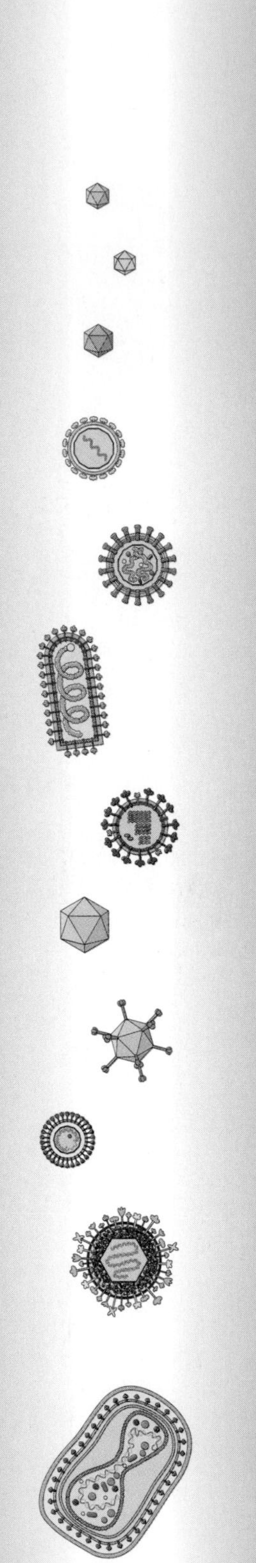

2

Virus Cultivation, Detection, and Genetics

You know my methods, Watson.
Sir Arthur Conan Doyle

Introduction

The knowledge of virus replication and pathogenesis discussed in subsequent chapters of this book had its origin in many research and diagnostic virology laboratories. A variety of experimental techniques have been employed to test hypotheses and to formulate new models of virus replication. The purpose of this chapter is to explain how some of these methods work. Knowledge of how experiments are performed not only is necessary for understanding why certain conclusions are made, but also gives the reader the tools to formulate new approaches to unsolved problems in virology. The technology for manipulating viral genomes, made available by the development of recombinant DNA methods, will also be explored. These techniques facilitate the study of viral gene function, and have paved the way for using viruses as vectors for the delivery of genes into cells and organisms. This chapter should be read in sequence with the remainder of the book and should also be consulted as a reference source for other chapters.

Cultivation of Viruses

Cell Culture

Types of Cell Culture

Although human and other animal cells were first cultured in the early 1900s, contamination with bacteria, mycoplasmas, and fungi initially made routine work with such cultures extremely difficult. For this reason, most viruses were grown in laboratory animals. In 1949, John Enders, Thomas Weller, and Frederick Robbins made the discovery that poliovirus could multiply in cultured cells not of neuronal origin. As noted in Chapter 1, this revolutionary finding, for which these three investigators were awarded the Nobel Prize in physiology or medicine in 1954, led the way to the propagation of many other viruses in cultured cells, the discovery of new viruses, and the development of viral vaccines such as those against poliomyelitis, measles, and rubella. The ability to in-

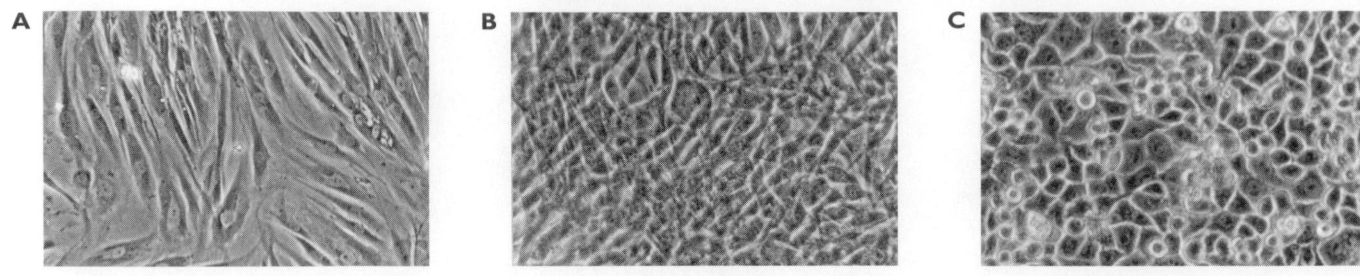

Figure 2.1 Different types of cell culture used in virology. Confluent cell monolayers photographed by low-power light microscopy. (A) Primary human foreskin fibroblasts; (B) established line of mouse fibroblasts (3T3); (C) continuous line of human epithelial cells (HeLa [Box 2.1]). Note the differences in morphology among the primary (human foreskin fibroblasts), established (i.e., immortalized [3T3]), and transformed (HeLa) cells, as well as the ability of transformed HeLa cells to overgrow one another. This property is the result of a loss of contact inhibition. Courtesy of R. Gonzalez, Princeton University.

fect cultured cells synchronously permitted studies of the biochemistry and molecular biology of viral replication. The growth and purification of viruses on a large scale allowed studies of the composition of virus particles, leading to the solution of high-resolution, three-dimensional structures of some viruses, as discussed in Chapter 4.

Cell culture is still the most common method for the propagation of viruses. To prepare a cell culture, tissues are dissociated into a single-cell suspension by mechanical disruption followed by treatment with proteolytic enzymes. The cells are then suspended in culture medium and placed in plastic flasks or covered plates. As the cells divide, they cover the plastic surface. Epithelioid and fibroblastic cells attach to the plastic and form a **monolayer**, whereas blood cells such as lymphocytes settle, but do not adhere. The cells are grown in a chemically defined medium consisting of an isotonic solution of salts, glucose, vitamins, coenzymes, and amino acids buffered to a pH between 7.2 and 7.4, and often supplemented with antibiotics to inhibit bacterial growth. Serum is usually added to the cell culture medium to provide a source of growth factors required by the cells. Commonly used cell lines double in 24 to 48 h in such media. Most cells retain viability after being frozen at low temperatures (–70 to –196°C).

There are three main kinds of cell cultures (Fig. 2.1). **Primary cell cultures** are prepared from animal tissues as described above. They include several cell types and have a limited life span, usually no more than 5 to 20 cell divisions.

BOX 2.1

The cells of Henrietta Lacks

The most widely used continuous cell line in virology, HeLa cells, was derived from Henrietta Lacks. In 1951, the 31-year-old mother of five visited a physician at Johns Hopkins Hospital in Baltimore and found that she had a malignant tumor of the cervix. A sample of the tumor was taken and given to George Gey, head of tissue culture research at Hopkins. Dr. Gey had been attempting for years, without success, to produce a line of human cells that would live indefinitely. When placed in culture, Henrietta Lacks's cells propagated as no other cells had before. On the day in October that Henrietta Lacks died, George Gey appeared on national television with a vial of Henrietta's cells, which he called HeLa cells. He said, "It is possible that, from a fundamental study such as this, we will be able to learn a way by which cancer can be completely wiped out." Soon after, HeLa cells were used to propagate poliovirus, which was causing poliomyelitis throughout the world, and they played an important role in the development of poliovirus vaccines. Henrietta Lacks's HeLa cells started a medical revolution: not only was it possible to propagate many different viruses in these cells, but also continuous cell lines could be produced from many human tissues. Sadly, the family of Henrietta Lacks did not learn about HeLa cells, or the revolution they started, until 24 years after her death. The family was shocked that cells from Henrietta lived in so many laboratories, and hurt that they had not been told that any cells were taken from her. The story of HeLa cells is a sad commentary on the lack of informed consent that pervaded medical research in the 1950s. Since then, biomedical ethics have changed greatly, and now there are regulations about informed consent: physicians may not take samples from patients without permission.

For additional information, see http://www.jhu.edu/~jhumag/0400web/01.html.

The most commonly used primary cell cultures are derived from monkey kidney, human embryonic amnion, kidney, or foreskin, and chicken or mouse embryos. Such cells are used for experimental virology when the state of cell differentiation is important or when appropriate cell lines are not available. They are also used in vaccine production: for example, live attenuated poliovirus vaccine strains may be propagated in primary monkey kidney cells. Primary cell cultures were mandated for the growth of viruses to be used as human vaccines to avoid contamination of the product with potentially oncogenic DNA from continuous cell lines (see below). Some viral vaccines are now prepared in **diploid cell strains**, which consist of a homogeneous population of a single type and can divide up to 100 times before dying. Despite the numerous divisions, these cell strains retain the diploid chromosome number. The most widely used diploid cells are those established from human embryos, such as the WI-38 strain derived from human embryonic lung. **Continuous cell lines** consist of a single cell type that can be propagated indefinitely in culture. These immortal lines are usually derived from tumor tissue or by treating a primary cell culture or a diploid strain with a mutagenic chemical or a tumor virus. Such cell lines often do not resemble the cell of origin; they are less differentiated (having lost the morphology and biochemical features that they possessed in the organ), often abnormal in chromosome morphology and number (**aneuploid**), and can be tumorigenic (they produce tumors when inoculated into nude mice). Examples of commonly used continuous cell lines include HEp-2 (human epithelial) and HeLa (Henrietta Lacks) cells (Box 2.1), derived from human carcinomas; Vero cells, from African green monkey kidneys; L and 3T3 cells, derived from mice; and BHK-21 (baby hamster kidney) cells, derived from hamster kidneys. Continuous cell lines provide a uniform population of cells that can be infected synchronously for growth curve analyses (see "The One-Step Growth Cycle" below) or biochemical studies of virus replication.

In contrast to cells that grow in monolayers on plastic dishes, others can be maintained in **suspension cultures**, in which a spinning magnet continuously stirs the cells. The advantage of suspension culture is that a large number of cells can be grown in a relatively small volume. This culture method is well suited for applications that require large quantities of virus particles, such as X-ray crystallography.

Initially, it was believed that viruses, as obligatory intracellular parasites, could not replicate in any cell-free medium, no matter how complex. This dictum was nullified in 1991 by the demonstration that infectious poliovirus could be produced in a cell-free extract of human cells incubated with viral RNA. Despite this finding, most work on viruses is done in vivo, using cultured cells, embryonated eggs, or laboratory animals (Box 2.2).

BOX 2.2

In vitro and in vivo

The terms "in vitro" and "in vivo" are common in the virology literature. In vitro means "in glass," and refers to experiments carried out in an artificial environment, such as a glass test tube. Unfortunately, the phrase "experiments performed in vitro" is used to designate not only work done in the cell-free environment of a test tube but also work done within cultured cells. The use of the phrase in vitro to describe living cultured cells leads to confusion and is inappropriate.

In this textbook, experiments being carried out in vitro will signify the absence of cells, e.g., in vitro translation. Since vivo means living, work done with cultured cells, or with animals, receives the appellation in vivo. When intact organ pieces are placed in culture for study, the work is done ex vivo.

Evidence of Viral Growth in Cultured Cells

Some viruses kill the cells in which they replicate, and the infected cells may eventually detach from the cell culture plate. As more cells are infected, the changes become visible and are called **cytopathic effects** (Table 2.1).

Table 2.1 Some examples of cytopathic effects of viral infection of animal cells

Cytopathic effect(s)	Virus(es)
Morphological alterations	
Nuclear shrinking (pyknosis), proliferation of membrane	Picornaviruses
Proliferation of nuclear membrane	Alphaviruses, herpesviruses
Vacuoles in cytoplasm	Polyomaviruses
Syncytia (cell fusion)	Paramyxoviruses, coronaviruses
Margination and breaking of chromosomes	Herpesviruses
Rounding up and detachment of cultured cells	Herpesviruses, rhabdoviruses, adenoviruses, picornaviruses
Inclusion bodies	
Virions in nucleus	Adenovirus
Virions in the cytoplasm (Negri bodies)	Rabies virus
"Factories" in the cytoplasm (Guarnieri bodies)	Poxviruses
Clumps of ribosomes in virions	Arenaviruses
Clumps of chromatin in nucleus	Herpesviruses

Many types of cytopathic effect can be seen with a simple light or phase-contrast microscope at low power, without fixing or staining the cells. These changes include the rounding up and detachment of cells from the culture dish, cell lysis, swelling of nuclei, and, sometimes, formation of fused cells called **syncytia** (Fig. 2.2). Other cytopathic effects require observation by high-power microscopy. These include the development of intracellular masses of virions or unassembled viral components in the nucleus and/or cytoplasm (**inclusion bodies**), formation of crystalline arrays of virus proteins, membrane blebbing, duplication of membranes, and fragmentation of organelles. The time required for the development of cytopathology varies greatly among animal viruses. For example, depending on the size of the inoculum, enteroviruses and herpes simplex virus can cause cytopathic effects in 1 to 2 days and destroy the cell monolayer in 3 days. In contrast, cytomegalovirus, rubella virus, and some adenoviruses may not produce such effects for several weeks.

Figure 2.2 Development of cytopathic effect. (A) Cell rounding and lysis during poliovirus infection. Cells were infected with poliovirus type 1 and incubated at 37°C. (Upper left) Uninfected cells. (Upper right) At 5 1/2 h after infection some cells have begun to "round up." (Lower left) At 8 h after infection, more cells have rounded up and detached from the culture dish. (Lower right) At 24 h after infection all cells have detached and formed clumps as they lyse. (B) Syncytium formation induced by murine leukemia virus. The field shows a mixture of individual small cells and syncytia, indicated by arrows, which are large, multinucleate cells. Courtesy of R. Compans, Emory University School of Medicine.

A

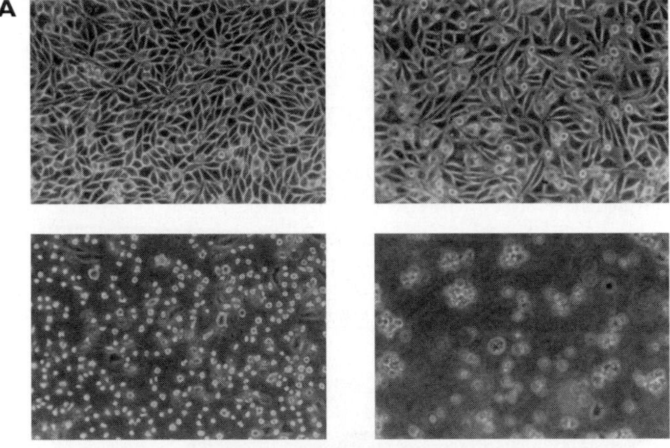

B

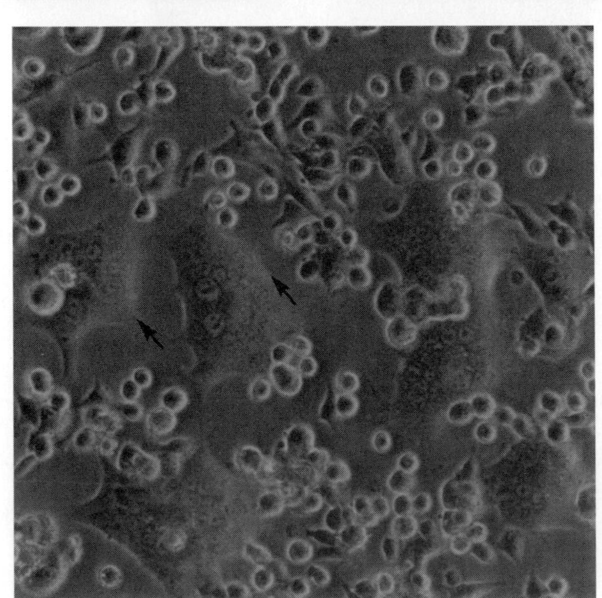

The development of cytopathic effects in infected cell cultures is frequently monitored in diagnostic virology during isolation of viruses from specimens obtained from infected patients or animals. However, cytopathic effect is also of value in the research laboratory: it can be used to monitor the progress of an infection, and it is often one of the phenotypic traits by which mutant viruses are characterized.

Some viruses multiply in cells without causing obvious cytopathic effects. For example, many members of the families *Arenaviridae*, *Paramyxoviridae*, and *Retroviridae* do not cause obvious damage to cultured cells. The growth of such viruses in cells must therefore be assayed using alternative methods, as described in "Detection of Viruses in the Host" below.

Embryonated Eggs

Before the advent of cell culture, many viruses were propagated in embryonated chicken eggs (Fig. 2.3). Five to 14 days after fertilization, a hole is drilled in the shell and virus is injected into the site appropriate for its replication. This method of virus propagation is now routine only for influenza viruses. The robust yield of this virus from chicken eggs has led to their widespread use in research laboratories and for vaccine production.

Laboratory Animals

In the early 1900s, when viruses were first isolated, freezers and cell cultures were not available and it was necessary to maintain virus stocks by continuous passage of virus from animal to animal. This practice was not only inconvenient but, as we shall see in Chapter 20, led to the selection of viral mutants. For example, monkey-to-monkey intracerebral passage of poliovirus selected a mutant that could no longer infect chimpanzees by the oral route, the natural means of infection. Cell culture has largely supplanted the use of animals for propagating viruses, but some viruses cannot be grown in this way. Many viruses that cause human gastroenteritis, such as Norwalk virus, cannot be grown in cell culture. Similarly, hepatitis C virus

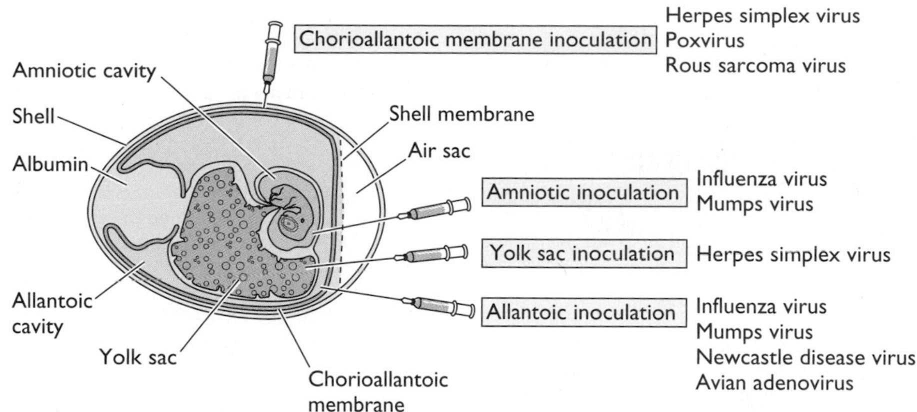

Figure 2.3 Growth of viruses in embryonated eggs. The cutaway view of an embryonated chicken egg shows the different compartments in which viruses may grow. The different routes by which viruses are inoculated into eggs are indicated. Adapted from F. Fenner et al., *The Biology of Animal Viruses* (Academic Press, New York, N.Y., 1974), with permission.

produces infectious particles only in humans and chimpanzees.

Experimental infection of laboratory animals has always been, and will continue to be, obligatory for studying the processes by which viruses cause disease, a field of study known as **viral pathogenesis**. The studies in monkeys of poliomyelitis, the paralytic disease caused by poliovirus, led to an understanding of the basis of this disease and was instrumental in the development of a successful vaccine. Similarly, the development of vaccines against hepatitis B virus would not have been possible without experimental studies with chimpanzees. Understanding how the immune system or any complex organ reacts to a virus cannot be achieved without research on living animals. The development of viral vaccines, antiviral drugs, and diagnostic tests for veterinary medicine has also benefited from research on viral diseases in laboratory animals.

Detection of Viruses in the Host

The ability to recognize viral infection in the host is of prime importance in a research or clinical virology laboratory. The approaches for detecting viruses fall into three categories. Because viruses were first recognized by their **infectivity,** the earliest assays focused on this most sensitive and informative property. In viral **serology**, the specific interaction of viral antigens with antibodies is used to detect viruses and the immune response against them. Viral proteins in infected cells or tissues can be detected with antibodies; alternatively, the assay can be designed to identify antibodies produced in the host in response to viral infection. More recently, **molecular diagnostics,** which allow the detection of viral nucleic acid sequences in infected cells or tissues, have been developed. Examples of each class of assay are discussed in this section.

Measurement of Infectious Units

One of the most important procedures in virology is measuring the concentration of a virus in a sample, or the **virus titer.** This parameter is determined by inoculating serial dilutions of virus into host cell cultures, chick embryos, or laboratory animals, and monitoring for evidence of virus multiplication. The response may be quantitative (as in assays for plaques, fluorescent foci, infectious centers, or transformation) or all-or-none, in which the presence or absence of infection is measured (as in an endpoint dilution assay).

Plaque Assay

In 1952, Renato Dulbecco modified the plaque assay developed to determine the titers of bacteriophage stocks for use in animal virology. The plaque assay was adopted rapidly for reliable determination of the titers of a wide variety of viruses. In this assay, monolayers of cultured cells are incubated with a preparation of virus to allow adsorption to cells. After removal of the inoculum, the cells are covered with nutrient medium containing a supplement, most commonly agar, that results in the formation of a gel. When the original infected cells release new progeny viruses, their spread to neighboring uninfected cells is restricted by the gel. As a result, each infectious particle produces a circular zone of infected cells, or **plaque**. If the infected cells are damaged, the plaque can be distinguished from the surrounding monolayer. In time, the plaque becomes large enough to see with the

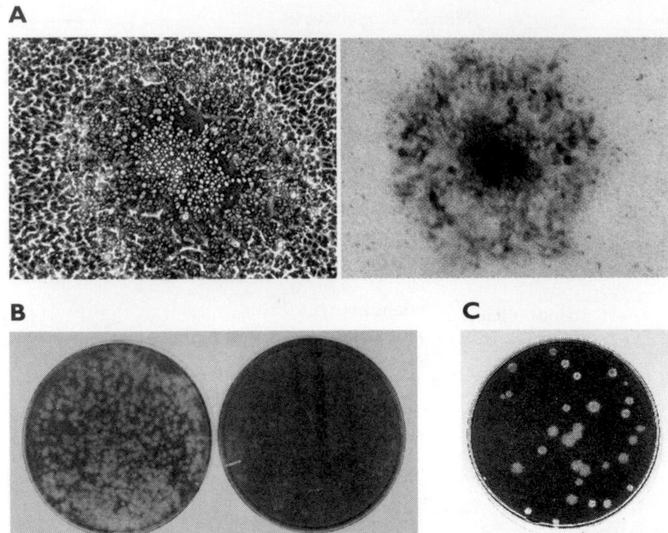
A
B
C

Figure 2.4 Plaques formed by different animal viruses. Plaque sizes reflect the life cycle of a virus in a particular cell type. (A) Photomicrograph of a single plaque formed by pseudorabies virus in Georgia bovine kidney cells. (Left) Unstained cells. Detached cells and syncytia are evident. (Right) Cells were stained with the chromogenic substrate X-Gal. The virus used carries the *lacZ* gene. Its product converts X-Gal to a blue compound. Infected and uninfected cells can be distinguished clearly. Courtesy of B. Banfield, Princeton University. (B) Different plaque morphology of influenza C virus strains. Monolayers were stained with crystal violet. One virus produces small, clear plaques (left), and the other produces small, turbid plaques that are difficult to see (right). (C) Plaques formed by poliovirus on human HeLa cells stained with crystal violet. Note the well-defined plaques.

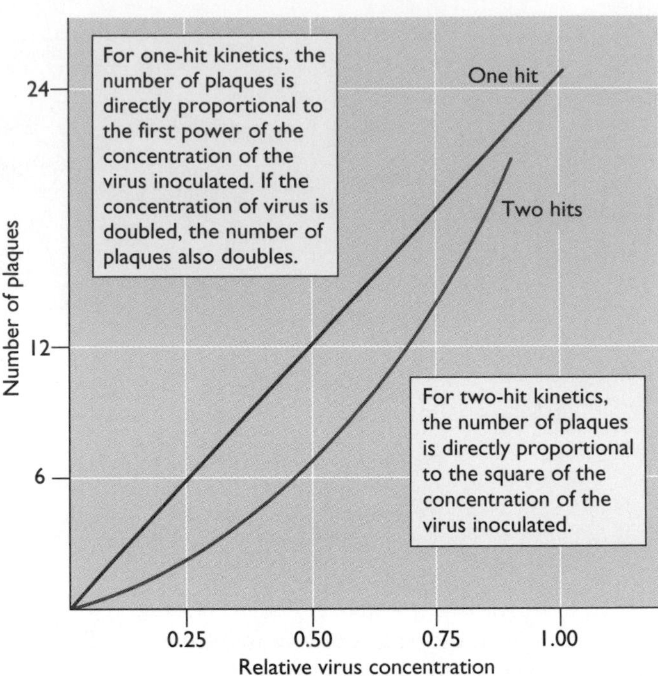
For one-hit kinetics, the number of plaques is directly proportional to the first power of the concentration of the virus inoculated. If the concentration of virus is doubled, the number of plaques also doubles.
One hit
Two hits
For two-hit kinetics, the number of plaques is directly proportional to the square of the concentration of the virus inoculated.
Number of plaques
24
12
6
0.25
0.50
0.75
1.00
Relative virus concentration

Figure 2.5 The dose-response curve of the plaque assay. The number of plaques produced by a virus with one-hit kinetics (red) or two-hit kinetics (blue) is plotted versus the relative concentration of the virus. In two-hit kinetics, there are two classes of uninfected cells, those receiving one particle and those receiving none. The Poisson distribution can be used to determine the proportion of cells in each class: they are e^{-m} and me^{-m} (Box 2.4). Because one particle is not sufficient for infection, $P(0) = e^{-m}(1 + m)$. At a very low multiplicity of infection, this equation becomes $P(i) = (1/2)m^2$, which gives a parabolic curve. Adapted from B. D. Davis et al., *Microbiology* (J. B. Lippincott Co., Philadelphia, Pa., 1980), with permission.

naked eye (Fig. 2.4B and C). Only viruses that cause visible damage of cultured cells can be assayed in this way.

In some cases, viral plaques can be seen without further manipulation of the monolayer. To enhance the contrast between the plaque and the surrounding monolayer, the cells may be stained with a vital dye, neutral red or crystal violet. Living cells absorb the stain, and plaques appear clear against a red (neutral red) or purple (crystal violet) background of healthy cells (Fig. 2.4B and C). When crystal violet is used, the agar overlay is first removed. However, neutral red staining, which is not toxic and can be done through the agar, allows the recovery of infectious virus from a plaque (see below).

For the majority of animal viruses, there is a linear relationship between the number of infectious virus particles and the plaque count (Fig. 2.5). One infectious particle is therefore sufficient to initiate infection, and the virus is said to infect cells with **one-hit kinetics**. Some examples of **two-hit kinetics**, in which two different types of virus particle must infect a cell to ensure replication, have been recognized. For example, the genomes of some (+) strand RNA viruses of plants consist of two RNA molecules that are separately encapsidated. Both RNAs are required for infectivity. The dose-response curve in plaque assays for these viruses is parabolic rather than linear (Fig. 2.5).

The titer of a virus stock can be calculated in **plaque-forming units (PFU) per milliliter**, from the dilution of the sample and the number of plaques observed. To minimize error in calculating the virus titer, a series of dilutions is used for the plaque assay, and only those plates containing between 20 and 100 plaques are counted, depending on the area of the cell culture vessel. According to statistical principles, when 100 plaques are counted, the sample titer varies ±10%. Plates with more than 100 plaques are generally not counted because the plaques may overlap, causing inaccuracies.

When one infectious virus particle initiates a plaque, the viral progeny within the plaque are clones, and virus stocks prepared from a single plaque are known as **plaque purified**. The tip of a small pipette is plunged into the overlay above the plaque, and the plug of agar containing the virus is recovered. The virus within the agar plug is eluted into buffer and used to prepare virus stocks. To ensure purity, this process is usually repeated at least one more time. Plaque purification is employed widely in virology to establish clonal virus stocks.

Fluorescent-Focus Assay

The fluorescent-focus assay, a modification of the plaque assay, is useful in determining the titers of viruses that do not kill cells. The initial procedure is the same as in the plaque assay. However, after a period sufficient for adsorption and gene expression, cells are permeabilized with methanol or acetone and incubated with an antibody raised against a viral protein. A second antibody that recognizes the first is then added. This second antibody is usually conjugated to an indicator, such as fluorescein. The cells are then examined under a microscope with ultraviolet illumination; foci of infected cells fluoresce against the dark background. The titer of the virus stock is expressed in fluorescent-focus-forming units per milliliter.

Infectious Center Assay

Another modification of the plaque assay, the infectious center assay, is used to determine the fraction of cells in a culture that are infected with a virus. Monolayers of infected cells are suspended before progeny viruses are produced. Dilutions of a known number of infected cells are then plated on monolayers of susceptible cells, which are covered with an agar overlay. The number of plaques that form on the indicator cells is a measure of the number of cells infected in the original population. The fraction of infected cells can, therefore, be determined. A typical use of the infectious center assay is to measure the proportion of infected cells in persistently infected cultures.

Transformation Assay

The transformation assay is useful for determining titers of some retroviruses that do not form plaques. For example, Rous sarcoma virus transforms chick embryo cells. As a result, the cells lose their contact inhibition (the property that governs whether cultured cells grow as a single monolayer [see Chapter 18]) and become heaped upon one another. The transformed cells form small piles, or **foci**, that can be distinguished easily from the rest of the monolayer (Fig. 2.6). Infectivity is expressed in focus-forming units per milliliter.

Endpoint Dilution Assay

The endpoint dilution assay provided a measure of virus titer before the development of the plaque assay. It is still used for certain viruses that do not form plaques or for determining the virulence of a virus in animals. Serial dilutions of a virus stock are inoculated into replicate test units (typically 8 to 10), which can be cell cultures, eggs, or animals. The number of test units that have become infected is then determined for each virus dilution. When cell culture is used, infection can be gauged by the devel-

Figure 2.6 Transformation assay. Chicken cells transformed by two different strains of Rous sarcoma virus. Loss of contact inhibition causes cells to pile up, rather than grow as a monolayer. One focus is seen in panel A, and three foci are seen in panel B at the same magnification. Courtesy of H. Hanafusa, Rockefeller University.

A

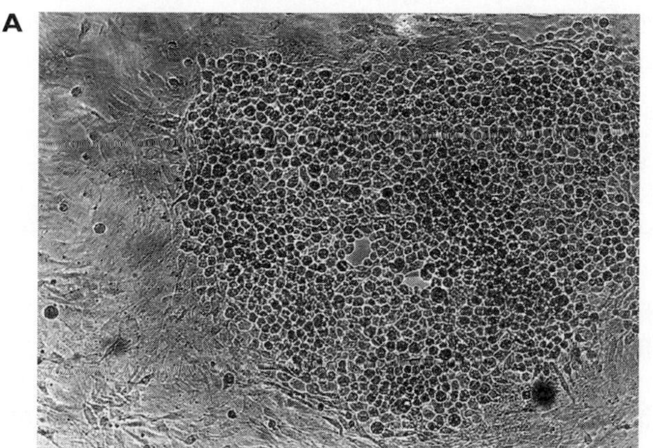

B

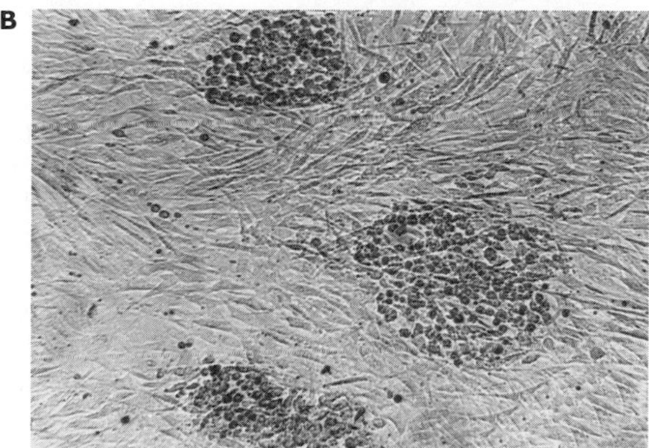

BOX 2.3

Endpoint dilution assays

In the first example, 10 monolayer cell cultures were infected with each virus dilution. After the incubation period, plates that displayed cytopathic effect were scored +. Fifty percent of the cell cultures displayed cytopathic effect at the 10^{-5} dilution, and therefore the virus stock contains 10^5 ID_{50} units.

Virus dilution	Cytopathic effect									
10^{-2}	+	+	+	+	+	+	+	+	+	+
10^{-3}	+	+	+	+	+	+	+	+	+	+
10^{-4}	+	+	–	+	+	+	+	+	+	+
10^{-5}	–	+	+	–	+	–	–	+	–	+
10^{-6}	–	–	–	–	–	–	+	–	–	–
10^{-7}	–	–	–	–	–	–	–	–	–	–

In most cases the 50% endpoint does not fall on a dilution tested as shown in the example; for this reason, various statistical procedures have been developed to calculate the endpoint of the titration. In one popular method, the dilution containing the ID_{50} is identified by interpolation between two dilutions on either side of this value. The assumption is made that the location of the 50% endpoint varies linearly with the log of the dilution. Because the number of test units used at each dilution is usually small, the accuracy of this method is relatively low. For example, if six test units are used at each 10-fold dilution, differences in virus titer of only 50-fold or more can be reliably detected. The method is illustrated in the following example, in which the lethality of poliovirus in mice is the endpoint. Eight mice were inoculated per dilution.

Dilution	Alive	Dead	Total alive	Total dead	Mortality ratio	Mortality (%)
1×10^{-2}	0	8	0	40	0/40	100
1×10^{-3}	0	8	0	32	0/32	100
1×10^{-4}	1	7	1	24	1/25	96
1×10^{-5}	0	8	1	17	1/18	94
1×10^{-6}	2	6	3	9	3/12	75
1×10^{-7}	5	3	8	3	8/11	27

In the method of Reed and Muench, the results are pooled, as shown in the table. The interpolated value of the 50% endpoint, known as *I*, which in this case falls between the fifth and sixth dilutions, is calculated by the following formula:

$$I = h\left(\frac{\text{\% of animals affected at dilution above 50\%} - 50\%}{\text{\% of animals affected at dilution above 50\%} - \text{\% of animals affected at dilution below 50\%}}\right)$$

where *h* is the log of the dilution factor, in this case –1. For this example *I* is –0.5. The endpoint is then calculated as follows:

$$\text{50\% endpoint titer} = 10^{\text{log total dilution above 50\%} - (I \times h)}$$

which in our example is $10^{-6-(-0.5 \times -1)} = 10^{-6.5}$. The virus sample therefore contains $10^{6.5}$ LD_{50} doses. The LD_{50} may also be calculated as the concentration of the stock virus in PFU per milliliter (1×10^9) times the 50% endpoint titer. In the example shown, the LD_{50} is 3×10^2 PFU.

opment of cytopathic effect; in eggs or animals, infection is gauged by death or disease. An example of an endpoint dilution assay using cell cultures is shown in Box 2.3. At high dilutions, none of the cell cultures are infected because no infectious particles are delivered to the cells; at low dilutions, every culture is infected. The endpoint is the dilution of virus that affects 50% of the test units. This number can be calculated from the data and expressed as 50% infectious dose (ID_{50}) per milliliter. The first preparation illustrated in Box 2.3 contains 10^5 ID_{50} per ml.

When the endpoint dilution assay is used to assess the virulence of a virus or its capacity to cause disease, as defined in Chapter 14, the result of the assay can be expressed in terms of 50% lethal dose (LD_{50}) per milliliter or 50% paralytic dose (PD_{50}), endpoints of death and paralysis, respectively. If the virus titer can be determined separately by plaque assay, the 50% endpoint determined in an animal host can be related to this parameter. In this way, the effects of route of inoculation or specific mutations on viral virulence can be quantified.

Efficiency of Plating

The term **relative efficiency of plating** was coined in studies of bacteriophages to assign a value to the plaque count determined with the same bacteriophage for different strains of bacteria. The value is a ratio of bacteriophage titers obtained on two different bacterial hosts. This number may be more or less than 1, depending on how well the bacteriophage grows in the different hosts. A very different value is the **absolute efficiency of plating**, which is defined as the plaque titer divided by the number of virus particles in the sample. The **particle-to-PFU ratio,** a term more commonly used today, is the inverse value (Table 2.2). For many bacteriophages, the particle-to-PFU ratio approaches 1, the lowest value that can be obtained. However, for animal viruses this value can be much higher, ranging from 1 to 10,000. These high values have complicated the study of animal viruses. For example, when the particle-to-PFU ratio is high, it is never certain whether properties measured biochemically are in fact those of the infectious particle or those of the noninfectious component.

Table 2.2 Particle-to-plaque-forming-unit ratios of some animal viruses

Virus	Particle-to-PFU ratio
Adenoviridae	20–100
Alphaviridae	
Semliki Forest virus	1–2
Herpesviridae	
Herpes simplex virus	50–200
Orthomyxoviridae	
Influenza virus	20–50
Papillomaviridae	
Papillomavirus	10,000
Picornaviridae	
Poliovirus	30–1,000
Polyomaviridae	
Polyomavirus	38–50
Simian virus 40	100–200
Poxviridae	1–100
Reoviridae	
Reovirus	10

Although the linear nature of the dose-response curve indicates that a single particle is capable of initiating an infection (one-hit kinetics) (Fig. 2.5), the high particle-to-PFU ratio of many viruses demonstrates that not all virions are successful. The high particle-to-PFU ratio of animal viruses may sometimes be caused by the presence of noninfectious particles with genomes that harbor lethal mutations or which have been damaged during growth or purification. An alternative explanation is that all viruses in a preparation are in fact capable of initiating infection, but that all do not succeed because of the complexity of the infectious cycle. Failure at any one step in the cycle prevents completion.

Measurement of Virus Particles and Their Components

Although the numbers of virus particles and infectious units are often not equal, assays for particle number are frequently used to approximate the quantity of virus present in a sample. Physical assays are usually more rapid and easier to carry out than assays for infectivity, which may be slow, cumbersome, or not available. Assays for subviral components also provide information on particle number if the stoichiometry of these components in the virus particle is known.

Electron Microscopy

With few exceptions, virus particles are too small to be observed by light microscopy. However, they can be seen readily in the electron microscope. If a sample contains only one type of virus, the particle count can be determined. First, a virus preparation is mixed with a known concentration of latex beads. The numbers of virus particles and beads are then counted, allowing the concentration of the virus in the sample to be determined by comparison.

Hemagglutination

Members of the *Adenoviridae*, *Orthomyxoviridae*, and *Paramyxoviridae*, among others, contain proteins that can bind to erythrocytes (red blood cells); these viruses can link multiple cells, resulting in a lattice. This property is called **hemagglutination**. For example, influenza viruses contain an envelope glycoprotein called hemagglutinin, which binds to *N*-acetylneuraminic acid-containing glycoproteins on erythrocytes. In practice, twofold serial dilutions of the virus stock are prepared, mixed with a defined quantity of red blood cells, and added to the small wells of a plastic tray (Fig. 2.7). Unadsorbed red blood cells tumble to the bottom of the well and form a sharp dot or button. In contrast, agglutinated red blood cells form a diffuse lattice that coats the well. Because the assay is rapid (30 min), it is often used as a quick indicator for the relative

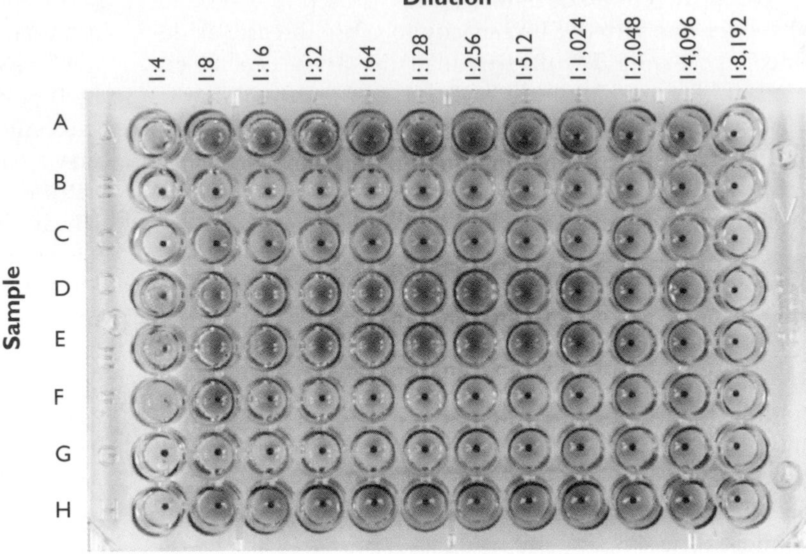

Figure 2.7 Hemagglutination assay. Samples of different influenza viruses were diluted, and a portion of each dilution was mixed with a suspension of chicken red blood cells and added to the wells. After 30 min at 4°C, the wells were photographed. Sample A causes hemagglutination until a dilution of 1:256 and therefore has a hemagglutination titer of 256. Many viruses that hemagglutinate red blood cells also contain a neuraminidase in the virus particle. The presence of this enzyme, which cleaves *N*-acetylneuraminic acid from glycoprotein receptors, elutes bound viruses from red blood cells. As a result, the lattice formed by hemagglutination is reversed. An example of elution of the virus from red blood cells can be seen in column 1, rows D and E. To minimize elution, hemagglutination assays should not be incubated for more than 30 min at 4°C. Courtesy of C. Basler and P. Palese, Mount Sinai School of Medicine of the City University of New York.

quantities of virus particles. However, it is not sufficiently sensitive to determine small numbers of particles.

Measurement of Viral Enzyme Activity

Some animal virus particles contain nucleic acid polymerases, which can be assayed by mixing permeabilized particles with radioactively labeled precursors and measuring the incorporation of radioactivity into nucleic acid. Assays for nucleic acid polymerase are most frequently used for retroviruses, many of which do not transform cells or form plaques. In one example, the reverse transcriptase incorporated into the virus particle is assayed by mixing cell culture supernatants with a mild detergent (to permeabilize the viral envelope), a poly(rC) template, an oligo(dG) primer, and [α-^{32}P]dGTP. If reverse transcriptase is present, a radioactive product will be produced by priming on the poly(rC) template. This product can be detected by precipitation or bound to a filter and quantified. Because enzymatic activity is proportional to particle number, the assay allows rapid tracking of virus production in the course of an infection.

Serological Methods

Many virological techniques are based on the specificity of the antibody-antigen reaction. Some of the techniques, such as immunostaining, immunoprecipitation, immunoblotting, and enzyme-linked immunosorbent assay, are by no means limited to the detection of viruses and viral proteins. All these approaches have been used extensively to study the structures and functions of cellular proteins.

Virus neutralization. When a virus preparation is inoculated into an animal, an array of antibodies is produced. These antibodies can bind to virus particles, but not all of them can block their infectivity (**neutralize**), as discussed in Chapter 15. Virus neutralization assays are usually conducted by mixing dilutions of antibodies with virus, incubating them, and assaying for remaining infectivity in cultured cells, eggs, or animals. The endpoint is defined as the highest dilution of antibody that inhibits the development of cytopathic effect in cells or virus replication in eggs or animals.

Some neutralizing antibodies define **type-specific antigens** on the virus particle. For example, the three **serotypes** of poliovirus are distinguished on the basis of neutralization tests; type 1 poliovirus is neutralized by antibodies to type 1 virus but not by antibodies to type 2 or type 3 poliovirus, and so forth. Neutralization tests have therefore been valuable for virus classification.

Knowledge of the antigenic structure of a virus is useful in understanding the immune response to these agents and in designing new vaccination strategies. The use of **monoclonal antibodies** (antibodies of a single specificity made by a clone of antibody-producing cells) in neutralization assays permits mapping of antigenic sites on a virus particle or of the amino acid sequences that are recognized by neutralizing antibodies. Each monoclonal antibody binds specifically to a short amino acid sequence (8 to 12 residues) that fits into the antibody-combining site. This amino acid sequence, which may be linear or nonlinear, is known as an **epitope**. In contrast, **polyclonal antibodies** comprise the repertoire produced in an animal against the many epitopes of an antigen. Antigenic sites may be identified by cross-linking the monoclonal antibody to the virus and determining which protein is the target of the antibody. The abilities of monoclonal antibodies to bind synthetic peptides representing viral protein sequences may also be assessed. When the monoclonal antibody recognizes a linear epitope, it may react with the protein in Western blot analysis (see

"Immunoprecipitation and immunoblotting" below), facilitating direct identification of the viral protein harboring the antigenic site. The most elegant understanding of antigenic structures has come from the isolation and study of variant viruses that are resistant to neutralization with specific monoclonal antibodies (called **monoclonal antibody-resistant** or ***mar*** **variants**). By identifying the amino acid change responsible for the *mar* phenotype, the antibody-binding site can be located and, together with three-dimensional structural information, can provide detailed information on the nature of antigenic sites that are recognized by neutralizing antibodies.

Hemagglutination inhibition. Antibodies against viral proteins with hemagglutination activity can block the ability of virus to bind red blood cells. In this assay, dilutions of antibodies are incubated with virus, and erythrocytes are added as outlined above. After incubation, the hemagglutination inhibition titer is read as the highest dilution of antibody that inhibits hemagglutination. This test is sensitive, simple, inexpensive, and rapid, and is the method of choice for assaying antibodies to any virus that causes hemagglutination. This assay can be used to detect antibodies to viral hemagglutinin in animal and human sera, or to identify the origin of the hemagglutinin of influenza viruses produced in cells coinfected with two parent viruses (see "Mapping Mutations" below).

Complement fixation. The complement fixation assay can be used to determine if antibodies against a virus are present in serum. The interaction of viral antigen and antibody can cause complement fixation, which leads to membrane lysis, as discussed in Chapter 15. Red blood cells are used as targets, because lysis of their membranes is readily observed. In the assay for complement-fixing antibodies, samples of patients' sera are heated to inactivate endogenous complement, and then incubated with preparations of viral antigen and a standardized quantity of guinea pig complement. If an antibody-antigen reaction takes place, complement fixation will occur. The latter is detected by adding sheep red blood cells that have been coated with rabbit anti-red blood cell antibodies. If complement has been fixed (complexed) by binding of viral antigen to antibody, the red blood cells will remain intact; if complement is not fixed, the red blood cells will be lysed by the action of free complement.

Because crude preparations of viral antigens are often used for the complement fixation test, it is not highly specific. The antigen preparations may include proteins or specific epitopes shared by related groups of viruses. As a result, the test detects most serotypes within a given group of viruses and is therefore said to recognize group-specific antigens. For example, the three serotypes of poliovirus share a complement-fixing, group-specific antigen. Complement fixation is not as sensitive as neutralization or hemagglutination inhibition. Nevertheless, it is often the first assay performed on sera from infected patients to identify the group to which the infecting virus belongs.

Protein detection. Antibodies can be used to visualize viral proteins in infected cells or tissues. In **direct immunostaining**, an antibody that recognizes a viral antigen is coupled directly to an indicator such as a fluorescent dye or an enzyme (Fig. 2.8). A more sensitive approach is

Figure 2.8 Direct and indirect methods for antigen detection. (A) The sample (tissue section, smear, or bound to a solid phase) is incubated with a virus-specific antibody (Ab). In direct immunostaining, the antibody is linked to an indicator such as fluorescein. In indirect immunostaining, a second antibody which recognizes a general epitope on the virus-specific antibody is coupled to the indicator. MAb, monoclonal antibody. (B) Use of immunofluorescence to demonstrate the nuclear location of herpes simplex virus protein VP22. Virus-infected cells were stained with a rabbit polyclonal antibody against VP22, and a mouse monoclonal antibody against α-tubulin, followed by fluorescein isothiocyanate-conjugated anti-rabbit immunoglobulin and Texas Red-conjugated anti-mouse IgG. VP22 (green) is located in the cell nucleus. Viral infection leads to rearrangement of microtubules (red), with the loss of microtubule organizing centers. Adapted from L. E. Pomeranz and J. A. Blaho, *J. Virol.* **73**:6769–6781, 1999, with permission. Courtesy of J. A. Blaho, Mount Sinai School of Medicine of the City University of New York.

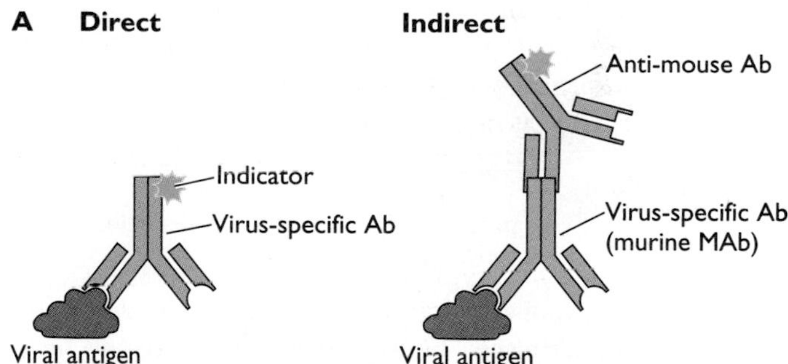

B

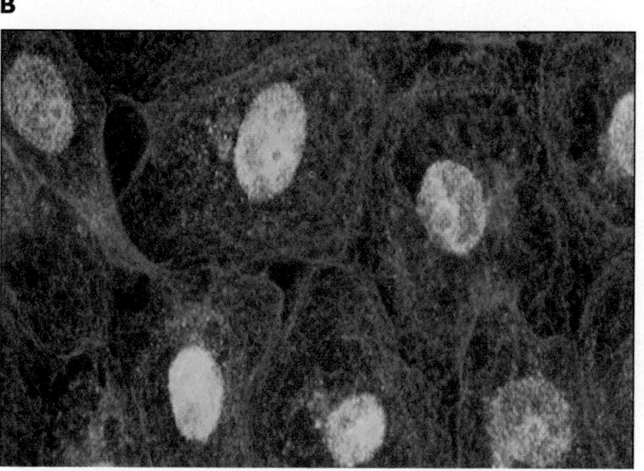

indirect immunostaining, in which a second antibody is coupled to the indicator. The second antibody recognizes a common epitope on the virus-specific antibody, for example, an anti-mouse immunoglobulin antibody prepared in goats. Multiple second-antibody molecules bind to the first antibody, resulting in an increased signal from the indicator compared with that obtained with direct immunostaining. Furthermore, a single indicator-coupled second antibody can be used in many assays, avoiding the need to purify and couple an indicator to multiple first antibodies.

In practice, virus-infected cells are fixed with acetone, methanol, or paraformaldehyde, and incubated with polyclonal or monoclonal antibodies directed against viral antigen. Excess antibody is washed away, and in direct immunostaining, cells are examined for the indicator by microscopy. For indirect immunostaining, the second antibody is added before examination of the cells by microscopy. Commonly used indicators include fluorescein and rhodamine, which fluoresce on exposure to ultraviolet light. Filters are placed between the specimen and the eyepiece to remove blue and ultraviolet light so that the field is dark, except for cells to which the antibody has bound, which emit green (fluorescein) or red (rhodamine) light (Fig. 2.8). Antibodies can also be coupled to molecules other than fluorescent indicators, including enzymes such as alkaline phosphatase, horseradish peroxidase, and β-galactosidase, a bacterial enzyme that in a test system converts the chromogenic substrate X-Gal (5-bromo-4-chloro-3-indolyl-β-D-galactopyranoside) to a blue product. After excess antibody is washed away, a suitable chromogenic substrate is added, and the presence of the indicator antibody is revealed by development of a color that can be visualized.

Immunostaining has been applied widely in the research laboratory for determining the subcellular localization of proteins in cells (Fig. 2.8), monitoring the synthesis of viral proteins, determining the effects of mutation on protein production, and investigating the sites of virus replication in animal hosts. It is the basis of the fluorescent-focus assay.

Immunostaining of viral antigens in smears of clinical specimens is also used to diagnose viral infections. For example, direct and indirect immunofluorescence assays with nasal swabs or washes are routine for diagnosis of infections with respiratory syncytial virus, influenza virus, parainfluenza virus, measles virus, and adenovirus.

Immunoprecipitation and immunoblotting. Immunoprecipitation depends on the interaction of specific antiviral antibodies with viral proteins in solubilized extracts of infected cells or tissues (Fig. 2.9). The antibody-protein complexes are isolated, and the viral proteins, which can be metabolically labeled with a radioactive amino acid, are dissociated from the complex and fractionated by electrophoresis in polyacrylamide gels. Immunoprecipitation can establish many properties of viral proteins, including the types present in virions and infected cells, molecular mass, kinetics of synthesis, subcellular location, and whether they associate with other viral or cellular proteins.

Immunoblotting (also known as Western blot analysis) depends on the reaction of an antibody with a viral

Figure 2.9 Visualization of proteins by immunoprecipitation. In the example shown, cell proteins are labeled during translation with radioactive amino acids. Cells are lysed with a detergent to solubilize the proteins. Antibodies that have been coupled to beads are added to the cell lysate. The beads are removed by centrifugation and washed free of proteins not bound by the antibody. The bound proteins can then be fractionated by gel electrophoresis and visualized by autoradiography. The numbers next to the gel on the right are molecular masses in kilodaltons. SDS-PAGE, sodium dodecyl sulfate-polyacrylamide gel electrophoresis.

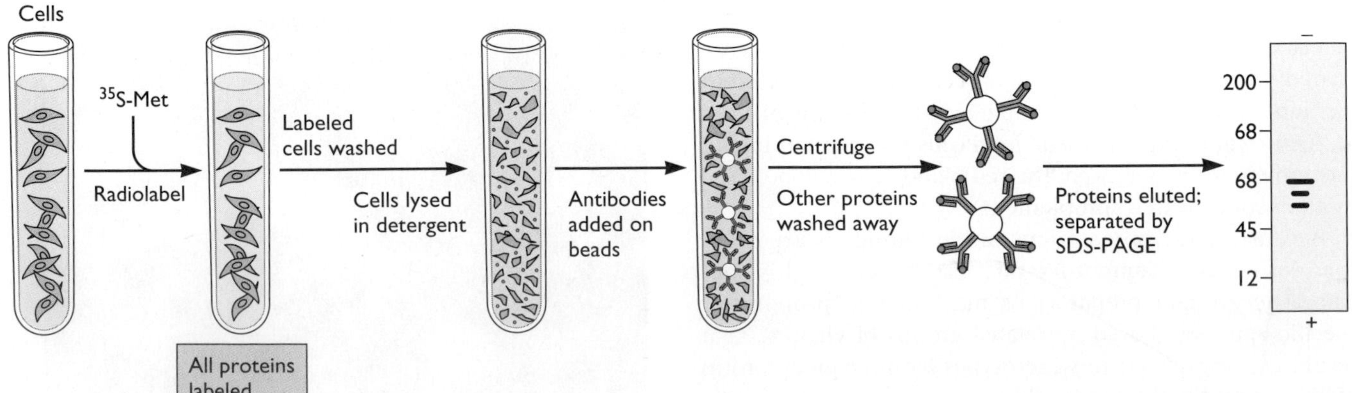

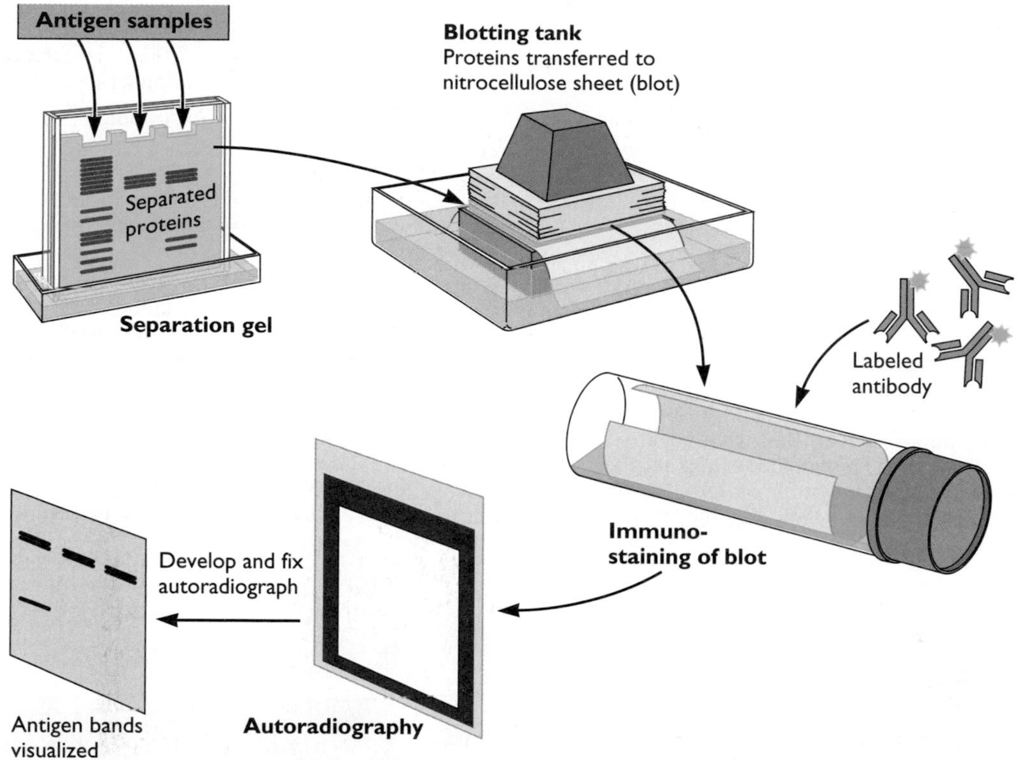

Figure 2.10 Visualization of proteins by immunoblotting (Western blot analysis). The sample, which can be cells, tissues, virus, or partially purified protein, is solubilized with detergent. Proteins are fractionated by electrophoresis in a polyacrylamide gel. A thin membrane is placed over the gel, and the fractionated proteins are transferred. In the illustration, protein transfer is accomplished by capillary action in a blotting tank; transfer may also be achieved by an electric current. The membrane, to which the proteins are tightly bound, is incubated with a specific antibody coupled to an enzyme such as horseradish peroxidase. Bound antibody is detected by the addition of a substrate that is converted to a luminescent compound after reaction with the enzyme, followed by autoradiography.

protein but in a different format, immobilized on a thin membrane (Fig. 2.10). Proteins are fractionated by electrophoresis in a polyacrylamide gel and then transferred to a thin, synthetic membrane that has a strong affinity for proteins. Proteins bound to the membrane are detected by immunostaining. Because proteins bound to the membrane are denatured, antibodies that recognize nonlinear determinants are usually not suitable for their detection. The main advantage of this technique is that it does not require labeling of proteins and therefore can be applied to tissues and organs as well as cultured cells. However, this assay does not provide information about the kinetics of protein synthesis. When used alone, immunoblotting cannot reveal whether a viral protein associates with other proteins. However, such information can be obtained if proteins are first isolated by immunoprecipitation and then subjected to immunoblotting with different antibodies capable of identifying the putative interacting protein(s).

A variation of immunoblotting is used in the clinical laboratory to identify antibodies to human immunodeficiency virus in clinical specimens and in donated blood. The sample is incubated with membranes to which viral proteins separated by gel electrophoresis have been transferred. If antibodies against human immunodeficiency virus are present, one or more of the viral proteins on the membrane will be stained.

Enzyme-linked immunosorbent assay (ELISA). Detection of viral antigen or antibody can be accomplished by solid-phase methods, in which viral antibody or protein is adsorbed to a plastic surface. To detect viral antigens in serum or clinical samples, a "capture" antibody, directed against the virus, is linked to a solid support, a plastic dish

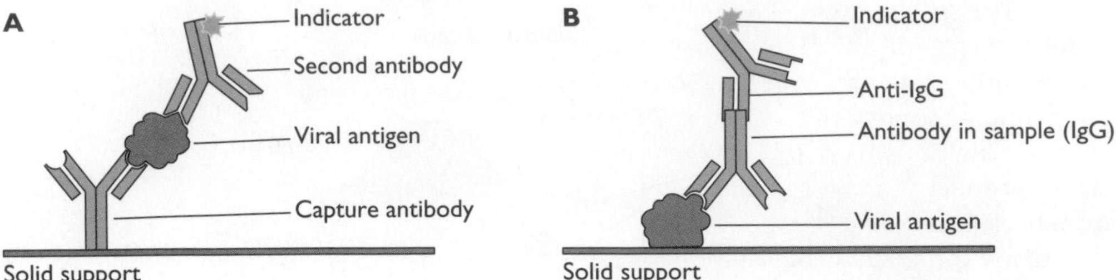

Figure 2.11 Detection of viral antigen or antibodies against viruses by ELISA. (A) To detect viral proteins in a sample, antibodies specific for the virus are immobilized on a solid support such as a plastic well. The sample is placed in the well, and viral proteins are "captured" by the immobilized antibody. After washing to remove unbound proteins, a second antibody against the virus is added, which is linked to an indicator. Another wash is done to remove unbound second antibody. If viral antigen has been captured by the first antibody, the second antibody will bind, and the complex will be detected by the indicator. (B) To detect antibodies to a virus in a sample, viral antigen is immobilized on a solid support. The sample is placed in the well, and viral antibodies bind the immobilized antigen. After washing to remove unbound antibodies, a second antibody, directed against a general epitope on the first antibody, is added. Another wash removes unbound second antibody. If viral antibodies are bound by the immobilized antigen, the second antibody will bind, and the complex will be detected by the indicator. IgG, immunoglobulin G.

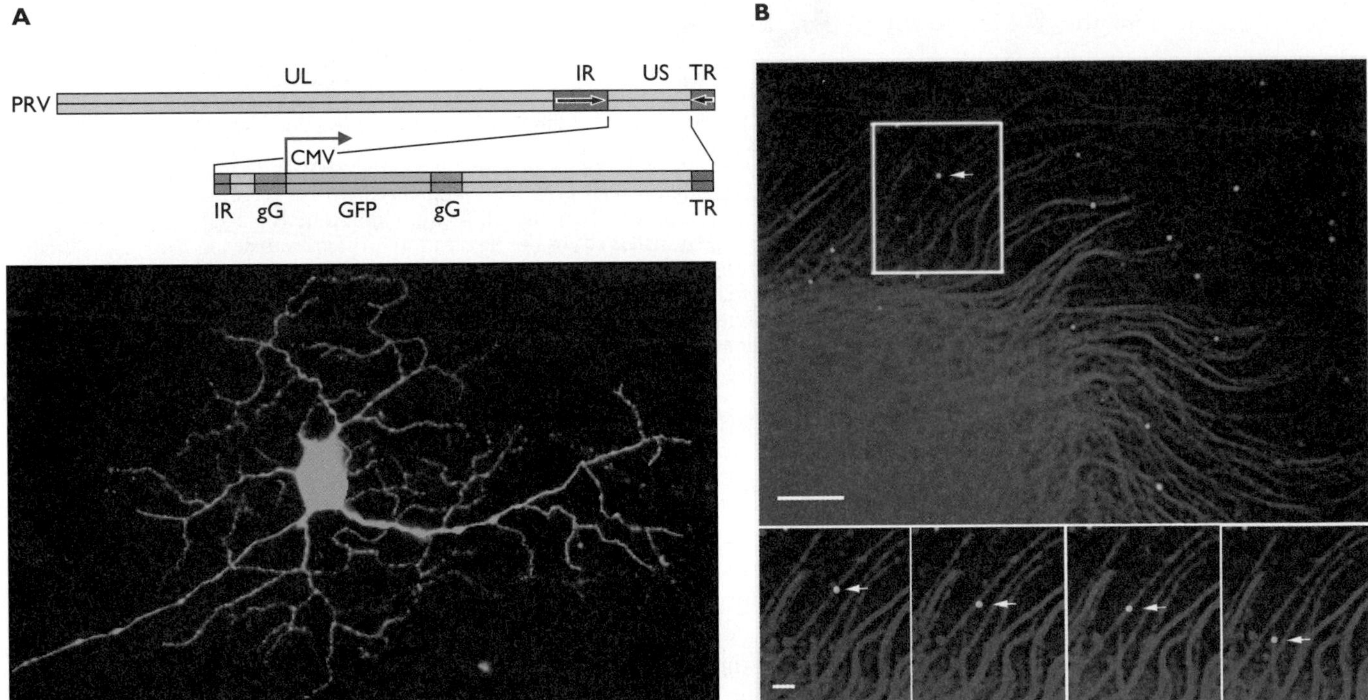

Figure 2.12 The use of green fluorescent protein. (A) Imaging of neurons. (Top) A diagram of the pseudorabies virus (PRV) DNA genome, with the right end expanded to show the disruption of a viral gene with DNA encoding green fluorescent protein (GFP). In infected cells, transcription of the green fluorescent protein gene is directed by a cytomegalovirus promoter (CMV). IR, inverted repeat; TR, terminal repeat; UL, unique sequences of long region; US, unique sequences of short region; gG, glycoprotein G. (Bottom) Retinal ganglion cells from hamsters infected intraocularly with the pseudorabies virus described above. After infection, the retina was removed and viewed under a microscope with ultraviolet illumination. Green fluorescent protein has diffused throughout each infected neuron. Adapted from G. E. Pickard et al., *J. Neurosci.* **22:**2701–2710, 2002. (B) Single-virus-particle imaging with green fluorescent protein illustrates microtubule-dependent movement of human immunodeficiency virus type 1 particles in cells. Rhodamine-tubulin was injected into cells to label microtubules (red). The cells were infected with virus particles that contain a fusion of green fluorescent protein with Vpr. Virus particles can be seen as green dots. Upper image shows the cell at the beginning time point. The particle in the box is shown below at 30-s intervals moving along the microtubule network. Bars, 5 μm (top) and 2 μm (bottom). Courtesy of David McDonald, University of Illinois.

or bead (Fig. 2.11A). The specimen is added to the plastic support, and if viral antigens are present, they will be captured by the bound antibody. Bound viral antigen is detected by using a second antibody linked to an enzyme. A chromogenic molecule that is converted by the enzyme to an easily detectable product is then added. The enzyme amplifies the signal because a single catalytic enzyme molecule can generate many product molecules. To detect immunoglobulin G (IgG) antibodies to viruses, viral antigen is first linked to the plastic support, and then the specimen is added (Fig. 2.11B). If antibodies against the virus are present in the specimen, they will bind to the immobilized antibody. The bound antibodies are then detected by using a second antibody directed against a common epitope on the first antibody. Like other detection methods, ELISAs are used in both diagnostic and experimental virology.

Green Fluorescent Protein

The discovery of green fluorescent protein revolutionized the study of the cell biology of virus infection. This protein, isolated from the jellyfish *Aequorea victoria*, is a convenient reporter for monitoring transcription or translation, because it is directly visible in living cells without the need for fixation, substrates, or coenzymes. In one approach, the coding sequence for green fluorescent protein is inserted into the viral genome. Upon infection of cells with the recombinant virus, green fluorescent protein is produced. This approach has been used to trace neuronal connections using herpesviruses (Fig. 2.12A). Many viral proteins retain their functions when fused with green fluorescent protein, permitting studies of subcellular location. The use of green fluorescent protein to track movement of individual virus particles in cells is also gaining momentum (Fig. 2.12B). Using this approach, entry, uncoating, replication, assembly, and egress of single particles can all theoretically be observed in living cells.

Nucleic Acid Detection

The development of methods to detect viral nucleic acids among the vast excess of cellular sequences was a watershed in the study of viruses. In **Southern blot hybridization** (developed by Edward Southern), DNA preparations are digested with a restriction endonuclease, fractionated by gel electrophoresis, denatured within the gel, and transferred to a membrane to which they bind strongly. The bound DNAs are then detected by hybridization, originally with an isotopically labeled, viral nucleic acid sequence (Fig. 2.13). However, radioactive nucleic acids are now used less frequently because of the short half-lives and the potential hazards posed by particle emissions. Other methods for labeling nucleic acids have been developed. In one version, the nucleic acid hybridization probes contain biotin. The biotinylated nucleic acids are detected by incubation with streptavidin, which binds with high affinity to biotin and which has been linked to an indicator such as β-galactosidase. In enhanced chemiluminescence, nucleic acid hybridization probes are linked to perioxidase. This enzyme reduces hydrogen peroxide to O_2^-, which in turn oxidizes luminol, producing a blue light.

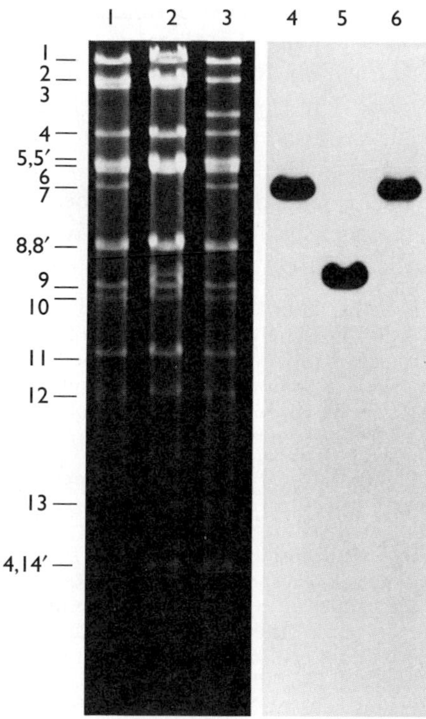

Figure 2.13 Southern blot hybridization. DNAs from wild-type pseudorabies virus (lanes 1 and 4), and from mutants deficient in glycoproteins gE/I (lanes 2 and 5) or gM (lanes 3 and 6), were cleaved with *Bam*HI and fractionated on agarose gels. Lanes 1 to 3, ethidium bromide-stained gel showing the DNA fragments. The locations of the *Bam*HI fragments of pseudorabies virus wild-type DNA are on the left. Lanes 4 to 6, DNAs transferred to a membrane filter that was then hybridized with ^{32}P-labeled *Bam*HI fragment 7 produced from wild-type viral DNA. The DNA probe hybridizes with only one DNA fragment in each lane. Adapted from A. R. Brack et al., *J. Virol.* **74:**4004–4016, 2000, with permission.

In one variation of Southern blot analysis, DNAs are concentrated onto a small region of the filter, generally by a vacuum. This method, called slot blot or dot blot analysis, depending on the shape of the manifold wells, provides a rapid means for measuring relative concentrations of different nucleic acids in the same sample or detecting a specific viral nucleic acid in different samples. **Northern blot hybridization** is used to detect RNA in a similar manner. The name of this technique, like that of Western blot analysis, was not derived from the inventor but is a play on Dr. Southern's name. Nucleic acid hybridization probes can also be used to locate viral genomes or transcripts in sections of cells, tissues, or even whole animals by a technique known as **in situ hybridization.**

The **polymerase chain reaction** has also found wide application in the detection of viral nucleic acids. Specific oligonucleotides are used to amplify viral DNA sequences from infected cells or tissues; in situ methods based on the polymerase chain reaction have also been developed. Polymerase chain reaction techniques have also allowed the identification of new viruses associated with human diseases, as described in Chapter 20. Clinical virology laboratories also employ these methods to diagnose viral infections.

DNA microarrays provide an important new method for studying the gene expression profile of a cell in response to virus infection. In this method, hundreds or thousands of unique DNA sequences are fixed on a glass slide in aligned rows (Fig. 2.14A). The resulting slide is called a DNA microarray chip. By using differentially labeled hybridization probes made from uninfected and virus-infected cell messenger RNAs (mRNAs), it is possible to determine whether production of specific mRNAs is induced, repressed, or unchanged after virus infection (Fig. 2.14B). Such information can then be used to identify cellular genes that play a role in viral replication and pathogenesis.

Figure 2.14 DNA microarray technology. (A) Experimental procedure. In this experiment, the effects of virus infection on the expression of cellular RNAs are compared. RNA is extracted from virus-infected or uninfected cells, and complementary DNA (cDNA) is produced by reverse transcription with fluorescently labeled nucleoside triphosphates. cDNAs from virus-infected and uninfected cells are labeled with a red and a green dye, respectively. The cDNA probes are then mixed and hybridized to DNA microarrays, which are glass slides on which have been printed thousands of DNAs of unique sequence, representing individual cellular genes. After hybridization, the slide is exposed to ultraviolet light to produce red or green emission. The red and green images are superimposed. Red dots identify genes whose expression has increased after virus infection, and green dots identify genes whose expression level has decreased. Yellow dots (red + green) represent genes whose expression has not changed after virus infection. (B) Changes in gene expression in mammalian cells after infection with rotavirus. (Left) Each horizontal row represents a single cDNA, and each vertical column represents a single microarray hybridization. Each square represents the ratio of hybridization signal of labeled cDNA prepared from mRNA of virus-infected or mock-infected cells relative to mock-infected cells. Red squares indicate genes whose expression has increased, and green squares denote genes whose expression has decreased. Black squares represent genes whose expression has not changed after virus infection. The scale at the bottom shows colors associated with the amount of increased or decreased gene expression. Columns labeled "control" are control hybridizations (mock infected versus mock infected or virus infected versus virus infected), and the lanes labeled "infection" are experimental hybridizations (virus infected versus mock infected). Genes shown are those for which mRNA levels varied more than twofold in at least two of the three experimental comparisons. (Right) Some regions are amplified to show the name and expression profile of selected genes. Panel B is reprinted from M. A. Cuadras et al., *J. Virol.* **76:**4467–4482, 2002, with permission.

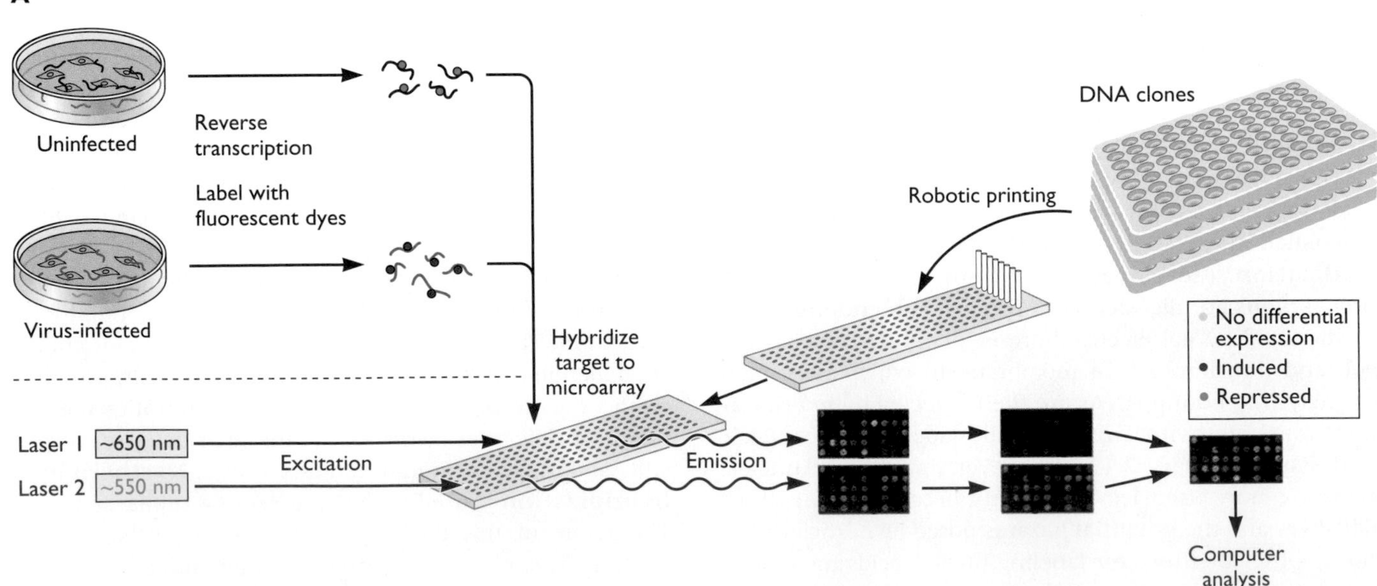

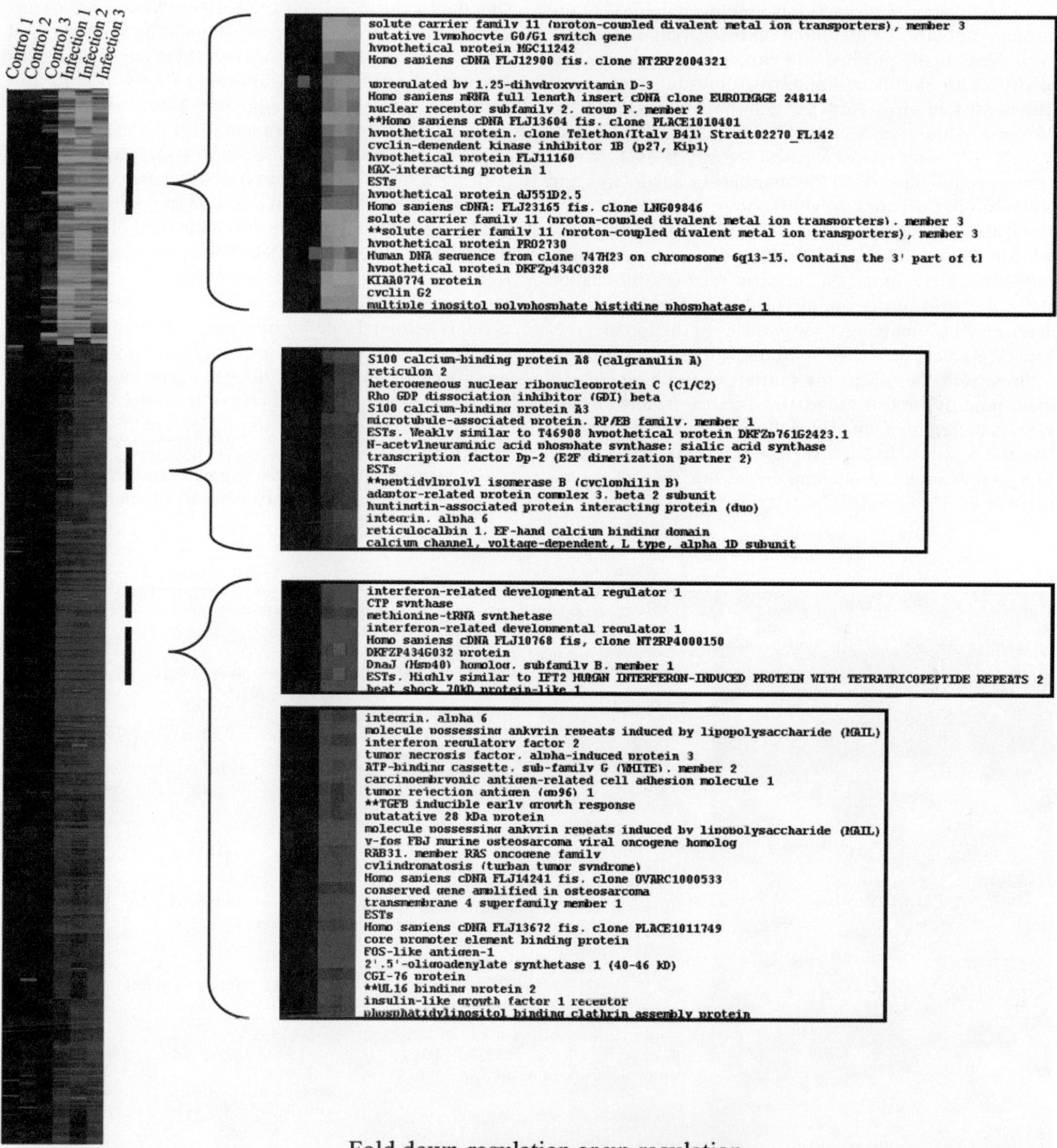
B
Control 1
Control 2
Control 3
Infection 1
Infection 2
Infection 3
solute carrier family 11 (proton-coupled divalent metal ion transporters), member 3
putative lymphocyte G0/G1 switch gene
hypothetical protein MGC11242
Homo sapiens cDNA FLJ12900 fis, clone NT2RP2004321
upregulated by 1,25-dihydroxyvitamin D-3
Homo sapiens mRNA full length insert cDNA clone EUROIMAGE 248114
nuclear receptor subfamily 2, group F, member 2
**Homo sapiens cDNA FLJ13604 fis, clone PLACE1010401
hypothetical protein, clone Telethon(Italy B41) Strait02270_FL142
cyclin-dependent kinase inhibitor 1B (p27, Kip1)
hypothetical protein FLJ11160
MAX-interacting protein 1
ESTs
hypothetical protein dJ551D2.5
Homo sapiens cDNA: FLJ23165 fis, clone LNG09846
solute carrier family 11 (proton-coupled divalent metal ion transporters), member 3
**solute carrier family 11 (proton-coupled divalent metal ion transporters), member 3
hypothetical protein PRO2730
Human DNA sequence from clone 747H23 on chromosome 6q13-15. Contains the 3' part of tl
hypothetical protein DKFZp434C0328
KIAA0774 protein
cyclin G2
multiple inositol polyphosphate histidine phosphatase, 1
S100 calcium-binding protein A8 (calgranulin A)
reticulon 2
heterogeneous nuclear ribonucleoprotein C (C1/C2)
Rho GDP dissociation inhibitor (GDI) beta
S100 calcium-binding protein A3
microtubule-associated protein, RP/EB family, member 1
ESTs, Weakly similar to T46908 hypothetical protein DKFZp761G2423.1
N-acetylneuraminic acid phosphate synthase; sialic acid synthase
transcription factor Dp-2 (E2F dimerization partner 2)
ESTs
**peptidylprolyl isomerase B (cyclophilin B)
adaptor-related protein complex 3, beta 2 subunit
huntingtin-associated protein interacting protein (duo)
integrin, alpha 6
reticulocalbin 1, EF-hand calcium binding domain
calcium channel, voltage-dependent, L type, alpha 1D subunit
interferon-related developmental regulator 1
CTP synthase
methionine-tRNA synthetase
interferon-related developmental regulator 1
Homo sapiens cDNA FLJ10768 fis, clone NT2RP4000150
DKFZP434G032 protein
DnaJ (Hsp40) homolog, subfamily B, member 1
ESTs, Highly similar to IFT2 HUMAN INTERFERON-INDUCED PROTEIN WITH TETRATRICOPEPTIDE REPEATS 2
heat shock 70kD protein-like 1
integrin, alpha 6
molecule possessing ankyrin repeats induced by lipopolysaccharide (MAIL)
interferon regulatory factor 2
tumor necrosis factor, alpha-induced protein 3
ATP-binding cassette, sub-family G (WHITE), member 2
carcinoembryonic antigen-related cell adhesion molecule 1
tumor rejection antigen (gp96) 1
**TGFB inducible early growth response
putatative 28 kDa protein
molecule possessing ankyrin repeats induced by lipopolysaccharide (MAIL)
v-fos FBJ murine osteosarcoma viral oncogene homolog
RAB31, member RAS oncogene family
cylindromatosis (turban tumor syndrome)
Homo sapiens cDNA FLJ14241 fis, clone OVARC1000533
conserved gene amplified in osteosarcoma
transmembrane 4 superfamily member 1
ESTs
Homo sapiens cDNA FLJ13672 fis, clone PLACE1011749
core promoter element binding protein
FOS-like antigen-1
2',5'-oligoadenylate synthetase 1 (40-46 kD)
CGI-76 protein
**UL16 binding protein 2
insulin-like growth factor 1 receptor
phosphatidylinositol binding clathrin assembly protein
Fold down-regulation or up-regulation
>5.65
1:1
>5.65

Centrifugation

Centrifugation is a commonly employed method in the virology laboratory. Differential centrifugation can be used to prepare highly purified and concentrated preparations of viruses for biochemical and structural studies. The relative density of virus particles, determined by centrifugation in a solution of cesium chloride (CsCl), is a constant feature of closely related viruses. Density is expressed in grams per milliliter (e.g., the densities of adenovirus and poliovirus in CsCl are 1.34 g/ml) and provides information about the proportion of protein and nucleic acid in the particle. The sedimentation coefficient, measured in units called Svedberg units (S), imparts relative information about the sizes of macromolecules. Viral S values can be determined by applying virus particles to the top of a preformed gradient of sucrose solution, with a density that increases from the top to the bottom of the tube. After a fixed time in the centrifuge, the distance traveled by the virus is determined, and this information is used to calculate the S value. In such sucrose gradients, poliovirus is said to sediment at 160S, and eukaryotic ribosomes sediment at 80S.

The One-Step Growth Cycle

One important objective of research in virology is to understand how viruses enter individual cells, replicate, and assemble new infectious particles. These studies are usually carried out with cell cultures rather than with animals, because cell cultures provide a much simpler and more homogeneous experimental system. More important, cell cultures can be infected in such a way as to ensure that a single replication cycle occurs synchronously in every infected cell. The idea that one-step growth analysis can be used to study the single-cell life cycle of viruses was first set forth by Ellis and Delbrück in 1939, in their studies with bacteriophages described in Chapter 1. Synchronous infection, the key to the one-step growth cycle, is accomplished by infecting cells with a sufficient number of virus particles to ensure that most of the cells are infected rapidly. Most one-step growth experiments are conducted at a **multiplicity of infection** (Box 2.4) of 5 to 10 PFU per cell to ensure that almost all cells receive at least 1 infectious unit.

One-step growth analysis begins with removal of the medium from the cell monolayer and addition of virus in

BOX 2.4

Multiplicity of infection (MOI)

Infection depends on the random collision of cells and virus particles. When susceptible cells are mixed with a suspension of virus, some cells are uninfected and other cells receive one, two, three, etc., particles. The distribution of virus particles per cell is best described by the Poisson distribution:

$$P(k) = \frac{e^{-m}m^k}{k!}$$

In this equation, $P(k)$ is the fraction of cells infected by k virus particles. The multiplicity of infection, m, is calculated from the proportion of uninfected cells, $P(0)$, which can be determined experimentally. If k is made 0 in the above equation, then

$$P(0) = e^{-m} \text{ and } m = -\ln P(0)$$

The fraction of cells receiving 0, 1, and more than one virus particle in a culture of 10^6 cells infected with an MOI of 10 can be determined as follows.

Fraction of cells that receive 0 particles:

$$P(0) = e^{-10} = 4.5 \times 10^{-5}$$

and in a culture of 10^6 cells this equals 45 uninfected cells.

Fraction of cells that receive 1 particle:

$$P(1) = 10 \times 4.5 \times 10^{-5} = 4.5 \times 10^{-4}$$

and in a culture of 10^6 cells, 450 cells receive 1 particle.

Fraction of cells that receive >1 particle:

$P(>1) = 1 - e^{-m}(m + 1)$* = 99.95% of cells receive more than 1 particle

For MOI = 0.001:

$P(0) = 99.99\%$
$P(1) = 0.0999\%$ (for 10^6 cells, 10^4 are infected)
$P(>1) = 10^{-6}$

The MOI required to infect 99% of the cells in a cell culture dish:

$P(0) = 1\% = 0.01$
$m = -\ln (0.01) = 4.6$ PFU per cell

*Obtained by subtracting from 1 (the sum of all probabilities for any value of k) the probabilities $P(0)$ and $P(1)$.

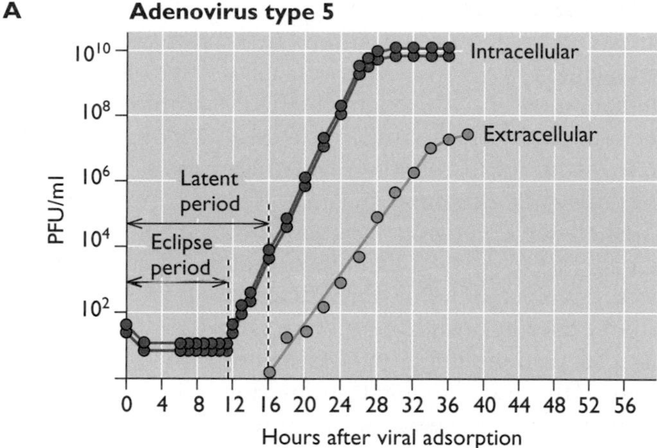

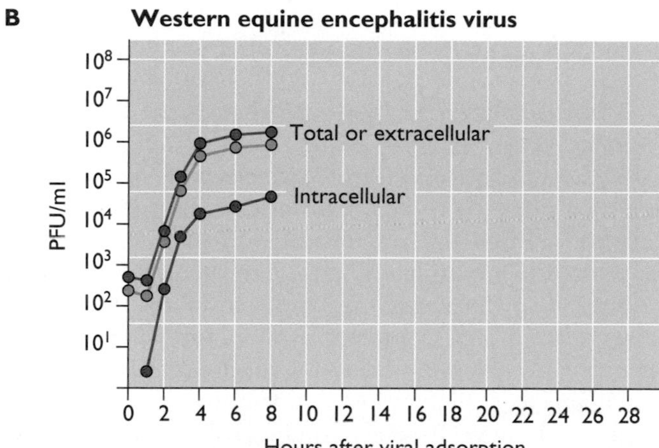

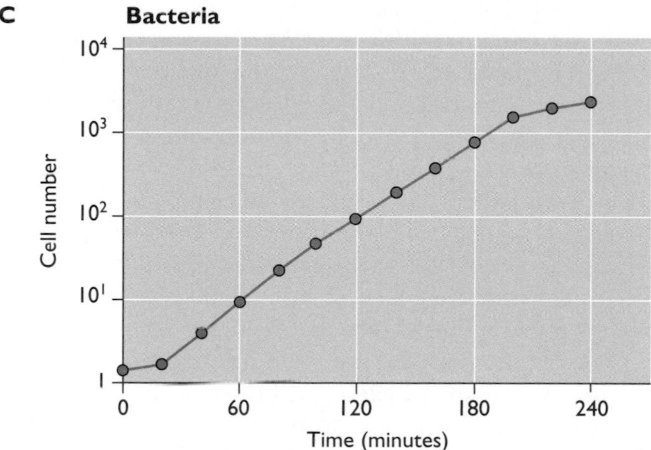

Figure 2.15 One-step growth curves. (A) Growth of a nonenveloped virus, adenovirus type 5. The titers of extracellular virus (yellow), intracellular virus (red), and the sum of both (purple) are plotted as a function of the number of hours after adsorption. The yield of infectious virus per cell can be calculated by subtracting the residual infectivity observed during the eclipse period from the total number of infectious viruses produced and dividing this number by the number of cells in the culture. (B) Growth of an enveloped virus, western equine encephalitis virus, a member of the *Togaviridae*. The total replication cycle is short, as are the eclipse, latent, and synthetic periods. This virus acquires infectivity after maturation at the plasma membrane, and therefore little intracellular virus can be detected. The small amounts observed at each time point probably represent released virus contaminating the cell extract. (C) Growth curve for a bacterium. The number of bacteria is plotted as a function of time. One bacterium is added to the culture at time zero; after a brief lag, the bacterium begins to divide. The number of bacteria doubles every 20 min until nutrients in the medium are depleted and the growth rate decreases. (A and B) Adapted from B. D. Davis et al., *Microbiology* (J. B. Lippincott Co., Philadelphia, Pa., 1980), with permission. (C) Adapted from B. Voyles, *The Biology of Viruses* (McGraw-Hill, New York, N.Y., 1993), with permission.

a small volume to promote rapid adsorption. After approximately 1 h, the unadsorbed inoculum is removed, the cells are washed, and fresh medium is added. At different times after infection, samples of the cell culture supernatant are collected, and the virus titer is determined.

The kinetics of intracellular virus production can be monitored by removing the medium containing extracellular particles, scraping the cells into fresh medium, and lysing them by repeated cycles of freeze-thawing. A cell-free extract is prepared after removal of cellular debris by centrifugation, and the virus titer in the extract is measured.

When the results of a one-step growth experiment are plotted graphically, a number of important features about viral replication are revealed. In the example shown in Fig. 2.15A, the first 11 h after infection constitute the **eclipse period**, during which the viral nucleic acid is uncoated from its protective shell and no infectious virus can be detected inside cells. The low level of infectivity detected during this period probably results from adsorbed virus that was not uncoated. Beginning at 12 h after adsorption, the quantity of intracellular infectious virus begins to increase, marking the onset of the synthetic phase, during which new virus particles are assembled. During the **latent period** no extracellular virus can be detected. At 18 h after adsorption, virus is released from cells and found in the extracellular medium. Ultimately, virus production plateaus as the cells become metabolically and structurally incapable of supporting additional replication.

The yield of infectious virus per cell can be calculated from the data collected during a one-step growth experiment (Fig. 2.15). This value varies widely among different viruses and with different virus-host cell combinations. For many viruses, increasing the multiplicity of infection above a certain point does not increase the yield. Cells have a finite capacity to produce new virus particles.

The nature of the one-step growth curve can vary dramatically among different viruses. For example, enveloped viruses that mature by budding from the plasma membrane, as discussed in Chapter 13, generally become infectious only as they leave the cell, and therefore little intracellular infectious virus can be detected (Fig. 2.15B). The first one-step growth curves of viruses were prepared for bacteriophages, and the results surprised scientists who had expected that they would resemble the growth curves of bacteria or cultured cells. After a short lag, bacterial cell growth becomes exponential (i.e., each progeny cell is capable of dividing) and follows a straight line (Fig. 2.15C). Exponential growth continues until the nutrients in the medium are exhausted. The one-step growth curves of viruses are very different: they begin with a lag period (eclipse) during which no virus growth is observed, followed by the sudden appearance of new infectious virus. We now know that during the eclipse period the components of new virus particles are being synthesized and assembled. The curve shown in Fig. 2.15A illustrates the pattern observed for a DNA virus with the long latent and synthetic phases typical of many DNA viruses, some retroviruses, and reovirus. For small RNA viruses, the entire growth curve is complete within 6 to 8 h, and the latent and synthetic phases are correspondingly shorter. Counterintuitively, polyomavirus, with one of the smallest genomes of the DNA viruses, has a very long latent period. The basis for these differences is related to the various strategies of gene expression and genome replication, to be discussed in Chapter 3.

One-step growth curve analysis can provide quantitative information about different virus-host systems. It is frequently employed to study mutant viruses to determine what parts of the replication cycle are affected by a particular genetic lesion. It is also valuable for studying the multiplication of a new virus or viral replication in a new virus-host cell combination.

When cells are infected at a low multiplicity of infection, several cycles of viral replication may occur. Growth curves established under these conditions can also provide useful information. For example, if a mutation fails to have an obvious effect on viral replication in a one-step growth curve, the defect may become obvious following a low-multiplicity infection. Because the effect of a mutation in each cycle is multiplied over several cycles, a small effect can be amplified. Defects in the ability of viruses to spread from cell to cell may also be revealed when multiple cycles of replication occur.

Genetic Analysis of Viruses

The application of genetic methods to study the structure and function of animal viral genes and proteins began with development of the plaque assay by Dulbecco in 1952. This assay permitted the preparation of clonal stocks of virus, the measurement of virus titers, and a convenient system for studying viruses with conditional lethal mutations. Although a limited repertoire of classical genetic methods was available, the mutants that were isolated were invaluable in elucidating many aspects of infectious cycles and of cell transformation. Contemporary methods of genetic analysis based on recombinant DNA technology confer an essentially unlimited scope for genetic manipulation; in principle, any viral gene of interest can be mutated, and the precise nature of the mutation can be predetermined by the investigator. Much of the large body of information about viruses and their lifestyles that we now possess can be attributed to the power of these methods.

Classical Genetic Methods

Spontaneous and Induced Mutations

In the early days of experimental virology, mutant viruses could be isolated only by screening stocks for interesting phenotypes, for none of the tools that we now take for granted, such as restriction endonucleases, efficient DNA sequencing methods, or molecular cloning procedures, were developed until the mid- to late 1970s. RNA virus stocks usually contain a high proportion of mutants, and it is only a matter of devising the appropriate selection conditions (e.g., high or low temperature or drugs that inhibit viral growth) to select mutants with the desired phenotype from the total population. For example, the live attenuated poliovirus vaccine strains developed by Albert Sabin are mutants that were selected from a virulent virus stock (Fig. 19.6). RNA virus mutants resistant to neutralization with monoclonal antibodies are often isolated from stocks at a frequency of 1 in 100,000 PFU. Mutation frequencies of RNA virus genomes are on the order of 1 misincorporation in 10^4 to 10^5 nucleotides polymerized, compared with 1 misincorporation in 10^8 to 10^{11} nucleotides incorporated for DNA virus genomes. This difference has been attributed to a lack of proofreading and error-correcting abilities in enzymes that replicate RNA.

The low spontaneous mutation rate of DNA viruses necessitated random mutagenesis of the virus by exposure to a chemical **mutagen**. Mutagens such as nitrous acid, hydroxylamine, and alkylating agents chemically modify the nucleic acid in preparations of virus particles, resulting in changes in base pairing during subsequent replication, and the substitution of an incorrect nucleotide. Mutagens such as base analogs, intercalating agents, and ultraviolet light are applied to the infected cell and cause changes in the viral genome during replication. Such agents introduce mutations more or less at random. Some mutations are

BOX 2.5

What is wild type?

Terminology can be confusing. Virologists often use terms such as strains, variants, and mutants to designate a virus that differs in some heritable way from a parental or wild-type virus. In conventional usage, the wild type is defined as the original (often laboratory-adapted) virus from which mutants are selected and which is used as the basis for comparison. A wild-type virus may **not** be synonymous with a virus isolated from nature. In fact, the genome of such a virus may include numerous mutations accumulated during propagation in the laboratory. For example, the first isolate of poliovirus obtained in 1909 probably is very different from the virus we call wild type today.

We distinguish carefully between laboratory wild types and new virus isolates from the natural host. The latter are called field isolates or clinical isolates.

lethal under all conditions, while others have no effect and are said to be silent. To increase the chances of obtaining viruses with a single genetic change, selective screens were applied after exposure to the lowest concentration of mutagen that produced a useful number of mutants. Under such conditions, many of the virus particles in the mutagenized population contain **wild-type** genomes (Box 2.5), making identification of mutants laborious. Some investigators have used chemicals to mutagenize RNA genomes, despite their high mutation rate.

To facilitate identification of mutants, the population must be screened for a phenotype that can be identified easily in a plaque assay. One such phenotype is temperature-sensitive growth of the virus. Virus mutants with this phenotype reproduce well at low temperatures, but poorly or not at all at high temperatures. The permissive and nonpermissive temperatures are typically 39 and 33°C, respectively, for viruses that replicate in mammalian cells. Other commonly sought phenotypes are changes in plaque size or morphology, cold-sensitive growth, drug resistance, antibody escape, and host range (that is, loss of the ability to multiply in certain hosts or host cells). The nomenclature used to identify viral and cellular genes, essential for describing mutations, is explained in Box 2.6.

Mapping Mutations

Before the advent of recombinant DNA technology, it was extremely difficult for investigators to determine the locations of mutations in viral genomes. The **marker rescue** technique (described in "Introducing Mutations into the Viral Genome" below) was a solution to this problem, but before it was developed other, less satisfactory approaches were exploited.

Recombination mapping can be applied to both DNA and RNA viruses. Recombination results in genetic exchange between genomes within the infected cell. For viruses with unimolecular genomes, the recombination frequency between two mutations increases with the physical distance separating them. In practice, cells are coinfected with two mutants, and the frequency of recombination is calculated by dividing the titer of phenotypically wild-type virus obtained under restrictive conditions (e.g., high temperature) by the titer obtained under permissive

BOX 2.6

Genetic nomenclature

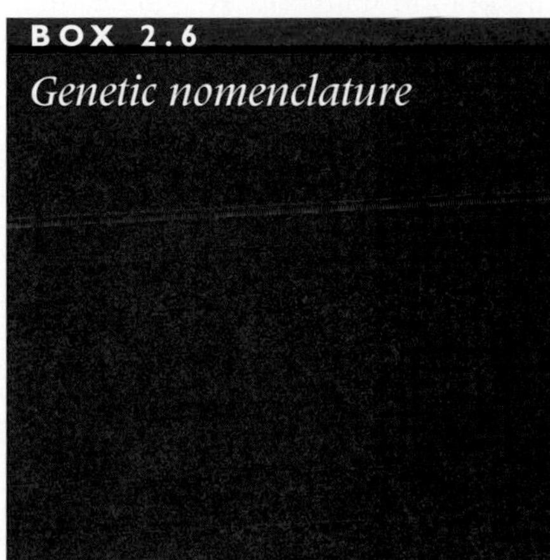

From the earliest days of genetic analysis, it has been customary to use abbreviations to designate gene names. Unfortunately, no standard nomenclature has been established for the genes of mammalian cells and their viruses. To facilitate our discussion of genes and mutations, we shall use the following conventions. Abbreviations for cellular genes will be given in lowercase, italicized letters (e.g., *pvr* for poliovirus receptor gene). In abbreviations for proteins, the first letter will be capitalized, and the remainder will be lowercase (e.g., Pvr for poliovirus receptor protein). We will use the names of some cellular proteins, and all viral genes and proteins, that do not conform to these conventions: they have become firmly entrenched in the scientific literature, and have come to have a life of their own. This nomenclature is not consistent among viruses, and in some cases the gene and protein names do not relate to one another (e.g., the VP1 capsid protein of poliovirus is encoded by the 1A gene, and the membrane glycoprotein gD of herpes simplex virus is encoded by the US6 gene). The reader should consult Appendix A to avoid confusion about such names.

conditions (e.g., low temperature). The recombination frequency is determined between pairs of mutants, allowing the mutations to be placed on contiguous maps. Although a location can be assigned for each mutation relative to others, this approach does not result in a physical map of the actual location of the base change in the genome.

For RNA viruses with segmented genomes, the technique of **reassortment** allowed the assignment of mutations to specific genome segments. When cells are coinfected with both mutant and wild-type viruses, the progeny includes **reassortants** that inherit RNA segments from either parental virus. The origins of the RNA segments can be deduced from their migration patterns after gel electrophoresis (Fig. 2.16) or by nucleic acid hybridization. By analyzing a panel of such reassortants, the segment responsible for the phenotype can be identified. When the protein products of each RNA segment are later identified, the mutation can be assigned unambiguously.

Functional Analysis

The term **complementation** describes the ability of gene products from two different nonreplicating mutant viruses to interact functionally in the same cell, permitting viral replication. If the mutations are in separate genes, each virus is able to supply a functional gene product, allowing both viruses to replicate. If both viruses carry mutations in the same gene, no replication will occur. In this way, the members of collections of mutants obtained by chemical mutagenesis were initially organized into complementation groups defining separate viral functions. In theory, there can be as many complementation groups as genes.

Figure 2.16 Reassortment of influenza virus RNA segments. (A) Schematic diagram of influenza virus reassortment. Cells are coinfected with two influenza virus strains, L and M. In infected cells the eight (–) strand RNAs of both viruses replicate. Progeny viruses include both parents and viruses that derived RNA segments from them. Recombinant R3 has inherited segment 2 from the L strain and the remaining seven segments from the M strain. (B) ^{32}P-labeled influenza virus RNAs were fractionated in a polyacrylamide gel and detected by autoradiography. Migration differences of parental viral RNAs (M and L) permitted identification of the origin of RNA segments in the progeny virus R3. Solid arrowheads next to the RNAs of recombinant viruses indicate RNA segments derived from the M parent virus, and the open arrowhead indicates an RNA segment derived from the L parent virus. Panel B is reprinted from V. R. Racaniello and P. Palese, *J. Virol.* **29:**361–373, 1979, with permission.

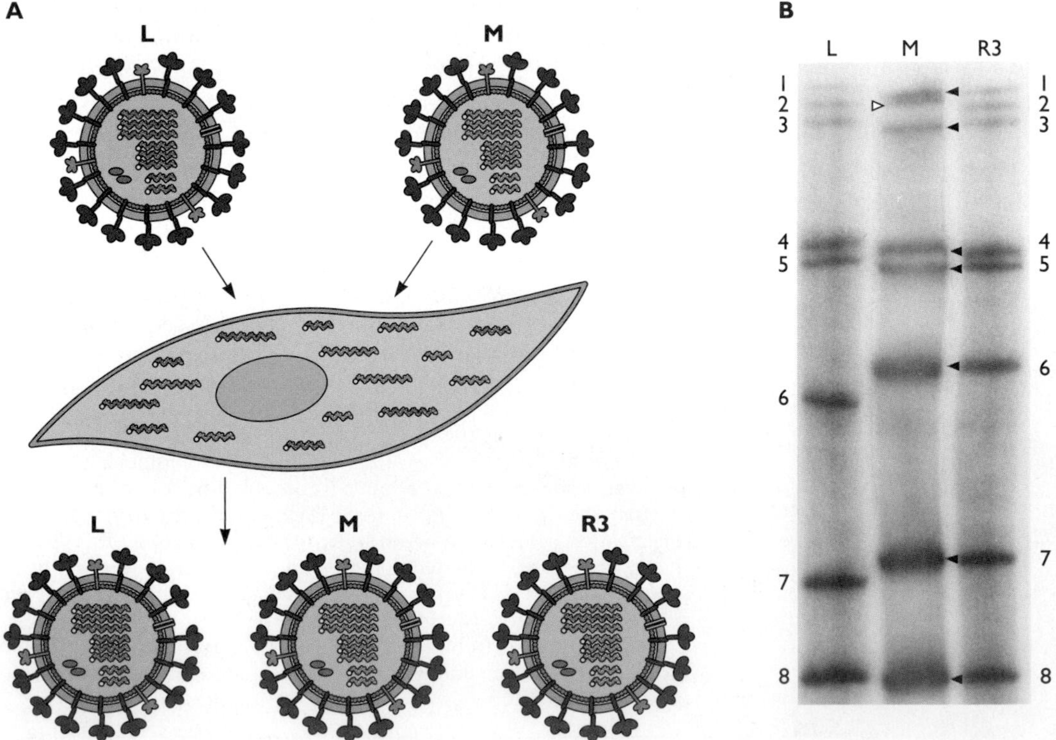

Complementation can be distinguished from recombination or reassortment by examining the progeny produced by coinfected cells. True complementation yields both parental mutants, while wild-type genomes result from recombination or reassortment.

Engineering Mutations into Viral Genomes

Infectious DNA Clones

Recombinant DNA techniques have made it possible to introduce any kind of mutation anywhere in the genome of most animal viruses, whether that genome comprises DNA or RNA. The holy grail of virology today is the **infectious DNA clone,** a double-stranded DNA copy of the viral genome that is carried on a bacterial plasmid. Infectious DNA clones, or in vitro transcripts derived from them, can be introduced into cultured cells by **transfection** (Box 2.7) to recover infectious virus. This approach is a modern validation of the Hershey-Chase experiment described in Chapter 1. The availability of site-specific bacterial restriction endonucleases, DNA ligases, and an array of methods for mutagenesis has made it possible to manipulate these infectious clones at will. Infectious DNA clones also provide a stable repository of the virus, which is particularly important for vaccine strains.

DNA viruses. Current genetic methods for the study of most viruses with DNA genomes are based on the infectivity of viral DNA. When deproteinized viral DNA molecules are introduced into permissive cells by transfection, they generally initiate a complete infectious cycle, although the infectivity (i.e., plaques per microgram of DNA) may be low. For example, the infectivity of deproteinized human adenoviral DNA is between 10 and 100 PFU per mg. When the genome is isolated by procedures that do not degrade the covalently attached terminal protein, infectivity is increased by 2 orders of magnitude, probably because this protein participates in the assembly of initiation complexes on the viral origins of replication.

The complete genomes of polyomaviruses, papillomaviruses, and adenoviruses can be cloned in plasmid vectors, and such DNA is infectious under appropriate conditions. The DNA genomes of herpesviruses and poxviruses are too large to insert into conventional bacterial plasmid vectors, but they can be cloned in vectors that can accept larger insertions (e.g., cosmids and bacterial artificial chromosomes). The plasmids containing these cloned herpesvirus genomes are infectious. Poxvirus DNA is not infectious unless early functions are provided by a helper virus.

BOX 2.7

DNA-mediated transformation and transfection

The introduction of foreign DNA into cells is called DNA-mediated transformation to distinguish it from oncogenic transformation of cells caused by tumor viruses and other insults. The term transfection (transformation-infection) was coined to describe the production of infectious virus after transformation of cells by viral DNA, first demonstrated with bacteriophage lambda. Unfortunately, the term transfection is now routinely used to describe the introduction of any DNA or RNA into cells. In this textbook, we use the correct nomenclature. Usage of the term transfection is restricted to the introduction of viral DNA or RNA into cells with the goal of obtaining virus replication.

RNA viruses. *(+) strand RNA viruses.* The genomic RNA of retroviruses is copied into a double-stranded DNA form by the viral enzyme reverse transcriptase early during infection, a process described in Chapter 7. Such viral DNA is infectious when introduced into cells, as are molecularly cloned forms inserted into bacterial plasmids.

Introduction of a plasmid containing cloned poliovirus DNA into cultured mammalian cells results in the production of progeny virus (Fig. 2.17A). The mechanism by which cloned poliovirus DNA initiates infection is not known, but it has been suggested that the DNA enters the nucleus, where it is transcribed by cellular DNA-dependent RNA polymerase from cryptic, promoter-like sequences on the plasmid. The resulting (+) strand RNA transcripts initiate an infectious cycle. During replication, the extra terminal nucleotide sequences must be removed or ignored, because the viruses that are produced contain RNA with the authentic 5′ and 3′ termini.

The genomic RNA of poliovirus has a higher specific infectivity (10^6 PFU per μg) than cloned DNA (10^3 PFU per μg). By incorporating promoters for bacteriophage T7 DNA-dependent RNA polymerase in plasmids containing poliovirus DNA, full-length (+) strand RNA transcripts can be synthesized in vitro. The specific infectivity of such RNA transcripts resembles that of genomic RNA. Infectious DNA clones have been constructed for many (but not all) (+) strand RNA viruses, including members of the *Arteriviridae, Caliciviridae, Coronaviridae, Flaviviridae, Picornaviridae,* and *Togaviridae.*

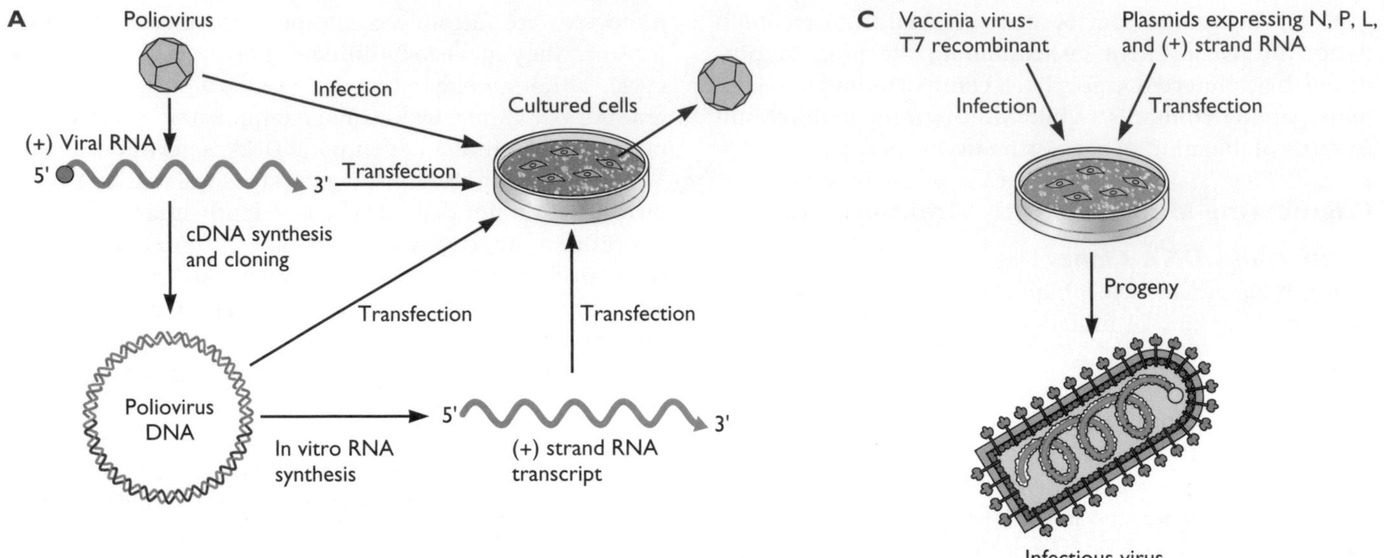

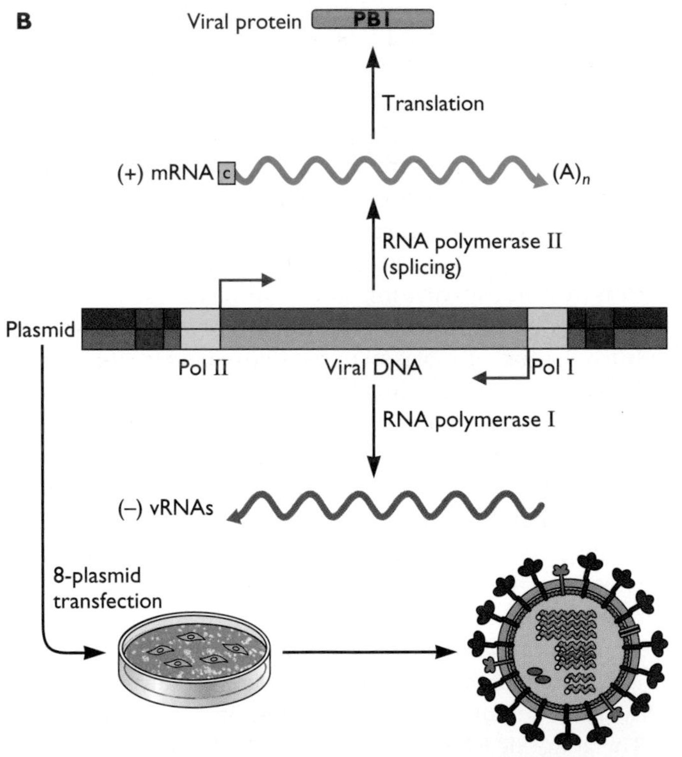

Figure 2.17 Genetic manipulation of (+) and (–) strand RNA viruses. (A) Recovery of infectivity from cloned DNA of (+) strand RNA genomes as exemplified by genomic RNA of poliovirus, which is infectious when introduced into cultured cells by transfection. A complete DNA clone of the viral RNA, carried in a plasmid, is also infectious, as are RNAs derived by in vitro transcription of the full-length DNA. cDNA, complementary DNA. (B) Recovery of influenza viruses by transfection of cells with eight plasmids. Cloned DNA of each of the eight influenza virus RNA segments is inserted between an RNA polymerase I promoter (Pol I) and terminator. This transcription unit is flanked by an RNA polymerase II promoter (Pol II) and a polyadenylation signal. When the plasmids are introduced into mammalian cells, (–) strand viral RNA (vRNA) molecules are synthesized from the RNA polymerase I promoter, and mRNAs are produced by transcription from the RNA polymerase II promoter. The mRNAs contain 5′ cap structures and 3′ poly(A) tails, and are translated into viral proteins. The result is production of infectious virus from the transfected cells. For clarity, only one cloned viral RNA is shown. Adapted from E. G. Hoffmann et al., *Proc. Natl. Acad. Sci. USA* **97:**6108–6113, 2000, with permission. (C) Recovery of infectivity from cloned DNA of viruses with a (–) strand RNA genome. Cells are infected with a vaccinia virus recombinant that expresses T7 RNA polymerase and transformed with plasmids that express (+) strand RNA and proteins required for viral RNA synthesis. The (+) strand RNA is copied into (–) strands, which in turn are used as templates for mRNA synthesis and genome replication. Infectious progeny viruses are produced. The example shown is for viruses with a single (–) strand RNA genome (e.g., rhabdoviruses and paramyxoviruses). A similar approach has been demonstrated for Bunyamwera virus, with a genome comprising three (–) strand RNAs.

(–) strand RNA viruses. Genomic RNA of (–) strand RNA viruses is not infectious because it can be neither translated nor copied into (+) strand RNA by host cell RNA polymerases, as discussed in Chapter 6. Three different experimental approaches have been developed using cloned DNA to enable genetic manipulation of these viruses (Fig. 2.17B and C).

The first system for isolation of influenza viruses containing RNA segments produced from genetically modified DNA depended on reconstitution of functional ribonucleoprotein complexes. Although this method permitted the introduction of defined mutations into influenza viruses, it could not be used to reconstitute infectious viruses entirely from cloned DNA. This limitation has been overcome by an expression system in which cloned DNA copies of the eight RNA segments of the viral genome are inserted between RNA polymerase I promoter and terminator sequences. The RNA polymerase I transcription unit is flanked by an RNA polymerase II promoter and a polyadenylation site. The orientation of the two transcription units leads to the synthesis of (–) strand viral RNA and mRNA from one DNA template. When eight plasmids carrying DNA for each viral RNA segment are introduced into cells, infectious influenza virus is produced (Fig. 2.17B).

For (–) strand RNA viruses with a nonsegmented genome, such as vesicular stomatitis virus (a rhabdovirus), the approach to obtaining an infectious DNA has been to transcribe a DNA clone of the entire genome in vitro to produce a full-length (+) strand RNA (Fig. 2.17C). This RNA is introduced into cells that synthesize the vesicular stomatitis virus nucleocapsid protein, phosphoprotein, and polymerase from separate plasmids. The (+) strand RNA is copied into (–) strand RNAs, which initiate an infectious cycle leading to the production of new virus particles. For unknown reasons, introduction of a full-length (–) strand RNA into cells that produce these viral proteins fails to initiate an infectious cycle.

Types of Mutation

Recombinant DNA techniques allow the introduction of many kinds of mutation at any desired site in cloned DNA (Box 2.8). Indeed, provided that the sequence of the segment of the viral genome to be mutated is known, there is little restriction on the type of mutation that can be introduced. **Deletion mutations** can be used to remove an entire gene to assess its role in replication, to produce truncated gene products, or to assess the functions of specific segments of a coding sequence. Noncoding regions can be deleted to identify and characterize regulatory sequences such as promoters. **Insertion mutations** can be made by the addition of unrelated sequences or sequences derived from a closely related virus. **Substitution mutations**, which can correspond to one or more nucleotides, are often made in coding or noncoding regions. Included in the latter class are **nonsense mutations**, in which a termination codon is introduced, and **missense mutations**, in which a single nucleotide or a codon is changed, resulting in the production of a protein with a single amino acid substitution. The introduction of a termination codon is frequently exploited to cause truncation of a membrane protein so that it is secreted, or to eliminate the synthesis of a protein without changing the size of the viral genome or mRNA. Substitutions are used to assess the roles of specific nucleotides in regulatory sequences or of single amino acids in protein function, such as polymerase activity or binding of a viral protein to a cell receptor. An example of a longer substitution is the exchange of nucleic acid sequences encoding a specific protein sequence among virus strains that differ in host range. The construction of such hybrid viruses can be used to assign a biological function to a specific gene product, and to determine whether it functions independently of the viral genetic background.

BOX 2.8

Operations on DNA and on protein

A mutation is a change in DNA or RNA comprising base changes, nucleotide additions, deletions, and rearrangements. When mutations occur in open reading frames, they can be manifested as changes in the synthesized proteins. For example, one or more base changes in a specific codon may produce a single amino acid substitution, a truncated protein, or no protein. The terms mutation and deletion are often used incorrectly or ambiguously to describe alterations in proteins. In this textbook, these terms are used to describe genetic changes, and the terms amino acid substitution and truncation are used to describe protein alterations.

A variety of experimental methods can be used to construct deletion, insertion, and substitution mutations, and it is beyond the scope of this book to cover all of them. Much of what can be achieved with mutagenesis today depends on the development of two important technologies: low-cost chemical synthesis of specific **oligonucleotides** and the polymerase chain reaction for amplifying selected regions of DNA. Figure 2.18 illustrates how the polymerase chain reaction can be used to introduce mutations into DNA.

Introducing Mutations into the Viral Genome

Mutations can be introduced rapidly into a viral genome when it is cloned in its entirety. Mutagenesis is usually carried out on cloned subfragments of the viral genome, which are then substituted into full-length cloned DNA using recombinant DNA techniques. The final step is introduction of the mutagenized DNA into cultured cells by transfection. This approach has been applied to cloned DNA copies of RNA viral genomes and smaller viral DNA genomes, and has recently been extended to the genomes of human adenoviruses, herpesviruses, and poxviruses.

For the large DNA viruses such as adenoviruses, herpesviruses, and poxviruses, it is not always practical to replace cloned and mutagenized subfragments of the viral

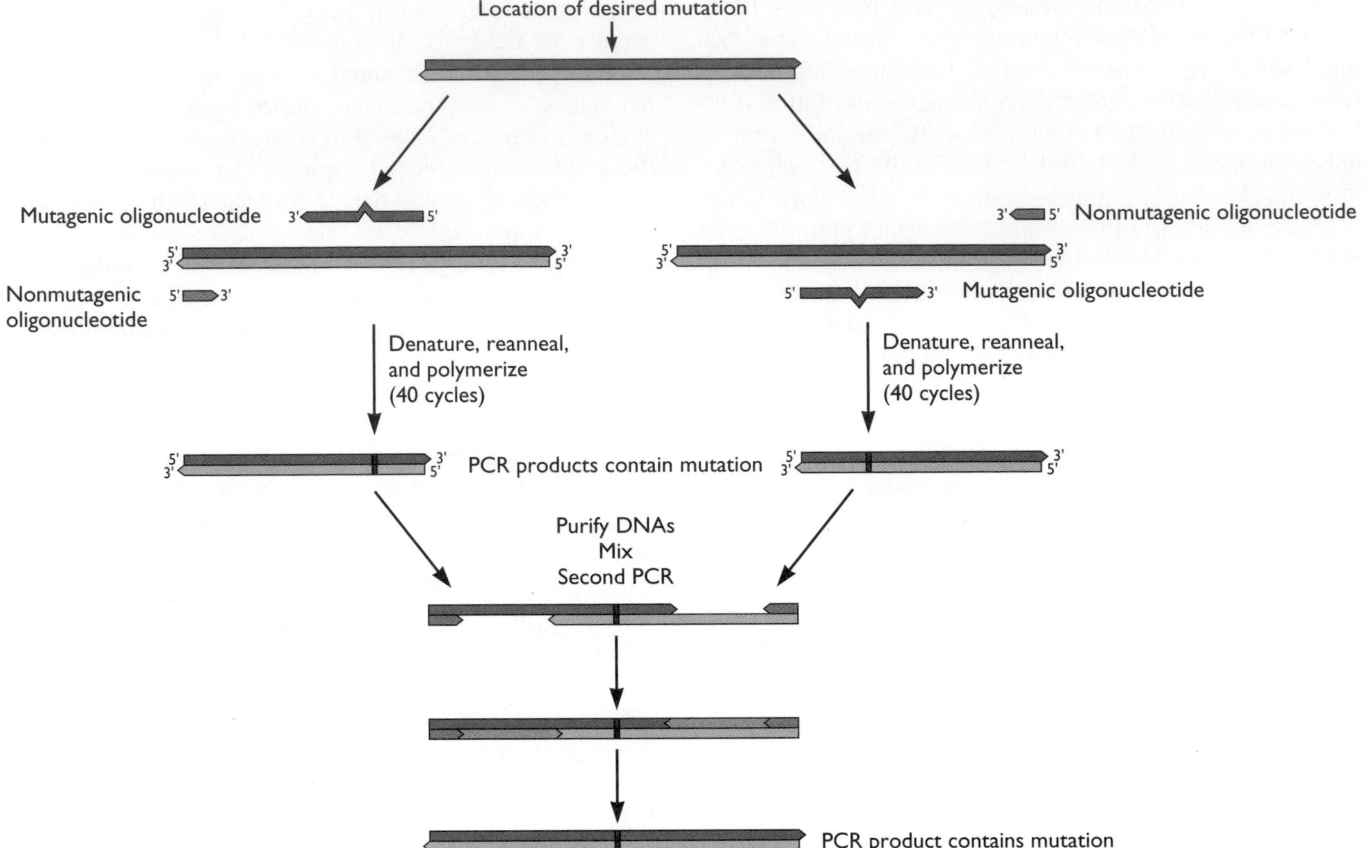

Figure 2.18 Mutagenesis by polymerase chain reaction (PCR). To introduce a mutation into DNA, two PCR mixtures are assembled with oligonucleotides complementary to either the left or right ends of the DNA to be mutagenized (which may be part of a larger sequence) and a mutagenic oligonucleotide. The products of the first set of reactions are two fragments that overlap in the area to be mutagenized. Next, the two DNAs are combined for polymerase chain reaction with both the left- and right-end oligonucleotides. The double-stranded DNAs are denatured, and a fraction anneals such that the two different products overlap in the area of the mutation. Oligonucleotide-primed DNA synthesis results in the formation of completely double-stranded molecules containing the mutation at the desired site. Depending on the sequence of the oligonucleotide, this strategy may be used to introduce point mutations, deletions, or insertions into DNA.

DNA because of the lack of convenient, unique restriction enzyme sites. A less direct, but effective, method for introducing mutations into these viral DNA genomes is **marker transfer** (a mutation is marked or traced by its phenotype). This method was originally developed to locate mutations within the viral genome in mutant viruses isolated by classical methods. For DNA viruses, site-specific bacterial restriction endonucleases were used to generate physical maps of viral DNA on which the reference points are the sites at which such enzymes cleave. These endonucleases also allowed the preparation of discrete subgenomic fragments of viral DNAs. The discrete fragments representing the complete genome of the virus are individually introduced into host cells together with a mutant virus. Only recombination with the wild-type fragment that spans the site of the mutation generates wild-type progeny virus, allowing the mutation to be placed within a specific interval on the physical map. This method, called marker rescue, is now largely of historical interest, as mutations can be introduced at predetermined sites within the viral genome. However, this technique is widely used to introduce mutations into viral genomes and is now known as marker transfer.

The marker transfer method is applicable to any DNA virus provided the target sequence is available as a DNA fragment in a vector that allows cloning and amplification in *Escherichia coli*. As in the marker rescue method, cells are transfected with the viral genome together with a DNA fragment containing the desired mutation. Recombination between the DNAs produces viral genomes containing the desired mutation. The use of marker transfer to introduce mutations into the genome of adenoviruses or herpesviruses is shown in Fig. 2.19 and 2.20. In these examples, recombination is carried out in *E. coli* rather than in mammalian cells. Marker transfer in mammalian cells is used to introduce mutations into the genome of poxviruses (Fig. 19.10).

Introduction of mutagenized viral nucleic acid into cultured cells by transfection may produce one of several possible results. The mutation may have no effect on viral replication; it may have a subtle effect that is discovered only on subsequent study; it may impart a readily detectable phenotype, which could be constitutive or conditionally lethal; or it may prevent the production of infectious virus. In the last-mentioned case, it is necessary to complement the defect in cell lines that have been engineered to produce specific viral proteins.

When the sequences of interest control viral gene expression or replication, complementation is not possible. This limitation is not as prohibitive as it might appear. One approach that can be used to study control regions is to introduce mutations that impair but do not eliminate virus replication. For example, certain mutations introduced into the major late promoter of human adenoviruses reduce virus yield more than 30-fold. An alternative approach is to study the effects of lethal mutations in cloned copies of the viral genome carried in plasmids or bacterial artificial chromosomes.

Mutations introduced into a specific gene might have unanticipated **polar** effects on other genes that are not al-

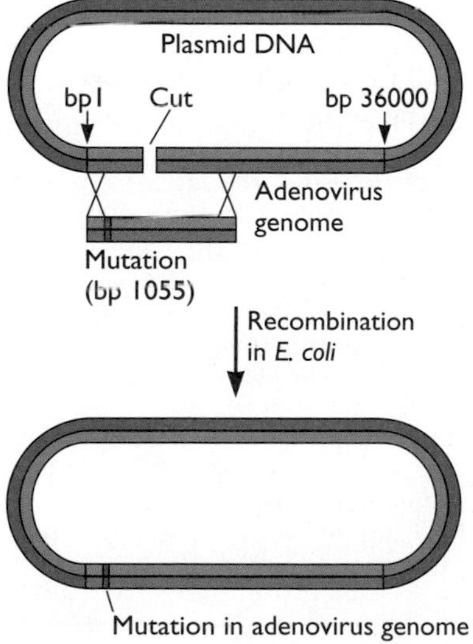

Figure 2.19 Introduction of mutations into the adenoviral genome. In the example shown, a mutation is introduced into the E1A region. A plasmid carrying the entire adenoviral DNA genome is cleaved with a suitable restriction endonuclease to produce a single cut in the area of the desired mutation. This cleaved plasmid is introduced into *E. coli* cells together with a shorter DNA containing the desired mutation. The cleaved plasmid is unable to replicate in *E. coli*. Only plasmids produced by homologous recombination between the adenoviral sequences in the cleaved plasmid and on the DNA fragment will be propagated in *E. coli*. The resulting recombinant plasmids contain a full-length copy of the adenoviral DNA genome with the desired mutation.

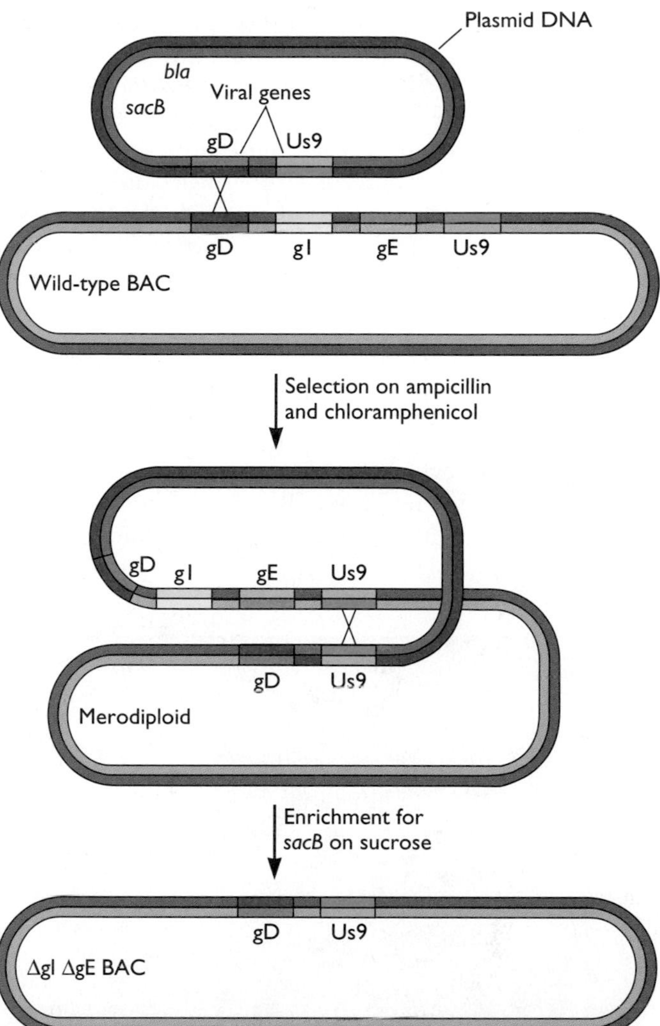

Figure 2.20 Introduction of mutations into the herpesviral genome. The viral DNA genome is cloned into a bacterial artificial chromosome (BAC). Mutations are introduced into viral genes cloned in plasmids, and substituted into the viral genome by homologous recombination. The region of the viral genome to be mutagenized is cloned into a bacterial plasmid, which carries genes encoding ampicillin resistance (*bla*, β-lactamase gene) and levansucrase (*sacB*) for selection of recombinants. Shown is the sequence of steps in recombination of the plasmid with the bacterial artificial chromosome. Only one of two possible crossover events is illustrated for each recombination step. Courtesy of G. A. Smith and L. W. Enquist, Princeton University.

tered in sequence. Insertions of the *lacZ* gene into the US4 gene that encodes the gG of pseudorabies virus impaired cell-to-cell spread of the virus. It was concluded that this phenotype is the result of loss of gG gene expression. However, introduction of a nonsense mutation that prevents gG synthesis produced viruses with no defects in cell-to-cell spread. It was then discovered that large DNA insertions in US4 cause dramatic reductions in proteins encoded in an upstream gene, US3. These findings emphasize the risk of unanticipated changes in gene expression when inserting new sequences into the viral genome.

Reversion Analysis

The phenotypes caused by mutation can **revert** in one of two ways: by reversion of the original mutation to the wild-type sequence or by acquisition of a mutation at a second site, either in the same gene or in a different gene. Phenotypic reversion caused by second-site mutation is known as **suppression**, or **pseudoreversion**, to distinguish it from reversion at the original site of mutation. Reversion has been studied since the beginnings of classical genetic analysis (Box 2.9). When suspected pseudorevertants are crossed with wild-type viruses, some viruses with the mutant phenotype should be observed in the progeny, as a result of segregation of the original mutation from the suppressor mutation. In the modern era of genetics, cloning and sequencing techniques can be used to demonstrate suppression and to identify the nature of the suppressor mutation (see below). The identification of suppressor mutations is a powerful tool for studying protein-protein and protein-nucleic acid interactions. Some suppressor mutations complement changes made at several sites, whereas **allele-specific** suppressors complement only a specific change. The allele specificity of second-site

BOX 2.9

The raison d'être of reversion analysis: is the observed phenotype due to the mutation?

The Problem

The requirement for construction and analysis of revertants is well understood in genetic analysis. The assertion that a phenotype arises from the mutation and the conclusion that the mutated gene mediates the function cannot stand without careful reversion analysis.

In genetic analysis of viruses, mutations are made in vitro by a variety of techniques, all of which can introduce unexpected mutations. Errors can be introduced during cloning, during polymerase chain reaction, and by making incorrect assumptions about sequences. Unexpected mutations often arise during the introduction of viral DNA and plasmid DNA into the eukaryotic cell. The process of recombination when viral DNA meets plasmid DNA in the eukaryotic cell is not error-proof. Tens of thousands of DNA copies are forced into a cell (transfection), and only a tiny portion give rise to the desired genome. Viral DNA isolated from cells or virions contains nicks, gaps and even RNA. Isolation of large, linear viral DNA with standard micropipetting techniques usually shears the DNA into a population of subgenomic fragments that must be repaired and rejoined in the eukaryotic cells. Less than perfect repair fidelity is a potential source of undesired, second-site mutations. Indeed, linear DNA in eukaryotic cells can be attacked at the ends, and the single-stranded DNA that results from such an attack can invade at nicked sites in linear DNA. In addition, the transfection process itself induces a variety of stress responses that might induce repair pathways.

Some Solutions

- Construct more than one mutant using totally different sources of DNA. It is unlikely that an unlinked mutation with the same phenotype would occur twice.
- Marker rescue. All local DNA around the mutation is replaced with parental DNA. If the mutation indeed causes the phenotype, the wild-type phenotype should be restored in the rescued virus.
- Ectopic expression of the wild-type protein in the mutant background. If the wild-type phenotype is restored (complemented), then the probability is high that the phenotype arises from the mutation. The merit of this method over marker rescue is that the latter only shows that unlinked mutations are unlikely to be the cause of the phenotype.

All of these approaches have limitations, and it is prudent to use more than one.

Reversion analysis is an integral part of genetics: when confidence is high that the phenotype observed is due to the known mutation, research progresses rapidly.

mutations provides evidence for physical interactions among proteins and nucleic acids.

Phenotypic revertants can be isolated either by propagating the mutant virus under restrictive conditions or, with mutants exhibiting nonconditional phenotypes, by searching for wild-type properties. For DNA viruses, chemical mutagenesis may be required to produce revertants, but this is not necessary for RNA viruses, which spawn mutants at a higher frequency. Next, nucleotide sequence analysis is used to determine if the original mutation is still present in the genome of the revertant. The presence of the original mutation indicates that reversion has occurred by second-site mutation. Even in the smallest viral genomes, it is not practical to sequence the entire nucleic acid to identify the suppressor mutation. Therefore, a form of marker rescue is used to identify the region that includes the suppressor. For DNA viruses, restriction endonuclease fragments derived from different parts of the genome of phenotypic revertants can be substituted in the original mutant viral genome by ligation in vitro or recombination in vivo. The objective is to identify a fragment that contains the suppressor mutation and is small enough (~1 kilobase [kb]) so that its nucleotide sequence can be determined. A similar analysis can be carried out for RNA viruses, except that DNA clones must first be derived from the genomes of phenotypic suppressors. The final step is introduction of the suspected suppressor mutation into the genome of the original mutant virus to confirm its effect. Several specific examples of suppressor analysis are provided below.

Some mutations within the origin of replication (Ori) of simian virus 40 reduce viral DNA replication, and induce formation of small viral plaques. Pseudorevertants of Ori mutants were isolated by random mutagenesis of mutant viral DNA, followed by introduction into cultured cells and selection of viruses that form large plaques. The second-site mutations that suppressed the replication defects were localized to a specific region within the gene for large T antigen by in vitro recombination or marker rescue experiments. These results indicated that a specific domain of large T antigen interacts with the Ori sequence during viral replication.

The 5′ untranslated region of the poliovirus genome contains extensive RNA secondary-structure features that are important for RNA translation and replication, as discussed in Chapters 6 and 11. Disruption of such secondary structure by substitution of an eight-nucleotide sequence produces a virus that is poorly infectious and readily gives rise to pseudorevertants (Fig. 2.21). Nucleotide sequence analysis of two pseudorevertants demonstrated that they contain base changes that restore the disrupted secondary structure. These results confirm that the RNA secondary structure is important for the biological activity of this untranslated region.

Mutations in the viral genome may also suppress defects in a host cell protein. An example of such extragenic

Figure 2.21 Effect of second-site suppressor mutations on predicted secondary structure in the 5′ untranslated region of poliovirus (+) strand RNA. A diagram of the region between nucleotides 468 and 534 is shown, which corresponds to stem-loop V (Chapter 11). Shown are the sequences of wild-type poliovirus type 1, a mutant containing the nucleotide changes highlighted in orange, and two phenotypic revertants. Two C:G base pairs present in the wild-type parent and destroyed by the mutation are restored by second-site reversion (blue shading). Adapted from A. A. Haller et al., *J. Virol.* **70:**1467–1474, 1996, with permission.

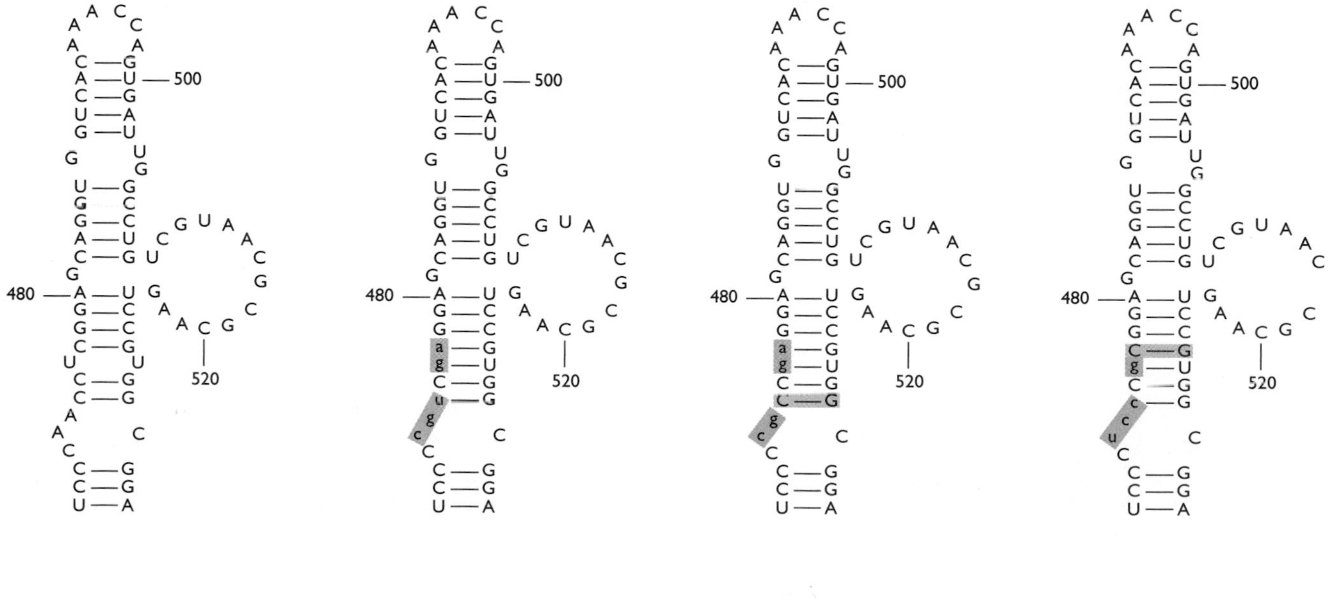

suppression is the selection of virus mutants that recognize altered cell receptors (Fig. 2.22). Amino acid changes were introduced into the poliovirus receptor. Wild-type virus cannot bind to cells that display the altered receptors. Nevertheless, viruses that bind to the altered receptors can be selected. Some of the selected viruses contain allele-specific suppressor mutations that allow the virus to bind to a receptor containing a single, specific amino acid change. These capsid protein alterations define regions of the viral capsid that bind to the cell receptor.

Genetic Interference by Double-Stranded RNA

RNA interference (**RNAi**) has become a powerful and widely used tool for analyzing gene function. In this technique, duplexes of 21-nucleotide-long RNA molecules, called short interfering RNAs (**siRNAs**), corresponding to the gene to be silenced are synthesized chemically or by transcription reactions. siRNAs or plasmids which encode them are introduced into cultured cells by transformation, and efficiently inhibit the production of specific proteins by causing sequence-specific mRNA degradation. The functions of specific viral or cellular proteins during infection can therefore be studied by using this technique (Fig. 2.23).

mRNA degradation mediated by siRNA is carried out by an endonuclease composed of multiple protein subunits. The siRNA binds to the endonuclease and is unwound, and the (–) strand guides the endonuclease to the complementary sequence on the mRNA. Endonucleolytic cleavage of the target mRNA occurs once, in the center of the duplex region formed by the siRNA and the mRNA.

Figure 2.22 Viral mutations that suppress host cell receptor defects. Cell lines that make wild-type and altered versions of the poliovirus receptor, Pvr, are established. Poliovirus cannot bind efficiently to altered Pvr molecules, and as a result viral replication is impaired. The titer of poliovirus on cells displaying wild-type Pvr is approximately 10^8 PFU per ml, and that on cells with altered Pvr is 10^3 PFU per ml. Viruses selected after growth on cells with altered Pvr have similar titers (10^8 PFU per ml) on both cell lines. These selected viruses contain mutations that allow them to bind to the defective receptor (in addition to the wild-type receptor), restoring their ability to grow on cells with mutant Pvr. Some of the selected viruses contain allele-specific suppressors that allow the virus to utilize receptors with a specific amino acid change. Others contain mutations that can suppress the defects in all altered Pvr molecules.

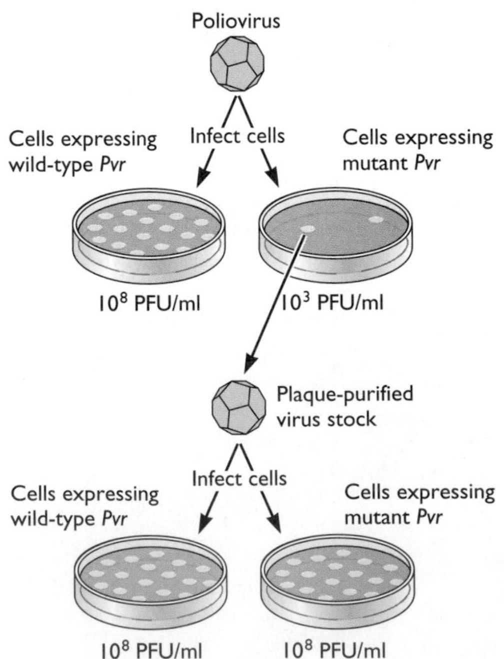

Engineering Viral Genomes: Viral Vectors

Naked DNA can be introduced into cultured animal cells as complexes with calcium phosphate or lipid-based reagents or by electroporation. Such DNA can support expression of its gene products transiently or stably from an integrated or episomal copy. Introduction of DNA into cells is a routine method in virological research and is also employed for certain clinical applications, such as the production of a therapeutic protein or a vaccine. However, this approach is not suitable for certain applications. For example, one goal of gene therapy is to deliver a gene to patients who either lack the gene or carry defective versions of it (Table 2.3). An approach to this problem is to create a cell line that synthesizes the gene product. After infusion into patients, the cells can become permanently established in tissues. If the primary cells to be used are limiting in a culture (e.g., stem cells), it is not practical to select and amplify the rare cells that receive naked DNA. Recombinant viruses carrying foreign genes can infect a greater percentage of cells and thus facilitate generation of the desired cell lines. These viral vectors have also found widespread use in the research laboratory. A complete understanding of the structure and function of viral vectors requires knowledge of viral genome replication, which is discussed in subsequent chapters for selected viruses and is summarized in Appendix A.

The design requirements for viral vectors include the inclusion of an appropriate promoter, maintenance of genome size within the packaging limit of the particle, and elimination of viral virulence, the capacity of the virus to cause disease. Expression of foreign genes from viral vectors may be controlled by homologous or heterologous promoters and enhancers chosen to support efficient transcription (e.g., the human cytomegalovirus immediate-early transcriptional control region), depending on the goals of the experiment. Such genes can be built directly into the viral genome or introduced by recombination in

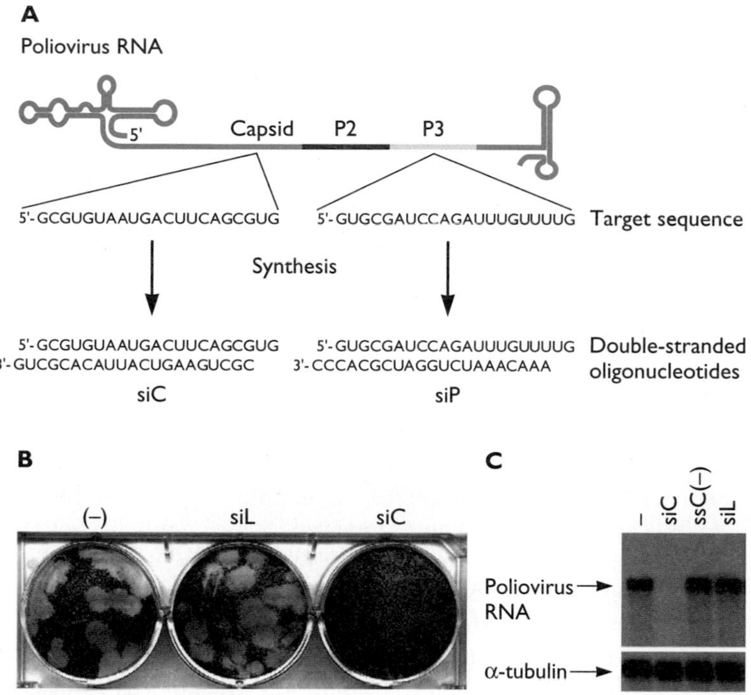

Figure 2.23 Inhibition of poliovirus replication by siRNA. siRNAs were introduced into cells by transformation, followed by poliovirus infection. (A) Location of siRNAs siC and siP on a map of the poliovirus RNA genome. (B) Inhibition of plaque formation by siRNA. Forty-eight hours after poliovirus infection, cell monolayers were stained with crystal violet to visualize plaques. The number of plaques is not reduced in untreated cells (–) or when siRNA from *Renilla* luciferase is used (siL). No plaques form in cells treated with siC. A similar effect was observed with siP (not shown). (C) Northern blot analysis of RNA from poliovirus-infected cells 6 h after infection. Poliovirus RNA replication is blocked by siC but not by the (–) strand of siC RNA, ssC(–), or siL. The blot was rehybridized with a DNA probe directed against α-tubulin to ensure that all lanes contained equal amounts of RNA. Adapted from L. Gitlin et al., *Nature* **418:** 430–434, 2002, with permission.

cells, as described previously. However, the recipient viral genome generally carries deletions and sometimes additional mutations. Deletion of some viral sequences is often required to overcome the limitations on the size of viral genomes that can be packaged in virions. For example, adenoviral DNA molecules more than 105% normal length are packaged very poorly. As this limitation would allow only 1.8 kb of exogenous DNA to be inserted into the genome, adenovirus vectors often include deletions of the E3 gene (which is not essential for growth in cells in culture) and of the E1A and E1B transcription units, which encode proteins that can be provided by complementing cell lines.

When viral vectors are designed for therapeutic purposes, it is essential to prevent replication and destruction of target host cells. The deletions necessary to accommodate a foreign gene may contribute to such disabling of the vector. For example, the E1A protein-coding sequences that are invariably deleted from adenovirus vectors are necessary for efficient transcription of viral early genes; in

Table 2.3 Some genetic diseases that might be treated by using viral vectors containing human genes

Disease	Defect	Incidence	Target cell
Severe combined immunodeficiency	Adenosine deaminase (25% of patients)	Rare	Bone marrow cells or T lymphocytes
Hemophilia A	Factor VII deficiency	1 in 10,000 males	Liver, muscle, fibroblasts, bone marrow cells
Hemophilia B	Factor IX deficiency	1 in 30,000 males	Liver, muscle, fibroblasts, bone marrow cells
Familial hypercholesterolemia	Deficiency of low-density lipoprotein receptor	1 in 1,000,000	Liver
Cystic fibrosis	Defective salt transport in lung epithelium	1 in 3,000 whites	Lung airways
Hemoglobinopathies and thalassemias	Defects in α or β globin gene	1 in 600 in specific ethnic groups	Bone marrow precursors of red blood cells
Gaucher's disease	Defect in glucocerebrosidase	1 in 450 Ashkenazi Jews	Bone marrow cells, macrophages
α_1-Antitrypsin deficiency, inherited emphysema	α_1-Antitrypsin not produced	1 in 3,500	Lung or liver cells
Duchenne muscular dystrophy	Dystrophin not produced	1 in 3,000 males	Muscle cells

their absence, viral yields from cells in culture are reduced by about 3 to 6 orders of magnitude (depending on the cell type). Removal of E1A coding sequences from adenovirus vectors is therefore doubly beneficial, although not sufficient to ensure that the vector cannot replicate or induce damage in a host animal. As discussed in detail in Chapter 14, producing viruses that do not cause disease can be more difficult to achieve.

A summary of viral vectors is presented in Table 2.4, and examples are discussed below.

DNA Virus Vectors

One goal of gene therapy is to introduce genes into terminally differentiated cells. Such cells normally do not divide, they cannot be cultured in vitro, and the organs of which they are a part cannot be infused with virus-infected cells. DNA virus vectors have been developed to overcome some of these problems.

Adenovirus vectors were originally developed for the treatment of cystic fibrosis because of the tropism of the virus for the respiratory epithelium. Adenovirus can infect terminally differentiated cells, but only transient expression is provided, as this viral DNA is not integrated into host cell DNA. Adenoviruses carrying the cystic fibrosis transmembrane conductance regulator gene, which is defective in patients with this disease, have been used in clinical trials. Many other gene products with therapeutic potential have been synthesized in a wide variety of cell types. In the first adenovirus vectors that were designed, foreign genes were inserted into the E1 and/or E3 regions. As these vectors had limited capacity, vectors that contain minimal adenovirus sequences were designed (Fig. 2.24). This strategy allows 27 to 38 kb of foreign sequence to be introduced into the vector. In addition, elimination of most viral genes reduces the host immune response against viral proteins, simplifying multiple immunizations.

Adenovirus-associated virus has attracted much attention as a vector for gene therapy, because the virus integrates into the host genome. Genes packaged into recombinant adenovirus-associated virus particles integrate at numerous sites in the human genome and persist, in some cases with high levels of expression, in many different tissues. There has been increasing interest in recombinant adenovirus-associated virus vectors to target therapeutic genes to smooth muscle, a tissue that is highly susceptible and supports sustained high-level expression of foreign genes. Although first-generation adenovirus-associated virus vectors were limited in the size of inserts that could be transferred, a two-plasmid delivery system that overcomes the limited genetic capacity has been developed (Fig. 2.25).

Table 2.4 Viral vectors

Virus	Insert size	Integration	Duration of expression	Advantages	Disadvantages
Adeno-associated virus	~4.5–9 (?) kb	Low efficiency	Long	Nonpathogenic, episomal, infects nondividing cells	Immunogenic, toxicity
Adenovirus	2–38 kb	No	Short	Efficient gene delivery	Transient, immunogenic
Alphavirus	~5 kb	No	Short	Broad host range, high-level expression	Virulence
Herpes simplex virus	~30 kb	No	Long in central nervous system, short elsewhere	Neurotropic, large capacity	Virulence, persistence in neurons
Influenza virus	Unknown	No	Short	Strong immune response	Virulence
Lentivirus	7–18 kb	Yes	Long	Stable integration; infects nondividing and terminally differentiated mammalian cells	Insertional mutagenesis
Poliovirus	~300 bp for helper-free virus; ~3 kb for defective virus	No	Short	Excellent mucosal immunity	Limited capacity, reversion to neurovirulence
Retrovirus	1–7.5 kb	Yes	Shorter than formerly believed	Stable integration	May rearrange genome, insertional mutagenesis, require cell division
Rhabdovirus	Unknown	No	Short	High-level expression, rapid cell killing	Virulence, highly cytopathic
Vaccinia virus	At least ~25 kb, probably ~75–100 kb	No	Short	Wide host range, ease of isolation, large capacity, high-level expression	Transient, immunogenic

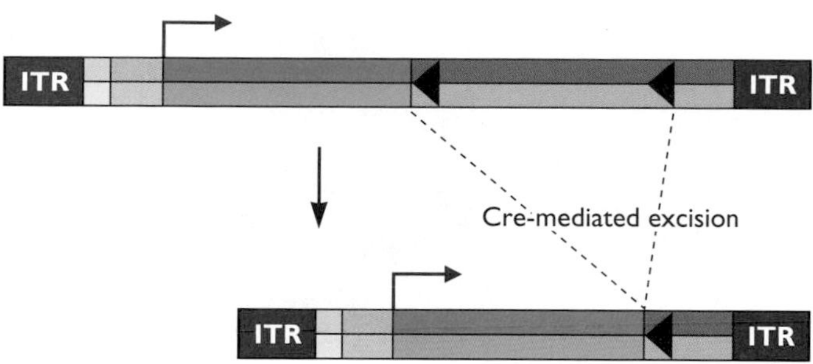

Figure 2.24 Adenovirus vectors. High-capacity adenovirus vectors are produced by inserting a foreign gene and promoter into the viral E1 region, which has been deleted. The E3 region also has been deleted. Two *loxP* sites have been introduced into the adenoviral genome (black arrowheads). These are sites for cleavage by the Cre recombinase. Infection of cells that produce this recombinase leads to excision of sequences flanked by the *loxP* sites. The result is a "gutless" vector that contains only the origin of replication-containing inverted terminal repeats (ITR), the packaging signal (yellow), the viral E4 transcription unit (orange), and the transgene with its promoter (green). Adapted from A. Pfeifer and I. M. Verma, *in* D. M. Knipe et al. (ed.), *Fields Virology*, 4th ed. (Lippincott Williams & Wilkins, Philadelphia, Pa., 2001), with permission.

Vaccinia virus vectors have the advantages of a wide host range, a genome that accepts very large fragments, high expression, and relative ease of preparation. Foreign DNA is usually inserted into the viral genome by homologous recombination, with an approach similar to that described for marker transfer (Fig. 19.10). Because of the relatively low pathogenicity of the virus, vaccinia virus recombinants have been considered candidates for human and animal vaccines.

RNA Virus Vectors

A number of RNA viruses have also been developed as vectors for foreign gene expression (Table 2.4). Because poliovirus is such an excellent mucosal immunogen, it has been an attractive candidate for the delivery of antigens to these surfaces. Alphavirus vectors have been shown to be capable of expressing foreign genes. These (+) strand RNA viruses produce subgenomic mRNAs from which efficient gene expression can be obtained. In one type of vector, the foreign gene replaces those encoding the viral structural proteins. As a result, this vector requires the presence of a helper virus to provide the missing proteins, so that the vector genome can be packaged into virus particles. An alternative arrangement is to place the foreign gene under

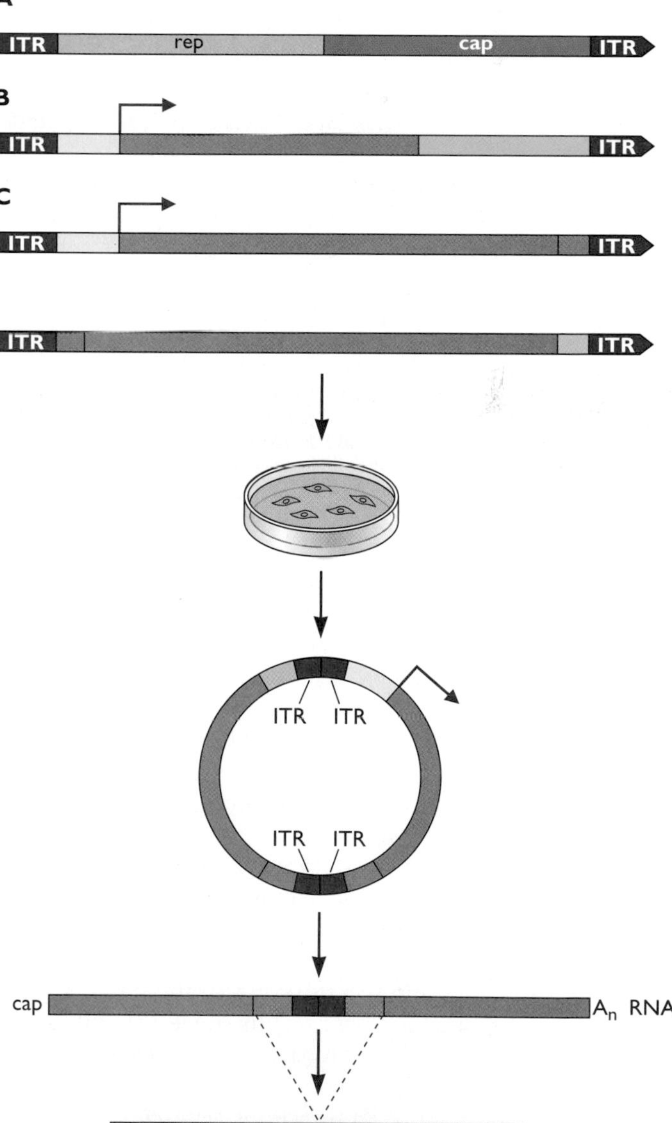

Figure 2.25 Adenovirus-associated vectors. (A) Map of the genome of wild-type adenovirus-associated virus. The viral DNA is single stranded and flanked by two inverted terminal repeats (ITR), and encodes capsid (blue) and nonstructural (orange) proteins. (B) In one type of vector, the viral genes are replaced with the transgene (pink) and its promoter (yellow) and a poly(A) addition signal (green). These DNAs are introduced into cells which produce capsid proteins, and the vector genome is encapsidated into virus particles. A limitation of this vector structure is that only 4.1 to 4.9 kb of foreign DNA can be packaged efficiently. To overcome this limitation, a two-plasmid expression system was developed. (C) In this example, the transgene of interest (pink) is divided between two adenovirus-associated virus vectors. One plasmid contains the promoter, the 5′ part of the transgene, and a 5′ splice site (purple), and the second plasmid contains a 3′ splice site, the remainder of the transgene, and the polyadenylation site. Upon introduction into cells, the recombinant adenovirus-associated virus genomes form circular multimers that bring the divided transgene together in a head-to-tail orientation. A long RNA is then produced that undergoes splicing to form a functional mRNA that can be translated into protein. Adapted from A. Pfeifer and I. M. Verma, *in* D. M. Knipe et al. (ed.), *Fields Virology*, 4th ed. (Lippincott Williams & Wilkins, Philadelphia, Pa., 2001), and Z. Yan et al., *Proc. Natl. Acad. Sci. USA* **97:**6716–6721, 2000, with permission.

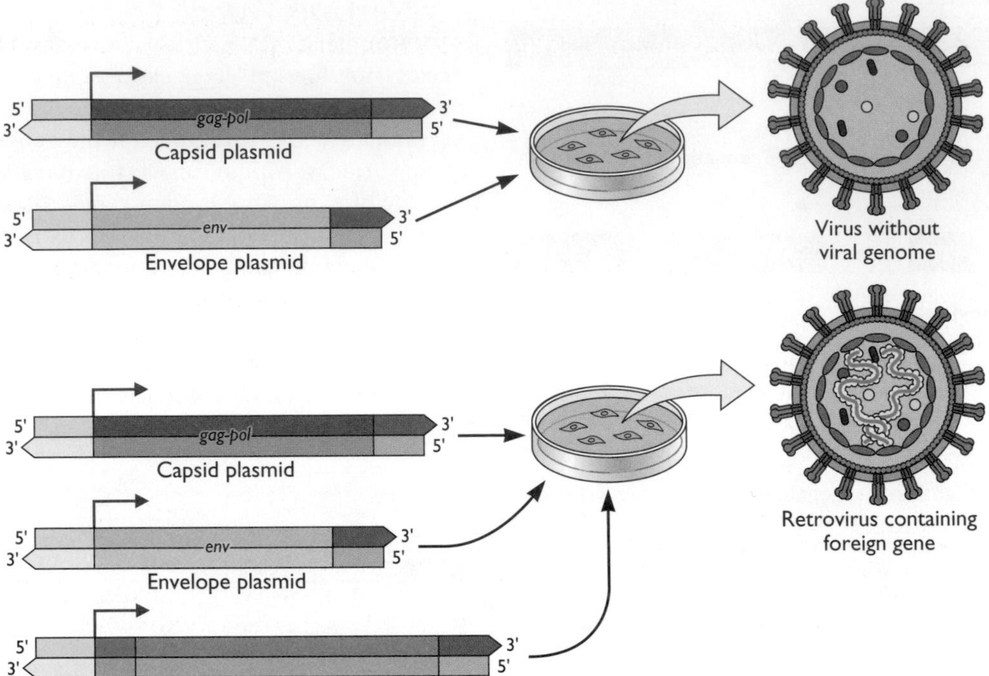

Figure 2.26 Retroviral vectors. The minimal viral sequences required for retroviral vectors are 5′- and 3′-terminal sequences (yellow and blue) that control gene expression and packaging of the RNA genome. The foreign gene (pink) and promoter (green) are inserted between the viral sequences. To package this DNA into viral particles, it is introduced into cultured cells with plasmids that encode viral proteins required for encapsidation. No wild-type viral RNA is present in these cells. If only these plasmids are introduced into cells, virus particles that do not contain viral genomes are produced. When all three plasmids are introduced into cells, the result is the formation of retrovirus particles that contain only the recombinant vector genome and no wild-type particles. The host range of the recombinant vector can be controlled by the type of envelope protein. Envelope protein from amphotropic retroviruses allows the recombinant virus to infect human and mouse cells. The vesicular stomatitis virus glycoprotein G allows infection of a broad range of cell types in many species, and also permits concentration with simple methods. Adapted from A. Pfeifer and I. M. Verma, *in* D. M. Knipe et al. (ed.), *Fields Virology,* 4th ed. (Lippincott Williams & Wilkins, Philadelphia, Pa., 2001), with permission.

the control of a second, subgenomic RNA promoter, leaving intact the promoter that controls synthesis of the structural proteins. This vector replicates without a helper and produces infectious particles.

The ability to manipulate the genome of influenza virus has been exploited for vaccine development. For example, immunization of mice with a recombinant influenza virus that produces specific antigens of *Plasmodium yoelii* resulted in a protective immune response against the malaria parasite *Plasmodium falciparum*. Similarly, mice immunized with recombinant influenza viruses that produce epitopes from human immunodeficiency virus synthesized virus-specific antibodies and secretory IgA in the nasal, vaginal, and intestinal mucosa. These results suggest that influenza viruses may be engineered for use as live vaccines for protection against other pathogens.

The development of an infectious DNA clone of vesicular stomatitis virus, a (–) strand RNA virus, has led to an approach to therapy for human immunodeficiency virus type 1 infection. Recombinant vesicular stomatitis virus particles that lack the viral glycoprotein G gene used for attachment to host cells have been produced. The G gene is replaced by genes encoding two cellular proteins required for the attachment of human immunodeficiency virus type 1 to cells (CD4 and CXCR4) (Chapter 5). The recombinant vesicular stomatitis viruses cannot infect normal cells. However, they can infect and lyse cells infected with human immunodeficiency virus type 1, because these cells synthesize viral glycoproteins on the surface. This approach has the potential of limiting the number of infected cells in the host.

Retroviruses have enjoyed great popularity as vectors because their replication cycles include the integration of a double-stranded DNA copy of viral RNA into the cell genome, a topic of Chapter 7. The integrated provirus remains permanently in the cell's genome and is passed on to progeny cells during cell division. This feature of retro-

Table 2.5 Some Nobel Prizes awarded for the development of new technologies

Year	Prize	Recipient	Discovery
1993	Chemistry	Kary Mullis	Polymerase chain reaction
		Michael Smith	Oligonucleotide-directed mutagenesis
1984	Medicine	Niels K. Jerne, Georges Köhler, Cesar Milstein	Monoclonal antibodies
1982	Chemistry	Aaron Klug	Crystallographic electron microscopy
1980	Chemistry	Paul Berg	Recombinant DNA
		Walter Gilbert, Frederick Sanger	DNA sequencing
1978	Medicine	Werner Arber, Daniel Nathans, Hamilton Smith	Restriction endonucleases
1977	Medicine	Rosalyn Yalow	Radioimmunoassay
1975	Medicine	David Baltimore, Howard Temin	Reverse transcriptase
1958	Chemistry	Frederick Sanger	Protein sequencing
1954	Medicine	John Enders, Thomas Weller, Frederick Robbins	Growth of poliovirus in cell culture

viral vectors results in permanent modification of the genome of the infected cell. The minimal viral sequences required for retroviral vectors are 5′- and 3′-terminal sequences required for gene expression and packaging of the RNA genome and integration of the DNA (Fig. 2.26). The foreign gene is cloned between the terminal sequences, and the DNA is introduced into cells that synthesize viral proteins. As no viral RNA genomes are present, the result is the formation of retrovirus particles containing only the recombinant vector genome.

One problem with the use of retroviruses in correcting genetic deficiencies is that only a few cell types can be infected by the commonly used murine retroviral vectors, and these can integrate their DNA efficiently only in actively dividing cells. Often the cells that are targets of gene therapy—bone marrow stem cells, hepatocytes, and muscle cells—do not divide. This problem can be circumvented if ways can be found to induce such cells to divide before infection with the retrovirus. Another important limitation of the murine retrovirus vectors is the phenomenon of gene silencing, which represses foreign gene expression in many cells. An alternative approach is to use viral vectors that contain sequences from human immunodeficiency virus type 1, which can infect nondividing cells and are less severely affected by gene silencing.

Perspectives

The development of new technologies is essential for scientific progress. Methods are not developed for their own sake, but rather to drive a field forward and open new vistas. For example, the development of methods for propagating animal viruses in cell culture in the 1950s led to an explosion of activity in the field of viral genetics. The techniques of DNA sequencing, molecular cloning, site-directed mutagenesis, and monoclonal antibody production introduced in the 1970s and 1980s made it possible to isolate, characterize, mutagenize, and express individual genes, the foundations of today's genetic analyses. In recognition of the importance of technology to scientific progress, Nobel Prizes have been bestowed for the development of new methods (Table 2.5). Just when it appeared that it would be difficult to top recombinant DNA methodology, the polymerase chain reaction was developed, and it is now used in nearly every molecular biology laboratory. The techniques of DNA microarray analysis and RNAi are the latest power tools in the virologists' toolkit. We can only guess what the next blockbuster method might be. The beauty of a great method is not only its universal utility, but also its ability to invoke in all of us the same thought: "I knew that would work!"

References

Textbooks

Ausubel, F. M., R. Brent, R. E. Kingston, D. D. Moore, J. G. Seidman, J. A. Smith, and K. Struhl. 1998. *Current Protocols in Molecular Biology.* John Wiley & Sons, Inc., New York, N.Y.

Coligan, J. E., B. M. Dunn, H. C. Ploegh, D. W. Speicher, and P. T. Wingfeld. 1998. *Current Protocols in Protein Science.* John Wiley & Sons, Inc., New York, N.Y.

Griffiths, A. J. F., W. M. Gelbart, R. C. Lewontin, and J. H. Miller. 2002. *Modern Genetic Analysis.* W. H. Freeman & Co., New York, N.Y.

Harlow, E., and D. Lane. 1999. *Using Antibodies: a Laboratory Manual.* Cold Spring Harbor Laboratory Press, Cold Spring Harbor, N.Y.

Sambrook, J., and D. Russell. 2001. *Molecular Cloning: a Laboratory Manual,* 3rd ed. Cold Spring Harbor Laboratory Press, Cold Spring Harbor, N.Y.

Review Articles

Conzelmann, K.-K., and G. Meyers. 1996. Genetic engineering of animal RNA viruses. *Trends Microbiol.* **4:**386–393.

Maramorosch, K., and H. Koprowski. *Methods in Virology.* Academic Press, New York, N.Y. A series of volumes begun in the 1960s that contain review articles on virological methods.

Papers of Special Interest

Almazan, F., J. M. Gonzalez, Z. Penzes, A. Izeta, E. Calvo, J. Plana-Duran, and L. Enjuanes. 2000. Engineering the largest RNA virus genome as an infectious bacterial artificial chromosome. *Proc. Natl. Acad. Sci. USA* **97:**5516–5521.

Bridgen, A., and R. M. Elliott. 1996. Rescue of a segmented negative-strand RNA virus entirely from cloned complementary DNAs. *Proc. Natl. Acad. Sci. USA* **93:**15400–15404.

Demmin, G. L., A. C. Clase, J. A. Randall, L. W. Enquist, and B. W. Banfield. 2001. Insertions in the gG gene of pseudorabies virus reduce expression of the upstream Us3 protein and inhibit cell-to-cell spread of virus infection. *J. Virol.* **75:**10856–10869.

Elbashir, S. M., J. Harborth, W. Lendeckel, A. Yalcin, K. Weber, and T. Tuschl. 2001. Duplexes of 21-nucleotide RNAs mediate RNA interference in cultured mammalian cells. *Nature* **411:**494–498.

Goff, S. P., and P. Berg. 1976. Construction of hybrid viruses containing SV40 and lambda phage DNA segments and their propagation in cultured monkey cells. *Cell* **9:**695–705.

Graham, F. L., J. Rudy, and P. Brinkley. 1989. Infectious circular DNA of human adenovirus type 5: regeneration of viral DNA termini from molecules lacking terminal sequences. *EMBO J.* **8:**2077–2085.

Hoffmann, E., G. Neumann, Y. Kawaoka, G. Hobom, and R. G. Webster. 2000. A DNA transfection system for generation of influenza A virus from eight plasmids. *Proc. Natl. Acad. Sci. USA* **97:**6108–6113.

Luytjes, W., M. Krystal, M. Enami, J. D. Pavin, and P. Palese. 1989. Amplification, expression, and packaging of a foreign gene by influenza virus. *Cell* **59:**1107–1113.

Mackett, M., G. L. Smith, and B. Moss. 1982. Vaccinia virus: a selectable eukaryotic cloning and expression vector. *Proc. Natl. Acad. Sci. USA* **79:**7415–7419.

Palese, P., and J. L. Schulman. 1976. Mapping of the influenza virus genome: identification of the hemagglutinin and the neuraminidase genes. *Proc. Natl. Acad. Sci. USA* **73:**2142–2146.

Post, L. E., and B. Roizman. 1981. A generalized technique for deletion of specific genes in large genomes: alpha gene 22 of herpes simplex virus 1 is not essential for growth. *Cell* **25:**227–232.

Racaniello, V. R., and D. Baltimore. 1981. Cloned poliovirus complementary DNA is infectious in mammalian cells. *Science* **214:**916–919.

Saiki, R. K., D. H. Gelfand, S. Stoffel, S. J. Scharf, R. Higuchi, G. T. Horn, K. B. Mullis, and H. A. Erlich. 1988. Primer-directed enzymatic amplification of DNA with a thermostable DNA polymerase. *Science* **239:**487–491.

Schnell, M. J., T. Mebatsion, and K. K. Conzelmann. 1994. Infectious rabies viruses from cloned cDNA. *EMBO J.* **13:**4195–4203.

Smith, G. A., and L. W. Enquist. 2000. A self-recombining bacterial artificial chromosome and its application for analysis of herpesvirus pathogenesis. *Proc. Natl. Acad. Sci. USA* **97:**4873–4878.

van Dinten, L. C., J. A. den Boon, A. L. M. Wassenaar, W. J. M. Spaan, and E. J. Snijder. 1997. An infectious arterivirus cDNA clone: identification of a replicase point mutation that abolishes discontinuous mRNA transcription. *Proc. Natl. Acad. Sci. USA* **94:**991–996.

Yan, Z., Y. Zhang, D. Duan, and J. F. Engelhardt. 2000. Trans-splicing vectors expand the utility of adeno-associated virus for gene therapy. *Proc. Natl. Acad. Sci. USA* **97:**6716–6721.

Journals

BioTechniques (http://www.biotechniques.com)

Trends in Biotechnology (http://www.trends.com/tibtech/)

II Molecular Biology

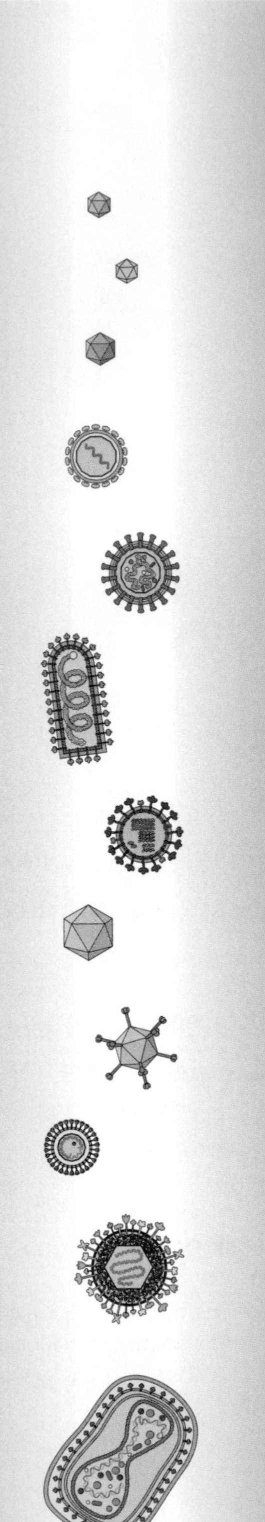

3

Genomes

> *... everywhere an interplay between nucleic acids and proteins; a spinning wheel in which the thread makes the spindle and the spindle the thread.*
>
> ERWIN CHARGAFF
> *as quoted in A. N. H. Creager, The Life of a Virus, University of Chicago Press, 2002*

Introduction

The world abounds with uncountable numbers of viruses of apparently overwhelming diversity. Fortunately, as discussed in Chapter 1, taxonomists have devised methods of classifying viruses. As a result, the number of identifiable groups becomes rather manageable (Table 3.1). One of the contributions of molecular biology has been a detailed analysis of the genetic material of representatives of these main virus families. What emerged from these studies was the principle that the **viral genome** is the nucleic acid-based repository of the information needed to build, replicate, and transmit a virus (Box 3.1). These analyses further revealed that the structures of viral genomes are far less complex than is implied by the existence of thousands of distinct entities defined by classical taxonomic methods. In fact, it is possible to organize the known viruses into seven groups, based on the structures of their genomes.

Genome Principles and the Baltimore System

One universal function of viral genomes once inside a cell is to specify proteins. However, viral genomes do not encode the machinery needed to carry out protein synthesis. One important principle is that all viral genomes must be copied to produce messenger RNAs (mRNAs) that can be read by host ribosomes. Literally, all viruses are parasites of their host cells' mRNA translation system. A second principle is that there is unity in diversity; even with immeasurable time, evolution has led to the formation of only seven major types of viral genome. The Baltimore classification system integrates these two principles to construct an elegant molecular algorithm for virologists (Fig. 1.10). When the bewildering array of viruses are classified by this system, we find fewer than 10 pathways to mRNA. The elegance of the Baltimore system is that by knowing only the nature of the viral genome, one can deduce the basic steps that must take place to produce mRNA. Perhaps more pragmatically, the system simplifies comprehension of the extraordinary life cycles of viruses.

Table 3.1 DNA and RNA viruses according to the report of the International Committee on Taxonomy of Viruses (2000)[a]

Type	Genome	No. of families or unassigned genera[b]	Morphology			
			Isometric[c]		Other[d]	
			Enveloped	Naked	Enveloped	Naked
DNA	ds	24	1	9	11	3
	ss	6	0	5	0	1
RNA	ds	7	0	4	1	2
	(+)ss	44	4	18	1	21
	(−)ss	10	0	0	8	2
Subtotal			5	36	21	29
Total		91	41		50	

[a] The known viruses have been grouped into 1,550 species (probably 30,000–40,000 total virus isolates), 56 families, 9 subfamilies, 233 genera, subviral agents (satellites, viroids, and prions), and unclassified viruses.

[b] Many genera of viruses have not been assigned to families.

[c] Particles likely to be built with icosahedral symmetry; the latter can only be determined by structural studies, which have not been done for all viruses.

[d] This category includes nonisometric forms, including tailed bacteriophages, pleomorphic, rod shaped, lemon shaped, droplet shaped, spherical, bacilliform, reniform, and filamentous.

The Baltimore system omits the second universal function of viral genomes, to serve as a template for synthesis of progeny genomes. There is a finite number of nucleic acid template copying strategies, each with unique primer, template, and termination requirements. We shall combine this principle with that embodied in the Baltimore system to define seven classes of genomes based on mRNA synthesis **and** genome replication.

BOX 3.1

What information is encoded in a viral genome?

Gene products and regulatory signals required for

- replication of the viral genome
- assembly and packaging of the genome
- regulation and timing of the replication cycle
- evasion of host defenses
- spread to other cells and hosts

Information **not** contained in viral genomes:

- no genes or evolutionary relics of genes encoding protein synthesis machinery (e.g., no ribosomal RNA and no ribosomal or translation proteins). Note: the genomes of bacteriophage T4 and phycodnaviruses encode some transfer RNA genes.
- no genes encoding proteins of energy metabolism or membrane biosynthesis
- no telomeres (to maintain genomes) or centromeres (to ensure segregation of genomes)

For most viruses with DNA genomes, replication and mRNA synthesis present no obvious challenges, as all cells use DNA-based mechanisms. In contrast, animal cells possess no known mechanisms to copy RNA templates and to produce mRNA from them. For RNA viruses to survive, their RNA genomes must, by definition, encode a nucleic acid polymerase for genome replication and mRNA synthesis.

Structure and Complexity of Viral Genomes

Despite the simplicity of expression strategies, the composition and structures of viral genomes are more varied than those seen in the entire bacterial, plant, or animal kingdoms. Nearly every possible method for encoding information in nucleic acid can be found in viruses. Viral genomes can be

- DNA or RNA
- DNA with short segments of RNA
- DNA or RNA with covalently attached protein
- single stranded (+) strand, (−) strand, or ambisense (Box 3.2)
- double stranded
- linear
- circular
- segmented
- gapped

What is the significance of such diversity in genome composition and structure? Obviously, viral genomes are survivors of uncounted selections, but can we deduce something from their existence? Is one configuration more advantageous? These are difficult questions to answer. It is

BOX 3.2

Important conventions: plus (+) and minus (−) strands

mRNA is defined as the positive (+) strand, because it contains immediately translatable information. A strand of DNA of the equivalent polarity is also designated as a (+) strand; i.e., if it were mRNA, it would be translated into protein.

The RNA and DNA complement of the (+) strand is the (−) strand. The (−) strand cannot be translated into the appropriate protein; it must first be copied to make the (+) strand.

clear that the nature of the genome reflects the method of replication and packaging, but there is little information about selective advantages of one type over another. Virus evolution will be discussed further in Chapter 20.

Figures 3.1 through 3.7 illustrate the seven classes of viral genome configuration. For each class, a panel contains the following essential information:

- genome configuration
- degree of dependence on host for replication
- gene expression strategies
- noteworthy features of the interaction with the host

Examples of specific viruses in each class are provided, with diagrams of their genomes.

Figure 3.1

A dsDNA genome: *Polyomaviridae, Adenoviridae, Herpesviridae, Poxviridae*

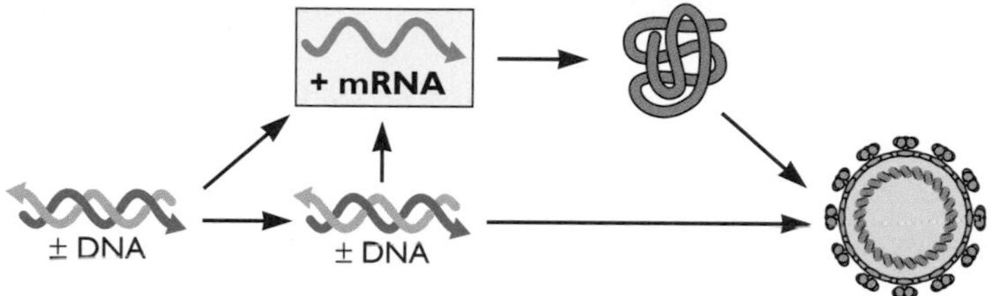

Figures 3.1 to 3.7 Seven classes of viral genomes based on mRNA synthesis and genome replication. Abbreviations: ITR, inverted terminal repeat; TP, terminal protein; poly(A), polyadenylate; UTR, untranslated region; IRES, internal ribosome entry site.

B *Polyomaviridae* (5 kbp)

Genome configuration

- Circular DNA genomes, <10 open reading frames

Degree of dependence on host for replication

- Dependent on cell for both replication and gene expression
- No viral DNA polymerase
- RNA primers and replication fork to replicate DNA, analog of the cellular mechanism

Gene expression strategies

- Temporal control of early and late gene expression
- Splicing is a major regulatory process
- Other decoding tactics: overlapping genes, leaky scanning

Noteworthy features of the interaction with the host

- Polyomavirus and simian virus 40 encode tumor (T) antigens that modulate cell cycle control

C *Adenoviridae* (36–48 kbp)

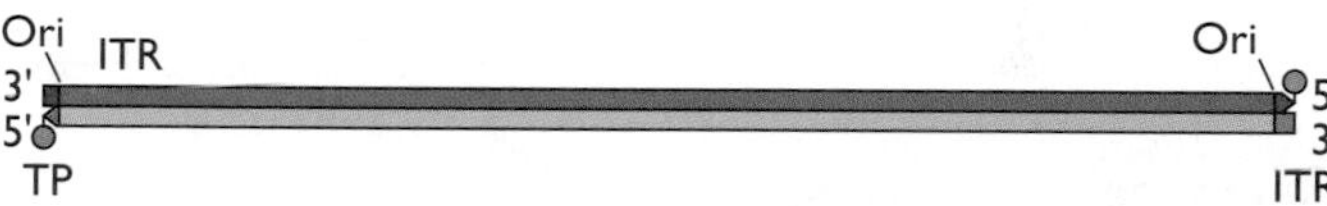

Genome configuration

- Linear DNA, short inverted terminal repeat sequence, protein bound to 5' termini

Degree of dependence on host for replication

- Encodes DNA replication system including DNA polymerase
- Protein priming with unusual strand displacement mechanism

Gene expression strategies

- Temporal control (immediate-early, early, intermediate, and late) of gene expression
- 8 pre-mRNAs that are differentially polyadenylated and/or spliced give rise to many mRNAs translated into viral proteins
- Other decoding tactics: overlapping coding and/or control sequences
- Small viral RNAs produced by host RNA polymerase III (VA RNAs)

Noteworthy features of the interaction with the host

- E1A proteins are powerful transcriptional activators and regulators of the cell cycle

Figure 3.1 ***(continued)***

D ***Herpesviridae*** **(120–220 kbp)**

L
S
TRL
UL
IRL
IRS
US
TRS
OriL
OriS
OriS

Genome configuration

- Linear DNA, some of the biggest DNA genomes
- Genome isomers

Degree of dependence on host for replication

- Genomes encode a complete replication system, including a viral DNA polymerase, and enzymes that synthesize DNA precursors
- Genome replication with RNA primers and replication fork

Gene expression strategies

- Cascade regulation; extent of splicing variable among different subfamilies of herpesviruses
- Encode a variety of transcriptional and posttranscriptional regulators
- Package activators of transcription and other regulator proteins in the virion (tegument)
- Production of many mRNAs from multiple promoters (nested set tactics)

Noteworthy features of the interaction with the host

- Latent infection: hallmark of all herpesviruses

E ***Poxviridae*** **(130–375 kbp)**

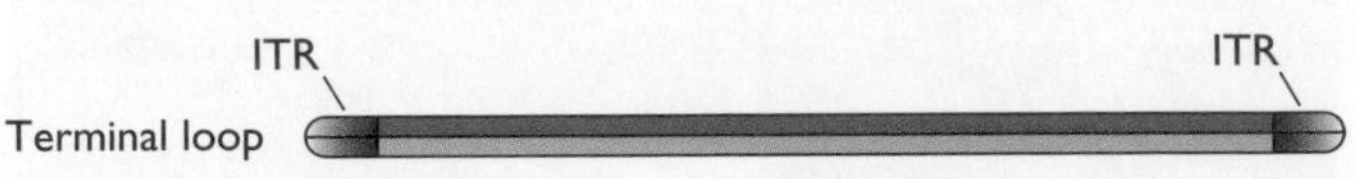

Genome configuration

- Large DNA genome, cross-linked ends
- No known origins

Degree of dependence on host for replication

- Replicates in cytoplasm; nucleus barely used at all
- Genome encodes DNA replication system
- Strand displacement DNA synthesis; no RNA primers

Gene expression strategies

- Genome encodes a complete mRNA synthesis system including RNA polymerase, capping enzyme, and polyadenylation system
- Cascade regulation
- Many viral enzymes packaged in virus particle

Noteworthy features of the interaction with the host

- Encode many immune modulation genes

Figure 3.2

A Gapped, circular, dsDNA genome: ***Hepadnaviridae***

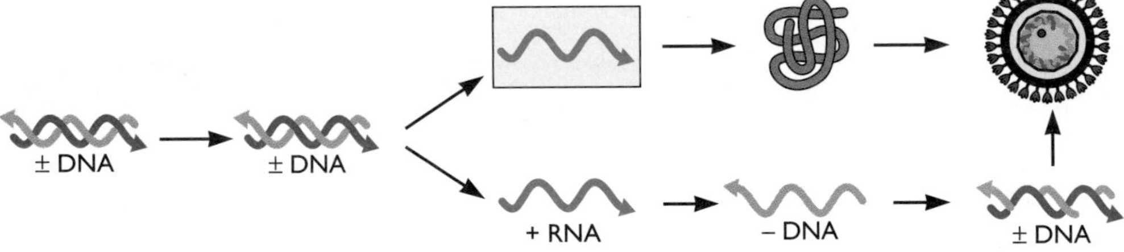

B ***Hepadnaviridae*** **(3.4 kbp)**

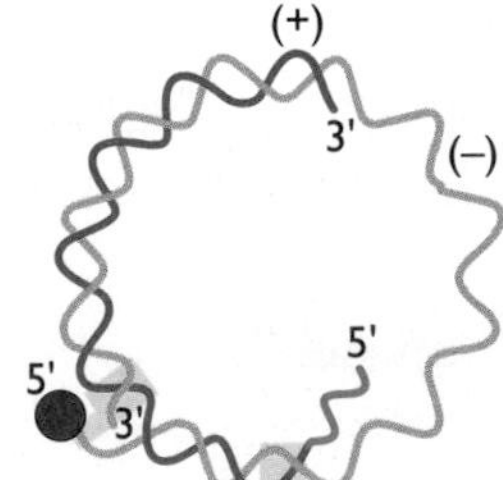

Genome configuration

- Gapped DNA genome; full length but nicked (+) and partial (–) strand
- Reverse transcriptase bound to 5' end of (–) strand; RNA bound to 5' end of (+) strand

Degree of dependence on host for replication

- Gapped DNA genome is repaired in nucleus by host enzyme to form covalently closed circular DNA
- Longer-than-genome-length copy of mRNA is produced by host RNA polymerase II, and the viral reverse transcriptase copies it to make the gapped dsDNA genome

Gene expression strategies

- Splicing to encode seven proteins from four open reading frames (hepatitis B virus)
- Genome information is compressed by overlapping reading frames

Noteworthy features of the interaction with the host

- The X gene product of mammalian members is a potent transcriptional regulator

Figure 3.3

A ssDNA genome: *Circoviridae, Parvoviridae*

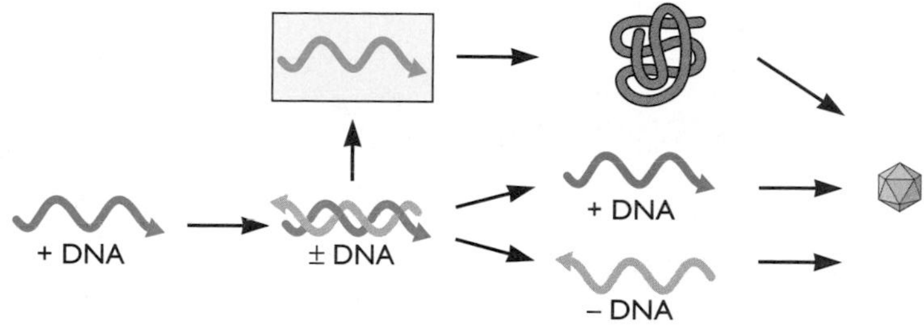

B *Circoviridae* (1.7–2.2 kb)

Genome configuration

- Circular ssDNA genome, antisense and ambisense (TT virus is first human virus with circular ssDNA genome)

Degree of dependence on host for replication

- Highly cell-dependent; genome replication by host DNA replication system

Gene expression strategies

- Two open reading frames (encoding replication and capsid proteins)
- Differential mRNA splicing and translation lead to two forms of replication protein

Noteworthy features of the interaction with the host

- Likely replicate in nucleus, although this is unproven

C *Parvoviridae* (4–6 kb)

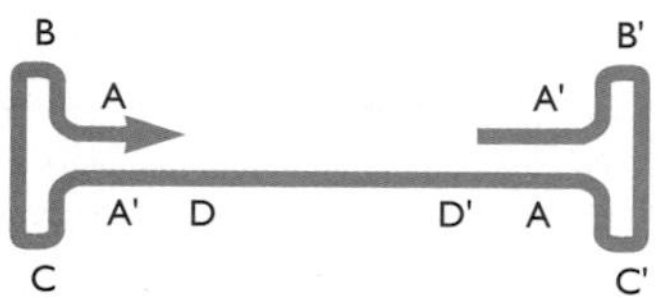

Genome configuration

- Genomes are linear with 3' and 5' hairpins; can be both (–) and (+) strands
- 3' hairpin primes DNA synthesis
- NS1 protein covalently attached to 5' end (autonomous parvoviruses)

Degree of dependence on host for replication

- Highly cell-dependent; genome replication by host DNA replication system

Gene expression strategies

- Two open reading frames (encoding replication and capsid proteins)
- Multiple splicing of pre-RNAs

Noteworthy features of the interaction with the host

- Replicate only in cells in S phase (autonomous parvoviruses) or cells infected by helper virus (dependoviruses)

DNA Genomes

The strategy of having DNA as a viral genome appears at first glance to be simplicity itself: the host genetic system is based on DNA, so the viral genome replication and expression could simply emulate the host's system. Many surprises await those who believe that this is all such a strategy entails.

Double-Stranded DNA (dsDNA) (Fig. 3.1)

Viral genomes may consist of double-stranded or partially double-stranded nucleic acids. There are 22 families of viruses with dsDNA genomes; those that include mammalian viruses are the *Adenoviridae, Herpesviridae, Papillomaviridae, Polyomaviridae,* and *Poxviridae*. These dsDNA genomes may be linear or circular. mRNA is produced by copying of the genome by host or viral DNA-dependent RNA polymerase.

Gapped DNA (Fig. 3.2)

For partially double-stranded DNA, such as the gapped DNA of *Hepadnaviridae*, the gaps must be filled to produce perfect duplexes. This repair process must precede mRNA synthesis. The unusual gapped DNA genome is produced from an RNA template by a virus-encoded reverse transcriptase homologous to that of the retroviruses.

Single-Stranded DNA (ssDNA) (Fig. 3.3)

Five families and one genus of viruses containing ssDNA genomes have been recognized; the *Circoviridae*

Figure 3.4

A dsRNA genome: ***Reoviridae***

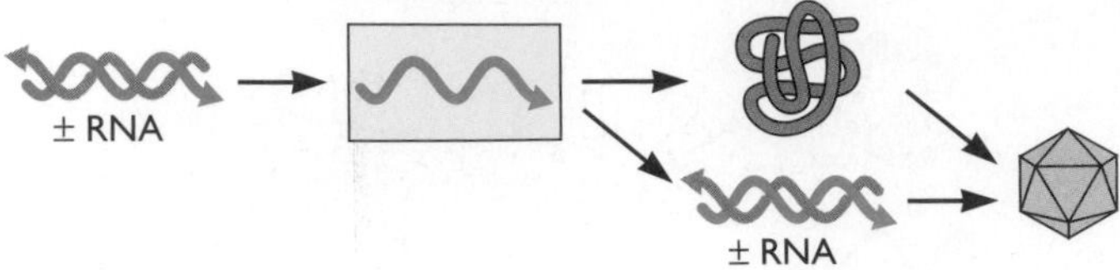

B ***Reoviridae*** **(19–32 kbp in 11 dsRNA segments)**

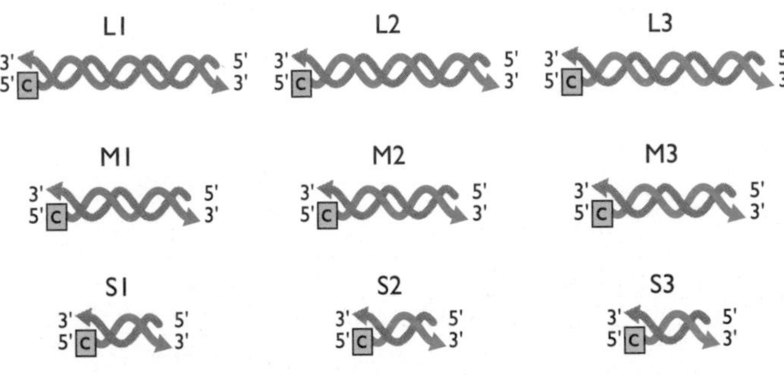

Genome configuration

- Segmented RNA genome

Degree of dependence on host for replication

- Viral RNA polymerase and other enzymes needed for mRNA synthesis are packaged in the particle

Gene expression strategies

- Separate monocistronic mRNA copied from each RNA segment
- mRNAs are capped, lack poly(A), and leave the particle to be translated
- Template remains inside particle
- mRNA incorporated into assembling particle
- Newly assembling particle contains the RNA polymerase that copies the mRNA faithfully to produce the (–) strands to make the dsRNA genome

Noteworthy features of the interaction with the host

- Proteolytic activation in late endosomes and lysosomes and penetration into cytoplasm

and *Parvoviridae* include viruses that infect mammals. ssDNA must be copied into mRNA before proteins can be produced. However, as RNA can only be made from a double-stranded DNA template, no matter what sense the single-stranded DNA is, DNA synthesis **must** precede mRNA production in the replication cycles of these viruses. The single-stranded genome is produced by cellular DNA polymerases.

RNA Genomes

Cells have no RNA-dependent RNA polymerase that can replicate the genomes of RNA viruses or make mRNA from RNA. The simple solution that has evolved is that

Figure 3.5

A ss (+) RNA: ***Coronaviridae, Flaviviridae, Picornaviridae, Togaviridae***

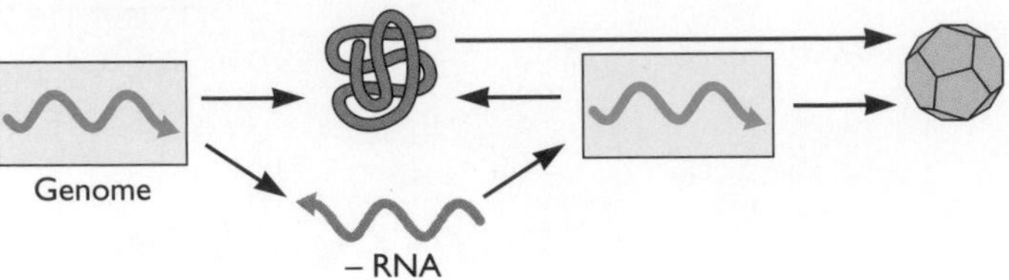

B ***Coronaviridae*** **(28–33 kb)**

Flaviviridae **(10–12 kb)**

Picornaviridae **(7–8.5 kb)**

Togaviridae **(10–13 kb)**

Genome configuration

- Genomes are single molecules of (+) strand RNA
- Picornaviruses: VPg protein, covalently attached to the 5' end, is primer for RNA synthesis
- No 3'-poly(A) in genomes of flaviviruses

Degree of dependence on host for replication

- Encode viral RNA polymerase to replicate genomes and synthesize mRNAs
- Capping enzyme is virus-encoded (all except picornaviruses)

Gene expression strategies

- Genome is mRNA and is translated immediately after infection
- Genome is copied to make (–) sense RNA, which is copied to make more mRNA and more genomes
- Picornaviruses and some flaviviruses: translation from an IRES
- Decoding tactics:

  Production of a polyprotein that is cleaved by proteases into individual proteins

  Production of subgenomic mRNAs (all except *Picornaviridae*)

  Nested set of subgenomic mRNAs with identical nontranslated leader sequence at their 5' ends and coincident 3' ends (*Coronaviridae*)

  Ribosomal frameshifting (*Coronaviridae*)

  Nonsense suppression (*Togaviridae*)

Figure 3.6

A ss (+) RNA with DNA intermediate: *Retroviridae*

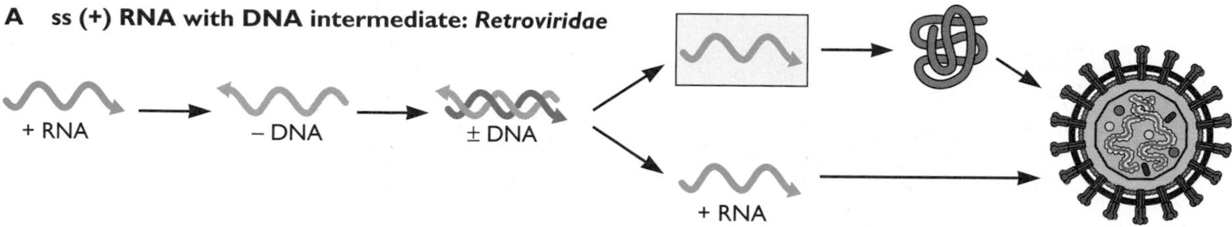

B *Retroviridae* (7–10 kb)

Genome configuration

- Virions contain two copies of the (+) strand RNA genome, each bound with a tRNA primer
- Virions contain reverse transcriptase

Degree of dependence on host for replication

- RNA genome is not mRNA upon infection, even though it is (+) sense, capped, and polyadenylated
- The genome is the template for reverse transcription into dsDNA, which is integrated into the host genome
- Host RNA polymerase II transcribes the DNA into the genomic RNA

Gene expression strategies

- Splicing, mRNA transport, translational regulation, and posttranslational regulation
- Posttranslational tactics: polyproteins, frameshifting, suppression of termination

Noteworthy features of the interaction with the host

- May lead to oncogenic transformation of cells

RNA virus genomes encode novel RNA-dependent RNA polymerases which produce RNA (both genomes and mRNA) from RNA templates. The mRNA produced is readable by host ribosomes. One notable exception is retrovirus mRNAs, which are produced from DNA templates by host RNA polymerase II.

dsRNA (Fig. 3.4)

There are seven families of viruses with dsRNA genomes. The number of dsRNA segments ranges from 1 (*Totiviridae* and *Hypoviridae*, viruses of fungi, protozoans, and plants) to 10 to 12 (*Reoviridae*, viruses of mammals, fish, and plants). While dsRNA contains a (+) strand, it cannot be translated as part of a duplex. The (–) strand RNAs are first copied into mRNAs by a viral RNA-dependent RNA polymerase to produce viral proteins. Newly synthesized mRNAs are encapsidated and then copied to produce dsRNAs.

(+) Strand RNA (Fig. 3.5)

The (+) strand RNA viruses are the most plentiful on this planet; 22 families and 22 genera have been recognized. The *Arteriviridae*, *Astroviridae*, *Caliciviridae*, *Coronaviridae*, *Flaviviridae*, *Picornaviridae*, *Retroviridae*, and *Togaviridae* include viruses that infect mammals. (+) strand RNA genomes usually can be translated directly into protein by host ribosomes. The genome is replicated in two steps. First, the (+) strand genome is copied into a full-length (–) strand. The (–) strand is then copied into full-length (+) strand genomes. A subgenomic mRNA is produced in cells infected with some viruses of this class.

(+) Strand RNA with DNA Intermediate (Fig. 3.6)

The retroviruses are unique (+) strand RNA viruses because a dsDNA intermediate is produced from genomic RNA by an unusual viral RNA-dependent DNA polymerase called reverse transcriptase. This DNA then serves as the template for viral mRNA and genome RNA synthesis by cellular enzymes.

(–) Strand RNA (Fig. 3.7)

Viruses with (–) strand RNA genomes are found in seven families and three genera. Viruses of this class that can infect mammals are found in the *Bornaviridae*, *Filoviridae*, *Orthomyxoviridae*, *Paramyxoviridae*, and *Rhabdoviridae*. By definition, (–) strand RNA genomes cannot be translated directly into protein but first must be copied to make (+) strand mRNA by a viral RNA-dependent RNA polymerase. There are no enzymes in the cell that can produce mRNAs from the RNA genomes of (–) strand RNA viruses. These virus particles therefore contain virus-encoded, RNA-dependent RNA polymerases that produce mRNAs from the (–) strand genome. This strand is also the template for the synthesis of full-length (+) strands, which in turn are copied to produce (–) strand genomes. (–) strand RNA viral genomes can be either single molecules (nonsegmented) or segmented.

A curious strategy of certain (–) strand RNA viruses (e.g., *Arenaviridae* and *Bunyaviridae*) is the mixture of (+) and (–) strand information on a single strand of RNA

Figure 3.7

A ss (–) RNA: *Orthomyxoviridae, Paramyxoviridae, Rhabdoviridae*

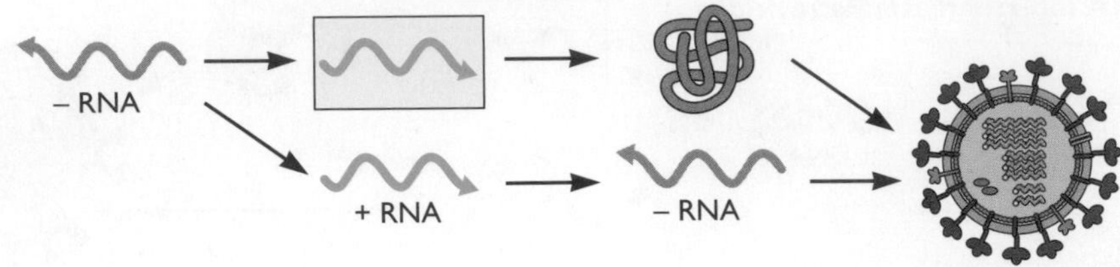

B Segmented genomes: *Orthomyxoviridae* (10–15 kb in 6–8 RNAs)

(–) strand RNA segments

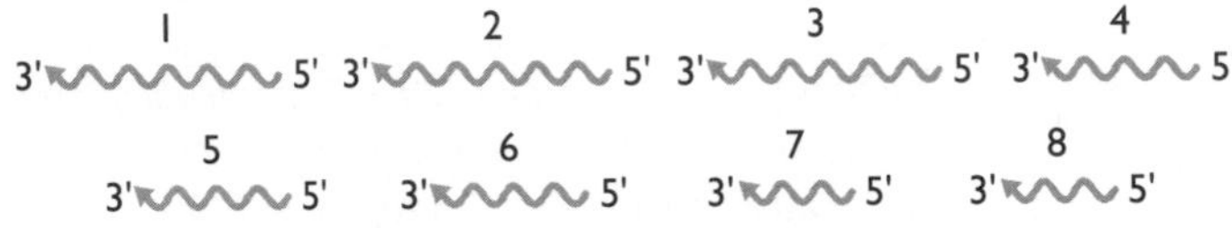

Nonsegmented genomes: *Paramyxoviridae* (15–16 kb)

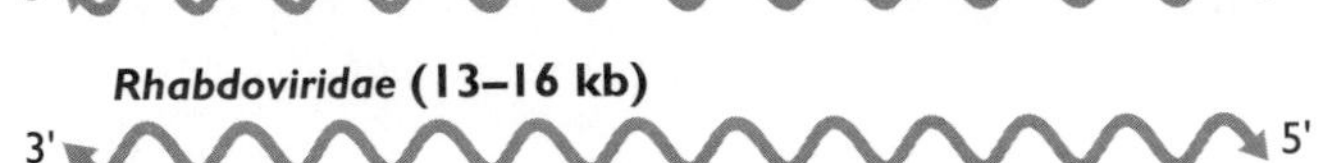

C Ambisense (–) strand RNA
***Arenaviridae* (11 kb in 2 RNAs)**
***Bunyaviridae* (12–23 kb in 3 RNAs)**

Genome configuration

- Genomes are segmented or nonsegmented RNA

Degree of dependence on host for replication

- Cells cannot copy or translate (–) strand RNA
  RNA-dependent RNA polymerase in virus particle

Gene expression strategies

- Capped, polyadenylated mRNAs produced by viral polymerase
- Cap-stealing mechanism for priming transcription: caps from nuclear pre-mRNA
- Nonsegmented genomes: protein level is regulated by transcriptional polarity (more 3' mRNA than 5' mRNA)
- RNA editing in *Paramyxoviridae* (P gene) and *Filoviridae* (membrane protein gene)
- Replication requires synthesis of full-length (+) sense antigenome as template for (–) sense genome
- Monocistronic mRNAs are translated into individual proteins
- Splicing of some mRNAs occurs in nucleus, enabling production of two proteins from one segment (*Orthomyxoviridae*)
- Leaky scanning and reinitiation (*Orthomyxoviridae*)
- RNA transport mechanisms required for export of unspliced mRNA and genome segments into the cytoplasm

Noteworthy features of the interaction with the host

- *Orthomyxoviridae* replicate in the nucleus

Genome configuration

- Arenaviral and bunyaviral genomes are segmented and ambisense; part (–) and part (+) strand

Degree of dependence on host for replication

- Cells cannot copy or translate (–) strand RNA
- RNA-dependent RNA polymerase in virus particle

Gene expression strategies

- Primer for mRNA synthesis is derived from host cytoplasmic mRNAs

Noteworthy features of the interaction with the host

- Bunyavirus virions form intracellularly by budding in the Golgi apparatus

(Fig. 3.7C). Usually the (+) sense information is translated upon entry of the viral RNA genome into cells. The RNA genome is then copied, and the (+) sense information in the copy of the genome is translated. A viral RNA polymerase copies the genome to make a full-length template that is, in turn, copied again to produce the viral genome.

What Do Viral Genomes Look Like in the Virus?

As we have seen, viral genomes exhibit considerable diversity in the form in which they are packaged in virus particles and function in infected cells. Some small RNA and DNA genomes enter cells from virus particles as naked molecules of nucleic acid, whereas others are always associated with specialized viral nucleic acid binding proteins. A fundamental difference between the genomes of viruses and those of hosts is that while viral genomes often are covered with proteins, they are not bound by histones (polyomaviral genomes are a remarkable exception).

While viral genomes are all nucleic acids, they should **not** be thought of as one-dimensional structures. Virology textbooks (this one included) often draw genomes as straight, one-dimensional lines, but this notation is for illustrative purposes only; physical reality is certain to be dramatically different. Genomes have the potential to adopt amazing secondary and tertiary structures (Fig. 3.8). The sequences and structures at the ends of viral genomes are often indispensable for replication, mRNA synthesis, and packaging of the genome into virions (Fig. 3.9).

Coding Strategies

In general, the compact genome of most viruses renders the "one gene, one mRNA" dogma inaccurate. Extraordinary tactics have evolved for information retrieval from viral genomes (Fig. 3.10). In general, the smaller the genome, the greater the compression of genetic information.

What Can Viral Sequences Tell Us?

Knowledge about the physical nature of genomes and coding strategies was first obtained by study of the nucleic acids of viruses. Indeed, DNA sequencing technology was perfected on viral genomes. The first genome of any kind to be sequenced was that of the *Escherichia coli* bacteriophage ϕX174, a circular single-stranded DNA of 5,386 nucleotides. dsDNA genomes of the largest viruses, such as herpesviruses and poxviruses (vaccinia virus), were completely sequenced by the 1990s. Since then, the sequences of over 1,100 different viral genomes have been determined. Published viral genome sequences can be found at the website http://www.ncbi.nlm.nih.gov/PMGifs/Genomes/10239.html.

Viral genome sequences have a variety of important uses, including classification of viruses. Sequence analysis has identified many relationships among diverse viral genomes, providing considerable insight into the origin of viruses. In outbreaks or epidemics of viral infection, even partial genome sequences can provide information on the origin of the virus and the movement of the virus in different popu-

Figure 3.8 Genome structures in cartoons and in real life. (Top) Linear representation of the poliovirus genome. (Bottom) Computer-predicted model of the genome of poliovirus by optimal and suboptimal minimum free energy folding algorithms. The arrow points to the 5′ end of the RNA. It is not known whether the viral RNA folds in this configuration. Genetic and biochemical methods can be used to determine whether the structures actually exist. Adapted from A. C. Palmenberg and J.-Y. Sgro, *Semin. Virol.* **8**:231–241, 1997, with permission. Courtesy of J.-Y. Sgro and A. C. Palmenberg, University of Wisconsin, Madison.

Linear (+) strand RNA genome of a picornavirus

Computer-produced structure of a picornavirus genome

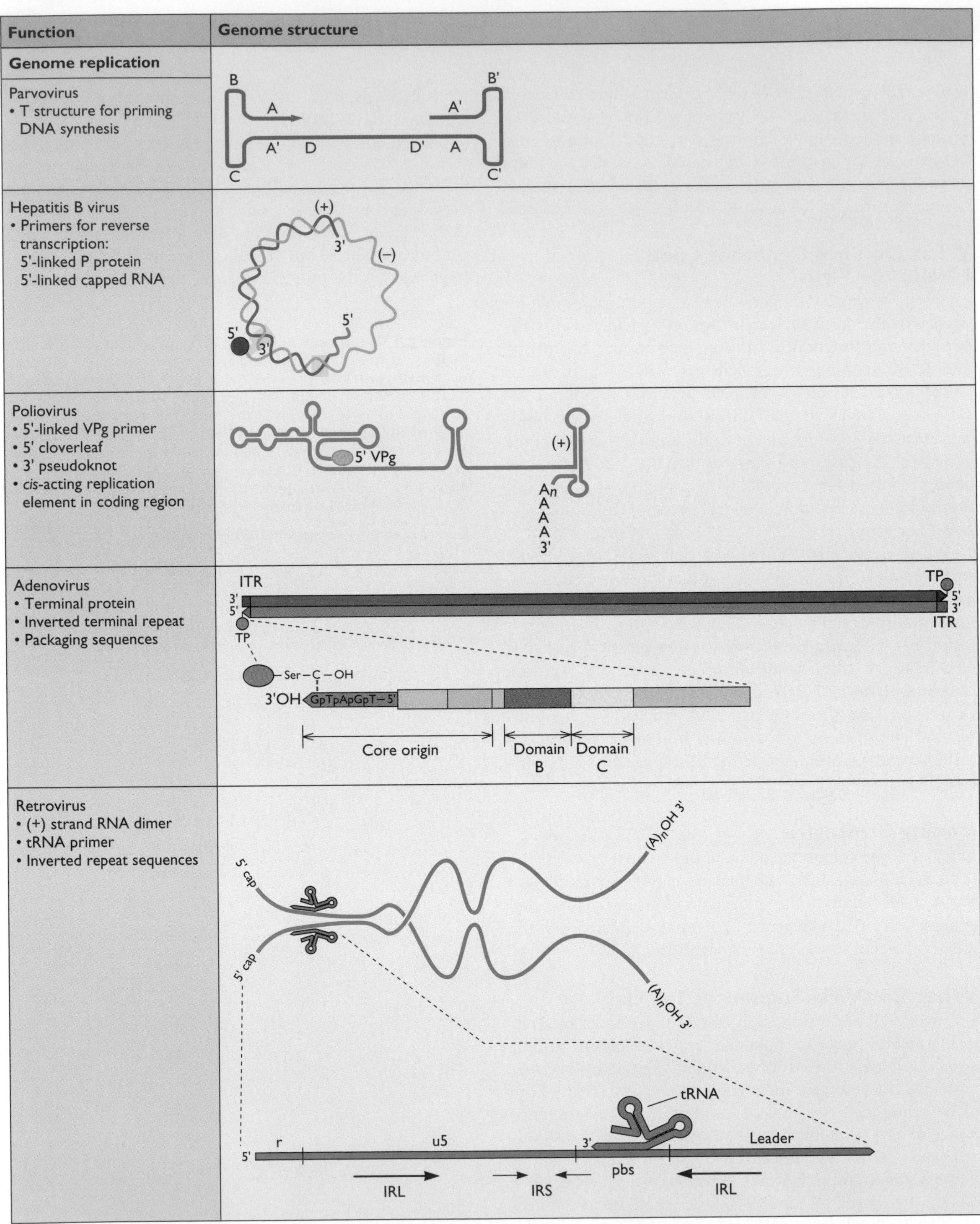

Figure 3.9 Genome structures critical for function. Depicted are unusual structures at the ends of viral DNA and RNA genomes that are important for DNA replication, transcription, packaging, and translation. Abbreviations: ITR, inverted terminal repeat; TP, terminal protein; IRES, internal ribosome entry site; PBS, primer-binding site; DIS, dimerization initiation site; DLS, dimer linkage structure.

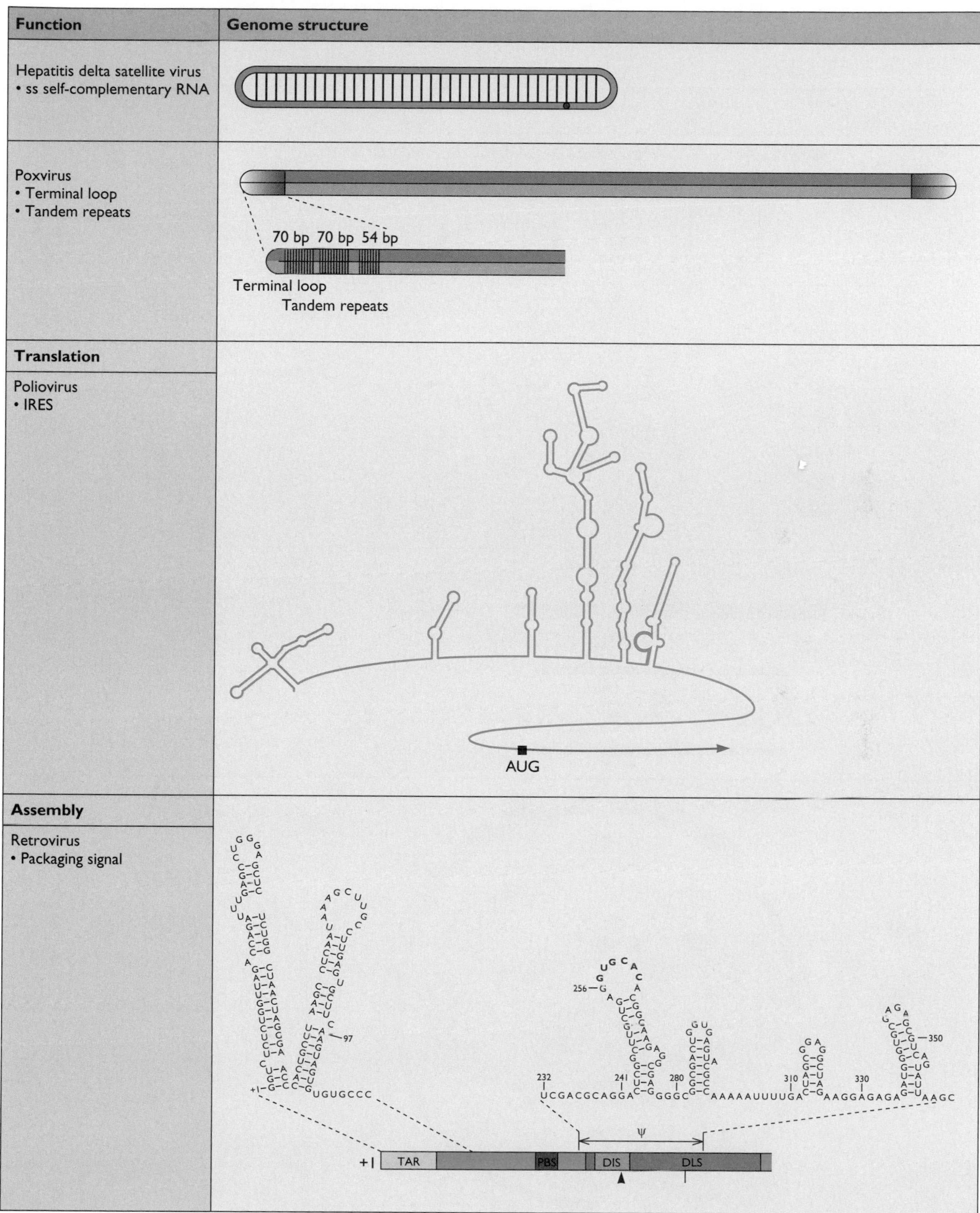
Function
Genome structure
Hepatitis delta satellite virus
• ss self-complementary RNA
Poxvirus
• Terminal loop
• Tandem repeats
70 bp 70 bp 54 bp
Terminal loop
Tandem repeats
Translation
Poliovirus
• IRES
AUG
Assembly
Retrovirus
• Packaging signal
+1
TAR
PBS
DIS
DLS
Ψ

Figure 3.9 ***(continued)***

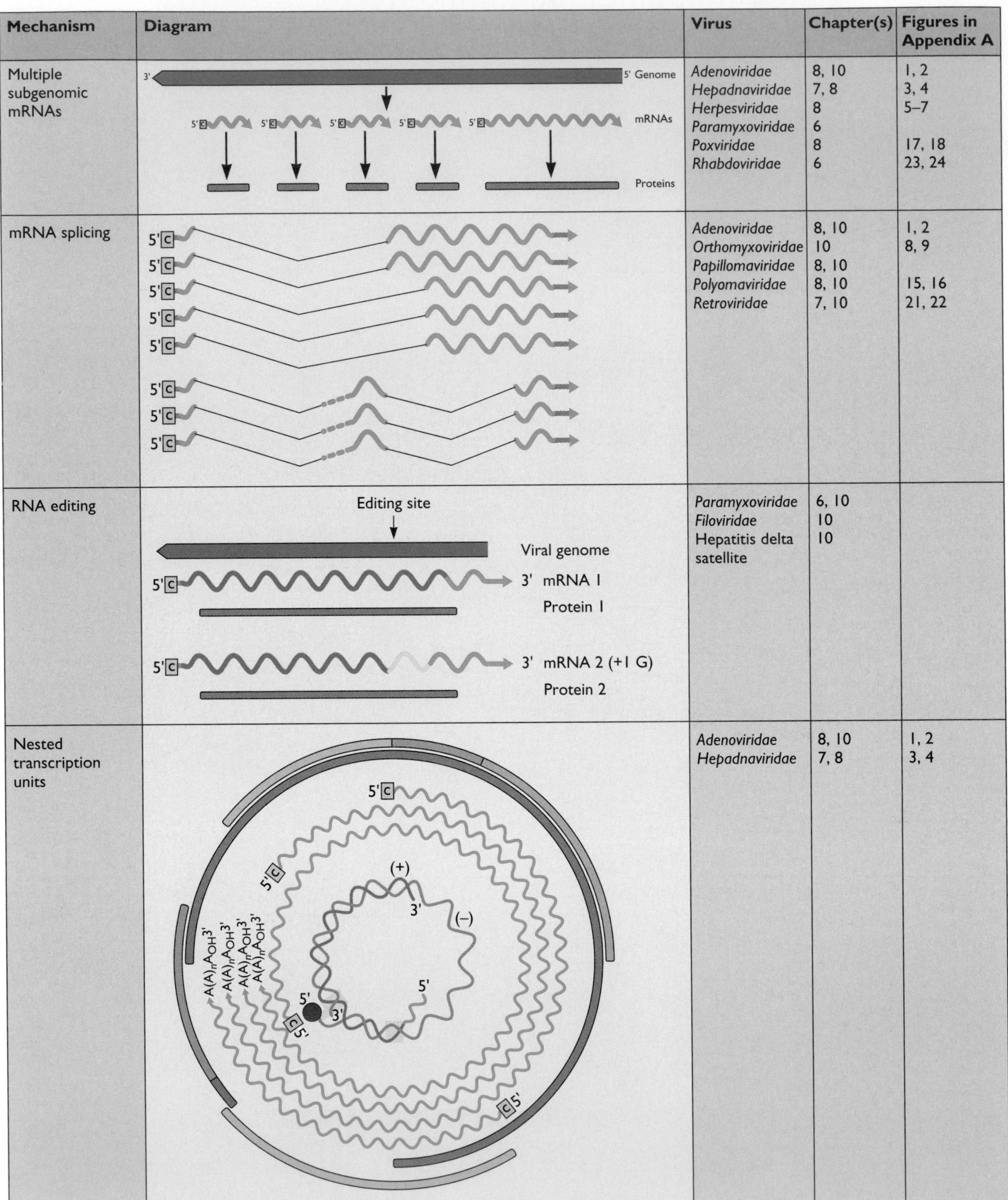

Mechanism	Diagram	Virus	Chapter(s)	Figures in Appendix A
Multiple subgenomic mRNAs	3' 5' Genome; 5' C mRNAs; Proteins	*Adenoviridae* *Hepadnaviridae* *Herpesviridae* *Paramyxoviridae* *Poxviridae* *Rhabdoviridae*	8, 10 7, 8 8 6 8 6	1, 2 3, 4 5–7 17, 18 23, 24
mRNA splicing	5' C	*Adenoviridae* *Orthomyxoviridae* *Papillomaviridae* *Polyomaviridae* *Retroviridae*	8, 10 10 8, 10 8, 10 7, 10	1, 2 8, 9 15, 16 21, 22
RNA editing	Editing site; Viral genome; 5' C 3' mRNA 1; Protein 1; 5' C 3' mRNA 2 (+1 G); Protein 2	*Paramyxoviridae* *Filoviridae* Hepatitis delta satellite	6, 10 10 10	
Nested transcription units	5' C; (+); (–); 3'; 5'; $A(A)_nA_{OH}3'$	*Adenoviridae* *Hepadnaviridae*	8, 10 7, 8	1, 2 3, 4

Figure 3.10 Information retrieval from viral genomes. Different strategies for decoding the information in viral genomes are depicted. IRES, internal ribosome entry site.

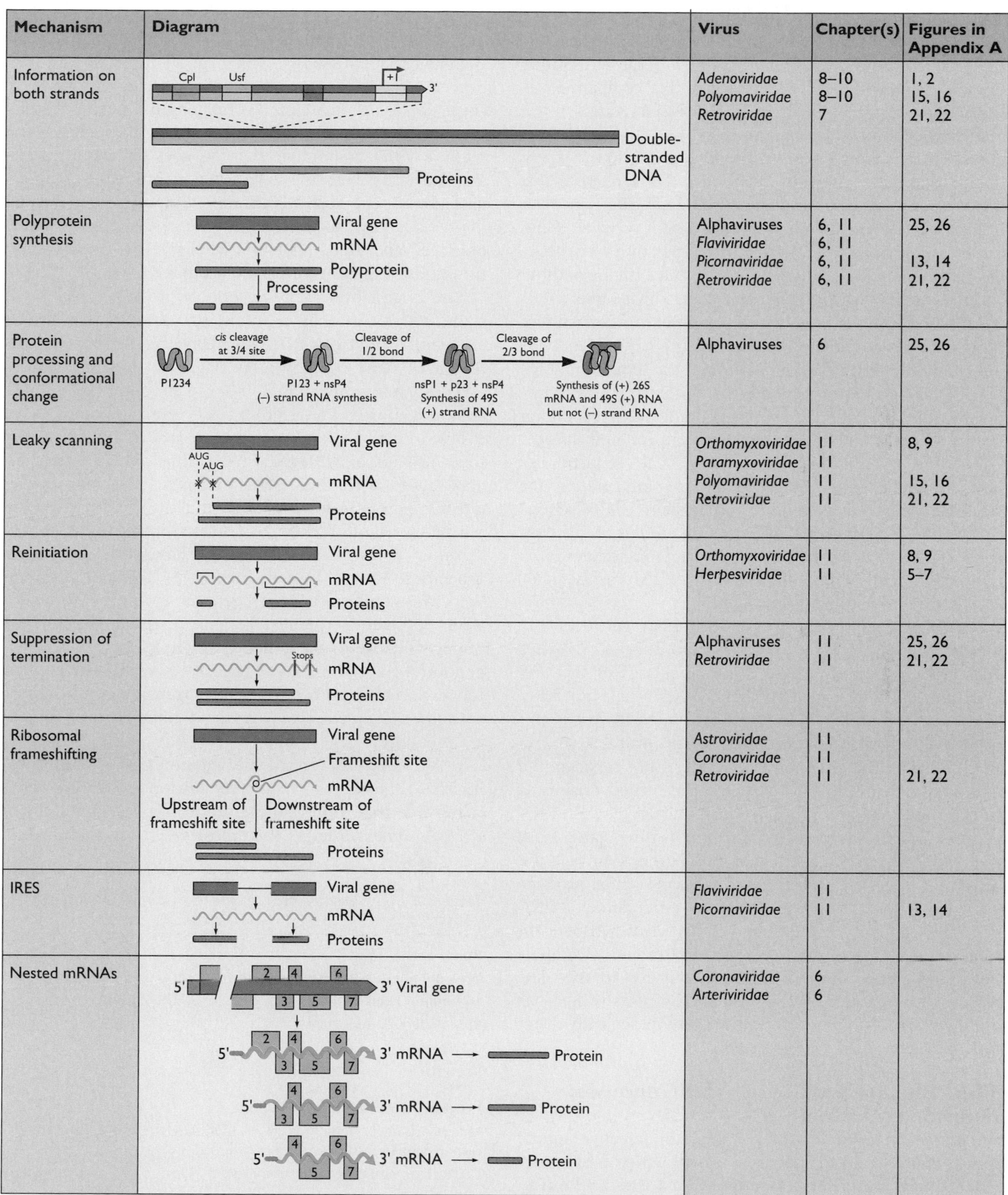

Mechanism	Diagram	Virus	Chapter(s)	Figures in Appendix A
Information on both strands		*Adenoviridae* *Polyomaviridae* *Retroviridae*	8–10 8–10 7	1, 2 15, 16 21, 22
Polyprotein synthesis		Alphaviruses *Flaviviridae* *Picornaviridae* *Retroviridae*	6, 11 6, 11 6, 11 6, 11	25, 26 13, 14 21, 22
Protein processing and conformational change		Alphaviruses	6	25, 26
Leaky scanning		*Orthomyxoviridae* *Paramyxoviridae* *Polyomaviridae* *Retroviridae*	11 11 11 11	8, 9 15, 16 21, 22
Reinitiation		*Orthomyxoviridae* *Herpesviridae*	11 11	8, 9 5–7
Suppression of termination		Alphaviruses *Retroviridae*	11 11	25, 26 21, 22
Ribosomal frameshifting		*Astroviridae* *Coronaviridae* *Retroviridae*	11 11 11	 21, 22
IRES		*Flaviviridae* *Picornaviridae*	11 11	 13, 14
Nested mRNAs		*Coronaviridae* *Arteriviridae*	6 6	

Figure 3.10 ***(continued)***

lations. New viral nucleic acid sequences can be associated with disease and characterized in the absence of standard virological techniques (Chapter 20). A new herpesvirus called human herpesvirus 8 was identified by comparing sequences present in diseased and nondiseased tissues.

However, it quickly became clear that a complete understanding of how viruses reproduce is not found solely in the genome sequence or structure. In retrospect, this fact should have been anticipated. The genome sequence of a virus is at best a biological parts list: it provides some information about the intrinsic properties of a virus (e.g., viral proteins and their composition), but little or nothing about how the virus interacts with cells, hosts, and populations. Because genomes are so complex, understanding them requires a reductionist analysis, the study of individual components in isolation. Although the reductionist approach is often experimentally the simplest, it is also important to understand how the genome behaves among others (population biology) and how the genome changes with time (evolution). Nevertheless, the reductionists have provided much needed detailed information for tractable virus-host systems. These systems allow genetic and biochemical analyses, and provide models of infection in vivo and in vitro. One problem is that viruses and hosts that are difficult or impossible to manipulate in the laboratory are understudied or ignored.

Why are there both viruses with DNA genomes and viruses with RNA genomes? Why aren't there just viruses with DNA genomes? There is no final answer to these questions, but one possibility is that RNA viruses are relics of the "RNA world," a period during which RNA was **the** information **and** catalytic molecule (no proteins yet existed). During this time, possibly billions of years ago, life might have evolved from RNA, and the earliest organisms might have had RNA genomes. Viruses with RNA genomes might have evolved during this time. Later, DNA replaced RNA as the genetic material, perhaps through the action of reverse transcriptases, enzymes that produce DNA from RNA templates. With the emergence of DNA genomes came the evolution of DNA viruses. However, those with RNA genomes were still evolutionarily competitive, and hence they continue to survive to this day. Although the origins of RNA and DNA viruses will probably never be known, we do know that viruses with either type of nucleic acid are successful.

The "Big and Small" of Viral Genomes: Does Size Matter?

Currently, the prize for the smallest nondefective animal virus genome goes to the circoviruses, which possess circular, single-stranded DNA genomes of 1.7 to 2.3 kb (Fig. 3.3B). Members of the *Circoviridae* include chicken anemia virus, beak and feather disease virus of chickens, and TT virus, a ubiquitous human virus of no known consequence. The consolation prize goes to the *Hepadnaviridae*, such as hepatitis B virus, which causes hepatitis and liver cancer in millions of people. Its genome contains 3.4 kb of "unusual" DNA: one strand is full length, but the complementary strand is incomplete (Fig. 3.2). Although it is not really a virus, honorable mention goes to hepatitis delta satellite virus, with a 1.7-kb, single-stranded, but highly base-paired, circular RNA genome. This agent depends on hepatitis B virus to provide an envelope for transmission of its genome. Hepatitis B and delta satellite viruses infect hundreds of millions of people around the world.

The largest known virus genome, a DNA molecule of some 800 kb, is that of mimivirus, a virus that infects amoebae. The largest RNA virus genome, 31 kb, is characteristic of some coronaviruses (Fig. 3.5). Despite detailed analyses, nothing stands to indicate that one size is more advantageous than another. All viral genomes have evolved under relentless selection, so extremes of size must provide advantages. One feature distinguishing big genomes from smaller ones is the presence of many genes encoding proteins for viral replication, nucleic acid metabolism, and evasion of the host defense systems (Table 3.2). In other words, these large viruses have sufficient coding capacity to escape some restrictions of host cell biochemistry. The smallest viable cell genome is thought to be less than 300 genes (predicted from bacterial genome sequences), close to the genetic content of large viral DNA genomes. Nevertheless, the big viruses are **not** cells; their replication absolutely requires host translation machinery, as well as host cell systems to make membranes and generate energy.

What limits the size of viral genomes? It is worth noting that thousands of viruses remain undiscovered, and it is possible that among these are examples with DNA or RNA genomes far larger than those we know of today. But

Table 3.2 Some viral proteins encoded by almost all large DNA viruses

Enzymes
DNA polymerase plus accessory replication proteins
Thymidine kinase
Ribonucleotide reductase
dUTPase
Exonucleases
Proteins that facilitate survival
Many "virulence" gene products to modulate the host's immune response or to foster invasion and spread
Cytokine homologs and receptors
Apoptosis inhibitors
Immune defense system modulators

how do we account for the limitations that we know of? Cellular DNA and RNA molecules which are much longer than those found in virus particles exist. One limitation might be the length of time needed to replicate the viral genome. Long DNA and RNA viral genomes might require unacceptably long periods to replicate, placing the virus at an evolutionary disadvantage. This idea is probably valid only for RNA viruses. There are multiple origins of replication in DNA viral genomes, permitting rapid copying of even very large templates.

The capsid (Box 3.3) might also limit the size of viral genomes. There is a penalty inherent in having a large genome: a huge particle must be provided, and this is not a simple proposition. The 150-kb herpes simplex virus genome resides in a $T = 16$ icosahedral capsid, built from a total of 2,760 protein molecules. In addition, the virion carries multiple copies of 20 or 30 tegument proteins and more than 15 membrane proteins in the viral envelope. Not evident in the inventory of the particle itself are the gene products required to assemble large complicated capsids. In the case of herpes simplex virus, 50 to 60 gene products are needed to build the final particle to house the genome, yet there are only 84 known open reading frames. In other words, 75% of the viral genetic information is dedicated to building the capsid. The largest known DNA viral genomes, those of mimivirus (800 kb) and phycodnaviruses (560 kb) are housed in the biggest capsids constructed with icosahedral symmetry. Although the principles of icosahedral symmetry are quite flexible in allowing a wide range of capsid sizes, perhaps building a very large and stable capsid that can also come apart to release the viral genome is beyond the intrinsic properties of macromolecules.

One solution to the capsid size problem is exemplified by the baculoviruses, which are 250 to 300 nm long and 30 to 60 nm in diameter. The ends of these particles are isometric, while the "middle" is helical, composed of ring-like subunits stacked one on top of another. Particles built with helical symmetry can in principle accommodate very large genomes, but in practice the longest are those of the filoviruses [19-kb (–) strand RNA] and the plant closteroviruses [15- to 19-kb (+) strand RNA]. Longer helical particles are not known, probably because they are easily sheared.

In cells, DNA molecules are much longer than RNA molecules. RNA molecules are less stable than DNA, but in the cell much of the RNA is meant to be used for the synthesis of proteins and therefore need not exceed the size needed to specify the largest polypeptide. However, this constraint does not apply to viral genomes. Yet, the largest viral single-molecule RNA genomes, the 27- to 31-kb (+) strand RNAs of the coronaviruses, are dwarfed by the largest (800-kb) DNA virus genomes. It seems likely that the susceptibility of RNA to nuclease attack limits the size of viral RNA genomes. A solution to RNase susceptibility is to make the RNA segmented [orthomyxoviruses, (–) strand segmented RNA, 10 to 15 kb], or segmented and double-stranded (reoviruses, segmented dsRNA, 19 to 32 kb). However, even these genomes are far below the theoretical size that could be encapsidated, for example, in an icosahedral shell the size of that of a mimivirus.

RNA genome sizes may be limited because there are no enzymes that can correct errors in RNA. RNA polymerases, like their DNA counterparts, make mistakes. DNA polymerases can eliminate errors during polymerization, a process known as proofreading, and the errors can also be corrected after synthesis is complete. Such processes are not available during RNA synthesis. The average error frequencies for RNA genomes are about one misincorporation in 10^4 or 10^5 nucleotides polymerized. In an RNA viral genome of 10 kb, a mutation frequency of 1 in 10^4 nucleotides polymerized would produce about one mutation in every replicated genome. Hence, very long viral RNA genomes, perhaps longer than 32 kb, would sustain too many lethal mutations. Even the 7.5-kb genome of poliovirus exists at the edge of viability: treatment of the virus with the RNA mutagen ribavirin causes a >99% loss of infectivity after a single round of replication.

BOX 3.3

Some curious observations

- Most RNA genomes in the world are in **capsids** (no cells have RNA genomes).
- There are no known animal viruses with "tailed" capsids like those of the bacteriophages.
- Most plant viruses have RNA genomes. DNA viruses of plants exist, of course (e.g., curious geminiviruses and the strange cauliflower mosaic virus).
- Most RNA viruses can cross species barriers; multiple hosts are common. For example, West Nile virus's hosts include birds, horses, and humans.
- DNA viruses tend to be more species specific than RNA viruses.
- Circular RNA viral genomes are rare, but circular DNA genomes abound.

How To Use This New Classification

The information in this chapter can be used as a "road map" for negotiating this book and for planning a virology course. Figures 3.1 through 3.7 serve as the points of departure for detailed analyses of the principles of virology. These figures illustrate the seven classes of viruses, based on mRNA synthesis and genome replication, with references to chapters containing information about the replication and pathogenesis of members of each class. The material in this chapter can be used to structure individual reading, or to design a virology course based on specific viruses or groups of viruses while adhering to the overall organization of this textbook by function. Refer to this chapter and the figures to find answers to questions about specific virus families or individual species. For example, Fig. 3.5 provides information about (+) strand RNA viruses, and Fig. 3.10 indicates specific chapters in which these viruses are discussed.

References

Books

Casjens, S. 1985. *Virus Structure and Assembly*. Jones and Bartlett Publishers, Inc., Boston, Mass.

Reviews

Baltimore, D. 1971. Expression of animal virus genomes. *Bacteriol. Rev.* **35:**235–241.

Domingo, E. 2000. Viruses at the edge of adaptation. *Virology* **270:**251–253.

Drake, J. W., and J. J. Holland. 1999. Mutation rates among RNA viruses. *Proc. Natl. Acad. Sci. USA* **96:**13910–13913.

Hino, S. 2002. TTV, a new human virus with single stranded circular DNA genome. *Rev. Med. Virol.* **12:**151–158.

Joyce, G. F. 2002. The antiquity of RNA-based evolution. *Nature* **418:**214–221.

La Scola, B., S. Audic, C. Robert, L. Jungang, X. de Lamballerie, M. Drancourt, R. Birtles, J.-M. Claverie, and D. Raoult. 2003. A giant virus in amoebae. *Science* **299:**2033.

McGeoch, D. J., and A. J. Davison. 1995. Origins of DNA viruses, p. 67–75. *In* A. J. Giffs, C. H. Calisher, and F. Garcia-Arenal (ed.), *Molecular Basis of Virus Evolution.* Cambridge University Press, Cambridge, England.

Tidona, C. A., and G. Darai. 2000. Iridovirus homologues of cellular genes—implications for the molecular evolution of large DNA viruses. *Virus Genes* **21:**77–81.

Van Etten, J. L., M. V. Graves, D. G. Müller, W. Boland, and N. Delaroque. 2002. Phycodnaviridae—large DNA algal viruses. *Arch. Virol.* **147:**1479–1516.

Papers of Special Interest

Chang, Y., E. Cesarman, M. S. Pessin, F. Lee, J. Culpepper, D. M. Knowles, and P. S. Moore. 1994. Identification of herpesvirus-like DNA sequences in AIDS-associated Kaposi's sarcoma. *Science* **266:**1865–1869.

Crotty, S., C. E. Cameron, and R. Andino. 2001. RNA virus error catastrophe: direct molecular test by using ribavirin. *Proc. Natl. Acad. Sci. USA* **98:**6895–6900.

Palmenberg, A. C, and J. Y. Sgro. 1997. Topological organization of picornaviral genomes: statistical prediction of RNA structural signals. *Semin. Virol.* **8:**231–241.

Sanger, F., A. R. Coulson, T. Friedmann, G. M. Air, B. G. Barrell, N. L. Brown, J. C. Fiddes, C. A. Hutchison III, P. M. Slocombe, and M. Smith. 1978. The nucleotide sequence of bacteriophage phiX174. *J. Mol. Biol.* **25:**225–246.

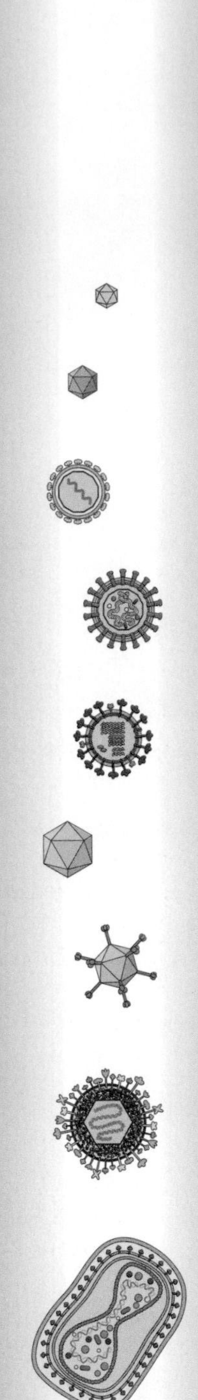

4

Structure

In order to create something that functions properly—a container, a chair, a house—its essence has to be explored, for it should serve its purpose to perfection, i.e. it should fulfill its function practically and should be durable, inexpensive and beautiful.

WALTER GROPIUS
Neue Arbeiten der Bauhaus Werk Stäten, Bauhaus Books, no. 7

Introduction

Virus particles are elegant assemblies of viral, and occasionally cellular, macromolecules. They are marvelous examples of architecture on the molecular scale, with forms beautifully adapted to their functions. Virus particles come in many sizes and shapes (see Fig. 1.10) and vary enormously in the number and nature of the molecules from which they are built. Nevertheless, they fulfill common functions and are built according to general principles that apply to them all. These properties are described in subsequent sections, in which we also discuss some examples of the architectural detail characteristic of different virus families.

Functions of the Virion

Virus particles are designed for effective transmission of the nucleic acid genome from one host cell to another within a single animal, or among host organisms. A primary function of the **virion**, an infectious virus particle, is therefore protection of the genome, which can be damaged irreversibly by a break in the nucleic acid or by mutation during passage through hostile environments. During its travels, a virus particle may encounter a variety of potentially lethal chemical and physical agents, including proteolytic and nucleolytic enzymes, extremes of pH or temperature, and various forms of natural radiation. In all virus particles, the nucleic acid is sequestered within a sturdy barrier formed by extensive interactions among the viral proteins comprising the protein coat. Such protein-protein interactions maintain surprisingly stable capsids: many virus particles composed of only protein and nucleic acid survive exposure to large variations in the temperature, pH, or chemical composition of their environment. Some, such as certain picornaviruses, are even resistant to strong detergents like sodium dodecyl sulfate. The highly folded nature of coat proteins and their dense packing within virions render them largely inaccessible to proteolytic enzymes. Some viruses also possess an **envelope**, derived from cellular membranes, into which viral glycoproteins have been inserted. The envelope adds not only a protective lipid membrane, but also an external layer of

Table 4.1 Functions of virion proteins

Protection of the genome
Assembly of a stable, protective protein shell
Specific recognition and packaging of the nucleic acid genome
In many virus particles, interaction with host cell membranes to form the envelope
Delivery of the genome
Specific binding to external receptors of the host cell
Transmission of specific signals that induce uncoating of the genome
Induction of fusion with host cell membranes
Interaction with internal components of the infected cell to direct transport of the genome to the appropriate site
Other interactions with the host
With cell host components to ensure an efficient infectious cycle
With cellular components for transport to intracellular sites of assembly
With the host immune system

protein and sugars formed by the glycoproteins. Like the cellular membranes from which they are derived, viral envelopes are impermeable and block entry of chemicals or enzymes in aqueous solution into virus particles.

To protect the nucleic acid genome, virus particles must be stable structures. However, virions must also attach to an appropriate host cell and deliver the genome to the interior of that cell, where the particle is at least partially disassembled. The functions of virus particles therefore depend on very stable intermolecular interactions among their components during assembly, egress from the virus-producing cell, and transmission. But these interactions must be reversed readily during entry and uncoating in a new host cell. The molecular mechanisms by which these apparently paradoxical requirements are met are understood in only a few cases. Nevertheless, it is clear that contact of a virion with the appropriate cell-surface receptor, or exposure to a specific intracellular environment, can trigger substantial conformational changes. Virus particles are therefore **metastable structures** that have not yet attained the minimum free energy conformation. The latter state can be gained only when an unfavorable energy barrier is surmounted, following induction of the irreversible conformational transitions associated with attachment and entry. Virions are not simply inert structures. Rather, they are molecular machines that play an active role in delivery of the nucleic acid genome to the appropriate host cell and initiation of the reproductive cycle.

Specific functions of virion proteins associated with protection and delivery of the genome are summarized in Table 4.1.

Nomenclature

Virus architecture is described in terms of **structural units** of increasing complexity, from the smallest biochemical unit (the polypeptide chain) to the infectious particle (or virion). These terms, which will be used throughout this text, are defined in Table 4.2. Although virus particles are complex assemblies of macromolecules exquisitely suited for protection and delivery of viral genomes, they are constructed according to the general principles of biochemistry and protein structure. In this and subsequent chapters, we describe structural features of viral proteins using the nomenclature summarized in Box 4.1.

Methods for Studying Virus Structure

The method most widely used in the examination of virus structure and morphology is electron microscopy. This technique, which has been applied to viruses since the middle decades of the last century, traditionally relied on negative staining of purified virus particles (or of sections of infected cells) with an electron-dense material, such as uranyl acetate or phosphotungstate. It can yield quite detailed and often beautiful images (Fig. 1.7; see Appendix A) and provided the first rational basis for the classification of viruses. The greatest contrast between virus particle and stain occurs where portions of the folded protein chain pro-

Table 4.2 Nomenclature used in description of virus structure

Term	Synonym	Definition
Subunit (protein subunit)	None	Single, folded polypeptide chain
Structural unit	Protomer, asymmetric unit	Unit from which capsids or nucleocapsids are built; may comprise one protein subunit or multiple, different protein subunits
Morphological unit	Capsomere	Surface structures (e.g., knobs, projections, clusters) seen by electron microscopy. Because these structures do not necessarily correspond to structural units (see text), the term is restricted to descriptions of electron micrographs of viruses.
Capsid	Coat	Protein shell surrounding the nucleic acid genome
Nucleocapsid	Core	Nucleic acid-protein assembly packaged within the virion; used when this complex is a discrete substructure of a particle
Envelope	Viral membrane	Host cell-derived lipid bilayer carrying viral glycoproteins
Virion	None	Infectious virus particle

BOX 4.1

The description of protein structure

The **primary structure** (A) of a polypeptide chain is its amino acid sequence, conventionally written in the N- to C-terminal direction. Throughout this text, amino acid sequences are given in the one-letter code, summarized at the right. The **secondary structure** (B) describes regular, local structures formed in proteins. As illustrated, these are stabilized by numerous hydrogen bonds. The α-helix (left) and β-sheet (right) are by far the most common secondary structures. The former is formed by amino acids that are consecutive in the primary sequence. However, individual β-strands in a β-sheet are connected by sharp turns or larger loops, and a β-sheet can therefore be formed with amino acids that are not close neighbors in the primary sequence. β-sheets may be antiparallel (shown) or parallel, when all strands possess the same orientation. In the simplified diagrams of protein structure that have become the convention in the field, α-helices are represented by cylinders or coiled ribbons, and the individual strands of a β-sheet are shown as arrows drawn in the N- to C-terminal direction of the strand. These allow the **tertiary structure** (C), the folded structure of the polypeptide chain, to be summarized in a clear, accessible fashion. The large number of high-resolution protein structures that has now been obtained has established that tertiary structure is based on a finite number of structural **motifs**. These are combinations of a limited number of secondary-structure units in a specific geometric arrangement, as illustrated with the Greek key motif. This motif is represented in a two-dimensional **topology diagram**, which summarizes the orientations of and connections among secondary-structure units. In general, several such motifs interact in the folded protein to form globular structures termed **domains.** A domain is a part of a protein that can fold into a stable tertiary structure independently of other parts of the protein. For example, the β-barrel jelly roll domain, which is found in many viral capsid proteins, is formed from two Greek key motifs linked to one another by four connections across both the top and bottom of the barrel-shaped structure formed by the β-sheets. Simple proteins contain a single domain, but many proteins are multidomain structures in which each domain is an independent structural unit and performs a specific function. Because of these properties, it is often possible to study the functions of individual domains of complex

(continued)

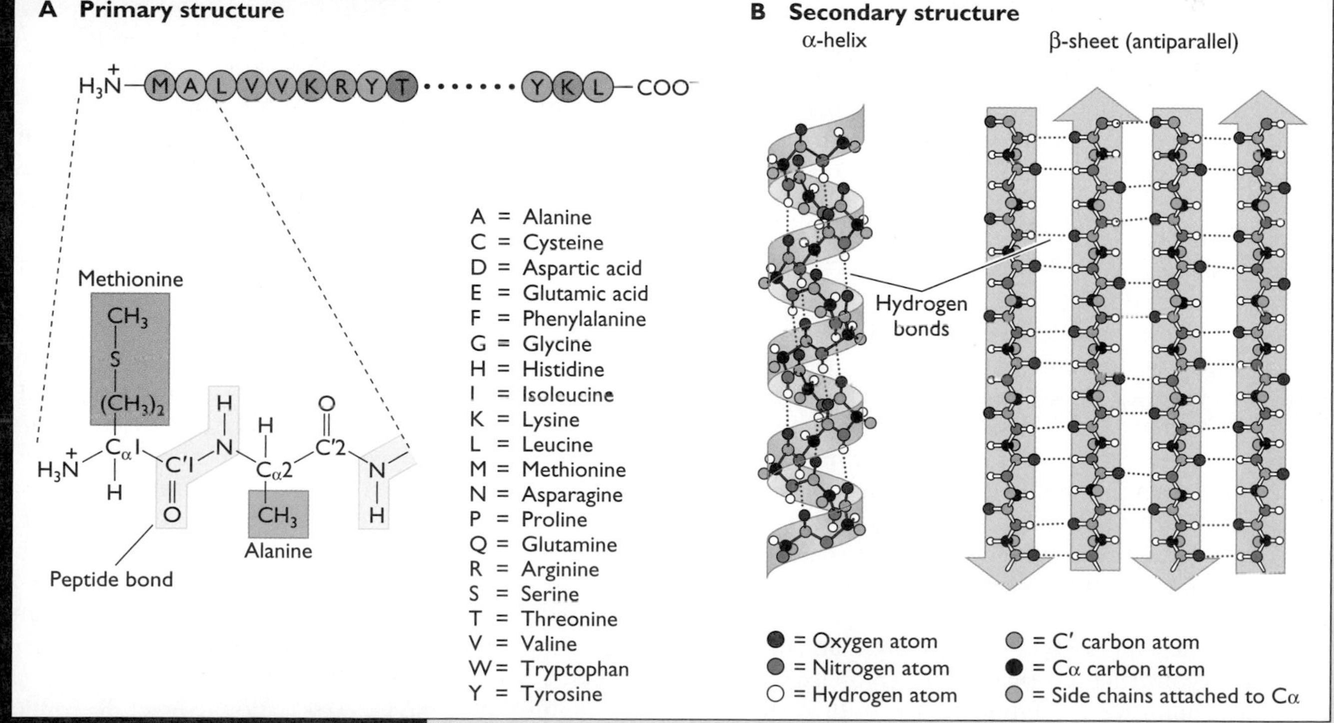

BOX 4.1 (continued)

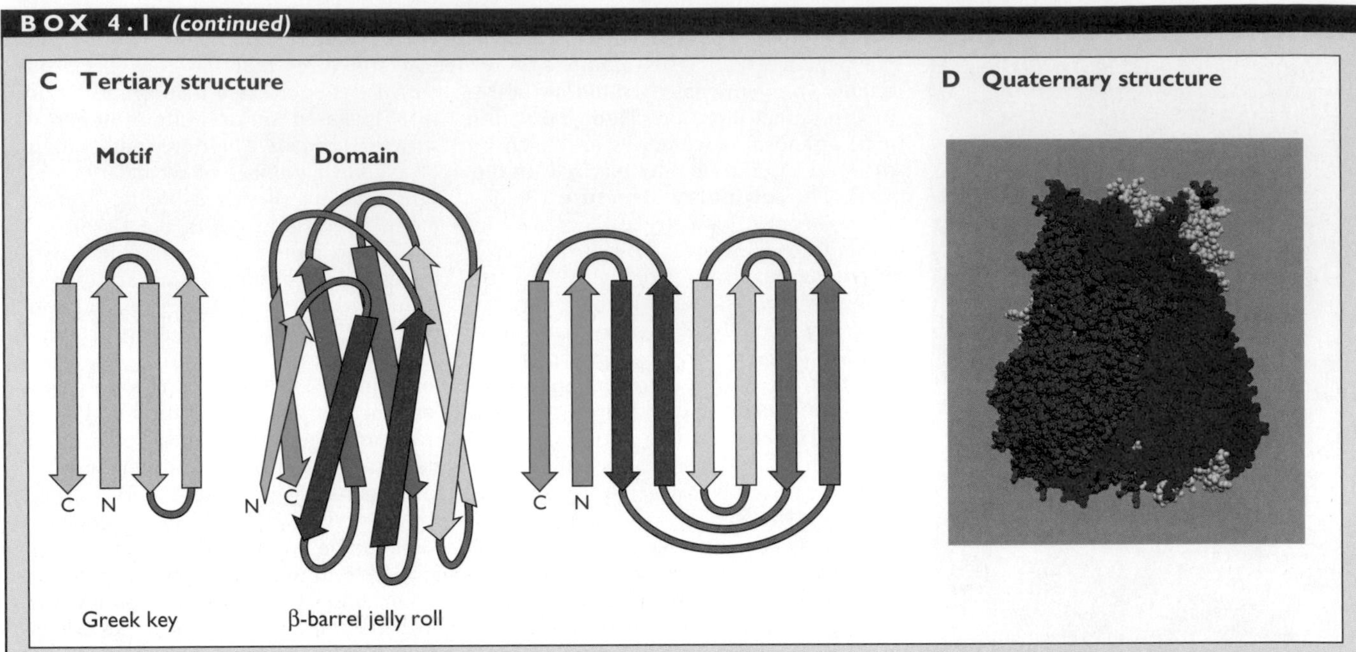

proteins. Many functional proteins comprise a single folded polypeptide chain. However, many others function only as multimeric structures containing several copies of the same polypeptide chain (**homomeric proteins**) or two or more different polypeptide chains (**heteromeric proteins**). The **quaternary structure** (D) of a protein describes the arrangement of and interactions among the individual folded proteins (**subunits**) in such multimeric proteins. Shown is the human adenovirus type 2 hexon, the major structural unit of the capsid, which is a homotrimer of adenoviral protein II. Each of the three molecules of protein II in the hexon (shown in different colors) contains two β-barrel jelly roll domains. Most other virion proteins described in this chapter are also multimeric. Panel C is adapted from C. Branden and J. Tooze, *An Introduction to Protein Structure* (Garland Publishing, Inc., New York, N.Y., 1991), with permission.

trude from the virion surface. Consequently, surface knobs or projections, termed **morphological units**, are the main features identified by negative-contrast electron microscopy. However, these structures are often formed by protein domains from neighboring structural units in the virion, and so their organization does not necessarily correspond to that of the individual proteins that form the capsid shell. Even when virus structure is well preserved and a high degree of contrast can be achieved, the minimal size of an object that can be distinguished by classical electron microscopy, its **resolution**, is limited to 50 to 75 Å. Detailed structural interpretation of images of negatively stained virus particles, which possess dimensions on a scale of a few hundred to a few thousand angstroms, is therefore impossible. Cryo-electron microscopy, in which samples are rapidly frozen and examined at very low temperatures in a hydrated, vitrified (noncrystalline, glasslike) state, preserves native structure, and allows direct visualization of the contrast inherent in the virus particle rather than that between virus and stain. When combined with computerized mathematical methods of image analysis and three-dimensional reconstruction (Fig. 4.1), cryo-electron microscopy can improve resolution to 10 to 20 Å. Indeed, sufficient resolution (less than 10 Å) was attained in recent applications of these techniques to identify secondary-structure features of the core protein of hepatitis B virus, and of transmembrane domains of viral glycoproteins. Within the past decade, cryo-electron microscopy has become a standard tool of structural biology. Its application to virus particles has provided a wealth of previously inaccessible information about the external and internal structures of multiple members of at least twenty virus families.

The diameter of an α-helix in a protein is on the order of 10 Å, so even the highest resolution attained by these sophisticated techniques cannot reveal the molecular interactions that cement structural units in the virion. Such information can be obtained by X-ray crystallography (Fig. 4.2), provided that the virus yields crystals suitable for X-ray diffraction. It has been known for more than 60 years that simple plant viruses can be crystallized, and the first high-resolution virus structure determined was that of tomato bushy stunt virus. Since this feat was accomplished in 1978, high-resolution structures of a number of increasingly larger animal viruses have been determined, placing our understanding of the principles of capsid structure on a firm molecular foundation. Although they cannot capture dynamic processes, atomic-level descriptions of specific viruses, or of individual viral proteins, have also

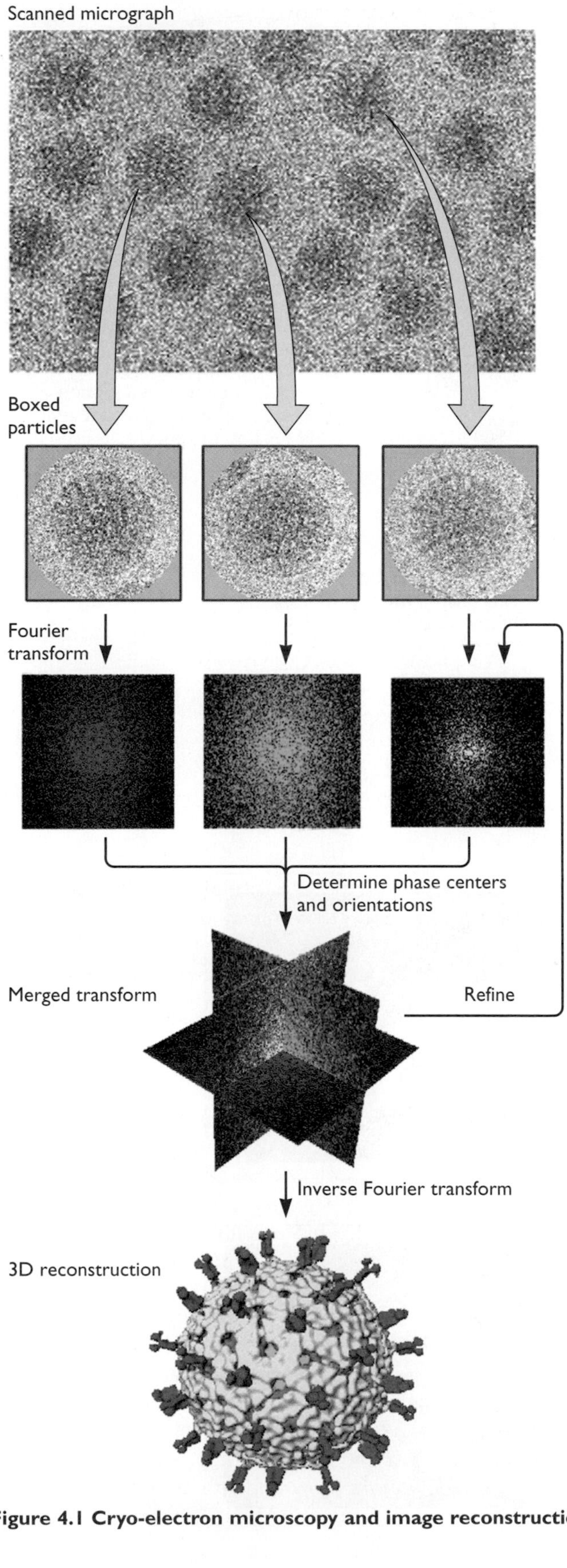

Scanned micrograph

Boxed particles

Fourier transform

Determine phase centers and orientations

Merged transform

Refine

Inverse Fourier transform

3D reconstruction

Concentrated preparations of purified virus particles are prepared for cryo-electron microscopy by rapid freezing on an electron microscope grid so that a glasslike, noncrystalline (vitrified) water layer is produced. This procedure avoids damage to the sample that can be caused by crystallization of the water as well as alterations that may be caused by chemical modification or dehydration during conventional negative-contrast electron microscopy. The sample is maintained at or below −160°C during all subsequent operations, including microscopy at low electron doses and recording of images. Fields containing sufficient numbers of vitrified virus particles are identified by transmission electron microscopy at low magnification (to minimize sample damage from the electron beam) and photographed at high resolution. Because the inherent contrast of unstained virus particles is low, defocusing of the objective lens of the microscope is used to generate phase contrast. Images may be recorded at different levels of defocus to enhance features of different sizes in the sample.

These electron micrographs can be treated as two-dimensional projections (Fourier transforms) of the particles. Three-dimensional structures can be reconstructed from such two-dimensional projections by mathematically combining the information given by different views of the particles, which may be provided by particles lying in different orientations or by tilting the sample in the electron microscope. For the purpose of reconstruction, the images of different particles are therefore treated as different views of the same structure.

For reconstruction, micrographs are first digitized for computer processing. Each particle to be analyzed is then framed and centered inside a box, and its orientation is determined by application of programs that orient the particle on the basis of its icosahedral symmetry. The parameters that define the orientation of the particle must be determined with a high degree of accuracy, for example, to within 1° for even a low-resolution reconstruction (>40 Å). These parameters are improved in accuracy (**refined**) by comparison of different views (particles) to identify common data.

Once the orientations of a number of particles sufficient to represent all parts of the asymmetric unit of the icosahedron have been determined, a low-resolution three-dimensional reconstruction is calculated from the initial set of two-dimensional projections by using computerized algorithms.

This reconstruction is refined by including data from additional views (particles). The number of views required depends on the size of the particle and the resolution sought. The reconstruction is initially interpreted in terms of the external features of the virus particle. Various computational and computer graphics procedures have been developed to facilitate interpretation of internal features. Courtesy of B. V. V. Prasad, Baylor College of Medicine.

And is it not true that even the small step of a glimpse through the microscope reveals to us images that we should deem fantastic and over-imaginative if we were to see them somewhere accidentally, and lacked the sense to understand them.
Paul Klee, *On Modern Art*, translated by Paul Findlay (London, United Kingdom, 1948)

Figure 4.1 Cryo-electron microscopy and image reconstruction illustrated with images of rotavirus.

A

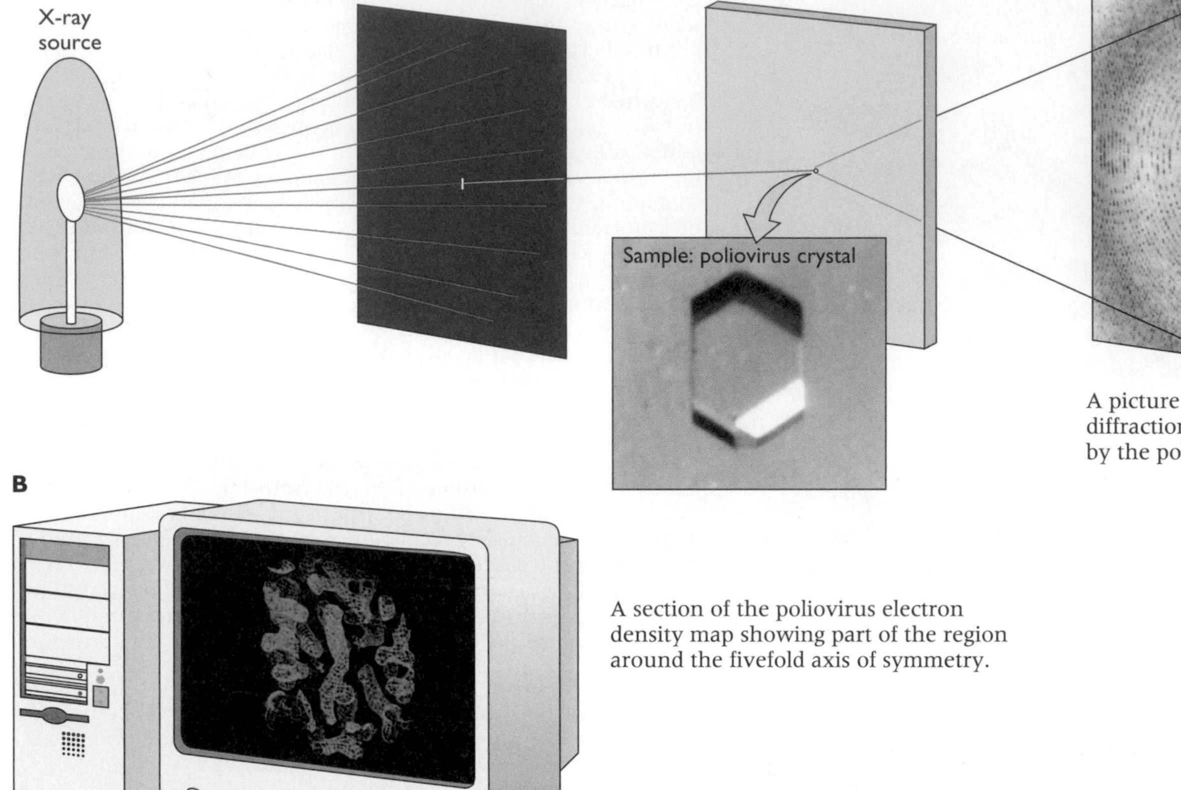

A picture of a section of the diffraction pattern generated by the poliovirus crystal.

B

A section of the poliovirus electron density map showing part of the region around the fivefold axis of symmetry.

C

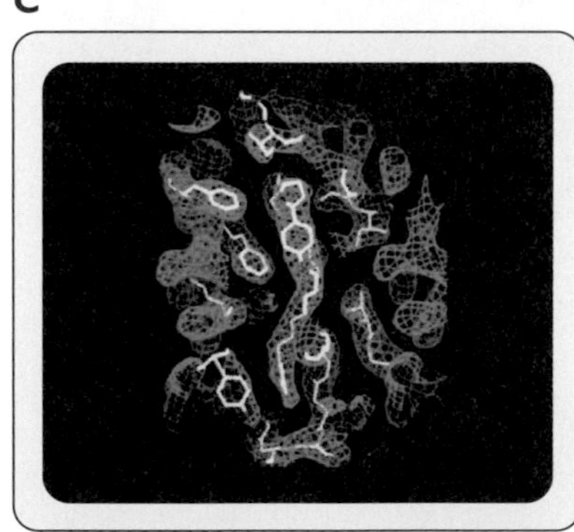

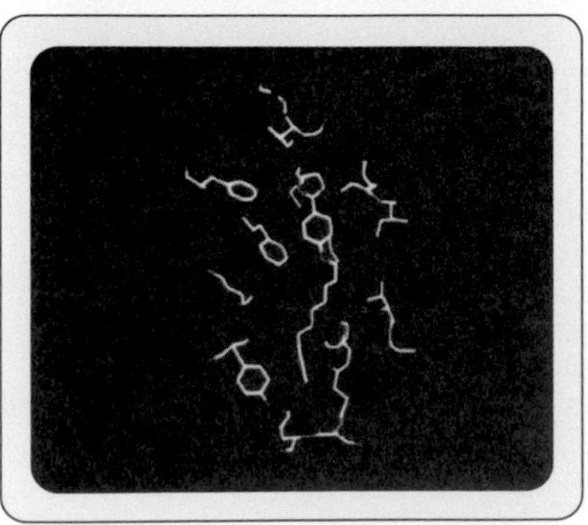

The same section of the map through one plane of the virus particle with segments of the structure built to fit the electron density (left). The built structure without the electron density is shown below. The C–C bonds and atoms are white, blue indicates nitrogen, and red indicates oxygen. The double-ring structure with the long aliphatic chain is an antiviral drug that is bound to the poliovirus particles in the crystal.

D

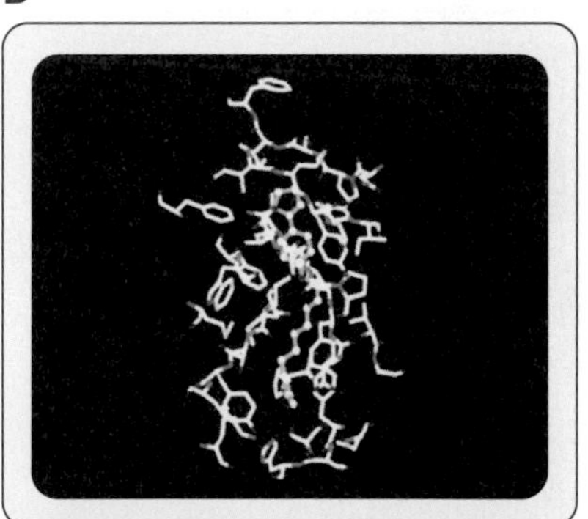

A portion of the completely built virus structure as a ball-and-stick model. The region displayed is that surrounding the drug binding pocket. A portion of this structure corresponds to the electron density shown in panel C. The antiviral drug is shown in green.

E

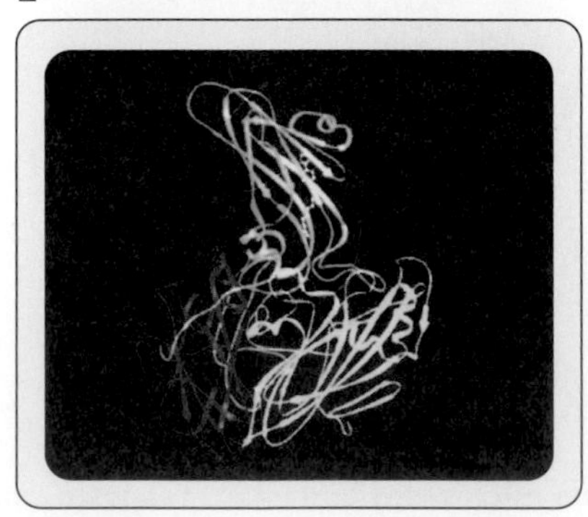

A portion of the virus structure shown as a ribbon diagram, with the three proteins that form the surface, VP1, VP2, and VP3, colored blue, yellow, and red, respectively.

Figure 4.2 Determination of virus structure by X-ray diffraction. A virus crystal is composed of virus particles arranged in a well-ordered three-dimensional lattice. When the crystal is bombarded with a monochromatic X-ray beam, each atom within the virus particle scatters the radiation. Interactions of the scattered rays with one another form a diffraction pattern that is recorded (A). Each spot contains information about the position and the identity of all atoms in the crystal. The locations and intensities of the spots are stored electronically. Determination of the three-dimensional structure of the virus from the diffraction pattern requires information that is lost in the X-ray diffraction experiment. This missing information (the phases of the diffracted rays) can be retrieved by collecting the diffraction information from otherwise identical (isomorphous) crystals, in which the phases have been systematically perturbed by the introduction of heavy metal atoms at known positions; heavy metal atoms scatter X rays more strongly because of their high electron density. Comparison of the two diffraction patterns, by using computer programs to calculate how spot intensities at each position have been altered by the presence of the heavy metal, yields the phases. This process is called multiple isomorphous replacement. Alternatively, if the structure of a related molecule is known, the diffraction pattern collected from the crystal can be interpreted by using the phases from the known structure as a starting point and subsequently using computer algorithms to iteratively calculate the actual values of the phases. This method is known as molecular replacement. Once the phases are known, the intensities and spot positions from the diffraction pattern are used to calculate the locations of the atoms within the crystal, again by using computer programs. The product of this mathematical analysis is an electron density map (B), which is a map of the location of electron-dense atoms (typically C, N, O, and S), as well as of the covalent bonds within and between amino acids in the virus. A model of the peptide backbone linking amino acids within proteins is built by tracing the electron density through each section of the map and specifying the location of each atom (C and D). The amino acids are identified by the shapes of the electron density of the side chains and by using the primary sequence of the proteins. When the entire structure is built, the completed images can be reassembled and visualized in various representations (D and E). Courtesy of J. Hogle, Harvard Medical School, and M. Chow, University of Arkansas for Medical Sciences.

greatly improved our knowledge of mechanisms of attachment and entry of virions and of virus assembly. They have also provided new opportunities for the design of antiviral drugs.

More complex viruses cannot yet be examined directly by X-ray crystallography: they generally do not form suitable crystals, and the larger viruses lie beyond the power of the current procedures by which X-ray diffraction spots are converted into a structural model (Fig. 4.2). Nevertheless, their component proteins can be examined by this method (Box 4.2), an approach that has been particularly important in illuminating mechanisms of attachment and entry of en-

BOX 4.2

Human immunodeficiency virus type 1, a virus understood in great structural detail

This causative agent of acquired immunodeficiency syndrome (AIDS) was identified about 20 years ago. Yet, as illustrated in the figure, the structures of most of the proteins present in the complex virions have been solved at high resolution, by one or more of the methods described in the text. The need to develop effective drugs for the treatment of AIDS has provided an important impetus for structural studies of human immunodeficiency virus type 1 proteins.

Human immunodeficiency virus type 1. The virion is depicted at the center, with the structures of the component proteins illustrated on either side in surface representation. RNA is green. Adapted from H. Berman et al., *Am. Sci.* **90:**350–359, 2002, with permission.

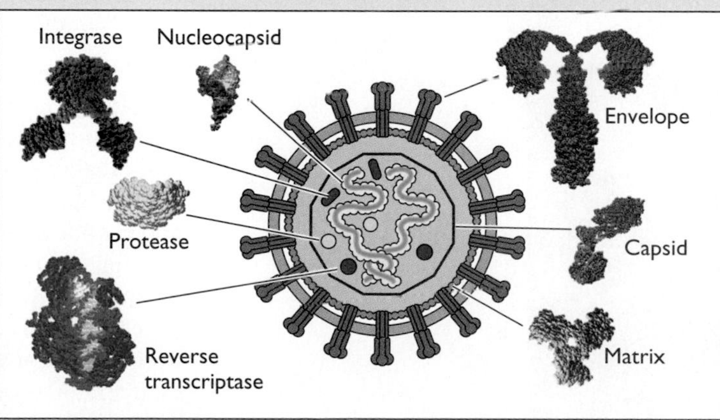

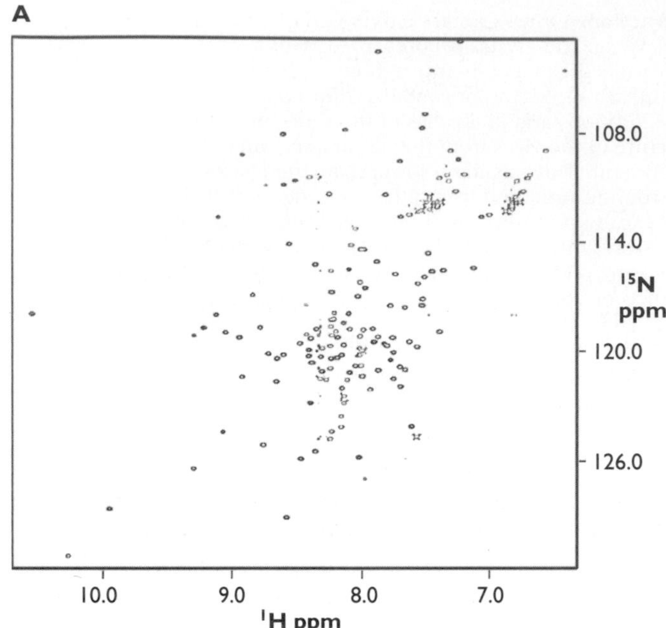

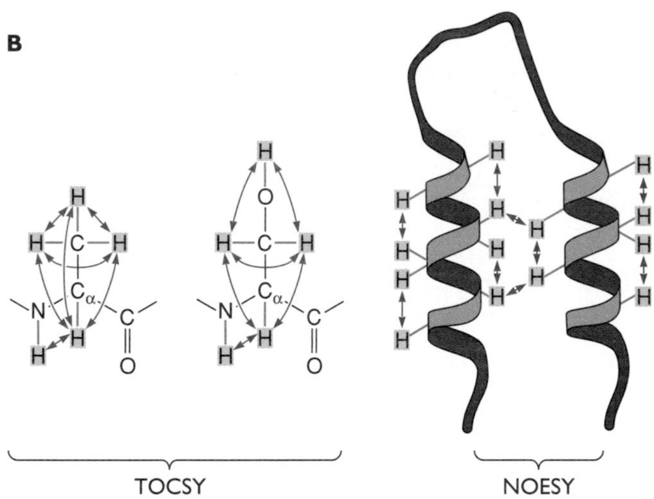

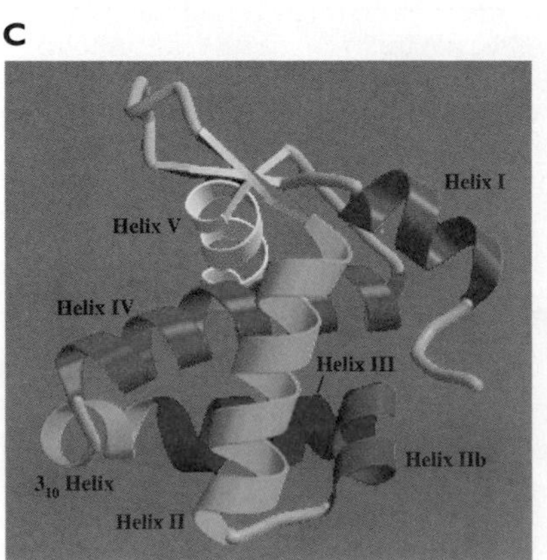

Figure 4.3 Information obtained from multidimensional nuclear magnetic resonance (NMR) spectroscopy. (A) Example of a two-dimensional $^{1}H^{15}N$ heteronuclear, single quantum coherence (HSQC) NMR spectrum of the human immunodeficiency virus type 1 matrix (MA) protein. The horizontal and vertical axes show the radiation absorbed by covalently linked protons and nitrogen atoms, respectively, expressed relative to reference signals and termed the chemical shifts. These signals are observed when the protein is placed in a strong magnetic field and exposed to radiofrequency pulses. The chemical shift is different for each atom and is determined in part by the molecular environment of its nucleus. The different atoms in a complex molecule like the MA protein therefore exhibit different chemical shifts in this kind of experiment. Modern multidimensional NMR spectroscopic techniques allow structural determinations of moderate-sized proteins (up to ~30 kDa), although proteins at the upper end of this size range typically require enrichment with the "NMR-active" atomic nuclei ^{15}N and ^{13}C (in addition to protons). The initial problem is to resolve and assign the very large number of signals generated by even a small protein (~1,000 for a protein of 15 kDa). The assignment is typically made by using multidimensional experiments, which allow individual protons to be resolved on the basis of several different chemical shift indices (^{15}N and ^{1}H in the spectrum shown). (B) Proton-proton correlations obtained from $^{1}H,^{1}H$ total correlation spectroscopy (TOCSY) and nuclear Overhauser effect spectroscopy (NOESY) NMR spectra. TOCSY reveals correlations between protons that are covalently connected via only one or two additional atoms. A $^{1}H,^{1}H$ TOCSY experiment identifies interactions among protons within the same amino acid residue in a protein. Hydrogen atoms in adjacent amino acids are connected by at least three other atoms (Box 4.1). A TOCSY experiment therefore generates a characteristic set of linked signals associated with each amino acid, which differ for the different amino acids because each bears a unique side chain. NOESY identifies correlations between protons that are closer than 5 Å in space, regardless of whether they are closely linked in the primary sequence of the protein. Secondary-structure elements, such as α-helices, generate characteristic sets of NOE signals. Because NOE signals require only that protons be close in space, they also provide information about the tertiary structure of the protein. NOE signals place constraints on the distances between specific pairs of hydrogen atoms in the protein. When sufficient NOE constraints have been collected and assigned, it is possible to produce structural models of the protein (C) that are consistent with these constraints, as shown here for the human immunodeficiency virus type 1 MA protein. (A and C) Courtesy of M. F. Summers, University of Maryland, Baltimore County, and W. I. Sundquist, University of Utah. (B) Adapted from C. Branden and J. Tooze, *An Introduction to Protein Structure* (Garland Publishing, Inc., New York, N.Y., 1991), with permission.

veloped viruses. The structures of individual viral proteins in solution can also be determined by multi-dimensional nuclear magnetic resonance (NMR) methods. These techniques allow structural models to be constructed from knowledge of the distances between specific atoms in a polypeptide chain (Fig. 4.3). At present, NMR methods can be applied to only relatively small proteins (20 to 30 kilodaltons [kDa]), but their power is being expanded rapidly. When combined with information collected by cryo-electron microscopy of the intact virus, the high-resolution structure of a single virion protein can reveal much detailed information about the architecture of virus particles.

Building a Protective Coat

Regardless of their structural complexity, all virions contain at least one protein coat, the capsid or nucleocapsid (Table 4.2), that encases and protects the nucleic acid genome. As discussed in Chapter 1, genetic economy dictates that such structures be built from identical copies of a small number of viral proteins with structural properties that permit regular and repetitive interactions among them. These protein molecules are arranged to provide maximal contact and non-covalent bonding among subunits and structural units. The repetition of such interactions among a limited number of proteins results in a regular structure, with symmetry that is determined by the spatial patterns of the interactions. The protein coats of all but the most complex animal viruses display **helical** or **icosahedral symmetry**.

Helical Structures

The **nucleocapsids** of some enveloped animal viruses, as well as certain plant viruses and bacteriophages, are rod-like or filamentous structures with helical symmetry. Helical symmetry is described by the number of structural units per turn of the helix, μ, the axial rise per unit, ρ, and the pitch of the helix, P, given by the formula

$$P = \mu \times \rho$$

A characteristic feature of a helical structure is that any volume can be enclosed simply by varying the length of the helix. Such a structure is said to be **open.** In contrast, capsids with icosahedral symmetry (described below) are **closed** structures of fixed internal volume.

From a structural point of view, the best understood helical nucleocapsid is that of tobacco mosaic virus, the first virus to be identified. The virus particle comprises a single molecule of (+) strand RNA, about 6.4 kb in length, enclosed within a helical protein coat (Fig. 4.4A and 1.7). The coat is built from a single protein that folds into an extended structure shaped like a Dutch clog. Repetitive in-

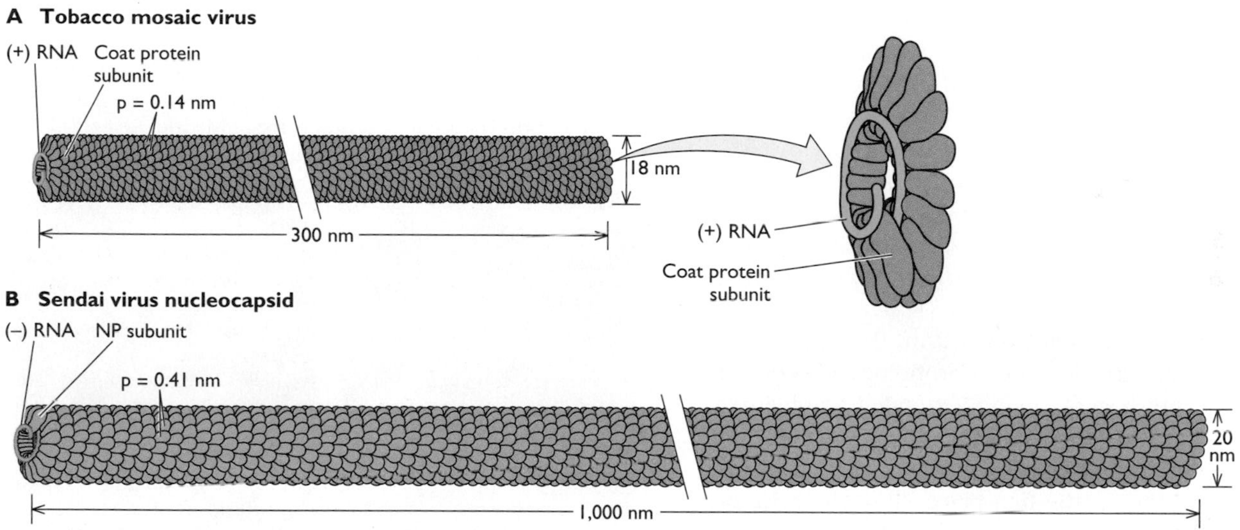

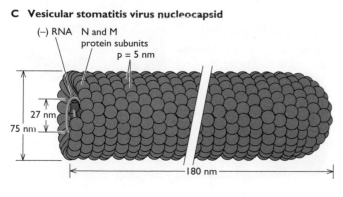

Figure 4.4 Virus structures with helical symmetry. (A) Tobacco mosaic virus. The structure of this virus has been determined at high resolution by X-ray diffraction of fibers or oriented gels of the particles. The single genomic RNA molecule and the single coat protein form an extended right-handed helix with $16\frac{1}{3}$ protein subunits per turn, μ, and an axial rise per residue, ρ, of 0.14 nm, for a pitch, P, of 2.3 nm. Each subunit engages in the same interactions with its neighbors, and each binds to three nucleotides in the RNA. The broken disk-shaped organization of the proteins in a single turn of the helix can be seen in the inset. (B and C) The structures of Sendai virus (a paramyxovirus) and vesicular stomatitis virus (a rhabdovirus) nucleocapsids are based on electron microscopy of virions or nucleocapsids released from them. The Sendai virus nucleocapsid resembles the tobacco mosaic virus particle in diameter and presence of a hollow core, but is in the form of a left-handed helix with distinctive helical parameters ($\mu = 13$, $\rho = 0.41$ nm, $P = 5.3$ nm). The vesicular stomatitis virus nucleocapsid is also hollow and comprises about 30 turns of uniform diameter, followed by 5 or 6 turns of decreasing diameter. Helical parameters have not been determined.

teractions among coat protein subunits form disks that have been likened to lock washers, which in turn assemble as a long, rodlike, right-handed helix with 16.3 coat protein molecules per turn. In the interior of the helix, each coat protein molecule binds three nucleotides of the RNA genome. The coat protein molecules therefore engage in **identical,** equivalent interactions with one another and with the genome, allowing construction of a large, stable structure from a single protein subunit.

The virions of several families of animal viruses with (–) strand RNA genomes, including paramyxoviruses, rhabdoviruses, and orthomyxoviruses, contain internal structures with helical symmetry encased within an envelope. In all cases, these structures contain an RNA molecule, many copies of an RNA packaging protein (designated NP or N), and the viral RNA polymerase and associated enzymes responsible for synthesis of messenger RNA (mRNA). Despite common helical symmetry and similar composition, the internal components of these (–) strand RNA viruses exhibit considerable diversity in morphology and organization. Like tobacco mosaic virus particles, the nucleocapsids of paramyxoviruses and rhabdoviruses contain a single molecule of RNA (of about 15 and 11 kb, respectively) tightly associated with the nucleocapsid protein. Nucleocapsids of paramyxoviruses, such as Sendai virus, are long, filamentous structures in which the RNA and NP protein form a left-handed helix with a hollow core (Fig. 4.4B). Nucleocapsids of vesicular stomatitis virus (and other rhabdoviruses) are quite different in shape (Fig. 4.4C). Furthermore, an additional virion protein is essential to maintain their organization. Vesicular stomatitis virus nucleocapsids released from within the virion envelope retain the dimensions and morphology observed in intact particles, but become highly extended and filamentous once the matrix (M) protein is also removed.

The internal components of influenza A virus particles differ more radically. In the first place, they comprise not a single nucleocapsid but rather multiple ribonucleoproteins, one for each molecule of the segmented RNA genome present in the virion (Appendix A, Fig. 8). The viral NP protein organizes each RNA into a helical structure analogous to the nucleocapsids of viruses with nonsegmented (–) strand RNA genomes. However, each influenza A virus ribonucleoprotein is further folded into a compact, circular conformation as a result of the binding of the virion enzyme complex (P proteins) to specific sequences conserved at the 5′ and 3′ ends of each (–) strand RNA molecule (Fig. 4.5A). Finally, apart from the terminal sequences bound by P proteins, the RNA in these ribonucleoproteins is fully accessible to solvent. This property suggests that the RNA is wound around a helical core formed by the NP protein, rather than sequestered within a helical structure. Natural ribonucleoproteins have not

Figure 4.5 Structure of influenza A virus ribonucleoproteins. (A) Circular conformation maintained by the P proteins. The electron micrographs show negatively stained ribonucleoproteins as purified from virions (top and middle) or following removal of P proteins by exposure to detergent and density gradient centrifugation (bottom). The samples shown in the middle and bottom panels were incubated in low salt prior to electron microscopy, a treatment that unwinds the compact native ribonucleoproteins (top) to circular structures (middle). In the absence of the P proteins, only linear structures are observed upon unwinding (bottom). Therefore, the P proteins, which protect both 5′- and 3′-terminal sequences of RNA segments against chemical modification, hold the virion ribonucleoproteins in a compact, circular conformation. From K. Klumpp et al., *EMBO J.* **16:**1248–1257, 1997, with permission. Courtesy of R. W. H. Ruigrok, European Molecular Biology Laboratory Grenoble Outstation, Grenoble, France. (B) Three-dimensional reconstruction of a recombinant "mini"-ribonucleoprotein, showing perspective, top, and bottom views. Ribonucleoproteins assembled on a 248-nucleotide influenza virus genomic-like RNA in cells synthesizing the viral NP and P proteins were examined by electron microscopy of negatively stained preparations and image reconstruction. The nine NP monomers present interact more extensively at the base than at the top of the ribonucleoprotein. The perimeter of the NP ring is less than half the length of the RNA present, indicating that the RNA must be wound in some manner on NP monomers. The circular form is observed because the RNA is too short to form a helical supercoil, the conformation exhibited by natural and recombinant ribonucleoproteins containing RNA molecules of more than 400 nucleotides. The P protein complex associates with two adjacent NP monomers. Bar = 50 Å. Adapted from J. Martin-Renito et al., *EMBO Rep.* **2:**313–317, 2001, with permission. Courtesy of J. Ortin, Centro Nacional de Biotecnología, Madrid, Spain.

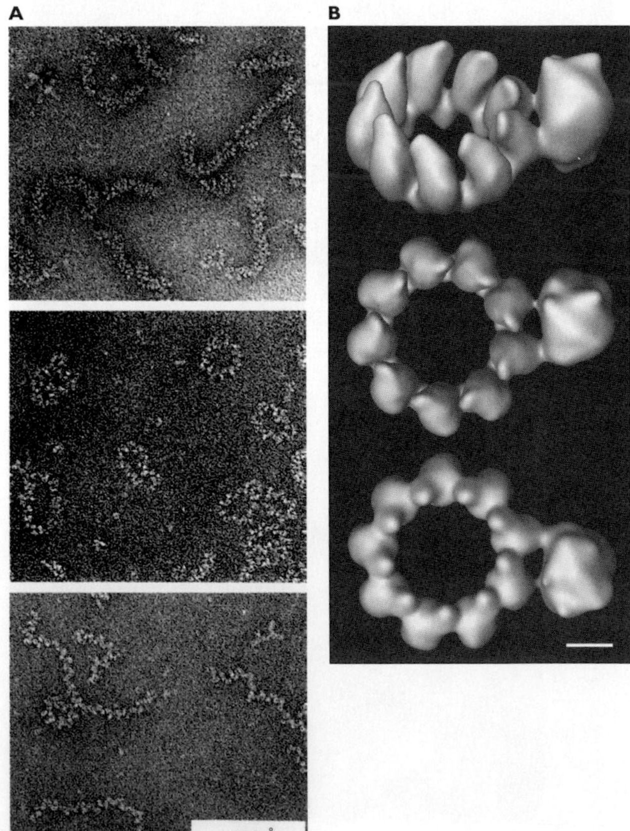

been examined at high resolution. However, the dimensions and organization of influenza viral "mini"-ribonucleoproteins determined by cryo-electron microscopy (Fig. 4.5B) are consistent with this model.

Capsids or Nucleocapsids with Icosahedral Symmetry

General Principles

Icosahedral symmetry. An icosahedron is a solid with 20 triangular faces and 12 vertices related by two-, three-, and fivefold axes of rotational symmetry (Fig. 1.12). The protein shells (capsids or nucleocapsids) of many virions are closed structures based on icosahedral symmetry. Such particles do not always appear icosahedral, because the most prominent surface structures or the viral glycoproteins in the envelope do not conform to the underlying symmetry of the capsid shell. Nevertheless, the symmetry with which the structural units interact is that of an icosahedron. Icosahedral symmetry allows formation of a closed shell with the smallest number (60) of identical subunits. In such an arrangement, these subunits are related to one another by the two-, three-, and fivefold rotational axes that define icosahedral symmetry (Fig. 4.6A). Such well-defined symmetry poses a number of interesting questions about the organization and assembly of icosahedral viruses. This property may also prove to have great practical value (Box 4.3).

Large capsids and quasiequivalent bonding. In the simple icosahedral packing arrangement shown in Fig. 4.6A, each of the 60 subunits (or asymmetric units) consists of a single molecule in a structurally identical environment. Consequently, all subunits interact with their neighbors in an identical (or **equivalent**) manner, just like the subunits of a helical nucleocapsid such as that of

Figure 4.6 Icosahedral packing in simple structures. A comma represents a single protein molecule. In the simplest and ideal case, $T = 1$ (A), the protein molecule forms the structural unit, and each of the 60 molecules is related to its neighbors by the two-, three-, and fivefold rotational axes that define a structure with icosahedral symmetry. At a twofold rotational axis, rotation of an object by 180° results in an appearance identical to that at the starting position; at a threefold axis, rotations of 120° and 240° restore the starting appearance; and rotations of 72°, 144°, 216°, and 288° around a fivefold axis restore identity. In such a simple icosahedral structure, the interactions of all molecules with their neighbors are identical. In the $T = 3$ structure (B) with 180 identical protein subunits, there are three modes of packing of a subunit shown in orange, yellow, and purple because the structural unit (outlined in blue) is now the asymmetric unit, which, when replicated according to 60-fold icosahedral symmetry, generates the complete structure. The orange subunits are present in pentamers, formed by tail-to-tail interactions, and interact in rings of three (back to back) with purple and yellow subunits, and in pairs (head to head) with a purple or a yellow subunit. The purple and yellow subunits are arranged in rings of six molecules (also by tail-to-tail interactions) that alternate in the particle. Despite these packing differences, the bonding interactions in which each subunit engages are similar, that is, **quasiequivalent:** for example, all engage in tail-to-tail and head-to-head interactions. This type of packing arrangement is seen in small plant RNA viruses, such as tomato bushy stunt virus, in which the variation in subunit-subunit interactions needed for the three types of packing is accommodated by whether an N-terminal arm on each molecule adopts an ordered (purple subunits) or a disordered (orange and yellow subunits) conformation. Adapted from S. C. Harrison et al., *in* B. N. Fields et al. (ed.), *Fundamental Virology* (Lippincott-Raven, New York, N.Y., 1995), with permission.

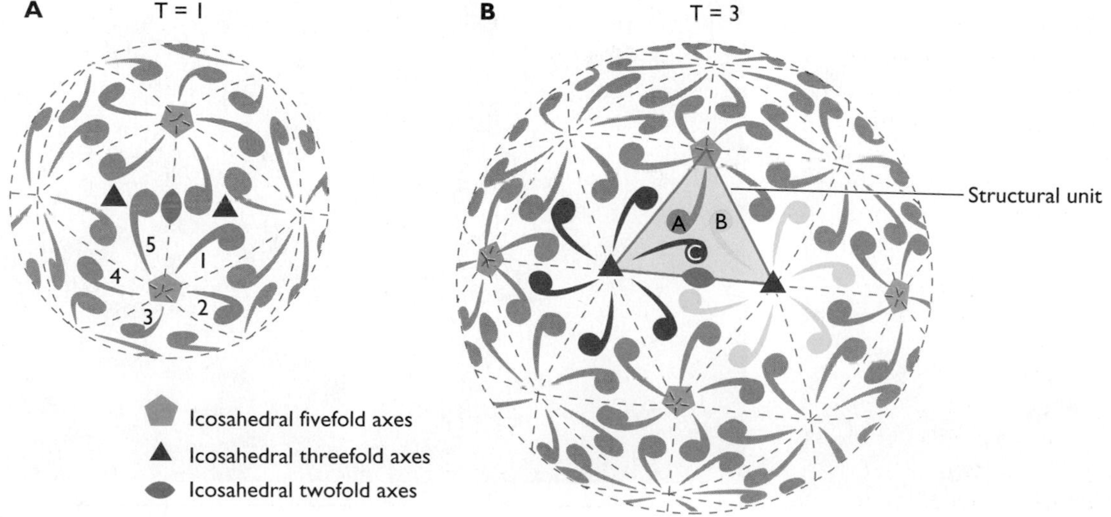

BOX 4.3

Nanoconstruction with virus particles

Nanochemistry is the synthesis and study of well-defined structures with dimensions of 1 to 100 nanometers. Nanobuilding blocks span the size range between molecules and materials. Molecular biologists study nanochemistry, nanostructures, and molecular machines including the ribosome, the photosynthetic center, and membrane-bound signaling complexes. Icosahedral virus particles are proving to be precision building blocks for nanochemistry. The $T = 1$ icosahedral cowpea mosaic virus particle is 30 nm in diameter and its atomic structure is known in detail. Grams of virions can be prepared easily from kilograms of infected leaves. Insertional mutagenesis is straightforward and precise amino acid changes can be introduced. As illustrated in the figure, cysteine residues inserted in the capsid protein provide functional groups for chemical attachment of 60 precisely placed molecules.

High local concentrations of the attached chemical agent coupled with precise placement enable rather remarkable nano-construction. For example, virus surfaces can be patterned with metal nano-particles that may function as a conducting "wire" for electronics. In addition, the propensity of virions for self-organization into two- and three-dimensional lattices leads to well-ordered arrays of 10^{13} particles in a 1-mm^3 crystal.

Viruses are not just for infections any more! They will provide a rich source of building blocks for applications spanning the worlds of molecular biology and materials science.

Wang, Q., T. Lin, L. Tang, J. E. Johnson, and M. G. Finn. 2002. Icosahedral virus particles as addressable nanoscale building blocks. *Angew. Chem. Int. Ed. Engl.* **41**:459–462.

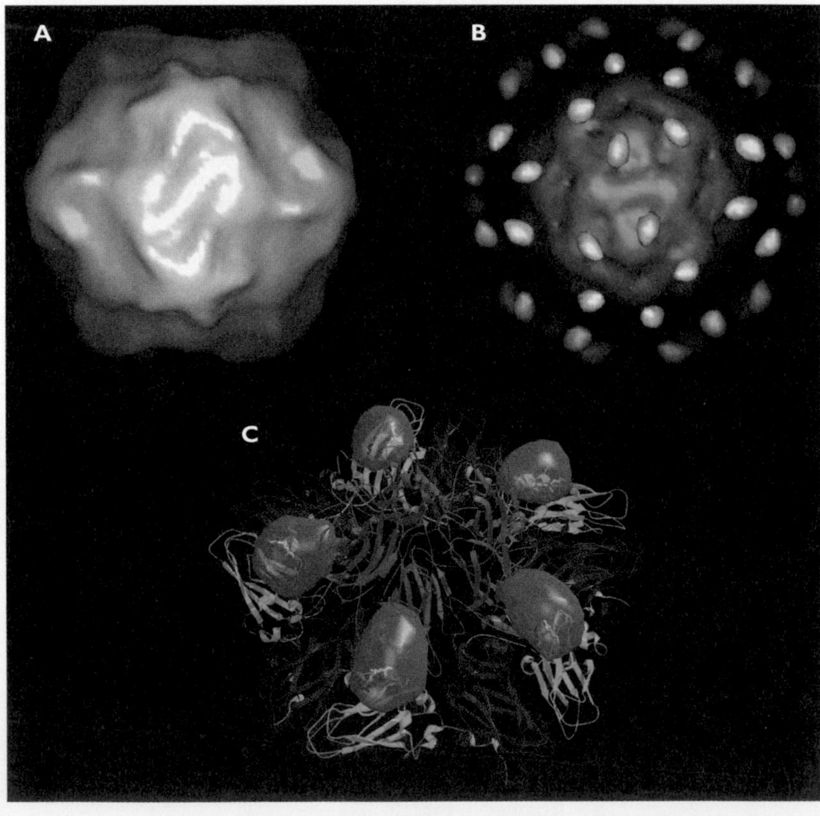

Cowpea mosaic virus. Cryo-electron microscopy of derivatized cowpea mosaic virus with a cysteine residue inserted on the surface of each of the 60 subunits and to which nanogold particles with a diameter of 1.4 nm were chemically linked. (A) Three-dimensional reconstruction of such derivatized virus particles at 29-Å resolution. (B) Difference electron density map obtained by subtracting the density of unaltered cowpea mosaic virus at 29 Å from the density map shown in panel A. This procedure reveals both the genome (which was not included in the density of the native particle), shown in green, and the gold nanoparticles. (C) A section of the difference map imposed on the atomic model of cowpea mosaic virus. The positions of the gold indicate that it is attached at the sites of the cysteine insertion. Courtesy of M. G. Finn and J. Johnson, The Scripps Research Institute.

tobacco mosaic virus. However, construction of a protein shell according to such a simple icosahedral design severely restricts the size of the genome that can be packaged. In fact, the capsids or nucleocapsids of the majority of animal viruses are built from structural units composed of several copies of the same protein, or of several different proteins, and can accommodate quite large genomes. In 1962, Donald Caspar and Aaron Klug developed a theoretical framework accounting for the structural properties of larger particles with icosahedral symmetry. This theory has had enormous influence on the way virus structures are described and interpreted.

A cornerstone of the theory developed by Caspar and Klug was the proposition that when a capsid contains more than 60 subunits, each subunit occupies a **quasi-equivalent position**; that is, the non-covalent bonding properties of subunits in different structural environments are **similar** (but not identical, as is the case for the simplest, 60-subunit structure). This property is illustrated in Fig. 4.6B for a particle with 180 identical subunits. In the small, 60-subunit structure, 5 subunits make fivefold symmetric contact at each of the 12 vertices (Fig. 4.6A). In the larger structure with 180 subunits, this arrangement is retained at the 12 vertices, but the additional subunits, arranged with sixfold symmetry, are interposed between the five-fold symmetric clusters (Fig. 4.6B). In such a structure, each subunit can be present in one of three **different** structural environments, designated A, B, or C in Fig. 4.6B. Nevertheless, all subunits bond to their neighbors in similar ways, for example, via head-to-head and tail-to-tail interactions (Fig. 4.6B).

The triangulation number, *T*. A second important idea introduced by Caspar and Klug was that of **triangulation**, the description of the triangular face of a large icosahedral structure in terms of its subdivision into smaller triangles termed facets. As shown in Fig. 4.7, combining several triangular facets allows assembly of a larger face of an icosahedrally symmetric structure from the same structural unit, in this case a homotrimer. This process is described by the triangulation number, T, which gives the number of structural units per face. The total number of subunits in the structure is $60T$. As in the $T = 3$ structure illustrated in Fig. 4.6B, the packing interactions of the 240 (60×4) protein subunits in the $T = 4$ structure shown in Fig. 4.7 cannot be identical, but according to the original theory of Caspar and Klug they are very similar, or quasiequivalent.

Icosahedrally symmetric structures can be formed from quasiequivalent subunits for only certain values of T (Box 4.4). Virus structures corresponding to various values of T, some very large, have been described (Table 4.3). The capsids of a number of small (+) strand RNA viruses, such as

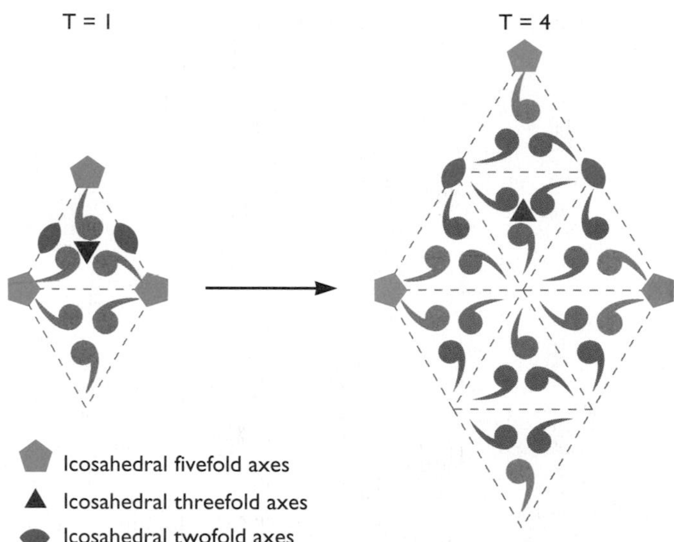

Figure 4.7 $T = 4$ triangulation of an icosahedral face. Two faces of a $T = 1$ icosahedron, each composed of three protein subunits, are shown flattened into a single plane on the left. The two-, three-, and fivefold rotational axes that define icosahedral symmetry are indicated. The combination of four such identical faces (now facets) to form a $T = 4$ face is shown at the right. The icosahedral symmetry axes are in the same relative positions as in the $T = 1$ face. In such a $T = 4$ structure, the size of the face, which is defined by the length of one of its sides, is twice that of the $T = 1$ structure. Adapted from S. Casjens, *in* S. Casjens (ed.), *Virus Structure and Assembly* (Jones and Bartlett Publishers Inc., Boston, Mass., 1985), with permission.

that of tomato bushy stunt virus ($T = 3$), show quasiequivalent bonding of their identical subunits: the local contacts made by the A, B, and C monomers that occupy distinctive packing environments (Fig. 4.6B) are largely similar, and the corresponding parts of each monomer engage in similar interactions. However, an important determinant of which type of packing interaction each subunit adopts is whether a flexible arm of the protein is ordered or disordered. This property indicates that interactions among chemically identical protein subunits must be regulated during assembly, with appropriate conformational switching of the flexible segment. Such large conformational differences between small regions of chemically identical (and otherwise quasiequivalent) subunits were not anticipated in early considerations of virus structure. This omission is not surprising, for these principles were formulated when little was known about the structural properties of proteins.

The triangulation number and quasiequivalent bonding among subunits describe the structural properties of many simple viruses with icosahedral symmetry. However, as we discuss in the next sections, the structural properties of both small and more complex viruses can depart radically from the constraints imposed by quasiequivalent bonding.

BOX 4.4

The triangulation number, T, and how it is determined

In developing their theories about virus structure, Caspar and Klug used graphic illustrations of capsid subunits, such as the net of flat hexagons shown at the top of (A) and (B). Each hexagon represents a hexamer, with identical subunits shown as equilateral triangles. When all subunits assemble into such hexamers, the result is a flat sheet, or lattice, which can never form a closed structure. However, curvature can be introduced into the hexagonal net by converting hexamers to pentamers. As an icosahedron has 12 axes of fivefold symmetry, 12 pentamers must be introduced to form a closed structure with icosahedral symmetry. If 12 adjacent hexamers are converted to pentamers (A), an icosahedron of the minimal size possible for the net is formed. This structure is built from 60 equilateral triangle asymmetric units and corresponds to a $T = 1$ icosahedron. Larger structures with icosahedral symmetry are built by converting 12 **nonadjacent** hexamers to pentamers at precisely spaced, and regular, intervals. To illustrate this, we use nets in which an origin (O) is fixed, and the positions of all other hexamers are defined by the coordinates along the axes labeled h and k, where h and k are any integers greater than or equal to zero. The hexamer (h, k) is therefore defined as that reached from the origin (O) by h steps in the direction of the h axis and k steps in the direction of the k axis. In the $T = 1$ structure, $h = 1$ and $k = 0$, or $h = 0$ and $k = 1$, and adjacent hexamers are converted to pentamers (A, center). To construct a model of an icosahedron when $h = 1$ and $k = 1$ (B), we generate one of its faces in the net by conversion of the origin (O) and (1, 1) hexamers to pentamers. The lattice point (1, 1) shown is reached by one step in each direction. The third hexamer replaced (–1, 1) is that identified by threefold symmetry to complete the large, marked equilateral triangle (black) representing the face of an icosahedron formed when $h = 1$ and $k = 1$. The resulting quasiequivalent lattice (B, center) and the $T = 1$ lattice (A, center) can be folded to form the icosahedra shown, by excision from the sheet and sequential joining of the edges marked A to D with those marked A′ to D′, respectively.

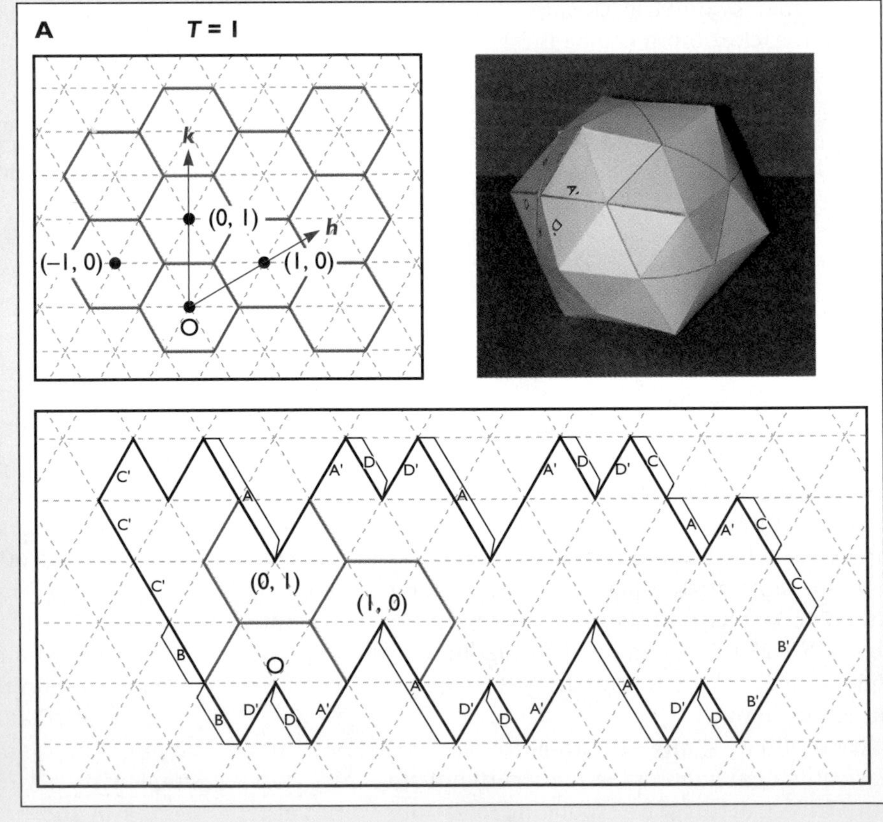

BOX 4.4 *(continued)*

The triangulation number, T, is the number of asymmetric units per face of icosahedra constructed in this way. A geometric derivation of T in terms of h and k, developed by Caspar, is shown in panel C. In the general case, it is the points (h, k) in the hexamer lattice that are converted to pentamers to create the new face, represented by the large equilateral triangle (purple) with sides of length S. Because each small equilateral triangle represents a subunit of the smallest icosahedron, to calculate T we need to know how many such small triangles can be accommodated in the large triangle of sides S that represents the face of the larger (h, k) icosahedron. Therefore,

$$T = \frac{\text{area of equilateral triangle of sides } S \text{ (the new face)}}{\text{area of small equilateral triangle representing 1 subunit}}$$

The area of an equilateral triangle with sides of length S is $(\sqrt{3}/4)S^2$. By definition, the small triangle, which represents one subunit, has sides of unit length.

$$\therefore\ T = \frac{(\sqrt{3}/4)S^2}{(\sqrt{3}/4)(1)^2}$$

As shown on the right, we can draw additional triangles on the lattice to derive S in terms of h and k. Here k is the hypotenuse of the 30°-60°-90° triangle that connects the lattice point (h, k) to the h axis. The other sides of this triangle are therefore $\sqrt{3}k/2$ and $k/2$, as shown. S is therefore the hypotenuse of a right-angled triangle with sides of length $h + k/2$ and $\sqrt{3}k/2$. By the Pythagorean theorem,

$$S^2 = \left(h + \frac{k}{2}\right)^2 + \left(\frac{\sqrt{3}k}{2}\right)^2$$

$$= h^2 + hk + k^2$$

$$\therefore\ T = h^2 + hk + k^2$$

Therefore, when both h and k are 1 (B), $T = 3$, and each face of the icosahedron contains three asymmetric units. The total number of units, which must be $60T$ (see text), is 180.

The integers h and k describe the spacing and spatial relationships of pentamers in the lattice, and of fivefold vertices in the corresponding icosahedron. For example, when $h = 1$ and $k = 0$, or when $h = 0$ and $k = 1$, pentamers are in direct contact with other pentamers (e.g., Fig. 4.6A), but when $h = 1$ and $k = 1$, each pentamer is separated

(continued)

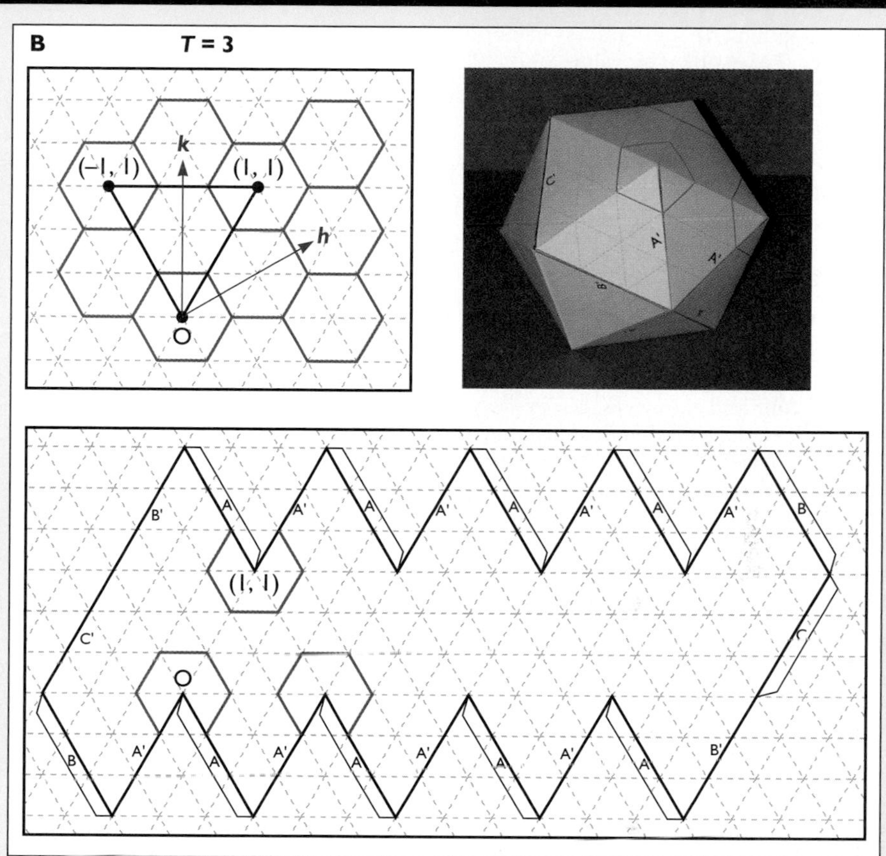

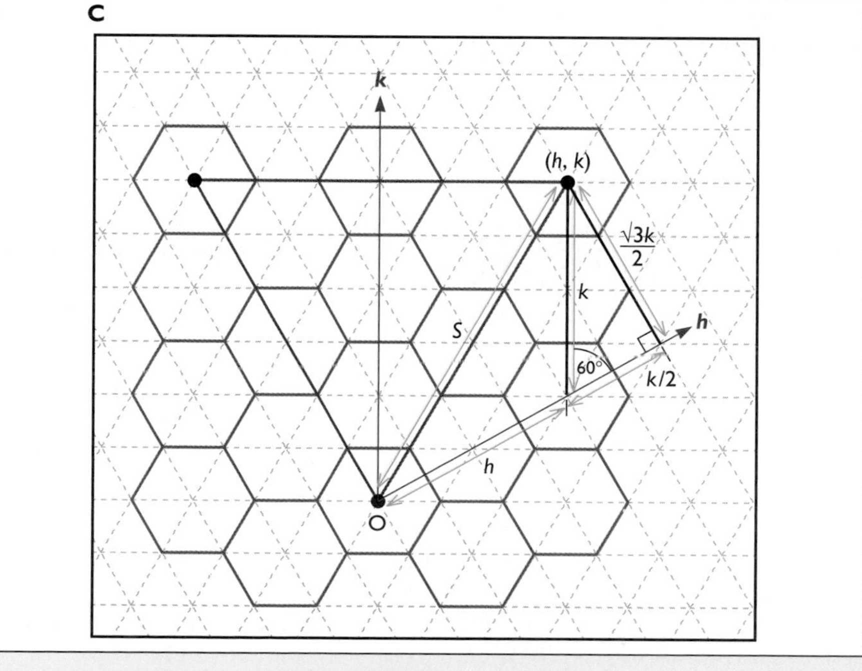

BOX 4.4 (continued)

from neighboring pentamers by one hexamer (Fig. 4.6B). The values of h and k are therefore determined by inspection of electron micrographs of virus particles or their constituents (D). For example, in the bacteriophage p22 capsid (top), one pentamer is separated from another by two steps (from one structural unit to the next) along the h axis, and one step along the k axis, as illustrated for the bottom left pentamer shown. Hence, $h = 2$, $k = 1$, and $T = h^2 + hk + k^2 = 7$. In contrast, pentamers of the herpes simplex virus type 1 (HSV-1) nucleocapsid (bottom) are separated by four and zero steps, respectively, in the directions of the h and k axes. Therefore, $h = 4$, $k = 0$, and $T = 16$. Cryo-electron micrographs of bacteriophage P22 and herpes simplex virus type 1 courtesy of B. V. V. Prasad and W. Chiu (Baylor College of Medicine), respectively. (A and B) Adapted from Fig. 2 of J. E. Johnson and A. J. Fisher, *in* R. G. Webster and A. Granoff (ed.), *Encyclopedia of Virology*, 3rd ed. (Academic Press, London, England, 1994), with permission. (C) Adapted from S. Casjens, *in* S. Casjens (ed.), *Virus Structure and Assembly* (Jones and Bartlett Publishers, Inc., Boston, Mass., 1985), with permission

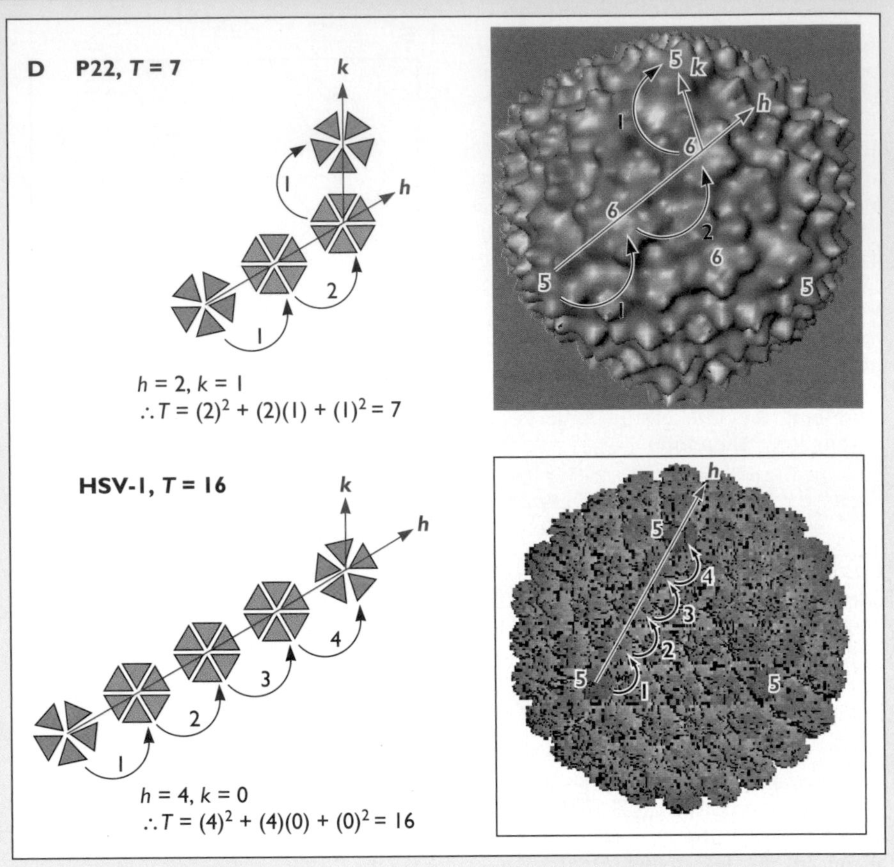

For example, the capsid of the small polyomavirus simian virus 40 is built from 360 subunits, corresponding to the $T = 6$ triangulation number excluded by the rules formulated by Caspar and Klug (Box 4.4). Moreover, a capsid stabilized by covalent joining of subunits to form viral "chain mail" has been described recently (Box 4.5). Our current view of icosahedrally symmetric virus structures is therefore one that includes greater diversity in the mechanisms by which stable capsids can be formed than was anticipated by the pioneers in this field.

Table 4.3 T numbers of representative viruses

Triangulation number (T)[a]	Family	Member(s)	60T	Protein(s) that forms the capsid or nucleocapsid shell
1	*Parvoviridae*	Canine parvovirus	60	60 copies of VP2
3	*Nodaviridae*	Black beetle virus	180	180 copies of coat protein
Pseudo-3	*Picornaviridae*	Poliovirus, human rhinovirus	180	60 copies each of VP1, VP2, and VP3
4	*Alphaviridae*	Ross River virus	240	240 copies of C protein
7	*Polyomaviridae*	Simian virus 40, polyomavirus	420	360 copies of VP1
13	*Reoviridae*	Reovirus (outer shell)	780	60 copies of λ2, 600 copies of μ1
16	*Herpesviridae*	Herpes simplex virus type 1	960	960 copies of VP5
25	*Adenoviridae*	Adenovirus type 2	1,500	720 copies of protein II, 60 copies of protein III
189–217	*Iridoviridae*	Ranavirus	11,340; 13,020	?

[a]The triangulation number, T, is a description of the face of an icosahedron indicating the number of equilateral triangles into which each face is divided following the laws of solid geometry (Box 4.4). The total number of subunits is equal to $60T$, because a minimum of 60 subunits is required to form a closed shell with icosahedral symmetry. Thus, the simplest assembly is $T = 1$. A $T = 3$ icosahedron requires 180 subunits in sets of three; a $T = 4$ icosahdron requires 240 subunits; a $T = 7$ icosahdron requires 420 subunits; and so forth. Each of the 60 structural units of picornavirus capsids contains three different polypeptides (VP1, VP2, and VP3). Hence, the capsid is described as a pseudo $T = 3$ structure. Note that polyomaviruses, reoviruses, and adenoviruses do not contain the number of subunits predicted.

BOX 4.5

Viral chain mail: stabilization of the bacteriophage HK97 capsid by formation of covalently bonded, interlinked subunit rings

The mature capsid of the tailed, double-stranded DNA bacteriophage HK97 is a $T = 7$ structure built from hexamers and pentamers of a single viral protein, Gp5. It is formed in a multistep assembly and maturation pathway, like many other DNA bacteriophage (e.g., lambda, T4) and some animal viruses (e.g., herpesviruses). The first hints of the remarkable and unprecedented mechanism of stabilization of this structure came from biochemical experiments, which showed that

- a previously unknown covalent protein-protein linkage forms in the final reaction in the assembly of the HK97 capsid: the side chain of a lysine in every Gp5 subunit forms a covalent bond with an asparagine in an adjacent subunit;
- this reaction is **autocatalytic,** depending only on Gp5 subunits organized in a particular conformational state: the capsid is enzyme, substrate, and product;
- HK97 mature particles are extraordinarily stable and cannot be disassembled into individual subunits by boiling in sodium dodecyl sulfate: it was therefore proposed that the cross-linking also interlinked the subunits from adjacent structural units to catenate rings of hexamers and pentamers.

The determination of the structure of the HK97 capsid to 3.6-Å resolution by X-ray crystallography has confirmed the formation of such capsid "chain mail" (see figure).

The HK97 capsid is the first example of a protein catenane. This unique structure has been shown to increase the stability of the virus particle, and may be of particular advantage as the capsid shell is very thin. The delivery of the DNA genome to host cells via the tail of the particle obviates the need for capsid disassembly and reversal of its covalent subunit-subunit bonds.

Duda, R. L. 1998. Protein chainmail: catenated protein in viral capsids. *Cell* **94:**55–60.

Wikoff, W. R., L. Liljas, R. L. Duda, H. Tsuruta, R. W. Hendrix, and J. E. Johnson. 2000. Topologically linked protein rings in the bacteriophage HK97 capsid. *Science* **289:** 2129–2133.

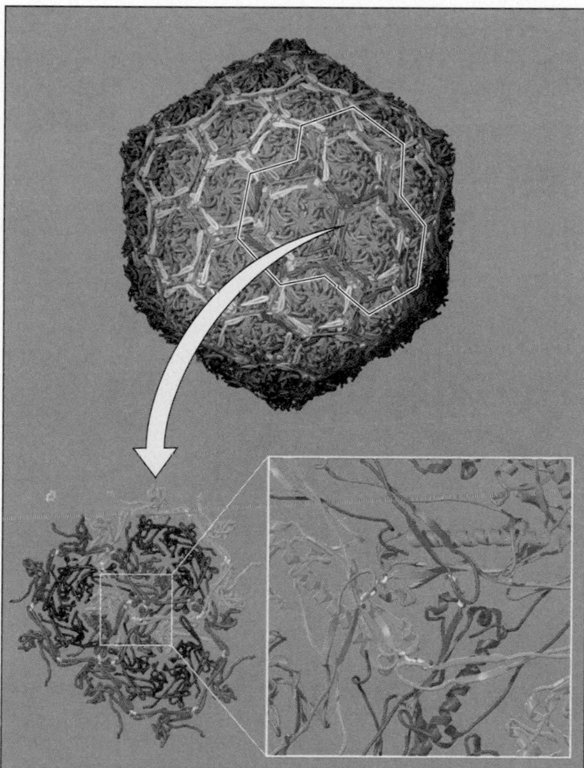

Chain mail in the bacteriophage HK97 capsid. The exterior of the HK97 capsid is shown at the top, with hexamers and pentamers of the Gp5 protein in gray. Subunits that are cross-linked into rings are colored the same, with only the segment between cross-links colored to illustrate the formation of catenated rings of subunits. The cross-linking is shown in the more detailed view (below) down a quasithreefold axis with three pairs of cross-linked subunits. The K-N isopeptide bonds are shown in yellow. The cross-linked monomers shown in blue loop over a second pair of covalently joined subunits (green), which in turn cross over a third pair (magenta). Adapted from W. R. Wikof et al., *Science* **289:**2129–2133, 2000, with permission. Courtesy of J. Johnson, The Scripps Research Institute.

Structurally Simple Capsids

Several nonenveloped animal viruses are small enough to be amenable to high-resolution structural studies by X-ray crystallography. We have chosen three examples, the nodavirus Nodamura virus, the picornavirus poliovirus, and the polyomavirus simian virus 40, to illustrate molecular foundations of icosahedral architecture.

Structure of Nodamura virus. The nodaviruses are small (290 to 320 Å in diameter) nonenveloped viruses that infect mammals, fish, and insects. The capsid of Nodamura virus (Fig. 4.8) (and other members of the family) encases the (+) strand RNA genome, which comprises two molecules of RNA. It is built from 180 copies of a single coat protein, organized according to a $T = 3$, quasi-equivalent design. The 60 structural units contain one copy of each of three types of coat protein subunit that are defined by occupancy of structurally distinct environments (A, B, and C [Fig 4.8B]). The A subunits are arranged as pentamers around the 12 axes of fivefold rotational symmetry, whereas the B and C subunits alternate in hexameric rings around the threefold symmetry axes (Fig. 4.8B; compare with Fig. 4.6B). Despite the differences in subunit packing, interactions among A, B, and C subunits within a structural unit are closely similar. However, subunits engage in either flat or bent contacts at the bases or sides, respectively, between asymmetric units (Fig. 4.8B). This difference is the result of the presence of a short, ordered protein segment only in the C subunits, and of a duplex segment of the RNA genome at the flat contacts. These structural features hold the neighboring subunits further apart than at the bent contacts, so that the asymmetric units form a flat, diamond-shaped structure (Fig. 4.8B). Whether internal surfaces of adjacent subunits interact with ordered segments of the RNA genome is therefore a crucial determinant of the quasiequivalent, icosahedral architecture of Nodamura virus (and other members of the *Nodaviridae*).

Structure of poliovirus. As their name implies, the picornaviruses are among the smallest and simplest of animal viruses. The approximately 300-Å-diameter poliovirus particle is composed of 60 copies of each of four virally encoded structural proteins, VP1, VP2, VP3, and VP4 (VP = virion protein). These proteins form a closed capsid encasing the (+) strand RNA genome of about 7.5 kb and its covalently attached 5′-terminal protein, VPg (Appendix A, Fig. 13). Our understanding of the structure of the *Picornaviridae* took a quantum leap in 1985 with the determination of high-resolution structures of human rhinovirus 14 (genus *Rhinovirus*) and poliovirus (genus *Enterovirus*).

The heteromeric structural unit of the poliovirus capsid contains one copy each of VP1, VP2, VP3, and VP4 (Fig. 4.9A). VP4 lies on the inner surface of the protein shell formed by VP1, VP2, and VP3. Although the three latter proteins are not related in amino acid sequence, all con-

Figure 4.8 Structure of Nodamura virus. The structure of the Nodamura virus particle determined by X-ray crystallography is shown in panel A and summarized schematically in panel B. The coat protein subunits that occupy A, B, and C structural environments are colored blue, red, and green, respectively. In panel B, the structural (asymmetric) unit is outlined in blue. The interactions between subunits at the sides of the structural units are extensive and closely similar, and the asymmetric units make contact at an angle of 144°. In contrast, flat contacts between B and C subunits are made between the bases of asymmetric units. This difference is the result of interaction of an ordered segment of RNA with the internal surfaces of the subunits, and the ordering of a specific N-terminal segment of the protein only in the C subunits. Panel A courtesy of P. Natarajan and J. Johnson, The Scripps Research Institute and Virus Particle Explorer (VIPER). See V. S. Reddy et al., *J. Virol.* **75**:11943–11947, 2001.

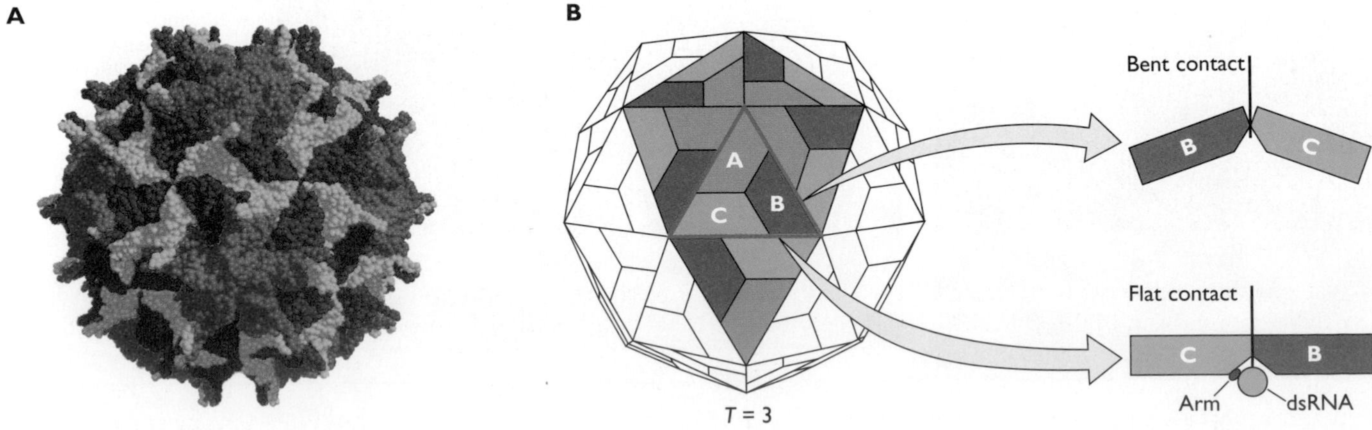

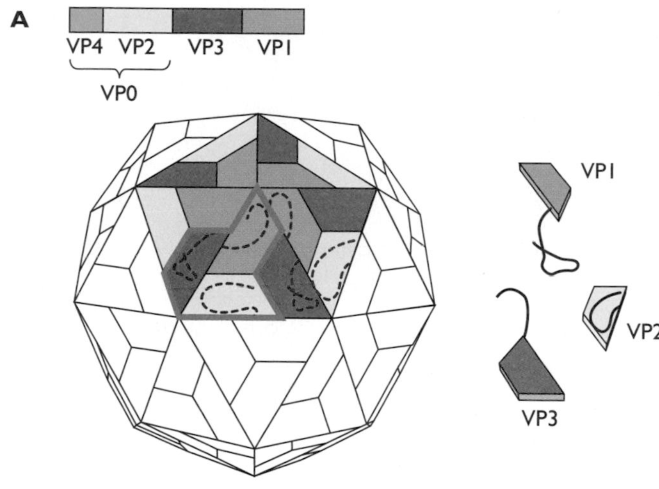

Figure 4.9 Packing and structures of poliovirus proteins. (A) The packing of the 60 VP1-VP2-VP3 structural units, represented by wedge-shaped blocks corresponding to their β-barrel domains. Note that the structural unit (outlined in blue) contributes to two adjacent faces of an icosahedron, rather than corresponding to a facet. The wedge-shaped β-barrels of VP1, VP2, and VP3 facilitate tight packing of the structural units, with VP1 proteins from five adjacent structural units clustered around the 12 fivefold axes of the virus particle and the thin ends of VP2 and VP3 alternating around the threefold axes (see also Fig. 4.10). (B) The topology of the polypeptide chain in a β-barrel jelly roll is shown at the top left. The β-strands, indicated by arrows, form two antiparallel sheets juxtaposed in a wedgelike structure. The two α-helices (purple cylinders) that surround the open end of the wedge are also conserved in location and orientation in these proteins. The VP1, VP2, and VP3 proteins each contain a central β-barrel jelly roll domain. However, the loops that connect the β-strands in this domain of the three proteins vary considerably in length and conformation, particularly at the top of the β-barrel, which, as represented here, corresponds to the outer surface of the capsid. The N- and C-terminal segments of the proteins, which extend from their β-barrel cores, also vary in length and structure. The very long N-terminal extension of VP3 has been truncated in this representation. Adapted from J. M. Hogle et al., *Science* **229:**1358–1365, 1985, with permission.

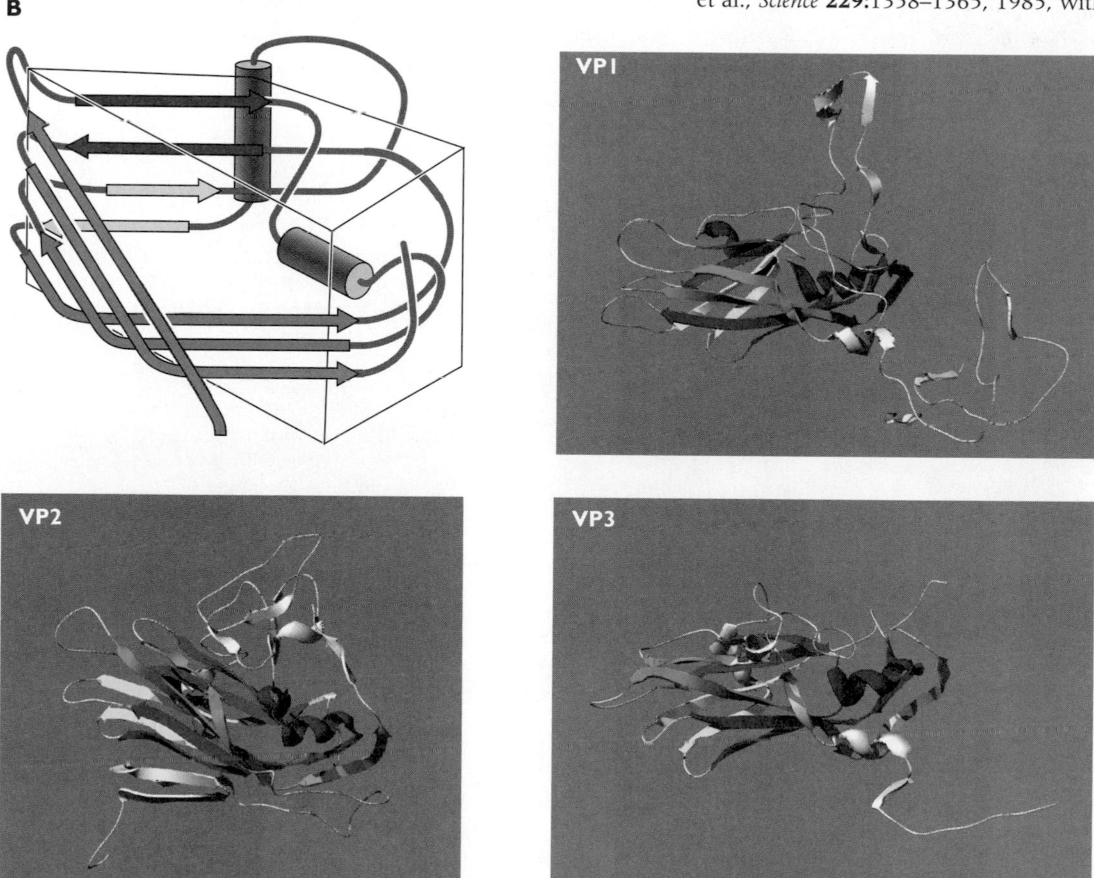

tain a central β-sheet structure termed a **β-barrel jelly roll**. The organization of this β-barrel domain is illustrated schematically in Fig. 4.9B, for comparison with the actual structures of VP1, VP2, and VP3. It is a wedge-shaped structure comprising two anti-parallel β-sheets. One of the sheets forms one wall of the wedge, while the second, sharply twisted β-sheet forms both the second wall and the floor (Fig. 4.9B). The β-barrel domains of VP1, VP2, and VP3 are folded in the same way; that is, they possess the same **topology**. The differences among these proteins

BOX 4.6

Ancient evolutionary relationships deduced from structural comparisons

The specific and unusual properties of the major capsid proteins of human adenovirus and the bacteriophage PRD1 make a compelling case for common ancestry of these viruses. The adenovirus hexons form the faces of the icosahedral capsid (Fig. 4.12). This structural unit is formed not by six monomers arranged with hexagonal symmetry, but is a trimer. However, each hexon monomer contains two β-barrel jelly roll domains, labeled P1 and P2 in the figure, such that the hexon exhibits pseudohexagonal symmetry (see Fig. 4.12). As the figure below illustrates, the same arrangement is seen in the major capsid protein, P3, of bacteriophage PRD1. Moreover, the hexon and P3 monomers have similar connections within and between the β-barrel domains. These human and bacterial viruses also have in common $T = 25$ capsids, an arrangement not seen in any other virus family, and a structural unit built from distinct proteins at the positions of fivefold symmetry, from which project proteins specialized for attachment. They also share features of their genome organization and mechanism of viral DNA synthesis. It is difficult to escape the conclusion that these modern viruses evolved from an ancient common ancestor.

Benson, S. D., J. K. H. Bamford, D. H. Bamford, and R. M. Burnett. 1999. Viral evolution revealed by bacteriophage PRD1 and human adenovirus coat protein structures. *Cell* **98**:825–833.

Hendrix, R. W. 1999. Evolution: the long evolutionary reach of viruses. *Curr. Biol.* **9**:R914–R917.

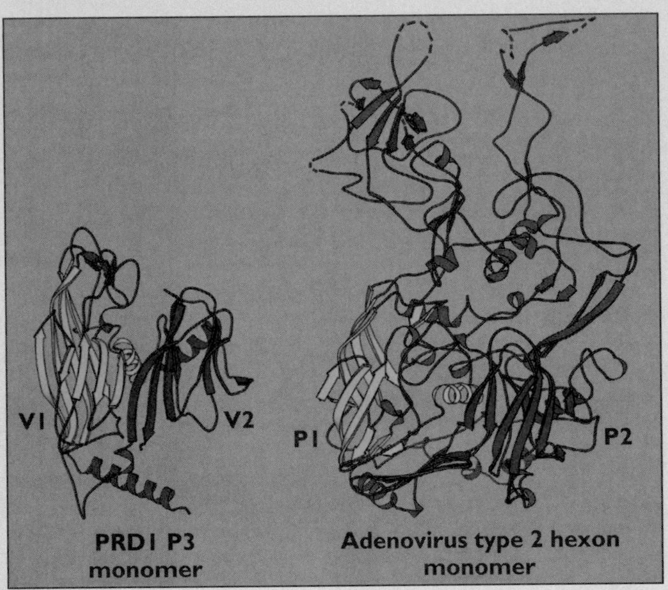

β-barrel jelly rolls. The human adenovirus type 2 hexon (right) and bacteriophage protein P3 (left) monomers are shown, with the two β-barrel jelly rolls present in each colored green and blue. Adapted from R. W. Hendrix, *Curr. Biol.* **9**:R914–R917, 1999, with permission. Courtesy of R. Burnett, The Wistar Institute.

are therefore largely restricted to the loops that connect β-strands and to the N- and C-terminal segments that extend from the central β-barrel domains (Fig. 4.9B).

The β-barrel jelly rolls of these picornaviral proteins are similar in structure to the core domains of capsid proteins of a number of plant, insect, and vertebrate (+) strand RNA viruses, such as tomato bushy stunt virus and Nodamura virus, described previously. This property was entirely unanticipated. Even more remarkably, this relationship is not restricted to small RNA viruses: the major capsid proteins of the DNA-containing polyomaviruses and adenoviruses also contain such β-barrel domains (see below). In fact, the β-barrel jelly roll continues to be the structural motif most commonly encountered as the structures of additional capsid proteins are determined. It is well established that the three-dimensional structures of cellular proteins have been highly conserved during evolution, even though there may be very little amino acid sequence identity. For example, all globins possess a common three-dimensional structure based on eight α-helices, even though

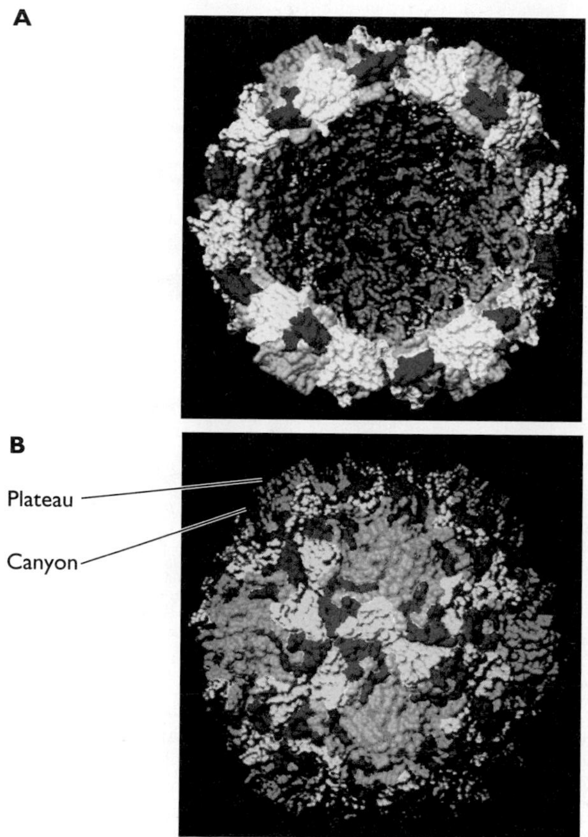

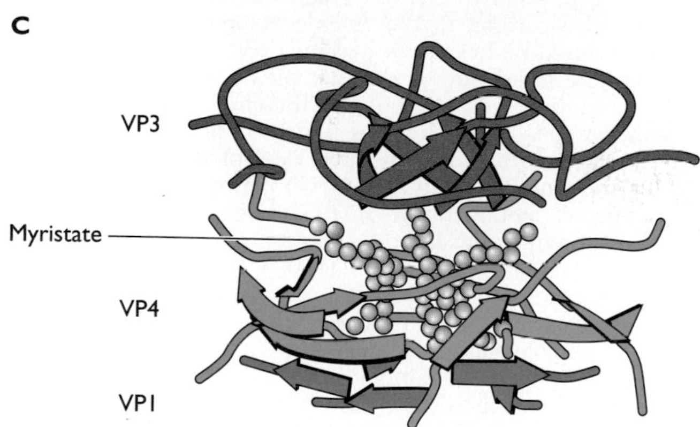

Figure 4.10 Interactions among the proteins of the poliovirus capsid. (A) Space-filling representation of the particle, with four pentamers removed from the capsid shell and VP1 in blue, VP2 in yellow, VP3 in red, and VP4 in green. Note the large central cavity in which the RNA genome resides, the dense protein shell formed by packing of the VP1, VP2, and VP3 β-barrel domains, and the interior location of VP4, which decorates the inner surface of the capsid shell. (B) Space-filling representation of the exterior surface showing the packing of the β-barrel domains of VP1, VP2, and VP3. Interactions among the loops connecting the upper surface of the β-barrel domains of these proteins create the surface features of the virion, the plateaus at the fivefold axes, and the knobs and ridges surrounding the threefold axes. For example, the outward tilt of the upper surface of each VP1 β-barrel domain and of the loops that connect the β-strands at this surface (Fig. 4.9B) of each of the five VP1 molecules around the fivefold axes forms the plateaus, which are encircled by a deep cleft or canyon. The virion is also stabilized by numerous interactions among the proteins on the inner side of the capsid. As illustrated in panel C, these internal contacts are most extensive around the fivefold axes, where the N termini of five VP3 molecules are arranged in a tube-like, parallel β-sheet. The N termini of VP4 molecules carry chains of the fatty acid myristate (gray), which are added to the protein posttranslationally. The lipids mediate interaction of the β-sheet formed by VP3 N termini with a second β-strand structure, containing strands contributed by both VP4 (green) and VP1 (blue) molecules. This internal structure is not completed until the final stages of, or after, assembly of virus particles, when proteolytic processing liberates VP2 and VP4 from their precursor, VP0. This reaction therefore stabilizes the capsid, and also primes it for entry into a new host cell. (A and B) From J. M. Hogle et al., *Science* **229:**1358–1365, 1985, with permission. Courtesy of M. Chow, University of Arkansas for Medical Sciences, and J. M. Hogle, Harvard Medical School.

the only residues conserved are those important for function. One interpretation of the common occurrence of the β-barrel jelly roll domain in viral capsid proteins is therefore that seemingly unrelated modern viruses (e.g., picornaviruses and adenoviruses) share some portion of their evolutionary history. However, as the origins and evolution of viruses are not well understood (see Chapter 20), it is also possible that this structural domain represents one of a limited number commensurate with packing of proteins to form a sphere, and is an example of convergent evolution. The case for descent of viruses with capsid proteins built on the β-barrel jelly roll motif from a distant common ancestor is considerably strengthened by remarkable similarities in the detailed structures of the capsid proteins of human adenoviruses and the bacteriophage PRD1 (Box 4.6).

The overall similarity in shape of the β-barrel domains of poliovirus VP1, VP2, and VP3 facilitates both their interaction with one another to form the 60 structural units of the capsid and the packing of these structural units in the capsid. How well these interactions are tailored to form a protective shell is illustrated by the space-filling model of the capsid shown in Fig. 4.10A and B: the extensive interactions among the β-barrel domains of adjacent proteins form a dense, rigid protein shell around a central cavity in which the genome resides. The packing of the β-barrel domains is reinforced by a network of protein-protein contacts on the inside of the capsid. These interactions are particularly extensive about the fivefold axes (Fig. 4.10C).

One of several important lessons learned from high-resolution structures of picornaviruses is that their design does not strictly conform to the principle of **quasiequivalence**. For example, despite the topological identity and geometric similarity of the central domains of the poliovirus proteins that form the capsid shell, the subunits do not engage in quasiequivalent bonding: interactions among VP1 molecules around the fivefold axes are not chemically or structurally equivalent to those in which

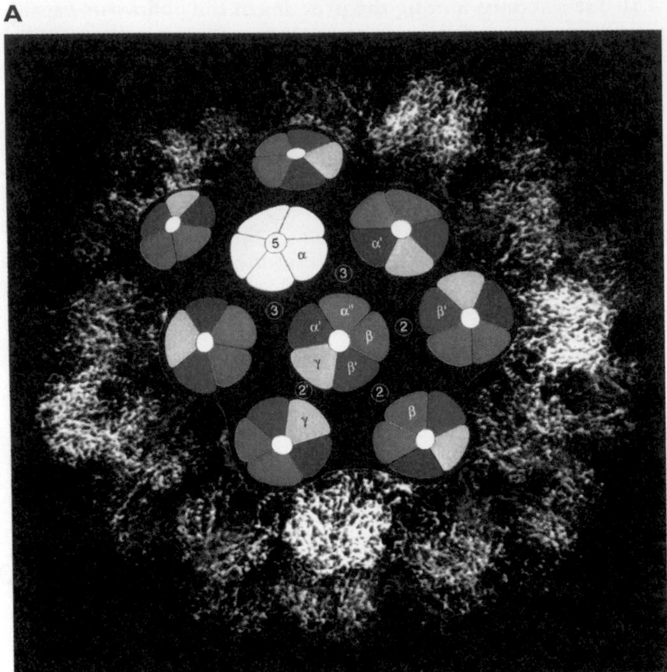

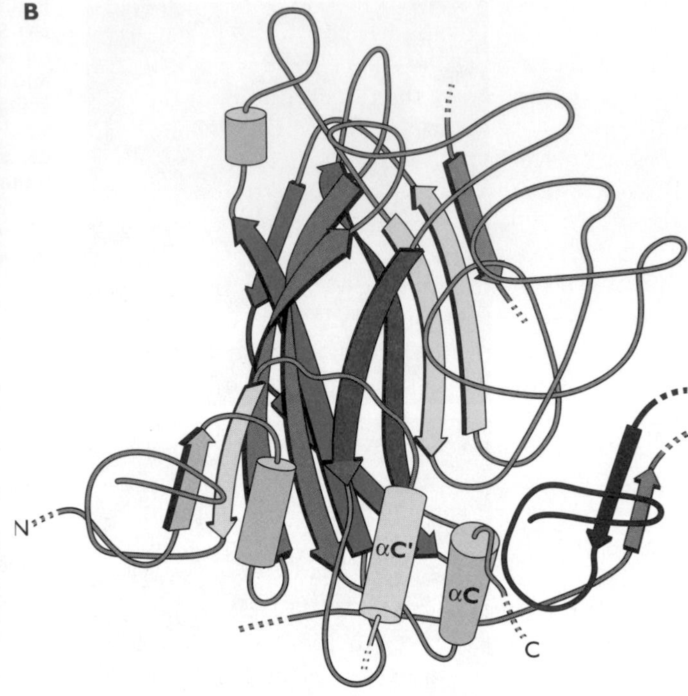

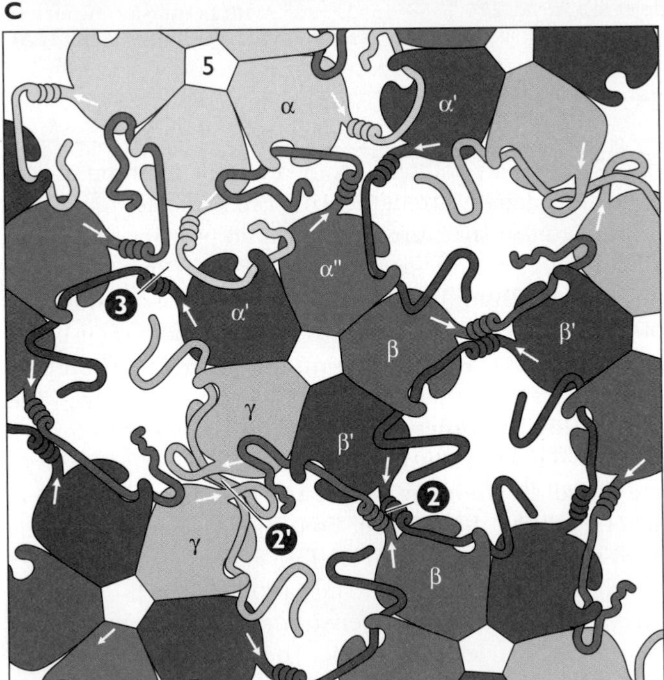

Figure 4.11 Structural features of the simian virus 40 virion. The structure of simian virus 40 (and other polyomaviruses) was long viewed as an enigma, because organization of these particles cannot be explained by the theoretical principles elaborated by Caspar and Klug. In the first place, the 360 subunits of the virion would correspond to $T = 6$, a triangulation number that is not permitted according to the rules developed by these authors. Furthermore, the hexameric packing of 60 VP1 pentamers represents a direct violation of the principle of quasiequivalence. The high-resolution view shown here therefore made an important contribution to advancing our understanding of the molecular principles of virus structure. (A) View of the simian virus 40 virion showing the organization of VP1 pentamers. The twelve 5-coordinated pentamers are shown in white, and the 60 pentamers present in hexameric arrays are colored. The three types of interpentameric clustering that allow 5-coordinated and 6-coordinated association of pentamers are shown in the schematic overlay. The VP1 subunits shown in white, purple, and green form a threefold cluster, designated 3. Those shown in red (β) and blue (β′) engage in one kind of twofold interaction, designated 2, and the yellow subunits (γ) form a second kind of twofold cluster, labeled 2′. (B) The topology of the VP1 protein shown in a ribbon diagram. The strands of the β-barrel jelly roll are shown as in Fig. 4.9B. This β-barrel domain is radial to the capsid surface, rather than organized tangentially as in poliovirus VP1, VP2, and VP3 proteins. Two C-terminal arms, which include αC-helices, are shown in orange and light gray. The former is the C-terminal arm of the VP1 subunit shown, which invades a neighboring pentamer (not shown). The second is the invading arm from a different neighboring pentamer (not shown), which is clamped in place by extensive interactions of its β-strand with the N-terminal segment of the subunit shown. The subunit shown also interacts with the N-terminal arm from its anticlockwise neighbor in the same pentamer (dark gray) and with the C-terminal arm of the pentamer that invades that neighbor (black).

(continued on next page)

VP2 or VP3 engage (Fig. 4.10A and B). These differences account for the characteristic features of the surface of the capsid. The interaction of five VP1 molecules unique to the fivefold axes results in a prominent protrusion extending to about 25 Å from the capsid shell (Fig. 4.10A and B). The resulting structure appears as a steep plateau encircled by a valley or cleft. In the capsids of many picornaviruses, these depressions are so deep that they have been termed canyons, and they are the receptor binding sites.

Structure of simian virus 40. The capsids of the small DNA polyomaviruses, simian virus 40 and polyomavirus, about 500 Å in diameter, are organized according to a rather different design. The structural unit is a pentamer of the major structural protein, VP1 (Fig. 4.11A). The capsid is built from 72 such structural units engaged in one of two kinds of interaction, which are described by the number of neighbors that surround any particular pentamer. Twelve structural units occupy the 12 positions of fivefold rotational symmetry, in which each is surrounded by five neighbors. Each of the remaining 60 structural units is surrounded by six neighbors at positions of sixfold rotational symmetry in the capsid (Fig. 4.11A). In contrast to the 60 structural units of poliovirus, the 72 pentamers of simian virus 40 do not engage in identical interactions. Rather, they occupy a number of different local environments in the capsid, because of differences in packing around the five- and sixfold axes (Fig. 4.11A and C).

Like the three poliovirus proteins that form the capsid shell, simian virus 40 VP1 contains a large central β-barrel jelly roll domain, in this case with an N-terminal arm and a long C-terminal extension. However, the arrangement and packing of VP1 molecules bear little resemblance to the organization of poliovirus capsid proteins. In the first place, the VP1 β-barrels in each pentamer project outward from the surface of the capsid to a distance of about 50 Å (Fig. 4.11A), in sharp contrast to those of the poliovirus capsid proteins, which tilt along the surface of the capsid shell. As a result, the surface of simian virus 40 is much more "bumpy" than that of poliovirus (compare Fig. 4.11A and 4.10B). Furthermore, the VP1 molecules present in adjacent pentamers in the simian virus 40 capsid do not make extensive contacts via the surfaces of their β-barrel domains. Rather, stable interactions among pentamers are mediated by their N- and C-terminal arms (Fig. 4.11B and C). The packing of VP1 pentamers in both pentameric and hexameric arrays in the simian virus 40 capsid requires different contacts among these structural units, depending on their local environment. In fact, just three kinds of interpentamer contact are needed (Fig. 4.11C). The different contacts are the result of alternative conformations and arrangements of the long C-terminal arms of VP1 molecules (Fig. 4.11B and C).

Along its long axis, each VP1 pentamer contains a tapering cavity that is narrowest toward the outer surface of the virion. This aperture contains a common C-terminal sequence of either VP2 or VP3, which associates tightly with the VP1 pentamer. Both these proteins are internal and make no contribution to the outer surface of the virion. VP2 plays an important role during entry, but why VP3 is present in virions is not known. This protein is identical in sequence to the C-terminal segment of the larger VP2 protein (Appendix A, Fig. 15). It is possible that the virus particle is simply too small to accommodate 72 molecules of VP2 (as well as the DNA genome) in its interior. VP3 could serve the same structural function as the C-terminal portion of VP2, presumably stabilizing VP1 pentamers, but occupy less internal space.

Simian virus 40 and poliovirus capsids differ in their surface appearance, in the number of structural units, and in the ways in which the structural units interact. Despite such differences, these virions share important features, including modular organization of proteins that form the capsid shell and a common β-barrel domain as the capsid building block. Neither poliovirus nor simian virus 40 capsids conform to strict quasiequivalent construction: all contacts made by all protein subunits are not similar, and in the case of simian virus 40, the majority of VP1 **pentamers** are packed in **hexameric** arrays. Nevertheless, close packing with icosahedral symmetry is achieved by limited variations of the contacts, either among topologically and geometrically similar, but chemically distinct, surfaces (poliovirus) or made by a flexible arm (simian virus 40).

The capsids of nonenveloped viruses such as picornaviruses and polyomaviruses must perform both protective

Figure 4.11 *(continued)* Consequently, the pentamers interact in the capsid shell via extensive contacts among N- and C-terminal arms of VP1 subunits in the same and in neighboring pentamers. (C) The interpentamer contacts of C-terminal arms, with pentamers and their subunits represented and color coded as in panel A, with C-terminal extensions shown as lines and with coils representing the αC-helices shown in panel B. The three types of interpentamer clustering shown in panel A, 3, 2, and 2′, are accommodated by variation in interpentamer contacts made by the C-terminal extensions. For example, in the threefold cluster, the C-terminal α-helices from the three interacting VP1 molecules form a three-stranded α-helical bundle. A two-chain α-helical bundle is formed at one kind of twofold cluster (2, between red and blue subunits), whereas at the second kind of twofold cluster (2′, yellow subunits) the subunits are packed so closely that there is no space to accommodate an α-helical structure. (A and B) From R. C. Liddington et al., *Nature* **354:**278–284, 1991, with permission. Courtesy of S. C. Harrison, Harvard University.

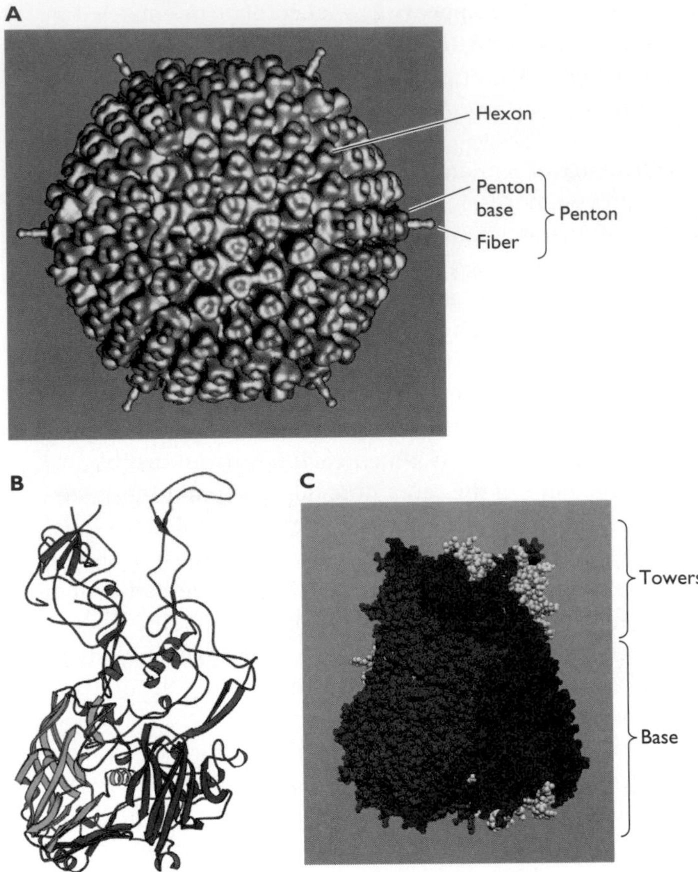

Figure 4.12 Structural features of adenovirus particles. (A) Surface view of the adenovirus type 2 virion, obtained by cryo-electron microscopy and image reconstruction. This view is oriented along an icosahedral threefold axis and has a nominal resolution of 30 Å. This is sufficient to visualize individual hexons, as well as their general shape, and to distinguish pentons. Only the virion-proximal portions of the projecting fibers are observed. From P. L. Stewart et al., *Cell* **67:**145–154, 1991, with permission. (B and C) Structure of the hexon. The monomer (B) is shown as a ribbon diagram, with gaps indicating regions that were not defined in the X-ray crystal structure at 2.9 Å resolution, and the trimer (C) is shown as a space-filling model with each monomer in a different color. The monomer contains two β-barrel jelly roll domains colored green and blue. These form the base of the hexon trimer (C), from which rise three towers formed by loops emanating from the β-barrel domains of each monomer. The trimers are stabilized by extensive interactions within both the base and the towers. The shape of the base facilitates packing of hexons in hexameric arrays to form the capsid shell. From M. M. Roberts et al., *Science* **232:**1148–1151, 1986, and F. K. Athappilly et al., *J. Mol. Biol.* **242:**430–455, 1994, with permission. Courtesy of J. Rux, S. Benson, and R. M. Burnett, The Wistar Institute.

and delivery functions. As mentioned above and described in detail in Chapter 5, their surfaces contain binding sites for specific cellular receptors. Moreover, their tightly knit protective protein shells (Fig. 4.10A and 4.11B) are designed for disassembly when appropriate conditions are encountered.

Structurally simple icosahedral capsids or nucleocapsids in more complex virions. Several viruses that are structurally more sophisticated than the picornaviruses or polyomaviruses nevertheless possess simple protein coats built from one or a few structural proteins. The complexity comes from the additional protein and lipid layers in which the capsid is enclosed. For example, togaviruses, such as Semliki Forest virus and Ross River virus (genus *Alphavirus*), contain a $T = 4$ icosahedral capsid built from a single protein (Table 4.3) within an envelope.

Table 4.4 Specialization of adenovirus type 2 structural proteins[a]

Protein	Molecular mass (kDa)	No. of copies	Location	Function
II	109,677	720	Hexon (trimer)	Formation of capsid shell
III	63,296	60	Penton base (pentamer)	Formation of capsid shell; entry
IV	61,960	36	Fiber (trimer)	Attachment to host cell
IIIa	63,287	74 ± 1	Edges between hexon faces	Stabilization of capsid
VI	23,449	342 ± 4	Hexon associated	Stabilization of capsid
VIII	14,539	211 ± 2	Hexon associated	Stabilization of capsid
IX	14,339	247 ± 2	Outer surface of groups-of-nine hexons	Stabilization of capsid
V	41,631	157 ± 1	Core, outer surface	Packaging of DNA genome
VII	19,412	835 ± 20	Core, bound to DNA	Packaging of DNA genome
μ	2,441	120 ± 15	Core	Packaging of DNA genome

[a]In addition to the structural proteins listed, the adenovirus virion contains two copies of pTP, one covalently linked to each 5′ end of the viral genome, the viral L3 protease, and several small proteins generated upon cleavage of proteins that enter the virion as precursors. Only one of these, the C-terminal extension from pVI, which activates the L3 protease, has been ascribed a function.

Structurally Complex Capsids

Some naked viruses are considerably larger and more elaborate than the small RNA and DNA viruses described in the previous section. The characteristic feature of such virus particles is the presence of proteins devoted to specialized structural or functional roles. Despite such complexity, reasonably detailed pictures of the organization of this type of virion can be constructed by using combinations of biochemical, structural, and genetic methods. The well-studied human adenovirus type 2 and members of the *Reoviridae* exemplify this approach.

Adenovirus. The most striking morphological features of the adenovirus particle (maximum diameter, 1,500 Å) are the well-defined icosahedral appearance of the capsid and the presence of long projections, the fibers, at the 12 vertices (Fig. 4.12A). Each fiber, which terminates in a distal knob that binds to the cell surface adenoviral receptor, is attached to one of the 12 penton bases located at positions of fivefold symmetry in the capsid. The remainder of the shell is built from 240 additional subunits, the hexons, each of which is a trimer of viral protein II (Table 4.4). Formation of this capsid therefore depends on nonequivalent interactions among subunits: the hexons that surround pentons occupy a different bonding environment than those surrounded entirely by other hexons. Furthermore, the 780 total protein subunits (12 × 5 in pentons plus 3 × 240 in hexons) that form the closed capsid are not arranged to correspond to a $T = 13$ (780/60) icosahedron. Instead, determination of the triangulation number of the adenovirus capsid by electron microscopy (Box 4.4, figure panel D) yielded a value of $T = 25$.

The adenovirus particle contains seven additional structural proteins (Table 4.4), as well as several small polypeptides produced by proteolytic cleavage of precursor proteins during maturation of newly assembled virus particles. The large size of the virion and the presence of so many proteins have made elucidation of adenovirus structure a challenging problem. One approach that has proved generally useful in the study of complex viruses is the isolation and characterization of discrete subviral particles. For example, adenovirus particles can be dissociated into a core structure that contains the DNA genome, groups-of-nine hexons, and pentons (Fig. 4.13). Analysis of the composition of

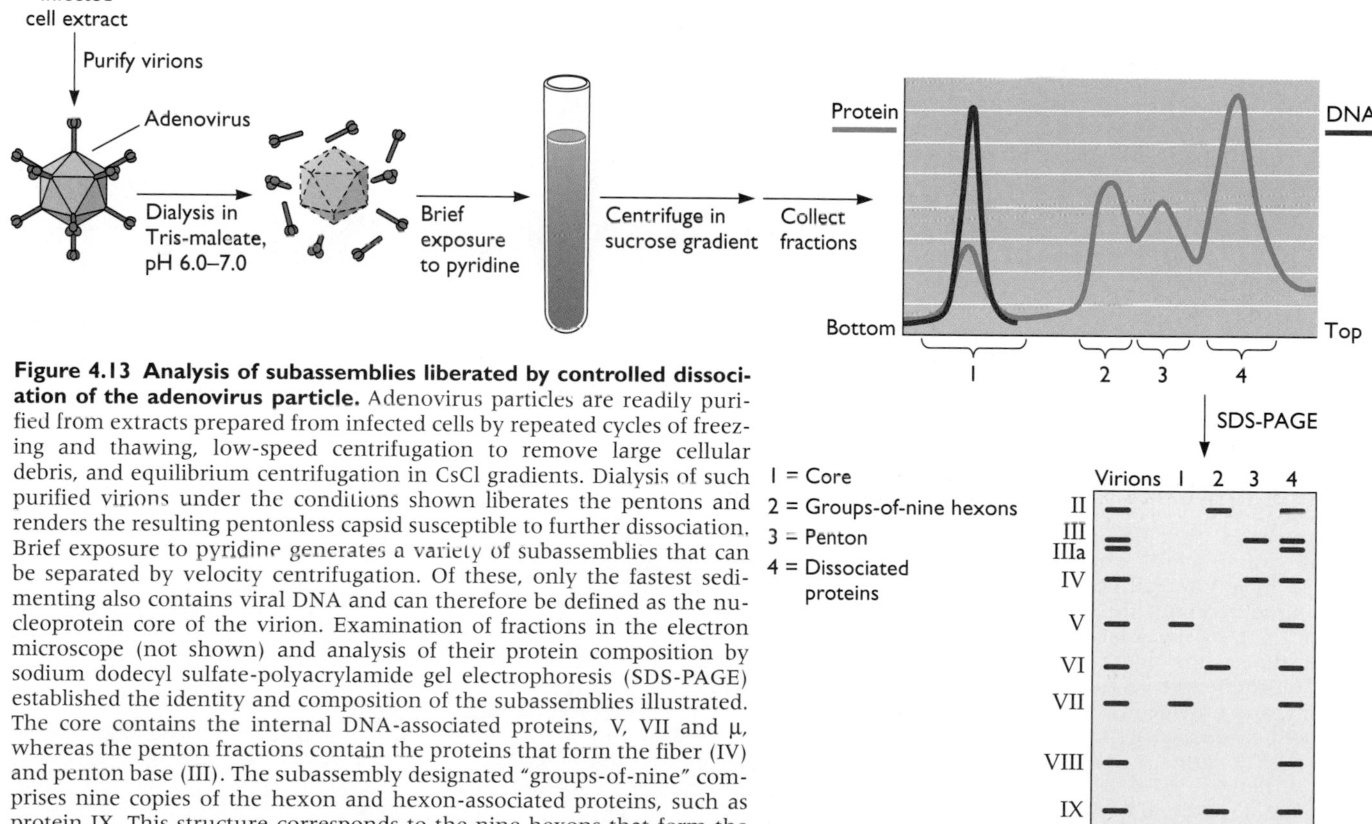

Figure 4.13 Analysis of subassemblies liberated by controlled dissociation of the adenovirus particle. Adenovirus particles are readily purified from extracts prepared from infected cells by repeated cycles of freezing and thawing, low-speed centrifugation to remove large cellular debris, and equilibrium centrifugation in CsCl gradients. Dialysis of such purified virions under the conditions shown liberates the pentons and renders the resulting pentonless capsid susceptible to further dissociation. Brief exposure to pyridine generates a variety of subassemblies that can be separated by velocity centrifugation. Of these, only the fastest sedimenting also contains viral DNA and can therefore be defined as the nucleoprotein core of the virion. Examination of fractions in the electron microscope (not shown) and analysis of their protein composition by sodium dodecyl sulfate-polyacrylamide gel electrophoresis (SDS-PAGE) established the identity and composition of the subassemblies illustrated. The core contains the internal DNA-associated proteins, V, VII and μ, whereas the penton fractions contain the proteins that form the fiber (IV) and penton base (III). The subassembly designated "groups-of-nine" comprises nine copies of the hexon and hexon-associated proteins, such as protein IX. This structure corresponds to the nine hexons that form the center of each of the 20 faces of the icosahedral capsid. The separation of these structures is idealized in the figure.

such subassemblies, together with identification of virion proteins that contact one another (by cross-linking methods), identified two classes of virion proteins in addition to the major capsid proteins described above. One comprises the proteins present in the core, such as protein VII, the major DNA-binding protein (Table 4.4). The remaining proteins are associated with either individual hexons or the groups of hexons that form an icosahedral face of the capsid (Table 4.4), suggesting that they stabilize the structure. Protein IX has been clearly identified as capsid "cement": a mutant virus that lacks the protein IX coding sequence produces the typical yield of virions, but these particles are much less heat stable than wild-type virions. During assembly, interactions among hexons and other major structural proteins must be relatively weak, so that incorrect associations can be reversed and corrected. However, the assembled virion must be stable enough to survive passage from one host to another. It has been proposed that the incorporation of stabilizing proteins like protein IX allows these paradoxical requirements to be met.

The intact adenovirus particle has not been studied by X-ray crystallography, but high-resolution structures have been determined for the knob of the fiber and the hexon. Each hexon subunit (Table 4.4) contains two β-barrel domains, each with the topology of the β-barrels of the simpler RNA and DNA viruses described in the previous section (Fig. 4.12B). As the two β-barrels are very similar in structure, the hexon trimer exhibits pseudohexagonal symmetry, a property that facilitates its close packing in the capsid. In the trimer, the β-barrel domains of the three monomers are packed together in a hollow base from which rise three towers formed by intertwining loops from each monomer (Fig. 4.12C). The interactions among monomers are very extensive, particularly in the tower. Consequently, once the trimer has formed, the monomers cannot be dissociated easily, and the hexon is an extremely stable structure.

The high-resolution structure of the hexon has allowed the resolution gap between X-ray crystallography and electron microscopy to be bridged by the technique of **difference imaging**. This procedure subtracts an image obtained by X-ray crystallography (at appropriate resolution) from an electron microscopic image. When applied to the adenoviral groups-of-nine hexons this procedure revealed the 12 molecules of protein IX present in this subassembly (Fig. 4.14). Each protein IX monomer extends along a

A Groups-of-nine hexons

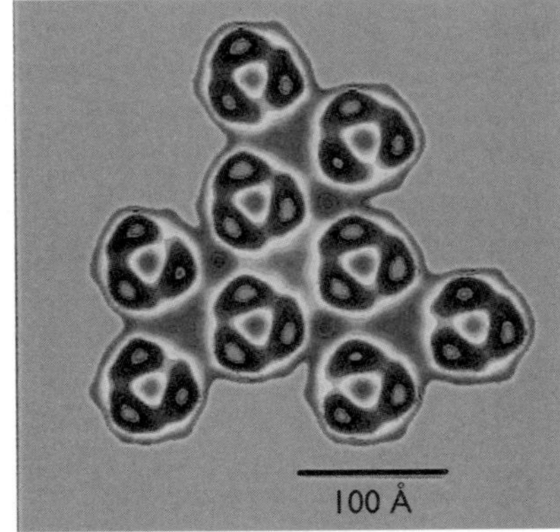

B Single hexon

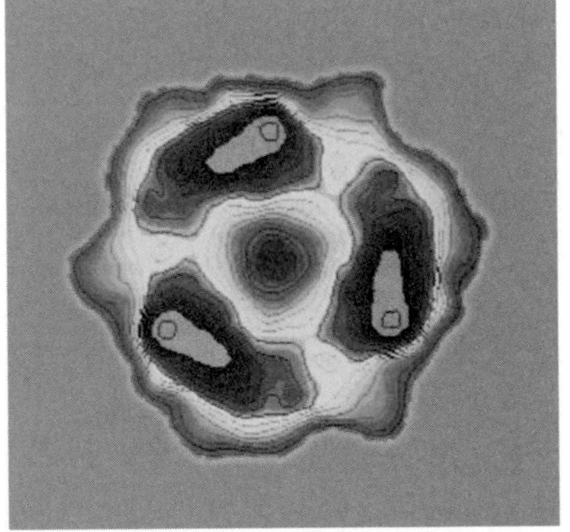

C Artificial image of 9 hexons placed as in A

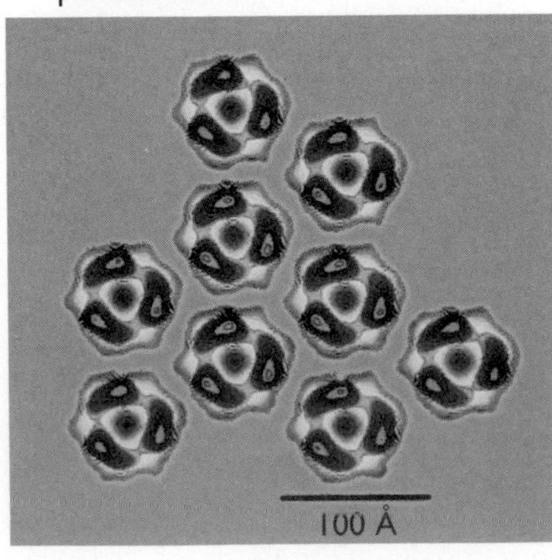

D Difference image, A minus C

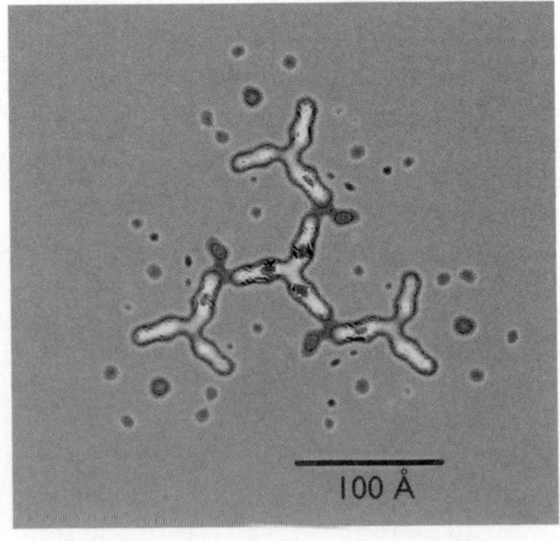

Figure 4.14 Difference imaging of adenovirus groups-of-nine hexons. The images used in this process are shown as two-dimensional contour plots. (A) The average of 57 images of groups-of-nine hexons obtained by scanning transmission electron microscopy (STEM) with a resolution of 15 to 17 Å. (B) Two-dimensional image of a single hexon trimer at the same resolution, constructed from the high-resolution hexon structure determined by X-ray diffraction. (C) Artificial image of the groups-of-nine hexons created by rotational and translational alignment of nine copies of the X-ray-based hexon image (B) with the STEM image of the assembly (A). The very good agreement of hexon separation in this artificial image with that observed by electron microscopy established the accuracy of the image shown. (D) Difference image formed by subtracting the artificial image of the placement of hexons in groups of nine (C) from the STEM image of natural groups of nine containing both hexons and protein IX (A). This procedure reveals 12 molecules of protein IX arranged in four clusters of three molecules in the hexon-hexon interfaces. More recently, three-dimensional difference imaging of the intact particle has located protein IIIa to the edges between icosahedral faces, consistent with its necessary role in virion assembly, and provided insights into the packing of trimeric hexons around the penton base pentamer at each vertex. From P. S. Furcinitti et al., *EMBO J.* **8**:3563–3570, 1989, with permission. Courtesy of J. Rux and R. M. Burnett, The Wistar Institute.

hexon-hexon interface, like mortar between bricks in a wall, an arrangement that neatly accounts for its stabilizing function. Difference imaging has yielded previously inaccessible structural information about adenoviruses, and its application to other complex viruses is proving equally valuable (see Fig. 4.17, for example).

Reoviruses. Reoviruses are naked $T = 13$ icosahedral particles, 700 to 900 Å in diameter, containing the 10 to 12 segments of the double-stranded genome and the enzymatic machinery to synthesize viral mRNA. Like adenoviruses, reoviruses are built with specialized proteins. However, reovirus particles exhibit a very different architecture: they contain multiple protein shells. The particles of human reovirus (genus *Orthoreovirus*) contain eight proteins organized in two concentric shells, with spikes projecting from the inner layer through and beyond the outer layer at each of the 12 vertices (Fig. 4.15A). Members of the genus *Rotavirus*, which includes the leading causes of severe infantile gastroenteritis in humans (Appendix B, Fig. 21), contain three nested protein layers, with 60 spikes projecting from the surface of the outer layer (Fig. 4.15A). Although differing in architectural detail, reovirus particles share common structural features, including an unusual design of the innermost protein shell. Removal of the outermost protein layer, a process believed to occur during entry into a host cell (Chapter 5), yields an inner core structure, comprising one shell (orthoreoviruses) or two (rotaviruses and members of the genus *Orbivirus*, such as bluetongue virus). These structures also contain the genome and virion enzymes and synthesize viral mRNAs under appropriate conditions in vitro. High-resolution structures have been obtained for bluetongue virus and human reovirus cores, the largest viral assemblies yet to be examined by X-ray crystallography.

The thin inner layer contains 120 copies of a single protein, termed VP3 in bluetongue virus and λ1 in human reovirus. These proteins are not related in their primary sequences, but nevertheless have similar topological features and the same plate-like shape (Fig. 4.15B and C, right). Moreover, in both cases, the protein monomers occupy one of two different structural environments and to do so adopt one of two distinct conformational states (A and B in Fig. 4.15B and C, right). Because of this arrangement, the A and B monomers are **not** quasiequivalent: virtually all contacts in which A and B monomers engage are very different. In the cores of human reovirus, there are also strikingly nonequivalent interactions between the λ1 subunits and the 150 copies of the stabilizing protein of the inner shell, σ2 (Fig. 4.15A). Bluetongue virus lacks a σ2 equivalent, and the VP3 inner shell abuts directly to the inner surface of the middle layer, which comprises trimers of a single protein (VP7) organized into a classical $T = 13$ lattice (Fig. 4.15B, left). A large number of different (nonequivalent) contacts between VP3 and VP7 structural units weld the two layers together. These properties of reoviruses illustrate that a quasiequivalent structure is not the **only** solution to the problem of building large viral particles: viral proteins that interact with each other and with other proteins in multiple ways can provide an effective alternative.

Complex Viruses

Some large viruses contain no structural elements that conform to helical or icosahedral symmetry. For example, the cores of the complex poxviruses, such as vaccinia virus, appear as smooth rectangles encased within a membrane and surface protein layer (Fig. 4.16). This shape may be the result of the combination of icosahedrally symmetric, terminal caps with a tubular structure with helical symmetry, as in the elongated heads of certain bacteriophages.

Packaging the Nucleic Acid Genome

A definitive property of a virion is the presence of a nucleic acid genome. Incorporation of the genome requires its discrimination from a large population of cellular nucleic acid and its packaging. Three mechanisms for condensing, and presumably organizing, nucleic acid molecules within capsids or nucleocapsids can be distinguished (Table 4.5).

Direct Contact of the Genome with a Protein Shell

In the simplest arrangement, the viral nucleic acid makes direct contact with the protein(s) that forms a protective shell of the particle (Table 4.5). The structural units of the icosahedral capsids of many small RNA viruses, including poliovirus and other picornaviruses, contain proteins that interact with the viral genome at the inner surface of the capsid shell. As we have seen, the interior surface of the poliovirus capsid can be described in detail. Nevertheless, we possess no structural information about the arrangement of the RNA genome, for the nucleic acid is not visible in the X-ray structure. This property indicates that the RNA genome lacks the symmetry of the virion, and does not adopt the identical conformation in every virus particle. In contrast, in other small viruses with icosahedral symmetry, segments of the RNA or DNA genomes are highly ordered. For example, in the $T = 3$ nodavirus Nodamura virus (Fig. 4.8), double-stranded segments of the RNA genome bind to subunits of adjacent structural units at the icosahedral twofold axes.

Use of the same protein or proteins both to package the genome and to build a capsid allows efficient utilization of limited genetic capacity. It is therefore an advantageous arrangement for viruses with small genomes. However, this mode of genome packing is also characteristic of some

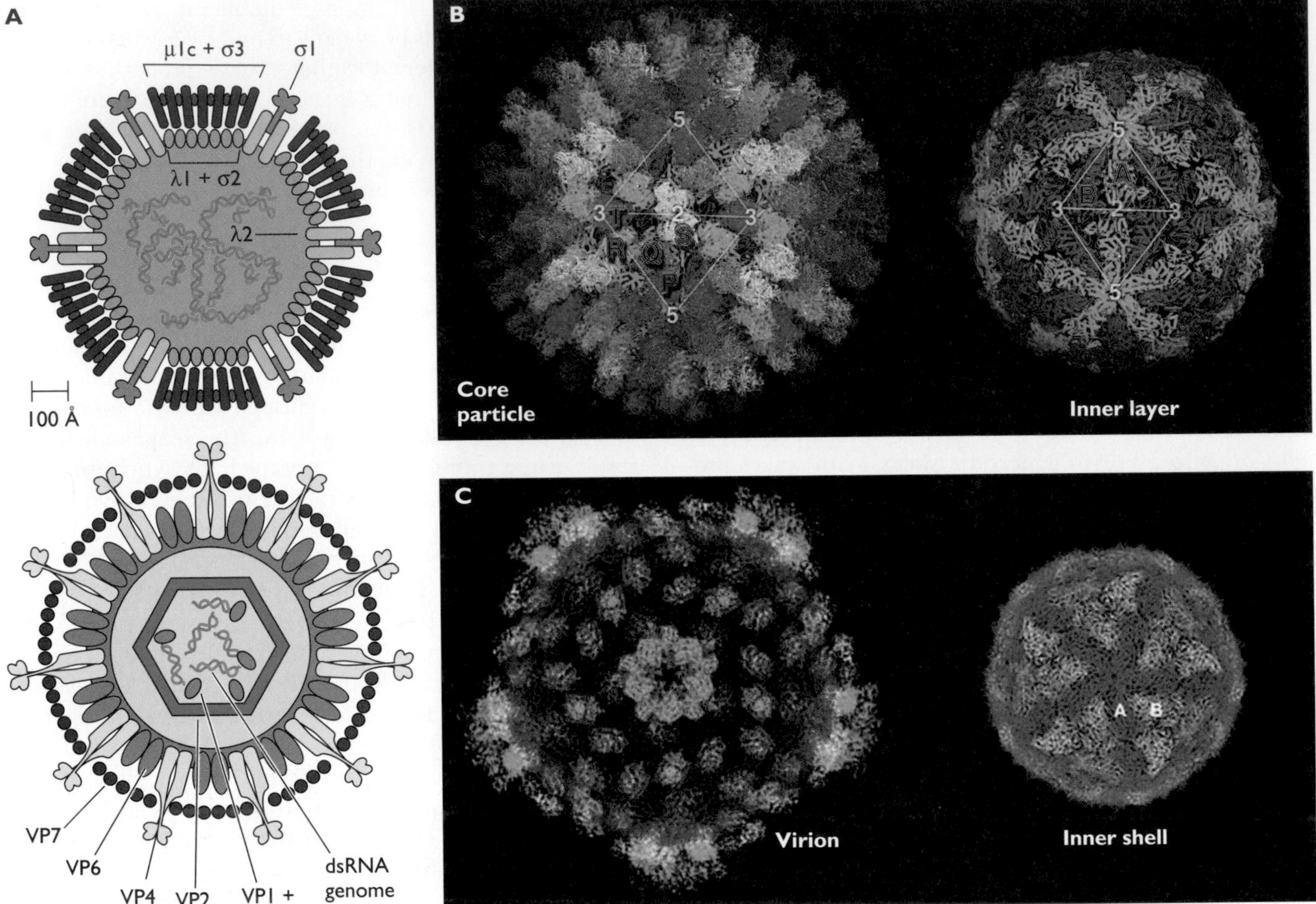

Figure 4.15 Structures of members of the *Reoviridae*. (A) Organization of mammalian reovirus and rotavirus particles is shown schematically to indicate the locations of proteins, deduced from the protein composition of virions and of subviral particles that can be readily isolated from them. (B) X-ray crystal structure of the core of bluetongue virus, a member of the *Orbivirus* genus of the *Reoviridae*, showing the core particle (left) and the inner scaffold (right). The VP7 trimers project radially from the outer layer. As indicated, each icosahedral asymmetric unit, the triangular areas defined by the icosahedral symmetry axes, contains 13 copies of VP7 arranged as five trimers designated P, Q, R, S, and T and colored red, orange, green, yellow, and blue, respectively. Trimer T, which is located at the threefold axes of icosahedral symmetry, contributes only a monomer to an asymmetric unit. The outer layer is organized with classical $T = 13$ icosahedral symmetry. As shown on the right, the inner layer is built from VP3 monomers that occupy one of two completely different structural environments, designated A and B and colored green and red, respectively. A-type monomers span the icosahedral twofold axes and interact around the icosahedral fivefold axes. In contrast, B-type monomers are organized as triangular "plugs" around the threefold axes. Differences in the interactions among monomers at different positions allow close packing to form the closed shell, the equivalent of a $T = 2$ lattice. As might be anticipated, VP7 trimers in pentameric or hexameric arrays in the outer layer make completely different contacts with the two classes of VP3 monomer in the inner layer. Nevertheless, each type of interaction is extensive and, in total, these contacts compensate for the symmetry mismatch between the two layers of the core. The details of these contacts suggest that the inner shell both defines the size of the virus particle and provides a template for assembly of the outer $T = 13$ structure. From J. M. Grimes et al., *Nature* **395:**470–478, 1998, with permission. Courtesy of D. I. Stuart, University of Oxford. (C) X-ray crystal structure of the core of human reovirus (genus *Orthoreovirus*), showing the inner icosahedral shell (right) and the complete particle (left), represented by α-carbon traces of the protein subunits. Note the similar organization of the A and B conformations of the λ1 protein of the inner shell to that of VP3 A and B monomers of bluetongue virus. There are three binding sites for the stabilizing protein σ2 within each icosahedral asymmetric unit of the λ1 shell, and the protein-protein contacts at these sites are nonequivalent. The core is completed by 60 copies of the λ2 protein (blue) arranged in pentameric, turret-like structures at each axis of fivefold symmetry. This protein has all the enzymatic activities necessary to add a 5′ blocking group (the cap) to viral mRNA molecules as they pass through the turret. From K. M. Reinish et al., *Nature* **404:**960–969, 2000, with permission. Courtesy of S. C. Harrison, Harvard University.

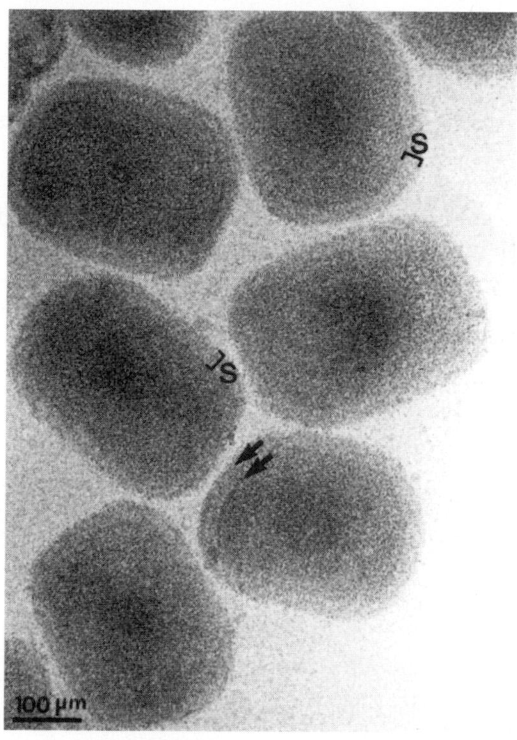

Figure 4.16 Cryo-electron micrograph of mature vaccinia virus particles. These particles comprise a 300-Å-thick surface domain (S) bounded by the membranes marked by the arrows and surrounding a homogeneous core. The latter is studded with 200-Å spikes and has the shape of a rounded rectangle. From J. Dubochet et al., *J. Virol.* **68:**1935–1941, 1994, with permission. Courtesy of M. Adrian and J. Dubochet, Université de Lausanne.

more complex viruses, notably rotaviruses and herpesviruses. The genome of rotaviruses comprises 11 segments of double-stranded RNA located within the innermost of the three protein shells of the virion. About 25% of the RNA (more than 4,000 base pairs [bp]) is highly ordered, forming a dodecahedral structure in which RNA helices are in close contact with the interior surface of the inner nucleocapsid (Fig. 4.17).

One of the most surprising properties of the large herpesviral nucleocapsid (described in "Large Enveloped Viruses with Numerous Internal Proteins" below) is the absence of internal proteins associated with viral DNA: despite intense efforts, no such core proteins have been identified. It is therefore probable that the double-stranded DNA genome is packaged in a liquid crystalline state, in which the DNA is present at a high concentration (typical of crystals) but not organized into a specific structure (like the molecules in a liquid). Such packing must require neutralization of the negative charges of the sugar-phosphate backbone, either by proteins that form the inner surface of the nucleocapsid, or by incorporation of small, positively charged, cellular molecules like spermine and spermidine.

Table 4.5 Mechanisms of packaging the viral genome

Mechanism	Example	Protein(s) contacting nucleic acid
By direct contact with nucleocapsid or capsid proteins	Alphavirus	C
	Calicivirus	Capsid?
	Herpesvirus	?
	Parvovirus	Ca
	Picornavirus	VP4
	Reovirus	λ1 (orthoreovirus), VP2 (rotavirus)
By specialized viral nucleic acid-binding proteins	Adenovirus	VII, V, μ
	Coronavirus	N
	Filovirus	NP
	Orthomyxovirus	NP
	Paramyxovirus	NP
	Poxvirus	L4R, A10L, A3L, A4L, F17R, I1, I6?
	Retrovirus	NC
	Rhadovirus	N
By cellular DNA-binding proteins	Papillomavirus, polyomavirus	Histones H2A, H2B, H3, and H4

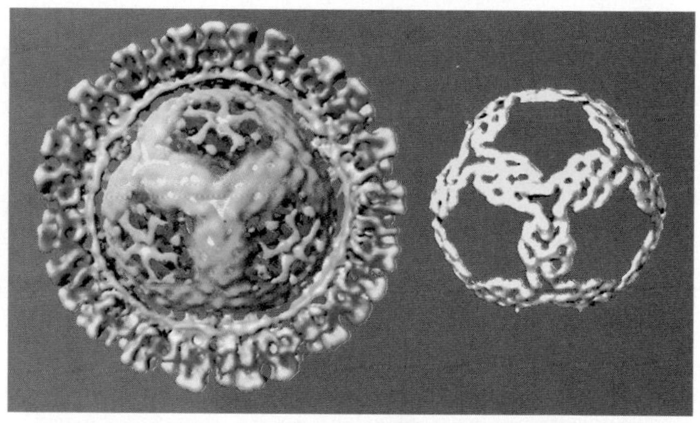

Figure 4.17 Structural organization of double-stranded genomic RNA within rotavirus. This structure was determined by cryo-electron microscopy, image reconstruction, and difference imaging of infectious double-layered particles and various virus-like particles. The 19-Å-resolution structure of the double-layered particle is shown at the left, with the outer VP6 capsid of the double-layered particle (middle capsid in the intact virion) in blue and the inner VP2 capsid in green. A complex of the virion RNA-dependent RNA polymerase and RNA capping enzyme (VP1 and VP3), shown in red, is associated with the inner surface of the VP2 nucleocapsid at the axes of fivefold rotational symmetry. The double-stranded RNA, which is packed around the structures formed by VP1 and VP3, is shown in yellow. The dodecahedral shell of ordered RNA, in which each strand is about 20 Å in diameter as expected for RNA in double-stranded helices, is shown separately at the right. From B. V. V. Prasad et al., *Nature* **382:**471–473, 1996, with permission. Courtesy of B. V. V. Prasad, Baylor College of Medicine.

Packaging by Specialized Virion Proteins

In many other virus particles, the genome is associated with specialized nucleic acid binding proteins, such as the nucleocapsid proteins of (–) strand RNA viruses, and retroviruses, or the core proteins of adenoviruses (Table 4.5). Although these viral nucleic acid-binding proteins were recognized many years ago, relatively few details of the nucleoprotein structures they form have been established. As described above, the ribonucleoproteins of (–) strand RNA viruses are helical structures. Electron microscopy of cores released from adenovirus particles (Fig. 4.13) suggested that the internal nucleoprotein is also organized in some regular fashion. However, this structure has proved difficult to study in detail, in part because the core is not a stable structure once released from virions, nor have structures of core proteins been determined. The fundamental DNA packaging unit is a multimer of protein VII (Table 4.5), which appears as beads on a string of adenoviral DNA when other core proteins are removed. Protein V probably occupies a more external position in the core, where it makes contact with capsid proteins. These, and the third core protein, μ, are basic, as would be expected for proteins that bind to a negatively charged DNA molecule without sequence specificity.

Retrovirus virions contain approximately 2,000 molecules of the nucleocapsid (NC) protein that bind to the two copies of the encapsidated (+) strand RNA genome (Appendix A, Fig. 21). The NC protein is also responsible for recognition of a specific packaging signal in the RNA during assembly and therefore binds both specifically and nonspecifically to the genome. Like other viral genome binding proteins, NC is positively charged, and in most retroviruses contains at least one copy of a well-characterized nucleic acid-binding motif. As discussed in more detail in Chapter 13, specific binding of NC to the RNA packaging signal is mediated by this motif. The structures of human immunodeficiency virus type 1 NC and of NC bound to the RNA packaging signal determined by NMR methods indicate that a long, N-terminal helix rich in basic residues interacts nonspecifically with the RNA. Whether this interaction is responsible for coating the RNA in the virion, and how NC molecules condense the genome, are not yet known. Retroviral ribonucleoproteins are encased within a protein shell built from the capsid (CA) protein to form an internal core. Although they contain the same components, these cores vary considerably in morphology (Fig. 4.18). Studies of the structures of the human immunodeficiency virus type 1 CA protein and the assemblies it forms in vitro (Box 4.7), and the aberrant shape of cores induced by alterations in the CA sequence, indicate that this protein determines the conical shape of the core.

Packaging by Cellular Proteins

The final mechanism of packaging the viral genome, by cellular proteins, is unique to polyomaviruses, such as simian virus 40, and papillomaviruses. The circular, double-stranded DNA genomes of these viruses are organized into nucleosomes that contain the four cellular core his-

Figure 4.18 Morphology of retroviruses. (A) Cryo-electron micrograph of mature human immunodeficiency virus type 1 showing the elongated internal cores of these particles. Courtesy of T. Wilk, European Molecular Biology Laboratory. (B) Variations in the morphology of retroviruses shown schematically. Although retrovirus particles are assembled from the same components, some contain roughly spherical cores, whereas the cores of lentiviruses like human immunodeficiency virus type 1 are elongated and conical.

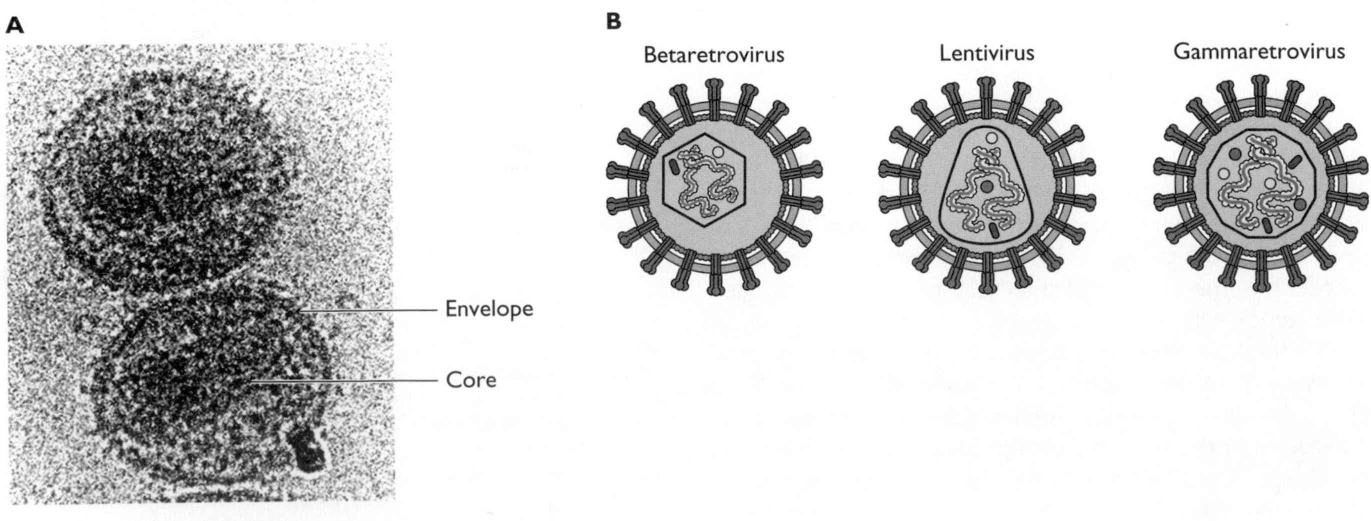

BOX 4.7

A fullerene cone model of the human immunodeficiency virus type 1 core

Purified human immunodeficiency virus type 1 CA-NC protein self-assembles into cylinders and cones when incubated with a segment of the viral RNA genome in vitro. Although the RNA facilitates formation of these structures, it is not essential, and specific viral sequences or structures are not required. The cones assembled in vitro are capped at both ends, and many appear very similar in dimensions and morphology to cores isolated from viral particles (panel A).

The very regular appearance of the synthetic CA-NC cones suggested that, despite their asymmetry, they are constructed from a regular, underlying lattice (analogous to the lattices that describe structures with icosahedral symmetry discussed in Box 4.4). In fact, these human immunodeficiency virus type 1 cores can be modeled using the geometric principles that describe cones formed by carbon. Such elemental carbon cones comprise helices of hexamers closed at each end by caps of buckminsterfullerene, which are structures that contain pentamers surrounded by hexamers. As in structures with icosahedral symmetry (Box 4.4), the positions of pentamers determine the geometry of cones. However,

(continued)

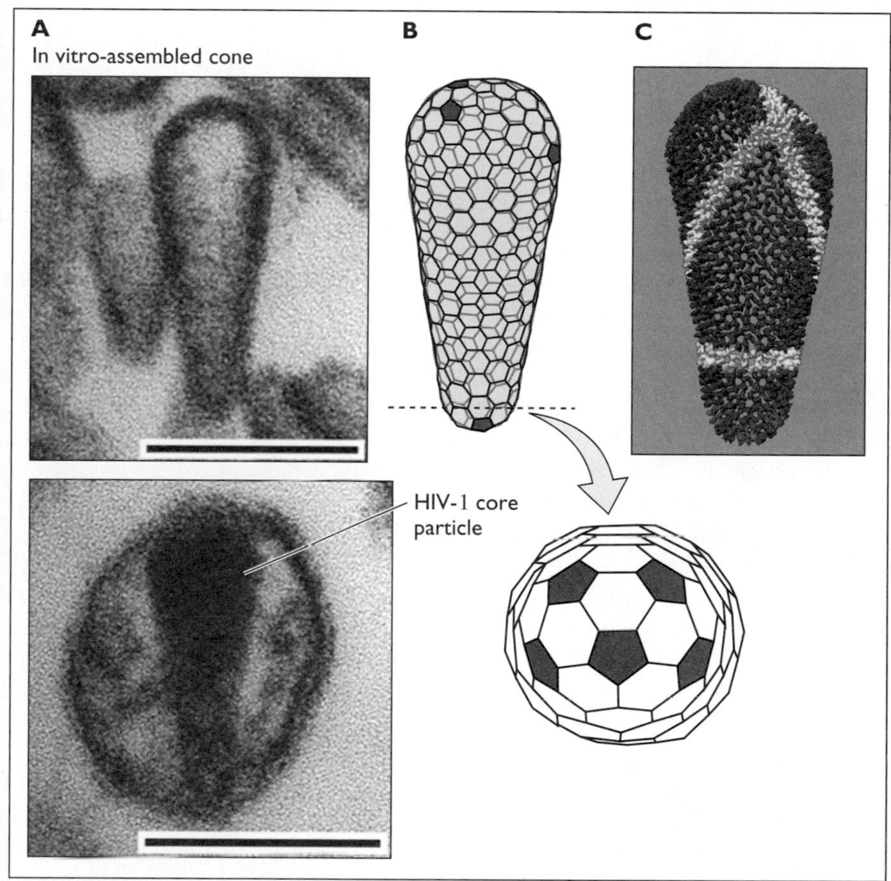

(A) Comparison of a synthetic CA-NC/RNA cone assembled in vitro (top) and core of the human immunodeficiency virus type 1 virion (bottom) by thin-section electron microscopy. The scale bars are 100 nm. (B) View of a fullerene cone closed with 5 pentamers (red) at the narrow end, and an end-on view of a cap built from 5 pentamers. (C) Model of the viral CA capsid, based on the CA-NC/RNA and CA structures assembled in vitro. Pentamers are shown in red, and a continuous line of CA hexamers is indicated in gold to illustrate the changes in a CA spiral. The model was created by placing molecular images of CA hexameric rings obtained by cryo-electron microscopy and image reconstruction onto an idealized fullerene cone (P = 5 narrow end) containing 1,572 CA subunits. Courtesy of Wes Sundquist, University of Utah.

BOX 4.7 *(continued)*

in cones, pentamers are present **only** in the terminal caps. The human immunodeficiency virus type 1 cones formed in vitro can be modeled as a fullerene cone assembling on a curved hexagonal lattice with 5 pentamers at the narrow end of the cone (panel B). The wide end would be closed by an additional 7 pentamers (because 12 pentamers are required to form a closed structure from a hexagonal lattice). In this type of structure, the cone angle at the narrow end can adopt only one of five allowed angles, determined by the number of pentamers. A narrow cap with 5 pentamers, as in the model shown in panel B, should exhibit a cone angle of 19.2°. Approximately 90% of all the synthetic CA-NC cores examined met this prediction, consistent with the fullerene cone model (B). Furthermore, subsequent cryo-electron microscopy and image reconstruction of tubes formed by purified CA protein in vitro established that these tubes are also constructed from a curved lattice of hexameric rings. The N-terminal domain of the CA protein, which is essential for capsid assembly, forms the hexameric rings, and adjacent hexamers interact via the C-terminal dimerization domain. In tubes, the planes of hexameric rings are parallel to the helix axis. However, as illustrated in panel C, to form cones the planes of the hexameric rings must be tilted with respect to the cone axis. Consequently, hexamers are organized in spirals that change gradually in pitch, diameter, and eventually helical handedness (panel C). This model requires that CA hexameric rings adopt many different orientations and therefore possess considerable structural plasticity. Just such conformational plasticity, imparted by both a flexible hinge between the N- and C-terminal domains of each CA monomer and loose packing of adjacent hexameric rings, was observed in the collection of CA structures analyzed in these studies.

Ganser, B. K., S. Li, V. Y. Klishko, J. T. Finch, and W. I. Sundquist. 1999. Assembly and analysis of conical models for the HIV-1 core. *Science* **283**:80–83.

Li, S., C. P. Hill, W. I. Sundquist, and J. T. Finch. 2000. Image constructions of helical assemblies of the HIV-1 CA protein. *Nature* **407**:409–413.

tones H2A, H2B, H3, and H4. This viral genome is organized within the virion (and in infected cells) like cellular DNA in chromatin to form a minichromosome. This packaging mechanism is elegant, with two major advantages: none of the limited viral genetic information needs to be devoted to DNA-binding proteins, and the viral genome, which is transcribed by cellular RNA polymerase II, enters the infected cell nucleus as a nucleoprotein closely resembling the cellular templates for this enzyme. In each simian virus 40 particle, the 20 or so nucleosomes, which package all sequences of the viral genome, condense the DNA by a factor of approximately 7. Within the virion, the minichromosome must be further compacted, presumably as a result of its interactions with the internal proteins of the capsid, VP2 and VP3, and perhaps the N-terminal arms of VP1 that lie on the interior surface. Both VP2 and VP3 can bind nonspecifically to DNA as well as to cellular histones, and all three proteins associate with the minichromosome during virion assembly. Neither the structure of this internal nucleoprotein nor how it interacts with capsid proteins has been determined.

Although three different ways of condensing and organizing genomic nucleic acids within virions can be distinguished readily (Table 4.5), none of these packaging arrangements is understood in detail. High-resolution structural descriptions of virion interiors, or of the nucleoproteins that reside within them, would undoubtedly provide important insights into mechanisms of condensation of the nucleic acid. Such structural information might also improve our understanding of the advantages conferred by the various packaging mechanisms described above.

Viruses with Envelopes

Many types of animal virus contain structural elements in addition to the capsids or nucleocapsids described previously. All such virus particles possess an **envelope** formed by a viral glycoprotein-containing membrane derived from the host cell, but they vary considerably in size, morphology, complexity, and nature of the membrane. Typical features of viral envelopes and their proteins are described in the next section, to set the stage for consideration of the structures of envelope proteins and the mechanisms by which they interact with internal components of the virion.

Viral Envelope Components

The foundation of all viral envelopes is a lipid membrane acquired from the host cell during assembly. This membrane consists of a bilayer of phospholipids, molecules containing negatively charged head groups attached to long chains of hydrophobic fatty acids. As discussed in more detail in Chapter 5, the apolar fatty acyl chains are arranged in the interior, with the polar head groups on the outside, facing the aqueous environment. However, the precise lipid composition is variable, for viral envelopes can be derived from different kinds of cellular membranes. Embedded in the membrane are viral proteins, the great majority of which are **glycoproteins** that carry covalently linked sugar chains, or **oligosaccharides** (Fig. 4.19). Sugars are added to the proteins posttranslationally, during transport to the cellular membrane at which progeny virions assemble. Intra- or interchain disulfide bonds, another common chemical feature of these proteins, are also acquired during transport to assembly sites. These covalent

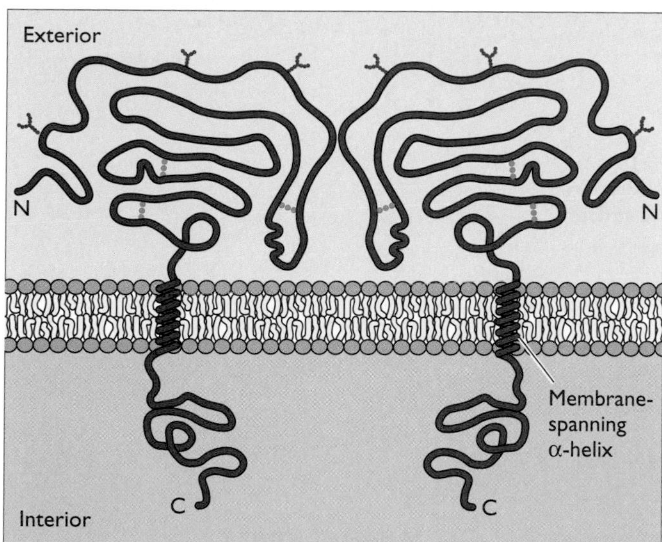

Figure 4.19 Structural and chemical features of a typical viral envelope glycoprotein shown schematically. The protein is inserted into the lipid bilayer via a single membrane-spanning domain. This segment separates a larger external domain, decorated with N-linked oligosaccharides (purple) and containing disulfide bonds (green), from a smaller internal domain. Viral glycoproteins are oligomers (Table 4.6), with external portions that form the spikes or other projections characteristic of the outer surfaces of enveloped viruses.

Table 4.6 Oligomeric structures of some viral membrane proteins[a]

Virus	Protein(s)	Quaternary structure
Alphavirus		
Semliki Forest virus	E1, E2, E3	$(E1E2E3)_3$
Ross River virus	E1, E2	$(E1E2)_3$
Herpesvirus		
Herpes simplex type 1	gH, gL	$(gHgL)_?$
Orthomyxovirus		
Influenza virus	HA	$(HA1\text{-}HA2)_3$
	NA	$(NA\text{-}NA)_2$
	M2	$(M2\text{-}M2)_2$
Retrovirus		
Avian sarcoma virus	Env	$(SU\text{-}TM)_3$
Rhabdovirus		
Vesicular stomatitis virus	G	$(G)_3$

[a]The best-predicted oligomeric structures of the viral membrane proteins listed are shown in column 3, with hyphens indicating disulfide-bonded protein chains. The proteins that comprise many of the heteromeric subunits listed (e.g., influenza virus HA, retroviral Env) are produced by proteolytic processing of precursors during transport to the cell surface.

bonds stabilize the tertiary or quaternary structures of viral glycoproteins (Table 4.6).

Envelope Glycoproteins

Viral glycoproteins are **integral membrane proteins** firmly embedded in the lipid bilayer by a short **membrane-spanning domain** (Fig. 4.19), or less often by two or more transmembrane segments. Membrane-spanning domains of both viral and cellular proteins are hydrophobic α-helices of sufficient length to span the lipid bilayer. The membrane-spanning α-helices of viral glycoproteins generally separate large external domains, the regions decorated with oligosaccharides, from smaller internal domains (Fig. 4.19). The former contain binding sites for cell surface virus receptors, major antigenic determinants, and sequences that mediate fusion of viral with cellular membranes during entry. Internal domains, which make contact with other components of the virion, are often essential for virus assembly.

With few if any exceptions, the structures formed by viral membrane glycoproteins are oligomeric. These oligomers vary considerably in composition. Some comprise multiple copies of a single protein, but in many cases each subunit contains two or more protein chains (Table 4.6). The subunits are generally held together by noncovalent interactions. On the exterior of the virion, these oligomers form spikes or other surface projections. Because of their critical roles in initiating infection, the structures of several viral glycoproteins have been determined. Indeed, the hemagglutinin (HA) protein of human influenza A virus was the first glycoprotein for which high-resolution structural information was obtained.

The HA protein is a trimer of disulfide-linked HA1 and HA2 molecules. Its structure is illustrated in Fig. 4.20 and described in detail in Chapter 5, in the context of the well-understood attachment and entry functions of this protein. The HA protein contains a globular head with a top surface that is projected about 135 Å from the viral membrane by a long stem (Fig. 4.20). The latter is formed by the coiling of α-helices present in each monomer, the major interaction that stabilizes the trimer. The membrane-distal globular domain contains the binding site for the virus receptor. This important functional region is therefore located more than 100 Å away from the lipid membrane of influenza virus particles. Other viral glycoproteins that mediate cell attachment and entry, such as the E protein of the flavivirus tick-borne encephalitis virus, are quite different in structure; the external domain of E protein is a flat, elongated dimer that would lie on the surface of the viral membrane rather than projecting from it (Fig. 4.20). Despite their lack of common structural features, both the HA protein and the E protein are primed for dramatic conformational change to allow entry of internal virion components into a host cell.

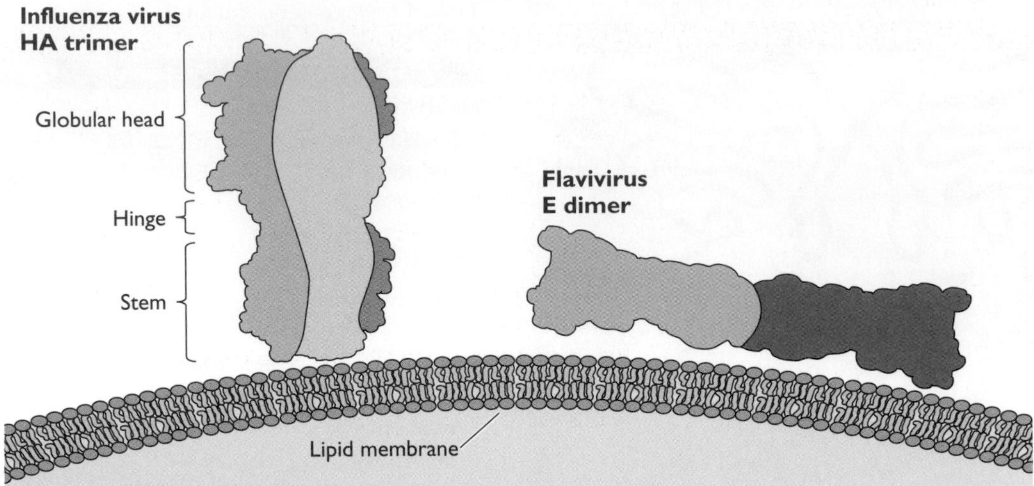

Figure 4.20 Structures of extracellular domains of viral glycoproteins, illustrated schematically. The influenza virus HA trimer and the tick-borne encephalitis virus E protein dimer are the envelope proteins that mediate attachment to and entry of these virus particles into a host cell. Nevertheless, these proteins are quite different in their tertiary and quaternary structures and orientation with respect to the membrane of the viral envelope.

The particles of the majority of other enveloped viruses contain one or two glycoproteins with typical properties. For example, the surfaces of all retroviruses are covered by a dense array of spikes (Fig. 4.18) formed by the Env proteins TM (transmembrane) and SU (surface unit). However, some enveloped viruses contain a larger collection of glycoproteins. One of the several remarkable features of the virions of herpesviruses is the large number of envelope proteins: so far, more than 10 viral glycoproteins have been identified (Table 4.7). As expected, some of them are important for attachment of the virus to host cells and for entry.

The high-resolution viral glycoprotein structures mentioned above are those of the large external domains of the proteins cleaved from the viral envelope by proteases. This treatment facilitated crystallization, but of course precluded determination of the structure of membrane-spanning or internal segments of the proteins, both of which also play important structural or functional roles. Membrane-spanning domains can contribute to the stability of oligomeric glycoproteins, as in influenza virus HA, and transmit signals from the exterior to the interior of the virion. Internal domains can participate in anchoring the envelope to internal virion structures. Recent improvements in resolution achieved by application of cryo-electron microscopy and image reconstruction (Fig. 4.1) have allowed visualization of these segments of glycoproteins of some enveloped viruses. This important advance has provided much previously inaccessible information, as discussed in "Simple Enveloped Viruses: Direct Contact of External Proteins with the Capsid or Nucleocapsid" below.

Other Envelope Proteins

The envelopes of some more complex viruses, including orthomyxoviruses, herpesviruses and poxviruses, contain integral membrane proteins that lack large external domains or possess multiple membrane-spanning segments (Table 4.7). Among the best characterized of these is the influenza A virus M2 protein. This small (97-amino-acid) protein is a minor component of virions. Estimates of its concentration range from 14 to 68 copies per particle. In the viral membrane, two disulfide-linked M2 dimers associate to form a noncovalent tetramer that functions as an ion channel. The M2 ion channel is the target of the influenza virus inhibitor amantadine (Fig. 19.23). The effects of this drug, as well as of mutations in the M2 coding sequence, indicate that M2 can play important roles in entry and assembly, by controlling the pH of the local environments in which virus particles and HA molecules are found.

Simple Enveloped Viruses: Direct Contact of External Proteins with the Capsid or Nucleocapsid

In the simplest enveloped viruses, exemplified by (+) strand RNA alphaviruses such as Semliki Forest and Ross River viruses, the envelope directly abuts an inner nucle-

Table 4.7 Some herpes simplex virus type 1 virion proteins

Location	Protein	Gene	Function or property
Nucleocapsid	VP5	UL19	Major capsid protein; forms both hexamers and pentamers
	VP19C/VP23	UL38/UL18	The heterotrimer ("triplex") connecting VP5 subunits on surface
	VP24	UL26	Protease; synthesized as a precursor linked to scaffolding protein VP21
	VP26	UL35	Caps VP5 hexons, but not pentons
	Portal	UL6	Present at one vertex; required for entry of DNA
Tegument	VP1-3	UL36	Very large, ≈273 kDa
	VP16	UL48	Abundent structural protein; activator of immediate-early gene transcription
	VP18.8	UL13	Protein kinase
	VP22	UL49	Hyperacetylation and stabilization of microtubules
	Vhs	UL41	Host shutoff factor
	US11	US11	Mysristylated protein; envelopment and transport of nascent virions
Envelope	gB	UL27	Fusion; binds heparan sulfate; binds tegument proteins, e.g., VP16
	gC	UL44	Attachment protein; binds heparan sulfate; also binds complement component C3b
	gD	US6	Binds cell surface receptors; entry
	gE/gI	US8/US7	Heterodimer; binds Fc domain of immunoglobulin G; cell-cell spread
	gH/gL	UL22/UL1	Heterodimer; fusion
	gG	US4	Not known
	gJ	US5	Not known
	gK	UL53	Virus-induced cell fusion; egress
	gM/gN	UL10/UL49.5	Heterodimer; function not known
	US9	US9	Function not known for herpes simplex virus type 1; no large external domain
		UL20	Egress; probably contains multiple membrane-spanning domains

ocapsid containing the (+) strand RNA genome. This inner protein layer is a T = 4 icosahedral shell built from 240 copies of a single capsid (C) protein arranged as hexamers and pentamers. The outer glycoprotein layer contains 240 copies of the envelope proteins (Table 4.6). They cover the surface of the particle, such that the lipid membrane is not exposed on the exterior. The glycoproteins are also organized into a T = 4 icosahedral shell (Fig. 4.21A), as a result of binding of their internal domains to the C-protein subunits of the underlying nucleocapsid. The structure of Semliki Forest virus has been determined by cryo-electron microscopy and image reconstruction to 9-Å resolution, the highest yet achieved for an enveloped virus. Structures of the E1 and C proteins have been solved at high resolution. The organization of this enveloped virus, including the transmembrane anchoring of the outer glycoprotein layer to structural units of the nucleocapsid, can therefore be described with unprecedented precision (Fig. 4.21). The transmembrane segments of the E1 and E2 glycoproteins form a pair of tightly associated α-helices, with the cytoplasmic domain of E2 in close opposition to a cleft in the capsid protein (Fig. 4.21D). On the outer surface of the membrane, the external portions of these glycoproteins, together with the E3 protein, form an unexpectedly elaborate structure: a thin icosahedral protein layer (called the skirt) covers most of the membrane and supports the spikes, which are hollow, three-lobed projections (Fig. 4.21B and C).

Enveloped Viruses with One or Two Additional Protein Layers

Virions of several enveloped viruses contain additional proteins that mediate interactions of the innermost, genome-containing structure with the viral envelope. In the simplest case, a single viral structural protein, termed the matrix protein, welds an internal ribonucleoprotein to the envelope. This arrangement is found in members of several groups of (–) strand RNA viruses (Appendix A, Fig. 8 and 23). Retrovirus particles also contain an analogous, membrane-associated matrix protein, but this protein makes contact with an internal capsid in which the viral ribonucleoprotein is encased. Because the internal

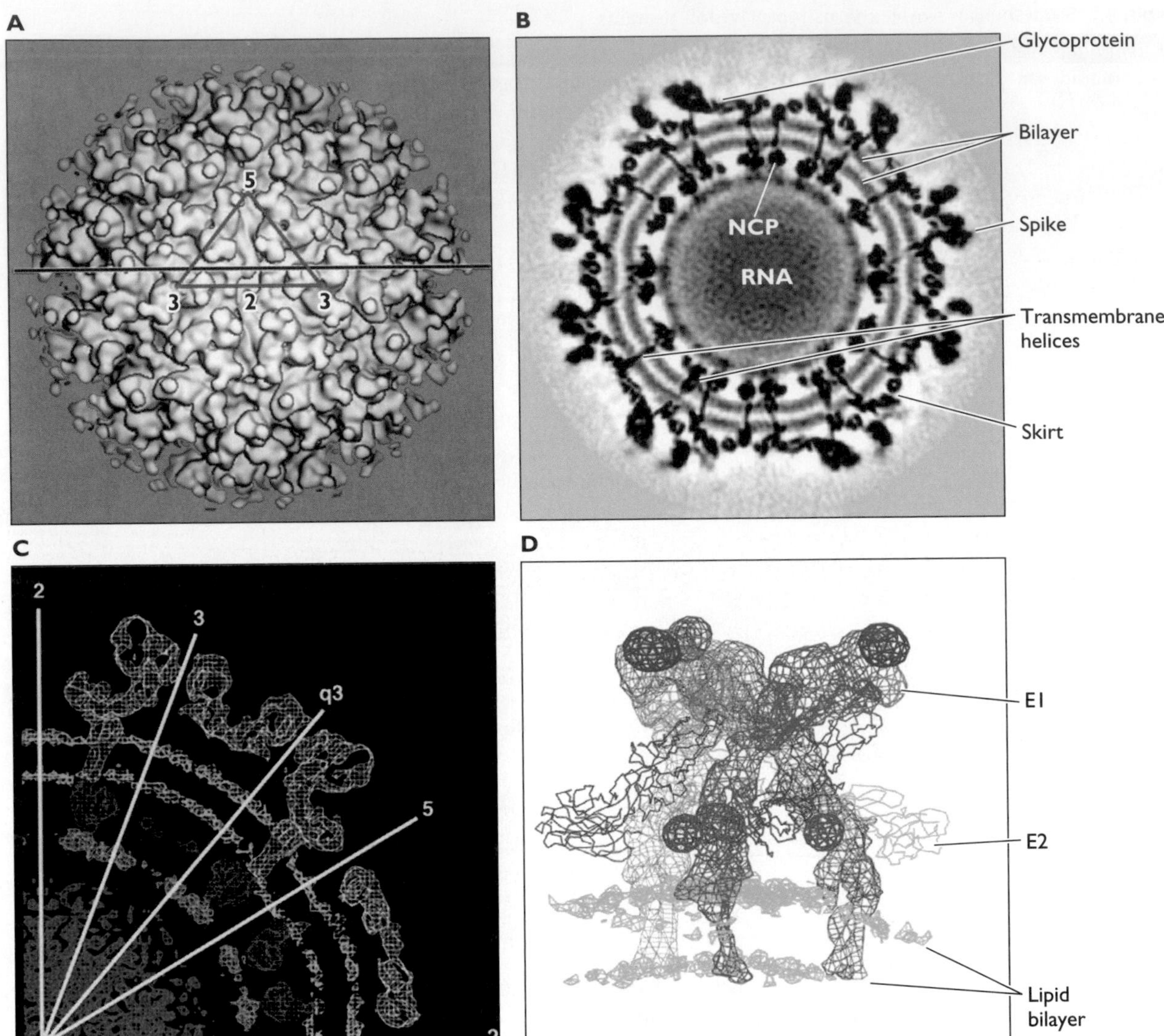

Figure 4.21 Structure of a simple enveloped virus, Sindbis virus. (A) The surface structure of Sindbis virus, a member of the alphavirus genus of the *Togaviridae*, at 20-Å resolution was determined by cryo-electron microscopy. The boundaries of the asymmetric unit are demarcated by the red triangle, on which the icosahedral five-, three-, and twofold axes of rotational symmetry are indicated. This outer surface is organized as a $T = 4$ icosahedral shell studded with 80 spikes, each built from three copies of each of the transmembrane glycoproteins E1 and E2. These spikes are connected by a thin, external protein layer, termed the skirt. (B) Cross section through the density map at 11-Å resolution, along the black line shown in panel A. The lipid bilayer of the viral envelope is clearly defined at this resolution, as are the transmembrane domains of the glycoproteins. (C) Different layers of the particle, based on the fitting of a high-resolution structure of the E1 glycoprotein into a 9-Å reconstruction of the virus particle. The nucleocapsid (NC) surrounds the genomic (+) strand RNA. The RNA is the least well-ordered feature in the reconstruction, although segments lying just below the C protein appear to be ordered by interaction with this protein. The capsid protein penetrates the inner leaflet of the lipid membrane, where it interacts with the cytoplasmic domain of the E2 glycoprotein (that of E1 contains only two amino acids). The membrane is penetrated by rodlike structures that are connected to the skirt by short stems. These transmembrane contacts account for the congruent, $T = 4$ icosahedral symmetries of the nucleocapsid and outer glycoprotein shells. (D) The structure of the E1 and E2 glycoproteins, obtained by fitting the crystal structure of the closely related Semliki Forest virus E1 glycoprotein into the 11-Å density map and assigning density unaccounted for to the E2 glycoprotein. The view shown is around a quasithreefold symmetry axis (q3 in panel C); the three E2 glycoprotein molecules are in a trimeric spike colored light blue, brown, and purple, and the E1 molecules are shown as Cα backbones colored red, green, and dark blue. The E1 glycoprotein is largely tangential to the surface of the particle. The portions of the proteins that cross the lipid bilayer are helical, twisting around one another in a left-handed coiled coil. Adapted from W. Zhang et al., *J. Virol.* **76:**11645–11658, 2002, with permission.

capsids or nucleocapsids of these more complex enveloped viruses are not in direct contact with the envelope, the organization and symmetry of the internal structure are not necessarily evident from the external appearance of the surface glycoprotein layer. For example, the outer surface of all retroviruses appears as a dense, roughly spherical array of projecting knobs or spikes, regardless of whether the internal core is spherical or cone shaped (Fig. 4.18). Likewise, influenza virus particles, which contain helical nucleocapsids, are generally roughly spherical particles 800 to 1,200 Å in diameter (Appendix A, Fig. 8) although long, filamentous forms are common in clinical isolates.

In general, neither the interior architecture of these enveloped viruses nor the molecular connections among internal structures and envelope components can be described in detail. There is good evidence, from both mutational analyses and in vitro assays, for binding of the proteins that abut the envelope to internal domains of envelope glycoproteins and for their direct association with the lipid membrane. These interactions are exemplified by the binding of the vesicular stomatitis virus matrix (M) protein to the internal domain of the envelope glycoprotein (G) and of the matrix (MA) proteins of some simple retroviruses to the inner surface of the lipid bilayer. Deletion of the coding sequence for the internal domain of the TM proteins of such retroviruses perturbs neither virion integrity nor incorporation of the glycoprotein (or any other component) into virions. The matrix proteins of several viruses, including influenza A virus and human immunodeficiency virus type 1, interact with both glycoprotein and lipid components of the envelope.

Internal proteins that mediate contact with the viral envelope are not embedded within the lipid bilayer, but rather bind to its inner face. Such viral proteins are targeted to, and interact with, membranes by means of specific signals, which are described in more detail in Chapter 12. For example, a posttranslationally added fatty acid chain is important for membrane binding of the MA proteins of most retroviruses. The matrix proteins and their Gag polyprotein precursors (Appendix A, Fig. 21) carry the 14-carbon, saturated, fatty acid myristate covalently linked to N-terminal glycine residues. In some cases, this lipid is necessary and sufficient for binding of MA to the membrane. However, the human immunodeficiency virus type 1 MA protein contains a bipartite membrane-binding signal, comprising the myristoylated N-terminal segment and a highly basic sequence located a short distance nearby. This MA protein was the first viral peripheral membrane protein for which a high-resolution structure was determined, initially by NMR methods (Fig. 4.3). Subsequent analysis by X-ray crystallography established that MA is a trimer (Fig. 4.22). Each MA molecule comprises a compact, globular domain of α-helices capped by a β-sheet that contains the positively charged amino acids necessary for membrane binding. As illustrated in the model of MA oriented on a membrane shown in Fig. 4.22, the basic residues form a positively charged surface, positioned for interaction with phospholipid head groups on the inner surface of the envelope.

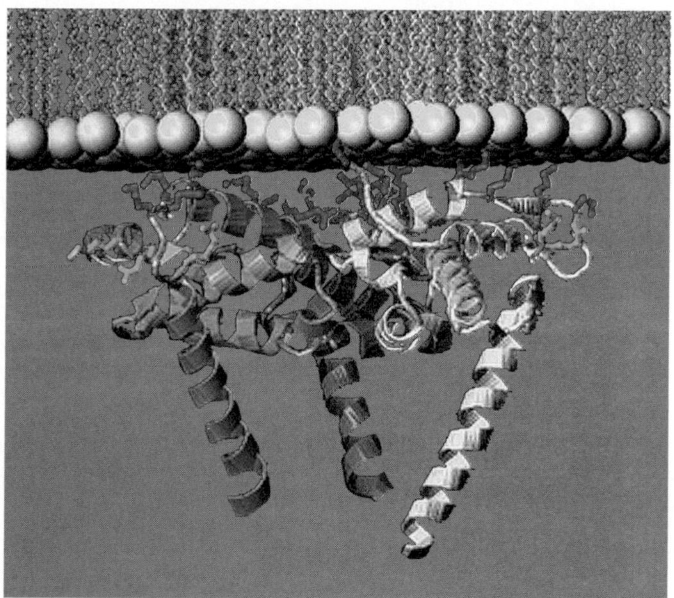

Figure 4.22 Model of the interaction of human immunodeficiency virus type 1 MA protein with the membrane. This model is based on the X-ray crystal structure of recombinant MA protein synthesized in *Escherichia coli* and consequently lacking the N-terminal myristate chain normally added in human cells. The three monomers in the MA trimer are shown in different colors. Basic residues in the β-sheet that caps the globular α-helical domain are magenta or green. Substitution of residues shown in magenta impairs replication of the virus in cells in culture. The position of the myristate (red) was modeled schematically. From C. P. Hill et al., *Proc. Natl. Acad. Sci. USA* **93**:3099–3104, with permission. Courtesy of C. P. Hill and W. I. Sundquist, University of Utah.

Large Enveloped Viruses with Numerous Internal Proteins

Members of the *Herpesviridae* contain many more proteins than any virus described in previous sections, and exhibit a number of unusual architectural features. Over half of the more than 80 genes of herpes simplex virus type 1 encode proteins found in virus particles (Table 4.7), which are correspondingly large, about 2,000 Å in diameter. These proteins are components of the envelope from which glycoprotein spikes project or of two distinct internal structures. The latter are the nucleocapsid surrounding the DNA genome and the protein layer encasing this structure, called the **tegument** (Fig. 4.23A). As noted previously, the envelope contains at least 13 viral proteins (Table 4.7).

A single protein (VP5) forms both the hexons and the pentons of the T = 16 icosahedral nucleocapsid of herpes simplex virus type 1 (Fig. 4.23B). Like the structural units of the smaller simian virus 40 capsid, the VP5-containing assemblies make direct contact with one another. However, the large (~1,500-Å-diameter) herpesviral nucleocapsid is stabilized by additional proteins, VP19C and

Figure 4.23 Structural features of herpesvirus particles. (A) Electron micrograph of a frozen, hydrated herpes simplex virus type 1 virion showing the surface glycoproteins, lipid membrane, tegument, and nucleocapsid. Note the regular appearance not only of the capsid but also of the tegument and envelope; the traditional view of the tegument as an amorphous, irregular structure was the result of distortion during negative staining. From F. J. Rixon, *Semin. Virol.* **4:**135–144, 1993, with permission. Courtesy of F. Rixon, Institute of Virology, Glasgow, United Kingdom, and W. Chiu, Baylor College of Medicine. (B) Reconstruction of the herpes simplex virus type 1 nucleocapsid (8.5-Å resolution), with VP5 hexamers and pentamers colored blue and red, respectively, and the triplexes that reinforce the connections among these structural units in green. VP5 hexamers, but not pentamers, are capped by a hexameric ring of VP26 protein molecules. Adapted from Z. H. Zhou et al., *Science* **288:**877–880, 2000, with permission. Courtesy of W. Chiu, Baylor College of Medicine. (C) The single portal of herpes simplex virus type 1 nucleocapsids was visualized by staining with an antibody specific for the viral UL6 protein conjugated to gold beads. The gold beads are electron-dense and appear as dark spots in the electron microscope. Gold beads are present at a single vertex in each nucleocapsid, which therefore contains one portal. The inset shows a model of the UL6 protein portal based on electron microscopy of negatively stained preparations of the protein. Adapted from W. W. Newcomb et al., *J. Virol.* **75:**10923–10932, 2001, with permission. Courtesy of J. Brown, University of Virginia. (D) Interactions of two tegument proteins with the simian cytomegalovirus nucleocapsid. Tegument proteins that bind to hexons plus pentons and to triplexes are shown in blue and red, respectively. These proteins were visualized by cryo-electron microscopy, image reconstruction (to 22-Å resolution), and difference mapping of nucleocapsids purified from the nucleus and cytoplasm of virus-infected cells. The latter carry the tegument, but the former do not. Adapted from B. L. Trus et al., *J. Virol.* **73:**2181–2192, 1999, with permission. Courtesy of A. C. Steven, National Institutes of Health.

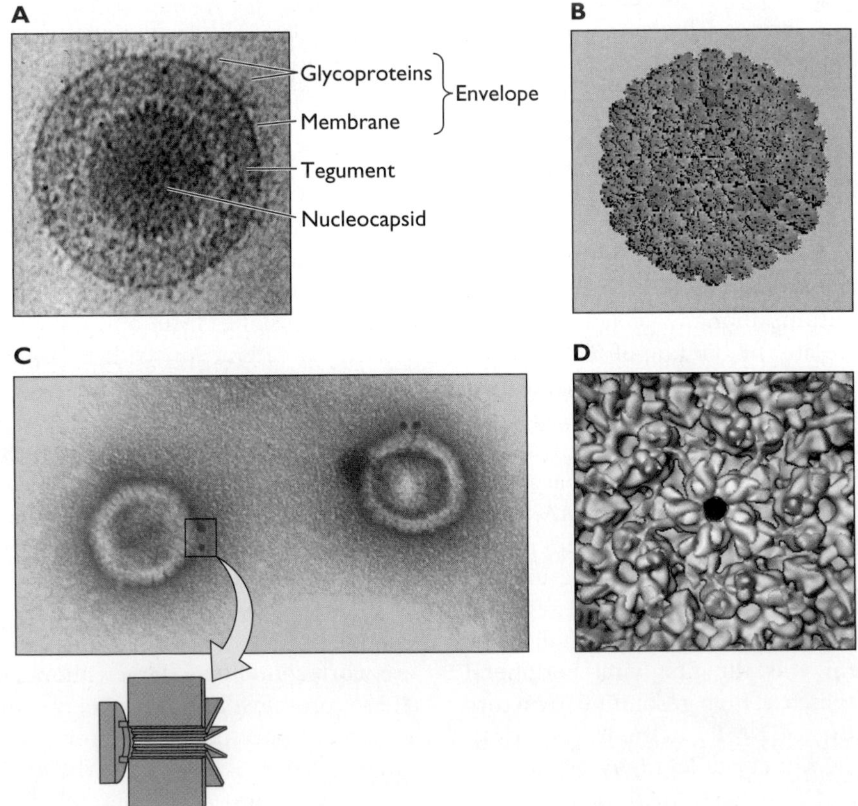

VP23. These two proteins form triplexes that link the major structural units (Fig. 4.23B). Although apparently a typical and quite simple icosahedral shell, this viral nucleocapsid is in fact an asymmetric structure: in infectious virions, one of the 12 vertices is occupied not by a VP5 penton, but by a unique structure termed the **portal** (Fig. 4.23C). The portal is so named because it is the structure through which the viral DNA genome enters the nucleocapsid. It comprises 12 copies of the UL6 protein (Table 4.7). In the electron microscope, the purified protein is seen in ring- and Y-shaped structures, suggesting that the UL6 portal may be goblet-shaped, with a tapering channel through which DNA can pass (Fig. 4.23C). The asymmetry of the herpesviral nucleocapsid and the incorporation of the portal have important implications for the mechanism of assembly of this structure.

The tegument, which contains at least eight viral proteins, appears as a regular spherical structure (Fig. 4.23A). Cryo-electron microscopy and difference imaging of herpesvirus particles with and without teguments have established that specific proteins of this layer are icosahedrally ordered, as a result of direct contacts with the structural units of the nucleocapsid (Fig. 4.23D). The tegument proteins that make contact with the inner surface of the envelope have not yet been identified. However, herpesvirus particles completely lacking certain tegument proteins appear morphologically normal and are fully infectious, suggesting that this layer may be structurally plastic.

The common functions of envelope glycoproteins in attachment and entry of the virus at the beginning of an infectious cycle are well established. High-resolution structural descriptions of these proteins are appearing at an increasing rate. Nevertheless, we possess few detailed views of the structure, or organization within virus particles, of internal proteins that bind to viral envelopes. Consequently, interactions among internal and envelope components, which must impart stability to the virion and influence its morphology, are well understood only for the simplest enveloped viruses (Fig. 4.21). It is hoped that the methods of image reconstruction that have generated such clear views of direct anchoring of the envelope to the internal protein shell will be equally informative when applied to more complex enveloped viruses. However, the lack of symmetry characteristic of some of these viruses (e.g., influenza virus) imposes a huge hurdle for such methods.

Other Components of Virions

Some virus particles comprise only the nucleic acid genome and structural proteins, or proteins plus a lipid membrane, necessary for protection and delivery into a host cell. However, many contain additional viral proteins or other components, which are generally present at much lower concentrations, but play essential or important roles in establishing an efficient infectious cycle (Table 4.8).

Virion Enzymes

Many types of virus particle contain enzymes necessary for synthesis of viral nucleic acids within an infected cell. Such enzymes generally catalyze synthetic reactions unique to viruses, such as synthesis of viral mRNA from an RNA template or of viral DNA from an RNA template. These reactions, which have no cellular counterparts, are catalyzed by RNA-dependent RNA polymerases present in the nucleocapsids of (–) strand RNA viruses and retroviral reverse transcriptase, respectively. In contrast, virions of the vaccinia virus contain a DNA-dependent RNA polymerase, analogous to cellular RNA polymerases, as well as several enzymes that modify viral RNA transcripts (Table 4.8). This complement of enzymes is necessary because transcription of the viral double-stranded DNA genome takes place in the cytoplasm of infected cells, whereas cellular DNA-dependent RNA polymerases and the RNA-processing machinery are restricted to the nucleus. Other types of enzyme present in virions include integrase, cap-dependent endonuclease, and proteases (Table 4.8). The latter, which eliminate covalent connections among specific proteins from which virions assemble, are necessary for the production of infectious particles.

Other Viral Proteins

More complex virions may also contain additional viral proteins that are not enzymes, but nonetheless are important for an efficient infectious cycle (Table 4.8). Among the best characterized examples of this class are several tegument proteins of herpesviruses (Table 4.8). The VP16 protein activates transcription of viral immediate-early genes to initiate the viral program of gene expression. In contrast to the majority of viral proteins discussed in this section, VP16 is present at high concentration, perhaps because it is necessary for virion assembly. Other herpesvirus tegument proteins induce degradation of cellular mRNA or block the cellular mechanism by which viral proteins are presented to the host's immune system. Complex retroviruses like human immunodeficiency virus type 1 also contain additional proteins required for efficient viral replication in certain cell types. At least one such protein, Vpr, is present in the internal core. This protein is discussed in Chapter 17.

Nongenomic Viral Nucleic Acid

The **viral** nucleic acids present in virions have, by definition, been considered to be genomes. It was therefore quite a surprise to find other, functional viral nucleic acids within virus particles. This unexpected property was first

Table 4.8 Some virion enzymes and other components

Virus	Component	Function(s)
Viral enzymes		
Adenovirus (human adenovirus type 2)	L3 23k	Protease, production of infectious virions; uncoating
Herpesvirus (herpes simplex virus type 1)	VP24	Protease, capsid maturation for genome encapsidation
	UL13 protein	Protein kinase
Orthomyxovirus (influenza A virus)	P proteins	RNA-dependent RNA polymerase, synthesis of viral mRNA and genome RNA; cap-dependent endonuclease, initiation of mRNA synthesis.
Poxvirus (vaccinia virus[a])	Eight subunits of DNA-dependent RNA polymerase	Synthesis of viral mRNA
	Poly(A) polymerase (two proteins)	Synthesis of poly(A) on viral mRNA
	Capping enzyme (two proteins)	Addition of 5′ caps to viral pre-mRNA
	DNA topoisomerase	Sequence-specific nicking of viral DNA
Reovirus (reovirus type 1)	λ2	Guanyl transferase
	λ3 + ?	Double-stranded RNA-dependent RNA polymerase
Retrovirus (avian sarcoma virus, human immunodeficiency virus type 1)	Pol	Reverse transcriptase, proviral DNA synthesis
	IN	Integrase, integration of proviral DNA into cellular genome
	PR	Protease, production of infectious virions
Rhabdovirus (vesicular stomatitis virus)	L	RNA-dependent RNA polymerase, synthesis of viral mRNA and genome RNA
Other viral proteins		
Herpesviruses		
Herpes simplex virus type 1	VP16	Structural protein, activation of immediate-early gene transcription
	Vhs	Induction of degradation of cellular and viral mRNA
Human cytomegalovirus	pp65	Inhibition of presentation of viral immediate-early proteins as antigens
Poxvirus (vaccinia virus)	VETF	Binds to early promoters, essential for their transcription
	RAP94	Associated with virion RNA polymerase, specificity factor
Retrovirus (human immunodeficiency virus type 1)	Vpr, Nef	Required for efficient infection in some cell types
Other viral nucleic acids		
Herpesvirus (human cytomegalovirus)	5 late mRNAs	?
Cellular components		
Polyomavirus (simian virus 40)	Histones H2A, H2B, H3, and H4	Packaging of circular, double-stranded DNA genome
Retroviruses		
Avian sarcoma virus, human immunodeficiency virus type 1	tRNA	Primer for synthesis of first proviral DNA strand
Human immunodeficiency virus type 1	Cyclophilin A	?; an early step in the infectious cycle
	Uracil DNA glycosylase	?; a DNA repair enzyme
	Icam-1	?; can facilitate attachment and entry

[a]Vaccinia virions contain at least 11 enzymes, only a few of which are listed. ?, function unknown.

reported for human cytomegalovirus, a betaherpesvirus with one of the largest DNA genomes known and an important human pathogen (Appendix B, Fig. 9). As discussed in more detail below, it can be difficult to distinguish macromolecules that are specifically assembled into enveloped virus particles from those incorporated nonspecifically. However, of the 200 or so viral mRNAs synthesized in human cytomegalovirus-infected cells, only 4 can be readily detected in purified virions. Moreover, these same viral mRNAs were detected within cells very soon after infection in the presence of a drug that prevents synthesis of **all** mRNAs. These viral mRNAs must therefore be packaged specifically. The virion-delivered mRNAs, which probably reside within the tegument, are translated in the cytoplasm of newly infected cells. Neither the functions of the viral proteins made from these viral mRNAs, nor the mechanism by which the mRNAs enter assembling virus particles, are yet understood.

Cellular Macromolecules

Virus particles can also contain cellular macromolecules that play important roles during the infectious cycle, such as the cellular histones that package polyomaviral and papillomaviral DNAs. Because they are formed by budding, enveloped viruses can readily incorporate cellular proteins and other macromolecules. Cellular glycoproteins may not be excluded from the membrane from which the viral envelope is derived. Moreover, as a bud enlarges and pinches off during virus assembly, internal cellular components may be trapped within it. In addition, enveloped viruses are generally more difficult to purify than naked viruses. As a result, preparations of these viruses may be contaminated with vesicles formed from cellular membranes during extraction of the virus from infected cells. Consequently, it can be difficult to distinguish cellular components specifically incorporated into enveloped virus particles from those trapped randomly, or copurifying with the virus. Nevertheless, in some cases it is clear that cellular molecules are important components of virus particles: these molecules are reproducibly observed in virions at a specific stoichiometry, and can be shown to play essential or important roles in the infectious cycle. The cellular components captured in retrovirus particles have been particularly well characterized.

The primer for initiation of synthesis of the first strand of proviral DNA during reverse transcription is invariably a cellular transfer RNA (tRNA). This RNA is incorporated into virions by virtue of its binding to a specific sequence in the RNA genome and to reverse transcriptase. More radical is the incorporation of cellular proteins into some retroviral particles. One of the most unusual properties of human immunodeficiency virus type 1 is the presence of cellular cyclophilin A, a **chaperone** that assists or catalyzes protein folding. This protein is a member of a ubiquitous family of peptidyl-prolyl isomerases, which catalyze the intrinsically slow *cis-trans* isomerization of peptide bonds preceding proline residues in newly synthesized proteins. Cyclophilin A is the major cytoplasmic member of this family. It is incorporated within human immunodeficiency virus type 1 particles (but not those of even closely related retroviruses) via specific interactions with the central portion of CA in multimers of the Gag polyprotein. Although incorporation of cyclophilin A is not a prerequisite for assembly, particles that lack this cellular chaperone have reduced infectivity, because they fail to complete some early step in the replication cycle. Cellular membrane proteins, such as Icam-1 (Chapter 5) and a specific major histocompatibility complex class II protein (Chapter 15), can also be incorporated in the viral envelope, and can contribute to attachment and entry of human immunodeficiency virus type 1 particles.

It is clear from these examples that, despite the limited coding capacity of their genomes and the finite, small volumes of their particles, viruses contain a surprisingly broad repertoire of molecular functions that are delivered to their host cells (Table 4.8). This repertoire is undoubtedly larger than we presently appreciate, for the precise molecular functions and/or roles in the infectious cycles of many of these virion components have yet to be established.

Perspectives

Much has been learned about the architecture of viruses and the molecular principles of their design since investigators first considered the problem of constructing these particles from a limited number of components. Despite great variety in size and morphology, all virus particles contain at least one protein shell that encases and protects the nucleic acid genome. With few exceptions, these capsids or nucleocapsids are built from a small number of protein subunits arranged with helical or icosahedral symmetry. The detailed views of nonenveloped virions provided by X-ray crystallography emphasize just how well such protein shells are designed for protection of the genome during passage from one host cell or organism to another. They have also identified several mechanisms by which identical or nonidentical subunits can interact to form icosahedrally symmetric structures. As we shall see, atomic-level descriptions of virus particles, or of the individual proteins that contact surface components of the host cell, can provide critical insights into the mechanisms of attachment and delivery.

More complex virus particles, which may contain additional protein layers, a lipid envelope carrying viral proteins, and enzymes or other proteins necessary to initiate the infectious cycle, cannot be described in the exquisite detail provided by a high-resolution structure. For many years we possessed only schematic views of their structures, deduced from negative-contrast electron microscopy and biochemical or genetic methods of analysis. Within the past decade, the development and refinement of cryo-electron microscopy and techniques of image reconstruction have revolutionized structural studies of larger and more complex viruses. These methods have yielded much new information about the structural features and the roles of individual proteins of both naked and enveloped viruses, as well as remarkable views of genome organization, or interactions among envelope and internal proteins. The recent improvement in the resolution that can be achieved by these techniques to <10 Å, the power of difference imaging methods, and the now routine ability to produce large quantities of specific structures from recombinant viral proteins promise many new insights into the structure, molecular organization, and function of virus particles. Such information will improve our understanding, presently quite limited, of mechanisms of packaging of nucleic acid genomes and of tethering envelopes (when present) to internal structures. Some surprises are undoubtedly in store. But we can predict with some confidence that future studies will reinforce and elaborate the general principles of virus structure described here.

References

Chapters in Books

Baker, T. S., and J. E. Johnson. 1997. Principles of virus structure determination, p. 38–79. *In* W. Chiu, R. M. Burnett, and R. L. Garcea (ed.), *Structural Biology of Viruses*. Oxford University Press, New York, N.Y.

Chow, M., R. Basavappa, and J. M. Hogle. 1997. The role of conformational transitions in poliovirus pathogenesis, p. 157–187. *In* W. Chiu, R. M. Burnett, and R. L. Garcea (ed.), *Structural Biology of Viruses*. Oxford University Press, New York, N.Y.

Garcea, R. L., and R. C. Liddington. 1997. Structural biology of polyomaviruses, p. 187–208. *In* W. Chiu, R. M. Burnett, and R. L. Garcea (ed.), *Structural Biology of Viruses*. Oxford University Press, New York, N.Y.

Harrison, S. C., J. J. Skehel, and D. C. Wiley. 1996. Virus structure, p. 59–99. *In* B. N. Fields, D. M. Knipe, and P. K. Howley (ed.), *Fundamental Virology*. Lippincott-Raven, New York, N.Y.

Reviews

Baker, T. S., N. H. Olson, and S. D. Fuller. 1999. Adding the third dimension to virus life cycles: three-dimensional reconstruction of icosahedral viruses from cryo-electron micrographs. *Microbiol. Mol. Biol. Rev.* **63:**862–922.

Chapman, M. S., V. L. Giranda, and M. G. Rossmann. 1990. The structures of human rhinovirus and mengo virus: relevance to function and drug design. *Semin. Virol.* **1:**413–427.

Chiu, W., and F. J. Rixon. 2002. High resolution structural studies of complex icosahedral viruses: a brief overview. *Virus Res.* **82:**9–17.

Stabbs, G. 1990. Molecular structures of viruses from the tobacco mosaic virus group. *Semin. Virol.* **1:**405–512.

Stewart, P. L., and R. M. Burnett. 1990. The structure of adenovirus. *Semin. Virol.* **1:**477–487.

Strauss, J. H., and E. G. Strauss. 2001. Virus evolution: how does an enveloped virus make a regular structure? *Cell* **105:**5–8.

Wilk, T., and S. D. Fuller. 1999. Towards the structure of human immunodeficiency virus: divide and conquer? *Curr. Opin. Struct. Biol.* **9:**231–243.

Papers of Special Interest

Theoretical Foundations

Caspar, D. L. D., and A. Klug. 1962. Physical principles in the construction of regular viruses. *Cold Spring Harbor Symp. Quant. Biol.* **27:**1–22.

Crick, F. H. C., and J. D. Watson. 1956. Structure of small viruses. *Nature* **177:**473–475.

Structures of Nonenveloped Viruses

Brenner, S., and R. W. Horne. 1959. A negative staining method for high resolution electron microscopy of viruses. *Biochim. Biophys. Acta* **34:**103–110.

Cheng, R. H., R. G. Kuhn, N. H. Olson, M. G. Rossmann, H.-K. Choi, T. J. Smith, and T. S. Baker. 1995. Nucleocapsid and glycoprotein organization in an enveloped virus. *Cell* **80:**621–630.

Harrison, S. C., A. Olson, C. E. Schutt, F. K. Winkler, and G. Bricogne. 1978. Tomato bushy stunt virus at 2.9 Å resolution. *Nature* **276:**368–373.

Hogle, J. M., M. Chow, and D. J. Filman. 1985. Three-dimensional structure of poliovirus at 2.9 Å resolution. *Science* **229:**1358–1365.

Liddington, R. C., Y. Yan, H. C. Zhao, R. Sahli, T. L. Benjamin, and S. C. Harrison. 1991. Structure of simian virus 40 at 3.8 Å resolution. *Nature* **354:**278–284.

Prasad, B. V. V., R. Rothnagel, C. Q.-Y. Zeng, J. Jakana, J. A. Lawton, W. Chiu, and M. K. Estes. 1996. Visualization of ordered genomic RNA and localization of transcriptional complexes in rotavirus. *Nature* **382:**471–473.

Reinisch, K. M., M. L. Nibert, and S. C. Harrison. 2000. Structure of the reovirus core at 3.6 Å resolution. *Nature* **404:**960–967.

Rossman, M. G., E. Arnold, and J. W. Erickson. 1985. Structure of a common cold virus and functional relationship to other picornaviruses. *Nature* **317:**145–153.

Structures of Enveloped Viruses

Mancini, E. J., M. Clarke, B. E. Gowen, T. Rulten, and S. D. Fuller. 2000. Cryo-electron microscopy reveals the functional organization of an enveloped virus, Semliki Forest Virus. *Mol. Cell* **5:**255–266.

Newcomb, W. W., R. M. Juhas, D. R. Thomsen, F. L. Homa, A. D. Burch, S. K. Weller, and J. C. Brown. 2001. The UL6 gene product forms the portal for entry of DNA into the herpes simplex virus capsid. *J. Virol.* **75:**10923–10932.

Trus, B. L., W. Gibson, N. Cheng, and A. C. Steven. 1999. Capsid structure of simian cytomegalovirus from cryoelectron microscopy: evidence for tegument attachment sites. *J. Virol.* **73:**2181–2192.

Wynne, S. A., R. A. Crowther, and A. G. W. Leslie. 1999. The crystal structure of the human hepatitis B virus capsid. *Mol. Cell* **3:**771–780.

Zhou, Z. H., M. Dougherty, J. Jakana, J. He, F. J. Rixon, and W. Chiu. 2000. Seeing the herpesvirus capsid at 8.5 Å resolution. *Science* **288:**877–880.

Structures of Virion Proteins

Böttcher, B., S. A. Wynne, and R. A. Crowther. 1997. Determination of the fold of the core protein of hepatitis B virus by electron cryomicroscopy. *Nature* **386:**88–91.

Chen, X. S., T. Stehe, and S. Harrison. 1998. Interaction of the polyomavirus internal protein VP2 with the capsid protein VP1 and implications for participation of VP2 in viral entry. *EMBO J.* **37:**3233–3240.

Lescar, J., A. Roussel, M. W. Wien, J. Navaza, S. D. Fuller, G. Wengler, G. Wengler, and F. A. Rey. 2001. The fusion glycoprotein shell of Semliki Forest virus: an icosahedral assembly primed for fusogenic activation at endosomal pH. *Cell* **105:**137–148.

Malashkevich, V. N., B. J. Schneider, M. L. McNally, M. A. Milhollen, J. X. Pang, and P. S. Kim. 1999. Core structure of the envelope glycoprotein GP2 from Ebola virus at 1.9 Å resolution. *Proc. Natl. Acad. Sci. USA* **96:**2662–2667.

Massiah, M. A., M. R. Starick, C. Paschall, M. F. Summers, A. M. Christensen, and W. I. Sundquist. 1994. Three-dimensional structure of the human immunodeficiency virus type 1 matrix protein. *J. Mol. Biol.* **244:**198–223.

Rey, F. A., F. X. Heinz, C. Mandl, C. Kunz, and S. C. Harrison. 1995. The envelope glycoprotein from tick-borne encephalitis virus at 2 Å resolution. *Nature* **375:**291–298.

Wilson, J. A., T. S. Skehel, and D. C. Wiley. 1981. Structure of the haemagglutinin membrane glycoprotein of influenza virus at 3 Å resolution. *Nature* **289:**366–373.

Other Components of Virions

Bresnahan, W. A., and T. Shenk. 2000. A subset of viral transcripts packaged within human cytomegalovirus particles. *Science* **288:**2373–2376.

Poon, D. T. K., L. V. Coren, and D. E. Ott. 2000. Efficient incorporation of HLA class II onto human immunodeficiency virus type 1 requires envelope glycoprotein packaging. *J. Virol.* **74:**3918–3923.

Rizzuto, C. D., and J. G. Sodroski. 1997. Contribution of virion ICAM-1 to human immunodeficiency virus infectivity and sensitivity to neutralization. *J. Virol.* **71:**4847–4851.

Thali, M., A. Bukovsky, E. Kondo, B. Rosenwirth, C. T. Walsh, J. Sodroski, and H. G. Göttlinger. 1994. Functional association of cyclophilin A with HIV-1 virions. *Nature* **372:**363–365.

Websites

http://mmtsb.scripps.edu/viper/ *Virus Particle Explorer*

http://www.virology.net/Big_Virology/BVHomePage.html *The Big Picture Book of Viruses*

http://rhino.bocklabs.wisc.edu/virusworld *Computer Visualization of Viruses*

http://icb.usp.br/~mlracz/animations/bock/ap.htm *Figures and animations from the* Encyclopedia of Virology

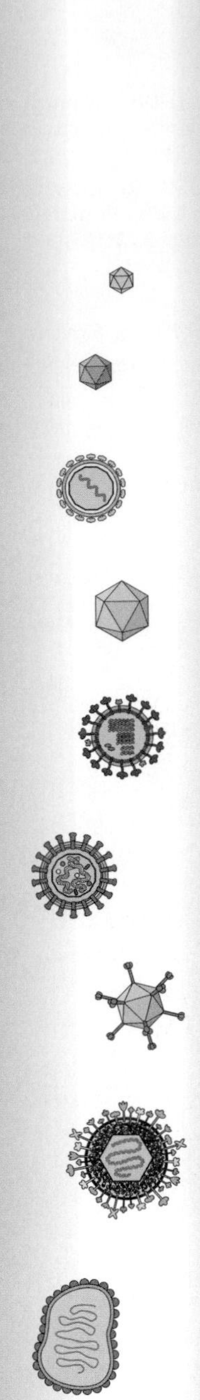

5

Attachment and Entry

Who hath deceived thee so often as thyself?
BENJAMIN FRANKLIN

Introduction

To initiate replication in a host cell, most viruses must first adhere to the plasma membrane, an interaction mediated by binding to a specific **receptor**, a cell surface molecule. Exceptions include viruses of yeasts and fungi, which have no extracellular phases, and plant viruses, which are believed to enter cells through openings produced by mechanical damage, such as those caused by farm machinery or insects. For some viruses, the receptor is the only cell surface molecule required for entry into cells. For others, binding to a cellular receptor is not sufficient for infection: an additional cell surface molecule, or **coreceptor**, is required for entry. Until the 1980s, our understanding of attachment lagged far behind that of other steps in viral replication, because technical limitations prevented scientists from identifying viral receptors. As a result of the development of monoclonal antibody and recombinant DNA methods, this field exploded with the identification of receptors for a number of medically important viruses, including human immunodeficiency virus type 1 and poliovirus. With the isolation of viral receptors came the ability to understand the virus-receptor interaction at a molecular level, to learn how the interaction leads to uncoating of the viral genome, to design novel antiviral therapies, and to produce new transgenic mouse models for studying viral diseases and disease prevention.

Our understanding of the earliest interactions of viruses with cells comes almost exclusively from analysis of synchronously infected cells in culture with a one-step growth curve. In our discussion, we will for the moment ignore the many steps that occur before attachment, including how viruses enter the environment, the initial contact with a host organism, and the bypassing of host defense mechanisms. These steps are very important parts of the total picture of viral infection and are discussed in Chapters 14 to 16.

Once a virus has attached to a cellular receptor, it must enter the host cell so that genome expression and replication can begin. Early studies of virus entry into the host cell, from the 1950s until the late 1970s, led to the view that viruses enter by an entirely passive process: virus particles attach to the cell surface, are taken up into the cell, release their genomes, and begin to replicate. No active role for the receptor in release of the viral genome into the cell, a process called **uncoating**, was envisioned. In these studies, investigation of entry largely comprised photographing sections of virus-infected cells at high magnification. While these findings showed that viruses did indeed cross the plasma membrane, the mechanisms of virus entry remained unknown. Indeed, in a popular virology textbook published in 1974, virus entry was described in only a few paragraphs. Beginning in the 1980s, the techniques of cellular, molecular, and structural biology were applied to elucidate the earliest events in viral infection. It is now understood that virus entry into cells is **not** a passive process but rather relies on viral usurpation of normal cellular processes, including endocytosis, membrane fusion, vesicular trafficking, and transport into the nucleus.

Protection and Release of Genetic Material

Successful entry of a virus into a host cell requires that the virus cross the plasma membrane, and in some cases the nuclear membrane as well, that the virus particle disassemble to make the viral genome accessible in the cytoplasm, and that the nucleic acid be targeted to the correct cellular compartment. These are not simple processes. Cell membranes are not permeable to particles as large as viruses. Viruses are brought across such barriers by the specific transport pathways by which large molecules normally enter cells. To survive in the extracellular environment, the viral genome must be encapsidated in a protective coat that will shield viral nucleic acid from the variety of potentially harsh conditions that a virion may meet in the environment. For example, mechanical shearing, ultraviolet (UV) irradiation (from sunlight), extremes of pH (in the gastrointestinal tract), dehydration (in the air), and enzymatic attack (in body fluids) are all capable of damaging viral nucleic acids. However, once in the host cell, the protective structures must become sufficiently unstable to release the genetic material. Virus particles cannot be viewed only as passive vehicles: they must be able to undergo structural transformations that are important for attachment and entry into a new host cell and for the subsequent disassembly required for viral replication. Viral disassembly often begins with conformational changes in the viral capsid or nucleocapsid or **integral membrane proteins**. In some cases, interaction of the virus particle with its cell receptor initiates conformational change that primes the capsid for uncoating. In others, the cell receptor is nothing more than a tether that holds the virus to the cell surface until it can be taken into the cell. Conformational changes are then triggered by low pH or by the action of proteases. Finally, the viral genome must be transported to the correct cellular compartment, either the nucleus or the cytoplasm. Viruses that replicate in the nucleus include all DNA-containing viruses (except poxviruses), RNA-containing retroviruses, influenza viruses, and Borna disease virus. All other RNA-containing viruses replicate in the cytoplasm. Because all viruses first enter the cytoplasm, those destined for the nucleus must undergo an additional trafficking step.

Finding the Right Cells To Infect

Viral infection is initiated by a collision between the virus particle and the cell, a process that is governed by chance. Therefore, a higher concentration of virus particles increases the probability of infection. However, a virus is not able to infect every cell it encounters. It must come in contact with the cells and tissues in which it can replicate. Such cells are normally recognized by means of a specific virion-cell surface receptor interaction, a process that can be either promiscuous or highly selective, depending on the virus and the distribution of the cell receptor. The presence of such receptors determines whether the cell will be **susceptible** to the virus. However, whether a cell is **permissive** for the replication of a particular virus depends on other, intracellular components found only in certain cell types. Cells must be both susceptible and permissive if an infection is to be successful.

In general, viruses have no means of locomotion, but their small size facilitates their diffusion driven by Brownian movement. Their propagation is dependent on essentially random encounters with potential hosts and host cells. In this sense, they can be thought of as tiny, opportunistic "Darwinian machines." For such machines, features that increase the probability of favorable encounters are very important. Consequently, viral propagation is critically dependent on the production of large numbers of progeny virions with surfaces composed of many copies of structures that enable the virions to attach to susceptible cells.

The Architecture of Cell Surfaces

The initiation and spread of viral infection most often follow encounters at the epithelial surfaces of the body that are exposed to the environment. Cells cover these surfaces, and the part of these cells exposed to the environment is called the **apical surface**. Conversely, the **basolateral surfaces** of such cells are in contact with adjacent or underlying cells or tissues. These cells exhibit a differential (**polar**) distribution of proteins and lipids in the plasma membranes that creates the two distinct surface domains. As illustrated in Fig. 5.1, these cell layers can differ in thickness and organization. Movement of macromolecules between the

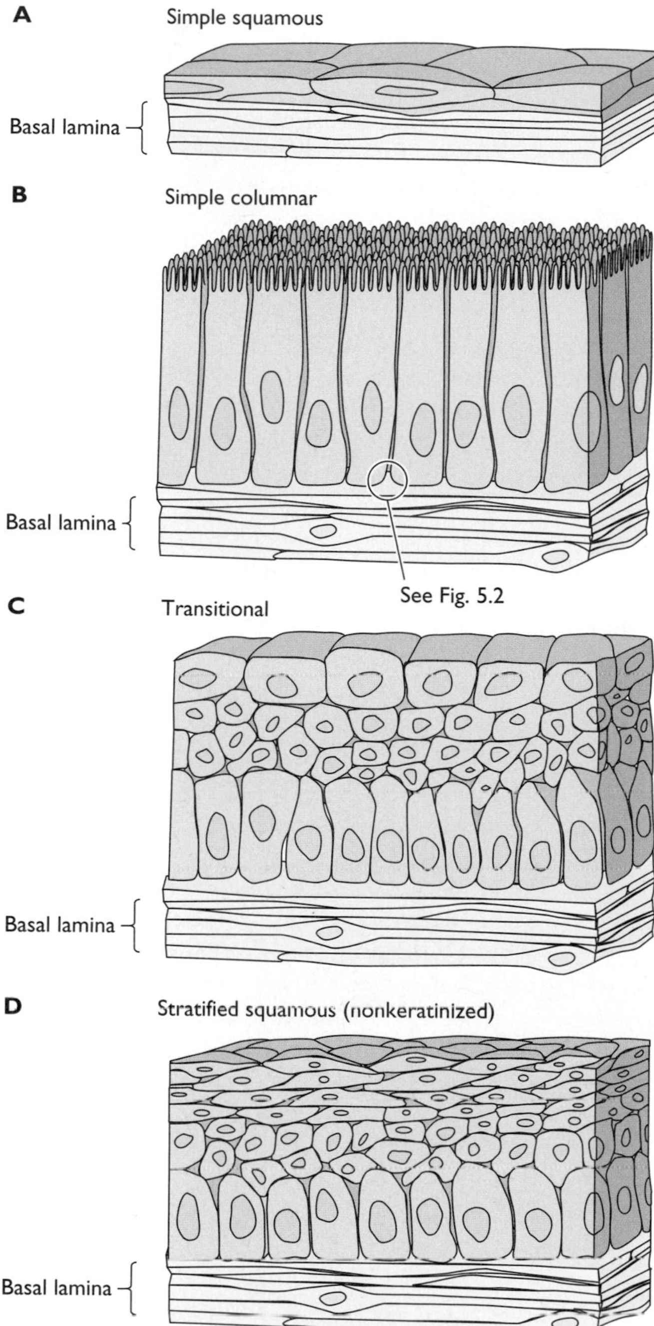

Figure 5.1 Major types of epithelia. (A) Simple squamous epithelium made up of thin cells such as those lining blood vessels and many body cavities. (B) Simple columnar epithelium. In the lining of the stomach and the cervical tract, this layer includes mucus-secreting cells. In the lining of the small intestine, it includes absorptive cells. (C) Transitional epithelium is composed of several layers with cells of different shapes. This layer lines cavities, such as the urinary bladder, that are subject to expansion and contraction. (D) Stratified, nonkeratinized epithelium lining surfaces such as the mouth and vagina. These layers resist abrasion. Adapted from H. Lodish et al., *Molecular Cell Biology*, 3rd ed. (W. H. Freeman & Co., New York, N.Y., 1995), with permission.

cells in the epithelium is prevented by **tight junctions**, which circumscribe the cells at the apical edges of their lateral membranes. Many viral infections are initiated upon entry of epithelial or endothelial cells at their exposed apical surfaces, often by attaching to cell surface molecules specific for this domain. Viruses that both enter and are released at apical membranes can be transmitted laterally from cell to cell without ever transversing the epithelial or endothelial layers; they generally cause localized infections. In other cases, progeny virions are transported to the basolateral surface and released into the underlying cells and tissues, a process that facilitates viral spread to other sites of replication.

There are also more specialized pathways by which viruses reach susceptible cells. For example, some epithelial tissues contain M cells, specialized cells that overlie the collections of lymphoid cells in the gut known as Peyer's patches. M cells function in the transport of intestinal contents to Peyer's patches. Certain viruses, such as poliovirus and human immunodeficiency virus type 1, can also be transported through them by a mechanism called **transcytosis** to gain access to underlying tissues. Such specialized pathways of invasion will be considered in Chapter 14. Below we describe briefly the structures that surround cells and tissues, the membrane components that mediate the interaction of cells with their environments, and the ways in which viruses attach to cells by means of these components.

Extracellular Matrix: Components and Biological Importance

Extracellular matrices hold the cells and tissues of the body together. The extracellular matrix is made up of two main classes of macromolecules (Fig. 5.2). The first class comprises glycosaminoglycans (such as heparan sulfate and chondroitin sulfate), which are unbranched polysaccharides made of repeating disaccharides. Glycosaminoglycans are usually linked to proteins to form **proteoglycans**. The second class of macromolecules in the extracellular matrix consists of fibrous proteins with structural (**collagen** and **elastin**) or adhesive (**fibronectin** and **laminin**) functions. The proteoglycan molecules in the matrix form hydrated gels in which the fibrous proteins are embedded, providing strength and resilience to the matrix. The gel provides resistance to compression and allows the diffusion of nutrients between blood and tissue cells. The extracellular matrix of each cell type is specialized for the particular function required, varying in thickness, strength, elasticity, and degree of adhesion.

Most organized groups of cells, like epithelial cells of the skin (Fig. 5.1 and 5.3), are bound tightly on their basal surface to a thin layer of extracellular matrix called

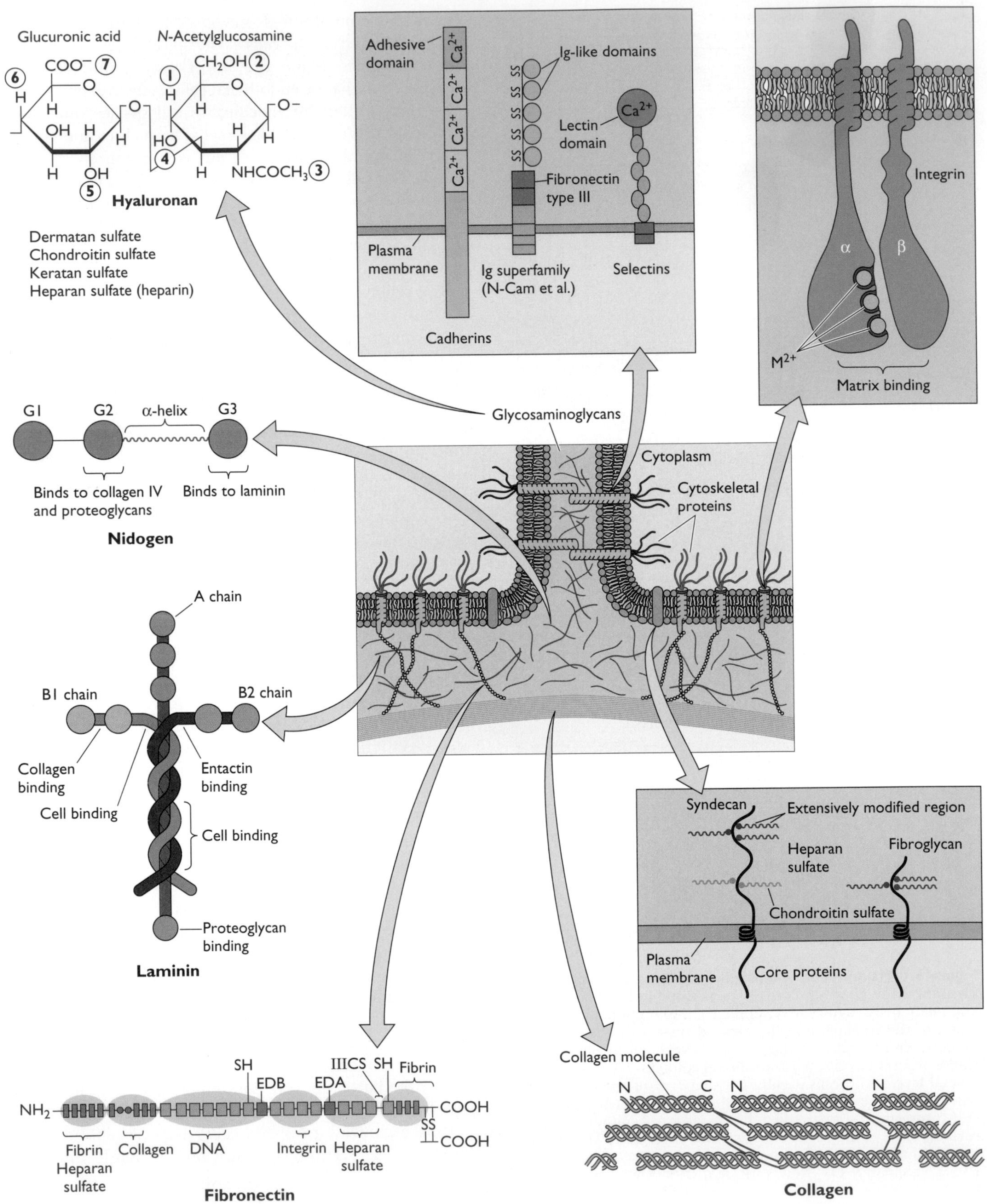

Glucuronic acid
N-Acetylglucosamine
Hyaluronan
Dermatan sulfate
Chondroitin sulfate
Keratan sulfate
Heparan sulfate (heparin)
Adhesive domain
Ig-like domains
Lectin domain
Fibronectin type III
Plasma membrane
Ig superfamily (N-Cam et al.)
Selectins
Cadherins
Integrin
Matrix binding
G1
G2
α-helix
G3
Binds to collagen IV and proteoglycans
Binds to laminin
Nidogen
Glycosaminoglycans
Cytoplasm
Cytoskeletal proteins
A chain
B1 chain
B2 chain
Collagen binding
Entactin binding
Cell binding
Cell binding
Proteoglycan binding
Laminin
Syndecan
Extensively modified region
Heparan sulfate
Fibroglycan
Chondroitin sulfate
Plasma membrane
Core proteins
Fibrin
Collagen
DNA
Integrin
Heparan sulfate
Fibrin Heparan sulfate
Fibronectin
Collagen molecule
Collagen

Figure 5.2 Cell adhesion molecules and components of the extracellular matrix. The diagram in the center (expanded from Fig. 5.1B) illustrates the variety of cell surface components that contribute to cell-cell adhesion and attachment to the extracellular matrix. Additional details are provided for each major molecular class. Different glycosaminoglycans such as dermatan sulfate and chondroitin sulfate are produced by covalent linkage of SO_3 at the numbered positions on the hyaluronan molecule. M^{2+}, divalent cation. Adapted from H. Lodish et al., *Molecular Cell Biology,* 3rd ed. (W. H. Freeman & Co., New York, N.Y. 1995), and G. M. Cooper, *The Cell: a Molecular Approach* (ASM Press, Washington, D.C., and Sinauer Associates, Sunderland, Mass., 1997), with permission.

the **basal lamina**. This matrix is linked to the basolateral membrane by specific receptor proteins called integrins (which are discussed in "Cell Membrane Proteins" below). Integrins are anchored to the intracellular structural network (the **cytoskeleton**) at the inner surface of the cell membrane. The basal lamina is attached to collagen and other material in the underlying loose connective tissue found in many organs of the body (Fig. 5.3). Capillaries, glands, and specialized cells are embedded in this connective tissue. Several viruses gain access to susceptible cells by attaching specifically to components of the extracellular matrix, including some cell adhesion proteins, fibronectins, proteoglycans, and integrins (Table 5.1).

Properties of the Plasma Membrane

The plasma membrane of every cell type is composed of a similar phospholipid/glycolipid bilayer, but different sets of membrane proteins and glycolipids allow the cells of different tissues to carry out their specialized functions. The lipid bilayer consists of molecules that possess both hydrophilic and hydrophobic portions; they are known as

Figure 5.3 Cross section through skin. In this diagram of skin from a pig, the precursor epidermal cells rest on a thin layer of extracellular matrix called the basal lamina. Underneath is loose connective tissue consisting mostly of extracellular matrix. Fibroblasts in the connective tissue synthesize the connective tissue proteins, hyaluronan, and proteoglycans. Blood and lymph capillaries are also located in the loose connective tissue layer. Adapted from H. Lodish et al., *Molecular Cell Biology,* 3rd ed. (W. H. Freeman & Co., New York, N.Y., 1995), with permission.

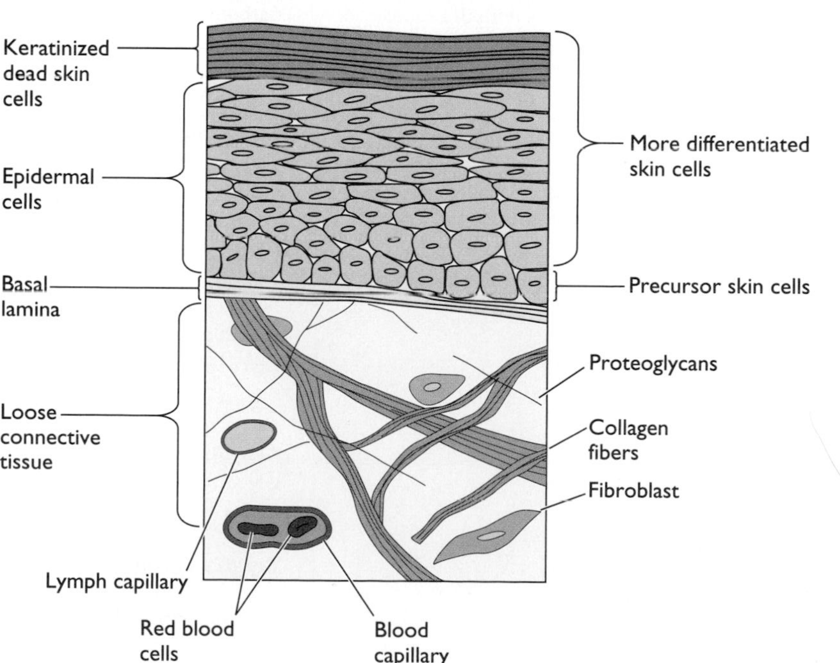

Table 5.1 Viral receptors and coreceptors[a]

Virus	Receptor	Type of molecule	Coreceptor
Adenoviridae			
Adenovirus subgroups A, C, D E, F	Car (coxsackievirus-adenovirus receptor)	Ig-like	α_v integrins
Adenovirus type 5 (subgroup C)	Major histocompatibility class II molecule	Ig-like	α_v integrins
Adenovirus type 2 (subgroup C)	$\alpha_M\beta_2$	Integrin	α_v integrins
Adenovirus type 9 (subgroup D)	α_v integrins	Integrin	
Bunyaviridae			
Hantavirus	β_3 integrins	Integrin	
Coronaviridae			
Mouse hepatitis virus	Bgp (biliary glycoprotein)	Ig-like	
Human coronavirus, 229 E	Amniopeptidase N	Protease	
Transmissable gastroenteritis virus	Amniopeptidase N	Protease	
Human coronavirus OC43	Sialic acid	Carbohydrate	
Bovine coronavirus			
Herpesviridae			
Herpes simplex virus type 1	Heparan sulfate	Glycosaminoglycan	HveA, Prr1
	3-O-sulfated heparan sulfate	Glycosaminoglycan	None
Herpes simplex virus type 2	Heparan sulfate	Glycosaminoglycan	HveA, Prr1, Prr2
Pseudorabies virus	Heparan sulfate	Glycosaminoglycan	Pvr, Prr1, Prr2
Bovine herpesvirus	Heparan sulfate	Glycosaminoglycan	Pvr, Prr1
Human herpesvirus 7	CD4	Ig-like	
Human herpesvirus 8	$\alpha_3\beta_1$, $\alpha_2\beta_1$	Integrin	
Epstein-Barr virus	Complement receptor Cr2 (CD21)	SCR-like (complement cascade)	
Human cytomegalovirus	Heparan sulfate	Glycosaminoglycan	Aminopeptidase N (CD13)
Orthomyxoviridae			
Influenza A and B viruses	Sialic acids (*N*-acetylneuraminic acid)	Carbohydrate	
Influenza C virus	Sialic acids (9-*O*-acetylneuraminic acid)	Carbohydrate	
Paramyxoviridae			
Measles virus	Membrane cofactor protein, CD46	Complement-regulating protein	
	Signaling lymphocyte-activation molecule (SLAM, CD150)	Ig-like	
Sendai virus	Sialic acid	Carbohydrate	
	Asialoglycoprotein receptor GP-2	Transport protein (receptor-mediated endocytosis)	
Respiratory syncytial virus	Heparan sulfate	Glycosaminoglycan	
Parvoviridae			
Bovine parvovirus	Sialic acids	Carbohydrate	
Adeno-associated virus type 2	Heparan sulfate	Glycosaminoglycan	$\alpha_V\beta_5$ integrin
Picornaviridae			
Foot-and-mouth disease virus (cell culture adapted)	Heparan sulfate	Glycosaminoglycan	
Foot-and-mouth disease virus	$\alpha_V\beta_3$ (vitronectin receptor)	Integrin	

Table 5.1 *(continued)*

Virus	Receptor	Type of molecule	Coreceptor
Encephalomyocarditis virus	Vcam-1	Ig-like	
	Sialylated glycophorin A (for hemagglutination only)	Carbohydrate	
Poliovirus types 1–3	Pvr (CD155)	Ig-like	
Coxsackievirus A13, A18, A21	Icam-1	Ig-like	
Coxsackievirus A21	Decay-accelerating factor (CD55)	SCR-like (complement cascade)	Icam-1
Coxsackievirus A9	$\alpha_V\beta_3$	Integrin	
Coxsackievirus B1 to B6	Car	Ig-like	
Coxsackievirus B1, B3, B5	CD55	SCR-like (complement cascade)	$\alpha_V\beta_6$ integrin
Echovirus 1 and 8	$\alpha_2\beta_1$ (Vla-2)	Integrin	β_2 microglobulin
Echovirus 22	$\alpha_V\beta_3$ (vitronectin receptor)	Integrin	
Echovirus 3, 6, 7, 11 to 13, 20, 21, 24, 29, 33	CD55	SCR-like (complement cascade)	β_2 microglobulin
Enterovirus 70	CD55	SCR-like (complement cascade)	
Bovine enterovirus	Sialic acid	Carbohydrate	
Hepatitis A virus	HAVcr-1	Ig-like, mucin-like	
Major group rhinovirus (91 serotypes)	Icam-1	Ig-like	
Minor group rhinovirus (10 serotypes)	Low-density lipoprotein receptor protein family	Signaling receptor	
Rhinovirus 87	Sialic acid	Carbohydrate	
Polyomaviridae			
Simian virus 40	Major histocompatability class I molecule	Ig-like	
Poxviridae			
Vaccinia virus	Heparan sulfate	Glycosaminoglycan	
	Epidermal growth factor receptor	Signaling receptor	
Reoviridae			
Orthoreovirus	Sialic acids	Carbohydrate	
	Junction adhesion molecule	Ig-like	
Group A porcine rotavirus	Sialic acids	Carbohydrate	
Human rotavirus	$\alpha_2\beta_1$, $\alpha_4\beta_1$	Integrin	$\alpha_V\beta_3$
Retroviridae			
Human immunodeficiency virus type 1	CD4	Ig-like	Chemokine receptors (CCr5, CXCr4, CCr3)
	Galactosylceramide	Glycolipid	
Human immunodeficiency virus type 2	CD4	Ig-like	Chemokine receptors
	CXCr4	7-transmembrane superfamily	
Simian immunodeficiency virus	CD4	Ig-like	Chemokine receptors
Feline immunodeficiency virus	CXCr4	7-transmembrane superfamily	
Gibbon ape leukemia virus	Pit1	Sodium-dependent phosphate transport protein	
Feline leukemia virus B	Pit1, Pit2	Sodium-dependent phosphate transport protein	
Feline leukemia virus C	Flvcr	Organic anion transport protein	
Feline leukemia virus T	Pit1	Sodium-dependent phosphate transport protein	Felix (Env-like protein)

(continued)

Table 5.1 Viral receptors and coreceptors[a] *(continued)*

Virus	Receptor	Type of molecule	Coreceptor
Amphotropic murine leukemia virus	Pit2	Sodium-dependent phosphate transport protein	
Ecotropic murine leukemia virus	Cat	Basic amino acid transport protein	
Xenotropic and polytropic murine leukemia virus	XPr1	G protein-coupled signaling or transport	
Subgroup A avian leukosis and sarcoma virus	Tva	Low-density lipoprotein receptor protein family	
Subgroup B, D, and E avian leukosis and sarcoma viruses	Tvb	Fas/Tnfr-like receptor	
Bovine leukemia virus	Blvr	Unknown	
Mouse mammary tumor virus	Transferrin receptor 1	Iron transporter	
RD114	RDr	Neutral amino acid transport	
	RDr2	Glutamate, neutral amino acid transport	
Porcine endogenous retrovirus A	Par-1, Par-2	G-coupled receptor	
Rhabdoviridae			
Rabies virus	Nicotinic acetylcholine receptor	Neurotransmitter receptor	
	Neural cell adhesion molecule CD56	Ig-like	
	Low-affinity nerve growth factor receptor	Tnfr protein superfamily	
Togaviridae			
Semliki Forest virus	Major histocompatibility class I molecule	Ig-like	
Sindbis virus	High-affinity laminin receptor	Integrin	
	Heparan sulfate	Glycosaminoglycan	
Dengue virus	Heparan sulfate	Glycosaminoglycan	

[a]The name of the receptor and the type of molecule are listed for selected viruses. When coreceptors have been identified, they are listed; a blank in the coreceptor column indicates that none have been identified to date. Abbreviations: Car, coxsackievirus and adenovirus receptor; Hve, herpesvirus entry mediator; Pvr, poliovirus receptor; Prr1, Prr2, Pvr-related proteins 1 and 2; Par, porcine endogenous retrovirus A receptor; Vcam, vascular cell adhesion molecule; SCR, short consensus repeat; Ig, immunoglobulin; Pit, phosphate (Pi) transport; Felix, feline leukemia virus infection accessory protein; Cat, cationic amino acid transporter; Tnfr, tumor necrosis factor receptor; Icam-1, intercellular adhesion molecule 1.

amphipathic molecules, from the Greek word *amphi* (on both sides) (Fig. 5.4). They form a sheetlike structure in which polar head groups face the aqueous environment of the cell's cytoplasm (inner surface) or the surrounding environment (outer surface). The polar heads of the inner and outer leaflets bear side chains with different lipid compositions. The fatty acyl side chains form a continuous hydrophobic interior about 3 nm thick. Hydrophobic interactions are the driving force for formation of the bilayer. However, hydrogen bonding and electrostatic interactions among the polar groups and water molecules or membrane proteins also stabilize the structure.

Thermal energy permits the phospholipid and glycolipid molecules comprising natural cell membranes to rotate freely around their long axes and diffuse laterally. If unencumbered, a lipid molecule can diffuse the length of an animal cell in only 20 s at 37°C. In most cases, phospholipids and glycolipids do not flip-flop from one side of a bilayer to the other, and the outer and inner leaflets of the bilayer remain separate. However, membrane proteins not anchored to the extracellular matrix and/or the underlying structural network of the cell can diffuse rapidly, moving laterally like icebergs in this fluid bilayer. In this way, certain membrane proteins can form functional aggregates. Intracellular organelles such as the nucleus, endoplasmic reticulum, and lysosomes are also enclosed in lipid bilayers, although their composition and physical properties differ.

For many years the plasma membrane was viewed as a uniform and fluid sea, in which lipid and protein compo-

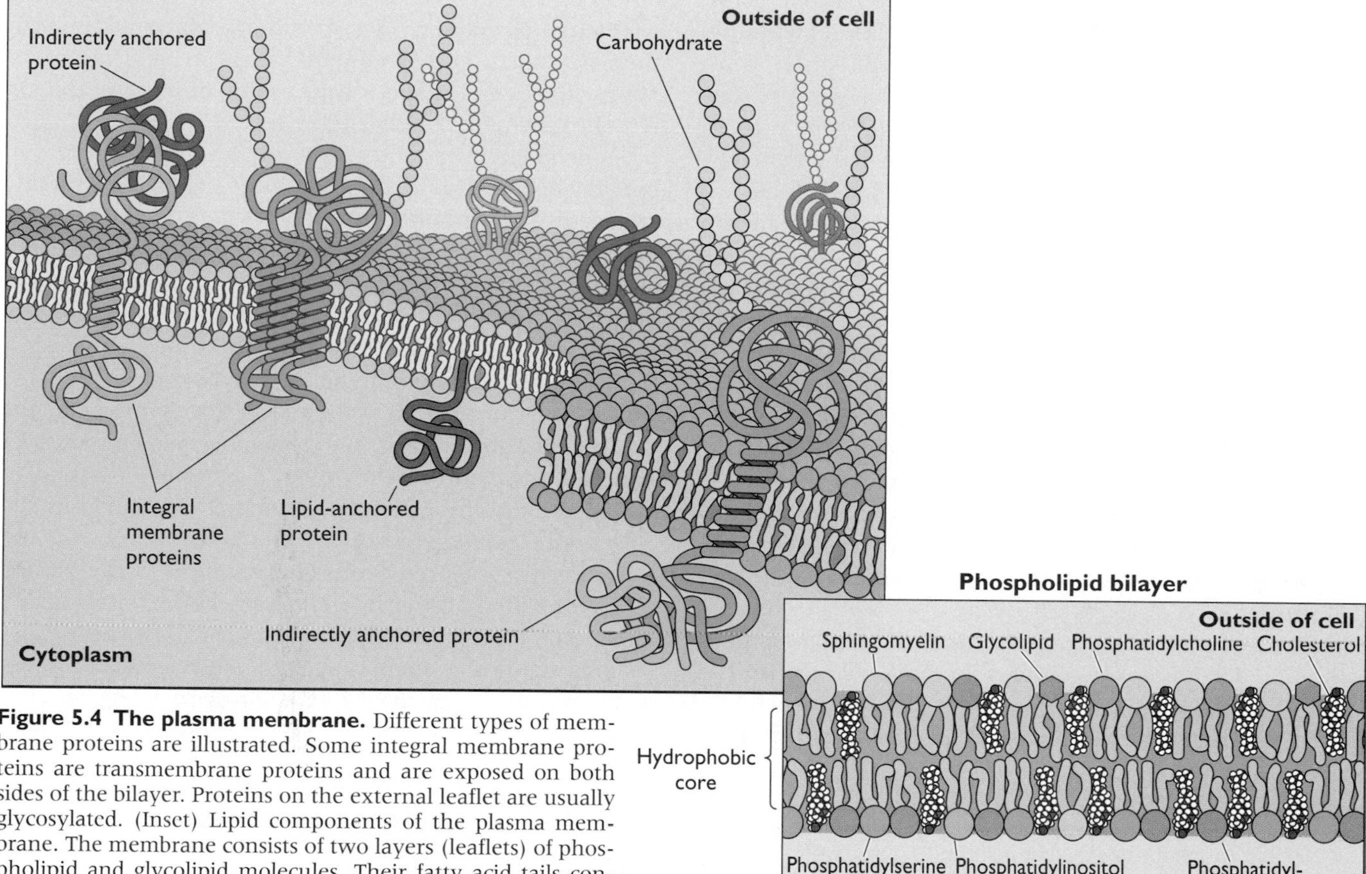

Figure 5.4 The plasma membrane. Different types of membrane proteins are illustrated. Some integral membrane proteins are transmembrane proteins and are exposed on both sides of the bilayer. Proteins on the external leaflet are usually glycosylated. (Inset) Lipid components of the plasma membrane. The membrane consists of two layers (leaflets) of phospholipid and glycolipid molecules. Their fatty acid tails converge to form the hydrophobic interior of the bilayer; the polar hydrophilic head groups (shown as balls) line both surfaces. Major components of the outer leaflet include phosphatidylcholine, sphingomyelin, and glycolipids; the inner leaflet is made up largely of phosphatidylethanolamine, phosphatidylserine, and phosphatidylinositol. Cholesterol is found in both leaflets. The head groups of phosphatidylserine and phosphatidylinositol have a net negative charge. Adapted from G. M. Cooper, *The Cell: a Molecular Approach* (ASM Press, Washington, D.C., and Sinauer Associates, Sunderland, Mass., 1997), with permission.

nents diffused randomly in the plane of the membrane. This simplistic model has been dispelled by experimental findings of the past 10 years. It is now recognized that plasma membranes comprise **microdomains,** regions with distinct lipid and protein composition. The **lipid raft** is one type of microdomain that is important for virus replication. Lipid rafts are enriched in cholesterol and saturated fatty acids and consequently are more densely packed and less fluid than other regions of the membrane. The assembly of a variety of viruses takes place at lipid rafts (Chapter 13). Furthermore, the entry of some viruses requires lipid rafts. For example, virions of human immunodeficiency virus 1, Ebola virus, and echovirus 11 enter cells at lipid rafts, and virus entry is blocked by treatment of cells with compounds that disrupt these microdomains. One explanation for this requirement might be that cell membrane proteins required for entry are present only in lipid rafts: receptors and coreceptors for human immunodeficiency virus are preferentially located in lipid rafts.

Cell Membrane Proteins

Membrane proteins are classified into two broad categories, **integral membrane proteins** and **indirectly anchored proteins,** names that describe the nature of their interactions with the plasma membrane (Fig. 5.4).

Integral proteins are embedded in the lipid bilayer, because many contain one or more **membrane-spanning domains,** as well as portions that protrude out into the exterior and interior of the cell (Fig. 5.4). A membrane-spanning domain consists of an α-helix typically 3.7 nm long

(Fig. 5.5). It includes 20 to 25 generally hydrophobic or uncharged residues embedded in the membrane, with the hydrophobic side chains protruding outward to make contacts with the fatty acyl chains of the lipid bilayer. The first and last residues are often positively charged amino acids (lysine or arginine) that can interact with the negatively charged polar head groups of the phospholipids to stabilize the membrane-spanning domain. Proteins with membrane-spanning domains enable the cell to respond to signals from outside the cell. Such membrane proteins are designed to bind external ligands (e.g., hormones, cytokines, or membrane proteins on the same cell or on other cells) and to signal the occurrence of such interactions to molecules in the interior of the cell. Some proteins with multiple membrane-spanning domains (Fig. 5.5) form critical components of molecular pores or pumps, which mediate the internalization of required nutrients, or expulsion of undesirable material from the cell, or maintain homeostasis with respect to cell volume, pH, and ion concentration.

In many cases, the external portions of membrane proteins are decorated by complex or branched **carbohydrate chains** linked to the peptide backbone through nitrogen (**N linked**) in the side chain of asparagine residues or through oxygen (**O linked**) in the side chains of serine or threonine residues. Such membrane **glycoproteins,** as they are called, quite frequently serve as viral receptors. Many viruses attach specifically to one or more of the external components of glycoproteins.

Some membrane proteins do not span the lipid bilayer, but are anchored in the inner or outer leaflet by covalently attached hydrocarbon chains. These lipid-bound proteins fall into three classes according to the nature of the anchoring hydrocarbons (Fig. 5.6). They include amide-linked **myristoyl** and thioether-linked **farnesyl** chains by which proteins may be bound to the inner cytoplasmic face of the plasma membrane. A myristoyl group anchors the proteins of several viruses (e.g., retroviruses) to the cell membrane during virion assembly. A third class of nonembedded, integral membrane proteins is bound to the outer portion of the cell membrane via amide-linked glycosylphosphatidylinositol.

Indirectly anchored proteins are bound to the plasma membrane lipid bilayer by interacting either with integral membrane proteins or with the charged sugars of the glycolipids within the membrane lipid. Fibronectin, a protein in the extracellular matrix that binds to integrins (Fig. 5.2), is an example.

Figure 5.5 Membrane-spanning domains. (A) Amino acid sequence of a typical single membrane-spanning domain, in this case for the CD4 receptor of human immunodeficiency virus type 1 and several other primate lentiviruses. Essential features include an α-helical stretch of 25 hydrophobic residues bounded on the C-terminal edge by a hydrophilic positively charged (R, Arg) residue. The polar carbonyl (C=O) and amide (NH) groups are contained within the helix. (B) Different arrangements of N and C termini and membrane-spanning domains in a variety of integral membrane proteins. LDL, low-density lipoprotein. Adapted from L. H. Stryer, *Biochemistry,* 4th ed. (W. H. Freeman & Co., New York, N.Y., 1995), with permission.

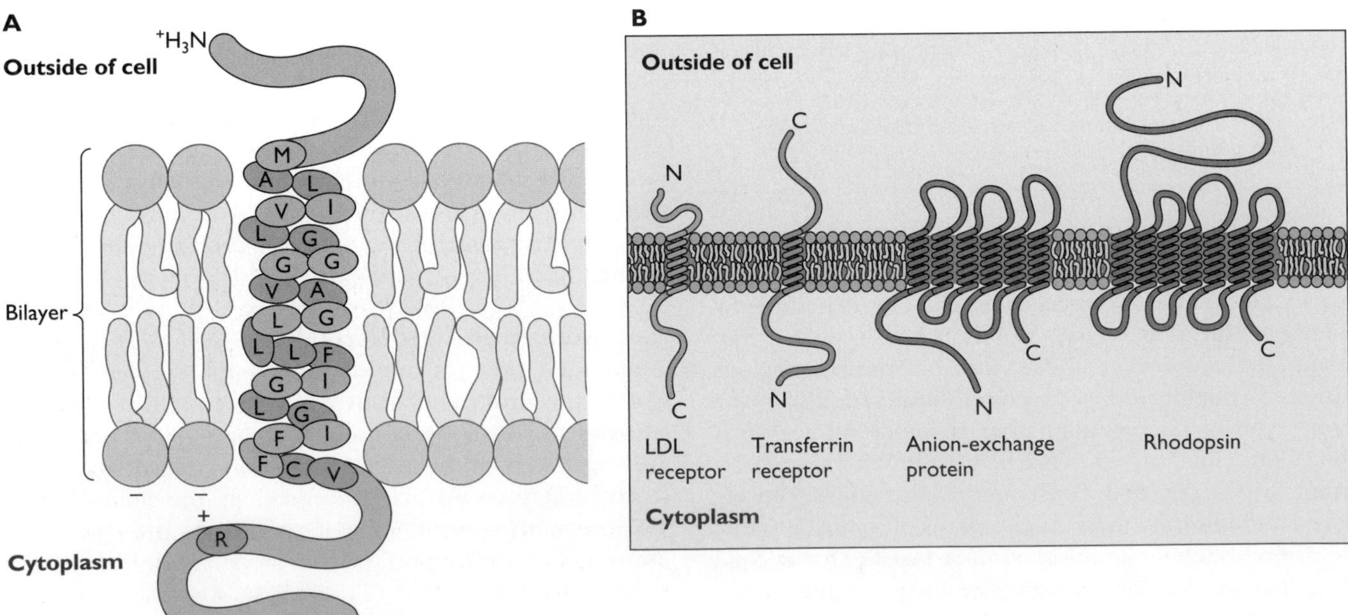

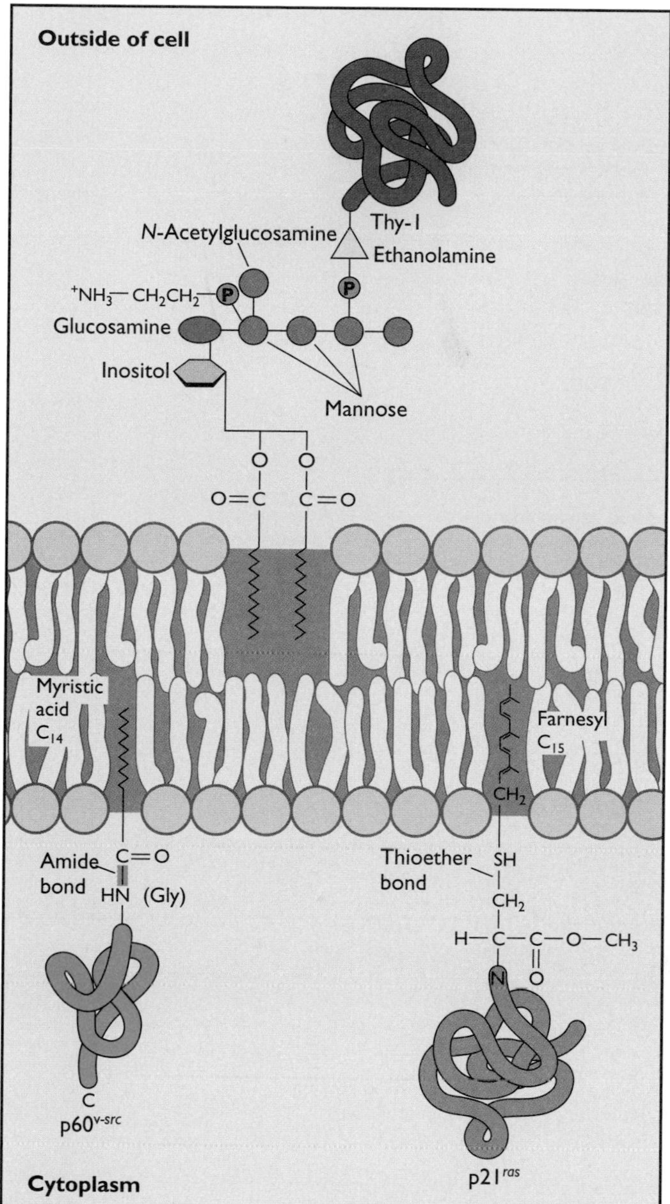

Figure 5.6 Lipid anchors for some integral membrane proteins. Myristyl- and farnesyl-anchored proteins are found only on the cytoplasmic face of the plasma membrane. They include various proteins involved in cell growth, as well as a number of viral proteins. Glycosylphosphatidylinositol-anchored proteins are found only on the extracellular face.

Interaction of Viruses with Cell Receptors

General Principles

Many different cell surface molecules can serve as receptors for the attachment of viruses (Table 5.1). Some viruses attach to more than one receptor, and some receptors are shared by viruses of different families.

The cell receptor may determine the **host range** of a virus, its ability to infect a particular animal or cell culture. For example, poliovirus infects primates and cultured primate cells but not mice or mouse cell cultures. Mouse cells synthesize a protein that is homologous to the poliovirus receptor but is sufficiently different so that poliovirus cannot attach to it. In this example, the poliovirus receptor is **the** determinant of poliovirus host range. However, production of the virus receptor in a particular cell type does not ensure that virus replication will occur. Some primate cell cultures produce the poliovirus receptor but cannot be infected. The restricted host range of the virus in such cells is most likely due to a block in viral replication beyond the attachment step. Cell receptors can also be determinants of tissue **tropism**, the predilection of a virus to invade and replicate in a particular cell type. However, there are many other determinants of tissue tropism. For example, the sialic acid residues on membrane glycoproteins or glycolipids, which are receptors for influenza virus, are found on many tissues, yet viral replication in the host is restricted. The basis for such restriction is discussed in Chapter 14.

Cell culture systems have made it possible to study the general properties of attachment. The initial interactions between virions and cells are probably via electrostatic forces, as they are sensitive to low pH or high concentrations of salt. Subsequent high-affinity binding relies mainly on hydrophobic and other short-range forces whose strength and specificity are governed primarily by the conformations of the interacting viral and cellular interfaces. Virion binding can usually occur at 4°C (even though entry does not) as well as at body temperature (e.g., 37°C). Infection of cultured cells can therefore be synchronized by allowing binding to take place at a low temperature and then shifting the cells to a physiological temperature to allow the initiation of subsequent steps.

The first steps in virus attachment are governed largely by the probability that a virion and a cell will collide, and therefore by the concentrations of free virions and host cells. The rate of attachment can be described by the equation $dA/dt = k[V][H]$, where $[V]$ and $[H]$ are the concentrations of virions and host cells, respectively, and k is a rate constant. Values of k for animal viruses vary greatly from a maximal value that represents the limits of diffusion to one that is as much as 5 orders of magnitude lower. It can be seen from this equation that if a mixture of viruses and cells is diluted after virions have been allowed to attach, subsequent binding will be greatly reduced. For example, a 100-fold dilution of the mixture will reduce the attachment rate 10,000-fold (i.e., $1/100 \times 1/100$). Dilution can be used to prevent subsequent virus adsorption and hence to synchronize an infection.

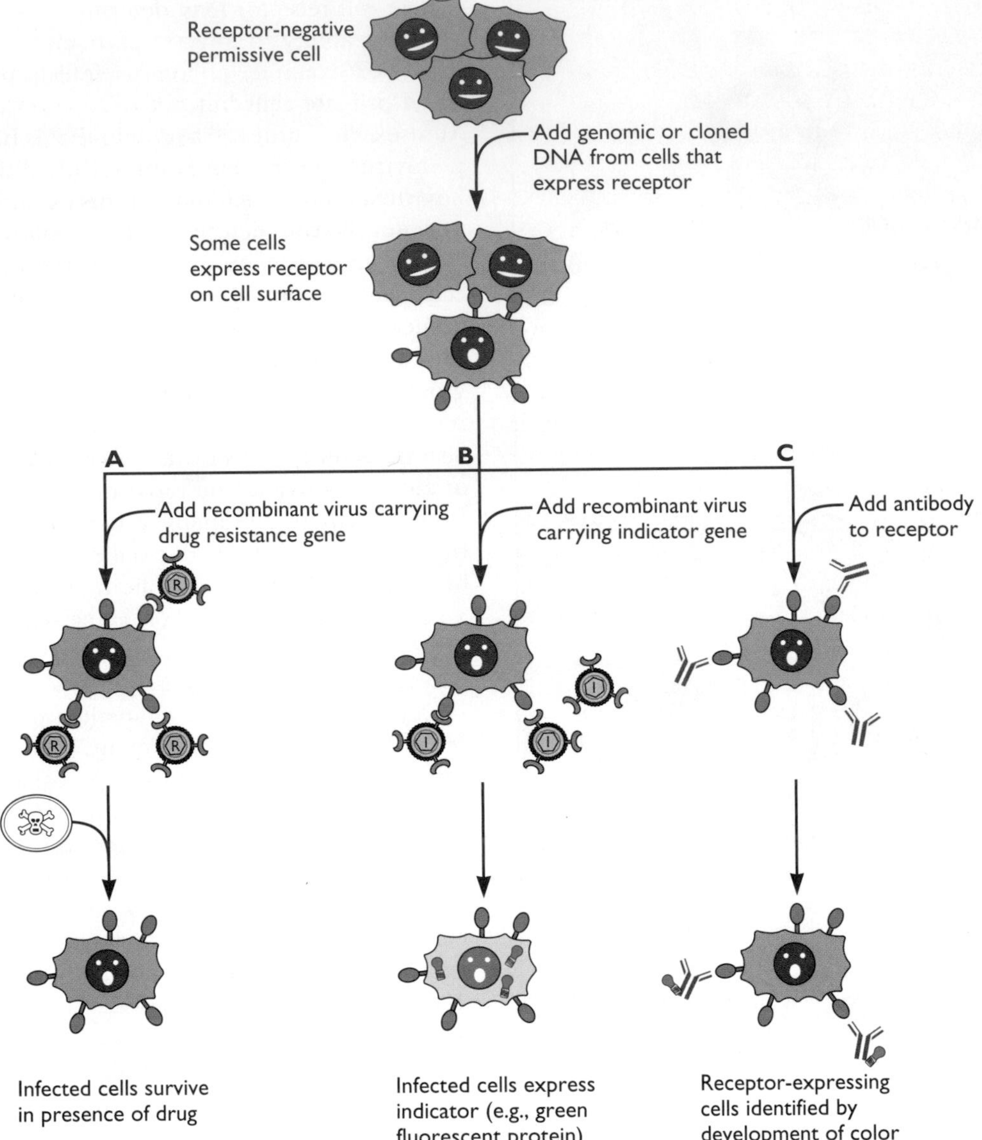

Figure 5.7 Experimental strategies for identification and isolation of cellular genes encoding viral receptors. Genomic DNA or pools of DNA clones (e.g., prepared by reverse transcription of messenger RNAs [mRNAs]) from cells known to synthesize the receptor are introduced into receptor-negative permissive cells. A small number of recipient cells express the receptor. Three different strategies for identifying such rare receptor-expressing cells are outlined. (A) The cells are infected with a virus that has been engineered so that it carries a gene encoding drug resistance. Cells that express the receptor will become resistant to the drug. The receptor DNA can then be isolated from these cells. This strategy works only for viruses that persist in cells without killing them. (B) For lytic viruses, an alternative is to engineer the virus to express an indicator, such as green fluorescent protein or β-galactosidase. Cells that make the correct receptor and become infected with such viruses can be distinguished by a color change, such as green in the case of green fluorescent protein. (C) The third approach depends on the availability of an antibody directed against the receptor, which binds to cells that express the receptor gene. Bound antibodies can be detected by an indicator molecule. After one of these assays has been conducted, one of two general approaches is used to isolate receptor DNA. When the donor DNA is genomic DNA, the presence of species-specific repetitive sequences allows it to be distinguished from the DNA of the recipient cell. Genomic DNA from the susceptible cell is cloned into an appropriate vector, and the repetitive sequence can be used as a hybridization probe to identify the cloned receptor DNA. When complementary DNA (cDNA) cloned in a plasmid is used as the donor DNA, pools of individual clones (usually 10,000 clones per pool) are prepared and introduced individually into cells. The specific DNA pool that yields receptor-expressing cells is then subdivided, and the screening process is repeated until a single receptor-encoding DNA is identified.

Identification of Viral Receptors on Cells

Experimental Approaches

Early investigations of viral receptors made use of a variety of enzymes to characterize cell surface components that are required for virus attachment. The first viral receptor discovered, sialic acid, which binds influenza virus, was identified because the enzyme neuraminidase removes this carbohydrate from cells and blocks virus attachment. In a similar way, experiments with proteases showed that many receptors are proteins. These types of analysis provided the first clues concerning the chemical nature of cell surface components to which virions become attached. It was also possible to determine whether receptors are shared by different viruses, by determining whether saturating cells with one virus prevented binding of a second. If significant competition was observed, then it was surmised that both viruses bound the same receptor molecule. This method was first used to establish different viral subgroups among avian retroviruses and to type cells from individual animals for the presence of genetic determinants of susceptibility to each subgroup.

Despite these approaches, identification of cell receptors for viruses languished because biochemical purification of these molecules proved difficult. As late as 1985, only one viral receptor, the sialic acid receptor of influenza viruses, had been identified unequivocally. The development of three crucial technologies rapidly changed this situation. The first, production of monoclonal antibodies, provided a powerful means for isolating and characterizing individual cell surface proteins. In practice, hybridoma cell lines that secrete monoclonal antibodies that block virus attachment are obtained after immunizing mice with intact cells. Such antibodies can be used to purify the receptor protein by immunoprecipitation or affinity chromatography.

A second technology that advanced the viral receptor field was the development of DNA-mediated transformation. This method was crucial for isolating receptor genes following introduction of DNA from susceptible cells into nonsusceptible cells (Fig. 5.7). Cells that acquire DNA encoding the receptor and carry the corresponding protein on their surface will be able to bind virus specifically. Clones of such cells are recognized and selected, for example, by the binding of receptor-specific monoclonal antibodies. The receptor genes can then be cloned from these selected cells using a third technology, molecular cloning.

The availability of cloned receptor genes has made it possible to investigate the details of receptor interaction with viruses by site-directed mutagenesis. Receptor proteins can be synthesized in heterologous systems and purified, and their properties can be studied in vitro, while animal cells producing altered receptor proteins can be used to test the effects of mutations on virus attachment. Because of their hydrophobic membrane-spanning domains, many of these cell surface receptor proteins are relatively insoluble and difficult to work with. In a number of cases, soluble forms have been obtained by the expression of genes encoding truncated receptor proteins lacking the membrane-spanning domain. Soluble forms of receptors have been essential for structural studies of receptor-virus interactions. Cloned receptor genes have also been used to produce transgenic mice that synthesize receptor proteins. Such transgenic animals can serve as useful models in the study of human viral diseases.

Although receptors for many viruses have now been identified (Table 5.1), many more remain to be discovered. Criteria for identifying receptors are discussed in Box 5.1. Despite their obvious importance to our understanding of viral biology and pathogenesis, the normal

BOX 5.1

Criteria for identifying virus receptors on cells

The combination of monoclonal antibodies, molecular cloning, and DNA-mediated transformation provides a powerful approach for identifying cell proteins that are receptors for viruses. Each method has associated uncertainties. For example, a monoclonal antibody that blocks virus attachment might recognize not the receptor, but a closely associated membrane protein (see "The poliovirus receptor"). To prove that the protein recognized by the monoclonal antibody **is** a receptor, the protein must be isolated and its DNA must be cloned and introduced into cells to determine that it can confer virus-binding activity. Any of the approaches outlined in Fig. 5.7 can result in identification of a cellular gene that encodes a putative receptor. However, the encoded protein might not be a receptor, but may modify another cellular protein so that it can serve as a viral receptor. Proof that the DNA codes for a viral receptor would come from the identification of a monoclonal antibody that blocks virus attachment and is directed against the encoded protein.

For some viruses, synthesis of the receptor on cells leads to binding but not infection. In such cases, a coreceptor is required, either for internalization or for membrane fusion. The techniques of molecular cloning can also be used to identify coreceptors. For example, production of CD4 on mouse cells leads to binding of human immunodeficiency virus type 1 but not infection, because fusion of viral and cell membranes does not occur. To identify the coreceptor, a DNA clone that allowed membrane fusion catalyzed by viral SU protein in mouse cells expressing CD4 was isolated from human cells.

functions of few of the receptors are known. We know even less about the molecular mechanisms that mediate virus attachment and entry. Among the best-studied receptors are those for members of the picorna-, orthomyxo-, and retrovirus families.

A Single Receptor Molecule Mediates Virus Binding and Entry

The rhinovirus receptor, intercellular adhesion molecule 1. Members of the rhinovirus genus of the *Picornaviridae* are unstable below pH 5 to 6 and multiply primarily in the upper respiratory tract. Up to 50% of all human common colds are caused by members of the major group of rhinoviruses. The cell surface receptor for these rhinoviruses (~90 serotypes) was identified by screening monoclonal antibodies for their ability to block rhinovirus infection. When such a monoclonal antibody was identified, it was used to isolate a 95-kDa cell surface glycoprotein by affinity chromatography. Amino acid sequence analysis of the purified protein, which bound to rhinovirus in vitro, identified it as the integral membrane protein **intercellular adhesion molecule 1 (Icam-1).**

Icam-1 (Fig. 5.8 and 5.9) is a member of the immunoglobulin (Ig) superfamily. It has five domains that are homologous to one another and to the constant domains of antibody molecules. Each domain is stabilized by intrachain disulfide bonds. Icam-1 is found on the surface of many cell types, including those of the nasal epithelium, the normal entry site for rhinoviruses. The natural ligand of Icam-1 is another integral membrane glycoprotein, the integrin known as **lymphocyte function-associated antigen 1 (Lfa-1).** The normal function of Icam-1 is to bind Lfa-1 on the surface of lymphocytes and promote a wide variety of immunological and inflammatory responses. Mediators of inflammation induce increased synthesis of Icam-1. Therefore, the initial reaction of host defenses to rhinoviruses, which leads to the production of such mediators, might actually induce the appearance of its receptor in nearby cells, and thereby enhance subsequent spread of the virus.

Mutational analyses of cloned Icam-1 DNA established that the binding sites for rhinovirus and Lfa-1 are located in domains 1 and 2, with critical contact points in both cases mapping to domain 1, the most membrane distal and accessible (Fig. 5.9). Although the binding sites for rhinovirus and Lfa-1 partially overlap, they are distinct. The three-dimensional atomic structure of the two N-terminal domains of a soluble form of Icam-1 has been determined (Fig. 5.9). This structure was then modeled into a reconstruction of the rhinovirus–Icam-1 complex made by cryo-electron microscopy. The residues of Icam-1 that contact the virus are located in the BC, DE, and FG loops of the first N-terminal domain (Fig. 5.9). Amino acids identified by mutational analyses as crucial for virus binding are located in these loops.

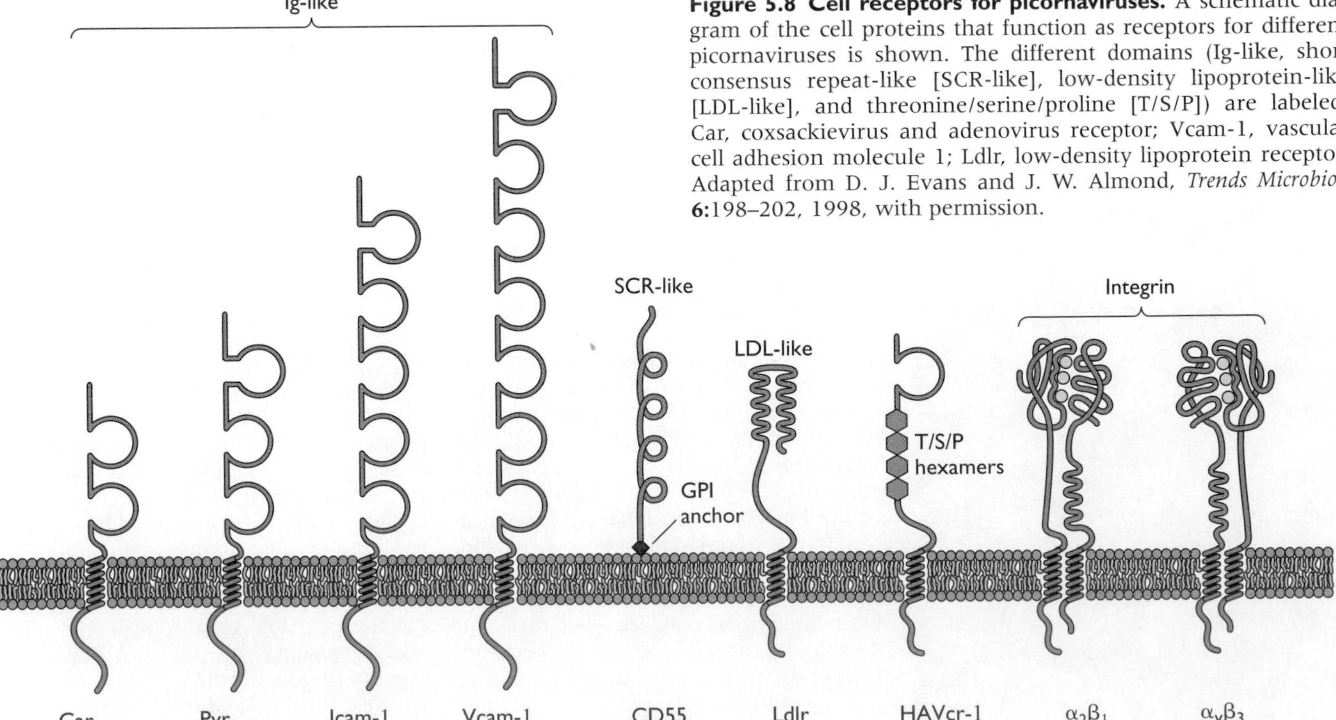

Figure 5.8 Cell receptors for picornaviruses. A schematic diagram of the cell proteins that function as receptors for different picornaviruses is shown. The different domains (Ig-like, short consensus repeat-like [SCR-like], low-density lipoprotein-like [LDL-like], and threonine/serine/proline [T/S/P]) are labeled. Car, coxsackievirus and adenovirus receptor; Vcam-1, vascular cell adhesion molecule 1; Ldlr, low-density lipoprotein receptor. Adapted from D. J. Evans and J. W. Almond, *Trends Microbiol.* **6:**198–202, 1998, with permission.

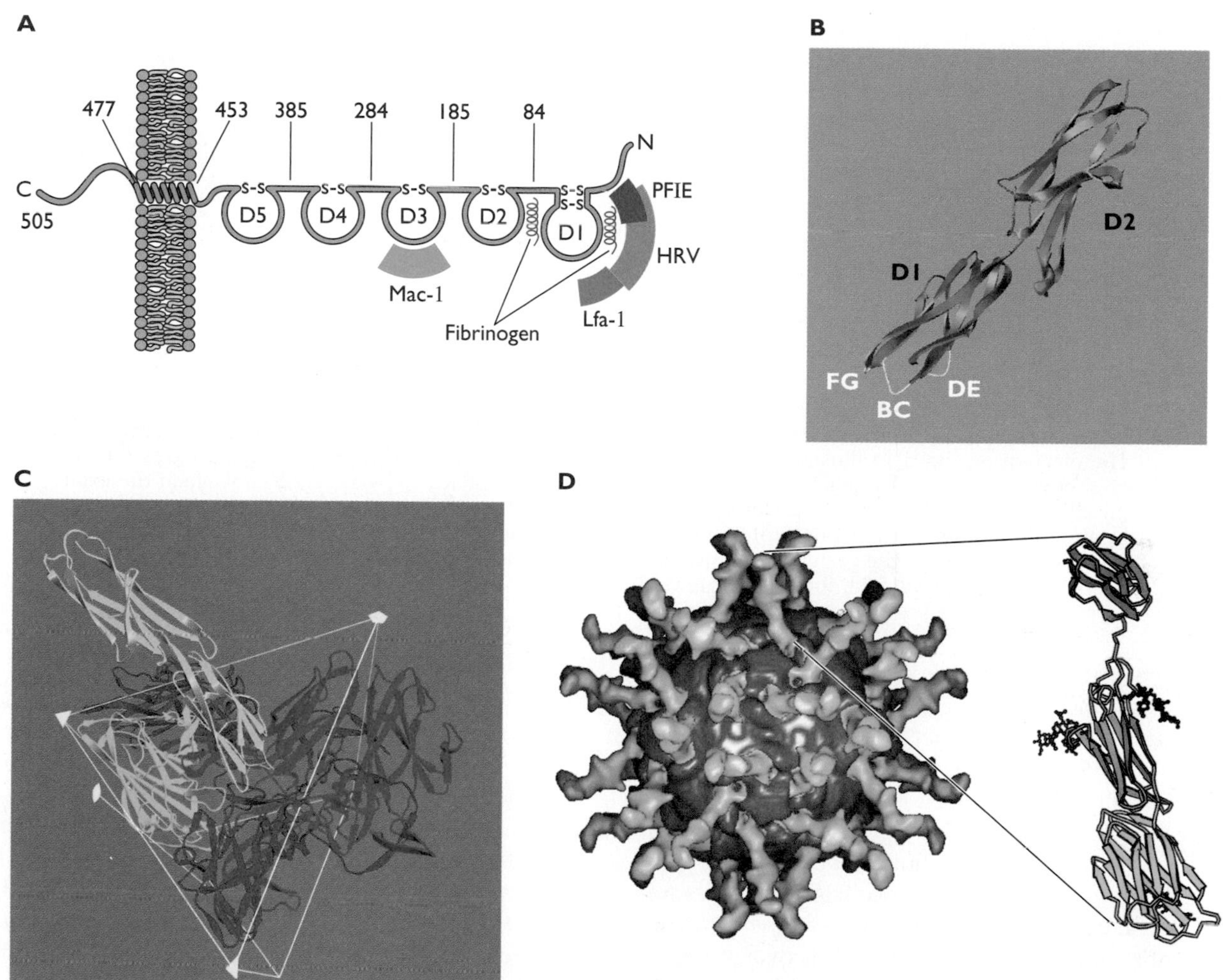

Figure 5.9 Structures of Icam-1 and Pvr. (A) Schematic diagram of Icam-1, showing the five Ig-like domains and binding sites for Lfa-1, Mac-1 (macrophage-1 antigen), rhinoviruses (HRV), fibrinogen, and malaria-infected erythrocytes (PFIE). (B) Model of domains 1 and 2 of Icam-1 derived from X-ray crystallographic data. The loops of domain 1 that contact rhinovirus are white and are labeled BC, DE, and FG. (C) Interaction of Icam-1 (yellow) in the canyon of rhinovirus 16. Viral proteins VP1, VP2, and VP3 are blue, green, and red. Two symmetry-related copies of VP1 and VP3 are included. Five-, three-, and twofold axes of symmetry are indicated; white lines indicate the boundary of an icosahedral symmetric unit. (D) Structure of poliovirus (red) bound to a soluble form of Pvr (green), derived by cryo-electron microscopy and image reconstruction. There are 60 Pvr-binding sites on capsid. At the right is a model of Pvr, with each Ig-like domain in a different color. The first Ig-like domain of Pvr (blue) binds in the canyon of the viral capsid. (A to C) Adapted from J. Bella et al., *Proc. Natl. Acad. Sci. USA* **95**:4140–4145, 1998, with permission. (C) Courtesy of Michael Rossmann, Purdue University. (D) Courtesy of David Belnap and Alasdair Steven, National Institutes of Health.

The results of competition assays had indicated that coxsackievirus A21 and human rhinovirus 14 bind to the same cellular receptor. It was subsequently found that coxsackieviruses A13, A18, and A21 could bind to mouse cells that synthesize human Icam-1. Icam-1 serves as a receptor for these coxsackieviruses and the major group rhinoviruses (Table 5.1).

The poliovirus receptor. The *Picornaviridae* genus *Enterovirus* includes human polioviruses, coxsackieviruses, echoviruses, and enteroviruses. These viruses are distinguished from rhinoviruses by their stability at acid pH and their ability to multiply in the gastrointestinal tract. However, they also can replicate in other tissues, such as nerve, heart, and muscle. The poliovirus receptor (Pvr) was

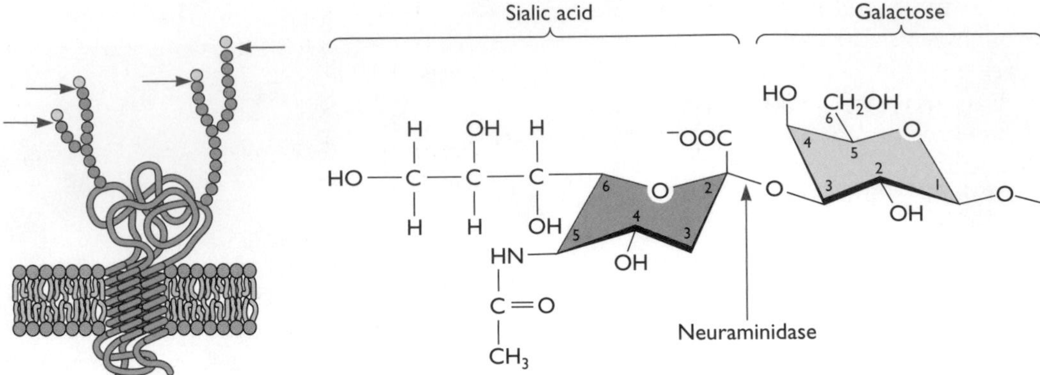

Figure 5.10 Sialic acid receptors for influenza viruses. The diagram on the left represents an integral membrane glycoprotein; the arrows point to terminal sialic acid units that are attachment sites for influenza virus. The diagram on the right illustrates the structure of a terminal sialic acid moiety that is recognized by the viral envelope protein hemagglutinin. Sialic acid is attached to galactose by an $\alpha(2,3)$ linkage in the example shown; certain HA subtypes preferentially bind to molecules with an $\alpha(2,6)$ linkage. The site of cleavage by the influenza virus envelope protein neuraminidase is indicated. The sialic acid shown is *N*-acetylneuraminic acid, which is the preferred receptor for influenza A and B viruses. These viruses do not bind to 9-*O*-acetyl-*N*-neuraminic acid, the receptor for influenza C viruses.

identified at about the same time as the rhinovirus receptor with the DNA transformation and cloning strategy outlined in Fig. 5.7. It was well known that mouse cells cannot be infected with poliovirus because they do not produce the viral receptor. Introduction of infectious poliovirus RNA into mouse tissue culture cells leads to poliovirus replication, indicating that there is no intracellular block to virus multiplication. Cloning of the human gene from such receptor-positive mouse cells established that the Pvr glycoprotein is also a member of the Ig superfamily, in this case possessing three Ig-like domains (Fig. 5.8). The first of these domains is essential and sufficient for receptor function, but efficient (wild-type) binding requires all three. A structure of Pvr bound to poliovirus has been determined (Fig. 5.9D). The structure confirms the results of mutational studies, which indicate that at least two β-strands and adjacent loops of Pvr contact poliovirus. A different region has been mapped for contacts with rhinovirus in domain 1 of Icam-1 (Fig. 5.9B). These findings demonstrate that interactions may differ even among structurally related viruses and receptors.

As mouse cells are permissive for poliovirus replication, and susceptibility appeared to be limited **only** by the absence of Pvr, a small-animal model for the disease was developed by producing transgenic mice that synthesize this receptor. Inoculation of Pvr transgenic mice with poliovirus by various routes produces paralysis, tremors, and death, as is observed in human poliomyelitis. These Pvr-synthesizing mice represented the first new animal model for the study of viral disease created by transgenic technology. As new receptors are identified, similar approaches should lead to animal models for a variety of viral diseases.

The observation that even some human cells that make Pvr are resistant to poliovirus infection prompted the search for a second cellular protein that might regulate receptor activity. One candidate was the lymphocyte homing receptor, CD44. This molecule was thought to be a coreceptor for poliovirus, because a monoclonal antibody against it blocks poliovirus binding to cells. This multifunctional 100-kDa membrane glycoprotein normally helps direct the migration of lymphocytes to the lymph nodes and regulates lymphocyte adhesion and other functions. However, CD44 is not a receptor for poliovirus, nor is it required for poliovirus infection of cells that produce Pvr. It is believed that Pvr and CD44 are associated in the plasma membrane and that anti-CD44 antibodies may block poliovirus attachment by sterically hindering the poliovirus-binding site on Pvr. These results emphasize that monoclonal antibodies that block virus binding do not necessarily identify the virus receptor on cells. Why certain cells that carry Pvr are resistant to poliovirus infection remains unknown.

The fact that Ig superfamily members are cell surface receptors for the major group rhinoviruses, poliovirus, and group B coxsackieviruses might lead one to believe that attachment to such proteins is a general property of the family *Picornaviridae*. However, this notion is quickly dispelled by inspection of Fig. 5.8 and Table 5.1. The receptor for the minor group of human rhinoviruses, Ldlr, is different, and various members of the other genera in the picornavirus family bind specifically to yet other types of cell surface molecule. For example, specific integrins are receptors for some picornaviruses. Most members of the integrin superfamily are known to recognize a specific amino acid sequence, RGD, common in a variety of extracellular matrix proteins that are the normal ligands of integrins. As we shall see in our discussion of viral attachment sites, this motif is also found in the capsid proteins of both picornaviruses and adenoviruses that attach to integrins.

The influenza virus receptor, sialic acid. The family *Orthomyxoviridae* comprises the three genera of influenza viruses, A, B, and C. Influenza A viruses are the best-studied members of this family. Influenza A virus binds to negatively charged, terminal sialic acid moieties present in oligosaccharide chains that are covalently attached to cell surface glycoproteins or glycolipids. The presence of sialic acid on most cell surfaces accounts for the ability of influenza virus to attach to many types of cell. The interaction of influenza virus with individual sialic acid moieties is of low affinity, but the opportunity for multiple interactions among the numerous hemagglutinin (HA) molecules on the surface of the virion and multiple sialic acid residues on cellular glycoproteins and glycolipids results in a high overall avidity of the virus for the cell surface. The surfaces of influenza viruses were shown in the early 1940s to contain an enzyme that, paradoxically, removes the receptors for attachment from the surface of cells. Later this enzyme was identified as the virus-encoded envelope protein neuraminidase, which cleaves the glycoside linkages of sialic acids (Fig. 5.10).

Receptors and Coreceptors Required for Infection

Human immunodeficiency virus type 1. Animal retroviruses have long been of interest because of their ability to cause a variety of serious diseases, especially cancers (caused by oncogenic retroviruses) and neurological disorders (caused by lentiviruses). When retroviruses infect a cell, they exhibit a phenomenon called **interference,** a block to subsequent infection by a related retrovirus that binds the same receptor. The mechanism of interference is not well understood but is presumed to be the result of occupation of receptor molecules by the envelope proteins produced in the infected cells. A study of receptor interference by 20 different mammalian retroviruses that can infect human cells in culture indicated that these viruses recognize no fewer than eight different receptors (Table 5.2).

Table 5.2 Receptor interference groups of mammalian retroviruses that can affect human cells[a]

Receptor group	Virus[b]	Receptor[c]	Type	Species from which isolated	Origin[d]
1	RD114	RDr or RDr2	C	Domestic cat	Endogenous
	BaEV		C	Baboon	Endogenous
	MPMV (SRV-3)		D	Rhesus monkey	Exogenous
2	MLV-A	GLVr2/rPit2	C	Wild mouse	Exogenous
3	MLV-X	XPr1	C	Laboratory mouse	Endogenous
4	FeLV-C	Flvcr	C	Domestic cat	Exogenous
5	FeLV-B	GLVr1/hPit1	C	Domestic cat	Exogenous
	SSAV		C	Woolly monkey	Exogenous
	GALV		C	Gibbon	Exogenous
6	BLV	Blvr	C (complex)	Cow	Exogenous
7	HTLV-1	Unknown	C (complex)	Human	Exogenous
	HTLV-2		C (complex)	Human	Exogenous
8	HIV-1	CD4	Lentivirus	Human	Exogenous
	HIV-2		Lentivirus	Human	Exogenous
	SIV_{mac}		Lentivirus	Rhesus monkey	Exogenous
	SIV_{smm}		Lentivirus	Sooty mangabey	Exogenous
	SIV_{agm}		Lentivirus	Green monkey	Exogenous

[a]Data from M. A. Sommerfelt and R. A. Weiss, *Virology* **176**:58–69, 1990, with permission. The family *Retroviridae* has been reorganized into seven genera, based on knowledge of molecular structure and probable evolutionary relatedness. The first four of these include known oncoretroviruses with simple genomes (Appendix A). They comprise mammalian type B, mammalian type C, and mammalian type D viruses that were originally distinguished by the type of morphology that characterized their virion cores and their sites of assembly; members of the fourth genus, avian type C, do not replicate in human cells and were not included in the survey. *Retroviridae* genera that comprise viruses with more complex genomes include bovine (bovine lentivirus) and human (human T-cell leukemia virus) retroviruses, which have cores with type C morphology; lentiviruses, which have cone-shaped cores and produce diseases with long or intermittent clinical latency; and spumaviruses (not included in this survey), which cause no disease. Receptors for members of five of these genera have been identified.

[b]RD114, feline rhabdomyosarcoma virus 114; BaEV, baboon endogenous virus; MPMV, Mason-Pfizer monkey virus; SRV-3, simian retrovirus 3; MLV-A, murine leukemia virus, amphotropic; MLV-X, murine leukemia virus, xenotropic; FeLV-C and -B, feline leukemia virus subgroups C and B; SSAV, simian sarcoma-associated virus; GALV, gibbon ape leukemia virus; BLV, bovine leukemia virus; HTLV-1 and -2, human T-lymphotropic virus types 1 and 2, HIV-1 and -2, human immunodeficiency virus types 1 and 2; SIV_{mac}, simian immunodeficiency virus, macaque; SIV_{smm}, simian immunodeficiency virus, sooty mangabey; SIV_{agm}, simian immunodeficiency virus, African green monkey.

[c]RDr, RD114 receptor; GLVr, gibbon ape leukemia virus receptor; rPit, rat phosphate (Pi) transport protein; XPr1, human glycoprotein with eight or nine transmembrane domains; Flvcr, feline leukemia virus cell receptor; hPit, human phosphate (Pi) transport protein; Blvr, bovine leukemia virus receptor.

[d]Endogenous origin refers to proviruses that are integrated into the germ line cells of the indicated species.

A B

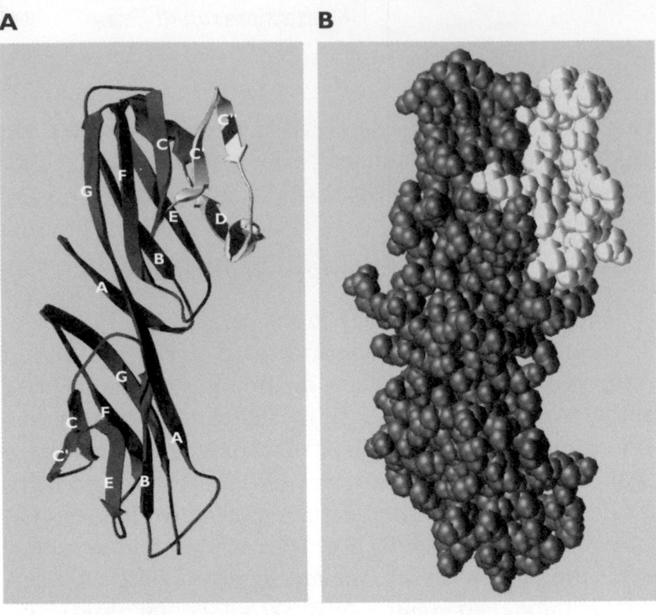

G
F
C'
C''
E
D
B
A
G
C
F
C'
E
B
A

C

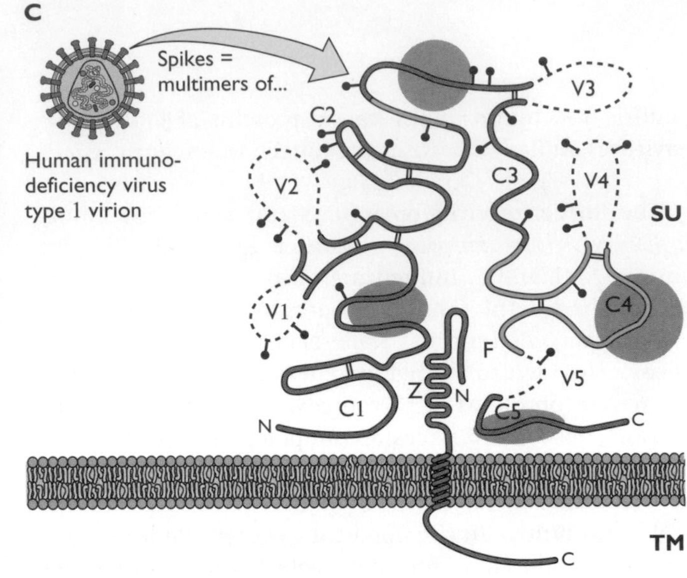

Spikes =
multimers of...
Human immuno-
deficiency virus
type 1 virion
V3
C2
V2
C3
V4
SU
V1
C4
F
V5
Z
N
C1
N
C5
C
TM
C

D

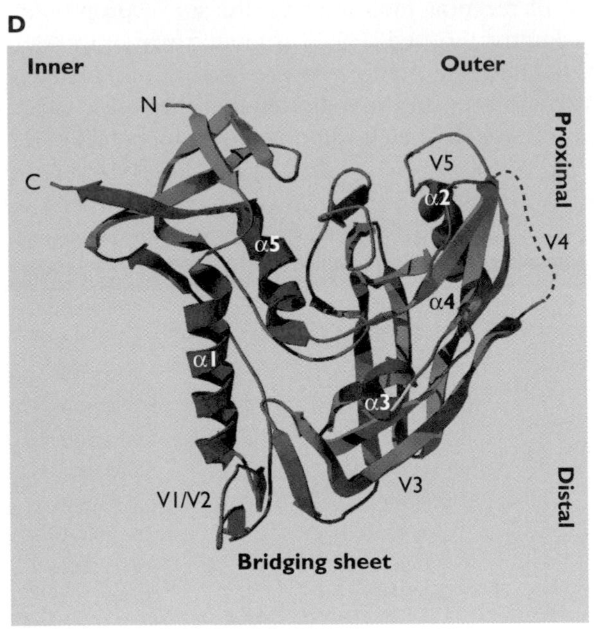

Inner
Outer
N
C
V5
α2
Proximal
V4
α5
α4
α1
α3
V1/V2
V3
Distal
Bridging sheet

E

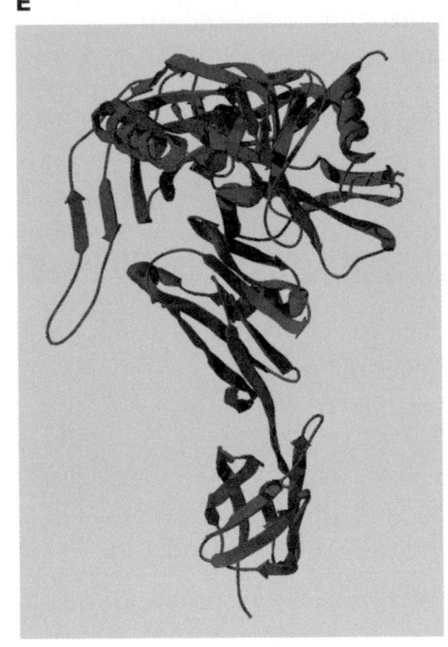

F

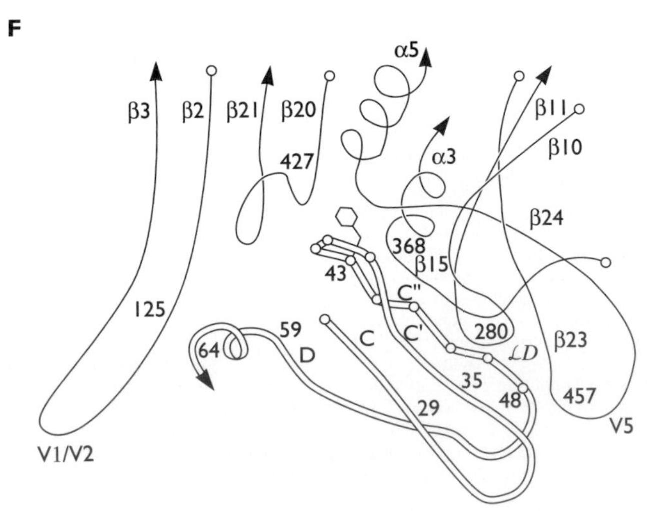

β3
β2
β21
β20
α5
β11
β10
427
α3
β24
368
β15
43
C''
125
59
C
C'
280
∠D
β23
64
D
35
48
457
29
V5
V1/V2

G

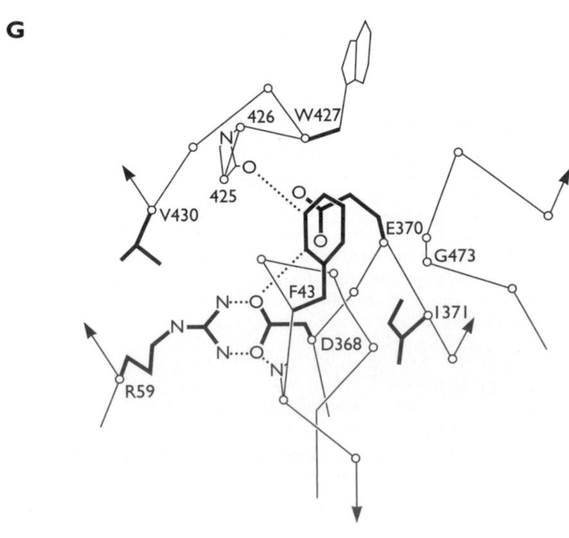

426
W427
N
O
425
V430
O
E370
G473
F43
N
O
I371
D368
R59
N

Although a seemingly large number, it is still considerably smaller than the number of retroviral species. Receptors for seven of the groups listed in Table 5.2 have been identified. At least one is the receptor for members of another viral family (e.g., CD4 and human herpesvirus 7 [Table 5.1]).

The worldwide acquired immunodeficiency syndrome (AIDS) epidemic has focused enormous attention on the lentivirus human immunodeficiency virus type 1 and its close relatives. Although this group (group 8 in Table 5.2) comprises the most recently discovered retroviruses, their interactions with cell surface receptors have been among the most intensively studied and currently are the best understood.

Early immunological studies showed that sensitivity to human immunodeficiency virus type 1 correlated with the presence of the T-helper cell surface marker CD4 and suggested that CD4 might be important for virus attachment. Subsequently, the observation that certain monoclonal antibodies against CD4 could block infection and cell fusion (syncytium formation) provided strong evidence that this integral membrane glycoprotein was the major cell receptor for the virus. The demonstration that nonsusceptible human cells engineered to produce CD4 became susceptible to infection confirmed the importance of this receptor for human immunodeficiency virus type 1. The CD4 protein, a 55-kDa rodlike molecule, is a member of the Ig superfamily that contains four Ig-like domains. A variety of techniques have been used to identify the site of interaction with human immunodeficiency virus type 1, including site-directed mutagenesis, monoclonal antibody studies, and X-ray crystallographic studies of a complex of CD4 and the viral attachment protein SU (Fig. 5.11). Residues of CD4 that contact SU are located between amino acids 25 and 64.

The interaction site for SU in domain 1 of CD4 is in a region analogous to the site in Pvr that binds to poliovirus. Because two viruses with entirely different virion architectures bind to analogous surfaces of these Ig-like domains, it can be hypothesized that some feature of this region is especially advantageous for virus attachment.

One of the first strategies to be considered for the treatment of AIDS was the use of soluble CD4 protein, which lacks the membrane-spanning domain, to inhibit viral infection. The rationale for such treatment was that soluble CD4 should bind virus particles and block their attachment to CD4 on host cell surfaces. Although inhibition of viral infectivity could be demonstrated in cell culture experiments, clinical trials gave disappointing results. This failure can be attributed in part to the fact that each viral envelope includes many copies (~30) of the structure that binds CD4 such that relatively high serum concentrations of soluble CD4 are required in order to block all of them. This problem was further compounded by the short half-life of soluble CD4 in the blood. Furthermore, human immunodeficiency virus can also be spread by cell fusion, a process that is not readily blocked by circulating, soluble CD4.

Figure 5.11 Interaction of human immunodeficiency virus type 1 SU with its cell receptor, CD4. (A) Ribbon model of the backbone carbons of domains 1 and 2 (residues 1 to 182), derived from X-ray crystallography. The β-strands are labeled with letters; strand directions are indicated by arrowheads, and connecting loops are shown as strings. (B) A space-filling model of CD4 shows the positions of residues (yellow) that interact with human immunodeficiency virus type 1 during attachment, as revealed by mutagenesis. (C) Schematic diagram of a monomer of human immunodeficiency virus type 1 envelope glycoprotein. Conserved (C) and variable (V) regions of SU are indicated. Lines show the position of disulfide bridges, and lollipops show the positions of glycosylation. F is the fusogenic peptide, and Z is a leucine zipper-like domain in the transmembrane protein, TM. Regions in the C domains of SU that are known to contact the CD4 receptor are highlighted in gray. (D) Ribbon diagram of SU, derived from X-ray crystallographic data. The structure was determined for a core of SU lacking 52 and 19 amino acids from the N and C termini, respectively, and containing Gly-Ala-Gly in place of 67 V1/V2 loop residues and 32 V3 loop residues. This modified SU binds CD4 with affinity comparable to that of the native protein. α-Helices are red, β-strands are magenta, and β-strand 15 is yellow. The last forms an antiparallel β-sheet with strand C″of CD4. Loops are gray, except for loop V4 (dashed line), which is disordered. Selected features of the molecule are labeled. (E) Ribbon diagram depicting the SU-CD4 interaction, derived from X-ray crystallographic data. SU is red, and CD4 is brown. (F) Schematic diagram of the interface between SU and CD4. Solid black lines represent the six segments of SU that interact with CD4, which is shown by double lines. Secondary structure elements are labeled, and some residues of SU that interact with CD4 are indicated. The directions of the main chains are indicated by arrows. An important residue of CD4 identified by mutagenesis, Phe43, is shown penetrating into SU. Phe43 makes 23% of the interatomic contacts between CD4 and SU. This residue is at the center of the interface and appears to stabilize the entire complex. (G) Contacts in the area of CD4 amino acids Phe43 and Arg59. SU residues in direct contact with these two CD4 amino acids are labeled. Electrostatic interactions are shown as dashed lines. There are hydrophobic interactions between Phe43 of CD4 and Trp427, Glu370, Gly473, and Ile371 of SU and between Arg59 of CD4 and Val430 of SU. Adapted from J. Wang et al., *Nature* **348:**411–418, 1990 (A and B); B. N. Fields et al. (ed.), *Fields Virology,* 3rd ed. (Lippincott-Raven Publishers, Philadelphia, Pa., 1996) (C); and P. D. Kwong et al., *Nature* **393:**648–659, 1998 (D to G), with permission.

Although the presence of CD4 is sufficient to render human cells susceptible to virus infection, synthesis of CD4 on the surface of mouse cells allows the virus to attach but not to enter. This property suggested that one or more human coreceptors are required for steps in infection subsequent to attachment. The experimental strategies used to isolate these coreceptors, their identification as chemokine receptors, and their role in promoting cell fusion are described below. Certain isolates of human immunodeficiency virus type 2 and simian immunodeficiency virus enter cells independently of CD4 via chemokine receptors. Given the large number of members of the chemokine receptor family and the ability of human immunodeficiency virus type 1 to interact with these proteins, it is possible that CD4-independent, chemokine receptor-mediated infection may occur with some frequency. If so, then the host range of the virus would be broader than previously suspected, thereby changing our view of the pathogenesis of AIDS.

It is noteworthy that both CD4 and human immunodeficiency virus type 1 coreceptor molecules are coupled via their cytoplasmic domains to intracellular signaling pathways. For example, the normal role of CD4 is to participate with the T-cell receptor in the recognition of major histocompatibility class II molecules during the presentation of antigens (Fig. 15.20) and to signal such interactions by contact with a protein kinase ($p56^{lck}$) attached to the inner leaflet of the plasma membrane. The chemokine receptors also signal interaction with their ligand, affecting cellular gene expression. One might speculate that contact with viral envelope proteins could activate these signaling pathways and produce cellular responses that enhance virus propagation and/or affect pathogenesis. Indeed, interactions between SU and CD4 have been implicated in virus-induced cell killing. One example is the binding of SU to human $CD4^+$ T cells, followed by signaling through the T-cell receptor and induction of apoptosis. It has also been reported that interactions between SU and chemokine receptors on neuronal cells induce apoptosis. The destruction of cytotoxic T cells by macrophages has also been attributed to such interactions. Such effects could help to explain the neurological disorders and loss of cytotoxic T cells that are symptoms of AIDS.

Picornaviruses. Decay-accelerating protein (CD55), a member of the complement cascade, has been identified as the cell receptor for many enteroviruses (Fig. 5.8; Table 5.1). For most of these viruses, binding to decay-accelerating protein is not sufficient for infection, which requires the presence of a coreceptor. For example, coxsackievirus A21 can bind to cell surface decay-accelerating protein, but this interaction does not lead to infection unless Icam-1 is also present. Virus binding to Icam-1 alone is sufficient for infection. However, when both Icam-1 and decay-accelerating protein are present in the same cell, complete blockade of virus binding and infection requires the use of antibodies against both receptors. In such cells, decay-accelerating protein and Icam-1 form a receptor-coreceptor complex required for infection. It is believed that virions first bind to decay-accelerating protein and then interact with Icam-1 to allow cell entry. A similar process, with different receptors and coreceptors, is likely to occur during infection with other enteroviruses (Table 5.1).

Depending on the cell type, decay-accelerating protein, Icam-1, or a complex of both molecules may serve as a receptor for coxsackievirus A21. When Icam-1 is not present, decay-accelerating protein can mediate infection with coxsackievirus A21 if the receptor is first bound by antibodies. In host cells, binding of a cellular ligand may cause cross-linking of decay-accelerating protein. This interaction might alter the structure of the receptor, exposing a binding site for the virus.

Adenoviruses. The results of competition experiments (see "Experimental Approaches" above) indicated that members of two different virus families, group B coxsackieviruses and adenoviruses, share a cell receptor. This receptor has been identified as a 46-kDa member of the Ig superfamily and is called Car (coxsackievirus and adenovirus receptor) (Table 5.1). Adenoviruses of subgroups A, C, D, E, and F bind to Car. This protein is located on the basolateral surface of cells lining the respiratory tract, where it is inaccessible to virus. Adenovirus infections probably occur when transient breaks in the epithelium allow virus particles to reach sites of Car expression.

An alternative receptor has been identified for adenovirus types 2 and 5, members of subgroup C. Adenovirus type 2 can bind to major histocompatibility class I molecules on the surface of human epithelial and B lymphoblastoid cells, while adenovirus type 5 binds to the integrin $\alpha_M\beta_2$.

Binding to Car, major histocompatibility class I molecules, or $\alpha_M\beta_2$ is not sufficient for infection by most adenoviruses. Interaction with a coreceptor, the α_v integrin $\alpha_v\beta_3$ or $\alpha_v\beta_5$, is required for uptake of the capsid into the cell by receptor-mediated endocytosis. An exception is adenovirus type 9, which apparently can infect hematopoietic cells after binding directly to α_v integrins.

Alphaherpesviruses. The alphaherpesvirus subfamily of the *Herpesviridae* includes herpes simplex virus types 1 and 2, pseudorabies virus, and bovine herpesvirus. Initial contact of these viruses with the cell surface is made by low-affinity binding to glycosaminoglycans (preferentially heparan sulfate), abundant components of the extracellular matrix (Fig. 5.2). This interaction concentrates virus particles near the cell surface and facilitates subsequent attachment to a membrane coreceptor, which is required for entry into the cell. Members of at least two different protein families serve as entry coreceptors for alphaherpesviruses (Table 5.1). One of these families comprises the poliovirus receptor Pvr and related proteins, yet another example of receptors shared by different viruses. When members of these two protein families are not present, 3-O-sulfated heparan sulfate can serve as an entry receptor for alphaherpesviruses.

Alternative Receptors

Some examples of the use of alternative cell surface molecules as receptors for the same virus have already been discussed. Two additional examples illustrate how receptor usage depends on the nature of the virus isolate or the cell line. Infection with foot-and-mouth disease virus type A12 requires the RGD-binding integrin $\alpha_v\beta_3$ (Table 5.1). However, the receptor for the O strain of foot-and-mouth disease virus, which has been extensively passaged in cell culture, is not integrin $\alpha_v\beta_3$ but cell surface heparan sulfate. High concentrations of heparin (a derivative of heparan sulfate proteoglycan [Fig. 5.2]) adsorb to virus particles and block their attachment to cells. On the other hand, the type A12 strain cannot infect cells that lack integrin $\alpha_v\beta_3$, even if heparan sulfate is present. In a similar way, adaptation of Sindbis virus to cultured cells has led to the selection of variants that bind heparan sulfate. Laboratory adaptation of viruses to cells that express low levels of the virus receptor can clearly lead to the use of alternative receptors.

How Virions Attach to Receptors

Animal viruses have multiple receptor binding sites on their surfaces. Of necessity, one or more of the capsid proteins of nonenveloped viruses specifically interacts with the cell receptor. Receptor binding sites for enveloped viruses are surface glycoproteins that have been incorporated into their cell-derived membranes. Some receptors are "hooks" that hold virus particles to cells, while uncoating is triggered by low pH or proteases. Other receptors are "unzippers"; their binding produces dramatic conformational changes in the structures of virions that facilitate subsequent steps of infection. Although the details vary among viruses, most virus-receptor interactions follow one of several mechanisms as illustrated by the best-studied examples described below.

Nonenveloped Viruses

Attachment via surface features: canyons and loops. The RNA genomes of picornaviruses are protected by capsids made up of four virus-encoded proteins, VP1, VP2, VP3, and VP4, arranged with icosahedral symmetry (Fig. 4.9A). Three-dimensional structures have been determined by X-ray crystallography for at least one member of all five picornavirus genera. The picornavirus capsid is built from 60 subunits arranged as 12 pentamers (Fig. 4.9A). Each subunit contains the four capsid proteins in an identical arrangement, with portions of the first three exposed on the surface. Although similar in arrangement, the surface architecture of the three exposed proteins varies among the family members, a property that accounts for the different serotypes and the different modes of interaction that can take place with cell receptors. For example, the capsids of rhinoviruses and some enteroviruses like poliovirus have deep canyons surrounding the 12 fivefold axes of symmetry in the pentamers (Fig. 5.12A), whereas cardioviruses and aphthoviruses lack this feature.

The canyons in the capsids of some rhinoviruses and enteroviruses are the sites of interaction with cell surface receptors (Fig. 5.12A). Consistent with this view, amino acids that line the canyons are more conserved than any others on the viral surface, and their substitution can alter the affinity of binding to cells. A human rhinovirus bound to a receptor fragment containing Icam-1 domains 1 and 2 has been visualized in reconstructed images from cryo-electron microscopy. The results indicate that Icam-1 binds to the central portion of the canyon in an orientation perpendicular to the surface of the virion. The BC, DE, and FG loops of the first Icam-1 domain penetrate deep into the canyon, while the CD loop lies against one wall (Fig. 5.9C). There are extensive charge-charge interactions between rhinovirus and Icam-1. Amino acids of the rhinovirus capsid that are crucial for receptor binding (previously identified by mutagenesis) all fall within the footprint of the receptor. Comparison of the amino acid sequences of major and minor group rhinoviruses (~10 serotypes) in the areas that contact Icam-1 do not reveal why the latter bind a different receptor (Table 5.1). In fact, it appears that both major and minor group rhinoviruses should be able to bind Icam-1.

The canyons of picornaviruses were first believed to be too deep and narrow to permit entry of antibody molecules with adjacent Ig domains. It was hypothesized that this physical barrier would allow amino acids crucial for receptor interactions to be hidden from the immune system. However, X-ray crystallographic analyses of a specific rhinovirus-antibody complex have shown that the antibody penetrates deep into the canyon, as does Icam-1 (Fig. 5.12A): the shape of the canyon is not likely to play a role in immune escape.

Although canyons are present in the capsid of rhinovirus type 2, they are not the binding sites for the cellular receptor. Rhinovirus type 2 is one of 10 viral serotypes that bind low-density lipoprotein receptor. The binding site on the capsid for this receptor is located on the star-shaped plateau at the fivefold axis of symmetry (Fig.

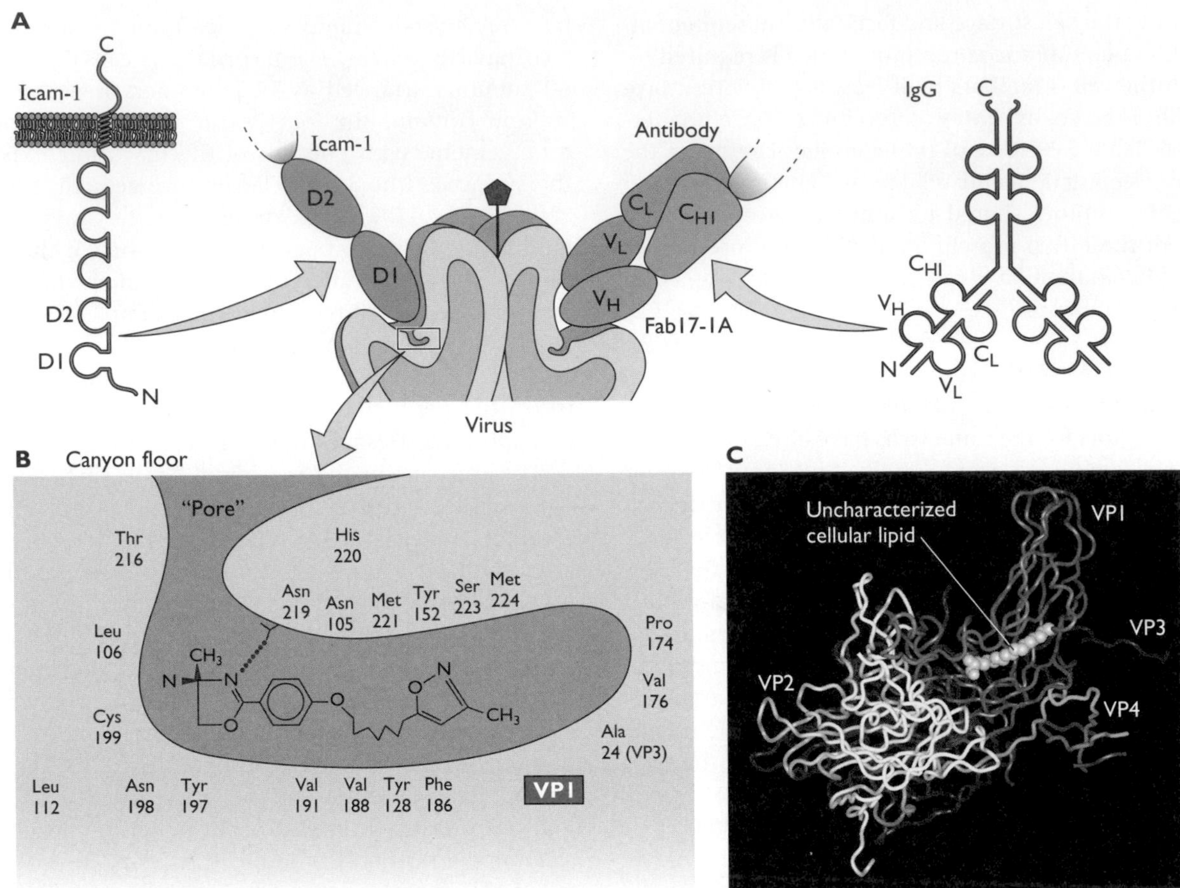

Figure 5.12 Receptor, antibody, and drug binding to the picornavirus capsid. (A) Schematic diagram of the canyon in the human rhinovirus capsid. The domain structure of the cell receptor Icam-1 is illustrated at the left, and the model in the center shows the tip of domain 1 dipping into the canyon. An antibody molecule is illustrated on the right. The Fab section contacts a good deal of the canyon, but not residues at the deepest regions. Antibodies that bind to the virus in this manner neutralize viral infectivity by blocking entry of receptor into the canyon. (B) Location of a WIN compound in a hydrophobic pocket below the canyon floor. The drug displaces the lipid and prevents the uncoating of picornaviruses. The attachment of certain rhinovirus serotypes is also blocked by the presence of a WIN compound. Adapted from T. Smith et al., *Nature* **383:**350–354, 1996 (A), and J. Badger et al., *Proc. Natl. Acad. Sci. USA* **85:**3304–3308, 1988 (B), with permission. (C) Location of lipid, possibly sphingosine, in the capsid of poliovirus type 1. Shown is a protomer consisting of one copy of VP1 (blue), VP2 (yellow), VP3 (red), and VP4 (green). The lipid, shown as gray spheres, is bound in the hydrophobic tunnel beneath the canyon floor. Crystallographic analyses of picornaviruses show that most contain an uncharacterized lipid in the canyon pocket. This lipid can be displaced by antiviral WIN compounds.

5.12A). The difference in the receptor-binding site among major and minor group rhinoviruses may explain differences in mechanisms of virus entry. Binding of Icam-1 to the capsids of major group rhinoviruses is sufficient to trigger uncoating. In contrast, uncoating of minor group rhinoviruses requires internalization into acidic compartments of the cell (see "Mechanisms of Entry" below).

For picornaviruses with capsids that do not have prominent canyons, including coxsackievirus A and foot-and-mouth disease virus, attachment is to VP1 surface loops that include RGD motifs recognized by their integrin receptors. Alteration of the RGD sequence in VP1 of foot-and-mouth disease virus blocks virus binding. However, changes in the same sequence of coxsackievirus A9 do not abolish binding of virus particles to cells, suggesting that viruses can bind to an alternative receptor.

Attachment via protruding fibers. The nonenveloped DNA-containing adenoviruses are much larger than picornaviruses, and their icosahedral capsids are more complex, comprising at least 10 different polypeptides. By electron microscopy adenovirus particles can be seen to have fibers protruding from each pentamer (Fig. 4.12A). The fibers are composed of homotrimers of the adenovirus fiber protein and are anchored in the pentameric penton base protein; both proteins have roles to play in virus attachment and uptake.

For many adenovirus serotypes, attachment via the fibers is necessary but not sufficient for infection. A region comprising the N-terminal 40 amino acids of each subunit of the fiber protein is bound noncovalently to the penton base. The central shaft region is composed of repeating motifs of approximately 15 amino acids; the length of the shaft in different serotypes is determined by the number of these repeats. The three constituent shaft regions appear to form a rigid triple-helical structure in the trimeric fiber. The C-terminal 180 amino acids of each subunit interact to form a terminal knob. Genetic analyses and competition experiments indicate that determinants for the initial, specific attachment to host cell receptors reside in this knob. The crystal structure of this receptor-binding domain bound to Car receptor reveals that surface loops of the knob contact one face of Car (Fig. 5.13).

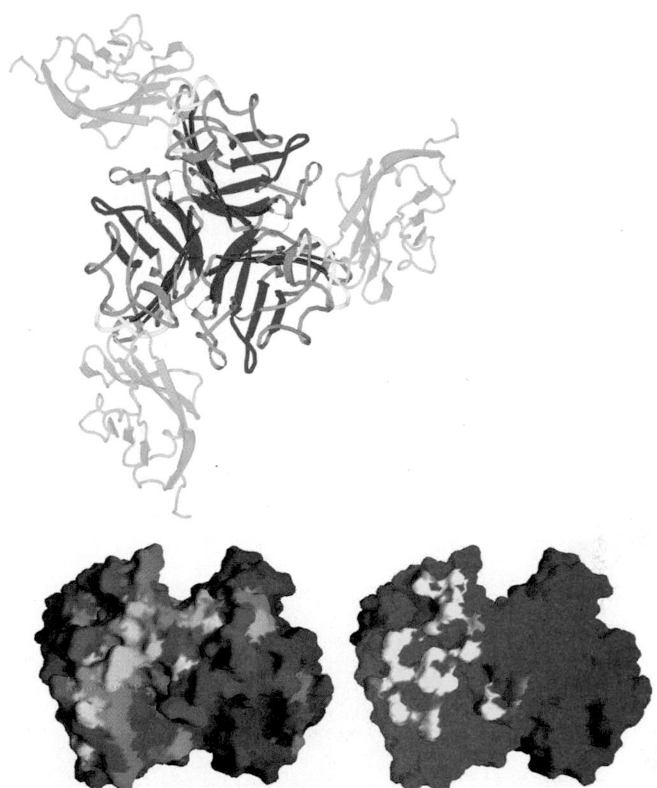

Figure 5.13 Structure of the adenovirus 12 knob complexed with the Car receptor. (Top) Ribbon diagram of the knob-Car complex as viewed down the axis of the viral fiber. The trimeric knob is in the center. The AB loop of the knob protein, which contacts Car, is in yellow. The first Ig-like domains of three Car molecules bound to knob are colored blue. (Bottom) Surface models of the interface between the knob and Car domain 1. Both models depict two knob monomers and are viewed looking down at the interface with Car. (Left) Conservation of amino acids among all known adenovirus knob protein sequences is represented by a color scale from white (conserved amino acids) to red (nonconserved amino acids). The white strip of conserved amino acids is covered when Car is bound. (Right) Amino acids are colored according to whether they contact Car (yellow) or not (red). From M. C. Bewley et al., *Science* **286:**1579–1583, 1999, with permission.

The observation that soluble fiber protein can bind to cells, but is not endocytosed, suggested that adenoviruses require a coreceptor for uptake into the cell. A second interaction between the penton base and cellular integrins has been identified in what should now be recognized as a recurrent theme: as in certain members of the picornaviruses (echoviruses, group A coxsackieviruses, and foot-and-mouth disease virus) (Table 5.1), attachment is mediated by RGD sequences that mimic the normal ligands of cell surface integrins. The same RGD sequences in each of the five subunits of the adenovirus penton base bind to the $\alpha_v\beta_3$ and $\alpha_v\beta_5$ vitronectin-binding integrins and perhaps other integrin family members. This interaction is critical, because adenovirus infection is blocked by competition with RGD-containing peptides or by substitutions in the integrin-binding region of the penton base protein. However, with the exception of adenovirus type 9 (see "Adenoviruses" above), while necessary, in most cases the integrin interaction is not sufficient to support infection in the absence of specific attachment via the fiber protein. Attachment and uptake is therefore hypothesized to be a two-step process, with the presumed role of the second receptor being to trigger receptor-mediated endocytosis of virus captured by interaction with Car.

Enveloped Viruses Attach to Receptors by Means of Viral Membrane Glycoproteins

As noted above, the lipid membranes of enveloped viruses originate from the cells they infect. The process of virion assembly includes insertion into membranes of specific viral proteins that carry membrane-spanning domains analogous to those of cellular integral membrane proteins. Attachment sites (i.e., viral ligands) on one or more of these envelope proteins bind to specific cell receptors. The two best-studied examples of enveloped virus attachment and its consequences are provided by the interactions of influenza A virus and the retrovirus human immunodeficiency virus type 1 with their cell receptors.

The HA of influenza virus. Influenza virus HA is the viral glycoprotein that binds to the cell receptor sialic acid. The HA monomer is synthesized as a precursor that is glycosylated and subsequently cleaved to form HA1 and HA2 subunits. Each HA monomer consists of a long, helical stalk anchored in the membrane by HA2 and topped by a large HA1 globule, which includes the sialic acid-binding pocket (Fig. 5.14). While attachment of all influenza A virus strains requires sialic acid, strains vary in their affinities for different sialyloligosaccharides. For example, avian virus strains bind preferentially to sialic acids attached to galactose with an α(2,3) linkage (Fig. 5.10), while human virus strains are preferentially bound by sialic acids attached to galactose with an α(2,6) linkage. Amino acid differences at positions 226 and 228 of certain HA subtypes account for these differences in sugar linkage specificity. Changes of these residues, which are in the receptor-binding pocket in HA (Fig. 5.14), influence its contact with sialic acid. These amino acids can also influence the host range of the virus. Avian strains of influenza virus replicate in the intestine, while human strains do not. A recombinant duck virus, containing a particular human HA gene (H3), does not replicate in duck intestine. However, introduction of two mutations into the H3 HA gene at codons 226 and 228 allows the recombinant virus to replicate in duck intestine.

The envelope glycoprotein of human immunodeficiency virus type 1. When examined by electron microscopy, the envelopes of human immunodeficiency virus type 1 and other retroviruses appear to be studded with "spikes" (Fig. 5.11). These structures are composed of trimers of the single viral envelope glycoprotein. The

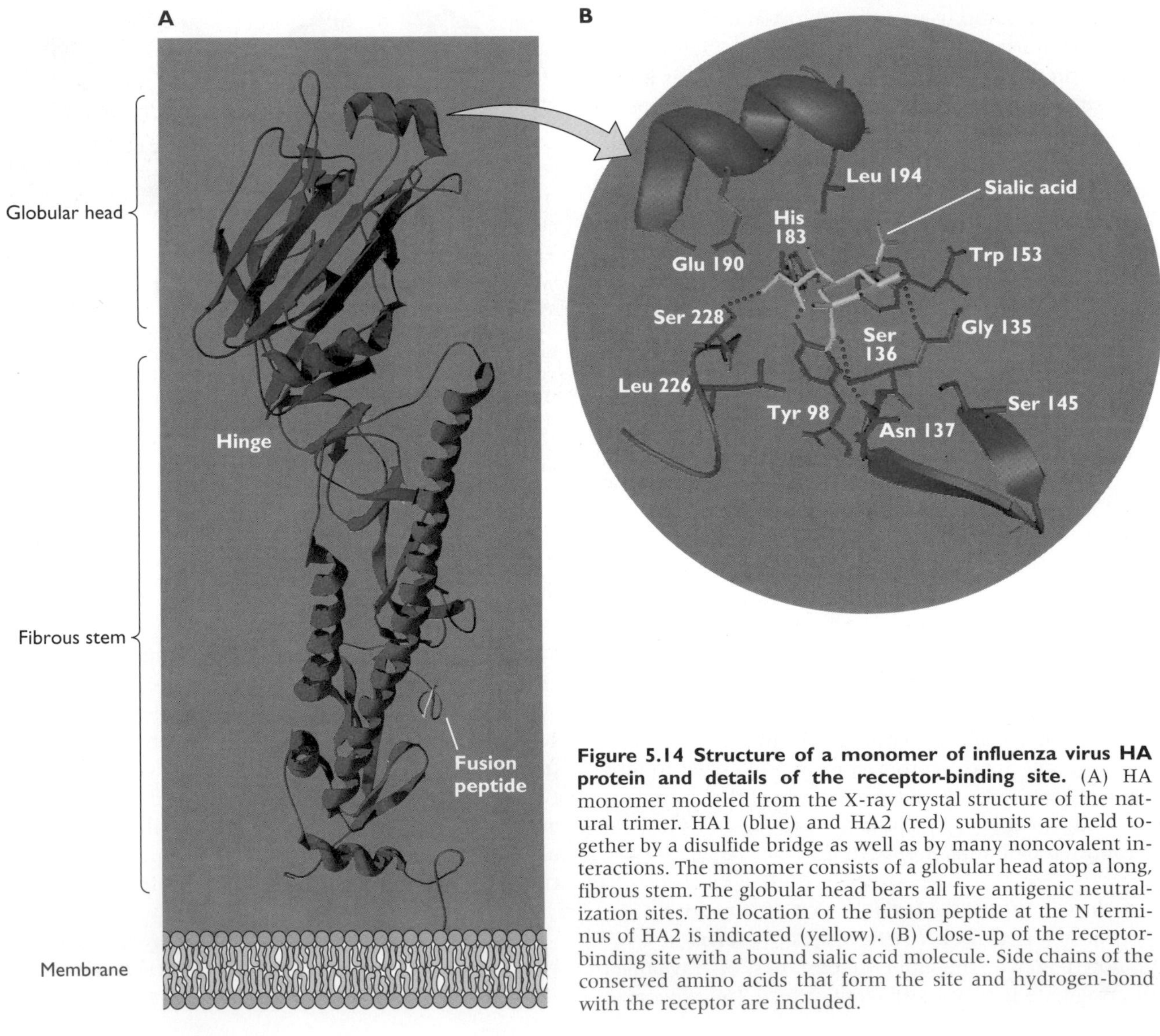

Figure 5.14 Structure of a monomer of influenza virus HA protein and details of the receptor-binding site. (A) HA monomer modeled from the X-ray crystal structure of the natural trimer. HA1 (blue) and HA2 (red) subunits are held together by a disulfide bridge as well as by many noncovalent interactions. The monomer consists of a globular head atop a long, fibrous stem. The globular head bears all five antigenic neutralization sites. The location of the fusion peptide at the N terminus of HA2 is indicated (yellow). (B) Close-up of the receptor-binding site with a bound sialic acid molecule. Side chains of the conserved amino acids that form the site and hydrogen-bond with the receptor are included.

spikes bind the cell receptor CD4 (Fig. 5.11). The monomers of the spike protein are synthesized as heavily glycosylated precursors that are cleaved by a cellular protease to form SU and TM. The latter is anchored in the envelope by a single membrane-spanning domain and remains bound to SU by numerous noncovalent bonds. The extensive antigenic variability among different isolates of human immunodeficiency virus type 1 maps to five discrete regions called V1 to V5. They are separated by relatively conserved regions, C1 to C5. Sequences important for binding to the major receptor, CD4, are included in regions C2 to C4.

The atomic structure of a complex of human immunodeficiency virus type 1 SU, a two-domain fragment of CD4, and a neutralizing antibody against SU has been determined by X-ray crystallography (Fig. 5.11). The polypeptide of SU is folded into an inner and an outer domain linked by an antiparallel four-stranded "bridging sheet." A depression at the interface of the outer and inner domains and the bridging sheet forms the binding site for CD4. Interatomic contacts are made between 22 amino acids of CD4 and 26 amino acids of SU. Although the structure of SU in the absence of CD4 is not known, certain features of the complex indicate that receptor binding induces a conformational change in SU. For example, CD4 amino acid F43 is located in an SU cavity that is lined with hydrophobic amino acids, but it is not likely that these residues would be exposed in the absence of CD4. The conformational changes induced in SU by CD4 binding expose binding sites on SU for the chemokine receptors, the interaction required for fusion of viral and cell membranes (see "Virus-Cell Membrane Fusion That Requires a Cellular Coreceptor" below).

Uptake of Macromolecules by Cells

After attaching to a cellular receptor, a virus must enter the host cell so that genome replication can begin. Viruses enter cells by the same pathways by which cells take up macromolecules. The plasma membrane, the limiting membrane of the cell, permits nutrient molecules to enter and waste molecules to leave, thereby ensuring an appropriate internal environment. Water, gases, and small hydrophobic molecules such as ethanol can freely traverse the lipid bilayer, but most metabolites, including glucose, ATP, nucleosides, amino acids and proteins, and certain ions (Ca^{2+}, H^+, K^+, and Na^+), cannot diffuse through the membrane. These essential components enter the cell by specific transport processes. Integral membrane proteins known as permeases are responsible for the transport of ions, sugars, and amino acids, while proteins and large particles are taken into the cell by phagocytosis or endocytosis. The former process (Fig. 5.15) is nonspecific, which means that any particle or molecule can be taken into the cell. Specific molecules are selectively taken into cells from the extracellular fluid by **receptor-mediated endocytosis** (Fig. 5.15 and 5.16), a mechanism of entry of some

Figure 5.15 Mechanisms for the uptake of macromolecules from extracellular fluid. During phagocytosis, large particles such as bacteria or cell fragments that come in contact with the cell surface are engulfed by extensions of the plasma membrane. The extensions fuse to form a 1- to 2-mm-diameter closed vesicle, or phagosome, containing the particle. Phagosomes ultimately fuse with lysosomes, resulting in degradation of the material within the vesicle. Macrophages use phagocytosis to ingest bacteria and destroy them. Endocytosis comprises the invagination and pinching off of small regions of the plasma membrane, resulting in the nonspecific internalization of molecules (pinocytosis or fluid-phase endocytosis) or the specific uptake of molecules bound to cell surface receptors (receptor-mediated endocytosis). Adapted from J. Darnell et al., *Molecular Cell Biology* (Scientific American Books, New York, N.Y., 1986), with permission.

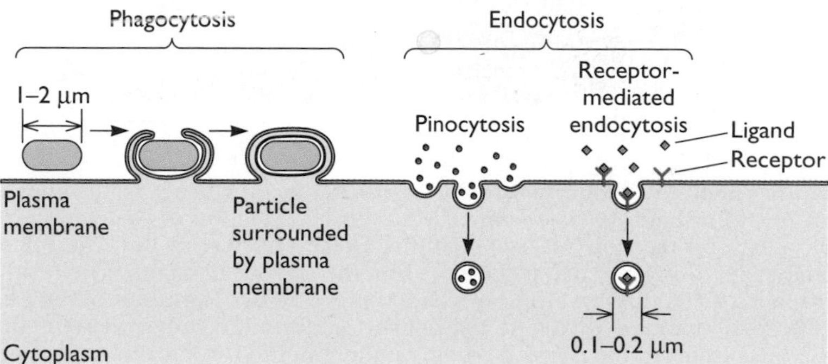

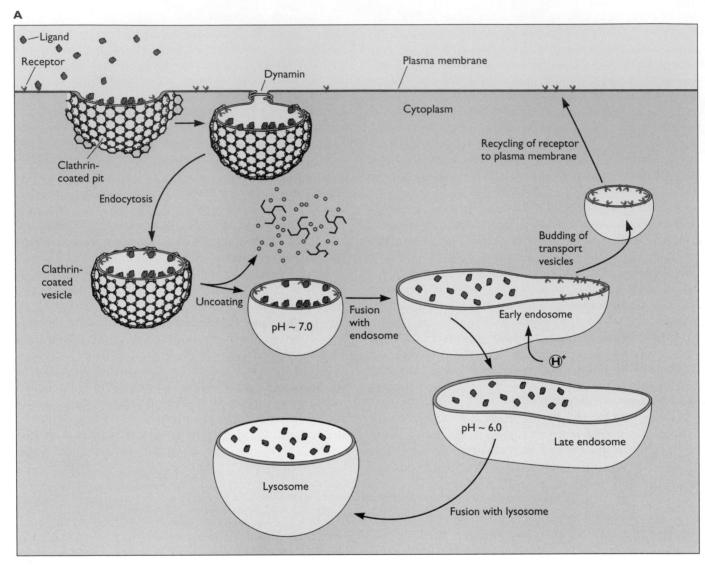

A

Ligand
Receptor
Dynamin
Plasma membrane
Cytoplasm
Clathrin-coated pit
Recycling of receptor to plasma membrane
Endocytosis
Budding of transport vesicles
Clathrin-coated vesicle
Uncoating
pH ~ 7.0
Fusion with endosome
Early endosome
H+
pH ~ 6.0
Late endosome
Lysosome
Fusion with lysosome

B

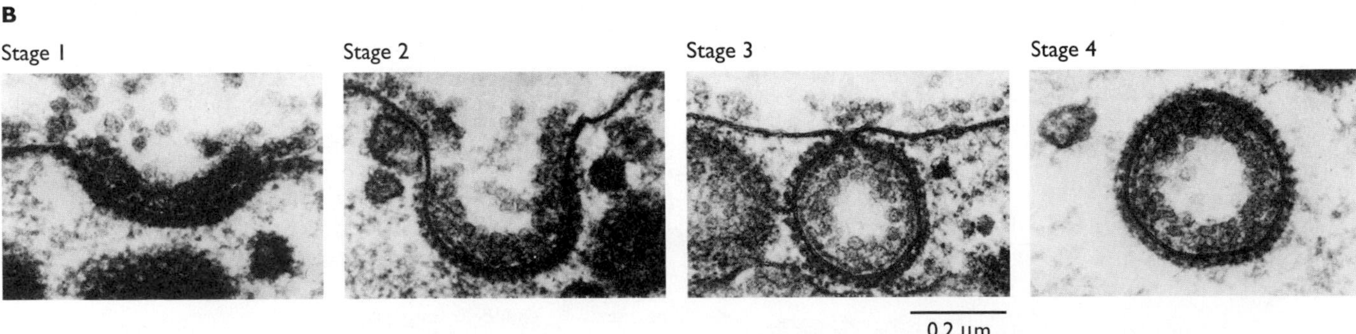

Figure 5.16 Receptor-mediated endocytosis. (A) This process begins when a ligand binds a specific cell surface receptor. The ligand-receptor complex diffuses into an invagination of the plasma membrane coated with the protein clathrin on the cytosolic side (clathrin-coated pits; see below). The coated pit further invaginates and pinches off, a process that is facilitated by the GTPase dynamin. The resulting coated vesicle then fuses with an early endosome. The early endosome is acidic, as a result of the activity of vacuolar proton ATPases. Receptors may be recycled to the cell surface from the early endosome by the budding off of transport vesicles. Transport of the contents of early endosomes to the late endosomes occurs by a microtubule-mediated process. The late endosomes then fuse with lysosomes. Whether a receptor is recycled from

(continued on next page)

viruses. Ligands in the extracellular medium bind to cells via specific plasma membrane receptor proteins. The receptor-ligand complex diffuses along the membrane until it reaches an invagination that is coated on its cytoplasmic surface by a cagelike lattice that is composed of the fibrous protein clathrin and is referred to as a clathrin-coated pit (Fig. 5.16). Clathrin-coated pits can comprise as much as 2% of the surface area of a cell, and some receptors are clustered over these areas even in the absence of their ligands. Following the accumulation of receptor-ligand complexes, the clathrin-coated pit invaginates and then pinches off, forming a clathrin-coated vesicle containing the ligand-receptor complex. Within a few seconds, the clathrin coat is lost and the vesicles fuse with small, smooth-walled vesicles located near the cell surface, called early **endosomes**. The contents of the early endosome are then transported to late endosomes located close to the nucleus. The lumen of late endosomes is acidic, a result of energy-dependent transport of protons into the interior of the vesicles by a proton pump in the endosomal membrane. Some ligands dissociate from their receptors in the acidic environment of the endosome, and the receptors are recycled to the cell surface by transport vesicles that bud from the endosome and fuse with the plasma membrane. Late endosomes in turn fuse with **lysosomes**, which are vesicles that contain a variety of enzymes that degrade sugars, proteins, nucleic acids, and lipids. Ligands that reach the lysosomes are degraded by enzymes for further use of their constituents. In some cases, the entire ligand-receptor complex travels to the lysosomal compartment, where it is degraded.

Cell Membranes Fuse during Endocytosis

The formation of vesicles during the process of endocytosis requires the fusion of cell membranes. For example, during endocytosis, fusion produces the intracellular vesicle following invagination of a small region of the plasma membrane (Fig. 5.16). Membrane fusion also takes place during many other cellular processes, such as cell division, myoblast fusion, and exocytosis.

Cell Membrane Fusion

Membrane fusion must be regulated in order to maintain the integrity of the cell and its intracellular compartments. Therefore, membrane fusion does not occur spontaneously but proceeds by specialized mechanisms mediated by proteins. The two membranes must first come into close proximity. This reaction is mediated by interactions of integral membrane proteins that protrude from the lipid bilayer, a targeting protein on one membrane and a docking protein on the other. The next step, fusion, requires an even closer approach of the membranes to within 1.5 nm of each other. This step depends on the removal of water molecules from the membrane surfaces, an energetically unfavorable process. In mammalian cells, a multisubunit protein complex is thought to provide the energy required for such a close approach. The complex formed by targeting and docking proteins recruits additional proteins that lead to fusion of the two membranes. After fusion occurs, the complex dissociates until needed once again. As individual components of the complex lack fusion activity, fusion can be regulated by assembly and disassembly.

Viral Membrane Fusion Proteins

The precise mechanism by which lipid bilayers fuse is not well characterized, but the action of fusion proteins is believed to result in the formation of an opening called a **fusion pore**, allowing exchange of material across the membranes. Much of our understanding about membrane fusion reactions comes from studies using viral fusion proteins. Individual viral proteins that promote membrane fusion have been identified. Membrane fusion reactions catalyzed by these viral proteins appear to be less complex than those mediated by cellular proteins, for in most cases a single viral gene product is sufficient. The sites at which viruses enter cells, the plasma membrane and the endosomal membrane, are relatively homogeneous. As a result, it is not necessary to accommodate extensive diversity in the fusion process. In contrast, fusion mediated by cellular proteins occurs in many different cellular compartments, necessitating a more flexible, complex fusion machinery. In addition, viral fusion proteins need not be recycled, as abundant quantities will be produced in the virus-infected cell. Cell fusion proteins are far less abundant and must be recycled, a requirement that is best accomplished by assembling and disassembling a multisubunit protein complex.

Figure 5.16 *(continued)* early endosomes or is degraded in the lysosome varies according to the particular function of the receptor. For example, the low-density lipoprotein (Ldl) receptor is dissociated from its ligand, Ldl, in the early endosome and is recycled to the plasma membrane. Ldl is then carried to lysosomes. In contrast, the epidermal growth factor (Egf)-Egf receptor complex is transported to lysosomes, where both are degraded. (B) Clathrin-coated pits and formation of clathrin-coated vesicles. Electron micrographs of four sequential stages in the formation of clathrin-coated vesicles from clathrin-coated pits are shown. Adapted from G. M. Cooper, *The Cell: a Molecular Approach* (ASM Press, Washington, D.C., and Sinauer Associates, Sunderland, Mass., 1997), with permission.

The membranes of enveloped viruses fuse with those of the cell as a first step in delivery of the viral nucleic acid into a host cell. Viral fusion may occur either at the plasma membrane or from within an endosome. The membranes of the virus and the cell are first brought into close contact by interaction of a viral glycoprotein with a cell receptor. The same viral glycoprotein, or a different viral integral membrane protein, then catalyzes the fusion of the juxtaposed membranes. As we will see in the following sections, virus-mediated fusion must be regulated to prevent viruses from aggregating, or to ensure that fusion does not occur in the incorrect cellular compartment. In some cases, fusogenic potential is masked until the fusion protein interacts with other integral membrane proteins. In others, low pH is required to expose fusion domains. The activity of fusion proteins may also be regulated by cleavage of a precursor. This requirement probably prevents activation of fusion potential during virus assembly: if fusion proteins were to become cleaved before assembly is complete, the mildly acidic compartments of the transport pathway might induce aberrant fusion reactions. Cleavage also generates the metastable states of viral glycoproteins that can subsequently undergo the conformational rearrangements required for fusion activity.

The Architecture of the Cytoskeleton

Viruses and subviral particles move within the host cell during entry and egress (Chapters 12 and 13). However, movement of molecules larger than 500 kDa does not occur by passive diffusion because the cytoplasm is crowded with organelles, high concentrations of proteins, and the cytoskeleton (Box 5.2). Rather, viruses and their

BOX 5.2

Passive diffusion cannot account for intracellular movement of virion components

In aqueous solutions, molecules can move rapidly by diffusion, a process of random motion produced by collision with other molecules in the solution. Under ideal conditions, diffusion coefficients typically range from 10^{-6} cm^2/s to 10^{-8} cm^2/s. However, the intracellular milieu is far from such an ideal: the very high intracellular protein concentrations (up to 300 mg/ml), the presence of numerous organelles, and the cytoskeletal networks (see figure) severely restrict diffusion of molecules with molecular mass greater than 500 kDa. Measurements of diffusion coefficients of beads microinjected into cells, of cytoplasmic vesicles, and of DNA molecules indicate that these values are from 5- to 1,000-fold lower in the cytoplasm than in aqueous solution. As shown in the table below, such estimates indicate that viral particles (or the components to be assembled into progeny virions) could not reach the appropriate intracellular destinations by passive diffusion within even a few years, let alone the few hours or days that comprise infectious cycles.

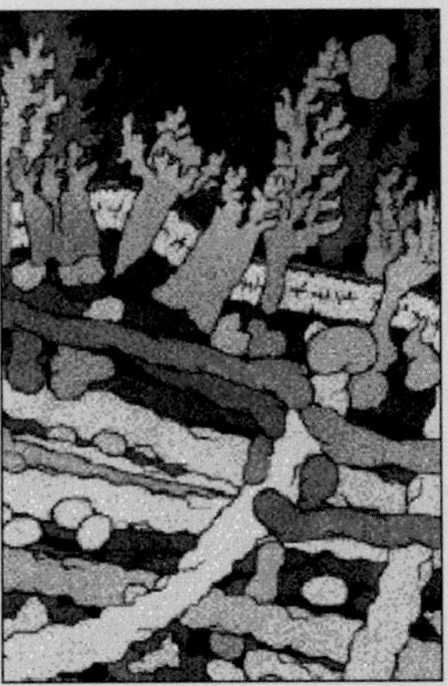

The crowded cytoplasm of a cell. The image shows the interior of a eukaryotic cell, starting at the cell surface, which is studded with membrane proteins. A small portion of the cytoplasm is shown, illustrating the profusion of cytoskeletal elements, ribosomes, and other small molecules. Adapted from David S. Goodsell (http://www.scripps.edu/pub/goodsell/gallery/patterson.html), with permission.

Estimated rates of transport of viral components by diffusion[a]

Viral component	Time to travel 10 μm[b]	
	In H_2O (s)	In cytoplasm (h)
Poliovirus capsid	3.85[c]	0.5
Herpes simplex virus nucleocapsid	14.6	2.0
Vaccinia virus intracellular mature virion	35.0	4.9

[a]Adapted from B. Sodeik, *Trends Microbiol.* **8:**465–472, 2000, with permission.
[b]The length of a typical human cell. Note the different time scales for H_2O and cytoplasm.
[c]Diffusion constants were calculated by a formula that considers the radius of the virus particle and the viscosity of water at room temperature. The assumption was made that diffusion constants in the cytoplasm would be 500 times lower than in water.

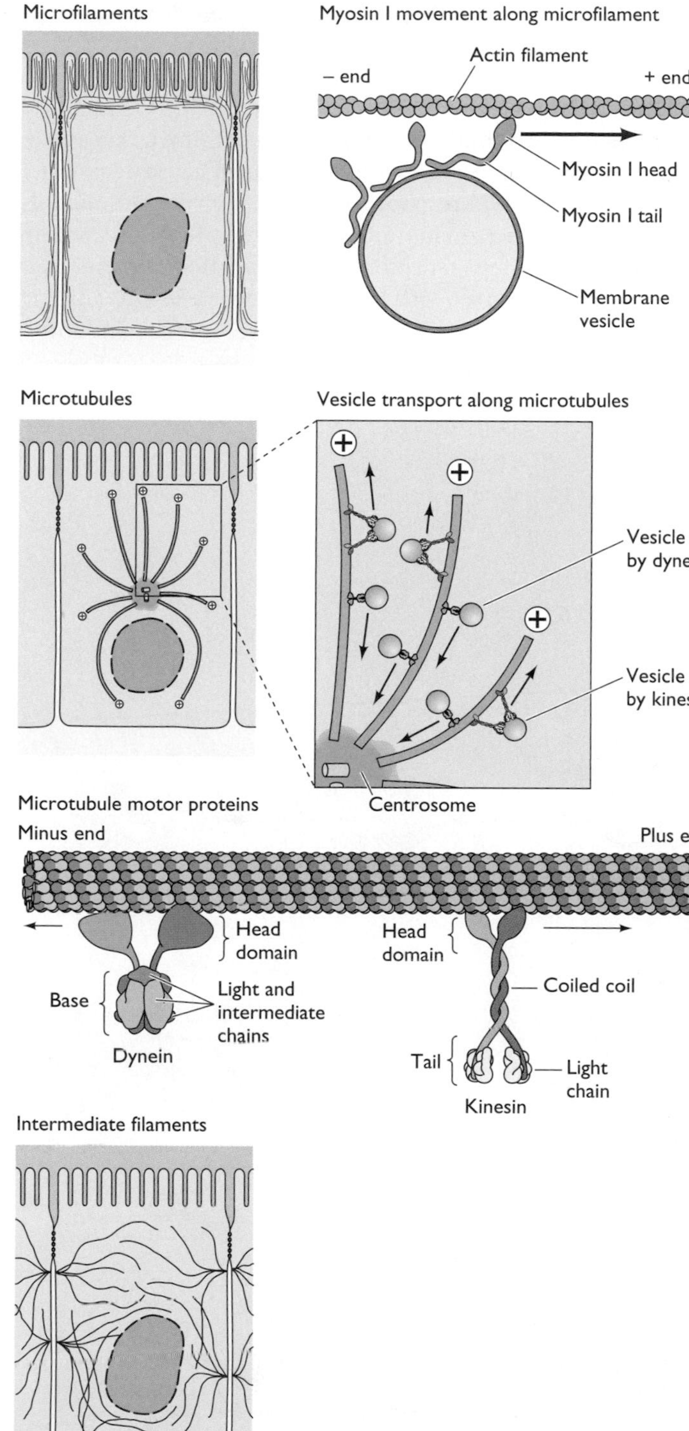

Figure 5.17 Architecture of the cytoskeleton. The cytoskeleton is composed of three types of filament—microfilaments, microtubules, and intermediate filaments. Microfilaments are two-stranded helical polymers of the ATPase actin with a fast-growing plus end and a slow-growing minus end. They are dispersed throughout the cell but most highly concentrated beneath the plasma membrane, where they are connected via integrins and other proteins to the extracellular matrix. Microfilaments are organized into linear bundles, two-dimensional networks, and three-dimensional gels. Transport along microfilaments is accomplished by myosin motors. Microtubules are 25-nm hollow cylinders made of the GTPase tubulin with a fast-growing plus end and a less dynamic minus end. They radiate from the **centrosome** to the cell periphery. Microtubules are organized in a polarized manner, with minus ends situated at the microtubule organizing center, and plus ends situated at the cell periphery. Movement on microtubules is carried out by kinesin and dynein motors. Intermediate filaments are 8- to 10-nm fibers made up of a family of related proteins called intermediate filament proteins. They surround the nucleus and extend to the cell periphery, but are not polarized and do not provide directionality for transport. Adapted from G. M. Cooper, *The Cell: a Molecular Approach* (ASM Press, Washington, D.C., and Sinauer Associates, Sunderland, Mass., 1997), with permission.

components are transported via the actin and microtubule cytoskeletons.

The cytoskeleton is a dynamic network of protein filaments that extends throughout the cytoplasm. It is composed of three types of filament—microtubules, intermediate filaments, and microfilaments (Fig. 5.17). Microtubules are organized in a polarized manner, with minus ends situated at the microtubule organizing center, and plus ends situated at the cell periphery. This arrangement permits directed movement of cellular and viral components. Microfilaments have a fast-growing plus end and a slow-growing minus end. Transport along microfilaments is accomplished by myosin motors. Intermediate filaments surround the nucleus and extend to the cell periphery. In contrast to microtubules or microfilaments, intermediate filaments are not polarized and do not provide directionality for transport.

Transport along microfilaments is accomplished by myosin motors, and movement on microtubules is carried out by kinesin and dynein motors (Fig. 5.17). Hydrolysis of ATP provides the energy for the motors to move their cargo along cytoskeletal tracks. Dyneins and kinesins participate in movement of viral components during both entry (see "Mechanisms of Entry" below) and egress (Chapters 12 and 13). In some cases, the actin cytoskeleton is remodeled during entry and egress, for example, when viruses bud from the plasma membrane.

Mechanisms of Entry

Uncoating is the release of viral nucleic acid from its protective protein coat or lipid envelope, although in most cases the liberated nucleic acid is still complexed with viral proteins. In some cases, uncoating is relatively simple and is accomplished upon fusion of the viral membrane with the plasma membrane (Fig. 5.18). For other viruses, uncoating is a multistep process that may include passage through the endocytic pathway and even docking at the nuclear membrane (Fig. 5.18).

Uncoating at the Plasma Membrane

The particles of many enveloped viruses, including members of the family *Paramyxoviridae*, such as Sendai virus, measles virus, and respiratory syncytial virus, fuse directly with the plasma membrane at neutral pH. These virions bind to cell surface receptors via a viral integral membrane protein (Fig. 5.19). Once the viral and cell membranes have been closely juxtaposed by this receptor-ligand inter-

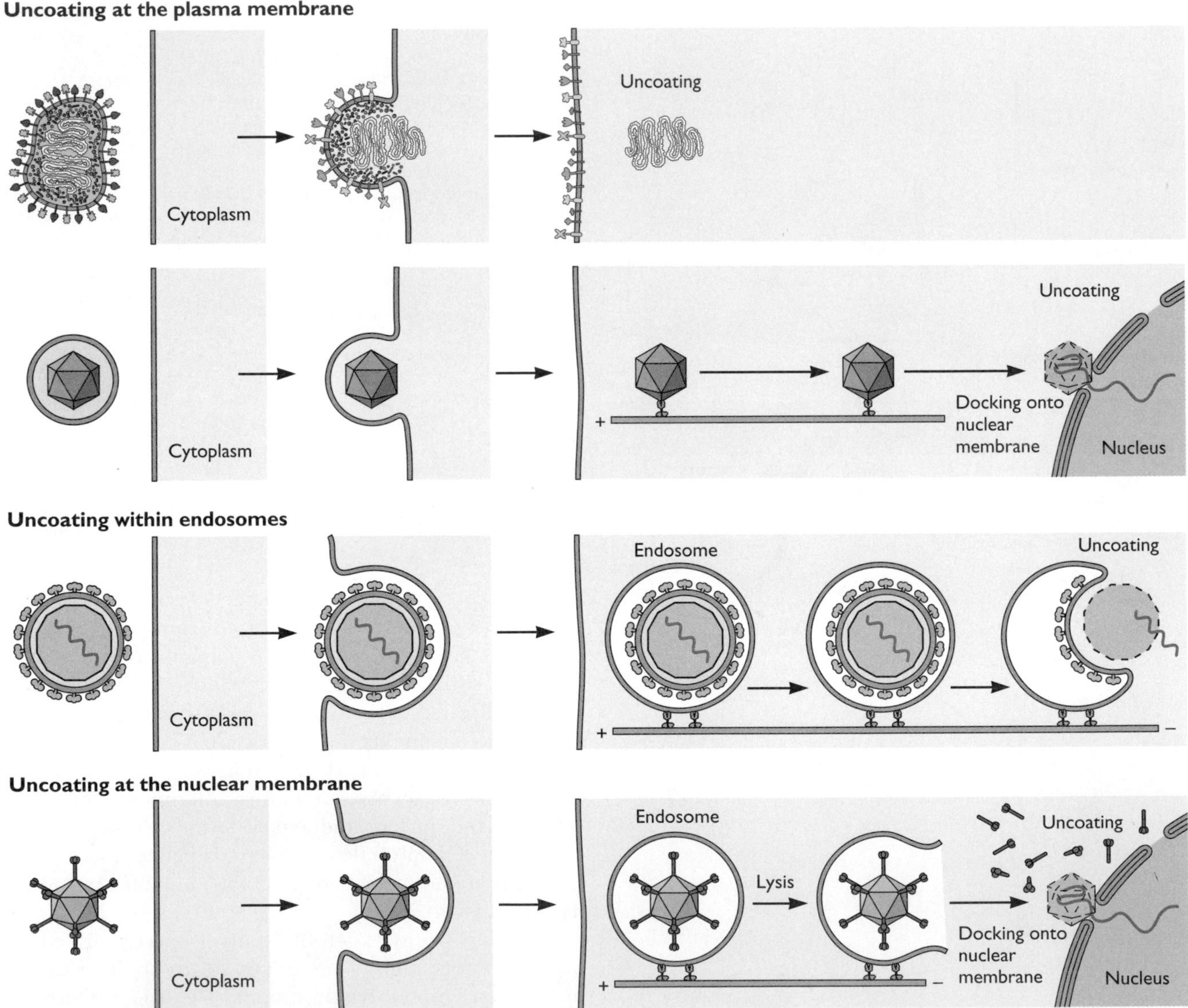

Figure 5.18 Three uncoating strategies. Examples of genome uncoating at the plasma membrane (top), from within endosomes (middle), and at the nuclear membrane (bottom) are illustrated. In the first two cases, the viral genome is released into the cytoplasm, and in the last case, the genome is released into the nucleus. Subviral particles, or virions within endosomes, are transported within the cytoplasm on microtubules. Not shown are viruses such as reovirus that penetrate the cytoplasm at the plasma membrane, after activation by proteases.

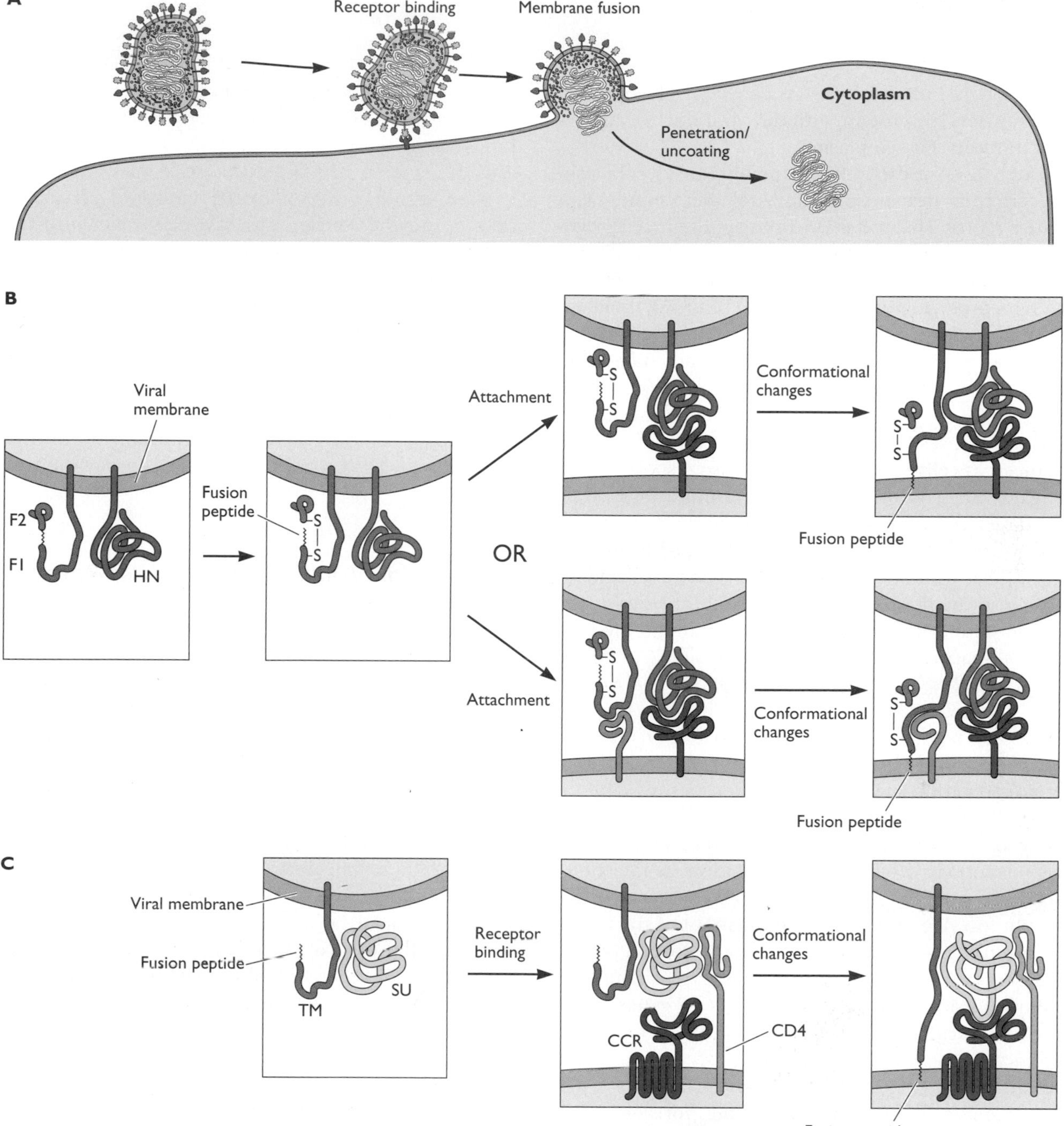

Figure 5.19 Penetration and uncoating at the plasma membrane. (A) Overview. Shown is entry of a member of the *Paramyxoviridae*, which bind to cell surface receptors via HN, H, or G glycoprotein. The fusion protein (F) then catalyzes membrane fusion at the cell surface at neutral pH. The viral nucleocapsid, a ribonucleoprotein complex, is released into the cytoplasm, where RNA synthesis begins. The mechanism by which contacts between the viral nucleocapsid and the M protein, which forms a shell beneath the lipid bilayer, are broken to facilitate release of the nucleocapsid into the cytoplasm is not known. (B) Models for F-protein-mediated membrane fusion. In one model, binding of HN to the cell receptor (red) induces conformational changes in HN that in turn induce conformational changes in the F protein, moving the fusion peptide from a buried position nearer to the cell membrane. In another model, interaction of HN with a cell receptor leads to the binding of F to a putative receptor (orange), which induces conformational changes in F, exposing the fusion peptide. (C) Model of the role of chemokine receptors in human immunodeficiency virus type 1 fusion at the plasma membrane. For simplicity, the envelope glycoprotein is shown as a monomer, although trimer and tetramer forms have been reported. Binding of SU to CD4 exposes a high-affinity chemokine receptor-binding site on SU. The SU-chemokine receptor interaction leads to conformational changes in TM that expose the fusion peptide and permit it to insert into the cell membrane, catalyzing fusion in a manner similar to that proposed for influenza virus (cf. Fig. 5.20 and 5.21).

action, fusion is induced by a second viral glycoprotein known as fusion (F) protein, and the viral nucleocapsid is released into the cell cytoplasm.

F protein is a type I integral membrane glycoprotein (the N terminus lies outside the viral membrane) with similarities to influenza virus HA in its synthesis and structure. F protein is synthesized as a precursor called F0 that is cleaved during transit to the cell surface by a host cell protease to produce two subunits, F1 and F2, held together by disulfide bonds. The newly formed N-terminal 20 amino acids of the F1 subunit, which are highly hydrophobic, form a region called the **fusion peptide** because it is believed to insert into target membranes to initiate fusion. Viruses with the uncleaved F0 precursor can be produced in cells that lack the protease responsible for its cleavage. Such virus particles are noninfectious; they bind to target cells, but the viral genome does not enter. Cleavage of the F0 precursor is necessary for fusion, presumably because the fusion peptide is made available for insertion into the plasma membrane, and to generate the metastable state of the protein that can undergo the conformational rearrangements needed for fusion (see "Acid-Catalyzed Membrane Fusion" below).

Because cleaved F-protein-mediated fusion can occur at neutral pH, it must be controlled, both to ensure that virus particles fuse with only the appropriate cell and to prevent aggregation of newly assembled virions. The fusion peptide of F1 is probably buried in the native protein. Conformational changes in F protein might move the fusion peptide from a buried position in the molecule to an exposed one from which it can contact the target membrane (Fig. 5.19). Such movement of the fusion peptide in influenza virus HA in response to acid pH has been described in atomic detail (see "Acid-Catalyzed Membrane Fusion" below). The results of experiments in which hemagglutinin-neuraminidase (HN) and F glycoprotein genes are synthesized in cultured mammalian cells indicate that F protein lacks fusion activity if HN glycoprotein is not present. It has therefore been hypothesized that an interaction between HN and F proteins is essential for fusion. It is believed that binding of HN protein to its cellular receptor induces conformational changes, which in turn trigger conformational change in F protein, exposing the fusion peptide and making the protein fusion competent (Fig. 5.19). The requirement for HN protein in F fusion activity has been observed only with certain paramyxoviruses, including Newcastle disease virus, human parainfluenza virus type 3, and mumps virus. In contrast, F protein of the paramyxovirus simian virus 5 apparently does not require HN protein for fusion activity. Conformational changes in simian virus 5 F glycoprotein needed to expose the fusion peptide might occur as the protein contacts the cell membrane, or as it interacts with an as yet unidentified F protein receptor on the plasma membrane (Fig. 5.19).

As a result of fusion of the viral and plasma membranes, the viral nucleocapsid, which is a ribonucleoprotein (RNP) consisting of the (–) strand viral RNA genome associated with the viral proteins L, NP, and P, is released into the cytoplasm (Fig. 5.19). Once in the cytoplasm, L, NP, and P proteins begin the synthesis of viral messenger RNAs (mRNAs), a process that will be discussed in Chapter 6. Because members of the *Paramyxoviridae* replicate in the cytoplasm, fusion of the viral and plasma membranes achieves uncoating and transport of the viral genome to the correct cellular compartment in a single step.

Virus-Cell Membrane Fusion That Requires a Cellular Coreceptor

Fusion of human immunodeficiency virus type 1 with the plasma membrane requires participation not only of the cell receptor CD4, but also of an additional cellular protein. When human CD4 protein is synthesized in cultured mouse cells, human immunodeficiency virus type 1 binds to the receptor but cannot fuse with the plasma membrane, because mouse cells lack a cellular coreceptor required for fusion. The experimental strategies used to identify the first of these coreceptors started with cultured mouse cells engineered to produce CD4; DNA that included candidate coreceptor genes was then introduced. Cells that make both CD4 and the coreceptor were selected on the basis of their ability to fuse (form syncytia) with other cells that carried the viral envelope protein on their surface. The coreceptor gene was then cloned from the selected cells. Other, related human immunodeficiency virus type 1 coreceptors were identified and cloned within a few weeks of this first report. All the coreceptors are cell surface receptors for small molecules that many cells can produce to attract and stimulate cells of the immune defense system at sites of infection; hence these small molecules are called **chemotactic cytokines** or **chemokines**. The chemokine receptors on such cells comprise a large family of proteins with seven membrane-spanning domains and are coupled to intracellular signal transduction pathways.

The predicted structure of the first coreceptor for human immunodeficiency virus type 1 to be isolated resembles that of receptors for CXC chemokines (i.e., chemokines characterized as having their first two cysteines separated by a single amino acid). However, when first identified, the coreceptor was considered an orphan receptor, as its natural ligand was unknown. This ligand was subsequently identified as stromal cell-derived protein, and the coreceptor was then given the "family" name of CXC chemokine receptor 4 (CXCr4). This molecule appears to be a specific coreceptor for human immunodeficiency virus type 1 strains that preferentially infect T cells.

Hints about the identity of a second coreceptor came from earlier observations that a specific subgroup of CC chemokines appeared to inhibit infection by macrophage-tropic strains of the virus (CC chemokines are distinguished from CXC chemokines by having the first two cysteines adjacent). As a chemokine receptor that responded to this same chemokine subgroup was already known, namely, the receptor CCr5, its cloned DNA was tested directly and shown to encode a coreceptor for the macrophage-tropic strains of the virus. The chemokines that bind to this receptor activate both T cells and macrophages, and the receptor is found on both types of cell. Interestingly, individuals who are homozygous for deletions in the CCr5 gene and produce nonfunctional coreceptors appear to be resistant to infection with human immunodeficiency virus type 1, but have no discernible immune function abnormality. Even heterozygous individuals seem to be somewhat resistant to the virus. Other members of the CC chemokine receptor family (i.e., CCr2b and CCr3) were subsequently found to serve as coreceptors for the virus.

Clues about how the chemokine receptors facilitate virus fusion come from experiments demonstrating that SU can bind to CCr5, but that the binding affinity is far greater when SU is first bound to CD4. Attachment to CD4 appears to create a high-affinity binding site on SU for CCr5. The atomic structure of SU bound to CD4 revealed that binding of CD4 induces conformational changes that expose binding sites for chemokine receptors

Figure 5.20 Influenza virus entry. The globular heads of native HA mediate binding of the virus to sialic acid-containing cell receptors. The virus-receptor complex is endocytosed, and import of H^+ ions into the endosome acidifies the interior. Upon acidification, the viral HA undergoes a conformational rearrangement that produces a fusogenic protein. The loop region of native HA (yellow) becomes a coiled coil, moving the fusion peptides (red) to the top of the molecule near the cell membrane. At the viral membrane, the long α-helix (purple) packs against the trimer core, pulling the globular heads to the side. The long coiled coil splays into the cell membrane, bringing it closer to the viral membrane so that fusion can occur. Not shown is the tilting of HA that occurs. To allow release of vRNP into the cytoplasm, the H^+ ions in the acidic endosome are pumped into the virion interior by the M2 ion channel. As a result, vRNP is primed to dissociate from M1 after fusion of the viral and endosome membranes. The released vRNPs are imported into the nucleus through the nuclear pore complex via a nuclear localization signal-dependent mechanism (see "Import of Influenza Virus RNP"). Adapted from C. M. Carr and P. S. Kim, *Science* **266:**234–236, 1994, with permission.

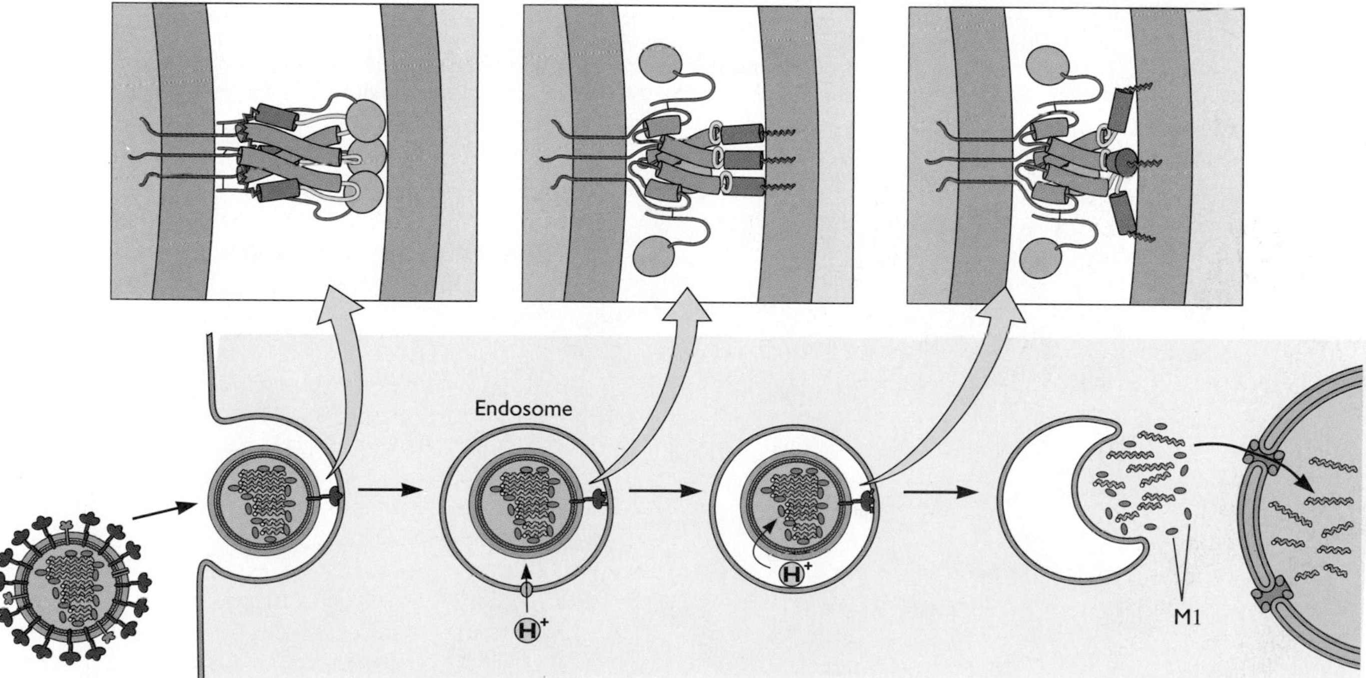

on SU (Fig. 5.11). Studies of CCr5 have shown that the first N-terminal extracellular domain is crucial for coreceptor function, suggesting that this sequence might interact with SU.

Human immunodeficiency virus type 1 TM is likely to mediate envelope fusion with the cell membrane. The high-affinity SU-CCr5 interaction may induce conformational changes in TM to expose the fusion peptide, placing it near the cell membrane, where it can catalyze fusion (Fig. 5.19). Such changes might be similar to those that influenza virus HA undergoes upon exposure to low pH. X-ray crystallographic analysis of fusion-active human immunodeficiency virus type 1 TM revealed that its structure is strikingly similar to that of the low-pH fusogenic form of HA (Fig. 5.22) (see "Acid-Catalyzed Membrane Fusion" below).

Entry via the Endocytic Pathway

Acid-Catalyzed Membrane Fusion

In contrast to fusion at plasma membranes, which is characteristic of many viral infections, other enveloped viruses undergo fusion within an endosomal compartment. The entry of influenza virus from the endosomal pathway is one of the best-understood viral entry mechanisms. At the cell surface, the virus attaches to sialic acid-containing receptors via the viral HA glycoprotein (Fig. 5.14 and 5.20). The virus-receptor complex is then internalized by the clathrin-dependent receptor-mediated endocytic pathway. When the vesicular pH reaches approximately 5.0, in the late endosome, HA undergoes an acid-catalyzed conformational rearrangement, exposing a fusion peptide. The viral and endosomal membranes then

Figure 5.21 Cleavage- and low-pH-induced structural changes in the influenza virus HA. (Left) Structure of the uncleaved HA0 precursor. HA1 subunits are blue, HA2 subunits are red, residues 323 of HA1 to 12 of HA2 are yellow, and locations of some of the N and C termini are indicated. The viral membrane is at the bottom, and the globular heads are at the top. The transmembrane segment is not present because the HA was cleaved from cells producing it with the protease trypsin. The cleavage site between HA1 and HA2 is in a loop adjacent to a deep cavity. (Middle) Structure of the HA0 trimer at neutral pH. The transmembrane segment is not present because the HA was cleaved from the virus with the protease bromelain. Cleavage of HA0 generates new N and C termini, which are separated by 20 Å. The N and C termini visible in this model are labeled. The cavity is now filled with residues 1 to 10 of HA2, part of the fusion peptide. (Right) Structure of the low-pH trimer. The protein used for crystallization was prepared from bromelain HA by further treatment with thermolysin and trypsin, and therefore the HA1 subunit and the fusion peptide are not present. This treatment is necessary to prevent aggregation of HA at low pH. At neutral pH the fusion peptide is close to the viral membrane, linked to a short α-helix, and at acid pH this α-helix is reoriented toward the cell membrane, carrying with it the fusion peptide. The structures are aligned on a central α-helix that is unaffected by the conformational change. Adapted from J. Chen et al., *Cell* **95**:409–417, 1998, with permission.

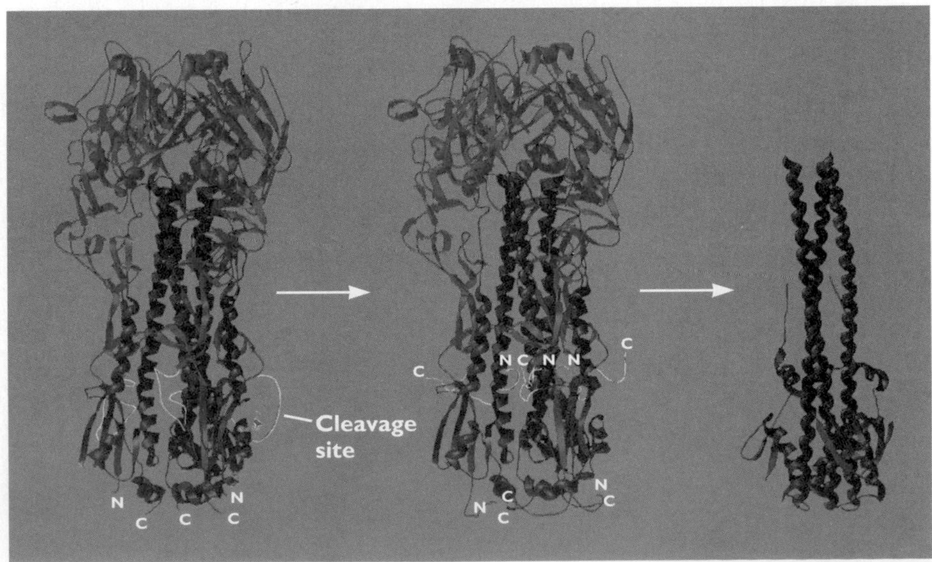

fuse, allowing penetration of the viral ribonucleoprotein into the cytoplasm.

The fusion reaction mediated by the influenza virus HA protein is a remarkable event when viewed at atomic resolution (Fig. 5.21). In native HA, the fusion peptide is joined to the three-stranded coiled-coil core by which the HA monomers interact by a 28-amino-acid sequence that forms an extended loop structure buried deep inside the molecule, about 100 Å from the globular head. In contrast, in the low-pH HA structure, this loop region is transformed into a three-stranded coiled coil. In addition, the long α-helices of the coiled coil bend upward and away from the viral membrane. The result is that the fusion peptide has moved a great distance toward the endosomal membrane (Fig. 5.21). Despite these dramatic changes, HA remains trimeric, and the globular heads can still bind sialic acid. In this conformation, HA holds the viral and endosome membranes 100 Å apart, too distant

Figure 5.22 Structures of influenza virus HA, simian virus 5 F protein, Ebola virus Gp2, human immunodeficiency virus type 1 TM, and Moloney murine leukemia virus TM. (Top) View from the top of the structures. (Bottom) Side view. The structure shown for HA is the low-pH, or fusogenic, form. The structure of simian virus 5 F protein is of peptides from the N- and C-terminal heptad repeats. Structures of retrovirus TM proteins are derived from interacting human immunodeficiency virus type 1 peptides and a peptide from Moloney murine leukemia virus and are presumed to represent the fusogenic forms because of structural similarity to HA. In all three molecules, fusion peptides would be located at the membrane-distal portion (the tops of the molecules in the bottom view). Despite the differences in the mechanism of membrane fusion among the three viruses (HA-mediated fusion is acid catalyzed, that of human immunodeficiency virus type 1 requires a cellular coreceptor, and that of Moloney murine leukemia virus occurs at the plasma membrane), they all present fusion peptides to cells on top of a central three-stranded coiled coil supported by C-terminal structures. Adapted from K. A. Baker et al., *Mol. Cell* **3**:309–319, 1999, with permission.

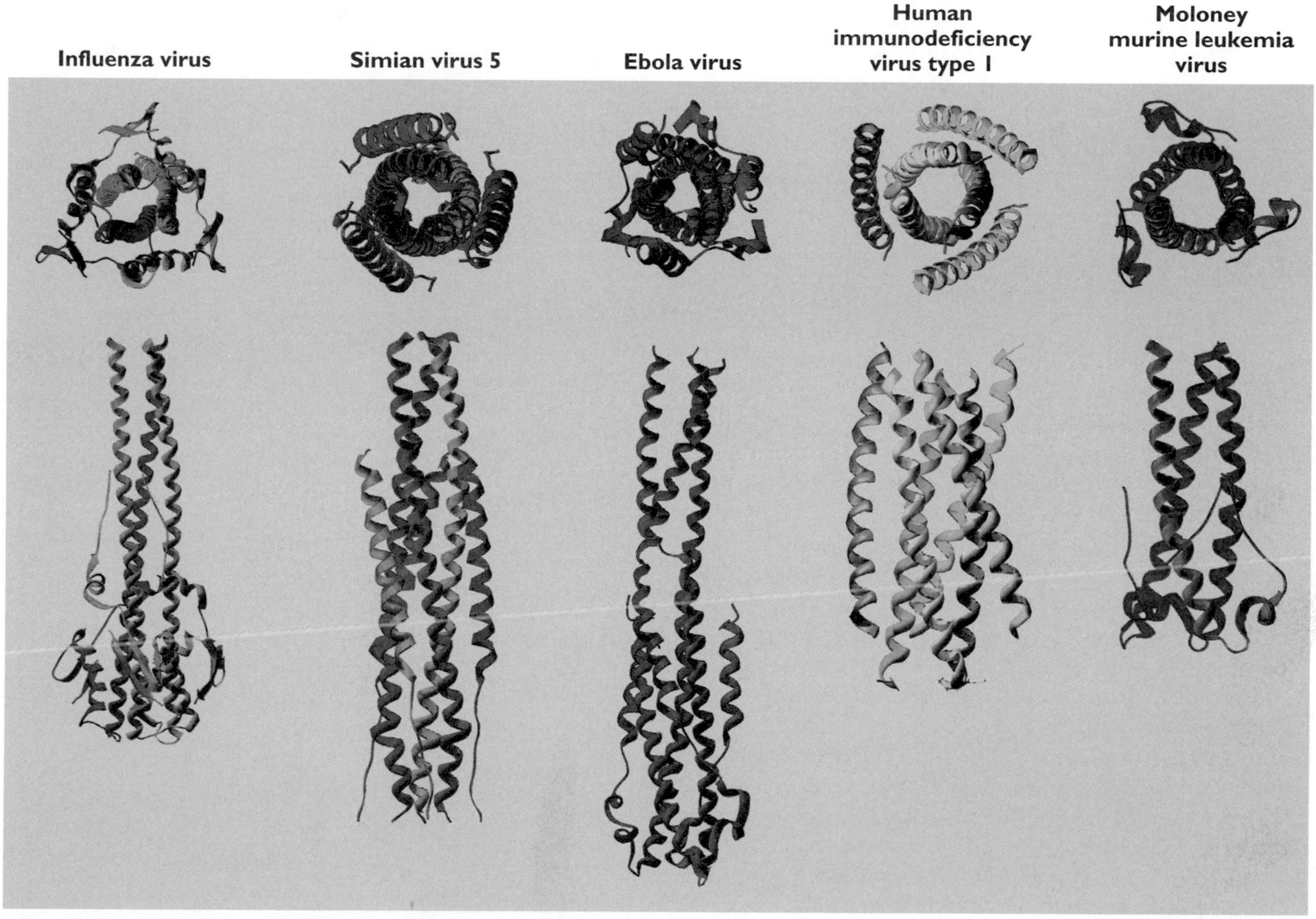

Influenza virus | **Simian virus 5** | **Ebola virus** | **Human immunodeficiency virus type 1** | **Moloney murine leukemia virus**

for the fusion reaction to occur. To bring the viral and cellular membranes closer, the top of the acid-induced, coiled coil is believed to splay apart, spreading into the lipid bilayer (Fig. 5.20). The stems of the HA probably tilt with respect to the membrane, further facilitating close contact of the membranes.

Cleavage of HA is a prerequisite for the low-pH-induced conformational change, because it not only produces the free N terminus of HA2 containing the fusion peptide, but also converts HA to a metastable state that can undergo the structural rearrangements that promote fusion. In contrast to cleaved HA, the precursor HA0 is stable at low pH and cannot undergo structural changes. How does cleavage of HA produce a metastable protein? Cleavage of the covalent bond between HA1 and HA2 might simply allow movement of the fusion peptide, which is restricted in the uncleaved molecule. Another possibility is suggested by the observation that cleavage of HA is accompanied by movement of the fusion peptide into the cavity in HA (Fig. 5.21). This movement buries ionizable residues of the fusion peptide, perhaps setting the low-pH "trigger." It should be emphasized that after cleavage, the N terminus of HA2 is tucked into the hydrophobic interior of the trimer (Fig. 5.21). This rearrangement presumably buries the fusion peptide so that newly synthesized virions do not aggregate and lose infectivity.

When the core structure of influenza virus HA is compared with those of the TM proteins of two retroviruses (human immunodeficiency virus type 1 and Moloney murine leukemia virus), the F protein of simian virus 5, and Gp2 of Ebola virus, remarkable similarities become apparent, despite the diversity in mechanisms of fusion (Fig. 5.22). In all five cases, the fusion peptides are presented to membranes on top of a three-stranded coiled coil. Such a scaffold might be a common feature of viral membrane fusion proteins: many have a region of high α-helical content and a 4-3 heptad repeat of hydrophobic amino acids, characteristic of coiled coils, next to the N-terminal fusion peptide.

The envelope proteins of alphaviruses and flaviviruses exemplify a different class of viral fusion protein. These viral fusion proteins contain an internal fusion peptide and are tightly associated with a second viral protein. Proteolytic cleavage of the second protein converts the fusion protein to a metastable state that can undergo structural rearrangements at low pH that promote fusion. In contrast, the fusion peptide of the influenza virus HA is adjacent to the cleavage point and becomes the N terminus of the mature fusion protein. The envelope proteins of alphaviruses and flaviviruses do not form coiled coils, as do fusion proteins like the influenza virus HA. Rather they contain predominantly β-barrels that are believed to tilt toward the membrane at low pH, thereby exposing the fusion peptide (Fig. 5.23).

Membrane Fusion Proceeds through a Hemifusion Intermediate

Because its fusogenic activity can be controlled by pH, the HA polypeptide has become a model for studying how proteins mediate membrane fusion. Fusion is believed to proceed through a hemifusion intermediate in which the outer leaflets of two opposing bilayers fuse (Fig. 5.24), followed by fusion of the inner leaflets and the formation of a fusion pore. Direct evidence that fusion proceeds via a hemifusion intermediate has been obtained with influenza virus HA (see Fig. 5.24 for details). This experiment also demonstrates that the transmembrane domain of the HA polypeptide plays a role in the fusion process. The extracellular portion of the HA is believed to move horizontally in the cell membrane, perhaps in response to low pH (Fig. 5.24). The transmembrane domain also reorients, and this movement generates stress within the single lipid bilayer that has resulted from hemifusion. The stress is then relieved by the formation of a fusion pore. When HA attached to the membrane by linkage with glycosylphosphatidylinositol, instead of its normal transmembrane

Figure 5.23 Models for low-pH-induced movement of alphavirus and flavivirus glycoproteins. Low pH causes conformational changes in the viral glycoproteins to produce the fusion-active forms. In alphavirus virions (A), the fusion peptide in E1 is masked by E2. Low pH leads to disruption of E1-E2 dimers, exposing the fusion peptide. In flavivirus virions (B), the fusion peptide is buried in dimers of the fusion glycoprotein E. At low pH, the dimers are disrupted, the proteins rotate to form trimers, and the fusion peptide is directed toward the cell membrane. Adapted from R. J. Kuhn et al., *Cell* **108:**717–725, 2002, with permission.

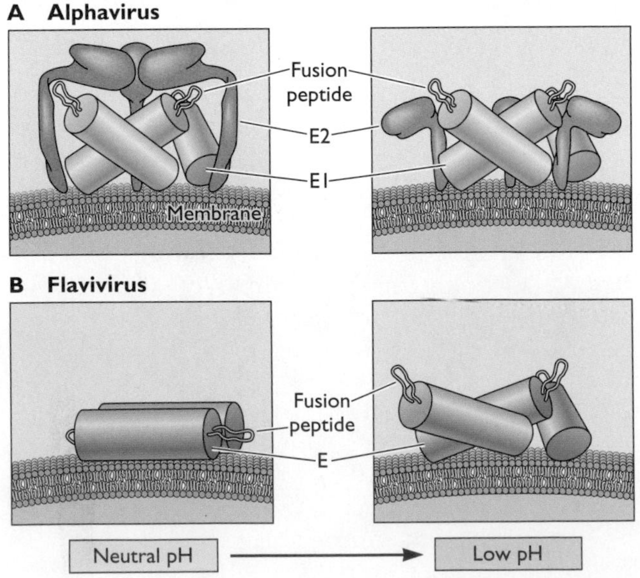

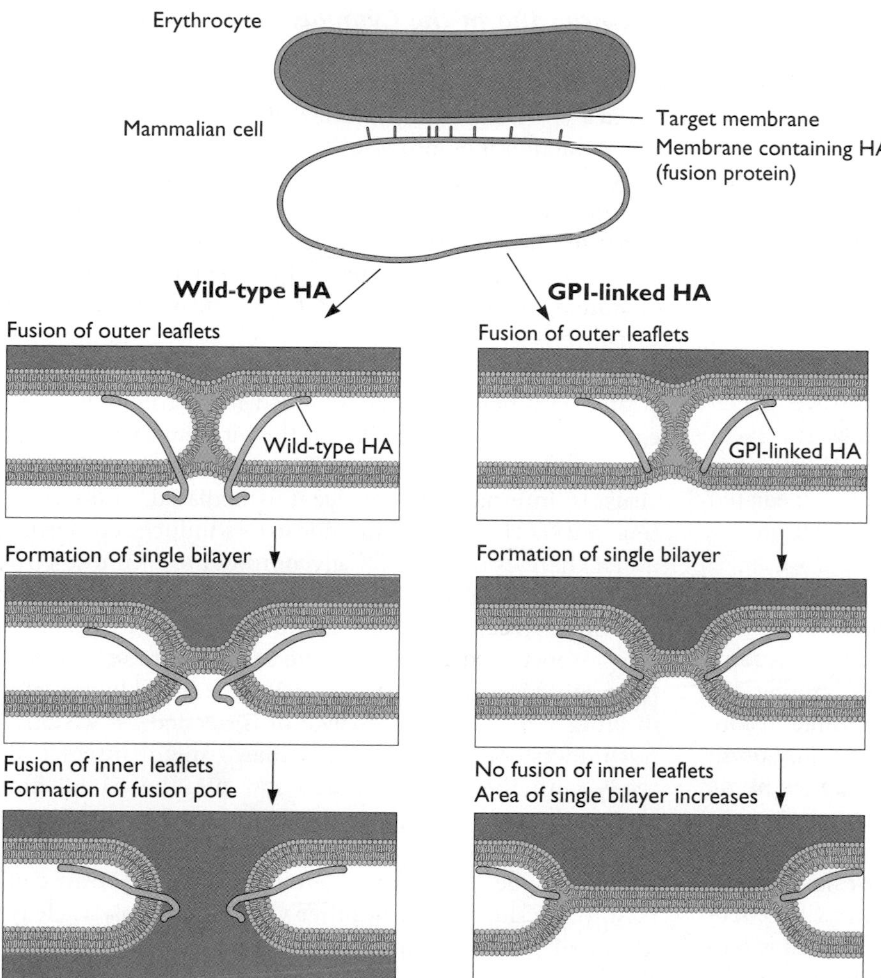

Figure 5.24 Glycosylphosphatidylinositol-anchored influenza virus HA induces hemifusion. (Left) Model of the steps of fusion mediated by wild-type HA. Cultured mammalian cells expressing wild-type HA are fused with erythrocytes containing two different types of fluorescent dye, one in the cytoplasm and one in the lipid membrane. Upon exposure to low pH, HA undergoes conformational change and the fusion peptide is inserted into the erythrocyte membrane. The green dye is transferred from the lipid bilayer of the erythrocyte to the bilayer of the cultured cell. The HA trimers tilt, causing reorientation of the transmembrane domain and generating stress within the hemifusion diaphragm. Fusion pore formation relieves the stress. The red dye within the cytoplasm of the erythrocyte is then transferred to the cytoplasm of the cultured cell. (Right) An altered form of HA was produced lacking the transmembrane and cytoplasmic domains, with membrane anchoring provided by linkage to a glycosylphosphatidylinositol (GPI) moiety. Upon exposure to low pH, the HA fusion peptide is inserted into the erythrocyte membrane, and green dye is transferred to the membranes of the mammalian cell. When the HA trimers tilt, no stress is transmitted to the hemifusion diaphragm because no transmembrane domain is present, and the diaphragm becomes larger. Fusion pores do not form, and there is no mixing of the contents of the cytoplasm, indicating that complete membrane fusion has not occurred. These results prove that hemifusion, or fusion of only the inner leaflet of the bilayer, can occur among whole cells. Adapted from G. B. Melikyan et al., *J. Cell Biol.* **131**:679–691, 1995, with permission.

segment, is exposed to low pH, no stress occurs, and the area of the single lipid bilayer expands (Fig. 5.24).

Release of Viral Ribonucleoprotein

The genomes of many enveloped RNA viruses are present as viral ribonucleoprotein (vRNP) in the virus particle. One mechanism for release of vRNP from the viral membrane during virus entry has been identified by studies of influenza virus. Fusion of the influenza virus envelope with the endosomal membrane exposes vRNPs to the cytoplasm. Each influenza virus vRNP is composed of a segment of the RNA genome complexed with nucleoprotein (NP) molecules at about 10- to 15-nucleotide intervals and the virion RNA polymerase. This complex interacts with viral M1 protein, an abundant virion protein that underlies the viral envelope and provides rigidity (Fig. 5.20). The M1 protein also contacts the internal tails of HA and neuraminidase proteins in the viral envelope. This arrangement presents two problems. Unless M1-vRNP interactions are disrupted, vRNPs might not be released into the cytoplasm. Furthermore, the vRNPs must enter the nucleus, where transcription and replication occur. However, vRNP cannot enter the nucleus if M1 protein remains bound, because this protein masks a nuclear localization signal (see "Import of Influenza Virus RNP" below).

The influenza virus M2 protein, the first viral protein discovered to be an ion channel, probably provides the solution to both problems. The virion envelope contains a small number (14 to 68) of molecules of M2 protein, which form a homotetramer. When made in *Xenopus laevis* oocytes or in cultured mammalian cells, the M2 protein can transport monovalent cations into the cell interior, a property of proteins with ion channel activity. Furthermore, when purified M2 was reconstituted into synthetic lipid bilayers, ion channel activity was also observed, indicating that this property requires only M2 protein. The M2 protein channel is structurally much simpler than other ion channels, which generally possess 24 membrane-spanning domains, and is the smallest channel discovered to date.

The M2 ion channel is activated by the low pH of the endosome before HA-catalyzed membrane fusion occurs. As a result, protons enter the interior of the virus particle. It has been suggested that the reduced pH of the virion interior leads to conformational changes in the M1 protein, thereby disrupting M1-vRNP interactions. When fusion between the viral envelope and the endosomal membrane subsequently occurs, vRNPs are released into the cytoplasm free of M1 and can then be imported into the nucleus (Fig. 5.20). Support for this model comes from studies with the anti-influenza virus drug **amantadine,** which specifically inhibits M2 ion channel activity (Fig. 19.23). In the presence of this drug, influenza virus particles can bind to cells, enter endosomes, and undergo HA-mediated membrane fusion, but vRNPs are not released from the endosomal membrane.

Receptor Priming for Low-pH Fusion

The entry of avian leukosis virus into cells demonstrates a recently recognized principle, that receptor binding can prime the viral fusion protein for low-pH-activated fusion. It was believed that avian leukosis virus, like many other simple retroviruses, entered cells at the plasma membrane in a pH-independent mechanism resembling that of members of the *Paramyxoviridae* (Fig. 5.19). Members of the subgroup A avian sarcoma and leukosis viruses bind to the cellular receptor Tva via the SU subunit of the viral envelope glycoprotein (Env-A). The TM subunit, which is embedded in the viral envelope at its C terminus, contains a fusion peptide at its N terminus. Both F protein of paramyxoviruses and Env-A protein of avian sarcoma and leukosis viruses must undergo proteolytic cleavage so that conformational rearrangements can occur, leading to exposure of the fusion peptide. The cleavage of Env-A into SU and TM subunits, which occurs in the *trans*-Golgi network during virus maturation, is required for fusion. The fusion peptide is probably buried within the Env-A molecule. Originally it was believed that binding of Env-A to Tva led to conformational rearrangements of the glycoprotein and exposure of the fusion peptide, which then catalyzed fusion of the viral and plasma membranes. Recent findings indicate that this model is incorrect. Rather, binding of Env-A to Tva induces conformational rearrangements that convert Env-A from a native metastable state that is insensitive to low pH, to a second metastable state. In this state, exposure of Env-A to low pH, within the endosomal compartment, leads to membrane fusion and release of the viral capsid. Whether other retroviruses enter cells by a similar mechanism remains to be determined.

Uncoating in the Cytoplasm by Ribosomes

By allowing protons to enter virus particles as they pass through acidic endosomes, the M2 ion channel of influenza viruses promotes release of vRNPs from the endosomal membrane into the cytoplasm. Other enveloped RNA-containing viruses, such as Semliki Forest virus, contain nucleocapsids that must be disassembled in the cytoplasm to release viral RNA. The nucleocapsid of this virus is an icosahedral shell composed of a single viral protein, C protein, which encloses the (+)-strand viral RNA. This structure is surrounded by an envelope containing viral glycoproteins called E1 and E2. These glycoproteins are arranged as heterodimers clustered into groups of three, each cluster forming a spike on the virus surface.

After binding to its cell surface receptor, Semliki Forest virus is internalized by receptor-mediated endocytosis (Fig. 5.25). The low pH in endosomes induces conformational changes in the spike glycoprotein to promote virus fusion. Semliki Forest virus-mediated fusion exhibits several unusual features. For example, this process requires the presence of cholesterol in the cell membrane, whereas this steroid is not required for fusion mediated by proteins of other viruses. Why cholesterol is needed for fusion is not understood. One notion is that it might interact directly with the E1 subunit. At low pH, the E1-E2 heterodimer spikes present in the viral envelope dissociate, and the E1 protein molecules reassociate as trimers, which may be the fusion-active form (Fig. 5.23). In contrast to the situation with other viruses, proteolytic cleavage of E1 is not required to produce a fusogenic protein. However, protein processing may control fusion potential in another way. In the endoplasmic reticulum, E1 protein is associated with the precursor of E2, called p62. In this heterodimeric form, p62-E1, E1 protein cannot be activated for fusion by mildly acidic conditions. Only after p62 has been cleaved to E2 can low pH induce disruption of E1-E2 heterodimers and formation of fusion-active E1 homotrimers.

Fusion of the viral envelope and endosomal membrane exposes the nucleocapsid to the cytoplasm (Fig. 5.25). The viral RNA within this structure is sensitive to digestion with RNase, suggesting that the nucleocapsid is permeable. Crystallographic studies of the nucleocapsid of Sindbis virus, a closely related alphavirus, confirm the presence of holes ranging in diameter from 30 to 60 Å. These holes permit the entry of small proteins such as RNase (25 to 40 Å in diameter) into the nucleocapsid, but not of ribosomes, whose individual subunits are approximately 150 Å in diameter. To begin translation of (+) strand viral RNA, the nucleocapsid must be disassembled, a process mediated by an abundant cellular component—the ribosome. Each ribosome binds from three to six molecules of C pro-

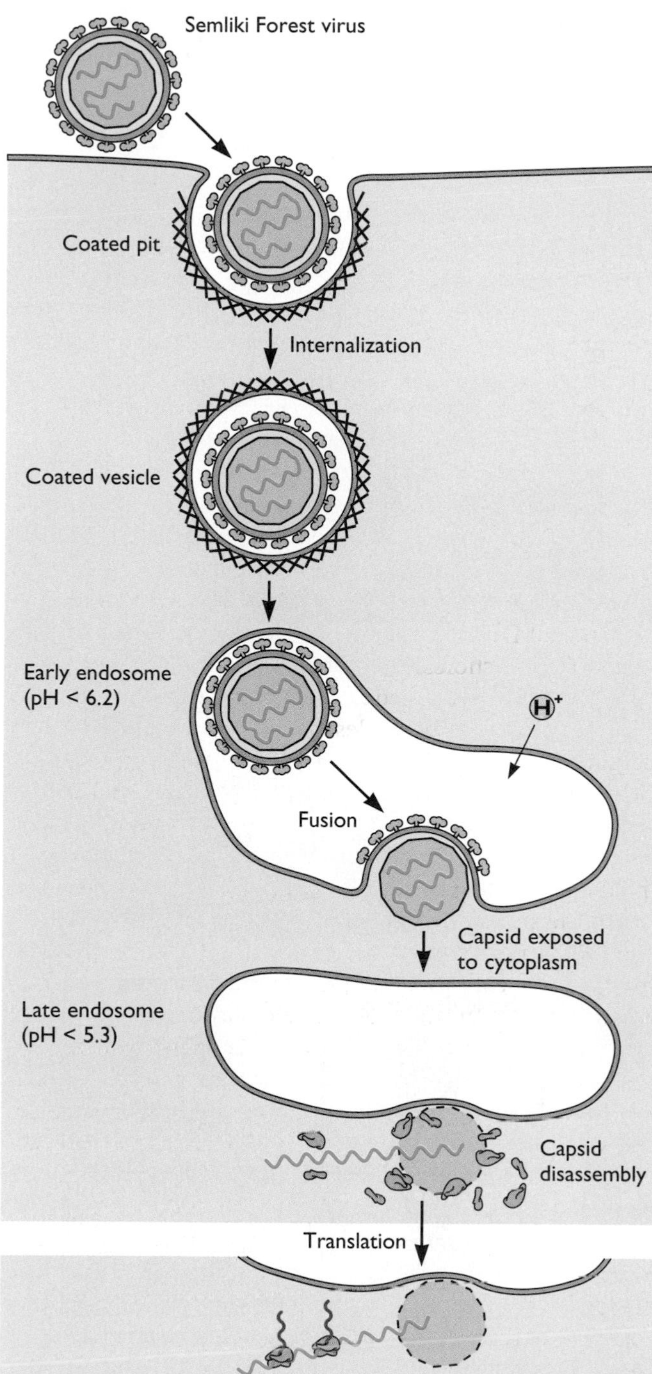

Figure 5.25 Entry of Semliki Forest virus into cells. Semliki Forest virus enters cells by clathrin-dependent receptor-mediated endocytosis, and membrane fusion is catalyzed by acidification of endosomes. Fusion results in exposure of the viral nucleocapsid to the cytoplasm, although the nucleocapsid remains attached to the cytosolic side of the endosome membrane. Cellular ribosomes then bind the capsid, disassembling it and distributing the capsid protein throughout the cytoplasm. The viral RNA is then accessible to ribosomes, which initiate translation. Adapted from M. Marsh and A. Helenius, *Adv. Virus Res.* **36:**107–151, 1989, with permission.

tein, causing them to detach from the nucleocapsid. This process occurs while the nucleocapsid is attached to the cytoplasmic side of the endosomal membrane (Fig. 5.25) and ultimately results in disassembly. The uncoated viral RNA remains associated with cellular membranes, at which translation and replication begin. The mechanism by which the replication complex is anchored to the vesicle membrane is not known.

Entry of Nonenveloped Viruses into Cells

Enveloped virus components enter cells by crossing cellular membranes when viral and cellular membranes fuse. How nonenveloped viruses enter cells is more difficult to envision because it is not obvious how naked viral capsids can cross the plasma membrane. Information about this step obtained from study of the members of several virus families has established that naked viruses gain entry to the cell by one of several mechanisms.

Disrupting the Endosomal Membrane

Adenoviruses are composed of a double-stranded DNA genome packaged in an icosahedral capsid made up of at least 10 structural proteins, as described in Chapter 4. Internalization of most adenovirus serotypes by receptor-mediated endocytosis requires attachment of fiber to an integrin or Ig-like cell surface receptor and binding of the penton base to a second cell receptor, the integrins $\alpha_v\beta_3$ and $\alpha_v\beta_5$. As the virus particle is transported via the endosomes from the cell surface toward the nuclear membrane, it undergoes multiple uncoating steps by which structural proteins are sequentially removed (Fig. 5.26). In the endosome, the fiber, protein IIIa, and protein VIII are dissociated from the capsid. As the endosome becomes acidified, the low pH causes release of the penton base and the facet-stabilizing protein IX. The penton base is believed to mediate lysis of the endosomal membrane, thereby releasing the remainder of the virus into the cytoplasm. Protein VI is proteolytically degraded in the acidic endosome, probably by the L3 viral protease. This protease, which is present in the core of the virus particle, is inactive in the virion, possibly because of intrachain disulfide bonds that form in the oxidizing extracellular environment. The protease is activated on entry into the reducing environment of the endosome. Furthermore, interaction of the penton base with integrin receptors is believed to lead to a conformational change in the penton that exposes L3 protease cleavage sites on protein VI. What remains of the capsid docks onto the nuclear pore complex through interactions with the filament protein CAN/Nup214 (Fig. 5.26). Small amounts of histone H1 from the nucleus bind to hexon proteins on the nuclear side of the viral capsid. The H1 import proteins importin β and importin 7 recognize H1

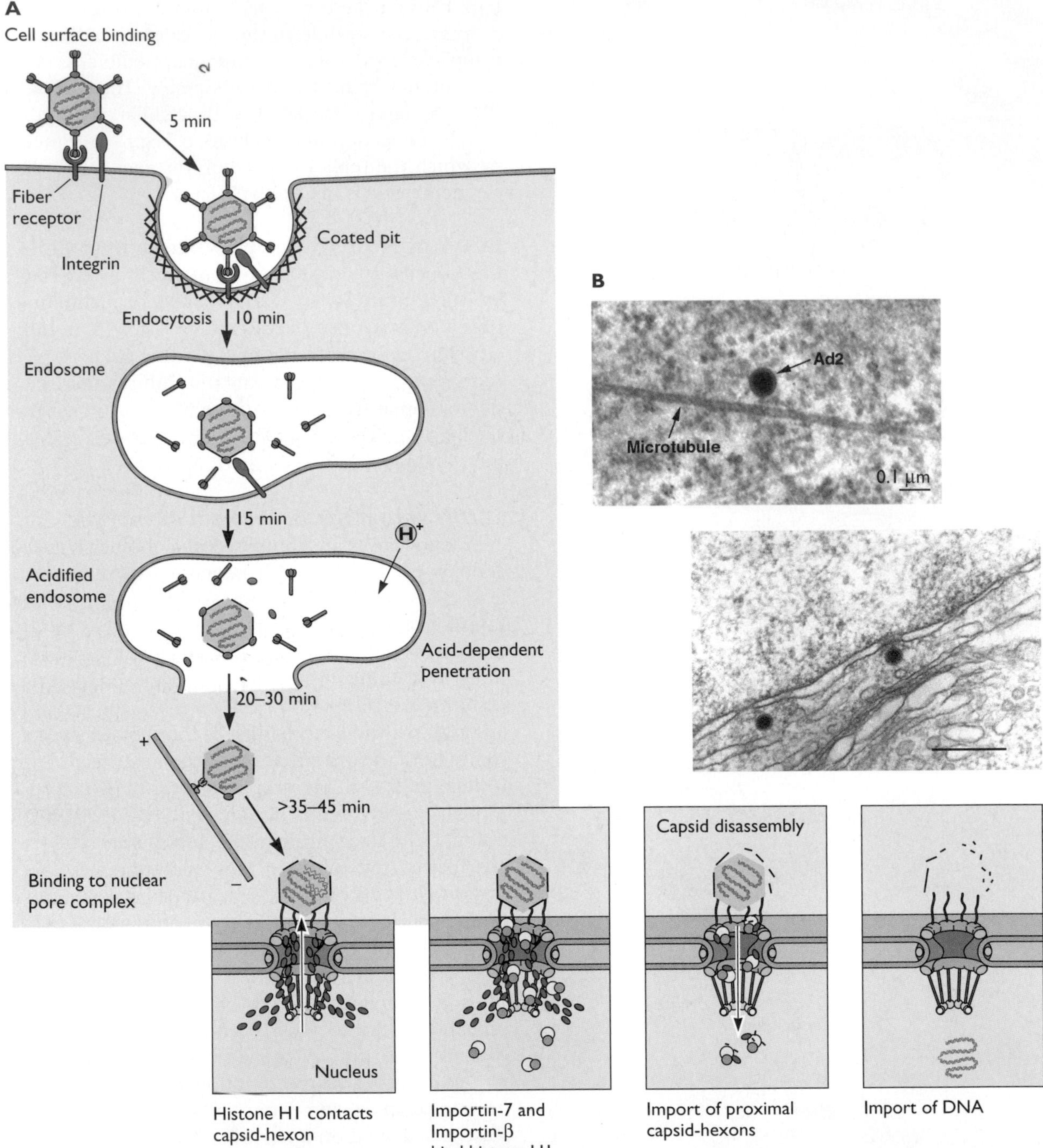

Figure 5.26 Stepwise uncoating of adenovirus during cell entry. (A) Adenoviruses bind the cell receptor via the fiber protein. Interaction of the penton base with an integrin receptor leads to internalization by endocytosis. Viral capsid proteins are sequentially removed in the endosome and the acidified endosome, leading to penetration of the capsid into the cytoplasm. The capsid is transported in the cytoplasm along microtubules and docks onto the nuclear pore complex-filament protein CAN/Nup214. Histone H1 from the nucleus (green ovals) bind to hexon. Importin β and importin 7 bind histone H1, and the import of this complex into the nucleus initiates further disassembly of the capsid. Once the capsid is sufficiently dismantled, the viral DNA is delivered into the nucleus. Adapted from U. F. Greber et al., *Cell* **75**:477–486, 1993, and L. C. Trotman et al., *Nat. Cell Biol.* **3**:1092–1100, 2001, with permission. (B) Electron micrograph of adenovirus type 2 bound to a microtubule (top) and bound to the cytoplasmic face of the nuclear pore complex (bottom). Bar in bottom panel = 200 nm. Reprinted from U. F. Greber et al., *Trends Microbiol.* **2**:52–56, 1994, with permission. Courtesy of Ari Helenius, Urs Greber, and Paul Webster, University of Zurich.

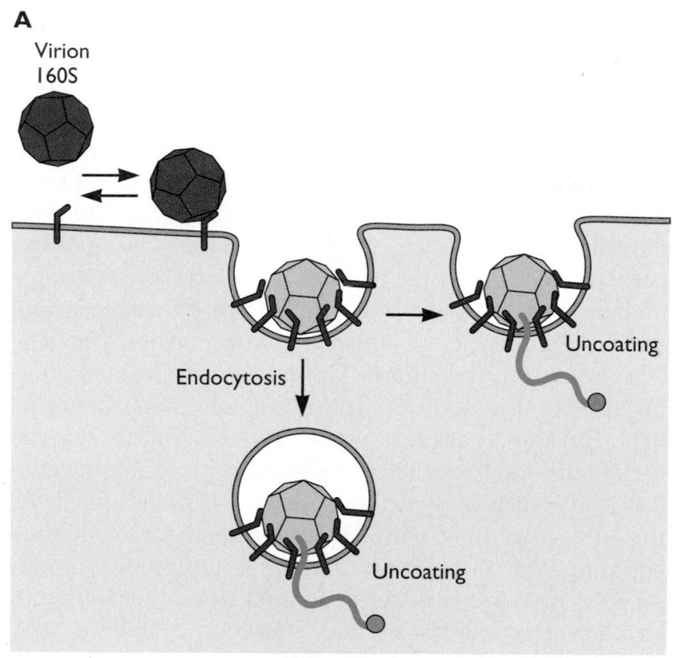

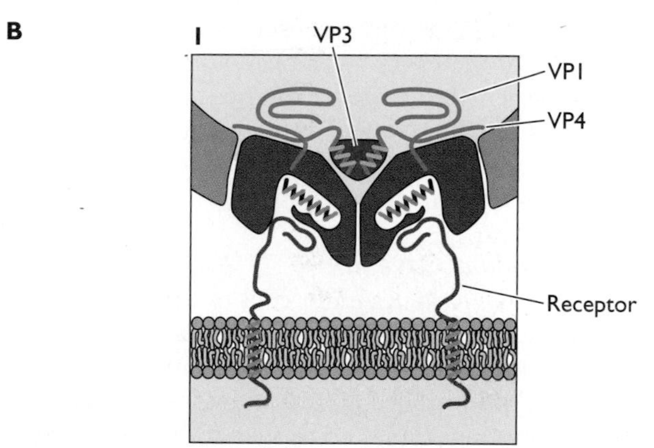

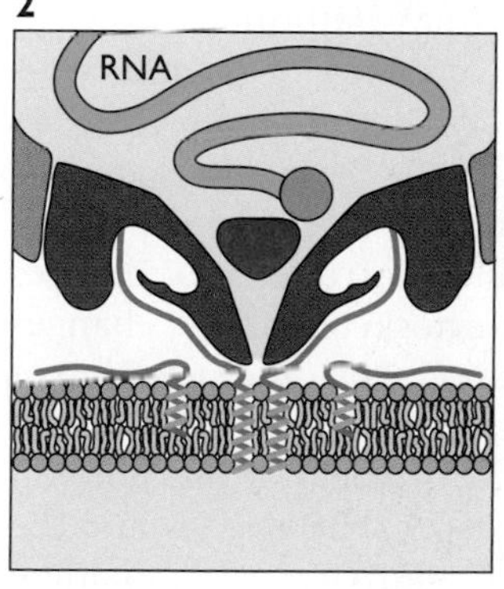

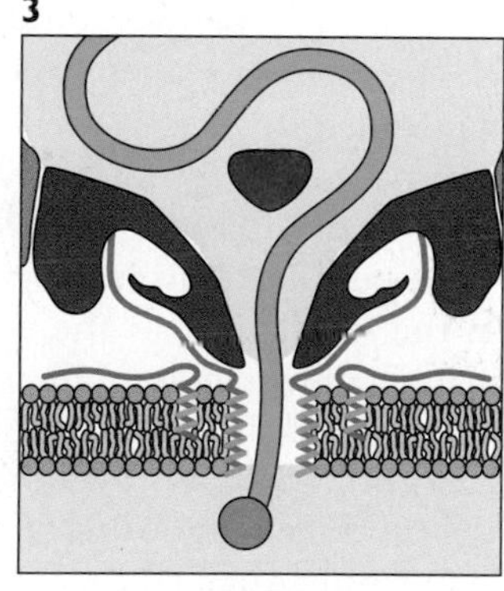

Figure 5.27 Model for poliovirus entry into cells. (A) Overview. The native virion (160S) binds to its cell receptor, Pvr, and at temperatures higher than 33°C undergoes a receptor-mediated conformational transition resulting in the formation of altered (A) particles. The A particles have lost the internal viral protein VP4, and the hydrophobic N terminus of VP1, also normally internal, is displaced to the virion surface. The A particle may be an intermediate in the uncoating of the viral RNA. It is not known whether the viral RNA, shown as a curved green line, leaves the capsid at the plasma membrane or from within endosomes. (B) Model of the formation of a pore in the cell membrane after poliovirus binding. 1, Poliovirus (shown in cross section, with capsid proteins purple) binds to Pvr (brown). 2, A conformational change leads to displacement of the pocket lipid (black) The pocket may be occupied by sphingosine in the capsid of poliovirus type 1. The hydrophobic N termini of VP1 (blue) are extruded and insert into the plasma membrane. 3, A pore is formed in the cell membrane by the VP1 N termini, through which the RNA is released from the capsid into the cytosol. Adapted from J. M. Hogle and V. R. Racaniello, p. 71–83, *in* B. L. Semler and E. Wimmer (ed.), *Molecular Biology of Picornaviruses* (ASM Press, Washington, D.C., 2002), with permission.

bound to hexon and import the protein into the nucleus, triggering capsid disassembly. The viral DNA and associated proteins are released into the nucleus, where viral transcription begins. The entire process of adenovirus entry, from the beginning of endocytosis to the release of DNA into the nucleus, takes approximately 30 min at 37°C (Fig. 5.26).

Forming a Pore in the Cell Membrane

The genome of the nonenveloped picornaviruses is transferred across the cell membrane by a different mechanism. Structural information at the atomic level, and complementary genetic and biochemical data obtained from studies of cell entry, is available for these viruses. The interaction of poliovirus with its immunoglobulinlike cell receptor, Pvr, leads to major conformational rearrangements in the virus particle (Fig. 5.27A). These altered (A) particles are missing the internal capsid protein VP4, and the N terminus of capsid protein VP1 is on the surface of the capsid rather than on the interior. Because of the latter change, altered particles are hydrophobic and possess an increased affinity for membranes compared to the native virus particle. It is believed that the exposed lipophilic N terminus of VP1 inserts into the cell membrane, forming a pore that allows transport of viral RNA into the cytoplasm (Fig. 5.27B). In support of this model, ion channel activity can be detected when A particles are added to lipid bilayers.

The fate of VP4 is not known, but the study of a virus with an amino acid change in VP4 indicates that this protein is required for an early stage of cell entry. Mutant virus particles can bind to target cells and convert to altered particles, but are blocked at a subsequent, unidentified step in entry. During poliovirus assembly, VP4 and VP2 are part of the precursor VP0, which remains uncleaved until the viral RNA has been encapsidated. The cleavage of VP0 during poliovirus assembly therefore primes the capsid for uncoating by separating VP4 from VP2.

Whether the altered particle is an obligatory uncoating intermediate (Fig. 5.27A) has been called into question by the observation that poliovirus can replicate at 25°C, a temperature at which conversion to altered particles cannot be detected. These findings suggest the possibility that the altered particle detected at higher temperatures may be a stable end product that is an exaggerated form of the uncoating intermediate. The latter may be short-lived and consequently difficult to detect.

It is not known if the release of poliovirus RNA occurs at the plasma membrane or from within endosomes (Fig. 5.27A). Bafilomycin A1, a drug that prevents the acidification of endosomes by inhibiting the transport of protons into the vesicles, has no effect on poliovirus infection. In contrast, this drug completely inhibits infection of cells with influenza virus, Semliki Forest virus, and vesicular stomatitis virus, all of which enter the cytoplasm from acidified endosomes. Furthermore, arrest of the clathrin-dependent endocytic pathway interferes with infection by Semliki Forest virus but not poliovirus. If poliovirus does enter the cell by endocytosis, this pathway is not sufficient to trigger uncoating. For example, antibody-poliovirus complexes can bind to cells that produce Fc receptors but cannot infect them. As the Fc receptor is known to be endocytosed, these results suggest that interaction of poliovirus with Pvr is required to induce conformational changes in the particle that are required for uncoating.

Some picornaviruses enter cells by a pH-dependent pathway, but precisely how acidic conditions lead to uncoating of these viruses in the cytoplasm is not understood. For example, foot-and-mouth disease virus is believed to enter cells by receptor-mediated endocytosis. Uncoating is then triggered by acidification of the endosome. Consistent with this entry mechanism, antibody-coated foot-and-mouth disease virus can bind to and infect cells that carry Fc receptors, in contrast to findings with poliovirus. This result suggests that the cell receptor for foot-and-mouth disease virus is a "hook": it serves only to tether the virus to the cell and bring it into the endocytic pathway and does not induce uncoating-related changes in the virus particle.

Lysosomal Uncoating

Most viruses that enter cells by receptor-mediated endocytosis leave the pathway before the vesicles reach the lysosomal compartment. This departure is not surprising, for lysosomes contain proteases and nucleases that would degrade virus particles. However, these enzymes play an important role during the uncoating of members of the *Reoviridae*, an event that takes place in lysosomes.

Orthoreoviruses are naked icosahedral viruses containing a double-stranded RNA genome of 10 segments. The viral capsid is a double-shelled structure composed of eight different structural proteins. These viruses bind to cell receptors via protein σ1 and are internalized into cells via endocytosis (Fig. 5.28A). Infection of cells by reoviruses is sensitive to bafilomycin A1, indicating that acidification of endosomes is required to enter cells. Low pH activates lysosomal proteases, which then modify several virion proteins, enabling the virus to cross the vesicle membrane. One viral outer capsid protein is cleaved and another is removed from the particle, producing an infectious subviral particle. These subviral particles penetrate the lysosome membrane and escape into the cytosol by a mechanism that is not yet understood. Isolated infectious subviral particles cause cell membranes to become permeable to toxins and produce pores in artificial membranes. These particles can initiate an infection by penetrating the plasma membrane, entering the cytoplasm directly. Their infectivity is not sensitive to bafilomycin A1, further supporting the idea that these particles are primed for membrane entry and do not require further acidification for this process. The core particles generated from infectious subviral particles after penetration into the cytoplasm carry out viral mRNA synthesis.

Cryo-electron microscopy has established the structural basis for some of the entry steps of reovirus (Fig. 5.28A). This virus is designed to undergo a series of sequential disassembly steps in which layers of the capsid are removed, ultimately resulting in entry of the core into the cytoplasm and initiation of viral mRNA synthesis.

Endocytosis via Caveolae

Simian virus 40 particles enter cells by endocytosis into small, uncoated vesicles called **caveolae**, instead of clathrin-coated vesicles. Caveolae are membranous vesicles that pinch off from the plasma membrane and participate in transcytosis, signal transduction, and uptake of membrane components and extracellular ligands. Caveolae can be distinguished from clathrin-coated vesicles by their size, shape, and the presence of a marker protein called caveolin. Furthermore, ligand binding triggers internalization through caveolae. An example is the induction of internalization by the binding of simian virus 40 to caveolae. This interaction leads to a series of changes in the vesicle and in the actin cytoskeleton. The changes include disassembly of filamentous actin, and recruitment of actin and dynamin II to the cytosolic surface of caveolae. Both actin and dynamin are needed for pinching off the caveolae, and for the transport of the vesicles into the cytoplasm. After endocytosis, simian virus 40-containing caveolae fuse with larger organelles called caveosomes. From the latter the virus particles are sorted to the endoplasmic reticulum.

An interesting question is how simian virus 40 virus particles travel from the endoplasmic reticulum to the nu-

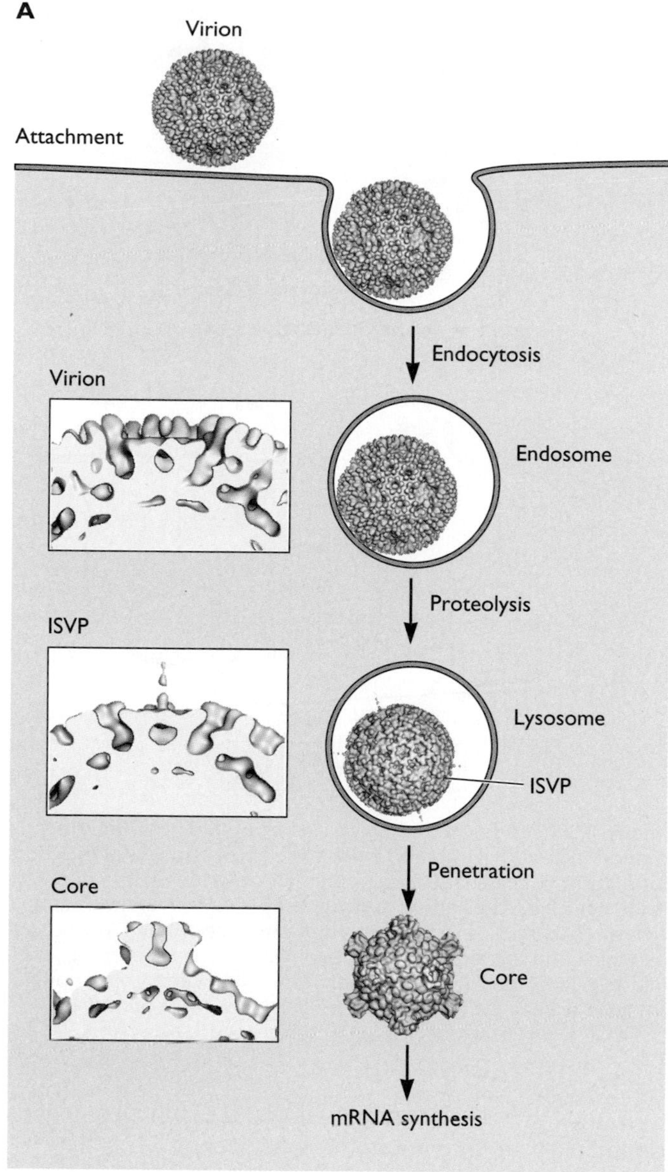

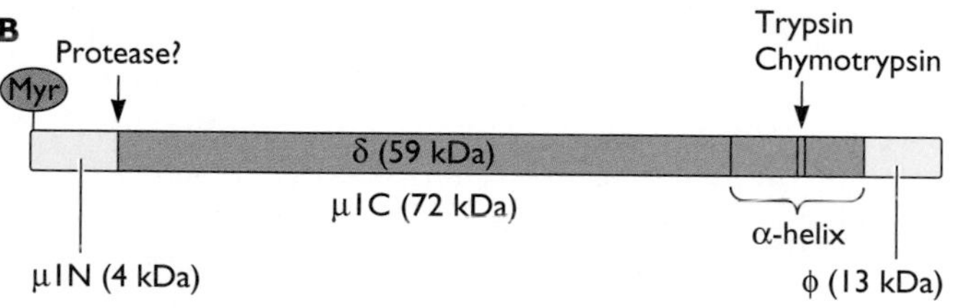

Figure 5.28 Entry of reovirus into cells. (A) The different stages in cell entry of reovirus. After the attachment of σ1 protein to the cell receptor, the virus particle enters the cell by receptor-mediated endocytosis. Proteolysis in the late endosome produces the infectious subviral particle (ISVP), which may then cross the lysosome membrane and enter the cytoplasm as a core particle. The intact virion is composed of two concentric, icosahedrally organized protein capsids. The outer capsid is made up largely of σ3 and μ1 and stabilized by σ3-μ1, λ2-σ3, and λ2-μ1 interactions. The dense core shell is formed mainly by λ1 and σ2. In the ISVP, 600 σ3 subunits have been released by proteolysis, and the σ1 protein changes from a compact form to an extended flexible fiber, possibly as a result of the release of σ3. The μ1 protein is believed to mediate interaction of the ISVP with membranes. This protein is present in virions as two cleaved fragments, μ1N and μ1C (see schematic of μ1 in panel B). The N terminus of μ1N is modified with myristate, suggesting that the protein functions in the penetration of membranes. A pair of amphipathic α-helices flank a C-terminal trypsin/chymotrypsin cleavage site at which μ1C is cleaved by lysosomal proteases. Such cleavage may release the helices to facilitate membrane penetration. The membrane-penetrating potential of μ1C in the virion may be masked by σ3; release of the σ3 in ISVPs might then allow μ1C to interact with membranes. The core is produced by the release of 12 σ1 fibers and 600 μ1 subunits. In the transition from ISVP to core, domains of λ2 rotate upward and outward to form a turretlike structure. (Insets) Close-up views of the emerging turretlike structure as the virus progresses through the ISVP and core stages. This structure may facilitate the entry of nucleotides into the core and the exit of newly synthesized viral mRNAs. (B) Schematic of the μ1 protein, showing locations of myristate, protease cleavage sites, and amphipathic α-helices. Virus images reprinted from K. A. Dryden et al., *J. Cell Biol.* **122:**1023–1041, 1993, with permission. Courtesy of Norm Olson and Tim Baker, Purdue University.

cleus, the site of replication of this virus. In the endoplasmic reticulum, virus particles would not have access to cytoplasmic import proteins, or the cytoplasmic side of the nuclear pore complex (see "The Cellular Pathway of Protein Import into the Nucleus" below). There may be a specific mechanism by which the virus crosses either the membrane of the endoplasmic reticulum or the outer nuclear membrane to enter the cytoplasm, so that it can gain access to the nuclear import machinery. Alternatively, virus particles might cross the inner nuclear membrane and enter the nucleoplasm directly. This mechanism would be unique.

Other viruses that are believed to enter cells via caveolae include Ebola virus, Marburg virus, echovirus type 1, and Hantaan virus.

Cytoplasmic Transport during Entry

There are two basic ways for viruses to travel within the cell—within a membrane vesicle, which interacts with the cytoskeletal transport machinery, or directly in the cytoplasm (Fig. 5.18). In the latter case, a virus particle must bind directly to the transport machinery. For example, the cytoplasmic domain of Pvr, the cellular receptor for poliovirus, binds the light chain of the motor protein dynein. This interaction might target endocytic vesicles containing Pvr, to the microtubule network, allowing transport of the viral capsid in the cytoplasm. Herpes simplex virus type 1 fuses with the plasma membrane, and the capsids then move from the cell surface to the nucleus by microtubule-based retrograde transport. Similarly, subviral particles of human immunodeficiency virus 1 are propelled along microtubules toward the nucleus. After leaving endosomes, the capsids of adenoviruses and parvoviruses are transported along microtubules to the nucleus. Although adenovirus has an overall net movement toward the nucleus, capsids exhibit bidirectional plus- and minus-end-directed microtubule movement. Adenovirus binding to cells activates two different signal transduction pathways that increase the net velocity of minus-end-directed capsid motility. The upregulated signaling pathways are therefore required for efficient delivery of the viral genome to the nucleus. An important, but unanswered, question is how viral capsids are released from the microtubules to move to the nuclear pore complex, where the viral genomes are injected into the nucleus.

It is not known if microtubule-based retrograde transport carries all viruses to their sites of replication after entry. Furthermore, the dependence on microtubules is not always absolute. For example, treatment of cells with nocodazole, a drug that depolymerizes microtubules, blocks adenovirus accumulation at microtubule organizing centers, but the viral genome still reaches the nucleus. The actin cytoskeleton may be an alternative transport pathway, as has been found for an insect virus. Although actin plays a role in the endocytosis of some viruses, there is yet no direct evidence for a role of actin in transport of mammalian viruses during entry.

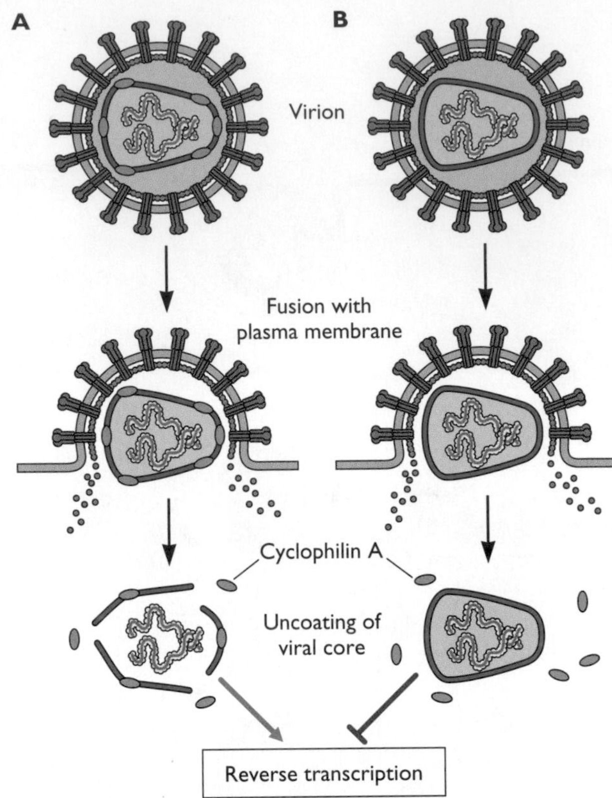

Figure 5.29 Model of the role of cyclophilin A in human immunodeficiency virus type 1 uncoating. The pathways of infection for mature virus particles containing (A) or lacking (B) cyclophilin A are depicted. After fusion at the plasma membrane, the viral nucleocapsid is released into the cytoplasm. Cyclophilin A (orange) packaged into the virus particle destabilizes the nucleocapsid, leading to release of CA and the initiation of reverse transcription. Adapted from J. L. Luban, *Cell* **87:**1–20, 1996, with permission.

Cellular Molecules That Participate in Uncoating

Uncoating of Human Immunodeficiency Virus Type 1 Requires Cyclophilin A

There is evidence that members of the cyclophilin family of proteins play an important role in the changes in the human immunodeficiency virus type 1 core that occur after it is released at the plasma membrane. Members of this family are chaperones, facilitating protein folding and assembly, and possess peptidylprolyl isomerase activity. Cyclophilin A is incorporated into human immunodeficiency virus type 1 virions at a stoichiometry of approximately 1 molecule to 10 molecules of the capsid protein CA. When cyclophilin A is excluded from virions, for example, by changing the sequence of CA that binds to cyclophilin A, the virus particles made are morphologically and biochemically normal and can enter cells. However, they are not infectious because of a block early in replication before reverse transcription takes place (Fig. 5.29). The presence of cyclophilin A in infected cells does not rescue infectivity of the mutant virus.

A model has been proposed for the function of cyclophilin A in human immunodeficiency virus type 1 replication. Shortly after the virus fuses with the plasma membrane to release the core into the cytoplasm, this structure may need to expand to facilitate reverse transcription. This expansion may be caused by the release of some of the CA subunits. It has been suggested that cyclophilin A acts as a chaperone, a protein that facilitates the folding of other proteins, and assists in the release of CA. When cyclophilin A is not present in the virus particle, the core does not expand, a property that may explain why reverse transcription cannot begin (Fig. 5.29). If correct, this mechanism would identify an unexpected function for chaperones, which are best known for their roles in facilitating protein folding or assembly, not disassembly.

Examination of the crystal structure of cyclophilin bound to the N-terminal 151 residues of the CA protein (called CA_{151}) has identified features consistent with a function for cyclophilin in uncoating of the viral core. Human immunodeficiency virus type 1 CA_{151} is pyramidal with a high content of α-helices (Fig. 5.30). One of the few parts of CA_{151} that do not form an α-helix is a loop believed to be exposed on the surface of the virion core. This loop contains a proline-rich region required for binding cyclophilin A and is the only part of CA protein that contacts cyclophilin A in cocrystals of the two molecules. Substitution of residues G89 and P90 of CA abrogates binding to cyclophilin A. These amino acids are located at the top of the loop in CA_{151}, buried deep within the hydrophobic pocket of cyclophilin A. A model for the organization of CA within the virion core suggests that "strips" composed of two rows of CA molecules interact side to side to form a two-dimensional surface (Fig. 5.30). In this model, the N termini of all the CA molecules point upward toward the viral membrane. The presence of a molecule of cyclophilin A for each 10 molecules of CA protein might introduce minor dislocations between neighboring strips of CA molecules, thereby destabilizing the core and facilitating disassembly. This mechanism of disassembly would be specific for human immunodeficiency virus type 1, because other retroviruses do not contain cyclophilin A.

Cellular Lipids Regulate Poliovirus Uncoating

A critical regulator of the receptor-induced structural transitions of poliovirus appears to be a hydrophobic tunnel located below the surface of each structural unit (Fig. 5.12). The tunnel opens at the base of the canyon and extends toward the fivefold axis of symmetry. In poliovirus type 1, each tunnel is occupied by a natural ligand believed to be a molecule of sphingosine. Rhinoviruses 1A and 16 each also contain a lipid, although they are shorter than the sphingosine of poliovirus. Because of the symmetry of the capsid, each virion may contain up to 60 lipid molecules.

The lipids are believed to contribute to the stability of the native virus particle, possibly by locking the capsid in a stable conformation. Removal of the lipid is probably necessary to endow the particle with sufficient flexibility to permit the RNA to leave the shell. These conclusions come from the study of antiviral drugs known as WIN compounds (named after Sterling-Winthrop, the pharmaceutical company at which they were discovered) that displace the lipid and fit tightly in the hydrophobic tunnel (Fig. 5.12). Polioviruses containing bound WIN compounds can bind to the cell receptor, but altered particles are not produced. WIN compounds may therefore inhibit

Figure 5.30 Location of cyclophilin A in the capsid of human immunodeficiency virus type 1. (A) Schematic of the arrangements of MA, CA, NC, and the viral RNA. The viral membrane is shown at the top. The structure of CA151 is shown as a ribbon diagram; the remainder of the molecule is labeled C. The cyclophilin A binding site is located in the loop between helices IV and V of CA. (B) Higher-order interaction of CA molecules (yellow and black) to form a shell beneath the viral envelope. The arrow indicates the possible side-to-side interaction between two strips of CA molecules. The cyclophilin A binding loop of CA projects upward from the page. Adapted from T. R. Gamble et al., *Cell* **87**:1285-1294, 1996, with permission.

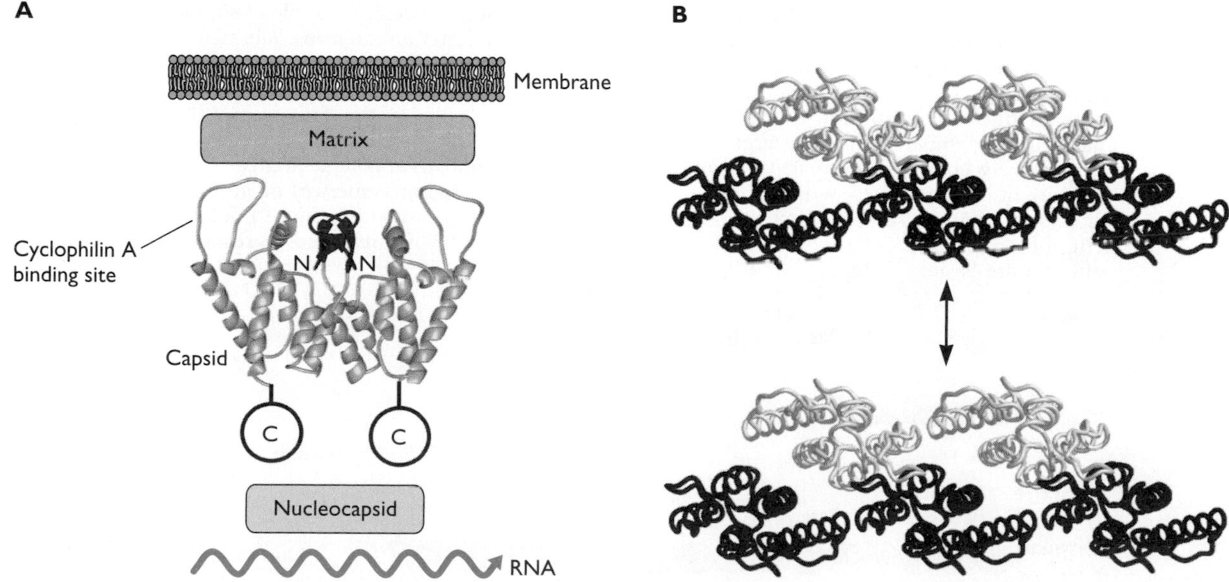

poliovirus infectivity by preventing the receptor-mediated conformational alterations required for uncoating. The properties of poliovirus mutants that cannot replicate in the absence of WIN compounds further support the role of the lipids in uncoating. These drug-dependent mutants spontaneously convert to altered particles at 37°C, in the absence of the cell receptor, probably because they do not contain lipid in the hydrophobic pocket. The lipids are therefore viewed as switches, because their presence or absence determines whether the virus is stable or will be uncoated. The interaction of the virus particle with its receptor probably initiates structural changes in the virion that lead to the release of lipid. Consistent with this hypothesis is the observation that Pvr docks onto the poliovirus capsid just above the hydrophobic pocket (see "Attachment via surface features: canyons and loops" above).

Transport of the Viral Genome into the Nucleus

The replication of DNA viruses, and RNA viruses such as influenza viruses, Borna disease virus, and hepatitis delta satellite virus, begins in the cell nucleus. The genomes of these viruses, as well as the DNA synthesized from retroviral RNA in the cytoplasm, must therefore be transported from the cytoplasm across the nuclear membrane. This movement is accomplished via the cellular pathway for protein import into the nucleus.

Although the discussions in this chapter concern use of the nuclear pore complex to enter the nucleus, it should be noted that this portal is also used for export of cellular and viral mRNAs. For viruses that replicate in the cell nucleus, the nuclear pore complexes mediate the transport of nucleoproteins into the cytoplasm for the final steps of assembly. These events are discussed in Chapters 10 and 12.

The Cellular Pathway of Protein Import into the Nucleus

Nuclear Localization Signals

Proteins that reside within the nucleus are characterized by the presence of specific nuclear targeting sequences. Such **nuclear localization signals** are both necessary for nuclear localization of the proteins in which they are present and sufficient to direct heterologous, nonnuclear proteins to enter this organelle. Nuclear localization signals identified by these criteria share a number of common properties: they are generally fewer than 20 amino acids in length; they are not found at the same position within all nuclear proteins, they are not removed after entry of the protein into the nucleus, and they are usually rich in basic amino acids. Despite these similarities, no consensus nuclear localization sequence can be defined.

Most nuclear localization signals belong to one of two classes, simple or bipartite sequences (Fig. 5.31). A particularly well characterized example of a simple nuclear localization signal is that of simian virus 40 large T antigen, which comprises five contiguous basic residues flanked by a single hydrophobic amino acid (Fig. 5.31). This se-

Figure 5.31 Nuclear localization signals. The general form and a specific example of simple and bipartite nuclear localization signals are shown in the one-letter amino acid code, where X is any residue. The sequences shown were defined by the criteria described in the text. Furthermore, specific substitutions, such as that of K129 of the simian virus 40 (SV40) T-antigen nuclear localization signal, prevent nuclear accumulation of the protein. As the name implies, bipartite nuclear targeting signals are defined by the presence of two clusters of positively charged amino acids separated by a spacer region of variable sequence. Both clusters of basic residues, which often resemble the simple targeting sequences of proteins like SV40 T antigen, are required for efficient import of the proteins in which they are found. The subscript indicates either length (3–7) or composition (e.g., 3/5 means at least 3 residues out of 5 are basic).

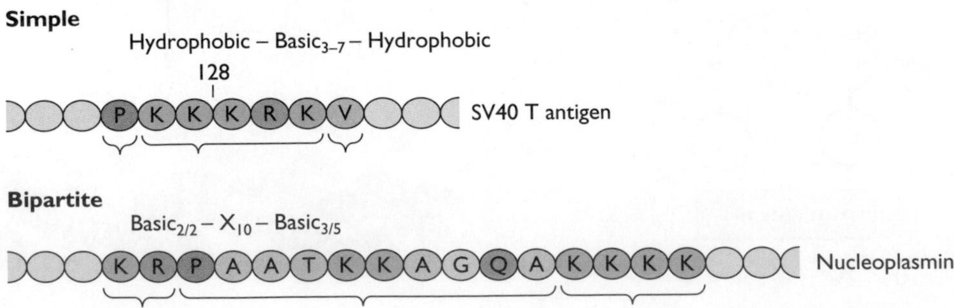

quence is sufficient to relocate the enzyme pyruvate kinase, normally found in the cytoplasm, to the nucleus. Many other viral and cellular nuclear proteins contain short, basic, nuclear localization signals, but these signals are not identical in primary sequence to the T-antigen signal (Table 5.3). The presence of a nuclear localization signal is all that is needed to target a macromolecular substrate for import into the nucleus. Even gold particles with diameters as large as 26 nm are readily imported following their microinjection into the cytoplasm, as long as they are coated with proteins or peptides containing a nuclear localization signal.

The Nuclear Pore Complex

The nuclear envelope is composed of two typical lipid bilayers separated by a luminal space (Fig. 5.32A). Like all other cellular membranes, it is impermeable to macromolecules such as proteins. However, the nuclear pore complexes that stud the nuclear envelopes of all eukaryotic cells provide aqueous channels that span both the inner and outer nuclear membranes for exchange of small molecules, macromolecules, and macromolecular assemblies between nuclear and cytoplasmic compartments. Numerous experimental techniques, including direct visualization of gold particles attached to proteins or RNA molecules as they are transported, have established that nuclear proteins enter and RNA molecules exit the nucleus by transport through the nuclear pore complex. The functions of the nuclear pore complex in both protein import and RNA export are far from completely understood, not least because this important cellular machine is large (molecular mass, approximately 124×10^3 kDa in vertebrates), built from many different proteins, and architecturally complex (Fig. 5.32C). In comparison, ribosomes, which consist of ~82 proteins and four RNA molecules, have a molecular mass of 4.2×10^3 kDa.

The nuclear pore complex allows passage of cargo in and out of the nucleus by either passive diffusion or facilitated translocation. Passive diffusion does not require interaction between the cargo and components of the nuclear pore complex, and becomes inefficient as molecules approach 20 to 40 kDa in mass. Objects as large as several megadaltons can pass through nuclear pore complexes by facilitated translocation. This process requires specific interactions between the cargo and components of the nuclear pore complex and is therefore highly selective.

The Nuclear Import Pathway

Import of a protein into the nucleus via nuclear localization signals occurs in two distinct, and experimentally separable, steps. A protein containing such a signal first binds to a soluble cytoplasmic receptor protein. This complex then engages with the cytoplasmic surface of the nuclear pore complex, in a reaction often called docking, and

Table 5.3 Nuclear localization signals of some viral proteins

Virus	Protein	Nuclear targeting signal	
		Position (residues)[a]	Sequence
Structural proteins			
Simian virus 40	VP1	1–11	APTKRK[6]
	VP2[b]	316–324	GPNKKKRKL
Adenovirus type 2	Fiber	1–11	MKRARPSEDTF
	Pre-TP	362–373	RLPVRRRRRRVP
Influenza virus	M1	101–105	RKLKR
	NP	327–445	AAFEDLRVRS[345]
Nonstructural proteins			
Simian virus 40	Large T antigen	126–132	PKKKRKV
Adenovirus type 2	E1A 13S	285–289	KRPRP
	DNA polymerase	1–48	Bipartite[c]
Epstein-Barr virus	EBNA-1	379–386	LKRPRSPSS
Human immunodeficiency virus type 1	Tat	48–52	GRKKR
	Rev	38–45	RRNRRRRW
Influenza virus	NS1		Two functionally independent sequences

[a]Position indicates the region of the protein shown to contain a nuclear localization signal. When the sequences listed in column 4 are part of a large region shown in column 3, they were generally identified by similarity to another nuclear localization signal, rather than actually demonstrated to be a nuclear localization signal.

[b]This sequence is also present in the VP3 protein.

[c]Bipartite nuclear localization signals contain two basic sequences (see Fig. 5.31).

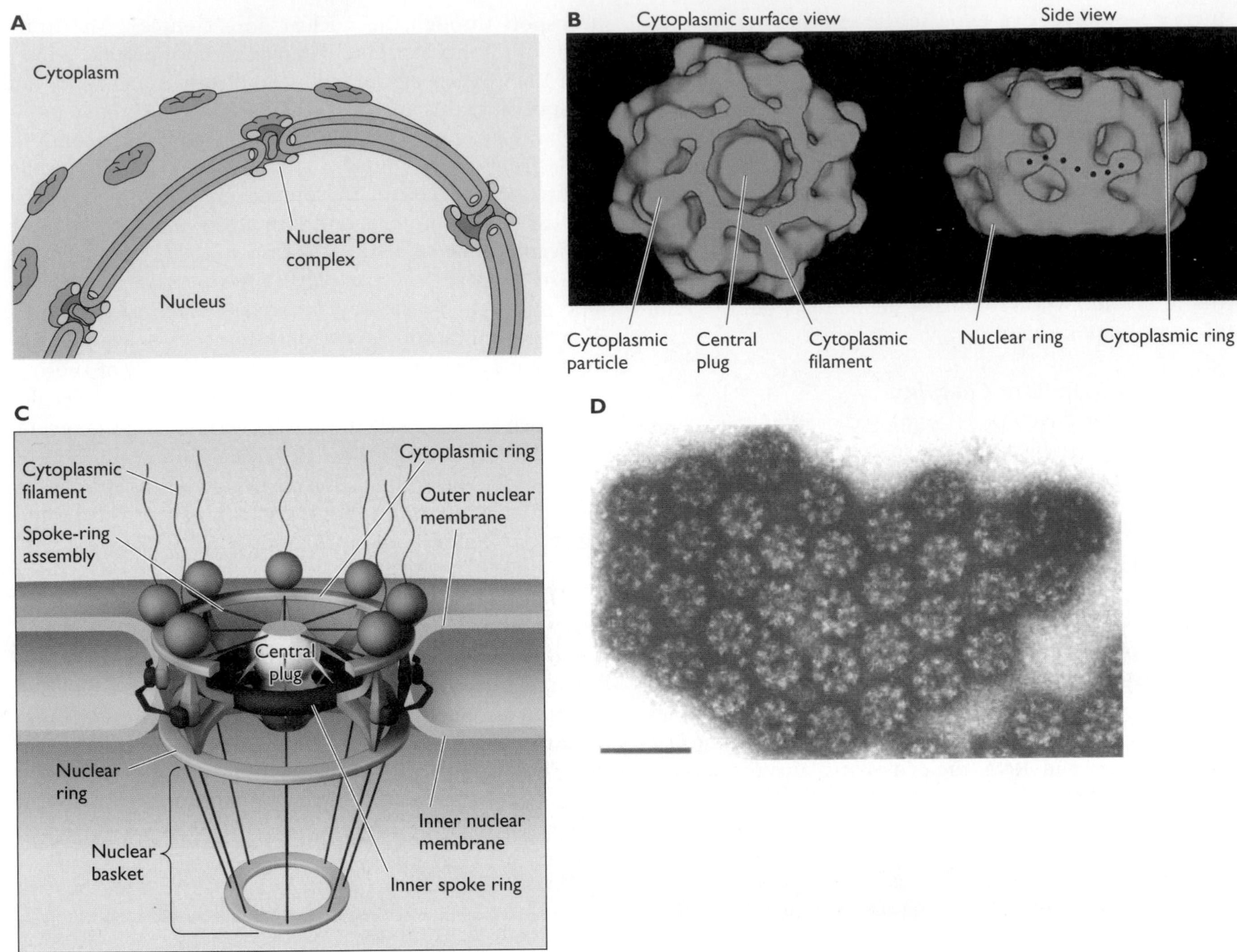

Figure 5.32 Structural model of the nuclear pore complex. (A) Overview of the nuclear membrane, showing the topology of the nuclear pore complexes. (B) Three-dimensional image reconstructions from cryo-electron microscopy of a cytoplasmic surface view and a side view of the vertebrate nuclear pore complex. Reprinted from Q. Yang, M. P. Rout, and C. W. Akey, *Mol. Cell* **1**:223–234, 1998, with permission. (C) Schematic drawing of the nuclear pore complex, showing the spoke-ring assembly at the waist of the nuclear complex and its attachment to cytoplasmic filaments and the nuclear basket. The latter comprises eight filaments, extending 50 to 100 nm from the central structure and terminating in a distal annulus. The nuclear pore channel is shown containing the transporter. This model is based on electron microscopy (D) and image analysis of nuclear pore complexes (B), generally of *Xenopus* oocytes, in which the nuclear envelope is packed with a paracrystalline array of nuclear pore complexes. It represents a composite of features observed in different experiments, for no single method of sample preparation preserves all the structural elements shown in the figure. Adapted from Q. Yang, M. P. Rout, and C. W. Akey, *Mol. Cell* **1**:223–234, 1998, with permission. (D) Face-on view of negatively stained nuclear pore complexes in an electron microscope. The very abundant nuclear complexes, which are about 150 nm in diameter, clearly exhibit eightfold symmetry. A central channel is visible and in some cases lacks a plug. Bar, 0.2 μm. Courtesy of Ron Milligan (Scripps Research Institute).

is translocated through the nuclear pore complex into the nucleus. In the nucleus, the complex is disassembled, releasing the protein "cargo." Our current, but still incomplete, model of the nuclear import pathway and its components is summarized in Fig. 5.33.

Different groups of proteins are imported into the nucleus by specific receptor systems. In what is known as the "classical system" of import, cargo proteins contain basic nuclear localization signals. The best-studied cytoplasmic nuclear localization signal receptor protein is im-

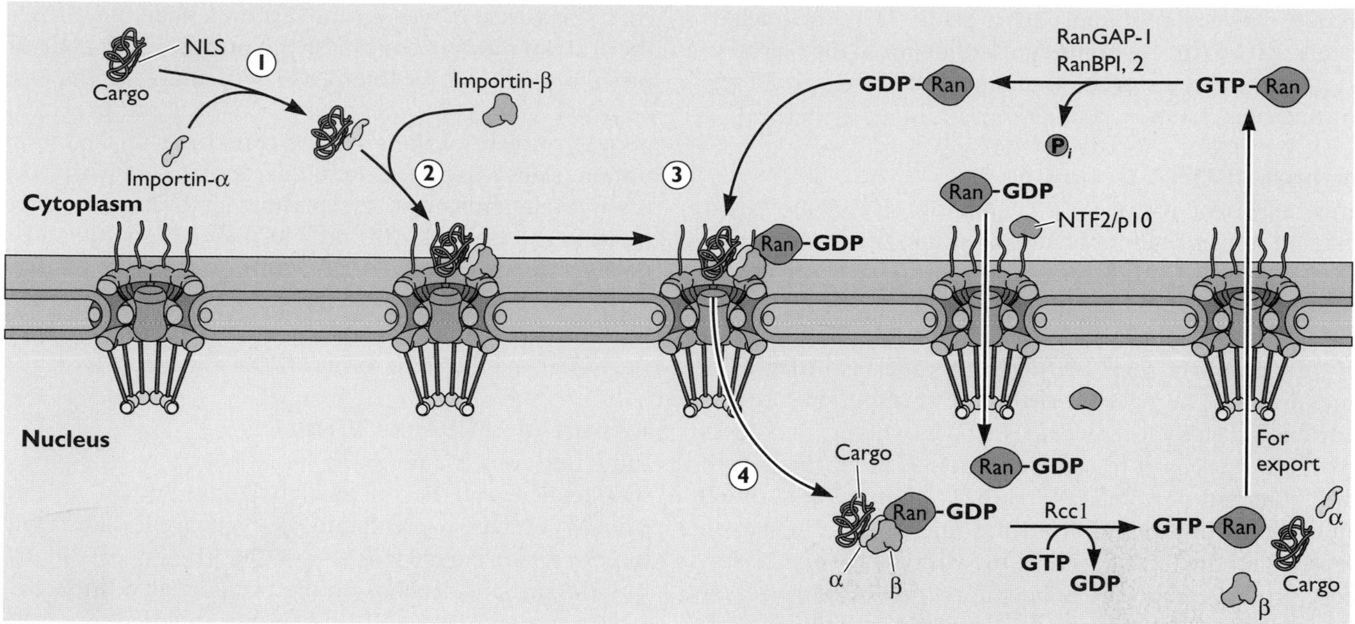

Figure 5.33 The protein import pathway. This pathway is illustrated schematically from left to right. Cytoplasmic and nuclear compartments are shown separated by the nuclear envelope studded with nuclear pore complexes. In step 1, a nuclear localization signal on the cargo (red) is recognized by importin α. Although the simple simian virus 40 and some bipartite nuclear localization signals (Table 5.2) bind to the same nuclear localization signal (NLS) receptor, metazoan cells contain multiple related, but distinct, importin α proteins which do not appear to fulfill identical functions. In step 2, importin β binds the cargo-importin α complex and docks onto the nucleus, probably by associating initially with nucleoporins present in the cytoplasmic filaments of the nuclear pore complex. As shown, this interaction would bring the nuclear localization signal-containing protein to the cytoplasmic face of the nuclear pore complex. Translocation of the substrate into the nucleus (step 4) requires additional soluble proteins, including the guanine nucleotide-binding protein Ran (step 3). A Ran-specific guanine nucleotide exchange protein (Rcc1) and a Ran-GTPase-activating protein (RanGAP-1) are localized in the nucleus and cytoplasm, respectively. The action of RanGAP-1, with the accessory proteins RanBP1 and RanBP2, maintains cytoplasmic Ran in the GDP-bound form. When Ran is in the GTP-bound form, nuclear import cannot occur. Following import, the complexes are dissociated when Ran-GDP is converted to Ran-GTP by Rcc1. Ran-GTP participates in export from the nucleus. When Ran is in the GDP-bound form, nuclear export cannot occur. The nuclear pool of Ran-GDP is replenished by the action of the transporter NTF2/p10, which efficiently transports Ran-GDP from the cytoplasm to the nucleus. Hydrolysis of Ran-GTP in the cytoplasm and GTP-GDP exchange in the nucleus therefore maintain a gradient of Ran-GTP/Ran-GDP. The asymmetric distribution of RanGAP-1 and Rcc1 allows for the formation of such a gradient. This gradient provides the driving force and directionality for nuclear transport.

portin α (Fig. 5.33). This complex then binds importin β, which mediates docking with the nuclear pore complex by binding to members of a family of nucleoporins. Some of these nucleoporins are found in the cytoplasmic filaments of the nuclear pore complex (Fig. 5.32C), which associate with import substrates as seen by electron microscopy. These properties suggest that cytoplasmic filaments are the sites of the initial association of receptor-substrate complexes with the nuclear pore complex. A monomeric receptor called transportin mediates import of heterogeneous nuclear RNA-binding proteins that contain glycine- and arginine-rich nuclear localization signals. Transportin is related to importin β, as are other monomeric receptors that mediate nuclear import of ribosomal proteins.

Translocation requires additional transport proteins, in most cases including a small GTP-binding protein termed Ran (Fig. 5.33) and nucleoporins (Table 5.3). How these components work together to move the import substrate through the channel of the nuclear pore complex, a distance of more than 100 nm, is not yet understood. One model proposes that nucleoporins, which contain characteristic repeats of hydrophobic amino acids, form a permeability "sieve" that can be entered only by cargo bound to importin receptors. It is clear that a single translocation through the nuclear pore complex does not require energy consumption. However, maintenance of a gradient of the guanosine nucleotide bound forms of Ran, with Ran-GDP and Ran-GTP concentrated in the cytoplasm and nucleus, respectively, is essential for continued transport. For example, conversion of Ran-GDP to Ran-GTP in the nucleus catalyzed by the guanine nucleotide exchange protein Rcc-1 promotes dissociation of imported proteins from importins (Fig. 5.33).

Import of DNA Genomes

The capsids of many DNA-containing viruses are larger than 26 nm and cannot be imported into the nucleus from the cytoplasm. A different mechanism has evolved for crossing the nuclear membrane in which capsids are released into the cytoplasm, dock onto the nuclear pore complex, and deliver their DNA into the nucleus. Adenoviral and probably herpesviral DNAs are transported into the nucleus via this mechanism. The adenovirus capsid docks onto the nuclear pore complex through interactions with the filament protein CAN/Nup214 (Fig. 5.26). Small amounts of histone H1 from the nucleus bind to hexon proteins on the nuclear side of the viral capsid. The H1 import proteins importin β and importin 7 recognize H1 bound to hexon and import the viral DNA into the nucleus.

Import of Retroviral Genomes

As we have seen, fusion of retroviral and plasma membranes releases the viral core into the cytoplasm. The retroviral core consists of the viral RNA genome, coated with NC protein, and the enzymes RT and IN, enclosed by CA protein. Retroviral DNA synthesis commences in the cytoplasm, within the nucleocapsid core, and after 4 to 8 h of DNA synthesis the preintegration complex, comprising viral DNA, IN, and perhaps other proteins, localizes to the nucleus. There the viral DNA is integrated into a cellular chromosome, and viral transcription begins. The mechanism of nuclear import of the preintegration complex is poorly understood, but it is quite clear that this structure is too large (ca. 60S) to pass through the nuclear pore complex. The betaretrovirus Moloney murine leukemia virus can infect only dividing cells. These and other observations suggest that the breakdown of the nuclear membrane that occurs during mitosis is essential to allow efficient entry of the preintegration complex of this retrovirus into the nucleus.

In contrast to Moloney murine leukemia virus, human immunodeficiency virus type 1 can replicate in nondividing cells. The preintegration complex of this virus, and probably other lentiviruses, must therefore be actively transported into the nucleus. Some evidence suggests that Vpr and a minor, phosphorylated form of the matrix protein direct nuclear import of the viral preintegration complex. Nuclear localization signals may also contribute to such import.

Avian sarcoma and leukosis viruses, like Moloney murine leukemia virus, do not replicate in nondividing cells. However, it was recently shown that the DNA of these avian retroviruses can be integrated in cell cycle-arrested cells and during interphase in cycling cells, implying an active, mitosis-independent mechanism of nuclear import. IN protein of these viruses contains a nuclear localization signal in the C-terminal domain which, when fused to heterologous, cytoplasmic proteins, can direct them to the nucleus. This protein may have a role in the nuclear import of the preintegration complex of these avian retroviruses. In any case, the block to replication of these viruses in nondividing cells is likely to occur at some step other than nuclear localization.

Import of Influenza Virus RNP

Influenza virus is among the few RNA-containing viruses that replicate in the cell nucleus. After vRNPs separate from M1 and are released into the cytosol, they are rapidly imported into the nucleus (Fig. 5.20). The import of vRNP into the nucleus depends on the presence of a nuclear localization signal in the NP protein, a component of vRNPs. Naked viral RNA does not dock onto the nuclear pore complex nor is it taken up into the nucleus, but in the presence of NP the viral RNA can enter this organelle. The two nuclear localization signals identified in NP protein are called nonconventional because they do not resemble "classical" import sequences.

Perspectives

The receptors and viral attachment proteins discussed in this chapter represent only a small subset of those that have been studied to some degree or are still to be explored. Receptors for fewer than 100 viruses are listed in Table 5.1, a small fraction of the ~4,000 known viruses. Although many more viral receptors await identification, studies of a small number have yielded fundamental principles about how receptors function during virus attachment and entry. A general conclusion that can be drawn is that the interactions governing virus attachment are complex, consisting of reciprocal changes in the virion and the cell. For some viruses, such as poliovirus, rhinovirus, and influenza virus, a single type of cellular molecule is sufficient for virus binding. In other cases, including human immunodeficiency virus type 1, adenovirus, and alphaherpesviruses, one type of cellular molecule is required for attachment, and a coreceptor is necessary for virus entry. There is a growing list of viruses that can attach to multiple receptors; which one is bound depends on the virus strain, the type of cultured cell, the host species, and even the site of replication within the host. The notion that only one receptor is needed for virus entry into cells, influenced by early studies with influenza viruses, has been shattered.

Perusal of Table 5.1 yields two striking conclusions about viral receptors. These molecules are a small subset of the large number of cell surface components. It is also surprising that many different viruses share the same cell receptor. Perhaps as additional receptors for viruses are identified, these patterns will no longer be so obvious. However, it seems more likely that viruses attach to only certain types of molecules because this interaction induces cellular responses that are beneficial for replication. Possible consequences of virus-receptor interactions include enhanced virus entry, more efficient replication, recruitment of susceptible cells to the site of infection, or inhibition of cellular defense mechanisms. Induction of cell death by receptor interaction may serve to facilitate release of virus particles from cells. The triggering of such events by virus binding to receptors may mimic interaction of the receptor with its cellular ligand. Future research is likely to focus more intensely on the question of whether virus-receptor interactions lead to intracellular signaling. The results should clarify why viruses bind certain receptors, and may also lead to new ways to prevent viral disease.

Although virus particles are structurally diverse, the mechanisms by which they enter cells appear to be surprisingly limited. Depending on whether the virus is enveloped, and whether the viral genome must reach the cytoplasm or the nucleus, common entry themes are evident. For example, all enveloped viruses must fuse with a membrane, either at the cell surface or from within an intracellular vesicle. Studies of this entry step have established that fusion is catalyzed by a viral glycoprotein with fusogenic potential that may be regulated by proteolytic cleavage. Fusion may be triggered by a reduction in pH or by interaction with coreceptors, which cause structural changes in the viral glycoprotein and expose the fusion peptide. Although the structures of some protein intermediates in the fusion reaction have been solved, "snapshots" of others will be required to understand this process. We know even less about the precise mechanism by which lipid bilayers fuse.

Once fusion has occurred, the genome can enter the cytoplasm. Release of the influenza virus RNP requires disruption of interactions with the M1 protein. A similar step may be required during uncoating of other (–) strand viruses, but few have been studied. In some enveloped virions, the genome is surrounded by a capsid or nucleocapsid, which must be disassembled to allow the RNA access to the cytoplasm. At least one togavirus nucleocapsid is disassembled in the cytoplasm by ribosomes; whether this is a universal mechanism for this family is not yet known.

There are large gaps in our understanding of how nonenveloped viruses enter cells. One strategy of entry is exemplified by adenoviruses and reoviruses, which become structurally altered as they move along the endocytic pathway, resulting in particles that can escape from endosomes. Poliovirus may form a pore at the cell surface, but crucial aspects of this hypothesis must be tested. Structural studies of intermediates in the entry of these three viruses should identify the underlying molecular mechanisms.

The genomes of many viruses replicate in the nucleus. Incoming viral genomes enter this cellular compartment by transport through the nuclear pore complex. Two distinct themes have emerged from the study of this step in infection. In one theme, exemplified by influenza virus, vRNP, which has been released into the cytoplasm, is imported into the nucleus. In the other theme, seen with DNA viruses, the viral capsid docks onto the nuclear pore complex, allowing the genome into the nucleus. There are many unanswered questions about nuclear import, e.g., whether components of the nuclear pore complex play a role in capsid disassembly and how molecules pass through the nuclear pore. Studies of viral entry into cells have been invaluable in clarifying aspects of normal cellular processes such as membrane fusion and endocytosis, and similar benefits are likely to come from studying how viral macromolecules enter the nucleus.

References

Reviews

Duncan, M. J., J. S. Shin, and S. N. Abraham. 2002. Microbial entry through caveolae: variations on a theme. *Cell. Microbiol.* **4:**783–791.

Dutch, R. E., T. S. Jardetzky, and R. A. Lamb. 2000. Virus membrane fusion proteins: biological machines that undergo a metamorphosis. *Biosci. Rep.* **20:**597–612.

Hogle, J. M. 2002. Poliovirus cell entry: common structural themes in viral cell entry pathways. *Annu. Rev. Microbiol.* **56:**677–702.

Pelkmans, L., and A. Helenius. 2002. Endocytosis via caveolae. *Traffic* **3:**311–320.

Ploubidou, A., and M. Way. 2001. Viral transport and the cytoskeleton. *Curr. Opin. Cell Biol.* **13:**97–105.

Poranen, M. M., R. Daugelavicius, and D. H. Bamford. 2002. Common principles in viral entry. *Annu. Rev. Microbiol.* **56:**521–538.

Schneider-Schaulies, J. 2000. Cellular receptors for viruses: links to tropism and pathogenesis. *J. Gen. Virol.* **81:**1413–1429.

Sieczkarski, S. B., and G. R. Whittaker. 2002. Dissecting virus entry via endocytosis. *J. Gen. Virol.* **83:**1535–1545.

Smith, G. A., and L. W. Enquist. 2002. Break ins and break outs: viral interactions with the cytoskeleton of mammalian cells. *Annu. Rev. Cell Dev. Biol.* **18:**135–161.

Sodeik, B. 2000. Mechanisms of viral transport in the cytoplasm. *Trends Microbiol.* **8:**465–472.

Weis, K. 2003. Regulating access to the genome. Nucleocytoplasmic transport throughout the cell cycle. *Cell* **112:**441–451.

Whittaker, G. 1998. Squeezing through pores: how microorganisms get into and out of the nucleus. *Trends Microbiol.* **6:**178–179.

Papers of Special Interest

Adenoviruses

Bergelson, J. M., J. A. Cunningham, G. Droguett, E. A. Kurt-Jones, A. Krithivas, J. S. Hong, M. S. Horwitz, R. L. Crowell, and R. W. Finberg. 1997. Isolation of a common receptor for coxsackie B viruses and adenoviruses 2 and 5. *Science* **275:**1320–1323.

Bewley, M. C., K. Springer, Y. B. Zhang, P. Freimuth, and J. M. Flanagan. 1999. Structural analysis of the mechanism of adenovirus binding to its human cellular receptor, CAR. *Science* **286:**1579–1583.

Mabit, H., M. Y. Nakano, U. Prank, B. Saam, K. Dohner, B. Sodeik, and U. F. Greber. 2002. Intact microtubules support adenovirus and herpes simplex virus infections. *J. Virol.* **76:**9962–9971.

Suomalainen, M., M. Y. Nakano, K. Boucke, S. Keller, and U. F. Greber. 2001. Adenovirus-activated PKA and p38/MAPK pathways boost microtubule-mediated nuclear targeting of virus. *EMBO J.* **20:**1310–1319.

Trotman, L. C., N. Mosberger, M. Fornerod, R. P. Stidwill, and U. F. Greber. 2001. Import of adenovirus DNA involves the nuclear pore complex receptor CAN/Nup214 and histone H1. *Nat. Cell Biol.* **3:**1092–1100.

Walters, R. W., P. Freimuth, T. O. Moninger, I. Ganske, J. Zabner, and M. J. Welsh. 2002. Adenovirus fiber disrupts CAR-mediated intercellular adhesion allowing virus escape. *Cell* **110:**789–799.

Wickham, T. J., P. Mathias, D. A. Cheresh, and G. R. Nemerow. 1993. Integrins alpha v beta 3 and alpha v beta 5 promote adenovirus internalization but not virus attachment. *Cell* **73:**309–319.

Arenaviruses

Cao, W., M. D. Henry, P. Borrow, H. Yamada, J. H. Elder, E. V. Ravkov, S. T. Nichol, R. W. Compans, K. P. Campbell, and M. B. A. Oldstone. 1998. Identification of α-dystroglycan as a receptor for lymphocytic choriomeningitis virus and Lassa fever virus. *Science* **282:**2079–2081.

Coronaviruses

Dveksler, G. S., C. W. Dieffenbach, C. B. Cardellichio, K. McCuaig, M. N. Pensiero, G. S. Jiang, N. Beauchemin, and K. V. Holmes. 1993. Several members of the mouse carcinoembryonic antigen-related glycoprotein family are functional receptors for the coronavirus mouse hepatitis virus-A59. *J. Virol.* **67:**1–8.

Yeager, C. L., R. A. Ashmun, R. K. Williams, C. B. Cardellichio, L. H. Shapiro, A. T. Look, and K. V. Holmes. 1992. Human aminopeptidase N is a receptor for human coronavirus 229E. *Nature* **357:**420–422.

Herpesviruses

Dohner, K., A. Wolfstein, U. Prank, C. Echeverri, D. Dujardin, R. Vallee, and B. Sodeik. 2002. Function of dynein and dynactin in herpes simplex virus capsid transport. *Mol. Biol. Cell* **13:**2795–2809.

Geraghty, R. J., C. Krummenacher, G. H. Cohen, R. J. Eisenberg, and P. G. Spear. 1998. Entry of alphaherpesviruses mediated by poliovirus receptor-related protein 1 and poliovirus receptor. *Science* **280:**1618–1620.

Montgomery, R. I., M. S. Warner, B. J. Lum, and P. G. Spear. 1996. Herpes simplex virus-1 entry into cells mediated by a novel member of the TNF/NGF receptor family. *Cell* **87:**427–436.

Ojala, P. M., B. Sodeik, M. W. Ebersold, U. Kutay, and A. Helenius. 2000. Herpes simplex virus type 1 entry into host cells: reconstitution of capsid binding and uncoating at the nuclear pore complex in vitro. *Mol. Cell. Biol.* **20:**4922–4931.

Shieh, M. T., D. WuDunn, R. I. Montgomery, J. D. Esko, and P. G. Spear. 1992. Cell surface receptors for herpes simplex virus are heparan sulfate proteoglycans. *J. Cell Biol.* **116:**1273–1281.

Shukla, D., J. Liu, P. Blaiklock, N. W. Shworak, X. Bai, J. D. Esko, G. H. Cohen, R. J. Eisenberg, R. D. Rosenberg, and P. G. Spear. 1999. A novel role for 3-O-sulfated heparan sulfate in herpes simplex virus 1 entry. *Cell* **99:**13–22.

Influenza Virus

Carr, C. M., C. Chaudhry, and P. S. Kim. 1997. Influenza hemagglutinin is spring-loaded by a metastable native conformation. *Proc. Natl. Acad. Sci. USA* **94:**14306–14313.

Chen, J., K. H. Lee, D. A. Steinhauer, D. J. Stevens, J. J. Skehel, and D. C. Wiley. 1998. Structure of the hemagglutinin precursor cleavage site, a determinant of influenza pathogenicity and the origin of the labile conformation. *Cell* **95:**409–417.

Han, X., J. H. Bushweller, D. S. Cafiso, and L. K. Tamm. 2001. Membrane structure and fusion-triggering conformational change of the fusion domain from influenza hemagglutinin. *Nat. Struct. Biol.* **8:**715–720.

Rogers, G. N., G. Herrler, J. C. Paulson, and H. D. Klenk. 1986. Influenza C virus uses 9-O-acetyl-*N*-acetylneuraminic acid as a high affinity receptor determinant for attachment to cells. *J. Biol. Chem.* **261:**5947–5951.

Vines, A., K. Wells, M. Matrosovich, M. R. Castrucci, T. Ito, and Y. Kawaoka. 1998. The role of influenza A virus hemagglutinin residues 226 and 228 in receptor specificity and host range restriction. *J. Virol.* **72:**7626–7631.

Weis, W., J. H. Brown, S. Cusack, J. C. Paulson, J. J. Skehel, and D. C. Wiley. 1988. Structure of the influenza virus hemagglutinin complexed with its receptor, sialic acid. *Nature* **333:**426–431.

Polyomavirus

Breau, W. C., W. J. Atwood, and L. C. Norkin. 1992. Class I major histocompatibility proteins are an essential component of the simian virus 40 receptor. *J. Virol.* **66:**2037–2045.

Pelkmans, L., J. Kartenbeck, and A. Helenius. 2001. Caveolar endocytosis of simian virus 40 reveals a new two-step vesicular-transport pathway to the ER. *Nat. Cell Biol.* **3:**473–483.

Pelkmans, L., D. Puntener, and A. Helenius. 2002. Local actin polymerization and dynamin recruitment in SV40-induced internalization of caveolae. *Science* **296:**535–539.

Paramyxoviruses

Baker, K. A., R. E. Dutch, R. A. Lamb, and T. S. Jardetzky. 1999. Structural basis for paramyxovirus-mediated membrane fusion. *Mol. Cell* **3:**309–319.

Dorig, R. E., A. Marcil, A. Chopra, and C. D. Richardson. 1993. The human CD46 molecule is a receptor for measles virus (Edmonston strain). *Cell* **75:**295–305.

Erlenhofer, C., W. P. Duprex, B. K. Rima, V. ter Meulen, and J. Schneider-Schaulies. 2002. Analysis of receptor (CD46, CD150) usage by measles virus. *J. Gen. Virol.* **83:**1431–1436.

Naniche, D., G. Varior-Krishnan, F. Cervoni, T. F. Wild, B. Rossi, C. Rabourdin-Combe, and D. Gerlier. 1993. Human membrane cofactor protein (CD46) acts as a cellular receptor for measles virus. *J. Virol.* **67:**6025–6032.

Russell, C. J., T. S. Jardetzky, and R. A. Lamb. 2001. Membrane fusion machines of paramyxoviruses: capture of intermediates of fusion. *EMBO J.* **20:**4024–4034.

Tatsuo, H., N. Ono, K. Tanaka, and Y. Yanagi. 2000. SLAM (CDw150) is a cellular receptor for measles virus. *Nature* **406:**893–897.

Parvoviruses

Chipman, P. R., M. Agbandje-McKenna, S. Kajigaya, K. E. Brown, N. S. Young, T. S. Baker, and M. G. Rossmann. 1996. Cryo-electron microscopy studies of empty capsids of human parvovirus B19 complexed with its cellular receptor. *Proc. Natl. Acad. Sci. USA* **93:**7502–7506.

Hueffer, K., J. S. Parker, W. S. Weichert, R. E. Geisel, J. Y. Sgro, and C. R. Parrish. 2003. The natural host range shift and subsequent evolution of canine parvovirus resulted from virus-specific binding to the canine transferrin receptor. *J. Virol.* **77:**1718–1726.

Summerford, C., and R. J. Samulski. 1998. Membrane-associated heparan sulfate proteoglycan is a receptor for adeno-associated virus type 2 virions. *J. Virol.* **72:**1438–1445.

Summerford, C., J. S. Bartlett, and R. J. Samulski. 1999. a_vb_5 integrin: a coreceptor for adeno-associated virus type 2 infection. *Nat. Med.* 5:78–82.

Poxviruses

Chung, C. S., J. C. Hsiao, Y. S. Chang, and W. Chang. 1998. A27L protein mediates vaccinia virus interaction with cell surface heparan sulfate. *J. Virol.* **72:**1577–1585.

Lalani, A. S., J. Masters, W. Zeng, J. Barrett, R. Pannu, H. Everett, C. W. Arendt, and G. McFadden. 1999. Use of chemokine receptors by poxviruses. *Science* **286:**1968–1971.

Rietdorf, J., A. Ploubidou, I. Reckmann, A. Holmstrom, F. Frischknecht, M. Zettl, T. Zimmermann, and M. Way. 2001. Kinesin-dependent movement on microtubules precedes actin-based motility of vaccinia virus. *Nat. Cell Biol.* **3:**992–1000.

Picornaviruses

Belnap, D. M., D. J. Filman, B. L. Trus, N. Cheng, F. P. Booy, J. F. Conway, S. Curry, C. N. Hiremath, S. K. Tsang, A. C. Steven, and J. M. Hogle. 2000. Molecular tectonic model of virus structural transitions: the putative cell entry states of poliovirus. *J. Virol.* **74:**1342–1354.

Belnap, D. M., B. M. McDermott, Jr., D. J. Filman, N. Cheng, B. L. Trus, H. J. Zuccola, V. R. Racaniello, J. M. Hogle, and A. C. Steven. 2000. Three-dimensional structure of poliovirus receptor bound to poliovirus. *Proc. Natl. Acad. Sci. USA* **97:**73–78.

Bergelson, J. M., J. A. Cunningham, G. Droguett, E. A. Kurt-Jones, A. Krithivas, J. S. Hong, M. S. Horwitz, R. L. Crowell, and R. W. Finberg. 1997. Isolation of a common receptor for coxsackie B viruses and adenoviruses 2 and 5. *Science* **275:**1320–1323.

He, Y., V. D. Bowman, S. Mueller, C. M. Bator, J. Bella, X. Peng, T. S. Baker, E. Wimmer, R. J. Kuhn, and M. G. Rossmann. 2000. Interaction of the poliovirus receptor with poliovirus. *Proc. Natl. Acad. Sci. USA* **97:**79–84.

Hewat, E. A., E. Neumann, and D. Blaas. 2002. The concerted conformational changes during human rhinovirus 2 uncoating. *Mol. Cell* **10:**317–326.

Hewat, E. A., E. Neumann, J. F. Conway, R. Moser, B. Ronacher, T. C. Marlovits, and D. Blaas. 2000. The cellular receptor to human rhinovirus 2 binds around the 5-fold axis and not in the canyon: a structural view. *EMBO J.* **19:**6317–6325.

Jackson, T., A. P. Mould, D. Sheppard, and A. M. King. 2002. Integrin alphavbeta1 is a receptor for foot-and-mouth disease virus. *J. Virol.* **76:**935–941.

Mendelsohn, C. L., E. Wimmer, and V. R. Racaniello. 1989. Cellular receptor for poliovirus: molecular cloning, nucleotide sequence and expression of a new member of the immunoglobulin superfamily. *Cell* **56:**855–865.

Mueller, S., X. Cao, R. Welker, and E. Wimmer. 2002. Interaction of the poliovirus receptor CD155 with the dynein light chain Tctex-1 and its implication for poliovirus pathogenesis. *J. Biol. Chem.* **277:**7897–7904.

Shafren, D. R. 1998. Viral cell entry induced by cross-linked decay-accelerating factor. *J. Virol.* **72:**9407–9412.

Xing, L., K. Tjarnlund, B. Lindqvist, G. G. Kaplan, D. Feigelstock, R. H. Cheng, and J. M. Casasnovas. 2000. Distinct cellular receptor interactions in poliovirus and rhinoviruses. *EMBO J.* **19:**1207–1216.

Reovirus

Barton, E. S., J. C. Forrest, J. L. Connolly, J. D. Chappell, Y. Liu, F. J. Schnell, A. Nusrat, C. A. Parkos, and T. S. Dermody. 2001. Junction adhesion molecule is a receptor for reovirus. *Cell* **104:**441–451.

Chandran, K., D. L. Farsetta, and M. L. Nibert. 2002. Strategy for nonenveloped virus entry: a hydrophobic conformer of the reovirus membrane penetration protein micro 1 mediates membrane disruption. *J. Virol.* **76:**9920–9933.

Ebert, D. H., J. Deussing, C. Peters, and T. S. Dermody. 2002. Cathepsin L and cathepsin B mediate reovirus disassembly in murine fibroblast cells. *J. Biol. Chem.* **277:**24609–24617.

Guerrero, C. A., E. Mendez, S. Zarate, P. Isa, S. Lopez, and C. F. Arias. 2000. Integrin alpha(v)beta(3) mediates rotavirus cell entry. *Proc. Natl. Acad. Sci. USA* **97:**14644–14649.

Retroviruses

Human immunodeficiency virus

Choe, H., M. Farzan, Y. Sun, N. Sullivan, B. Rollins, P. D. Ponath, L. Wu, C. R. Mackay, G. LaRosa, W. Newman, N. Gerard, C. Gerard, and J. Sodroski. 1996. The β-chemokine receptors CCR3 and CCR5 facilitate infection by primary HIV-1 isolates. *Cell* **85:**1135–1148.

Deng, H., R. Liu, W. Ellmeier, S. Choe, D. Unutmaz, M. Burkhart, P. DiMarzio, S. Marmon, R. E. Sutton, C. M. Hill, C. B. Davis, S. C. Peiper, T. J. Schall, D. R. Littman, and N. R. Landau. 1996. Identification of a major co-receptor for primary isolates of HIV-1. *Nature* **381:** 661–666.

Doranz, B. J., J. Rucker, Y. Yi, R. J. Smyth, M. Samson, S. C. Peiper, M. Parmentier, R. G. Collman, and R. W. Doms. 1996. A dual-tropic primary HIV-1 isolate that uses fusin and the β-chemokine receptors CKR-5, CKR-3 and CKR-2b as fusion cofactors. *Cell* **85:**1149–1158.

Dragic, T., V. Litwin, G. P. Allaway, S. R. Martin, Y. Huang, K. A. Nagashima, C. Cayanan, P. J. Maddon, R. A. Koup, J. P. Moore, and W. A. Paxton. 1996. HIV-1 entry into $CD4^+$ cells is mediated by the chemokine receptor CC-CKR-5. *Nature* **381:**667–673.

Feng, Y., C. C. Broder, P. E. Kennedy, and E. A. Berger. 1996. Human immunodeficiency virus type 1 entry cofactor: functional cDNA cloning of a seven-transmembrane, G protein-coupled receptor. *Science* **272:**872–877.

Herbein, G., U. Mahlknecht, F. Batliwalla, P. Gregersen, T. Pappas, J. Butler, W. A. O'Brien, and E. Verdin. 1998. Apoptosis of CD8+ T cells is mediated by macrophages through the interaction of human immunodeficiency virus gp120 with chemokine receptor CXCR4. *Nature* **395:**189–194.

Hesselgesser, J., D. Taub, P. Basker, M. Greenberg, J. Hoxie, D. L. Kolson, and R. Horuk. 1998. Neuronal apoptosis induced by human immunodeficiency virus-1 gp120 and the chemokine Sdf-1a is mediated by the chemokine receptor CXCR4. *Curr. Biol.* **8:**595–598.

Kwong, P. D., R. Wyatt, J. Robinson, R. W. Sweet, J. Sodroski, and W. A. Hendrickson. 1998. Structure of an HIV gp120 envelope glycoprotein in complex with the CD4 receptor and a neutralizing human antibody. *Nature* **393:**648–659.

Maddon, P. J., A. G. Dalgleish, J. S. McDougal, P. R. Clapham, R. A. Weiss, and R. Axel. 1986. The T4 gene encodes the AIDS virus receptor and is expressed in the immune system and the brain. *Cell* **47:**333–348.

McDonald, D., M. A. Vodicka, G. Lucero, T. M. Svitkina, G. G. Borisy, M. Emerman, and T. J. Hope. 2002. Visualization of the intracellular behavior of HIV in living cells. *J. Cell Biol.* **159:**441–452.

Ryu, S.-E., P. D. Kwong, A. Truneh, T. G. Porter, J. Arthos, M. Rosenberg, X. Dai, N.-H. Xuong, R. Axel, R. W. Sweet, and W. A. Hendrickson. 1990. Crystal structure of an HIV-binding recombinant fragment of human CD4. *Nature* **348:**419–426.

Wang, J., Y. Yan, T. P. J. Garrett, J. Liu, D. W. Rodgers, R. L. Garlick, G. E. Tarr, Y. Husain, E. L. Reinherz, and S. C. Harrison. 1990. Atomic structure of a fragment of human CD4 containing two immunoglobulin-like domains. *Nature* **348:**411–418.

Other retroviruses

Albritton, L. M., L. Tseng, D. Scadden, and J. M. Cunningham. 1989. A putative murine ecotropic retrovirus receptor gene encodes a multiple membrane-spanning protein and confers susceptibility to virus infection. *Cell* **57:**659–666.

Bates, P., J. A. Young, and H. E. Varmus. 1993. A receptor for subgroup A Rous sarcoma virus is related to the low density lipoprotein receptor. *Cell* **74:**1043–1051.

Brojatsch, J., J. Naughton, M. M. Rolls, K. Zingler, and J. A. Young. 1996. Carl, a TNFR-related protein, is a cellular receptor for cytopathic avian leukosis-sarcoma viruses and mediates apoptosis. *Cell* **87:**845–855.

Follis, K. E., S. J. Larson, M. Lu, and J. H. Nunberg. 2002. Genetic evidence that interhelical packing interactions in the gp41 core are critical for transition of the human immunodeficiency virus type 1 envelope glycoprotein to the fusion-active state. *J. Virol.* **76:**7356–7362.

Mothes, W., A. L. Boerger, S. Narayan, J. M. Cunningham, and J. A. Young. 2000. Retroviral entry mediated by receptor priming and low pH triggering of an envelope glycoprotein. *Cell* **103:**679–689.

Rhabdoviruses

Jeetendra, E., C. S. Robison, L. M. Albritton, and M. A. Whitt. 2002. The membrane-proximal domain of vesicular stomatitis virus G protein functions as a membrane fusion potentiator and can induce hemifusion. *J. Virol.* **76:**12300–12311.

Thoulouze, M. I., M. Lafage, M. Schachner, U. Hartmann, H. Cremer, and M. Lafon. 1998. The neural cell adhesion molecule is a receptor for rabies virus. *J. Virol.* **72:**7181–7190.

Tuffereau, C., J. Benejean, D. Blondel, B. Kieffer, and A. Flamand. 1998. Low-affinity nerve-growth factor receptor (P75NTR) can serve as a receptor for rabies virus. *EMBO J.* **17:**7250–7259.

Togaviruses

Klimstra, W. B., K. D. Ryman, and R. E. Johnston. 1998. Adaptation of Sindbis virus to BHK cells selects for use of heparan sulfate as an attachment receptor. *J. Virol.* **72:**7357–7366.

Kuhn, R. J., W. Zhang, M. G. Rossmann, S. V. Pletnev, J. Corver, E. Lenches, C. T. Jones, S. Mukhopadhyay, P. R. Chipman, E. G. Strauss, T. S. Baker, and J. H. Strauss. 2002. Structure of dengue virus: implications for flavivirus organization, maturation, and fusion. *Cell* **108:**717–725.

Lescar, J., A. Roussel, M. W. Wien, J. Navaza, S. D. Fuller, G. Wengler, G. Wengler, and F. A. Rey. 2001. The fusion glycoprotein shell of Semliki Forest virus: an icosahedral assembly primed for fusogenic activation at endosomal pH. *Cell* **105:**137–148.

Wang, K. S., R. J. Kuhn, E. G. Strauss, S. Ou, and J. H. Strauss. 1992. High-affinity laminin receptor is a receptor for Sindbis virus in mammalian cells. *J. Virol.* **66:**4992–5001.

RNA Virus Genome Replication and mRNA Production

Truth is ever to be found in the simplicity, and not in the multiplicity and confusion of things.
SIR ISAAC NEWTON

Introduction

The genomes of RNA viruses come in a number of conformations, including unimolecular or segmented, single stranded of (+) or (–) polarity, double stranded, and circular (Table 6.1). These structurally diverse viral RNA genomes share a common requirement: they must be efficiently copied within the infected cell to provide both genomes for assembly into progeny virions and messenger RNAs (mRNAs) for the synthesis of viral proteins. The synthesis of these RNA molecules by RNA viruses is a unique process that has no parallel in the cell. The genomes of all RNA viruses except retroviruses (see below) encode an **RNA-dependent RNA polymerase** (Box 6.1) to catalyze the synthesis of new genomes and mRNAs.

The virions of RNA viruses with (–) strand and double-stranded RNA genomes must contain the RNA polymerase, because the incoming viral RNA can be neither translated nor copied by the cellular machinery. The deproteinized RNA genomes of (–) strand and double-stranded RNA viruses are therefore noninfectious. In contrast, viral particles containing a (+) strand RNA genome lack a virion polymerase; the deproteinized RNAs of these viruses are infectious because they are translated in cells to produce, among other viral proteins, the viral RNA polymerase.

The mechanisms by which viral mRNA is made and the RNA genome is replicated in cells infected by RNA viruses (Fig. 6.1) appear even more diverse than the structure and organization of viral RNA genomes (Table 6.1). For example, the genomes of both picorna- and alphaviruses are single molecules of (+) strand genomic RNA, but the strategies for the production of viral RNA are quite different (Fig. 6.1). Nevertheless, each of the mechanisms of viral RNA synthesis summarized in Fig. 6.1 meets two essential requirements common to the infectious cycles of all these viruses. During replication, the RNA genome must be copied from one end to the other with no loss of nucleotide sequence. The second requirement is for the production of viral mRNAs that can be translated efficiently by the cellular protein synthetic machinery.

Table 6.1 Proteins that participate in RNA synthesis of representative RNA-containing viruses

Virus	Genome structure	RNA polymerase[a]	Viral accessory proteins involved RNA synthesis	Host proteins
Poliovirus	Single-stranded linear (+) strand	3D^pol^	3AB: membrane-bound VPg donor; cofactor for 3D^pol^ 3B (VPg): primer for RNA synthesis; linked to 5′ end of (+) and (-) strand RNAs 3CD^pro^: viral protease; binds cloverleaf in the (+) strand 3C^pro^: viral protease that cleaves 3CD and 3AB 2C: NTPase; possible helicase; anchors RNA polymerase to cellular membranes	Poly(rC)-binding protein Poly(A)-binding protein 1
Reovirus	Double-stranded linear, 10–11 segments	λ3	λ2: guanylyltransferase, methyltransferase λ1: NTPase, RNA helicase; role in mRNA synthesis not known μ2: role in mRNA synthesis not known	None identified
Sindbis virus	Single-stranded linear (+) strand	nsP4	nsP1: methyltransferase nsP2: NTPase, possible RNA helicase; protease that cleaves nsP1234 nsP3: unknown function nsP1, 2, 3, 4: form a multisubunit polymerase	Required but not identified
Influenza virus	Single-stranded linear (–) strand, 7–8 segments	PB1	PA: role in genome RNA synthesis PB2: cap-binding subunit NP: nucleocapsid protein	Hsp90 UAP56/BAT1 p36
Vesicular stomatitis virus	Single-stranded linear (–) strand	L	N: nucleocapsid protein P: phosphoprotein, binds L, required for polymerase activity	Translation elongation protein EF-1
Arenavirus	Single-stranded linear (–) strand, 2 segments	L	Z: regulation of mRNA synthesis NP: nucleocapsid protein	None identified
Hepatitis delta virus	Single-stranded circular, rodlike (–) strand	Cellular DNA-dependent RNA polymerase II	Delta antigen	Cellular general initiation proteins?

[a]The catalytic subunit; accessory proteins may be required for enzyme activity.

BOX 6.1

What should we call RNA polymerases and the processes they catalyze?

Viral RNA polymerases are frequently multisubunit assemblies consisting of several viral gene products. One protein is the catalytic subunit that polymerizes the RNA chain, while accessory proteins have other functions. In this book, the term **RNA-dependent RNA polymerase** refers to the protein assembly required to carry out RNA synthesis, not just the polypeptide that catalyzes chain elongation.

Historically, viral RNA-dependent RNA polymerases were given two different names depending on their activities during infection. The term **replicase** was used to describe the enzyme that copies the viral RNA to produce additional genomes, while the enzyme that produces mRNA was called **transcriptase.** In some cases this terminology indicates true differences in the enzymes that carry out synthesis of functionally different RNAs. However, these terms can also be inaccurate and misleading, and so they will not be used here. For example, for some RNA viruses, genomic replication and mRNA synthesis are the **same** reaction. For double-stranded RNA viruses, mRNA synthesis produces templates that can also be used for genomic replication.

The production of mRNAs from viral RNA templates is often designated **transcription.** However, this term refers to a specific process, the copying of genetic information carried in DNA into RNA. Consequently, it will not be used here to describe synthesis of the mRNAs of viruses with RNA genomes. Similarly, we will reserve use of the term **promoter** to designate sequences controlling transcription of DNA.

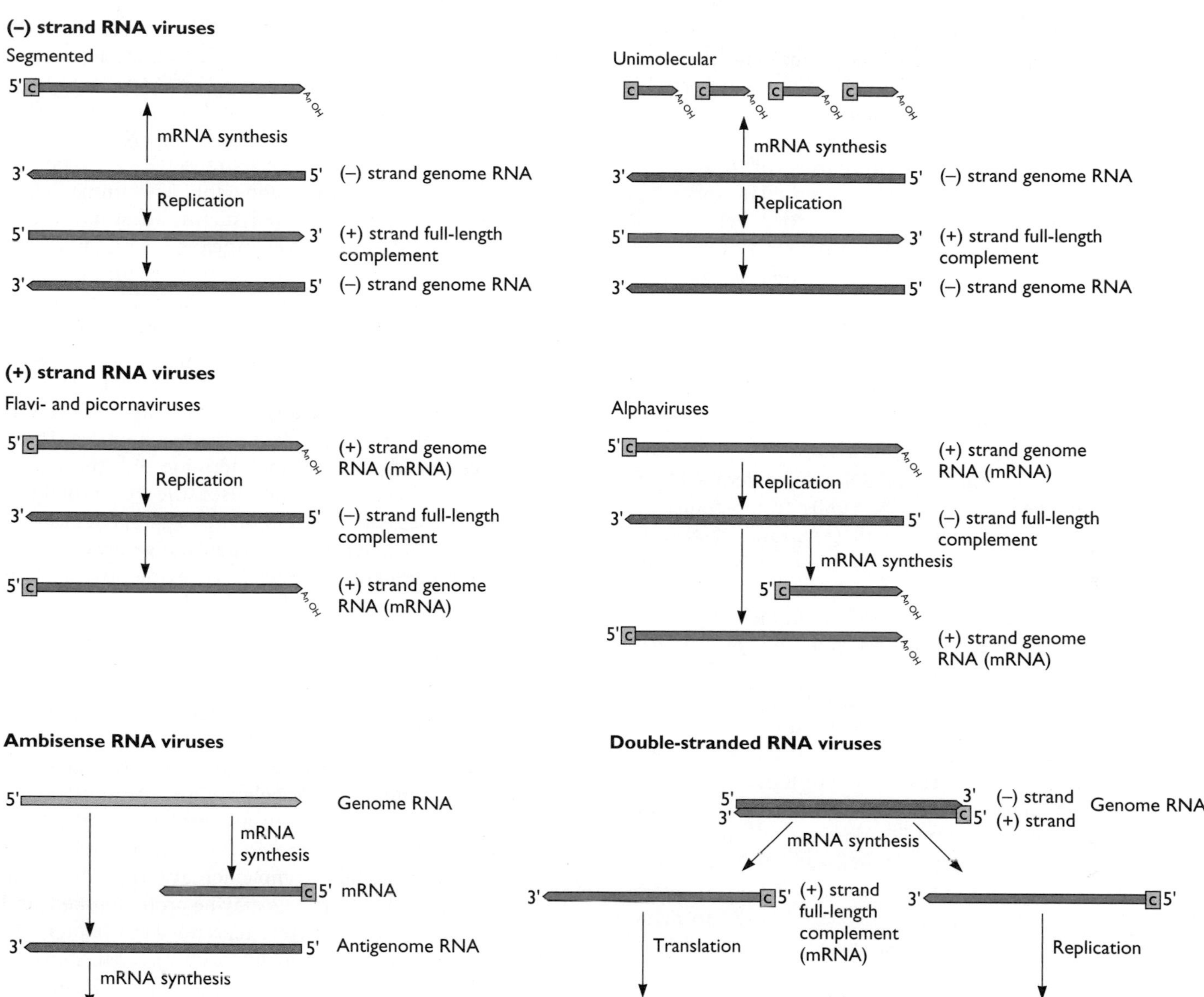

Figure 6.1 Strategies for replication and mRNA synthesis of RNA virus genomes. Subgenomic mRNAs are synthesized during infection by (–) strand viruses: one or two mRNAs for segmented RNA viruses or multiple mRNAs for viruses with unimolecular genomes. As these mRNAs are incomplete copies of the genomic RNA, they cannot be used as replicative intermediates. Antitermination mechanisms result in the synthesis of full-length (+) strands, which can then be copied to produce additional (–) genomic RNAs. For (+) strand viruses, the genomic RNA is also an mRNA. A subgenomic mRNA is also produced in alphavirus-infected cells. The segmented genomes of ambisense RNA viruses are copied to form one mRNA; a second mRNA is synthesized from the complement, or antigenome. In cells infected with double-stranded RNA viruses, the mRNAs first synthesized can either be translated into protein or serve as templates for the synthesis of (–) strands, resulting in double-stranded genomic RNA. Polioviral genomic RNA is linked to VPg at the 5′ end. The (+) genomic RNA of some flaviviruses does not contain poly(A).

In this chapter we will consider the mechanisms of viral RNA synthesis, the mechanism for switching from mRNA production to genome replication, and how the process of RNA-directed RNA synthesis leads to genetic diversity. Much of our understanding of viral RNA synthesis comes from experiments done with purified components. Because it is possible that events proceed differently in infected cells, the results of these in vitro studies are used to build hypothetical models for the different steps in RNA synthesis. While many models exist for each reaction, those presented in this chapter were selected because they are consistent with experimental results obtained in different laboratories.

The general principles of RNA synthesis will be illustrated with a few viruses, including influenza virus, vesicular stomatitis virus, poliovirus, alphaviruses, and arenaviruses, as examples. Summaries of the genome organization of these viruses are presented in Appendix A. Members of another family of RNA viruses, the *Retroviridae,* will be discussed in Chapter 7. Retroviruses encode an RNA-dependent DNA polymerase, and therefore mRNA synthesis and genome replication are very different from those of other RNA viruses.

Mechanisms of Viral RNA Synthesis

RNA-Dependent RNA Polymerases

Identification

The first evidence for a viral RNA-dependent RNA polymerase emerged in the early 1960s from studies with mengovirus and poliovirus, both (+) strand RNA viruses. In these experiments, extracts were prepared from virus-infected cells and incubated with the four ribonucleoside triphosphates (ATP, UTP, CTP, and GTP), one of which was radioactively labeled. The incorporation of nucleoside monophosphate into RNA was then measured. Infection with mengovirus or poliovirus caused the appearance of a cytoplasmic enzyme that could synthesize viral RNA in the presence of actinomycin D, a drug that was known to inhibit cellular DNA-directed RNA synthesis. Lack of sensitivity of the enzyme activity to actinomycin D suggested that it was virus specific and could copy RNA from an RNA template and not from a DNA template. This enzyme was presumed to be an RNA-dependent RNA polymerase. Several years later, similar assays were used to demonstrate that the virions of (–) strand viruses and of double-stranded RNA viruses contain an RNA-dependent RNA polymerase that synthesizes mRNAs from the (–) strand RNA present in the virus particles.

The initial discovery of a putative RNA polymerase in poliovirus-infected cells was followed by attempts to purify the enzyme and show that it can copy viral RNA. Because polioviral genomic RNA contains a 3′ poly(A) sequence, polymerase activity was measured with a poly(A) template and an oligo(U) **primer**. After several fractionation steps, a poly(U) polymerase that could copy polioviral genomic RNA in the presence of an oligo(U) primer was purified from infected cells. Poly(U) polymerase activity coincided with a single polypeptide, now known to be the polioviral RNA polymerase $3D^{pol}$ (see Appendix A, Fig. 13, and Chapter 11 for a description of this nomenclature). The $3D^{pol}$ RNA polymerase produced in infected human cells, bacteria, or insect cells cannot copy polioviral genomic RNA in the absence of a primer. The $3D^{pol}$ is therefore a template- and primer-dependent enzyme.

Because many (+) strand viral RNA-dependent RNA polymerases are tightly associated with membranes or nucleocapsids in virus-infected cells, these enzymes have been difficult to purify and study. Indeed, polioviral $3D^{pol}$, vesicular stomatitis virus L, hepatitis C virus NS5b protein, reovirus λ3, and retroviral reverse transcriptase (to be discussed in Chapter 7) are among the few animal virus RNA-dependent nucleic acid polymerases that have been purified and characterized. Nevertheless, many viral RNA polymerases have been identified. Assays for the RNA polymerase activity described above have been used to demonstrate virus-specific enzymes in virions or in extracts of infected cells. Amino acid sequence alignments (see "Sequence Relationships among RNA Polymerases" below) can be used to identify viral proteins with motifs characteristic of RNA-dependent RNA polymerases. These approaches have been used to determine that the L proteins of paramyxo- and bunyaviruses, the PB1 protein of influenza viruses, and nsP4 protein of alphaviruses are likely to be RNA polymerases. When the genes encoding these polymerases are expressed in cells, the proteins that are produced can copy viral RNA templates.

RNA-directed RNA synthesis follows a set of universal rules that differ slightly from those followed by DNA-dependent DNA polymerases. RNA synthesis initiates and terminates at specific sites in the template and is catalyzed by virus-encoded polymerases, but viral accessory proteins and even host cell proteins may also be required. Like cellular DNA-dependent RNA polymerases, most RNA-dependent RNA polymerases can initiate RNA synthesis de novo. However, some require a primer with a free 3′-OH end to which nucleotides complementary to the template strand are added. RNA primers may be protein linked, and others contain a 5′ cap structure (described in Chapter 10). RNA is usually synthesized by template-directed, stepwise incorporation of ribodeoxynucleoside monophosphates (NMPs) into the 3′-OH end of the growing RNA chain, which undergoes **elongation** in the 5′ → 3′ direction. However, some examples of nontemplated synthesis of viral RNA have been described in cells infected with viruses of the *Paramyxoviridae* and *Filoviridae,* or with hepatitis delta satellite virus (described in Chapter 10).

Viral RNA synthesis is highly efficient: a single molecule of polioviral RNA is copied to approximately 50,000 progeny during the course of an 8-h infection.

Sequence Relationships among RNA Polymerases

The amino acid sequences of viral RNA polymerases have been compared to identify conserved regions essential for polymerase function and to provide information about the evolution of these enzymes. Despite wide variation in genomic conformation, and very different amino acid se-

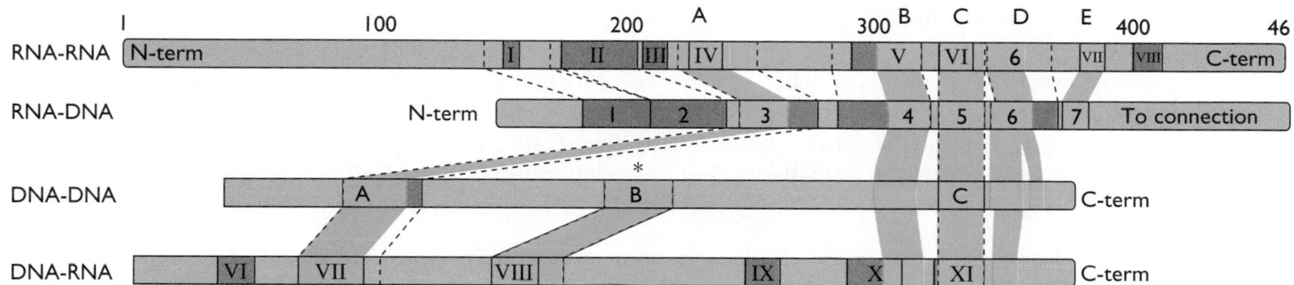

Figure 6.2 Protein domain alignments for the four categories of nucleic acid polymerases. Numbers at the top are from the 3D^{pol} amino acid sequence. Sequence motifs in each polymerase category are enclosed with solid lines and labeled, and the alignments between the different polymerases are shown with dashed lines. Motifs A to E are found in RNA-dependent RNA and DNA polymerases; of these, only motifs A and C are also found in DNA-dependent DNA and RNA polymerases. Motif B* found in DNA-dependent DNA and RNA polymerases is not the same as motif B in the other two enzymes. The structure and sequence motifs that resulted from comparing the crystal structure of 3D^{pol} are labeled A to E and are shown in green. This comparison allowed the identification of motifs B and D in DNA-dependent DNA and RNA polymerases. Adapted from J. L. Hansen et al., *Structure* **5**:1109–1127, 1997, with permission. Courtesy of S. Schultz, University of Colorado, Boulder.

quences, four common motifs (A to D) have been identified in the sequences of all polymerases (Fig. 6.2). One of these motifs (C) includes a Gly-Asp-Asp sequence conserved in the RNA polymerases of most (+) strand RNA viruses that have been studied. It was suggested that this sequence is part of the active site of the enzyme. In support of this hypothesis, changing the first Asp in polioviral 3D^{pol} produces an inactive polymerase. Evidence that a viral protein is an RNA polymerase is considerably strengthened when this three-amino-acid sequence is found.

The Asp-Asp sequence is also conserved in RNA-dependent **DNA** polymerases of retroviruses and in RNA polymerases of double-stranded RNA and segmented (–) strand viruses. The RNA polymerases of nonsegmented (–) strand viruses contain Gly-Asp-Asn instead of Gly-Asp-Asp. Mutational studies have shown that this sequence in the L protein of vesicular stomatitis virus is essential for RNA synthesis.

When the sequences of polymerases that copy DNA templates were included in comparisons, it became evident that two of the motifs (A and C) identified in RNA-dependent RNA polymerases were present (Fig. 6.2). Furthermore, the entire polymerase domain of DNA-dependent DNA and RNA-dependent DNA polymerases of retroviruses could be aligned. These comparisons suggested that all four classes of nucleic acid polymerases might share a common topology.

Three-Dimensional Structure of 3D^{pol}

The crystal structures of three of the four types of polymerases—DNA-dependent DNA polymerase, DNA-dependent RNA polymerase, and RNA-dependent DNA polymerase (reverse transcriptase)—are often characterized as being analogous to a right hand consisting of a palm, fingers, and a thumb, with the active site of the enzyme located in the palm subdomain. The first structure of an RNA-dependent RNA polymerase obtained was that of polioviral 3D^{pol}, which was determined by X-ray crystallography (Fig. 6.3). The enzyme has the same overall shape as

Figure 6.3 Structure of polioviral 3D^{pol}. The thumb, palm, and fingers subdomains are labeled. Structure/sequence motifs are labeled and colored as follows: A, red; B, green; C, yellow; D, purple; E, dark blue. The top of the fingers subdomain is disordered in the crystals; the two N-terminal regions that are ordered are in white, and amino acids flanking disordered regions are numbered. The two white regions have no counterpart in other categories of polymerases. Because the region connecting the two white sequences (residues 33 to 66) is disordered, it is not known how they are connected. It seems unlikely that residues 33 to 66 stretch across the active cleft of the polymerase, and therefore it has been suggested that residues 14 to 35 are from a separate molecule. Such interactions could lead to oligomerization of 3D^{pol} (see Fig. 6.6). Adapted from J. L. Hansen et al., *Structure* **5**:1109–1127, 1997, with permission. Courtesy of S. Schultz, University of Colorado, Boulder.

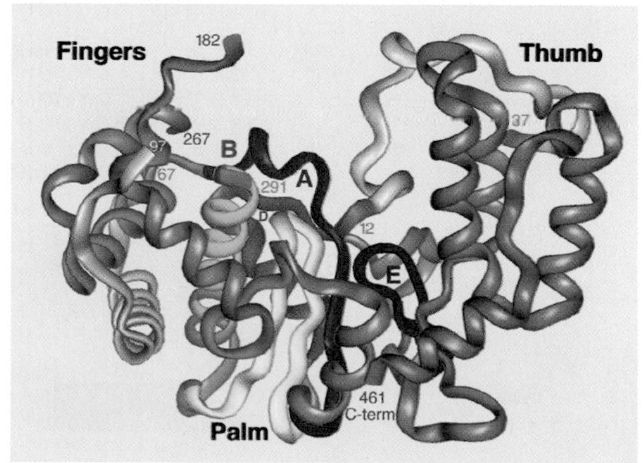

other polymerases, although the fingers and thumb are distinctive.

The palm subdomain (A to D) of $3D^{pol}$ is structurally similar to that of other polymerases and contains the four motifs conserved in RNA-dependent polymerases (Fig. 6.3 and 6.4A). The fifth motif, which is present in RNA-dependent, but not in DNA-dependent, polymerases, lies between the palm and thumb subdomains. Determination of the $3D^{pol}$ structure has allowed the realignment of the amino acid sequences of the five motifs for all four classes of polymerases (Fig. 6.4B). This realignment has revealed new functional relationships among individual amino acids. For example,

Figure 6.4 Polymerase structure and sequence motifs in representative structures of each of the four types of nucleic acid polymerases. (A) Ribbon diagrams of the polymerase domain of the large (Klenow) fragment of *Escherichia coli* DNA polymerase I, a DNA-dependent DNA polymerase; T7 RNA polymerase (T7 RNAP), a DNA-dependent RNA polymerase; human immunodeficiency virus type 1 reverse transcriptase (HIV-1 RT), an RNA-dependent DNA polymerase; and polioviral $3D^{pol}$, an RNA-dependent RNA polymerase. The thumb subdomain is at the right, and the fingers subdomain is at the left. The structure/sequence motifs A, B, C, D, and E are red, green, yellow, purple, and dark blue, respectively. (B) Sequence alignments based on the structures of $3D^{pol}$, HIV-1 RT, the large (Klenow) fragment of DNA polymerase I, and T7 RNA polymerase. Motif B is the green portion in panel A, not motif B* of DNA-dependent polymerases (Fig. 6.2). α-helical and β-strand regions are labeled. The boxes, which are color coded to match the motifs in panel A, indicate the extent of structural agreement. The most conserved amino acids are in bold type. Double lines are regions in which structures differ. Abbreviations: C-term, C terminus; h, helix; T, turn. Adapted from J. L. Hansen et al., *Structure* **5:**1109–1127, 1997, with permission. Courtesy of S. Schultz, University of Colorado, Boulder.

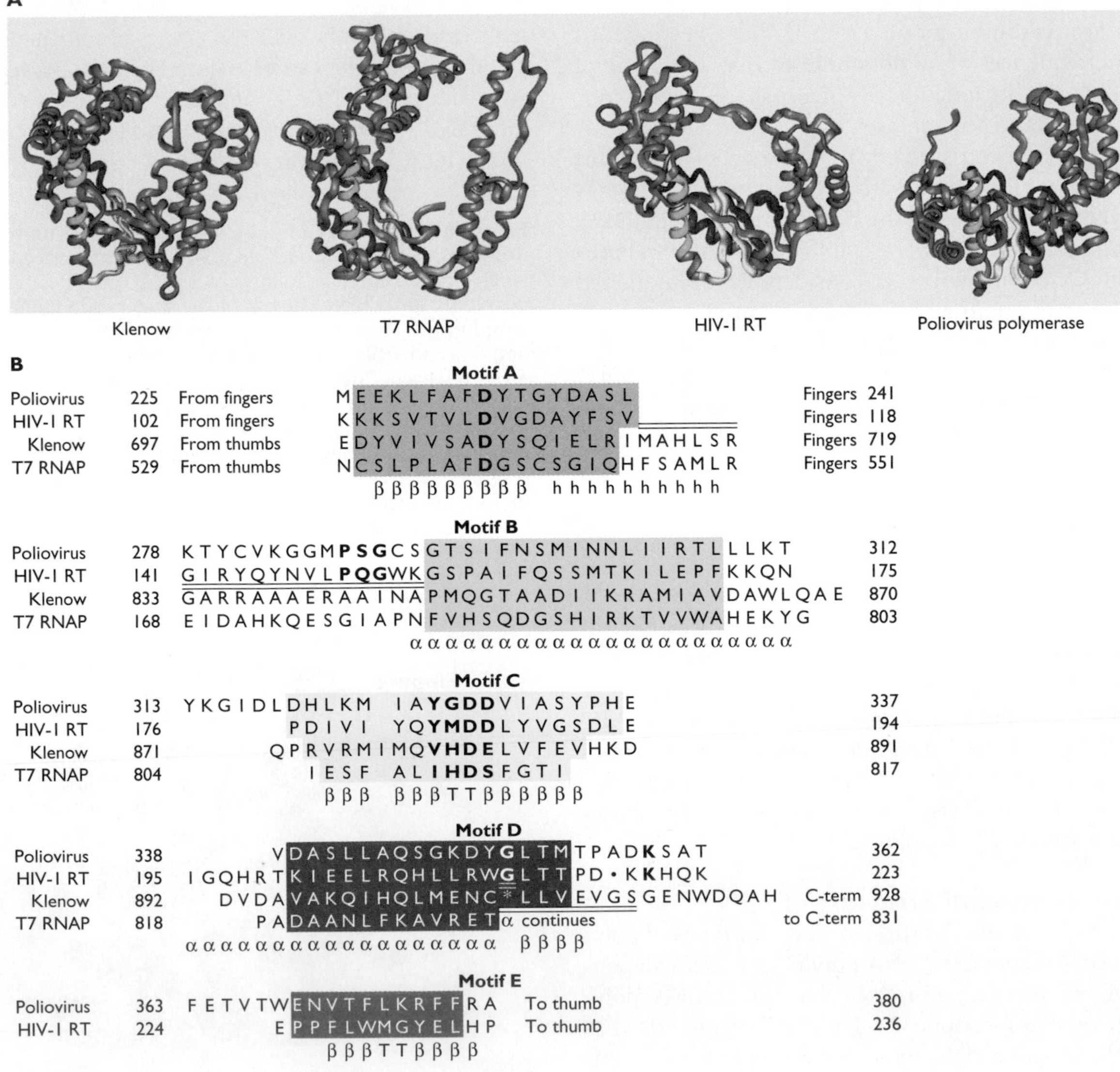

in motif A, Tyr115 of human immunodeficiency virus 1 reverse transcriptase is aligned with Asp238 of $3D^{pol}$. The two residues are conserved, and it has been suggested that they control the discrimination between ribonucleoside triphosphates and deoxyribonucleoside triphosphates (dNTPs), a major functional difference between the two enzymes. Motif C of $3D^{pol}$ contains the Asp-Asp sequence conserved in RNA-dependent polymerases; the first Asp is also conserved in DNA-dependent polymerases. All nucleic acid polymerases require a metal ion cofactor for activity, and the two Asp residues of motif C and the conserved Asp of motif A form a cluster that may coordinate such metals.

Based on the similarities between polioviral $3D^{pol}$ and members of the other classes of polymerases, it is likely that the structures of RNA polymerases of other RNA viruses also resemble a right hand. This expectation was confirmed by resolution of the structure of the RNA-dependent RNA polymerase of hepatitis C virus, NS5b. The structure of NS5b reveals additional information about template and substrate specificity. As discussed above, Asp238 of $3D^{pol}$ is believed to control substrate discrimination. The corresponding residue in NS5b, Asp225, is located in a shallow hydrophilic pocket. This pocket is well suited to bind NTPs, in which the ribose is bulkier and more polar than that in deoxyribonucleoside triphosphates. This property, together with hydrogen bonding of the ribose with Asp225, probably contributes to substrate selectivity. The structure of NS5b also revealed that the fingers domain, which had not been visible in the structure of $3D^{pol}$, is divided into two subdomains called α-fingers and β-fingers because they are rich in α-helices and β-strands, respectively. The α-fingers subdomain may be responsible for discriminating between an RNA and a DNA template.

The Nature of the RNA Template

Naked or Nucleocapsid RNA

The genomes of (–) strand viruses are organized into nucleocapsids in which protein molecules, including the RNA-dependent RNA polymerase and accessory proteins, are bound to the genomic RNAs at regular intervals. These tightly wound ribonucleoprotein complexes are very stable and resistant to RNase. The RNA polymerases of (–) strand viruses copy viral RNAs **only** when they are present in the nucleocapsid and not when they are naked RNAs. For example, vesicular stomatitis virus genomic RNA is a template for RNA polymerase only when it is bound to the nucleocapsid protein N. In contrast, the genomes of (+) strand RNA viruses are not coated with viral proteins in the virion. The main reason for this structural difference may be that the genomes of (–) strand RNA viruses must be ready to begin mRNA synthesis upon cell entry, whereas the genomes of (+) strand RNA viruses are translated. The exception is the (+) strand RNA genome of retroviruses, which is not translated, but copied into DNA (Chapter 7).

The viral nucleoproteins are cooperative, single-stranded RNA-binding proteins, as are the single-stranded nucleic acid-binding proteins required during DNA-directed DNA and RNA synthesis. Their function during replication is to keep the RNA single stranded and prevent base pairing between the template and product, so that additional rounds of RNA synthesis can occur. Many (+) strand RNA viruses encode helicases that serve a similar function (see below). In addition to its enzymatic activity, polioviral $3D^{pol}$ is a cooperative single-stranded RNA-binding protein and can unwind RNA duplexes without the hydrolysis of ATP characteristic of helicase-mediated unwinding. Polioviral RNA polymerase is therefore functionally similar to the RNA-binding nucleoproteins of (–) strand viruses. Consistent with this hypothesis, two regions of polymerase-polymerase interaction have been observed in $3D^{pol}$ protein crystals (Fig. 6.5). Interface I is ex-

Figure 6.5 Interactions between $3D^{pol}$ molecules in crystals of the polymerase. Four $3D^{pol}$ molecules are shown; the structure and sequence motifs and N-terminal regions are color coded as in Fig. 6.3. The two interfaces between polymerase molecules, called interface I and interface II, are labeled. Interface I is formed by interactions between the front of the thumb subdomain of one molecule and the back of the palm subdomain of another. This interaction leads to head-to-tail fibers of polymerase. Interface II is formed by interactions between the two N-terminal regions of the polymerase, residues 12 to 37 and residues 67 to 97 (shown in white). Interface II is shown formed by the contact of polymerase molecules in neighboring fibers formed by interface I. Because of the lack of order at the N-terminal regions, the origin of these regions among interacting $3D^{pol}$ molecules is not known. However, it has been suggested that residues 14 to 35 might derive from an adjacent polymerase molecule. The consequences of such a model are shown in Fig. 6.6. Adapted from J. L. Hansen et al., *Structure* **5:**1109–1127, 1997, with permission. Courtesy of S. Schultz, University of Colorado, Boulder.

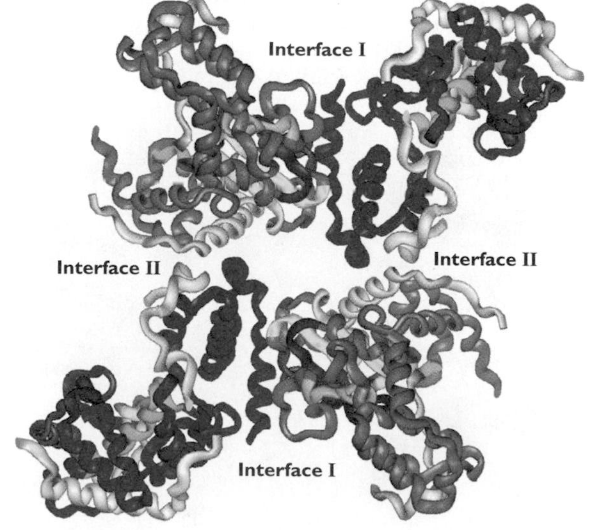

tensive and consists of more than 23 amino acid side chains on two different surfaces of the protein. This interaction occurs in a head-to-tail fashion and results in long fibers of polymerase molecules. When RNA is modeled into the $3D^{pol}$ structure according to the location of DNA in the human immunodeficiency virus type 1 reverse transcriptase-DNA cocrystal, it falls along the fibers formed by interface I interactions (Fig. 6.6). Interface II is formed by N-terminal polypeptide segments, perhaps from different polymerase molecules, which, in combination with interactions at interface I, might lead to a network of polymerase fibers. Because the N terminus of $3D^{pol}$ is required for enzyme activity, interface II interactions seem likely to be functionally important. Purified $3D^{pol}$ forms large sheets of planar arrays, hundreds of molecules of polymerase in length and width. Furthermore, alterations of $3D^{pol}$ that disrupt interfaces I or II prevent the formation of these ordered structures. Such polymerase arrays are present on the surfaces of membranes in virus-infected cells and are the sites of RNA polymerization (see "Cellular Sites of Viral RNA Synthesis" below).

Secondary Structures in Viral RNA

RNA molecules are not simple linear chains but can form secondary structures that have important functions. Viral RNA genomes contain secondary-structure elements such as base-paired **stem regions**, **hairpin loops**, **bulge loops**, **interior loops**, and **multibranched loops** (Fig. 6.7). An **RNA pseudoknot** is formed when a single-stranded loop region base pairs with a complementary sequence outside the loop (Fig. 6.7). These structures are important for RNA synthesis, translation, and assembly.

The first step in identifying a structural feature in RNA is to scan the nucleotide sequence with computer programs designed to fold the RNA into energetically stable structures. Comparative sequence analysis can provide evidence for RNA secondary structures. For example, comparison of the RNA sequences of several related viruses might reveal that the structure, but not the sequence, of a stem-loop is conserved. The best evidence for RNA structure comes from experiments in which RNAs are treated with enzymes or chemicals that attack single- or double-stranded regions specifically. The results of such analyses confirm that predicted stem regions are base paired, while loop regions are unpaired.

Figure 6.6 Model of oligomerization of the polioviral RNA-dependent RNA polymerase $3D^{pol}$. Two fibers formed by interface I interactions are shown; each contains four molecules of $3D^{pol}$ (labeled 1, 2, 3, 4, and 1′, 2′, 3′, 4′). The two fibers are shown interacting along interface II. The fibers cross at 120° angles; as a result, 1 and 1′ come out of the page, and 4 and 4′ go into the page. The axis of each fiber is shown in black. Shown in white are the N-terminal strands of the thumb subdomains that might be contributed by a molecule in the other fiber. Only one set of interface II interactions is shown. Adapted from J. L. Hansen et al., *Structure* **5**:1109–1127, 1997, with permission. Courtesy of S. Schultz, University of Colorado, Boulder.

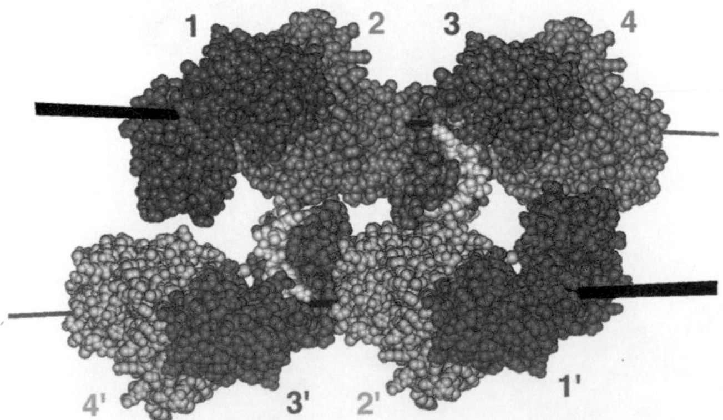

Accessory Proteins in RNA-Dependent RNA Synthesis

Viral Proteins

In addition to the RNA-dependent RNA polymerase, other viral proteins participate in RNA synthesis. Such accessory proteins can play several roles: they may direct the polymerase to the correct intracellular site, facilitate recognition of the initiation site on the RNA template, or stimulate the activity of the polymerase. The genomes of many (+) strand RNA viruses encode a helicase, an enzyme that unwinds double-stranded RNA structures by disrupting the hydrogen bonds between the two strands.

Directing RNA polymerase to the correct intracellular site. Viral RNA synthesis does not occur at random locations in the cell, but takes place either in the nucleus or on specific cytoplasmic assemblies (see "Cellular Sites of Viral RNA Synthesis" below). Because influenza virus RNA synthesis occurs in the cell nucleus, the viral RNA polymerase must be transported there from the cytoplasm. As described in Chapter 5, nuclear import of influenza virus RNA polymerase is mediated by nuclear localization sequences on the NP protein. Polioviral RNA synthesis occurs on membranous vesicles, and two polioviral polypeptides, 2C and 3AB, bring the RNA polymerase to these membranes. Low concentrations of guanidine hydrochloride inhibit polioviral RNA synthesis, and it is believed that this compound disrupts the interaction between the RNA polymerase and cell membranes. Mutations responsible for resistance of poliovirus to guanidine hydrochloride map to the coding sequence for the 2C protein, which contains an RNA-binding domain. It therefore appears that protein 2C anchors viral RNA to membranes during replication. The viral protein

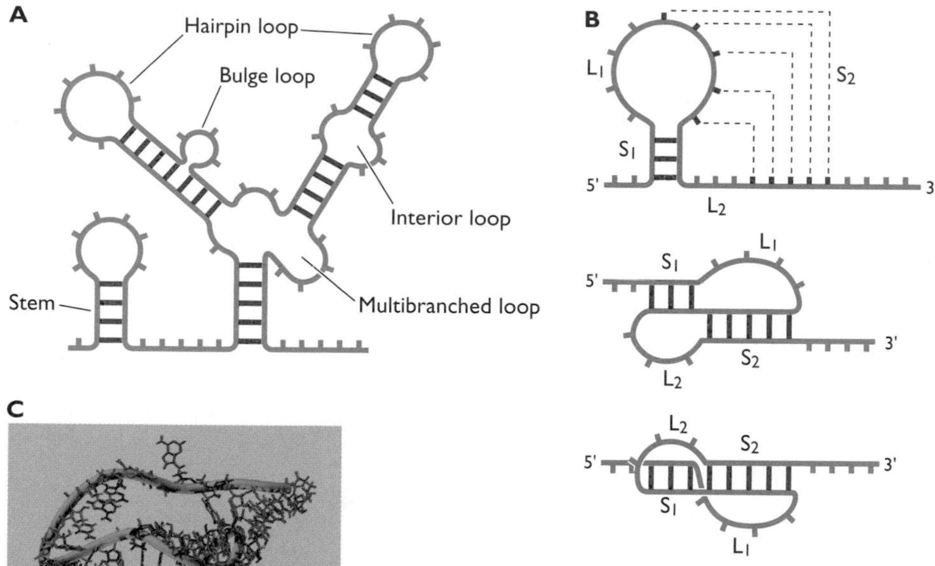

Figure 6.7 RNA secondary structure. (A) Schematic of different structural motifs in RNA. Red bars indicate base pairs; green bars indicate unpaired nucleotides. (B) Schematic of a pseudoknot. (Top) Stem 1 (S_1) is formed by base pairing in the stem-loop structure, and stem 2 (S_2) is formed by base pairing of nucleotides in the loop with nucleotides outside the loop. (Middle) A different view of the formation of stems S_1 and S_2. (Bottom) Coaxial stacking of S_1 and S_2 resulting in a quasicontinuous double helix. (C) Structure of a pseudoknot as determined by nuclear magnetic resonance. The sugar backbone is highlighted with a tube. Stacking of the bases in the areas of S_1 and S_2 can be seen. Adapted from C. W. Pleij, *Trends Biochem. Sci.* **15:**143–147, 1990, with permission.

3AB anchors the viral protein primer for RNA synthesis, VPg, in cell membranes (Fig. 6.8B). Polioviral 3D^{pol} cannot by itself associate with membranes, but is brought to the replication complex by binding to protein 3AB.

Targeting RNA polymerase to the correct initiation site on RNA templates. How does the polymerase know where to start copying the RNA template? Such template specificity may reside in part in the recognition of structures at the 5′ and 3′ ends of viral RNAs by viral proteins. The 3′ noncoding region of polioviral genomic RNA contains an RNA pseudoknot structure that is conserved among picornaviruses (Fig. 6.8B). Protein 3AB-3CD binds the 3′-terminal sequence of polioviral RNA and may direct the polymerase to that site for the initiation of (–) strand RNA synthesis. Polioviral 3CD^{pro} protein plays an important role in viral RNA synthesis by participating in the formation of a ribonucleoprotein at the 5′ end of the (+) strand RNA. The 3CD protein is a precursor of the 3C^{pro} protease and 3D^{pol}. Protein 3CD, together with a cellular protein known as poly(rC)-binding protein 2, binds to a cloverleaf structure in the viral RNA (Fig. 6.8B). The RNA-binding domain of 3CD is contained within the 3C portion of the protein, and alterations within this domain abolish complex formation and RNA synthesis without affecting viral protein processing.

Internal RNA sequences may also confer initiation specificity to RNA polymerases. The coding regions of several picornaviruses contain short RNA sequences that are required for RNA synthesis. *cis*-acting replication elements (cre) have been identified in the coding sequence of poliovirus protein 2C (cre2C) and rhinovirus capsid protein VP1 (creVP1). These sequences are binding sites for 3CD^{pro} and serve as a template for the priming of viral RNA synthesis by VPg protein (Fig. 6.8B).

Protein-protein interactions also play roles in directing the RNA polymerase to the RNA template. The vesicular stomatitis virus RNA polymerase for mRNA synthesis consists of the P protein and the L protein, the catalytic subunit. The P protein binds both the L protein and the complex of N and the (–) strand RNA. In this way the P protein brings the L polymerase to the RNA template. Cellular general initiation proteins have a similar function in bringing RNA polymerase II to the correct site to initiate transcription of DNA templates.

RNA helicases. Base-paired regions in viral RNA must be disrupted to permit copying by RNA-dependent RNA polymerase. RNA helicases, which are encoded in the genomes of many RNA viruses (see Table 6.1 for examples), are believed to unwind the genomes of double stranded RNA viruses and the secondary structures in template RNAs. They also prevent extensive base pairing between template RNA and the nascent complementary strand. The RNA helicase of the flavivirus hepatitis C virus, an important human pathogen (Appendix B, Fig. 7), has been studied extensively because the protein is a potential target for chemotherapeutic intervention. To facilitate the development of new agents that inhibit the hepatitis C virus helicase, its three-dimensional structure has been determined

A

Cleavage

5'

VPg

Poliovirus genome RNA

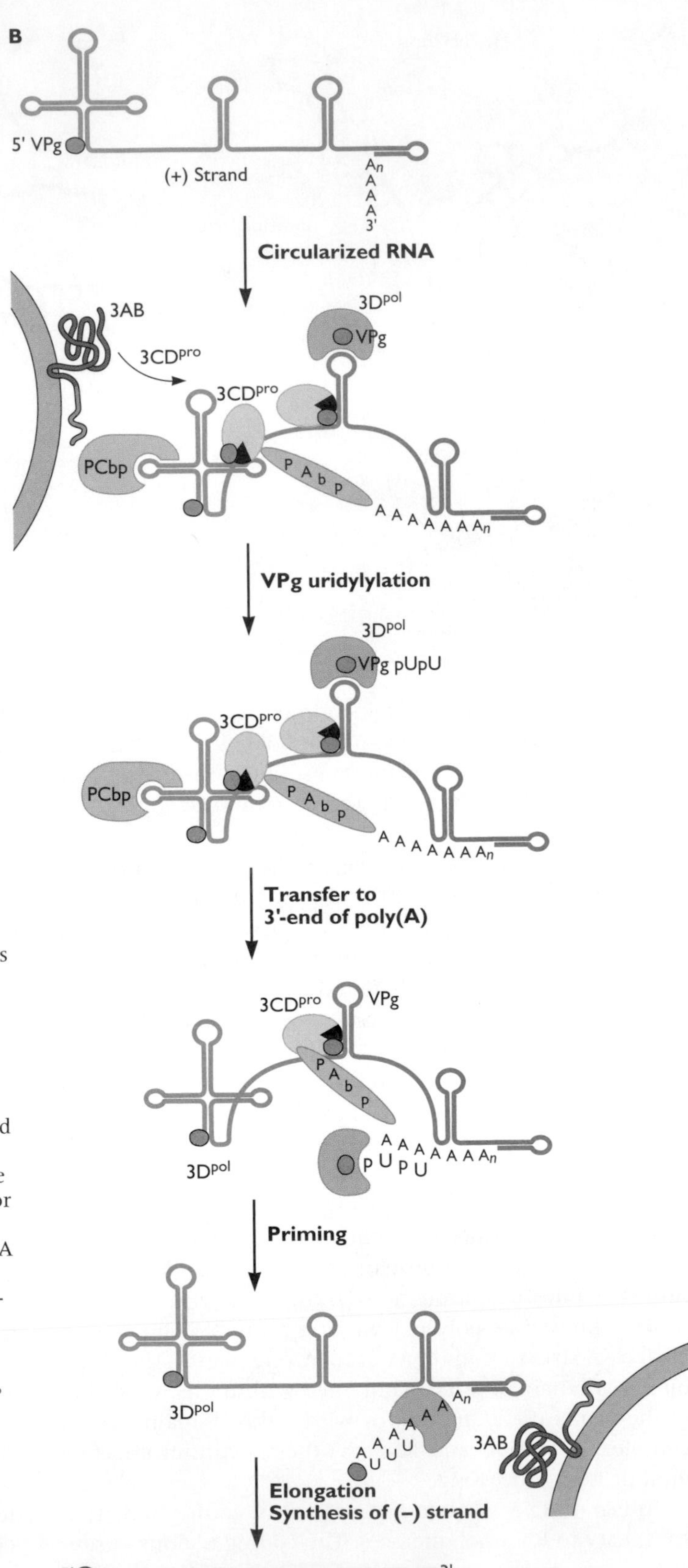

Figure 6.8 Poliovirus RNA synthesis. (A) Linkage of VPg to polioviral genomic RNA. Polioviral RNA is linked to the 22-amino-acid-long VPg (orange) via an O4-(5′-uridylyl)-tyrosine linkage. This phosphodiester bond is cleaved at the indicated site by a cellular enzyme to produce the viral mRNA containing a 5′-terminal Up. (B) Model of poliovirus (–) strand RNA synthesis. The (+) strand template is at top, showing the 5′ cloverleaf structure, the internal cre (*cis*-acting replication elements) sequence, and the 3′ pseudoknot. A precursor of VPg, known as 3AB, is a membrane-bound polypeptide and therefore an ideal candidate to act as a VPg donor in membranous replication complexes. The 3AB protein contains a hydrophobic domain. A virus carrying a mutation that reduces the hydrophobicity of this 3AB domain is defective in initiation of RNA synthesis, in vitro uridylylation of VPg, and in vivo synthesis of (+) strand viral RNAs. A ribonucleoprotein complex is formed when poly(rC)-binding protein 2 (PCbp2) and $3CD^{pro}$ bind the cloverleaf structure located within the first 108 nucleotides of (+) strand RNA. The ribonucleoprotein complex interacts with poly(A)-binding protein 1 (PAbp1), which is bound to the 3′-poly(A) sequence, producing a circular genome. Protease $3CD^{pro}$ cleaves membrane-bound 3AB to produce VPg and 3A. The cre sequence binds $3D^{pol}$, $3CD^{pro}$, and VPg. VPg-pUpU is synthesized by $3D^{pol}$ using the sequence AAACA of cre as a template. The complex is transferred to the 3′ end of the genome, and $3D^{pol}$ uses VPg-pUpU as a primer for RNA synthesis. The priming reaction may be stimulated by membrane-bound 3AB. (C) Model of polioviral (+) strand RNA synthesis. The replicative form (RF), produced by (–) strand RNA synthesis, is shown at top. The strands of the RF are separated by 2C, which binds to the cloverleaf in the (–) strand, and by PCbp2, $3CD^{pro}$, and 3AB, which bind to the cloverleaf in the (+) strand. Membrane-bound 3AB is cleaved by $3CD^{pro}$ to form 3A and VPg, and $3CD^{pro}$ cleaves itself to produce $3C^{pro}$ and $3D^{pol}$. VPg-pUpU is synthesized by $3D^{pol}$, using the 3′-terminal A residues of (–) strand RNA as template. VPg-pUpU is then elongated by $3D^{pol}$ to form (+) strands in a reaction that may be stimulated by membrane-bound 3AB. Multiple initiation reactions can occur on the (+) strand RNA while (–) strands are being elongated. (B and C) Modified from A. V. Paul, p. 227–246, *in* B. L. Semler and E. Wimmer (ed.), *Molecular Biology of Picornaviruses* (ASM Press, Washington, D.C., 2002), with permission.

Figure 6.8 *(continued)*

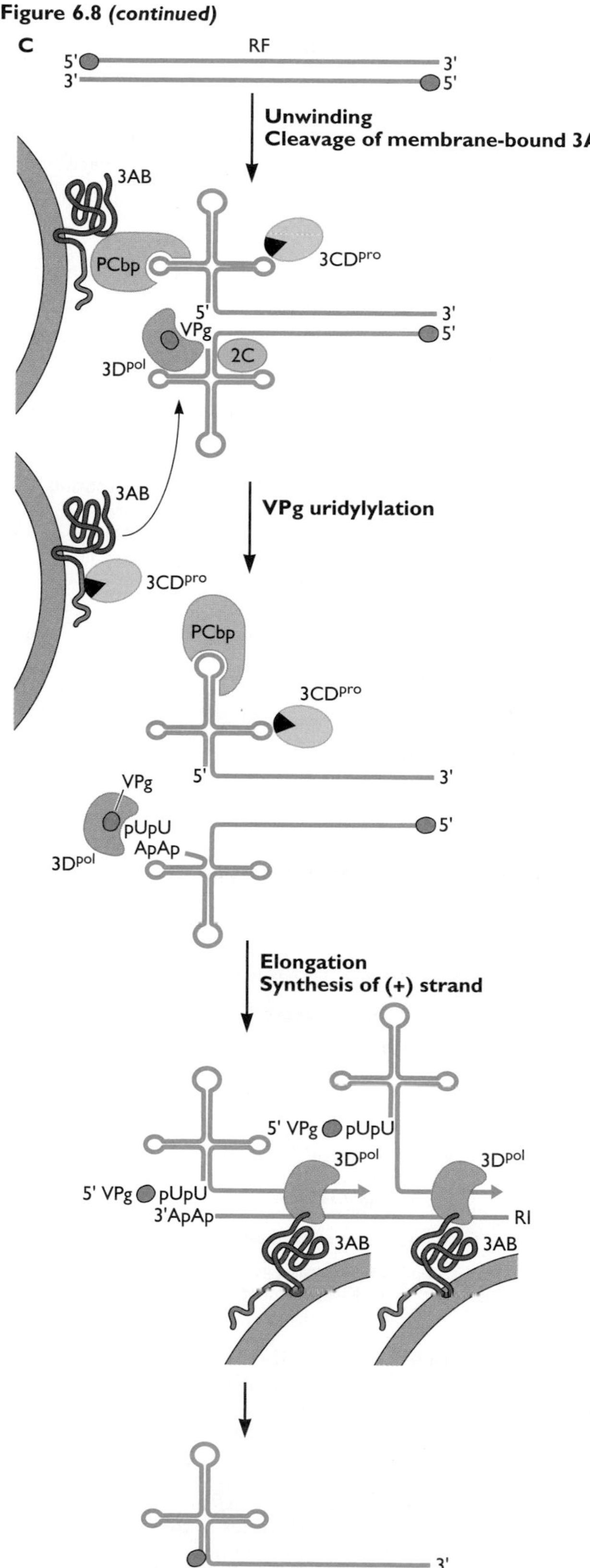

by X-ray crystallography. The molecule is Y shaped and comprises three domains: an NTPase domain, an RNA-binding domain, and a helical domain. Between the RNA-binding domain and the other two domains is a cleft that is large enough to accommodate single-stranded RNA but not double-stranded RNA. Unwinding of double-stranded RNA probably occurs as one strand of RNA passes through the cleft and another passes outside of the molecule.

Cellular Proteins in Viral RNA Synthesis

It has been known for over 35 years that host cell components are required for viral RNA synthesis. Such components were initially called host factors because nothing was known about their chemical composition. Evidence that host cell proteins are essential components of a viral RNA polymerase first came from studies of the bacteriophage Qβ. The RNA-dependent RNA polymerase of this virus is a multisubunit enzyme, consisting of a 65-kDa virus-encoded protein and three host proteins: ribosomal protein S1 and the translation elongation proteins EF-Tu and EF-Ts. Proteins S1 and EF-Tu contain RNA-binding sites that enable the RNA polymerase to recognize the viral RNA template. The 65-kDa viral protein exhibits no RNA polymerase activity in the absence of the host proteins, but has sequence similarity with known RNA-dependent RNA polymerases.

Polioviral RNA synthesis also requires host cell proteins. When purified polioviral RNA is incubated with a cytoplasmic extract prepared from uninfected permissive cells, the genomic RNA is translated and the viral RNA polymerase is made. If guanidine hydrochloride is included in the reaction, the polymerase assembles on the viral genome, but RNA synthesis does not initiate. The RNA polymerase-template assembly can be isolated free of guanidine, but RNA synthesis cannot begin unless a new cytoplasmic extract is added, indicating that soluble cellular proteins are required for initiation. A similar conclusion comes from studies in which polioviral RNA was injected into oocytes derived from the African clawed toad *Xenopus laevis*. Polioviral RNA cannot replicate in *Xenopus* oocytes unless it is coinjected with a cytoplasmic extract from human cells. These observations can be explained by the requirement of the viral RNA polymerase for a mammalian host cell protein that is absent in toad oocytes.

Host cell poly(rC)-binding protein 2 is required for polioviral RNA synthesis. This protein binds to a cloverleaf structure that forms in the first 108 nucleotides of (+) strand RNA (Fig. 6.8B). Formation of a ribonucleoprotein composed of the 5′ cloverleaf, 3CD, and poly(rC)-binding protein 2 is essential for the initiation of viral RNA synthesis. Interaction of poly(rC)-binding protein 2 with the cloverleaf facilitates the binding of viral protein 3CD to

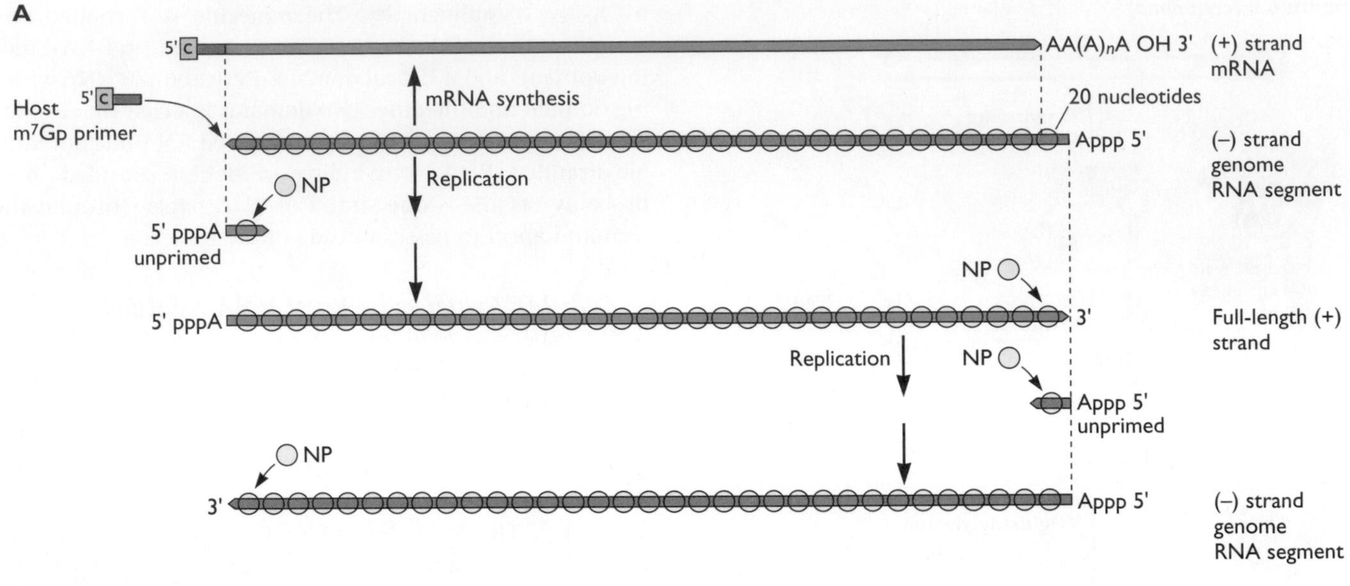

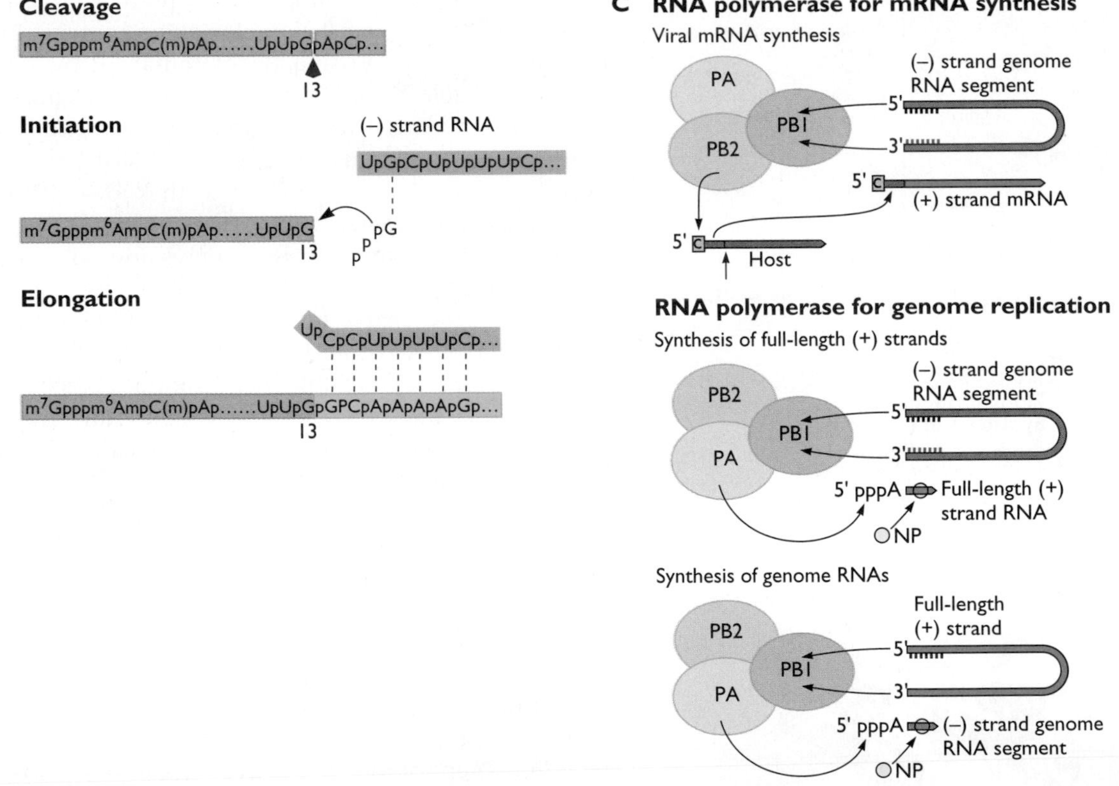

Figure 6.9 Influenza virus RNA synthesis. (A) Viral (–) strand genomes are templates for the production of either subgenomic mRNAs or full-length (+) strand RNAs. The switch from viral mRNA synthesis to genomic RNA replication is regulated by both the number of nucleocapsid (NP) protein molecules and the acquisition by the viral RNA polymerase of the ability to catalyze initiation without a primer. Binding of the NP protein to elongating (+) strands enables the polymerase to read all the way to the 5′ end of genomic RNA. (B) Mechanism of cap snatching: capped RNA-primed initiation of influenza virus mRNA synthesis. The first step is the cleavage of capped RNA fragments from the 5′ ends of cellular nuclear RNAs. These fragments then serve as primers for viral mRNA synthesis. The 10 to 13 nucleotides in these primers do not need to hydrogen-bond to the common sequence found at the 3′ ends of the influenza virus genomic RNA segments. The first nucleotide added to the primer is a G residue templated by the

(continued on next page)

the opposite side of the same cloverleaf. A model of how these interactions lead to viral RNA synthesis is presented in Fig. 6.8.

Another candidate for a host protein that is essential for polioviral RNA synthesis is poly(A)-binding protein 1. This protein interacts with poly(rC)-binding protein 2, 3CD^{pro}, and the 3′-poly(A) tail of poliovirus RNA, circularizing the viral genome (Fig. 6.8B). Formation of this circular ribonucleoprotein complex is required for negative-strand RNA synthesis. It is not known if genome circularization occurs in infected cells.

Host cell cytoskeletal proteins participate in paramyxoviral RNA synthesis. Measles virus and Sendai virus mRNA synthesis in vitro is stimulated by tubulin, the major structural component of microtubules of the cell's cytoskeleton, and is inhibited by anti-β-tubulin antibodies. In contrast, mRNA synthesis of human parainfluenza virus type 3 and respiratory syncytial virus requires cellular actin, which forms the microfilaments of the cytoskeletal network. Actin and tubulin might serve as anchoring sites on the cytoskeleton for the viral RNA polymerase. In support of this hypothesis, it has been shown that the incoming viral ribonucleoproteins associate with the cytoskeletal framework, where they are active in RNA synthesis. Assembling the RNA polymerase on the cytoskeleton may ensure high local concentrations of replication components and hence increase the rates or efficiencies of replication reactions.

Initiation and Elongation

Initiation Mechanisms

The requirement for a primer in the initiation step of nucleic acid synthesis varies among the different classes of polymerases. All DNA polymerases are primer-dependent enzymes, while DNA-dependent RNA polymerases can initiate RNA synthesis de novo. Most of the RNA-dependent RNA polymerases also initiate RNA synthesis without a primer. There are three notable exceptions: the enzymes of picornaviruses, influenza viruses, and bunyaviruses. Nucleic acid synthesis by these RNA polymerases is initiated by one of two unusual priming mechanisms.

Protein priming. Polioviral RNA replication is primed by a genome-linked protein, a mechanism that occurs during adenoviral DNA replication. As we have seen, polioviral 3D^{pol} is a primer-dependent enzyme that will not copy polioviral RNA in vitro without an oligo(U) primer. Polioviral genomic RNA, as well as newly synthesized (+) and (–) strand RNAs, are covalently linked at their 5′ ends to the 22-amino-acid-long protein VPg (Fig. 6.8A). This finding led to the suggestion that VPg might function as a primer for RNA synthesis. This hypothesis was supported by the discovery of a uridylylated form of the protein, VPg-pUpU, in infected cells. The template for uridylylation of VPg is an RNA hairpin, the *cis*-acting replication element, located in the coding region of picornaviruses. A model of VPg priming of polioviral RNA synthesis is depicted in Fig. 6.8B and C.

Priming by capped RNA fragments. Influenza virus mRNA synthesis is blocked by treatment of cells with the fungal toxin α-amanitin at concentrations that inhibit cellular DNA-dependent RNA polymerase II. In contrast, the toxin does not affect viral mRNA synthesis in vitro. This surprising finding demonstrated that the viral RNA polymerase is dependent on a host nuclear function, a consequence of the requirement by the viral RNA polymerase for capped RNA primers derived by cleavage of host cell RNA polymerase II transcripts.

We now understand why viral mRNA synthesis is inhibited by α-amanitin. In a process called **cap snatching**, nuclear RNA polymerase II transcripts are cleaved by a virus-encoded, cap-dependent endonuclease that is part of the RNA polymerase (Fig. 6.9B). The resulting 10- to 13-nucleotide-long capped fragments serve as primers for the initiation of viral mRNA synthesis. Inhibition by α-amanitin is explained by the requirement for a continuous supply of newly synthesized cellular RNA polymerase II transcripts to provide primers for viral mRNA synthesis.

Bunyaviral mRNA synthesis is also primed with capped fragments of cellular RNAs. In contrast to that of influenza virus, bunyaviral mRNA synthesis occurs in the cytoplasm,

Figure 6.9 ***(continued)*** penultimate C residue of the genomic RNA segment, followed by elongation of the mRNA chains. The terminal U residue of the genomic RNA segment does not direct the incorporation of an A residue. The 5′ ends of the viral mRNAs therefore comprise 10 to 13 nucleotides plus a cap structure snatched from host nuclear RNAs. Adapted from S. J. Plotch et al., *Cell* **23:**847–858, 1981, with permission. (C) Functionally different forms of the viral RNA polymerase carry out mRNA synthesis and genomic replication. During mRNA synthesis, the PB2 cap-binding protein is required for the initiation of mRNA chains by host cell-capped RNA primers, and the PA protein subunit does not have a role. In contrast, during replication the PB2 protein would have no role, and the PA protein is postulated to be required for initiation of both (+) and (–) strands without a primer. The three virus-specific RNA sequences highlighted in different colors activate the polymerase for mRNA synthesis and genomic replication. One of the virus-specific sequences (shown in red) binds to the RNA polymerase for mRNA synthesis or genomic replication and is found in both viral mRNAs and full-length (+) strand RNAs.

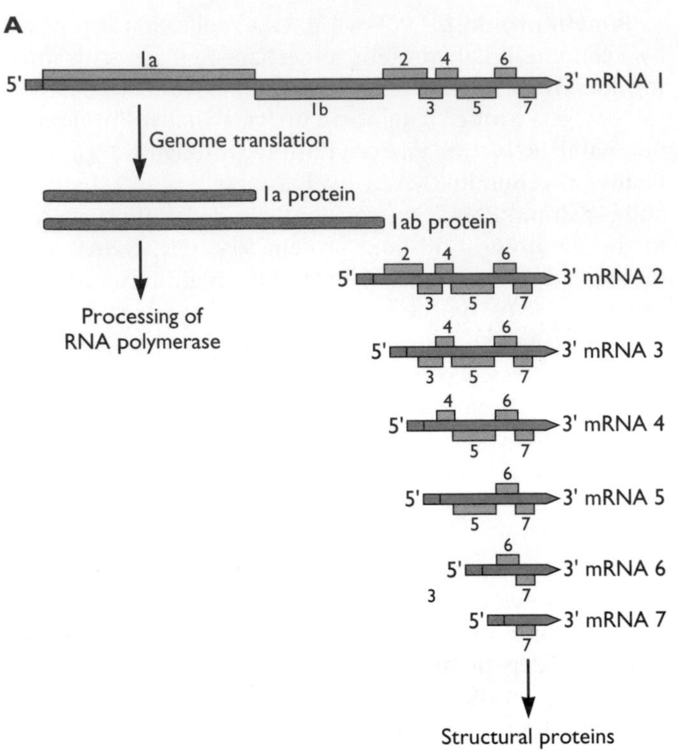

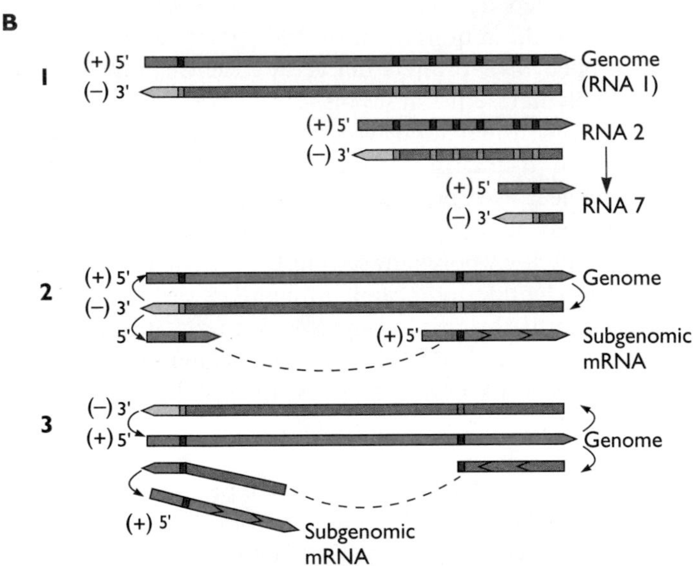

Figure 6.10 Nidoviral genome organization and expression. (A) Organization of open reading frames. The (+) strand viral RNA is shown at the top, with open reading frames as boxes. The genomic RNA is translated to form proteins 1a and 1ab, which are processed to form the RNA polymerase. Structural proteins are encoded by six nested mRNAs that share a common 5′ leader sequence (orange box). (B) Models of the synthesis of nested mRNAs. (1) Overview of (+) and (–) strand RNAs. (2) Discontinuous transcription was proposed to occur during (+) strand RNA synthesis by leader priming on a full-length (–) strand RNA template. Leader RNA (orange box) is joined to the mRNAs by a junction sequence also found adjacent to the leader RNA at the 5′ end of the viral RNA. The (+) strand junction sequence at the 3′ end of the leader can base pair with the (–) strand junction sequence of the region that encodes each mRNA. (3) After a nested set of subgenomic replicative intermediates was discovered in infected cells [i.e., containing both (+) and (–) strand RNAs], a new model was proposed in which a (–) strand is synthesized on the (+) strand genomic RNA. Most of the (+) strand template is not copied, probably because it loops out as the polymerase completes synthesis of the leader RNA. The resulting (–) strand RNAs, with leader sequences at the 3′ ends, are then copied to form mRNAs. The nested set of mRNAs would be produced by translocation signaled by the individual junction sequences between mRNA coding regions. Adapted from E. J. Snijder et al., *J. Gen. Virol.* **79:**961–979, 1998, with permission.

where there are many capped cellular RNAs, and is not inhibited by α-amanitin. Active RNA polymerase II is not required to provide a continuous supply of capped RNA primers.

Synthesis of Nested Subgenomic mRNAs

An unusual pattern of mRNA synthesis occurs in cells infected with members of the families *Coronaviridae* and *Arteriviridae,* in which subgenomic mRNAs that form a 3′-coterminal nested set with the viral genome are synthesized (Fig. 6.10A). These viral families were recently combined into the order *Nidovirales* to reflect this property (*nidus* is Latin for nest).

The subgenomic mRNAs of these viruses are composed of a leader and a body that are synthesized from noncontiguous sequences in the viral (+) strand genome. The leader is encoded in the 5′ end of the viral genome, and the mRNA bodies are encoded in the 3′ end. The leader and body are separated by a conserved junction sequence encoded both in the 3′ end of the leader and in the 5′ end of the mRNA body. The complementarity between the (+) strand junction sequence at the 3′ end of the leader and the (–) strand junction sequence at the start of each mRNA body led to the leader-primed model for mRNA synthesis (Fig. 6.10B). This model was initially supported by the failure to detect subgenome-length (–) strands, either free or as part of replicative intermediates. Recently, both these structures have been detected in infected cells, leading to a new model of discontinuous mRNA synthesis (Fig. 6.10B). (–) strand subgenome-length RNAs serve as templates for mRNA synthesis.

Conformational Changes and Regulation of RNA Synthesis

Alterations of viral RNA polymerases such as proteolytic processing and conformational changes may influence template recognition and initiation. The cap-binding and endonuclease functions of the influenza virus RNA polymerase are activated, probably by conformational changes, via sequential interactions with two viral RNA sequences (Fig. 6.11). When cosynthesized, the three P proteins form a multisubunit assembly that can neither bind to capped primers nor synthesize mRNAs. Addition of a sequence corresponding to the 5′-terminal 11 nucleotides of the viral RNA, which is highly conserved in all eight genome segments, activates the cap-binding activity of the P proteins. The PB1 protein binds this RNA sequence and activates the cap-binding PB2 subunit, probably by conformational change. Concomitantly with activation of cap binding, the P proteins acquire the ability to bind to a conserved sequence at the 3′ ends of genomic

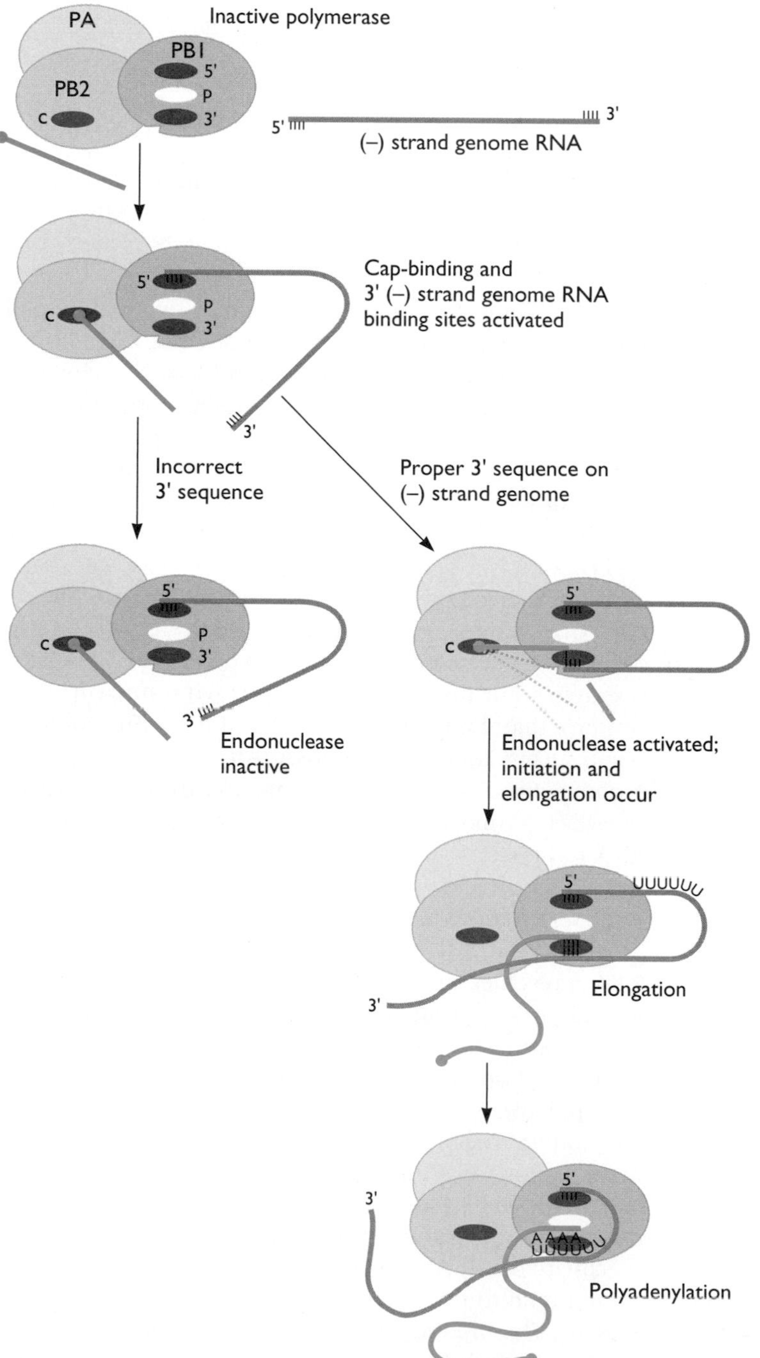

Figure 6.11 Activation of the influenza virus RNA polymerase by specific virion RNA sequences. The RNA polymerase, comprising the three P proteins, lacks activity until it binds to the 5′-terminal sequence of genomic RNA. Binding of the PB1 protein to this RNA sequence induces the PB2 protein to bind to the cap of a cellular RNA. When the 3′-terminal sequence of genomic RNA binds to a second amino acid sequence in the PB1 protein, the polymerase acquires the activity to cleave the capped cellular RNA 10 to 13 nucleotides from the cap. The RNA polymerase can then carry out initiation and elongation of mRNAs. Cleavage of capped cellular mRNAs occurs at a site (not shown) on PB1 that is distinct from the polymerase active site, p. 5′ and 3′ indicate the binding sites for the 5′ and 3′ ends, respectively, of (–) strand genomic RNA. Capped RNA is green; the 5′ cap structure is a circle at one end; p is the polymerase active site. Blue indicates an inactive site, and red indicates an active site. The p site is not colored because it is not known if it is constitutively active or requires activation. The mature polymerase is bound to both the 5′ and 3′ ends of the genomic RNA, with the capped RNA primer associated with the PB2 protein. During RNA synthesis, the polymerase remains bound to the 5′ end of the genomic RNA, and the 3′ end of the genomic RNA is threaded through (or along the surface of) the polymerase as the PB1 protein catalyzes each nucleotide addition to the growing mRNA chain. This threading process continues until the mRNA reaches a position on the genomic RNA that is close to the binding site of the polymerase. At this point the polymerase itself blocks further mRNA synthesis, and reiterative copying of the adjacent U7 tract occurs. After about 150 A residues are added to the 3′ end of the mRNA, mRNA synthesis terminates. The polymerase specifically binds to the sequence AGCAAAGCAGG, which is found in all viral mRNAs near their 5′ ends, immediately downstream of the sequence snatched from host cell capped RNAs. Such binding blocks access of the viral cap-dependent endonuclease to the viral mRNAs. The 5′ ends of host cell mRNAs are not protected in this way and are therefore efficiently cleaved by the viral endonuclease. (Right) Activated endonuclease; (left) influenza virus mRNA bound to the RNA polymerase and protected from cleavage by the endonuclease. Adapted from D. M. Knipe et al. (ed.), *Fields Virology*, 4th ed. (Lippincott Williams & Wilkins, Philadelphia, Pa., 2001); P. Rao et al., *EMBO J.* **22:**1188–1198, 2003; and S. R. Shih and R. M. Krug, *Virology* **226:**430–436, 1996, with permission.

RNA segments. This second interaction activates the endonuclease that cleaves capped host cell RNAs, enabling the P proteins to synthesize viral mRNA.

Imbalanced (–) and (+) Strand Synthesis

One of the unexplained curiosities of RNA virus infectious cycles is the imbalance in the synthesis of genomes and their full-length complements in infected cells. In cells infected with poliovirus, genomic RNA is produced at 100-fold higher concentrations than its complement. Assuming that RNA genomes and their complementary strands have similar stabilities, the simplest explanation for this difference is that the two strands are synthesized by different mechanisms that vary in efficiency. Viral (–) strand RNA is approximately 20 to 50 times more abundant than (+) strand RNAs in cells infected with vesicular stomatitis virus. The asymmetry was believed to occur because initiation of RNA synthesis at the 3′ end of (+) strand RNA is more efficient than initiation at the 3′ end of (–) strand RNA. An elegant proof of this hypothesis came from the construction and study of a rabies virus genome with identical initiation sites for RNA synthesis at the 3′ ends of both (–) and (+) strand RNAs. In cells infected with this virus, the ratio of (–) to (+) strands is 1:1.

In alphavirus-infected cells, the abundance of genomic RNA is explained by the observation that (–) strand RNAs are synthesized only for a short time early in infection. The RNA polymerase that catalyzes (–) strand RNA synthesis is produced only during this time. The synthesis of (+) strands continues for much longer and leads to accumulation of mRNA and (+) strand genomic RNA.

The Problem of Ribosomes Encountered by Viral Polymerases on (+) Strand RNAs

The genomic RNA of (+) strand viruses can be translated in the cell, and the translation products include the viral RNA polymerase. At a certain point in infection, the RNA polymerase copies the RNA in a 3′ → 5′ direction, while ribosomes traverse it in a 5′ → 3′ direction (Fig. 6.12), raising the question of how the viral polymerase avoids collisions with ribosomes. When ribosomes are frozen on polioviral RNA by using inhibitors of protein synthesis, replication is inhibited. In contrast, when ribosomes are released, replication of the RNA increases. These results suggest that ribosomes must be cleared from viral RNA before it can serve as a template for (–) strand RNA synthesis; in other words, replication and translation cannot occur simultaneously.

The interactions of viral and cellular proteins with the cloverleaf structure in the polioviral 5′ untranslated region might determine whether the RNA genome is translated or replicated. In this model, interaction of cellular

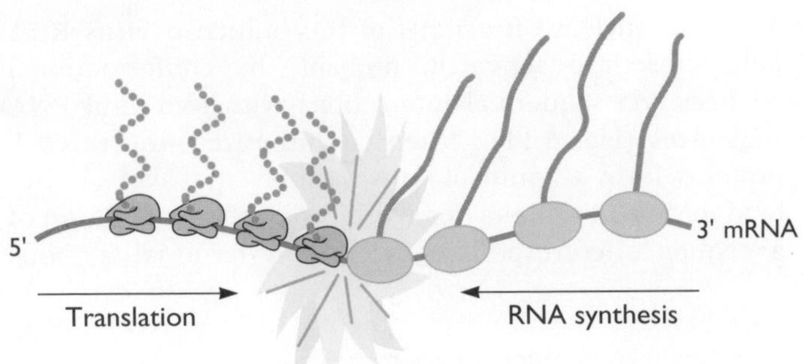

Figure 6.12 Ribosome-RNA polymerase collisions. A strand of viral RNA is shown, with ribosomes translating in the 5′ → 3′ direction and RNA polymerase copying the RNA chains in the 3′ → 5′ direction. Ribosome-polymerase collisions would occur in cells infected with (+) strand RNA viruses unless mechanisms exist to avoid simultaneous translation and replication.

poly(rC)-binding protein 2 with the cloverleaf initially stimulates translation in infected cells. Once the 3CD precursor has been synthesized, it binds to the cloverleaf structure and sequesters poly(rC)-binding protein, an interaction which represses translation and promotes RNA synthesis (Fig. 6.8B). However, this model cannot account for the observation that viral RNA translation continues in infected cells for several hours after the onset of RNA replication.

Even if mechanisms exist for controlling whether the polioviral genome is translated or replicated, some ribosome-RNA polymerase collisions do occur. This conclusion is drawn from the isolation of a polioviral mutant with a genome that contains an insertion of a 15-nucleotide sequence from 28S ribosomal RNA. After colliding with a ribosome, the RNA polymerase apparently copied 15 nucleotides of ribosomal RNA before returning to the viral RNA template.

Discrimination of Viral and Cellular RNAs

How do viral RNA polymerases selectively copy viral RNA molecules in the midst of hundreds of thousands of polyadenylated cellular mRNAs? Conserved RNA structural features in coding and noncoding regions of viral RNAs are responsible for the specific recognition by viral RNA polymerases. The *cis*-acting RNA elements located in the coding region of picornaviruses, which direct the uridylylation of VPg, are binding sites for $3CD^{pro}$ (Fig. 6.8B). The RNA pseudoknot that is important for the initiation of polioviral (–) strand synthesis plays a role in the specificity of copying by $3D^{pol}$. The $3D^{pol}$ and 3CD proteins cannot bind to the 3′ end of polioviral RNA unless 3AB is present. The interaction of 3AB and $3D^{pol}$ may determine the specificity of binding to the 3′ pseudoknot.

The isolation of infectious polioviruses without the entire 3′ noncoding region of the viral RNA has led to the suggestion that the template specificity imparted by terminal structures of RNA might be of greater importance early in infection. Early in infection, the 3′ pseudoknot structure might facilitate template selection when few viral polymerase molecules are available and membrane association has not yet provided highly concentrated areas of replication components. Later in infection, determinants of template selection by the polymerase might include the membrane association of the RNA polymerase, and the fact that translation of $3D^{pol}$, the most 3′-terminal gene, would position the polymerase at the 3′ end of the genome, ready for initiation.

The specific RNA-binding activity of the influenza virus P proteins explains the selective snatching of caps only from cellular RNAs. If the 5′ ends of newly synthesized viral mRNAs were also cleaved and used as primers, then no net synthesis of viral mRNAs would occur. The PB2 protein binds not only to the cap of a nascent viral mRNA, but also to a (+) strand RNA sequence located immediately downstream of the capped cellular RNA primer (Fig. 6.11). Such binding to two sites in the nascent viral mRNA blocks access of a second P protein and selectively protects newly synthesized viral mRNA from endonucleolytic cleavage by P proteins.

Synthesis of Poly(A)

The mRNAs synthesized during infection by most RNA viruses contain a 3′-poly(A) sequence, as do the vast majority of cellular mRNAs (exceptions are arena- and reoviruses). The poly(A) sequence is encoded in the viral RNA of (+) strand viruses. For example, polioviral (+) strand RNAs contain a 3′ stretch of poly(A), approximately 62 nucleotides in length, which is required for infectivity. The (–) strand RNA contains a 5′ stretch of poly(U), which is copied to form poly(A) of the (+) strand.

Another mechanism for poly(A) addition is reiterative copying of, or "stuttering" at, a short U sequence in the (–) strand template that leads to poly(A) addition. After initiation, vesicular stomatitis virus mRNAs are elongated until the RNA polymerase reaches a conserved stop-polyadenylation signal (3′-AUACU7-5′) located in each intergenic region (Fig. 6.13D). Poly(A) (approximately 150 nucleotides) is added by reiterative copying of the U stretch, followed by termination. Polyadenylation is achieved by a similar mechanism during influenza virus mRNA synthesis (Fig. 6.11).

Conventional models for RNA synthesis invoke a stationary template and moving polymerase. In a model for influenza virus mRNA synthesis with a stationary enzyme and moving template, a mechanism is apparent for reiterative copying of the five to seven U residues that produces the poly(A) sequence (Fig. 6.11). The RNA polymerase specifically binds the 5′ end of (–) strand RNA and remains bound at this site of each genomic RNA segment throughout mRNA synthesis. The genomic RNAs would be threaded through the polymerase in a 3′ → 5′ direction as mRNA synthesis proceeds. Eventually the template would be unable to move, causing reiterative copying of the U residues.

Cellular Sites of Viral RNA Synthesis

Cytoplasm

Genome and mRNA synthesis of most RNA viruses occurs in the cytoplasm of the cell, invariably in specific structures such as the nucleocapsids of (–) strand RNA viruses, subviral particles of double-stranded RNA viruses, and membrane-bound replication complexes for (+) strand RNA viruses. Membrane-bound replication complexes of different viruses are morphologically diverse, and the membranes originate from various cellular compartments. Alphaviral RNA synthesis occurs on the cytoplasmic surface of endosomes and lysosomes, and polioviral RNA polymerase is located on the surfaces of small, membranous vesicles that assemble into higher-order structures.

The mechanisms for the production of membranous replication complexes in virus-infected cells are beginning to be unraveled. The membrane vesicles observed early in poliovirus-infected cells originate from the endoplasmic reticulum. The production of these vesicles is regulated by proteins of the coat protein complex II (CopII) (Chapter 12). Unlike vesicles produced from the endoplasmic reticulum in uninfected cells, those in poliovirus-infected cells do not fuse with the Golgi and therefore accumulate in the cytoplasm. Membrane vesicles similar to those observed in poliovirus-infected cells are induced in cells by synthesis of polioviral protein 2BC. This viral protein might stimulate vesicle production by interacting with the CopII complex, and could prevent fusion with the Golgi by stabilizing CopII coats on the vesicles. In cells infected with poliovirus, membrane vesicle transport from the endoplasmic reticulum to the cell surface is inhibited, a consequence of the action of viral proteins 3A and 2B. This inhibition is likely to contribute to the accumulation of membrane vesicles during infection. Poliovirus infection therefore induces formation of the membrane vesicles important for replication by inhibition of an intracellular protein transport pathway. Because poliovirus is nonenveloped, its replication does not require the protein glycosylation and transport functions of the endoplasmic reticulum and Golgi apparatus.

Viral RNA polymerases are recruited to membranous replication complexes in different ways. Polioviral $3D^{pol}$ is

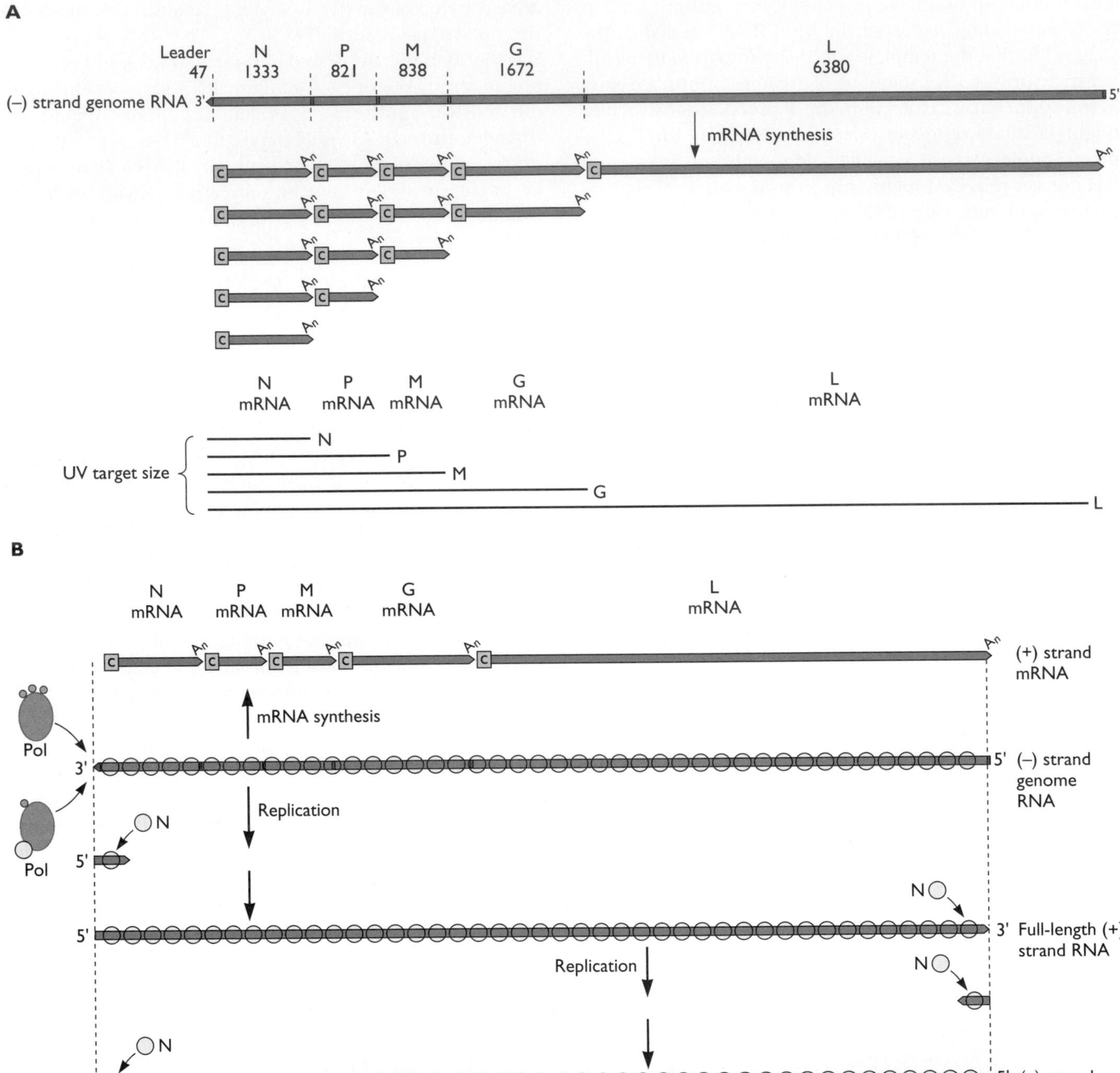

Figure 6.13 Vesicular stomatitis virus RNA synthesis. (A) Vesicular stomatitis virus mRNA map and UV map. The genome is shown as a dark green line at the top, and the N, P, M, G, and L genes and their relative sizes are indicated. The 47-nucleotide leader RNA is encoded at the 3′ end of the genomic RNA. The leader and intergenic regions are shown in orange. The RNAs encoded at the 3′ end of the genome are made in larger quantities than the RNAs encoded at the 5′ end of the genome. UV irradiation experiments determined the size of the vesicular stomatitis virus genome (UV target size) required for synthesis of each of the viral mRNAs. The UV target size of each viral mRNA corresponded to the size of the genomic RNA sequence encoding the mRNA plus all of the genomic sequence 3′ to this coding

(continued on next page)

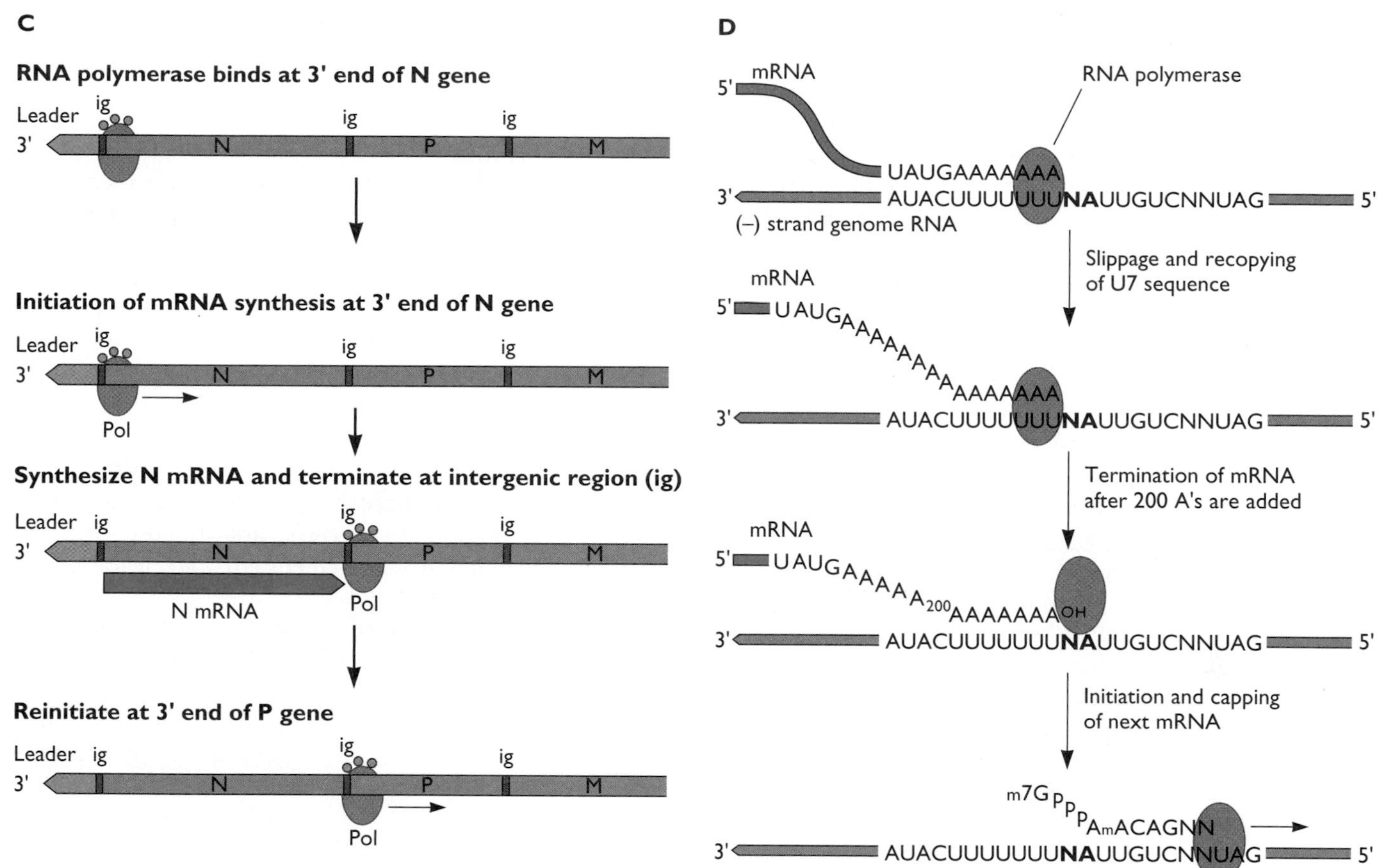

Figure 6.13 ***(continued)*** sequence. (B) Viral (–) strand genomes are templates for the production of either subgenomic mRNAs or full-length (+) strand RNAs. The switch from mRNA synthesis to genomic RNA replication is regulated by two different RNA polymerases, and by the N protein. Viral mRNA synthesis is carried out by an RNA polymerase that consists of 1 molecule of L protein complexed with 3 molecules of P protein. mRNA synthesis initiates at the beginning of the N gene, near the 3′ end of the viral genome. Poly(A) addition is a result of reiterative copying of a sequence of seven U residues present in each intergenic region. Chain termination and release occur after approximately 150 A residues have been added to the mRNA. The RNA polymerase then initiates synthesis of the next mRNA at the conserved start site 3′UUGUC . . . 5′. This process is repeated for all five viral genes. The transition from reiterative copying and termination to initiation is not perfect, and only about 70 to 80% of the polymerase molecules accomplish this transition at each intergenic region. Such inefficiency accounts for the observation that 3′-proximal mRNAs are more abundant than 5′-proximal mRNAs. Synthesis of the full-length (+) strand begins at the exact 3′ end of the viral genome, and is carried out by the RNA polymerase L-N-P. The (+) strand RNA is bound by the viral nucleocapsid (N) protein, which is complexed with the P protein in a 1:1 molar ratio. The N-P complexes bind to the nascent (+) strand RNA, allowing the RNA polymerase to read through the intergenic junctions (ig) at which polyadenylation and termination take place during mRNA synthesis. (C) Stop-start model of mRNA synthesis. The RNA polymerase (Pol) initiates RNA synthesis at the 3′ end of the N gene. After synthesis of the N mRNA, RNA synthesis terminates at the intergenic region, followed by reinitiation at the 3′ end of the P gene. This process continues until all five mRNAs are synthesized. Reinitiation does not occur after the last mRNA (the L mRNA) is synthesized, and as a consequence the 59 5′-terminal nucleotides of the vesicular stomatitis virus genomic RNA are not copied. Only a fraction of the polymerase molecules successfully make the transition from termination to reinitiation of mRNA synthesis at each intergenic region.
(D) Poly(A) addition and termination at an intergenic region. After the sequence of 7 U residues at the end of the mRNA-encoding sequence has been copied, the resulting seven A residues in the mRNA slip off the genomic sequence, which is then recopied. This process continues until approximately 200 A residues are added to the 3′ end of the mRNA. Termination then occurs, completing the synthesis of the mRNA, followed by initiation and capping of the next mRNA. The dinucleotide NA in the genomic RNA is not copied. The genomic RNA being copied is bound to viral N protein molecules at regular intervals (15 to 20 nucleotides).

targeted to membrane vesicles by binding to the viral membrane protein 3AB. In contrast, a C-terminal, transmembrane segment of the hepatitis C virus RNA polymerase, NS5b, is responsible for attachment of the enzyme to cellular membrane replication complexes.

Why viral replication complexes should be membrane associated is not known. Tight membrane association may ensure high local concentrations of replication components and hence increase the rates or efficiencies of replication reactions. As we have seen, this property may contribute to the specificity of polioviral $3D^{pol}$ for viral RNA templates. Membrane association may also have other roles, such as allowing efficient packaging of progeny RNA into virions, or providing lipid components or physical support to the replication complex. The surfaces of membranous replication complexes isolated from poliovirus-infected cells appear to be coated with two-dimensional arrays of polymerase similar to those depicted in Fig. 6.6. The surfaces of host cytoplasmic membranes may therefore be viewed as foundations for two-dimensional polymerase arrays. Surface catalysis is known to have several advantages, including a higher probability of collision among reactants, increase in substrate affinity from clustering of multiple binding sites, and retention of reaction products. The latter property would facilitate multiple rounds of copying (+) and (–) strand RNA templates.

The replication of some RNA viruses, such as reovirus, occurs in the cytoplasm in **inclusion bodies** that are not bound by membranes. These bodies contain viral RNA, proteins, and subviral particles and probably serve the same role as membranes, to concentrate the replication machinery in discrete locations in the cell.

Nucleus

It was initially surprising to find that mRNA synthesis and genome replication of some RNA viruses, such as influenza virus and Borna disease virus, a paramyxovirus with a nonsegmented (–) strand RNA genome, take place in the cell nucleus. For influenza virus, the requirement for the nucleus cannot be dictated by the need to produce capped primers from cellular mRNAs. Such mRNAs are found in the cytoplasm and are used to provide primers for bunyavirus mRNA synthesis. Rather, influenza virus RNA synthesis may take place in the nucleus to allow production of spliced viral mRNAs. The mRNA and genomic RNAs of the viral satellite agent hepatitis delta virus are also made in the nucleus, where these processes are carried out by host cell DNA-dependent RNA polymerase II, a nuclear enzyme.

Unique Mechanisms of mRNA and Genome Synthesis of the Hepatitis Delta Satellite Virus

The strategy for synthesis of the hepatitis delta satellite virus genome is apparently unique among animal viruses. The viral RNAs are synthesized by host cell DNA-dependent RNA polymerase II, and the hepatitis delta virus RNAs are RNA catalysts, or **ribozymes** (Box 6.2). The genome of hepatitis delta virus is a 1,700-nucleotide (–) strand circular RNA, the only RNA with this structure that has been found in animal cells. As approximately 70% of the nucleotides are base paired, the viral RNA is folded into a rodlike structure (Fig. 6.14A).

BOX 6.2

Ribozymes

A **ribozyme** is an enzyme in which RNA, not protein, carries out the catalytic activity. The first ribozyme discovered was the group I intron of the ciliate *Tetrahymena thermophila*. Other ribozymes have since been discovered, including RNase P of bacteria, group II self-splicing introns, hammerhead RNAs of viroids and satellite RNAs, and the ribozyme of hepatitis delta virus. Ribozymes are very diverse in size, sequence, and the mechanism for catalysis. For example, the hepatitis delta satellite virus ribozyme catalyzes a transesterification reaction that produces products with 2′,3′-cyclic phosphate and 5′-OH termini. Only an 85-nucleotide sequence is required for activity of this ribozyme, which can cleave optimally with as little as a single nucleotide 5′ to the site of cleavage.

By joining the 85-nucleotide fragment to other upstream sequences, accurate 3′ ends of heterologous RNA transcripts synthesized in vitro can be obtained. This property proved crucial for producing infectious RNAs from cloned DNA copies of the genomes of (–) strand RNA viruses. Ribozymes can also be used to target the cleavage of other RNAs and are therefore being widely explored as antiviral and antitumor agents.

Kruger, K., P. J. Grabowski, A. J. Zaug, J. Sands, D. E. Gottschling, and T. R. Cech. 1982. Self-splicing RNA: autoexcision and autocyclization of the ribosomal RNA intervening sequence of Tetrahymena. *Cell* **31:**147–157.

Westhof, E., and F. Michel. 1998. Ribozyme architectural diversity made visible. *Science* **282:**251–252.

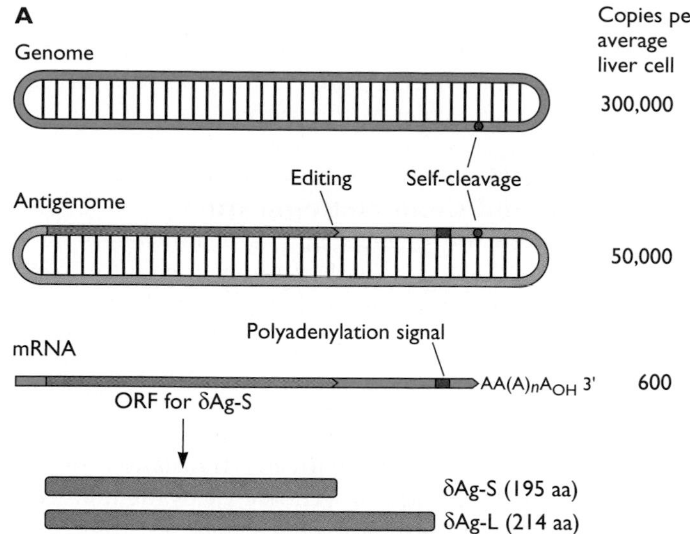

Figure 6.14 Hepatitis delta satellite virus RNA synthesis. (A) Schematic of the forms of hepatitis delta virus RNA and δ antigen found in infected cells. aa, amino acids. (B) Overview of hepatitis delta virus mRNA and genomic RNA synthesis. In steps 1 to 3, RNA synthesis is initiated, most likely by host RNA polymerase II, at the indicated position on the (–) strand genomic RNA. The polymerase passes the poly(A) signal (purple box) and the self-cleavage domain (red circle). In steps 4 and 5, the 5′ portion of this RNA is processed by cellular enzymes to produce delta mRNA with a 3′-poly(A) tail, while RNA synthesis continues beyond the cleavage site and the RNA undergoes self-cleavage (step 6). RNA synthesis continues until at least one unit of the (–) genomic RNA template is copied. The poly(A) signal is ignored in this full-length (+) strand. In steps 7 to 10, after self-cleavage to release a full-length (+) strand, self-ligation produces a (+) strand circular RNA. In steps 11 to 20, mRNA synthesis initiates on the full-length (+) strand to produce (–) strands by a rolling-circle mechanism. Unit-length genomes are released by the viral ribozyme (step 15) and self-ligated to form (–) strand circular genomic RNAs. Adapted from J. M. Taylor, *Curr. Top. Microbiol. Immunol.* **239:**107–122, 1999, with permission.

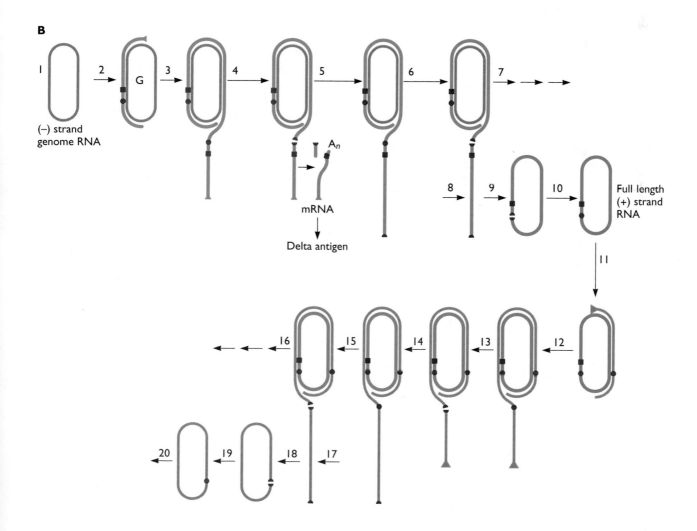

Surprisingly, all hepatitis delta virus-specific RNA synthesis is catalyzed by cellular DNA-dependent RNA polymerase II, which has not been known to utilize RNA as a template. Genomic RNA replication requires reactions catalyzed by ribozymes residing in both the hepatitis delta virus genomic (–) RNA and its full-length (+) RNA copy. Both these hepatitis delta virus RNAs catalyze self-cleavage in the absence of any protein and may also catalyze self-ligation.

All hepatitis delta satellite virus RNAs are synthesized in the nucleus by RNA polymerase II (Fig. 6.14B). The switch from mRNA synthesis to the production of full-length (+) RNA is controlled by suppression of a poly(A) signal (see "Suppression of Polyadenylation" below). Full-length (–) and (+) RNAs are copied by a rolling-circle mechanism, and ribozyme self-cleavage releases linear monomers. Subsequent ligation of the two termini by the same ribozyme produces a monomeric circular RNA. The hepatitis delta virus ribozymes are therefore needed to process the intermediates of rolling-circle RNA replication.

The Switch from mRNA Production to Genome RNA Synthesis

Exact replicas of the RNA genome must be made for assembly of infectious viral particles. However, the mRNAs of most RNA viruses are not complete copies of the viral RNA. The replication cycle of these viruses must therefore include a switch from mRNA synthesis to the production of full-length genomes. In this section we will consider examples of different mechanisms for the switch from mRNA synthesis to the production of full-length genomes. As might be expected, the majority of these mechanisms regulate either the initiation or termination of RNA synthesis.

No switch from mRNA to genomic RNA synthesis is needed when the only viral mRNA made in infected cells is identical to the (+) strand genome, as occurs in cells infected with picornaviruses (Fig. 6.1). Newly synthesized polioviral (+) strand RNA molecules can serve as RNA templates for further genomic RNA replication, as mRNAs for the synthesis of viral proteins, or as genomic RNAs to be packaged into progeny virions. Because picornaviral mRNA is identical in sequence to the viral RNA genome, all RNAs needed for the reproduction of these viruses can be made by a simple set of RNA synthesis reactions (Fig. 6.1). Such simplicity comes at a price, because there can be no regulation of individual gene products at the level of mRNA synthesis. However, polioviral gene expression can be controlled by the rate and extent of polyprotein processing. For example, $3CD^{pro}$, the precursor of the viral polymerase $3D^{pol}$, cannot produce RNA. Rather, $3CD^{pro}$ is a protease that cleaves at certain Gln-Gly amino acid pairs in the polyprotein. The concentration of RNA polymerase can be controlled by regulating the processing of 3CD.

Different RNA Polymerases for mRNA Synthesis and Genome Replication

The mechanism of mRNA synthesis of other (+) strand RNA viruses allows structural and nonstructural proteins (generally needed in greater and lesser quantities, respectively) to be produced separately. The latter are produced from full-length (+) strand (genomic) RNA, while structural proteins are translated from subgenomic mRNA(s). This strategy is a feature of the replication cycles of corona-, calici-, and alphaviruses. Translation of the Sindbis virus (+) strand RNA genome yields the nonstructural proteins that synthesize a full-length (–) strand (Fig. 6.15A). Such RNA molecules contain not only a 3′-terminal sequence for initiation of (+) strand RNA synthesis, but also an internal initiation site. The initiation of (+) strand synthesis at this internal site produces a 26S subgenomic mRNA that is translated to form the precursor of the structural proteins.

Alphaviral genome and mRNA synthesis is regulated by the sequential production of three RNA polymerases with different template specificities. All three enzymes are derived from the nonstructural polyprotein P1234 and contain the complete amino acid sequence of this precursor (Fig. 6.15A). The covalent connections among the segments (P1, P2, P3, P4) of the polyprotein are successively broken, with ensuing alterations in RNA polymerase specificity (Fig. 6.15B). It seems likely that each proteolytic cleavage induces a conformational change in the polymerase that alters its template specificity.

The genes of RNA viruses with a nonsegmented (–) strand RNA genome are expressed by the production of subgenomic mRNAs in infected cells. However, replication requires synthesis of a full-length (+) strand copy of the genome. The switch from subgenomic mRNA to full-length (+) strand synthesis may also be accomplished by producing RNA polymerases with different specificities.

Two different RNA polymerases are produced during the influenza virus infectious cycle (Fig. 6.9C). The RNA polymerase comprising the PA, PB1, and PB2 proteins synthesizes only subgenomic polyadenylated mRNAs. As discussed previously, the cap-binding subunit of the RNA polymerase is essential during influenza virus mRNA synthesis because it produces RNA primers for initiation. In contrast, the enzyme responsible for replication initiates the synthesis of full-length (+) and (–) strand RNAs de novo. In the presence of NP protein, this polymerase reads

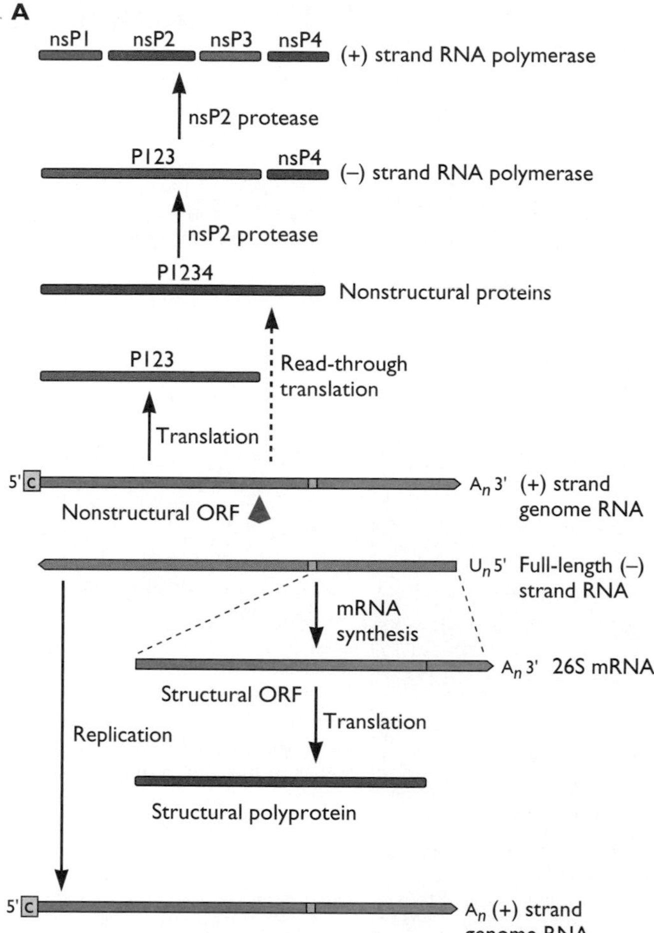

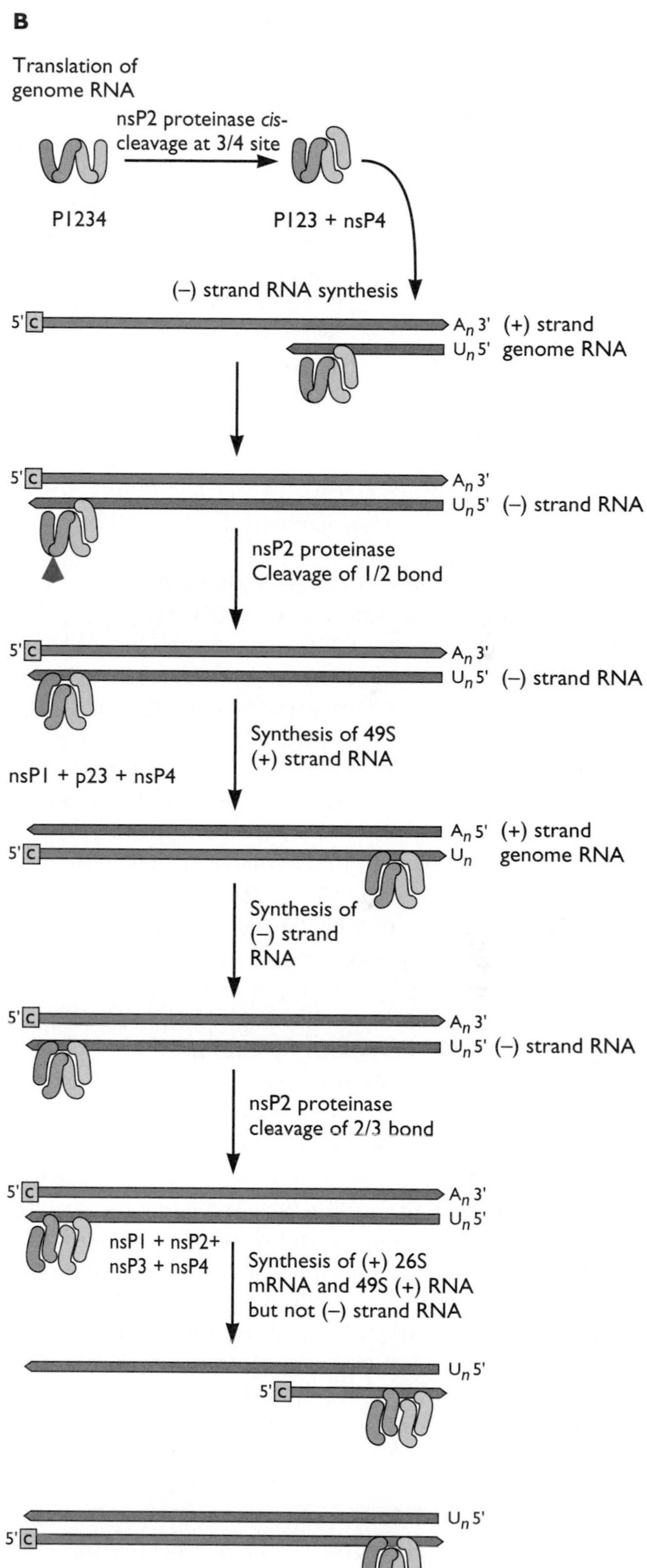

Figure 6.15 RNA replication of an alphavirus, Sindbis virus. (A) Genome structure and expression. The 11,703-nucleotide Sindbis virus genome contains a 5′-terminal cap structure and a 3′-poly(A) tail. A conserved RNA secondary structure at the 3′ end of (+) strand genomic RNA is believed to control the initiation of (–) strand RNA synthesis. At early times after infection, the 5′ region of the genomic RNA (nonstructural open reading frame [ORF]) is translated to produce two nonstructural polyproteins: P123, whose synthesis is terminated at the first translational stop codon (indicated by the red diamond), and P1234, produced by an occasional (15%) readthrough of this stop codon. The P1234 polyprotein is proteolytically cleaved to produce the enzymes that catalyze the various steps in genomic RNA replication: the synthesis of a full-length (–) strand RNA, which serves as the template for (+) strand synthesis, and either full-length genomic RNA or subgenomic 26S mRNA. The 26S mRNA, shown in expanded form, is translated into a structural polyprotein (p130) that undergoes proteolytic cleavage to produce the virion structural proteins. The 26S RNA is not copied into a (–) strand because a functional initiation site fails to form at the 3′ end. (B) Three RNA polymerases with distinct specificities in alphavirus-infected cells. These RNA polymerases contain the entire sequence of the P1234 polyprotein and differ only in the number of proteolytic cleavages in this sequence. Adapted from J. H. Strauss and E. G. Strauss, *Microbiol. Rev.* **58:**491–562, 1994, with permission.

through the polyadenylation and termination signals used for mRNA production (Fig. 6.9A). It is believed that the PA subunit of the RNA polymerase, which plays no known role in mRNA synthesis, is important for replication, but exactly how the initiation properties of the enzyme are regulated is not yet clear.

The RNA polymerases that carry out vesicular stomatitis virus mRNA synthesis and genome replication are also different (Fig. 6.13B). A polymerase composed of 1 molecule of L protein complexed with 3 molecules of P protein might carry out mRNA synthesis, while genome replication would be accomplished by an L-(N-P) complex. The ratio of the two polymerases at different times after infection would therefore be regulated both by viral protein synthesis and by the phosphorylation state of the P protein, which controls its ability to form oligomers.

Suppression of Intergenic Stop-Start Reactions by Nucleocapsid Protein

The transition from mRNA to genome RNA synthesis in cells infected with vesicular stomatitis virus requires not only different RNA polymerases, but also the viral nucleocapsid (N) protein. Individual mRNAs are produced by a series of initiation and termination reactions as the RNA polymerase moves down the viral genome. This start-stop mechanism accounts for the observation that 3′-proximal genes must be copied before downstream genes (Box 6.3); the viral RNA polymerase is unable to initiate synthesis of each mRNA independently. To produce a full-length (+) strand RNA, the stop-start reactions at intergenic regions are suppressed (Fig. 6.13). Suppression depends on the synthesis of the N and P proteins, which are complexed in a 1:1 molar ratio. The N-P complexes bind to leader RNA and cause antitermination, signaling the polymerase to begin processive RNA synthesis. Additional N protein molecules then associate with the (+) RNA as it is elongated and eventually bind to the seven A bases in the intergenic region. This interaction prevents the seven A bases from slipping backward along the genomic RNA template and therefore blocks reiterative copying of the seven U bases in the genome by the L-P protein. Consequently, RNA synthesis continues through the intergenic region. The number of N-P protein complexes in infected cells therefore regulates the relative efficiencies of mRNA synthesis and genome RNA replication. Approximately 2,600 N monomers are required to coat the viral genome.

The copying of full-length (+) strand RNAs to (–) strand genomic RNAs also requires the binding of N-P protein complexes to elongating RNA molecules. Newly synthesized (–) strand RNAs are produced as nucleocapsids that can be readily packaged into progeny viral particles.

The (–) strand RNA genome of paramyxoviruses is copied efficiently only when its length in nucleotides is a multiple of 6. This requirement, called the **rule of six**, is probably a consequence of the association of each N monomer with exactly six nucleotides. Assembly of the nucleocapsid begins with the first nucleotide at the 5′ end of the RNA and continues until the 3′ end is reached. If the genome length is not a multiple of six, then the 3′ end of the genome will not be precisely aligned with the last N monomer. Such misalignment reduces the efficiency of initiation of RNA synthesis at the 3′ end.

The influenza virus NP protein also regulates the switch from viral mRNA to full-length (+) strand synthe-

BOX 6.3

Mapping gene order by ultraviolet irradiation

The effects of ultraviolet (UV) irradiation provided insight into the mechanism of vesicular stomatitis virus mRNA synthesis. In these experiments, vesicular stomatitis virus particles were irradiated with UV light, and the effect on the synthesis of individual mRNAs was determined. In principle, larger genes require less UV irradiation to inactivate mRNA synthesis and have a greater **target size.** The dosage of UV irradiation needed to inactivate synthesis of the N mRNA corresponded to the predicted size of the N gene, but this was not the case for the other viral mRNAs. The target size of each other mRNA was the sum of its size plus the size of other genes located 3′ of the gene. For example, the UV target size of the L mRNA is the size of the entire vesicular stomatitis virus genome (Fig. 6.13A). These results indicate that these mRNAs are synthesized sequentially, in the 3′ → 5′ order in which their genes are arranged in the viral genome: N-P-M-G-L.

Ball, L. A., and C. N. White. 1976. Order of transcription of genes of vesicular stomatitis virus. *Proc. Natl. Acad. Sci. USA* **73:**442–446.

sis (Fig. 6.9A). The RNA polymerase for genome replication reads through the polyadenylation and termination signals used for mRNA production only if NP is present. The NP protein is believed to bind nascent (+) strand transcripts and block poly(A) addition by a mechanism analogous to that described for vesicular stomatitis virus N protein. Copying of (+) strand RNAs into (–) strand RNAs also requires NP protein. Intracellular concentrations of NP protein are therefore an important determinant of whether mRNAs or full-length (+) strands are synthesized.

Suppression of Termination Induced by a Stem-Loop Structure

The subgenomic mRNAs of certain (–) strand RNA viruses, such as arenaviruses, are produced when the RNA polymerase terminates at a stem-loop structure in the viral RNA template. This suppression results in the synthesis of full-length (+) RNAs.

Although arenaviruses are considered (–) strand RNA viruses, their genomic RNA is in fact **ambisense**: mRNAs are produced both from (–) strand genomic RNA and from complementary full-length (+) strands. The arenavirus genome comprises two RNA segments, S (small) and L (large) (Fig. 6.16). Shortly after infection, a virion-associated RNA polymerase synthesizes mRNAs from the 3′ region of the S and L genomic RNA segments. Synthesis of each mRNA terminates at a stem-loop structure (Fig. 6.16). These mRNAs, which are translated to produce the nucleocapsid (N) protein and RNA polymerase (L) protein, respectively, are the only viral RNAs made during the first several hours of infection. Later in infection, the block imposed by the stem-loop structure is overcome by the viral N protein, permitting the synthesis of full-length S and L (+) RNAs. The mechanism by which N protein allows the RNA polymerase to bypass the transcription termination signal is not known. A stem-loop structure also restricts mRNA synthesis to the 3′ ends of (+) strand RNAs (Fig. 6.16).

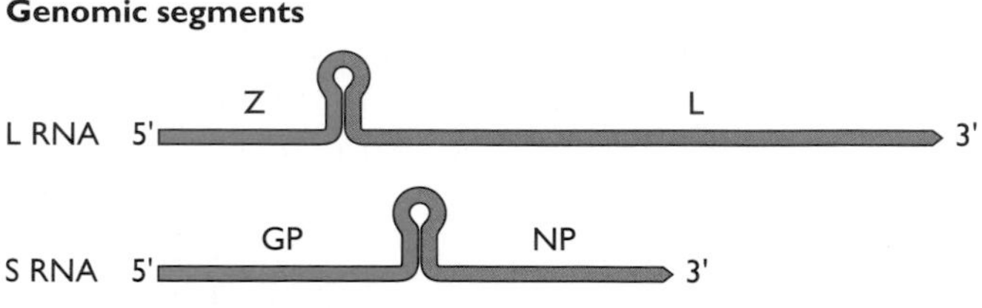

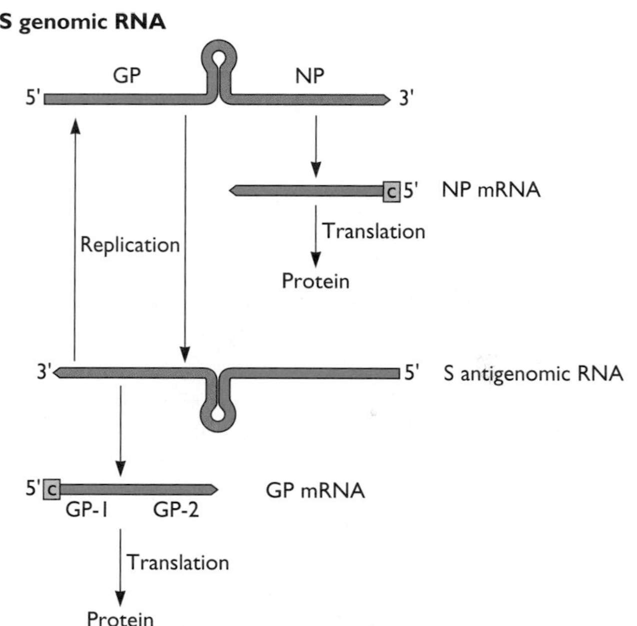

Figure 6.16 Arenavirus RNA synthesis. Arenaviruses contain two genomic RNA segments, L (large) and S (small) (top). At early times after infection, only the 3′ region of each of these segments is copied to form mRNA: the N mRNA from the S genomic RNA and the L mRNA from the L genomic RNA. Copying of the remainder of the S and L genomic RNAs may be blocked by a stem-loop structure in the genomic RNAs. After the S and L genomic RNAs are copied into full-length (+) strands, the 3′ regions of the (+) RNAs are copied to produce mRNAs: the glycoprotein precursor (GP) mRNA from S (+) RNA and the Z mRNA (encoding an inhibitor of viral RNA synthesis) from the L (+) RNA. Termination of mRNA synthesis on the (+) strands may also occur at a stem-loop structure. Only RNA synthesis from the S RNA is shown in detail. Adapted from D. M. Knipe et al. (ed.), *Fields Virology*, 4th ed. (Lippincott Williams & Wilkins, Philadelphia, Pa., 2001), with permission.

Different Templates Used for mRNA Synthesis and Genome Replication

A distinctive feature of the infectious cycle of double-stranded RNA viruses such as reovirus is the production of mRNAs and genomic RNAs from different templates in different viral particles. Because the viral genomes are double stranded, they cannot be translated by the cell. Therefore, the first step in infection is the production of mRNAs from each viral RNA segment by the virion-associated RNA polymerase (Fig. 6.17). Reoviral mRNAs contain 5′ cap structures but lack 3′ poly(A) sequences.

The viral mRNAs are extruded from the viral particle into the cytoplasm. Examination of the structure of actively transcribing rotavirus, a member of the *Reoviridae*,

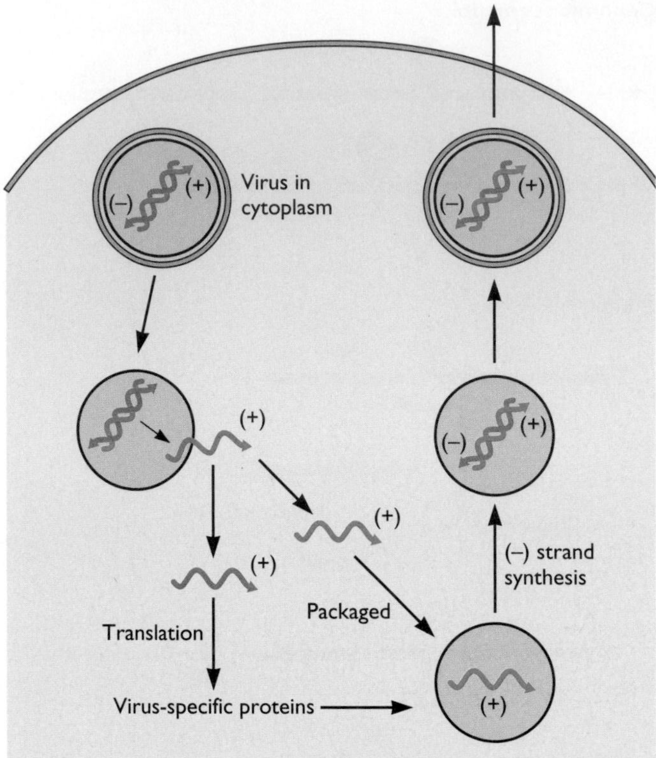

Figure 6.17 mRNA synthesis and replication of double-stranded RNA genomes. These processes occur in subviral particles containing the RNA templates and required enzymes. During cell entry, the virus passes through the lysosomal compartment, and proteolysis of viral capsid proteins activates the RNA synthetic machinery. Single-stranded (+) viral mRNAs, which are synthesized in parental subviral particles, are extruded into the cytoplasm, where they serve either as mRNAs or as templates for the synthesis of (–) RNA strands. In the latter case, viral mRNAs are first packaged into newly assembled subviral particles in which the synthesis of (–) RNAs to produce double-stranded RNAs occurs. These subviral particles become infectious particles. Only one double-stranded RNA is shown, whereas the reoviral genome consists of 10 to 12 double-stranded RNA segments.

has allowed a three-dimensional visualization of how mRNAs are released from the particle (Box 6.4). Viral (+) strand RNAs that will serve as templates for (–) strand RNA synthesis are first packaged into newly assembled subviral particles (Fig. 6.17). Each (+) strand RNA is then copied just once within the subviral particle to produce double-stranded RNA. This two-step replication process of double-stranded RNA bears little resemblance to the simultaneous copying of both strands that generally occurs during the replication of double-stranded DNA.

Suppression of Polyadenylation

Another mechanism for switching from mRNA synthesis to genomic replication is to suppress a poly(A)-addition signal; an example occurs in hepatitis delta satellite virus-infected cells. Host RNA polymerase II initiates viral mRNA synthesis at a position on the genome near the beginning of the delta antigen-coding region (Fig. 6.14B). Once the polymerase has moved past the polyadenylation signal and the self-cleavage domain, the 3′-poly(A) end of the mRNA is produced by host cell enzymes. The RNA downstream of the poly(A) site is not degraded, in contrast to that of other mRNA precursors made by RNA polymerase II, but is elongated until a complete full-length (+) strand is made. The poly(A) addition site in this full-length (+) RNA is not used. The delta antigen bound to the rodlike RNA may block access of cellular enzymes to the poly(A) signal, thereby inhibiting polyadenylation.

Origins of Diversity in RNA Viruses

Misincorporation of Nucleotides

All nucleic acid polymerases insert incorrect nucleotides during chain elongation. DNA-directed DNA polymerases have **proofreading** capabilities in the form of exonuclease activities that can correct such mistakes. RNA-dependent RNA polymerases do not possess proofreading ability. The result is that error frequencies in RNA replication can be as high as one misincorporation per 10^3 to 10^4 nucleotides polymerized, whereas the frequency of errors in DNA replication is about 1,000-fold lower. Many of these polymerization errors cause amino acid changes, and the viruses may not be infectious, while other mutations may appear in the genomes of infectious virions. This phenomenon has led to the realization that RNA virus populations are **quasispecies**, or mixtures of many different genome sequences. The errors introduced during RNA replication have important consequences for viral pathogenesis and evolution, as discussed in Chapter 20. Because RNA virus stocks are mixtures of genotypically different viruses, specific viral mutants may be isolated directly from such stocks. For example, live attenuated poliovirus vaccine strains are viral mutants that were isolated from an unmutagenized stock of wild-type virus.

Segment Reassortment and RNA Recombination

Reassortment is the exchange of entire RNA molecules between genetically related viruses with segmented genomes. In cells coinfected with two different influenza viruses, the eight genome segments of each virus replicate. When new progeny viruses are assembled, they can package RNA segments from either parental virus. Because reassortment involves the simple exchange of RNA segments, it can occur at high frequencies.

BOX 6.4

Release of mRNA from rotavirus particles

Rotaviruses, the most important cause of gastroenteritis in children, are large icosahedral viruses made of a three-shelled capsid containing 11 double-stranded RNA segments. The structure of this virus reveals that a large portion of the viral genome (~25%) is ordered within the particle and forms a dodecahedral structure (Fig. 4.17). In this structure, the RNAs interact with the inner capsid layer and pack around the RNA polymerase located at the fivefold axis of symmetry. Further analysis of rotavirus particles in the process of synthesizing mRNA has shown that newly synthesized molecules are extruded through the capsid through several channels located at the fivefold axes (see figure). Multiple mRNAs are released at the same time from such particles. On the basis of these observations, it has been suggested that each double-stranded genomic RNA segment is copied by an RNA polymerase located at a fivefold axis of symmetry. This model may explain why no double-stranded RNA virus with more than 12 genomic segments—the maximum number of fivefold axes—has ever been identified.

Lawton, J. A., M. K. Estes, and B. V. Prasad. 1997. Three-dimensional visualization of mRNA release from actively transcribing rotavirus particles. *Nat. Struct. Biol.* **4:**118–121.

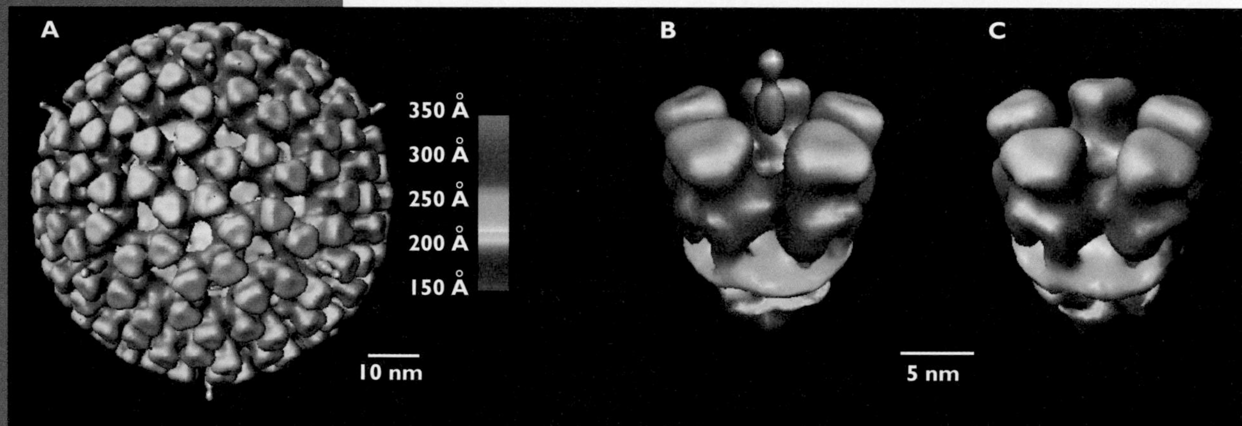

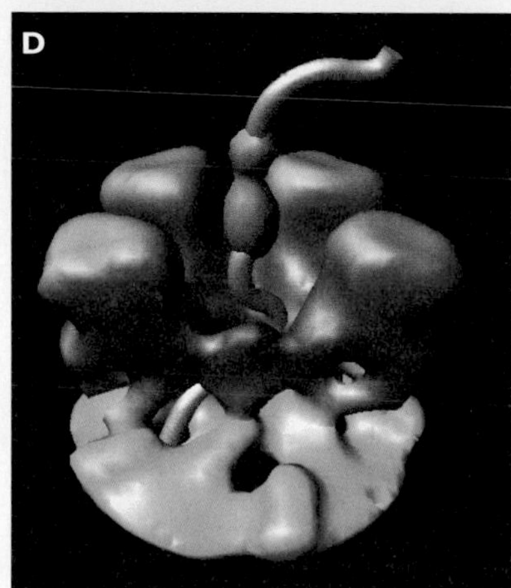

Three-dimensional visualization of mRNA release from rotavirus particles synthesizing mRNA. (A) Three-dimensional structure of rotavirus in the process of synthesizing mRNA. The capsid is depth-cued according to the color chart. Parts of newly synthesized mRNA that are ordered, and therefore structurally visible, are shown in pink at the fivefold axes of symmetry. (B) Close-up view of the channel at the fivefold axis and the visible mRNA. The mRNA is surrounded by five trimers of capsid protein VP6. (C) Close-up view of the channel at the fivefold axis of a particle not in the process of synthesizing mRNA. (D) Model of the pathway of mRNA transit through the capsid. One VP6 trimer has been omitted for clarity. The green protein is VP2, and the mRNA visible in the structure is shown in pink. The gray tube represents the possible path of an mRNA molecule passing through the VP2 and VP6 layers through the channel. Courtesy of B. V. V. Prasad, Baylor College of Medicine. Reprinted from J. A. Lawton et al., *Nat. Struct. Biol.* **4:**118–121, 1997, with permission.

In contrast to reassortment, recombination is the exchange of nucleotide sequences among different genomic RNA molecules (Fig. 6.18, top). Recombination, a feature of many RNA viruses, is an important mechanism for producing new genomes with selective growth advantages. This process has shaped the RNA virus world by rearranging genomes or moving functional parts of RNA molecules among different viruses. RNA recombination was first discovered in cells infected with poliovirus and was subsequently found to occur with other (+) and (–) strand

RNA recombination

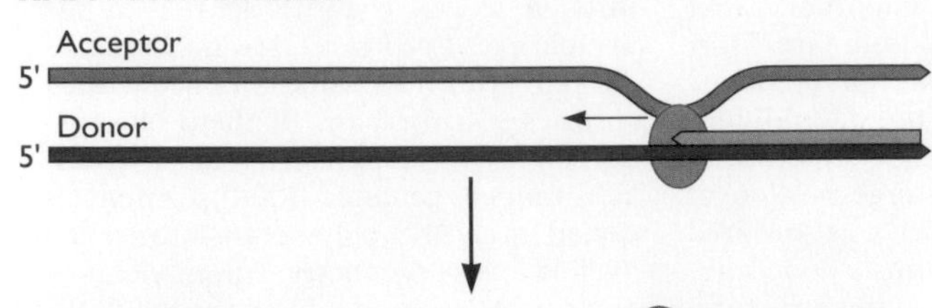

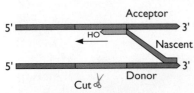

Figure 6.18 RNA recombination. (Top) Schematic representation of RNA recombination occurring during template switching by RNA polymerase, or copy choice. Two parental genomes are shown. The RNA polymerase (purple oval) has copied the 3′ end of the donor genome and is switching to the acceptor genome. The resulting recombinant molecule is shown. (Bottom) Three classes of RNA recombination. All three classes involve RNA polymerase-mediated template exchange. Events that occur after the exchange are shown by an arrow. The hairpin symbolizes various RNA features required for class 2 and 3 recombination. In base-pairing-dependent recombination, substantial sequence similarity between parental RNAs is required and is the major determinant of recombination. In base-pairing-independent recombination, sequence similarity is not required but may be present. Recombination may be determined by other RNA features, such as RNA polymerase-binding sites, secondary structures, and heteroduplex formation between parental RNAs. Base-pairing-assisted recombination combines features of class 1 and class 2 recombination. Sequence similarity influences the frequency or site of recombination, but additional RNA features are required. Adapted from P. D. Nagy and A. E. Simon, *Virology* **235**:1–9, 1997, with permission.

Class 1: Base pairing dependent

Acceptor
5' 3'
HO
Nascent
5' 3'
Donor
Cut

RNA-dependent RNA-polymerase internal pausing/termination or RNA breakage

Class 2: Base pairing independent

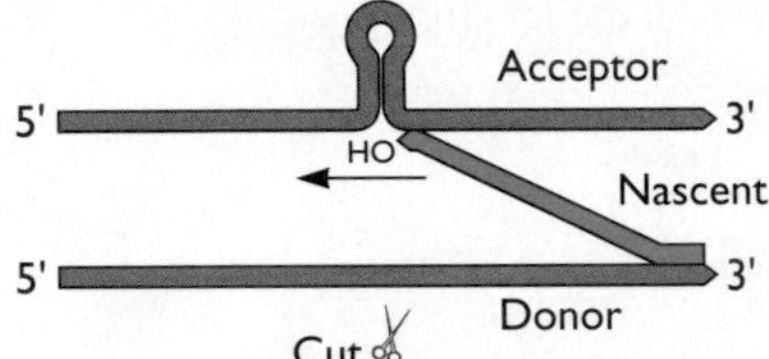

Class 3: Base pairing assisted

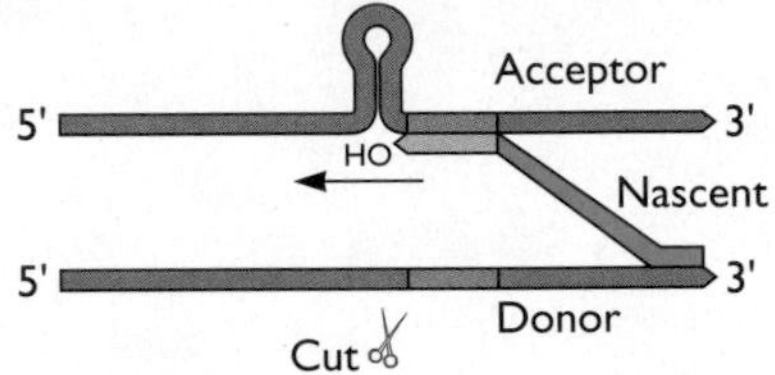

RNA viruses. The frequency of recombination can be relatively high: it has been estimated that 10 to 20% of polioviral genomic RNAs recombine in one growth cycle. RNA recombination is not limited to the research laboratory. For example, recombinants among the three serotypes of poliovirus are readily isolated from the intestines of individuals immunized with the live (Sabin) vaccine. These recombinants may possess an improved ability to replicate in the human alimentary tract and have a selective advantage over the parental viruses. Recombination may also lead to the production of pathogenic viruses (Box 6.5).

Polioviral recombination is predominantly **base pairing dependent**: it occurs between nucleotide sequences that have a high percentage of nucleotide identity (Fig. 6.18, bottom). Other viral genomes undergo **base-pairing-independent** recombination between very different nucleotide sequences. RNA recombination is coupled with the process of genomic RNA replication: it occurs by template exchange during (–) strand synthesis, as first demonstrated in poliovirus-infected cells. The RNA polymerase first copies the 3′ end of one parental (+) strand and then exchanges one template for another at the corresponding position on a second parental (+) strand (Fig. 6.18, bottom). Template exchange in poliovirus-infected cells occurs predominantly during (–) strand synthesis presumably because the concentration of (+) strands available for template exchange is 100-fold higher than that of (–) strands. This template exchange mechanism of recombination is also known as **copy choice**. The exact mechanism of template exchange is not known, but it might be triggered by pausing of the polymerase during chain elongation. In the **breakage-induced template exchange** model, it is postulated that breakage of the RNA template that is being copied promotes exchange of one template for another. Recombinant RNAs may be formed by a breakage-and-ligation mechanism similar to that which occurs during some DNA recombination. Although these last two models are plausible, they have not yet been demonstrated experimentally.

If the RNA polymerase skips sequences during template switching, deletions may occur. These RNAs can replicate if they contain the appropriate signals for the initiation of RNA synthesis. Subgenomic RNAs replicate more rapidly

BOX 6.5

RNA recombination leading to the production of pathogenic viruses

A remarkable property of pestiviruses, members of the *Flaviviridae*, is that RNA recombination produces viruses that cause disease. Bovine viral diarrhea virus causes a usually fatal gastrointestinal disease. Infection of a fetus with this virus during the first trimester is noncytopathic, but RNA recombination produces a cytopathic virus that causes severe gastrointestinal disease after the animal is born.

Pathogenicity of bovine viral diarrhea virus is associated with the synthesis of a nonstructural protein, NS3, encoded by the recombinant cytopathic virus (see figure). The NS3 protein cannot be produced in cells infected by the noncytopathic parental virus because its precursor, the NS2-3 protein, is not proteolytically processed. In contrast, NS3 is synthesized in cells infected by cytopathic bovine viral diarrhea virus because base-pairing-independent RNA recombination reactions add an extra protease cleavage site in the viral polyprotein, precisely at the N terminus of the NS3 protein (see figure). This cleavage site can be created in several ways. One of the most frequent is insertion of a cellular RNA sequence coding for **ubiquitin,** which targets cellular proteins to a degradative pathway. Insertion of ubiquitin at the N terminus of NS3 permits cleavage of NS2-3 by any member of a widespread family of cellular proteases. This recombination event provides a selective advantage because pathogenic viruses outgrow nonpathogenic viruses. Why cytopathogenicity is associated with release of the NS3 protein, which is thought to be part of the enzymatic system for genomic RNA replication, is not known.

Retroviruses acquire cellular genes by recombination, and the resulting viruses have lethal disease potential, as described in Chapter 18.

Pathogenicity of bovine viral diarrhea virus is associated with production of the NS3 nonstructural protein by a recombinant cytopathic virus. During infection of cells with noncytopathic bovine viral diarrhea virus, the NS3 protein is not produced but remains part of the precursor polypeptide NS2-3. During replication, RNA recombination leads to precise insertion of the cellular ubiquitin gene into the polyprotein. Cleavage of the polyprotein at the N terminus of the NS3 protein is catalyzed by any member of a widespread family of cellular proteases that cleave ubiquitin in uninfected cells. Because of the release of this NS3 protein, the resulting viral strains are pathogenic. Two cytopathic viruses, Osloss and CP1, in which the ubiquitin sequence (UCH) has been inserted at different sites, are shown. In Osloss, UCH has been inserted into the NS2-3 precursor, and NS3 is produced. In CP1, a duplication has also occurred such that an additional copy of NS3 is present after the UCH sequence. Adapted from D. M. Knipe et al. (ed.), *Fields Virology*, 4th ed. (Lippincott Williams & Wilkins, Philadelphia, Pa., 2001), with permission.

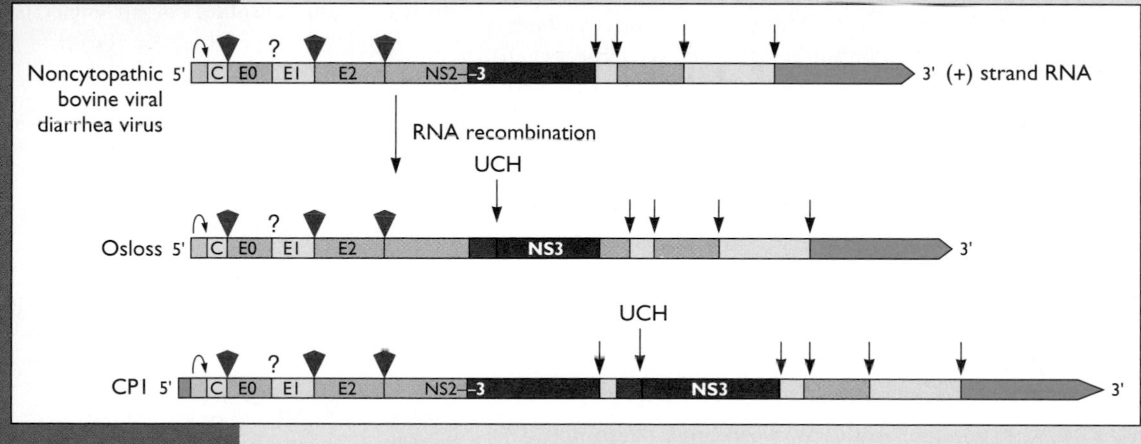

than full-length RNA and therefore compete for the components of the RNA synthesis machinery. Because of this property they are called **defective-interfering RNAs** because they lack parts of the genome and interfere with the replication of full-length RNAs. Such RNAs are packaged into viral particles only in the presence of a **helper virus** that provides viral proteins.

Defective-interfering viruses accumulate during the replication of both (+) and (–) strand RNA viruses. Production of these viruses requires a high multiplicity of infection, a condition that is achieved readily in the laboratory but rarely in nature. It is not known whether defective-interfering viruses generally play a role in viral pathogenesis. However, some recombination reactions that lead to the appearance of cytopathic bovine viral diarrhea viruses delete viral RNA sequences rather than inserting cellular RNA sequences (Box 6.5). Such deletions create a new protease cleavage site at the N terminus of the NS3 protein, and the defective-interfering viruses also cause severe gastrointestinal disease in livestock.

RNA Editing

Diversity in RNA viral genomes is also achieved by RNA editing. Viral mRNAs can be edited either by insertion of a nontemplated nucleotide during synthesis or by altering the base after synthesis. Examples of RNA editing have been documented in members of the *Paramyxoviridae* and *Filoviridae* and in hepatitis delta satellite virus. This process is described in Chapter 10.

Perspectives

Our understanding of genomic replication and mRNA synthesis of RNA viruses has expanded considerably since the discovery of RNA-dependent RNA polymerase in the early 1960s. From early studies of infected cells and of partially purified enzymes, the field has progressed enormously with the development of recombinant DNA methods. These procedures have allowed the production and purification of RNA polymerases and their associated proteins, and the production of altered proteins and RNA templates, to better understand the mechanisms of RNA synthesis. As outlined in this chapter, many questions still remain. Our knowledge is rudimentary compared with that on reverse transcriptase, to which the entire next chapter is devoted. We expect that future research will provide insight into the nature and function of host proteins that are required for viral RNA synthesis. Other unanswered questions include why RNA synthesis occurs in certain cellular compartments, and how translation and replication are regulated on individual RNA molecules. The crystal structure of polioviral 3D^{pol} hints at the beautiful pictures of other viral RNA polymerases and their associated proteins that will be forthcoming.

References

Reviews

Barr, J. N., S. P. Whelan, and G. W. Wertz. 2002. Transcriptional control of the RNA-dependent RNA polymerase of vesicular stomatitis virus. *Biochim. Biophys. Acta* **1577:**337–353.

Deiman, B. A. L. M., and C. W. A. Pleij. 1997. Pseudoknots: a vital feature in viral RNA. *Semin. Virol.* **8:**166–175.

Horikami, S. M., and S. A. Moyer. 1995. Structure, transcription, and replication of measles virus. *Curr. Top. Microbiol. Immunol.* **191:**35–50.

Nagy, P. D., and A. E. Simon. 1997. New insights into the mechanisms of RNA recombination. *Virology* **235:**1–9.

Nowakowski, J., and I. Tinoco. 1997. RNA structure and stability. *Semin. Virol.* **8:**153–165.

Paul, A. V. 2002. Possible unifying mechanism of picornavirus genome replication, p. 227–246. *In* B. L. Semler and E. Wimmer (ed.), *Molecular Biology of Picornaviruses.* ASM Press, Washington, D.C.

Sawicki, D. L., and S. G. Sawicki. 1998. Role of the nonstructural polyproteins in alphavirus RNA synthesis. *Adv. Exp. Med. Biol.* **440:**187–198.

Sawicki, S. G., and D. L. Sawicki. 1998. A new model for coronavirus transcription. *Adv. Exp. Med. Biol.* **440:**215–219.

Strauss, J. H., and E. G. Strauss. 1994. The alphaviruses: gene expression, replication, and evolution. *Microbiol. Rev.* **58:**491–562.

Vulliémoz, D., and L. Roux. 2001. "Rule of six": how does the Sendai virus RNA polymerase keep count? *J. Virol.* **75:**4506–4518.

Papers of Special Interest

Alphaviruses

Fata, C. L., S. G. Sawicki, and D. L. Sawicki. 2002. Alphavirus minus-strand RNA synthesis: identification of a role for Arg183 of the nsP4 polymerase. *J. Virol.* **76:**8632–8640.

Lemm, J. A., A. Bergqvist, C. M. Read, and C. M. Rice. 1998. Template-dependent initiation of Sindbis virus RNA replication in vitro. *J. Virol.* **72:**6546–6553.

Lemm, J. A., T. Rumenapf, E. G. Strauss, J. H. Strauss, and C. M. Rice. 1994. Polypeptide requirements for assembly of functional Sindbis virus replication complexes: a model for the temporal regulation of minus- and plus-strand RNA synthesis. *EMBO J.* **13:**2925–2934.

Sawicki, D. L., and S. G. Sawicki. 1994. Alphavirus positive and negative strand RNA synthesis and the role of polyproteins in formation of viral replication complexes. *Arch. Virol. Suppl.* **9:**393–405.

Wang, Y. F., S. G. Sawicki, and D. L. Sawicki. 1994. Alphavirus nsP3 functions to form replication complexes transcribing negative-strand RNA. *J. Virol.* **68:**6466–6475.

Bacteriophage

Brown, D., and L. Gold. 1996. RNA replication by Q beta replicase: a working model. *Proc. Natl. Acad. Sci. USA* **93:**11558–11562.

Flavivirus

Ago, H., T. Adachi, A. Yoshida, M. Yamamoto, N. Habuka, K. Yatsunami, and M. Miyano. 1999. Crystal structure of the RNA-dependent RNA polymerase of hepatitis C virus. *Structure Fold Des.* **7:**1417–1426.

Khromykh, A. A., H. Meka, K. J. Guyatt, and E. G. Westaway. 2001. Essential role of cyclization sequences in flavivirus RNA replication. *J. Virol.* **75:**6719–6728.

Lohmann, V., F. Korner, J. Koch, U. Herian, L. Theilmann, and R. Bartenschlager. 1999. Replication of subgenomic hepatitis C virus RNAs in a hepatoma cell line. *Science* **285:**110–113.

Influenza Virus

Beaton, A. R., and R. M. Krug. 1986. Transcription antitermination during influenza viral template RNA synthesis requires the nucleocapsid protein and the absence of a 5′ capped end. *Proc. Natl. Acad. Sci. USA* **83:**6282–6286.

Cianci, C., L. Tiley, and M. Krystal. 1995. Differential activation of the influenza virus polymerase via template RNA binding. *J. Virol.* **69:**3995–3999.

Hagen, M., T. D. Chung, J. A. Butcher, and M. Krystal. 1994. Recombinant influenza virus polymerase: requirement of both 5′ and 3′ viral ends for endonuclease activity. *J. Virol.* **68:**1509–1515.

Klumpp, K., R. W. Ruigrok, and F. Baudin. 1997. Roles of the influenza virus polymerase and nucleoprotein in forming a functional RNP structure. *EMBO J.* **16:**1248–1257.

Li, M. L., B. C. Ramirez, and R. M. Krug. 1998. RNA-dependent activation of primer RNA production by influenza virus polymerase: different regions of the same protein subunit constitute the two required RNA-binding sites. *EMBO J.* **17:**5844–5852.

Li, M. L, P. Rao, and R. M. Krug. 2001. The active sites of the influenza cap-dependent endonuclease are on different polymerase subunits. *EMBO J.* **20:**2078–2086.

Momose, F., T. Naito, K. Yano, S. Sugimoto, Y. Morikawa, and K. Nagata. 2002. Identification of Hsp90 as a stimulatory host factor involved in influenza virus RNA synthesis. *J. Biol. Chem.* **277:**45306–45314.

Palese, P., M. B. Ritchey, and J. L. Schulman. 1977. P1 and P3 proteins of influenza virus are required for complementary RNA synthesis. *J. Virol.* **21:**1187–1195.

Plotch, S. J., M. Bouloy, I. Ulmanen, and R. M. Krug. 1981. A unique cap(m7GpppXm)-dependent influenza virion endonuclease cleaves capped RNAs to generate the primers that initiate viral RNA transcription. *Cell* **23:**847–858.

Pritlove, D. C., L. L. Poon, E. Fodor, J. Sharps, and G. G. Brownlee. 1998. Polyadenylation of influenza virus mRNA transcribed in vitro from model virion RNA templates: requirement for 5′ conserved sequences. *J. Virol.* **72:**1280–1286.

Rao, P., W. Yuan, and R. M. Krug. 2003. Crucial role of CA cleavage sites in the cap-snatching mechanism for initiating viral mRNA synthesis. *EMBO J.* **22:**1188–1198.

Paramyxoviruses

Barik, S. 1992. Transcription of human respiratory syncytial virus genome RNA in vitro: requirement of cellular factor(s). *J. Virol.* **66:**6813–6818.

Grosfeld, H., M. G. Hill, and P. L. Collins. 1995. RNA replication by respiratory syncytial virus (RSV) is directed by the N, P, and L proteins; transcription also occurs under these conditions but requires RSV superinfection for efficient synthesis of full-length mRNA. *J. Virol.* **69:**5677–5686.

Gupta, S., B. P. De, J. A. Drazba, and A. K. Banerjee. 1998. Involvement of actin microfilaments in the replication of human parainfluenza virus type 3. *J. Virol.* **72:**2655–2662.

Holmes, D. E., and S. A. Moyer. 2002. The phosphoprotein (P) binding site resides in the N terminus of the L polymerase subunit of sendai virus. *J. Virol.* **76:**3078–3083.

Horikami, S. M., S. Smallwood, and S. A. Moyer. 1996. The Sendai virus V protein interacts with the NP protein to regulate viral genome RNA replication. *Virology* **222:**383–390.

Poliovirus

Andino, R., G. E. Rieckhof, P. L. Achacoso, and D. Baltimore. 1993. Poliovirus RNA synthesis utilizes an RNP complex formed around the 5′-end of viral RNA. *EMBO J.* **12:**3587–3598.

Baltimore, D., and R. M. Franklin. 1963. Properties of the mengovirus and poliovirus RNA polymerases. *Cold Spring Harbor Symp. Quant. Biol.* **28:**105–108.

Barton, D. J., and J. B. Flanegan. 1997. Synchronous replication of poliovirus RNA: initiation of negative-strand RNA synthesis requires the guanidine-inhibited activity of protein 2C. *J. Virol.* **71:**8482–8489.

Barton, D. J., B. J. Morasco, and J. B. Flanegan. 1999. Translating ribosomes inhibit poliovirus negative-strand RNA synthesis. *J. Virol.* **73:**10104–10112.

Charini, W. A., S. Todd, G. A. Gutman, and B. L. Semler. 1994. Transduction of a human RNA sequence by poliovirus. *J. Virol.* **68:**6547–6552.

Cuconati, A., A. Molla, and E. Wimmer. 1998. Brefeldin A inhibits cell-free, de novo synthesis of poliovirus. *J. Virol.* **72:**6456–6464.

Dasgupta, A., M. H. Baron, and D. Baltimore. 1979. Poliovirus replicase: a soluble enzyme able to initiate copying of poliovirus RNA. *Proc. Natl. Acad. Sci. USA* **76:**2679–2683.

Gamarnik, A. V., and R. Andino. 1998. Switch from translation to RNA replication in a positive-stranded RNA virus. *Genes Dev.* **12:**2293–2304.

Gamarnik, A. V., and R. Andino. 1996. Replication of poliovirus in *Xenopus* oocytes requires two human factors. *EMBO J.* **15:**5988–5998.

Giachetti, C., and B. L. Semler. 1991. Role of a viral membrane polypeptide in strand-specific initiation of poliovirus RNA synthesis. *J. Virol.* **65:**2647–2654.

Hansen, J. L., A. M. Long, and S. C. Schultz. 1997. Structure of the RNA-dependent RNA polymerase of poliovirus. *Structure* **5:**1109–1122.

Herold, J., and R. Andino. 2001. Poliovirus RNA replication requires genome circularization through a protein-protein bridge. *Mol. Cell* **7:**581–591.

Kirkegaard, K., and D. Baltimore. 1986. The mechanism of RNA recombination in poliovirus. *Cell* **47:**433–443.

Lyle, J. M., E. Bullitt, K. Bienz, and K. Kirkegaard. 2002. Visualization and functional analysis of RNA-dependent RNA polymerase lattices. *Science* **296:**2218–2222.

McKnight, K. L., and S. M. Lemon. 1998. The rhinovirus type 14 genome contains an internally located RNA structure that is required for viral replication. *RNA* **12:**1569–1584.

Molla, A., A. V. Paul, and E. Wimmer. 1991. Cell-free, de novo synthesis of poliovirus. *Science* **254:**1647–1651.

Parsley, T. B., J. S. Towner, L. B. Blyn, E. Ehrenfeld, and B. L. Semler. 1997. Poly(rC) binding protein 2 forms a ternary complex with the 5′-terminal sequences of poliovirus RNA and the viral 3CD proteinase. *RNA* **3:**1124–1134.

Paul, A. V., J. H. van Boom, D. Filippov, and E. Wimmer. 1998. Protein-primed RNA synthesis by purified poliovirus RNA polymerase. *Nature* **393:**280–284.

Rust, R. C., L. Landmann, R. Gosert, B. L. Tang, W. Hong, H. P. Hauri, D. Egger, and K. Bienz. 2001. Cellular COPII proteins are involved in production of the vesicles that form the poliovirus replication complex. *J. Virol.* **75:**9808–9818.

Teterina, N. L., D. Egger, K. Bienz, D. M. Brown, B. L. Semler, and E. Ehrenfeld. 2001. Requirements for assembly of poliovirus replication complexes and negative-strand RNA synthesis. *J. Virol.* **75:**3841–3850.

Todd, S., J. S. Towner, D. M. Brown, and B. L. Semler. 1997. Replication-competent picornaviruses with complete genomic RNA 3′ noncoding region deletions. *J. Virol.* **71:**8868–8874.

Tolskaya, E. A., L. I. Romanova, M. S. Kolesnikova, A. P. Gmyl, A. E. Gorbalenya, and V. I. Agol. 1994. Genetic studies on the poliovirus 2C protein, an NTPase: a plausible mechanism of guanidine effect on the 2C function and evidence for the importance of 2C oligomerization. *J. Mol. Biol.* **236:**1310–1323.

Towner, J. S., T. V. Ho, and B. L. Semler. 1996. Determinants of membrane association for poliovirus protein 3AB. *J. Biol Chem.* **271:**26810–26818.

Walter, B. L., T. B. Parsley, E. Ehrenfeld, and B. L. Semler. 2002. Distinct poly(rC) binding protein KH domain determinants for poliovirus translation initiation and viral RNA replication. *J Virol.* **76:**12008–12022.

Reovirus

Barro, M., P. Mandiola, D. Chen, J. T. Patton, and E. Spencer. 2001. Identification of sequences in rotavirus mRNAs important for minus strand synthesis using antisense oligonucleotides. *Virology* **288:**71–80.

Lawton, J. A., M. K. Estes, and B. V. Prasad. 1997. Three-dimensional visualization of mRNA release from actively transcribing rotavirus particles. *Nat. Struct. Biol.* **4:**118–121.

Lawton, J. A., C. Q. Zeng, S. K. Mukherjee, J. Cohen, M. K. Estes, and B. V. Prasad. 1997. Three-dimensional structural analysis of recombinant rotavirus-like particles with intact and amino-terminal-deleted VP2: implications for the architecture of the VP2 capsid layer. *J. Virol.* **71:**7353–7360.

Li, T., and A. K. Pattnaik. 1999. Overlapping signals for transcription and replication at the 3′ terminus of the vesicular stomatitis virus genome. *J. Virol.* **73:**444–452.

Prasad, B. V., R. Rothnagel, C. Q. Zeng, J. Jakana, J. A. Lawton, W. Chiu, and M. K. Estes. 1996. Visualization of ordered genomic RNA and localization of transcriptional complexes in rotavirus. *Nature* **382:**471–473.

Tao, Y., D. L. Farsetta, M. L. Nibert, and S. C. Harrison. 2002. RNA synthesis in a cage—structural studies of reovirus polymerase lambda3. *Cell* **111:**733–745.

Rhabdoviruses

Das, T., M. Mathur, A. K. Gupta, G. M. Janssen, and A. K. Banerjee. 1998. RNA polymerase of vesicular stomatitis virus specifically associates with translation elongation factor-1 alphabetagamma for its activity. *Proc. Natl. Acad. Sci. USA* **95:**1449–1454.

Finke, S., and K. K. Conzelmann. 1997. Ambisense gene expression from recombinant rabies virus: random packaging of positive- and negative-strand ribonucleoprotein complexes into rabies virions. *J. Virol.* **71:**7281–7288.

Gupta, A. K., D. Shaji, and A. K. Banerjee. 2003. Identification of a novel tripartite complex involved in replication of vesicular stomatitis virus genome RNA. *J. Virol.* **77:**732–738.

Pattnaik, A. K., L. Hwang, T. Li, N. Englund, M. Mathur, T. Das, and A. K. Banerjee. 1997. Phosphorylation within the amino-terminal acidic domain I of the phosphoprotein of vesicular stomatitis virus is required for transcription but not for replication. *J. Virol.* **71:**8167–8175.

Stillman, E. A., and M. A. Whitt. 1998. The length and sequence composition of vesicular stomatitis virus intergenic regions affect mRNA levels and the site of transcript initiation. *J. Virol.* **72:**5565–5572.

Testa, D., P. K. Chanda, and A. K. Banerjee. 1980. Unique mode of transcription in vitro by vesicular stomatitis virus. *Cell* **21:**267–275.

Whelan, S. P., and G. W. Wertz. 2002. Transcription and replication initiate at separate sites on the vesicular stomatitis virus genome. *Proc. Natl. Acad. Sci. USA* **99:**9178–9183.

7

Reverse Transcription and Integration

> *"One can't believe impossible things," said Alice.*
>
> *"I dare say you haven't had much practice," said the Queen. "Why, sometimes I've believed as many as six impossible things before breakfast."*
>
> LEWIS CARROLL
> *Alice in Wonderland*

Retroviral Reverse Transcription

Discovery

In 1970, back-to-back reports in the scientific journal *Nature* from the laboratories of Howard Temin and David Baltimore provided the first concrete evidence for the existence of an RNA-directed DNA polymerase activity in retrovirus particles. The pathways to this discovery were quite different in the two laboratories. In Temin's case, the discovery came about through attempts to understand how this group of RNA-containing viruses could permanently alter the heredity of cells, as they do in the process of oncogenic transformation. Temin proposed that retroviral RNA genomes become integrated into the host cell's chromatin in a DNA form, an idea supported by the observation that cellular DNA polymerases can use RNA as a template under certain in vitro conditions. Furthermore, studies of bacterial viruses such as bacteriophage lambda had established a precedent for viral DNA integration into host DNA. However, with the technology available at the time, it was a difficult hypothesis to test, and attempts by Temin and others to demonstrate such a phenomenon in infected cells were generally met with skepticism. Baltimore's entrée into the problem of reverse transcription came from his interest in virion-associated polymerases, in particular one that he had just discovered to be present in vesicular stomatitis virus, a virus with a (–) strand RNA genome. It occurred to Baltimore and Temin independently that retroviruses might also contain the sought-after RNA-dependent DNA polymerase. As subsequent experiments showed, this was indeed the case, and the retrovirus particles themselves yielded up the enzyme activity that had earlier eluded Temin. In 1975, Temin and Baltimore were awarded the Nobel Prize in physiology or medicine for their independent discoveries of retroviral reverse transcriptase (RT).

Impact

The immediate impact of the discovery of RT was to amend the then-accepted central dogma of molecular biology, that the transfer of genetic information is unidirectional: DNA → RNA → protein. It was now apparent that there could also be a "retrograde" flow of information from RNA to DNA, and the name **retroviruses** eventually came to replace the earlier designation of RNA tumor viruses. In the years following this revision of dogma, many additional reverse transcription reactions have been discovered. Furthermore, as Temin hypothesized, study of RT has contributed to our understanding of cancer. As described in Chapter 18, the study of oncogenic retroviruses has provided a framework for current concepts of the genetic basis of this disease. Study of reverse transcription (and integration) has also allowed us to understand the persistence of retroviral infections and other aspects of the pathogenesis of acquired immunodeficiency syndrome (AIDS). Finally, RT itself, first purified from virions and now synthesized in bacteria, has become an indispensable tool in molecular biology, allowing experimentalists to capture cellular messenger RNAs (mRNAs) as complementary DNAs (cDNAs), which can then be amplified, cloned, and expressed by well-established methodologies. For such reasons, we devote an entire chapter to these very important reactions.

The Pathways of Reverse Transcription

Significant insight into the mechanism of reverse transcription can be obtained by comparing the amino acid sequences of RTs with those of other enzymes that catalyze similar reactions. For example, RTs share certain sequence motifs with the RNA and DNA polymerases of both prokaryotes and eukaryotes, in regions known to include critical active site residues (Fig. 6.4). It is hypothesized, therefore, that these enzymes employ similar catalytic mechanisms for nucleic acid polymerization reactions. Like DNA polymerases, viral RTs cannot initiate DNA synthesis de novo, but require a specific primer. In this chapter, we provide a detailed description of priming and reverse transcription for retroviruses and hepadnaviruses. But it should be noted that even as arcane and distinct from each other as these two systems may appear, they do not exhaust the repertoire for reverse transcription reactions that have evolved in nature. A wide variety of primers as well as sites and modes of initiation are used by the RTs of other retroelements.

Much of what has been learned about reverse transcription in retroviruses comes from the analysis of intermediates in the reaction pathway as it occurs in infected cells. Reverse transcription intermediates have also been analyzed from **endogenous reactions**, which take place within purified virus particles, using the encapsidated viral RNA template. It was amazing to discover that intermediates and products virtually identical to those made in infected cells can actually be synthesized in purified virions; all that is required is the application of mild detergents to permeabilize the envelope and addition of the metal cofactor and deoxyribonucleoside triphosphate substrates. The efficiency of the endogenous reaction suggests that the reverse transcription system is poised for action as soon as the virus enters the cell. Indeed, small amounts of viral DNA can be detected in purified human immunodeficiency virus type 1 virions, presumably having been synthesized using substrates picked up from the infected cell during budding. Retroviral reverse transcription intermediates have also been analyzed in totally reconstituted reactions with purified enzymes and model RNA templates.

Retroviral RT is the only protein required to accomplish all the diverse steps in the pathway described below. However, as the reactions that take place inside cells are significantly more efficient than those observed in either endogenous or reconstituted systems, not all the essential features have been reproduced.

Essential Components

Genomic RNA. Retrovirus particles contain two copies of the RNA genome held together by multiple regions of base pairing. Such RNA sediments in a 70S complex, as expected for a dimer of 35S genomes. Partial denaturation and electron microscopic analyses of the 70S complex indicate that the most stable pairing is via sequences located near the 5′ ends of the two genomes (Fig. 7.1). Retroviral genomes can be thought of as being annealed head to head, an arrangement that may discourage the encapsidation of multiples larger than two (i.e., concatemers). The 70S RNA complex also includes two molecules of a specific cellular transfer RNA (tRNA) that serves as a primer for the initiation of reverse transcription (discussed below).

Despite the fact that two genomes are encapsidated, generally only one integrated copy of the viral DNA is detected after infection with single virion. Therefore, retroviral virions are said to be **pseudodiploid**. Why should such a fea-

Figure 7.1 The diploid retroviral genome. The diploid genome includes the following, from 5′ to 3′: the m^7Gppp capping group, the primer transfer RNA (tRNA), the coding regions for viral structural proteins and enzymes, *gag*, *pol*, and *env*, and the 3′-poly(A) sequence. From J. M. Coffin, p. 1767–1848, *in* B. N. Fields et al. (ed.), *Fields Virology*, 3rd ed. (Lippincott-Raven, Philadelphia, Pa., 1996), with permission.

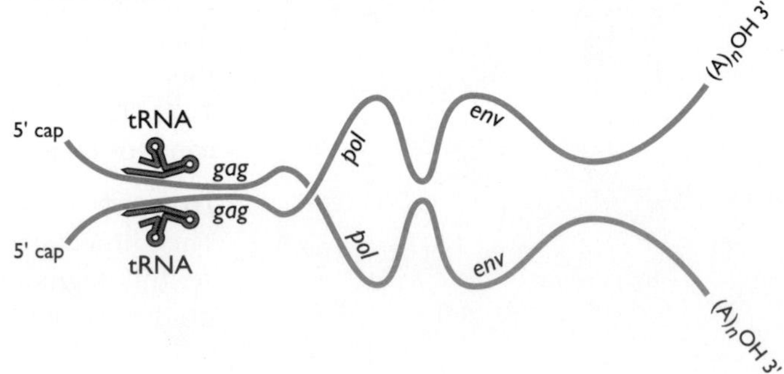

ture have been selected during evolution? One popular notion is that the availability of two RNA templates can help retroviruses survive extensive damage to their genomes. At least parts of both genomes can be, and typically are, used during the reverse transcription process, accounting for the high rates of genetic recombination in these viruses. Presumably, being able to patch together one complete DNA provirus from two randomly damaged RNA genomes would provide survival value, as might the genetic recombination produced during the process. Nevertheless, genetic experiments have shown that the use of two RNA templates is not an essential feature of the reverse transcription process; one genome can suffice for all the steps known.

Like the genomes of (–) strand RNA viruses, the retroviral genome is coated along its length by a viral nucleocapsid protein (NC), with approximately one molecule for every 10 nucleotides. This small basic protein can bind nonspecifically to both RNA and DNA and can promote the annealing of nucleic acids. Biochemical experiments suggest that NC may perform a role in reverse transcription similar to that of prokaryotic single-stranded binding protein. In the synthesis of DNA catalyzed by bacterial DNA polymerases, this protein enhances processivity and facilitates template exchanges. The ability of NC first to organize RNA genomes within the virion, and then to facilitate reverse transcription within the infected cell, may account for some of the differences in efficiency observed when comparing reactions reconstituted in vitro with those that take place in a natural infection.

Primer tRNA. It was surprising to find that retroviral virions contain a collection of cellular RNAs in addition to the viral genome. These include approximately 100 copies of a nonrandom collection of tRNAs, some 5S ribosomal RNA (rRNA), 7S RNA, and traces of cellular mRNAs. We do not know how most of these cellular RNAs become incorporated into virus particles, and most have no obvious function. However, one particular tRNA molecule does have a critical role, that of serving as primer for the initiation of reverse transcription. The tRNA primer is positioned on the template genome during virus assembly and maturation via interactions with both RT and viral RNA, probably facilitated by NC. The primer tRNA is partially unwound and hydrogen-bonded near the 5′ end of each RNA genome (Fig. 7.1 and 7.2) in a region called the **primer-binding site (pbs).** The reverse transcriptases of all retroviruses studied to date are primed by one of only a few classes of cellular tRNAs. For most mammalian retroviral RTs, these include $tRNA^{Pro}$, $tRNA^{Lys3}$, or $tRNA^{Lys1,2}$.

In addition to the 3′-terminal 18 nucleotides that anneal to the pbs, other regions in the tRNA primer contact the RNA template and modulate reverse transcription. The template-primer interaction has been studied extensively in reconstituted reactions with RNA and RT of avian sarcoma/leukosis virus. In these in vitro analyses, the ability of the viral RNA to form stem-loop structures, and specific interactions between the primer $tRNA^{Trp}$ and one of these loops, appears to be critical for reverse transcription (Fig. 7.2). Similar interactions have been reported for human

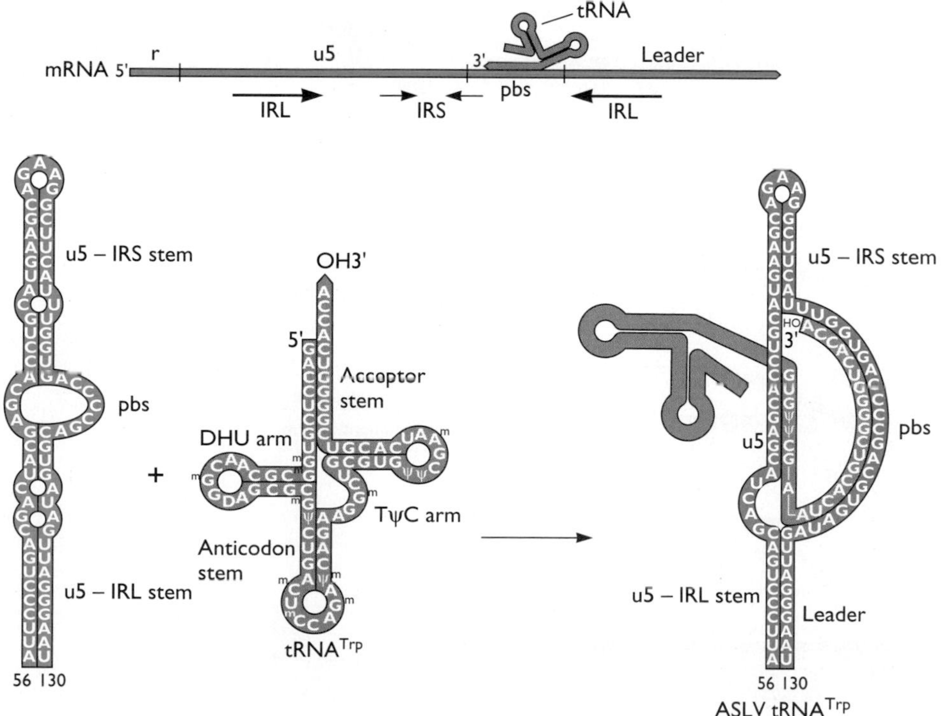

Figure 7.2 Primer tRNA binding to the retroviral RNA genome. (Top) Linear representation of the 5′ terminus of retroviral RNA. The r, u5, and leader regions are indicated. A tRNA primer is shown schematically annealed to the pbs. Two inverted repeat (IR) sequences that flank the pbs are represented by arrows. (Bottom) Avian sarcoma/leukosis virus RNA can form an extended hairpin structure around the pbs in the absence of primer tRNA (left). (Middle) Primer $tRNA^{Trp}$ in the cloverleaf structure. Modified bases are indicated. (Right) Viral RNA annealed to $tRNA^{Trp}$. The u5-leader and u5-IR stem structures are indicated. An interaction with the TψC arm of the primer and u5 RNA is also shown. (Bottom diagram) From J. Leis et al., p. 33–47, *in* A. M. Skalka and S. P. Goff (ed.), *Reverse Transcriptase* (Cold Spring Harbor Laboratory Press, Cold Spring Harbor, N.Y., 1993), with permission.

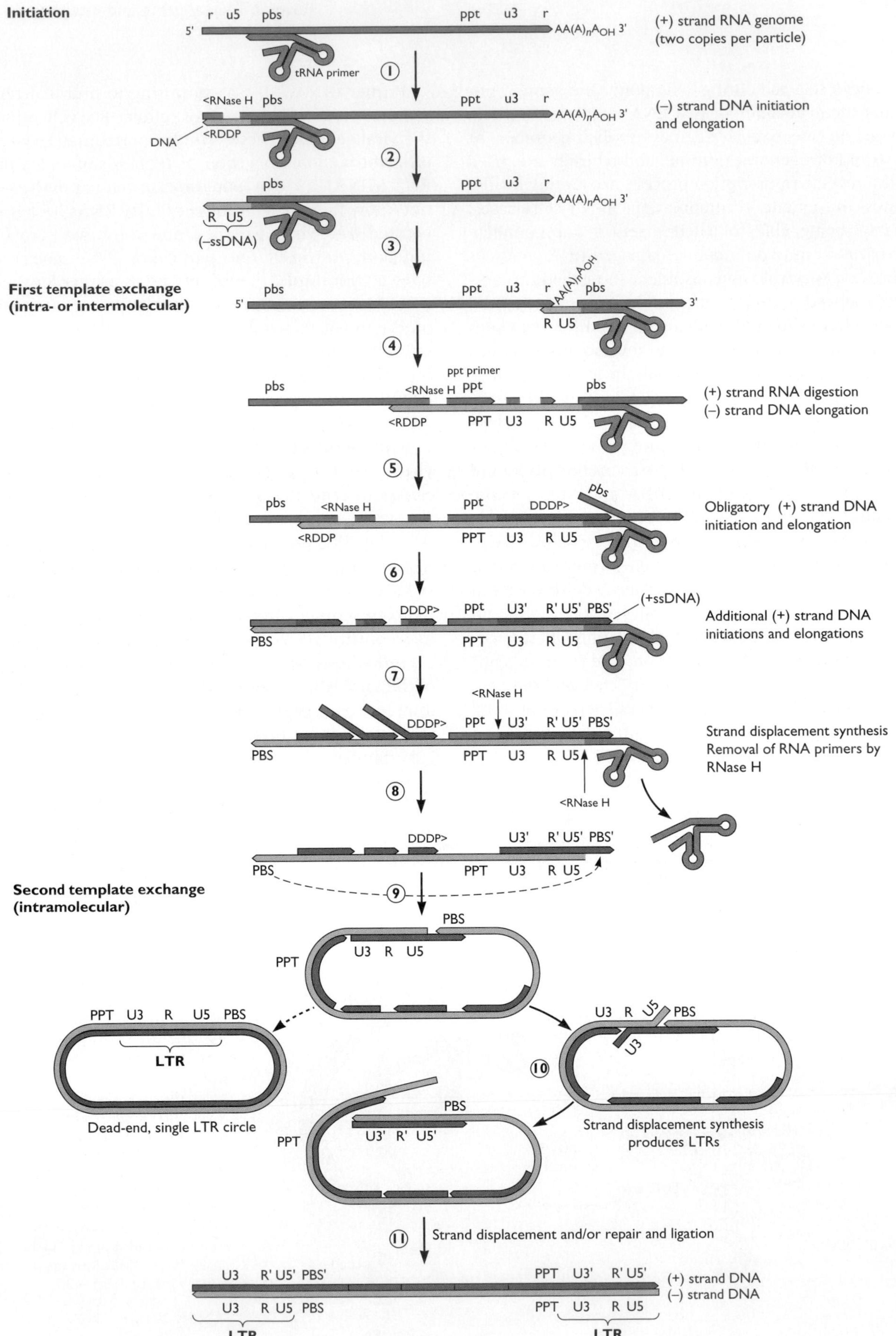
Initiation
r u5 pbs ppt u3 r
5' AA(A)nAOH 3'
(+) strand RNA genome (two copies per particle)
tRNA primer
1
<RNase H pbs ppt u3 r
DNA <RDDP
(–) strand DNA initiation and elongation
2
pbs ppt u3 r
R U5
(–ssDNA)
3
First template exchange (intra- or intermolecular)
pbs ppt u3 r pbs
R U5
4
ppt primer
pbs <RNase H ppt pbs
<RDDP PPT U3 R U5
(+) strand RNA digestion
(–) strand DNA elongation
5
pbs <RNase H ppt DDDP> pbs
<RDDP PPT U3 R U5
Obligatory (+) strand DNA initiation and elongation
6
DDDP> ppt U3' R' U5' PBS' (+ssDNA)
PBS PPT U3 R U5
Additional (+) strand DNA initiations and elongations
7
<RNase H
DDDP> ppt U3' R' U5' PBS'
PBS PPT U3 R U5
<RNase H
Strand displacement synthesis
Removal of RNA primers by RNase H
8
DDDP> U3' R' U5' PBS'
PBS PPT U3 R U5
Second template exchange (intramolecular)
9
PBS
PPT U3 R U5
PPT U3 R U5 PBS
LTR
Dead-end, single LTR circle
U3 R U5 PBS
U3
10
Strand displacement synthesis produces LTRs
PBS
PPT U3' R' U5'
11
Strand displacement and/or repair and ligation
U3 R' U5' PBS' PPT U3' R' U5'
(+) strand DNA
(–) strand DNA
U3 R U5 PBS PPT U3 R U5
LTR LTR

immunodeficiency virus RNA and its primer. Although the interactions are likely to be significant biologically, we do not yet know how RTs recognize structural features in these template-primer complexes.

Reverse transcriptase. Retroviral particles contain 50 to 100 molecules of RT. Viral RTs probably function as dimers, but the number of dimers in each virion that are actually engaged in reverse transcription is not known. As noted above, results from studies of endogenous reverse transcription in purified virions suggest that viral DNA synthesis can begin as soon as the viral envelope is removed and deoxynucleoside triphosphates are made available. With the three orthoretroviruses studied most extensively (avian sarcoma/leukosis virus, murine leukemia virus, and human immunodeficiency virus type 1), reverse transcription takes place mainly in the cytoplasm, before viral DNA enters the nucleus. Enzymes of these three retroviruses will be used as examples throughout this chapter.

Retroviral RTs are complex molecular machines with moving parts and multiple activities. Four distinct catalytic activities brought into play at various stages in the pathway of reverse transcription are illustrated in Fig. 7.3. These include RNA-directed and DNA-directed DNA polymerization, DNA unwinding (helicase), and the hydrolysis of RNA in RNA·DNA heteroduplexes by RNase H. In many other cases these catalytic activities would reside in separate proteins, but in RT they are all combined, the first three within a single polymerase domain, with the RNase H comprising a separate domain. In digesting heteroduplexed RNA, the RNase H can function as both an exonuclease and an endonuclease, producing oligoribonucleotides 2 to 15 nucleotides in length. These activities of the RNase H domain of RT degrade the genomic RNA after it has been copied into cDNA, form the primer for (+) strand DNA synthesis from the genomic RNA, and, finally, remove this primer and the tRNA primer from the 5′ ends of the viral DNA duplex.

Critical Reactions in Reverse Transcription

RNA priming and template strand exchange. Based on our understanding of DNA synthesis, the simplest way of copying an RNA template to produce DNA would be to start at its 3′ end and finish at its 5′ end. It was therefore somewhat of a shock to early researchers to discover that reverse transcription in fact started near the 5′ end—only to run out of template after little more than about 100 nucleotides, at which point most DNA synthesis stopped. As we will see later, this counterintuitive strategy for initiation of DNA synthesis allows the duplication and translocation of critical transcription and integration signals encoded in both the 5′ and 3′ ends of the genomic RNA (called u5 and u3, respectively). (Lowercase designations are used throughout this chapter to refer to RNA se-

Figure 7.3 The reverse transcription process. Two copies of the capped, 7- to 10-kb RNA genome with a poly(A) tail are present in each retrovirus particle. Critical *cis*-acting sequences at the ends of the RNA genome are indicated in lowercase letters. They are represented in uppercase letters in the DNA products. (–) strand DNA synthesis starts near the 5′ end of the (+) strand RNA genome with a specific host tRNA as primer. The partially unfolded tRNA is annealed via at least 18 nucleotides at its 3′ end to a site on the RNA genome called the primer-binding site (pbs). Step 1: Synthesis proceeds to the 5′ end of the RNA genome through the u5 region (ca. 100 nucleotides), ending after the r region, which is terminally redundant in the RNA; this comprises the (–) strand strong-stop DNA (–ssDNA). Step 2: The RNA portion of the RNA·DNA hybrid product is digested by the RNase H of RT, resulting in a single-strand DNA product, attached to the tRNA primer. Step 3: This exposure of single-stranded DNA facilitates hybridization with the r region at the 3′ end of the same or the second RNA genome, a reaction that produces the first template exchange for RT, and allows continuation of RNA-dependent DNA polymerization (RDDP). Step 4: When (–) strand RDDP passes a specific region near the 3′ end of the RNA genome, characterized by a polypurine tract (ppt), the complementary sequence of the RNA template escapes digestion by RNase H and serves as a primer for the synthesis of (+) strands in DNA-dependent DNA polymerization (DDDP). (+) Strand DDDP then continues back to the U5 region using the (–) strand DNA as a template. Step 5: Meanwhile, (–) strand RDDP continues along the (+) strand RNA template, during which the RNA template is removed via the RT RNase H. Step 6: The RNase H digestion products formed can, in some cases, provide additional primers for (+) strand DDDP at internal locations along the (–) strand DNA. Step 7: ppt-initiated (+) strand DNA synthesis stops after the annealed portion of the tRNA has been copied, producing the (+) strand DNA form of the PBS, and forming the (+) strand strong-stop product, denoted +ssDNA. The tRNA and ppt primers are removed by RNase H activity of RT. Step 8: The single-stranded PBS sequence exposed by removal of the tRNA primer can now anneal to its complement on the (–) strand DNA, facilitating the second template exchange (step 9). A circular DNA intermediate is formed. Repair of this intermediate by host cell enzymes can produce a dead-end product, a circular DNA molecule with a single long terminal repeat (LTR) (left, dashed arrow). In the productive pathway, strand displacement synthesis by RT to the PBS and PPT ends (step 10) produces a linear molecule with LTRs (step 11). Internal discontinuities in the (+) strand, represented by a broken line, are probably repaired by host cell enzymes. Adapted from R. A. Katz and A. M. Skalka, *Annu. Rev. Biochem.* **63:**133–173, 1994, with permission.

quences; uppercase designations identify the same or complementary sequences in DNA.) The short (ca. 100-nucleotide) product of this first reaction (Fig. 7.3, steps 1 and 2) accumulates in large quantities in the endogenous and reconstituted systems and, therefore, is called **(–) strong-stop DNA.** In the next step, the 3′ end of the same or the second RNA genome is engaged as template so that DNA synthesis can continue. This reaction, which we call the **first template exchange,** corresponds to the substitution of one end of the RNA for another to be copied by the RT "machine" (Box 7.1).

The first template exchange (Fig. 7.3, middle) requires the presence of direct repeats, r, at the ends of the viral RNA. After it is copied to form (–) strong-stop DNA, the 5′ end of the genome RNA is digested by the RNase H of RT. The R complement in (–) strong-stop DNA can then anneal to the complementary r sequence upstream of the poly(A) tail at the 3′ end of either the same or the second RNA genome in the virion (Fig. 7.3, steps 3 and 4). It has been suggested that transient melting, facilitated by NC or by RT helicase, may allow this exchange to take place without complete digestion of the RNA, and that some special properties of the u3-r junction may facilitate the reaction. As (–) strong-stop DNA is barely detectable in infected cells, this first template exchange must be efficient in vivo.

Once the 3′ end of the genome RNA is engaged, the RNA-directed DNA polymerase activity of RT can proceed with copying all the way to the 5′ end of that template (Fig. 7.3, steps 4 to 7). This (–) strand DNA product (cDNA) terminates at the primer binding site at the 5′ end of the RNA template (Fig. 7.3, step 7). It comprises an entire genome's equivalent of DNA, but in permuted order. This description is, of course, idealized. In actuality, analyses of reaction intermediates show that the enzyme pauses along the way, presumably at some sequences (or possibly breaks) that impede copying. If a break is encountered in the RNA template, synthesis can be completed by utilization of the second RNA genome. Such internal template exchanges, known to occur even in the absence of such breaks, probably proceed via the same steps outlined for the first template exchange. Internal exchanges that take place during RNA-directed DNA synthesis are estimated to be the source of at least half of the genetic recombination that occurs in retroviruses; the mechanism is known as **copy choice** (see Fig. 7.5).

Production of primers for (+) strand DNA synthesis by RNase H digestion. The viral RNA template is digested by the RNase H of RT, right after it has been copied during (–) strand DNA synthesis (as illustrated in Fig. 7.12). To initiate synthesis of the second, (+) strand of DNA and commence DNA-directed DNA polymerization, another primer is needed by RT. In contrast to the tRNA primer for (–) strand DNA synthesis, this second primer is a specific product of RNase H digestion of the viral RNA genome.

The (+) strand RNA primer comprises a **polypurine tract** approximately 11 nucleotides long. Recent studies suggest that this sequence adopts abnormal, partially unwound and incomplete hydrogen bonding in an RNA·DNA hybrid. Such departure from Watson-Crick base pairing may account for resistance of the polypurine tract to digestion during (–) strand DNA synthesis. The DNA-directed DNA polymerase begins as soon as the polypurine tract primer is formed, usually before (–) strand DNA synthesis is completed. Synthesis of (+) strand DNA proceeds to the nearby end of the (–) strand DNA template and terminates after copying the first 18 nucleotides of the primer tRNA, when it encounters a modified base that cannot be copied by RT. This product is called (+) strong-stop DNA, and after removal of the tRNA primer by the RNase H, its 3′ end becomes available to anneal to the PBS sequence at the 3′ end of the (–) strand DNA (Fig. 7.3, step 9). This process is called the **second template exchange.**

Other oligoribonucleotides formed by RNase H digestion of the RNA template can also serve as primers for DNA-directed DNA polymerization by RT. Human immunodeficiency virus contains a second, specific internal polypurine tract for priming (+) strand synthesis located near the center of the genome. With both the human virus

BOX 7.1

Enzymes can't jump!

The exchange of one template for another to be copied by either DNA or RNA polymerases is sometimes referred to as enzyme "jumping." This inappropriate term comes from a too literal reading of cartoon illustrations of the process, in which the templates to be exchanged may be opposite ends of the nucleic acid or different nucleic acid molecules. In actuality, such enzyme movement is quite improbable, and use of this terminology can cloud thinking about these processes. In almost all cases, these polymerases are components of large complexes with architecture designed to bring different parts of the template, or different templates, close to each other. Although some dynamic changes must occur for the proteins to accommodate template exchanges, it seems likely that most of the "movement" is done by the flexible nucleic acid molecules.

and avian sarcoma/leukosis viruses, (+) strand DNA synthesis can also be initiated at several additional locations; this is the situation illustrated in Fig. 7.3 (steps 6 and 7). On the other hand, there is little or no evidence of (+) strand initiation at sites other than the 3′ polypurine tract during murine leukemia virus reverse transcription.

Strand displacement synthesis and formation of long terminal repeats. The product of the second template exchange is a circular DNA molecule, formed by annealing of the complementary PBS sequences at the 3′ ends of the (+) and (–) strands of newly synthesized DNA (Fig. 7.3, step 9). Were reverse transcription to stop at this point, repair enzymes in the nucleus of the cell could produce a covalently closed, circular DNA molecule with a single equivalent of the genome. Such molecules have been isolated from infected cells and are most likely by-products of incomplete reverse transcription produced by such repair. However, owing perhaps to its helicase activity, RT displaces the DNA strand annealed upstream, as synthesis continues (Fig. 7.3, step 10). DNA synthesis is eventually terminated by breaks in the template strands, at the sites from which the (–) and (+) strand RNA primers were removed. The position of these sites demarcates the U3-R-U5 region. The termination of DNA synthesis at these breaks leads to production of a blunt-ended, linear DNA copy of the viral genome that also includes **long terminal repeats (LTRs)** (Fig. 7.3, step 11). Sequences at the ends of the LTRs are essential for DNA integration, the next step in the viral replication cycle.

Retroviral reverse transcription has been called "destructive replication," as there is no net gain of genomes, but rather a substitution of one double-stranded DNA for two molecules of single-stranded RNA. However, by this rather complex but elegant pathway, RT not only makes a linear DNA copy of the retroviral genome to be integrated but also produces the LTRs that include signals necessary for transcription of the integrated DNA, which is called the **provirus.** Integration ensures subsequent replication of the provirus via the host's DNA synthesis machinery as the cell divides. In addition, the promoter in the upstream LTR is now in the appropriate location for synthesis of progeny RNA genomes and viral mRNAs by host cell RNA polymerase II.

Structure of retroviral DNA products. Unintegrated linear retroviral DNA extracted from infected cells appears to have continuous (–) strands and discontinuous (+) strands. For example, the (+) strands of murine retroviruses (mouse mammary tumor virus and murine leukemia virus) contain large gaps. In contrast, avian sarcoma/leukosis virus (+) strands contain single-stranded tails formed by strand displacement synthesis through internal sites of initiation on the DNA template (Fig. 7.4).

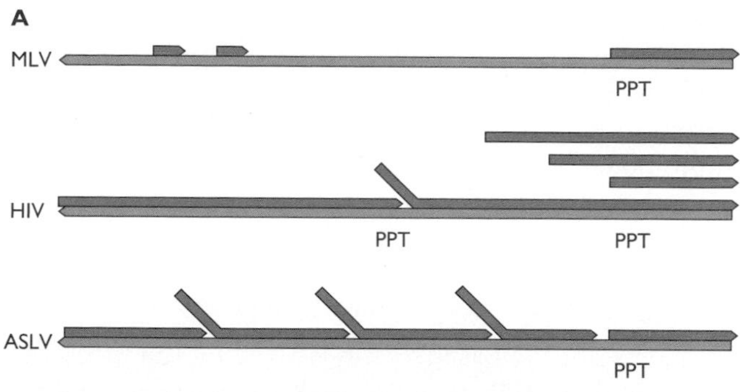

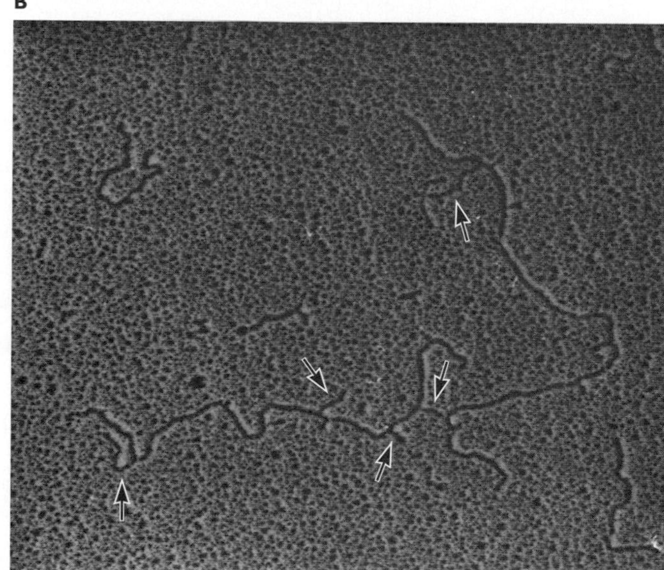

Figure 7.4 Differences in the structure of retroviral DNAs extracted from infected cells. (A) Schematic representations. Like all retroviral DNAs, most unintegrated murine leukemia virus (MLV) DNA molecules extracted from infected cells (top) have a full-length, continuous DNA (–) strand but only short fragments of the (+) strand. Human immunodeficiency virus type 1 (HIV-1) DNA (middle) has two classes of (+) strands. In the left half of the molecules (+) strands are continuous to a point just beyond the internal polypurine tract (PPT), where they appear to encounter a block to further elongation. In the right half there appear to be many possible initiation sites, producing a collection of (+) strands of varying length. The (+) strands of avian sarcoma/leukosis virus (ASLV) DNA can be initiated at several (apparently random) locations. These are elongated past adjacent initiation sites to generate long, single-stranded tails. (B) An electron micrograph of ASLV DNA formed in an endogenous reaction in purified virions. Arrows point to single-stranded tails (about three or four per molecule). The image shows two molecules of full-length DNA joined by a single-stranded tail as proposed in the strand displacement-assimilation recombination model. From R. P. Junghans et al., *Cell* **30:**53–62, 1982, with permission.

Human immunodeficiency virus DNA also has a uniquely positioned tail generated by strand displacement synthesis through the internal polypurine tract (Fig. 7.4A). This apparent abundance of single-stranded DNA, opposite gaps and in tails, could provide additional opportunities for recombination. It is estimated that approximately half of the genetic recombination that occurs during retroviral reverse transcription takes place during (+) strand synthesis. One proposed mechanism, the strand displacement-assimilation model, is illustrated in Fig. 7.5.

It is not known how or when linear DNA products with (+) strand discontinuities or branches are processed to complete the covalent joining that produces an integrated provirus. From in vitro studies it appears unlikely that strand displacement synthesis by RT can completely displace all downstream (+) strand segments. It may be that cellular repair enzymes in the nucleus participate in the final process. Theoretically, such repair could occur even after the viral DNA is joined to the host DNA, as such joining can proceed as soon as the duplex LTR ends are formed. Indeed, studies with murine leukemia virus show that the first step in the integration process, processing of the blunt ends of the LTRs, takes place in the cytoplasm, before the viral DNA enters the nucleus. Furthermore, nucleoprotein complexes isolated from the cytoplasmic fraction of human

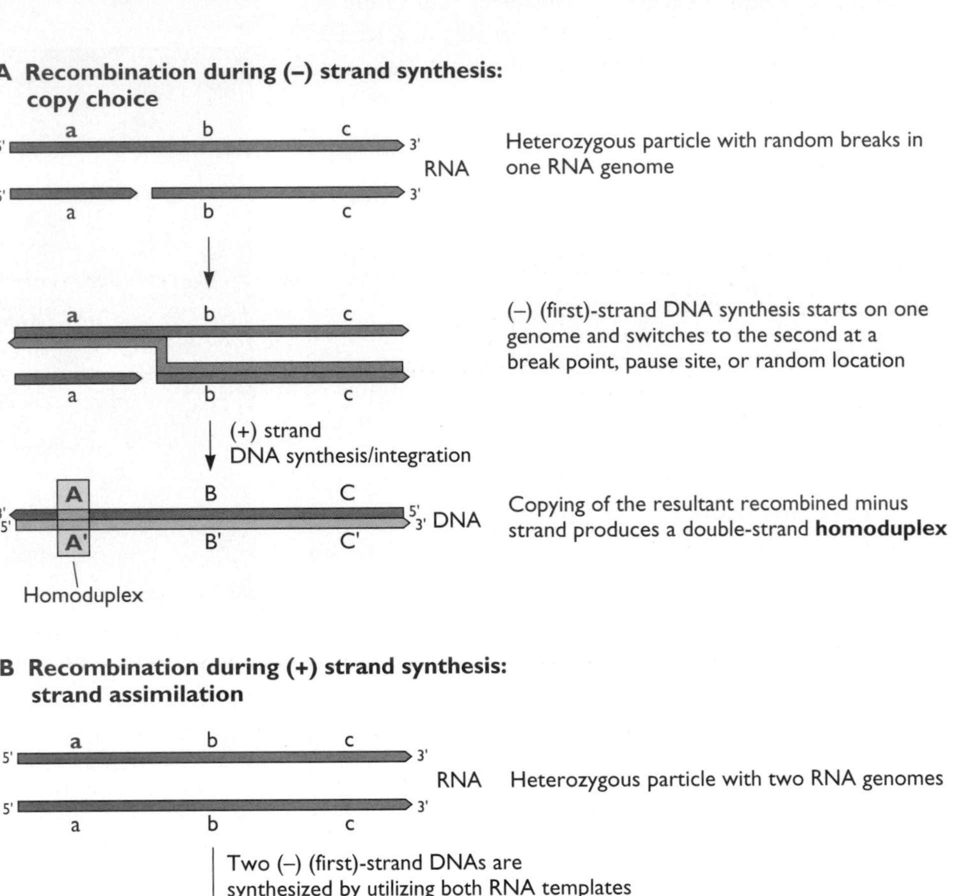
A Recombination during (–) strand synthesis: copy choice
a
b
c
5'
3'
RNA
3'
Heterozygous particle with random breaks in one RNA genome
(–) (first)-strand DNA synthesis starts on one genome and switches to the second at a break point, pause site, or random location
(+) strand DNA synthesis/integration
A
B
C
A'
B'
C'
DNA
Homoduplex
Copying of the resultant recombined minus strand produces a double-strand homoduplex
B Recombination during (+) strand synthesis: strand assimilation
Heterozygous particle with two RNA genomes
Two (–) (first)-strand DNAs are synthesized by utilizing both RNA templates
(+) (second)-strand DNA synthesis accompanied by strand displacement and assimilation of single-strand DNA tails onto DNA from the second genome
Resolution/repair/integration
Heteroduplex
Segregation of wt and mutant strands
The assimilated strand in the DNA of the second genome produces a double strand with heteroduplex regions

Figure 7.5 Two models for recombination during reverse transcription. Virtually all retroviral recombination occurs between coencapsidated genomes at the time of reverse transcription. The copy choice model (A) postulates a mechanism for genetic recombination during first, or (–) strand, DNA synthesis; the strand displacement-assimilation model (B) proposes a mechanism for recombination during (+) strand DNA synthesis. These two models are not mutually exclusive, and there is experimental support for both. Viral genetic markers, arbitrarily labeled *a*, *b*, and *c*, are shown to illustrate recombination. Single recombinations are shown for simplicity, focusing on the *a* allele, with the mutant form in red. However, multiple crossover events are frequently observed. The boxes highlight the region of recombination. (A) Recombination during (–) strand synthesis. A break in the RNA is represented as a gap. Exchanges between RNA templates may occur at RNA breaks, as shown, but such breaks may not be required. Introducing intentional breaks, e.g., by gamma irradiation, does not seem to increase the frequency of such recombination. The copy choice mechanism predicts only one DNA homoduplex product from two RNA molecules. (B) Recombination during (+) strand synthesis. (+) Strand DNA synthesis initiates at internal sites with RNA primers produced by partial RNase H digestion. The strand assimilation model requires that (+) strand DNA synthesis take place on two templates and predicts formation of a heteroduplex as a consequence of recombination of (+) strands. One DNA molecule is shown for simplicity. Adapted from R. Katz and A. M. Skalka, *Annu. Rev. Genet.* **24:** 409–445, 1990, with permission.

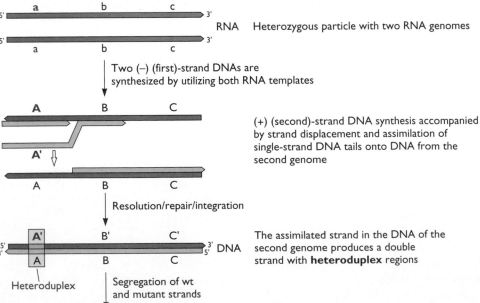

immunodeficiency virus-infected cells are competent to join viral DNA to exogenously supplied target DNA, despite the fact that their (+) strands are not yet continuous.

Beginning at approximately 4 to 8 h after infection of cells in culture, a portion of the newly synthesized viral DNA enters the nucleus; at about the same time, integration events are also first detected. Nuclear viral DNA comprises both linear molecules and a small, but variable, proportion of circular molecules that contain one or two LTRs. It is now clear that linear viral DNA molecules are the substrates for integration (see below). The circles appear to be dead-end products. However, as joined LTRs are easy to detect by polymerase chain reaction techniques, the two LTR circles have been used as convenient markers for the nuclear transport of viral DNA.

General Properties and Structure of Retroviral Reverse Transcriptases

Domain Structure and Subunit Organization

The RTs of retroviruses are encoded in the *pol* gene. Despite the sequence homologies and similar organization of coding sequences in their *pol* regions, there is considerable diversity in the subunit composition and organization of RTs from different retroviruses (Fig. 7.6). For example, each of the three enzymes studied most extensively—those of avian sarcoma/leukosis, murine leukemia, and human immunodeficiency viruses—has a distinctive arrangement. Little is known about the subunit compositions of RTs from most other retroviruses. In the absence of more structural data, it is difficult to gauge the significance of this apparent structural diversity. However, it seems likely that in the confines of the subviral nucleoprotein complex in which reverse transcription takes place in vivo, Pol-derived proteins all work together and in similar ways.

Catalytic Properties

DNA polymerization is slow. The biochemical properties of RT have been studied with enzymes purified from virions or synthesized in bacteria, and model templates and primers. Kinetic analyses have revealed an ordered reaction pathway for DNA polymerization similar to that of other polymerases. Like cellular polymerases and nucleases, RTs require divalent cations (Mg^{2+} or Mn^{2+}) as cofactors. The in vitro rate of elongation by RT on natural RNA templates is 1 to 1.5 nucleotides per second, approximately 1/10 the rate of other eukaryotic DNA polymerases. The long time required to generate a genome's equivalent of retroviral DNA after infection, approximately 4 h for ca. 9,000 nucleotides, supports the view that even in vivo reverse transcription is a relatively slow process.

In reactions in vitro, the rate of enzyme-template·primer dissociation decreases considerably after addition of the first nucleotide, suggesting that initiation and elongation are distinct steps in reverse transcription, as is the case during DNA synthesis by DNA-dependent DNA polymerases. However, in contrast to such polymerases, retroviral RTs do not remain attached to their template·primers during a large number of successive deoxyribonucleotide additions; this property is described as "poor processivity."

Fidelity is low. Retroviral genomes, like those of other RNA viruses, accumulate mutations at much higher rates than cellular genes. RTs lack an editing activity (resembling the 3′ → 5′ exonuclease of *Escherichia coli* DNA polymerase I, which is capable of excising mispaired nucleotides) and have been shown to be error-prone in vitro. RTs are presumed therefore to contribute to the high in vivo mutation rate of retroviruses.

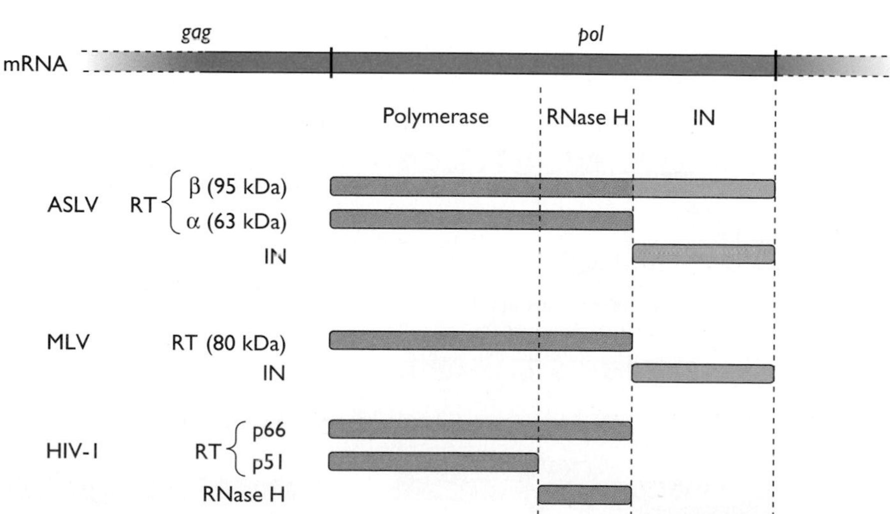

Figure 7.6 Domain and subunit relationships of RTs from different retroviruses. The RT of avian sarcoma/leukosis virus (ASLV) includes sequences corresponding to integrase in its larger subunit, β. Both subunits of this enzyme include RNase H sequences. Murine leukemia virus (MLV) RT functions as a homodimer; both subunits contain the RNase H domain, but neither includes IN sequences. Like that of ASLV, the HIV-1 RT is a heterodimer. However, in this case only one subunit includes the RNase H domain, and neither includes IN sequences. Adapted from R. A. Katz and A. M. Skalka, *Annu. Rev. Biochem.* **63:**133–173, 1994, with permission.

In vitro, RT errors have been shown to include not only misincorporations but also rearrangements such as deletions and additions (Fig. 7.7). Misincorporations by human immunodeficiency virus type 1 RT can occur as frequently as 1 per 70 copies at some positions, and as infrequently as 1 per 10^6 copies at others. Many types of genetic experiments have been conducted in attempts to determine the in vivo error rates of RTs in a single replication cycle. The general conclusion is that such rates are also quite high, with reported misincorporations in the range of 1 per 10^4 to 1 per 10^6 nucleotides polymerized, in contrast to 1 per 10^7 to 1 per 10^{11} for cellular DNA replication. As retroviral genomes are approximately 10^4 nucleotides in length, this rate can translate to approximately one lesion per retroviral genome per replication cycle. This high mutation rate explains, in part, the difficulties inherent in treating AIDS patients with inhibitors of RT or other viral proteins; a large population of mutant viruses preexist in every chronically infected individual, some encoding drug-resistant proteins. These mutants can replicate in the presence of a drug and quickly comprise the bulk of the population (see Chapters 17 and 19).

Several unique activities of RTs are also likely to contribute to their high error rates. The avian sarcoma/leukosis virus and human immunodeficiency virus enzymes are both proficient at extending mismatched terminal base pairs, such as those that result from nontemplated addition (Fig. 7.7). This process facilitates incorporation of the mismatched nucleotide into the RT product. RTs also seem to allow a certain type of slippage within homopolymeric runs, in which one or more bases are extruded on the template strand (Fig. 7.7B). Mispairing occurs after the next deoxyribonucleotide is added and the product strands attempt to realign with the template. Deletions can also be produced by this mechanism (Fig. 7.7D). Both deletions and insertions are also known to occur when template exchanges are made in vivo, apparently because these exchanges can take place within short sequence repeats (e.g., four or five nucleotides) that are not in homologous locations on the two RNA templates. This mechanism is of major importance in the capture of oncogenes.

RNase H. The RNase H of RT also requires a divalent cation, Mg^{2+} or Mn^{2+}. Like other RNase H enzymes, present in all prokaryotic and eukaryotic cells, the RNase H of RT digests only RNA that is annealed to cDNA. RNase H cleaves phosphodiester bonds to produce 5′-PO_4 and 3′-OH ends; the latter can be extended directly by the RT.

Structure of RT

Although RTs from avian and murine retroviruses have been studied extensively in vitro, the potential of human immunodeficiency virus type 1 RT as a target for drugs to treat AIDS has focused intense interest and resources on this enzyme. As a consequence, we know more about the human virus protein than about any other RT. Figure 7.8 shows a linear map of RT, which indicates the evolutionarily conserved regions in both the polymerase and RNase H domains, and the acidic amino acids that are required for catalysis. The aspartic acid residues highlighted in the polymerase region are included in conserved motifs in a large number of polymerases (Fig. 6.4) and are believed to coordinate the required metal ions and contribute to binding deoxyribonucleoside triphosphates and catalysis. As indicated by the arrows at the bottom of Fig. 7.8, the primary sequence of the p51 subunit is the same as that of p66, minus the RNase H domain. However, as discussed below, the analogous portions of these subunits are arranged quite differently in RT.

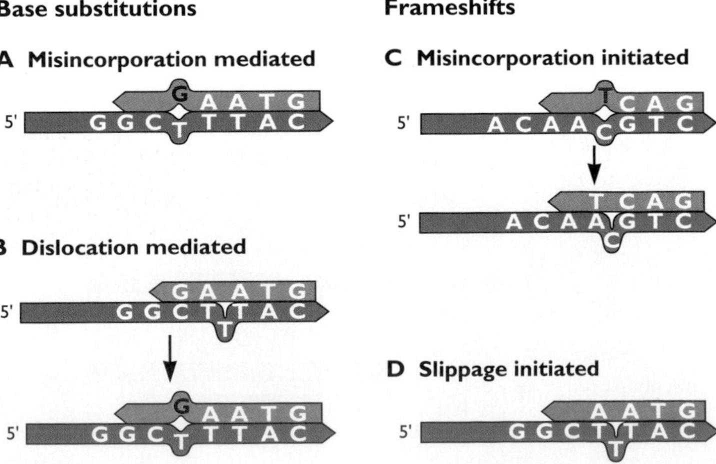

Figure 7.7 Mutational intermediates for base substitution and frameshift errors. Misincorporated deoxynucleotides are indicated in red. Slippage and dislocations are presumed to be mediated by looping out of nucleotides in the template. Only single-nucleotide dislocations are shown here, but large dislocations leading to deletions are also possible. From K. Bebenek and T. A. Kunkel, p. 85–102, *in* A. M. Skalka and S. P. Goff (ed.), *Reverse Transcriptase* (Cold Spring Harbor Laboratory Press, Cold Spring Harbor, N.Y., 1993), with permission.

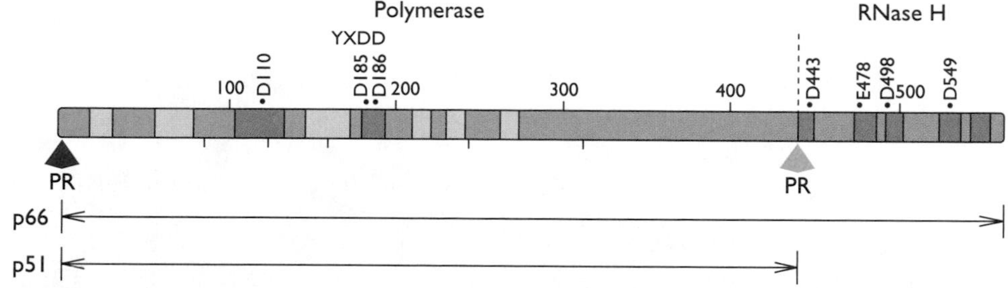

Figure 7.8 Linear map showing conserved residues in HIV-1 RT. The most highly conserved residues, which are thought to be components of the two active sites, are indicated by dots. The sequences included in the two subunits, p66 and p51, are shown by the lines below. In the polymerase domain, the colored blocks in the blue background indicate regions of homology within the RT family. The more intensely colored areas indicate regions of homology for all polymerases. In the RNase H domain, the intensely colored areas indicate regions with homology to *E. coli* RNase H. The arrowheads show sites of complete (red) or partial (pink) processing by the retroviral protease (PR). The dashed line indicates partial processing. Adapted from R. A. Katz and A. M. Skalka, *Annu. Rev. Biochem.* **63:**133–173, 1994, with permission.

The first portion of human immunodeficiency virus type 1 RT to be crystallized was the independently folded RNase H domain. As illustrated by the ribbon models in Fig. 7.9, the general structure of this domain is almost superimposable on that of *E. coli* RNase H. Included in the regions of structural homology are seven residues that are components of conserved motifs present in all retroviral and bacterial RNases H, clustered at what is believed to be the site at which the metal ion cofactors are bound. The results of site-directed mutagenesis have confirmed that many of the conserved residues are important for the activities of both *E. coli* and human retroviral RNase H.

The first high-resolution structure of the intact human immunodeficiency virus type 1 RT heterodimer was obtained for complexes with the nonnucleoside inhibitor nevirapine. Subsequently, several additional structures were determined, including RT with no bound inhibitors, and RT complexed with a short RNA·DNA duplex or a

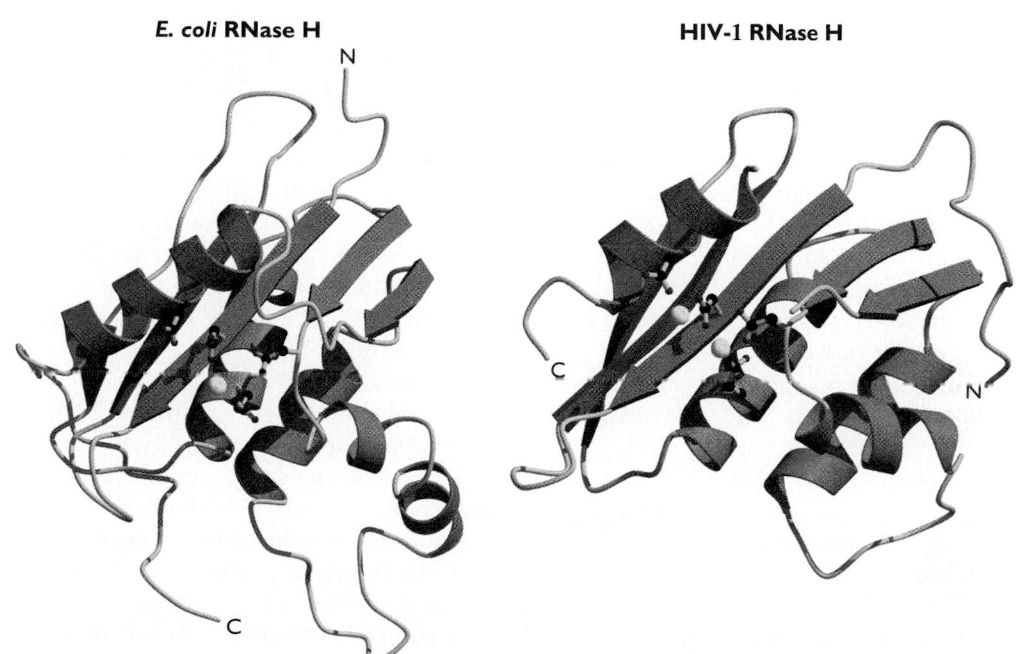

Figure 7.9 Ribbon diagrams based on the crystal structures of *E. coli* and HIV-1 RNase H. The highly conserved acidic residues are shown as ball-and-stick models, with metal ion complexes indicated by the large yellow spheres. Courtesy of Mark Andrake, Fox Chase Cancer Center.

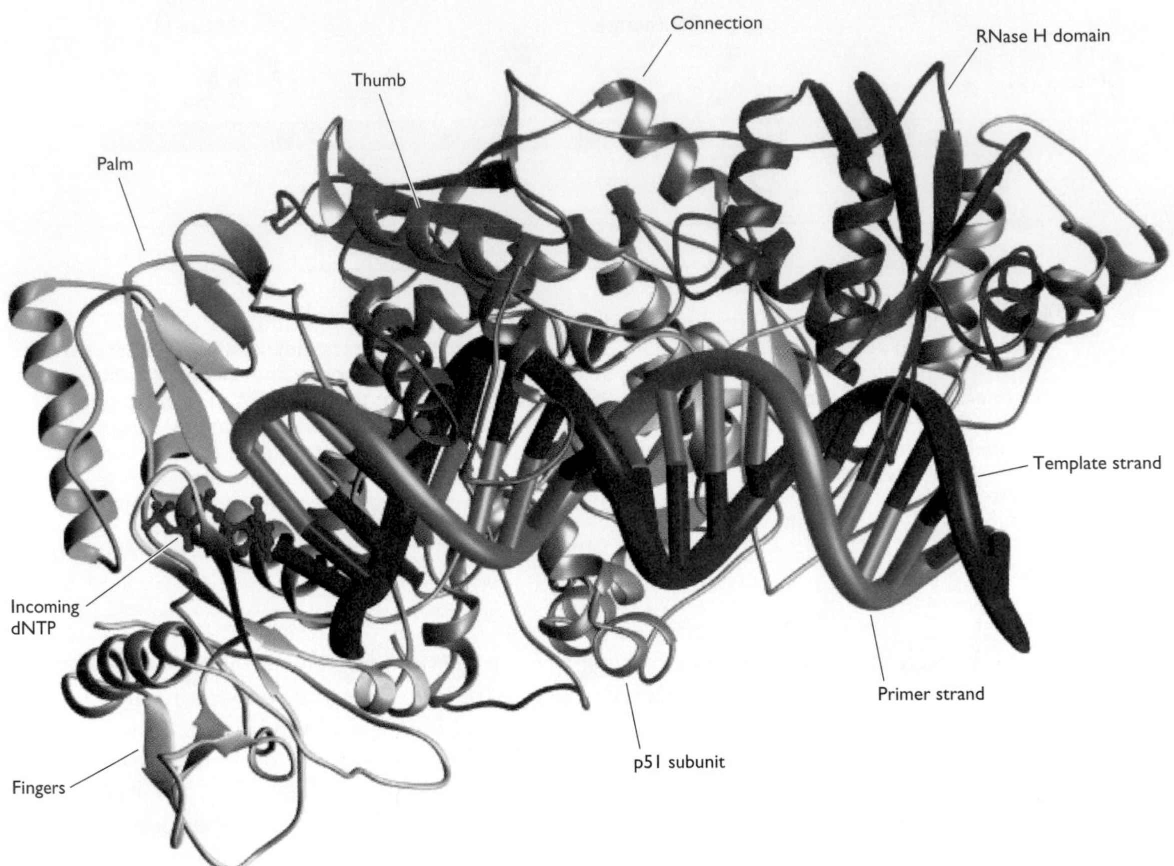

Figure 7.10 A ribbon representation of HIV-1 reverse transcriptase in complex with DNA template/primer. dNTP, deoxyribonucleoside triphosphate. Courtesy of Rajiv Chopra and Stephen Harrison, Harvard University.

DNA duplex and a deoxynucleotide substrate (Fig. 7.10). The most surprising finding to come from the analysis of the first RT crystals was the **structural asymmetry** in the subunits. Results of biochemical studies showed that the two subunits also perform different functions in the heterodimeric enzyme. The catalytic functions are contributed by the larger subunit, p66, whereas the role of p51 appears to be mainly structural. In the crystal structures, the two subunits are nestled one on top of the other, with an extensive subunit interface. The p66 polymerase domain was divided into three subdomains denoted "finger," "palm," and "thumb" by analogy with the convention used for describing the topology of the *E. coli* DNA polymerase I Klenow fragment, described in Chapter 6 (Fig. 6.3 and 6.4). A fourth subdomain lies between the remainder of the polymerase and the RNase H domain, and is therefore called the "connection" domain. It contains the major contacts between the two subunits of RT. The extended thumb of p51 contacts the RNase H domain of p66, an interaction that appears to be required for RNase H activity. Not only are human immunodeficiency virus type 1 RT and *E. coli* DNA polymerase similar topologically, but this retroviral RT can actually substitute for the bacterial enzyme in *E. coli* cells that carry a temperature-sensitive mutation in their DNA polymerase I gene. Such genetic complementation may be used to develop a screen in *E. coli* for nucleoside analogs that block the human retroviral RT activity.

Analysis of the various structural models predicts highly dynamic interactions among human immunodeficiency virus type 1 RT, its template·primer, and deoxyribonucleoside triphosphate substrates. The substrates are bound in a defined order: the template·primer first and then the deoxyribonucleoside triphosphate to be added. Upon binding the template·primer, the thumb is moved away from the fingers, allowing contacts between the fingers and the 5′ extension of the template. Another conformational change takes place when the deoxyribonucleoside triphos-

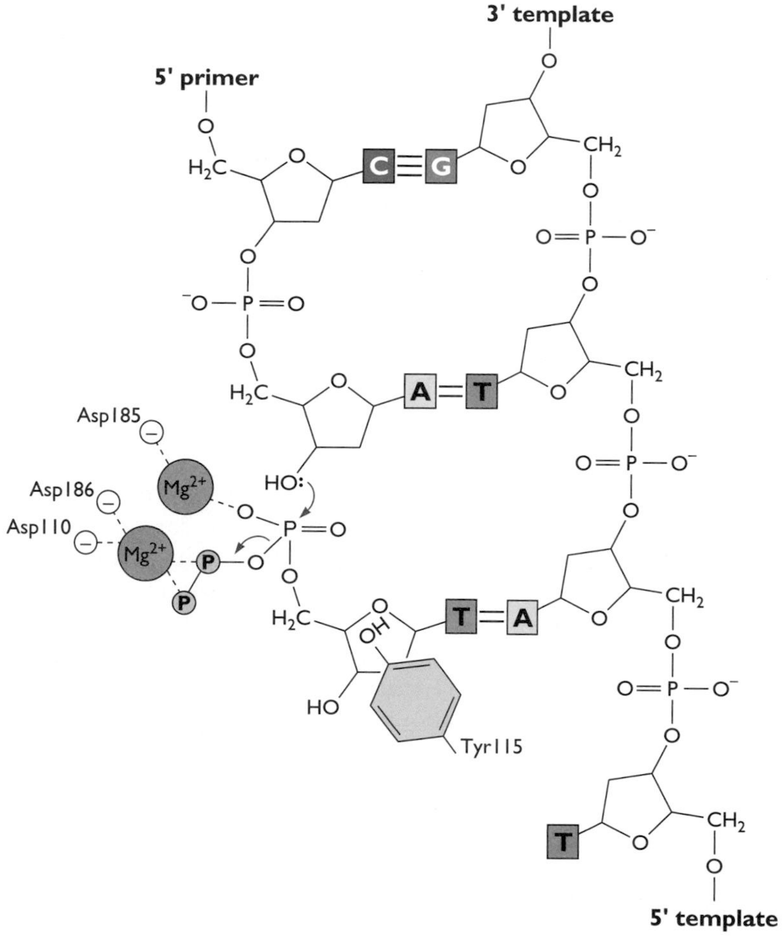

Figure 7.11 Schematic representation of the proposed mechanism of phosphodiester bond formation by HIV-1 RT. The chemical mechanism proposed for the polymerase reaction of both *E. coli* DNA polymerase I and the human retroviral RT (and RNase H) is based on the two-metal mechanism of catalysis deduced for the 3′ → 5′ exonuclease reaction of the DNA polymerase I Klenow fragment. The reaction is proposed to be metal-mediated, requiring two divalent cations separated by ~4 Å (Mg^{2+} in the model can also be replaced by Mn^{2+} or Zn^{2+}). One of the metal ions is shown coordinating Asp-110 and Asp-186 with the β- and γ-phosphates of the incoming deoxynucleoside triphosphate, while the other ion is chelated by Asp-185 and the α-phosphate. The second metal ion is in position to participate directly in the nucleophilic attack step by contributing to the electropositive character of the α-phosphate. The resulting PP_i may bind transiently to Asp-110 and Asp-186 via Mg^{2+} coordination prior to release. Tyr-115 is believed to stabilize binding of the incoming deoxyribonucleotide by base-stacking interactions. Red arrows indicate the net movement of electrons. From P. H. Patel et al., *Biochemistry* **34:** 5351–5363, 1995, with permission.

phate is bound. This substrate interacts directly with two fingertip residues; the interaction may induce closure of the binding pocket. This conformational change facilitates attack of the 3′ OH of the primer on the α-phosphate of the incoming deoxyribonucleotide. After this addition, the fingertips may resume their open position, allowing the diphosphate product to be released and the template·primer to be translocated, so that the next deoxyribonucleoside triphosphate can be accepted. Such translocation may be driven in part by the energy released upon hydrolysis of the previous deoxyribonucleoside triphosphate substrate.

A predicted mechanism for formation of new phosphodiester bonds at the catalytic center of the polymerase domain is shown in Fig. 7.11; the presumed roles of residues known to be critical from both structural and biochemical data are indicated. Figure 7.12 is a schematic drawing of

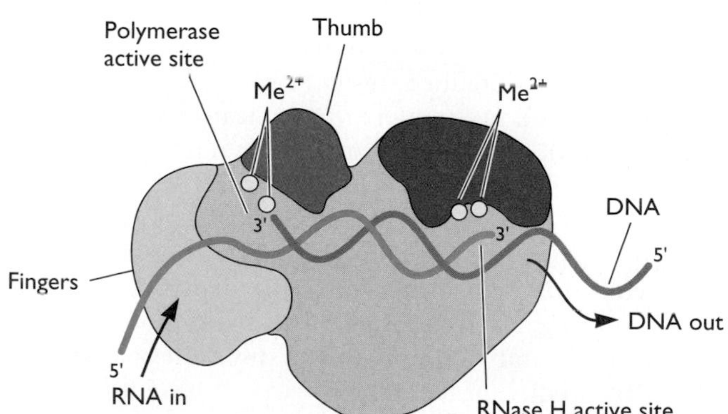

Figure 7.12 Model for a DNA·RNA hybrid bound to HIV-1 RT. The RNA template·DNA product duplex is shown lying in a cleft. The polymerase active site and the putative RNase H active site are indicated. Me^{2+} signifies a divalent metal ion. As illustrated, the RNA template enters from the left, and is degraded at the RNase H active site after being copied into DNA at the polymerase active site. The newly synthesized DNA strand exits to the right. Adapted from L. A. Kohlstaedt et al., p. 223–250, *in* A. M. Skalka and S. P. Goff (ed.), *Reverse Transcriptase* (Cold Spring Harbor Laboratory Press, Cold Spring Harbor, N.Y., 1996), with permission.

an RNA·DNA heteroduplex bound in the cleft region of the human immunodeficiency virus type 1 RT. As illustrated in the structure shown in Fig. 7.10, the cleft is flanked on one end by the polymerase active site and on the other end by the RNase H active site. In this model the RNA strand is fed into the polymerizing site from the left. As a new DNA chain is synthesized by addition of deoxyribonucleotides to the primer, the template RNA is translocated in stepwise fashion to the RNase H site at the right, where it can be degraded. This model is consistent with biochemical evidence for polymerase-coupled RNase H activity. The distance between the polymerizing and RNase H sites can account for the length (ca. 18 nucleotides) of the terminal RNase H oligoribonucleotide product.

Production of the p51 subunit, which possesses identical amino acid sequences but has both structure and function that are distinct from the p66 subunit, appears to be a unique example of viral genetic economy (Fig 7.8). The C terminus of the p51 subunit, at the end of the connection domain, is buried within the N-terminal β-sheet of the RNase H domain of p66. This arrangement suggests a model for proteolytic processing in which a p66 homodimer intermediate in the reaction is also arranged asymmetrically and the RNase H domain of the subunit destined to become p51 is unfolded. This mechanism would account for asymmetric cleavage by the viral protease. Results of cross-linking experiments suggest that the p51 subunit may perform a unique function in the RT heterodimer, that of binding the tRNA primer. Without comparable structural data for other retroviral RTs, we can only wonder if an analogous processing strategy and similar economy of coding apply to other retroviruses. The variability in subunit composition of the three best-studied systems makes this an interesting question (Fig. 7.6).

Other Examples of Reverse Transcription

When it was first discovered, RT was thought to be a peculiarity of retroviruses. We now know that other animal viruses, such as the hepadnaviruses, and some plant viruses, the caulimoviruses, also replicate by producing duplex genomic DNA via an mRNA intermediate. All are classified therefore as **retroid viruses**. In fact, the discovery of RT activity in some strains of myxobacteria and *E. coli* places the evolutionary origin of this enzyme before the separation of prokaryotes and eukaryotes. As it is now widely held that the evolving biological world was initially based on RNA molecules as both catalysts and genomes, the development of the modern (DNA) stage of evolution would have required an RT activity. If so, the retroid viruses may also be viewed as living fossils, shining the first dim light into an ancient evolutionary passageway from the primordial world (Chapter 20).

Since the discovery of RT in retroviruses, additional RT-related sequences have been found in cellular genomes. Some of these sequences are contained in cellular constituents known generally as **retroelements**, which are derived from retroviruses. During the retroviral life cycle (Appendix A, Fig. 22), the double-stranded DNA molecule synthesized by reverse transcription is integrated into the genomes of infected animal cells by a second retroviral enzyme, integrase (discussed in the following section). In some cases, retroviral DNA may be integrated into the DNA of germ line cells in a host organism. These integrated DNAs are then passed on to future generations in Mendelian fashion as **endogenous proviruses**. Endogenous proviruses are often replication-defective, a property that may facilitate coexistence with their hosts. A surprisingly large fraction of the mammalian genome (an estimated 10%) comprises endogenous proviruses and other retroelements that have been accumulated during evolution. Three other types of reverse-transcribed elements that are common residents in the genomes of animal cells include retrotransposons, retroposons, and processed pseudogenes (Box 7.2). Chromosomal telomeres are also formed via reverse transcription by an enzyme known as telomerase. Comparisons of the predicted RT-related amino acid sequences of representative retroelements provide clues about their relatedness and hints to their possible origin (Fig. 7.13).

Retroviral DNA Integration

The **integrase (IN)** of retroviruses and the related retrotransposons catalyzes specific and efficient splicing of the DNA product of RT into the host cell DNA. This activity is unique in the eukaryotic world. Some retroposons also contain IN-related sequences (Fig. 7.13). Establishment of an integrated copy of the genome is a critical step in the life cycle of retroviruses, as this reaction ensures stable association of viral DNA with the host cell's genome. The integrated **proviral DNA** is transcribed by cellular RNA polymerase II to produce the viral RNA genome and the mRNAs required to complete the replication cycle.

IN is encoded in the 3′ region of the retroviral *pol* gene (Fig. 7.6), and the mature protein is made, in most cases, by viral protease (PR)-mediated processing of the Gag-Pol polyprotein precursor. During progeny virus assembly, all three viral enzymes (PR, RT, and IN) are incorporated into the virion core. Virus particles contain equimolar concentrations of RT and IN (ca. 50 to 100 molecules per virion). The DNA product of RT is the direct substrate for IN, and genetic and biochemical studies indicate that these en-

BOX 7.2
Retroelements

Retrotransposons are dispersed widely in nature. Their gene content and arrangements are similar to those of retroviruses. Like retroviral proviruses, integrated retrotransposon DNAs have LTRs that include signals for transcription by cellular RNA polymerase II and short direct repeats of host DNA at their borders. RT sequence comparisons (Fig. 7.13) have allowed classification of these elements into two families. Most retrotransposons are distinguished from retroviruses by lack of an extracellular phase; they have no *env* gene, and hence the virus-like particles formed intracellularly are noninfectious. However, in the family *Metaviridae*, members of one genus do include open reading frames corresponding to *env*, and at least one of these elements, the *Drosophila* gypsy, produces infectious extracellular particles. Because of these similarities, retrotransposons may be thought of as retroviral progenitors. An alternative possibility, but with less phylogenetic support, is that retrotransposons are degenerate forms of retroviruses.

Retroposons, also called non-LTR retrotransposons, are dispersed widely in the DNA of human cells. Like retroviral proviruses and retrotransposons, they are flanked by direct repeats of host DNA at their boundaries. Although they lack LTRs, some contain internal promoters for transcription by cellular RNA polymerase III. All retroposons have A-rich stretches at one terminus, presumed to be derived by reverse transcription of the 3′-poly(A) tails in their RNA intermediates. Retroposons comprise fairly long stretches of related sequences (up to 6 kb). These long interspersed repeat elements are commonly called LINEs. Most LINEs encode RT-related sequences, but they often contain large deletions and translational stop codons and are therefore defective. Recently, LINEs that are capable of retrotransposition and encode an active RT have been discovered in the human genome.

Retrosequences, another type of reverse-transcribed sequence, are found as short interspersed repeat elements on the order of 300 bp long. These sequences, called SINEs, are even more abundant than LINEs. SINEs have no known open reading frames. The mechanism by which they are retrotransposed is not yet known.

Processed pseudogenes have no introns (hence "processed"), and their sequences are related to exons in functional genes that map elsewhere in the genome. Like retroposons, they include long A-rich stretches at one end and lack LTRs. However, they contain no promoter for transcription, and no RT sequences have been found associated with processed pseudogenes. Generally, they are thought to arise from rare reverse transcription of cellular mRNAs catalyzed by the RTs of retroviruses or nondefective LINEs. The table summarizes the defining characteristics of the major classes of retroelements.

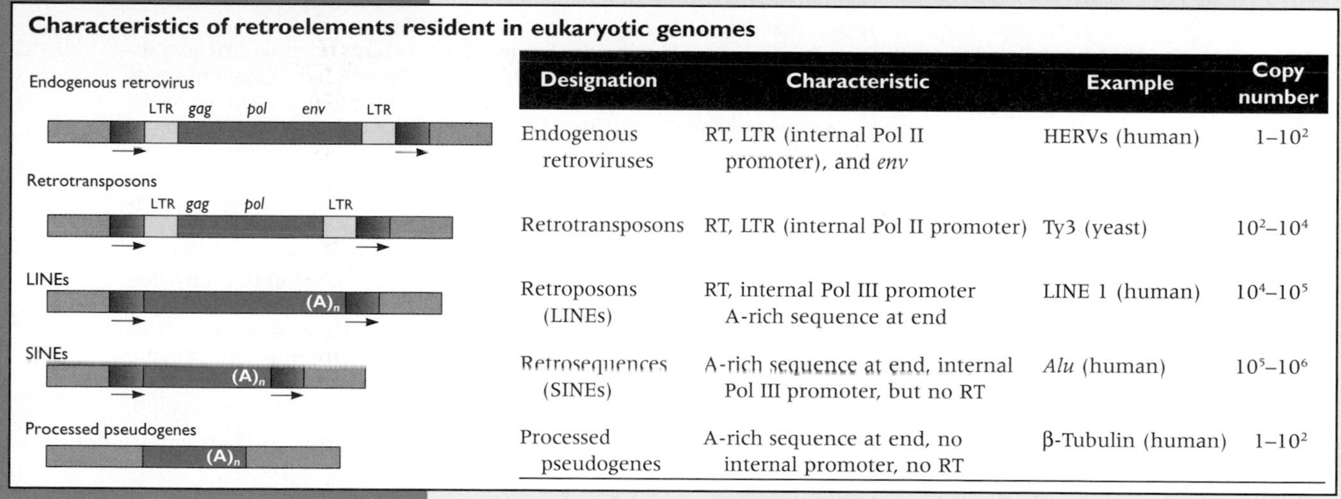

Designation	Characteristic	Example	Copy number
Endogenous retroviruses	RT, LTR (internal Pol II promoter), and *env*	HERVs (human)	$1–10^2$
Retrotransposons	RT, LTR (internal Pol II promoter)	Ty3 (yeast)	$10^2–10^4$
Retroposons (LINEs)	RT, internal Pol III promoter A-rich sequence at end	LINE 1 (human)	$10^4–10^5$
Retrosequences (SINEs)	A-rich sequence at end, internal Pol III promoter, but no RT	*Alu* (human)	$10^5–10^6$
Processed pseudogenes	A-rich sequence at end, no internal promoter, no RT	β-Tubulin (human)	$1–10^2$

zymes function in concert within infecting particles. As already noted, IN sequences are actually included in one of the subunits of avian sarcoma/leukosis virus RT, and gentle extraction of murine leukemia virus particles yields RT-IN complexes. However, as with RT, virtually nothing is known about the molecular organization of IN within virions.

The first insights into the mechanism of the integration process came in the early 1980s, when it was established that proviral DNA is flanked by LTRs and is colinear with

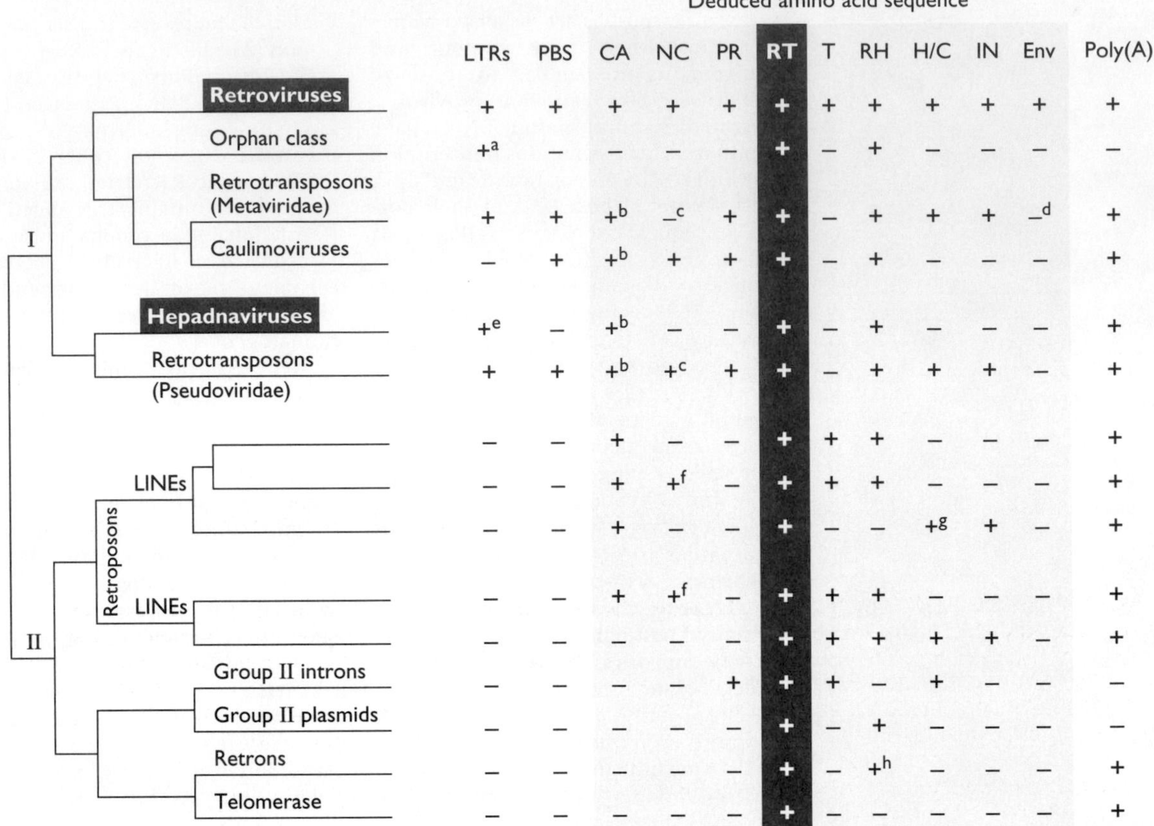

Figure 7.13 Schematic representation of the eukaryotic retroid family phylogeny. The unrooted family tree derived from comparison of deduced RT amino acid sequences is displayed at the left. It is separated into two major branches (I and II), based primarily on the presence of long terminal repeats (LTRs) that contain important regulatory sequences in the DNA versions of some of its members. The table compares DNA sequences or deduced amino acid sequences encoded in all family members with those sequences that are characteristic of retroviruses. The region that defines membership in the retroid family, the RT, is highlighted in red. Superscript letters in the figure indicate the following. (a) Some members have inverted LTR-like sequences. (b) Members of the retrotransposon, caulimovirus, and hepadnavirus lineages have capsid proteins (CA), but convincing similarity to retroviral capsid sequences has not been demonstrated. (c) The presence of nucleocapsid protein (NC) varies among different members of the retrotransposon lineages. (d) Some members of this retrotransposon lineage have a region that encodes an envelope-like protein (Env). (e) The direct repeats, DR1 and DR2, of hepadnaviruses are not long terminal repeats. (f) Among the retroposons, the NC relationship consists only of alignment to the essential conserved $CX_2CX_4HX_4C$ motif of the retroviral NC. This Zn^{2+}-binding motif, found in an upstream open reading frame, is variable in copy number and position. (g) In one of the LINEs (R2Bm), this region is highly divergent. (h) Among the retrons, RNase H activity and sequence similarity are variable. Gene key: LTRs, long terminal repeats flanking the retroviral genome; PBS, tRNA primer-binding site; CA, capsid protein; NC, nucleocapsid; PR, protease; RT, reverse transcriptase domain; T, "tether" region connecting the RT and RNase H domains; RH, RNase H domain; H/C, histidine and cysteine motif in the N-terminal portion of the integrase; IN, integrase; Env, envelope protein; poly(A), 3′-poly(A) tract in RNA or DNA form. From M. A. McClure, *in* A. M. Skalka and S. P. Goff (ed.), *Reverse Transcriptase* (Cold Spring Harbor Laboratory Press, Cold Spring Harbor, N.Y., 1996), with permission.

the unintegrated viral DNA and the RNA genome (Fig. 7.14). Nucleotide sequencing of cloned retroviral DNAs and host-viral DNA junctions revealed several unique features of the process. Both viral and cellular DNAs undergo characteristic changes; viral DNA is cropped, usually by 2 bp from each end, and a short (4- to 6-bp) duplication of host DNA flanks the provirus on either end. Finally, the proviral ends of all retroviruses comprise the same dinucleotide: 5′-TG . . . CA-3′. This dinucleotide is often embedded in an extended, imperfect inverted repeat that can

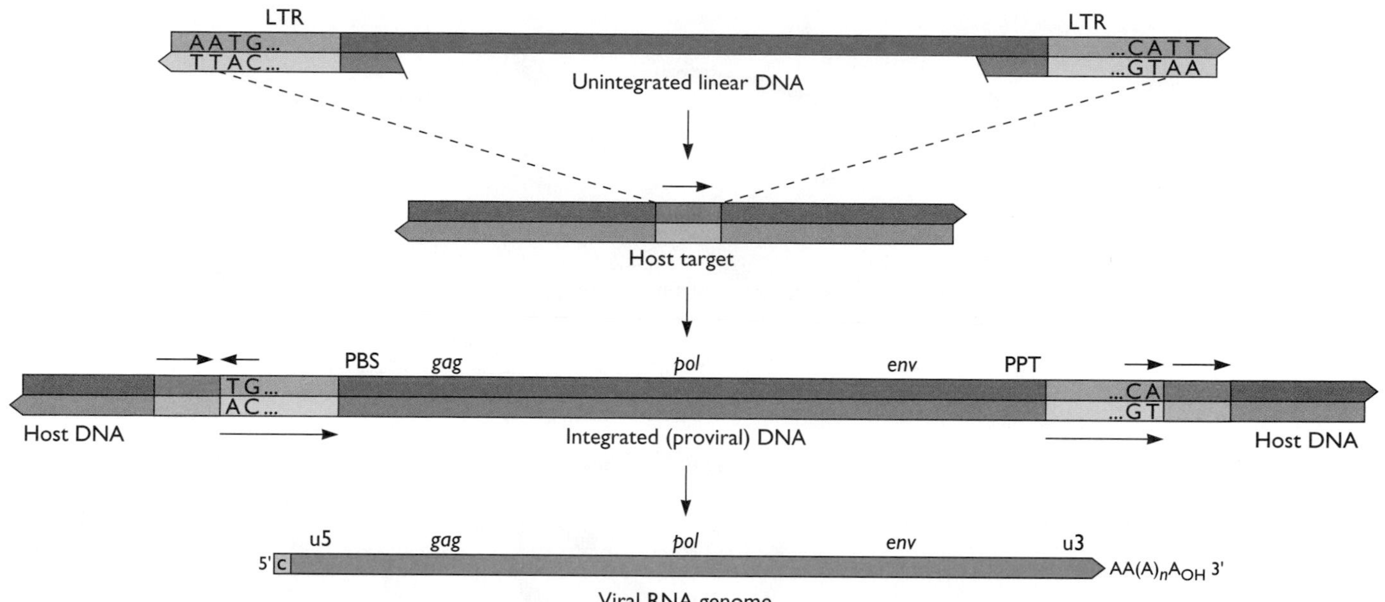

Figure 7.14 Characteristic features of retroviral integration. Unintegrated linear DNA of the avian retrovirus avian leukosis virus (top) after reverse transcription has produced two blunt-ended LTRs (Fig. 7.3). The break in the bottom strand indicates that this strand can include discontinuities, whereas the top strand is continuous (Fig. 7.3). Two base pairs (AA·TT) are removed from both termini on integration. A 6-bp "target site" in host DNA (indicated by an arrow) is duplicated on either side of the proviral DNA. The integrated proviral DNA (middle) includes 15-bp imperfect, inverted repeats at its termini, which end with the conserved 5′-TG . . . CA-3′ sequence; these inverted repeats are embedded in the LTR, which is itself a direct repeat. The gene order is identical in unintegrated and proviral DNA and is colinear with that in the viral RNA genome (bottom), for which a provirus serves as a template (described more fully in Chapter 8).

be as long as 20 bp for some viruses. The fact that the length of the host cell DNA duplication is characteristic of the virus provided the first clue that a viral protein must play a critical role in the process.

The inverted terminal repeat, conserved terminal dinucleotide sequence, and flanking direct repeats of host DNA were strikingly reminiscent of features observed in a number of bacterial insertion sequences and the *E. coli* bacteriophage, Mu (for mutator). Predicted amino acid sequences homologous to a portion of the retroviral IN were also found in the transposases of certain bacterial insertion sequences such as Tn*5*, suggesting that, like RT, IN probably evolved before the divergence of prokaryotes and eukaryotes. These similarities suggest common mechanisms for retroviral DNA integration and DNA transposition.

IN-Catalyzed Steps in the Integration Process

A generally accepted model for the IN-catalyzed reactions has been developed on the basis of results from many different types of experiment, including in vivo analyses and studies with reconstituted systems. The two ends of viral DNA are recognized specifically and nicked, and the new ends are then joined covalently to the host DNA in a sequence-independent manner, at staggered nicks also introduced by IN. The enzyme accomplishes these multiple activities as a multimer in a large nucleoprotein complex, probably assisted by other viral components and perhaps by cellular proteins. As illustrated in Fig. 7.15, IN catalysis occurs in two biochemically and temporally distinct steps.

The first step catalyzed by IN is a processing reaction, which requires duplex ends and therefore can take place only when synthesis of the ends of the unintegrated linear viral DNA is complete. This requirement prevents the insertion of (defective) molecules with imperfect ends into the host's genome. A specific sequence in the duplex DNA and close proximity of this sequence to a terminus are critical (Fig. 7.15), but sequences upstream also affect the efficiency of the reaction (Fig. 7.16). It has been shown that the processing reaction can take place in the cytoplasm of an infected cell, before viral DNA enters the nucleus within a subviral structure commonly referred to as the preintegration complex (described below). Although there is strong evidence of sequence specificity for processing in vivo, attempts to demonstrate sequence-specific binding of purified IN protein to retroviral DNA in vitro have generally not been fruitful. This failure may be a manifestation of the need for IN to recognize host DNA in a sequence-independent manner in the joining step.

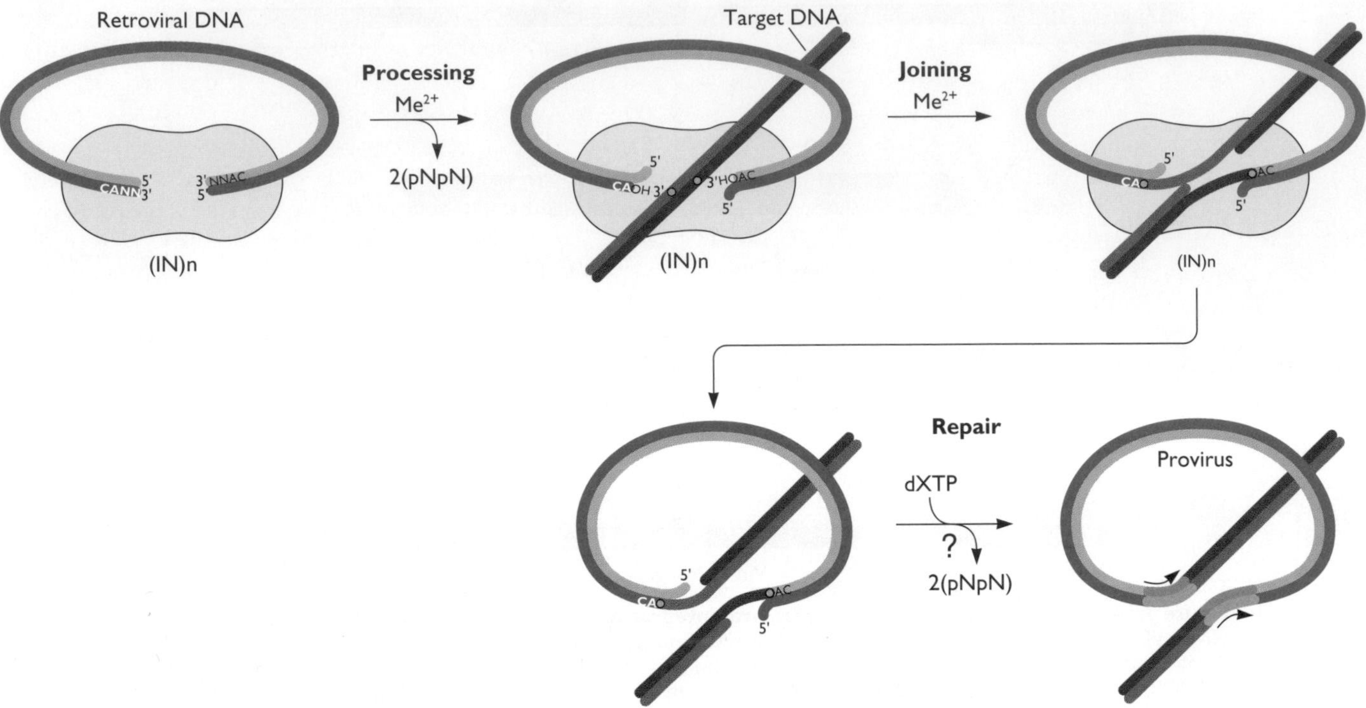

Figure 7.15 Three steps in the retroviral DNA integration process. Endonucleolytic nicking adjacent to the conserved dinucleotide near each DNA end results in the removal of a terminal dinucleotide, with formation of a new, recessed 5′ . . . CA_{OH}-3′ end that will be joined to target DNA in the second step of the IN-catalyzed reaction. The region shaded in beige represents an IN multimer [(IN)n]. Results of site-directed mutagenesis of viral DNA ends have shown that the conserved 5′ . . . CA_{OH}-3′ dinucleotide is essential for correct and efficient integration. Indeed, mutation of this sequence at only one DNA terminus is sufficient to impair processing of both viral DNA ends, implying that these reactions are coordinated within the preintegration complex. The small gold circles represent the phosphodiester bonds cleaved in the joining reaction. The final step in the integration process is a repair reaction that utilizes host enzymes. Adapted from N. D. Grindley and A. E. Leschziner, *Cell* **83:**1063–1066, 1995, with permission.

The second step catalyzed by IN is a concerted cleavage and ligation reaction in which the two newly processed 3′ viral DNA ends are joined to staggered (4- to 6-bp) phosphates at the target site in host DNA (Fig. 7.15). The product of the joining step is a **gapped intermediate** in which the 5′-PO_4 ends of the viral DNA are not linked to the 3′-OH ends of host DNA. These gaps are subsequently repaired, most likely by cellular enzymes, producing the characteristic flanking direct repeats of host DNA and a net loss of 2 bp from the ends of the viral DNA.

Figure 7.16 Nucleotide sequences at the termini of retroviral DNAs. Sequences are arranged with juxtaposed termini. Red-shaded sections are important for integration. The most intensely shaded regions appear to be most critical, and the conserved 5′ . . . CA . . . 3′ is essential for efficient and accurate integration. Red arrowheads indicate the location of the nick produced during processing by IN.

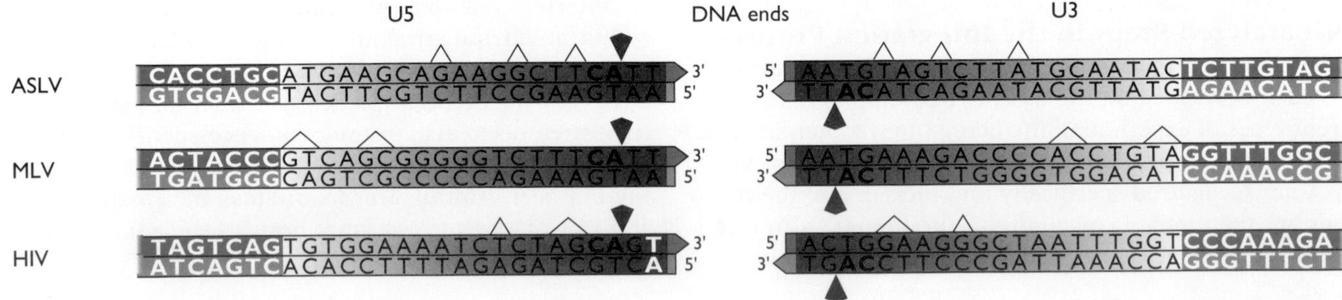

The Preintegration Complex and Nuclear Entry

Rapidly sedimenting nucleoprotein complexes have been isolated from the cytoplasm of avian sarcoma/leukosis virus-, murine leukemia virus-, and human immunodeficiency virus type 1-infected cells. These complexes appear to be compact, but the viral DNA associated with them is accessible to nucleases. The nucleoprotein complexes contain IN and viral DNA in a form that can be joined to exogenously provided plasmid or bacteriophage DNA. Such ex vivo integrants exhibit all the features expected for products of authentic integration. Other viral proteins, among them RT, capsid, and nucleocapsid proteins, have also been reported to be present in the preintegration complexes of murine leukemia and human immunodeficiency viruses, but no clearly defined role in the integration process has yet been assigned to any of these proteins. How nucleoprotein preintegration complexes gain access to target sites in host DNA is largely unknown. Some recent models are described in Chapter 5.

Specificity of the IN-Catalyzed Reactions

Integration is site-specific with respect to viral DNA. IN proteins recognize a specific sequence in their cognate viral DNA ends that includes a conserved 5′ . . . CA-3′ dinucleotide and a short stretch of adjacent sequences that are often part of an imperfect inverted repeat (Fig. 7.16). While specificity for viral DNA end sequences can be demonstrated with certain DNA substrates in vitro, such recognition is clearly less efficient than that which occurs in vivo. It seems likely therefore that some structural features or interactions among components within the preintegration complex help to place IN at its site of action near the viral DNA termini.

Many sites in host DNA can be targets for retroviral DNA joining. Furthermore, comparison of cellular sequences that flank a number of independently integrated viral DNA have identified no clear consensus target sequence. In vitro studies with purified enzymes and DNA substrates have failed to show any sequence specificity for target sites, although the patterns exhibited by different retroviral INs with the same DNA target are not identical. In the absence of discernible sequence or gene specificity, should we expect any selectivity for target sites in host chromosomes? The available data suggest yes. The IN protein of the Ty3 yeast retrotransposon appears to interact with RNA polymerase III transcription initiation proteins during target site selection. A specific interaction has also been observed between human immunodeficiency virus type 1 IN and a cellular protein (called IN-interacting protein 1 [Ini-1]), which is a component of a large complex that binds to chromatin and activates transcription. In vitro studies have established a preference for IN-catalyzed joining of viral DNA end sequences into target DNA sequences that are bent, owing to intrinsic properties, or underwound, as a consequence of being wrapped around a nucleosome. Furthermore, the binding of proteins to DNA can prevent this reaction by blocking access to the DNA, or stimulate it by distorting the DNA structure. These results suggest that interactions with particular chromatin-associated proteins may guide the preintegration complexes to sites where the DNA conformation favors joining.

Repair of the Integration Intermediate

This final step in the retroviral DNA integration process (Fig. 7.15) is likely to depend on host cell enzymes. Indeed, recent studies have shown that host proteins that participate in the cellular response to DNA damage are required for stable retroviral DNA integration. As with the evolutionarily related recombination of cellular immunoglobulin genes (V[D]J recombination), the nonhomologous end-joining DNA repair pathway has been implicated in the retroviral DNA integration process. It is likely that other host proteins participate in repair of the lesions introduced into host DNA by the joining step, repair of any remaining discontinuities in the (+) strand of the viral DNA, and reconstitution of chromatin structure at the site of integration.

Model Reactions In Vitro

The development of simple in vitro assays for the processing and joining steps catalyzed by IN marked an important turning point for investigation of the biochemistry of these reactions (Box 7.3). With such simple in vitro assays it was discovered that retroviral IN protein is both necessary and sufficient for catalysis, that no source of energy (ATP or an ATP-generating system) is needed, and that the only required cofactor is a divalent cation, Mn^{2+} or Mg^{2+}. Use of simple substrates and other biochemical analyses with purified IN protein helped to delineate the sequence and structural requirements for DNA recognition. They also showed that while the processing, joining, and disintegration reactions produce different products, their underlying chemistries are the same.

Assays with single-viral-end model substrates have been invaluable in elucidation of the catalytic mechanisms of IN. Nonetheless, they are limited in that the major products represent "half-reactions" in which only one viral end is processed and joined to a target. More recently, conditions for efficient, concerted processing and joining of two viral DNA ends have been described, with a variety of specially designed, "miniviral" model DNA substrates. These studies demonstrate that IN alone is also sufficient to accomplish such concerted reactions, an impressive feat for this relatively small protein, even acting as a multimer.

BOX 7.3

Principles of simple in vitro assays for IN activity with oligodeoxynucleotides

The substrates in these assays comprise short duplex oligodeoxynucleotides (15 to 25 bp) whose sequences correspond to one retroviral DNA terminus. The use of such model viral DNA substrates showed that IN can also catalyze an apparent reversal of the joining reaction, which has been called disintegration. These IN-catalyzed reactions all comprise a nucleophilic attack on a phosphorus atom by the oxygen in an OH group, and result in cleavage of a phosphodiester bond in the DNA backbone. In **processing,** the OH comes from a water molecule. In **joining,** the OH is derived from the processed 3′ end of the viral DNA, and the result is a direct transesterification. In **disintegration,** also a direct transesterification, a 3′-OH end in the interrupted duplex attacks an adjacent phosphorus atom, forming a new phosphodiester bond and releasing the overlapping DNA.

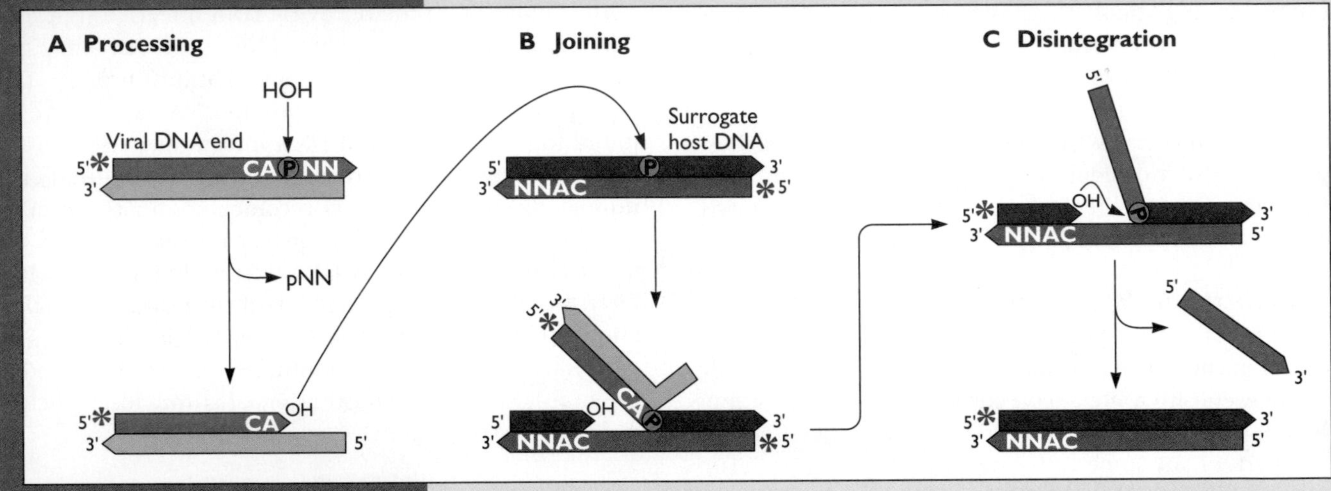

(A) In the simplest form of the oligodeoxynucleotide assay, the 5′ end of the viral DNA end sequence to be cleaved is labeled radioactively with ^{32}P, shown here as a red asterisk. The processed product is then detected on a denaturing sequencing gel, as a single chain two nucleotides shorter than the substrate. Production of the other product (pNN) can also be monitored if the 3′-terminal phosphate in the substrate strand is labeled. (B) In the joining reaction, a collection of products that are longer than the substrate is generated. These molecules can also be detected in a denaturing gel. If the target DNA is a much larger molecule, such as a bacterial plasmid or phage DNA, the products can be detected by electrophoretic separation in agarose gels. (C) Only the simplest version of the disintegration assay is shown here. A variety of disintegration substrates, in which all strands are duplex and some are continuous, have been used.

Integrase Structure and Mechanism

IN Proteins Are Composed of Three Structural Domains

Retroviral IN proteins are approximately 300 amino acids in length, comprising three domains (Fig. 7.17). Attempts to determine the three-dimensional structure of a full-length retroviral IN by X-ray crystallography have so far proved unsuccessful, perhaps because the linkages between the domains are too flexible. However, considerable insight has been obtained from analyses of single- and two-domain polypeptides.

The **N-terminal domain,** comprising approximately the first 50 amino acids, is characterized by two pairs of invariant, Zn^{2+}-chelating histidine and cysteine residues. The isolated domain forms a dimer in solution. A Zn^{2+} ion bound by each monomer stabilizes a helix-turn-helix structural motif (Fig. 7.17, left) that is almost identical in topology to the DNA-binding domain of the bacterial *trp* repressor protein. The **catalytic core domain** (Fig. 7.17, center), included in a central region of approximately 150 amino acids, is characterized by a constellation of three invariant acidic amino acids, the last two separated by 35 residues, the D,D(35)E motif. Results of site-directed mutagenesis experiments have shown that these acidic amino acids are required for all catalytic functions of IN. This finding indicates that there is one catalytic center for both processing and joining. Solution of the crystal structures of the catalytic core domains of the human and avian viral IN proteins was a significant milestone in the study of integration. In overall topology, the core domains of these two retroviral INs are quite similar; both are closely knit dimers, with an extensive subunit interface. These core domain structures established

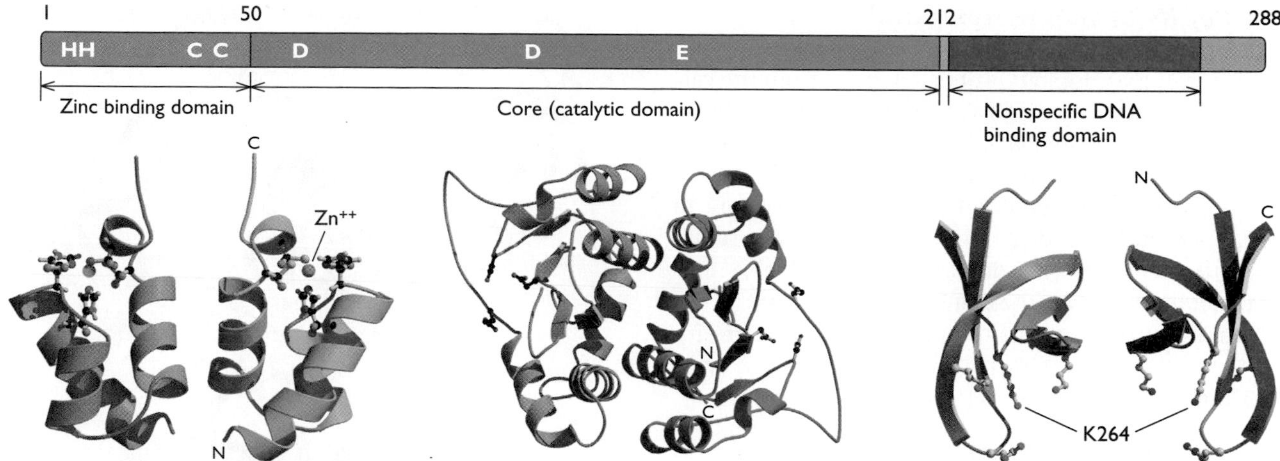

Figure 7.17 Linear map of human immunodeficiency virus type 1 IN showing the three independently folding domains. Numbers at the top indicate amino acid residues starting with 1 at the N terminus. Evolutionarily conserved amino acids are indicated in the single-letter code. (Left) Ribbon diagram showing the nuclear magnetic resonance (NMR) structure of a dimer of the N-terminal domain of HIV-1 IN (residues 1 to 46). The structure of the single subunit is shown in green, the zinc atom is displayed as a pink ball, and the coordinating cysteines and histidine residues are shown in ball-and-stick representations. Coordinates were kindly provided by Marius Clore, National Institutes of Health. (Center) Ribbon model of the structure of a dimer of the catalytic core domain determined by X-ray crystallography. The structure of the single subunit is shown in blue. Side chains of conserved acidic residues of the active site are shown in a ball-and-stick representation. The primary contacts between subunits at this interface are between α-helix 1 of one subunit and α-helix 5′ of the other subunit. Coordinates were kindly provided by Alex Wlodawer, National Cancer Institute-Frederick Cancer Research and Development Center. (Right) Ribbon model of the NMR-determined structure of a dimer of the C-terminal domain (residues 220 to 270). The structure of a single subunit is shown in red. The positions of several basic amino acids that are proposed to bind DNA are indicated in ball-and-stick representation. Coordinates were kindly provided by Marius Clore; the figure was prepared by Mark Andrake, Fox Chase Cancer Center.

that IN proteins are members of a large superfamily of nucleases and recombinases that includes the RNase H domain of the human virus RT (Fig. 7.18). It is quite striking that this same superfamily structure is represented twice in retroviral *pol*. The sequence of the 80- to 100-amino-acid **C-terminal domain** is the least conserved among IN proteins from different retroviral genera (Fig. 7.17, right). Nevertheless, this domain contains critical DNA-binding activity and multimerization determinants. The isolated C-terminal domain also forms a dimer in solution.

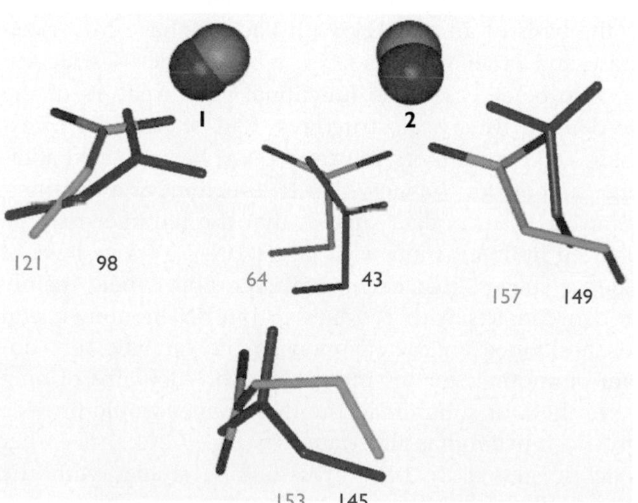

Figure 7.18 Metal complexes in the HIV-1 RNase H and the avian sarcoma/leukosis virus IN catalytic core domain. The orientation of the side chains and the positions of the metals in the human immunodeficiency virus type 1 and avian sarcoma virus IN proteins are superimposable on those of metal ion complexes obtained with the RNase H domain of the human immunodeficiency virus type 1 RT (Fig. 7.9). In the comparison of the avian viral IN and the human viral RNase H shown here, the side chains of critical acidic amino acids form a tripod that complexes with two divalent cations in almost superimposable positions. The IN catalytic core also binds a single Mn^{2+} or Mg^{2+} ion in position 1 with no change in the position of the side chains of D64 and D121. The binding of two metal ions at the active sites of the viral RNase H and IN proteins is consistent with a two-metal mechanism of catalysis by this superfamily, as proposed for the human retroviral RT polymerase activity (Fig. 7.11).

Model of IN Structure from Analysis of Two-Domain Polypeptides

Structures of two-domain proteins derived from the catalytic core and either the N- or C-terminal domains of several retroviral IN proteins have been solved by X-ray crystallography. All show the same dimerization interactions in the catalytic core as observed with the isolated domain. In contrast, either different or no dimer interactions were observed with the other two domains. A model of a full-length IN dimer, based on these structures, is shown in Fig. 7.19. Although no C-terminal domain interactions are indicated in this model, previous biochemical studies indicate that this domain contributes to formation of a tetramer.

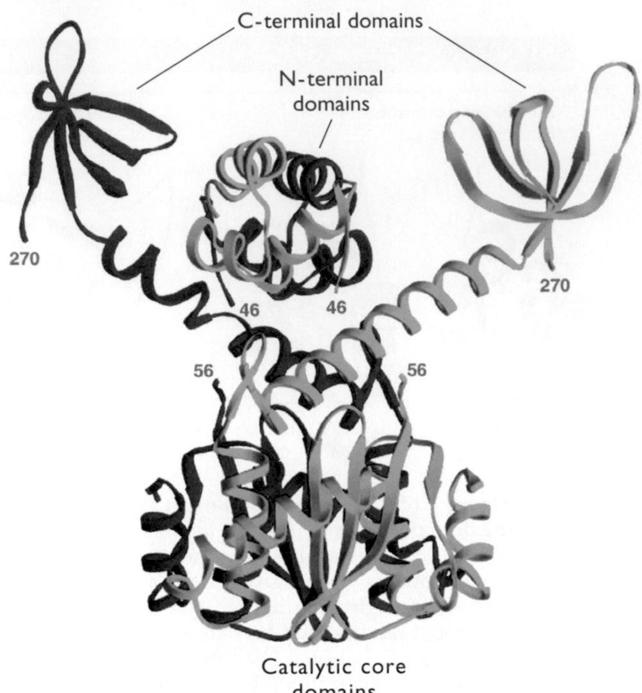

Figure 7.19 Model of a dimer of HIV-1 integrase. This ribbon model is derived from the superposition of the nearly identical core domain dimers in the crystal structures of the core together with N-terminal domain (Protein Data Bank code 1K6Y), and the core together with C-terminal domain (Protein Data Bank code 1EX4). The domains comprising the individual monomers are distinguished by coloring in green and blue. The catalytic core dimer interface is identical to that of the isolated domain in Fig. 7.17, but is shown here in a different orientation. However, the N-terminal dimer interface is different than that formed by the isolated domain, and the C-terminal domains do not contact each other in this dimer model, although they may do so in formation of a tetramer. Courtesy of Robert Craigie, National Institutes of Health.

Higher-Order Structure in the Preintegration Complex

The joining of viral to host DNA comprises a dynamic reaction in which two separate DNA duplexes are bound to an IN complex, cut, and recombined. We have yet to understand completely how the substrates and protein are organized and acted upon in any recombination reaction. Details of the higher-order structure of the preintegration complex are important to the development of antiviral drugs that target the integration process.

DNA organization. In most models of the IN-catalyzed reactions, viral DNA ends are adjacent to one another, as presumed to be necessary for their concerted processing and joining to a target site. The presence of inverted repeats at these ends (Fig. 7.16) suggests that they are recognized by a multimeric protein with a twofold axis of symmetry. In vitro experiments indicate that the invariant dinucleotides at each viral DNA end are bound to IN at a fixed distance from each other during the concerted processing reaction. The optimal distance between them is related to the distance between the staggered cuts in target DNA made during joining. These and other experiments also indicate that the viral DNA ends do not maintain B-DNA conformation when bound at the active site of the enzyme, but that the two strands are partially unwound and distorted. As noted above, partially unwound, bent DNA is a preferred target for joining. During the joining reaction, therefore, the viral DNA ends and target DNA appear to assume similar conformations, independent of their sequence.

A multimeric form of IN is necessary for activity. The results of biochemical analyses and in vitro complementation indicate that IN functions as a multimer. A reversible equilibrium among monomeric, dimeric, and tetrameric forms of IN is observed in the absence of DNA substrate. A conservative estimate of the intracapsid concentration of IN gives a value (~150 μM) high enough for most of the IN protein in virions to be in the form of dimers and tetramers. On the basis of an analogy with bacteriophage Mu transposase and other recombinases, a tetrameric IN structure seems to be a reasonable functional unit. Analysis of the two-domain IN crystal structures, and of bacterial transposase Tn*5* bound to its substrate DNA, has provided additional support for the view that IN functions as a tetramer; recent biochemical data suggest that the tetramer may be stabilized by interaction with target DNA. Various lines of evidence suggest that each viral DNA end is held mainly through contacts with residues in one IN monomer, and then positioned and acted upon by the catalytic core domain of another. In this model, the core domains of only two of the four subunits in the IN tetramer would provide catalytic function (as illustrated in Fig. 7.20). An earlier model suggested by DNA cross-linking studies with the human retrovirus IN proposed that at least eight monomers

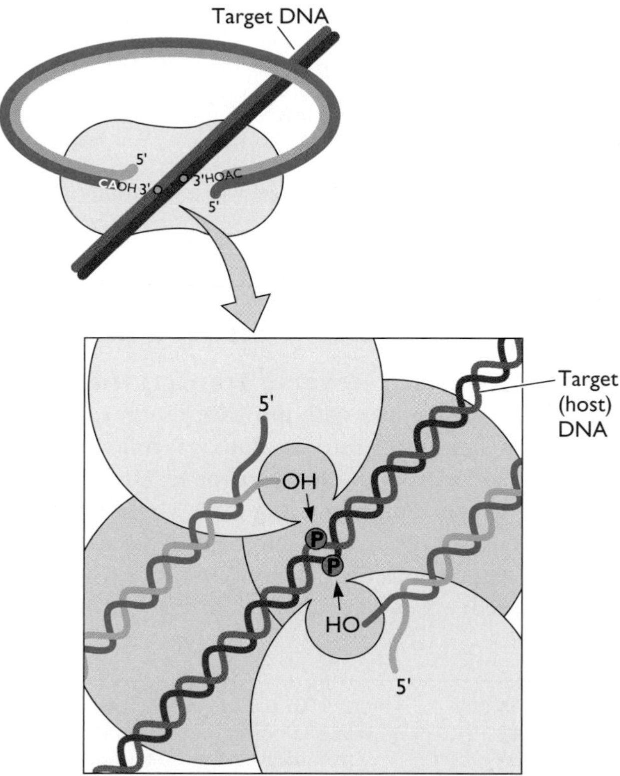

Figure 7.20 Tetramer model for functional integrase. The active sites of only two IN monomers participate in catalysis during the concerted joining reaction illustrated. The other two subunits likely contribute to the stability of the complex. Binding of the target DNA to basic residues at the dimer-dimer interface may help to stabilize the tetramer.

are required for concerted integration of two viral DNA ends into a target DNA. But in this case as well, only two of the subunits function catalytically; the other subunits help to position the substrates and stabilize the complex.

DNA footprinting experiments with preintegration complexes isolated from Moloney murine leukemia virus-infected cells show that several hundred base pairs of the viral DNA ends are protected by association with integrase, suggesting that many multimers may be bound. Furthermore, mutation at one end of the viral DNA, which renders it incompetent for integration, affects the footprint at the other viral DNA end. It seems certain that IN molecules "talk" to each other by protein-protein interactions in some large complex. But their organization and mechanism of "communication" have yet to be elucidated.

Host Proteins That May Regulate the Integration Reaction

The viral DNA within the preintegration complex is not itself a target for joining; such autointegration reactions would be suicidal. In murine leukemia virus preintegration complexes, this restriction was shown to depend on the presence of an 89-amino-acid cellular protein called **barrier-to-autointegration protein.** This small protein forms dimers in solution, binds to DNA, and can produce intermolecular bridges that compact the DNA. It has been proposed that such compaction prevents autointegration. As purified virions do not contain this cellular protein, it must be acquired from the cytoplasm of a newly infected host cell (Fig. 7.21). Recent evidence suggests that a homologous human protein blocks autointegration in human immunodeficiency virus preintegration complexes.

In the best-understood prokaryotic recombination systems, such as integration of bacteriophage lambda DNA and transposition of bacteriophage Mu DNA, cellular DNA-bending proteins are required to establish the proper configuration of DNA substrates. The same may be true for retroviral integration. High-mobility-group proteins, which are nonhistone, DNA-bending chromosomal proteins, stimulate the concerted joining of two viral DNA ends to a target DNA by purified IN proteins. Furthermore, one such high-mobility-group protein, Hmg I(Y), was found to be present in preintegration complexes isolated from cells infected with the human immunodeficiency virus type 1. Salt extraction of such complexes abolished integration of the viral DNA into exogenously provided target DNA, but this activity was restored partially by the addition of purified Hmg I(Y). Like the barrier-to-integration protein, the high-

Figure 7.21 Host proteins may regulate the integration process. The barrier-to-autointegration protein (BAF) binds to newly synthesized viral DNA, causing it to condense. Such compacted DNA may be an unsuitable target for autointegration. The high-mobility-group (Hmg) protein I(Y) also enters the preintegration complex and may facilitate the subsequent joining reaction via interaction with the viral DNA ends.

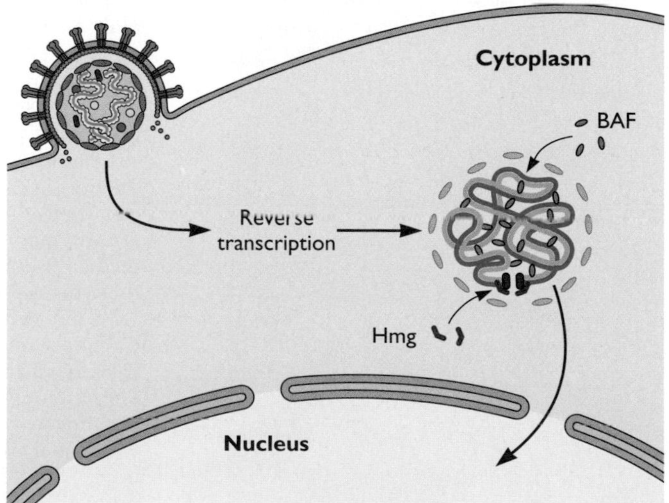

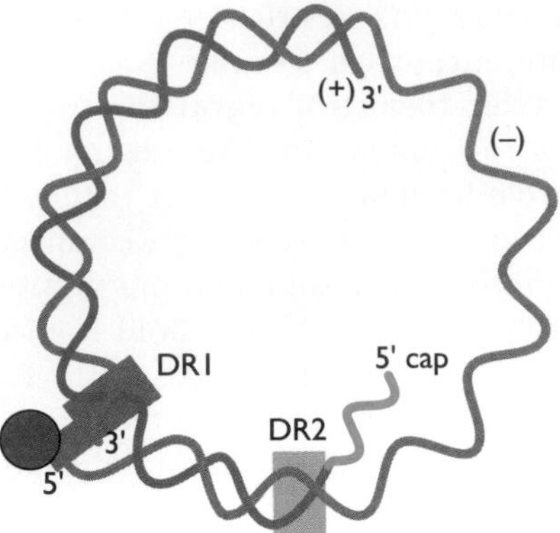

Figure 7.22 Hepadnaviral DNA. The DNA in extracellular hepadnavirions is a relaxed circular molecule of approximately 3 kb with circularity that is maintained by overlapping 5′ ends. The (–) strand is slightly longer than unit length, and the polymerase, shown as a blue ball, is attached to its 5′ end. The (+) strand has a capped RNA of 18 nucleotides at its 5′ end, and is less than unit length. The 5′ ends are near or in (10- to 12-bp) direct repeats called DR1 and DR2. As in retroviruses, these repeat sequences play an important role in facilitating critical template transfers during reverse transcription. Details of the genetic content are provided in Fig. 4 of Appendix A.

mobility-group protein is presumed to be acquired after the virion enters the cell (Fig. 7.21). It is not yet clear exactly how this occurs, whether the activities of the proteins are related, or how important each is to the organization of functional viral DNA-IN complexes in vivo.

Hepadnaviral Reverse Transcription

A DNA Virus with Reverse Transcriptase?

The concept that a virus with an RNA genome can replicate by means of a DNA intermediate was followed, about a decade later, with another big surprise: RNA as an intermediate in the replication of a DNA virus. Early hints that a mechanism other than semiconservative DNA synthesis was responsible for hepadnaviral replication came from the discovery of asymmetries in the genomic DNA

BOX 7.4

A retrovirus with a DNA genome?

The *Spumavirinae* comprise a subfamily of retroviruses isolated from primate, feline, and bovine species, among others. Spumaviruses are commonly called **foamy viruses** because they cause vacuolization and formation of syncytia in cultured cells. These viruses exhibit no known pathogenesis and received little attention from virologists until recent years. However, these recent studies have revealed that the foamy viruses are most **unconventional** retroviruses, with many properties that seem more similar to those of hepadnaviruses than of other retroviral family members.

For example:

- Although the arrangement of genes and the mechanism of reverse transcription are the same as that in other retroviruses, the human foamy virus reverse transcriptase is not synthesized as part of a Gag-Pol precursor, but rather from translation of a separate *pol* mRNA, as is also the case for the hepatitis B virus RT.
- Mature foamy virus particles do not include the usual processed retroviral structural proteins (MA, CA, and NC), but instead contain two large Gag proteins that differ only by a 3-kDa extension at the C terminus. These Gag proteins contain glycine-arginine-rich domains that bind with equal affinity to RNA and DNA, much like the hepadnaviral core protein.
- As with the hepadnaviruses, foamy virus budding requires both Gag and Env proteins, and most budding occurs into the endoplasmic reticulum.
- Most foamy virus particles remain within the infected cell. This property probably accounts for the large quantities of intracellular viral DNA, and might explain why persistently infected cells contain numerous integrated proviruses. It is possible that, as with the hepadnaviruses, foamy virus DNA integration occurs through an intracellular recycling pathway of progeny virions.
- Reverse transcription appears to be a late event in viral morphogenesis, and is largely complete before extracellular virus infects new host cells. Furthermore, although they contain both RNA and DNA, the genome-length DNA extracted from such virions can account entirely for virus infectivity. Like other retroviral family members, foamy virus genome replication requires an RNA intermediate, but the functional nucleic acid in extracellular virions appears to be DNA.

The foamy viruses therefore may represent an evolutionary link between the retroviruses and the hepadnaviruses, with genomes that are replicated by reverse transcription.

Linial, M. A. 1999. Foamy viruses are unconventional retroviruses. *J. Virol.* **73:**1747–1755.

and in the endogenous DNA polymerase reaction in isolated virions. The viral DNA comprises one full-length (–) strand and an incomplete, complementary (+) strand (Fig. 7.22), and the endogenous polymerase reaction can extend only the (+) strand. The replication intermediates isolated from infected cells were also unusual, comprising mainly (–) strands of less than unit length, few of which were associated with (+) strands. To Jesse Summers and William Mason, all this seemed suspiciously like retroviral reverse transcription, and in 1982 they published the results of their landmark studies with the duck hepatitis B virus, revealing the unique features of hepadnaviral replication. Unlike the endogenous reaction typical of extracellular virions, newly formed intracellular "core" particles were found to incorporate deoxynucleoside triphosphates into both strands. As with RNA-dependent DNA polymerization in retroviruses, synthesis of the (–) strand was resistant to the DNA-intercalating drug actinomycin D, whereas synthesis of the (+) strand was inhibited by this compound. Furthermore, a portion of the newly synthesized (–) strand DNA sedimented with the density of RNA·DNA hybrids. These and related findings marked an important turning point in our understanding of hepadnaviruses and extended our knowledge of reverse transcription. (See Box 7.4 for yet another, more recent, surprise.)

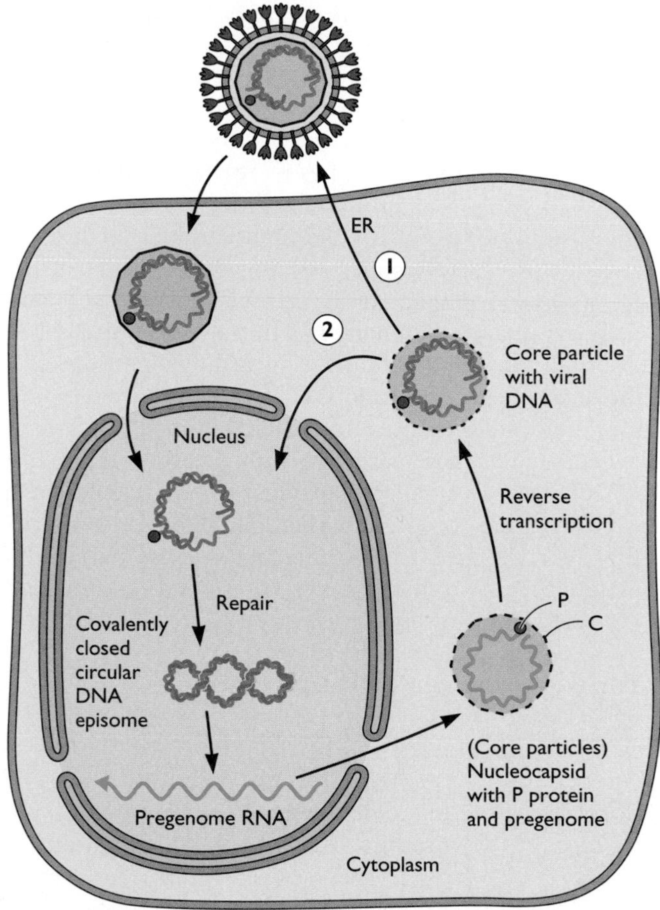

Figure 7.23 Single-cell replication cycle for hepadnaviruses. Pathway 2 provides additional copies of covalently closed circular episomal DNA. ER, endoplasmic reticulum.

Role of Reverse Transcription in the Hepadnaviral Life Cycle

An abbreviated model for the single-cell replication cycle of hepadnaviruses is shown in Fig. 7.23. The gapped DNA of an entering virus particle is imported into the nucleus and repaired, producing a covalently closed, circular molecule by mechanisms that are still undefined. Unlike retroviral DNA, hepadnaviral DNA is not normally integrated into the host's genome; the hepadnavirus genome encodes no integrase. However, the covalently closed circular DNA persists in the nucleus as an autonomous episome from which viral RNAs are transcribed by the host cell RNA polymerase II. The 3.5-kb **pregenomic mRNA** is exported to the cytoplasm, where it serves as the template for reverse transcription. This process takes place in a newly formed subviral particle that includes the core and polymerase proteins (products of the C and P genes). P protein provides all the activities required for reverse transcription. Following this process, the DNA-containing, nascent core particles can follow one of two pathways. Late in infection, when the cisternae of the endoplasmic reticulum contain an abundance of viral envelope glycoprotein, they can bud into the endoplasmic reticulum and eventually be secreted as progeny virions (Chapter 13). Conversely, if they do not become enveloped, the core particles can be directed to the nucleus, where their DNA is converted to additional copies of covalently closed circular DNA. This second pathway predominates at early times after infection when there is little envelope protein available. Eventually, at least for duck hepatitis B virus, ca. 10 to 20 covalently closed circles accumulate in the nucleus, and a steady-state balance of DNA and cytoplasmic virion protein components is maintained. It seems reasonable to assume that similar regulation occurs during the replication of mammalian hepadnaviruses, but this has not yet been demonstrated.

Covalently closed circular hepadnaviral DNA is not replicated by the host's DNA synthesis machinery; **all** hepadnaviral DNA is produced by reverse transcription. This situation contrasts with that of retroviruses, in which the integrated nuclear form, the provirus, is replicated with the host DNA. Both retroviruses and hepadnaviruses cause persistent infections by maintaining nuclear DNAs for the life of their host. But they do so in quite different ways.

The analysis of hepadnaviral reverse transcription has been difficult for a number of technical reasons. For one, suitable tissue culture systems were not available until 1986, when hepatoma cell lines were identified that were permissive for viral replication following transfection with cloned viral DNA. Furthermore, mutational studies are confounded by the compact coding organization of the DNA. The tiny genome (ca. 3.2 kb) is organized very efficiently, with more than half of its nucleotides translated in more than one reading frame. This arrangement makes it difficult to produce mutations that change only one protein. Finally, although reverse transcription takes place in newly assembled core particles, until quite recently it had not been possible to prepare enzymatically active P protein to study the reaction. Despite these difficulties, details currently available reveal fascinating analogies, but also striking differences, in the reverse transcription of hepadnaviruses and retroviruses.

Pathway of Reverse Transcription

Essential Components

Pregenomic RNA. The **pregenomic RNA** molecule that serves as the template for production of hepadnaviral genomic DNA is capped and polyadenylated and is also the mRNA for both C and P proteins. The transcription of pregenomic RNA from covalently closed circular DNA in the nucleus is initiated at a position approximately 6 bp upstream of one copy of a short, direct repeat of 11 to 12 nucleotides called DR1 (Fig. 7.22). Transcription then proceeds along the entire DNA molecule, past the initiation site, to terminate after a polyadenylation signal just downstream of DR1. Consequently, pregenomic RNA is longer than its template DNA. Because the region from the transcription initiation site to the polyadenylation site is copied twice, there is a longer direct repeat of ca. 200 nucleotides (r) at either end of the RNA. This longer repeat includes dr1 and a structural element of ca. 100 nucleotides called **epsilon** (ε) (Fig. 7.24 and 7.27). Deletion of ε within the 3′ copy of r has no impact on viral replication. In contrast, ε at the 5′ end provides both the site for initiation of (–) strand synthesis and the signal for encapsidation into core particles. Although all viral transcripts have ε at their 3′ ends, only the pregenomic RNA has a copy of ε at its 5′ end.

There is a marked preference for reverse transcription of the pregenomic RNA from which P protein is translated. The basis of such *cis*-selectivity is unknown; C protein, which is also translated from this RNA, and has nucleic acid-binding properties, appears to function perfectly well in *trans*. It is possible that the nascent P polypeptide binds

Figure 7.24 *cis*-acting signals on pregenomic RNA. (A) The RNA bears terminal repetitions of ca. 200 nucleotides (r) that contain copies of the packaging signal (ε), but only the 5′ copy has functional activity in vivo. Indicated are positions of initiation of the 5′ ends of (–) and (+) strand DNAs, and the 5′-UUAC-3′ motifs in duck hepatitis B virus within ε and at dr1, that are important for (–) strand DNA synthesis. Both the structural features of ε and the specific sequence in the loop are critical for its function. Adapted from C. Seeger and W. S. Mason, p. 815–832, *in* M. L. DePamphilis (ed.), *DNA Replication in Eukaryotic Cells* (Cold Spring Harbor Laboratory Press, Cold Spring Harbor, N.Y., 1996), with permission. (B) Sequences and proposed base pairing in ε of duck hepatitis B virus (DHBV) and human hepatitis B virus (HBV). From J. R. Pollack and D. Ganem, *J. Virol.* **68:**5579–5587, 1994, with permission.

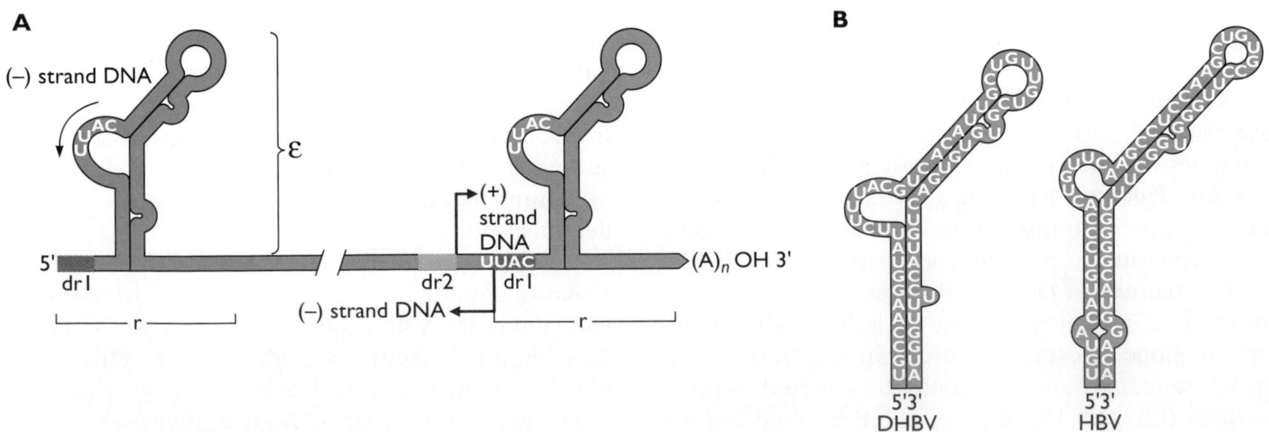

to its own mRNA cotranslationally. An attendant benefit for such a mechanism would be the selection for genomes that express functional P protein.

The newly formed hepadnaviral core in which reverse transcription occurs is believed to contain only one pregenomic RNA (Table 7.1). This selectivity is determined, in part, by the presence of the encapsidation signal(s) at the 5′ end of the pregenomic RNA. Quantitation of P protein in cytoplasmic core particles suggests a molar ratio of approximately 1:1 DNA molecule, implying that hepadnaviruses contain one copy of the viral genome per virion.

Primers. The primers for hepadnaviral RT remain attached to the 5′ ends of the viral DNA strands. They are for (−) strand synthesis, the P protein itself, and for (+) strand synthesis, a capped RNA fragment derived from the 5′ end of pregenomic RNA. A protein priming mechanism (Chapter 9) was first described for adenovirus DNA replication and later for the bacteriophage ϕ29. Priming by a viral protein, VPg, is also a feature of poliovirus RNA synthesis. Hepadnaviral reverse transcription is distinguished by the fact that the protein primer and the polymerase are components of the same polypeptide.

P protein, a self-priming reverse transcriptase. P protein is translated from an internal AUG in the pregenomic RNA, approximately 550 to 650 nucleotides downstream from the start of the overlapping C protein reading frame (Appendix A, Fig. 3). The translation product is a 90-kDa protein with C-terminal enzymatic domains that were first identified by amino acid sequence alignment with the retroviral RTs (Fig. 7.25). The highly conserved residues in the homologous domains are essential for hepadnaviral reverse transcription. Hepadnaviral P protein also contains an N-terminal domain separated from the RT region by a spacer, so called because it is refractory to mutation and not necessary for any known activity of the protein. The N-terminal domain, referred to as the **terminal protein region,** includes a tyrosine residue utilized for priming (−) strand DNA synthesis. In addition to its other functions, P protein is also required for encapsidation of viral RNA, a departure from the retroviral scheme, in which the NC protein sequences in the Gag polyprotein are used for this purpose (Table 7.1).

Table 7.1 Comparison of retroviral and hepadnaviral reverse transcription

Parameter	Retroviruses[a]	Hepadnaviruses[a]
Viral genome	RNA (pseudodipoid)	DNA (incomplete duplex)
Template RNA also:	Genome RNA, *mRNA (gag, pol)*	pregenome RNA, *mRNA* (C and P proteins)
DNA intermediate	*Circular DNA with 5′ overlaps*	*Circular DNA with 5′ overlaps*
Virus-encoded enzyme	RT	P protein
Molecules per core	50–100	1
Functions	*DNA polymerase, RNase H,* helicase (strand displacement)	*DNA polymerase, RNase H,* protein priming, template RNA encapsidation
Primer, first (−) DNA strand	tRNA (host)	Viral P protein
Site of initiation	*Near 5′ end of genome*	*Near 5′ end of pregenome*
First DNA product	(−) strong-stop DNA, ca. 100 nucleotides	Four nucleotides copied from bulge in 5′ ε
First template exchange	*To complementary sequence in repeated sequence, r, at 3′ end of template RNA*	*To complementary sequence in repeated sequence, R, at 3′ end of template RNA*
Primer, second (+) DNA strand	*Derived from template RNA,* internal RNase H product (ppt)[b]	*Derived from template RNA,* 5′ cap, terminal RNase H product
Site of initiation	*Near 5′ end of (−) strand DNA*	*Near 5′ end of (−) strand DNA*
Time of initiation	Before completion of (−) strand	After completion of (−) strand
Type of priming	Priming in situ	Primer translocated
Second template exchange	*To the 3′ end of (−) strand DNA via complementary PBS[c] sequence*	*To the 3′ end of (−) strand DNA via complementary sequence*
Reverse transcribing nucleoprotein complex	*Subviral "core" particles,* deposited in the *cytoplasm* on viral entry	*Nascent subviral "cores," cytoplasmic* intermediates in viral assembly
Final product(s)	Double-stranded linear DNA	Circular viral DNA or covalently closed episomal DNA
DNA maintained in the nucleus	Integrated into host genome, proviral DNA	Nonintegrated episome in host nucleus

[a]Italics indicate similarities.

[b]ppt, polypurine tract.

[c]PBS, primer-binding site.

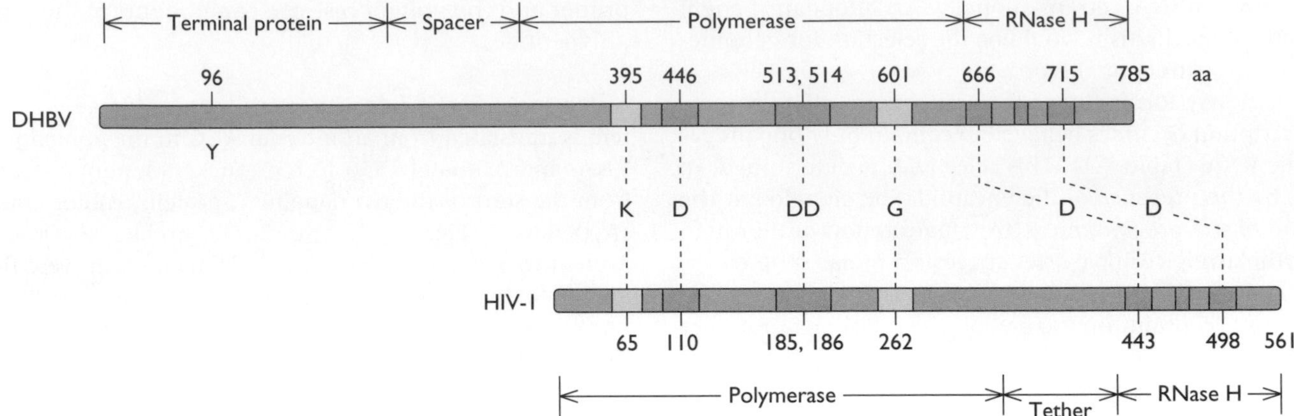

Figure 7.25 Comparison of hepadnaviral and retroviral RT proteins. Linear maps of the duck hepatitis B virus (DHBV) and human immunodeficiency virus type 1 (HIV-1) *pol* gene products. The maps were aligned relative to amino acids that are generally conserved among all RTs. Position 1 on the DHBV protein corresponds to the AUG codon at position 170 on the DHBV genome. See also Fig. 7.8. Adapted from C. Seeger and W. S. Mason, p. 815–832, *in* M. L. DePamphilis (ed.), *DNA Replication in Eukaryotic Cells* (Cold Spring Harbor Laboratory Press, Cold Spring Harbor, N.Y., 1996), with permission.

An important breakthrough in the study of hepadnaviral reverse transcription was achieved with the demonstration that enzymatically active P protein can be produced upon translation of P mRNA from duck hepatitis B virus in a cell-free rabbit reticulocyte lysate. Essential components in this reaction include the ε sequence and host proteins that assist in the formation of an active nucleoprotein complex. The results of subsequent studies showed that a short, three- or four-nucleotide DNA fragment copied from the bulge in ε was covalently attached to a tyrosine residue in the newly synthesized polymerase. As described below, this short DNA product serves as a primer for further RNA-dependent DNA synthesis following the first template exchange, and P protein remains covalently attached to this DNA during **all** subsequent steps in reverse transcription and in the virion (see Fig. 7.27).

Host proteins may facilitate P protein folding. P protein is the only viral protein required for initiation of DNA synthesis. Binding to ε may be required for the P protein to fold into an active conformation. If the ε region is not present during synthesis of P protein in yeast, the enzyme is inactive, even if the ε region is supplied later. Furthermore, P protein synthesized in the presence of the ε region is more resistant to proteolysis. Recent studies suggest that host cell proteins may also affect P-protein folding. Synthesis of active P protein in the cell-free system requires the presence of cellular chaperone proteins and a source of energy (ATP). Furthermore, incorporation of these host cell proteins into viral cores appears to require the P protein polymerase activity. It has been proposed that chaperones are needed to maintain viral P protein in a conformation that is competent to bind to ε and prime DNA synthesis, and to interact with assembling core subunits (Fig. 7.26).

Critical Steps in Reverse Transcription

Initiation and the first template exchange. Synthesis of the (–) strand of duck hepatitis B virus DNA is initiated by the polymerization of three or four nucleotides primed by the OH group of a tyrosine residue in the terminal protein domain of P protein. The template for this reaction is a specific sequence in the bulge of ε at the 5′ end of the pregenomic RNA. Why this copy and not the ε at the 3′ end is normally chosen for initiation by P protein remains a mystery. This initial synthesis is followed by a template exchange in which the enzyme-bound, four-nucleotide product anneals to another CAUU at the edge of dr1 at the 3′ end of the pregenomic RNA (Fig. 7.24 and 7.27, step 2). Although the sequence at this end is complementary to the short, initial product, it is not unique in the pregenomic RNA (Fig. 7.24). Furthermore, mutation of the normal acceptor sequence leads to the synthesis of (–) strands with 5′ ends that map to other sites in the vicinity of the 3′ dr1 that apparently can serve as alternative acceptors. It seems reasonable therefore to assume that selection of the normal site is guided by specific organization of pregenomic RNA in core particles, which places the acceptor sequence close to the site of initiation. P protein remains covalently attached to the 5′ end of the (–) strand during the

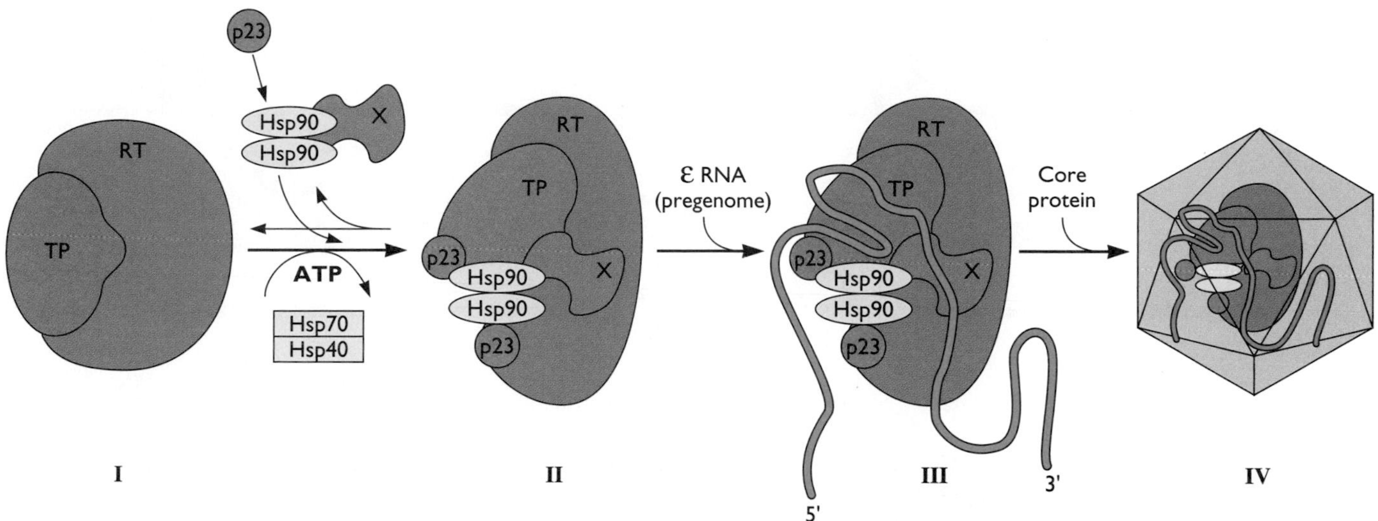

Figure 7.26 Model for the assembly of nucleocapsids in hepadnaviruses. P protein is translated to form an apostructure (I) that provides a substrate for the heat shock protein 90 (Hsp90) pathway. TP denotes the N-terminal domain of the P protein (the so-called terminal protein), which is unique to hepadnaviral RTs and harbors the primer tyrosine residue for the initiation of reverse transcription. RT indicates the polymerase plus RNase H domains which are conserved among all known RTs. The Hsp90 complex is depicted together with the chaperone partner p23 and possible additional proteins (X) that may be components of the complex. The function of Hsp90 also depends on the ATPase Hsp70 and its partner Hsp40. Interaction of the polymerase with the chaperone complex induces a conformational change (II) that allows binding of P protein to ε RNA (III), which in turn provides the signal for nucleocapsid assembly and initiation of viral DNA synthesis (IV). From J. Hu et al., *EMBO J.* **16:**59–68, 1997, with permission.

first template exchange and, as noted previously, **all** subsequent steps.

Elongation and RNase H digestion of the RNA template. Following the first template exchange, (–) strand DNA synthesis continues all the way to the 5′ end of the pregenomic RNA template (Fig. 7.27, steps 3 and 4). Because synthesis is initiated in the 3′ dr1, a short repeat of 7 to 8 nucleotides (3′R) is produced at the end of this elongation step when the 5′ dr1 sequence is copied (Fig. 7.27, step 4). The reader should note this fact, as it will become important in a later step of viral DNA synthesis. The RNA template is degraded by the RNase H activity of P protein as (–) strand synthesis proceeds. Unlike the retroviral RNase H products, none of these hepadnaviral RNAs are used as primers for (+) strand DNA synthesis in situ, and (+) strand DNA synthesis does not start until the (–) strand has been completed (Table 7.1). The final product of RNase H digestion is a short RNA molecule, corresponding to the capped end of the pregenomic RNA, which includes the 5′ dr1. This 15- to 18-nucleotide product is similar in length to the limit product of polymerase-associated RNase H digestion by human immunodeficiency virus type 1 RT. In the human retroviral protein, such a product provides a measure for the distance between the polymerase and RNase H active sites on the enzyme (Fig. 7.12); a similar relationship may apply to hepadnaviral P protein (Table 7.1).

The short, capped RNA could serve as a primer in situ for synthesis of the (+) strand of DNA (Fig. 7.27, step 5a). Indeed, in 5 to 10% of reverse transcription reactions, such priming does take place, and double-stranded linear genomes are produced. These linear DNAs can be converted to covalently closed circular DNA. However, most of these circular molecules are formed by base-pairing-independent recombination of their linear ends, a process which is mutagenic. Consequently, infectious virus particles would generally not be produced by this mechanism. What happens most of the time is a translocation of the capped RNA primer to another location on the (–) strand DNA template, DR2.

Translocation of the primer for (+) strand DNA synthesis. Translocation of the primer for (+) strand DNA synthesis is likely to be facilitated by the homology between DR1 and DR2 (Fig. 7.27, step 5). The capped RNA primer, which includes dr1 sequences, can anneal to both. How the primer is induced to dissociate from DR1 and associate with DR2 is another mystery. A small hairpin structure

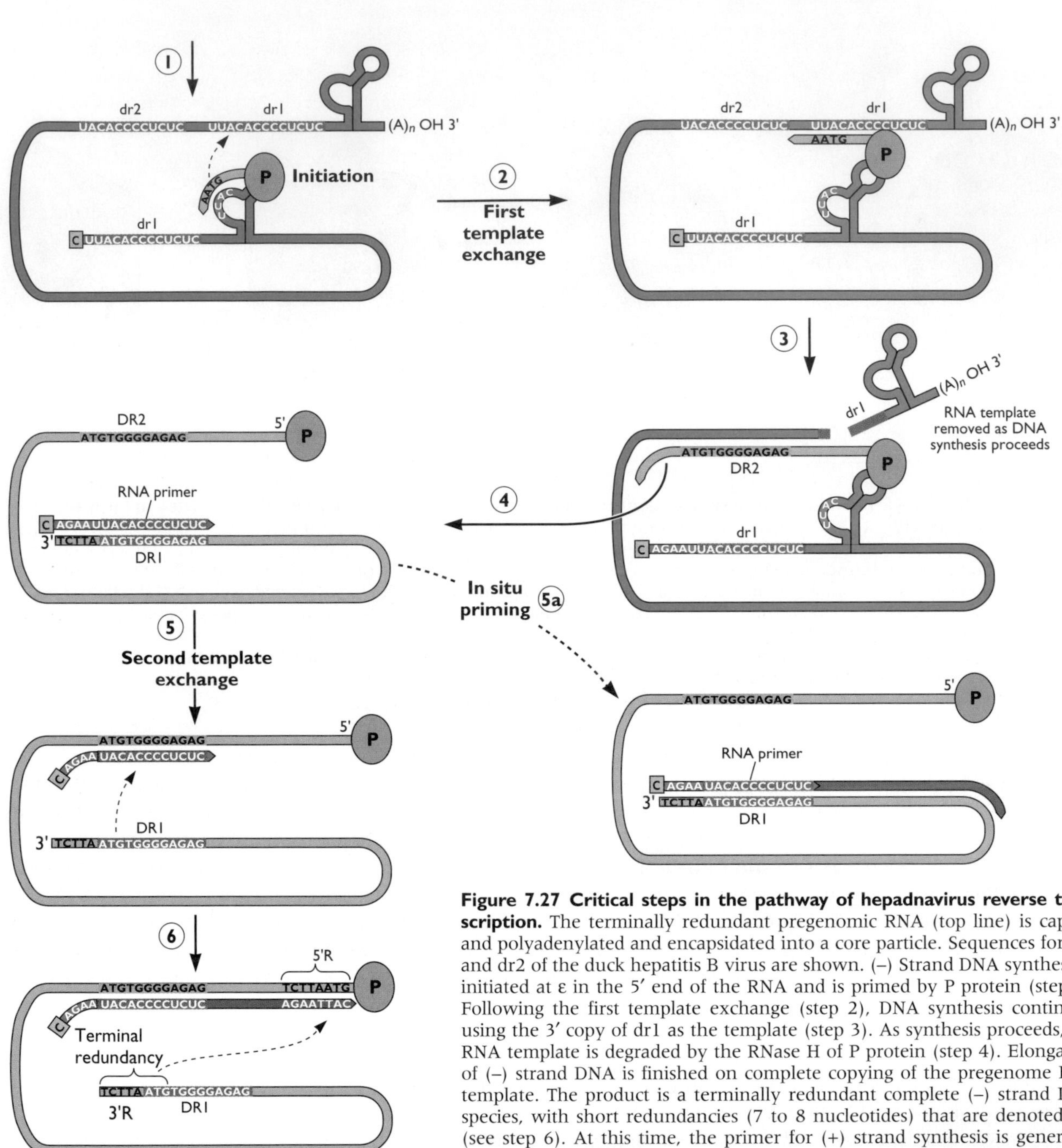

Figure 7.27 Critical steps in the pathway of hepadnavirus reverse transcription. The terminally redundant pregenomic RNA (top line) is capped and polyadenylated and encapsidated into a core particle. Sequences for dr1 and dr2 of the duck hepatitis B virus are shown. (–) Strand DNA synthesis is initiated at ε in the 5′ end of the RNA and is primed by P protein (step 1). Following the first template exchange (step 2), DNA synthesis continues, using the 3′ copy of dr1 as the template (step 3). As synthesis proceeds, the RNA template is degraded by the RNase H of P protein (step 4). Elongation of (–) strand DNA is finished on complete copying of the pregenome RNA template. The product is a terminally redundant complete (–) strand DNA species, with short redundancies (7 to 8 nucleotides) that are denoted 3′R (see step 6). At this time, the primer for (+) strand synthesis is generated from the 5′-terminal 15 to 18 nucleotides of the pregenomic RNA, which remain as the limit product of RNase H digestion. The primer is capped and includes the short sequence 3′ of dr1. At a low frequency (5 to 10%), instead of being translocated, the (+) strand primer is extended in situ (the structure set off by a dashed arrow) (step 5a). Elongation of this (+) strand results in a duplex linear genome. In the majority of cases, the primer is translocated to base pair with the DR2 sequence near the 5′ end of (–) strand DNA (step 5). (+) Strand synthesis is initiated, and then elongation begins (step 6). On reaching the 5′ end of (–) strand DNA, an intramolecular template exchange occurs, resulting in a circular DNA genome (step 7). This exchange is facilitated by the short terminal redundancy, 5′R, in (–) strand DNA. (+) strand DNA synthesis then continues for a variable distance, resulting in the relaxed circular form of the genome found in mature virions. Adapted from Fig. 1 of J. W. Habig and D. D. Loeb, *J. Virol.* **76:**980–989, 2002, with permission.

that includes the 5′ end of DR1 in the (–) strand of duck hepatitis B virus DNA appears to contribute to the translocation by inhibiting in situ priming and, perhaps, facilitating annealing of the capped RNA fragment with the complementary sequence in DR2 (Fig. 7.28). As in the first template exchange, specific organization of the template in the core particles is thought to be required for this translocation to occur. Furthermore, if the potential for the primer to hybridize with DR2 is disrupted by mutation, the pathway leading to formation of linear duplex DNA molecules predominates.

The (+) strand synthesis primed by the translocated capped hepadnaviral RNA primer is similar to that which produces the strong-stop DNAs in retroviral reverse transcription. The (+) strand DNA synthesis begins near the 5′ end at DR2 and soon runs out of (–) strand template. As in the retroviral case, this problem is solved by a second template exchange, in this instance facilitated by the short repeat, 5′R, produced during synthesis of the (–) strand (Fig. 7.27, step 6).

The second template exchange forms a circle. The structural requirements for the next step in hepadnaviral reverse transcription must be complicated, as P protein is still attached to the 5′ end of the (–) strand, which must be displaced to allow a template transfer. Studies in which genomes of heron and duck hepatitis viruses were exchanged indicate that, in addition to DR1 and DR2, *cis* interactions among other sequences, at the ends and in a central region of (–) strand DNA, are important in this final step. It has been suggested that the simultaneous interaction of the central region with both ends may hold the ends in a position that facilitates both (+) strand primer translocation as well as the second template exchange. Even with such "help," it is difficult to envision how a single protein accommodates all three DNA ends at once and catalyzes polymerization while still attached to one of them. Nevertheless, this exchange does occur with high efficiency in vivo, and subsequent elongation of the (+) strand produces the partially duplex circle that comprises virion DNA (Fig. 7.27, step 7).

Figure 7.28 Model for (+) strand priming. Formation of a putative hairpin in the (–) strand DNA template displaces the 3′ end of the capped RNA fragment, preventing in situ priming and facilitating annealing with the homologous sequence in DR2. The ensuing translocation of the RNA primer allows initiation of (+) strand DNA synthesis. Adapted from Fig. 2 of J. W. Habig and D. D. Loeb, *J. Virol.* **76:**980–989, 2002, with permission.

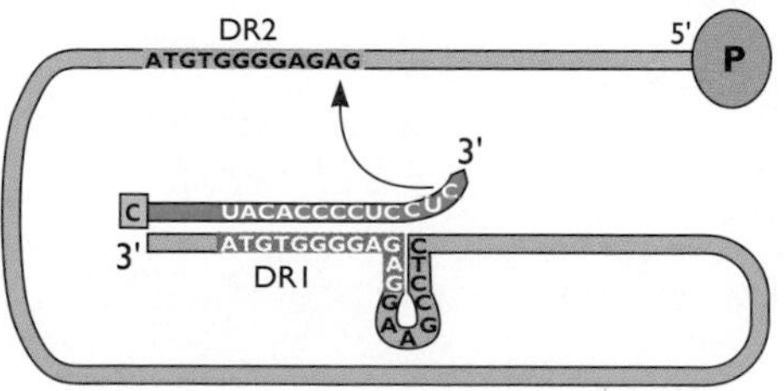

It is not clear what causes premature termination of the (+) strand of hepadnaviral DNA. Synthesis of (+) strand DNA is affected by mutations in C protein, consistent with the idea that this protein is somehow involved in this reaction. It has been proposed that DNA synthesis induces a change in the outer surface of the core, and that envelopment is regulated by interaction of the envelope proteins with the altered core. Once the cores are enveloped, DNA synthesis stops, presumably because deoxyribonucleoside triphosphate substrates can no longer enter the particle.

In mammalian hepadnaviruses, the (+) strand comprises only about half of the genome, but in avian hepadnaviruses, it is nearly full-length. However, even in avian viral DNA the capped RNA primer remains attached to the (–) strand. The primer is not removed by the P protein RNase H, as is the case during retroviral replication, and strand displacement synthesis, an essential activity of retroviral RT, is not observed with P protein (Table 7.1).

Perspectives

Comparison of Systems

The description of critical steps in hepadnaviral reverse transcription reveals interesting points of similarity and contrast with retroviral systems, as summarized in Table 7.1. Amino acid sequences and functions are conserved among retroviral RT and hepadnaviral P proteins, and both enzymes use terminal nucleic acid repeats to mediate template exchanges. However, the mechanisms by which their templates are reverse transcribed are quite distinct. Differences in the form and function of the final products of the two pathways are especially striking. A relaxed DNA circle with overlapping 5′ ends is an intermediate in the formation of the final product of retroviral reverse transcription, a linear duplex DNA. Repair of this relaxed-circle intermediate is an aberrant reaction, and the covalently closed circle formed is a dead-end product. In contrast, linear DNA is an aberrant product of hepadnaviral reverse transcription, and the covalently closed circle is the functional form.

The single-cell replication cycles of retroviruses and hepadnaviruses are in a sense permutations of one another. In comparing them and the unconventional foamy viruses (Box 7.4), it is also instructive to consider cauliflower mosaic virus, a plant retroid virus that seems to combine some features of both animal viruses during reverse transcription (Fig. 7.29). This plant virus has a relaxed circular DNA genome and directs synthesis of a covalently closed episomal form, but its reverse transcription and priming mechanisms are quite analogous to those of retroviruses and retrotransposons. On the other hand, as with hepadnaviruses, all, or remnants of, the RNA primers remain attached to the 5′ ends of cauliflower mosaic virus DNA. As suggested by the family tree of retroelements (Fig. 7.13), retroid viruses represent a continuum in evolution and remind us of the varied combinations of strategies that exist in nature for replicating viral genomes and related genetic elements.

Figure 7.29 Replication cycles of hepadnaviruses, cauliflower mosaic virus, and retroviruses. The double-stranded DNA circle found in cauliflower mosaic virus particles contains three interruptions. At each interruption there is a short 5′ overlap of DNA as if formed by strand displacement synthesis. Ribonucleotides are often found attached to the 5′ ends. The (–) strand starts with either a ribo- or a deoxyriboadenosine. The 5′ ends of the (+) strand each contain 8 to 10 purine-rich matches to the viral DNA at the same location, suggesting a primer function. r, short sequence at both ends of viral RNA; R, same sequence in DNA. Shaded boxes show the nucleic acid (genomes) encapsidated in the virions of each virus, which represent different components in analogous pathways. Updated from H. E. Varmus, *Nature* **304:**116–117, 1983, with permission.

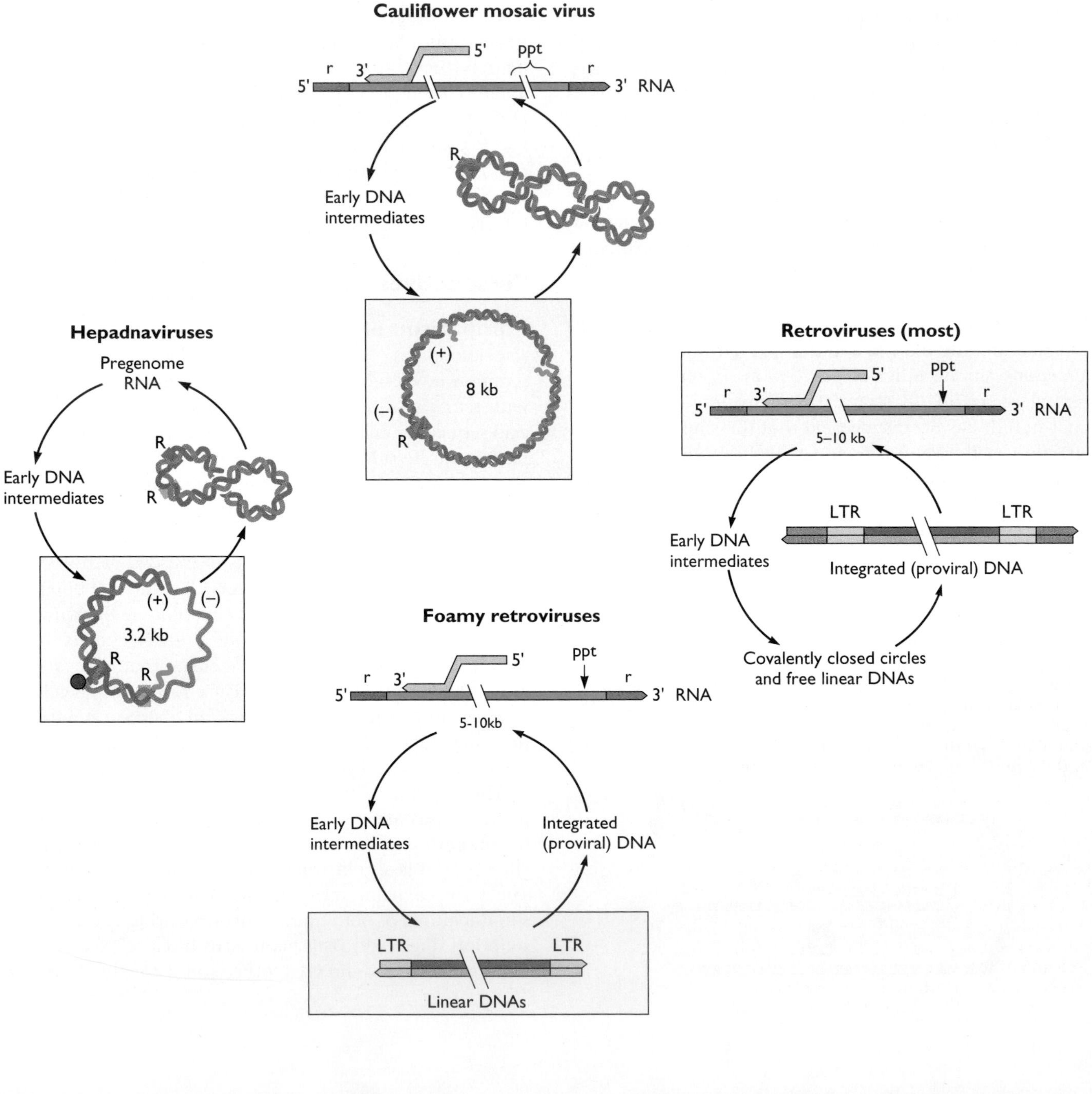

References

Books

Cooper, G. M., R. G. Temin, and W. Sugden (ed.). 1995. *The DNA Provirus: Howard Temin's Scientific Legacy.* ASM Press, Washington D.C.

Reviews

Arts, E. J., and S. F. Le Grice. 1998. Interaction of retroviral reverse transcriptase with template-primer duplexes during replication. *Prog. Nucleic Acid Res. Mol. Biol.* **58:**339–393.

Brown, P. O. 1997. Integration, p. 161–204. *In* J. M. Coffin, S. H. Hughes, and H. E. Varmus (ed.), *Retroviruses* Cold Spring Harbor Laboratory Press, Cold Spring Harbor, N.Y.

Ganem, D., and R. J. Schneider. 2001. *Hepadnaviridae*: the viruses and their replication, p. 2923–2969. *In* D. M. Knipe, P. M. Howley, and D. F. Griffin (ed.), *Fields Virology,* 4th ed. Lippincott Williams & Wilkins, Philadelphia, Pa.

Linial, M. L. 1999. Foamy viruses are unconventional retroviruses. *J. Virol.* **73:**1747–1755.

Seeger, C., and J. Hu. 1997. Why are hepadnaviruses DNA and not RNA viruses? *Trends Microbiol.* **5:**447–450.

Seeger, C., and W. S. Mason. 1996. Replication of the hepatitis virus genome, p. 815–832. *In* M. L. DePamphilis (ed.), *DNA Replication in Eukaryotic Cells* (Cold Spring Harbor Laboratory Press, Cold Spring Harbor, N.Y.

Skalka, A. M. (ed.). 1999. *Advances in Virus Research,* vol. 52. *Seminars in Virology: Retroviral DNA Integration.* Academic Press, Inc., New York, N.Y.

Skalka, A. M., and S. Goff (ed.). 1993. *Reverse Transcriptase.* Cold Spring Harbor Laboratory Press, Cold Spring Harbor, N.Y.

Steitz, T. A. 1993. DNA- and RNA-dependent DNA polymerases. *Curr. Opin. Struct. Biol.* **3:**31–38.

Telesnitsky, A., and S. P. Goff. 1997. Reverse transcriptase and the generation of retroviral DNA, p.121–160. *In* J. M. Coffin, S. H. Hughes, and H. E. Varmus (ed.), *Retroviruses.* Cold Spring Harbor Laboratory Press, Cold Spring Harbor, N.Y.

Yang, W., and T. A. Steitz. 1995. Recombining the structures of HIV integrase, RuvC and RNase H. *Structure* **3:**131–134.

Papers of Special Interest

Retroviral Reverse Transcription

Landmark papers

Baltimore, D. 1970. RNA-dependent DNA polymerase in virions of RNA tumour viruses. *Nature* **226:**1209–1211.

Temin, H. M., and S. Mizutani. 1970. RNA-dependent DNA polymerase in virions of Rous sarcoma virus. *Nature* **226:**1211–1213.

Recent work

Beerens, N., F. Groot, and B. Berkhout. 2001. Initiation of HIV-1 reverse transcription is regulated by a primer activation signal. *J. Biol. Chem.* **276:**31247–31256.

Davies, J. F., Jr., Z. Hostomska, Z. Hostomsky, S. R. Jordan, and D. A. Matthews. 1991. Crystal structure of the ribonuclease H domain of HIV-1 reverse transcriptase. *Science* **252:**88–95.

Hsiou, Y., J. Ding, K. Das, A. D. Clark, Jr., S. H. Hughes, and E. Arnold. 1996. Structure of unliganded HIV-1 reverse transcriptase at 2.7 Å resolution: implications of conformational changes for polymerization and inhibition mechanisms. *Structure* **4:**853–860.

Huang, H., R. Chopra, G. L. Verdine, and S. Harrison. 1998. Structure of a covalently trapped catalytic complex of HIV-1 reverse transcriptase: implications for drug resistance. *Science* **282:**1669–1675.

Jacobo-Molina, A., J. Ding, R. G. Nanni, A. D. Clark, Jr., X. Lu, C. Tantillo, R. L. Williams, G. Kamer, A. L. Ferris, P. Clark, A. Hizi, S. H. Hughes, and E. Arnold. 1993. Crystal structure of human immunodeficiency virus type 1 reverse transcriptase complexed with double-stranded DNA at 3.0 Å resolution shows bent DNA. *Proc. Natl. Acad. Sci. USA* **90:**6320–6324.

Kim, B., and L. Loeb. 1995. Human immunodeficiency virus reverse transcriptase substitutes for DNA polymerase I in *Escherichia coli*. *Proc. Natl. Acad. Sci. USA* **92:**684–688.

Kohlstaedt, L. A., J. Wang, J. M. Friedman, P. A. Rice, and T. A. Steitz. 1992. Crystal structure at 3.5Å resolution of HIV-1 reverse transcriptase complexed with an inhibitor. *Science* **256:**1783–1790.

Morris, S., M. Johnson, E. Stavnezer, and J. Leis. 2002. Replication of avian sarcoma virus in vivo requires an interaction between the viral RNA and the TPsiC loop of the $tRNA^{Trp}$ primer. *J. Virol.* **76:**7571–7577.

Patel, P. H., A. Jacobo-Molina, J. Ding, C. Tantillo, A. D. Clark, Jr., R. Raag, R. G. Nanni, S. H. Hughes, and E. Arnold. 1995. Insights into DNA polymerization mechanisms from structure and function analysis of HIV-1 reverse transcriptase. *Biochemistry* **34:**5351–5363.

Rodgers, D. W., S. J. Gamblin, B. A. Harris, S. Ray, J. S. Culp, B. Hellmig, D. J. Woolf, C. Debouck, and S. C. Harrison. 1995. The structure of unliganded reverse transcriptase from the human immunodeficiency virus type 1. *Proc. Natl. Acad. Sci. USA* **92:**1222–1226.

Sarafianos, S. G., K. Das, C. Tantillo, A. D. Clark, Jr., J. Ding, J. M. Whitcomb, P. L. Boyer, S. H. Hughes, and E. Arnold. 2001. Crystal structure of HIV-1 reverse transcriptase in complex with a polypurine tract RNA:DNA. *EMBO J.* **20:**1449–1461.

Topping, R., M.-A. Demoitie, N. H. Shin, and A. Telesnitsky. 1998. *cis*-acting elements required for strong stop acceptor template selection during Moloney murine leukemia virus reverse transcription. *J. Mol. Biol.* **281:**1–15.

Retroviral DNA Integration

Landmark papers

Bujacz, G., M. Jaskólski, J. Alexandratos, A. Wlodawer, G. Merkel, R. A. Katz, and A. M. Skalka. 1995. High-resolution structure of the catalytic domain of the avian sarcoma virus integrase. *J. Mol. Biol.* **253:**333–346.

Craigie, R., T. Fujiwara, and F. Bushman. 1990. The IN protein of Moloney murine leukemia virus processes in the viral DNA ends and accomplishes their integration in vitro. *Cell* **62:**829–837.

Dyda, F., A. B. Hickman, T. M. Jenkins, A. Engelman, R. Craigie, and D. R. Davies. 1994. Crystal structure of the catalytic domain of HIV-1 integrase: similarity to other polynucleotidyl transferases. *Science* **266:**1981–1986.

Katz, R. A., G. Merkel, J. Kulkosky, J. Leis, and A. M. Skalka. 1990. The avian retroviral IN protein is both necessary and sufficient for integrative recombination in vitro. *Cell* **63:**87–95.

Murphy, J. E., T. De Los Santos, and S. P. Goff. 1993. Mutational analysis of the sequences at the termini of the Moloney murine leukemia virus DNA required for integration. *Virology* **195:**432–440.

Recent work

Aiyar, A., P. Hindmarsh, A. M. Skalka, and J. Leis. 1996. Concerted integration of linear retroviral DNA by the avian sarcoma virus integrase in vitro: dependence on both long terminal repeat termini. *J. Virol.* **70:**3571–3580.

Bao, K. K., H. Wang, J. K. Miller, D. A. Erie, A. M. Skalka, and I. Wong. 2003. Functional oligomeric state of avian sarcoma virus integrase. *J. Biol. Chem.* **278:**1323–1327.

Eijkelenboom, A.P.A.M., F. M. I. van den Ent, R. Wechselberger, R. H. A. Plasterk, R. Kaptein, and R. Boelens.

2000. Refined solution structure of the dimeric N-terminal HHCC domain of HIV-2 integrase. *J. Biomol. NMR* **18:**119–128.

Eijkelenboom, A.P.A.M., R. A. Puras Lutzke, R. Boelems, R. H. A. Plasterk, R. Kaptein, and K. Hård. 1995. The DNA-binding domain of HIV-1 integrase has an SH3-like fold. *Nat. Struct. Biol.* **2:**807–810.

Farnet, C. M., and F. D. Bushman. 1997. HIV-1 cDNA integration: requirement of HMG I(Y) protein for function of preintegration complexes in vitro. *Cell* **88:**483–492.

Heuer, T. S., and P. O. Brown. 1998. Cross-linking studies suggest a model for the architecture of an active human immunodeficiency virus type 1 integrase-DNA complex. *Biochemistry* **37:**6667–6678.

Jones, J. S., R. W. Allan, and H. W. Temin. 1994. One retroviral RNA is sufficient for synthesis of viral DNA. *J. Virol.* **68:**207–216.

Jordan A., P. Defechereux, and E. Verdin. 2001. The site of HIV-1 integration in the human genome determines basal transcriptional activity and response to Tat transactivation. *EMBO J.* **20:**1726–1738.

Lai, L., H. Liu, X. Wu, and J. C. Kappes. 2001. Moloney murine leukemia virus integrase protein augments viral DNA synthesis in infected cells. *J. Virol.* **75:**11365–11372.

Lee, M. S., and R. Craigie. 1998. A previously unidentified host protein protects retroviral DNA from autointegration. *Proc. Natl. Acad. Sci. USA* **95:**1528–1533.

Lodi, P. J., J. A. Ernst, J. Kuszewski, A. B. Hickman, A. Engelman, R. Craigie, G. M. Clore, and A. M. Gronenborn. 1995. Solution structure of the DNA binding domain of HIV-1 integrase. *Biochemistry* **34:**9826–9833.

Schröder, A. R. W., P. Schinn, H. Chen, C. Berry, J. R. Ecker, and F. Bushman. 2002. HIV-1 integration in the human genome favors active genes and local hotspots. *Cell* **110:**521–529.

Sinha, S., M. H. Pursley, and D. P. Grandgenett. 2002. Efficient concerted integration by recombinant human immunodeficiency virus type 1 integrase without cellular or viral cofactors. *J. Virol.* **76:**3105–3113.

Turelli, P., V. Doucas, E. Craig, B. Mangeat, N. Klages, R. Evans, G. Kalpana, and D. Trono. 2001. Cytoplasmic recruitment of INI1 and PML on incoming HIV preintegration complexes: interference with early steps of viral replication. *Mol. Cell* **7:**1245–1254.

Wang, J.-Y., H. Ling, W. Yang, and R. Craigie. 2001. Structure of a two-domain fragment of HIV-1 integrase: implications for domain organization in the intact protein. *EMBO J.* **20:**7333–7343.

Wei, S. Q., K. Mizuuchi, and R. Craigie. 1998. Footprints on viral DNA ends in Moloney murine leukemia virus preintegration complexes reflect a specific association with integrase. *Proc. Natl. Acad. Sci. USA* **95:**10535–10540.

Withers-Ward, E. S., Y. Kitamura, J. P. Barnes, and J. M. Coffin. 1994. Distribution of targets for avian retrovirus DNA integration in vivo. *Genes Dev.* **8:**1473–1487.

Hepadnaviral Reverse Transcription

Landmark papers

Summers, J., and W. S. Mason. 1982. Replication of the genome of a hepatitis B-like virus by reverse transcription of an RNA intermediate. *Cell* **29:**403–415.

Recent work

Habig, J. W., and D. D. Loeb. 2002. Small DNA hairpin negatively regulates in situ priming during duck hepatitis B virus reverse transcription. *J. Virol.* **76:**980–989.

Hu, J., D. O. Toft, and C. Seeger. 1997. Hepadnavirus assembly and reverse transcription require a multi-component chaperone complex which is incorporated into nucleocapsids. *EMBO J.* **16:**59–68.

Mueller-Hill, K., and D. D. Loeb. 2002. *cis*-acting sequences 5E, M, and 3E interact to contribute to primer translocation and circularization during reverse transcription of avian hepadnavirus DNA. *J. Virol.* **76:**4260–4266.

Wang, G.-H., and C. Seeger. 1992. The reverse transcriptase of hepatitis B virus acts as a protein primer for viral DNA synthesis. *Cell* **71:**663–670.

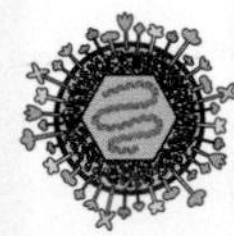
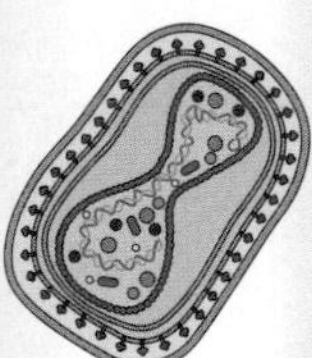

8

Transcription Strategies: DNA Templates

It is possible that nature invented DNA for the purpose of achieving regulation at the transcriptional rather than at the translational level. The level of control in small DNA viruses and large RNA viruses should be interesting to study.

A. CAMPBELL
1967

Introduction

During replication of viruses with DNA genomes, synthesis of viral messenger RNA (mRNA) must precede production of proteins. In cells infected by the majority of DNA viruses, this step is accomplished by the cellular enzyme that produces cellular mRNA, RNA polymerase II. The proviral DNA of retroviruses is also transcribed by this cellular enzyme. The signals that control expression of the genes of these viruses are therefore closely similar to those of cellular genes. In fact, much of our current understanding of the mechanisms of cellular transcription stems from study of the transcription of viral DNA templates. In contrast, the genomes of poxviruses are transcribed by a viral DNA-dependent RNA polymerase and accessory viral proteins, which control the recognition of viral promoters. However, this strategy is an exception, not the rule.

A hallmark of cells infected by DNA viruses is the expression of viral genes in a strictly defined, reproducible sequence. In general, viral enzymes and regulatory proteins are made during the initial period of infection, whereas structural proteins from which virus particles are built are made only after viral DNA synthesis begins. Such orderly expression of viral genes is primarily the result of transcriptional regulation by viral proteins. This pattern of gene expression is quite different from the continual expression of all viral genes characteristic of the infectious cycles of many RNA viruses (Chapter 6). The molecular strategies that ensure sequential transcription of the genes of these DNA viruses therefore lay the foundations for the infectious cycle. As discussed in this chapter, the elucidation of these strategies, one of the major goals of studies of DNA viruses, has identified a number of common mechanisms executed in virus-specific fashion. An unanticipated dividend has been insights into the cellular mechanisms that control progression through the cell cycle.

Properties of Cellular Transcription Systems

Cellular RNA Polymerases That Transcribe Viral DNA

The Three Transcriptional Systems of Eukaryotes

A general feature of eukaryotic cells is the division of transcriptional labor among three DNA-dependent RNA polymerases: these enzymes, designated RNA polymerases I, II, and III, synthesize different kinds of cellular RNA (Table 8.1). RNA polymerase II makes mRNA species or their precursors, while RNA polymerases I and III produce stable RNAs, such as ribosomal RNAs (rRNAs) and transfer RNAs (tRNAs), respectively. Synthesis of the latter "housekeeping" RNAs must be adjusted to match the rates of cell growth and division. But regulation of mRNA synthesis is crucial for orderly development and differentiation in eukaryotes, as well as for the responses of cells to their environment. The evolution of RNA polymerases with distinct transcriptional responsibilities appears to be a device for maximizing opportunities for regulation of mRNA synthesis, while maintaining a constant and abundant supply of the RNA species essential for the metabolism of all cells.

The three eukaryotic RNA polymerases transcribe different classes of genes and must meet different regulatory requirements. Nevertheless, several of the 12 to 16 subunits of these large enzymes are identical, and others are related in sequence. Such conservation of sequence, which extends to the proteins of prokaryotic and viral DNA-dependent RNA polymerases, can be attributed to the common biochemical capabilities of the enzymes. These activities include binding of ribonucleoside triphosphate substrates, binding to template DNA and to product RNA, and catalysis of phosphodiester bond formation. The recently described structure of a 10-subunit form of yeast RNA polymerase II revealed that the organization of its active centers is similar to that of smaller DNA-dependent RNA polymerases, as well as of enzymes that make DNA from DNA or RNA templates (Fig. 6.2).

Table 8.1 Eukaryotic RNA polymerases synthesize different classes of cellular and viral RNA

Enzyme	RNAs synthesized[a]	
	Cellular	Viral
RNA polymerase I	Pre-rRNA	None known
RNA polymerase II	Pre-mRNA, snRNAs	Pre-mRNA, mRNA
RNA polymerase III	Pre-tRNAs, 5S rRNA, U6 snRNA	Ad2 VA-RNAs, EBV EBER RNAs

[a]rRNA, ribosomal RNA; snRNA, small nuclear RNA; tRNA, transfer RNA; Ad2, adenovirus type 2; EBV, Epstein-Barr virus.

Transcription of cellular and viral genes by either cellular or viral RNA polymerases requires not only template-directed synthesis of RNA, but also correct interpretation of DNA punctuation signals that mark the sites at which transcription must start and stop. The point at which copying of the DNA should begin, the initiation site, must be identified by the transcriptional machinery. Such recognition together with synthesis of the first few phosphodiester bonds of RNA completes initiation. During the **elongation** phase, nucleotides are added rapidly to the 3′ end of the nascent RNA as the transcriptional machinery reads the sequence of a gene. When termination sites are encountered in the DNA, both the RNA product and the RNA polymerase are released from the template. Purified RNA polymerases I, II, and III perform the elongation reactions in vitro but, despite their complexity, are incapable of specific **initiation of transcription** without the assistance of additional, polymerase-specific proteins. This requirement for additional proteins during initiation is a universal property of DNA-dependent RNA polymerases, including those encoded in viral genomes.

Transcription of Viral Templates by Cellular RNA Polymerases II and III

Because prokaryotic RNA polymerases transcribe many bacteriophage genes, the transcription of DNA viral genomes in eukaryotic cells by RNA polymerase II was not a surprise. In contrast, the discovery that some of these viral DNAs also contain transcription units recognized by RNA polymerase III was not anticipated. This phenomenon was initially observed in human cells infected by adenovirus, but RNA polymerase III transcription units are also present in the genomes of other viruses. The cellular transcriptional machinery can also contribute to viral reproduction in unique ways. The acquisition of 5′ caps and primers for viral mRNA synthesis from cellular pre-mRNAs or mRNAs synthesized by RNA polymerase II in cells infected by certain (–) strand RNA viruses (Chapter 6) is a dramatic example of such virus-specific mechanisms.

Because the host cells' RNA polymerase II transcriptional machinery is responsible for transcription of many viral DNA templates, this cellular transcription system is described in some detail below. We then consider the regulatory mechanisms that ensure appropriate expression of genetic information of DNA viruses and retroviruses by this cellular system.

Production of Viral DNA Templates Transcribed by Cellular RNA Polymerases

All viral DNA molecules that are transcribed by cellular RNA polymerases must enter the infected cell nucleus. However, there is considerable variation in the reactions needed to produce templates that can be recognized by the cellular transcriptional machinery from genomic nucleic acids (Table 8.2). Some of these viral genomes, like that of the host cell, are double-stranded DNA molecules that can be transcribed as soon as they reach the nucleus. In the case of papillomaviruses and polyomaviruses, the resemblance to cellular templates for transcription is even greater, because the circular viral genomes enter cells as "minichromosomes," in which the DNA is packaged by cellular nucleosomes. Transcription of specific genes is therefore the first biosynthetic reaction in cells infected by adenoviruses, herpesviruses, papillomaviruses, and polyomaviruses. Although viral DNA synthesis is not required prior to transcription, this process dramatically increases the concentration of templates and, as we shall see, is crucial for transcription of specific viral genes.

Other viral DNA genomes must be converted from the form in which they enter the cell to double-stranded molecules that serve as transcriptional templates (Table 8.2). The hepadnaviral genome, an incomplete circular DNA molecule with a large gap in one strand (Appendix A, Fig. 3B) enters the nucleus, where it is repaired (probably by cellular enzymes) to form a fully double-stranded DNA molecule. During the initial period of the hepadnaviral infectious cycle, the supply of templates is supplemented, when pregenomic RNA transcribed by RNA polymerase II is copied into double-stranded DNA by the viral DNA polymerase. The prerequisites for expression of retroviral genetic information are even more demanding, for the (+) strand RNA genome must be both converted into viral DNA and integrated into the cellular genome, following import into the nucleus. Reverse transcription creates both an appropriate double-stranded DNA template and the signals needed for its recognition by components of the cellular transcriptional machinery (Chapter 7). Once the viral DNA becomes integrated into cellular DNA, it is indistinguishable from cellular genes in all its properties, including intranuclear location and organization in chromatin.

Transcription by RNA Polymerase II

Accurate initiation of transcription by RNA polymerase II is directed by specific DNA sequences located near the site of initiation, called the **promoter** (Fig. 8.1). The promoter and the set of DNA sequences that control transcription make up the **transcriptional control region** (Fig. 8.1). Transcriptional control regions of DNA viruses and retroviruses were among the first to be examined experimentally. For example, the human adenovirus type 2 major late promoter was the first from which accurate initiation of transcription was reconstituted in vitro (Box 8.1). Subsequently, the study of viral transcription yielded fundamental information about control signals and the mechanisms by which RNA polymerase II transcription is initiated and regulated.

Biochemical studies of model transcriptional control regions, such as the adenoviral major late promoter, established that initiation of transcription by RNA polymerase II is a multistep process. The initiation reactions include promoter recognition, formation of an open initiation complex in which the two strands of the DNA template in the

Table 8.2 Prerequisites for transcription of viral genes by RNA polymerase II

Requirement	Virus	Form of the intracellular template
Entry of genome into nucleus	Adenovirus	
	Human adenovirus type 2	Linear double-stranded DNA associated with viral protein VII
	Polyomaviruses	
	Simian virus 40	Closed, circular, double-stranded DNA organized by cellular nucleosomes
Entry of viral genome plus a virion protein (e.g., herpesviral VP16) into the nucleus	Herpesviruses	
	Herpes simplex virus type 1	Double-stranded DNA
Production of double-stranded DNA from genomic DNA and entry into nucleus	Hepadnaviruses	
	Hepatitis B virus	Closed circular DNA synthesized from genomic molecules with a gap in the (+) strand and a nick in the (–) strand
	Parvoviruses	
	Adeno-associated virus	Linear double-strand DNA synthesized from single-stranded DNA genome during replication
Reverse transcription, entry into nucleus and integration	Retroviruses	
	Avian sarcoma virus	Linear double-stranded DNA integrated into the cellular genome, produced by action of the viral reverse transcriptase and integrase

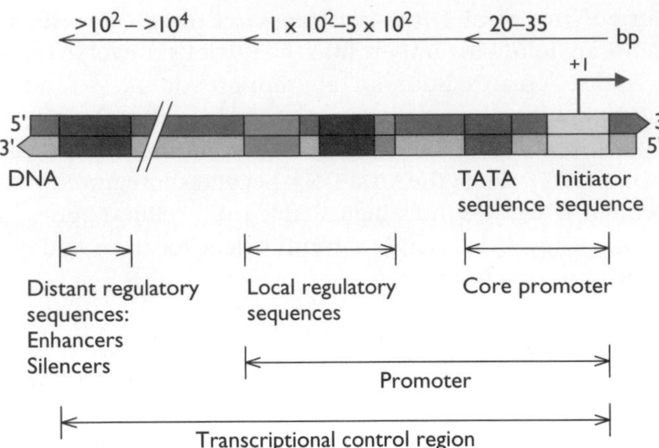

Figure 8.1 RNA polymerase II transcriptional control elements. The core promoter comprises the minimal sequence necessary to specify accurate initiation of transcription. The site of initiation is represented by the jointed red arrow drawn in the direction of transcription on the nontranscribed DNA strand, a convention used throughout this chapter. The TATA sequence is the binding site for TfIId (Box 8.3) and conforms to a well-conserved consensus sequence. The core promoter often also contains a sequence sufficient to specify initiation at a unique site. Such autonomous initiator sequences are more variable than TATA sequences, do not conform to a single consensus, and are recognized by a variety of different proteins. Although TATA sequences are considered a characteristic feature of RNA polymerase II promoters, they are absent from many promoters (e.g., Fig. 8.4). The activity of the core promoter is modulated by the local regulatory sequences found within a few hundred base pairs of the initiation site. A constellation of local regulatory sequences upstream of the TATA sequence, as shown, is a common arrangement. However, such sequences can also lie within transcribed sequences (e.g., Fig. 8.4). Distant regulatory sequences that stimulate (enhancers) or repress (silencers) transcription are present in a large number of RNA polymerase II transcriptional control regions. Such sequences can function over distances of greater than 10,000 bp in the DNA and when located upstream of the promoter, within an intron, or downstream of the gene. Enhancers play critical roles in regulation of transcription during development and differentiation or in response to environmental stimuli. Both local regulatory and enhancer sequences are binding sites for sequence-specific DNA-binding proteins.

BOX 8.1

Mapping of a human adenovirus type 2 initiation site and accurate transcription in vitro

When cellular RNA polymerase II was identified in 1969, investigators had access to only preparations of total cellular DNA. Furthermore, nothing was known about the organization of eukaryotic transcription units. The genomes of smaller DNA viruses such as simian virus 40 and human adenovirus type 2 were therefore perceived as a valuable resource for investigation of mechanisms of transcription, a view that inspired many investigators to enter virology. Indeed, it was detailed information about a particular adenoviral transcription unit that finally allowed biochemical studies of the mechanism of initiation by this enzyme. In 1978, the site at which major late transcription begins was mapped precisely, by determining the sequence of the 5′ end of the RNA transcript. This knowledge was exploited to develop a simple assay for accurate initiation of transcription, the "runoff" assay shown in Fig. 8.3. Purified RNA polymerase II produced no specific transcripts in this assay, but unfractionated nuclear extracts of human cells were shown to contain all the components necessary for accurate initiation of transcription from the adenoviral major late promoter.

Weil, P. A., D. S. Luse, J. Segall, and R. G. Roeder. 1979. Selective and accurate initiation of transcription at the Ad2 major late promotor in a soluble system dependent on purified RNA polymerase II and DNA. *Cell* **18:**469–484.

Ziff, E. B., and R. M. Evans. 1978. Coincidence of the promoter and capped 5′ terminus of RNA from the adenovirus 2 major late transcription unit. *Cell* **15:**1463–1475.

vicinity of the initiation site are unwound, and promoter clearance, movement of the transcribing complex away from the promoter (Fig. 8.2). At least 40 proteins, which comprise RNA polymerase II itself and auxiliary initiation proteins, are needed to complete the intricate process of initiation. Our understanding of the functions of these proteins and of the DNA sequences that control initiation of transcription is based largely on in vitro systems or simple assays for detecting gene expression within cells (Fig. 8.3). Application of these methods has identified a very large number of transcriptional control sequences. Fortunately, all of them can be assigned to one of the three functionally distinct regions identified in Fig. 8.1.

Core promoters of viral and cellular genes contain all the information necessary for recognition of the site of initiation. They direct assembly of an RNA polymerase II-con-

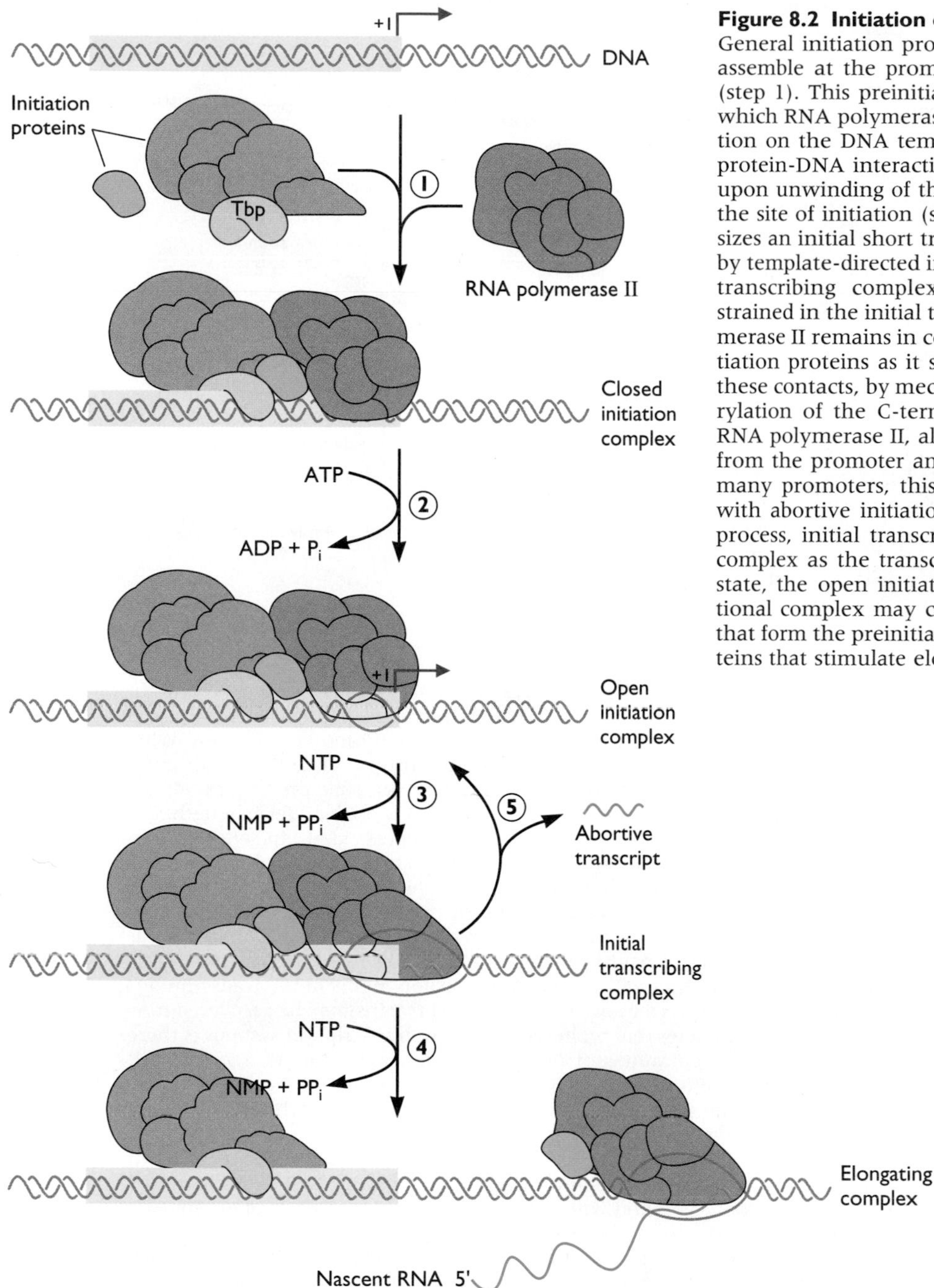

Figure 8.2 Initiation of transcription by RNA polymerase II. General initiation proteins (Box 8.3) and RNA polymerase II assemble at the promoter into the closed initiation complex (step 1). This preinitiation complex is a very stable entity, in which RNA polymerase II is firmly anchored at a specific position on the DNA template by numerous protein-protein and protein-DNA interactions. The open initiation complex forms upon unwinding of the DNA template in the region spanning the site of initiation (step 2). RNA polymerase II then synthesizes an initial short transcript (less than 10 to 15 nucleotides) by template-directed incorporation of nucleotides (step 3). The transcribing complex is believed to be conformationally strained in the initial transcribing complex, because RNA polymerase II remains in contact with promoter-bound general initiation proteins as it synthesizes short RNAs. The severing of these contacts, by mechanisms that probably include phosphorylation of the C-terminal domain of the largest subunit of RNA polymerase II, allows the transcribing complex to escape from the promoter and proceed with elongation (step 4). On many promoters, this promoter clearance step is inefficient, with abortive initiation (step 5) predominating. In the latter process, initial transcripts are cleaved and released from the complex as the transcribing machinery returns to its ground state, the open initiation complex. The elongating transcriptional complex may contain some but not all of the proteins that form the preinitiation complex, as well as a variety of proteins that stimulate elongation.

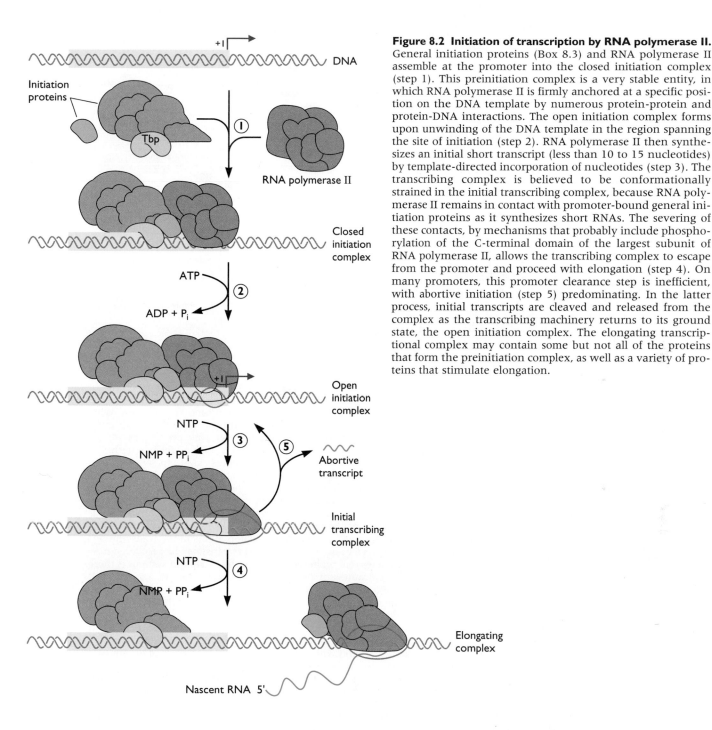

taining complex competent to begin transcription when supplied with ribonucleoside triphosphates. Such **preinitiation complexes** (Fig. 8.2) contain a common set of general initiation proteins and RNA polymerase II in an assembly precisely organized to support initiation at the appropriate site(s). A hallmark of core RNA polymerase II promoters is the presence of a TA-rich TATA sequence 20 to 35 bp upstream of the site of initiation (Fig. 8.1 and 8.4A and B). Short sequences, termed **initiators,** which specify accurate (but inefficient) initiation of transcription in the absence of any other promoter sequences, are also commonly found (Fig. 8.1 and 8.4). Either an initiator or an ap-

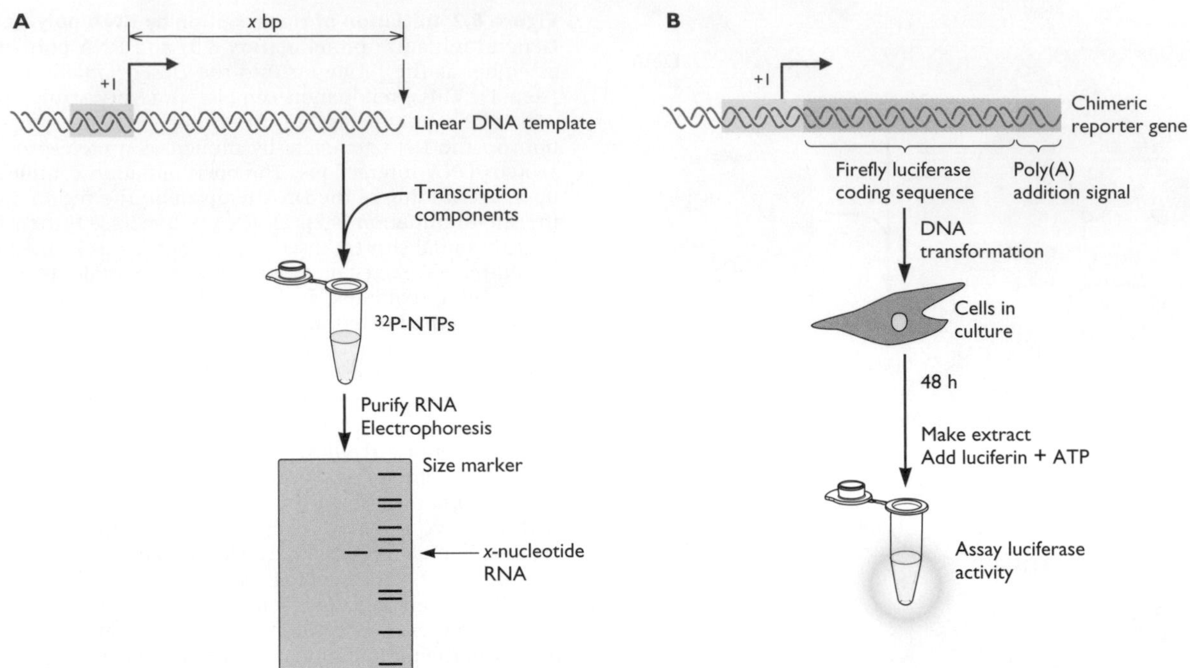

Figure 8.3 Assays for the activity of RNA polymerase II promoters. (A) In vitro transcription assay. In this simple assay, linear DNA templates are prepared by restriction endonuclease cleavage (black arrow at right end of plasmid DNAs), a known distance, *x* bp, downstream of position +1 of the promoter of interest. When the template is incubated with the transcriptional machinery and nucleoside triphosphate (NTP) substrates, transcription initiated at position +1 continues until the transcribing complex "runs off" the linear template. Specific transcription is therefore assayed as the production of ^{32}P-labeled RNA *x* nucleotides in length. This runoff transcription assay is convenient and can be used to assess both specificity and efficiency of transcription. However, it can be applied only to linear DNA templates and often suffers from high background, for example, "end-to-end" transcription of the linear DNA template. These problems can be circumvented by the use of circular templates and an indirect assay for specific transcripts. Such in vitro transcription assays differ from transcription under normal intracellular conditions in several important ways. In in vitro assays, transcription templates are typically provided as naked circular or linear DNA molecules, physical states that may not resemble those of cellular or viral genes within cells. When components of the transcriptional machinery are extracted from cells, some of the enzymes or other essential proteins may be recovered more efficiently than others. Another important parameter that is altered in in vitro transcription systems is therefore the relative, as well as the absolute, concentrations of the proteins necessary for transcription. (B) Transient-expression assay. For transient-expression assays, a segment of DNA containing the transcriptional control region of interest is ligated to the coding sequence (orange) of an enzyme not synthesized in the recipient cells to be used, and RNA processing signals, such as those specifying polyadenylation, shown by the green box at the end of the coding sequence. In this example, the firefly luciferase coding sequence is shown as the reporter gene. Plasmids containing such chimeric reporter genes are introduced into cells in culture by any one of several methods, including coprecipitation with calcium phosphate and electroporation. The proportion of cells that will take up foreign DNA varies with a number of parameters, including cell type, but is generally low: uptake of DNA by 10% of cells is considered excellent in this kind of assay. Within a cell that does take up the reporter gene, the DNA enters the nucleus, where the transcriptional control region directs transcription of chimeric RNA. The RNA is exported from the nucleus following processing and translated on cytoplasmic polyribosomes. The activity of the luciferase enzyme is then assayed, generally 48 h after introduction of the reporter gene, as a measure of the activity of the transcriptional control region. Note that this indirect measure of transcription assumes that it is only the activity of the transcriptional control region that determines the intracellular level of the enzyme whose activity is measured. Alternatively, the concentration of the chimeric reporter RNA can be measured.

propriately positioned TATA element can be sufficient to establish specific initiation, but both may be present (Fig. 8.4).

In vitro studies of formation of preinitiation complexes on core promoters elucidated a stepwise assembly pathway in which initiation proteins (nomenclature explained in Box 8.2) and RNA polymerase II enter the assembling complex in a defined sequence (Box 8.3). In the cell, many of the interactions among components of the transcriptional machinery take place before a promoter is encountered: RNA polymerase II is present in extremely large assemblies

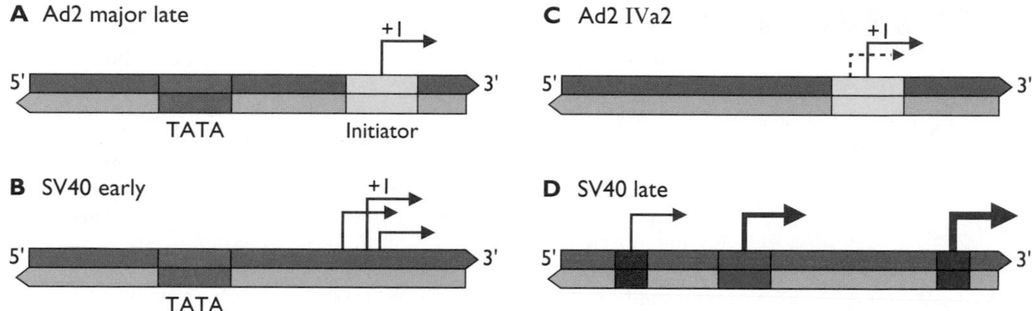

Figure 8.4 Variations in core RNA polymerase II promoter architecture. Some of the variability in the DNA sequence elements of RNA polymerase II core promoters is illustrated with four viral promoters represented as in Fig. 8.1. The TATA or initiator sequences of the different promoters are not identical in DNA sequence. In the case of the simian virus 40 (SV40) late transcription unit (D), each of the several sites of initiation is included within a DNA sequence resembling an initiator of another promoter. It has not been shown experimentally that all actually function as autonomous initiator sequences. The relative frequency with which different initiation sites are used in a single promoter is indicated by the thickness of the arrow. Ad2, adenovirus type 2.

BOX 8.2

Nomenclature of general initiation proteins

The idealized profile of elution of human (HeLa cell) general initiation proteins from a phosphocellulose column illustrates the origin of the nomenclature of these proteins. The fractions essential for transcription by purified RNA polymerase II in vitro were designated A to D in order of their elution, and the initiation proteins present in them were termed transcription factor (Tf) IIa, IIb, and so on. The fraction designated C in the original experiments proved not to be an essential component. Further chromatography showed that some fractions contained more than one essential protein, as summarized in the figure. Homologs of all the general initiation proteins shown are present in the yeast *Saccharomyces cerevisiae*. They are therefore considered universal components of eukaryotic cells.

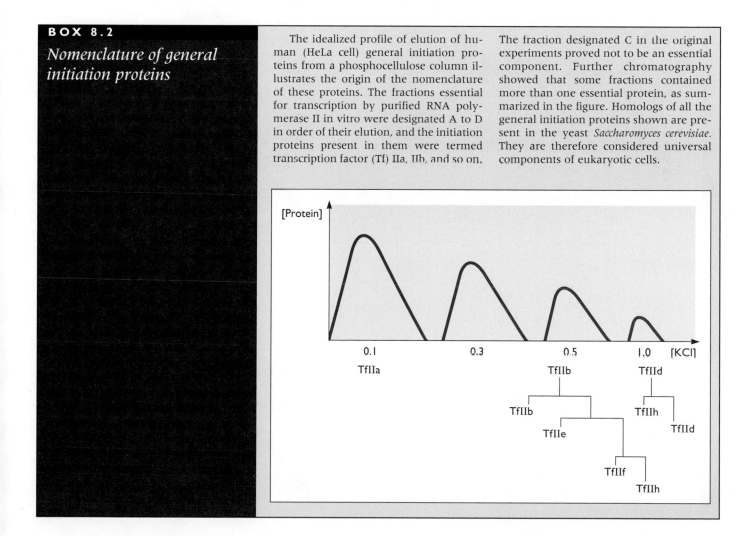

[Protein]

0.1 0.3 0.5 1.0 [KCl]

TfIIa TfIIb TfIId

TfIIb TfIIh TfIId

TfIIe

TfIIf TfIIh

BOX 8.3

Stepwise assembly of the RNA polymerase II closed initiation complex

The assembly of general initiation proteins and RNA polymerase II into the closed initiation complex (Fig. 8.2) is shown for a promoter that contains both a TATA and an initiator sequence (e.g., the adenovirus major late promoter). The first protein to bind, TfIId (step 1), contains a subunit that recognizes the TATA sequence (TATA-binding protein [Tbp]) and 8 to 10 additional subunits, termed Tbp-associated proteins (Tafs). X-ray crystal structures of DNA-bound Tbp, such as that of *Arabidopsis thaliana* Tbp bound to the adenoviral major late TATA sequence shown in the inset, revealed that this protein induces sharp bending of the DNA. One popular hypothesis is that such bending facilitates interaction among proteins bound to local regulatory sequences located upstream of the TATA sequence and the basal transcriptional machinery. TfIId is required for transcription from all RNA polymerase II promoters. It can recognize those that lack TATA sequences (e.g., Fig. 8.4) by binding of a Taf to an initiator or an internal sequence, or it can be recruited to the promoter by interactions with proteins bound to specific sequences

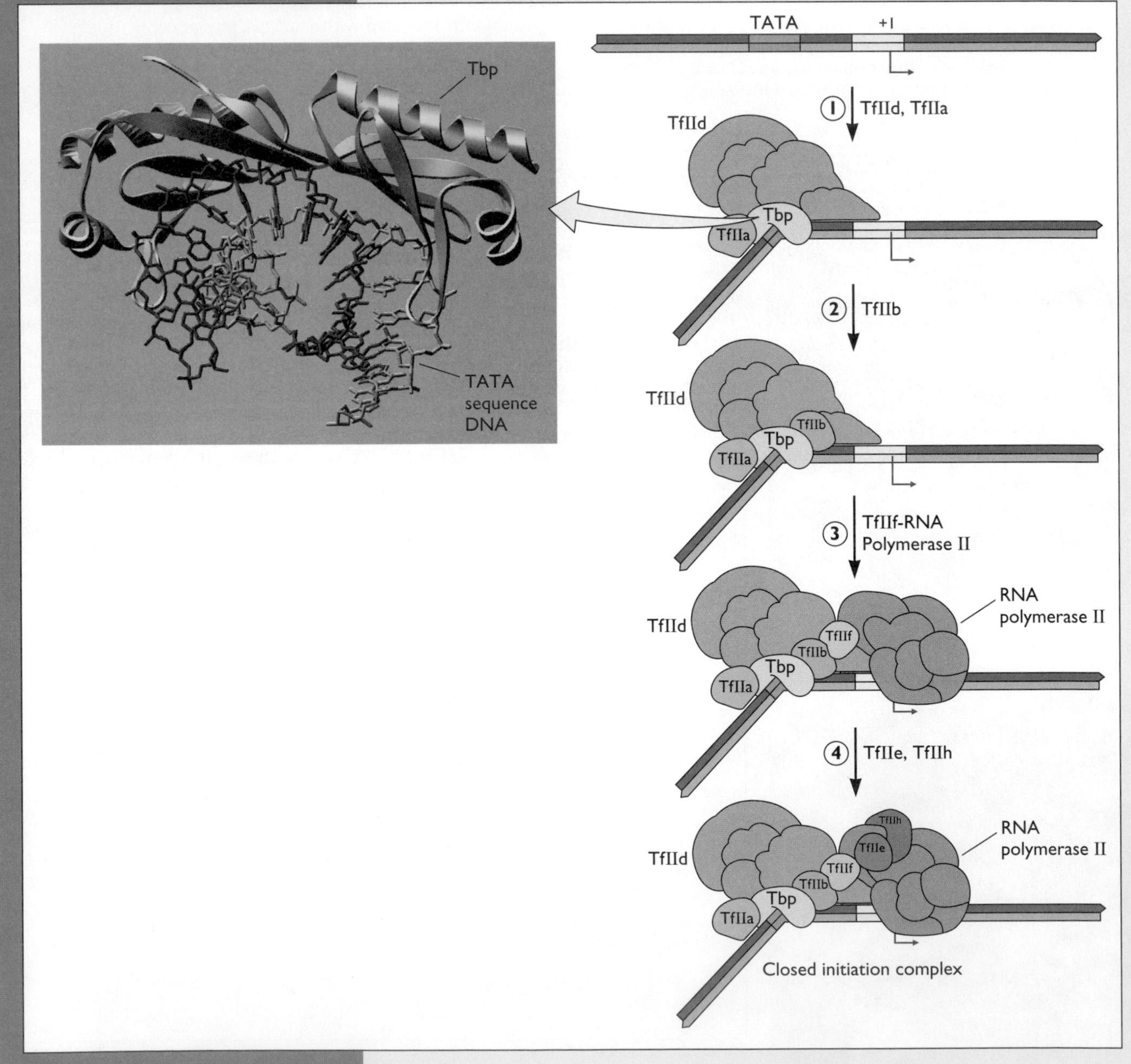

BOX 8.3 (continued)

near the initiation site. The TfIIa protein is shown entering the complex at step 1, because it can facilitate binding of Tbp to DNA and bind directly to one of the Tbp-associated proteins present in TfIId. Binding of TfIIb to the TfIId-promoter complex (step 2) is necessary for delivery of RNA polymerase II by TfIIf. TfIIb binds directly to Tbp, RNA polymerase II, and TfIIf, and TfIIf also interacts with the enzyme. The complex containing TfIId, TfIIb, and TfIIf RNA polymerase II is competent to initiate transcription when the DNA template is supercoiled, a state that can be considered energetically activated. Such a complex therefore contains the minimal set of components necessary for promoter recognition. The assembly process is completed by the sequential association of TfIIe and TfIIh. Protein components of TfIIe and TfIIh interact with one another and with RNA polymerase II. TfIIh contains at least nine different proteins. Its largest subunit, which also plays an important role in nucleotide excision repair of damaged DNA, supplies DNA-dependent ATPase and helicase activities essential for transcription. Two others are components of a kinase that can phosphorylate the C-terminal segment of the largest subunit of RNA polymerase II. Although assembly of initiation complexes within cells is probably via a much-truncated pathway (see text), elucidation of this in vitro pathway identified protein-DNA and protein-protein interactions important in the assembly and stabilization of the closed initiation complex, as well as reactions that can be regulated.

Buratowski, S., S. Hahn, L. Guarente, and P. A. Sharp. 1989. Five intermediate complexes in transcription initiation by RNA polymerase II. *Cell* **56:**549–561.

Kim, J. L., D. B. Nikolov, and S. K. Burley. 1993. Co-crystal structure of TBP recognizing the minor groove of a TATA element. *Nature* **365:**520–527.

that contain initiation proteins, as well as other proteins that are essential for transcription or that participate in its regulation. Such assemblies, termed **holoenzymes**, appear to be poised to initiate transcription as soon as they are recruited to a promoter, for example, by binding of one or more constituent proteins to DNA-bound TfIId (Box 8.3).

Regulation of RNA Polymerase II Transcription

Patterns of transcriptional regulation. Numerous patterns of gene expression are necessary for eukaryotic life: some RNA polymerase II transcription units must be expressed in all cells, whereas others are transcribed only during specific developmental stages or in specialized differentiated cells. Many others must be maintained in a silent ground state from which they can be activated rapidly in response to specific stimuli, and to which they can be returned readily. Transcription of viral genes is also regulated during the single-cell life cycles of most of the viruses considered in this chapter. Large quantities of viral proteins for assembly of progeny virions must be made within a finite (and often short) infectious cycle. Consequently, some viral genes must be transcribed at high rates. As discussed previously, the genes comprising the majority of viral DNA templates are transcribed in a specific temporal sequence. Such regulated transcription is achieved in part by means of cellular control mechanisms, for example, signal transduction cascades that transmit specific environmental stimuli to the transcriptional machinery, or cellular proteins that repress transcription. However, viral proteins are generally critical components of the circuits that establish orderly transcription of viral genes.

Local and distant regulatory sequences and their recognition. Both local and distant sequences (Fig. 8.1) can control transcription from core promoters. However, in many cases, local sequences are sufficient for proper transcriptional regulation. These local regulatory sequences are recognized by sequence-specific DNA-binding proteins (Fig. 8.5), a

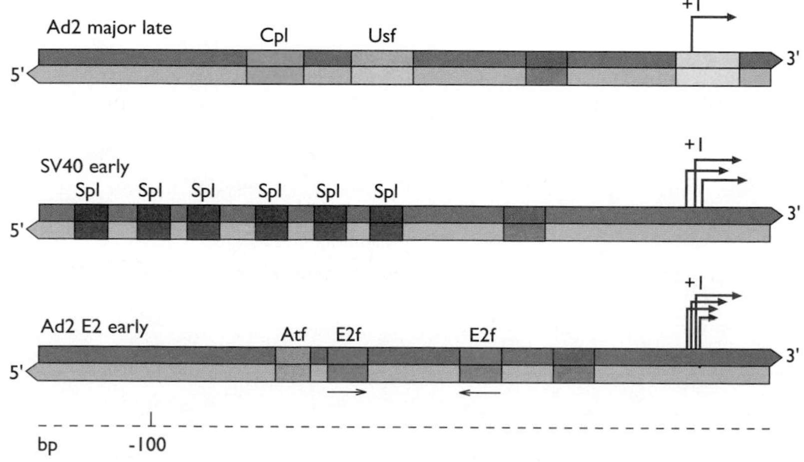

Figure 8.5 Local regulatory sequences of three viral transcriptional control regions. The TATA sequences, initiator sequences, and sites of transcription initiation of the human adenovirus type 2 (Ad2) major late and E2 early and the simian virus 40 (SV40) early transcriptional control regions are depicted as in Fig. 8.4. The local regulatory sequences of each promoter, which are recognized by the cellular DNA-binding proteins listed, are drawn to the scale shown at the bottom.

Table 8.3 Some cellular transcriptional regulators that recognize specific DNA sequences

Abbreviation	Name	Characteristics	Viral or cellular promoters/ enhancers recognized
Ap-1	Activator protein 1	Dimers of various basic-leucine zipper proteins, including c-Fos and c-Jun; regulated by dimerization and phosphorylation; mediate transcriptional response to a variety of extracellular stimuli	Simian virus 40 enhancer
Atf-2	Activating transcription factor 2	Member of large Atf/Creb family; basic-leucine zipper proteins, related to Ap-1 family; mediate transcriptional responses to second messenger cyclic AMP	Adenovirus type 2 E2E, E3, and E4 promoters; virus-responsive enhancer of human β-interferon gene
C/Ebps, e.g., A1/Ebp, C/Ebp-α	CCAAT/enhancer-binding proteins	Members of basic leucine-zipper family; C/Ebp-α enriched in hepatic cells	Rous sarcoma virus enhancer; hepatitis B virus enhancer
Cp-1	CCAAT-binding protein 1	Heterodimer	Adenovirus type 2 major late promoter
Creb	CAMP-response element (CRE)-binding protein	Member of the large Atf/Creb family (see above)	Human cytomegalovirus major immediate-early enhancer
E2f	E2 factor	Member of family of dimeric activators; activity regulated by association with Rb and Rb-related proteins; important in cell cycle regulation	Adenovirus type 2 E2E promoter
Ets-1	E twenty-six 1	Defined by conserved DNA-binding domain; expressed in T cells; expression and activity regulated by multiple mechanisms	Human immunodeficiency virus type 1 enhancers; herpes simplex virus type 1 ICP4 promoter
Gaf	Gamma interferon (IFN-γ) activated factor	Dimer of Stat-1; formed when Stat-1 phosphorylation is induced by IFN-γ	Many IFN-γ-responsive cellular promoters
Gata-3	GATA (binding protein) 3	Member of family defined by conserved Zn finger DNA-binding domain; expressed in T cells and specific neurons of the central nervous system	Human immunodeficiency virus type 1 enhancers
Ibp	Initiator binding protein	Member of nuclear hormone receptor superfamily; binds via Zn fingers as dimers	Simian virus 40 major late promoter
Irf-1, -9	Interferon-regulatory factors 1, 9	Members of family defined by conserved DNA-binding domain; synthesis regulated by virus infection and mitogens	Human interferon gene enhancers
Isgf3	Interferon-stimulated gene factor 3	Heterotrimer of Stat-1, Stat-2, and Irf-9; formed when Stat-2 phosphorylation is induced by IFN	Many IFN-responsive promoters
Nf-κb	Nuclear factor-κb	Dimer of p50-p65 Rel family proteins; activity regulated by sequestration in cytoplasm (Fig. 8.6)	Human immunodeficiency virus type 1 core enhancer; simian virus 40 enhancer
NfI	Nuclear factor I	Binds as dimer via novel DNA-binding domain to viral and cellular promoters	BK virus late promoter; mouse mammary tumor virus enhancer; adenovirus type 2 origins of replication (Chapter 9)
Nf-IL6	Nuclear factor for IL-6	Member of basic-leucine zipper protein family	Rous sarcoma virus and human immunodeficiency virus type 1 enhancers
Opbs	Octamer-binding proteins	Defined by binding to an octameric DNA sequence and POU-homeodomain DNA-binding motif; some ubiquitous, some restricted in expression	Simian virus 40 enhancer

Table 8.3 *(continued)*

Abbreviation	Name	Characteristics	Viral or cellular promoters/ enhancers recognized
Oct-1	Octamer binding-protein 1	Ubiquitously expressed member of Obp family	Herpes simplex virus type 1 immediate-early promoters, in complex with VP16 and Hcf (see text)
Rxr	9-*cis*-Retinoic acid receptor	Heterodimeric member of the nuclear receptor superfamily, which also includes Gr; can activate or repress transcription by binding to various coactivators or corepressors	Human immunodeficiency virus type 1 core enhancer; human cytomegalovirus immediate-early proximal enhancer
Sp1	Stimulatory protein 1	Binds as monomer via Zn-finger DNA-binding domain; first sequence-specific transcription factor identified	Simian virus 40 early promoter
Srf	Serum response factor	Members of Mads box DNA-binding domain family; mediates transcriptional response to serum growth factors	Rous sarcoma virus enhancer/promoter; interacts with human T-cell leukemia virus type 1 Tax
Stat1–Stat9	Signal transducer and activator of transcription 1–9	Members of family of regulators activated in response to interferon; dimerization and nuclear localization regulated by phosphorylation	Many IFN-responsive cellular promoters
Tef-1	Transcriptional enhancer factor 1	Complex DNA-binding domains; no independent activation domain identified; requires limiting coactivator	Simian virus 40 enhancer; human papillomavirus type 16 E6/E7 promoter
Tef-2	Transcription enhancer factor 2	Ubiquitous	Simian virus 40 enhancer; human papillomavirus type 16 enhancer
Usf	Upstream stimulatory factor	Contains basic, helix-loop-helix, and leucine zipper; domains; binds as dimer	Adenovirus type 2 major late promoter

property first demonstrated with the simian virus 40 early promoter. An enormous number of sequence-specific proteins that regulate transcription are now known, many first identified through analyses of viral promoters (Fig. 8.5). Unfortunately, the nomenclature applied to these regulatory proteins presents serious difficulties for both writer and reader, for it is unsystematic and whimsical. The names of some transcriptional regulators and the DNA sequences they recognize indicate their function (e.g., the glucocorticoid receptor [Gr] and the glucocorticoid response element [GRE]). Many other names, however, originate from the promoters in which the protein's binding sites were first identified (e.g., adenovirus E2 transcription factor [E2f]) or report some very general property (e.g., upstream stimulatory factor [Usf]). The universal use of abbreviations and the subsequent recognition that many "factors" in fact comprise families of closely related proteins compound this difficulty. A summary of the sequence-specific regulatory proteins described in this and other chapters is therefore provided in Table 8.3.

Proper regulation of transcription of many viral and cellular genes also requires more distant regulatory sequences in the DNA template, which possess properties that were entirely unanticipated (Fig. 8.1). The first example was discovered in the genome of simian virus 40 and was termed an **enhancer**, because it stimulated transcription to a large degree. Enhancers are defined by their position- and orientation-independent stimulation of transcription of homologous and heterologous genes over distances as great as 10,000 bp in the genome. Despite these unusual properties, enhancers are built with binding sites for the proteins that recognize local promoter sequences.

All viral templates transcribed by cellular RNA polymerase II contain transcriptional control regions that are directly recognized by cellular regulatory proteins (e.g., Fig. 8.5). Indeed, the transcriptional programs of retroviruses with simple genomes are executed by the cellular transcriptional machinery alone. However, many viruses also encode transcriptional regulatory proteins, some of which resemble cellular proteins that recognize local promoter or more distant enhancer sequences. Cellular, sequence-specific, transcriptional regulatory proteins therefore play pivotal roles in transcription from viral promoters and in the regulation of viral gene expression. These roles are illustrated in Fig. 8.6 by the many ways in which the cellular regulator called nuclear factor κb (Nf-κb) contributes to viral gene expression and is regulated in virus-infected cells.

Properties of sequence-specific proteins that regulate transcription. The cellular DNA-binding proteins necessary for transcription from viral DNA templates, as well as viral proteins that bind to specific transcriptional control sequences, share a number of common properties. Their most characteristic feature is modular organization: they

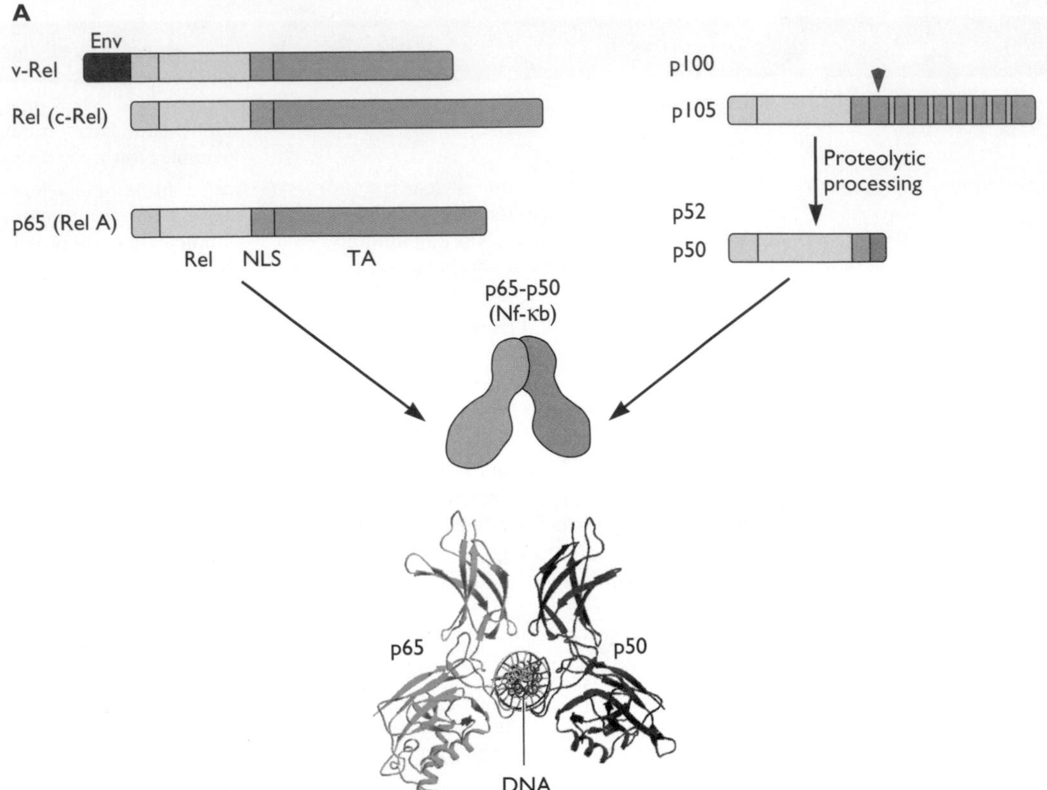

Figure 8.6 The cellular transcriptional regulator Nf-κb and its participation in viral transcription. (A) Nf-κb and the Rel protein family. The members of this family are defined by the presence of the Rel homology region (orange segment), which contains DNA-binding and dimerization motifs and a nuclear localization signal (pink). The p50-p65 Nf-κb heterodimer contains proteins from each of the two groups into which the Rel family is divided. One group (right) contains p50 (Nf-κb1) and p52 (Nf-κb2). As illustrated for p50, they are synthesized as precursors, which include a C-terminal segment closely related in sequence to members of the Iκb family of inhibitory proteins. Proteolytic processing of such precursors (e.g., p105) at the site indicated by the vertical arrow produces mature p50. The members of the second group of Rel proteins, which includes c-Rel, the *Drosophila* protein Dorsal, and p65 (Rel A), also contain an acidic activation domain (blue) at their C termini. The v-*rel* gene of the avian retrovirus reticuloendotheliosis virus strain T is a transduced version (see Chapter 18) of c-*rel* fused to the viral *env* gene. Its product is responsible for the oncogenic properties of this retrovirus. The X-ray crystal structure of a p50-p65 heterodimer bound specifically to DNA is shown below. The structure is viewed down the helical axis of DNA with the two strands in purple and yellow, and the p50 and p65 subunits are shown in green and red, respectively. The dimer makes extensive contact with DNA via protein loops, and the DNA-binding residues of the protein and those mediating contact between the dimers are in proximity. From F. E. Chen et al., *Nature* **391:**410–413, 1998. Courtesy of G. Ghosh, University of California, San Diego. (B) Activation of Nf-κb. Two forms of inactive Nf-κb are found in the cytoplasm (e.g., of unstimulated T cells). One consists of mature Rel heterodimers, p50-p65, associated with an inhibitory protein such as Iκb-α (left). Other members of the inhibitor family, which are characterized by the presence of a C-terminal region containing seven ankyrin repeats, include Iκb-β, the Bcl-3 oncogene product, and the *Drosophila* protein Cactus. The second inactive form of Nf-κb is the p65 protein dimerized with a Rel precursor, e.g., the p105 precursor of p50 (right). The C-terminal segment of p105 functions like Iκb, with which it shares sequences, to block nuclear localization signals and retain Nf-κb in the cytoplasm. Exposure of the cells to any of several growth factors, indicated by tumor necrosis factor α (Tnf-α), interleukin-1 (IL-1), and phorbol esters, results in activation (green arrows) of protein kinases that phosphorylate specific residues of Iκb or p105. Upon phosphorylation, Iκb dissociates from the complex and is recognized by the system of enzymes that adds branched chains of ubiquitin (Ub) to proteins targeted for degradation. It is then degraded by the proteasome, a multicatalytic protease that degrades polyubiquitinated proteins. Specific p105 cleavage (A) produces the p50-p65 dimer Nf-κb. Free Nf-κb generated by either mechanism can translocate to the nucleus because its nuclear localization signals (A) are now accessible. In the nucleus of uninfected cells, Nf-κb binds to specific promoter sequences to stimulate transcription via the p65 activation domain (A). Viral transcriptional control regions to which Nf-κb binds and some viral proteins that induce activation (green arrows) of Nf-κb are indicated. Binding of Nf-κb to the Epstein-Barr virus regulator Zta blocks the ability of Zta to stimulate transcription (red bar). HCMV, human cytomegalovirus; HIV-1, human immunodeficiency virus type 1; HTLV-1, human T-lymphotropic virus type 1; SV40, simian virus 40; EBV, Epstein-Barr virus.

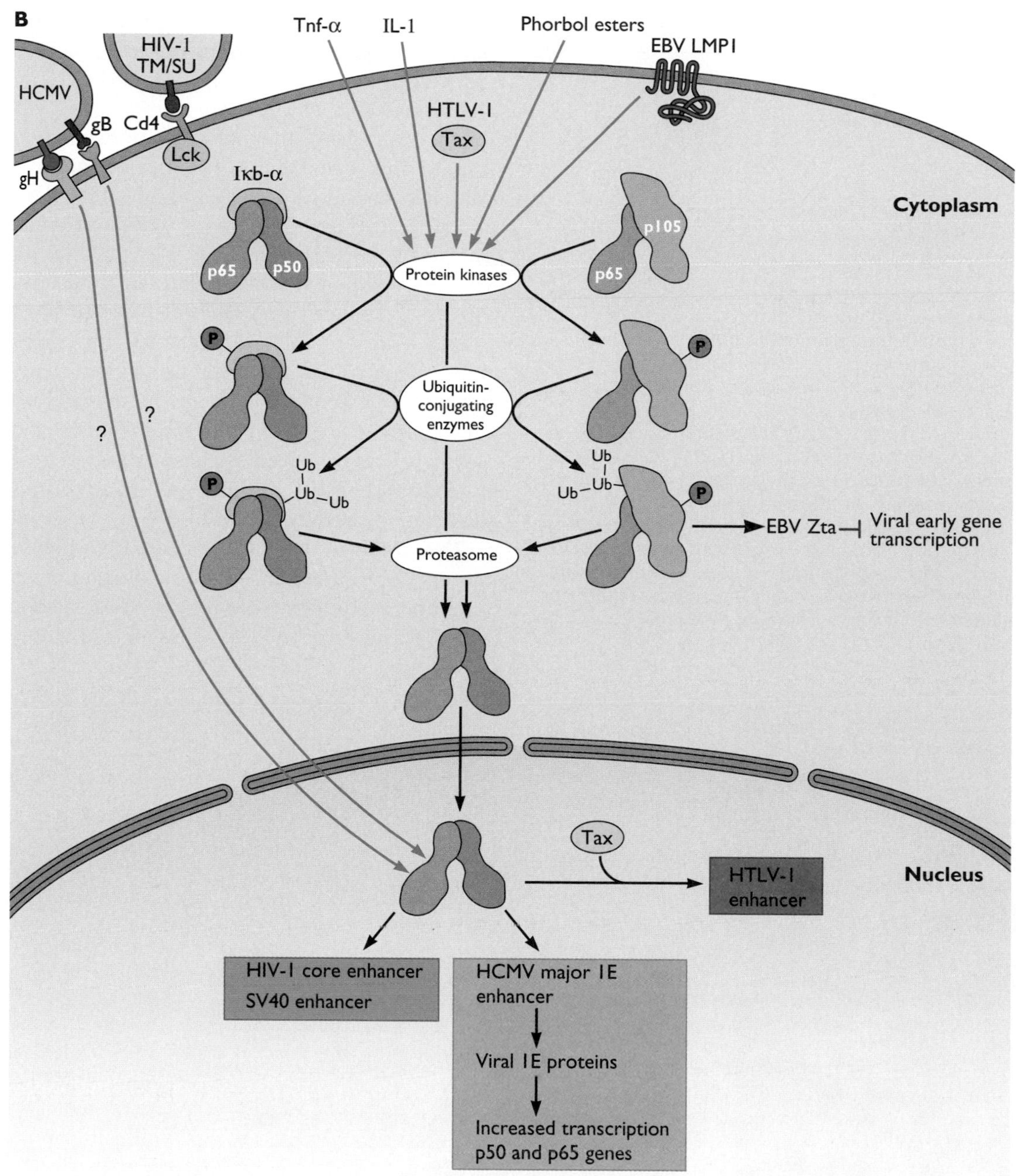

Figure 8.6 ***(continued)***

are built from discrete structural and functional domains (Fig. 8.7). The basic modules are a DNA-binding domain and an activation domain, which function as independent units. Other common properties of these regulators of transcription are binding to DNA as dimers, and membership in families of related proteins that share the same types of DNA-binding and dimerization domains (Fig. 8.7).

Once a regulatory protein is bound to a specific sequence in a transcriptional control region, it must effect a change in the rate of initiation (the speed at which the reactions summarized in Fig. 8.2 are completed), in the number of initiation cycles, or in other transcription reactions. Although the molecular mechanisms responsible for regulation of transcription are not fully understood, it is clear that even a single regulatory protein can act by multiple mechanisms. This property is discussed in more detail below.

Regulation of transcription by sequence-specific DNA-binding proteins usually requires additional proteins termed **coactivators** (or corepressors) and mediators. In general, these proteins cannot bind specifically to DNA, nor can they

DNA binding	Dimer formation	Activation
Zn finger Helix-turn-helix Basic	Leucine zipper	Acidic Glutamine rich Proline rich Isoleucine rich
Rel homology region		

Figure 8.7 Modular organization of sequence-specific transcriptional activators. Common functional domains of eukaryotic proteins that regulate transcription by binding specifically to regulatory regions of promoters are listed. Some of the types of each domain are listed below. DNA-binding and activation domains are defined by their structure (e.g., Zn finger or helix-turn-helix) and chemical makeup (e.g., acidic, glutamine-rich), respectively. As discussed in the text, such DNA-binding and activation domains can function as independent units. For the experimenter, this arrangement affords the opportunity to build artificial transcriptional regulators by mixing and matching DNA-binding and activation domains from different proteins. Transcriptional activators are often more complex than illustrated here. They can contain more than a single activation domain and regulatory domains, such as ligand-binding domains.

modulate transcription on their own. However, they dramatically augment (or damp) the transcriptional responses induced by sequence-specific regulators. Among the best-characterized coactivators are the Swi/Snf complex (named for the products of the yeast mating type switch [*swi*] and sucrose nonfermentable [*snf*] genes present in the first such complex identified) and the highly related proteins p300 and Cbp (cyclic AMP response element-binding protein). The former was discovered by virtue of its binding to adenoviral E1A proteins, which stimulate transcription of both viral and cellular genes. Each of these coactivators can cooperate with multiple, sequence-specific transcriptional activators and stimulate transcription from many promoters, but neither is required globally. Many questions about the mechanism of action of coactivators remain to be addressed. However, it is now well established that many can alter the organization of the chromatin template to stimulate multiple reactions in transcription.

The variety in the nature of core RNA polymerase II promoters and in the constellations of sequence-specific proteins and coactivators that regulate their activity contribute to the plasticity of the RNA polymerase II system. Equally

BOX 8.4

Synergistic stimulation of RNA polymerase II transcription by promoter-bound proteins

The synergistic stimulation of transcription is illustrated for a hypothetical case in which a promoter, with a basal transcription rate defined as 1.0, contains binding sites for the two stimulatory proteins X and Y. Binding of X alone or Y alone stimulates the rate of transcription 10- or 5-fold, respectively. If X and Y targeted different reactions in the initiation sequence and functioned independently of one another, the 10-fold stimulation by protein X would be amplified a further 5-fold by protein Y, for a total 50-fold activation. **Synergism** is the ability of one protein to assist the action of the second when both are bound to the promoter, in this case resulting in a significantly greater-than-50-fold increase in the rate of transcription. The synergistic action of two proteins is therefore defined as an increase in transcription when both are present that is greater than the product of the increases induced by each protein on its own.

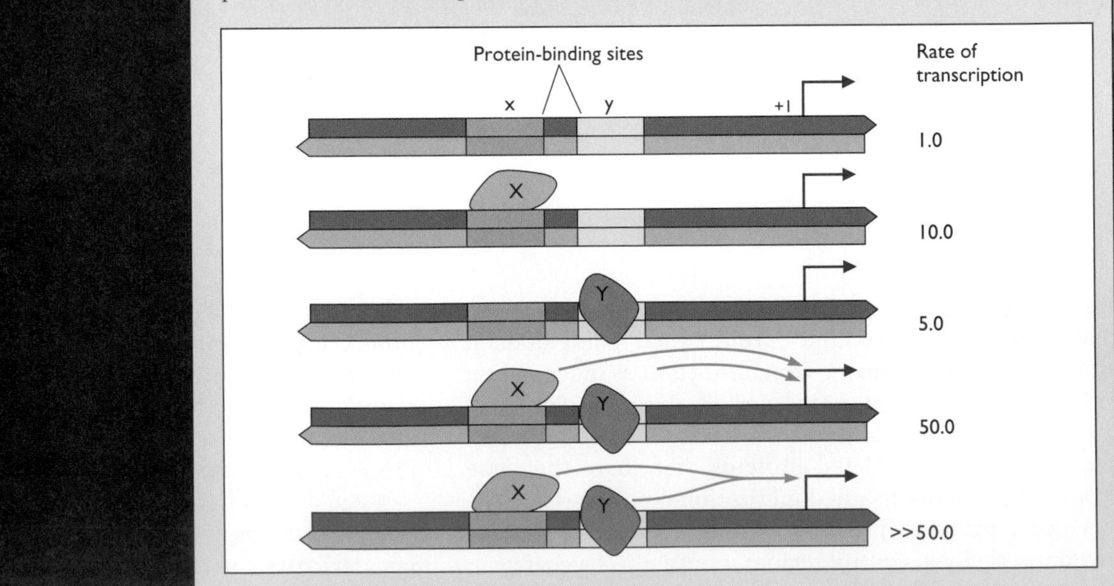

important is the power of the transcriptional machinery to integrate signals from multiple, promoter-bound regulators (Box 8.4). The transcriptional machinery must also be able to sense and respond to developmental and environmental cues. The proteins that control transcription are therefore frequently regulated by mechanisms that determine their activity, availability, or intracellular concentration. These mechanisms include regulation of the phosphorylation of specific amino acids, which can determine how well a protein binds to DNA, its oligomerization state, or the properties of its regulatory domain(s). In some cases, the intracellular location of a sequence-specific DNA-binding protein or its association with inhibitory proteins within the nucleus is controlled. Autoregulation of transcription of the genes encoding transcriptional regulators is also common. This brief summary illustrates the varied repertoire of mechanisms available for regulation of transcription of viral templates by RNA polymerase II. Not surprisingly, virus-infected cells provide examples of all items on this menu, with the added zest of virus-specific mechanisms.

Transcription of Viral DNA Templates by the Cellular Machinery Alone

In cells infected by many retroviruses, the components of the cellular transcriptional machinery described in the previous section complete the viral transcriptional program without the assistance of viral proteins. The proviral DNA created by reverse transcription and integration (Table 8.2) comprises a single RNA polymerase II transcription unit organized into chromatin like the cellular templates for transcription.

Integrated proviral DNA is a permanent resident in the cellular genome. Because the genomes of these retroviruses do not encode transcriptional regulators, the rate at which such proviral DNA is transcribed is determined by the constellation of cellular transcription proteins present in an infected cell. This rate may be influenced by the nature and growth state of the infected cell, as well as by the organization of cellular chromatin around the site of proviral DNA integration. Transcription of viral genetic information can occur throughout the lifetime of the host cell, indeed even in descendants of the cell initially infected. Retroviral transcription is characterized by production of a single viral transcript, which serves as both the genome for assembly of progeny virions and the source of viral mRNA species. This strategy for transcription of viral DNA is exemplified by avian sarcoma and leukosis viruses, such as Rous-associated viruses.

The long terminal repeat (LTR) of these proviral DNAs contains a compact enhancer located immediately upstream of the viral promoter (Fig. 8.8). Although enhancers can stimulate transcription over large distances, and generally do so when regulating the activity of cellular genes, the close proximity of the avian proviral LTR en-

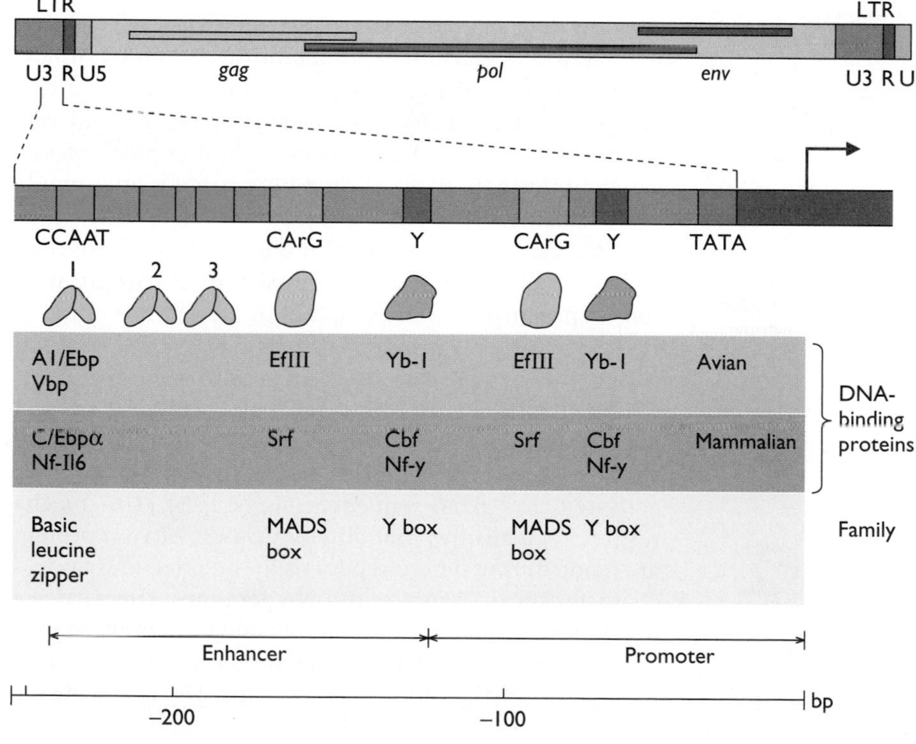

Figure 8.8 The transcriptional control region of an avian retrovirus. The proviral DNA of an avian leukosis virus is shown at the top. The enhancer and promoter present in the U3 regions of the LTRs are drawn to scale below. Each of the multiple CCAAT, CArG, and Y-box sequences present in this transcriptional control region, which are required for maximally efficient transcription, is recognized by both avian and mammalian proteins. All those recognizing the CCAAT enhancer sequences are members of the basic-leucine zipper family of proteins described in the text (see also Fig. 8.9). The chicken protein EfIII that binds to the CArG sequences appears to be the avian homolog of mammalian serum response factor (Srf). This protein plays an important role in the activation of transcription in response to serum growth factors, hence its name. These proteins are members of a large, widespread family defined by a conserved sequence motif within the DNA-binding domain (the Mads box) and named for four of the originally identified members.

hancer to promoter sequences is typical of viral genomes (see, for example, Fig. 8.11 and 8.16). The avian and mammalian serum response proteins that bind to the enhancer also bind to a specific sequence in the promoter (Fig. 8.8). This arrangement emphasizes the fact that enhancer and promoter sequences cannot be distinguished by the kinds of protein that recognize them. The other proteins that bind to this viral enhancer are all members of the basic-leucine zipper family of proteins (Fig. 8.7). Such proteins share a "leucine zipper" dimerization motif located immediately adjacent to a DNA-binding domain rich in basic residues (Fig. 8.9). As with many other transcriptional regulators, dimerization is essential for DNA binding, in this case to align the adjacent basic regions of the monomers to form the DNA-binding surface (Fig. 8.9).

The most remarkable property of the avian retroviral transcriptional control region is that it is active in many different cell types of both the natural avian hosts and mammals. This unusual feature can be explained by the widespread distribution of the cellular proteins that bind to it. These enhancer- and promoter-binding proteins, presumably assembled on the LTR transcriptional control region in different combinations in different cell types, allow efficient transcription of the provirus in both avian and mammalian cells. This property of the LTR enhancers/promoters of avian retroviruses has been exploited in the development of viral vectors used both in the laboratory and for gene therapy.

Figure 8.9 The structure of a basic-leucine zipper domain bound specifically to DNA. The model of the leucine zipper and adjacent basic region of the Jun-Fos heterodimer (Ap1) bound to a specific recognition sequence in DNA is based on the structure determined by X-ray crystallography. The leucine zipper forms an α-helical coiled coil, both in crystals and in solution. In the protein-DNA complex, the basic (DNA-binding) region is also α-helical, but in solution it is disordered: DNA binding induces a major conformational change in proteins of this class. The angle between the two leucine zipper region α-helices in the dimer increases slightly as the coiled coil approaches the DNA. As a result, the α-helical regions containing the basic amino acids that make specific contact with DNA fit snugly into successive major grooves. Consequently, the basic-leucine zipper dimerization and DNA-binding domains have been likened to DNA "forceps." Adapted from J. N. M. Glover and S. C. Harrison, *Nature* **373:**257–261, 1995, with permission.

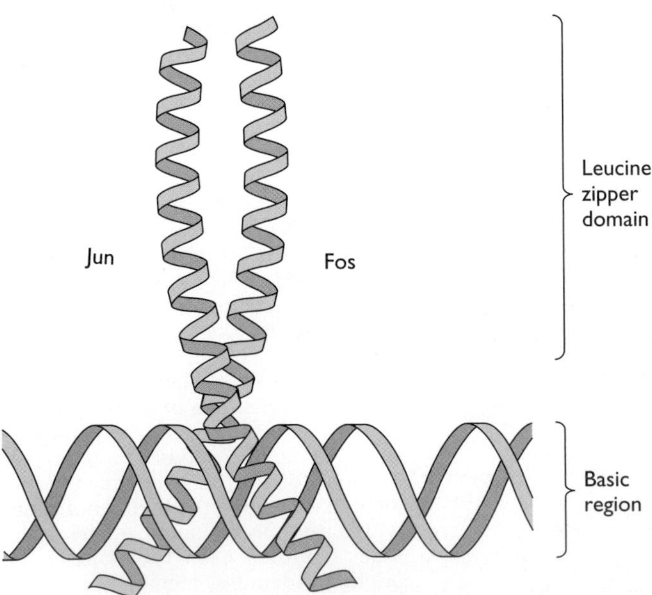

Because the LTRs are direct repeats of one another (Fig. 8.8), transcription directed by the 3′ LTR cannot contribute to the expression of retroviral genetic information. In fact, the transcriptional control region of the 3′ LTR is normally inactivated by a process called **promoter occlusion:** the passage of transcribing complexes initiating at the 5′ LTR through the 3′ LTR prevents recognition of the latter transcriptional control signals by enhancer- and promoter-binding proteins.

Absolute dependence on cellular components for the production of viral transcripts avoids the need to devote limited viral genetic information to transcriptional regulatory proteins. Nevertheless, such a strategy is not the rule, even among retroviruses.

Viral Proteins That Regulate Transcription by RNA Polymerase II

Patterns of Regulation

Transcription of many viral DNA templates by the RNA polymerase II machinery results in the synthesis of large quantities of viral transcripts (in some cases, more than 10^5 copies of individual mRNA species per cell) in relatively short periods. Such bursts of transcription of viral genetic information are elicited by transcriptional regulatory proteins. Viral proteins that stimulate RNA polymerase II transcription establish one of two kinds of regulatory circuit. The first is a **positive autoregulatory loop,** epitomized by transcription of human immunodeficiency virus type 1 proviral DNA (Fig. 8.10A). A viral activating protein stimulates the rate of transcription of proviral DNA, but does not alter the nature of viral proteins made in infected cells. The second is a **transcriptional cascade,** in which different viral transcription units are activated in a fixed sequence (Fig. 8.10B). This mechanism, which ensures that different classes of viral proteins are made during different periods in the infectious cycle, is characteristic of viruses with DNA genomes. The participation of viral regulatory proteins presumably confers a measure of control lacking when the transcriptional program is executed solely by cellular components. The following sec-

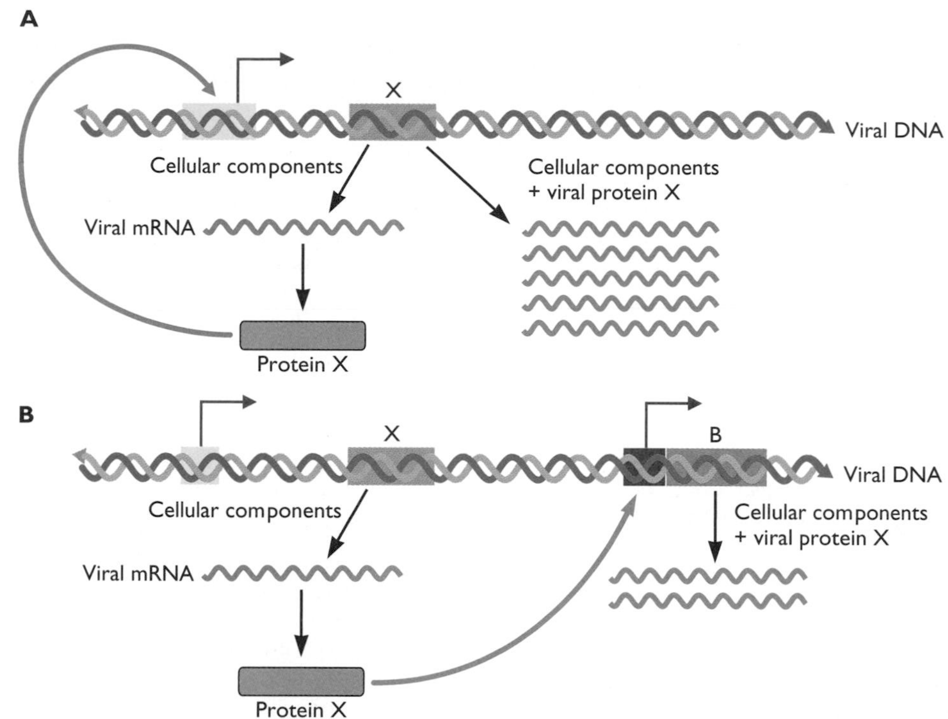

Figure 8.10 Mechanisms of stimulation of transcription by viral proteins. Cellular transcriptional components acting alone transcribe the viral gene for viral protein X. Once synthesized and returned to the nucleus, viral protein X can stimulate transcription either of the same transcription unit (A) or of a different viral transcription unit (B). In either case, viral protein X acts in concert with components of the cellular transcriptional machinery. The autoregulatory strategy (A) is exemplified by complex retroviruses such as human immunodeficiency virus type 1. The transcriptional programs of viruses with DNA genomes are variations of the sequential transcriptional cascade (B).

tions describe some well-studied examples of regulatory circuits established by viral proteins.

Autoregulation by the Human Immunodeficiency Virus Type 1 Tat Protein

Like those of their simpler cousins, the proteins of human immunodeficiency virus type 1 and other retroviruses with complex genomes are encoded in a single proviral transcription unit controlled by an LTR enhancer and promoter. However, in addition to the common structural proteins and enzymes, genomes of these retroviruses encode auxiliary proteins, including transcriptional regulators. Some of these regulatory proteins, such as the Tax protein of human T-lymphotropic virus type 1, resemble activators of other virus families and stimulate transcription from a wide variety of viral and cellular promoters (Table 8.4). Others, exemplified by the transactivator of transcription (Tat) of human immunodeficiency virus type 1, are unusual, sequence-specific activators of transcription that recognize an RNA element in nascent transcripts.

In outline, the positive feedback loop (Fig. 8.10A) that is established once a sufficient concentration of Tat has accumulated in an infected cell is simplicity itself. Cellular proteins initially direct transcription of the proviral DNA in infected cells at some basal rate. Among the processed products of viral primary transcripts are spliced mRNA species from which the Tat protein is synthesized. This protein is imported into the nucleus, where it stimulates transcription of the proviral template upon binding to its RNA recognition site in nascent viral transcripts. However, the molecular mechanisms that establish this autostimulatory loop are sophisticated and unusual. Their elucidation has been an important area of research, because the Tat protein is essential for virus propagation and represents a good target for antiviral therapy.

Cellular Proteins and the Human Immunodeficiency Virus Type 1 LTR Transcriptional Control Region

The LTR enhancer and promoter support some proviral transcription before Tat is made in infected cells, and are also necessary for Tat-dependent transcription. In contrast to avian retroviruses, human immunodeficiency virus type 1 propagates efficiently in only a few cell types, notably $CD4^+$ T lymphocytes and cells of the monocytic lineage. Viral reproduction (i.e., transcription) in infected T cells in culture is stimulated by T cell growth factors, indicating that viral transcription requires cellular components available only in such stimulated T cells. Indeed, the failure of the virus to propagate efficiently in unstimulated T cells correlates with the absence of active forms of specific enhancer-binding proteins. The distribution of cellular enhancer-binding proteins is therefore an important determinant of the host range of retroviruses with both simple and complex genomes. The difference is that the transcription of

Table 8.4 Viral enhancers and their recognition

Virus	Enhancer(s), location(s)	Enhancer-binding proteins		Enhancer properties and functions
		Cellular	Viral	
Adenovirus				
Human adenovirus type 2	Enhancer 1, sequences repeated at –300 and –200 of the E1A transcriptional control region	E2f, Ets family members	None	Broad cell-type specificity; active in a variety of cell types; negatively regulated in undifferentiated cells by Zn-finger containing cellular repressor
	Enhancer II, between repeated sequences of enhancer 1	?	None	Stimulates transcription from all immediate-early and early promoters, independently of enhancer 1
Hepadnavirus				
Hepatitis B virus	Enhancer I, adjacent to X gene promoter	Nf-κb, Ap-1, Nf-1, bZip protein, C/Ebp, hepatocyte nuclear factor 3 (Hnf3), Hnf4	None	Strong specificity for hepatic cells, because C/Ebp, Hnf3, and Hnf4 are specific for or enriched in these cells; some activity in nonhepatic cells; stimulates transcription from all viral promoters in infected cells; activity may be increased by the viral X protein
	Enhancer II, adjacent to pregenomic RNA promoter	Hnf1, Hnf3, Hnf4, C/Ebp, nuclear receptor superfamily member human B1-binding factor, B1f	None	Hepatic cell specific; stimulates transcription from pregenomic and pre-S promoters in infected cells
Herpesvirus				
Human cytomegalovirus	Immediate-early proximal enhancer, –613 to –70 of major immediate-early transcription unit	Nf-κb, Ap-1, Creb, Srf, Ets family member Elk1, nuclear receptor superfamily member Rxr	None	Can function as a strong basal enhancer in established cell lines in culture and an inducible enhancer activated by signal transduction pathways that converge on Nf-κb, Creb, and Srf + Ets; activity believed to be crucial for entry into lytic cycle and reactivation from latency
Papillomavirus				
Human papillomavirus type 16	"Constitutive" enhancer in the long control region between coding sequences for early and late proteins	Nf-κb, Ap-1, Tef-1, Tef-2, Oct-1, Nf-1	None	Epithelial cell specific, at least in part because of the presence of a specific subset of Nf-1 family members in these cells; Ap-1 binding sites confer responsiveness to epidermal growth factor
	E2 protein-dependent enhancers formed by pairs of the E2-binding sites present in the long control region	None	E2 protein dimers	Active only in cells synthesizing the viral E2 protein; activates transcription from all early genes
Polyomavirus				
Simian virus 40	Between early and late promoters; tandem copies of a 72-bp repeat sequence	Nf-κb, Ap-1, Tef-1, octamer family members	None	Broad cell-type specificity; active in many mammalian cell types; activated by signal transduction pathways that converge on Ap-1

Table 8.4 *(continued)*

Virus	Enhancer(s), location(s)	Enhancer-binding proteins: Cellular	Enhancer-binding proteins: Viral	Enhancer properties and functions
Retrovirus				
Human immunodeficiency virus type 1	Core enhancer, –95 to –50 of viral transcription unit	Nf-κb, Ets-1	None	Active only in cells in which Nf-κb is activated (e.g., T cells exposed to various growth factors); necessary for efficient HIV-1 transcription in vivo
	Upstream enhancer region, 5′ to –100 of viral transcription unit	Ets-1, Gata-3, Lef, Nf-IL6	None	Active in T cells and hematopoietic cells that contain Ets-1 and Gata-3 (i.e., narrow cell-type specificity)
Rous sarcoma virus	LTR enhancer, –250 to –130 of the viral transcription unit	bZip proteins, Srf, Y-box-binding proteins	None	Broad cell-type specificity; active in many avian and mammalian cells because recognized by ubiquitous cellular proteins

proviral DNA of avian retroviruses depends on proteins that are widely distributed, whereas human immunodeficiency virus type 1 transcription requires proteins that are found in only a few cell types, or that are active only under specific conditions.

Within the LTR, the promoter is immediately preceded by two important regulatory regions (Fig. 8.11). The promoter-proximal enhancer core is essential for enhancer function in transient-expression assays (Fig. 8.3B) and virus reproduction in T cells. The upstream enhancer region is also necessary for efficient viral transcription in both peripheral blood lymphocytes and certain T-cell lines. Both the core and the upstream enhancers are densely packed with binding sites for cellular proteins, many of which are enriched in the types of cell in which the virus can propagate. For example, the human Gata-3 and Ets-1 proteins, which can bind within the upstream enhancer region (Fig. 8.11) and stimulate viral transcription in tran-

Figure 8.11 Protein-binding sites in the human immunodeficiency virus type 1 transcriptional control region. The organization of the U3 region of the viral LTR is shown to scale. The proteins that bind to the core enhancer or promoter (discussed in the text) are shown below the DNA sequence. As shown above the line, the core enhancer also contains binding sites for other proteins. The upstream enhancer region, which stimulates human immunodeficiency virus type 1 transcription in infected cells, also contains binding sites for several proteins that are enriched in cells in which the virus reproduces. For example, the Lef protein is lymphocyte specific. Upon binding, this protein induces a large (130°) bend in DNA. Similarly, Ets-1 and Gata-3 are enriched in and restricted to, respectively, T lymphocytes. The cellular enhancer-binding proteins listed have been shown to modulate the efficiency of transcription from the viral promoter in experimental situations (usually in transient-expression assays). However, with the exception of Nf-κb (see text), their contributions to human immunodeficiency virus type 1 transcription in infected cells are generally not known.

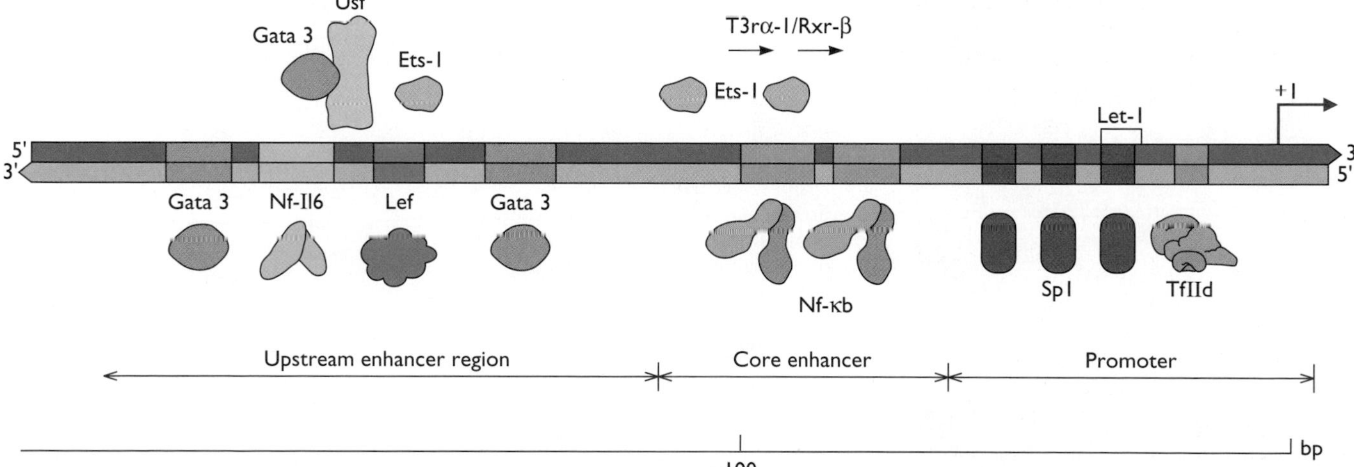

sient-expression assays, are restricted to cells of the T-lymphocyte and hematopoietic lineages, respectively. However, it is not known whether these proteins are important for transcription of the provirus in infected cells. Transient-expression assays in which a reporter gene is introduced together with a sequence encoding the protein of interest (Gata 3 or Ets-1 in this case) do not reproduce physiological conditions (Box 8.5). Consequently, a positive result in this kind of assay establishes only that a certain protein **can** stimulate transcription, not that it normally does so.

The mechanisms by which viral transcription is regulated by cellular pathways are illustrated by the cellular transcriptional activator Nf-κb, which plays a critical role in human immunodeficiency virus type 1 propagation in T cells. Unstimulated T cells display no Nf-κb activity, because the protein is retained in inactive form in the cytoplasm by binding of inhibitory proteins of the Iκb family (Fig. 8.6B). This interaction masks the nuclear localization signals of Nf-κb, preventing its translocation to the nucleus. Upon T-cell activation by treatment with any one of several growth factors, Iκb proteins rapidly disappear from the cytoplasm, with the concomitant appearance of active Nf-kb in the nucleus. Phosphorylation of Iκb at specific sites, the result of signaling pathways activated by growth factors at the cell surface, targets the inhibitor for degradation by the cytoplasmic multiprotease complex (the **proteasome**). Consequently, Nf-κb is free for transit to the nucleus, where it can bind to its recognition sites within the viral LTR core enhancer (Fig. 8.11). This pathway can account for the induction of human immunodeficiency virus type 1 transcription observed when T cells are stimulated. The importance of activation of Nf-κb in the infectious cycle of this virus is emphasized by the severe, or complete, inhibition of virus reproduction (transcription) in normal human $CD4^+$ lymphocytes induced by mutations that prevent binding of this cellular protein to the enhancer. This cellular protein is also indispensable for viral transcription in macrophages, which, when differentiated, contain nuclear pools of constitutively active Nf-κb. Nevertheless, Nf-κb and the other cellular proteins that act via LTR enhancer or promoter binding sites do not support efficient expression of human immunodeficiency virus type 1 genes. This process depends on synthesis of the viral Tat protein.

Regulation of Transcription by the Tat Protein

Recognition of an RNA sequence by Tat. Stimulation of human immunodeficiency virus type 1 transcription by Tat requires an LTR sequence, termed the **trans-activation response (TAR)** element, which lies within the transcription unit (Fig. 8.12A). This sequence is ac-

BOX 8.5

Caution: transient-expression assays do not reproduce conditions within virus-infected cells

In transient-expression assays, the transcriptional control region of interest linked to a reporter gene is introduced into cells in culture with a vector that directs synthesis of a viral (or cellular) regulator of transcription. Such assays provide a powerful, efficient way to investigate control of transcription. Advantages include the following:

- simplicity and sensitivity of assays for reporter gene activity (Fig. 8.3B)
- ready analysis of mutated promoters to identify DNA sequences needed for the action of the regulatory protein
- application with chimeric fusion proteins and synthetic promoters to avoid transcriptional responses due to endogenous cellular proteins
- simplification of complex transcriptional regulatory circuits to focus on the activity of a single protein

Despite these advantages, transient-expression assays do not necessarily tell us how transcription is regulated in virus-infected cells, for they do not reproduce intracellular conditions. Important differences include the following:

- association of transcriptional templates with different proteins: the exogenous DNA may associate with cellular histones, but viral DNA may be packaged by virus-specific proteins.
- abnormally high concentrations of exogenous template DNA: concentrations of reporter genes as high as 10^6 copies per cell are not unusual, a value significantly greater than even maximal concentrations of viral DNA molecules attained toward the end of an infectious cycle.
- abnormally high concentrations of the regulatory protein as a result of its deliberate overproduction
- because of these high concentrations of template and protein, potential interactions of the viral regulator with template or components of the cellular transcriptional machinery that would not take place in an infected cell
- absence of viral components that might negatively or positively modulate the activity of the protein under investigation

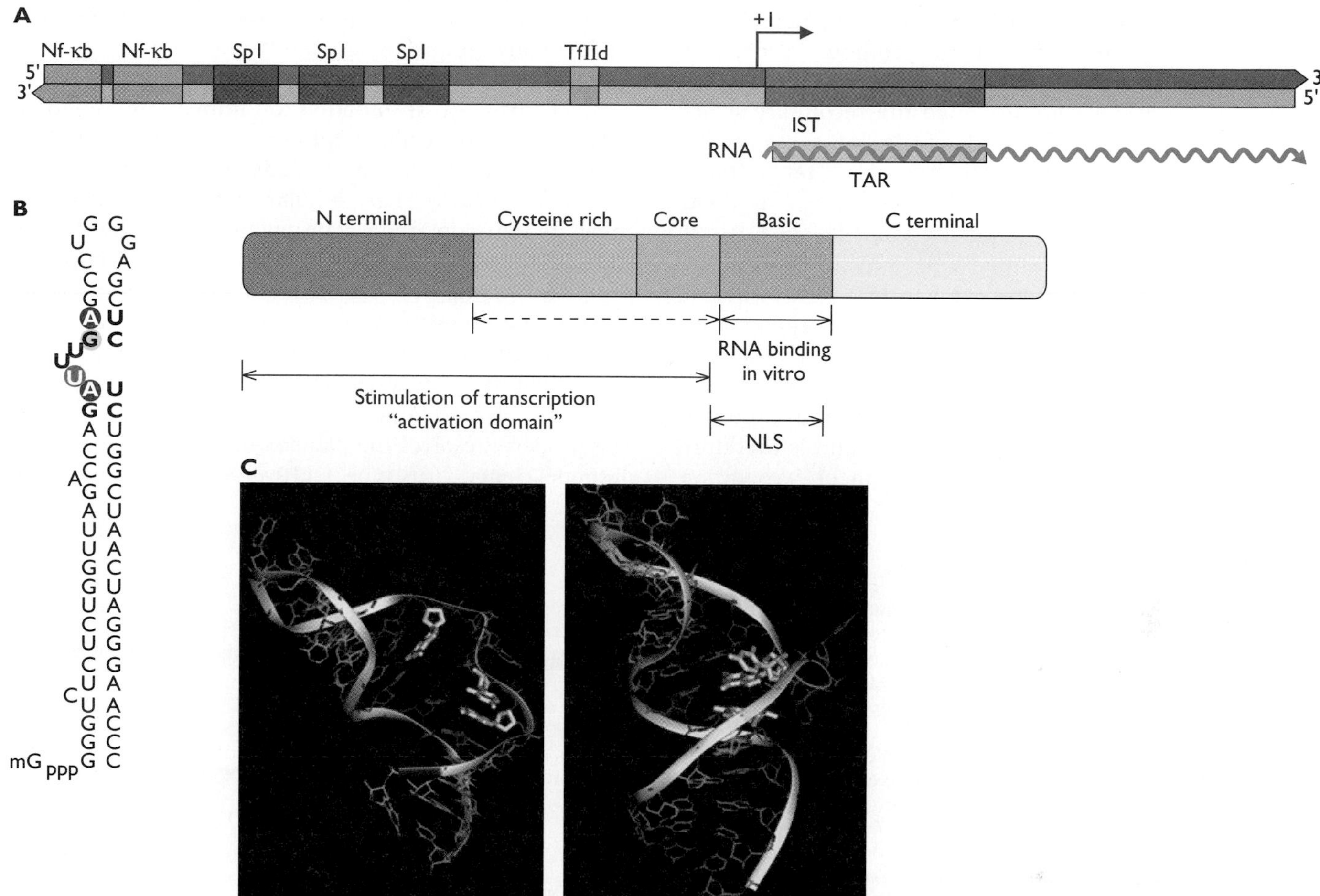

Figure 8.12 Human immunodeficiency virus type 1 TAR and the Tat protein. (A) Locations of regulatory sequences. The region of the viral genome spanning the site of transcriptional initiation is drawn to scale, with the core enhancer and promoter depicted as in Fig. 8.11. The DNA sequence lying just downstream of the initiation site (pink) controls synthesis of short transcripts from the human immunodeficiency virus type 1 promoter and has therefore been termed the inducer of short transcripts (IST). Transcription of the proviral DNA produces nascent transcripts that contain the TAR sequence (pale orange box). (B) TAR RNA and the Tat protein. The TAR RNA hairpin (left) extends from position 11 to position 159 in viral nascent RNA and includes the minimal TAR sequence (119 to 142) necessary and sufficient to support Tat-dependent stimulation of transcription in vivo. Sequences important for recognition of TAR RNA by the Tat protein are colored. Optimal stimulation of transcription by Tat requires not only this binding site in TAR but also the terminal loop. The Tat protein (right) is made from several different, multiply spliced mRNAs (Appendix A, Fig. 21B) and therefore varies in length (at its C terminus) from 82 to 101 amino acids. The regions of the protein are named for the nature of their sequences (Basic, Cysteine rich) or greatest conservation among lentiviral Tat proteins (Core). Experiments with fusion proteins containing various segments of Tat and a heterologous RNA-binding domain identified the N-terminal segment as sufficient to stimulate transcription. The basic region contains the nuclear localization signal (NLS). Analysis of the RNA-binding properties of short peptides indicated that the basic region, which is very rich in arginine residues, can bind specifically to RNA containing the bulge characteristic of TAR RNA (left). However, high-affinity binding, effective discrimination of wild-type TAR from mutated sequences in vitro, and RNA-dependent stimulation of transcription within cells require additional, N-terminal regions of the protein, shown by the dashed arrow. The sequences required for the stimulation of transcription and specific recognition of TAR RNA in cells do not form discrete domains of the protein. (C) Conformational change in TAR RNA induced by binding of the Tat peptide that contains residues from the core and basic regions of the protein. Major groove views of structures of a free TAR RNA corresponding to the apical stem and loop regions (but with a truncated stem) (left) and of the same RNA when bound to the Tat peptide (right) were determined by nuclear magnetic resonance methods. The residues shown in yellow are A22, U23, and G26 (from bottom to top in the left panel), which are colored red, blue, and yellow, respectively, in panel B. Note the substantial conformational change in the trinucleotide bulge region on binding of the Tat peptide. From F. Aboul-ela et al., *J. Mol. Biol.* **253:**313–332, 1995, and F. Aboul-ela et al., *Nucleic Acids Res.* **24:**3974–3981, 1996, with permission. Courtesy of M. Afshar, RiboTargets, and J. Karn, MRC Laboratory of Molecular Biology.

tive only in the sense orientation and only when located close to the initiation site of the promoter, properties that distinguish it from enhancer elements. The observation that mutations that disrupted the secondary structure predicted for the RNA copy of the TAR sequence inhibited stimulation of transcription by Tat initially indicated that the TAR element is recognized as RNA. Indeed, the Tat protein binds specifically to a trinucleotide bulge and adjacent base pairs in the stem of the TAR RNA stem-loop structure (Fig. 8.12B). Such specific binding requires an arginine-rich basic region and adjacent sequences of Tat (Fig. 8.12B). Binding of Tat to this region of TAR induces a local conformational rearrangement in the RNA, resulting in formation of a more stable and compact structure (Fig. 8.12C), and is therefore energetically favorable. This mechanism of recognition of a transcriptional control sequence by a regulatory protein remains unique to Tat.

Mechanism of stimulation of human immunodeficiency virus type 1 transcription. Binding of Tat to TAR RNA stimulates production of viral RNA as much as 100-fold. In contrast to many cellular and viral proteins that stimulate transcription by RNA polymerase II, the Tat protein has little effect on initiation. Rather, it greatly improves elongation. Complexes that initiate transcription in the absence of Tat elongate poorly, and many terminate transcription within 60 bp of the initiation site (Fig. 8.13). Consequently, in the absence of Tat, full-length transcripts of proviral DNA account for no more than 10% of the total. This property resolves the paradox of why the human immunodeficiency virus type 1 LTR enhancer and promoter

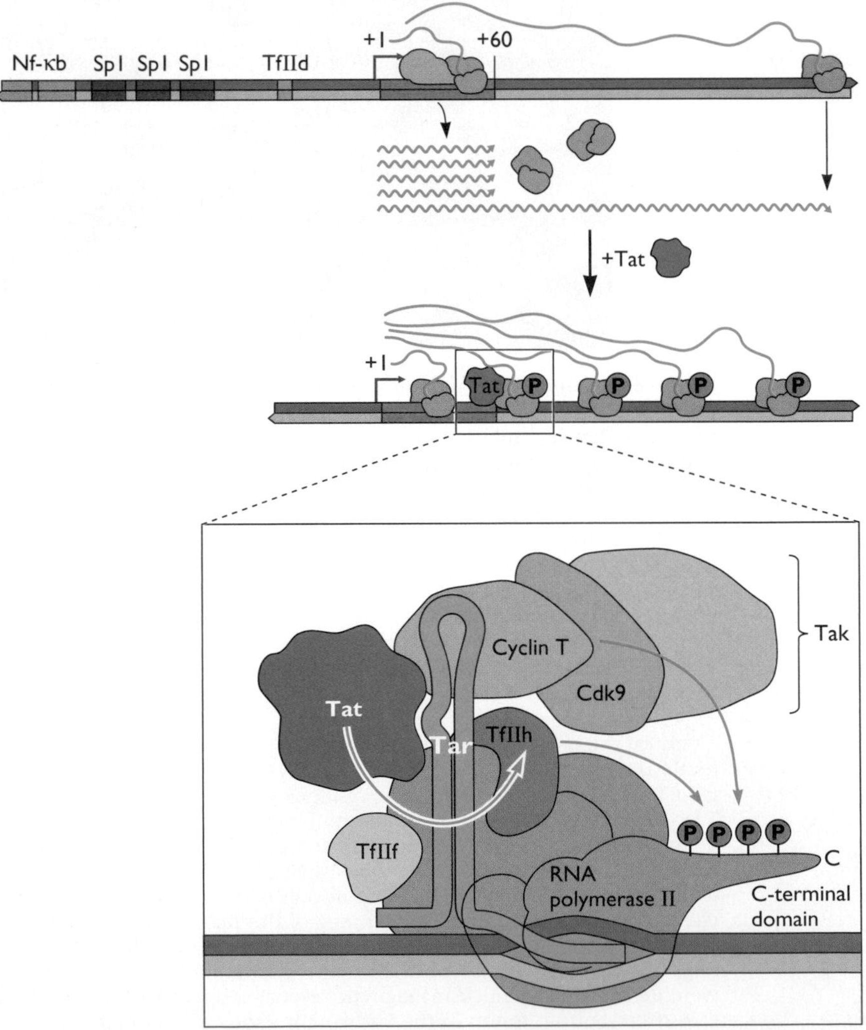

Figure 8.13 Mechanism of stimulation of transcription by the human immunodeficiency virus type 1 Tat protein. The regulatory sequences flanking the site of initiation of transcription are depicted as in Fig. 8.11. Initiation of transcription by the RNA polymerase II machinery results in the synthesis of short transcripts containing the TAR sequence: in the absence of Tat, transcription complexes are poorly processive, and the great majority (9 out of 10) terminate transcription within 60 bp of the initiation site, releasing transcription components and short transcripts. Synthesis of full-length human immunodeficiency virus type 1 RNA is therefore inefficient. Production of the Tat protein upon translation of mRNAs spliced from rare, full-length transcripts allows all transcriptional complexes to pass through the elongation blocks to synthesize full-length viral RNA. Binding of Tat to TAR in conjunction with the cyclin T subunit of Tak (see text) leads to stimulation of phosphorylation (P) of the C-terminal domain of the largest subunit of RNA polymerase II by the TfIIh-associated kinase and allows phosphorylation of this domain by Tak. It is believed that the two kinases act sequentially (see text). As a result of modification of this C-terminal domain, transcriptional complexes become competent to carry out highly processive transcription. The Tat protein also binds to histone acetylases, including p300, pCaf, and hGcn5, and is itself acetylated at Lys 28 and Lys 50, and Lys 51, within the activation and TAR RNA-binding domains, respectively (Fig. 8.12). These modifications stimulate Tat-dependent transcription from the viral transcriptional control region. A second elongation protein, Tat-Sf1, is also essential for stimulation of transcription by Tat, but its mechanism of action is not yet known.

do not support efficient viral RNA synthesis: they direct efficient initiation, but the transcription complexes formed cannot carry out sustained transcription of the proviral genome over long distances. The Tat protein overcomes such poor **processivity** of elongating complexes to allow efficient production of full-length viral transcripts.

A search for cellular proteins that stimulate viral transcription when bound to the N-terminal region of Tat (Fig 8.12B) identified a human Ser/Thr kinase called Tat-associated kinase (Tak). This protein, which proved to correspond to a protein that stimulates elongation (positive-acting transcription factor b), is essential for Tat-dependent stimulation of processive viral transcription both in vitro and in vivo. The Tat-associated kinase comprises a regulatory subunit termed a **cyclin** (because the first members of the family to be identified accumulate during specific periods of the cell cycle) and a cyclin-dependent kinase (Cdk9). It phosphorylates an unusual domain at the C terminus of the largest subunit of RNA polymerase II, which is essential for Tat-dependent stimulation of viral transcription. This domain comprises multiple, tandem copies of a heptapeptide sequence rich in Ser and Thr residues. The C-terminal domain of RNA polymerase II present in preinitiation complexes is hypophosphorylated, and its hyperphosphorylation has been implicated in successful promoter clearance and transcriptional elongation.

Although Tat makes no contact with residues in the TAR RNA loop (Fig. 8.12B), these sequences are crucial for the stimulation of transcription by the viral protein. The activation domain of Tat binds directly to the cyclin subunit of the Tat-associated kinase. The viral-cellular protein complex makes contact with the RNA loop and binds to TAR RNA with higher affinity and greater specificity than does Tat alone. It appears that Tat, and presumably the kinase bound to it, then becomes associated with transcription complexes (Fig. 8.13).

In this mechanism, the sole function of Tat is to ensure association of the C-terminal domain kinase (Tak) with the transcriptional complex. However, the viral protein stimulates the activity of a second, C-terminal domain kinase present in the general initiation protein TfIIh, and inhibition of the TfIIh kinase impairs the ability of Tat to stimulate processive viral transcription. As we have seen, TfIIh is a component of the preinitiation complex and acts during initiation and promoter clearance. However, it is released from the transcriptional complex when nascent RNA chains of 30 to 50 nucleotides have been made. In contrast, the Tat-associated kinase modulates the activity of transcriptional complexes at a later stage, when nascent RNAs are 15 to about 500 nucleotides in length. It is therefore believed that phosphorylation of the C-terminal domain of RNA polymerase II is a two-step process, with TfIIh and Tat-associated kinase acting sequentially (Fig. 8.13).

At this juncture, it is difficult to appreciate the value of the intricate transcriptional program of human immunodeficiency virus type 1. Those of many other viruses are executed successfully by transcriptional regulators that operate, directly or indirectly, via specific DNA sequences, and Tat can stimulate transcription effectively when made to bind to DNA experimentally. Why, then, is efficient transcription of the proviruses of human immunodeficiency virus type 1 and its relatives mediated by the unique, RNA-dependent mechanism of bringing a transcriptional activator to the promoter? Binding of Tat to nascent viral RNA close to the site at which many transcriptional complexes pause or stall (Fig. 18.13) could provide a particularly effective way to recruit the cellular proteins that stimulate processive transcription. Or it may be that regulation of transcription via an RNA sequence is a legacy from retroviral ancestors that replicated in an RNA world.

It is clear that the efficient production of progeny human immunodeficiency virus type 1 genomes by transcription of the provirus ultimately depends on the constellation or activation state of cellular enhancer-binding proteins. Transcription of even small quantities of full-length viral transcripts in response to these cellular proteins allows the synthesis of Tat mRNA and protein, and consequently activation of the positive, autoregulatory loop. The provirus can be considered dormant in cells that do not contain the necessary enhancer-binding proteins or in which these proteins are inactive. However, it is important to keep in mind that such transcriptional inactivity is **not** the cause of the clinical latency characteristic of human immunodeficiency virus type 1 infections. In clinical latency, few, if any, symptoms are manifested. Nevertheless, virus is produced continuously (Chapter 17), because the positive, autoregulatory circuit is triggered whenever infected cells can support LTR enhancer-dependent transcription of proviral DNA.

The Transcriptional Cascades of DNA Viruses

The transcriptional strategies of viruses with DNA genomes exhibit a number of common features. The most striking is the transcription of viral genes in a reproducible and precise sequence. Prior to initiation of viral DNA synthesis, during **immediate-early** and **early** phases in the reproductive cycle, infected cells are devoted to the production of viral proteins necessary for viral DNA synthesis, efficient expression of viral genes, or other regulatory functions. Transcription of the **late** genes, most of which encode structural proteins, requires viral DNA synthesis

(Fig. 8.14). This property ensures coordinated production of the DNA genomes and the structural proteins from which progeny virus particles are assembled. Another common feature is the control of the transitions from one transcriptional stage to the next by both viral proteins and replication of viral DNA (Fig. 8.14).

The transcriptional activating proteins encoded by different viruses exhibit distinctive molecular properties. However, all cooperate with cellular transcriptional components to stimulate transcription of specific sets of viral genes expressed inefficiently in their absence. In addition, many (including simian virus 40 large T antigen and the herpes simplex virus type 1 ICP4 protein) can repress transcription of their own genes. Such autoregulation presumably circumvents unfavorable consequences of too-high concentrations of these viral proteins, for example, unnecessary use of the resources of the infected cell, or inappropriate transcription of cellular genes. These viral transcriptional programs closely resemble those regulating many developmental processes in animals, in both the transcription of individual genes in a predetermined sequence, and the sequential action of viral proteins that regulate the transcription of different sets of viral genes.

The Transcriptional Programs of DNA Viruses

Sequential transcription of viral genes. The viral transcriptional programs established in cells infected by viruses with small DNA genomes are quite simple. For example, the genome of simian virus 40 contains only two transcription units (Appendix A, Fig. 15B), each of which encodes more than one protein. This type of organization reduces the genetic information that must be devoted to transcription punctuation marks and regulatory sequences, a significant advantage when genome size is limited by packaging constraints. The price for such a transcriptional strategy is heavy dependence on the host cell's RNA processing systems to generate multiple mRNAs by differential polyadenylation and/or splicing of a single primary transcript (Chapter 10). Expression of the simian virus 40 early transcription unit by the cellular enhancer- and promoter-binding proteins described below leads to the synthesis of large T antigen in infected cells. This multifunctional viral protein induces initiation of viral DNA synthesis and activates late transcription (Fig. 8.14). Consequently, T-antigen synthesis in a permissive host cell induces progression to the late phase of infection, when the segment of the viral genome encoding structural proteins is transcribed.

Adenoviral genomes also contain a minimal number of transcriptional regulatory sequences, for the sequences encoding the more than 40 viral proteins are organized into only eight RNA polymerase II transcription units (Appendix A, Fig. 1B). In contrast to that of polyomaviruses, the initial period of the adenoviral infectious cycle comprises two distinct transcriptional phases (Fig. 8.14). Upon entry of the viral genome into the nucleus, a single viral gene (E1A) is transcribed under the control of typical enhancers. If func-

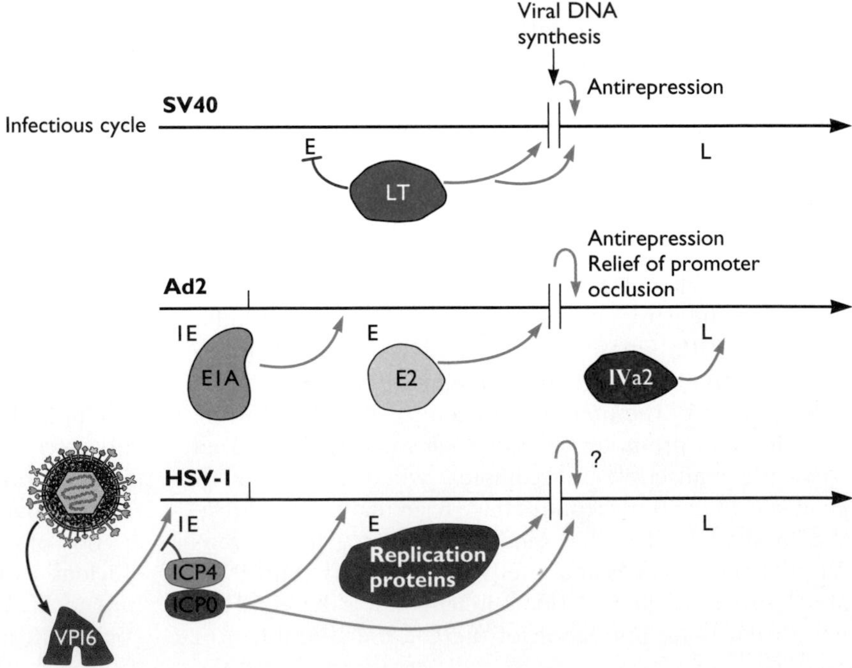

Figure 8.14 The simian virus 40 (SV40), human adenovirus type 2 (Ad2), and herpes simplex virus type 1 (HSV-1) transcriptional programs. The transcriptional programs of these three viruses are depicted by the horizontal time lines, on which the onset of viral DNA synthesis is indicated by vertical lines. For comparative purposes only, the three reproductive cycles are represented by lines of equal length. The immediate-early (IE), early (E), and late (L) transcriptional phases are indicated, as are viral proteins that participate in regulation of transcription. Stimulation of transcription by these proteins and effects contingent on viral DNA synthesis in infected cells are indicated by green and orange arrows, respectively. Red bars indicate negative regulation of transcription.

tional E1A proteins cannot be made, progression of the infectious cycle beyond this immediate-early phase is severely impaired. The E1A proteins, which regulate transcription by multiple mechanisms, are necessary for efficient transcription of all viral early transcription units (Fig. 8.14). Among this set is the E2 gene, which encodes the proteins required for viral DNA synthesis and entry into the late phase of infection (Fig 8.14).

The adenoviral late phase is marked not only by transcription of late phase-specific genes, but also by an increased rate of initiation from the major late promoter that is active during the early phase of infection. Late genes of the more complex DNA viruses are therefore defined as those that attain their maximal rates of transcription following viral DNA synthesis. However, no known activator of late transcription is included among the adenoviral early gene products. Instead, the synthesis of progeny adenoviral DNA molecules is indirectly coordinated with production of the protein components that will encapsidate them in virions (Fig. 8.14). In this respect, adenoviruses can be distinguished from both simpler polyomaviruses and more complex herpesviruses, such as herpes simplex virus type 1.

The organization of the herpes simplex virus type 1 genome demands an entirely different transcriptional strategy. The more than 80 known genes of this virus are, with few exceptions, expressed as individual transcription units. Furthermore, splicing of herpes simplex virus type 1 primary transcripts is the exception, rather than the rule. In contrast to those of simian virus 40 or human adenoviruses, herpes simplex virus type 1 genes that encode proteins that participate in the same process are not clustered, but are scattered throughout the genome (Appendix A, Fig. 5B). The basic distinction of early and late phases is maintained in the herpes viral transcriptional program (Fig. 8.14), but temporal control of the activity of more than 80 viral promoters is obviously more complicated. In fact, the potential for finely tuned regulation is much greater when the viral genome comprises a large number of independent transcription units.

Initial expression of both polyomaviral and adenoviral genes is directed by enhancers that operate efficiently with cellular proteins. In contrast, a viral activating protein is imported into cells infected by herpes simplex virus type 1 (Fig. 8.14). This virion structural protein, VP16, is necessary for efficient transcription of viral immediate-early genes. Although this simple device might seem to guarantee transcription of these genes in **all** infected cells, this is not necessarily the case: VP16 can function only in conjunction with specific cellular proteins. Stimulation of immediate-early gene transcription by VP16 is of considerable importance for the success of the viral reproductive cycle. For example, viruses encoding VP16 that is defective for activation of transcription grow well only at high multiplicities of infection. Mutant adenoviruses lacking E1A coding sequences exhibit the same phenotype. Neither of these transcriptional regulatory proteins is absolutely essential for virus reproduction, at least in cells in culture.

Herpesviral immediate-early gene products resemble adenoviral E1A proteins in performing a number of regulatory functions. The regulatory scheme of herpes simplex virus type 1 is, however, considerably more complex, for this viral genome carries five such genes (Appendix A, Fig. 5B). The products of at least two, ICP4 and ICP0, are transcriptional regulators. The ICP4 protein, which is essential for progression beyond the immediate-early phase of infection, is regarded as the major transcriptional activator. It stimulates transcription of both early and late genes, and also acts as a repressor of immediate-early gene transcription. As for the simpler DNA viruses, transcription of herpesviral late genes is governed by viral DNA replication (Fig. 8.14). Herpes simplex virus type 1 late genes can be divided into those that are not transcribed prior to viral DNA synthesis, and those that are transcribed at maximal rates following initiation of such synthesis. More subtle distinctions among the large number of late genes may be made as their transcriptional regulation becomes better understood.

This summary of DNA virus transcriptional programs illustrates the increasing sophistication of the circuits that control transcription with increasing genome size, a feature of little surprise. Underlying such diversity are the common themes of the central role of virus-encoded transcriptional regulators and the coordination of transcriptional control with viral DNA synthesis. Regulation of initiation is the mainstay of the transcriptional control programs of these (and other) DNA viruses. Nevertheless, expression of some genes is controlled at other steps in the transcription cycle, such as termination.

Entry into one of two alternative transcriptional programs. The herpesviruses are unique among the viruses described in this section in their ability to establish **latent infections** in specific cell types. This state is characterized by both lack of efficient expression of all viral genes that are transcribed during productive infection, and activation of a unique, latent-phase transcriptional program. For example, during latent infection of neurons by herpes simplex virus type 1, a set of specific RNAs, termed the **latency-associated transcripts** (LATs) (Fig. 8.15), are the only viral products synthesized in large quantities. The 2.0- and 1.5-kb latency-associated transcripts (Fig. 8.15) accumulate to 40,000 to 100,000 copies within the nucleus of a latently infected neuron, lack poly(A) tails, and are not linear mole-

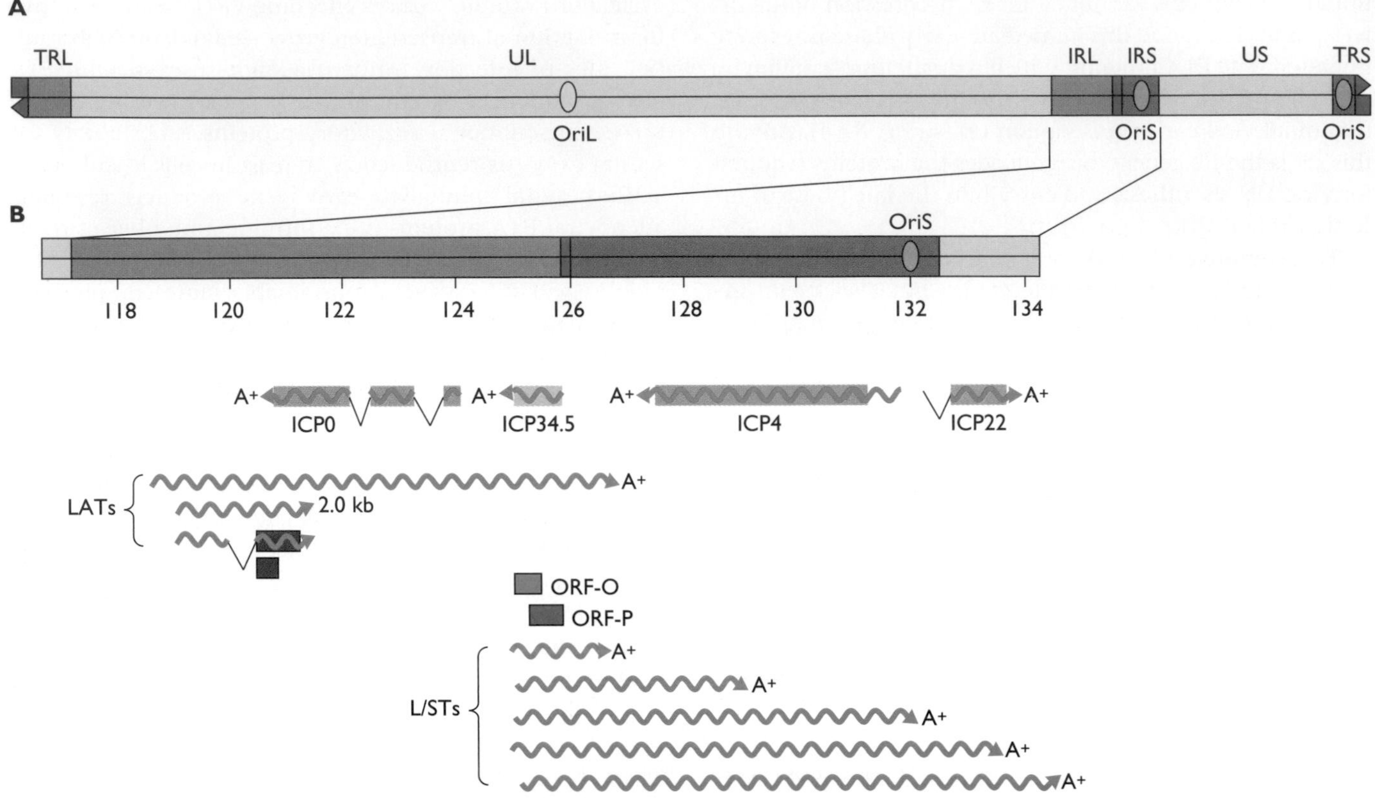

Figure 8.15 The latency-associated transcripts of herpes simplex virus type 1. (A) Diagram of the herpes simplex virus type 1 genome, showing the unique long and short segments, U_L and U_S, respectively, the terminal repeat (TRL, TRS) and internal repeat (IRL, IRS) sequences; and the origins of replication, OriL and OriS. (B) Expanded map of the IRL region, with the scale in kilobase pairs. This region encodes immediate-early proteins ICP0, ICP4, and ICP22, which play important roles in establishing a productive infection. Below are shown the locations of sequences encoding the latency-associated transcripts (LATs) and the transcripts named L/STs. The arrow indicates the direction of transcription, and A+ indicates 3′ poly(A) sequences where known. Open reading frames, including ORF-O and ORF-P, are shown as boxes. The evidence currently available suggests that ORF-P (and perhaps ORF-O) may control LAT RNA synthesis.

cules. Indeed, all properties established to date indicate that they are stable introns produced by splicing of precursor RNA, as discussed in Chapter 10. Regulatory mechanisms that determine whether the transcriptional cascade described above is activated or repressed and that control transcription of LAT genes must be of paramount importance in determining the outcome of herpes simplex virus infection. Reentry into the productive cycle from the latent state, a process called **reactivation** (see Chapter 16), must also be governed by transcriptional regulatory mechanisms.

The mechanisms that control whether productive or latent infections are established, and that mediate reactivation, are by no means fully understood. One viral transcriptional regulator likely to be important in these processes is the immediate-early protein ICP0. This protein is a powerful activator of transcription of viral and cellular genes in simplified experimental systems, and is required for efficient viral replication in cells infected at low multiplicity. In its absence, productive infection of cultured cells by herpes simplex virus type 1 is very inefficient, as is reactivation from latently infected murine ganglia. However, recent studies indicate that ICP0 modulates transcription of viral genes by an unusual, indirect mechanism. This protein localizes to specific nuclear structures called **promyelocytic leukemia bodies** (described in more detail in Chapter 9) and induces their reorganization, concomitant with degradation of some protein components by the proteasome. Mutations that compromise this activity of ICP0 also impair progression through the productive cycle. Furthermore, inhibitors of the proteasome block both the productive cycle in cells infected by wild-type virus, and the ability of ICP0 to support reactivation from latency. These properties indicate that reorganization of promyelocytic leukemia bodies induced by the ICP0 protein overcomes repression of transcription of viral genes early in the productive cycle, or during reactivation from latency. The ICP0 protein induces the degradation of several cellular proteins, but which of these proteins repress viral transcription is not yet known.

The latency-associated transcripts themselves also appear to be important for establishing a latent infection. Stable production of the abundant latency-associated transcripts in neuronal cells suppresses replication of the virus and production of the immediate-early gene products needed for progression through the productive infectious cycle. The intriguing question of how viral RNAs with the properties of introns excised from pre-mRNAs might repress expression of these viral genes cannot yet be answered. However, the 2.0-kb LAT RNA has been reported to be associated with cellular ribosomes and splicing proteins, suggesting that it might alter production of cellular proteins that modulate viral gene expression. The products of other RNAs made in smaller quantities in latently infected neurons, such as the L/ST RNAs (Fig. 8.16), can block the binding of the immediate-early ICP4 protein to promoter sequences. These observations suggest that once latency-associated genes are expressed in an infected neuron, the transcriptional cascade characteristic of productive infection is repressed. Much less clear at present are the mechanisms that control transcription of these genes. As discussed below, the regulatory circuits that govern entry into the productive or latent transcriptional programs of other herpesviruses, notably the gammaherpesvirus Epstein-Barr virus, appear to be considerably simpler and are better understood.

The Simian Virus 40 Enhancer: a Model for Viral and Cellular Enhancers

The majority of viral DNA templates described in this chapter contain enhancers of transcription that are recognized by cellular DNA-binding proteins. The simian virus 40 enhancer has been studied intensively, in part because it was the first such regulatory sequence to be identified. Its properties and mechanism of action are considered characteristic of many enhancers, whether of viral or cellular origin.

Figure 8.16 Organization of the simian virus 40 enhancer. The positions of the 72-bp repeat region of the simian virus 40 genome containing the enhancer elements relative to the early promoter are shown at the top. Shown to scale below are functional DNA sequence units, and the proteins that bind to them, of the early promoter-distal 72-bp repeat and its 5′-flanking sequence, which forms part of enhancer element B. All the protein-binding sites shown between the expansion lines are repeated in the promoter-proximal 72-bp repeat. The complete enhancer contains one copy of the enhancer element B1 and two directly repeated copies of the enhancer elements B2 and A. The definition of enhancer elements is discussed in the text. Some enhancer elements are built from repeated binding sites for a single sequence-specific protein. The Sph-II sequences, which are recognized by the cellular protein Tef-1, are representative of this class, as formation of a functional enhancer element requires cooperative binding of Tef-1 to the two Sph-II sequences. Such cooperative binding, which renders enhancer activity sensitive to small changes in the concentration of a single protein, represents a regulatory device of great potential sensitivity. A second class of enhancer elements comprises sequences bound by two different proteins, as illustrated by enhancer element B. Although the binding of Tef-1 and Tef-2 to the GT-IIC and GT-I sequences, respectively, is not cooperative, these proteins interact once bound to DNA to form an active enhancer element. A third kind of enhancer element contains neither a precisely spaced duplicate (to permit cooperative binding) nor a second protein-binding site. Rather, oligomerization of a single sequence is sufficient to form an active enhancer element. The TC-II sequence recognized by Nf-κb functions in this manner.

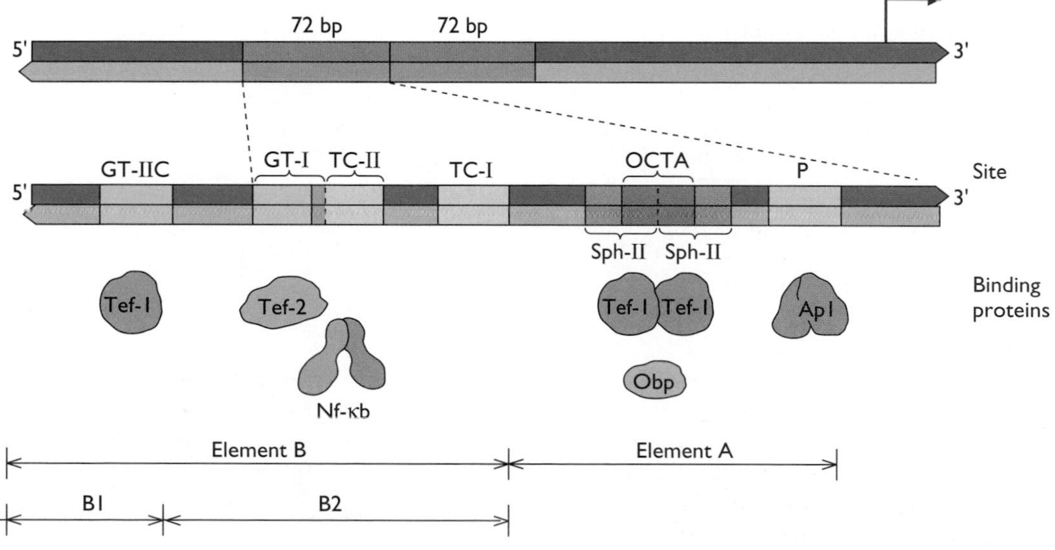

BOX 8.6

Typical properties of the simian virus 40 enhancer

An enhancer is composed of multiple units, termed enhancer elements (e.g., the B1, B2, and A elements of the simian virus 40 enhancer) (Fig. 8.16).

Enhancer elements operate synergistically: a single element has little activity, but the complete set stimulates transcription more than 100-fold.

Multimerization of a single inactive enhancer element can generate an active enhancer.

Enhancer activity is relatively insensitive to the orientation or position of individual elements: stimulation of transcription generally depends on the presence of multiple elements, rather than on the exact way in which these are arranged.

Enhancer elements comprise multiple sequences recognized by sequence-specific DNA-binding proteins that can also bind to promoter sequences (Fig. 8.8, 8.11, and 8.16).

The simian virus 40 enhancer displays the hierarchical organization typical of these regulatory regions (Box 8.6). It is built from three units, termed enhancer elements, which are subdivided into smaller sequences recognized by DNA-binding proteins (Fig. 8.16). The DNA-binding proteins that interact with this viral enhancer are differentially produced in different cell types. For example, Nf-κb and certain members of the octamer-binding protein (Obp) family (Fig. 8.16) are specific to cells of lymphoid origin, and their binding sites are necessary for enhancer activity

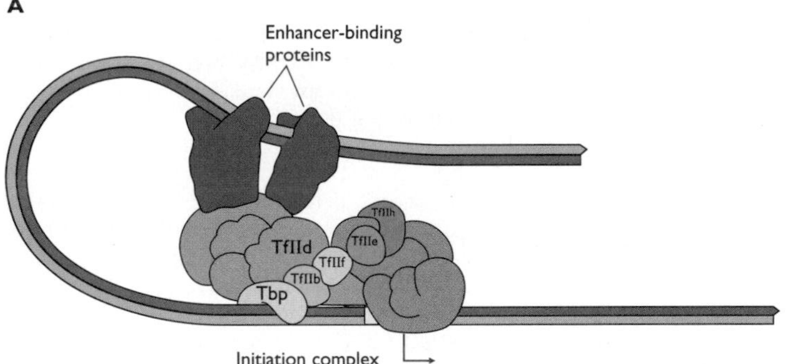

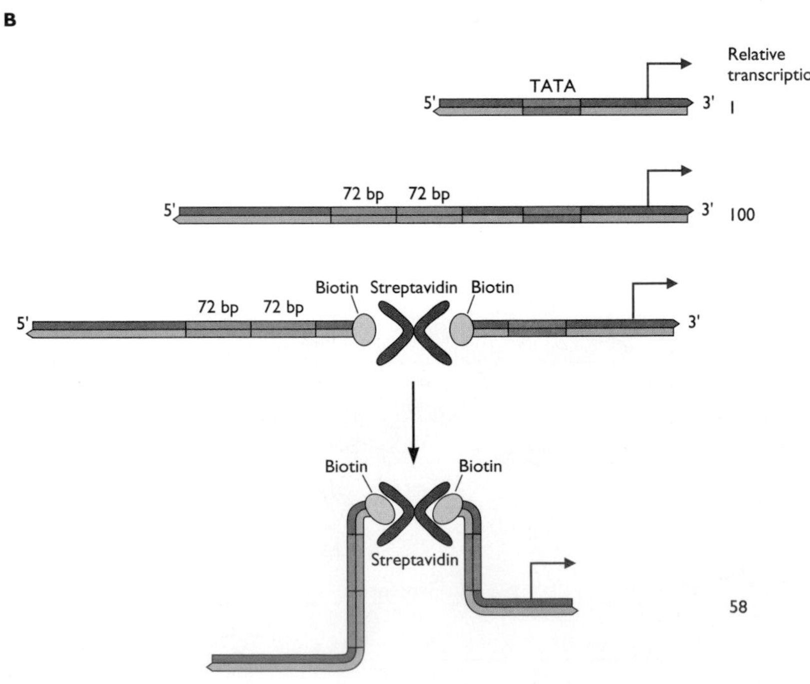

Figure 8.17 Mechanisms of enhancer action. (A) The DNA looping model postulates that proteins bound to a distant enhancer, here shown upstream of a gene, interact directly with components of the transcription initiation complex, with the intervening DNA looped out. These interactions of proteins bound to a distal enhancer are analogous to those that can take place when the enhancer is located close to the promoter. Such interactions might stabilize the initiation complex and therefore stimulate transcription. (B) An enhancer noncovalently linked to a promoter via a protein bridge is functional. When placed immediately upstream of the rabbit β-globin gene promoter, the simian virus 40 enhancer stimulates specific in vitro transcription from circular plasmids by a factor of 100. In the experiment summarized here, the enhancer and promoter were separated by restriction endonuclease cleavage. Under this condition, the enhancer cannot stimulate transcription (not shown). The protein biotin was added to the ends of each DNA fragment by incorporation of biotinylated UTP. Biotin binds the protein streptavidin noncovalently, but with extremely high affinity (K_d, 10^{-15} M). Because streptavidin can bind four molecules of biotin, its addition to the biotinylated DNA fragments allows formation of a noncovalent protein "bridge" linking the enhancer and the promoter. Under these conditions, the viral enhancer stimulates in vitro transcription from the rabbit β-globin promoter almost as efficiently as when present in the same DNA molecule, as summarized at the right. Because this result indicates that an enhancer can stimulate transcription when present in a separate DNA molecule (i.e., in *trans*), it rules out models in which enhancers are proposed to serve as entry sites for RNA polymerase II (or other components of the transcription machinery): such a mechanism requires that RNA polymerase II slide along the DNA from the enhancer to the promoter, a passage that would be blocked by the protein bridge. Rather, this result suggests that enhancer function requires close proximity (but not covalent linkage) to the promoter. The results of this experiment are therefore consistent with the looping model shown in panel A.

in these cells. Other elements of the simian virus 40 enhancer, such as the activator protein 1 (Ap-1) binding sites, confer responsiveness to cellular signaling pathways. The combination of DNA sequences ensures the activity of the enhancer, and therefore initiation of the viral infectious cycle, in many different cellular environments. This property is exhibited by several other viral enhancers, including those of avian retroviruses discussed previously (Table 8.4). In contrast, some viral templates for RNA polymerase II transcription contain enhancers that are active only in a specific tissue (for example, hepatitis B virus DNA) or only in the presence of inducers or viral proteins (Table 8.4).

The simian virus 40 enhancer is located within 200 bp of the transcription initiation site. More typically, enhancers are found hundreds, thousands, or tens of thousands of base pairs up- or downstream of the promoters that they regulate. The most popular model of the mechanism by which these sequences exert remote control of transcription, the DNA looping model, invokes interactions among enhancer-bound proteins and the transcriptional components assembled at the promoter, with the intervening DNA looped out (Fig. 8.17A). Compelling evidence in favor of this model has been collected by using the simian virus 40 enhancer (Fig. 8.17B). These regulatory sequences can also facilitate access of the transcriptional machinery to chromatin templates. For example, the simian virus 40 enhancer lies within, and contains DNA sequences necessary for formation of, a nucleosome-free region of the viral genome in infected cells. Enhancers can, therefore, stimulate RNA polymerase II transcription by multiple molecular mechanisms. The primary effect of these mechanisms is to increase the probability that the gene to which an enhancer is linked will be transcribed.

Transcriptional Regulators Encoded in Viral Genomes

Viral proteins that stimulate transcription are linchpins of the transcriptional cascades illustrated in Fig. 8.10. Consequently, they have received much attention. In this section, we focus on a few selected viral regulators to illustrate general principles of their operation, or fundamental insights into cellular processes that have been gained through their study.

Some viral transcriptional regulators (Table 8.5) are close relatives of well-studied cellular proteins that bind to spe-

Table 8.5 Some viral transcriptional regulators that bind to specific DNA sequences

Virus	Protein	Properties	Functions
Adenovirus			
Human adenovirus type 2	IVa2	Binds intragenic sequence in the major late promoter; late-phase-specific	In conjunction with a second, infected cell-specific protein that binds intragenic sequences, stimulates transcription from major late promoter
Baculovirus			
Autographa californica nuclear polyhedrosis virus	IE-1	Typical domain organization; binds as dimer to 28-bp palindromic sequences	Stimulates transcription from early promoters, including its own, by binding to enhancer elements
Herpesvirus			
Herpes simplex virus type 1	ICP4	Typical domain organization; one domain related in sequence to a domain of the cellular coactivator p15; binds to a degenerate consensus sequence	Stimulates transcription from early and late promoters by a mechanism that is not well understood, but may not require binding to a specific DNA sequence; represses transcription from its own and other IE promoters by binding to promoter sequences
Epstein-Barr virus	Zta	bZip protein; synthesis and activity regulated by multiple mechanisms (see text)	Essential activator of early gene transcription; commits to lytic infection (see text)
Papillomavirus			
Bovine papillomavirus type 1	E2	Typical domain organization; binds as dimers; truncated forms lack activation domains; can bind TfIIb and TfIId	Stimulation of transcription from several promoters as enhancer-binding protein; truncated forms compete for DNA binding and inhibit transcriptional stimulation by full-length E2 protein; replication of viral DNA in vivo
Poxvirus			
Vaccinia virus	VETF	Binds to TA-rich early promoter sequences as heterodimer; DNA-dependent ATPase	Essential for recognition of, and transcription from, early promoters by the viral RNA polymerase (Fig. 8.25)

Table 8.6 Some viral transcriptional regulators that lack sequence-specific DNA-binding activity

Virus	Protein	Functions and properties
Adenovirus		
Human adenovirus type 2	E1A proteins	Larger stimulates early gene transcription in infected cells via CR3; both bind to the Rb protein releasing cellular E2f, and to the cellular coactivator p300 (see text); both repress enhancer-dependent transcription
Hepadnavirus		
Hepatitis B virus	X	Required for efficient virus reproduction in vivo; promiscuous activator of transcription by RNA polymerases II and III in experimental assays; stimulation of RNA polymerase II transcription depends on cellular, sequence-specific activators; may be a general coactivator; may also stimulate transcription indirectly via activation of cellular signaling pathways
Herpesvirus		
Herpes simplex virus type 1	VP16	Stimulates transcription from immediate-early promoters via potent acidic activation domain; achieves promoter specificity by interaction with the cellular Oct-1 protein (see text)
	ICP0	Cooperates with ICP4 to stimulate transcription of E and L genes in infected cells; important for reactivation from latency (see text); stimulates transcription from many promoters in experimental assays; can stimulate transcription synergistically with ICP4 in such assays and bind to ICP4
Epstein-Barr virus	EBNA-2	Essential for B-cell transformation; induces transcription of viral LMP-1 and LMP-2 genes and of some cellular genes via specific DNA sequence; sequence recognition achieved by interaction with the cellular recombination binding protein Jκ; contains an acidic activation domain that binds to Tbp and a Tbp-associated protein of TfIId
Polyomavirus		
Simian virus 40	Large T antigen[a]	Stimulates late gene transcription; can stimulate transcription from many simple promoters in experimental assays; can bind to Tbp, TfIIb, and Tef-1 via different segments; contains no activation domain; may stabilize initiation complexes
Poxvirus		
Vaccinia virus	VTF1	Subunit of viral RNA polymerase with sequence homology to cellular RNA polymerase II elongation protein TfIIs; required for transcription of viral intermediate genes
	VTL1, -2, and -3	Necessary and sufficient for transcription from viral late promoters by the viral RNA polymerase
Retrovirus		
Human T-lymphotropic virus type 1	Tax	Stimulates transcription from the viral LTR transcriptional control region; in vitro, binds directly to the basic region of bZIP proteins to stabilize bZIP protein dimers and increase their affinity for DNA; binds to free and DNA-bound Nf-κb in vitro; also regulates transcription indirectly, by inducing degradation of IκB

[a] The sequence-specific DNA binding activity of large T antigen (see Fig. 9.3) is not required for stimulation of transcription by this protein.

cific DNA sequences in promoters or enhancers. Not every viral DNA-binding protein listed in Table 8.5 has been studied in detail, but all their known properties emphasize this resemblance. These viral proteins possess discrete DNA-binding domains, some with sequences that conform to motifs characteristic of cellular DNA-binding proteins, and activation domains that interact with general initiation proteins (Table 8.5). These properties are described in more detail for one such viral protein, the Epstein-Barr virus Zta protein, in the next section. Sequence-specific DNA-binding proteins play ubiquitous roles in the transcription of cellular genes by RNA polymerase II, so it is hardly surprising that the genomes of DNA viruses transcribed by this enzyme contain coding sequences for analogous proteins. However, viral transcriptional regulators that possess no intrinsic ability to bind specifically to DNA are equally common (Table 8.6). The preponderance of such viral proteins, quite unexpected when they were first characterized, was a strong indication that host cells also contain proteins that control transcription without themselves binding to DNA. Many such proteins (e.g., coactivators or mediators) have now been recognized. Two examples, the herpes simplex virus type 1 VP16 and adenovirus E1A proteins, illustrate the diversity of mechanisms by which this class of viral proteins regulates transcription.

Multiple mechanisms of regulation of the activity and availability of the Epstein-Barr virus Zta protein. When the gammaherpesvirus Epstein-Barr virus infects B lymphocytes, a latent state characterized by expression of a limited number of viral genes is established. The products of these genes maintain the viral genome via replication from a latent-phase-specific origin of replication, modulate the immune system, and alter the growth properties of the cells. Treatment of such latently infected lymphocytic cells in culture with a variety of agents induces the viral productive cycle. Epstein-Barr virus also enters this cycle upon infection of epithelial cells. Virus reproduction begins with synthesis of three viral proteins that regulate gene expression. However, just one of these, the transcriptional regulator Zta (also known as ZEBRA, Z, or EB-1), is sufficient to interrupt latency and induce entry into the productive cycle.

The Zta protein exhibits many properties characteristic of the cellular proteins that recognize local promoter sequences. It is a modular, sequence-specific DNA-binding protein that belongs to the basic-leucine zipper family (Table 8.5). This dimerization and DNA-binding domain mediates binding of Zta to viral promoters that contain its recognition sequence. Its discrete activation domain, which can bind directly to cellular initiation proteins, such as subunits of TfIIId, is believed to facilitate the assembly of preinitiation complexes. The Zta protein acts synergistically (Box 8.4) with a second viral activating protein (Rta), also made at the beginning of the productive cycle, to stimulate transcription from viral early promoters. The availability or activity of Zta is regulated by numerous mechanisms. Expression of the viral gene encoding Zta is controlled by several cellular DNA-binding proteins that stimulate or repress its transcription (Fig. 8.18). Many of the cellular proteins that stimulate transcription from the Zta promoter are targets of signal transduction cascades that induce entry of Epstein-Barr virus into the productive cycle in latently infected B cells (Fig. 8.18). In addition, the Zta protein is itself a positive autoregulator of transcription. The availability of Zta mRNA for translation is also regulated, in part, by annealing of Zta pre-mRNA to the complementary transcripts of the viral EBNA-1 gene (Fig. 8.18). Finally, phosphorylation of Zta modulates its DNA-binding activity, whereas binding of cellular proteins, including the p53 tumor suppressor protein and the p65 subunit of Nf-κb (Fig. 8.6), prevents Zta from stimulating transcription.

The net effect of these regulatory mechanisms, which depends on the type, growth state, differentiation state, and environment of the Epstein-Barr virus-infected cell, determines whether active Zta protein is available, and consequently whether infection leads to latent or productive replication. Entry into the productive cycle appears to be an inevitable consequence of production of active Zta. This protein stimulates transcription from the promoter of its own gene and also that of the viral gene encoding the Rta protein, with which Zta cooperates to activate early gene transcription. Furthermore, Zta plays an important role in replication from the lytic origins.

Sequence-specific transcriptional regulation via cellular DNA-binding proteins: the herpes simplex virus type 1 VP16 protein. The herpesviral VP16 protein, which enters infected cells in the virion (Fig. 8.14), has taught us much about mechanisms by which transcription by RNA polymerase II can be stimulated. Furthermore, its unusual mode of promoter recognition illustrates the importance of conformational change during formation of DNA-bound complexes.

Although it lacks a DNA-binding domain, the herpes simplex virus type 1 VP16 protein has been used widely to investigate mechanisms of stimulation of transcription. Its C-terminal, acidic activation domain is one of the most potent known. Chimeric proteins in which this acidic activation domain are fused to heterologous DNA-binding domains strongly stimulate transcription from promoters that contain the appropriate binding sites. When part of

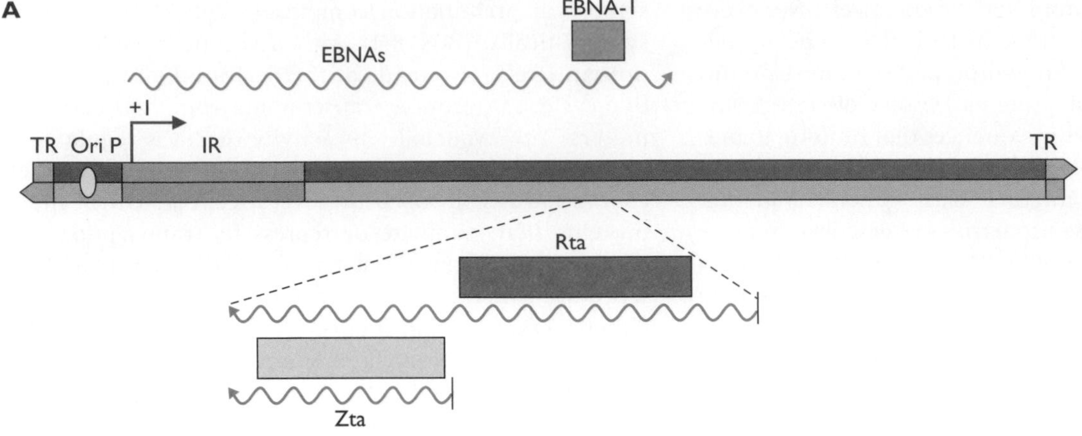

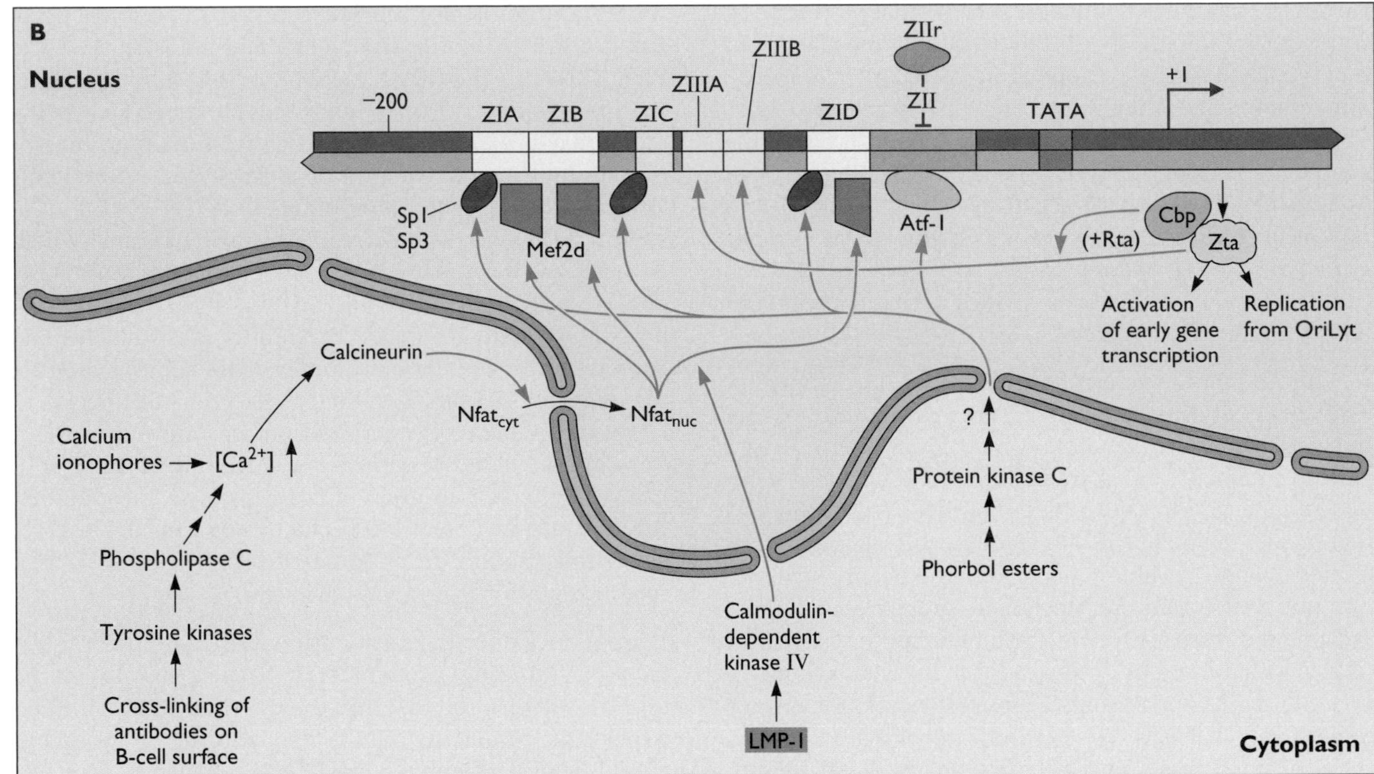

Figure 8.18 Organization and regulation of the Epstein-Barr virus Zta gene promoter Zp. (A) Organization of the transcription units that contain the coding sequence for the Epstein-Barr virus nuclear antigen (EBNA) proteins (an ~100-kb transcription unit), Zta, and Rta. The locations of the genomic terminal (TR) and internal (IR) repeat sequences, the origin of replication for plasmid maintenance (OriP), and the coding sequences for the EBNA-1, Zta, and Rta proteins are indicated. The last two proteins are synthesized from spliced mRNAs processed from the primary transcripts shown. (B) Sequences that regulate transcription from the Zp promoter, to scale, and the cellular or viral proteins that operate through them. The Atf-1 and Mef2d proteins are members of the basic-leucine zipper and Mads families, respectively. The activities of the cellular transcriptional regulators are controlled by several pathways that transmit signals from the surface of the cell to the transcriptional machinery. Such signals, which include cross-linking of antibodies on the B-cell surface by anti-immunoglobulin antibodies and phorbol esters, induce entry into the lytic cycle, presumably as a result of activation of transcription from the Zp promoter. Expression of viral latent membrane protein 1 (LMP-1) can also activate transcription, as does synthesis of Zta itself. The viral protein binds to the ZIII sites in the promoter and stimulates transcription in conjunction with the cellular coactivator Cbp.

(continued on next page)

such fusion proteins, the VP16 acidic activation domain can stimulate several initiation reactions, consistent with its ability to bind directly to multiple general initiation proteins (Fig. 8.19A). It can also increase the rate of transcriptional elongation and stimulate transcription from chromatin templates (Fig. 8.19B). These properties established that a single protein can regulate RNA polymerase II transcription by multiple molecular mechanisms.

The herpes simplex virus type 1 VP16 protein is the founding member of a class of viral regulators that activate transcription from promoters that contain a specific consensus sequence, yet possess no sequence-specific DNA-binding activity. The 5′-flanking regions of viral immediate-early genes contain at least one copy of a consensus sequence, 5′TAATGARAT3′ (where R is a purine), which is necessary for VP16-dependent activation of their transcription. This sequence is bound by VP16 only in association with at least two cellular proteins, octamer-binding protein 1 (Oct-1) and host cell factor (Hcf) (Fig. 8.20). The Oct-1 protein is a ubiquitous transcriptional activator named for its recognition of a consensus DNA sequence termed the octamer motif. This protein and VP16 can associate to form a ternary (three-component) complex on the 5′TAATGARAT3′ sequence, but the second cellular protein, Hcf, is necessary for stable, high-affinity binding. The VP16 protein and Hcf form a heteromeric complex in the absence of Oct-1 or DNA. The primary function of Hcf is therefore believed to be induction of a conformational change in VP16, to allow its stable binding to Oct-1 on the immediate-early promoters (Fig. 8.20).

One of the most remarkable features of the mechanism by which the VP16 protein is recruited to viral immediate-early promoters is its specificity for Oct-1. This protein is a member of a family of related transcriptional regulators defined by a common DNA-binding motif called the POU-homeodomain. The VP16 protein distinguishes Oct-1 from all other members of this family, including Oct-2, which binds to exactly the same DNA sequence as Oct-1. In fact, VP16 detects a single amino acid difference in the exposed surfaces of DNA-bound Oct-1 and Oct-2 homeodomains. Such discrimination is not only remarkable but also of biological importance, because only the Oct-1 protein is synthesized ubiquitously.

The GARAT segment of TAATGARAT-containing elements of immediate-early genes is indispensable for assembly of VP16 into the DNA-bound complex and therefore defines promoters that are recognized by the viral protein. Nevertheless, the DNA-binding domain of Oct-1, not VP16, recognizes the GARAT motif (Fig. 8.20). This DNA sequence is therefore the crucial effector of assembly of the quaternary complex. The mechanism by which VP16, in conjunction with Hcf and Oct-1, recognizes herpesviral immediate-early promoters illustrates the important contributions of induced conformational changes to transcriptional specificity.

The VP16 protein interacts with cellular Hcf and Oct-1 proteins via its N-terminal region. Its C-terminal region contains the acidic activation domain discussed previously, which can act from sites far from a promoter when fused to a heterologous DNA-binding domain. In contrast, when present in the ternary complex with cellular proteins, VP16 stimulates transcription only if the complex is bound close to the promoter. This difference indicates that the activation domain does not function in the same manner in natural complexes and artificial chimeras. Which, if any, of the biochemical activities exhibited by the VP16 acidic activation domain when fused to a heterologous DNA-binding domain (e.g., Fig. 8.19) pertain to the mechanism of stimulation of immediate-early gene transcription in infected cells has yet to be determined.

The incorporation of the VP16 protein into virions at the end of one infectious cycle is an effective way to ensure transcription of viral genes and initiation of a new cycle, once a new host cell has been infected. Nevertheless, some features of this mechanism are not fully appreciated, in particular the benefits conferred by the indirect mechanism by which VP16 recognizes immediate-early promoters. One advantage over direct DNA binding may be the opportunity to monitor the growth state of the host cell that is provided by the requirement for binding to Hcf, a protein that plays a role in cell cycle progression in uninfected cells. And the strictly local action of VP16 imposed by its association with Oct-1 and Hcf (see above) might well be necessary to prevent untimely transcription of other viral genes in response to the potent acidic activation domain of this viral transcriptional regulator. The dependence on Hcf may also contribute to the establishment of latent infections in neurons. In contrast to many cell types in which Hcf is present in the nucleus, this protein is sequestered in the cytoplasm in neurons.

Figure 8.18 *(continued)* Transcription from the Zp promoter is negatively regulated by the ZII sequence, to which the cellular repressor ZIIr binds, and by several additional sequences upstream of position 2220. Transcripts from the upstream promoter of the Rta gene (A) yield processed bicistronic mRNAs that encode Zta and Rta. Zta, a sequence-specific DNA-binding protein of the basic-leucine zipper family, activates both transcription of viral genes and replication from the lytic origin, OriLyt.

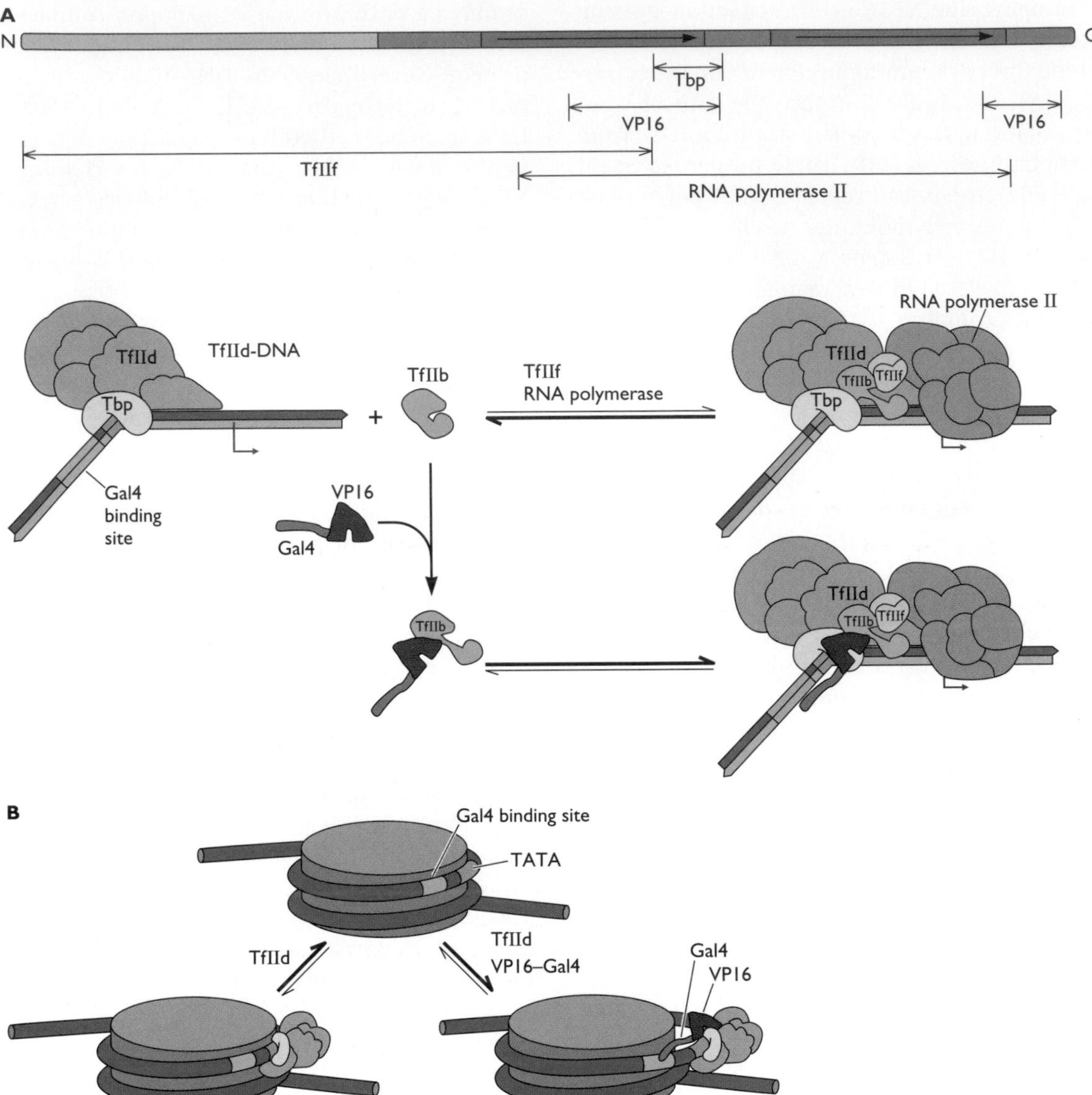

Figure 8.19 Models for transcriptional activation by the acidic activation domain of herpes simplex virus type 1 VP16 protein. (A) Induction of conformational change in TfIIb. The C-terminal domain of human TfIIb protein (top) contains two directly repeated sequences and sequences required for binding to Tbp, the acidic activation domain of VP16 and RNA polymerase II. The N-terminal domain is necessary for the binding of TfIIf to TfIIb. In native TfIIb, the N- and C-terminal domains interact with one another such that the segments of the protein that interact with TfIIf-RNA polymerase II are blocked. Entry of TfIIb into an assembling initiation complex and recruitment of TfIIf-RNA polymerase II to the complex are therefore unfavorable reactions that require conformational change in TfIIb. Binding of the acidic activation domain of VP16, as a chimera with the DNA-binding domain of the yeast protein Gal4, disrupts the intermolecular association of the N- and C-terminal domains of TfIIb, exposing its binding sites for TfIIf and RNA polymerase II. Consequently, formation of the preinitiation complex containing TfIIf and RNA polymerase II is now more favorable. When the VP16 activation domain is bound to the promoter, here through interaction of the Gal4 DNA-binding domain with its recognition site, its binding to TfIIb can also facilitate interaction of this protein with the TfIId-DNA complex. Two sequential reactions in the assembly sequence (Box 8.3) are therefore driven in the forward direction by interaction of the VP16 acidic activation domain with TfIIb. However, this activation domain can also bind directly to proteins of TfIId. These interactions, or those with TfIIf or TfIIh, may also favor formation of the closed initiation complex by directly stabilizing assembling complexes or by inducing favorable conformational changes. Or they may be related to the ability of the VP16 acidic activation domain to stimulate other reactions, such as promoter clearance or passage of the transcriptional machinery through sequences in the DNA template that can cause premature termination of transcription.

(continued on next page)

Figure 8.19 *(continued)* Other acidic activating domains have also been shown to facilitate formation of the open initiation complex. (B) Alleviation of transcriptional repression by nucleosomes. Many activators, including the acidic activation domain of VP16, stimulate transcription from nucleosomal DNA templates to a much greater degree than they do transcription from naked DNA. This property is the result of their ability to alleviate repression of transcription by nucleosomes, an activity therefore termed antirepression. The association of DNA with a nucleosome, an octamer of two copies each of the histones H2A, H2B, H3, and H4, can block access of proteins to their DNA-binding sites, as illustrated for binding of TfIId to a TATA sequence (left). Interaction of the DNA template with the histone proteins may directly prevent binding of the protein, or the conformation adopted by the DNA to wrap around the nucleosome may be incompatible with its recognition by the protein. Binding of the acidic activation domain of VP16 to the template, in this example via the heterologous Gal4 DNA-binding domain, alters the interaction of the DNA with the nucleosome to allow TfIId access to the TATA sequence (right). The mechanism by which VP16 and other activators might remodel nucleosomal DNA has not been established, but might involve their interactions with nucleosomal histones or with certain coactivators. These coactivators are multicomponent complexes that, in ATP-dependent reactions, dramatically alter the association of nucleosomes with DNA (but do not displace them) and in so doing stimulate the binding of Tbp and other DNA-binding proteins. Other coactivators stimulate acetylation of nucleosomal histones, a modification long known to be associated with transcriptionally active chromatin. Although many activators alter nucleosomal DNA templates in this way, some can prevent binding of nucleosomes to promoters once bound to their recognition sequences.

Viral proteins with multiple regulatory activities: adenoviral E1A proteins. Two E1A proteins, synthesized from differentially spliced mRNAs (Fig. 8.21), are synthesized within an hour or two of adenovirus infection. These two proteins share all sequences except for an internal segment (conserved region 3 [CR3]) that is unique to the larger protein. Nevertheless, they differ considerably in their regulatory potential, because the CR3 segment is primarily responsible for stimulation of transcription. As the larger E1A protein neither binds specifically to DNA nor

Figure 8.20 Conformational changes are necessary for Oct-1 protein-dependent recruitment of VP16 to herpes simplex virus type 1 promoters. The VP16 protein, which enters herpes simplex virus type 1-infected cells in the virion tegument, must interact with the host cell protein Hcf to adopt a conformation that allows recognition of the cellular Oct-1 protein bound to DNA. The homeodomain segment of the bipartite DNA-binding domain of Oct-1 also undergoes a conformational change necessary for its recognition by VP16. When the binding site (Octamer-GARAT) contains the GARAT sequence, the conformation of the DNA-bound homeodomain of the Oct-1 protein is altered from the structure observed when the protein is bound to the octamer sequence alone, which has been determined by X-ray crystallography. The unique conformation adopted on binding sites that contain the GARAT sequence has not been determined and is illustrated schematically at the left. The conformationally altered VP16 protein can then recognize the unique form of the Oct-1 homeodomain bound to a site containing the GARAT sequence. This mechanism ensures that the VP16 protein is recruited only to promoters that contain the GARAT sequence, viral immediate-early promoters. As a result, binding of VP16 to the many cellular promoters that contain the octamer sequence, interactions that would be of little benefit to the virus, is avoided.

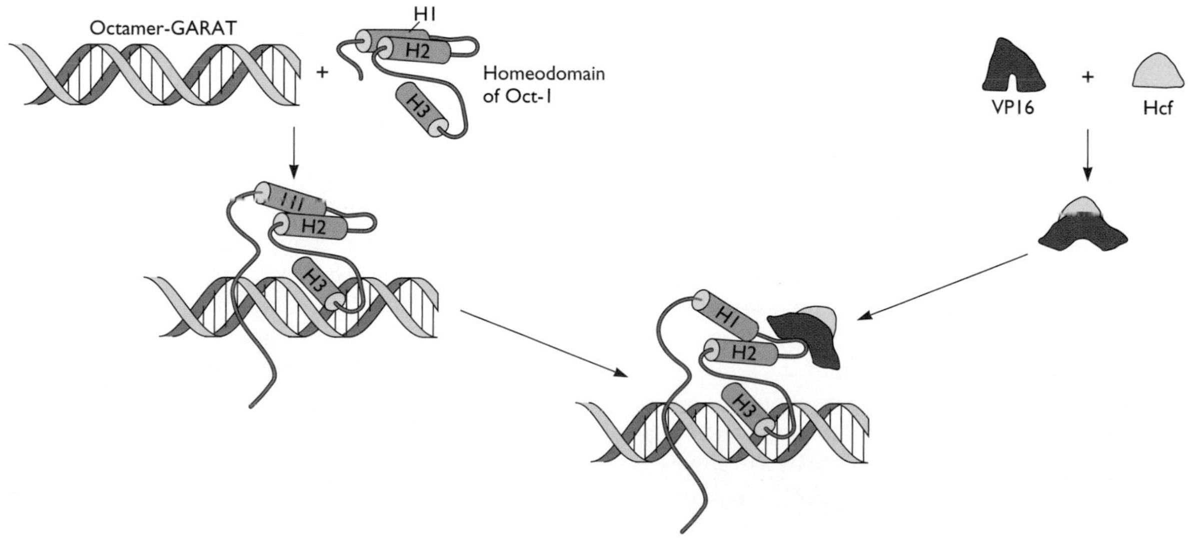

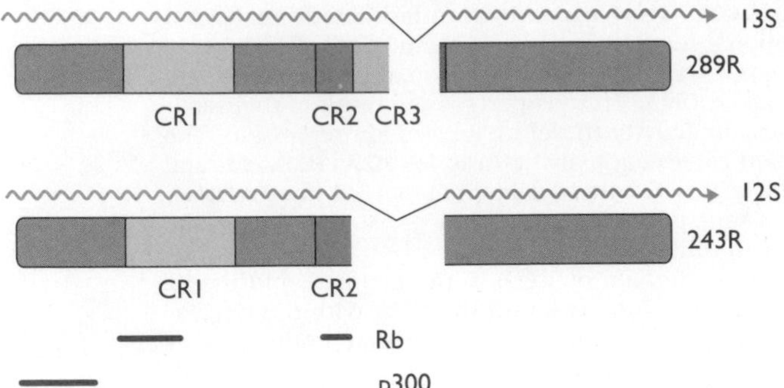

Figure 8.21 Structural and functional features of the adenovirus type 2 E1A proteins. Primary transcripts of the immediate-early E1A gene are alternatively spliced to produce the abundant 13S and 12S mRNAs. As such splicing does not change the translational reading frame, the E1A proteins synthesized from these mRNAs are identical, except for an internal segment of 46 amino acids unique to the larger protein. Comparison of the E1A sequences of different human adenovirus serotypes identified the three most highly conserved regions, designated CR1, CR2, and CR3. The CR3 segment includes the unique sequence of the larger E1A protein, which is required for stimulation of transcription of viral early genes in infected cells. Conserved regions CR1 and CR2 present in both E1A proteins are necessary for the mitogenic activity of E1A gene products and for transformation (see Chapter 18). A large number of cellular proteins bind to these regions, as well as to N-terminal sequences, of E1A proteins. The regions of the E1A proteins necessary for interaction with two of these, the Rb protein and p300, are indicated.

depends on a specific promoter sequence, it is often considered the prototypical example of viral proteins that stimulate transcription by indirect mechanisms, some of which are listed in Table 8.6.

In transient-expression assays (Fig. 8.3B), stimulation of transcription by the larger E1A protein is remarkably insensitive to the nature of the promoter. The protein can stimulate transcription from the great majority of the many RNA polymerase II promoters that have been tested, as well as transcription by RNA polymerase III. The CR3 segment can interact physically with RNA polymerase II initiation proteins, such as TfIId subunits, and bind to and/or stimulate transcription via a large variety of sequence-specific transcriptional regulators (Fig 8.22A). Such interactions among the viral and cellular proteins are required for stimulation of transcription in simplified experimental systems. However, the functional significance of these interactions during the adenoviral infectious cycle has not been established. Indeed, when human adenovirus type 2 infects a permissive cell, the larger E1A protein does not stimulate transcription globally, as might be expected from the properties just described. The one exception is binding of the CR3 region to Sur2, a subunit of a cellular coactivator complex: CR3-dependent stimulation of viral E2 transcription is observed in mouse embryonic stem cells, but not in mutant cells, homozygous for deletion of the *sur2* gene.

Adenoviral E1A proteins activate transcription by a second mechanism, which was discovered more recently but is better understood. This function, mediated by the conserved regions of the N-terminal exon, CR1 and CR2 (Fig. 8.21), was elucidated through studies of transformation and immortalization by E1A gene products. The CR1 or CR2 segments interact with several cellular proteins. Two of these, Rb and p300 (Fig. 8.21), are of special relevance to transcriptional regulation. The Rb protein is the product of the cellular retinoblastoma susceptibility gene, a tumor suppressor that plays a crucial role in cell cycle progression (Chapter 18). The discovery that Rb binds to adenoviral E1A proteins, as well as to transforming proteins of papillomaviruses and polyomaviruses, led to elucidation of the mechanism of action of this important cellular protein, and of one mechanism by which E1A proteins regulate transcription.

In uninfected cells, Rb binds to cellular E2f proteins, sequence-specific transcriptional activators originally discovered because they bind to the human adenovirus type 2 E2 early promoter (Fig. 8.5). Such E2f-Rb complexes possess the specific DNA-binding activity characteristic of E2f, but Rb represses transcription (Fig. 8.22B). The CR1 and CR2 regions of the E1A proteins bind to the same segments of Rb as does E2f. Binding of the E1A proteins to Rb therefore releases E2f proteins from association with this repressor to allow transcription from E2f-dependent promoters. During the early phase of infection, E2f proteins are essential for efficient transcription of the E2 gene, which encodes the proteins required for viral DNA synthesis. Sequestration of Rb by the E1A proteins therefore ensures production of the E2 replication proteins and progression into the late phase of the infectious cycle (Fig. 8.14).

The N-terminal sequences common to the two E1A proteins also bind to a second cellular protein that participates in regulation of transcription, the p300 protein (Fig. 8.21). This protein is a member of a family of coactivators that bind to, and mediate stimulation of transcription by, the basic-leucine zipper protein Creb and steroid/thyroid hormone receptors. The p300 protein is a histone acetylase and it binds to other histone acetylases. Modification of histones by these enzymes alters the structure of transcriptionally active chromatin and stimulates multiple reactions in transcription. The E1A proteins, which bind directly to p300, both disrupt interaction of p300 with Creb and compete for binding of p300 to the histone acetylase. This activity of E1A proteins has been implicated in repression of

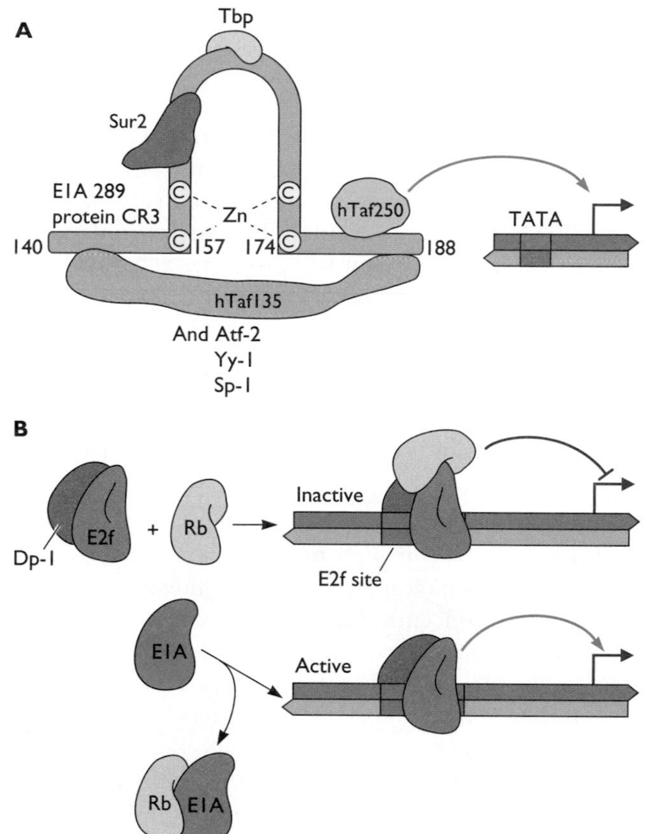

Figure 8.22 Indirect stimulation of transcription by adenoviral E1A proteins. (A) Interactions of the E1A CR3 sequences with components of the RNA polymerase II transcriptional machinery. The CR3 sequence, which is present in the larger E1A protein, binds Zn in vitro. The putative Zn-finger motif is shown with the Zn-coordinating cysteines indicated. This segment of CR3 is required for binding to Tbp and the Sur2 subunit of the coactivator called mediator. In contrast, specific Tbp-associated proteins (Tafs) can bind to sequences flanking the Zn-finger motif. The sequence-specific DNA-binding activators listed below can also bind to CR3, and have been implicated in CR3-dependent stimulation of transcription from specific viral or cellular promoters. (B) Model of competition between E1A proteins and E2f for binding to Rb protein. The E2f transcriptional activators are heterodimers of a member of the E2f protein family (described in Chapter 18) and a related Dp protein (differentiation-regulated transcription factor protein), such as Dp-1. The binding of E2f to its recognition sites in specific promoters is not inhibited by association with the Rb protein. However, when present in a ternary complex, Rb can repress transcription from E2f-dependent promoters (top). Adenoviral E1A proteins made in infected (or transformed) cells bind to Rb and disrupt the E2f-Rb interaction. Binding of E1A proteins to Rb releases E2f, which can then stimulate transcription.

enhancer-dependent transcription, and is required for induction of cell proliferation and transformation.

The multiplicity of mechanisms by which the E1A proteins engage with components of the cellular transcriptional machinery is one of their most interesting features. Regulation by multiple mechanisms may prove to be a definitive property of viral proteins that cannot bind directly to viral promoters (Table 8.6). For example, the human T-lymphotropic virus type 1 Tax protein stimulates transcription by both binding to specific cellular members of the basic-leucine zipper family and activating Nf-κb. The latter mechanism includes activation of the signaling cascade that induces degradation of Iκb and translocation of Nf-κb from the cytoplasm to the nucleus (Fig. 8.6), a pathway illustrating the potential for intervention of viral regulators in cellular pathways well upstream of any transcriptional activator.

Coordination of Transcription of Late Genes with Viral DNA Synthesis

The structural proteins of viruses must be made in quantities sufficient to ensure assembly of the maximal number of progeny. Such a high concentration of foreign (viral) proteins is undoubtedly injurious to the host cell and could therefore interfere with cellular processes needed for virus reproduction. In the case of the viruses under consideration here, this potential problem is avoided, because structural proteins are synthesized only during the late phase of infection. As replication of viral DNA genomes is in full swing by this time, virion assembly is the only step then necessary to complete the reproductive cycle. This pattern of viral structural protein synthesis results from the dependence of late gene transcription on viral DNA replication: drugs or mutations that inhibit viral DNA synthesis in infected cells also block expression of late genes. Despite their importance, the mechanisms by which activation of transcription can be integrated with viral DNA synthesis remain incompletely understood.

Regulation of transcription initiation. The most obvious consequence of genome replication in cells infected by any DNA virus is the large increase in intranuclear concentration of viral DNA molecules. Even in experimental situations, infected cells contain a relatively small number of copies of the viral genome during the early phase of infection, typically 5 to 100 copies per cell depending on the multiplicity of infection. As soon as viral DNA synthesis begins, this number rises geometrically to values as high as hundreds of thousands of viral DNA molecules per infected cell nucleus. At such high concentrations, viral promoters can compete effectively for components of the cellular transcription machinery. The increase in DNA template concentration also titrates cellular repressing proteins that bind to specific sequences of certain viral late promoters. For example, the simian virus 40 major late promoter contains multiple binding sites for a cellular repressor that belongs to the steroid/thyroid hormone receptor superfamily (Fig. 8.23A). Viral DNA replication increases the concentration of the late promoter above that at which every copy can be bound by the cellular repressor, and therefore allows this promoter to become active (Fig. 8.23B).

Maximally efficient transcription from the simian virus 40 major late promoter also requires the viral early gene product large T antigen. This protein controls simian virus 40 late transcription both directly, as an activating protein (Fig. 8.14; Table 8.6), and indirectly, as a result of its essential functions in viral DNA synthesis. Among the more complex DNA viruses, the molecular coupling of replication to transcription of late genes can be more indirect. For example, late phase-specific transcription of the adenoviral IVa2 gene, like simian virus 40 major late transcription, is controlled by viral DNA synthesis-dependent titration of a cellular repressor. However, the IVa2 protein is itself a sequence-specific activator of transcription. Once synthesized in infected cells, it cooperates with a second, as yet unidentified, protein to stimulate the rate of initiation of transcription from the major late promoter at least 20-fold. Synthesis of adenoviral DNA therefore initiates a transcriptional cascade in which late promoters are activated sequentially (Fig. 8.14).

Figure 8.23 Regulation of the activity of the simian virus 40 late promoter by cellular repressors. (A) The sequence surrounding the simian virus 40 major late initiation site (the thickest arrow in Fig. 8.4D) contains three binding sites for the cellular protein termed initiator-binding protein (Ibp), which contains members of the steroid/thyroid receptor superfamily. (B) "Antirepression" resulting from synthesis of viral DNA synthesis in infected cells. During the early phase of infection, the concentration of Ibp relative to that of the viral major late promoter is sufficiently high to allow all Ibp-binding sites in the viral genomes to be occupied. Consequently, the late promoter is transcriptionally silent, because Ibp is a repressor of transcription. The concentration of Ibp does not change during the course of simian virus 40 infection. However, as viral DNA synthesis takes place in the infected cell, the concentration of the major late promoter becomes sufficiently high that not all Ibp-binding sites can be occupied. Consequently, the major late promoter becomes accessible to cellular transcription components. Therefore, although we generally speak of "activation" of late gene transcription, this DNA replication-dependent mechanism is one of escape from repression.

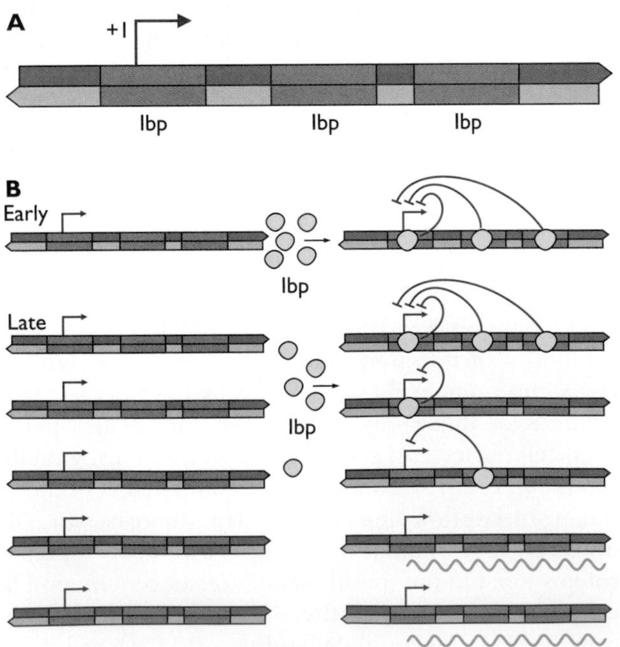

The increase in intracellular concentration of viral DNA can also alleviate promoter occlusion. The pIX gene of adenovirus is completely contained within the early E1B transcription unit (Appendix A, Fig. 1B), and transcription from the pIX promoter is initially inhibited by transcription from the upstream E1B promoter. As the concentration of viral DNA molecules increases following viral DNA synthesis, the concentration of E1B promoters exceeds that of the cellular transcriptional components necessary for initiation of transcription. Consequently, some viral DNA molecules carry transcriptionally inactive E1B genes, in which the pIX promoter is accessible.

Newly replicated DNA molecules must be partitioned for such fates as entry into additional replication cycles, service as templates for transcription, or assembly into virions. This process must be regulated so that different fates predominate at different times in the infectious cycle. Although transcription of all viral DNA molecules made in infected cells seems to be a simple mechanism to ensure efficient transcription of late genes, no more than 5 to 10% of the large numbers that accumulate are transcriptionally active. In the case of simian virus 40, transcriptional activity can be ascribed to establishment of the open chromatin region in minichromosomes described above. An analogous mechanism cannot be operating in cells infected by adenoviruses or herpesviruses (Box 8.7), and it is not known whether a subset of these viral DNA molecules is also marked for transcriptional activity by some unknown molecular device. However, one important parameter governing the concentration of transcriptional templates must be the relative concentrations of viral DNA molecules and the proteins that package them for assembly into virions, because packaging and transcription are mutually exclusive.

Regulation of termination. Viral DNA synthesis can also affect transcriptional termination, a regulatory device illustrated by the human adenovirus type 2 major late transcription unit. During the early phase of infection, major late transcription terminates within a region in the middle

BOX 8.7

Unusual viral templates for RNA polymerase II transcription?

In contrast to the genomes of papillomaviruses and polyomaviruses, adenoviral and herpesviral DNAs are not packaged by cellular nucleosomes in either virions or infected cells. The significance of the unusual nature of these transcriptional templates is not fully appreciated, but there are hints that this property might contribute to regulation of viral gene expression.

In virions, adenoviral DNA is packaged by specialized viral DNA-binding proteins (Appendix A, Fig. 1A), at least one of which is retained when the genome enters the infected cell nucleus. Because these proteins are products of late genes, viral DNA synthesis must necessarily be accompanied by reorganization of the viral nucleoproteins. Indeed, the templates for major late transcription during the late phase of infection appear protein-free in the electron microscope. It is therefore possible that production of protein-free viral DNA molecules contributes to alterations in termination of major late transcription (see "Coordination of Transcription of Late Genes with Viral DNA Synthesis") that accompany entry into the late phase.

Herpes simplex virus DNA is not associated with dedicated viral packaging proteins in virions, nor is it bound by nucleosomes in productively infected cells. Whether viral proteins are associated with the viral templates for transcription during any, or all, phases of the productive cycle is not known. However, there is accumulating evidence that changes in the nucleoprotein organization of herpesviral DNA accompany establishment of latency, and reactivation from this state. In contrast to viral DNA in productively infected cells, that in latently infected murine neurons has been reported to exhibit a regular, nucleosomal organization. Furthermore, one of the promyelocytic leukemia-associated proteins that are degraded when the viral ICO protein is made (see "Entry into one of two alternative transcriptional programs") forms a complex with a heterochromatin-associated protein called HP-1. These observations suggest that cellular proteins that form and organize chromatin are important for the repression of herpesviral transcription during latent infection. This hypothesis implies that **lack** of association of viral DNA with such proteins is equally important for efficient transcription of viral immediate-early genes, and entry into the productive cycle.

of the transcription unit, such that no elongating complexes reach the L4 poly(A) addition site (Fig. 8.24). As discussed in Chapter 10, such restricted transcription is coupled with preferential utilization of specific RNA processing signals to produce a single major late mRNA and protein during the early phase. Viral DNA synthesis is necessary to induce transcription of distal segments of the transcription unit to a termination site close to the right-hand end of the viral genome (Fig. 8.24). The inefficient termination of early transcription at many sites spread over about 12 kb of DNA and the fact that only replicated viral DNA molecules can support complete transcription are unusual features that suggest a novel regulatory mechanism. One hypothesis is that alterations in template structure upon viral DNA synthesis (Box 8.7) may contribute to regulation of termination of major late transcription.

A Viral DNA-Dependent RNA Polymerase

The DNA genomes of viruses considered in preceding sections are replicated in the nucleus of infected cells, where the cellular transcriptional machinery resides. In contrast, poxviruses such as vaccinia virus are reproduced exclusively in the cytoplasm of their host cells. This feat is pos-

Figure 8.24 DNA replication-dependent alterations in termination of human adenovirus type 2 major late transcription. The segment of the viral genome from 15 to 100 map units containing the major late transcription unit is shown to scale (top). The five sites at which primary major late transcripts are polyadenylated, designated L1 to L5, are indicated by the vertical arrows. During the early phase of infection, major late transcripts synthesized in response to the stimulatory activity of the larger E1A protein terminate at multiple sites beyond the L1 site, as indicated by the dashed line. No transcript extends beyond the position indicated by the arrowhead. The mechanisms responsible for this termination, which takes place within a DNA segment of 11 to 12 kb, are not understood. Following synthesis of both viral DNA and late proteins, all regions of the major late transcription unit to termination sites near the right-hand end of the genome are transcribed with equal efficiency.

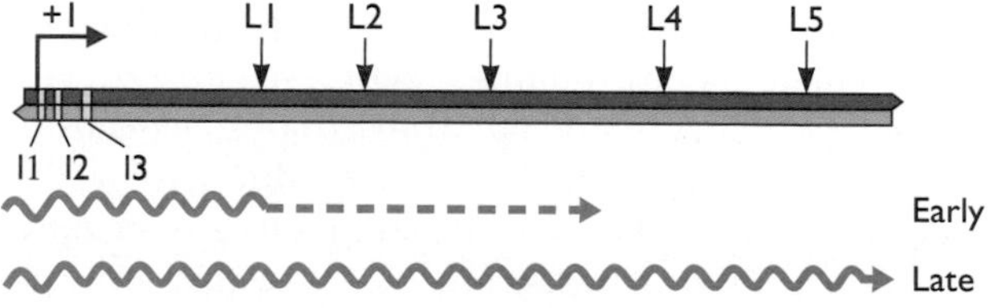

sible because the genomes of these viruses encode the components of a transcription and RNA processing system that produces viral mRNAs with all the hallmarks of cellular mRNAs [5′ caps and 3′ poly(A) tails]. This system, which is carried into infected cells within virions, includes a DNA-dependent RNA polymerase responsible for transcription of all vaccinia virus genes. A striking feature of this viral enzyme is its structural and functional resemblance to cellular RNA polymerases.

The vaccinia virus RNA polymerase, like RNA polymerase II and the other eukaryotic RNA polymerases, is a complex, multisubunit enzyme built from the products of at least eight genes. The amino acid sequences of several of these subunits, including the two largest and the smallest, are clearly related to subunits of RNA polymerase II. The viral enzyme transcribes all classes of vaccinia virus genes expressed at different times in the infectious cycle (early, intermediate, and late). Like the cellular enzymes, vaccinia virus RNA polymerase recognizes promoters by cooperation with additional proteins. For example, formation of transcriptionally competent initiation complexes on vaccinia virus early promoters is mediated by two viral proteins, VETF and RAP94, which are responsible for the recognition of promoter sequences and recruitment of the RNA polymerase, respectively (Fig. 8.25). These viral proteins are functional analogs of the cellular RNA polymerase II general initiation proteins TfIId and TfIIf (Box 8.3). However, the vaccinia virus transcriptional machine is not analogous to its cellular counterpart in every respect. Transcription of the majority of viral early genes terminates at discrete sites, 20 to 50 bp downstream of specific T-rich sequences in the template. Termination requires the viral termination protein, which is also the viral mRNA capping enzyme. The 3′ ends of the viral mRNAs correspond to sites of transcription termination. In contrast, cellular RNA polymerase II can transcribe far beyond the sites at which the 3′ ends of mature cellular or viral mRNAs are produced by processing of the primary transcript, and does not terminate transcription at simple, discrete sites.

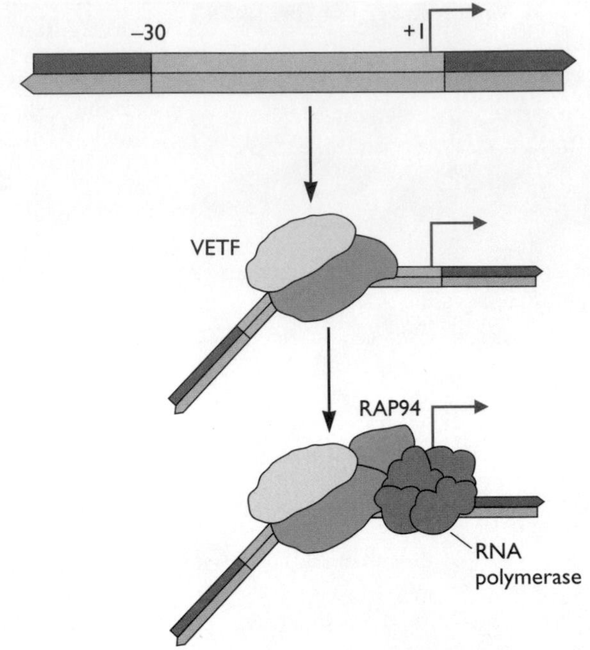

Figure 8.25 Assembly of an initiation complex on a vaccinia virus early promoter. Vaccinia virus early promoters are characterized by the presence of an AT-rich sequence (tan) immediately upstream of the site of initiation of transcription. Vaccinia virus RNA polymerase cannot recognize these (or any other) viral promoters in the absence of other viral proteins. One of these, the vaccinia virus early transcription protein (VETF), is necessary for early promoter recognition and must bind before the viral RNA polymerase. This heteromeric protein is composed of subunits of 77 and 82 kDa and binds specifically to early promoters and induces DNA bending; it also possesses DNA-dependent ATPase activity. This protein, and the second protein necessary for early promoter specificity, RAP94, enters infected cells in virions. Some 40% of the virion RNA polymerase molecules are associated with RAP94, and it is only this RNA polymerase population that can transcribe early promoters in vitro. The RAP94-RNA polymerase complex associates with early promoter-bound VETF, an interaction that probably includes direct binding of RAP94 to VETF, to form a functional initiation complex. Assembly of these vaccinia virus initiation complexes is therefore analogous to, although simpler than, formation of RNA polymerase II initiation complexes (Box 8.3).

The viral RNA polymerase and proteins necessary for transcription of early genes enter host cells within vaccinia virus particles. Subsequently, viral gene expression is controlled by the ordered synthesis of viral transcriptional regulatory proteins that permit sequential recognition of intermediate and late promoters. For example, transcription of intermediate genes requires synthesis of the viral RPO30 gene product, while late transcription depends on synthesis of several intermediate gene products. Activation of intermediate and late gene transcription also requires viral DNA replication. Transcription of vaccinia virus genetic information is therefore regulated by mechanisms similar to those operating in cells infected by other DNA viruses, even though the transcriptional machinery is viral in origin.

Surprisingly, the vaccinia virus transcription system is not entirely self-contained: a cellular protein is essential for transcription of viral intermediate genes. This protein, termed Vitf2, is located in the nucleus of uninfected cells, but is present in both the cytoplasm and the nucleus of infected cells. As a significant number of vaccinia virus genes encode proteins necessary for transcription, such dependence on a cellular protein must confer some special advantage. An attractive possibility is that interaction of the viral transcriptional machinery with a cellular protein serves to integrate the viral reproductive cycle with the growth state of its host cell. The molecular identity of Vitf2 and elucidation of its function in uninfected cells should shed light on the physiological significance of the limited, but essential, participation of cellular components in the transcription of vaccinia virus genetic information.

Transcription of Viral Genes by RNA Polymerase III

Several of the viruses considered in this chapter contain RNA polymerase III transcription units in their genomes (Table 8.7). The first, and still best understood, example of such a viral gene is that encoding the major RNA polymerase III transcription product of human adenovirus type 2, virus-associated RNA I (VA-RNA I). The VA-RNA I gene specifies an RNA product that ameliorates the effects of a host cell defense mechanism (Chapter 15). It contains a typical, intragenic promoter that has been widely used in studies of the mechanism by which RNA polymerase III initiates transcription.

Transcription of the Adenoviral VA-RNA Genes

The VA-RNA I Promoter

The human adenovirus type 2 genome contains two VA-RNA genes located very close to one another (Appendix A, Fig. 1B). The VA-RNA I promoter is described here, for it has been more thoroughly characterized than the closely related VA-RNA II promoter. Transcription of the VA-RNA I gene depends on two intragenic sequences, the A and B boxes (Fig. 8.26A). Sequences upstream of sites of initiation are not necessary for promoter activity, but do modulate the specificity or efficiency of initiation. As in the RNA polymerase II system, the essential promoter sequences are binding sites for accessory proteins necessary for promoter recognition. The

Table 8.7 Viral RNA polymerase III transcription units

Virus	RNA polymerase III transcript	Function
Adenovirus		
Human adenovirus type 2	VA-RNA I	Prevention of activation of RNA-dependent protein kinase
	VA-RNA II	Not known
Herpesvirus		
Epstein-Barr virus	EBER 1, EBER 2	Not known
Herpesvirus saimiri	HSVR 1–5	Degradation of certain cellular mRNAs
Retrovirus		
Moloney murine leukemia virus	Let	Stimulation of transcription of specific cellular genes (e.g., CD4 and major histocompatibility complex class I genes)

Figure 8.26 Organization and recognition of the human adenovirus type 2 VA-RNA I RNA polymerase III promoter. (A) The VA-RNA I gene is depicted to scale, in base pairs. The intragenic A and B box sequences are essential for efficient VA-RNA I transcription by RNA polymerase III. The sequences of the A and B boxes of the VA-RNA gene are closely related to the consensus A and B sequences of cellular tRNA genes. The VA-RNA termination site sequences are also typical of those of cellular genes transcribed by RNA polymerase III. (B) Assembly of an RNA polymerase III initiation complex on the VA-RNA I promoter is initiated by binding of TfIIIc2 to the intragenic promoter sequences. This reaction is stimulated by TfIIIc1, which makes contact with both the A box and sequences spanning the termination site. The TfIIIc-DNA complex is recognized by TfIIIb, which in turn allows recruitment of RNA polymerase III to the promoter. Although TfIIIb does not possess sequence-specific DNA-binding activity, it does contact DNA in the vicinity of the site of initiation. All mammalian RNA polymerase III initiation proteins are large, multiprotein complexes. The TATA-binding protein is an essential component of TfIIIb.

A

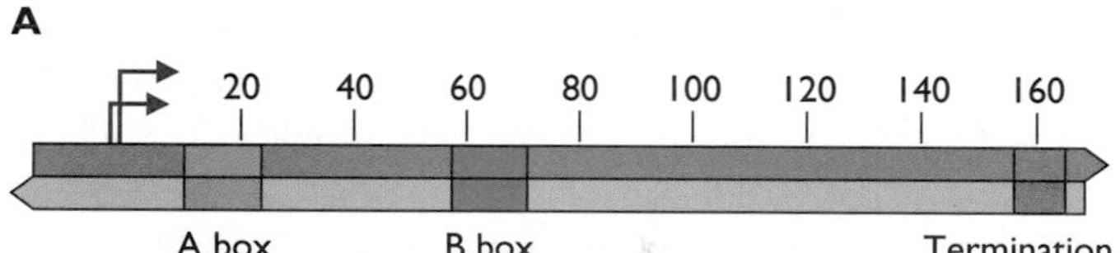

B

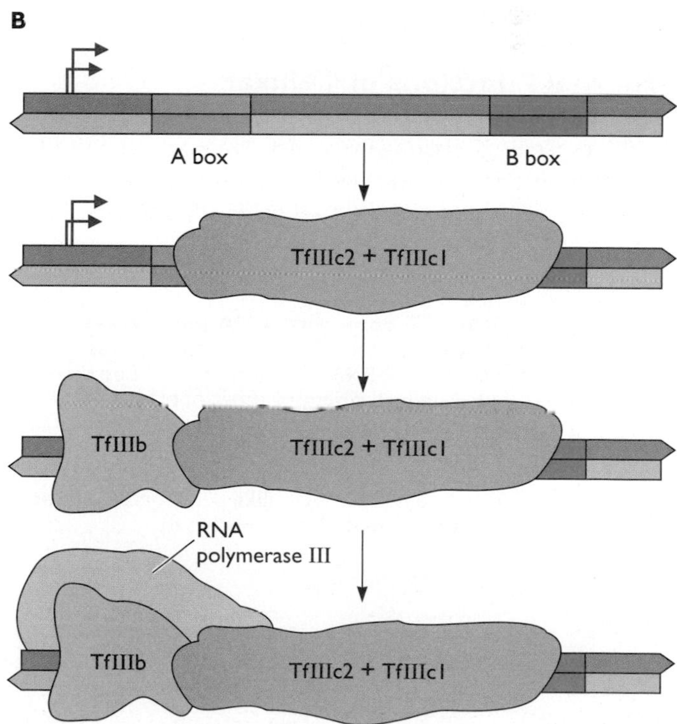

internal sequences are specifically recognized by the RNA polymerase III-specific initiation protein TfIIIc, which binds to the promoter to begin assembly of an initiation complex that also contains TfIIIb and the enzyme (Fig. 8.26B). The pathway of initiation of VA-RNA I transcription by RNA polymerase III shown in Fig. 8.26B was elucidated by using in vitro assays. We can be reasonably confident that this same pathway operates in adenovirus-infected cells, for there is excellent agreement between the effects of A and B box mutations on VA-RNA I synthesis in vitro and in mutant virus-infected cells.

Regulation of VA-RNA Gene Transcription

The two VA-RNA genes are initially transcribed at similar rates, but during the late phase of infection, production of VA-RNA I, but not of VA-RNA II, is greatly accelerated. Such preferential transcription is the result of competition of the VA-RNA I promoter with the intrinsically much weaker transcriptional control region of the VA-RNA II gene for a limiting component of the RNA polymerase III transcriptional machinery. Repression of VA-RNA I transcription may account for the similar rates at which the two genes are transcribed during the early phase. The control of transcription of VA-RNA genes during the adenoviral infectious cycle emphasizes the fact that transcription by RNA polymerase III can, and must, be regulated, although the mechanisms are probably less elaborate than those that govern transcription by RNA polymerase II. The properties of other viral RNA polymerase III transcription units illustrate the kinship of the RNA polymerase II and III systems (Fig. 8.27).

Unusual Functions of Cellular Transcription Components

In the preceding sections, we concentrated on the similarities among the mechanisms by which viral and cellular DNA are transcribed. Even though every mechanism of regulation of expression of viral genes by the host cell's RNA polymerase II or RNA polymerase III cannot be described in molecular detail, it seems clear that the majority are not unique to viral systems. It is therefore an axiom of molecular virology that every mechanism by which viral transcription units are expressed by cellular components, or by which their activity is regulated, will prove to have a normal cellular counterpart. However, virus-infected cells also provide examples of novel functions or activities of cellular proteins that mediate transcription.

One example of the latter kind of mechanism is the production of hepatitis delta satellite virus RNA from an **RNA** template by RNA polymerase II, described in Chapter 6. The RNA of viroids, infectious agents of plants, is synthesized in the same manner (Chapter 20). Such RNA-dependent RNA synthesis by RNA polymerase II from a specific RNA promoter is one of the most remarkable interactions of a viral genome with the cellular transcriptional machinery. No cellular analog of this reaction is yet known. Even more divergent functions of cellular transcriptional components in virus-infected cells are illustrated by the participation of the RNA polymerase III initiation proteins TfIIIb and TfIIIc in integration of the yeast retroid element Ty3 (see Chapter 7). Given the large repertoire of molecular and biochemical activities displayed by components of the eukaryotic transcriptional machinery, it seems likely that other unusual activities of these cellular proteins will be discovered in virus-infected cells.

Inhibition of the Cellular Transcriptional Machinery in Virus-Infected Cells

Inhibition of cellular transcription in virus-infected cells offers several advantages. Cellular resources, such as substrates for RNA synthesis, can be devoted to the exclusive production of viral mRNAs (and, in many cases, RNA genomes), and competition between viral and cellular mRNAs for components of the translational machinery is

Figure 8.27 Some viral RNA polymerase III genes contain both external and internal promoter elements. The 5′ end of the Epstein-Barr virus EBER-2 transcription unit is shown to scale. This gene, which is transcribed by RNA polymerase III in Epstein-Barr virus-infected cells, contains intragenic A and B box sequences, which resemble those of other genes transcribed by this enzyme in both their sequences and their positions. However, efficient transcription by RNA polymerase III also depends on the 5′-flanking sequence, which includes binding sites for the RNA polymerase II stimulatory proteins Sp1 and Atf. It has not been determined whether these proteins stimulate RNA polymerase III transcription from the EBER-2 promoter in infected cells. The TATA-like sequence is important for efficient transcription and essential for specifying transcription by RNA polymerase III.

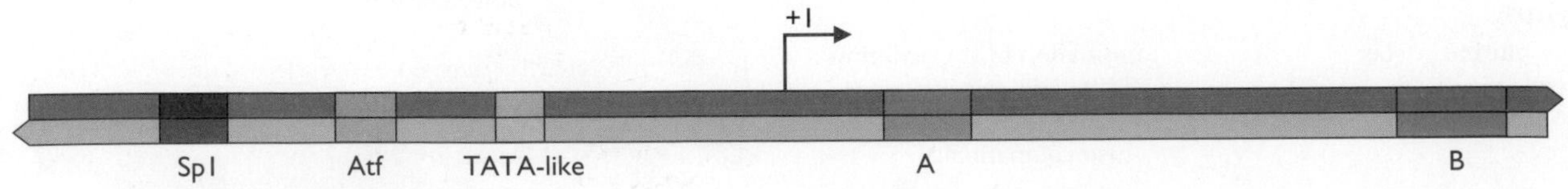

minimized. The essential participation of cellular transcriptional systems in the infectious cycles of most viruses considered in this chapter precludes inactivation of this transcriptional machinery. However, posttranscriptional mechanisms allow selective expression of adenoviral and herpesviral genes (Chapter 10). Furthermore, herpes simplex virus type 1 infection induces inhibition of transcription of the majority of cellular genes by RNA polymerase II. Selective transcription of viral genes has been correlated with loss from infected cells of RNA polymerase II in which the C-terminal domain of the largest subunit is hyperphosphorylated, and the concomitant appearance of an alternatively phosphorylated form of the cellular enzyme unique to herpesvirus-infected cells. The latter change requires the ICP22 protein and a protein kinase encoded by the viral UL3 gene, and is accompanied by loss of the initiation protein TfIIe from the RNA polymerase II holoenzyme. How these properties alter the specificity of transcription by the cellular enzyme is not yet known. The promoters of viral late genes possess a particularly simple organization, comprising TATA and **initiator sequences**, but lacking upstream binding sites for cellular transcriptional activators. This unusual property may facilitate selective initiation of transcription from these promoters by the modified transcriptional machinery of the host cell. The poxviruses induce rapid inhibition of synthesis of all classes of cellular RNA in infected cells. Such inhibition requires viral proteins, but these have not been identified.

Replication of the majority of viruses with RNA genomes requires neither the cellular transcriptional machinery nor its RNA products, and is accompanied by inhibition of cellular mRNA synthesis. Among the best-characterized examples is the inhibition of transcription by RNA polymerase II characteristic of poliovirus-infected cells. The poliovirus protease $3C^{pro}$ is both necessary and sufficient for such inactivation of this cellular transcription system. The defect in RNA polymerase II transcription in extracts of poliovirus-infected cells can be explained by the fact that $3C^{pro}$ cleaves the Tbp subunit of TfIId at several sites. This modification eliminates the DNA-binding activity of Tbp and hence transcription by RNA polymerase II. The TATA-binding protein is also a subunit of initiation proteins that function with RNA polymerase III (TfIIIb) and RNA polymerase I, and is the only protein known to be common to the three cellular transcription systems. Its cleavage by $3C^{pro}$ in poliovirus-infected cells therefore appears to be a very efficient way to inhibit all cellular transcriptional activity.

Two gene products of the rhabdovirus vesicular stomatitis virus have been implicated in the inhibition of RNA polymerase II transcription. The first is the viral leader RNA described in Chapter 6, which inhibits transcription by both RNA polymerase II and RNA polymerase III in vitro and is primarily responsible for the rapid inhibition of cellular RNA synthesis in infected cells. Following synthesis in the cytoplasm, the leader RNA enters the nucleus. The question of how short RNA molecules inhibit DNA-dependent RNA transcription cannot yet be answered, although in vitro experiments suggest that binding of a cellular protein to specific sequences within the RNA may be important. The viral M protein is also a potent inhibitor of transcription by RNA polymerase II, even in the absence of other viral gene products. This activity may become important later in infection, when replication of genome RNA predominates over transcription such that less leader RNA is produced.

Perspectives

It is difficult to exaggerate the contributions of viral systems to the elucidation of mechanisms of transcription and its regulation in eukaryotic cells. The organization of RNA polymerase II promoters considered typical was first described for viral transcriptional control regions, enhancers were first discovered in viral genomes, and many important cellular regulators of transcription (e.g., Table 8.3) were discovered by virtue of their specific binding to viral transcriptional control regions. Perhaps even more important, efforts to elucidate the molecular basis of transcriptional regulatory circuits crucial to viral infectious cycles have established general principles of eukaryotic transcriptional regulation. These include the importance of proteins that do not recognize transcriptional control regions directly, and the ability of a single transcriptional regulator to modulate multiple components of the transcriptional machinery. The insights into regulation of elongation by RNA polymerase II gained from studies of the human immunodeficiency virus type 1 Tat protein emphasize the intimate relationship of viral regulatory proteins with cellular transcriptional components that make viral systems such rich resources for the investigation of eukaryotic transcription.

The identification of cellular and viral proteins necessary for transcription of specific viral genes has allowed many regulatory circuits to be traced. For example, the tissue distribution or the availability of particular cellular activators that bind to specific viral DNA sequences can account for the tropism of individual viruses, or conditions under which different transcriptional programs (e.g., latent or lytic) can be established. Furthermore, the mechanisms that allow sequential expression of viral genes are quite well established. Regardless of whether regulatory circuits are constructed of largely cellular or mostly viral proteins, these transcriptional cascades share such mechanistic features as sequential production of viral activators in infected cells, and integration of transcription of late genes with synthesis of viral DNA. Our ability to describe these regulatory mechanisms in some detail represents enormous progress. Nevertheless, it is important to appreciate that greater challenges lie ahead.

In very few cases have the models for individual regulatory processes that were developed in convenient and powerful experimental systems actually been tested in virus-infected cells. This dearth must be remedied if we are to gain a better understanding of the regulation of transcription of viral DNA templates. It is clear that such experimental systems (e.g., in vitro transcription reactions and transient-expression assays) do not reproduce all features characteristic of infected cells, even in the simplest cases, and cannot address such issues as how transcription of specific genes can be coupled with replication of the viral genome. It is more difficult to elucidate the molecular functions and mechanisms of action of transcriptional components in vivo. Many viral regulatory proteins perform multiple functions, a property that can confound genetic analysis. Moreover, the study of individual reactions, such as binding of a protein to a specific promoter sequence, or formation of preinitiation complexes, is technically more demanding. Nevertheless, viral *cis*-acting sequences and regulatory proteins remain more amenable to genetic analyses of their function in the natural context than do their cellular counterparts. In conjunction with increasingly powerful and sensitive methods for examining intracellular processes, continued efforts to exploit such genetic malleability will eventually establish how transcription of viral DNA templates is mediated and regulated within infected cells. Such information will not only address outstanding virological questions, but should also allow current models of the fundamental cellular process of transcription to be assessed and refined.

References

Books

McKnight, S. L., and K. Yamamoto (ed.). 1992. *Transcriptional Regulation*. Cold Spring Harbor Laboratory Press, Cold Spring Harbor, N.Y.

Book Chapters

Cole, C. N. 2001. Polyomaviridae: the viruses and their replication, p. 2141–2174. *In* D. M. Knipe and P. M. Howley (ed.), *Fields Virology*, 4th ed. Lippincott Williams and Wilkins, Philadelphia, Pa.

Kieff, E. 2001. Epstein-Barr virus and its replication, p. 2511–2574. *In* B. N. Fields, D. M. Knipe, P. M. Howley, R. M. Chanock, J. L. Melnick, T. P. Monath, B. Roizman, and S. E. Straus (ed.), *Fundamental Virology*. Lippincott-Raven, Philadelphia, Pa.

Roizman, B., and D. M. Knipe. 2001. Herpes simplex viruses and their replication, p. 2461–2510. *In* B. N. Fields, D. M. Knipe, P. M. Howley, R. M. Chanock, J. L. Melnick, T. P. Monath, B. Roizman, and S. E. Straus (ed.), *Fundamental Virology*. Lippincott-Raven, Philadelphia, Pa.

Shenk, T. 2001. Adenoviridae: the viruses and their replication, p. 2301–2326. *In* B. N. Fields, D. M. Knipe, P. M. Howley, R. M. Chanock, J. L. Melnick, T. P. Monath, B. Roizman, and S. E. Straus (ed.), *Fundamental Virology*. Lippincott-Raven, Philadelphia, Pa.

Reviews

Block, T. M., and J. M. Hill. 1997. The latency-associated transcripts (LAT) of herpes simplex virus: still no end in sight. *J. Neurovirol.* **3:**313–321.

Everett, R. D. 1999. A surprising role for the proteasome in the regulation of herpesvirus infection. *Trends Biochem. Sci.* **24:**293–295.

Flint, J., and T. Shenk. 1997. Viral transactivating proteins. *Annu. Rev. Genet.* **31:**177–212.

Greenblatt, J. 1997. RNA polymerase II holoenzyme and transcriptional activation. *Curr. Opin. Cell Biol.* **9:**310–319.

Hassan, A. H., K. E. Neely, M. Vignali, J. C. Reese, and J. L. Workman. 2001. Promoter targeting of chromatin-modifying complexes. *Front. Biosci.* **6:**D1054–1064.

Hiscott, J., L. Petropoulos, and J. Lacoste. 1995. Molecular interactions between HTLV-1 Tax protein and the Nf-κB/IκB complex. *Virology* **214:**3–11.

Karn, J. 1999. Tackling Tat. *J. Mol. Biol.* **293:**235–254.

Khoury, G., and P. Grüss. 1983. Enhancer elements. *Cell* **33:**313–314.

Maniatis, T., S. Goodbourn, and J. F. Fischer. 1987. Regulation of inducible and tissue-specific gene expression. *Science* **236:**1237–1245.

Moss, B., B.-Y. Ahn, B. Amegadzie, P.-D. Gershon, and J. G. Keck. 1991. Cytoplasmic transcription system encoded by vaccinia virus. *J. Biol. Chem.* **266:**1355–1358.

Naar, A. M., B. D. Lemon, and R. Tjian. 2001. Transcriptional coactivator complexes. *Annu. Rev. Biochem.* **70:**475–501.

O'Hare, P. 1993. The virion transactivator of herpes simplex virus. *Semin. Virol.* **4:**145–155.

Preston, C. M. 2000. Repression of viral transcription during herpes simplex virus latency. *J. Gen. Virol.* **81:**1–19.

Roeder, R. G. 1996. The role of general initiation factors in transcription by RNA polymerase II. *Trends Biochem. Sci.* **21:**327–335.

Ruddell, A. 1995. Transcription regulatory elements of the avian retroviral long terminal repeat. *Virology* **206:**1–7.

Urnov, F. D., and A. P. Wolffe. 2001. Chromatin remodeling and transcriptional activation: the cast (in order of appearance). *Oncogene* **20:**2991–3006.

Papers of Special Interest

Viral RNA Polymerase II Promoters and the Cellular Transcriptional Machinery

Cramer, P., D. A. Bushnell, and R. D. Kornberg. 2001. Structural basis of transcription: RNA polymerase II at 2.8 Ångstrom resolution. *Science* **292:**1863–1876.

Dynan, W. S., and R. Tjian. 1983. The promoter-specific transcription factor Sp1 binds to upstream sequences in the SV40 early promoter. *Cell* **35:**79–87.

Grüss, P., R. Dhar, and G. Khoury. 1981. Simian virus 40 tandem repeated sequences as an element of the early promoter. *Proc. Natl. Acad. Sci. USA* **78:**943–947.

Hu, S. L., and J. L. Manley. 1981. DNA sequences required for initiation of transcription in vitro from the major late promoter of adenovirus 2. *Proc. Natl. Acad. Sci. USA* **78:**820–824.

Matsui, T., J. Segall, P. A. Weil, and R. G. Roeder. 1980. Multiple factors required for accurate initiation of transcription by purified RNA polymerase II. *J. Biol. Chem.* **255:**11992–11996.

Yamamoto, T., B. de Crombrugghe, and I. Pastan. 1980. Identification of a functional promoter in the long terminal repeat of Rous sarcoma virus. *Cell* **22:**787–797.

Viral Transcriptional Enhancers

Banerji, J., S. Rusconi, and W. Schaffner. 1981. Expression of the α-globin gene is enhanced by remote SV40 DNA sequences. *Cell* **27:**299–308.

Fromental, C., M. Kanno, H. Nomiyama, and P. Chambon. 1988. Cooperativity and hierarchical levels of functional organization in the SV40 enhancer. *Cell* **54:**943–953.

Moreau, P., R. Hen, B. Wasylyk, R. Everett, M. P. Gaub, and P. Chambon. 1981. The SV40 72 base repeat has a striking effect on gene expression both in SV40 and other chimeric recombinants. *Nucleic Acids Res.* **9:**6047–6068.

Müller, H. P., J. M. Sogo, and W. Schaffner. 1989. An enhancer stimulates transcription in trans when attached to the promoter via a protein bridge. *Cell* **58:**767–777.

Nabel, G. J., and D. Baltimore. 1987. An inducible transcription factor that activates expression of human immunodeficiency virus in T cells. *Nature* **326:**711–713.

Spalholz, P. A., Y.-C. Yang, and P. M. Howley. 1985. Transactivation of a bovine papillomavirus transcriptional regulatory element by the E2 gene product. *Cell* **42:**183–191.

Yee, J. 1989. A liver-specific enhancer in the core promoter of human hepatitis B virus. *Science* **246:**658–670.

Regulation of Transcription of DNA Genomes by Viral Proteins

Bandara, L. R., and N. B. La Thangue. 1991. Adenovirus E1A prevents the retinoblastoma gene product from complexing with a cellular transcription factor. *Nature* **351:**494–497.

Batterson, W., and B. Roizman. 1983. Characterization of the herpes simplex virion-associated factor responsible for the induction of alpha genes. *J. Virol.* **46:**371–377.

Berk, A. J., F. Lee, T. Harrison, J. F. Williams, and P. A. Sharp. 1979. Pre-early adenovirus 5 gene product regulates synthesis of early viral messenger RNAs. *Cell* **17:**935–944.

Chellappan, S. P., S. Hiebert, M. Mudryj, J. M. Horowitz, and J. R. Nevins. 1991. The E2F transcription factor is a cellular target for the Rb protein. *Cell* **65:**1053–1061.

Chi, T., and M. Carey. 1993. The ZEBRA activation domain: modular organization and mechanism of action. *Mol. Cell. Biol.* **13:**7045–7055.

Jones, N., and T. Shenk. 1979. An adenovirus type 5 early gene function regulates expression of other early viral genes. *Proc. Natl. Acad. Sci. USA* **76:**3665–3669.

Keller, J. M., and J. C. Alwine. 1984. Activation of the SV40 late promoter: direct effects of T antigen in the absence of viral DNA replication. *Cell* **36:**381–389.

Lai, J.-S., M. A. Cleary, and W. Herr. 1992. A single amino acid exchange transfers VP16-induced positive control from the Oct-1 to the Oct-2 homeo domain. *Genes Dev.* **6:**2058–2065.

Lillie, J. W., and M. R. Green. 1989. Transcription activation by the adenovirus E1A protein. *Nature* **338:**39–44.

Mador, N., D. Goldenberg, O. Cohen, A. Panet, and I. Steiner. 1998. Herpes simplex virus type 1 latency-associated transcripts suppress viral replication and reduce immediate-early gene mRNA levels in a neuronal cell line. *J. Virol.* **72:**5067–5075.

Tribouley, C., P. Lutz, A. Staub, and C. Kédinger. 1994. The product of the adenovirus intermediate gene IVa_2 is a transcriptional activator of the major late promoter. *J. Virol.* **68:**4450–4457.

Vales, L. D., and J. E. Darnell. 1989. Promoter occlusion prevents transcription of adenovirus polypeptide IX mRNA until after DNA replication. *Genes Dev.* **3:**49–59.

Walker, S., S. Hayes, and P. O'Hare. 1994. Site-specific conformational alteration of the Oct-1 POU domain-DNA complex as the basis for differential recognition by Vmw65 (VP16). *Cell* **79:**841–852.

Wiley, S. R., R. J. Kraus, F. Zuo, E. E. Murray, K. Loritz, and J. E. Mertz. 1993. SV40 early-to-late switch involves titration of cellular transcriptional repressors. *Genes Dev.* **7:**2206–2219.

Wilson, A. C., R. W. Freeman, H. Goto, T. Nishimoto, and W. Herr. 1997. VP16 targets an amino-terminal domain of HCF involved in cell cycle progression. *Mol. Cell. Biol.* **17:**6139–6146.

RNA-Dependent Stimulation of Human Immunodeficiency Virus Type 1 Transcription by the Tat Protein

Dingwall, C., J. Ernberg, M. J. Gait, S. M. Green, S. Heaphy, J. Karn, M. Singh, and J. J. Shinner. 1990. HIV-1 Tat protein stimulates transcription by binding to a U-rich bulge in the stem of the TAR RNA structure. *EMBO J.* **9:**4145–4153.

Feinberg, M. B., D. Baltimore, and A. D. Frankel. 1991. The role of Tat in the human immunodeficiency virus life cycle indicates a primary effect on transcriptional elongation. *Proc. Natl. Acad. Sci. USA* **88:**4045–4049.

Fisher, A. G., M. B. Feinberg, S. F. Josephs, M. E. Harper, L. M. Marselle, G. Reyes, M. A. Gonda, A. Aldovini, C. Debouk, R. C. Gallo, and F. Wong-Staal. 1986. The trans-activator gene of HTLV-III is essential for virus replication. *Nature* **320:**367–371.

Mancebo, H. S. Y., G. Lee, J. Flygare, J. Tomassini, P. Luu, Y. Zhu, J. Peng, C. Blau, D. Hazuda, D. Price, and O. Flores. 1997. P-TEFb kinase is required for HIV Tat transcriptional activation in vivo and in vitro. *Genes Dev.* **11:**2633–2644.

Parada, C. A., and R. G. Roeder. 1996. Enhanced processivity of RNA polymerase II triggered by Tat-induced phosphorylation of its carboxy-terminal domain. *Nature* **384:**375–378.

Wei, P., M. F. Garber, S.-M. Fang, W. H. Fischer, and K. A. Jones. 1998. A novel CDK9-associated C-type cyclin interacts directly with HIV-1 Tat and mediates its high-affinity, loop-specific binding to TAR RNA. *Cell* **92:**451–462.

The Poxviral Transcriptional System

Hagler, J., and S. Shuman. 1992. A freeze-frame view of eukaryotic transcription during elongation and capping of nascent RNA. *Science* **255:**983–986.

Kates, J. R., and B. R. McAuslan. 1967. Poxvirus DNA-dependent RNA polymerase. *Proc. Natl. Acad. Sci. USA* **58:**134–141.

Lu, J., and S. S. Broyles. 1993. Recruitment of vaccinia virus RNA polymerase to an early gene promoter by the viral early transcription factor. *J. Biol. Chem.* **268:**2773–2780.

Rosales, R., G. Sutter, and B. Moss. 1994. A cellular factor is required for transcription of vaccinia viral intermediate-stage genes. *Proc. Natl. Acad. Sci. USA* **91:**3794–3798.

Transcription of Viral Genes by RNA Polymerase III

Fowlkes, D. M., and T. Shenk. 1980. Transcriptional control regions of the adenovirus VA1 RNA gene. *Cell* **22:**405–413.

Hoeffler, W. K., R. Kovelman, and R. G. Roeder. 1988. Activation of transcription factor IIIC by the adenovirus E1A protein. *Cell* **53:**907–920.

Howe, E., and M.-D. Shu. 1993. Upstream basal promoter element important for exclusive RNA polymerase III transcription of EBER2 gene. *Mol. Cell. Biol.* **13:**2656–2665.

Huang, W., R. Pruzan, and S. J. Flint. 1994. In vivo transcription from the adenovirus E2 early promoter by RNA polymerase III. *Proc. Natl. Acad. Sci. USA* **91:**1265–1269.

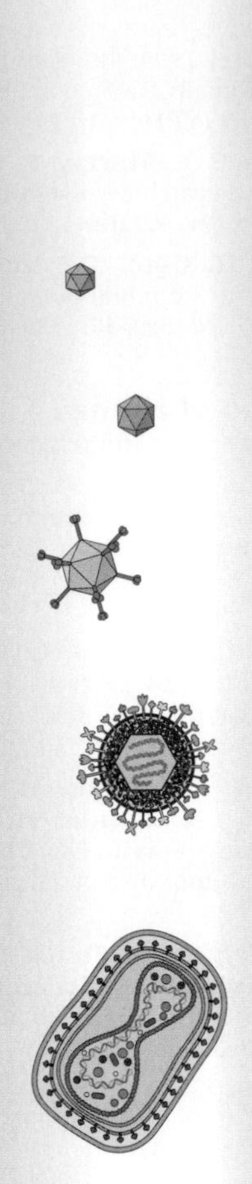

9

Genome Replication Strategies: DNA Viruses

The more the merrier.
ANONYMOUS

Introduction

The genomes of DNA viruses come in a considerable variety of sizes and shapes, from small molecules of single-stranded DNA to large double-stranded molecules that may be linear or circular (Table 9.1). Whatever their physical nature, viral DNA molecules must be replicated efficiently within an infected cell to provide genomes for assembly into progeny virions. Such replication of DNA genomes invariably requires the synthesis of at least one viral protein and often expression of several viral genes. Consequently, viral DNA synthesis cannot begin immediately upon arrival of the genome at the appropriate intracellular site, but rather is delayed until viral replication proteins have attained a sufficient concentration. Initiation of viral DNA synthesis typically leads to many cycles of replication and the accumulation of very large numbers of newly synthesized viral DNA molecules. However, more long-lasting virus-host cell interactions, such as latent infections, are also common, both in nature and in the laboratory. In these circumstances, the number of viral DNA molecules made is strictly controlled.

Replication of DNA in all organisms, from the simplest virus to the most complex vertebrate cell, follows a set of universal rules: (1) DNA is always synthesized by template-directed, stepwise incorporation of deoxynucleoside monophosphates (dNMPs) from deoxynucleoside triphosphate (dNTP) substrates into the 3′-OH end of the growing DNA chain, which is elongated in the $5' \rightarrow 3'$ direction; (2) each parental strand of a duplex DNA template is copied by base pairing to produce two daughter molecules identical to one another and to their parent (**semiconservative replication**); (3) replication of DNA begins and ends at specific sites in the template, termed **origins** and **termini**, respectively; and (4) DNA synthesis is catalyzed by DNA-dependent DNA polymerases, but many accessory proteins are required for initiation or **elongation**. In contrast to all DNA-dependent, and many RNA-dependent, RNA polymerases, no DNA polymerase can initiate template-directed DNA synthesis de novo. All require a

Table 9.1 Cellular and viral proteins used in synthesis of viral DNA

Virus	Genome	Viral origin-binding protein(s)	DNA polymerase(s)	Other replication proteins	Effects on cellular DNA synthesis
Adenovirus					
Human adenovirus type 2	35,937 bp, linear, double stranded	DNA polymerase preterminal protein complex	Viral	Cellular transcriptional activators (Nf-1, Oct-1) and topoisomerase; viral preterminal protein primer; viral single-stranded DNA protein	Induction of entry into S phase in quiescent cells; inhibition of cellular DNA synthesis in actively growing cells
Circovirus					
Chicken anemia virus	2,298 bp, circular (−) strand	Rep, Rep′	Cellular, DNA polymerases α and δ	Presumably other cellular replication proteins; no viral	None known
Herpesviruses					
Epstein-Barr virus	172 kb, linear, double stranded				
OriP		EBNA-1	Cellular	Many cellular, no viral	Immortalization of latently infected B cells
OriLyt		Zta	Viral	Several viral, analogous to HSV-1 replication proteins	Induction of cell cycle arrest and inhibition of cellular DNA synthesis by Zta
Herpes simplex virus type 1	~150 kb, linear, double stranded, 4 isomers	UL9	Viral	Cellular protein requirement not known; at least 7 viral, including a primase complex	Inhibition of cellular DNA synthesis
Parvovirus					
Adeno-associated virus type 2	4,680 bp, linear, single (−) or (+) strands	Rep 78, Rep 68	Cellular, DNA polymerases δ and ε?	Many cellular; no viral, but Rep 78/68 supplies ATP-dependent helicase activity	None
Polyomavirus					
Simian virus 40	5,243 bp, closed, double-stranded circle	LT antigen	Cellular, DNA polymerases α and δ	Many cellular; no viral, but LT supplies ATP-dependent helicase activity	Induction of entry into S phase and cellular DNA synthesis
Poxvirus					
Vaccinia virus	~200 kb, linear, double stranded with closed terminal loops	None	Viral	Several viral, no cellular	Inhibition of cellular DNA synthesis

BOX 9.1

The two mechanisms of synthesis of double-stranded viral DNA molecules

Replication of double-stranded nucleic acids proceeds either by synthesis of daughter strands from a replication fork, or by a strand displacement mechanism. No other replication mechanisms are known.

Among viral genomes, only those of certain double-stranded DNA viruses are synthesized via a replication fork mechanism. Replication of viral double-stranded RNAs **never** proceeds via this mechanism.

DNA synthesis by a replication fork mechanism is **always** initiated from an RNA primer. In contrast, strand displacement synthesis of viral DNA **never** requires an RNA primer.

Fork	Displacement (primer)
Papillomaviruses	Adenoviruses (protein)
Polyomaviruses	Parvoviruses (DNA hairpin)
Herpesviruses	Poxviruses (DNA hairpin)

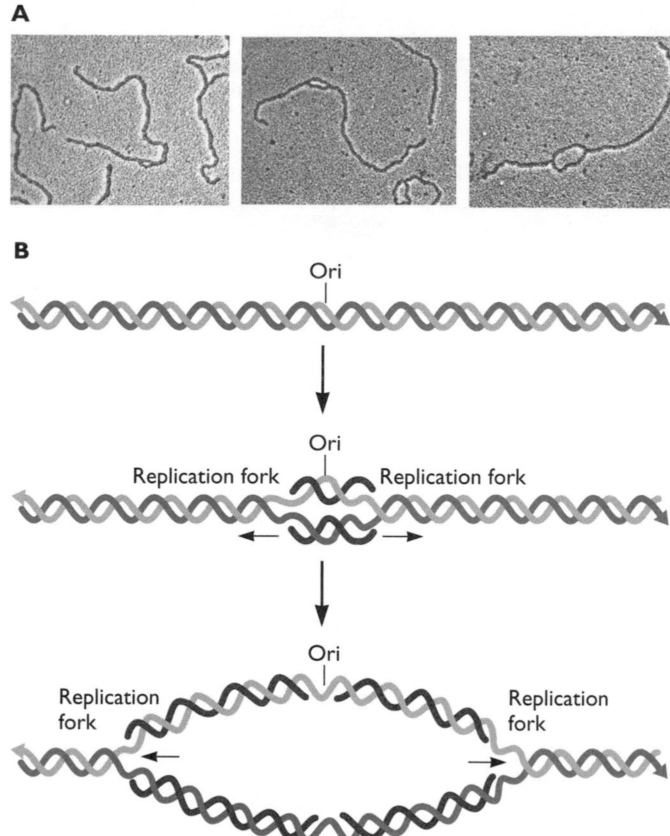

Figure 9.1 Properties of replicons. (A) Electron micrographs of replicating simian virus 40 DNA, showing the "bubbles" of replicating DNA, in which the two strands of the template are unwound. These linear DNA molecules were obtained by *Eco*RI cleavage of viral DNA that had replicated to different degrees in infected cells. They are arranged in order of increasing degree of replication to illustrate the progressive movement of the two replication forks from a single origin of replication, which was established by measurement of the lengths of the unreplicated double-stranded "arms." From G. C. Fareed et al., *J. Virol.* **10:** 484–491, 1972, with permission. (B) Bidirectional replication from an origin. This is the most common pattern at viral and cellular origins and can also be visualized by autoradiography of DNA labeled in cells exposed to [^{3}H]thymidine.

primer with a free 3′-OH end to which dNMPs complementary to those of the template strand are added.

The genomes of RNA viruses must encode enzymes that catalyze RNA-dependent RNA or DNA synthesis. In contrast, those of DNA viruses can be replicated by the cellular machinery. Indeed, replication of the smaller DNA viruses, such as parvoviruses and polyomaviruses (Table 9.1), requires but a single viral replication protein, and the majority of reactions are carried out by cellular proteins. This strategy avoids the need to devote limited viral genetic coding capacity to enzymes and other proteins needed for DNA synthesis. In contrast, the genomes of all larger DNA viruses encode DNA polymerases and additional replication proteins (Table 9.1). In the extreme case, exemplified by poxviruses, the viral genome encodes a complete DNA synthesis system and is replicated in the cytoplasm of its host cells.

There is also considerable variety in the mechanism of priming of viral DNA synthesis. In some cases, DNA synthesis is initiated with RNA primers, the mechanism by which cellular genomes are replicated. In others, unusual structural features of the genome or viral proteins serve as primers. Despite such distinctions, the replication strategies of different viral DNAs are based on the only two mechanisms of double-stranded DNA synthesis known (Box 9.1) and on common molecular principles. For example, the genomes of polyomaviruses and herpesviruses, which are quite different in size and structure (Table 9.1), are replicated by the cellular replication machinery and viral replication proteins, respectively. Nevertheless, synthesis of these two DNAs is initiated by the same priming mechanism, and the virally encoded replication machinery of herpesviruses carries out the same biochemical reactions as the host proteins that participate in synthesis of simian virus 40 DNA.

DNA Synthesis by the Cellular Replication Machinery: Lessons from Simian Virus 40

Our current understanding of the intricate reactions by which both strands of a typical double-stranded DNA template are completely copied is based on in vitro studies of simian virus 40 DNA synthesis. In this section, we focus on the cellular replication machinery and the molecular functions of its components that have been established by using this viral system. However, we first briefly discuss general features of eukaryotic DNA replication and why simian virus 40 proved to be a priceless resource for those seeking to understand this process.

Eukaryotic Replicons

General Features

The complete replication of large eukaryotic genomes within the lifetime of an actively growing cell depends on their organization into smaller units of replication termed **replicons** (Fig. 9.1): at the maximal rate of replication observed in vivo, a typical human chromosome could not be replicated as a single unit in less than 10 days! Each chromosome therefore contains many replicons, ranging in length from about 20 to 300 kb. With rare exceptions, no viral DNA genome is larger than the upper limit reported for such cellular replicons. Nevertheless, all but the smallest are replicated from two or three origins (see "Viral Replication Origins and Their Recognition" below).

Each replicon contains a single origin at which replication begins. The sites at which nascent DNA chains are being synthesized, the ends of "bubbles" seen in the electron microscope (Fig. 9.1A), are termed **replication forks**. In bidirectional replication, two replication forks are established at a single origin and move away from it as the new DNA strands are synthesized (Fig. 9.1B). However, as DNA must be synthesized in the 5′ → 3′ direction, only one of the two parental strands can be copied continuously from a primer deposited at the origin. The long-standing conundrum of how the second strand is synthesized was solved with elucidation of the discontinuous mechanism of synthesis (Fig. 9.2A). Primers for DNA synthesis are synthesized at multiple sites, such that this new DNA strand is initially made as short, discontinuous segments.

The discontinuous mechanism of DNA synthesis creates a special problem at the ends of linear DNAs, where excision of the terminal primer creates a gap at the 5′ end of the daughter DNA molecules (Fig. 9.2B). In the absence of a mechanism for completing synthesis of termini, discontinuous DNA synthesis would lead to an intolerable loss of genetic information. In chromosomal DNA, specialized elements, called **telomeres**, at the ends of each chromosome prevent loss of end sequences. These structures comprise simple, repeated sequences maintained by reverse transcription of an RNA template, which is an essential component of the ribonucleoprotein enzyme **telomerase**. Complete replication of all sequences of linear viral DNA genomes (Table 9.1) is achieved by a variety of elegant mechanisms.

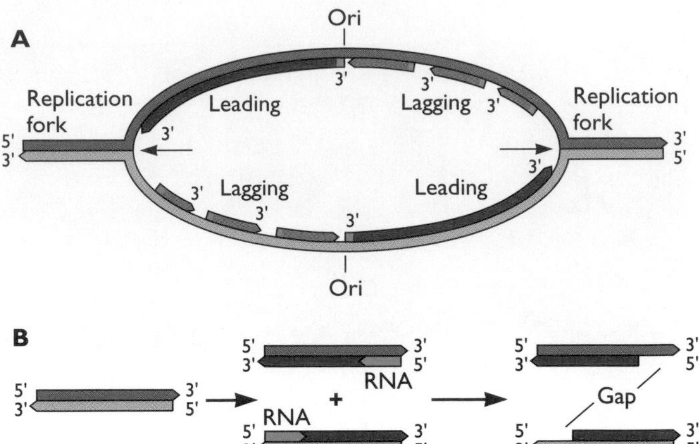

Figure 9.2 Semidiscontinuous DNA synthesis from a bidirectional origin. (A) Semidiscontinuous synthesis of the daughter strands. Synthesis of the RNA primers (green) at the origin allows initiation of continuous copying of one of the two strands on either side of the origin in the replication bubble. The second strand cannot be made in the same way, because DNA can be synthesized only in the 5′ → 3′ direction. The nascent DNA population, identified by pulse-labeling with [^{3}H]thymidine, contains many small molecules termed **Okazaki fragments** in honor of the investigator who first described them. The presence of short segments of RNA at the 5′ ends of Okazaki fragments indicated that the primers necessary for DNA synthesis are molecules of RNA. With increasing time of replication, these small fragments are incorporated into long DNA molecules, indicating that they are precursors. It was therefore deduced that the second nascent DNA strand is synthesized discontinuously, also in the 5′ → 3′ direction. Because synthesis of this strand cannot begin until the replication fork has moved some distance from the origin, it is called the **lagging strand,** while the strand synthesized continuously is termed the **leading strand.** Because the lagging strand is initially made as short pieces, its complete replication requires not only an enzyme to make the RNA primer (a primase) and a DNA polymerase, but also enzymes that can remove RNA primers, repair the gaps thus created, and ligate the individual DNA fragments to produce a continuous copy of the template strand. (B) Incomplete synthesis of the lagging strand. When a DNA molecule is linear, synthesis of the leading strand can be completed simply by copying of its template to the end of the molecule. In contrast, removal of the terminal RNA primer from the 5′ end of the lagging strand creates a gap that cannot be repaired by any DNA polymerase. In the absence of specialized mechanisms for maintaining the ends of linear DNA molecules, successive cycles of replication would therefore progressively eliminate genetic information.

Origins of Cellular Replication

Identification of replication origins of eukaryotes (other than budding yeasts) has proved to be one of the most frustrating endeavors of modern molecular biology. A large body of evidence indicates that replication initiates at specific sites in vivo. With the availability of cloned genomic DNA segments, these origins can be identified and mapped with some precision. The difficulty lies in their functional identification. *Saccharomyces cerevisiae* origins can be assayed readily, because they support replication of small plasmids and hence their maintenance as episomes. All yeast origins behave as such **autonomously replicating sequences.** In contrast, this simple functional assay has generally failed to identify analogous mammalian sequences, even when applied to DNA segments containing origins mapped in vivo. Conversely, many long DNA fragments (including prokaryotic DNAs) support replication of plasmids in mammalian cells or *Xenopus* eggs, but apparently direct random initiation of DNA replication. The paradox of why DNA sequences that support plasmid replication are so difficult to find when replication initiates at specific sites in vivo has yet to be fully resolved. However, there is accumulating evidence that metazoan DNA replication initiates within long (several kilobase pairs or more) DNA segments, and that their function can depend on far more distant sequences. Despite the considerable differences in size and complexity (and presumably regulation) of fungal and mammalian origins of replication, the origin recognition machinery (described in "Regulation of Replication via Different Viral Origins: Epstein-Barr Virus" below) is highly conserved among eukaryotes.

The Origin of Simian Virus 40 DNA Replication

Because the identification of functional origins of most eukaryotic genomes has been so difficult, progress in elucidation of the mechanisms of origin-dependent DNA synthesis came through study of viral origins, notably that of simian virus 40. This origin, which supports bidirectional replication, was the first viral control sequence to be located on a physical map in which the reference points are restriction endonuclease cleavage sites (Fig. 9.1; Box 9.2). We now possess a detailed picture of this origin (Fig. 9.3) and of the binding sites for the simian virus 40 origin recognition protein, large T antigen (LT).

A 64-bp sequence, the **core origin,** which lies between the initiation sites of early and late transcription, is sufficient for initiation of simian virus 40 DNA synthesis in infected cells. This sequence contains four copies of a pentanucleotide-binding site for LT, flanked by an AT-rich element and a 10-bp imperfect palindrome (Fig. 9.3). Additional DNA sequences within this busy control region of the viral genome increase the efficiency of initiation of DNA synthesis from the core origin.

Cellular Replication Proteins and Their Functions during Simian Virus 40 DNA Synthesis

Eukaryotic DNA Polymerases

It has been known for about 45 years that eukaryotic cells contain DNA-dependent DNA polymerases. Mammalian cells contain several such nuclear enzymes, which are distinguished by such properties as sensitivities to var-

BOX 9.2

Mapping of the simian virus 40 origin of replication by analysis of the distribution of radioactivity in pulse-labeled DNA molecules that completed replication during a brief labeling period

As illustrated in panel A (left), exposure of simian virus 40-infected monkey cells to [^{3}H]thymidine ([^{3}H]dT) for a period less than the time required to complete one round of replication (e.g., 5 min) results in labeling of the growing points of replicating DNA. If replication proceeds from a specific origin (Ori) to a specific termination site (T), the DNA replicated last will be preferentially labeled in the population of completely replicated molecules (panel A, right). The distribution of [^{3}H]thymidine among the fragments of completely replicated viral DNA generated by cleavage with *Hin*dII and *Hin*dIII is shown in panel B. The circular simian virus 40 genome is represented as cleaved within the G fragment, and relative distances are given with respect to the junction of the A and C fragments. The observation of two decreasing gradients of labeling that can be extrapolated (dashed lines) to the same region of the genome confirmed that simian virus 40 replication is bidirectional (Fig. 9.1A) and allowed location of the origin on the physical map of the viral genome. Modified from K. J. Danna and D. Nathans, *Proc. Natl. Acad. Sci. USA* **69:**3097–3100, 1972, with permission.

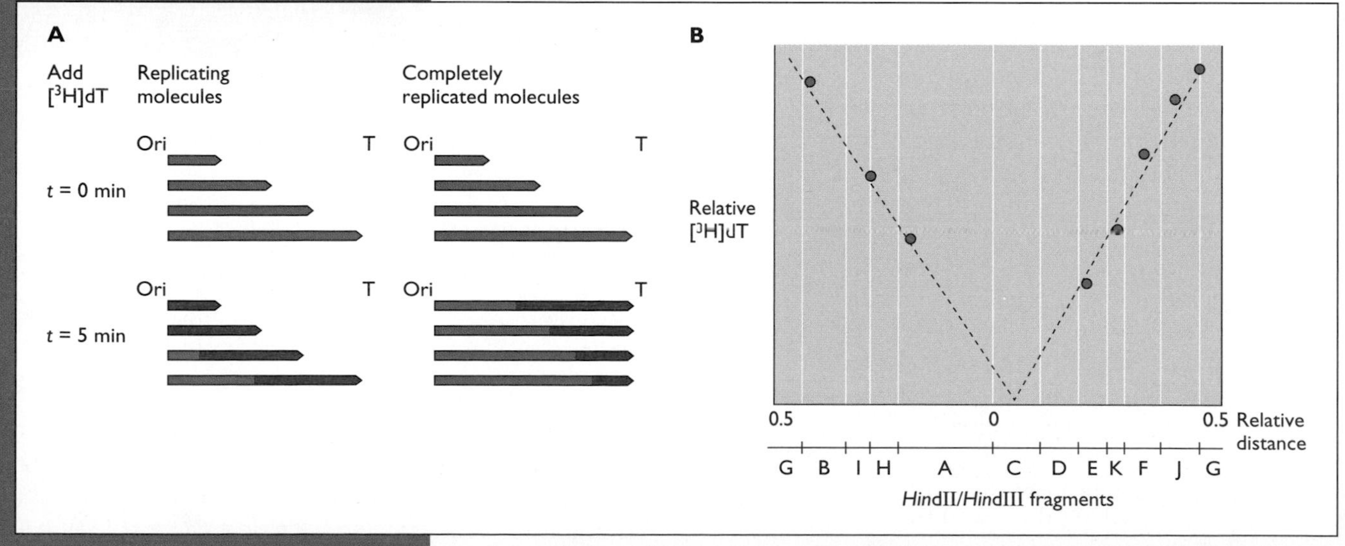

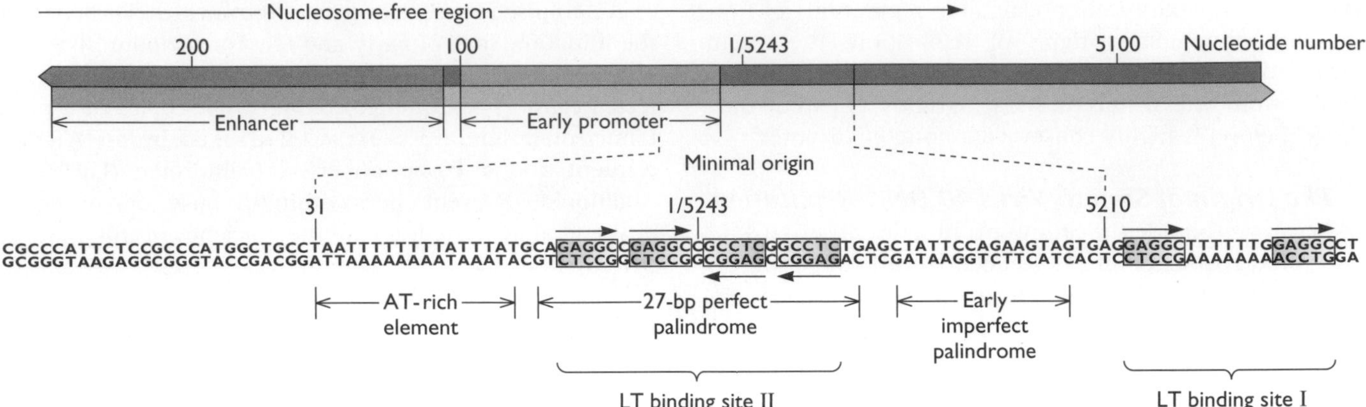

Figure 9.3 The origin of simian virus 40 DNA replication. The positions of the minimal origin necessary for simian virus 40 DNA replication in vivo and in vitro and of the enhancer and early promoter (Chapter 8) in the viral double-stranded DNA genome are indicated. The sequence of the minimal origin is shown below, with the LT-binding sites ($G^{A}_{C}GGC$) in yellow. The AT-rich element and early imperfect palindrome, as well as LT-binding site II, are essential for replication. A second LT-binding site (site I) stimulates in vivo replication modestly. Other sequences, including the enhancer and Sp1 binding sites of the early promoter, increase the efficiency of viral DNA replication at least 10-fold. The transcriptional regulators that bind to the enhancer or promoter might facilitate initiation of replication by various mechanisms. One possibility is that the activation domains of these proteins (Chapter 8) help recruit essential replication proteins to the origin. This model is consistent with popular ideas about how activators stimulate transcription. Alternatively, the binding of transcriptional activators might induce remodeling of the chromatin structure in the vicinity of the origin. This possibility is particularly attractive in the case of simian virus 40, because, as indicated, the region of the genome containing the origin and transcriptional control regions is nucleosome-free in a significant fraction (about 25%) of minichromosomes in infected cells (but not in virions).

ious inhibitors and their degree of **processivity,** the number of nucleotides incorporated into a nascent DNA chain per initiation from a primer carrying a 3′-OH group (Table 9.2). These characteristics can be readily assayed in in vitro reactions with artificial template-primers, such as gapped or nicked DNA molecules. However, distinguishing their roles in DNA replication was impossible until simian virus 40 origin-dependent DNA synthesis was reproduced in 1984 (see next section). The requirements for viral DNA synthesis in vitro and genetic analyses (performed largely with yeasts) identified DNA polymerases α, δ, and ε as replication enzymes. Only DNA polymerase α is associated with priming activity. The heteromeric **primase** is tightly bound to DNA polymerase α and copurifies with it.

One of the most striking properties of these replicative DNA polymerases is their obvious evolutionary relationships to prokaryotic and viral enzymes. Six regions are highly conserved in both primary amino acid sequence and relative position in a wide variety of DNA polymerases, including human DNA polymerase α and the enzymes encoded by several DNA viruses. As discussed in Chapters 6 and 7, all template-directed nucleic acid polymerases share several sequence motifs and probably a similar core structure, indicating that important features of the mechanisms by which they catalyze synthesis are also common to all these enzymes.

Identification of Other Replication Proteins by Using In Vitro Replication Systems

Studies of eukaryotic DNA replication took a quantum leap forward with the development of a cell-free system for synthesis of adenoviral DNA from an exogenous tem-

Table 9.2 Properties of mammalian nuclear DNA polymerases

Properties	DNA polymerase			
	α	δ	ε	β
Mass (kDa)	>250	170	256	36–38
Associated functions				
3′→ 5′ Exonuclease	Cryptic	Yes	Yes	No
Primase	Yes	No	No	No
Characteristics				
Response to Pcna	No	Yes	Yes	No
Processivity	Low	High[a]	High[a]	Low
Fidelity	High	High	High	Low
Inhibitors				
Aphidicolin	Strong	Strong	Strong	No
Dideoxy-NTPs	None	Weak	Weak	Strong
Replication	Yes	Yes	Yes	No

[a]In the presence of Pcna.

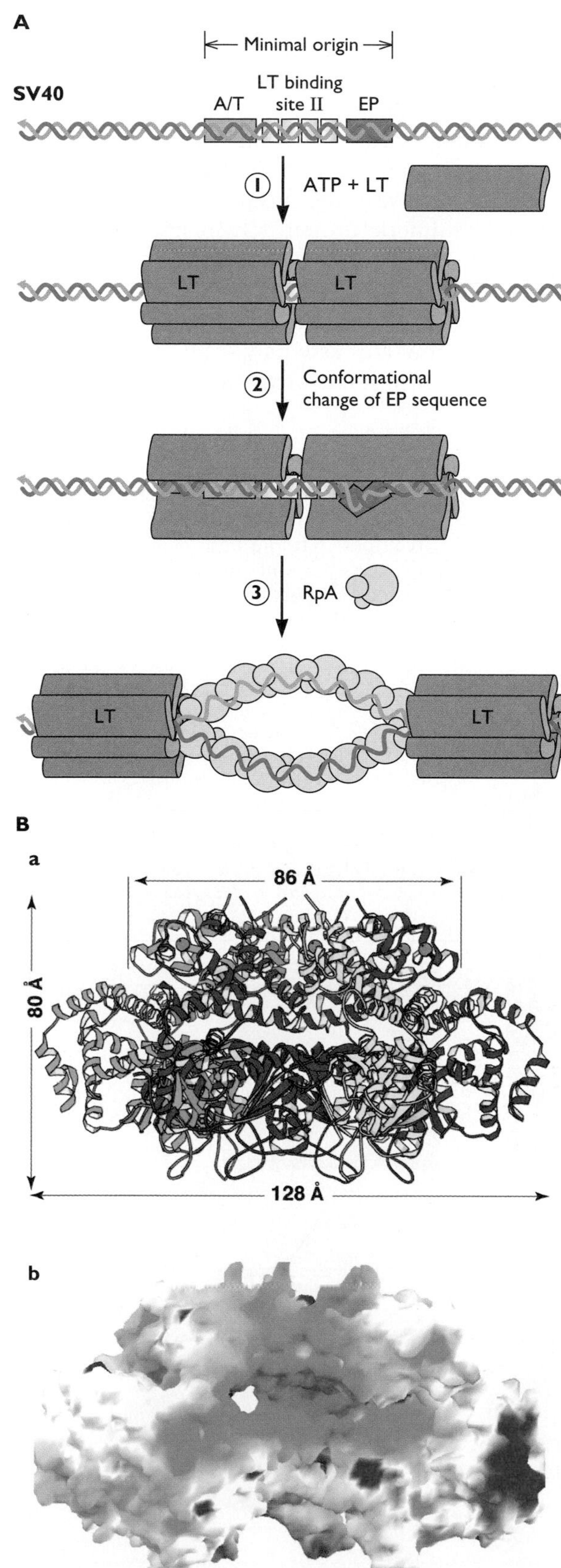

Figure 9.4 Recognition and unwinding of the simian virus 40 (SV40) origin. (A) Schematic model. In the presence of ATP, the viral early protein LT assembles into hexamers. Two hexamers bind to the origin via the pentanucleotide LT-binding sites in the center of the minimal origin (step 1). Binding of LT hexamers protects the flanking AT-rich (A/T) and early palindrome (EP) sequences of the minimal origin from DNase I digestion and induces structural changes in them, in particular, distortion of the early palindrome (step 2). However, extensive, stable unwinding of the origin requires the cellular, single-stranded-DNA-binding protein replication protein A (Rp-A). This heterotrimeric protein binds to the single strands of the unwound region to stabilize the unwound DNA, but appears to fulfill additional roles in simian virus 40 DNA synthesis. Rp-A binds to LT via its smaller subunits (Table 9.3). LT helicase activity in concert with Rp-A and topoisomerase I progressively unwinds the origin (step 3). For simplicity, the two LT hexamers are shown moving apart, but they probably remain in contact (Box 9.4). (B) Structural features of the LT helicase. The structure of residues 251 to 627 of simian virus 40 LT, a fragment that possesses helicase activity and binds to both single- and double-stranded DNA, was determined by X-ray crystallography. (a) The side view of the hexamer formed by this LT fragment in the absence of ATP is shown in ribbon form, with each monomer in a different color. This ring comprises two tiers of different diameters around a central channel. Both hydrophobic (in the smaller tier) and hydrophilic (in the larger tier) interactions stabilize the hexamer. (b) A surface representation of the interior of the central channel is shown, with positively charged, negatively charged, and uncharged residues in blue, red, and white, respectively. Two monomers have been removed from the front of the hexamer. Consistent with the DNA binding activity of this LT fragment, the central channel is lined with positively charged residues. This view also shows the large chamber that is formed in the middle of the central channel. This chamber is wide enough (~67-Å diameter) to accommodate DNA strands separated during DNA unwinding. The diameters of the channel on either side of this chamber are sufficient to accommodate double-stranded DNA in the smaller tier (~25 Å), but only single-stranded DNA (~15 Å) in the larger tier. The sizes of these channels were observed to differ significantly in different crystal forms of the LT hexamer and under different buffer conditions. This property, and structural features of LT monomers and the interfaces that form the two-tier hexamer, has led to the proposal that during DNA unwinding, the central channel might expand and constrict like the diaphragm (iris) of a camera. Adapted from D. Li et al., *Nature* **423:**512–518, 2003, with permission. Courtesy of Xiaojiang Chen, University of Colorado Health Science Center.

plate. This breakthrough was soon followed by origin-dependent replication of simian virus 40 DNA in vitro. Because cellular components are largely responsible for simian virus 40 DNA synthesis, this system proved to be the watershed in the investigation of eukaryotic DNA replication, allowing the identification of previously unknown cellular replication proteins and elucidation of the mechanism of DNA synthesis.

Mechanism of Simian Virus 40 DNA Synthesis

Origin recognition and unwinding. The first step in simian virus 40 DNA synthesis is the recognition of the origin by LT, the major early gene product of the virus. In the presence of ATP, LT binds cooperatively to the core origin (Fig. 9.3) as two hexamers that encase the entire DNA sequence (Fig. 9.4A). Each hexamer comprises a two-tiered ring enclosing a central channel of varying diameter (Fig 9.4B). The LT hexamers undergo large conformational changes upon binding to DNA (Box 9.3) and, in turn, induce structural transitions in the AT-rich and the early imperfect palindrome sequences that flank the LT-binding sites (Fig. 9.4A). The next reaction, LT-induced unwinding of the origin, requires cellular replication protein A (Rp-A), which possesses single-stranded-DNA-binding activity and binds specifically to LT (Table 9.3). In concert with Rp-A and cellular topoisomerase I (see "Termination and resolution" below), the intrinsic $3' \rightarrow 5'$ helicase activity of LT unwinds DNA bidirectionally from the core origin (Fig. 9.4A). The high-resolution structure of the helicase domain of LT suggests that DNA distortion and unwinding may be accomplished by twisting and untwisting of the two tiers of the hexamer relative to one another (Fig. 9.4B). Formation of the LT complex at the simian virus 40 origin resembles assembly reactions at well-characterized prokaryotic replication origins, such as *Escherichia coli* OriC or the origin of phage λ, in which multimeric protein structures form on AT-rich sequences.

Leading-strand synthesis. Binding of DNA polymerase α-primase to both LT and Rp-A at the simian virus 40 origin completes assembly of the **presynthesis complex** and sets the stage for the initiation of leading-

BOX 9.3

DNA binding induces conformational changes in simian virus 40 LT antigen

The simian virus 40 LT antigen is both the viral origin recognition protein and the ATP-dependent helicase that unwinds parental DNA in both directions from the origin. In the presence of ATP, it binds cooperatively to the origin to form a double hexamer, arranged in a head-to-head orientation. Electron microscopy and image reconstruction of LT double hexamers assembled in the presence of ATP and of these structures bound to a model replication fork revealed substantial conformational change induced by DNA binding. This interaction does not alter the overall shape or dimensions of the double hexamer (maximal diameter of ~130 Å and height of ~100 Å). Rather, specific LT domains, for example, the C-terminal domain, become reoriented (compare left and right panels in the figure). As a result, the central channel is widened for better accommodation of the DNA replication fork.

(Top) Averages of images of free LT double hexamers (left) and of double hexamers formed in the presence of forked DNA (right). Bar = 100 Å. (Bottom) Three-dimensional reconstructions of the free double hexamer (left) and the double-hexamer formed in the presence of forked DNA (right). The rotation of N- and C-terminal domains upon DNA binding widens the central channel. Adapted from M. S. van Loock et al., *Curr. Biol.* **12:**472–476, 2002, with permission.

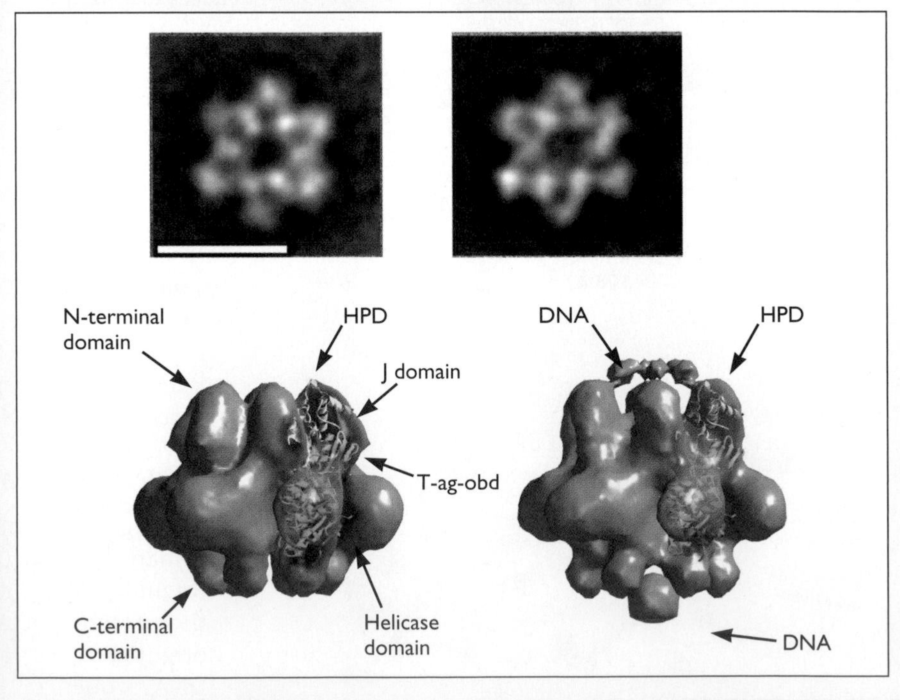

Table 9.3 Cellular proteins required for simian virus 40 DNA replication

Protein[a]	Synonym(s)	Subunits (kDa)	Contact(s)	Functions
Rp-A	Rf-A, ssB	70 32 11	Primase LT LT	Binds to single-stranded DNA; origin unwinding in cooperation with LT
DNA polymerase α-primase	Polα-primase	165–180, 72, 58, 48	LT	Synthesis of RNA primers and Okazaki fragments on leading and lagging strands
Rf-C	Activator 1	140, 40, 38, 37, 36	Pcna	ATP-dependent clamp loading; also required for release of Pcna from DNA; largest subunit contacts Pcna and DNA
Pcna		34	Rf-C	Sliding clamp
DNA polymerase δ		125, 48	Pcna	Processive synthesis of leading and lagging strands when bound to Pcna
RNase H1		32	Unknown	Endonucleolytic degradation of RNA base paired with DNA and attached to the 5′ end of a DNA strand; removal of RNA primers
Fen1		48	Pcna	5′ → 3′ exonuclease; removal of RNA primers
DNA ligase I		98	Pcna	Sealing daughter DNA fragments

[a]Rf-A, replication factor A; ssB, single-stranded-DNA-binding protein B; Fen1, Flap endonuclease 1. See the text for definitions of other protein designations.

strand synthesis (Fig. 9.5; Table 9.3). The primase synthesizes the RNA primers of the leading strand at each replication fork, while DNA polymerase α extends them to produce short **Okazaki fragments**. The 3′-OH ends of these small DNA fragments are then bound by cellular replication factor C (Rf-C), proliferating cell nuclear antigen (Pcna), and DNA polymerase δ (Fig. 9.5; Table 9.3). Pcna, which is highly conserved, is the processivity factor for DNA polymerase δ: it is required for synthesis of long DNA chains from a single primer and essential for simian virus 40 DNA replication in vitro. This mammalian protein is the functional analog of the β subunit of *E. coli* DNA polymerase III and phage T4 gene 45 product. These remarkable proteins are **sliding clamps,** which track along the DNA template and serve as movable platforms for DNA polymerases (Fig. 9.6). As sliding clamp proteins are closed rings, Pcna cannot load onto DNA molecules that lack ends, such as the circular simian virus 40 genome or chromosomal domains of genomic DNA. This essential, ATP-dependent step is carried out by the clamp-loading protein Rf-C, which presumably induces transient opening of the Pcna ring. Because Rf-C binds to the 3′-OH ends of Okazaki fragments, it places the processivity protein at the replication forks (Fig. 9.5). Binding of these two proteins inhibits DNA polymerase α, and induces dissociation of the DNA polymerase α-primase complex. Subsequent binding of DNA polymerase δ completes assembly of a multiprotein structure capable of leading-strand synthesis by continuous copying of the parental template strand.

DNA replication can neither initiate nor proceed without the action of enzymes that unwind the strands of the DNA template. In the simian virus 40 in vitro replication system, LT is the helicase responsible for unwinding of the origin, and remains associated with each of the two replication forks, unwinding the template during elongation (Fig. 9.5). LT translocates along DNA in the 3′ → 5′ direction of each parental strand and must therefore be bound to the leading strand in a replication fork.

Lagging-strand synthesis. The first primer and Okazaki fragment of the lagging strand are synthesized by the DNA polymerase α-primase complex (Fig. 9.5, step 4). The lagging strand is also synthesized by DNA polymerase δ, at least in the in vitro system, and transfer of the 3′ end of the first Okazaki fragment to this enzyme is believed to proceed as on the leading strand. The lagging-strand template is then copied toward the **origin of replication**. However, synthesis of the lagging strand requires initiation by DNA polymerase α-primase at multiple sites progressively further from the origin (Fig. 9.5). The mechanisms by which leading- and lagging-strand syntheses are coordinated are not fully understood. If the replication machinery tracked along an immobile DNA template, the complexes responsible for leading- and lagging-strand syntheses would have to move in opposite directions (Fig. 9.5). Furthermore, lagging-strand synthesis would require repeated assembly of priming complexes at sites that do not contain LT-binding sequences. Therefore, a more attractive alternative is that the DNA template is spooled

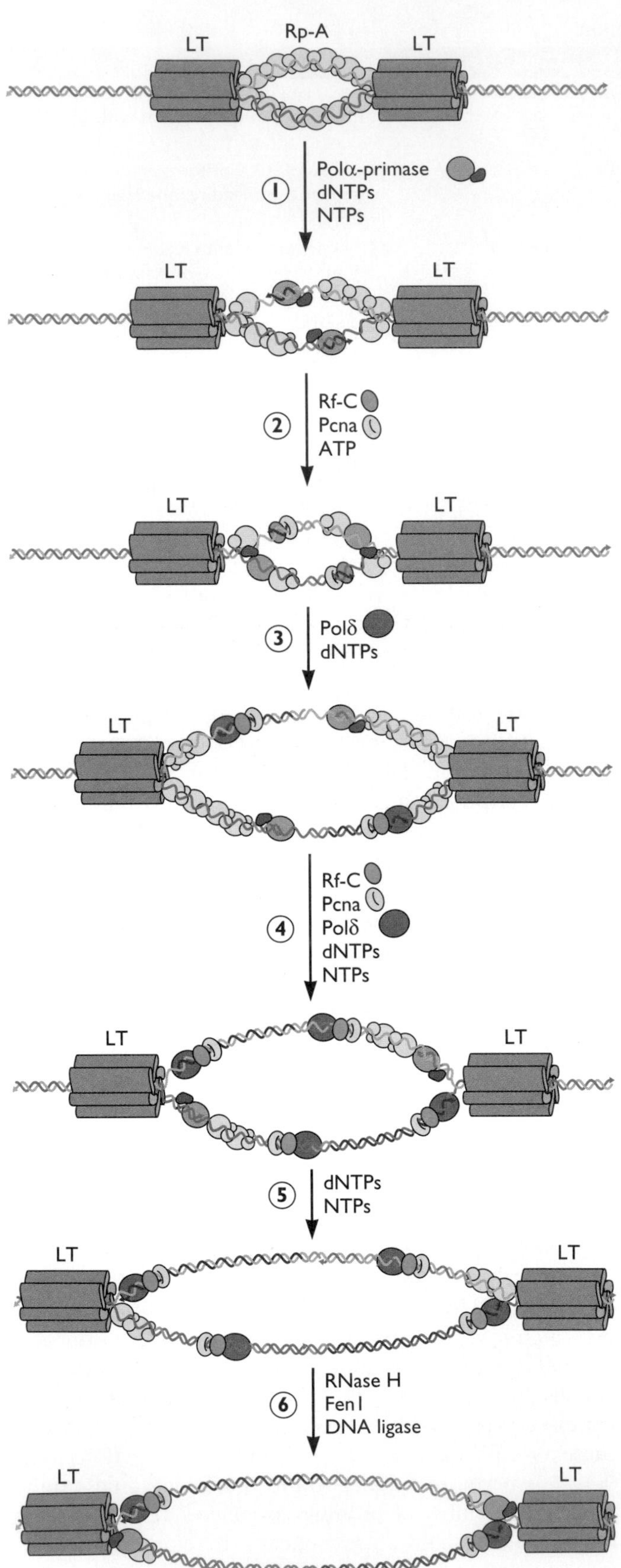

Figure 9.5 Synthesis of leading and lagging strands. The DNA polymerase α-primase responsible for the synthesis of Okazaki fragments binds specifically to both the Rp-A and LT proteins assembled at the origin in the presynthesis complex. Once bound, this enzyme complex initiates synthesis of the leading strands with production of RNA primers that are then extended as DNA by the polymerase (step 1). The 3′ OH group of the nascent RNA-DNA fragment (about 30 nucleotides in total length) is then bound by replication factor C (Rf-C) in a reaction that, like assembly of LT at the origin, requires ATP but not its hydrolysis. Rf-C allows ATP-dependent loading of proliferating cell nuclear antigen (Pcna) onto the template (step 2). The conformation of Pcna, a closed ring (Fig. 9.6), must be substantially altered during the loading reaction to allow passage of the replicating DNA into its central channel. DNA polymerase δ then binds to the Pcna/Rf-C complex (step 3). This replication complex is competent for continuous and highly processive synthesis of the leading strands (steps 4 and 5). Although DNA polymerase δ is sufficient for leading-strand synthesis during simian virus 40 DNA synthesis in vitro, the second Pcna-responsive enzyme, DNA polymerase ε (Table 9.2), appears to contribute to chromosomal replication in vivo. Lagging-strand synthesis begins with synthesis of the first Okazaki fragment by DNA polymerase α-primase (step 3). While there is evidence indicating that initiation of lagging-strand synthesis takes place after formation of the replication complex on the leading strand, it is not known whether DNA polymerase α-primase transfers directly from the leading to the lagging strands. Transfer of the 3′-OH terminus of the Okazaki fragment to the Rf-C/Pcna/DNA polymerase, as in steps 2 and 3, allows synthesis of a lagging-strand segment (step 4). The multiple DNA fragments produced by discontinuous lagging-strand synthesis (step 5) are sealed by removal of the primers by RNase H (an enzyme that specifically degrades RNA hybridized to DNA) and the 5′ → 3′ exonuclease Fen1, repair of the resulting gaps by DNA polymerase δ, and joining of the DNA fragments by DNA ligase I (step 6).

through an immobile replication complex that contains all proteins necessary for synthesis of both daughter strands. This mechanism would allow simultaneous copying of the template strands in opposite directions by a single replication machine present at each fork (Fig. 9.7). Replication of chromosomal DNA occurs at fixed sites in the nucleus, as would be predicted by a DNA spooling mechanism. Moreover, structures indicative of DNA spooling have been observed in the electron microscope during the initial, LT-dependent unwinding of the simian virus 40 origin (Box 9.4). While a model invoking an immobile replication machine possesses the virtues of elegance and parsimony, it remains to be firmly established.

Termination and resolution. Because the circular simian virus 40 DNA genome possesses no termini, its replication does not lead to gaps in the strands made discontinuously. Nevertheless, additional cellular proteins are needed for the production of two daughter molecules from the circular template. These essential components of the simian virus 40 replication system are cellular enzymes

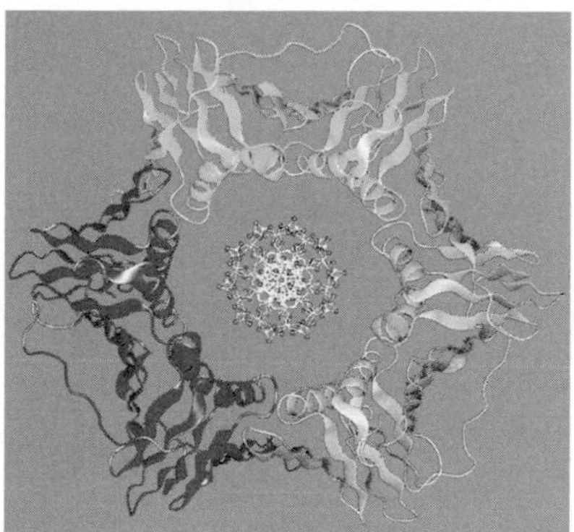

Figure 9.6 Structure of yeast proliferating cell nuclear antigen. The protein trimer is represented schematically in ribbon form, with the three subunits colored red, yellow, and orange. Each monomer contains two topologically identical domains. A model of double-stranded B form DNA is shown in the geometric center of the structure to illustrate how the closed ring formed by tight association of the three subunits might encircle DNA. The α-helices from the three subunits line the central channel, forming a surface of positively charged residues favorable for interaction with the negatively charged phosphodiester backbone of DNA. The nonspecificity of such interactions and the orientation of the internal α-helices appear ideal for strong, but nonspecific, contact with the phosphate groups of the backbone of the DNA threaded through the channel, as modeled in the figure. From T. S. R. Krishna et al., *Cell* **79:**1233–1243, 1994, with permission. Courtesy of J. Kuriyan, Rockefeller University.

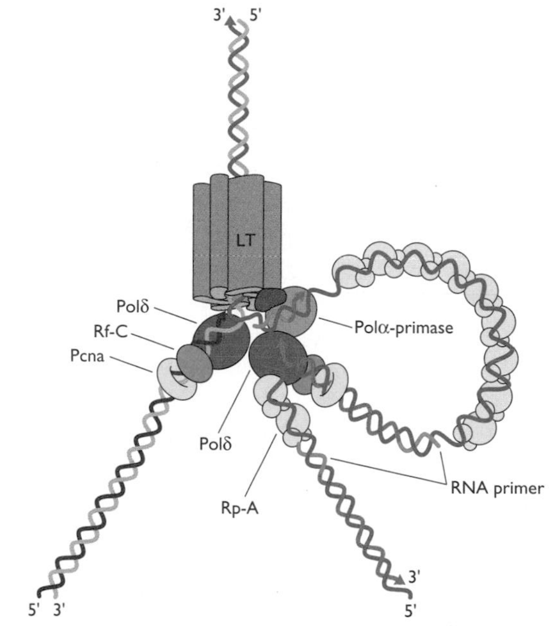

Figure 9.7 A hypothetical simian virus 40 replication machine. A replication machine containing all proteins necessary for both continuous synthesis of the leading strand and discontinuous synthesis of the lagging strand would assemble at each replication fork. Spooling of a loop of the template DNA strand for discontinuous synthesis would allow the single complex to copy the two strands in opposite directions. Although there is indirect evidence for such a model, formation of this kind of replication complex remains to be demonstrated directly.

BOX 9.4

Spooling of DNA during unwinding of the simian virus 40 origin

Visualization by electron microscopy of structures formed during LT-dependent unwinding from the simian virus 40 origin in vitro suggests that the two hexamers remain in contact as DNA is unwound. (A) LT bound to the origin, as a characteristic bilobed structure (the double hexamer shown in Fig. 9.4); (B) unwinding intermediates; (C) the intermediate at the bottom right in panel B at higher magnification. This intermediate contains a bilobed LT complex connecting the two replication forks so the single-stranded DNA (ssDNA) is looped out as "rabbit ears." The formation of such structures containing a dimer of the LT hexamer, in which each monomer is bound to a replication fork, stimulates the helicase activity of LT. This property supports the view that the DNA template is spooled through an immobile replication machine (see text). dsDNA, double-stranded DNA. From R. Wessel et al., *J. Virol.* **66:** 804–815, 1992, with permission. Courtesy of H. Stahl, Universität des Saarlandes.

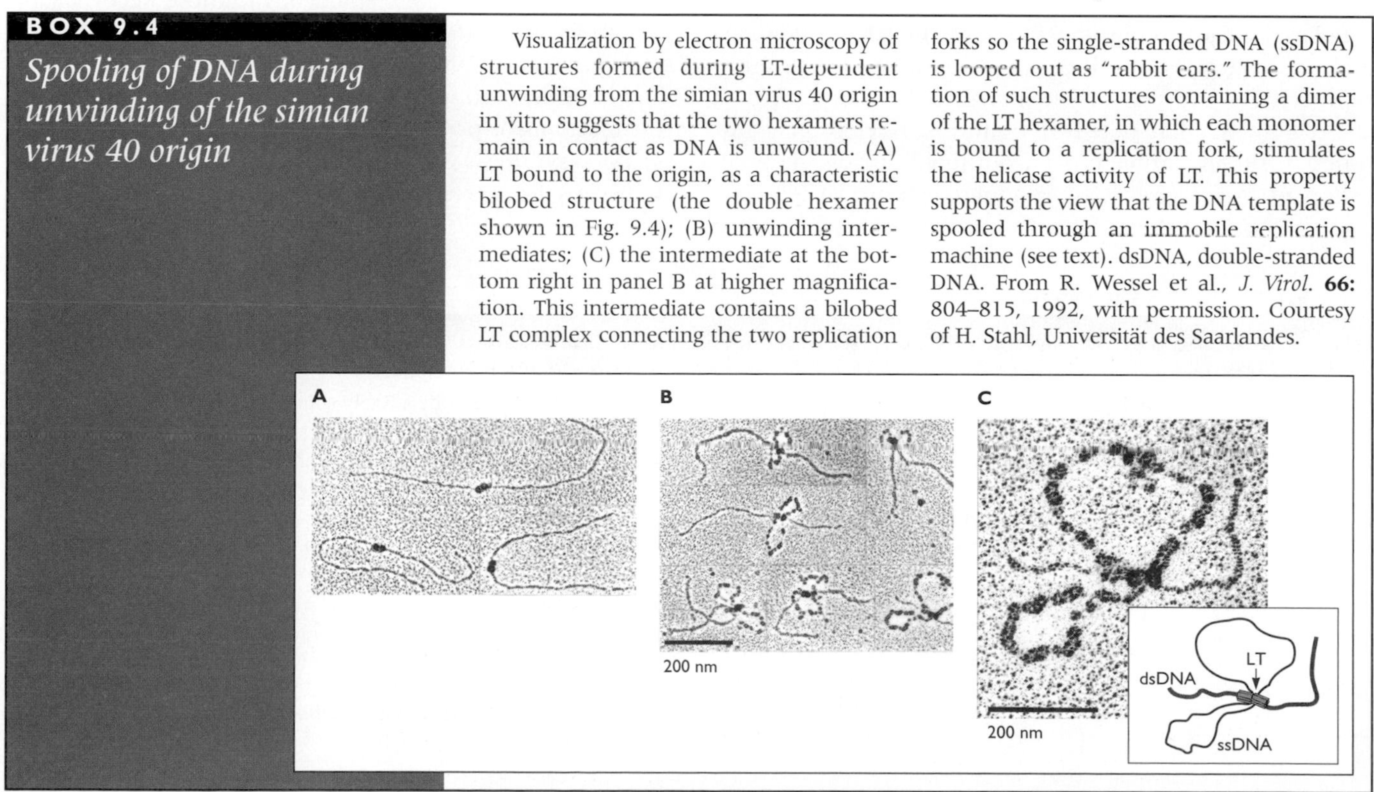

that alter the topology of DNA, topoisomerases I and II (Table 9.4). These enzymes, which differ in their physical properties, catalytic mechanisms, and functions in the cell, reverse the winding of one duplex DNA strand around another (**supercoiling**). Because they remove supercoils, topoisomerases are said to **relax** DNA. In a closed circular DNA molecule like the simian virus 40 genome, the unwinding of duplex DNA at the origin and subsequently at the replication forks is necessarily accompanied by supercoiling of the remainder of the DNA (Fig. 9.8A). If not released, the torsional stress so introduced would act as a brake on movement of the replication forks, eventually bringing them to a complete halt. Both topoisomerases I and II remove such torsional stress during simian virus 40 DNA replication in vitro to allow movement of the replication forks. These enzymes play an analogous role during replication of chromosomal DNA in vivo (at least in yeasts). Topoisomerase II is also specifically required for the separation of the viral daughter molecules from late replication intermediates. A cycle of simian virus 40 DNA synthesis produces two locked (catenated) circular DNA molecules that can be separated only when one DNA molecule is passed through a double-strand break in the other, and the break is then re-sealed (Fig. 9.8B). Topoisomerase II catalyzes this series of reactions (Fig. 9.8B; Table 9.4).

Replication of chromatin templates. The simian virus 40 genome is associated with cellular nucleosomes both in the virion and in infected cell nuclei. It is therefore replicated as a minichromosome, in which the DNA is wrapped around nucleosomes. This arrangement raises the question of how the replication machinery rapidly copies a DNA template that is tightly bound to nucleosomal histones. A similar problem is encountered during the replication of many viral RNA genomes, when the template RNA is packaged by viral RNA-binding proteins in a ribonucleoprotein. The mechanisms by which replication complexes circumvent the nucleosomal barriers to movement are not understood in detail. Nevertheless, there is considerable evidence from simple model systems that enzymes that copy DNA can displace nucleosomes to a new site on the template. The organization of the simian virus 40 genome into a minichromosome also implies that viral DNA replication must be coordinated with binding of newly synthesized DNA to cellular nucleosomes. Various mechanisms can deposit nucleosomes onto newly replicated DNA, including coordination of histone production with replication-dependent nucleosome assembly by a chromatin assembly protein.

Summary. Analysis of simian virus 40 replication in vitro has identified essential cellular replication proteins, led to molecular descriptions of crucial reactions in the complex process of DNA synthesis, and provided new insights into chromatin assembly. The detailed understanding of the reactions completed by the cellular DNA replication machinery laid the foundation for elucidation of both the mechanisms by which other animal viral DNA genomes are replicated and some of the intricate circuits that regulate DNA synthesis and its initiation.

Mechanisms of Viral DNA Synthesis

The replication of all viral DNA genomes within infected cells comprises reactions analogous to those necessary for simian virus 40 DNA synthesis, namely, origin recognition and priming of DNA synthesis, elongation, termination, and often resolution of the replication products. However, the mechanistic problems associated with each of the steps in DNA synthesis are solved by a considerable variety of virus-specific mechanisms.

Table 9.4 Cellular DNA topoisomerases

Topoisomerase I	Topoisomerase II
90- to 135-kDa monomer	150- to 180-kDa homodimers
Relaxes positively and negatively supercoiled DNA in ATP-independent reactions	Relaxes positively and negatively supercoiled DNA in ATP- and Mg^{2+}-dependent reactions
Catenates and decatenates double-stranded circles, provided one molecule carries a nick or gap	Catenates, decatenates, knots, and unknots double-stranded DNA molecules without nicks or gaps
Forms intermediate in relaxation, in which one strand of the DNA is cleaved and covalently bound via 3′ phosphate to the protein; changes linking number in steps of one	Cleaves both strands of DNA to form a 4-bp staggered cut; 5′ ends of each strand covalently bound to one subunit of the enzyme; alters linking number in steps of two
Cleaves at preferred sequences	Modest sequence specificity
Supports replication fork movement in vitro; participates in transcription elongation and recombination; not essential for DNA replication in vivo	Supports replication fork movement; essential in vivo, because required for separation of daughter chromosomes; decatenates replication products in vitro; participates in recombination; a component of chromosome scaffolds and matrices

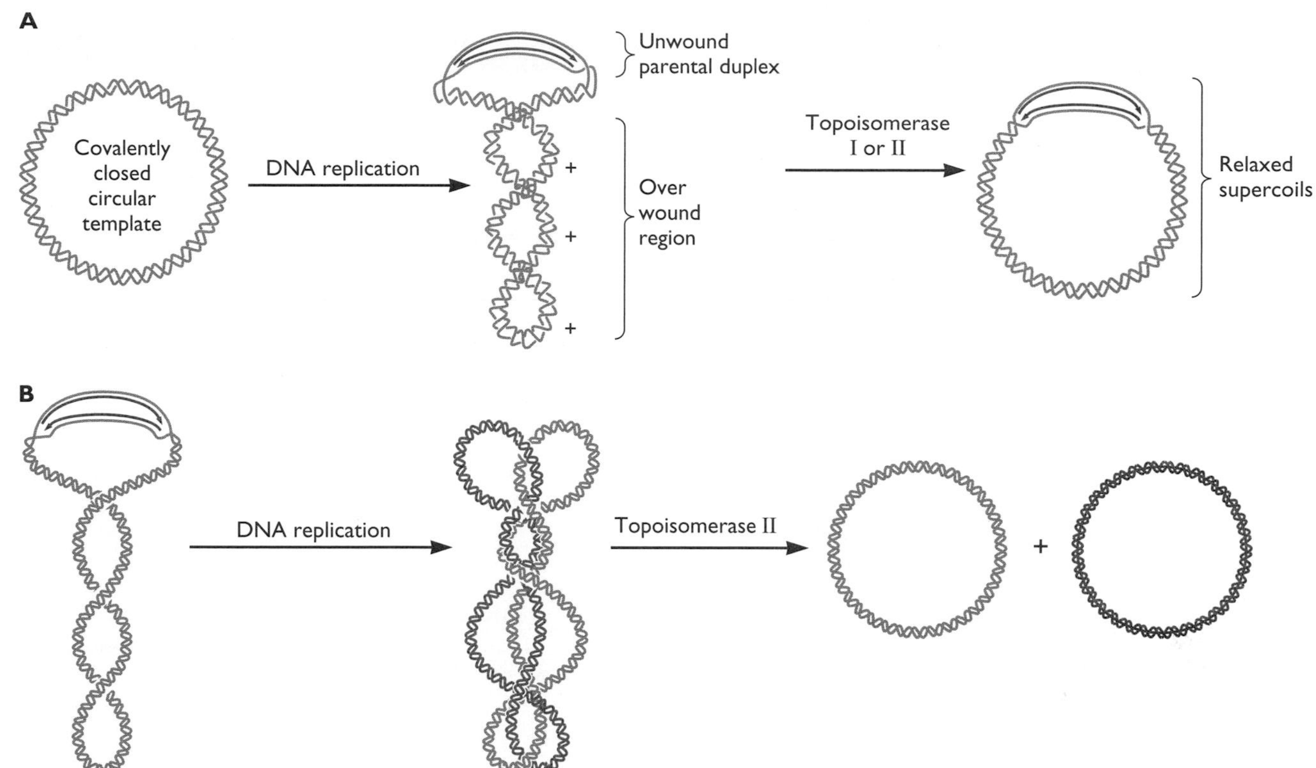

Figure 9.8 Function of topoisomerases during simian virus 40 DNA replication. (A) Relief from overwinding. In a closed, double-stranded DNA circle like the simian virus 40 genome, unwinding of the template DNA at the origin and two replication forks is necessarily accompanied by overwinding (positive supercoiling) of the DNA ahead of the replication forks (middle). Either topoisomerase I or topoisomerase II can act as a swivel, removing the supercoils to allow continued movement of the replication fork. (B) Decatenation of replication products. The products of a cycle of simian virus 40 DNA replication are two interlocked daughter molecules. Their separation to release daughter molecules identical to the parental genome requires topoisomerase II, which makes a double-strand break in DNA, passes one double strand over the other to unwind one turn, and reseals the DNA in reactions that require hydrolysis of ATP. During chromosomal replication, this enzyme is likewise essential for separation of daughter chromosomes during mitosis.

Priming and Elongation

Synthesis of viral DNA molecules is initiated by a number of unusual mechanisms in which not only RNA but also DNA and even protein molecules function as primers. Because each of the viral priming strategies has profound consequences for the mechanism of elongation, priming and elongation are considered together.

Synthesis of Viral RNA Primers by Cellular or Viral Enzymes

The standard method of priming is synthesis of a short RNA molecule by a specialized primase. As we have seen, cellular DNA polymerase α-primase synthesizes all RNA primers needed for replication of both template strands of polyomaviral genomes. A similar mechanism operates at certain origins of some herpesviruses, such as that directing replication of the episomal Epstein-Barr viral genome in latently infected cells (see "Regulation of Replication via Different Viral Origins: Epstein-Barr Virus" below). The integrated proviral genomes of retroviruses are also replicated via RNA primers synthesized by the cellular primase at the origins of the cellular replicon into which the provirus is inserted. In actively dividing cells, proviral DNA is therefore replicated once per cell cycle by the cellular replication machinery. Such proviral DNA replication makes no contribution to the production of progeny retrovirus particles in a single-cell infectious cycle. However, it does provide an effective means of expanding the population of infected cells, for all descendants of the original infected cell will carry the integrated provirus.

Less common among DNA viruses is synthesis of RNA primers by viral proteins. However, this mechanism is

Table 9.5 Viral proteins that participate in genome replication[a]

Virus	Protein	Properties and functions
Proteins needed for DNA synthesis		
Adenovirus		
Human adenovirus type 2	DNA polymerase	Initiates replication as complex with preterminal protein by covalent linkage of dCMP to the protein primer; completes synthesis of all daughter strands
	Preterminal protein	Protein primer for DNA synthesis
	Terminal protein	Covalently bound to 5′ termini of the parental genome; facilitates origin unwinding
	Single-stranded DNA-binding protein	Binds displaced single-stranded DNA; confers processivity to viral DNA polymerase
Herpesvirus		
Herpes simplex virus type 1	UL9 protein	Origin recognition, contains consensus ATP-binding site; can bind to UL42 proteins
	UL30 protein (DNA polymerase)	Synthesis of viral DNA; associated 3′ → 5′ exonuclease for proofreading; target of several antiherpesviral drugs
	UL5, UL8, and UL52 proteins	Form heteromeric complex with primase and helicase activitites; UL8 protein binds to the viral DNA polymerase
	UL42 protein	Processivity protein for viral DNA polymerase
	UL29 protein (ICP8)	Single-stranded DNA-binding protein essential for viral DNA synthesis; binds DNA cooperatively; binds to the UL42 protein
Poxvirus		
Vaccinia virus	DNA polymerase	Synthesis of vaccinia virus DNA; associated 3′ → 5′ exonuclease activity for proofreading
	A2OR protein	Component of a proactive DNA polymerase complex
	D5R protein[b]	Nucleoside triphosphatase
	D4R protein[b]	Uracil DNA glycosylase
Enzymes of nucleic acid metabolism		
Herpesvirus		
Herpes simplex virus type 1	Thymidine kinase (UL23 protein, ICP36)	Phosphorylates thymidine and other nucleosides; essential for efficient replication in animal hosts
	Ribonucleotide reductase; $\alpha_2\beta_2$ dimer of UL39 and U40 proteins	Reduces ribose to deoxyribose in ribonucleotides; essential in nondividing cells
	dUTPase (UL50 protein)	Hydrolyzes dUTP to dUMP, preventing incorporation of dUTP into DNA and providing dUMP for conversion to dTMP; not essential for growth in tissue culture
	Uracil DNA glycosylase (UL2 protein)	Corrects insertion of dUTP or deamination of C in viral DNA; presumed to be important in DNA repair
	Alkaline nuclease (UL12 protein)	Required for efficient virus reproduction and production of progeny virions; processes newly replicated DNA to remove branched structures
Poxvirus		
Vaccinia virus	Thymidine kinase	Phosphorylates thymidine; required for efficient virus reproduction in animal hosts
	Thymidylate kinase	Phosphorylates TMP
	Ribonucleotide reductase, dimer	Reduces ribose to deoxyribose in ribonucleotides; essential in nondividing cells
	dUTPase	Hydrolyzes dUTP to dUMP (see above)
	ATP-dependent DNA ligase	Not essential for viral replication in cells in culture
	DNase	Has nicking-joining activity; present in virion cores
	DNA type I topoisomerase	Essential; present in virions; structurally and functionally similar to cellular counterpart

[a]Predicted from coding sequence.
[b]Whether these proteins participate directly in viral DNA synthesis or function indirectly is not known.

characteristic of herpesviral replication in productively infected cells. Such productive replication has been best characterized in cells infected by herpes simplex virus type 1 when a viral primase synthesizes RNA primers. This primase is a heterotrimer of the products of the viral UL5, UL8, and UL25 genes (Table 9.5).

An inevitable consequence of DNA synthesis from RNA primers by either cellular or viral DNA polymerases is that one of the two parental DNA strands must be copied discontinuously, as described in detail for simian virus 40. When the template is circular, no special mechanism is needed to complete copying of both strands. In contrast, the ends of the lagging strands of linear templates such as herpesviral DNAs cannot be copied by the discontinuous mechanism. During productive infection, the incoming herpesviral genome remains linear, and the way in which complete copying of both strands is achieved is not known. The properties of both herpes simplex virus type 1 **replication intermediates**, molecules that contain nascent DNA, and the products of replication, which are long concatemers containing multiple copies of the genome (see "Resolution and Processing of Viral Replication Products" below), suggest that recombination is likely to contribute.

Regardless of whether cellular or viral proteins are used to make primers, the replication of DNA viruses via RNA priming is subject to the difficulties intrinsic to this mechanism (i.e., how to copy ends of lagging strands). In contrast, other DNA viral genomes are replicated by means of alternative priming mechanisms that eliminate the need for discontinuous synthesis.

Priming via DNA: Specialized Structures in Viral Genomes

Self-priming of viral DNA synthesis via specialized structures in the viral genome is a hallmark of all *Parvoviridae*, among the smallest DNA viruses that replicate in animal cells. This virus family includes the dependoviruses, such as adeno-associated viruses, and the autonomous parvoviruses, such as minute virus of mice. The dependoviruses can reproduce only in cells coinfected by a helper adenovirus or herpesvirus, which encode proteins needed for efficient expression of parvoviral genes. Replication of the autonomous parvoviruses requires passage of an infected cell through S phase. Nevertheless, synthesis of all parvoviral DNAs exhibits a number of unusual features, the most striking being self-priming. This mechanism is illustrated here with adeno-associated virus.

The adeno-associated virus genome is a small molecule (<5 kb) of single-stranded, linear DNA that carries **inverted terminal repetitions** (ITRs). Genomic DNA is of both (+) and (–) polarity, for both strands are encapsidated, but in separate virus particles. As illustrated in Fig. 9.9A, palindromic sequences within the central 125 nucleotides of the inverted terminal repetition base pair to form T-shaped structures. Formation of this structure at the 3′ end of either single strand of viral DNA provides an ideal template-primer for initiation of the first cycle of viral DNA synthesis (Fig. 9.9B). Experimental evidence for such **self-priming** includes the dependence of adeno-associated virus DNA synthesis on self-complementary sequences within the ITR. Following recognition of the viral DNA primer, the single template strand of an infecting genome can be copied by a continuous mechanism, analogous to leading-strand synthesis during replication of double-stranded DNA templates. In subsequent cycles of replication, the same 3′-terminal priming structures form in the duplex replication intermediate produced in the initial round of synthesis (Fig. 9.9B). Adeno-associated virus DNA synthesis is therefore always continuous, and the DNA polymerase α-primase is not required for replication of parvoviral genomes.

On the other hand, a specialized mechanism **is** necessary to complete replication, by copying of the sequences that form the priming structure: the initial product retains the priming hairpin and is largely duplex DNA in which parental and daughter strands are covalently connected (Fig 9.9.B, step 2). This step is achieved by nicking of the intermediate within the parental DNA strand at a specific site, between nucleotides 124 and 125. The new 3′ OH end liberated in this way primes continuous synthesis to the end of the DNA molecule (Fig. 9.9B). Consequently, this part of the 3′ ITR of the viral template is replaced by its complementary sequence during each cycle of viral DNA synthesis (Fig. 9.9B; Appendix A, Fig. 10B). The nick is introduced by the related viral proteins Rep 78 and Rep 68 (Rep 78/68) (Fig. 9.9C). These proteins are site- and strand-specific endonucleases, which bind to, and cut at, specific sequences within the ITR. During this **terminal resolution** process, Rep 68 or Rep 78 becomes covalently linked to the cleaved DNA at the site that will become the 5′ terminus of the fully replicated molecule (Fig. 9.9B). Following the synthesis of a duplex of the genomic DNA molecule (the **replication intermediate**), formation of the same 3′-terminal priming hairpin allows continuous synthesis of single-stranded genomes by a strand displacement mechanism, with re-formation of the replication intermediate (Fig. 9.9B, steps 6 and 7).

Rep 78 and Rep 68 are similar to simian virus 40 LT in several respects, and can be considered origin recognition proteins (Table 9.6). They are the only viral gene products necessary for parvoviral DNA synthesis. In addition to recognizing the terminal resolution site, these proteins provide the ATP-dependent, 3′ → 5′ helicase activity needed for both unwinding of double-stranded DNA templates and formation of the priming hairpin.

Priming via self-complementary terminal sequences of the viral genome eliminates the need for proteins that

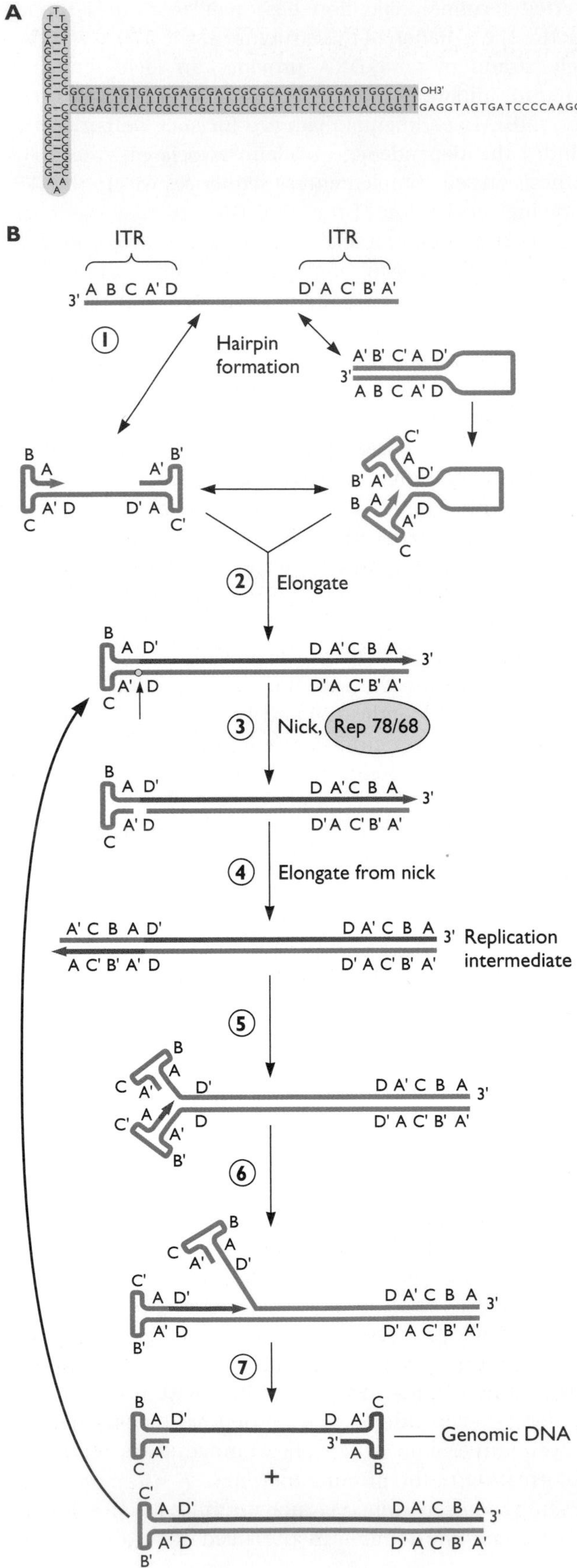

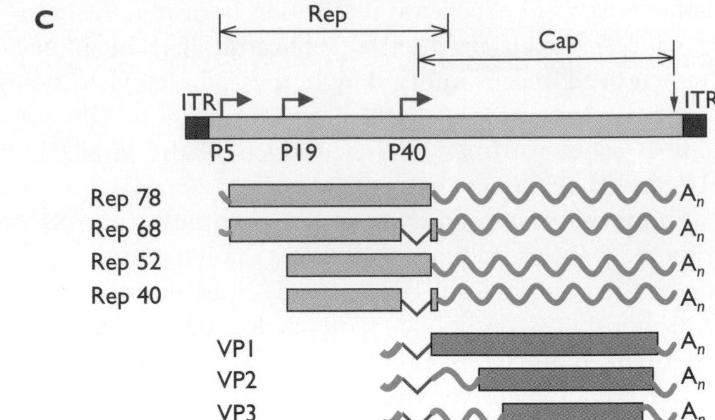

Figure 9.9 Replication of parvoviral DNA. (A) Sequence and secondary structure of the adeno-associated virus type 2 inverted terminal repetition (ITR). A central palindrome (tan background) is flanked by a longer palindrome (light blue background) within the central 125 bp of the 145-bp ITR. When these bases pair at the 3′ end of the genome, a T-shaped structure in which the internal duplex stem terminates in a free 3′-OH group is formed. This structure is an ideal template-primer structure for cellular DNA polymerases. (B) Model of adeno-associated virus DNA replication. The ITRs are represented by 3′ABCA′D5′ and 5′A′B′C′AD′3′. Formation of the 3′-terminal hairpin provides a primer-template for continuous DNA synthesis to the 5′ end of the parental single-stranded DNA (step 1, left). However, it is possible that the self-complementary terminal repeat sequences first base pair to form a "panhandle," a single-stranded circle stabilized by the duplex terminal sequence (step 1, right). Such a structure could explain the repair of deletions or mutations within one ITR observed when the other is intact. In either case, elongation from the 3′-OH group of the hairpin allows continuous synthesis (red) to the 5′ end of the parental strand (step 2). Complete copying of the parental strand requires introduction of a nick to generate a new 3′ OH (step 3). Such nicking takes place at a specific site (the **terminal resolution site**) marked by the arrow, and is performed by the viral Rep 78 and Rep 68 proteins, which are site-specific endonucleases. Elongation from the nick results in copying of sequences that initially formed the self-priming hairpin to form the double-stranded replication intermediate (step 4). However, the parental strand then contains newly replicated DNA (red) at its 3′ end. As a result, the ITR of the parental strand is no longer the initial sequence but rather its complement. This palindromic sequence is therefore heterogeneous in populations of adeno-associated virus DNA molecules, being present in one of the two orientations listed above. Indeed, such sequence heterogeneity provided an important clue for elucidation of the mechanism of viral DNA synthesis. The newly replicated 3′ end of the replication intermediate can form the same terminal hairpin structure (step 5) to prime a new cycle of DNA synthesis (step 6). One product is a molecule of single-stranded genomic DNA, displaced on copying to the end of the template strand (step 7). The second, which corresponds to the incompletely replicated molecule initially produced, can undergo additional cycles of replication following terminal resolution as in steps 3 and 4. (C) Organization of coding sequences in the adeno-associated virus genome. The viral genome is depicted at the top, with the locations of the ITR, the promoters for transcription by RNA polymerase II (P5, P19, and P40), and the single site of polyadenylation (vertical red arrow) indicated. The two coding regions within the viral genome are termed Rep (replication) and Cap (capsid) for the functions of the proteins they encode. As shown below, each is expressed as multiple related proteins, which are synthesized from differentially spliced mRNAs transcribed from the different promoters. The two smaller products of the Rep region, Rep 52 and Rep 40, are not required for viral DNA synthesis, but appear to facilitate encapsidation of progeny single-stranded genomic DNA molecules during assembly of virus particles.

Table 9.6 Viral origin recognition proteins[a]

Virus	Protein(s)	DNA-binding properties	Other activities and functions
Adenovirus			
Human adenovirus type 2	pTP, DNA polymerase	Binds to origins as complex, stimulated by cellular transcriptional activators Nf-1 and Oct-1	Priming of DNA synthesis via addition of dCMP to pTP by the DNA polymerase; continuous synthesis of both strands of viral genome
Herpesvirus			
Herpes simplex virus type 1	UL9	Binds cooperatively to specific sites in viral origins; distorts DNA to which it binds	ATPase and 3′ → 5′ helicase; binds UL29 protein, UL8 subunit of viral primase, and UL42 processivity protein
Epstein-Barr virus	EBNA-1	Binds as dimer to multiple sites in two clusters (FR and DS) in viral OriP; may use different DNA-binding motifs to form complexes in which DNA is distorted	Required for maintenance of episomal viral genomes; binding to FR sequences stimulates transcription from specific viral promoters; binds to cellular origin recognition proteins
Papillomavirus			
Bovine papillomavirus type 5	E1	Binds origin with low affinity; binds strongly and cooperatively in presence of E2 protein	Binding to E2 essential for viral DNA replication; DNA-dependent ATPase and helicase; helicase active upon assembly of a hexamer; binds catalytic subunit of Polα-primase
	E2	Binds transiently as dimer to specific sequences in origin	Regulates transcription by binding to viral enhancers; loads E1 onto origin
Parvovirus			
Adeno-associated virus	Rep78/68	Binds to specific sequences in ITR; Rep 78 binds as hexamer	Site- and strand-specific endonuclease; ATPase and ATP-dependent 3′ → 5′ helicase; transcriptional regulator; binding to single-stranded DNA stimulates ATPase activity and inhibits endonuclease activity
Polyomavirus			
Simian virus 40	LT	Binds cooperatively to origin site II to form double hexamer; distorts origin; binds to two other nearby sites (I, III) in genome; DNA-binding activity regulated by phosphorylation at specific sites	DNA-dependent ATPase and 3′ → 5′ helicase; binds to cellular Rp-A and Polα-primase; autoregulates early transcription; activates late transcription; binds cellular Rb protein to induce reentry into, or progression through, the cell cycle; binds cellular p53

[a]pTP, preterminal protein; Polα-primase, DNA polymerase α-primase complex.

make RNA primers, or are themselves primers for DNA synthesis (see next section). Such a mechanism would therefore seem especially advantageous for viruses with small genomes, such as the parvoviruses. Nevertheless, a similar self-priming mechanism is believed to initiate replication of the extremely large double-stranded DNA genomes of poxviruses such as vaccinia virus.

Protein Priming

Judged by the criterion of simplicity, the most effective mechanism of initiation is via a protein primer. Nevertheless, this mechanism is rare, restricted to hepadnaviruses and adenoviruses among DNA viruses that infect animal cells. The replication of some viral RNA genomes is also initiated from a protein primer, notably the VPg protein of poliovirus discussed in Chapter 6. In a template-dependent reaction, the adenoviral DNA polymerase covalently links the α-phosphoryl group of dCMP to the hydroxyl group of a specific serine residue in the viral preterminal protein (Fig. 9.10). The 3′-OH group of the protein-linked dCMP then primes synthesis of daughter viral DNA strands by this viral enzyme. The nucleotide is added to the preterminal protein only when the protein primer is assembled into preinitiation complexes at the

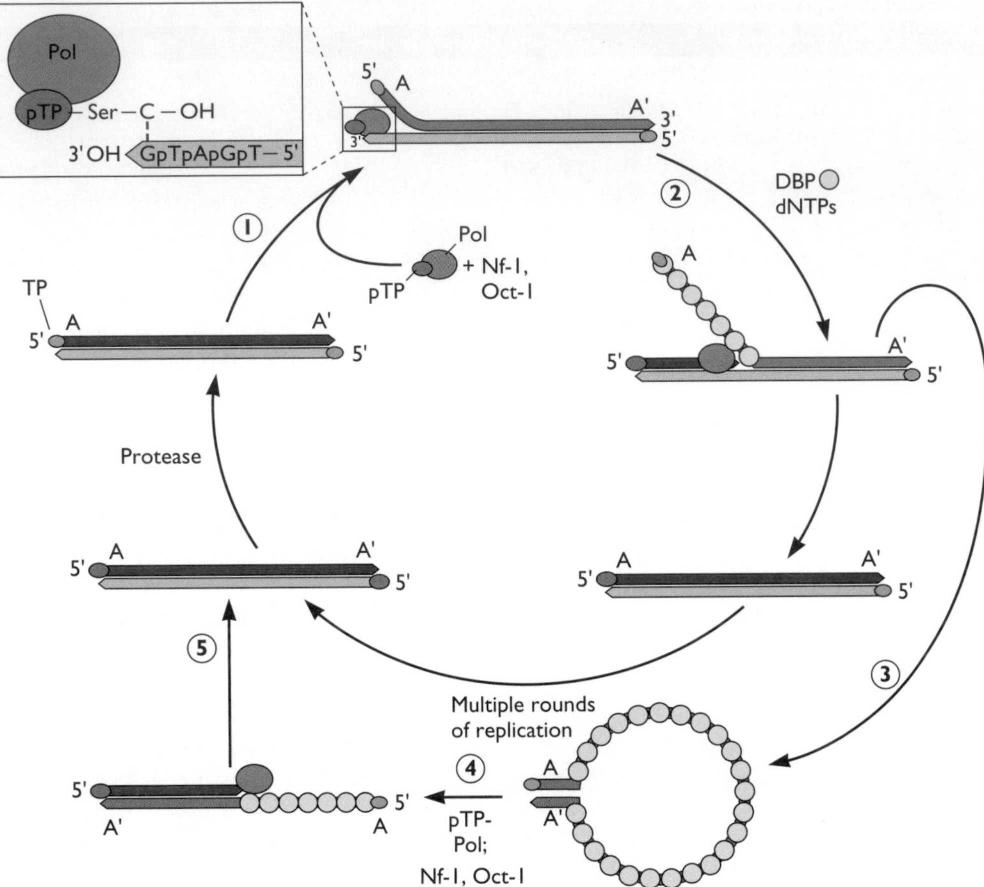

Figure 9.10 Replication of adenoviral DNA. Assembly of the viral preterminal protein (pTP) and DNA polymerase (Pol), both early proteins encoded by the E2 transcription unit (Appendix A, Fig. 1B), into a preinitiation complex at each terminal origin of replication activates covalent linkage of dCMP to a specific serine residue in pTP by the DNA polymerase (step 1). The cellular transcriptional activators Nf-1 and Oct-1 bind specifically adjacent to the core origin sequence (Fig. 9.12) and facilitate assembly of the preinitiation complex by binding to the viral replication proteins. The free 3′-OH group of preterminal protein-dCMP primes continuous synthesis of viral DNA in the 5′ → 3′ direction by Pol (step 2). This reaction also requires the viral E2 single-stranded-DNA-binding protein (DBP), which coats the displaced second strand of the template DNA molecule, and a cellular topoisomerase. As the terminal segments of the viral genome comprise an inverted repeat sequence (A and A′), there is an origin at each end, and both parental strands can be replicated by this displacement mechanism. Reannealing of the complementary terminal sequences of the parental strand displaced in single-stranded form during the initial step forms a short duplex stem identical to the terminus of the double-stranded genome (step 3). The origin re-formed in this way directs a new cycle of protein priming and continuous DNA synthesis (steps 4 and 5).

origins of replication. As the origins lie at the ends of the linear genome, each template strand is copied continuously from one end to the other (Fig. 9.10). The parental template strand initially displaced is copied by the same mechanism, following formation of a duplex stem upon annealing of an inverted terminal repeat sequence. This unusual strand displacement mechanism of replication therefore results in semiconservative replication, even though the two parental strands of viral DNA are not copied at the same replication fork.

Viral Replication Origins and Their Recognition

As we have seen, origins of replication contain the sites at which viral DNA synthesis begins and can be defined experimentally as the minimal DNA segment necessary for initiation of DNA synthesis in cells or in in vitro reactions. Viral origins of replication support initiation of DNA synthesis by a variety of mechanisms, including some with no counterpart in cellular DNA synthesis. Nevertheless, the majority are discrete DNA segments that contain sequences to which viral origin recognition proteins bind specifically to seed assembly of multiprotein complexes. When initiation is by self-priming, DNA sequences essential for replication include those needed to form and maintain a specific secondary structure in the template, as well as the sequence at which replication intermediates are cleaved for complete copying of the parental strand (Fig. 9.9A). Even though viral origins may be unconventional (compared with a cellular origin) and differ in number and location in viral genomes, they exhibit a number of common features, as do the proteins that recognize them.

Number of Origins

In contrast to papillomaviral and polyomaviral DNAs, the genomes of the larger DNA viruses contain not one, but two or three origins. The two identical adenoviral origins at the ends of the linear genome are the sites of assembly of complexes containing the viral DNA polymerase and protein primer (Fig. 9.10). The genomes of herpesviruses such as Epstein-Barr virus and herpes simplex virus type 1 contain three origins of replication. Different functions can be ascribed to the different Epstein-Barr virus origins: a single origin (OriP) allows maintenance of episomal genomes in latently infected cells, while two others (OriLyt) support replication of the viral genome during productive infection. The advantage of three origins—two copies of OriS and one copy of OriL (Fig. 9.11)—to herpes simplex viruses is less clear, for a full complement of these origins is not necessary for efficient viral DNA synthesis, at least in cells in culture. The two types of origin possess considerable nucleotide sequence similarity, but differ in their organization, and can be functionally distinguished. For example, OriL is activated when differentiated neuronal cells are exposed to a glucocorticoid hormone, but OriS is repressed. As glucocorticoids are induced by stress, a condition that induces reactivation of herpes simplex virus type 1 from latency, it has been suggested that replication from OriL may be particularly important during the transition from a latent to a productive infection. The presence of three origins might contribute to rapid replication of the viral genome during productive infection, or be necessary for the mechanism by which herpes simplex virus DNA is synthesized.

Sequence Features of Viral Replication Origins

Even though the origins of replication of double-stranded DNA viruses are recognized by different proteins and support different mechanisms of initiation, they exhibit a number of common features (Fig. 9.12). The most prominent of these is the presence of AT-rich sequences. In general, AT base pairs contain only two hydrogen bonds, whereas GC pairs interact via three such noncovalent bonds. The less stable AT-rich sequence blocks (Fig. 9.12) are therefore believed to facilitate the unwinding of origins that is necessary for initiation of viral DNA

Figure 9.11 Features of the herpes simplex virus type 1 genome. The long (L) and short (S) regions of the viral genome that are inverted with respect to one another in the four genome isomers are indicated at the top. Each segment comprises a unique sequence (UL or US) flanked by internal and terminal repeated sequences. The locations of the two identical copies of OriS, in repeated sequences, and of the single copy of OriL are indicated. The orientation shown at the top is defined as the prototype (P) genome isomer. The other three isomers differ, with respect to the P form, in the orientation of S (IS), in the orientation of L (IL), or in both S and L (ISIL). These differences are illustrated using the *Hin*dIII fragments of herpes simplex virus type 1 DNA. The unusual isomerization of this viral genome was deduced from the presence of fragments that span the terminal or internal inverted repeat sequences at 0.5 and 0.25 M concentrations, respectively, in such *Hin*dIII digests, and examination of partially denatured DNA in the electron microscope.

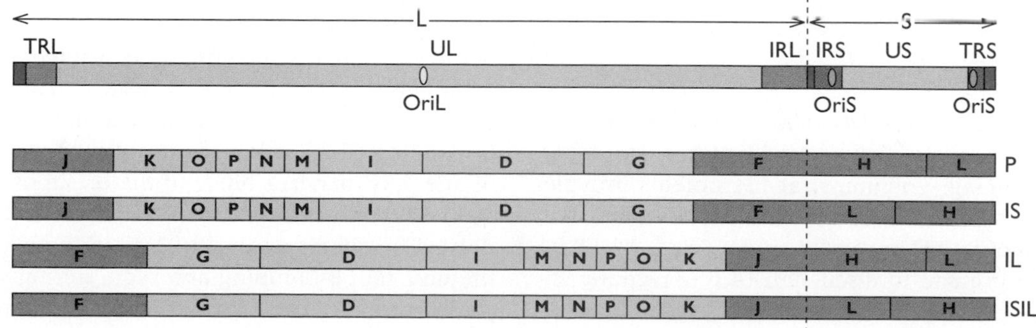

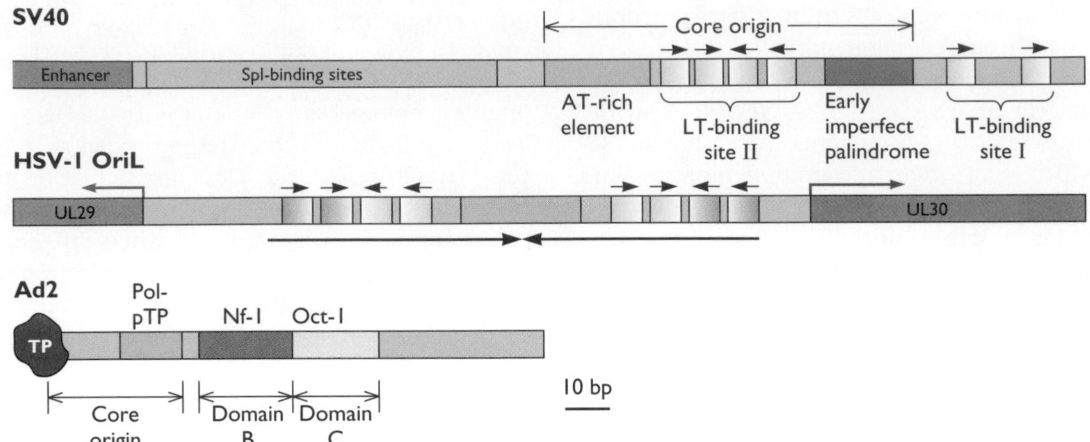

Figure 9.12 Common features of viral origins of DNA replication. The single simian virus 40 (SV40) origin, herpes simplex virus type 1 (HSV-1) OriL (Fig. 9.11), and the adenovirus type 2 (Ad2) origin, which is repeated at each end of the viral genome (Fig. 9.10), are illustrated to scale, emphasizing common features. Binding sites for origin recognition proteins and AT-rich sequences are indicated in yellow and green, respectively. Herpes simplex virus type 1 OriL contains two types of binding site for the viral origin recognition protein (which is encoded by the UL9 gene) distinguished by the two shades of yellow. This origin, which is a perfectly symmetrical palindrome as indicated by the arrows, is located within the transcriptional control regions of the divergently transcribed UL29 and UL30 genes, which encode the DNA-binding protein ICP8 and the viral DNA polymerase, respectively. The two copies of OriS (Fig. 9.11) are very similar in sequence to OriL, but lack the rightmost copy of the UL9 protein-binding site. The small size and relatively simple organization of the adenoviral origin probably reflect the unusual mechanism of initiation. The terminal sequence designated the core origin functions inefficiently in the absence of the adjacent binding site for the transcriptional activator nuclear factor 1 (Nf-1).

synthesis from either RNA or protein primers. The best-characterized viral origins of DNA replication (e.g., those of the simian virus 40 and human adenovirus type 2 genomes) comprise a minimal essential core origin flanked by sequences that are dispensable, but nonetheless significantly increase replication efficiency (Fig. 9.12). These stimulatory sequences contain binding sites for cellular transcriptional activators. Yet other viral origins, those of papillomaviruses and OriLyt of Epstein-Barr virus, include binding sites for viral proteins that are both transcriptional regulators and essential replication proteins. And all three herpes simplex virus type 1 origins lie between sites at which transcription of viral genes is initiated (Fig. 9.12). A close relationship between origin sequences and those that regulate transcription therefore appears to be a general feature.

Recognition of Viral Replication Origins

Because they were the first to be identified and have been studied in detail, polyomaviral LT proteins provide the paradigm for viral origin recognition proteins. We therefore describe the properties of simian virus 40 LT in more detail as a prelude to discussion of other viral proteins with similar functions.

Properties of simian virus 40 LT. *Functions and organization.* The LT proteins of polyomaviruses are remarkable proteins that provide essential replication functions and play other important roles in the infectious cycle. As we have seen, simian virus 40 LT is both necessary and sufficient for recognition of the viral origin. This viral protein also supplies the helicase activity necessary for origin unwinding and movement of the replication fork. The LT proteins make a major contribution to the species specificity of polyomaviruses. Although the genomes of simian virus 40 and mouse polyomavirus are closely related in organization and sequence, they replicate only in simian and murine cells, respectively. Such host specificity in viral DNA replication is largely the result of species-specific binding of LT to the largest subunit of DNA polymerase α of the host cell in which the virus will replicate (Table 9.3). Although the precise mechanism remains to be determined, assembly of preinitiation complexes competent for unwinding of the origin does not take place when the LT of one polyomavirus binds to the origin of another.

LT proteins also ensure that the cellular components needed for simian virus 40 DNA synthesis are available in the host cell. By binding and sequestering specific cellular proteins, LT perturbs mechanisms that control cell proliferation and can induce infected cells to enter S phase when they would not normally do so (see "Induction of Synthesis of Cellular Replication Proteins by Viral Gene Products" below). In addition, LT carries out a number of transcriptional regulatory functions during the infectious cycle, including autoregulation of its own synthesis and activation of late gene expression.

Sequences of simian virus 40 LT necessary for its numerous activities have been mapped by analysis of the effects of specific alterations in the protein on virus replication in infected cells, DNA synthesis in vitro, or the individual biochemical activities of the protein. The properties of such altered proteins suggest that LT comprises discrete structural and functional domains, such as the minimal domain for specific binding to sites I and II at the viral origin or the discrete sequences by which LT binds to cellular replication proteins (Fig. 9.13). However, the activities of such functional regions defined by genetic and biochemical methods may be influenced by distant sites, as discussed in the next section.

Regulation of LT synthesis and activity. As the simian virus 40 early enhancer and promoter are active in many cell types (Chapter 8), the viral early gene encoding LT is efficiently transcribed as soon as the viral chromosome enters the nucleus. The spliced early messenger RNA (mRNA) encoding LT is the predominant product of processing of these early transcripts. Although production of LT is not regulated during the early phase of infection, at least until the protein attains concentrations sufficient to repress early gene transcription, its activity is tightly controlled.

Simian virus 40 LT is phosphorylated at multiple Ser and Thr residues, most of which are located within one of two clusters near the N and C termini of the protein (Fig. 9.13). Although the significance of phosphorylation at every site is not known, it is clear that specific modifications regulate the ability of LT to support viral DNA synthesis. For example, phosphorylation of Thr124 is absolutely required. This modification specifically stimulates binding of LT to origin site II (Fig. 9.3) and promotes co-

Figure 9.13 Functional organization of simian virus 40 LT. The minimal segments of LT required for binding to the DNA polymerase α-primase complex (Polα), to the cellular chaperone Hsc70, to the cellular retinoblastoma (Rb) and p53 proteins that negatively regulate cell growth and proliferation, to sites I and II in the origin of replication (origin DNA binding), and to single-stranded (ss) DNA binding, as well as for the helicase and ATPase activities of the protein and hexamer assembly at the origin, are indicated on a linear representation of the 708-amino-acid protein. The nuclear localization signal (NLS) and a C-terminal sequence necessary for virion production but not viral DNA synthesis (Host range) are also shown. The region that binds to Hsc70 lies within an N-terminal segment (amino acids 1 to 82) which is termed the J domain because it shares sequences and functional properties with the *E. coli* protein DnaJ, a chaperone that assists the folding and assembly of proteins and is required during replication of bacteriophage λ. The chaperone functions of the J domain and Hsc70 are essential for replication, and seem likely to assist assembly or rearrangement of the preinitiation complex. Although the three-dimensional structure of full-length LT has not been determined, several observations indicate that it is organized into discrete structural and functional domains. For example, specific protease-resistant fragments that retain DNA binding or ATPase activity have been identified, and the structure of the helicase domain has been determined (Fig. 9.4B). Furthermore, subsegments of the protein that span the DNA-binding region can be synthesized as stable, active proteins in *E. coli*, indicating that this region of LT independently adopts a stable, folded structure. Below are shown the two regions of the protein in which sites of phosphorylation are clustered, indicating modifications that have been shown to inhibit (red) or activate (green) the replication activity of LT.

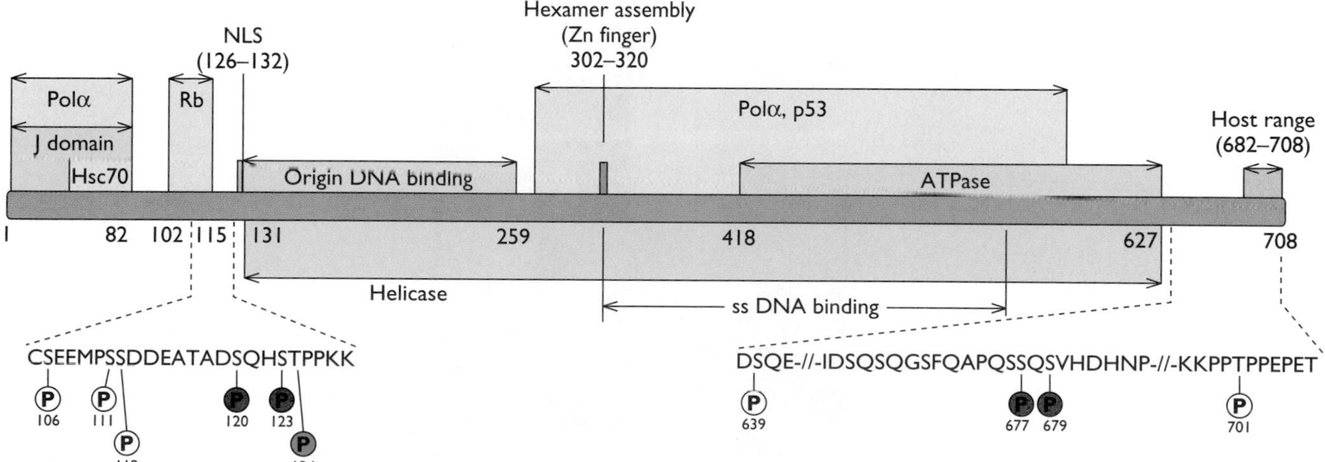

BOX 9.5

Binding of the Epstein-Barr virus EBNA-1 protein to OriP: a mechanism for origin distortion

(A) Organization of binding sites for EBNA-1 dimers in OriP shown to scale. Initiation of replication within the dyad symmetry (DS) sequence, which comprises four EBNA-1 binding sites (1 to 4), is indicated by the vertical arrow. The activity of the DS sequence origin is stimulated by the family of repeat (FR) sequence, which is also necessary for maintenance of episomal viral DNA in latently infected B cells. In the absence of the DS origin, DNA synthesis can initiate at EBNA-1 binding sites within the FR sequence, suggesting that OriP is functionally complex. Exactly how the FR sequence stimulates replication is not clear, but the ability of EBNA-1 to bind to both DS and FR sequences, which are about 1 kb apart, looping out the intervening DNA, appears to be important. Within the DS sequence, sites 1 and 2 function as a minimal origin with activity that is critically dependent on the 3-bp spacing of these EBNA-1 binding sites. (B) Three-dimensional structure of the EBNA-1 DNA-binding and dimerization domain dimer bound to DNA. The core domain of the dimer is shown in blue with the flanking

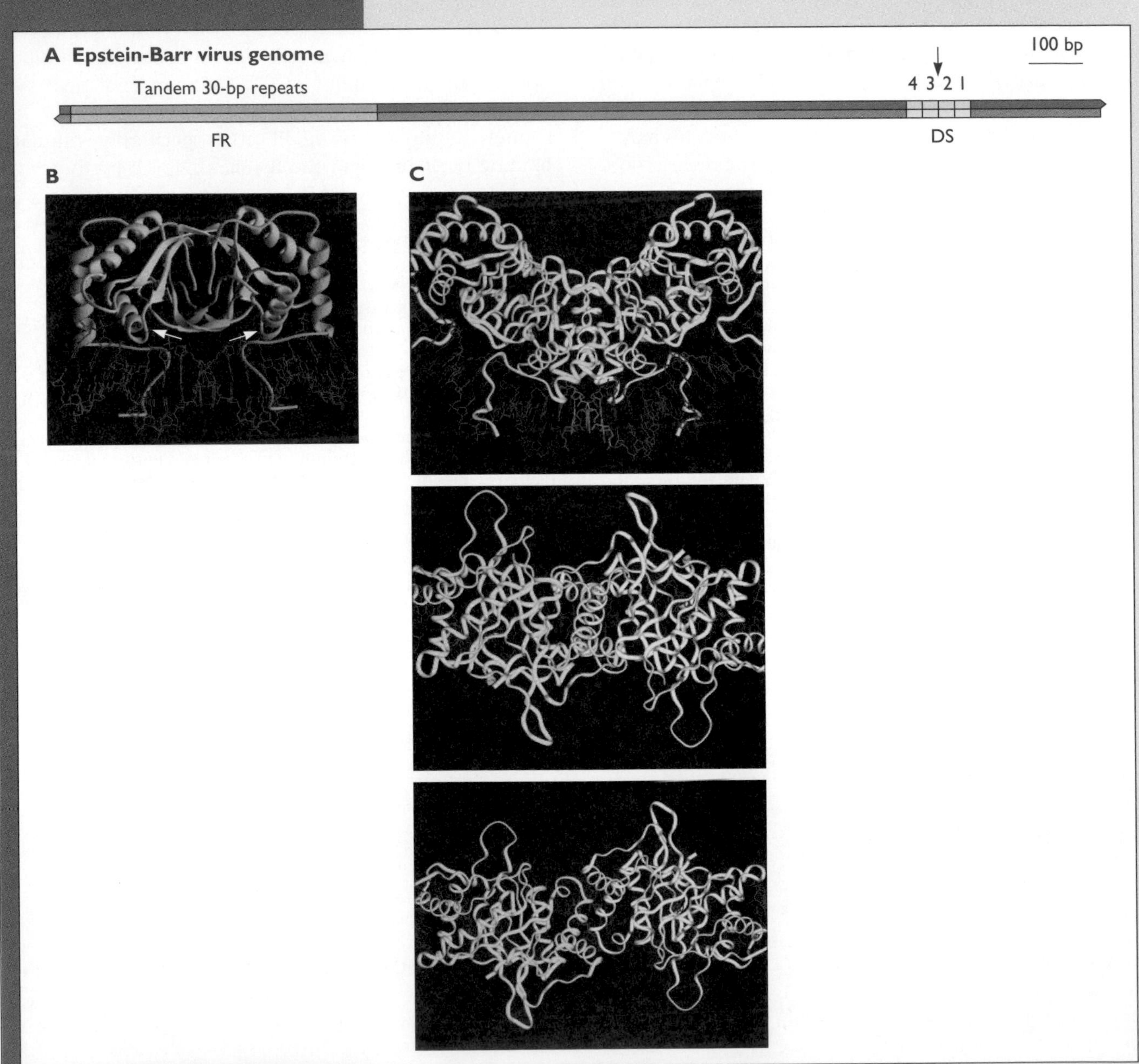

BOX 9.5 *(continued)*

domains in orange. This crystal structure of EBNA-1 bound to a single recognition site yielded a number of interesting surprises. In the first place, the structure of the central domain of the protein is essentially the same as that of the corresponding portion of the bovine papillomavirus E2 protein dimer. As the E2 protein is not the origin recognition protein of this virus (see text), such structural identity was not anticipated. Furthermore, despite their structural identity, EBNA-1 and E2 dimers contact DNA in different ways. The EBNA-1 dimer interacts with DNA via its flanking domains (orange), but also possesses two α-helices (white arrows) that are structurally analogous to the DNA recognition helices of the papillomavirus E2 protein. When EBNA-1 is bound to a single site, these helices are too far from the DNA to make contact with it, as shown. However, it seems likely that when EBNA-1 dimers bind to adjacent sites in a functional origin, the α-helices do indeed make contact with DNA, to facilitate distortion (and ultimately unwinding) of the origin as modeled in panel C. (C) Model of the binding of two EBNA-1 dimers to a minimal OriP. The minimal origin comprises EBNA-1 binding sites 1 and 2 (A), separated by 3 bp, to which two EBNA-1 dimers bind cooperatively. The close spacing of these binding sites in the minimal origin precludes binding of two dimers in the same way as in the single dimer-DNA complex shown in panel B: two EBNA-1 dimers collide when modeled on two DNA-binding sites separated by 3 bp (top and middle). The two dimers are blue and yellow, and the model of 3-bp connection between the DNA-bound dimers is white. The model is shown in the same view as in panel B (top) and viewed from above the DNA (middle). EBNA-1 binding to OriP is known to distort the DNA. Specific distortion of the DNA and twisting of one EBNA-1 dimer relative to the other would relieve the steric clash when dimers bind to the minimal origin (bottom). The potential α-helical DNA-binding motifs of EBNA-1 dimers are postulated to facilitate such distortion, which presumably favors subsequent unwinding of the origin. Additional contacts between the core domain α-helices (white arrows in panel B) and the specifically distorted DNA would stabilize the distorted origin. (B and C) From A. Bochkarev et al., *Cell* **84**:791–800, 1996, with permission. Courtesy of A. Bochkarev, University of Oklahoma Health Sciences Center, and L. Frappier, University of Toronto.

operative interactions among LT molecules bound to the core origin and hence assembly of the double-hexamer complex (Fig. 9.4). Moreover, it is essential for unwinding of DNA from the origin by LT. As Thr124 does not lie within the minimal origin-binding domain of LT (Fig. 9.13), such regulation of DNA-binding activity is believed to be the result of conformational change induced by phosphorylation at this site. The best candidate for the protein kinase that phosphorylates Thr124 is cyclin-dependent kinase 2 (Cdk2) associated with cyclin A. The accumulation and activity of this complex are strictly regulated as growing cells traverse the cell cycle, but the mechanisms by which LT ensures that cellular replication proteins are made in simian virus 40-infected cells also result in production of active Cdk2 (see "Induction of Synthesis of Cellular Replication Proteins by Viral Gene Products" below).

Origin recognition proteins of other viruses. Other viral origin recognition proteins share with simian virus 40 LT the ability to bind specifically to DNA sequences within the appropriate origin of replication. They also bind to other replication proteins (although these may be viral or cellular), and several possess the biochemical activities exhibited by LT (Table 9.6). For example, the herpes simplex virus type 1 UL9 protein, which recruits viral rather than cellular replication proteins, performs functions analogous to those of simian virus 40 LT at the viral origins of replication. This protein binds cooperatively to specific sites that flank the AT-rich sequences of these origins (Fig. 9.12), which are then distorted, and it possesses an ATP-dependent helicase activity that unwinds DNA in the 3′ → 5′ direction (Table 9.6). A single viral protein, EBNA-1, is also responsible for recognition of the origin of plasmid maintenance, OriP, of Epstein-Barr virus. The structure of the DNA-binding and dimerization domains of EBNA-1 bound specifically to DNA has provided new insights into mechanisms of distortion of origin DNA (Box 9.5).

Although recognition of viral origins of replication by a single viral protein is common, it is not universal (Table 9.6). For example, although the bovine papillomavirus type 1 E1 protein possesses the same activities as simian virus 40 LT (Table 9.6), to which it is related in sequence, it cannot support papillomaviral DNA replication in infected cells: a second viral protein, the E2 transcriptional regulator, is also necessary. Essential sequences of the minimal origin of replication of papillomaviral genomes include adjacent binding sites for the E1 and E2 proteins (Fig 9.14). The E1 protein binds to DNA, but with only low specificity for origin sequences. In contrast, the E1 and E2 proteins together bind cooperatively to the origin, and the specificity and the affinity of the E1-DNA interaction are increased significantly. However, the E2 protein functions like clamp-loading proteins and is required only transiently: once a specific E1-E2 protein complex has assembled and induced a bend in the DNA, the E2 protein dissociates from the complex, and additional molecules of E1 then associate (Fig. 9.14). The final product is a hexamer of E1 assembled on, and probably closed around, single-stranded DNA at the origin. The E2 protein therefore serves to load the papillomaviral initiation protein.

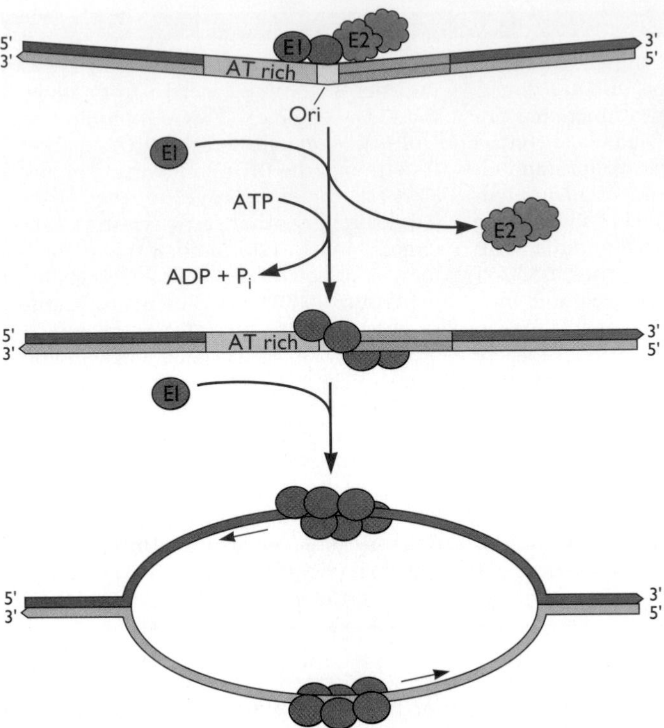

Figure 9.14 Origin loading of the papillomaviral E1 initiation protein by the viral E2 protein. The sequence features of the minimal origin of replication of bovine papillomavirus type 1 are depicted as in Fig. 9.12. This origin contains an essential binding site for the viral E2 protein, a sequence-specific transcriptional regulator. The model of the origin loading of the viral E1 by the E2 protein is based on in vitro studies of the interactions of these proteins with the origin. The E1 and E2 proteins, which are both homodimers, bind cooperatively to the viral origin, with specificity and affinity far greater than that exhibited by the E1 protein alone. Such cooperativity is the result of interactions between the DNA-binding domains of the two proteins that induce a sharp bend in the DNA, thereby permitting the activation domains of the E2 proteins to bind to the E1 helicase domains. When ATP is hydrolyzed (presumably by the ATPase of the E1 protein), the $(E1)_2(E2)_2$-Ori complex is destabilized, the E2 dimers are displaced, and additional E1 molecules bind to the complex. In the resulting E1-Ori complex, the DNA is distorted and becomes partially single stranded. Finally, a larger E1 structure (probably a ring-like hexamer) is assembled on single-stranded DNA.

The adenoviral origins of replication are also recognized by a complex of viral proteins, the preterminal protein and viral DNA polymerase. These proteins associate as they are synthesized in the cytoplasm of infected cells and, once within the nucleus, bind specifically to a conserved sequence within the minimal origin of replication (Fig. 9.12). Assembly of the viral proteins into a preinitiation complex at the origin is stimulated by their direct interaction with two cellular transcriptional regulators that bind to adjacent sequences (Fig. 9.12). These viral and cellular proteins are sufficient to reconstitute initiation of origin-dependent adenoviral DNA synthesis from templates that carry the mature terminal protein covalently linked to their 5′ ends, as in viral chromosomes that enter infected cell nuclei (Fig. 9.10). The terminal protein present on such natural adenoviral DNA templates facilitates unwinding of the duplex termini early in the initiation reaction, and may be important for directing viral DNA molecules to the specialized nuclear sites at which replication takes place (see "Localization of Replicating Viral DNA to Specialized Intracellular Sites" below).

Viral DNA Synthesis Machinery

In addition to origin recognition proteins, larger viral DNA genomes encode DNA polymerases and other essential replication proteins (Table 9.5). The simplest viral replication apparatus, and consequently the best understood, is that of adenoviruses, which comprises the preterminal protein primer and DNA polymerase and only one other protein, a single-stranded-DNA-binding protein. The latter protein stimulates initiation and is essential during elongation, when it coats the displaced strands of the template DNA molecule (Fig. 9.10). It binds single-stranded DNA cooperatively to stimulate the activity of the viral DNA polymerase as much as 100-fold and to induce highly processive DNA synthesis by the viral DNA polymerase. The most interesting property of the DNA-binding protein is that it supports movement of the replication fork over quite large distances without hydrolysis of ATP. Consequently, no helicase activity is needed during adenoviral DNA synthesis. The crystal structure of a large segment of the DNA-binding protein indicates that this protein multimerizes via a C-terminal hook (Fig. 9.15). The formation of long protein chains results in cooperative, high-affinity binding to single-stranded DNA, and is the driving force for ATP-independent unwinding of the duplex template. Other single-stranded DNA-binding proteins, such as the herpes simplex virus type 1 UL8 protein or cellular replication protein A, may destabilize double-stranded DNA helices by a similar mechanism.

Complete copying of adenoviral DNA templates also requires a cellular topoisomerase, such as topoisomerase I (Table 9.4), which is presumed to relieve overwinding of the template upon extensive replication, as during simian virus 40 DNA synthesis. The simplicity of the adenoviral DNA synthesis machinery stands in stark contrast to the complexity of the cellular replication complex, assembled, for example, at simian virus 40 replication forks. This property can be attributed to the protein priming mechanism, in which single molecules of the preterminal protein-dCMP primer support continuous synthesis of full-length adenoviral daughter DNA strands (Fig. 9.10).

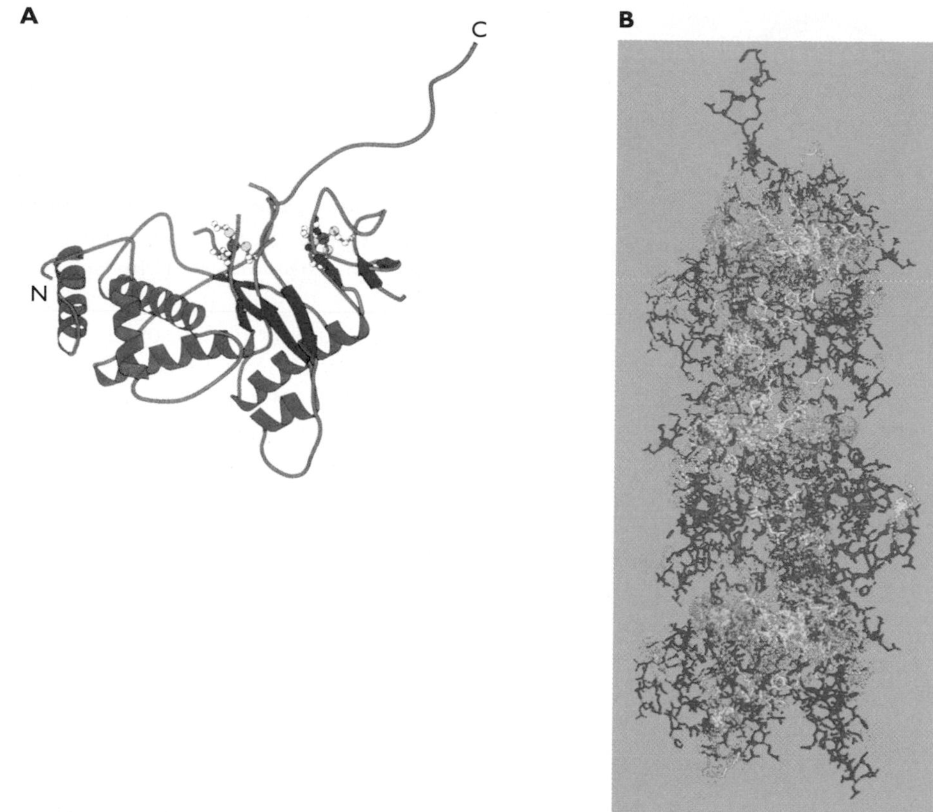

Figure 9.15 Crystal structure of the adenoviral single-stranded-DNA-binding protein. (A) Ribbon diagram of the C-terminal nucleic acid-binding domain (amino acids 176 to 529) of the human adenovirus type 5 single-stranded-DNA-binding protein, showing the two sites of Zn^{2+} (red atom) coordination. The most prominent feature is the long (~40-Å) C-terminal extension. This C-terminal hook of one protein molecule invades a cleft between two α-helices in its neighbor in the infinite protein array formed in the crystal. Deletion of the C-terminal 17 amino acids of the DNA-binding protein fragment eliminates cooperative binding of the protein to DNA, indicating that the interaction of one molecule with another via the C-terminal hook is responsible for cooperativity in DNA binding. (B) Likely DNA-binding surface in a chain of three DNA-binding protein molecules. This surface was identified both by the locations of substitutions that inhibit the DNA-binding and replication functions of the protein, and as a region of high positive potential (yellow) for interaction with the phosphate backbone of DNA. The regions colored cyan in the model are the ends of the polypeptide chain of each monomer. The organization of the positively charged surface and the locations of residues implicated in DNA binding suggest that single-stranded DNA is wound around the DNA-binding protein chain. From P. A. Tucker et al., *EMBO J.* **13:**2994–3002, 1994, with permission. Courtesy of P. C. van der Vliet, Utrecht University.

Other viral replication systems include a larger number of accessory replication proteins (Table 9.5). Herpes simplex virus type 1 genes encoding essential replication proteins have been identified by both genetic methods and a DNA-mediated transformation assay for the gene products necessary for plasmid replication directed by a viral origin (Fig. 9.16). Replication from a herpes simplex virus type 1 origin requires not only the viral DNA polymerase and origin recognition protein, but also five other viral proteins (Table 9.5). These proteins are functional analogs of essential components of the cellular replication machinery. The products of the UL5, UL8, and UL52 genes form a viral primase, which also functions as a helicase, and the UL42 protein is a processivity factor. Although the herpesviral proteins listed in Table 9.5 provide a large repertoire of replication functions, they do not support viral DNA synthesis in vitro. The additional viral and/or cellular proteins needed to reconstitute herpesviral DNA synthesis have not yet been identified. One may be cellular topoisomerase II, which is essential for replication in infected cells.

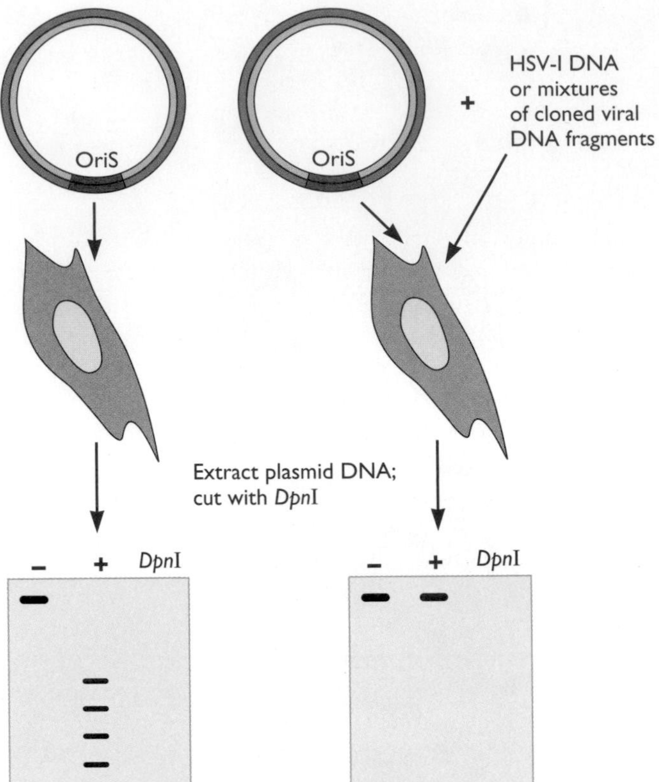

Figure 9.16 DNA-mediated transformation assay for essential herpes simplex virus type 1 replication proteins. A plasmid carrying a viral DNA fragment spanning OriS is introduced into monkey cells permissive for herpesvirus replication. In the absence of viral proteins, the plasmid DNA is not replicated and retains the methyl groups added to A residues in a specific sequence, which are introduced in *E. coli* by the *dam* methylation system. These sequences include the recognition site for the restriction endonuclease *Dpn*I, which cleaves only such methylated DNA. The unreplicated plasmid DNA therefore remains sensitive to *Dpn*I cleavage. However, when all viral genes encoding proteins required for OriS-dependent replication are also introduced into the cells (right), the plasmid is replicated. Because the newly replicated DNA is not methylated at *Dpn*I sites, it cannot be cleaved by this enzyme. The sensitivity of the plasmid to *Dpn*I cleavage therefore provides a simple assay for plasmid replication, and hence for the identification of viral proteins required for replication from OriS.

Resolution and Processing of Viral Replication Products

Several of the viral DNA replication mechanisms described in preceding sections yield products that do not correspond to the parental viral genome. For example, replication of simian virus 40 DNA yields two interlocked, double-stranded, circular DNA molecules that must be separated by cellular topoisomerase II. Such resolution is required whenever circular templates (e.g., episomal Epstein-Barr virus DNA) are replicated as monomers. In other cases, replication yields multimeric DNA molecules, from which linear genomes of fixed length and sequence must be processed for packaging into virions. This situation is exemplified by the herpes simplex virus type 1 genome.

Exactly how head-to-tail concatemers containing multiple copies of the herpes simplex virus type 1 genome are made from viral DNA templates in infected cells is not clear, nor is the origin of the nonlinear, branched herpes simplex virus type 1 replication intermediates. Conversion of the genome from one of its four isomers (Fig. 9.11) to another occurs by the time that newly replicated DNA can first be detected in infected cells. Such isomerization is the result of recombination between repeated viral DNA sequences (see "Recombination of Viral Genomes" below). These properties suggest that recombination may be an essential reaction during herpes simplex virus type 1 DNA synthesis and responsible for production of concatemers.

Linear herpes simplex virus type 1 DNA molecules with termini identical to those of the infecting genome are liberated from concatemeric replication products by cleavage at specific sites within the *a* repeats (Fig. 9.11). As described in Chapter 13, such cleavage is coupled with encapsidation of viral DNA molecules during virion assembly.

The products of adenoviral DNA synthesis are unit-length copies of the linear viral genome that require no processing (Fig. 9.10). However, recent studies indicate that accumulation of these viral DNA molecules requires inactivation of a cellular DNA repair system. In the absence of either the viral E4 Orf 3 or Orf 6 proteins (Appendix A, Fig. 1B), newly synthesized viral DNA forms concatemers far too large to be packaged into progeny virions. Accumulation of such multimeric DNA molecules depends on a cellular DNA repair complex that normally functions in double-stranded break repair (and telomere maintenance) and accumulates at sites of DNA damage. In adenovirus-infected cells, the protein components of this complex become redistributed within nuclei and are then degraded by the proteasome, alterations induced by the Orf 3 and Orf 6 proteins, respectively. When the E4 proteins cannot be made, the cellular repair proteins accumulate in viral replication centers. These E4 gene products therefore appear to prevent triggering of a double-strand-break repair mechanism by the accumulation of linear viral DNA genomes in infected cell nuclei.

Mechanisms of Exponential Viral DNA Replication

The details of the mechanisms by which complete replication of viral DNA genomes is achieved vary considerably from one virus to another. Nevertheless, each of these strategies generally ensures efficient replication of viral

DNA. Production of 10^4 to 10^5 viral genomes, or more, per infected cell is not uncommon, as the products of one cycle of replication are recruited as templates for the next. Such **exponential** viral DNA synthesis sets the stage for assembly of a large burst of progeny virions. In this section, we discuss regulatory mechanisms that ensure efficient viral DNA synthesis. These regulatory circuits impinge primarily on the expression of cellular or viral genes encoding the proteins that carry out viral DNA synthesis.

Induction of Synthesis of Cellular Replication Proteins by Viral Gene Products

With few exceptions, virus reproduction is studied in established cell lines that are permissive for the virus of interest. Such immortal or transformed cell lines grow and divide indefinitely and differ markedly from the cells in which viruses reproduce in nature. For example, highly differentiated cells, such as neurons or the outer cells of an epithelium, do not divide and are permanently in a specialized resting state, termed the **G_0 state.** Many other cells in an organism divide only rarely, or only in response to specific stimuli, and therefore spend much of their lives in the G_0 state. Such cells do not contain many of the components of the cellular replication machinery and are characterized by generally slow rates of synthesis of RNAs and proteins. The G_0 state would not seem to provide a hospitable environment for virus reproduction, which entails the synthesis of large quantities of viral nucleic acids and proteins, often at a rapid rate. Nevertheless, viruses often reproduce successfully within cells infected when they are in G_0. In some cases, such as replication of several herpesviruses in neurons, the replication machinery is encoded within the viral genome. Infection by other viruses stimulates resting or slowly growing cells to abnormal activity, by disruption of cellular circuits that restrain cell proliferation. This strategy is characteristic of polyomaviruses and adenoviruses. The discovery that infection by these viruses disrupts the same cellular cell growth control circuits was quite unanticipated, and of the greatest importance in elucidating the roles of critical regulators of cell proliferation, such as the cellular retinoblastoma (Rb) protein.

Functional Inactivation of the Rb Protein

Loss or mutation of both copies of the cellular retinoblastoma (*rb*) gene is associated with the development of tumors of the retina in children and young adults. Because it is the **loss** of normal function that leads to tumor formation, *rb* is defined as a **tumor suppressor gene**. As discussed in Chapter 18, the Rb protein is an important component of the regulatory program that ensures that cells grow, duplicate their DNA, and divide in an orderly manner. In particular, the Rb protein controls entry into the period of the cell cycle in which DNA is synthesized, the **S phase,** from the preceding G_1 phase. Our current appreciation of the critical participation of this protein in the control of cell cycle progression and of the mechanism by which it operates stems from the discovery that Rb binds directly to the two adenoviral E1A proteins (see Chapter 8).

In uninfected cells in the G_1 phase, the Rb protein is bound to cellular transcriptional activators of the E2f family. These complexes, which bind to specific promoters via the DNA-binding activity of E2f, function as repressors of transcription (Fig. 9.17A). Binding of adenoviral E1A proteins (or of simian virus 40 LT or E7 proteins of highly oncogenic human papillomaviruses) to Rb releases E2f from this association and sequesters Rb. The E2f protein therefore becomes available to stimulate transcription of cellular genes that encode proteins that participate directly or indirectly in DNA synthesis, or in control of cell cycle progression (Fig. 9.17A; see Chapter 18).

Although the polyomaviral LT and adenoviral E1A proteins both release E2f from inhibitory associations with the cellular Rb protein (Fig. 9.17A), the DNA genomes of these viruses are replicated by entirely different mechanisms. Consequently, the benefits conferred by activation of E2f when a polyomavirus or an adenovirus infects a cell that is growing slowly, or is withdrawn from the cell cycle, are virus specific. While the advantages for viral DNA replication can be deduced, it is important to stress that essentially all studies of the cell growth control functions of LT, E7, or E1A proteins have been directed toward elucidation of mechanisms of transformation. For example, it has been shown that the smaller adenoviral E1A protein is necessary for efficient viral DNA synthesis in quiescent human cells. However, it has not been established that this is because the E1A protein plays a critical role in activation of cellular E2f, as assumed in the model described below.

E2f and simian virus 40 DNA synthesis. Polyomaviral DNA synthesis requires the cellular replication machinery, and therefore cannot take place unless these proteins are present in the infected cell. Expression of the cellular genes encoding DNA polymerases, accessory replication proteins, histones, and enzymes that synthesize substrates for DNA synthesis is tightly controlled during the cell cycle, such that these proteins are made only just before they are needed in S phase. The transcriptional control regions of many of these genes contain E2f-binding sites, which are thought to be required for their efficient transcription in response to E2f. Synthesis of simian virus 40 LT is therefore believed to activate production of the cellular proteins necessary for viral DNA synthesis, and hence to

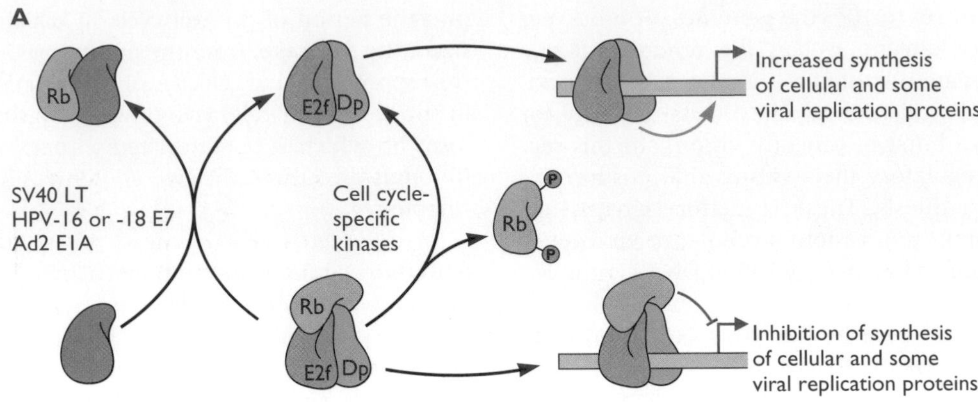

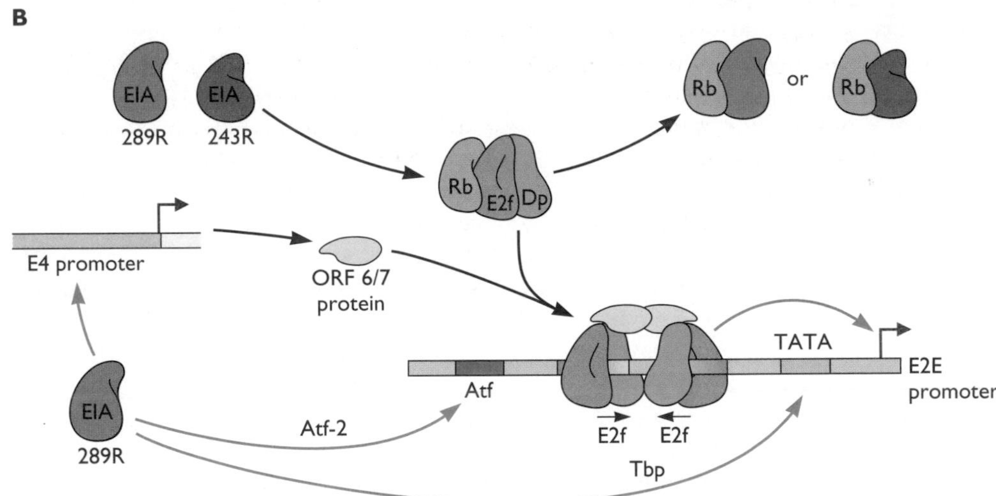

Figure 9.17 Regulation of production of adenoviral replication proteins. (A) Model for the abrogation of the function of the Rb protein by adenoviral proteins. The members of the E2f family of transcriptional regulators are heterodimeric proteins, each containing one E2f and one Dp subunit. To date, six E2f (E2f-1 to E2f-6) and three Dp (Dp1 to Dp3) genes have been identified. As discussed in Chapter 18, there is accumulating evidence that these E2f proteins fulfill different functions. On binding to specific promoter sequences, E2f dimers stimulate transcription of cellular genes encoding replication proteins, histones, and proteins that allow passage through the cell cycle (green arrow). Binding of Rb protein does not prevent promoter recognition by E2f. However, Rb protein present in the complex represses transcription (red bar). Phosphorylation of Rb protein at specific sites, a modification that normally takes place to allow cells to enter S phase (see Chapter 18), induces its dissociation from E2f and activates transcription of cellular genes expressed in S phase. The adenoviral E1A proteins, simian virus 40 LT, and the E7 proteins of certain human papillomaviruses (types 16 and 18) bind to the region of Rb protein that contacts E2f. These proteins can therefore disrupt Rb-E2f complexes and sequester Rb protein to activate E2f-dependent transcription. (B) Stimulation of transcription from the adenoviral E2 early promoter by E1A proteins. The E2E promoter-binding sites for the cellular Atf, E2f, and TfIId (TATA sequence) proteins are necessary for E2E transcription in infected cells. The inversion of the two E2f sites and their precise spacing are essential for assembly of an E2f-DNA complex unique to adenovirus-infected cells, in which binding of an E2f heterodimer to each of the two sites is stabilized by binding of two molecules of the E4 ORF6/7 protein. Such binding of the E4 protein greatly promotes cooperative binding of E2f to these sites and increases the lifetime of E2f-DNA complexes. The availability of the cellular E2f and viral E4 ORF6/7 proteins is a result of the action of immediate-early E1A proteins: either the 243R or 289R protein can sequester unphosphorylated Rb to release active E2f from Rb-E2f complexes in infected cells (A), and the 289R protein stimulates transcription from the E4 promoter. This larger E1A protein can also stimulate transcription from the E2E promoter directly, probably via components of the general initiation protein TfIId, such as Tbp, which binds to the TATA element, and specific members of the Atf family (Chapter 8).

ensure efficient replication of the viral genome regardless of the growth state of the host cell. The ability of simian virus 40 LT both to induce synthesis of components of the cellular replication machinery and to initiate viral DNA synthesis seems likely to coordinate viral replication with the induced entry of the host cell into S phase. Indeed, as discussed previously, LT replication functions are regulated by phosphorylation, probably by the cyclin A–cyclin-dependent kinase 2 that accumulates as cells enter S phase.

E2f and adenoviral DNA synthesis. Activation of E2f in adenovirus-infected cells is believed to promote viral DNA synthesis in two ways: stimulation of transcription of cellular genes that encode enzymes that make substrates for DNA synthesis, such as thymidine kinase and dihydrofolate reductase, and activation of production of all three viral replication proteins. The viral DNA polymerase, preterminal protein primer, and DNA-binding protein are encoded within the E2 gene, which is transcribed from an early promoter that contains two binding sites for E2f (Fig. 9.17B). In fact, these critical cellular regulators derive their name from these E2 binding sites, which are necessary for efficient E2 transcription during the early phase of adenovirus. As discussed previously, the viral E1A proteins disrupt Rb-E2f complexes and sequester Rb. They also regulate transcription from the E2 promoter by two other mechanisms (Fig. 9.17B). These E1A-dependent regulatory mechanisms presumably operate synergistically, to allow synthesis of the viral mRNAs encoding replication proteins in quantities sufficient to support numerous cycles of viral DNA synthesis. A post-transcriptional regulatory mechanism may allow production of the appropriate concentrations of the three E2 replication proteins (Box 9.6).

BOX 9.6

Production of appropriate quantities of the adenoviral replication proteins by differential polyadenylation?

The three adenoviral replication proteins, Pol, pTP, and DBP, are all encoded within the E2 transcription unit of the viral genome. However, they are needed in very different quantities during viral DNA synthesis. The replication of one molecule of viral DNA requires two molecules each of Pol and the pTP primer (Fig. 9.10). Once the daughter strands have been synthesized, Pol dissociates and can catalyze additional replication cycles. This protein is therefore required at **catalytic** concentrations. In contrast, the pTP remains covalently attached to the 5′ end of each newly synthesized DNA strand and is incorporated into progeny virus particles. This protein is therefore required at a greater concentration than Pol. The DBP coats the single-stranded DNA displaced during viral DNA synthesis (Fig. 9.10). This protein, which interacts with some 12 nucleotides, binds cooperatively to such DNA to form a long chain (Fig. 9.15). Thousands of molecules of DBP are therefore needed during replication of a single molecule of the viral genome.

As shown in the figure, E2 primary transcripts can be polyadenylated at either a promoter-proximal (DBP mRNA) or a promoter-distal (Pol and pTP mRNAs) site. The former appears to be utilized much more frequently than the latter: DBP mRNA accumulates to 10- to 20-fold-higher concentrations than the Pol and pTP mRNAs. In this way, the replication proteins can be made in the appropriate relative concentrations, even though they are encoded in the same transcription unit.

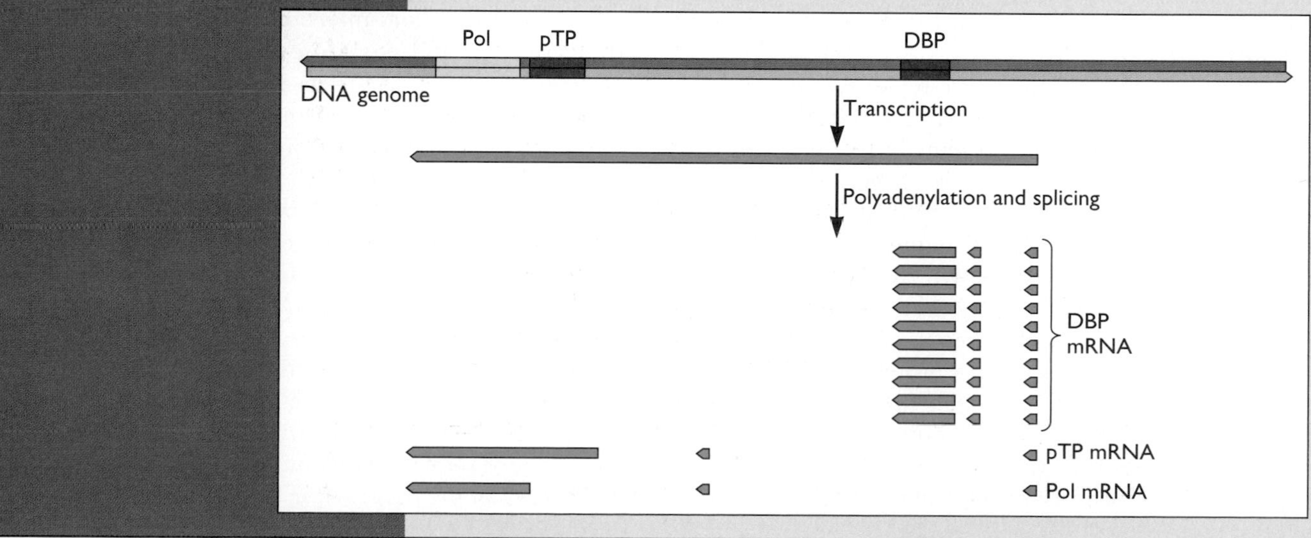

Synthesis of Large Sets of Viral Replication Proteins

The DNA genomes of several viruses, exemplified by that of herpes simplex virus type 1, encode large cohorts of proteins that participate directly or indirectly in viral DNA synthesis. Two classes of such replication proteins are synthesized during productive infection. The first comprises the viral DNA polymerase and other proteins responsible for viral DNA synthesis discussed previously (Table 9.5). The viral proteins in the second class are enzymes, analogous to host cell enzymes, that catalyze reactions by which dNTPs are synthesized (Table 9.5). For example, viral thymidine kinase and ribonucleotide reductase are synthesized in herpesvirus-infected cells. In general, members of this second class of proteins are dispensable for replication in proliferating cells in culture, but are essential in animal hosts. Herpes simplex viruses that lack thymidine kinase or ribonucleotide reductase genes cannot replicate in neurons, because such terminally differentiated cells are permanently withdrawn from the cell cycle, and do not make enzymes that produce substrates for DNA synthesis.

Efficient synthesis of all herpes simplex virus type 1 replication proteins is primarily the result of the viral transcriptional cascade described in Chapter 8. Expression of the genes encoding these viral proteins, which are early (β) genes, is regulated by products of immediate-early genes. These immediate-early proteins operate transcriptionally (e.g., ICP0 and ICP4) or posttranscriptionally (e.g., ICP27) to induce synthesis of viral replication proteins at concentrations sufficient to support exponential replication of viral DNA. Whether the multiple origins of herpes simplex virus type 1 and other herpesviruses increase the rate of replication will not be clear until the viral replication mechanism has been better characterized.

Viral DNA Replication Independent of Cellular Proteins

One method guaranteed to ensure replicative success of a DNA virus, regardless of the growth state of the host cell, is encoding of **all** necessary proteins in the viral genome. On the other hand, this mechanism is genetically expensive, which may be the reason why it is restricted to the viruses with the largest DNA genomes, such as vaccinia virus. The genome of this virus, which is replicated in the cytoplasm at specialized foci sometimes called **viral factories,** encodes all the proteins needed for viral DNA synthesis. The current catalog of such proteins includes a DNA polymerase and several accessory replication proteins (Table 9.5). The vaccinia virus genome also encodes several other enzymes that would be expected to participate in DNA synthesis or in resolution of replication products, including a type I topoisomerase and a DNase with endonucleolytic activity, as well as several enzymes for synthesis of dNTPs (Table 9.5). None of the latter appear to be essential for virus reproduction in actively growing cells in culture. However, several enzymes, such as the thymidine kinase, are necessary for efficient virus growth in quiescent cells or animal hosts, where they presumably increase less than optimal substrate pools. Much remains to be learned about the mechanism by which the vaccinia virus genome is replicated and about the functions of viral replication proteins. Nevertheless, it is clear that no cellular proteins are needed, because vaccinia virus replicates efficiently in enucleated cells.

Delayed Synthesis of Virion Structural Proteins

During productive infection by DNA viruses, each cycle of replication increases the number of DNA molecules that can be copied in the subsequent cycle. This increase in the pool of replication templates, a doubling in each cycle of the simpler viruses like simian virus 40 and adenovirus, undoubtedly makes an important contribution to the rapid amplification of genomes that follows initiation of viral DNA synthesis. Progeny viral DNA molecules are eventually encapsidated during the assembly of new virus particles, and consequently become unavailable to serve as templates for DNA synthesis. However, assembly and sequestration of the genome are delayed with respect to initiation of viral DNA synthesis, because transcription of late genes encoding virion proteins cannot begin until viral DNA has been replicated.

Inhibition of Cellular DNA Synthesis

When viral DNA synthesis is carried out largely by viral proteins, cellular DNA synthesis is often inhibited, presumably to eliminate competition for substrates from cellular DNA replication. Infection by the larger DNA viruses (herpesviruses and poxviruses) induces severe inhibition of synthesis of cellular DNA during productive infection. Cellular DNA synthesis is also blocked when adenoviruses infect actively growing cells in culture, despite the elaborate mechanisms by which these viruses induce quiescent host cells to reenter the cell cycle. Although inhibition of cellular DNA synthesis in cells infected by these DNA viruses was described in some of the earliest studies of their infectious cycles, very little is known about the mechanisms that shut down this cellular process. It appears that inhibition depends on specific mechanisms, rather than on passive competition between viral and cellular DNA polymerases for the finite pools of dNTP substrates. For example, infection of actively growing cells by adenovirus or the herpesvirus human cytomegalovirus has been reported to induce cell cycle arrest. Synthesis of the Epstein-Barr virus Zta protein, a sequence-specific transcriptional regulator and origin-binding protein, also arrests cells at a point in the cell cycle prior to S phase, such that cellular DNA synthesis is precluded. In this case, arrest is the result of increased concentrations of cellular proteins that negatively regulate progression through the cell cycle, such as the Rb protein.

Localization of Replicating Viral DNA to Specialized Intracellular Sites

A common, probably universal, feature of cells infected by viruses with DNA genomes is the presence of virus-specific structures that are the sites of viral DNA synthesis. Vaccinia virus DNA replication takes place in discrete cytoplasmic, viral factories that contain both the DNA templates and viral replication proteins. The replication of other viral DNA genomes within infected cell nuclei also takes place in specialized compartments, which can be visualized as distinctive, infected cell-specific structures containing viral proteins (Fig. 9.18). Such structures, known as **replication centers** or **replication compartments,** have been best characterized in human cells infected by adenovirus or herpes simplex virus type 1. They contain newly synthesized viral DNA and the viral proteins necessary for viral DNA synthesis (Table 9.5), as well as other viral and cellular proteins. Among the cellular proteins associated with herpes simplex virus type 1 replication centers are several that catalyze DNA transactions, such as DNA polymerases α and γ, DNA ligase, and topoisomerase II. Their presence in replication centers might be because they participate in viral DNA synthesis, or may be a consequence of the assembly of viral replication centers at sites of initiation of cellular DNA synthesis. The localization of both the templates for viral DNA synthesis and the replication proteins at a limited number of sites undoubtedly facilitates exponential viral DNA replication. This arrangement increases the local concentrations of proteins that must interact with one another, or with viral origin sequences or replication forks, favoring such intermolecular interactions by the law of mass action. In addition, the high local concentrations of replication templates and proteins appear ideally suited for efficient recruitment of the products of one replication cycle as templates for the next.

Viral replication centers also serve as foci for viral gene expression, presumably in part by concentrating templates for transcription with the proteins that carry out or regulate this process. For example, the herpes simplex virus type 1 immediate-early ICP4 and ICP27 proteins, as well as the host cell's RNA polymerase II, are recruited to these nuclear sites. Similarly, an adenoviral early protein complex necessary for selective export of viral late mRNAs from the nucleus is associated with the periphery of viral replication centers, as is nascent viral RNA.

Viral replication centers do not assemble at random sites, but rather are formed by viral colonization of specialized niches within mammalian cell nuclei. When they

Figure 9.18 Nuclear replication compartments of herpes simplex virus type 1-infected monkey cells. The locations of the UL29 single-stranded DNA-binding protein (A) and newly replicated DNA (B) were visualized by immunofluorescence with antibodies against UL29 protein and bromodeoxyuridine incorporated into DNA, respectively. Newly synthesized viral DNA and the UL29 protein colocalize to a limited number of globular patches, which contain all seven known viral replication proteins. These viral replication compartments are found at the peripheries of nuclear domains containing cellular promyelocytic leukemia protein (see text). The infecting viral DNA is found associated with these structures, but there is subsequent reorganization following expression of the viral ICP0 protein. From C. I. Lukonis and S. K. Weller, *J. Virol.* **71:**2390–2399, 1997, with permission. Courtesy of S. Weller, University of Connecticut Health Sciences Center.

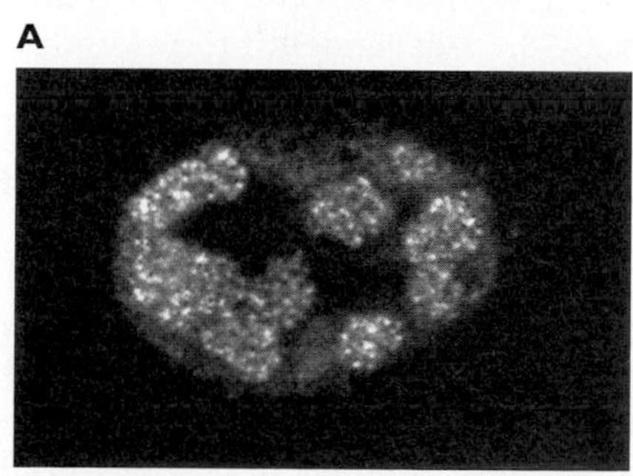

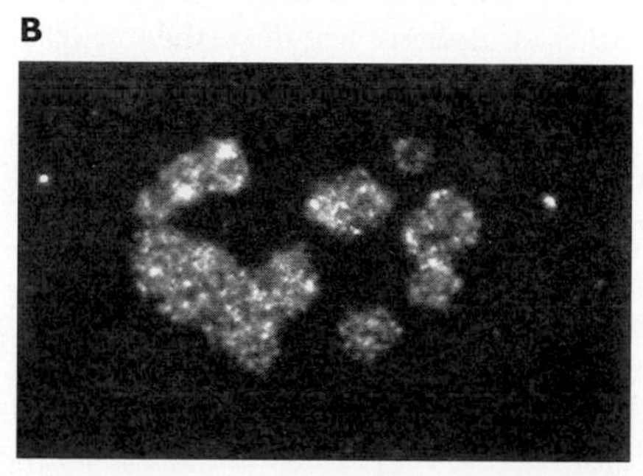

enter the nucleus, infecting adenoviral or herpes simplex virus type 1 genomes localize to preexisting nuclear bodies that contain a specific cellular protein called the promyelocytic leukemia protein (Pml). These structures are therefore called **Pml bodies,** or nuclear domain 10 structures, a name derived from the average number present in most cells. Viral proteins then induce reorganization of Pml bodies as viral replication centers are established. For example, the herpes simplex virus type 1 ICP0 protein causes disruption of these cellular structures and degradation of some of their constituent proteins by the cellular proteasome. Others, including specific forms of the Pml protein, become associated with viral replication centers. Recruitment of specific proteins present in Pml bodies of host cell nuclei to viral replication centers also occurs in adenovirus-infected cells, and requires the E4 Orf 3 protein. The association of replication centers of different DNA viruses with constituents of the same intranuclear structures suggests that some feature of the organization of their host cell nuclei facilitates viral DNA synthesis. The discovery that viral DNA genomes home to Pml bodies has stimulated characterization of their constituents, but much remains to be learned about the molecular functions of these nuclear structures. One hypothesis is that Pml bodies represent a form of intrinsic antiviral defense (Chapter 15). For example, when disruption of these structures is blocked in herpes simplex virus-infected cells, because of alterations in the ICP0 protein or addition of proteasome inhibitors, the yield of progeny virions is decreased by several orders of magnitude. There can be little doubt that further characterization of viral replication centers will shed new light on a poorly understood aspect of eukaryotic cell biology, the structural and functional organization of the nucleus.

Limited Replication of Viral DNA

Exponential replication of viral DNA is the typical pattern when the majority of DNA viruses infect cells in culture. Nevertheless, several can establish long-term relationships with their hosts and host cells, in which the number of genomes produced is limited. Various mechanisms that effect such copy number control are described in this section.

Replication as Part of the Cellular Genome: Integrated Parvoviral DNA

As mentioned previously, the adeno-associated viruses reproduce only in cells coinfected with a helper adeno- or herpesvirus. Although the latter viruses are widespread in hosts infected by adeno-associated viruses, the chances that a particular host cell will be simultaneously infected by two viruses are very low. The strategy of exploiting other viruses to provide functions for efficient expression of the genetic information of adeno-associated viruses would therefore appear to be potentially lethal for individual virus particles. In fact, this is not the case, for adeno-associated virus can survive by means of an alternative mechanism: when the virus infects a cell that contains no helper virus, its genome becomes integrated into that of the host cell, and is replicated as part of a cellular replicon.

Activation of this program for long-term survival of the adeno-associated virus genome depends on expression of its regulatory region (Rep) (Fig. 9.9C). The two larger proteins encoded by this region, Rep 78/68, are multifunctional and control all phases of the viral life cycle (Table 9.6). When helper virus proteins, such as adenoviral E1A, E1B, and E4 proteins, allow synthesis of large quantities of Rep 78/68, adeno-associated virus DNA is replicated by the mechanism described previously. In the absence of helper functions, only very small quantities of Rep 78/68 are made, even in healthy, dividing cells. Consequently, there is little viral DNA synthesis, and the genome is integrated into that of the host cell. The latter reaction is mediated by Rep 78/68 and depends on the sequence at which viral DNA is nicked during productive replication. One of the most unusual features of this integration reaction is that it occurs only near one end of human chromosome 19. Both the viral and cellular sequences required for integration include binding sites for Rep 78/68 and adjacent sequences at which these endonucleases can nick the DNA. This protein can bind simultaneously to both viral and human chromosomal 19 DNA, at least in vitro. The current model of integration therefore proposes that its specificity is the result of such simultaneous binding to the two DNA molecules by Rep 78/68.

The integrated viral genome is preserved within the genome of the host cell and its descendants, until a helper virus infects an adeno-associated virus-carrying cell. Following activation of transcription of Rep sequences, the viral genome is excised from that of the host cell, in a Rep 78/68-dependent process, for entry into the productive cycle.

Regulation of Replication via Different Viral Origins: Epstein-Barr Virus

During herpesviral latent infections, the viral genome is stably maintained at low concentrations, often for long periods (Chapter 16). Furthermore, replication of viral and cellular genomes can be coordinated. This pattern is characteristic of human B cells latently infected by Epstein-

Barr virus. Many cell lines latently infected with this virus have been established from patients with Burkitt's lymphoma, and this state is the usual outcome of Epstein-Barr virus infection of B cells in culture. Characteristic features of latent Epstein-Barr virus infection include expression of only a small number of viral genes, the presence of a finite number of viral genomes, and replication from a specialized origin. Because replication from this origin, which is not active in lytically infected cells, is responsible for maintenance of episomal viral genomes in latently infected cells, it is termed the **origin for plasmid maintenance** (OriP).

The Epstein-Barr virus genome is maintained in nuclei of latently infected cells as a stable circular **episome**, present at 10 to 50 copies per cell. For example, one Burkitt's lymphoma cell line (Raji) has carried about 50 copies per cell of episomal viral DNA for over 35 years in culture. When Epstein-Barr virus infects a B cell, the linear viral genome circularizes by a mechanism that is not well understood. The circular viral DNA is then amplified during S phase of the host cell to the final concentration listed above. Such replication is by the cellular DNA polymerases and accessory proteins that synthesize simian virus 40 DNA. However, it also requires the viral origin for latent viral DNA replication, OriP, and the viral protein that binds specifically to it, EBNA-1 (Table 9.6), which is invariably synthesized in latently infected cells. In contrast to the productive replication of this and other viruses, such amplification of the episomal viral genome is limited to a few cycles, by an unknown mechanism. Following such limited amplification, the viral DNA is duplicated once per cell cycle, in S phase, such that its concentration is maintained as the host lymphocyte cell divides. The EBNA-1 protein and OriP are sufficient for both such once-per-cell cycle replication of episomal viral genomes and their orderly segregation to daughter cells. The latter process requires the binding of EBNA-1 to its high-affinity sites in the family of repeat sequence of OriP (Box 9.6). The mechanism by which these sequences prevent loss of episomal genomes is not known, but may be related to their binding to the nuclear matrix in cells that contain EBNA-1.

The availability of cellular replication proteins only in late G_1 and S can account for the timing of Epstein-Barr virus replication in latently infected cells. However, this property **cannot** explain why each genome is replicated only once in each cell cycle. The controlled initiation of replication of episomal viral DNA molecules is analogous to the tight control of initiation of replication from cellular origins, each of which also fires once and only once in each S phase. The mechanisms that control such once-per-cycle firing of eukaryotic origins, a process termed **replication licensing**, were initially elucidated in budding yeasts, which, as noted previously, contain compact origins of replication. Mammalian homologs of the yeast origin recognition complex and proteins that regulate initiation of DNA synthesis, such as Mcm and Cdc6, have been identified in all other eukaryotes examined. The human origin recognition complex (Orc) proteins are associated with OriP and can bind to EBNA-1. Experimental manipulations that reduce the concentrations of specific Orc proteins severely inhibit OriP-dependent replication. Human Mcm proteins are also associated with OriP during G_1 phase. However, they are not present at OriP during G_2, as would be expected if replication from OriP were licensed by the mechanism shown in Fig. 9.19. The inhibition of OriP-dependent replication by overproduction of a protein that prevents recruitment of Mcm to the replication complex (Fig. 9.19) provides strong support for the conclusion that synthesis of viral DNA genomes in latently infected cells is governed by the mechanisms that ensure once-per-cell-cycle firing of cellular origins.

In addition to EBNA-1-binding sites, OriP contains three copies of a nonameric sequence that resemble repeated sequences present in telomeres, the specialized DNA sequences responsible for maintenance of the ends of linear chromosomal DNA. In the presence of EBNA-1, several cellular proteins bind to the repeats. These cellular proteins include telomerase-associated poly(ADP-ribose) polymerase and telomere repeat-binding protein 2. Mutations of the nonamer sequence impair maintenance of plasmids containing OriP in human cells, as do compounds that inhibit poly(ADP-ribose) polymerase. These observations indicate that cellular proteins that normally regulate the length of telomeres facilitate maintenance of the viral episome, although how they do so is not yet clear.

As a latent infection is established, the Epstein-Barr virus genome becomes increasingly methylated at C residues present in CG dinucleotides, although sequences that must function in latently infected cells, such as OriP, generally escape this modification. Such DNA methylation is associated with repression of transcription, and contributes to inhibition of expression of viral genes. The viral genome also becomes packaged by cellular nucleosomes and is therefore replicated as a circular minichromosome, much like that of simian virus 40. Replication of the Epstein-Barr virus genome once per cell cycle persists unless conditions that induce entry into the viral productive cycle are encountered. The critical step for this transition is activation of transcription of the viral genes encoding the transcriptional activators Zta and Rta (Chapter 8). These

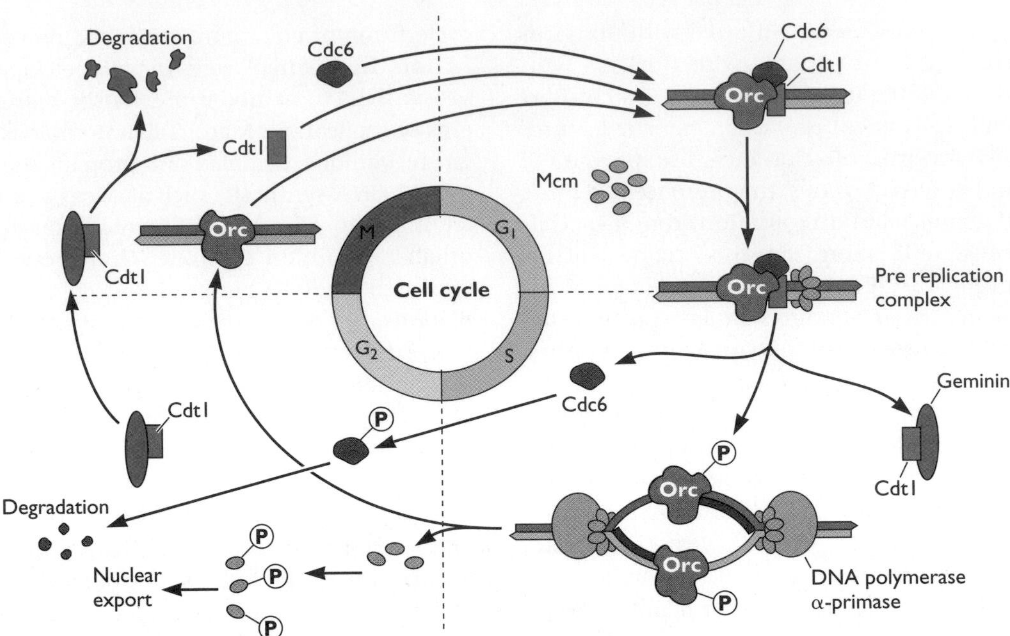

Figure 9.19 Licensing of replication. The multiprotein origin recognition complex (Orc) is present throughout the cell cycle and is associated with replication origins. However, initiation of DNA synthesis requires loading of the hexameric minichromosome maintenance complex (Mcm), which is believed to remodel chromatin and therefore provide the enzymes that synthesize DNA access to the template. It is the recruitment of Mcm that is controlled during the cell cycle to set the timing of the initiation of DNA synthesis in S phase. This reaction requires two proteins, Cdc6 and Cdt1. The concentrations and activities of both are tightly controlled during the cell cycle, and therefore so in loading of Mcm onto the Orc-DNA complex. As cells complete mitosis and enter G_1, Cdc6 and Cdt1 accumulate in the nucleus, where they can associate with DNA-bound Orc. These interactions permit loading of Mcm at the G_1-to-S-phase transition, and subsequently of components of the DNA synthesis machinery, such as Rpa and the DNA polymerase α (Polα)-primase. The latter step requires phosphorylation of specific components of the prereplication complex by cyclin-dependent kinases that accumulate during the G_1-to-S-phase transition (Chapter 18). Reinitiation of DNA synthesis is prevented by several mechanisms. A cyclin-dependent kinase that accumulates during the G_2 and M phases phosphorylates both Mcm proteins and Cdc6. This modification induces nuclear export of the former and degradation of the latter. In addition, the protein called geminin is present in the nucleus from S until M phase (when it is degraded). This protein binds to Cdt1, sequestering it from interaction with Cdc6 and Orc. As a consequence of such regulatory mechanisms, the prereplication complex can form **only** in the G_1 phase, ensuring firing of the origin once per cell cycle.

proteins induce expression of the viral early genes that encode the viral DNA polymerase and other proteins necessary for replication from one of the two copies of OriLyt, and also bind to a sequence of these origins. Indeed, Zta appears to be the OriLyt recognition protein. Consequently, once this protein is made in an Epstein-Barr virus-infected cell, its indirect and direct effects on viral DNA synthesis ensure a switch from OriP-dependent to OriLyt-dependent replication, and progression through the infectious cycle.

Controlled and Exponential Replication from a Single Origin: the Papillomaviruses

Three different modes of viral DNA replication are associated with papillomavirus infection (Fig. 9.20). Entry of a papillomaviral genome into the nucleus of a host cell initiates a period of amplification of the circular genome, just as during the early stages of latent infection by Epstein-Barr virus. Replication continues until a moderate number of viral genomes (~50 to 100) accumulates in the cell. A maintenance replication pattern, in which the complement of viral episomes is duplicated on average once per cell cycle, is then established. The mechanism that governs the switch from amplification to maintenance replication is not known. In natural papillomavirus infections, these two types of replication take place in the proliferating basal cells of an epithelium. They can also be reproduced in cells in culture transformed by these viruses.

The single viral origin and the viral E1 and E2 proteins that bind to specific origin sequences are necessary for both the initial amplification of the bovine papillomavirus genome and its maintenance for long periods at a more-or-less constant concentration. However, such maintenance

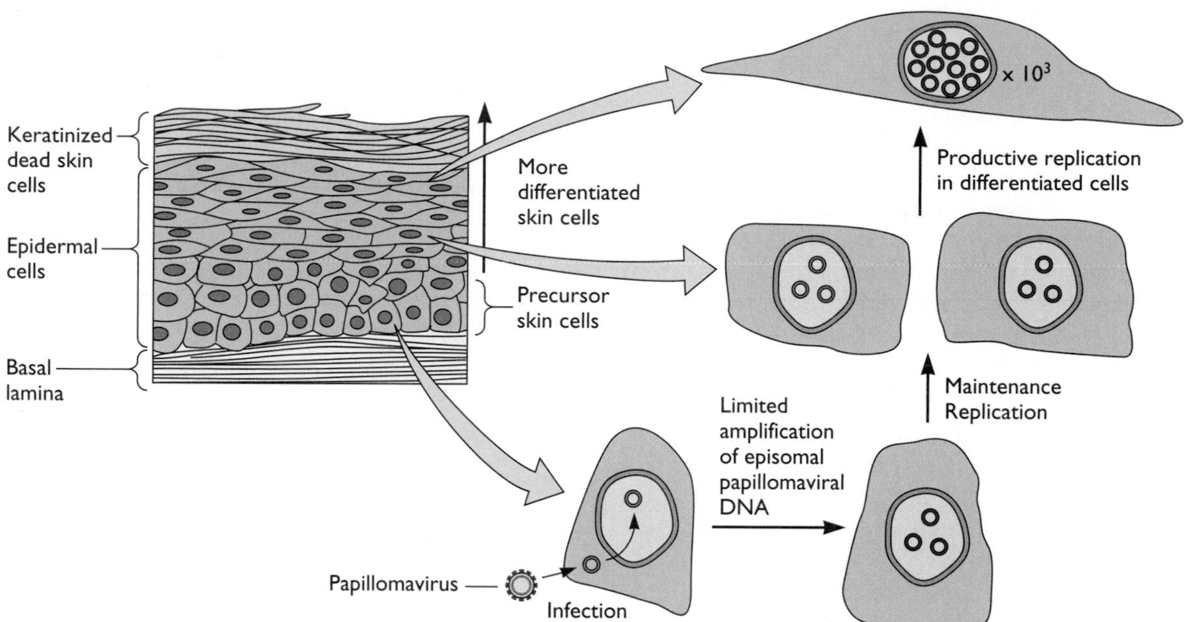

Figure 9.20 Regulation of papillomaviral DNA replication in epithelial cells. The outer layers of the skin are shown as depicted in Fig. 5.3. The virus infects proliferating basal epithelial cells, to which it probably gains access after wounding. The double-stranded, circular viral DNA genome is imported into the infected cell nucleus and initially amplified to a concentration of 50 to 100 copies per cell. This concentration of viral DNA episomes is maintained by further limited replication as the basal and parabasal cells of the epithelium divide **(maintenance replication).** As cells move to the outer layers of the epidermis and differentiate, productive replication of the viral genome to thousands of copies per cell takes place. The mechanism by which differentiation induces the switch from maintenance to productive replication of papillomaviral DNA is not fully understood.

replication does not appear to be the result of strict, once-per-cell-cycle replication of viral DNA. Rather, replication of individual viral episomes appears to occur at random, taking place on average once per cell cycle. Stable maintenance of the viral genome requires an additional sequence, called the **minichromosome maintenance element,** which is composed of multiple binding sites for the E2 protein. The mechanism by which the minichromosome maintenance element acts has not been established, but the most attractive possibility is that, like the family of repeat sequences of Epstein-Barr virus OriP, it promotes orderly partitioning of viral genomes during host cell division.

The final stage of papillomaviral DNA replication—production of high concentrations of the viral genome for assembly into progeny virions—is restricted to differentiated epithelial cells, such as terminally differentiated keratinocytes formed as basal cells move toward the surface of an epithelium (Fig. 9.20). This so-called productive replication is not yet fully understood, because cell culture systems that support this process have been developed only recently. The shift from maintenance to productive replication entails a switch from a simian virus 40-like to a rolling-circle mechanism of replication. The viral E7 protein is necessary for productive replication and induction of synthesis of the cellular replication proteins needed for viral DNA synthesis, such as DNA polymerase α and Pcna, by the mechanism shown in Fig. 9.17A. This protein perturbs the program of keratinocyte differentiation. The viral E5 protein also contributes to maintenance of proliferative competence in infected keratinocytes. The functions of these viral proteins therefore appear to be to establish a cellular environment favorable for productive replication of the papillomaviral genome.

Origins of Genetic Diversity in DNA Viruses

Fidelity of Replication by Viral DNA Polymerases

Proofreading Mechanisms

Cellular DNA replication is a high-fidelity process with an error rate of only about one mistake in every 10^9 base pairs synthesized. Such fidelity, which is essential to maintain the integrity of the genome, is based on the accurate pairing of substrate and template deoxyribonucleotide bases prior to synthesis of each phosphodiester bond.

Nonstandard base pairs can form quite readily. However, initiation of DNA synthesis requires a primer that is base paired perfectly with the template strand, and DNA synthesis does not begin if the terminal nucleotide, or the preceding region of the primer-template, is mismatched. In such circumstances, the mismatched base in the primer strand is excised by a 3′ → 5′ exonuclease present in all replicative DNA polymerases until a perfectly base-paired primer-template is created (Fig. 9.21). Replicative DNA polymerases are therefore self-correcting enzymes, removing errors made in newly synthesized DNA as replication continues (Fig. 9.21).

The cellular DNA polymerases (α, δ, and perhaps ε) that replicate small viral DNA genomes possess such proofreading exonucleases. Infection by these viruses does not result in inhibition of cellular protein synthesis, and indeed may induce expression of cellular replication proteins. The cellular mechanisms of mismatch repair (Fig. 9.22), which correct errors in the daughter strand of newly replicated DNA missed during proofreading by DNA polymerases, are therefore available to operate on progeny viral genomes. The replication of the genomes of these small DNAs viruses is therefore likely to be as accurate as that of the genome of their host cells.

Figure 9.21 Proofreading during DNA synthesis. If permanently fixed into the genome, mispaired bases would result in mutation. However, the majority are removed by the proofreading activity of replicative DNA polymerases. A mismatch at the 3′-OH terminus of the primer-template during DNA synthesis by such an enzyme (step 1) activates the 3′ → 5′ exonuclease activity of all replicative DNA polymerases, which excises the mismatched region to create a perfect duplex (step 2) for further extension (step 3). In the best-characterized case, DNA polymerase I of *E. coli*, the rate of extension from a mismatched nucleotide is much slower than when a correct base pair is formed at the 3′ terminus of the nascent strand. This slow rate of extension allows time for spontaneous unwinding (breathing) of the new duplex region of the DNA and transfer of the 3′ end to the 3′ → 5′ exonuclease site for removal of the mismatched nucleotide. Because preferential excision of mismatched nucleotides is the result of differences in the rate at which the polymerase can add the next nucleotide, this mechanism is called **kinetic proofreading.**

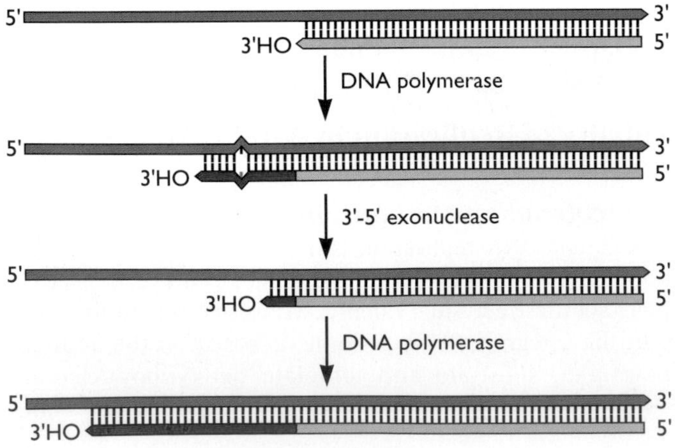

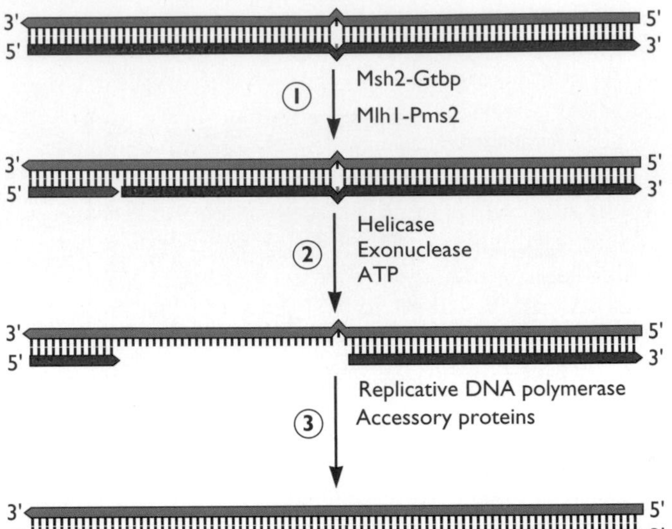

Figure 9.22 Mismatch repair in newly synthesized DNA. This activity requires recognition of the newly synthesized strand containing a misincorporated nucleotide and introduction of a nick (step 1), exonucleolytic degradation of the new strand from the nick as it is unwound (step 2), and resynthesis of DNA to repair the gap (step 3). The long-patch repair system is responsible for removal of mismatches introduced during DNA synthesis in eukaryotic cells, although not all components have been identified. Those listed, the mammalian Msh2-Gtbp and Mlh1-Pms2 heterodimers, are composed of homologs of the well-characterized *E. coli* mismatch repair proteins MutS and MutL, respectively. This relationship suggests that they participate in steps 1 and 2. The MutS homologs bind to mismatched base pairs in duplex DNA, as well as to mismatches resulting from small insertions or deletions with high specificity, and possess weak ATPase and DNA helicase activities. The biochemical functions of the mammalian MutL homologs are not well understood, nor is it yet clear how the newly synthesized DNA strand is distinguished. The importance of this repair system is illustrated by the predisposition to tumor development induced by mutations in the human genes encoding MutS or MutL homologs.

Proofreading by Viral DNA Polymerases

The question of how accurately viral DNA is replicated by viral DNA polymerases, such as those of adenoviruses, herpesviruses, and poxviruses, has received relatively little attention. However, each of these viral DNA polymerases possesses an intrinsic 3′ → 5′ exonuclease that preferentially excises mismatched nucleotides from duplex DNAs in vitro. These viral enzymes contain short, conserved sequences related to those that flank the catalytically active exonuclease residues of well-characterized cellular DNA polymerases. Indeed, recent experiments have shown that mutations that impair the exonuclease activity of the herpes simplex virus type 1 DNA polymerase greatly increase the rate of mutation of the viral genome.

Expression of cellular genes and cellular DNA synthesis are inhibited in cells infected by the larger DNA viruses. It is therefore possible that the cellular mismatch repair proteins that normally back up proofreading by DNA polymerases are not present in the concentrations necessary for effective surveillance and repair of newly synthesized viral DNA. However, neither the concentrations nor activities of these cellular repair proteins have been examined in cells infected by adenoviruses, herpesviruses, or poxviruses. More detailed information about the rates at which viral DNA polymerases introduce errors during DNA synthesis in vitro, and the rates of mutation of viral DNA genomes when replicated in infected cells, would help establish the role of cellular repair systems in maintaining the integrity of these viral genomes. Similarly, the contributions of viral enzymes that could prevent or repair DNA damage, such as the dUTPase and uracil DNA glycosylases of herpesviruses and poxviruses (Table 9.5), remain to be established.

Recombination of Viral Genomes

General Mechanisms of Recombination

Genetic recombination, the rearrangement of DNA sequences within a genome or between two genomes, is an important reaction in the life cycles of many DNA viruses. Much of our understanding of the mechanisms of recombination is based on studies of bacterial viruses, such as bacteriophage lambda. Similar principles apply to recombination of DNA genomes of animal viruses. Two types of recombination are generally recognized: site specific and homologous. In **site-specific recombination,** exchange of DNA takes place at short DNA sequences, specifically recognized by proteins that catalyze recombination. These sequences may be present in only one or both of the DNA sequences that are recombined in this way. Much more common during reproduction of DNA viruses is **homologous recombination,** the exchange of genetic information between **any** pair of related DNA sequences. Such exchange is generally initiated by introduction of a nick into one DNA strand, which then invades the homologous duplex and base pairs with it (Fig. 9.23). These reactions are

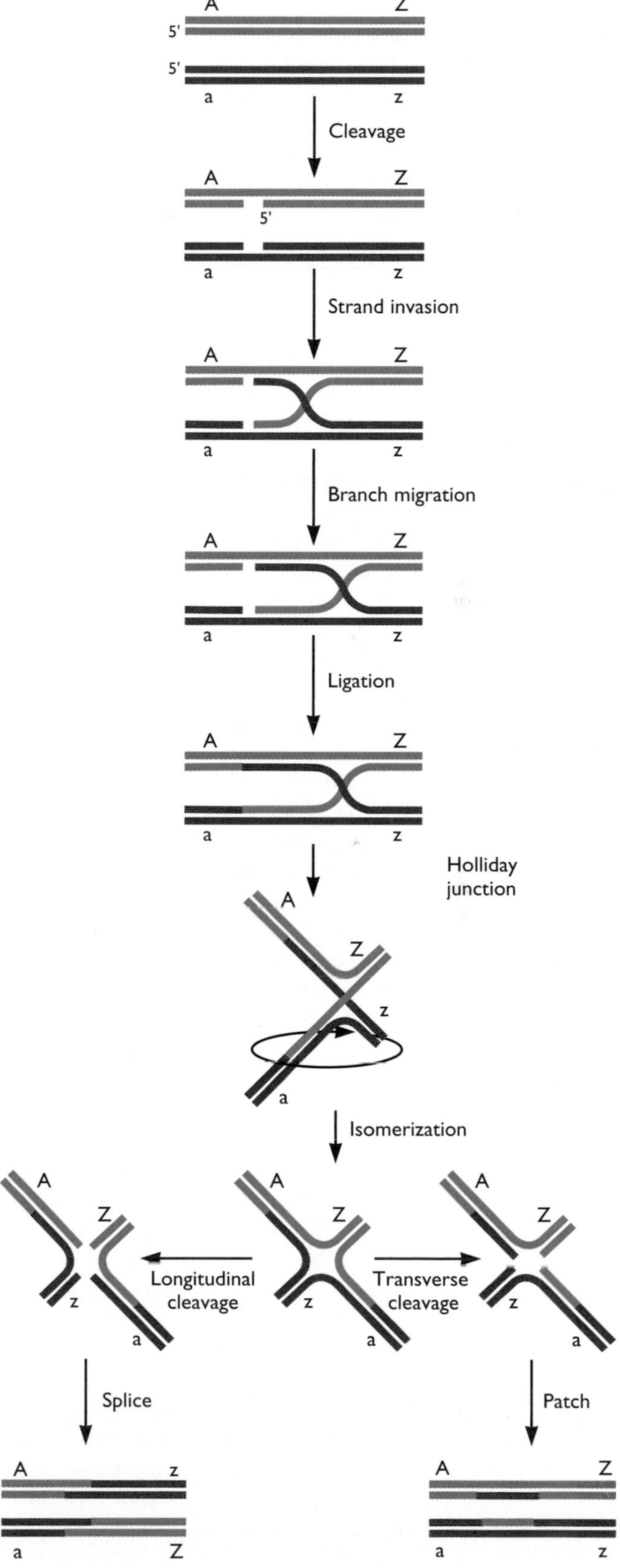

Figure 9.23 General homologous recombination. Formation of a Holliday junction (or cross-strand exchange) on alignment of homologous duplexes is initiated by cleavage in one strand of each duplex, which then invades its partner. These reactions and subsequent branch migration are catalyzed by specialized recombination proteins. Homologous recombination can also be initiated by single-stranded DNA by other mechanisms. For example, if a single strand from only one of the two duplexes invaded its partner, DNA synthesis would be necessary to repair the gap left by the invading strand. As illustrated, the sites at which the Holliday junction is cleaved during resolution following its isomerization determine whether short segments of DNA on one strand are exchanged (patch) or whether the duplexes are recombined (splice).

facilitated by single-stranded DNA-binding proteins, as well as by specialized recombination proteins that bind to both single- and double-stranded DNA to promote the invasion reaction (Fig. 9.23). The specialized proteins then catalyze branch migration to extend the length of the heteroduplex in which the strands from the two DNA duplexes are paired. The initial invasion reactions are followed by cross-strand exchange, a process that can be achieved in several ways. As illustrated in the generic mechanism shown in Fig. 9.23, isomerization of the structure held together by the exchange of two of the four strands of the recombining DNA molecules (termed a **Holliday junction** or **cross-strand exchange**), followed by cleavage of the crossing strands and ligation, results in exchange of DNA between the two original duplexes.

Viral Genome Recombination

The integration of adeno-associated virus DNA into the cellular genome, and its excision when conditions are appropriate, is the result of site-specific recombination reactions mediated by the Rep 78/68 viral proteins, which bind to specific sequences in both DNA molecules. In contrast, integration of retroviral DNA (Chapter 7) is site specific only for the viral DNA.

All viral DNA genomes undergo homologous recombination. These genomes do not generally encode specific recombination proteins, although the herpes simplex virus type 1 alkaline nuclease (Table 9.5) may participate in resolution of branched DNA molecules formed by replication. It is therefore believed that general homologous recombination of viral genomes is catalyzed by host cell enzymes. Nevertheless, viral recombination is favored by the large numbers of viral DNA molecules present in productively infected cells, and the concentration of replicating DNA within specialized replication centers in the nucleus or cytoplasm: the initial step in recombination, pairing of homologous sequences with one another (Fig. 9.23), depends on random collision and is therefore concentration dependent. Furthermore, the structures of replication intermediates, or the nicking of viral DNA during replication or packaging that yields DNA ends, can facilitate recombination among viral DNA molecules. Conversely, recombination may be necessary for productive replication of some viral genomes (those of herpesviruses) or stimulate viral DNA synthesis, for example, from the nicks that initiate recombination. Furthermore, recent evidence indicates that recombination can rescue replication when this process has stalled at unfavorable sequences (or chromosomal sites) in the template. The ease with which viral DNA sequences can recombine is an important factor in the evolution of these viruses. It is also of great benefit to the experimenter, facilitating introduction of specific mutations into the viral genome or construction of viral vectors (see Chapter 2).

Although recombination among animal viral DNA sequences has been widely exploited in the laboratory, the mechanisms have not received much attention. One exception is adenovirus recombination. Recombination between markers in two adenovirus genomes, such as restriction endonuclease sites or temperature-sensitive mutations, exhibits many properties typical of this process, such as the dependence of recombination frequency on the distance between the markers. However, several features suggest that recombination is mechanistically coupled with viral DNA synthesis, because the initial invasion is by the single strands of viral DNA displaced during replication of the adenoviral genome (Fig. 9.10). In particular, this mechanism accounts for the **polarity** of adenovirus recombination, the decreasing gradient of recombination frequency with distance of the recombining sequences from the ends of the viral genome. Another important exception is the homologous recombination of DNA sequences of some herpesviruses, including herpes simplex virus type 1, which is responsible for isomerization of the genome.

Populations of viral DNA molecules purified from herpes simplex virus type 1 virions contain four isomers of the genome, defined by the relative orientations of the two unique sequence segments (L and S) with respect to one another (Fig. 9.11). These unique sequences are flanked by several inverted repeats, including the conserved *a* sequence (Fig. 9.11). The viral DNA population isolated from a single plaque contains all four isomers at equimolar concentrations, suggesting that a single virus particle containing just one genome isomer gives rise to all four by recombination between repeated DNA sequences. It was initially believed that the *a* sequences mediate these recombination reactions by a site-specific mechanism, for example, because insertion of an *a* sequence at an ectopic site induces additional inversions. However, it is now clear that an *a* sequence can act as a hot spot for recombination simply because it contains the viral packaging signal. Double-stranded DNA breaks that promote recombination are made within this sequence during cleavage of replication products for packaging of the viral genome. In fact, the *a* sequences are dispensable for production of all four genome isomers at the normal frequency, and recombination between **any** of the inverted repeat sequences in the viral genome promotes inversion of the L and S segments. Such homologous recombination takes place during viral DNA synthesis and requires the viral replication machinery.

Despite some 25 years of study, the function of the unusual isomerization of the genome of herpes simplex virus type 1 and certain other herpesviruses remains enigmatic. Isomerization is not absolutely essential for virus replication in cells in culture, because viruses "frozen" as a single isomer by deletion of internal inverted repeats are viable. On the other hand, the reduced yield of such viruses and the presence of the inverted repeat sequences in **all** strains of herpes simplex virus type 1 examined emphasize the importance of the repeated sequences. It may be that these sequences themselves fulfill some beneficial function (as yet unknown). Recombinational isomerization would then be a secondary result of the presence of multiple, inverted copies of these sequences in the viral genome. Alternatively, isomerization might be a consequence of an essential role for recombination in replication of the viral genome. The value of the unusual isomerization of the herpes simplex virus type 1 genome may become clearer as understanding of the mechanism of viral DNA synthesis improves.

Perspectives

Our understanding of mammalian replication proteins and the intricate reactions they carry out during DNA synthesis would still be rudimentary were it not for the simian virus 40 origin of replication. This relatively simple viral DNA sequence, which was initially well characterized by genetic methods, supports origin-dependent replication in vitro when cellular proteins are supplemented with a single viral protein, LT. Many origins have been defined in mammalian cell genomes, and some have been characterized in considerable detail, but origin-dependent DNA synthesis from them still cannot be reproduced in experimental systems.

The mechanism of synthesis of this small viral DNA genome has also provided the conceptual framework within which to appraise the considerable diversity in replication of viral DNA genomes. One parameter that varies considerably is the degree of dependence on the host cell's replication machinery. In contrast to those of parvoviruses, papillomaviruses, and polyomaviruses, the genomes of the larger DNA viruses (herpesviruses and poxviruses) encode the components of a complete DNA synthesis system as well as accessory enzymes responsible for the production of dNTP substrates. Nevertheless, replication of **all** viral DNA genomes requires proteins that carry out the reactions first described for simian virus 40 DNA synthesis, notably an origin recognition protein(s), one or more DNA polymerases, proteins that confer processive DNA synthesis, origin-unwinding and helicase proteins, and, usually, proteins that synthesize, or serve as, primers.

Replication of viral DNA genomes ranges from simple, continuous synthesis of both strands of a linear, double-stranded DNA template (adenovirus) to baroque (and not well understood) mechanisms that produce DNA concatemers (herpesviruses). These various replication strategies represent alternative mechanisms for circumventing the inability of all known DNA-dependent DNA polymerases to initiate DNA synthesis de novo. In some cases, initiation of viral DNA synthesis requires RNA primers and the lagging strand is synthesized discontinuously, but in others the priming mechanism leads to continuous synthesis of all daughter DNA strands from protein or DNA sequence primers.

Efficient reproduction of DNA viruses requires the rapid production of very large numbers of progeny viral DNA molecules for assembly of viral particles. One factor contributing to such exponential replication is the efficient production of the proteins that mediate or support DNA synthesis, be they viral or cellular in origin. However, it is also likely that viral DNA replication at specialized intracellular sites, a common feature of cells infected by these viruses, contributes to efficient viral DNA synthesis. Further exploration of this recently recognized and incompletely understood phenomenon should shed new light on host cell biology, in particular the structural and functional compartmentalization of the nucleus. Also far from well understood are the cues that set the stage for the alternative mode of limited replication characteristic of some DNA viruses, when the number of replication cycles and their timing with respect to the host cell cycle are governed by the nature (e.g., alphaherpesviruses) or differentiation state (papillomaviruses) of the host cell. Elucidation of the mechanisms that result in such close integration of viral DNA synthesis with the physiological state of the host cell seems certain to provide important insights into both host cell control mechanisms and the long-term relationships these viruses can establish with their hosts.

References

Books

Kornberg, A., and T. Baker. 1992. *DNA Replication*, 2nd ed. W. H. Freeman and Company, New York, N.Y.

Chapters in Books

Muzyczka, N., and V. I. Berns. 2001. Parvoviridae: the viruses and their replication, p. 2327–2359. *In* D. M. Knipe and P. M. Howley (ed.), *Fields Virology*, 4th ed. Lippincott, Williams and Wilkins, Philadelphia, Pa.

Weller, S. K. 1995. Herpes simplex virus DNA replication and genome maturation, p. 189–213. *In* G. M. Cooper, R. Greenberg Temin, and B. Sugden (ed.), *The DNA Provirus: Howard Temin's Scientific Legacy*. American Society for Microbiology, Washington, D.C.

Reviews

Beaud, G. 1995. Vaccinia virus DNA replication: a short review. *Biochimie* **77**:774–779.

Bell, S. P. 2002. The origin recognition complex: from simple origins to complex functions. *Genes Dev.* **16:**659–672.

Boehmer, P. E., and I. R. Lehman. 1997. Herpes simplex virus DNA replication. *Annu. Rev. Biochem.* **66:**347–384.

Borowiec, J. A., F. B. Dean, P. A. Bullock, and J. Hurwitz. 1990. Binding and unwinding—how T antigen engages the SV40 origin of DNA replication. *Cell* **60:**181–184.

Chow, L. T., and T. R. Broker. 1994. Papillomavirus DNA replication. *Intervirology* **37:**150–158.

Everett, R. D. 2001. DNA viruses and viral proteins that interact with PML nuclear bodies. *Oncogene* **20:**7266–7273.

Fanning, E. 1994. Control of SV40 DNA replication by protein phosphorylation: a model for cellular DNA replication? *Trends Cell. Biol.* **4:**250–255.

Gilbert, D. M. 2001. Making sense of eukaryotic DNA replication origins. *Science* **294:**96–100.

Kolodner, R. D., and G. T. Marsischky. 1999. Eukaryotic DNA mismatch repair. *Curr. Opin. Genet. Dev.* **9:**89–96.

Lindahl, T., and R. D. Wood. 1999. Quality control by DNA repair. *Science* **286:**1897–1905.

Marintcheva, B., and S. K. Weller. 2001. A tale of two HSV-1 helicases: roles of phage and animal virus helicases in DNA replication and recombination. *Prog. Nucleic Acid Res. Mol. Biol.* **70:**77–118.

Quintana, D. G., and A. Dutta. 1999. The metazoan origin recognition complex. *Front. Biosci.* **4:**D805–815.

Sugden, B. 2002. In the beginning: a viral origin exploits the cell. *Trends Biochem. Sci.* **27:**1–3.

Van der Vliet, P. C. 1995. Adenovirus DNA replication. *Curr. Top. Microbiol. Immunol.* **199**(Pt. 2):1–30.

Waga, S., and B. Stillman. 1998. The DNA replication fork in eukaryotic cells. *Annu. Rev. Biochem.* **67:**721–751.

Weinberg, R. A. 1995. The retinoblastoma protein and cell cycle control. *Cell* **81:**323–330.

Papers of Special Interest

Viral Replication Mechanisms and Replication Proteins

Challberg, M. D. 1986. A method for identifying the viral genes required for herpesvirus DNA replication. *Proc. Natl. Acad. Sci. USA* **83:**9094–9098.

Challberg, M. D., S. V. Desidero, and T. J. Kelly. 1980. Adenovirus DNA replication in vitro: characterization of a protein covalently linked to nascent DNA strands. *Proc. Natl. Acad. Sci. USA* **77:**5105–5109.

Danna, K. J., and D. Nathans. 1972. Bi-directional replication of simian virus 40 DNA. *Proc. Natl. Acad. Sci. USA* **69:**3097–3100.

Earl, P. L., E. V. Jones, and B. Moss. 1986. Homology between DNA polymerases of poxviruses, herpesviruses and adenoviruses: nucleotide sequence of the vaccinia virus DNA polymerase gene. *Proc. Natl. Acad. Sci. USA* **83:**3659–3663.

Efstathiou, S., S. Kemp, G. Darby, and A. C. Minson. 1989. The role of herpes simplex virus type 1 thymidine kinase in pathogenesis. *J. Gen. Virol.* **70:**869–879.

Fixman, E. D., G. S. Hayward, and S. D. Hayward. 1995. Replication of Epstein-Barr virus OriLyt: lack of a dedicated virally encoded origin-binding protein and dependence on Zta in co-transfection assays. *J. Virol.* **69:**2998–3006.

Im, D. S., and N. Muzyczka. 1990. The AAV origin-binding protein Rep68 is an ATP-dependent site-specific endonuclease with DNA helicase activity. *Cell* **61:**447–457.

Lechner, R. L., and T. J. Kelly. 1977. The structure of replicating adenovirus 2 DNA molecules. *Cell* **12:**1007–1020.

Li, J. J., and T. J. Kelly. 1984. Simian virus 40 DNA replication in vitro. *Proc. Natl. Acad. Sci. USA* **81:**6973–6977.

Li, J. J., K. W. Peden, R. A. Dixon, and T. J. Kelly. 1986. Functional organization of the simian virus 40 origin of DNA replication. *Mol. Cell. Biol.* **6:**1117–1128.

Rekosh, D., W. C. Russell, A. J. D. Bellett, and A. J. Robinson. 1977. Identification of a protein linked to the ends of adenovirus. *Cell* **11:**283–295.

Sanders, C. M., and A. Stenlund. 1998. Recruitment and loading of the E1 initiator protein: an ATP-dependent process catalysed by a transcription factor. *EMBO J.* **17:**7044–7055.

Stow, N. D. 1982. Localization of an origin of DNA replication within the TRS/IRS repeated region of the herpes simplex virus type 1 genome. *EMBO J.* **1:**863–867.

Straus, S. E., E. D. Sebring, and J. A. Rose. 1976. Concatemers of alternating plus and minus strands are intermediates in adenovirus-associated virus DNA synthesis. *Proc. Natl. Acad. Sci. USA* **73:**742–746.

Waga, S., and B. Stillman. 1994. Anatomy of a DNA replication fork revealed by reconstitution of SV40 DNA replication in vitro. *Nature* **369:**207–212.

Weller, S. K., A. Spadaro, J. E. Schaffer, A. W. Murray, A. M. Maxam, and P. A. Schaffer. 1985. Cloning, sequencing and functional analysis of OriL, a herpes simplex virus type 1 origin of DNA synthesis. *Mol. Cell. Biol.* **5:**930–942.

Control of Viral DNA Replication

Burkham J., D. M. Coen, C. B. Hwang, and S. K. Weller. 2001. Interactions of herpes simplex virus type 1 with ND10 and recruitment of PML to replication compartments. *J. Virol.* **75:**2353–2367.

Deng, Z., L. Lezina, C. J. Chen, S. Shtivelband, W. So, and P. M. Lieberman. 2002. Telomeric proteins regulate episomal maintenance of Epstein-Barr virus origin of plasmid replication. *Mol. Cell* **9:**493–503.

Dhar, S. K., K. Yoshida, Y. Machida, P. Khaira, B. Chaudhuri, J. A. Wohlschlegel, M. Leffak, J. Yates, and A. Dutta. 2001. Replication from oriP of Epstein-Barr virus requires human ORC and is inhibited by geminin. *Cell* **106:**287–296.

Doucas, V., A. M. Ishov, A. Romo, H. Juguilon, M. D. Weitzman, R. M. Evans, and G. G. Maul. 1996. Adenovirus replication is coupled with the dynamic properties of the PML nuclear structure. *Genes Dev.* **10:**194–207.

Flores, E. R., and P. F. Lambert. 1997. Evidence for a switch in the mode of human papillomavirus type 11 DNA replication during the viral life cycle. *J. Virol.* **71:**7167–7179.

Gilbert, D. M., and S. N. Cohen. 1987. Bovine papilloma virus plasmids replicate randomly in mouse fibroblasts throughout S phase of the cell cycle. *Cell* **50:**59–68.

Kotin, R. M., M. Siniscalco, R. J. Samulski, X. D. Zhu, L. Hunter, C. A. Laughlin, S. McLaughlin, N. Muzyczka, M. Rocchi, and K. I. Berns. 1990. Site-specific integration by adeno-associated virus. *Proc. Natl. Acad. Sci. USA* **87:**2211–2215.

Reisman, D., J. L. Yates, and B. Sugden. 1985. A putative origin of replication of plasmids derived from Epstein-Barr virus is composed of two cis-acting components. *Mol. Cell. Biol.* **5:**1822–1832.

Spindler, K. R., C. Y. Eng, and A. J. Berk. 1985. An adenovirus early region 1A protein is required for maximal viral DNA replication in growth-arrested human cells. *J. Virol.* **53:**742–750.

Whyte, P., J. J. Buchkovich, J. M. Horowitz, S. H. Friend, M. Raybuck, R. A. Weinberg, and E. Harlow. 1988. Association between an oncogene and an anti-oncogene: the adenovirus E1A proteins bind to the retinoblastoma gene product. *Nature* **334:**124–129.

Repair and Recombination of Viral DNA

Hayward, G. S., R. J. Jacob, R. C. Wadsworth, and B. Roizman. 1975. Anatomy of herpes simplex virus DNA: evidence for four populations of molecules that differ in the relative orientations of their long and short segments. *Proc. Natl. Acad. Sci. USA* **72:**4243–4247.

Hwang, Y. T., B.-Y. Liu, D. M. Coen, and C. B. C. Hwang. 1997. Effects of mutations in the ExoIII motif of the herpes simplex virus DNA polymerase gene on enzyme activities, viral replication and replication fidelity. *J. Virol.* **71:**7791–7798.

Linden, R. M., E. Winocour, and K. I. Berns. 1996. The recombination signals for adeno-associated virus site-specific integration. *Proc. Natl. Acad. Sci. USA* **93:**7966–7972.

Martin, D. W., and P. C. Weber. 1996. The *a* sequence is dispensable for isomerization of the herpes simplex virus type 1 genome. *J. Virol.* **70:**8801–8812.

Mocarski, E. S., L. E. Post, and B. Roizman. 1980. Molecular engineering of the herpes simplex virus genome: insertion of a second LS junction causes additional genome inversions. *Cell* **22:**243–245.

Stracker, T. H., C. T. Carson, and M. D. Weitzman. 2002. Adenovirus oncoproteins inactivate the Mre11-Rad50-NBS1 DNA repair complex. *Nature* **418:**348–352.

Weitzman, M. D., S. R. M. Kyöstiö, R. M. Kotin, and R. A. Owens. 1994. Adeno-associated virus (AAV) rep proteins mediate complex formation between AAV DNA and its integration site in human DNA. *Proc. Natl. Acad. Sci. USA* **91:**5808–5812.

Processing of Viral Pre-mRNA

All is flux, nothing stays still.
HERACLITUS

Introduction

Viral mRNAs are synthesized by either viral or cellular enzymes and may be made in the nucleus or the cytoplasm of an infected cell. Regardless of how and where they are made, all viral mRNAs must be translated by the protein-synthesizing machinery of the host cell. Consequently, they must conform to the requirements of the cell's translational system. A series of covalent modifications, collectively known as **RNA processing** (Fig. 10.1), endow mRNAs with the molecular features needed for recognition by the protein synthesis machinery and translation of the protein-coding sequence by cellular ribosomes. Most RNA-processing reactions were discovered in viral systems, primarily because virus-infected cells provide large quantities of specific mRNAs for analysis.

Two modifications important for efficient translation are the addition of m^7GpppN to the 5′ end (**capping**) and the addition of multiple A residues to the 3′ end (**polyadenylation**) (Fig. 10.1). The enzymes that perform these chemical additions (Box 10.1) may be encoded by viral or cellular genes. When an RNA is produced in the nucleus, another chemical rearrangement, called **splicing**, is possible. During splicing, short blocks of noncontiguous coding sequences (**exons**) are joined precisely to create a complete protein-coding sequence for translation, while the intervening sequences (**introns**) are discarded (Fig. 10.1). Splicing therefore dramatically alters the precursor mRNA (**pre-mRNA**) initially synthesized. As no viral genome encodes even part of the complex machinery needed to catalyze splicing reactions, splicing of viral pre-mRNAs is accomplished entirely by cellular gene products. Some viral pre-mRNAs undergo a different type of internal chemical change, in which a single nucleotide is replaced by another, or one or more nucleotides are inserted at specific positions (Box 10.1). Such **RNA-editing** reactions introduce nucleotides that are not encoded in the genome, and therefore may change the sequence of the encoded protein.

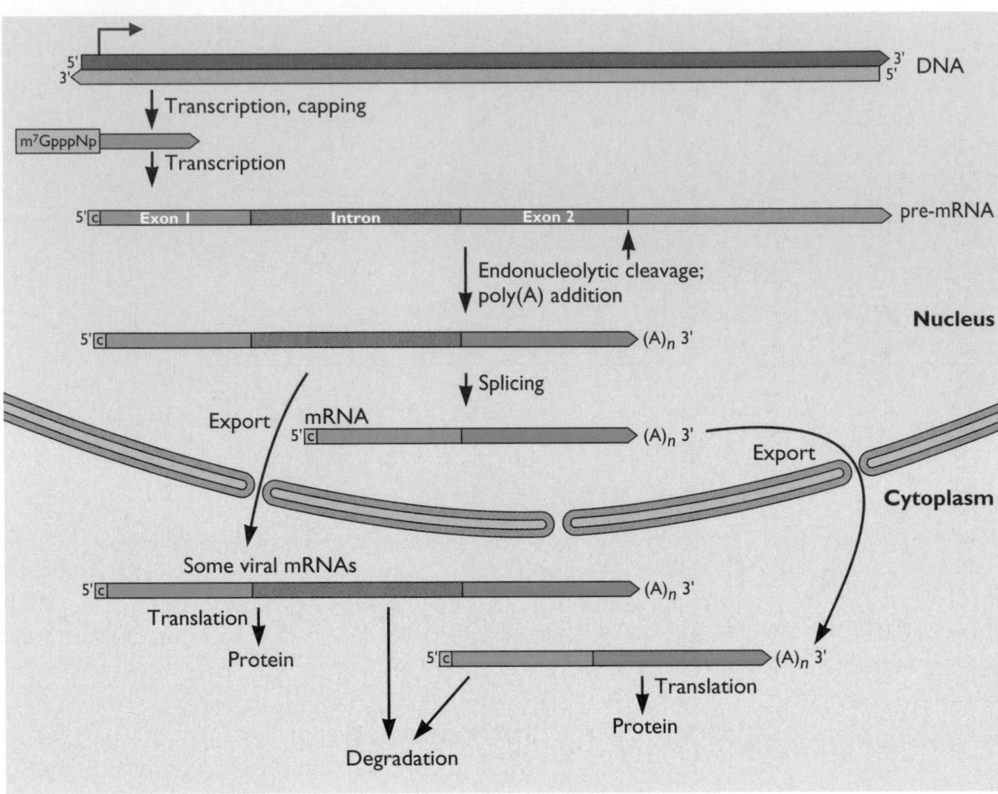

Figure 10.1 Processing of a viral or cellular mRNA synthesized by RNA polymerase II. The reactions by which mature mRNA is made from a typical RNA polymerase II transcript are shown. The first such reaction, capping, takes place cotranscriptionally. For clarity, the exons of a hypothetical, partially processed (i.e., polyadenylated but unspliced) pre-mRNA are depicted, even though polyadenylation and splicing are often coupled, and in long primary transcripts, splicing of cap-proximal exons occurs before the sequences encoding 3′ exons and polyadenylation signals have been transcribed. Consequently, many splicing reactions are cotranscriptional. Most cellular and viral pre-mRNAs synthesized by RNA polymerase II are processed by this pathway. However, some viral mRNAs that are polyadenylated but not spliced, or are incompletely spliced, are exported to the cytoplasm.

When a viral pre-mRNA is produced and processed in the nucleus, it must be exported to the cytoplasm for translation (Fig. 10.1). Such export of mature viral and cellular mRNAs is considered to be part of mRNA processing, even though the RNA is not known to undergo any chemical change during transport. Viral mRNAs invariably leave the nucleus by cellular pathways, but nuclear export mechanisms may be altered in activity or specificity in cells infected by some viruses. Once within the cytoplasm, an mRNA has a finite lifetime before it is recognized and degraded by ribonucleases. The susceptibilities of individual mRNA species to attack by these destructive enzymes vary greatly, and can be modified in virus-infected cells.

BOX 10.1

Types of change during RNA processing

Addition of sequences
- 5′ caps
- 3′ poly(A) tails
- One or a few internal nucleotides by editing

Removal of sequences
- Introns during splicing

Substitution of sequences
- Editing

Relocation
- Export from nucleus to cytoplasm
- Transport within the cytoplasm

The RNA-processing reactions summarized in Fig. 10.1 not only produce functional mRNAs but also provide numerous opportunities for posttranscriptional control of gene expression. Regulation of RNA processing can increase the coding capacity of the viral genome, determine when specific viral proteins are made during the infectious cycle, and facilitate selective expression of viral genetic information. In this chapter, we focus on these RNA-processing reactions to illustrate both crucial viral regulatory mechanisms and the seminal contributions of viral systems to the elucidation of essential cellular processes.

Covalent Modification during Viral Pre-mRNA Processing

Capping the 5′ Ends of Viral mRNA

The first mRNAs shown to carry the 5′-terminal structure termed the **cap** were those of reovirus and vaccinia virus (Box 10.2). These viral mRNAs are made and processed by virus-encoded enzymes (see "Synthesis of Viral 5′ Cap Structures by Viral Enzymes" below), but sub-

BOX 10.2

Identification of 5′ cap structures and 3′ poly(A) sequences on viral mRNAs

The first clues that the termini of mRNAs made in eukaryotic cells possess special structures came when viral mRNAs did not behave as predicted from the known structure of prokaryotic mRNAs.

In the figure, panel A shows identification of 5′ cap structures. The 5′ end of reoviral (or vaccinia virus) mRNA, in contrast to that of a prokaryotic mRNA, could not be labeled by polynucleotide kinase and [γ-^{32}P]ATP after alkaline phosphatase treatment. This property established that the 5′ end did not carry a simple phosphate group, but rather was blocked. As indicated below, the structure of the 5′ blocking group (termed the **cap**) was elucidated by differential labeling of specific groups of the viral mRNA as indicated by colors in the figure, followed by digestion of the mRNA with nucleases with different specificities. In addition to the expected product, 5′pN$_{OH}$ 3′, cleavage with P1 nuclease liberated a structure that corresponded to the 5′ terminus of the mRNA (the only position at which β or γ phosphates from NTPs are retained in an mRNA). This structure also contained all the methyl-^{3}H label, indicating that it contained methyl groups. Its digestion with nucleotide pyrophosphatase produced free phosphate with all the ^{32}P label and two methyl-^{3}H-labeled products. Because the terminal structure was known to contain G (from incorporation of ^{32}P from [β,γ^{32}P]GTP), these nucleotides were identified as m^{7}Gp and 2′-*O*-methylguanosine monophosphate (i.e., methylated at the 2′ position of the sugar ring) by comigration with these compounds. Panel B shows identification of 3′ poly(A) sequences. Polyadenylation of

(continued)

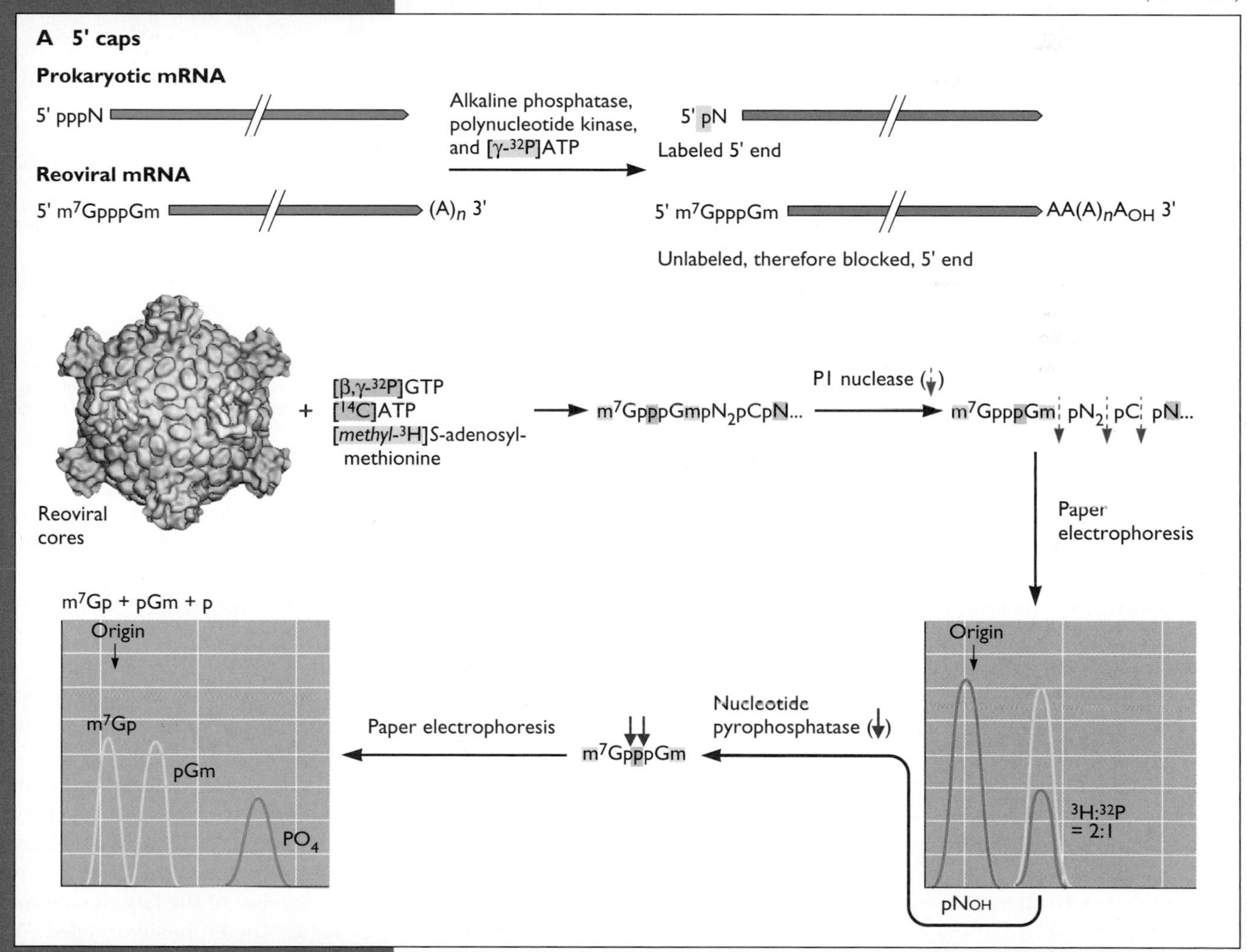

BOX 10.2 *(continued)*

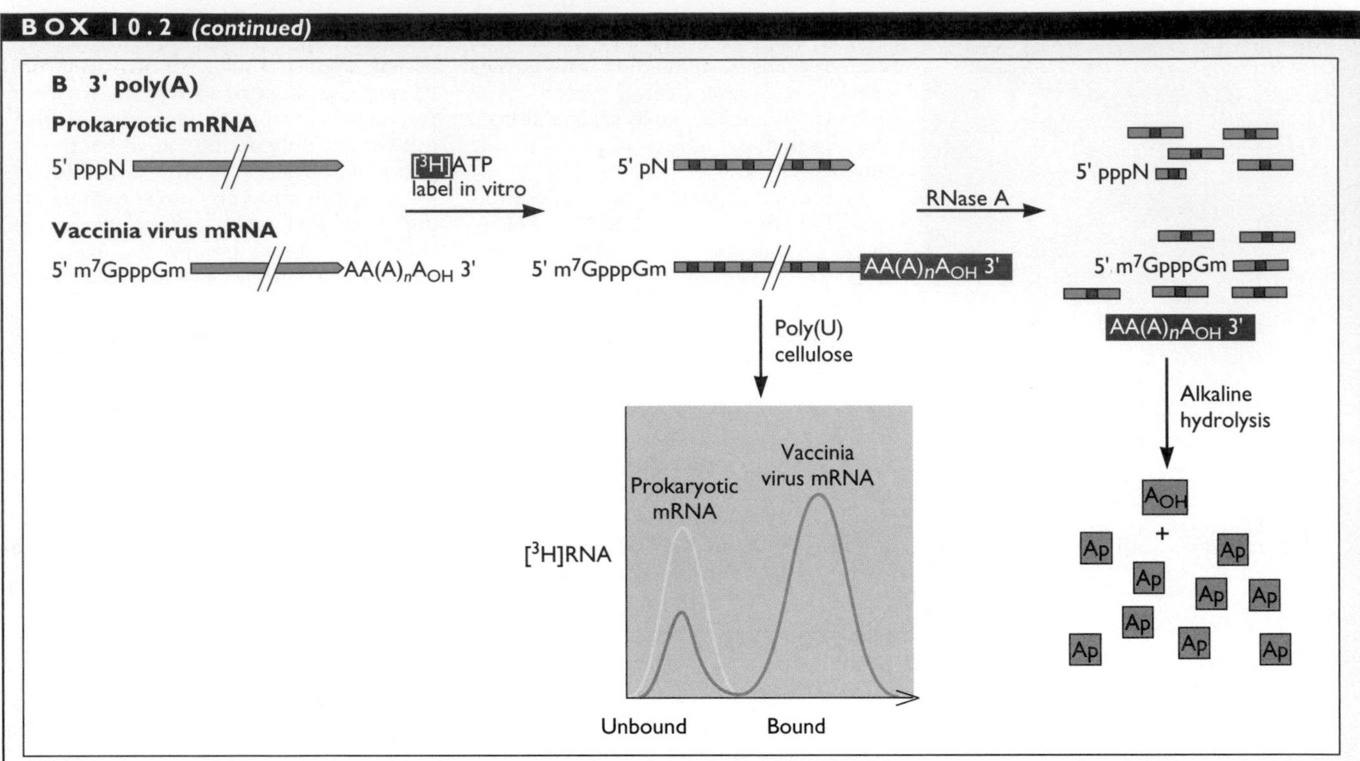

viral mRNAs was first indicated by the observation that an RNA chain resistant to digestion by RNase A, which cleaves after U and C, was produced when vaccinia virus mRNA, but not prokaryotic mRNA, was treated with this enzyme following labeling with [^{3}H]ATP. The presence of a tract of poly(A) was confirmed by the specific binding of vaccinia virus mRNA, but not of prokaryotic mRNA, to poly(U)-Sepharose under conditions that allowed annealing of complementary nucleic acids. The position of the poly(A) sequence in viral mRNA was determined by analysis of the products of alkaline hydrolysis, when phosphodiester bonds are broken to produce nucleotides with 5′ hydroxyl and 3′ phosphate (Ap) groups. The liberation of A_{OH} 3′ by this treatment indicated that the poly(A) was located at the 3′ end of the mRNA.

Furuichi, Y., M. Morgan, S. Muthukrishnan, and A. J. Shatkin. 1975. Reovirus messenger RNA contains a methylated, blocked 5′-terminal structure: m^{7}G(5′)ppp(5′)GpCp. *Proc. Natl. Acad. Sci. USA* **72:**362–366.

Kates, J. 1970. Transcription of the vaccinia virus genome and occurrence of polyriboadenylic acid sequences in messenger RNA. *Cold Spring Harbor Symp. Quant. Biol.* **35:**743–752.

Miura, K., K. Watanabe, M. Sugiura, and A. J. Shatkin. 1974. The 5′-terminal nucleotide sequences of the double-stranded RNA of human reovirus. *Proc. Natl. Acad. Sci. USA* **71:**3979–3983.

sequent research established that virtually all viral and cellular mRNAs possess the same cap structure, m^{7}GpppN, where N = any nucleotide. This structure protects mRNAs from 5′ exonucleolytic attack and allows recognition during splicing of the 5′-terminal exons of pre-mRNAs made by cellular RNA polymerase II. In addition, the cap structure is essential for the efficient translation of most viral and cellular mRNAs, for it is recognized by proteins that mediate initiation of translation. The principal exceptions are the uncapped mRNAs of certain (+) strand viruses, notably picornaviruses and the flavivirus hepatitis C virus, which are translated by the cap-independent mechanism described in Chapter 11.

Although most other viral mRNAs carry a 5′-terminal cap structure, there is considerable variation in how this modification is made. Three mechanisms can be distinguished: de novo synthesis by cellular enzymes, synthesis by viral enzymes, and acquisition of preformed 5′ cap structures from cellular pre-mRNA or mRNA (Table 10.1).

Synthesis of Viral 5′ Cap Structures by Cellular Enzymes

Viral pre-mRNA substrates for the cellular capping enzyme described below are invariably made in the infected cell nucleus by cellular RNA polymerase II (Table 10.1). The formation of cap structures on the 5′ ends of viral and cellular RNA polymerase II transcripts, the first step in their processing, is a cotranscriptional reaction that takes place when the nascent RNA is only 20 to 30 nucleotides in length. At this early stage, the transcriptional complex pauses and the C-terminal domain of the largest subunit of RNA polymerase II becomes hyperphosphorylated. This modification is the signal for binding of the cellular capping enzyme to this unique domain of RNA polymerase II and

Table 10.1 Mechanisms of synthesis of 5′-terminal cap structures of viral mRNAs

Mechanism	Virus family	Enzyme synthesizing pre-mRNA
Synthesis by host cell enzymes	*Adenoviridae, Hepadnaviridae, Herpesviridae, Papillomaviridae, Parvoviridae, Polyomaviridae, Retroviridae*	Cellular DNA-dependent RNA polymerase II
Synthesis by viral enzymes	*Reoviridae, Rhabdoviridae, Togaviridae*	Viral RNA-dependent RNA polymerase
	Poxviridae	Viral DNA-dependent RNA polymerase
Acquisition from cellular pre-mRNA or mRNA	*Bunyaviridae, Orthomyxoviridae*	Viral RNA-dependent RNA polymerase

capping of the nascent RNA. The intimate relationship between the cellular capping enzyme and RNA polymerase II ensures that all transcripts made by this enzyme are capped at their 5′ ends. It also explains why RNA species synthesized by other cellular RNA polymerases are not modified in this way: these enzymes do not carry an equivalent of the C-terminal domain of RNA polymerase II.

The 5′ cap structure is assembled by the action of several enzymes (Fig. 10.2). The 5′ ends of nascent RNA polymerase II transcripts carry a triphosphate group from the

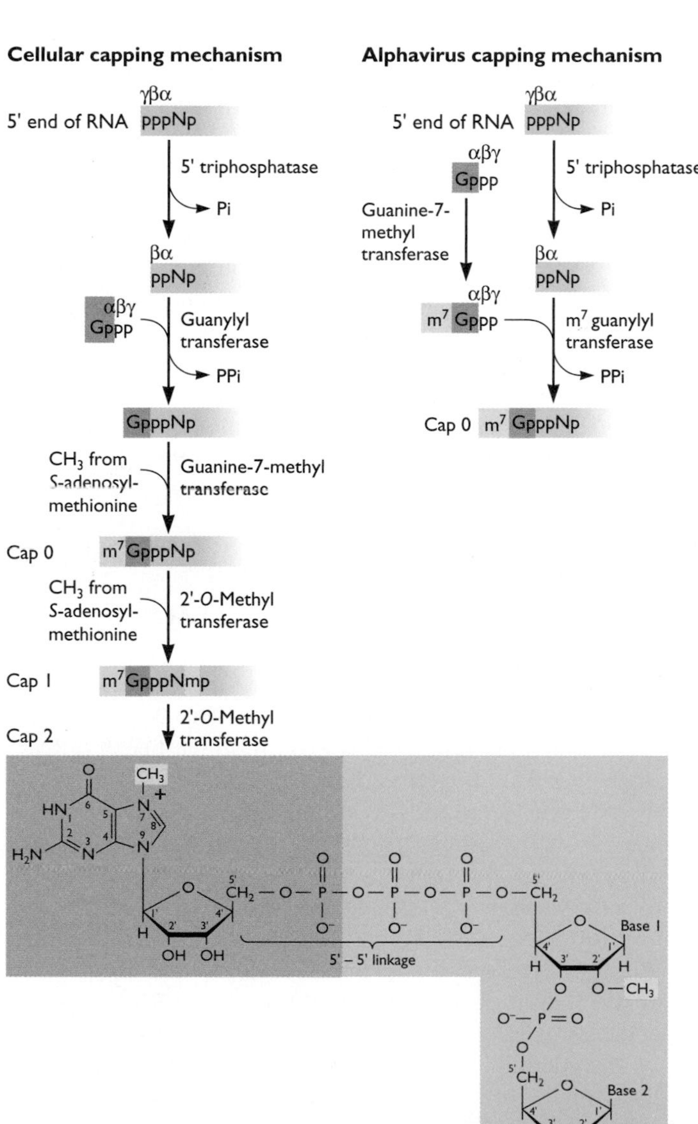

Figure 10.2 The 5′ cap structure and its synthesis by cellular or viral enzymes. The cap structure shown in full, cap 2, in which the sugars of the two transcribed nucleotides adjacent to the terminal m^{7}G contain 2′-*O*-methyl groups, is found in higher eukaryotes. The first and second nucleotides synthesized are methylated in the nucleus and in the cytoplasm, respectively. The enzymes and reactions by which this cap is synthesized by cellular enzymes (left) are compared with the synthesis of the caps of Semliki Forest virus (a togavirus) mRNAs by viral enzymes in the cytoplasm of infected cells (right).

initiating nucleotide. The terminal phosphate is removed by a 5′ triphosphatase to form the substrate for a guanylyltransferase. This enzyme transfers GMP from GTP to the 5′ end of the nascent transcript to form the protective GpppN structure (Fig. 10.2), which is resistant to nucleases that cleave 5′-3′ phosphodiester bonds. In mammalian cells, a single protein, commonly called **capping enzyme,** contains both the 5′ triphosphatase and the guanylyltransferase. Following the action of capping enzyme, the terminal G residue is modified by methylation at specific positions (Fig. 10.2). The cap 1 structure, m^7GpppNm, is common in viral and mammalian mRNAs. However, the sugar of the second nucleotide can also be methylated by a cytoplasmic enzyme, to form the cap 2 structure (Fig. 10.2).

Synthesis of Viral 5′ Cap Structures by Viral Enzymes

When viral mRNAs are made in the cytoplasm of infected cells, their 5′ cap structures are, of necessity, synthesized by viral enzymes. These enzymes form cap structures typical of those present on cellular mRNA, although with some variations in the mechanism of cap construction. For example, during synthesis of the caps of vesicular stomatitis virus mRNAs, only the α-phosphate group of the initiating nucleotide is retained and GDP is added. The capping enzymes of alphaviruses (members of the *Togaviridae*) methylate the N^7 position of GTP prior to transfer of GMP to the 5′ end of viral pre-mRNA, and the resulting cap structure is not modified further (Fig. 10.2). Alphaviral mRNAs therefore carry the cap 0 structure, which is rarely found on mRNAs made in mammalian cells.

Like their cellular counterparts, viral capping enzymes are intimately associated with the viral RNA polymerases responsible for mRNA synthesis. Indeed, in the simplest case, the several enzymatic activities required for synthesis of a 5′ cap structure are supplied by the viral RNA polymerase. This mechanism is exemplified by the vesicular stomatitis virus L protein, which carries out all reactions necessary for synthesis and capping of viral mRNAs. Viruses with more complex genomes encode dedicated capping enzymes, such as the λ-2 protein of reovirus particles. This protein, which is the guanylyltransferase and probably also the *N*-methyltransferase, interacts with the viral RNA-dependent RNA polymerase in the core of the virion. One of the first capping enzymes to be analyzed in detail was the vaccinia virus enzyme. This protein binds directly to the viral RNA polymerase described in Chapter 8 and adds 5′ cap structures cotranscriptionally to nascent viral transcripts that are approximately 30 nucleotides in length. The capping enzyme of vaccinia virus therefore displays striking functional similarities to its host cell counterpart, and has provided an important model for the cellular capping machinery and its mechanism of action.

With the exception of that of vaccinia virus, viral capping enzymes cooperate with viral RNA-dependent RNA polymerases that can synthesize both (–) and (+) strand RNAs, but only (+) strand RNAs become capped. The activities of these enzymes must therefore be regulated. The mechanisms that coordinate capping activity with viral mRNA synthesis are not well understood. However, in the case of alphaviruses, such as Sindbis and Semliki Forest viruses, activation of capping enzymes may be the result of proteolytic processing. As discussed in Chapter 6, the viral P1234 polyprotein is responsible for the initial synthesis of (–) strand RNA from the (+) stand viral genome. This polyprotein includes the sequences of the RNA polymerase and the capping enzyme (Nsp2 and Nsp1, respectively), but the latter is inactive. Cleavage of the polyprotein is necessary for synthesis of viral mRNAs (see Fig. 6.15), and this processing reaction may also activate the capping enzyme.

Acquisition of Viral 5′ Cap Structures from Cellular RNAs

The 5′ cap structures of orthomyxoviral and bunyaviral mRNAs are produced by cellular capping enzymes, but in a unique manner. The 5′ caps of these viral mRNAs are acquired when viral, cap-dependent endonucleases cleave cellular transcripts to produce the primers needed for viral mRNA synthesis, a process called **cap snatching** (see Fig. 6.9). The 5′-terminal segments and caps of influenza virus mRNAs are obtained from cellular pre-mRNA in the nucleus. On the other hand, bunyaviral mRNA synthesis is primed with 5′-terminal fragments cleaved from mature cellular mRNAs in the cytoplasm.

Synthesis of 3′ Poly(A) Segments of Viral mRNA

Like the 5′ cap structure, a 3′ poly(A) segment was first identified in a viral mRNA (Box 10.2). This 3′-end modification was soon found to be a common feature of mRNAs made in eukaryotic cells. Like the 5′ cap, the 3′ poly(A) sequence stabilizes mRNA, and it also increases the efficiency of translation. Therefore, it is not surprising that viral mRNAs generally carry a 3′ poly(A) tail. Those that do not, such as reoviral and arenaviral mRNAs, may survive by virtue of a 3′-terminal stem-loop structure that blocks nucleolytic attack. Such structures are also present at the 3′ ends of cellular, poly(A)-lacking mRNAs that encode histones.

The addition of 3′ poly(A) segments to viral pre-mRNAs, like capping of their 5′ ends, can be carried out by either cellular or viral enzymes (Table 10.2). However, cellular and viral polyadenylation mechanisms differ markedly.

Table 10.2 Mechanisms of addition of poly(A) to viral RNAs

Mechanism	Enzyme	Viruses
During mRNA synthesis		
Reiterative copying of short U stretches in the template RNA	Viral	Orthomyxoviruses, paramyxoviruses, rhabdoviruses
Copying of a long 5′-terminal U stretch in the template RNA	Viral	Picornaviruses, togaviruses
Posttranscriptional		
Cleavage of pre-mRNA followed by polyadenylation	Cellular	Adenovirus, hepadnaviruses, hepatitis delta satellite virus, herpesviruses, polyomaviruses, retroviruses
Transcription termination followed by polyadenylation	Viral	Poxviruses

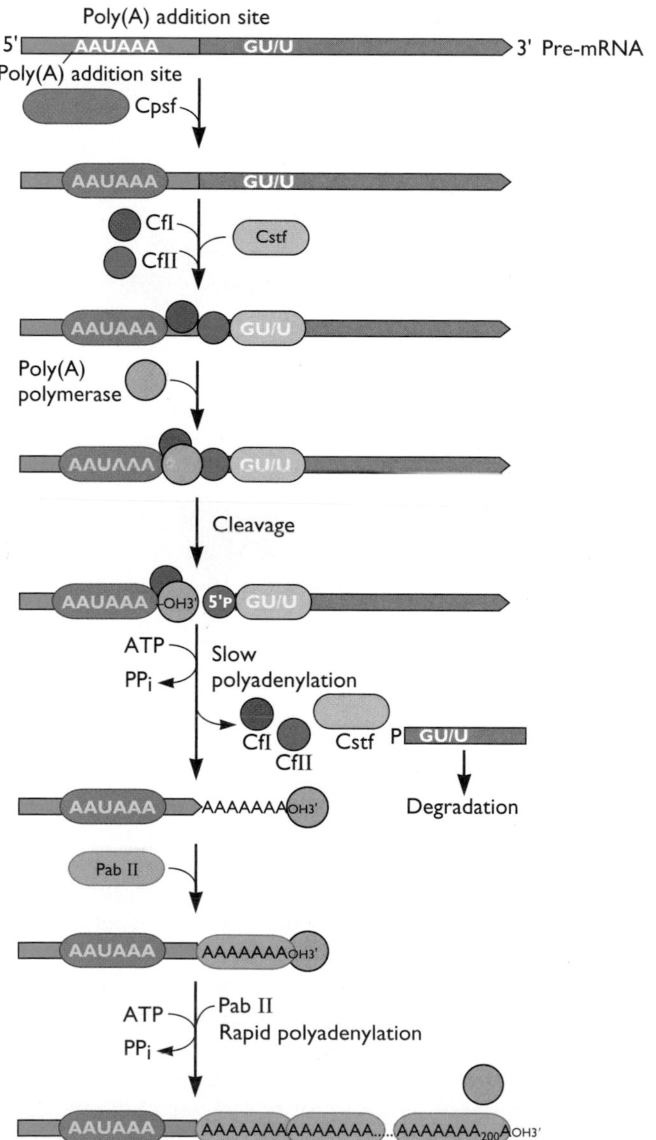

Figure 10.3 Cleavage and polyadenylation of vertebrate pre-mRNAs. The cleavage and polyadenylation specificity protein (Cpsf), which contains four subunits, binds to the 5′AAUAAA3′ poly(A) addition signal that lies upstream of the site at which poly(A) will be added. Cleavage-stimulatory protein (Cstf) then interacts with the downstream U/GU-rich sequence to stabilize a complex that also contains the two cleavage proteins CfI and CfII. Binding of poly(A) polymerase is followed by cleavage at the poly(A) addition site, and CfI, CfII, Cstf, and the downstream RNA cleavage product are then released. The enzyme slowly adds 10 to 20 A residues to the 3′-OH terminus produced by the cleavage reaction. Poly(A)-binding protein II (Pab II) then binds to this short poly(A) sequence and, in conjunction with Cpsf, tethers poly(A) polymerase to the poly(A) sequence. This association facilitates rapid and processive addition of A residues until a poly(A) chain of about 200 residues has been synthesized.

Polyadenylation of Viral Pre-mRNAs by Cellular Enzymes

Viral pre-mRNAs synthesized in infected cell nuclei by RNA polymerase II are invariably polyadenylated by cellular enzymes (Table 10.2). Transcription of a viral or cellular gene by RNA polymerase II proceeds beyond the site at which poly(A) will be added. The 3′ end of the mRNA is determined by endonucleolytic cleavage of its pre-mRNA at a specific position. Poly(A) is then added to the new 3′ terminus, while the RNA downstream of the cleavage site is degraded (Fig. 10.3). Cleavage and polyadenylation sites are identified by specific sequences, first characterized in simian virus 40 and adenoviral pre-mRNAs. The 3′ end of mature mRNA is formed 10 to 30 nucleotides downstream of a highly conserved and essential polyadenylation signal, the sequence 5′AAUAAA3′. However, this sequence is not sufficient to specify poly(A) addition. For example, it is found within mRNAs at internal positions that are never used as polyadenylation sites. Sequences at the 3′ side of the cleavage site, notably a U- or GU-rich sequence located 5 to 20 nucleotides downstream, are also required. In

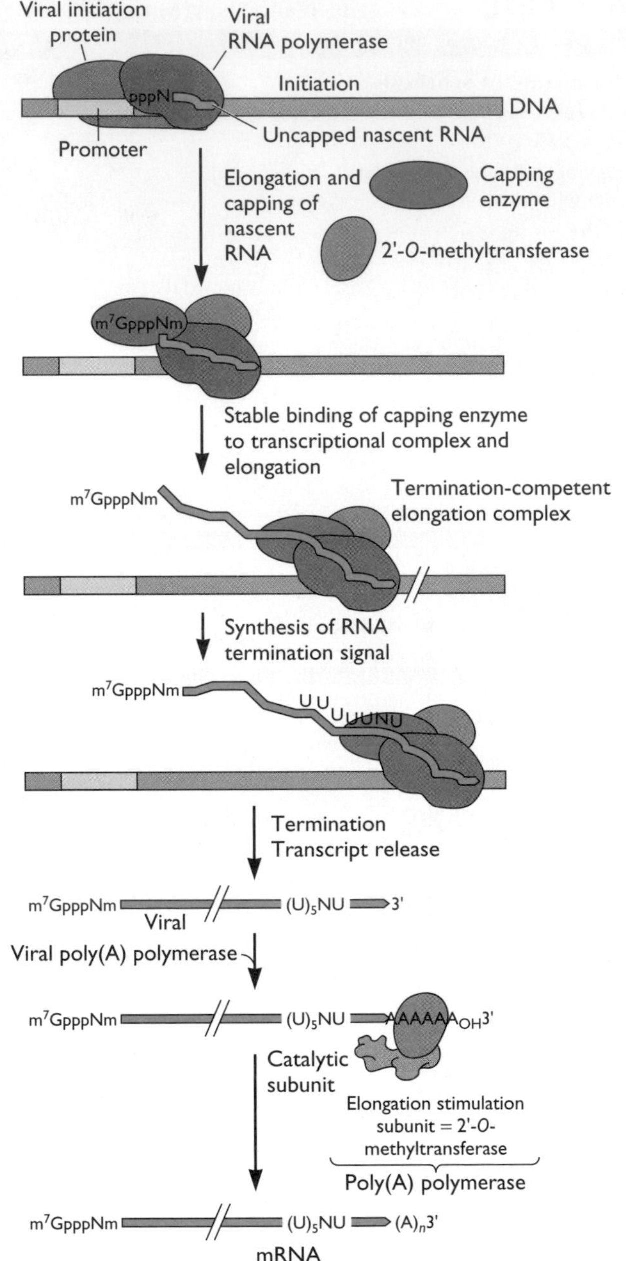

Figure 10.4 The vaccinia virus capping enzyme and 2′-O-methyltransferase process both the 5′ and 3′ ends of vaccinia virus mRNAs. After capping the 5′ ends of nascent viral mRNA chains about 30 nucleotides in length, the capping enzyme remains bound to the nascent RNA chain and to the RNA polymerase as the latter enzyme transcribes the template DNA. The viral 2′-*O*-methytransferase, which produces a cap 1 structure, also binds to the viral RNA polymerase and stimulates elongation during transcription of viral intermediate and late genes. This protein is also a subunit of the viral poly(A) polymerase. Termination of transcription, which takes place 30 to 50 nucleotides downstream of the RNA sequence 5′UUUUUNU3′, is mediated by the termination protein/capping enzyme and the viral nucleoside triphosphate phosphohydrolase I, which is a single-stranded DNA-dependent ATPase. A fraction of the 2′-*O*-methyltransferase molecules act as an elongation stimulation protein for the viral poly(A) polymerase, analogous to cellular poly(A)-binding protein II. This viral enzyme, like its cellular counterpart (Fig. 10.3), adds poly(A) to the 3′ ends of the mRNA in a two-step process.

many mRNAs (particularly viral mRNAs) additional sequences 5′ to the cleavage site are also important.

The first reaction in polyadenylation is recognition of the 5′AAUAAA3′ sequence by the protein termed Cpsf (cleavage and polyadenylation specificity protein) (Fig. 10.3). This interaction is stabilized by other proteins, including the cleavage-stimulatory protein (Cstf), which binds to the downstream U/GU-rich sequence. Poly(A) polymerase is then recruited to the complex prior to cleavage of the pre-mRNA. The enzyme initially adds A residues at a low rate to synthesize a short poly(A) segment of 10 to 20 residues. This structure is then bound by poly(A)-binding protein, which induces rapid and processive addition of more A residues until a long tail, typically 200 to 250 nucleotides, has been made (Fig. 10.3).

Components of the polyadenylation machinery, like capping enzymes, are associated with RNA polymerase II, and the polymerase is required for efficient polyadenylation in in vitro reactions. Both Cpsf and Cstf bind to the C-terminal domain of the largest subunit of RNA polymerase II. Such interactions are essential for polyadenylation in vivo, and are believed to coordinate synthesis of pre-mRNAs with maturation of the 3′ ends. They may also be required for termination of transcription by RNA polymerase II: polyadenylation signals are necessary for this process, even though the enzyme can continue synthesis for long distances beyond sites of polyadenylation.

Polyadenylation of Viral Pre-mRNAs by Viral Enzymes

Formation of 3′ ends by termination of transcription. The 3′ ends of vaccinia virus early mRNAs are formed by termination of transcription by the viral DNA-dependent RNA polymerase at specific sites (Fig. 10.4), a mechanism with no known counterpart in cellular or other viral mRNA synthesis systems. Unexpectedly, the vaccinia virus capping enzyme was identified as one protein that is required for termination of transcription at specific positions. As discussed previously, the capping enzyme binds to the viral RNA polymerase. It is believed to remain with the transcriptional complex until the termination signal

5′UUUUUNU3′ is encountered in the nascent RNA. Termination of transcription, which also requires a specific subunit of the early form of the RNA polymerase and a single-stranded DNA-dependent ATPase, take place 30 to 50 nucleotides downstream of this signal. The viral poly(A) polymerase adds poly(A) to the 3′ end thus formed, by a two-step process that is remarkably similar to the mechanism of synthesis of poly(A) by the cellular machinery (compare Fig. 10.4 and 10.3). The regulatory subunit of the viral poly(A) polymerase that induces processive synthesis of full-length poly(A) after the initial production of short chains is the viral 2′-*O*-methyltransferase. All components of the vaccinia virus capping machinery therefore also participate in formation of the 3′ ends of viral mRNAs. This property may be a manifestation of efficient utilization of viral genetic information or may facilitate coordination of the reactions by which viral mRNAs are synthesized.

Polyadenylation during viral mRNA synthesis. The synthesis of the poly(A) segments of vaccinia virus early mRNAs resembles the cellular mechanism in several respects. In contrast, the poly(A) sequences of mRNAs made by other viral RNA polymerases are produced by quite different mechanisms, during synthesis of the mRNA. In the simplest case, exemplified by (+) strand picornaviruses, a poly(U) sequence present at the 5′ end of the (–) strand RNA template is copied directly into a poly(A) sequence of equivalent length. The mRNAs of (–) strand RNA viruses such as vesicular stomatitis virus and influenza virus are polyadenylated by reiterative copying of short stretches of U residues in the (–) strand RNA template, a mechanism described in Chapter 6.

Splicing of Viral Pre-mRNA

Discovery of Splicing

Between 1960 and the mid-1970s, the study of putative nuclear precursors of mammalian mRNAs established that these RNAs are larger than the mature mRNAs translated in the cytoplasm, and heterogeneous in size. They were therefore named **heterogeneous nuclear RNAs** (hnRNAs). Such hnRNAs were shown to carry both 5′-terminal cap structures and 3′ poly(A) sequences, leading to the conclusion that both ends of the hnRNA were preserved in the smaller, mature mRNA. Investigators were faced with the dilemma of deducing how smaller mRNAs could be produced from larger hnRNAs while both ends of the hnRNA were retained.

The puzzle was solved by two groups of investigators, led by Phillip Sharp and Richard Roberts, who shared the 1993 Nobel Prize in physiology or medicine for this work. These investigators showed that adenoviral major late mRNAs are encoded by four **separate** genomic sequences (Box 10.3). The distribution of the mRNA-coding sequences into four separate blocks in the genome, in conjunction with the large size of major late mRNA precursors, implied that these mRNAs were synthesized by excision of noncoding sequences from primary transcripts, with precise joining of coding sequences. The demonstration that nuclear major late transcripts contain the noncoding sequences (introns) confirmed that the mature mRNAs are formed by **splicing** of noncontiguous coding sequences in the pre-mRNA.

This mechanism of mRNA synthesis had great appeal, because it could account readily for the puzzling properties of hnRNA. Indeed, it was shown within a matter of months that splicing of pre-mRNA is not an obscure, virus-specific device. Splicing occurs in all eukaryotic cells, although it is less prevalent in unicellular organisms such as the yeast *Saccharomyces cerevisiae*. The great majority of mammalian pre-mRNAs, like the adenoviral major late mRNAs, comprise exons interspersed among introns. In all cases, the capped 5′ end and sequences immediately adjacent to the 3′ poly(A) tail are conserved in mature mRNA.

The organization of protein-coding sequences into exons separated by introns has profound implications for the evolution of the genes of eukaryotes and their viruses. Introns are generally much longer than exons, and only short sequences at the ends of introns are necessary for accurate splicing (see "Mechanism of Splicing" below). Consequently, introns provide numerous sites at which DNA sequences can be broken and rejoined without loss of coding information. Therefore, they greatly increase the frequency with which random recombination reactions can create new functional genes by rearrangement of exons. Evidence of such "exon shuffling" can be seen in the modular organization of many modern proteins. Such proteins comprise combinations of a finite set of structural and functional domains or motifs, or multiple repeats of a single protein domain, each often encoded by a single exon. In similar fashion, the presence of introns is believed to have facilitated the transfer (by recombination) of cellular genetic information into the genomes of DNA viruses and retroviruses.

Mechanism of Splicing

Sequencing of DNA copies of a large number of cellular and viral mRNAs and of the genes that encode them identified short consensus sequences at the 5′ and 3′ **splice sites,** which are joined to each other in mature mRNA

BOX 10.3

Discovery of the spliced structure of adenoviral major late mRNAs

The demonstration that several different adenoviral major late mRNAs carry the same capped 11-nucleotide sequence at their 5′ ends (figure panel A) was both surprising and puzzling. Hybridization studies indicated that these 5′ ends were not encoded adjacent to the main segments of major late mRNAs. Direct visualization of such mRNAs hybridized to viral DNA provided convincing proof that their coding sequences are dispersed in the viral genome. A schematic diagram of one major late mRNA (hexon mRNA), hybridized to a complementary adenoviral DNA fragment extending from the left end to a point within the hexon coding sequence, is shown in panel B. Three loops of unhybridized DNA (thin lines), designated A, B, and C, bounded or separated by three short segments (1, 2, and 3) and one long segment (hexon mRNA) of DNA-RNA hybrid (thick lines), were observed. Other adenoviral late mRNAs examined yielded the same sets of hybridized and unhybridized viral DNA sequences at their 5′ ends, but differed in the length of loop C and the length and location of the 3′-terminal RNA-DNA hybrid. It was therefore concluded that the major late mRNAs contain a common 5′-terminal sequence (segments 1, 2, and 3) built from sequences encoded at three different sites in the viral genome, and termed the tripartite leader sequence. This sequence is joined to the mRNA body, a long sequence complementary to part of the hexon coding sequence in the viral genome in the example shown. (B) Adapted from S. M. Berget et al., *Proc. Natl. Acad. Sci. USA* **74:**3171–3175, 1977, with permission.

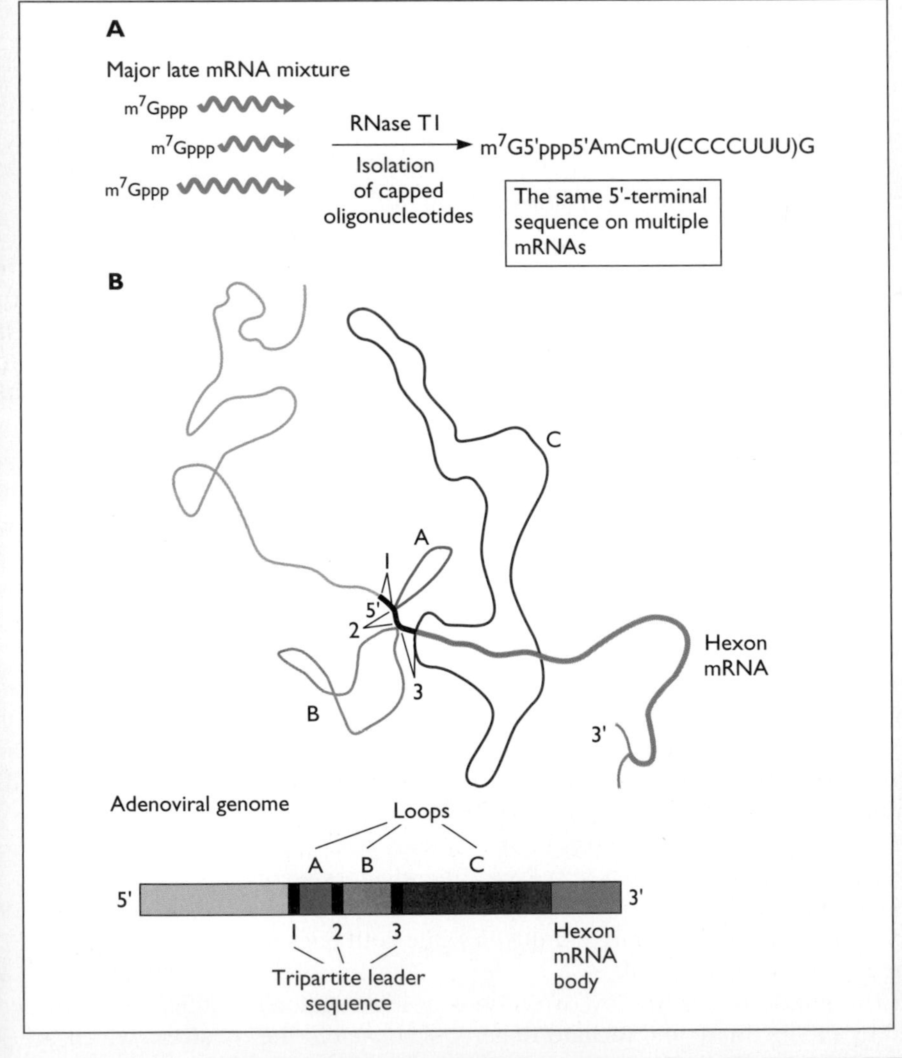

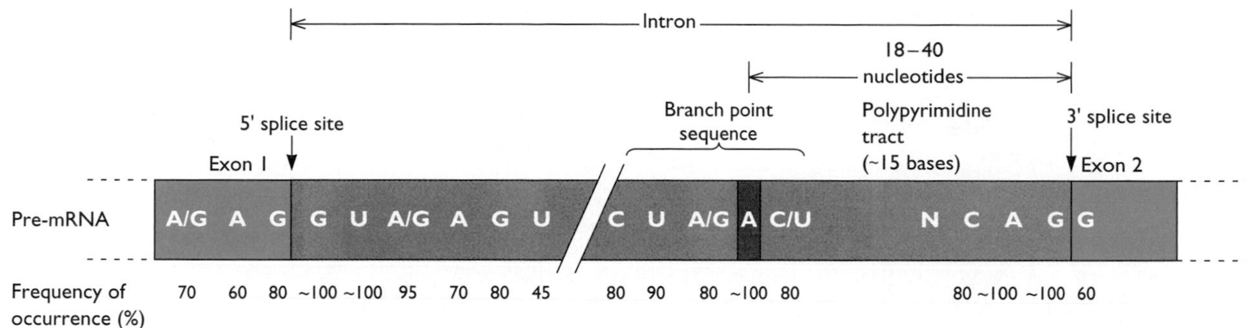

Figure 10.5 Consensus splicing signals in cellular and viral pre-mRNAs. Conserved sequences are found at the 5′ and 3′ splice sites at the junctions of exons (green) and introns (pink) and at the 3′ ends of introns. The intronic 5′GU3′ and 5′AG3′ dinucleotides at the 5′ and 3′ ends, respectively, of introns and branch point A (highlighted) are present in all but rare mRNAs made in higher eukaryotes.

(Fig. 10.5). The conserved sequences lie largely within the introns. The dinucleotides GU and AG are found at the 5′ and 3′ ends, respectively, of almost all introns. Mutation of any one of these four nucleotides eliminates splicing, indicating that all are essential. Elucidation of the mechanism of splicing came with the development of in vitro systems in which model pre-mRNAs (initially of viral origin) are accurately spliced. Numerous components needed for splicing have now been identified by both biochemical studies based on such systems, and genetic analyses in organisms such as yeasts and the fruit fly *Drosophila melanogaster.*

Pre-mRNA splicing occurs by two transesterification reactions, in which one phosphodiester bond is exchanged for another without the need for an external supply of energy (Fig. 10.6). The first reaction yields two products, the 5′ exon and the intron-3′ exon **lariat**. In the second transesterification reaction, the newly formed hydroxyl group at the end of the 5′ exon attacks the phosphodiester bond at the 3′ splice site, yielding the spliced exon product and the intron lariat (Fig. 10.6).

From a chemical point of view, the splicing of pre-mRNA is a simple process. However, each splicing reaction must be completed with a high degree of accuracy to

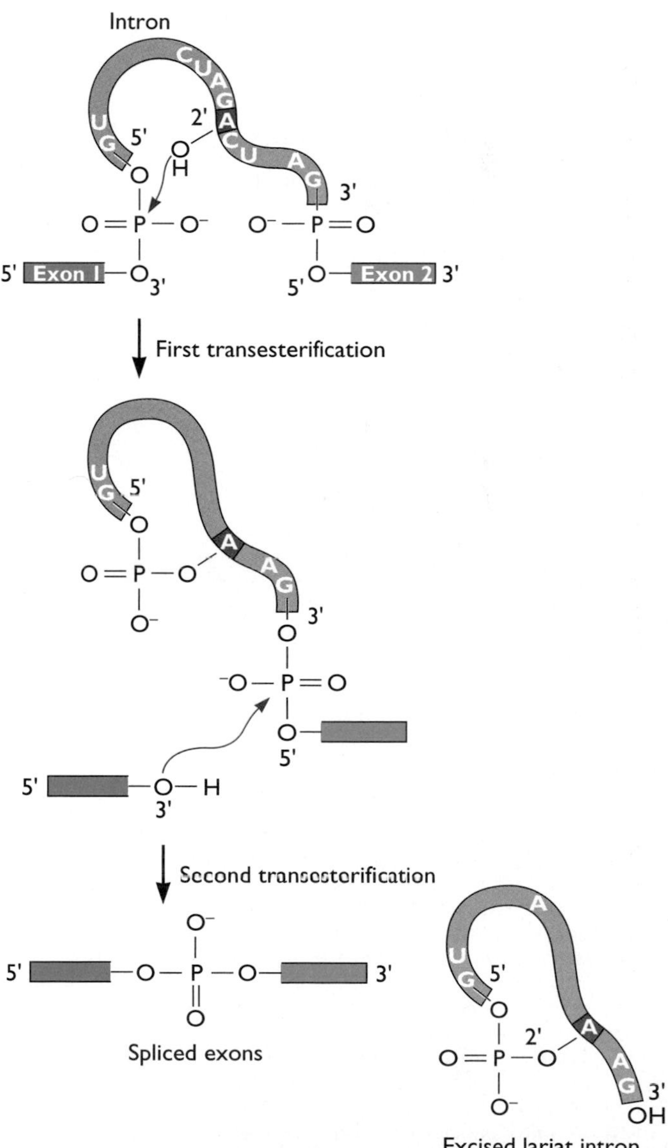

Figure 10.6 The two transesterification reactions of pre-mRNA splicing. In the first reaction, the 2′ hydroxyl group of the conserved A residue in the intronic branch point sequence, which is located 18 to 40 nucleotides upstream of the 3′ splice site (Fig. 10.5), makes a nucleophilic attack on the phosphodiester bond at the 5′ side of the GU dinucleotide at the 5′ splice site. In the intron-3′ exon lariat product, this attacking A residue is linked via 3′-5′ and 2′-5′ phosphodiester bonds to two different nucleotides. The second product is the 5′ exon. A second nucleophilic attack by the newly formed 3′ hydroxyl group of the 5′ exon on the phosphodiester bond at the 3′ splice site then yields the spliced exons and the intron lariat.

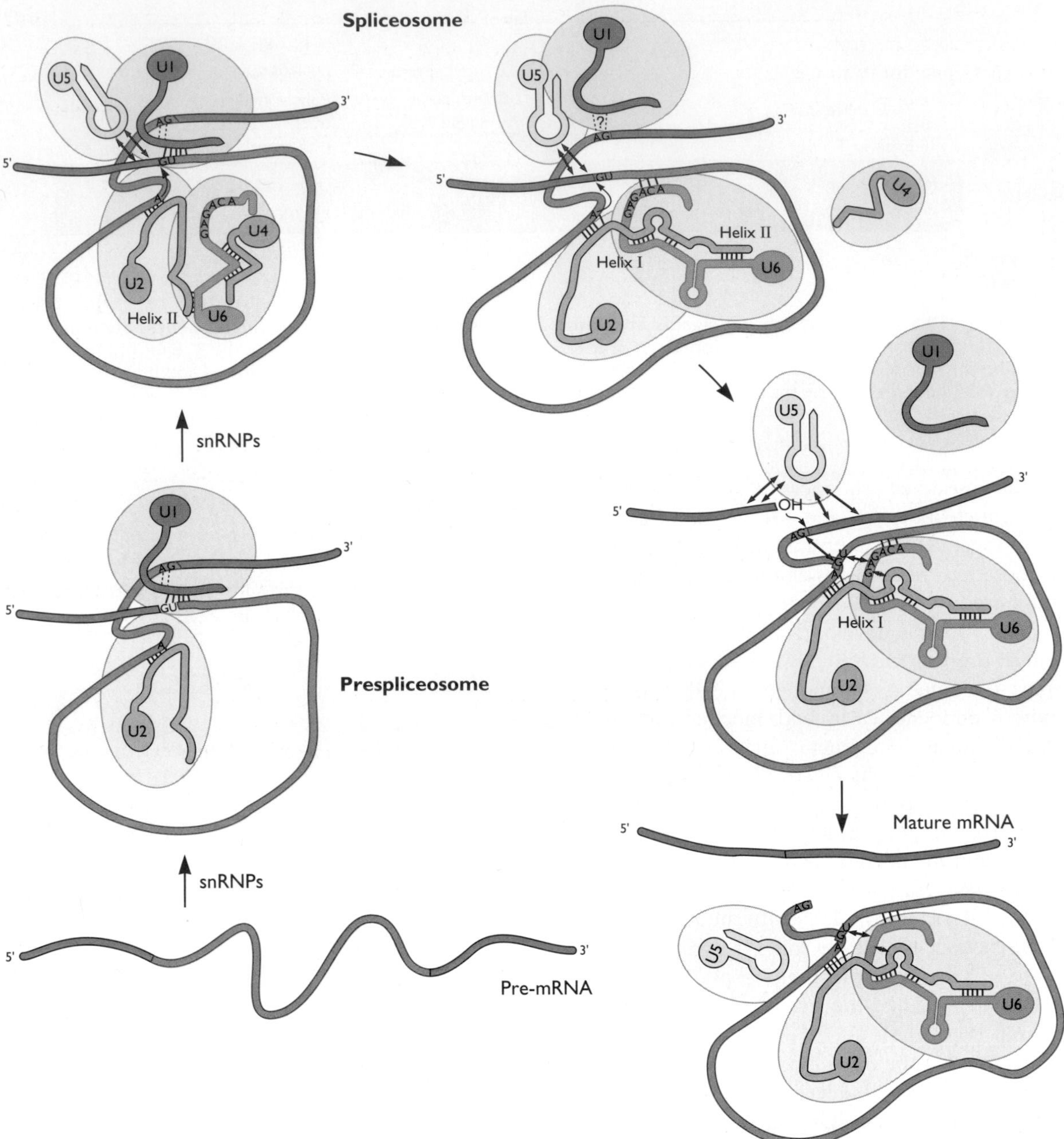

Figure 10.7 Interactions among snRNAs or snRNAs and pre-mRNA during formation of active spliceosomes. Base pairs are indicated by dashes and experimentally observed or presumed contacts among the RNA molecules by the two-headed arrows. The U1 and U2 snRNAs initially base pair with the 5′ splice site and branch point sequence, respectively, in the pre-mRNA. Usually only three base pairs between U2 snRNA and the pre-mRNA are formed at the branch point sequence. The U2-associated protein facilitates and stabilizes this interaction. The other snRNPs then enter the assembling spliceosome. The U4 and U6 snRNAs, which are present in a single snRNP, are base paired with one another over an extended complementary region. This snRNP binds to the U5 snRNP, and the snRNP complex associates with that containing the pre-mRNA and U1 and U2 snRNAs to form the spliceosome. RNA rearrangements then activate spliceosomes for catalysis of splicing. U4 snRNA dissociates from U6 snRNA, which forms hydrogen bonds with both U2 snRNA and the pre-mRNA. The interaction of U6 snRNA with the 5′ splice site displaces U1 snRNA, which plays no further role in splicing. One of the U2 sequences hydrogen-bonded to U6 snRNA (helix I) is adjacent to the U2 snRNA sequence that is base paired with the pre-mRNA branch point region. The interactions of U2 and U6 snRNAs with each other and with the pre-mRNA therefore juxtapose the branch point and 5′ splice site sequences for the first transesterification reaction (Fig. 10.6). The U5 snRNA base pairs to sequences in both the 5′ and 3′ exons and may align them for the second transesterification reaction. The many proteins that participate in spliceosome assembly and activation or that package the pre-mRNA and snRNAs (hnRNP and snRNP proteins, respectively) (see text) are not shown. Adapted from T. Nilsen, *Cell* **78:**1–4, 1994, with permission.

ensure that coding information is not lost or altered. The chemically active hydroxyl groups (Fig. 10.6) must also be brought into close proximity to the phosphodiester bonds they will attack. Furthermore, some genes contain 50 or more exons separated by **much** larger introns, which must be spliced in the correct order. It is presumably for such reasons that pre-mRNA splicing requires both many proteins and several RNAs, which associate with the pre-mRNA to form the large structure called the **spliceosome**.

Five small nuclear RNAs (snRNAs) participate in splicing: the U1, U2, U4, U5, and U6 snRNAs. In vertebrate cells, these RNAs vary in length from 100 to 200 nucleotides and are associated with proteins to form **small nuclear ribonucleoproteins (snRNPs).** The RNA components of the snRNPs recognize splice sites and other sequences in cellular and viral pre-mRNAs. Indeed, they participate in multiple interactions with the pre-mRNA and with each other during the three steps necessary to complete splicing. In the first, assembly of the spliceosome, the snRNA components of the U1 and U2 snRNPs base pair with the 5′ splice site and the intronic branchpoint, respectively, and the other snRNPs then enter the spliceosome (Fig. 10.7). At this stage, the U4 and U6 snRNAs are extensively hydrogen-bonded to one another and present in a single snRNP. The next step is dramatic rearrangement of the RNA-RNA interactions within the spliceosome. The end result is juxtaposition of the 5′ splice site and the branch point for the first transesterification reaction, as the result of interactions of U6 snRNA with the pre-mRNA and with U2 snRNA (Fig. 10.7). The RNA of the U5 snRNP may juxtapose the 5′ and 3′ exons to facilitate catalysis of their joining by the U2-U6 RNA complex. However, RNAs of the snRNPs do much more than simply organize the pre-mRNA sequences into a geometry suitable for transesterification. It has long been suspected that the spliceosome might be an RNA enzyme (or **ribozyme**), and recent experiments have provided direct evidence for catalysis by U6 and U2 snRNAs during splicing (Box 10.4).

Although the RNAs of snRNPs play essential roles in splicing as both guides and catalysts, the spliceosome also contains ~150 non-snRNP proteins. One class comprises proteins that package the pre-mRNA substrate, hnRNP proteins. These proteins bind to all pre-mRNAs made by RNA polymerase II, probably as they are synthesized, and

BOX 10.4

Evidence for RNA catalysis of the first transesterification reaction in pre-mRNA splicing

Coordination of a Crucial Metal Ion by U6 snRNA

By the time a spliceosome competent to catalyze splicing has assembled, the snRNPs containing U1 and U4 snRNAs have been released (Fig. 10.7). Of the remaining snRNAs, U6 has long been considered likely to participate directly in catalysis, for example, because it acts at (or very near) the active site of the spliceosome. This snRNA was recently shown to bind to a metal ion essential for splicing. Substitution of either of the two nonbonded oxygen atoms of the phosphodiester bond of residue 80 of the RNA of yeast U6 snRNP by sulfur completely inhibited pre-mRNA splicing in vitro. None of the various precatalytic assembly reactions were defective. Splicing normally requires Mg^{2+} ions, which can bind to phosphoryl oxygen atoms, but not to sulfur, suggesting that the substituted residue coordinates an essential Mg^{2+} ion. This inference was confirmed by the demonstration that the sulfur-substituted RNA of U6 snRNP supported efficient splicing in the presence of Mn^{2+} ions, which can bind sulfur atoms. These experiments therefore established that metal ion coordination by the RNA of U6 snRNP is crucial for splicing, but could not distinguish whether this role is catalytic or noncatalytic.

RNA Catalysis by U2 and U6 snRNAs in the Absence of Any Proteins

In the absence of proteins, U2 and U6 snRNAs synthesized in vitro base pair to form a stable structure with a three-dimensional conformation similar to that believed to be present at the active site of the spliceosome. This U2-U6 snRNA complex binds specifically to a short, synthetic RNA molecule containing the sequence of an intron branch point. The branch point adenosine is activated when the substrate RNA is bound to the snRNA complex, and its 2′-OH group attacks a specific phosphodiester bond in U6 RNA to form an unusual phosphotriester bond. This reaction is analogous to the first transesterification during pre-mRNA splicing by the spliceosome. It requires RNA sequences necessary for authentic splicing, including a specific base-pairing interaction between U2 and U6 snRNAs and a particular U6 RNA sequence. These observations therefore provide direct evidence for RNA catalysis of the first splicing reaction.

Valadkhan, S., and J. L. Manley. 2001. Splicing-related catalysis by protein-free snRNAs. *Nature* **413:**701–707.

Yean, S.-L., G. Wuenschell, J. Termini, and R.-J. Lin. 2000. Metal-ion coordination by U6 small nuclear RNA contributes to catalysis in the spliceosome. *Nature* **408:**881–884.

play important structural roles. There is some evidence that certain members of this family may participate more directly in splicing. Many other splicing proteins contain both RNA-binding and protein-protein interaction domains. Such proteins bind to pre-mRNA sequences within or adjacent to exons, and act positively or negatively to regulate splicing or exon recognition. Members of one family of such proteins, called the SR proteins because they contain a domain rich in serine (S) and arginine (R), act at early stages of splicing to recruit snRNPs to the spliceosome. For example, the SR protein named splicing factor 2 (Sf2, or alternative splicing factor) facilitates binding of U1 snRNP to 5′ splice sites and stabilizes the prespliceosomal complex (Fig. 10.7). Members of the SR protein family can also recognize exon sequences that stimulate splicing of upstream introns, termed **splicing enhancers,** and can assist juxtaposition of 5′ and 3′ splice sites. Other proteins important for splicing are RNA-dependent helicases, which are thought to catalyze the multiple rearrangements of hydrogen bonding among different snRNAs and the pre-mRNA substrate (Fig. 10.7). Such helicases are generally ATP dependent, and spliceosome assembly and rearrangement depend on energy supplied by ATP hydrolysis.

Splicing of pre-mRNAs is commonly cotranscriptional, and components of the splicing machinery may associate with RNA polymerase II. The hyperphosphorylated form of the C-terminal domain of the largest subunit of this enzyme can bind to both SR proteins and spliceosomal snRNPs. Peptide mimics of the C-terminal domain or antibodies raised against it inhibit pre-mRNA splicing in vivo or in vitro. Furthermore, nontranscribed sequences within an RNA polymerase II promoter can dictate whether a particular exon is retained or removed during splicing. As we have seen, association of components of the 5′ capping and 3′ polyadenylation systems with the C-terminal domain of RNA polymerase is necessary for these processing reactions. The synthesis of pre-mRNA and complete processing are therefore coordinated as a result of association of specific proteins needed for each processing reaction with the C-terminal domain of RNA polymerase II. Such a transcription and RNA-processing machine is analogous to, but much more complex than, that of vaccinia virus described above.

Production of Stable Viral Introns

Cells infected by certain herpesviruses provide a remarkable exception to the rule that viral intron-containing pre-mRNAs are spliced just like their cellular counterparts. As discussed in Chapter 8, the only abundant viral RNAs detected in neurons latently infected by herpes simplex viruses are the major latency-associated transcripts. These nuclear RNA species not only lack 3′ poly(A) sequences, but also exhibit the properties of excised introns: they are not linear molecules and contain the branch points characteristic of the intron lariats excised from pre-mRNAs during splicing. In contrast to typical introns removed from other viral and cellular pre-mRNAs, these herpesviral RNAs are very stable, and accumulate to high concentrations in the nuclei of latently infected cells. Neither the reason for such unusual stability, nor the molecular functions of the latency-associated transcripts, are yet well understood.

Alternative Splicing of Viral Pre-mRNA

Many viral and cellular pre-mRNAs contain multiple exons. Splicing of the majority removes all introns and joins all exons in the order in which they are present in the substrate pre-mRNA. Such **constitutive splicing** (Fig. 10.8A) produces a single mature mRNA. However, some cellular and many viral pre-mRNAs yield more than one mRNA as a result of the splicing of different combinations of exons, a process termed **alternative splicing**. Several different types of alternative splicing can be defined, depending on whether the splicing machinery skips an entire exon or recognizes alternative 5′ or 3′ splice sites (Fig. 10.8B). Alternative splicing generally leads to the synthesis of mRNAs that differ in their protein coding sequences.

The most obvious advantage of this mechanism is that it can expand the limited coding capacity of viral genomes. For example, the early genes of simian virus 40 and polyomavirus, as well as the adenoviral E1A and E1B genes, each specify two or more proteins as a result of splicing of primary transcripts at alternative 5′ or 3′ splice sites. Alternative splicing can also be important for temporal regulation of viral gene expression, or the control of a crucial balance in the production of spliced and unspliced mRNAs. In many cases, alternative splicing of viral pre-mRNAs is coupled with other RNA-processing reactions or regulated by viral proteins. These more complex phenomena are considered in later sections.

Examples of Alternative Splicing of Viral Pre-mRNAs

Identification of cellular proteins that control alternative splicing: simian virus 40 early pre-mRNA. The simian virus 40 early pre-mRNA is spliced at one of two suboptimal 5′ splice sites to form mRNAs encoding either large T (LT) or small T (sT) antigen (Fig. 10.9). The longer mRNA encodes sT, because it includes a translation termination codon that is removed from LT mRNA by splicing. The two proteins therefore contain a common N-terminal

A Constitutive splicing

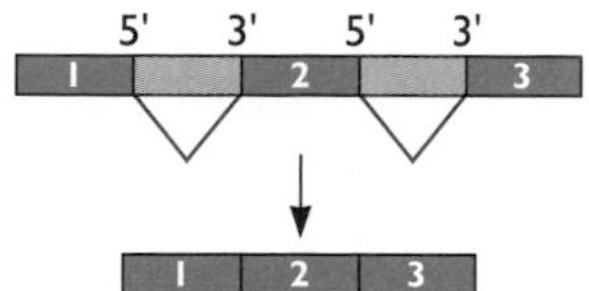

B Alternative splicing

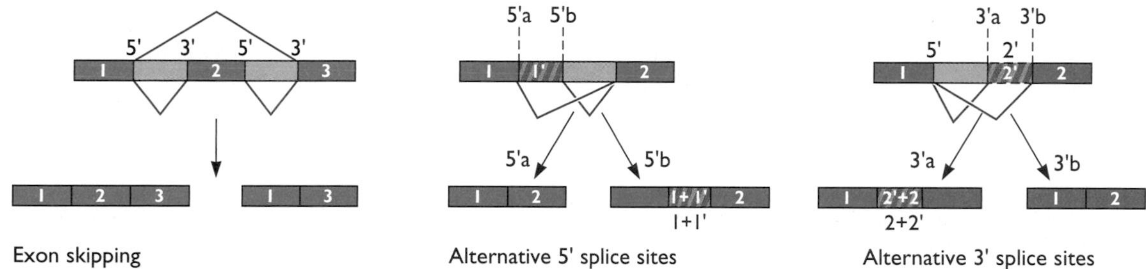

Figure 10.8 Constitutive and alternative splicing. In constitutive splicing (A), all exons (green) are joined sequentially and all introns (pink) are excised. Alternative splicing (B) occurs by several mechanisms. In exon skipping, the 3′ splice site of exon 2 is sometimes ignored, so that this exon is not included in some fraction of the spliced mRNA molecules. Alternatively, one of two 5′ splice sites (5′a and 5′b) in exon 1 or one of two 3′ splice sites (3′a and 3′b) in exon 2 are recognized. Recognition of different 5′ and 3′ splice sites produces alternatively spliced simian virus 40 early mRNAs (Fig. 10.9) and adenoviral major late mRNAs (Fig. 10.12), respectively.

Figure 10.9 Regulation of alternative splicing of simian virus 40 early pre-mRNA by splicing factor 2 (Sf2). At low concentrations of Sf2, also known as alternative splicing factor (Asf), the 5′ splice site for LT mRNA is recognized predominantly in in vitro splicing reactions supplied with simian virus 40 early pre-mRNA. At high concentrations of this cellular protein, the 5′ splice site that produces sT mRNA is activated. Sf2, an SR protein, can facilitate binding of U1 snRNP to 5′ splice sites and promote interaction of the U1 and U2 snRNPs. High concentrations might therefore improve recognition of the weak sT 5′ splice site, or circumvent steric constraints imposed by the very small size of the sT intron.

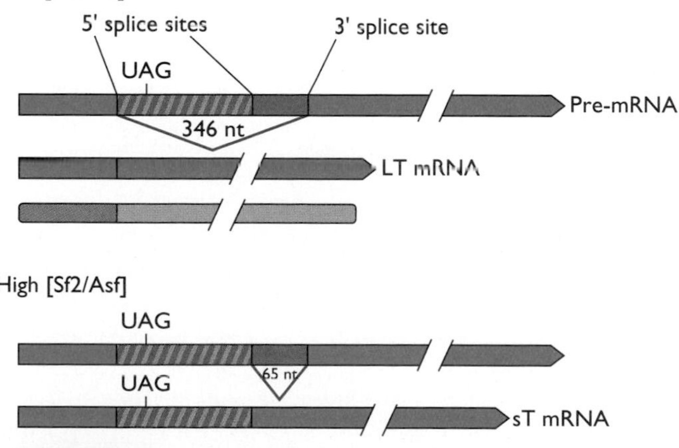

sequence, but are unrelated at their C termini. Study of the alternative splicing of simian virus 40 early pre-mRNA provided the first demonstration that specific proteins function in both constitutive and alternative splicing. The SR protein Sf2 regulates the recognition of the two 5′ splice sites of this viral pre-mRNA in vitro. As the concentration of this protein is increased above that which supports constitutive splicing, the splicing machinery switches from the LT to the sT 5′ splice site (Fig. 10.9). These in vitro observations set the stage for many other studies of these proteins, which are important regulators of splicing in both vertebrates and invertebrates.

Regulation of splicing and virus production by cellular differentiation: papillomaviral late pre-mRNAs. The late proteins of bovine papillomavirus type 1 are synthesized efficiently only in highly differentiated keratinocytes. Productive replication of viral DNA, and assembly and release of virions, is also restricted to these outer cells in an epithelium. Alterations in splicing of the late pre-mRNA are crucial for production of the mRNA encoding the major capsid protein, the L1 mRNA (Fig. 10.10A). In situ hybridization studies have shown that this mRNA is made only in fully differentiated cells (Fig. 10.10B). Production of the L1 mRNA requires activation of an alternative 3′ splice site that is not recognized in incompletely differentiated cells. Several *cis*-acting sequences

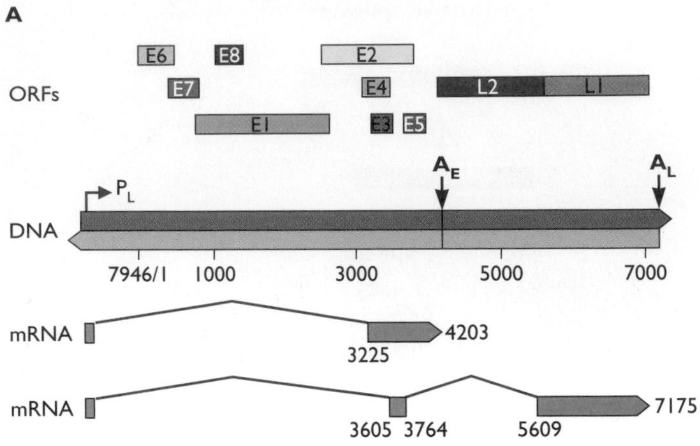

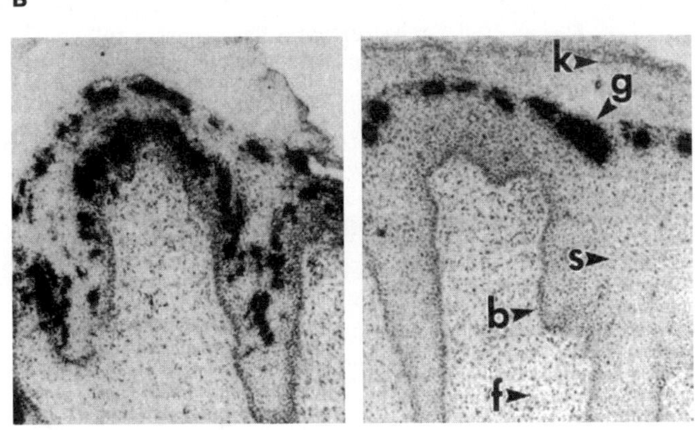

Figure 10.10 Control of the production of bovine papillomavirus type 1 late mRNAs by alternative splicing and alternative polyadenylation. (A) The circular bovine papillomavirus type 1 genome is represented in linear form, with open reading frames (ORFs) shown above. Two of the many mRNAs made from transcripts from the late promoter (P_L) are shown to illustrate the changes in recognition of splice sites and of poly(A) addition sites necessary to produce the L1 mRNA. Synthesis of this mRNA depends on recognition of a 3′ splice site at position 3605, rather than that at 3225, which is used during the early phase of infection. Polyadenylation of pre-mRNAs must also switch from the early (A_E) to the late (A_L) polyadenylation sites. (B) In situ hybridization of bovine fibropapillomas to probes that specifically detect mRNAs spliced at the 3225 3′ splice site (left) or at the 3605 site (right). The cell layers of the fibropapilloma are indicated in the right panel. Abbreviations: k, keratin horn; g, granular cell layer; s, spinous cell layer; b, basal cell layer; f, fibroma. Note the production of late mRNA spliced at the 3605 3′ splice site only in the outermost layer (g) of fully differentiated cells. From S. K. Barksdale and C. C. Baker, *J. Virol.* **69:**6553–6556, 1995, with permission. Courtesy of C. C. Baker, National Institutes of Health. (C) Mechanisms that regulate processing reactions to allow production of L1 mRNA, which are specific to highly differentiated keratinocytes of the granular cell layer. As shown at the top, the sequences that control alternative splicing at the 3225 and 3605 3′ splice sites are located between these splice sites. The two splicing enhancers, SE1 and SE2, are recognized by cellular SR proteins. The former and the adjacent sequence that inhibits splicing at the 3605 3′ splice site, termed the exonic splicing suppressor (ESS), are believed to facilitate recruitment of U2-associated protein (U2af) and recognition of the branch point sequence upstream of the 3225 3′ splice site. Splicing enhancer 2 is located very close to the 3605 3′ splice site and may block access to the branch point for splicing at this site until keratinocytes differentiate. However, the mechanisms by which this sequence and the exonic splicing suppressor function remain to be determined. Inhibition of polyadenylation at the A_E site by the binding of U1 snRNP to a pseudo-5′ splice site located nearby in the primary transcript (see text) is depicted below. Such inhibition may be the result of binding of a U1 snRNP protein to poly(A) polymerase.

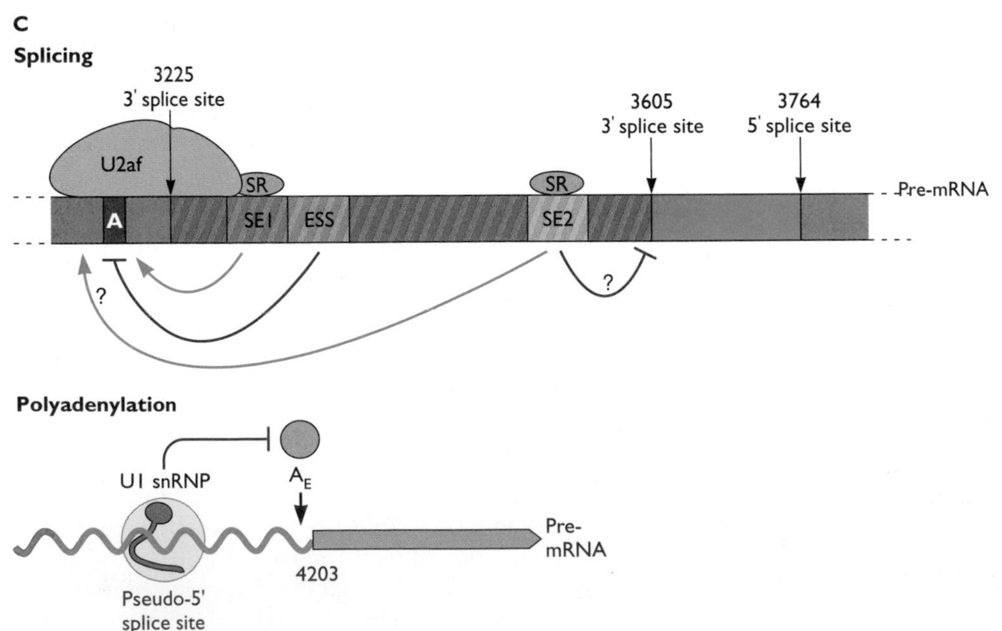

in the viral late pre-mRNA govern the choice of 3′ splice sites (Fig. 10.10C). These sequences are bound by cellular splicing proteins, including SR proteins. It has therefore been proposed that terminal differentiation of keratinocytes is accompanied by changes in the activity or abundance of these cellular proteins, such that L1 mRNA can be synthesized.

Production of spliced and unspliced RNAs essential for virus replication. The expression of certain coding sequences in the genomes of retroviruses (Gag and Pol) (Appendix A, Fig. 21B) and orthomyxoviruses (M1 and NS1) (Appendix A, Fig. 8B) depends on an unusual form of alternative splicing that produces both spliced and unspliced mRNAs: normally all pre-mRNAs that contain introns are fully spliced in mammalian cell nuclei. This phenomenon has been well studied in retrovirus-infected cells, in which splicing of less than 100% of viral transcripts is essential for production of the viral genome.

In cells infected by retroviruses with simple genomes, such as avian leukosis virus, a full-length unspliced transcript of proviral DNA serves as both the genome and the mRNA for the capsid proteins and viral enzymes, while a singly spliced mRNA specifies the viral envelope protein (Fig. 10.11A). Retrovirus production depends rather critically on the maintenance of a proper balance between the proportions of unspliced and spliced RNAs. Modest changes in splicing efficiencies cause replication defects (Fig. 10.11A). This phenomenon has been used as a genetic tool to select for mutations that affect splicing control. Such mutations arise in different splicing signals at the 3′ splice site, and alter the efficiency of either the first or second step in the splicing reaction. Features that maintain the proper splicing balance include a suboptimal 3′ splice site and a splicing enhancer in the adjacent exon. A negative regulatory sequence located more than 4000 nucleotides upstream of the 3′ splice site is also important. Specific sequences within exons that inhibit splicing (**exonic splicing suppressors**) have also been

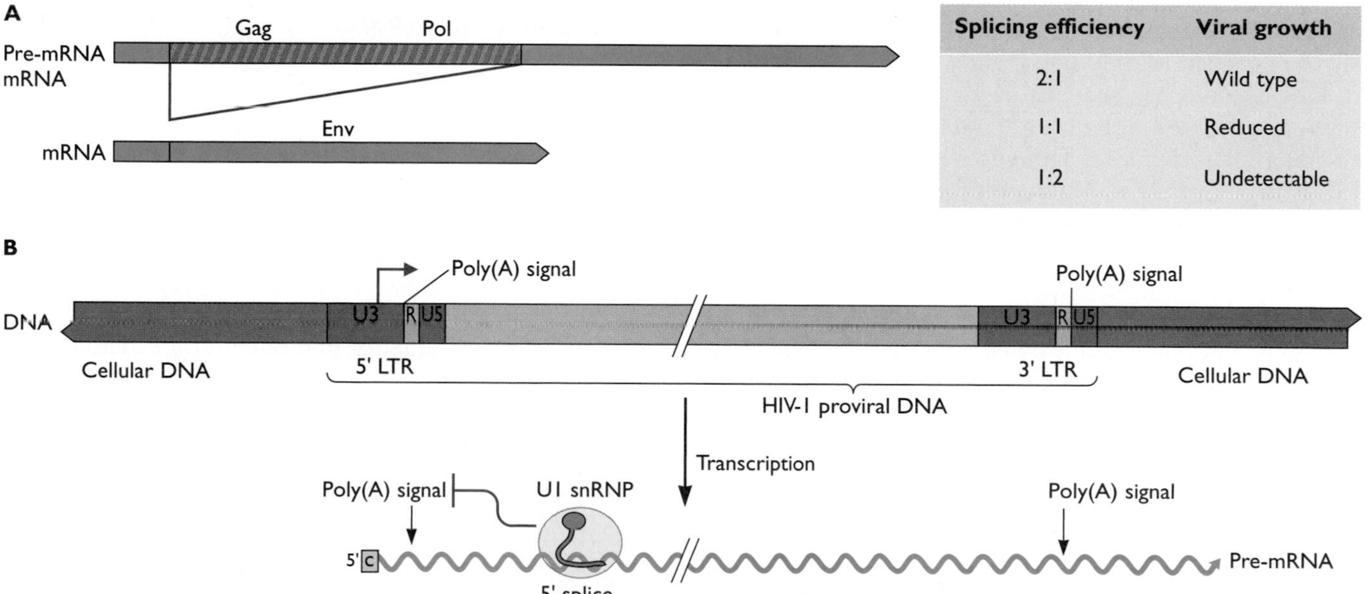

Splicing efficiency	Viral growth
2:1	Wild type
1:1	Reduced
1:2	Undetectable

Figure 10.11 Control of RNA-processing reactions during retroviral gene expression. (A) Balanced production of spliced and unspliced mRNAs is illustrated for avian leukosis virus. A single 3′ splice site is recognized in about one-third of the primary transcripts to produce spliced mRNA encoding the Env protein. The Gag and Pol proteins are synthesized from unspliced transcripts. Even a twofold reduction in the ratio of unspliced to spliced mRNA impairs virus reproduction (right). (B) Suppression of poly(A) site recognition. Polyadenylation signals are present in the long terminal repeats (LTRs) of retroviral proviruses, and the primary transcripts of many proviral DNAs contain identical poly(A) addition signals near both their 5′ and 3′ ends. In the case of human immunodeficiency virus type 1 (HIV-1), utilization of the 5′ polyadenylation site is inhibited by binding of U1 snRNP to the major 5′ splice site located 195 nucleotides downstream. This snRNP might block access to the 5′ poly(A) addition signal directly, or indirectly via proteins that mediate interaction of the cap structure with U1 snRNP. Alternatively, the U1 snRNP might inhibit the activity of one or more proteins needed for polyadenylation. This possibility is suggested by the ability of a protein specific to the U1 snRNP, the U1A protein, to bind to both poly(A) polymerase and the cleavage polyadenylation specificity protein.

identified. Two such sequences are present in Tat-encoding exons of human immunodeficiency virus type 1 pre-mRNA: the mutation of either increases the efficiency of splicing both in vitro and in vivo.

Coordination between Polyadenylation and Splicing

Although described separately in previous sections, capping, polyadenylation, and splicing of a pre-mRNA by cellular components are not independent reactions. Rather, one processing reaction governs the efficiency or specificity of another. For example, interaction of the nuclear cap-binding protein with the 5′ end of a pre-mRNA is necessary for both efficient removal of the intron closest to the 5′ end and maximally efficient cleavage at the 3′ poly(A) addition site. Similarly, the presence of a 3′ poly(A) addition signal generally stimulates removal of the intron closest to it. Conversely, the 3′ splice site at the beginning of the terminal exon increases the efficiency of polyadenylation. However, splice sites at other positions can suppress recognition of polyadenylation signals. Inhibition of polyadenylation at specific sites is essential for expression of the genes of some retroviruses and is an important device for the regulation of adenoviral gene expression.

The proviral DNAs of retroviruses contain long terminal repeat (LTR) sequences at either end, both of which contain a poly(A) addition signal (Fig. 10.11B). Transcription of some proviral DNAs, such as that of Rous sarcoma virus, initiates downstream of the LTR sequence encoding the essential 5′AAUAAA3′ polyadenylation sequence. Consequently, a functional poly(A) addition site is present only at the 3′ ends of these retroviral pre-mRNAs. However, many other retroviral transcripts carry complete signals for this modification at both their 5′ and 3′ ends. Nevertheless, poly(A) is added to only the 3′ ends of these pre-mRNAs. At least two mechanisms ensure that the correct poly(A) addition signal of human immunodeficiency virus type 1 pre-mRNA is recognized. Sequences present only at the 3′ end of the pre-mRNA stimulate polyadenylation in vitro and in vivo by facilitating binding of Cpsf to the nearby 5′AAUAAA3′ sequence [poly(A) signal]. In addition, recognition of the 5′ poly(A) signal is suppressed by the 5′ splice site lying immediately downstream (Fig. 10.11B). Such inhibition is believed to be mediated by U1 snRNP, which can bind to the 5′ splice site. A related mechanism, binding of U1 snRNP to a pseudo 5′ splice site, appears to suppress recognition of the poly(A) addition site for production of bovine papillomavirus type 1 late mRNAs until the keratinocyte is fully differentiated (Fig. 10.10C).

The production of adenoviral major late mRNAs exemplifies complex alternative splicing and polyadenylation at multiple sites in a pre-mRNA. The major late pre-mRNA contains the sequences of at least 15 different mRNAs. These mRNAs fall into five families (L1 to L5) defined by which of five polyadenylation sites in the pre-mRNA is recognized (Fig. 10.12). The frequency with which each site is used must therefore be regulated to allow production of all major late mRNAs. For example, high-efficiency polyadenylation at the L1 site would prevent efficient synthesis of L2 to L5 mRNAs, just as efficient splicing of retroviral mRNAs would preclude production of the essential, unspliced transcripts. During the late phase of adenovirus infection each of the five polyadenylation sites directs 3′-end formation with approximately the same efficiency. The mechanism responsible for such balanced recognition of multiple poly(A) addition sites is not fully understood. However, the activities of cellular polyadenylation proteins are altered as infection proceeds (see "Posttranscriptional Regulation of Viral or Cellular Gene Expression by Viral Proteins" below).

As discussed previously, all major late mRNAs contain the 5′-terminal tripartite leader sequence. The splicing reactions that produce this sequence from three small exons can take place before polyadenylation of the primary transcript. The final splicing reaction joins the tripartite leader sequence to one of many mRNA sequences (Fig. 10.12). Each primary transcript therefore yields only a **single** mRNA, even though it contains the sequences for many, and most of its sequence is discarded (Fig. 10.12). It remains something of a mystery why the majority of adenoviral late mRNAs are made by this bizarre mechanism. However, one contributing factor may be that it ensures that each major late mRNA molecule carries the 5′-terminal tripartite leader sequence, which is important for efficient translation late in the adenoviral infectious cycle.

Editing of Viral mRNAs

The term **RNA editing** was first coined in 1980 to describe the origin of uridine residues that are not encoded in the DNA template in a mitochondrial mRNA of trypanosomes. Since this modification was discovered, RNA editing has been identified in many different eukaryotes, as well as in some important viral systems.

Viral mRNAs are edited by either insertion of nucleotides not directly specified in the template during synthesis or alteration of a base in situ. Both mechanisms result in an mRNA sequence that differs from that encoded in the viral genome, altering the sequence and function of the protein specified by edited mRNA. Consequently, RNA editing has the potential to make an important contribution to regulation of viral gene expression.

Editing during Pre-mRNA Synthesis

The mRNAs of the *Paramyxoviridae* (e.g., measles and mumps viruses) and *Filoviridae* (e.g., Ebola virus) are edited during their synthesis. Among paramyxoviruses,

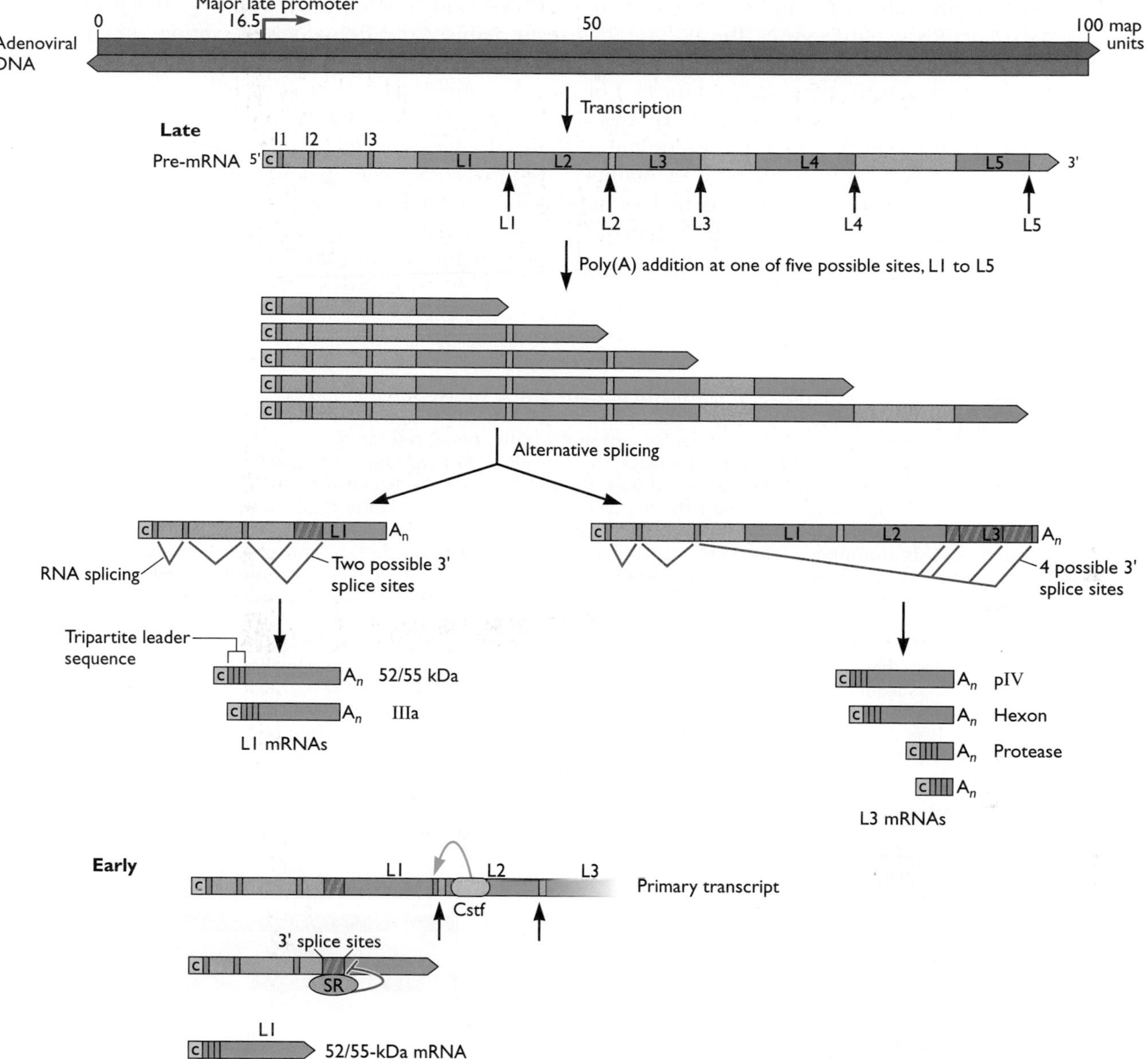

Figure 10.12 Alternative polyadenylation and splicing of adenoviral major late transcripts. During the late phase of adenoviral infection (top), major late primary transcripts extend from the major late promoter almost to the right end of the genome. They contain the sequences for at least 15 mRNAs and are polyadenylated at one of five sites, L1 to L5, each of which is used with approximately equal frequency. The tripartite leader sequence, present at the 5′ ends of all late mRNAs, is assembled by the splicing of three short exons, l1, l2, and l3. This sequence is then ligated to alternative 3′ splice sites. Such joining of the spliced tripartite leader sequence to an mRNA sequence has been reported to take place after polyadenylation of pre-mRNA. Polyadenylation therefore appears to determine which 3′ splice sites can be utilized during the final splicing reaction. During the early phase of infection (below), transcription from the major late promoter terminates upstream of map unit 70 (see Chapter 8). Although the transcripts contain the L1, L2, and L3 poly(A) addition sites, only the L1 site is recognized, because its downstream U/GU-rich sequence provides a strong binding site for cleavage stimulatory protein (Fig. 10.3). The decrease in activity of this protein as infection continues (see text) allows recognition of the L2 to L5 polyadenylation sites, which are less dependent on this protein. During the early phase of infection, only the L1 3′ splice site for the 52/55-kDa protein is utilized, because binding of SR proteins to the pre-mRNA blocks recognition of the 3′ splice site for production of the mRNA for protein IIIa. An E4 protein induces dephosphorylation of these cellular proteins by protein phosphatase 2. This modification inhibits binding of the SR proteins to the pre-mRNA and reverses suppression of recognition of the 3′ splice site for production of mRNA encoding protein IIIa.

this modification occurs only in the mRNAs that encode the RNA-dependent RNA polymerase, the P protein. During mRNA synthesis, the viral RNA polymerase inserts one or two G residues at specific positions in a fraction of the RNA molecules. The genomic RNA template contains a polypyrimidine run at the RNA-editing site. Insertion of G residues is therefore believed to occur by a reiterative copying mechanism (Fig. 10.13A), analogous to that by which the viral RNA polymerases synthesize 3′ poly(A) tails. The observation that increased stability of the nascent RNA-template RNA duplex just upstream of the editing site reduces the frequency of editing, and vice versa, is consistent with this mechanism.

Because insertional editing of both measles and mumps mRNAs alters the translational reading frame, paramyxoviral P genes encode two distinct proteins (Fig. 10.13B). Similarly, RNA editing determines whether an Ebola virus gene is expressed as a full-length glycoprotein that is required for virus entry, or as a truncated secreted protein, and modulates the cytotoxicity of the virus (Box 10.5).

Figure 10.13 Cotranscriptional editing of paramyxoviral mRNAs. (A) Proposed mechanism of cotranscriptional editing of measles and mumps virus mRNAs. The viral polymerase pauses near a junction of U_n and C_n sequences in the template virion RNA after two C residues of the template have been copied into G residues in the nascent mRNA. As a result of this pausing, some fraction of viral RNA polymerase molecules and their attached nascent mRNA chains slip backwards, such that additional nucleotides are incorporated when RNA synthesis resumes. Because of differences in the sequences of the virion RNA templates, the most stable structures (boxed in light green) are formed when the measles virus and mumps virus RNA polymerases slip backward by one (–1) and two (–2) positions, respectively. Consequently, one and two additional G residues, respectively, are incorporated into measles virus and mumps virus mRNAs when RNA synthesis resumes. (B) The P genes of mumps and measles viruses. The unedited measles virus mRNA contains a continuous open reading frame (pink) for the P protein. The addition of one G residue at the editing site changes the translational reading frame to one that contains a termination codon. The edited mRNA therefore specifies the V protein, which differs from the P protein in its C-terminal sequence (orange). In contrast, the unedited mumps virus mRNA specifies the V protein. Insertion of two G residues at the editing site changes the reading frame to fuse the V open reading frame with the second, C-terminal open reading frame (orange) and produce the continuous P protein sequence.

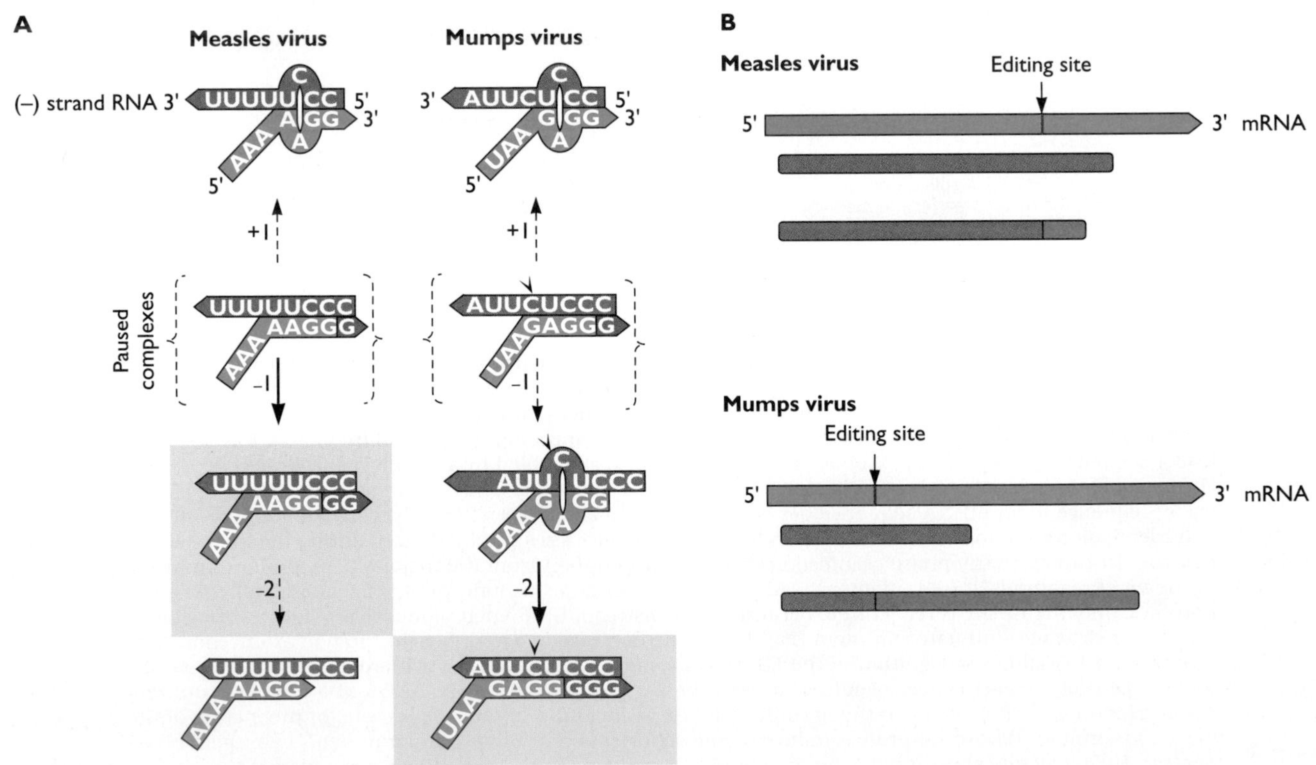

BOX 10.5

RNA editing regulates the cytotoxicity of Ebola viruses

The ~19,000 nucleotide (–) strand RNA genome of Ebola virus (a member of the *Filoviridae*) contains an editing site within the coding sequence for glycoprotein (GP) mRNA. This site comprises seven consecutive U residues and resembles the viral polyadenylation signal. Editing is therefore believed to take place cotranscriptionally, by reiterative copying, as during synthesis of paramyxoviral mRNA (Fig. 10.13).

As shown in the figure, the products of the edited and unedited GP mRNAs share an N-terminal sequence but carry different C-terminal sequences, because introduction of an additional A residue by editing changes the reading frame of the mRNA. The protein specified by edited mRNA, GP, is the viral envelope glycoprotein and localizes to the plasma membrane of infected cells. In contrast, the protein synthesized from the unedited mRNA (sGP) is secreted.

When the editing site was eliminated by mutation of cloned DNA from which infectious virus can be recovered (see Chapter 2), sGP was not made, as was expected. The concentration of GP increased, also as anticipated, but most of the protein accumulated as an immature precursor in the endoplasmic reticulum. Furthermore, and unexpectedly, these alterations in production of the GP proteins were accompanied by severe cytotoxicity: the mutant virus formed small plaques because of the earlier than normal death of infected cells.

These observations indicate that maintenance of a proper ratio of the secreted and membrane-associated forms of the viral glycoprotein is necessary to limit the cytotoxicity of Ebola virus. The secreted glycoprotein is therefore likely to make an important contribution to pathogenicity by indirectly increasing virus reproduction and spread.

Volchkov, V. E., V. A. Volchkova, E. Mühlberger, L. V. Kolesnikova, M. Weik, O. Dolnik, and H.-D. Klenk. 2001. Recovery of infectious Ebola virus from complementary DNA: RNA editing of the GP gene and viral cytotoxicity. *Science* **291:**1965–1969.

The editing site in the GP gene of the viral RNA genome is shown at the top, and the differences between the sGP and GP proteins specified by unedited and edited mRNAs, respectively, are illustrated below.

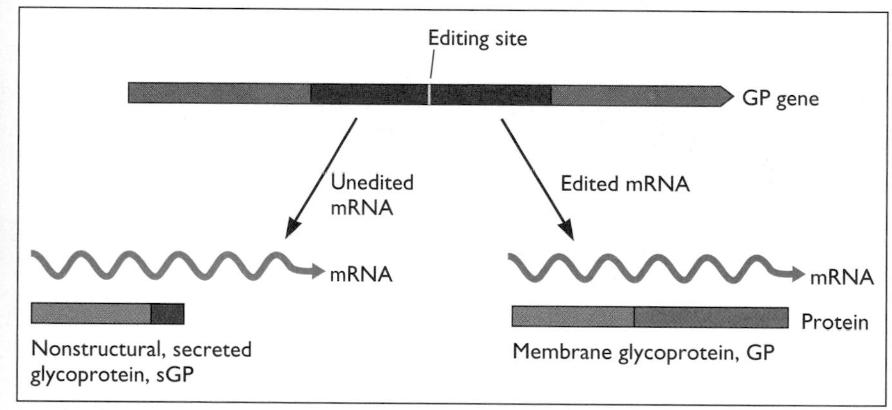

Editing following mRNA Synthesis

Posttranscriptional editing of mRNAs is accomplished by cellular enzymes that deaminate adenine bases in double-stranded RNA regions to form inosine (I). One such double-stranded RNA adenosine deaminase has been implicated in editing of the mRNA of hepatitis delta satellite virus. Two forms of the protein encoded in the genome of this agent, called hepatitis delta antigen, are synthesized in infected cells. Both are necessary for satellite virus reproduction. The large delta antigen, which is required for assembly and inhibits RNA replication, is made when editing at a specific site converts a UAG termination codon in the mRNA to a codon for tryptophan, UGG (Fig. 10.14). As many as 50% of satellite virus mRNA molecules are modified at this site, but few other sequences in the viral RNA are edited. The mechanisms that restrict action of the adenosine deaminase, which exhibits little specificity in vitro, have not been elucidated. The larger form of hepatitis delta antigen can inhibit editing in vivo. This property is important for synthesis of the essential, smaller form of the protein, which is translated from unedited mRNA molecules (Fig. 10.14). It also prevents

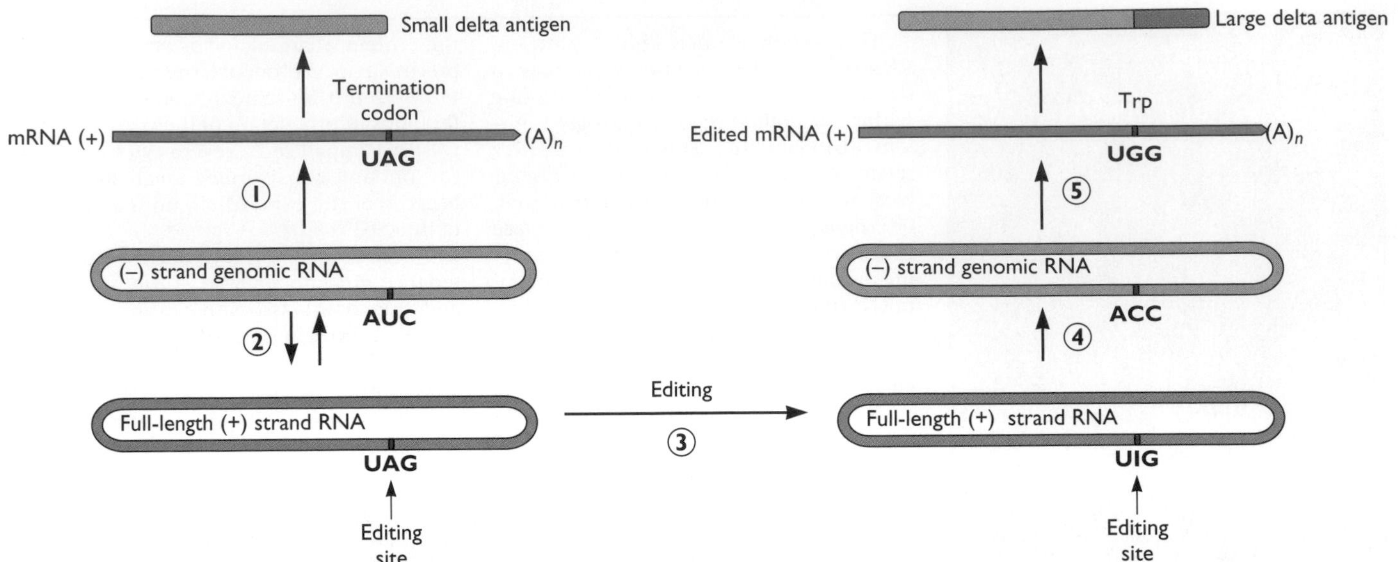

Figure 10.14 Editing of hepatitis delta satellite virus RNA by double-stranded RNA adenosine deaminase. The mRNA synthesized from genomic (–) strand RNA specifies the small delta antigen (step 1), which is required for replication of the genome (2). Double-stranded RNA adenosine deaminase acts on the full-length (+) strand RNA to convert a specific A residue to I (3). Because I base pairs with C, the genome (–) strands copied from edited full-length (+) strands contain a C residue (4). Such edited (–) strand RNA is therefore copied into (+) mRNA that contains a UGG codon (tryptophan) at the editing site (5) rather than the UAG stop codon. As a result, the mRNA made from edited RNA specifies the large delta antigen, which contains a 19-amino-acid C-terminal extension (purple). This protein inhibits genomic RNA replication and is needed for association of the hepatitis delta satellite virus genome with envelope proteins of its helper virus, hepatitis B virus.

production of too high a concentration of large delta antigen and consequently inhibition of replication of the satellite virus: overproduction of an adenosine deaminase in human cells containing hepatitis delta virus RNA greatly increased the efficiency of editing of the viral RNA and strongly inhibited its replication.

Double-stranded RNA adenosine deaminases may play a broader role in the biology of RNA viruses. Before the discovery of these enzymes, it was assumed that nucleotide changes in viral RNA genomes arise solely by the introduction of the wrong nucleotide by RNA polymerases, or by RNA recombination (Chapter 6). However, several changes can now be attributed to editing. For example, many of the genomic RNAs of defective measles viruses isolated from the brains of patients who died of subacute sclerosing panencephalitis appear to have been edited by cellular adenosine deaminases; the relationship of such mRNAs to this pathology is unknown.

Export of RNAs from the Nucleus

Any mRNA made in the nucleus must be transported to the cytoplasm for translation. Other classes of RNA, including small viral RNAs made by RNA polymerase III, must also enter the cytoplasm permanently (e.g., tRNAs and rRNAs) or transiently (snRNAs). The export of viral mRNAs and small RNAs is mediated by the host cell machinery and, in most cases, is indistinguishable from export of analogous cellular RNAs. However, transport to the cytoplasm of unusual mRNA molecules is a critical step in some viral life cycles. In this section, we describe the cellular export machinery and the mechanisms that ensure export of such viral mRNA substrates.

The Cellular Export Machinery

The substrates for mRNA export are not naked RNA molecules, but rather nucleic acid-protein complexes. Some of these proteins travel to the cytoplasm with the mRNA, but are then displaced and return to the nucleus. Consequently, such hnRNP proteins shuttle between the nucleus and cytoplasm (Box 10.6). One such protein, the abundant hnRNP-A1 protein, contains a short sequence, termed M9, that directs both export and import. This sequence was the first nuclear export signal to be identified. A different kind of nuclear export signal is found in proteins that direct export of certain viral mRNAs and small cellular RNAs (see "Export of Viral mRNA" below). In fact, export of RNA mol-

BOX 10.6

The interspecies heterokaryon assay for shuttling of proteins between the nucleus and the cytoplasm

Human cells synthesizing the protein of interest, e.g., the human immunodeficiency virus type 1 Rev protein, are fused with mouse cells that do not express the protein to form heterokaryons containing multiple nuclei (figure panel C). In this experiment, human (hu) and mouse (mo) nuclei were distinguished by immunofluorescent staining with antibodies specific for the human hnRNP C protein (A). They can also be identified by their distinctive patterns of DNA staining. The location of the protein is then determined by immunofluorescence with antibodies specific for the protein of interest, Rev. When a protein can shuttle between the nucleus and the cytoplasm, it is exported from human cell nuclei and reimported into all nuclei present in the heterokaryon. Consequently, it is detected in mouse as well as human nuclei (B). Although illustrated here with the Rev protein, this assay was originally developed to examine the shuttling of cellular proteins (nucleolar proteins and the hnRNP-A1 protein) between the nucleus and the cytoplasm. Adapted from B. E. Meyer and M. H. Malim, *Genes Dev.* **8:**1538–1547, 1994, with permission. Courtesy of M. H. Malim, University of Pennsylvania Medical Center.

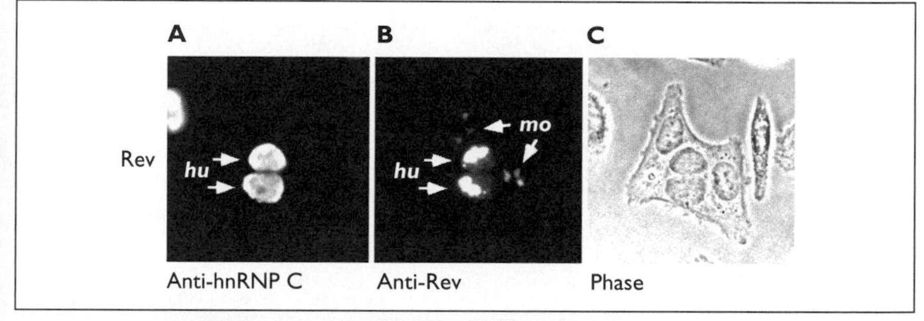

ecules (with the exception of tRNAs) is directed by sequences present in the proteins associated with them.

Like proteins entering the nucleus, RNA molecules travel between nuclear and cytoplasmic compartments via the nuclear pore complexes described in Chapter 5. Numerous genetic, biochemical, and immunocytochemical studies have demonstrated that specific nucleoporins (the proteins from which nuclear pore complexes are built) participate in nuclear export. Export of RNA molecules also shares several mechanistic features with import of proteins into the nucleus. Substrates for nuclear export or import are identified by specific protein signals, and some soluble proteins, including the small guanosine nucleotide-binding protein Ran, function in both import and export. And RNA export, like protein import, is mediated by receptors that recognize nuclear export signals and direct the proteins and RNA-protein complexes that contain them to and through nuclear pore complexes.

Export of Viral mRNA

All viral mRNAs made in infected cell nuclei carry the same 5′- and 3′-terminal modifications as cellular mRNAs that are exported. Furthermore, many viral mRNAs, like their cellular counterparts, are produced by splicing of intron-containing precursors. Cellular pre-mRNAs that contain introns and splice sites are ordinarily retained in the nucleus, at least in part because they remain associated with spliceosomes. Furthermore, a protein complex that marks mature mRNAs for export is assembled on the RNA only during splicing. These pre-mRNAs therefore are either spliced to completion and exported, or degraded within the nucleus. However, replication of retroviruses, herpesviruses, and orthomyxoviruses requires production of mRNAs that are incompletely spliced, or not spliced at all, and their export from the nucleus for translation. Efforts to address the question of how the normal restrictions on cellular mRNA export are circumvented in cells infected by these viruses have provided important insights into the complex process of export of macromolecules from the nucleus.

Export of Intron-Containing mRNAs by the Human Immunodeficiency Virus Type 1 Rev Protein

The human immunodeficiency virus type 1 Rev protein is by far the best understood of the viral proteins that modulate mRNA export from the nucleus. This protein and related proteins of other lentiviruses promote export of the

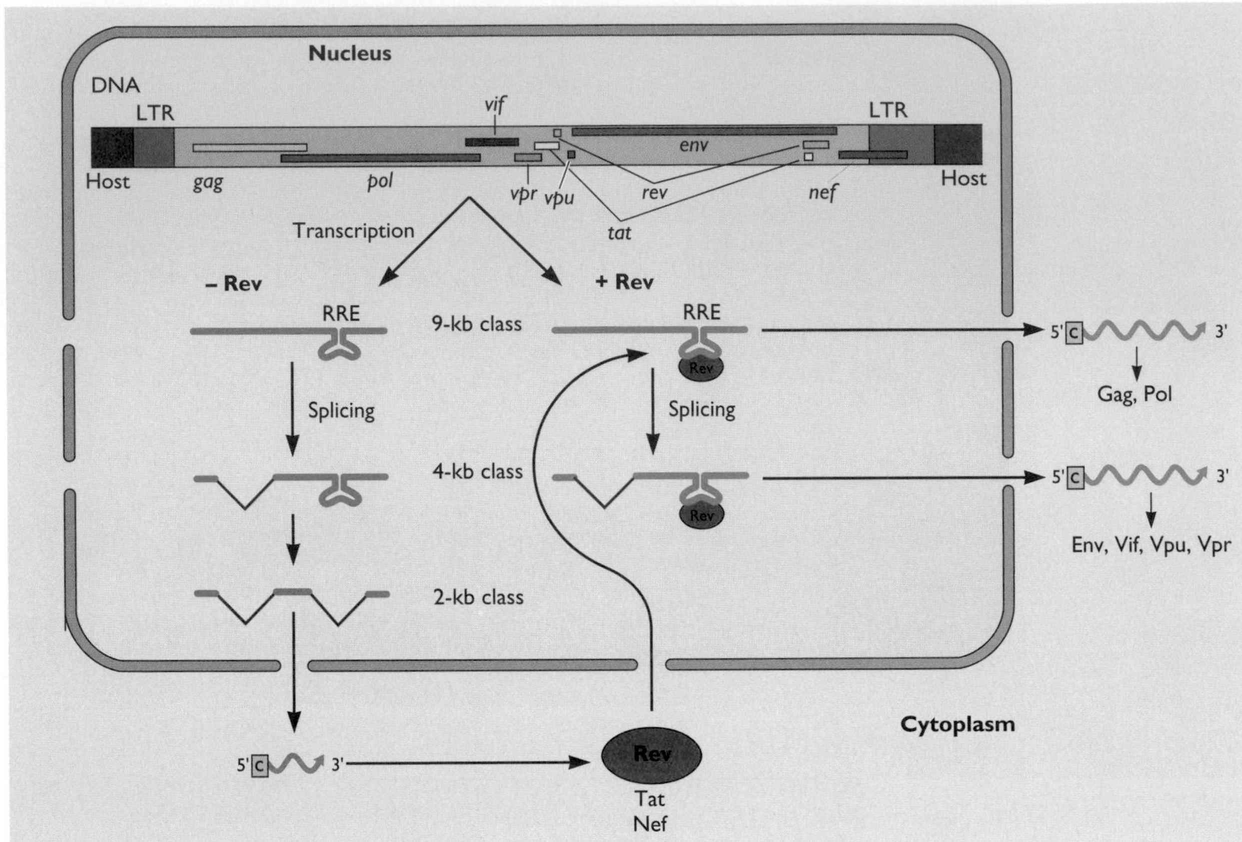

Figure 10.15 Regulation of export of human immunodeficiency virus type 1 mRNAs by the viral Rev protein. Before the synthesis of Rev protein in the infected cell, only fully spliced (2-kb class) viral mRNAs are exported to the cytoplasm (left). These mRNAs specify viral regulatory proteins, including Rev. The Rev protein enters the nucleus, where it binds to an RNA structure, the Rev-responsive element (RRE) present in unspliced (9-kb class) and singly spliced (4-kb class) viral mRNAs. This interaction induces export to the cytoplasm of the RRE-containing mRNAs, from which virion structural proteins and enzymes are made (right). The Rev protein therefore alters the pattern of viral gene expression as the infectious cycle progresses.

unspliced and partially spliced viral mRNAs. Rev binds specifically to an RNA sequence termed the **Rev-responsive element (RRE)** that lies within an alternatively spliced intron of viral pre-mRNA (Fig. 10.15). The Rev-responsive element is some 250 nucleotides in length and contains several stem-loops (Fig. 10.16A). One such structure contains a high-affinity binding site for the arginine-rich RNA-binding domain of Rev (Fig. 10.16B and C). Export of RNAs that contain the Rev-responsive element depends on the formation of Rev-protein oligomers on this sequence and a nuclear export signal present in Rev.

The C-terminal domain of Rev contains a short nuclear export signal (Fig. 10.16B) that is sufficient to induce export of heterologous proteins. In the oligomeric Rev-RNA complex, the nuclear export signals of the protein become organized on the surface. One cellular protein that binds to the leucine-rich nuclear export signal of Rev is exportin-1 (also known as Crm-1). Exportin-1 binds simultaneously to Rev and the GTP-bound form of Ran. This protein is the **receptor** for Rev-dependent export of the human immunodeficiency virus type 1 RNAs bound to it from the nucleus (Fig. 10.17). The viral protein therefore functions as an **adapter,** directing viral intron-containing mRNAs to a preexisting cellular export receptor. Translocation of the complex containing viral RNAs, Rev, and cellular proteins through the nuclear pore complex to the cytoplasm requires specific nucleoporins. In the cytoplasm, hydrolysis of GTP bound to Ran by a Ran-specific GTPase activating protein present only in the cytoplasm induces dissociation of the export complex. Rev then shut-

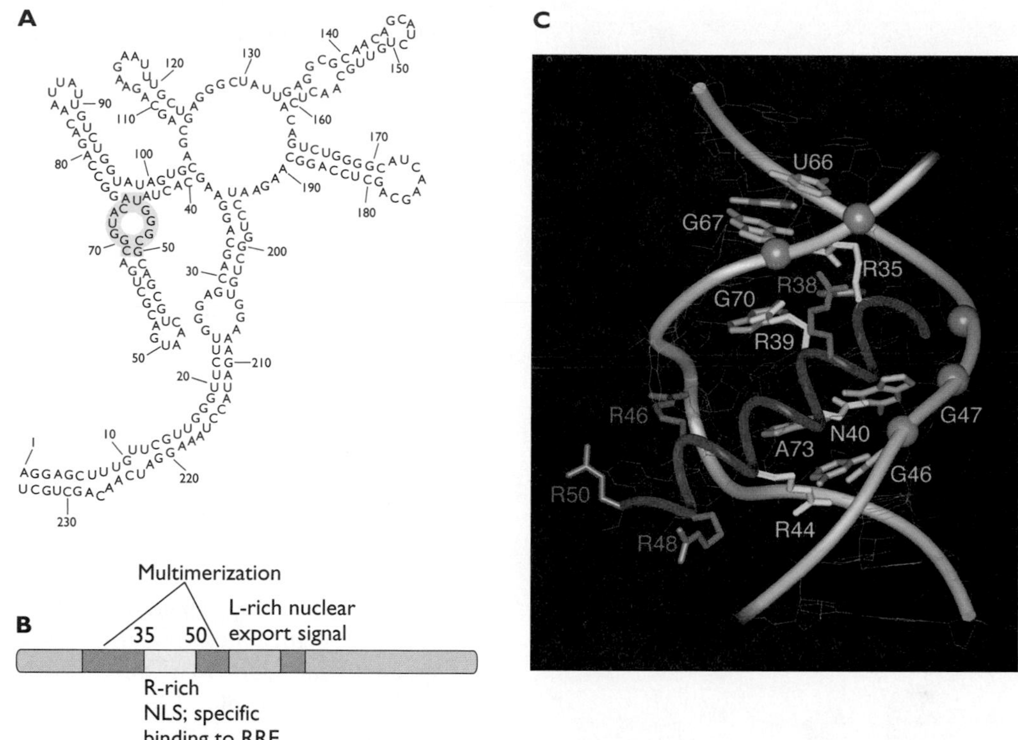

Figure 10.16 Binding of the Rev protein to the Rev-responsive element. (A) Predicted secondary structure of the 234-nucleotide Rev-responsive element, with the high-affinity binding site for Rev shaded in yellow. (B) The functional organization of the Rev protein is shown to scale. (C) Structure of a Rev peptide bound to the high-affinity binding site of the Rev-responsive element, determined by nuclear magnetic resonance methods. Comparison of the high-resolution structures of a Rev peptide bound to the high-affinity site and free Rev has provided important mechanistic insights into how specific RNA sequences and structural features are recognized. Binding of the protein induces substantial conformational change in the RNA to allow extensive interaction of the α-helical RNA-binding domain of Rev with the major groove in the RNA. The Rev peptide corresponding to amino acids 34 to 50 of the full-length protein forms an α-helix (red) that binds to the major groove of the RNA (blue), with phosphates that upon chemical modification impair Rev binding shown as spheres. The bases shown are invariant in RNA molecules selected from a large population for high-affinity binding to Rev in vitro. The major groove of an A-form helix is too narrow to accommodate an α-helix. However, in the Rev-RNA complex, the groove is widened by two purine-purine base pairs and distortion of the RNA backbone. The purine-purine pairs are not observed in the solution structure of the RNA alone, indicating that binding of Rev induces a substantial conformational change. The Rev protein penetrates deeply into the major groove and interacts extensively with the RNA over three to four turns of the α-helix. These interactions include base-specific contacts, for example, via the three arginine and one asparagine residues shown in yellow, and numerous contacts with the phosphate backbone of the RNA, via residues colored orange. From J. L. Battiste et al., *Science* **273:**1547–1551, 1996, with permission. Courtesy of J. R. Williamson, The Scripps Research Institute.

tles back into the nucleus via a typical nuclear localization signal (Fig. 10.16B), where it can pick up another cargo RNA molecule.

Perhaps the most interesting aspect of Rev-dependent RNA export is the exit of mRNAs by a pathway that normally does not handle such cargo, but rather exports small RNA species of the host cell. The Rev nuclear export signal is similar in sequence to, and can be functionally replaced by, that of the cellular protein TfIIIa. This protein binds specifically to 5S rRNA molecules, and it is required for export of this cellular RNA from the nucleus. Peptides containing the Rev nuclear export signal inhibit export of 5S rRNA but do not impair transport of cellular mRNAs from the nucleus to cytoplasm. The human immunodeficiency virus type 1 Rev protein therefore circumvents the normal restriction on the export of intron-containing pre-mRNAs from the host cell nucleus by diverting such viral mRNAs to a cellular pathway that handles unspliced RNAs.

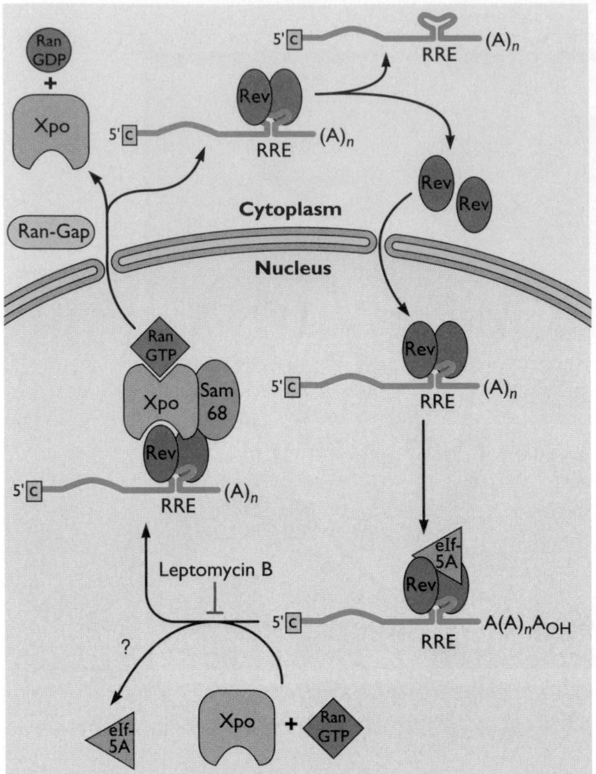

Figure 10.17 Model of export of human immunodeficiency virus type 1 mRNA containing introns mediated by the Rev protein. All cellular proteins shown in the nucleus (eIF-5A, exportin-1 [Xpo], the GTP-bound form of Ran [Ran-GTP], and the 68-kDa Src-associated protein in mitosis [Sam68]) have been implicated in Rev-dependent mRNA export. In the presence of Ran-GTP, Rev binds to exportin-1. This protein is related to the import receptors described in Chapter 5 and interacts with nucleoporins. The inhibition of both formation of the Rev–exportin-1–Ran-GTP ternary complex and Rev-dependent export by leptomycin B indicates that the quaternary complex, with Rev-responsive element mRNA bound to Rev oligomers, is translocated through the nuclear pore complex to the cytoplasm via interactions of exportin-1 with nucleoporins, such as Can/Nup14 and Nup98. The eukaryotic initiation factor 5A protein (eIF-5A), which binds to both Rev and ribosomal protein L5, is therefore depicted functioning upstream of the ternary complex to direct Rev to exportin-1. However, the sequence of interactions and the molecular functions of the eIF-5A protein in export remain to be determined. The Sam68 protein can bind to the Rev nuclear export signal, but does not appear to shuttle between nucleus and cytoplasm. It may therefore act prior to docking of the viral RRE-containing RNA complex at the nuclear pore complex. Hydrolysis of GTP bound to Ran to GDP induced by the cytoplasmic Ran GTPase activating protein (Ran-Gap) is presumed to dissociate the export complex, releasing viral mRNA for translation or assembly, and Ran and exportin-1 for reentry into the nucleus.

RNA Signals Mediating Export of Intron-Containing Viral mRNAs by Cellular Proteins

The genomes of many retroviruses do not encode regulatory proteins analogous to Rev, even though unspliced viral mRNAs must reach the cytoplasm. These unspliced viral mRNAs contain specific sequences that promote export (Table 10.3). Because they must function by means of cellular proteins, such sequences were termed **constitutive transport elements** (CTE). The first such sequence was found in the 3′ untranslated region of the genome of Mason-Pfizer monkey virus.

Even low concentrations of RNA containing the Mason-Pfizer monkey virus CTE inhibit export of mature mRNAs when microinjected into *Xenopus* oocyte nuclei, but this RNA does not compete with Rev-dependent export. This observation indicates that this retroviral RNA sequence is recognized by components of the cellular mRNA export pathway. A search for such proteins led to the first identification of a mammalian protein mediating mRNA export: the human Tap protein binds specifically to the CTE and is essential for export from the nucleus of the unspliced viral RNAs and spliced cellular mRNAs. Tap proved to be the human homolog of a yeast protein that is essential for mRNA export.

The pathway of Tap-dependent mRNA export has not yet been fully elucidated, but the Ran-GTP protein does not appear to participate. The direct and specific binding of Tap to the CTE of unspliced retroviral RNAs may bypass a cellular process that ensures that export is normally strictly coupled with splicing (Fig. 10.18). The Tap protein can bind only nonspecifically and with low affinity to cellular pre-mRNAs, but it does interact with components of a protein complex that is deposited on cellular mRNAs as they are spliced, such as the Ref proteins. Indirect recruitment of Tap to an mRNA, which couples export to splicing, is circumvented in the case of retroviral pre-mRNAs containing CTEs recognized directly by Tap, allowing export of unspliced RNAs from the nucleus.

The Balance of Incomplete Splicing of Viral Pre-mRNAs with Nuclear Export

A finely tuned balance in the synthesis of fully spliced and unspliced (or incompletely spliced) mRNA is crucial for the successful replication of retroviruses and orthomyxoviruses. This balance is primarily maintained by the relative efficiencies of splicing and export. On the one hand, splicing of viral pre-mRNA must be inefficient to allow export of the essential, intron-containing mRNAs (Fig. 10.11A). Indeed, increasing the efficiency of splicing of human immunodeficiency virus type 1 pre-mRNA, by replacing the natural, suboptimal splice sites with efficient ones, leads to complete splicing of all pre-mRNA molecules before Rev can recognize and export the unspliced mRNA to the cytoplasm. On the other hand, when unspliced mRNAs remain in the nucleus (e.g., before Rev is made in infected cells), they are eventually spliced to completion. Efficient export is therefore required to place unspliced mRNA into the cytoplasm for translation or incorporation into virus particles.

Table 10.3 Viral RNA sequences that direct export of unspliced mRNAs by cellular proteins

Virus	Sequence	Properties[a]	Functions
Hepadnaviruses			
Hepatitis B virus	Posttranscriptional regulatory element	433-nucleotide sequence within viral mRNAs; comprises two subelements, each with low activity, that function cooperatively; partially overlaps a transcriptional enhancer in the viral genome	Required for export to cytoplasm of viral mRNA; can substitute for Rev/RRE to promote export of unspliced human immunodeficiency virus type 1 mRNA; can increase cytoplasmic concentrations of synthetic mRNAs that lack introns; activity is exportin 1-independent; bound by hnRNP I protein, which stimulates export of viral RNA
Woodchuck hepatitis virus	Posttranscriptional regulatory element	Includes the two conserved subelements and an additional subelement (PRE-γ); functions more efficiently than hepatitis B virus PRE	Required for export to cytoplasm of viral mRNA; possesses both exportin-1-dependent and exportin-1-independent activities that may function cooperatively
Retroviruses			
Mason-Pfizer monkey virus	Constitutive transport element	154-nucleotide sequence in 3′ untranslated region; forms a single double-strand stem with two internal loops; contains degenerate repeat sequences; sequences and structure conserved in RNA of other simian retroviruses	Required for export of unspliced viral mRNA; can substitute for Rev/RRE to promote export of human immunodeficiency virus type 1 intron-containing mRNA; bound by cellular Tap protein; accesses cellular mRNA export pathway
Rous sarcoma virus	Direct repeat sequence	~100-nucleotide sequence flanking Src gene with secondary-structure features similar to constitutive transport elements of simian retroviruses	Required for efficient cytoplasmic accumulation of unspliced mRNA

[a]PRE, posttranscriptional regulatory element.

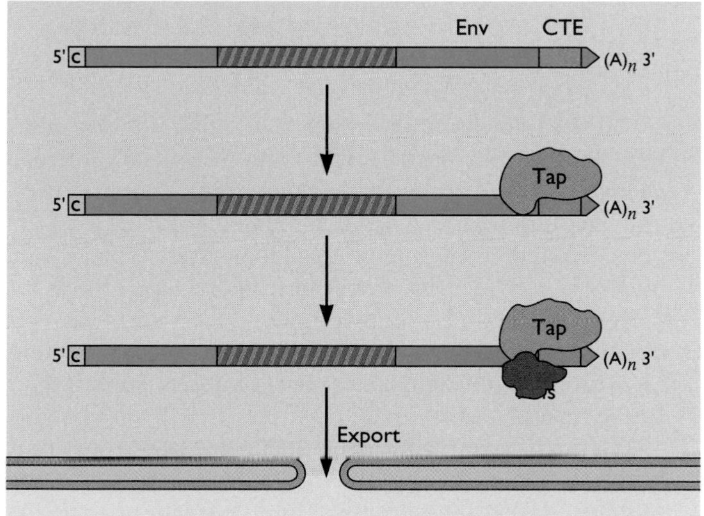

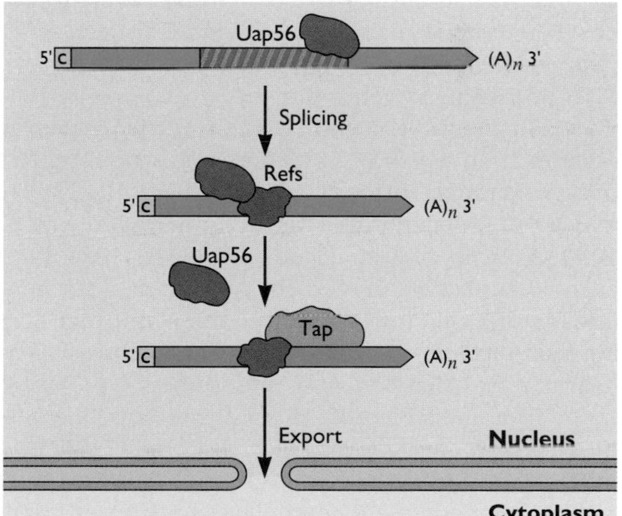

Figure 10.18 Model of export of unspliced RNA of retroviruses with simple genomes from the nucleus. Export of unspliced, primary transcripts of many retroviruses depends on the constitutive transport element (CTE) in the RNA. This sequence is recognized by the cellular protein Tap, which is then bound by proteins that mark mRNAs as appropriate substrates for export, such as Ref. A variety of experimental approaches, including genetic studies with yeasts, have indicated that Tap is an essential component of a cellular mRNA export pathway. However, Tap does not bind to cellular mRNAs with high affinity, but rather becomes associated with them via interactions with Ref proteins. The latter is deposited on the cellular mRNA during splicing, and therefore marks spliced mRNAs for export. The direct interaction of Tap with retroviral CTEs therefore appears to bypass the mechanism that couples the splicing of cellular mRNAs with their export.

Export of Single-Exon Viral mRNAs

Most of the viral mRNAs made in cells infected by hepadnaviruses, herpesviruses, or orthomyxoviruses are not spliced. In contrast to the retroviral mRNAs described in previous sections, these viral mRNAs do not retain introns. Rather, the viral genes that encode them contain no introns and the RNAs **cannot** be spliced. We therefore designate such mRNAs **single-exon mRNAs** to distinguish them from those that retain introns.

As such viral mRNAs do not contain splice sites, they should not become trapped in the nucleus by a spliceosome retention mechanism. On the other hand, they have few counterparts in uninfected mammalian cells growing under normal conditions, a property that raises the question of how viral single-exon mRNAs are transported efficiently to the cytoplasm. Recent experiments have identified both viral proteins and RNA sequences that promote export of such viral mRNA, analogous to the retroviral Rev protein and constitutive transport elements, respectively.

Most viral mRNAs made in herpes simplex virus-infected cells lack introns. The viral early mRNA encoding thymidine kinase contains a sequence that directs export by cellular components. This sequence is sufficient to induce efficient, cytoplasmic accumulation of unspliced β-globin mRNA when fused to it, and its recognition by the cellular hnRNP L protein correlates with RNA accumulation in the cytoplasm. It may therefore direct the viral mRNA to a cellular pathway for export of the rare, cellular mRNAs that are not spliced. This same pathway may be responsible for export of hepadnaviral mRNAs. These viral mRNAs are made by cellular transcriptional and RNA-processing systems, but the great majority lack introns (Appendix A, Fig. 3B). Their efficient cytoplasmic accumulation depends on a conserved sequence, termed the **posttranscriptional regulatory element** (PRE), which can function in the absence of viral proteins (Table 10.3). Functionally analogous sequences have recently been identified in one of the uncommon mammalian mRNAs that lack introns (a histone 2A mRNA), suggesting that hepadnaviral mRNAs leave the nucleus via a cellular pathway that handles a minor fraction of the mRNA traffic in uninfected cells. The posttranscriptional regulatory element of other hepadnaviruses may facilitate viral mRNA export from the nucleus by promoting cooperation among components of multiple cellular export systems (Table 10.3).

The efficient export of the majority of herpesviral single-exon mRNAs requires the viral ICP27 protein. This protein, like the products of all immediate-early genes, is made at the beginning of the infectious cycle. It is a nuclear phosphoprotein that contains a leucine-rich nuclear export signal resembling that of the human immunodeficiency virus Rev protein and shuttles between the nucleus and the cytoplasm. ICP27 specifically facilitates the export of viral single-exon mRNAs: these exon mRNAs do not enter the cytoplasm efficiently in cells infected by mutant viruses that do not direct the synthesis of ICP27, but export of viral mRNAs spliced from intron-containing precursors continues normally. This viral protein has been reported to bind to viral RNAs in infected cells. It contains different types of RNA-binding motifs in its N-terminal and C-terminal portions, both of which are necessary for optimal replication of the virus. A considerable body of genetic evidence indicates that the RNA-binding motifs of ICP27 contribute to efficient cytoplasmic accumulation of viral single-exon mRNAs. On the other hand, the identity of the cellular export pathway, or pathways, to which ICP27 directs these mRNAs remains a matter of debate: observations consistent with export of herpesviral mRNAs by both exportin-1- and Tap-dependent mechanisms have been reported (Box 10.7).

Posttranscriptional Regulation of Viral or Cellular Gene Expression by Viral Proteins

The production of viral mRNAs in cells infected by a variety of viruses depends critically on the cellular RNA processing machinery. The genomes of several of these viruses encode proteins that regulate one or more RNA processing reactions (Table 10.4). These proteins play crucial roles in temporal regulation of viral gene expression, or in inhibition of the production of cellular mRNAs (Table 10.4).

Temporal Control of Viral Gene Expression

Regulation of Alternative Polyadenylation by Viral Proteins

The frequencies with which alternative poly(A) addition sites within specific viral pre-mRNAs are utilized change during infection by adenoviruses, herpesviruses, and papillomaviruses. As discussed previously, the polyadenylation of bovine papillomavirus type 1 late mRNA is activated by a specific complement of cellular proteins found only in fully differentiated cells of the epidermis. In contrast, viral proteins have been implicated in regulation of polyadenylation in cells infected by the larger DNA viruses.

Despite its name, the adenoviral major late promoter is active during the early phase of infection, prior to the onset of viral DNA synthesis. The major late pre-mRNAs made during this period are polyadenylated predominantly at the L1 mRNA site (Fig. 10.12), even though they also contain the L2 and L3 3′ processing sites. Such selective recognition of the L1 site depends on the cellular cleavage-stimulatory protein (Cstf), which binds to the U/GU-rich sequence 3′ to the cleavage site (Fig. 10.3). As infection continues, the activity of this cellular protein decreases. The mechanism responsible for this change in

BOX 10.7

Does the herpes simplex virus type 1 ICP27 protein engage multiple cellular RNA export pathways?

The nucleocytoplasmic shuttling of the ICP27 protein under the direction of its leucine-rich nuclear export signal originally suggested that this herpesviral protein, like Rev, engages the exportin-1-mediated export pathway of the host cell. The observation that leptomycin B, a specific inhibitor of exportin-1, impairs the cytoplasmic accumulation of viral single-exon mRNAs is consistent with this mechanism.

In contrast, ICP27-dependent export of herpesviral mRNAs from microinjected *Xenopus* oocyte nuclei was found to require not exportin-1, but proteins that mediate mRNA export, Ref and Tap. In this system, export of herpesviral mRNAs is inhibited by the simian retroviral constitutive transport element (Table 10.3) and depends on binding of ICP27 to Ref.

Is it possible that ICP27 can induce export of viral mRNAs from the nucleus by accessing either the exportin-1 or the Tap-containing cellular pathways? Several observations are consistent with this hypothesis.

When herpesvirus-infected cells are treated with leptomycin B, the cytoplasmic accumulation of some viral, single-exon mRNAs (e.g., UL17 and gC mRNAs) is reduced severely, but that of others is unaffected (e.g., VP22 mRNA).

Deletion of the sequence encoding the leucine-rich nuclear export signal from the ICP27 gene resulted in restriction of the protein to the nucleus. However, the mutant displayed only 100-fold reduction in replication, and much less severe defects in viral late protein synthesis or accumulation of viral mRNAs in the cytoplasm, than cells infected by an ICP27 null mutant. As the altered ICP27 retained the sequence by which it binds to Ref, it may continue to direct viral mRNAs to this pathway.

This viral protein therefore may be an example of an export adapter that can direct mRNA to more than one export receptor, a mechanism established for the cellular HuR protein.

Koffa, M. D., J. B. Clements, E. Izzauralde, S. Wadd, S. A. Wilson, I. W. Mattaj, and S. Kursten. 2001. Herpes simplex virus ICP27 protein provides viral mRNAs with access to the cellular mRNA export pathway. *EMBO J.* **20:**5769–5778.

Lenggel, J., C. Guy, V. Leong, S. Borge, and S. A. Rice. 2002. Mapping of functional regions in the amino-terminal portion of the herpes simplex virus ICP27 regulatory protein: importance of the leucine-rich nuclear export signal and RGG box RNA-binding domain. *J. Virol.* **76:**11866–11879.

Soliman, T. M., and S. J. Silverstein. 2000. Herpesvirus mRNAs are sorted for export via Crm1-dependent and -independent pathways. *J. Virol.* **74:**2814–2825.

Table 10.4 Viral proteins that regulate RNA processing reactions

Virus	Protein(s)	Properties and functions
Adenovirus		
Human adenovirus type 2	E4 ORF4	Induces dephosphorylation of cellular SR proteins by protein phosphatase 2A; relieves inhibition of L1 pre-mRNA splicing at the IIIa 3′ splice site by phosphorylated SR proteins present early in infection
	E1B 55 kDa-E4 ORF6	The complex of the two proteins inhibits export of fully processed cellular mRNAs to induce selective export of viral late mRNAs
Herpesvirus		
Herpes simplex virus type 1	ICP27	Stimulates polyadenylation of some viral late mRNAs with suboptimal U/GU-rich sequences at their polyadenylation sites; inhibits splicing of intron-containing cellular and probably viral mRNAs; promotes export of viral single-exon mRNAs; contains RNA-binding domains and a leucine-rich nuclear export signal, recognized by exportin-1
Retrovirus		
Human immunodeficiency virus type 1	Rev	Mediates export of unspliced and incompletely spliced viral mRNAs; binds to the Rev-responsive element in these mRNAs as an oligomer and directs them to the cellular pathway for export of unspliced small RNAs via interactions with cellular exportin-1 proteins
Orthomyxovirus		
Influenza A virus	P proteins	Inhibit splicing of M1 mRNA at the M3 5′ splice site upon binding to 5′ end of the RNA; allows binding of cellular Sf2 to a purine-rich sequence in M1 and activation of the suboptimal M2 mRNA 5′ splice site

activity has not been elucidated, but synthesis of viral proteins is required. The recognition of the other four polyadenylation sites present in major late pre-mRNA synthesized during the late phase (Fig. 10.12) is much less dependent on Cstf. It is therefore believed that these poly(A) addition signals compete more effectively with the L1 site for components of the polyadenylation machinery later in the infectious cycle.

Some herpesviral late pre-mRNAs contain poly(A) addition signals that function poorly in the absence of the ICP27 protein, at least in part because the positions or sequences of the downstream U/GU-rich sequences are not optimal. This viral protein appears to stimulate binding of Cstf to such suboptimal sites, and consequently increases the rate of polyadenylation of these mRNAs.

Regulation of Splicing

Control of alternative splicing. Some viral proteins that regulate pre-mRNA splicing alter the balance among alternative splicing reactions at specific points in the infectious cycle. For example, the ratios of alternatively spliced mRNA products of several adenoviral pre-mRNAs change with the transition into the late phase of infection. This phenomenon has been studied most extensively using the L1 mRNAs. The L1 pre-mRNA, the product of major late transcription and polyadenylation during the early phase of infection, can be spliced at one of two alternative 3′ splice sites (Fig. 10.12). However, only the L1 mRNA that specifies the 52/55-kDa protein is made prior to the onset of viral DNA synthesis. At such early times in the infectious cycle, binding of cellular SR proteins to a negative regulatory sequence located immediately upstream of the branchpoint for the L1 IIIa mRNA blocks its recognition (Fig. 10.12). A viral early protein encoded within the E4 transcription unit induces dephosphorylation of the SR proteins, which are highly phosphorylated in uninfected cells. Such dephosphorylated SR proteins lose the ability to repress recognition of the 3′ splice site of the L1 IIIa mRNA. Overproduction of the SR protein Sf2 in adenovirus-infected cells impairs not only synthesis of the L1 IIIa mRNA, but also the temporal shift in splicing of early E1A and E1B mRNAs, as well as viral replication. This observation indicates that dephosphorylation of cellular SR proteins makes a major contribution to posttranscriptional regulation of adenoviral gene expression. However, other mechanisms are also important, at least in some cases. For example, the L1 pre-mRNA sequence required for suppression of recognition of the IIIa mRNA splice site prior to SR protein phosphorylation also acts as a potent, late phase-specific enhancer of splicing at this site.

A cellular SR protein (Sf2) is also important for temporal regulation of synthesis of the spliced mRNA that encodes the M2 ion channel protein of influenza A viruses. In this case, the SR protein acts positively, after binding of viral proteins to the splicing substrate (Fig. 10.19).

Regulation of mRNA Export

Even though all are encoded within a single proviral transcription unit, the regulatory and virion proteins of human immunodeficiency virus type 1 are made sequentially in infected cells as a result of regulation of mRNA export by the Rev protein. The viral pre-mRNAs initially made are all multiply spliced prior to export from the nucleus to the cytoplasm (Fig. 10.15). These mRNAs encode the viral regulatory proteins such as Tat and Rev. The Rev protein enters the nucleus, and mediates export of unspliced and partially spliced viral mRNAs that contain the Rev-responsive element. These mRNAs specify the major structural proteins and enzymes of the virion (Fig. 10.15). The Rev protein therefore regulates a switch in viral gene expression from early production of viral regulatory proteins to a later phase, in which virion components are made.

Viral proteins that modulate mRNA export may play subsidiary, but nonetheless crucial, roles in temporal regulation of viral gene expression. For example, the transcriptional program described in Chapter 8 results in efficient transcription of herpes simplex virus late genes only following initiation of viral DNA synthesis in infected cells. Because all but one of the late mRNAs comprises a single exon, their efficient entry into the cytoplasm and the synthesis of viral late proteins requires ICP27. Consequently, this posttranscriptional regulator is essential for putting the viral transcriptional program into effect.

Inhibition of Cellular mRNA Production by Viral Proteins

All viral mRNAs are translated by the protein synthesis machinery of the host cell. Inhibition of production of cellular mRNAs can therefore favor this essential step in viral replication. Several mechanisms of selective inhibition of cellular RNA processing operate in virus-infected cells.

Inhibition of Polyadenylation and Splicing

The influenza virus NS1 protein can inhibit both polyadenylation and splicing of cellular pre-mRNAs. This viral protein comprises an N-terminal RNA-binding domain and a C-terminal segment that is required for inhibition of polyadenylation. This C-terminal domain contains binding sites for both Cpsf and PabII (Fig. 10.3). The former interaction inhibits polyadenylation of cellular mRNAs in experimental systems, and the NS1 protein is required for inhibition of cellular mRNA polyadenylation and cytoplasmic accumulation in virus-infected cells. These properties

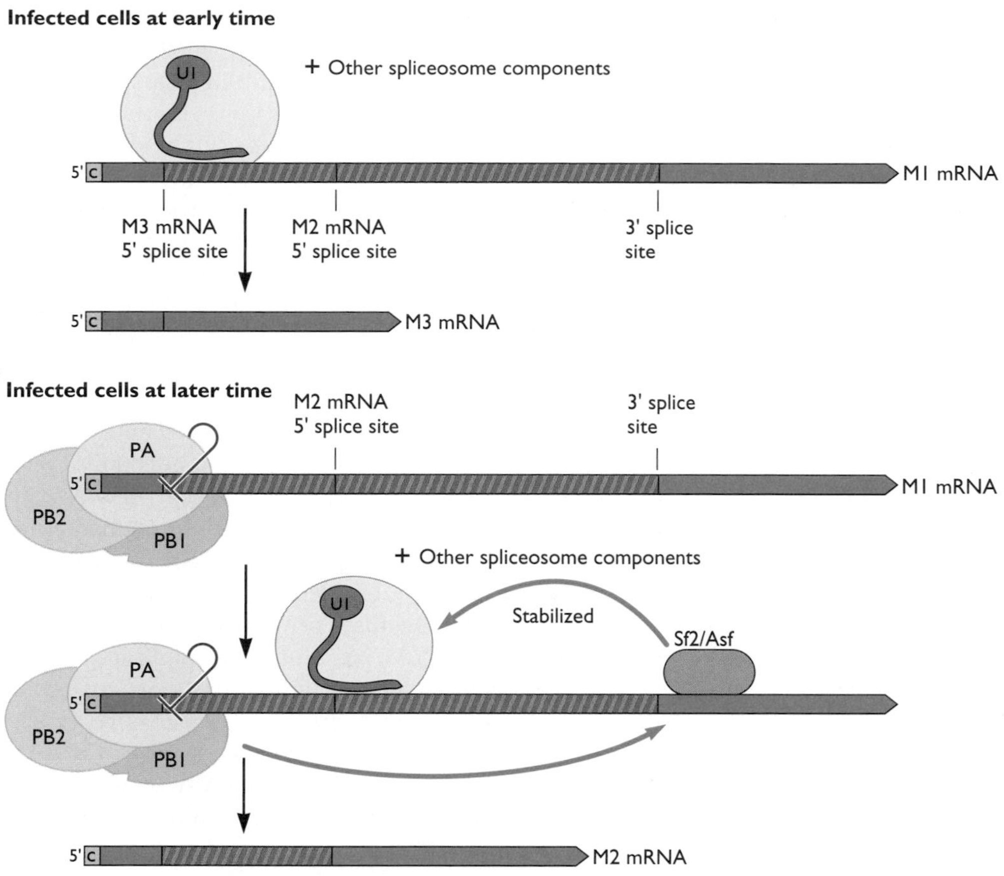

Figure 10.19 Model of regulation of splicing of the influenza A virus M1 mRNA. The mRNA synthesized from the M genomic RNA segment encodes the M1 protein but can also be spliced to produce the mRNA specifying the M2 protein. However, at early times of infection (top), before the synthesis of P proteins, the strong 5′ splice site for a third RNA, M3 mRNA, is used exclusively and only M3 mRNA is made. At later times (below), the P proteins bind to a specific sequence near the 5′ end of the influenza A virus mRNAs. Such binding blocks access to the M3 mRNA 5′ splice site, allowing unspliced M1 mRNA to be synthesized in infected cells. The 5′ splice site for the M2 mRNA is suboptimal, and its recognition depends on binding of the cellular SR protein Sf2 to a purine-rich splicing enhancer in the 3′ exon of M1 mRNA. Such binding requires interaction of the P proteins with the 5′ end of M1 mRNA, suggesting that the conformation of the RNA is altered upon binding of P proteins.

suggest that NS1 could increase the intranuclear concentrations of cellular pre-mRNAs, facilitating cap-snatching (Chapter 6) and indirectly inhibiting translation of cellular mRNAs. However, cellular protein synthesis is inhibited effectively when NS1 is not made in infected cells. Under these circumstances, infected cells produce larger quantities of interferon mRNAs and of other cellular mRNAs encoding proteins with antiviral activities than do wild-type virus-infected cells. Inhibition of processing of such cellular mRNAs may therefore contribute to the circumvention of host cellular defenses, a critical function of the NS1 protein (see Chapters 11 and 15).

In addition to its other activities, the herpes simplex virus protein ICP27 inhibits splicing of cellular pre-mRNAs. This protein induces redistribution of spliceosomal snRNPs from a pattern of diffuse speckles in the nucleus to a limited number of sites with a punctate pattern. The viral protein colocalizes with snRNPs at the latter sites, and may bind directly to them. Such abnormal intranuclear distribution of essential splicing components might contribute to the inhibition of splicing, but ICP27 also acts more directly: extracts prepared from infected cells are defective for splicing in vitro, but only when ICP27 is made in the cells. Direct interaction of the viral protein with components of the spliceosome inhibits splicing at an early step in assembly of the functional complex. Genetic analyses have shown that disruption of cellular RNA processing by ICP27 leads to inhibition of cellular protein synthesis. Moreover, this function is genetically separable from the role of this protein in efficient production of viral late mRNAs. Because herpesviral genes generally lack introns, inhibition of splicing is an effective strategy for the selective inhibition of cellular gene expression.

Inhibition of Cellular mRNA Export by Viral Proteins

In contrast to the other viruses considered in this section, adenovirus infection disrupts cellular gene expression by direct inhibition of export of cellular mRNAs from the nucleus. During the late phase of infection, the great majority of newly synthesized mRNAs entering the cytoplasm are viral in origin. Transcription and processing of cellular pre-mRNAs are unaffected, but these mRNAs are not exported and are degraded within the nucleus. Viral mRNAs are selectively exported, a phenomenon that is important for efficient synthesis of viral late proteins. When selective viral mRNA export is prevented by mutations in the viral genome, both the quantities of late proteins made in infected cells and virus yield are reduced substantially. These same phenotypes are seen in herpes simplex virus-infected cells synthesizing an altered ICP27 protein that is defective only for the inhibition of pre-mRNA splicing. These properties emphasize the importance of posttranscriptional inhibition of cellular mRNA production for efficient virus reproduction.

The preferential export of viral late mRNAs in adenovirus-infected cells requires two viral early proteins, the E1B 55-kDa and the E4 ORF6 proteins. These proteins are found both free and in association with one another in infected cell nuclei. The complex is responsible for the regulation of mRNA export. Like human immunodeficiency virus type 1 Rev and herpesviral ICP27, both the E1B and the E4 proteins contain leucine-rich nuclear export signals and can shuttle between the nucleus and cytoplasm. However, whether shuttling of either viral protein is necessary for regulation of mRNA export is currently a matter of debate. The selectivity of mRNA export is especially puzzling, because the viral mRNAs possess all the characteristic features of cellular mRNAs and are made in the same way. One hypothesis is that the viral early protein complex recruits nuclear proteins needed for export of mRNA to the specialized sites within the nucleus at which the adenoviral genome is replicated and transcribed. As a result of such sequestration, export of cellular mRNAs would be inhibited.

Regulation of Turnover of Viral and Cellular mRNAs in the Cytoplasm

Individual mRNAs may differ in the rate at which they are translated, and also in such properties as cytoplasmic location and stability. Indeed, the intrinsic lifetime of an mRNA can be a critical parameter in the regulation of gene expression.

In the cytoplasm of mammalian cells, the lifetimes of specific mRNAs can differ by as much as 100-fold. This property is described in terms of the time required for 50% of the population of the mRNAs to be degraded under conditions in which replenishment of the cytoplasmic pool is blocked, the **half-life** of the mRNA. Many mRNAs are very stable, with half-lives exceeding 12 h. As might be anticipated, such mRNAs encode proteins needed in large quantities throughout the lives of all cells, such as structural proteins (e.g., actin) and ribosomal proteins. At the other extreme are very unstable mRNAs with half-lives of less than 30 min. This class includes mRNAs specifying regulatory proteins that are synthesized in a strictly controlled manner in response to cues from external or internal environments of the cell, such as cytokines, and proteins that regulate cell proliferation. The short lifetimes of these mRNAs ensure that synthesis of their products can be shut down effectively once they are no longer needed. Specific sequences that signal the rapid turnover of the mRNA in which they reside, such as a 50- to 100-nucleotide AU-rich sequence within the 3′ untranslated regions, have been identified, as have several mechanisms of mRNA degradation (Fig. 10.20).

The stabilities of viral mRNAs have not been examined in much detail, in part because many viral infectious cycles are completed within the normal range of mRNA half-lives. Nevertheless, it is clear that a viral protein that induces RNA degradation plays an important role in selective expression of viral genes in herpes simplex virus-infected cells. Regulation of the stability of specific viral or cellular mRNAs has also been implicated in the permanent changes in cell growth properties (transformation) induced by some viruses. Furthermore, RNA-mediated induction of degradation of specific mRNAs, a widespread phenomenon known as **RNA interference**, is believed to contribute to host antiviral defense mechanisms (Box 10.8).

Figure 10.20 Destabilization of cellular mRNAs by 5′AUUUA3′ sequences, repeated in the 3′ untranslated regions. Such sequences, which are present in short-lived mRNAs encoding cytokines and proteins that regulate cell growth and division, are specifically recognized by one or more proteins. Some proteins stabilize mRNAs upon binding to 5′AUUUA3′ sequences, but binding of others, depicted by the blue oval, appears to induce shortening of the poly(A) tail. This reaction is followed by degradation of the mRNA, in one mechanism via decapping and 5′-3′ exoribonucleolytic attack.

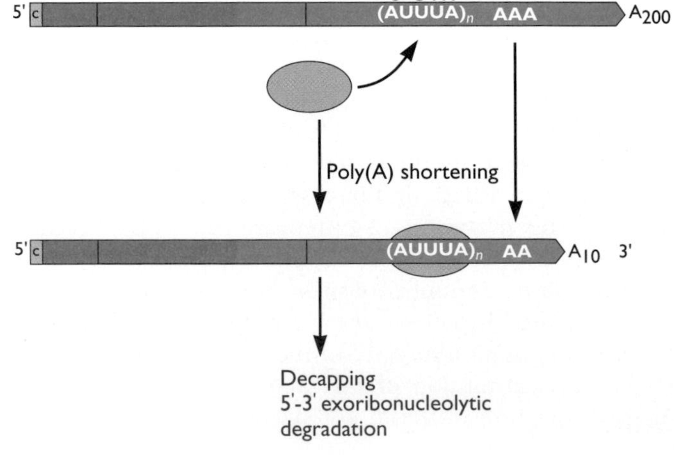

BOX 10.8

RNA interference: an ancient cellular defense mechanism

In the early 1990s, attempts to produce more vividly purple petunias by creation of transgenic plants carrying an additional copy of the enzyme that makes the purple pigment often resulted in white flowers. It is now clear that this seemingly esoteric observation represented the first example of a previously unknown mechanism of posttranscriptional regulation of gene expression called **RNA interference.** In this process, which is ancient and conserved among eukaryotes, small RNA molecules containing sequences specified in cellular genes direct degradation of the mRNAs synthesized from the corresponding gene.

As illustrated in the figure, in *Drosophila* and human cells, small interfering RNAs (siRNAs) are 21- to 23-nucleotides-long, double-stranded RNAs with overhangs of two unpaired nucleotides at their 3′ ends. Introduction of synthetic siRNAs into cells is proving a valuable experimental tool with which to inhibit the synthesis of specific gene products (Chapter 2). Naturally, siRNAs are produced by cleavage of longer RNA molecules by double-stranded RNA-specific endonucleases, called Dicer after the *Drosophila* enzyme. Following their production, single-stranded siRNAs became associated with a second, and distinct, endoribonuclease, the RNA-induced silencing complex (Risc). This enzyme complex cleaves mRNA molecules containing a sequence complementary to the siRNA that is bound to it. In this way, the siRNA targets the ribonuclease to degrade a specific mRNA (see figure).

In some organisms, including plants and the nematode *Caenorhabditis elegans*, RNA interference requires amplification of target mRNA sequences by RNA-dependent RNA polymerases. In contrast, there is no evidence for such an amplification step in *Drosophila* or mammalian cells. The reason for this difference is not yet understood, but one hypothesis is that siRNAs are more stable in the latter organisms and therefore attain a sufficiently high concentration without amplification. Also incompletely understood are the natural signals and mechanisms that trigger RNA interference.

(continued)

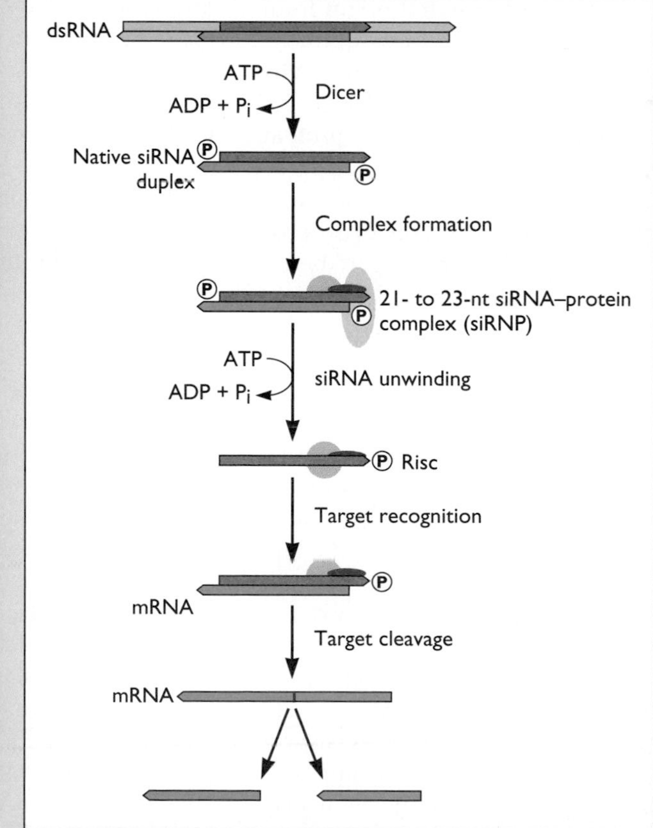

Model for production and action of siRNAs, based on studies with *Drosophila* and human cells. Adapted from P. D. Zamore, *Science* **296:**1265–1269, 2002, with permission.

BOX 10.8 (continued)

One well-established trigger is double-stranded RNA, a form of nucleic acid that is produced at higher than normal concentrations in virus-infected cells, where it activates various host immune defense mechanisms (Chapter 15). There is accumulating evidence that RNA interference impairs replication of (+) strand RNA viruses in plants and animal cells and that viral countermeasures to this newly recognized defense have evolved.

Ahlquist, P. 2002. RNA-dependent RNA polymerases, viruses, and RNA silencing. *Science* **296:**1270–1273.

Zamore, P. D. 2002. Ancient pathways programmed by small RNAs. *Science* **296:** 1265–1269.

Regulation of mRNA Stability by a Viral Protein

The virion host shutoff protein (Vhs) of herpes simplex virus reduces the stability of mRNAs in infected cells. As its name implies, Vhs is a structural protein of the virion. It is present at low concentrations in the tegument and therefore delivered to infected cells at the start of the infectious cycle, before viral gene expression begins. It remains in the cytoplasm where it mediates degradation of cellular mRNAs. The Vhs protein is an RNase that cleaves mRNA endonucleolytically. The removal of cellular mRNAs at the start of infection facilitates viral gene expression, presumably by reducing or eliminating competition from cellular mRNAs during translation.

The Vhs protein specifically targets mRNAs, probably by virtue of its binding to the translation initiation protein eIF4H, but it cannot distinguish those of viral origin from their cellular counterparts. It therefore induces degradation of both viral and cellular mRNAs. Although more Vhs protein is made in infected cells once its coding sequence is expressed during the late phase of infection, the protein is sequestered in the teguments of assembling virus particles by interaction with the viral VP16 protein. As a result, the activity of Vhs decreases as the infection cycle progresses. This mechanism presumably contributes to the efficient synthesis of viral proteins characteristic of the late phase of infection.

Regulation of mRNA Stability in Transformation

Stabilization of specific viral mRNAs appears to be important in the development of cervical carcinoma associated with infection by high-risk human papillomaviruses, such as types 16 and 18. As discussed in Chapter 18, the E6 and E7 proteins of these viruses induce abnormal proliferation of the cells in which they are made. In benign lesions, the circular human papillomavirus genome is not integrated (Fig. 10.21). The E6 and E7 mRNAs that are synthesized from such templates contain destabilizing, AU-rich sequences in their 3′ untranslated regions and possess short lifetimes. This property, as well as repression of transcription from the viral promoter from which the viral pre-mRNAs are transcribed, results in low, steady-state concentrations of the E6 and E7 mRNAs. In cervical carcinoma cells, the viral DNA is integrated into the cellular genome. Such reorganization of viral DNA frequently disrupts the sequences encoding the E6 and E7 mRNAs, such that their 3′ untranslated regions are copied from cellular DNA sequences (Fig. 10.21). These hybrid mRNAs lack the destabilizing AU-rich sequences and are more stable. The increase in the stability of the viral mRNAs encoding the papillomaviral transforming proteins accounts, at least in part, for their higher concentrations in tumor cells.

The regulation of mRNA stability may also play an important role in the alteration of cell growth by herpesvirus saimiri. This gammaherpesvirus causes lymphomas or leukemias in certain species of New World monkeys, and immortalizes T cells. Three small viral RNAs produced in large quantities in immortalized cells have been implicated in such transformation. These viral RNAs share several properties with spliceosomal snRNAs, including binding to many of the same proteins. However, they appear to alter cell growth by increasing the stability of cellular mRNAs that contain multiple copies of the sequence 5′AUUUA3′ in their 3′ untranslated regions. Binding of one or more proteins to this sequence in an mRNA leads to shortening of the poly(A) sequence and degradation of the mRNA (Fig. 10.20). Some of the herpesvirus saimiri small RNAs also contain multiple copies of the 5′AUUUA3′ sequence and compete for binding to at least one protein that recognizes this sequence in unstable mRNAs. It has therefore been proposed that the viral small RNAs act as decoys for proteins that initiate degradation of cellular mRNAs marked by the presence of 5′AUUUA3′ sequences. The resulting increase in concentration of these mRNAs would lead to overproduction of cellular proteins that regulate cellular proliferation, many of which are specified by such unstable mRNAs.

Perspectives

Many of the molecular processes required for replication of animal viruses, including such virus-specific reactions as synthesis of genomic and messenger RNAs from an RNA template, were foretold by the properties of the bacteriophages that parasitize prokaryotes. In contrast, the covalent modifications necessary to produce functional mRNAs in eukaryotic cells were without precedent when discovered in viral systems. Study of the processing of viral RNAs

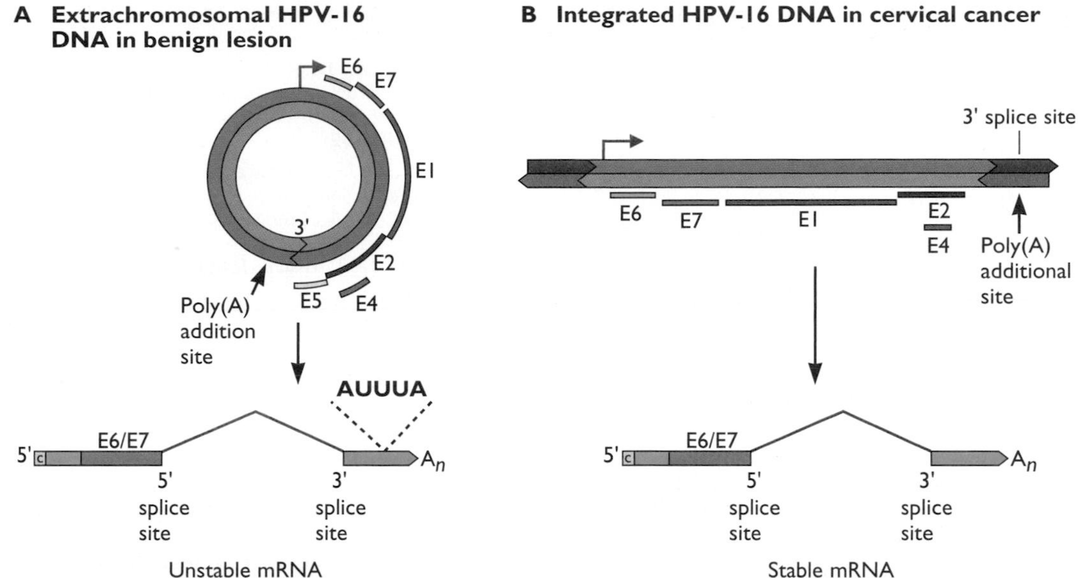

Figure 10.21 Stabilization of human papillomavirus type 16 (HPV-16) mRNAs upon integration of the viral genome into cellular DNA. (A) In benign lesions, the viral genome is maintained as an extrachromosomal, circular episome. Transcription of such viral DNA and pre-mRNA processing produce various alternatively spliced mRNAs containing the E6 and/or E7 protein-coding sequences, but, as illustrated, all contain destabilizing 5′AUUUA3′ sequences in their 3′ untranslated regions. (B) In cervical carcinoma cells, the viral genome is integrated into cellular DNA (purple) such that the viral genome is disrupted upstream of the E6/E7 mRNA 3′ splice site. The mRNAs encoding these viral proteins are therefore made by using 3′ splice and polyadenylation sites transcribed from adjacent cellular DNA, and they lack the destabilizing sequence.

has yielded much fundamental information about the mechanisms of capping, polyadenylation, and splicing. More recently, viral systems have provided equally important insights into export of mRNA from the nucleus to the cytoplasm. Perhaps the most significant lesson learned from the study of viral mRNA processing is the importance of these reactions in the regulation of gene expression.

Regulation of viral RNA processing can be the result of differences in the concentrations or activities of specific cellular components in different cell types, or actively induced by viral gene products. Several mechanisms by which viral proteins or RNAs can regulate or inhibit polyadenylation or splicing reactions, export of mRNA from the nucleus, or mRNA stability have been quite well characterized. However, our understanding of regulation of viral gene expression via RNA-processing reactions is far from complete; the mechanisms of action of several crucial viral regulatory proteins have not been fully elucidated, and many of the specific mechanisms deduced by using experimental systems have yet to be confirmed in virus-infected cells. Among the greatest challenges for the future is a full understanding of the recently identified physical and functional coupling among the reactions that produce mRNAs in eukaryotic cell nuclei. The intimate relationships among synthesis, capping, polyadenylation, and splicing of pre-mRNAs suggest that the transcription, RNA-processing, and export machineries are organized within the nucleus to optimize all reactions in the production of a functional mRNA. Further study of viral mRNA production via these processes seems likely to be as valuable in addressing such complex issues as many viral systems were in the initial elucidation of RNA-processing reactions.

References

Books

Hauber, J., and P. K. Vogt (ed.). 2001. *Nuclear Export of Viral RNAs.* Springer-Verlag, Berlin, Germany.

Reviews

Banks, J. D., K. L. Beemon, and M. F. Linial. 1997. RNA regulatory elements in the genomes of simple retroviruses. *Semin. Virol.* **8:**194–204.

Colgan, D. F., and J. L. Manley. 1997. Mechanism and regulation of mRNA polyadenylation. *Genes Dev.* **11:**2755–2766.

Cullen, B. R. 1998. Retroviruses as model systems for the study of nuclear RNA export pathways. *Virology* **249:**203–210.

Emerman, M., and M. H. Malim. 1998. HIV-1 regulatory/accessory genes: keys to unraveling viral and host cell biology. *Science* **280:**1880–1884.

Grate, D., and C. Wilson. 1997. Role REVersal: understanding how RRE RNA binds its peptide ligand. *Structure* **5:**7–11.

Lei, E. P., and P. A. Silver. 2002. Protein and RNA export from the nucleus. *Dev. Cell* **2:**261–272.

Maniatis, T., and R. Reed. 2002. An extensive network of coupling among gene expression machines. *Nature* **416:**499–506.

Misteli, T., and D. L. Spector. 1998. The cellular organization of gene expression. *Curr. Opin. Cell Biol.* **10:**323–331.

Mitchell, P., and D. Tollervey. 2000. mRNA stability in eukaryotes. *Curr. Opin. Genet. Dev.* **10:**193–198.

Nigg, E. A. 1997. Nucleocytoplasmic transport: signals, mechanisms and regulation. *Nature* **386:**779–787.

Nilsen, T. W. 1994. RNA–RNA interactions in the spliceosome: unraveling the ties that bind. *Cell* **78:**1–4.

Proudfoot, N. J., A. Furger, and M. J. Dye. 2002. Integrating mRNA processing with transcription. *Cell* **208:**501–512.

Shuman, S. 1995. Capping enzyme in eukaryotic mRNA synthesis. *Prog. Nucleic Acid Res. Mol. Biol.* **50:**101–129.

Smith, H. C., J. M. Gottand, and M. R. Hanson. 1997. A guide to RNA editing. *RNA* **3:**1105–1123.

Staley, J. P., and C. Guthrie. 1998. Mechanical devices of the spliceosome: motors, clocks, springs, and things. *Cell* **92:**315–326.

Varani, G. 1997. A cap for all occasions. *Structure* **5:**855–858.

Wickens, M., P. Anderson, and R. J. Jackson. 1997. Life and death in the cytoplasm: messages from the 3′ end. *Curr. Opin. Genet. Dev.* **7:**220–232.

Papers of Special Interest

Capping of Viral mRNAs or Pre-mRNAs

Furuichi, Y., M. Morgan, S. Muthukrishnan, and A. J. Shatkin. 1975. Reovirus messenger RNA contains a methylated, blocked 5′ terminal structure, m^{7}G(5′)ppp(5′)GpCp. *Proc. Natl. Acad. Sci. USA* **72:**362–366.

Hagler, J., and S. Shuman. 1992. A freeze-frame view of eukaryotic transcription during elongation and capping of nascent mRNA. *Science* **255:**983–986.

Lou, Y., X. Mao, L. Deng, P. Cong, and S. Shuman. 1995. The D1 and D12 subunits are both essential for the transcription termination activity of vaccinia virus capping enzyme. *J. Virol.* **69:**3852–3856.

Plotch, S. J., M. Bouloy, I. Ulmanen, and R. M. Krug. 1981. A unique cap (m^{7}GpppXm)-dependent influenza virion endonuclease cleaves capped RNAs to generate the primers that initiate viral RNA transcription. *Cell* **23:**847–858.

Salditt-Georgieff, M., M. Harpold, S. Chen-Kiang, and J. E. Darnell. 1980. The addition of the 5′ cap structure occurs early in hnRNA synthesis and prematurely terminated molecules are capped. *Cell* **19:**69–78.

Wei, C. M., and B. Moss. 1975. Methylated nucleotides block 5′ terminus of vaccinia virus messenger RNA. *Proc. Natl. Acad. Sci. USA* **72:**318–322.

Polyadenylation of Viral mRNA or Pre-mRNA

Ashe, M. L. P., L. H. Pearson, and N. J. Proudfoot. 1997. The HIV-1 5′ LTR poly(A) site is inactivated by U1 snRNP interaction with the major splice donor site. *EMBO J.* **16:**5752–5763.

Fitzgerald, M., and T. Shenk. 1981. The sequence 5′AAUAAA3′ forms part of the recognition site for polyadenylation of late SV40 mRNAs. *Cell* **24:**251–260.

Gilmartin, G. M., S.-L. Hung, J. D. DeZazzo, E. S. Fleming, and M. J. Imperiale. 1996. Sequences regulating poly(A) site selection within the adenovirus major late transcription unit influence the interaction of constitutive processing factors with the pre-mRNA. *J. Virol.* **70:**1775–1783.

McGregor, F., A. Phelan, J. Dunlop, and J. B. Clements. 1996. Regulation of herpes simplex virus poly(A) site usage and the action of immediate-early protein IE63 in the early-late switch. *J. Virol.* **70:**1931–1940.

Mohamed, R. M., D. R. Latner, R. C. Condit, and E. G. Niles. 2001. Interaction between the J3R subunit of vaccinia virus poly(A) polymerase and the H4L subunit of the viral RNA polymerase. *Virology* **280:**143–152.

Sheldon, R., C. Jurale, and J. Kates. 1972. Detection of polyadenylic acid sequences in viral and cellular eukaryotic RNA. *Proc. Natl. Acad. Sci. USA* **69:**417–421.

Splicing of Viral Pre-mRNA

Barksdale, S. K., and C. C. Baker. 1995. Differentiation-specific alternative splicing of bovine papillomavirus late mRNAs. *J. Virol.* **69:**6553–6556.

Berget, S. M., C. Moore, and P. A. Sharp. 1977. Spliced segments at the 5′ terminus of adenovirus 2 late mRNA. *Proc. Natl. Acad. Sci. USA* **74:**3171–3175.

Bouck, J., X.-D. Fu, A. M. Skalka, and R. A. Katz. 1995. Genetic selection for balanced retroviral splicing: novel regulation involving the second step can be mediated by transitions in the polypyrimidine tract. *Mol. Cell. Biol.* **15:**2663–2671.

Chow, L. T., R. E. Gelinas, T. R. Broker, and R. J. Roberts. 1977. An amazing sequence arrangement at the 5′ ends of adenovirus 2 messenger RNA. *Cell* **12:**1–8.

Ge, H., and J. L. Manley. 1990. A protein factor, ASF, controls cell-specific alternative splicing of SV40 early pre-mRNA in vitro. *Cell* **62:**25–34.

Kanopka, A., O. Mühlemann, S. Petersen-Mahrt, C. Estmer, C. Öhrmalm, and G. Akusjärvi. 1998. Regulation of adenovirus alternative RNA splicing by dephosphorylation of SR proteins. *Nature* **393:**185–187.

Lamb, R. A., and C. J. Lai. 1980. Sequence of interrupted and uninterrupted mRNAs and cloned DNA coding for the two overlapping non-structural proteins of influenza virus. *Cell* **21:**475–485.

Zabolotny, J. M., C. Krummenacher, and N. W. Fraser. 1997. The herpes simplex virus type 1 2.0-kilobase latency-associated transcript is a stable intron which branches at a guanosine. *J. Virol.* **71:**4199–4208.

Editing of Viral RNA

Polson, A. G., B. L. Bass, and J. L. Casey. 1996. RNA editing of hepatitis delta virus antigenome by dsRNA–adenosine deaminase. *Nature* **380:**454–456.

Polson, A. G., H. J. Ley III, B. L. Bass, and J. L. Casey. 1998. Hepatitis delta virus RNA editing is highly specific for amber/W site and is suppressed by hepatitis delta antigen. *Mol. Cell. Biol.* **18:**1919–1926.

Sanchez, A., S. G. Trappier, B. W. Mahy, C. J. Peters, and S. T. Nichol. 1996. The virion glycoproteins of Ebola viruses are encoded in two reading frames and are expressed through transcriptional editing. *Proc. Natl. Acad. Sci. USA* **93:**3602–3607.

Thomas, S. M., R. A. Lamb, and R. G. Paterson. 1988. Two mRNAs that differ by two nontemplated nucleotides encode the amino coterminal proteins P and V of the paramyxovirus SV5. *Cell* **54:**891–902.

Export of Viral mRNAs from the Nucleus

Ernst, R. K., M. Bray, D. Rekosh, and M.-L. Hammarskjöld. 1997. A structured retroviral RNA element that mediates nucleocytoplasmic export of intron-containing RNA. *Mol. Cell. Biol.* **17:**135–144.

Fischer, U., J. Huber, W. C. Boelens, I. W. Mattaj, and R. Lührmann. 1995. The HIV-1 Rev activation domain is a nuclear export signal that accesses an export pathway used by specific cellular RNAs. *Cell* **82:**475–483.

Huang, Z. M., and T. S. Yen. 1995. Role of the hepatitis B virus post-transcriptional regulatory element in export of intronless transcripts. *Mol. Cell. Biol.* **15:**3864–3869.

Meyer, B. E., and M. H. Malim. 1994. The HIV-1 Rev trans-activator shuttles between the nucleus and the cytoplasm. *Genes Dev.* **8:**1538–1547.

Pasquinelli, A. E., R. K. Ernst, E. Lund, C. Grimm, M. L. Zapp, D. Rekosh, M.-L. Hammarskjöld, and J. E. Dahlberg. 1997. The constitutive transport element (CTE) of Mason Pfizer monkey virus (MPMV) accesses a cellular mRNA export pathway. *EMBO J.* **16:**7500–7510.

Popa, I., M. E. Harris, J. E. Donello, and T. J. Hope. 2002. CRM1-dependent function of a *cis*-acting RNA export element. *Mol. Cell. Biol.* **22:**2057–2067.

Zang, W.-Q., B. Li, P.-Y. Huang, M. M. C. Lai, and T. S. B. Yen. 2001. Role of polypyrimidine tract binding protein in the function of the hepatitis B virus posttranscriptional regulatory element. *J. Virol.* **75:**10779–10786.

Regulation of RNA Processing by Viral Proteins

Gonzalez, R. A., and S. J. Flint. 2002. Effects of mutations in the adenoviral E1B 55-kilodalton protein coding sequence on viral late mRNA metabolism. *J. Virol.* **76:**4507–4519.

McGregor, F., A. Phelan, J. Dunlop, and J. B. Clements. 1996. Regulation of herpes simplex virus poly(A) site usage and the action of immediate early protein IE63 in the early-late switch. *J. Virol.* **70:**1931–1940.

Molin, M., and G. Akusjärvi. 2000. Overexpression of essential splicing factor ASF/SF2 blocks the temporal shift in adenovirus pre-mRNA splicing and reduces virus progeny formation. *J. Virol.* **74:**9002–9009.

Noah, D. L., K. Y. Twu, and R. M. Krug. 2003. Cellular antiviral responses against influenza virus are countered at the post-transcriptional level by the viral NS1A protein via its binding to a cellular protein required for the 3′ end processing of cellular pre-mRNAs. *Virology* **307:**386–395.

Pilder, S., M. Moore, J. Logan, and T. Shenk. 1986. The adenovirus E1B 55K transforming polypeptide modulates transport or cytoplasmic stabilization of viral and host cell mRNAs. *Mol. Cell. Biol.* **6:**470–476.

Shih, S. R., M. Nemeroff, and R. M. Krug. 1995. The choice of alternative 5′ splice sites in influenza virus M1 mRNA is regulated by the viral polymerase complex. *Proc. Natl. Acad. Sci. USA* **92:**6324–6328.

Stability of Viral and Cellular mRNA

Everly, D. N., Jr., P. Feng, I. S. Mian, and G. S. Read. 2002. mRNA degradation by the virion host shutoff (Vhs) protein of herpes simplex virus: genetic and biochemical evidence that Vhs is a nuclease. *J. Virol.* **76:**8560–8571.

Jeon, S., and P. F. Lambert. 1995. Integration of human papillomavirus type 16 DNA into the human genome leads to increased stability of E6 and E7 mRNAs: implications for cervical carcinogenesis. *Proc. Natl. Acad. Sci. USA* **92:**1654–1658.

Lam, Q., C. A. Smibert, K. E. Koop, C. Lavery, J. P. Capone, S. P. Weinheimer, and J. R. Smiley. 1996. Herpes simplex virus VP16 rescues viral mRNA from destruction by the virion shutoff function. *EMBO J.* **15:**2575–2581.

Myer, V. E., S. I. Lee, and J. A. Steitz. 1992. Viral small nuclear ribonucleoproteins bind a protein implicated in messenger RNA destabilization. *Proc. Natl. Acad. Sci. USA* **89:**1296–1300.

Zelus, B. D., R. S. Stewart, and J. Ross. 1996. The virion host shutoff protein of herpes simplex virus type 1: messenger ribonucleolytic activity in vitro. *J. Virol.* **70:**2411–2419.

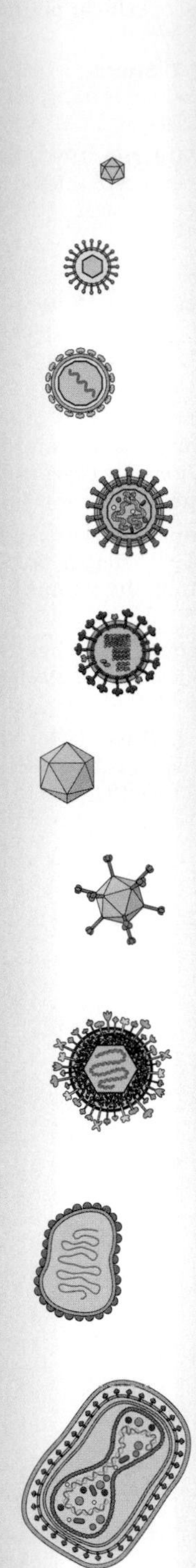

11

Control of Translation

The difficulty lies, not in the new ideas, but in escaping old ones....
JOHN MAYNARD KEYNES

Introduction

Viruses are obligate intracellular parasites; their replication is dependent on components of the host cell. Some aspects of viral multiplication depend more on cellular contributions than do others. For example, all viral genomes encode at least one protein needed for viral nucleic acid synthesis. However, viruses are completely dependent on the host cell for the translation of their mRNAs because most viral genomes do not encode any part of the translational machinery. Virus infection often results in modification of the host's translational apparatus so that viral mRNAs can be translated selectively.

Studies of viruses have contributed much to our understanding of translation and its regulation. Before the advent of recombinant DNA technology, viruses were a ready source of large quantities of relatively pure mRNA for in vitro studies of protein synthesis. The 5′ cap structure was identified on a viral RNA, and new translation initiation mechanisms, such as internal ribosomal entry, were discovered during studies of virus-infected cells. Our understanding of how the activity of the multisubunit cap-binding complex can be regulated originated from the finding that one of its subunits is cleaved in virus-infected cells.

Translation is a universal process in which proteins are produced from mRNA templates read in the 5′ → 3′ direction, and the growing polypeptide chain is synthesized from the amino to the carboxy terminus. Each amino acid is specified by a genetic code consisting of three bases, a **codon**, in the mRNA. Translation takes place on **ribosomes,** and **transfer RNAs (tRNAs)** are the adapter molecules that link specific amino acids with individual codons in the mRNA. In this chapter, we discuss the basic mechanisms by which translation occurs in eukaryotic cells. A consideration of the many ways that viral mRNAs are translated to provide optimal coding capacity in genomes of limited size is followed by a discussion of how translation is regulated in virus-infected cells.

Mechanisms of Eukaryotic Protein Synthesis

General Structure of Eukaryotic mRNA

Most eukaryotic mRNAs, with the exception of organelle mRNAs and certain viral mRNAs, begin with a 5′ 7-methylguanosine (m^7G) **cap structure** (Fig. 11.1; see also Fig. 10.2). It is joined to the second nucleotide by a 5′-5′ phosphodiester linkage, in contrast to the 5′-3′ linkages found in the remainder of the mRNA. The unique cap structure directs pre-mRNAs to processing and transport pathways, regulates mRNA turnover, and is required for efficient translation by the 5′-end-dependent mechanism (see "5′-End-Dependent Initiation" below). Most eukaryotic mRNAs contain **5′ untranslated regions,** which may vary in length from 3 to more than 1,000 nucleotides, although these regions are typically 50 to 70 nucleotides in length. Such 5′ untranslated regions often contain secondary structures (e.g., hairpin loops [see Fig. 6.7]) formed by base pairing of the RNA. These double-helical regions must be unwound to allow passage of ribosomal 40S subunits during translation.

Translation begins at an **initiation codon** and ends at a **termination codon.** The termination codon is followed by a **3′ untranslated region,** which can regulate initiation, translation efficiency, and mRNA stability. At the very 3′ end of the mRNA is a stretch of adenylate residues known as the **poly(A) tail,** which is added to nascent pre-mRNA. The poly(A) tail is necessary for efficient translation and may promote interactions among proteins that bind both ends of the mRNA (see "5′-End-Dependent Initiation" below).

Most prokaryotic mRNAs are **polycistronic**: they encode several proteins, and each open reading frame is separated from the next by an untranslated spacer region. The vast majority of eukaryotic mRNAs are **monocistronic**, i.e., they encode only a single protein (Fig. 11.1). A small number of eukaryotic mRNAs are functionally polycistronic, and different strategies have evolved for synthesizing multiple proteins from a single mRNA. Members of the virus family *Dicistroviridae* are unique because they have true bicistronic mRNAs (see Fig. 11.9).

Figure 11.1 Structure of eukaryotic and prokaryotic mRNAs. Both mRNAs contain 5′ and 3′ untranslated regions (UTR). Eukaryotic mRNAs carry a 5′ 7-methylguanosine cap and a 3′ poly(A) tail. Most prokaryotic mRNAs are polycistronic: they encode multiple proteins, and each open reading frame is separated from the next one by an untranslated region. In contrast, the vast majority of eukaryotic mRNAs are monocistronic: they encode only one protein. However, some eukaryotic mRNAs are polycistronic, as discussed in the text. Adapted from G. M. Cooper, *The Cell: a Molecular Approach* (ASM Press, Washington, D.C., and Sinauer Associates, Sunderland, Mass., 1997), with permission.

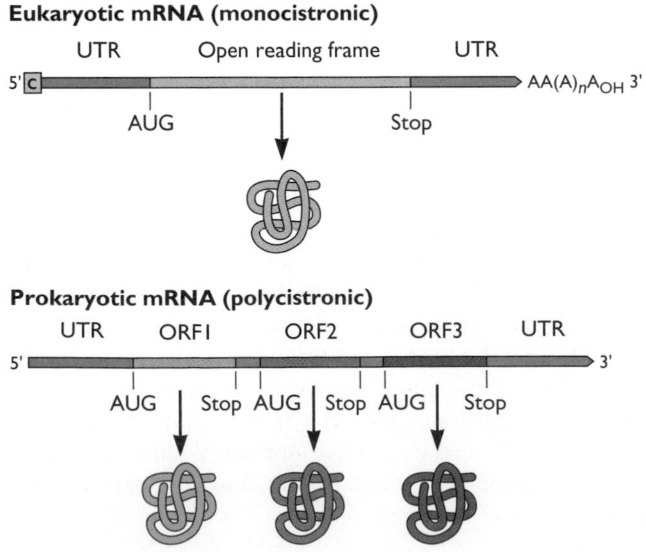

The Translation Machinery

Ribosomes

Mammalian ribosomes, the sites of translation, are composed of two subunits designated according to their sedimentation coefficients, 40S and 60S (Fig. 11.2A). The 40S subunit is made up of an 18S rRNA molecule and 33 proteins, while the 60S subunit contains three RNAs (5S, 5.8S, and 28S rRNAs) and 47 proteins. Actively growing mammalian cells may contain approximately 10 million ribosomes.

rRNAs are believed to play a catalytic role in protein synthesis. After removal of 95% of the ribosomal proteins, the 60S ribosomal subunit can still catalyze the formation of peptide bonds. The protein components of ribosomes are now believed to help fold the rRNAs properly, so that they can fulfill their catalytic function, and to position the tRNAs.

tRNAs

tRNAs are adapter molecules that align each amino acid with its corresponding codon on the mRNA. Each tRNA is 70 to 80 nucleotides in length and folds into a highly base-paired L-shaped structure (Fig. 11.2B). This shape is believed to be required for the appropriate interaction between tRNA and the ribosome during translation. The adapter function of tRNAs is carried out by two distinct regions of the molecule. At their 3′ ends, all tRNAs have the sequence 5′-CCA-3′, to which amino acids are covalently linked by **aminoacyl-tRNA synthetase.** Each of these enzymes recognizes a single amino acid and the correct tRNA. At the opposite end of the tRNA structure is the **anticodon loop,** which base pairs with the mRNA template.

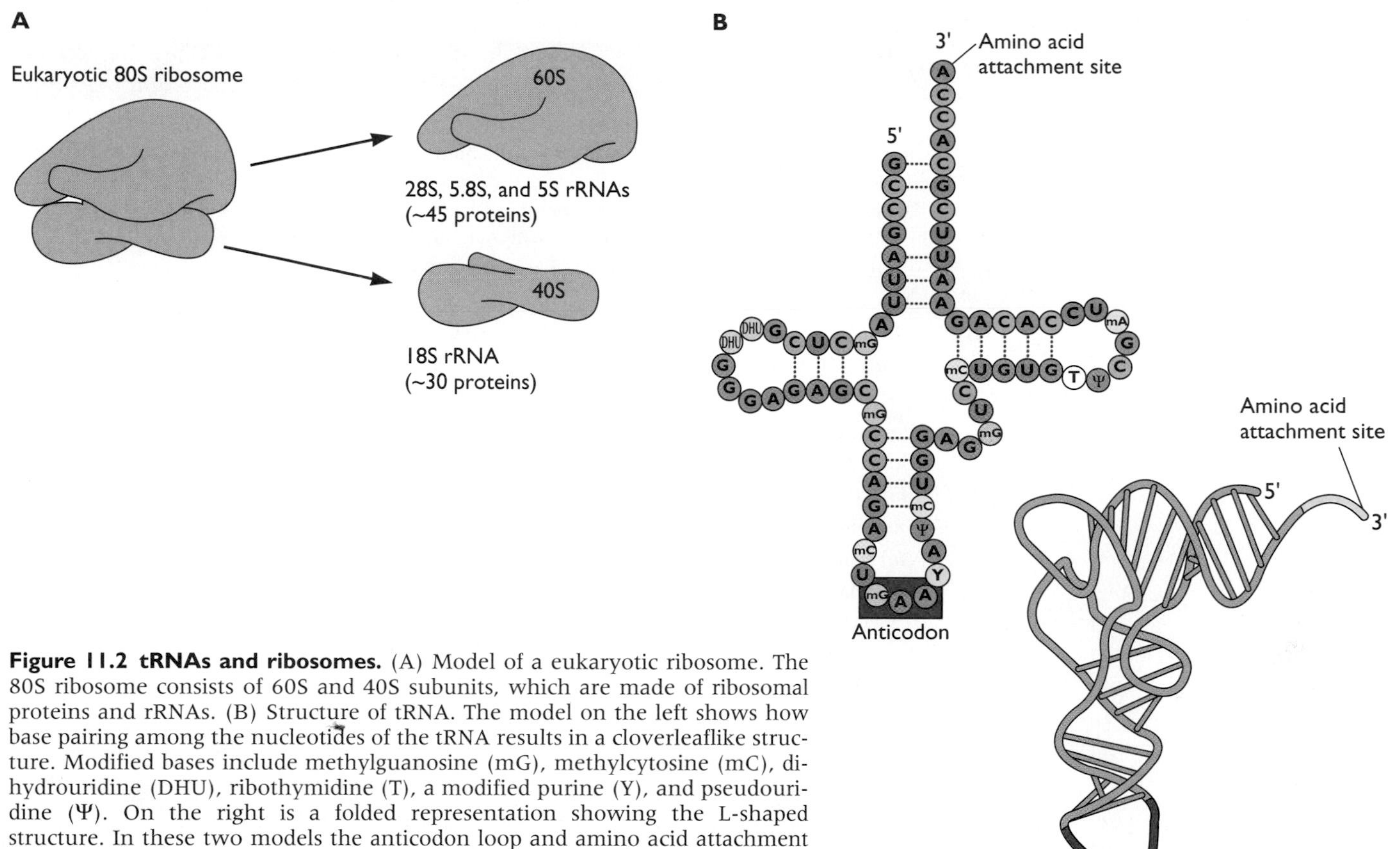

Figure 11.2 tRNAs and ribosomes. (A) Model of a eukaryotic ribosome. The 80S ribosome consists of 60S and 40S subunits, which are made of ribosomal proteins and rRNAs. (B) Structure of tRNA. The model on the left shows how base pairing among the nucleotides of the tRNA results in a cloverleaflike structure. Modified bases include methylguanosine (mG), methylcytosine (mC), dihydrouridine (DHU), ribothymidine (T), a modified purine (Y), and pseudouridine (Ψ). On the right is a folded representation showing the L-shaped structure. In these two models the anticodon loop and amino acid attachment sites are labeled. Adapted from G. M. Cooper, *The Cell: a Molecular Approach* (ASM Press, Washington, D.C., and Sinauer Associates, Sunderland, Mass., 1997), with permission.

Faithful incorporation of amino acids therefore depends on the specificity of codon-anticodon base pairing, as well as on the correct attachment of amino acids to tRNAs by aminoacyl-tRNA synthetases. The accuracy of protein synthesis is therefore maintained by two different proofreading mechanisms.

Translation Proteins

Many nonribosomal proteins are required for eukaryotic translation (Table 11.1). Some form multisubunit complexes containing as many as eight different proteins, while others function as monomers. Translation can be separated experimentally into three distinct stages: initiation, elongation, and termination. The proteins that participate at each stage are named eukaryotic initiation, elongation, and termination proteins. These proteins are designated in the same way as their prokaryotic counterparts, with the prefix "e" to distinguish them. The amino acid sequences of these proteins are conserved from mammals to yeasts, indicating that the mechanisms of translation are similar throughout eukaryotes.

Initiation

The majority of regulatory mechanisms function during initiation because this is the rate-limiting step in the translation of most mRNAs (see "Regulation of Translation during Viral Infection" below). At least 11 initiation proteins participate in this energy-dependent process. The end result is formation of a complex containing the mRNA, the ribosome, and the initiator Met-$tRNA_i$ in which the reading frame of the mRNA has been set. The 80S ribosome, which is the predominant species in cells, must be dissociated because it is the 40S subunit that participates in initiation. Three initiation proteins, eIF1A, eIF3, and eIF6 (Table 11.1) promote the dissociation of 80S ribosomes into 60S and 40S subunits.

There are two mechanisms by which ribosomes bind to mRNA in eukaryotes. In 5′-end-dependent initiation, by which the majority of mRNAs are translated, the initiation complex binds to the 5′ cap structure and moves, or scans, in a 3′ direction until the initiating AUG codon is encountered. In internal ribosome entry, the initiation complex binds at, or just upstream of, the initiation codon. Internal

Table 11.1 Mammalian translation proteins[a]

Name	Subunits	Molecular mass (kDa)	Function
Initiation proteins			
eIF1		12.6	Enhances initiation complex formation at the appropriate AUG initiation codon; destabilizes aberrant 48S complexes
eIF1A		16.5	Enhances initiation complex formation at the appropriate AUG initiation codon; affects ribosome dissociation; promotes Met-tRNA$_i$ binding to 40S; promotes scanning
eIF2		125	Binds Met-tRNA$_i$ and GTP; associates with eIF3, eIF5
	α	36.2	Affects eIF2B binding by phosphorylation
	β	39.0	Binds Met-tRNA$_i$ and associates with eIF2B, eIF5
	γ	51.8	Binds GTP, Met-tRNA$_i$; GTPase
eIF2B		270	eIF2 recycling
	α	33.7	Recognition of phosphorylated eIF2
	β	39.0	Binds GTP; recognition of phosphorylated eIF2
	δ	57.8	Binds ATP; recognition of phosphorylated eIF2
	γ	50.4	Guanine nucleotide exchange
	ε	80.2	Guanine nucleotide exchange
eIF3		650	Dissociates 80S ribosomes; promotes Met-tRNA$_i$ and mRNA binding to 40S ribosomal subunit
	p28	25.1	
	p35	29.0	
	p36	36.5	
	p40	39.9	
	p44	35.4	Binds mRNA, eIF4B
	p47	37.6	
	p48	52.2	
	p66	64.0	Binds mRNA
	p110	105.3	
	p116	98.9	Binds eIF1, eIF5
	p170	166.5	Binds RNA
eIF4AI		44.4	ATPase, RNA helicase
eIF4AII		46.3	ATPase, RNA helicase
eIF4B		69.2	Binds RNA; promotes mRNA-40S ribosomal subunit interaction; promotes helicase activity of eIF4A
eIF4E		25.1	Bind m^{7}G cap of mRNA
eIF4GI		171.6	Binds eIF4E, eIF4A, eIF3, Mnk1, Pab1p, Paip-1, RNA
eIF4GII		176.5	Binds eIF4E, eIF4A, eIF3, Mnk1, Pab1p, Paip-1, RNA
eIF4H		25	Binds RNA; stimulates ATPase and helicase of eIF4A
eIF5		48.9	Promotes GTPase activity of eIF2
eIF5B		139.0	Ribosome-dependent GTPase; required for 60S ribosomal subunit joining
eIF6		26	Binds to 60S ribosomal subunit; promotes 80S ribosome dissociation
Elongation proteins			
eEF1A		50.1	GTP-dependent binding of aminoacyl-tRNAs; GTPase
eEF1B		114	Guanine nucleotide exchange on eEF1A
	α	30	Guanine nucleotide exchange activity
	β	36	Guanine nucleotide exchange activity
	γ	48	
eEF2		95	GTPase; promotes translocation of peptidyl-tRNA from A site to P site
Termination proteins			
eRF1		55	Recognizes termination codons; promotes ribosome-catalyzed peptidyl-tRNA hydrolysis; interacts with peptidyl transferase of 60S ribosomal subunit and A site of ribosome
eRF3		55.8	GTPase; associates with eRF1 and Pab1p

[a]Modified from N. Sonenberg et al., *Translational Control of Gene Expression* (Cold Spring Harbor Press, Cold Spring Harbor, N.Y.), 2000, with permission.

ribosome entry sites were first discovered with picornavirus mRNAs, and are now known to be present in some cellular mRNAs.

5′-End-Dependent Initiation

How ribosomes assemble at the proper end of mRNA. The first step in this pathway is recognition of the m^7G cap by the cap-binding protein, eIF4E (Fig. 11.3). eIF4G acts as a scaffold between the cap structure and the 40S subunit, which associates with the mRNA through an interaction of eIF3 with the C-terminal domain of eIF4G. This important adapter molecule was first discovered as the target of proteolytic cleavage in poliovirus-infected cells that results in the abrupt termination of host protein synthesis (see "Regulation of Translation during Viral Infection" below). The 40S ribosomal subunit then moves in a 3′ direction on the mRNA in a process called **scanning.** When the 40S subunit reaches the AUG initiation codon, initiation proteins are released and GTP is hydrolyzed, allowing the 60S ribosomal subunit to associate with the 40S subunit, forming the 80S initiation complex.

The presence of a poly(A) tail can stimulate mRNA translation. This surprising effect is a consequence of interactions between proteins associated with the 5′ and 3′ ends of the mRNA, promoting 40S subunit recruitment. Such interactions were first demonstrated in the yeast *Saccharomyces cerevisiae,* in which poly(A)-tail-binding protein, Pab1p, is required for efficient mRNA translation. Stimulation of translation by poly(A), which requires Pab1p, occurs by enhancing the binding of 40S ribosomal subunits to mRNA. Pab1p has been shown to interact with the N terminus of eIF4G (Fig. 11.3C). Alteration of the Pab1p-binding site on eIF4G destroys stimulation of translation by poly(A). These results have led to a model in which Pab1p, bound to the poly(A) tail, interacts with eIF4G bound to the 5′ cap, perhaps stabilizing the interaction and assisting in recruiting 40S subunits (Fig. 11.3C). A consequence of these interactions is that the 5′ and 3′ ends of the mRNA are brought in close proximity.

Many viral mRNAs, such as the mRNA of barley yellow dwarf luteovirus, lack a 5′-terminal cap and 3′-poly(A) sequence. Nevertheless, the ends of these mRNAs are brought together by base pairing between discrete sequences in the 5′ and 3′ untranslated regions. Translation of mRNA of the flavivirus dengue virus, which has a 5′ cap structure but lacks a 3′-poly(A) sequence, may also depend on complementarity between the untranslated regions.

The juxtaposition of mRNA ends might be a mechanism to ensure that only intact mRNAs [containing 5′ cap and poly(A)] are translated. Such structures might also stabilize mRNA by preserving the interaction among the translation initiation proteins associated with both ends. Translation reinitiation might also be stimulated by such an arrangement. Once the ribosome terminates translation, it might be rapidly repositioned at the AUG initiation codon rather than dissociate into the two subunits. Regardless of the terminal structures, the 5′ and 3′ ends of mRNAs are brought together. The conservation of this feature implies that a closed loop may be necessary for all translation.

The role of RNA secondary structure in mRNA translation. Translation efficiency is reduced by the presence of stable secondary structure in the mRNA 5′ untranslated region, particularly when the structure is near the 5′ terminus. There are at least two reasons for this effect. If RNA secondary structure is adjacent to the 5′ cap, it can inhibit binding of the 40S ribosomal subunit. In addition, the presence of secondary structure blocks ribosome movement toward the initiation codon.

The eIF4A protein is an ATP-dependent RNA helicase that is enhanced by eIF4B and eIF4H (Fig. 11.3). Such activity is believed to unwind regions of double-stranded RNA (dsRNA) near the 5′ end of the mRNA, allowing the 43S preinitiation complex to bind. The helicase may also migrate in a 3′ direction, unwinding dsRNA and enabling ribosomes to scan. mRNAs with less secondary structure in the 5′ untranslated region have a reduced requirement for RNA helicase activity during mRNA translation, and hence are less dependent on the cap structure through which the helicase is brought to the mRNA. Dependence of translation on the cap structure can be measured experimentally by determining the effect on protein synthesis of cap analogs such as m^7GDP and m^7GTP. These compounds competitively inhibit 5′-end-dependent initiation by binding to eIF4E and preventing it from associating with capped mRNAs. For example, the 5′ untranslated region of alfalfa mosaic virus RNA segment 4 is largely free of secondary structure, and translation of this mRNA is quite resistant to inhibition by cap analogs.

Choosing the initiation codon. For over 90% of mRNAs, translation initiation occurs at the 5′-proximal AUG codon. The efficiency of initiation is influenced by the nucleotide sequence surrounding this codon. Studies of the effects of mutating these sequences have shown that the consensus sequence 5′-GCC**A**CCAUGG-3′ is recognized most efficiently in mammalian cells; the presence of a purine at the –3 position (boldface) is most important for high levels of translation. However, only 5% of eukaryotic mRNAs contain this ideal consensus sequence; most have suboptimal sequences that result in less efficient translation. This finding indicates that not all mRNAs must be translated maximally, only at levels appropriate

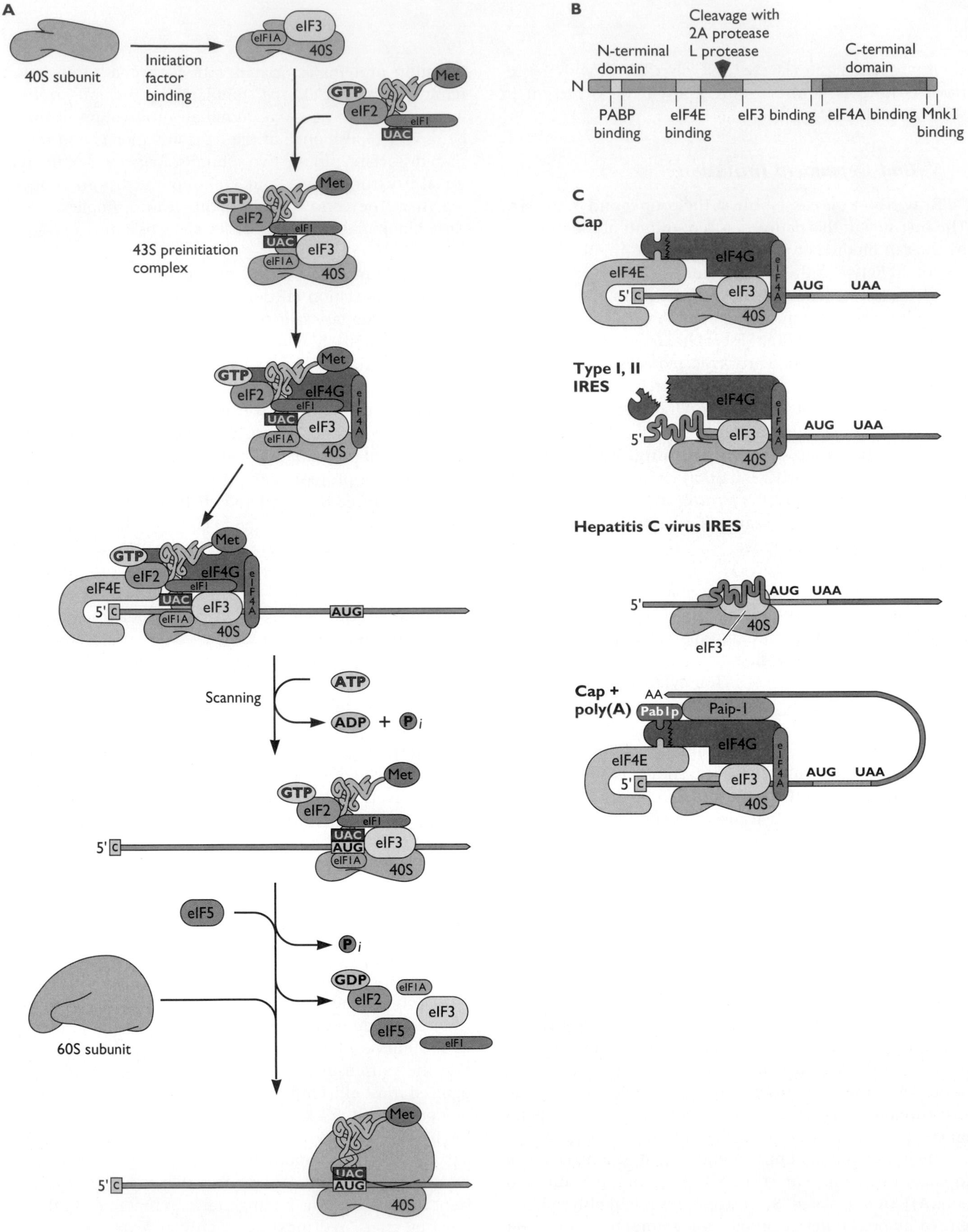

A

40S subunit

Initiation factor binding

eIF1A eIF3 40S

GTP eIF2 Met eIF1 UAC

43S preinitiation complex

eIF4G eIF4A

eIF4E 5' C AUG

Scanning

ATP

ADP + P_i

5' C UAC AUG

eIF5

P_i

GDP eIF2 eIF1A eIF3 eIF5 eIF1

60S subunit

5' C UAC AUG 40S

B

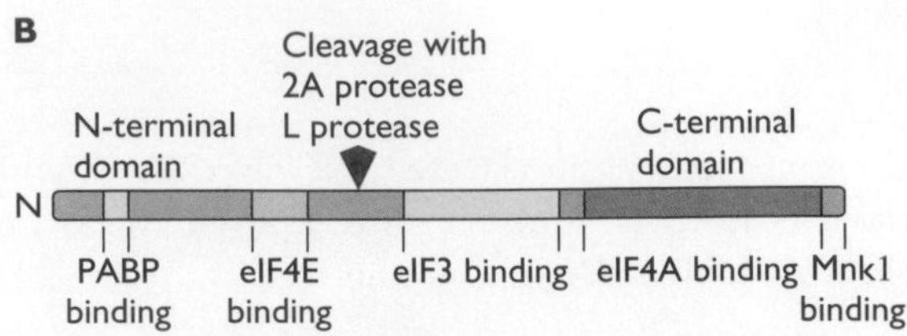

C

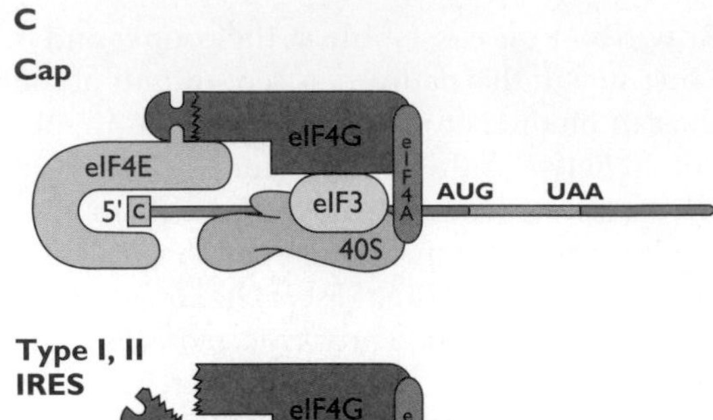

Hepatitis C virus IRES

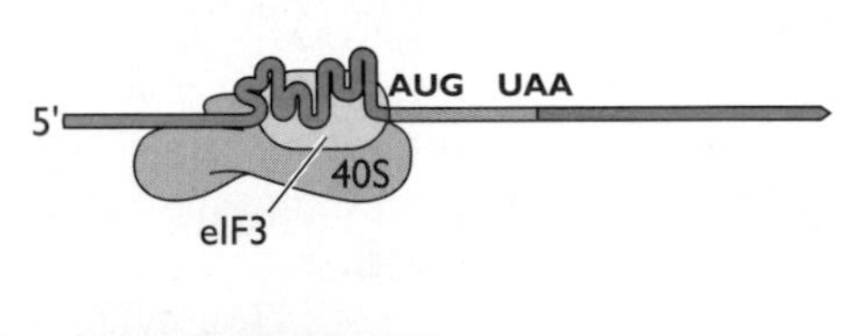

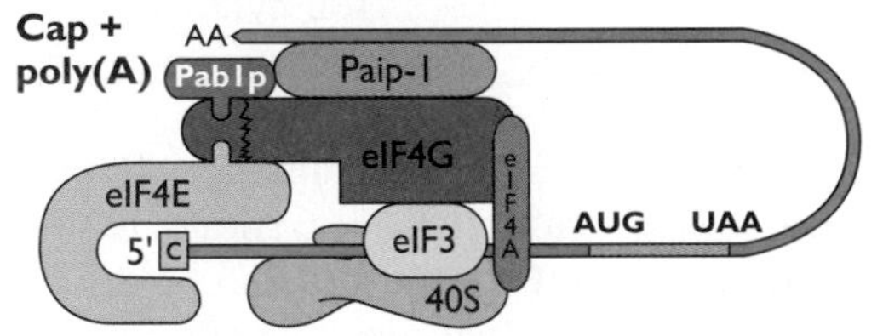

Figure 11.3 The initiation complex. (A) Model of initiation complex assembly. Initiation proteins eIF3 and eIF1A bind to free 40S subunits to prevent their association with the 60S subunit, while interaction of eIF6 (not shown) with the larger subunit prevents it from associating with the 40S subunit. eIF4F, which consists of three proteins—eIF4A, eIF4E, and eIF4G—binds the cap via the eIF4E subunit, and the ribosome binds a ternary complex containing eIF2, GTP, and Met-tRNA$_i$, forming a 43S preinitiation complex. The ribosome then binds eIF4G via eIF3. Two functional forms of eIF4G, called eIF4GI and eIF4GII, are present in cells. Alternatively, eIF4G may first join the 43S preinitiation complex and then bind the mRNA via eIF4E bound to the cap. The 40S subunit then scans down the mRNA until the AUG initiation codon is reached. eIF1 and eIF1A are required for selection of the correct AUG initiation codon. eIF5 triggers GTP hydrolysis, eIF2 bound to GDP is released along with other initiation proteins, and the 60S ribosomal subunit joins the complex. In a variation of this model, the helicases eIF4A/eIF4B scan the mRNA and stop at the initiating AUG, at which point the 40S subunit binds the mRNA. Adapted from G. M. Cooper, *The Cell: a Molecular Approach* (ASM Press, Washington, D.C., and Sinauer Associates, Sunderland, Mass., 1997), with permission. (B) Schematic of eIF4G. Sites for binding of eIF4E, eIF3, eIF4A, Pab1p, and Mnk1 are shown, as are the cleavage sites for viral proteases. Adapted from S. J. Morley et al., *RNA* **3**:1085–1104, 1997, with permission. (C) Hypothetical models of initiation complex assembly involving the 5′ cap structure, internal ribosome entry site (IRES) elements, and poly(A). (Top) In 5′-end-dependent initiation, eIF4F is brought to the mRNA 5′ end by interaction of eIF4E with the cap structure. The N terminus of eIF4G binds eIF4E, and the C terminus binds eIF4A. The 40S ribosomal subunit binds to eIF4G indirectly via eIF3. (Upper middle) Initiation in the IRES does not depend on the presence of a cap structure but requires the C-terminal fragment of eIF4G to recruit the 40S ribosomal subunit via its interaction with eIF3. eIF4G probably binds directly to the IRES. (Lower middle) The ribosomal 40S subunit binds to the hepatitis C virus IRES without the need for translation initiation proteins. eIF3 also binds the IRES and is believed to be necessary for recruitment of the 60S ribosomal subunit. (Bottom) 5′-end-dependent initiation is stimulated by the poly(A)-binding protein Pab1p, which interacts with eIF4G. This interaction may bring the mRNA ends together and facilitate formation of the initiation complex at the 5′ end. Paip-1 binds Pab1p and eIF4G, facilitating juxtaposition of mRNA ends and stimulating translation. Adapted from M. W. Hentze, *Science* **275**:500–501, 1997, with permission.

for the function of the protein product. If a very poor match to this consensus sequence is present, the AUG codon may be passed over by the ribosome and initiation may occur further downstream (see "The Diversity of Viral Translation Strategies" below). If the 5′-proximal AUG codon is mutated so that it cannot serve as an initiation codon, translation will start at the next downstream AUG. Insertion of an AUG codon upstream of the initiating codon causes initiation at the more 5′-proximal site.

The results of genetic and biochemical experiments demonstrate that eIF1 plays a role in discrimination during initiation codon selection. In the absence of eIF1, the 43S complex is unable to discriminate between AUG and AUU or GUG codons. Certain alterations of eIF1 of yeasts enhance initiation at a UUG triplet. eIF1 might influence codon-anticodon pairing, either directly or by modulating the conformation of the 43S complex.

Although AUG is the initiation codon for most proteins, synthesis may also begin at ACG, GUG, and CUG codons in viral and cellular mRNAs, although far less frequently. In all cases, the first amino acid of the protein is a methionine. Precisely how ribosomes recognize non-AUG codons so that methionine is inserted is not known. When translation begins at a non-AUG initiator codon, it is always the 5′-proximal start site and in a good sequence context for initiation. Because the efficiency of initiation at these sites is always low, non-AUG initiation might be another mechanism for regulating translation efficiency. Not all mRNAs support initiation from a non-AUG site, suggesting that there may be structural elements in the mRNA that regulate expression from such a codon.

Methionine-independent initiation. The structural proteins of some viruses begin not with methionine but with glutamine (CAA), proline (CCU), or alanine (GCU or GCA). Translation initiation of these viral proteins does not require initiator tRNA methionine or the ternary complex because the viral mRNA mimics the structure of tRNA (Fig. 11.4A). The tRNA-like structure occupies the P site of the ribosome, allowing initiation to take place within the A site. These mRNAs require no translation initiation proteins, and can bind ribosomes and induce them to enter the elongation phase of translation. Methionine-independent initiation of the mRNA of turnip yellow mosaic virus is accomplished in a similar way, except that the tRNA-like structure is located in the 3′ untranslated region of the viral RNA (Fig. 11.4B). The tRNA-like structure is aminoacylated with valine, which is incorporated as the first amino acid of the viral polyprotein.

Does the 40S subunit move on the mRNA? Translation of most mRNAs begins at the 5′-proximal AUG codon. How does the ribosome reach this location? In one popular model, the 40S ribosomal subunit migrates

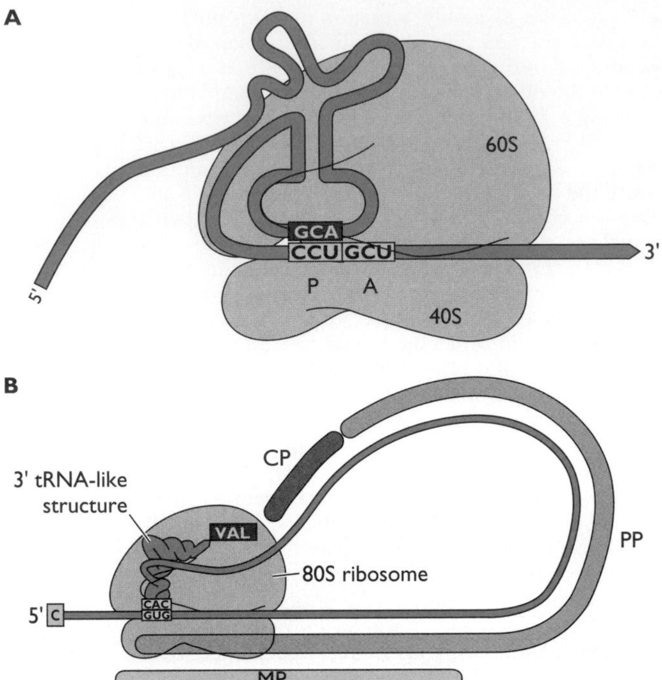

Figure 11.4 Two mechanisms of methionine-independent initiation. (A) The viral mRNA of picornavirus-like viruses of insects mimics the structure of tRNA, which occupies the P site of the ribosome, allowing initiation to take place within the A site. Adapted from M. Bushell and P. Sarnow, *J. Cell Biol.* **158:**395–399, 2002, with permission. (B) A tRNA-like structure in the 3′ untranslated region of turnip yellow mosaic virus RNA occupies the P site of the ribosome. The tRNA-like structure is aminoacylated with valine, which is incorporated as the first amino acid of the viral polyprotein. Adapted from S. Barends et al., *Cell* **112:** 123–129, 2003, with permission.

through the 5′ untranslated region until it reaches the AUG initiation codon. Experimental proof for this long-standing hypothesis is still lacking. Hydrolysis of ATP is necessary for movement of the 43S complex on RNA containing weak secondary structures. However, energy is required by the ATP-dependent RNA helicase, not for ribosome movement. Movement of ribosomes on RNA lacking secondary structure does not require ATP hydrolysis. In an alternative model, the ribosome does not travel to the initiation codon, but remains at the 5′ cap structure. The mRNA is threaded through the ribosome, eventually bringing the initiation codon to the P site.

Ribosome shunting. Stable RNA secondary structures in 5′ untranslated regions may inhibit scanning of 40S ribosomes. In some RNAs, such hairpin structures are not inhibitory because ribosomes bypass or shunt over them. This process, called **ribosome shunting**, may be dependent or independent of viral proteins. Shunting on the 35S cauliflower mosaic virus RNA requires translation of the very short upstream open reading frame on the same viral RNA. In contrast, shunting on adenovirus major late mRNAs occurs in the absence of viral protein, although its efficiency is increased by the viral 100-kDa polypeptide. The mechanism of ribosome shunting remains to be elucidated.

5′-End-Independent Initiation

The internal ribosome entry site. The mRNAs of picornaviruses differ from most host cell mRNAs. They lack the 5′-terminal cap structure: the VPg protein, which is linked to the 5′ end, is removed as the RNA enters the cell. In addition, the 5′ untranslated regions are highly structured and contain multiple AUG codons. Infection of host cells by many picornaviruses results in the inhibition of host cell translation. These observations led to the hypothesis that translation of the mRNA of (+) strand picornaviruses was initiated by an unusual mechanism. It was suggested that the ribosome bound internally rather than at the mRNA 5′ end. In a key experiment, the 5′ untranslated region of poliovirus mRNA was shown to promote internal binding of the 40S ribosomal subunit, and was termed the **internal ribosome entry site** (IRES) (Box 11.1).

An IRES has been identified in the mRNAs of all picornaviruses; in other viral mRNAs, including those of pestiviruses and hepatitis C virus; and in some cellular mRNAs. There is very little nucleotide sequence conservation among these IRESs, with the exception of an oligopyrimidine tract 25 nucleotides upstream of the 3′ end of the IRES. Viral IRESs contain extensive regions of RNA secondary structure (Fig. 11.5A and B). Although the RNA secondary structure is not strictly conserved, it is of extreme importance for ribosome binding.

Viral IRESs have been placed in five groups, depending on a variety of criteria, including primary sequence and secondary structure conservation, the location of the initiation codon, and activity in different cell types. In the type 1 IRES, characteristic of enteroviruses and rhinoviruses, and the type 3 IRES (hepatitis A virus), the initiation codon is located 50 to 100 nucleotides past the 3′ end of the IRES, while it is at the 3′ end of a type 2 IRES (cardio- and aphthoviruses). The intergenic IRES of picornavirus-related viruses of insects (type 5) also ends at the initiating codon, although it is not an AUG codon (Fig. 11.5D). The 3′ end of the hepatitis C virus IRES (type 4) extends beyond the AUG initiation codon (Fig. 11.5C).

The discovery of the IRES makes even more puzzling the rarity of eukaryotic mRNAs that contain several open reading frames reminiscent of prokaryotic mRNAs (Fig. 11.1). In principle, all the open reading frames of a poly-

BOX 11.1

Key experiment: discovery of the IRES

The hypothesis that poliovirus mRNA is translated by internal ribosome binding was first tested by examining the translation of mRNAs containing two open reading frames (ORFs) separated by the poliovirus 5′ untranslated region (figure panel A). These bicistronic mRNAs directed the synthesis of two proteins, but the second ORF was efficiently translated only if it was preceded by the picornavirus 5′ untranslated region. It was concluded that ribosomes bind within the picornavirus 5′ untranslated region, thereby permitting translation of the second ORF. The region of the 5′ untranslated region that directs internal ribosome entry was called the IRES.

It had long been known that covalently closed circular mRNAs cannot be translated by 5′-end-dependent initiation. Translation by internal ribosome binding, however, should not require a free 5′ end. To test this hypothesis, circular mRNAs with and without an IRES were created. The circular mRNA was translated only if an IRES was present (panel B). This experiment formally proved that translation initiation directed by an IRES occurs by internal binding of ribosomes and does not require a free 5′ end.

Chen, C. Y., and P. Sarnow. 1995. Initiation of protein synthesis by the eukaryotic translational apparatus on circular RNAs. *Science* **268:** 415–417.

Pelletier, J., and N. Sonenberg. 1988. Internal initiation of translation of eukaryotic mRNA directed by a sequence derived from poliovirus RNA. *Nature* **334:**320–325.

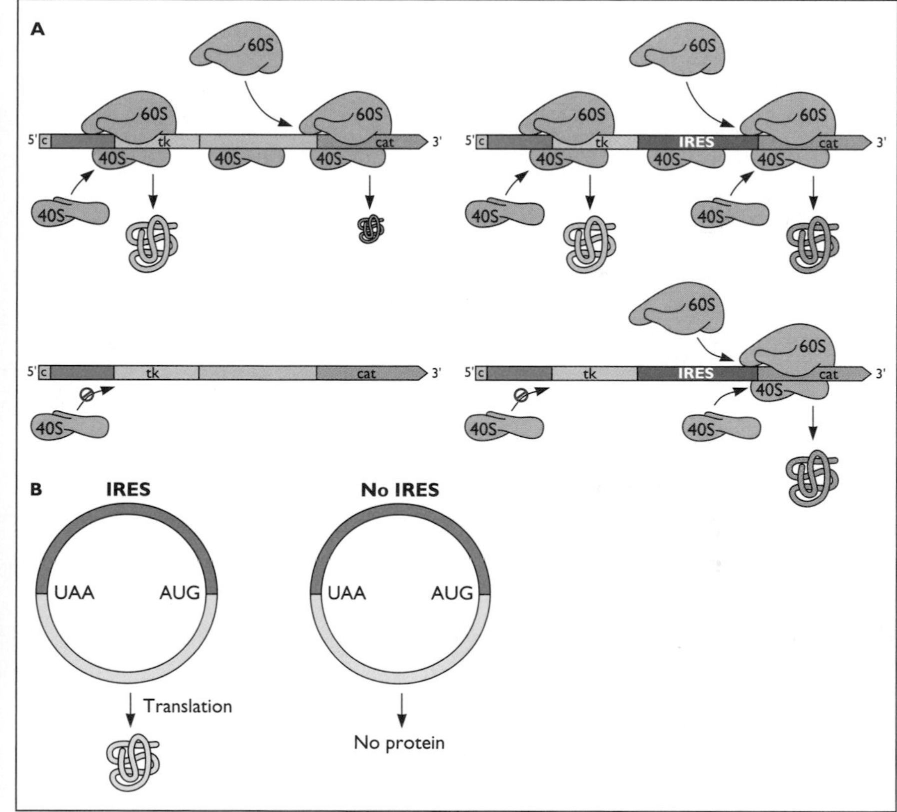

Assays for an IRES. (A) Bicistronic mRNA assay for an IRES. Plasmids were constructed that encode bicistronic mRNAs encoding the thymidine kinase (tk) and chloramphenicol acetyltransferase (cat) proteins separated by a spacer (light green) or a poliovirus IRES (dark green). Plasmids were introduced into mammalian cells by transformation. In uninfected cells containing either plasmid (top lines), both tk and cat proteins were detected, although without an IRES, cat synthesis was inefficient. Translation of cat from this plasmid likely occurs by reinitiation. In poliovirus-infected cells, 5′-end-dependent initiation is blocked, and no proteins are observed without an IRES. cat protein is detected in infected cells when the IRES is present, demonstrating internal ribosome binding. Adapted from J. Pelletier and N. Sonenberg, *Nature* **344:**320–325, 1988, with permission. (B) Circular mRNA assay for an IRES. Circular mRNAs containing an ORF (yellow) were produced and translated in vitro in cell extracts. No protein product was observed unless an IRES was included in the circular mRNA, demonstrating that 5′-end dependent translation requires a free 5′ end, while translation initiation by ribosome binding does not. Adapted from C. Y. Chen and P. Sarnow, *Science* **268:**415–417, 1995, with permission.

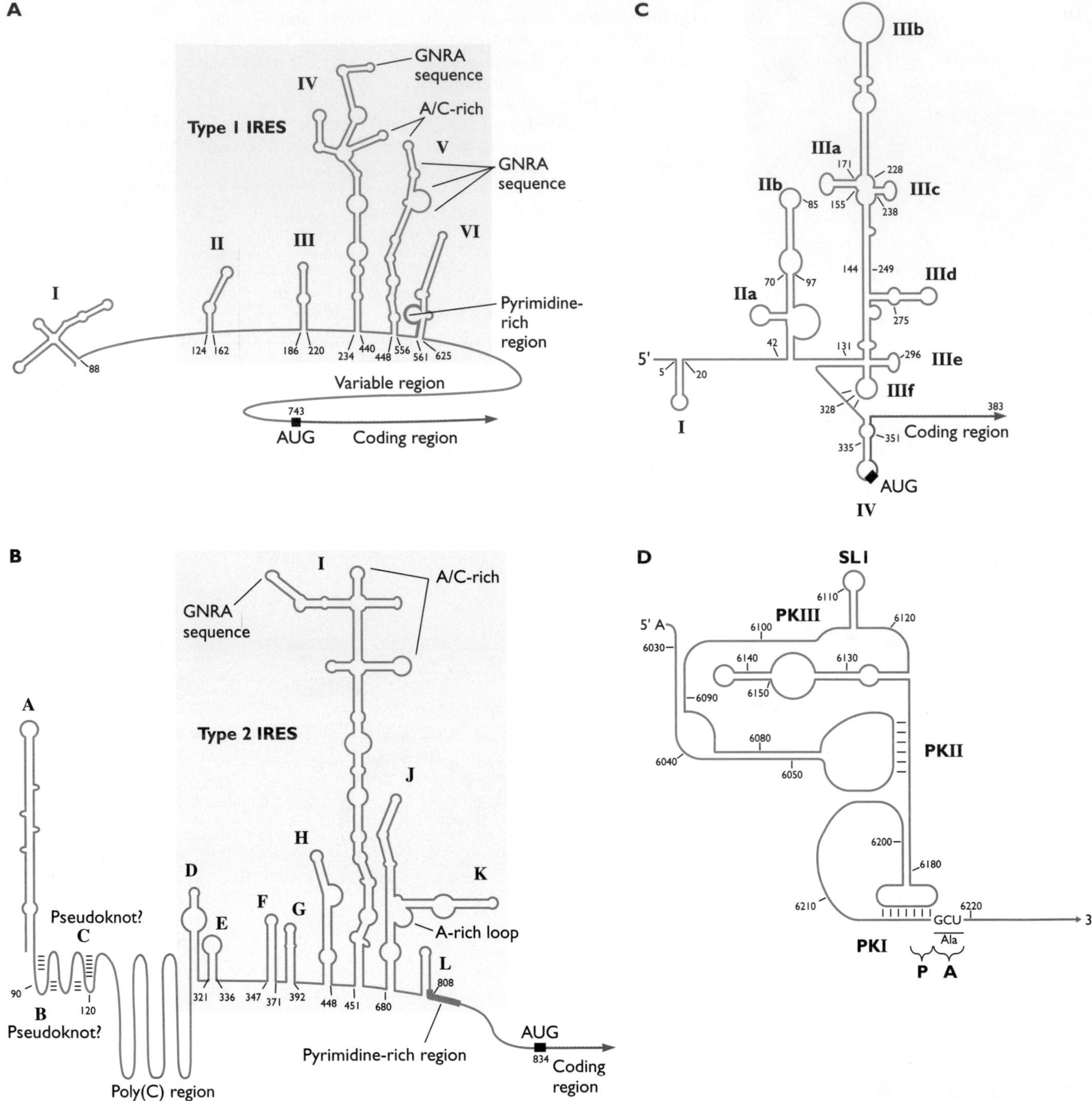

Figure 11.5 Four types of IRES. The 5′ untranslated regions from poliovirus (A), encephalomyocarditis virus (B), hepatitis C virus (C), and cricket paralysis virus (D) are shown. The IRES in panels A and B is indicated by yellow shading. Predicted secondary and tertiary RNA structures (RNA pseudoknots) are shown. The poliovirus IRES is a type 1 IRES, which is found in the genomes of enteroviruses and rhinoviruses. The ribosome probably enters the IRES at domains V and VI, and scans to the AUG initiation codon. The type 2 IRES is found in the genomes of aphthoviruses and cardioviruses. Conserved sequence features include the GNRA- and A/C-rich loops and a pyrimidine-rich region. Ribosome binding at the AUG initiation codon of the type 2 IRES requires eIF2, eIF3, eIF4G, and eIF4A. eIF4G interacts with multiple regions of the type 2 IRES, including the J and K structures. Binding of eIF4G/eIF4A induces ATP-dependent conformational

(continued on next page)

cistronic mRNA can be translated in a eukaryotic cell as long as each frame is preceded by an IRES. Nevertheless, only one such naturally occurring polycistronic mRNA has been identified in eukaryotes, and it is not known if an IRES is present. Bicistronic mRNAs produced in the laboratory have been used in the expression and cloning of genes (Box 11.2).

The mechanism of internal initiation. Different sets of translation initiation proteins are required for the function of various IRESs. Internal ribosome binding on the hepatitis A virus IRES requires all of the initiation proteins, including eIF4E. At the other extreme, the intergenic IRES of cricket paralysis virus requires none of the translation initiation proteins. A subset of translation initiation proteins is required for the activity of most IRESs.

Translation initiation via a type 1 IRES comprises binding of the 40S ribosomal subunit to the IRES, followed by scanning of the subunit to the initiation codon. The 40S subunit may bind directly to the RNA, or could be recruited to the IRES by means of interaction with translation initiation proteins (Fig. 11.3). In poliovirus-infected cells, the scaffold eIF4G is cleaved, inactivating the translation of most cellular mRNAs. It was therefore believed that this initiation protein was not required for function of the poliovirus IRES. However, the C-terminal fragment of eIF4G, which contains binding sites for eIF3 and eIF4A (Fig. 11.3B and C), stimulates translation directed by the poliovirus IRES. These results have led to a model in which the 40S ribosomal subunit is recruited to the IRES via interaction with eIF3 bound to the C-terminal domain of eIF4G (Fig. 11.3C). It has been reported that IRES function is markedly enhanced in cells in which the poliovirus protease $2A^{pro}$ is synthesized. Because $2A^{pro}$ is responsible for cleavage of eIF4G, such stimulation of translation may be due to production of the C-terminal proteolytic fragment of eIF4G. The 40S ribosomal subunit binds at or near the AUG initiation codon of the type 2 IRES, and no scanning occurs.

The IRESs of hepatitis C virus (Fig. 11.5C) and pestiviruses such as bovine viral diarrhea virus and classical swine fever virus function very differently from those of picornaviruses. The formation of the 48S complex on the mRNA is independent of eIF4A, eIF4B, and eIF4F, and is highly sensitive to secondary RNA structure around and downstream of the AUG initiation codon. Purified 40S ribosomal subunits bind directly to stem-loop IIId of the hepatitis C virus IRES, and single point mutations in this structure abolish the interaction and block internal initiation. Addition of only Met-tRNA$_i$, eIF2, and GTP is required to form 48S complexes. A dramatic conformational change in the 40S ribosomal subunit occurs when it binds the hepatitis C virus IRES (Fig. 11.6), clamping the mRNA in place and setting the AUG initiation codon within the P site of the ribosome. The IRES also contacts the E site of the ribosome, where the deacylated tRNA is harbored after translocation of the 80S ribosome. Initiation of translation from the IRES of hepatitis C virus and related viruses therefore resembles translation initiation of prokaryotic mRNAs.

The intergenic IRESs of picornavirus-like viruses of insects (*Plautia stali* intestine virus, *Rhopalosiphum padi* virus, and cricket paralysis virus) form complexes with the 40S ribosome independent of initiation proteins, and translation begins at other than an AUG initiation codon. Genetic experiments done in yeasts demonstrate that initiation from these IRESs is inhibited by ternary complex (Met-tRNA$_i$–eIF2–GTP) and a high concentration of Met-tRNA$_i$. In cells infected with these viruses, recycling of eIF2-GDP is blocked and the concentration of ternary complex is low, inhibiting cellular mRNA translation, but activity of the intergenic IRES is not reduced. The secondary structure of the IRES of these viruses mimics an uncharged tRNA, and mutations that destabilize the fold abrogate translation. The tRNA-like structure is recognized and bound by the 40S ribosomal subunit, placing the initiation codon within the A site instead of the P site. Initiation is therefore dependent on elongation proteins eEF1A and eEF2 and the appropriate aminoacylated tRNA. Like the IRESs of hepatitis C and pestiviruses, IRESs of the picornavirus-like viruses of insects also occupy the E site of the 80S ribosome.

As discussed above, translation of cellular mRNAs is enhanced by the juxtaposition of mRNA ends (Fig. 11.3C). Translation of viral mRNAs by internal initiation is also stimulated by this arrangement, which may be established in a number of different ways. Translation from picornavi-

Figure 11.5 *(continued)* changes in the type 2 IRES that are believed to prepare the area surrounding the AUG initiation codon for the arrival of the ribosome. The initiator AUG codon (black box) is part of the IRES of hepatitis C virus. The IRES of cricket paralysis virus mimics a tRNA and occupies the P site in the 40S ribosomal subunit. Translation initiates with a non-AUG codon from the A site. (A and B) Adapted from S. R. Stewart and B. L. Semler, *Semin. Virol.* **8**:242–255, 1997, with permission. (C) Adapted from S. M. Lemon and M. Honda, *Semin. Virol.* **8**:274–288, 1997, with permission. (D) Adapted from E. Jan and P. Sarnow, *J. Mol. Biol.* **324**:889–902, 2002, with permission.

BOX 11.2

Use of the IRES in cloning and expression vectors

The IRES has been used widely in the expression and cloning of foreign genes in eukaryotes. One strategy for the expression of genes in mammalian cells is to produce mRNAs in the cytoplasm by a bacteriophage DNA-dependent RNA polymerase, such as T7 RNA polymerase. Such mRNAs are poorly translated because they are not capped; inclusion of an IRES in the 5′ untranslated region allows them to be translated efficiently.

Another application of the IRES is in the functional cloning of new genes (figure, top panel). A DNA library is made by using a cloning vector that produces a bicistronic mRNA encoding both the desired gene and a selectable marker. This library is introduced into cells that are then screened for expression of the desired protein. The use of a selectable marker on the same mRNA increases the efficiency of screening because most transformants also express the encoded gene.

IRESs have also been used in the isolation of mutant mice by homologous recombination in embryonic stem cells. Bicistronic vectors have been designed to produce mRNA encoding the altered protein and the *lacZ* gene, separated by an IRES (figure, bottom panel). Cells in which this DNA is integrated into the wild-type gene are selected; these cells express a mutant form of the protein and the β-galactosidase enzyme. The latter can be readily assayed in tissue sections by staining with a chromogenic substrate, and since it is encoded on the same mRNA as the targeted gene product, it serves as a marker for mRNA expression.

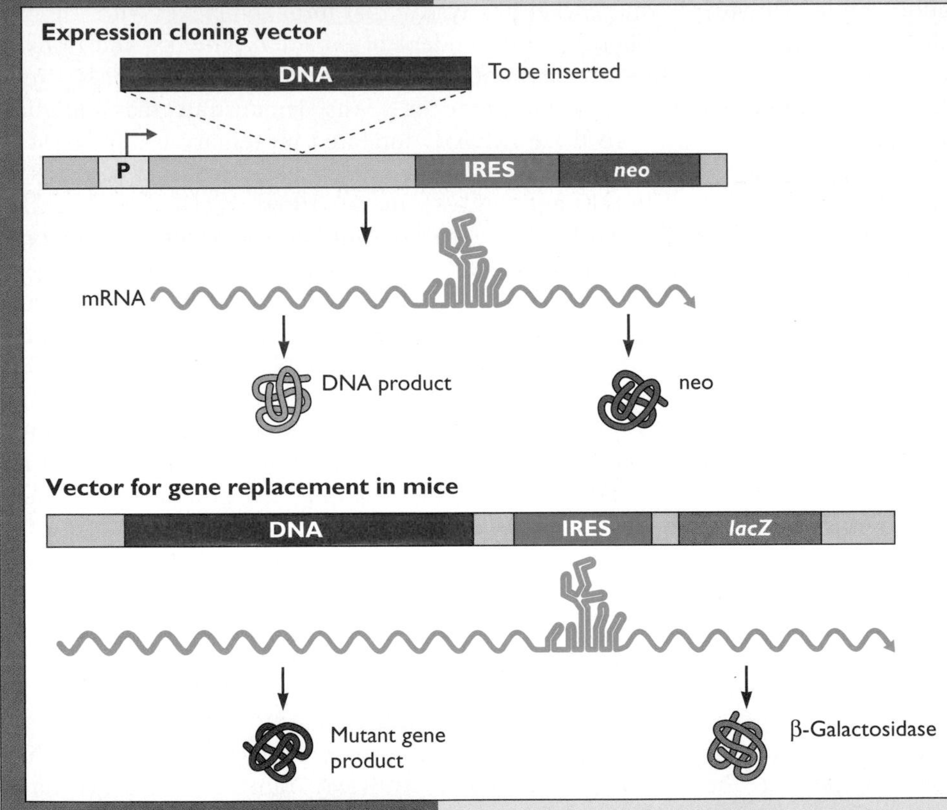

Uses of the IRES in molecular biology. (Top) Design of plasmids for expression cloning. The expression library is created by inserting DNA into a site on the plasmid downstream of a promoter. The foreign DNA is followed by an IRES linked to a selectable marker such as neomycin resistance. The mRNA produced from this plasmid DNA, after introduction into mammalian cells, encodes the cloned DNA product and the protein conferring neomycin resistance. (Bottom) Vector for gene replacement in mice. In this example, the goal is to replace the gene with a mutant version. The targeting plasmid consists of mutant DNA followed by an IRES and the *lacZ* gene. The flanking light blue bars represent sequences from the mouse gene that will mediate homologous recombination. After replacement of the endogenous gene with this synthetic version, mRNA will be produced that encodes the mutant gene product as well as the β-galactosidase protein. The latter can be detected in tissues by staining with the chromogenic substrate X-Gal (5-bromo-4-chloro-3-indolyl-β-D-galactopyranoside).

ral IRESs is stimulated by poly(A), Pab1p, and eIF4G. However, in poliovirus-infected cells the mRNA ends must be brought together in another way, because cleavage of eIF4G removes the binding site for Pab1p (see "Regulation of Poly(A)-Binding Protein Activity" below). The mRNA of hepatitis C virus is not polyadenylated and therefore cannot bind Pab1p. The interaction between molecules of polypyrimidine-tract-binding protein that bind both the viral 3′ untranslated region and the IRES may bring together the 5′ and 3′ ends of the mRNA.

Other host cell proteins that contribute to IRES function. The poliovirus IRES functions poorly in reticulocyte

A

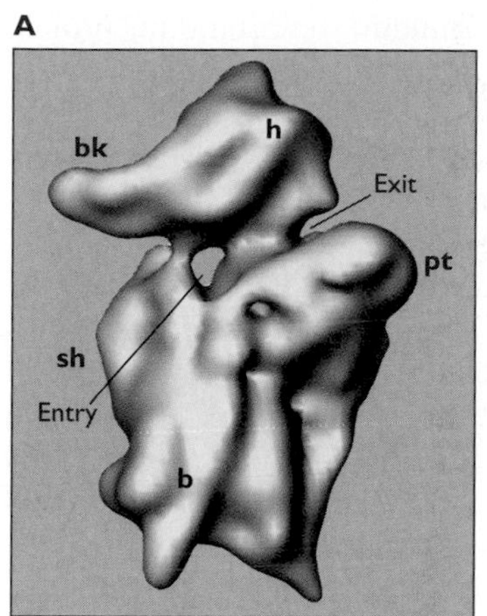

B

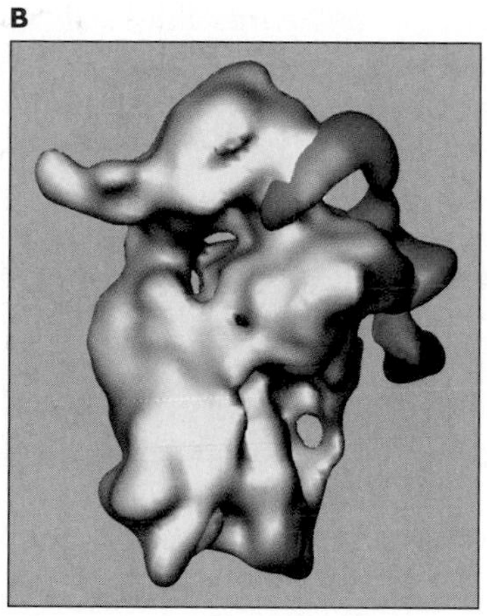

C

AUGAGCACAAA...

IIb

IIId/e/f

IIIa/b/c

Figure 11.6 Complex of the 40S ribosomal subunit with the hepatitis C virus IRES. (A) Structure of 40S ribosomal subunit from reticulocytes, showing structural features: bk, beak; h, head; pt, platform; sh, shoulder; b, body. (B) Structure of 40S ribosomal subunit complexed with 350-nucleotide RNA encompassing the IRES of hepatitis C virus. Binding of the RNA induces major conformational changes in the 40S subunit. The head clamps down on the mRNA and touches the platform. The asterisk marks a conformational change in the body of the ribosome. (C) Enlarged side view of hepatitis C virus IRES bound to the 40S ribosomal subunit. The viral RNA lies within the cleft. Domains of the IRES are identified by boxes, and their sequences are shown below. The AUG initiation codon is in bold. The arrow indicates the mRNA-binding groove, where the coding region of the viral RNA may be located. Structures were obtained by cryo-electron microscopy and image reconstruction. Adapted from C. M. Spahn, J. S. Kieft, R. A. Grassucci, P. A. Penczek, K. Zhou, J. A. Doudna, and J. Frank, *Science* **291:**1959–1962, 2001, with permission.

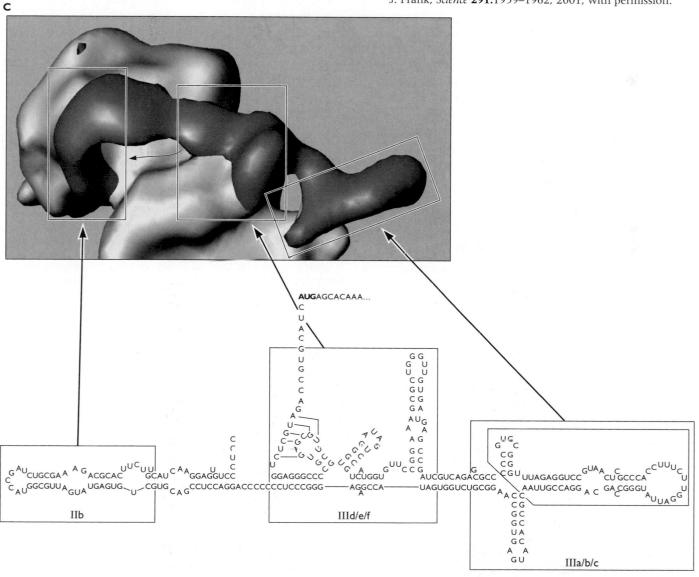

lysates, in which most capped mRNAs are translated efficiently (Box 11.3). Addition of a cytoplasmic extract to reticulocyte lysates restores efficient translation from this IRES. These observations led to the suggestion that ribosome binding to the IRES requires more than translation initiation proteins. Such proteins have been identified by their ability to bind to the IRES and to restore its function in the reticulocyte lysate. One host protein identified by this approach is the La protein, which binds to the 3′ end of the poliovirus IRES. This protein is associated with the 3′ termini of newly synthesized small RNAs, including all nascent transcripts of cellular RNA polymerase III. While predominantly nuclear, La protein is localized to the cytoplasm in poliovirus-infected cells. La protein is present in low concentrations in reticulocyte lysates, and when added to them it stimulates the function of the poliovirus IRES.

Polypyrimidine-tract-binding protein (Ptb), or heterogeneous nuclear ribonucleoprotein I (hnRNPI), was also found to bind the poliovirus IRES. This predominantly nuclear protein, a negative regulator of alternative pre-mRNA splicing, is redistributed to the cytoplasm of the cell during poliovirus infection. Polypyrimidine-tract-binding protein binds to sequences upstream of the pyrimidine-rich sequence of the poliovirus IRES, and to both the 5′ and 3′ untranslated regions of hepatitis C virus RNA. If this protein is removed from a cell extract, the activity of the poliovirus and hepatitis C virus IRESs is eliminated. However, neither IRES activity is restored when purified Ptb is added to the depleted extracts. These findings suggest that another protein(s) associated with polypyrimidine-tract-binding protein might also be required for IRES function.

Another cellular protein necessary for the activity of the enterovirus and rhinovirus IRES is the cytoplasmic RNA-binding protein, poly(rC)-binding protein 2, originally identified by its ability to bind stem-loop IV of the poliovirus IRES (Fig. 11.5A). Mutations in the poliovirus 5′ untranslated region that abolish binding of poly(rC)-binding proteins cause decreased translation in vitro. Depletion of poly(rC)-binding proteins from human translation extracts inhibits translation dependent on the IRESs of poliovirus, coxsackievirus B, and rhinovirus, but not on those of encephalomyocarditis virus or foot-and-mouth

BOX 11.3

Translation in vitro: the reticulocyte lysate and wheat germ extract

Our present understanding of the fundamentals of translation initiation, elongation, and termination, as well as the translation strategies of different viruses, would not be possible without the technique of in vitro translation in cell-free extracts. In this method, cells are lysed and the nuclei are removed by centrifugation. The mRNA is added to the lysate, and the mixture is incubated in the presence of an isotopically labeled amino acid precursor which is incorporated into the translation product.

The ideal extract for in vitro translation has two important properties: high translation efficiency and low levels of protein synthesis in the absence of added mRNA. By the early 1970s, cell-free extracts prepared from Krebs II ascites tumor cells or rabbit reticulocytes were found to translate protein with high efficiency, but the presence of endogenous mRNAs that were also translated complicated the analysis of proteins made from added mRNA. In 1973 a cell-free extract from commercial wheat germ that had low background levels of protein synthesis, and in which exogenous mRNAs were translated very efficiently, was developed. A few years later, the background in a reticulocyte lysate was eliminated by treatment with micrococcal nuclease, which destroyed the endogenous mRNA. This nuclease requires calcium for its activity, and it was therefore a simple matter of adding EGTA [ethylene glycol-bis(β-aminoethylether)-*N,N,N′,N′*-tetraacetic acid], a calcium chelator, to the reaction to prevent the degradation of exogenously added mRNA.

Wheat germ extract and reticulocyte lysate are still widely used in translation studies because the cells are abundant and inexpensive and are excellent sources of initiation proteins, leading to high translation efficiencies. Micrococcal nuclease followed by calcium chelation has been successfully used to make mRNA-dependent extracts from many mammalian cell types, although the translation efficiency of such systems does not approach that of wheat germ or reticulocyte lysates. Unfortunately, it has not been possible to consistently prepare translation extracts from normal mammalian tissues, which has hampered the study of tissue-specific translation regulation in virus-infected and uninfected cells.

Pelham, H. R. B., and R. J. Jackson. 1976. An efficient mRNA-dependent translation system from reticulocyte lysates. *Eur. J. Biochem.* **67**:247–256.

Roberts, B. E., and B. M. Patterson. 1973. Efficient translation of tobacco mosaic virus RNA and rabbit globin 9S RNA in a cell-free system from commercial wheat germ. *Proc. Natl. Acad. Sci. USA* **70**:2330–2334.

disease virus. Translation activity of the IRESs was restored by addition of purified poly(rC)-binding protein 2.

The use of binding assays to identify proteins that are required for activity of the IRES can be misleading. There are several examples of cellular proteins that specifically bind an IRES yet have no role in its function. Functional translation assays are required to prove that cellular proteins are essential for internal initiation. The poor activity of the rhinovirus IRES in reticulocyte lysates can be overcome with the addition of an extract from HeLa cells. Two cellular proteins, polypyrimidine-tract-binding protein and Unr, a second protein of unknown function, account for the increased activity. Addition of both proteins to a reticulocyte lysate stimulates the activity of the rhinovirus IRES to the same extent as the addition of an extract from HeLa cells.

The restricted tropism of some viruses may be governed in part by the distribution of cellular proteins required for optimal IRES activity. There has been little evidence in support of this hypothesis; the cell proteins discussed above are ubiquitously expressed. An exception may be murine proliferation-associated protein 1 (Mpp-1), which is required for function of the IRES of foot-and-mouth disease virus. This protein binds to a central domain of the foot-and-mouth disease virus IRES and acts synergistically with Ptb to increase the binding of eIF4F. It has been suggested that Mpp-1 may determine the tissues in which the IRES functions. To assess this possibility, a recombinant virus was constructed by replacing the IRES of Theiler's murine encephalomyocarditis virus with that of foot-and-mouth disease virus. Theiler's murine encephalomyelitis virus replicates in the brain, but the recombinant virus cannot. The inability of the recombinant virus to replicate in the brain may be due to the dearth of Mpp-1.

No single host cell protein that is essential for the function of all viral IRESs has been identified. It has been suggested that La protein, Ptb protein, poly(rC)-binding protein, Unr protein, and Mpp-1 act as RNA chaperones, maintaining the IRES in its appropriate three-dimensional structure for binding to ribosomes and translation initiation proteins. In support of this hypothesis is the observation that these are RNA-binding proteins that can form multimers with the potential to contact the IRES at multiple points.

Elongation and Termination

During elongation, the ribosome selects aminoacylated tRNA according to the sequence of the mRNA codon, and catalyzes the formation of a peptide bond between the polypeptide and the incoming amino acid. Elongation is assisted by three proteins that act as catalysts and maintain the speed and accuracy of translation. In the 80S initiation complex, the Met-tRNA$_i$ is bound to the peptidyl (P) site of the ribosome (Fig. 11.7). Elongation of the peptide chain begins with addition of the next amino acid. An important

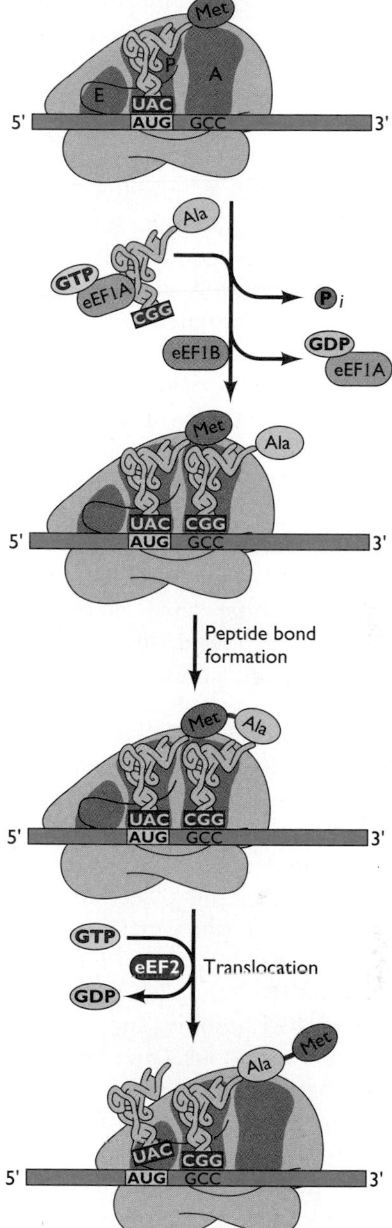

Figure 11.7 Translation elongation. There are three tRNA-binding sites on the ribosome, called peptidyl (P), aminoacyl (A), and exit (E). After the initiating Met-tRNA$_i$ is positioned at the P site, the second aminoacyl-tRNA (alanyl-tRNA is shown) is brought to the A site by eEF1A complexed with GTP. After GTP hydrolysis, eEF1A is released. The guanine nucleotide exchange protein eEF1B exchanges GDP of eEF1A-GDP with GTP, allowing eEF1A to interact with a tRNA synthetase and bind a newly aminoacylated tRNA. The peptide bond is then formed, followed by movement of the ribosome three nucleotides along the mRNA, a reaction that requires GTP hydrolysis and eEF2. The peptidyl (Met-Ala) tRNA moves to the P site, and the uncharged tRNA moves to the E site. The A site is now empty, ready for another aminoacyl-tRNA. Adapted from G. M. Cooper, *The Cell: a Molecular Approach* (ASM Press, Washington, D.C., and Sinauer Associates, Sunderland, Mass., 1997), with permission.

component of this process is elongation factor eEF1A, which is bound to aminoacylated tRNA, a molecule of GTP, and the nucleotide exchange protein eEF1B. Interaction between the codon and the anticodon leads to a conformational change in the ribosome called accommodation, and the hydrolysis of GTP and the release of eEF1A-GDP. Accommodation preserves the fidelity of translation, because it can only occur on proper codon-anticodon base pairing, and is required for GTP hydrolysis. If an incorrect tRNA enters the A site, accommodation does not occur and the aminoacylated tRNA is rejected. The large ribosomal subunit catalyzes the formation of a peptide bond between the amino acids occupying the P and A sites. The 80S ribosome then moves three nucleotides along the mRNA. Translocation is dependent upon eEF2 and hydrolysis of GTP. This motion moves the uncharged tRNA to the exit (E) site and the peptidyl-tRNA to the P site, allowing a new aminoacylated tRNA to enter the A site and release of the uncharged tRNA. This cycle is repeated until the ribosome encounters a stop codon. mRNAs are usually bound by many ribosomes (**polysomes**), with each ribosome separated from its neighbors by approximately 100 to 200 nucleotides, synthesizing a polypeptide chain.

Termination is a modification of the elongation process: once the stop codon enters the A site of the ribosome, it is recognized by the 40S subunit, and the 60S subunit cleaves the ester bond between the protein chain and the last tRNA. Recognition of the three stop codons (UAA, UAG, or UGA) by the 40S ribosomal subunit is facilitated by the **release proteins** eRF1 and eRF3. The structure of eRF1 mimics that of tRNA, allowing the release protein to occupy the A site of the ribosome. The N terminus of eRF1 recognizes all three stop codons. Once bound in the A site, eRF1 induces a rearrangement of the 80S ribosome that causes release of the polypeptide. The interaction between eRF1 and the ribosome stimulates the GTPase activity of eRF3, which is bound to the C terminus of eRF1. GTP hydrolysis is required for release of the protein.

Although stop codons are the major determinants of translation termination, other sequences can affect the efficiency of this process. The nucleotide immediately downstream of the stop codon can influence chain termination and ribosome dissociation. In eukaryotes, the preferred termination signals are UAA(A/G) and UGA(A/G).

Exactly when ribosomes are released from the mRNA is not clear. This reaction may occur at the termination codon, or ribosomes may continue to transit through the 3′ noncoding region to the poly(A) tail and even beyond. Because eRF3 binds both eRF1 and Pab1p, the mRNA ends may be brought together by the ribosome-eRF1-eRF3-Pab1p-eIF4G interaction (Fig. 11.8C). As a result, ribosomes may shuttle from the 3′ end of the mRNA back to the 5′ end, beginning the synthesis of another molecule of the protein.

Once ribosomes are released from the mRNA, they dissociate into 40S and 60S subunits. This disassembly is favored by the binding of eIF1A, eIF3, and eIF6.

The Diversity of Viral Translation Strategies

A variety of unusual translation strategies have evolved that expand the coding capacity of viral genomes, which are relatively small, and allow the synthesis of multiple polypeptides from a single RNA genome (Fig. 11.9). All were discovered in virus-infected cells and subsequently shown to operate also during translation of cellular mRNAs. Nontranslational solutions for maximizing the number of proteins encoded in viral genomes are discussed in other chapters and include the synthesis of multiple subgenomic mRNAs, mRNA splicing, and RNA editing.

Polyprotein Synthesis

One strategy allowing the synthesis of multiple proteins from a single RNA genome is to synthesize from a single mRNA a polyprotein precursor, which is then proteolytically processed to form functional viral proteins.

A dramatic example of protein processing occurs in picornavirus-infected cells: nearly the entire (+) strand RNA is translated into a single large polyprotein (Fig. 11.10A). Processing of this precursor is carried out by two virus-encoded proteases, $2A^{pro}$ and $3C^{pro}$, which cleave between Tyr and Gly and between Gln and Gly, respectively. In both cases, flanking amino acid residues control the efficiency of cleavage so that not all Tyr-Gly and Gln-Gly pairs in the polyprotein are processed. These two proteases are active in the nascent polypeptide and release themselves from the polyprotein by self-cleavage. Consequently, the polyprotein is not observed in infected cells because it is processed as soon as the protease coding sequences have been translated. After the proteases have been released, they cleave the polyprotein in *trans*.

Protein production can be controlled by the rate and extent of polyprotein processing. In addition, alternative utilization of cleavage sites can produce proteins with different activities. For example, the poliovirus protease $3C^{pro}$ does not efficiently process the capsid protein precursor P1. Rather, the $3C^{pro}$ precursor, 3CD, is required for processing of P1. By regulating the amount of 3CD produced, the extent of capsid protein processing can be controlled. Because 3CD and $3C^{pro}$ process Gln-Gly pairs in the remainder of the polyprotein with the same efficiency, an interesting question is why 3CD, which also contains $3D^{pol}$

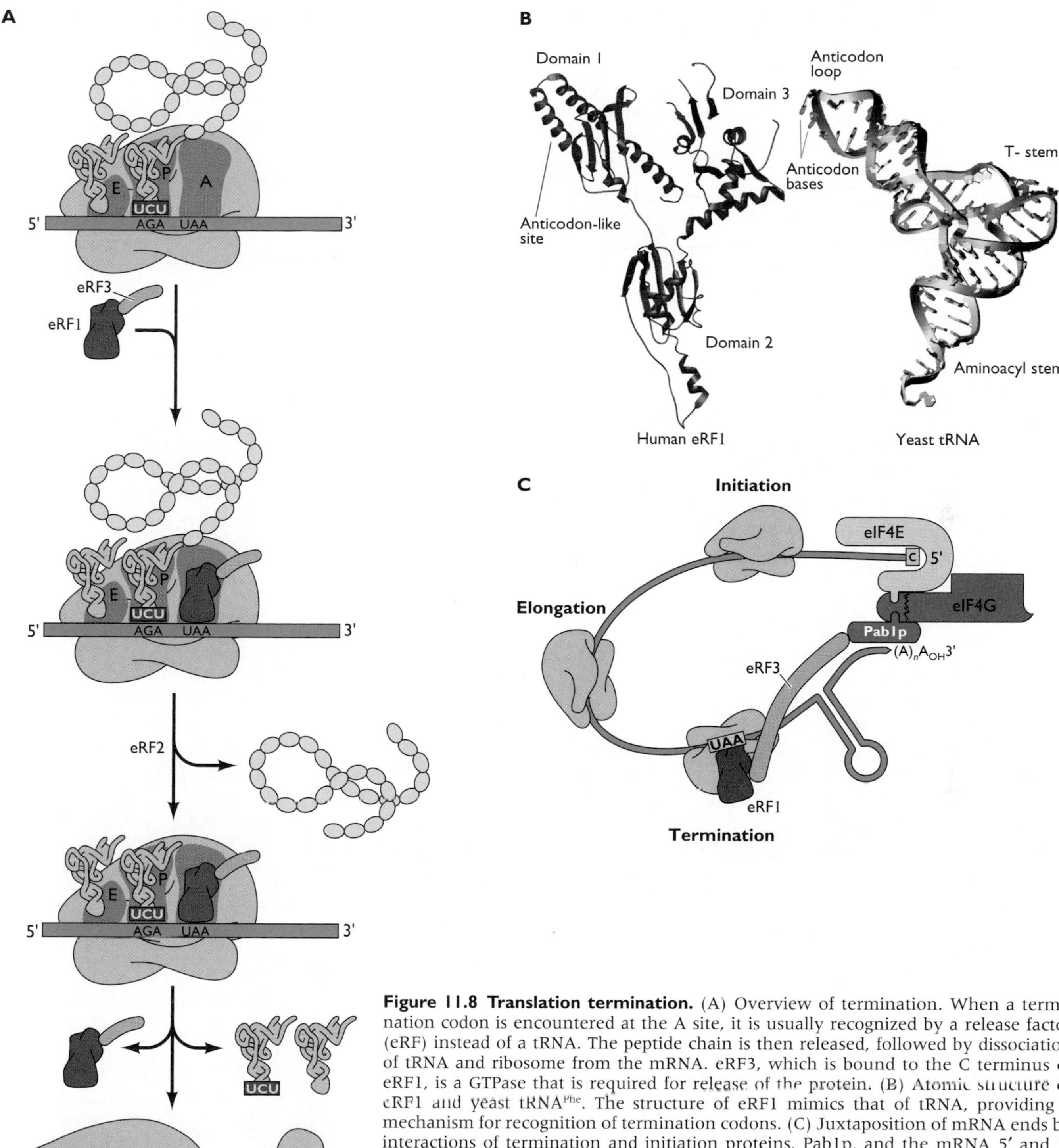

Figure 11.8 Translation termination. (A) Overview of termination. When a termination codon is encountered at the A site, it is usually recognized by a release factor (eRF) instead of a tRNA. The peptide chain is then released, followed by dissociation of tRNA and ribosome from the mRNA. eRF3, which is bound to the C terminus of eRF1, is a GTPase that is required for release of the protein. (B) Atomic structure of eRF1 and yeast $tRNA^{Phe}$. The structure of eRF1 mimics that of tRNA, providing a mechanism for recognition of termination codons. (C) Juxtaposition of mRNA ends by interactions of termination and initiation proteins, Pab1p, and the mRNA 5′ and 3′ ends. eRF3 binds both eRF1 and Pab1p. (A) Adapted from G. M. Cooper, *The Cell: a Molecular Approach* (ASM Press, Washington, D.C., and Sinauer Associates, Sunderland, Mass., 1997), with permission. (B) Adapted from H. Song, P. Mugnier, A. K. Das, H. M. Webb, D. R. Evans, M. F. Tuite, B. A. Hemmings, and D. Barford, *Cell* **100:**311–321, 1998, with permission. (C) Adapted from N. Uchida, S. Hoshino, H. Imataka, N. Sonenberg, and T. Katada, *J. Biol. Chem.* **277:**50286–50292, 2002, with permission.

Mechanism of translation	Examples	
Polyprotein synthesis	Picornaviruses Flaviviruses Alphaviruses Retroviruses	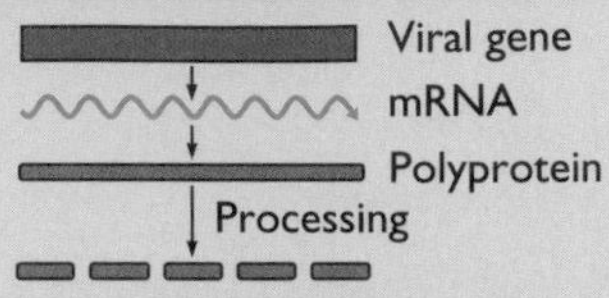 Viral gene mRNA Polyprotein Processing
Leaky scanning	Sendai virus P/C mRNA Influenza B virus RNA 6 Human immunodeficiency virus type 1 Env/Vpu Human T-cell leukemia virus Tax, Rex Simian virus 40 VP2, VP3 Simian virus 40 agnoprotein	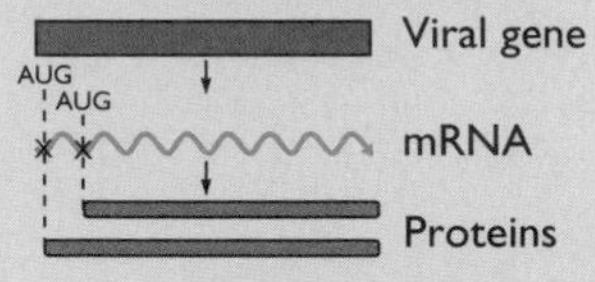 Viral gene AUG AUG mRNA Proteins
Reinitiation	Influenza B virus RNA 7 Cytomegalovirus gp48 mRNA	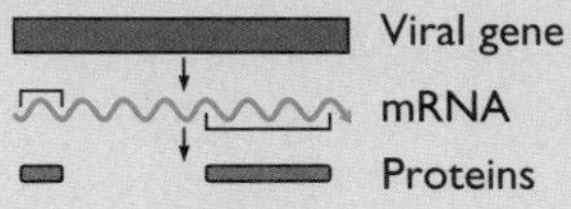 Viral gene mRNA Proteins
Suppression of termination	Alphavirus nsP4 Retrovirus Gag-Pol	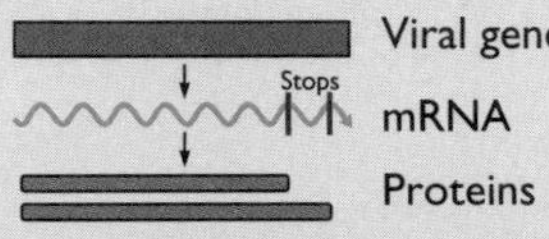 Viral gene Stops mRNA Proteins
Ribosomal frameshifting	Coronavirus orf1a-orf1b Human astrovirus type 1 orf1a-orf1b Retrovirus Gag-Pol	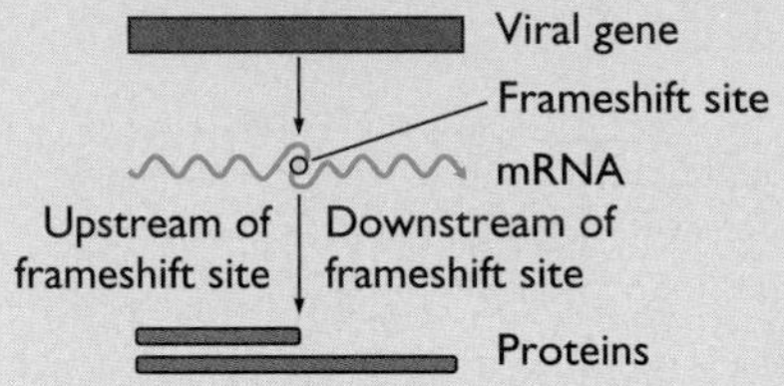 Viral gene Frameshift site mRNA Upstream of frameshift site Downstream of frameshift site Proteins
Internal ribosome entry	Picornaviruses Flaviviruses	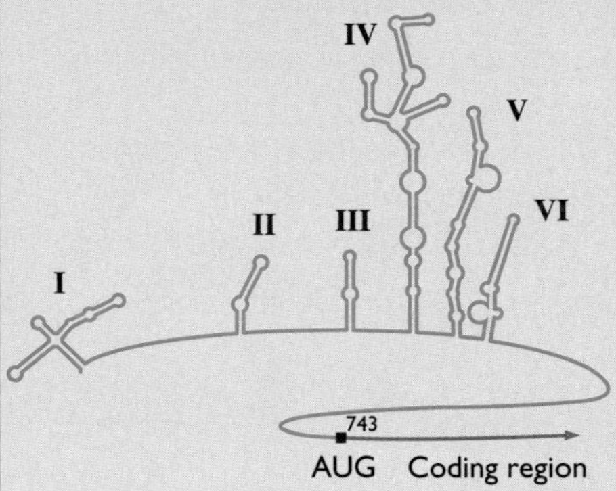 IV V II III VI I 743 AUG Coding region
Ribosome shunting	Adenovirus Cauliflower mosaic virus	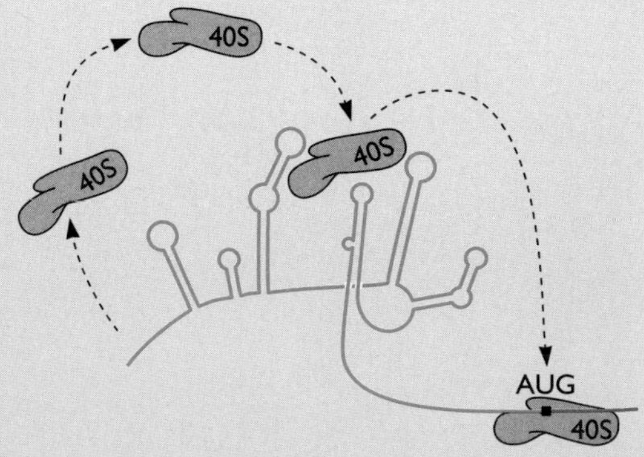 40S 40S 40S AUG 40S

Figure 11.9 The diversity of viral translation strategies. Examples of different mechanisms of translation that occur in virus-infected cells are listed. On the right are schematic diagrams of the strategies used for making multiple proteins from one gene, and diverse initiation mechanisms.

Figure 11.9 *(continued)*

Mechanism of translation	Examples	
Internal initiation mediated by tRNA-like structure in the 3' untranslated region	Turnip yellow mosaic virus	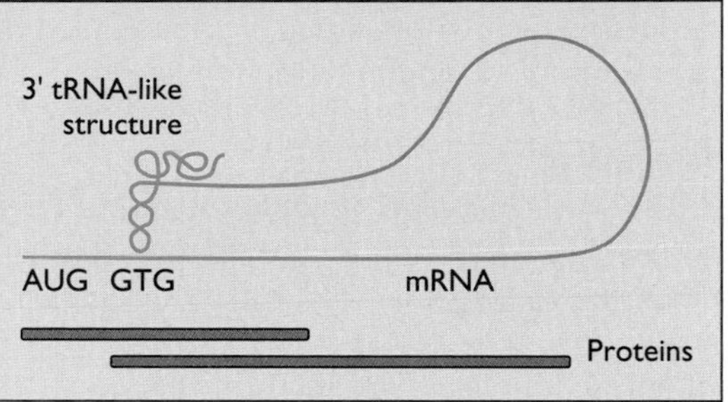 3' tRNA-like structure; AUG GTG; mRNA; Proteins
Dicistronic mRNAs	Cricket paralysis virus *Rhopalosiphum padi* virus	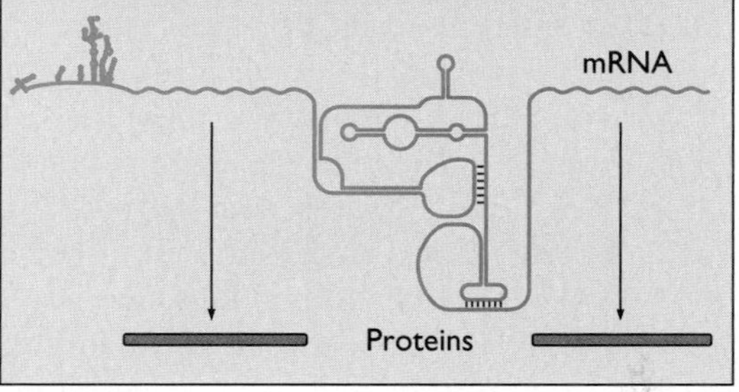 mRNA; Proteins

protein, is further processed to produce $3C^{pro}$ (Fig. 11.10A). The answer is that 3CD protein, while proteolytically active, does not possess RNA polymerase activity and therefore some of it must be cleaved to allow RNA replication.

Some viral precursor proteins are processed by cellular proteases. The genome of flaviviruses encodes an open reading frame of more than 10,000 bases (Fig. 11.10B). This mRNA is translated into a polyprotein precursor that is processed by a viral serine protease and by host signal

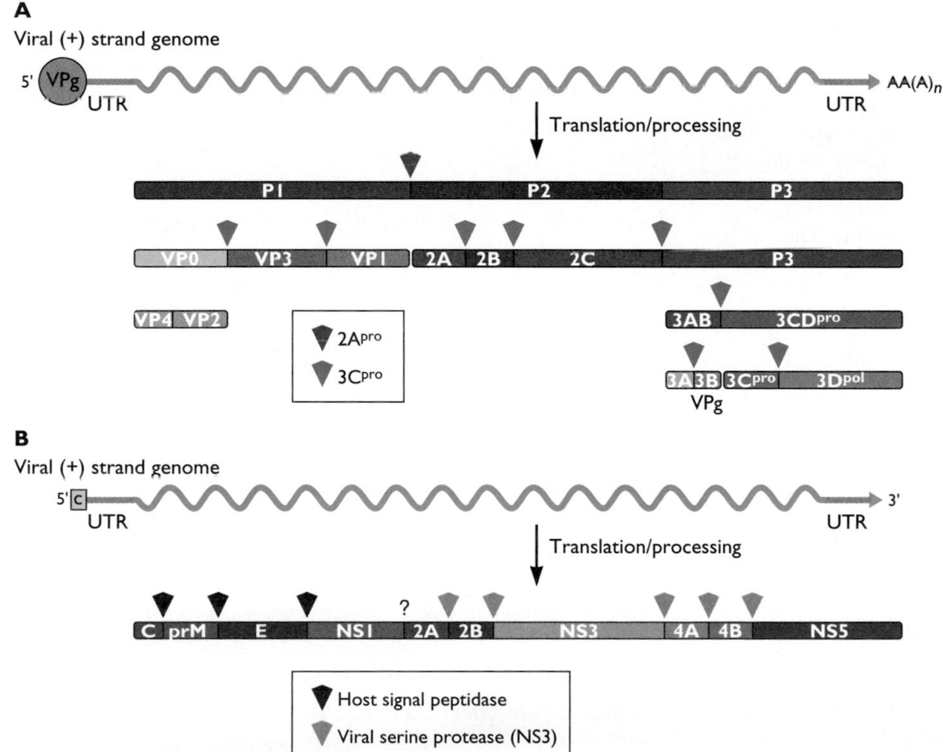

Figure 11.10 Polyprotein processing of enteroviruses and flaviviruses. (A) Processing map of protein encoded by the poliovirus genome. The viral RNA is translated into a long precursor polyprotein that is processed by two viral proteases, $2A^{pro}$ and $3C^{pro}$, to form viral proteins. Cleavage sites for each protease are shown. (B) Cleavage map of protein encoded in the flavivirus genome. Processing of the flavivirus precursor polyprotein is carried out either by the host signal peptidase, which occurs in the lumen of the endoplasmic reticulum, or by the viral protease NS3.

peptidase. The latter enzyme is located in the endoplasmic reticulum, where it removes the signal sequence from proteins translocated into the lumen (Chapter 12). The viral proteins processed by the host signal peptidase must therefore be inserted into the endoplasmic reticulum.

Leaky Scanning

Although the vast majority of eukaryotic mRNAs are monocistronic (Fig. 11.1), some viral mRNAs that encode two proteins in overlapping reading frames have been identified. The P/C gene of Sendai virus is the model for genes that encode mRNAs with such translational flexibility (Fig. 11.11). P protein is translated from an open reading frame beginning with an AUG codon at nucleotide 104. C proteins are produced from a different reading frame, which begins at nucleotide 81, and are completely different from P proteins. No less than four C proteins (called C′, C, Y1, and Y2) are produced by translation beginning at four in-frame initiation codons. The first start site is at an unusual ACG codon, and the third, fourth, and fifth are at AUG codons; the result is a nested set of proteins with a common C terminus. mRNA editing has no effect on these proteins because their coding sequence terminates before the editing site.

The first three initiation sites on P/C mRNA are likely to be arranged to permit translation by **leaky scanning**. The first start site, $ACG^{81/C'}$, is surrounded by a good initiation context but is inefficient because of the unusual start codon. The second initiator is in a poor initiation context (CGCAUGG), while the third is in a better context (AAGAUGC). Because initiation at the ACG codon is inefficient, some ribosomes can bypass it and initiate at the second; as the context at this AUG is poor, some ribosomes can also find their way to the third initiation codon, which has a better initiation context. Consistent with this hypothesis, mutagenesis of $ACG^{81/C'}$ to AUG abolishes translation at $AUG^{104/P}$ and $AUG^{114/C}$. When successive initiation codons are used in leaky scanning, they are increasingly efficient as start sites.

The last two C-protein initiation codons, $AUG^{183/Y1}$ and $AUG^{201/Y2}$, are not likely to be translated by leaky scanning because they are in the poorest contexts of all five. Furthermore, mutagenesis of $ACG^{81/C'}$ to AUG has no effect on the synthesis of Y1 and Y2 proteins. Rather, Y1 and Y2 proteins are believed to be translated by ribosome shunting. An interesting question is how the different translation strategies of P/C mRNA are coordinated such that, for example, shunting does not dominate at the expense of translation of upstream AUG codons. The answer to this question is not known, but Y protein synthesis relative to that of the other C proteins varies in different cell lines. This result suggests that cellular proteins might regulate ribosome shunting on P/C mRNA, although no such protein has been identified.

Translation of overlapping reading frames also occurs in other viral mRNAs (Fig. 11.9). Influenza viruses are classified into three types, A, B, and C; most of the previous discussion in this textbook has concerned work on influenza A virus. The mRNA synthesized from influenza B virus RNA segment 6 encodes two different proteins, NB and NA, in overlapping reading frames. NB protein translation is initiated at the 5′-proximal AUG codon, while NA protein is initiated from an AUG codon four nucleotides downstream.

Figure 11.11 Leaky scanning and mRNA editing in the Sendai virus P/C gene. Open reading frames are shown as boxes. The P protein open reading frame is shown as a brown horizontal box, and the C protein open reading frame is shown as a blue box. An enlargement of the 5′ end of the mRNA is shown below, indicating the different start sites for four of the C proteins. The site at which G residues are added during mRNA synthesis is shown above; addition of one or two G residues produces the V or W C terminus of P protein. Adapted from J. Curran et al., *Semin. Virol.* **8:**351–357, 1997, with permission.

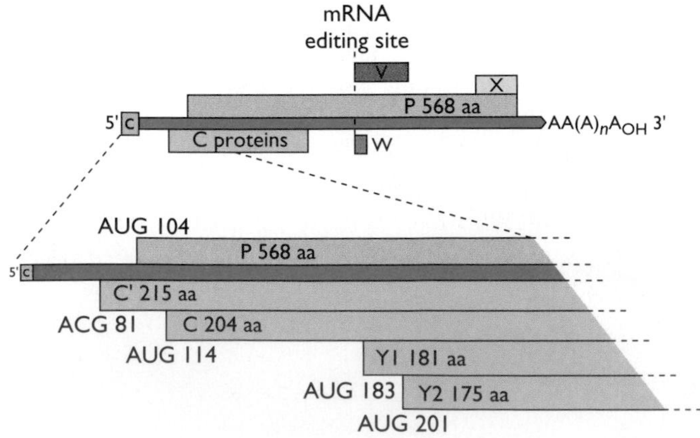

Reinitiation

Reinitiation is another strategy for producing two proteins from a single mRNA. About 10% of eukaryotic mRNAs contain upstream AUG codons that are followed by short open

Figure 11.12 Reinitiation of translation. (Top) Some mRNAs contain one or more short upstream open reading frames (uORFs) that may be translated. Expression of the longer, downstream ORF depends on reinitiation. (Bottom) mRNA produced from influenza B virus RNA segment 7 encodes two proteins, M1 and BM2. The initiation AUG codon for BM2 overlaps the termination codon of M1. Synthesis of BM2 occurs by reinitiation.

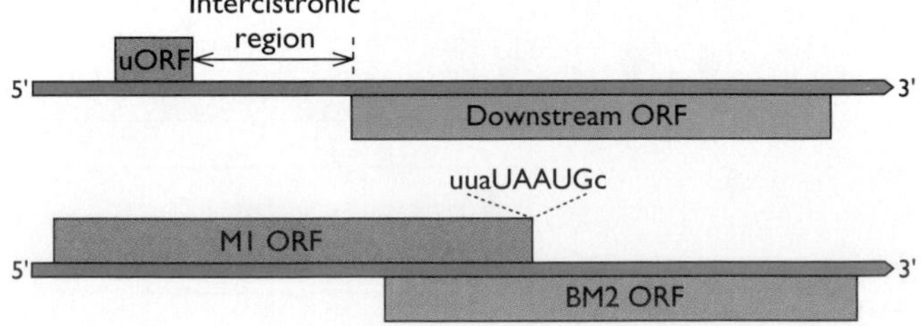

BOX 11.4

Translational control of yeast GCN4 by upstream open reading frames

The best-characterized example of reinitiation is on *GCN4* mRNA of the yeast *S. cerevisiae*. The *GCN4* gene product activates the transcription of at least 40 different genes that encode biosynthetic enzymes. When nutrients are abundant, four short upstream open reading frames in the 5′ untranslated region of this mRNA act in *cis* to prevent ribosomes from initiating at the downstream start site. Under conditions of nutrient deprivation, the upstream open reading frames are unable to prevent ribosomes from translating at the *GCN4* AUG codon. The mechanism of action involves the phosphorylation state of eIF2α (see figure).

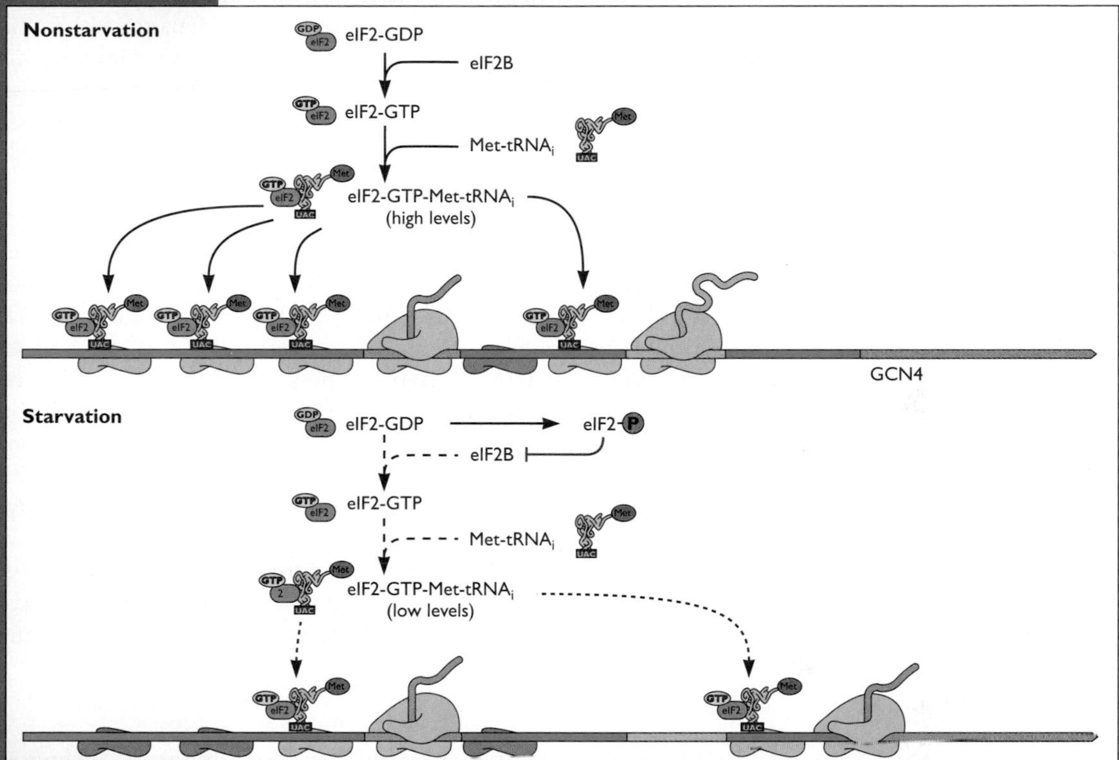

Model for translational control of yeast *GCN4*. A schematic of *GCN4* mRNA is shown, with upstream open reading frames 1 (uORF1) (blue) and 4 (uORF4) (orange) and the GCN4 coding sequence (pink). The gray 40S ribosomal subunits are associated with eIF2, GTP, and Met-tRNA$_i$ and are competent to reinitiate; olive subunits are not associated with this ternary complex and cannot reinitiate. Under all conditions, all ribosomes translate uORF1; half of them continue scanning as 40S ribosomal subunits. Under nonstarvation conditions, eIF2α is not phosphorylated; high levels of eIF2 are available to form the ternary complex, which reassembles with 40S ribosomes after scanning uORF1, allowing reinitiation at the second, third, and fourth upstream open reading frames, after which they dissociate and fail to reach the *GCN4* start site. Under starvation conditions, eIF2α is phosphorylated, leading to limiting amounts of eIF2-GTP and reducing the level of eIF2–GTP–Met-tRNA$_i$. As a result, 40S subunits scan the mRNA longer before they bind to a Met-tRNA$_i$, increasing the likelihood that they will reach the *GCN4* start site before initiating translation. Adapted from A. G. Hinnebusch, Fig. 1, p. 202, *in* J. W. B. Hershey et al., *Translational Control* (Cold Spring Harbor Laboratory Press, Cold Spring Harbor, N.Y., 1996), with permission.

reading frames (Fig. 11.12). These open reading frames may be translated, with reinitiation occurring at the downstream AUG. For example, a 22-amino-acid peptide is synthesized from an open reading frame in the 5′ region of human cytomegalovirus gp48 mRNA. Other examples are found in mRNAs of retroviruses and in cellular mRNAs such as those encoding *S*-adenosylmethionine decarboxylase and fibroblast growth factor 5. The presence of upstream open reading frames can affect translation of downstream open reading frames (Box 11.4). The extent of regulation depends on many factors, such as sequence context of the upstream open reading frame AUG initiation codon, the presence of RNA secondary structure, and the distance between the upstream and downstream open reading frames.

In most cases, translation reinitiation involves short upstream open reading frames that precede the main open reading frame. Reinitiation of longer, overlapping reading frames occurs on mRNA of influenza B virus RNA 7, which encodes two proteins, M1 protein and BM2 protein (Fig. 11.12). M1 protein is translated from the 5′-proximal AUG codon, while the BM2 protein AUG initiation codon is part of the termination codon for M1 protein (UAAUG). Translation of the BM2 open reading frame is dependent on the synthesis of M1 protein, as deletion of the M1 AUG codon abrogates BM2 synthesis.

Suppression of Termination

Suppression of termination occurs during translation of many viral mRNAs as a means of producing a second protein with an extended C terminus. The Gag and Pol genes of Moloney murine leukemia virus are encoded in a single mRNA and separated by an amber termination codon, UAG (Fig. 11.13). Infected cells contain a polyprotein precursor called Gag-Pol. This precursor is synthesized by translational suppression of the amber termination codon. The efficiency of suppression is about 4 to 10%. The Gag-Pol precursor is subsequently processed proteolytically to liberate the Gag and Pol proteins. Without this suppression strategy, the viral enzymes RT and IN would not be produced. In a similar way, translational suppression of a different termination codon, UGA, is required for the synthesis of nsP4 of alphaviruses (Fig. 11.13). In this case, the efficiency of suppression is about 10% of the synthesis of the normally terminated nsP3 protein. Because nsP4 encodes the RNA-dependent RNA polymerase, suppression is essential for viral RNA replication. Translational suppression in eukaryotic mRNAs is extremely rare.

Most translational suppression occurs when normal tRNAs misread termination codons, but the efficiency is usually low, e.g., 10%. The misreading of the amber codon in Moloney murine leukemia virus Gag protein for a Gln codon is an example. More rare are suppressor tRNAs that can recognize termination codons and insert a specific amino acid. One example is a suppressor tRNA that inserts selenocysteine, the 21st amino acid, in place of a UGA codon.

The nucleotide sequence 3′ of the termination codon plays an important role in the efficiency of translational suppression. In prokaryotes, the nucleotide next to this codon is highly influential. In eukaryotes, the signals range from very simple to complex. In Sindbis virus, efficient suppression of the UGA codon requires only a single C residue 3′ of the termination codon. The effect of this nucleotide on Sindbis virus suppression may be explained in three ways. The 3′ nucleotide might influence the recognition of termination codons by release proteins; it might affect the interaction of misreading tRNAs with the codon by increasing the stacking energy; or the misreading tRNA and a tRNA recognizing the next codon might interact, affecting the efficiency of elongation and suppression.

Figure 11.13 Suppression of termination codons of alphaviruses and retroviruses. (A) Suppression of termination during the synthesis of alphavirus P123 to produce nsP4, the RNA-dependent RNA polymerase. The termination codon is shown by a red arrowhead. Adapted from J. H. Strauss and E. G. Strauss, *Microbiol. Rev.* **58**:491–562, 1994, with permission. (B) Structure of the termination site between Gag and Pol of Moloney murine leukemia virus. The stop codon that terminates synthesis of Gag is underlined; it is followed by a pseudoknot that is important for suppression of termination.

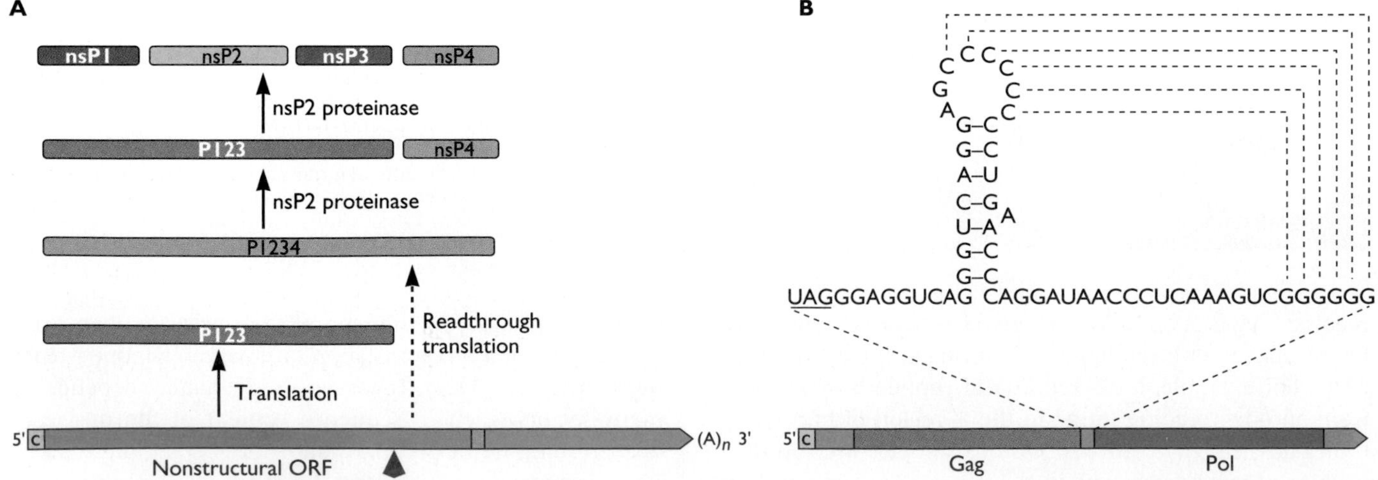

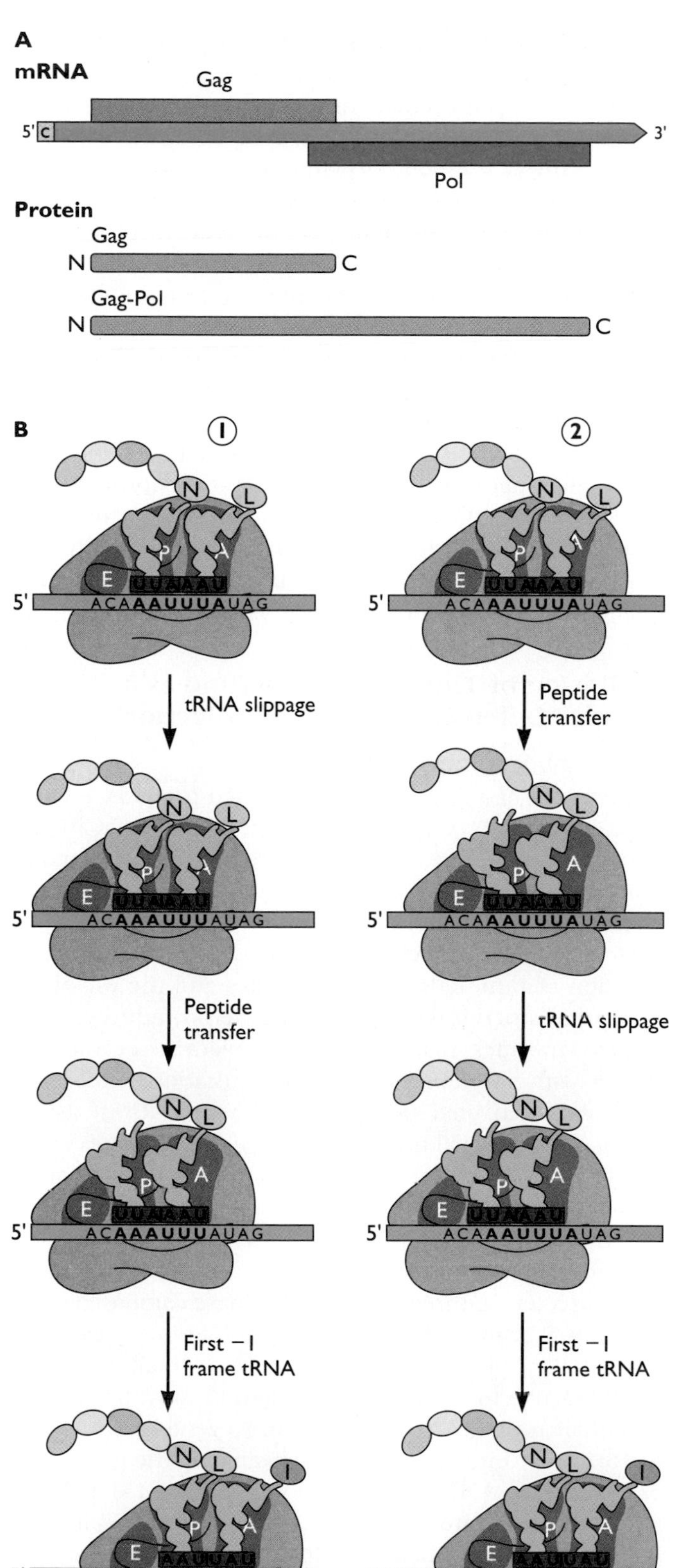

Figure 11.14 Frameshifting on a retroviral mRNA. (A) Structure of open reading frames. Rous sarcoma virus mRNA encodes Gag and Pol proteins in reading frames that overlap by −1. Normal translation and termination produce the Gag protein; ribosomal frameshifting to the −1 frame results in the synthesis of a Gag-Pol fusion protein. (B) Tandem shift model for −1 frameshifting. Two different models are illustrated. In model 1, slippage of the two tRNAs occurs after aminoacyl-tRNA enters the A site but before peptide transfer. In model 2, slippage occurs after peptide transfer, when tRNAs occupy the hybrid E-P and P-A sites. In both cases, slippage allows the tRNAs to form only two base pairs with the mRNA. The site shown is that of Rous sarcoma virus. One-letter amino acid codes are used. Adapted from P. J. Farabaugh, *Microbiol. Rev.* **60:**103–134, 1996, with permission.

In contrast, readthrough of the UAG codon in Moloney murine leukemia virus mRNA requires a purine-rich sequence 3′ of the termination codon, as well as a pseudoknot structure further downstream (see Chapter 6 for a description of pseudoknots). Because of these differences, it is likely that readthrough can be mediated by different mechanisms. It has been suggested that the pseudoknot of Moloney murine leukemia virus RNA causes the ribosome to pause and, together with the eight-nucleotide, purine-rich segment of the suppression signal, allows the suppressor tRNA to compete with release factor at the suppression site.

Suppression of termination is far more prevalent during translation of viral mRNAs than of cellular mRNAs. The RNA sequences and structures required for suppression are not found in most cellular mRNAs. For example, there is a strong bias against cytidine residues at the 3′ end of UGA termination codons in cellular mRNAs.

Ribosomal Frameshifting

Ribosomal frameshifting is a process by which, in response to signals in mRNA, ribosomes move into a different reading frame and continue translation in that new frame. It was discovered in cells infected with Rous sarcoma virus and has since been described for many other viruses, including additional retroviruses, eukaryotic (+) strand RNA viruses, and herpes simplex virus. There is only one known example of frameshifting in a mammalian mRNA.

In the genome of retroviruses, the *gag* and *pol* genes may be separated by a stop codon (Fig. 11.13), or they may be in different reading frames, with *pol* overlapping *gag* in the −1 direction (Fig. 11.14A). During synthesis of Rous sarcoma virus Gag protein, ribosomes frameshift before reaching the Gag stop codon and continue translating Pol, such that a Gag-Pol fusion is produced at about 5% of the frequency of Gag protein. Studies on the requirements for frameshifting in retroviruses and coronaviruses have identified two essential components: a "slippery" homopolymeric sequence, which is a heptanucleotide stretch with two homopolymeric triplets of the form X-XXY-YYZ (e.g., in Rous sarcoma virus A-AAU-UUA), and an RNA secondary structure, usually a pseudoknot, five to eight nucleotides downstream. These observations led to the proposal of the tandem shift model for frameshifting, in which two tRNAs in the zero reading frame (X-XXY-YYZ) slip back by one nucleotide during the frameshift to the –1 phase (XXX-YYY). Each tRNA base pairs with the mRNA in the first two nucleotides of each codon (Fig. 11.14B). The peptidyl-tRNA is transferred to the P site, the –1 frame *pol* codon is decoded, and translation continues to produce the fusion protein. In this model, slippage occurs before peptide transfer, with the peptidyl- and aminoacyl-tRNAs bound to the P and A sites. However, it is possible that the shift occurs after peptide transfer but before translocation of the tRNAs (Fig. 11.14B). These two models cannot be distinguished by mutagenesis or by the sequence of the protein products.

The function of the pseudoknot is not known, but it is believed to cause the ribosome to pause over the slippery sequence, increasing the probability that realignment to the –1 reading frame will occur. Some viral mRNAs contain simple stem-loop structures rather than pseudoknots in this position, but the highest frequencies of frameshifting are associated with the presence of the latter structure. However, when the pseudoknot of a coronavirus is replaced with a stem-loop structure, frameshifting is abolished but ribosomal pausing still occurs. The pseudoknot might not simply act as a barrier to ribosome movement; perhaps structural features of the pseudoknot are also important to promote frameshifting. It has been suggested that cellular proteins might interact with the pseudoknot to bring about frameshifting, but such proteins have not been identified.

Bicistronic mRNAs

The mRNAs of members of the *Dicistroviridae*, including cricket paralysis virus and *Rhopalosiphum padi* virus, are bicistronic (Fig. 11.9). The upstream open reading frame begins with an AUG codon and is preceded by an IRES similar to those of picornaviruses. The downstream open reading frame, which encodes the viral capsid proteins, is independently translated from a completely different IRES. The 40S ribosomal subunit binds directly to the intergenic region that is partially folded to mimic a tRNA. The tRNA-like structure occupies the P site of the ribosome, and initiation occurs from the A site at a nonmethionine codon. Translation of this cistron is therefore dependent on ribosomes, elongation, and termination proteins. Because initiation proteins are not required, translation regulation may occur at the stages of elongation and termination.

Regulation of Translation during Viral Infection

Alterations in the cellular translation apparatus are commonplace in virus-infected cells. As part of the antiviral defense, the cell responds to viral infection with measures designed to inhibit protein synthesis and thereby limit virus production. Many viral genomes encode proteins or nucleic acids that neutralize this response and are targeted to restoring translation and maximizing virus replication. In addition, many viral gene products modify the host translation apparatus to favor synthesis of viral proteins over those of the cell. As a result, the entire synthetic capability of the cell is turned to the production of new virus particles, which should enhance virus yield and perhaps accelerate replication. This supposition is supported by the growth defects of different viral mutants that cannot inhibit cellular translation. These cellular and viral modifications of the translation apparatus are usually aimed at the initiation stages, which are rate-limiting.

Inhibition of Translation Initiation as a Cellular Defense against Viral Infection

Phosphorylation of eIF2α

Members of a large family of secreted proteins, including interferons, are produced as part of the rapid innate immune response of vertebrates in response to viral infection (discussed in Chapter 15). Interferons diffuse to neighboring cells, bind to cell surface receptors, and activate a signal transduction pathway that results in transcription of hundreds of cellular genes and the establishment of an **antiviral state**. Two interferon-induced genes encode enzymes that effectively prevent association of mRNA with polysomes and hence translation: protein kinase, RNA-activated (**Pkr**), and **RNase L.** RNase L degrades RNA and will not be considered further here, while Pkr phosphorylates eIF2α, thereby inhibiting translation initiation. Because the block to translation is global, the infected cell may die, but by slowing down viral replication the organism may be spared.

Pkr is a serine-threonine protein kinase composed of an N-terminal regulatory domain and a C-terminal catalytic domain (Fig. 11.15; see also Chapter 15). Small quantities of an inactive form of Pkr are present in most uninfected mammalian tissues; transcription of its gene is induced 5- to 10-fold by interferon. Pkr is activated by the binding of dsRNA to two dsRNA-binding motifs (Fig. 11.15) at the N terminus of the protein. Such dsRNA is produced in cells infected by either DNA or RNA viruses (Chapter 15). Activation is accompanied by autophosphorylation of Pkr. Low concentrations of dsRNA activate Pkr, but high concentrations are inhibitory, leading to the suggestion that one molecule of Pkr phosphorylates another while bound

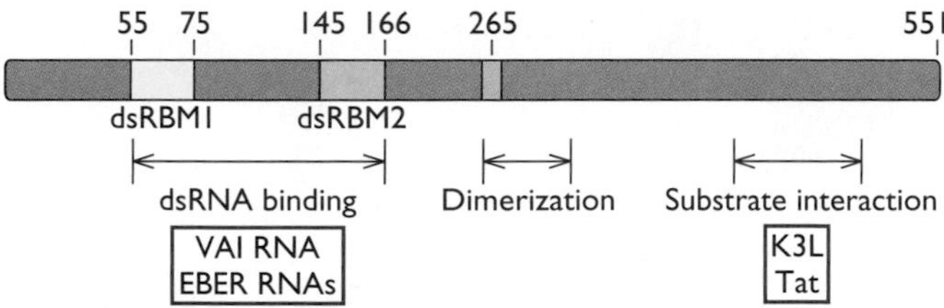

Figure 11.15 Schematic structure of dsRNA-activated protein kinase. A schematic diagram of the protein is shown, with the locations of the regulatory domain, dsRNA-binding sites, and the catalytic domain indicated. Sites where viral gene products interfere with Pkr are shown. These include antagonism of dsRNA binding (adenovirus VA-RNA I [VAI] and Epstein-Barr virus EBERs) and substrate interaction (vaccinia virus K3L and human immunodeficiency virus type 1 Tat). Adapted from M. Gale and M. G. Katze, *Pharmacol. Ther.* **78:**29–46, 1998, with permission.

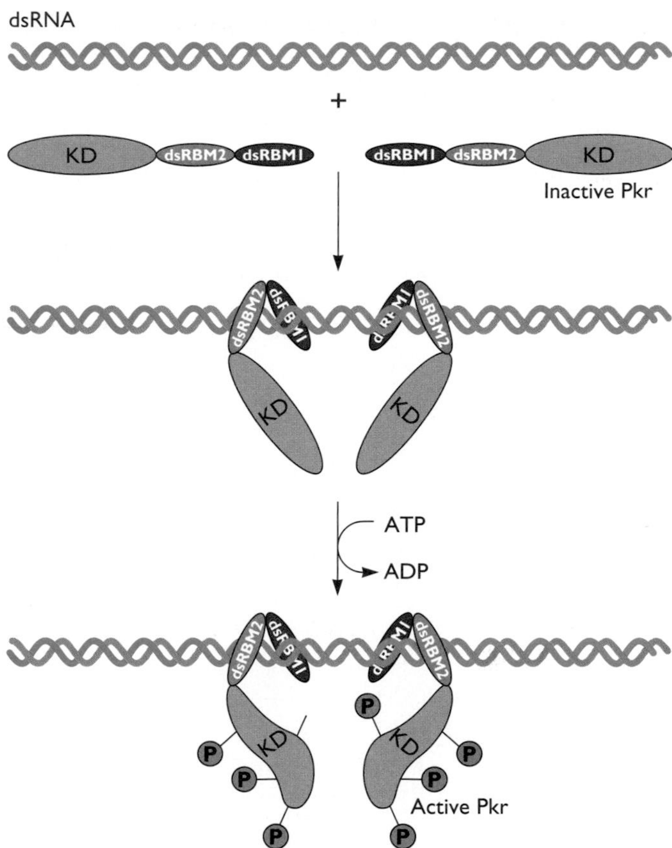

Figure 11.16 Model of activation of Pkr. In the absence of dsRNA, Pkr is an inactive monomer. When two or more molecules of inactive Pkr bind to one dsRNA molecule, cross-phosphorylation occurs through the physical proximity of the molecules. Phosphorylation is believed to cause a conformational change in the catalytic domain (KD) to allow phosphorylation of other substrates, including eIF2α. Adapted from J. W. B. Hershey et al. (ed.), *Translational Control* (Cold Spring Harbor Laboratory Press, Cold Spring Harbor, N.Y., 1996), with permission.

to the same molecule of dsRNA (Fig. 11.16). Phosphorylation is believed to induce conformational changes in the enzyme, thereby rendering the kinase active without further need for dsRNA. Pkr molecules may also interact via a dimerization region (Fig. 11.15), which could lead to activation of the kinase independently of dsRNA.

Activated Pkr can phosphorylate many substrates, but our discussion will focus on eIF2α, which is modified on serine 51. This initiation protein is part of the complex that also contains GTP and Met-tRNA$_i$ (Fig. 11.3). After GTP hydrolysis, the bound GDP moiety must be exchanged for GTP to permit the binding of another Met-tRNA$_i$. This exchange is carried out by eIF2B (Fig. 11.17). When the alpha subunit of eIF2 is phosphorylated, eIF2-GDP binds eIF2B with such high affinity that it is effectively trapped; recycling of eIF2 stops, and ternary complexes are depleted. eIF2B is less abundant than eIF2, and phosphorylation of about 10 to 40% of eIF2 (depending on the cell type and the relative concentrations of eIF2 and eIF2B) results in the complete sequestration of eIF2B, leading to a complete block in protein synthesis. As viral translation is also impaired, the synthesis of new virus particles is diminished.

Viral Regulation of Pkr

Most viral infections induce activation of Pkr and consequent phosphorylation of eIF2α. Global inhibition of translation is clearly a threat to successful viral replication. At least five different viral mechanisms to block Pkr activation or to stop activated Pkr from inhibiting translation can be distinguished (Table 11.2).

Inhibition of dsRNA binding. The 166-nucleotide adenovirus VA-RNA I, which accumulates to massive concentrations (up to 10^9 copies per cell) late in infection following transcription of the viral gene by RNA polymerase III, is a potent inhibitor of Pkr. An adenovirus mutant that cannot express the VA-RNA I gene grows poorly. In cells infected with this mutant virus, eIF2α becomes extensively phosphorylated, causing global translational inhibition. VA-RNA I binds the dsRNA-binding region of Pkr and blocks activation. It has been suggested that binding of VA-RNA I to Pkr blocks interaction with authentic dsRNA and hence prevents activation of the kinase. In another model, translation of viral mRNAs is not impaired because bound VA-RNA I inactivates Pkr in the vicinity. However, cellular mRNAs have no such guardian, and their translation ceases because of local inhibition of eIF2α recycling. Epstein-Barr virus and human immunodeficiency virus type 1 genomes encode small RNAs that inhibit Pkr activation in vitro, but whether they function in a similar manner in infected cells is not clear.

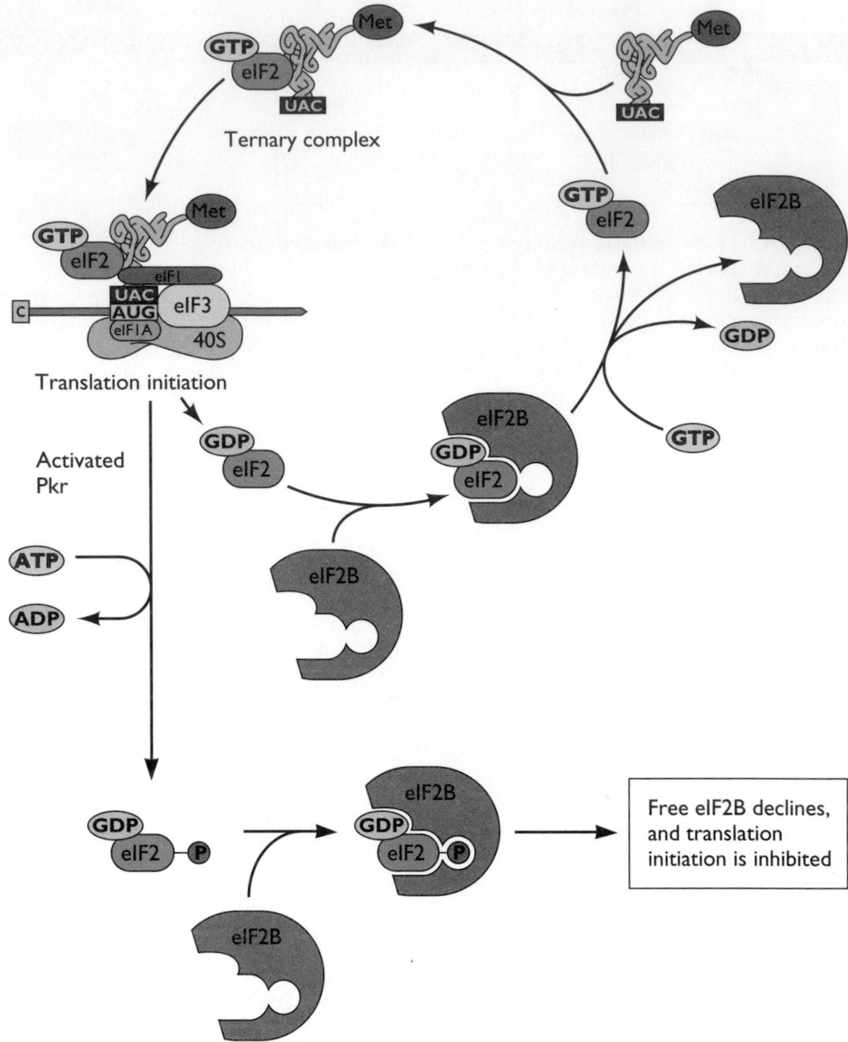

Figure 11.17 Effect of eIF2α phosphorylation on catalytic recycling. eIF2-GTP and tRNA-Met$_i$ form the ternary complex required for translation initiation. During initiation, GTP is hydrolyzed to GDP, and in order for initiation to continue, eIF2 must be recharged with GTP. This recycling is accomplished by eIF2B, which exchanges GTP for GDP on eIF2. When eIF2 is phosphorylated on the alpha subunit, it binds irreversibly to eIF2B, preventing the latter from carrying out its role in recycling GDP. As a result, the concentration of eIF2-GTP declines and translation initiation is inhibited.

While adenovirus VA-RNA I binds Pkr and blocks activation by dsRNA, the vaccinia virus genome encodes a protein that sequesters dsRNA. The viral E3L protein contains the same dsRNA-binding region as Pkr; it binds dsRNA and prevents it from activating the kinase. Deletion of the gene encoding the E3L protein renders the virus more sensitive to interferon and results in larger quantities of active Pkr in infected cells. The influenza virus NS1 protein and the reovirus σ3 protein also sequester dsRNA.

Inhibitors of kinase function. The genomes of several viruses encode proteins that directly inhibit the kinase activity of Pkr, and some do so by acting as pseudosubstrates. For example, vaccinia virus K3L protein has amino acid homology with the N terminus of eIF2α but lacks Ser51, which is phosphorylated by Pkr. The protein binds tightly to Pkr within the catalytic cleft and blocks its ability to phosphorylate Pkr. The growth of vaccinia virus mutants lacking the K3L gene is severely impaired by interferon. Human immunodeficiency virus type 1 Tat protein may inhibit Pkr by a similar mechanism. Pkr is degraded in poliovirus-infected cells by a mechanism that has not been elucidated.

In uninfected cells, Pkr is complexed with the chaperone proteins Hsp40 and Hsp70. Influenza virus infection induces the release of Hsp40 from this complex. Such dissociation may allow refolding of Pkr by Hsp70, leading to inactivation of the enzyme.

Dephosphorylation of eIF2α. Another mechanism for reversing the consequences of Pkr activation is dephosphorylation of its target. In herpes simplex virus-infected cells, Pkr is activated, but eIF2α is not phosphorylated. During infection with viruses lacking the viral

Table 11.2 Targets and mechanisms of viral Pkr inhibitors

Target	Virus	Inhibitor	Mechanism
dsRNA	Herpes simplex virus	US11	Binds and sequesters dsRNA
	Influenza virus	NS1	Binds and sequesters dsRNA
	Reovirus	σ3	Binds and sequesters dsRNA
	Vaccinia virus	E3L	Binds and sequesters dsRNA
dsRBM[a]	Adenovirus	VAI RNA	Binds Pkr, blocks activation by dsRNA
	Epstein-Barr virus	EBER	Binds Pkr, blocks activation by dsRNA
	Human immunodeficiency virus type 1	TAR RNA	Binds Pkr, blocks activation by dsRNA
Dimerization	Baculovirus	PK2	Inhibits dimerization
	Hepatitis C virus	NS5a	Inhibits dimerization
Substrate interaction	Hepatitis C virus	E2	Pseudosubstrate, blocks Pkr-eIF2α interaction
	Human immunodeficiency virus type 1	Tat	Pseudosubstrate, blocks Pkr-eIF2α interaction
	Vaccinia virus	K3L	Pseudosubstrate, blocks Pkr-eIF2α interaction
Synthesis and stability	Human immunodeficiency virus type 1	Tat	Reduces Pkr expression
	Poliovirus	Unknown	Degrades Pkr
Regulation of eIF2α	Herpes simplex virus	γ34.5	Binds phosphatase, directs to eIF2α
	Simian virus 40	T antigen	Downstream of eIF2α?

[a]RBM, RNA-binding motif.

ICP34.5 gene, Pkr is activated and eIF2α becomes phosphorylated, causing global inhibition of protein synthesis. This viral protein product associates with a type 1a protein phosphatase and acts as a regulatory subunit, redirecting the phosphatase to dephosphorylate eIF2α. The effects of activated Pkr are reversed, ensuring continued protein synthesis.

Regulation of eIF4F Activity

The eIF4F protein plays several important roles during 5′-end-dependent initiation, including recognition of the cap, recruitment of the 40S ribosomal subunit, and unwinding of RNA secondary structure. It is not surprising, therefore, that several viral proteins modify the activity of this protein. The cap-binding subunit eIF4E is frequently a target, probably because its activity can be modulated in at least two ways and because it is present in limiting quantities in cells. The cap-binding complex can also be inactivated by cleavage of eIF4G.

Cleavage of eIF4G

Poliovirus infection of cultured mammalian cells results in a dramatic inhibition of cellular protein synthesis. By 2 h after infection, polyribosomes are disrupted, and translation of nearly all cellular mRNAs ceases (Fig. 11.18). Translationally competent extracts from infected cells can

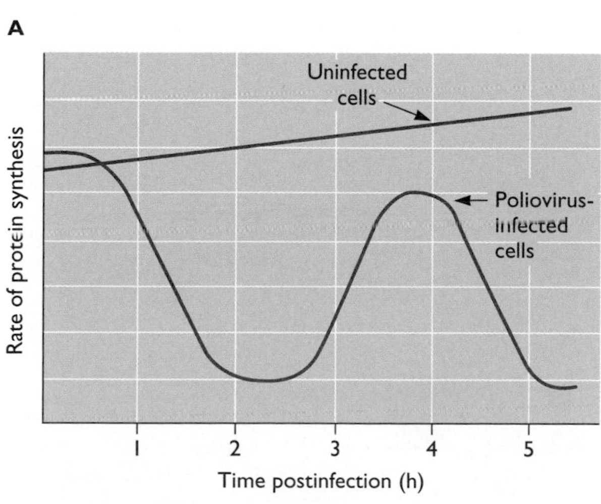

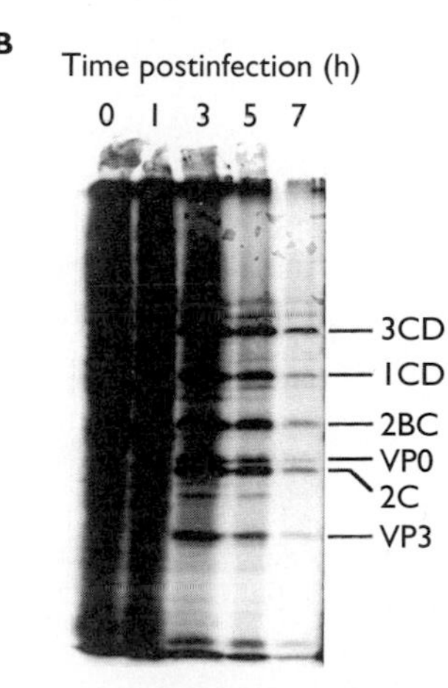

Figure 11.18 Inhibition of translation in poliovirus-infected cells. (A) Rate of protein synthesis in poliovirus-infected and uninfected cells. During poliovirus infection, host cell translation is inhibited by 2 h after infection and is replaced by translation of viral proteins. Adapted from H. Fraenkel-Conrat and R. R. Wagner (ed.), *Comprehensive Virology* (Plenum Press, New York, N.Y., 1984), with permission. (B) Polyacrylamide gel electrophoresis of [^{35}S]methionine-labeled proteins at different times after poliovirus infection. In this experiment, host translation is shut off by 5 h postinfection and is replaced by the synthesis of viral proteins, some of which are labeled at the right.

readily translate poliovirus mRNA, but not capped mRNAs. Studies of these extracts demonstrated that they are defective in the translation initiation protein, eIF4F. Specifically, in poliovirus-infected cells, both isoforms of eIF4G are proteolytically cleaved. As the N-terminal domain of eIF4G binds eIF4E, which in turn binds the 5′ cap of cellular mRNAs, such cleavage effectively prevents eIF4F from recruiting 40S ribosomal subunits to capped mRNAs (Fig. 11.19). Because poliovirus mRNA is uncapped and is translated by internal ribosome binding, a process that does not require eIF4E, viral mRNA translation is unaffected. However, IRES function appears to require the C-terminal fragment of eIF4G, which, as discussed above (see "The internal ribosome entry site"), is necessary to recruit 40S ribosomal subunits to the IRES. Consequently, cleavage of eIF4G not only inhibits cellular mRNA translation but also is a strategy for stimulating IRES-dependent translation. Although both eIF4GI and eIF4GII are cleaved in poliovirus- and rhinovirus-infected cells, the kinetics of shutoff of host translation correlates only with cleavage of eIF4GII.

Cleavage of eIF4G is carried out by viral proteases such as $2A^{pro}$ of poliovirus, rhinovirus, and coxsackievirus, and the L protease of foot-and-mouth disease virus. Purified $2A^{pro}$ of rhinovirus cleaves eIF4G directly in vitro, although very inefficiently unless eIF4G is bound to eIF4E. The eIF4F complex appears to be the target of $2A^{pro}$ cleavage. The binding of eIF4E to eIF4G might induce conformational changes in the latter protein that make it a more efficient substrate for the protease. Poliovirus $2A^{pro}$ efficiently cleaves eIF4GI but not eIF4GII, consistent with the differential processing of these proteins during viral infection.

Modulation of eIF4E Activity by Phosphorylation

The regulated phosphorylation of eIF4E at Ser209 has been recognized for many years. Inhibition of cellular translation during mitosis and heat shock correlates with reduced phosphorylation of eIF4E. Two protein kinases which are also associated with eIF4G phosphorylate Ser209 of eIF4E, Mnk1, and Mnk2. However, the effect of phosphorylation on the function of eIF4E is unclear. It has

Figure 11.19 Regulation of eIF4F activity. The activity of eIF4F can be regulated in at least three ways: phosphorylation of eIF4E, interaction with two eIF4E-binding proteins, and proteolytic cleavage of eIF4G. Dephosphorylation of eIF4E, which occurs in cells infected with adenovirus and influenza virus, causes eIF4E to dissociate from eIF4F. The proteins 4E-bp1 and 4E-bp2 bind eIF4E and block its ability to bind eIF4G. Dephosphorylation of 4E-bp1, which occurs in cells infected with poliovirus and encephalomyocarditis virus, allows it to bind eIF4E. Cleavage of eIF4G occurs in cells infected with poliovirus and foot-and-mouth disease virus. In all cases, the effect is to prevent the interaction of eIF4E and eIF4F, rendering the complex unable to recruit ribosomes to the mRNA via the 5′ cap structure. As a result, translation by 5′-end-dependent initiation is inhibited.

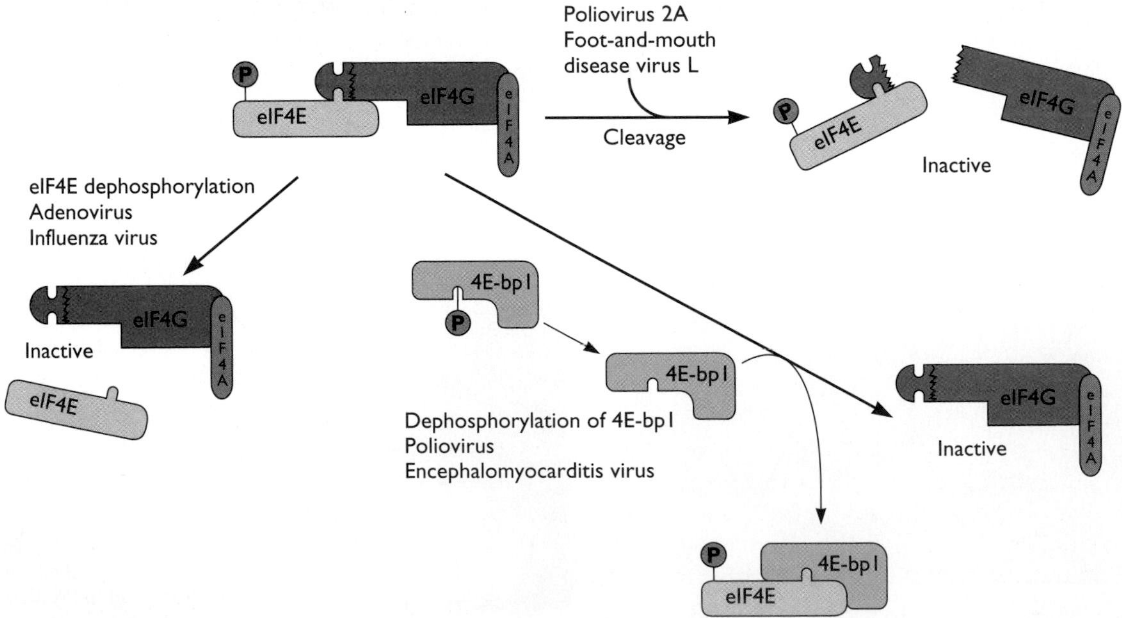

been suggested that phosphorylation of eIF4E allows tighter binding to the 5′-terminal cap structure.

A decrease in eIF4E phosphorylation has been suggested to be responsible for the inhibition of mRNA translation in cells infected with several viruses. For example, cellular protein synthesis is inhibited at late times in cells infected with adenovirus, a result of virus-induced underphosphorylation of eIF4E. The viral L4 100-kDa protein binds to the C terminus of eIF4G, preventing binding of Mnk1, presumably blocking phosphorylation of eIF4E. Adenoviral late mRNAs continue to be translated because they possess a reduced requirement for eIF4E. The majority of these viral mRNAs contain the tripartite leader (Fig. 10.12), a common 5′ noncoding region that mediates translation by ribosome shunting. Initiation by this mechanism is less dependent on eIF4F, presumably because the shunting of part of the 5′ untranslated region reduces the requirement for RNA unwinding (helicase) activity associated with initiation by cap binding and scanning. Furthermore, adenovirus late mRNAs efficiently recruit the small amounts of phosphorylated eIF4E present late in infection, a feature of mRNAs with little RNA secondary structure near the 5′ cap. The tripartite leader therefore confers selective translation of viral over host mRNAs under conditions in which eIF4E is underphosphorylated. Adenovirus-induced translation inhibition not only boosts viral late mRNA translation but also enhances cytopathic effects and consequently release of virus from cells.

Despite the correlation between reduced phosphorylation of eIF4E and translation inhibition in virus-infected cells, it now appears that phosphorylation of eIF4E markedly reduces its affinity for the 5′-terminal cap structure. A new model for the role of eIF4E phosphorylation has therefore been suggested. Phosphorylation of eIF4E occurs after it is bound to eIF4G and the mRNA. The consequent reduction in affinity of eIF4E for the mRNA 5′-terminal cap structure allows the eIF4F complex to be released, facilitating movement of the 40S ribosomal subunit.

Modulation of eIF4E Activity by Two Specific Binding Proteins

Two related low-molecular-weight cellular proteins, 4E-bp1 and 4E-bp2, can bind to eIF4E and inhibit translation following 5′-end-dependent scanning, but not by internal ribosome entry (Fig. 11.19). The former was found to be identical to a previously described protein, called phosphorylated heat- and acid-stable protein regulated by insulin (Phas-I). This protein was known to be an important phosphorylation substrate in cells treated with insulin and growth factors. Phosphorylation of 4E-bp1 in vitro prevents it from associating with eIF4E. It was subsequently shown that binding of either 4E-bp1 or 4E-bp2 to eIF4E does not prevent it from interacting with the cap, but rather inhibits binding to eIF4G. As a result, active eIF4F is not formed. eIF4G and 4E-bp proteins carry a common sequence motif that binds eIF4E. Treatment of cells with hormones and growth factors leads, through signal transduction pathways, to the phosphorylation of 4E-bp1 and its release from eIF4E. Translation of mRNAs with extensive secondary structure in the 5′ untranslated region is preferentially sensitive to the phosphorylation state of 4E-bp1. Cellular translation may be independently regulated by the phosphorylation state of eIF4E and/or 4E-bp proteins.

Phosphorylation of 4E-bp1 is inhibited by the drug **rapamycin.** As expected, treatment of cells with this drug inhibits translation initiation by 5′-end-dependent scanning but not by internal ribosome entry. Rapamycin is an immunosuppressant that binds to an immunophilin protein (Fkbp). The latter binds to members of the Fkbp-rapamycin-associated family of protein kinases in humans. Clearly it will be important to identify the reactions that lead to phosphorylation of 4E-bp proteins to establish links between cell surface binding of growth factors and translation initiation.

Infection with several different picornaviruses results in alteration of the phosphorylation state of 4E-bp1 and 4E-bp2 (Fig. 11.19). In contrast to the shutoff that occurs in poliovirus-infected cells, inhibition of cellular protein synthesis in encephalomyocarditis virus-infected cells occurs late in infection, and is not mediated by cleavage of eIF4G. Rather, encephalomyocarditis virus infection induces dephosphorylation of 4E-bp1. As a result, translation of cellular mRNAs is inhibited, but, because the viral mRNA contains an IRES, its translation is unaffected.

Regulation of Poly(A)-Binding Protein Activity

The poly(A)-binding protein Pab1p plays a crucial role in mRNA translation, bringing together the ends of the mRNA (Fig. 11.3 and 11.8C). Two mammalian proteins that bind to Pab1p and regulate its function have been identified. Poly(A)-binding protein interacting protein-1 (Paip-1) is similar to the central portion of eIF4G and can bind to eIF4A. Overexpression of Paip-1 in mammalian cells stimulates translation, perhaps by acting as a scaffold between Pab1p and eIF4G, juxtaposing the mRNA ends (Fig. 11.3C). Paip-2, which is unrelated to eIF4G, inhibits translation by blocking the function of Pab1p: this protein decreases the affinity of Pab1p for the poly(A) sequence, and competes with Paip-1 for Pab1p binding.

In cells infected with rotaviruses, inhibition of host translation is a consequence of blocking the function of Pab1p (Fig. 11.20). The 3′ ends of rotaviral mRNAs are not polyadenylated and therefore cannot interact with this protein. Instead, these 3′ untranslated regions contain a conserved sequence that binds the viral protein nsP3. This viral protein associates with eIF4G, bringing together the viral mRNA ends. Therefore nsP3 assumes the function of Pab1p in translation of rotavirus mRNAs. The nsP3 protein occupies the Pab1p binding site of eIF4G, thereby evicting Pab1p and preventing juxtaposition of the mRNA ends. The binding of nsP3 to eIF4G is the molecular basis for rotavirus inhibition of host cell translation.

In cells infected with poliovirus, Pab1p is cleaved by the viral proteases 2A^{pro} and 3C^{pro}, which may contribute to the inhibition of cellular mRNA translation. Because translation from picornaviral IRESs is stimulated when the mRNA ends are brought together, this might seem counterintuitive. However, cleavage of Pab1p does not abolish binding of the protein to the poly(A) tail. In cells infected with these viruses, the mRNA ends might be brought together by the interaction between cleaved Pab1p, poly(rC)-binding protein 2, and viral protein 3CD (Chapter 6). Alternatively, Paip-1 might form a scaffold with cleaved Pab1p and the central domain of eIF4G (Fig. 11.3C).

Perspectives

Molecular mimicry is a common theme in translation. During translation termination, stop codons are recognized by release protein eRF1, which mimics the structure of tRNA and occupies the ribosomal A site. The intergenic IRES of cricket paralysis virus resembles a tRNA, allowing it to enter the P site of the ribosome and mimic the presence of initiator tRNA. The 3′ tRNA-like structure of turnip yellow mosaic virus is actually charged with valine, and this molecular mimic is placed in the P site of the ribosome as initiation commences at the 5′ end of the viral mRNA. The myxoma virus protein M156R is a structural mimic of eIF2α, and competes with Pkr for phosphorylation of the initiation protein. The art of molecular mimicry is just now being appreciated and is surely more extensive than we know.

Evidence is accumulating that despite variations in the structure of mRNA ends, they must be brought together for efficient translation. Such juxtaposition may be achieved by RNA-RNA, protein-protein, or RNA-protein interactions. When this process is prevented by virus infection, as in poliovirus-mediated cleavage of eIF4G and Pab1p, other means of bringing viral mRNA ends together have evolved. Precisely how mRNA translation is stimu-

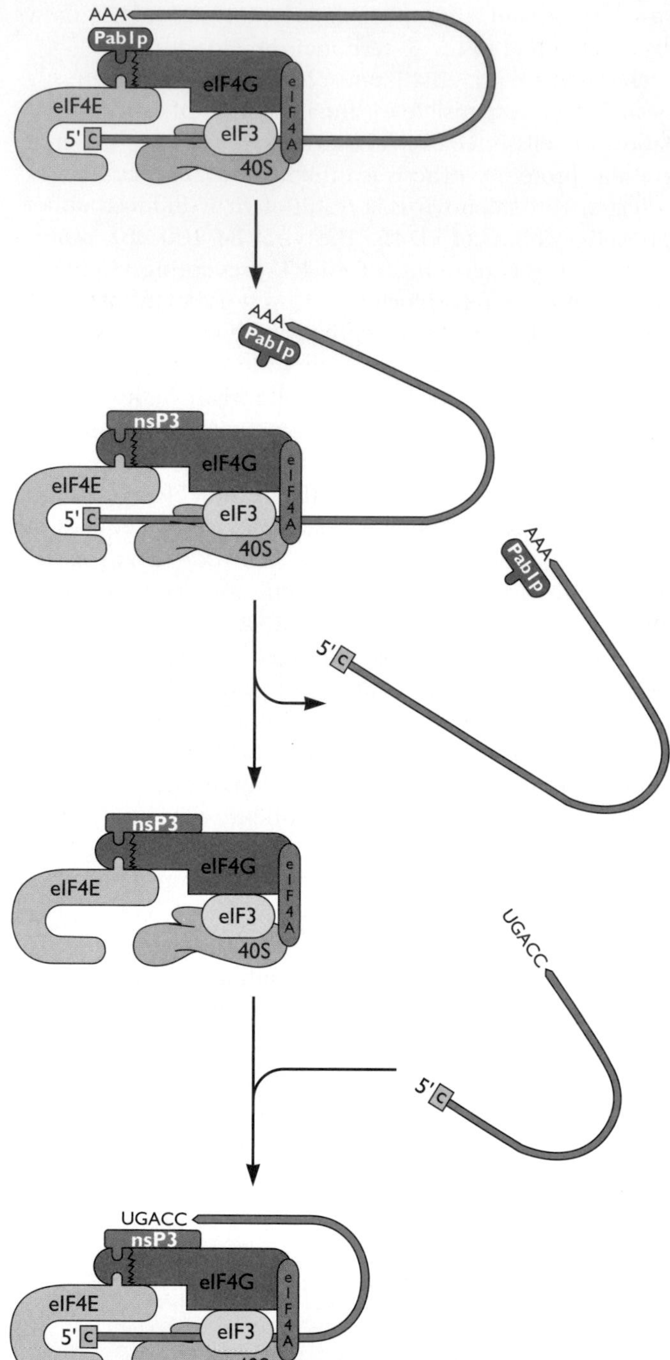

Figure 11.20 Eviction of Pab1p from eIF4G by rotavirus nsP3. Top panel shows the ends of host cell mRNA brought together by the interaction of Pab1p with poly(A) and eIF4G. Rotavirus nsP3 associates with eIF4G at the Pab1p binding site, and also binds the 3′ untranslated region of the viral mRNA. As a result, host mRNAs are replaced by viral mRNAs in initiation complexes, and translation occurs in a 5′-end-dependent manner. Adapted from M. Piron, P. Vende, J. Cohen, and D. Poncet, *EMBO J.* **17**:5811–5821, 1998.

lated by bringing the ends of the mRNA together remains to be understood.

Despite our detailed understanding of how viral proteins regulate eIF4F activity, there are many examples of virus-mediated translation regulation that are poorly understood. For example, in the late phase in cells infected with herpes simplex virus, mRNA encoding the viral glycoprotein gD accumulates in the cytoplasm, but only small quantities of the protein are found. How translation of this and many other mRNAs is controlled has not been elucidated. Understanding how viruses regulate translation remains an area of intense research activity. Such studies will not only enhance our understanding of how viruses efficiently infect cells, but will also contribute to our knowledge of translation during normal and abnormal cell growth.

Our discussion has been restricted to regulation of the initiation of mRNA translation. With the discovery that viral mRNA translation can proceed without initiation proteins, the potential for regulation during elongation and termination must be considered. Translation elongation rates can be modulated by phosphorylation of elongation protein eEF2 in response to growth-promoting stimuli. The activity of this protein is also modified by ADP-ribosylation caused by the toxin produced by the pathogenic bacterium *Clostridium diphtheriae*. Given the enormous complexity of the translation apparatus, it seems likely that many other regulatory points remain to be discovered.

References

Monographs

Harford, J. B., and D. R. Morris. 1997. *mRNA Metabolism and Post-Transcriptional Gene Regulation*. Wiley-Liss, New York, N.Y.

Sonenberg, N., J. W. B. Hershey, and M. B. Mathews (ed.). 2000. *Translational Control of Gene Expression*. Cold Spring Harbor Laboratory Press, Cold Spring Harbor, N.Y.

Review Articles

Curran, J., P. Latorre, and D. Kolakofsky. 1997. Translational gymnastics on the Sendai virus P/C mRNA. *Semin. Virol.* **8:**351–357.

Ehrenfeld, E., and N. L. Teterina. 2002. Initiation of translation of picornavirus RNAs: structure and function of the internal ribosome entry site, p. 159–170. *In* B. L. Semler and E. Wimmer (ed.), *Molecular Biology of Picornaviruses*. ASM Press, Washington, D.C.

Gale, M., and M. G. Katze. 1998. Molecular mechanisms of interferon resistance mediated by viral-directed inhibition of Pkr, the interferon-induced protein kinase. *Pharmacol. Ther.* **78:**29–46.

Harger, J. W., A. Meskauskas, and J. D. Dinman. 2002. An "integrated model" of programmed ribosomal frameshifting. *Trends Biochem. Sci.* **27:**448–454.

Jackson, R. J. 2002. Proteins involved in the function of picornavirus internal ribosome entry sites, p. 171–186. *In* B. L. Semler and E. Wimmer (ed.), *Molecular Biology of Picornaviruses*. ASM Press, Washington, D.C.

Lemon, S. M., and M. Honda. 1997. Internal ribosome entry sites within the RNA genomes of hepatitis C virus and other flaviviruses. *Semin. Virol.* **8:**274–288.

Morley, S. J., P. S. Curtis, and V. M. Pain. 1997. eIF4G: translation's mystery factor begins to yield its secrets. *RNA* **3:**1085–1104.

Mountford, P. S., and A. G. Smith. 1995. Internal ribosome entry sites and dicistronic RNAs in mammalian transgenesis. *Trends Genet.* **11:**179–184.

Sachs, A. B., P. Sarnow, and M. W. Hentze. 1997. Starting at the beginning, middle and end: translation initiation in eukaryotes. *Cell* **89:**831–838.

Papers of Special Interest

Juxtaposition of mRNA Ends

Craig, A. W. B., A. Haghighat, A. T. K. Yu, and N. Sonenberg. 1998. Interaction of polyadenylate-binding protein with the eIF4G homologue Paip enhances translation. *Nature* **392:**520–523.

Guo, L., E. M. Allen, and W. A. Miller. 2001. Base-pairing between untranslated regions facilitates translation of uncapped, non-polyadenylated viral RNA. *Mol. Cell* **7:**1103–1109.

Tarun, S. Z., and A. B. Sachs. 1996. Association of the yeast poly(A) tail binding protein with translation initiation factor eIF4G. *EMBO J.* **15:**7168–7177.

Tarun, S. Z., S. E. Wells, J. A. Deardorff, and A. B. Sachs. 1997. Translation initiation factor eIF4G mediates in vitro poly(A) tail-dependent translation. *Proc. Natl. Acad. Sci. USA* **94:**9046–9051.

Uchida, N., S. Hoshino, H. Imataka, N. Sonenberg, and T. Katada. 2002. A novel role of the mammalian GSPT/eRF3 associating with poly(A)-binding protein in cap/poly(A)-dependent translation. *J. Biol. Chem.* **277:**50286–50292.

Ribosome Shunting

Pooggin, M. M., J. Futterer, K. G. Skryabin, and T. Hohn. 2001. Ribosome shunt is essential for infectivity of cauliflower mosaic virus. *Proc. Natl. Acad. Sci. USA* **98:**886–891.

Yueh, A., and R. J. Schneider. 1996. Selective translation initiation by ribosome jumping in adenovirus infected and heat shocked cells. *Genes Dev.* **10:**1557–1567.

The IRES

Blyn, L. B., J. S. Towner, B. L. Semler, and E. Ehrenfeld. 1997. Requirement of poly(C) binding protein for translation of poliovirus RNA. *J. Virol.* **71:**6243–6246.

Elroy-Stein, O., T. R. Fuerst, and B. Moss. 1989. Cap-independent translation of mRNA conferred by encephalomyocarditis virus 5′ sequence improves the performance of the vaccinia virus/bacteriophage T7 hybrid expression system. *Proc. Natl. Acad. Sci. USA* **86:**6126–6130.

Jang, S. K., H.-G. Kräusslich, M. J. H. Nicklin, G. M. Duke, A. C. Palmenberg, and E. Wimmer. 1988. A segment of the 5′ nontranslated region of encephalomyocarditis virus RNA directs internal entry of ribosomes during in vitro translation. *J. Virol.* **62:**2636–2643.

Kaminski, A., S. L. Hunt, J. G. Patton, and R. J. Jackson. 1995. Direct evidence that polypyrimidine tract binding protein (PTB) is essential for internal initiation of translation of encephalomyocarditis virus RNA. *RNA* **1:**924–938.

Kolupaeva, V. G., T. V. Pestova, C. U. Hellen, and I. N. Shatsky. 1998. Translation eukaryotic initiation factor 4G recognizes a specific structural element within the internal ribosome entry site of encephalomyocarditis virus RNA. *J. Biol. Chem.* **273:**18599–18604.

Macejak, D. G., and P. Sarnow. 1991. Internal initiation of translation mediated by the 5′ leader of a cellular mRNA. *Nature* **353:**90–94.

Meerovitch, K., Y. V. Svitkin, H. S. Lee, F. Lejbkowicz, D. J. Kenan, E. K. L. Chan, V. I. Agol, J. D. Keene, and N. Sonenberg. 1993. La autoantigen enhances and corrects aberrant translation of poliovirus RNA in reticulocyte lysate. *J. Virol.* **67:**3798–3807.

Pestova, T. V., I. N. Shatsky, S. P. Fletcher, R. J. Jackson, and C. U. Hellen. 1998. A prokaryotic-like mode of cytoplasmic eukaryotic ribosome binding to the initiation codon during internal translation initiation of hepatitis C and classical swine fever virus RNAs. *Genes Dev.* **12:**67–83.

Non-AUG Initiation

Barends, S., H. H. Bink, S. H. van den Worm, C. W. Pleij, and B. Kraal. 2003. Entrapping ribosomes for viral translation: tRNA mimicry as a molecular Trojan horse. *Cell* **112:**123–129.

Sasaki, J., and N. Nakashima. 2000. Methionine-independent initiation of translation in the capsid protein of an insect RNA virus. *Proc. Natl. Acad. Sci. USA* **97:**1512–1525.

Wilson, J. E., T. V. Pestova, C. U. Hellen, and P. Sarnow. 2000. Initiation of protein synthesis from the A site of the ribosome. *Cell* **102:**511–520.

Leaky Scanning

Shin, B. S., D. Maag, A. Roll-Mecak, M. S. Arefin, S. K. Burley, J. R. Lorsch, and T. E. Dever. 2002. Uncoupling of initiation factor eIF5B/IF2 GTPase and translational activities by mutations that lower ribosome affinity. *Cell* **111:**1015–1025.

Williams, M. A., and R. A. Lamb. 1989. Effect of mutations and deletions in a bicistronic mRNA on the synthesis of influenza B virus NB and NA glycoproteins. *J. Virol.* **63:**28–35.

Reinitiation

Horvath, K. M., M. A. Williams, and R. A. Lamb. 1990. Eukaryotic coupled translation of tandem cistrons: identification of the influenza B virus BM2 polypeptide. *EMBO J.* **9:**2639–2647.

Translational Suppression

Feng, Y.-X., H. Yuan, A. Rein, and J. G. Levin. 1992. Bipartite signal for read-through suppression in murine leukemia virus mRNA: an eight-nucleotide purine-rich sequence immediately downstream of the gag termination codon followed by an RNA pseudoknot. *J. Virol.* **66:**5127–5132.

Li, G., and C. M. Rice. 1993. The signal for translational readthrough of a UGA codon in Sindbis virus RNA involves a single cytidine residue immediately downstream of the termination codon. *J. Virol.* **67:**5062–5067.

Yoshinaka, Y., I. Katoh, T. D. Copeland, and S. Oroszlan. 1985. Murine leukemia virus protease is encoded by the gag-pol gene and is synthesized through suppression of an amber termination codon. *Proc. Natl. Acad. Sci. USA* **82:**1618–1622.

Frameshifting

Chen, X., M. Chamorro, S. I. Lee, L. X. Shen, J. V. Hines, I. Tinoco, and H. E. Varmus. 1995. Structural and functional studies of retroviral RNA pseudoknots involved in ribosomal frameshifting: nucleotides at the junction of the two stems are important for efficient ribosomal frameshifting. *EMBO J.* **14:**842–852.

Hwang, C. B. C., B. Horsburgh, E. Pelosi, S. Roberts, P. Digard, and D. M. Coen. 1994. A net –1 frameshift permits synthesis of thymidine kinase from a drug-resistant herpes simplex virus mutant. *Proc. Natl. Acad. Sci. USA* **91:**5461–5465.

Jacks, T., H. D. Madhani, F. R. Masiarz, and H. E. Varmus. 1988. Signals for ribosomal frameshifting in the Rous sarcoma virus gag-pol region. *Cell* **55:**447–458.

Kim, Y. G., L. Su, S. Maas, A. O'Neill, and A. Rich. 1999. Specific mutations in a viral RNA pseudoknot drastically change ribosomal frameshifting efficiency. *Proc. Natl. Acad. Sci. USA* **96:**14234–14239.

Pkr

He, B., M. Gross, and B. Roizman. 1997. The γ(1)34.5 protein of herpes simplex virus 1 complexes with protein phosphatase 1alpha to dephosphorylate the alpha subunit of the eukaryotic translation initiation factor 2 and preclude the shutoff of protein synthesis by double-stranded RNA-activated protein kinase. *Proc. Natl. Acad. Sci. USA* **94:**843–848.

Kitajewski, J., R. J. Schneider, B. Safer, S. M. Munemitsu, C. E. Samuel, B. Thimmappaya, and T. Shenk. 1986. Adenovirus VAI RNA antagonizes the antiviral action of the interferon-induced eIF2 alpha kinase. *Cell* **45:**195–200.

Leib, D. A., M. A. Machalek, B. R. Williams, R. H. Silverman, and H. W. Virgin. 2000. Specific phenotypic restoration of an attenuated virus by knockout of a host resistance gene. *Proc. Natl. Acad. Sci. USA* **97:**6097–6101.

Polyak, S. J., N. Tang, M. Wambach, G. N. Barber, and M. G. Katze. 1996. The p59 cellular inhibitor complexes with the interferon-induced, double-stranded RNA-dependent protein kinase, Pkr, to regulate its autophosphorylation and activity. *J. Biol. Chem.* **271:**1702–1707.

Control of eIF4F Activity

Cuesta, R., Q. Xi, and R. J. Schneider. 2000. Adenovirus-specific translation by displacement of kinase Mnk1 from cap-initiation complex eIF4F. *EMBO J.* **19:**3465–3474.

Feigenblum, D., and R. J. Schneider. 1993. Modification of eukaryotic initiation factor 4F during infection by influenza virus. *J. Virol.* **67:**3027–3035.

Feigenblum, D., and R. J. Schneider. 1996. Cap-binding protein (eukaryotic initiation factor 4E) and 4E-inactivating protein BP-1 independently regulate cap-dependent translation. *Mol. Cell. Biol.* **16:**5450–5457.

Gingras, A.-C., and N. Sonenberg. 1997. Adenovirus infection inactivates the translational inhibitors 4E-bp1 and 4E-bp2. *Virology* **237:**182–186.

Gingras, A.-C., Y. Svitkin, G. J. Belsham, A. Pause, and N. Sonenberg. 1996. Activation of the translational suppressor 4E-bp1 following infection with encephalomyocarditis virus and poliovirus. *Proc. Natl. Acad. Sci. USA* **93:**5578–5583.

Gradi, A., Y. V. Svitkin, H. Imataka, and N. Sonenberg. 1998. Proteolysis of human eukaryotic translation initiation factor eIF4GII, but not eIF4GI, coincides with the shutoff of host protein synthesis after poliovirus infection. *Proc. Natl. Acad. Sci. USA* **95:**11089–11094.

Haghighat, A., Y. Svitkin, I. Novoa, E. Kuechler, T. Skern, and N. Sonenberg. 1996. The eIF4G-eIF4E complex is the target for direct cleavage by the rhinovirus 2A protease. *J. Virol.* **70:**8444–8450.

Ohlmann, T., M. Rau, V. M. Pain, and S. J. Morley. 1996. The C-terminal domain of eukaryotic protein synthesis initiation factor eIF4G is sufficient to support cap-independent translation in the absence of eIF4E. *EMBO J.* **15:**1371–1382.

Zhang, Y., D. Feigenblum, and R. J. Schneider. 1994. A late adenovirus factor induces eIF4E dephosphorylation and inhibition of cell protein synthesis. *J. Virol.* **68:**7040–7050.

Other Mechanisms of Translational Regulation

Geballe, A. P., F. S. Leach, and E. S. Mocarski. 1986. Regulation of cytomegalovirus late gene expression: γ genes are controlled by posttranscriptional events. *J. Virol.* **57:**864–874.

Johnson, D. C., and P. G. Spear. 1984. Evidence for translational regulation of herpes simplex virus type 1 gD expression. *J. Virol.* **51:**389–394.

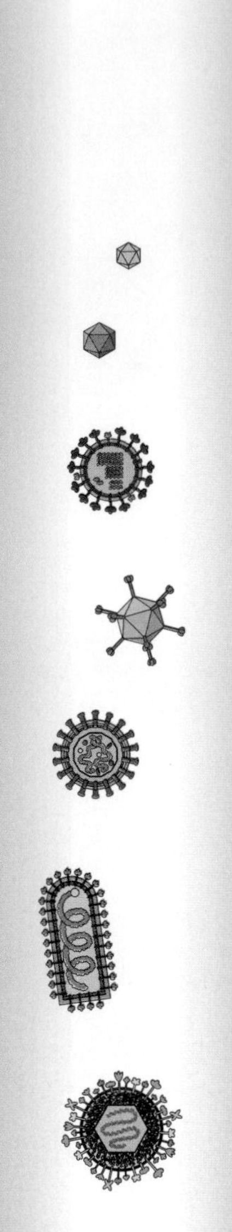

12

Intracellular Trafficking

Anatomy is destiny.
SIGMUND FREUD
Collected Writings, vol. 5, 1924

Introduction

Successful viral reproduction requires the intracellular assembly of progeny virions from their protein, nucleic acid, and, in many cases, membrane components. In preceding chapters, we have considered molecular mechanisms that ensure the synthesis of the macromolecules from which virions are constructed in the host cell. Because of the structural and functional compartmentalization of eukaryotic cells, virion components are generally produced at different intracellular locations and must be brought together for assembly. Intracellular trafficking, or sorting, of viral nucleic acids, proteins, and glycoproteins to the appropriate intracellular sites is therefore an essential prelude to the assembly of all animal viruses.

From our point of view, animal cells are very small, with typical diameters of 10 to 30 μm. However, in the microscopic world inhabited by viruses, an animal cell is large: the distances over which virion components must be transported within a cell are roughly equivalent to a mile on the macroscopic, human scale. The properties of the intracellular milieu indicate that viral particles, genomes, or subassemblies could not reach the appropriate intracellular destinations during entry or egress within reasonable periods simply by diffusion (Boxes 5.2 and 12.1). Their movement therefore requires transport systems and a considerable expenditure of energy, supplied by the host cell. The cellular highways most commonly used for movement of viral components for assembly are those formed by microtubules (as is also true during entry). These filaments are both polarized and highly organized within the cell, with (–) ends at the centrosome (near the nucleus) and (+) ends at the cell periphery. They are traveled by cellular (–) end- and (+) end-directed motor proteins that convert the chemical energy of ATP into kinetic energy, and carry cargo.

The intracellular trafficking of viral macromolecules must be appropriately directed, so that individual virion components are delivered to the correct assembly site. Virion assembly can occur at any one of several intracellular addresses, determined largely by the presence of an **envelope**, the nature of the

BOX 12.1

Getting from point A to point B in heavy traffic

Major problems in cell biology are directional movement and coordination of such movement in space and time. Concentrations of high-molecular-weight reactants and products are rarely controlled by diffusion, as they are in vitro. Indeed, the inside of a cell is so tightly packed with organelles and cytoskeletal structures (panel A in figure) that it is simply inappropriate to think of the contents of the cytoplasm, the nucleus, or organelle lumen as a "gel" or a "suspension."

Directional movement in cells is achieved by two general processes (panel B). For short distances, movement across membranes, or in and out of capsids, is measured in angstroms to nanometers and is accomplished primarily via protein channels.

- Common channels are transporters, translocons, pores, and portals
- Movement requires energy
- Diffusion in the classical sense contributes little to the process

For long distances, movement of proteins, viral particles or their components, and organelles inside cells is measured in micrometers to meters.

- Requires molecular motors moving on cytoskeletal tracks
- Requires energy
- Myosin motors move cargo on actin fibers
- Dynein and kinesin motors move cargo on microtubules

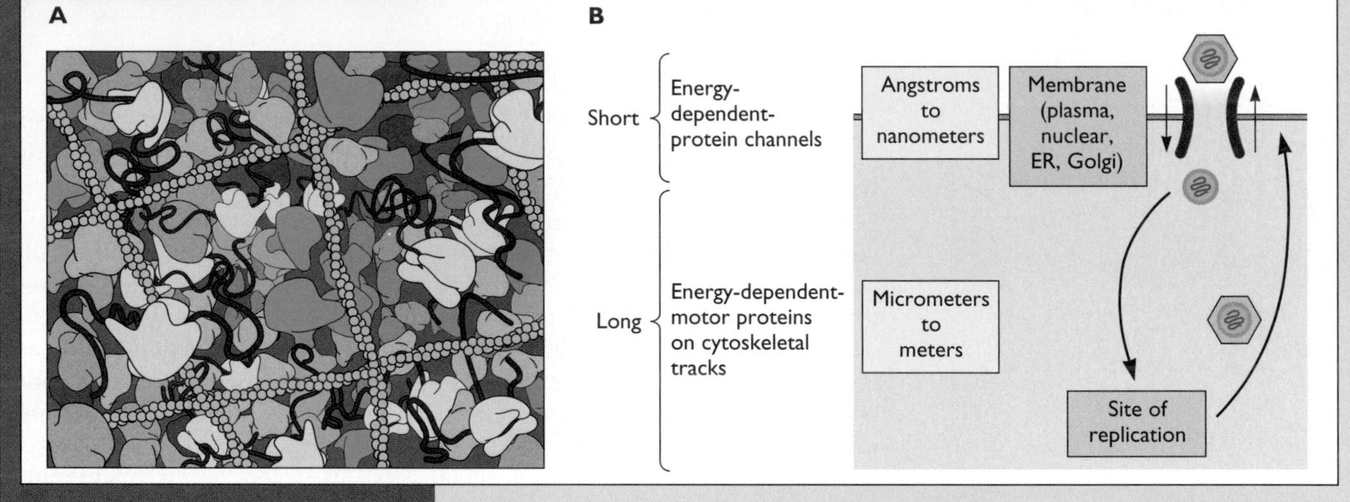

nucleic acid genome, and the mechanism of genome replication (Table 12.1). All viral envelopes are derived from one of the host cell's membranes, which are modified by insertion of viral proteins. Many virus particles assemble at the plasma membrane, but some envelopes are derived from membranes of internal compartments. Assembly of enveloped viruses therefore requires delivery of some viral proteins to the appropriate membrane, as well as transport of other virion proteins and the nucleic acid genome to the modified membrane. Another common assembly site is the cell nucleus. In this case, virion structural proteins made in the cytoplasm must be imported into the nucleus. Yet other viruses complete assembly within the cytoplasm, where all virion components are also made. These strategies impose less-complex trafficking problems than assembly of enveloped viruses at membrane sites, but additional mechanisms may be required to ensure egress of progeny virions from the cell. In some cases, genome-containing nucleocapsids are formed in infected cell nuclei, but virion assembly is completed at a cellular membrane. Such spatial and temporal separation of assembly reactions depends on appropriate coordination among multiple transport processes. For example, specific viral proteins must be imported into the nucleus for assembly of nucleocapsids, but exported once these structures are formed.

The need for movement of proteins and nucleic acids from one cellular compartment to another, or for insertion of proteins into specific membranes, is not unique to viruses. The majority of cellular proteins are made by translation of messenger RNAs (mRNAs) on cytoplasmic polyribosomes and must then be transported to their sites of operation. Similarly, most cellular RNA species are exported from the nucleus, in which transcription takes place. Eukaryotic cells are therefore constantly engaged in

Table 12.1 Intracellular trafficking requirements for virus assembly

Assembly site(s)	Viruses	Trafficking requirements
Within the nucleus	Adenovirus, polyomavirus	Transport of structural proteins from cytoplasm to nucleus
Within the cytoplasm	Picornavirus	Transport of structural proteins to specialized vesicles in which genome replication and assembly take place
At the plasma membrane	Alphavirus, retrovirus, rhabdovirus	Transport of viral glycoproteins to the plasma membrane; transport of other virion proteins made in the cytoplasm to the plasma membrane; transport of viral RNA genomes from nuclear or cytoplasmic sites of synthesis to the plasma membrane
At an internal cellular membrane	Bunyavirus, coronavirus, poxvirus	Transport and sorting of viral glycoproteins to the appropriate internal membrane; transport of other virion proteins and genomes to the internal membrane
Within the nucleus and at a cellular membrane	Herpesvirus, orthomyxovirus	Transport of structural proteins to the nucleus for assembly of the nucleocapsid; transport of the nucleocapsid from the nucleus to the membrane site of assembly; transport of internal virion proteins from cytoplasmic sites of synthesis to the membrane assembly site; transport and sorting of viral glycoproteins to the membrane assembly site

transport of macromolecules among their compartments via intracellular trafficking systems. Cellular systems that sort macromolecules to each of many possible intracellular sites are therefore just as indispensable for viral replication as the cellular biosynthetic machinery responsible for transcription, DNA synthesis, or translation. Indeed, the rapid advances in our understanding of cellular trafficking mechanisms made in recent years can be traced to initial studies with viral membrane or nuclear proteins. In the following sections, the cellular transport pathways required during viral reproduction are described in the context of the site at which virion assembly takes place.

Assembly within the Nucleus

Assembly of the majority of viruses with DNA genomes, including polyomaviruses and adenoviruses, takes place within infected cell nuclei, the site of viral DNA synthesis. All structural proteins of these nonenveloped viruses are imported into the nucleus following synthesis in the infected cell cytoplasm (Fig. 12.1), allowing complete assembly within this organelle (Table 12.1). In contrast, assembly of the more complex herpesviruses, which contain a DNA-containing nucleocapsid assembled within the nucleus, is completed at extranuclear sites (Table 12.1), as is that of some enveloped RNA viruses, such as orthomyxoviruses. In these cases, a subset of virion proteins must be imported into the nucleus.

As far as we know, all viral structural proteins that enter the nucleus do so via the normal cellular pathway of nuclear protein import. As discussed in Chapter 5, these same pathways are responsible for import of both viral genomes (or nucleoproteins) and viral, nonstructural proteins that function in the nucleus early in the infectious cycle. Proteins destined for the nucleus are so labeled by the presence of nuclear localization signals (Fig. 5.31). These signals are recognized by components of the cellular nuclear import machinery for subsequent transport into the nucleus.

Import of Viral Proteins for Assembly

Viral proteins that are known to be localized to the nucleus generally contain specific sequences that target them for import (for examples, see Table 5.3). The nuclear localization signals of viral proteins cannot be distinguished from those of cellular proteins. Indeed, many of the efforts to identify the cellular receptors that initiate nuclear import employed the well-characterized nuclear localization signal of simian virus 40 large T antigen. It is therefore believed that viral proteins that enter the nucleus contain typical nuclear localization signals to ensure their recognition by the cellular import machinery. Nevertheless, such signals have been verified for only a small fraction of the large number of viral proteins that are imported into the nucleus. Furthermore, there are a large number of cellular, nuclear localization signal receptors, some of which are expressed in tissue-specific fashion. These receptors may recognize different types of viral (and cellular) nuclear localization signals.

The cellular components that mediate import of viral proteins into the nucleus are present at finite concentrations. A typical mammalian cell contains on the order of 3,000 to 4,000 nuclear pore complexes, each with a very high translocation capacity (a mass flow of up to 80 MDa per second with 10^3 translocation events per second).

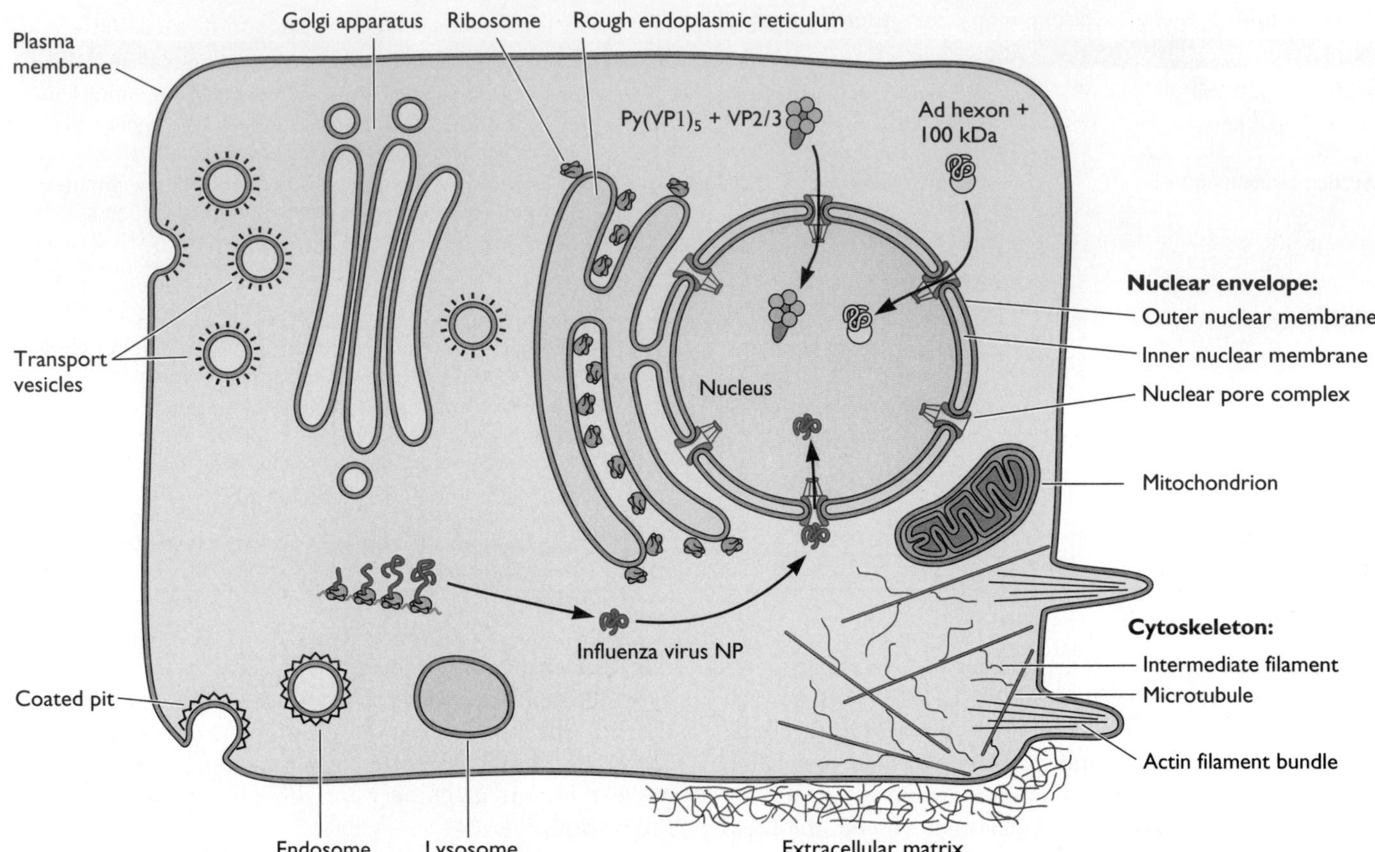

Figure 12.1 Localization of viral proteins to the nucleus. The nucleus and major membrane-bound compartments of the cytoplasm, as well as components of the cytoskeleton, are illustrated schematically and not to scale. Transport vesicles are components of the secretory pathway by which proteins travel to the plasma membrane or are secreted from the cell, whereas clathrin-coated vesicles and endosomes mediate uptake of material into the cell. Viral proteins destined for the nucleus are synthesized on cytoplasmic polyribosomes, as illustrated for the influenza virus NP protein. They engage with the cytoplasmic face of the nuclear pore complex and are translocated into the nucleus by the protein import machinery of the host cell. Some viral structural proteins enter the nucleus as preassembled structural units (polyomaviral [Py] VP1 pentamers associated with one molecule of either VP2 or VP3) or in association with a viral chaperone (adenoviral [Ad] hexon monomers bound to the L4 100-kDa protein).

However, nuclear import also depends on the limited supply of soluble transport proteins. As very large quantities of viral structural proteins must enter the nucleus prior to assembly, there is potential for competition among viral and cellular proteins for access to receptors, the nuclear pore complex, or transport proteins. Such competition is minimized in cells infected by the larger DNA viruses, such as adenoviruses and herpesviruses: by the time structural proteins are made during the late phase of infection, cellular protein synthesis is severely inhibited. The proteins of viruses that do not induce inhibition of cellular protein synthesis, such as those of the polyomaviruses, must enter the nucleus despite continual transport of cellular proteins. Whether import of viral proteins is favored in such circumstances, for example, by the presence of particularly effective nuclear localization signals, is not known.

Many viral structural proteins that enter infected cell nuclei form multimeric capsid components. In some cases, structural units of the virion are assembled in the cytoplasm prior to import into the nucleus. Pentamers of the major capsid protein (VP1) of simian virus 40 and polyomavirus specifically bind a monomer of either VP2 or VP3, the minor virion proteins, which share C-terminal sequences (Appendix A, Fig. 15B). Such heteromeric assemblies are the substrates for import into the nucleus. Indeed, efficient nuclear localization of polyomavirus VP2 and VP3 proteins can occur only in cells in which VP1 is also made. Assembly of the heteromeric complex therefore facilitates import of the minor structural proteins, even though each contains a nuclear localization signal. The increased density of these signals may allow more effective utilization of essential components of the import pathway, or the nuclear localization signals may be more accessible in the complex.

Despite such potential advantages as increased efficiency of import of viral proteins and transport of the structural proteins in the appropriate stoichiometry, import of preassembled capsid components is not universal. For example, adenoviral hexons, which are trimers of viral protein II, are found only in the nucleus of the infected cell. Association of newly synthesized hexon monomers with a late, nonstructural protein, the L4 100-kDa protein (Appendix A, Fig. 1B), is essential for both the entry of hexon subunits into the nucleus and their assembly into trimers. The monomeric hexon-L4 protein complex therefore must be the import substrate. The viral L4 100-kDa protein might supply the nuclear localization signal for hexon import, or promote nuclear retention, by facilitating assembly of hexon trimers within the nucleus.

Assembly at the Plasma Membrane

Assembly of enveloped viruses frequently takes place at the plasma membrane of infected cells (Table 12.1). Before such virions can form, viral integral membrane proteins must be transported to this cellular membrane (Fig. 12.2). The first stages of the pathway by which viral and cellular proteins are delivered to the plasma membrane were identified more than 30 years ago, and the process is now quite well understood. Viruses with envelopes derived from the plasma membrane also contain internal proteins, which may be membrane associated, and of course nucleic acid genomes. These internal components must therefore be sorted to appropriate plasma membrane sites for assembly (Fig. 12.2).

Figure 12.2 Localization of viral proteins to the plasma membrane. Viral envelope glycoproteins (red) are cotranslationally translocated into the ER lumen and folded and assembled within that compartment. They travel to and through the Golgi apparatus and from the Golgi apparatus to the plasma membrane. The internal proteins of the particle (purple) and the genome (green) are also directed to plasma membrane sites of assembly.

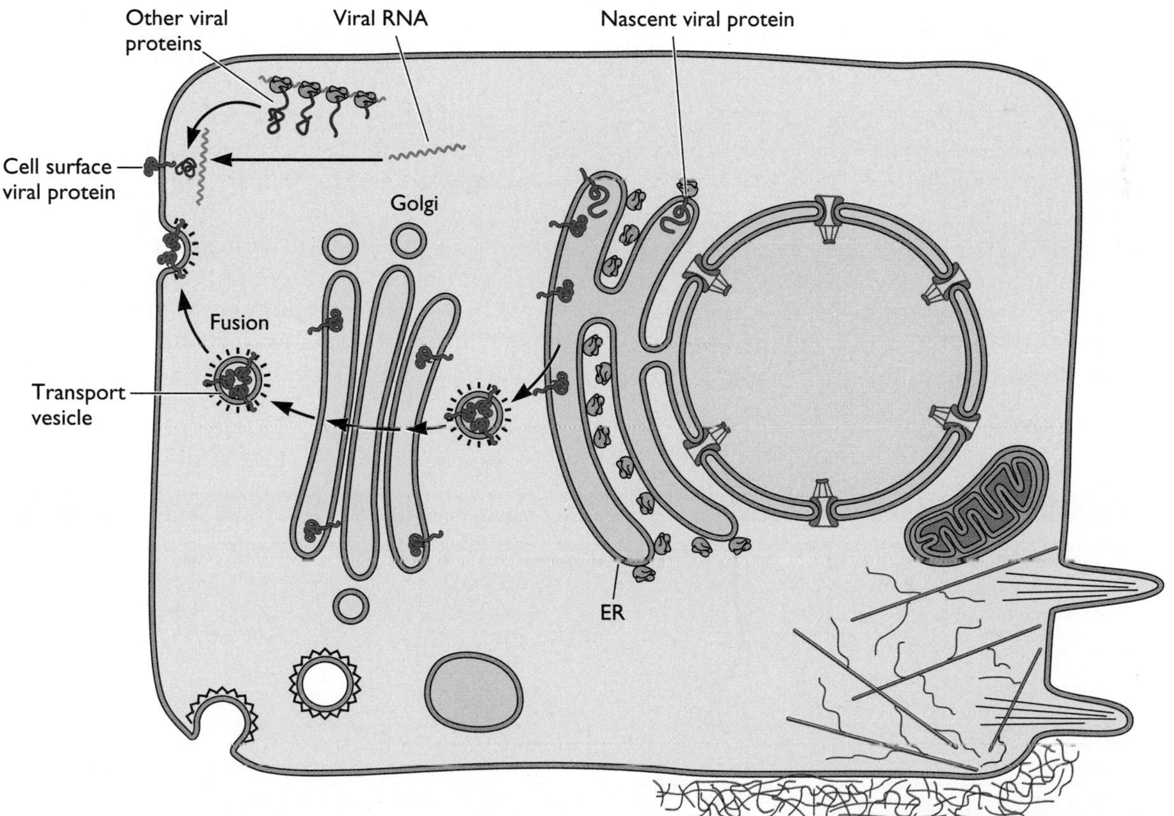

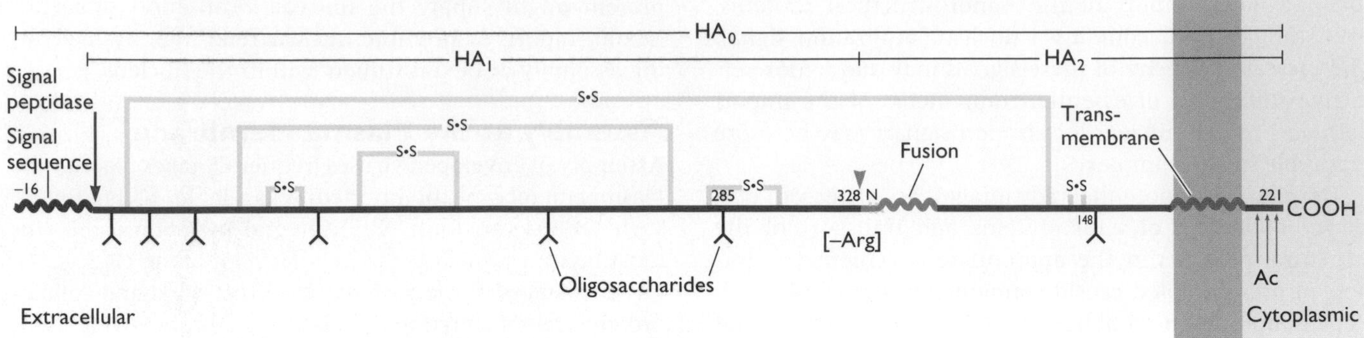

Figure 12.3 Primary sequence features and covalent modifications of the influenza virus HA protein. The primary sequence of the HA0 protein is depicted in the center, with the orange arrowhead indicating the site of the proteolytic cleavages that produce the HA1 and HA2 subunits. Residues of the HA1 and HA2 proteins are numbered from each N terminus. The fusion peptide, the N-terminal signal sequence which is removed by signal peptidase in the ER, and the C-terminal transmembrane domain are hydrophobic. Disulfide bonds, one of which maintains covalent linkage between the HA1 and HA2 proteins following HA0 cleavage, are indicated, as are sites of N-linked glycosylation (oligosaccharides) and palmitoylation (Ac).

Transport of Viral Membrane Proteins to the Plasma Membrane

Viral membrane proteins reach their destinations by the highly conserved, cellular **secretory pathway**. Many of the steps in the pathway have been studied by using viral membrane glycoproteins, such as the vesicular stomatitis virus G and influenza virus hemagglutinin (HA) proteins. Such viral proteins offer several experimental advantages: they are frequently synthesized in large quantities, their synthesis is initiated in a controlled fashion following infection, and their transport can be studied readily by both genetic and biochemical methods.

Entry into the first staging post of the secretory pathway, the endoplasmic reticulum (ER), is accompanied by membrane insertion of integral membrane proteins. Viral envelope proteins generally span the cellular membrane, into which they are inserted only once and therefore contain a single transmembrane domain. In viral proteins, transmembrane segments (described in Chapter 5) usually separate large extracellular domains from smaller cytoplasmic domains (Fig. 12.3; see also Fig. 4.19). The former include the binding sites for cellular receptors, crucial for initiation of the infectious cycle, whereas the latter are important in virus assembly. As discussed in Chapter 4 and summarized in Table 12.2, viral membrane proteins are usually oligomers. Most interactions among the subunits of viral membrane proteins are noncovalent, but some examples of association via covalent, interchain disulfide bonds are known (Table 12.2). Oligomer assembly takes place during transit from the cytoplasm to the cell surface, as does the proteolytic processing necessary to produce some mature (functional) envelope glycoproteins from the precursors that enter the secretory pathway. For example, the influenza virus HA0 precursor is cleaved into HA1 and HA2 subunits (Fig. 12.3). Viral (and cellular) proteins that travel the secretory pathway also

Table 12.2 Properties of some viral membrane proteins[a]

Virus	Protein(s)	Oligomeric structure	Efficiency of transport
Alphavirus (Semliki Forest virus)	E1, E2, E3	$(E1\text{-}E2\text{-}E3)_3$	High
Herpesvirus (herpes simplex virus type 1)	gH, gL	$(gHgL)_?$	High, but zero when only one protein is synthesized
Orthomyxovirus (influenza A virus)	HA	$(HA1\text{-}HA2)_3$	>90%
	NA	$(NA\text{-}NA)_2$	>60%
	M2	$(M2)_4$	High
Retrovirus (Rous sarcoma virus)	Env	$(SU\text{-}TM)_3$	High
Rhabdovirus (vesicular stomatitis virus)	G	$(G)_3$	>90%

[a]The best-predicted oligomeric structures of the viral membrane proteins listed are shown in column 3, with hyphens indicating disulfide-bonded subunits. The efficiency with which these proteins reach their final destinations is listed in column 4. Based on Table 1 of R. W. Doms et al., *Virology* **193:**545–562, 1993.

possess distinctive structural features, including disulfide bonds and covalently linked oligosaccharide chains (Fig 12.3). As illustrated in Fig. 12.4 for the influenza A virus HA0 protein, these characteristic covalent modifications (as well as oligomerization) take place as proteins travel through a series of specialized compartments that provide the chemical environments and enzymatic machinery necessary for their maturation. The first such compartment, the ER, is encountered by viral membrane proteins as they are synthesized.

Figure 12.4 Maturation of influenza virus HA0 protein during transit along the secretory pathway. The covalent modifications that occur during transit of the influenza virus HA0 protein through the various compartments of the secretory pathway are listed or illustrated. In the ER, these are translocation and signal peptide cleavage (1), disulfide bond formation, and addition of N-linked core oligosaccharides (2), as the protein folds (3). The cytoplasmic domain acquires palmitate (green) while the protein travels to the plasma membrane, but it has not been established when this modification takes place. For simplicity, the protein is depicted as a monomer, although oligomerization also takes place in the ER lumen, and the protein travels most of the pathway as a trimer. Note that the protein domain initially introduced into the ER lumen, in this case the N-terminal portion of the protein (type I orientation), corresponds to the extracellular domain of the cell surface protein.

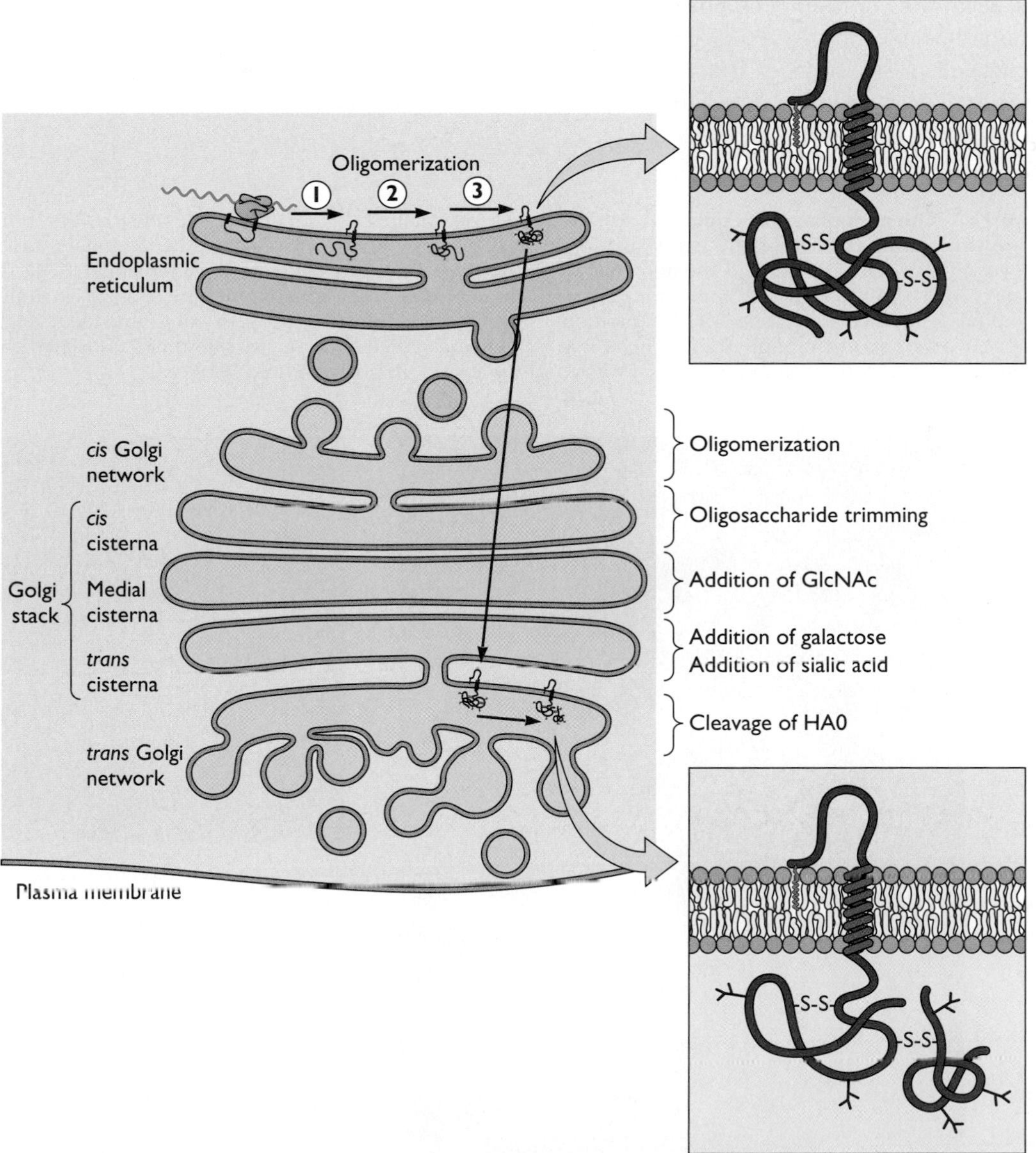

Translocation of Viral Membrane Proteins into the Endoplasmic Reticulum

All proteins destined for insertion into the plasma membrane, or the membranes of such intracellular organelles as the Golgi apparatus, enter the ER as they are translated (Fig. 12.2). This membranous structure appears as a basketwork of tubules and sacs extending throughout the cytoplasm (Fig. 12.5A). The ER membrane demarcates a geometrically convoluted but continuous internal space, the **ER lumen**, from the remainder of the cytoplasm. The ER lumen has a chemically distinctive environment, topologically equivalent to the outside of the cell. Proteins that enter the ER during their synthesis are therefore sequestered from the cytoplasmic environment as they are made.

Most proteins, regardless of their final ports of call, are synthesized on polyribosomes in the cytoplasm. However, polyribosomes engaged in synthesis of proteins that will enter the secretory pathway become associated with the cytoplasmic face of the ER membrane soon after translation begins. Those areas of the ER to which polyribosomes are bound form the **rough ER** (Fig. 12.5B). The association of polyribosomes with the ER membrane is directed by a short sequence in the nascent protein termed the **signal peptide**. It is now taken for granted that the primary sequences of proteins include "zip codes" specifying the cellular addresses at which the proteins must reside to fulfill their functions, such as the nuclear localization signals discussed in the previous section. The signal peptides of proteins that enter the ER lumen were the first such zip codes to be identified and established this paradigm some 25 years ago. Signal peptides are commonly found at the N termini of proteins destined for the secretory pathway. They are usually about 20 amino acids in length and contain a core of 15 hydrophobic residues. Signal peptides are often transient structures that are removed enzymatically during protein translocation into the ER by a protease located in the lumen, **signal peptidase**.

Figure 12.5 The endoplasmic reticulum. (A) The ER of a mammalian cell in culture. The reticular ER, which extends throughout the cytoplasm, was visualized by fluorescence microscopy of fixed African green monkey kidney epithelial cells stained with the lipophilic fluorescent dye 3,3′-dihexyloxacarbocyanine iodide. This dye also stains mitochondria. The ER membrane accounts for over half of the total membrane of a typical animal cell and possesses a characteristic lipid composition. Courtesy of M. Terasaki, University of Connecticut Health Center. (B) Electron micrograph of the rough ER in rat hepatocytes. Note the many ribosomes associated with the cytoplasmic surface of the membrane. From R. A. Rodewald, Biological Photo Service.

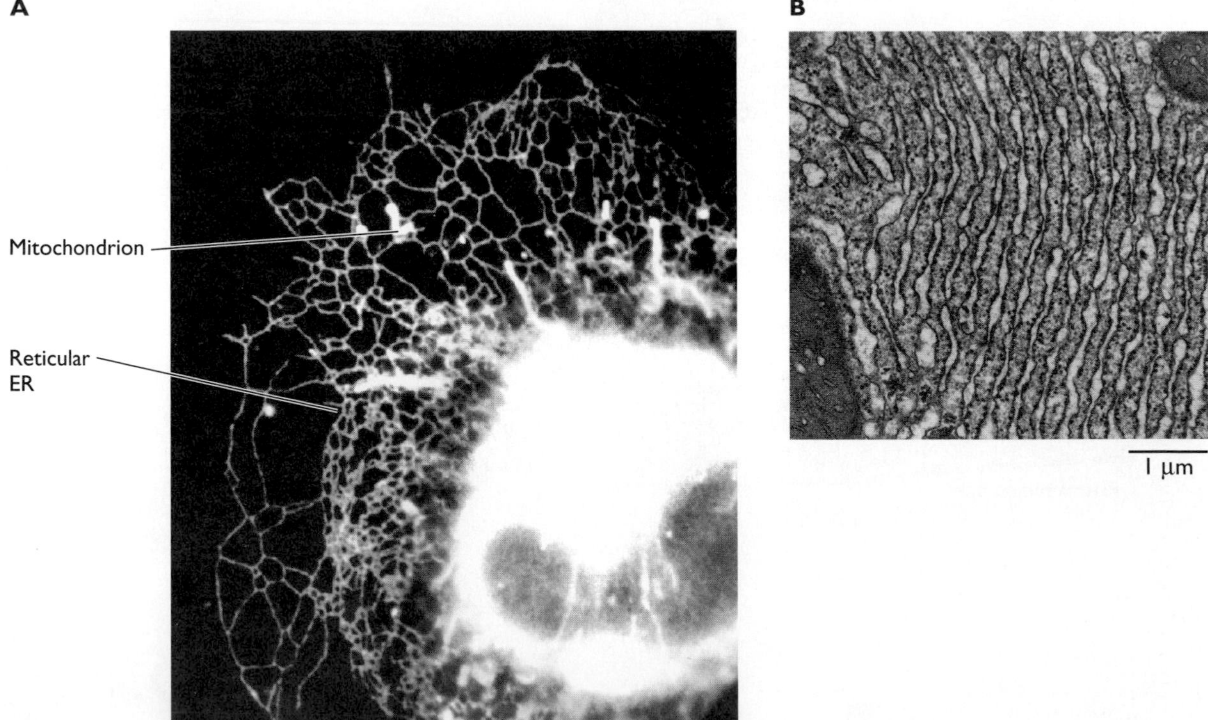

Translation of a protein that will enter the ER begins in the normal fashion and continues until the signal peptide emerges from the ribosome (Fig. 12.6). This signal then directs binding of the translation machinery to the ER membrane by means of two components: the signal peptide is recognized by the **signal recognition particle (SRP),** which in turn binds to the cytoplasmic domain of an integral ER membrane protein termed the **SRP receptor**. Binding of the signal recognition particle to the ribosome temporarily halts translation, to allow the stalled translation complex to bind to the ER membrane via the SRP receptor. Both the signal recognition particle and its receptor contain subunits that bind GTP, and efficient targeting requires the presence of this nucleotide. Following the initial

Figure 12.6 Targeting of a nascent protein to the ER membrane. Translation of an mRNA encoding a protein that will enter the ER lumen proceeds until the signal peptide (purple) emerges from the ribosome. The signal recognition particle (SRP), which contains a small RNA molecule and several proteins, binds to both the signal peptide and the ribosome to halt or pause translation, upon binding of GTP to one of the protein subunits (step 1). The nascent polypeptide-SRP-ribosome complex then binds to the SRP receptor in the ER membrane (step 2). This interaction triggers hydrolysis of GTP bound to both the SRP and its receptor, release of the SRP (step 3), and close association of the ribosome with the ER membrane and the multisubunit protein translocation channel (step 4). The lumenal end of the translocation channel is also initially closed. During translocation, the latter structure forms an aqueous pore 40 to 60 Å in diameter, the largest "hole" yet observed in a membrane. The tight association of the ribosome with the membrane blocks the cytoplasmic end of the aqueous channel and maintains the barrier between the cytoplasm and the ER lumen. The hydrolysis of GTP coordinates the release of SRP with binding of the signal peptide to the translocation channel. Translation is then resumed when the seal maintained at the lumenal end of the channel early in translocation is reversed. The growing polypeptide chain is then transferred through the membrane as its translation continues (step 5). Signal peptidase removes the signal peptide cotranslationally (step 6). The walls of the channel must open within the membrane for lateral transfer of the transmembrane domain of translocated proteins into the ER membrane. The interaction of the ribosome with the translocation channel appears to be important for this process, but the mechanism is not yet well understood.

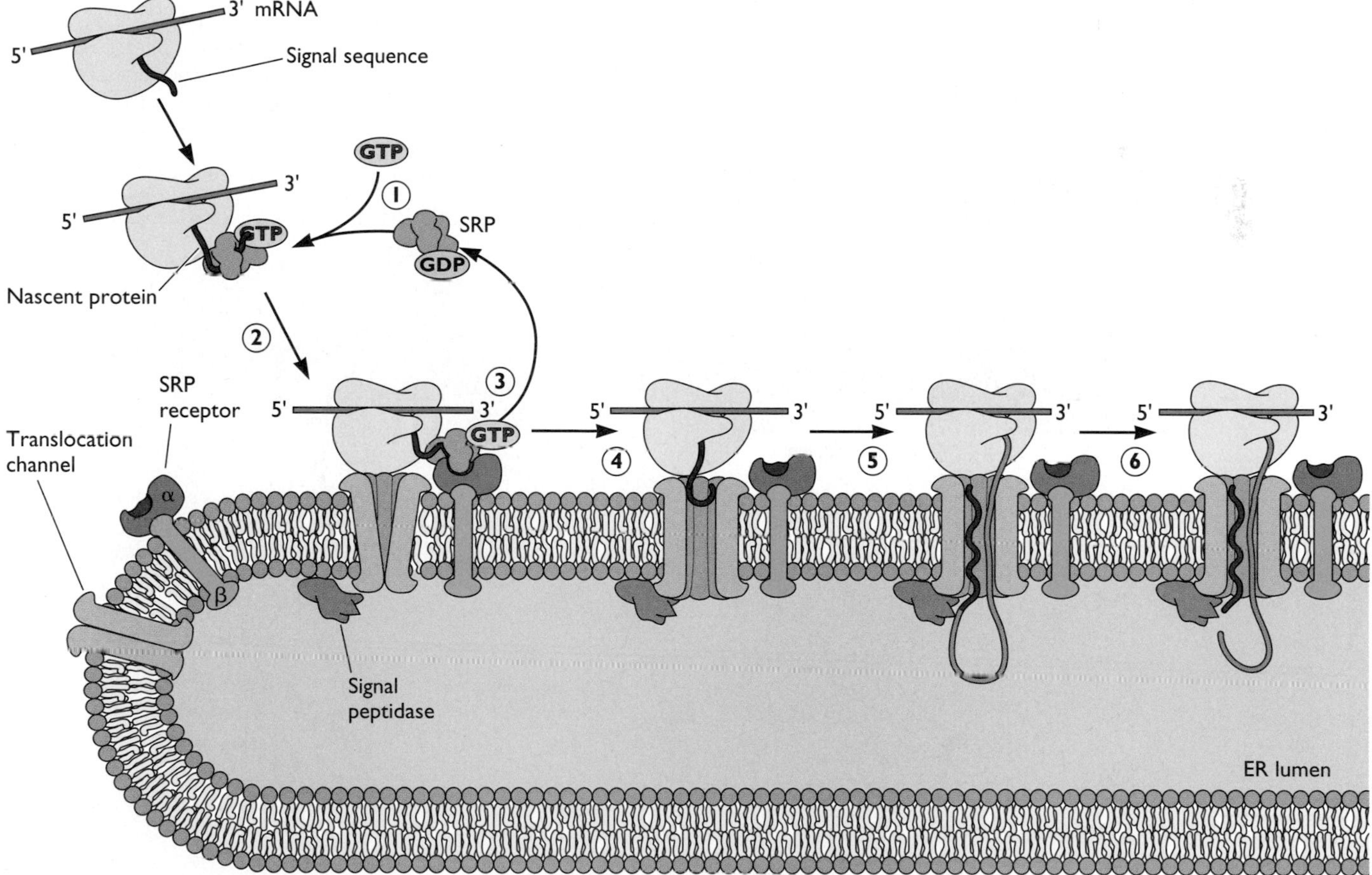

docking of the complex at the membrane, the ribosome becomes tightly bound to the membrane and engaged with a protein translocation channel, which forms a gated, aqueous pore through the ER membrane. This interaction is coordinated with release of the signal recognition particle, association of the signal peptide with the translocation channel, and resumption of translation. Because the ribosome remains bound to the membrane upon release of the signal recognition particle, continued translation facilitates movement of the growing polypeptide chain through the membrane. Such coupling of translation and translocation ensures that the protein crosses the membrane as an unfolded chain that can be accommodated by the translocation channel.

When the protein entering the ER is destined for secretion from the cell, translocation continues until the entire polypeptide chain enters the lumen. During translocation, the signal peptide is proteolytically removed by signal peptidase, releasing the soluble protein into the ER. In contrast, translocation of integral membrane proteins, such as viral envelope proteins, halts when a hydrophobic **stop transfer signal** is encountered in the nascent protein. This sequence may be the signal peptide itself or a second, internal hydrophobic sequence. The number, location, and orientation of such sequences within a protein determine the topology with which it is organized in the ER membrane (Box 12.2). The programming of insertion of proteins into the ER membrane by signals built into their pri-

BOX 12.2

Establishing the topology of integral membrane proteins that contain a single transmembrane domain

Many proteins with a single transmembrane domain, including the influenza virus HA0 protein, contain not only an N-terminal signal peptide, but also a second hydrophobic sequence. The signal peptide initiates cotranslational translocation of the nascent polypeptide through the protein translocation channel, as shown in more detail in Fig. 12.6. During translation, the resident ER enzyme signal peptidase cleaves off the signal peptide, forming the N terminus of the mature protein. Translocation continues until the second hydrophobic sequence is encountered. This sequence acts as a stop transfer signal and halts translocation, with lateral discharge of the protein into the ER membrane by a multistep process that is not well understood. This mechanism results in type I orientation (i.e., N terminus within the lumen), as in the influenza virus HA0 protein (Fig. 12.4). The position of the stop transfer sequence within the polypeptide determines the sizes of its lumenal (extracellular) and cytoplasmic domains. In proteins that contain multiple hydrophobic sequences of this kind, the first one encountered by the signal recognition particle starts the transfer. Consequently, it establishes whether subsequent hydrophobic sequences will stop or start transfer.

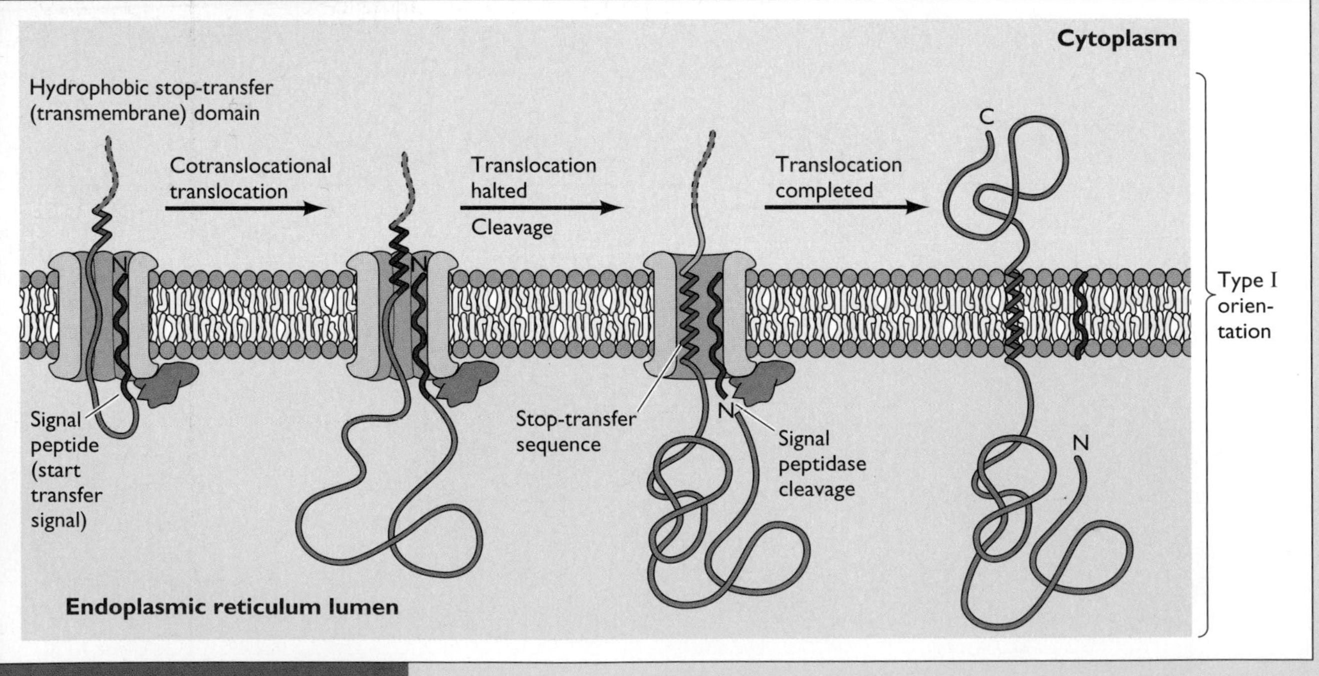

mary sequences ensures that every molecule of a particular protein adopts the identical topology in the membrane. As this topology is maintained during the several membrane budding and fusion reactions by which proteins reach the cell surface, the way in which a protein is inserted into the ER membrane determines its orientation in the plasma membrane (Fig. 12.4).

Reactions within the ER

The folding and initial posttranslational modification of proteins that enter the secretory pathway take place within the ER. The lumen contains many of the enzymes that catalyze chemical modifications, such as disulfide bond formation and the initial steps in the assembly of oligosaccharide chains (glycosylation), or that promote folding and oligomerization (Table 12.3).

Glycosylation. Viral envelope proteins, like cellular proteins that travel the secretory pathway, are generally modified by the addition of oligosaccharides to either asparagine (N-linked glycosylation) or serine or threonine (O-linked glycosylation). The presence of oligosaccharides on a protein can be detected by the incorporation of radioactively labeled monosaccharides or, more usually, as changes in the protein's electrophoretic mobility, following exposure of cells to inhibitors of glycosylation, or of the protein to enzymes that cleave the oligosaccharide (Fig. 12.7A). The first steps in N-linked glycosylation take place as a polypeptide chain emerges into the ER lumen. Oligosaccharides rich in mannose preassembled on a lipid carrier are added to asparagine residues by an oligosaccharyltransferase (Fig. 12.7B). Subsequently, several sugar residues are trimmed from N-linked core oligosaccharides in preparation for additional modifications that take place as the protein travels from the ER to the plasma membrane. As discussed below, cycles of trimming and readdition of the terminal glucose residues couple proper folding of glycoproteins to their native conformations with ER retention or release. Sites of N-linked glycosylation are characterized by the sequence NXS/T (where X = any amino acid except proline), but not every potential glycosylation site is modified. Even a single specific site within a protein is not necessarily modified with 100% efficiency. Each glycoprotein therefore exists as a heterogeneous mixture of **glycoforms**, varying in whether a particular site is glycosylated, as well as in the composition and structure of the oligosaccharide present at each site. As many viral and cellular proteins contain a large number of potential N-linked glycosylation sites, single proteins can exist in an extremely large number of glycoforms. This property complicates investigation of the physiological functions of oligosaccharide chains present on glycoproteins. Nevertheless, this modification has been assigned a wide variety of roles.

As essential components of receptors and ligands, oligosaccharides participate in many molecular recognition reactions. These processes include binding of certain hormones to their cell surface receptors, interactions of cells with one another, and binding of viruses, such as influenza A virus and herpesviruses, to their host cells. Glycosylation at specific sites can also protect viral envelope proteins from recognition by components of the host's immune defense system. Some sugar units serve as signals, targeting proteins to specific locations, in particular to lysosomes. Glycosylation can also regulate the activity of proteins, for example, of certain hormones. This modification has also been suggested to fulfill more general functions, such as protecting proteins (and viruses) that circulate in body fluids from degradation. Many pro-

Table 12.3 Some ER enzymes and chaperones that participate in quality control

Protein	Properties and functions
Calnexin (Cnx)	Integral membrane lectin that binds to almost all glycoproteins in the ER; promotes proper folding; retains incompletely folded glycoproteins in the ER; directs misfolded proteins for degradation
Calreticulin (Crt)	Soluble homolog of Cnx present in the ER lumen; similar but not identical substrate specificity to Cnx
Erp57	Soluble thioloxidoreductase; as cochaperone binds to and functions with Cnx and Crt
Erp72	Thioloxidoreductase; forms disulfide bonds within incompletely folded or assembled proteins
Glucosidase I and II	Catalyze removal of terminal glucose residues from high-mannose, N-linked oligosaccharides; glucosidase II promotes release of substrates from Cnx and Crt
Grp78 (Bip)	Lumenal protein; binds to exposed hydrophobic surfaces on incompletely folded proteins; retains such proteins in the ER
Protein disulfide isomerase (Pdi)	Catalyzes formation and isomerization of disulfide bonds; retains incompletely folded or assembled proteins in the ER; appears to operate with Grp78/BiP
UDP-glucose-glycoprotein transferase (Gt)	Lumenal ER enzyme that specifically recognizes unfolded proteins; adds glucose residues to their N-linked oligosaccharides; promotes recognition by Cnx and Crt and thus ER retention

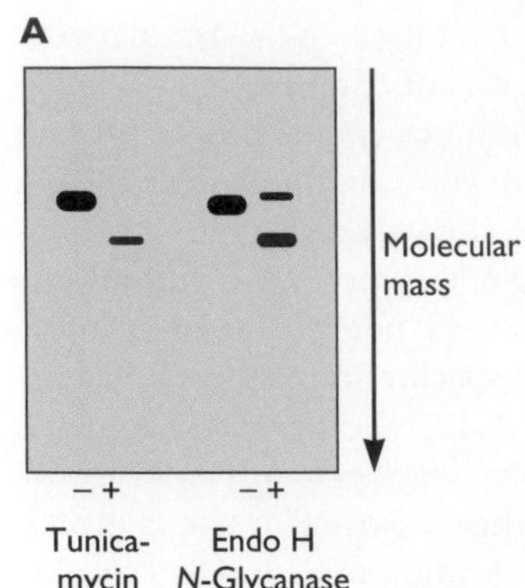

Figure 12.7 Detection and structure of N-linked oligosaccharides. (A) Detection of N-linked oligosaccharides. Glycoproteins carrying N-linked oligosaccharides can be distinguished from nonglycosylated proteins by several methods. These include their labeling in cells supplied with radioactive sugars and their specific binding to lectins, proteins that bind to sugar residues. Alternatively, addition to cells of tunicamycin, an inhibitor of the first step in synthesis of the oligosaccharide precursor, prevents N-linked glycosylation, so that the mobility of glycoproteins is altered (left). In vitro treatment of glycoproteins with enzymes (glycosidases) that cleave within the oligosaccharide, such as endoglycosidase H (endo H) or *N*-glycanase, can also alter glycoprotein mobility (right). Complex-type N-linked oligosaccharides are not sensitive to the former enzyme. (B) The branched, mannose-rich oligosaccharide added via an N-glycosidic bond to asparagine residues of proteins is shown at the left. This common precursor is transferred to N-linked glycosylation sites as proteins are translocated into the ER, from the lipid carrier dolichol upon which the oligosaccharide is built, one sugar unit at a time. While still within the ER, three glucose residues and one mannose residue are trimmed from the core oligosaccharide. Trimming continues, and additional sugars are added as the protein travels through the Golgi apparatus to form the three types of mature N-linked oligosaccharide chains (see Fig. 12.12).

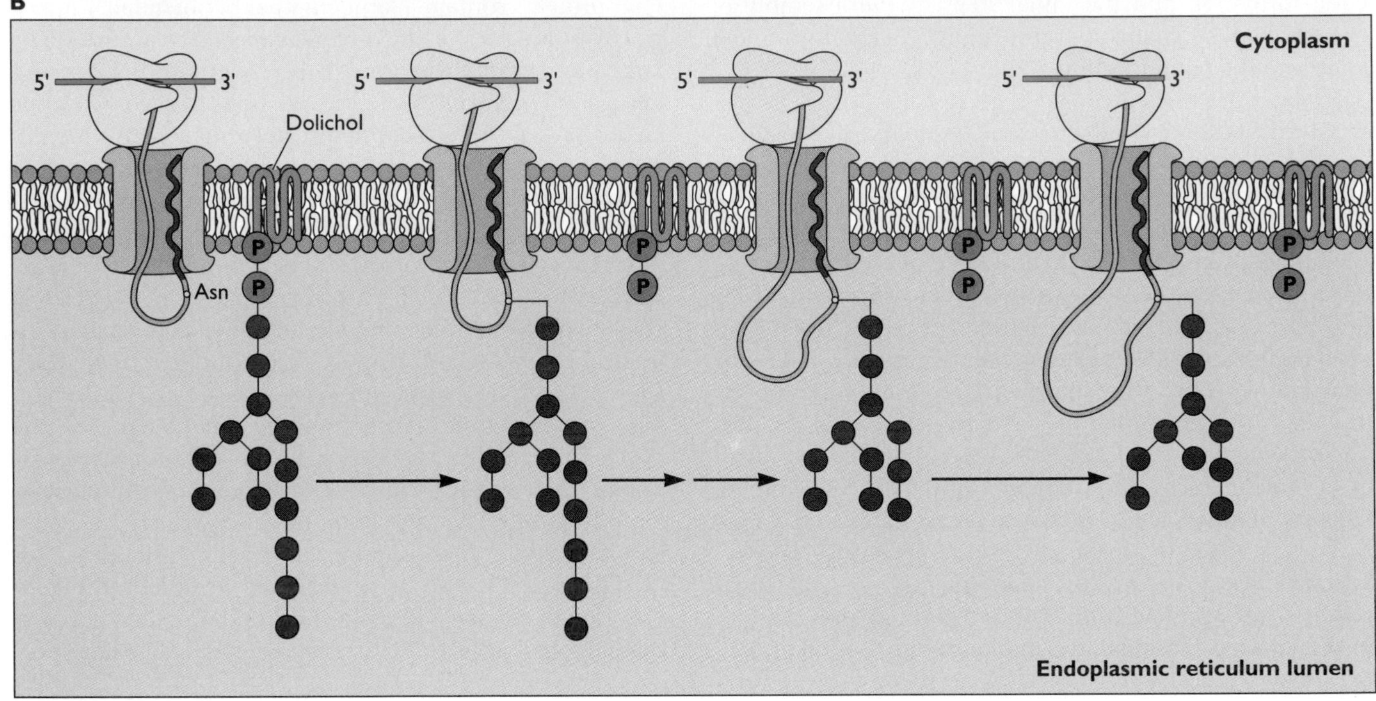

teins contain such a large number of glycosylation sites that carbohydrate can contribute more than 50% of the mass of the mature protein, for example, the poliovirus receptor and the respiratory syncytial virus G protein. The hydrophilic oligosaccharides are present on the surface of such proteins, where they can form a sugar "shell," masking much of the proteins' surfaces.

Studies of viral glycoproteins have established that glycosylation is absolutely required for the proper folding of some. Elimination (by mutagenesis) of all sites at which the vesicular stomatitis virus G or influenza virus HA proteins are glycosylated blocks the folding of these proteins and their exit from the ER (see "Protein folding and quality control" below). Before a protein folds, its hydrophobic amino acids, which are ultimately buried in the interior, are exposed. Such exposed hydrophobic patches on individual unfolded polypeptide chains tend to interact with one another nonspecifically, leading to aggregation. The hydrophilic oligosaccharide chains are believed to counter this tendency.

Disulfide bond formation. A second chemical modification that generally is restricted to proteins entering the secretory pathway, and essential for the correct folding of many, is the formation of intramolecular disulfide bonds between pairs of cysteine residues (Fig. 12.3). These bonds can make important contributions to the stability of a folded protein. However, they rarely form in the reducing environment of the cytoplasm. The more oxidizing ER lumen provides an appropriate chemical environment for disulfide bond formation. This compartment contains high concentrations of protein disulfide isomerase and other thioloxidoreductases, enzymes that catalyze the formation, reshuffling, or even the breakage of disulfide bonds under appropriate redox conditions (Table 12.3). As formation of the full and correct complement of disulfide bonds in a protein is often the rate-limiting step in its folding, these enzymes are important catalysts of this process.

Protein folding and quality control. A number of other cellular proteins assist the folding of the extracellular domains of viral membrane glycoproteins as they enter the lumen of the ER (Table 12.3). In contrast to the enzymes described above, these proteins do not alter the covalent structures of proteins. Rather, their primary function is to facilitate folding by preventing improper associations among unfolded, or incompletely folded, polypeptide chains, such as the nonspecific, hydrophobic interactions described above. Such **molecular chaperones** play essential roles in the folding of individual polypeptides and in the oligomerization of proteins. The ER chaperones, which include Grp78 and calnexin, are also crucial for **quality control**, processes that determine the sorting and fate of newly synthesized proteins translocated into the ER.

The Grp78 protein is a member of the Hsp70 family of stress response proteins. It associates transiently with incompletely folded viral and cellular proteins. Binding of this chaperone, generally at multiple sites in a single nascent protein molecule, is believed to protect against misfolding and aggregation by sequestering sequences prone to nonspecific interaction, such as hydrophobic patches. The binding to, and release from, unfolded proteins of Grp78 is controlled by the hydrolysis of ATP bound to the chaperone. Multiple cycles of association with, and dissociation from, Grp78 probably take place as a protein folds. Once the sequences to which the chaperone binds are buried in the interior of the protein, such interactions cease. For example, molecules of vesicular stomatitis virus G, Semliki Forest E1, or influenza virus HA0 proteins that have acquired the full complement of correct disulfide bonds can no longer associate with Grp78.

Calnexin is an integral membrane protein of the ER that also binds transiently to immature proteins, as does a soluble relative of this protein, calreticulin, present in the ER lumen. In contrast to Grp78, which recognizes protein sequences directly, calnexin and calrecticulin distinguish newly synthesized glycoproteins by binding to immature oligosaccharide chains. For example, the vesicular stomatitis virus G and influenza virus HA0 proteins bind to calnexin only when their oligosaccharide chains retain terminal glucose residues (Fig 12.7B). In fact, formation of the mature oligosaccharide is intimately coupled with folding of glycoproteins and their retention within the ER (Fig. 12.8). Proteins with monoglucosylated sugars are recognized by calnexin (or calreticulin), which forms a complex with an ER thioloxidoreductase, but released upon removal of the glucose by the enzyme glucosidase II (Table 12.3). An enzyme that readds terminal glucose (UDP-glucose-glycoprotein transferase [Table 12.3]) appears to be the "sensor" of the folded state of the glycoprotein: it recognizes and specifically reglucosylates incompletely folded proteins, controlling cycles of substrate binding and release from calnexin or calreticulin (Fig. 12.8). This specificity ensures that only fully folded proteins can escape these chaperones and travel along the secretory pathway.

The ER contains numerous other folding catalysts and chaperones, many specific for particular proteins. Relatively little is yet known about the parameters that determine the chaperone(s) to which a newly synthesized protein will bind, and the order in which chaperones operate. However, studies of specific viral glycoproteins in living cells indicate that the positions of oligosaccharides within the protein chain are one important determinant of chaperone selection (Box 12.3).

Proteins that are misfolded or not modified correctly cannot escape covalent or noncovalent associations with ER enzymes or molecular chaperones (Table 12.3). For example, a temperature-sensitive vesicular stomatitis virus G protein remains bound to calnexin, and hence to the ER membrane, at a restrictive temperature. Consequently, egress of nonfunctional proteins from the ER to subsequent compartments in the secretory pathway is prevented. These interactions also target misfolded (and misassembled) proteins for degradation. The first step in this process is back-translocation of such proteins through the translocation channel into the cytoplasm. The mechanisms responsible for specific recognition of misfolded or incompletely folded proteins, and induction of transport from ER to cytoplasm, are currently receiving intense scrutiny. Once the proteins re-enter the cytoplasm, oligosaccharides are removed and multiple copies of the protein ubiquitin (polyubiquitin) are added. Such modification targets the proteins for degradation by the proteasome. The quality control functions of resident ER chaperones therefore ensure that nonfunctional proteins are cleared from the secretory pathway at an early step.

Oligomerization. Most viral membrane proteins are oligomers (Table 12.2) that must assemble as their constituent protein chains are folded and covalently modified.

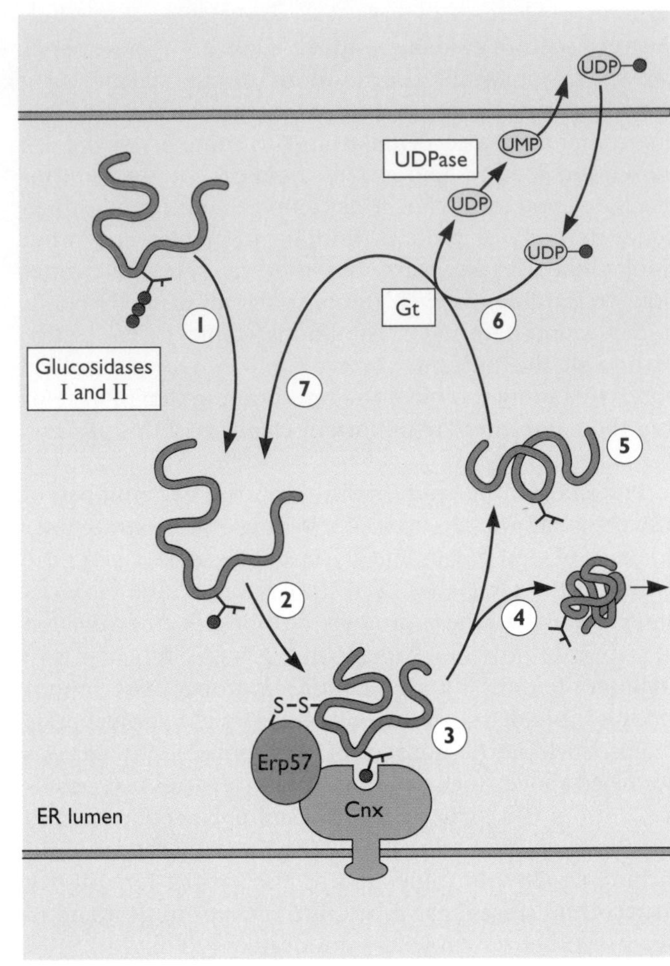

Figure 12.8 Integration of folding and glycosylation in the ER. The model illustrates the coordination of ER retention by calnexin (Cnx) with glycosylation and folding of a newly synthesized glycoprotein (red ribbon) containing an N-linked oligosaccharide, depicted as in Fig. 12.3. Trimming of two of the three terminal glucose residues by glucosidases I and II (1) yields a monoglucosylated chain, to which Cnx (or calreticulin) binds (2). Because the thioloxidoreductase Erp57 forms a complex with Cnx, the newly synthesized protein is brought into contact with Erp57, with which transient intermolecular disulfide bonds (-S-S-) can form. When the remaining glucose is removed by glucosidase II (3), the protein dissociates from the Cnx/Erp57 complex. If it has attained its native structure, the protein can leave the ER (4). However, if it is incompletely (or incorrectly) folded (5), the protein is specifically recognized by UDP-glucose glycoprotein transferase (Gt) (6), which readds terminal glucose residues to the oligosaccharide (7) and therefore allows rebinding to Cnx. Cycles of binding and modification are repeated until the protein is either folded properly or targeted for degradation. Although the signals and mechanisms that lead to degradation are not fully understood, removal of terminal mannose residues has been implicated. Adapted from L. Ellgaard et al., *Science* **286:**1882, 1999, with permission.

Such assembly generally begins in the ER, as the surfaces that mediate interactions among protein subunits adopt the correct conformation. For many proteins, these reactions are completed within the ER. For example, influenza virus HA0 protein monomers are restricted to the ER lumen, whereas trimers are found in this and all subsequent compartments of the secretory pathway. Indeed, several viral and some cellular heteromeric membrane proteins must oligomerize to exit the ER, for folding of one subunit depends on association with the other(s). This requirement has been characterized in some detail for the

BOX 12.3

Selectivity of chaperones for viral glycoproteins entering the ER

The parameters that determine which of the many ER chaperones operate on individual proteins are not well understood. However, analysis of the folding and chaperone association of viral glycoproteins suggests that the position of glycosylation sites can determine chaperone selection. This conclusion is based on the following observations.

The Semliki Forest virus E1 and p62 (pre-E2) glycoproteins enter the ER cotranslationally, and are cleaved from a precursor by signal peptidase. However, E1, which folds via three intermediates differing in their disulfide bonding, initially associates with Grp78, whereas nascent p62 molecules bind to calnexin.

This difference is not a trivial result of a lack of Grp78-binding sites in p62: when access to calnexin was blocked, this protein did bind to Grp78.

One major difference between E1 and p62 is that the latter contains glycosylation sites close to the N terminus. Addition of oligosaccharides at such sites could allow recognition of nascent p62 by calnexin and preclude association with Grp78.

This hypothesis was tested by elimination of N-linked glycosylation sites of the influenza virus HA0 protein, either close to the N terminus or in more C-terminal positions. The wild-type protein does not bind to Grp78, but inhibition of glycosylation at N-terminal sites (positions 8, 22, and 38) led to association with this chaperone.

These observations show that nascent proteins that carry N-linked glycosylation sites close to the N terminus (p62, HA0) enter the calnexin folding pathway directly. Others, such as the Semliki Forest virus E1, the vesicular stomatitis virus G, and the Uukuniemi virus (bunyavirus) G1 and G2 proteins, associate initially with Grp78 and protein disulfide isomerase and are transferred to the calnexin pathway as they mature.

Molinari, M., and A. Helenius. 2000. Chaperone selection during glycoprotein translocation into the endoplasmic reticulum. *Science* **288:**331–333.

glycoproteins of alphaviruses, such as Sindbis virus. These proteins are translated from the 26S subgenomic mRNA (Appendix A, Fig. 25B) as a precursor that contains the E1 and E2 protein sequences, but are liberated from the precursor by signal peptidase cleavage as they enter the ER. The E1 protein associates with Grp78 and folds via three disulfide-bonded intermediates. Folding beyond the second such intermediate, and release from chaperone association, requires dimerization of the E1 with the E2 protein. Furthermore, the E2 protein misfolds when synthesized in the absence of E1 protein. The association of these two envelope proteins within the ER therefore appears to be essential for the productive folding of both. Similarly, the herpes simplex virus type 1 envelope glycoproteins gH and gL must interact with one another for the transport of either from the ER, and in the absence of gL protein, the gH protein cannot fold correctly.

Assembly of other viral membrane proteins is completed following exit from the ER: disulfide-linked dimers of the hepatitis B virus surface antigen form higher-order complexes in the next compartment in the pathway, and oligomers of the human immunodeficiency virus type 1 Env protein can be detected only in the Golgi apparatus. At present, we can discern no simple rules describing the relationship of oligomer assembly and transport of membrane proteins from the ER. Nevertheless, oligomerization begins, and in some cases must be completed, within the ER, where it can be facilitated by the folding catalysts and chaperones characteristic of this compartment.

Vesicular Transport to the Cell Surface

Mechanism of vesicular transport. Viral membrane proteins, like their cellular counterparts, travel to the cell surface through a series of membrane-bound compartments and vesicles. The first step in this pathway, illustrated schematically in Fig. 12.9, is transport of the folded protein from the ER to the Golgi apparatus. Within the Golgi apparatus (Fig. 12.10), proteins are sorted according to the addresses specified in their primary sequences or covalent modifications. **Transport vesicles**, which bud from one compartment and move to the next, provide an

Figure 12.9 Compartments in the secretory pathway. The lumen of each membrane-bound compartment shown is topologically equivalent to the exterior of the cell. Proteins destined for secretion or for the plasma membrane travel from the ER to the cell surface via the Golgi apparatus, as indicated by the red arrows. However, proteins can be diverted from this pathway to lysosomes, or to secretory granules that carry proteins to the cell surface for regulated release. The return of proteins from the Golgi apparatus to the ER is indicated (purple arrow). The endocytic pathway (blue arrows), discussed in Chapter 5, and the secretory pathway intersect in endosomes and the Golgi apparatus.

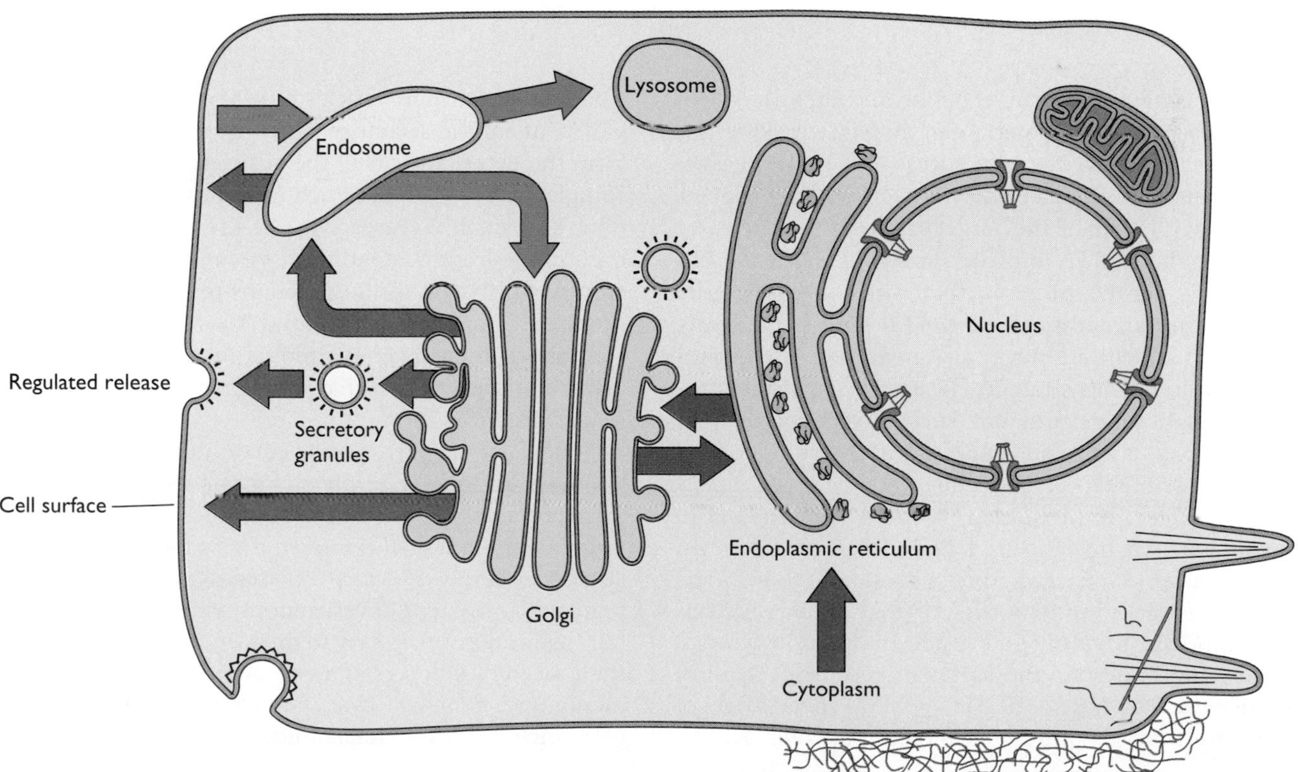

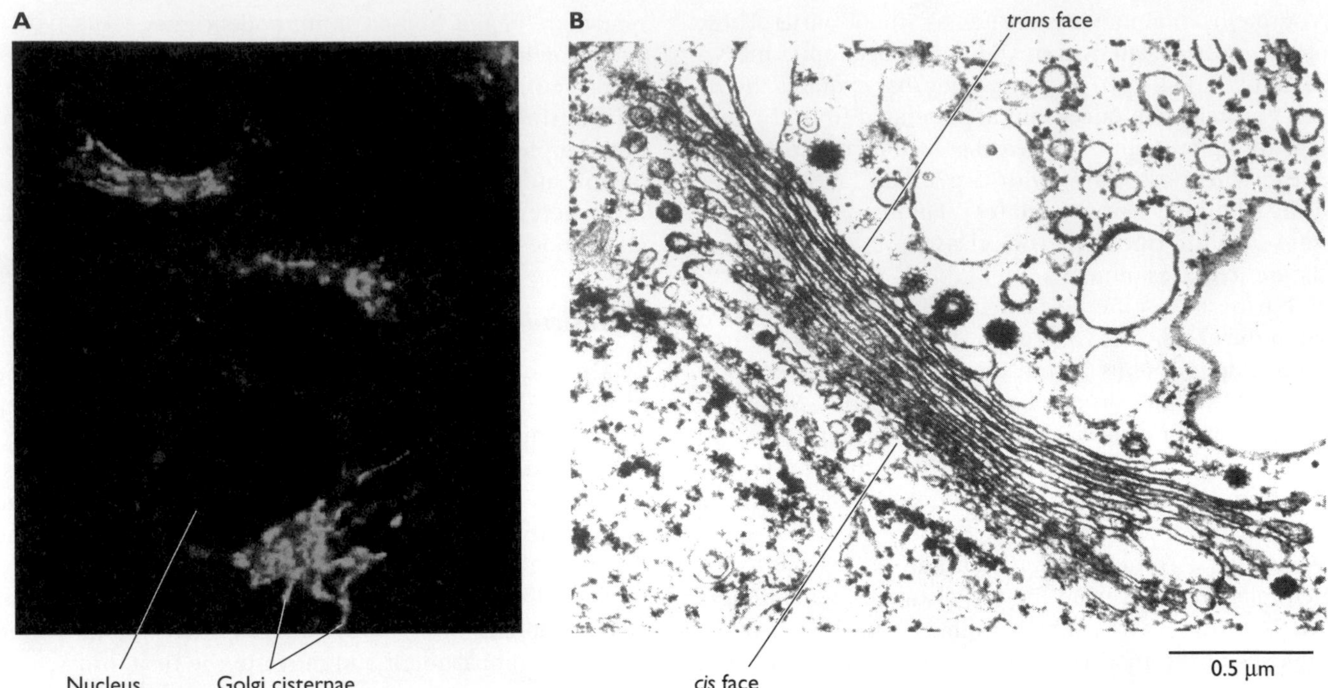

Figure 12.10 The Golgi apparatus. (A) Immunofluorescence of pig kidney epithelial cells following staining with antibody against a Golgi protein (p115). Courtesy of A. Brideau, Princeton University. (B) Electron micrograph of negatively stained sections of the alga *Ochromonas danica*. From G. T. Cole, Biological Photo Service. As can be seen, this organelle is usually located near the nucleus. It is composed of flattened membranous sacs, termed cisternae, bordered on either side by more convoluted structures of connected tubules and sacs, the *cis*- and *trans*-Golgi networks. Because the cisternae lie on top of one another, they are often called Golgi stacks. The number of cisternae varies considerably from one cell type to another and under different physiological conditions.

important mechanism for travel between compartments of the secretory pathway. Proteins can also move along this pathway in larger vesicular structures (Box 12.4). Regardless of the nature of the transport vehicle, fusion of its membrane with that of the target compartment releases the cargo of the vesicle into the lumen of that compartment. Consequently, proteins that enter the secretory pathway upon translocation into the ER (and are correctly folded and modified) are never again exposed to the cytoplasm of the cell. This strategy effectively sequesters proteins that might be detrimental, such as secreted or lysosomal proteases, and avoids exposure of disulfide-bonded proteins to a reducing environment.

Many soluble and membrane proteins participating in vesicular transport have been identified and characterized by biochemical, molecular, and genetic methods. The properties of these proteins suggest that similar mechanisms control the budding and fusion of different types of transport vesicle, despite the stringent specificity requirements of vesicular transport. This general mechanism of vesicular transport is quite well understood (Fig. 12.11). Budding of transport vesicles from the membranes of compartments of the secretory pathway requires proteins that form the external coats of the vesicles, such as the protein complex called COP II, which mediates ER-to-Golgi transport, and small GTPases (Fig. 12.11A). Transport vesicles then move from compartment to compartment of the secretory pathway, shuttling proteins toward their final destinations. Such movement may be by passive diffusion over short distances, but there is increasing evidence for active transport of vesicular structures over both short and longer distances.

When a transport vesicle encounters its target membrane, it docks as a result of specific interactions among Snare proteins (defined below) present in the vesicle and target membranes. Docking requires a member of the very large Rab family of small GTPases, which are thought to regulate many steps in vesicular transport. A complex containing proteins necessary to prepare for membrane fusion then assembles in a regulated manner and juxtaposes the membranes that will fuse. The components of this complex include *N*-ethylmaleimide-sensitive fusion protein

BOX 12.4

ER-to-Golgi transport in living cells

The vesicular stomatitis virus G protein made in cells infected with the mutant virus *ts*045 misfolds and is retained in the ER at high temperature (40°C). It refolds and is transported to the Golgi apparatus when the temperature is reduced to 32°C. This temperature-sensitive protein has therefore been used extensively to study transport through the secretory pathway. For examination of this process in living cells, the green fluorescent protein was attached to the cytoplasmic tail of the viral G protein. Control experiments established that this modification did not alter the temperature-sensitive folding or transport of the G protein. Time-lapse fluorescence microscopy of cells shifted from high to low temperature (see figure) demonstrated that the chimeric G protein rapidly left the ER at multiple peripheral sites. The protein appeared in membranous structures, which were often larger than typical transport vesicles. These structures moved rapidly toward the Golgi in a stop-start manner, with maximal velocities of 1.4 μm/s. Such transport, but not formation of post-ER structures, was blocked when microtubules were depolymerized by treatment with nocodazole, or when the (–) end-directed microtubule motor dynein was inhibited. It was therefore concluded that vesicles and other membrane-bound structures that emerge from the ER at peripheral sites are actively transported to the Golgi complex.

Presley, J. F., N. B. Cole, T. A. Schroer, K. Hirschberg, K. J. M. Zaal, and J. Lippincott-Schwartz. 1997. *Nature* **384:**81–85.

Shown are images of the vesicular stomatitis virus *ts*045 G protein-green fluorescent protein observed in cells 5 min after a shift to 32°C from 40°C (A) and after increasing periods of transfer to 32°C (B). Red arrows indicate the positions of G protein-green fluorescent protein. (C) Fluorescence associated with the Golgi complex was photobleached upon the shift to 32°C to allow visualization of delivery of the fluorescent protein from intermediates to the Golgi. By 396 s, significant quantities of the fusion protein have reached the Golgi complex. (D) The path of such movement is shown schematically.

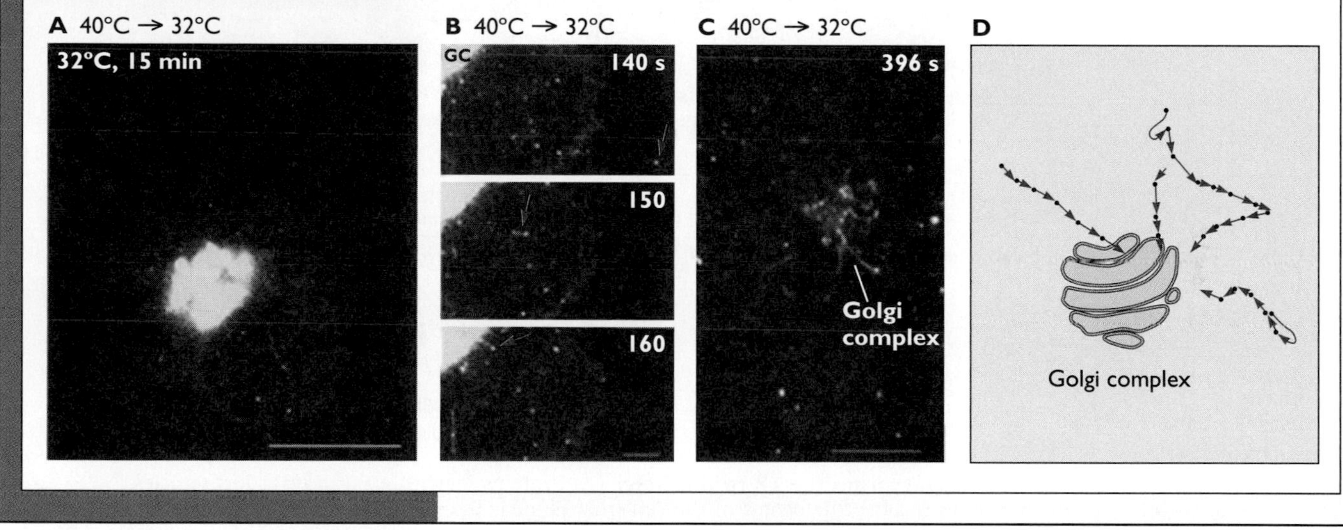

(Nsf) and soluble Nsf attachment proteins (Snaps), which shuttle between assembled complexes and the cytoplasm (Fig. 12.11B). The Nsf protein associates with α-Snap, which in turn binds to Snap receptors (the **Snares**) present in the membranes of the transport vesicle and of the compartment with which it will fuse. The final reaction of vesicular transport, membrane fusion, has not been well characterized, and it is not known whether this process requires proteins in addition to those present in the prefusion complex. At present, the best-understood examples of membrane fusion are those between enveloped viruses and the membranes of their host cells, such as the HA protein-mediated fusion process described in Chapter 5.

The high density of intracellular protein traffic requires considerable specificity during vesicle formation and fusion. For example, vesicles that transport proteins from the ER to the first compartment of the Golgi apparatus, the *cis*-Golgi network (Fig. 12.9), must take up only the appropriate

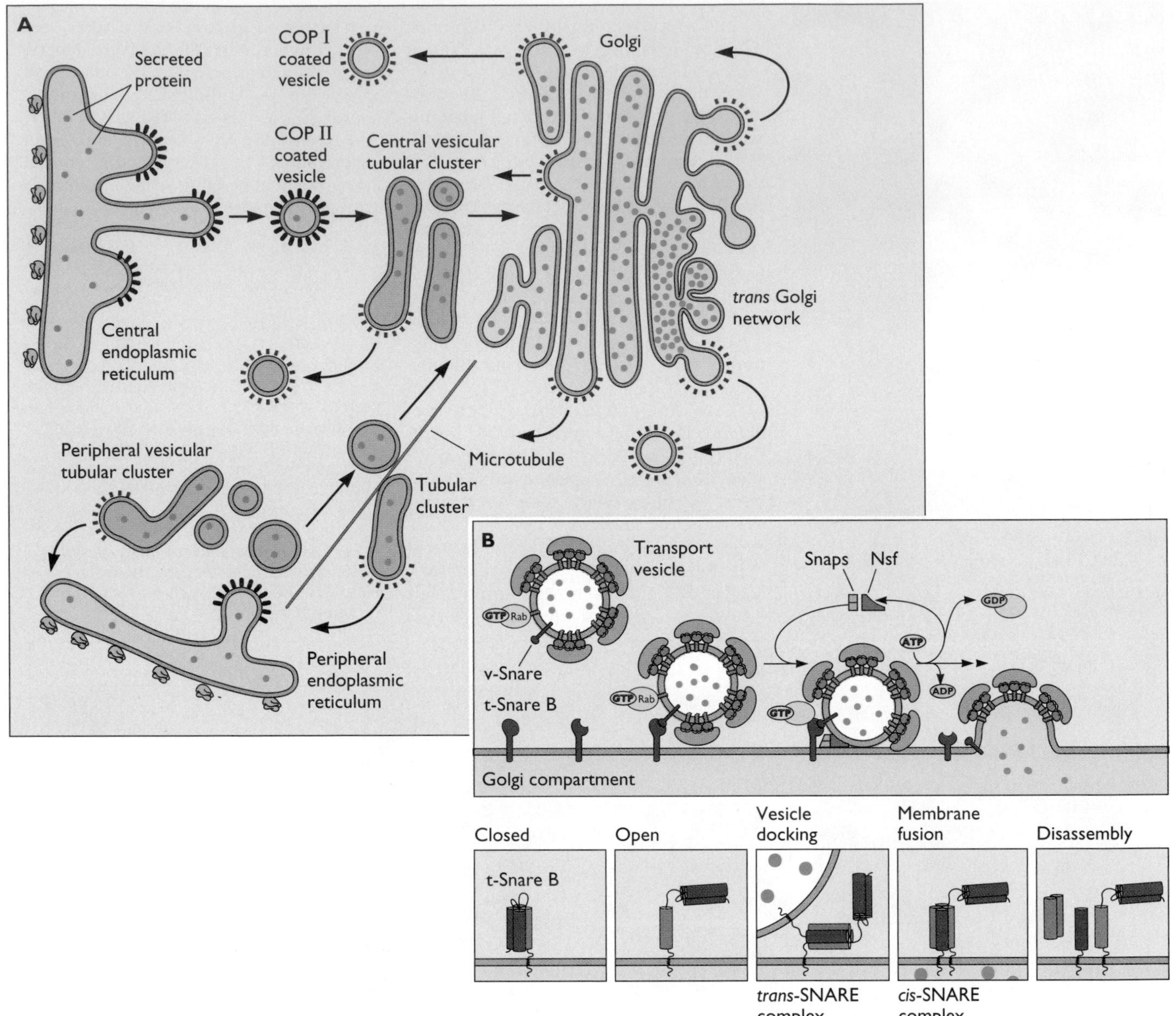

Figure 12.11 Protein transport from the ER to the Golgi. (A) Proteins leave the ER in transport vesicles at regions free of ribosomes in both the center of the cell (that is, near the Golgi) and peripheral regions. Vesicle formation is initiated by binding of the cytoplasmic coat protein complex II (COP II), which contains a small GTPase (Sar1) and several other proteins, to the membrane. The vesicle membranes also carry proteins that direct them to appropriate destinations, notably members of the Snare family such as syntaxin 5 and membrin. The COP II coat induces budding and pinching off of vesicles. The mechanisms of cargo loading are not fully understood: soluble proteins may be included nonselectively in vesicles leaving the ER (bulk-flow), but there is evidence that membrane proteins are specifically recruited into COP II-coated vesicles. After formation, COP II vesicles lose their coats and fuse to form vesicular tubular clusters located both in the center of the cell near the *cis*-Golgi network and at peripheral sites. These clusters are also known as the ER-Golgi intermediate compartment. As illustrated in the enlargement in panel B, the specificity of membrane fusion is determined, at least in part, by the assembly of specific v-Snare and t-Snare complexes. This process can be regulated via conformational change in the t-Snare syntaxin. Once this protein is induced to adopt the open form, it can form a *trans*-Snare bundle with a v-Snare (red) and a second t-Snare (purple). Such assembly, via formation of a bundle of four α-helices from the different Snares, docks the vesicle at the target membrane. Docking also requires members of the Rab family of small GTPases, which are

(continued on next page)

proteins when budding from the ER and must fuse only with the membrane of the *cis*-Golgi network. The mechanisms that ensure specificity of cargo loading during formation of transport vesicles are not fully understood. However, both direct association of cargo proteins with certain of the proteins forming the vesicular coat and indirect interactions via "cargo carrier" proteins have been implicated. The Snare proteins resident in the membranes of transport vesicles (v-Snares) and in those of the target compartment (t-Snares) are believed to be important determinants of the specificity of vesicular transport (Fig. 12.11B). Mammalian cells contain a large number of these proteins, and individual Snares are both restricted to specific compartments of the secretory pathway and form specific associations with one another in vivo. It is now well established that assembly of complexes containing v- and t-Snares is regulated, in part, by induction of conformational change in specific Snare proteins. Such regulatory mechanisms, as well as directed movement of transport vesicles to the target compartment and actions of the Rab proteins, seem likely to contribute to the fidelity of vesicular transport.

Reactions within the Golgi apparatus. One of the most important staging posts in the secretory pathway is the Golgi apparatus, which is composed of a series of membrane-bound compartments and located near the cell nucleus. Proteins enter the Golgi apparatus from the ER via the *cis*-Golgi network, which is composed of connected tubules and sacs (Figs. 12.9 and 12.10). A similar structure, the *trans*-Golgi network, forms the exit face of this organelle. The *cis*- and *trans*-Golgi networks are separated by a variable number of cisternae termed the *cis*, medial, and *trans* compartments. Each of the compartments of the Golgi apparatus, which can comprise multiple cisternae, is the site of specific processing reactions. As illustrated in Fig. 12.12, the N-linked oligosaccharides attached to proteins in the ER are further processed as the proteins move through the Golgi stacks. The enzymes responsible for early reactions in maturation of these oligosaccharides are located in *cis* cisternae, whereas those that carry out later reactions are present in the medial and *trans* compartments. Such spatial separation ensures that oligosaccharide processing follows a strict sequence as proteins pass through the compartments of the Golgi apparatus. This property has provided an important experimental tool with which to distinguish among, and hence localize other reactions to, the various Golgi compartments. Synthesis of O-linked oligosaccharides by glycosyltransferases, which add one sugar unit at a time to certain serine or threonine residues in a protein, also takes place in the Golgi apparatus.

A number of viral envelope glycoproteins are also proteolytically processed by cellular enzymes resident in late Golgi compartments. Retroviral Env glycoproteins are cleaved in the *trans*-Golgi network to produce the TM (transmembrane) and SU (surface unit) subunits from the Env polyprotein precursor (Fig. 12.13). The larger SU protein associates with TM noncovalently and in many simple retroviruses also via a disulfide bond (Fig. 12.13B). Similarly, the HA0 protein of certain avian influenza A viruses is cleaved into the HA1 and HA2 chains in the *trans*-Golgi network (Fig. 12.4). These and other viral

Figure 12.11 *(continued)* specifically localized to different membranes of the cell and appear to regulate the specificity of docking and therefore of membrane fusion. These Rab GTPases are believed to contribute to the specificity of vesicular transport by regulating the tethering of vesicles to target membranes. This reaction precedes formation of the *trans*-Snare complex, and is mediated by specifically located peripheral membrane proteins that are generally extended, coiled-coil proteins (e.g., p115 and giantin). Docking is followed by assembly of a membrane fusion complex via recruitment of two soluble components, *N*-ethylmaleimide-sensitive factor (Nsf) and soluble Nsf attachment proteins (Snaps, α or β, and γ). Nsf, which is an ATPase, and Snaps were originally identified in studies that used in vitro vesicle fusion assays. It is not known if this complex mediates fusion or if additional proteins are needed. However, the targeting Snare proteins may represent a minimal membrane fusion machine, for recombinant v- and t-Snare proteins promote docking and at least partial fusion of synthetic lipid vesicles in vitro. When formed at sites far from the Golgi, the vesicular-tubular compartments move to this organelle along microtubules (Box 12.4). Vesicles coated by the COP I coat bud off from these compartments and mediate transport of membrane proteins and escaped proteins back to the ER. This process is believed to make an important contribution to sorting proteins for forward transport. Whether vesicular-tubular compartments fuse with a fixed *cis*-Golgi network (stable Golgi compartment model) or form *cis*-cisternae upon fusion (cisternal maturation model) is a question currently receiving intense study. There is also considerable debate about the mechanisms by which proteins travel through the Golgi stack. Evidence supporting forward transport via both COP I transport vesicles and by maturation of Golgi cisternae has been reported, and it may be that different cargoes move by different mechanisms. Recent studies of transport of the vesicular stomatitis virus G protein indicate that this plasma membrane-destined protein passes rapidly through the Golgi without entering vesicles, whereas resident Golgi enzymes are found in COP I vesicles.

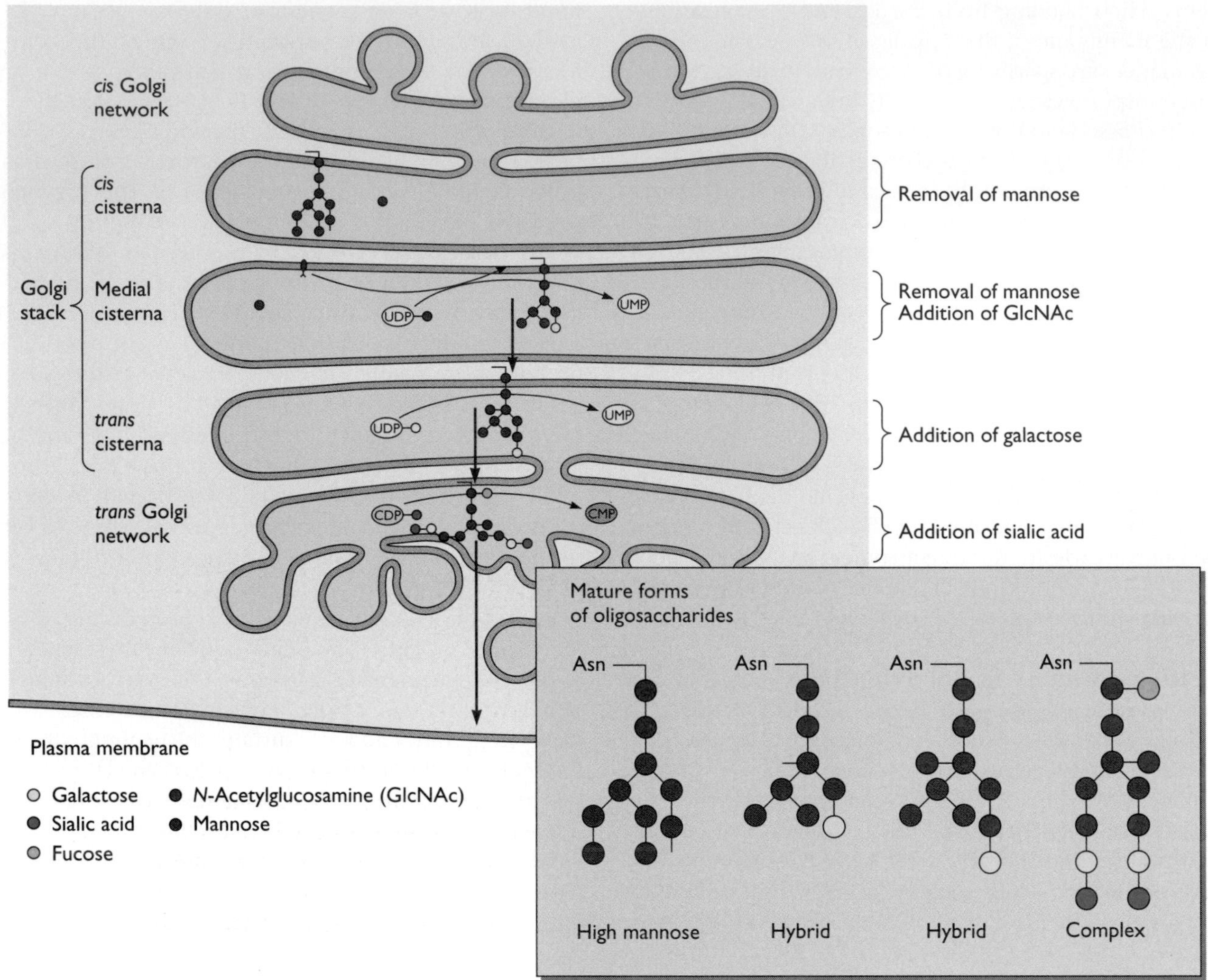

Figure 12.12 Compartmentalization of processing of N-linked oligosaccharides. The reactions by which mature N-linked oligosaccharide chains are produced from the high-mannose core precursor added in the ER (Fig. 12.7B) are shown in the Golgi compartment in which they take place. Trimming of terminal glucose and mannose residues of the common core precedes stepwise addition of the sugars found in the mature chains. These additions, which require activated sugar-nucleotide substrates, such as UDP-*N*-acetylglucosamine, are catalyzed by compartment-concentrated transferases.

membrane proteins (Table 12.4) are processed by members of a family of resident Golgi proteases that cleave after pairs of basic amino acids. The members of this family, which in animal cells includes furin and at least five other proteins found in the *trans*-Golgi network, are serine proteases related to the bacterial enzyme subtilisin. Various furin family members have been shown by genetic and molecular methods to process viral glycoproteins; their normal function is to process cellular polyproteins such as certain hormone precursors. These proteolytic cleavages are generally essential for formation of infectious particles, although they are not necessary for assembly. For example, proteolytic processing of envelope proteins of retroviruses and alphaviruses is necessary for infectivity, probably because sequences important for fusion and entry become accessible. It is well established that proteolytic processing of the influenza virus HA0 protein liberates the fusion peptide as the N terminus of HA2 (Fig. 12.3A). If HA0 is not cleaved, the fusion peptide is not available to initiate the infectious cycle. As discussed in more detail in Chapter 14, virulent strains of avian influenza A virus encode HA0 proteins that can be processed by the ubiquitous furin family proteases, such that virus particles carrying fusion-active HA protein are released. It seems likely that the common dependence on furin family proteases (Table 12.4), which act on proteins relatively late in the secretory

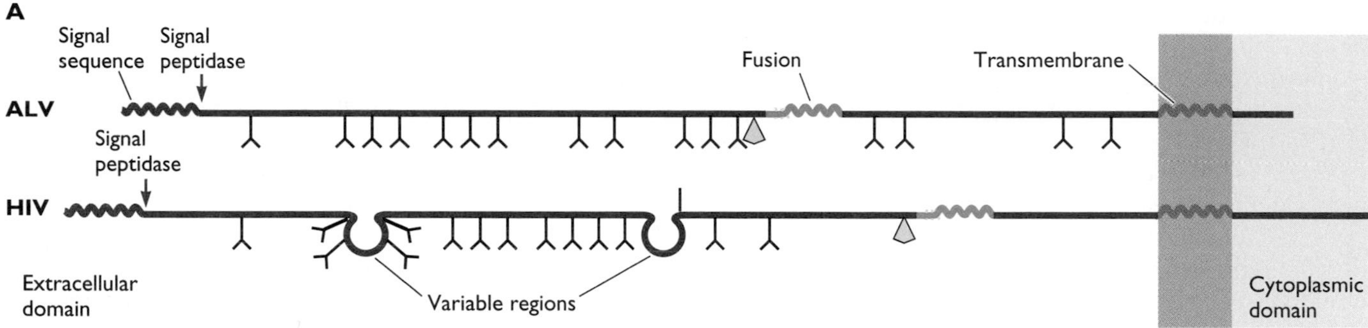

B

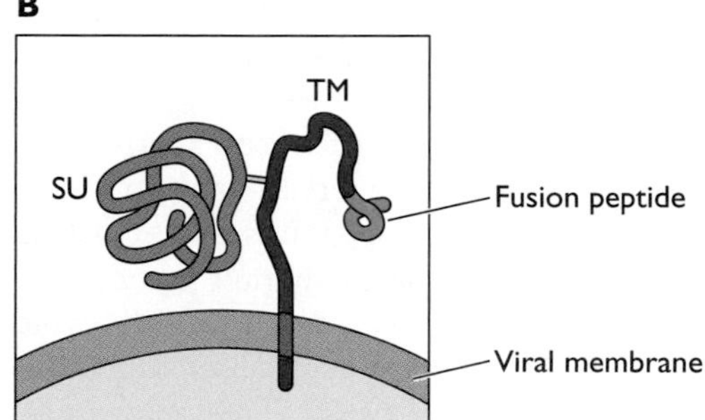

Figure 12.13 Modification and processing of retroviral Env polyproteins. Sequence features and modifications of the Env proteins of avian leukosis virus (ALV) and human immunodeficiency virus type 1 (HIV) (A) are depicted as in Fig. 12.3. The variable regions of human immunodeficiency virus type 1 Env differ greatly in sequence among viral isolates. The translocation products shown here are cleaved by the ER signal peptidase (red arrows) and by furin family proteases in the *trans*-Golgi network (yellow arrowheads). The latter liberates the transmembrane (TM) and surface unit (SU) subunits from the Env precursor. The interaction between the subunits has not been examined by structural methods. Both noncovalent interactions and a disulfide bond (yellow line) can maintain the association of the SU and TM proteins (B).

pathway, helps minimize complications that would arise if viral glycoproteins were initially synthesized with their fusion peptides in an active conformation.

For at least some influenza A viruses, such as fowl plague virus, the ion channel activity of the viral M2 protein helps to maintain HA in a fusion-incompetent conformation, because its HA protein switches to the fusion-competent conformation at a pH higher than that required by HA proteins of human influenza A viruses. The M2 protein, which forms a cation-selective ion channel, is found in infected cells at quite high concentrations in the membranes of secretory pathway compartments. By increasing the pH of normally acidic compartments, such as those of the *trans*-Golgi network, this protein prevents premature switching of proteolytically processed HA to the fusion-active conformation described in Chapter 5.

Although all the envelope proteins of viruses that assemble at the plasma membrane travel via the cellular

Table 12.4 Viral envelope glycoprotein precursors processed by signal peptidase or furin family proteases

Virus family	Precursor glycoprotein	Membrane-associated cleavage products
Signal peptidase		
Alphavirus	Envelope polyprotein precursor	E1, pE2
Bunyavirus	Translation product of M mRNA	G1, G2
Furin family proteases		
Alphavirus	pE2	E2
Flavivirus	PrM	M
Hepadnavirus	preC	C antigen[a,b]
Herpesvirus[b,c]	pre-gB	gB
Orthomyxovirus[d]	HA0	HA1, HA2
Paramyxovirus	F0	F1, F2
Retrovirus	Env	TM, SU

[a]This cleavage product is largely secreted into the extracellular medium, but is also associated with the plasma membrane of infected cells.
[b]Cleavage is not necessary for production of infectious virus particles in cells in tissue culture.
[c]Some alphaherpesviruses (e.g., varicella-zoster virus), and all known betaherpesviruses.
[d]Virulent strains of avian influenza A virus.

secretory pathway, there is considerable variation in the rate and efficiency of their transport (Table 12.2). A champion is the influenza virus HA protein, which folds and assembles with a half time of only 7 min, with more than 90% of the newly synthesized HA molecules reaching the cell surface. Many other viral proteins do considerably less well. Parameters determining the rate and efficiency of transport may include the complexity of the protein and the inherent asynchrony of protein folding. With few exceptions (see "Inhibition of Transport of Cellular Proteins by Viral Infection" below), cellular proteins continue to enter and traverse the secretory pathway as enveloped viruses assemble at the plasma membrane. Competition among viral and cellular proteins, which may vary with the nature and physiological state of the host cell, is also likely to affect the transport of viral proteins to the cell surface.

We have focused our discussion of viral envelope proteins that travel the cellular secretory pathway on the well-understood maturation of their extracellular domains. However, the cytoplasmic domains of these proteins are also frequently modified. Many are **acylated** by the covalent linkage of the fatty acid palmitate to cysteine residues in their cytoplasmic domains (Table 12.5). This modification can be necessary for optimally efficient production of progeny virions. For example, inhibition of palmitoylation of the Sindbis virus E2 glycoprotein or of the human immunodeficiency virus type 1 Env protein impairs virus assembly and budding. The bulky fatty acid chains attached to the short cytoplasmic tails may regulate envelope protein conformation or association with specific membrane domains (see "Signal Sequence-Independent Transport of Viral Proteins to the Plasma Membrane" below). However, the functions of acylation are not well established in other cases.

Sorting of Viral Proteins in Polarized Cells

Proteins that are not specifically targeted to an intracellular address travel from the Golgi apparatus to the plasma membrane. However, the plasma membrane is not uniform in all animal cells: differentiated cells often devote different parts of their surfaces to specialized functions,

Table 12.5 Examples of acylated or isoprenylated viral proteins

Virus	Protein	Lipid	Probable function
Envelope proteins			
Alphavirus			
Sindbis virus	E2	Palmitate	Required for efficient budding of virus particles
Hepadnavirus			
Hepatitis B virus	L (pre-S1)	Myristate	Initiation of infection
Orthomyxovirus			
Influenza A virus	HA	Palmitate	May facilitate fusion activity
	M2	Palmitate	Unknown
Retroviruses			
Human immunodeficiency virus type 1, Moloney murine leukemia virus	Env, TM	Palmitate	Virion budding
Rhabdovirus			
Vesicular stomatitis virus	G	Palmitate	Unknown
Other viral proteins			
Alphavirus			
Semliki Forest virus	Nsp1	Palmitate	Tight association with plasma membrane; movement of Nsp1 (capping enzyme) to filopodia
Hepatitis delta satellite virus	Large delta antigen	Geranylgeranol	Interaction with hepatitis B virus L protein; assembly; inhibition of hepatitis D virus RNA replication
Picornavirus			
Poliovirus	VP0, VP4	Myristate	Virion assembly; uncoating
Polyomavirus			
Simian virus 40	VP2	Myristate	Virion assembly
Retroviruses			
Human immunodeficiency virus type 1, murine leukemia virus	Gag, MA	Myristate	Membrane association, assembly and budding
Rous sarcoma virus	$pp60^{src}$	Myristate	Membrane association, tranformation (Chapter 18)
Harvey sarcoma virus	$p21^{ras}$	Farnesol	Membrane localization

and the plasma membranes of such **polarized cells** are divided into correspondingly distinct regions. During infection by many enveloped viruses, the asymmetric surfaces of such cells are distinguished during entry, and by the sorting of virion components to a specific plasma membrane region. In this section, we describe the final steps in the transport of proteins to specialized plasma membrane regions in two types of polarized cell in which animal viruses often replicate, epithelial cells and neurons (Fig. 12.14).

Epithelial Cells

As discussed in Chapter 5, epithelial cells, which cover the external surfaces of vertebrates and line all their internal cavities (such as the respiratory and gastrointestinal tracts), are primary targets of virus infection. The cells of an epithelium are organized into close-knit sheets both by the tight contacts they make with one another and by their interactions with the underlying basal lamina, a thin layer of extracellular matrix (Fig. 5.1). Within the best-characterized epithelia, such as those that line the intestine, each cell is divided into a highly folded **apical domain** exposed to the outside world and a **basolateral domain** (Figs. 5.1 and 12.14). The former performs more specialized functions, whereas the latter is associated with cellular housekeeping. These two domains differ in their protein and lipid contents, in part because they are separated by specialized cell-cell junctions (tight junctions) (Fig. 5.1), which prevent free diffusion and mixing of components in the outer leaflet of the lipid bilayer. However, such physical separation does not explain how the polarized distribution of plasma membrane proteins is established and maintained.

Figure 12.14 Polarized epithelial cells and neurons.

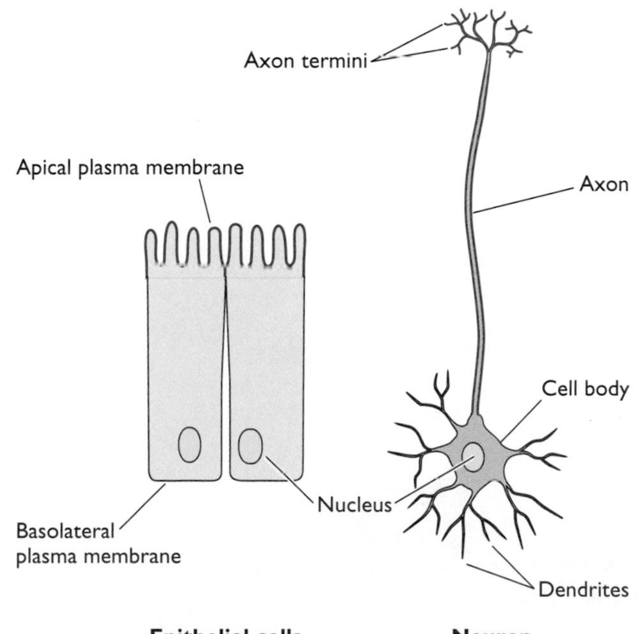

Viruses have been important tools in efforts to elucidate the molecular mechanisms responsible for the polarity of typical epithelial cells, because certain enveloped viruses bud asymmetrically. For example, in the epithelial cells of all organs studied, influenza A virus buds apically and vesicular stomatitis virus buds basolaterally. The specific advantages of polarized assembly and release of viruses are not fully appreciated. However, as discussed in Chapter 14, this process can facilitate virus spread within or among host organisms. The polarity of virus budding is generally the result of accumulation of envelope proteins, such as HA and G, in the apical and basolateral domains, respectively. The most common mechanism for selective localization appears to be signal-dependent sorting of proteins in the *trans*-Golgi network, in which proteins destined for apical or basolateral surfaces are distinguished and packaged into appropriately targeted transport vesicles. Signals necessary for basolateral targeting comprise short amino acid sequences located in the cytoplasmic domains of membrane proteins. Many basolateral targeting signals overlap with those for the direction of proteins to clathrin-coated pits, the first staging post of the endocytic pathway. Indeed, in polarized epithelial cells, certain proteins are transferred through endosomes from one membrane domain to the other, a process termed transcytosis (see Fig. 15.23). The sorting of viral glycoproteins to basolateral membrane domains can also be governed by additional viral proteins. The two envelope proteins of measles virus (F and H) possess signals that direct them to the basolateral membrane. However, when the viral matrix protein binds to the cytoplasmic tails of F and H, these proteins accumulate at the apical surface of epithelial cells.

The signals that direct proteins to the apical membrane are currently a matter of debate, but both N-glycosylation and association of proteins with specific lipids may be important. The apical membranes of epithelial cells are enriched in sphingolipids and cholesterol, which form specific microdomains called **lipid rafts**. There is now considerable evidence indicating that recruitment of proteins to lipid rafts within the *trans*-Golgi is an important determinant of transport to the apical membrane. Such rafts are dynamic assemblies that can selectively incorporate specific proteins, and were initially shown to mediate apical transport of glycosylphosphatidylinositol-anchored proteins. The influenza virus HA and NA proteins bind specifically to lipid rafts via their transmembrane domains, which determine apical sorting. Cellular proteins known to participate in apical trafficking of viral glycoproteins, such as caveolin-1, annexin XIIIb, myelin, and lymphocyte-binding protein, are also associated with these rafts: inhibition of the activity or synthesis of these proteins disrupts transport of the influenza virus HA protein (and other proteins) from the Golgi complex to the apical membrane.

The role of lipid rafts has been recognized only quite recently, and many questions about the signals that load cargo and induce formation of apical transport vesicles remain to be addressed. Lipid rafts seem likely to be more generally important in targeting of viral membrane proteins and assembly: measles virus glycoproteins are selectively enriched in lipid rafts in nonpolarized cells, and association of the human immunodeficiency virus type 1 Gag polyprotein with these membrane domains promotes production of virus particles.

Neurons

Neurons are probably the most dramatically specialized among the many polarized cells of vertebrates. The axon is typically long and unbranched, whereas the dendrites form an extensive branched network of projections (Fig. 12.14). Axons are specialized for the transmission of nerve impulses, ultimately via the formation of synaptic vesicles and release of their contents. In contrast, dendrites provide a large surface area for the receipt of signals from other neurons. The nucleus, the rough ER, and the Golgi apparatus are also located in the dendritic region and the cell body of a neuron. Although axonal and dendritic surfaces are not separated by tight junctions, proteins must be distributed asymmetrically in neurons. Several mechanisms contribute to the establishment and maintenance of neuronal polarity, including transport of vesicles in specific directions along the highly organized microtubules of the axon (**axonal transport**) and transport of mRNA to dendrites. Nevertheless, it is believed that sorting and targeting of membrane proteins for delivery to axonal or dendritic surfaces is also essential for polarization.

The directional movement of vesicles and many cellular organelles in neurons is dependent on polarized microtubules and motor molecules that travel toward either their – or their + ends. Such motors therefore mediate transport both toward the cell body from axons and dendrites and away from the cell body (Fig. 12.15). Viruses that infect neurons depend on these mechanisms during infection, assembly, and egress of particles. An important example is provided by the neurotropic alphaherpesviruses, a group that includes herpes simplex virus type 1 and varicella-zoster virus. Upon infection of sensory neurons, herpesvirus capsids are transported to the nucleus by microtubule-based transport, probably mediated by (–) end-directed motors such as dynein (Fig. 12.15). For many years it was believed that, after replication and assembly, enveloped virus particles were transported within axons. However, direct visualization of virion and nucleocapsid transport at the beginning and end, respectively, of the infectious cycle (Box 12.5) established that the DNA-containing nucleocapsid and viral envelope

Figure 12.15 Axonal transport in neurons. Vesicles containing cargo destined for synaptic vesicles (e.g., neurotransmitters) become associated with molecular motors of the kinesin family, and are transported down the axon on the tracks formed by axonal microtubules. The force generated by kinesin family motors allows transport at rates of 1 to 2 μm/s. Transport toward the cell body, for example, of degenerating mitochondria, is believed to be mediated by the cytoplasmic motor dynein. In infected neurons, herpes simplex virus type 1 (HSV-1) components are transported in opposite directions at the beginning and end of the infectious cycle. Newly synthesized internal components and envelope proteins travel the axon separately.

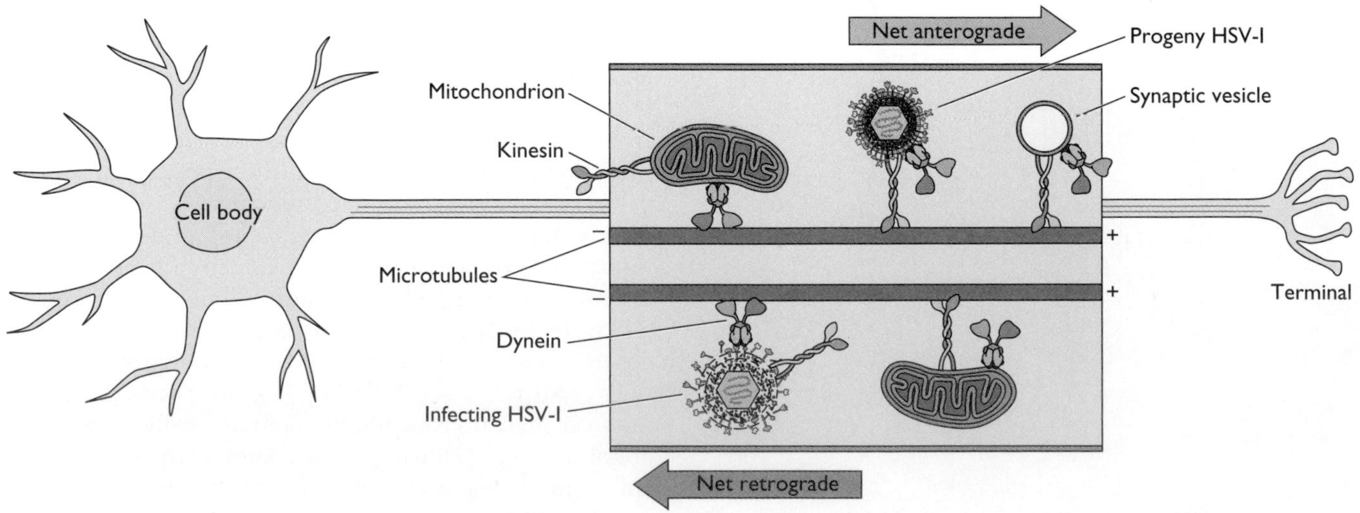

BOX 12.5

The complexities of axonal transport of alphaherpesviral components

The transport of the pseudorabies virus nucleocapsid has been directly visualized in living neurons by fusion of the VP26 capsid protein coding sequence in the viral genome to that for the green fluorescent protein. Such capsids appear as discrete dots (figure panel A), and their movement can be recorded for analysis using laser-scanning, time-lapse confocal microscopy. One important finding was that nucleocapsids and viral envelope proteins newly synthesized in the cell body are transported separately to axon termini (panel A). Kinetic studies of nucleocapsid movement during entry and egress led to the unexpected conclusion that movement is saltatory (that is, characterized by stops, starts, and changes of direction) (panel B). However, while the properties of retrograde movement are the same during entry and egress, anterograde movement exhibits both a higher velocity and much longer runs during egress than during entry (panel B). This difference accounts for whether net nucleocapsid movement is retrograde or anterograde. It implies that interactions of nucleocapsids with the anterograde motor(s) or motor function must be regulated.

Smith, G. A., S. Gross, and L. W. Enquist. 2001. *Proc. Natl. Acad. Sci. USA* **98:**3466–3470.

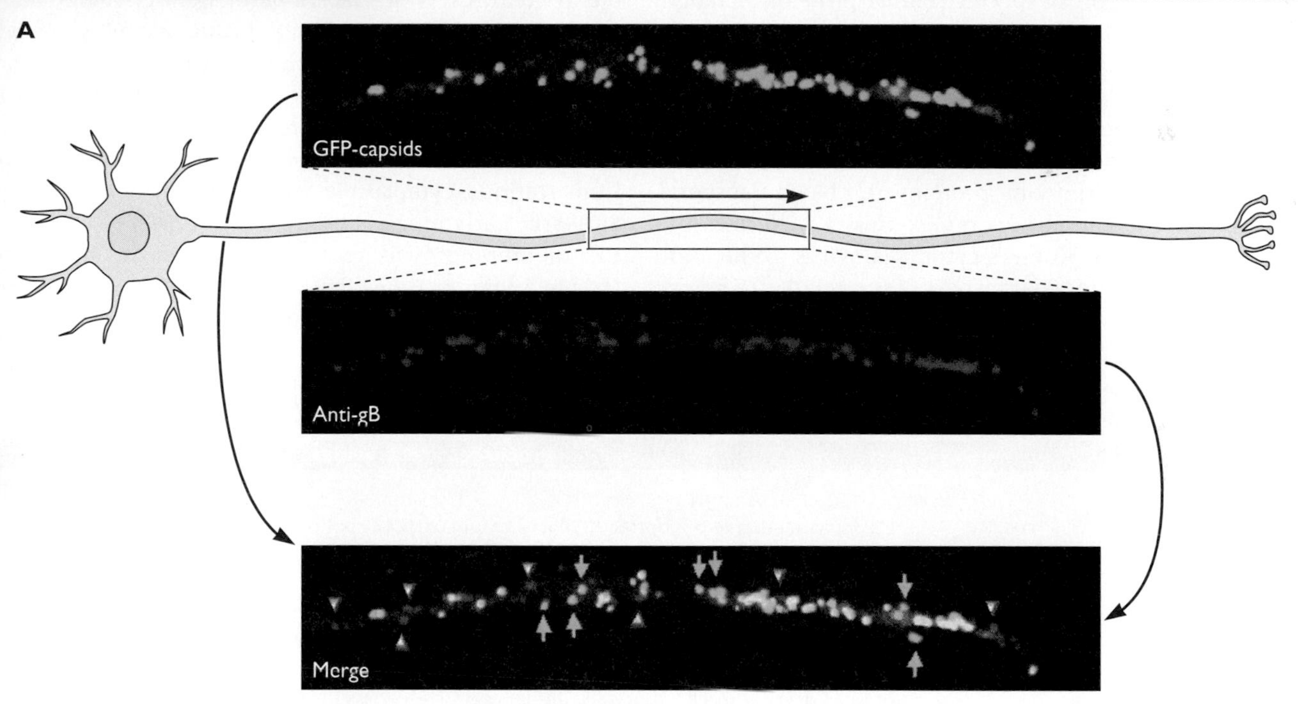

B Capsid motion: egress versus entry in neurons

Direction	Rate (μm/s)	Run length (μm)
Egress		
Anterograde	1.97	13.1
Retrograde	1.28	6.8
Entry		
Anterograde	0.85	0.5
Retrograde	1.33	8.1

(A) Chicken embryo dorsal root ganglia infected with the pseudorabies viruses described above were isolated 15 h after infection and examined by confocal microscopy. Nucleocapsids are detected by autofluorescence of the VP26-green fluorescent protein (green), and the envelope gB protein (red) was detected by indirect immunofluorescence. Comparison of the locations of the nucleocapsid and gB along the axon (merge) indicates that these virion components are transported separately. (B) The kinetic properties of nucleocapsid transport during entry and egress are summarized. One hypothesis for the differences in anterograde transport during entry and egress is that nucleocapsids are associated with different populations of tegument proteins.

proteins are transported separately along axons. Exactly how these virion components are moved is not yet well understood, although (+) end-directed motors, such as members of the kinesin family, seem likely to participate. Another remarkable finding is that the spread of herpesviruses from neuron to neuron occurs at or near sites of synaptic contact, indicating that virus particles must be targeted to specific areas within neurons for egress. As discussed in Chapter 14, this attribute can be exploited to define neuronal connections in a living animal by using the virus as a self-amplifying tracer.

Inhibition of Transport of Cellular Proteins by Viral Infection

The genomes of several viruses encode proteins that interfere with the transport of specific cellular proteins, notably major histocompatibility class I molecules (Tables 12.6 and 15.3). For example, the adenovirus E3 glycoprotein gp19 (Appendix A, Fig. 1B) binds to these important components of immune defense in the ER and prevents their exit from this compartment, probably by inhibiting folding and assembly. Several herpesviral proteins also block transport of major histocompatibility complex class I molecules to the cell surface. The human cytomegalovirus US11 and US2 gene products induce transport of the cellular proteins from the ER to the cytoplasm for rapid degradation by the cellular proteasome. When this activity was first discovered, such back transport from the ER to the cytoplasm appeared very unusual. However, as discussed previously, it is now clear that this is a normal activity of uninfected cells. The human immunodeficiency virus type 1 Vpu protein (Fig 17.4; Appendix A, Fig. 21), a transmembrane phosphoprotein, also induces selective degradation of newly synthesized major histocompatibility complex class I proteins, and of CD4, by a similar mechanism (Table 12.6). Such degradation of CD4, the major receptor for this virus, is important for assembly and release: tight binding of this cellular protein to the viral Env glycoprotein in the ER prevents transit of both proteins to the cell surface.

Proteins encoded by certain other viruses exert more drastic effects on the cellular secretory pathway. For example, rotaviruses, which lack a permanent envelope but are transiently membrane enclosed during assembly, encode a protein that disrupts the ER membrane. This protein is believed to allow removal of the temporary envelope formed during virion assembly. The replication of most (+) strand RNA viruses takes place in membranous structures derived from various cytoplasmic membranes of the host cell. Similarly, early in infection by the poxvirus vaccinia virus, ER membranes are reorganized to form membrane-wrapped viral factories that are required for efficient viral DNA synthesis. Such remodeling of cytoplasmic mem-

Table 12.6 Viral proteins that inhibit transport of cellular proteins through the secretory pathway

Viral protein	Properties and activities
Adenovirus	
Ad2 E3gp19	ER transmembrane glycoprotein; blocks exit of MHC[a] class I proteins from the ER; probably inhibits folding and assembly
Herpesvirus	
Human cytomegalovirus US2 and US11	ER transmembrane proteins; induce back transport of MHC class I proteins from ER to cytoplasm for proteasomal degradation
Varicella-zoster virus ORF66	Induces decreased cell surface accumulation of MHC class I proteins, probably by their retention in Golgi compartments
Picornavirus	
Poliovirus	
2BC, 3A	Inhibit vesicular transport from ER; induce accumulation of infected cell-specific vesicles derived from ER and disassembly of the Golgi complex; associated with membranes of the viral replication complex
Retrovirus	
Human immunodeficiency virus type 1	
Vpu	Transmembrane protein with phosphorylated cytoplasmic domain; induces degradation of MHC class I and CD4 proteins; Vpu-induced degradation of CD4 is by the cytoplasmic proteasome; this process releases Env from tight association with CD4 and allows its transport to the plasma membrane
Nef	Induces decreased cell surface accumulation of MHC class I proteins and retention in intracellular compartments; increases rate of endocytosis of CD4

[a]MHC, major histocompatibility complex.

branes of the host cell is best understood for poliovirus and other enteroviruses, which induce a dramatic reorganization of the compartments of the secretory pathway.

Virus-induced membranous vesicles, 40 to 50 nm in diameter, appear during the poliovirus infectious cycle and eventually fill the cytoplasm of the infected cell (Fig. 12.16A). These vesicles are derived from the membranes of the compartments of the secretory pathway, and late in infection no Golgi apparatus can be seen in poliovirus-infected cells. Not surprisingly, such dismantling of the cellular secretory pathway results in the inhibition of protein traffic to the cell surface. Because poliovirus particles lack an envelope, this cellular system is not needed for virus reproduction. However, formation of membranous vesicles is essential for replication of the viral RNA genome (Chapter 6). Poliovirus replication complexes are assembled from individual poliovirus-induced vesicles. These infected-cell-specific structures are also the sites of assembly of progeny virions and might participate in the nonlytic release of virus particles described in Chapter 13.

A

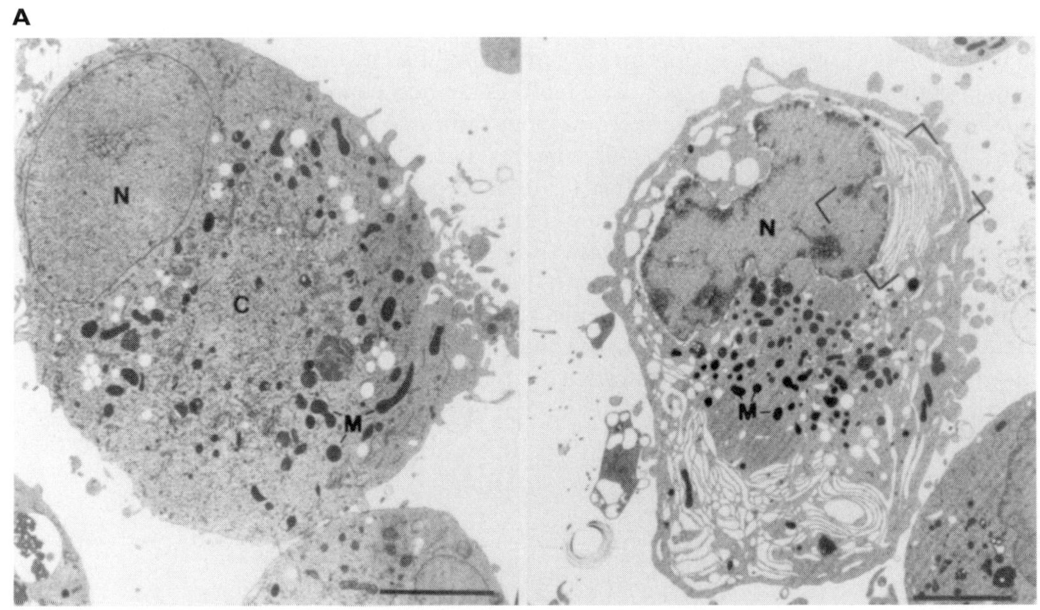

B

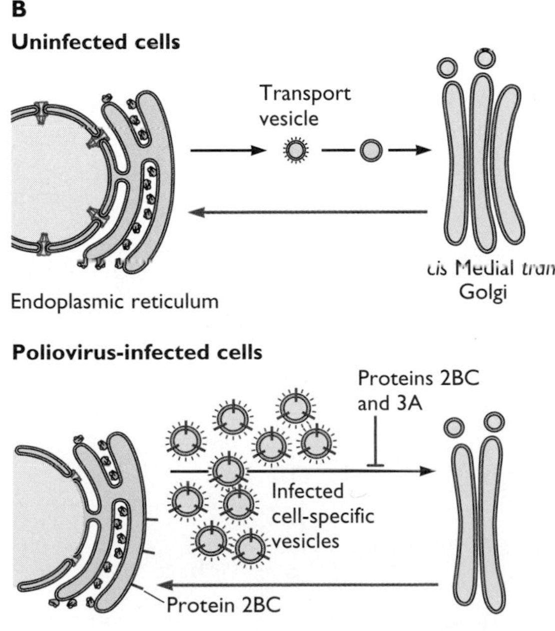

Figure 12.16 Inhibition of the cellular secretory pathway in poliovirus-infected cells. (A) Electron micrographs of monkey cells following introduction of a control plasmid (left) and a plasmid from which the poliovirus 3A protein is expressed (right). Frozen cells were fixed and negatively stained at low temperature prior to sectioning and electron microscopy. Bar = 4 μm. Note the high concentration of membranes of the ER and Golgi apparatus in the center of the cell (C) that does not express the 3A protein (left) and their replacement by dilated membranes with tubular and whorled morphologies when the 3A protein is synthesized (right). N, nucleus; M, mitochondria. From J. R. Doedens et al., *J. Virol.* **71**:9054–9064, 1997, with permission. Courtesy of K. Kirkegaard, Stanford University School of Medicine. (B) Model for formation of poliovirus-induced vesicles and inhibition of vesicular transport from the ER to the cell surface. The poliovirus 2BC protein accumulates in patches on the cytoplasmic face of the ER membrane and recruits the COP II coat protein to induce formation of vesicles by the normal cellular budding mechanism. These infected-cell-specific vesicles retain COP II coats, and do not fuse with *cis*-Golgi cisternae. They assemble to form poliovirus replication complexes. The redirection of ER membranes inhibits transport to the cell surface via the secretory pathway.

Each of the poliovirus 2BC, 2B, and 3A proteins can inhibit vesicular transport from the ER, when synthesized in mammalian cells in the absence of other viral proteins. All these viral proteins are found associated with the cytoplasmic membranes of the infected-cell-specific vesicles. However, only the 2BC protein induces formation of vesicles with the physical properties of those from virus-infected cells. Such vesicles, like those produced early during poliovirus infection, lack resident membrane proteins of the ER and Golgi, but carry COP II coat proteins on their surfaces, as well as the viral 2BC protein. This viral protein, which is bound to the ER membrane before vesicle budding begins, is believed to recruit COP II proteins and induce vesicle formation (Fig. 12.16B). Although formed by the normal cellular mechanism, such poliovirus-induced vesicles do not fuse with Golgi cisternae, perhaps because they do not lose their COP II coats. Later in infection, additional cytoplasmic membranes may contribute to vesicles of the poliovirus replication complex, which have been reported to acquire properties of autophagic vesicles formed in uninfected cells in response to nitrogen or amino acid starvation.

Signal Sequence-Independent Transport of Viral Proteins to the Plasma Membrane

As discussed in Chapter 4, many enveloped viruses contain proteins lying between, and making contact with, the inner surface of the membrane of the particle and the capsid or nucleocapsid. In contrast to the integral membrane proteins of enveloped viruses, such internal virion proteins do not enter the secretory pathway, but are synthesized in the cytoplasm of an infected cell and directed to membrane assembly sites by specific signals. The membrane-targeting signals of some viral proteins are built into their primary sequences and are therefore analogous to the nuclear localization signals or signal sequences described in previous sections. However, in many cases, posttranslationally added lipid chains are important components of membrane-targeting signals.

Lipid-Plus-Protein Signals

It has been known for many years that cytoplasmic proteins can be modified by the covalent addition of lipid chains (Table 12.5). Best characterized are the addition of the 14-carbon, saturated fatty acid myristate to N-terminal glycine residues, or of unsaturated polyisoprenes, such as farnesol (C_{15}) or geranylgeranol (C_{20}), to a specific C-terminal sequence (Fig. 12.17). Palmitate is also added to some viral proteins that do not enter the secretory pathway. The discovery that transforming proteins of oncogenic retroviruses, the Src and Ras proteins, are myristoylated and isoprenylated, respectively, led to a resurgence of interest in these modifications. As discussed in Chapter 18, lipid addition is essential for the function of these transforming proteins. In this section,

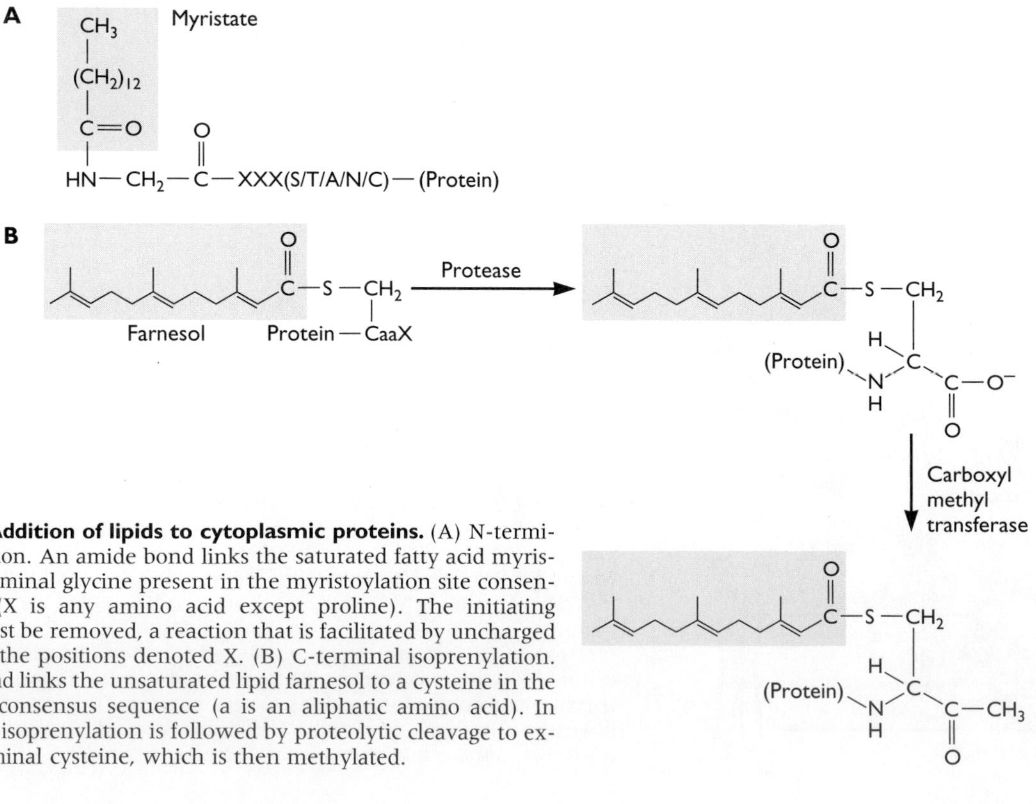

Figure 12.17 Addition of lipids to cytoplasmic proteins. (A) N-terminal myristoylation. An amide bond links the saturated fatty acid myristate to an N-terminal glycine present in the myristoylation site consensus sequence (X is any amino acid except proline). The initiating methionine must be removed, a reaction that is facilitated by uncharged amino acids in the positions denoted X. (B) C-terminal isoprenylation. A thioether bond links the unsaturated lipid farnesol to a cysteine in the isoprenylation consensus sequence (a is an aliphatic amino acid). In many proteins, isoprenylation is followed by proteolytic cleavage to expose the C-terminal cysteine, which is then methylated.

we focus on myristoylation and isoprenylation of viral structural proteins.

Myristoylation of the cytoplasmic Gag proteins of retroviruses and its consequences have been examined in most detail. The internal structural proteins of these viruses, MA (matrix), CA (capsid), and NC (nucleocapsid), are produced by proteolytic cleavage of the Gag polyprotein following virus assembly. The Gag proteins of the majority of retroviruses are myristoylated at their N-terminal glycines. Mutations that alter the sequence at which murine leukemia virus or human immunodeficiency virus type 1 Gag proteins are myristoylated prevent interaction of the protein with the cytoplasmic face of the plasma membrane, induce cytoplasmic accumulation of Gag, and inhibit virus assembly and budding. In the case of the human immunodeficiency virus type 1 Gag protein, the myristoylated N-terminal segment and a highly basic sequence located a short distance downstream (Fig. 12.18) form a bipartite signal, which allows membrane binding in vitro and virus assembly and budding in vivo.

The hepatitis B virus large surface (L) protein is also myristoylated at its N terminus. However, in contrast to retroviral Gag, the L protein is present in the envelope of the virion. Modification of its N terminus must therefore occur while it traverses the secretory pathway. In this case, myristoylation is not necessary for assembly or release of virions, but is required for infection of primary hepatocytes, presumably because it contributes to the initial interaction of the virus with, or its entry into, the host cell.

More surprising is the myristoylation of structural proteins of poliovirus (VP4) and polyomavirus (VP2): although these virions do not contain an envelope, such modification is necessary for efficient assembly of both viruses. In mature poliovirus particles, the myristate chain at the N terminus of VP4, which is processed from VP0, interacts with the VP3 protein on the inside of the capsid (Fig. 4.10B). The hydrophobic lipid chain must therefore facilitate protein-protein interactions necessary for the assembly of virions. The fatty acid is also important during entry into cells of poliovirus particles and their uncoating at the beginning of an infectious cycle.

Among viral structural proteins, only the large delta protein of the hepatitis delta satellite virus has been found to be isoprenylated. Formation of the particles of this satellite virus depends on structural protein provided by the helper virus, hepatitis B virus. The isoprenylation of large delta protein is necessary, but not sufficient, for its binding of the hepatitis B virus S protein during assembly of the satellite. This hydrophobic tail of large delta protein seems likely to facilitate interaction with the plasma membrane adjacent to the helper envelope of S protein in cells infected by the two viruses.

The addition of lipid chains allows the association of virion proteins with the cytoplasmic face of the plasma membrane, providing opportunities for interactions with viral membrane glycoproteins. Therefore, we can now explain (at least in principle) how cytoplasmic structural proteins, such as the Gag and Gag-Pol polyproteins of most retroviruses, associate with the plasma membrane. It is also clear that such lipid chains can do more than simply provide a "greasy foot" for membrane binding: they can stabilize virions, facilitate virus-host cell interactions, and participate in uncoating reactions.

Figure 12.18 Human immunodeficiency virus type 1 Gag proteins and their targeting signals. The locations of the internal structural proteins MA (matrix), CA (capsid), and NC (nucleocapsid) in the Gag polyprotein are shown at the top. Sequence features, localization signals (MA), and the RNA-binding domain (NC) are shown below. The lengths of the MA and NC proteins are listed as approximate because of the variation among virus isolates. Specific amino acids are given in the single-letter code in the boxes, and a plus sign indicates a basic amino acid. The basic region of MA of simple retroviruses, such as avian sarcoma virus, is not required for membrane binding. The CH boxes of NC contain three cysteines and one histidine, and each coordinates one atom of Zn. CH box I is conserved among retroviruses, but CH box II is not.

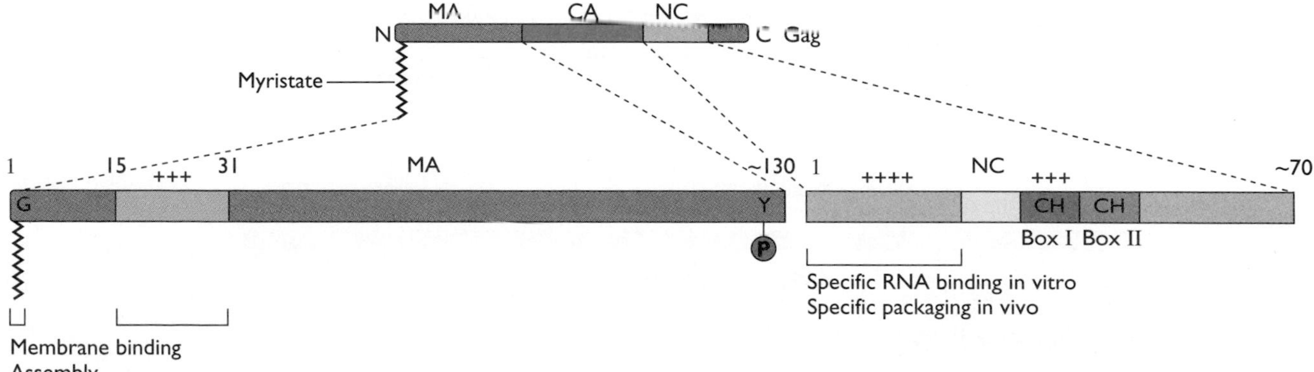

Protein Sequence Signals

The matrix proteins of several families of (–) strand RNA viruses lie between the nucleocapsid and the envelope and are essential for correct localization and packaging of RNA genomes. During assembly, matrix proteins, such as M of vesicular stomatitis virus and M1 of influenza A virus, must bind to the plasma membrane of infected cells. These proteins are produced in the cytoplasm but, as far as is known, receive no lipid after translation. Direct membrane binding in vivo appears to be an intrinsic property of matrix proteins. For example, when the influenza virus M1 protein is synthesized in host cells in the absence of any other viral proteins, a significant fraction is tightly associated with cellular membranes. Both this protein and the vesicular stomatitis virus M protein contain specific sequences that are necessary for their interaction with the plasma membrane in vivo or with lipid vesicles in vitro. This region of the influenza A virus M1 protein contains two hydrophobic sequences (Fig. 12.19A), which might form a hydrophobic surface structure in the folded protein. In addition to an internal hydrophobic segment, membrane association of the vesicular stomatitis virus M protein requires a basic N-terminal sequence (Fig. 12.19B). This latter segment might participate directly in membrane binding, like the basic sequence of the human immunodeficiency virus type 1 Gag membrane-targeting signal, or it might stabilize a conformation of the internal sequence favorable for interaction with the membrane.

Assembly of (–) strand RNA viruses requires association of their matrix proteins with the plasma membrane. In some cases, such as the vesicular stomatitis virus M protein, specificity for the plasma membrane is an intrinsic property. However, binding of matrix proteins to the cytoplasmic tails of viral envelope glycoproteins can also be an important determinant of membrane association. The cytoplasmic domains of both the NA and HA proteins of influenza A virus stimulate membrane binding by M1 protein. Similarly, membrane binding by the matrix protein of Sendai virus (a paramyxovirus) is independently stimulated by the presence of either of the two viral glycoproteins (F or HN) in the membrane.

Interactions with Internal Cellular Membranes

The envelopes of a variety of viruses are acquired from internal membranes of the infected cell, rather than from the plasma membrane. The majority assemble at the cytoplasmic faces of compartments of the secretory pathway (Table 12.7). Although a single budding reaction is typical, the more complex herpesviruses and poxviruses interact with multiple internal membranes of the cell during assembly and exocytosis. Even some viruses with mature forms that lack an envelope, such as rotaviruses, can enter the secretory pathway transiently.

The diversity of the internal membranes with which these viruses associate during assembly and exocytosis is the result of variations on a single mechanistic theme: the site of assembly is determined by the intracellular location

Figure 12.19 Targeting signals of matrix proteins of influenza virus (A) and vesicular stomatitis virus (VSV) (B). Sequence features of specific segments of the proteins and the boundaries of targeting and RNP-binding domains are indicated. Amino acids are indicated in the one letter code, and a plus sign indicates a basic amino acid. NLS, nuclear localization signal.

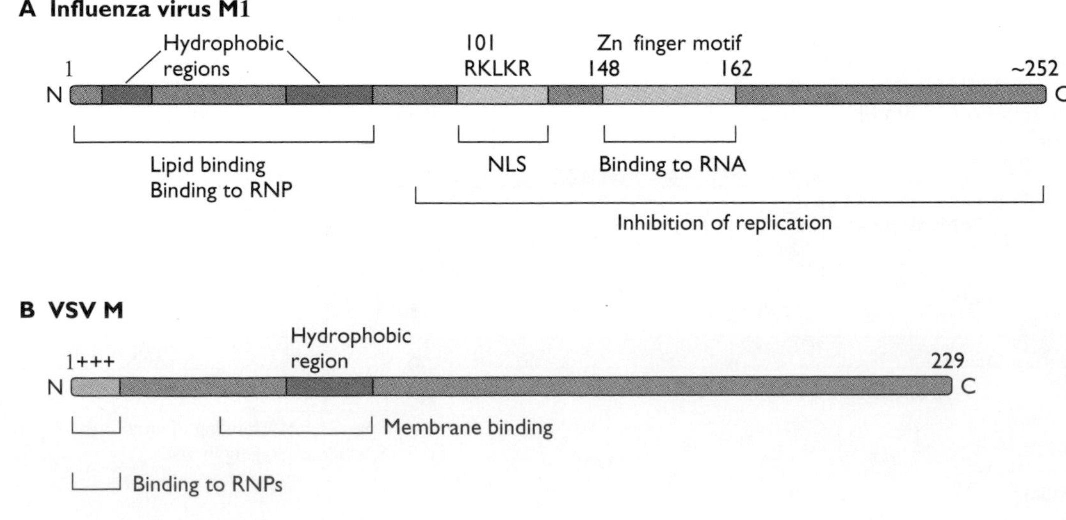

Table 12.7 Interactions of viruses with internal cellular membranes

Virus family	Example	Integral membrane protein(s)	Intracellular membrane(s)	Mechanism of envelopment
Bunyaviruses	Uukuniemi virus, Hantaan virus	G1, G2	*cis*- and medial Golgi cisternae	Budding into Golgi cisternae
Coronavirus	Mouse hepatitis virus	M, S	ER, *cis*-Golgi network	Budding into ER and *cis*-Golgi network
Hepadnavirus	Hepatitis B virus	L, M, S	ER and other compartments	Budding into ER?
Herpesvirus	Herpes simplex virus type 1	UL31, UL34	Nuclear membrane	Primary envelopment; budding of capsids from inner nuclear membrane
		gE-gI	*trans*-Golgi cisternae	Budding at *trans*-Golgi membrane
Poxvirus	Vaccinia virus, immature	A14L, A13L, A17L	*cis*-Golgi cisternae	Engulfment of particle
	Vaccinia virus, intracellular mature	A56R (HA), F13L, B5R	Late *trans*-Golgi cisternae and post-Golgi vesicles	Engulfment to form structure with four membranes
Rotavirus	Simian rotavirus	VP7, NS28	ER	Budding into ER
Togavirus	Rubella virus	E1-E2	Medial and *trans*-Golgi cisternae	Budding into Golgi cisternae

of viral envelope proteins (Fig. 12.20), just as assembly at the plasma membrane is the result of transport of such proteins to that site. Assembly of viruses at internal membranes therefore requires transport of envelope proteins to, and their retention within, appropriate intracellular compartments.

Localization of Viral Proteins to Compartments of the Secretory Pathway

The bunyaviruses, a family that includes Uukuniemi virus and Hantaan virus, are among the best-studied viruses that assemble by budding into compartments of the secretory pathway. Bunyavirus particles contain two integral

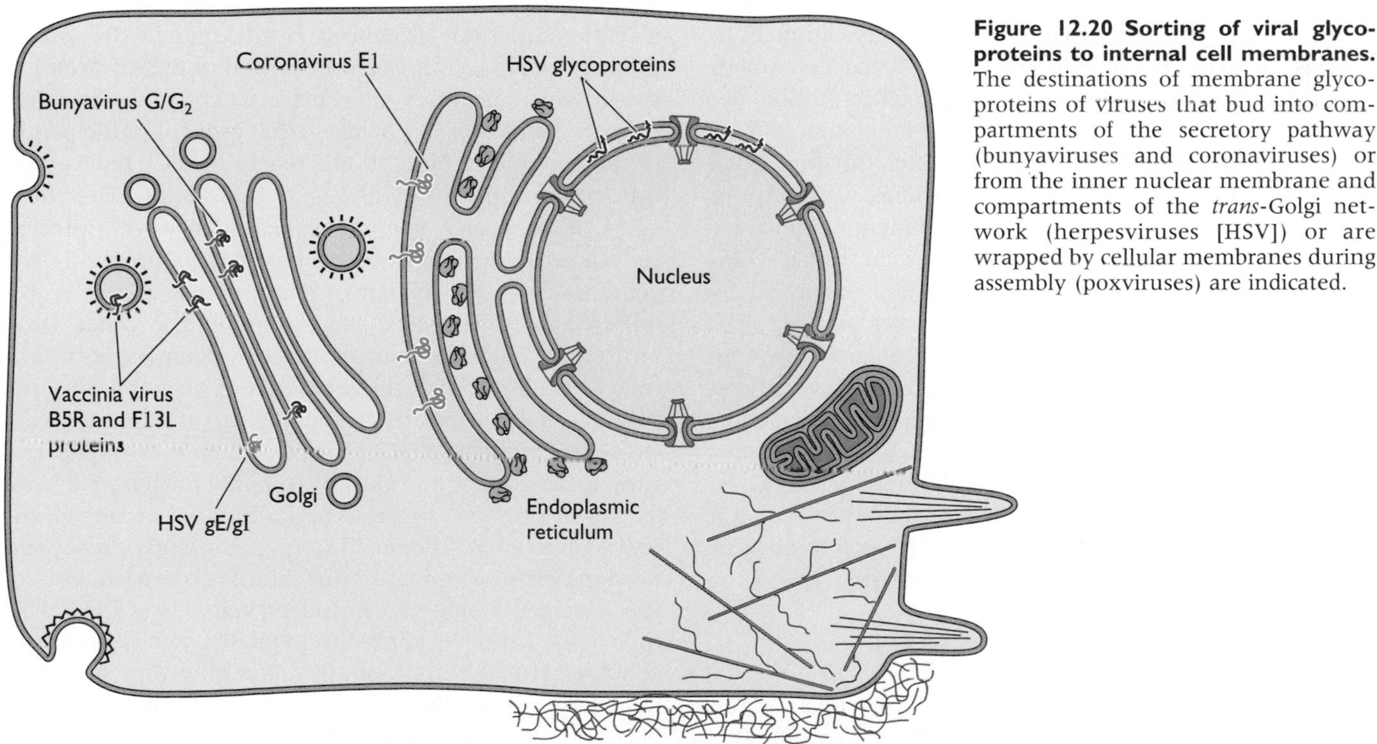

Figure 12.20 Sorting of viral glycoproteins to internal cell membranes. The destinations of membrane glycoproteins of viruses that bud into compartments of the secretory pathway (bunyaviruses and coronaviruses) or from the inner nuclear membrane and compartments of the *trans*-Golgi network (herpesviruses [HSV]) or are wrapped by cellular membranes during assembly (poxviruses) are indicated.

membrane glycoproteins, called G1 and G2, which are encoded within a single open reading frame of the M genomic RNA segment. Like alphaviral envelope proteins, the bunyaviral polyprotein containing G1 and G2 is processed cotranslationally by signal peptidase as the precursor enters the lumen of the ER. However, association of G1 with G2 is required for transport to, and retention in, Golgi compartments. For example, when synthesized alone, the G1 protein accumulates in the Golgi complex in normal fashion, but the G2 protein fails to leave the ER. When both glycoproteins are made, the G2 protein is now transported to the Golgi complex. The G1 protein may therefore contain signals necessary for transport along the secretory pathway, or it may be required for correct folding of the G2 protein and its exit from the ER. Both visualization of the intracellular distribution of the G1 and G2 proteins, and the structures of their N-linked oligosaccharides, indicate that they do not reach the *trans*-Golgi network, but accumulate in *cis* or medial compartments. The signals that specify such locations have not been identified precisely, but are included within the transmembrane domain and the adjacent segment of the cytoplasmic domain of the G1 protein.

Golgi cisternae are by no means the only compartments of the secretory pathway at which virus budding can occur. For example, rotaviruses assemble at and bud into the ER, but they do not retain the envelope, whereas coronaviruses bud into both the ER and the *cis*-Golgi network. In both cases, it is the presence of viral membrane glycoproteins in specific cellular membranes (Table 12.7) that determines the site of assembly and budding. As discussed in more detail in Chapter 13, vaccinia virus acquires membranes from both the *cis*-Golgi network and a late Golgi or post-Golgi compartment (Table 12.7). This virus does not bud into these cisternae, but rather induces engulfment of immature particles by their membranes, a process often called wrapping. Consequently, unlike enveloped viruses formed by budding, which acquire a typical lipid bilayer, vaccinia virus particles acquire two membranes at each wrapping step. These reactions are not fully understood, but several viral membrane proteins that accumulate in specific Golgi compartments and are required for the first or second membrane-wrapping reaction have been identified (Table 12.7). As for simpler viruses, the presence of these proteins in a specific intracellular membrane is the first step in acquisition of that membrane by the assembling virus.

Localization of Viral Proteins to the Nuclear Membrane

Herpesviruses such as herpes simplex virus type 1 are the only enveloped viruses that are known to assemble initially within, and bud from, the nucleus. The first association of an assembling herpesvirus with a cellular membrane is therefore budding of the nucleocapsid through the inner membrane of the nuclear envelope. The dense network of protein filaments (the nuclear lamina) that underlies the inner nuclear membrane is disrupted in herpes simplex virus type 1-infected cells, presumably to allow close association of assembled nucleocapsids with the membrane. Budding from this nuclear membrane takes place at morphologically distinct sites, which appear as thickened, curved regions of the membrane. As the outer nuclear membrane is continuous with that of the ER (Fig. 12.20), the nucleocapsid enters the ER lumen upon budding through the inner nuclear membrane. This reaction requires the viral UL31 and UL34 proteins, both of which are present in intracellular particles, but absent from mature virions. The UL34 gene product accumulates in the inner and outer nuclear membranes, a property specified by as yet unidentified signals, and is necessary to localize the UL31 phosphoprotein to the nuclear membrane. In the absence of either the UL34 or the UL31 protein, nucleocapsids are trapped within the nucleus, probably in part because these viral proteins induce disruption of the network of lamin proteins that lines the inner nuclear membrane (Box 12.6). Whether other viral proteins participate in this unusual process is not yet known.

Transport of Viral Genomes to Assembly Sites

Like the structural proteins and enzymes of the virion, progeny genomes must be transported to the intracellular site at which assembly takes place. Such transport is frequently coupled to mechanisms that prevent participation of the genome in biosynthetic reactions, i.e., replication, transcription, mRNA synthesis, or translation. The packaging of most DNA genomes takes place in the infected cell nucleus. In these cases, no transport of progeny DNA genomes from one cellular compartment to another is required to set the stage for assembly. For other DNA viruses, as well as for many with RNA genomes, both synthesis of the nucleic acid genome and assembly of virus particles take place in the cytoplasm of the host cell. However, many of these viruses assemble at membrane sites, necessitating transport of genomic nucleic acid from its site of synthesis to the cytoplasmic face of the appropriate membrane. The membranes of both poxviruses and hepadnaviruses originate from internal compartments of the host cell (Table 12.7), and proteins associated with their DNA genomes appear to mediate association of the genome with the appropriate membrane. For example, the hepadnaviral DNA genome is synthesized in the cytoplasm from the pregenomic RNA within cores containing the viral C protein. When cells infected by duck hepatitis

BOX 12.6

Disruption of the nuclear lamina: a general mechanism for escape of herpesviral capsids from the nucleus

The nuclear lamina is a dense protein network (some 50 nm thick) that abuts and stabilizes the inner nuclear envelope. It is built from intermediate filament-like proteins termed lamins, which bind to integral membrane proteins of the inner nuclear membrane. This structure, which is normally disassembled only during mitosis upon phosphorylation of lamins, represents a major obstacle during transport of herpesviral particles from the nucleus to the cytoplasm.

The murine cytomegalovirus counterparts of the herpes simplex virus type 1 UL34 and UL31 proteins (see text), called M50/p35 and M53/p38, respectively, are sufficient to induce the alterations in the nuclear envelope characteristic of virus-infected cells. These changes include an irregular staining pattern for the lamins. The M50/p35 viral protein induces both association of protein kinase C with the inner nuclear membrane and phosphorylation of lamins A, B and C. Herpes simplex virus type 1 infection has been shown to lead to increased solubility of lamins, and alterations in the lamin B receptor (although whether the UL34 and UL31 proteins induce these changes is not yet known). These observations suggest that disruption of the nuclear lamina and hence modification of the inner nuclear membrane is a general mechanism that facilitates primary envelopment of herpesviral nucleocapsids.

Muranyi, W., et al. 2002. Cytomegalovirus recruitment of cellular kinases to dissolve the nuclear lamina. *Science* **297:**854–857.

Scott, E. S., and P. O'Hare. 2001. Fate of the inner nuclear membrane protein lamin B receptor and nuclear lamins in herpes simplex virus type 1 infection. *J. Virol.* **75:**8818–8830.

B virus contain a sufficient concentration of the large surface (L) glycoprotein in the ER membrane, core and L proteins can bind to one another, bringing the genome to the appropriate cell membrane and allowing budding of particles into the ER. Similarly, many RNA genomes made in the cytoplasm must be transported to the plasma membrane for assembly of virions. Yet other RNA genomes must travel even further: both influenza virus and retroviral genomic RNAs are synthesized within the infected cell nucleus, but progeny virions bud from the plasma membrane.

Transport of Genomic and Pregenomic RNA from the Nucleus to the Cytoplasm

The transcripts of integrated retroviral proviruses are processed in the infected cell nucleus to produce both unspliced and spliced mRNA species. The unspliced RNA is also the retroviral genome. As discussed in Chapter 10, the inefficient export of unspliced mRNAs characteristic of host cells must therefore be circumvented. In the case of complex retroviruses, sequence-specific viral RNA-binding proteins that promote export of unspliced RNA, such as human immunodeficiency virus type 1 Rev protein, overcome this limitation. Simple retroviruses also rely on specific sequences for the export of genomic RNA, but these are recognized by cellular proteins. The pregenomic RNA of hepadnaviruses is also synthesized by cellular RNA polymerase II within infected cell nuclei. The export of this unspliced viral RNA is also directed by RNA signals that are most likely recognized by cellular proteins. Despite these different strategies for efficient transport of unspliced genomic RNA to the cytoplasm, subsequent transport of the RNA to the plasma membrane (described below) occurs by similar mechanisms.

Perhaps the most elaborate requirements for transport of viral RNA species between nucleus and cytoplasm are found in influenza virus-infected cells, in which both the direction of transport of genomic RNA and the nature of the viral RNA exported from the nucleus change as the infectious cycle progresses. When the cycle is initiated, viral genomic ribonucleoproteins (RNPs) released into the cytoplasm enter the nucleus under the direction of the nuclear localization signal of the NP protein. The mechanisms that ensure export for translation of viral (+) strand mRNAs are not fully understood. With the switch to replication, genomic (–) strand RNA segments are synthesized in infected cell nuclei and accumulate as vRNPs containing the NP protein and the three P proteins. These RNPs must be exported to allow virus assembly and completion of the infectious cycle, a reaction that requires the viral M1 and NEP proteins. The former is the most abundant protein of the virus particle and enters the nucleus by means of a typical nuclear localization signal (Table 5.3). Binding of M1 to genomic RNPs (Fig. 12.19A) both inhibits RNA synthesis and promotes genomic RNP export. The NEP, which contains an N-terminal nuclear export signal, appears to direct genomic RNPs to components of the export machinery. The NEP protein possesses no intrinsic RNA-binding activity, but includes a C-terminal M1-binding domain. This domain is believed to allow recognition of RNPs to which the M1 protein is bound. Because the M1 protein is not available in large quantities until the later stages of infection, such an indirect recognition mechanism probably ensures that genomic RNPs are not exported from the nucleus before sufficient quantities of viral mRNA and progeny genomic RNA have been synthesized. Furthermore, the RNPs enter the cytoplasm only when they have associated with the protein (M1) that is necessary for guiding them to the plasma membrane. The reimport of vRNPs into the nucleus via the NLS of the NP protein appears to be prevented by tight binding of the NP protein to actin filaments.

Transport of Genomes from the Cytoplasm to the Plasma Membrane

Like nuclear export, transport of the RNA genomes of enveloped viruses to the appropriate cellular membrane depends on signals present in viral proteins bound to the RNA. As we have seen, the influenza virus M1 protein contains an N-terminal sequence that allows it to bind to lipids (Fig. 12.19A). This membrane-binding domain of M1 allows association of the genomic RNPs with the plasma membrane. Although more limited in its RNA transport functions, the M protein of vesicular stomatitis virus shares several properties with the influenza virus M1 protein.

Replication of vesicular stomatitis virus RNA takes place in the cytoplasm, where newly synthesized genomic RNA molecules assemble with the N, L, and NS proteins to form helical RNPs. Genomic RNA molecules within RNPs can serve as templates for additional cycles of replication or for mRNA synthesis. However, these RNPs must eventually travel to the plasma membrane for association with the G protein and incorporation into virions. Entry into the latter pathway is determined by the viral M protein, which associates with RNPs containing genomic RNA (Fig. 12.19B). This interaction induces formation of a tightly coiled RNA-protein "skeleton," the final structure of the rhabdoviral nucleocapsid (Fig. 12.21). Immunoelectron microscopy suggests that the nucleocapsid is wound around a core built from the M protein. Formation of this structure precludes replication and mRNA synthesis and directs RNPs to the plasma membrane, by means of the membrane-binding domains of the M protein described above.

The retroviral proteins that mediate membrane association of genomic RNA are similar to the matrix proteins of these (–) strand RNA viruses in several respects. Once within the cytoplasm, unspliced retroviral RNA is translated on polyribosomes into the Gag and, at low frequency, Gag-Pol polyproteins. The functions of Gag (Fig. 12.18) include transport of unspliced RNA molecules to membrane assembly sites and packaging of the RNA into assembling capsids. Consequently, whether unspliced genomic RNA molecules continue to be translated, or are redirected for virion assembly, is controlled by the cytoplasmic concentration of Gag. Accumulation of Gag may

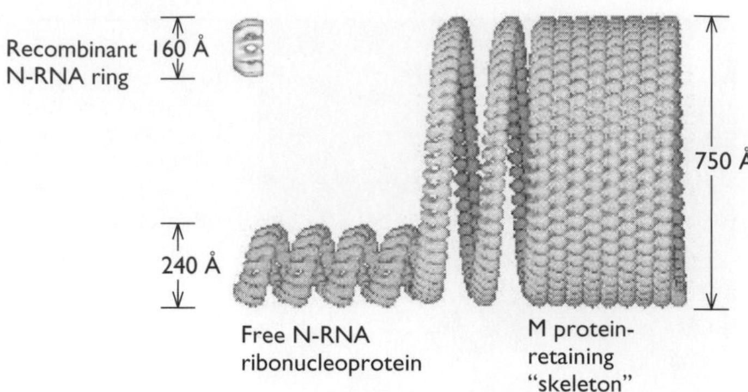

Figure 12.21 Models of the rabies virus nucleocapsid, showing the free nucleocapsid and the nucleocapsid present in virions. The models are based on cryo-electron microscopy and image reconstruction of the two forms of the nucleocapsid, as well as of rings of 9 or 10 molecules of the viral N protein and RNA assembled when the protein is produced in insect cells. The free nucleocapsid, which is the template for viral RNA synthesis, is a loosely coiled helix with a variable pitch and diameter of 240 Å. In contrast, the nucleocapsid helix incorporated into virus particles is tightly wound, with a small pitch and a much larger diameter. These structural transitions are induced by binding of the M protein to the free nucleocapsid. Adapted from G. Schoehn et al., *J. Virol.* **75**:490–498, 2001, with permission.

inhibit translation of the viral RNA genome directly: binding of the MA segment to the translational elongation protein Ef1a inhibits translation in vitro.

The NC (nucleocapsid) segment of Gag (Fig. 12.18) contains an RNA-binding domain required for specific recognition of the RNA-packaging sequences described in Chapter 13. The NC region, which functions as an independent protein in mature virions, is very basic and contains at least one copy of a zinc-binding motif (Fig. 12.18). This "zinc finger" domain makes a major contribution to the specificity with which Gag or NC proteins bind to unspliced retroviral RNA and, in conjunction with basic amino acids located nearby, is responsible for the RNA-packaging activity of Gag. However, the N-terminal MA portion of Gag contains the signals described above that target the polyprotein to the plasma membrane. Binding of Gag to unspliced retroviral RNA therefore allows delivery of the genome to assembly sites at the plasma membrane, just as binding of matrix proteins to influenza virus or vesicular stomatitis virus genomic RNPs directs these structures to the membrane.

The association of Gag polyprotein molecules with one another appears to be an additional prerequisite for transport to the membrane. Several regions in human immunodeficiency virus type 1 Gag, including the spacer between CA and NC and sequences within MA, CA, and NC (Fig. 12.18), are necessary for efficient binding of Gag molecules to one another. Altered proteins that lack such sequences are not assembled into virus particles, and accumulate at cytoplasmic sites. The rescue of such mutants by wild-type Gag indicates that interaction among Gag molecules is necessary not only for virus assembly but also for transport of the protein and its associated RNA to the plasma membrane.

The mechanisms by which the various viral RNPs containing genomic RNA species, RNA-packaging proteins, and membrane-targeting proteins travel to the plasma membrane are not well established. In principle, such transport could be the result of passive, random diffusion and retention of the RNPs at membrane sites, or of an active process in which the RNPs are carried along microtubules to the membrane by motor proteins. The former mechanism seems unlikely (Boxes 5.2 and 12.1), and there is accumulating evidence for active transport of structures containing Gag proteins. The Gag proteins of several retroviruses bind to a (+) end-directed microtubule motor of the kinesin family in vivo and/or in vitro.

The influenza virus M1, vesicular stomatitis virus M, and retroviral Gag proteins each possess the ability to bind both to RNPs containing genomic RNA and to membranes. Such interactions commit the genomic RNA to the assembly pathway and prevent its participation in biosynthetic processes. Despite their different sequences and mechanisms of membrane association, these three proteins share the ability to direct genomic RNA to the plasma membrane and to promote interactions among internal and envelope components of virions. These properties are essential at the end of an infectious cycle, when the primary task is assembly of progeny virions. On the other hand, they would be disastrous if the interactions could not be reversed before or at the beginning of a new cycle, when the infecting genome must reach nuclear (influenza virus) or cytoplasmic (vesicular stomatitis virus and retroviruses) sites distant from the plasma membrane. In the case of (–) strand viruses, matrix proteins are removed during virus entry. Matrix-free RNPs can then enter the nucleus for mRNA synthesis (influenza virus) or begin this process in the cytoplasm (vesicular stomatitis virus). Simple retroviruses exhibit a more elegant mechanism: following virus assembly and budding, Gag (and Gag-Pol) polyproteins are processed by the viral protease to the individual structural proteins listed in Fig. 12.18. Such cleavages place the RNA-binding domain of NC in protein molecules separate from membrane-binding signals of MA, so that matrix-free core RNPs can be released into the cell to initiate a new infectious cycle. Membrane association of internal components of human immunodeficiency virus type 1 appears to be controlled by a more subtle mechanism. The N-terminal signal responsible for association of the Gag polyprotein with the plasma membrane (Fig. 12.18) does not support membrane binding by mature MA. A C-terminal α-helix of mature MA sequesters the N-terminal myristate chain. Conformational changes in MA induced when the Gag polyprotein is cleaved therefore appear to release this protein from its membrane association.

Perspectives

The cellular trafficking systems described in this chapter are just as crucial for virus reproduction as the host cell's biosynthetic capabilities. The trafficking requirements during the infectious cycle can be quite complex, with transport of viral macromolecules (or structures built from them) over large distances, or in opposite directions during different periods. Assembly of progeny particles of all viruses depends on the prior sorting of virion components by at least one cellular trafficking system.

This property has provided important tools (viral proteins or nucleic acids synthesized in large quantities in infected cells) with which to study these essential cellular processes. Indeed, the fundamental principle of protein sorting, that a protein's final destination is dictated by specific signals within its amino acid sequence and/or covalently attached sugars or lipids, was established by analyses of viral proteins. Furthermore, the study of viral proteins that enter the secretory pathway has provided much of what we know of the reactions by which proteins are folded and processed within the ER, as well as those that clear misfolded proteins from the pathway. It therefore seems certain that viral systems will provide equally important insights into the signals and sorting mechanisms that are presently less well characterized, such as those responsible for the direction of proteins to specialized membrane regions of polarized cells or to the inner nuclear membrane.

One of the greatest current challenges in this field is elucidation of the mechanics of the movement of proteins, nucleic acids, nucleoproteins, or transport vesicles from one cellular compartment or site to another. With the exception of axonal transport in neurons, we really know very little about these processes. It is to be hoped that continued technical advances, such as the use of fluorescent proteins to visualize transport in living cells in real time, will lead to a clearer view of transport mechanics.

Some viruses interfere with the normal transport of specific cellular proteins to the cell surface or completely disrupt the secretory pathway. Such inhibitory consequences of virus infection can facilitate evasion of the host's immune defense systems, for example, because the display of viral antigens bound to major histocompatibility complex class I proteins on the surface of the infected cell is impaired or prevented. However, in most cases, transport of the components from which virus particles are built to assembly sites appears to have little detrimental effect on the host cell, despite the large quantities of viral cargo that must be handled. In all cases, the most important consequence of such transport is the formation within infected cells of microenvironments containing high concentrations of viral structural proteins and the nucleic acid genome, ideal niches for the assembly of progeny virions from their multiple parts.

References

Chapters in Books

Alberts, B., A. Johnson, J. Lewis, M. Raff, K. Roberts, and P. Walter. 2002. *Molecular Biology of the Cell,* 4th ed., p. 659–710. Garland Publishing Inc., New York, N.Y.

Alberts, B., A. Johnson, J. Lewis, M. Raff, K. Roberts, and P. Walter. 2002. *Molecular Biology of the Cell,* 4th ed., p. 711–766. Garland Publishing Inc., New York, N.Y.

Cooper, G. M. 2000. *The Cell: a Molecular Approach,* 2nd ed., p. 347–386. ASM Press, Washington, D.C., and Sinauer Associates, Sunderland, Mass.

Reviews

Allan, B. A., and W. E. Balch. 1999. Protein sorting by directed maturation of Golgi compartments. *Science* **285:**63–66.

Braakman, I., and E. van Anken. 2000. Folding of viral envelope viral glycoproteins in the endoplasmic reticulum. *Traffic* **1:**533–539.

Chen, Y. A., and R. H. Scheller. 2001. SNARE-mediated membrane fusion. *Nat. Rev. Mol. Cell. Biol.* **2:**98–106.

Drubin, D. G., and W. J. Nelson. 1996. Origins of cell polarity. *Cell* **84:**335–344.

Ellgard, L., M. Molinari, and A. Helenius. 1999. Setting the standards: quality control in the secretory pathway. *Science* **286:**1882–1888.

Gahmberg, C. G., and M. Tolvanen. 1996. Why mammalian cell surface proteins are glycoproteins. *Trends Biochem. Sci.* **21:**308–311.

Griffiths, G., and P. Rottier. 1993. Cell biology of viruses that assemble along the biosynthetic pathway. *Semin. Cell Biol.* **3:**367–381.

Hruby, D. E., and C. A. Franke. 1993. Viral acylproteins: greasing the wheels of assembly. *Trends Microbiol.* **1:**20–25.

Ikonen, E. 2001. Roles of lipid rafts in membrane transport. *Curr. Opin. Cell Biol.* **13:**470–477.

Kluperman, J. 2000. Transport between ER and Golgi. *Curr. Opin. Cell Biol.* **12:**445–449.

Matlack, K. E., W. Mothes, and T. A. Rappaport. 1998. Protein translocation: tunnel vision. *Cell* **92:**381–390.

Mostov, K. E., M. Verges, and Y. Altschuler. 2000. Membrane traffic in polarized epithelial cells. *Curr. Opin. Cell Biol.* **12:**483–490.

Pelham, H. R. B. 2001. Traffic through the Golgi apparatus. *J. Cell Biol.* **155:**1099–1101.

Pelham, H. R. B., and J. E. Rothman. 2000. The debate about transport in the Golgi—two sides of the same coin? *Cell* **102:**713–719.

Ploubidou, A., and M. Way. 2001. Viral transport and the cytoskeleton. *Curr. Opin. Cell Biol.* **13:**97–105.

Rothman, J. E., and T. H. Sollner. 1997. Throttles and dampers: controlling the engine of membrane fusion. *Science* **276:**1212–1213.

Schatz, G., and B. Dobberstein. 1996. Common principles of protein translocation across membranes. *Science* **271:**1519–1528.

Tomishima, M. J., G. A. Smith, and L. W. Enquist. 2001. Sorting and transport of alpha herpesviruses in axons. *Traffic* **2:**429–436.

Traub, L. M., and S. Kornfeld. 1997. The trans-Golgi network: a late secretory sorting station. *Curr. Opin. Cell Biol.* **9:**527–533.

Papers of Special Interest

Import of Viral Proteins into the Nucleus

Forstova, J., N. Krauzewicz, S. Wallace, A. J. Street, S. M. Dilworth, S. Beard, and B. E. Griffin. 1993. Cooperation of structural proteins during late events in the life cycle of polyomavirus. *J. Virol.* **67:**1405–1413.

Lombardo, E., J. C. Ramirez, M. Agbandje-McKenna, and J. M. Almendral. 2000. A beta-stranded motif drives capsid protein oligomers of the parvovirus minute virus of mice into the nucleus for viral assembly. *J. Virol.* **74:**3804–3814.

Zhao, L. J., and R. Padmanabhan. 1988. Nuclear transport of adenovirus DNA polymerase is facilitated by interaction with preterminal protein. *Cell* **55:**1005–1015.

Transport of Viral and Cellular Proteins via the Secretory Pathway

Balch, W. E., W. G. Dunphy, W. A. Braell, and J. E. Rothman. 1984. Reconstitution of the transport of proteins between successive compartments of the Golgi measured by the coupled incorporation of N-acetyl glucosamine. *Cell* **39:**405–416.

Chackerjan, B., L. M. Rudensey, and J. Overbaugh. 1997. Specific N-linked and O-linked glycosylation modifications in the envelope V1 domain of simian immunodeficiency virus variants that evolve in the host alter recognition by neutralizing antibodies. *J. Virol.* **71:**7719–7727.

Hammond, C., and A. Helenius. 1994. Folding of VSV G protein: sequential interaction with BiP and calnexin. *Science* **266:**456–458.

Horimoto, T., K. Nakayama, S. P. Smeekens, and Y. Kawaoka. 1994. Preprotein-processing endoproteases PC6 and furin both activate hemagglutinin of virulent avian influenza viruses. *J. Virol.* **68:**6074–6078.

Li, Z., A. Pinter, and S. C. Kayman. 1997. The critical N-linked glycan of murine leukemia virus envelope protein promotes both folding of the C-terminal domains of the precursor polypeptide and stability of the postcleavage envelope complex. *J. Virol.* **71:**7012–7019.

Manié, S. N., S. Debreyne, S. Vincent, and D. Gerlier. 2000. Measles virus structural components are enriched into lipid raft microdomains: a potential cellular location for virus assembly. *J. Virol.* **74:**305–311.

Rothman, J. E., and H. F. Lodish. 1977. Synchronized transmembrane insertion and glycosylation of a nascent membrane protein. *Nature* **269:**775–780.

Rousso, I., M. B. Mixon, B. K. Chen, and P. S. Kim. 2000. Palmitoylation of the HIV-1 envelope glycoprotein is critical for viral infectivity. *Proc. Natl. Acad. Sci. USA* **97:**13523–13525.

Sapin, S., O. Colard, O. Delmas, C. Tessier, M. Breton, V. Enouf, S. Chwetzoff, J. Ouanich, J. Cohen, C. Wolf, and G. Trugnan. 2002. Rafts promote assembly and atypical targeting of a nonenveloped virus, Rotavirus, in Caco-2 cells. *J. Virol.* **76:** 4591–4602.

Scheiffele, P., M. G. Roth, and K. Simons. 1997. Interaction of influenza virus hemagglutinin with sphingolipid-cholesterol membrane domains via its transmembrane domain. *EMBO J.* **18:**5501–5508.

Takeuchi, K., and R. A. Lamb. 1994. Influenza virus M2 protein ion channel activity stabilizes the native form of fowl plague virus hemagglutinin during intracellular transport. *J. Virol.* **65:**911–919.

Transport of Viral Proteins to Intracellular Membranes

Alconada, A., U. B. B. Sodeik, and B. Hoflack. 1999. Intracellular traffic of herpes simplex virus glycoprotein gE: characterization of the sorting signals required for its *trans*-Golgi network localization. *J. Virol.* **73:**377–387.

Hobman, T. C., H. F. Lemon, and K. Jewell. 1997. Characterization of an endoplasmic reticulum retention signal in the rubella virus E1 glycoprotein. *J. Virol.* **71:**7670–7680.

Husain, M., and B. Moss. 2001. Vaccinia virus F13L protein with a conserved phospholipase catalytic motif induces colocalization of the B5R envelope glycoprotein in post-Golgi vesicles. *J. Virol.* **75:**7528–7542.

Law, L. M. J., R. Duncan, A. Esmaili, H. L. Nakhasi, and T. C. Hobman. 2001. Rubella virus E2 signal peptide is required for perinuclear localization of capsid protein and virus assembly. *J. Virol.* **75:**1978–1983.

Melin, L., R. Persson, A. Andersson, A. Bergstrom, R. Ronnholm, and R. F. Pettersson. 1995. The membrane glycoprotein G1 of Uukuniemi virus contains a signal for localization to the Golgi complex. *Virus Res.* **36:**49–66.

Salmons, T., A. Kuhn, F. Wylie, S. Schleich, J. R. Rodriguez, D. Rodriguez, M. Esteban, G. Griffiths, and J. K. Locker. 1997. Vaccinia virus membrane proteins p8 and p16 are cotranslationally inserted into the rough endoplasmic reticulum and retained in the intermediate compartment. *J. Virol.* **71:**7404–7420.

Virus-Induced Inhibition of Transport via the Secretory Pathway

Burgert, H. G., and S. Kvist. 1985. An adenovirus type 2 glycoprotein blocks cell surface expression of human histocompatibility class I antigens. *Cell* **41:**987–997.

Cho, M. W., N. Teterina, D. Egger, K. Bienz, and E. Ehrenfeld. 1994. Membrane rearrangement and vesicle induction by recombinant poliovirus 2C and 2BC proteins in human cells. *Virology* **202:**129–145.

Rust, R. C., L. Landmann, R. Gosert, B. L. Tang, W. Hong, H.-P. Hauri, D. Egger, and K. Bienz. 2001. Cellular COPII proteins are involved in production of the vesicles that form the poliovirus replication complex. *J. Virol.* **75:**9808–9818.

Schubert, U., L. C. Antón, I. Bačík, J. H. Cox, S. Bour, J. R. Bennink, M. Orlowski, K. Strebel, and J. W. Yewdell. 1998. CD4 glycoprotein degradation induced by human immunodeficiency virus type 1 Vpu protein requires the function of proteasomes and the ubiquitin-conjugating pathway. *J. Virol.* **72:**2280–2288.

Suhy, D. A., T. H. Giddings, Jr., and K. Kirkegaard. 2000. Remodeling the endoplasmic reticulum by poliovirus infection and by individual viral proteins: an autophagy-like origin for virus-induced vesicles. *J. Virol.* **74:**8953–8965.

Wiertz, E. J. H., T. R. Jones, L. Sun, M. Bogyo, H. J. Gueze, and H. L. Ploegh. 1996. The human cytomegalovirus US11 gene product dislocates MHC class I heavy chains from the endoplasmic reticulum to the cytosol. *Cell* **84:**709–779.

Transport of Viral RNA Genomes to Sites of Assembly

Chong, L. D., and J. K. Rose. 1994. Interactions of normal and mutant vesicular stomatitis virus matrix proteins with the plasma membrane and nucleocapsids. *J. Virol.* **68:**441–447.

Enami, M., and K. Enami. 1996. Influenza virus hemagglutinin and neuraminidase stimulate the membrane association of the matrix protein. *J. Virol.* **70:**6653–6657.

Martin, K., and A. Helenius. 1991. Nuclear transport of influenza virus ribonucleoproteins: the viral matrix protein (M1) promotes export and inhibits import. *Cell* **67:**117–130.

O'Neill, R. E., J. Tulon, and P. Palese. 1998. The influenza virus NEP (NS2 protein) mediates the nuclear export of viral ribonucleoproteins. *EMBO J.* **17:**288–296.

Tang, Y., U. Winkler, E. O. Freed, T. A. Torrey, W. Kim, H. Li, S. P. Goff, and H. C. Morse III. 1999. Cellular motor protein KIF-4 associates with retroviral Gag. *J. Virol.* **73:**10508–10513.

Ye, Z. P., R. Pal, W. Fox, and R. R. Wagner. 1987. Functional and antigenic domains of the matrix (M1) protein of influenza A virus. *J. Virol.* **61:**239–246.

Zhou, W., L. J. Parent, J. W. Wills, and M. D. Resh. 1994. Identification of a membrane-binding domain within the amino-terminal region of human immunodeficiency virus type 1 Gag protein which interacts with acidic phospholipids. *J. Virol.* **68:**2556–2569.

13

Assembly, Exit, and Maturation

The probability of formation of a highly complex structure from its elements is increased, or the number of possible ways of doing it diminished, if the structure in question can be broken down in a finite series of successively smaller substrates!

J. D. Bernal

The Origins of Life on Earth (A. I. Oparin, ed.), Pergamon, Oxford, United Kingdom, 1959

Introduction

Virus particles exhibit considerable diversity in size, composition, and structural sophistication, ranging from those comprising a single nucleic acid molecule and one structural protein to complex structures built from many different proteins and other components. Nevertheless, because of their unique mode of replication, all viruses must complete a common set of de novo assembly reactions to ensure reproductive success. These processes include formation of the structural units of the protective protein coat from individual protein molecules, assembly of the coat by interaction among the structural units, incorporation of the nucleic acid genome, and release of newly assembled progeny virions (Fig. 13.1). In many cases, formation of internal virion structures must be coordinated with acquisition of a cellular membrane into which viral proteins have been inserted, or additional maturation steps must be completed to produce infectious particles. Assembly of even the simplest viruses is therefore a remarkable process requiring considerable specificity in, and coordination among, each of multiple reactions. Furthermore, virus reproduction is successful only if each of the assembly reactions proceeds with reasonable efficiency, and if the overall pathway is irreversible. The diverse mechanisms by which viruses assemble represent powerful solutions to the problems imposed by de novo assembly. Indeed, infectious virus particles are produced in prodigious numbers with great specificity and efficiency.

The structure of a virus particle determines the nature of the reactions by which it is formed (Fig. 13.1), as well as the mechanisms by which it enters a new host cell, reproduces, and promotes pathogenesis within a host animal. For example, the exceptionally stable poliovirus survives passage through the stomach to replicate in the gut. Despite such virus-specific variations in structure and biological properties, all viruses must be metastable structures well suited for protection of the nucleic acid genome in extracellular environments. They must also be built in a way that allows their ready disassembly during

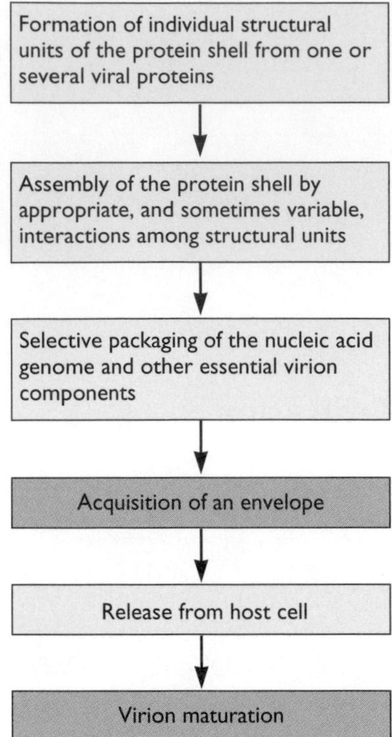

Figure 13.1 Hypothetical pathway of virion assembly and release. Reactions common to all viruses are shown in yellow, and those common to many viruses are shown in blue. The structural units that are often the first assembly intermediates are the homo- or hetero-oligomers of viral structural proteins from which virus particles are built (see Table 4.1). The arrows indicate a general sequence that strictly applies to only some viruses (see text). Packaging of the genome can be coordinated with assembly of the capsid or nucleocapsid and, for enveloped viruses, assembly of internal components can be coordinated with acquisition of the envelope.

entry into a new host cell. A number of elegant mechanisms resolve the apparently paradoxical requirements for very stable associations among virion components during assembly and transmission, but facile reversal of these interactions when appropriate signals are encountered upon infection of a host cell.

Like synthesis of viral nucleic acids and proteins, production of virus particles depends on host cell components, such as the cellular proteins that catalyze or assist the folding of individual protein molecules. Furthermore, the components from which virions are built are transported to the appropriate assembly site by cellular pathways (Chapter 12). Such localization of virion components to a specific intracellular compartment or region undoubtedly facilitates virus production: the concentration of virion components at an assembly site must increase the rates of reactions, for example, those by which structural units and protein shells are formed. The establishment of specialized microenvironments for assembly also seems likely to restrict the number of interactions in which a particular virion component can engage, thereby increasing the specificity of assembly reactions.

A successful virus-host interaction, the survival and propagation of a virus in a host population, generally requires dissemination of the virus beyond the cells infected upon initial contact. Progeny virus particles assembled during the later stages of the infectious cycle must therefore escape from the infected cell for transmission to new host cells within the same animal or to new host animals. The majority of viruses leave an infected cell by one of two general mechanisms: they are released into the external environment (upon budding from, or lysis of, the cell), or they are transferred directly into a new host cell.

Methods of Studying Virus Assembly and Egress

Mechanisms of virus assembly and release can be understood only with the integration of information obtained by structural, biochemical, and genetic methods of analysis. These methods are introduced briefly in this section.

Structural Studies of Virus Particles

The mechanisms by which viruses form within, and leave, their host cells are intimately related to their structural properties. Our understanding of these processes therefore improves dramatically whenever the structure of a virus particle is determined. An atomic-level description of the contacts among the structural units that maintain the integrity of the virus particle identifies the interactions that mediate assembly, and the ways in which these interactions must be regulated. For example, the X-ray crystal structure of the polyomavirus simian virus 40 particle described in Chapter 4 solved the enigma of how VP1 pentamers could be packed in hexameric arrays. Three distinct modes of interpentamer contact, mediated by conformationally flexible N- and C-terminal arms of VP1 subunits, were also identified. Assembly of the simian virus 40 capsid therefore must require specific variations in the way in which pentamers associate, depending on their position in the capsid shell. Such subtle, yet sophisticated, regulation of association of structural units during assembly was certainly not anticipated and could be revealed only by high-resolution structural information. Such information can also provide important insights into specific features of individual assembly pathways or the mechanisms that ensure that assembly is irreversible, as discussed for picornaviruses in subsequent sections.

The atomic structures of larger naked particles and enveloped viruses cannot yet be determined by X-ray crystallography. However, the newer methods for structural analysis, notably, cryo-electron microscopy and image analysis, are rapidly improving our understanding of the mechanisms by which these more complex virions assemble.

Visualization of Assembly and Exit by Microscopy

While high-resolution structural studies of purified virions or virion proteins provide a molecular foundation for describing virus assembly, they offer no clues about how assembly (or exit) actually proceeds in an infected cell. Electron microscopy can be applied to investigation of these processes. Examination of thin sections of cells infected by a wide variety of viruses has provided important information about intracellular sites of assembly, the nature of assembly intermediates, and mechanisms of **envelope** acquisition and release of particles. This approach can be particularly useful when combined with immunocytochemical methods for identification of individual viral proteins, or the structures they form, via binding of specific antibodies detached to electron-dense particles of gold.

The labeling of viral proteins by fusion with the green fluorescent protein (Chapter 2) allows direct visualization of assembly and egress, an approach inconceivable even a few years ago. Such chimeric proteins and virus particles containing them can be observed in living cells, and their associations and movements can therefore be recorded in real time by video microscopy. Consequently, these techniques overcome the limitations associated with traditional microscopic methods, which provide only static views of populations of proteins or virus particles.

Biochemical and Genetic Analysis of Assembly Intermediates

Although of great value, the information provided by X-ray crystallography or microscopy is not sufficient to describe the dynamic processes of virus assembly and release. An understanding of virus assembly requires identification of the intermediates in the pathway by which individual viral proteins and other virion components are converted to a mature infectious virion.

When extracts are prepared from the appropriate compartment of infected cells under conditions that preserve protein-protein interactions, a variety of viral assemblies can often be detected by techniques that separate on the basis of mass and conformation (velocity sedimentation in sucrose gradients or gel filtration) or of density (equilibrium centrifugation). These assemblies range from structural units of the capsid or nucleocapsid (see Table 4.1 for the definition of structural units) to empty capsids and mature virions. Similar methods have identified various subcomplexes formed by viral structural proteins in in vitro reactions. Furthermore, such structures can be organized into a sequence logical for assembly, from the least to the most complex. On the other hand, it is often quite difficult to **prove** that structures identified by these approaches, such as empty capsids, are true intermediates in the pathway.

By definition, the intermediates in any pathway do not accumulate, unless the next reaction is rate limiting. For this reason, assembly intermediates are generally present within infected cells at low concentrations against a high background of the starting material (mono- or oligomeric structural proteins) and the final product (virus particles). This property makes it difficult to establish precursor-product relationships by standard pulse-chase experiments: the large pools of structural proteins initially labeled are converted only slowly and inefficiently into subsequent intermediates in the pathway. Genetic methods of analysis provide one powerful solution to this problem. Mutations that confer temperature-sensitive or other phenotypes and block a specific reaction have been invaluable in the elucidation of assembly pathways. A specific intermediate may accumulate in mutant virus-infected cells, and can often be purified and characterized more readily. The judicious use of temperature-sensitive mutants can allow the reactions in a pathway to be ordered, and second-site suppressors of such mutations can identify proteins that interact with one another. Of even greater value is the combination of genetics with biochemistry, an elegant approach pioneered more than 30 years ago during studies of assembly of the complex bacteriophage T4 with the development of in vitro complementation (Box 13.1).

The difficulties inherent in kinetic analyses are compounded by the potential for formation of dead-end products and the unstable nature of some assembly intermediates. Dead-end assembly products are those that form by off-pathway (side) reactions. Because they are not true intermediates, they may accumulate in infected cells and be identified incorrectly as components in the pathway. By definition, such dead-end products differ from true intermediates in some structural property that prevents them from completing assembly. Furthermore, some authentic intermediates may be fragile structures, because they lack the complete set of intermolecular interactions that stabilize the virus particle. Less obvious, and recognized more recently, is the conformational instability of some intermediates: such structures do not fall apart during isolation and purification, but rather undergo irreversible conformational changes so that the structures studied experi-

BOX 13.1

Late steps in T4 assembly

As illustrated, the head, tail, and tail fibers of this morphologically complex bacteriophage first form separately and then assemble with one another. The many genes encoding products that participate in building the T4 particle are listed by the reaction for which they are required. These gene products and the order in which they act were identified by genetic methods, including identification of second-site suppressors of specific mutations (see Chapter 2). The development of in vitro systems in which specific reactions were reconstituted was also of the greatest importance, allowing the development of in vitro complementation. For example, noninfectious T4 particles lacking tail fibers accumulate in infected cells when the tail fiber pathway (right part of figure) is blocked by mutation. These incomplete particles can be converted to infectious phage when mixed in vitro with extracts prepared from cells infected with T4 mutated in the gene encoding the major "head" protein. The fact that the phages formed in vitro in this way were infectious established that assembly was accurate. This type of system was used to identify the genes encoding proteins required for assembly of heads or tails, as well as scaffolding proteins essential for assembly but not present in the virus particle. Adapted from W. B. Wood, *Harvey Lect.* **73**:203–223, 1978, and W. B. Wood et al., *Fed. Proc.* **27**:1160–1166, 1968, with permission.

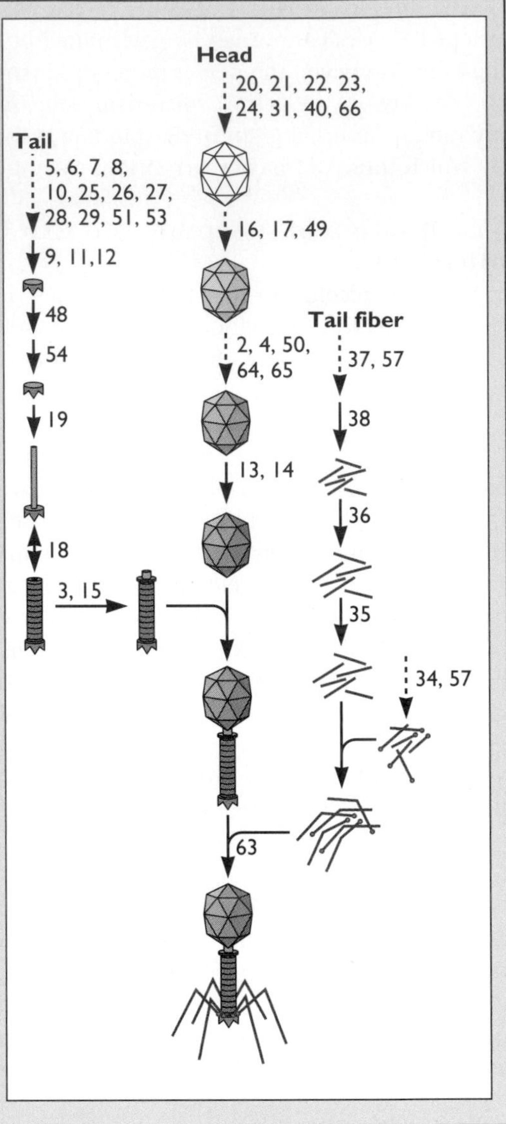

mentally do not correspond to **any** present in the infected cell. This kind of instability is not easy to detect because monoclonal antibodies that distinguish specific structural features, rather than a simple linear sequence of amino acids, are needed. Consequently, such conformational change may well escape notice, as was initially the case for poliovirus empty capsids.

Methods Based on Recombinant DNA Technology

Modern methods of molecular biology and the application of recombinant DNA technology have greatly facilitated the study of virus assembly. Especially valuable is the simplification of this complex process that can be achieved by the synthesis of an individual viral protein or small sets of proteins in the absence of other viral components (Box 13.2).

Assembly of Protein Shells

Although virus particles are far simpler in structure than any cell, they are built from multiple components, such as a capsid (or nucleocapsid), a nucleoprotein core containing the genome, and a lipid envelope carrying viral glycoproteins. The first steps in assembly are therefore the formation of the various components of the virion from their parts. Such intermediates must then associate in ordered fashion, in some cases following transport to the appropriate intracellular site, to complete construc-

BOX 13.2

Assembly of herpes simplex virus type 1 nucleocapsids in a simplified system

The assembly and egress of herpesviruses from infected cells are complex processes comprising multiple steps (Fig. 13.5 and 13.18). To facilitate analysis of the initial reactions that lead to assembly of the nucleocapsid, viral genes encoding its constituent proteins were cloned in baculovirus vectors. Formation of the nucleocapsid was examined by electron microscopy in insect cells infected with various combinations of the recombinant baculoviruses. Nucleocapsids indistinguishable from those formed in herpes simplex virus type 1-infected cells were observed when the six viral genes encoding VP5 (the major nucleocapsid protein that forms the hexons and pentons), VP19C and VP23 (which form triplexes linking the VP5 structural units), VP21, VP22a, VP24, and VP26 were expressed together. By omission of individual recombinant baculoviruses, it was shown that VP26 is not necessary for nucleocapsid assembly. Furthermore, only partial or deformed structures assemble in the absence of VP24, VP21, and VP22a (the protease and scaffolding proteins; see "Viral Scaffolding Proteins: Chaperones for Assembly"). Synthesis of a subset of viral proteins is also widely used to identify and characterize protein-protein and protein-membrane interactions critical for assembly.

Tatman, J. D., V. G. Preston, P. Nicholson, R. M. Elliot, and F. J. Rixon. 1994. Assembly of herpes simplex virus type 1 capsids using a panel of recombinant baculoviruses. *J. Gen. Virol.* **75:**1101–1113.

tion of the virus particle. Application of the techniques described in the previous section has allowed us to sketch the pathways by which many viruses are assembled, and describe some specific reactions in exquisite detail. In this section, we draw on this large body of information to illustrate mechanisms for the efficient assembly of protective protein coats for genomes, the first reactions listed in Fig. 13.1.

Formation of Structural Units

For some viruses, fabrication of a protein shell is coordinated with binding of structural proteins to the viral genome, as during assembly of the ribonucleoproteins of (–) strand RNA viruses. Consequently, structures built entirely from proteins do not accumulate. In other cases, the first assembly reaction is formation of the protein structural units from which the protein shell is constructed (Fig. 13.1). Compared to many other steps, this process is relatively simple: individual structural units contain a small number of protein molecules, typically two to six, that must associate appropriately following their synthesis. Nevertheless, several mechanisms have evolved for formation of structural units, and in some cases additional proteins are required to assist the reactions (Fig. 13.2).

Assembly from Individual Proteins

The structural units of some protein shells, including the VP1 pentamers of simian virus 40, assemble from their individual protein components (Table 13.1; Fig. 13.2A). This straightforward mechanism is analogous to formation of cellular structures containing multiple proteins, such as nucleosomes. In some cases, exemplified by adenoval pentons, assembly is a two-step process (Fig. 13.2A). In this kind of reaction, the surfaces of individual protein molecules that contact either other molecules of the same protein or a different protein (Table 13.1) are formed prior to assembly of the structural unit. This arrangement facilitates specific binding when appropriate protein molecules encounter one another. No energetically costly conformational change is required, and subunits that come into contact can simply interlock. Production of these structural units generally can be reconstituted in vitro or in cells that synthesize the component proteins. Such experiments confirm that all information necessary for accurate assembly is contained within the primary sequence, and hence the folded structure, of the protein subunits. On the other hand, the individual protein subunits must find one another in an intracellular environment, in which the concentration of irrelevant (cellular) proteins is as high as that attained in protein crystals (20 to 40 mg/ml). Such a milieu offers uncountable opportunities for nonspecific binding of viral proteins to unrelated cellular proteins. This problem may account for the synthesis of viral structural proteins in quantities far in excess of those incorporated in virus particles, a common feature of virus-infected cells. Such high concentrations must facilitate the encounter of viral proteins with one another by random diffusion and would provide a sufficient reservoir to compensate for any loss by nonspecific binding to cellular components. Another benefit of high protein concentration is that formation of structural units proceeds efficiently (Fig. 13.2A), driving the assembly pathway in the productive direction.

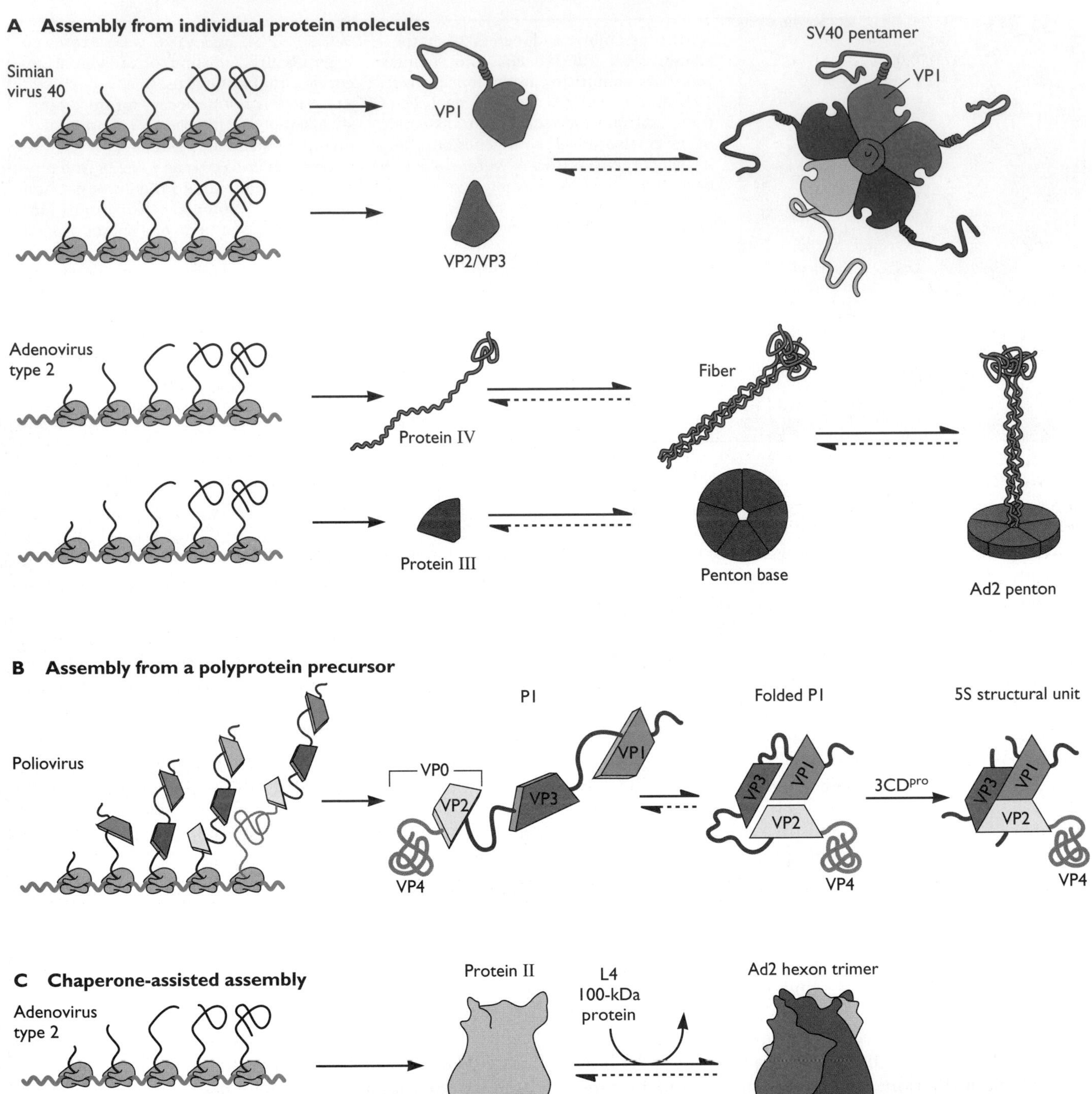

Figure 13.2 Mechanisms of assembly of the structural units of virion protein shells. (A) Assembly from folded protein monomers, illustrated with simian virus 40 (SV40) VP1 pentamers and adenovirus pentons. The assembly reactions shown are the result of specific interactions among the proteins that form structural units. Some, notably those among VP1 molecules in the simian virus 40 pentamer, have been described at atomic resolution (Chapter 4). In other cases, parts of the protein required for assembly have been identified by genetic analysis. For example, formation of the trimeric adenovirus fiber requires both the entire head of each monomer and the region of the shaft immediately adjacent to it. These assembly

(continued on next page)

Figure 13.2 ***(continued)*** reactions are driven in a forward direction by the high concentrations of protein subunits synthesized in infected cells, as indicated by the solid arrows. (B) Assembly from a polyprotein precursor, illustrated with the poliovirus polyprotein that contains the four proteins that form the heteromeric structural unit. The latter proteins are synthesized as part of the single polyprotein precursor from which all viral proteins are produced by proteolytic processing. For simplicity, only the P1 capsid protein precursor and its cleavage by the viral 3CD protease following folding and assembly of the proteins of the immature structural unit (VP0, VP3, and VP1) are shown. The covalent connections between VP1 and VP3 and between VP3 and VP0 in the P1 precursor, which are flexible regions exaggerated for clarity in the figure, are severed by the protease to form the 5S structural unit. However, VP4 remains covalently linked to VP2 in VP0 until assembly is completed (see text). (C) Assisted assembly. Some structural units are assembled only with the assistance of viral chaperones, such as the adenoviral L4 100-kDa protein, which is required for formation of the hexon trimer.

Assembly from Polyproteins

An alternative mechanism for production of structural units is assembly while covalently linked in a polyprotein precursor (Fig. 13.2B). This mechanism circumvents the need for protein subunits to meet by random diffusion, and avoids competition from nonspecific binding reactions. The structural units of several viruses are assembled by this mechanism (Table 13.1), which is exemplified by formation of picornaviral capsids. The first poliovirus intermediate, which sediments as a 5S particle (Fig. 13.3), is the immature structural unit that contains one copy each of VP0, VP3, and VP1 (Fig. 13.2B). These three proteins are liberated from the capsid protein precursor P1 upon cleavage by the viral 3CD^{pro} protease. However, it is believed that folding of their central β-barrel domains (Fig. 4.9) takes place during synthesis of P1. The poliovirus structural unit can then form by intramolecular interactions among the surfaces of these β-barrel domains before the covalent connections that link the proteins are severed.

Retrovirus assembly illustrates elegant and effective variation on the polyprotein theme. Mature retrovirus particles contain three protein layers. An inner coat of NC protein, which packages the dimeric RNA genome, is enclosed within the capsid built from the CA protein. The capsid is in turn surrounded by the MA protein, which lies beneath, and is in intimate contact with, the inner surface of the viral envelope (Appendix A, Fig. 21). These three structural proteins are synthesized as the Gag polyprotein precursor, which contains their sequences in order of the protein layers they form in virus particles, with MA at the N terminus (Fig. 13.4). Retrovirus particles assemble from such Gag polyprotein molecules by a unique mechanism that allows orderly construction of the three protein layers and, as we shall see, coordination of this reaction with encapsidation of the genome and acquisition of the envelope.

Table 13.1 Mechanisms of assembly of structural units

Mechanism	Virus	Structural unit
Association of individual protein molecules	Adenovirus (adenovirus type 2)	Protein IV trimer (fiber), protein III pentamer (penton base) that form penton
	Hepadnavirus (hepatitis B virus)	C (capsid) protein dimers
	Polyomavirus (simian virus 40)	VP1 pentamer, with one molecule of VP2 or VP3 in its central cavity
	Reovirus (reovirus type 1)	λ, σ2 (inner capsid protein) homo-oligomers; σ3-μ, (outer capsid protein) hetero-oligomers
Assisted assembly of protein subunits	Adenovirus (adenovirus type 2)	Hexon trimers of protein II, formed with assistance of the L4 100-kDa protein
	Herpesviruses (herpes simplex virus type 1)	VP5 pentamers and hexamers, formed with assistance of VP22a
Assembly from polyprotein precursors	Alphavirus (Sindbis virus)	Capsid (C) protein folds in, and cleaves itself from, a nascent polyprotein also containing glycoprotein sequences
	Picornavirus (poliovirus)	Immature 5S structural units, VP0-VP3-VP1
	Retrovirus (avian sarcoma virus)	NC, CA, and MA protein shells assembled via Gag polyprotein

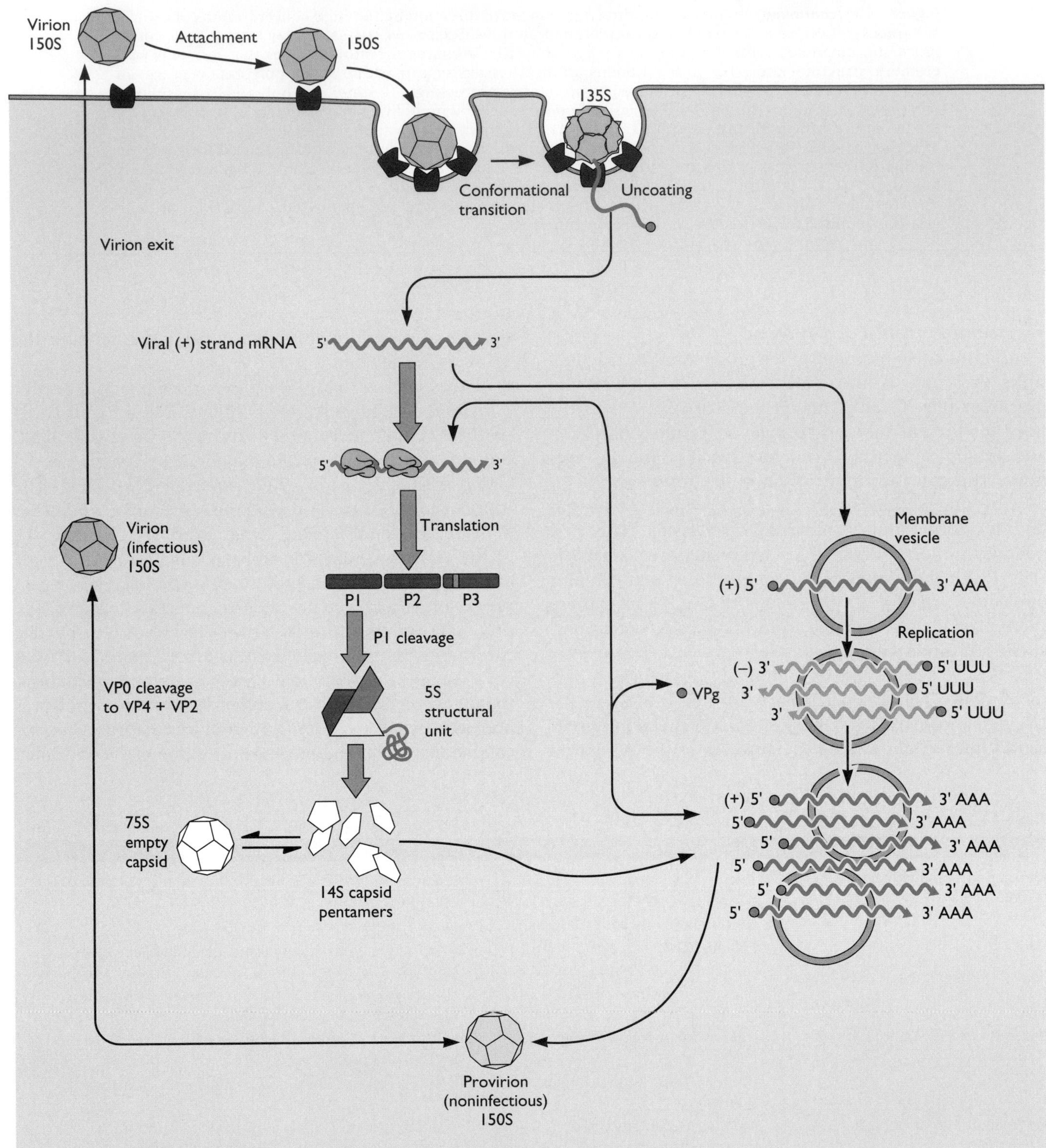

Figure 13.3 Assembly of poliovirus in the cytoplasm of an infected cell. Most of the assembly reactions are essentially irreversible, either because of proteolytic cleavage (formation of 5S structural units and mature virions), or because of extensive stabilizing interactions in the assembled structure (formation of 14S pentamers and of provirions). Stable, empty capsids, originally considered the precursors of provirions and termed procapsids, do not possess the same conformation as the mature virion, as symbolized by the white

(continued on next page)

Figure 13.3 *(continued)* color. They are dead-end products, or a storage form of 14S pentamers. The entering viral genome is replicated when associated with the membranes of the infected-cell-specific vesicles described in Chapter 12. Formation of the capsid shell from 14S pentamers is coordinated with genome encapsidation, and requires replication of genomic RNA. The conformational transition upon attachment to the poliovirus receptor, for which the virion is primed by cleavage of VP0 to VP2 and VP4, is also illustrated.

Participation of Cellular and Viral Chaperones

The assembly of viral proteins into structural units is undoubtedly assisted by cellular **chaperones**. These specialized proteins facilitate protein folding by preventing nonspecific, improper associations among exposed, sticky patches on nascent and newly synthesized proteins. The first chaperone to be identified, the product of the *Escherichia coli groEL* gene, was discovered because it is essential for reproduction of bacteriophages T4 and lambda. As discussed in Chapter 12, the participation of chaperones resident in the lumen of the endoplasmic reticulum (ER) in folding and assembly of oligomeric viral glycoproteins is well established. Cytoplasmic and nuclear chaperones seem likely to play equally important roles in the formation of structural units. A number of viral structural proteins have been shown to interact with one or more cellular chaperones (Table 13.2). In most cases, a role for these catalysts of protein folding in viral assembly is presently based on "guilt by association." However, some cellular chaperones have been directly implicated in viral assembly reactions. Alterations in the Gag protein of the betaretrovirus Mason-Pfizer monkey virus that prevent binding to a cytoplasmic chaperone (Table 13.2) reduce the accumulation of stable Gag, and of capsids, in cells transiently synthesizing the viral protein. These observations, and the dependence of association of Gag molecules with one another on the chaperone, indicate that the latter protein facilitates proper folding of the Gag polyprotein.

Chaperones are abundant in all cells and some accumulate to concentrations even greater than that of ribosomes. Nevertheless, the genomes of several viruses encode proteins with chaperone activity, some with sequences and functions homologous to those of cellular proteins (Table 13.2). In some cases, such viral chaperones are essential participants in the reactions by which structural units are formed. Production of adenoviral hexon trimers, which form the faces of the icosahedral capsid, depends on such an accessory protein, the viral L4 100-kDa protein (Fig. 13.2C). The latter protein facilitates folding of monomeric hexon subunits or their assembly into trimers, although its biochemical activity has not been identified.

Figure 13.4 Radial organization of the Gag polyprotein in immature human immunodeficiency virus type 1 particles. The model for the arrangement of the Gag polyprotein shown below a cryo-electron micrograph of a virus-like particle assembled from Gag was deduced from radial density measurements of digitized images of the particles. The plot shows density as a function of distance from the particle center, in angstroms. Courtesy of T. Wilk, European Molecular Biology Laboratory.

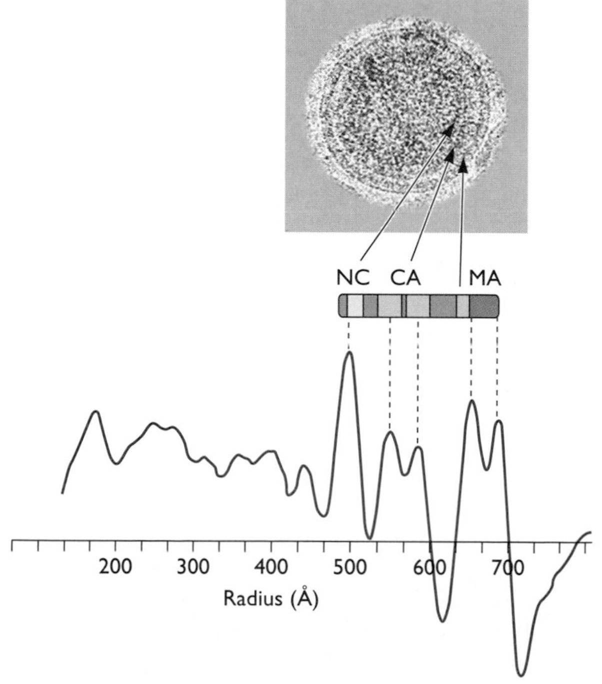

Capsid and Nucleocapsid Assembly

The accumulation of virion structural units within the appropriate compartment of an infected cell sets the stage for the assembly of more complex capsids or nucleocapsids (see Table 4.1 for nomenclature). For reasons discussed previously, the reactions by which these structures are formed are often not understood in detail. Nevertheless, several different mechanisms for their assembly can be distinguished.

Intermediates in Assembly

A striking feature of well-characterized pathways of bacteriophage assembly (Box 13.1) is the sequential formation of progressively more complex structures: heads, tails, and tail fibers are each assembled in stepwise fashion via defined intermediates. Such an assembly line mechanism appears ideally suited for orderly formation of virus particles, which can be large and morphologically complex. Discrete intermediates also form during assembly of some icosahedral animal viruses. A stepwise assembly mechanism has been well characterized for poliovirus: the 5S structural unit described

Table 13.2 Cellular and viral proteins implicated in viral assembly reactions

Chaperone	Properties and function(s)	Viral protein(s) bound
Cellular chaperones		
Bacterial		
Chaperonins GroEL and GroES	GroEL comprises two rings of 8 identical subunits to which the single heptameric GroES ring binds and dissociates, regulated by the ATPase of GroEL; nonnative proteins enter the GroEL cavity where they fold protected from aggregation with other unfolded proteins; required for assembly of phage T4 and λ heads.	Phage λ B protein, phage T4 gene 31 protein
Mammalian		
Chaperonin TriC	Large, double-ring complexes of ~800 kDa surrounding a central cavity; encapsulates substrates upon ATP binding; most abundant substrates are actin and tubulin.	Mason-Pfizer monkey virus Gag, an interaction implicated in productive folding of Gag
Hsp70 proteins	Cytoplasmic proteins synthesized constitutively (Hsc70) and in response to stress; in conjunction with cochaperones of the Hsp40 family, participate in ATP-dependent cycles of binding to, and release from, nascent proteins to prevent misfolding or nonspecific aggregation	Adenovirus protein IV Hepatitis B virus L protein Human immunodeficiency virus type 1 Gag (present in virions) Poliovirus P1 polyprotein Simian virus 40 VP1
Hsp68	Contains ATP binding motifs and an epitope present in a subunit of TriC; RNase L inhibitor	Human immunodeficiency virus type 1 Gag; facilitates a late reaction in assembly of Gag capsids in vitro
Viral chaperones		
Adenovirus type 2 L4 100-kDa protein	Required for formation of hexon trimers; mechanism of action not known	Hexon protein II
African swine fever virus CAP80	Required for productive folding of the major capsid protein p73	Capsid protein p73
Herpes simplex virus type 1 VP22a	Required for formation of VP5 pentamers	VP5
Simian virus 40 LT antigen	N-terminal J-domain necessary for assembly of virions; binds to and stimulates activity of cellular Hsc70 proteins	None known
Viral scaffolding proteins		
Adenovirus type 2		
L1 52/55-kDa proteins	Necessary for formation of capsids; present in immature particles but not virions; may be required for encapsidation rather than to form a scaffold	?
L4 33-kDa protein	Required for formation of any particles, even empty capsids; molecular function(s) unknown	?
Herpes simplex virus type 1 VP22a	Self-associates to form a scaffold-like structure that organizes assembly of the empty nucleocapsid; removed by the viral VP24 protease prior to encapsidation of viral DNA	VP5 VP19-VP23 triplexes

in the previous section is the immediate precursor of a 14S pentamer, which in turn is efficiently incorporated into virus particles (Fig. 13.3). The pentamer is stabilized by extensive protein-protein contacts and by interactions mediated by the myristate chains present on the five VP0 N termini (see Fig. 4.10C). The contribution of the lipids to pentamer stability is so great that this structure does not form at all when myristoylation of VP0 is prevented. The extensive interactions among its components result in molecular interlocking of the five structural units of the pentamer and impart great stability. Consequently, formation of the 14S assembly intermediate is irreversible under normal conditions in an infected cell, a property that imposes the appropriate directionality on the entire assembly pathway (Fig. 13.3).

Despite the apparent advantages of a stepwise assembly mechanism, for many viruses discrete assembly intermediates like the poliovirus pentamer have been difficult to identify. In some cases, the absence of discrete intermediates can be attributed to coordination of assembly of protein shells with binding of the structural proteins to the nucleic acid genome. This mode of assembly is exemplified by the ribonucleoproteins of (–) strand RNA viruses, which assemble as genomic RNA is synthesized. Nucleocapsid formation depends on interactions of the protein components with both the nascent RNA and other protein molecules previously bound to the RNA. Because the synthesis of genomic RNA molecules is an all-or-none process, so too is the assembly of ribonucleoproteins in infected cells.

Methods that permit synthesis of subsets of virion proteins have begun to provide insights into how such ribonucleoproteins assemble. The vesicular stomatitis virus N protein, which is a dimer in the helical nucleocapsid, aggregates when synthesized alone in *E. coli.* However, when the viral N and P proteins are made, aggregation does not occur, and discrete, disc-like oligomers assemble. The assembly contains 10 molecules of the N protein, 5 molecules of the P protein, and an RNA molecule of some 90 nucleotides. The disc-like oligomer is equivalent to one turn of the ribonucleoprotein helix formed in vesicular stomatitis virus-infected cells, and the RNA (presumably of bacterial origin) is of the length required to bind to 10 molecules of the N protein. As no further assembly takes place in bacterial cells, even though they contain numerous, long RNA molecules, it has been proposed that assembly of the ribonucleoprotein containing genomic RNA requires a viral or mammalian cell assembly chaperone.

For many enveloped viruses, including retroviruses, assembly of a protein shell is coordinated with binding of structural proteins to a cellular membrane. This property makes isolation of intermediates a technically demanding task. Nevertheless, new methods for separation of intermediates make it possible to examine assembly reactions of these viruses. Specific assembly reactions can also be studied in simplified experimental systems. For example, when synthesized in a cell-free transcription-translation system, the human immunodeficiency virus type 1 Gag protein multimerizes through a series of discrete intermediates to form 750S particles. These structures resemble virus-like particles released when Gag is the only viral gene expressed in mammalian cells. Conversion of a 500S late intermediate to the 750S particle requires ATP and a previously unrecognized cellular chaperone (Table 3.2). This observation illustrates the power of simplified approaches to the study of virus assembly. An important caveat is that such experimental systems must faithfully reproduce reactions that take place within infected cells. There is good reason to believe that the in vitro assembly of Gag particles meets this crucial criterion: the intermediates formed in vitro can also be detected when Gag is made in mammalian cells. Furthermore, assembly phenotypes exhibited by altered Gag proteins in vitro correspond closely to those observed in vivo.

The general dearth of structures simpler than empty capsids in cells infected by nonenveloped viruses might simply be a consequence of the properties that make such intermediates difficult to detect, notably low concentration and instability. In addition, the formation of assembly intermediates may be rate limiting, allowing stable structural units to be stockpiled before the final assembly reactions begin. While these possibilities cannot be excluded, it is more likely that assembly of many capsids is a highly cooperative, all-or-none process. Both simple and more complex icosahedral capsids are built by the repetition of interactions among multiple copies of one or a small number of structural units. Consequently, once the first few structural units were associated in the correct manner, assembly of the capsid would proceed rapidly to completion.

Self-Assembly and Assisted Assembly Reactions

The primary sequences of virion proteins contain sufficient information to specify assembly, including complex reactions like the alternative five- and sixfold packing of VP1 pentamers in the simian virus 40 capsid. When synthesized in *E. coli,* VP1 is isolated as pentamers that assemble into capsid-like structures in vitro. While self-assembly of structural proteins is the primary mechanism for formation of virus particles, other viral components or cellular proteins can assist assembly.

Viral and Cellular Components That Regulate Self-Assembly

Self-assembly of viral structural proteins may be the mortar for construction of virus particles, but other components of the virion often provide an essential foundation or the blueprint for correct assembly. As we have seen, assembly of the nucleocapsids of (–) strand RNA viruses is both coordinated with, and dependent on, synthesis of genomic RNA. The RNA must serve as a template for productive and repetitive binding of nucleocapsid proteins to one another. RNA may also serve as a scaffold during retrovirus assembly (Box 13.3). In other cases, the viral genome plays a more subtle yet equally important role, ensuring that the interactions among structural units are those necessary for infectivity. For example, poliovirus empty capsids lack internal structural features characteristic of the mature virion (Fig. 4.10) because VP0 is not cleaved to VP4 and VP2. The RNA genome is believed to participate in the autocatalytic cleavage of this precursor, which is essential for the production of infectious particles.

BOX 13.3

A scaffolding function for RNA?

When synthesized in the absence of any other viral component, retroviral Gag polyproteins direct assembly and release of virus-like particles (Fig 13.14). It has therefore been believed for many years that this protein contains all information necessary and sufficient for assembly of particles. However, more recent experiments suggest that RNA acts as a scaffold during Gag assembly.

In vitro studies of the ability of altered Gag proteins to multimerize with the full-length protein initially underscored the importance of the NC RNA-binding domain for efficient assembly. Association of Gag with RNA was also required for multimerization in this system. The apparent contradiction between these findings and efficient assembly of Gag in mammalian cells in the absence of genomic RNA has recently been resolved: virus-like particles formed in cells infected by a Moloney murine leukemia virus mutant with a deletion in the signal that directs packaging of the RNA genome (Fig 13.11) contain cellular mRNAs. Furthermore, when the Gag coding sequence was expressed from an alphavirus vector, which directs very efficient synthesis of genomic and subgenomic viral RNAs, these alphaviral mRNAs could be easily detected in the Gag virus-like particles. Finally, RNase digestion of cores assembled from Gag in wild-type Moloney murine leukemia virus-infected cells was shown to dissociate these structures, indicating that interaction of Gag molecules with RNA, as well as with one another, maintains particle stability.

Muriaux, D., J. Mirro, D. Harvin, and A. Rein. 2001. RNA is a structural element in retrovirus particles. *Proc. Natl. Acad. Sci. USA* **98**:5246–5251.

Association of structural proteins with a cellular membrane is essential for the assembly of some protein shells, a situation exemplified by many retroviruses: the sequences of MA that specify Gag myristoylation and binding to the cytoplasmic surface of the plasma membrane (described in Chapter 12) are also required for assembly of the core.

Binding of structural proteins to the genome or to a cellular membrane might simply raise their local concentration sufficiently to drive self-assembly, might organize the proteins in such a way that their interactions become cooperative, or might induce conformational changes necessary for productive association of structural units. These mechanisms, which are not mutually exclusive, have not been distinguished experimentally, but there is evidence for induction of conformational transitions in specific cases. We do not understand adequately the molecular mechanisms by which binding of structural proteins to other virion components directs or regulates particle assembly. However, such a requirement offers the important advantage of integration of assembly of protein shells with acquisition of other essential parts of the virion.

Cellular components can also modulate the fidelity with which viral structural proteins bind to one another. The capsid-like structures assembled when simian virus 40 VP1 is made in insect or mammalian cells are much more regular in appearance than those formed in vitro by bacterially synthesized VP1. Modification of VP1 (by acetylation and phosphorylation) or the participation of nuclear chaperones must therefore improve the accuracy with which VP1 pentamers associate to form capsids. Similarly, in vitro self-assembly of poliovirus structural proteins is very slow, proceeding at a rate at least 2 orders of magnitude lower than that observed in infected cells. Furthermore, the empty capsids that form have the altered conformation described previously, unless the reaction is seeded by 14S pentamers isolated from infected cells. This property indicates that the appropriate folding, modification, and/or interaction of the viral proteins that form the pentamer are critical for subsequent assembly reactions to proceed productively. Within infected cells, these crucial reactions are likely to be modulated by cellular chaperones, such as Hsp70, which is associated with the polyprotein during its folding to form 5S structural units. It is clear from such examples that host cells provide a hospitable environment for productive virus assembly, one that is not necessarily well reproduced when viral structural proteins are made and assemble in vitro.

Viral Scaffolding Proteins: Chaperones for Assembly

Accurate assembly of some large icosahedral protein shells, such as those of adenoviruses and herpesviruses, is mediated by proteins that are not components of mature virions. Because these proteins participate in reactions by which the capsid or nucleocapsid is constructed but are then removed, they are termed **scaffolding proteins**. Among the best characterized of such proteins is the precursor of the herpes simplex virus type 1 VP22a protein.

This protein is the major component of an interior core present in assembling nucleocapsids (Fig. 13.5). In the absence of other viral proteins, it forms specific scaffold-like structures and in immature nucleocapsids isolated from infected cells appears as an ordered sphere that lacks the icosahedral symmetry of the mature nucleocapsid. Self-association of pre-VP22a stimulates binding of the scaffolding protein to VP5, the protein that forms the hexameric structural units of the nucleocapsid. The VP5 and pre-

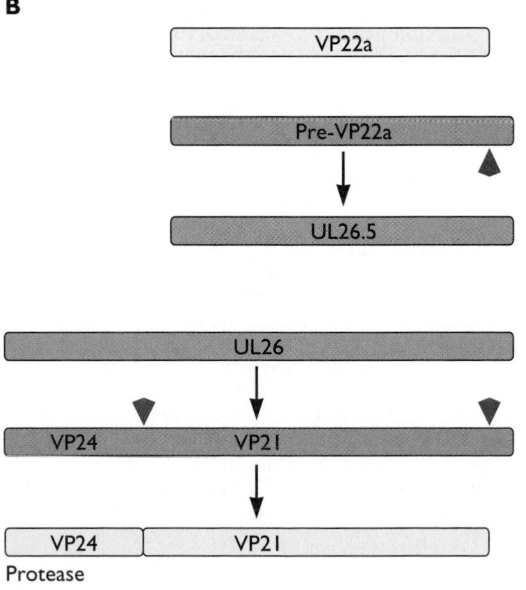

Figure 13.5 Assembly of herpes simplex virus type 1 nucleocapsids. (A) Assembly begins as soon as nucleocapsid proteins accumulate to sufficient concentrations in the infected cell nucleus. Intermediates include pentamers and hexamers of the major capsid protein VP5, which form pentons and hexons in the capsid, and triplexes of the minor proteins VP23 and VP19C. Whether structural units assemble prior to transport into the nucleus is not clear. Viral proteins essential for assembly of the nucleocapsid but not present in mature virions, the scaffolding protein (pre-VP22a) and the viral protease precursor (VP24-VP21), must also enter the nucleus. Assembly of nucleocapsids depends on the formation of an internal scaffold around which the protein shell assembles. Subsequent reactions require the herpes simplex virus type 1 protease to remove the scaffolding protein, allowing entry of the DNA genome. As discussed in the text, encapsidation is concurrent with cleavage of the concatemeric products of herpesviral DNA replication. (B) Overlapping sequences of herpes simplex virus type 1 scaffolding proteins. The UL26 and UL26.5 reading frames are shown in purple, and their primary translation products are light brown. The initiating methionine of VP22a protein is within the larger reading frame that encodes the VP24-VP21 polyprotein. VP21 and VP22a consequently are identical in sequence, except that the former contains a unique N-terminal segment. All proteolytic cleavages, at the sites indicated by the red arrowheads, including those that liberate the protease itself from the VP24-VP21 precursor, are carried out by the VP24 protease. The cleavage at the C-terminal site in VP22a disengages the scaffolding from the capsid proteins.

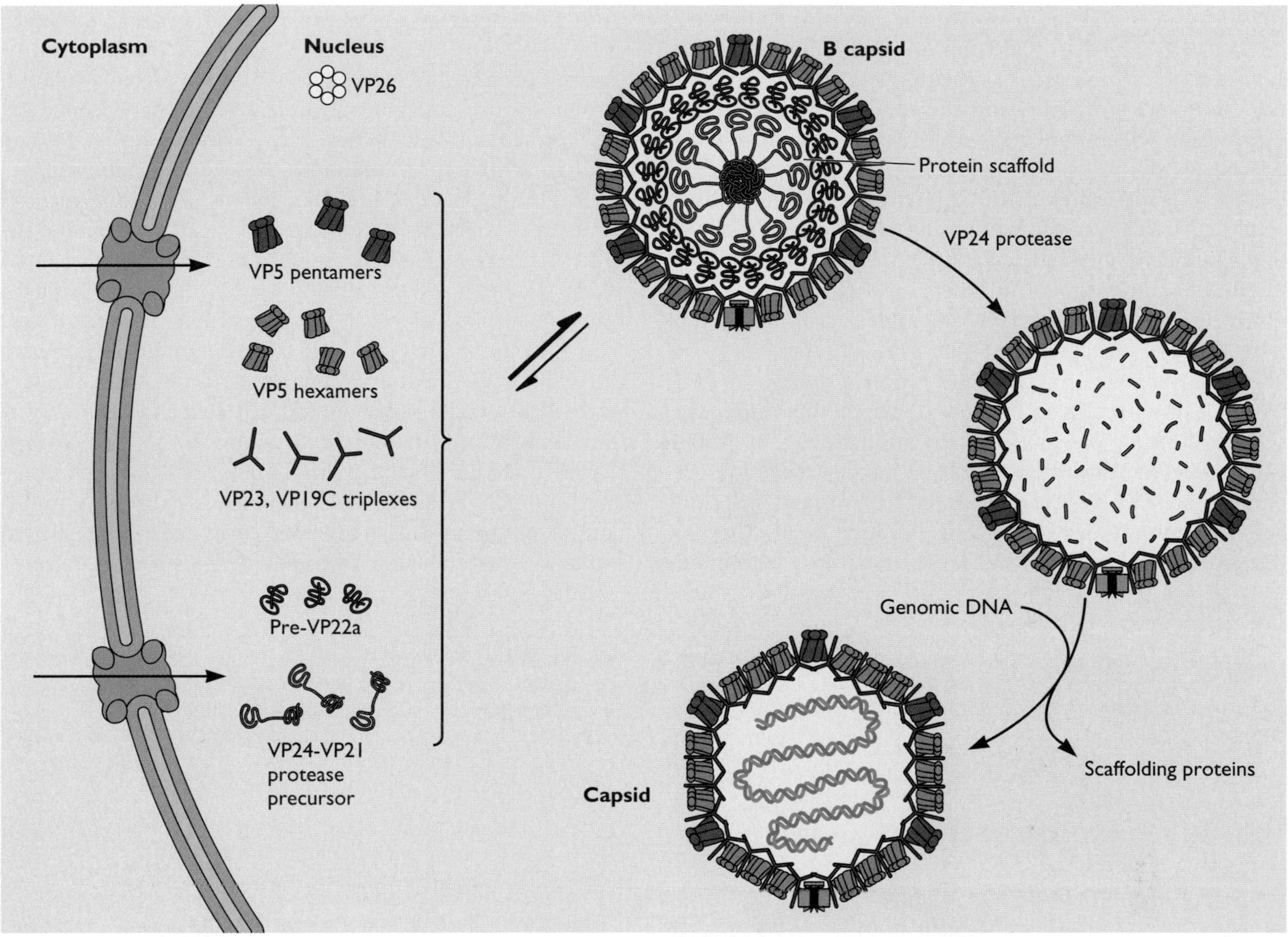

VP22a proteins form a core, to which are added additional VP5 hexamers and the triplex structures formed by VP19 and VP23 (Table 4.7). The latter are also required for capsid assembly, which occurs by sequential formation of partial dome-like structures and the spherical immature nucleocapsid. The interactions of VP5 with the scaffolding protein guide and regulate the intrinsic capacity of VP5 hexamers (and other nucleocapsid proteins) for self-assembly: omission of the scaffolding protein from a simplified assembly system (Box 13.2) leads to the production of unclosed and deformed nucleocapsid shells.

Once nucleocapsids have assembled, scaffolding proteins must be disposed of so that viral genomes can be accommodated (Fig. 13.5). The virion protease (VP24), which is also present in the core of assembling nucleocapsids, is essential for such DNA encapsidation. This protein is incorporated into the assembling nucleocapsid as a precursor (Fig. 13.5B). The protease precursor possesses some activity and initiates cleavage to produce VP24. The protease also cleaves a short C-terminal sequence from the scaffolding protein. Because this sequence is required for formation of sealed nucleocapsids with a regular structure and binding of the scaffolding protein to VP5, such processing presumably disengages scaffolding from structural proteins, once assembly of the nucleocapsid is complete. The protease is also believed to degrade the scaffolding protein so that encapsidation of the genome can begin.

The protease cleavages that liberate the VP5 structural units from their association with the scaffold may also induce major changes in the organization of the nucleocapsid shell. Assembly of the nucleocapsid from its constituent proteins in vitro proceeds via a short-lived, spherical precursor to the mature structure. This intermediate possesses the $T = 16$ icosahedral symmetry characteristic of the mature nucleocapsid. However, it is unstable and dissociates at low temperature, because the strong interactions among VP5 structural units that form the floor of the nucleocapsid shell are absent. Similarly cold-sensitive particles accumulate at nonpermissive temperatures in cells infected by a mutant virus encoding a temperature-sensitive viral pro-

BOX 13.4

Visualization of structural transitions during assembly of DNA viruses

The assembly of viruses that package double-stranded DNA genomes into a preformed protein shell exhibits several common features, regardless of the host organism. These include the presence of a portal for DNA entry in the capsid or nucleocapsid precursor and probably the mechanism of DNA packaging (see text). In addition, as illustrated below for bacteriophage lambda and herpes simplex virus type 1, assembly of DNA-containing structures is accompanied by major reorganizations of the protein shell.

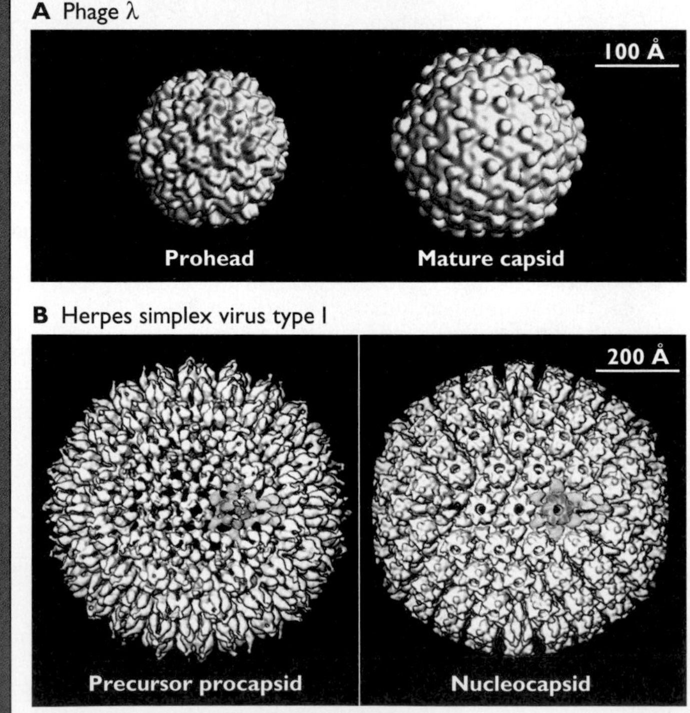

(A) Cryo-electron micrographs of the phage lambda prohead and the DNA-containing mature capsid. The former comprises hexamers and pentamers of the capsid protein gpE organized with $T = 7$ icosahedral symmetry, and is assembled prior to encapsidation of the DNA genome. It is smaller than the mature capsid (270 Å and 315 Å in diameter, respectively), but its protein shell is considerably thicker. Packaging of the DNA genome leads to an expansion of the capsid shell, as a result of reorganization of gpE hexamers. This change is accompanied by binding of the gpD protein, which contributes to capsid stabilization. Adapted from T. Dokland and H. Murialdo, *J. Mol. Biol.* **233**:682–694, 1993, with permission.
(B) Cryo-electron micrographs of herpes simplex virus type 1 nucleocapsid precursor and mature nucleocapsid, viewed along a twofold axis of icosahedral symmetry. Some copies of the proteins that form the particles' surfaces are colored as follows: VP5 hexons, red; VP5 pentons, yellow; triplexes containing one molecule of VP19C and two of VP23, green. The precursor nucleocapsid is spherical and less angular than the mature, DNA-containing structure, and its protein shell is thicker. Furthermore, the VP5 hexamers are not organized in a highly regular, symmetric manner in the precursor, resulting in a more open protein shell. The precursor nucleocapsid also lacks the VP26 protein, which binds to the external surface of VP5 hexamers, but not pentamers, in the mature nucleocapsid. B. L. Trus et al., *J. Mol. Biol.* **203**:447–462, 1996. Adapted from A. C. Steven et al., *FASEB J.* **10**:733–742, 1997, with permission.

tease, and can form infectious virions following shift to a permissive temperature. These properties suggest that the open structures initially assembled in vitro may correspond to bona fide, but short-lived, intermediates in herpesvirus assembly in vivo, analogous to the well-characterized **procapsids** formed during assembly of certain DNA-containing bacteriophage (Box 13.4). Because the protease is essential for production of virions, it is a target for development of anti-herpesviral compounds (Chapter 19).

The adenoviral L1 52/55-kDa proteins, which are differentially phosphorylated forms of a single translation product, exhibit properties expected of scaffolding proteins: they are necessary for assembly of adenovirus particles and are present in all capsid-like structures that can be isolated from infected cells, but are not found in mature virions (Fig. 13.6). Nevertheless, their primary function is in stable encapsidation of the viral DNA genome.

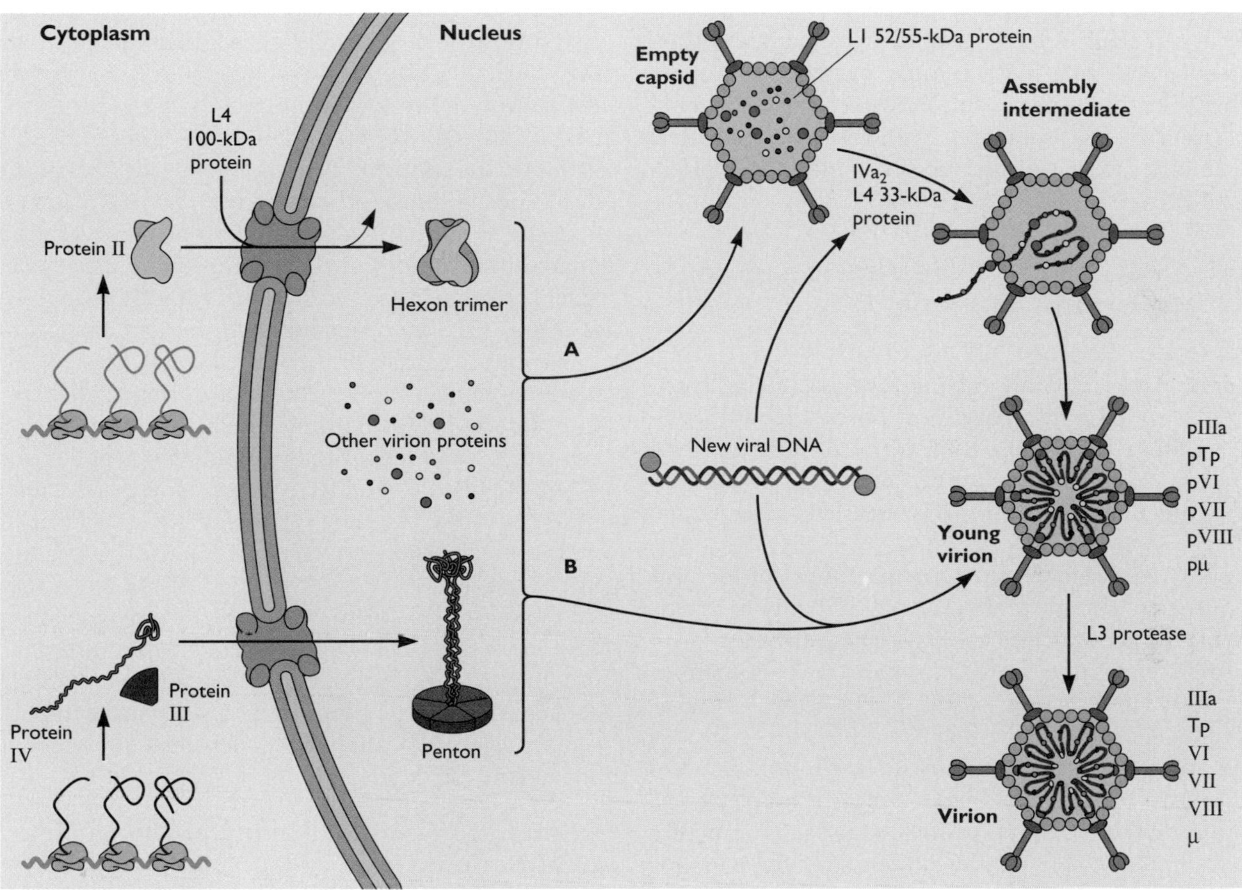

Figure 13.6 Adenovirus assembly pathways. Synthesis and assembly of hexons and pentons and their transport into the nucleus set the stage for assembly. The L4 100-kDa protein is required for formation of hexons, but its molecular function is not known. In the pathway originally proposed (A), these structural units and the proteins that stabilize the capsid (see Table 4.4) assemble into empty capsids. The L1 52/55-kDa proteins are necessary for the formation of structures that can complete assembly, and decrease in concentration as assembly proceeds. The DNA is then inserted into this structure via the packaging signal located near the left end of the genome. The viral IVa2 protein binds specifically to this sequence in vitro and is required for assembly in infected cells. Premature breakage of DNA in the process of insertion would yield the structure designated "Assembly intermediate," in which an immature capsid is associated with a DNA fragment derived from the left end of the viral genome. Core proteins are encapsidated along with the viral genome to yield noninfectious young virions. Mature virions are produced upon cleavage of the precursor proteins listed to the right of the young virion. The alternative pathway (B) is based on the failure of any capsid-like structures to assemble in cells infected by viruses with mutations that eliminate the packaging signal, prevent synthesis of the IVa2 protein, or remove the C-terminal segment of the L4 33-kDa protein. In this model, capsid assembly and encapsidation of the genome are concerted reactions. Empty capsids and the assembly intermediate would then be viewed as dead-end products or artifacts of the methods by which particles are extracted from infected cells. For example, the structure designated "Assembly intermediate" in pathway A could represent unstable structures with DNA genomes that were partially extruded and broken during extraction of intermediates.

Assembly of simpler protein shells can also depend critically on a viral protein. In addition to its many other functions, the simian virus 40 large T antigen (LT) participates in virion assembly. This protein does not form a scaffold, but an N-terminal domain of LT appears to be essential to organize the capsid shell: alterations within this domain block production of virions and induce accumulation of an incomplete structure, not normally observed during virus assembly. This abnormal assembly contains both the viral chromatin and VP1. The N-terminal segment of LT possesses chaperone activity (Table 13.2). It is similar in sequence to a specific domain of cellular chaperones of the DnaJ family and, like these cellular proteins, stimulates the activity of Hsp70 chaperones. The chaperone activity of LT may ensure the productive binding of VP1 pentamers to one another, and to other components of the virion during assembly.

Selective Packaging of the Viral Genome and Other Virion Components

Concerted or Sequential Assembly

Incorporation of the viral genome into assembling particles is often called **packaging**. This process requires specific recognition of genomic RNA or DNA molecules (see below). It is clear that all viral genomes are packaged by one of two general mechanisms, concerted or sequential assembly.

In concerted assembly, the structural units of the protective protein shell assemble productively only in association with the genomic nucleic acid. The nucleocapsids of (−) strand RNA viruses form by a concerted mechanism (Fig. 13.7), as do retrovirus particles (Fig. 13.8). In the alternative mechanism, sequential assembly, the genome is inserted into a preformed protein shell. The formation of herpesviral nucleocapsids provides a clear example of this type of assembly (Fig. 13.5). A nucleocapsid-like structure that lacks viral DNA but contains internal scaffolding proteins is an early intermediate in the assembly of these complex viruses. Neither mutations that inhibit viral DNA synthesis nor those that prevent DNA packaging block assembly of these capsids. The DNA genome must enter preformed capsids, when the proteins that assist capsid assembly are lost. In contrast to concerted assembly, encapsidation of the genome in a preformed structure requires specialized mechanisms to pull or push the genome into the capsid, as well as to maintain or open a portal for entry of the nucleic acid. The herpesviral portal protein UL6, which is present at only one of the 12 vertices of the nucleocapsid (Fig. 13.5 and 4.23), fulfills this crucial function. The identity of this portal was established only recently, and the mechanisms that ensure incorporation of but one portal into nucleocapsids are not yet understood.

Despite the clear differences between concerted and sequential assembly pathways, it can be quite difficult to decide which mechanism applies to some viruses with icosahedral symmetry. A classic case in point is poliovirus assembly, which has been studied for more than 25 years. In the first scheme to be proposed, an empty capsid containing 60 copies of the VP0-VP3-VP1 structural unit but lacking the RNA genome was viewed as the precursor of the **provirion** (Fig. 13.3), the RNA-containing but immature virion. Such 75S empty capsids, initially called procapsids, are readily detected in infected cell extracts and also form by self-assembly of 14S pentamers in vitro. In the alternative pathway, pentamers are proposed to condense around the RNA genome (Fig. 13.3). In this mechanism, capsid assembly and packaging of the RNA genome are coupled, and empty capsids are considered dead-end products, or a storage form of 14S pentamers. The concerted pathway is supported by several experimental observations, including the ability of radioactively labeled pentamers to form virions without the appearance of a procapsid intermediate, and the demonstration that stable empty capsids are produced by irreversible, conformational changes that take place during their extraction from infected cells. Furthermore, in a recently developed cell-free system for synthesis of infectious poliovirus particles, exogenously added 14S pentamers assemble with newly synthesized viral (+) strand RNA to form virions with antigenic sites characteristic of virions produced in infected cells. In contrast, exogenously added empty capsids undergo no further assembly, even when genomic RNA is synthesized, confirming that they are dead-end products. In this case, there is now strong evidence for the concerted pathway, but in others, such as adenovirus assembly (Fig. 13.6), there is still debate about which mechanism is used.

Recognition and Packaging of the Nucleic Acid Genome

The assembly mechanisms described in previous sections are fruitless unless the viral genome is packaged within progeny virions. Despite diversity in size, composition, and morphology, specific incorporation of the genome during assembly of virus particles is achieved by a limited repertoire of mechanisms, which are discussed in the next section. The special problems imposed by segmented genomes are considered subsequently. At the outset, it is important to appreciate that particles released from an infected cell do not necessarily contain a genome. Because synthesis of the viral genome is often rate limiting for assembly, empty particles may be produced by "off-pathway" reactions among structural proteins. Or they may form as a result of failure of genome encapsidation.

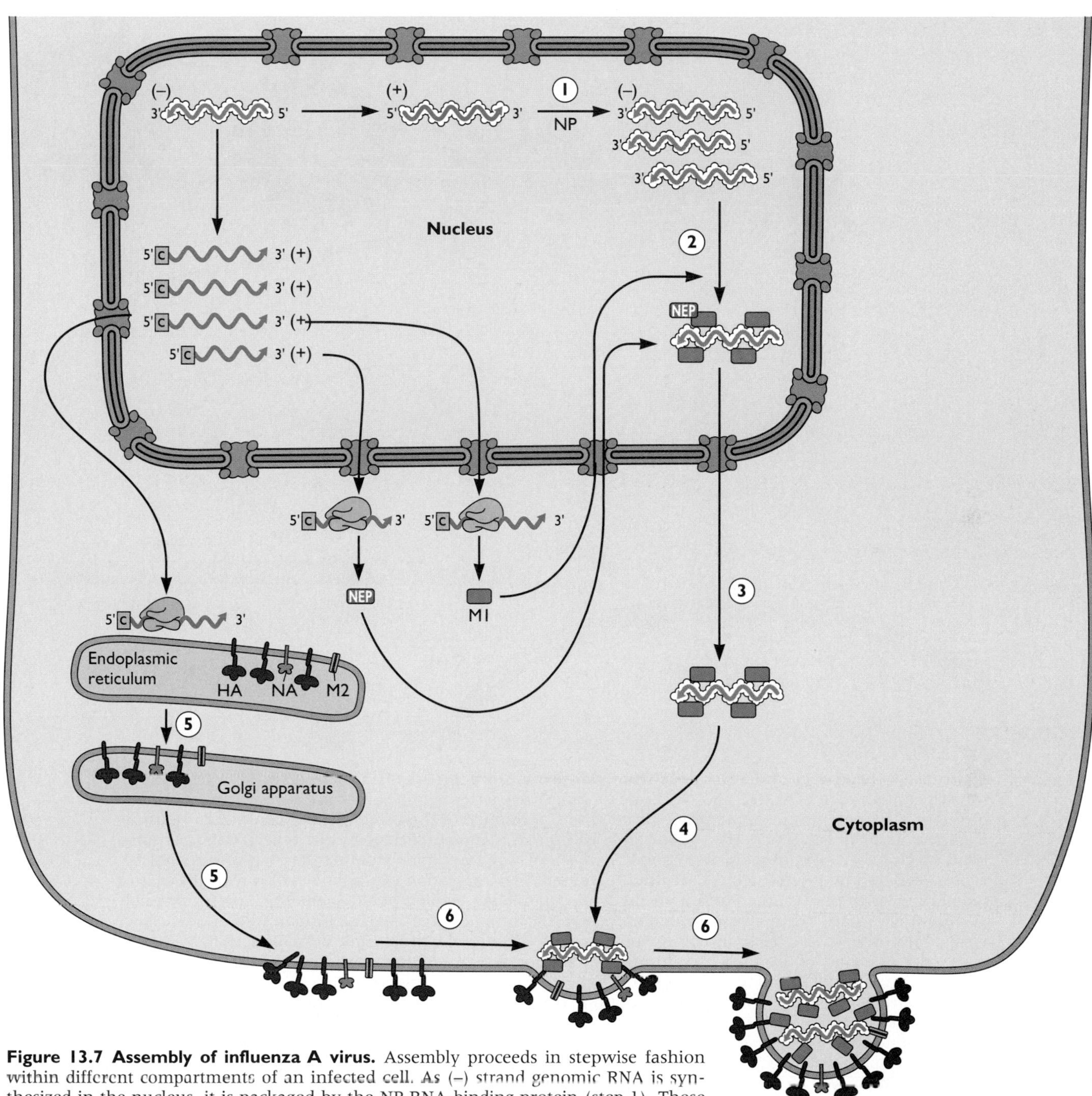

Figure 13.7 Assembly of influenza A virus. Assembly proceeds in stepwise fashion within different compartments of an infected cell. As (–) strand genomic RNA is synthesized in the nucleus, it is packaged by the NP RNA-binding protein (step 1). These ribonucleoproteins may serve as templates for transcription, participate in further cycles of replication, or bind the M1 protein (step 2). The latter interaction prevents further transcription or replication and allows binding of NEP and export of the nucleocapsid to the cytoplasm (step 3). The M1 protein also binds to the cytoplasmic face of the plasma membrane via specific sequences and directs the nucleocapsid to the plasma membrane (step 4). The sequences via which M1 binds to the membrane and to nucleocapsids have not been mapped at high resolution, nor is it known whether a specific sequence is necessary for assembly. The plasma membrane carries the viral HA, NA, and M2 proteins, which reach this site via the cellular secretory pathway (step 5). The M1 protein probably controls budding (step 6). The mechanism by which membranes fuse to release the enveloped particle (step 7) is not known. Only two of the eight genome segments are illustrated for clarity.

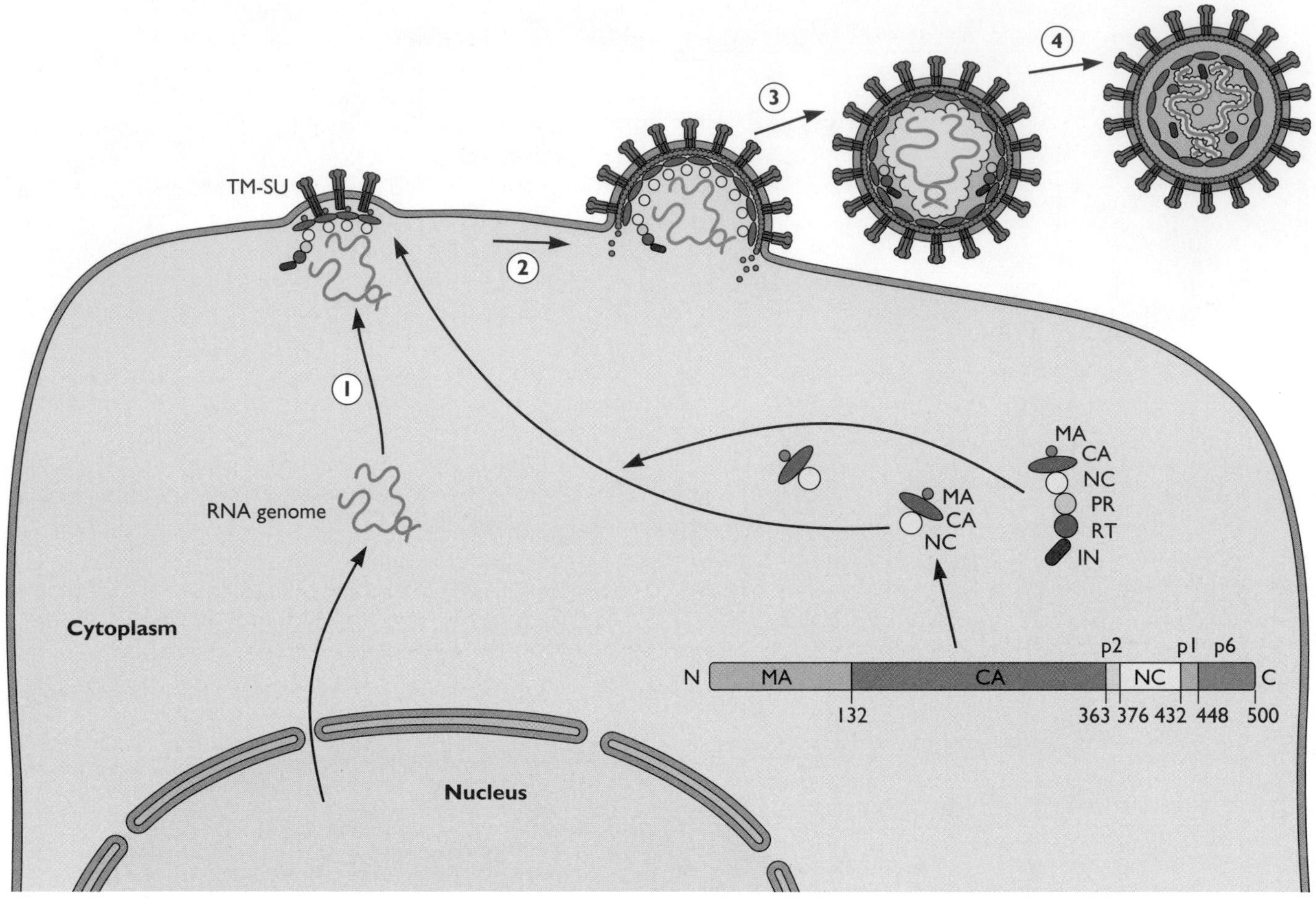

Figure 13.8 Assembly of a retrovirus from polyprotein precursors. The Gag polyprotein of all retroviruses contains the MA, CA, and NC proteins linked by spacer peptides that are variable in length and position. The proteins are in the order (from N to C terminus) of the protein shells of the virus particle, from the outer to the inner. The organization of human immunodeficiency virus type 1 Gag is summarized on the right. A minor fraction, about 1 in 10, of Gag translation products carry the retroviral enzymes, denoted by PR, RT, and IN, at their C termini. The association of Gag molecules with the plasma membrane, with one another, and with the RNA genome via binding of NC segments initiates assembly at the inner surface of the plasma membrane (step 1). In some cases, such as human immunodeficiency virus type 1, the MA segment also binds specifically to the internal cytoplasmic domain of the TM-SU glycoprotein. Assembly of the particle continues by incorporation of additional molecules of Gag (step 2). This pathway is typical of many retroviruses, but some (e.g., betaretroviruses) complete assembly of the core in the interior of the cell prior to its association with the plasma membrane. The dimensions of the assembling particle are determined by interactions among Gag polyproteins. Eventually, fusion of the membrane around the budding particle (step 3) releases the immature noninfectious particle. Cleavage of Gag and Gag-Pol polyproteins by the viral protease (PR) produces infectious particles (step 4) with a morphologically distinct core (see Fig. 13.19).

Mechanisms of Nucleic Acid Recognition and Packaging

During encapsidation, viral nucleic acid genomes must be distinguished from the cellular DNA or RNA molecules present in the compartment in which assembly takes place. This process requires a high degree of discrimination among similar nucleic acid molecules. For example, retroviral genomic RNA constitutes much less than 1% of an infected cell's cytoplasmic RNA population and bears all the hallmarks of cellular messenger RNAs (mRNAs). Yet it is **the** RNA packaged in the great majority of retrovirus particles. Such discrimination is the result of specific recognition of sequences or structures unique to the viral genome, termed **packaging signals**. These can be defined by genetic analysis as the sequences necessary for incorporation of the nucleic acid into the assembling virion, or sufficient to direct incorporation of foreign nucleic acid. The organization of the packaging signals of several viruses is therefore quite well understood.

Nucleic acid packaging signals. *DNA signals.* The products of polyoma- or adenoviral DNA synthesis are genomic DNA molecules that can be incorporated into assembling virus particles without further modification. These DNA genomes contain discrete packaging signals with several common properties. The signals comprise repeats of short sequences, some of which are also part of viral promoters or enhancers; they are positioned close to an origin of replication, and their ability to direct DNA encapsidation depends on this location. However, the molecular mechanisms of packaging are not well understood.

The simian virus 40 DNA packaging signal is located in the regulatory region of the genome containing the origin of replication, the enhancer, and early and late promoters. Multiple sequences within this region contribute to the encapsidation signal, and binding sites for the cellular transcriptional regulator Sp1 are required. Neither the internal proteins of the simian virus 40 capsid (VP2/3) nor the major capsid protein (VP1) binds specifically to DNA. In contrast, they recognize Sp1 bound to its six tandem binding sites present in the viral DNA packaging signal with high affinity and specificity. This cellular protein stimulates the in vitro assembly of infectious virus particles by an order of magnitude, consistent with a role in mediating indirect recognition of the packaging signal by capsid proteins. Although the cellular genome contains numerous binding sites for Sp1, the particular arrangement of sequences recognized by Sp1 in the packaging signal is unique to the viral genome.

The encapsidation signal of the adenoviral genome, which is located close to the left inverted repeat sequence and origin, likewise comprises a set of repeated sequences. Several of these sequences overlap enhancers that stimulate transcription of viral genes (Fig. 13.9) and are bound specifically in vitro by cellular proteins. In this case, however, the packaging signal is recognized by a viral protein during assembly. The viral IVa2 protein, a sequence-specific activator of transcription from the major late promoter (Chapter 8), binds specifically to this DNA sequence in vitro and is absolutely required for production of any particles in infected cells.

The replication of herpesviruses produces not genomic DNA molecules, but rather concatemers containing many head-to-tail copies of the viral genome. Individual genomes must therefore be liberated from such concatemers. The herpes simplex virus type 1 packaging signals *pac1* and *pac2,* which lie within the terminal *a* repeats of the genome, are necessary for both recognition of the viral DNA and its cleavage within the adjacent DR repeats (Fig. 13.10A). The proteins that recognize the *pac* sequences must therefore include a specific endonuclease that releases unit-length DNA molecules from concatemers. It is generally believed that cleavage is concomitant with genome encapsidation. In one model (Fig. 13.10B), it is proposed that a protein complex formed on the unique short *pac* sequence (Fig. 13.10A) is recognized by a nucleocapsid-associated protein (presumably the portal protein). Following the first DNA cleavage, a unit-length genome is then reeled into the nucleocapsid prior to the second DNA cleavage. This mechanism is analogous to that by which concatemeric DNA products of bacteriophage T4 replication are cleaved and packaged by a "terminase" complex, which associates transiently with a specific portal protein of a preformed capsid. The encapsidation of both T4 and herpesviral genomes requires ATP. The products of at least seven herpesviral genes are required specifically for cleavage and packaging of the viral genome (but not for DNA replication). The product of one (UL15) exhibits sequence similarities to an ATPase present in the T4 terminase complex, including similarity to an ATP-binding site essential for its function. Mutations that alter the putative ATP binding site of the UL15 protein prevent viral DNA cleavage and packaging. This protein forms a complex with two others necessary

Figure 13.9 Packaging signal of human adenovirus type 5. The location of the repeated sequences (blue arrows) of the packaging signal relative to the left inverted terminal repeat (ITR), the origin of replication (Ori), and the E1A transcription unit is indicated. These repeated sequences are AT rich and functionally redundant. The positions of transcriptional enhancers within this region are also shown. Enhancer 1 specifically stimulates transcription of the immediate-early E1A gene, whereas enhancer II increases the efficiency of transcription of all viral genes.

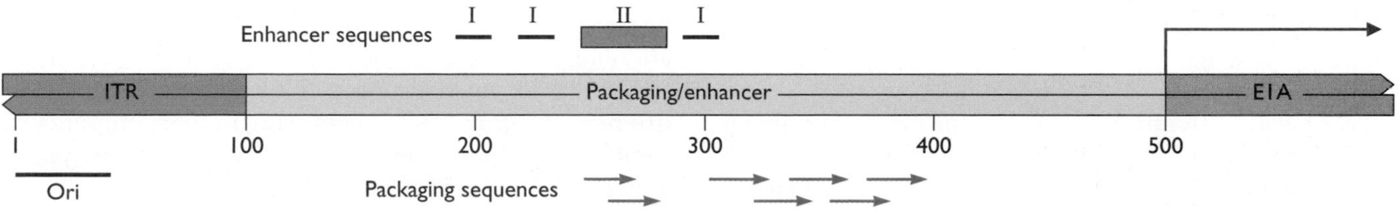

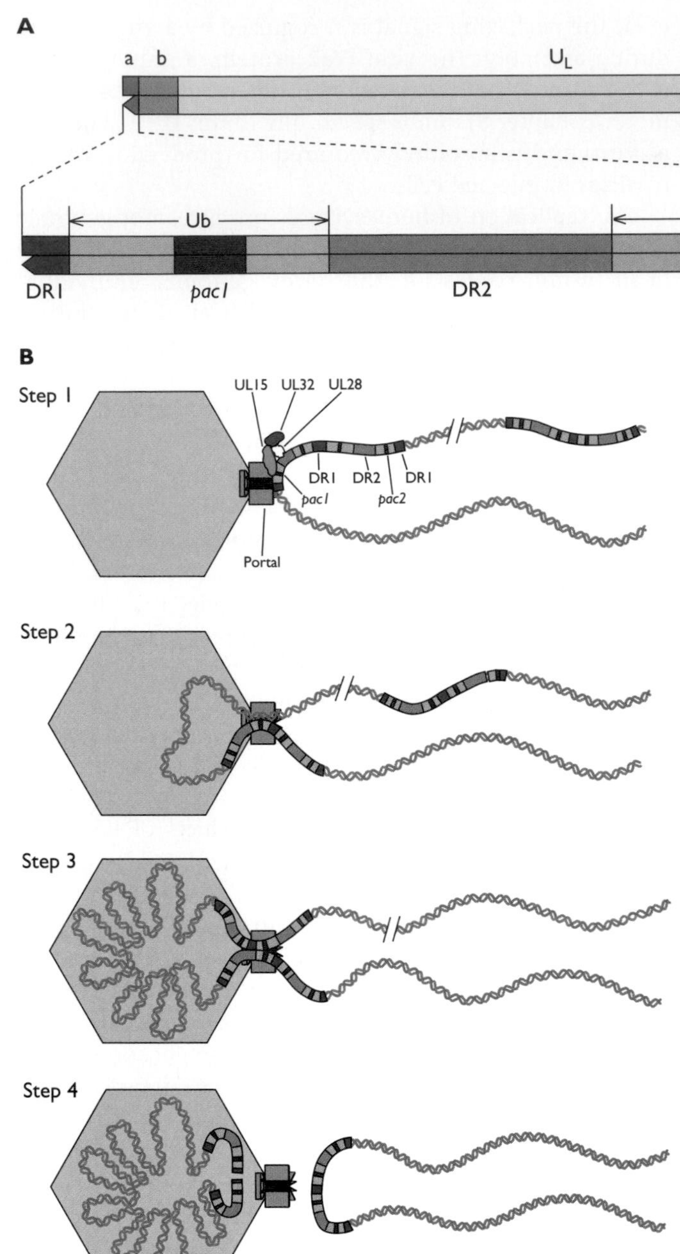

Figure 13.10 Packaging of herpes simplex virus type 1 DNA. (A) The organization of the *a* repeats of the viral genome, showing the location of the *pac1* and *pac2* sequences within the nonrepeated sequences Ub and Uc and relative to the flanking and internal direct repeats DR1 and DR2. One to several copies of the *a* sequence are present at the end of the unique long (U_L) segment and at the internal L-S junction, but only one copy lies at the end of the U_S region. (B) Model of herpes simplex virus type 1 DNA packaging in which encapsidation is initiated by formation of a terminase complex, which includes the proteins indicated, on the packaging sequence. This protein-DNA complex is oriented to interact with the portal of the nucleocapsid (step 1). The DNA is then reeled into the capsid (steps 2 and 3) until a headful threshold is reached and an *a* sequence in the same orientation (i.e., one genome equivalent) is encountered (step 3), when cleavage in DR1 sequences takes place (step 4). When *a* sequences are tandemly repeated, adjacent copies share a single intervening DR1 sequence. The unit-length DNA genomes packaged in virions are therefore assumed to be released from the concatemeric products of viral DNA synthesis by cleavage at a specific site within shared DR1 sequences.

for encapsidation of the genome, including the UL28 gene product that binds specifically to an unusual DNA structure formed by the *pac* sequence. Although evidence for a herpesviral terminase complex is accumulating, much remains to be learned about the biochemical properties and functions of the proteins that mediate cleavage and packaging of unit-length genomes.

RNA signals. Because it is also an mRNA, the retroviral genome must be distinguished during encapsidation from both cellular mRNA and subgenomic viral mRNA. In addition, two genomic RNA molecules must interact with one another, for the retroviral genome is packaged as a dimer. This unusual property is believed to facilitate reverse transcription of the genome into proviral DNA at the start of the infectious cycle. In virions, the dimeric genome is in the form of a 70S complex held together by many noncovalent interactions between the RNA molecules. However, most attention has focused on sequences that allow formation of stable dimers in vitro, termed the **dimer linkage sequence**. In vitro experiments with human immunodeficiency virus type 1 have provided evidence for base pairing between loop sequences of a specific hairpin (SL1) within the dimer linkage sequence (Fig. 13.11A) and the forma-

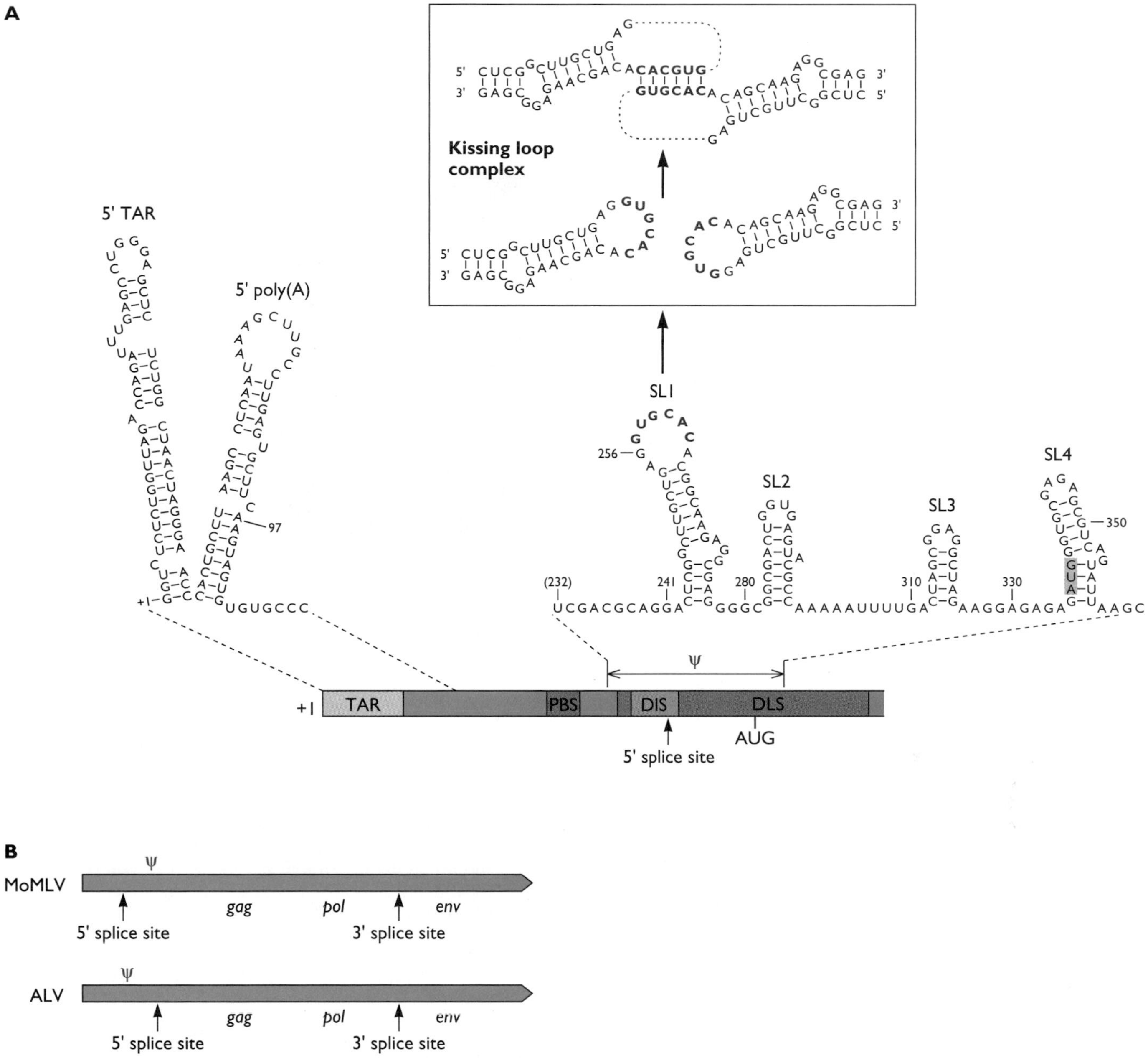

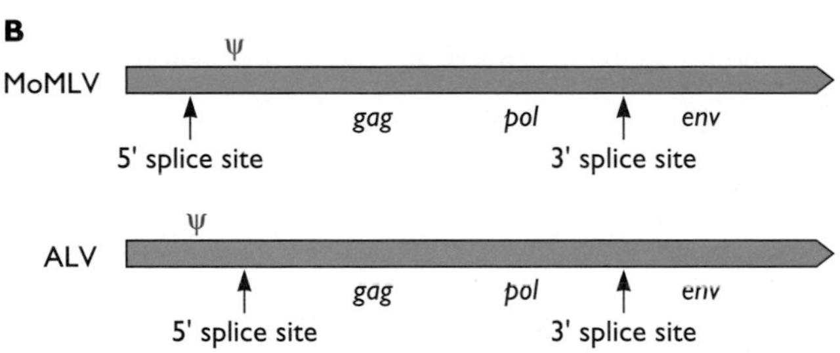

Figure 13.11 Sequences important in packaging of retroviral genomes. (A) The 5′ end of the human immunodeficiency virus type 1 genome is shown to scale at the bottom, indicating the positions of TAR, the tRNA primer-binding site (PBS), the 5′ splice site, a packaging signal designated ψ, the sequence that forms the dimer linkage structure (DLS), and the dimerization initiation site (DIS), which can initiate dimerization in vitro. The hairpins formed by TAR and the adjacent sequence containing a poly(A)-addition signal, as well as the four hairpins (SL1 to SL4) present in the ψ segment, are shown above. The SL1 hairpin is also the dimer initiation sequence. The loop-loop "kissing" complex proposed to form when two genomic RNA molecules dimerize via the self-complementary sequence shown in red is depicted at the top. The ψ sequence, which includes intronic sequences and therefore is present only in unspliced RNA, appears to be necessary but not sufficient for encapsidation of genomic RNA. The hairpin containing the poly(A)-addition signal near the 5′ end of the RNA and the bottom of the TAR hairpin, as well as more distantly located sequences, may also be required. (B) Locations within the RNA genomes of sequences necessary for encapsidation of Moloney murine leukemia virus (MoMLV) and avian leukosis virus (ALV) RNAs, designated ψ. The latter ψ signal resides only upstream of the 5′ splice site. Even though both genomic and subgenomic RNA contain this sequence, spliced mRNA molecules are not encapsidated efficiently. There must be some additional mechanism to prevent entry of subgenomic avian leukosis virus RNA into assembling particles.

tion of an intermolecular four-stranded helical structure (known as a G tetrad or G quartet). The effects of mutations in this sequence indicate that it mediates formation of genomic RNA dimers in vivo. For example, duplications of the sequence in the viral genome induce the appearance of monomeric RNA in virions, without significantly decreasing the efficiency with which viral RNA is packaged.

Sequences necessary for packaging of retroviral genomes, termed Psi (ψ), vary considerably in their complexity and location. In some cases, exemplified by Moloney murine leukemia virus, a contiguous ψ sequence of about 350 nucleotides (Fig. 13.11B) is both necessary and sufficient for RNA encapsidation. As this sequence lies downstream of the 5′ splice site, only unspliced genomic RNA molecules are recognized for packaging. The human immunodeficiency virus type 1 genome also contains a primary RNA-packaging sequence (Fig. 13.11A) that distinguishes the full-length genome from spliced viral RNA molecules. However, this sequence fails to direct packaging of heterologous RNA species into retrovirus particles, indicating that it is not sufficient. Additional sequences required for genomic RNA encapsidation lie within the TAR sequence and adjacent poly(A)hairpin (Fig. 13.11A) and at more distant locations. The genomes of this and some other retroviruses therefore appear to contain packaging signals that are dispersed in the primary sequence, but presumably form a distinctive structural feature in folded RNA molecules.

The NC domain of Gag, a basic region that contains at least one copy of a zinc-binding motif (Fig. 13.12A), mediates selective and efficient encapsidation of genomic RNA during retroviral assembly. The central region of NC containing the zinc-binding motif(s) and adjacent basic sequences binds specifically to RNAs that contain ψ sequences in vitro and is necessary for selective packaging of the genome in infected cells. Because even the simplest retroviral packaging signals comprise long RNA sequences, a complete picture of how NC recognizes ψ sequences is not yet available. However, the structure of human immunodeficiency virus type 1 NC bound to a highly conserved hairpin within the ψ sequence (SL3 in Fig. 13.11A) has been determined by nuclear magnetic resonance methods. Residues within the RNA-binding region of NC that are highly conserved among human immunodeficiency virus type 1 isolates, or important for specific RNA binding and selective packaging, make numerous specific contacts with the RNA (Fig. 13.12B). The NC protein is inherently flexible. It therefore seems likely that the NC regions of additional Gag molecules bind with analogous specificity (but via distinctive intramolecular interactions) to the other RNA sequences needed for packaging (Fig. 13.11A).

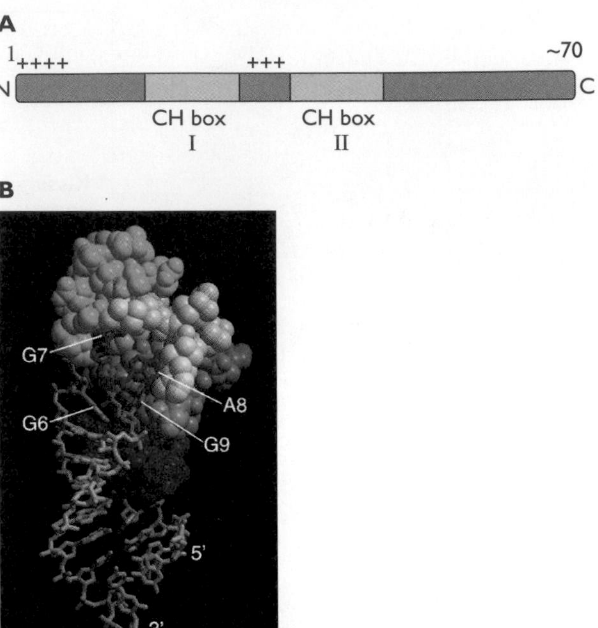

Figure 13.12 The retroviral NC RNA-packaging protein. (A) Schematic illustration of the NC protein, showing the locations of the Cys-His (CH) boxes and basic regions necessary for both nonspecific binding to RNA and specific recognition of the packaging signal. (B) Structural model of human immunodeficiency virus type 1 NC protein bound to the SL3 hairpin of the ψ sequence, determined by nuclear magnetic resonance methods, shown as a space-filling model. RNA is shown in gray except for specific G and A residues contacted by the protein, which are shown in pink (G7), violet (A8), and orange (G9). The N- and C-terminal zinc fingers (light blue and green, respectively) and residues within the basic region between them (yellow) make specific contacts with these exposed bases. For example, the N-terminal zinc finger binds to both G9 and A8 via hydrophobic interactions and hydrogen bonds from the protein backbone. The N-terminal segment (royal blue) preceding the first zinc finger includes a helix that binds within the major groove of the double-stranded RNA stem. From R. N. De Guzman et al., *Science* **279**:384–388, 1998, with permission. Courtesy of R. N. De Guzman and M. F. Summers, University of Maryland, Baltimore County.

The approximately 2,000 molecules of processed NC present in a mature virion also bind nonspecifically to genomic RNA. Whether specific recognition of the packaging signal by NC-embedded in Gag and nonspecific binding take place coordinately or sequentially during assembly is not clear. As discussed in Chapter 7, the NC protein may also facilitate dimerization of retroviral genomic RNA.

Other parameters that govern genome encapsidation. Specific signals may be required to mark a viral genome for encapsidation, but their presence does not guarantee packaging. The fixed dimensions of the closed icosahedral capsids or nucleocapsids of many viruses impose an upper limit on the size of viral nucleic acid that can be accommodated. Consequently, nucleic acids that are more than 5 to 10% larger than the wild-type genome cannot be encapsidated, even when they contain appropriate packaging signals. This property has important implications for the development of viral vectors. In some cases, the length of the DNA that can be accommodated in the particle (a "headful") is a critical parameter. This mechanism is exemplified by the coupled cleavage and encapsidation of genomic herpes simplex virus type 1 DNA molecules from the concatemeric products of replication, when both specific sequences and a headful of DNA are recognized. Indeed, the packaging of some viral DNA genomes, such as T4 DNA, depends solely on the latter parameter.

The coupling of encapsidation of a viral nucleic acid with its synthesis may also contribute to the specificity with which the viral genome is incorporated into assembling structures. As mentioned previously, such coordination is typical of the assembly of (–) strand RNA viruses (e.g., Fig. 13.7). However, the mechanisms by which nascent genomic RNA molecules are recognized by RNA-binding proteins are not well understood. Coordination of replication and encapsidation may also contribute to the great specificity with which picornaviral genomes are packaged: not only abundant cytoplasmic cellular RNA species (tRNAs, rRNAs, and mRNAs), but also (–) strand poliovirus RNA and poliovirus mRNA lacking VPg are excluded from virions. No packaging signal has yet been identified in the poliovirus genome. As discussed in Chapter 12, poliovirus infection induces the formation of cytoplasmic vesicles, with which both viral replication and virion assembly are associated. Such sequestration of genomic RNA molecules with the viral proteins that must bind to them clearly could make a major contribution to packaging specificity, by reducing competition from cytoplasmic cellular RNAs. There is accumulating evidence that encapsidation of the genome is coupled with RNA replication and that this step in assembly is assisted by the poliovirus 2C protein. This protein is a nucleoside triphosphatase that might facilitate release of RNA from membrane-bound replication complexes. As packaging of flavivirus RNA also depends on replication of genomic RNA, coincident genome synthesis and assembly may be a general feature of (+) strand RNA viruses.

Packaging of Segmented Genomes

Segmented genomes pose an intriguing packaging problem. The best-studied example among animal viruses is the influenza A virus genome, which comprises eight molecules of RNA. It has been appreciated for many years that production of an infectious virus particle requires incorporation of at least one copy of each of the eight genomic segments. Nevertheless, random packaging still cannot be distinguished from a selective mechanism for inclusion of a full complement of genomic RNAs.

Packaging of the bacteriophage ϕ6 genome provides clear precedent for a selective mechanism. The genome of this bacteriophage comprises one copy of each of three double-stranded RNA segments designated S, M, and L. The (+) strand of each segment is packaged prior to synthesis of complementary (–) strands within particles, a mechanism analogous to synthesis of the double-stranded RNA segments of the reovirus genome. The particle-to-plaque-forming-unit ratio of ϕ6 is close to 1, indicating that essentially all particles contain a complete complement of genome segments. Such precise packaging appears to be the result of the serial dependence of packaging of the (+) strand RNA segments. In in vitro reactions, the S segment packages alone, but entry of M RNA requires the presence of S RNA within particles, and packaging of the L segment is dependent on prior entry of both S and M RNAs.

A random packaging mechanism, in which any eight RNA segments of the influenza genome were incorporated into virions, would yield a maximum of 1 infectious particle for every 400 or so assembled ($8!/8^8$). This ratio might seem impossibly low, but is within the range of ratios of noninfectious to infectious particles found in virus preparations. Furthermore, if packaging of more than eight RNA segments were possible, the proportion of infectious particles would increase significantly. For example, with 12 RNA molecules per virion, 10% of the particles would contain the complete viral genome. Particles containing more than eight RNA segments have been isolated, consistent with random packaging. In a plasmid-based system for production of infectious influenza A virus particles, differentially marked versions of three of the genomic RNA segments competed nonspecifically for packaging. Furthermore, some of the virions produced contained two copies of the **same** RNA segment, a result that can be explained by a random, but not by a specific, packaging mechanism. On the other hand, there is also evidence for selective packaging: RNA segment 1 competes only with packaging of distinguishable segment 1 RNA.

Incorporation of Virion Enzymes and Other Nonstructural Proteins

In many cases, the production of infectious particles requires incorporation into the assembling virion of essential viral enzymes, or other proteins that are important in establishing an efficient infectious cycle. Some of these proteins are also structural proteins of the virion, needing no special mechanism to ensure that they enter the assembling virion. For example, the herpes simplex virus type 1 VP16 protein is both a major component of the virion tegument and the activator of transcription of viral immediate-early genes. However, the majority of virion enzymes and other nonstructural proteins must be incorporated by specific mechanisms.

Among the best understood is the simple, yet elegant, mechanism that ensures entry of retroviral enzymes (PR, RT, and IN) into the assembling core. In most cases, these enzymes are synthesized as C-terminal extensions of the Gag polyprotein. The organization and complement of these translation products, here designated Gag-Pol, varies among retroviruses, but the important point is that they contain not only Pol but also the sequences specifying Gag-Gag association described below. These are presumed to direct incorporation of Gag-Pol molecules into assembling particles (Fig. 13.8). The low efficiency with which Gag-Pol polyproteins are translated determines their concentrations relative to Gag in the cell and in virions (1:9). A low Gag-Pol/Gag ratio may be necessary for particle assembly: Gag-Pol molecules cannot direct assembly in the absence of Gag, presumably because they are too large to be packed into a protein shell of the size dictated by the interactions among Gag domains. All retroviral cores also contain the cellular tRNA primer for reverse transcription, brought into particles by its base pairing with a specific sequence in the RNA genome and by specific binding to RT.

The enzymes present in other virus particles, such as the RNA-dependent RNA polymerases of (–) strand RNA viruses (see Table 4.8), are synthesized as individual molecules and therefore must enter assembling particles by noncovalent binding to the genome or to structural proteins.

Acquisition of an Envelope

Formation of many types of virus particle requires envelopment of capsids or nucleocapsids by a lipid membrane carrying viral proteins. Most such enveloped viruses assemble by virtue of specific interactions among virion components at a cellular membrane before budding and pinching off of a new virus particle. However, they vary in the intracellular site at which the envelope is acquired and therefore in the relationship of envelopment to release of the virus particle. As discussed in Chapter 12, whether particles assemble at the plasma membrane or internal membranes is determined by the destination of viral proteins that enter the cellular secretory pathway. Enveloped viruses assemble by one of two mechanisms, distinguished by whether acquisition of the envelope follows assembly of internal structures, or whether these processes take place simultaneously.

Sequential Assembly of Internal Components and Budding from a Cellular Membrane

For most enveloped viruses, the assembly of internal structures of the virion and their interaction with a cellular membrane modified by insertion of viral proteins are spatially and temporally separated. This class of assembly pathways is exemplified by (–) strand RNA viruses such as influenza A virus (Fig. 13.7) and vesicular stomatitis virus. Influenza A virus ribonucleoproteins containing individual genomic RNA segments, NP protein, and the polymerase proteins are assembled in the infected cell nucleus as genomic RNA segments are synthesized. They are then transported to the cytoplasm in the M1- and NEP-dependent reactions described in Chapter 12. The viral glycoproteins HA and NA and the M2 membrane protein travel separately to the plasma membrane via the cellular secretory pathway (Fig. 13.7). The M1 protein interacts with both viral nucleocapsids and the inner surface of the plasma membrane to direct assembly of progeny particles at that membrane. Vesicular stomatitis virus assembles in a similar fashion, although no transport of the ribonucleoprotein from nucleus to cytoplasm is required. The matrix proteins of these (–) strand RNA viruses therefore provide the links among ribonucleoproteins and the modified cellular membrane necessary for assembly and budding. One or more tegument proteins of the more complex herpesviruses (see Fig. 4.23) must function like these matrix proteins, mediating binding of the preassembled nucleocapsid to the modified cellular membrane from which the virion derives its envelope.

The cellular membranes destined to form the envelopes of virus particles contain viral integral membrane proteins that play essential roles in the attachment of virus particles to, and their entry into, host cells. In simple enveloped alphaviruses, direct binding of the cytoplasmic portions of the viral glycoproteins to the single nucleocapsid protein (see Fig. 4.21) is necessary for acquisition of the envelope during budding from the plasma membrane. The crucial role and specificity of these interactions in the final steps in assembly are illustrated by the failure of a chimeric Sindbis virus containing the coding sequence for the E1 glycoprotein of Ross River virus, a second togavirus, to bud efficiently. The pE2 and E1 glycoproteins of the chimeric virus form heterodimers that are correctly processed (by cleavage of pE2 to E2) and transported to the plasma membrane. However, such chimeric glycoproteins exhibit an altered conformation and fail to bind to nucleocapsids at the plasma membrane. Binding of viral

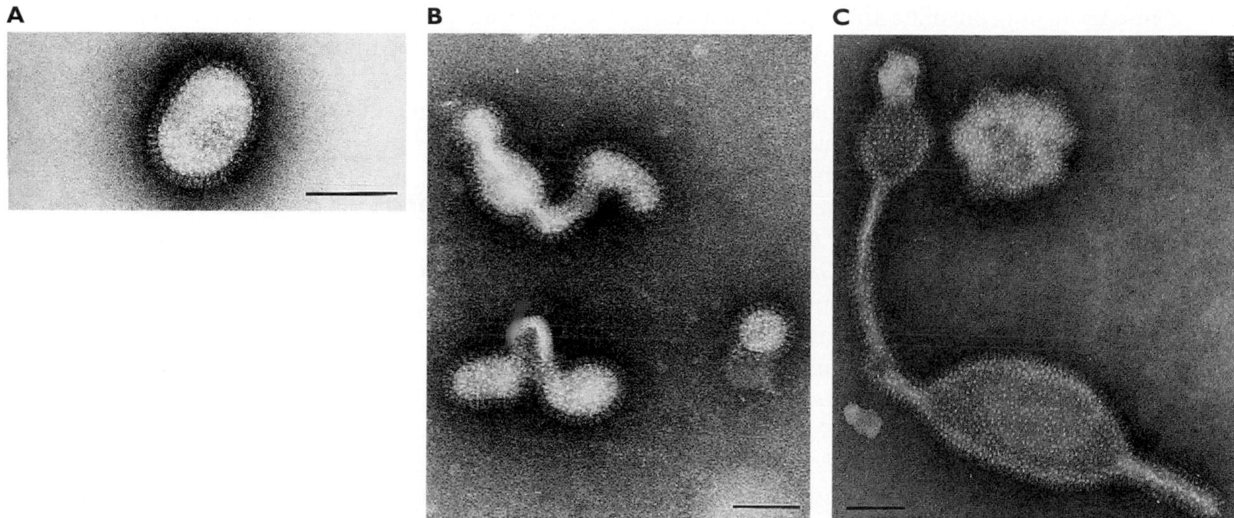

Figure 13.13 Deformed influenza A virus particles. These particles assemble when the cytoplasmic (internal) domains of the NA glycoprotein (B) or both NA and HA glycoproteins (C) are truncated. Purified wild-type (A) and mutant virions were negatively stained with phosphotungstic acid and examined by electron microscopy. Bars = 100 μm. The mutant particles are also considerably larger than wild-type virus particles. It is therefore believed that binding of the internal portions of HA, and particularly of NA, to internal virion components (M1 protein and/or the ribonucleoproteins) determines the spherical shape and characteristic size of these virus particles, as well as the efficiency of budding. From H. Jin et al., *EMBO J.* **16:**1236–1247, 1997, with permission. Courtesy of R. A. Lamb, Northwestern University.

glycoproteins to internal components also appears to be important for production of the more complex enveloped viruses. Interactions between the influenza virus M1 protein and the cytoplasmic tails of the HA and NA glycoproteins are necessary for formation of virus particles with normal size and morphology (Fig. 13.13), and with the appropriate concentration of genomic RNA.

Coordination of the Assembly of Internal Structures with the Acquisition of the Envelope

The alternative pathway of acquiring an envelope, in which assembly of internal structures and budding from a cellular membrane are coincident in space and time, is exemplified by many retroviruses. Assembling cores of the majority first appear as crescent-shaped patches at the inner surface of the plasma membrane. These structures extend to form a closed sphere as the plasma membrane wraps around and eventually pinches off the assembling particle (Fig. 13.8). Formation of the virion is mediated by interaction of Gag polyprotein molecules with one another to form the protein core, with the RNA genome via the NC portion, and with the plasma membrane via the MA segment. Indeed, when synthesized alone, Gag directs the budding of virus-like particles (Fig. 13.14), indicating that it contains all the information necessary for assembly and release.

Figure 13.14 Formation of virus-like particles by a retroviral Gag polyprotein synthesized in the absence of other viral proteins in human T cells. The electron micrograph shows a thin section (fixed and stained) of a human T cell synthesizing the viral Gag-Pol protein. Prior to electron microscopy, viral particles (arrowhead) were labeled with polyclonal antibodies (attached to gold beads) recognizing the CA protein. Bar = 1.0 μm. N, nucleus; M, mitochondrion. Courtesy of J. J. Wang, Institute of Biomedical Sciences, Academica Sinica, Taipei, Taiwan, and B. Horton and L. Ratner, Washington University School of Medicine.

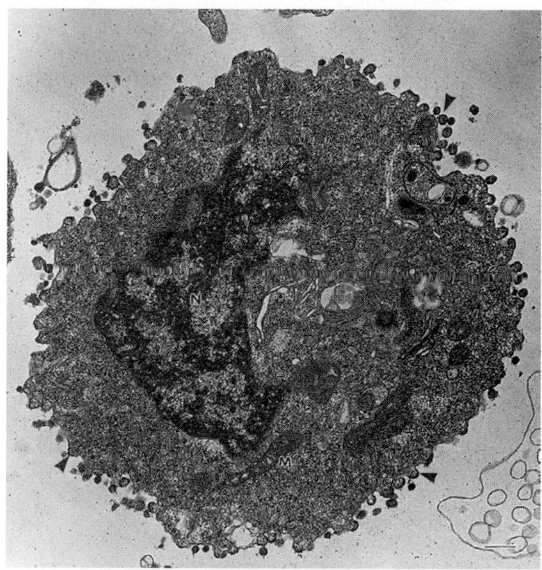

Specific segments of Gag mediate the orderly association of polyprotein molecules with one another and are required for proper assembly. These sequences include an essential C-terminal multimerization domain of the CA segment. Removal of certain sequences present only in the polyprotein results in the assembly of misshapen particles. Clearly, interactions among these segments of Gag polyproteins are important in determining the size and shape of retroviral virions. Association among Gag polyprotein molecules must coincide with their binding to the plasma membrane via signals in the MA segment of the polyprotein described in Chapter 12. Particle formation also requires sequences predicted to lie at the interfaces of the MA trimers formed in crystals (see Fig. 4.22). Such binding of Gag molecules to the inner surface of the plasma membrane is essential for productive assembly, probably because Gag adopts an assembly-competent conformation only when the MA segment is firmly anchored in the plasma membrane. This property provides an effective mechanism for ensuring that the assembly of internal protein shells is coordinated with the binding of Gag to the membrane. Conversely, efficient membrane binding of Gag depends on sequences other than the membrane-binding region of MA, such as a sequence in the N-terminal portion of NC. Because this sequence is not required for production of stable Gag or its transport to the plasma membrane, it may promote Gag-Gag or Gag-RNA interactions that lead to cooperative and stable binding of Gag molecules to the membrane.

In some cases, the MA segment of Gag also binds to the cytoplasmic tail of the viral envelope glycoprotein. For example, association of the assembling human immunodeficiency virus type 1 core with the TM-SU glycoprotein requires the N-terminal 100 amino acids of MA. Such Gag-Env interactions ensure specific incorporation of viral glycoproteins into virions. Nonetheless, they do not appear to be universal: glycoprotein-containing virions are produced even when the C-terminal tails of TM of other retroviruses (e.g., avian sarcoma virus) are deleted. Nor can a model based solely on Gag-Env interactions account for the ease with which "foreign" viral and cellular glycoproteins are included in the envelopes of all retroviruses. It is therefore believed that as a particle assembles at the plasma membrane, Gag-membrane interactions displace cellular membrane proteins connected to internal components of the cell. Such displacement would allow lateral diffusion of viral (and cellular) glycoproteins that are not connected in this way into these regions of the cellular membrane, and hence their passive incorporation into assembling virus particles.

As discussed previously, the dimensions and morphology of budding retroviral structures are set by interactions among Gag polyproteins. The final reaction, fusion of membrane regions juxtaposed as the particle assembles (Fig. 13.8), is shared with other viruses that assemble at the plasma membrane. This process is considered in the next section.

Release of Virus Particles

Many enveloped viruses assemble at, and bud from, the plasma membrane. Consequently, the final assembly reaction, fusion of the bud membrane, releases the newly formed virus particle into the extracellular milieu. When the envelope is derived from an intracellular membrane, the final step in assembly, budding, is also the first step in egress, which must be followed by transport of the particles to the cell surface. The assembly of enveloped viruses is therefore both mechanistically coupled and coincident with (or at least shortly followed by) their exit from the host cell. The egress of some viruses without envelopes from certain types of host cell is also by specific mechanisms. However, replication of such viruses more commonly results in destruction (lysis) of the host cell. Large quantities of assembled virions may accumulate within infected cells for hours, or even days, prior to release of progeny on cell lysis. The release of many enveloped viruses also destroys the cells that support their replication. However, in some cases nondestructive budding permits a long-lasting relationship with the host cell. The progeny of many simple retroviruses are released throughout the lifetime of an infected cell, which is not harmed (but may be permanently altered; see Chapter 18).

Release of Nonenveloped Viruses

The most usual fate of host cells permissive for reproduction of nonenveloped viruses is death and destruction (but see Chapter 16). In natural infections, the host defenses are an important cause of infected-cell destruction. However, infection by these viruses destroys host cells more directly: they are cytopathic to cells in culture. In general, we are remarkably ignorant about the mechanisms by which replication of nonenveloped viruses induces death and lysis of host cells.

Infection by many viruses, including poliovirus and adenovirus, leads to inhibition of expression of cellular genetic information by specific effects on cellular transcription, RNA export from the nucleus, or translation. In the case of adenovirus, with a one-step infectious cycle that is a relatively long 1 to 2 days, the shutdown of production of cellular proteins during the late phase of infection makes an important contribution to the eventual destruction of the infected cell. However, more specific mechanisms have also been implicated in the release of adenovirus and must be essential for exit of poliovirus:

infection by the latter virus induces host cell lysis more rapidly (within approximately 8 h) than can be explained by inhibition of synthesis of cellular macromolecules. One or more poliovirus nonstructural proteins have been implicated in cell lysis. For example, protein 3A increases membrane permeability. A number of adenoviral proteins may contribute to virus release by altering cellular structures. Synthesis of the early E1B 19-kDa protein induces both the collapse of some intermediate filaments and disruption of the nuclear lamina. Later in the infectious cycle, other intermediate filament components, specific cytokeratins, are cleaved by the L3 protease into polypeptides that can no longer polymerize. Such disruption of intermediate filament networks, which requires inhibition of cellular protein synthesis so that the network cannot be rebuilt, seems likely to damage the structural integrity of the cell and facilitate virus release. In addition, a small E3-encoded glycoprotein is necessary for efficient nuclear disruption and lysis of cells in culture. This viral glycoprotein is made in large quantities late in the infectious cycle and accumulates in the nuclear envelope, but its mechanism of action is not yet known.

While cell lysis is the most common means of escape of naked viruses, there is evidence for the release of some without any cytopathic effect. Under certain conditions, polioviruses and other picornaviruses are released without lysis of the infected cell. When poliovirus replicates in polarized epithelial cells resembling those lining the gastrointestinal tract (a natural site of infection), progeny virions are released exclusively from the apical surface by a nondestructive mechanism. It is unlikely that such directed, nonlytic release requires the cellular system for sorting components in polarized cells, because poliovirus induces dramatic rearrangement of the cells' internal membranes and inhibition of the secretory pathway. In contrast, simian virus 40 is released from permissive cells before induction of cytopathic effects and leaves polarized epithelial cells via their apical surfaces, via the secretory transport pathway of a cell. It will be of considerable interest to learn how such nonenveloped virions enter the membrane-bound compartments of this pathway.

Assembly at the Plasma Membrane: Budding of Viral Particles

The release of enveloped virus particles from the plasma membrane is a complex process that comprises induction of membrane curvature by viral components (bud formation), bud growth and fusion of the bud membrane. Although budding has been visualized repeatedly, often in striking images, the mechanisms of this crucial process are still not well understood.

For many years, it was generally believed that bud formation was driven by interactions among viral envelope glycoproteins and internal components of viral particles, a mechanism exemplified by alphaviruses such as Sindbis virus. Such a mechanism (Fig. 13.15) can account

Figure 13.15 Interaction of viral proteins responsible for budding at the plasma membrane. Four distinct budding strategies have been identified. In type I budding, exemplified by alphaviruses such as Sindbis virus, both the envelope glycoproteins and the internal capsid are essential. Quite detailed structural pictures of alphaviruses are now available (Chapter 4). Certain altered or chimeric envelope proteins that reach the membrane normally do not support budding. These properties indicate that lateral interactions among the envelope heterodimers, as well as those of the heterodimers with the capsid, cooperate to drive budding. Type II budding, such as Gag-dependent budding of retroviruses, requires only internal matrix or capsid proteins. Conversely, budding can be driven solely by envelope proteins (Type III), a mechanism exemplified by the envelope proteins of the coronavirus mouse hepatitis virus. Type IV budding is driven by matrix proteins, but its proper functioning depends on these additional components. For example, in the case of rhabdoviruses and orthomyxoviruses, internal matrix proteins alone can drive budding. However, this process is inefficient, or results in deformed or incomplete particles in the absence of envelope glycoproteins or the internal ribonucleoprotein. Adapted from H. Garoff et al., *Mol. Microbiol. Rev.* **62:**1171–1190, 1998, with permission.

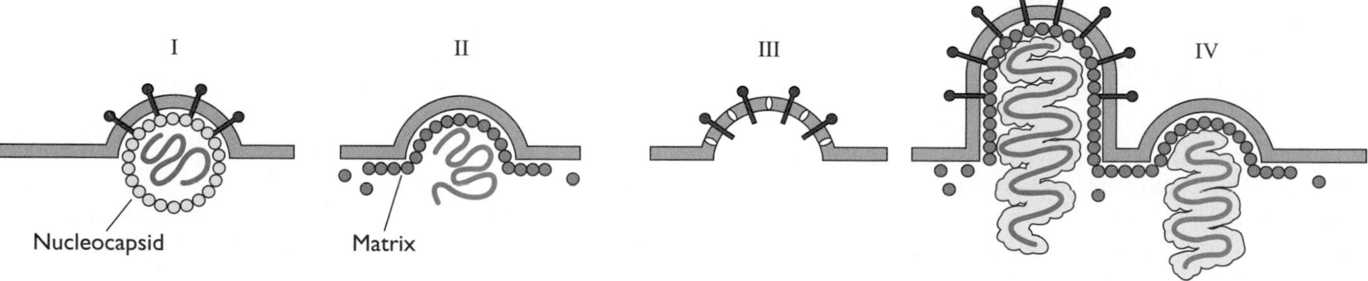

nicely for how a bud site is determined, by initial binding of the internal nucleocapsid to a cluster of the heterodimeric E1-E2 glycoproteins. The bud then expands by cooperative, lateral interactions among glycoprotein spikes and between their internal segments and the capsid protein. Nevertheless, as described in the previous section and summarized in Fig. 13.15, it is now clear that there is greater variety in the viral proteins required for budding than originally appreciated. How interactions among such proteins, for example, those among retroviral Gag proteins or alphaviral envelope and capsid proteins, provide the force for membrane bending remains a mystery.

The membrane fusion reactions that pinch off the bud are even less well characterized than those responsible for bud formation. However, recent studies have underscored the crucial participation of host cell components. Regions of retroviral Gag proteins removed during their proteolytic processing are essential for fusion: their alteration leads to the accumulation of immature particles that remain attached to the host cell by a thin membrane stalk. The segments necessary for membrane fusion during assembly of avian sarcoma virus (p26) and human immunodeficiency virus type 1 (p6) are located at different positions within Gag, and are unrelated in essential amino acid sequence. Nevertheless, they can substitute for one another. Such sequences, which have been termed L (late) domains, may be a general feature of viral proteins that direct assembly and release of enveloped particles from the plasma membrane: the matrix proteins of vesicular stomatitis virus and Ebola virus (a filovirus) can substitute for the L domains of retroviral Gag proteins to direct particle release.

The amino acid motifs of these viral L domains are known to mediate protein-protein interactions. For example, that of human immunodeficiency virus type 1 Gag conforms to the PXXP motif recognized by Src homology 3 domains (defined in Chapter 18). In fact, these L domains do interact with host cell components, specific ubiquitin ligases. Such enzymes are part of the machinery that adds mono- or polyubiquitin chains to proteins. The latter modification marks proteins for destruction by the proteasome, while the former has been ascribed various functions. Several observations indicate that binding of viral protein L domains to ubiquitin ligases promotes release of viral particles. Addition of a sequence encoding an L domain to an MA-CA protein stimulates both monoubiquitination of the polyprotein and release of virus-like particles. Conversely, depletion of the cellular pool of ubiquitin inhibits retrovirus and rhabdovirus budding. Exactly how monoubiquitination promotes membrane fusion and release of virus particles is not yet clear. This modification has been implicated in endocytosis, suggesting that release of these enveloped viruses might require the cellular machinery that takes vesicles into cells. Indeed, the interaction of the L domain of human immunodeficiency virus type 1 Gag with a second cellular protein that participates in endocytic trafficking, the product of tumor susceptibility gene 101 (Tsg101), is required for budding and release of virus particles. The Tsg101 protein and ubiquitination have been implicated in the sorting of endocytic cargo to structures called multivesicular bodies, which form from late endosomes and fuse with lysosomes. Exactly how these cellular proteins function in budding of virus particles and the mechanisms(s) by which viral components convert machinery of the endocytic pathway into a system for exocytosis remain to be determined.

Completion of assembly at the plasma membrane can also depend on viral proteins other than major structural proteins. For example, efficient release of human immunodeficiency virus type 1 requires the viral Vpu protein. This function of Vpu is distinct from the induction of proteolysis of CD4 in the ER described in Chapter 12. In the absence of Vpu, particles accumulate in intracellular vacuoles, or tethered to the infected cell surface. The Vpu protein, which lacks a large extracellular domain, has recently been shown to form an ion channel. Although the ion channel activity of Vpu has been implicated in promoting the release of virus particles, the mechanism has not yet been elucidated.

Assembly at Internal Membranes: the Problem of Exocytosis

Compartments of the Secretory Pathway

Several enveloped viruses are assembled at the cytoplasmic surfaces of membranes of internal compartments of the secretory pathway under the direction of specifically located viral glycoproteins (Table 12.4). The final step in virion production is budding of the particle into the lumen of one of these compartments, for example, of bunyaviruses into Golgi cisternae. These particles therefore lie within membrane-bound organelles. It is generally assumed that such virus particles must be packaged within cellular transport vesicles for travel along the secretory pathway to the cell surface, as illustrated for herpesviral particles (see Fig. 13.18). Despite almost universal acceptance of this mechanism, we have little direct evidence for this mode of transport.

The budding of virus particles into internal compartments of the secretory pathway is initiated by interactions among the cytoplasmic domains of viral membrane proteins and internal components of the particle. Consequently, this process generally begins as soon as the integral membrane and cytoplasmic viral proteins attain sufficient concentrations in the infected cell. For example, the concentration of viral membrane proteins (surface proteins) determines the fate of hepadnaviral cores, which contain the capsid (C) protein (Table 13.1), a DNA copy of the pregenomic RNA, and the viral polymerase (Appendix A, Fig. 3). Early in infection, the concentration of the large surface protein in ER membranes is too low for efficient envelopment of cores, which therefore enter the nucleus, where they contribute to the pool of viral DNA templates for transcription. As the concentration of the viral membrane protein increases, it interacts with cores, and enveloped particles form and bud into the ER. The ability of hepadnaviral cores to bind to the viral glycoprotein is also regulated by the nature of the nucleic acid they contain. Nascent cores that contain the pregenomic RNA do not enter the ER or leave the cell. Synthesis of DNA within the core must therefore induce some change in its exterior surface that allows association with large surface protein molecules within the ER membrane. It is clear that production of (–) strand DNA activates envelopment, but the nature of the signal that regulates acquisition of the hepadnaviral envelope remains to be determined.

An intriguing question is why some viruses formed by a budding mechanism assemble at intracellular membranes rather than at the plasma membrane, where budding ensures release of the virion from the host cell. One possible advantage of intracellular budding is that the concentration of viral glycoproteins exposed on the surface of the infected cell is reduced. This property would decrease the likelihood that the infected cell would be recognized by other components of the immune system, before the maximal number of progeny virions were assembled and released. Alternatively, the simpler cytoplasmic surfaces of internal membranes, which are not burdened with cytoskeletal structures and the proteins that attach them to the extracellular matrix, may make for more facile assembly or budding reactions. Or the distinctive lipid composition of internal membranes may confer some special property (as yet unknown) advantageous to these viruses.

The interaction of vaccinia virus with internal cellular membranes is most unusual. The DNA and all proteins of these complex particles are synthesized and assembled in the cytoplasm of the infected cell, and all steps in the infectious cycle appear to take place in association with cellular membranes. The viral DNA associates preferentially with the cytoplasmic face of the ER membrane upon release from the cores that enter the host cell. A remarkable feature of vaccinia virus reproduction is the assembly of two **different** infectious particles, the intracellular mature virus and the extracellular enveloped virion. Furthermore, infectious particles can leave the host cell by at least three distinct routes. Although many of the mechanisms of assembly and release of these particles are only now being worked out, it is clear that assembly begins with formation of viral crescents, curved segments of membrane with spikes on their convex side associated with structures containing newly replicated viral DNA (Fig 13.16A). These structures mature to the spherical immature virus covered by two membranes, which become clearly visible when the immature virus forms the brick-shaped, intracellular mature virus (Fig. 13.16). The origin of these two membranes has been much debated, but most evidence indicates that they are derived from the membranes of the intermediate compartment by a process of wrapping (see Chapter 12). The mechanism by which vaccinia virus proteins induce wrapping of these membranes during assembly of the immature virus is not well understood, although viral proteins required for specific reactions in assembly are being identified at an increasing rate (e.g., Table 12.7; Fig 13.16B).

The intracellular mature virus is released only upon lysis of the infected cell. However, a variable proportion of this population becomes engulfed by the membranes of a second intracellular compartment, probably a post-Golgi or endocytic compartment, to form the intracellular enveloped virus (Fig. 13.16B). This particle can be released from the cell as the three-membrane-containing extracellular enveloped virion following transport to the cell surface and fusion of its outer membrane with the plasma membrane (Fig. 13.16B). As the intracellular mature virion and the extracellular enveloped virion bind to different cell surface receptors, the release of two types of infectious particle may increase the range of cell types that can be infected.

The exit of intracellular enveloped vaccinia virus from the host cell, an unusually well understood mechanism of viral egress, is truly amazing. These particles initially travel from perinuclear sites of assembly to the plasma membrane on microtubules, carried by the cellular motor protein kinesin. Fusion of intracellular envelope virions with the plasma membrane releases particles that remain cell associated (cell-associated virions) because of a remarkable activity: they induce a dramatic reorganization of the actin cytoskeleton just below the site of fusion. The number of typical actin stress fibers is significantly decreased, because the virus induces the formation of new filamentous

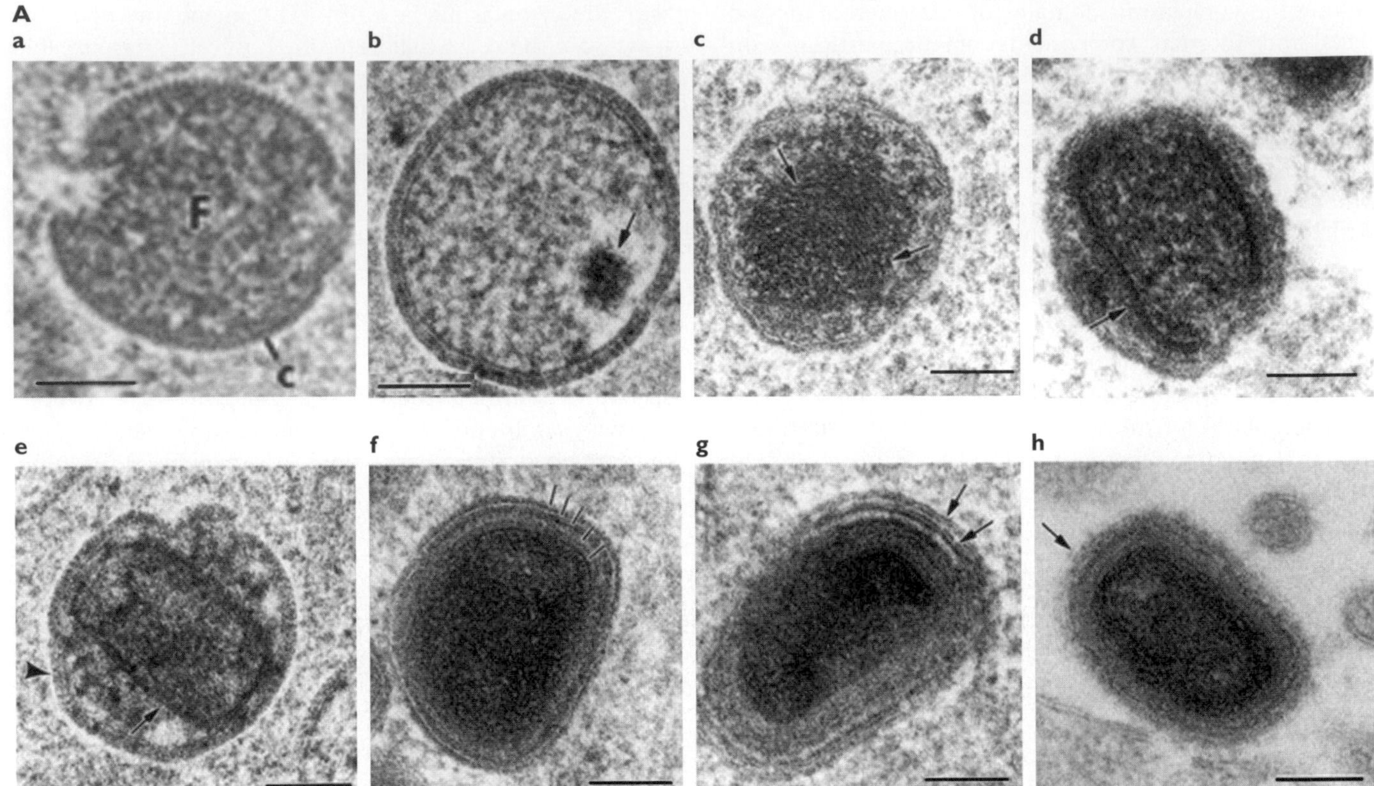

Figure 13.16 Vaccinia virus assembly and exocytosis. (A) Viral structures observed in infected cells. HeLa cells infected with vaccinia virus for 10 or 24 h were prepared for electron microscopy by quick freezing and negative staining while frozen. These procedures preserve fine structural detail. The following structures are shown. (a) Viral factory comprising a viral crescent (c) around a viroplasm focus (F). (b) Immature virions, in this example associated with DNA (arrows) entering via a pore in the particle. (c) Spherical, dense particles containing DNA-like material (arrows). (d and e) Potential intermediates in the assembly of the internal core (arrows) and the intracellular mature virion. These structures retain the IV membrane (arrowhead in e) around the assembling core. (f) Intracellular mature virion in which at least five layers (short lines) can be distinguished. (g) Intracellular enveloped virion with the additional double membrane (arrows). (h) Extracellular enveloped virion, which carries an external fuzzy layer (arrow). Bars = 100 nm. Adapted from C. Risco et al., *J. Virol.* **76:**1839–1855, 2000, with permission. (B) Schematic model of assembly and exocytosis. The model shows the formation of immature virus by the acquisition of two cellular membranes from a compartment of the *cis*-Golgi network and its maturation to the infectious intracellular mature virion. These cellular membranes are modified by the incorporation of viral proteins (Table 12.7) to form crescent-shaped structures on the periphery of the specific cytoplasmic areas in which viral DNA synthesis takes place (viral factories). The altered membranes coalesce into spherical structures containing granular material acquired from the viral factories and then mature to form the intracellular mature virion. The intracellular mature virion is released from infected cells upon lysis. However, a small proportion of these structures undergo a second wrapping, in membranes derived from a late or post-Golgi compartment, to form the intracellular enveloped virion. This particle is transported to the plasma membrane, where it induces formation of actin tails (Fig. 13.17) following fusion with this cellular membrane to form cell-associated enveloped virus. Viral proteins that are required for specific steps in assembly and egress are indicated. The upper and lower insets show cryo-electron micrographs of sections through the intracellular mature virion and the intracellular enveloped virion, respectively. Adapted from B. Sodeik and J. Krijnse-Locker, *Trends Microbiol.* **10:**15–24, 2002, with permission. Courtesy of J. Krijnse-Locker, European Laboratory of Molecular Biology, Heidelberg, Germany.

actin-containing structures (Fig. 13.17A and B). Each of these, which are termed actin tails, is in contact with a single virus particle. Their formation requires viral genes that encode envelope proteins of the intracellular enveloped virion, such as A36R. Viral particles attached to the tips of actin tails (Fig. 13.17C) are propelled at an average speed of 0.18 nm/s, by polymerization of actin at the front end of the tail and its depolymerization at the back end. As the infection progresses, they can be seen on large microvilli induced by the actin tails (Fig. 13.17C).

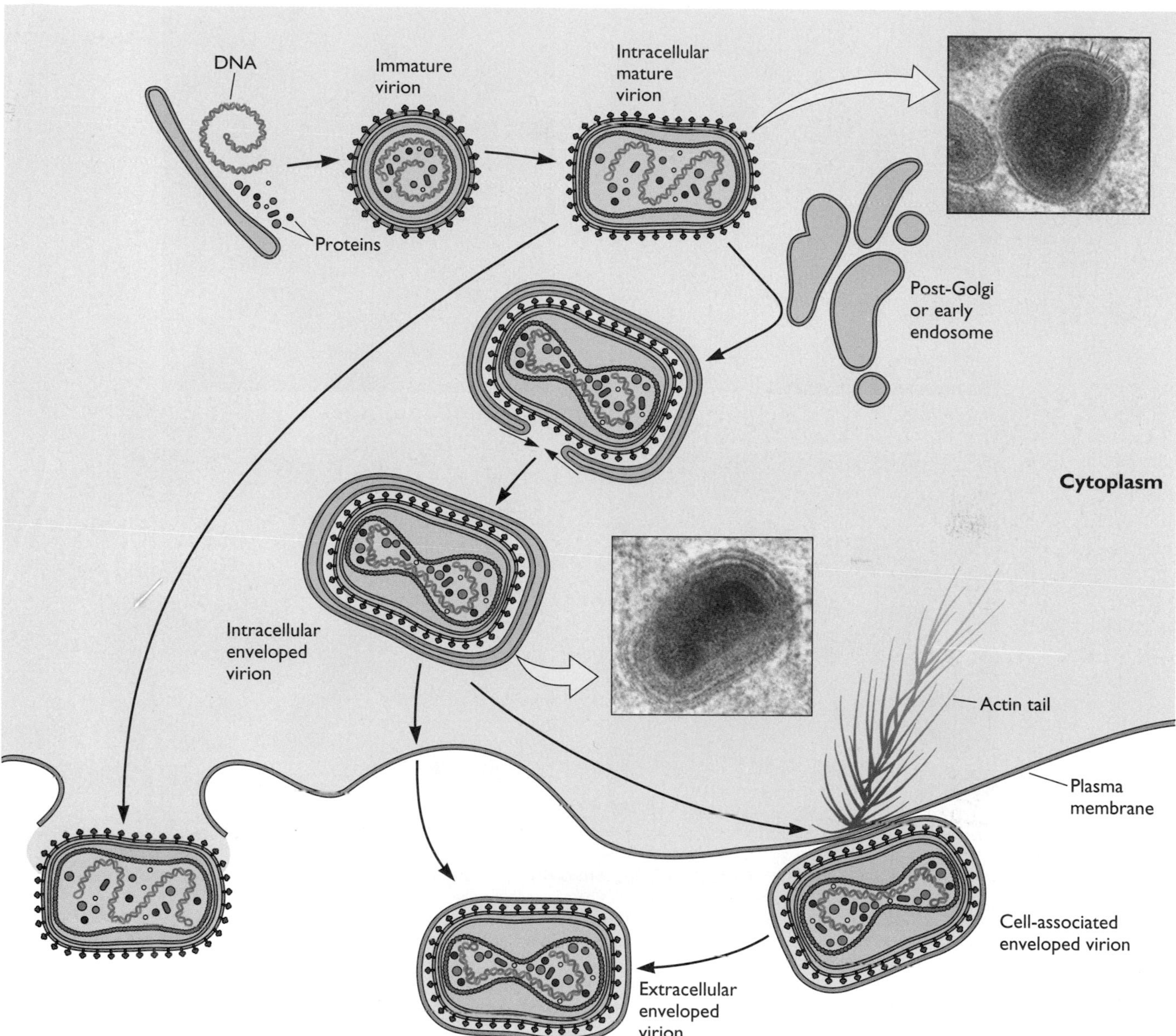

Figure 13.16 ***(continued)***

Formation of vaccinia-actin tails is the result of modulation of a cellular mechanism that controls the organization of the actin cytoskeleton. This process depends on host cell Src family protein kinases, which phosphorylate the viral A36R transmembrane protein that is required for induction of actin tails. This modification in turn initiates a signaling cascade that controls cellular proteins with actin-nucleating activity.

The formation of vaccinia-actin tails is necessary for efficient spread of the virus, for mutants that cannot induce these structures form only small plaques on cultured cells. The viral particles attached to the outer surface of cellular projections containing actin tails (Fig 13.17C) can dissociate to become extracellular enveloped virions. However, such projections can extend from infected cells into neighboring uninfected cells, suggesting that they may also facilitate direct cell-to-cell spread of infectious particles.

Intranuclear Assembly

The problem of egress is especially acute for the enveloped herpesviruses, because the nucleocapsids assemble in the nucleus. For years, the pathway by which the

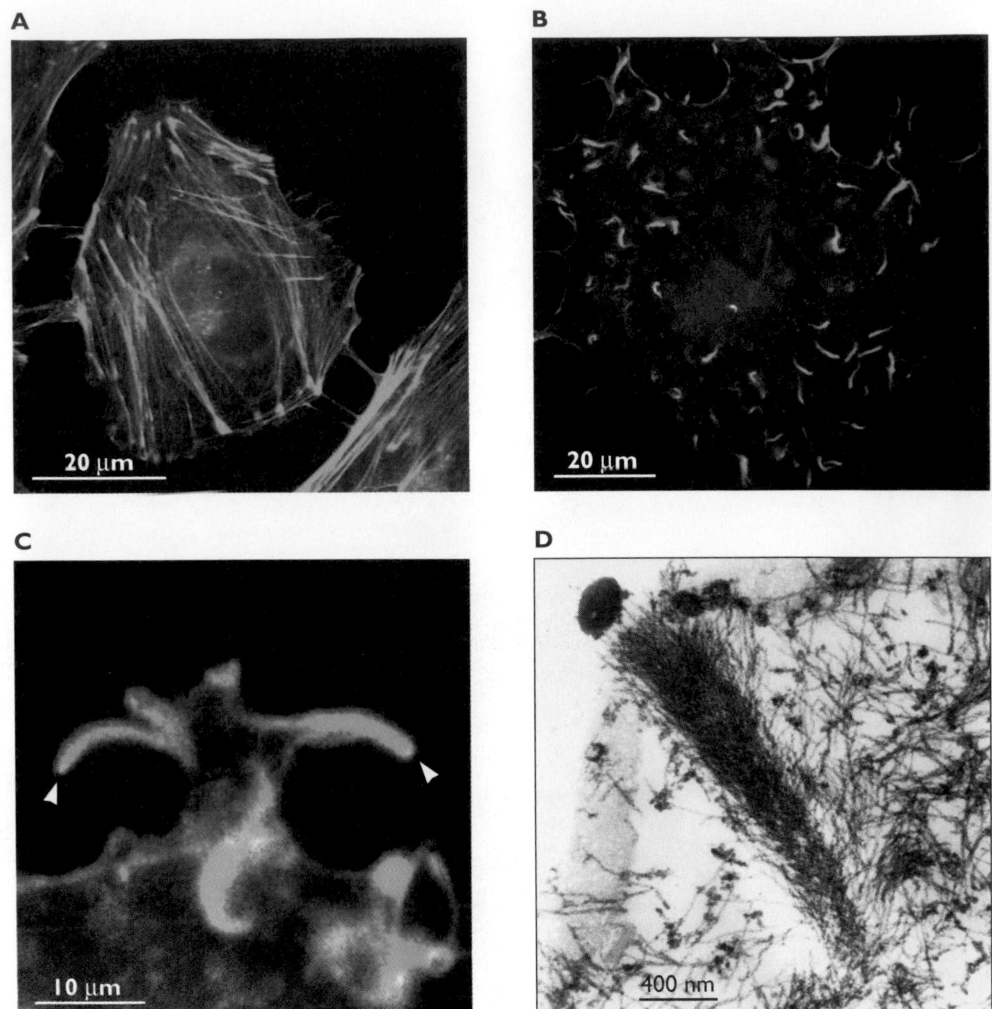

Figure 13.17 Movement of vaccinia virus on actin tails. In the immunofluorescence micrographs (A to C), actin is green and vaccinia virus particles are red. The dramatic reorganization of the actin cytoskeleton of an uninfected HeLa cell (A) induced by the virus can be seen (B). (C) A close-up of virus particles at the ends of the cell surface projections containing actin tails. The coincidence of the tips of the projecting actin and viral particles gives yellow-orange signals (marked by white arrowheads), indicating that the particles are projected from the cell surfaces on the tips of actin tails. When infected cells are plated with uninfected cells, such actin-containing structures to which virus particles are attached can be seen extending from the former into the latter. (D) An electron micrograph of a virus particle attached to an actin tail. Adapted from S. Cudmore et al., *Nature* **378:**636–638, 1995 (C and D); F. Frischknecht et al., *Curr. Biol.* **9:**89–92, 1999 (A); and C. M. Sanderson et al., *J. Gen. Virol.* **79:**1415–1425, 1998 (B), with permission. Courtesy of F. Frischknecht, S. Cudmore, and M. Way, European Molecular Biology Laboratory.

virus leaves the cell was a topic of fierce controversy. There is complete agreement that assembled nucleocapsids bud through the inner nuclear membrane, a process described in Chapter 12. The controversy concerned the mechanism by which herpesviral nucleocapsids acquire the tegument and envelope prior to exit from the host. It was believed for many years that the entire tegument is deposited around the nucleocapsid, prior to budding of the particle through the inner nuclear membrane, and that this envelope is retained as the particle leaves the nucleus, enters the secretory pathway, and travels to the cell surface. However, it is now generally accepted that the assembled nucleocapsid bound by only a subset of tegument proteins initially buds through the inner nuclear membrane. Moreover, the nuclear membrane acquired in this process is lost upon fusion of the immature particle with the outer nuclear membrane to release the nucleocapsid into the cytoplasm (Fig. 13.18). Additional tegument proteins assemble with this structure at the site of final envelopment, a membrane surface of a late Golgi compartment or a late endosome. Some evidence suggests that the tegument proteins bind to cytoplasmic tails of viral membrane proteins at the site of envelopment. An alternative idea is that some tegument proteins bind to the cytoplasmic nucleocapsids prior to engaging the cytoplasmic tails of the viral membrane proteins. It may be that some tegument proteins function like the matrix proteins of enveloped RNA viruses to facilitate budding and assembly. Further work is necessary to understand the complex problem of tegument acquisition and secondary envelopment of herpesviruses.

Maturation of Progeny Virions

Proteolytic Processing of Virion Proteins

The products of assembly of several viruses are noninfectious particles, often called **immature virions.** In all cases, proteolytic processing of specific proteins with which the particles are initially built converts them to infectious virions. Such maturation cleavages should not be confused with the proteolytic processing of some viral glycoproteins that takes place during their transit to cellular membranes, and that can also be necessary for production of infectious particles. The maturation reactions are carried out by virus-encoded enzymes and take place late in assembly of particles or following release of immature virions from the host cell. Proteolytic cleavage of virion proteins introduces an irreversible reaction into the assembly pathway, driving it in a forward direction. This modification can also make

Figure 13.18 Pathway of herpesvirus assembly and egress. The mature nucleocapsid assembled within the nucleus (Fig. 13.5 and 13.10) initially acquires an envelope by budding through the inner nuclear membrane. Upon fusion with the outer nuclear membrane, this membrane is lost as unenveloped nucleocapsids are released into the cytoplasm. The addition of tegument proteins, a process that is not well understood, is illustrated schematically. The final envelope is acquired upon budding of tegument-containing structures into a late compartment of the secretory pathway. Viral gene products implicated in specific reactions are indicated. The reactions are illustrated in the corresponding electron micrographs of cells infected by the alphaherpesvirus pseudorabies virus. Bar = 150 nm. Adapted from T. C. Mettenleiter, *J. Virol.* **76:**1537–1547, 2002, with permission. Courtesy of T. C. Mettenleiter, Federal Research Center for Virus Diseases of Animals, Insel Riems, Germany.

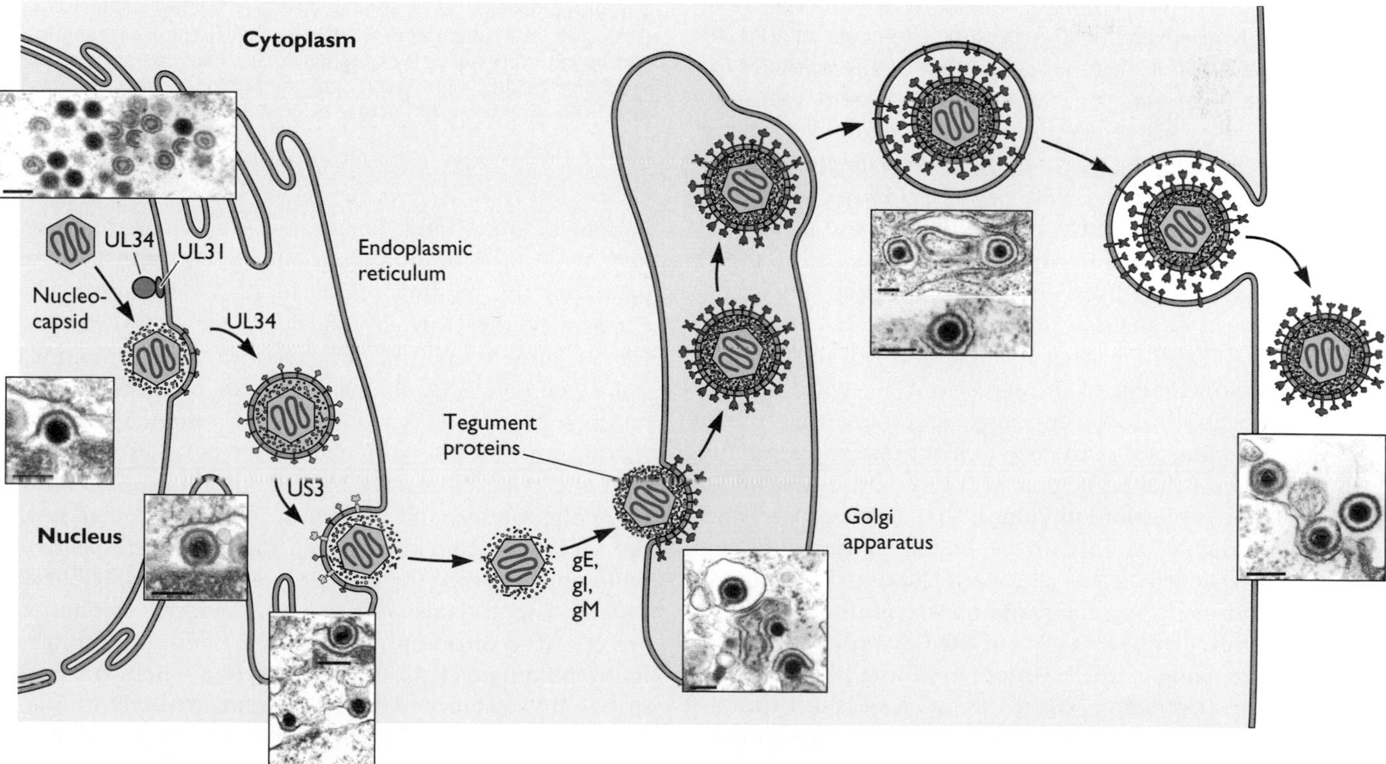

an important contribution to resolving the contradictory requirements of assembly and virus entry. One consequence of proteolytic processing is the exchange of covalent linkages between specific protein sequences for much weaker noncovalent interactions, which can be disrupted in a subsequent infection. A second is the liberation of a new N terminus and a new C terminus at each cleavage site, and hence opportunities for additional protein-protein contacts. Such changes in chemical bonding among virion proteins clearly facilitate virus entry, for the proteolytic cleavages that introduce them are necessary for infectivity. Accordingly, viral proteases and the structural consequences of their action are of considerable interest. Moreover, these enzymes are excellent targets for antiviral drugs, a property exemplified by the success of therapeutic agents that inhibit the human immunodeficiency virus type 1 protease.

The alterations in the structure of the virus particle and their functional correlates are best understood for small RNA viruses such as poliovirus. A single cleavage to liberate VP4 and VP2 from VP0 converts noninfectious provirions to mature infectious virions (Fig. 13.3). As the viral proteases are not incorporated into particles, VP0 cleavage may be catalyzed by a specific feature of the virion itself with internal genomic RNA participating in the reaction. The structural changes induced by such maturation cleavage can be described in great detail, for the structures of mature virions and empty particles in which VP0 have been determined at high resolution. Cleavage of VP0 allows the extensive internal structures of the particle, including the network of numerous intra- and interpentamer contacts via segments of VP4 and the N-terminal arms of VP2 and VP1 (Fig. 4.10), to be established in its final form. Cleavage of VP0 therefore allows additional protein-protein interactions that make important contributions to the stability of the virion.

Cleavage of VP0 to VP4 and VP2 is also necessary for release of the RNA genome into a new host cell. The conformational transitions that mediate entry of the genome following attachment of the virus to its receptor are not fully understood. However, many alterations that impair receptor binding and entry map to just those regions of the capsid proteins that participate in the structures that adopt their final organization only upon VP0 cleavage. Cleavage of VP0 therefore not only further stabilizes the virion, but also "spring-loads" it for the conformational transitions that take place during entry and release of the genome.

Upon proteolytic processing of the Gag polyprotein, immature retroviral particles undergo substantial morphological rearrangements (Fig. 13.19) as a result of conformational reorganization (Box 13.5). Conformational changes in segments of Gag during assembly may regulate

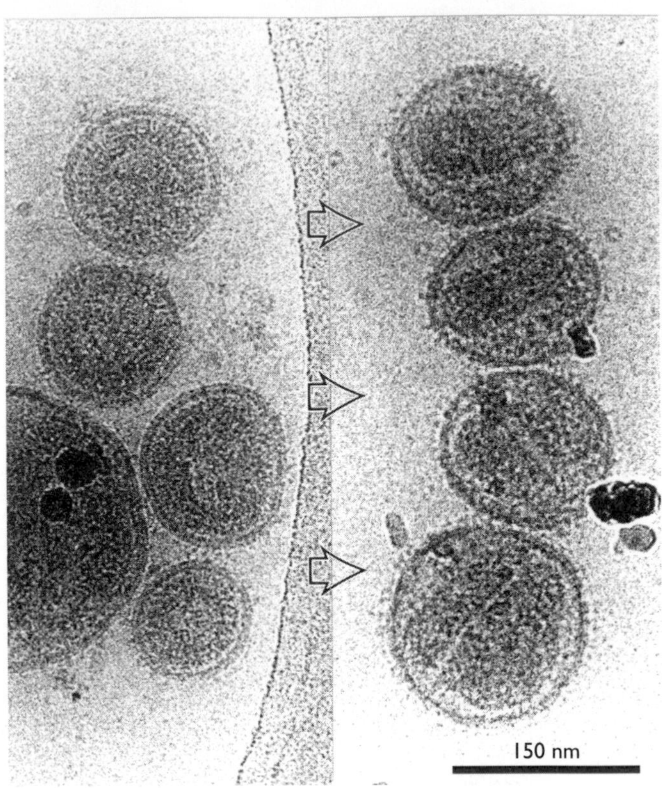

Figure 13.19 Morphological rearrangement of retrovirus particles upon proteolytic processing of the Gag polyprotein. These two cryo-electron micrographs show the maturation of human immunodeficiency virus type 1 virions. The immature particles (left) contain a Gag polyprotein layer below the viral membrane and its external spikes. Processing of Gag converts such particles to mature virions (right) with elongated internal capsids. Courtesy of T. Wilk, European Molecular Biology Laboratory.

proteolytic processing. For example, efficient cleavage to liberate human immunodeficiency virus type 1 NC depends on the binding of NC to RNA, at least in vitro. Conversely, the cleavage that liberates the N terminus of the NC protein appears to be necessary for formation of stable genomic RNA dimers within the particle.

These processing reactions play an essential part in the mechanisms by which infectious retroviruses are assembled and released. As we have seen, interactions among Gag polyproteins, and of their NC and MA domains with the viral RNA and the plasma membrane, respectively, build and organize the assembling particle. Efficient and orderly assembly also depends on "spacer" peptides that are removed during proteolysis. Furthermore, the membrane-binding signal of MA is exposed when MA is part of Gag, but is blocked by a C-terminal α-helix of MA in the mature protein. It is therefore very unlikely that particles could be correctly constructed from mature Gag

BOX 13.5

Model for refolding of the human immunodeficiency virus type 1 CA protein on proteolytic processing of Gag

The model for the radial organization of the human immunodeficiency virus type 1 Gag polyprotein (left) is based on cryo-electron micrographs like that in Fig. 13.4. The three-dimensional structures of the processed proteins, MA (green), CA (blue), and NC (violet) bound to the SL3 packaging signal (red), shown on the right are derived from high-resolution structures discussed in this and preceding chapters.

In the X-ray crystal structure of the N-terminal portion of mature CA (right), the charged N terminus is folded back into the protein by a β hairpin formed by amino acids 1 to 13 and forms a buried salt bridge with the carboxylate of Asp51. The lack of a charged N terminus prior to cleavage of CA from MA, and the steric difficulties of burying the N terminus of CA attached to an MA extension (left), indicate that the β hairpin and buried salt bridge can form only after proteolytic cleavage. Furthermore, the viral protease recognizes the cleavage site between MA and CA in an extended conformation. As the N-terminal β hairpin of mature CA forms a CA-CA interface, it has been proposed that proteolytic cleavage and the consequent refolding of the N terminus of CA facilitate the rearrangements to form the conical core during maturation of virus particles. The inhibition of core assembly and formation of infectious viral particles in vivo by alteration of amino acids in this interface support this model. Figure courtesy of T. L. Stemmler and W. Sundquist, University of Utah.

von Schwedler, U. K., T. L. Stemmler, V. Y. Klishko, S. Li, K. H. Albertine, D. R. Davis, and W. I. Sundquist. 1998. Proteolytic refolding of the HIV-1 capsid protein amino-terminus to facilitate viral core assembly. *EMBO J.* **17:**1555–1568.

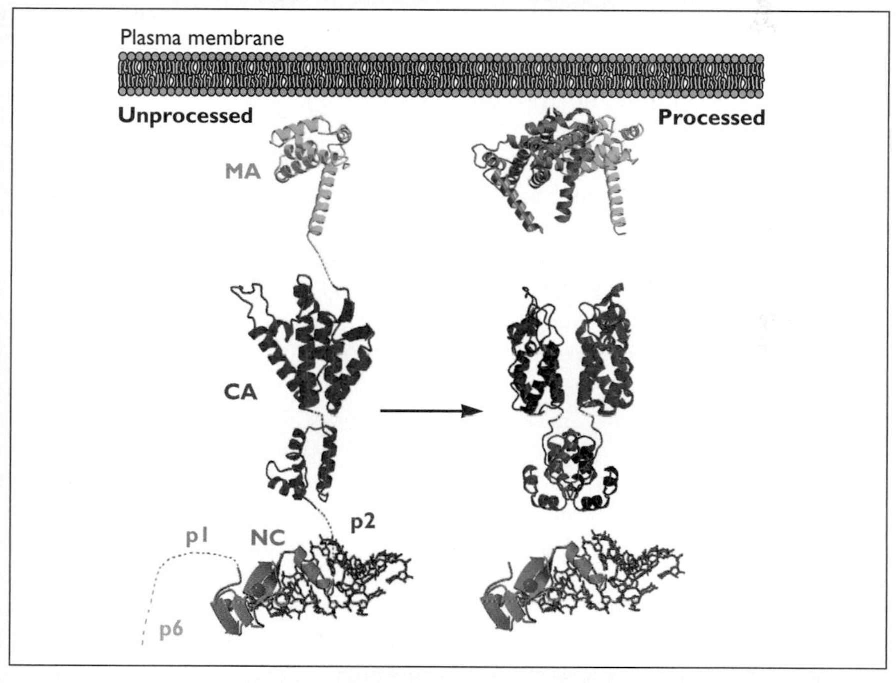

proteins. Indeed, alterations that increase the catalytic activity of the retroviral protease inhibit budding and production of infectious particles, indicating that premature processing of the polyproteins is detrimental to assembly. On the other hand, the covalent connection of the virion proteins that is so necessary during assembly is incompatible with release of the virion core following fusion of the viral envelope with the membrane of a new host cell. Such covalent linkage also precludes efficient activity of virion enzymes, which are incorporated as Gag-Pol proteins. In some retroviruses, including Moloney murine leukemia virus, the protease also removes a short C-terminal segment of the cytoplasmic tail of the TM envelope protein to activate the fusionogenic activity of TM. The retroviral proteases that sever such connections therefore are absolutely necessary for production of infectious virions, even though they are not required for formation of particles.

The retroviral proteases belong to a large family of enzymes with two aspartic acid residues at the active site (aspartic proteases). The viral and cellular members of this family are similar in sequence, particularly around the active site, and are also similar in three-dimensional structure (see Fig. 19.17). All aspartic proteases contain an active site formed between the two lobes of the protein, each of which contributes a catalytic aspartic acid. The retroviral proteases are homodimers, in which each monomer corresponds to a single lobe of their cellular cousins. Consequently, the active site is formed only upon dimerization of two identical subunits. This property undoubtedly helps avoid premature activity of the protease within infected cells, in which the low concentration of their polyprotein precursors mitigates against dimerization. Indeed, dimerization of the protease appears to be rate limiting for maturation of virions. Fusion of the protease to the NC domain of Gag also inhibits dimerization. Consequently, synthesis of the protease as part of a polyprotein precursor not only allows incorporation of the enzyme into assembling virions but also contributes to regulation of its activity. These properties raise the question of how the protease is activated, a step that requires its cleavage from the polyprotein. Polyproteins containing the protease (e.g., made in bacteria) possess some activity, sufficient to liberate fully active enzyme at a very low rate in vitro. It is therefore believed that such activity of the polyproteins initially releases protease molecules within the particle. The high local concentrations of protease molecules within the assembling particle would facilitate the initial *trans* cleavages, as well as dimerization of protease molecules to form the fully active enzyme.

Like its retroviral counterpart, the adenoviral protease converts noninfectious particles to infectious virions, in this case by cleavage at multiple sites within six virion proteins (Fig. 13.6). Although the adenoviral enzyme does not process polyprotein precursors, the cleavage of so many proteins also appears to alter protein-protein interactions necessary for virion assembly, in preparation for early steps in the next infectious cycle. The enzyme is a cysteine protease containing an active-site cysteine and two additional cysteines, all highly conserved. One mechanism by which its activity is regulated is by interaction with a small peptide, a product of cleavage of the virion protein pVI, or with pVI itself (Fig. 13.20). The pVI peptide binds covalently via a disulfide bond to the proteases both in vitro and in virions to increase the catalytic efficiency of the enzyme over 1,000-fold. Its binding site is not located close to the active site of the protease, suggesting that pVI peptide induces or stabilizes an active-site conformation that is optimal for catalysis.

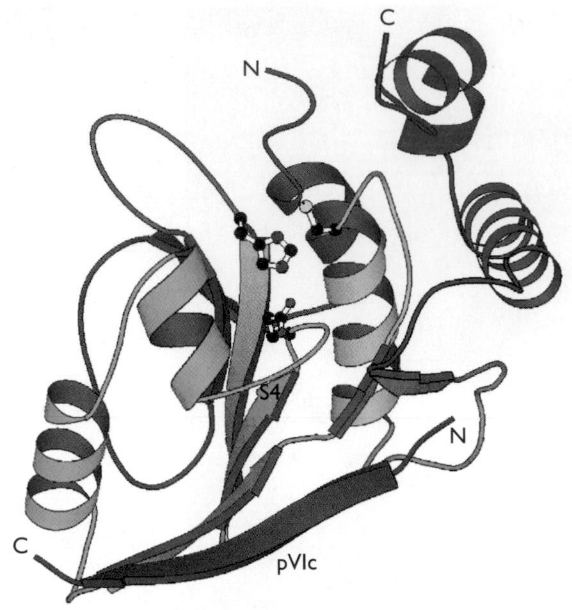

Figure 13.20 Three-dimensional structure of the human adenovirus type 2 protease bound to the pVI peptide. This structure, determined by X-ray crystallography of the complex and shown as a ribbon diagram, allowed identification of the active site of the enzyme (at which side chains are shown) by comparison with the secondary and tertiary structures of cellular cysteine protease. This structure closely resembles that of papain, suggesting that the adenoviral protease employs the same catalytic mechanism. The β-strand formed by the activating pVIc peptide (deep red) is covalently linked to the enzyme by a disulfide bond, but also makes many noncovalent interactions. Examination of the structure suggests that the cofactor might activate the enzyme by stabilizing the conformation of the active site optimal for catalysis, or by altering the conformation of the protease active site. Adapted from J. Ding et al., *EMBO J.* **15:**1778–1783, 1995, with permission. Courtesy of W. Mangel, Brookhaven National Laboratory.

Other Maturation Reactions

Newly assembled virus particles appear to undergo few covalent modifications other than proteolytic processing reactions. One important exception is the removal of terminal sialic acid residues from the complex oligosaccharides added to the envelope HA and NA glycoproteins of influenza A virus during their transit to the plasma membrane. The influenza A virus receptor is sialic acid, which is specifically recognized by the HA protein. Consequently, newly synthesized virus particles have the potential to aggregate with one another, and with the surface of the host cell, by binding of an HA molecule on one particle to a sialic acid present in an envelope protein of another particle or on cell surface proteins. Such aggregation is observed when the influenza virus neuraminidase is inactivated. The activity of this enzyme, which removes terminal sialic acid residues from oligosaccharide chains, is essential for effective release of progeny virus particles from the surface of a host cell. Presumably, the neuraminidase eliminates such binding of newly synthesized virions to one another and to cell surface proteins.

Cell-to-Cell Spread

The raison d'être of all progeny virions is to infect a new host cell in which the infectious cycle can be repeated. Many viruses are released as free particles by the mechanisms described in preceding sections and must travel within the host until they encounter a susceptible cell. The new host cell may be an immediate neighbor of that originally infected or a distant cell reached via the circulatory or nervous systems of the host. Virions are designed to withstand such intercellular passage, but they are susceptible to several host defense mechanisms that can destroy virus particles (Chapter 15). Localized release of virus particles only at points of contact between an infected cell and its uninfected neighbor(s) can minimize exposure to these host defense mechanisms. Furthermore, some viruses can spread from one cell to another by mechanisms that circumvent the need for release of progeny virions into the extracellular environment.

In some cases, viruses can be transferred directly from an infected cell to its neighbors (Box 13.6), a strategy that avoids exposure to host defense mechanisms targeted against extracellular virus. Such cell-to-cell spread, which is defined operationally as infection that still occurs when released virus particles are neutralized by addition of antibodies, depends on the viral fusion machinery. In the case of herpes simplex virus type 1, the glycoproteins gD, gB, gH, and gL are required, while additional glycoproteins, gE and gL, facilitate this process. Mutant viruses that lack the gE or gI genes form only small plaques when transfer of free virus particles from one cell to another is prevented. They are also defective for both lateral spread of infection in polarized epithelial cells and spread of infection from an axon terminal to an uninfected neuron in animals. Such cell-to-cell spread of herpesviruses is believed to occur at specialized junctions, such as tight junctions of epithelial cells, sites of synaptic contact between individual neurons, and adhesion contacts between fibroblasts in culture. Experiments with herpes simplex virus gE mutants in polarized cells demonstrate that in the absence of gE, virions do not accumulate at lateral surfaces with tight junctions. The junctions between diverse cell types may therefore have more in common than we presently appreciate. Exactly how gE/gI glycoprotein oligomers promote cell-to-cell spread via these specialized junctions is not yet clear.

Persistent measles virus (a paramyxovirus) infection of the brain is associated with subacute sclerosing panencephalitis (Appendix B, Fig. 14). Spread of this virus between neurons occurs by a mechanism different from that in nonneuronal tissue: little infectious virus can be recovered from brain tissue of patients with this disease, although the genomic RNA and viral proteins are present. Indeed, budding of virus particles does not take place from the surfaces of infected mouse or human neurons in culture, which contain nucleocapsids accumulating at presynaptic membranes. Nor does spread of measles virus between cultured neurons require the viral receptors. Rather, cell-cell contact and the fusion protein are necessary. In neurons, measles virus therefore spreads without release and attachment to infected cells of free virus particles, but rather from cell to cell, perhaps through synapses.

BOX 13.6

Extracellular and cell-to-cell spread

Many viruses spread from one host cell to another as extracellular virions released from an infected cell. (A) Such extracellular dissemination is necessary to infect another naive host. Some viruses, notably alphaherpesviruses, can also spread from cell to cell without passage through the extracellular environment (B) and therefore can use both mechanisms (C).

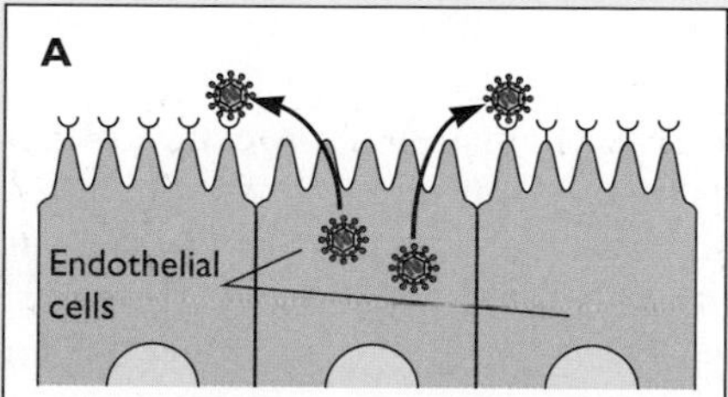

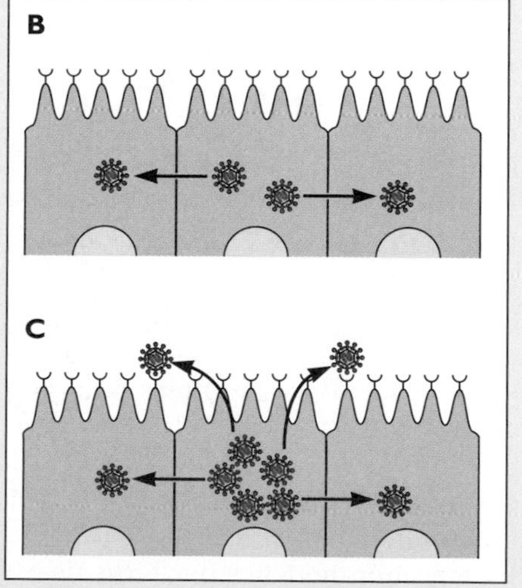

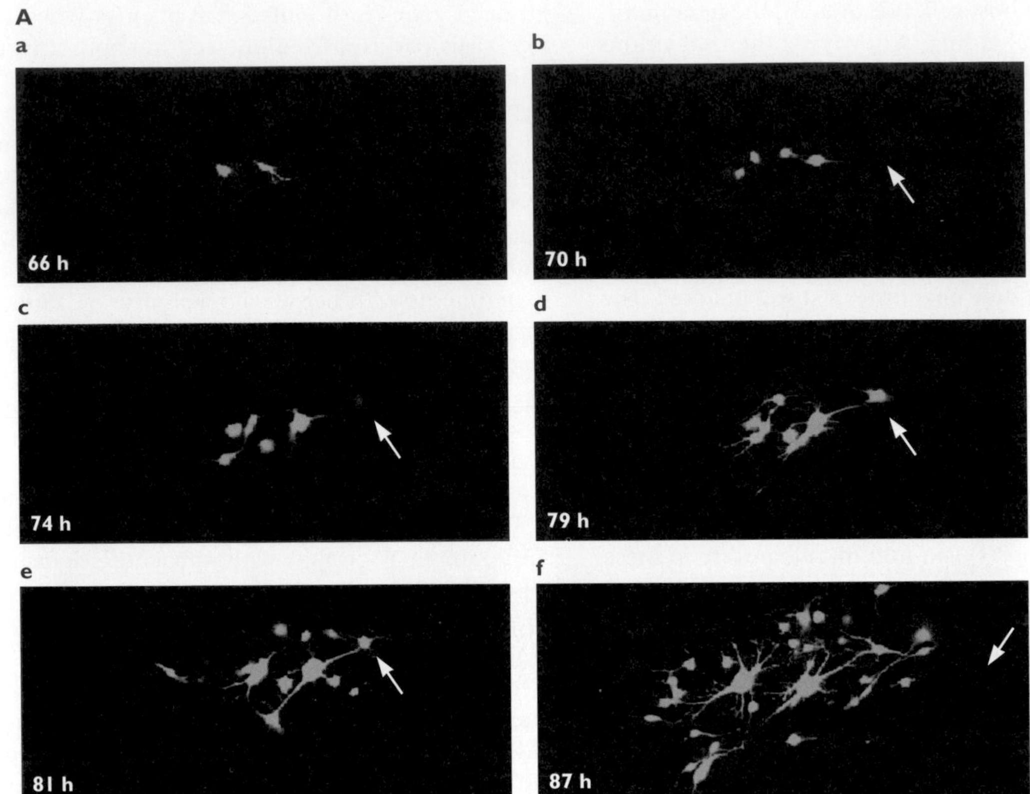

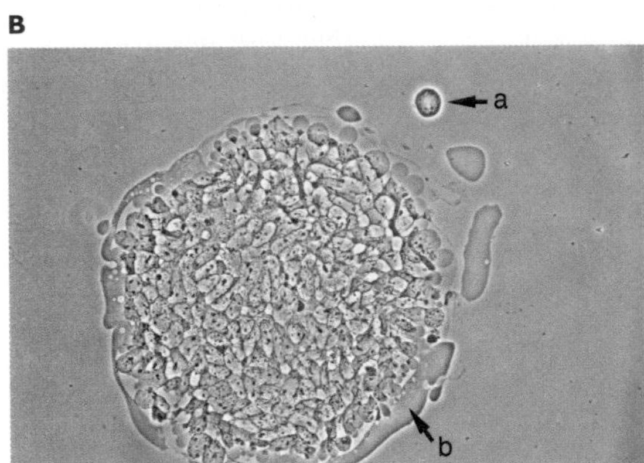

Figure 13.21 Cell-to-cell spread via formation of syncytia. (A) Cell-to-cell spread by measles virus. Human astrocytoma cells were infected at low multiplicity with a recombinant measles virus carrying a green fluorescent protein coding sequence in its genome. The autofluorescence of this protein identifies infected cells (a) and allows spread of the virus to be monitored in living cells. With increasing time, the virus spreads to cells neighboring those initially infected and can be clearly seen in the processes connecting the cells that become infected (b to f). The arrows point to an extended astrocyte process of a newly infected cell (b), the weak autofluorescence of a nucleus of a cell in a very early phase of infection (c), the nucleus from panel c 5 and 7 h later (d and e, respectively), and an extended astrocytic process issuing from the cell shown in panels d and e (f). From W. P. Duprex et al., *J. Virol.* **73:** 9568–9575, 1999. Courtesy of W. D. Duprex, Queen's University, Belfast, United Kingdom. (B) Cell-to-cell spread by human immunodeficiency virus type 1 in a T-cell line. The photograph shows a large syncytium of SupT1 cells (a $CD4^+$ T-lymphotropic cell line) that had been infected with a viral vector that expresses the *env* gene from a dualtropic strain of the virus. Large quantities of this Env protein accumulate at the cell surfaces mediating fusion. Arrow a shows a single cell for comparison. Arrow b indicates the cytoplasm surrounding the multiple nuclei present in the large syncytium. Courtesy of Matthias Schnell, Philip McKenna, and Joseph Kulkosky, Thomas Jefferson University School of Medicine, Philadelphia, Pa.

There are other examples of more radical mechanisms of transfer. In astrocyctes (supporting cells of the central nervous system), measles virus spreads by inducing the formation of syncytia, sheets of neighboring cells fused to one another (Fig. 13.21A). Certain cell types infected by human immunodeficiency virus type 1 also form syncytia (Fig. 13.21B) when they would not normally do so. This property gives the virus ready access to many cell nuclei, in which proviral DNA integration and transcription take place, without the need to travel from one host cell to another. Perhaps most unusual is the direct transfer of vaccinia virus from an infected cell to its neighbors by the novel actin-containing structures induced by the virus, as described above.

The production of "decoys," noninfectious particles released in large quantities, is one alternative strategy to avoid host defense mechanisms during transmission. By far the majority of particles detected in hepatitis B virus-infected humans are empty. Another strategy would be to disguise virus particles with normal products of a host cell. Some viral envelopes retain cellular proteins, such as major histocompatibility complex class II proteins and the adhesion receptor Icam-1, in their membranes. The latter protein substantially increases the infectivity of human immunodeficiency virus type 1 particles. However, the importance of such a "wolf in sheep's clothing" strategy for the spread of a virus from one cell to another in the host has yet to be documented.

Perspectives

The assembly of even the simplest virus is a complex process in which multiple reactions must be completed in the correct sequence and coordinated in such a way that the overall pathway is irreversible. These requirements for efficient production and release of stable structures must also be balanced with the fabrication of virus particles primed for ready disassembly at the start of a new infectious cycle. The integration of information collected by the application of structural, biochemical, and genetic methods of analysis has allowed an outline of the dynamic processes of assembly, release, and maturation for many viruses. Despite the considerable structural diversity of virions, the repertoire of mechanisms for successful completion of the individual reactions (Fig. 13.1) is limited. Furthermore, we can identify common mechanisms that ensure that assembly proceeds efficiently and irreversibly, or that resolve the apparent paradox of great particle stability during assembly and release but facile disassembly at the start of the next infectious cycle. These mechanisms include high concentrations of virion components at specific sites within the infected cell, and proteolytic cleavage of virion proteins at one or more steps in the production of infectious particles. Indeed, for some smaller viruses, the structural changes that accompany the production of infectious virions from noninfectious precursor particles can be described in atomic detail. Such information has revealed unanticipated relationships between structures that stabilize virions and interactions that prime them for conformational rearrangements during entry.

On the other hand, the pathways for production and release of even the simplest virions cannot be fully described. These reactions arc difficult to study in infected cells, and even the simplest proved more difficult to reconstitute in vitro than originally anticipated. The latter property emphasizes the crucial contributions to virus assembly that can be made by cellular proteins that assist protein folding and oligomerization (chaperones), or that covalently modify virion proteins. Historically, assembly reactions have received less attention than mechanisms of viral gene expression or replication of viral genomes. However, the development of new structural methods, coupled with the experimental power and flexibility provided by modern molecular biology, has revitalized investigation of the crucial processes of assembly, release, and maturation of virus particles. This renaissance has been further stimulated by the success of therapeutic agents designed to inhibit virus-specific reactions crucial for the production of infectious particles. Consequently, the advances in our understanding of assembly and egress illustrated in this chapter seem certain to continue at an accelerated pace.

References

Chapters in Books

Hellen, C. U. T., and E. Wimmer. 1995. Enterovirus structure and assembly, p. 155–170. *In* H. A. Rotbart (ed.), *Human Enterovirus Infections*. American Society for Microbiology, Washington, D.C.

Weldon, R. A., and E. Hunter. 1997. Molecular requirements for retrovirus assembly, p. 381–410. *In* W. Chiu, R. M. Burnett, and R. L. Garcia (ed.), *Structural Biology of Viruses*. Oxford University Press, New York, N.Y.

Wood, W. B., and J. King. 1979. Genetic control of complex bacteriophage assembly, p. 581–633. *In* H. Frankel-Conrat and R. R. Wagner (ed.), *Comprehensive Virology*, vol. 13. Plenum Press, New York, N.Y.

Reviews

Ansardi, D. C., D. C. Porter, M. J. Anderson, and C. D. Morrow. 1996. Poliovirus assembly and encapsidation of genomic RNA. *Adv. Virus Res.* **46:**1–68.

Carter, C. A. 2002. Tsg101: HIV-1's ticket to ride. *Trends Microbiol.* **10:**203–205.

D'Halluin, J. C. 1995. Virus assembly. *Curr. Top. Microbiol. Immunol.* **199:**47–66.

Enquist, L. W., P. J. Husak, B. W. Banfield, and G. A. Smith. 1999. Infection and spread of alphaherpesviruses in the nervous system. *Adv. Virus Res.* **51:**237–347.

Freed, E. O. 2002. Viral late domains. *J. Virol.* **76:**4679–4687.

Garoff, H., R. Hewson, and D. J. E. Opstelten. 1998. Virus maturation by budding. *Microbiol. Mol. Biol. Rev.* **199:**227–235.

Harrison, S. C. 1995. Virus structures and conformational rearrangements. *Curr. Opin. Struct. Biol.* **15:**157–164.

Jewell, N. A., and L. M. Mansky. 2000. In the beginning: genome recognition, RNA encapsidation and the initiation of complex retrovirus assembly. *J. Gen. Virol.* **81:**1889–1899.

Johnson, D. C., and M. T. Huber. 2002. Directed egress of animal viruses promotes cell-to-cell spread. *J. Virol.* **76:**1–8.

Johnson, J. E., and W. Chiu. 2000. Structures of virus and virus-like particles. *Curr. Opin. Struct. Biol.* **10:**229–235.

Knopf, C. W. 2000. Molecular mechanisms of replication of herpes simplex virus 1. *Acta Virol.* **44:**289–307.

Mettenleiter, T. C. 2002. Herpesvirus assembly and egress. *J. Virol.* **76:**1537–1547.

Moss, B., and B. M. Ward. 2001. High-speed mass transit for poxviruses on microtubules. *Nat. Cell. Biol.* **3:**E245–E246.

Seeger, C., and J. Hu. 1997. Why are hepadnaviruses DNA and not RNA viruses? *Trends Microbiol.* **5:**447–450.

Sodeik, B., and J. Krijnse-Locker. 2002. Assembly of vaccinia virus revisited: de novo membrane synthesis or acquisition from the host? *Trends Microbiol.* **10:**15–24.

Strauss, J. H., E. G. Strauss, and R. J. Kuhn. 1995. Budding of alphaviruses. *Trends Microbiol.* **3:**346–350.

Sullivan, C. S., and J. M. Pipas. 2001. The virus-chaperone connection. *Virology* **287:**1–8.

Weber, J. 1995. The adenovirus endopeptidase and its role in virus infection. *Curr. Top. Microbiol. Immunol.* **199:**227–235.

Wien, M. W., M. Chow, and J. M. Hogle. 1996. Poliovirus: new insights from an old paradigm. *Structure* **4:**763–767.

Papers of Special Interest

Assembly of Protein Shells

Basavappa, R., R. Syed, O. Flore, J. P. Icenogle, D. J. Filman, and J. M. Hogle. 1994. Role and mechanism of the maturation cleavage of VP0 in poliovirus assembly: structure of the empty capsid assembly intermediates at 2.9Å resolution. *Protein Sci.* **3:**1651–1669.

Desai, P., N. A. DeLuca, J. C. Glorioso, and S. Person. 1993. Mutations in herpes simplex virus type 1 genes encoding VP5 and VP23 abrogate capsid formation and cleavage of replicated DNA. *J. Virol.* **67:**1357–1364.

Edvardsson, B., E. Everitt, E. Jörnvall, L. Prage, and L. Philipson. 1976. Intermediates in adenovirus assembly. *J. Virol.* **19:**533–547.

Khromykh, A. A., A. N. Varnavski, P. L. Sedlak, and E. G. Westaway. 2001. Coupling between replication and packaging of flavivirus RNA: evidence derived from the use of DNA-based full-length cDNA clones of Kunjin virus. *J. Virol.* **75:**4633–4640.

Ng, S. C., and M. Bina. 1984. Temperature-sensitive BC mutants of SV40: block in virion assembly and accumulation of capsid-chromatin complexes. *J. Virol.* **50:**471–477.

Reicin, E. S., A. Ohagen, L. Yin, S. Hoglund, and S. P. Goff. 1996. The role of Gag in human immunodeficiency virus type 1 virion morphogenesis and early steps of the viral life cycle. *J. Virol.* **70:**8645–8652.

Verlinden, Y., A. Cuconati, E. Wimmer, and B. Rombaut. 2000. Cell-free synthesis of poliovirus: 14S subunits are the key intermediates in the encapsidation of poliovirus RNA. *J. Gen. Virol.* **81:**2751–2754.

Yuen, L. K. C., and R. A. Consigli. 1985. Identification and protein analysis of polyomavirus assembly intermediates from infected primary mouse embryo cells. *Virology* **144:**127–136.

Assembly Chaperones and Scaffolds

Cepko, C. L., and P. A. Sharp. 1982. Assembly of adenovirus major capsid protein is mediated by a non-virion protein. *Cell* **31:**407–415.

Desai, P., S. C. Watkins, and S. Person. 1994. The size and symmetry of B capsids of herpes simplex virus type 1 are determined by the gene products of the UL26 open reading frame. *J. Virol.* **68:**5365–5374.

Dokland, T., R. McKenna, L. L. Hag, B. R. Bowman, N. L. Incardona, B. A. Fane, and M. G. Rossman. 1997. Structure of a viral procapsid with molecular scaffolding. *Nature* **389:**308–313.

Gao, M., L. Matusick-Kumar, W. Hurlburt, S. F. DiTusa, W. W. Newcomb, J. C. Brown, P. C. McCann III, I. Deckma, and R. J. Colonno. 1994. The protease of herpes simplex virus type 1 is essential for functional capsid formation and viral growth. *J. Virol.* **68:**3702–3712.

Gustin, K. E., and M. J. Imperiale. 1998. Encapsidation of viral DNA requires the adenovirus L1 52/55-kilodalton protein. *J. Virol.* **72:**7860–7870.

Hasson, T. B., P. D. Soloway, D. A. Ornelles, W. Doerfler, and T. Shenk. 1989. Adenovirus L1 52- and 55-kilodalton proteins are required for assembly of virions. *J. Virol.* **63:**3612–3621.

Zhou, Z. H., S. J. Macnab, J. Jakana, R. Scott, W. Chiu, and F. J. Rixon. 1998. Identification of the sites of interaction between the scaffold and outer shell in herpes simplex virus type 1 capsids by difference imaging. *Proc. Natl. Acad. Sci. USA* **95:**2778–2783.

Zimmerman, C., K. C. Klein, P. K. Kiser, A. R. Singh, B. L. Firestein, S. C. Riba, and J. R. Lingappa. 2002. Identification of a host protein essential for assembly of immature HIV-1 capsids. *Nature* **415:**88–92

Packaging the Viral Genome

Adelman, K., B. Salmon, and J. D. Baines. 2001. Herpes simplex virus DNA packaging sequences adopt novel structures that are specifically recognized by a component of the cleavage and packaging machinery. *Proc. Natl. Acad. Sci. USA* **98:**3086–3091.

Ansardi, D. C., and C. D. Marrow. 1993. Poliovirus capsid proteins derived from P1 precursors with glutamine-valine sites have defects in assembly and RNA encapsidation. *J. Virol.* **67:**7284–7297.

Bancroft, C. T., and T. G. Parslow. 2002. Evidence for segment-nonspecific packaging of the influenza A virus genome. *J. Virol.* **76:**7133–7139.

Beard, P. M., N. S. Taus, and J. D. Baines. 2002. DNA cleavage and packaging proteins encoded by genes U(L)28, U(L)15, and U(L)33 of herpes simplex virus type 1 form a complex in infected cells. *J. Virol.* **76:**4785–4791.

De Guzman, R. N., Z. R. Wu, C. C. Stalling, L. Pappalardo, P. W. Boner, and M. E. Summer. 1998. Structure of the HIV-1 nucleocapsid protein bound to the SL3 C-RNA recognition element. *Science* **279:**384–388.

Duhaut, S. D., and J. W. McCauley. 1996. Defective RNAs inhibit the assembly of influenza virus genome segments in a segment-specific manner. *Virology* **216:**326–337.

Frilander, M., and D. H. Bamford. 1995. In vitro packaging of the single-stranded RNA genomic precursors of the segmented double-stranded RNA bacteriophage f6: the three segments modulate each other's packaging efficiency. *J. Mol. Biol.* **246:**418–428.

Gordon-Shaag, A., O. Ben-Nun-Shaul, V. Roitman, Y. Yosef, and A. Oppenheim. 2002. Cellular transcription factor Sp1 recruits simian virus 40 capsid proteins to the viral packaging signal, *ses. J. Virol.* **76:**5915–5924.

Nugent, C. I., K. L. Johnson, P. Sarnow, and K. Kirkegaard. 1999. Functional coupling between replication and packaging of poliovirus replicon RNA. *J. Virol.* **73:**427–435

Sakuragi, J., T. Shioda, and A. T. Panganiban. 2001. Duplication of the primary encapsidation and dimer linkage region of human immunodeficiency virus type 1 RNA results in the appearance of monomeric RNA in virions. *J. Virol.* **75:**2557–2565.

Schmid, S. I., and P. Hearing. 1997. Bipartite structure and functional independence of adenovirus type 5 packaging elements. *J. Virol.* **71:**3375–3384.

Zhang, J., G. P. Leser, A. Pekosz, and R. A. Lamb. 2000. The cytoplasmic tails of the influenza virus spike glycoproteins are required for normal genome packaging. *Virology* **269:**325–334.

Zhang, W., and M. J. Imperiale. 2000. Interaction of the adenovirus IVa2 protein with viral packaging sequences. *J. Virol.* **74:**2687–2693.

Acquisition of an Envelope

Bruss, V., and D. Ganem. 1991. The role of envelope proteins in hepatitis B virus assembly. *Proc. Natl. Acad. Sci. USA* **88:**1059–1063.

Freed, E. O., and M. A. Martin. 1996. Domains of the human immunodeficiency virus type 1 matrix and gp41 cytoplasmic tail required for envelope incorporation into virions. *J. Virol.* **70:**341–351.

Justice, P. A., W. Sun, Y. Li, Z. Ye, P. R. Grigera, and R. R. Wagner. 1995. Membrane vesiculation function and exocytosis of wild-type and mutant matrix proteins of vesicular stomatitis virus. *J. Virol.* **69:**3156–3160.

Lenthoff, R., and J. Summers. 1994. Coordinate regulation of replication and virus assembly by the large envelope protein of an avian hepadnavirus. *J. Virol.* **68:**4565–4571.

Vangenderen, I. L., R. Brandimarti, M. R. Torrisi, G. Campadelli, and G. van Meer. 1994. The phospholipid-composition of extracellular herpes-simplex virions differs from that of host-cell nuclei. *Virology* **200:**831–836.

Yao, E., E. G. Strauss, and J. H. Strauss. 1998. Molecular genetic study of the interaction of Sindbis virus E2 with Ross River virus E1 for virus budding. *J. Virol.* **72:**1418–1423.

Virion Maturation

Bernstein, H., D. Bizub, and A. M. Skalka. 1991. Assembly and processing of avian retroviral Gag polyproteins containing linked protease dimers. *J. Virol.* **65:**6165–6172.

Kohl, N. E., E. A. Emini, W. A. Schleif, J. L. Davis, J. C. Heimbach, R. A. Dixon, E. M. Scolnick, and J. S. Sigal. 1988. Active human immunodeficiency virus protease is required for viral infectivity. *Proc. Natl. Acad. Sci. USA* **85:**4686–4690.

Lee, W.-M., S. S. Monroe, and R. R. Rueckert. 1993. Role of maturation cleavage in infectivity of picornaviruses: activation of an infectosome. *J. Virol.* **67:**2110–2122.

Webster, A., R. T. Hay, and G. Kemp. 1993. The adenovirus protease is activated by a virus-coded disulfide-linked peptide. *Cell* **72:**97–104.

Release of Virions

Clayson, E. T., L. V. Brando, and R. W. Compans. 1989. Release of simian virus 40 virions from epithelial cells is polarized and occurs without cell lysis. *J. Virol.* **63:**2278–2288.

Johnson, D. C., M. Webb, T. W. Wisner, and C. Brunetti. 2001. Herpes simplex virus gE/gI sorts nascent virions to epithelial cell junctions, promoting virus spread. *J. Virol.* **75:**821–833.

Roper, R. L., E. J. Wolfe, A. Weisberg, and B. Moss. 1998. The envelope protein encoded by the A33R gene is required for formation of actin-containing microvilli and efficient cell-to-cell spread of vaccinia virus. *J. Virol.* **72:**4192–4204.

Strack, B., A. Calistri, M. A. Accola, G. Palu, and H. G. Gottlinger. 2000. A role for ubiquitin ligase recruitment in retrovirus release. *Proc. Natl. Acad. Sci. USA* **97:**13063–13068.

Tollefson, A. E., A. Scaria, T. W. Hermiston, J. S. Ryerse, L. H. Wold, and W. S. Wold. 1996. The adenovirus death protein (E3-11.6K) is required at very late stages of infection for efficient cell lysis and release of adenovirus from infected cells. *J. Virol.* **70:**2296–2306.

Zhang, Y., and R. J. Schneider. 1995. Adenovirus inhibition of cell translation facilitates release of virus particles and enhances degradation of the cytokeratin network. *J. Virol.* **68:**2544–2555.

Pathogenesis

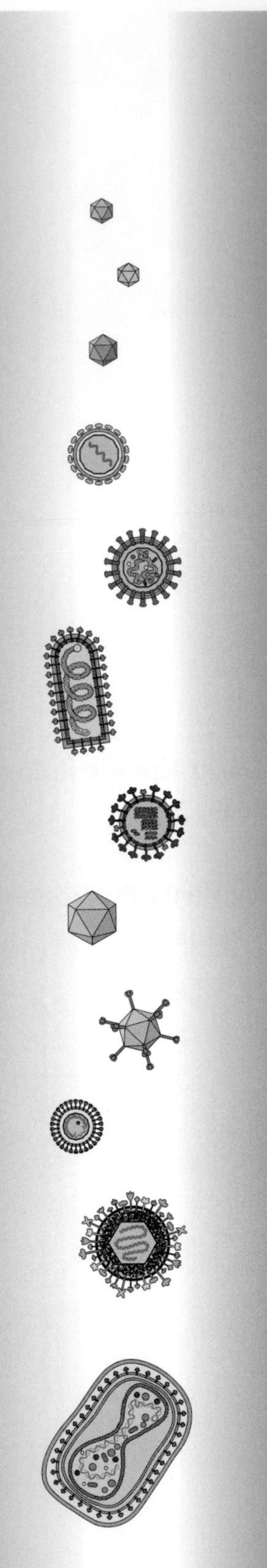

14

Dissemination, Virulence, and Epidemiology

Before I came here I was confused about this subject. Having listened to your lecture, I am still confused—but at a higher level.
ENRICO FERMI

Introduction

There is no easy answer to the seemingly straightforward question, how does a viral infection cause disease in its host? This chapter will provide a framework for dissection of this question, by defining and discussing the basic principles of **viral pathogenesis** (Table 14.1), the entire process by which viruses cause **disease** (Table 14.2). The term virulence (synonymous with pathogenicity) describes the **capacity** of a virus to cause disease. Viral disease is a sum of the effects on the host of virus replication and of the immune response.

Interest in viral pathogenesis stems in large part from the desire to treat or eliminate viral diseases that affect humans. However, progress in understanding the molecular basis of viral pathogenesis has come not from experiments with human subjects, but from studies of animal models. The mouse has become a particularly fruitful host for studying viral pathogenesis because the genome of this animal can be manipulated readily (Box 14.1). In some cases, nonhuman hosts can be infected with the same viruses that infect humans, but close relatives of human viruses must often be used. While the knowledge obtained from animal models is essential for understanding how viruses cause disease in humans, the results of such studies must be interpreted with caution. What is true for a mouse may not be true for a human. For example, although transgenic mice that express the human poliovirus receptor gene develop paralytic disease after inoculation with poliovirus, they cannot be infected by ingestion of virus, the principal route of infection in humans.

Because animals are complex organisms, the study of viral pathogenesis encompasses a broad range of issues (Table 14.3). These questions are the topics of this chapter.

This chapter is organized according to the individual steps that comprise viral pathogenesis. For information on diseases caused by specific viruses, the reader should consult Appendix B.

Table 14.1 Determinants of viral pathogenesis

Interaction of virus with target tissue
Access of virus to target tissue
Stability of virus in body
Temperature
Acid and bile of gastrointestinal tract
Capacity to establish viremia
Capacity to spread through the reticuloendothelial system
Target tissue
Presence of viral receptors
Ability of infection to kill cells (cause cytopathology)
Efficiency of viral replication in the cell
Best temperature for replication
Cell permissivity
Cytotoxic viral proteins
Inhibition of macromolecular synthesis
Production of viral proteins and structures (inclusion bodies)
Altered cell metabolism
Host response to infection
Innate immune response
Acquired immune response
Viral immune escape mechanisms
Immunopathology
Interferon: systemic symptoms
T-cell responses: delayed-type hypersensitivity
Antibody: complement, antibody-dependent cellular cytotoxicity, immune complexes

Table 14.2 Determinants of viral disease

Nature of the disease
Target tissue
Where virus enters the body
Ability of virus to gain access to target tissue
Viral tropism
Permissivity of cells
Strain of virus
Severity of disease
Ability of infection to kill cells (cytopathic effect)
Immunity to virus
Intact immune response
Immunopathology
Quantity of virus inoculated
Duration of infection
General health of the host
Host nutritional status
Other infections which might affect immune response
Host genotype
Age of host

Initiating an Infection

Basic Requirements

Three requirements must be satisfied to ensure successful infection in an individual host: sufficient virus must be available to initiate infection; the cells at the site of infection must be accessible, susceptible, and permissive for the virus; and the local host antiviral defense systems must be absent or at least initially ineffective.

The first requirement erects a substantial barrier to any infection and forms a significant weak link in the transmission of infection from host to host. Free virus particles face both a harsh environment and rapid dilution that can reduce their concentration. Viruses spread by contaminated water and sewage must be stable in the presence of osmotic shock, pH changes, and sunlight, and must not adsorb irreversibly to debris. Aerosol-dispersed viruses must stay hydrated and highly concentrated to infect the next host. Such viruses do best in populations in which individuals are in close contact. In contrast, viruses that are spread by biting insects, contact of mucosal surfaces, or other means of direct contact, including contaminated needles, have little environmental exposure.

Even if one virus particle survives the passage from one host to another, infection may fail simply because the concentration is not sufficient. In principle, a single virus particle should be able to initiate an infection, but host physical and immune defenses, coupled with the complexity of the infection process itself, demand the presence of many particles. How many particles are required to initiate and maintain an infection? Unfortunately, there is no general answer to this question, because the success of an infection depends on the particular virus, the site of infection, and the physiology and age of the host. However, some basic facts help to guide us.

Statistical analysis of infections in cultured cells demonstrates that on average a single virus particle can initiate an infection, but that many perfectly competent virions fail to do so. Such failure can be explained in part by the complexity of the infectious cycle: there are many distinct reactions, and the probability of a single virus particle completing any one is not 100%. For example, virus particles face many potentially nonproductive interactions with debris and extracellular material during their initial encounter with the cell surface. Even if a virus attaches successfully to a permissive cell, it may be delivered to a digestive lysosome upon entry. Many of these false starts or inappropriate interactions are irreversible, aborting infection by the virion.

In addition, populations of viruses often contain particles that are not capable of completing an infectious cycle. For example, defective particles can be produced by mistakes during virus replication or from interaction with

BOX 14.1

Transgenic and knockout mice for studying viral pathogenesis

Mice have always played an important role in the study of viral pathogenesis (see figure). Now that it is possible to manipulate this animal genetically, we can expect a wealth of new information about how viruses cause disease. Introducing a gene into the mouse germ line to produce a transgenic mouse and ablating specific genes (gene knockouts) both have wide use in virology.

New mouse models for viral diseases have been established for poliomyelitis and measles by producing transgenic mice that synthesize the human viral receptor. When viral receptors have not been identified, or are not sufficient for infection, an alternative approach is to express either the entire viral genome or a selected viral gene in mice. For example, transgenic mice expressing the hepatitis B virus genome have been used to study virus-immune response interactions. Transgenic mice have also been produced that express T-cell-receptor transgenes or genes encoding soluble immune mediators. Such mice have been used to study the effect of immune cells on virus clearance, and the protective and deleterious effects of cytokines.

Mice lacking specific components of the immune response have proven invaluable for studying immunity and immunopathogenesis. For example, mice lacking the gene encoding perforin, a molecule essential for the ability of cytotoxic T lymphocytes to lyse target cells, cannot clear infection with lymphocytic choriomeningitis virus, despite the presence of an otherwise intact immune response. Studies of mice with disruptions in genes encoding components of the immune response have led to the identification of cells that are important for mediating recovery from a variety of viral infections, including measles, influenza, and lymphocytic choriomeningitis.

Rall, G. F., D. M. P. Lawrence, and C. E. Patterson. 2000. The application of transgenic and knockout mouse technology for the study of viral pathogenesis. *Virology* **271:**220–226.

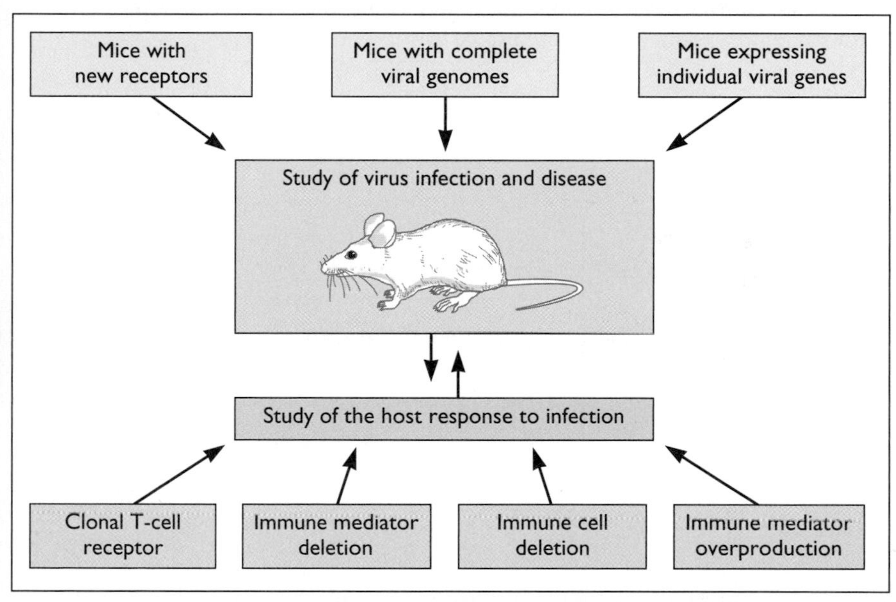

Table 14.3 Fundamental questions of viral pathogenesis

How does a virion enter the host?
What is the initial host response?
Where does primary replication occur?
How does the infection spread in the host?
What organs and tissues are infected?
How is the infection transmitted to other hosts?
Is the infection cleared from the host or is a persistent infection established?

inhibitory compounds in the environment. In the laboratory, a quantitative measure of the proportion of infectious viruses is the particle-to-plaque-forming-unit (PFU) ratio. As described in Chapter 2, the number of physical particles in a given preparation are counted, usually with an electron microscope, and compared with the number of infectious units, or PFU per unit volume. This ratio is a useful indicator of the quality of a virus preparation, as it should be constant for a given virus prepared by identical or comparable procedures.

Successful infection also depends on whether cells at the site of infection are physically accessible by the virus, susceptible (bear receptors for entry), and permissive (contain intracellular gene products needed for viral replication). The crucial role of antiviral defense systems in initiation of infection is the topic of Chapter 15.

Dissemination in the Host

Viral Entry

In general, virions must first enter cells at a body surface. Common sites of entry include the mucosal linings of the respiratory, alimentary, and urogenital tracts, the outer surface of the eye (conjunctival membranes or cornea), and the skin (Fig. 14.1).

Figure 14.1 Sites of viral entry into the host. A representation of the human host is shown with sites of virus entry and shedding indicated. The body is covered with skin, which has a relatively impermeable (dead) outer layer. However, layers of living cells are present to absorb food, exchange gases, and release urine and other fluids. These layers offer easier pathways for the entry of viruses than the skin. Virions can also be introduced through the skin by a scratch or injury, a vector bite, or inoculation with a needle.

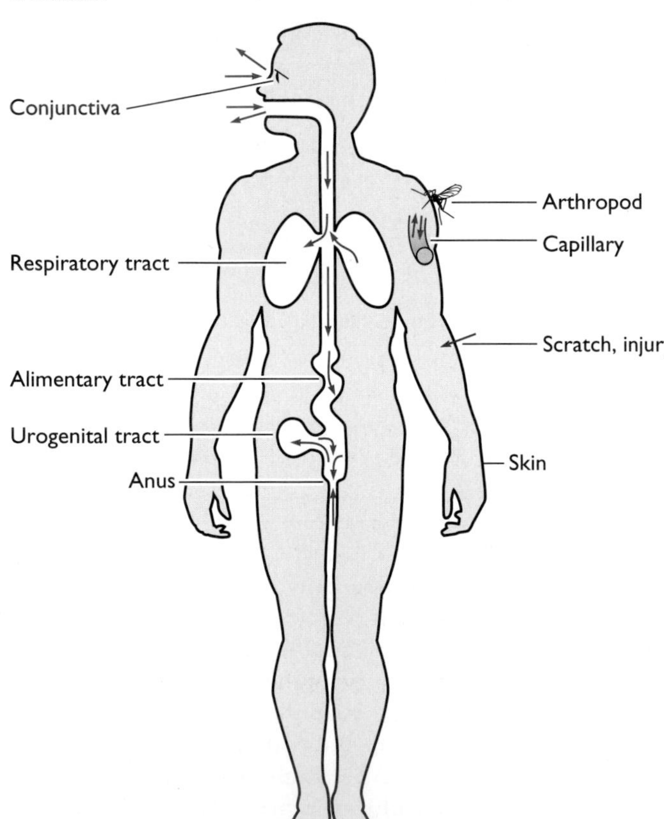

Respiratory Tract

Probably the most common route of viral entry is through the respiratory tract. In a human lung there are about 300 million terminal sacs, called alveoli, that function in gaseous exchange between inspired air and the blood. To accomplish this function, each sac is in close contact with capillary and lymphatic vessels. The combined absorptive area of the human lung is almost 140 m^2. Humans have a resting ventilation rate of 6 liters of air per minute, which introduces large numbers of foreign particles and aerosolized droplets into the lungs with every breath. Many of these particles and droplets contain viruses. Fortunately, there are numerous host defense mechanisms to block respiratory tract infection. Mechanical barriers play a significant role in antiviral defense. For example, the tract is lined with a mucociliary blanket consisting of ciliated cells, mucus-secreting goblet cells, and subepithelial mucus-secreting glands (Fig. 14.2). Foreign particles deposited in the nasal cavity or upper respiratory tract are trapped in mucus, carried to the back of the throat, and swallowed. In the lower respiratory tract, particles trapped in mucus are brought up from the lungs to the throat by ciliary action. The lowest portions of the tract, the alveoli, lack cilia or mucus, but macrophages lining the alveoli are responsible for ingesting and destroying particles. Other cellular and humoral immune responses also intervene.

Many viruses enter the respiratory tract in the form of aerosolized droplets expelled by an infected individual by coughing or sneezing (Table 14.4). Infection can also spread through contact with saliva from an infected individual. Larger virus-containing droplets are deposited in the nose, while smaller droplets find their way into the airways or the alveoli. To infect the respiratory tract successfully, viruses must not be swept away by mucus, neutralized by antibody, or destroyed by alveolar macrophages.

Alimentary Tract

The alimentary tract, the tube that connects the oral cavity to the anus, is always in motion. Eating, drinking, and some social activities routinely place viruses in the alimentary tract. It is designed to mix, digest, and absorb food, providing a good opportunity for viruses to encounter a susceptible cell and to interact with cells of the circulatory, lymphatic, and immune systems. The alimentary tract is a common route of infection and dispersal (Table 14.4). It is an extremely hostile environment for virions. The stomach is acidic, the intestine is alkaline, digestive enzymes and bile detergents abound, mucus lines the epithelium, and the lumenal surfaces of intestines contain antibodies and phagocytic cells.

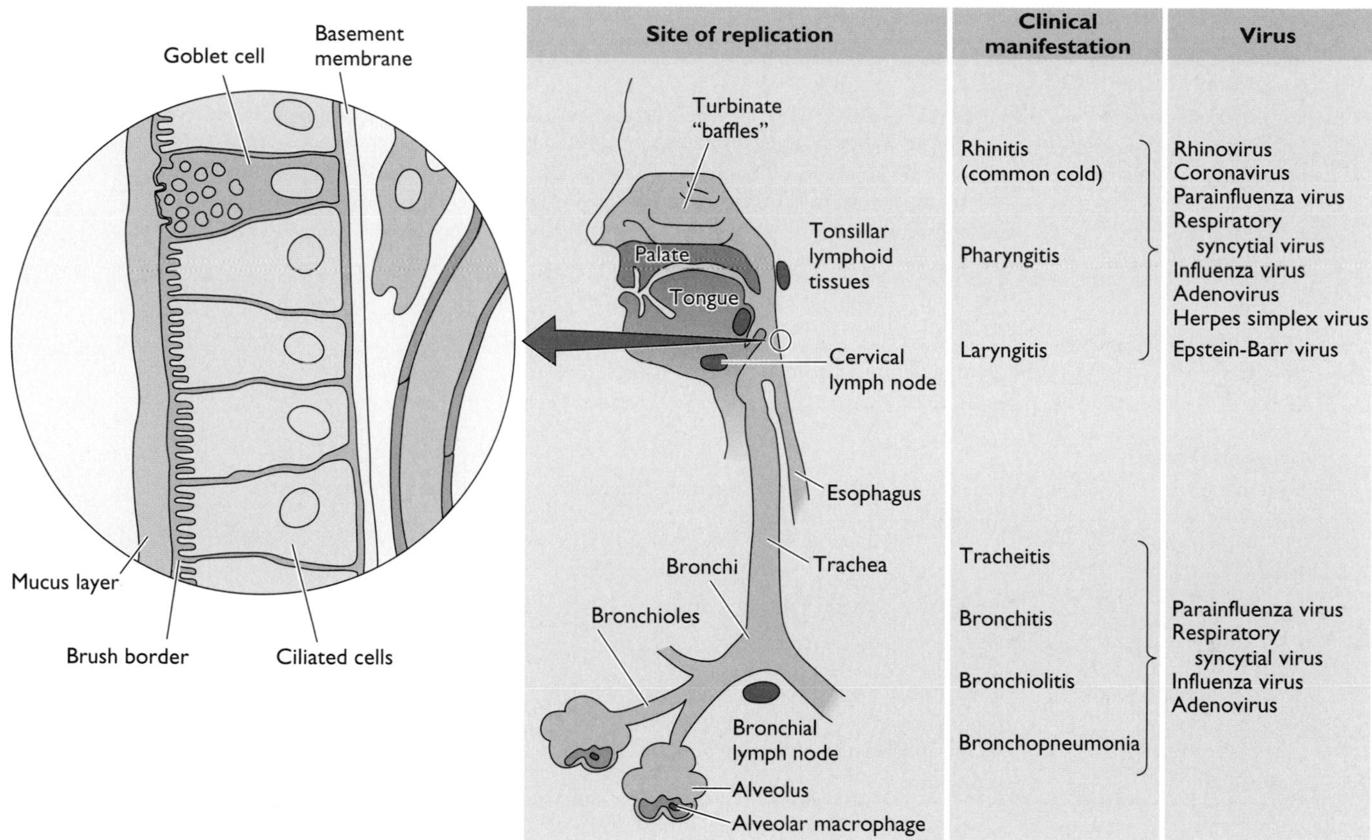

Figure 14.2 Sites of viral entry in the respiratory tract. Viruses that replicate at different levels of the respiratory tract are indicated on the right, along with the associated clinical syndromes. A detailed view of the respiratory epithelium is shown. A layer of mucus, produced by goblet cells, is a formidable barrier to virus attachment. Virions that pass through this layer may multiply in the ciliated cells or pass between them, reaching another physical barrier, the basement membrane. Beyond this membrane are tissue fluids from which particles may be taken into lymphatic capillaries and reach the blood. Local macrophages patrol the tissue fluids in search of foreign particles. Adapted from C. A. Mims et al., *Mims' Pathogenesis of Infectious Disease* (Academic Press, Orlando, Fla., 1995), with permission.

Virions that infect by the intestinal route must, at a minimum, be resistant to extremes of pH, proteases, and bile detergents. Indeed, virions that lack these features are destroyed when exposed to the alimentary tract, and must infect at other sites. The family *Picornaviridae* comprises both acid-labile viruses (e.g., rhinoviruses) and acid-resistant viruses (e.g., poliovirus) (Box 14.2). Rhinoviruses are respiratory pathogens and cannot infect the upper intestine. They spread in a population by means of inhalant aerosols. In contrast, poliovirus can survive ingestion and establish an infection of the upper intestine, and spreads by the fecal-oral route. The hostile environment of the alimentary tract actually facilitates infection by some viruses. For example, reovirus particles are converted by host proteases in the intestinal lumen into infectious subviral particles, the forms that subsequently infect intestinal cells. As might be expected, most enveloped viruses do not initiate infection in the alimentary tract, because viral envelopes are susceptible to dissociation by detergents such as bile salts. Enteric coronaviruses are notable exceptions, but it is not known why these enveloped viruses can withstand the harsh conditions in the alimentary tract.

Nearly all of the intestinal surface is covered with columnar villous epithelial cells with apical surfaces that are densely packed with microvilli (Fig. 14.3). This brush border, together with a surface coat of glycoproteins and glycolipids, and the overlying mucus layer, is permeable to electrolytes and nutrients, but presents a formidable barrier to microorganisms. Nevertheless, viruses such as enteric adenoviruses and Norwalk virus, a calicivirus, replicate extensively in intestinal epithelial cells. The mechanisms by which they bypass the physical barriers and enter susceptible cells are not well understood. Scattered throughout the intestinal mucosa are lymphoid

Table 14.4 Different routes of viral entry in the host

Location	Virus(es)
Respiratory tract	
Localized upper tract	Rhinovirus; coxsackievirus; coronavirus; arenaviruses; hantavirus; parainfluenza virus types 1–4; respiratory syncytial virus; influenza A and B viruses; human adenovirus types 1–7, 14, 21
Localized lower tract	Respiratory syncytial virus; parainfluenza virus types 1–3; influenza A and B viruses; human adenovirus types 1–7, 14, 21; infectious bovine rhinotracheitis virus (herpesvirus)
Entry via respiratory tract followed by systemic spread	Foot-and-mouth disease virus, rubella virus, arenaviruses, hantavirus, mumps virus, measles virus, varicella-zoster virus, poxviruses
Alimentary tract	
Systemic	Enterovirus, reovirus, adenovirus
Localized	Coronavirus, rotavirus
Urogenital tract	
Systemic	Human immunodeficiency virus type 1, hepatitis B virus, herpes simplex virus
Localized	Papillomavirus
Eyes	
Systemic	Enterovirus 70, herpes simplex virus
Skin	
Arthropod bite	Bunyavirus, flavivirus, poxvirus, reovirus, togavirus
Needle puncture, sexual contact	Hepatitis C and D viruses, cytomegalovirus, Epstein-Barr virus, hepatitis B virus, human immunodeficiency virus, papillomavirus (localized)
Animal bite	Rhabdovirus

follicles that are covered on the luminal side with a specialized follicle-associated epithelium consisting mainly of columnar absorptive cells and M (membranous epithelial) cells. The M cell cytoplasm is very thin, resulting in a membrane-like bridge that separates the lumen from the subepithelial space. As discussed in Chapter 15, M cells ingest and deliver antigens to the underlying lymphoid tissue by **transcytosis**. In this process, material taken up on the luminal side of the M cell traverses the cytoplasm virtually intact, and is delivered to the underlying basal membranes and extracellular space (Fig. 14.3). It is thought that M cell transcytosis provides the mechanism by which some enteric viruses gain entry to deeper tissues of the host from the intestinal lumen. After crossing the mucosal epithelium in this manner, a virus could enter lymphatic vessels and capillaries of the circulatory system,

BOX 14.2

Why is rhinovirus but not poliovirus sensitive to low pH?

Rhinoviruses are sensitive to low pH, which causes irreversible disassembly of the viral capsid. In contrast, poliovirus is not inactivated by low pH; the changes in the capsid that occur under acidic conditions are fully reversible. Although high-resolution crystallographic structures of rhinovirus and poliovirus have been determined, they do not provide information about the basis for the difference in pH sensitivity. Acid-resistant mutants of rhinovirus contain amino acid changes near the fivefold axes of symmetry. Rhinovirus acid-resistant mutants behave similarly to poliovirus, in that low pH induces reversible conformational changes in the capsid. The mutations responsible for acid resistance probably result in stabilization of the particle, so that it does not dissociate at low pH. Such mutations are not selected during natural infections, because they are not necessary for the virus to replicate in the respiratory tract.

Giranda, V. L., B. A. Heinz, M. A. Oliveira, I. Minor, K. H. Kim, P. R. Kolatkar, M. G. Rossmann, and R. R. Rueckert. 1992. Acid-induced structural changes in human rhinovirus 14: possible role in uncoating. *Proc. Natl. Acad. Sci. USA* **89:**10213–10217.

Skern, T., H. Torgersen, H. Auer, E. Kuechler, and D. Blaas. 1991. Human rhinovirus mutants resistant to low pH. *Virology* **183:**757–763.

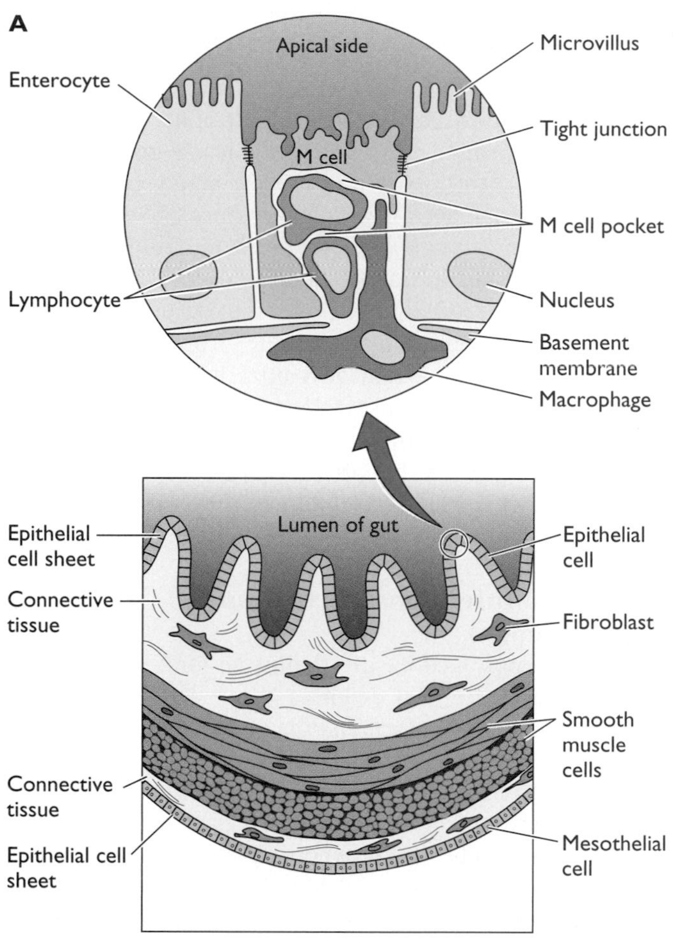

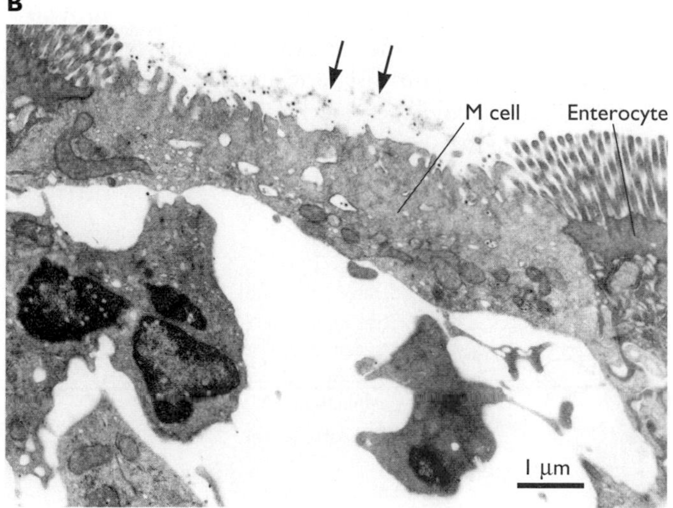

Figure 14.3 Viral entry in the intestine through M cells. (A) Schematic drawing of the intestinal wall. This organ is made up of epithelial, connective, and muscle tissues. Each is formed by different cell types that are organized by cell-cell adhesion within an extracellular matrix. A section of the epithelium has been enlarged, and a typical M cell is shown, surrounded by two enterocytes. Lymphocytes and macrophages move in and out of invaginations on the basolateral side of the M cell. Adapted from A. Siebers and B. B. Finlay, *Trends Microbiol.* **4**:22–28, 1996, and B. Alberts et al., *Molecular Biology of the Cell* (Garland Publishing, New York, N.Y., 1994), with permission. (B) Reovirus attached to, and within vesicles of, an M cell. An electron micrograph of the gut epithelium shows reovirus (small black arrows) attached to the surface of an M cell and also within intracellular vesicles. Reprinted from J. L. Wolf and W. A. Bye, *Annu. Rev. Med.* **35**:95–112, 1984, with permission. Photo courtesy of R. Finberg, Harvard Medical School.

facilitating spread within the host. A particularly well studied example is transcytosis of reovirus. After attaching to the M cell surface, reovirus particles are transported to cells underlying the lymphoid follicle, where they replicate and spread to other tissues. Rather than spread by transcytosis across the M cell, some viruses actively replicate in them and do not spread to underlying tissues. For example, human rotavirus and the coronavirus transmissible gastroenteritis virus destroy M cells, resulting in mucosal inflammation and diarrhea.

It is possible for virions to enter the body through the lower gastrointestinal tract without passing through the upper tract and its defensive barriers. For example, human immunodeficiency virus can be introduced into the body by anal intercourse. As M cells abound in the lower colon, these cells are likely to provide a portal of entry for this virus into susceptible lymphocytes in the underlying lymphoid follicles. Once in the follicle, the virus can infect migratory lymphoid cells and spread throughout the body.

Urogenital Tract

Some viruses enter the urogenital tract as a result of sexual activities (Table 14.4). The urogenital tract is well protected by physical barriers, including mucus and low pH (in the case of the vagina). Normal sexual activity can result in minute tears or abrasions in the vaginal epithelium or the urethra, allowing virions to enter. Some viruses infect the epithelium and produce local lesions (e.g., certain human papillomaviruses, which cause genital warts). Other viruses gain access to cells in the underlying tissues and infect cells of the immune system (e.g., human immunodeficiency virus type 1), or sensory and autonomic neurons (in the case of herpes simplex viruses). These two viruses invariably spread from the urogenital tract to establish lifelong persistent or latent infections, respectively.

Eyes

The epithelium covering the exposed part of the sclera (the outer fibrocollagenous coat of the globe of the eye) and the inner surfaces of the eyelids (conjunctivae) is the route of entry for several viruses. Every few seconds the eyelid passes over the sclera, bathing it in secretions that wash away foreign particles. There is usually little opportunity for viral infection of the eye, unless it is injured by abrasion. Direct inoculation into the eye may occur during ophthalmologic procedures or from environmental contamination (e.g., improperly sanitized swimming pools and hot tubs). In most cases, replication is localized and results in inflammation of the conjunctiva, a condition called conjunctivitis. Systemic spread of the virus from the eye is rare, although it does occur (e.g., paralytic illness after enterovirus 70 conjunctivitis). Herpesviruses, in particular herpes simplex virus type 1, can also infect the cornea, mainly at the site of a scratch or other injury. Such an infection may lead to immune destruction of the cornea and blindness. Inevitably, herpes simplex virus infection of the cornea is followed by spread of the virus to sensory neurons, and then to neuronal cell bodies in the sensory ganglia (e.g., the trigeminal ganglia), where a latent infection is established.

Skin

The skin protects the body yet provides sensory contact with the environment. The external surface of skin, or **epidermis**, is composed of several layers, including a basal germinal layer of proliferating cells, a granular layer of dying cells, and an outer layer of dead, keratinized cells (Fig. 14.4). The skin of most animals is an effective barrier against viral infections, as the dead outer layer cannot support viral growth. Entry through this organ occurs primarily when its integrity is breached by breaks or punctures. Examples of viruses that can gain entry in this manner are some human papillomaviruses and certain poxviruses (e.g., myxoma virus) that are transmitted mechanically by insect vectors such as the arthropod depicted in Fig. 14.1 (Table 14.4). Replication is usually limited to the site of entry because the epidermis is devoid of blood or lymphatic vessels that could provide pathways for further spread. The epidermis is supported by the highly vascularized **dermis**. Other viruses can gain entry to the dermis through the bites of arthropod vectors such as mosquitoes, mites, ticks, and sand flies. Even deeper inoculation, into the tissue and muscle below the dermis, can occur by hypodermic needle punctures, body piercing or tattooing, or sexual contact when body fluids are mingled through skin abrasions or ulcerations (Table 14.4). Animal bites introduce rabies virus into tissue and muscle rich with nerve endings by which the virus can invade the peripheral nervous system. In contrast to the strictly localized replication of viruses in the epidermis, viruses that initiate infection in dermal or subdermal tissues can reach nearby blood vessels, lymphatic tissues, and cells of the nervous system. As a consequence, they may spread to other sites in the body.

Successful Infections Must Evade Host Defenses

To initiate any infection, there must be viral mechanisms for countering the host defenses by an active or passive mechanism, or by a combination of the two. The actual pattern of infection that ensues is determined by the kinetics of virus replication in the face of host defenses. The interplay between virus offense and host defense is dynamic: there will be different consequences of a fast-replicating virus or a slow-replicating virus, depending on the strength of host defenses (Fig. 14.5).

Active avoidance of host defenses requires synthesis of viral gene products. It is likely that evolution has provided all viruses with active mechanisms for bypassing, modulating, or disarming intrinsic and immune host defenses. We will discuss many such mechanisms in Chapter 15.

Some host defenses may be overcome passively by an overwhelming virus inoculum. Single droplets found in

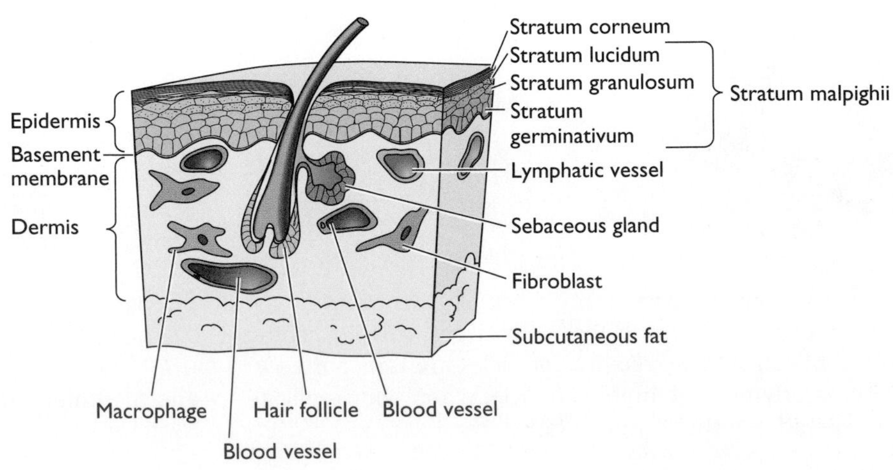

Figure 14.4 Schematic diagram of the skin. The epidermis consists of a layer of dead, keratinized cells (stratum corneum) over the stratum malpighii. The latter may have two layers of cells with increasing amounts of keratin granules (stratum granulosum and stratum lucidum) and a basal layer of dividing epidermal cells (stratum germinativum). Below this is the basement membrane. In the dermis are blood vessels, lymphatic vessels, fibroblasts, and macrophages. A hair follicle and a sebaceous gland are shown. Adapted from F. Fenner et al., *The Biology of Animal Viruses* (Academic Press, New York, N.Y., 1974), with permission.

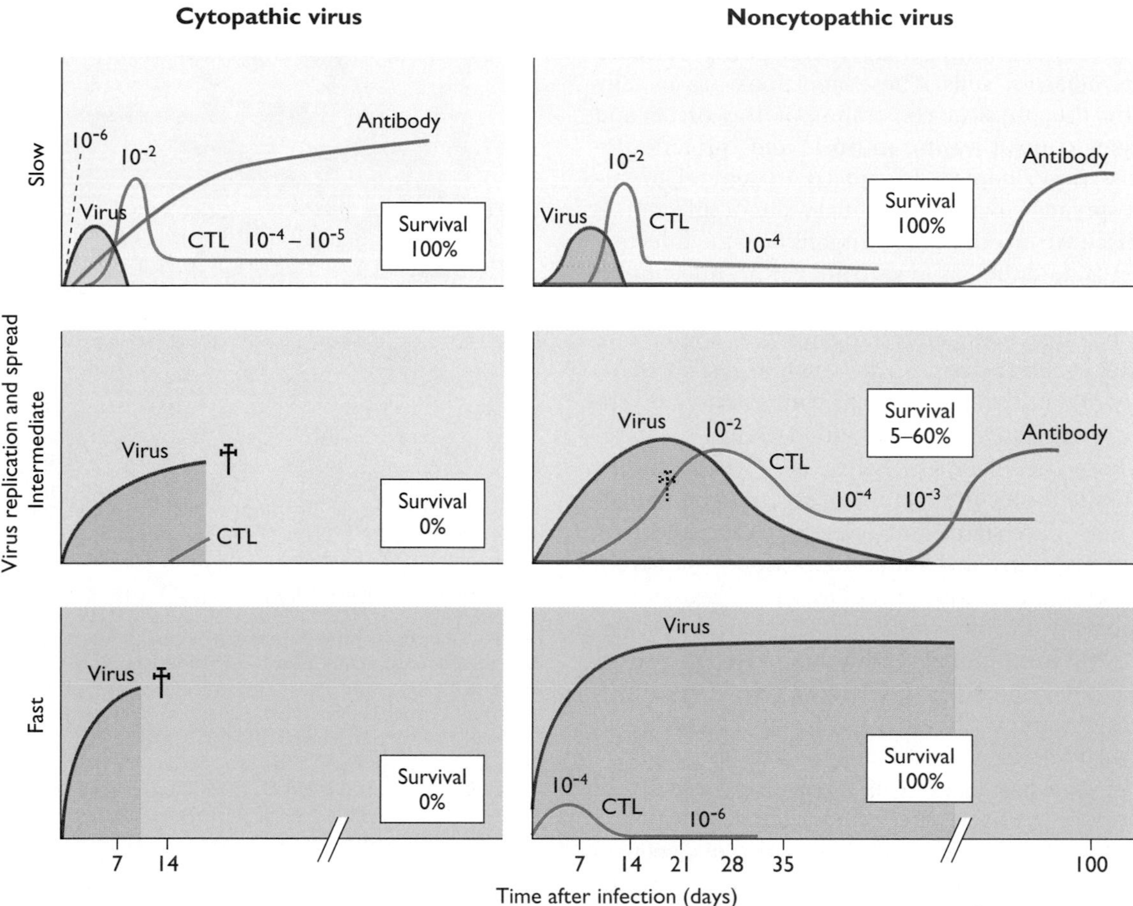

Figure 14.5 General schema illustrating kinetics of viral replication and immune responses. Three infections by two types of viruses are illustrated. The measurement of viral replication and spread on the *y* axes with respect to time after infection on the *x* axes is illustrated. The three infections are characterized as slow, localized spread (top); intermediate spread (middle); and fast, disseminated spread (bottom). A cytopathic virus (left) is compared with a noncytopathic virus (right). The percentage of animals that survive each infection is indicated in the box. The relative virus titers are depicted by the red lines, and the shaded areas under these curves indicate the period of disease. The immune responses as measured by CD8+ T cells (CTL; green lines) or neutralizing antibody titers (blue lines) are superimposed on the graphs. The frequencies of CTL precursors in the spleen are noted at 10^{-2} to 10^{-6} values, where 10^{-2} indicates that there was 1 CTL in 100 spleen cells. The death of animals is indicated by a cross. The point to be noted is that the kinetics of viral replication and the kinetics of the defensive response both affect the outcome. For example, in this hypothetical scheme only the virus spread kinetics were changed from slow to fast, but the overall interaction between virus and immune response can also vary similarly if immune responses shift from fast and disseminated to slow and localized. Adapted from R. M. Zinkernagel, *Science* **271:**173–178, 1996, with permission.

the aerosol produced by sneezing can contain as many as 100 million rhinovirus particles; a similarly high number of hepatitis B virus particles can be found in 1 μl of blood from a hepatitis patient. At these concentrations, it may be impossible for physical and innate defenses to block every infecting virus particle. Free passage of virus through the primary physical barriers of skin and mucus layers made possible by a cut, abrasion, or needle stick may also allow passive evasion of defenses. Some viruses (e.g., herpesviruses, papillomaviruses, and rabies virus) establish unique infections because they infect organs or cells not exposed to antibodies or cytotoxic lymphocytes. A more egregious breach of both primary and secondary defenses may occur during organ transplantation, which places viruses in direct contact with potentially susceptible cells in immunosuppressed patients.

Viral Spread

Following replication at the site of entry, virus particles can remain localized (rhinovirus infections that are restricted to the respiratory epithelium) or can spread to other tissues (spread of measles virus from the respiratory tract to other tissues) (Table 14.4). Local spread of the infection in the epithelium occurs when newly released virus infects adjacent cells. These infections are usually contained by the physical constraints of the tissue and brought under control by the intrinsic and immune defenses discussed in Chapters 15 and 16. In general, an infection that spreads beyond the primary site of infection is said to be **disseminated**. If many organs become infected, the infection is described as **systemic**. For an infection to spread beyond the primary site, physical and immune barriers must be breached. After crossing the epithelium, virus particles reach the basement membrane (Fig. 14.3). The integrity of that structure may be compromised by epithelial cell destruction and inflammation. Below the basement membrane are subepithelial tissues, where virions encounter tissue fluids, the lymphatic system, and phagocytes. All three play significant roles in clearing foreign particles, but also may disseminate infectious virus from the primary site of infection.

One important mechanism for avoiding local host defenses and facilitating spread within the body is the directional release of virions from polarized cells at the mucosal surface (Chapter 12). Virions can be released from the apical surface, from the basolateral surface, or from both (Fig. 14.6). After replication, particles released from the apical surface are back where they started, outside the host. Such directional release facilitates the dispersal of many newly replicated enteric viruses in the feces (e.g., poliovirus). In contrast, virus particles released from the basolateral surfaces of polarized epithelial cells have been moved away from the defenses of the luminal surface. Directional release is therefore a major determinant of the infection pattern. In general, virions released at apical membranes establish a localized or limited infection. In this case, local lateral spread from cell to cell occurs in the infected epithelium, but virions rarely invade the underlying lymphatic and circulatory vessels. On the other hand, release of particles at the basal membrane provides

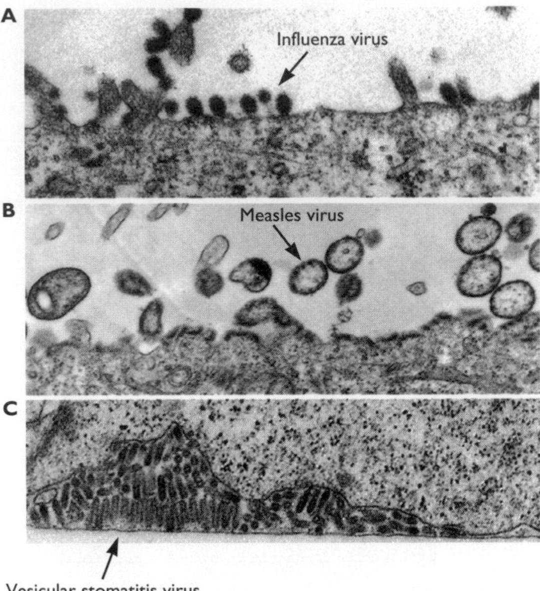

Figure 14.6 Polarized release of viruses from cultured epithelial cells visualized by electron microscopy. (A) Influenza virus released by budding from the apical surface of MDCK cells. (B) Budding of measles virus on the apical surface of Caco-2 cells. (C) Release of vesicular stomatitis virus at the basolateral surface of MDCK cells. Arrows indicate virus particles. Magnification, ×324,000. Reprinted from D. M. Blau and R. W. Compans, *Semin. Virol.* **7**:245–253, 1996, with permission. Courtesy of D. M. Blau and R. W. Compans, Emory University School of Medicine.

Table 14.5 Cell-associated and plasma viremia[a]

Cell type	Virus	Duration of viremia
Monocyte/macrophage	Dengue virus	Acute
	Rubella virus	Persistent
	Measles virus	Acute
	Lymphocytic choriomeningitis virus	Persistent
	Human immunodeficiency virus	Persistent
	Cytomegalovirus	Acute
	Mousepox virus	Acute
B lymphocyte	Murine leukemia virus	Persistent
	Epstein-Barr virus	Persistent
T lymphocyte	Human immunodeficiency virus	Persistent
	Human T-lymphotropic virus type 1	Persistent
	Human herpesviruses 6 and 7	Acute
Erythrocyte	Colorado tick fever virus	Acute
Neutrophil	Influenza virus	Acute
Free in plasma	Poliovirus	Acute
	Yellow fever virus	Acute
	Hepatitis B virus	Persistent

[a]The cell associations may not play a role in pathogenesis.

access to the underlying tissues and may facilitate systemic spread.

The consequences of directional release are striking. This assertion is underscored by studies of Sendai virus, which is normally released from the apical surfaces of polarized epithelial cells. Consequently, only a localized infection of the respiratory tract occurs. In stark contrast to infection by wild-type virus, a mutant virus, which is released at both apical and basal surfaces, is disseminated.

Hematogenous Spread

Viruses that escape from local defenses to produce a disseminated infection often do so by entering the bloodstream (**hematogenous spread**) (Table 14.5). Virus particles may enter the blood directly through capillaries, by replicating in endothelial cells, or through inoculation by a vector bite. Once in the blood, virions may have access to almost every tissue in the host (Fig. 14.7). Hematogenous spread begins when newly replicated particles produced at the entry site

Figure 14.7 Entry, dissemination, and shedding of blood-borne viruses. Shown are the target organs for selected viruses that enter at epithelial surfaces and spread via the blood. The sites of virus shedding (red arrows), which may lead to transmission to other hosts, are shown. Adapted from N. Nathanson (ed.), *Viral Pathogenesis* (Lippincott-Raven Publishers, Philadelphia, Pa., 1997), with permission.

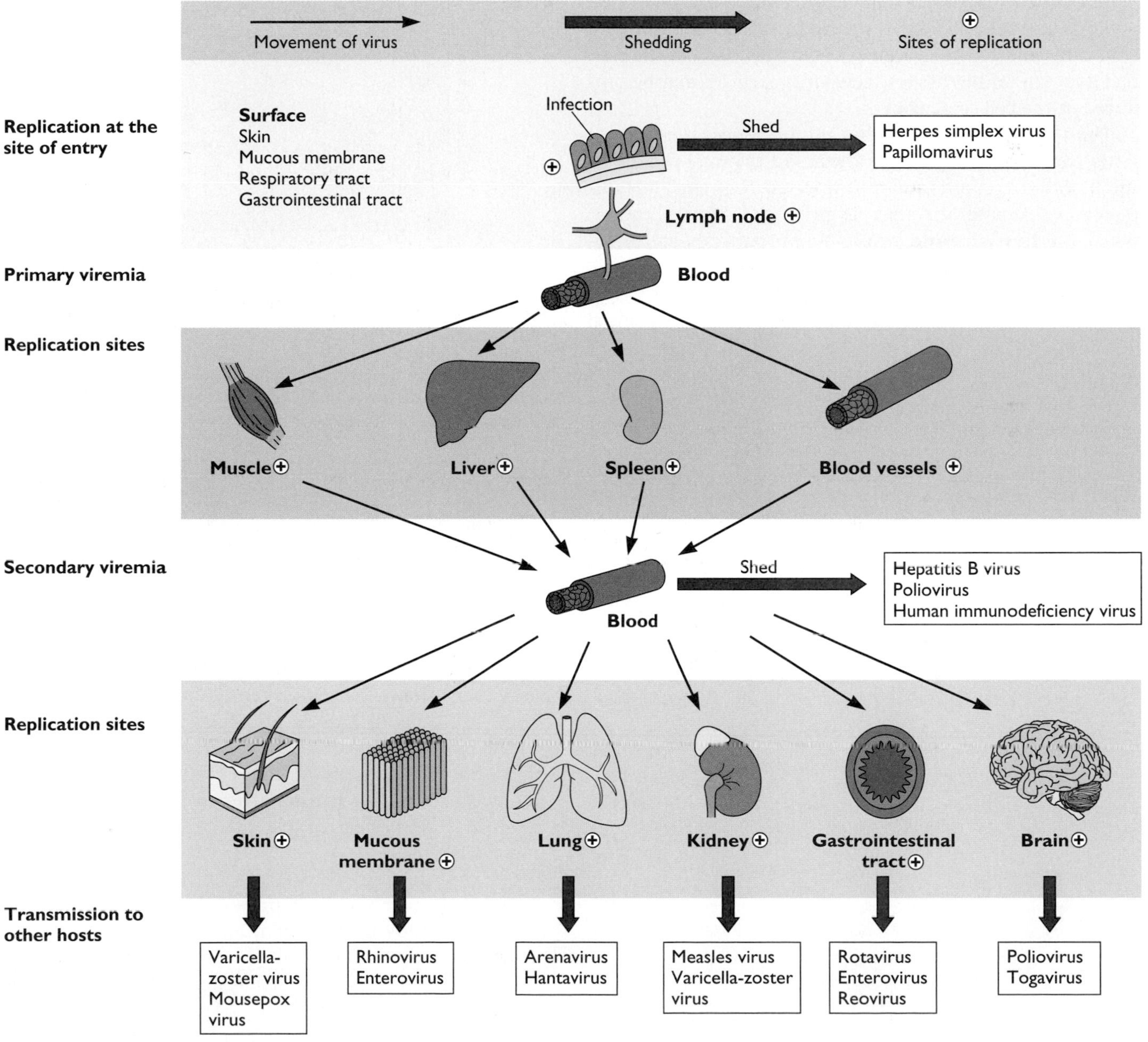

are released into the extracellular fluids, which can be taken up by the local lymphatic vascular system (Fig. 14.8). Lymphatic capillaries are considerably more permeable than circulatory system capillaries, facilitating virus entry. As the lymphatic vessels ultimately join with the venous system, virus particles in lymph have free access to the bloodstream. In the lymphatic system, virions pass through lymph nodes, where they encounter migratory cells of the immune system. Viral pathogenesis resulting from the direct infection of immune system cells (e.g., human immunodeficiency virus and measles virus) is initiated in this fashion. Some viruses replicate in the infected lymphoid cells, and progeny are released into the blood plasma. The infected lymphoid cell may also migrate away from the local lymph node to distant parts of the circulatory system. There may be little viral replication while the cell is in the bloodstream. However, when the infected cell receives chemotactic signals that direct it to enter other tissues, new virus particles may be produced in the cell so activated.

The term **viremia** describes the presence of infectious virus particles in the blood. These virions may be free in the blood or contained within infected cells such as lymphocytes. **Active viremia** is produced by replication, while **passive viremia** results when particles are introduced into the blood without viral replication at the site of entry (injection of a virion suspension into a vein) (Fig. 14.9). Progeny virions released into the blood after initial replication at the site of entry constitute **primary viremia**. The concentration of particles during primary viremia is usually low. However, the subsequent disseminated infections that result are often extensive, releasing considerably more virions. Such delayed appearance of a high concentration of infectious virus in the blood is termed **secondary viremia**. The two phases of viremia were first described in classic studies of mousepox (Fig. 14.10).

The concentration of virions in blood is determined by the rates of their synthesis in permissive tissues, and by how quickly they are released into, and removed from, the blood. Circulating particles are removed by phagocytic cells of the reticuloendothelial system in the liver, lung,

Figure 14.8 The lymphatic system. Pathways for virus spread from the epithelium to the blood via the lymphatic system are shown. Lymphocytes flow from the blood into the lymph node through postcapillary venules. Adapted from C. A. Mims et al., *Mims' Pathogenesis of Infectious Disease* (Academic Press, Orlando, Fla., 1995), with permission.

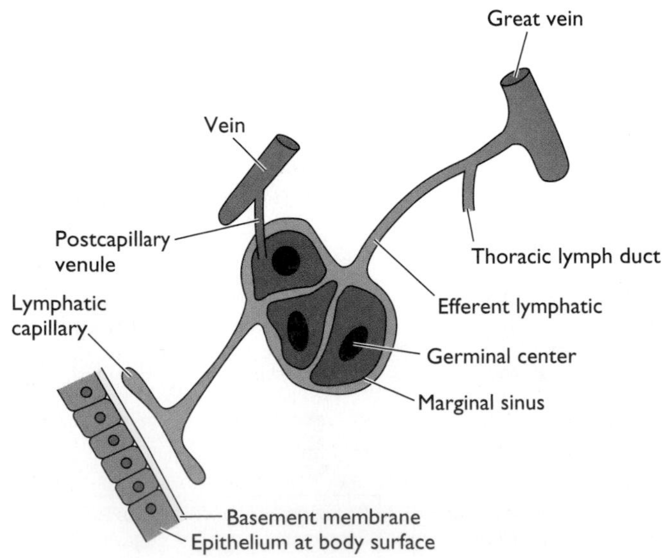

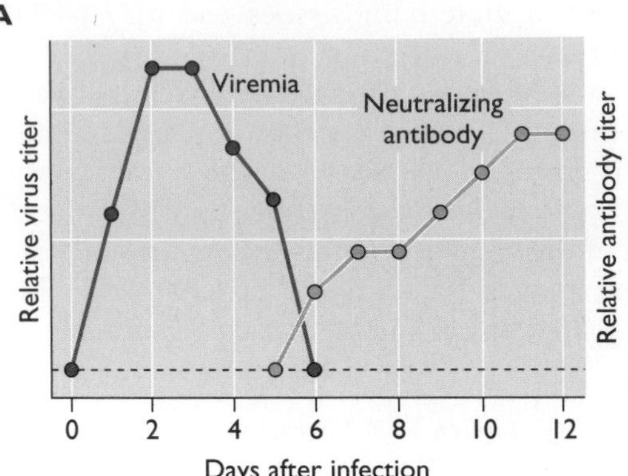

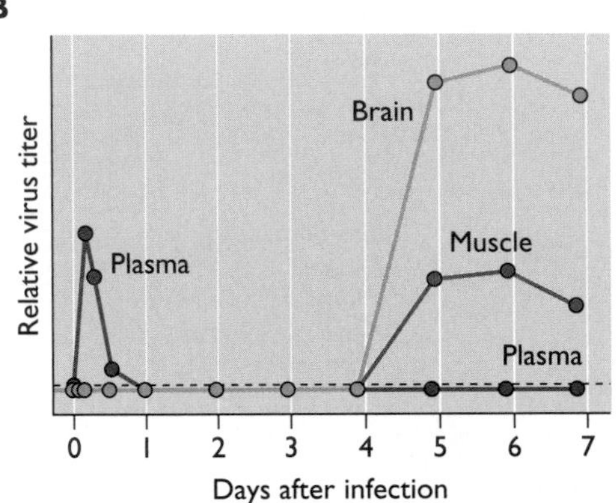

Figure 14.9 Characteristics of active and passive viremia. (A) Active viremia. Poliovirus was inoculated intramuscularly into monkeys, and the presence of virus and neutralizing antibody in the blood was determined at different times after infection. Viremia ceases just as serum antibody appears. (B) Passive viremia. La Crosse virus, a bunyavirus, was injected into weanling mice, and virus titers in plasma, brain, and muscle were determined at different times afterward. No virus can be detected in the blood after 1 day; however, beginning at 5 days after infection, virions can be found in the brain and muscle. The latter is viral progeny produced by multiplication of the inoculum in the target organs. Adapted from N. Nathanson (ed.), *Viral Pathogenesis* (Lippincott-Raven Publishers, Philadelphia, Pa., 1997), with permission.

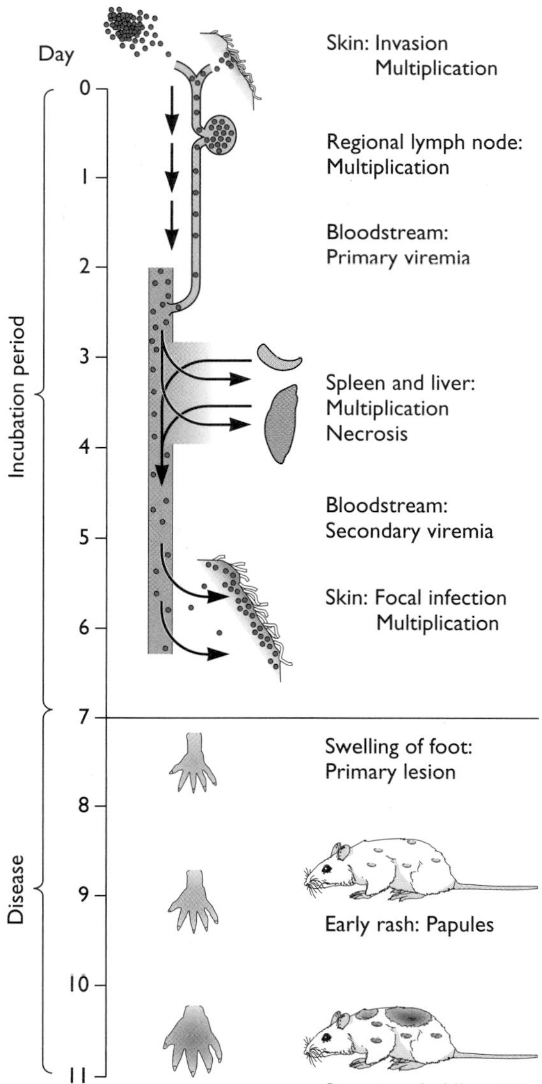

Figure 14.10 Pathogenesis of mousepox. Sequence of events in the pathogenesis of mousepox after inoculation of virus into the footpad. The figure is based on the classic studies of Frank Fenner, which were the first to demonstrate how disseminated viral infections develop from local multiplication to primary and secondary viremia. In the case of mousepox, after local multiplication in the foot, the host response leads to swelling at the site of inoculation, and after viremia, the host response to replication in the skin results in a rash. Adapted from F. Fenner et al., *The Biology of Animal Viruses* (Academic Press, New York, N.Y., 1974), with permission.

spleen, and lymph nodes. When serum antibodies appear, virions in the blood may bind these antibodies and be neutralized, as described in Chapter 15. Formation of a complex of antibodies and virus particles facilitates uptake by Fc receptors carried by macrophages lining the circulatory vessels. Virus-antibody complexes can be sequestered in significant quantities in the kidney, spleen, and liver. The time a virion is present in the blood usually varies from 1 to 60 min, depending on such parameters as the physiology of the host (e.g., age and health) and the size of the virus (larger particles are cleared more rapidly than small particles).

While viremias are usually of short duration, some viral infections are noteworthy for the long-lasting presence of infectious particles in the blood. Hosts infected with hepatitis B and C viruses or lymphocytic choriomeningitis virus may have viremias that last for years. One source of prolonged viremia is viral replication in macrophages (Table 14.5). Such infected macrophages often cannot perform normally, leading to delays in virus clearance and exacerbation of disease symptoms.

Viremias obviously are of diagnostic value and can be used to monitor the course of infection, but they also present practical problems. Infections can be spread inadvertently in the population when pooled blood from thousands of individuals is used directly for therapeutic purposes (transfusions) or as a source of therapeutic proteins (e.g., gamma globulin or blood-clotting factors). We have learned from unfortunate experience that hepatitis and acquired immunodeficiency syndrome (AIDS) can be spread by contaminated blood and blood products. Obviously, sensitive detection methods and stringent purification protocols are required to protect those who use and dispense these products. Frequently, it may be difficult, or technically impossible, to quantify infectious particles in the blood, as is certainly the case for hepatitis B virus. Accordingly, the presence of characteristic viral proteins provides surrogate markers of the viremia. Immunoblotting, hemagglutination, and complement fixation are techniques frequently used to detect viral proteins in blood (Chapter 2).

Neural Spread

Many viruses spread from the primary site of infection by entering local nerve endings (Table 14.6). For certain viruses (e.g., rabies virus and alphaherpesviruses), neuronal spread is the definitive characteristic of their pathogenesis. For other viruses (e.g., poliovirus and reovirus), invasion of the nervous system is a less frequent diversion from their site of replication and destination. Some viruses (e.g., mumps virus, human immunodeficiency virus, and measles virus) may replicate in the brain, but spread is by the hematogenous route. The molecular mechanisms that dictate spread by neural or hematogenous pathways are not well understood for most viruses. Clues may be found through the study of strains of the same virus that differ in their ability to spread by these two routes (Table 14.7).

Table 14.6 Steps required to establish an infection of neurons

1. *Enter* the neuron at the axon, sensory terminal, or cell body, depending on site of infection.
2. *Transport* the virus particle or subviral particle (e.g., nucleocapsid) toward the cell body of the neuron where replication occurs.
3. *Replicate* the genome.
4. Assemble virus particles that *egress* from the infected neuron in a directional manner (see Box 14.3).

Table 14.7 Viral spread to the central nervous system

Pathway	Viruses
Neural	Poliovirus, yellow fever virus, mouse hepatitis virus, Venezuelan encephalitis virus, rabies virus, reovirus (type 3 only; type 1 spread by viremia), herpes simplex virus types 1 and 2, pseudorabies virus
Olfactory	Poliovirus (experimental), herpes simplex virus, coronavirus
Hematogenous	Poliovirus, coxsackievirus, arenavirus, mumps virus, measles virus, herpes simplex virus, cytomegalovirus

Neurons are polarized cells with functionally and structurally distinct processes (axons and dendrites) that can be separated by enormous distances. For example, in adult humans the axon terminals of motor neurons that control stomach muscles can be 50 cm away from the cell bodies and dendrites in the brain stem. We currently have a limited understanding of how viral particles move in and among cells of the nervous system. It is likely that virions enter neurons by the same mechanisms used to enter other cells. Because protein synthesis does not occur in the extended processes of neuronal cells, virus particles must be transported over relatively long distances to the site of viral replication. All evidence indicates that virions are carried in the infected neuron by cellular systems, but viral proteins may facilitate the direction of spread.

Directionality of movement within a neuron is most likely mediated by microtubules and their attendant motor proteins, including kinesin and dynein (Box 14.3). For example, drugs such as colchicine that disrupt microtubules efficiently block the spread of many neurotropic viruses

BOX 14.3

Which direction, anterograde or retrograde?

Those who study virus spread in the nervous system often use the words **retrograde** and **anterograde** to describe direction. Unfortunately, confusion arises because the terms can be used to describe directional movement of virus inside a cell as well as spread of virus between synaptically connected neurons. Spread from the primary neuron to the second-order neuron in the direction of the nerve impulse is said to be anterograde spread (see figure). Spread in the opposite direction is said to be retrograde spread. Inside a neuron, spread is defined by microtubule polarity. Transport on microtubules that goes from (–) to (+) ends is said to be anterograde, while transport on microtubules that goes from (+) to (–) ends is said to be retrograde.

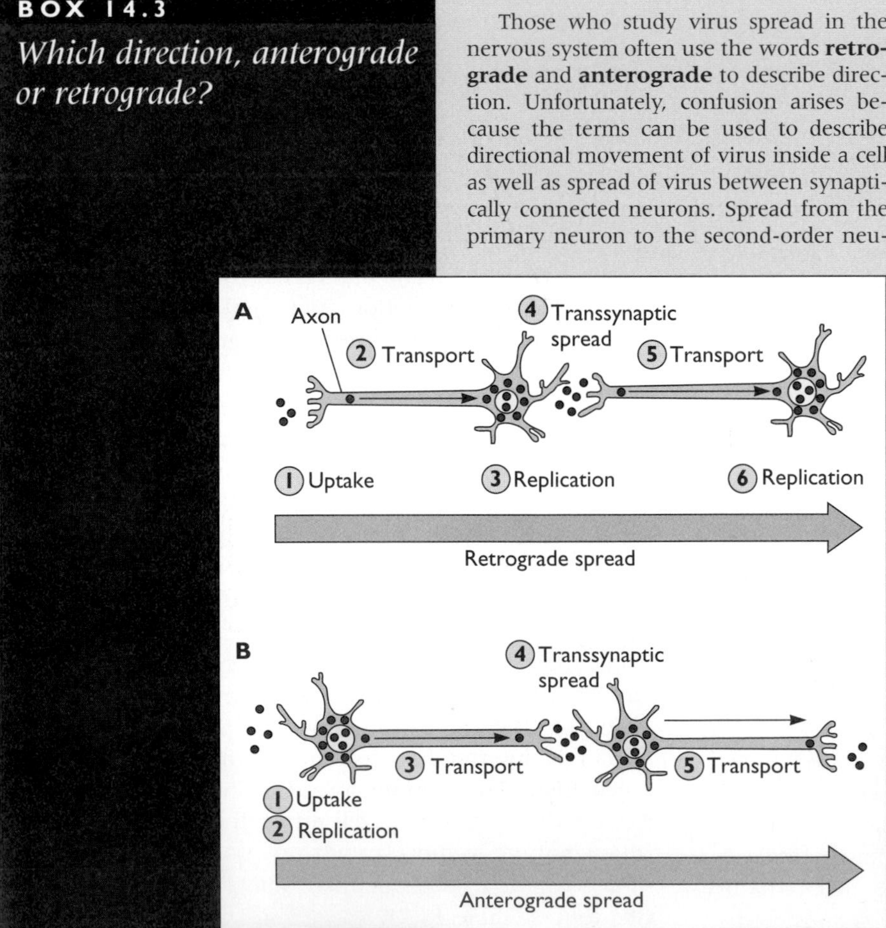

Retrograde and anterograde spread of virus in nerves. (A) Retrograde spread of infection. Virus invades at axon terminals and spreads to the cell body, where replication ensues. Newly replicated virus particles spread to a neuron at sites of synaptic contact. Particles enter the axon terminal of the second neuron to initiate a second cycle of replication and spread. (B) Anterograde spread of infection. Virus invades at dendrites or cell bodies and replicates. Virus particles then spread to axon terminals, where virions cross synaptic contacts to invade dendrites or cell bodies of the second neuron.

BOX 14.4

Infection of the nervous system: definitions and distinctions

A **neurotropic** virus can infect neural cells; infection may occur by neural or hematogenous spread initiating from a peripheral site. A **neuroinvasive** virus can enter the central nervous system (spinal cord and brain) after infection of a peripheral site. A **neurovirulent** virus can cause disease of nervous tissue, manifested by neurological symptoms and often death.

Examples:

Herpes simplex virus has low neuroinvasiveness of the central nervous system, but high neurovirulence. It always enters the peripheral nervous system but rarely enters the central nervous system. When it does, the consequences are almost always severe, if not fatal.

Mumps virus has high neuroinvasiveness but low neurovirulence. Most infections lead to invasion of the central nervous system, but neurological disease is mild.

Rabies virus has high neuroinvasiveness and high neurovirulence. It readily infects the peripheral nervous system and spreads to the central nervous system with 100% lethality unless antiviral therapy is administered shortly after infection.

from the site of peripheral inoculation to the central nervous system. The precise intracellular form of any virus that spreads in the nervous system has not been established. Both mature virions and nucleocapsids have been observed in axons and dendrites of animals infected by rabies virus and alphaherpesviruses.

While viruses that infect the nervous system are often said to be neurotropic (Box 14.4), they are generally capable of infecting a variety of cell types. Viral replication usually occurs first in nonneuronal cells, with virions subsequently spreading into afferent (e.g., sensory) or efferent (e.g., motor) nerve fibers innervating the infected tissue (Fig. 14.11). Concurrent with the primary infection in nonneuronal cells at the epithelial surface, viruses that invade neurons must complete four general processes to establish the neuronal infection (Table 14.6). Under some circumstances, virions can enter neurons directly with no prior replication in nonneuronal cells. However, this event is probably infrequent, as nerve endings are rarely exposed directly to the environment. Unless virus particles are injected directly into the brain or spinal cord, enter olfactory neurons (Fig. 14.12), or infect retinal ganglion cells, the first neurons to be infected are constituents of the peripheral nervous system. These neurons represent the first cells in circuits connecting the innervated peripheral tissue with the spinal cord and brain. Once in the peripheral nervous system, alphaherpesviruses, some rhabdoviruses (e.g., rabies virus), and some coronaviruses can spread among neurons connected by synapses (Box 14.5). Infection of nonneuronal support cells and satellite cells in ganglia may also occur. Virus spread by this mode can continue through chains of connected neurons of the peripheral nervous system to the spinal cord and brain, with devastating results (Fig. 14.13).

An important component of neuronal infection is movement and release of infectious particles in polarized neurons. As with polarized cells discussed earlier, direc-

Figure 14.11 Possible pathways for the spread of infection in nerves. Virus particles may enter the sensory or motor neuron endings. They may be transported within axons, in which case viruses taken up at sensory endings reach dorsal root ganglion cells. Those taken up at motor endings reach motor neurons. Viruses may also travel in the endoneural space, in perineural lymphatics, or in infected Schwann cells. Directional transport of virus inside the sensory neuron is defined as anterograde [movement from the (–) to the (+) ends of microtubules] or retrograde [movement from the (+) to the (–) ends of microtubules]. Adapted from R. T. Johnson, *Viral Infections of the Nervous System* (Raven Press, New York, N.Y., 1982), with permission.

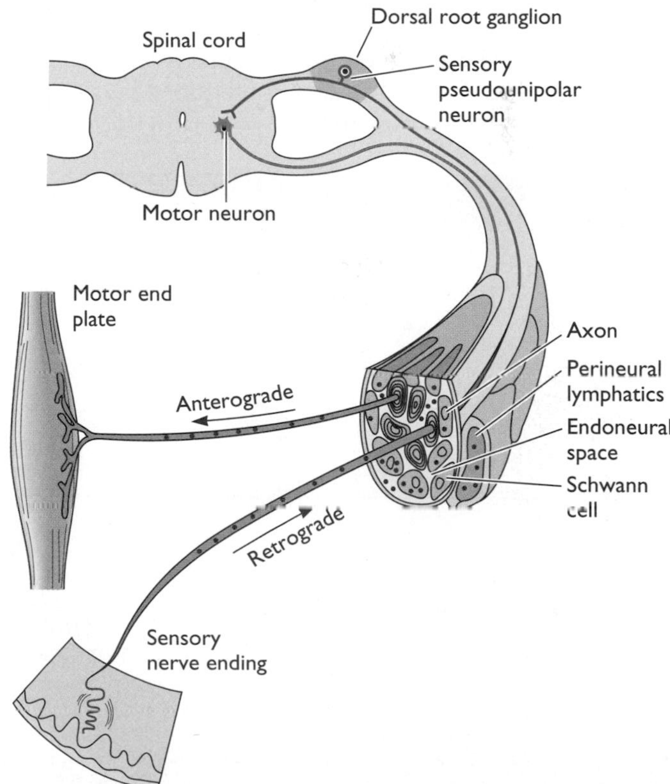

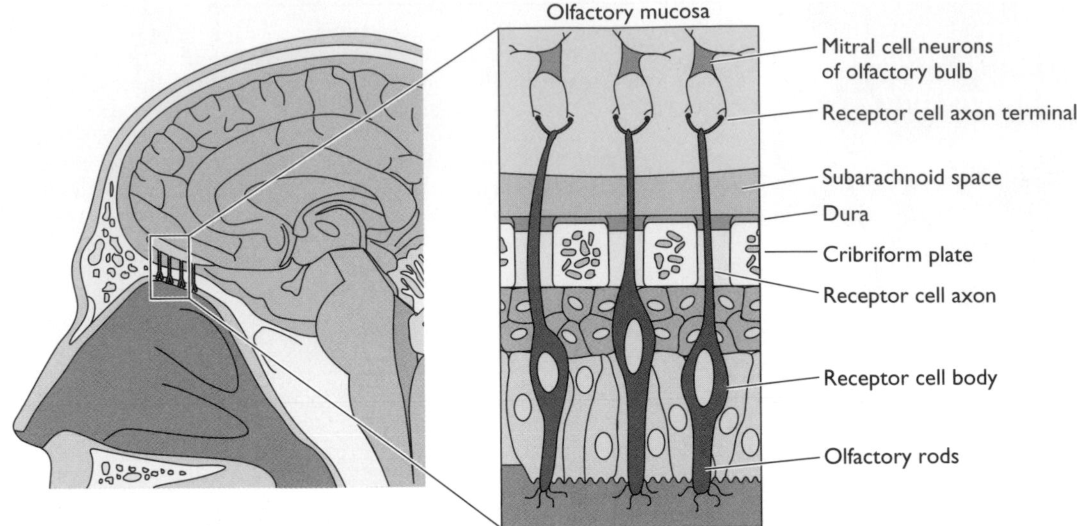

Figure 14.12 Pathways for viral entry into the central nervous system by olfactory routes. Olfactory neurons are unusual in that their cell bodies are present in the olfactory epithelia and their axon termini are in synaptic contact with olfactory bulb neurons. The olfactory nerve fiber passes through the skull via an opening called the arachnoid. Adapted from R. T. Johnson, *Viral Infections of the Nervous System* (Raven Press, New York, N.Y., 1982), with permission.

tional release of infectious virus from neurons affects the outcome of infection. For example, after primary infection, alphaherpesviruses become latent in peripheral neurons that innervate the site of infection (Fig. 14.13). Reactivation from the latent state results in viral replication in the primary neuron and subsequent transport of progeny viruses from the neuron cell body back to the innervated peripheral tissue. The virus can then spread from the peripheral to the central nervous system, or it can spread back to the peripheral site serviced by that particular group of neurons in the ganglion (Fig. 14.13). The direction taken is the difference between a minor

BOX 14.5

Tracing neuronal connections in the nervous system with viruses

The identification and characterization of synaptically linked multineuronal pathways in the brain are important in understanding the functional organization of neuronal circuits. Conventional tracing methodologies have relied on the use of markers such as wheat germ agglutinin-horseradish peroxidase or fluorochrome dyes. The main limitations of these tracers are their low specificity and sensitivity. During experimental manipulation, it is difficult to restrict the diffusion of tracers to a particular cell group or nucleus, and so uptake occurs in neighboring neurons or adjacent nonconnected axon fibers, producing false-positive labeling of a circuit. Neurons located one or more synapses away from the injection site receive progressively less label, because the tracer is diluted at each stage of transneuronal transfer.

Some alphaherpesviruses and rhabdoviruses have considerable promise as self-amplifying tracers of synaptically connected neurons. Under proper conditions, second- and third-order neurons show the same labeling intensity as first-order neurons. Moreover, the specific pattern of infected neurons observed in tracing studies is consistent with transsynaptic passage of virus rather than lytic spread through the extracellular space.

The detection of viruses typically involves immunohistochemical localization of viral antigens by light microscopy. More recently, reporter genes such as the green fluorescent protein gene from *Aequorea victoria* have been introduced into the genomes of neurotropic viruses for simpler visualization of viral infection.

Enquist, L. W., P. J. Husak, B. W. Banfield, and G. A. Smith. 1999. Infection and spread of alphaherpesviruses in the nervous system. *Adv. Virus Res.* **51:**237–347.

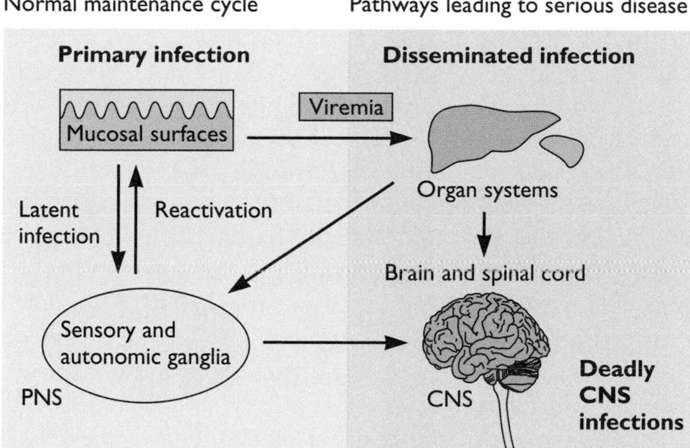

Figure 14.13 Outline of the spread of alphaherpesviruses and relationship to disease. Abbreviations: PNS, peripheral nervous system; CNS, central nervous system.

local infection (a cold sore) or a lethal viral encephalitis. Luckily, spread back to the peripheral site is by far more common.

Organ Invasion

Once virions enter the blood and are dispersed from the primary site, any subsequent replication requires invasion of new cells and tissues. There are three main types of blood vessel-tissue junctions that provide routes of tissue invasion (Fig. 14.14). In some tissues the endothelial cells are continuous with a dense basement membrane. At other sites the endothelium contains gaps, and at still others there may be **sinusoids** in which macrophages form part of the blood-tissue junction.

Figure 14.14 Blood-tissue junction in a capillary, venule, and sinusoid. (Left) Continuous endothelium and basement membrane found in the central nervous system, connective tissue, skeletal and cardiac muscle, skin, and lungs. (Center) Fenestrated endothelium found in the choroid plexus, villi of the intestine, renal glomerulus, pancreas, and endocrine glands. (Right) Sinusoid, lined with macrophages of the reticuloendothelial system, as found in the adrenal glands, liver, spleen, and bone marrow. Adapted from C. A. Mims et al., *Mims' Pathogenesis of Infectious Disease* (Academic Press, Orlando, Fla., 1995), with permission.

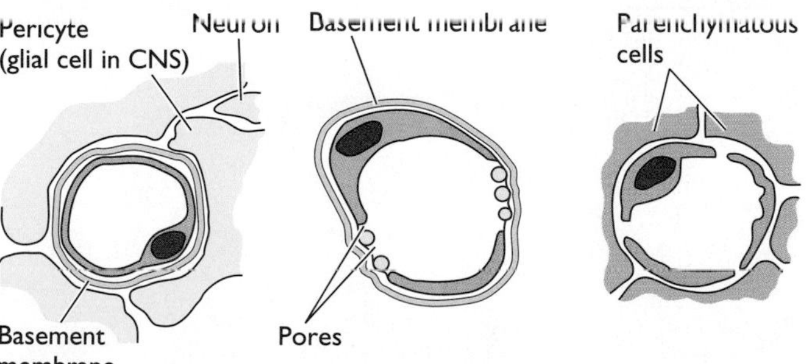

Skin

In a number of systemic viral infections, rashes are produced when virions leave blood vessels (Table 14.8). Different types of skin lesions are recognized. **Macules** and **papules** develop when inflammation occurs in the dermis, with the infection confined in or near the vascular bed. **Vesicles** and **pustules** occur when viruses spread from the capillaries to the superficial layers of the skin. Destruction of cells by virus replication is the primary cause of lesions.

Similar considerations govern the formation of lesions in mucosal tissues such as those in the mouth and throat. Because these surfaces are wet, vesicles break down more rapidly than on the skin. During measles infection, vesicles in the mouth become ulcers before the appearance of skin lesions. Such Koplik spots are diagnostic for measles virus infection. In the respiratory tract, after virions leave the subepithelial capillaries, only a single layer of cells must be traversed before particles reach the exterior. Hence, during infections with measles virus and varicella-zoster virus, virions appear in respiratory tract secretions before the skin rash appears. By the time that the infection is recognized from the skin rash, viral transmission to others may already have occurred.

The Liver, Spleen, Bone Marrow, and Adrenal Glands

The liver, spleen, bone marrow, and adrenal glands are characterized by the presence of sinusoids lined with macrophages. Such macrophages, known as the reticuloendothelial system, function to filter the blood and remove foreign particles. They often provide a portal of entry into various tissues. For example, viruses that infect the liver, the major filtering and detoxifying organ of the body, usually enter from the blood. The presence of virus

Table 14.8 Viruses that cause skin rashes in humans

Virus	Disease	Feature
Coxsackievirus A16	Hand-foot-and-mouth disease	Maculopapular rash
Echoviruses 4, 6, 9, 16; coxsackieviruses A9, 16, 23	Rash (not distinguishable)	Maculopapular rash
Measles virus	Measles	Maculopapular rash
Parvovirus	Erythema infectiosum	Maculopapular rash
Rubella virus	German measles	Maculopapular rash
Varicella-zoster virus	Chickenpox, zoster	Vesicular rash

particles in the blood invariably leads to the infection of **Kupffer cells**, the macrophages that line the liver sinusoids (Fig. 14.15). Virions may be transcytosed across Kupffer and endothelial cells without replication to reach the underlying hepatic cells. Alternatively, viruses may multiply in Kupffer and endothelial cells and then infect underlying hepatocytes. Either mechanism may induce inflammation and necrosis of liver tissue, a condition termed **hepatitis**.

Central Nervous System, Connective Tissue, and Skeletal and Cardiac Muscle

Coxsackieviruses, members of the *Picornaviridae*, multiply in skeletal and cardiac muscle, and rubella virus, a togavirus, infects connective tissue. The mechanisms by which these viruses enter muscles from the blood have not been adequately studied. It is known that certain coxsackieviruses can replicate in cultured endothelial cells. Virions may be released into muscle cells after multiplying within cells of the capillary walls. Some coxsackieviruses infect B lymphocytes, which might carry the infection from the blood into muscle.

In the central nervous system, connective tissue, and skeletal and cardiac muscle, capillary endothelial cells are generally not fenestrated and are backed by a dense basement membrane (Fig. 14.14 and 14.16). In the central nervous system the basement membrane is the foundation of the blood-brain barrier. However, it is not so much a barrier as a selective permeability system. Much study has been devoted to determining how infection spreads to the brain from the blood. Routes of invasion into the central nervous system are summarized in Fig. 14.17.

In several well-defined parts of the brain, the capillary epithelium is fenestrated and the basement membrane is sparse. These highly vascularized sites include the choroid plexus, a tissue that produces more than 70% of the cerebrospinal fluid that bathes the spinal cord and ventricles of the brain. Some viruses (e.g., mumps virus and certain togaviruses) pass through the capillary endothelium and enter the stroma of the choroid plexus, where they may cross the epithelium into the cerebrospinal fluid either by transcytosis or by replication and directed release. Once in the cerebrospinal fluid, infection spreads to the ependymal cells lining the ventricles and the underlying brain tissue. Other viruses may directly infect, or be transported across,

Figure 14.15 Routes of viral entry into the liver. Two layers of hepatocytes are shown, with the sinusoid at the center lined with Kupffer cells. Endothelial cells are not shown. Adapted from C. A. Mims et al., *Mims' Pathogenesis of Infectious Disease* (Academic Press, Orlando, Fla., 1995), with permission.

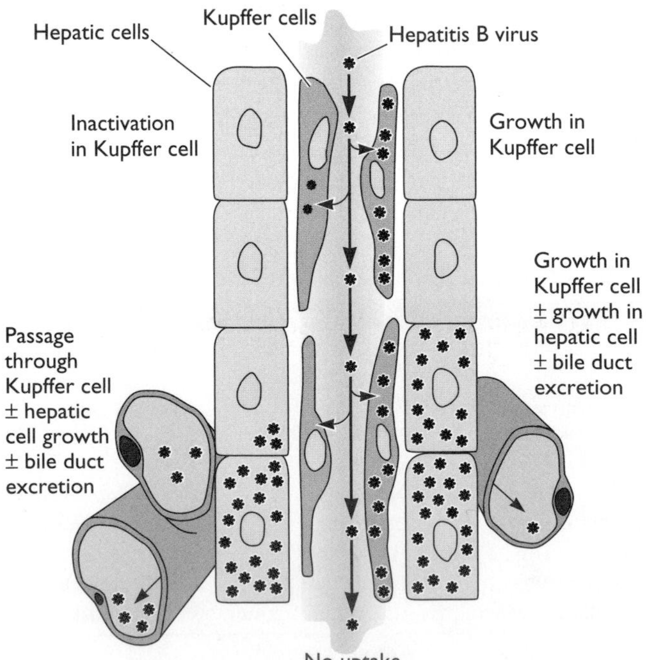

Figure 14.16 The blood-brain junction. An electron micrograph of an arteriole in the cortex of an adult mouse is shown. The open arrows indicate parts of three endothelial cells sealed by tight junctions. The filled arrows show the dense basement membrane. Magnification, ×36,850. Reprinted from R. T. Johnson, *Viral Infections of the Nervous System* (Raven Press, New York, N.Y., 1982), with permission. Courtesy of R. T. Johnson, Johns Hopkins University School of Medicine.

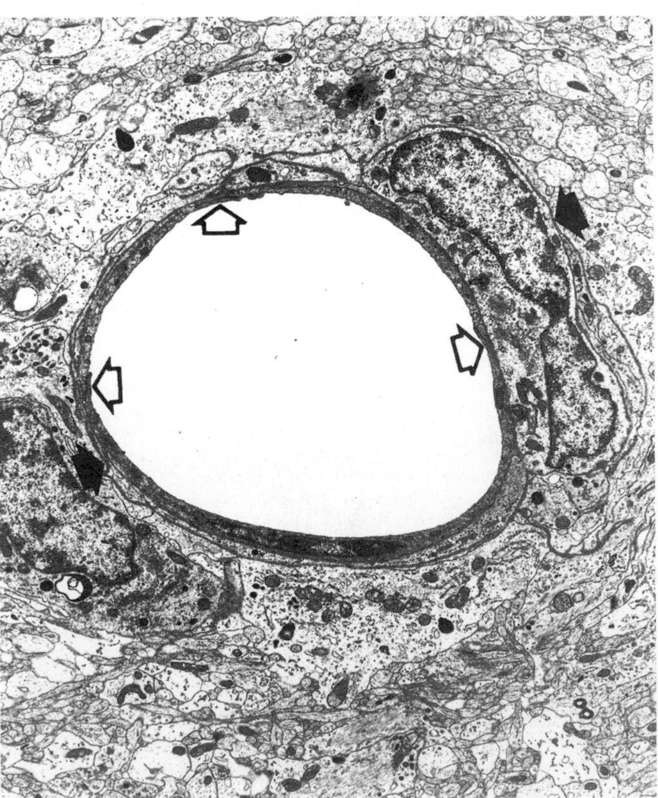

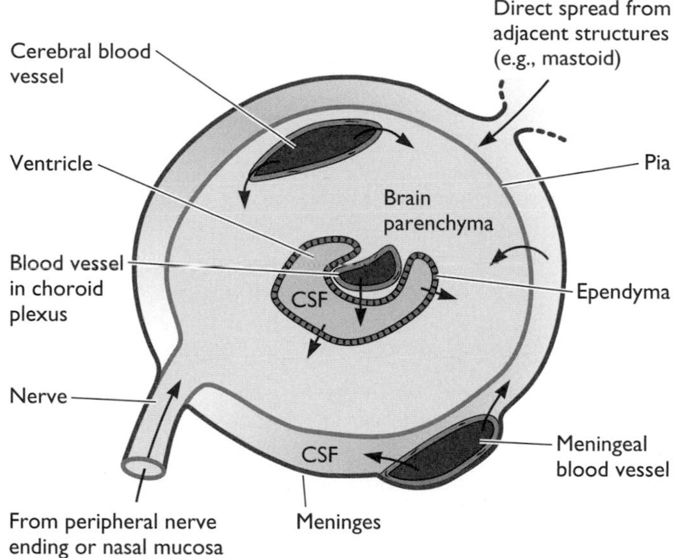

Figure 14.17 How viruses gain access to the central nervous system. A summary of the mechanisms by which viruses can enter the brain is shown. CSF, cerebrospinal fluid. Adapted from C. A. Mims et al., *Mims' Pathogenesis of Infectious Disease* (Academic Press, Orlando, Fla., 1995), with permission.

the capillary endothelium (e.g., picornaviruses and togaviruses). Some viruses cross the endothelium within infected monocytes or lymphocytes (e.g., human immunodeficiency virus and measles virus). Increased local permeability of the capillary endothelium, caused, for example, by certain hormones, may also permit virus entry into the brain and spinal cord.

The Renal Glomerulus, Pancreas, Ileum, and Colon

To enter tissues that lack sinusoids (Fig. 14.14), virions must first adhere to the endothelial cells lining capillaries or venules, where the blood flow is slowest and the walls are thinnest. Endothelial cells are not highly phagocytic, and therefore adhesion of virus particles to these cells requires the presence of cellular receptors. To increase the chances of adhesion, virions must be present in a high concentration and circulate for a sufficient length of time. Clearly there is a race between adhesion and removal of virions by the reticuloendothelial system. Once blood-borne viruses have adhered to the vessel wall, they can readily invade the renal glomerulus, pancreas, ileum, or colon because the endothelial cells that make up the capillaries are fenestrated (with "windows" between cells; loosely joined together), permitting virions or virus-infected cells to cross into the underlying tissues (e.g., poliovirus). Some viruses (e.g., herpes simplex, yellow fever, and measles viruses) cross the endothelium while being carried by infected monocytes or lymphocytes in a process called **diapedesis**.

The Fetus

In a pregnant female, viremia may result in infection of the developing fetus (Table 14.9). The risk of fetal infection in infants whose mothers have been infected with rubella virus during the first trimester is approximately 80%. Similarly, intrauterine transmission of human cytomegalovirus occurs in approximately 40% of pregnant women with primary infection. Human immunodeficiency virus can certainly be transmitted from mother to infant; about 1 in 5 infants born to infected mothers are infected in utero. The basement membrane is less well developed in the fetus, and infection can occur by invasion of the placental tissues and subsequent invasion of the fetal tissue. Infected circulating cells such as monocytes may enter the fetal bloodstream directly. There is ample evidence that virus may be transmitted to the baby during delivery or breast-feeding.

Tropism

Most viruses do not infect all the cells of a host but are restricted to specific cell types of certain organs. Tropism is a predilection of viruses to infect certain tissues and not others. For example, an **enterotropic** virus replicates in the

Table 14.9 Some congenital viral infections

Syndrome	Virus	Fetus
Fetal death and abortion	Smallpox virus	Human
	Parvovirus	Human
	Various alphaherpesviruses	Swine, horses, cattle
Congenital defects	Cytomegalovirus	Human
	Rubella virus	Human
Immunodeficiency	Human immunodeficiency virus type 1	Human
Inapparent (lifelong carrier)	Lymphocytic choriomeningitis virus	Mouse
	Noncytopathic bovine viral diarrhea virus	Cattle
	Murine leukemia virus	Mouse
	Avian leukosis virus	Chicken

Table 14.10 Viral receptors and viral tropism

Virus	Tropism	Receptor (distribution)
Major group rhinovirus	Respiratory tract	Icam-1 (ubiquitous)
Influenza virus	Respiratory tract	Sialic acid (ubiquitous)
Poliovirus	Oropharyngeal and intestinal mucosa, motor neurons	Pvr (most organs)
Herpes simplex virus	Mucoepithelial cells, neurons	Glycosaminoglycans (ubiquitous)

gut, whereas a **neurotropic** virus replicates in cells of the nervous system. Some viruses are **pantropic**, infecting and replicating in many cell types and tissues. Tropism is governed by at least four parameters. It can be determined by the distribution of receptors for entry (**susceptibility**), or by a requirement of the virus for differentially expressed intracellular gene products to complete the infection (**permissivity**). However, even if the cell is permissive and susceptible, infection may not occur because the virus is physically prevented from interacting with the tissue (**accessibility**). Finally, an infection may not occur even when the tissue is accessible and the cells are susceptible and permissive because of the local intrinsic and innate immune defenses. In most cases, tropism is determined by a combination of two or more of these parameters.

Tropism influences the pattern of infection, pathogenesis, and long-term virus survival. Human herpes simplex virus is often said to be neurotropic because of its noteworthy ability to infect, and be reactivated from, the nervous system. But in fact, herpes simplex virus is pantropic and replicates in many cells and tissues in the host. By infecting neurons, it establishes a stable latent infection (see "Two Viruses That Produce Latent Infections" in Chapter 16), but, because it is pantropic, it spreads to other tissues and host cells. One serious consequence is that, if the virus escapes host defenses at the site of infection, it will spread widely, causing disseminated disease, as can occur when herpes simplex virus infects babies and immunocompromised adults (Fig. 14.13). Herpes simplex virus neurotropism leads to yet another serious result of infection. On rare occasions, this virus can enter the central nervous system and cause often-fatal encephalitis. This lethal excursion is not due to a breach in immune defenses because immune-deficient individuals experience the same low rate of brain infections as immune-competent individuals do.

Cell Receptors for Viruses

A cell may be susceptible to infection if the viral receptor(s) is present and functional. If the proper form of viral receptor is not made, the tissue cannot become infected. The location of the receptor might also be a determinant of cellular susceptibility. If the cellular receptor is present only on the basal cell membrane of polarized epithelial cells, a virus cannot replicate unless it first traverses that location by some other means. In some cases, virus particles bound to antibodies can be taken up by Fc receptors on nonsusceptible cells (see "Immunopathological lesions associated with B cells," Chapter 15).

As cellular receptors for viruses have been identified and studied, it has become evident that they can be determinants of tropism. Despite the simplicity of this mechanism, the distribution of cellular receptors in host tissues generally is more widespread than the observed tropism of the virus (Table 14.10). Clearly, receptors are necessary for infection but are not sufficient to explain viral tropism.

Cellular Proteins That Regulate Viral Transcription

Sequences in viral genomes that control transcription of viral genes such as enhancer regions may be determinants of viral tropism. In the brain, JC polyomavirus replicates only in oligodendrocytes (Box 14.6). The results of in vitro

BOX 14.6

JC virus, a ubiquitous human polyomavirus

JC virus is widespread in the human population. Most humans experience inapparent childhood infections, but the virus then persists for life in the kidneys or brain with little consequence. If the immune system is compromised by pregnancy or chemotherapy, virus often reactivates from kidney tissues, and infectious virus can be found in the urine. On rare occasions, JC virus reactivates in the brain, causing more serious problems. The disease is called progressive multifocal leukoencephalopathy, a demyelinating disease affecting oligodendrocytes. This disease is often seen in AIDS patients and after immunosuppressive therapy for organ transplant procedures. Given the rarity of the disease and the lack of suitable animal models, it has been difficult to determine how the permissive infection is maintained and reactivated.

and in vivo experiments indicate that the JC virus enhancer is active only in this cell type. Similar results have been obtained with the liver-specific enhancer of hepatitis B virus, the keratinocyte-specific enhancer of human papillomavirus type 11, and the enhancers in the long terminal repeat of certain retroviruses.

Cellular Proteases

Cellular proteases are often required to cleave viral proteins to form the mature infectious virus particle. A cellular protease cleaves the influenza virus HA precursor into two subunits so that fusion of the viral envelope and cell membrane can proceed. In mammals, the replication of influenza virus is restricted to the epithelial cells of the upper and lower respiratory tract. The tropism of this virus is thought to be influenced by the limited expression of the protease that processes HA. This serine protease, called tryptase Clara, is secreted by nonciliated Clara cells of the bronchial and bronchiolar epithelia (Fig. 14.18). The purified enzyme can cleave HA and activate HA in virions in vitro. Alteration of the HA cleavage site so that it can be recognized by other cellular proteases dramatically influences the tropism of the virus and its pathogenicity: some highly virulent avian influenza virus strains contain an insertion of multiple basic amino acids at the cleavage site of HA. This new sequence permits processing by ubiquitous intracellular proteases such as furins. As a result, these variant viruses are released in an active configuration and are able to infect many organs of the bird, including the spleen, liver, lung, kidney, and brain. Naturally occurring mutants of this type cause high mortality in poultry farms. Avian influenza viruses isolated from 16 people in Hong Kong contained similar amino acid substitutions at the HA cleavage site. Indeed, many of these individuals had gastrointestinal, hepatic, and renal, as well as respiratory symptoms. A virus with such an HA site alteration had not been previously identified in humans, and its isolation led to fears that an influenza pandemic was imminent. To prevent the virus from spreading, all the chickens in Hong Kong were slaughtered. The concern was based on the knowledge that the HA of a pantropic influenza virus strain isolated over 50 years ago is processed by ubiquitous proteases (Box 14.7).

Figure 14.18 Cleavage of influenza virus HA0 by tryptase Clara. Influenza viruses replicate in respiratory epithelial cells in humans. These virus particles contain the uncleaved form of HA (HA0) and are noninfectious. Clara cells secrete the protease tryptase Clara, which cleaves the HA0 of extracellular particles, thereby rendering them infectious. Adapted from M. Tashiro and R. Rott, *Semin. Virol.* **7**:237–243, 1996, with permission.

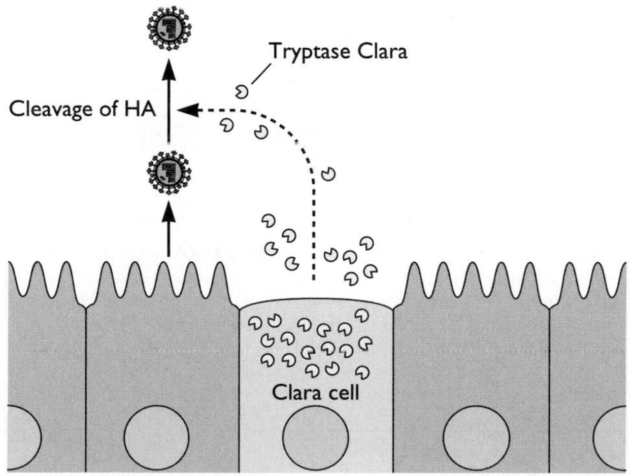

The Site of Entry Often Establishes the Pathway of Spread

For viruses that spread by neural pathways, the innervation at the primary site of inoculation determines the neuronal circuits that become infected. The only areas in the brain or spinal cord that become infected by herpes simplex virus are those that contain neurons with axon terminals or dendrites connected to the site of inoculation. After peripheral infection, poliovirus never reaches certain areas of the spinal cord and brain. However, replication occurs if the virus is placed directly into these sites. The conclusion is that accessibility of susceptible cells, not a lack of susceptibility or permissiveness, can determine the tropism of infection.

As we have discussed, the brain can be infected either by hematogenous or neural spread. When the neurotropic NWS strain of influenza virus is inoculated intraperitoneally into mice, it is disseminated to the brain by hematogenous spread and replicates in the meninges, choroid plexus, and ependymal cells lining the brain and spinal cord. However, when the virus inoculum is placed in the nose, there is no viremia. Instead, virus enters the brain by neural spread. It first replicates in olfactory epithelia and then enters the sensory neurons that richly endow the nasopharyngeal cavity (Fig. 14.12). Virions then can be found in the trigeminal ganglia, neurons in the brain stem that are in contact with trigeminal neurons, and the olfactory bulb in the brain. The experiment with the NWS influenza virus strain makes a significant point: the site of inoculation can determine the pathway of spread of a particular virus.

Virus Shedding and Transmission

The release of infectious viruses from an infected host is called **shedding**. Shedding is usually an absolute requirement for viral propagation in the host population. The exceptions are direct transmission of viral genomes in the germ lines of their hosts, and infections transmitted by cannibalism, such as the prion diseases kuru in humans

BOX 14.7

A mechanism for expanding the tropism of influenza virus is revealed by analyzing infections that occurred in 1940

Until the isolation of the H5N1 virus from 16 persons in Hong Kong, the HA cleavage site mutation that permits cleavage by intracellular furin proteases had not been found in humans. However, the WSN/33 strain of influenza virus, produced in 1940 by passage of a human isolate in mouse brain, is pantropic in mice. Unlike most human influenza virus strains, WSN/33 can replicate in cell culture in the absence of added trypsin, because its HA can be cleaved by serum plasmin. It was subsequently found that the NA of WSN/33 is necessary for HA cleavage by serum plasmin. This NA contains an in-frame deletion that removes a glycosylation site at residue 46. Recently it was found that the mutant NA can bind plasminogen, sequestering it on the cell surface, where it can be converted to the active form, plasmin (see figure). Plasmin then cleaves HA into HA1 and HA2. A mutation in NA, not in HA, allows cleavage of HA by a ubiquitous cellular protease. This property may, in part, explain the pantropic nature of WSN/33.

WSN/33 can replicate in the mouse brain, causing encephalitis. Interestingly, recombinant viruses containing only NA from WSN/33 did not produce encephalitis; replication in the brain requires the presence of the M and NS genes in addition to the NA from WSN/33. The virulence of WSN/33 is polygenic and cannot be explained solely by the expanded tropism of this strain.

Goto, H., and Y. Kawaoka. 1998. A novel mechanism for the acquisition of virulence by a human influenza A virus. *Proc. Natl. Acad. Sci. USA* **95:**10224–10228.

Taubenberger, J. K. 1998. Influenza virus hemagglutinin cleavage into HA1 and HA2: no laughing matter. *Proc. Natl. Acad. Sci. USA* **95:**9713–9715.

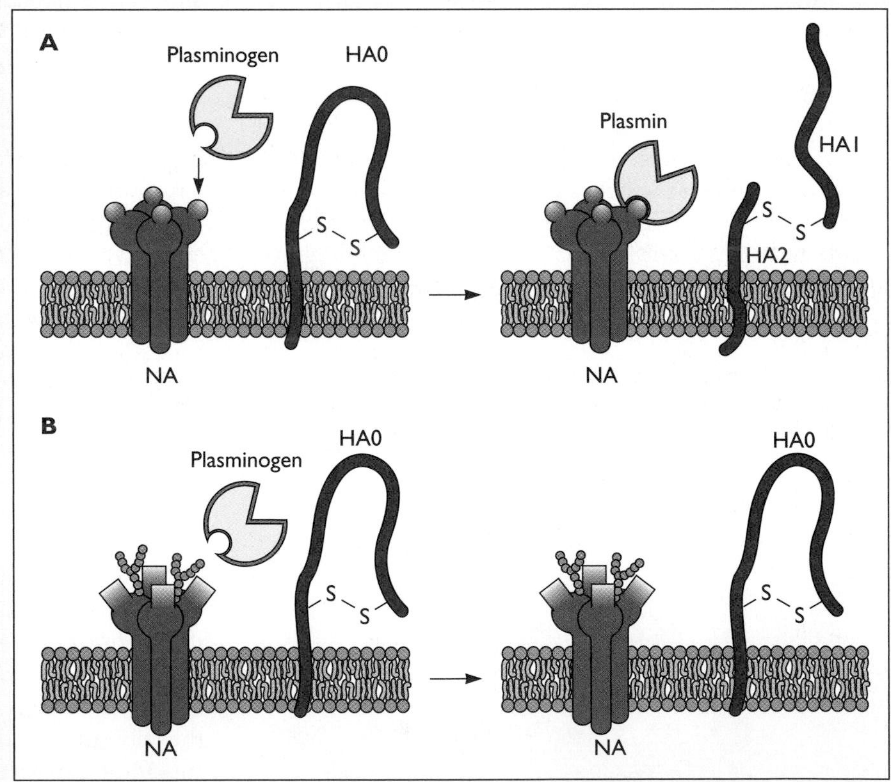

Proposed mechanism for activation of plasminogen and cleavage of HA. (A) Plasminogen binds to NA, which has a lysine at the carboxyl terminus. A cellular protein converts plasminogen to the active form, plasmin. Plasmin then cleaves HA0 into HA1 and HA2. (B) When NA does not contain a lysine at the carboxyl terminus, plasminogen cannot interact with NA and is not activated to plasmin. Therefore, HA is not cleaved. Adapted from H. Goto and Y. Kawaoka, *Proc. Natl. Acad. Sci. USA* **95:**10224–10228, 1998, with permission.

and bovine spongiform encephalopathy ("mad cow disease") in cattle (see Chapter 20). During localized infections, shedding takes place from the primary site of replication at one of the body openings. In contrast, release of virions that cause disseminated infections can occur at many sites. Effective transmission of virions from one host to another depends directly on the concentration of released particles and the mechanisms by which the virions are introduced into the next host. The shedding of small quantities of virus particles may be irrelevant to transmission, while the shedding of high concentrations of infectious virus may facilitate transmission even in minute quantities of tissue or fluid. For example, the concentration of hepatitis B virus in blood can be so high that as little as a few microliters can be sufficient to initiate an infection. The extent of virion survival in the environment also influences the efficiency of transmission.

Respiratory Secretions

Respiratory transmission depends on the production of airborne particles, or aerosols, that contain viruses. Aerosols are produced during speaking, singing, and normal breathing, while coughing produces even more forceful expulsion. Shedding from the nasal cavity requires sneezing and is much more effective if infection induces the production of nasal secretions. A sneeze produces up to 20,000 droplets (in contrast to several hundred expelled by coughing), and all may contain rhinovirus if the individual has a common cold. The largest droplets fall to the ground within a few meters. Many viruses are inactivated by drying (e.g., measles virus, influenza virus, and rhinovirus), and therefore close proximity is required for transmission. The remaining droplets travel a distance related to their size. Droplets 1 to 4 μm in diameter may remain suspended indefinitely, because air is continually in motion. Such particles may reach the lower respiratory tract. Nasal secretions also frequently contaminate hands or tissues. The infection may be transmitted when these objects contact another person's fingers and that person in turn touches his or her nose or conjunctiva. In today's crowded society, the physical proximity of people may select for those viruses that spread efficiently by this route.

Saliva

Some viruses that replicate in the lungs, nasal mucosa, or salivary glands are shed into the oral cavity. Transmission may occur through aerosols, as discussed above, or by contaminated fingers, kissing, or spitting. Animals that lick, nibble, and groom may also transmit infections in saliva. Human cytomegalovirus, mumps virus, and some retroviruses are known to be transmitted by this route.

Feces

Enteric and hepatic virus particles are shed in the feces and are generally more resistant to inactivation by environmental conditions than those released from other sites. An important exception is hepatitis B virus, which is shed in bile into the intestine, but is inactivated as a consequence and not transmitted by feces. Instead, hepatitis B virus is transmitted through the blood. Viruses transmitted by fecal spread usually survive dilution in water, as well as drying. Inefficient or no sewage treatment, contaminated irrigation systems, and the use of animal manures are prime sources of fecal contamination of food, water supplies, and living areas. Any one of these conditions provides an efficient mode for continual reentry of these viruses into the alimentary canal of their hosts. Two hundred years ago, such contamination was inevitable in most of the world. Disposal of human feces in the streets was a common practice. With modern sanitation, the flow of feces into human mouths has been largely interrupted in developed countries, but still is common throughout the rest of the world. Even so, the closing of beaches because of high coliform counts and the contamination of clam and oyster beds by sewage outflows provide regular reminders of our continuous exposure to enteric and hepatic viruses.

Blood

Viremias are a common feature of many viral infections, and viremic blood is a prime means of virus transmission. Arthropods acquire virions when they bite viremic hosts. Although hepatitis and AIDS are not spread this way, they can be transmitted by virus-laden blood during transfusions and injections. Infections may be transmitted from viremic blood during coitus or childbirth, and eating raw meat may place viremic blood in contact with the alimentary and respiratory tracts. Health care workers, dentists, and emergency rescue workers are exposed routinely to viremic blood. Indeed, for many of the fatal hemorrhagic fevers caused by viruses (such as members of the *Bunyaviridae* and *Filoviridae*), their only mode of transmission to humans is by blood and body fluids. Consequently, health care workers often are the first people to die in an outbreak of such viral diseases.

Urine, Semen, and Milk

Virus-containing urine is a common contaminant of food and water supplies. The presence of virus particles in the urine is called **viruria**. Hantaviruses and arenaviruses that infect rodents cause persistent viruria. Consequently, humans may be infected by exposure to dust that contains dried urine of infected rodents. A variety of human viruses replicate in the kidney and are shed in urine.

However, for most viruses, viruria is not important for transmission of infection.

Some retroviruses, including human immunodeficiency virus type 1, herpesviruses, and hepatitis B virus, are shed in semen and transmitted during coitus. Herpesviruses and papillomaviruses that infect the genital mucosa are shed from lesions and transmitted by genital secretions.

Mouse mammary tumor virus is spread primarily by milk, as are some tick-borne encephalitis viruses. Mumps virus and cytomegalovirus are shed in human milk but are probably not often transmitted by this route.

Skin Lesions

Many viruses replicate in the skin, and the lesions that form contain infectious virus particles that can be transmitted to other hosts. Spread of infection is usually by direct body contact. Viruses that are transmitted from skin lesions include poxviruses, herpes simplex virus, varicella-zoster virus, papillomaviruses, and Ebola virus.

Viral Virulence

Once infected, the host may develop a wide range of disease symptoms, as discussed in Chapter 16. **Virulence** refers to the capacity of infection to cause disease. It is a quantitative statement of the degree or extent of pathogenesis. In general, a **virulent** virus causes significant disease, whereas an **avirulent** or **attenuated** virus causes no or reduced disease, respectively.

From the earliest days of experimental virology, it was recognized that viral strains often differ in virulence. Virologists believed that the study of attenuated strains would answer the question of how viruses cause disease. This belief has stood the test of time; indeed, the study of attenuated viruses is still a common strategy. We have learned how to alter viral virulence by direct and indirect methods and have produced viruses of such limited virulence that they can be used as live vaccines (see Chapter 19). Today, the methods of recombinant DNA technology permit a more systematic and unbiased analysis, by introducing defined mutations into viral genomes so that virulence genes can be identified. The goal of these studies is to understand the mechanisms by which viral and cellular gene products control viral virulence.

Measuring Viral Virulence

Virulence can be quantified in a number of ways. One approach is to determine the concentration of virus that causes death or disease in 50% of the infected animals. This parameter is called the 50% lethal dose (LD_{50}), the 50% paralytic dose (PD_{50}), or the 50% infectious dose (ID_{50}), depending on the parameter that is measured. Other measurements of virulence include mean time to death or appearance of symptoms, and measurement of fever or weight loss. Virus-induced tissue damage can be measured directly by examining histological sections or blood. The safety of live attenuated poliovirus vaccine is determined by assessing the extent of pathological lesions in the central nervous system in experimentally inoculated monkeys. The reduction in blood concentration of $CD4^+$ lymphocytes caused by human immunodeficiency virus type 1 infection is another example. Indirect measures of virulence include assays for liver enzymes (alanine or aspartate aminotransferases) that are released into the blood as a result of virus-induced liver damage.

It is important to recognize that the virulence of a single virus strain may vary dramatically depending on the dose and the route of infection, as well as on the species, age, gender, and susceptibility of the host. The effect of inoculation route on virulence is illustrated in Table 14.11. Clearly, virulence is a relative property. Consequently, when the levels of virulence of two very similar viruses are compared, the assays must be conducted in identical fashion. Quantitative terms such as LD_{50} cannot be used to compare virulence among different viruses.

Genetic Determinants of Virulence

A major goal of animal virology is to identify viral and host genes that control virulence. Once such genes have been identified, defined mutations can be made, gene products can be purified, and hypotheses can be tested. If the molecular mechanisms by which viruses cause disease are known, drugs that block these processes may be synthesized. In addition, the role of virulence genes in estab-

Table 14.11 Effect of route of inoculation on viral virulence

Virus	PFU/LD_{50}			
	Suckling mice		Adult mice	
	Intracerebral infection	Subcutaneous infection	Intracerebral infection	Subcutaneous infection
Wild-type La Crosse virus	~1	~1	~1	~10
Attenuated La Crosse virus mutant	~1	$>10^5$	$>10^6$	$>10^7$

lishing the characteristic patterns of viral infections can be determined. It may be possible to test the hypothesis that virulence influences the long-term survival of viruses in nature.

Alteration of Viral Virulence

To identify viral virulence genes, it is necessary to compare viruses that differ only in their degree of virulence. Before the era of modern virology, several approaches were used to attain this goal. Occasionally, natural isolates of some avirulent viruses were identified. Although wild-type strains of poliovirus type 2 readily cause paralysis after intracerebral inoculation into monkeys, a type 2 isolate from the feces of healthy children was shown to be completely avirulent after inoculation by the same route. Serial passage of virulent viruses either in animal hosts or in cell culture yielded viruses with reduced virulence (Fig. 19.6). With the development of cell cultures for use in virology, serial passage of viruses in such cells became a common method for isolating attenuated viruses.

Although these approaches were useful, they were laborious. To facilitate the isolation of attenuated viruses, viral genomes often were mutagenized, as described in Chapter 2, and the altered viruses were assayed for virulence in animals. However, controlling the degree of mutagenesis was difficult, and multiple mutations were often introduced. Until the advent of recombinant DNA technology, the ability to introduce mutations in a specific gene was limited. More recently, rapid sequencing of entire viral genomes, polymerase chain reaction amplification of selected genomic segments, and site-directed mutagenesis have become routine procedures in the quest to identify viral virulence genes and their products.

Viral Virulence Genes

Despite modern technological advances, the identification and analysis of virulence genes in a systematic way have not been straightforward. Part of the problem is that there are few simple tissue culture assays for virulence. Many of the pathogenic effects promoted by viruses are a result of action of the intrinsic defense and innate and adaptive immune systems, and it is not possible to reproduce the complex actions of the immune system in a tissue culture dish. Another problem confronting investigators is the simple fact that it is not obvious a priori by inspection of viral genomes what comprises a virulence gene. Consequently, most studies begin with the premise that a virus that causes reduced or no disease in an animal host harbors a defective virulence gene. The genomes of attenuated viruses obtained by empirical efforts (see Chapter 19) were often found to harbor multiple mutations, and the contributions of the individual mutations to the attenuation phenotype could be difficult to ascertain. However, by reversion of point mutations, repairing of deletions, and crossing of mutants and wild-type viruses, many relevant defects were identified. Some mutations reduced, eliminated, or augmented protein function, while others affected the binding of transcription, translation, or replication proteins.

A significant drawback to studies of virulence phenotypes of viruses that infect humans is that relevant animal models of disease are not always readily available. Nevertheless, considerable progress has been made in recent years. In the following sections, we will discuss examples of viral virulence genes that can be placed in one of four general classes (Box 14.8). It should be understood, however, that the vast majority of such genes have not been studied sufficiently to place them in one of these categories.

BOX 14.8

Four classes of viral virulence genes

The viral genes affecting virulence can be sorted into four general classes (and some may be included in more than one). The genes in these classes specify proteins that

- affect the ability of the virus to replicate
- modify the host's defense mechanisms
- facilitate virus spread in and among hosts
- are directly toxic

As might be expected, mutations in these genes often have minimal or no effect on replication in cell culture and, as a consequence, often are called "nonessential genes," an exceedingly misleading appellation. Virulence genes require careful definition, as exemplified by the first general class listed above (ability of the virus to replicate). Any defect that impairs virus reproduction or propagation often results in reduced virulence. In many cases, this observation is not particularly insightful or useful. The difficulty in distinguishing an indirect effect due to inefficient replication from an effect directly relevant to disease has plagued the study of viral pathogenesis for years. There is an adage in genetics that says, you always get what you select, but you may not get what you want.

Although this discussion focuses on producing viruses that are less virulent, the opposite approach, producing viruses that are **more** virulent, has also been informative. Perhaps the best example is the production of a recombinant ectromelia virus containing the gene encoding interleukin-4 (IL-4) (Box 14.9).

Gene products that alter the ability of the virus to replicate. Genes that encode proteins affecting both viral replication and virulence can be placed in one of two subclasses (Fig. 14.19). Classification is accomplished by analyzing mutants defective in the gene of interest. Viral mutants of one subclass exhibit reduced or no replication in the animal host and in many cultured cell types. Reduced virulence results from failure to produce sufficient numbers of virus particles to cause disease. Such a phenotype may be caused by mutations in any viral gene.

Mutants of the second subclass exhibit impaired virulence in animals, but no replication defects in cells in culture (except perhaps in cell types representative of the tissue in which disease develops). Such host-range mutants should provide valuable insight into the basis of viral virulence, because they identify genes specifically required for disease. Host-range mutations in a wide variety of viral genes encoding proteins that participate in many of the steps in viral replication have been described. We will discuss three gene families found in alphaherpesviruses as specific examples of this type of mutant.

A primary requirement for the replication of DNA viruses is access to large pools of deoxyribonucleoside triphosphates. This need poses a significant problem for viruses that replicate in terminally differentiated, nonreplicating cells such as neurons. The genomes of many small DNA viruses encode proteins that alter the cell cycle such that the cellular substrates for DNA synthesis are produced. Another solution, exemplified by alphaherpesviruses, is to encode enzymes that function in nucleotide metabolism and DNA synthesis, such as thymidine kinase and ribonucleotide reductase. Mutations in these genes often reduce the neurovirulence of herpes simplex virus because the mutants cannot replicate in neurons, or in any cell unable to complement the deficiency.

Several attenuated herpesvirus strains harbor viral DNA polymerase gene mutations that alter virus replication in neurons but not in other cell types. Such mutants are attenuated after direct inoculation into the brains of mice, yet replicate well if introduced in the periphery. Furthermore, transfer of the DNA polymerase gene from a neuroinvasive herpes simplex virus type 1 strain to a nonneuroinvasive herpes simplex virus type 2 strain

BOX 14.9

Creation of a killer poxvirus

Australia had a wild rodent infestation, and scientists were attempting to attack this problem with a genetically engineered ectromelia virus, a member of the *Poxviridae*. The idea was to introduce the gene for the mouse egg shell protein zona pellucida 3 into a recombinant ectromelia virus. When the virus infects mice, the animals would trigger an antibody response that would destroy eggs in female mice. Unfortunately, the strategy did not work in all the mouse strains that were tested. It was decided to incorporate the gene for IL-4 into the recombinant virus. This strategy was based on the previous observation that incorporation of the IL-4 gene into vaccinia virus boosts antibody production in mice. The presence of IL-4 was therefore expected to increase the immune response against zona pellucida 3.

To the researchers' great surprise, the recombinant virus replicated out of control in inoculated mice, destroying their livers and killing them. Mice that were vaccinated against ectromelia could not survive infection with the recombinant virus; half of them died. "This was a complete shock to us," said one researcher. Essentially, they had shown that a common laboratory technique could be used to overcome the host immune response.

Those who conducted this work debated whether to publish their findings, but eventually did so. Their findings raised alarms about whether such technology could be used to produce biological weapons, and the incident was widely reported in the press. Although the result was a surprise to the investigators, analysis of previously published data suggests that increased virulence of the recombinant virus could have been predicted. This incident emphasizes the need to carefully consider experimental design, and to be aware of possible dangers that might arise from the inappropriate use of genetic engineering.

Jackson, R. J., A. J. Ramsay, C. D. Christensen, S. Beaton, D. F. Hall, and I. A. Ramshaw. 2001. Expression of mouse interleukin-4 by a recombinant ectromelia virus suppresses cytolytic lymphocyte responses and overcomes genetic resistance to mousepox. *J. Virol.* **75:**1205–1210.

Müllbacher, A., and M. Lobigs. 2001. Creation of killer poxvirus could have been predicted. *J. Virol.* **75:**8353–8355.

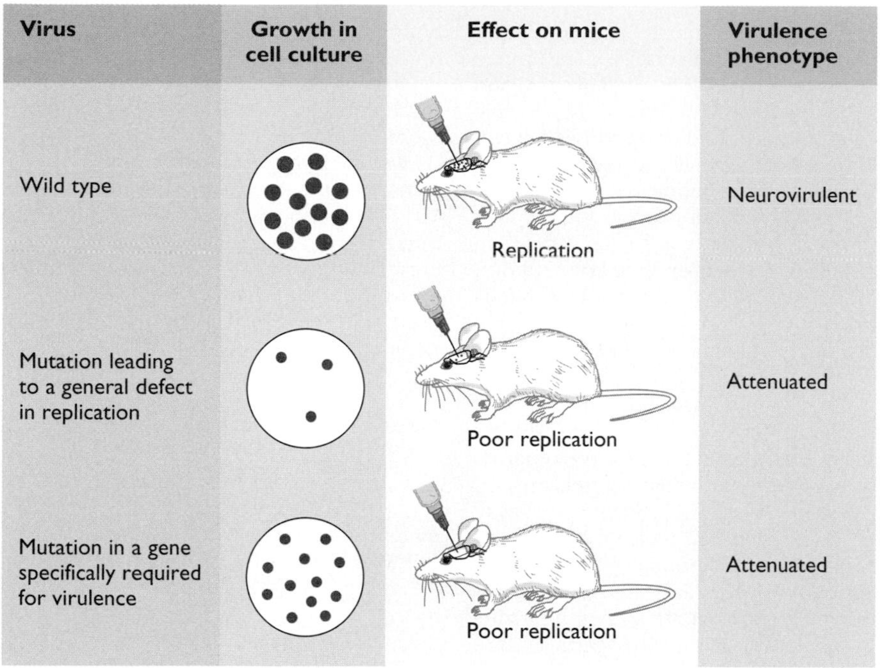

Virus	Growth in cell culture	Effect on mice	Virulence phenotype
Wild type		Replication	Neurovirulent
Mutation leading to a general defect in replication		Poor replication	Attenuated
Mutation in a gene specifically required for virulence		Poor replication	Attenuated

Figure 14.19 Different types of virulence genes. Examples of virulence genes that affect viral growth, using intracerebral neurovirulence in adult mice as an example. Wild-type viruses grow well in cell culture; after inoculation into the mouse brain, they replicate and are virulent. Mutants with defects in replication do not grow well in cultured cells, or in mouse brain, and are attenuated. Mutants with a defect in a gene specifically required for virulence replicate well in certain cultured cells, but not in the mouse brain, and are attenuated. Adapted from N. Nathanson (ed.), *Viral Pathogenesis* (Lippincott-Raven Publishers, Philadelphia, Pa., 1997), with permission.

yielded a recombinant virus that spread readily to the brain after eye infection in mice. In another example, point mutations that affect the helicase activity of the helicase-primase complex result in a virus that cannot replicate in neurons, and is attenuated. These observations imply that neuron-specific proteins cooperate with the viral DNA replication machinery to promote DNA synthesis. Analogous attenuating mutations are also found in the genomes of members of other virus families.

Deletion of the herpes simplex virus gene encoding ICP34.5 protein produces a mutant virus so dramatically attenuated that it is difficult to determine an LD_{50}, even when it is injected directly into the brain. Such mutants can replicate in some, but not all, cell types in culture and in the brain (Box 14.10). Notably, they are unable to grow in postmitotic neurons. The molecular basis for the cell type specificity of ICP34.5 mutants has yet to be determined. Their lack of virulence is related to the interaction of this protein with components of the interferon-activated Pkr pathway described in Chapters 11 and 15.

Noncoding sequences that affect the ability of the virus to replicate. The attenuated strains used for the live Sabin poliovirus vaccine are examples of viruses with mutations that are not in protein-coding sequences (Fig. 19.6). Each of the three serotypes in the vaccine contains a mutation in the 5′ noncoding region of the viral RNA that impairs replication in the brain. They also reduce the translation of viral messenger RNA (mRNA) in cultured cells of neuronal origin, but not in certain other cell types. An interesting finding is that attenuated viruses bearing these mutations apparently do not replicate efficiently at the primary site of infection in the gut. Consequently, many fewer virus particles are available for hematogenous or neural spread to the brain. Mutations in the 5′ noncoding regions of other picornaviruses also affect virulence in animal models. Deletions within the long poly(C) tract within the 5′ noncoding region of mengovirus reduce virulence in mice without affecting viral replication in cell culture.

Gene products that modify the host's defense mechanisms. The study of viral virulence genes has identified a diverse array of viral proteins that sabotage the body's intrinsic defenses and innate and adaptive systems. Some of these viral proteins are called **virokines** (secreted proteins that mimic cytokines, growth factors, or similar extracellular immune regulators) or **viroceptors** (homologs of host receptors). Mutations in genes encoding either class of protein affect virulence, but these genes are **not** required for growth in cell culture (Box 14.11). Most virokines and viroceptors have been discovered in the genomes of large DNA viruses (see Fig. 15.5). Their mechanisms of action will be discussed in detail in Chapters 15 and 16.

As will be discussed in Chapter 15, many viruses induce apoptosis, an intrinsic response that contributes significantly to viral pathogenesis. For example, the pattern of

BOX 14.10

The use of attenuated herpes simplex viruses to clear human brain tumors

Malignant glioma, a common brain tumor, is almost universally fatal, despite advances in surgery, radiation, and chemotherapy. Patients rarely survive longer than a year after diagnosis. Several groups have proposed the use of cell-specific replication mutants of herpes simplex virus to kill glioma cells in situ. One such virus under study carries a deletion of the ICP34.5 gene and the large subunit of ribonucleotide reductase. These mutant viruses replicate well in dividing cells, such as glioma cells, but not in nondividing cells, such as neurons. The theory is that attenuated virus injected into the glioma will replicate and kill the dividing tumor cells, but will not replicate or spread in the nondividing neurons.

This idea works in principle: studies of mice have indicated that direct injection of this virus into human gliomas transplanted into mice causes clearing of the tumor. The virus is attenuated and safe: injection of 1 billion virus particles into the brain of *Aotus nancymai* (a monkey highly sensitive to herpes simplex virus brain infections) had no pathogenic effect on the animal. This degree of attenuation is remarkable, because the virus replicates to reasonably high titers in nonneural monkey and human cells.

Several human trials are in progress to test safety and dosage. In one study, up to 10^5 PFU of virus was inoculated directly into the brain tumors of nine patients. No encephalitis, adverse clinical symptoms, or reactivation of latent herpes simplex virus was observed. Higher concentrations will be used until a therapeutic effect is attained.

Mineta, T., S. D. Rabkin, T. Yazaki, W. D. Hunter, and R. L. Martuza. 1995. Attenuated multi-mutated herpes simplex virus-1 for the treatment of malignant gliomas. *Nat. Med.* **1:**938–943.

Rampling, R., G. Cruickshank, V. Papanastassiou, J. Nicoll, D. Hadley, D. Brennan, R. Petty, A. MacLean, J. Harland, E. McKie, R. Mabbs, and M. Brown. 2000. Toxicity evaluation of replication-competent herpes simplex virus (ICP 34.5 null mutant 1716) in patients with recurrent malignant glioma. *Gene Ther.* **7:**859–866.

infection of Sindbis virus changes from nonlethal persistent to lethal acute, depending on whether the infected cell can mount an apoptotic response. Interestingly, after infection, apoptosis can be either indirect (uninfected cells die) or direct (infected cells die), and both responses influence subsequent pathogenesis. African swine fever, a highly contagious disease of pigs caused by a double-stranded DNA virus transmitted by insects, is an example of a disease caused by indirect apoptosis. The severe lesions and hemorrhages are striking, but intense destruction of lymphoid tissue is characteristic of the disease. This destruction is caused by apoptosis of uninfected lymphocytes induced by cytokines and apoptotic mediators released from infected macrophages. Direct pathogenic effects of virus-induced apoptosis have been suggested in a model of herpes simplex virus-mediated fulminant hepatitis, in which massive hepatocyte death occurs within 24 h of virus injection into the bloodstream. In this disease model, virions in the blood are removed rapidly by cells of the reticuloendothelial system, primarily those found in

BOX 14.11

Variola virus virulence: a highly efficient inhibitor of complement encoded in the genome

Variola virus, which causes the human disease smallpox, is the most virulent member of the *Orthopoxvirus* genus. The prototype poxvirus, vaccinia virus, does not cause disease in immunocompetent humans, and is used to vaccinate against smallpox. Both viral genomes encode inhibitors of the complement pathway. The vaccinia virus complement control protein is secreted from infected cells and functions as a cofactor for the serine protease factor I. The variola virus homolog, called smallpox inhibitor of complement, differs from the vaccinia virus complement control protein by 11 amino acid substitutions. Because the variola virus protein had not been studied, it was produced by mutagenesis of DNA encoding the vaccinia virus homolog. The variola virus protein was found to be 100 times more potent than the vaccinia virus protein at inactivating human complement. In this way, the variola virus protein interferes with complement-mediated virus clearance.

These findings suggest that the virulence of variola virus, and the avirulence of vaccinia virus, might be controlled in part by complement inhibitors encoded in the viral genome. Furthermore, if smallpox should reemerge as a consequence of bioterrorism, the smallpox inhibitor of complement might be a useful therapeutic target.

Rosengard, A. M., Y. Liu, N. Zhiping, and R. Jimenez. 2002. Variola virus immune evasion design: expression of a highly efficient inhibitor of human complement. *Proc. Natl. Acad. Sci. USA* **99:**8808–8813.

the liver. If the phagocytic activity of macrophages is impaired, virus replication ensues in the liver followed by massive apoptotic death throughout the organ.

Genes that enable the virus to spread in the host. The mutation of some viral genes disrupts spread from peripheral sites of inoculation to the organ where disease occurs. For example, after intramuscular inoculation in mice, reovirus type 1 spreads to the central nervous system through the blood, while type 3 spreads by neural routes. Studies of viral recombinants between types 1 and 3 indicate that the gene encoding the viral outer capsid protein σ1, which recognizes the cell receptor, determines the route of spread. Only one PFU of the bunyavirus La Crosse virus causes lethal encephalitis after intracerebral injection into mice. However, subcutaneous inoculation of over 10^7 PFU causes no disease because no viremia is produced (Table 14.11).

Viral membrane proteins have been implicated in neuroinvasiveness. For example, the change of a single amino acid in the gD glycoprotein of herpes simplex virus type 1 blocks spread to the central nervous system via nerves after footpad inoculation. Similarly, studies of neuroinvasive and nonneuroinvasive strains of bunyaviruses indicate that the G1 glycoprotein is an important determinant of entry into the central nervous system from the periphery. Although it is tempting to speculate that these viral glycoproteins, which participate in entry of the virus into cells, facilitate direct entry into nerve termini, the mechanisms by which they govern neuroinvasiveness are unknown. They may also influence the ability of the host's humoral immune system to clear virus from the primary site of infection.

Toxic viral proteins. Some viral gene products cause cell injury directly, and alterations in these genes reduce viral virulence. Evidence of their intrinsic activity is usually obtained by adding purified proteins to cultured cells, or by synthesis of the proteins from plasmids or viral vectors. The most convincing example of a viral protein with intrinsic toxicity relevant to the viral disease is the nsP4 protein of rotaviruses, which cause gastroenteritis and diarrhea. nsP4 is a nonstructural glycoprotein that participates in the formation of a transient envelope as the particles bud into the endoplasmic reticulum. When this protein is produced in insect cells, it causes an increase in intracellular calcium concentration. When fed to young mice, nsP4 causes diarrhea by potentiating chloride secretion. nsP4 acts as a viral enterotoxin and triggers a signal transduction pathway in the intestinal mucosa.

The SU and TM glycoproteins of human immunodeficiency virus are toxic to cultured cells. TM causes death of cultured cells most likely through alterations in membrane permeability. The addition of SU to cells results in a high influx of calcium. Unfortunately, the contribution of this toxicity to pathogenesis in humans remains untested: the genes encoding these glycoproteins cannot be deleted without destroying viral infectivity, and there is no suitable animal model for AIDS.

Cellular Virulence Genes

Disruption of cellular genes that encode components of the host's immune response—proteins required for T- and B-cell function and cytokines—may have enormous effects on viral infection. In some cases viral disease becomes more severe, while in others disease severity is lessened. Such cellular genes can therefore be considered virulence determinants. A consideration of immune response genes can be found in Chapter 15.

The host proteins required for viral translation, genome replication, and mRNA synthesis are prime candidates for virulence determinants. Tissue-specific differences in such proteins could in principle influence virus replication. A common approach in studying viral pathogenesis is to construct mice with specific gene disruptions (Box 14.1). It may be possible to derive mice lacking such proteins to study viral virulence. Few studies have been done in this largely uncharted aspect of animal virology, an area that may be one of the most interesting yet to be explored in viral pathogenesis.

Epidemiology

While pathogenesis refers to the development of disease in a single individual, epidemiology is the study of the occurrence of disease in a population. The study of epidemiology includes the consideration of mechanisms of viral transmission, risk factors for infection, size of the population needed for virus transmission, geography, season, and means of control (Table 14.12).

Viral Transmission

Viruses survive only if they spread from one susceptible host to another (transmission). There are two general patterns of viral transmission: (i) the perpetuation of infection in one species and (ii) alternate infection of insect and vertebrate hosts. Some viruses such as rabies virus and influenza viruses spread across species, but transmission within each species is relatively self-contained. Most human viruses are transmitted from human to human, although there are viruses for which an extrahuman cycle is needed for maintenance. Measles virus and hepatitis A virus are transmitted from human to human and are maintained solely in the human population. Humans are therefore the **reservoir** for these viruses. In contrast, rabies virus is transmitted from animal to human, but the virus is maintained in animal-to-animal cycles. Some arboviruses (dengue virus and urban yellow fever virus) are transmitted and maintained in vector-to-human cycles, while oth-

Table 14.12 Viral epidemiology

Mechanisms of transmission
Aerosol
Food and water
Fomites
Body secretions
Sexual activity
Birth
Transfusion or transplant
Zoonoses (animals, insects)

Factors that promote transmission
Virus stability
Virus in aerosols and secretions
Asymptomatic shedding
Ineffective immune response

Geography and season
Vector ecology (habitat and season)
School year
Home heating season

Risk factors
Age
Health
Immunity
Occupation
Travel
Lifestyle
Children (schools, day care centers)
Sexual activity

Critical population size
Numbers of seronegative susceptible individuals

Means of control
Quarantine
Vector elimination
Immunization
Antivirals

Table 14.13 Viral transmission

Route of transmission	Examples
Respiratory	Paramyxoviruses, influenza viruses, picornaviruses, varicella-zoster virus, parvovirus B19
Fecal-oral	Picornaviruses, rotavirus, adenovirus
Contact: lesions, saliva, fomites	Herpes simplex virus, rhinovirus, poxvirus, adenovirus
Zoonoses: insects, animals	Togaviruses (arthropod bite), flaviviruses (arthropod bite), bunyaviruses (urine, arthropod bite), arenaviruses (urine), rabies virus (animal bite)
Blood	Human immunodeficiency virus, human T-cell lymphotropic virus, hepatitis B virus, hepatitis C virus, cytomegalovirus
Sexual contact	Herpes simplex virus, human papillomavirus
Maternal-neonatal	Rubella virus, cytomegalovirus, echovirus, herpes simplex virus, varicella-zoster virus
Genetic	Prions, retroviruses

ers (St. Louis encephalitis virus and western equine encephalitis virus) are maintained in vector-to-vertebrate cycles. Viral diseases shared by humans and animals or insects are called **zoonoses**.

Viruses are transmitted among hosts in specific ways (Table 14.13). The route of transmission is determined by the site of virus excretion and the physical stability of the virus. The presence or absence of a lipid envelope is a major determinant of the mode of transmission. Enveloped viruses are fragile and sensitive to low pH. Hence, they are most often transmitted by aerosols or secretions, injection, or organ transplantation. In contrast, nonenveloped viruses can withstand drying, detergents, low pH, and higher temperatures. These viruses can be transmitted by the respiratory and fecal-oral routes, and are often acquired via contaminated objects or **fomites**. **Iatrogenic** transmission occurs when some activity of a health care worker leads to infection of the patient. Such transmission can occur when nonsterile instruments and needles are used, or if a health care worker is infectious. **Nosocomial** transmission occurs when an individual is infected while in a hospital or health care facility. Viruses that cause acute infections of short duration must be efficiently transmitted to new hosts. These viruses are usually transmitted by the respiratory or fecal-oral route, and infections lead to excretion of large numbers of virus particles. Viruses that cause persistent infections need not be efficiently transmitted because virions are excreted for many years.

Vertical transmission refers to the transfer of viruses between parent and offspring, while **horizontal transmission** includes all other forms. In **germ line transmission** the agent is transmitted as part of the genome. Vertical transmission may occur during gestation, when virus crosses the placenta, during birth, or by passage of virus from mother to child in breast milk or through close physical contact (Table 14.9). Most maternal infections have little effect on the fetus, but some viruses that spread by viremia can infect the placenta and even reach fetal circulation. As a result, the fetus may die and be aborted. Less well understood effects of normally nonlethal infections are congenital (teratogenic or birth) defects, including deafness, blindness, and abnormalities of the heart and nervous system. Other symptoms are obvious upon birth. For example, the "congenital rubella syndrome" is recognized in the newborn by liver and spleen enlargement, jaundice, and skin discoloration.

Most arthropod-borne viruses are maintained by cycling between insect and vertebrate hosts. Most viruses have a major or exclusive vector, because they replicate in one or a few related species of insects. These insects in turn have a preference for feeding on a few vertebrate species. Dengue and urban yellow fever viruses are the only two arthropod-borne viruses for which humans are the major vertebrate host. In most other infections caused by arthropod-borne viruses, humans are accidentally infected during vector bites. Normally the vectors prefer to feed on other vertebrates such as birds or woodland rodents. In such arbovirus infections the humans are dead-end hosts and do not maintain the virus (Chapter 20). Other examples are rabies and Korean hemorrhagic fever.

Geography and Season

Some viruses are found only in specific geographical locations. Such restrictions may be due to the requirement for a specific vector or animal reservoir. The availability of an immunologically naive and susceptible population may also explain why a virus is geographically limited. Before global travel was possible, the distribution of viruses was limited. Now, viruses are transported routinely around the globe in airplanes and ships.

Most acute viral infections have a striking seasonal variation in incidence (Fig. 14.20). Respiratory virus infections are more frequent in winter months, and enteric virus infections predominate in the summer. Seasonal differences in diseases caused by arthropod-borne viruses clearly re-

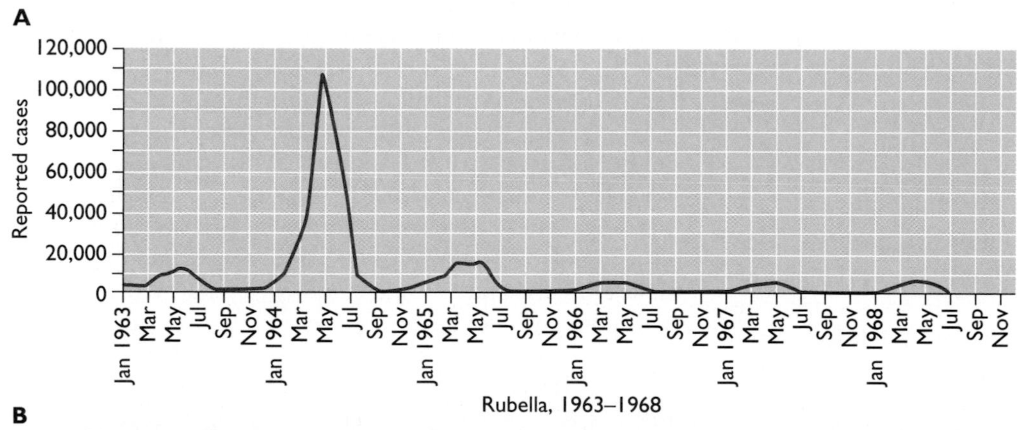

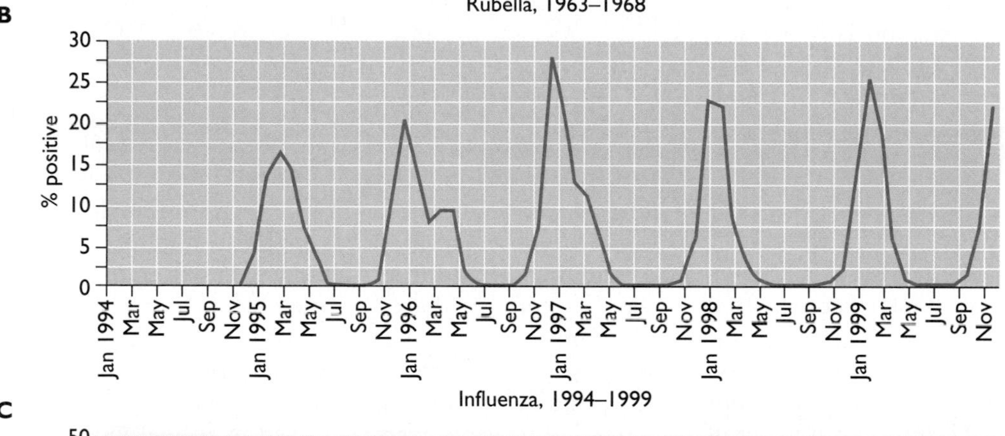

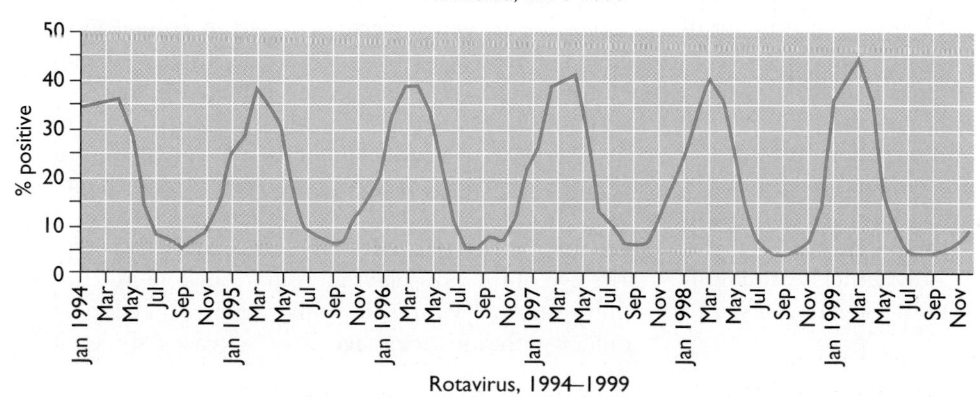

Figure 14.20 Seasonal variation in disease caused by three human pathogens in the United States. (A) Annual cycles of rubella between larger epidemics, which occurred every 6 to 9 years. (B) Percentage of specimens testing positive for influenza viruses. (C) Pattern of rotavirus infection. Adapted from S. F. Dowell, *Emerg. Infect. Dis.* **7**:369–374, 2001, with permission.

flect the life cycle of the vector or the animal reservoir. However, the basis for the seasonal nature of non-arthropod-borne virus infections is unclear. Variations in disease incidence correlate with changes in climate. For example, poliomyelitis was seasonal in New England but not in Hawaii. It has been suggested that seasonality of viruses is due to differences in sensitivity to humidity. According to this hypothesis, during winter months of low humidity, poliovirus is inactivated, but influenza virus remains infectious. Annual variations in viral disease may also result from changes in the susceptibility of the host. Such changes might be linked to the light-dark cycle, and could be governed by alterations in mucosal surfaces, epithelial receptors, immune cell numbers and responsiveness, or other aspects of the intrinsic and immune defense systems. If annual changes in host resistance contribute to the seasonality of viral disease, then the genes encoding such changes should be identified to provide therapeutic targets for intervention.

Host Susceptibility to Viral Disease

When viruses are introduced into populations of humans or animals, many different responses are possible. Some hosts may be highly resistant to infection, some may become infected, and still others may fall in the spectrum between the two extremes. Those who become infected may develop disease ranging from asymptomatic to fatal. Susceptibility to infection and disease vary independently. Understanding the basis for such variation is important, because it may suggest methods for preventing viral disease. The results of such studies demonstrate that both host genes and nongenetic parameters control how populations respond to viral infections.

Immunity

Epidemiologists divide populations into two groups: susceptible and immune. Individuals who have been infected in the past are immune and are not likely to transmit infection. Susceptible individuals can develop disease and spread the virus to others. Persistence of a virus in a population depends on the presence of a sufficient number of susceptible individuals. How efficiently infection occurs determines this number. Immunization against viruses, through natural infection or vaccination, reduces the number of susceptible people, and therefore limits viral persistence and spread. Epidemics of poliomyelitis were self-limiting because the asymptomatic spread of the virus immunized the population. The competence of the immune response also determines how quickly and efficiently the infection is resolved, and the severity of symptoms (Fig. 14.5).

Genetic Determinants of Susceptibility

The genes that control the humoral and cellular immune responses are determinants of susceptibility to viral infections. The class I and class II major histocompatibility complex proteins, for which there are many genes, control the presentation of foreign peptides to T cells. The ability of these proteins to interact with peptides derived from viral proteins determines, in part, how efficiently the infection is cleared. Diversity in major histocompatibility protein genes makes it more likely that there will be a suitable molecule to present peptides for a given infectious agent. Populations from isolated islands have less polymorphism in these genes, a property that may account for their greater susceptibility to infection.

Genes other than those that control the immune response are likely to govern susceptibility to viral infections. Several examples have been studied in animal systems. The best characterized is the *mx* gene of mice, which confers resistance to infection with influenza A virus. The *mx* gene encodes a GTPase that inhibits influenza virus replication, as discussed in Chapter 15. Resistance of mice to flavivirus disease has been mapped to the *flv* locus. Although these viruses replicate in resistant mice, the virus titers are 1,000 to 10,000-fold lower than in susceptible animals, and the infection is cleared before disease symptoms develop. The product of the *flv* locus is 2′-5′-oligo(A) synthetase, an interferon-induced enzyme that activates RNase L, leading to degradation of host and viral mRNAs (Chapter 15). The enzyme from susceptible mice lacks 30% of the C-terminal sequence, compared with the protein from resistant mice. Expression of the enzyme from resistant mice in susceptible embryo fibroblasts renders the cells partially resistant to flavivirus replication. Host susceptibility to flavivirus infection in mice is therefore determined by a gene in the interferon-mediated antiviral response. The *fv-1* locus confers resistance of mice to infection with murine leukemia virus. This gene has been shown to encode the Gag protein of a defective endogenous provirus. It is believed that the *fv-1* gene product in some way blocks infection by interfering at a stage after entry and before integration of the provirus.

mx and *flv* genes were identified in animal models, but have not been shown to play roles in human infections. The best examples of human genes influencing susceptibility to viral infections are those encoding the chemokine receptors that serve as cofactors for entry of human immunodeficiency virus type 1 into cells, as discussed in Chapter 18.

Other Determinants of Susceptibility

The age of the host plays an important role in determining the result of viral infections. Some infections become milder with advancing age. The increased susceptibility of infants and young children is likely explained by the immaturity of their immune responses. Sindbis virus causes lethal encephalitis in newborn but not older mice. This virus kills cells by inducing apoptosis, which can be blocked by the oncogene *bcl-2*. Higher levels of this protein may explain the resistance of older mice. Although young animals may have an immature immune response and become infected more frequently, they also have greater freedom from immunopathology. Intracerebral inoculation of lymphocytic choriomeningitis virus in adult mice is lethal because of the T-cell response, while infant mice survive because of their weaker response. In infants, infections with enteric coronaviruses are severe because the alimentary canal is not fully active. In addition, the pH tends toward neutrality, and digestive enzymes are not available.

Other physiological and physical differences may explain age-dependent variation in susceptibility. As animals age, their alveoli become less elastic, the respiratory muscles weaken, and the cough reflex is diminished. These changes may explain in part why elderly people have increased susceptibility to respiratory infections. Respiratory syncytial virus causes severe lower respiratory tract infections in infants, but only mild upper tract infections in adults. It is not known if the difference is due to host defenses or to variations in the susceptibility of cells to viral infection.

Some viral infections are less severe in children than in adults. An example is poliovirus infection, which was more frequently asymptomatic in childhood. Paralysis was more commonly observed when infection occurred later in life (Chapter 20). The basis for this difference is not known.

Males are more susceptible to viral infections than females, but the difference is very slight and the reasons are not understood. Hormonal differences, which affect the immune system, may be partly responsible. Pregnant women are more susceptible to infectious disease than nonpregnant women, probably for similar reasons. Hepatitis A, B, and E are more lethal, and paralytic poliomyelitis was more common, in pregnant women than in others.

Malnutrition increases susceptibility to infection because the physical barriers to infection, as well as the immune response, are compromised. An example is the increased susceptibility to measles in children with protein deficiency. For this reason, the disease is 300 times more lethal in developing countries than in Europe and North America. When children are malnourished, Koplik spots become massive ulcers, the skin rash is much worse, and lethality may approach 10 to 50% (Chapter 15). Such severe measles infections are observed in children in tropical Africa and in aboriginal children in Australia.

Many other parameters influence susceptibility to infection. Corticosteroid hormones have large effects because they are essential for the body's response to the stress of infection. These hormones have an anti-inflammatory effect which is believed to limit tissue damage. Cigarette smoking is believed to increase susceptibility to respiratory infections, but the data are conflicting. Increased susceptibility is correlated with a poor mental state, such as occurs in stressful life situations (e.g., a death in the family, injury, or the loss of a job). Our mothers told us that exposure to changes in temperature increases susceptibility to infection. Fortunately, the results of studies with rhinovirus have failed to reveal any relationship between exposure to low temperature and the common cold.

Perspectives

Most of this chapter has been concerned with descriptions: how viruses enter the host, how they spread, and how they cause disease. Our intention is to stress the importance of recognizing and understanding the consequences of viral infections. Without this knowledge, it is impossible to define the underlying molecular mechanisms, the quest of the molecular virologist. While the field of viral pathogenesis is almost 100 years old, the viral and host genes that control this process are only now being enumerated. Even though many genes that affect pathogenesis have been identified, this information rarely provides more than a glimpse of the molecular mechanisms responsible. For example, knowing that a gene controls the spread of infection by the blood or by the nervous system does not answer the question of how the gene product determines which route will be taken. Although it is difficult to obtain such mechanistic knowledge, many investigators believe that viral pathogenesis is the most exciting field in virology today, simply because more fundamental questions remain to be answered than in any other area. The challenge is to build on our understanding of viral replication embodied in the previous chapters of this book to provide a comprehensive molecular description of how viruses cause disease. One obvious consequence of such knowledge will be new approaches for preventing, treating, and curing viral disease.

References

Monographs

Becker, Y., and G. Darai (ed.). 1994. *Frontiers of Virology,* vol. 3. *Pathogenicity of Human Herpesviruses due to Specific Pathogenicity Genes.* Springer-Verlag, Berlin, Germany.

Johnson, R. T. 1982. *Viral Infections of the Nervous System.* Raven Press, New York, N.Y.

Mims, C. A., A. Nash, and J. Stephen. 2001. *Mims' Pathogenesis of Infectious Disease,* 5th ed. Academic Press, Orlando, Fla.

Nathanson, N. (ed.). 1997. *Viral Pathogenesis.* Lippincott-Raven Publishers, Philadelphia, Pa.

Notkins, A. L., and M. B. A. Oldstone (ed.). 1984. *Concepts in Viral Pathogenesis.* Springer-Verlag, New York, N.Y.

Richman, D. D., R. J. Whitley, and F. G. Hayden (ed.). 2002. *Clinical Virology*, 2nd ed. ASM Press, Washington, D.C.

Roizman, B., R. J. Whitley, and C. Lopez (ed.). 1993. *The Human Herpesviruses.* Raven Press, New York, N.Y.

Scheld, W. M., R. J. Whitley, and D. T. Durack (ed.). 1991. *Infections of the Central Nervous System.* Raven Press, New York, N.Y.

Wright, P. F. 1996. *Seminars in Virology,* vol. 7. *Viral Pathogenesis.* The W. B. Saunders Co., Philadelphia, Pa.

Review Articles

Biondi, M., and L. Zannino. 1997. Psychological stress neuroimmunomodulation and susceptibility to infectious diseases in animals and man: a review. *Psychother. Psychosom.* **66:**3–26.

Enquist, L. W., P. J. Husak, B. W. Banfield, and G. A. Smith. 1999. Infection and spread of alphaherpesviruses in the nervous system. *Adv. Virus Res.* **51:**237–347.

Graham, B. S., J. A. Rutigliano, and T. R. Johnson. 2002. Respiratory syncytial virus immunobiology and pathogenesis. *Virology* **297:**1–7.

Kennedy, P. G. E. 1992. Molecular studies of viral pathogenesis in the central nervous system; the Linacre lecture 1991. *J. R. Coll. Physicians Lond.* **26:**204–214.

Ohka, S., and A. Nomoto. 2001. Recent insights into poliovirus pathogenesis. *Trends Microbiol.* **9:**501–506.

Oldstone, M. B. A. 1996. Principles of viral pathogenesis. *Cell* **87:**799–801.

Rall, G. F., D. M. P. Lawrence, and C. E. Patterson. 2000. The application of transgenic and knockout mouse technology for the study of viral pathogenesis. *Virology* **271:**220–226.

Takada, A., and Y. Kawaoka. 2002. The pathogenesis of Ebola hemorrhagic fever. *Trends Microbiol.* **9:**506–511.

Tracy, S., N. M. Chapman, J. Romero, and A. I. Ramsingh. 1996. Genetics of coxsackievirus B cardiovirulence and inflammatory heart muscle disease. *Trends Microbiol.* **4:**175–179.

Virgin, H. W. 2002. Host and viral genes that control herpesvirus vasculitis. *Clevel. Clin. J. Med.* **69:**SII7–SII12.

Papers of Special Interest

Chou, J., E. R. Kern, R. J. Whitley, and B. Roizman. 1990. Mapping of herpes simplex virus-1 neurovirulence to $\gamma_1$34.5, a gene nonessential for growth in culture. *Science* **250:**1262–1266.

Dong, Y., C. Q.-Y. Zeng, J. M. Ball, M. K. Estes, and A. P. Morris. 1997. The rotavirus enterotoxin NSP4 mobilizes intracellular calcium in human intestinal cells by stimulating phospholipase C-mediated inositol 1,4,5-trisphosphate production. *Proc. Natl. Acad. Sci. USA* **94:**3960–3965.

Dropulic, L. K., J. M. Hardwick, and D. E. Griffin. 1997. A single amino acid change in E2 glycoprotein of Sindbis virus confers neurovirulence by altering an early step of virus replication. *J. Virol.* **71:**6100–6105.

Haller, B. L., M. L. Barkon, G. P. Vogler, and H. W. Virgin IV. 1995. Genetic mapping of reovirus virulence and organ tropism in severe combined immunodeficient mice: organ-specific virulence genes. *J. Virol.* **69:**357–364.

Leib, D. A., M. A. Machalek, B. R. Williams, R. H. Silverman, and H. W. Virgin. 2000. Specific phenotypic restoration of an attenuated virus by knockout of a host resistance gene. *Proc. Natl. Acad. Sci. USA* **97:**6097–6101.

Levine, B., J. E. Goldman, H. H. Jian, D. E. Griffin, and J. M. Hardwick. 1996. Bcl-2 protects mice against fatal alphavirus encephalitis. *Proc. Natl. Acad. Sci. USA* **93:**4810–4815.

Morimoto, K., D. C. Hooper, H. Carbaugh, Z. F. Fu, H. Koprowski, and B. Dietzschold. 1998. Rabies virus quasispecies: implications for pathogenesis. *Proc. Natl. Acad. Sci. USA* **95:**3152–3156.

Morrison, L. A., and B. N. Fields. 1991. Parallel mechanisms in neuropathogenesis of enteric virus infections. *J. Virol.* **65:**2767–2772.

Perelygin, A. A., S. V. Scherbik, I. B. Zhulin, B. M. Stockman, Y. Li, and M. A. Brinton. 2002. Positional cloning of the murine flavivirus resistance gene. *Proc. Natl. Acad. Sci. USA* **99:**9322–9327.

Ren, R., and V. Racaniello. 1992. Human poliovirus receptor gene expression and poliovirus tissue tropism in transgenic mice. *J. Virol.* **66:**296–304.

Smith, G. A., S. P. Gross, and L. W. Enquist. 2001. Herpesviruses use bidirectional fast-axonal transport to spread in sensory neurons. *Proc. Natl. Acad. Sci. USA* **98:**3466–3470.

Tufariello, J., S. Cho, and M. S. Horwitz. 1994. The adenovirus E3 14.7-kilodalton protein which inhibits cytolysis by tumor necrosis factor increases the virulence of vaccinia virus in a murine pneumonia model. *J. Virol.* **68:**453–462.

Tyler, K. L., H. W. Virgin IV, R. Bassel-Duby, and B. N. Fields. 1989. Antibody inhibits defined stages in the pathogenesis of reovirus serotype-3 infection of the central nervous system. *J. Exp. Med.* **170:**887–900.

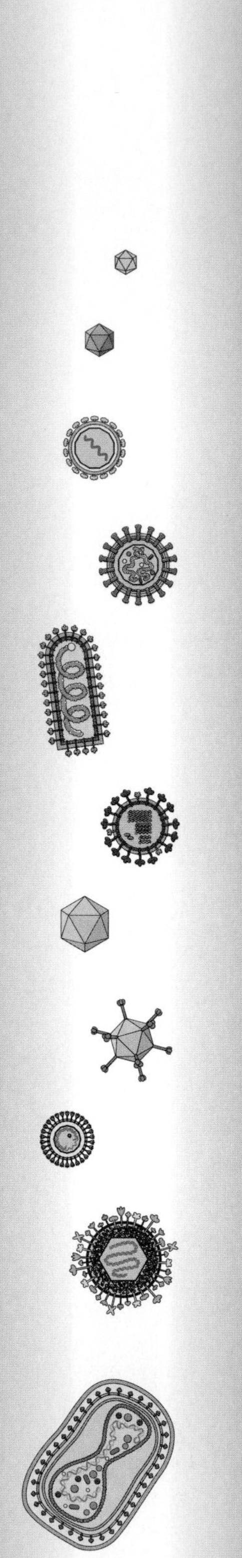

15

Virus Offense Meets Host Defense

The trouble with facts is that there are so many of them.
ANONYMOUS

I hear you knockin' but you can't come in!
LITTLE RICHARD

The Host Defense against Viral Infections

We live and prosper in a literal cloud of viruses. The numbers of potentially infectious particles that impinge on us daily are astronomical. Seasonal "colds," "flu," childhood rashes, measles, chicken pox, and mumps, as well as acquired immunodeficiency syndrome (AIDS) and Ebola fever, all serve notice of our vulnerability. In this chapter, we consider the powerful primary and secondary defense systems that ensure our survival.

The labyrinthine complexity of the host defense system is bewildering and the details initially can be overwhelming. However, the subject of host defense against viral infection provides both insight and focus: the devil and the delight are in the details. Defense mechanisms, even in healthy hosts, are imperfect despite millions of years of evolution, in large part because the genomes of successful pathogens encode gene products to modify, redirect, or block every step of host defense. The take-home message is that for every host defense, there is a viral offense (Box 15.1).

Primary Physical and Chemical Defenses

The primary, but often underappreciated, host defenses are physical and chemical barriers: the skin, surface coatings such as mucous secretions, tears, acid pH, and surface-cleansing mechanisms (discussed in more detail in Chapter 14). The skin is the largest organ of the body, weighing more than 5 kg in an average adult, and is a strong barrier to infection. Unless broken by cuts, abrasions, or punctures (e.g., insect bites and needle sticks), it is impervious. Many virus particles that land on the skin are inactivated by desiccation, by acids, or by other inhibitors formed by indigenous commensal microorganisms. Surfaces exposed to the environment but not covered by skin are lined by living cells and therefore depend on other primary defenses for immediate protection. These surfaces are at risk for infection despite the remarkable and continuous action of cleansing mechanisms.

BOX 15.1

Is it evasion or modulation?

From the online *Merriam-Webster Dictionary:*

Evade: to elude by dexterity or stratagem

Modulate: to adjust to or keep in proper measure or proportion

The phrase "immune evasion" is popular in the virology literature. It is meant to describe the viral mechanisms that thwart the host immune defense systems. However, in many cases, the phrase can be inaccurate, imprecise, and even misleading. The term "evasion" implies that host defenses are completely ineffective. In reality, defense and offense are matters of degree, not absolutes. If viruses really could evade the immune system, we might not be here discussing semantic issues.

Perhaps a more accurate term to describe viral gene products that engage immune defenses is "immune modulators."

Intrinsic Cellular Defenses

All cells have genetic programs that respond to various stresses, such as starvation, irradiation, and infection. We define these processes as **intrinsic cellular defenses** to distinguish them from **immune defenses,** which depend on cytokines and white blood cells. Intrinsic cellular defenses are highly conserved and arose very early in evolution. Below, we consider several that are relevant to viral infection.

Apoptosis (Programmed Cell Death)

Many cells die quickly after viral infection. When the biochemical alterations initiated by infection are detected within a cell, a process of self-destruction called **apoptosis** may be induced (Fig. 15.1 and 15.2). A primary function of apoptosis is to eliminate particular cells during development and differentiation. Cells in organs are in numerical equilibrium despite rather large changes in cell birth and death depending on physiological stimuli. Accordingly, apoptotic pathways are controlled by a variety of signals that monitor the processes of growth regulation, cell cycle progression, and metabolism. Many diseases are associated with malfunctions in regulation of apoptosis (including cancer). Survival signals from the cell's environment, and internal signals reporting on cell integrity, normally keep the apoptotic response in check.

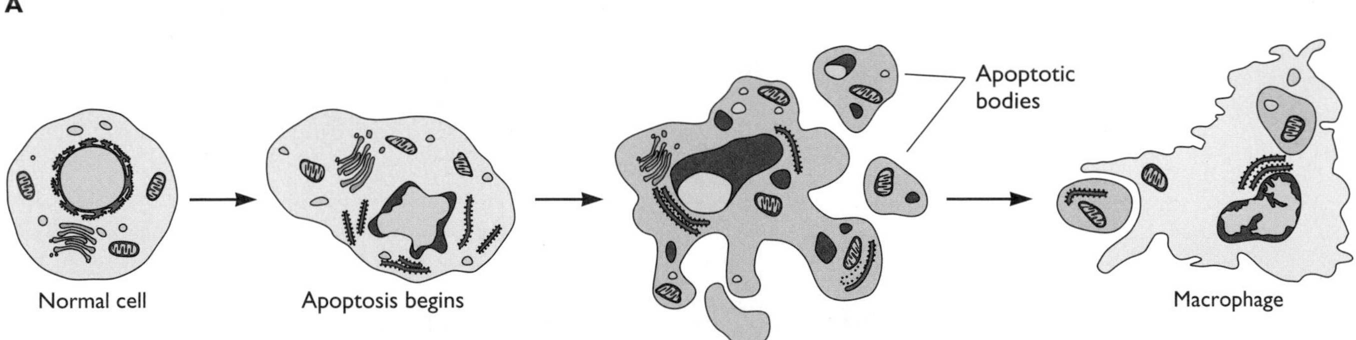

Figure 15.1 Apoptosis, the process of programmed cell death. (A) Apoptosis can be recognized by several distinct changes in cell structure. A normal cell is shown at the left. When programmed cell death is initiated, as indicated by the second cell, the first visible event is the compaction and segregation of chromatin into sharply delineated masses that accumulate at the nuclear envelope (dark blue shading around periphery of nucleus). The cytoplasm also condenses and the outline of the cell and nuclear membranes changes, often dramatically. The process can be rapid: within minutes the nucleus fragments and the cell surface convolutes, giving rise to the characteristic "blebs" and stalked protuberances illustrated. These blebs then separate from the dying cell and are called apoptotic bodies. Macrophages (the cell at the right) engulf and destroy these apoptotic bodies. Reprinted from J. A. Levy, *HIV and the Pathogenesis of AIDS,* 2nd ed. (ASM Press, Washington, D.C., 1998), with permission. (B) Apoptosis in human epithelial cells infected with herpes simplex virus type 1. Infection by herpes simplex virus type 1 initiates apoptosis, but de novo synthesis of infected cell proteins prevents cell death. The viral immediate-early ICP27 protein is one of several proteins involved in this inhibition: infection by a mutant virus with a deletion of the ICP27 gene (Δ27) induces apoptosis but does not block cell death. (a) Fluorescence microscopy of the nuclei of infected cells stained

(continued on next page)

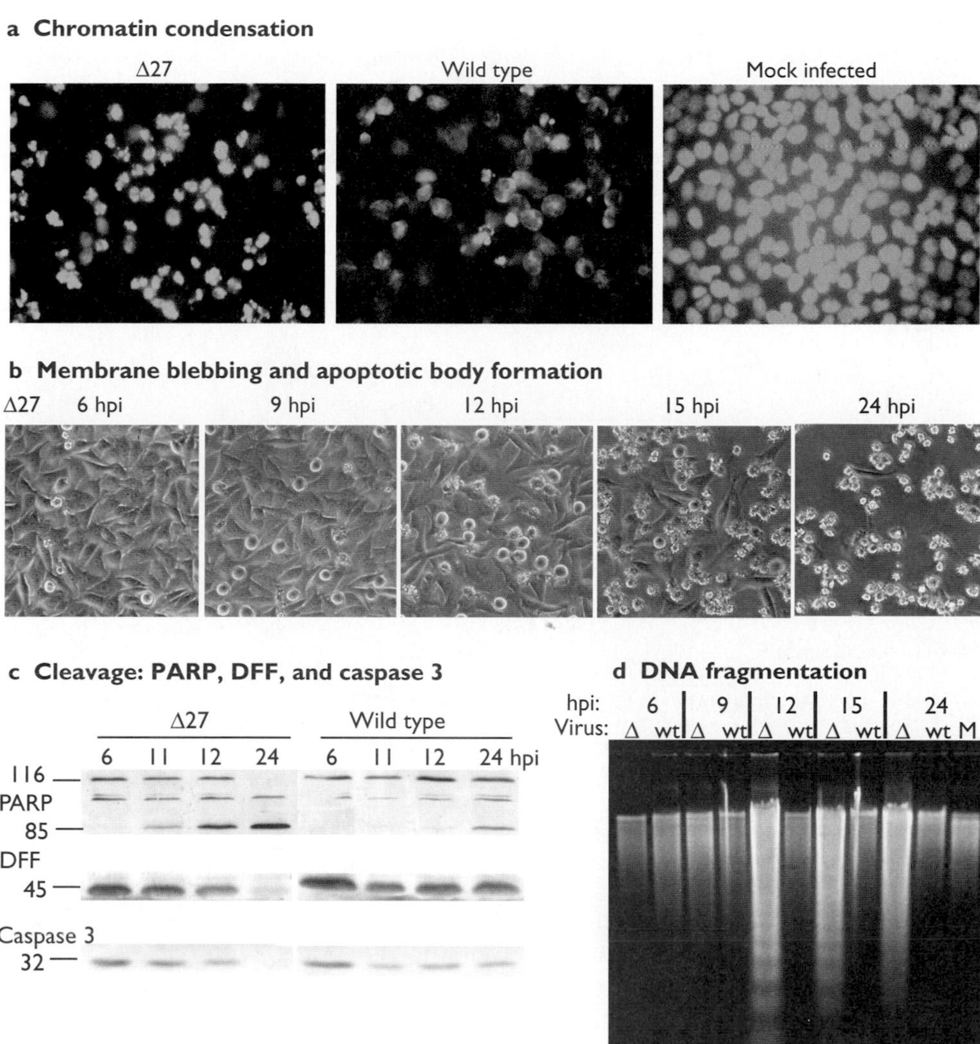

Figure 15.1 *(continued)* with Hoechst 33258. Chromatin condensation, typical of apoptosis, is visible in cells infected with Δ27, but not in mock-infected cells or cells infected with wild-type virus. (b) Phase-contrast microscopy of human epithelial cells infected with Δ27. Membrane blebbing and apoptotic body formation typical of apoptosis increase with time after infection. hpi, hours postinfection. (c) Determination of apoptosis induction after infection by monitoring proteolytic processing of three proteins: DNA fragmentation factor (DFF), poly(ADP-ribose) polymerase (PARP), and caspase 3. Proteins are detected with specific antibodies. The processing of PARP, a 116-kDa protein, produces an 85-kDa product. Apoptosis-induced processing of DFF (45 kDa) and caspase 3 (32 kDa) results in the loss of reactivity with the anti-DFF and anti-caspase 3 antibodies. (d) DNA fragmentation after infection. Appearance of a ladder of short DNA fragments is indicative of apoptosis. The time course of DNA laddering in cells infected with Δ27 or wild-type (wt) herpes simplex virus type 1 is shown. Reprinted from M. Aubert and J. A. Blaho, *Microbes Infect.* **3:**859–866, 2001, with permission.

When these signals are perturbed, cell death invariably ensues.

Apoptosis can be activated by a large variety of both external and internal stimuli. Regardless of the nature of the initiation signal, all converge on common effectors, the **caspases**. Caspases are members of a family of cysteine proteases that specifically cleave after aspartate residues. These proteases are first synthesized as precursors with little or no activity. A mature caspase with full activity is produced after cleavage by another protease (often another caspase).

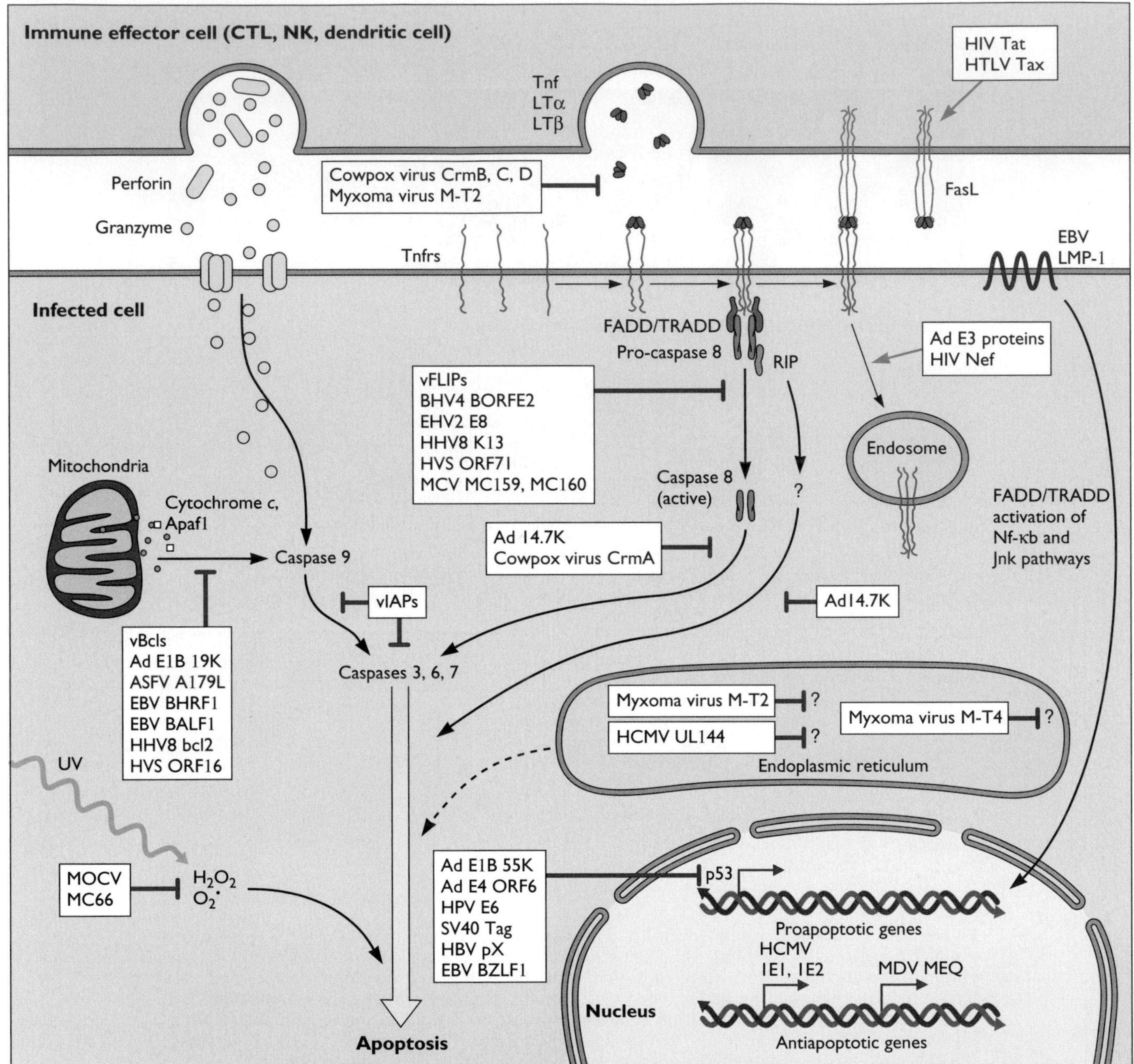

Figure 15.2 Modulation of apoptosis by selected viral gene products. A flow chart of the major processes and selected signals in the pathways mediating programmed death is shown. Many host proteins involved in the process have been omitted for clarity. The extrinsic pathway (death receptor-mediated) initiates when immune effector cells (top) detect infected cells (e.g., via Toll-like receptor signaling) and release cytokines. These cytokines include members of the Tnf family (Tnf; lymphotoxin alpha [LTα], also called Tnf-β; lymphotoxin beta [LTβ]) that bind to Tnf death receptors (e.g., Tnfr1, TRAILR2, and Fas). Later in infection, Fas ligand, produced by antigen-specific CTLs, induces apoptosis in target cells that produce Tnf receptors. Apoptosis of infected cells also is induced when CTLs or NK cells release perforin and granzyme proteins. The intrinsic pathway (dependent on mitochondria) is stimulated when internal sensor proteins respond to cellular stresses such as viral infection and DNA damage. The intrinsic pathway can lead to activation of p53 and the resulting increased transcription of proapoptotic genes. In general, loss of mitochondrial integrity is a hallmark action of the intrinsic pathway. The apoptotic pathways converge to activate one or more cytosolic caspases, cysteine proteases that act in cascade to carry out the final stages of apoptosis. At least 11 caspases

(continued on next page)

Figure 15.2 ***(continued)*** are known; we illustrate only 4 here. Caspases cleave a variety of substrates, including lamins and pronucleases. As a result of caspase action, the cell architecture is altered and cellular DNA is degraded. Viral proteins can modulate essentially every step in the intrinsic and extrinsic pathways. Selected viral proteins are indicated in the white boxes. Green arrows indicate induction, stimulation, or activation; red bars indicate inhibition. Dotted arrow from endoplasmic reticulum indicates an unknown pathway that leads to caspase activation. Viral proteins that inhibit this pathway are indicated in the lumen of the endoplasmic reticulum. Abbreviations: Ad, adenovirus; ASFV, African swine fever virus; BHV4, bovine herpesvirus 4; EBV, Epstein-Barr virus; EHV2, equine herpesvirus 2; FADD, Fas-associated death domain protein; HBV, hepatitis B virus; HCMV, human cytomegalovirus; HHV8, human herpesvirus 8 (Kaposi's sarcoma herpesvirus); HIV, human immunodeficiency virus; HPV, human papillomavirus; HTLV, human T-lymphotropic virus type 1; HVS, herpesvirus saimiri; MCV, murine cytomegalovirus; MDV, Marek's disease virus; MOCV, molluscum contagiosum virus; SV40, simian virus 40; Tag, large T antigen; TRADD, Tnfr1-associated death domain protein; UV, ultraviolet light stimulation of reactive oxygen species; vBcls, viral mimics of host Bcl function; vFLIPs, viral FLICE inhibitory proteins (FLICE is another name for caspase 8); vIAPs, viral inhibitors of apoptosis.

Alternatively, increasing the concentration of some caspase precursors results in cleavage-independent activation. These protease cascades are not unlike blood clotting or the complement cascade (see "Complement" below). The principle is similar: a modest initial signal can be amplified significantly, culminating in an all-or-none response.

Two convergent caspase cascades are known: the extrinsic and intrinsic pathways. The **extrinsic pathway** begins when a cell surface receptor binds a proapoptotic ligand (e.g., the cytokine tumor necrosis factor binding to its receptor). Binding changes the cytoplasmic domain of the receptor so that death-inducing signaling proteins are recruited. This complex of proteins attracts pro-caspase 8, which is activated on binding. Caspase 8 cleaves and activates pro-caspase 3, the terminal enzyme for both extrinsic and intrinsic pathways. The **intrinsic pathway** is often called the mitochondrial pathway and integrates stress responses as well as internal developmental cues. Common intracellular initiators include DNA damage and ribonucleotide depletion. In these situations, the cell cycle regulatory protein p53 is activated (see Chapter 18) and apoptosis ensues. The cascade is triggered when a proapoptotic protein such as Bax moves to the mitochondria. As a result of this interaction, cytochrome *c* is released into the cytoplasm where it binds to Apaf-1 and, in the presence of dATP or ATP, oligomerizes. This oligomeric assembly binds and cleaves pro-caspase 9, which in turn activates pro-caspase 3. The extrinsic and intrinsic signaling pathways can converge in other ways. For example, after cell death receptors bind external ligands (extrinsic pathway), mature caspase 8 may cleave a proapoptotic protein called Bid that translocates to the mitochondria to trigger the intrinsic pathway. There is ample evidence to suggest that the intrinsic pathway can serve to amplify the extrinsic pathway.

Once caspase 3 is activated (no matter what the initial activating signal), the end results are always the same: cell and organelle dismantling, vesicle and membrane bleb formation, phosphatidylserine exposure on the cell surface, and DNA cleavage to nucleosome-sized fragments (Fig. 15.2).

It is best to think of apoptosis as a default pathway held in check by the continuous action of a variety of regulatory molecules. Cells synthesize several suppressors including members of the Bcl-2 family. Inducers of apoptosis (e.g., Bax and Bad) must be synthesized for the maintenance of important aspects of normal cell physiology and homeostasis, such as the regulation of cell numbers in development and proper functioning of the immune system. Cellular debris resulting from apoptosis is taken up by macrophages and dendritic cells and presented to T cells in lymph nodes.

Apoptosis and defense against viral infection. Because virions engage cell receptors on entry, and because viral replication engages all or part of the host's transcription, translation, and replication machinery, a variety of signals may activate the extrinsic and intrinsic pathways (Box 15.2). In many infections, the target cell is quiescent and, hence, unable to provide the enzymes and other proteins needed by the infecting virus. Consequently, viral proteins induce the cell to leave the resting state. However, cell cycle checkpoint proteins then respond to this unscheduled event by inducing apoptosis. Typically, viral early proteins activate the cell cycle (e.g., adenoviral E1A proteins or simian virus 40 large T protein). Not surprisingly, most viral infections trigger apoptosis. Accordingly, these viral genomes often encode proteins that modulate this potentially lethal process.

Viral proteins that inhibit apoptosis. Viral proteins that modulate the apoptotic pathway have proved valuable in dissecting the complex pathways and regulatory circuits in normal cells (Fig. 15.1 and Table 15.1). As we noted above, apoptosis is normally held in check by regulatory proteins called inhibitors of apoptosis, or IAPs. The prototype IAP gene was described in baculovirus genomes

BOX 15.2

The many ways by which virus infections perturb apoptotic pathways

At the Cell Surface

Stimulation of apoptosis-inducing cytokines and receptors (e.g., ligand and receptor mimicry)
Alteration of membrane organization (e.g., virion passage into the cytoplasm via receptors)

In the Cytoplasm

Production of metabolic inhibitors (e.g., arrest of host translation)
Modification of cytoskeleton (e.g., disruption of actin microfilaments)
Disruption of signal transduction pathways (e.g., death domain proteins and kinase- and phosphatase-binding proteins)

In the Nucleus

Degradation of DNA
Alteration of gene expression (e.g., increased expression of heat shock genes)
Disruption of the cell cycle (e.g., inactivation of p53 or pRb/p300)

Hay, S., and G. Kannourakis. 2002. A time to kill: viral manipulation of the cell death program. *J. Gen. Virol.* **83:**1547–1564.

Miller, L. K., and E. White (ed.). 1998. Apoptosis in viral infections. *Semin. Virol.* **8:**443–523.

by the late Lois Miller and colleagues in 1993. This seminal work led to the discovery of cellular orthologs in yeasts, worms, flies, and humans. Mutant viruses unable to inhibit apoptosis were detected originally because the host DNA of infected cells was unstable, the cells lysed prematurely, and, as a consequence, viral yields were reduced, resulting in small plaques. Table 15.1 provides many examples of viral proteins that modulate every possible point in the apoptosis cascade.

Other Intrinsic Defenses

At least three other widely conserved cellular processes may function in cellular defense against viral infections. These processes are autophagy, ND10-mediated repression (see "Promyelocytic leukemia protein" below), and RNA interference. **Autophagy** is the process by which cells are induced to degrade the bulk of their contents by formation of specialized membrane compartments related to lysosomes. This process is evoked by stress such as nutrient starvation or viral infection. Stress-induced alterations in translation are modulated by eIF2α kinases. It has been proposed that phosphorylated eIF2α triggers autophagy, which in turn leads to engulfment and digestion of cytoplasmic virions. Two lines of experimentation are consistent with the proposal. Sindbis virus replication in neurons is inhibited after induction of autophagy. Herpes simplex virus infection transiently activates Pkr-mediated autophagy, but a viral protein (ICP34.5) blocks the process by reversing the phosphorylation of eIF2α.

Specialized complexes of proteins in the nucleus called **nuclear domain 10 (ND10) bodies** have been suggested to prevent transcription of foreign DNA that enters the nucleus. Such activity may prove to be an intrinsic defense against DNA viruses that replicate in the nucleus. Interferon stimulates synthesis of proteins that comprise ND10 bodies, implicating them in antiviral defense (see "Promyelocytic leukemia protein" below). The global repression of ND10-bound DNA can be relieved by viral proteins such as the ICP0 protein of herpes simplex virus type 1. The ICP0 protein accumulates at ND10 bodies and induces the proteasome-mediated degradation of several ND10 proteins. The human cytomegalovirus IE1 protein, the Epstein-Barr EBNA5 protein, and the adenovirus E4 ORF3 protein all affect ND10 protein localization or synthesis. How the components of ND10 domains bind DNA and repress transcription are active areas of study (see Chapters 8 and 9).

RNA silencing is a mechanism of sequence-specific degradation of RNA observed among diverse plants and animals. It is likely to have arisen early in the evolution of eukaryotes to detect and destroy foreign nucleic acids. RNA silencing is related to a process called **RNA interference** (RNAi) that was first described to occur in *Caenorhabditis elegans* and subsequently in fungi and algae. In both processes, double-stranded RNA (dsRNA) produced by the invading nucleic acid (virus, plasmid, or transposon) serves as the initial trigger and, after recognition by binding proteins, is processed by a ribonuclease (RNase) called Dicer into short fragments of 21 nucleotides (primary silencing RNAs). These fragments are incorporated into a dsRNA-induced silencing complex that binds target messenger RNAs (mRNAs) that contain sequences complementary to the 21-nucleotide primary silencing RNA. These complementary sequences are then amplified by a polymerase to yield secondary silencing RNAs. Primary or secondary silencing RNAs then direct repeated cycles of target mRNA degradation. We know very little about RNA interference as an antiviral defense, but at least one animal viral genome encodes an inhibitor of the process, and other examples are certain to be discovered.

Table 15.1 Selected viral regulators of apoptosis

Cellular target	Virus	Gene	Function
Bcl-2	Adenovirus	E1B 19kDa	Bcl-2 homolog
	African swine fever virus	A179L	Bcl-2 homolog
	Epstein-Barr virus	BHRF1	Bcl-2 homolog
		BALF1	Bcl-2 homolog
		LMP-1	Increases synthesis of Bcl-2; mimics CD40/Tnf receptor signaling
	Herpesvirus saimiri	ORF16	Bcl-2 homolog
	Human herpesvirus 8	Ksbcl-2/ORF16	Bcl-2 homolog
Caspases	Adenovirus	E3 14.7kDa	Inactivates caspase 8
	African swine fever virus	A224L/5 HL	IAP homolog
	Baculovirus	p35, IAP	Inhibits multiple caspases
	Cowpox virus	CrmA/SPI-2	Serpin; inhibits caspases 1 and 8 and granzyme B
	Vaccinia virus	SPI-2/B13R2	Serpin; CrmA homolog
Cell cycle	Adenovirus	E1B 55kDa	Binds/inactivates p53
		E4 Orf6	Binds/inactivates p53
	Hepatitis B virus	pX	Blocks p53-mediated apoptosis
	Human papillomavirus	E6	Targets p53 degradation
	Simian virus 40	Large T	Binds/inactivates p53
Fas/Tnf receptors	Adenovirus	E3 10.4/14.5kDa	Internalization of Fas
	Cowpox virus	CrmB	Neutralizes Tnf and LTα
		CrmC	Neutralizes Tnf but not LTα
		CrmD	Neutralizes Tnf and LTα
	Human cytomegalovirus	UL144	Function not known
	Myxoma virus	M-T2	Secreted Tnf receptor homolog
vFLIPs; DED box-containing proteins	Bovine herpesvirus 4	BORFE2	Blocks activation of caspases by death receptors
	Equine herpesvirus 2	E8	Blocks activation of caspases by death receptors
	Murine cytomegalovirus	MC159, MC160	Block activation of caspases by death receptors
	Herpesvirus saimiri	ORF71	Blocks activation of caspases by death receptors
	Human herpesvirus 8	K13	Blocks activation of caspases by death receptors
Oxidative stress	Molluscum contagiosum virus	MC66	Inhibits UV- and peroxide-induced apoptosis; homologous to human glutathione peroxidase
Transcription	Human cytomegalovirus	IE1, IE2	Inhibit Tnf-α but not UV-induced apoptosis
	Marek's disease virus	MEQ	Inhibits Tnf-α, ceramide, UV, and serum withdrawal induced apoptosis
Unclassified	Myxoma virus	M-T2	Blocks apoptosis in infected T cells
	Herpes simplex virus	US3	Blocks apoptosis in many infected cells; serine/threonine kinase
		Latency-associated transcript (LAT)	Unknown mechanism
		ICP27	Unknown mechanism

Immune Defenses

Occasionally, the intrinsic cellular defenses are unable to contain virions at the site of entry, and viral genomes begin to replicate. Almost immediately, a remarkable defense, called the **immune response**, begins. This highly coordinated response depends on the interplay of secreted proteins, receptor-mediated signaling, and intimate cell-cell communication. Several remarkable life-or-death decisions are made quickly: the nature of the invader is established (e.g., is it an intracellular virus or an extracellular bacterium?), and an appropriate mixture of soluble proteins and white blood cells is mobilized to remove the invader. Amazingly, infections of just a few cells can be detected in the context of billions of uninfected cells.

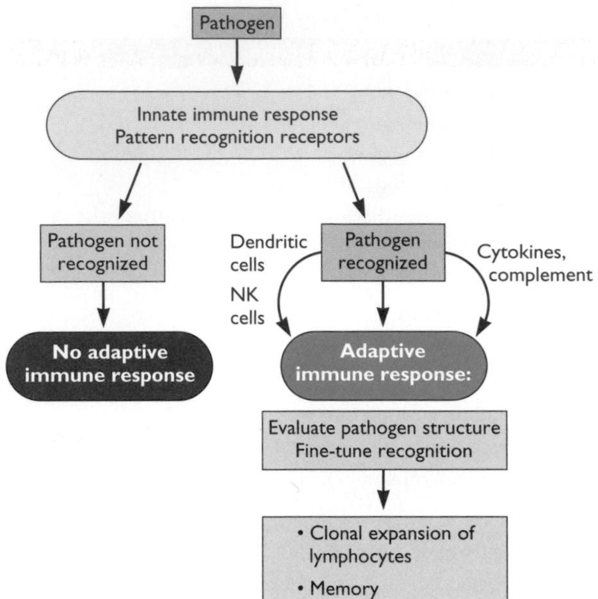

Figure 15.3 Integration of innate and adaptive immune recognition. An invading pathogen is first detected by molecular interactions that depend on pattern recognition receptors of the innate immune system. In this initial and rapid process of molecular recognition and response, molecules of microbial origin are usually detected, resulting in a variety of responses including new cytokine production and release of stimulators of inflammation. Migratory sentinel cells (e.g., dendritic cells) take packets of ingested proteins to lymph nodes, where they contact and stimulate cells of the adaptive immune system. Molecules of the invading microbe and the effector molecules of the innate immune system then interact further with the adaptive immune system, often causing clonal expansion of distinct classes of lymphocytes. Highly specific effectors such as antibodies produced by B cells and cytotoxic T cells are released into the circulation to recognize the invading microbe and the infected cells. This adaptive response enables recognition of antigens with a high degree of structural specificity. The innate immune system is the front line, not only for immediate deployment but also for reconnaissance and transfer of information to the adaptive immune system. Adapted from D. T. Fearon and R. M. Locksley, *Science* **272:**50–54, 1996, with permission.

Three critical steps in immune defense are recognition, amplification, and control. A viral infection must be recognized early, defenses must be activated quickly, and amplified if needed, and then all responses must be turned off when the infection has ceased. Pathologies can result if any of these processes is defective, modulated, or bypassed. The paramount importance of the immune system in antiviral defense is amply documented by the devastating viral infections observed in children who lack normal immune function, as well as in patients with AIDS.

Innate and Adaptive Immune Defenses

The immune response to viral infection consists of an innate (nonspecific) and an adaptive (specific) defense (Fig. 15.3). The **innate response** is the **first line of immune defense,** because it functions continually in a normal host without any prior exposure to the invading virus (Box 15.3). Indeed, most viral invasions are repelled by the innate immune system before viral replication outpaces host defense. However, once this threshold is passed, second-line defenses (the **adaptive response**) must be mobilized if the host is to survive.

The **adaptive defense** consists of the antibody response and the lymphocyte-mediated response, often called the **humoral** and **cell-mediated responses**, respectively. This system is called "adaptive" because it not only differentiates infected self from noninfected self, but also is tailored on the spot to the particular foreign invader such that an appropriate combination of soluble molecules (antibodies and cytokines) and lymphocytes (Table 15.2) participate in the action. The tailoring of the specific adaptive response requires close communication with cells that participate in the innate response. Such communication occurs by binding of cytokines and intimate cell-cell interactions among dendritic cells and lymphocytes in the lymph nodes. Considerable evidence indicates that the cells and cytokines of the innate response provide essential information to the adaptive immune system about the nature of the potential hazard confronting the host. In fact, it is possible to demonstrate that the adaptive response cannot be established without the innate immune system.

A defining feature of the adaptive defense system is **memory**; subsequent infections by the same agent are met almost immediately with a robust and highly specific response that usually stops the infection as soon as it starts, with minimal reliance on the innate defenses.

BOX 15.3

Innate defense stands alone

- It is the **only** immune defense available for the first few **days** after viral infection.
- It is the only system that can discern the general nature of the invader (viruses, bacteria, protozoans, fungi, or worms).
- It is the only system that can inform the adaptive response when infection reaches a dangerous threshold.

While the innate immune system responds to essentially any infection, the term "nonspecific" does a great disservice to its incredible discriminatory power.

Table 15.2 Important white blood cells that participate in the innate and adaptive defense systems

Cell	Source/function
Lymphocytes	Responsible for specificity of immune responses; recognize and bind to foreign antigens; derived from bone marrow
T cells	Differentiate into T-helper cells (Th cells), which secrete cytokines, and cytolytic T cells (CTLs). Regulatory or suppressor T cells are a type of Th cell. All have T-cell receptors.
B cells	Produce antibodies
NK cells	Natural killer cells; large granular cytolytic cells; most do not have T-cell receptors
Mononuclear phagocytes	Responsible for phagocytosis and antigen presentation; derived from bone marrow; monocyte lineage; macrophages, Kupffer cells in the liver, alveolar macrophages in the lung
Dendritic cells	Responsible for induction of immune response; antigen presentation to T-helper cells; migratory cells found in every tissue except the brain
Interdigitating dendritic cells	Bone marrow-derived; present in most organs, lymph nodes, and spleen; Langerhans cells in the skin
Follicular dendritic cells	Not bone marrow-derived; present in germinal layers of lymphoid follicles in spleen, lymph nodes, mucosal lymphoid tissue; trap antigens and antigen/antibody complexes for display to B cells in lymphoid tissue
Plasmacytoid dendritic cells	Immature dendritic cells found in the blood and T-cell zones of lymph nodes; capable of synthesizing large amounts of interferon to protect immune cells from viral infection
Granulocytes	Contain abundant cytoplasmic granules; inflammatory cells
Neutrophils	Polymorphonuclear leukocytes; respond to chemotactic signals, phagocytose foreign particles; major leukocyte in the inflammatory response
Eosinophils	Function in defense against certain pathogens (such as worms) that induce IgE antibody; not critical in antiviral defense; involved in hypersensitivity (allergic) reactions
Basophils	Circulating counterparts to tissue mast cells; mediate hypersensitivity caused by IgE antibody repsonse

The agents of the innate and adaptive immune responses are the **myelomonocytes** (monocytes, macrophages, dendritic cells, and a variety of granulocytes) and the **lymphocytes** (natural killer, T and B cells [Table 15.2]). One fascinating aspect of the immune system is that its effector cells are dispersed throughout the body, yet the response can be directed quickly to the focal point of infection. It is the powerful cytokines that coordinate the activity of this dispersed cellular defense system.

The Innate Immune Response

General Features

The innate defense system comprises **cytokines** (soluble proteins such as interferons released from infected cells), local **sentinel cells** (dendritic cells and macrophages), a complex collection of serum proteins termed **complement**, and cytolytic lymphocytes called **natural killer cells (NK cells)**. Neutrophils and other granulocytic white blood cells also play important roles in innate defense in response to the initial burst of cytokines from dendritic cells, macrophages, and infected cells.

Innate defense relies on unusual receptors that reside on surfaces of sentinel cells. These **pattern recognition receptors** can detect "nonself" targets; they bind certain molecular motifs found only in microbial macromolecules. It is worth emphasizing this crucial feature once more: pattern recognition receptors do not, and **must** not, recognize any self structure. Members of a second class of receptors, found mainly on NK T cells, recognize "missing self" or "altered self." Innate defense immediately unleashes potent molecules and effector cells targeted to destroy the invader and cells harboring it. Failure of this system, for example, via a pattern recognition receptor that bound to a self ligand, could lead to death of the organism. Similarly, any false-positive registered by the missing-self system is potentially lethal. As a result, pattern recognition receptors and missing-self receptor systems have been selected over evolutionary time to be highly pathogen selective. Our first insights into the nature of pattern recognition receptors came from *Drosophila* developmental genetics (Box 15.4). We now understand that all innate defense systems arose early in the evolution of multicellular organisms and remain absolutely essential for survival.

The innate immune response is crucial in antiviral defense because it can be activated quickly and can begin functioning within minutes to hours of infection. Such rapid action contrasts with the activation of the adaptive response, which is orders of magnitude slower than the

BOX 15.4

Toll receptors: pattern recognition by the innate immune system

- The Toll signaling pathway was defined initially as being essential for the establishment of the dorsal-ventral axis in *Drosophila* embryos. Eric Wieschaus and Christiane Nusslein-Volhard discovered the first Toll mutants. When Wieschaus showed the unusual mutant *Drosophila* embryos to Nusslein-Volhard, she exclaimed "Toll!" (a German slang term comparable to "crazy" or "far out").
- Toll signaling also initiates the response of larval and adult *Drosophila* to microbial infections.
- Toll-like receptors bind a variety of microbe-specific components and trigger a defensive reaction in both flies and mammals via signal transduction pathways and activation of new gene expression.

Anderson, K. V. 2000. Toll signaling pathways in the innate immune response. *Curr. Opin. Immunol.* **12:**13–19.

Mushegian, A., and R. Medzhitov. 2001. Evolutionary perspective on innate immune recognition. *J. Cell Biol.* **155:**705–710.

replication cycles of some viruses. It takes days to weeks to orchestrate the effective response of antibodies and activated lymphocytes specifically tailored for the infecting virus. While the rapidity of the innate response is important, this response must also be transient because its continued activity is damaging to the host. Below, we discuss six important players in the innate response. Despite their apparent diversity, every component initiates an immediate action that, in turn, is detected and amplified by cells of the adaptive immune system. Viral genomes encode a surprising variety of gene products that modulate every step of this innate defense.

Cytokines

The presence of cytokines in the blood is one of the first indications that the host has been infected. In general, the cytokines can be divided into three functional groups: the proinflammatory cytokines, the anti-inflammatory cytokines, and the chemokines (Table 15.3).

Table 15.3 Three primary classes of cytokines[a]

Functional group	Selected members	Activity
Proinflammatory	IL-1, Tnf, IL-6, IL-12	Promote leukocyte activation
Anti-inflammatory	IL-10, IL-4, Tgf-β	Suppress activity of proinflammatory cytokines; return system to basal "circulate and wait" state
Chemokines	IL-8	Recruit immune cells during early stages of immune response

[a]The terms "lymphokines" and "monokines" were originally used to denote secreted proteins produced by activated lymphocytes or monocytes, repectively. Similarly, the term "interleukin" identified proteins such as IL-2 that communicated signals between different populations of white blood cells and other nonhematopoietic cells.

Cytokines participate in almost every phase of the host response to viral infection, including control of inflammation, induction of an antiviral state, and regulation of the adaptive response. They constitute a major means of communication between the innate and adaptive arms of the immune system.

The initial response to viral infection is the production of potent cytokines by the infected cell. These cytokines engage receptors on sentinel dendritic cells, macrophages, and adjacent uninfected cells, which then synthesize a burst of cytokines, amplifying the initial response. The first cytokines to appear in high concentrations are alpha interferon (IFN-α) and IFN-β, followed by tumor necrosis factor alpha (Tnf-α), interleukin 6 (IL-6), IL-12, and IFN-γ. The action of Tnf-α induces marked changes in nearby capillaries that attract and facilitate entry of circulating white blood cells to the site of infection (see "Inflammatory Response" below).

Tnf-α is a multifunctional cytokine made primarily by activated monocytes and macrophages. It can induce an antiviral response when it binds to receptors on infected cells. Within seconds, the combination of infection and binding of Tnf-α to its receptor initiates a signal transduction cascade that activates caspases and apoptosis. Viral proteins that modulate the function of Tnf-α are well known (Fig. 15.4).

Infected cells make cytokines after stimulation by a variety of initiating cues (Fig. 15.5). All stimulated pathways converge, often by means of Nf-κb activation, to the same regulatory regions of cytokine genes to promote their transcription (Fig. 15.6). In general, several overlapping positive and negative regulatory sequences reside upstream of cytokine genes. In unstimulated cells, inhibitory proteins bind and repress transcription. In infected cells, proteins that increase transcription replace these repressors.

While cytokines function locally in antiviral defense, they also evoke global responses that are both essential and

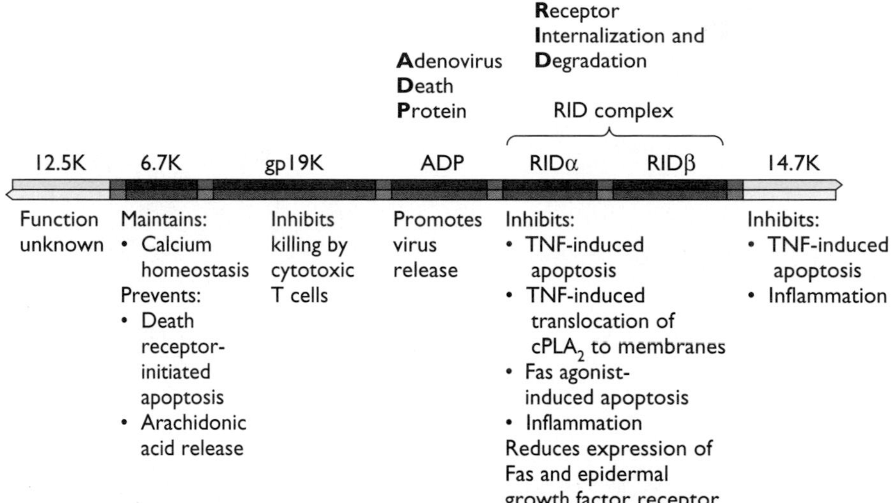

Figure 15.4 The adenovirus type 2 E3 region, a cluster of seven genes encoding proteins that mediate host defense modulation. The proteins were named initially according to their apparent molecular masses (e.g., 14.7K for 14.7 kDa). Recently, some of the proteins have been given names that reflect their known functions. RID is an acronym for "receptor internalization and degradation." The RID protein complex (previously called E3-10.4K/14.5K) is composed of RIDα (10.4 kDa) and RIDβ (14.5 kDa). RID proteins have multiple functions, as indicated. These functions may or may not represent the same molecular mechanism. ADP is an abbreviation for "adenovirus death protein." Even though the ADP gene is in the E3 cluster, it is expressed as a late gene by alternative splicing from the major late promoter. ADP promotes cell lysis and virus release after virus replication is complete. Integral membrane proteins are indicated by reddish shading. gp19K is a glycoprotein that reduces major histocompatibility complex class I protein synthesis, and inhibits killing by cytotoxic T cells. The 14.7K protein inhibits Tnf-induced apoptosis. The 6.7K protein maintains calcium homeostasis and prevents death-receptor-initiated apoptosis and arachidonic acid release. The functions of the 12.5K protein have not been identified. $cPLA_2$, cytoplasmic phospholipase A2. Adapted from W. S. M. Wold and A. E. Tollefson, *Semin. Virol.* **8:**515–523, 1998, with permission.

familiar to all who experience viral infections (Fig. 15.7). For example, they act directly on cells of the nervous system to produce many of the characteristic behaviors and responses exhibited by infected individuals, including sleepiness (somnolence), lethargy, muscle pain (myalgia), appetite suppression, and nausea. Proinflammatory cytokines (Table 15.3) stimulate the liver to synthesize characteristic **acute-phase proteins,** many of which are required to repair tissue damage and to clear the infection. Members of the colony-stimulating factor class of cytokines, which are made in the bone marrow after an inflammatory response, control the growth and maturation of lymphocytes and other cells essential in antiviral defense.

Cytokines bind to specific cell surface receptors and activate gene expression. They are potent molecules, capable of inducing a response at nanogram-per-milliliter concentrations. More than 80 cytokines are known, and the list is growing. Some of the more important in antiviral defense are listed in Tables 15.4 and 15.5.

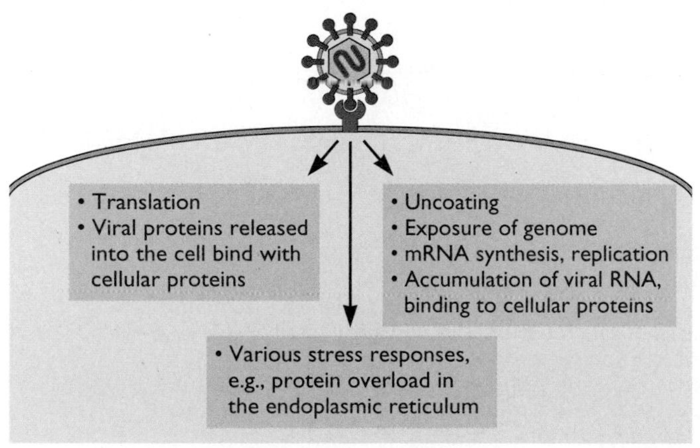

Figure 15.5 Activation of cellular signal transduction and gene expression by viral infection. An infected cell invariably responds by altered gene expression and the secretion of cytokines. Changes in transcription are stimulated by several general pathways common to all viral infections. First, virus particles attach to their receptors and may, as a result, activate signal transduction pathways before any viral gene expression occurs. Second, uncoating of the virus particle to release the genome, mRNA synthesis, and replication activate signal transduction pathways. Third, synthesis of viral proteins may activate signal transduction directly or indirectly. Fourth, viral membrane protein synthesis may overload the secretory system and stimulate stress response pathways.

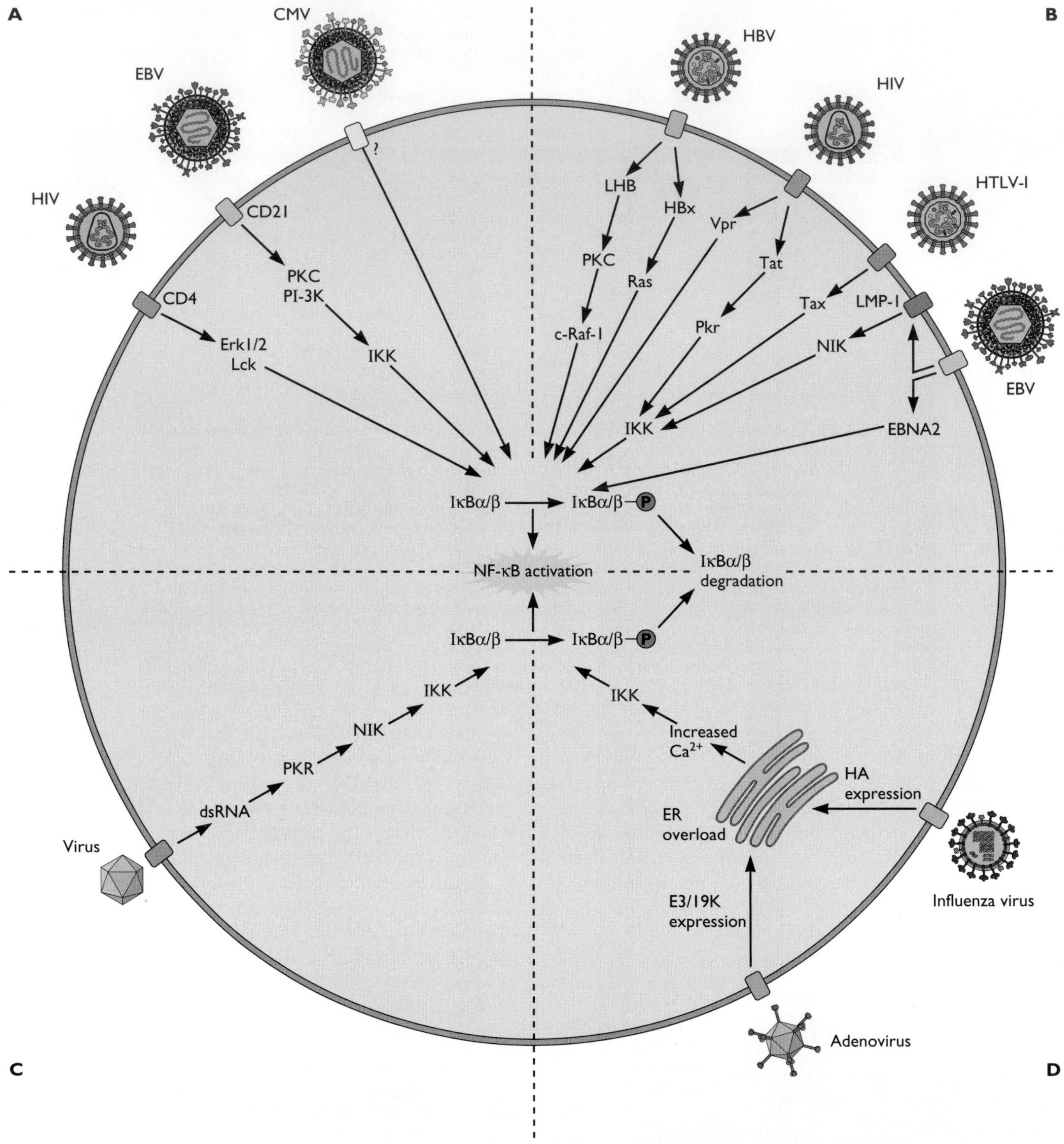

Figure 15.6 Activation of the transcription regulator Nf-κb by viral infection. Nf-κb is an important transcription control protein in the response to viral infection. Four diverse mechanisms that result in activation of Nf-κb, corresponding to the pathways described in the legend to Fig. 15.5, are illustrated. (A) Activation of signal transduction pathways on binding of a virus particle to its receptor; (B) activation of Nf-κb by viral proteins synthesized in the infected cell; (C) activation of Pkr and Nf-κb by dsRNA; (D) activation of Nf-κb by overproduction of viral proteins in the ER due to calcium release. Abbreviations: HIV, human immunodeficiency virus; EBV, Epstein-Barr virus; CMV, cytomegalovirus; HBV, hepatitis B virus; HTLV-1, human T-lymphotropic virus type 1; PKC, protein kinase C; HA, hemagglutinin.

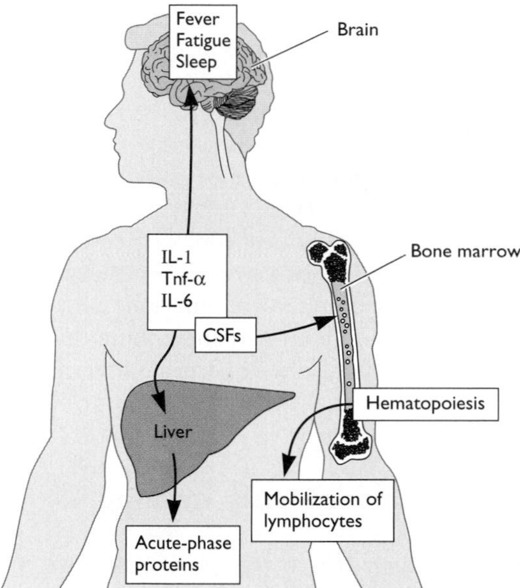

Figure 15.7 Systemic effects of cytokines in inflammation. A localized viral infection often produces global effects, including fever and lethargy, lymphocyte mobilization (swollen glands), and production of new proteins in the blood. The proinflammatory cytokines IL-1, IL-6, and Tnf all act on the brain (particularly the hypothalamus) to produce a variety of effects, including these typical responses to viral infection. These cytokines also act in the liver to cause the release of iron, zinc, and acute-phase proteins including mannose-binding protein, fibrinogen, C-reactive protein, and serum amyloid protein. Acute-phase proteins have innate immune defense capabilities: e.g., C-reactive protein binds phosphorylcholine on microbial surfaces and activates complement. The colony-stimulating factors (CSFs) activated by an inflammatory response have long-range effects in the bone marrow on hematopoiesis and lymphocyte mobilization. Adapted from A. S. Hamblin, *Cytokines and Cytokine Receptors* (IRL Press, Oxford, United Kingdom, 1993), with permission.

Table 15.4 Some cytokines that function in the immune response to viral infection

Cytokine	Source	Target/action
IFN-γ	T cell/NK cell	Activates macrophages; promotes adhesion of Th cells to vascular endothelium; inhibits IL-6 effects; induces antiviral state
IFN-α/β	Immature dendritic cells; many types	Induces antiviral state; inhibits cell proliferation; stimulates growth and cytolytic funtion of NK cells; increases expression of MHC class I and decreases expression of MHC class II molecules
Tnf-α	T cell/macrophage	Activates neutrophils; induces inflammatory response, fever, and initiates catabolism of muscle and fat (cachexia); induces adhesion molecules on vascular endothelial cells; potentiates lysis of some virus infected cells
Tgf-β	T cell/macrophage	Negative regulator of immune responses; inhibits T-cell proliferation; inhibits macrophage activation; induces IgA
Gm-csf	T cell, macrophage, fibroblast, endothelial cell	Induces myelomonocytic cell growth and differentiation; important maturation protein for dendritic cells
Mcp-1	Macrophage, monocyte, T cell, endothelium, smooth muscle, keratinocytes, fibroblast	Induces monocyte chemotaxis
Mip-1α	Macrophage, T cell, B cell, neutrophil, Langerhans cell	Chemoattractant for T cells, B cells, monocytes, and eosinophils
Mip-1β	T cell, macrophage, B cell	Chemoattractant for T cells and monocytes
Rantes	T cell, platelet	Chemoattractant for T cells, monocytes, and eosinophils
IL-1	Macrophage, T cell, B cell, epithelial cells	Costimulator of T cells; initiates inflammatory response; affects brain to produce fever; causes metabolic wasting (cachexia); induces acute-phase protein synthesis in the liver
IL-2	T cell	Induces proliferation of T cells; stimulates growth and cytolytic function of NK cells; induces antibody synthesis in B cells
IL-3	Th cell	Multilineage colony-stimulating factor; stimulates growth and differentiation of precursor cells in bone marrow; shares many activities with IL-4
IL-4	Mast cell, bone marrow	Induces B cell proliferation and differentiation; required for production of IgE; inhibits macrophage function and blocks effects of IFN-γ on macrophages; stimulates growth and differentiation of Th2 cells; stimulates synthesis of adhesion molecules on vascular endothelium
IL-5	Th2 cell	Induces B-cell differentiation with IL-2 and IL-4; increases production of eosinophil killing functions
IL-6	Macrophage, T cells; vascular endothelial cells, fibroblasts	Made in response to IL-1 and Tnf; stimulates B-cell growth; costimulator of T cells; stimulates hepatocytes to synthesize acute-phase proteins
IL-7	Marrow stromal cells	Stimulates differentiation of B lymphocytes
IL-10	T and B cells	Promotes growth and differentiation of B cells; inhibits macrophage function
IL-12	Macrophage, monocyte	Potent stimulator of NK cell growth and killing activity; promotes differentiation of Th cells to Th1 subset; stimulates differentiation of immature $CD8^+$ T cells to functionally active CTLs; stimulates expansion and activation of autoreactive $CD4^+$ T cells
IL-15	Many non-T cells, glial cells in central nervous system	T cell, NK cell, and intestinal epithelial cell growth factor; IL-15-deficient mice lack NK cells and memory $CD8^+$ T cells

Table 15.5 Chemokine receptors and their ligands[a]

Receptor[b]	Chemokine ligand
CCr1	Mip-1α, Rantes, Mcp-3, Hcc-1, Mpif-1
CCr2	Mcp-1, -3, -4
CCr3	Eotaxin, Mcp-2, -3, -4, -5, Rantes, Mpif-2
CCr4	Tarc
CCr5	Rantes, Mip-1α, -β
CCr6	Larc/Mip-3α
CCr7	Elc, Mip-3β
CCr8	I309
CXCr1	IL-8, Gcp-2
CXCr2	IL-8, Groα, -β, -γ, Nap-2, Ena-78, Gcp-2
CXCr3	IP-10, Mig, I-Tac
CXCr4	Sdf-1
CXCr5	Bca-1/Blc

[a]Data from C. R. Mackay, *Curr. Biol.* **7:**R384–R386, 1997.

[b]The four families of chemokine receptors are distinguished by the pattern of cysteine residues near the amino terminus and are abbreviated as CXC, CC, C, and CX3C. Only two types are listed in this table. The CXC family has an amino acid between two cysteines; the CC family has none; the C family has only one cysteine; and the CX3C family has three amino acids between two cysteines. Subfamilies of these major four groups also exist.

Many viral gene products can modulate cytokine responses (Tables 15.6 and 15.7). These proteins have been called **virokines** if they mimic host cytokines, or **viroceptors** if they mimic host cytokine receptors. The arsenal includes remarkable proteins such as soluble IL-1 receptors, a variety of chemokine antagonists, and functional homologs of IL-10 and IL-17. DNA virus genomes encode most of the well-known virokines and viroceptors, but RNA virus genomes contain some surprises. For example, the envelope protein of respiratory syncytial virus is a mimic of fractalkine, the only known chemokine that is a membrane protein. The viral envelope protein competes with fractalkine binding to the chemokine receptor, which also functions as a viral receptor.

Interferons, Cytokines of Early Warning and Action

The **interferons**, which are cytokines synthesized by mammals, birds, reptiles, and fish, are critical signaling proteins of the host frontline defense (Box 15.5). Inter-

Table 15.6 Virokines[a]

Virus	Protein	Function/homolog
Herpesviridae		
Epstein-Barr virus	BCRF-1	Viral IL-10
Equine herpesvirus 2	EHVIL-10	Viral IL-10
Herpes simplex virus	Glycoprotein D	Binds Tnf-β receptor; mimics LIGHT ligand
Human cytomegalovirus	UL111A	Viral IL-10
	UL146	Viral IL-8
Human herpesvirus 6	U83	CC or CX3 chemokine
Human herpesvirus 6	vIL-6	Increases angiogenesis and hematopoiesis
Human herpesvirus 8	Viral MIP-1	CCr8 agonist; Th2 chemoattractant
Human herpesvirus 8	Viral MIP-2	Th2 chemoattractant; chemokine receptor antagonist
Human herpesvirus 8	K2Orf	Viral IL-6
Herpesvirus saimiri	HVS13	Viral IL-17; T-cell mitogen
Marek's disease virus	vIL-8	Viral IL-8
Murine cytomegalovirus	m131/m129	Chemokine homolog; promotes viral dissemination
Poxviridae		
Cowpox virus	38K	Inhibits IL-1β convertase
Molluscum contagiosum virus	MC148R	Blocks Mip-1α
Myxoma virus	MGF	Mimics Egf, Tgf-α
Myxoma virus	SERP-1	Secreted serine protease
Myxoma virus	SERP-2	Inhibits IL-1β convertase
Orf virus	vIL-10	Viral IL-10
Shope fibroma virus	SFGF	Mimics Egf
Vaccinia virus	19-kDa protein	Mimics Egf
Vaccinia virus	A39R	Binds semaphorin receptor Vespr; regulates monocyte Icam-1 production
Paramyxoviridae		
Respiratory syncytial virus	G protein	Mimics fractalkine
Retroviridae		
Simian sarcoma virus	Sis	Pdgf homolog

[a]Data from http://www.copewithcytokines.de and P. M. Murphy, *Nat. Immunol.* **2:**116–122, 2001.

Table 15.7 Viroceptors[a]

Virus	Receptor	Ligand
Herpesviridae		
Epstein-Barr virus	BARF1	M-Csf
Herpesvirus saimiri	ECRF-3	Chemokines IL-8, Mgsa
Human cytomegalovirus	US28, US27, UL33, UL78	Various chemokines
Human herpesvirus 8	ORF74	Constitutively active chemoreceptor; IL-8
Human herpesviruses 6, 7	U12, U51	Chemokine receptors
Murine cytomegalovirus	R33, M33	Various chemokines
Poxviridae		
Capripox virus	Q2/3L	G protein-coupled chemokine receptor
Cowpox virus	CrmB	Tnf-α, Tnf-β
Cowpox virus	CrmD	Tnf-α, LTα
Myxoma virus	M11L	Various cytokines
Myxoma virus	M-T1	CC chemokines, Rantes
Myxoma virus	M-T7	IFN-γ
Myxoma virus	T2	Tnf-α, Tnf-β
Swinepox virus	K2R	Chemokine receptor
Tanapox virus	38-kDa protein	IFN-γ, IL-2, IL-5
Vaccinia virus	vCKBP	Various chemokines
Vaccinia virus	B15R	IL-1β
Vaccinia virus	B18R	IFN-α
Variola virus	vCCI	Chemokines, Mcp-1

[a]Data from http://www.copewithcytokines.de and P. M. Murphy, *Nat. Immunol.* **2:**116–122, 2001.

feron was discovered in 1957 when A. Isaacs and J. Lindenmann observed that chicken cells exposed to inactivated influenza virus contained a substance that interfered with the infection of other chicken cells by live influenza virus. Not only do most cells make IFN when infected, but also their released IFN inhibits replication of a wide spectrum of viruses. Remarkably, as originally noted by Isaacs and Lindenmann, virus replication is not required to induce IFN production. In some cases, virion binding to cells is sufficient. In fact, a variety of agents and

BOX 15.5

The IFN system is crucial for antiviral defense

Many steps in a viral life cycle can be inhibited by IFN, depending on the virus family and cell type. When this cytokine binds its receptor, more than 300 genes are transcribed. The proteins so produced can inhibit viral penetration and uncoating, synthesis of viral mRNAs or viral proteins, replication of the viral genome, and assembly and release of progeny virions. More than one of these steps in a virus life cycle can be inhibited, providing a strong cumulative effect.

The contribution of IFN can be demonstrated in animal models in which the IFN response is reduced by treatment with anti-IFN antibodies, or in mice harboring mutations that delete or inactivate IFN genes, IFN receptor genes, genes that regulate the IFN response, or genes that are induced by IFN.

Animals with a defective IFN response exhibit a reduced ability to contain viral infections, and often show an increased incidence of illness or death. When the IFN-α/β response is abrogated, there is a global increase in susceptibility to most viruses. When the IFN-γ response is blocked, viral pathogenesis is modestly affected, at best.

These observations suggest that IFN α/β is crucial as a general antiviral defense, whereas the IFN-γ response has other roles.

Huang, S., W. Hendriks, A. Althage, S. Hemmi, H. Bluethmann, R. Kamijo, J. Vilcek, R. M. Zinkernagel, and M. Aguet. 1993. Immune response in mice that lack the interferon-gamma receptor. *Science* **259:**1742–1745.

Ryman, K. D., W. B. Klimstra, K. B. Nguyen, C. A. Biron, and R. E. Johnston. 2000. Alpha/beta interferon protects adult mice from fatal Sindbis virus infection and is an important determinant of cell and tissue tropism. *J. Virol.* **74:**3366–3378.

Stojdl, D. F., N. Abraham, S. Knowles, R. Marius, A. Brasey, B. D. Lichty, E. G. Brown, N. Sonenberg, and J. C. Bell. 2000. The murine double-stranded RNA-dependent protein kinase Pkr is required for resistance to vesicular stomatitis virus. *J. Virol.* **74:**9580–9585.

Zhou, A., J. Paranjape, S. Der, B. Williams, and R. Silverman. 1999. Interferon action in triply deficient mice reveals the existence of alternative anti-viral pathways. *Virology* **258:**435–440.

Table 15.8 The interferons: antiviral cytokines

Interferon[a]	Producer cells	Inducers
IFN-α	Most, if not all, nucleated cells	Viral infection, dsRNA
IFN-β	Most, if not all, nucleated cells	Viral infection, dsRNA
IFN-γ	T cells, NK cells	Antigens, mitogens, IL-2, IL-12

[a]The molecular mass of each IFN is approximately 20 kDa. Human IFN-α is produced from more than 24 closely related intronless genes. Human IFN-β is produced from a single intronless gene with about 30 to 45% homology to IFN-α genes. IFN-γ is produced from a single-copy gene with three introns. There is little homology with IFN-α and IFN-β. IFN-α and IFN-β are sometimes called type I interferons. IFN-γ is often called type II or immune interferon.

treatments, including natural dsRNA, synthetic dsRNA, endotoxin, and even inhibitors of transcription (actinomycin D) or translation (cycloheximide) promote the synthesis of interferon.

There are three classes of interferons (Table 15.8). In following sections, we use the abbreviation IFN to mean both IFN-α and IFN-β (also called type I interferons). We will refer specifically to IFN-γ, which is induced only when certain lymphocytes are stimulated to replicate and divide after binding a foreign antigen. In contrast, IFN-α and IFN-β are induced directly by viral infection of almost any cell type.

IFN Is Made by Infected Cells and by Immature Dendritic Cells

Virus-infected cells produce IFN, but the signals that trigger its synthesis are not completely understood. One signal is accumulation of dsRNA, an inevitable product of RNA and DNA virus replication. Single-stranded RNA molecules, including viral mRNA, invariably contain stable, double-stranded structures formed by base pairing between complementary sequences. In cells infected by DNA viruses, dsRNA can also be formed by the pairing of RNA transcripts of complementary strands of DNA. Other signals unique to viral infection can also lead to IFN production. For example, structural proteins of some viruses stimulate IFN synthesis on binding of virions to cells. In other cases, virus-induced degradation of the inhibitor of Nf-κb (Iκbα) leads to activation of IFN gene transcription (Fig. 15.6).

Virus-infected cells invariably produce IFN, but the uninfected, local sentinel cells also make this cytokine in response to products released from infected cells. Such products include viral proteins, viral nucleic acids, and cellular stress proteins (e.g., heat shock proteins). The sentinel cell response is essential for amplification of the subsequent immune response and will be discussed in the next section. If the infection is not contained and spreads to more cells, large quantities of IFN may be synthesized by dendritic cell precursors in the blood called **plasmacytoid cells.** Such systemic production of IFN leads to many of the general symptoms of viral infection.

Production of IFN by infected cells and uninfected, immature dendritic cells at the site of infection is rapid, but transient; it occurs within hours of infection and declines in less than 10 h. Regulation of IFN synthesis and secretion is complex, and much remains to be discovered. For example, we know that the many *IFN-α* genes are differentially expressed after infection (Box 15.6). In addition, transcription of the human *IFN-β* gene is stimulated by infection, but only for a short period. The *IFN-β* enhancer possesses several remarkable properties that allow precise temporal control of transcription (Box 15.7). The biological significance of such a mix of α and β IFNs is unclear. Moreover, the quantity of IFN released from cells infected by different isolates of a given virus is astonishingly variable. In the case of vesicular stomatitis virus infection, the released IFN concentration can vary over a 10,000-fold range, depending on the serotype of the infecting virus. As discussed later, many viral proteins affect the quantity of IFN, as well as its action.

IFN Affects Only Cells with IFN Receptors

IFN functions only when it occupies its receptor on the surfaces of cells. A cell without IFN receptors may synthesize IFN, but cannot be affected by this cytokine. Binding of IFN to its receptor initiates a signal transduction cascade that culminates in increased transcription of many genes.

BOX 15.6

Differential induction of IFN-α genes by viral infection

- Most cells can produce IFN-α, and there are more than 20 genes, depending on the species.
- *IFN-α* genes are expressed differentially. Different members of this gene family are induced by specific viruses. For example, transcription of *IFN-A4* but not *IFN-A1* is induced by infection of mice with Newcastle disease virus. The function of such mixtures of potent IFNs is not understood and not well studied.
- Differential expression occurs as a result of specific binding of the transcriptional activator proteins Irf3 and Irf7 to specific *IFN-α* promoters.

Pitha, P. M., and W. C. Au. 1995. Induction of interferon-alpha gene expression, p. 151. *In* P. M. Pitha (ed.), *Interferon and Interferon Inducers.* Academic Press, London, United Kingdom.

BOX 15.7

Switching IFN-β *transcription on and off*

Viral infection activates transcription of the human *IFN-β* gene, but only for a short period. This on-off response is controlled by an enhancer located immediately upstream of the core promoter. Like other enhancers, this regulatory sequence contains binding sites for multiple transcriptional activators, including Nf-κb and members of the Ap-1 and Atf families (see figure). However, the *IFN-β* enhancer possesses several remarkable properties that allow precise temporal control of transcription.

(continued)

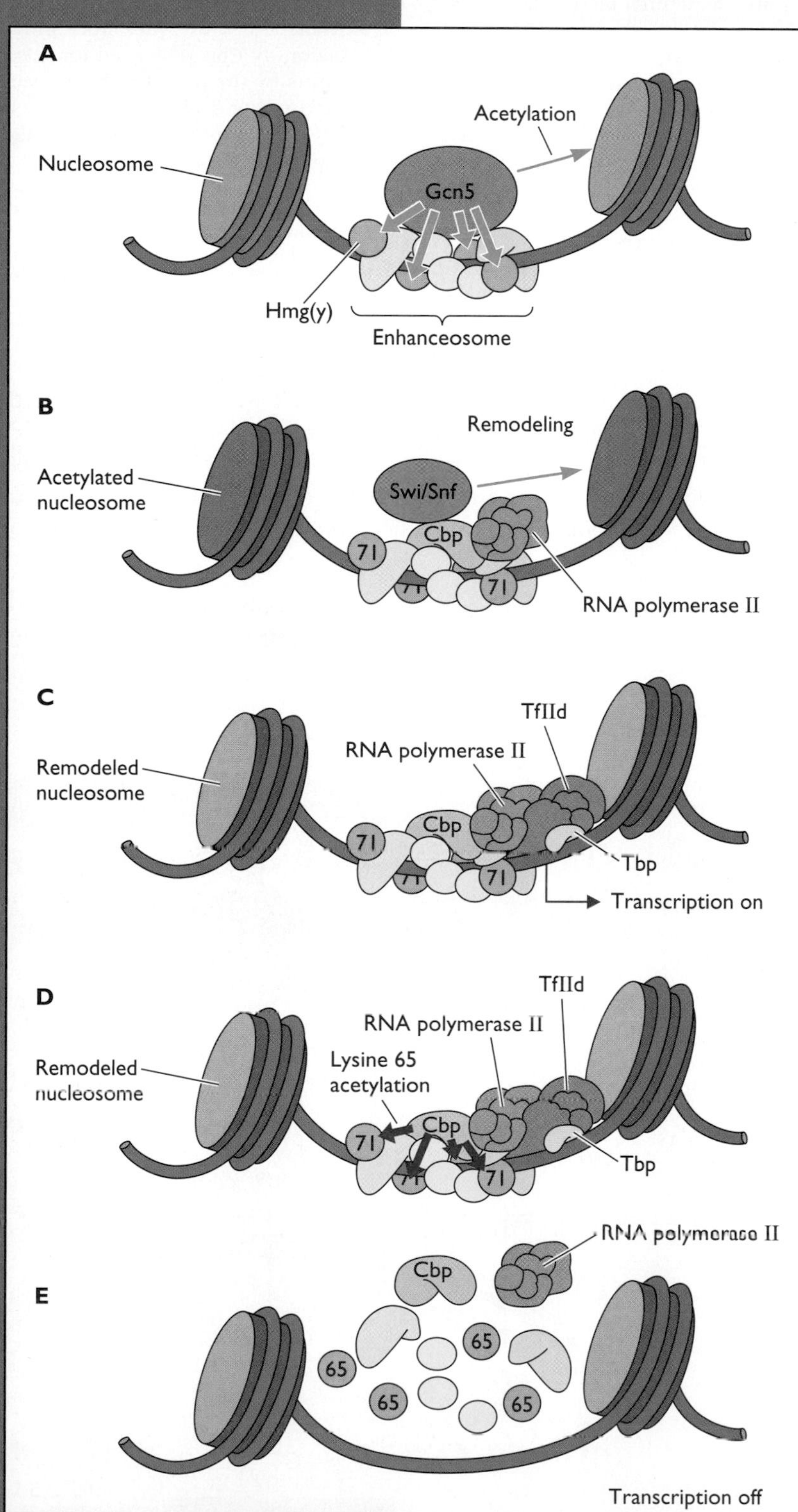

(A) Viral infection of human cells leads to assembly of a multiprotein complex on the *IFN-β* enhancer, which lies in a nucleosome-free region of the gene. The signals that direct binding of transcriptional activators (blue, yellow, and tan) and Hmg(y) (green) are not fully understood. The precisely organized surface of the complex allows binding of Gcn5, which acetylates both histones in nearby nucleosomes and Lys71 of Hmg(y) (red arrows). This modification stabilizes the enhanceosome. (B) A complex of Cbp and RNA polymerase II and the chromatin remodeling protein Swi/Snf bind sequentially to the stabilized complex. The latter alters the adjacent nucleosome that contains the core promoter DNA (red arrow). (C) Such alteration allows binding of TfIId and activation of transcription. Because Lys71 of Hmg(y) is acetylated, Cbp cannot acetylate Lys65. (D) Eventually, Cbp does acetylate Lys65 of Hmg(y) (red arrows), but how the inhibition induced by Lys71 acetylation is overcome is not yet clear. Regardless, Hmg(y) modification by Cbp disrupts the complex and switches off transcription. Adapted from K. Struhl, *Science* **293:**1054, 2001, with permission.

BOX 15.7 *(continued)*

- The enhancer also contains four binding sites for the architectural protein Hmg(y), which alters DNA conformation to direct assembly of a precisely organized nucleoprotein complex on the enhancer.
- In contrast to typical modular enhancers, all binding sites **and** their natural arrangement are essential for activation of *IFN-β* transcription specifically in response to viral infection.
- Formation of the complex takes place in stages, and is not complete until several hours after infection.
- Activation of transcription then requires sequential recruitment of the histone acetylase Gcn5, the coactivator Cbp and RNA polymerase II, and the chromatin remodeling complex Swi/Snf.
- In addition to modifying nucleosomes, Gcn5 acetylates Hmg(y) at Lys71. This modification stabilizes the complex.
- The Hmg(y) protein is also acetylated by Cbp at Lys65. However, **this** modification impairs DNA-binding activity and results in disruption of the complex and cessation of *IFN-β* transcription.
- Remarkably, this inhibitory modification by Cbp is blocked for several hours by the prior Gcn5 acetylation of Hmg(y). The "off" switch is delayed for a sufficient period to allow a burst of *IFN-β* transcription.

Munshi, N., T. Agalioti, S. Lomvardas, M. Merika, G. Chen, and D. Thanos. 2001. Coordination of a transcriptional switch by HMG1(Y) acetylation. *Science* **293:**1133–1136.

A simplified outline of this signal transduction cascade is shown in Fig. 15.8.

The Jak/Stat pathway contains a family of proteins that respond not only to binding of IFN but also to the binding of IL-6 and other cytokines. There are four known Jak kinases and seven structurally and functionally related *stat* genes. Their targeted disruption in mice has revealed much about their functions. For example, a mouse in which the *stat1* gene has been inactivated has no innate response to viral or bacterial infection, whereas deletion of

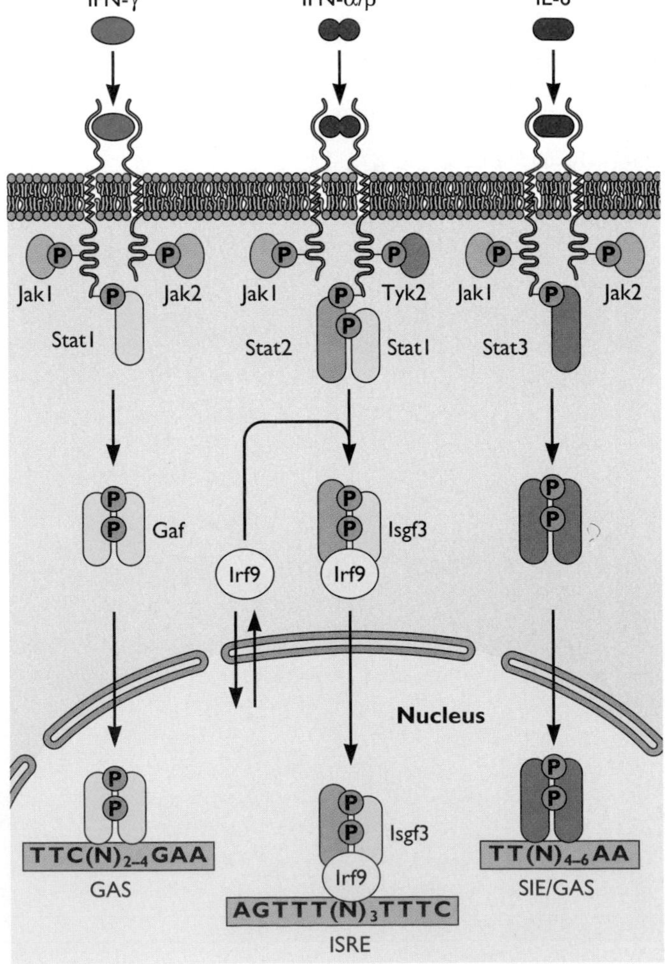

Figure 15.8 Overlapping signal transduction pathways for IFN-α/β, IFN-γ, and IL-6. IFN signals via the Jak/Stat pathway, characterized by a family of tyrosine kinases given the acronym Jak (Janus kinases; Janus, a Roman god, guardian of gates and doorways, is represented with two faces and therefore faces in two directions at once) and a set of transcription proteins named Stat (signal transduction and activators of transcription). The receptors for IFN-α/β, IFN-γ, and IL-6 are different, but all affect components of the Jak/Stat signal transduction pathway. All three IFN proteins and IL-6 bind to their receptors with high affinity (equilibrium dissociation constant [K_d] of about 10^{-10} M). Binding of IFN or IL-6 to the appropriate receptor leads to tyrosine phosphorylation of tyrosine kinases (Tyk2 and Jak1 for IFN-α/β and Jak1 and Jak2 for IFN-γ or IL-6) as well as of the receptor itself. These modifications are followed by tyrosine phosphorylation of the Stat proteins. In mammals there are seven Stat genes encoding Stat1, -2, -3, -4, -5A, -5B, and -6. The phosphorylated Stat proteins then form a variety of dimers that enter the nucleus. Within that organelle, Stat dimers bind, in some cases in conjunction with other proteins (e.g., Irf9), to specific transcriptional control sequences of IFN-α/β, IFN-γ, and IL-6-inducible genes called interferon-stimulated response elements (ISREs), gamma-activated site (GAS) elements, and Sis-inducible element (SIE), respectively. Later in the transcriptional response to IFN-α/β, a second transcriptional activator called Irf1 replaces Isgf3. For further information, see D. S. Aaronson and C. M. Horvath, *Science* **296:**1653–1655, 2002.

stat4 and *stat6* abrogates specific functions of the adaptive response. *stat* gene homologs are encoded in the genomes of *Drosophila melanogaster* and *Dictyostelium discoideum*, underscoring the ancient evolutionary origin of this signaling pathway.

Signaling via Jak/Stat activates transcription dependent on specific DNA sequences called **interferon-stimulated response elements (ISREs).** These sequences are found in the promoters of the more than 300 IFN-activated genes. Regulation is also exerted through suppressors of cytokine signaling proteins, phosphatases, and proteases.

IFN-γ Signaling

Unlike the common signaling cascades induced by IFN-α/β, IFN-γ signaling is mediated by several pathways (Box 15.8). The best-understood pathway leads to formation of the transcriptional activator called gamma-activated factor, Gaf, which binds to specific promoter sequences of IFN-γ activated genes. While cross talk between pathways does occur, the results of such mixed signals are neither predictable nor understood, even in principle. In the milieu of an infected tissue, multiple cytokines are produced and many signaling pathways are operating in both infected and uninfected cells. In theory, the orchestration of these multiple signaling pathways enables the whole (control of infection and appropriate host response) to be greater than the sum of its parts. How the integrated output is achieved remains a mystery.

IFN Action Produces an Antiviral State

As the name coined by Isaacs and Lindenmann so aptly indicates, IFN interferes with the replication of a wide variety of viruses in cultured cells and animals. Shortly after infection, newly made IFN released from infected cells and local immature dendritic cells can be found circulating in the body, but IFN concentration is highest at the site of infection, where it is bound by any cell with the appropriate receptor. Those cells that bind and respond to IFN are unable to propagate many different viruses; they are said to be in an **antiviral state.**

What does IFN do to a cell to make it inhospitable for the replication of almost any virus? We now know that IFNs can induce the synthesis of more than 300 cellular proteins, but their mix and concentrations vary according to cell type and specific IFN. Which subset of the hundreds of IFN-inducible proteins establishes the antiviral state for any given cell remains an open question. It is possible that many antiviral constellations exist depending on the cell type, virus, and cocktail of IFN and other cytokines sensed by that cell. Many of the products of IFN-inducible genes possess potent broad-spectrum antiviral activities, but the relevant molecular mechanisms of only a few are understood. IFN not only induces death of the infected cells, but also ensures that uninfected cells in the vicinity are induced to kill themselves should they become infected. Such a local cauterizing response has led some to characterize IFN action as a **firebreak** to infection.

The IFN-induced proteins are functionally diverse and participate in signal transduction, chemokine action, antigen presentation, regulation of transcription, the stress response, and control of apoptosis. Some of the proteins induced by IFN also are induced by other stimuli, including dsRNA, bacterial lipopolysaccharides, Tnf-α, or IL-1. For example, attachment of human cytomegalovirus to cells stimulates the transcription of many genes that also are induced by IFN.

IFN also functions as an important cytokine in individuals who are not infected with a virus. Interferon blocks cell proliferation, increases the ability of NK cells to lyse cells infected by other pathogens, reduces amino acid biosynthesis, and alters transcription of major histocompatibility complex (MHC) genes. Because IFN induces the expression of many deleterious gene products and is potentially lethal to any cell that has specific receptors (most cells in our bod-

BOX 15.8

IFN-γ production

- IFN-γ is produced only by NK cells and T cells, critical players in the innate and adaptive immune responses, respectively.
- Production of IFN-γ is stimulated by IL-1, IL-2, estrogen, and IFN-γ itself.
- IFN-γ synthesis is suppressed by glucocorticoids and the cytokines transforming growth factor β and IL-10.
- Transcription of *IFN-γ* is regulated by repressors or activators. A combination of positive and negative regulatory sequences in the *IFN-γ* promoter is essential for controlled expression, but the details remain to be elucidated.
- IFN-γ translation is also regulated in a remarkable fashion: a pseudoknot structure in the 5′ end of the mRNA activates Pkr, which, in turn, blocks translation of IFN-γ mRNA.

Ben-Asouli, Y., Y. Banai, Y. Pel-Or, A. Shir, and R. Kaempfer. 2002. Human IFN-γ mRNA autoregulates its translation through a pseudoknot that activates the interferon-inducible protein kinase Pkr. *Cell* **108:**221–232.

Goodbourn, S., L. Didcock, and R. E Randall. 2000. Interferons: cell signaling, immune modulation, anti-viral responses and virus countermeasures. *J. Gen. Virol.* **81:**2341–2364.

ies), the production of large quantities of IFN in an infected individual has dramatic physiological effects. These responses include such common symptoms as fever, chills, nausea, and malaise. All infections lead to IFN production, one reason why these "flulike" symptoms are so common.

Soon after its discovery, IFN was touted as the "magic bullet," a broad-spectrum antiviral drug. However, the IFN response is more of a blunderbuss than a sharpshooter's rifle. The side effects often are worse than the infection. Nevertheless, use of IFN has been effective in the treatment of some persistent infections, particularly those caused by hepatitis B and C viruses.

IFN-Induced Proteins and Their Antiviral Actions

dsRNA-activated protein kinase (Pkr). Often viral **and** cellular protein synthesis in infected cells stops abruptly. In many cases, this lethal defense is mediated by a cellular dsRNA-activated protein kinase (Pkr) (also described in Chapter 11). Establishment of the Pkr-mediated antiviral state is a two-step process in which IFN promotes the increased production and accumulation of an inactive protein that subsequently can become activated **only** when the cell is infected.

All mammalian cells contain low concentrations of inactive Pkr, a serine/threonine kinase that has not only antiviral properties, but also antiproliferative and antitumor activities. The signal transduction cascade initiated by IFN binding to its receptor leads to a dramatic increase in the concentration of inactive Pkr. If the cell is infected, this enzyme can be activated by binding viral dsRNA. The enzyme then phosphorylates the alpha subunit of the eIF2 translation initiation protein (eIF2α), effectively rendering it incapable of supporting new protein synthesis in the cell (see Chapter 11). Phosphorylated eIF2α also can trigger autophagy, an intrinsic cell defense, as discussed above.

Many viral genomes encode proteins that can block the lethal actions of Pkr (Table 15.9), but our understanding of the biological significance of this phenomenon is incomplete. An exception is the herpes simplex virus type 1 ICP34.5 protein, which redirects the cellular protein phosphatase 1 to dephosphorylate eIF2α after it has been inactivated by the Pkr kinase. While wild-type virus is fully virulent in mice, ICP34.5-null mutants are markedly attenuated. Significantly, the ICP34.5 mutant regains wild-type virulence in mice lacking the *pkr* gene. This experiment provides convincing evidence that Pkr mediates defense against herpes simplex virus infection in mice.

RNase L and 2′-5′ oligo(A) synthetase. Another well-studied antiviral system induced by IFN comprises two enzymes and dsRNA. RNase L is a nuclease that can degrade most cellular and viral RNA species. Its concentration increases 10- to 1,000-fold after IFN treatment, but the protein remains inactive unless a second enzyme is synthesized. This enzyme, 2′-5′-oligo(A) synthetase, makes oligomers of adenylic acid, but only when activated by dsRNA. These unusual nucleotide oligomers then induce formation of active RNase L, which in turn begins to degrade all host and viral mRNA. It is known from the study of mouse mutants defective in RNase L that this enzyme participates not only in IFN-mediated antiviral defense, but also apoptosis. As a death effector, RNase L contributes to the antiviral activity of IFN by eliminating infected cells, as well as by destroying viral RNA. The RNase L/2′-5′ oligo(A) system destabilizes IFN-induced mRNAs and therefore may contribute to attenuation of the interferon response.

Mx proteins. Unlike the broad-spectrum antiviral effects of Pkr and RNase L, at least one IFN-induced mouse protein and two related human proteins appear to be directed against specific viruses. Mouse strains that have an IFN-inducible gene called *mx1* are completely resistant to influenza virus infection. The Mx1 protein is part of a small family of IFN-inducible GTPases with potent activities against various (–) strand RNA viruses. After IFN induction, this protein accumulates in the nucleus and inhibits the unusual influenza virus "cap-snatching" mechanism (see Chapter 6). It is likely that the Mx1 protein interferes with the function of the viral polymerase subunit PB2, for if this viral protein is overproduced, the antiviral effect of Mx1 protein is overcome. The significance of the *mx1* gene in the biology of influenza virus or of mice is not at all clear, as influenza virus does not circulate among wild mice. Moreover, in these animals, about one-quarter of the population lacks a functional *mx1* gene with no clear consequences.

Two human genes related to the murine *mx1* gene are termed *mxA* and *mxB*. Expression of these genes is also induced by IFN, but unlike the murine protein, the human proteins reside in the cytoplasm. MxA, but not MxB, blocks replication of influenza virus. Interestingly, in contrast to murine Mx1, which inhibits only influenza virus, the human MxA protein also prevents replication of vesicular stomatitis virus, measles virus, human parainfluenza virus type 3, and some bunyaviruses. The Mx proteins are related to members of the dynamin superfamily of GTPases, which regulate endocytosis and vesicle transport, but how this fact relates to their antiviral activities is unknown. IFN also induces transcription of genes encoding other GTPases of unknown function (e.g., guanylate-binding proteins 1 and 2).

P200 proteins. IFN induces transcription of the P200 gene cluster that encodes several proteins, the best studied of which are P202 and P204. The P202 protein blocks cell proliferation by inhibiting an impressive number of cell cycle regulatory proteins, including Nf-κb, E2f, p53, c-Fos, c-Jun, MyoD, myogenin, c-Myc, and Rb. The P204 protein is found in the nucleolus, where it inhibits ribosomal RNA synthesis by binding to a protein essential for the transcription of ribosomal RNA genes.

Interferon regulatory proteins (Irf). Members of the Irf protein family are required for sustained IFN transcription after induction. Mice lacking the *irf1* gene are incapable of mounting an effective IFN response to viral infection. Other members of this gene family (*irf2* to *irf9*) were discovered because their protein products bound to the ISRE in promoters of IFN-regulated genes. The Irf2 protein is a repressor of transcription and cell growth. Irf4 is synthesized only in T and B cells, and Irf8 is made only in cells of the macrophage lineage. Mice defective for *irf8* gene expression are markedly more susceptible to infection and cannot synthesize proinflammatory cytokines. The protein Irf9 (also known as Isgf3γ or p48) is the DNA-binding component of the transcriptional regulator Isgf3 (Fig. 15.8). Several viral Irf-like proteins that block IFN action have been identified (Table 15.9).

Nitric oxide synthase. Nitric oxide synthase is an IFN-γ-inducible protein with important antiviral activities. This enzyme exists as several isoforms, each of which has a distinctive tissue distribution. Nitric oxide synthase produces nitric oxide during the conversion of arginine to citrulene. Nitric oxide exerts a variety of effects, including inhibition of poxvirus and herpesvirus replication. The nitric oxide made by IFN-γ-activated NK cells accounts for much of their antiviral activity (see "NK cells" below).

Promyelocytic leukemia protein (Pml). The promyelocytic leukemia protein is present in both the nucleoplasm and discrete multiprotein complexes known as nuclear bodies (Pml bodies, ND10 bodies, or PODs [discussed in Chapter 9]; see also "Intrinsic Cellular Defenses" above). These structures appear to be important in the intrinsic cellular response to infection because they bind foreign DNA that enters the cell. Importantly, the structures, and presumably their antiviral activity, are affected by IFN. For example, IFN does not cause apoptosis in Pml-deficient cells, and overproduction of Pml inhibits the replication of vesicular stomatitis virus, influenza virus, and human foamy virus. Pml and other proteins present in the complex are thought to exert their antiviral effects by transcriptional repression. Many viral infections promote disorganization of Pml bodies, in part as a measure to override global repression.

Ubiquitin-proteasome pathway components. The proteasome is a large multisubunit protease that degrades cytoplasmic and nuclear proteins targeted for degradation by polyubiquitination. This process is important for the destruction of abnormal or damaged proteins, in the turnover of short-lived regulatory proteins, and in the production of peptides for assembly of MHC-I complexes. All IFNs induce transcription of a number of genes encoding proteins of the ubiquitin-proteasome pathway. Increased protein degradation may contribute to the antiviral response against some viruses. For example, proteasome inhibitors block the anti-hepatitis B virus action of IFN and IFN-γ. For this virus, activation of the proteasome may be **the** major antiviral effect, because the results of other experiments demonstrate that Pkr and RNase L systems are completely ineffective. The hepatitis B virus X protein binds to various proteasomal subunits and may modulate protease activity.

Other IFN-induced proteins. The most strongly induced IFN gene, *isg56*, is one member of a gene family induced by IFN, and encoding the structurally related proteins p54, p56, p58, and p60. The p58 protein interacts with the p48 subunit of the translation initiation protein eIF3 and blocks the initiation of protein synthesis. Other proteins with antiviral effects certainly remain to be discovered among the 300-plus IFN-induced proteins. For example, the IFN response is required to clear human cytomegalovirus infections, but Pkr, Mx, and RNase L proteins are not. Similarly, uncharacterized IFN-induced proteins block penetration and uncoating of simian virus 40 and some retroviruses. Others impair the maturation, assembly, and release of vesicular stomatitis virus, herpes simplex virus, and some retroviruses.

Viral Gene Products That Counter the IFN Response

While the term "antiviral state" implies complete resistance to infection after the IFN response, it is misleading, because infections vary considerably in their sensitivity to the effects of this cytokine. The replication of some viruses, like vesicular stomatitis virus, is so sensitive to IFN that this property is used to titrate the cytokine. Ironically, influenza virus replication is rather resistant, even though the virus was instrumental in the discovery of IFN. We now know that there are numerous viral mechanisms for confounding IFN production and action (Table 15.9). For example, viral soluble IFN receptors function as decoys, and viral regulatory proteins alter or block IFN-stimulated transcription.

Many viral genomes encode dsRNA-binding proteins that interfere with IFN induction. For example, the reovirus σ3 protein, the multifunctional influenza virus NS1 protein, the porcine rotavirus nsP3 protein, and the hepatitis B virus core antigen are all well characterized dsRNA binding proteins with anti-IFN effects. The vaccinia virus

Table 15.9 Viral modulation of interferon response

Type of modulation	Representative virus	Viral protein	Mechanism of action
Inhibition of IFN synthesis	Epstein-Barr virus	BCRF-1	IL-10 homolog, inhibits production of IFN-γ
	Vaccinia virus	A18R	Regulates dsRNA production
	Human papillomavirus type 16	E6	Binds Irf3
		E7	Binds Irf1
	Influenza A virus	NS1	Mechanism under study
	Hepatitis B virus	OrfC and terminal protein	
	Measles virus	?	
	Foot-and-mouth disease virus	L	Host protein synthesis block
IFN receptor decoys	Myxoma virus	M-T7	Soluble IFN-γ decoy receptor
	Vaccinia virus	B18R	Soluble IFN-α/β decoy receptor
Inhibition of IFN signaling	Adenovirus	E1A	Decreases quantity of Stat1 and p48; blocks Isgf3 formation; interferes with Stat1 and Cbp/P300 interactions
	Epstein-Barr virus	EBNA-1, EBNA-2	?
	Herpes simplex virus type 1	?	Inhibits phosphorylation of Stats
	Vaccinia virus	VH1	Viral phosphatase reverses Stat1 activation
	Human papillomavirus type 16	E7	Binds p48
	Polyomavirus	Large T antigen	Binds and inactivates Jak1
	Human cytomegalovirus	?	Reduces levels of Jak1 and p48
	Hepatitis B virus	Terminal protein	?
	Hepatitis C virus	NS5a	Blocks formation of Isgf3 and Stat dimers
	Sendai virus	C proteins	Block Stat phosphorylation
	Simian virus 5	Protein V	Degrades Stat1
	Bovine respiratory syncytial virus	NS1, NS2	Block IFN induction
	Parainfluenza virus type 2	?	Degrades Stat2
	Parainfluenza virus type 3	?	Blocks Stat1 phosphorylation
	Nipah virus	V protein	Prevents Stat1 and Stat2 activation and nuclear accumulation
Block function of IFN-induced proteins	Adenovirus	VA-RNA I	Binds dsRNA, blocks Pkr
	Simian virus 40	?	Blocks RNase L
	Baculovirus	PK2	Blocks eIF2α kinases and Pkr
	Epstein-Barr virus	EBER RNA	Binds dsRNA, blocks Pkr
	Herpes simplex virus type 1	US11	Blocks Pkr activation
		ICP34.5	Redirects protein phosphatase 1α to dephosphorylate eIF2α; reverses Pkr action
		?	Produces 2′-5′ derivatives that antagonize 2′-5′ oligo(A) synthetase
		ICP0	Promotes degradation of Pml
	Human herpesvirus 8	Irf homolog	Blocks transcriptional responses to IFN and IFN-γ
	Vaccinia virus	E3L	Binds dsRNA and blocks Pkr
		K3L	Pkr pseudosubstrate, decoy
	Human immunodeficiency virus type 1	TAR RNA	Blocks activation of Pkr
		Tat	Pkr decoy
		?	Induces RNase L inhibitor
	Hepatitis C virus	NS5a	Blocks Pkr
		E2	Blocks Pkr
	Hepatitis B virus	Capsid protein	Inhibits MxA

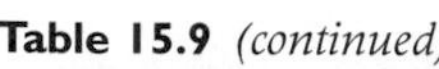

Table 15.9 *(continued)*

Type of modulation	Representative virus	Viral protein	Mech
Block function of IFN-induced proteins *(continued)*	Poliovirus	?	Degrades Pkr
	Influenza virus	NS1	Binds dsRNA and
		?	Induces p58 cellu
	Reovirus	σ3	Binds dsRNA, inh
	Rotavirus	nsP3	Binds dsRNA and inhibits Pkr and 2′-5′ oligo(A) synthetase

E3L protein and the herpes simplex virus type 1 US11 protein have dsRNA-binding properties that correlate with inhibition of IFN induction (Box 15.9). Interestingly, adenovirus VA-RNA I acts as a dsRNA decoy and blocks the activation of Pkr by directly binding to the enzyme.

An inescapable inference from the various countermeasures encoded by the genomes of diverse viruses is that IFN is an essential host defense component. But numerous questions remain. For example, some infections (e.g., Newcastle disease virus) are inhibited only by IFN-α, while others (e.g., herpes simplex virus type 1) are inhibited primarily by IFN-β. IFN is induced after infection by vaccine strains of measles virus, while little IFN is made after wild-type virus infection. When animals are infected, the IFN response varies depending on the route of infection (Box 15.10). It is clear that much remains to be learned about the antiviral effects of IFN and virus countermeasures.

Dendritic Cells

Dendritic cells and macrophages are crucial as **sentinel cells** in peripheral compartments (e.g., the skin, the mucosal membranes, the respiratory tract). Dendritic cells bind cytokines produced by infected cells and take up viral proteins from dead and dying cells (Fig. 15.9). Sentinel cells are specially equipped, not only to initiate immediate immune defense, but also to convey information of the attack to the adaptive immune system. This latter ability is emphasized by their common name of **professional antigen-presenting cells**. Even if a viral protein is introduced by injection into the skin or muscle, local dendritic cells and macrophages will most likely bind some of the molecules and stimulate an immune response to that protein. Indeed, most vaccines would not be effective without dendritic cells.

Dendritic cells play two major roles in antiviral responses: they directly inhibit viral replication at the onset of infection by producing large amounts of cytokines such as IFN, and they subsequently trigger adaptive, T-cell-mediated immunity appropriate to the infection. Dendritic cells exist in two functionally distinct states called **immature** and **mature** (Fig. 15.9). Immature dendritic cells are found in the periphery of the body, around body cavities, and under mucosal surfaces. Dendritic cell types can be identified by cell surface markers and probably differ in function. In general, dendritic cells are proficient at endocytosis and can synthesize copious quantities of cytokines

BOX 15.9

Three herpes simplex virus proteins modulate the IFN response in the cytoplasm and the nucleus

Herpes simplex virus is only marginally sensitive to IFN in cultured cells, yet this cytokine plays a major role in limiting the acute infection in animals. At least three viral proteins modulate the IFN response.

- ICP34.5 acts as a regulatory subunit of cellular protein phosphatase 1 and acts to reverse Pkr-induced eIF2α phosphorylation. Consequently, this viral protein prevents the translational block and autophagy induction imposed by IFN.
- US11 is an RNA binding protein that prevents Pkr activation.
- ICP0 functions in the nucleus to launch the replication cycle. This protein prevents the IFN-induced block to RNA polymerase II transcription.

The closely related human alphaherpesvirus varicella-zoster virus is similarly insensitive to IFN in cultured cells, but it has no counterparts to the ICP34.5 or US11 genes. It is unclear if the ICP0 homolog provides the only IFN defense for this common herpesvirus of humans.

Harle, P., B. Sainz, Jr., D. J. Carr, and W. P. Halford. 2002. The immediate-early protein, ICP0, is essential for the resistance of herpes simplex virus to interferon-alpha/beta. *Virology* **293:**295–304

Mossman, K., and J. R. Smiley. 2002. Herpes simplex ICP0 and ICP34.5 counteract distinct interferon-induced barriers to virus replication. *J. Virol.* **76:**1995–1998.

Most natural infections are acquired through respiratory or mucosal routes. For ease of experimentation, many investigators resort to rather unnatural routes, including injecting virus into the bloodstream or into the peritoneal cavity. Can we assume that the innate defenses activated by unnatural infections are similar to those following infection by a natural route? Infections of transgenic animals lacking innate defense genes have provided some insight.

Mice are relatively resistant to infection by vesicular stomatitis virus no matter the route of infection. However, Pkr-null mutant mice become highly susceptible to this virus, **but only** after respiratory infection.

These findings show that the Pkr response in the respiratory tract is a primary defense against the virus. Defense after injection of virions into the bloodstream or the peritoneal cavity is mediated by some other mechanism.

The lesson is to be wary of generalizations about the innate immune response. Not all routes of infection are defended in the same manner.

Levy, D. E. 2002. Whence interferon? Variety in the production of interferon in response to viral infection. *J. Exp. Med.* **195:**F15–F18.

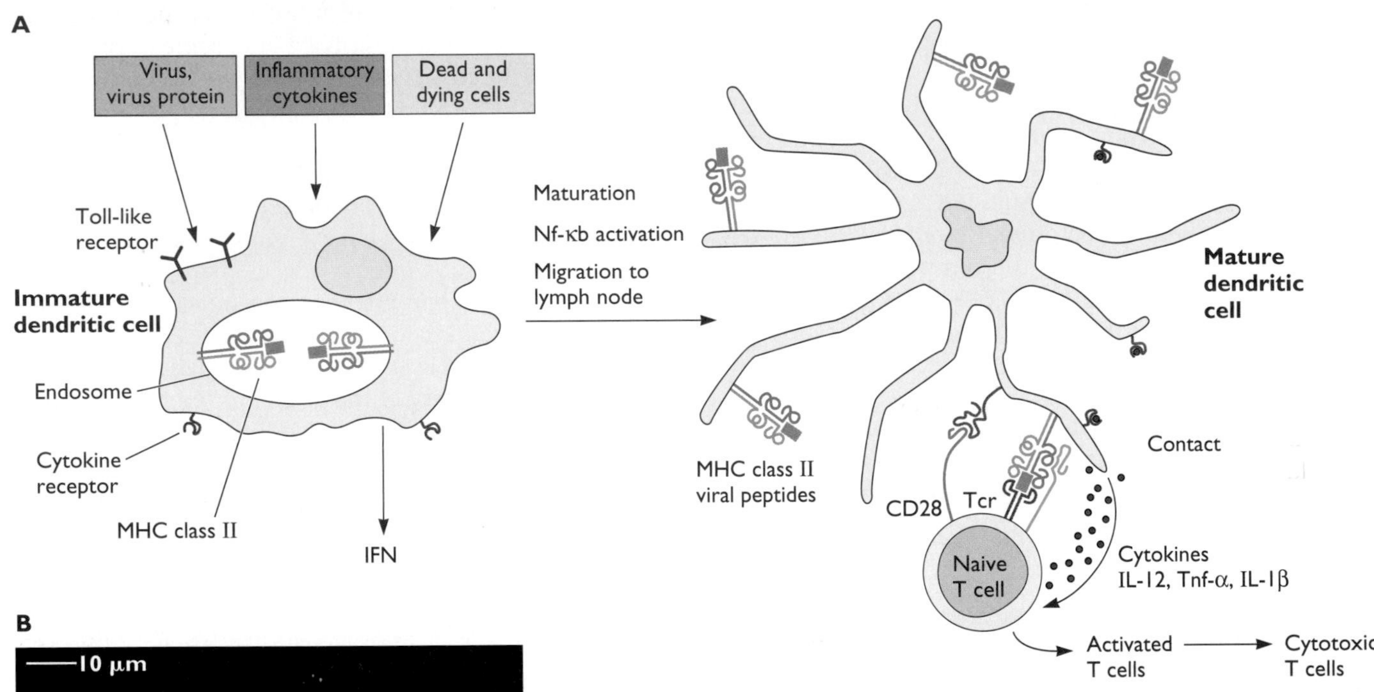

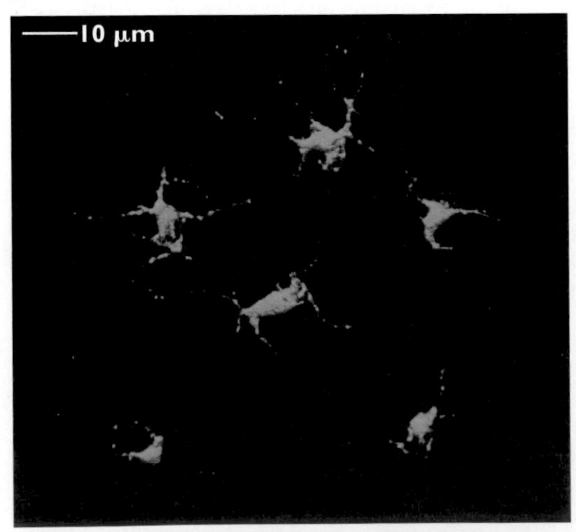

Figure 15.9 Dendritic cells provide cytokine signals and packets of protein information to naive T cells. (A) Immature dendritic cells are migratory cells that patrol the periphery and carry cytokine receptors and Toll-like receptors on their surfaces. Immature dendritic cells actively take up extracellular proteins by endocytosis and store the proteins internally. They do not carry MHC class II complexes on their surfaces. When microbial products bind to the Toll-like receptors or when proinflammatory cytokines bind their receptors, the immature dendritic cells differentiate into mature dendritic cells. The mature dendritic cells no longer have the capacity for endocytosis of proteins and display a new repertoire of cell surface receptors. Some of these are chemokine receptors that enable the dendritic cell to migrate to the local lymph nodes. The proteins ingested by the immature cell are now processed into peptides and loaded on to MHC class II proteins for subsequent transport to the cell surface. Mature cells extend long dendritic processes to increase surface area for binding of naive T cells in the lymph node. Mature dendritic cells release proinflammatory cytokines such as Tnf-α, IL-1β,

(continued on next page)

Figure 15.9 ***(continued)*** and IL-12 to stimulate T-cell differentiation. Naive but antigen-specific T cells bind to the MHC class II-peptide complexes via their T-cell receptors. The interaction is strengthened by the presence of increased costimulatory ligands on the mature dendritic cell. The T cell is activated, begins the maturation process into its final effector state, and moves out of the lymph node into the circulation. (B) Langerhans cells are abundant in mouse ear epithelium and are visualized here in a live tissue preparation by their production of an MHC class II-EGFP fusion protein. Ear snips were obtained from MHC class II-EGFP knock-in mice. Epidermal sheets were produced by separation of ear snips into dorsal and ventral halves and visualized by confocal microscopy. Figure provided by Marianne Boes, Jan Cerny, and Hidde Ploegh (Harvard Medical School, Boston, Mass.).

when stimulated. Soluble proteins are taken up avidly and retained in endosomes until the dendritic cell matures, which may be hours, if not days, after the actual uptake of viral proteins. Immature dendritic cells also capture viral proteins from dead or dying cells by taking up complexes containing heat shock proteins and unfolded proteins, as well as cellular debris and vesicles that are produced by apoptosis. As the cell matures, internalized proteins are processed into peptides and moved to the cell surface complexed to MHC class II molecules that can be recognized by lymphocytes of the adaptive immune system (see "Antigen Presentation and Activation of Immune Cells" below).

Immature dendritic cells carry important pattern recognition receptors on their surfaces (see "Pattern recognition receptors" below; see also Box 15.4), as well as receptors for various proinflammatory cytokines. When these receptors bind the appropriate ligands, most immature dendritic cells undergo dramatic morphological and functional changes, and differentiate. These mature cells no longer have the capacity for endocytosis and they display a new repertoire of cell surface receptors. Some of these are chemokine receptors that direct migration of the mature dendritic cell (loaded with stored viral proteins) to the local lymph nodes (often called **homing**). Other new receptors are T-cell adhesion receptors and T-cell costimulatory molecules essential for binding and activating naive T cells on arrival in the lymph node. A remarkable change in morphology also occurs as mature cells extend long eponymous dendritic processes that increase their surface area. Mature dendritic cells become potent mobile signaling centers releasing proinflammatory cytokines that act locally and at a distance.

Once within lymphoid tissue, mature dendritic cells instruct the adaptive immune response by directly engaging and stimulating naive, antigen-specific T lymphocytes. The density of mature dendritic cells and the cytokines they secrete in the lymph node dictates the type of adaptive response that will ensue (see "Th1 and Th2 cells" below). Dendritic cell maturation is an essential link between innate and adaptive immunity.

While dendritic cells function by sampling proteins and cytokines found in the area of infection, they may also be infected by the virus that they detect (Box 15.11). One obvious outcome is that the mobile, infected cells travel to the lymph node where the virus can be transmitted to T and B cells. Indeed, human immunodeficiency virus 1 virions can bind to a lectin (DC-SIGN [dendritic-cell-specific, Icam-3-grabbing nonintegrin]) on immature dendritic cells with grave consequences. This protein is essential for establishing contact of a mature dendritic cell with a naive T cell in lymphoid tissue. Virus particles so bound do not replicate in the dendritic cell, but rather are retained just below the cell surface during migration to the lymph node, where they subsequently infect $CD4^+$ T cells (see Chapter 17). Two mosquito-borne viruses, Venezuelan equine en-

BOX 15.11

Infection of the sentinels: dysfunctional immune modulation

When viruses infect the dendritic cells, the immune system's first command-and-control link is compromised. Some of the many possible consequences of sentinel cell infection, any one of which could suppress the immune response locally or systemically, include the following:

- reduction in the number of functional precursors
- interference with recruitment to peripheral sites of infection
- impairment of antigen uptake or processing
- infection and destruction of immature dendritic cells
- interference with maturation
- impairment of migration to lymphoid tissue
- interference with activation of T cells

Mellman, I., and R. M. Steinman. 2001. Dendritic cells: specialized and regulated antigen processing machines. *Cell* **106:**255–258.

Raftery, M. J., M. Schwab, S. M. Eibert, Y. Samstag, H. Walczak, and G. Schonrich. 2001. Targeting the function of mature dendritic cells by human cytomegalovirus: a multilayered viral defense strategy. *Immunity* **15:**997–1009.

cephalitis virus and dengue virus, replicate initially in immature dendritic cells at the site of inoculation. These infected cells then migrate to lymph nodes, where they propagate the infection with severe consequences to the mouse host. This property has inspired vaccine strategies in which attenuated versions of these viruses are used to deliver immunizing antigens directly to the dendritic cells.

As might be expected, viral mechanisms to modulate dendritic cell functions have evolved. For example, infection by either herpes simplex virus type 1 or vaccinia virus inhibits maturation of dendritic cells by blocking a signal transduction cascade or by interfering with cytokine stimulation of maturation, respectively. In contrast, dendritic cells infected with murine cytomegalovirus and measles virus are fully capable of maturing, but the mature cells are incapable of stimulating T cells. Such interference with a critical component of the innate response is likely to contribute to the profound immunosuppression that follows cytomegalovirus and measles virus infection. The contribution of dendritic cell infection to immunosuppression is a topic of considerable interest.

Pattern Recognition Receptors

A major conundrum for many years has been how the innate immune system recognizes microbial invaders and not "self." The problem in the case of viruses is obvious: the basic building materials are derived from self cells; the only difference being the way materials are put together. This property actually provides one solution to the conundrum. The innate immune system recognizes **patterns of macromolecules** that are unique to invaders (sometimes called PAMPS for "pathogen-associated molecular patterns"). **Toll-like receptors** (Box 15.4) are considered to be the prototypical pattern recognition molecules of the innate immune system. Other pattern recognition receptors include the complement C1q protein and certain NK cell receptors. Toll-like receptors are type I transmembrane proteins that are conserved from insects to humans. More than 10 members of this receptor family have been identified in mammals. Their ligands have been difficult to identify, because they are structurally diverse and vary among pathogens. However, some ligand/receptor pairs are known (Table 15.10). Ligands that might identify viral infections include CpG DNA and dsRNA, both recognized by particular Toll-like receptors. Unmethylated CpG tracts are present in bacterial and most viral DNA genomes, while dsRNA is commonly found in virus-infected cells. Like many cytokine receptors, after binding their unusual ligands, these Toll-like receptors usually engage the Nf-κb pathway of signal transduction to activate transcription of inflammatory cytokines and T-cell costimulatory molecules.

Table 15.10 Toll-like receptors recognize microbial macromolecular patterns[a]

Toll-like receptor	Pattern recognized
Tlr3	dsRNA
Tlr4	Lipopolysaccharide; integral component of the gram-negative bacterial outer membrane
Tlr2	Glycolipids and lipoproteins; proteins containing lipid covalently linked to N-terminal cysteines
Tlr5	Flagellin; 55-kDa component of bacterial flagella
Tlr9	CpG DNA; unmethylated CpG oligonucleotides; DNA-dependent protein kinase is activated by binding of CpG oligonucleotides to Tlr9

[a]Data from G. Barton and R. Medzhitov, *Curr. Opin. Biol.* **14:**380–383, 2002, and M. Schnare et al., *Nat. Immunol.* **10:**947–950, 2001.

Data from recent experiments indicate that Toll-like receptors bind not only to products on the surfaces of microbes, but also to endogenous ligands induced by contact with proinflammatory cytokines. For example, the cytoplasmic tails of the Toll-like receptors have domains similar to those of various IL-1 family receptors. Consequently, many downstream signaling steps in pattern recognition and inflammation are mediated by common components. Such ability of some Toll-like receptors to bind their ligands inside dendritic cells may be critical for recognizing viral infections. It is unlikely that ligands important in detecting infection, such as CpG-containing DNA fragments and dsRNA, will be found outside the dendritic cell. In fact, a good question is: where are these ligands produced? One idea is that they form inside the dendritic cell after it has taken up virions and fragments of dead and dying cells. Endocytosed proteins and virus particles end up in dendritic cell lysosomal compartments where they can be digested. Some Toll-like receptors, including Tlr3 and Tlr9, are located in endosomes and lysosomes, perfectly placed to bind these unusual viral ligands.

Toll-like receptors may be responsible for early recognition of viral pathogens. Respiratory syncytial virus persists longer in the lungs of infected *tlr4*-null mice than in wild-type mice. NK cell responses and IL-12 synthesis are also reduced after challenge of *tlr4*-null mice with this virus. One interpretation of these observations is that the Tlr4 protein is important for recognition of the infection and production of an antiviral response. An alternative idea is that virus propagation is dependent on signaling from this receptor in the dendritic cell. Other clues concerning the contribution of Toll-like receptors to viral pathogenesis come from the study of viruses that encode proteins with the potential to disrupt their functions. Two vaccinia virus proteins, A46R and A52R, are similar in sequence to segments in the cytoplasmic domain of Toll-like and IL-1 receptors. These two viral proteins can inhibit IL-1- and Tlr4-mediated signal transduction. Vaccinia virus may modulate host immune responses by competing with this domain-dependent intracellular signaling.

NK Cells

NK cells are in the immediate front line of innate defense: they recognize and kill virus-infected cells. Like dendritic cells, they recognize infected cells in the company of vast numbers of uninfected cells. However, the mechanism of recognition is completely different: NK cells recognize "missing self" or "altered self." NK cells are a distinct, abundant lymphocyte population (representing as much as 20% of the circulating lymphocyte population [Table 15.2]) that patrols the blood and lymphoid tissues. They are large, granular lymphocytes, distinguished from others by the absence of antigen receptors found on B and T cells. When an NK cell binds an infected target cell, it releases a mix of cytokines (notably IFN-γ and Tnf-α) that contribute to the inflammatory response and alert cells of the adaptive immune system. NK cells also participate later in adaptive defense by binding to infected cells coated with IgG (immunoglobulin G) antibody and killing them (see "Antibody-dependent cell-mediated cytotoxicity" below).

The number of NK cells increases quickly after viral infection and then declines as the acquired immune response is established. These cells are stimulated to divide whenever infected cells and sentinel dendritic cells make IFN. NK cells kill after contact with the target by releasing perforins and granzymes that perforate membranes and trigger caspase-mediated cell death, mechanisms similar to those used by cytotoxic T lymphocytes (see below). In humans, NK cells are particularly important in controlling primary infection by many herpesviruses, as patients with NK cell deficiencies suffer from severe infections with varicella-zoster virus, human cytomegalovirus, and herpes simplex virus. While a role for direct NK cell-mediated killing in antiviral defense is difficult to establish experimentally, NK cell production of IFN-γ clearly provides significant antiviral action.

NK Cell Recognition of Infected Cells: Detection of "Missing Self" or "Altered Self" Signals

As we will discuss below, MHC proteins are important receptors in the adaptive immune response. Accordingly, they are found on the surfaces of most cells of the body. The MHC molecules are the **self antigens** that, when missing, cause the NK cell to kill the target cell. A possible mechanism for detection of **missing self** is illustrated in Fig. 15.10. At least two receptor-binding interactions are required for such discrimination: one to activate the NK cell and the other to block this activation if the cell is not infected. The activating signal is delivered when an NK cell receptor binds a pathogen-specific ligand (e.g., virus-infected cells may present new glycoproteins on their surface). As a consequence, a signal transduction cascade is initiated and the NK cell is stimulated to secrete a burst of cytokines and kill the cell. However, a negative regulatory signal is produced when an MHC class I-specific receptor on the NK cell engages MHC class I molecules on the surface of the same target cell. Because many infected cells carry fewer MHC class I molecules on their surfaces (discussed in Chapter 16), they are prime NK cell targets. The

Figure 15.10 NK cells can distinguish normal healthy cells from aberrant, infected cells in the body by "missing self" receptors. Both positive (stimulating) and negative (inhibiting) signals may be received when an NK cell contacts a target cell. The converging signal transduction cascades from the two classes of receptor regulate NK cell cytotoxicity and release of cytokines. The inhibitory receptors dominate all interactions with normal, healthy cells. Their ligands are the MHC class I proteins. When NK cells contact MHC class I molecules on the surface of the target cell, signal transduction blocks the response of activating receptors. Virus-infected cells have altered MHC class I proteins on their surface which do not activate the NK-cell inhibitory receptor. As a consequence, the response of the activating receptors is not blocked when NK cells bind to infected target cells and the NK cell lyses the infected cell, produces IFN-γ, or both. The ligands on the target cell that bind the activating receptors are not well characterized, but the signal transduction cascades that ensue are fairly well understood.

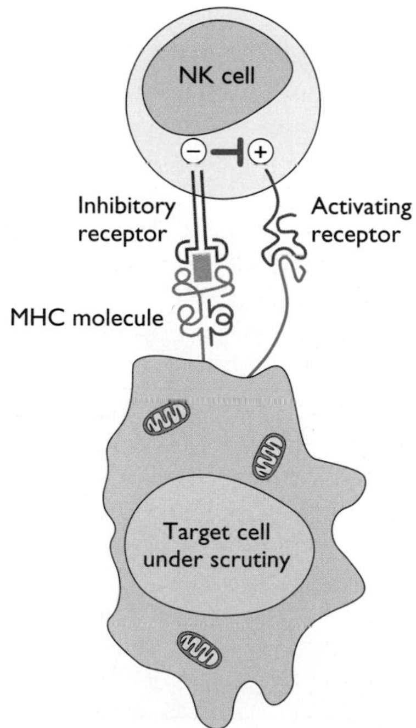

unusual two-receptor recognition system employed by NK cells ensures that normal cells that synthesize MHC class I proteins are not killed by NK cells, even if they have distinct ligands on their surfaces. Much work is in progress to test this idea.

NK MHC Class I Receptors Produce the Inhibitory Signals

Human NK cells synthesize two inhibitory MHC class I receptors of either the C-type lectin family or the immunoglobulin family (called killer cell immunoglobulin-like inhibitory receptors, or Kirs) (Fig. 15.10). Mouse NK cells produce inhibitory MHC receptors only from the Ly49 subfamily of C-type lectins. NK cells also can recognize and spare target cells carrying HLA-E, an unusual MHC class I protein that binds peptides derived from the signal sequences of other MHC class I molecules. The presence of HLA-E protein complexed with signal peptide informs the NK cell that MHC class I synthesis is normal. An intriguing finding is that infection by human cytomegalovirus induces synthesis of HLA-E protein, thereby diverting potential NK cell recognition and lysis.

Many viral genomes encode proteins that block or confound NK cell recognition and killing. For example, the hepatitis C virus E2 envelope protein binds to CD8, a protein on the surface of NK cells, and blocks activation signals. As a result, the NK cell no longer recognizes infected cells. Similarly, the M157 protein of murine cytomegalovirus is a ligand for the inhibitory NK cell receptor Ly49H, which is activated when bound to MHC class I on the target cell. The NK cell will not kill the infected cell if M157 is synthesized. Resistance of mice to infection by mousepox virus and herpes simplex virus type 1 maps to *nkc*, a locus encoding NK cell receptors. It is likely that these genomes encode ligands similar to M157.

NK T Cells

NK T cells are a subset of NK cells that have cell surface markers characteristic of NK cells as well as T-cell receptors. They synthesize an inhibitory receptor that binds to CD1, a distant cousin of the MHC class I proteins, which is complexed with **glycolipid** instead of peptides. NK T cells account for 20 to 30% of lymphocytes present in the liver, and are capable of releasing IFN-γ after infection.

Complement

The **complement system** was identified in 1890 as a heat-labile serum component that lysed bacteria in the presence of antibody (it "complemented" antibody). The complement system in the blood is a major primary defense and a clearance component for **both** the innate and adaptive immune responses (Box 15.12). There are three distinct complement pathways: the **classical**, **alternative**, and **mannan-binding** pathways. As part of the innate defense system, complement action can be initiated by direct recognition of a microbial invader by C1q or C3b proteins in the alternative pathway (Fig. 15.11). The mannan-binding lectin pathway triggers complement action by the binding of a lectin similar to C1q with mannose-containing carbohydrates on bacteria or viruses. Importantly,

BOX 15.12

The complement cascade has four major biological functions

Cytolysis

Cytolysis and lysis occur when specific activated complement components (C6, C7, C8, and C9) polymerize on a foreign cell or enveloped virus, forming pores or holes that disrupt the lipid bilayer and compromise its function. The cell or virus lyses by osmotic effects.

Activation of Inflammation

Inflammation is stimulated by several peptide products of proteolysis during establishment of the complement cascade. These peptides (C3a, C4a, and C5a) bind to vascular endothelial cells and various classes of lymphocytes to stimulate inflammation and to enhance responses to foreign antigens.

Opsonization

Complement proteins (typically C3b and C1q) can bind to virus particles so that phagocytic cells carrying appropriate receptors can then engulf the coated viruses and destroy them; this process is called opsonization. Complement receptors such as Cr1 present on phagocyte surfaces bind C3b-coated particles and stimulate their endocytosis.

Solubilization of Immune Complexes

Noncytopathic viral infections commonly result in pathological accumulations of antigen-antibody complexes in lymphoid organs and kidneys. Complement proteins can solubilize these complexes by binding both antigen and antibody, and facilitate their clearance from the circulatory system.

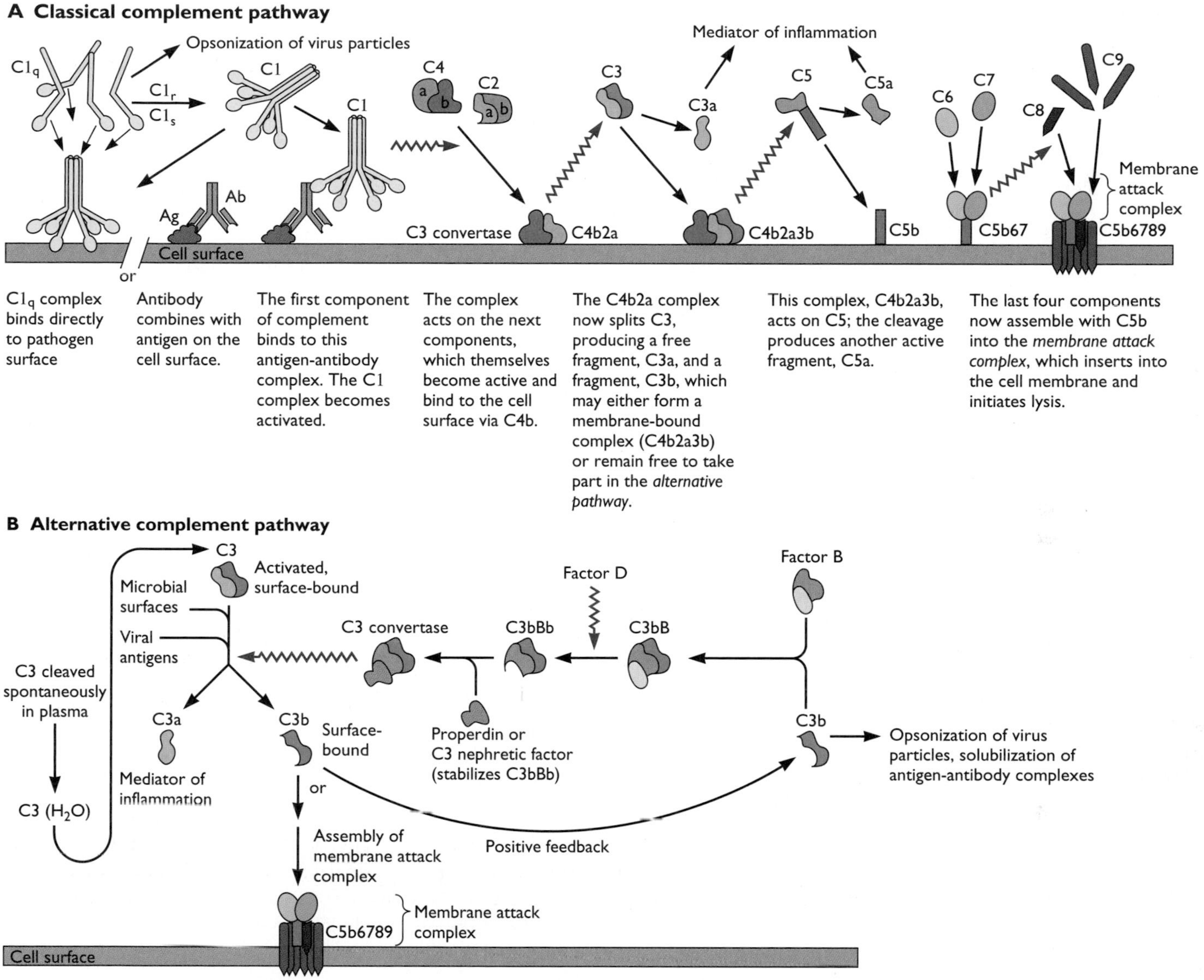

A Classical complement pathway

Opsonization of virus particles
Cl_q
Cl_r
Cl_s
C1
C1
C4
C2
C3
Mediator of inflammation
C3a
C5
C5a
C6
C7
C8
C9
Ag
Ab
Cell surface
C3 convertase
C4b2a
C4b2a3b
C5b
C5b67
Membrane attack complex
C5b6789
or

Cl_q complex binds directly to pathogen surface

Antibody combines with antigen on the cell surface.

The first component of complement binds to this antigen-antibody complex. The C1 complex becomes activated.

The complex acts on the next components, which themselves become active and bind to the cell surface via C4b.

The C4b2a complex now splits C3, producing a free fragment, C3a, and a fragment, C3b, which may either form a membrane-bound complex (C4b2a3b) or remain free to take part in the *alternative pathway*.

This complex, C4b2a3b, acts on C5; the cleavage produces another active fragment, C5a.

The last four components now assemble with C5b into the *membrane attack complex*, which inserts into the cell membrane and initiates lysis.

B Alternative complement pathway

C3
Activated, surface-bound
Microbial surfaces
Viral antigens
C3 cleaved spontaneously in plasma
C3 (H_2O)
C3a
Mediator of inflammation
C3b
Surface-bound
or
Assembly of membrane attack complex
C3 convertase
C3bBb
Factor D
C3bB
Factor B
Properdin or C3 nephretic factor (stabilizes C3bBb)
C3b
Opsonization of virus particles, solubilization of antigen-antibody complexes
Positive feedback
C5b6789
Membrane attack complex
Cell surface

Figure 15.11 The classical and alternative pathways for activation of complement. The C1q, mannan-binding lectin, and C3a initiator proteins are the recognition devices that, when bound to pathogens or infected cells, activate an unusual protease cascade using one or more members of a set of seven activating enzymes. (A) The classical complement cascade begins when initiator protein C1q binds to antigen-antibody complexes or pathogen surfaces. (B) The alternative pathway has a distinctive mode of initiation: the abundant C3 protein in plasma is spontaneously hydrolyzed to C3b, which then binds to essentially any membrane surface. When C3b binds to normal cell surfaces, complement regulatory proteins present in the membranes block further action. However, when C3b binds to pathogen membranes, the lack of complement regulatory proteins ensures that the cascade continues and amplifies. Both pathways converge in the formation of an enzyme activity called C3 convertase that cleaves complement component C3 into its active components. The C4b2a complex (classical pathway) and the C3bBbP complex (alternative pathway) are active C3 convertase enzymes. C3 convertase activity produces the membrane attack complex that forms holes in membranes of cells and viruses, and is instrumental in the formation of C3a, a potent mediator of inflammation. The complement cascade engages a substantial cast of players. Two membrane binding proteins and opsonins are critical primary products of C3 convertase (C4b and C3b), as well as at least three soluble peptide mediators of inflammation (C5a, C3a, and C4a). All three pathways converge to produce the membrane attack complex, which is built with five proteins (C5b, C6, C7, C8, and C9). In addition, five complement membrane protein receptors bind complement components, and at least nine complement regulatory proteins are required to keep the three complement pathways from spontaneously destroying normal cells.

complement can also function as an effector of the adaptive defense system by the binding of C1q to antibody-antigen complexes on the surface of an invader or infected cell (the classical pathway [Fig. 15.11]).

The Complement Cascade

Complement is not a single protein, but rather at least 18 distinct serum proteins and an almost equal number of membrane proteins that act sequentially (Fig. 15.11). Unfortunately, the nomenclature of the complement components is complicated because proteins were named in order of their discovery, and not in order of their function.

All three pathways follow a cascade of protease reactions to activate a central protease activity called **C3 convertase** (note that the three pathways yield the same enzyme activity, but the proteins comprising each convertase are different). A crucial property of C3 convertase enzymes is that they are bound covalently to the surface of the pathogen or the infected cell. The action of surface-bound C3 convertase on its substrate yields C3b, the primary effector of all three complement pathways, and C3a, a potent soluble mediator of inflammation. C3b remains on the pathogen's surface, where it binds more complement components to stimulate a protease cascade giving rise to many other bioactive proteins that stimulate inflammation, potentate the adaptive response, and kill infected cells.

More than 90% of complement components in the plasma are made in the liver. However, these components also are made elsewhere, including the major portals of pathogen entry. For example, the initiator complex C1 is synthesized mainly in the gut epithelium, and mannan-binding lectin is found in the respiratory tract. In addition, monocytes, macrophages, lymphocytes, fibroblasts, endothelial cells, and cells lining kidney glomeruli or synovial cavities all make most proteins of the complement system. Astrocytes in the brain can synthesize the full panoply of complement proteins when stimulated by inflammatory cytokines. As the brain is exquisitely sensitive to inflammation, it is not clear how the inflammatory action of peptides produced during the complement cascade is controlled.

One important consequence of complement cascade activation is the initiation of a local broad-spectrum defense (Box 15.12). Complement components released locally during the protease cascade aid in recruitment of monocytes and neutrophils (Fig. 15.11 and Table 15.11) to the site of infection, stimulate their activities, and also increase vascular permeability. The antiviral effects of complement are both direct and indirect. The membrane attack complex lyses infected cells and inactivates enveloped viruses, while phagocytes engulf and destroy virions coated with C3b protein. Complement components stimulate a local inflammatory response that can limit infection and convey the nature of the invader to the adaptive immune system.

Table 15.11 Biological activities of proteins and peptides released during the complement cascade

Substance	Biological activity
C5b, C6, C7, C8, and C9	Lytic membrane attack complex
C3a	Peptide mediator of inflammation, smooth muscle contraction; vascular permeability increase; degranulation of mast cells, eosinophils and basophils; histamine release; platelet aggregation
C3b	Opsonization of particles and solubilization of immune complexes; facilitation of phagocytosis
C3c	Neutrophil release from bone marrow; lyses leukocytes
C3dg	Molecular adjuvant; profoundly influences adaptive response
C4a	Smooth muscle contraction; vascular permeability increase
C4b	Opsonin for phagocytosis, processing, and clearance of antibody-antigen immune complexes
C5a	Peptide mediator of inflammation, smooth muscle contraction; vascular permeability increase; degranulation of mast cells, basophils, and eosinophils; histamine release; platelet aggregation; chemotaxis of basophils, eosinophils, neutrophils, and monocytes; hydrolytic enzyme release from neutrophils
Bb	Inhibition of migration and induction of monocyte and macrophage spreading
C1q	Opsonin for phagocytosis, clearance of apoptotic cells, processing and clearance of antibody-antigen immune complexes

The antibodies made during the Th1 response (see "Th1 and Th2 Cells" below) are predominantly of the immunoglobulin G2a (IgG2a) isotype and actively stimulate the complement cascade. In contrast, antibodies produced during the Th2 response typically are IgG1 (mice), IgG4 (humans), and IgE, which bind neither to complement nor to macrophage Fc receptors (Table 15.12). Many antibodies specific for viral envelope proteins cannot be recognized by C1q, a property likely to be a consequence of selective pressure to escape complement-mediated lysis.

"Natural Antibody" Protects against Infection

The classical complement pathway of humans and higher primates can be activated by a particular collection of antibodies present in serum prior to viral infection (historically called "natural antibody"). Synthesis of these antibodies is triggered by the antigen galactose α(1,3)-galactose (α-Gal) found as a terminal sugar on glycosylated cell surface proteins. Lower primates, most other animals, and

ing protein (a lectin) in the collectin family of proteins. Other collectins include soluble collagenous lectins such as the mannose-binding protein, conglutinin in plasma, and pulmonary surfactant proteins A and D. These proteins bind to polysaccharides on a wide variety of microbes and act as opsonins or activators of the complement cascade. Collectins bind the glycoproteins of a number of enveloped viruses, including human immunodeficiency virus type 1, herpes simplex virus, and influenza virus. They display antiviral activity in the laboratory, but their physiological contributions have not been well studied.

The Inflammatory Response

The release of cytokines and soluble mediators of the complement cascade at the site of infection initiates responses with far-reaching consequences. One very visible response is **inflammation**: not only does it control viral infections and contribute to the overt pathogenic effects that follow, but also, like most innate responses, it is essential for the initiation of adaptive immune defenses (Fig. 15.12). The four classical signs of inflammation are redness, heat, swelling, and pain. These symptoms result from increased blood flow, increased capillary permeability, influx of phagocytic cells, and tissue damage. Increased blood flow occurs when vessels that carry blood away constrict, resulting in engorgement of the capillary network in the area. This response produces redness (erythema) and an increase in tissue temperature. Capillary permeability increases, facilitating an efflux of fluid and cells from the engorged capillaries into the surrounding tissue. The fluid that accumulates has a high protein content, in contrast to that of normal fluid found in tissues, and contributes to the swelling. The cells that migrate into the damaged area are largely mononuclear phagocytes (Table 15.2). They are attracted by molecules synthesized by virus-infected cells, by cytokines elaborated by local defensive systems, and by subsequent secondary reactions that facilitate adherence of phagocytic cells to capillary walls near sites of damage. Neutrophils are found in abundance in the blood but normally are absent from tissues. They are the earliest phagocytic cells to be recruited to a site of infection and are classic cellular markers of the inflammatory response. Neutrophils also secrete a variety of cytokines and toxic products that can determine subsequent events. Some cytokines made by infected cells, dendritic cells, and macrophages, as well as soluble complement components are chemokines that direct the migration of monocytes and granular cells to regions of cell damage (Fig. 15.12). Monocytes are also important in the healing reactions that take place after the infection is cleared.

Noncytopathic viruses do not induce a strong inflammatory response and, as a consequence, have dramatically different interactions with the host. It is now understood

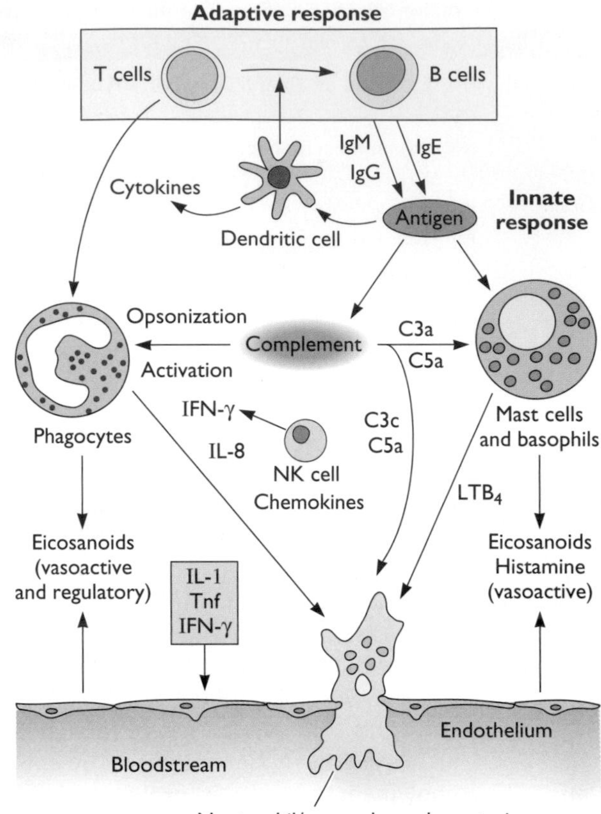

Figure 15.12 Inflammation provides integration and synergy with the main components of the immune system. Viral infections at entry sites in the body often trigger an inflammatory response. Inflammation reactions can be initiated in several ways, for example, by IFN released by immature dendritic cells as they detect infection, by the classical or alternative pathway of complement activation, or by mast cells that migrate to sites of cell damage responding to cytokine release, where they can be activated by IgE antibody and antigen. C3a, C3c, and C5a are protease digestion products of the complement cascade that stimulate the inflammatory response. C3a increases vascular permeability and activates mast cells and basophils, C3c stimulates neutrophil release, and C5a increases vascular permeability and chemotaxis of basophils, eosinophils, neutrophils, and monocytes and stimulates neutrophils. The cytokines IL-1, IFN-α, and Tnf-α act on the local endothelium to enhance leukocyte adhesion and migration. IL-8 and other chemokines promote lymphocyte and monocyte chemotaxis. IL-1 and Tnf-α bind to receptors on epithelial and mesenchymal cells to cause division and collagen synthesis and stimulate prostaglandin and leukotriene synthesis (eicosanoid compounds). LTB_4 is a particularly active leukotriene that is vasoactive and chemotactic. The activities of cells that enter an infected site where inflammation reactions are occurring are controlled by locally produced cytokines, particularly Tnf-α, IL-1, and IFN-γ. Adapted from D. Male et al., *Advanced Immunology,* 3rd ed. (Mosby, St. Louis, Mo., 1996).

that the early reactions in inflammation determine the type of immune response that will predominate, which in turn can influence the outcome of a viral infection.

Despite an incomplete understanding of the process, for many years scientists have induced inflammation deliberately by the use of adjuvants, such as Freund's adjuvant (killed mycobacterial cells in oil emulsion) or aluminum hydroxide gels. The adjuvant-induced inflammation mimics an infection to provide the environment for the induction of a strong immune response to injected viral proteins in vaccines. In fact, most vaccines would not work without adjuvants to stimulate the adaptive immune response (see Chapter 19 for a discussion of adjuvants and vaccines).

The nature and extent of the inflammatory response to viral infection depends on the tissue that is infected, as well as on the cytopathic nature of the virus. Many of the cells and proteins that participate in an inflammatory response come from the bloodstream and are directed by proinflammatory cytokines and soluble complement effectors to the site of infection. As a result, those tissues that have reduced access to the circulatory system (e.g., the brain and the interior of the eyeball) normally avoid the destructive effects of the inflammatory response. This affords protection of vital tissues. However, as a consequence of their so-called "privileged" state, the kinetics, extent, and final outcome of viral infections of these tissues can be markedly different from the outcome in tissues with more intimate access to the circulatory system.

Viral gene products that modulate the inflammatory response include soluble cytokine receptors expressed by poxviruses, the complement component-binding proteins of herpesviruses, and various cytokines and growth factors of beta- and gammaherpesviruses (Tables 15.6 and 15.7). Mutants that do not synthesize these proteins have markedly reduced virulence (virulence is discussed in Chapter 14). Such modulation can be critical for both host and virus. For example, if the normal inflammatory response induced by adenovirus infection is not suppressed by viral proteins in the E3 gene complex (Fig. 15. 4), severe damage and even death of the host can result.

The Adaptive Immune Response

General Features

The adaptive response comprises two complex actions, the **humoral response** (actions of antibody) and the **cell-mediated response** (actions of specific helper and effector cells) (Fig. 15.13 and Table 15.12). As we discuss the cells and actions that characterize each response, it is important to understand that both are essential in antiviral defense and function in concert, but that the relative contribution of each in any given infection varies with the nature of both the infecting virus and the host. In general, antibodies bind to virions in the bloodstream and at mucosal surfaces, decreasing the number of cells that could have been infected, whereas T cells recognize and kill infected cells as well as synthesize antiviral cytokines.

Like the innate response, the adaptive response to viral infection must distinguish infected self from uninfected self. This feat is accomplished in a markedly different fashion than in the innate immune system. Highly specific molecular recognition is mediated by two antigen receptors: membrane-bound antibody on B cells or the T-cell receptor, and one of two membrane glycoprotein oligomers that display fragments of internal cellular proteins on the cell surface. These latter proteins are members of the MHC protein family. MHC class I proteins display protein fragments on the surface of almost all cells, whereas MHC class II proteins generally are found only on the surfaces of mature dendritic cells, macrophages, and B cells (the professional antigen-presenting cells).

While B- and T-cell receptors both bind foreign antigens, they do so in very different ways. The B-cell receptor is an antibody and binds discrete **epitopes** (contiguous sequences or unique conformations) in intact proteins. In contrast, the T-cell receptor binds short, linear **peptides** derived from proteolytically processed proteins. The act of binding to an epitope or peptide by either receptor has profound effects on the cell bearing that receptor: it responds by producing cytokines, by replicating and dividing, by killing the cell that bears the foreign protein or peptide, or by synthesizing antibodies. The entire sequence of events initiated by the binding of foreign peptides or epitopes comprises the adaptive immune response.

As in all defense mechanisms, an uncontrolled or inappropriate adaptive response can be damaging. Therefore, precise initiation and rapid cessation of the response are as important as its amplification. Regulation is achieved largely by multiple interactions among the surfaces of lymphocytes and infected cells, the short life spans of the activated cells, and the short half-lives of the chemical mediators. When such regulation is disrupted actively or passively by large-scale viral infection, or when the cells of the immune system themselves are infected by viruses, severe consequences ensue. Many pathological effects associated with infection are manifestations of inappropriate reactions of the host immune defense system (see "Injury induced by viruses" below). On the other hand, all successful viruses have evolved mechanisms that modulate these host defenses. In some instances, viral infections can be long-lived, persistent, or latent in the face of an active, adaptive immune system (we will discuss these situations in Chapter 16).

Unlike the innate response, the adaptive response is tailored specifically to a particular invading organism or substance and requires more time (days to weeks) after exposure to become fully active. However, once a specific response has been established and the viral infection is subdued, the individual is immune to subsequent infection by

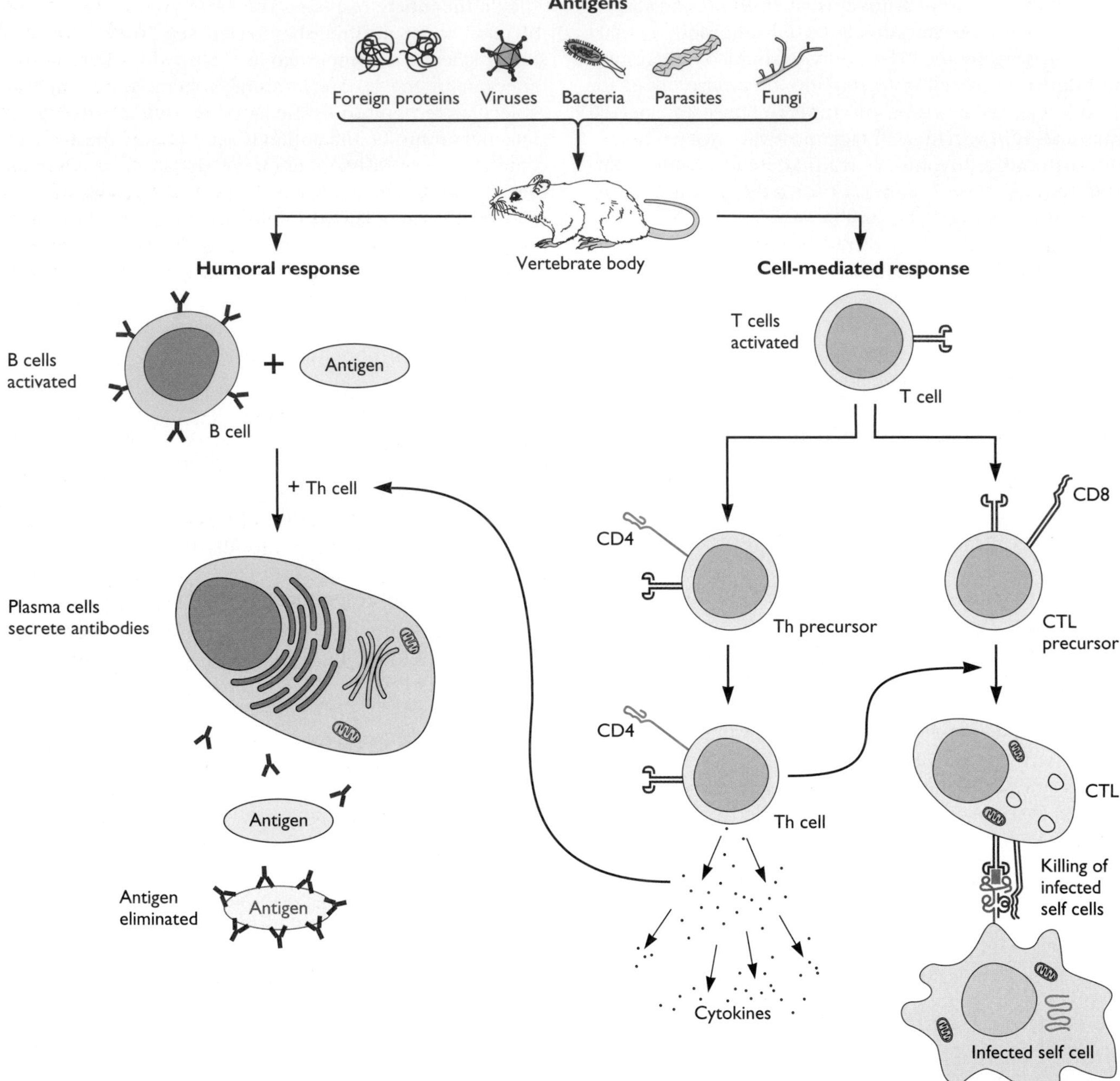

Figure 15.13 The humoral and cell-mediated branches of the adaptive immune system. The humoral branch comprises lymphocytes of the B-cell lineage. Antibodies are the important effector molecules produced by this response. The process begins with the interaction of a specific receptor on precursor B lymphocytes with antigens. Binding of antigen promotes differentiation into antibody-secreting cells (plasma cells). The cell-mediated branch comprises lymphocytes of the T-cell lineage. Immature cytotoxic T cells and T-helper cells (Th cells) are the critical effectors of this response. The process is initiated when T lymphocytes recognize antigens on the surface of "self" cells. The antigens on "self" cells can be recognized only by a receptor on the surface of T cells (the T-cell receptor) when they are bound to the MHC family of membrane proteins. Two subpopulations of T cells are illustrated. The Th cells recognize antigens bound to MHC class II molecules and produce powerful cytokines that "help" other lymphocytes (both B and T cells) by promoting or inhibiting cell division and gene activity. Once activated by Th cells, cytotoxic T-cell precursors differentiate into fully mature effector cytotoxic T cells (CTLs) capable of recognizing and killing virus-infected cells.

the same invader (Fig. 15.14). Such memory of previous infections is one of the most powerful features of the adaptive immune response and makes vaccines possible. While the primary response takes many days to reach its optimum, a subsequent encounter with an invader by an immune individual engenders a ferocious response that occurs within hours of the infection. The innate defenses also are stimulated during this secondary response to infection. The cellular and molecular mechanisms for maintenance of memory are the focus of considerable research and debate, but a subset of lymphocytes called **memory cells** is maintained after each encounter with a foreign antigen. These cells survive for years in the body and are ready to respond immediately to any subsequent encounter by rapid proliferation and efficient production of their protective products. Because such a secondary response is usually stronger than the primary one, childhood infection protects adults, and vaccination can last for years.

Figure 15.14 The specificity, self-limitation, and memory of the adaptive immune response. This general profile of a typical adaptive antibody response demonstrates the relative concentration of serum antibodies after time (weeks) of exposure to antigen A or a mixture of antigens A and B. The antibodies that recognize antigens A and B are given by the red and blue lines, respectively. The primary response to antigen A takes about 3 to 4 weeks to reach a maximum. When the animal is injected with a mixture of both antigens A and B at 7 weeks, the secondary response to antigen A is more rapid and more robust than the primary response. However, the primary response to antigen B takes about 3 to 4 weeks to reach a maximum and is less robust than the secondary response to antigen A. These properties demonstrate immunological memory. Antibody levels (also termed titers) decline with time after each immunization. This property is called self-limitation or resolution. From A. K. Abbas et al., *Cellular and Molecular Immunology* (The W. B. Saunders Co., Philadelphia, Pa., 1994), with permission.

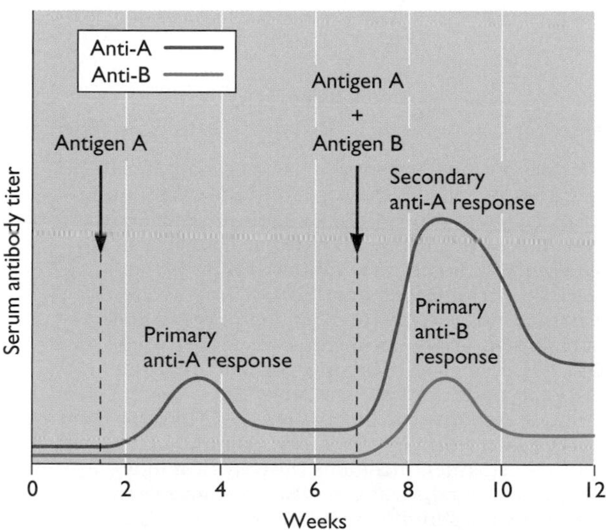

Whether memory cells are sequestered in various depots in the body for future use, or if they are constantly produced, is controversial. We know that the cytokine IL-15 is required for maintenance of some, but not all, memory T cells. One idea of how memory is maintained is that follicular dendritic cells in lymph nodes bind antibody-antigen complexes and keep such complexes on their surfaces for long periods. When the circulating antibody concentration drops, as happens after an infection is cleared, antigen is released gradually from the immune complexes. Such slow release of antigen would provide continual stimulation of the immune response.

Cells of the Adaptive Immune System

The adaptive response depends on two important cell groups: lymphocytes and antigen-presenting cells (Table 15.2). Lymphocytes are white blood cells produced by hematopoiesis in the bone marrow. They are migratory cells, leaving the bone marrow to circulate in the blood and lymphatic systems and often settling in various lymphoid organs throughout the body (Fig. 15.15). The small, bean-shaped organs called lymph nodes are essential for initiation of the adaptive immune response: in experimental models, when lymph nodes are removed, viral infections do not stimulate adaptive immunity. Lymphoid tissues are the collection centers and sites of communication for cells in the circulatory system.

T cells and **B cells** represent two major classes of lymphocytes. T cells (lymphocytes that mature in the thymus) and particular cytokines are the primary components of the cell-mediated response. Immature T cells differentiate into two critical effector T cells, the T-helper (Th) cell and the cytotoxic T cell (cytotoxic T lymphocyte, or CTL). Plasma cells, whose immediate precursors are B cells (lymphocytes derived from the bone marrow, the mammalian equivalent of the avian bursa), make antibodies that bind to foreign molecules and define the antibody response (also known as the humoral response).

There are two classes of **professional antigen-presenting cells.** The first comprises the dendritic cells discussed previously; the second includes the B cells and monocyte-derived cells. Migratory dendritic cells are a central part of the innate immune response, but are particularly important to initiate the adaptive response by presentation of viral antigens to naive, antigen-specific T cells. B cells and monocyte-derived cells present antigenic peptides on MHC molecules, but only to previously activated T cells. Many nonlymphoid cells could be called "nonprofessional" antigen-presenting cells because they synthesize MHC class I molecules carrying peptides that can be recognized by activated T cells.

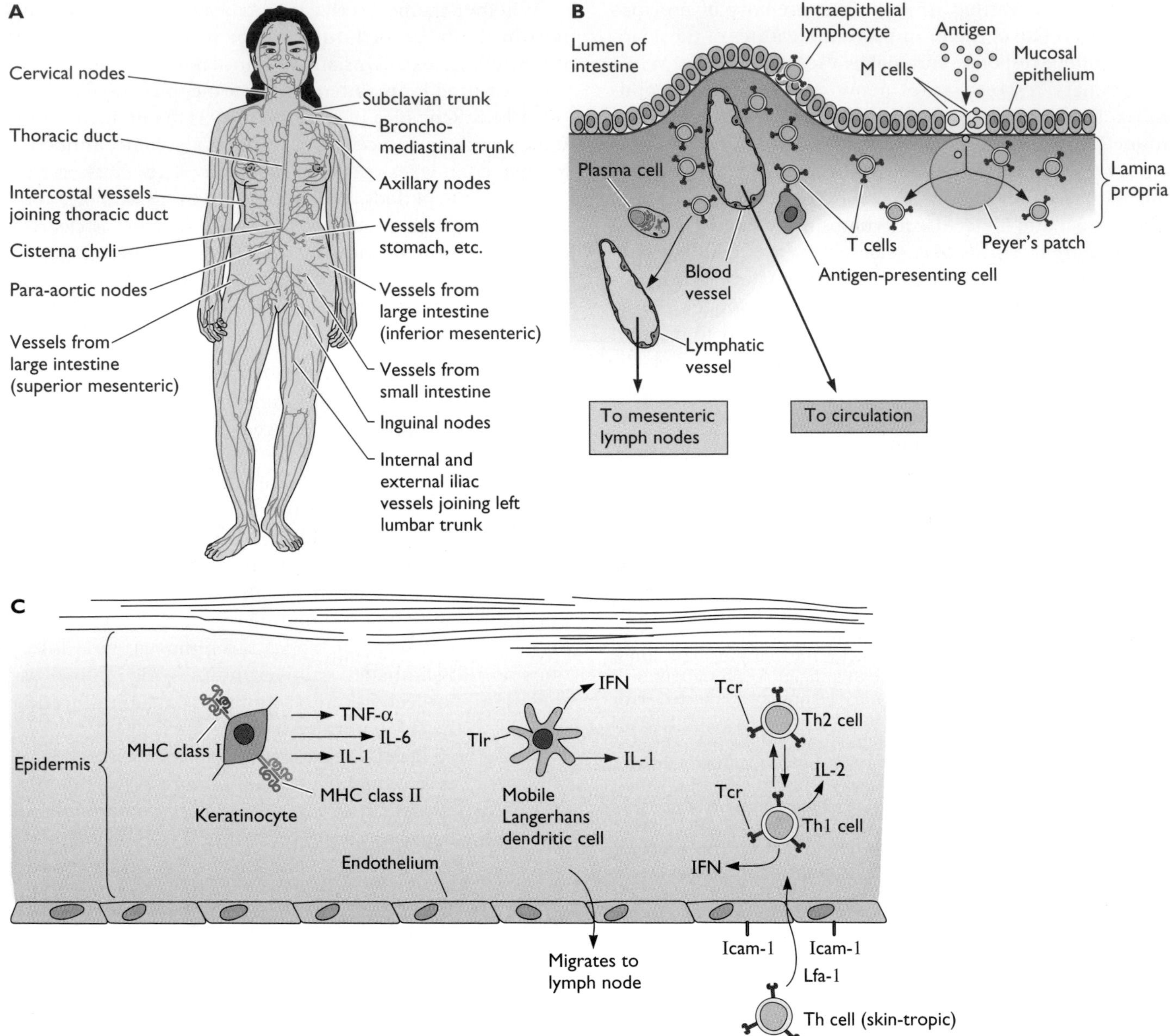

Figure 15.15 Components of the human lymphatic and mucosal immune systems. (A) The primary lymphatic system is illustrated in green. Many clusters of lymphatic tissue, called nodes, are found throughout the lymphatic system. Lymph nodes provide the interactive environment where mobile dendritic cells patrolling the local area exchange information with lymphocytes in the circulation. (B) Cellular components of the mucosal immune system in the gut (mucosa-associated lymphoid tissue). The lumen of the small intestine is at the top of the figure. The mucosal epithelial cells are shown with their basal surface oriented toward the dark purple line marking the start of the lamina propria. Cross sections of a lymphatic vessel and a capillary are shown, illustrating their juxtaposition to cells of the mucosal immune system. M cells (yellow) have large intraepithelial pockets filled with B and CD4$^+$ T lymphocytes, macrophages, and dendritic cells. M cells and intraepithelial lymphocytes are important in the transfer of antigen from the intestinal lumen to the lymphoid tissue in Peyer's patches, where an immune response can be initiated. (C) The cutaneous immune system (skin-associated lymphoid tissue) comprises three cell types: keratinocytes, Langerhans cells, and T cells. Keratinocytes actively secrete various cytokines, including Tnf-α, IL-1, and IL-6, and have phagocytic activity. They also synthesize both MHC class I and MHC class II proteins and can present antigens to T and B cells if stimulated by IFN-γ. Langerhans cells are migratory dendritic cells and

(continued on next page)

The Mucosal and Cutaneous Arms of the Immune System

At first glance, the adaptive immune system appears to exhibit considerable decentralization of important cells and tissues. However, every major organ and body surface harbors organized components to mount an adaptive immune response when signaled by the innate immune system. The **mucosal immune system** is usually the first adaptive defense to be engaged after infection. The lymphoid tissues below the mucosa of the gastrointestinal and respiratory tracts (often called mucosa-associated lymphoid tissue, or MALT) (Fig. 15.15B) are vital in antiviral defense. These clusters of lymphoid cells include the dome-shaped collection called Peyer's patches in the lamina propria of the small intestine, the tonsils in the pharynx, the submucosal follicles of the upper airways, and the appendix. A specialized epithelial cell of mucosal surfaces is the **M cell** (microfold or membranous epithelial cell), which samples and delivers antigens to the underlying lymphoid tissue by transcytosis. M cells have invaginations of their membranes (pockets) that harbor immature dendritic cells, B and $CD4^+$ T lymphocytes, and macrophages. The secreted antibody IgA (important in antiviral defense at mucosal surfaces; see below) is made by B cells that accumulate at adhesion sites in these M cell membrane pockets. As viral proteins transit through M cells, they emerge to be in intimate contact with all the appropriate immune cells. This process represents an essential step for the development of mucosal immune responses.

The skin, the largest organ of the body, possesses its own complex community of organized immune cells. Lymphocytes and Langerhans cells comprise the **cutaneous immune system** (also called skin-associated lymphoid tissue, or SALT) (Fig. 15.15C; see also Fig. 15.9). These cells are important in the initial response and resolution of viral infections of the skin. In particular, Langerhans cells, the predominant scavenger antigen-presenting cells of the epidermis, function as the sentinels or outposts of early warning and reaction. These abundant, mobile cells sample antigens and migrate to regional lymph nodes to transfer information to T cells, and to directly activate B lymphocytes. Certain T cells in the circulation have tropism for the skin and, after binding to the vascular endothelium, can enter the epidermis to interact with Langerhans cells and keratinocytes. These skin-tropic T cells play important roles in production of the virus-specific skin rashes and pox characteristic of measles virus and varicella-zoster virus infections.

Virus particles can interact with lymphoid cells associated with mucosal and cutaneous immune systems at the primary site of infection. Such events can suppress immune responses by killing or misregulation of immune cells. These interactions can govern the outcome of the primary infection and often will establish the pattern of infection characteristic of a given virus. The M cells in the mucosal epithelium have been implicated in the spread from the pharynx and the gut to the lymphoid system of a variety of viruses, including poliovirus, enteric adenoviruses, human immunodeficiency virus type 1, and reovirus. The antigen-presenting Langerhans cells of the cutaneous immune system are easily infected by Venezuelan equine encephalitis virus and dengue virus. These cells also have been suggested to be sites of persistent or latent infection for a number of other viruses, including herpes simplex virus.

The Immune System and the Brain

The nervous system coevolved with the intrinsic and innate defense systems over millions of years. Cells of the central nervous system (brain and spinal cord) can initiate a robust and transient innate defense, but, surprisingly, they are unable to mount an adaptive response. A primary reason for this deficit is that the central nervous system is devoid of lymphoid tissue and dendritic cells. The central nervous system of vertebrates also is separated from many cells and proteins of the bloodstream by tight endothelial cell junctions that comprise the so-called **blood-brain barrier.** As a consequence of these features, viral infections of the central nervous system can have unexpected outcomes. For example, if virus particles are injected directly into the ventricles or membranes covering the brain that are in contact with the bloodstream, the innate immune system is activated, a strong inflammatory response occurs, and the adaptive response ensues. In contrast, if

Figure 15.15 *(continued)* are the major antigen-presenting cells in the epidermis. When products of viral infections in the skin are detected, Langerhans cells secrete IFN and undergo maturation. Mature dendritic cells migrate to the local draining lymph node, where they present viral peptides on both MHC class I and MHC class II proteins to antigen-specific T cells. Special skin-tropic T cells can cross the endothelium to enter the epidermis where they can mature into Th1 or Th2 cells depending on the antigen and cytokine milieu. Activated T cells synthesize cytokines, including IFN-γ, that activate MHC expression from keratinocytes and Langerhans cells. Tcr, T-cell receptor. Adapted from A. K. Abbas et al., *Cellular and Molecular Immunology* (The W. B. Saunders Co., Philadelphia, Pa., 1994), with permission.

virus particles are injected directly into brain tissue, avoiding the blood vessels and ventricles, only a transient inflammatory response is found (the evolutionarily ancient innate defensive response). The adaptive response is not activated; antibodies and antigen-activated T cells are not made.

Although the central nervous system is unable to initiate an adaptive immune response, it is not isolated from the immune system. Indeed, the blood-brain barrier is open to entry of activated immune cells circulating in the periphery. Antigen-specific T cells regularly enter and travel through the brain, performing immune surveillance. Moreover, at least two glial cell types, astrocytes and microglia, have a variety of cell surface receptors that can engage these T cells. Astrocytes, the most numerous cell type in the central nervous system, respond to a variety of cytokines made by cells of the immune system. All natural brain infections begin in peripheral tissue, and any infection that begins outside the central nervous system activates the adaptive immune response. However, if the infection spreads to the brain, the resulting immune attack on this organ can be devastating. A careful scientist can inject virus particles experimentally into the brain without infecting peripheral tissue, thereby avoiding an adaptive response. On the other hand, if the animal is first immunized by injecting virus particles into a peripheral tissue, the adaptive response is activated as expected. Subsequent injection of an identical virion preparation into the brain of the immunized animal elicits massive immune attack on the brain: any virus-infected target in the brain is recognized and destroyed by the peripherally activated T cells. In both natural and experimental infections, the inflammatory response is not transient but sustained, resulting in capillary leakage, swelling, and cell death. Swelling of the brain in the closed confinement of the skull has many deleterious consequences and, when coupled with bleeding and cell death, the results are disastrous. It is clear that although the brain is "immunoprivileged" in some sense, it is not completely isolated and, when infected, is vulnerable to attack by T cells produced and stimulated by the immune system.

Adaptive Immunity Results from the Action of Lymphocytes That Carry Distinct Antigen Receptors

Unlike the germ line-encoded pattern recognition receptors of the innate immune system, the T-cell and B-cell antigen receptors are formed by somatic gene rearrangement during development of the organism. Their specificities are built originally without regard to any one target. Subsequently, those cells that can bind a particular antigen are killed if they recognize self antigens, or stimulated to divide and prosper if they recognize a foreign invader or infected cell. Lymphocytes are responsible for specificity, memory, and self-nonself discrimination (Fig. 15.13).

As noted above, each T or B cell carries on its surface a specific receptor that binds and responds to a particular peptide or epitope (Fig. 15.16). Antiviral defense is possible because T and B cells bearing these specific receptors have survived the normal selective process that eliminates cells that respond to self-peptides and self-epitopes. It is thought that the cells emerging from such selections enter the circulation and pass through the lymphatic system or are retained in various tissues in the body. They are said to be **naive,** because they are not completely differentiated, and are not armed to produce their ultimate immune effector response (e.g., antibody production or cell killing).

Figure 15.16 The antigen receptors on the surface of B and T cells. (A) B cells have about 100,000 molecules of a single membrane-bound receptor antibody per cell. Every receptor antibody on a given B cell has identical bivalent specificity for one antigen epitope. (B and C) T lymphocytes have about 100,000 T-cell receptors (Tcr) per cell, each with identical specificity. (B) T cells bearing the surface membrane protein CD4 always recognize peptide antigens bound to MHC class II proteins and generally function as T-helper (Th) cells. (C) T cells bearing the surface membrane protein CD8 always recognize peptide antigens bound to MHC class I proteins and generally function as cytotoxic T (Tc) cells.

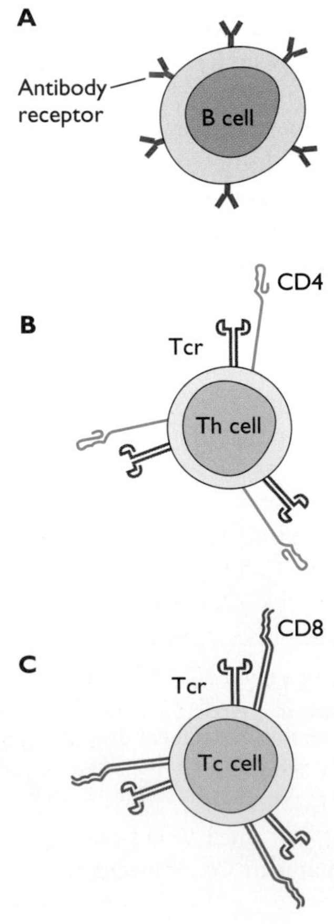

However, they are able to react to foreign (infected self) signals with remarkable diversity and specificity. Upon such binding, they rapidly acquire their specific immune effector activity.

The initial encounter with any foreign epitope, whether in lymphoid tissues or elsewhere in the body, involves only a few cells. For example, the frequency of lymphocytes that recognize infected cells on first exposure is as few as 1 in 10,000 to 1 in 100,000. So few cells certainly are not sufficient for protection against an infection that is spreading in the host. What makes the adaptive response so powerful is that the initial encounter can be amplified substantially during the ensuing 1 or 2 weeks. The number of virus-specific lymphocytes increases more than 1,000-fold because the original encounter stimulates previously uncommitted precursor lymphocytes to differentiate, divide, and produce antibodies or cytokines. The antigen-stimulated cell is said to be **activated.** It then becomes committed and undergoes numerous rounds of cell division such that each daughter cell has the same specific immune reactivity as the original parent.

B Cells

B cells are produced in the bone marrow. As they mature, each synthesizes an antigen receptor, which is a membrane-bound antibody (Fig. 15.13 and 15.16). When an antigen binds specifically to the membrane-bound antibody, a signal transduction cascade is initiated. As a consequence, new gene products are made and the cell begins to divide rapidly. The daughter cells produced by each division differentiate into memory B cells and effector cells called plasma cells. As their name implies, memory B cells, or their clonal progeny, are long-lived and continue to produce the parental membrane-bound antibody receptor. In contrast, plasma cells live for only a few days and no longer make membrane-bound antibody, but instead synthesize the same antibody in secreted form. A single plasma cell can secrete more than 2,000 antibody molecules per second.

T Cells

T-cell precursors are also produced in the bone marrow, but in contrast to a B-cell precursor, a T-cell precursor must migrate to the thymus gland to mature. The maturation process comprises two types of selection: positive selection for T cells that can bind appropriate surface molecules (see below) and negative selection that efficiently kills T cells that recognize cells that display self-peptides on their surfaces. As a result, only 1 to 2% of all immature T cells entering the thymus emerge as mature cells able to defend the host against viral infections. These mature but naive T cells are now able to respond to nonself antigens in lymphoid tissues. When they encounter such antigens presented to them by mature dendritic cells, they radically change their complement of cell surface proteins. Some differentiate into effector cells that can kill target cells when they leave the lymph node (CTLs), some differentiate into helper cells that can stimulate B cells to make antibody (Th cells), and others become memory cells that retain the capacity to differentiate into effector cells when they reencounter the stimulating antigen.

As noted above, T cells synthesize a membrane-bound receptor, the T-cell receptor, which is capable of binding antigens. The T-cell receptor is a disulfide-linked heterodimer composed of either alpha and beta (α/β complex) or gamma and delta (γ/δ complex) protein chains. The peptide-binding site of the T-cell receptor and the epitope-binding site of the B-cell receptor are very similar structures, formed by the folding of three regions in the amino-terminal domains of the proteins that participate in antigen recognition (the so-called hypervariable regions). However, unlike the B-cell receptor, which can recognize the epitope as part of an intact folded protein, the T-cell receptor can recognize **only** a peptide fragment produced by proteolysis. Furthermore, for recognition by the T-cell receptor, the peptide must be bound to MHC cell surface proteins (see below). When the T-cell receptor engages an MHC molecule carrying the appropriate antigenic peptide, a signal transduction cascade that leads to new gene expression is initiated. As a result, the stimulated T cell is capable of differentiating to form memory and various effector T cells.

Th Cells and CTLs

In general, lymphocytes can be distinguished by the presence on their surfaces of specific proteins called **cluster of differentiation (CD) markers** (e.g., CD3, CD4, CD8). The presence of these proteins can be detected with antibodies raised against them in heterologous organisms; they are often referred to as "CD antigens." Over 247 individual CD markers are known, and they are invaluable in identifying lymphocytes of a particular lineage or differentiation stage. Two well-known subpopulations of T cells are defined by the presence of either the CD4 or the CD8 surface proteins, which are coreceptors for MHC class II and MHC class I, respectively (Fig. 15.17). **$CD4^+$ T cells** are generally T-helper (Th) cells capable of interacting with B cells and antigen-presenting cells that have MHC class II proteins on their surfaces. After such interactions, $CD4^+$ Th cells mature into Th1 or Th2 cells (see below). Th cells synthesize cytokines and growth factors that stimulate (help) the specific classes of lymphocytes with which they interact. **$CD8^+$ T cells** mature into CTLs that can interact with almost all cells of the body expressing the more ubiquitous MHC class I proteins. CTLs recognize foreign peptides complexed with MHC class I proteins and, when productively engaged, actively destroy the cell presenting

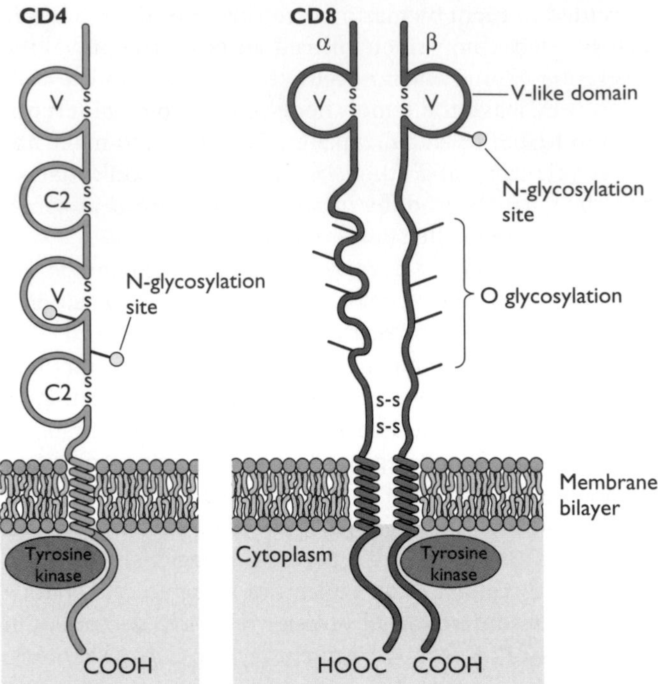

Figure 15.17 Simplified representations of CD4 and CD8 coreceptor molecules. These two molecules associate with the T-cell receptor on the surface of T cells. The CD4 molecule is a glycosylated type 1 membrane protein and exists as a monomer in membranes of T cells. It has four characteristic immunoglobulin-like domains labeled here as V and C2. The V domains are similar to the variable domain of immunoglobulin in the tertiary structure. The first two domains fold into a rigid rodlike structure linked to the other two domains by a flexible arm and are thought to be the domains for binding MHC class II proteins. The cytoplasmic domain forms a complex with specific tyrosine kinases, endowing CD4 with signal transduction properties. The CD8 molecule is a type I membrane protein with both N and O glycosylation. It is a heterodimer of an α chain and a β chain associated by disulfide bonds that interacts with MHC class I proteins. The two polypeptides are quite similar in sequence, each having an immunoglobulinlike V domain thought to exist in an extended conformation. Tyrosine kinases also associate with the CD8 cytoplasmic domain and participate in signal transduction events.

the peptides. Mature CTLs play important roles in eliminating virus-infected cells from the body by cell lysis and by production of IFN-γ and Tnf-α.

Th1 and Th2 Cells

When immature Th cells engage mature dendritic cells in lymphoid tissue, cytokines and receptor ligand interactions stimulate the naive T cell to differentiate into one of two Th cell types called Th1 and Th2 (Fig. 15 .15). These two cell types can be distinguished by the cytokines they produce (Table 15.2). Th1 cells are important for controlling most, but not all, viral infections. They promote the cell-mediated proinflammatory response by causing maturation of immature CTLs. They accomplish this, in part, by producing IL-2 and IFN-γ. In addition, Th1 cells provide stimulating cytokines to the antigen-presenting dendritic cell so that it can communicate with naive $CD8^+$ T cells. We know that if IL-12 is present at the time of antigen recognition, immature Th cells will differentiate into Th1 cells. IL-12 also stimulates NK and Th1 cells to secrete IFN-γ, a cytokine important in increasing the activity of inflammatory cells such as macrophages.

In the presence of IL-4, immature Th cells differentiate into Th2 cells that stimulate the antibody response rather than the proinflammatory response. The initial source of IL-4 may be the NK T cells discussed earlier. Th2 cells promote the antibody response by inducing maturation of immature B cells and resting macrophages. They also reduce the inflammatory response by producing IL-4, IL-6, and IL-10, but not IL-2 or IFN-γ. Th2 cells are more active after invasion by extracellular bacteria or multicellular parasites. Nevertheless, the Th2 response is critical for controlling infections that produce large quantities of virus particles in the blood.

While the Th1 response promotes CTL function, it is important to understand that antibodies also are produced by B cells as a result of the Th1 response. However, these are particular antibodies (isotypes). For example, one predominant Th1-stimulated antibody isotype is IgG2a, which actively promotes the complement cascade and binds to Fc receptors on macrophages to initiate antibody-dependent cell-mediated cytotoxicity (see below). Therefore, this antibody isotype is also part of a cytotoxic defense. In contrast, when stimulated by Th2 cells, B cells produce the IgG1 (mice), IgG4 (humans), and IgE isotypes, which bind neither complement nor macrophage Fc receptors.

In general, Th1 and Th2 responses have a yin-yang relationship; as one increases, the other decreases (Fig. 15.18). While IFN-γ turns up the Th1 response, it also inhibits the synthesis of IL-4 and IL-5 by Th2 cells, effectively dampening the latter response. On the other hand, production of Th2 cytokines is an important mechanism to shut off the proinflammatory and potentially dangerous Th1 response.

Viral infection induces the production of proinflammatory Th1 cytokines such as IL-12. In contrast, bacterial infection induces synthesis of cytokines, such as IL-4, that promote the Th2 response. What controls this discriminatory process? One idea is that mature dendritic cells automatically produce proinflammatory cytokines as their default pathway and will always activate a Th1 response unless appropriate Th2 signals are provided. A similar idea gaining credence is that when dendritic cells detect CpG sequences or dsRNA sequences via their Toll-like receptors, Th1 cytokine genes are transcribed via Nf-κb mediated pathways.

We know that many viral proteins modulate the Th1/Th2 balance in interesting ways. For example, infec-

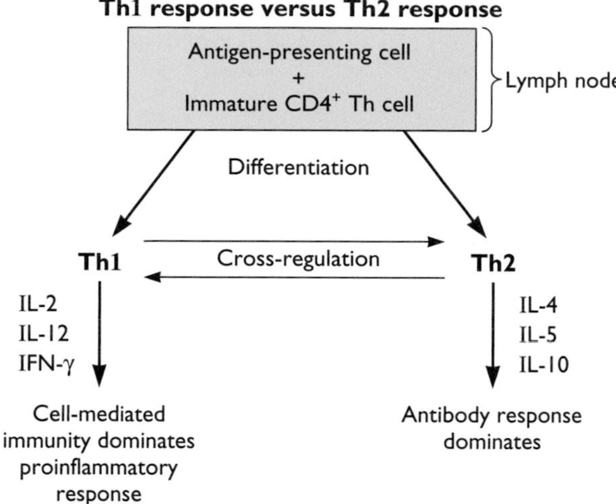

Immune cross-regulation by cytokines

Th1 response		Th2 response	
Enhance	Suppress	Enhance	Suppress
IL-2 IL-12 IFN-γ	IL-4 IL-10	IL-4 IL-5 IL-6 IL-10	IFN-γ IL-12

Figure 15.18 Th cells: the Th1 versus the Th2 response. Immature CD4+ Th cells differentiate into two general subtypes called Th1 and Th2, defined functionally according to the cytokines they secrete. Th1 cells produce cytokines that promote the inflammatory response and activity of cytotoxic T cells, and Th2 cells synthesize cytokines that stimulate the antibody response. The cytokines made by one class of T-helper cell tend to suppress expression of those of the other class.

tion of B cells by Epstein-Barr virus and equine herpesvirus type 2 should stimulate an active Th1 response. However, both viral genomes encode proteins homologous to IL-10, a regulatory cytokine produced by Th2 cells that represses the Th1 response. Viral IL-10 foils the Th1 antiviral defense that would kill infected B cells, while promoting differentiation into memory B cells that are important for long-term survival of the viral genome.

For most viral infections, a given Th response represents a spectrum of some Th1 and some Th2 cells, and consequently a mixture of cytokines, depending on the infectious agent and the nature of the infection. Establishment of the proper repertoire of Th cells therefore is an important early event in host defense; an inappropriate response has far-reaching consequences. For example, synthesis of the Th2 cytokine IL-4 by an attenuated mousepox virus recombinant resulted in lethal, uncontained spread of virus in an immune animal. How the fundamental Th1/Th2 decision is made is under much scrutiny in laboratories around the world. A practical reason for such interest is that the design of potent and effective vaccines depends on stimulating the appropriate spectrum of response.

Antigen Presentation and Activation of Immune Cells

Immature T cells engage the professional antigen-presenting cells by binding to MHC class II proteins (Fig. 15.9). This encounter results in differentiation of T cells into mature effector cells, the **CTLs.** Mature CTLs enter the circulation and are able to distinguish infected cells from uninfected cells by specific T-cell receptor interactions with MHC class I proteins (Box 15.13). MHC class I proteins are found on the surfaces of nearly all nucleated cells. Cells of the placenta,

BOX 15.13

Viruses provide Nobel Prize-winning insight: MHC restriction

In 1974 at the John Curtin School of Medical Research in Canberra, Australia, Rolf Zinkernagel and Peter Doherty performed a classic experiment that provided insight into how CTLs can recognize virus-infected cells. Initially, they teamed up to determine the mechanism of the lethal brain destruction observed when mice are infected with lymphocytic choriomeningitis virus, an arenavirus that does not directly kill the cells it infects. They anticipated that the brain damage was due to CTLs responding to replication of the noncytopathic virus in the brain.

When they infected mice of a selected MHC type with lymphocytic choriomeningitis virus and then isolated T cells capable of recognizing virus-infected cells, these T cells lysed virus-infected target cells in vitro only when the target cells and the T cells were of identical MHC haplotype. Uninfected target cells were not lysed even when they shared identical MHC alleles. This requirement for MHC matching was called **MHC restriction.**

Their Nobel Prize-winning insight was that a CTL must recognize two determinants present on a virus-infected cell: one specific for the virus (a short peptide derived from the virus) and one specific for the MHC of the host. CTLs recognize foreign antigens only when the T-cell receptor engages the foreign peptides bound to MHC class I proteins present on the surface of target cells.

Zinkernagel, R. M., and P. C. Doherty. 1974. Restriction of in vitro T-cell mediated cytotoxicity in lymphocytic choriomeningitis within a syngeneic or semiallogenetic system. *Nature* **248:**701–702.

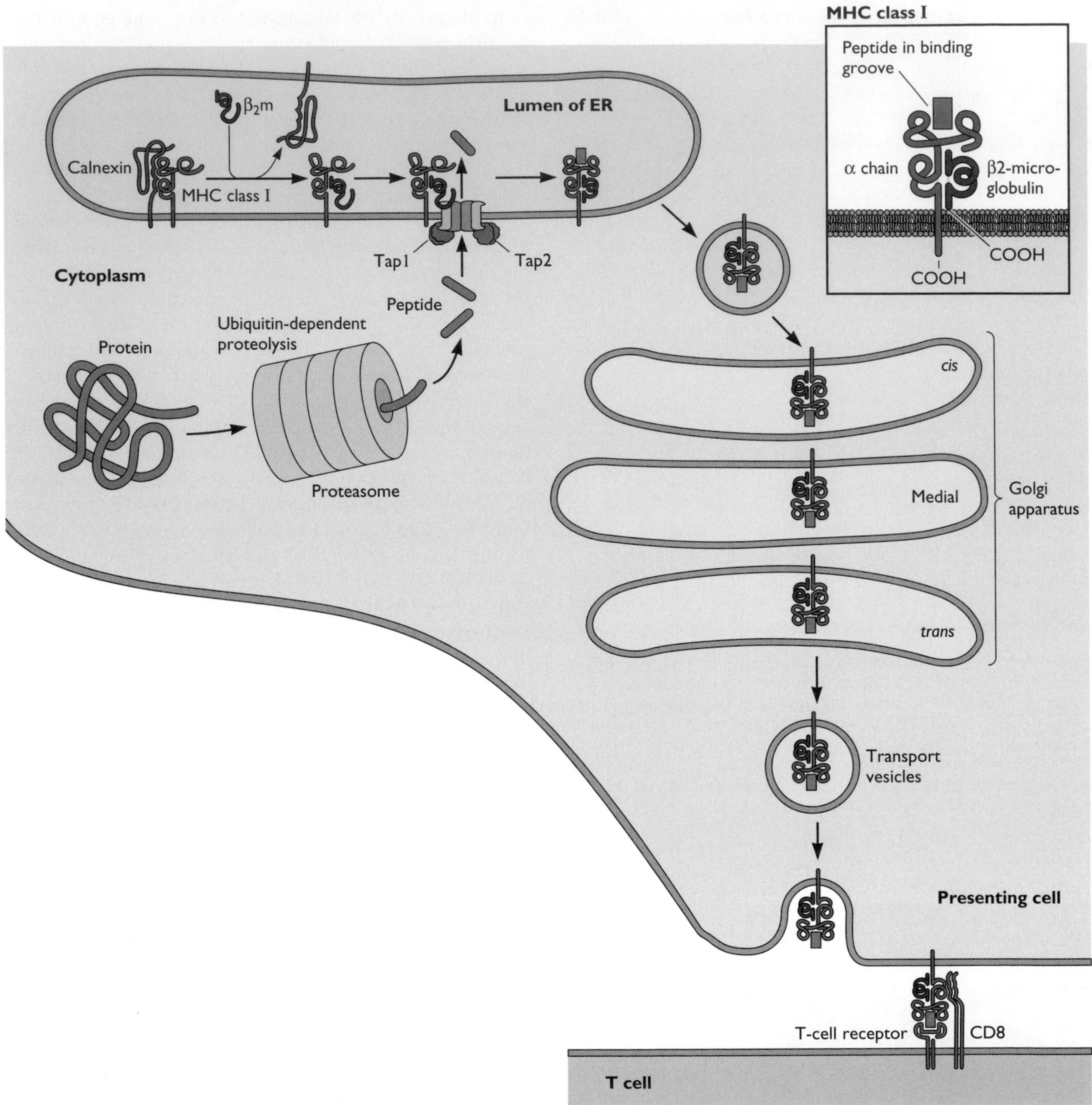

Figure 15.19 Endogenous antigen processing: the pathway for MHC class I peptide presentation. Intracellular proteins of both host cell and virus are degraded in the cytoplasm in a ubiquitin-dependent process. Proteins are marked for destruction by polyubiquitination. These proteins are then taken up and degraded by the proteasome. The resulting peptides, nine or more amino acids long, are transported into the ER lumen by the Tap1-Tap2 heterodimeric transporter in a reaction requiring ATP. Once in the ER lumen, the peptides associate with newly synthesized MHC class I molecules that bind weakly to the Tap complex. Assembly of the α chain and β_2-microglobulin of the MHC class I molecule is facilitated by the ER chaperone calnexin, but formation of the final native structure requires peptide loading. The MHC class I complex loaded with peptide is released from the ER to be transported via the Golgi compartments to the cell surface, where it is available for interaction with the T-cell receptor of a cytotoxic T cell carrying the CD8 coreceptor.

(continued on next page)

sperm cells at certain stages of differentiation, and neurons have been said to be devoid of cell surface MHC class I proteins. Recent evidence suggests that neurons do produce the MHC proteins to different extents during the life of an animal. The MHC class I complex comprises two protein subunits called the α chain (often called the heavy chain) and β_2-microglobulin (the light chain). Lymphocytes possess the highest concentration of MHC class I protein, with about 5×10^5 molecules per cell. In contrast, fibroblasts, muscle cells, and liver hepatocytes carry much lower quantities, sometimes 100 or fewer per cell. There are three MHC class I loci in humans (*A, B,* and *C*) and two in mice (*K* and *D*). Because there are many allelic forms of these genes in outbred populations, MHC class I genes are said to be **polymorphic**. For example, at the human MHC class I locus *hla-B,* more than 149 alleles with pairwise differences ranging from 1 to 49 amino acids have been identified. When cells bind cytokines such as IFN and IFN-γ, transcription of MHC class I α chains, β_2-microglobulin, and the linked genes for processing peptides (proteasome and peptide transporter genes) is markedly increased.

Synthesis of MHC class II proteins occurs primarily in the professional antigen-presenting cells (dendritic cells, macrophages, and B cells). Other cell types, including fibroblasts, pancreatic β cells, endothelial cells, and astrocytes, can produce MHC class II molecules, but only on exposure to IFN-γ. Human genomes contain three MHC class II loci (*DR, DP,* and *DQ*), and mouse genomes have two (*IA* and *IE*). MHC class II molecules are dimers comprising α and β subunits. The human *DR* cluster contains an extra β-chain gene, the protein product of which can pair with the *DR* α subunit. The three sets of human genes give rise to four types of MHC class II molecules. As with MHC class I, there are many alleles of MHC class II genes.

The defining property of both MHC class I and class II proteins is that they specialize in binding short peptides produced inside the cell and presenting them for recognition by lymphocytes with T-cell receptors on their surfaces. Both classes of MHC protein have a peptide-binding cleft that is sufficiently flexible to enable the binding of many peptides (Fig. 15.19 and 15.20), but not all possible peptides are bound. The ability of MHC molecules to bind and display peptides on the cell surface varies from individual to individual due to the many MHC alleles. As a result, a diversity of MHC alleles is needed to mount and sustain a strong immune defense. Indeed, inbred populations lose MHC diversity and have less robust immune defense.

In addition to MHC class I and class II proteins, mammals produce an additional family of MHC proteins called CD1. Humans carry five CD1 genes (*CD1a* through *CD1e*), and mice carry two (*mCD1.1* and *mCD1.2,* both related to human *CDd*). Like conventional MHC proteins, CD1 proteins are antigen-presenting molecules. However, they present not peptides, but rather lipids and glycolipids. CD1 proteins play important roles in immune regulation by interacting with a specialized subclass of $CD4^+$ T cells called NK1 T cells. When NK1 T cells engage CD1, they secrete cytokines, such as IL-4, that are known to control the differentiation of T lymphocytes. It is thought that CD1 lipid/glycolipid antigen presentation to the NK1 subset of $CD4^+$ T cells is an important part of the transition from innate to adaptive immune response. Many microbial invaders with distinctive membranes can be recognized by this lipid antigen-presenting system. Nevertheless, the contribution of the CD1 proteins to antiviral defense remains to be established.

T Cells Recognize Infected Cells by Using the MHC Class I System

The immune system must not only destroy virus-infected cells but also ignore uninfected cells. As viruses can infect many cell types, the recognition/destruction system certainly operates on all cells with high fidelity. As discussed below, recognition and selective killing are accomplished by CTLs that constantly patrol the body and scan the surface of essentially every cell. Virus-infected cells are identified, in part, because they display small viral peptides as well as cell peptides complexed to MHC class I proteins on their surfaces. The viral and cellular peptides are produced by the **endogenous pathway of antigen presentation** (Fig. 15.19).

In uninfected and infected cells, a fraction of most newly synthesized proteins is broken down in a controlled manner by the proteasome. The targeted protein is marked for destruction by the covalent attachment of multiple copies of a small protein called ubiquitin and, following ATP-dependent unfolding, the protein is degraded in the inner chamber of the proteasome. The peptide products are released and transported into the endoplasmic reticulum (ER) by a specific peptide transporter system and then assembled with MHC class I proteins. This interaction allows MHC class I molecules to fold into a conformation that can be transported via the secretory pathway to the cell surface. Hence, the MHC class I pathway displays an "inside-out" picture of the cell to the T cell.

Figure 15.19 ***(continued)*** (Inset) The MHC class I molecule is a heterodimer of the membrane-spanning type I glycoprotein α chain (43 kDa) and β_2-microglobulin (12 kDa) that does not span the membrane. The α chain folds into three domains, 1, 2, and 3. Domains 2 and 3 fold together to form the groove where peptide binds, and domain 1 folds into an immunoglobulinlike structure. Adapted from D. Male et al., *Advanced Immunology,* 3rd ed. (Mosby, St. Louis, Mo., 1996).

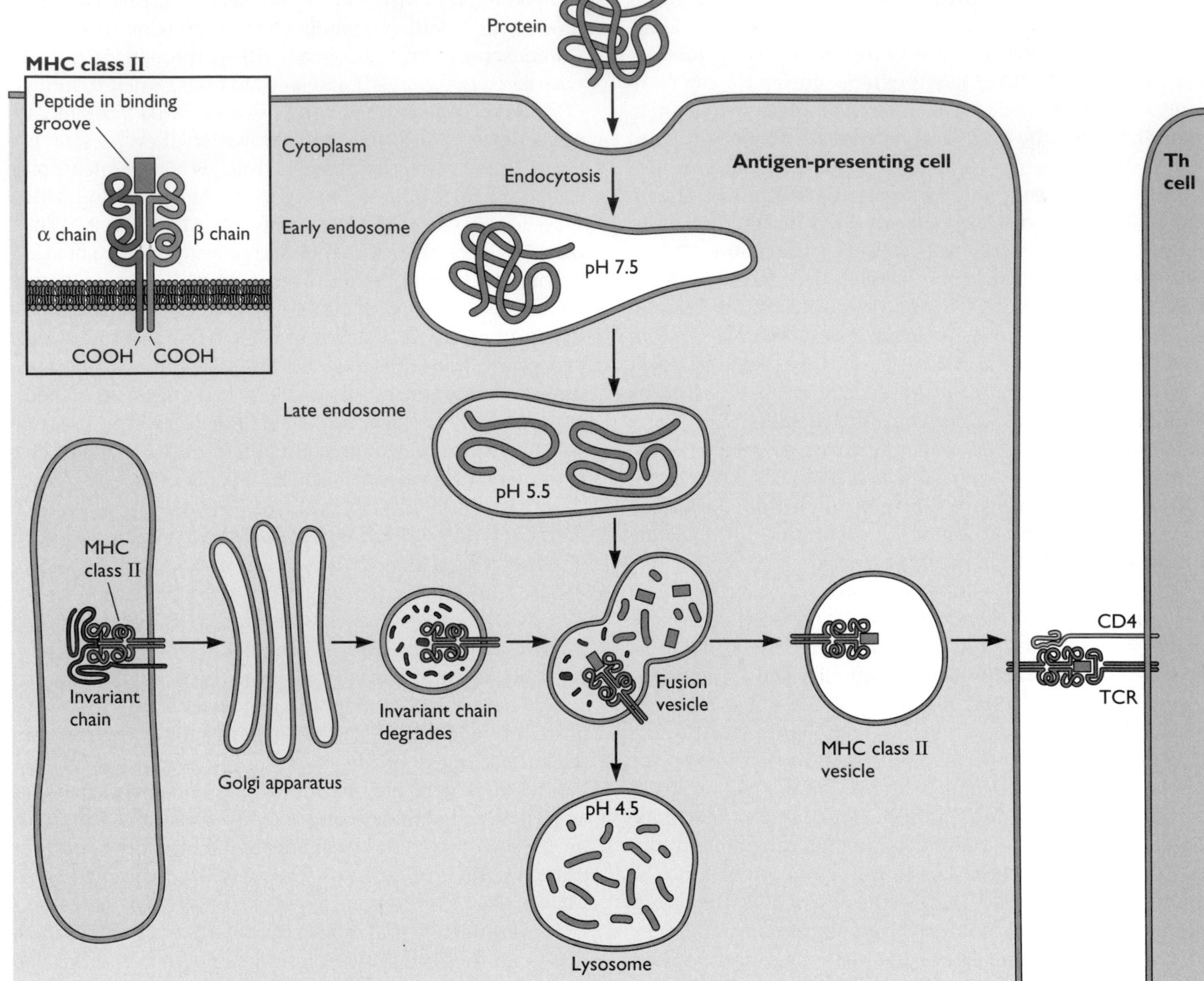

Figure 15.20 Exogenous antigen processing in the antigen-presenting cell; the pathway for MHC class II peptide presentation. (Left) Peptides in the ER lumen of the antigen-presenting cell are prevented from binding to the MHC class II peptide groove by association of a protein called the invariant chain with MHC class II molecules. The complex is transported through the Golgi compartments to a post-Golgi vesicle, where the invariant chain is removed by proteolysis. This reaction activates MHC class II molecules to accept peptides. The peptides are derived not from endogenous proteins but from extracellular proteins that enter the antigen-presenting cell. In some antigen-presenting cells, the protein enters by endocytosis (top). B cells capture antigens by the surface antibodies. In both cases, the antigens are internalized to early endosomes with neutral luminal pH. Endocytotic vesicles traveling to the lysosome via this pathway are characterized by a decrease in pH as they "mature" into late endosomes. The lower pH activates vesicle proteases that degrade the exogenous protein into peptides. Internalized endosomes with their peptides fuse at some point with the vesicles containing activated MHC class II. The newly formed peptide-MHC class II complex then becomes competent for transport to the cell surface, where it is available for interaction with the T-cell receptor (Tcr) of a Th cell carrying the CD4 coreceptor. (Inset) The MHC class II molecule is a heterodimer of the membrane-spanning type I α-chain (34 kDa) and β-chain (29 kDa) glycoproteins. Each chain folds into two domains, 1 and 2, and together the α and β chains fold into a structure similar to that of MHC class I. The two amino-terminal domains from α and β chains form the groove in which peptide binds. Unlike the closed MHC class I peptide groove, the MHC class II peptide-binding groove is open at both ends. The second domain of each chain folds into an immunoglobulinlike structure. Adapted from D. Male et al., *Advanced Immunology,* 3rd ed. (Mosby, St. Louis, Mo., 1996).

Patrolling CTLs move over the surface of potential target cells, engaging MHC class I peptide complexes by their T-cell receptors. Normal cells (self) are ignored, but when a viral peptide-MHC class I complex is bound by the T-cell receptor, a series of reactions is triggered such that the CTL becomes activated and kills the infected cell (see below). Surprisingly, the T-cell receptor has a low affinity, 1 μM or less, for its peptide-MHC class I ligand. How high-fidelity recognition is obtained from low-affinity binding has been a subject of intense research.

T Cells Recognize Professional Antigen-Presenting Cells by Using the MHC Class II System

Both antibody and CTL responses are controlled precisely by the cytokines produced by Th cells. Such precision is achieved by specific antigen recognition by the MHC class II proteins. A Th cell is activated only when the peptide antigen is presented on the surfaces of the professional antigen-presenting cells, such as dendritic cells. As dendritic cells mature, MHC class II glycoproteins loaded with peptides produced from their stores of endocytosed antigens appear on their surfaces. The mature antigen-presenting cells also carry high surface concentrations of costimulatory T-cell adhesion molecules that bind receptors on Th cells in the lymphoid tissue.

The process by which viral proteins are taken up from the outside of the cell and digested, and by which resulting peptides are loaded onto MHC class II molecules, is called **exogenous antigen presentation** (Fig. 15.20). In this case, the viral proteins are not produced inside the cell, as it is not infected, and their digestion takes place in endosomes rather than the proteasome. Furthermore, the peptides and MHC class II molecules are brought together by vesicular fusion. As with MHC class I, the complex is then transported to the surface of the antigen-presenting cell, where it is available to interact with appropriate T cells in the lymph node. Interaction of T-cell receptors on the naive $CD4^+$ Th cell with the MHC class II-peptide complex induces a series of membrane changes in the Th cell, leading to its activation and differentiation (Fig. 15.13). Full activation requires the interactions of other surface proteins and costimulatory molecules (Fig. 15.21).

A

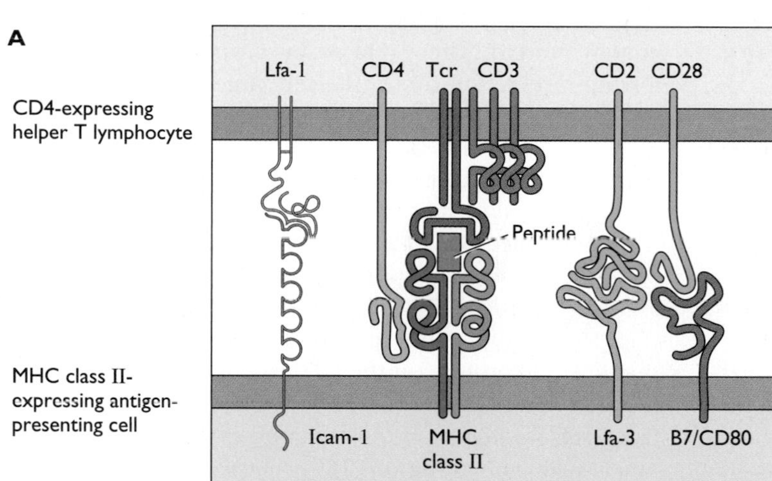

B

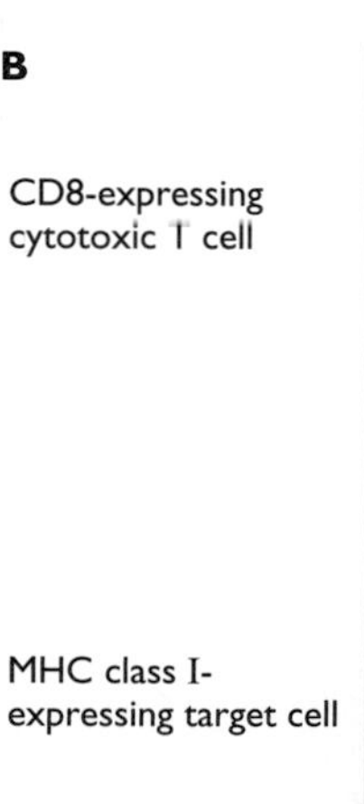

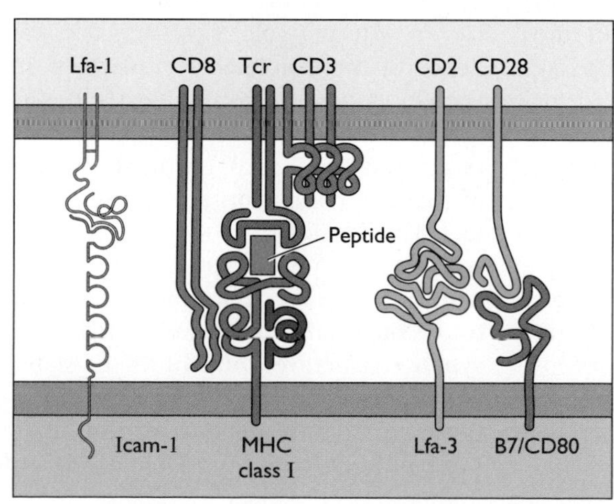

Figure 15.21 T-cell surface molecules and ligands. The interactions of these receptors and ligands are important for antigen recognition and initiation of signal transduction and other T-cell responses. (A) Interaction of a Th cell producing the CD4 coreceptor with an antigen-presenting cell. The antigen-presenting cell exhibits an MHC class II-peptide complex in addition to Icam-1, Lfa-3, and CD80 (B7) membrane proteins. These complexes all are capable of binding cognate receptors on the Th cell as illustrated. (B) Interaction of a CTL producing the CD8 coreceptor with its target cell. The target cell exhibits an MHC class I-peptide complex in addition to Icam-1, Lfa-3, and CD80 (B7) membrane proteins. These complexes all can be recognized and bound by receptors on the CTL as illustrated.

Th cells activated in this fashion produce IL-2, as well as a high-affinity receptor for this cytokine. The secreted IL-2 binds to the new receptors to induce autostimulation and proliferation of the Th cell. Such clonal expansion of specific Th1 or Th2 cells then promotes the activation of CTLs and B lymphocytes, which characterize the cell-mediated and antibody responses, respectively (Fig. 15.13).

While MHC class II proteins are definitive components of the professional antigen-presenting cell, some MHC class I molecules also can be loaded with peptides via the exogenous route in dendritic cells. This pathway may be a significant mechanism for activating the adaptive immune response. Virus-infected cells, no doubt, will harbor gene products that can confound or modulate the pathway.

The Cell-Mediated Adaptive Response

In general, the cell-mediated response facilitates recovery from a viral infection, because it eliminates virus-infected cells without damage to uninfected cells (Fig. 15.14). While the Th2-promoted antibody response is important for some infections in which virus particles spread in the blood, antibody alone is often unable to contain and clear an infection. Indeed, antibodies have little or no effect in natural infections that spread by cell-to-cell contact (e.g., neurotropic viruses such as alphaherpesviruses), or by viruses that infect circulating immune cells (e.g., lentiviruses and paramyxoviruses). These infections can be stopped only by CTL-produced antiviral cytokines and overt killing of infected cells.

CTLs

CTLs are superbly equipped to kill virus-infected cells, and once they complete one killing, they can detach and kill again. These lethal effector cells have a multistep maturation pathway that fully arms them for killing. They are not completely differentiated when they leave the thymus: they can recognize nonself antigens by their T-cell receptor, but they cannot lyse target cells. When the naive CD8$^+$ T cells encounter antigen-presenting cells and Th1 cells in the lymph node, they differentiate into CTLs capable of lysing target cells, and enter the circulatory system. However, for realization of their full killer potential, the cells require at least two additional reactions. These are the interaction of their T-cell receptor with foreign antigens presented by MHC class I molecules, and the binding of additional surface proteins (the coreceptors) on the potential CTL to their ligands on the infected cell. Signaling from the T-cell receptor when it engages the peptide antigen-MHC complex also requires clustering (aggregation) of a number of T-cell receptors in a particular structure and reorganization of the T-cell cytoskeleton. Only after these reactions have taken place does the CTL become capable of lysing an infected cell.

CTLs kill by two mechanisms: transfer of cytoplasmic granules from the CTL to the target cell, and induction of apoptosis. These killing systems are formed during the differentiation process. The maturing CTL fills with cytoplasmic granules that contain macromolecules required for target cell lysis, such as **perforin,** a membrane pore-forming protein, and **granzymes,** members of a family of serine proteases. Granules are released by CTLs in a directed fashion when in direct membrane contact with the target cell, and are taken up by that cell via receptor-mediated endocytosis. Perforin, as its name implies, makes holes in the endosomal membrane, allowing the release of granzymes that induce apoptosis of the infected cell. CTL killing by the perforin pathway is rapid, occurring in minutes after contact and recognition. Activated CTLs also can induce apoptotic cell death via binding of the Fas ligand on their surfaces to the Fas receptor (CD95) on target cells. Fas pathway killing is much slower than perforin-mediated killing. Many activated CTLs also secrete IFN-γ, which, as we have seen, is a potent inducer of both the antiviral state in neighboring cells and synthesis of MHC class I and II proteins. Activated CTLs also secrete powerful cytokines, such as IL-16, and chemokines such as Rantes (regulated on activation, normal T-cell expressed and secreted) protein. Their release by virus-specific CTLs following recognition of an infected target cell may assist in coordination of the antiviral response.

The primary interaction of the MHC-peptide complex with the T-cell receptor is often called **signal 1.** This interaction is not sufficient to activate the T cell. A second signal, often called **signal 2,** produced by binding of accessory or costimulatory molecules on the CTL to appropriate ligands on the infected cell, is required for efficient activation (Fig. 15.21). One required second interaction for signal 2 is that of CD28 protein on the CTL with the CD80 protein on the target cell. Examples of other interactions include those that facilitate adhesion of T cell and target cell to one another, such as binding of CD2 or leukocyte function antigen 1 (Lfa-1) on the CTL with Lfa-3 or Icam-1, respectively, on the infected cell. Viral infection can affect the abundance of these accessory molecules on cell surfaces and therefore alter CTL recognition and effector function.

Recent studies indicate that these proteins mediating target and T-cell recognition show an unexpected degree of spatial organization at the site of T-cell–target cell contact. This focal collection of membrane proteins and their respective binding partners has been called the "**immunological synapse,**" because of its functional analogy to the site of informational transfer between neurons. The synapse structure contributes to stabilizing signal transduction by the T-cell receptor for the prolonged periods of time required for gene activation. In addition, membrane proteins in the structure engage the underlying cytoskeleton and polarize the secretion apparatus so that a high local concentration of effector molecules is attained at the site of contact. Low numbers of peptide ligands complexed to MHC class I molecules apparently can stimulate a T cell because they serially engage a large total number of T-cell receptors on the opposing cell surface in the immunological synapse. Unengaged T-cell receptors subsequently entering this zone have an increased likelihood of binding specific ligand and signaling.

Typically, for a cytopathic virus infection, CTL activity appears within 3 to 5 days after infection, peaks in about a week, and declines thereafter. The magnitude of the CTL response depends on such variables as titer of infecting virions, route of infection, and age of the host. The critical contribution of CTLs to antiviral defense is demonstrated by **adoptive transfer** experiments in which virus-specific CTLs from an infected animal can be shown to confer protection to nonimmunized recipients (see also Box 15.13). However, CTLs can also cause direct harm by large-scale cell killing. Such immunopathology often follows infection by noncytopathic viruses, whereby cells can be infected yet still function. For example, the liver damage caused by hepatitis viruses is actually due to CTL killing of persistently infected liver cells. Other examples of immunopathology are discussed later in this chapter.

Control of CTL Proliferation

With new assays described in Box 15.14, several investigators have found a massive CTL precursor expansion after acute primary infections by viruses such as lymphocytic choriomeningitis virus (LCMV) and Epstein Barr virus. For example, more than 50% of CTLs from the spleen of an LCMV-infected mouse were specific for a single LCMV peptide. The response reached a maximum 8 days after infection, but up to 10% of specific T cells were still detectable after a year. Such results are in contrast to those for hepatitis B or human immunodeficiency virus infection: less than 1% of the CTLs from spleens of infected patients are specific for a single viral peptide.

These findings raise a number of questions. Why is there such a large primary expansion of particular CTL precursors for some infections and not others? Is expansion necessary or simply an overreaction? How is this rapid expansion regulated? Obviously, this response is not typical of all primary viral infections. What is the relationship of T-cell function (e.g., cell killing) to peptide specificity? Do antigen-specific cells reflect the antiviral potential of the expanded population? For now, these new assays have shed unexpected light and raised important questions concerning CTL responses to viral infection.

As we will discuss in Chapter 16, viral proteins can blunt the deadly CTL response with far-reaching effects ranging from rapid death of the host to long-lived, persistent infections. Many such proteins confound CTL recognition by disguising or reducing antigen presentation by MHC class I molecules (see Fig. 16.5). In the case of human immunodeficiency virus type 1, the viral genome encodes three proteins that interfere with CTL action: Nef and Tat induce Fas ligand production and subsequent Fas-mediated apoptosis of CTLs, while Env engages the CXCr4 chemokine receptor triggering death of the CTL. Human cytomegalovirus-infected cells contain at least six viral proteins that interfere with the MHC class I pathway and also evoke apoptosis of virus-specific CTLs by increasing synthesis of Fas ligand (see Table 16.3).

Noncytolytic Control of Infection by T Cells

It is widely assumed that complete clearance of intracellular viruses by the adaptive immune system depends on the destruction of infected cells by CTLs. However, the production of cytokines, such as IFN-γ and Tnf-α, by CTLs can lead to **purging** of viruses from infected cells without cell lysis. Such a mechanism requires that the infected cell retain the ability to activate antiviral pathways induced by binding of these cytokines to their receptors, and that viral replication be sensitive to the resulting antiviral response.

CTL killing of infected cells is likely to be an appropriate clearing response when relatively few cells in an organ or tissue are infected. However, as we have discussed, if vital organs such as brain or liver are infected, CTL killing can do more harm than good. In certain circumstances, such as infection of the liver by hepatitis B and C viruses, there are orders of magnitude more infected cells than there are virus-specific CTLs. When hepatitis B virus-specific CTLs are transferred to another animal (adoptive transfer), the IFN-γ and Tnf-α produced purge the infection from thousands of cells without their destruction. CTL production of powerful antiviral cytokines provides a simple explanation for how CTLs are able to control massive numbers of infected cells. The simple principle is that the antiviral effect of a single CTL is substantially greater than its cell lysis potential.

Noncytolytic clearing of infection by IFN-γ and Tnf-α produced by CTLs has now been documented for many viral infections, including those caused by arenaviruses, adenoviruses, coronaviruses, hepadnaviruses, and picornaviruses. Additional cytokines are likely to participate in viral clearance. For example, small antimicrobial peptides called **defensins** secreted by CTLs promote the noncytolytic clearing of human immunodeficiency virus.

$CD4^+$ T cells can also accomplish noncytolytic clearing of some infections with little involvement of CTLs. Such cases include infections by vaccinia virus, vesicular stomatitis virus, and Semliki Forest virus.

BOX 15.14

Measuring the antiviral cellular immune response

The Classic Assay: Limiting Dilution and Chromium Release

For the past 40 years, virologists have determined the presence of CTLs in blood, spleen, or lymphoid tissues of immune animals using the limiting dilution assay and chromium release from lysed target cells. This assay measures CD8$^+$ CTL precursors, or memory CTLs, based on two attributes: (i) the ability to replicate in the presence of the foreign protein or peptide they can recognize and (ii) the ability to lyse target cells.

Lymphocytes are obtained from an animal that has survived virus infection and are cultured for 1 to 2 weeks in the presence of whole inactivated virus, its proteins, or synthetic viral peptides. Under these conditions, virus-specific CTL precursors begin to replicate and divide (clonal expansion).

The expanded CTL population is then tested for its ability to destroy target cells that display viral peptides on MHC class I molecules. The MHC class I molecules on the target cell must be syngeneic with the CTLs (see Box 15.13) and the target cell must synthesize the viral peptide antigen. Accordingly, target cells can be infected with an innocuous virus vector (e.g., vaccinia virus, baculovirus, or adenovirus) encoding the desired viral protein. The target cells loaded with the viral antigen in question are then exposed to chromium 51, a radioactive isotope that binds to most intracellular proteins. After washing to remove external isotope, cultured CTLs are incubated with the target cells, and the release of chromium 51 into the supernatant is determined. Release of chromium 51 is used as an indicator that the CTLs have lysed the target cells.

Serial dilution before assay provides an estimate of the number of CD8$^+$ CTL precursors in the original cell suspension (providing the name of the assay: "limiting dilution"). This assay does provide a quantitative measure of cellular immunity, but it is time-consuming, technically demanding, and expensive.

Identifying and Counting Virus-Specific T Cells

The limiting dilution assay defines T cells by function, but not by their peptide specificity. Until recently, scientists have tried with little success to identify individual T cells based on their unusual antigen recognition properties. This failure has been attributed to the low affinity and fast "off" rates of the MHC-peptide complex and T-cell receptor. Without this information, it was difficult, if not impossible, to measure antigen-specific T-cell responses.

An important advance is the use of an **artificial MHC tetramer** as an antigen-specific, T-cell-staining reagent. The extracellular domains of MHC class I proteins are produced in *Escherichia coli*. These engineered MHC class I molecules have an unusual C-terminal 13-amino-acid sequence that enables them to be biotinylated. The truncated MHC class I proteins are folded in vitro with a synthetic peptide that will be recognized by a specific T cell. Biotinylated tetrameric complexes are purified and mixed with isolated T-cell pools from virus-infected animals. Individual T cells that bind the biotinylated MHC class I-peptide complex are detected by a variety of immunohistochemical techniques. Cell-sorting techniques can be used, and stained cells are viable.

Another assay for counting single T cells is the **enzyme-linked immunospot assay (ELISPot).** Fresh lymphoid cells are put into culture medium on plates that have been coated with antibody specific for the cytokine under study. These cells are stimulated with virus-specific peptides. Those CTLs that recognize the peptide secrete cytokines locally where they are bound by the antibody on the plate. After washing the plate, the foci of bound cytokine are stained and enumerated.

The **intracellular cytokine assay** is a relatively rapid method to count specific CTLs. Fresh lymphoid cells are treated with brefeldin A. This fungal metabolite blocks the secretory pathway and prevents the secretion of cytokines. The cells are then fixed with a mild cross-linking chemical that preserves protein, such as glutaraldehyde. Treated cells are permeabilized so that a specific antibody for a given cytokine can react with cytokines retained inside. Cells that react with the antibody can be quantified in a fluorescence-activated cell sorter. With appropriate software and calibration, the staining intensity corresponds to the level of cytokine expression, and the number of cells responding to a particular epitope and MHC class I molecule can be determined.

Measuring the Antiviral Antibody Response

Antibodies are the primary effector molecules of the humoral response. The methods to detect antibodies are wide and varied. However, a standard method in virology is the **neutralization assay.** Here, viral infectivity is determined in the presence and absence of antibody. Two varia-

BOX 15.14 *(continued)*

tions on this general theme include the **plaque reduction assay** and the **neutralization index.** In the plaque assay, a known number of plaque-forming units are exposed to serial dilutions of the antibody or serum in question. The highest dilution that will reduce the plaque count by 50% is taken as the plaque reduction titer of the serum or antibody. To compute a neutralization index, the titer of a virus stock is compared in the presence and absence of test antibody or serum. The index is calculated as the difference in viral titers. An obvious requirement for these assays is that the virus in question can be propagated in cultured cells, and that a measure of virus growth such as plaque formation is available.

Other important assays to determine binding of antibody to viral proteins are immunoprecipitation, Western blot, enzyme-linked immunosorbent assay, and hemagglutination inhibition. These assays are discussed in Chapter 2.

Rashes and Poxes: Examples of T-Cell-Mediated, Delayed-Type Hypersensitivity

Many infections, including those of measles virus, smallpox virus, and varicella-zoster virus, produce a characteristic rash or lesion over extensive areas of the body, even though the primary infection began at a distant mucosal surface. This phenomenon results when the primary infection escapes the local defenses and virions or infected cells spread in the circulation to initiate many foci of infected cells in the skin. Th1 cells and macrophages that were activated by the initial infection home in on these secondary sites and respond by aggressive synthesis of cytokines, including IL-2 and IFN-γ. Such cytokines then act locally to increase capillary permeability, a reaction partially responsible for a characteristic local reaction referred to as **delayed-type hypersensitivity**. This reaction is responsible for many virus-promoted rashes and lesions with fluid-filled vesicles.

The delayed-type hypersensitivity response typically is measured in a previously infected mouse by an ear-swelling assay. Viral antigen is injected intradermally on the dorsal pinna of a mouse ear, while the other ear is injected similarly with control antigen. Twenty-four hours later, the antigen-injected ear becomes inflamed and swells as a result of delayed-type hypersensitivity, but the control-injected ear does not. This assay can be used in the following adoptive transfer experiment to show that delayed-type hypersensitivity is due to T cells. When nonimmune animals were injected with purified, activated T cells from a virus-infected animal and subsequently given an intradermal inoculation of viral antigen, a lesion rich in macrophages and lymphocytes was seen 24 h later at the site of injection.

Antibody-Dependent Cell-Mediated Cytotoxicity: Specific Killing by Nonspecific Cells

The Th1 response produces a particular isotype of IgG that can bind to antibody receptors on macrophages and some NK cells. These receptors are specific for the carboxy-terminal, more conserved region of an antibody molecule, the Fc region (Fig. 15.22). If an antiviral antibody is bound

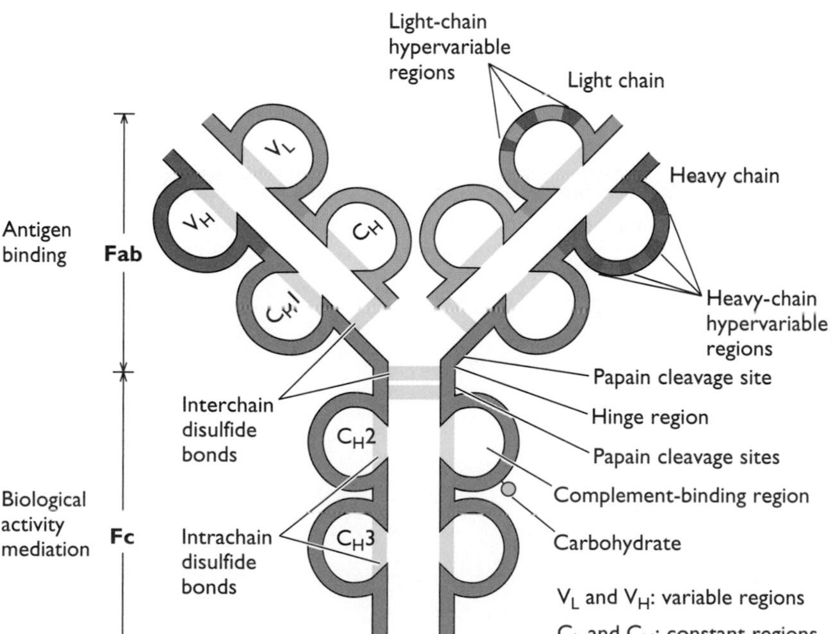

Figure 15.22 The structure of an antibody molecule. This is a schematic representation of an IgG molecule delineating the subunit and domain structures. The light and heavy chains are indicated in colors, as are the disulfide bonds that hold them together (yellow bars). The variable regions of the heavy (V_H) and light (V_L) chains, as well as the constant regions of the heavy (C_H) and light (C_L) chains, are indicated on the left part of the molecule. The hypervariable regions and invariable regions of the antigen-binding domain (Fab) are emphasized. The constant region (Fc) performs many important functions, including complement binding (activation of the classical pathway) and binding to Fc receptors found on macrophages and other cells. Clusters of papain protease cleavage sites are indicated, as this enzyme is used to define the Fab and Fc domains.

in this manner, the amino-terminal antigen-binding site is still free to bind viral antigen on the surface of the infected cell. In this way, the antiviral antibody can bring an NK cell or a macrophage to the infected cell for elimination. This process is called **antibody-dependent cell-mediated cytotoxicity** (often referred to as ADCC). The antibody provides the specificity for killing by a less discriminating NK cell. While this mechanism is well documented in cultured cells, its importance in controlling viral infections in animals is unknown.

The Antibody Response

Specific Antibodies Are Made by Activated B Cells Called Plasma Cells

When the B cell emerges from the bone marrow into the circulation and travels to lymph and lymphoid organs, it is unable to differentiate and synthesize antibody unless its surface antibody receptor is bound by the cognate antigen. The activating signal requires clustering of receptors complexed with antigen. Such receptor clustering activates a signal transduction cascade via Src family tyrosine kinases that associate with the cytoplasmic domains of the closely opposed receptors. B-cell coreceptors, such as CD19, CD21, and CD8, enhance signaling by recruiting tyrosine kinases to clustered antigen receptors and coreceptors. When antigen-binding and coreceptor interactions occur in the presence of appropriate cytokines made by Th2 cells and growth factors made by macrophages, the signal transduction cascade induces the immature B cell to produce a variety of new gene products and to differentiate into antibody-producing plasma cells and memory B cells (Fig. 15.14). Like dendritic cells, the B cell is an antigen-presenting cell that uses the MHC class II system and exogenous antigen processing (Fig. 15.20).

When the T-cell receptor of Th2 cells recognizes MHC class II-peptide complexes present on the B-cell surface, these Th2 cells produce a locally high concentration of stimulatory cytokines as well as CD40 ligand (a protein homologous to Tnf). The engagement of CD40 ligand with its receptor on the surface of B cells facilitates a local exchange of cytokines that further stimulates proliferation of the activated B cell and promotes its differentiation. Fully differentiated plasma cells produce prodigious amounts of specific antibodies: the rate of synthesis for IgG can be as high as 30 mg/kg of body weight/day.

Antibodies

Antibodies (immunoglobulins) have the common structural features illustrated in Fig. 15.22. Five classes of immunoglobulin—IgA, IgD, IgE, IgG, and IgM—are defined by their distinctive heavy chains—α, δ, ε, γ, and μ, respectively. Their properties are summarized in Table 15.13. IgG, IgA, and IgM are commonly produced after viral infection. During B-cell differentiation, "switching" of the constant region of heavy-chain genes occurs by somatic recombination and is regulated in part by specific cytokines. Consequently, after such maturation, each cell produces a specific type of antibody.

During the **primary antibody response,** which follows initial contact with antigen or viral infection, the production of antibodies follows a characteristic course. The IgM antibody appears first, followed by IgA on mucosal surfaces or IgG in the serum. The IgG antibody is the major antibody of the response and is remarkably stable, with a half-life of 7 to 21 days. Specific IgG molecules remain detectable for years, reflecting the presence of memory B cells. A subsequent challenge with the same antigen or viral infection promotes a rapid antibody response, the **secondary response** (Fig. 15.14 and Box 15.14).

Table 15.13 The five classes of immunoglobulins

Property	IgA	IgD	IgE	IgG	IgM
Function	Mucosal; secretory	Surface of B cell	Allergy; anaphylaxis; epithelial surfaces	Major systemic immunity; memory responses	Major systemic immunity; primary response; agglutination
Subclasses	2	1	1	4	1
Light chain	κ, λ	κ, λ	κ, λ	κ, λ	κ, λ
Heavy chain	α	δ	ε	γ	μ
Concentration in serum (mg/ml)	3.5	0.03	0.00005	13	1.5
Half-life (days)	6	2.8	2	25	5
Complement activation					
Classical	–	–	–	+	++
Alternative	–/+	–	+	–	–

Virus neutralization by antibodies. While the cell-mediated response clearly plays an important role in eliminating virus-infected cells, the antibody response is crucial for preventing many viral infections and may also contribute to resolution of infection (Box 15.14). Virions that infect mucosal surfaces are likely to encounter secretory IgA antibodies. Similarly, virions that spread in the blood are likely to encounter circulating IgG and IgM antibody molecules. In particular, antibodies are an important defense against infections caused by rabies virus, vesicular stomatitis virus, and enteroviruses. We also know that immunodeficient animals can be protected from some viral infections by injection with virus-specific antiserum or purified monoclonal antibodies. Additional confirmation that antibodies affect the outcome of an infection comes from studies on eastern and western equine encephalitis viruses (alphaviruses in the family *Togaviridae*). In these cases, the speed with which virus-specific antibodies appear after infection correlates well with the ability of patients to recover from encephalitis; those with low antibody titers are more likely to die. The best example of the importance of antibodies is the success of viral vaccines in preventing viral diseases.

The type of antibody produced can significantly influence the outcome of a viral infection. Poliovirus infection stimulates strong IgM and IgG responses in the blood, but it is mucosal IgA that is vital in defense. This isotype can neutralize poliovirus directly in the gut, the site of primary infection. The live attenuated Sabin poliovirus vaccine is effective because it elicits a strong mucosal IgA response.

Immunoglobulin A is synthesized by plasma cells that underlie the mucosal epithelium. This antibody is secreted from these cells as dimers of two conventional immunoglobulin subunits. The dimers then bind the polymeric immunoglobulin receptor on the basolateral surface of epithelial cells (Fig. 15.23). This complex is then internalized by endocytosis, and moved across the cell (**transcytosis**) to the apical surface. Protease cleavage of the receptor releases dimeric IgA into mucosal secretions, where it is free to interact with incoming virions.

It is widely assumed that the primary mechanism of antibody-mediated viral neutralization is via steric blocking of virion-receptor interaction (Fig. 15.24). While some antibodies do prevent virions from attaching to cell receptors, the vast majority of virus-specific antibodies are likely to interfere with the concerted structural changes that are required for entry. Antibodies also can promote aggregation of virions, and thereby reduce the effective concentration of infectious particles. Many enveloped viruses can be destroyed in a test tube when antiviral antibodies and serum complement disrupt membranes (the classical complement activation pathway). Nonneutralizing antibodies are also prevalent after infection; they bind specifically to virus particles, but do not interfere with infectivity. In some cases, such antibodies can even enhance infectivity. Antibody bound to virions is recognized by Fc receptors on macrophages, and the entire complex is brought into the cell by endocytosis (see Chapter 17).

Much of what we know about antibody neutralization comes from the isolation and characterization of "antibody escape" mutants or **monoclonal antibody-resistant mutants**. These mutants are selected after propagating virus in the presence of neutralizing antibody. The analysis of the mutant viruses allows a precise molecular definition, not only of antibody binding sites, but also of parts of viral proteins important for entry. Antigenic drift (see Chapters 16 and 20) is a consequence of selection and establishment of antibody escape mutants in viral populations.

IgA may also block viral replication inside infected mucosal epithelial cells (Fig. 15.23). Because IgA must pass through such a cell en route to secretion from the apical surface, it is available during transit for interaction with viral proteins produced within the cell. The antigen-binding domain of intracellular IgA lies in the lumen of the ER, the Golgi compartment, and any transport vesicles of the secretory pathway. It is therefore available to bind to the external domain of any type I viral membrane protein that has the cognate epitope of that IgA molecule. Such interactions have been demonstrated with Sendai virus and influenza virus during infection of cultured cells. In these experiments, antibodies colocalized with viral antigen only when the IgA could bind to the particular viral envelope protein. These studies suggest that clearing viral infection of mucosal surfaces need not be limited to the lymphoid cells of the adaptive immune system.

Antibodies can provoke other remarkable responses in virus-infected cells. For example, in a process analogous to CTL purging, antibodies that bind to the surface proteins of many enveloped viruses (e.g., alphaviruses, paramyxoviruses, arenaviruses) can clear these viruses from persistently infected cells. This process is noncytolytic and complement independent. In this case, antibodies act synergistically with IFN and other cytokines. Virus-specific antibodies bound to surfaces of infected cells can inhibit virion budding at the plasma membrane and can also reduce surface expression of viral membrane proteins by inducing endocytosis. Antiviral antibodies injected into the peritoneal cavities of mice can block the neurotropic transmission of reovirus, poliovirus, and herpesvirus to the brain, even when the virions are injected at sites far re-

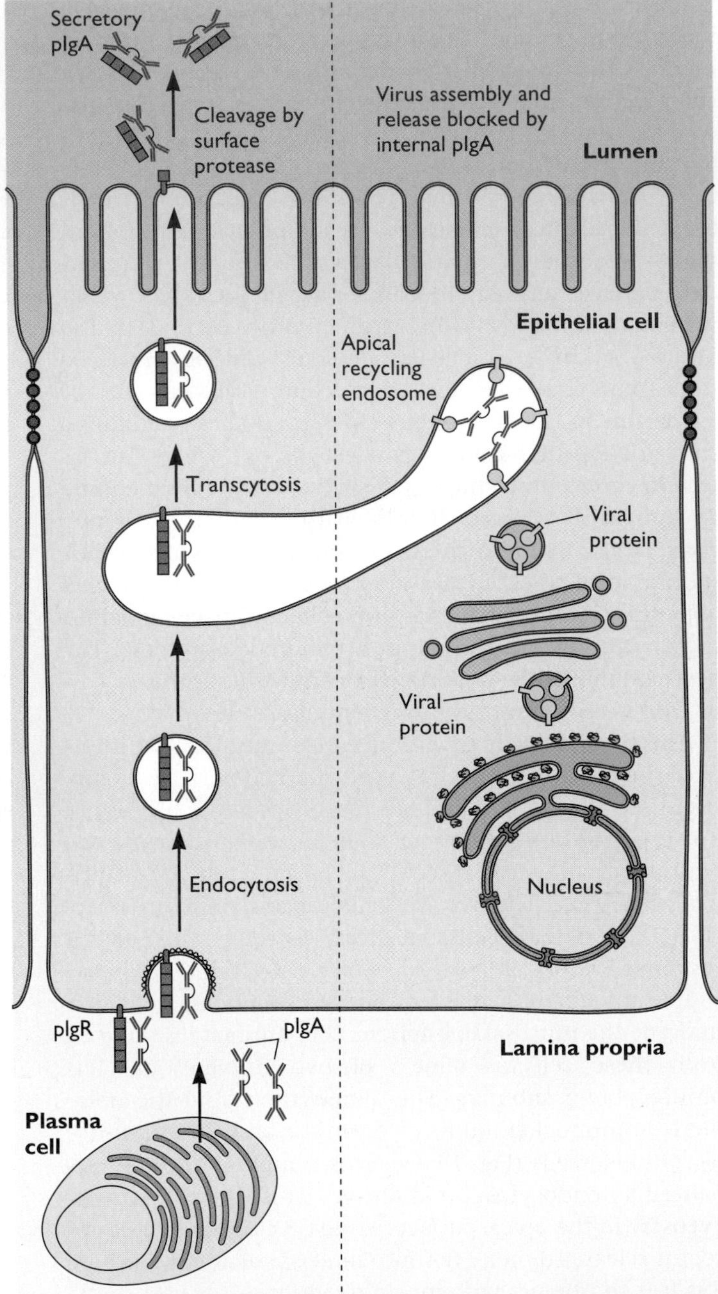

Figure 15.23 Secretory antibody, IgA, is critical for antiviral defense at mucosal surfaces. Antibody-producing B cells (plasma cells) in the lamina propria of a mucous membrane specifically secrete the IgA antibody (also called polymeric IgA [pIgA]). pIgA is a dimer, joined at its Fc ends. For IgA to be effective in defense, it must be moved to the surface of the epithelial cells that line body cavities. This process is called transcytosis and begins by the binding of pIgA to its receptor (pIgR) on the basolateral surface of an epithelial cell. The IgA-receptor complex is taken into the epithelial cell by endocytosis and then delivered to vesicles that transport the complex to the apical surface. Once on the apical surface, pIgA is released by cleavage of the receptor by a surface protease that releases IgA antibody into the lumen. A portion of the receptor remains bound to the dimeric IgA antibody (the so-called secretory component). The dimeric IgA antibody can bind antigens at many steps along this pathway. If antigen is bound before the IgA enters the epithelial cell, antigen-pIgA-pIgR complexes can be transported through the cell to the lumen. If virus is in the lumen, free IgA can bind and block infection at the apical surface. Interestingly, a virus particle infecting an epithelial cell potentially can be bound by internal IgA if virus components intersect with the IgA in the lumen of vesicles during transcytosis. This process is likely to occur for enveloped viruses, as their membrane proteins are processed in many of the same compartments as those mediating transcytosis. Adapted from M. E. Lamm, *Annu. Rev. Microbiol.* **51**:311–340, 1997, with permission.

moved from the peritoneum. The molecular bases for these intriguing effects remain elusive.

Injury Induced by Viral Infection

The clinical symptoms of viral disease in the host (e.g., fever, tissue damage, aches, pains, nausea) result primarily from the host response to infection. This response is initiated by cell injury caused by viral replication. Cell injury can result from the direct effects of viral replication on the cell, or from the consequences of the host's intrinsic, innate, and adaptive immune responses.

Direct Effects of Primary Infection

Infection of cultured cells may result in visible changes in the cells collectively called **cytopathic effect,** a topic of Chapter 2. Direct alteration of the cell can clearly account for some of the damage observed during infections in an animal host. For example, poliovirus induces cytopathic

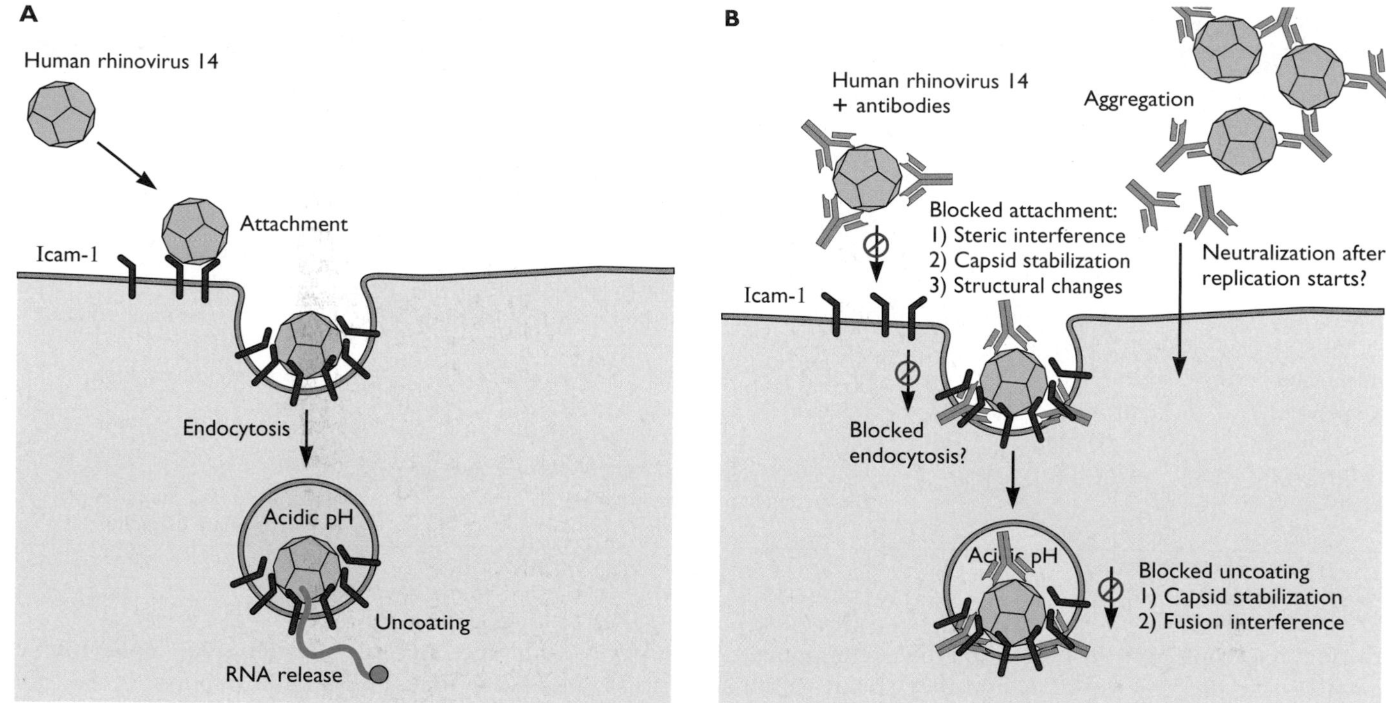

Figure 15.24 Human rhinovirus 14 infection and potential interactions with antibodies. (A) The normal route of infection. The virus attaches to the Icam-1 receptor and enters by endocytosis. As the internal pH of the endosome decreases, the particle uncoats and releases its RNA genome into the cytoplasm. (B) Possible mechanisms of neutralization of human rhinovirus 14 by antibodies. With well-characterized monoclonal antibodies, at least three modes of antibody neutralization have been proposed and are illustrated: (1) binding of antibody molecules to virus results in steric interference with virus-receptor binding; (2) antibodies bound to the particle fix the capsid in a stable conformation so that pH-dependent uncoating is not possible; (3) antibody molecules binding to the capsid can alter the capsid structure, affecting other steps in entry and uncoating. In addition, because all antibodies are divalent, they can aggregate virus particles (upper right), affecting concentration of virus particles and facilitating their destruction by phagocytes. Antibodies themselves may be taken up by endocytosis and interact with virus in the cell after infection starts. Adapted from T. J. Smith et al., *Semin. Virol.* **6**:233–242, 1995, with permission.

effects in cultured neuronal cells. The relevance of this finding may be that virus-induced killing of neurons in the central nervous system could account for the paralytic symptoms characteristic of poliomyelitis.

Because virus-induced cytopathic effects are clearly relevant to viral pathogenesis, much effort has been devoted to understanding how infection alters cell structure and metabolism. One major process that leads to visible cell damage is apoptosis. As we have discussed, many viral proteins both induce and block apoptosis, presumably to facilitate completion of the infectious cycle and the production of new progeny virions.

The consequences of viral infection may also include cessation of essential host processes such as translation, DNA and RNA synthesis, and vesicular transport (Table 15.14). A general outcome may be the increased permeability of cell membranes. Lysosomal contents would diffuse into the cytoplasm, resulting in autolytic digestion of the cell. Indeed, the 2B protein of some picornaviruses enhances membrane permeability and may cause such enzyme leakage from infected cells.

The host genome can be damaged directly by viral infection. For example, the retrovirus replication cycle requires insertion of a proviral DNA copy into random locations in the cell genome. Such insertion may affect the expression or integrity of a cellular gene, a process known as **insertional mutagenesis** (Chapters 7 and 18).

Some pathogenic effects are indirect: viral infection may not cause cell death, but may interfere with synthesis of a critical molecule needed for host survival. For example, when a strain of lymphocytic choriomeningitis virus is inoculated into newborn mice, the virus replicates in cells of the pituitary gland that produce growth hormone. The infection does not destroy these cells, but does lead to a

Table 15.14 Examples of virus-induced alteration of the host

Virus	Protein	Effect	Target
Poliovirus	$2A^{pro}$	Inhibition of 5′-end-dependent translation	eIF4G
Poliovirus	2B, 3A	Inhibition of ER-to-Golgi protein traffic	?
Poliovirus	2C, 2BC	Membranous vesicle accumulation	?
Poliovirus	?	Alteration of Map4 by unknown mechanism	Map4
Poliovirus	3C	Inhibition of transcription	Tbp, TfIIc complex
Sindbis virus	?	Increased plasma membrane permeability	Na,K ATPase?
Paramyxovirus	F protein	Fusion of cell membranes, syncytium formation	Plasma membrane
Adenovirus	E1B 55kDa, E4 34kDa	Block accumulation of host mRNA in cytoplasm	Cellular proteins involved in mRNA transport?
Adenovirus	?	Inhibition of 5′-end-dependent translation	eIF4E (dephosphorylation)
Herpesvirus	RNase encoded by the *vhs* gene	Disassembly of polysomes	Host mRNA
Herpes simplex virus	ICP27	Inhibition of transport and processing of host RNA	?
Many viruses	?	Depolymerization of cytoskeleton	Actin filaments, microtubules, intermediate filaments

twofold reduction in the synthesis of growth hormone. As a result, the mice fail to thrive, and they perish within a month. Genetic analysis establishes that two viral proteins are necessary. The viral membrane protein is required for entry into cells of the pituitary gland, while the viral nucleoprotein decreases the concentration of a cellular transcription protein specifically required for synthesis of growth hormone mRNA.

Immunopathology: Too Much of a Good Thing

Most symptoms and many diseases caused by viral infection are a consequence of the immune response (Table 15.15). Damage caused by the immune system is called **immunopathology,** and it may be the price paid by the host to eliminate a viral infection. For noncytolytic viruses it is likely that the immune response is the sole cause of disease. In fact, most viral infections with an immunopathological component are noncytolytic and persistent (see Chapter 16). Most immunopathology is induced by activated T cells, but there are examples of disease caused by antibodies or an excessive innate response (Box 15.15).

Table 15.15 Examples of virus-induced immunopathology

Proposed mechanism	Virus
$CD8^+$ T cell mediated	Coxsackievirus B
	Lymphocytic choriomeningitis virus
	Sin Nombre virus
	Human immunodeficiency virus type 1
	Hepatitis B virus
$CD4^+$ T cell mediated	
Th1	Theiler's virus
	Mouse coronavirus
	Semliki Forest virus
	Measles virus
	Visna virus
	Herpes simplex virus
Th2	Respiratory syncytial virus
Antibody mediated	Dengue virus
	Feline infectious peritonitis virus

Immunopathological Lesions

Lesions caused by CTLs. The best-characterized example of CTL-mediated immunopathology occurs during the infection of mice with lymphocytic choriomeningitis virus. Infection by this virus is not cytopathic and induces tissue damage only in immunocompetent animals. Experiments using adoptive transfer of T-cell subtypes, depletion of cells, and gene knockout and transgenic mice have clearly shown that tissue damage requires CTLs (Box 15.16). The mechanism by which these cells cause damage is not clear, but may be a result of CTL cytotoxicity. For example, knockout mice lacking perforin (the major cytolytic protein of CTLs) develop less severe disease after infection. CTLs may also release proteins that recruit inflammatory cells to the site of infection, which in turn elaborate proinflammatory cytokines.

Liver damage caused by hepatitis B virus also appears to depend on the action of CTLs. Production of the hepatitis B virus envelope proteins in transgenic mice has no effect on the animals. When the mice are injected with hepatitis B virus-specific CTLs, liver lesions develop that resemble those observed in acute human viral hepatitis. Cytotoxic lymphocytes first attach to hepatocytes and induce apop-

BOX 15.15

Defective viral vectors and lethal immunopathology

In September 1999, an 18-year-old man participated in a clinical trial to test the safety of a defective adenovirus designed as a gene delivery vector. It seemed like a routine procedure: normally, even replication-competent adenoviruses cause only mild respiratory disease. Most humans harbor adenoviruses as persistent colonizers of the respiratory tract and, indeed, produce antibodies against the virus. The young man was injected with a large dose of the viral vector. Four days after the injection, he died of multiple organ failure. What caused this devastating response to such an apparently benign virus?

Some relevant facts:

- Natural adenovirus infection never occurs by direct introduction of virions into the circulation. Most infections occur at mucosal surfaces with rather small numbers of infecting virions.
- Most humans have antibodies to the adenoviral vectors used for gene therapy.
- A large dose of virus was injected directly into his bloodstream.

One compelling idea is that most of the infecting defective virions were bound by antibody present in the young man's blood. As a consequence, the innate immune system, primarily complement proteins, responded to the antibody-virus complex resulting in massive activation of the complement cascade. The amplified complement cascade caused widespread inflammation in the vessel walls of the liver, lungs, and kidney, resulting in multiple organ failure.

Given the complexities of the immune response to infections, this idea is most certainly an oversimplification. Nevertheless, this fatal trial resulted in a large-scale reassessment of gene therapy methods and clinical protocols. Research is ongoing to find methods and tests to identify, reduce, or eliminate immunopathology so that the promise of gene therapy can be realized.

totic cell death. Next, cytokines released by these lymphocytes recruit neutrophils and monocytes, which cause even more extensive cell damage. Death from hepatitis can be prevented by the administration of antibody to IFN-γ or by depletion of macrophages. Cytotoxic lymphocytes are required for immunopathology, but tissue damage is largely mediated by nonspecific cytokines and cells recruited to the site of infection.

CTLs must have access to infected cells to promote damage. When hepatitis B virus proteins are produced in

BOX 15.16

Transgenic mouse models prove useful in defining the antiviral contribution of CTLs

The use of mice carrying mutations in genes that encode proteins required for CTL killing, B-cell function, and IFN responses has provided a wealth of information on the contributions of these gene products in immune defense. The only requirement is that the virus of interest be able to infect the transgenic mouse.

Genetic manipulation of the CTL response has been particularly informative. In the case of some noncytopathic viruses, perforin-mediated but not granzyme-mediated or Fas-mediated, killing is an important component of host CTL defense. When virus was not cleared in perforin-defective mice, the resulting persistent infection led to a lethal overproduction of cytokines, including Tnf-α and IFN-γ. In contrast, granzyme-deficient mice recover completely from viral infection. An important area of research is to understand how the various CTL killing proteins are coordinated to control other viral infections.

Transgenic models also demonstrate an unexpectedly limited contribution of the CTL killing functions in defense against certain cytopathic viruses. Vaccinia virus and vesicular stomatitis virus pathogenesis was not affected by lack of CTL killing systems, but rather recovery from primary infection depended largely on production of neutralizing IgG antibodies. In fact, cytokines secreted by $CD4^+$ Th cells and CTLs are required for early immune defense but not for recovery from infection by many different cytopathic viruses. One way to think about these observations is that recognition and lysis of infected cells by CTLs may be too slow to be effective as a primary defense against acute cytopathic infections: the rapid innate defenses and early cytokine responses of lymphocytes are the critical front-line systems.

Rall, G. F., D. M. P. Lawrence, and C. E. Patterson. 2000. The application of transgenic and knockout mouse technology for the study of viral pathogenesis. *Virology* **271:**220–226.

Yeung, R. S. M., J. Penninger, and T. W. Mak. 1994. T-cell development and function in gene knockout mice. *Curr. Opin. Immunol.* **6:**298–307.

other tissues, such as kidney or brain, injection of CTLs into the blood does not cause disease. However, disease will occur if the lymphocytes are inoculated directly into the infected organ. Accessibility of viral antigens to patrolling CTLs is an important determinant of immunopathology.

Myocarditis (inflammation of the heart muscle) caused by coxsackievirus B infection of mice also requires the presence of CTLs. In particular, perforin is a major determinant of myocarditis. Mice lacking the perforin gene develop a mild form of heart disease yet are still able to clear the infection. The disease is probably also exacerbated by the action of cytokines such as Tnf-α and IL-1β. Chemokines contribute to disease by controlling directional migration of lymphocytes into infected tissues. For example, mice lacking the chemokine macrophage inflammatory protein 1a do not develop myocarditis. Inflammation of heart muscle following infection is a result of the combined action of immune-mediated tissue damage and virus-induced cytopathology. The recently recognized acute and often fatal respiratory disease caused by hantaviruses is characterized by prominent infiltration of CTLs into the lung.

Lesions caused by CD4$^+$ T cells. CD4$^+$ T lymphocytes elaborate far more cytokines than do CTLs and recruit and activate many nonspecific effector cells. Such inflammatory reactions are usually called delayed-type hypersensitivity responses. Most of the recruited cells are neutrophils and mononuclear cells, which are protective and cause tissue damage. Immunopathology is the result of release of proteolytic enzymes, reactive free radicals such as peroxide and nitric oxide (see below), and cytokines such as Tnf-α. For noncytopathic persisting viruses, the CD4$^+$-mediated inflammatory reaction is largely immunopathological. Against cytopathic viruses, this response is protective, although there may be cases in which immunopathology occurs.

CD4$^+$ Th1 T cells. CD4$^+$ Th1 T cells are essential to provoke central nervous system disease of rodents infected with Theiler's murine encephalomyelitis virus (a picornavirus), mouse coronavirus, or Semliki Forest virus (a togavirus). These viruses persistently infect the central nervous system and cause inflammatory reactions leading to demyelination. In Theiler's virus-infected mice, CD4$^+$ Th1 cells are believed to cause inflammatory responses by the activation of monocytic phagocytes that mediate demyelination of neurons. It is not known how this process occurs, but perhaps the activated macrophages release superoxide and nitric oxide free radicals in addition to Tnf-α and IL-6. Tnf-β elaborated by Th1 cells may also destroy oligodendrocytes, which are the source of myelin. It is not clear whether similar mechanisms account for demyelination observed in humans after infection with many other viruses. That similar pathology is caused by so many different viral infections implies that an underlying immune-mediated mechanism is at work.

Herpes stromal keratitis, one of the most common causes of vision impairment in developed countries of the world, is an example of a cytolytic infection in which the pathology is mainly immune-mediated. In humans, herpes simplex virus infection of the eye induces lesions on the corneal epithelium, and repeated infections result in opacity and reduced vision. Studies of a mouse model for this disease have demonstrated the importance of CD4$^+$ Th1 T cells. Mice lacking T cells do not develop ocular disease following infection. Lymphocytes isolated from the corneas of mice with active disease are CD4$^+$ T cells that produce cytokines typical of the Th1 subset. Interestingly, viral replication occurs in the corneal epithelium, while inflammation is restricted to the stroma. Moreover, the expression of viral genes has ceased by the time that CD4$^+$ T cells attack the stroma. One interpretation of these findings is that herpes stromal keratitis represents an inflammatory reaction that results from activation of uninfected cells in the stroma by infection in the corneal epithelium (bystander cell activation).

CD4$^+$ Th2 cells. Respiratory syncytial virus disease may be mediated by CD4$^+$ Th2 cells. This noncytopathic virus is an important cause of lower respiratory tract disease in infants and the elderly. In immunosuppressed mice, lesions of the respiratory tract are minor, but become severe after transfer of viral antigen-specific T cells, particularly CD4$^+$ Th2 cells. How Th2 cells mediate tissue damage in the lung is not understood. Lesions in the respiratory tract contain many eosinophils, which may be responsible for pathology. One likely possibility is that products of CD4$^+$ Th2 T cells recruit these eosinophils.

The balance of Th1 and Th2 cells. An inappropriate Th1 or Th2 response can have pathogenic effects. Recent experiments indicate that it may be possible to alter this balance by vaccination. This concept is being exploited not only to produce safe vaccines, but also to facilitate analysis of immunopathogenic responses. For example, respiratory syncytial virus causes respiratory infections that can be prevented by vaccination in young children. Children who received a live virus vaccine were protected and suffered no ill effects. However, when children were vaccinated with a formalin-inactivated whole-virus vaccine, they were not protected from subsequent infection. Indeed, upon infection, they developed an atypically severe disease characterized by increased infiltration of eosinophils into the lungs. Such atypical augmented disease appears to be caused by CD4+ Th2 cells. Current vaccination efforts are based on preparations that induce only the Th1 response, which does not cause immunopathological disease.

Immunopathological lesions caused by B cells. Virus-antibody complexes accumulate to high concentrations when extensive viral replication occurs at sites inaccessible to the immune system or continues in the presence of an inadequate immune response. Such complexes are not efficiently cleared by the reticuloendothelial system and continue to circulate in the blood. They become deposited in the smallest capillaries and cause lesions that are exacerbated when the complement system is activated (Fig. 15.25). Deposition of these immune complexes in blood vessels, kidney, and brain may result in vasculitis, glomerulonephritis, and mental confusion, respectively. Such an immune complex disease was first described in mice infected with lymphocytic choriomeningitis virus. Although immune complexes have been demonstrated in humans, viral antigens have been found in the complexes only in hepatitis B virus infections.

Antibodies may also cause an enhancement of viral infection. This mechanism probably accounts for the pathogenesis of dengue hemorrhagic fever. This disease is transmitted by mosquitoes and is endemic in the Caribbean, Central and South America, Africa, and Southeast Asia, where billions of people are at risk. The primary infection is usually asymptomatic, but may result in an acute febrile illness with severe headache, back and limb pain, and a rash. It is normally self-limiting, and patients recover in 7 to 10 days. There are four viral serotypes, and antibodies to any one serotype do not protect against infection by another. After infection by another serotype of dengue virus, nonprotective antibodies bind virus particles and facilitate uptake into normally nonsusceptible peripheral blood monocytes carrying Fc receptors. Consequently, the infected monocytes produce proinflammatory cytokines, which in turn stimulate T cells to produce more cytokines. This vicious cycle results in high concentrations of cytokines and other chemical mediators that are believed to trigger the plasma leakage and hemorrhage characteristic of dengue hemorrhagic fever. There may be so much internal bleeding that the often fatal dengue shock syndrome results. Dengue hemorrhagic fever occurs in approximately 1 in 14,000 primary infections. However, after infection with a dengue virus of another serotype, the incidence of hemorrhagic fever increases dramatically to 1 in 90, and the shock syndrome is seen in as many as 1 in 50.

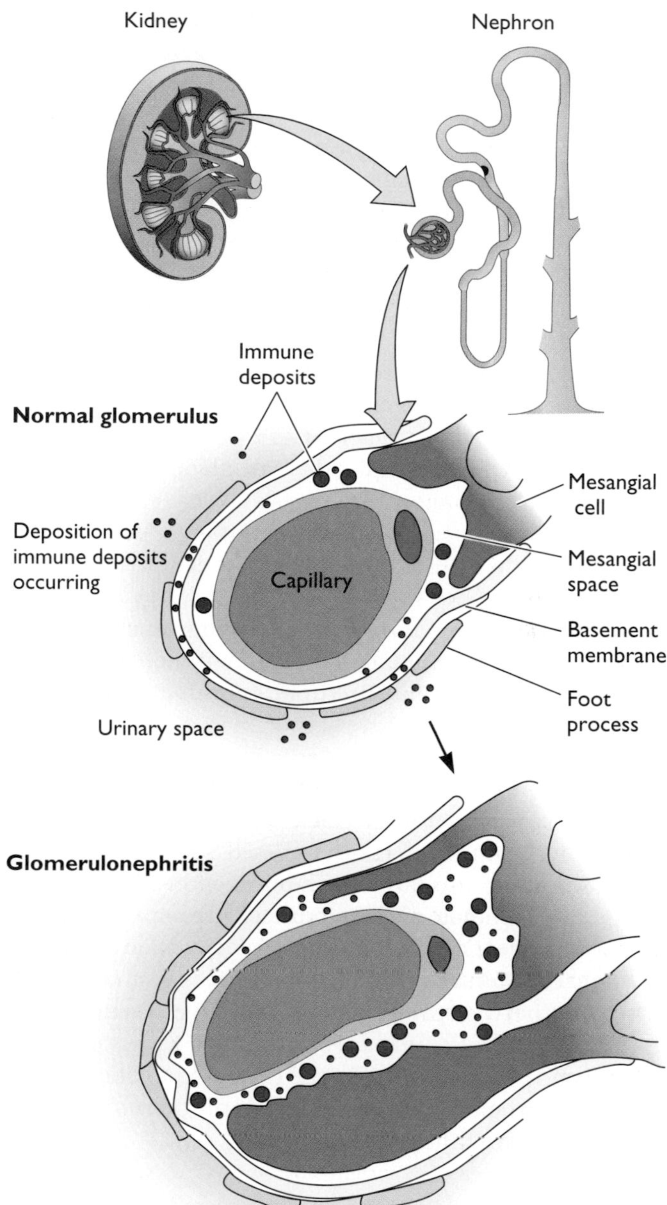

Figure 15.25 Deposition of immune complexes in the kidney leading to glomerulonephritis. (Top) Normal glomerulus and its location in the nephron and kidney. Red dots are immune complexes. The smaller complexes pass to the urine, and the larger ones are retained at the basement membrane. (Bottom) Glomerulonephritis. Complexes have been deposited in the mesangial space and around the endothelial cell. The function of the mesangial cell is to remove complexes from the kidney. In glomerulonephritis, the mesangial cells enlarge into the subepithelial space. This results in constriction of the glomerular capillary, and foot processes of the endothelial cells fuse. The basement membrane becomes leaky, filtering is blocked, and glomerular function becomes impaired, resulting in failure to produce urine. Adapted from C. A. Mims et al., *Mims' Pathogenesis of Infectious Disease* (Academic Press, Orlando, Fla., 1995), with permission.

Virus-induced immunosuppression. Modulation of the immune defenses by viral gene products can range from a mild and rather specific attenuation, to a marked global inhibition of the response (Table 15.16). The molecular mechanisms of immune modulation were discussed above. Immunosuppression by viral infection first was observed over 100 years ago because patients were unable to respond to a skin test for tuberculosis during and after measles infection. However, progress in understanding the phenomenon was slow until the human immunodeficiency virus epidemic and the well-known, devastating immunodeficiency syndrome stimulated unprecedented efforts. Immunosuppression by common human viruses such as rubella and measles viruses is also a serious public health concern. For example, the vast majority of the million children who die from measles virus each year in Third World countries succumb to other infections that arise during this transient immunosuppression.

Systemic inflammatory response syndrome. An important tenet of immune defense is that when viral replication exceeds a certain threshold, immune defenses are mobilized and amplified, resulting in a global response. How this threshold is determined is not known. Normally, the global response is well tolerated, and it quickly contains the infection. However, if the threshold is breached too rapidly, or if the immune response is not proportional to the infection, the large-scale production and systemic release of inflammatory cytokines and stress mediators can overwhelm and kill an infected host. Such a disastrous outcome often results if the host is very young, malnourished, or otherwise compromised. This type of pathogenesis is called the **systemic inflammatory response syndrome** and is sometimes referred to as a "cytokine storm." Similar syndromes (e.g., toxic shock and toxic sepsis) are triggered by other microbial pathogens.

Autoimmune Diseases

Human autoimmune disease is caused by an immune response directed against host tissues (often described as "breaking immune tolerance"). Some evidence suggests that viral infections can contribute to autoimmune disease. In laboratory animals, viral infection can trigger potent autoimmune responses, but it has been difficult to find evidence for any role for viral infections in human autoimmune disease. How can viral infections promote autoimmune disease in animals? One model posits that replication at an anatomically sequestered site leads to the release and subsequent recognition of self antigens. A modified version of this hypothesis proposes that infection leads to exposure of cellular self antigens normally hidden from the immune system. Cytokines, or even virus-antibody complexes that modulate the activity of proteases in antigen-presenting cells, might cause the unmasking of self antigens. Another possibility is that cytokines produced during infection may stimulate surface expression of host membrane proteins that are recognized by host defenses.

A popular hypothesis for virus-induced autoimmunity states that viral and host proteins share antigenic determinants. This idea is often called **molecular mimicry.** Infection leads to the elaboration of immune responses against such shared determinants, resulting in an autoreactive response. Although many peptide sequences are shared among viral and host proteins, direct evidence for this popular theory has been difficult to obtain. One reason is the long lag period between events that trigger human autoimmune diseases and the onset of clinical symptoms. To circumvent this problem, transgenic mouse models in which the products of foreign genes are expressed as self antigens have been established. This approach permits a precise study of how the autoreactive immune response is initiated and what cell types and soluble mediators are involved (Box 15.17).

Superantigens "Short-Circuit" the Immune System

Some viral proteins are extremely powerful T-cell mitogens known as **superantigens.** These proteins bind to MHC class II molecules on antigen-presenting cells and interact with the Vβ chain of the T-cell receptor. As approximately 2 to 20% of **all T cells** produce the particular Vβ chain that binds the superantigen, superantigens short-cir-

Table 15.16 Some mechanisms of immunosuppression by selected viruses

Mechanism	Representative virus	Degree of immunosuppression	Specific for infecting virus only
Infection of immune cells	Human immunodeficiency virus, canine distemper virus, lymphocytic choriomeningitis virus	Marked	No
Tolerance after infection of fetus	Rubella virus	Moderate	Yes
Disruption of cytokine defense pathways	Measles virus	Moderate	No
Virokines and viroceptors	Poxviruses, herpesviruses	Mild	Yes

BOX 15.17

Viral infections promote or protect against autoimmune disease

Promotion

- Transgenic mice that synthesize proteins of lymphocytic choriomeningitis virus in β cells of the pancreas or oligodendrocytes develop insulin-dependent diabetes mellitus or central nervous system demyelinating disease, respectively. The product of the viral transgene is present in the mouse throughout development, and therefore is a self antigen.
- Infection with virus stimulates an immune response in which the self antigen is recognized, leading to disease.
- The cross-reactive immune response by CTLs leads to inflammation, insulitis, and diabetes. Such virus-specific lymphocytes are present in the peripheries of uninfected mice, as determined from in vitro studies. The mechanisms that maintain unresponsiveness to self antigens and prevent autoimmunity can be elucidated with such transgenic models. Whether tolerance to self antigens is circumvented depends on the number of autoreactive T cells and their affinity for peptides presented by MHC class II proteins, the number of memory cells induced by infection, and the cytokine milieu.

Hypothesis: If an individual becomes infected in the first few months after birth with an organ-tropic virus that establishes a persistent infection, tolerance to the virus is established. Later in life, an infection with the same virus leads to an immune-mediated attack on virus-infected cells in the organ.

Protection

- Both the nonobese diabetic mouse strain and a particular strain of rat spontaneously develop type 1 diabetes.
- Infection of either animal with lymphocytic choriomeningitis virus reduces the incidence of diabetes.
- IL-12 is known to exacerbate autoimmune disease in these experimental models by promoting expansion and activation of autoreactive $CD4^+$ T cells. Moreover, production of IFN results in reduction in IL-12 production by monocytes and dendritic cells.

Hypothesis: Lymphocytic choriomeningitis virus infection induces synthesis of IFN that, in turn, protects against autoimmune disease by decreasing IL-12 expression.

Fujinami, R. S. 2001. Viruses and autoimmune disease—two sides of the same coin? *Trends Microbiol.* **9**:377–381.

von Herrath, M. G., and M. B. A. Oldstone. 1996. Virus-induced autoimmune disease. *Curr. Opin. Immunol.* **8**:878–885.

cuit the interaction of MHC class II-peptide complex and the T cell. Rather than activation of a small, specific subset of T cells (only 0.001 to 0.01% of T cells usually respond to a given antigen), **all subsets** of T cells producing the Vβ chain to which the superantigen binds are activated and proliferate. Superantigens clearly interfere with a coordinated immune response and divert the host's defenses.

All known superantigens are microbial products and many are produced after infection by viruses. The best-understood viral superantigen is encoded in the U3 region of the mouse mammary tumor virus long terminal repeat. This retrovirus is transmitted efficiently from mother to offspring via milk. However, the virus replicates poorly in most somatic tissues, so the question is, how do virions get into the milk? When B cells in the neonatal small intestine epithelium are infected, the viral superantigen is produced and recognized by T cells carrying the appropriate T-cell receptor Vβ chain. Consequently, extraordinarily large numbers of T cells are activated, and these cells produce growth factors and other molecules that stimulate the infected B cells. The infected B cells then carry the virus to the mammary gland. Infection of mice with mutants harboring a deletion of the superantigen gene results in limited viral replication and minimal transmission to offspring via milk. One function of the superantigen is therefore to stimulate the immune system to facilitate spread of infection from the gut to the mammary gland.

Mechanisms Mediated by Free Radicals

Two free radicals, superoxide ($O_2^{\cdot}$) and nitric oxide (NO), are produced during the inflammatory response and may play important roles in virus-induced pathology. Superoxide is produced by the enzyme xanthine oxidase present in phagocytes (Fig. 15.26). The production of $O_2^{\cdot}$ is significantly increased in the lungs of mice infected with influenza virus or cytomegalovirus. Inhibition of xanthine oxidase protects mice from virus-induced lethality. However, $O_2^{\cdot}$ is not toxic for some cells and viruses, and its effects might be the result of formation of peroxynitrite ($ONOO^{\cdot}$) through interaction with nitric oxide.

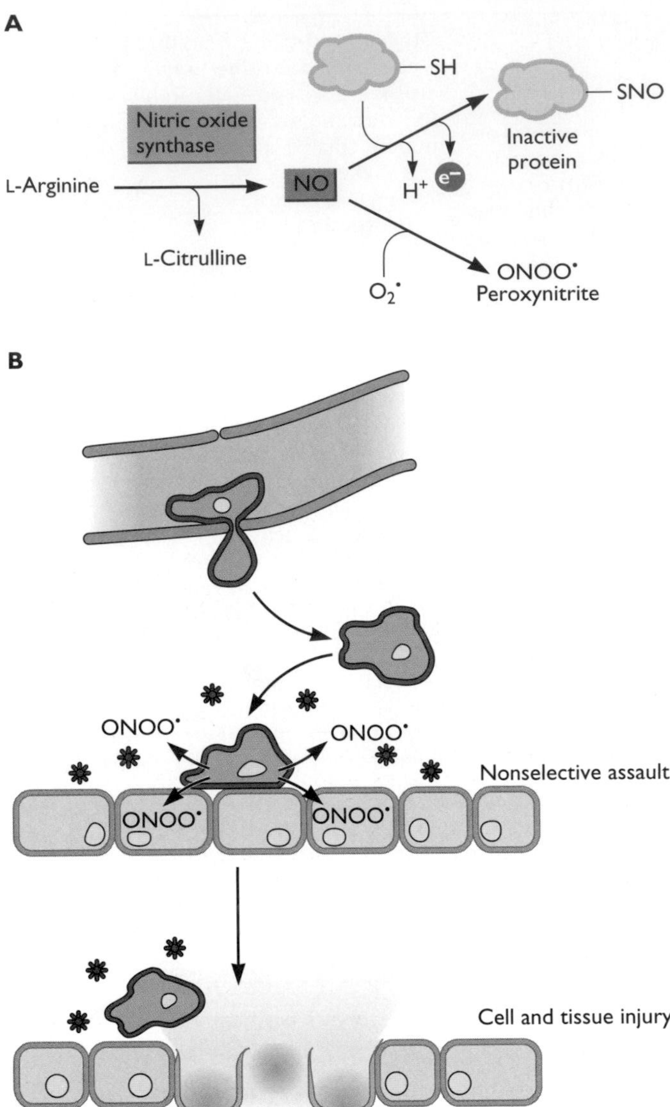

Figure 15.26 Consequences of nitric oxide production. (A) Formation of NO and ONOO˙. Nitric oxide, which is produced from L-arginine by nitric oxide synthase, reacts with proteins, inactivating them, or with $O_2^{\cdot}$, forming peroxynitrite, ONOO˙. Adapted from C. S. Reiss and T. Komatsu, *J. Virol.* **72:**4547–4551, 1998, with permission. (B) Biological effects of NO/$O_2^{\cdot}$/ONOO˙. Phagocytes move into the area of active virus replication and produce NO/$O_2^{\cdot}$/ONOO˙, which destroys cells, causing tissue injury but perhaps limiting virus yields. Adapted from T. Akaike et al., *Proc. Soc. Exp. Biol. Med.* **217:**64–73, 1998, with permission.

Nitric oxide is produced in abundance in virus-infected tissues during inflammation as part of the innate immune response. This gas has been shown to inhibit the replication of many viruses in cultured cells and in animal models. It probably acts intracellularly to inhibit viral replication, but the molecular sites of action are not well understood. Nitric oxide is produced by three different IFN-inducible isoforms of nitric oxide synthase. While low concentrations of NO have a protective effect, high concentrations or prolonged production have the potential to contribute to tissue damage. For example, treating infected animals with inhibitors of nitric oxide synthase prevents tissue damage. Although NO is relatively inert, it rapidly reacts with $O_2^{\cdot}$ to form peroxynitrite (ONOO˙), which is much more reactive than either radical and may be responsible for cytotoxic effects on cells (Fig. 15.26). In the central nervous system, NO is produced by activated astrocytes and microglia and may be directly toxic to oligodendrocytes and neurons.

Perspectives

The breadth and complexity of the host response to infection can be confusing, if not overwhelming. To illuminate the important principles, it is useful to consider a hypothetical viral infection that is cleared by the host response (Fig. 15.27).

To initiate the primary infection, physical barriers are breached and virus particles enter permissive cells. Replication begins and stimulates the infected cell to synthesize cytokines, such as IFN, as well as to initiate intrinsic defenses, such as apoptosis or autophagy. Local sentinel cells, the immature dendritic cells and macrophages, engage cytokines and internalize viral proteins produced by infected cells. The first response of the immature dendritic cell is to produce massive quantities of IFN and other cytokines. If viral anti-IFN or anti-apoptotic gene products are made, progeny virions are released. If the newly infected cells have already bound IFN, protein synthesis is inhibited when dsRNA is produced. Soon thereafter, NK cells can recognize the infected cells because of new surface antigens and a low or aberrant display of MHC class I proteins. The IFN produced by infected cells stimulates the NK cells to intensify their activities, which include target cell destruction and synthesis of IFN-γ. In some cases, serum complement can be activated, destroying enveloped viruses and infected cells. Some viral proteins sequester complement components and block the cascade. In general, the intrinsic and innate defenses bring most viral infections to an uneventful close before the adaptive response is required.

If viral replication outpaces the innate defense, a critical threshold is reached: increased IFN production by circulating immature dendritic cells elicits a more global host response, and flulike symptoms are experienced by the infected individual. As viral replication continues, viral antigens are delivered by mature dendritic cells to the local lymph nodes or spleen to establish sites of information exchange with T cells. T-cell recirculation is shut down be-

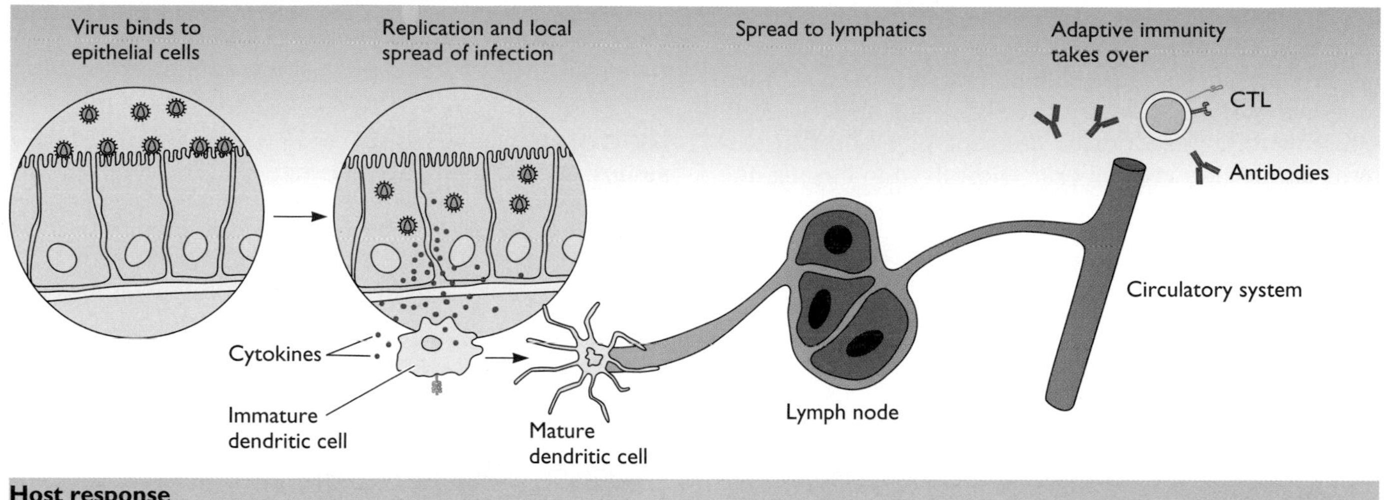
Virus binds to epithelial cells
Replication and local spread of infection
Spread to lymphatics
Adaptive immunity takes over
CTL
Antibodies
Circulatory system
Cytokines
Immature dendritic cell
Mature dendritic cell
Lymph node
Host response
Cells produce cytokines
Cytokines are released; Toll-like receptors on immature dendritic cells are activated; viral proteins are taken up by dendritic cells from dead and dying cells.
Dendritic cells mature, process viral proteins, and present peptides on MHC class II cell surface receptors and migrate to draining lymphatics; some viruses infect dendritic cells. Information exchange between dendritic cells and naive T cells takes place in lymphatics.
Most viral infections stimulate Th1 response. CTLs, antibody, and macrophages clear infection. Memory T cells are established.

Figure 15.27 The stages of viral infection and response of the host. Four states are illustrated for a general viral infection at an epithelial surface, a common site of viral entry. The virus must attach to epithelial cells and then initiate replication. Infected cells produce cytokines that signal local dendritic cells that respond appropriately. As a result, most often local defenses contain the infection. If virus replication is not stopped, the adaptive immune system is activated by contact with mature dendritic cells and cytokines. The adaptive response leads to production of activated effector T and B cells that are released into the circulation to clear the infection and provide precise memory of the particular invader.

cause of the massive recruitment of lymphocytes into lymphoid tissue. The swelling of lymph nodes so often characteristic of infection is a sign of this stage of immune action.

Within days, Th cells and CTLs appear; these cells are the first signs of activation of the adaptive immune response. Th cells produce cytokines that begin to direct the amplification of this response. The synthesis of antibodies, first of IgM and then of other isotypes, quickly follows. The relative concentrations of these isotypes are governed by the route of infection and the pattern of cytokines produced by the Th cells. As the immune response is amplified, CTLs kill or purge infected cells, and antibodies bind to virus particles and infected cells. Specific antibody-virus complexes can be recognized by macrophages and NK cells to induce antibody-dependent, cell-mediated cytotoxicity, and can also activate the classical complement pathway. Both of these processes lead to the directed killing of infected cells and enveloped viruses by macrophages and NK cells.

An inflammatory response often occurs as infected cells die, and innate and adaptive responses develop. Cytokines, chemotactic proteins, and vasodilators are released at the site of infection. These proteins, invading white blood cells, and various complement components all contribute to the swelling, redness, heat, and pain characteristic of the inflammatory response. Many viral proteins modulate this response and the subsequent activation of immune cells.

If infection spreads from the primary site, second and third rounds of replication can occur in other organs. T cells that were activated at the initial site of infection can cause delayed-type hypersensitivity (usually evident as a characteristic rash or lesion) at the sites of later replication. Immunopathology, particularly after infection by noncytopathic viruses, is caused by the cytokines, antibodies, and cells in response to the infection and can contribute to the severity of the disease.

Finally, the combination of innate and adaptive responses clears the infection, and the host is immune because of the presence of memory T and B cells and long-lived antibodies. The high concentrations of immune lymphocytes drop dramatically as these cells die by apoptosis, and the system returns to its normal, preinfection state. The adaptive response can be avoided completely, or in part, when organs or tissues that have poor immune responses or none are infected, when new viral variants are produced rapidly because of high mutation rates, or when progeny virions spread directly from cell to cell. Consequently, there can be dramatic differences in patterns of infection ranging from short-lived to lifelong, the topic of Chapter 16.

References

Textbooks

Janeway, C. A., P. Travers, M. Walport, and M. Shlomchik. 2001. *Immunobiology: the Immune System in Health and Disease,* 5th ed. Garland Publishing, Inc., New York, N.Y.

Male, D., M. Cooke, J. Owen, J. Trowsdale, and B. Champion. 1996. *Advanced Immunology,* 3rd ed. The C. V. Mosby Co., St. Louis, Mo.

Reviews

Apoptosis and Intrinsic Defenses

Ahlquist, P. 2002. RNA-dependent RNA polymerases, viruses, and RNA silencing. *Science* **296:**1270–1273.

Hay, S., and G. Kannourakis. 2002. A time to kill: viral manipulation of the cell death program. *J. Gen. Virol.* **83:**1547–1564.

Maul, G. 1998. Nuclear domain 10, the site of DNA virus transcription and replication. *Bioessays* **20:**660–667.

Plasterk, R. H. 2002. RNA silencing: the genome's immune system. *Science* **296:**1263–1265.

Redpath, S., A. Angulo, N. R. J. Gascoigne, and P. Ghazal. 2001. Immune checkpoints in viral latency. *Annu. Rev. Microbiol.* **55:**531–560.

Tyler, K. L., P. Clarke, R. L. DeBiasi, D. Kominsky, and G. J. Poggioli. 2001. Reoviruses and the host cell. *Trends Microbiol.* **9:**560–564.

Cytokines and Interferons

Goodbourn, S., Didcock, L., and R. E. Randall. 2000. Interferons: cell signaling, immune modulation, anti-viral responses and virus countermeasures. *J. Gen. Virol.* **81:**2341–2364.

Hatada, E. N., D. Krappmann, and C. Scheidereit. 2000. Nf-κb and the innate immune response. *Curr. Opin. Immunol.* **12:**52–58.

Kisseleva, T., S. Bhattacharya, J. Braunstein, and C. W. Schindler. 2002. Signaling through the JAK/Stat pathway, recent advances and future challenges. *Gene* **285:**1–24.

Mogensen, T. H., and S. R. Paludan. 2001. Molecular pathways in virus-induced cytokine production. *Microbiol. Mol. Biol. Rev.* **65:**131–150.

Sen, G. C. 2001. Viruses and interferons. *Annu. Rev. Microbiol.* **55:**255–281.

Silverman, R. H., and N. M. Cirino. 1997. RNA decay by the interferon-regulated 2-5A system as a host defense against viruses, p. 295–309. *In* J. B. Harford and D. R. Morris (ed.), *mRNA Metabolism and Post-Transcriptional Gene Regulation.* Wiley-Liss Inc., New York, N.Y.

Dendritic Cells and Toll-Like Receptors

Akira, S., K. Takeda, and T. Kaisho. 2001. Toll-like receptors: critical proteins linking innate and acquired immunity. *Nat. Immunol.* **2:**675–680.

Krieg, A. M. 2000. The role of CpG motifs in innate immunity. *Curr. Opin. Immunol.* **12:**35–43.

Mellman, R., and R. M. Steinman. 2001. Dendritic cells: specialized and regulated antigen processing machines. *Cell* **106:**255–258.

Rescigno, M., and P. Borrow. 2001. The host-pathogen interaction: new themes from dendritic cell biology. *Cell* **106:**267–270.

Steinman, R. M. 2000. DC-SIGN: a guide to some mysteries of dendritic cells. *Cell* **100:**491–494.

NK Cells

Biron, C. A., and L. Brossay. 2001. NK cells and NKT-cells in innate defense against viral infections. *Curr. Opin. Immunol.* **13:**458–464.

Campbell, K. S., and M. Colonna. 2001. Human natural killer cell receptors and signal transduction. *Int. Rev. Immunol.* **20:**325–362.

Lanier, L. L. 2000. Turning on natural killer cells. *J. Exp. Med.* **191:**1259–1262.

Moretta, A., C. Bottino, M. C. Mingari, R. Biassoni, and L. Moretta. 2002. What is a natural killer cell? *Nat. Immunol.* **3:**6–8.

Park, S.-H., and A. Bendelac. 2000. CD1-restricted T-cell responses and microbial infection. *Nature* **406:**788–792.

Complement and Inflammation

Lachman, P. J., and A. Davies. 1997. Complement and immunity to viruses. *Immunol. Rev.* **159:**69–77.

Weiss, R. A. 1998. Transgenic pigs and virus adaptation. *Nature* **391:**327–328.

Virus Immune Modulation

Levy, D. E., and A. Garcia-Sastre. 2001. The virus battles: IFN induction of the anti-viral state and mechanisms of viral evasion. *Cytokine Growth Factor Rev.* **12:**143–156.

Lorenzo, M. E., H. L. Ploegh, and R. S. Tirabassi. 2001. Viral immune evasion strategies and the underlying cell biology. *Semin. Immunol.* **13:**1–9.

Tortorella, D., B. E. Gewurz, M. H. Furman, D. J. Schust, and H. L. Ploegh. 2000. Viral subversion of the immune system. *Annu. Rev. Immunol.* **18:**861–926.

Adaptive Immunity

Delon, J., and R. N. Germain. 2000. Information transfer at the immunological synapse. *Curr. Biol.* **10:**R923–R933.

Dimmock, J. J. 1993. Neutralization of animal viruses. *Curr. Top. Microbiol. Immunol.* **183:**1–149.

Doherty, P. C., and J. P. Christensen. 2000. Accessing complexity: the dynamics of virus-specific T-cell responses. *Annu. Rev. Immunol.* **18:**561–592.

Guidotti, L. G., and F. V. Chisari. 2001. Non-cytolytic control of viral infections by the innate and adaptive immune response. *Annu. Rev. Immunol.* **19:**65–91.

Janeway, C. A., Jr. 2001. How the immune system works to protect the host from infection: a personal view. *Proc. Natl. Acad. Sci. USA* **98:**7461–7468.

Lamm, M. E. 1997. Interaction of antigens and antibodies at mucosal surfaces. *Annu. Rev. Microbiol.* **51:**311–340.

Matzinger, P., and E. J. Fuchs. 1996. Beyond "self" and "non-self": immunity is a conversation, not a war. *J. NIH Res.* **8:**35–39.

Central Nervous System Defense

Hickey, W. F. 2001. Basic principles of immunological surveillance of the normal central nervous system. *Glia* **36:**118–124.

Lowenstein, P. R. 2002. Immunology of viral-vector-mediated gene transfer into the brain: an evolutionary and developmental perspective. *Trends Immunol.* **23:**23–30.

Virus Injury to Cells

Davies, M., and P. Hagen. 1997. Systemic inflammatory response syndrome. *Br. J. Surg.* **84:**920–935.

Lane, T. E., and M. J. Buchmeier. 1997. Murine coronavirus infection: a paradigm for virus induced demyelinating disease. *Trends Microbiol.* **5:**9–14.

Reiss, C. S., and T. Komatsu. 1998. Does nitric oxide play a critical role in viral infections? *J. Virol.* **72:**4547–4551.

Rouse, B. T. 1996. Virus-induced immunopathology. *Adv. Virus Res.* **47:**353–376.

Simas, J. P., and S. Efstathiou. 1998. Murine gammaherpesvirus 68: a model for the study of gammaherpesvirus pathogenesis. *Trends Microbiol.* **7:**276–282.

Streilein, J. W., M. R. Dana, and B. R. Ksander. 1997. Immunity causing blindness: five different paths to herpes stromal keratitis. *Immunol. Today* **18:**443–449.

Classic Papers

Isaacs, A., and J. Lindenmann. 1957. Virus interference. I. The interferon. II. Some properties of interferon. *Proc. R. Soc. Lond. B* **147:**258–267, 268–273.

Marcus, P. I., and M. J. Sekellick. 1976. Cell killing by viruses. III. The interferon system and inhibition of cell killing by vesicular stomatitis virus. *Virology* **69:**378–393.

Zinkernagel, R. M., and P. C. Doherty. 1974. Restriction of *in vitro* T-cell-mediated cytotoxicity in lymphocytic choriomeningitis within a syngeneic or semiallogeneic system. *Nature* **248:**701–702.

Zinkernagel, R. M. 1996. Immunology taught by viruses. *Science* **271:**173–178.

Selected Papers

Apoptosis and Intrinsic Defenses

Bertin, J., R. C. Armstrong, S. Ottilie, D. A. Martin, Y. Wang, S. Banks, G.-H. Wang, T. G. Senkevich, E. S. Alnemri, B. Moss, M. J. Lenardo, K. J. Tomaselli, and J. I. Cohen. 1997. Death effector domain-containing herpesvirus and poxvirus proteins inhibit both Fas- and TNFR1-induced apoptosis. *Proc. Natl. Acad. Sci. USA* **94:**1172–1176.

Everett, R., M. Meredith, A. Orr, A. Cross, M. Kathoria, and J. Parkinson. 1997. A novel ubiquitin-specific protease is dynamically associated with the PML nuclear domain and binds to a herpesvirus regulatory protein. *EMBO J.* **16:**1519–1530.

Galvan, V., and B. Roizman. 1998. Herpes simplex virus type 1 induces and blocks apoptosis at multiple steps during infection and protects cells from exogenous inducers in a cell-type dependent manner. *Proc. Natl. Acad. Sci. USA* **95:**3931–3936.

Irie, H., H. Koyama, H. Kubo, A. Fukuda, K. Aita, T. Koike, A. Yoshimura, T. Yoshida, J. Shiga, and T. Hill. 1998. Herpes simplex virus hepatitis in macrophage-depleted mice: the role of massive, apoptotic cell death in pathogenesis. *J. Gen. Virol.* **79:**1225–1231.

Li, H., Li, W. X., and Ding, S. W. 2002. Induction and suppression of RNA silencing by an animal virus. *Science* **296:**1319–1321.

Regad, T., A. Saib, V. Lallemand-Breitenbach, P. P. Pandolfi, H. de The, and M. K. Chelbi-Alix. 2001. PML mediates the interferon-induced antiviral state against a complex retrovirus via its association with the viral transactivator. *EMBO J.* **20:**3495–3505.

Sawtell, N. M., and R. L. Thompson. 2001. Herpes simplex virus type 1 latency associated transcript gene promotes neuronal survival. *J. Virol.* **75:**6660–6676.

Talloczy, Z., W. Jiang, H. W. Virgin IV, D. A. Leib, D. Scheuner, R. J. Kaufman, E. L. Eskelinen, and B. Levine. 2002. Regulation of starvation and virus induced autophagy by the eIF2α kinase signaling pathway. *Proc. Natl. Acad. Sci. USA* **99:**190–195.

Taylor, J. L., D. Unverrich, W. O'Brien, and K. Wilcox. 2000. Interferon coordinately inhibits the disruption of PML-positive ND10 and immediate early gene expression by herpes simplex virus. *J. Interferon Cytokine Res.* **20:**805–815.

Thome, M., P. Schneider, K. Hofmann, H. Fickenscher, E. Meinl, F. Neipel, C. Mattmann, K. Burns, J.-L. Bodmer, M. Schroter, C. Scaffidi, P. H. Krammer, M. E. Peter, and J. Tschop. 1997. Viral FLICE-inhibitory proteins (FLIPs) prevent apoptosis induced by death receptors. *Nature* **386:**517–521.

Cytokines and Interferon

Balachandran, S., P. C. Roberts, L. E. Brown, H. Truong, A. K. Pattnaik, D. R. Archer, and G. N. Barber. 2000. Essential role of the dsRNA-dependent protein kinase Pkr in innate immunity to viral infection. *Immunity* **13:**129–141.

Ben-Asouli, Y., Y. Banai, Y. Pel-Or, A. Shir, and R. Kaempfer. 2002. Human interferon-gamma mRNA autoregulates its translation through a pseudoknot that activates the interferon-inducible protein kinase Pkr. *Cell* **108:**221–232.

Browne, E. P., B. Wing, D. Coleman, and T. Shenk. 2001. Altered cellular mRNA levels in human cytomegalovirus infected fibroblasts: viral block to the accumulation of anti-viral mRNAs. *J. Virol.* **75:**12319–12330.

Chinsangaram, J., M. E. Piccone, and M. J. Grubman. 1999. Ability of foot-and-mouth disease virus to form plaques in cell culture is associated with suppression of alpha/beta interferon. *J. Virol.* **73:**9891–9898.

Dokun, A. O., S. Kim, H. R. C. Smith, H.-S. P. Kang, D. T. Chu, and W. M. Yokoyama. 2001. Specific and nonspecific NK cell activation during virus infection. *Nat. Immunol.* **2:**951–956.

Garcin, D., J. Curran, and D. Kolakofsky. 2000. Sendai virus C proteins must interact directly with cellular components to interfere with interferon action. *J. Virol.* **74:**8823–8830.

Harle, P., V. Cull, M. P. Agbaga, R. Silverman, B. R. Williams, C. James, and D. J. Carr. 2002. Differential effect of murine alpha/beta interferon transgenes on antagonization of herpes simplex virus type 1 replication. *J. Virol.* **76:**6558–6567.

Ishov, A., and G. Maul. 1996. The periphery of nuclear domain 10 (ND10) as site of DNA virus deposition. *J. Cell Biol.* **134:**815–826.

Leib, D. A., T. E. Harrison, K. M. Laslo, M. A. Machalek, N. J. Moorman, and H. W. Virgin. 1999. Interferons regulate the phenotypes of wild-type and mutant herpes simplex viruses *in vivo*. *J. Exp. Med.* **189:**663–672.

Leib, D. A., M. A. Machalek, B. R. G. Williams, R. H. Silverman, and H. W. Virgin. 2000. Specific phenotypic restoration of an attenuated virus by knockout of a host resistance gene. *Proc. Natl. Acad. Sci. USA* **97:**6097–6101.

Li, X.-L., J. A. Blackford, C. S. Judge, M. Liu, W. Xiao, D. V. Kalvakolanu, and B. A. Hassel. 2000. RNase-L-dependent destabilization of interferon-induced mRNAs: a role for the 2-5A system in attenuation of the interferon response. *J. Biol. Chem.* **275:**8880–8888.

Marcus, P. I., L. L. Rodriguez, and M. J. Sekellick. 1998. Interferon induction as a quasispecies marker of vesicular stomatitis virus populations. *J. Virol.* **72:**542–549.

Mossman, K. L., and J. R. Smiley. 2002. Herpes simplex virus ICP0 and ICP34.5 counteract distinct interferon-induced barriers to virus replication. *J. Virol.* **76:**1995–1998.

Najarro, P., P. Traktman, and J. A. Lewis. 2001. Vaccinia virus blocks gamma interferon signal transduction: viral VH1 phosphatase reverses Stat1 activation. *J. Virol.* **75:**3185–3196.

Oura, C. A. L., P. P. Powell, and R. M. E. Parkhouse. 1998. African swine fever: a disease characterized by apoptosis. *J. Gen. Virol.* **79:**1427–1438.

Paludan, S. R., and S. C. Mogensen. 2001. Virus-cell interactions regulating induction of tumor necrosis factor alpha production in macrophages infected with herpes simplex virus. *J. Virol.* **75:**10170–10178.

Robek, M. D., S. F. Wieland, and F. V. Chisari. 2002. Inhibition of hepatitis B virus replication by interferon requires proteasome activity. *J. Virol.* **76:**3570–3574.

Thome, M., P. Schneider, K. Hofmann, H. Fickenscher, E. Meinl, F. Neipel, C. Mattmann, K. Burns, J.-L. Bodmer, M. Schroter, C. Scaffidi, P. H. Krammer, M. E. Peter, and J. Tschop. 1997. Viral FLICE-inhibitory proteins (FLIPs) prevent apoptosis induced by death receptors. *Nature* **386:**517–521.

Yokota, S., N. Yokosawa, T. Kubota, T. Suzutani, I. Yoshida, S. Miura, K. Jimbow, and N. Fujii. 2001. Herpes simplex virus type 1 suppresses the interferon signaling pathway by inhibiting phosphorylation of STATs and Janus kinases during an early infection stage. *Virology* **286:**119–124.

Young, D. F., L. Didcock, S. Goodbourn, and R. E. Randall. 2000. Paramyxoviridae use distinct virus-specific mechanisms to circumvent the interferon response. *Virology* **269:**383–390.

Dendritic Cells and Toll-Like Receptors

Haynes, L. M., D. D. Moore, E. A. Kurt-Jones, R. W. Finberg, L. J. Anderson, and R. A. Tripp. 2001. Involvement of toll-like receptor 4 in innate immunity to respiratory syncytial virus. *J. Virol.* **75:**10730–10737.

Kadowaki, N., S. Antonenko, J. Y.-N. Lau, and Y.-J. Liu. 2000. Natural interferon α/β-producing cells link innate and adaptive immunity. *J. Exp. Med.* **192:**219–225.

Ludewig, B., K. J. Maloy, C. Lopez-Macias, B. Odermatt, H. Hengartner, and R. M. Zinkernagel. 2000. Induction of optimal anti-viral neutralizing B-cell responses by dendritic cells requires transport and release of virus particles in secondary lymphoid organs. *Eur. J. Immunol.* **30:**185–196.

Trevejo, J. M., M. W. Marino, N. Philpott, R. Josien, E. C. Richards, K. B. Elkon, and E. Falck-Pedersen. 2001. TNF-α-dependent maturation of local dendritic cells is critical for activating the adaptive immune response to virus infection. *Proc. Natl. Acad. Sci. USA* **98:**12162–12167.

NK Cells

Brown, M. G., A. O. Dokun, J. W. Heusel, H. R. C. Smith, D. L. Beckman, E. A. Blattenberger, C. E. Dubbelde, L. R. Stone, A. A. Scalzo, and W. M. Yokoyama. 2001. Vital involvement of a natural killer cell activation receptor in resistance to viral infection. *Science* **292:**934–937.

Complement and Inflammation

Lubinski, J., L. Wang, D. Mastellos, A. Sahu, J. D. Lambris, and H. M. Friedman. 1999. *In vivo* role of complement interacting domains of herpes simplex virus type 1 glycoprotein gC. *J. Exp. Med.* **90:**1637–1646.

Rosengard, A., Y. Liu, Z. Nie, and R. Jimenez. 2002. Variola virus immune evasion design: expression of a highly efficient inhibitor of human complement. *Proc. Natl. Acad. Sci. USA* **99:**8808–8813.

Rother, R. P., and S. P. Squinto. 1996. The α-galactosyl epitope: a sugar coating that makes viruses and cells unpalatable. *Cell* **86:**185–188.

Sugano, N., W. Chen, M. L. Roberts, and N. R. Cooper. 1997. Epstein-Barr virus binding to CD21 activates the initial viral promoter via Nf-κb induction. *J. Exp. Med.* **186:**731–737.

Vanderplasschen, A., E. Mathew, M. Hollinshead, R. Sim, and G. Smith. 1998. Extracellular enveloped vaccinia virus is resistant to complement because of incorporation of host complement control proteins into its envelope. *Proc. Natl. Acad. Sci. USA* **95:**7544–7549.

Adaptive Immunity

Callan, M. F. C., L. Tan, N. Annels, G. S. Ogg, J. D. K. Wilson, C. A. O'Callaghan, N. Steven, A. J. McMichael, and A. B. Rickinson. 1998. Direct visualization of antigen-specific CD8+ T-cells during the primary immune response to Epstein-Barr virus *in vivo*. *J. Exp. Med.* **187:**1395–1402.

Nagashunmugam, T., J. Lubinski, L. Wang, L. T. Goldstein, B. Weeks, P. Sundaresan, E. Kang, G. Dubin, and H. M. Friedman. 1998. *In vivo* immune evasion mediated by herpes simplex virus type 1 IgG Fc receptor. *J. Virol.* **72:**5351–5359.

Posavad, C. M., J. J. Newton, and K. L. Rosenthal. 1994. Infection and inhibition of human cytotoxic T lymphocytes by herpes simplex virus. *J. Virol.* **68:**4072–4074.

Sigal, L. J., S. Crotty, R. Andino, and K. L. Rock. 1999. Cytotoxic T-cell immunity to virus-infected non-haematopoietic cells requires presentation of exogenous antigen. *Nature* **398:**77–80.

Central Nervous System Defense

Stevenson, P. G., J. M. Austyn, and S. Hawke. 2002 Uncoupling of virus-induced inflammation and anti-viral immunity in the brain parenchyma. *J. Gen. Virol.* **83:**1735–1743.

Stevenson, P. G., S. Hawke, D. J. Sloan, and C. R. Bangham. 1997. The immunogenicity of intracerebral virus infection depends on the anatomical site. *J. Virol.* **71:**145–151.

Stevenson, P. G., S. Freeman, C. R. Bangham, and S. Hawke. 1997. Virus dissemination through the brain parenchyma without immunologic control. *J. Immunol.* **159:**1876–1884.

Thomas, C., G. Schiedner, S. Kochaneck, M. G. Castro, and P. R. Lowenstein. 2000. Peripheral infection with adenovirus induces unexpected long term brain inflammation in animals injected intracranially with first generation, but not with high capacity adenovirus vectors: toward realistic long term neurological gene therapy for chronic diseases. *Proc. Natl. Acad. Sci. USA* **97:**7482–7487.

Virus Injury and Immunopathology

Deshpande, S. P., S. Lee, M. Zheng, B. Song, D. Knipe, J. A. Kapp, and B. T. Rouse. 2001. Herpes simplex virus-induced keratitis: evaluation of the role of molecular mimicry in lesion pathogenesis. *J. Virol.* **75:**3077–3088.

Evans, C. F. , M. S. Horwitz, M. V. Hobbs, and M. B. A. Oldstone. 1996. Viral infection of transgenic mice expressing a viral protein in oligodendrocytes leads to chronic central nervous system autoimmune disease. *J. Exp. Med.* **184:**2371–2384.

Gebhard, J. R., C. M. Perry, S. Harkins, T. Lane, I. Mena, V. C. Asensio, I. L. Campbell, and J. L. Whitton. 1998. Coxsackievirus B3-induced myocarditis: perforin exacerbates disease, but plays no detectable role in virus clearance. *Am. J. Pathol.* **153:**417–428.

Klimstra, W. B, K. D. Ryman, K. A. Bernard, K. B. Nguyen, C. A. Biron, and R. E. Johnston. 1999. Infection of neonatal mice with Sindbis virus results in a systemic inflammatory response syndrome. *J. Virol.* **73:**10387–10398.

Posavad, C. M., D. M. Koelle, M. F. Shaughnessy, and L. Corey. 1997. Severe genital herpes infections in human immunodeficiency virus-infected individuals with impaired herpes simplex virus-specific CD8+ cytotoxic T lymphocyte responses. *Proc. Natl. Acad. Sci. USA* **94:**10289–10294.

Ramsingh, A. I., W. T. Lee, D. N. Collins, and L. E. Armstrong. 1997. Differential recruitment of B and T-cells in coxsackievirus B4-induced pancreatitis is influenced by a capsid protein. *J. Virol.* **71:**8690–8697.

Tripp, R. A., A. M. Hamilton-Easton, R. D. Cardin, P. Nguyen, F. G. Behm, D. L. Woodland, P. C. Doherty, and M. A. Blackman. 1997. Pathogenesis of an infectious mononucleosis-like disease induced by a murine γ-herpesvirus: role for a viral superantigen? *J. Exp. Med.* **185:**1641–1650.

16

Patterns of Infection

> *You know something's happening, but you don't know what it is, do you, Mr. Jones?*
> R. ZIMMERMAN

Patterns of Infection

Viral infections of individuals in populations differ from infections of cultured cells in the laboratory. In the former, initiation of the infection and its eventual outcome rest upon complex variables such as host defenses, composition of the host population, and the environment. Despite such complexity and the plethora of viruses and hosts, common patterns of infection appear. In general, natural infections can be rapid and self-limiting (**acute infections**) or long-term (**persistent infections**). Variations and combinations of these two modes abound. While we can provide detailed descriptions of individual patterns of infection, we are in the early days of understanding the molecular mechanisms required to initiate or maintain any specific one.

Life Cycles

A cursory examination of the animal viruses that grow in cultured cells identifies many distinctive life cycles with common features. Some infections rapidly kill the cell, while producing a burst of new particles (caused by **cytopathic viruses**). Other infections yield virions without causing immediate host cell death (caused by **noncytopathic viruses**). Alternatively, some infections neither kill the cell nor produce any progeny. These apparently diverse patterns defined in cultured cells comprise the two primary patterns of infection in the host: acute and persistent infections (Fig. 16.1). Variations on these archetypes occur repeatedly. For example, a **latent infection** is an extreme version of a persistent infection. Similarly, **slow**, **abortive**, and **transforming infections** are more complicated variants of a persistent infection.

Once immediate host defenses have been breached and an infection is established, a cascade of reactions occurs in the host (see Chapter 15). Symptoms and disease may or may not be obvious, depending upon the virus, the infected tissue, and the host defenses. The initial period before the characteristic symptoms of a disease are obvious is called the **incubation period**. During this

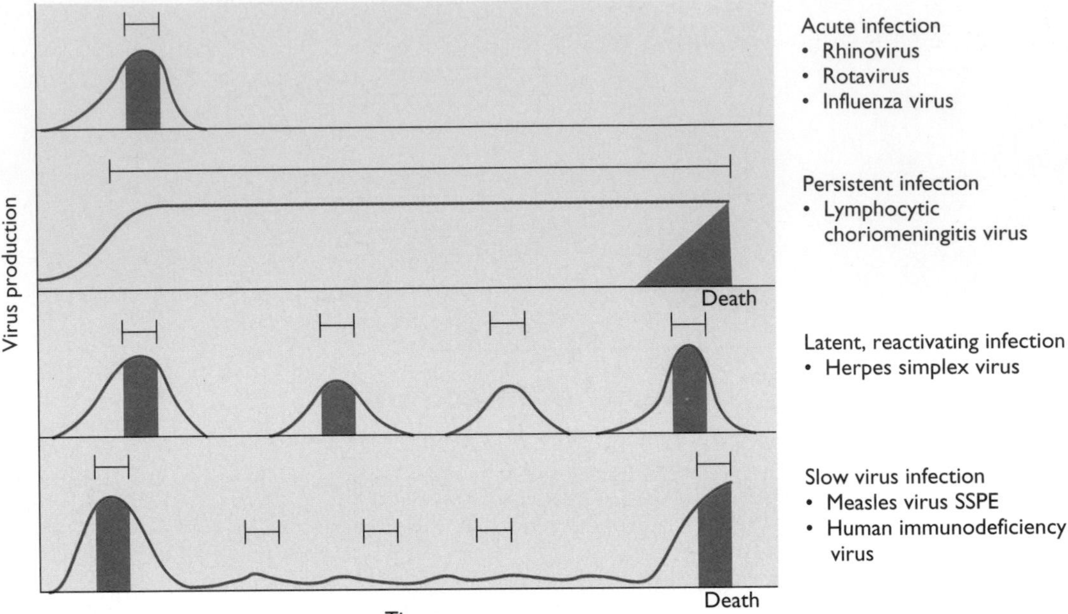

Figure 16.1 General patterns of infection. Relative virion production is plotted as a function of time after infection. The time when symptoms appear is indicated by the red shaded area, and the period in which infectious virus is released (available to infect other hosts) is indicated by the bracket. The top panel is the typical profile of an acute infection, in which virions are produced, symptoms appear, and the infection is cleared within 7 to 10 days after infection. The second panel is the typical profile of a persistent infection, in which virion production continues for the life of the host. Symptoms may or may not appear just before death, depending upon the virus. Infectious virions are usually produced throughout the infection. The bottom two panels are variations of the persistent infection. The third panel depicts a latent infection in which an initial acute infection is followed by a quiescent phase and repeated bouts of reactivation. Reactivation may or may not be accompanied by symptoms, but generally results in the production of infectious virions. The fourth panel depicts a slow virus infection, in which a period of years intervenes between a typical primary acute infection and the usually fatal appearance of symptoms. The production of infectious virions during the long period between primary infection and fatal outcome may be continuous (e.g., human immunodeficiency virus) or absent (e.g., measles virus SSPE). The brackets indicating infectious particle release are placed arbitrarily to indicate this phenomenon. Adapted from F. J. Fenner et al., *Veterinary Virology* (Academic Press, Inc., Orlando, Fla., 1993), with permission.

time, virions are replicating and the host is responding. Remarkably, incubation periods can vary from one or two days to years (Table 16.1). If infection at the primary site induces the characteristic symptoms of the disease, the incubation period usually is short. Long incubation times often indicate that the host response, or the tissue damage required to reveal the infection, is not the result of the primary infection. The events that occur during the incubation period must certainly dictate the resulting pattern of infection, but identifying and characterizing these processes remain major challenges for those studying pathogenesis. Meeting these challenges is of paramount importance if we are to design and implement useful diagnostic tools and treatments.

Acute Infections

Definition and Requirements

An acute infection is one of the best understood of all infection patterns because it is often easily studied, and also is characteristic of many infections of animals and cultured cells. By "acute," we mean rapid production of infectious virions, followed by rapid resolution and elimination of the infection by the host ("clearing" the infection). Acute infections can be established only when innate defenses are transiently bypassed. Acute infections are the typical, expected course for agents such as influenza virus and rhinovirus (Fig. 16.2). These infections usually are relatively brief, and in a healthy host, virions and virus-infected cells

Table 16.1 Incubation periods of some common viral infections

Disease	Incubation period (days)[a]
Influenza	1–2
Common cold	1–3
Bronchiolitis, croup	3–5
Acute respiratory disease (adenoviruses)	5–7
Dengue	5–8
Herpes simplex	5–8
Enterovirus disease	6–12
Poliomyelitis	5–20
Measles	9–12
Smallpox	12–14
Chickenpox	13–17
Mumps	16–20
Rubella	17–20
Mononucleosis	30–50
Hepatitis A	15–40
Hepatitis B and C	50–150
Rabies	30–100
Papilloma (warts)	50–150
AIDS	1–10 yr

[a]Until first appearance of prodromal symptoms.

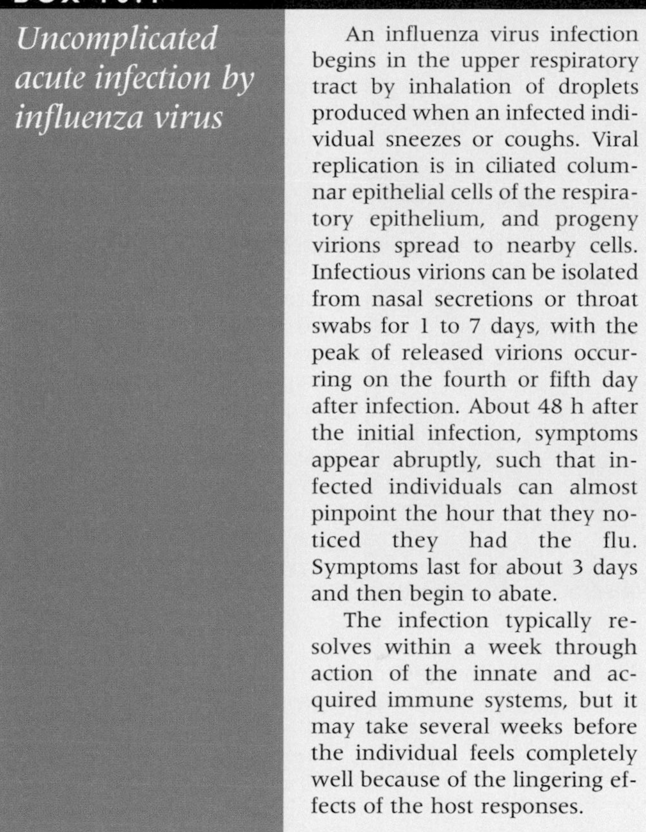

BOX 16.1

Uncomplicated acute infection by influenza virus

An influenza virus infection begins in the upper respiratory tract by inhalation of droplets produced when an infected individual sneezes or coughs. Viral replication is in ciliated columnar epithelial cells of the respiratory epithelium, and progeny virions spread to nearby cells. Infectious virions can be isolated from nasal secretions or throat swabs for 1 to 7 days, with the peak of released virions occurring on the fourth or fifth day after infection. About 48 h after the initial infection, symptoms appear abruptly, such that infected individuals can almost pinpoint the hour that they noticed they had the flu. Symptoms last for about 3 days and then begin to abate.

The infection typically resolves within a week through action of the innate and acquired immune systems, but it may take several weeks before the individual feels completely well because of the lingering effects of the host responses.

are cleared by the adaptive immune system within days. Nevertheless, some progeny are invariably available for spread to other hosts before the infection is resolved (Boxes 16.1 and 16.2). An initial acute infection may be followed by a second or third round of infection in the same animal. In these cases, virions or infected cells escape containment by local defenses and spread from the primary site to other tissues, where a second infection can occur (see Chapter 14). The pattern of multiple infections is characteristic of varicella-zoster virus, an alphaherpesvirus that causes the familiar childhood disease chickenpox (Fig. 16.3).

Acute infections need not produce disease. Indeed, **inapparent** (asymptomatic) acute infections are quite common and can be major sources of infections in populations

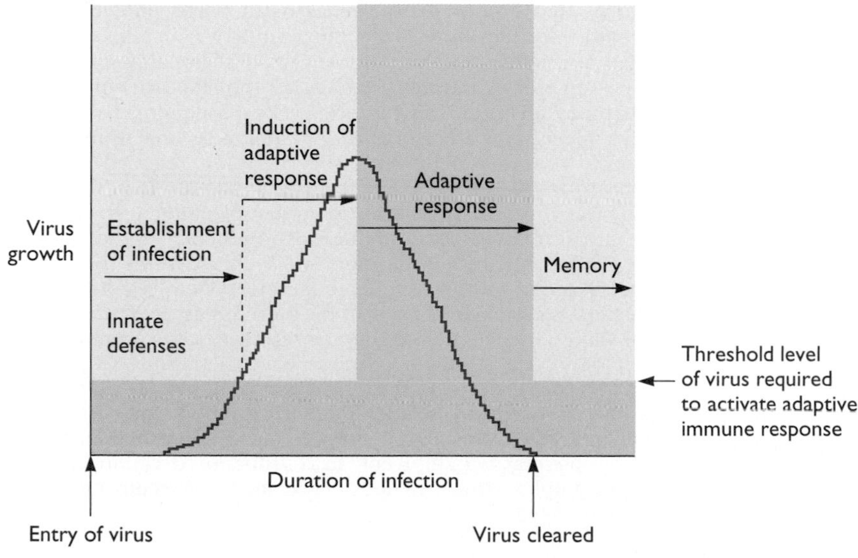

Figure 16.2 The course of a typical acute infection. Relative virus growth plotted as a function of time after infection. The concentration of infectious virions increases with time, as indicated by the jagged red line. During the establishment of infection, only the innate defenses are at work. If the infection reaches a certain threshold level characteristic of the virus and host (purple), the adaptive responses initiate. After 4 to 5 days, effector cells and molecules of the adaptive response begin to clear the infection by removing virus particles and infected cells (green). When virus particles are cleared and infected cells are eliminated, the adaptive response ceases. Antibodies, residual effector cells, and memory cells provide lasting protection, should the host be reinfected at a later date. Redrawn from C. A. Janeway, Jr., and P. Travers, *Immunobiology: the Immune System in Health and Disease* (Current Biology Ltd. and Garland Publishing, New York, N.Y., 1996), with permission.

BOX 16.2

Inapparent acute infections

It is important to distinguish an inapparent acute infection from an unsuccessful one. Inapparent infections are successful acute infections that produce no symptoms or disease. Sufficient virions are made to maintain the infection in the host population, but the quantity is below the threshold required to induce symptoms in infected individuals. The usual way an inapparent infection is detected is by a rise in antiviral antibody concentrations in an otherwise healthy individual. Well-adapted pathogens often follow this infection pattern, as demonstrated by poliovirus, in which more than 90% of infections are inapparent.

(Box 16.2). Such infections are recognized by the presence of virus-specific antibodies with no reported history of disease. For example, over 95% of the population of the United States has antibody to varicella-zoster virus, but less than half report that they have had chicken pox.

Defense against Acute Infections

The intrinsic and innate responses limit and contain most acute infections. As a result, such infections can be disastrous for individuals whose early defenses are compromised, primarily because viral particles do not remain localized to the primary site but spread widely. In a naive host, the adaptive immune response (antibody and activated cytotoxic T lymphocytes [CTLs]) does not influence viral replication for several days, but is essential for final clearance of virions and infected cells. The adaptive response also provides **memory** for defense against subsequent exposure.

Antigenic Variation, an Effective Mechanism for Maintenance of Some Viruses That Cause Acute Infections

Hosts that survive an acute virus infection are usually immune to infections by the same virus for the rest of their lives, as exemplified by poliovirus and measles virus. Nevertheless, acute infections caused by some viruses (e.g., rhinovirus, the common cold virus, influenza virus, and human immunodeficiency virus) occur repeatedly despite active immune clearance. The mechanisms responsible for reinfection are not completely understood, but the structural properties of virions and the ability of neutralizing antibodies to block infectivity are critical parameters.

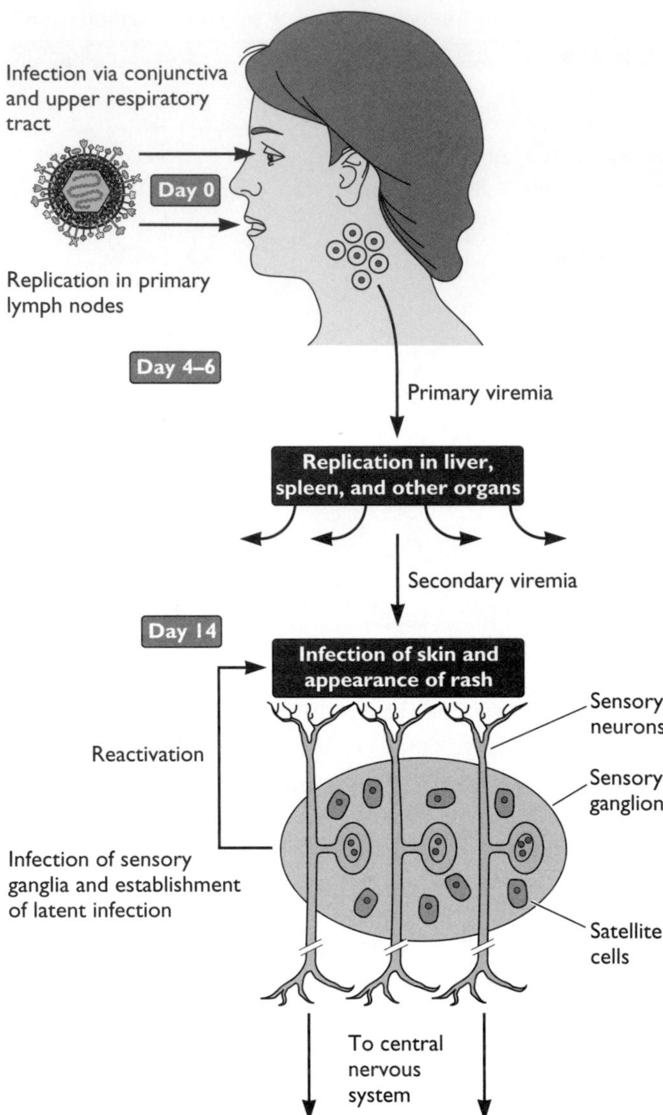

Figure 16.3 Model of varicella-zoster virus infection and spread. Infection initiates on the conjunctiva or mucosa of the upper respiratory tract and then moves to regional lymph nodes. After 4 to 6 days, infected T cells enter the bloodstream, causing a **primary viremia.** These infected cells subsequently invade the liver, spleen, and other organs, initiating a second round of infection. Virions and infected cells are then released into the bloodstream in a **secondary viremia,** subsequently infecting the skin. This third round of infection gives rise, after 2 weeks from the initial infection, to the characteristic vesicular rash of chicken pox. Following acute replication in the skin, virions then infect sensory ganglia of the peripheral nervous system, where a latent infection is established. Later in life, perhaps as the immune system wanes, the latent viral genome reactivates and another acute infectious cycle is initiated. Virions leave the peripheral neurons to infect the skin. The characteristic recurrent disease called shingles is often accompanied by a painful condition called postherpetic neuralgia. Normally, an infected individual experiences only one visible reactivation event. In contrast, reactivation of latent herpes simplex virus infections occurs many times during the life of the infected person.

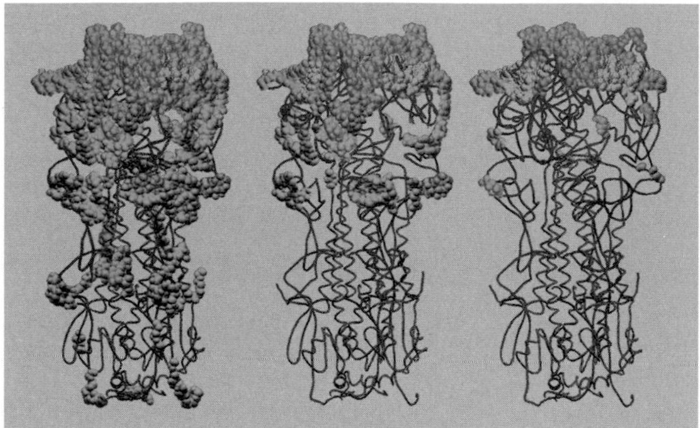

Figure 16.4 Antigenic drift: distribution of amino acid residue changes in hemagglutinins (HA) of viruses isolated during the Hong Kong pandemic era (1968 to 1995). The space-filling models represent the virus-receptor binding site (yellow) and the substituted amino acids (green). (Left) All substitutions in HAs of virions isolated between 1968 and 1995; (middle) amino acid substitutions between 1968 and 1995 that were retained in subsequent years; (right) amino acid substitutions detected in monoclonal antibody-selected variants of A/Hong Kong/68 HA. The α-carbon tracings of the HA trimers are colored blue to denote the HA1 chain and red to denote the HA2 chain. Adapted from T. Bizebard et al., *Curr. Top. Microbiol. Immunol.* **260**:55–64, 2001, with permission.

Virions that can tolerate many amino acid substitutions in their structural proteins and remain infectious are said to have **structural plasticity** (e.g., influenza virus, rhinovirus, human immunodeficiency virus). Antibody-resistant mutants are common among these viruses, and therefore reinfections of immune individuals are possible. In contrast, the virions of other viruses cannot tolerate wholesale amino acid change (e.g., poliovirus, measles virus, yellow fever virus). Accordingly, mutations in genes encoding virion proteins that result in reduced antibody binding are rare with these viruses.

The principles underlying the selection and maintenance of antibody-resistant virions in natural infections are not well understood. For example, the virions of rhinoviruses have remarkable structural plasticity, while those of poliovirus, a related picornavirus, do not. Over 100 different serotypes of rhinovirus are maintained in the population at all times, a property that accounts for the fact that individuals may contract more than one common cold each year. Why there are just three serotypes of poliovirus remains a mystery. Fortunately, this property ensures that poliovirus vaccine that was effective in the 1950s is just as potent in the 21st century. Similarly, the enveloped virions of influenza virus readily become resistant to antibodies, while the enveloped virions of measles virus and yellow fever virus exhibit little change in membrane protein amino acid sequence, and antibody-resistant variants rarely arise. Consequently, a flu vaccine is required every year, while a single measles virus vaccination lasts a lifetime.

Change of virion proteins in response to antibody selection is called **antigenic variation**. In an immunocompetent host, antigenic variation comes about by two distinct processes (Fig. 16.4). **Antigenic drift** is the appearance of virions with slightly altered surface proteins (antigen) following passage in the natural host. In contrast, **antigenic shift** is a major change in the surface protein of a virion as genes encoding completely new surface proteins are acquired. This dramatic shift in virion composition is the result of coinfection of a host with two viral serotypes. Virions with segmented genomes can exchange segments, or coreplicating genomes can produce recombinant genomes. The new reassortant and recombinant viruses have exchanged blocks of genetic information, and the resulting hybrid virions may temporarily avoid immune defenses (Box 16.3).

BOX 16.3

Recombination and antigenic shift during human immunodeficiency virus infections

While the contribution of antigenic shift to the dramatic influenza virus pandemics is well documented (see Box 20.4), the contribution of antigenic shift in human immunodeficiency virus pathogenesis is only now becoming appreciated. In this case, antigenic shift can only occur by recombination. The process requires packaging of two distinct genomes into a single particle for subsequent reverse transcription and copy-choice replication. At first glance, recombination appears to be an unlikely event. However, results of recent experiments indicate that recombination may be more frequent than expected. Splenocytes from human immunodeficiency virus-infected patients harbor as many as three or four distinct proviral genomes per cell, and give rise to huge numbers of recombinants. Antigenic shift arising from recombination is a double-edged sword for pathogen and host: virions can escape elimination by immune defenses with serious consequences to the host, but if the immune system is functional, CTLs and antibodies will recognize the new combinations and broaden the immune repertoire against the infected cell.

Jung, A., R. Maier, J.-P. Vartanian, G. Bocharov, V. Jung, U. Fischer, E. Meese, S. Wain-Hobson, and A. Meyerhans. 2002. Multiply infected spleen cells in HIV patients. *Nature* **418**:144.

Acute Infections Present Common Public Health Problems

An acute infection is most frequently associated with serious epidemics of disease affecting millions of individuals every year (e.g., polio, influenza, and measles). The nature of an acute infection presents difficult problems for physicians, epidemiologists, drug companies, and public health officials. The main problem is that by the time people feel ill, or mount a detectable immune response, most acute infections are essentially complete, and the infection has spread to the next host. Such infections can be difficult to diagnose retrospectively, or to control in large populations or in crowded environments (e.g., day care centers, military camps, college dormitories, nursing homes, schools, and offices). Effective antiviral drug therapy requires treatment early in the infection, often before symptoms are manifested, because by the time the patient feels ill, the infection has been resolved. Antiviral drugs can be given in anticipation of an infection, but this strategy demands that the drugs be safe and free of side effects. Moreover, as we discuss in Chapter 19, our arsenal of antiviral drugs is very small, and drugs effective for most common acute viral diseases simply do not exist.

Persistent Infections

Definition and Requirements

Persistent infections, like acute infections, begin when innate defenses are modulated or bypassed. However, unlike an acute infection, a persistent infection is not cleared efficiently by the adaptive immune response. Instead, virus particles or products continue to be produced for long periods. Virions may be produced continuously or intermittently for months or years (Fig. 16.1). In some instances, viral genomes remain long after viral proteins can no longer be detected. Distinctions have been made between persistent infections that are eventually cleared (**chronic infections**) and those that last the life of the host (latent infections or slow infections).

A surprising number of infections follow the persistent pattern (Table 16.2). For example, some arenaviruses, such as lymphocytic choriomeningitis virus, are inherently noncytopathic in their natural hosts and maintain a persistent infection if the host cannot clear the infected cells. Other life cycles, like those of Epstein-Barr virus, have alternative transcription and replication programs that maintain the viral genome in some cell types with no production of viral particles. Ubiquitous infections, such as those produced by adenoviruses, circoviruses, and human herpesvirus 7, persist uneventfully in most human populations by as yet uncharacterized mechanisms. Adenoviruses can be isolated from lymphoid tissue, including adenoids and tonsils, in most respiratory infections, but cultured lymphoid cells do not support efficient viral replication. It is possible that delayed kinetics of infection and replication observed in these cells contribute to the long-term maintenance of adenovirus in a fraction of lymphoid cells. What is clear from these examples is that no single mechanism is responsible for establishing the persistent infection. However, one common theme does emerge: when viral cytopathic effects and host defenses are reduced, a persistent infection is likely.

Table 16.2 Some persistent viral infections of humans

Virus	Site of persistence	Consequence
Adenovirus	Adenoids, tonsils, lymphocytes	None known
Epstein-Barr virus	B cells, nasopharyngeal epithelia	Lymphoma, carcinoma
Human cytomegalovirus	Kidney, salivary gland, lymphocytes?, macrophages?, stem cells?, stromal cells?	Pneumonia, retinitis
Hepatitis B virus	Liver, lymphocytes	Cirrhosis, hepatocellular carcinoma
Hepatitis C virus	Liver	Cirrhosis, hepatocellular carcinoma
Human immunodeficiency virus	$CD4^+$ T cells, macrophages, microglia	AIDS
Herpes simplex virus types 1 and 2	Sensory and autonomic ganglia	Cold sore, genital herpes
Human T-lymphotropic virus types 1 and 2	T cells	Leukemia, brain infections
Papillomavirus	Skin, epithelial cells	Papillomas, carcinomas
Polyomavirus BK	Kidney	Hemorrhagic cystitis
Polyomavirus JC	Kidney, central nervous system	Progressive multifocal leukoencephalopathy
Measles virus	Central nervous system	Subacute sclerosing panencephalitis, measles inclusion body encephalitis
Rubella virus	Central nervous system	Progressive rubella panencephalitis
Varicella-zoster virus	Sensory ganglia	Zoster (shingles), postherpetic neuralgia

An Ineffective Intrinsic or Innate Immune Response Can Result in Persistent Infection

Apoptosis is a common intrinsic cellular defense that can limit viral replication, but it can also affect patterns of infection. For example, in some vertebrate cell lines, Sindbis virus infection is acute and cytopathic because apoptosis is induced. When virus-induced apoptosis is blocked by synthesis of cellular proteins (e.g., Bcl-2), a persistent infection occurs. Similarly, a persistent infection is established in cultured, postmitotic neurons because these cells produce inhibitors of the cell death pathway. A corollary of the in vitro experiment can be observed in animals. When virions are injected into an adult mouse brain, a persistent noncytopathic infection is established. In contrast, when the same preparation is injected into neonatal mouse brains, the infection is cytopathic and lethal. Examination of cellular regulators of apoptosis indicates that Sindbis virus infection induces cell death in neonatal mouse brains because cellular inhibitors are not produced as they are in adult neurons.

The host IFN response plays unexpected roles in establishing patterns of infection. For example, bovine viral diarrhea virus, a pestivirus in the *Flaviviridae* family, establishes a lifelong persistent infection in the vast majority of cattle around the world. Remarkably, persistently infected animals have no detectable antibody or T-cell responses to viral antigens. Cytopathic and noncytopathic strains (called biotypes) have been useful in understanding how a persistent infection occurs. Infection of pregnant cattle with the noncytopathic biotype during the first 120 days of pregnancy results in birth of persistently infected offspring. In contrast, infection of the fetus by cytopathic virus is contained quickly and eliminated; no persistent infection is established. A mechanism to block infection by the cytopathic, but not the noncytopathic, biotype must exist. The fetal IFN response dictates the pattern of infection by both. One idea is that IFN in fetal tissue blocks establishment of the cytopathic infection and stimulates rapid adaptive immune system clearance. In contrast, noncytopathic infection of fetal tissue does not stimulate production of IFN, the adaptive immune system is not activated, and because infection does not kill cells, a persistent infection is established.

Perpetuating a Persistent Infection by Modulating the Adaptive Immune Response

Interference with Production and Function of MHC Proteins

The CTL response is one of the most powerful adaptive host defenses against viral infection. This response depends, in part, upon the ability of host T cells to detect viral antigens present on the surfaces of infected cells, and to kill them. Recognition requires the presentation of viral peptides by major histocompatibility complex (MHC) class I proteins. The pathway by which endogenous peptide antigens are produced and transported to the cell surface is discussed in Chapter 15 (see Fig. 15.19). Obviously, any mechanism that prevents viral peptides from appearing in MHC class I complexes, even transiently, provides an important selective advantage. Perhaps unexpectedly, MHC I production is modulated after some acute infections (Table 16.3). Presumably, the strategy is valuable late in the in-

Table 16.3 Viral regulation of MHC class I antigen synthesis

Virus	Observed effect or postulated mechanism
RNA	
Human immunodeficiency virus type 1	Tat protein-induced reduction in MHC class I gene transcription; Vpu interferes with an early step in synthesis of MHC class I; Nef promotes endocytosis of MHC class I from the cell surface
Mouse hepatitis virus	Decrease in transcription of specific MHC class I genes
Respiratory syncytial virus	Decrease in MHC class I gene transcription
Poliovirus	Protein 3A inhibits transport of vesicles containing MHC class I proteins
DNA	
Adenovirus	E3 gp19kDa retains MHC class I in endoplasmic reticulum (ER); reduced transcription of MHC class I genes
Epstein-Barr virus	EBNA-4 may block production of antigenic peptides or their transport from the cytosol to the ER; allele-specific decrease in MHC class I appearance on cell surface
Human cytomegalovirus	US3 retains MHC class I molecules in the ER; US6 inhibits peptide translocation by Tap (ER luminal domain); US11 and US2 dislocate MHC class I molecule from the ER lumen to the cytosol by different mechanisms; UL10 slows rate of MHC class I export from ER; UL83 blocks IE1 peptide presentation
Herpes simplex virus	ICP47 binds to Tap transporter and blocks import of peptides into the ER
Vaccinia virus	Lower abundance of MHC class I on cell surface by unknown mechanisms

fection to delay elimination by early-acting CTLs so that sufficient progeny can be disseminated. However, no direct test of this presumption has been provided.

Several viral proteins block MHC class I function at various points in the pathway (Fig. 16.5 and Table 16.3). The MHC class I pathway is of obvious importance in immunology, but the existence of many MHC-processing or regulatory steps was not known until these viral inhibitors were characterized. Peptide presentation by MHC class I proteins can be reduced by lowering expression of the

Figure 16.5 Viral proteins block cell surface antigen presentation by the MHC class I system. In almost every cell, a fraction of newly synthesized proteins translated in the cytoplasm is targeted to the proteasome, where the molecules are digested into peptide fragments (orange). These peptides are transported to the lumen of the endoplasmic reticulum (lavender) by the Tap transporters in the endoplasmic reticulum membrane (pink channel). Peptides then bind to a cleft in the newly synthesized MHC class I protein complex (blue) consisting of an MHC class I heavy chain and a β_2-microglobulin chain. The complete peptide-MHC class I complex moves into the Golgi apparatus (light blue) and then to the cell surface, where it can be recognized by T-cell receptors present on CD8+ T cells that are interacting with the infected cell. Specific viral gene products block (red bars) this process at almost every step along the pathway. Green arrows indicate stimulation. In the nucleus, transcription of MHC class I genes can be blocked by E1A or Tat at the promoter (P), as indicated. HSV, herpes simplex virus; hCMV, human cytomegalovirus; mCMV, mouse cytomegalovirus; HIV, human immunodeficiency virus.

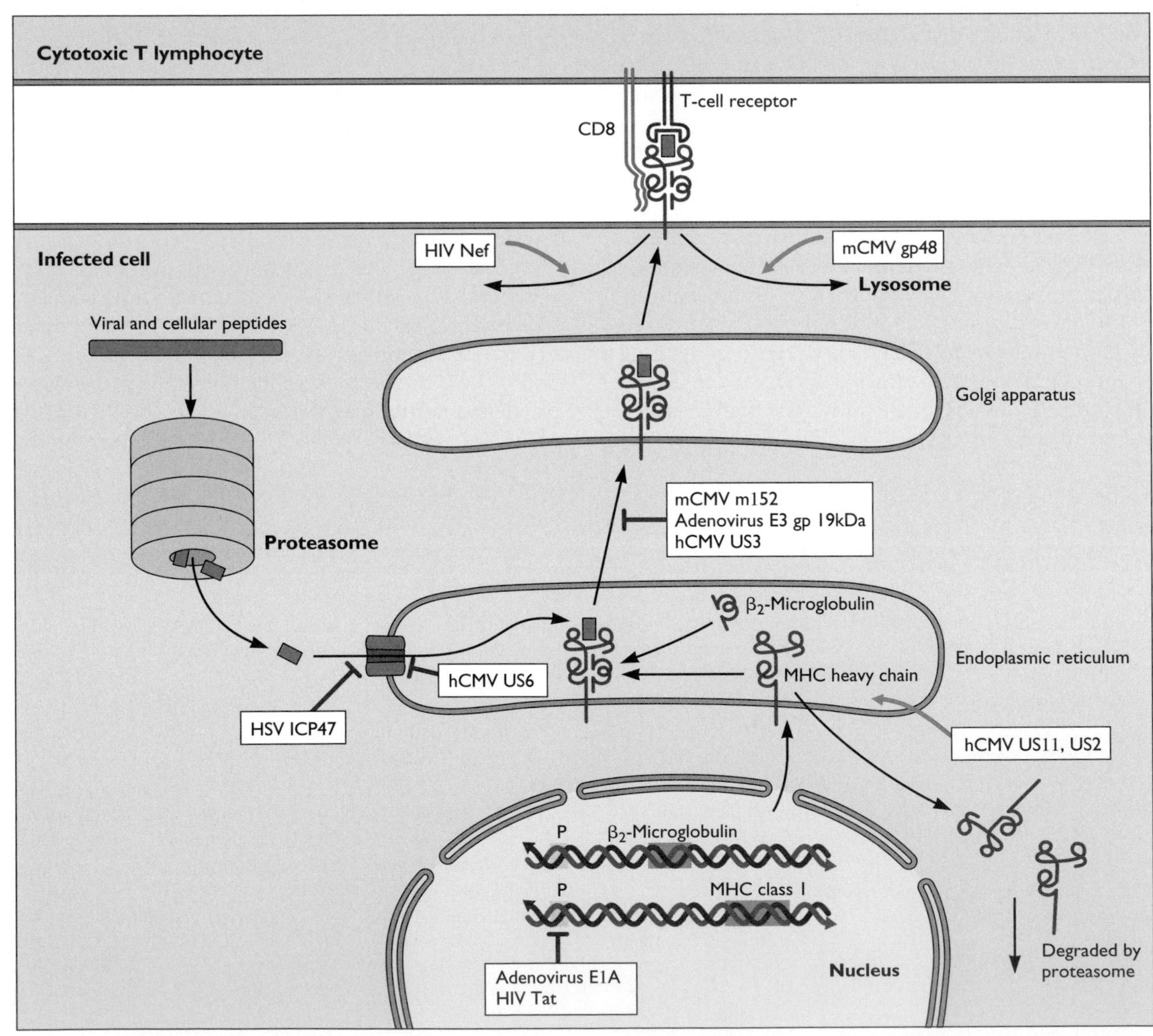

MHC genes directly, by blocking the production of peptides by the proteasome, or by interfering with subsequent assembly and transport of the MHC-peptide complex to the cell surface.

Human cytomegalovirus, a betaherpesvirus, causes a common childhood infection with inapparent to mild effects in healthy individuals (Appendix B). It is only when an infected person is immunosuppressed by drugs during transplantation, or by human immunodeficiency virus infections, that infections cause a life-threatening disease. A latent infection is established in cells of the monocyte/macrophage lineage, but mechanisms of establishment, maintenance, and **reactivation** remain to be discovered. Human cytomegalovirus deserves special mention because MHC class I presentation of viral antigens is inhibited by multiple mechanisms. The viral US6 protein inhibits peptide translocation into the endoplasmic reticulum lumen. The viral US3 protein binds to, and detains, MHC class I proteins in the endoplasmic reticulum, while the US11 and US2 proteins eject MHC class I molecules from that organelle into the cytoplasm, where they are degraded by the proteasome. Why human cytomegalovirus encodes so many proteins to block antigen presentation remains an open question. One possibility is that multiple gene products act additively or synergistically. Another idea is that concentrations of viral proteins may vary in different cell types.

Ubiquitination of proteins is an important regulatory mechanism that directs endocytosis, sorting, and degradation. The genomes of many gammaherpesviruses and poxviruses encode a class of zinc-binding RING finger proteins with E3 ubiquitin ligase activity, which disable the adaptive immune response, as well as stimulate replication and inhibit apoptosis. The K3 and K5 genes of human herpesvirus 8 and the MK3 gene of murine gammaherpesvirus 68 encode such proteins. The K5 and MK3 proteins are related type III transmembrane proteins, but act at surprisingly different steps in the MHC class I pathway. The K5 protein reduces the concentrations of MHC class I proteins, as well as costimulatory molecules, present on the surfaces of infected cells by adding ubiquitin to the cytoplasmic domains. This modification stimulates endocytosis of the marked proteins. The related MK3 protein is also an E3 ubiquitin ligase, but in this case, ubiquitination promotes proteasomal destruction of MHC class I proteins soon after they appear in the endoplasmic reticulum. The genome of myxoma virus encodes a similar RING finger E3 ligase called MV-LAP that also directs proteasomal destruction by a mechanism analogous to that used by the K5 protein. Importantly, while the effects of K5 protein upon human infections cannot be assessed, myxoma virus mutants lacking the MV-LAP gene are markedly attenuated in rabbits (the natural host).

Another strategy for avoidance of CTL destruction is exemplified by Epstein-Barr virus. Early observations indicated that Epstein-Barr virus-infected individuals do not produce CTLs capable of recognizing the viral protein EBNA-1. This phosphoprotein is found in the nuclei of latently infected cells, and is a protein regularly detected in malignancies associated with the virus (see Table 16.5). T cells specific for other Epstein-Barr virus proteins are made in abundance, indicating that EBNA-1 must possess some special features. Indeed, this protein contains an amino acid sequence with remarkable activities. It inhibits the host proteasome so that relevant EBNA-1 peptides are not produced at all. This inhibitory sequence can be fused to other proteins to inhibit their processing and subsequent presentation of peptide antigens normally produced from them. The biological relevance of this mechanism is evident after acute infection of B cells. The adaptive immune system kills all productively infected cells, sparing only those rare cells that produce EBNA-1. These cells harbor a latent viral genome.

MHC Class II Modulation after Infection

In the exogenous pathway of antigen presentation, proteins are internalized and degraded to peptides that can bind to MHC class II molecules (Fig. 15.20). These complexes are transported to the cell surface, where they can be recognized by the CD4$^+$ T-cell receptor. Activated CD4$^+$ T cells stimulate the development of CTLs and help coordinate an antiviral response against the pathogen. Any viral proteins that can modulate the MHC class II antigen presentation pathway would therefore interfere with Th-cell activation.

We have only begun to understand the different ways in which viral gene products modulate the MHC class II pathways. For example, the human cytomegalovirus US2 protein has been reported to promote proteasomal destruction of the class II DR-alpha and DM-alpha molecules. The Epstein-Barr virus BZLF2 protein interacts with intracellular and cell surface MHC class II molecules to block T-cell activation. Herpes simplex virus strain KOS infection results in removal of the MHC class II complex from the endocytic compartment. The human immunodeficiency virus Nef protein triggers the rapid internalization of CD4 and MHC molecules from the cell surface. In endosomal compartments, Nef may interfere with acidification, affecting the loading of antigenic peptides onto MHC class II proteins.

Bypassing Deadly CTLs by Mutation of Immunodominant Epitopes

In many viral infections, the Th cells and CTL populations found after replication are surprisingly limited, and the T cells respond to very few viral peptides. These peptides are said to be **immunodominant**. An extreme example of a limited CTL response is observed after infection of C57BL/6 mice with herpes simplex virus type 1. The virus-specific CTLs respond **almost entirely to a single peptide** in the viral envelope protein gB (the amino acid sequence of this peptide is SSIEFARL). Given that there are more than 85 open reading frames in the viral genome, it is remarkable that CTLs recognize only one peptide in this particular animal model of infection.

Focusing of the T-cell response upon a small repertoire of viral peptides provides a simple means to bypass T-cell recognition. A limited number of mutations in the coding sequence for these immunodominant peptides will render the infected cell invisible to the T-cell response produced early in the infection. Viruses with these mutations are called **CTL escape mutants** and are thought to contribute to progressive accumulation of virions because of decreased clearing of infected cells. For example, CTL escape mutants, which are of central importance in human immunodeficiency virus pathogenesis, arise because of error-prone replication and the constant exposure to an activated immune response. In some well-documented cases, the T-cell peptide sequence is completely deleted from the viral protein synthesized by the CTL escape mutant. Understanding how immunodominant peptides are selected, maintained, and bypassed is essential if effective vaccines against human immunodeficiency virus are to be developed. For example, a vaccine directed toward a dominant T-cell peptide that is part of a critical structural motif in a viral protein may have value, because CTL escape mutants will be less likely to survive and participate in subsequent spread of infection.

Immunodominant epitopes and CTL escape mutants play central roles in the increasingly common and dangerous infection caused by hepatitis C virus (more than 70 million people worldwide are infected [Appendix B]). The robust CTL response associated with acute infection is effective in less than 20 to 30% of individuals. An insidious persistent infection remains in the vast majority of cases. After several years, this persistent infection can lead to serious liver damage, and even fatal hepatocellular carcinoma. Persistently infected chimpanzees harbor viruses with CTL escape mutations in their genomes. In contrast, the viral population isolated from animals that resolved their acute infections rapidly included no such mutants. The conclusions from this work are clear. If CTL escape mutations occur early, a persistent infection is likely. If CTLs resolve the infection before escape mutants appear, no persistent infection is possible. This simple conclusion implies that early analysis of the viral population and CTL response during acute infection may be a valuable diagnostic tool to guide treatment.

The CTL epitope need not be deleted or mutated radically to escape CTL recognition. When a T cell specific for a given viral peptide engages a similar, **but not identical,** peptide complexed to MHC class I, the T cell may respond partially or not at all. In this case, both mutant and parent viruses are likely to be maintained in the population. Altered viral peptides of this kind have been identified in viral isolates from persistent hepatitis B infections.

Killing Activated T Cells

Sometimes, when the CTL engages an infected cell, the CTL dies instead of its target. This unexpected turn of events is a remarkable example of viral defense. Activated T cells produce a membrane receptor on their surfaces called Fas. Fas is related to the tumor necrosis factor family of membrane-associated cytokine receptors, and it binds a membrane protein appropriately called Fas ligand (FasL). When Fas on activated T cells binds FasL on target cells, the receptor trimerizes, triggering a signal transduction cascade that results in apoptosis of the T cell (see Fig. 15.1). If viral proteins increase the quantity of FasL on the cell surface, any engaging T cell (Th or CTL) will be killed. Such a mechanism has been suggested to explain the relatively high frequency of "spontaneous" T-cell apoptosis in human immunodeficiency virus-infected patients. The viral Nef, Tat, and SU proteins, human T-lymphotropic virus Tax protein, and an uncharacterized protein produced in human cytomegalovirus-infected cells have all been implicated in increased production of FasL and resulting T-cell apoptosis. This seemingly unusual mechanism to kill T cells is not unique to viral infections. If it were, evolution would have removed the Fas system long ago. Indeed, Fas-mediated CTL killing is a normal activity of most complex organisms: it removes T cells when they are no longer needed after infection, or when their presence in a tissue is detrimental. For example, certain delicate and irreplaceable tissues, such as the eye, remain free of potentially destructive T cells by maintaining a high concentration of FasL on cell surfaces.

Infections of Tissues with Reduced Immune Surveillance

Cells and organs of the body differ in the degree of their immune defense (see Box 15.10). When virus particles infect such tissues, a persistent infection may be established if the organ is not otherwise compromised. Tissues with surfaces exposed to the environment (e.g., skin, glands, bile ducts, and kidney tubules) are exposed routinely to foreign matter, and therefore have a higher threshold for activating immune defenses. Persistent infections by members of the *Papillomaviridae* and the *Betaherpesvirinae* are common in these tissues. By replicating in cells present on lumenal surfaces of glands and ducts with poor immune surveillance (kidney, salivary, and mammary glands), human cytomegalovirus particles are shed almost continually in secretions. Possibly the most extreme example of immune avoidance is represented by the papillomaviruses that cause skin warts. Productive replication of these infectious particles occurs only in the outer, terminally differentiated skin layer, where an immune response is impossible. Dry skin is continually flaking off, ensuring efficient spread of infection. One can verify this assertion by running a finger along a clean surface in the most hygienic hospital and noticing a white film, which is 70 to 80% keratin from human skin. Molecular biologists often discover this abundance of dried skin in the laboratory when examining silver-stained protein gels; the major band is often contaminating human keratin.

Certain compartments of the body, such as the central nervous system, vitreous humor of the eye, and areas of lymphoid drainage, are devoid of initiators and effectors of the inflammatory response, simply because these tissues can be damaged by the fluid accumulation, swelling, and ionic imbalances that characterize inflammation. In addition, because most neurons do not regenerate, immune defense by cell death is obviously detrimental. Accordingly, persistent infections of these tissues are common (Table 16.2).

Direct Infection of Cells of the Immune System

It is the rare virus that cannot infect a subset of cells in the immune system. In many cases, infected lymphocytes and monocytes migrate to the extremes of the body, providing easy transport of virions to new areas and hosts. In other cases, the function of infected lymphocytes and monocytes is compromised, resulting in some degree of immunosuppression directly proportional to the type and number of immune cells that are infected. Immunosuppression has far-reaching consequences for host and virus (see "Virus-induced immunosuppression," Chapter 15).

Human immunodeficiency virus provides a powerful reminder of how insidious infection of the immune system can be (see Chapter 17). The virus infects not only $CD4^+$ Th cells, but also monocytes and macrophages that can transport virions to the brain and other organs. Professional antigen-presenting cells, such as the dendritic cells in the spleen discussed in Chapter 15, bind virions and carry them to lymph nodes where T cells can be infected. Because of continuous replenishment of immune cells, an untreated infected individual continues to produce prodigious quantities of virions for years in the face of a highly engaged immune system. The early stages of disease are characterized by this persistent immune activation, and not by immunosuppression. It is only at the end stage of disease, when viral replication outpaces replenishment of immune cells and immune defenses, that massive immunosuppression occurs. Only then do the virus and other secondary infections spread in uncontrolled fashion, and the patient succumbs. As we will discuss below, some infections, such as those caused by measles virus, produce a transient immunosuppression that affects the progression of pathogenesis.

Two Viruses That Cause Persistent Infections

Measles Virus

Measles virus, a member of the family *Paramyxoviridae,* is a common human pathogen with no known animal reservoir (Appendix B). The genome organization and replication strategy are similar to those of the rhabdovirus vesicular stomatitis virus (Appendix A). Measles virus is a highly adapted human pathogen, persisting only in populations sufficient to produce a large number of new hosts (children). Population geneticists calculate that communities of 200,000 to 500,000 individuals are required in order to maintain measles virus. Measles is one of the most contagious human viruses, with about 40 million infections occurring worldwide each year, resulting in 1 to 2 million deaths (predominantly of children). Normally, a single infection protects the individual for life.

Cellular receptors for measles virus include the proteins CD46 and CD150. While CD150 is the receptor used by all tested strains, CD46 is the receptor used predominantly by vaccine- and laboratory-adapted viruses. Despite a rather broad distribution of these receptors, the site of primary infection is the respiratory tract. After primary replication, measles virus infects local monocytes and lymphoid cells that migrate to draining lymph nodes. After replication in these tissues, a small proportion of monocytes, B cells, and T cells are infected and enter the circulation. Secondary infections of lymph tissues result in a secondary viremia, and replication continues in the epithelial cells of the lung and the mouth. The characteristic mouth lesions, called Koplik spots, are caused by a delayed-type hypersensitivity reac-

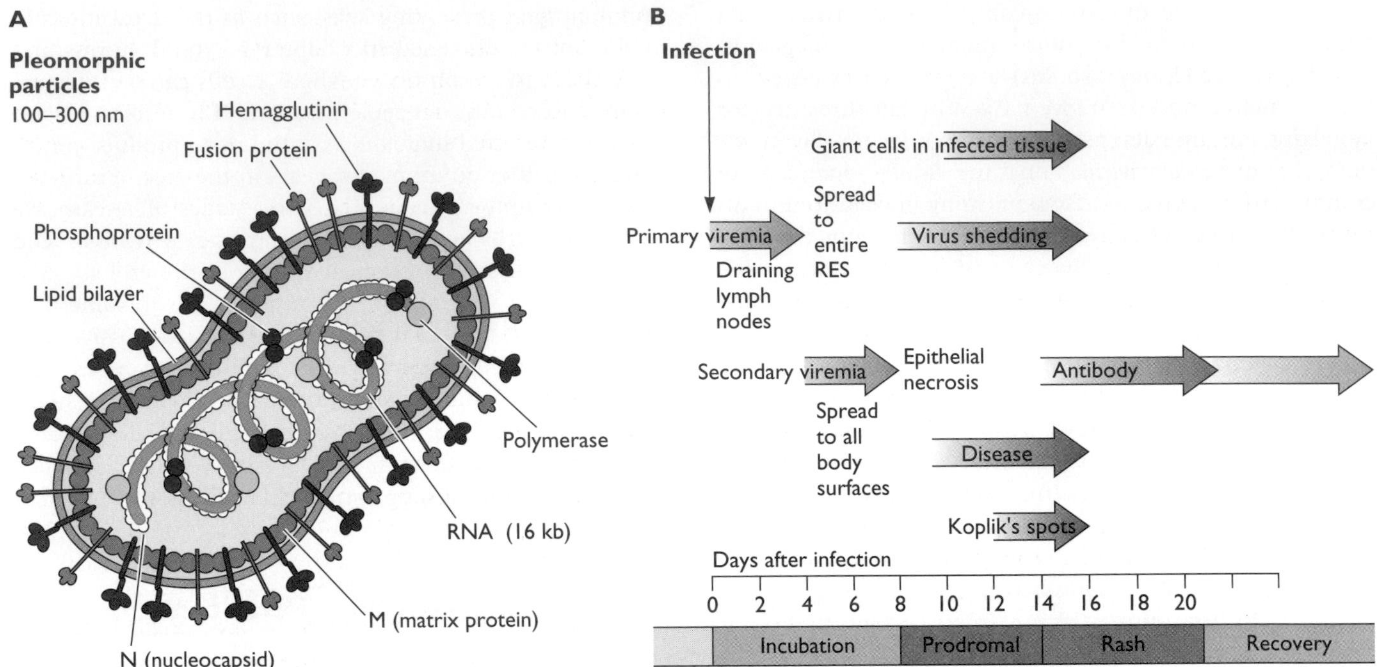

Figure 16.6 Infection by measles virus. (A) Diagrammatic representation of the structure of the pleomorphic measles virion. (B) Course of clinical measles infection and events occurring in the spread of the infection within the body. Four clinically defined temporal stages occur as infection proceeds (illustrated at the bottom). Characteristic symptoms appear as infection spreads by primary and secondary viremia from the lymph node to the entire reticuloendothelial system (RES) and finally to all body surfaces. The timing of typical reactions that correspond to the clinical stages is shown by the colored arrows. The telltale spots on the inside of the cheek (Koplik's spots) and the skin lesions of measles consist of pinhead-sized papules upon a reddened, raised area. They are typical of immunopathology produced in response to measles virus proteins. Redrawn from A. J. Zuckerman et al., *Principles and Practice of Clinical Virology*, 3rd ed. (John Wiley & Sons, Inc., New York, N.Y., 1994), with permission.

tion, analogous to that responsible for the typical measles skin rash. The course of acute infection—so-called uncomplicated measles—runs about 2 weeks (Fig. 16.6). An acute infection causes cough, fever, and conjunctivitis and confers lifelong immunity.

Upon closer examination, the hallmarks of acute measles infection are sinister and presage the subsequent, potentially serious problem of persistent infection. Such hallmarks include the following: measles virus kills cells by cell-cell plasma membrane fusion, not by shutting off host macromolecular synthesis; many tissues can be infected, including the nervous system and the immune system; the majority of acutely infected individuals exhibit an abnormal electroencephalograph indicating subtle neurological damage; and finally, an initial acute infection results in dissemination via viremia and, most important, a substantial immunosuppression.

The vast majority of measles victims have an uneventful recovery, but the immunosuppressive effect lasts for a week or two after the infection is resolved. Consequently, secondary infections during this period may be uncontested by host defenses with serious or fatal consequences if immediate intervention and care are not provided. The large number of children in the Third World who die after measles infection succumb to complications of secondary infections. The molecular basis for immunosuppression is not yet fully understood, but several mechanisms have been suggested (see also Chapter 15). Infected T and B cells, as well as macrophages, are arrested in the late G_1 phase of the cell cycle and cannot perform their normal functions. Uninfected lymphocytes can also be affected by direct contact with viral membrane proteins present on the surface of infected cells. In addition, when the virions bind to the CD46 receptor, interleukin-12 production is suppressed and the Th1 immune response is not efficiently activated.

On rare occasions, measles virus genomes and structural proteins are not cleared by the adaptive immune system, and may persist for years in an infected individual. The mechanisms responsible are only now being characterized. Some studies in humans infected with measles virus reported a correlation between the extent of antibody production and persistent infection. The significance of this correlation has been debated, but some insight has come from studying the effect of antibodies during infection of cultured cells. Measles virus-specific antibodies bind to viral membrane proteins present on cell surfaces to induce endocytosis and proteolysis of the antibody-protein complexes. Because one of these proteins, the viral fusion protein, is responsible for the cell-cell fusion that causes cell death, exposure to antibodies effectively blocks cell killing, and thereby allows viral persistence in cultured cells.

An important finding is that measles virus can enter the brain by infecting lymphocytes that traverse the body during the viremia following primary infection. Such a secondary infection of a tissue with reduced immune surveillance has a number of consequences. One is **acute postinfectious encephalitis,** which occurs in about 1 in 3,000 infections. The other is a rare, but delayed and often lethal, brain infection called **subacute sclerosing panencephalitis (SSPE)**. This disease is a manifestation of a slow infection, an unusual variation of a persistent infection (see "Slow Infections" below and Fig. 16.1). After young adults and children contract measles, about one in a million develop SSPE with a 6- to 8-year incubation period. SSPE is more likely to occur if children are infected in their first year or two of life than if they are infected later. This disease begins in rare infected cells of the brain. In these cells, viral gene expression, especially synthesis of envelope proteins, is reduced. In addition, fully assembled particles cannot be detected in brains of afflicted patients. An altered matrix (M) protein may be responsible for ineffective virion assembly. Even though assembly of mature virions is not observed, nucleoprotein complexes are produced, and infectious genomes spread between synaptically connected neurons. The mechanism of such spread in the absence of assembled virions is of some interest because it does require the viral fusion protein, but not viral receptors.

We lack testable hypotheses that relate these provocative findings to the mechanism of persistent infection. Important questions remain. Does transient immunosuppression during an acute infection facilitate infection of the brain? Are defects in M protein synthesis and particle assembly necessary and sufficient to cause disease, or are they effects of other selection processes in the brain? Can SSPE be produced in the absence of an acute infection? Are the defects in viral gene expression the cumulative results of selection after years of exposure to host defenses? Transgenic mice producing the human measles virus receptors are now available and should allow these important questions to be addressed in a rigorous and controlled fashion.

Lymphocytic Choriomeningitis Virus

Lymphocytic choriomeningitis virus, a member of the family *Arenaviridae,* was the first virus associated with aseptic meningitis in humans. Perhaps more important, in recent years its study has illuminated fundamental principles of immunology and viral pathogenesis, particularly those that underlie persistent infection and CTL recognition and killing. It was noted early on that the infection spreads from rodents (the natural host) to humans, in whom it can cause severe neurological and developmental damage. Infected rodents normally excrete large quantities of infectious virions in feces and urine throughout their lives without any apparent detrimental effect. These mice are called "carriers" because of such lifelong production of virions. The carrier state is established for two reasons: infection is not cytopathic, and, if mice are infected congenitally or immediately after birth, viral proteins are not recognized as "foreign" and the infection is not cleared by the immune system. However, if virions are injected into the brains of healthy adult mice, the mice die of acute immunopathologic encephalitis. This disease is similar to that contracted by humans who develop aseptic meningitis after lymphocytic choriomeningitis virus infection.

Results from the mouse model demonstrate that CTLs are required both for clearing virus and for the lethal response to intracerebral infections. If adult mice are depleted of CTLs, direct injection of virus particles into the brain is no longer fatal. Instead, the mice produce infectious virions throughout their lifetimes, precisely as seen in persistent infections of neonates. When virus-responsive CTLs are added back to neonates with a persistent infection, the infection is cleared after several weeks. In the case of neonatal persistent infection, the brain is not infected, so no encephalitis is promoted by the activated T cells. How the neonatal infection effectively "silences" the effector system from clearing the infection is currently under investigation. Recent experiments have implicated an active process of clonal deletion of T lymphocytes that are capable of recognizing the dominant lymphocytic choriomeningitis virus peptides.

Latent Infections

An Extreme Variant of the Persistent Infection

Latent infections can be characterized by three general properties: expression of productive cycle viral genes is absent or inefficient; immune detection of the cell harboring the latent genome is reduced or eliminated; and the viral genome itself persists intact so that at some later time a productive acute infection can be initiated to ensure spread of its progeny to a new host (Fig. 16.1). The latent genome can be maintained as a nonreplicating chromosome in a nondividing cell such as a neuron (e.g., herpes simplex virus, varicella-zoster virus), become an autonomous, self-replicating chromosome in a dividing cell (e.g., Epstein-Barr virus), or be integrated into a host chromosome (e.g., adeno-associated virus).

Such "long-term parking" of a viral genome in the latent infection is remarkable for its stability: a balance among the regulators of viral and cellular gene expression must be maintained. Generally, only a restricted set of viral gene products is made. Viral proteins required for productive replication may not be produced at all, a pattern exemplified by varicella-zoster virus (Fig. 16.3). This virus establishes a latent infection in neurons of the peripheral nervous system that can reactivate to produce infectious virions years after the primary infection. The disease produced after reactivation is called **shingles.** In contrast, several viral proteins may be required to maintain the latent infection, as is the case for Epstein-Barr virus, a herpesvirus that is widespread in human populations (Table 16.4). A latent infection is established in dividing B lymphocytes, and at least nine viral proteins needed to replicate the viral genome and modulate the host immune response are produced. Reactivation normally is not associated with disease.

If latency is to have any value as a survival strategy, there must exist a mechanism for reactivation so that infectious virions can spread to other hosts. Reactivation may be spontaneous, or follow trauma, stress, or other insults, conditions that may mark the host as unsuitable for continuing the latent infection. In the case of herpes simplex virus, reactivation can also provide a mechanism for reinfection and establishment of a latent infection in more neurons in the same individual. This is the so-called "round-trip strategy": the latent infection reactivates, an acute infection follows at mucosal surfaces, and the progeny enter local neuronal termini to reinfect the ganglia.

Two Examples of Latent Infections

Herpes Simplex Virus

Over three-quarters of all adults in the United States have antibodies to herpes simplex virus type 1 or type 2 and, therefore, harbor latent viral genomes in their peripheral nervous system. Approximately 40 million infected individuals will experience recurrent herpes disease due to reactivation of their own personal viruses sometime in their lifetimes. Many millions more carry latent viral genomes in their nervous systems, but never report reactivated infections. Herpes simplex virus is an example of a well-adapted pathogen, as demonstrated by its widespread prevalence in humans, its only known natural hosts. Such

Table 16.4 Epstein-Barr virus gene products synthesized in the latent infection

Gene product	Function(s)
EBNA-1	Maintains replication of the latent Epstein-Barr virus genome during S phase of the cell cycle. It is a sequence-specific DNA-binding protein and binds to a unique origin of replication called *oriP* that is distinct from the origin used in the productive replication cycle.
EBNA-2	A transcriptional regulator that coordinates Epstein-Barr virus and cell gene expression in the latent infection by activating the promoters for the LMP-1 gene and cellular genes such as CD23 (low-affinity immunoglobulin E Fc receptor) and CD21 (the Epstein-Barr virus receptor, CD23 ligand, and receptor for complement protein C3d)
EBNA-LP	Required for cyclin D2 induction in primary B cells in cooperation with EBNA-2
EBNA-3A and EBNA-3C	Play important roles early in the establishment of the latent infection
LMP-1	An integral membrane protein required to protect the latent infected B cell from the immune response. LMP-1 stimulates the synthesis of several surface adhesion molecules in B cells, a calcium-dependent protein kinase, and the apoptosis inhibitor Bcl-2
LMP-2	An integral membrane protein required to block the activation of the *src* family signal transduction cascade; an inhibitor of reactivation from latency. Two spliced forms exist: LMP-2A and LMP-2B; LMP-2B lacks a receptor-binding domain and may act to modulate LMP-2A.
EBER-1 and EBER-2	Nonpolyadenylated, small RNA molecules that do not encode proteins; transcribed by RNA polymerase III; 166 and 172 nucleotides in length, respectively

BOX 16.4

Why people vary in susceptibility to herpes simplex virus infection: the hygiene hypothesis

More than 80% of the adult population in the developed world harbor latent herpesviral genomes in their peripheral nervous system. Some individuals suffer from lesions after reactivation while some never report symptoms or lesions. What accounts for the high infectivity yet marked diversity in host response to infection?

Hypothesis: The effectiveness of innate immunity to stimulate appropriate adaptive immune responses is conditioned by the individual's exposure to microbes early in life. In the extreme case, life in a supersanitized environment fails to stimulate innate immunity, which in turn, fails to prime the adaptive immune system.

The hygiene hypothesis asserts that the rising incidence of allergy and asthma, as well as of herpes simplex virus infections in Western societies, results from "supersanitized" living conditions. Normally, exposure to common microbes early in life primes the adaptive immune system to mount a Th1-dominated immune response in the face of later infections. Hypersanitized living avoids the early microbial stimulation of innate immunity. The adaptive immune response is misdirected, such that subsequent reactivation of latent herpesvirus leads to more severe symptoms. Testing the hygiene hypothesis is not an easy matter; many observations that apparently support or refute the hypothesis are anecdotal or poorly controlled. Nevertheless, the idea is sure to stimulate research and debate.

Rouse, B. T., and M. Gierynska. 2001. Immunity to herpes simplex virus: a hypothesis. *Herpes* **8**(Suppl. 1):2A–5A.

Comparison of hypersanitized and undersanitized environments for infants

Hypersanitized	Undersanitized
No or reduced infection	Frequent infection
Sterile baby food	Microbes in food and water
Late vaccination and no tuberculosis vaccination (*Mycobacterium bovis* BCG vaccine)	BCG vaccination and natural mycobacterial infections
No or reduced maternal antibody to common viruses such as herpes simplex virus	Passive immunization provided by mother
Outcome	**Outcome**
No activation of innate immunity	Activation of innate immunity
Infection by virus results in Th2 response	Infection by virus results in Th1 response
Primary disease more severe	Mild primary disease
Frequent recurrences and inappropriate memory response	Rare recurrences and effective memory response

success reflects the high efficiency of both the productive and latent infections. However, why some people are more likely than others to be infected by this ubiquitous virus is poorly understood (Box 16.4). No animal reservoirs are known, although several laboratory animals, including rats, mice, guinea pigs, and rabbits, can be infected. The alphaherpesviruses, of which herpes simplex virus type 1 is the type species, are unique in establishing latent infections predominantly in terminally differentiated, nondividing neurons of the peripheral nervous system. Indeed, most other neurotropic viruses (e.g., rabies virus) initiate infections of the peripheral nervous system, but then continue on to the central nervous system to cause devastating disease or death.

In general, most neurons neither replicate their DNA nor divide, and so once established, a viral genome need not replicate to persist. The herpesvirus latent infection of peripheral neurons is an effective survival mechanism, as we have no vaccines or antivirals that can attack the latent infection. Consequently, once infected with these viruses, the host is infected for life—latency is absolute persistence.

Many details of the molecular aspects of herpes simplex virus latency await discovery and explanation, but the general pathway is well established. Neurons of sensory and autonomic ganglia are infected following primary rounds of replication in cells of mucosal or epidermal surfaces (Fig. 16.7). The general outline of possible regulatory steps necessary for the establishment, maintenance, and reactivation of a viral infection is shown in Fig. 16.8. A typical primary infection of a mouse, showing the time course of production of infectious virus and establishment

Table 15.12 The major cell-mediated and humoral immune responses to viral infections

Response	Effector	Activity
Cell mediated	IFN-γ secreted by Th and cytotoxic T lymphocytes (CTLs)	Induces antiviral state
	CTLs	Destroy virus-infected cells
	NK cells and macrophages	Destroy virus-infected cells directly or by antibody-dependent cell-mediated cytotoxicity
Humoral	Primarily secretory IgA	Inhibits virion-host attachment
	Primarily IgG	Inhibits fusion of enveloped viruses with host membrane
	IgG and IgM antibody	Enhances phagocytosis (opsonization) after binding to virions
	IgM antibody	Agglutinates virions
	Complement activated via IgG and IgM	Lyses enveloped viruses; opsonization by C3b-antibody complex

bacteria synthesize the enzyme galactosyltransferase that attaches α-Gal to membrane proteins. Importantly, humans and higher primates do not make this antigen, as they lack the enzyme. Because of constant exposure to bacteria producing α-Gal in the gut, human serum contains high levels of antibodies specific for this antigen; indeed, more than 2% of the IgM and IgG populations is directed against this sugar. It is this antibody that triggers the complement cascade and subsequent lysis of foreign cells and enveloped viruses bearing α-Gal antigens. The anti-α-Gal antibody-complement reaction is probably the primary reason why humans and higher primates are not infected by enveloped viruses of other animals, despite the ability of many of these viruses to infect human cells efficiently in culture. In support of this assertion, when such viruses are grown in nonhuman cells, they are sensitive to inactivation by fresh human serum, but when grown in human cells, they are resistant. Anti-α-Gal antibodies provide a link for cooperation of the adaptive immune system and the innate complement cascade to provide immediate, "uninstructed" action.

Regulation of the Complement Cascade

Any amplified host antiviral defense must be fail-safe and regulated with precision. In particular, complement is a double-edged sword with a capacity to harm as well as to heal. For example, the alternative pathway, which culminates in lysis of cells, is activated spontaneously. Tight control of the complement cascade is achieved in a variety of ways. Some regulation is intrinsic to the complement proteins themselves. Many are large and therefore cannot leave blood vessels to attack infected tissues unless there is local tissue damage that exposes cells directly to blood. Consequently, minor infections do not activate a substantial complement response. Many cascade intermediates are short-lived, with millisecond half-lives, and therefore do not exist long enough to diffuse far from the site of infection. Further control is maintained by complement-inhibitory proteins present in the serum and by integral membrane proteins found on the surface of many cells (e.g., the complement receptor type 1 protein [Cr1], decay-accelerating protein [decay-accelerating factor, Daf, or CD55], protectin [CD59], and membrane cofactor protein [CD46]). These proteins are the only regulators that can limit the alternative pathway cascade by binding complement components such as C3b and C4b (Fig. 15.11). Enveloped viruses that do not carry CD55 or Cr1 on their surfaces are susceptible to the action of complement, particularly via the alternative pathway. Others, such as human immunodeficiency virus type 1 and the extracellular form of vaccinia virus, incorporate CD46, CD55, and CD59 in their envelopes and are thereby protected from complement-mediated lysis.

Many viral genomes encode proteins that interfere with the complement cascade. For example, alphaherpesvirus glycoprotein C binds the C3b component, herpesvirus saimiri ORF4 protein is a homolog of complement control proteins CD46 and CD55, and several poxvirus proteins bind C3b and C4. The smallpox virus SPICE protein (smallpox inhibitor of complement enzymes) inactivates human C3b and C4b and is a major contributor to the high mortality of smallpox virus (see Box 14.10).

Several viral receptors, including those for measles virus and certain picornaviruses, are complement control proteins. Epstein-Barr virus particles bind to CD21 (the Cr2 complement receptor) with profound consequences for the host and virus. This interaction activates the Nf-κb pathway, which then allows transcription from an important viral promoter. Epstein-Barr virus binding to the complement receptor enables replication in resting B cells otherwise incapable of supporting viral transcription.

Pattern Recognition by C1q and the Collectins

The action of the complement initiator protein C1q exemplifies the definitive property of innate defense: C1q can recognize molecular patterns characteristic of pathogens and not of self. It is a calcium-dependent, sugar-bind-

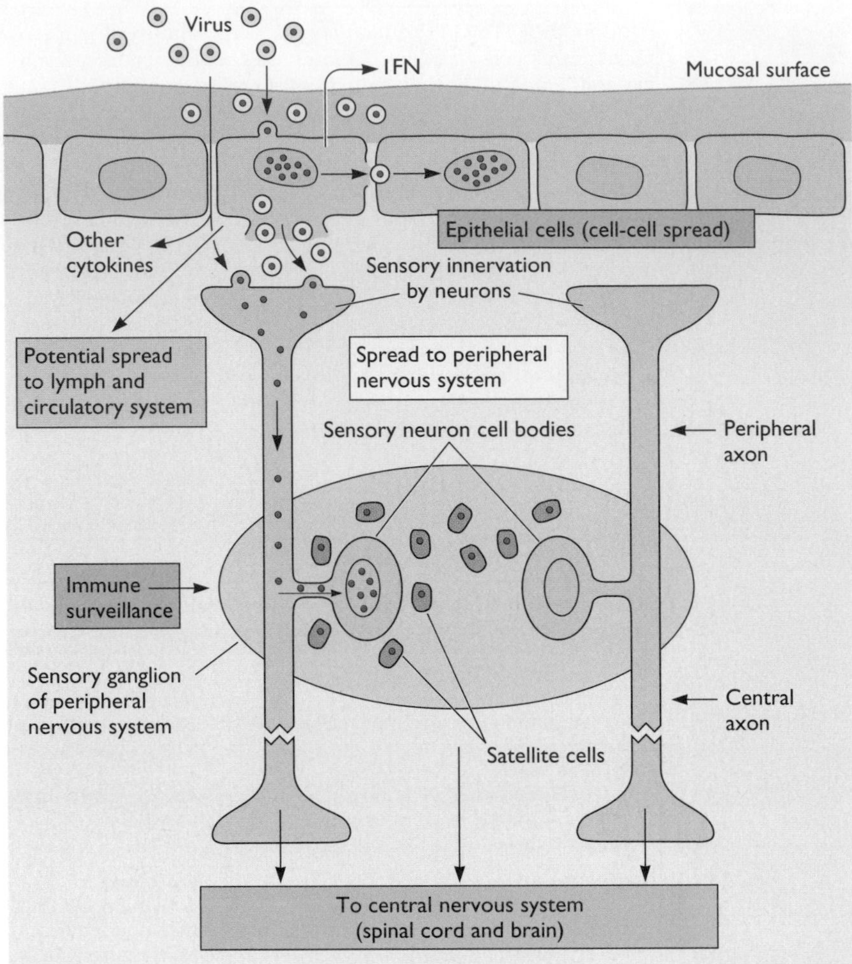

Figure 16.7 Herpes simplex virus primary infection of a sensory ganglion. Viral replication occurs at the site of infection (primary infection), usually mucosal surfaces, and the infection may or may not be apparent. Host innate defenses, including IFN and other cytokines, normally limit the spread of infection at this stage. Virions may infect local immune effector cells, including dendritic cells and infiltrating natural killer cells. The infection also spreads locally between epithelial cells and may enter lymph and the circulation. Virions will infect neurons that service the mucosal surfaces. Viral replication in epithelial cells may not be an absolute requirement for infection of neurons because replication-defective mutants can establish a latent infection in model systems. During the primary infection, virions attach to nerve endings of the local sensory and autonomic nerves innervating the area of infection. It is likely that the virion loses the envelope upon entry and the capsid is transported to the neuronal cell body by microtubule-based systems, where it delivers the viral DNA to the nucleus. In some neurons, many infectious virions are produced that then spread to other neurons in synaptic contact with the infected neuron (transsynaptic spread). Infection could spread among cells in the ganglion, as well as to other ganglia and the central nervous system (spinal cord and brain). Spread to the central nervous system from the peripheral nervous system is rare. Primary infection is normally confined to the ganglion. Viral proteins can also be detected in the nonneuronal satellite cells of the ganglion during the acute infection. Sensory and autonomic ganglia are in close contact with the bloodstream and are in direct contact with lymphocytes and humoral effectors of the immune system. Consequently, infected ganglia become inflamed and populated with lymphocytes and macrophages. Infection of the ganglion is usually resolved within 7 to 14 days after primary infection, virus particles are cleared, and a latent infection of some neurons in the ganglion is established (see also Fig. 16.9).

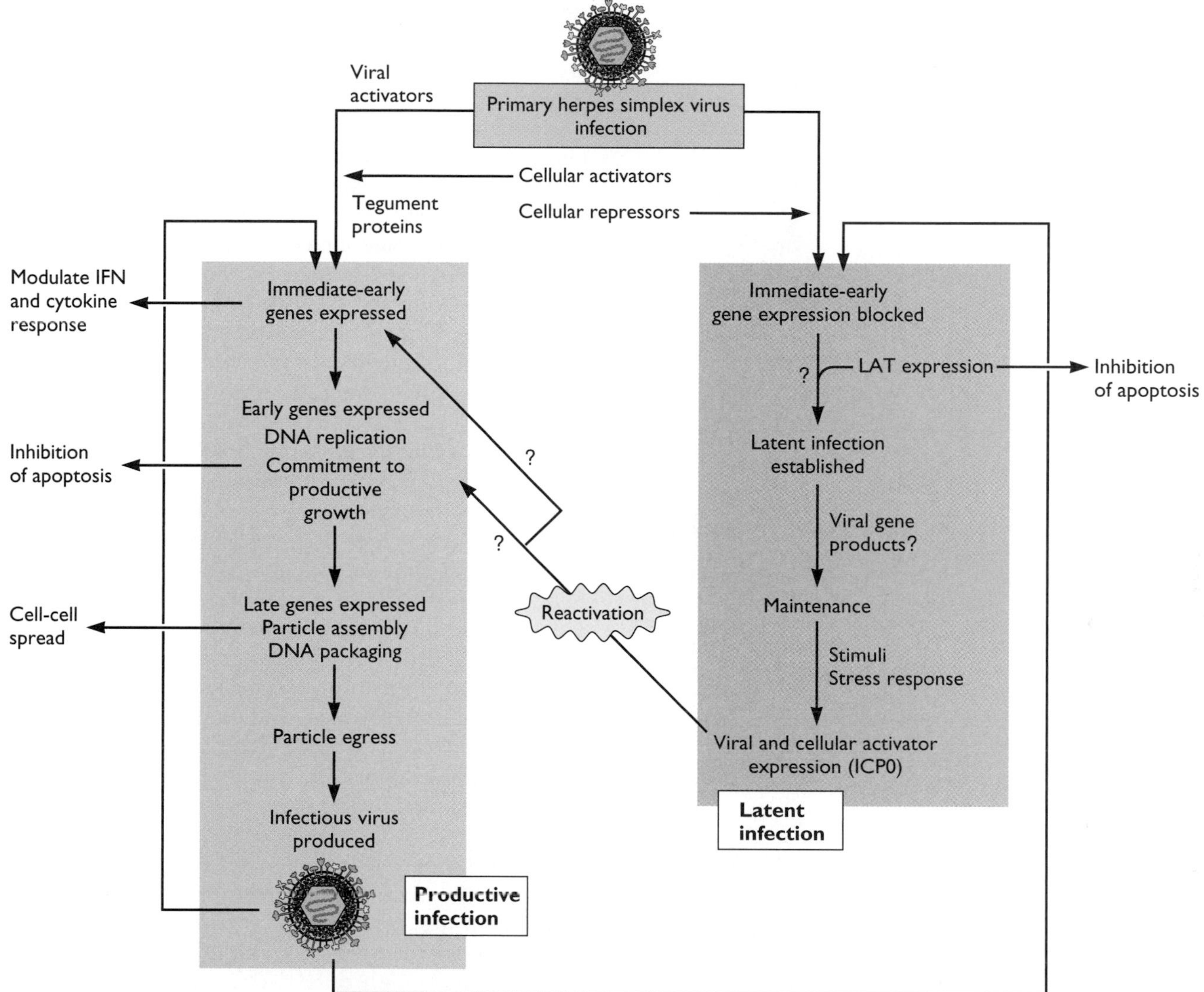

Figure 16.8 General flowchart for establishment, maintenance, and reactivation of a latent infection by herpes simplex virus. The green box at the top indicates the primary infection by virus particles at mucosal surfaces. The productive infection is shown by the pathway on the left, and the latent infection is indicated by the pathway on the right. The question marks indicate our lack of knowledge concerning synthesis and function of viral proteins at the indicated steps. Infectious particles produced by the productive pathway may infect other cells and enter either the productive or latent pathway as indicated. Infection can also spread from cell to cell without release of particles. Apoptosis induced by infection is inhibited by viral gene products. In addition, antiviral effects of IFN and other cytokines are modulated by viral gene products. The contribution of these processes in establishing the latent infection is not well understood. Reactivation is indicated by the diagonal arrow from the latent infection to the start of the productive infection. The question marks note the current controversy as to whether reactivation requires "going back to go" (immediate-early gene expression) or expression of early genes required for DNA replication. Experimental data indicate that synthesis of the immediate-early protein ICP0 is sufficient to activate the latent infection. Adapted from M. A. Garcia-Blanco and B. R. Cullen, *Science* **254:**815–820, 1991, with permission.

of a latent infection, is seen in Fig. 16.9. An often unappreciated fact is that peripheral ganglia undergo an acute infection with substantial production of virions followed by a strong inflammatory response. Nevertheless, 1 or 2 weeks after primary infection of the ganglion, infectious virions can no longer be isolated—the operational definition of an established latent infection. Inflammatory cells may persist in the infected ganglia for months. If the animal survives the primary infection, establishment of the latent infection is inevitable. The time frame for this

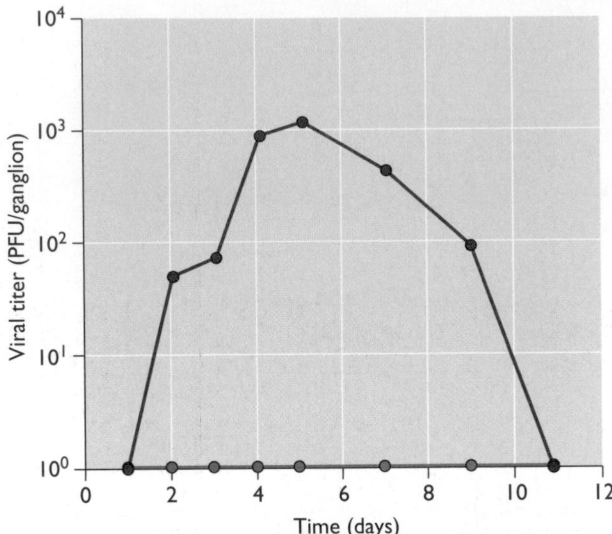

Figure 16.9 Replication of infectious herpes simplex virus type 1 in mouse trigeminal ganglia during acute infection. Mice were anesthetized and infected by a standard strain of herpes simplex virus by dropping approximately 10^5 plaque-forming units (PFU) onto the cornea of one eye that had been lightly scratched with a sterile needle. After a few minutes, the liquid was blotted and the animal was allowed to recover. At selected time points, animals were euthanized and the trigeminal ganglia were removed quickly and frozen. Each point on the graph (red line) represents the geometric mean titer in PFU from eight individual ganglia from two different experiments tested at the indicated time after infection. Uninfected animal controls are indicated by the blue line.

process varies depending on the animal species, the concentration and genotype of the infecting virus, and the site of primary infection. The latent genome persists in the nucleus as a nonintegrated, circular DNA molecule associated with nucleosomes, and is probably tethered at a specific site in the nucleus.

Many questions remain. For example, how do neurons in the ganglia survive the primary infection by a cytolytic virus? Is the lytic cycle gene expression pathway turned off after it has been started, or are some neurons productively infected and subsequently purged of infection? Many latently infected neurons synthesize RNA molecules termed **latency-associated transcripts** (LATs) (discussed in Chapter 8). Some researchers argue that all latently infected neurons synthesize LATs, while others report that only 5 to 30% do so. After infection of rabbits, mutants that do not synthesize LATs establish a latent infection, but spontaneous reactivation is markedly reduced. Identifying functions for the LATs continues to be a challenge. Two popular hypotheses are that the transcripts block apoptosis upon primary infection of neurons (or upon reactivation), or that they maintain the latent state through antisense inhibition of ICP0 translation (a crucial transcription activator). It is not clear if these hypotheses are mutually exclusive, or even if transcripts act directly as RNA molecules, or if proteins are produced.

Nonneuronal cells and the immune system play important roles in establishing the pattern of herpes simplex virus infection. For example, only 10% of the cells in a typical sensory ganglion are neurons; the remaining 90% are nonneuronal satellite cells and Schwann cells associated with a fibrocollagenous matrix. These nonneuronal cells are in intimate contact with ganglionic neurons. Some of the nonneuronal cells are infected during initial invasion of the ganglion and may be the major source of infectious virions isolated from infected ganglia. The peripheral nervous system is accessible to antibodies, complement, cytokines, and lymphocytes of the innate and adaptive immune system. In murine models, the immune response to productive infection in the ganglion actively influences the outcome of primary infection. For example, efficient establishment of latency occurs in vaccinated animals or animals that receive passive immunization with virus-specific antibodies prior to infection.

A latent infection in sensory ganglia is a particularly effective means of ensuring transmission of virions because mucosal contact is widespread among affectionate humans. An individual must be actively producing virions to transfer herpes simplex virus to another person, but production need only be transient. Indeed, local spread of reactivated virions is rapidly curtailed because the host is immunized during the primary infection. The spread of herpes simplex virus among mucosal epithelial cells after reactivation may be facilitated by action of the viral protein ICP47. This protein blocks MHC class I presentation of viral antigens to the T cells. Such activity may provide sufficient time for a few rounds of replication before elimination of the infected cell by activated CTLs.

Some individuals with latent herpes simplex virus experience reactivation every 2 to 3 weeks, while others have only rare or no episodes of reactivation. The signaling mechanisms that reactivate the latent infection are not well characterized, although sunburn, stress, nerve damage, depletion of nerve growth factor, steroids, heavy metals, and trauma (including dental surgery) all promote reactivation. Despite the apparent global nature of most reactivation stimuli, when reactivation does occur in animal models, only about 0.1% of those neurons containing the viral genome proceed to synthesize viral proteins and virions. Multiple levels of regulation must be operating, and the overwhelming thrust must be to maintain the latent state. The regulatory network employed may rely, in part, on nonuniformity within the latent population itself. Not only are different types of neurons infected in peripheral ganglia, but also the number of viral genomes in a given neuron varies dramatically (Box 16.5). It is likely that competency for reactivation is influenced by the number of viral genomes within a given neuron.

At first glance, the diversity of potential reactivation signals may be surprising, but we can imagine that the signals converge to activate specific cellular proteins needed for transcription of the herpes simplex virus immediate-early genes and consequently activate the productive transcriptional program. Indeed, all of these exogenous signals have the capacity to induce the synthesis of cell cycle and transcription regulatory proteins that may render neurons permissive for viral replication. It is known that synthesis of the viral immediate-early protein ICP0 is sufficient to reactivate a latent infection (Fig. 16.8). Reactivation may be an all-or-none process requiring but a single reaction to "flip the switch" that triggers the cascade of gene expression of the lytic pathway. Glucocorticoids are excellent examples of such activators, as they stimulate transcription rapidly and efficiently while inducing an immunosuppressive response. These properties explain the observation that clinical administration of glucocorticoids frequently results in reactivation of latent herpesvirus. An interesting idea is that spontaneous reactivation results from single reactions in individual neurons, each of which leads to a small burst of herpes simplex virus transcription. When the stimulus is strong enough, such sporadic transcription ultimately passes a threshold, resulting in replication and reactivation. This idea is consistent with the very low levels of immediate-early and early transcripts that can be detected in ganglia harboring a latent infection.

Epstein-Barr Virus

Epstein-Barr virus, which infects only humans, establishes latent infections in B lymphocytes (Fig. 16.10). About 90% of the world's population carry the latent viral genome in their peripheral blood lymphocytes. It is the only human herpesvirus consistently associated with human cancers (Table 16.5 and Appendix B). As we will learn in Chapter 18, such viral oncogenesis is a by-product of the mechanisms by which a latent infection is established. In contrast to the nonpathogenic latent state of herpes simplex virus, the latent state of Epstein-Barr virus is implicated in several important diseases, including **infectious mononucleosis** and several cancers (Table 16.5).

Epstein-Barr virus infects two major tissues: B lymphocytes and epithelia. Infected individuals produce virions, carry virus-specific CTLs, produce antibody specific for viral proteins, and yet harbor latently infected B cells. How latency is maintained in the face of an active immune response is an important question.

The primary site of Epstein-Barr virus infection is the epithelium of the oropharyngeal cavity. Children and teenagers are commonly afflicted, usually after oral contact (hence the name "kissing disease"). The acute infection requires expression of most viral genes and stimulates a strong immune response. Spread of infection to B cells in an individual with a normal immune system induces the infected B cells to divide, resulting in substantial immune and cytokine responses. The resulting disease is called infectious mononucleosis. The ensuing immune response destroys most infected cells, but approximately 1 in 100,000 infected B cells survive. They persist as small,

BOX 16.5

Neurons harboring latent herpes simplex virus often contain hundreds of viral genomes

The number of neurons in a ganglion that will ultimately harbor latent genomes following primary infection depends upon the host, the strain of virus, the concentration of infecting virions, and the conditions at the time of infection. It is possible to infect as few as 1% to as many as 50% of the neurons in a ganglion. In the mouse trigeminal ganglion (about 20,000 neurons per ganglion), this means that less than 200 or more than 10,000 neurons may carry latent genomes. In controlled experiments in mice, the number of latently infected neurons increases as the titer of infecting virions increases.

Interestingly, many infected neurons contain multiple copies of the latent viral genome, varying from fewer than 10 to more than 1,000; a small number have more than 10,000 copies. This variation in copy number has been enigmatic. Does it reflect multiple infections of a single neuron or is it the result of replication in a stimulated permissive neuron after infection by one particle? If it is the latter, how does the neuron recover from what should be an irreversible commitment to the productive cycle?

Recent experiments indicate that virions that cannot replicate, or whose replication is blocked by antiviral drugs, exhibit a significant reduction in the number of latently infected neurons with multiple genomes. It is likely that a single neuron can be infected by multiple viruses, each of which participates in the latent infection.

Sawtell, N. M. 1997. Comprehensive quantification of herpes simplex virus latency at the single-cell level. *J. Virol.* **71:**5423–5431.

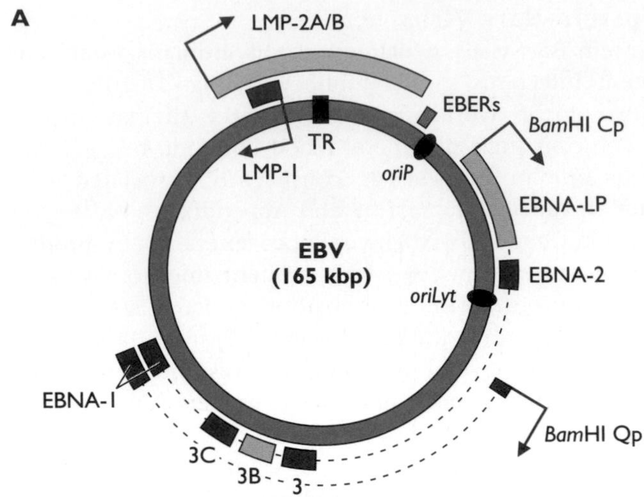

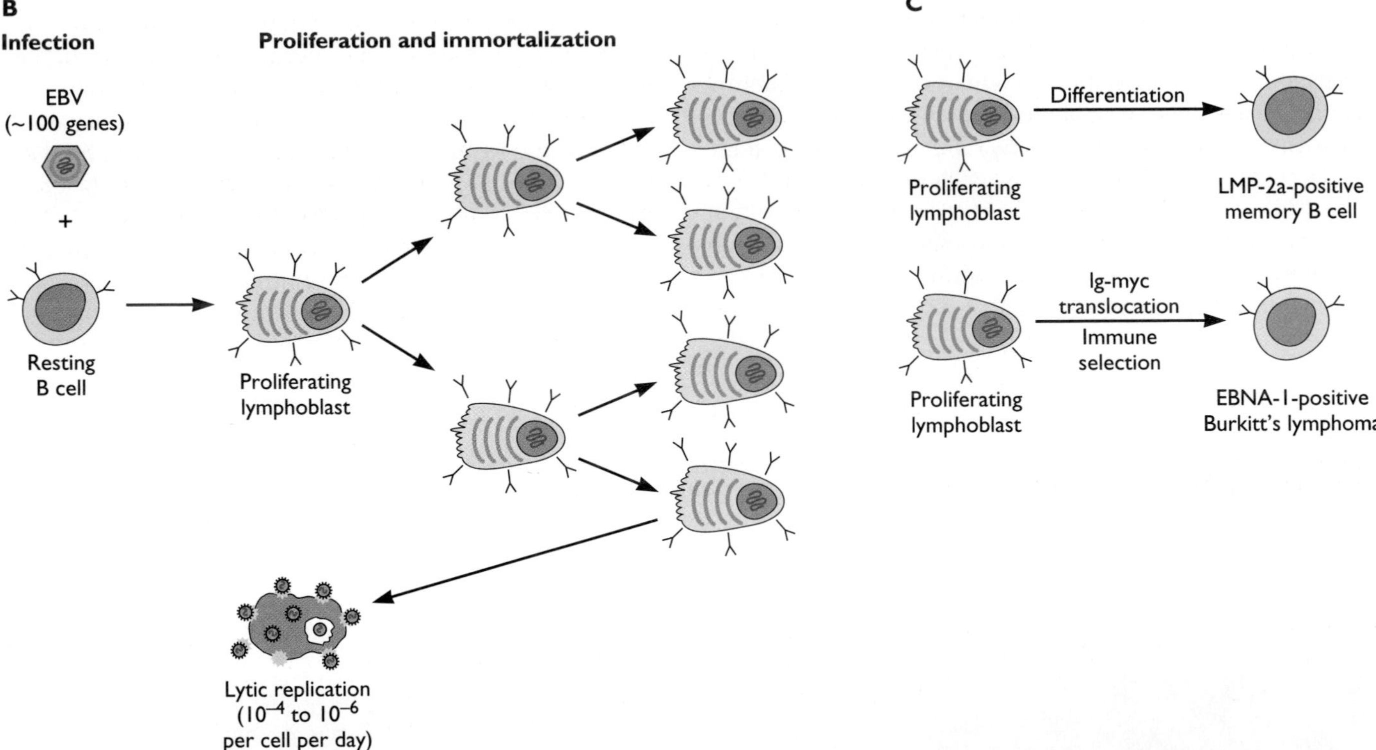

Figure 16.10 Epstein-Barr virus: patterns of infection in B cells. Two strains of Epstein-Barr virus are recognized that differ by their 0.5-kb terminal and 3-kb internal repeats as well as their production of nuclear antigens and the small RNAs (EBERs) during the latent infection. Epstein-Barr virus 1 is about 10 times more prevalent in the United States and Europe than is Epstein-Barr virus 2, while both strains are equally represented in Africa. (A) The Epstein-Barr virus genome. More than 90% of all viral genes are expressed during lytic replication. Genes expressed during the latent phase are in green or blue. Green represents proteins present in the cytoplasm and membranes; blue represents proteins that function in the nucleus; dark green and blue represent proteins required for induction and/or maintenance of proliferation; and light colors represent auxiliary functions. Promoters used during different forms of latency are illustrated in red; the black dashed lines represent the long introns; and the black boxes or ellipses represent critical *cis*-acting elements,

(continued on next page)

Figure 16.10 *(continued)* the terminal repeats (TR) at the ends of linear genomic DNA, *oriP* and *oriLyt,* the two origins of DNA synthesis used during the latent and lytic phases of the life cycle, respectively. (B) Proliferation and secretion of antibody begins soon after infection of resting B cells in vitro. Approximately 1 in 4 of the initially infected clones progresses to yield immortalized progeny cells after 70 to 90 cell generations. Approximately 1 in 10,000 to 1 in 1,000,000 of the proliferating B cells support productive infection per day and may be one source of infectious virus found in vivo. These events are thought to reflect what happens early in human infection. (C) Most infected B cells are killed by the adaptive immune response, but a few (approximately 1 in 100,000) persist in the blood as small, nonproliferating memory B cells that synthesize only LMP-2A mRNA. These memory B cells are presumably the long-term reservoir of Epstein-Barr virus in vivo and the source of infectious virus when peripheral blood cells are removed and cultured. A limited immune response toward these infected B cells leads to self-limiting proliferation, infectious mononucleosis, or unlimited proliferation (polyclonal B-cell lymphoma). A second outcome for the infected, proliferating B cells occurs in some children in regions of central Africa. A rare translocation that juxtaposes an immunoglobulin locus with that of c-*myc* leads to constitutive expression of c-*myc* and the ability of the B cells to proliferate in the absence of most viral latent gene expression. The host's residual immune response presumably selects for those cells that fail to produce viral proteins, and rare cells that synthesize only EBNA-1 grow as a Burkitt's lymphoma. Such an expression pattern is possible because EBNA-1 is transcribed from a distinct promoter (*Bam*HI Qp) such that other genes need not be expressed.

nonproliferating memory B cells that make only latent membrane protein 2A (LMP-2A) mRNA. They apparently home to lymphoid organs, where they are maintained. These cells do not produce the B7 coactivator receptor, and therefore are not killed by CTLs (see Chapter 15). These latently infected B cells proliferate indefinitely when stimulated or when propagated as cultured cells. They are the progenitors of the B-cell lines that grow out of peripheral blood from an infected patient. The latent viral genome is maintained as a circular episome that replicates via a program distinct from the one used during productive replication (Chapter 9).

When peripheral blood of an infected individual is cultured, latently infected B cells begin to replicate, while uninfected B cells die. It is important to understand that these cultured immortal lymphoblasts, most assuredly, are not the same as latently infected cells circulating in vivo. Nevertheless, this class of virus-infected B cells often yields immortalized progeny capable of being propagated indefinitely in the laboratory. Consequently, they are the best-understood models of Epstein-Barr virus latent infection. These cells synthesize a set of at least 10 gene products, including 6 nuclear proteins, termed EBNAs, 3 viral membrane proteins (LMP proteins) that are important in

Table 16.5 Diseases associated with Epstein-Barr virus infections[a]

Acute infection	
Infectious mononucleosis	The best known clinical presentation of infection; resolves in 1–2 weeks, fatigue symptoms may last longer
Oral hairy leukoplakia	Primary infection leading to a wartlike lesion of epithelial cells of the tongue seen in AIDS patients and transplant recipients
Abnormalities of latent infection; lymphoproliferative disorders and malignancies	
B-lymphoproliferative disease	Frequently observed in individuals experiencing a primary viral infection following tissue transplantation; initially benign; if untreated, can lead to B-cell lymphoma
X-linked lymphoproliferative syndrome	Certain males have X-linked mutations that lead to a severe immunodeficiency after primary viral infection.
Burkitt's lymphoma	The most common childhood cancer in equatorial Africa. Cells from Burkitt's B-cell lymphomas exhibit a reciprocal translocation involving the c-*myc* locus on the long arm of chromosome 8 and one of the immunoglobulin loci on chromosome 2, 14, or 22. Newly explanted B cells from tumors produce only EBNA-1.
Hodgkin's disease	Mixed cells in tumor; 1–2% are malignant and the remaining cells are infiltrating lymphocytes. Association of virus with Hodgkin's lymphomas varies with geography. Newly explanted B cells from tumors produce EBNA-1, LMP-1, and LMP-2A.
Nasopharyngeal carcinoma	A cancer of epithelial cells and one of the most common cancers in China. Tumor cells produce EBNA-1, LMP-1, LMP-2, but not EBNA-2.

[a]Data from G. C. Faulkner et al., *Trends Microbiol.* **8:**185–189, 2000.

altering the properties of the cells, and small RNA molecules called EBER-1 and EBER-2 (Table 16.4). While considerable effort has focused on the mechanism of transformation of B cells in culture, the viral genes required are not often expressed in human cancers associated with viral infection.

At least three distinct phenotypes of viral latency can be distinguished by the viral gene products made in an infected B cell. In the acute infection, infected B cells divide in response to viral proteins. These cells express all of the latency-associated genes (sometimes called type 3 latency) (Fig. 16.10 and Table 16.4). These viral proteins are required to establish the latent infection and to promote growth of the infected cell. This phenotype is often found in lymphomas of immunodeficient patients. A second phenotype is found in which only EBNA-1 protein is synthesized (sometimes called type 1 latency). Cells of this phenotype are found in Burkitt's lymphoma, but have been difficult to detect in virus-infected individuals. The third phenotype is defined by coproduction of EBNA-1 and LMP-1 and -2. Such a phenotype is characteristic of nasopharyngeal carcinoma, Hodgkin's disease, and T-cell lymphomas.

A remarkable feature of Epstein-Barr virus infection is the equilibrium established between active immune elimination of infected cells and viral persistence. Although immunocompetent individuals maintain CTLs directed against many of the viral proteins synthesized in latently infected B cells, these cells are not eliminated. Some viral proteins, such as LMP-1, inhibit apoptosis or immune recognition of the latently infected cells. Moreover, EBNA-1 peptides are not presented to T cells as discussed above. When the equilibrium between proliferation of latently infected B cells and the immune response that kills the infected cells is altered (e.g., after immunosuppression), the immortalized B cells can form lymphomas (Fig. 16.10; see also Chapter 18). It is a matter of debate if any viral protein is synthesized in the infected, nondividing B cell. Certainly, virus-infected proliferating cells produce viral proteins and are superb targets for the host's immune system. It is likely that the normal immune response selects nonproliferating B cells as the survivors of infection and ensures that the latent infection is benign in the majority of cases.

The signals that reactivate latent Epstein-Barr virus infection in vivo are not well understood, but considerable information has been obtained in studies with cultured cells. Certain signal transduction cascades or production of an essential viral transcriptional activator, Zta (Z or zebra protein; see Fig. 8.18 and below), induce the productive infection. Many signal transduction pathways efficiently reactivate Epstein-Barr virus from the latent state. Such pathways are activated by various interactions, including clustering of the B-cell antigen receptor CD21 after interaction with anti-immunoglobulin antibodies (activation of tyrosine kinases), binding of phorbol esters (stimulation of protein kinase C), and introduction of calcium ionophores. Therefore, it is surprising that latent infection is so stable. We now know that virus-encoded LMP-2A makes an important contribution to maintaining the latent infection by inhibiting tyrosine kinase signal transduction pathways. It is the first example of a viral protein that blocks reactivation of a latent infection. As reactivation occurs efficiently in vivo, a second signal transduction pathway that bypasses the LMP-2A block must exist, but such a pathway has not been found.

B cells harboring latent Epstein-Barr virus genomes can also be reactivated whenever the viral immediate-early gene product called Zta is made (see also Fig. 8.18). Indeed, as described in Chapter 8, Zta, like phorbol esters, increases transcription of viral lytic genes. This protein also represses the latency-associated promoters and is responsible for recognition of the viral lytic origin of replication.

Slow Infections: Sigurdsson's Legacy

Many fatal brain diseases, characterized by ataxia (movement disorders) or dementia (severe cognitive impairment), stem from another extreme variation of persistent infection, a pattern called slow infection (Fig. 16.1). It may be **years** from the time of initial contact of the infectious agent with the host until the appearance of recognizable symptoms. Such long incubation periods are remarkable taken in the context of other viral diseases (Table 16.1). Once symptoms appear, death usually follows quickly. Viruses such as measles virus, the polyomavirus JC virus, and retroviruses such as human immunodeficiency virus and human T-lymphotropic virus can establish slow infections with severe nervous system pathogenesis at the end stage of disease. In many cases, the persistent infection exists in peripheral compartments with no apparent effect, and only enters the brain after many years. Elucidating the molecular mechanisms responsible for an infectious disease process of such long duration is a formidable challenge. Experimental analysis of these unusual diseases began in the 1930s when a flock of Karakul sheep was imported from Germany to Iceland, where they infected the native sheep, causing a disease called maedi/visna. Thanks to the many years of careful work by Bjorn Sigurdsson and colleagues, we now know that the maedi/visna syndrome is caused by a lentivirus that is like human immunodeficiency virus. The striking feature that Sigurdsson discovered is the slow progression to disease after primary infection—often more than 10 years. He developed a framework of experimentation for studying the slow, relentless, usually progressive and fatal brain infections, including those now proposed to be caused by prions (see Chapter 20).

Abortive Infections

In an abortive infection, virions infect susceptible cells or hosts, but do not complete replication, usually because an essential viral or cellular gene is not expressed. Clearly, an abortive infection is nonproductive. Even so, it is not necessarily uneventful or benign for the infected host. Viral interactions at the cell surface and subsequent uncoating can initiate membrane damage, disrupt endosomes, or activate signaling pathways that cause apoptosis and cytokine production. In some instances, abortively infected cells may not be recognized by the immune system, and if they do not divide, the viral genome may persist as long as the cell survives. In some cases, an infection may proceed far enough so that the infected cell is recognized by CTLs. Such an infection would likely induce an IFN, as well as an inflammatory response that may damage the host if sufficient cells are involved. Recently, it was discovered that the human immunodeficiency virus structural protein Vpr can damage cells when it is associated with particles that contain noninfectious genomes. When these noninfectious particles bind to T cells, Vpr induces G_2 arrest. It has been suggested that, although most of the particles in an infected individual are noninfectious, because of the presence of virion Vpr they participate in immune suppression.

With the advent of modern viral genetics, virologists can construct defective viral genomes, which in the absence of a complementing gene product initiate an abortive infection. One popular idea is to use such defective genomes as vectors for gene therapy or as vaccines. To be effective, cytopathic genes of a prospective viral vector certainly must be eliminated. Many of the well-known, defective viral vectors lack essential genes and are designed to express only the therapeutic cloned gene. Care must be taken to ensure that the infection is truly noncytotoxic. Given that intrinsic and innate defenses can be activated by inactivated particles, prudence in assuming safety of viral vectors is essential (Box 15.16). Cytotoxicity and inflammatory host responses are of particular concern if the therapeutic gene is to be delivered to a substantial number of cells, a process that requires administration of many virus particles.

Transforming Infections

A transforming infection is a special class of persistent infection. A cell infected by certain DNA viruses or retroviruses may exhibit altered growth properties and begin to proliferate faster than uninfected cells. For some infections, this change is accompanied by integration of viral genetic information. For others, replication is in concert with the cell. Virus particles may no longer be produced, but some or all of their genetic material generally persists. We characterize this pattern of persistent infection as transforming because of the change in cell behavior. It is also considered oncogenic because transformed cells may cause cancer in animals. This important infection pattern is discussed in detail in Chapter 18.

Perspectives

Different patterns of infection occur primarily because host defenses are modulated passively or actively. It is likely that the genomes of all viruses encode gene products that engage and modulate intrinsic and immune defenses. Acute infections progress beyond physical, intrinsic, and innate defenses only to be blocked and cleared by the adaptive immune response. Remarkably, in persistent viral infections, essentially all defenses, including the adaptive immune system, are ineffective. These two primary patterns of infection confront us with several questions that challenge our basic understanding of virology. A particular infection pattern can be a defining characteristic of a virus family, yet why this should be so is not always obvious. Why has one particular pattern been selected over another? As we discuss in Chapter 20, new viral populations emerge as a consequence of selection pressures. What are the selective advantages or disadvantages of a given infection pattern? Is an acute infection more advantageous as a survival strategy than a persistent infection? Can we identify the correlates among the characteristic patterns of infection and pathogenesis?

One hypothesis is that successful viral populations establish symbiosis, neither helping nor harming the host. As suggested by Lewis Thomas, pathogenesis is an aberration of symbiosis, an overstepping of boundaries. Another hypothesis is that a successful viral population need not have a static or benign relationship with its host. Rather, the outcome of infection is in constant flux, and the successful relationship is better described as an approach to equilibrium (benign symbiosis may never be attained). In this relationship, some host organisms may be harmed in the short run to achieve long-term survival of the virus population. Accordingly, pathogenesis may be a necessary survival feature of the viral population, and would be selected during evolution of the relationship.

Selection may work both ways. Is it possible that some viral infections are advantageous to the host? Recent work with a mouse model suggests that memory T cells established after lymphocytic choriomeningitis virus infection can be activated by subsequent infection with vaccinia virus. Such vaccinia virus-infected mice exhibited decreased mortality and dramatic changes in pathology. This concept of a "good infection" certainly tests our current understanding of the complexities of host-virus interactions in the real world.

References

Books

Mims, C. A., A. Nash, and J. Stephen. 2001. *Mims' Pathogenesis of Infectious Disease,* 5th ed. Academic Press, San Diego, Calif.

Richman, D. D., R. J. Whitley, and F. G. Hayden (ed.). 2002. *Clinical Virology,* 2nd ed. ASM Press, Washington, D.C.

Reviews

Modulation of Host Defenses

Alcami, A., and U. Koszinowski. 2000. Viral mechanisms of immune evasion. *Trends Microbiol.* **8:**410–417.

Gewurz, B. E., R. Gaudet, D. Tortorella, E. W. Wang, and H. L. Ploegh. 2001. Virus subversion of immunity: a structural perspective. *Curr. Opin. Immunol.* **13:**442–450.

Neipel, F., J.-C. Albrecht, and B. Fleckenstein. 1997. Cell-homologous genes in the Kaposi's sarcoma-associated rhadinovirus human herpesvirus 8: determinants of its pathogenicity? *J. Virol.* **71:**4187–4192.

Smith, G. L. (ed.). 1998. Immunomodulation by virus infection. *Semin. Virol.* **8:**359–442.

Tortorella, D., B. E. Gewurz, M. H. Furman, D. J. Schust, and H. L. Ploegh. 2000. Viral subversion of the immune system. *Annu. Rev. Immunol.* **18:**861–926.

Weiss, R. A. 2002. Virulence and pathogenesis. *Trends Microbiol.* **10:**314–317.

Xu, X., G. R. Screaton, and A. J. McMichael. 2001. Virus infections: escape, resistance, and counterattack. *Immunity* **15:**867–870.

Measles Virus

Borrow, P., and M. B. A. Oldstone. 1995. Measles virus—mononuclear cell interactions. *Curr. Top. Microbiol. Immunol.* **191:**85–100.

Schneider-Schaulies, J., V. ter Meulen, and S. Schneider-Schaulies. 2001. Measles virus interactions with cellular receptors: consequences for viral pathogenesis. *J. Neurovirol.* **7:**391–399.

Schneider-Schaulies, S., S. Niewiesk, J. Schneider-Schaulies, and V. ter Meulen. 2001. Measles virus induced immunosuppression: targets and effector mechanisms. *Curr. Mol. Med.* **1:**163–181.

Herpes Latency

Nash, A. A. 2000. T-cells and the regulation of herpes simplex virus latency and reactivation. *J. Exp. Med.* **191:**1455–1457.

Redpath, S., A. Angulo, N. R. J. Gascoigne, and P. Ghazal. 2001. Immune checkpoints in viral latency. *Annu. Rev. Microbiol.* **55:**531–560.

Rickinson, A. 2002. Epstein-Barr virus. *Virus Res.* **82:**109–113.

Selected Papers

Modulation of Host Defenses

Bertin, J., R. C. Armstrong, S. Ottilie, D. A. Martin, Y. Wang, S. Banks, G.-H. Wang, T. G. Senkevich, E. S. Alnemri, B. Moss, M. J. Lenardo, K. J. Tomaselli, and J. I. Cohen. 1997. Death effector domain-containing herpesvirus and poxvirus proteins inhibit both Fas- and TNFR1-induced apoptosis. *Proc. Natl. Acad. Sci. USA* **94:**1172–1176.

Charleston, B., M. D. Fray, S. Baigent, B. V. Carr, and W. I. Morrison. 2001. Establishment of persistent infection with non-cytopathic bovine viral diarrhoea virus in cattle is associated with a failure to induce type I interferon. *J. Gen. Virol.* **82:**1893–1897.

Deitz, S. B., D. A. Dodd, S. Cooper, P. Parham, and K. Kirkegaard. 2000. MHC I dependent antigen presentation is inhibited by poliovirus protein 3A. *Proc. Natl. Acad. Sci. USA* **97:**13790–13795.

Erickson, A. L., Y. Kimura, S. Igarashi, J. Eichelberger, M. Houghton, J. Sidney, D. McKinney, A. Sette, A. L. Hughes, and C. M. Walker. 2001. The outcome of hepatitis C virus infection is predicted by escape mutations in epitopes targeted by cytotoxic T lymphocytes. *Immunity* **15:**883–895.

Griffith, T. S., T. Brunner, S. M. Fletcher, D. R. Green, and T. A. Ferguson. 1995. Fas ligand-induced apoptosis as a mechanism of immune privilege. *Science* **270:**1189–1192.

Guidotti, L. G., R. Rochford, J. Chung, M. Shapiro, R. Purcell, and F. V. Chisari. 1999. Viral clearance without destruction of infected cells during HBV infection. *Science* **284:**825–829.

Karp, C. L., M. Wysocka, L. M. Wahl, J. M. Ahearn, P. J. Duomo, B. Sherry, G. Trinchieri, and D. E. Griffin. 1996. Mechanism of suppression of cell-mediated immunity by measles virus. *Science* **273:**228–231.

Levitskaya, J., M. Coram, V. Levitsky, S. Imreh, P. M. Steigerwald-Mullen, G. Klein, M. G. Kurilla, and M. G. Masucci. 1995. Inhibition of antigen processing by the internal repeat region of the Epstein-Barr virus nuclear antigen-1. *Nature* **375:**685–688.

Reyburn, H. T., O. Mandelboim, M. Vales-Gomez, D. M. Davis, L. Pazmany, and J. L. Strominger. 1997. The class I MHC homologue of human cytomegalovirus inhibits attack by natural killer cells. *Nature* **386:**514–517.

Wiertz, E. J. H. J., T. R. Jones, L. Sun, M. Bogyo, H. J. Geuze, and H. L. Ploegh. 1996. The human cytomegalovirus US11 gene product dislocates MHC class I heavy chains from the endoplasmic reticulum to the cytosol. *Cell* **84:**769–779.

York, I. A., C. Roop, D. W. Andrews, S. R. Riddell, F. L. Graham, and D. C. Johnson. 1994. A cytosolic herpes simplex virus protein inhibits antigen presentation to CD81 T lymphocytes. *Cell* **77:**525–535.

Herpes Latency

Ahmed, M., M. Lock, C. G. Miller, and N. W. Fraser. 2002. Regions of the herpes simplex virus type 1 latency associated transcript that protect cells from apoptosis in vitro and protect neuronal cells in vivo. *J. Virol.* **76:**717–729.

Feldman, L. T., A. R. Ellison, C. C. Voytek, L. Yang, P. Krause, and T. P. Margolis. 2002. Spontaneous molecular reactivation of herpes simplex virus type 1 latency in mice. *Proc. Natl. Acad. Sci. USA* **99:**978–983.

Halford, W. P., and P. A. Schaffer. 2001. ICP0 is required for efficient reactivation of herpes simplex virus type 1 from neuronal latency. *J. Virol.* **75:**4240–3249.

Kramer, M. F., and D. Coen. 1995. Quantification of transcripts from the ICP4 and thymidine kinase genes in mouse ganglia latently infected with herpes simplex virus. *J. Virol.* **69:**1389–1399.

Miyashita, E. M., B. Yang, G. J. Babcock, and D. A. Thorley-Lawson. 1997. Identification of the site of Epstein-Barr virus persistence in vivo as a resting B-cell. *J. Virol.* **71:**4882–4891.

Perng, G. C., C. Jones, J. Ciacci-Zanella, M. Stone, G. Henderson, A. Yukht, S. M. Slanina, F. M. Hofman, H. Ghiasi, A. B. Nesburn, and S. L. Wechsler. 2000. Virus-induced neuronal apoptosis blocked by the herpes simplex virus latency associated transcript. *Science* **287:**1500–1503.

Sawtell, N. 1998. The probability of in vivo reactivation of herpes simplex virus type 1 increases with the number of latently infected neurons in ganglia. *J. Virol.* **72:**6888–6892.

Shimeld, C., J. L. Whiteland, N. A. Williams, D. L. Easty, and T. J. Hill. 1997. Cytokine production in the nervous system of mice during acute and latent infection with herpes simplex virus type 1. *J. Gen. Virol.* **78:**3317–3325.

Speck, P., and A. Simmons. 1998. Precipitous clearance of herpes simplex virus antigens from the peripheral nervous systems of experimentally infected C57BL/10 mice. *J. Gen. Virol.* **79:**561–564.

Zabolotny, J. M, C. Krummenacher, and N. W. Fraser. 1997. The herpes simplex virus type 1 2.0-kilobase latency-associated transcript is a stable intron which branches at a guanosine. *J. Virol.* **71:**4199–4208.

Slow Viruses

Iwasaki, Y. 1990. Pathology of chronic myelopathy associated with HTLV-1 infection (HAM/TSP). *J. Neurol. Sci.* **96:**103–123.

Sigurdsson, B. 1954. Rida, a chronic encephalitis of sheep with general remarks on infections which develop slowly and some of their special characteristics. *Br. Vet. J.* **110:**341–354.

Ter Meulen, V., J. R. Stephenson, and H. W. Kreth. 1983. Subacute sclerosing panencephalitis. *Comp. Virol.* **18:**105–159.

17

Human Immunodeficiency Virus Pathogenesis

Nature is not human-hearted.
LAO TZU
Tao Te Ching

Introduction

Worldwide Scope of the Problem

Acquired immunodeficiency syndrome (AIDS) is the name given to end-stage disease caused by infection with human immunodeficiency virus (HIV). By almost any criteria, HIV qualifies as one of the world's deadliest scourges. First recognized as a clinical entity in 1981, by 1992 AIDS had become the major cause of death in individuals 25 to 44 years of age in the United States. The current worldwide statistics are even more staggering, with the developing countries of Africa and parts of Asia being especially hard-hit (Table 17.1). As of December 2002, 20 million people in the world had already died of AIDS. An end-of-year report from the United Nations' AIDS program estimated the number of new HIV infections in 2002 to be 5 million, bringing the total number of infected people worldwide to 42 million. This number corresponds to approximately 1 in every 100 adults, ages 15 to 49, in the world's population. HIV/AIDS is now the leading cause of death in sub-Saharan Africa, with 2.4 million fatalities in 2002 alone. In certain parts of this region, 25 to 30% of the adult population have become infected. It is estimated that one-third of the children under 15 years of age in these areas have lost one or both parents to AIDS. In these places a whole generation of human beings has succumbed to this fatal disease.

HIV continues to spread faster than any known persistent infectious agent in the last half century. The recent clinical emergence is likely to be the consequence of a number of political, economic, and societal changes, including the breakdown of national borders, economic distress with the migration of large populations, and the ease and frequency of travel throughout the world.

HIV has also become the most intensely studied of any infectious agent. Research on the virus has contributed to our understanding of AIDS and related veterinary diseases, and has provided new insights into virology, cellular biology, and immunology. This chapter describes the many facets of HIV-

Table 17.1 Numbers of people infected with HIV[a]

Region	No. of infected persons
Sub-Saharan Africa	29,400,000
South and Southeast Asia	6,000,000
Latin America	1,500,000
North America	980,000
Western Europe	570,000

[a]Data from the Joint United Nations Programme on HIV/AIDS as of December 2002.

induced pathogenesis and what has been learned through its analysis. The complexities will illustrate the enormous scope of the challenges faced by biomedical researchers and physicians in their efforts to control this virus, which strikes at the very heart of the body's defense system.

HIV Is a Lentivirus

Discovery and Characterization

The first clue to the etiology of AIDS came in 1983 upon isolation of a retrovirus from the lymph node of a patient with lymphadenopathy at the Pasteur Institute in Paris. Although not fully appreciated initially, the significance of this finding became apparent in the following year with the isolation of a cytopathic, T-cell-tropic retrovirus from combined blood cells of AIDS patients by researchers at the U.S. National Institutes of Health and of a similar retrovirus from blood cells of an AIDS patient at the University of California, San Francisco. Although the National Institutes of Health isolate was later shown to be a contaminant of a sample received from the Pasteur Institute (Box 17.1), the virus isolated at the University of California, San Francisco, and subsequent isolates at the National Institutes of Health laboratory were

BOX 17.1

Lessons from discovery of the AIDS virus(es)

- The first AIDS virus was obtained from a patient with lymphadenopathy by Francoise Barré-Sinoussi in collaboration with Jean-Claude Chermann and Luc Montagnier at the Pasteur Institute (1983). The isolate, named Bru, grew only in primary cell cultures. We now know that Bru belonged to a class of slow-growing, low-titer viruses that are common in early-stage infection.
- Between 20 July and 3 August 1983, Bru-infected cultures at the Pasteur became contaminated with a second AIDS virus, called Lai, which had been isolated from a patient with full-blown AIDS and which belonged to a class of viruses that grow well in cell culture. HIV-1 Lai rapidly overtook the cultures.
- Unaware of this contamination, Pasteur scientists subsequently sent out virus samples from these cultures as "Bru" to several laboratories, including those of Robin Weiss in Britain and Malcolm Martin and Robert Gallo in the United States.
- Unlike earlier samples of Bru, this virus grew robustly in the laboratories to which it was distributed. Indeed, Lai was later discovered to have contaminated some AIDS patient "isolates" obtained by Weiss. In retrospect, such contamination is not surprising, as biological containment facilities were limited at the time, with the same incubators and hoods being used for maintaining HIV stocks and making new isolates.
- Lai also contaminated cultures of blood cells combined from several AIDS patients in the Gallo laboratory at the National Institutes of Health. Because the properties of this virus were found to be different from those described for Bru, Gallo and coworkers reported the discovery of a second type of AIDS virus, which they believed to have originated from one of their AIDS patients.
- This second claim, a race to develop blood screening tests, and the later revelation from DNA sequence analyses that the French and the Gallo viruses were one and the same (Lai) led to a much publicized scientific controversy with significant political overtones. Simon Wain-Hobson and colleagues at the Pasteur eventually sorted out the chain of events in 1991 by comparing nucleotide sequences of stored samples of the original stocks of Bru and Lai. The controversy has since subsided—what remains are important lessons in virology!

Goudsmit, J. 2002. Lots of peanut shells but no elephant. A controversial account of the discovery of HIV. *Nature* **416:**125–126. (Book review.)

Wain-Hobson, S. J. P. Vartanian, M. Henry, N. Chenciner, R. Cheynier, S. Delassus, L. P. Martins, M. Sala, M. T. Nugeyre, D. Guetard, et al. 1991. LAV revisited: origins of the early HIV-1 isolates from Institut Pasteur. *Science* **252:**961–965.

Weiss, R., and M. Martin. Personal communication.

unique. As commonly happens, each laboratory gave its isolate a different name: LAV, lymphadenopathy-associated virus; HTLV-III, human T-cell lymphotropic virus type III; and ARV, AIDS-associated retrovirus. Electron microscopic examination revealed that these viruses were morphologically similar to a known group of retroviruses, the lentiviruses, and further characterization confirmed this relationship. In 1986, the International Committee on Taxonomy of Viruses recommended the current name, human immunodeficiency virus.

Lentiviruses comprise a separate genus of the family *Retroviridae* (Table 17.2). The equine infectious anemia lentivirus was one of the first viruses to be identified. Discovered in 1904, this virus causes episodic autoimmune hemolytic anemia in horses. Lentiviruses of sheep (visna/maedi virus) and goats (caprine arthritis-encephalitis virus) have also been known for many years. Like the equine lentivirus, these viruses are associated with long incubation periods, and are therefore called **slow viruses** (Chapter 16). The discovery of HIV led to a search for additional lentiviruses and their subsequent isolation from cats (feline immunodeficiency virus) and a variety of nonhuman primates (simian immunodeficiency virus [SIV]). In 1986, a distinct type of HIV that is prevalent in certain regions of West Africa was discovered. It was called HIV type 2 (HIV-2) to distinguish it from the original type (HIV-1). Individuals infected with HIV-2 also develop AIDS, but with a longer incubation period and lower morbidity.

Many independent isolates of both HIV-1 and HIV-2 have been characterized over the last decade. Nucleotide sequence comparisons allow us to distinguish two major groups among HIV-1 isolates: group M includes most HIV-1 isolates, and group O represents what appear to be relatively rare "outliers" (Box 17.2). Ten distinct subtypes are currently recognized in group M (called **clades** A to K), each of which is prevalent in a different geographic area. For example, clade B is the most common subtype in North America and Europe. Five clades of HIV-2 have also been identified on the basis of similar criteria. A new group, N, was proposed for HIV-1 in 1998 based on a virus isolate, YBF30, obtained from an AIDS patient in Cameroon. The nucleotide sequence of this virus is more closely related to group M than to group O. Identification of related strains in Cameroon supports a three-pronged radiation of HIV-1 groups.

Figure 17.1 shows the phylogenetic relationships among the lentiviruses, based on sequences of their *pol* genes. The African monkey and ape isolates are endemic to each of the species from which they were obtained and do not appear to cause disease in their native hosts. However, a fatal AIDS-like disease is caused by infection of Asian macaques with virus originating from the African sooty mangabey (SIV_{smm}). Close contact between sooty mangabeys and humans is common, as these animals are hunted for food and kept as pets. Such interaction and the observation that several isolates of HIV-2 are nearly indistinguishable in nucleotide sequence from SIV_{smm} support the hypothesis that HIVs emerged via interspecies transmission between nonhuman primates and humans. This hypothesis is supported further by recent studies, which indicate that the known HIV-1 groups arose via at least three independent transmissions from chimpanzees (Box 17.2). The strains of SIV_{cpz} from the chimpanzee *Pan troglodytes troglodytes* are closest in sequence to HIV-1, implicating this subspecies as the origin of the human virus. In this chapter we will use the abbreviation HIV to describe properties shared by all the HIVs, and will specify the type when referring to one or the other.

As summarized in Table 17.2, lentiviruses cause immune deficiencies and disorders of the hematopoietic and central nervous systems and, sometimes, arthritis and autoimmunity. Lentiviral genomes are relatively large, with more genes than those of simple retroviruses (Fig. 17.2). In addition to the three structural polyproteins Gag, Pol, and Env, common to all retroviruses, lentiviral genomes encode a number of additional **auxiliary proteins**. Two HIV auxiliary proteins (Table 17.3), Tat and Rev, perform

Table 17.2 Lentiviruses[a]

Virus	Host infected	Primary cell type infected	Clinical disorder
Equine infectious anemia virus	Horse	Macrophages	Cyclic infection in the first year, autoimmune hemolytic anemia, sometimes encephalopathy
Visna/maedi virus	Sheep	Macrophages	Encephalopathy/pneumonitis
Caprine arthritis-encephalitis virus	Goat	Macrophages	Immune deficiency, arthritis, encephalopathy
Bovine immunodeficiency virus	Cattle	Macrophages	Lymphadenopathy, lymphocytosis, central nervous system disease (?)
Feline immunodeficiency virus	Cat	T lymphocytes	Immune deficiency
Simian immunodeficiency virus	Primate	T lymphocytes	Immune deficiency and encephalopathy
Human immunodeficiency virus	Human	T lymphocytes	Immune deficiency and encephalopathy

[a]Adapted from Table 1.1 of J. A. Levy, *HIV and the Pathogenesis of AIDS*, 2nd ed. (ASM Press, Washington, D.C., 1998), with permission.

BOX 17.2

The earliest record of HIV-1 infection

The earliest record of HIV-1 infection comes from a serum sample obtained in 1959 from a Bantu male in the city now known as Kinshasa, in the Democratic Republic of Congo. Phylogenetic analyses place the viral sequence (ZR59) near the ancestral node of clades B and D. As this is not at the base of the M group, this group must have originated earlier (red arrowhead near top of figure), and back calculations suggest that the M group of viruses arose via transspecies transmission from a chimpanzee into the African population around 1930. Its rapid evolution, giving rise to at least 10 subtypes (clades A to K), seems to have occurred near the end of or just after World War II. Separate transspecies transmissions (red arrowheads in bottom half of figure) account for the origin of the N and O groups.

Sharp, P. M. 2002. Origins of human virus diversity. *Cell* **108:**305–312.

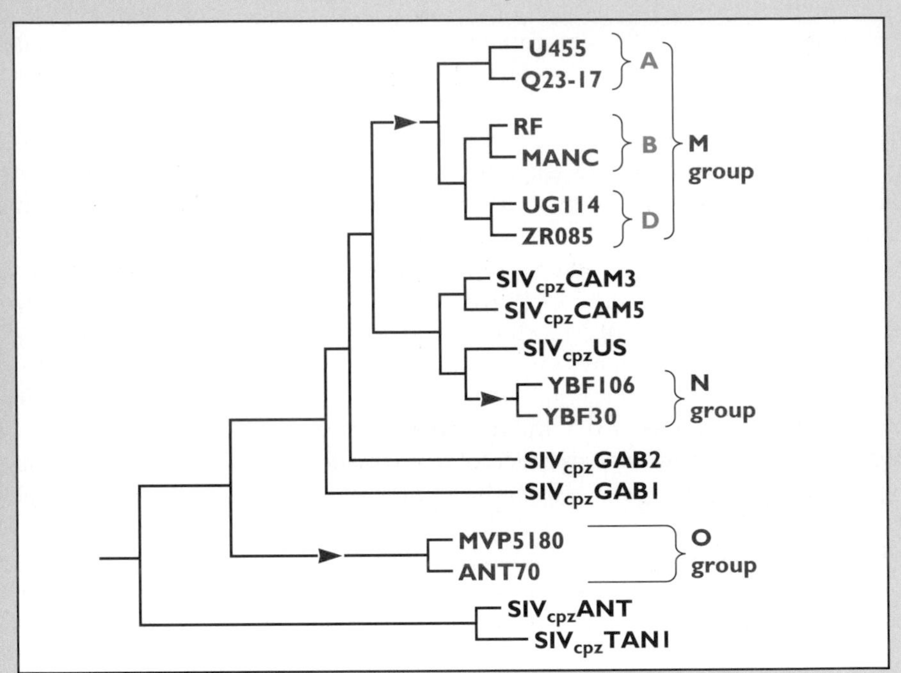

Figure 17.1 Phylogenetic relationships among lentiviruses. Representative lentiviruses compared by using *pol* gene nucleotide sequences for establishing phylogenetic relationships. Five groups of primate lentiviruses are shown: HIV-1, HIV-2, simian immunodeficiency virus from the sooty mangabey monkey (SIV_{smm}), Sykes' monkey (SIV_{syk}), chimpanzee (SIV_{cpz}), African green monkey (SIV_{agm}), and mandrill (SIV_{mnd}). Nonprimate lentiviruses are visna/maedi virus (VMV), caprine arthritis-encephalitis virus (CAEV), equine infectious anemia virus (EIAV), bovine immunodeficiency virus (BIV), and feline immunodeficiency virus (FIV). The scale indicates the percentage difference in nucleotide sequences in the *pol* gene. The branching order of the primate lentiviruses is controversial. Adapted from Fig. 2 of P. A. Luciw, p. 1881–1952, *in* B. N. Fields et al. (ed.), *Fields Virology*, 3rd ed. (Lippincott-Raven, Philadelphia, Pa., 1996), with permission.

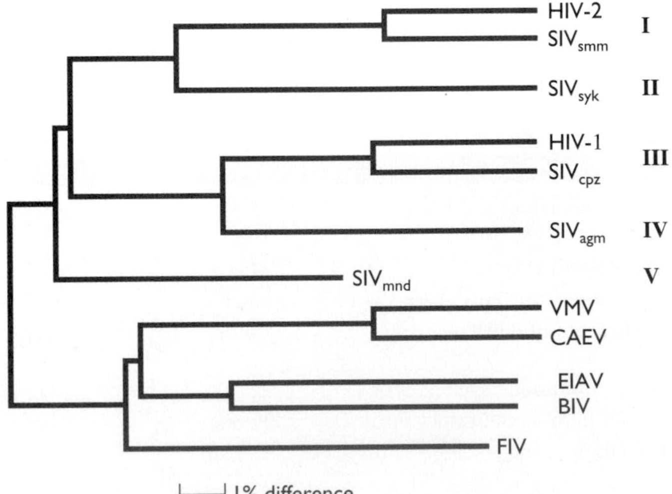

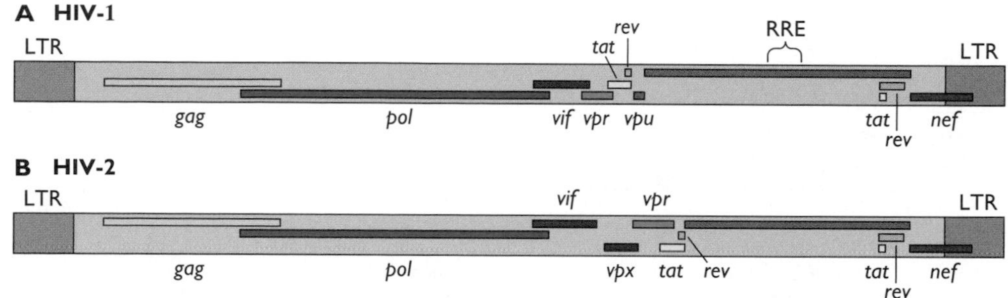

Figure 17.2 Genome organization of HIV-1 and HIV-2. Vertical positions denote each of the three different reading frames that encode viral proteins. The long terminal repeats (LTRs) contain sequences necessary for transcriptional initiation and termination, reverse transcription, integration, and binding of the viral transactivator Tat. The position of the RRE is indicated in the HIV-1 *env* gene. Adapted from Fig. 1 of M. Emerman, *Curr. Biol.* **6**:1096–1103, 1996, with permission.

regulatory functions that are essential for viral replication. The remaining four—Nef, Vif, Vpr, and Vpu—are not essential for viral reproduction in most immortalized T-cell lines and hence are known as **accessory** proteins. However, accessory proteins do modulate virus replication, and they are essential for efficient virus production in vivo.

Distinctive Features of the HIV Replication Cycle and the Roles of Auxiliary Proteins

cis-*Acting Regulatory Sequences and the Viral Regulatory Proteins Tat and Rev*

The long terminal repeat. As in all retroviruses, expression of integrated HIV DNA is regulated by sequences in the transcriptional control region of the viral long terminal repeat (LTR), which are recognized by the host cell's transcriptional machinery. The HIV-1 LTR functions as a promoter in a variety of cell types, but its basal level is very low. As described in Chapter 8 (Fig. 8.12), the LTR of HIV includes an enhancer sequence that binds a number of cell-type-specific transcriptional activators, for example, Nf-κb (Fig. 17.3). The release of Nf-κb from its cytoplasmic inhibitor in activated T cells may explain why HIV replication requires T-cell stimulation.

TAR and Tat. Just downstream of the site of initiation of transcription in the HIV LTR is a unique viral regulatory sequence, TAR (Fig. 17.3). As described in Chapter 8 (Fig. 8.13), TAR RNA forms a stable, bulged stem-loop structure that binds the 14-kDa viral regulatory protein Tat (Table 17.3). In the absence of Tat, viral transcription usually ter-

Table 17.3 HIV auxiliary proteins[a]

Protein[b]	Size (kDa)	Function	Location
Regulatory			
Tat	14	Stimulation of transcription; binds to TAR sequence in viral RNA to facilitate initiation and elongation of viral transcription	Primarily in cell nucleus
Rev	19	Regulation of viral mRNA production; binds RRE and facilitates nuclear export of unspliced or singly spliced viral RNAs	Primarily in cell nucleus
Accessory			
Nef	27	Pleiotropic, can increase or decrease virus replication; reduces expression of MHC class I, the CD4 receptor, and the CD8 receptor; affects T-cell activation; enhances virion infectivity	Cell cytoplasm, plasma membrane
Vif	23	Increases virus infectivity; affects virion assembly and/or viral DNA synthesis; antagonist of cellular protein CEM15/APOBEC3G	Cell cytoplasm
Vpr	15	Causes G_2 arrest; facilitates nuclear entry of preintegration complex	Virion
Vpu[c]	16	Affects virus release; disrupts Env-CD4 complex; CD4 degradation	Integral cell membrane protein
Vpx[d]	15	Nuclear entry of preintegration complexes	Virion

[a]Adapted from Table 1.5 of J. A. Levy, *HIV and the Pathogenesis of AIDS*, 2nd ed. (ASM Press, Washington, D.C., 1998), with permission.
[b]See Fig. 17.2 for location of the viral genes on the HIV genome.
[c]Present only with HIV-1. Expression appears to be regulated by Vpr.
[d]Encoded only by HIV-2. May be a duplication of Vpr.

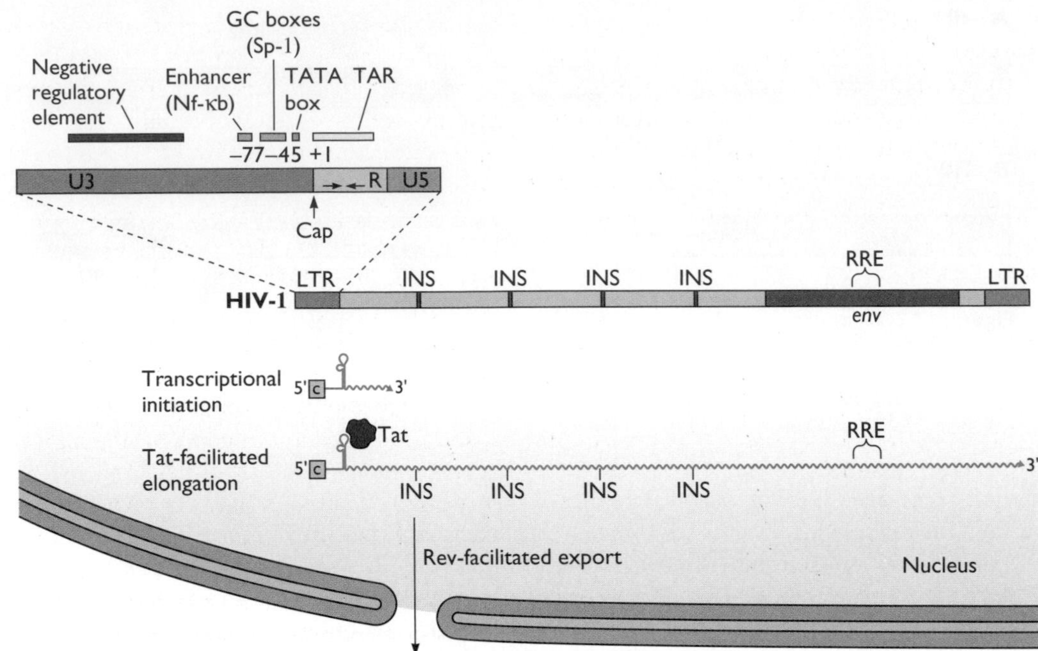

Figure 17.3 Mechanisms of Tat activation. Some regulatory sequences in the HIV LTR are depicted in the expanded section at the top. The numbers refer to positions relative to the site of initiation of transcription. The opposing arrows in R represent a palindromic sequence that folds into a stem-loop structure in the transcribed mRNA to which Tat binds (center). Tat is required for efficient elongation during HIV-1 RNA synthesis. The position of the RRE in the *env* transcript and the presence of *cis*-acting repressive sequences, also known as instability elements (INS), are also illustrated.

minates prematurely. The principal role of Tat is to enhance processivity and facilitate the elongation of viral RNA.

The Tat protein can be cytotoxic to cells in culture and is neurotoxic when inoculated intracerebrally into mice. The protein can also cause increased expression of adhesion molecules on endothelial cells, depolarization of cells, and degeneration of cell membranes, by mechanisms still unknown. It has been reported that transgenic expression of Tat in mice causes a disease that resembles Kaposi's sarcoma. Although the human disease is almost certainly caused by a herpesvirus, Tat contributes to the aggressive nature of this malignancy in AIDS patients by promoting the growth of spindle cells in Kaposi's sarcoma.

Multiple splice sites and the role of Rev. Unlike those of the simpler oncogenic retroviruses, the full-length HIV transcript contains numerous 5′ and 3′ splice sites. The regulatory proteins Tat and Rev and the accessory protein Nef are synthesized early in infection from multiply spliced mRNAs (Appendix A, Fig. 21). As Tat then stimulates transcription, these mRNAs are found in abundance at this early time. However, the accumulation of Rev protein brings about a change in the pattern of mRNAs, leading to a temporal shift in viral gene expression.

The Rev protein (Table 17.3) is an RNA-binding protein that recognizes a specific sequence within a structural element in *env* called the **Rev-responsive element (RRE)** (Fig. 17.3). As discussed in Chapter 10 (Fig. 10.15 to 10.17), Rev activates the nuclear export of any RRE-containing RNA. As the Rev concentration increases, unspliced or singly spliced transcripts containing the RRE are exported from the nucleus. In this way, Rev facilitates synthesis of the viral structural proteins and enzymes, and ensures the availability of full-length genomic RNA to be incorporated into new virus particles. The accessory proteins Vif, Vpr, and Vpu (for HIV-1) or Vpx (for HIV-2) are also expressed later in infection from singly spliced messenger RNAs (mRNAs) that are dependent on Rev for export to the cytoplasm (Appendix A, Fig. 22).

The dependence of HIV gene expression on Rev is due in part to *cis*-acting repressive sequences, also called **instability elements**, present in the unspliced or singly spliced transcripts. These sequences, some of which are characterized by a high A/U content, have been mapped to several locations within the regions encoding Gag and Pol. Mutations in these sequences have been shown to increase the stability, nuclear export, and translatability of the transcripts in the absence of Rev. The response to their presence appears to be cell dependent, but the mechanism(s) by which these sequences act, and exactly how Rev counteracts their effects, is not understood. This phenomenon does, however, provide an explanation for the puzzling failure of early attempts to express individual HIV-1 structural proteins and enzymes in primate or human cells from mRNAs that did not also encode Rev.

The Accessory Proteins

Nef protein. As noted above, Nef (which initially stood for "negative factor") is translated from multiply spliced early transcripts. The 5′ end of Nef mRNA includes two initiation codons and, as both are utilized, two forms of Nef are produced in infected cells. The apparent size of these proteins can vary from 27 to 37 kDa because of differences in posttranslational modification. Nef is myristylated posttranslationally at its N terminus (Fig. 17.4 and 17.5) and thereby anchored to the inner surface of the plasma membrane, probably in a complex with a cellular serine kinase. There are numerous potential threonine and serine phosphorylation sites in Nef, and the protein is phosphorylated, but the significance of such modification is as yet unknown. Nef includes a protein-protein (SH3) interaction domain, which mediates interactions with components of intracellular signaling pathways, eliciting a program of gene expression similar to that observed after T-cell activation. Such expression may provide an optimal environment for viral replication.

Most laboratory strains of HIV-1 that have been adapted to grow well in T-cell lines contain deletions or mutations in the *nef* gene. Restoration of *nef* reduces the efficiency of virus replication in these cells, hence the name "negative factor." Multiple functions have been attributed to Nef (Table 17.3 and Fig. 17.5), and it is now clear that Nef exerts pleiotropic effects on infected cells. The functions reported for the Nef protein vary for different strains of the virus and with different cell types. Nef has been shown to decrease cell surface expression of CD4 as a result of interaction with the cytoplasmic portion of this integral membrane protein and induction of its endocytosis (and lysosomal degradation). Nef-mediated decrease in surface expression of the immune costimulatory protein CD28 also involves the endocytosis pathway, but genetic studies indicate that the determinants for this activity are distinct from those for CD4 downregulation. Nef also decreases cell surface expression of MHC class I molecules in infected primary lymphocytes by a different, but as yet unknown, pathway. As a strong cytotoxic T-lymphocyte (CTL) response against viral infection requires recognition of viral epitopes presented by MHC class I molecules, this inhibitory activity of Nef may allow infected cells to evade lysis by CTLs and could be a major factor contributing to

Figure 17.4 Organization of known or presumed functional regions in HIV-1 accessory proteins. Locations of conserved and critical cysteine residues are indicated in Nef and Vif. Serines that require phosphorylation for protein function are noted in Vpu. Magenta regions in Vif are conserved but their functions are as yet unknown. Numbering corresponds to amino acid positions in proteins of the laboratory strain of HIV-1, NL43.

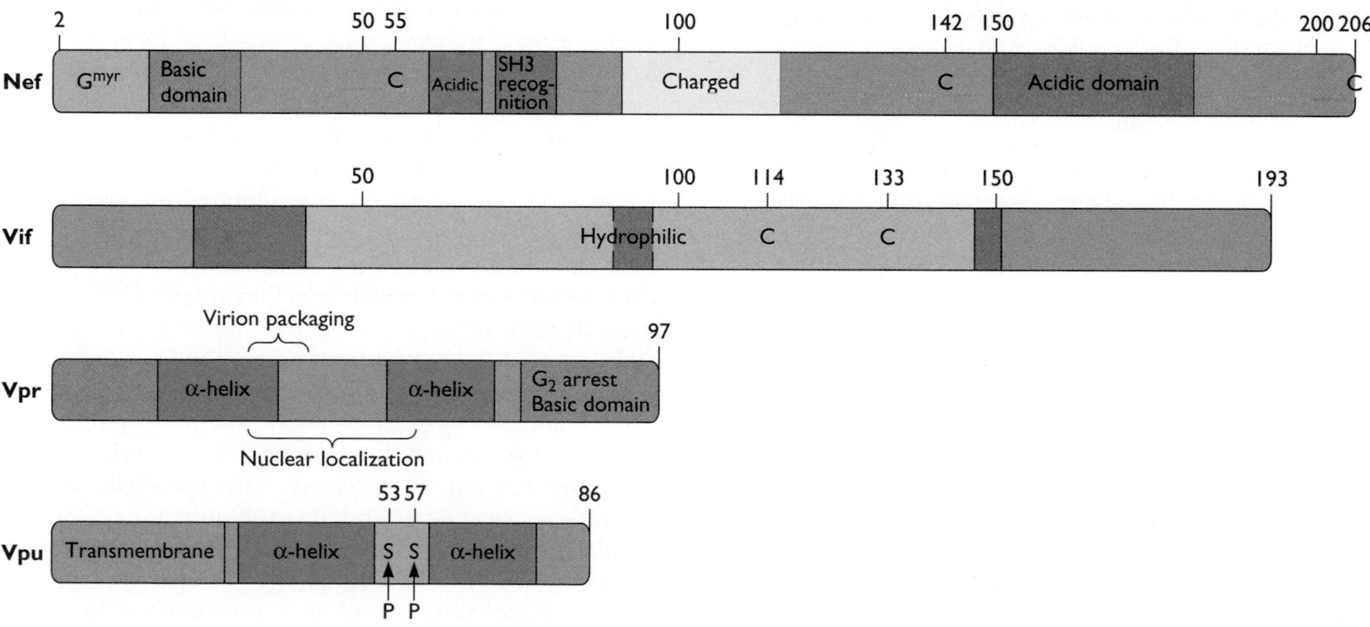

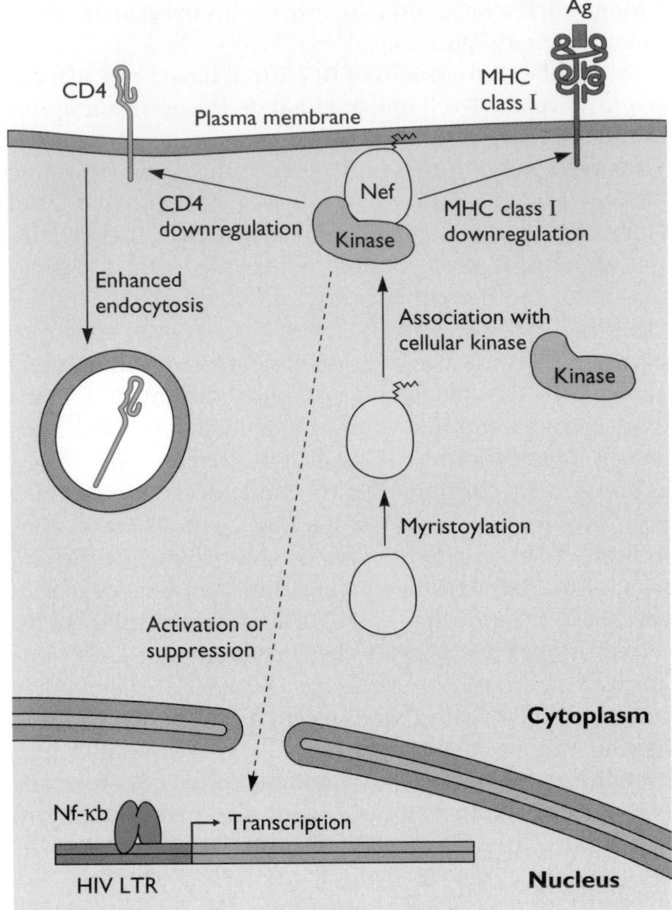

Figure 17.5 Intracellular functions attributed to Nef. Nef is myristoylated posttranslationally; the jagged protrusion represents myristic acid covalently linked to the glycine residue at position 2 in Nef. Myristoylation enables Nef to attach to cell membranes and bind to a cellular serine kinase and perhaps other cellular proteins. The cytoplasmic domain of CD4 contains sequences bound by Nef. Nef reduces cell surface expression of CD4 by enhancing endocytosis; MHC class I expression is also reduced. By an unknown mechanism, Nef increases the activity of the cellular transcriptional activator Nf-κb and perhaps other cellular transcription proteins, thereby augmenting transcription from the viral LTR. Nef has also been reported to suppress viral gene expression.

HIV-1 pathogenesis. Finally, Nef protein is packaged into virions, and its presence appears to enhance infectivity by facilitating the fusion of the viral envelope with the host cell membrane.

Although the initial cell culture experiments suggested a negative effect on virus production, subsequent experiments in animals indicated that Nef augments HIV pathogenesis significantly. Rhesus macaques inoculated with a Nef-defective mutant of SIV had low virus titers in their blood during early stages of infection, and the later appearance of high titers was associated with reversion of the mutation. More important, adult macaques inoculated with a virus strain containing a deletion of *nef* did not progress to clinical disease and were, in fact, immune to subsequent challenge with wild-type virus. The observation that *nef* had been deleted in HIV-1 isolates from some individuals who remained asymptomatic for long periods, and from transfusion recipients who did not develop AIDS, also suggests that this viral protein can contribute significantly to pathogenesis. Initial hopes that intentional deletion of *nef* might facilitate the development of a vaccine strain for humans were dashed when it was discovered that the humans infected with *nef* deletion mutants eventually developed AIDS and that newborns of the female macaques that had been immunized with the *nef*-minus strain of SIV developed an AIDS-like disease.

Vif protein. Vif stands for viral infectivity factor. This protein (Table 17.3 and Fig. 17.4) accumulates in the cytoplasm and at the plasma membrane of infected cells. Early studies showed that mutant viruses lacking the *vif* gene were approximately 1,000 times less infectious than the wild type in certain $CD4^+$ T-cell lines and peripheral blood lymphocytes and macrophages, although direct cell-to-cell transfer was only slightly lower than normal. Virions produced in the absence of Vif are therefore defective.

Production of Vif from a plasmid vector in susceptible host cells does not compensate for its absence in the cell that produces virions. Rather, it appears that Vif is needed at the time of virus assembly or maturation in the producing cells. Vif is an RNA-binding protein; small amounts can be detected in HIV particles, and also in heterologous retroviral particles produced by cells that have been transfected with a Vif-expressing plasmid. Virions produced from *vif*-defective HIV genomes contain the normal complement of progeny RNA, and they are able to enter susceptible cells and to initiate reverse transcription, but full-length double-stranded viral DNA is not detected. These observations indicate that Vif is required for a postentry step that is essential for completion of reverse transcription. The requirement for Vif is strikingly cell-type dependent. Experiments in which cells that are permissive for *vif* mutants were fused with cells that are nonpermissive established that the nonpermissive phenotype is dominant; the infectivity of virions produced in such heterokaryons was enhanced by Vif production. This observation suggested that Vif protein may suppress a host cell function that otherwise inhibits progeny virus infectivity. Recent studies indicate that Vif inhibits the antiviral action of a cellular cytidine deaminase, CEM15/ APOBEC3G, which is synthesized in nonpermissive cells. This enzyme catalyzes the deamination of deoxycytidine to form deoxyuridine (dU) in the first (–) strand of viral DNA to be synthesized by reverse transcriptase. The dU is a substrate for the cellular uracil-DNA glycosylase, and the abasic sites that are produced by its action would be targets for endonucleolytic digestion. Such destruction may account for the diminished infectivity of *vif* mutant viruses. If dU is not removed, the (+) strand complement of the deaminated (–) strand would contain deoxyadenosine in place of the normal deoxyguanosine at such sites. This mechanism may explain why G→A transitions are the most frequent point mutations in HIV genomes. The frequency of such transitions is abnormally high in the genomes of *vif*-defective virions produced in nonpermissive cells. This increased rate of mutagenesis is also incompatible with viral viability. It has been suggested that CEM15/APOBEC3G activity represents an ancestral mode of innate cellular defense against retroviruses.

Vpr protein. The viral protein R, or Vpr (Table 17.3 and Fig. 17.4), derives its name from the early observation that it affects the rapidity with which the virus replicates in, and destroys, T-cell lines. Most T-cell-adapted strains of HIV-1 carry mutations in *vpr*. The Vpr protein is encoded in an open reading frame lying between *vif* and *tat* in the genomes of primate lentiviruses (Fig. 17.2). The SIV and HIV-2 genomes include a second, related gene, *vpx*, which appears to have arisen as a duplication of *vpr*. The other lentiviruses do not contain sequences related to *vpr* but do include small open reading frames that might encode proteins with similar functions.

Vpr and Vpx, like Nef, are incorporated into virions. Vpr incorporation is dependent on specific interactions with a proline-rich domain (p6) at the C terminus of the Gag polyprotein. Although it was thought that Vpr might therefore be incorporated in amounts equimolar to Gag (1,500 to 2,000 copies per virions), only about 100 to 200 molecules of Vpr appear to be present in nucleocapsids. However, its presence in virions is consistent with the observation that Vpr function is required at some early stage in the virus replication cycle.

Two different functions have been recognized for HIV-1 Vpr. The protein causes a G_2 cell cycle arrest, and it promotes entry of viral nucleic acids into the nucleus. In HIV-2 these functions are segregated into Vpr and Vpx, respectively. Several studies have indicated that the G_2 arrest is a consequence of Vpr inhibition of the cyclin B/Cdc2 kinase complex that is required for cells to progress to mitosis. The biological advantage of preventing infected cells from entering mitosis is not clear, but the increased activity of the LTR promoters in the G_2 phase of the cell cycle may lead to enhanced virus production. Vpr has also been reported to stimulate transcription from the viral LTR by interaction with cellular transcription regulators. Vpr has been shown to bind to nuclear pore proteins; as noted in Chapter 5, these interactions may facilitate docking of the HIV-1 preintegration complex at the nuclear pore in preparation for import. Studies with SIV-infected macaques indicate that deletion of *vpr* attenuates, but does not eliminate, viral pathogenicity. However, deletion of both *vpr* and *vpx* does reduce virus replication in these animals. It seems likely therefore that HIV-1 Vpr, which combines the functions of the two SIV gene products, is crucial to HIV pathogenesis.

Vpu protein. This small protein is unique to HIV-1 and the related SIV_{cpz} (Fig. 17.1), hence the name viral protein U (Vpu). The predicted sequence of Vpu includes an N-terminal stretch of 27 hydrophobic amino acids that is likely to be a membrane-spanning domain (Fig. 17.4). Biochemical studies show that Vpu is an integral membrane protein that self-associates to form oligomeric complexes. In infected cells, the protein accumulates in the perinuclear region.

Expression of Vpu is required for the proper maturation and targeting of progeny virions and for their efficient release (Table 17.3). In its absence, virions containing multiple cores are produced, and budding is targeted to intracellular vacuolar compartments rather than to the plasma membrane. Vpu also appears to reduce the syncytium-mediated cytopathogenicity of HIV-1, perhaps because the efficient release of virions prevents the excessive accumulation of virion Env proteins at the cell surface.

The Vpu protein has structural and biochemical features similar to those of the influenza virus M2 protein. As noted in Chapter 5, M2 is an ion channel protein that modulates the pH in the Golgi compartment, thereby protecting the newly formed hemagglutinin protein from changing conformation prematurely in the secretory pathway. Vpu appears to oligomerize in lipid bilayers, forming channel-like structures, which are believed to play a role in virion release. In this connection, it is interesting to note that expression of Vpu also enhances the release of virions produced by other retroviruses, such as Moloney murine leukemia virus, and the lentiviruses visna/maedi virus and HIV-2.

A second function of Vpu is the degradation of CD4. Vpu traps this receptor in the endoplasmic reticulum via specific interactions with its cytoplasmic domain, preventing transport of CD4 to the cell surface and triggering its destruction by the proteasome. These two activities of Vpu are distinct, as the stimulation of virion release is independent of Env or CD4 expression.

Cellular Targets

As discussed in Chapters 4 and 5, virus attachment and entry into host cells are dependent on the interaction between viral proteins and cellular receptors. The major receptor for the HIV envelope protein, SU, is the cell surface CD4 molecule. The HIV envelope protein must also interact with a coreceptor to trigger fusion of the viral and cellular membranes and gain entry into the cytoplasm. The ability to bind to specific coreceptors is a critical determinant of the cell tropism of different HIV-1 strains. For example, binding to the α-chemokine receptor CXCr4 is a definitive feature of strains that infect T-cell lines. Infection with these strains also causes T cells to fuse, forming syncytia. Binding to the β-chemokine receptor CCr5 is characteristic of non-syncytium-inducing M (monocyte/macrophage)-tropic strains. However, some M-tropic strains can also infect T cells. Strains of HIV that bind to CXCr4 or CCr5 coreceptors are commonly referred to as X4 or R5 strains, respectively. The importance of these two chemokine receptors to HIV pathogenesis is demonstrated by two findings. People who carry a mutation in the gene encoding CCr5, and produce a defective receptor protein, are resistant to HIV-1 infection. So too are individuals who carry a mutation in the gene for a ligand of CXCr4 (Tables 15.2 and 15.3). The latter mutation may lead to increased availability of the ligand, which then blocks virus entry by competing for coreceptor binding. This idea is consistent with earlier studies showing that chemokine binding to these receptors inhibits the infectivity of specific strains of HIV in cell culture (Fig. 17.6). Cells of the hematopoietic lineage that bear CD4 and one or more of these chemokine receptors are the main targets of HIV infection, and they produce the highest titers of progeny virions.

Several additional coreceptors for HIV and SIV have been identified in cell culture experiments in various laboratories, but their roles in natural infection remain to be discovered. These additional coreceptors may allow the virus to enter a broader range of cells than first appreciated. Some of them are found on cells of the thymus gland and the brain, and they could play a role in infection in infancy or of cells in the nervous system. It has also been proposed that binding to these additional coreceptors may trigger signals that affect virus replication in target cells, or that harm nonpermissive cells, producing a "bystander" effect.

Cell culture studies have revealed additional mechanisms by which HIV may enter cells. For example, the virus can be transmitted through direct cell contact. In addition, cells may also be infected by virus particles that are endocytosed after binding to cell surface galactosyl ceramide or to Fc receptors (as antibody-virus complexes). HIV can infect many different types of human cells in culture and has been found in small quantities in several tissues of the body. As discussed below, infection of these cells and tissues is likely to be relevant to HIV-1 pathogenesis in humans.

Figure 17.6 Coreceptors for macrophage/monocyte- and T-cell-tropic strains of HIV-1. CXCr4 is the major coreceptor for T-cell-tropic strains; entry of such strains (denoted X4) is inhibited by the receptor's natural ligand, Sdf-1. CCr5 is the major coreceptor for macrophage/monocyte (M)-tropic strains (denoted R5), and their entry is inhibited by the receptor's natural ligands, Rantes, and the macrophage inflammatory proteins Mip-1α and Mip-1β. Primary T cells and monocytes produce both coreceptors; primary T cells are susceptible to both strains, but monocytes can be infected only by M-tropic strains for reasons that are not yet clear. Adapted from Fig. 3 of A. S. Fauci, *Nature* **384:**529–533, 1996, with permission.

T-cell-line-tropic strain of HIV-1

HIV (X4)
α-chemokine receptor (CXCr4)
CD4
Sdf-1
CD4+ target cell

Macrophage-tropic strain of HIV-1

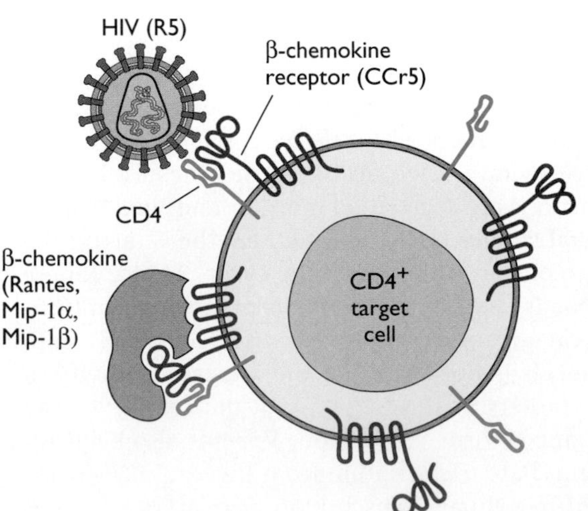

Routes of Transmission

Sources of Virus Infection

Even before HIV-1 was identified, epidemiologists had established the most likely routes of its transmission to be via sexual contact, via blood, and from mother to child. As might be anticipated, the efficiency of transmission is influenced greatly by the concentration of the virus in the body fluid to which an individual is exposed. Table 17.4 provides estimates of the percentage of infected cells and the concentration of HIV-1 in different body fluids. The highest values are observed in peripheral blood monocytes, in blood plasma, and in cerebrospinal fluid, but semen and female genital secretions also appear to be important sources of the virus.

Other routes of transmission have been shown to be relatively unimportant or nonexistent, at least for HIV-1; among these are casual nonsexual contact, exposure to saliva or urine from infected individuals, and exposure to blood-sucking insects. Fortunately, HIV-1 infectivity is reduced upon air drying (by 90 to 99% within 24 h), by heating (56 to 60°C for 30 min), by exposure to standard germicides (such as 10% bleach or 70% alcohol), or by exposure to pH extremes (e.g., <6 or >10 for 10 min). This information and results from epidemiology studies have been used to establish safety regulations to prevent transmission in the public sector and in the health care setting.

Table 17.4 Isolation of infectious HIV-1 from body fluids[a]

Fluid	Virus isolation[b]	Estimated quantity of virus[c]
Cell-free fluid		
Plasma	33/33	1–5,000[d]
Tears	2/5	<1
Ear secretions	1/8	5–10
Saliva	3/55	<1
Sweat	0/2	None detected
Feces	0/2	None detected
Urine	1/5	<1
Vaginal-cervical	5/16	<1
Semen	5/15	10–50
Milk	1/5	<1
Cerebrospinal fluid	21/40	10–10,000
Infected cells		
PBMC[e]	89/92	0.001–1%[d]
Saliva	4/11	<0.01%
Bronchial fluid	3/24	Not determined
Vaginal-cervical fluid	7/16	Not determined
Semen	11/28	0.01–5%

[a]Adapted from Table 2.1 of J. A. Levy (ed.), *HIV and the Pathogenesis of AIDS*, 2nd ed. (ASM Press, Washington, D.C., 1998), with permission.

[b]Number of samples positive/number analyzed.

[c]For cell-free fluid, units are infectious particles per milliliter; for infected cells, units are percentages of total cells capable of releasing virus. Results are from studies in the laboratory of J. A. Levy.

[d]High levels associated with acute infection and advanced disease (~5×10^6 PBMC per ml in blood).

[e]PBMC, peripheral blood mononuclear cells.

Modes of Transmission

Modes of HIV-1 transmission vary in different geographic locations (Fig. 17.7). In the United States, the major mode is via homosexual contact, whereas heterosexual transmission is most common in the rest of the world. Although a single contact can be sufficient for transmission of the virus, the likelihood of infection increases with the number of infected partners encountered. The presence of sexually transmitted diseases also increases the probability of HIV-1 transmission, presumably because infected inflam-

Figure 17.7 Modes of transmission of HIV worldwide and in the United States. From January 1998 data published by the Joint United Nations Programme on HIV/AIDS and the U.S. Centers for Disease Control and Prevention.

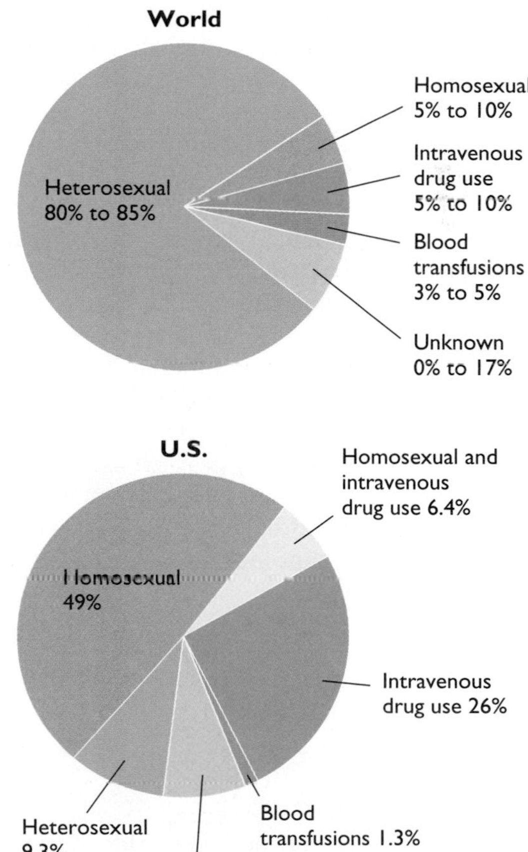

matory cells may be present in both seminal and vaginal fluids. Genital ulceration and consequent direct exposure to infected blood cells also increase the likelihood of transmission. In both homo- and heterosexual contact, the recipient partner is the one most at risk.

Intravenous drug use is the next most common route of transmission, owing to the common practice of sharing contaminated needles and other drug paraphernalia. Here again, the probability of transmission is a function of frequency of exposure and the prevalence of HIV-1 infection among a drug user's contacts. Of course, sexual partners of drug users are also at increased risk.

Until 1985, when routine HIV antibody testing of donated blood was established in the United States and other industrialized countries, individuals who received blood transfusions or certain blood products, such as clotting factors VIII and IX, were at high risk of becoming infected. Transfusion of a single unit (500 ml) of blood from an HIV-1-infected individual nearly always led to infection of the recipient. Appropriate heat treatment of clotting factor preparations and, more recently, their production by biotechnology has eliminated transmission from this source. This safeguard is small comfort to the many hemophiliacs who contracted the disease before this route was understood. Fortunately, other blood products, such as pooled immunoglobulin, albumin, and hepatitis B vaccine, were not implicated in HIV-1 transmission, presumably because their production methods include steps that destroy the virus.

Transmission of HIV from mother to child can occur across the placenta or, most frequently, at the time of delivery as a consequence of exposure to a contaminated genital tract. The virus can also be transmitted via infected cells in the mother's milk during breast-feeding. Rates of transmission from an infected mother to a child range from as low as 11% to as high as 60%, depending on the severity of infection (the concentration of virus present) in the mother and the prevalence of breast-feeding (the frequency of the infant's exposure). Antiviral drug therapy administered during pregnancy is an effective measure to reduce the amount of virus to which the newborn is exposed and, therefore, the frequency of transmission. Even a single treatment with a nonnucleoside reverse transcriptase inhibitor early in labor can reduce the incidence significantly. Unfortunately, because of the cost, this is not often an option in underdeveloped countries, where the risk is greatest. Current efforts are focused on discouraging breast-feeding by infected mothers. In 2002 alone, an estimated 800,000 children were newly infected worldwide.

Mechanics of Spread

Except in cases of direct needle sticks or blood transfusion, HIV enters the body through mucosal surfaces, as do most viruses (Chapter 14). In the case of sexual transmission, the initial target cells in the rectum or genital tract have not been identified. The most likely sources of transmission are virus-infected cells, as they can be present in much larger numbers than free infectious virus particles in vaginal or seminal fluids. Results of cell culture studies show that HIV-1 can be transferred directly to $CD4^-$ epithelial cells via cell-cell contact (Fig. 17.8). Whether such transfer is relevant to natural transmission is unknown. Results from analyses of tissue biopsies of bowel mucosae and cervical and uterine epithelia suggest that cells in these layers can be infected in the absence of any injury. As noted in the preceding section, this infection might occur by interaction of the virus with galactosyl ceramide or Fc receptors on the mucosal cells.

Figure 17.8 HIV-1-infected HUT 78 cell (top) cocultivated for 1 h with an ME1809 cervix-derived epithelial cell (bottom). Virus can be seen at the cell-cell interface. It is noteworthy that virus is produced by the T cell only at the point of contact with the epithelial cell. A role of cytokines in this induction of virus by localized release should be considered. Magnification, ×8,700. Reprinted from Fig. 2.6 of J. A. Levy, *HIV and the Pathogenesis of AIDS*, 2nd ed. (ASM Press, Washington, D.C., 1998), with permission.

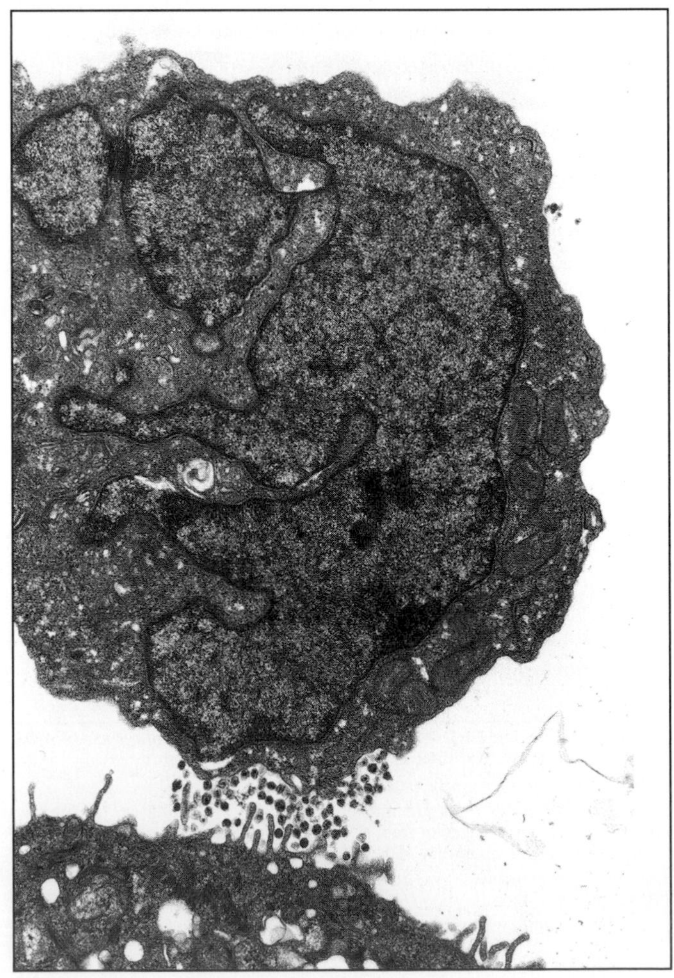

The main routes of initial infection include the acquisition of virus by cells of the mucosal and cutaneous immune system described in Chapter 15; these include M cells present in the bowel epithelium and the dendritic, antigen-presenting $CD4^+$ Langerhans cells in the vaginal and cervical epithelia. Dendritic cells express a glycoprotein on their surface, called DC-SIGN (dendritic-cell-specific, Icam-3-grabbing nonintegrin), that binds the HIV-1 envelope protein with high affinity and can stabilize the virus for several days until it encounters a susceptible T cell. Activated T cells, coincidental with genital infections that cause sores, are also likely targets. Although the insertive partner is at relatively low risk for infection, transmission to the male can occur through cells in the lining of the urethral canal of the penis, presumably from infected macrophages or Langerhans cells in the cervix or the intestinal mucosa of the infected partner. Uncircumcised males have a twofold-increased risk of infection, suggesting that the mucosal lining of the foreskin may be susceptible to HIV infection. Both male and female hormones appear to facilitate HIV transmission by stimulating cell-cell contact (prostaglandins) or erosion of the vaginal lining (progestin).

Free virus, virus attached to dendritic cells, or virus-infected cells enter draining lymph nodes or the circulatory system, where the next major targets are encountered, namely, susceptible cells that bear the CD4 receptor. As nonactivated peripheral blood mononuclear cells are not very permissive for infection, and few activated $CD4^+$ lymphocytes are circulating in the blood at any given time, the first $CD4^+$ T cells in the blood to be infected are probably macrophages. The macrophages are in a differentiated state, permissive for viral replication, and can pass progeny virions to activated lymphocytes in the lymph nodes. From this point the infection runs its protracted, but usually inevitable, course.

The Course of Infection

Patterns of Virus Appearance and Immune Cell Indicators of Infection

Pathological conditions associated with different phases in HIV-1 infection are summarized in Table 17.5.

The Acute Phase

In the first few days after infection, the virus is produced in large quantities by the activated lymphocytes in lymph nodes, sometimes causing the nodes to swell (lymphadenopathy) or producing flu-like symptoms. Virus released into the blood can be detected by infectivity with appropriate cell cultures or by screening directly for viral RNA or proteins (Fig. 17.9). As many as 5×10^3 infectious virions or 10^7 viral RNA molecules (i.e., ~5×10^6 particles) per ml of plasma can be found during this stage. Generally this viremia is greatly curtailed within a few weeks, most

Table 17.5 Pathological conditions associated with HIV-1 infection[a]

Acute phase

Mononucleosis-like syndrome: fever, malaise, pharyngitis, lymphadenopathy, headache, arthralgias, diarrhea, maculopapular rash, meningoencephalitis

Asymptomatic phase

Often none, but patients may present sporadically with one or more of the following symptoms: fatigue, mild weight loss, generalized lymphadenopathy, thrush, oral hairy leukoplakia, shingles

Symptomatic phase and AIDS

200–500 $CD4^+$ T cells per ml: generalized lymphadenopathy, oral lesions (thrush, hairy leukoplakia, aphthous ulcers), shingles, thrombocytopenia, molluscum contagiosum, basal cell carcinomas of the skin, headache, condyloma acuminatum, reactivation of latent *Mycobacterium tuberculosis*

Fewer than 200 $CD4^+$ T cells per ml

Protozoal infections: *Pneumocystis carinii, Toxoplasma gondii, Isospora belli,* cryptosporidia, microsporidia

Bacterial infections: *Mycobacterium avium-Mycobacterium intracellulare, Treponema pallidum*

Fungal infections: *Candida albicans, Cryptococcus neoformans, Histoplasma capsulatum*

Viral infections and malignancies: human cytomegalovirus, recurrent bouts of oral or genital herpes simplex virus, lymphoma (mostly Epstein-Barr virus, some human herpesvirus 8), Kaposi's sarcoma (human herpesvirus 8), anogenital carcinoma (human papillomavirus)

Neurological symptoms: aseptic meningitis; myelopathies, such as vacuolar myelopathy; pure sensory ataxia; paresthesia/dysesthesia; peripheral neuropathies, such as acute demyelinating polyneuropathy, mononeuritis multiplex, and distal symmetric polyneuropathy; myopathy; AIDS dementia complex

[a]Adapted from Table 1 of A. S. Fauci and R. C. Desrosiers, p. 587–635, *in* J. M. Coffin et al. (ed.), *The Retroviruses* (Cold Spring Harbor Laboratory Press, Plainview, N.Y., 1997), with permission.

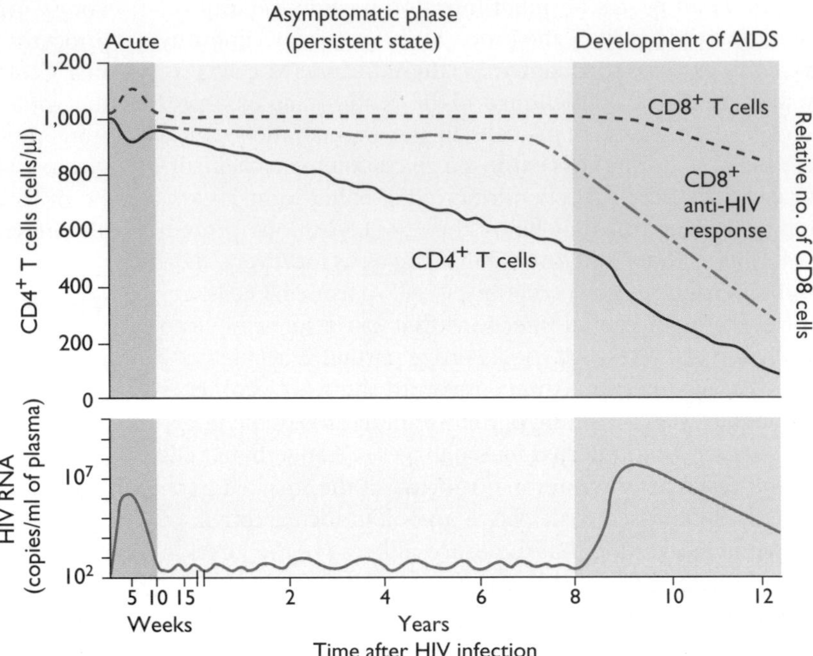

Figure 17.9 Schematic diagram of events occurring after HIV-1 infection. Adapted from Fig. 13.1 of J. A. Levy, *HIV and the Pathogenesis of AIDS*, 2nd ed. (ASM Press, Washington, D.C., 1998), with permission.

likely via the cell-mediated immune response, as the number of CTLs increases before neutralizing antibodies can be detected. The $CD4^+$ T-cell count decreases during this phase but then returns to near normal levels.

During the period of acute infection, the virus population is relatively homogeneous. In most cases, it appears that the predominant virus was a minor variant in the population present in the source of the infection. These early-stage isolates are generally macrophage tropic, but the reason for their apparent selective transmission is unknown.

The Asymptomatic Phase

By 3 to 4 months after infection, viremia is usually reduced to low levels, with small bursts of virus appearing from time to time. It is known that the degree of viremia at this stage of infection, the so-called virologic set point, is a direct predictor of how fast the disease will progress in a particular individual: the higher the viremia, the faster the progression. During this time, $CD4^+$ T-cell numbers decrease at a steady rate estimated to be approximately 60 cells/µl/year. The cause of this decline remains a mystery, although direct cytopathogenicity by the virus and/or apoptosis due to inappropriate cytokine production may be responsible. In this protracted asymptomatic period, which can last for years, the CTL level remains slightly elevated, but virus replication continues at a low rate, mainly in the lymph nodes. In lymphoid tissues a relatively large, stable pool of virions can be detected bound to the surface of follicular dendritic cells. A small number of infected T cells are also observed. During this asymptomatic phase of persistent infection, known also as **clinical latency**, only 1 in 300 to 400 infected cells in the lymph nodes may actually release virus. It is believed that, as in acute infection, virus propagation is suppressed at this stage by the action of antiviral CTLs. The number of these specific lymphocytes decreases toward the end of this stage. During the asymptomatic phase, the virus population becomes more heterogeneous, probably because of continual selection for specific mutations as a result of immunological pressures (Chapter 16).

The Symptomatic Phase and AIDS

The end stage of disease, when the infected individual develops symptoms of AIDS, is characterized by a $CD4^+$ T-cell count below 200 per ml and increased quantities of virus. The total CTL count also decreases, probably owing to the precipitous drop in the number specific for HIV. In the lymph nodes, virus replication increases with concomitant destruction of lymphoid cells and of the normal architecture of lymphoid tissue. The cause of this lymph node degeneration is not clear; it may be due directly to virus replication or an indirect effect of chronic immune stimulation.

In this last stage, the virus population again becomes relatively homogeneous and, generally, T cell tropic. Properties associated with increased virulence predominate, including an expanded cellular host range, ability to cause formation of syncytia, rapid replication kinetics, and $CD4^+$ T-cell cytopathogenicity. Late-emerging virus also appears to be less sensitive to neutralizing antibodies and more readily recognized by antibodies that enhance infectivity. In some cases, strains that have enhanced neurotropism or increased pathogenicity for other organ systems emerge. Where analyzed, these changes can be traced to specific mutations, for example, in the viral envelope gene or in a regulatory gene (e.g., *tat*).

Variability of Response to Infection

Studies of large cohorts of HIV-1-infected adults show that approximately 10% progress to AIDS within the first 2 to 3 years of infection. Over a period of 10 years, approximately 80% of infected adults will show evidence of disease progression and, of these, 50% will have developed AIDS. Of the remainder, 10 to 17% will be AIDS free for over 20 years; a very small percentage of these individuals will be completely free of symptoms, with no evidence of progression to disease. What are the parameters that contribute to such variability?

One parameter is the degree to which an individual's immune system may be stimulated by infection with other pathogens. HIV-1 replicates most efficiently in activated T cells, and it is known that virus quantities increase when the immune system is activated by opportunistic infections with other microorganisms. Such activation can explain the fact that HIV-1 disease is generally more aggressive in sub-Saharan Africa, where chronic infection by parasites and other pathogens is frequent. As might be expected for an outbred population, variations in an individual's genetic makeup can modulate immune response to infection and affect survival. As already noted, differences in chemokines or chemokine receptors and, probably, in any one of several components of the immune system can have an impact on the course of the disease.

Clearly, accumulation of mutations in the genomes of the virus will also influence the course of the disease. As noted above, some long-term survivors of HIV-1 infection harbor viruses with deletions in the *nef* gene. Others appear to be infected with otherwise attenuated strains that produce low titers in cultured cells and have restricted cell tropism. Low virus titers and the presence of viral neutralizing and not enhancing antibodies (Fig. 17.10) are other characteristics of these infections. Further study of the rare, fortunate individuals who are nonprogressors or long-term survivors of HIV-1 infection will lead to a better understanding of viral pathogenesis and suggest new strategies for effective prevention or therapy.

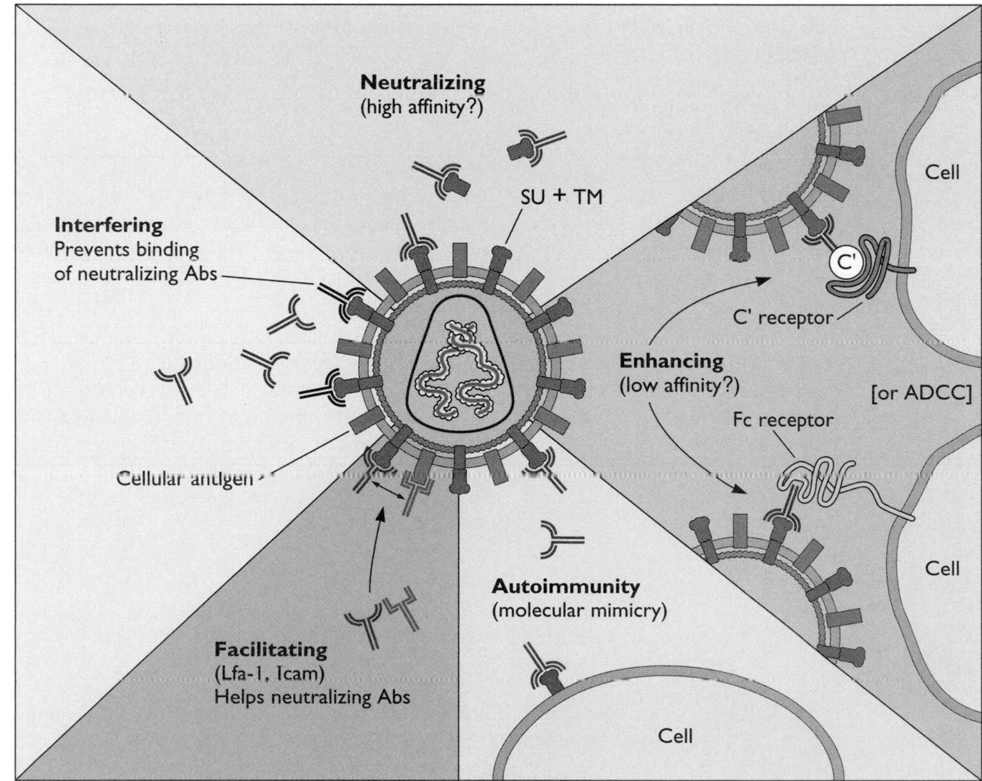

Figure 17.10 Antibody (Ab) responses to HIV infection. The figure summarizes the various responses described in the text. "Cellular antigen" refers to cellular membrane proteins that are incorporated in the virus envelope. One idea is that the relative affinities of the antibodies may be critical. According to this hypothesis, high-affinity antibodies neutralize the virus by binding tightly to SU, causing it to become detached from the virion; low-affinity antibodies bind to SU but not tightly enough to cause its detachment. Conformational changes in SU that might occur as a consequence of such low-affinity binding would then facilitate viral entry. C', complement; ADCC, antibody-dependent cellular cytotoxicity response.

Origins of Cellular Immune Dysfunction

The defining feature of HIV disease is impaired immune cell function. This defect is eventually devastating because these defenses are vital in the body's battle against this virus as well as other pathogens. At first the immune system appears to keep the HIV infection in check. However, the virus is not eliminated, and the infection that persists in the asymptomatic stage leads to increasing dysfunction among immune cells. In the end, most AIDS patients actually succumb to opportunistic infections with microorganisms that are little threat to individuals with healthy immune systems. The impairment in immune cell function typical of HIV-1 disease can result from direct attack of the virus on particular cell types, or it can come about indirectly as a response of uninfected cells to viral gene products or specific proteins made in infected or stimulated cells (Table 17.6).

$CD4^+$ T Lymphocytes

The major reservoir of HIV in the peripheral blood is the $CD4^+$ T cell. Even before the profound depletion of these cells that is a signal of end-stage disease, abnormalities in $CD4^+$ T-cell functions can be detected. These abnormalities include decreased ability to form colonies when grown in tissue culture, decreased expression of IL-2 and the IL-2 receptor, and reduced proliferative response to various antigens. The precise causes of these impairments are unknown. Some loss of function can be explained by direct killing of cells. In addition, as surface expression of CD4 is reduced by some viral proteins (SU, Vpu, and Nef), MHC class II-CD4 interactions become less frequent, leading to a decrease in antigen-specific responses. Trapping of infected $CD4^+$ lymphocytes in lymphoid organs may account for some of the decline. Destruction of infected $CD4^+$ lymphocytes in the circulation by anti-HIV-1 CTLs also reduces their total number. Noninfectious virions and viral proteins shed from infected cells can bind to or enter uninfected $CD4^+$ lymphocytes, triggering inappropriate responses. For example, HIV-1 envelope and Tat proteins have been reported to suppress proliferative responses of T cells to antigens and other mitogens. It is possible that SU bound to CD4 interferes sterically with its interaction with

Table 17.6 Immune cell dysfunction associated with the onset of AIDS

Affected cell type	Dysfunction	Known and/or postulated causes	
		Direct	Indirect
$CD4^+$ T cells	Colony formation ↓ Proliferative response to antigen ↓ Expression of IL-2 and IL-2 receptor ↓ Total number in circulation ↓	Direct killing by HIV-1 CD4 downregulated in infected cells (by SU, Vpu, Nef) Trapping of infected cells in lymphoid organs Destruction of infected cells by anti-HIV-1 $CD8^+$ cytotoxic lymphocytes	SU component of noninfectious virions binds to cell surface CD4, suppresses immune response, and causes inappropriate intracellular signaling
$CD8^+$ T cells	Abnormally high numbers following acute phase Loss of anti-HIV cytotoxic lymphocyte activity Loss in numbers toward end stage	Infection and killing of progenitor $CD4^+$/$CD8^+$ and $CD8^+$ immature thymocytes	Attempt of the immune system to establish homeostasis results in a higher production of $CD4^+$ **and** $CD8^+$ cells as $CD4^+$ cells are depleted Loss of IL-2 production as $CD4^+$ cell pool is depleted
Monocytes, macrophages, and dendritic cells	Defects in chemotaxis Monocyte-dependent T-cell proliferation ↓ Antigen-presenting cell activity ↓ Fc receptor function ↓ Complement C3 receptor-mediated clearance/oxidative burst ↓ Decrease in numbers	Only few circulatory cells are infected, but changes in cytokine production by such cells (e.g., Tnf-α ↓; IL-1 ↓) can cause apoptosis of $CD4^+$ cells and other abnormalities	Exposure to noninfectious virions (SU, TM), downregulates chemotactic ligand receptors and causes abnormal secretion of certain chemokines Decreased expression of costimulatory molecules (e.g., B7) Cells killed by eliciting cytotoxic response in cytotoxic lymphocytes
B cells	Abnormal proliferation Hypergammaglobulinemia Poor response to additional antigen signals Production of autoantibodies		Exposure to noninfectious virions (TM) causes polyclonal activation Loss of $CD4^+$ T-cell-helper function Molecular mimicry of host proteins by viral proteins
Natural killer cells	Cytotoxicity function ↓		Loss of IL-2 as $CD4^+$ cell pool is depleted

MHC class II molecules. Such binding may also disturb the normal signaling that must take place during antigen-specific responses. Exposure to HIV-1 SU protein also inhibits production of the lymphokine IL-2 and IL-2 receptor by uninfected T cells. Finally, changes in cytokine production by HIV-1-infected macrophages can trigger apoptosis in uninfected CD4+ T cells.

CTLs

As mentioned previously, the number of CTLs is abnormally high following the acute phase and decreases precipitously during the end stage of the disease. The early increase may be the result of an imbalance brought about as the immune system attempts to achieve homeostasis of CD8+ and CD4+ cells while CD4+ cells are being destroyed. The reduced numbers of anti-HIV-1 CTLs at late times can be explained in part by the direct infection and killing of their progenitors. Additionally, and most important, because CTL proliferation and function depend on inductive signals from CD4+ T cells (Fig. 15.13), the decline of the CD4+ population also contributes to CTL dysfunction.

Monocytes and Macrophages

HIV-1-infected macrophages can be detected readily in tissues throughout the body of an infected individual. However, as only a small proportion of monocytes/macrophages in the blood are infected with the virus, it seems likely that the functional impairment seen in this population of cells is due to indirect effects. Monocyte/macrophage abnormalities include defects in chemotaxis, inability to promote T-cell proliferation, and defects in Fc receptor function and complement-mediated clearance. Results of cell culture experiments suggest that some of these effects are caused by exposure to the viral envelope proteins.

B Cells

HIV-1-infected individuals produce abnormally large quantities of immunoglobulin G (IgG), IgA, and IgD. Such production is indicative of B-cell dysfunction that may result from increased proliferation of cells of the lymph nodes. Infected individuals also show poor responses to primary and secondary immunization. During end-stage disease, there may also be a decrease in the total number of B cells. Some of the B-cell loss may be attributable to the decline in CD4+ T-cell-helper function (Fig. 15.13). B cells isolated from infected individuals proliferate in culture without stimulation, and are also defective in their response to specific antigens or mitogens. Binding of viral proteins (e.g., TM) has been shown to induce polyclonal B-cell activation, a property that might explain such apparent spontaneous proliferation. Infection by Epstein-Barr virus and human cytomegalovirus, common in AIDS patients, may also contribute to abnormal B-cell function.

Natural Killer Cells

Impairment of natural killer (NK) cell function is observed throughout the course of HIV-1 infection, becoming more severe during end-stage disease. As NK cell cytotoxicity depends on IL-2, these abnormalities may be a consequence of impaired CD4+ T-cell function and the reduced production of this cytokine. This reduction in NK cell function cripples the innate immune response to infection by other microorganisms.

Autoimmunity

Because of the imbalance of the immune system, T-cell dysfunction, and abnormal B-cell activation described above, it is not surprising that immune disorders, such as a breakdown in the system's ability to distinguish self from nonself, accompany HIV-1 infection. In early studies, antibodies against platelets, T cells, and peripheral nerves were detected in AIDS patients. Subsequently, autoantibodies to a large number of normal cellular proteins have been found in infected individuals (Table 17.7). The specific reason for the appearance of such antibodies is not clear, but their production might be stimulated in part by cellular proteins on the surface of viral particles or by viral pro-

Table 17.7 Autoantibodies detected in HIV infection[a]

Antibodies to:	Associated clinical condition
Lymphocytes	Loss of CD4+, CD8+, and B lymphocytes
Platelets	Thrombocytopenia
Neutrophils	Neutropenia
Erythrocytes	Anemia
Nerves (myelin)	Peripheral neuropathy
Sperm, seminal plasma	Aspermia
Cellular components (Golgi complex, centriole, vimentin)	Immune disorder
Nuclear protein (antinuclear antibody)	Autoimmune symptoms
Lupus anticoagulant (phospholipid, cardiolipid)	Neurological disease (?), thrombosis (?)
Myelin basic protein	Dementia, demyelination
Collagen	Arthritis (?)
CD4	CD4+ cell loss
Hla	Lymphocyte depletion
Hydrocortisone	Addison's-like disease
Thymic hormone	Immune disorder
Thyroglobulin	Thyroid disease

[a]Adapted from Table 10.3 of J. A. Levy, *HIV and the Pathogenesis of AIDS*, 2nd ed. (ASM Press, Washington, D.C., 1998), with permission.

teins, regions of which may resemble cellular proteins (called molecular mimicry) (Chapter 15 and Table 17.8).

Immune Response to HIV

Humoral Responses

Antibodies to HIV-1 can be detected shortly after acute infection, sometimes as early as a few days after exposure to the virus but generally within 1 to 3 months. These antibodies are secreted into the blood and on mucosal surfaces of the body and can be detected in genital and other body fluids. This phenomenon has been exploited in the design of home kits for detecting anti-HIV antibodies in the blood or urine. Among the various isotypes, IgG1 antibodies are known to play a dominant role at all stages of infection, giving rise to an antibody-dependent cellular cytotoxicity response (known as ADCC), complement-dependent cytotoxicity, and neutralizing and blocking responses (Chapters 15 and 16). Levels of other classes of antibody may vary at different clinical times, but there is no known correlation between the isotype and the clinical stage of disease.

Neutralizing antibodies are likely to play a role in limiting viral replication during the early, asymptomatic stage of infection. However, the titers of these antibodies are generally very low. This property is consistent with recent structural analyses of the HIV-1 envelope protein which indicate that neutralizing epitopes are likely to be hidden from the immune system. The low titer may favor selection of resistant mutants. Indeed, many individuals produce antibodies that neutralize earlier virus isolates but not those present at the time of serum collection, suggesting effective immune "escape" by the virus (see Fig. 16.4). Some studies show loss of neutralizing antibodies with progression to AIDS, but the clinical relevance of such antibodies during the later stages of infection remains obscure.

Neutralizing antibodies generally bind to specific sites on the viral envelope complex, SU-TM. Variable region 3 (V3) in HIV-1 SU is one of the initial targets. Anti-V3 antibodies appear to block coreceptor interactions that occur after the virus attaches to the CD4 receptor. Because of the high sequence variation within the V3 loop (hence "variable"), neutralizing antibodies to this region are usually strain specific. Consequently, despite its relatively strong antigenicity, V3 is not a good target for the development of vaccines or broadly specific antiviral drugs. Members of another class of neutralizing antibodies block the binding of SU to the CD4 receptor. These antibodies bind to numerous sites on SU and are usually more broadly reactive than anti-V3 antibodies. Other conserved and variable regions on both SU and TM can be targets for neutralizing antibodies. Broad neutralizing activity against carbohydrate-containing regions of the viral envelope protein has also been detected, but it is still not clear whether such activity represents true anticarbohydrate reactions or conformational epitopes on SU. Even less well understood is the relative importance of antibodies to other molecules on the surface of virions, such as adhesion molecules Lfa-1 and Icam. Antibodies to these cell surface adhesion molecules may inhibit syncytium formation, because the cells are less likely to form clumps. In some instances, these antibodies seem to enhance neutralization of HIV-1 by antiviral antibodies (Fig. 17.10, bottom left).

Antibodies called **interfering antibodies** can bind to virions or infected cells and block interaction with neutralizing antibodies (Fig. 17.10, left). Others, called **enhancing antibodies**, can actually facilitate infection by allowing virions coated with them to enter susceptible cells (Fig. 17.10, right). In complement-mediated antibody enhancement, the complement receptors Cr1, Cr2, and Cr3 appear to play a critical role in attaching virion-antibody complexes to susceptible cells. In Fc-mediated enhancement, attachment is via Fc receptors that are abundant not only on monocytes/macrophages and NK cells but also on other human cell types. As HIV-1 has been shown to replicate in cells that lack CD4 but express an Fc receptor, binding to the CD4 receptor is probably not required for Fc-mediated enhancement.

Table 17.8 Regions of HIV that resemble normal cellular proteins[a]

Normal cellular protein	HIV protein(s)	Relation
Hla (major histocompatibility complex)	SU	Sequence
	TM	Serology
	MA	Sequence
IL-2	TM	Sequence
IL-2 receptor	Nef	Sequence
Thymosin	MA	Serology
Vasoactive intestinal polypeptide	SU (peptide T)	Sequence, serology
Neuroleukin (phosphohexose)	SU	Sequence, serology
Immunoglobulin	SU	Sequence
Neurotoxin	Ner, Tat, TM	Sequence
Protein kinase	Nef	Sequence
Epithelial cell protein	MA	Serology
Astrocyte protein	MA, TM	Serology
Platelet glycoprotein	SU	Serology
Platelet cell protein	CA	Serology
Brain cell protein	SU (V3 loop)	Serology
Fas	SU	Sequence
Nuclear export sequences	Rev	Sequence

[a]Adapted from Table 10.5 of J. A. Levy, *HIV and the Pathogenesis of AIDS*, 2nd ed. (ASM Press, Washington, D.C., 1998), with permission.

It is noteworthy that the same cellular receptors (for complement and Fc) are implicated in infection enhancement and the antibody-dependent cellular cytotoxicity response. In the case of enhancement, the receptors allow antibody-coated virions to enter susceptible cells bearing such receptors. In the case of antibody-dependent cellular cytotoxicity, such receptors on CTLs, NK cells, or monocytes/macrophages mediate the recognition and killing of antibody-coated infected cells.

Both neutralizing and enhancing antibodies recognize epitopes on SU and TM. Consequently it has been difficult to identify the features that specify either response. Indeed, polyclonal antibodies against SU and the SU V3 loop have been shown to possess both neutralizing and enhancing activities, and it has not been possible to decide how either effect might predominate. This idea and the distinctions between other classes of anti-HIV-1 antibodies are summarized in Fig. 17.10.

The clinical importance of antibody-dependent enhancement is uncertain, but the fact that circulating, infectious virus-antibody complexes have been described suggests that this phenomenon may contribute to HIV-1 pathogenesis. In addition, infectivity-enhancing antibodies have been demonstrated in individuals who progress to disease. Results from studies of other lentiviral infections, as well as infections with other viruses (dengue viruses, coronaviruses, and others), show a correlation between increased symptoms of disease and increased quantities of enhancing antibodies. These observations and the finding that enhancing antibodies have been found in individuals who were vaccinated with Env from a different HIV-1 strain certainly complicate strategies for the development of an effective vaccine. As it is increasingly clear that antibodies do not play a major role in controlling HIV infections, attention has been focused on the role of cellular responses.

The Cellular Immune Response

Antigen-specific cellular immune responses include activities of CTLs and T-helper cells. The great majority of CTLs are $CD8^+$, and their general role in limiting or suppressing viral replication is discussed in Chapter 15. CTLs can also block virus replication by production of cellular antiviral factor(s) (Caf), including chemokines and small peptides called defensins, that suppress transcription of viral genes.

CTLs programmed to recognize virtually all HIV-1 proteins have been detected in infected individuals. As noted previously, there is a direct correlation among a good CTL response, a low virus load, and slower disease progression. Furthermore, a broadly reactive response appears to correlate with a less fulminant course of disease. End-stage disease is characterized by a rapid drop in anti-HIV-1 CTLs. Results of animal studies in severe combined immunodeficient (SCID) mice that have been reconstituted with human lymphoid cells show that adoptive transfer of human anti-HIV-1 CTLs provides some protection against subsequent challenge with HIV. These findings all demonstrate a significant role for such lymphocytes in fighting HIV-1 disease. Why, then, do these CTLs eventually fail to control infection? While the answer to this most pressing question is not yet known, it seems quite likely that much of the failure is due to reduced cytokine expression from infected and dysfunctional $CD4^+$ T cells (Table 17.6). However, the T-cell receptor may not be able to recognize all viral peptides as the viral population becomes more diverse, or the CTLs may simply respond inadequately in the face of high rates of virus production in AIDS patients.

Summary: the Critical Balance

HIV-1 replication is controlled by what may be thought of as a finely balanced scale that can be tipped in either direction by a number of stimulatory or inhibitory host proteins. Among these, various cytokines can have important but opposing effects. Immune responses and the production of specific chemokines can inhibit virus replication, whereas immune cell activation and certain antibodies can be stimulatory (Fig. 17.11). The challenge of modern med-

Figure 17.11 Control of HIV-1 replication and the progress of HIV-induced disease by the balance of host proteins. Cellular activation and proinflammatory cytokines stimulate viral replication. These are counterbalanced by inhibitory proteins. Adapted from Fig. 4 of A. S. Fauci, *Nature* **384:**529–533, 1996, with permission.

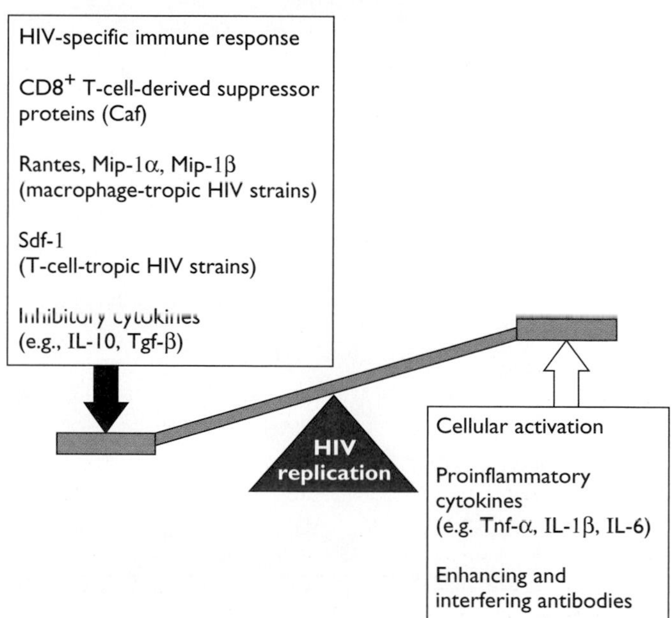

icine is to find practical interventions that will tip the balance in favor of patients' survival.

Dynamics of HIV-1 Replication in AIDS Patients

The availability of potent drugs that block HIV-1 replication by inhibiting the activity of the viral enzymes reverse transcriptase (RT) and protease (PR) made it possible to measure the dynamics of virus production in humans. Clinical studies performed with patients near end stage, whose $CD4^+$ T-cell counts are in decline (500 or fewer per ml), have revealed the magnitude of the battle between the virus and the immune system in HIV-1 disease. Within the first 2 weeks of treatment with a combination of these drugs, an exponential decline in viral RNA in the plasma is observed; this is followed by a second, slower rate of decline. The initial drop represents clearance of free virus and loss of virus-producing $CD4^+$ lymphocytes from the blood. The most important contributor to the second drop is presumed to be the loss of longer-lived infected cells, such as tissue macrophages and dendritic cells, with a minor contribution from the clearance of latently infected nonactivated T lymphocytes (Fig. 17.12).

At steady state, in the absence of drugs the rate of virus production must equal the rate of virus clearance. Mathematical analyses of the data from these clinical studies can therefore provide estimates for the rates of HIV-1 production in the blood and other compartments of the body, as well as for the rate of loss of virus and virus-infected cells. The results calculated for HIV-1 production are nothing less than astonishing. The minimal rate estimated for release into the blood is on the order of 10^{10} virions per day. This minimal number computes to approximately 1 cycle/infected cell/day. Continuous high-replication capacity is undoubtedly the principal engine that drives viral pathogenesis at this stage. Because of the high mutation rate of HIV-1, on average every possible mutation at every position in the genome is predicted to occur numerous times each day. Based on the high rates of mutation and virus propagation, it has been estimated that the genetic diversity of HIV produced in a single infected individual can be greater than the world diversity of influenza virus during an epidemic. This enormous variation and the con-

Figure 17.12 Summary of kinetics of HIV-1 production in the blood and other compartments. The percentages indicate the relative quantities of virus in blood plasma calculated to be produced by the various cell populations illustrated. The average time in days for 50% of the cells in each population to be destroyed or eliminated is indicated as the half-life ($t_{1/2}$). The average time in hours (h) for 50% of the virus to be eliminated from the plasma ($t_{1/2}$) is also shown. Adapted from Fig. 1 of D. D. Ho, *J. Clin. Investig.* **99:**2565–2567, 1997, with permission.

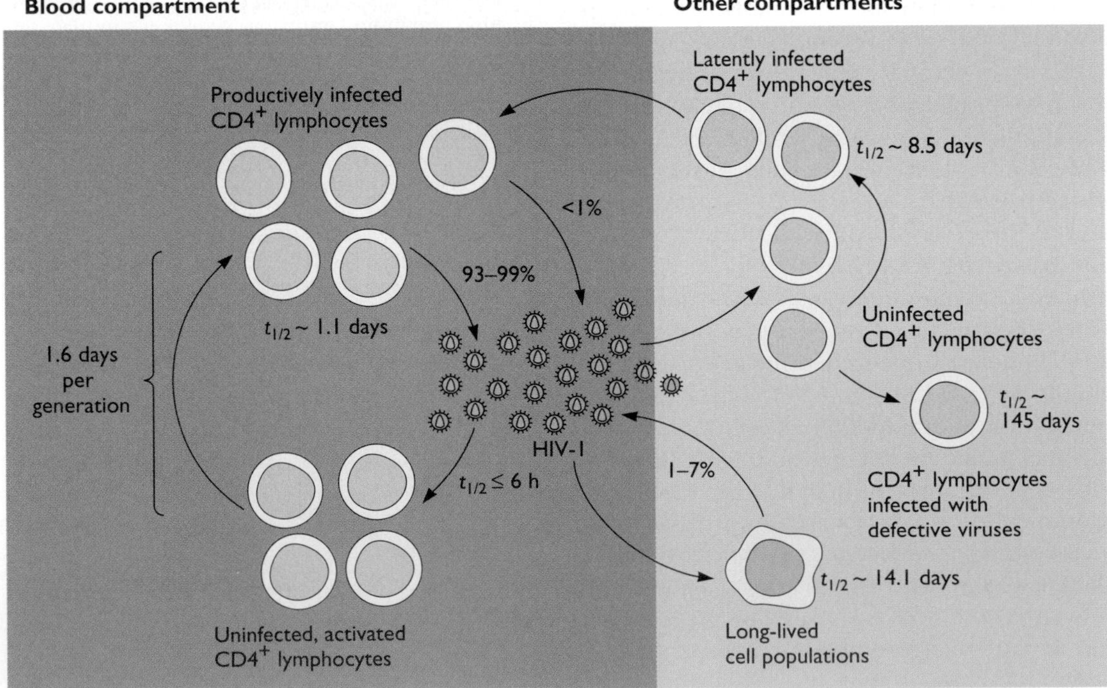

tinuous onslaught of infection must present a prodigious challenge to the immune system.

As illustrated in Fig. 17.12, more than 90% of the virus particles produced in the blood come from infected CD4+ lymphocytes with average half-lives of only ca. 1.1 days. A smaller percentage, approximately 1 to 7%, come from longer-lived cells in other compartments, with half-lives from 8.5 up to 145 days. If de novo synthesis of virus can be blocked completely by drug treatment, it could take approximately 3 to 5 years of treatment to eliminate virus from these longer-lived compartments. Sadly, this can be considered only a minimal estimate. Complete eradication may not be possible, as some virus may still lurk in undetected "sanctuary" compartments or at sites that are not accessible to some drugs, such as the brain. As all potent HIV-1 inhibitors produce undesirable side effects, most patients cannot tolerate such long-term treatment.

The other dramatic change that takes place in patients treated with antiviral drugs is a resurgence of the CD4+ lymphocyte count in the blood. From the initial rates of recovery, it has been calculated that during an ongoing infection, as many as 4×10^7 of these cells are replaced in the blood each day. Lymphocytes in the peripheral blood comprise a relatively small fraction (ca. 1/50) of the total in the body, and lymphocyte trafficking, homing, and recirculation are complex processes. It is still uncertain whether such CD4+ lymphocyte replacement following drug treatment represents new cells or a redistribution from other compartments. If one assumes that the increase in the circulation is proportional to the total, then as many as 2×10^9 new CD4+ T cells are produced each day. This estimate is controversial because it seems to exceed significantly the normal proliferative capacity of these cells. However, some studies suggest that HIV-1-infected individuals who are treated with potent drug combinations do produce new CD4+ T cells at rates higher than normal. Perhaps both increased rate of synthesis and redistribution contribute to the observed resurgence in CD4+ T cells in the periphery.

Long-term studies of patients treated with highly active antiretroviral therapies (HAART) should help to clarify some of these issues. From recent analyses it is unlikely that a simple inverse correlation exists between virus load and number of CD4+ cells. For example, in certain patients the virus reappears after 6 months, reaching levels near those recorded before treatment, even though CD4+ T-cell counts remain high. Whether this situation can be explained by a change in the genetic makeup of virus population or some sort of reactivation of the immune system remains to be determined.

Effects of HIV on Different Tissues and Organ Systems

Lymphoid Organs

The majority of the body's lymphocyte pool resides in lymphoid tissue (Chapter 15). Because these lymphocytes exist in a different environment than those in the periphery, their response to viral infection may also be different.

The function of these tissues is to filter invading pathogens, such as HIV, and to present them to immunocompetent cells. The majority of virus in lymph nodes is trapped within the germinal centers that comprise networks of follicular dendritic cells with long interdigitating processes which surround lymphocytes. These follicular dendritic cells also trap antibodies and complement and present antigens to B cells. The process of cellular activation that takes place in these tissues could make resident and recruited lymphocytes permissive for HIV-1 replication. Lymph nodes appear to contain a much larger percentage of virus-infected cells than the peripheral blood.

Early in infection the germinal centers in lymph nodes appear to remain intact, although there is some proliferation of activated immune cells (Fig. 17.13). Early in the asymptomatic stage, infected individuals often have palpable lymphadenopathy at two or more sites as a result of follicular dendritic cell hyperplasia and capillary endothelial cell proliferation. Later, during an intermediate stage of disease (CD4+ counts of 200 to 500 per ml), the nodes begin to deteriorate, there is evidence of cell death, and the trapping efficiency of the follicular dendritic cells declines. At a more advanced stage of disease (<200 CD4+ cells per ml), the architecture of the lymphoid tissue is almost completely destroyed and the follicular dendritic cells disappear. At this point, there is also a significant increase of HIV-1 in the peripheral blood. Finally, because the lymphoid tissue is no longer able to mediate an immune response, there is an increased incidence of opportunistic infections.

The Nervous System

Approximately one-third of AIDS patients are diagnosed with neurological disorders at some time during the course of their disease. HIV-1 enters the nervous system early after infection, either as free virions in plasma or via infected monocytes or lymphocytes from the bloodstream. The most prevalent disorder associated with infection of the brain is subacute encephalitis, also called AIDS dementia complex; nearly two-thirds of all HIV-1-infected individuals ultimately develop AIDS dementia. The disease progresses slowly over a period of up to 1 year, but mean survival time from the onset of severe symptoms is less than

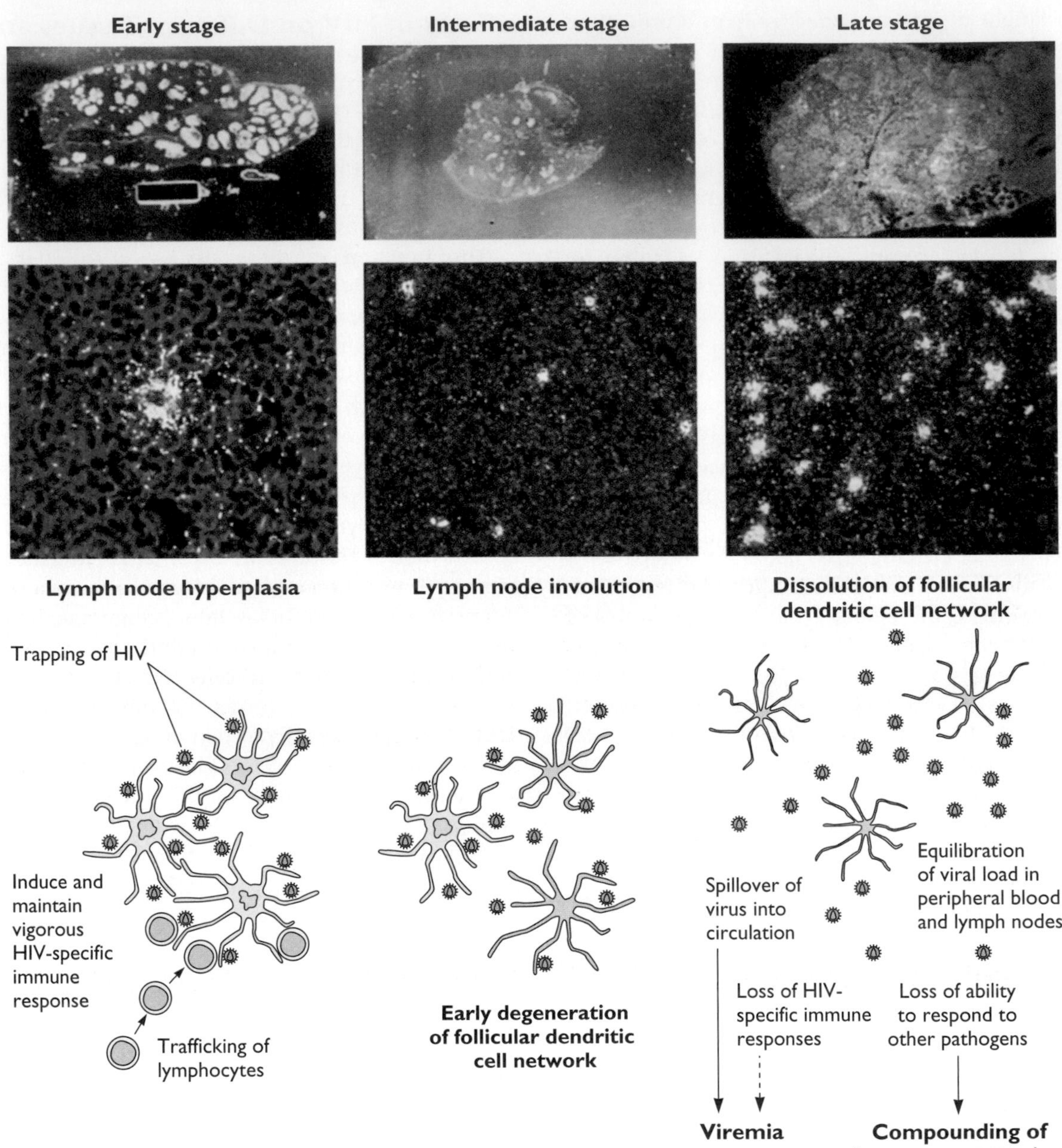

Figure 17.13 Effects of HIV-1 infection on lymphoid tissue. Top panels indicate the changes in lymph node germinal centers as determined by selective staining (above) and by the location of viral replication in lymph node tissue (below). HIV-1 replication was detected by polymerase chain reaction procedures. The examples illustrate conditions in the early, intermediate, and late stages of HIV infection. The bottom panels illustrate events that take place in lymph node germinal centers during various stages of HIV-1 disease. In early-stage HIV disease (left), the virus is trapped on the follicular dendritic cell network. $CD4^+$ T cells that migrate into the area to provide help to B cells in the germinal center are susceptible to infection by these virions. At this stage of disease, the follicular dendritic cell network is relatively intact. Over time, the lymphoid tissue begins to involute and early degeneration of the network is observed (middle). The trapping ability of the follicular dendritic cells is then compromised. In advanced-stage disease (right), there is dissolution of the network; virus can no longer be trapped effectively and spills over readily into the circulation. At this stage, there is loss of HIV-1-specific immune responses as well as loss of the ability to respond to other pathogens. Profound immunosuppression ensues. (Top) Reprinted from Color Plate 11 of J. A. Levy, *HIV and the Pathogenesis of AIDS*, 2nd ed. (ASM Press, Washington, D.C., 1998), with permission. (Bottom) Adapted from Fig. 10 of A. S. Fauci and R. C. Desrosiers, p. 587–635, *in* J. M. Coffin et al. (ed.), *The Retroviruses* (Cold Spring Harbor Laboratory Press, Cold Spring Harbor, N.Y., 1998), with permission.

6 months. Several AIDS-associated opportunistic infections can also produce neurological damage, and neurological abnormalities in infected individuals can be caused by lymphomas of the central nervous system (Table 17.9).

In general, brain-derived isolates of HIV-1 are macrophage tropic and distinct from blood-derived strains from the same individual. Results of viral genome sequencing suggest that the brain isolates represent a distinct lineage selected for neurotropism. The precise mechanisms by which these isolates cause neuronal abnormalities and damage are not known. In animal lentivirus infections, central nervous system disease is usually caused by direct infection of glial cells, astrocytes, and possibly brain macrophages called microglia. Infection of astrocytes and microglia also appears to be important in HIV-1 pathogenesis of the central nervous system. The virus can also be detected in brain capillary endothelial cells and has been shown to infect such cells in culture. Plausible hypotheses for how infection of the central nervous system is initiated are shown in Fig. 17.14A. As HIV-1 does not appear to infect neurons, direct replication is unlikely to explain their loss. It is more probable that the release of toxic cellular products and viral proteins (e.g., SU, TM, Nef, Rev, and Tat) from infected macrophages or microglial cells is responsible for the damage to neurons. There is ample evidence from cell culture experiments that these viral proteins could contribute to neuropathogenesis. In addition, transgenic mice expressing HIV-1 *env* under the control of neuronal promoters show abnormalities in astrocytes, dendrites, and neurons similar to those seen in HIV-1-infected individuals. A role for HIV-1 proteins in neuropathogenesis is further supported by the observation that antiviral therapy can reduce the symptoms of AIDS dementia significantly in some patients, at least for a period.

The Gastrointestinal System

The more advanced stages of HIV disease are often associated with damage to the gastrointestinal system. Diarrhea and chronic malabsorption, with consequent malnourishment and weight loss, are frequently observed. In Africa, this has been called "slim disease." In some cases, the disorders are associated with opportunistic infections with other microbial agents, including human cytomegalovirus and herpes simplex viruses. Nevertheless, in cases where no opportunistic agent can be identified, HIV-1 itself may be the primary cause of gastrointestinal pathogenesis. As in the central nervous system, the tissue macrophage appears to be a major target cell of HIV-1 replication in the gastrointestinal system. HIV-1-inflicted damage is thought to result from virus replication, toxic effects of viral proteins, and/or indirect destructive effects of certain cytokines.

Other Organ Systems

HIV has been found in the lungs of patients with pneumonia, in the hearts of some with heart muscle dysfunction (cardiomyopathy), in the kidneys of some with renal

Table 17.9 Major neurological disorders associated with HIV-1 infection[a]

Disorder	Prevalence (%)	Clinical features[b]
Opportunistic infections		
Toxoplasma gondii	10	Focal seizures, altered consciousness, CT ring-enhancing lesions
Cryptococcus neoformans	4–8	Headache, confusion, seizures, CSF cryptococcal antigen
Progressive multifocal leukoencephalopathy	1–4	Limb weakness, gait abnormalities, visual loss, altered mental state, CT-magnetic resonance imaging white matter lesions
Lymphoma	5–10	Diplopia, weakness, encephalopathy, CT contrast-enhancing lesions
HIV-1-related subacute encephalitis	90	Cognitive deficits, memory loss, psychomotor slowing, pyramidal tract signs, ataxia, weakness, depression, organic psychosis, incontinence, myoclonus, seizures
Peripheral neuropathies	10–50	
Chronic distal symmetric polyneuropathy		Painful dyesthesias, numbness, paresthesia, weakness, autonomic dysfunction
Chronic inflammatory myelinating polyneuropathy		Weakness, sensory deficits, mononeuropathy multiplex, cranial nerve palsies, hyporeflexia or areflexia, CSF pleocytosis
Vacuolar myelopathy	11–22	Gait ataxia, progressive spastic paraparesis, posterior column deficits, incontinence
Aseptic meningitis	5–10	Headache, fever, meningeal signs, cranial nerve palsies, CSF pleocytosis

[a]Adapted for Table 4 of M. S. Hirsch and J. Curran, p. 1953–1995, *in* B. N. Fields et al. (ed.), *Fields Virology*, 3rd ed. (Lippincott-Raven, Philadephia, Pa., 1996).
[b]CSF, cerebrospinal fluid; CT, computed tomography.

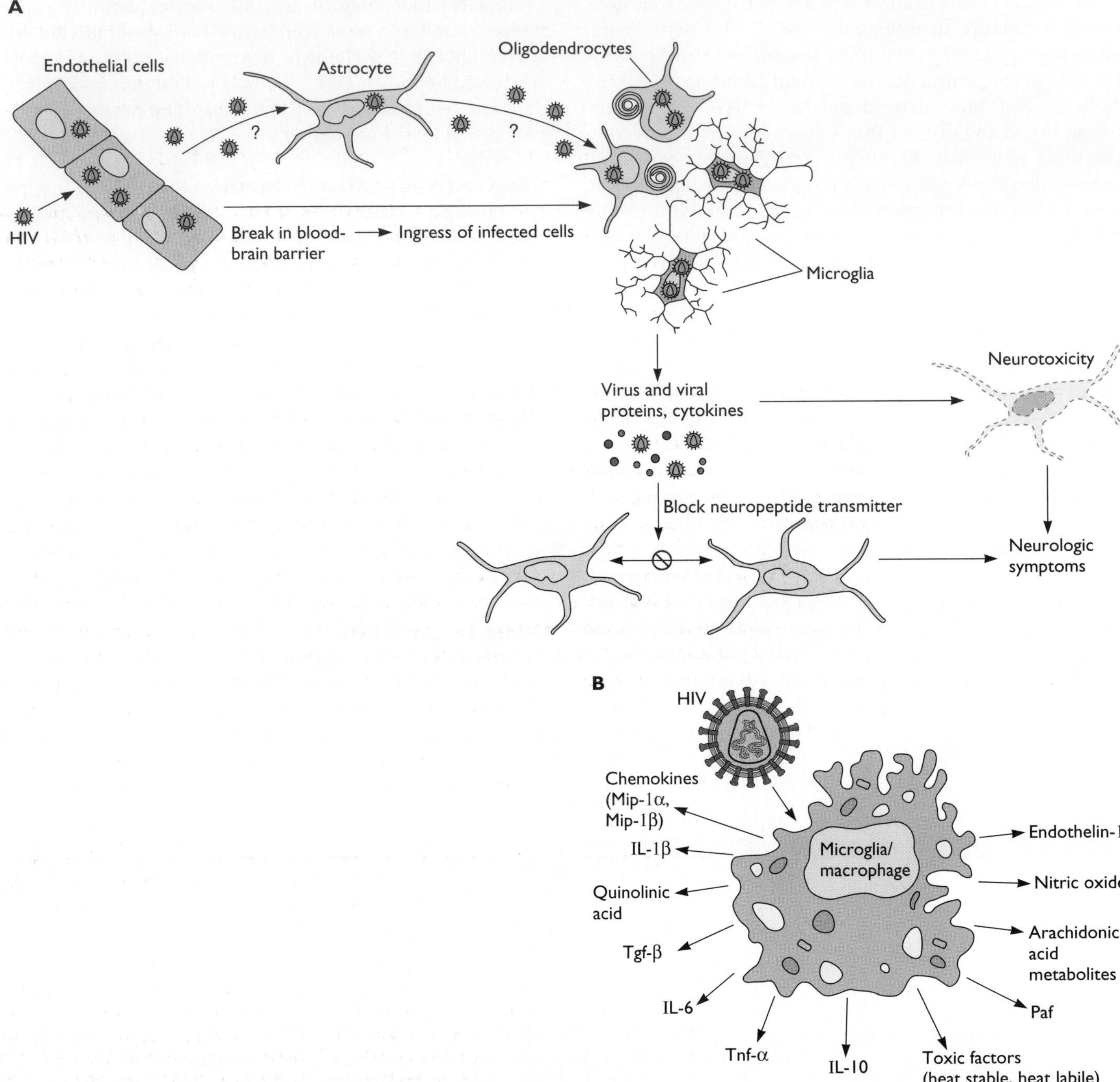

Figure 17.14 HIV-1 neuropathogenesis. (A) It is proposed that the virus infects capillary endothelial cells of the brain and passes into the brain via the basolateral surface of these cells. Infection of both these cells and, perhaps, astrocytes could lead to a breakdown in the blood-brain barrier and ingress of infected T cells and macrophages. The ultimate result is an infection of other brain cells (e.g., oligodendrocytes and microglia). Production by these cells of HIV-1 and viral proteins ensues, as does the release of various cytokines and other cellular products. These products could lead to an interruption of cell-to-cell transmission through a blockage of neurotropic factors. Direct infection of cells, as well as high levels of viral proteins (e.g., SU, TM, Tat, and Nef) and cytokines could lead to direct neurotoxicity through detrimental effects on the cell membrane. Viral replication in the endothelial cells and astrocytes causes damage to the blood-brain barrier, allowing access of infected $CD4^+$ lymphocytes, macrophages, or free virus from the blood. Alternatively, virus could pass through the endothelial layer to infect the adjacent astrocytes. Because astrocytes trigger

(continued on next page)

Figure 17.14 ***(continued)*** production of cytokines that help maintain the blood-brain barrier, as well as cytokines that are harmful to the brain, they could play a significant role in this pathology. Neurotropic viral mutants that replicate well in microglia might later trigger production of more virus, as well as harmful cytokines and other neurotoxic products. (B) The role of infected microglia/macrophages is illustrated. After exposure to HIV or to viral envelope proteins, microglia/macrophages may be induced to produce cytokines and the macrophage inflammatory chemokines that can be toxic to cells of the central nervous system. Production of endothelin-1, a monocyte-derived protein, can be increased on microglia/macrophage activation, and has potent vasoconstricting activity in the brain. Arachidonic acid metabolites and platelet-derived activating factor (Paf) produced by activated microglia/macrophages have also been implicated in toxicity to cells of the central nervous system. High levels of quinolinic acid also produced by such activated cells have been found in the spinal fluid and brain tissue of HIV-1-infected patients and are implicated in central nervous system diseases of infected children. Elevated concentrations of a monocyte-typical isoform of nitric oxide synthase have been detected in the central nervous systems of patients with dementia. Production of nitric oxide may explain the death of neuronal cells that come into direct contact with infected or activated microglia/macrophages. Other low-molecular-weight toxic products (heat stable and heat labile) produced by such activated cells have been detected in high concentrations in the blood of HIV-1-infected patients with dementia. Adapted from p. 216 and Fig. 9.11 of J. A. Levy, *HIV and the Pathogenesis of AIDS*, 2nd ed. (ASM Press, Washington, D.C., 1998), with permission.

injury, and in the joint fluid of patients with arthritis. The role of the virus in these pathologies is not clearly understood. Opportunistic infection of the lungs is common; *Pneumocystis carinii*, a protozoan parasite that is usually dormant in the host lung, causes pneumonia in approximately 50% of AIDS patients. Other microorganisms may also cause pulmonary infections, most notably *Mycobacterium tuberculosis, Mycobacterium avium*, fungi, and human cytomegalovirus. The known effects of SU on membrane permeability or other toxic effects of viral proteins might explain some of the lung damage or the electrophysiological abnormalities associated with heart disease. Direct infection of endothelial cells or other cells in the kidney has been proposed as a potential cause of tubular destruction in this organ. Deposition of antigen-antibody complexes could also account for some kidney damage. HIV also has been identified in the adrenal glands of infected individuals.

HIV and Cancer

It is not surprising that the immune dysfunction characteristic of HIV-1 infection leads to an increased incidence of neoplastic malignancies. Contributory factors might include the absence of proper immune surveillance directed against other (oncogenic) viruses or transformed cells. Also, high levels of cytokine production associated with HIV infection might induce inappropriate proliferation of cells, activate oncogenic-virus replication, and promote the generation of blood vessels (angiogenesis) in developing tumors. Indeed, cancers that develop in HIV-infected individuals generally are more aggressive than those in uninfected individuals. These malignancies can develop in a number of tissues and organs, but certain types, namely, Kaposi's sarcoma and B-cell lymphoma, are especially prevalent in AIDS patients. It may be important that in these cases the neoplastic cells are derived from the immune system. The endothelial cells thought to give rise to Kaposi's sarcoma can act as accessory cells in lymphocyte activation. One reasonable hypothesis is that proliferation of endothelial cells, B cells, and the epithelial cells that give rise to carcinomas may be promoted by cytokines produced by immune cells. As discussed in this section, many malignancies that develop in HIV-1-infected individuals are associated with infection by oncogenic viruses.

Kaposi's Sarcoma

Kaposi's sarcoma was first reported by the Hungarian physician Moritz Kaposi in 1872. It is a multifocal cancer that differs from more common tumors in that the lesions contain many cell types; the dominant type is a called a spindle cell, thought to be of endothelial origin. The tumors contain infiltrating inflammatory cells and many newly formed blood vessels. Kaposi's sarcoma was typically found in older men from the Mediterranean rim region and eastern Europe. In these areas, and in immunocompetent individuals, Kaposi's sarcoma normally appears in a nonaggressive (classical) form confined to the skin and extremities, and is rarely lethal. The classical form as well as a more aggressive, sometimes lethal form is found in sub-Saharan Africa, where there are more immunocompromised individuals. In HIV-1-infected men, Kaposi's sarcoma appears in the more aggressive form, affecting both mucocutaneous and visceral areas. This disease occurs in about 20% of HIV-1-infected homosexual men (Fig. 17.15) but in only about 2% of HIV-1-infected women, transfusion-infected recipients, and blood product-infected hemophiliacs in the United States. This distribution suggests a specific type of sexual transmission.

Spindle cell cultures established from Kaposi's sarcoma tumors are not fully transformed according to the criteria

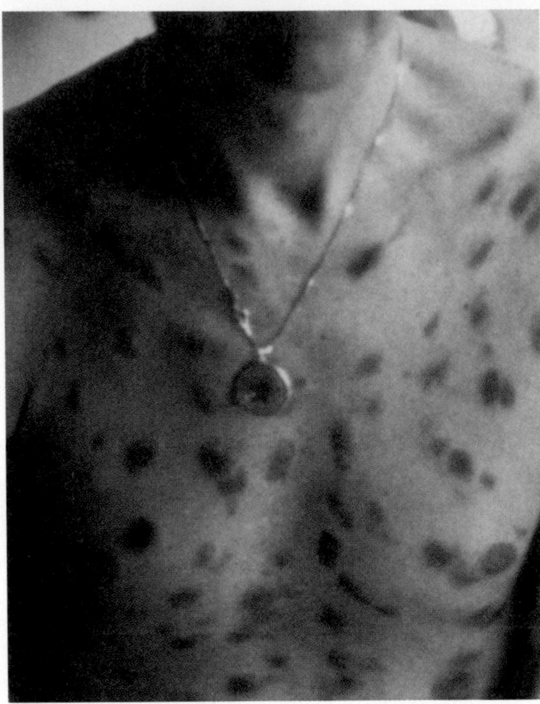

Figure 17.15 Kaposi's sarcoma in a young man infected with HIV-1. Note the distribution of the lesions, suggesting lymphatic involvement. Reprinted from Color Plate 13 of J. A. Levy, *HIV and the Pathogenesis of AIDS*, 2nd ed. (ASM Press, Washington, D.C., 1998), with permission. Photo courtesy of P. Volberding.

defined in Chapter 18, but they do produce a variety of proinflammatory and angiogenic proteins. It is thought that these products are responsible for recruiting the other cell types in these tumors. Spindle cells from AIDS patients are also not infected with HIV-1, and the disease occurs in individuals who are not infected with the virus. HIV cannot therefore be the sole factor in the development of Kaposi's sarcoma. Epidemiological studies suggest that another sexually transmitted virus was the inducing agent. Subsequently, a new member of the gammaherpesviruses was found to be associated with Kaposi's sarcoma (see Fig. 18.10). This new virus, called human herpesvirus 8, can infect spindle cells, and results of in situ hybridization studies show RNA transcripts from this virus in the vast majority of Kaposi's sarcoma lesions, irrespective of the presence or absence of HIV-1. Human herpesvirus 8 can also infect B cells and has been linked to certain AIDS-associated B-cell lymphomas (Fig. 17.16).

The evidence that infection with human herpesvirus 8 is necessary for the development of Kaposi's sarcoma is quite strong. Antibodies to this virus can be found in up to 6% of the general population but in 25% of the population in classic endemic areas, indicating widespread exposure. Nevertheless, very few people develop this disease. Other parameters are therefore likely to be important in its etiology. The immune deficiency associated with HIV-1 is certainly one explanation for its prevalence in AIDS patients, and synthesis of viral proteins, such as Tat, may be

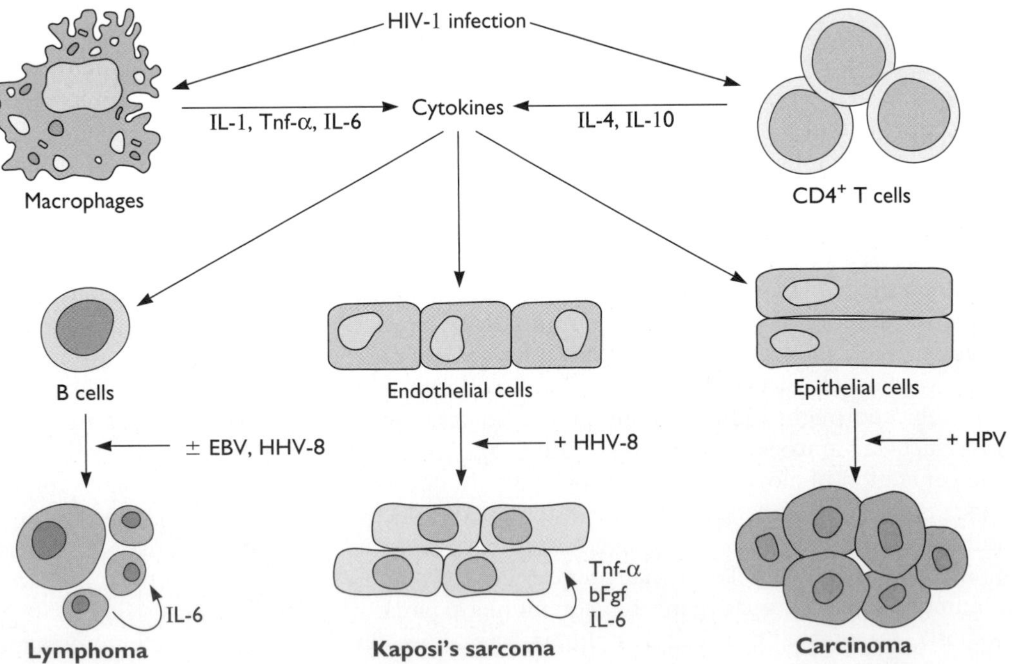

Figure 17.16 Induction of cancers in HIV-1 infection. Virus infection of macrophages, CD4$^+$ T cells, or other cells could lead to the production of cytokines that enhance the proliferation of certain other cells, such as B cells, endothelial cells, and epithelial cells. The enhanced replication of these cells, either through cytokine production or through subsequent viral infection, could lead to eventual development of the malignancies noted. In some cases, such as B-cell lymphomas and Kaposi's sarcoma, ongoing cytokine production by the tumor cells maintains the malignant state. Adapted from Fig. 12.11 of J. A. Levy, *HIV and the Pathogenesis of AIDS*, 2nd ed. (ASM Press, Washington, D.C., 1998), with permission.

another. The exact role of human herpesvirus 8 in Kaposi's sarcoma remains to be elucidated. The viral genome includes a number of open reading frames with sequence homology to genes encoding cellular proteins known to be important in growth control, cell signaling, and immunoregulation (see Fig. 18.10).

It is ironic that one of the first reports of AIDS in 1981 described seven patients with Kaposi's sarcoma. While HIV-1 was identified soon afterward, the discovery of human herpesvirus 8 did not occur until many years later even though both viruses were present and contributed to the pathology in these first patients.

B-Cell Lymphomas

B-cell lymphomas are 60 to 100 times more common in AIDS patients than in the general population. The incidence is especially high among patients whose survival has been prolonged by anti-HIV-1 drugs. Tumors can be found in many locations, including lymph nodes, the intestine, the central nervous system, and the liver. B-cell lymphomas in the peritoneal or other body cavities, called primary effusion lymphomas or body cavity-based lymphomas, are almost always associated with human herpesvirus 8. Both polyclonal and monoclonal B-cell lymphomas are found in the central nervous system, with monoclonal types being most frequent. Epstein-Barr virus is found in all AIDS-associated primary lymphomas in the brain. Why these two B-cell lymphoma-inducing herpesviruses show such site specificity remains unknown. On the other hand, approximately 60% of the tumors outside the brain show no evidence of infection with either herpesvirus, indicating that B-cell transformation in HIV-1-infected individuals does not require infection with such viruses.

Lymphomas may arise because of the destruction of germinal centers in the lymphatic system. Lysis of antigen-presenting follicular dendritic cells would render B cells less sensitive to normal apoptotic processes, allowing them to live longer and to replicate. Epstein-Barr virus latent membrane protein 1 (see Table 18.8 and Fig. 18.13) also inhibits apoptosis. Proliferation of B cells as a result of production of cytokines by macrophages or $CD4^+$ T cells (Fig. 17.16), or even some viral proteins (e.g., SU or TM), may also play a role in this process. Uncontrolled proliferation, by whatever mechanism, could lead to the chromosomal changes required for cell transformation and malignancy.

Anogenital Carcinomas

Anogenital carcinomas are two to three times more frequent in HIV-1-infected individuals than in the general population and are associated with human papillomavirus infections that are typically spread through sexual contact. The high-risk human serotypes 16 and 18 have been found to be associated with both cervical and anal carcinomas. Such cancers often arise in areas of squamous metaplasia near the glandular epithelium and reach more advanced stages in immunosuppressed individuals. The proposed mechanism of transformation is mutation in the p53 tumor suppressor gene in cases where a human papillomavirus is not involved, or binding and inactivation of the tumor suppressor proteins p53 and Rb by the viral proteins E6 and E7, as described in Chapter 18.

Prospects for Treatment and Prevention

Antiviral Drugs and Therapies

Antiviral drugs and therapies will be discussed in a broader context in Chapter 19. Some of the more clinically relevant details are summarized here. The first drug to be licensed in the United States for the treatment of AIDS was the nucleoside analog inhibitor of the viral reverse transcriptase zidovudine (AZT). AZT is converted to the triphosphate form after being taken up into cells. On the basis of laboratory experiments, it appears to act both as a competitor of normal nucleoside triphosphates and as a chain terminator when incorporated into viral DNA by reverse transcriptase. Effective treatment of HIV-infected patients with AZT was limited by the rapid appearance of resistant mutants. However, AZT, lamivudine (3TC, another nucleoside analog), or nevirapine (a nonnucleoside inhibitor) can be used as prophylactic treatment for pregnant woman to reduce infection of newborns. Indeed, a brief course of administration of such drugs has proven both effective and economical for this purpose in developing countries, where their availability is a major problem. Antiviral drugs have also been used after needle stick injuries and other types of acute exposure to the virus.

Highly Active Antiretroviral Therapy

The use of powerful antiviral drugs (Table 17.10) has had an enormous impact on AIDS in the developed world. The incidence of AIDS in the United State has declined since its high point in 1993, and the incidence of death has subsequently decreased fourfold. In addition, as treated patients experience a significant restoration of their immune cell function, it has been possible to withdraw prophylactic or suppressive therapy for a wide variety of opportunistic microorganisms. For maximum effectiveness, these antiviral drugs are given in combinations, as has proven advantageous in anticancer therapy. The most dramatic decrease in viral RNA levels in the blood of AIDS patients has been seen with triple-drug therapy in which two anti-reverse transcriptase and one anti-protease drug are combined

Table 17.10 Drugs approved for antiretroviral therapy

Reverse transcriptase inhibitors	Protease inhibitors
Nucleoside analogs	Amprenavir[a]
Abacavir (ABC)	Indinavir[a]
Didanosine (ddI)	Lopinavir[a]
Lamivudine (3TC)	Nelfinavir
Stavudine (d4T)	Ritonavir
Zalcitabine (ddC)	Saquinavir[a]
Zidovudine (AZT)	
Nonnucleoside inhibitors	
Efavirenz	
Delavirdine	
Nevirapine	

[a]Most often used with low-dose ritonavir for pharmacological enhancement.

(often called HAART). As is often the case in science and medicine, the development of inhibitors for both enzymes benefited partly from serendipity. AZT had been developed initially as a potential anticancer drug, and the earliest protease inhibitors were derivatives of renin inhibitors being developed as antihypertensive drugs. In addition, the firm base of both structural and biochemical knowledge available from research with these enzymes was a great advantage in developing this first group of drugs and is still being exploited to design the next generation of improved drugs.

Many patients treated with HAART sustain levels of fewer than 50 copies of HIV RNA per milliliter of plasma (the lower limit of detection) for years. But closer examination of tissues and blood shows that virus replication is not completely suppressed in such patients. Instead, they experience intermittent virus replication that is difficult to detect. In some patients, the emergence of drug-resistant virus hinders subsequent treatment. As drugs must be taken at least twice daily and have significant side effects, noncompliance is a large factor in treatment failure. Newly developed drugs and treatment regimens may help to minimize this problem in the future.

Other Approaches for Antiviral Agents

Many steps in the HIV-1 replication cycle offer opportunities for possible intervention (see Tables 19.10 and 19.11); a number have been or are currently being exploited. One obvious candidate is the third retroviral enzyme, integrase. Structural data and biochemical information about this protein have become available, and potent inhibitors of its activity have been identified. Drugs that disrupt the Zn^{2+} finger domain of the viral nucleocapsid protein, NC, may also have utility. Finally, inhibitors of HIV fusion that target either the viral envelope protein or the host cell chemokine coreceptors have been identified. The clinical utility of these inhibitors has yet to be tested. Most important, all future treatments must cope with the virus in inaccessible tissues or quiescent cells. For example, to accomplish a cure via drug therapy, it will be necessary to develop drugs that can penetrate the blood-brain barrier and enter the central nervous system to block replication of virus at this and other "privileged" sites. Quiescent, long-lived memory T cells can harbor latent HIV proviruses. As viral proteins are not expressed in such cells, these proviruses are invisible to the immune system, but they have the potential to produce infectious viral progeny upon activation of the memory cells.

Immune System-Based Therapies

Because immunization seems to be protective against subsequent SIV infection of rhesus macaques, treatment strategies that combine HAART with augmentation of HIV-specific immunity have been proposed. One approach, called structured treatment interruption, is based on the idea that temporary interruption of HAART will allow some virus replication and the release of HIV antigens that can provide autoimmunization. Several cycles of such interruption might augment antiviral immunity sufficiently to alter the course of the disease. Another approach is to deliver HIV immunogens while the patient is being maintained on HAART. Both approaches are currently under investigation.

Prophylactic Vaccine Development To Prevent Infection

The most potent defense against viral infection is a vaccine, a topic that is discussed in more detail in Chapter 19. For HIV-1, vaccine development presents unique challenges, not the least of which is the fact that the virus infects the immune system. In addition, the correlates of protective immunity against HIV have not been well defined (Box 17.3). Other confounding issues are the large genetic diversity and high mutation rates of the virus, and the danger of producing an inappropriate immune response that enhances infection. Finally, animal models of HIV that are faithful to the human disease are not available for testing vaccine candidates. The SIV-macaque model has been useful, and the discovery and cloning of HIV-1 coreceptors has made it possible to design promising transgenic mouse models. Ultimately, however, any protective vaccine must be tested in humans. Here there are many societal and ethical issues to be considered. At the present time, the best prophylaxis for people at risk is to adjust their behavior to avoid infection.

BOX 17.3

The correlates of protective immunity against HIV-1 have not been well defined

Vaccine development requires a deeper understanding of the specific immune responses essential for protection against HIV-1.

Is the humoral response essential? Must the antibodies be neutralizing or should they be able to activate killer cells (ADCC)? Is mucosal immunity in regions such as the rectum and vagina essential? Are T cells essential, and, if so, which kind?

The answers to many of these questions will dictate not only the nature of the vaccine but also the manner and site of its administration.

Possible requirements for protective immunity against HIV-1 infection[a]
Cytotoxic lymphocyte activity (CD4+ and CD8+ cells)
CD8+ cell noncytotoxic antiviral response
NK cell activity
Antibody-dependent cellular cytotoxicity (ADCC)
Neutralizing antibodies

[a]Adapted from Table 15.2 of J. A. Levy, *HIV and the Pathogenesis of AIDS*, 2nd ed. (ASM Press, Washington, D.C., 1998), with permission. The presence of these responses at mucosal sites is likely to be important. If attenuated virus is used, sufficient replication of the virus must occur to induce protective immune responses.

Despite the many problems, there is hope that an effective vaccine can be developed. Some encouragement can be drawn from the cross-protection against HIV-1 obtained in HIV-2-infected individuals, and from other lentiviral systems, in which inoculation with killed cell-free viruses has been shown to induce protection against high-dose virus challenges. Some individuals may have better innate defenses that abrogate any primary infection. Other genetic traits that confer protection are mutations in specific chemokines or chemokine coreceptors. In addition, as established by a study of a group of Kenyan and Gambian prostitutes who remained uninfected despite continuous exposure to HIV-1, particular MHC class I alleles are associated with protection. Such protection may signify that peptides presented by these MHC molecules specify strongly immunogenic regions of HIV-1 proteins. The identification of such peptides might present a fruitful approach for vaccine development.

In some instances in which prior infection with HIV appears to offer protection, the infecting virus may be defective or the dose of infecting virus may be such that the immune system is activated and the infection is cleared. Long-term protection against SIV in adult macaques was achieved by immunization with a live attenuated virus carrying a deletion of the *nef* gene. Unfortunately, as discussed above, immunized female animals were found to bear infected infants that succumbed to the attenuated virus. Whether a live attenuated vaccine will ever be acceptable for humans is unclear. It is also important to appreciate that an AIDS vaccine need not be 100% effective to be useful. With such a deadly disease, even partial protection that might spare 20 to 40% of potential victims would extend millions of lives and reduce transmission significantly.

There are several AIDS Vaccine Evaluation Units in the United States that can carry out initial vaccine trials. However, these studies are limited by the relatively small number of potential participants in this country. To determine efficacy properly, vaccines have to be tested in large, high-incidence populations. Such trials will require international collaboration, extensive testing in developing countries where such populations exist, and preparation of vaccines that target the predominant clade in the region of study. Such efforts are currently ongoing and it is hoped that some of these will be successful in the future.

Perspectives

> *Pneumocystis* Pneumonia—Los Angeles. In the period October 1980–May 1981, 5 young men, all active homosexuals, were treated for biopsy-confirmed *Pneumocystis carinii* pneumonia at three different hospitals in Los Angeles, California. Two of the patients died. . . .
>
> M. S. Gottlieb et al. (Centers for Disease Control)
> *Morb. Mortal. Wkly. Rep.* **30:**250–252, 1981

So began the first warning, soon echoed in large urban centers throughout the United States and Europe, where physicians were being confronted by a puzzling and ominous new disease that was killing young homosexual men. In a deceptively low-key editorial note with this report, it was observed that "*Pneumocystis* pneumonia is almost exclusively limited to severely immunosuppressed patients," that "the occurrence of the disease in these five previously healthy individuals is unusual," and that "the fact that these patients were all homosexuals suggests an association between some aspect of a homosexual lifestyle or disease acquired through sexual contacts. . . ." Soon the disease was to ravage this group and also the hemophiliac community, whose lives depended on blood products. This state of distress led to unprecedented social activism that demanded significant public investment for AIDS research and put patient advocates on scientific review panels for the first time.

Initial progress in AIDS research was impressive. The etiological agent of the disease was identified within 2 years, and screening assays to safeguard the blood supply were developed shortly thereafter. As a broad base of knowledge about retroviruses already existed, it seemed that treatment and a vaccine should soon be available. Unfortunately, more than 2 decades later, AIDS is still an emerging epidemic with a death toll that rises every year, with about 16,000 people newly infected each day. As is the case for most sexually transmitted diseases, efforts at prevention are fraught with difficulty and complicated by societal factors. Although a godsend to some, the treatments that have been developed to date are expensive, difficult to tolerate, and unavailable to most of the world's victims. And an effective vaccine is not yet on the horizon. Why has the initial optimism been so frustrated?

A major difficulty in combating HIV is the fact that the virus infects the immune system, destroying the delicate balance needed for the system to function properly. The pathology is complex, and the physician is forced to battle not only HIV but also many other infectious agents, some rarely seen in immunocompetent patients. Other problems include the enormous replicative capacity of the virus and the lack of fidelity of reverse transcriptase that produces vast populations of mutants. Drug and vaccine developers therefore face an ever-changing target, with few good animal models to help them. Furthermore, HIV is much more complicated than was first imagined, with several auxiliary genes and structural proteins that are themselves toxic to certain cell types. Perhaps, in the end, this very complexity will provide a key to effective and affordable treatment. There are numerous steps in the virus replication cycle, and several additional viral and host proteins may be good targets for such intervention. Although we already know a great deal about the biology of HIV, to exploit these targets, uncover others, and learn what is needed to develop an effective vaccine, much more knowledge will have to be acquired.

> What is it that we expected from our shamans, millennia ago, and still require from the contemporary masters of the profession? To do something, that's what.
>
> Lewis Thomas, *The Fragile Species*

References

Books

Levy, J. A. 1998. *HIV and the Pathogenesis of AIDS,* 2nd ed. ASM Press, Washington, D.C.

Reviews

Cohen, J. 2002. Confronting the limits of success. *Science* **296:**2320–2324.

Cohen, O. J., and A. S. Fauci. 2001. Pathogenesis and medical aspects of HIV-1 infection, p. 2043–2094. *In* D. M. Knipe and P. M. Howley (ed.), *Fields Virology,* 4th ed. Lippincott Williams & Wilkins, Philadelphia, Pa.

D'Souza, M. P., J. S. Cairns, and S. F. Plaeger. 2000. Current evidence and future directions for targeting HIV entry. *J. Am. Med. Assoc.* **284:**215–222.

Fauci, A. S., and R. C. Desrosiers. 1997. Pathogenesis of HIV and SIV, p. 587–636. *In* J. Coffin, H. Varmus, and S. Hughes (ed.), *Retroviruses.* Cold Spring Harbor Laboratory Press, Cold Spring Harbor, N.Y.

Freed, E. O., and M. A. Martin. 2001. HIVs and their replication, p. 1971–2042. *In* D. M. Knipe and P. M. Howley (ed.), *Fields Virology,* 4th ed. Lippincott Williams & Wilkins, Philadelphia, Pa.

Gaschen, B., J. Taylor, K. Yusim, B. Foley, F. Gao, D. Lang, V. Novitsky, B. Haynes, B. H. Hahn, T. Bhattacharya, and B. Korber. 2002. Diversity considerations in HIV-1 vaccine selection. *Science* **296:**2354–2360.

Kaul, M., G. A. Garden, and S. A. Lipton. 2001. Pathways to neuronal injury and apoptosis in HIV-associated dementia. *Nature* **410:**988–994.

Lederman, M. M., and H. Valdez. 2000. Immune restoration with antiretroviral therapies. *J. Am. Med. Assoc.* **284:**223–228.

Malim, M. H., and M. Emerman. 2001. HIV sequence variation: drift, shift, and attenuation. *Cell* **104:**469–472.

McCune, J. M. 2001. The dynamics of CD4$^+$ T-cell depletion in HIV disease. *Nature* **410:**974–979.

McMichael, A. J., and S. L. Rowland-Jones. 2001. Cellular immune responses to HIV. *Nature* **410:**980–987.

Nabel, G. J. 2001. Challenges and opportunities for development of an AIDS vaccine. *Nature* **410:**1002–1007.

O'Brien, S. J., and J. P. Moore. 2000. The effect of genetic variation in chemokines and their receptors on HIV transmission and progression to AIDS. *Immunol. Rev.* **177:**99–111.

Piot, P., M. Bartos, P. D. Ghys, N. Walker, and B. Schwartländer. 2001. The global impact of HIV/AIDS. *Nature* **410:**968–973.

Richman, D. D. 2001. HIV chemotherapy. *Nature* **410:**995–1001.

Rowland-Jones, S., S. Pinheiro, and R. Kaul. 2001. New insights into host factors in HIV-1 pathogenesis. *Cell* **104:**473–476.

Sharp, P. M. 2002. Origins of human virus diversity. *Cell* **18:**1–20.

Steffens, C. M., and T. J. Hope. 2001. Recent advances in the understanding of HIV accessory protein function. *AIDS* **15:**S21–S26.

Weiss, R. A. 2001. Gulliver's travels in HIVland. *Nature* **410:**963–967.

Research Articles of Special Interest

Discovery of the First AIDS Virus

Barré-Sinoussi, F., J. C. Chermann, F. Rey, M. T. Nugeyre, S. Chamaret, J. Gruest, C. Dauguet, C. Axler-Blin, F. Vézinet-Brun, C. Rouzioux, W. Rozenbaum, and L. Montagnier. 1983.

Isolation of a T-lymphotropic retrovirus from a patient at risk for acquired immune deficiency syndrome (AIDS). *Science* **220:**868–871.

HIV Origin

Gao, F., E. Baiels, D. L. Robertson, Y. Chen, C. M. Rodenburg, S. F. Michael, L. B. Cummins, L. O. Arthur, M. Peters, G. M. Shaw, P. M. Sharp, and B. H. Hahn. 1999. Origin of HIV-1 in the chimpanzee. *Nature* **397:**436–441.

Simon, F., P. Mauclére, P. Roques, I. Loussert-Ajaka, M. C. Müller-Trutwin, S. Saragosti, M. C. Georges-Courbot, F. Barré-Sinoussi, and F. Brun-Vézinet. 1998. Identification of a new human immunodeficiency virus type 1 distinct from group M and group O. *Nat. Med.* **4:**1032–1037.

Zhu, T., B. T. M. Korber, A. J. Nahmias, E. Hooper, P. M. Sharp, and D. D. Ho. 1998. An African HIV-1 sequence from 1959 and implications for the origin of the epidemic. *Nature* **391:**594–597.

Auxiliary Proteins

Harris, R. S., K. N. Bishop, A. M. Sheehy, H. M. Craig, S. K. Petersen-Mahrt, I. N. Watt, M. S. Neuberger, and M. H. Malim. 2003. DNA deamination mediates innate immunity to retroviral infection. *Cell* **113:**803–809.

Sheehy, A. M., N. C. Gaddis, J. D. Choi, and M. H. Malim. 2002. Isolation of a human gene that inhibits HIV-1 infection and is suppressed by the viral Vif protein. *Nature* **418:**646–650.

Dendritic Cells

Geijtenbeek, T. B. H., D. S. Kwon, R. Torensma, S. J. van Vliet, G. C. F. van Duijnhoven, J. Middel, I. L. M. H. A. Cornelissen, H. S. L. M. Nottet, V. N. KewalRamani, D. R. Littman, C. G. Figdor, and Y. van Kooyk. 2000. DC-SIGN, a dendritic cell-specific HIV-1-binding protein that enhances *trans*-infection of T cells. *Cell* **100:**587–597.

Structured Treatment Interruption

Lori, F., M. G. Lewis, J. Xu, G. Varga, D. E. Zinn, Jr., C. Crabbs, W. Wagner, J. Greenhouse, P. Silvera, J. Yalley-Ogunro, C. Tinelli, and J. Lisziewicz. 2000. Control of SIV rebound through structured treatment interruptions during early infection. *Science* **290:**1591–1593.

HIV-1 Inhibitors

Eckert, D. M., V. N. Malashkevich, L. H. Hong, P. A. Carr, and P. S. Kim. 1999. Inhibiting HIV-1 entry: discovery of D-peptide inhibitors that target the gp41 coiled-coil pocket. *Cell* **99:**103–115.

Root, M. J., M. S. Kay, and P. S. Kim. 2001. Protein design of an HIV-1 entry inhibitor. *Science* **291:**884–888.

T-Cell Depletion

Hellerstein, M., M. B. Hanley, D. Cesar, S. Siler, C. Papageorgopoulos, E. Wieder, D. Schmidt, R. Hoh, R. Neese, D. Macallan, S. Deeks, and J. M. McCune. 1999. Directly measured kinetics of circulating T lymphocytes in normal and HIV-1-infected humans. *Nat. Med.* **5:**83–89.

Vaccine Development

Amara, R. R., F. Villinger, J. D. Altman, S. L. Lydy, S. P. O'Neil, S. I. Staprans, D. C. Montefiori, Y. Xu, J. G. Herndon, L. S. Wyatt, M. A. Candido, N. Kozyr, P. L. Earl, J. M. Smith, H.-L. Ma, B. D. Grimm, M. L. Hulsey, J. Miller, H. M. McClure, J. M. McNicholl, B. Moss, and H. L. Robinson. 2001. Control of a mucosal challenge and prevention of AIDS by a multiprotein DNA/MVA vaccine. *Science* **292:**69–74.

Shiver, J. W., T.-M. Fu, L. Chen, D. R. Casimiro, M. E. Davies, R. K. Evans, Z.-Q. Zhang, A. J. Simon, W. L. Trigona, S. A. Dubey, L. Huang, V. A. Harris, R. S. Long, X. Liang, L. Handt, W. A. Schleif, L. Zhu, D. C. Freed, N. V. Persaud, L. Guan, K. S. Punt, A. Tang, M. Chen, K. A. Wilson, K. B. Collins, G. J. Heidecker, V. R. Fernandez, H. C. Perry, J. G. Joyce, K. M. Grimm, J. C. Cook, P. M. Keller, D. S. Kresock, H. Mach, R. D. Troutman, L. A. Isopi, D. M. Williams, Z. Xu, K. E. Bohannon, D. B. Volkin, D. C. Montefiori, A. Miura, G. R. Krivulka, M. A. Lifton, M. J. Kuroda, J. E. Schmitz, N. L. Letwin, M. J. Caulfield, A. J. Bett, R. Youil, D. C. Kaslow, and E. A. Emini. 2002. Replication-incompetent adenoviral vaccine vector elicits effective anti-immunodeficiency-virus immunity. *Nature* **415:**331–339.

Genetic Susceptibility

Martin, M.P., M. Dean, M. W. Smith, C. Winkler, B. Gerrard, N. L. Michael, B. Lee, R. W. Doms, J. Margolick, S. Buchbinder, J. J. Goedert, T. R. O'Brien, M. W. Hilgartner, D. Vlahov, S. J. O'Brien, and M. Carrington. 1998. Genetic acceleration of AIDS progression by a promoter variant of *CCR5*. *Science* **282:**1907–1911.

Smith, M. W., M. Dean, M. Carrington, C. Winkler, G. A. Huttley, D. A. Lomb, J. J. Goedert, T. R. O'Brien, L. P. Jacobson, R. Kaslow, S. Buchbinder, E. Vittinghoff, D. Vlahov, K. Hoots, M. W. Hilgartner, and S. J. O'Brien. 1997. Contrasting genetic influence of *CCR2* and *CCR5* variants on HIV 1 infection and disease progression. *Science* **277:**959–965.

Winkler, C., W. Modi, M. W. Smith, G. W. Nelson, X. Wu, M. Carrington, M. Dean, T. Honjo, K. Tashiro, D. Yabe, S. Buchbinder, E. Vittinghoff, J. J. Goedert, T. R. O'Brien, L. P. Jacobson, R. Detels, S. Donfield, A. Willoughby, E. Gomperts, D. Vlahov, J. Phair, and S. J. O'Brien. 1998. Genetic restriction of AIDS pathogenesis by an SDF-1 chemokine gene variant. *Science* **279:**389–393.

Website

http://www.unaids.org *UNAIDS, the Joint United Nations Programme on HIV/AIDS*

18

Transformation and Oncogenesis

> *Cause and effect, means and ends, seed and fruit, cannot be severed; for the effect already blooms in the cause, the end pre-exists in the means, the fruit in the seed.*
>
> RALPH WALDO EMERSON

Introduction

Cancer is a leading cause of death in developed countries: about 500,000 individuals succumb each year in the United States alone. Efforts to understand and control this deadly disease have therefore long been high priorities of public health institutions. Our general understanding of the mechanisms of **oncogenesis**, the development of cancer, as well as of normal cell growth, has improved enormously in the past 40 years. Such progress can in large part be traced to efforts to elucidate how members of several virus families cause cancer in animals. In fact, as we discuss in this chapter, study of such oncogenic viruses has led to a detailed understanding of the molecular basis of the development of cancer.

It is now clear that cancer (defined in Box 18.1) is a genetic disease: it results from the growth of successive populations of cells in which mutations have accumulated (Box 18.2). These mutations affect various steps in the regulatory pathways that control cell communication and proliferation, and lead to uncontrolled growth, increasing cellular disorganization and, ultimately, cancer. One or more mutations may be inherited (Box 18.2), or they may arise as a consequence of exposure to environmental carcinogens or infectious agents, including viruses. It is estimated that viruses are a contributing factor in approximately 20% of all human cancers. For some, such as liver and cervical cancer, they are the major cause. However, it is important to understand that the induction of malignancy is not a requirement for the propagation of any virus. Rather, this unfortunate outcome for the host is a side effect of either infection or the host's response to the presence of the virus. From this perspective, viruses can be thought of as cofactors or unwitting initiators of oncogenesis.

Understanding the development of cancer ultimately depends on knowledge of how individual cells behave within an animal. As described in Chapters 14 and 15, analysis of viral pathogenesis must encompass a consideration of the organism as a whole, especially the body's immune defenses.

Benign An adjective used to describe a growth that does not infiltrate into surrounding tissues; opposite of malignant

Cancer A malignant tumor; a growth that is not encapsulated and that infiltrates into surrounding tissues, replacing normal with abnormal cells; it is spread by the lymphatic vessels to other parts of the body; death is caused by destruction of organs to a degree incompatible with their function, by extreme debility and anemia, or by hemorrhage

Carcinogenesis The complex, multistage process by which a cancer develops

Carcinoma A cancer of epithelial tissue

Endothelioma Overproduction of erythrocytes

Fibroblast A cell derived from connective tissue

Fibropapilloma A solid tumor of cells of the connective tissue

Hepatocellular carcinoma A cancer of liver epithelial cells

Leukemia A cancer of white blood cells

Lymphoma A cancer of lymphoid tissue

Malignant An adjective applied to any disease of a progressive and fatal nature; opposite of benign

Neoplasm An abnormal new growth, i.e., a cancer

Oncogenic Causing a tumor

Retinoblastoma A cancer of cells of the retina

Sarcoma A cancer of fibroblasts

Tumor A swelling, caused by abnormal growth of tissue, not resulting from inflammation; may be benign or malignant

However, understanding how members of several virus families cause cancer in animals began with studies of cultured cells in the laboratory. In particular, early investigators noticed that the growth properties and morphologies of some cultured cells could be changed upon infection with certain viruses. We therefore consider such cells **transformed**. The advantages of these cell culture systems are many. The molecular virologist can focus attention on particular cell types, can manipulate their behavior in a controlled manner, and can easily distinguish effects specific to the virus. In many cases, cells transformed by viruses in culture can form tumors when implanted in animals. But tumors do not always form, and it is important to realize that transformed cultures are **not** tumors. The major benefit of cell culture systems is that they allow researchers to study the molecular events that establish an oncogenic potential in virus-infected cells.

Properties of Transformed Cells

Cellular Transformation

The multiplication of cells in the body is a strictly regulated process. In a young animal, total cell multiplication exceeds cell death as the animal grows to maturity. In an adult, the processes of cell multiplication and death are carefully balanced. For some cells, high rates of proliferation are required to maintain this balance. For example, human intestinal cells and white blood cells have half-lives of only a few days and need to be replaced rapidly. On the other hand, red blood cells live for over 100 days, and healthy neuronal cells rarely die. Occasionally, this carefully regulated process breaks down, and a particular cell begins to grow and divide even though the body has sufficient numbers of its type; such a cell behaves as if it were immortal. Acquisition of this property, **immortality**, is generally acknowledged to be an early step in oncogenesis. An immortalized cell may acquire one or more additional genetic changes to give rise to a clone of cells that is able to expand, ultimately forming a mass called a **tumor**. Some tumors are **benign**; they cease to grow in the body after reaching a certain size and generally do no great harm. Other tumor cells grow and divide indefinitely to form **malignant** tumors (see Box 18.1 for definition) that damage and impair the normal function of organs and tissues. Occasionally, some cells in a tumor acquire additional mutations (Box 18.2) that confer the ability to escape the boundary of the mass, to invade surrounding tissue, and to be disseminated to other parts of the body, where the cells take up residence. There they continue to grow and divide, giving rise to secondary tumors called **metastases**. Such cells cause the most serious and life-threatening disease.

Many studies of the molecular biology of oncogenic animal viruses employed primary cultures of normal cells, for example, rat or mouse embryo fibroblasts, obtained by

BOX 18.2

Genetic alterations associated with the development of colon carcinoma

The clinical stages in the development of human colon cancer, which is the fourth most common cancer in the United States, are particularly well defined. Furthermore, as shown, several genes that are frequently mutated to allow progression from one stage to the next have been identified. The early adenomas or polyps that initially form are benign lesions. Their conversion to malignant metastatic carcinomas correlates with the acquisition of additional mutations in the *p53* and *dcc* ("deleted in colon carcinoma") genes. Inherited mutations in the genes listed can greatly increase the risk that an individual will develop colon carcinoma. For example, patients with familial adenomatous polyposis can inherit defects in the *apc* (adenomatous polyposis coli) gene that result in the development of hundreds of adenomatous polyps. The large increase in the **number** of these benign lesions increases the chance that some will progress to malignant carcinomas. In contrast, patients with hereditary nonpolyposis colorectal cancer develop polyps at the same rate as the general population. However, polyps develop to carcinomas more frequently in these patients, because defects in genes encoding proteins that correct mismatched bases in DNA lead to a higher mutation rate. Consequently, the **likelihood** that an individual polyp will develop into a malignant lesion increases from 5 to 70%.

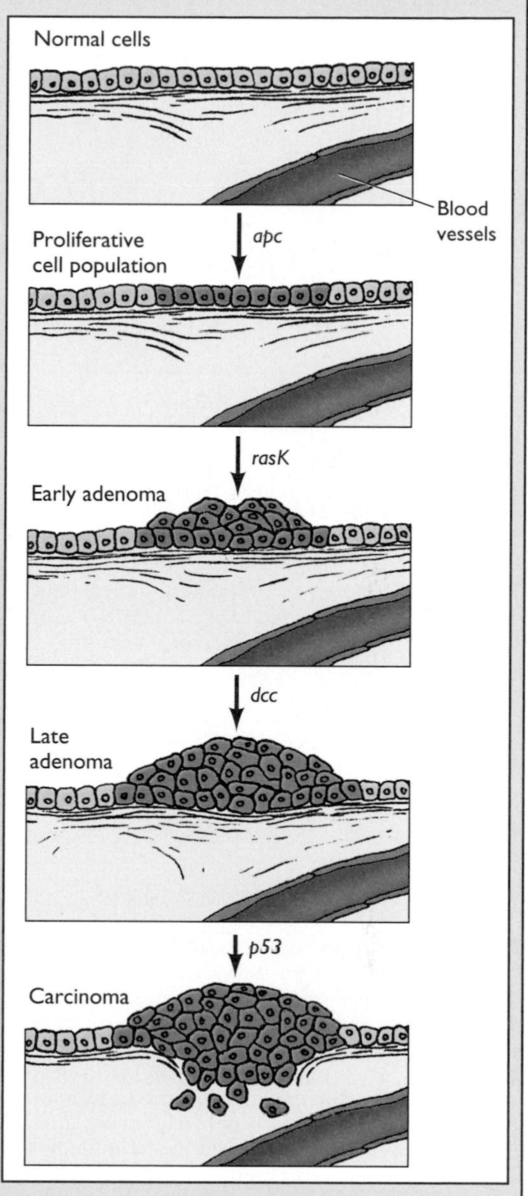

the methods described in Chapter 2. The advantage of these cells is that all the molecular changes necessary for the virus to convert them to the oncogenic state can be studied. On the other hand, cells of such primary cultures, like their normal counterparts in the body, have a finite capacity to grow and divide in culture. Cells from some animal species, such as rodents, undergo a spontaneous transformation when maintained in culture. Immortalized cells appear after a "crisis" period in which the great majority of the cells die (Fig. 18.1). As such cells are otherwise normal, and do not induce tumors when introduced into animals, they can be used to identify viral gene products needed for steps in transformation subsequent to immortalization. For reasons that are not fully understood, human and simian cells rarely undergo spontaneous transformation to immortality when passaged in culture. In fact, established lines of human cells generally can be derived only from tumors, or following exposure of primary cells to chemical carcinogens or to oncogenic RNA or DNA viruses (or their transforming genes). The realization that such transformed cells share a number of common properties, regardless of how they were obtained, provided a major impetus for the investigation of viral transformation.

A Human cells

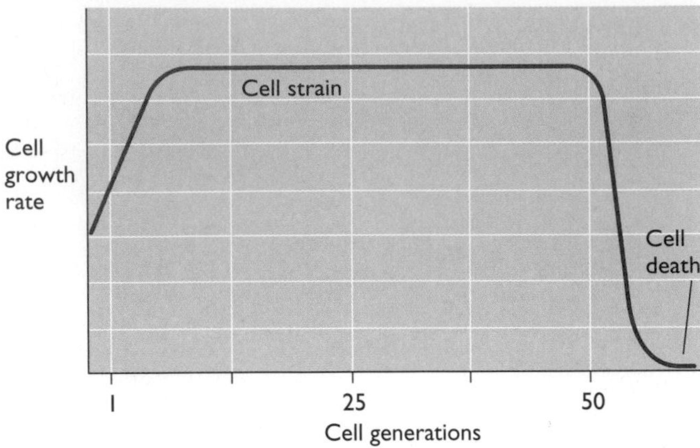

B Mouse cells

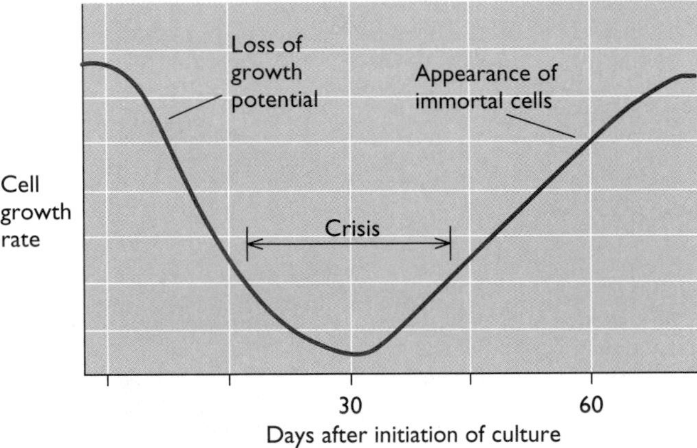

Figure 18.1 Stages in the establishment of a cell culture. (A) Human cells. When an initial explant is made (e.g., from foreskin), some cells die and others (mainly fibroblasts) start to grow; overall, the growth rate increases. If the surviving cells are diluted regularly, the cell strain grows at a constant rate for about 50 cell generations, after which growth begins to decrease. Eventually, all the cells die. The generation time of human cells varies from 24 to 72 h. (B) Mouse or other rodent cells. In a culture prepared from mouse embryo cells, cell death initially occurs, and then healthy growing cells emerge. As these cells are maintained in culture, they soon begin to lose growth potential and most cells die (the culture goes into crisis). Very rare cells do not die, but continue growth and division until their progeny overgrow the culture. These cells constitute a cell line, which will grow indefinitely if it is appropriately diluted and fed with nutrients: the cells are immortal.

Properties That Distinguish Transformed and Normal Cells

The definitive characteristic of transformed cells is their lack of response to the signals or conditions that normally control DNA replication and division. This property is illustrated by the list of growth parameters and behaviors provided in Table 18.1. As described above, transformed cells are immortal: they can grow and divide indefinitely, provided that they are diluted regularly into fresh medium. Production of **telomerase,** an enzyme that maintains telomeric DNA at the ends of chromosomes, has been implicated in immortalization. In addition, transformed cells typically exhibit a reduced requirement for the growth factors present in serum. Some transformed cells actually produce their own growth factors and the cognate receptors, providing themselves **autocrine growth stimulation**. Normal cells cease to grow and enter a quiescent state (called G_0, described in "Control of Cell Proliferation" below) when essential nutrient concentrations drop below a threshold level. Transformed cells are deficient in this capacity, and some may even kill themselves by trying to continue to grow in an inadequate environment.

Table 18.1 Growth parameters and behaviors of tranformed cells

- Immortal: can grow indefinitely
- Reduced requirement for serum growth factors
- Loss of capacity for growth arrest upon nutrient deprivation
- High saturation densities
- Loss of contact inhibition (can grow over one another or normal cells)
- Anchorage independent (can grow in soft agar)
- Altered morphology (appear rounded and refractile)
- Tumorigenic

Transformed cells grow to high densities. This characteristic is manifested by the cells' piling up, over, or under each other. They also grow on top of untransformed cells, forming visibly identifiable clumps called **foci** (Fig. 18.2). Transformed cells behave in this manner because they have lost **contact inhibition**, a response by which normal cells cease growth and movement when they sense the presence of their neighbors. Unlike normal cells, many transformed cells have also lost the need for a surface on which to adhere, and we describe them as being **anchorage independent**. Some anchorage-independent cells can also form isolated colonies in semisolid media (e.g., 0.6% agar). This property correlates well with the ability to form tumors in animals and often is used as an experimental surrogate for malignancy. Finally, transformed cells **look** different from normal cells; they are more rounded, with fewer processes, and as a result appear more refractile when observed under a microscope (Fig. 18.2).

There are other ways in which transformed cells can be distinguished from their normal counterparts. These properties include metabolic differences and characteristic changes in cell surface and cytoskeletal components. However, the list in Table 18.1 comprises the standard criteria used to judge whether cells have been transformed by viruses.

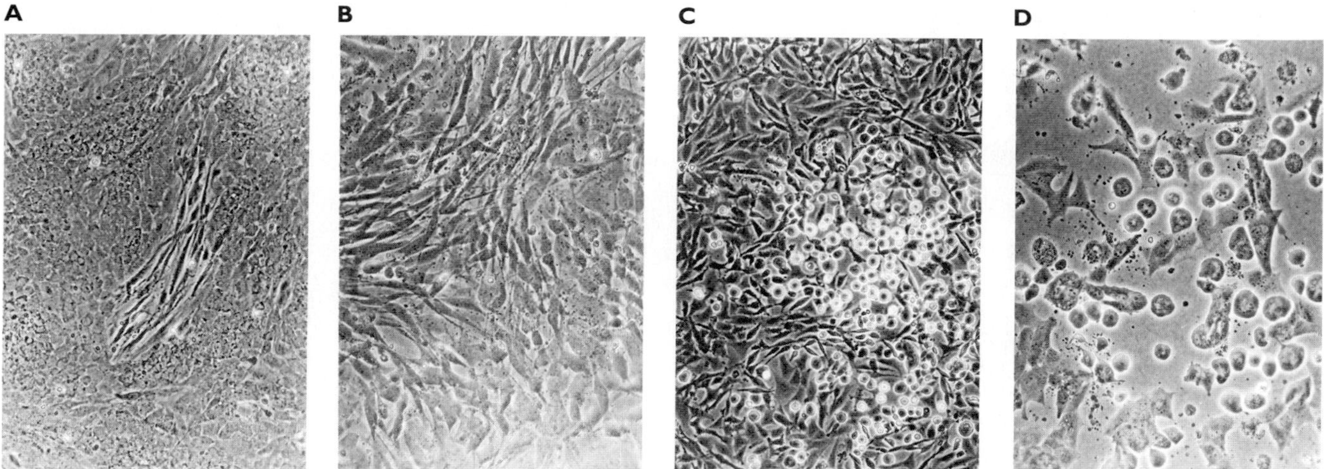

Figure 18.2 Foci formed by avian cells transformed with two strains of the Rous sarcoma virus. Differences in morphology are due to genetic differences in the transduced *src* oncogene. (A) A focus of infected cells with fusiform morphology shown on a background of flattened, contact-inhibited, uninfected cells. (B) Higher magnification of a fusiform focus showing lack of contact inhibition among the transformed cells. (C) A focus of highly refractile infected cells with rounded morphology and reduced adherence. (D) Higher magnification of rounded infected cells showing tightly adherent normal cells in the background. Courtesy of P. Vogt, The Scripps Research Institute.

Control of Cell Proliferation

Sensing the Environment

Normal cells possess elaborate pathways that receive and process growth-stimulatory or growth-inhibitory signals transmitted by other cells in the tissue or organism. Much of what we know about these pathways comes from study of the cellular genes transduced or activated by oncogenic retroviruses. Signaling often begins with the secretion of a growth factor by a specific type of cell. The growth factor may enter the circulatory system, as in the case of many hormones, or may simply diffuse through the spaces around cells inside a tissue. Growth factors bind to the external portion of specific receptor molecules on the surface of the same or other types of cells in the tissue (Chapter 5). Alternatively, signaling can be initiated by binding of a receptor on one cell to a specific protein (or proteins) present on the surface of another cell, or to components of the extracellular matrix. The binding of the ligand triggers a change, often via oligomerization of receptor molecules, that is transmitted to the cytoplasmic portion of the receptor. In the case illustrated in Fig. 18.3, the cytoplasmic domain of the receptor possesses protein tyrosine kinase activity, and interaction with the growth factor ligand triggers autophosphorylation. This modification sets off a **signal transduction cascade**, a chain of sequential physical interactions among, and biochemical modifications of, membrane-bound and cytoplasmic proteins. Membrane-associated guanine nucleotide-binding proteins with GTPase activity (**G proteins**) regulate the activities of some of the membrane-bound components of signaling pathways. A common biochemical modification of cytoplasmic protein components of a signaling cascade is phosphorylation of serine and/or threonine residues (e.g., via the mitogen-activated protein [Map] kinases).

As illustrated in Fig. 18.3, small molecules synthesized by some membrane-bound proteins in the cascade, such as cyclic nucleotides and lipids, act as diffusible **second messengers** in the signal relay. Changes in ion flux across the plasma membrane, or in membranes of the endoplasmic reticulum, may also contribute to transmission of signals.

Relay of the signal can terminate at cytoplasmic sites to alter metabolism, or cell morphology and adhesion. However, many signaling cascades culminate in the modification of transcriptional activators or repressors, and therefore alter the expression of specific cellular genes. The products of these genes either allow the cell to progress through another growth cycle or cause the cell to stop growing, to differentiate, or to die, whichever is appropriate to the situation. Errors in the signaling pathways that regulate these decisions can lead to cancer.

Phases of the Cell Cycle

The capacity of cells to grow and divide is regulated by a cellular timer. The timer comprises an assembly of proteins that integrate stimulatory and inhibitory signals received by, or produced within, the cell. Eukaryotic cells do not attempt to divide until all their chromosomes have

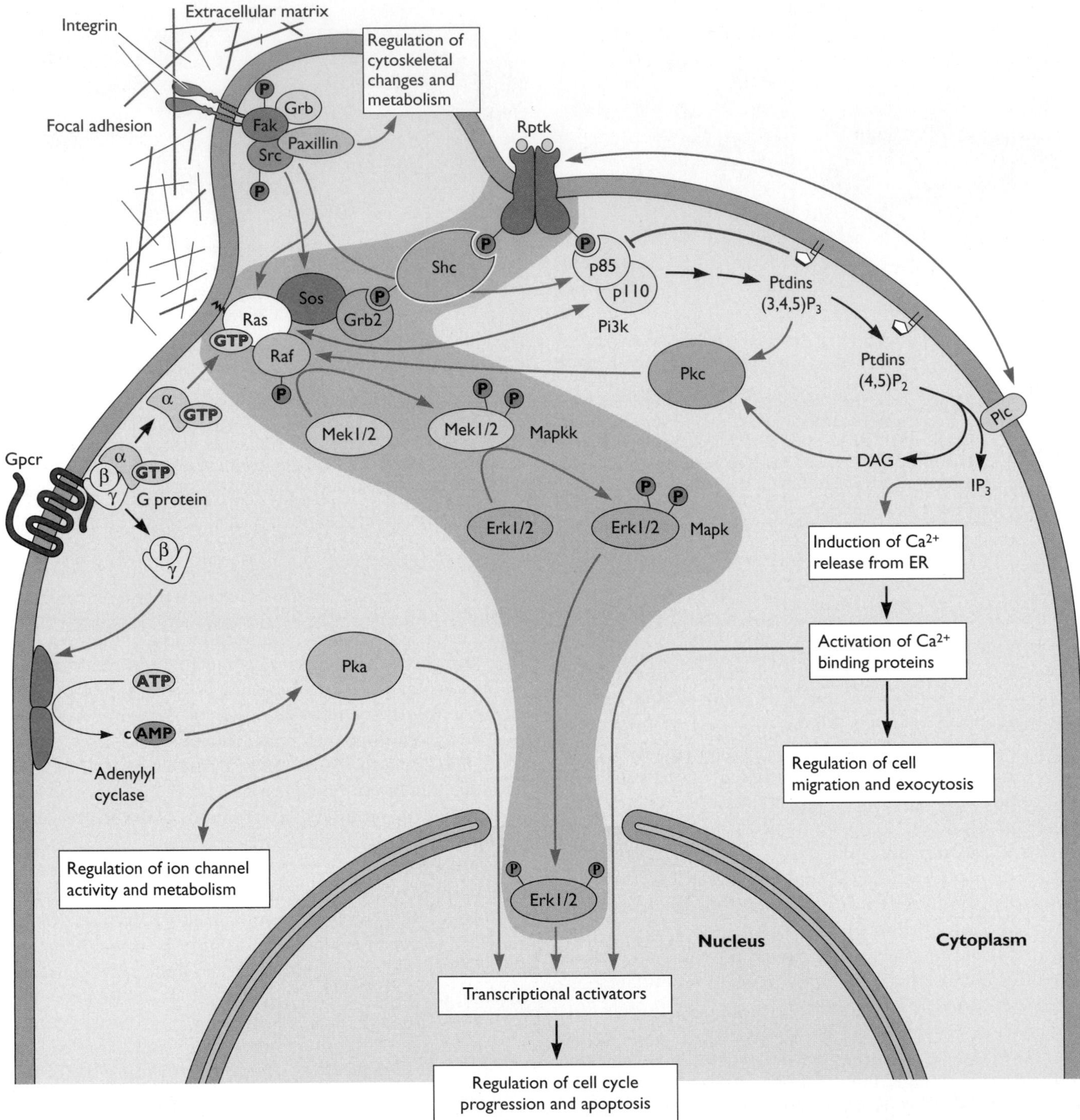

Figure 18.3 Some signal transduction pathways and their connections. The pathway extending from the cell surface to the nucleus (pink background) is the mitogen-activated protein (Map) kinase cascade, which is highly conserved among eukaryotes. Signal transduction is initiated by binding of ligand to the extracellular domain of a receptor protein tyrosine kinase (Rptk), for example, the receptor for epidermal growth factor or the receptor for platelet-derived growth factor. Binding of ligand induces receptor dimerization and autophosphorylation of tyrosine residues. Adapter proteins like Shc and the Grb2 component of the Grb2-Sos complex are recruited to the membrane by binding to phosphotyrosine-containing sequences in the

(continued on next page)

been duplicated and are precisely organized for separation into daughter cells. Nor do they initiate DNA synthesis and chromosome duplication until the previous cell division is complete, or unless the extra- and intracellular environments are appropriate. Consequently, the cellular timer controls a tightly ordered **cell cycle** comprising intervals, or **phases,** devoted to specific processes.

The duration of each of the separate phases in the cell cycle shown in Fig. 18.4 is typical of those of many mammalian cells growing actively in culture. However, there is considerable variation in the length of the cell cycle, largely reflecting differences in the **gap phases** (**G_1** and **G_2**). For example, early embryonic cells of animals, such as the fruit fly *Drosophila* and the toad *Xenopus laevis,* dispense with G_1 and G_2, do not increase in mass, and move immediately from the **DNA synthesis phase (S)** to **mitosis (M)** and again from M to S. Consequently, they possess extremely short cycles of 10 to 60 min. At the other extreme are cells that have ceased growth and division and have withdrawn from the cell cycle. The variability in duration of this specialized **resting state**, termed G_0, accounts for the large differences in the rates at which cells in multicellular organisms proliferate. As discussed in Chapter 9, many viruses replicate successfully in cells that spend all or most of their lives in G_0, a state that has been likened to "cell cycle sleep." In many cases, synthesis of viral proteins in such resting or slowly cycling cells induces them to reenter the cell cycle and grow and divide rapidly. In order to describe the mechanisms by which these viral proteins induce such abnormal activity and transform cells, we will first introduce the molecular mechanisms that control passage through the cell cycle.

The Cell Cycle Engine

The orderly progression of eukaryotic cells through periods of growth, chromosome duplication, and nuclear and cell division is driven by intricate regulatory circuits. The elucidation of these circuits must be considered a tour de force of contemporary biology. The first experimental hint that cells contain proteins that control transitions from one phase of the cell cycle to another came more than 30 years ago, when nuclei of slime mold (*Physarum polycephalum*) cells in early G_2 were found to enter mitosis immediately following fusion with cells in late G_2 or M. This crucial observation led to the conclusion that the latter cells must contain a **mitosis-promoting factor.** Subsequently, similar experiments with mammalian cells in culture identified an analogous S-phase-promoting factor. The convergence of many observations eventually led to the identification of the highly conserved cell cycle engine components (Fig.

Figure 18.3 ***(continued)*** cytoplasmic domains of activated receptor protein tyrosine kinases or to a substrate phosphorylated by the activated receptor, along with Ras. Juxtaposition of Sos and Ras at the membrane activates exchange of GDP for GTP bound to Ras by Sos, the guanine nucleotide exchange protein for Ras. The GTP-bound form of Ras binds to members of the Raf family, bringing these serine/threonine protein kinases to the plasma membrane. Raf is then activated by an unknown mechanism and becomes autophosphorylated. Such activation of Raf initiates the Map kinase (Mapk) cascade. The pathway shown contains Mek1 and Mek2 (Mek1/2) (Map kinase kinases [Mapkk]) and Erk1 and Erk2 (Erk1/2) (Map kinases). Activated Erk1/Erk2 can enter the nucleus, where they phosphorylate and activate transcriptional regulators. These kinases can also regulate transcription indirectly, by activation of other protein serine kinases. As shown, signals also feed into this pathway from other types of receptors. These include seven transmembrane domain receptors, which transmit signals by means of heterotrimeric G proteins, termed G protein-coupled receptors (Gpcr), or integrin receptors, which bind to components of the extracellular matrix at focal adhesions (sites at which the cell is in close contact with the extracellular matrix). The latter interaction induces binding of focal adhesion kinase (Fak) to the cytoplasmic domains of integrins, and autophosphorylation of Fak on tyrosine residues. Activated Fak then binds Src and other members of the Src family of protein tyrosine kinases, an interaction that leads to further phosphorylation of Fak, activation of the Src kinase, and binding and phosphorylation of adapter proteins that recognize sequences containing phosphotyrosine. The Fak-Src complex not only activates the Mapk pathway but also induces alterations in the cytoskeleton, in part via paxillin, and activation of phosphoinositol 3-kinase (Pi3k). As shown, this enzyme is also activated upon binding of its p85 subunit to activated receptor protein tyrosine kinases. The enzyme synthesizes phosphatidylinositol-3-phosphate [Ptdins(3)P], which is a substrate for synthesis of other phosphoinositides. Some of the autoregulatory effects and connections of these lipids to the Mapk pathway are depicted. The lipid products of activation of phospholipase C (Plc) act as second messengers. For example, inositol 1,4,5-triphosphate (IP3) stimulates release of Ca^{2+} from storage within the endoplasmic reticulum (ER), following binding to a receptor (Ip3r) in the ER membrane, while diacylglycerol (DAG) activates members of the protein kinase C (Pkc) family. Cyclic AMP (cAMP) produced upon activation of adenylyl cyclase by heterotrimeric G proteins is also a second messenger, which activates protein kinase A (Pka). Physical associations between proteins are shown directly or indicated by a blue arrow, with green arrows and red bars, indicating positive and negative effects, respectively, of the interaction.

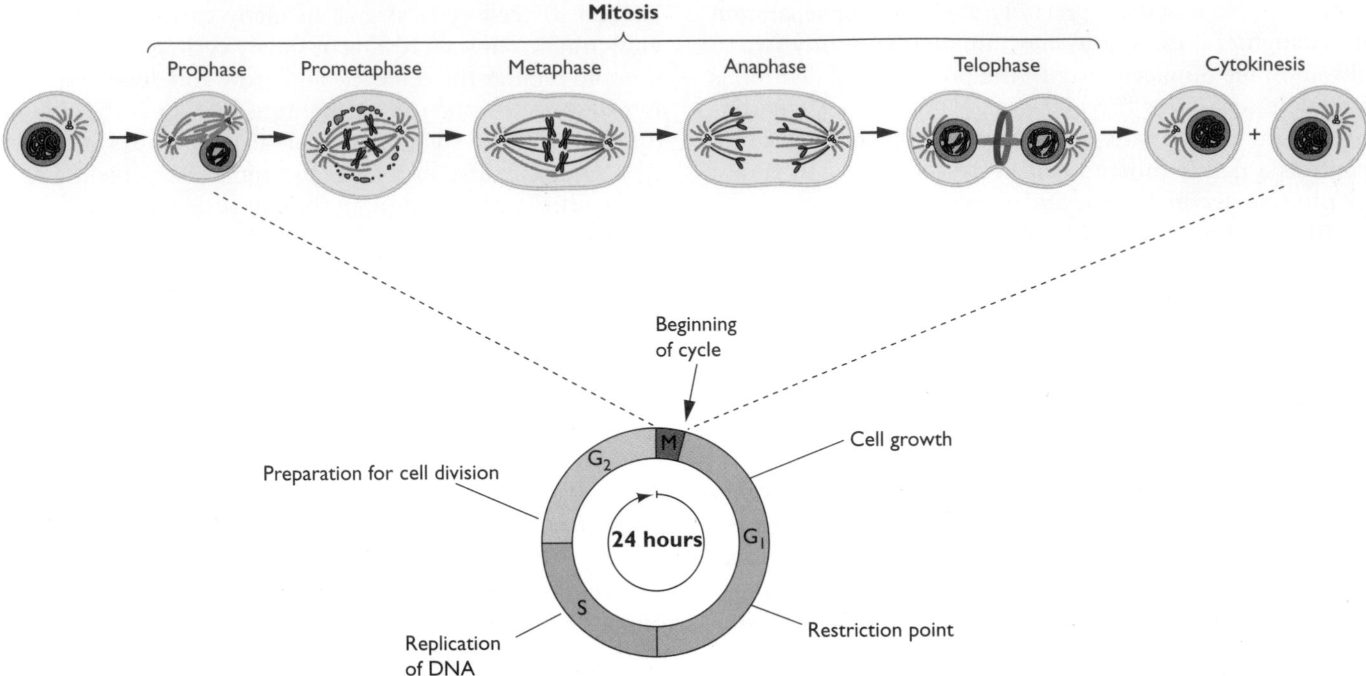

Figure 18.4 The phases of a eukaryotic cell cycle. The most obvious phase morphologically, and hence the first to be identified, is mitosis, or M phase, the process of nuclear division that precedes cell division. During this period, the nuclear envelope breaks down. Duplicated chromosomes become condensed and aligned on the mitotic spindle and are segregated to opposite poles of the cell, where nuclei reform upon chromosome decondensation (top). The end of M phase is marked by cytokinesis, the process in which the cell divides in two. Despite this remarkable reorganization and redistribution of cellular components, M phase occupies only a short period within the cell cycle. During the long interval from one mitosis to the next, interphase, cells grow continuously. Interphase was divided into three parts with the recognition that DNA synthesis takes place only during a specific period, the synthetic or S phase, beginning about the middle of interphase. The other two periods, which appeared as gaps between defined processes, are designated the G_1 and G_2 (for gap) phases.

18.5A). Mitosis-promoting factor purified from amphibian oocytes proved to be an unusual protein kinase: its catalytic subunit is activated by the binding of an unstable regulatory subunit. Furthermore, the concentration of the regulatory subunit was found to oscillate reproducibly during each and every cell cycle. The regulatory subunit was therefore given the descriptive name **cyclin,** and the associated protein kinase was termed **cyclin-dependent kinase (Cdk).** Similar proteins were implicated in cell cycle control in the yeast *Saccharomyces cerevisiae,* and it soon became clear that all eukaryotic cells contain multiple cyclins and Cdks, which operate in specific combinations to control progression through the cell cycle.

Figure 18.5A lists the various mammalian cyclin-Cdk complexes and shows the phases in the cell cycle in which they accumulate. The patterns of accumulation illustrate a critical feature of the cell cycle: individual cyclin-Cdk complexes, the active protein kinases, accumulate in successive waves. The concentration of each increases gradually during a specific period in the cycle, but decreases abruptly as the cyclin subunit is degraded. This orderly activation and deactivation of specific kinases govern passage through the cell cycle. For example, cyclin E synthesis is rate limiting for the transition from G_1 to S phase in mammalian cells, and the cyclin E-Cdk2 complex accumulates during late G_1. Soon after cells have entered S phase, cyclin E and the complex it forms with Cdk2 rapidly disappear from the cell; its task is completed until a new cycle begins.

While the oscillating waves of active Cdk accumulation and destruction are thought of as the ratchet that advances the cell cycle timer, it is important not to interpret this metaphor too literally. The orderly and reproducible sequence of DNA replication, chromosome segregation, and cytokinesis is not determined solely by the oscillating concentrations of individual cyclin-Cdk complexes. Rather, the cyclin-Cdk cycle serves as a device for integrating numerous signals from the exterior and interior of the cell into appropriate responses. The regulatory circuits that feed into and from the cycle are both many and complex (Fig. 18.5B). These regulatory signals ensure that the cell increases in mass and divides only when the environment is propitious or, in multicellular organisms, when the tim-

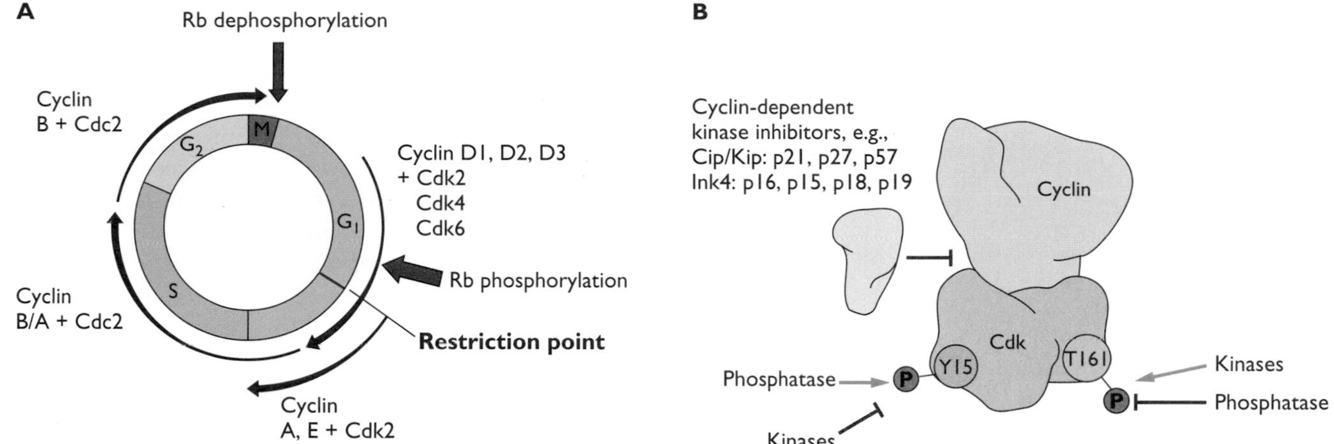

Figure 18.5 The mammalian cyclin-Cdk cell cycle engine. (A) The phases of the cell cycle are denoted on the inner circle. The progressive accumulation of specific cyclin-Cdk complexes and their abrupt disappearance from the cell are represented by the broadening arrows, with the arrowheads marking the time of disappearance. The cyclins are related in sequence to one another, and they share such properties as activation of cyclin-dependent kinases and controlled destruction by the proteasome. In early amphibian oocytes, the first system in which oscillation of a cyclin (cyclin B) was observed, such regulation of cyclin concentrations is posttranscriptional: the cyclins are made at more or less constant rates throughout the cell cycle, but their proteolysis is initiated abruptly at specific points in the cycle. In mammalian cells, proteolysis is also very important in resetting the concentrations of individual cyclins at specific points in the cycle, but production of cyclin mRNAs is also regulated. For example, the concentrations of mRNAs encoding cyclins E, A, and B, like those of the proteins, oscillate during the cell cycle. Regulation of nuclear export, which is inhibited by cyclin B phosphorylation at the G_2/M transition, makes an important contribution to nuclear accumulation of the B-type cyclins specifically in the M phase. The cyclin-dependent kinases are also related in sequence and activity. (B) The production, accumulation, and activities of both cyclins and cyclin-dependent kinases are regulated by numerous mechanisms. Activating and inhibitory reactions are indicated by green arrows and red bars, respectively. For example, activation of the kinases can require not only binding to the appropriate cyclin, but also phosphorylation at specific sites and removal of phosphate groups at other sites. The activities of the kinases are also controlled by association with members of two families of cyclin-dependent kinase-inhibitory proteins, which control the activities of only G_1 (Ink4 proteins) or all (Cip/Kip proteins) cyclin-Cdk complexes. Both types of inhibitor play crucial roles in cell cycle control. For example, the high concentration of $p27^{Kip1}$ characteristic of quiescent cells falls as they enter G_1, and inhibition of synthesis of this protein prevents cells from becoming quiescent.

ing is correct. Many signal transduction pathways that convey information about the local environment or the global state of the organism therefore converge on the cyclin-Cdk integrators. In addition, various surveillance mechanisms monitor such internal parameters as DNA damage, problems with DNA replication, and proper assembly and function of the mitotic spindle, to protect cells against potentially disastrous consequences of continuing a cell division cycle that would not be completed correctly. As we shall see, it is primarily these signaling and surveillance mechanisms that are compromised during transformation by oncogenic viruses.

Oncogenic Viruses

Oncogenic viruses cause cancer by inducing changes that affect the processes that control cell proliferation. Indeed, the study of the mechanisms of viral transformation and oncogenesis laid the foundation for our current understanding of cancer, for example, with the identification of oncogenes that are activated or captured by the retroviruses (originally known as RNA tumor viruses) and viral proteins that inactivate tumor suppressor gene products (Fig. 18.6). Specific members of a number of different virus families have been implicated in naturally occurring or experimentally induced cancers in animals (Table 18.2). It has been estimated that approximately 20% of all cases of human cancer are associated with infection by one of five viruses: Epstein-Barr virus, hepatitis B virus, hepatitis C virus, human T-lymphotropic virus type 1, and human papillomaviruses. In this section, we introduce oncogenic viruses and general features of their transforming interactions with host cells.

Discovery of Oncogenic Viruses

Retroviruses

Oncogenic viruses were discovered almost 100 years ago when Vilhelm Ellerman and Olaf Bang (1908) first showed that avian leukemia could be transmitted by filtered extracts (i.e., viruses) of leukemic cells or serum from in-

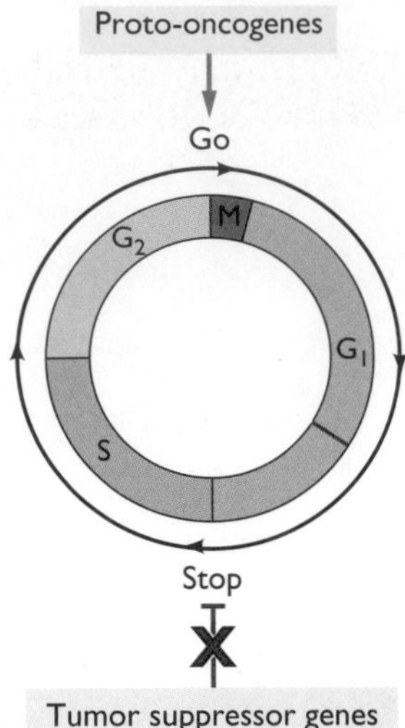

Figure 18.6 A genetic paradigm for cancer. The division cycle for all eukaryotic cells has an intrinsic pace that can be modulated both positively and negatively by different sets of gene products. Cancer arises from a combination of dominant, gain-of-function mutations in proto-oncogenes and recessive, loss-of-function mutations in tumor suppressor genes. The function of either type of gene product can be affected by oncogenic viruses.

fected birds. Because leukemia was not recognized as cancer in those days, the significance of this discovery was not generally appreciated. Shortly thereafter (in 1911), Peyton Rous demonstrated that solid tumors could be produced in chickens by using cell-free extracts from a transplantable sarcoma that had appeared spontaneously. Despite the viral etiology of this disease, the cancer viruses of chickens were thought to be oddities until similar murine malignancies, as well as mammary tumors, were found to be induced by viruses. These oncogenic viruses all proved to be members of the retrovirus family, which, fulfilling the criteria noted above, do not kill their host cells. We now know that retroviruses are endemic in many species, including mice and chickens. For example, most chickens in a flock will have been infected within a few months of hatching. In the vast majority of cases, infectious virus appears as a transient viremia and disease is rare. Animals can also be infected congenitally and thereafter become immune-tolerant to the virus. Such animals will then develop a chronic viremia and, in such cases, malignancy is more frequent. The appearance of a rapidly transforming and highly oncogenic strain of retroviruses, such as that isolated by Rous, is a very rare event but, as discussed below, one that has been much exploited by experimentalists.

Table 18.2 Oncogenic viruses and cancer

Family	Associated cancer(s)
RNA viruses	
Flaviviridae	
Hepatitis C virus	Hepatocellular carcinoma
Retroviridae	Hematopoietic cancers, sarcomas, carcinomas
DNA viruses	
Adenoviridae	Various solid tumors
Hepadnaviridae	Hepatocellular carcinoma
Herpesviridae	Lymphomas, carcinomas, and sarcomas
Papillomaviridae	Papillomas and carcinomas
Polyomaviridae	Various solid tumors
Poxviridae	Myxomas and fibromas

Early researchers classified the oncogenic retroviruses into two groups depending on the rapidity with which they caused neoplasia (Table 18.3). The first group comprises the rare, rapidly transforming **transducing oncogenic retroviruses**. These are all highly carcinogenic agents that cause malignancies in nearly 100% of infected animals in a matter of days. They were later discovered to have the ability to transform susceptible animal cells in culture. The second class, **nontransducing oncogenic retroviruses**, includes less carcinogenic agents. Not all animals infected with these viruses develop tumors, which appear only weeks or months after infection. In the late 1980s, a third type of oncogenic retrovirus, a **long-latency virus** for which tumorigenesis is very rare, and occurs months or even years after infection, was identified in humans.

Each group of oncogenic retroviruses induces tumors by a distinct mechanism. As their name implies, transducing retroviruses cause cancer because their genomes contain transduced cellular genes that become **oncogenes** (genes encoding proteins that cause transformation or tumorigenesis) when expressed in the viral context. The virally transduced versions of cellular genes are called **v-oncogenes**, and their normal cellular counterparts are called **c-oncogenes** or **proto-oncogenes**. The nontransducing retroviruses do not encode cell-derived oncogenes. Rather, the transcription of proto-oncogenes is inappropriately activated as a consequence of the nearby integration of a provirus in the host cell genome. In either situation, the oncogene products play no role in the reproductive cycle of the retroviruses themselves. The oncogenic potential of these viruses is, in a sense, an accident of their lifestyle. Nevertheless, the study of v-oncogenes and proto-oncogenes that are affected by retroviruses has been of great importance in advancing our understanding of the origins of cancer.

Oncogenic DNA Viruses

The first DNA virus to be associated with oncogenesis was the papillomavirus that causes warts (papillomas) in cottontail rabbits, isolated by Richard Shope in 1933.

Table 18.3 Oncogenic retroviruses

Property or characteristic	Transducing viruses	Nontransducing viruses	Nontransducing, long-latency viruses
Efficiency of tumor formation	High (~100% of infected animals)	High to intermediate	Very low (<5%)
Tumor latency	Short (days)	Intermediate (weeks, months)	Long (months, years)
Infecting viral genome	Viral-cellular recombinant; normally replication defective	Intact; replication competent	Intact; replication competent
Oncogenic element	Cell-derived oncogene carried in viral genome	Cellular oncogene activated in situ by a provirus	Virus-encoded regulatory protein controlling transcription?
Mechanism	Oncogene transduction	*cis*-acting provirus	*trans*-acting protein?
Ability to transform cells in culture	Yes	No	No

Important properties of the virus and its DNA genome were identified with remarkable speed given the techniques available at the time. But the lack of cell culture systems for papillomaviruses initially precluded their use as experimental models for oncogenesis. Other viruses (in particular polyomaviruses, such as simian virus 40, and human adenoviruses) proved much more tractable and soon dominated early studies of transformation and tumorigenesis by DNA viruses. It is important to note that neither simian virus 40 nor adenoviruses are associated with oncogenesis in their natural hosts. However, it was shown soon after their discovery that these viruses can induce tumors in rodents and transform cultured mammalian cells, and the possibility that simian virus 40 could have contributed to human cancers is currently under intense investigation (Box 18.3). Replication of these viruses destroys permissive primate host cells within a few days of infection. In contrast, rodent cells are nonpermissive for viral reproduction, or support only limited reproduction. Consequently, some infected cells survive infection and in rare cases become transformed (see "Viral Genetic Information in Transformed Cells").

In contrast to the v-oncogenes of retroviruses, the transforming genes of these DNA viruses **do** play important roles in virus reproduction. The proteins encoded by these genes contribute to oncogenesis or transformation by altering the activities of cellular products. In some cases, such cellular proteins are encoded by the same proto-oncogenes transduced or affected by retroviruses. This important discovery, initially made in studies of the early middle T protein (mT) of mouse polyomavirus in the early 1980s, provided the first indication that retroviruses and DNA viruses can transform cells by related mechanisms. Investigation of the biochemical properties of proteins encoded in other transforming genes of these DNA viruses led to equally important insights, notably the characterization of cellular proteins that can block cell cycle progression, the products of **tumor suppressor genes**.

Since the discovery of Epstein-Barr virus (Appendix B, Fig. 10) in cells derived from Burkitt's lymphoma in 1966, it has been appreciated that herpesviruses can be associated with the development of tumors in humans and other animals. Members of this family can immortalize or induce typical transformed phenotypes in susceptible cells in culture. Poxviruses can also induce cell proliferation, which may be prolonged or rapidly followed by cell death, depending on the virus. Indeed, some members of this family, such as Shope fibroma virus, cause tumors of the skin. However, poxviruses do not transform cells in culture, in part because they are highly cytotoxic. The large sizes of the herpesviral and poxviral genomes initially presented a major impediment to analysis of the transforming properties of these viruses. Nevertheless, we now know the molecular functions of many of the relevant viral proteins, and it appears that these viruses generally alter cell growth and proliferation by mechanisms related to those responsible for transformation by the smaller DNA viruses or retroviruses.

Contemporary identification of oncogenic viruses. Oncogenic viruses associated with human disease continue to be isolated with some regularity. The most recently discovered (in 1994) is a previously unknown member of the family *Herpesviridae,* human herpesvirus 8, which was isolated from tumor cells from Kaposi's sarcoma. Although this virus was identified only recently, it is already clear that its genome, like those of transducing retroviruses, contains homologs of cellular proto-oncogenes. Perhaps an even greater surprise was the realization that RNA viruses other than retroviruses can be associated with cancer: hepatitis C virus, a (+) strand RNA virus belonging to the family *Flaviviridae* identified in 1989, is associated with a high risk for hepatocellular carcinoma in 60 to 70% of persistently infected humans (Appendix B, Fig. 7). Humans are the only known natural hosts, and the virus cannot be propagated in tissue culture. However, a tissue culture system that can support early steps in hepatitis C virus replication has recently become available.

BOX 18.3

A debate reopened: has simian virus 40 contributed to human cancer?

In 1960, simian virus 40 was discovered in African green monkey kidney cells used to produce poliovirus vaccines and within 2 years was shown to be tumorigenic in hamsters. These were observations of great concern, because it was realized that many batches of the vaccines contained quite high concentrations of infectious simian virus 40. It has been estimated that 98 million people in the United States, and many more worldwide, were exposed to potentially contaminated poliovirus vaccines before screening to ensure preparation of simian virus 40-free vaccines was introduced in 1963. Ironically, monkey cells had been adopted for poliovirus vaccine production, because of the concern that human cells might contain then unknown human cancer viruses!

Epidemiological studies initiated in the 1960s and 1970s followed thousands of vaccine recipients for up to 20 years, with no evidence for increased cancer risk in this population. The initial alarm raised by the tumorigenicity of simian virus 40 in rodents therefore appeared to be laid to rest. However, during the past decade, reports that the DNA of this virus is present in human tumors, analogous to those induced by simian virus 40 in hamsters, have led some researchers to reconsider the contribution of this monkey virus to human cancer. Others remain skeptical. Some of the pros and cons of the current debate are listed below.

It is important to note that even if the debate is settled with a consensus that simian virus 40 DNA **is** present in human tumors, the question of whether it is a causal agent of cancer will remain to be addressed.

Garcea, R. L., and M. J. Imperiale. 2003. Simian virus 40 infection of humans. *J. Virol.* **77:**5039–5045.

Jasani, B, et al. 2001. Association of SV40 with human tumors. *Semin. Cancer Biol.* **11:**49–61G.

Pro	Con
Simian virus 40 DNA has been detected in several human tumors, including osteosarcoma, mesothelioma, and non-Hodgkin's lymphoma. The virus induces similar tumors in hamsters. In some studies, the viral DNA has been detected in tumors, but not in surrounding normal tissues.	Simian virus 40 DNA is not present in all samples of a particular cancer, and in some studies (e.g., of mesotheliomas) has not been detected in any.
Simian virus 40 DNA appears to be present in mesotheliomas only in countries in which contaminated poliovirus vaccine was used.	The viral DNA has been detected in tumors of individuals who could not have received potentially contaminated poliovirus vaccine.
A poliovirus vaccine produced in 1954 was found to be contaminated with a variant of simian virus 40 that can be distinguished from common lab strains. The same variant has now been found in three non-Hodgkin's lymphoma patients.	In one comparison of mesotheliomas and normal samples, simian virus 40 DNA was detected just as frequently in the normal tissues as in the tumors.

As hepatitis C virus is estimated to be the cause of about 30% of all acute liver disease in the United States, studies of cloned viral genes are being actively pursued in many academic and commercial laboratories.

Common properties of oncogenic viruses. Although they are members of different families (Table 18.2), these oncogenic viruses share several general features. In principle, any virus that encodes proteins that can engage and modulate the cell cycle or block programmed cell death (apoptosis) has the potential to transform cells and contribute to oncogenesis. The actual mechanism by which some of these viruses contribute to oncogenesis is poorly understood. However, one critical property appears to be the ability of the viruses to infect, but not kill, rapidly growing cells or stem cells. The capacity of some of the viruses to induce secretion of virally encoded or cellular molecules that stimulate the growth of uninfected cells (e.g., cytokines) and induce tissue hyperplasia, or to modulate immune system killing of the infected cell can also contribute to oncogenic potential.

In all cases that have been analyzed, transformation is observed to be a single-hit process (defined in Chapter 2) in the sense that infection of a susceptible cell with a single virus particle is sufficient to cause transformation. In addition, all or part of the viral genome is usually retained in the transformed cell. With few exceptions, cellular transformation is accompanied by the continuous expression of specific viral sequences. On the other hand, while specific viral genes are present and expressed, transformed cells need not and (except in the case of some retroviruses) **do not** produce infectious virus particles. Most important, viral transforming proteins alter cell proliferation by a limited repertoire of molecular mechanisms.

Viral Genetic Information in Transformed Cells

State of Viral DNA

Cells transformed by oncogenic viruses generally retain viral DNA in their nuclei. These DNA sequences correspond to all or part of the infecting DNA genome or the proviral DNA made in retrovirus-infected cells. In most cases, viral DNA sequences are maintained as a permanent addition to the cellular genome following their integration (Box 18.4).

As discussed in Chapter 7, integration of proviral DNA by the viral integrase is an essential step in the retroviral life cycle. Such proviral integration can occur at essentially any site in cellular DNA, but preserves a fixed order of viral genes and control sequences (see Fig. 7.14). When the provirus carries a v-oncogene, the site at which it is integrated into the cellular genome is of no importance (provided that the viral transcription unit is not silenced by integration into transcriptionally inert regions of cellular chromosomes). In contrast, integration of proviral DNA within specific regions of the cellular genome is a hallmark of, and essential for, the induction of tumors by nontransducing retroviruses. The proviral sequences present in every cell of a tumor induced by members of this class are found in the same chromosomal location, an indication that all arose from a single transformed cell. Such tumors

BOX 18.4

Detection and characterization of integrated viral DNA sequences in transformed cells

Integration of viral DNA into the cellular genome was initially demonstrated by the cosedimentation of viral DNA with high-molecular-mass cellular DNA under strongly denaturing conditions. The development of restriction endonucleases as molecular tools and of the Southern hybridization assay allowed direct proof of such integration as well as characterization of integrated viral DNA, as illustrated for a circular, double-stranded, viral DNA genome.

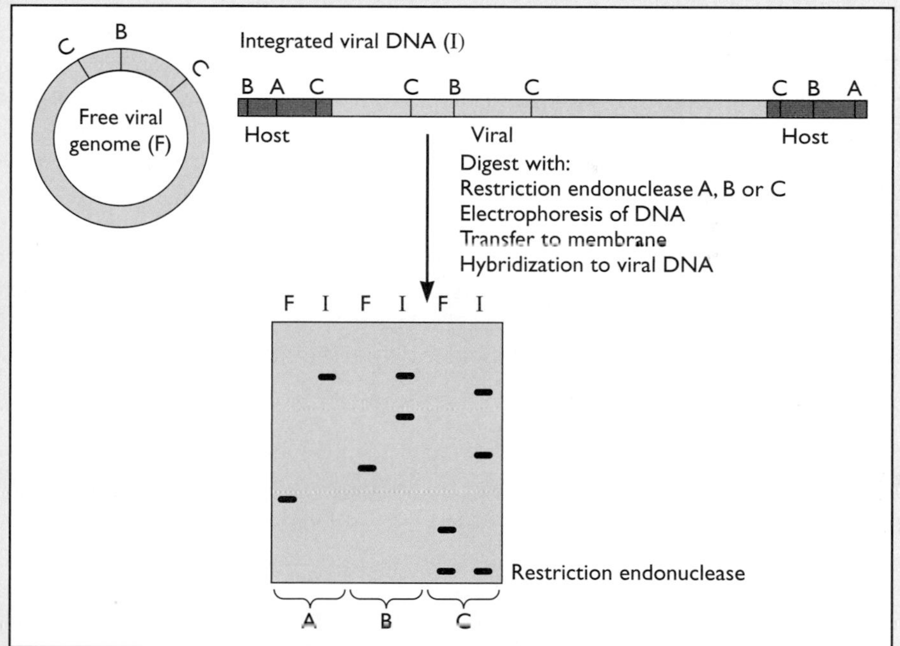

Cleavage of genomic DNA isolated from transformed cells with restriction endonuclease A (no sites in the viral genome) yields one high-molecular-mass fragment of lower mobility than free viral DNA for each site of integrated viral DNA. The number of integration sites can therefore be counted. When used with appropriate standards, the number of copies of viral sequences at each such site can also be estimated. Restriction endonuclease B (one site in the viral genome) generates two fragments differing in mobility from the free, linear viral DNA. The results obtained with enzymes A and B demonstrate that viral DNA must be covalently attached to cellular DNA. Restriction endonuclease C, with two cleavage sites in the viral genome, produces three fragments from transformed cell DNA. The comigration of one fragment with the smaller product of digestion of free viral DNA indicates that this segment of the viral genome is intact in the integrated viral DNA.

are, therefore, considered to be **monoclonal.** Although infection is initiated with a nondefective retrovirus, the proviruses in the resulting tumor cells have usually lost some or most of the proviral sequences. Nevertheless, at least one long terminal repeat (LTR) containing the transcriptional control region is always present. Viral transcription signals, but not protein-coding sequences, are therefore required for transformation by nontransducing retroviruses. The significance of these properties became apparent when it was discovered that proviruses are integrated in several tumors in the vicinity of some of the same cellular oncogenes that are captured by transducing retroviruses. The study of nontransducing oncogenic retroviruses has also led to the identification of some additional proto-oncogenes, as illustrated for two murine retroviruses in Table 18.4. As integration of retroviral DNA into the host genome is essentially random, the long latency for tumor induction by the nontransducing viruses can be explained in part by the limited probability that integration will occur in the vicinity of an oncogene.

Integration of viral DNA sequences is not a prerequisite for successful propagation of any oncogenic DNA viruses. Nevertheless, integration is the rule in adenovirus- or polyomavirus-transformed cells, which contain specific subsets of viral genes inserted into the cellular genome. Such integration is the result of rare recombination reactions (catalyzed by cellular enzymes) between generally unrelated cellular and viral DNA sequences. Integration can therefore occur at many sites in cellular genomes and, in contrast to proviral DNA integration, does not maintain specific viral DNA sequences at the junctions with cellular DNA. The low probability that viral DNA will become integrated into the cellular genome and the fact that only a fraction of these recombination reactions will maintain the integrity of viral transforming genes are major factors contributing to the low efficiencies of transformation by these viruses. In the case of simian virus 40, one focus-forming unit (see Chapter 2) corresponds to 10^2 to 10^4 plaque-forming units, while the ratio is even lower for human adenoviruses.

The great majority of cells transformed by adenoviruses or polyomaviruses retain only partial copies of the viral genome. The genomic sequences integrated can vary considerably among independent lines of cells transformed by the same virus, but a common, minimal set of genes is always integrated. Identification of this minimal set provided one of the first ways in which the genes of DNA viruses needed to maintain the transformed state were identified. From 1 copy to 10 to 20 copies of viral DNA can be integrated per cell and there can be multiple integration sites. However, there is little indication that such variation is associated with phenotypic differences among the cells transformed by these viruses.

A second mechanism by which viral DNA can persist in transformed cells is as a stable, extrachromosomal episome (Box 1.4). Such episomal viral genomes are a characteristic feature of B cells immortalized by Epstein-Barr virus, and they can also be found in cells transformed by papillomaviruses. The viral episomes are maintained at concentrations of tens to hundreds of copies per cell, by both replication of the viral genome in concert with cellular DNA synthesis and orderly segregation of viral DNA to daughter cells (see Chapter 9). Because of these requirements, permanent alteration of cell growth properties depends on the viral proteins necessary for the survival of viral episomes, in addition to the viral gene products that directly modulate cell growth and proliferation (Table 16.4).

Table 18.4 Some previously unknown loci targeted in tumors induced by two nontransducing, simple retroviruses[a]

Virus	Neoplasia	Cellular gene	Normal function
Moloney murine leukemia virus	Pre-B lymphoma	*Ahi-1*	Adaptor protein
	T-cell lymphoma	*cyclin D2/Vin-1*	G_1 cyclin
		Lck	Nonreceptor protein tyrosine kinase
		Notch-1/Mis-6	Transmembrane receptor that functions in development
		Pim-1	Serine/threonine kinase
		Tpl-2	Serine/threonine kinase
	B-cell lymphoma	*Pim-2*	Serine/threonine kinase
Mouse mammary tumor virus	Mammary tumors	*Int-1 (wnt-1)*	Secreted glycoprotein important in pattern formation in early embryonic development
		Int-2 (hst)	Secreted protein that may act as a growth factor
		Int-3	Protein presumed to function in development (*notch* family)
		Int-H/Int-5	Converts androgens to estrogens

[a]Adapted from J. M. Coffin, S. H. Hughes, and H. E. Varmus (ed.), p. 482–484, *Retroviruses* (Cold Spring Harbor Laboratory Press, Cold Spring Harbor, N.Y., 1997), with permission.

Identification and Properties of Viral Transforming Genes

Transforming genes of oncogenic viruses have been identified by classical genetic methods, characterization of the viral genes present and expressed in transformed cell lines, and analysis of the transforming activity of viral DNA fragments directly introduced into cells (Box 18.5). For example, analysis of transformation by temperature-sensitive mutants of mouse polyomavirus established as early as 1965 that the viral early transcription unit is necessary and sufficient to initiate and maintain transformation. Of even greater value were mutants of retroviruses, in particular two mutants of Rous sarcoma virus isolated in the early 1970s. The genome of one mutant carried a spontaneous deletion of approximately 20% of the viral genome; the mutant could no longer transform the cells it infected, but it could still replicate. The second mutant was temperature sensitive for transformation, but the virus could replicate at both temperatures. These properties of the mutants therefore showed unequivocally that cellular transformation and viral replication are distinct processes. More importantly, the deletion mutant allowed the preparation of the first nucleic acid probe specific for a v-oncogene. Complementary (–) strand DNA was prepared by reverse transcription of the Rous sarcoma virus RNA genome and then hybridized to (+) strand genomic RNA of the mutant. The nonhybridizing DNA, separated from the double-stranded hybrids, comprised the viral oncogenic (*src*) information. This *src*-specific probe was found to hybridize to cellular DNA, providing the first conclusive evidence that v-oncogenes are of cellular and not viral origin. This finding, for which J. Michael Bishop and Harold Varmus received the 1989 Nobel Prize in physiology or medicine, had far-reaching significance because it immediately suggested that such cellular genes might become oncogenes by means other than viral transduction.

The presence of cellular oncogenes in their genomes turned out to be the definitive characteristic of transducing retroviruses, as illustrated in Fig. 18.7. As noted earlier, the acquisition of these cellular sequences is a very rare event. In addition, with the exception of Rous sarcoma virus, the transducing retroviruses are all replication defective, having lost all or most of the viral coding sequences during oncogene capture. Such defective transducing viruses can, however, be propagated in mixed infections with replication-competent "helper" viruses, which provide all the necessary virion proteins. Such mixed viral stocks have been the source of many transducing oncogenic viruses. The majority carry a single v-oncogene in their genomes, but some include more than one oncogene (e.g., *erb*A and *erb*B in avian erythroblastosis virus ES4). In such cases, one is sufficient for transformation; the second enhances oncogenicity.

Viral and cellular protein-coding sequences are fused in many v-oncogenes (Fig. 18.7). The presence of viral sequences can enhance the efficiency of translation of the oncogene mRNA, stabilize the protein, or determine its lo-

BOX 18.5

Multiple lines of evidence identified the transforming proteins of the polyomavirus simian virus 40

Early gene products are necessary and sufficient to initiate transformation.

1. Viruses carrying temperature-sensitive mutations in the early transcription unit (*ts*A mutants), but no other region of the genome, fail to transform at a nonpermissive temperature.
2. Simian virus 40 DNA fragments containing only the early transcription unit transform cells in culture; DNA fragments containing other regions of the genome exhibit no activity.

Early gene products are necessary to maintain expression of the transformed phenotype.

1. Many lines of cells transformed by simian virus 40 *ts*A mutants at a permissive temperature revert to a normal phenotype when shifted to a nonpermissive temperature.
2. Integration of viral DNA sequences disrupts the late region of the viral genome but not the early transcription unit, and early gene products are synthesized in all transformed cell lines.

Both LT and sT contribute to transformation.

1. Simian virus 40 mutants carrying deletions of sequences expressed only in sT fail to transform rat cells to anchorage-independent growth.
2. Introduction and expression of LT complementary DNA are sufficient for induction of transformation, but expression of sT can stimulate transformation (especially at low LT concentrations), is necessary for expression of specific phenotypes in specific cells, and is required for transformation of resting cells.

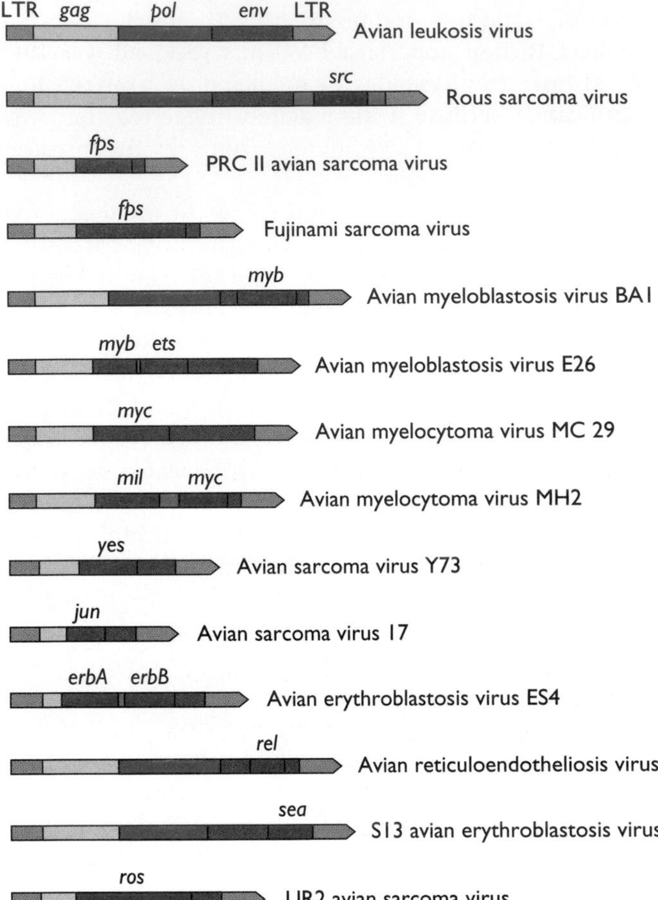

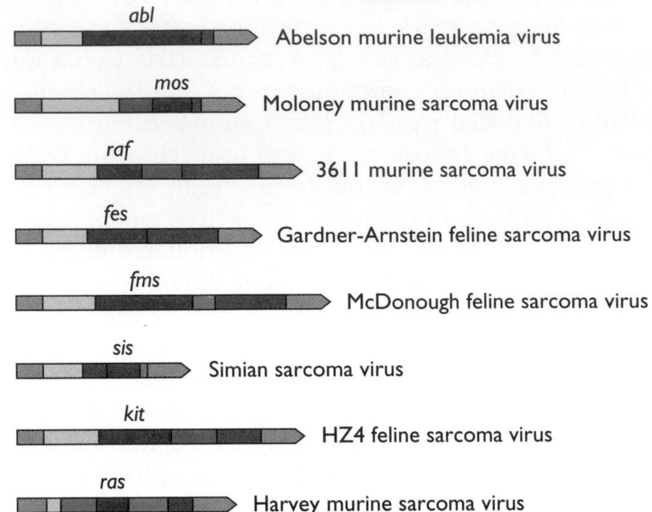

Figure 18.7 Genome maps of avian and mammalian transducing retroviruses. Avian leukosis virus (e.g., Rous-associated virus) is a prototypical retrovirus. Its genome contains the three major coding regions, *gag* (pink), *pol* (blue), and *env* (brown), and regulatory sequences that constitute the long terminal repeat (LTR) (lavender) of the provirus. In Rous sarcoma virus, the oncogene *src* is added to the complete viral genome. In all other avian and mammalian transducing retroviruses, some of the viral coding information is replaced by cell-derived oncogene sequences (red). Consequently, such transducing viruses are defective in replication. In some cases, additional cellular DNA sequences (green) were also captured in the viral genome. Adapted from T. Benjamin and P. Vogt, p. 317–367, *in* B. N. Fields et al. (ed.), *Fields Virology*, 2nd ed. (Raven Press, New York, N.Y., 1990), with permission.

cation in the cell. Unregulated expression or overexpression of the cellular sequence from the viral promoter is sufficient to cause transformation by some v-oncogenes (e.g., *myc* and *mos*). However, in most cases, the captured oncogenes have undergone additional changes that contribute to their transforming potential. These alterations, which include nucleotide changes, truncations at one or both ends, or other rearrangements, affect the normal function of the gene product.

DNA-mediated transformation allows assessment of the effects of individual viral genes (or DNA sequences encoding a single viral protein) on cell growth and proliferation. This method has been especially valuable in identification of transforming proteins encoded by DNA viruses. Transformation of primary cells by these viruses requires the products of two or more viral genes, as illustrated for adenoviruses, papillomaviruses, and polyomaviruses in Table 18.5. The majority of these genes exhibit some ability to alter the properties of the cells in which they are expressed in the absence of other viral proteins. However, some are required only for the induction of specific transformed phenotypes or only under certain conditions (e.g., simian virus 40 small T antigen [sT], bovine papillomavirus type 1 E7), and several exhibit no activity on their own (Table 18.5). A classic example of the latter phenomenon is provided by the adenoviral E1B gene: this gene, together with the E1A gene, was initially shown to be essential for transformation of rodent cells in culture, but it possesses no intrinsic ability to induce **any** transformed phenotype. This apparent paradox has been resolved with elucidation of the molecular functions of the viral gene products: E1A proteins induce apoptosis, but E1B proteins suppress this response. They therefore allow cells that synthesize E1A proteins to survive and express transformed phenotypes. A requirement for multiple viral genes is also the rule for transformation by the larger DNA viruses.

For some viruses, a complete protein-coding sequence is not necessary for oncogenesis or transformation: as we

Table 18.5 Tranforming gene products of adenoviruses, papillomaviruses, and polyomaviruses

Virus	Gene product(s)	Activities
Adenoviruses		
Human adenovirus type 2	E1A: 243R and 289R	Essential for transformation of established and primary rodent cells; cooperate with E1B products to transform primary cells; not sufficient for establishment of transformed cell lines
	E1B: 55kDa and 19kDa	Necessary for E1A-dependent transformation of primary and established cells; when expressed alone, cannot alter cell growth or proliferation; the two E1B proteins counter apoptosis, by different mechanisms
	E4: ORF6	Cooperates with E1A to transform primary cells; converts nontumorigenic E1A + E1B-transformed cells to cells tumorigenic in nude mice
	E4: ORF1	Transforms established cells; the adenovirus type 9 protein is necessary for induction of estrogen-dependent mammary tumors by this virus
Papillomaviruses		
Human papillomavirus types 16 and 18	E6	Required for efficient immortalization of primary human fibroblasts and keratinocytes
	E7	Transforms established rodent cells; cooperates with E6 to transform primary rodent cells; required for efficient immortalization of primary human fibroblasts or keratinocytes
	E5	Induces some transformed phenotypes in established cells; can increase the proliferative capacity of human keratinocytes
Bovine papillomavirus type 1	E5	Transforms bovine and rodent fibroblasts in culture; major viral protein synthesized in transformed cells
	E6	Required for expression of typical transformed phenotypes in primary mouse cells; can transform established cells
	E7	Required for expression of full transformed phenotype in primary mouse cells; no activity alone, but can cooperate with the E6 protein
Polyomaviruses		
Polyomavirus	LT	Immortalizes primary cells; required to induce but not to maintain transformation of primary cells
	mT	Transforms established cell lines; required both to induce and to maintain transformation of primary cells
	sT	See simian virus 40
Simian virus 40	LT	Immortalizes primary cells; required to induce and maintain transformation of primary and established cells; in some recipient cells, an N-terminal segment (amino acids 1–121) is sufficient, but in most cells, most LT sequences are required
	sT	Required under many conditions, depending on LT, genetic background of recipient cells, and transformation assay; necessary for transformation of resting cells

have seen, the minimal proviral sequences associated with oncogenesis by nontransducing retroviruses are those that comprise the transcriptional control sequences of the LTR.

The Origin and Nature of Viral Transforming Genes

Two classes of viral oncogenes can be distinguished on the basis of their similarity to cellular sequences. The oncogenes of transducing retroviruses and certain herpesviruses (e.g., human herpesvirus 8) are so closely related to cellular genes that it is clear that they were captured relatively recently from the genomes of cells in which these viruses replicate. Such acquisition of cellular genetic information must be a result of recombination between viral and cellular nucleic acids, a process that has been documented for transducing retroviruses. Retrovirus particles contain some cellular RNAs, and rare recombination reactions during reverse transcription can give rise to trans-

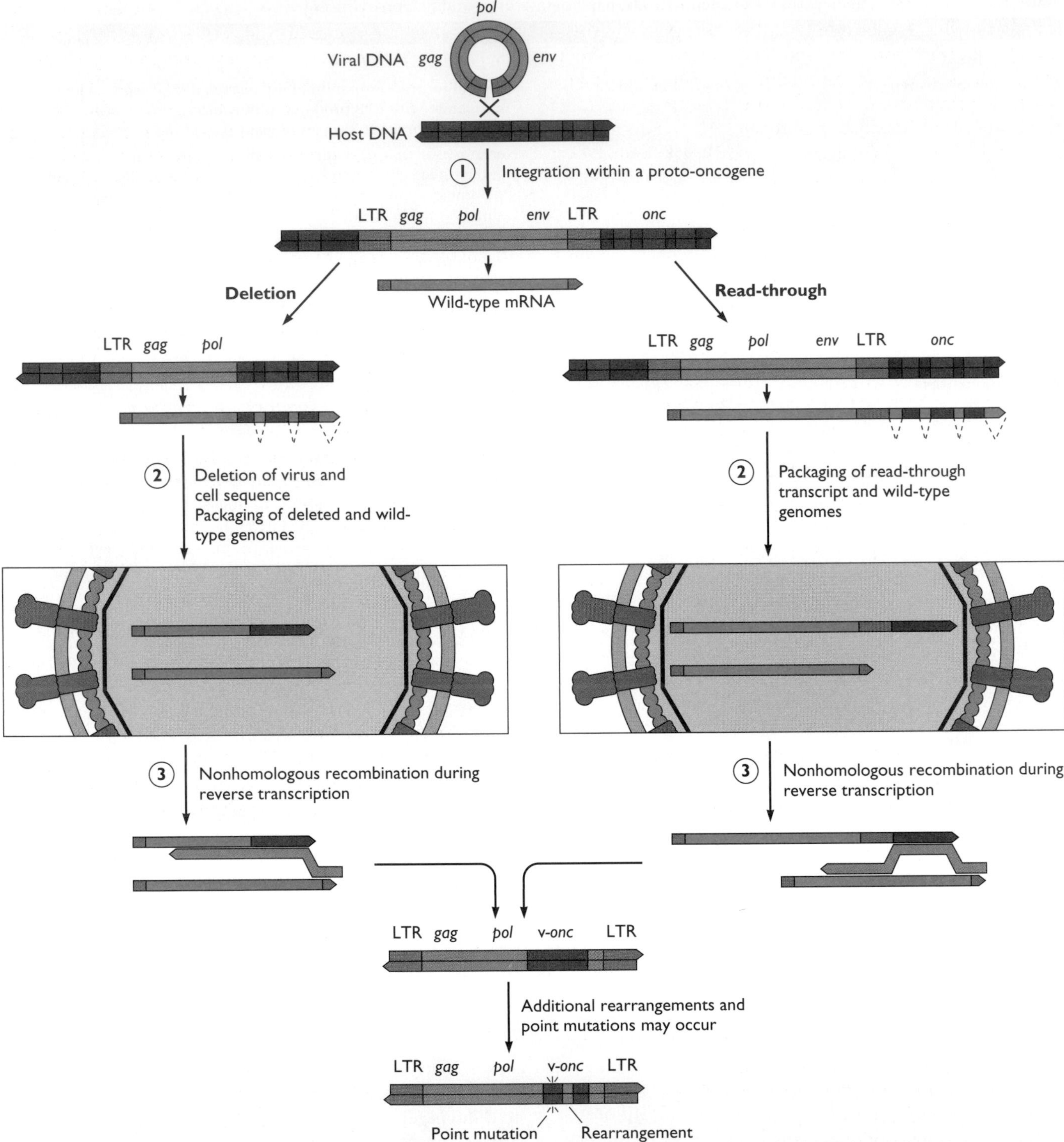

Figure 18.8 Possible mechanisms for oncogene capture by retroviruses. The first step in each of two mechanisms shown is integration of a provirus in or near a cellular gene. The deletion mechanism (left) requires removal of the right end of the provirus, thereby linking cellular sequences to the strong viral transcriptional control region in the left long terminal repeat (LTR). The first recombination step in this mechanism therefore takes place at the DNA level. It leads to synthesis of a chimeric RNA, in which viral sequences from the left end of the provirus are joined to cellular sequences. Should the chimeric RNA include the viral packaging

(continued on next page)

ducing retroviruses. The limiting factor appears to be the frequency with which cellular mRNA molecules are encapsidated into virions. Two mechanisms that can increase the likelihood of such encapsidation, and consequently increase the frequency of gene capture, have been described (Fig. 18.8). Both mechanisms depend on integration of a provirus in or near a cellular gene and final recombination reaction(s) between largely nonhomologous sequences. Model systems have provided experimental support for these more efficient mechanisms for producing transducing retroviruses, but it is still unclear exactly how these viruses arise in nature.

Many of the cellular proto-oncogenes from which v-oncogenes are derived have been highly conserved throughout evolution; many vertebrate examples have homologs in yeast. The products of such genes must therefore fulfill functions that are indispensable for a wide variety of eukaryotic cells. Furthermore, as single copies of v-oncogenes are sufficient to transform infected cells, their functions must override those of the resident, cognate proto-oncogenes. Accordingly, v-oncogenes function as **dominant** transforming genes.

Members of the second class of viral oncogenes are not obviously related to cellular genes. However, the products of these genes may contain short amino acid sequences also present in cellular proteins (for example, see Fig. 18.17). The precise origins of such oncogenes, like those of viruses themselves, therefore remain shrouded in mystery (see Chapter 20).

Functions of Viral Transforming Proteins

Many approaches have been employed to determine the functions of viral oncogene products. In some cases, the sequence of a viral transforming gene can immediately suggest the function of its protein product. For example, the genomes of certain herpes- and poxviruses include coding sequences that are closely related to cellular genes that encode growth factors, cytokines, and their receptors. Or the protein may contain predicted amino acid motifs characteristic of particular biochemical activities, such as tyrosine kinase or sequence-specific DNA binding. With some retroviral v-oncogene products, it has been possible to identify important biochemical properties, such as enzymatic activity, binding to a hormone or growth factor, or sequence-specific binding to nucleic acids. Table 18.6 includes a list of these retroviral v-oncogenes and the proteins they encode, organized according to the known or deduced functions of their normal cellular homologs.

The watershed in elucidation of the mechanisms of transformation by viral proteins encoded in the genomes of oncogenic DNA viruses came with mutational analyses that correlated their transforming activity with binding to specific cellular proteins. For example, two of the regions of adenoviral E1A proteins that are necessary for transformation mediate binding to the cellular retinoblastoma tumor suppressor protein, Rb. The similar relationship between Rb binding and the transforming activities of simian virus 40 large T antigen (LT) and the E7 protein of oncogenic human papillomaviruses rapidly established the general importance of interaction with Rb in transformation.

Viral transforming proteins exhibit great diversity in all their properties, from primary amino acid sequence to biochemical activity. They also differ in the number and nature of the cellular pathways they alter. Oncogenes transduced by retroviruses generally act on a limited number of cellular targets. In contrast, the transforming activities of the products of several oncogenic DNA viruses have been correlated with effects on multiple cellular pathways and proteins (Table 18.7). Despite such variation, the best-characterized mechanisms of transformation by viral proteins fall into one of two classes: permanent activation of cellular signal transduction cascades, or disruption of the circuits that regulate cell cycle progression.

Transformation by Activation of Cellular Signal Transduction Pathways

The products of transforming genes of both RNA and DNA viruses can alter cellular signal transduction cascades. The consequence is permanent activation of pathways that promote cell growth and proliferation. However, as dis-

Figure 18.8 *(continued)* signal, it could be incorporated efficiently into virions as a heterozygote with a wild-type genome produced from another provirus in the same cell. A second recombination reaction, during reverse transcription (as described in Chapter 7), is then required to add right-end viral sequences to the recombinant. At a minimum, these right-end sequences must include signals for subsequent integration of the recombinant viral DNA into the genome of the newly infected host cell, from which the transduced gene is then expressed. The read-through mechanism (right) does not require a chromosomal deletion. This hypothesis was suggested by the finding that viral transcription does not always terminate within the right LTR, with the result that cellular sequences downstream of an integrated provirus are occasionally incorporated into viral transcripts. Such chimeric transcripts can then be incorporated into virions together with the normal viral transcript. The cellular sequences can then be captured by recombination during reverse transcription, although a double crossover is required. In either case, important additional mutations and rearrangements probably occur during subsequent virus replication. Adapted from J. M. Coffin et al. (ed.), *Retroviruses* (Cold Spring Harbor Laboratory Press, Cold Spring Harbor, N.Y., 1997), with permission.

Table 18.6 Functional classes of oncogenes transduced by retroviruses[a]

Transduced oncogene	Viral oncoprotein[b]	Function of cellular homolog
Growth factors		
sis	p28$^{env\text{-}sis}$	Platelet-derived growth factor
Tyrosine kinase growth factor receptors		
*erb*B	gp65^{erbB}	Epithelial growth factor receptor
fms	gp180$^{gag\text{-}fms}$	Colony-stimulating factor 1 receptor
sea	gp160$^{env\text{-}sea}$	Receptor; ligand unknown
kit	gp80$^{gag\text{-}kit}$	Hematopoietic receptor; product of the mouse W locus
ros	p68$^{gag\text{-}ros}$	Receptor, ligand unknown
mpl	p31$^{env\text{-}mpl}$	Member of the hematopoietin receptor family
eyk	gp37^{eyk}	Receptor, ligand unknown
Hormone receptors		
*erb*A	p75$^{gag\text{-}erbA}$	Thyroid hormone receptor
G proteins		
H-*ras*	p21^{ras}	GTPase
K-*ras*	p21^{ras}	GTPase
Adaptor protein		
crk	p47$^{gag\text{-}crk}$	Signal transduction
Nonreceptor tyrosine kinases		
src	pp60^{src}	Signal transduction
abl	p460$^{gag\text{-}abl}$	Signal transduction
fps[c]	p130$^{gag\text{-}fps}$	Signal transduction
	p105$^{gag\text{-}fps}$	
fes[c]	p85$^{gag\text{-}fes}$	Signal transduction
fgr	p70$^{gag\text{-}actin\text{-}fgr}$	Signal transduction
yes	p90$^{gag\text{-}yes}$	Signal transduction
	p80$^{gag\text{-}yes}$	
Serine/threonine kinases		
mos	p37$^{env\text{-}mos}$	Required for germ cell maturation
raf[d]	p75$^{gag\text{-}raf}$	Signal transduction
mil[d]	p100$^{gag\text{-}mil}$	Signal transduction
akt	p86$^{gag\text{-}akt}$	Signal transduction
Nuclear proteins		
jun	p65$^{gag\text{-}jun}$	Transcriptional regulator (Ap-1 complex)
fos	p55^{fos}	Transcriptional regulator (Ap-1 complex)
myc	p100$^{gag\text{-}myc}$	
	p90$^{gag\text{-}myc}$	
	p200$^{gag\text{-}pol\text{-}myc}$	
	p59$^{gag\text{-}myc}$	
myb	p45^{myb}	Transcriptional regulator
	p135$^{gag\text{-}myb\text{-}ets}$	
ets	p135$^{gag\text{-}myb\text{-}ets}$	Transcriptional regulator
rel	p64^{rel}	Transcriptional regulator
maf	p100$^{gag\text{-}maf}$	Transcriptional regulator
ski	p110$^{gag\text{-}ski\text{-}pol}$	Transcriptional regulator
qin	p90$^{gag\text{-}qin}$	Transcriptional regulator of the forkhead/Hnk-3 family

[a]Adapted from J. Benjamin and P. Vogt, p. 331, *in* D. M. Knipe and P. M. Howley (ed.), *Fields Virology* (Lippincott-Raven Publishers, Philadelphia, Pa., 1996).

[b]Designations for viral proteins: p, protein; gp, glycoprotein; pp, phosphoprotein. The last is not applied consistently but is used mainly in conjuction with the *src* product. The numbers give the estimated molecular mass in kilodaltons, and the superscript lists the genes from which the coding information is derived in the 5′ → 3′ direction. The listing of more than one protein for an oncogene signifies its inclusion in independent virus isolates.

[c]*fps* and *fes* are the same oncogene derived from the avian and feline genomes, respectively.

[d]*raf* and *mil* are the same oncogene derived from the murine and avian genomes, respectively.

Table 18.7 Viral transforming proteins can bind to multiple cellular proteins

Viral protein	Cellular proteins that bind to sequences needed for transforming activity	Function of cellular protein
Adenovirus E1A	Rb family members (Rb, p107, and p130)	Nuclear phosphoproteins that regulate cell cycle progression, in part by binding E2f family members and repressing transcription; E1A proteins disrupt such Rb-E2f complexes and sequester Rb
	p300/Cbp	Nuclear coactivators required for stimulation of transcription by many sequence-specific activators, including p53; E1A protein binding inhibits their activity
	pCaf	Histone acetyltransferase that acetylates nucleosomal histones and can stimulate transcription; E1A binding via CR1 residues inhibits the latter activity
	$p27^{Kip1}$	Nuclear inhibitor of G_1/S phase cyclin-Cdks; E1A binding inhibits cell cycle arrest induced by the protein
	Ubc9	Ubiquitin-conjugating enzyme
Human papillomavirus type 16 or 18 E6	p53	Nuclear transcriptional regulator that induces G_1 arrest or apoptosis in response to DNA damage or stress; E6 binding stimulates ubiquitination and degradation
	p300/Cbp	See above
	Paxillin	Cytoplasmic protein that participates in signal transduction from focal adhesions to the cytoskeleton; E6 binding implicated in transformation
	Interferon regulatory factor 3 (Irf3)	Transcriptional regulator activated in response to interferon or virus infection; E6 binding inhibits induction of Irf3 mRNA

cussed in subsequent sections, these viral proteins can intervene at various points in the cascades that transmit signals from the exterior of the cell to nuclear components, and they operate in several different ways.

Viral Mimics of Cellular Signaling Molecules

The Transduced Cellular Genes of Acutely Transforming Retroviruses

The v-*src* paradigm. The protein product of v-*src* was the first retroviral transforming protein to be identified, when serum from rabbits bearing tumors induced by Rous sarcoma virus was shown to immunoprecipitate a 60-kDa phosphoprotein ($pp60^{src}$) (Table 18.6). The v-Src protein was soon found to possess protein tyrosine kinase activity, a property that provided the first clue that specific phosphorylation of cellular proteins is critical to oncogenesis. The discovery of this protein tyrosine kinase led to the identification of a large number of other proteins with similar enzymatic activity and important roles in cellular signaling. The Src protein contains a tyrosine kinase domain (SH1, for Src homology region 1) and two domains that mediate protein-protein interactions (Fig. 18.9). The first, the SH2 domain, binds to phosphotyrosine-containing sequences, whereas the SH3 domain has affinity for proline-rich sequences. Both domains are found in other proteins that participate in signal transduction pathways. One transforming protein, Crk, is made up of only an SH2 and an SH3 domain: it functions as an **adapter,** bringing other proteins in the signal transduction pathway together (Table 18.6). A fourth Src domain (SH4) includes the N-terminal myristoylation signal that directs Src to the plasma membrane. All four domains are required for transforming activity.

The Src kinase phosphorylates itself at specific tyrosine residues; these modifications regulate its enzymatic activity. For example, phosphorylation of Y416 in the kinase domain activates the enzyme, whereas phosphorylation of Y527 in the C-terminal segment inhibits activity. The crystal structures of cellular Src and another member of the Src family revealed the importance of the SH2 and SH3 domains in such regulation. As illustrated in Fig. 18.9, the phosphorylated C-terminal segment of the cellular Src protein is bound by the SH2 domain. This interaction brings a polyproline helix located between the SH2 and kinase domains into proximity to SH3. Binding of SH3 to the helix deforms the kinase domain, accounting for the inactivity of the Y527-phosphorylated form of the protein. Exchange of these intramolecular associations for binding of the SH2 or SH3 domains to phosphotyrosine- or polyproline-containing motifs, respectively, in other proteins initiates conformational changes that activate the kinase. This autoregulatory mechanism explains earlier findings that transduction and overproduction of the normal Src protein do not lead to cellular transformation, and that the oncogenic activity of v-*src* requires loss or mutation of the Y527 codon. Consequently, the v-Src protein is constitutively active.

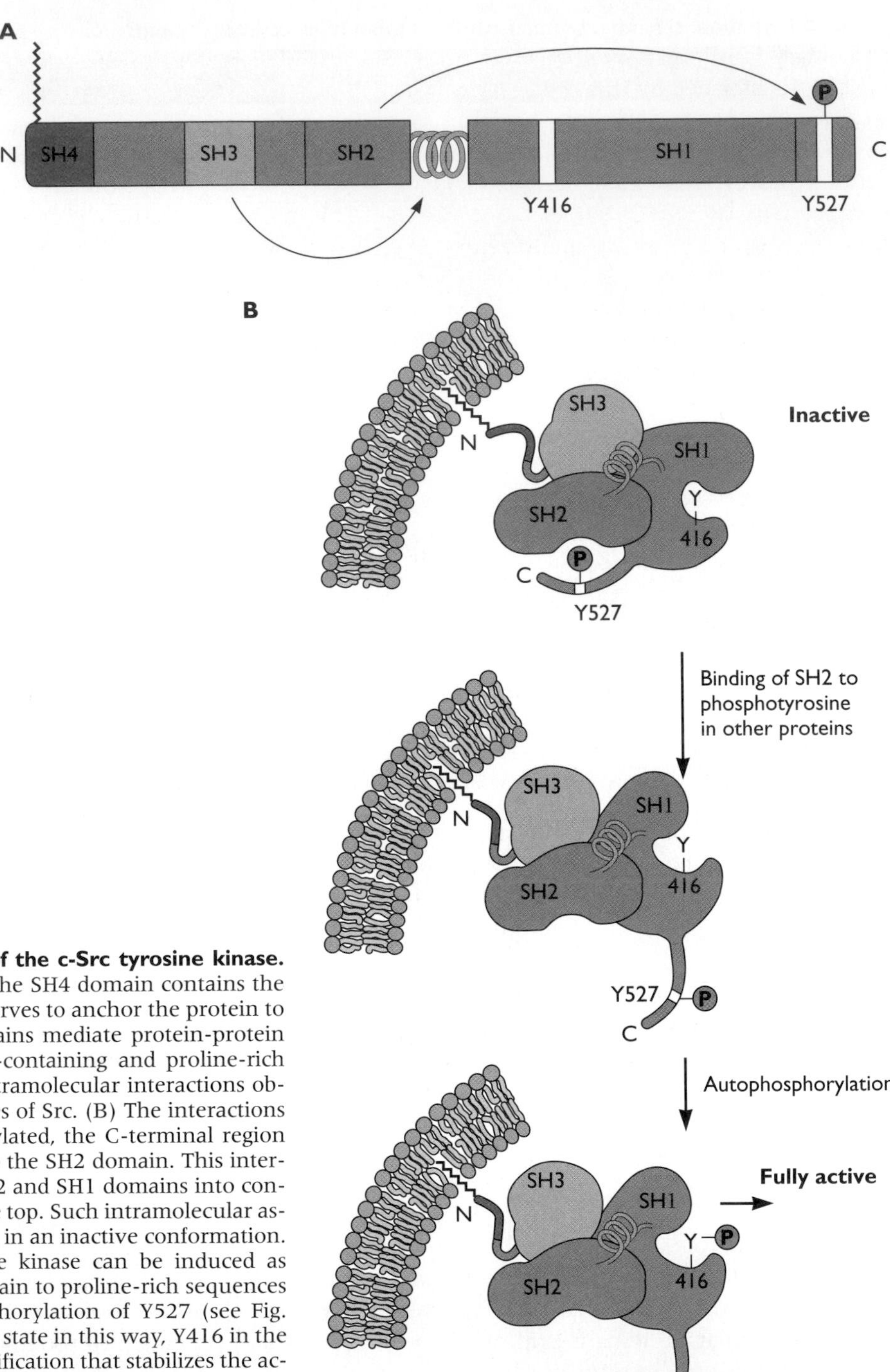

Figure 18.9 Organization and regulation of the c-Src tyrosine kinase. (A) The functional domains of the protein. The SH4 domain contains the site for addition of the myristate chain that serves to anchor the protein to the cell membrane. The SH2 and SH3 domains mediate protein-protein interactions by binding to phosphotyrosine-containing and proline-rich sequences, respectively. Arrows represent intramolecular interactions observed in the repressed-state crystal structures of Src. (B) The interactions and their reversal. When Y527 is phosphorylated, the C-terminal region of c-Src in which this residue lies is bound to the SH2 domain. This interaction brings a helix located between the SH2 and SH1 domains into contact with the SH3 domain, as illustrated at the top. Such intramolecular associations maintain the kinase domain (SH1) in an inactive conformation. A conformational change that activates the kinase can be induced as shown, as well as by binding of the SH3 domain to proline-rich sequences in other proteins and probably by dephosphorylation of Y527 (see Fig. 18.14). Once released from the autoinhibited state in this way, Y416 in the kinase domain is autophosphorylated, a modification that stabilizes the active conformation of the SH1 domain. The v-Src protein is not subject to such autoinhibition, because the sequence encoding the C-terminal regulatory region of c-Src was deleted during transduction of the cellular gene.

Soon after its kinase activity was first discovered, v-Src was shown to localize to focal adhesions, the areas where cells make contact with the extracellular matrix. This observation led to identification of a second, nonreceptor protein tyrosine kinase enriched in these areas as a protein exhibiting increased tyrosine phosphorylation in v-Src-transformed cells. This focal adhesion kinase (Fak) and Src family proteins turned out to be crucial components of a signal transduction cascade normally controlled by cell adhesion. This pathway modulates the properties of the actin cytoskeleton, and hence cell shape and adhesion, and also signals to the Ras-Map kinase pathway that controls cell proliferation (Fig. 18.3). Its constitutive activity in cells containing v-Src can therefore account for the properties of cells transformed by this oncogene product.

Other transduced oncogenes. The transduced oncogenes of retroviruses are homologs of cellular genes that encode many components of signal transduction cascades, from the external signaling molecules (e.g., v-Sis) and their receptors (v-ErbB and v-Kit) to the nuclear proteins at the end of the relay (v-Fos and v-Myc) (Fig. 18.3 and Table 18.6). It therefore seems likely that any positively acting protein in such a cascade has the potential to act as an oncoprotein. The oncogenic potential of such transduced oncogenes is realized by two nonexclusive mechanisms—genetic alterations, which lead to constitutive protein activity and inappropriate production, or overproduction of the protein. The former mechanism applies to most of the retroviral oncogenes listed in Table 18.6. For example, like other small guanine nucleotide-binding proteins, Ras normally cycles between active, GTP-bound and inactive, GDP-bound conformations. Such cycling is under the control of GTPase-activating and guanine nucleotide exchange proteins. The latter proteins (e.g., Sos) (Fig. 18.3) stimulate the release of GDP once bound GTP has been hydrolyzed. However, v-Ras proteins fail to hydrolyze GTP efficiently and therefore persist in the active, GTP-bound conformation that relays signals to downstream pathways, such as the Map kinase cascade (Fig. 18.3). Such constitutive activity is the result of mutations that lead to substitution of specific, single amino acids in the protein (at residues 12, 13, or 61) and render the protein refractory to the GTPase-activating protein. Analogous mutations are common in certain human tumors, such as colorectal cancers (Box 18.2), and were the first discrete genetic changes in a proto-oncogene linked to neoplastic disease in humans.

Less commonly, over- or misexpression of the transduced oncogene is sufficient to disrupt normal cell behavior. This type of mechanism is best characterized for *myc*. In normal cells, the expression of this gene is tightly regulated, such that the Myc protein is made only during a short G_1 period in the cell cycle and is not synthesized when cells withdraw from the cycle or differentiate. The production of even small quantities of Myc or Myc-fusion proteins specified by retroviruses, such as avian myelocytoma virus MH2, at an inappropriate time results in cellular transformation.

Other Viral Homologs of Cellular Genes

The genomes of some larger DNA viruses also contain coding sequences that are clearly related to cellular genes encoding signal transduction molecules. Human herpesvirus 8, a gammaherpesvirus related to Epstein-Barr virus, has been strongly implicated in the etiology of Kaposi's sarcoma, a malignancy common in AIDS patients. Its structural proteins and viral enzymes are closely related to those of other herpesviruses. The genome also contains several homologs of cellular genes that encode signaling proteins, many of which are clustered in a single region (Fig. 18.10). Among the best characterized is the gene

Figure 18.10 Some genes of human herpesvirus 8 and herpesvirus saimiri that are homologous to cellular genes. The two viral genomes are shown in orientations that align genes conserved among herpesviruses, the core gene blocks shown at the top. These conserved genes encode proteins needed for virus reproduction and assembly. These two herpesviral genomes, but not that of closely related Epstein-Barr virus, carry homologs of cellular genes (arrowheads) interspersed among the core gene blocks. Those shown in purple are related to cellular chemokines (v-IL-6, v-IL-17, and macrophage inflammatory factor [v-Mip] 1α or 1β), chemokine receptors (v-Gpcr; see text), or other signaling molecules (interferon-responsive protein [v-Irf] and an N-Cam family transmembrane protein that participates in intercellular signaling [v-Ox-2]). The human herpesvirus 8 v-IL-8 protein blocks the action of interferon (Table 15.8) and can also induce proliferation of B cells. Although this viral gene product may therefore contribute to hematopoietic tumors induced by the virus, it is not consistently made in Kaposi's sarcomas (endothelial cell tumors). Viral genes shown in red are related to cellular genes that encode proteins that regulate cell proliferation or apoptosis, cyclin D [v-cyclin; see text] and Bcl-2 [v-Bcl-2].

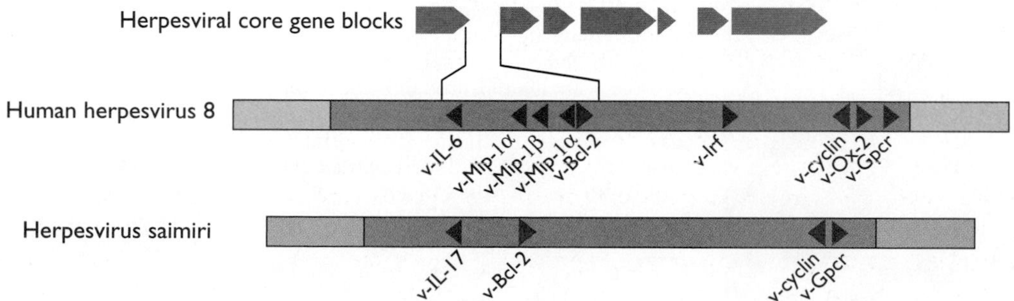

(v-*gpcr*) specifying a guanine nucleotide-binding, protein-coupled receptor that is most closely related to a cellular receptor for CXC chemokines. Members of this family of receptors bind chemokines released at sites of inflammation to activate signal transduction. In contrast, v-Gpcr is fully active in the absence of any ligand, and can trigger specific cellular signal transduction pathways, for example, a Map kinase cascade. When introduced into mouse NIH 3T3 cells in culture, the v-*gpcr* gene induces morphological transformation and secretion of vascular endothelial growth factor. A characteristic feature of Kaposi's sarcoma is extensive **angiogenesis,** the proliferation of new blood vessels, which is essential for tumor progression. Angiogenesis appears to be the result of the production of large quantities of vascular endothelial growth factor by spindle cells that proliferate in the tumor. These properties suggest that the v-*gpcr* gene plays a critical role in human herpesvirus 8 oncogenesis.

The functions of v-*gpcr* or any other homologs of cellular genes encoding signal transduction proteins (Fig. 18.10) in the viral infectious cycle have not yet been examined, for de novo infection of human cells in culture by this virus was reported only recently. For example, infection of normal endothelial cells by a primary isolate of human herpesvirus 8 was observed to induce formation of colonies of proliferating, spindle-shaped cells. Such colonies can be expanded into cultures consisting entirely of morphologically altered cells. Furthermore, the initial infection was propagated efficiently by passage of spindle cells with a large excess of uninfected cells, all of which acquired the spindle-shaped morphology. All such cells contained the viral protein (latent nuclear antigen) that is a characteristic marker of spindle cells present in late-stage Kaposi's sarcomas. This type of culture system should permit genetic analysis of persistent infection by human herpesvirus 8, and the mechanisms by which viral gene products modulate cell proliferation.

Viral homologs of cellular receptors and the ligands that bind to them are also encoded in the genomes of poxviruses. For example, vaccinia virus growth factor is a functional analog of cellular epidermal growth factor, and can bind to the appropriate receptor to activate signal transduction pathways that result in anchorage-independent growth. Closely related genes are present in the genomes of other poxviruses, including variola virus and Shope fibroma virus. Mutant vaccinia viruses lacking any growth factor gene fail to induce proliferation of ectodermal and endodermal cells of the chick chorioallantoic membrane, reproduce less efficiently, and are less pathogenic than their wild-type parents. There can therefore be little doubt that the poxviral homologs of cellular epidermal growth factor are important in the hyperplasia or development of tumors associated with infection by members of this family.

BOX 18.6

Inadvertent insertional activation of a cellular gene during gene therapy

Retroviruses have long been considered likely to be valuable vectors for gene therapy. One reason is that integration of the retroviral vector into the host genome results in permanent delivery of the therapeutic gene to all infected cells and their descendents. However, an outcome recently detected in one clinical trial indicates that this property is a double-edged sword.

A French trial was examining the potential of gene therapy using a vector based on mouse Moloney leukemia virus to treat children with severe combined immunodeficiency (SCID). This disease is caused by mutation in a single gene on the X chromosome, and the only therapies available are associated with severe, often fatal side effects. A trial with 10 children with the disease, who were identified and given gene therapy as early as possible, initially appeared to be very successful: in most cases, the immune system was restored without side effects. However, early in 2002, one patient was found to have developed a T-cell leukemia-like disease. The overproliferating T cells were monoclonal: all carried a provirus integrated into the same site on chromosome 11, in a gene (*lmo2*) that is abnormally expressed in a form of childhood acute lymphoblastic leukemia.

The monoclonal origin of the T cells that proliferated in this child indicates that proviral insertion contributed to the development of the disease. It initially seemed likely that other factors also did so: a predisposition to childhood cancers was evident in other members of the child's family. Sadly, however, in December 2002, a second child participating in the same trial was diagnosed with leukemia associated with insertion of the provirus in the same chromosome 11 site. Integration within the vicinity of the *lmo2* gene has also been detected in a third patient in the trial, although this child has not developed leukemia.

This outcome halted this and numerous other clinical trials of gene therapies using retroviral vectors in the United States and Europe. Trials for other conditions seem likely to resume once further safety reviews have been completed. However, the National Institutes of Health recently recommended that retroviral therapy for SCID be used only as a last resort.

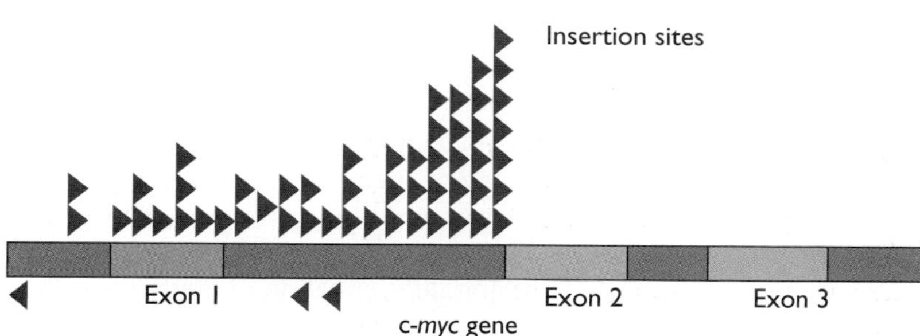

Figure 18.11 Insertional activation of c-*myc* by avian leukosis viruses. In avian cells derived from avian leukosis virus-induced B-cell lymphomas, individual proviral integration sites are clustered as shown (arrowheads) within noncoding exon 1 and intron 1 of the *myc* gene. Most integrated proviruses are oriented in the direction of *myc* transcription (arrowheads pointing to the right). Adapted from J. Nevins and P. Vogt, p. 301–343, *in* B. N. Fields et al. (ed.), *Fields Virology* (Lippincott-Raven Publishers, Philadelphia, Pa., 1996), with permission.

Alteration of the Expression or Activity of Cellular Signal Transduction Proteins

Insertional Activation by Nontransducing Retroviruses

Most tumors induced by nontransducing retroviruses arise as a result of increased transcription of cellular genes located in the vicinity of integrated proviruses. This mechanism of oncogenesis is known as **insertional activation**. It has recently been implicated in a leukemia-like disease developed by patients participating in a gene therapy trial (Box 18.6). As in the case of the transducing retroviruses, Rous sarcoma-derived avian viruses played a seminal role in delineating the mechanisms of insertional activation. The original stocks of viruses isolated by Rous included replication-competent leukosis viruses, called Rous-associated viruses 1 and 2. These viruses do not carry an oncogene, but in young chickens they induce B-cell tumors that originate in the bursa of Fabricius, the major lymphoid organ of these birds. A provirus was found integrated in the vicinity of the cellular *myc* gene in each of these tumors. Although the exact integration site varies from tumor to tumor, many integration sites lie in the intron between exon 1 (a noncoding exon) and exon 2 (Fig. 18.11). However, in some tumors proviruses are located upstream or downstream of the cellular *myc* gene. In this avian system, inappropriate synthesis of the cellular Myc protein is associated with lymphomagenesis; no changes in the protein are required. Analysis of the sites of proviral DNA integration and the gene products formed in these tumors provided the first evidence for two types of insertional activation, promoter insertion and enhancer insertion (Fig. 18.12).

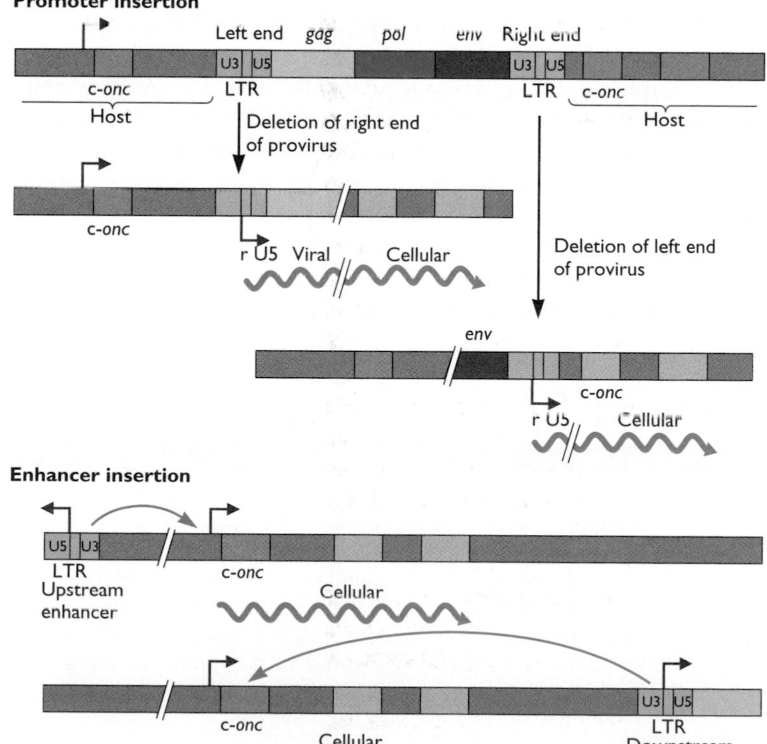

Figure 18.12 Mechanisms for insertional activation by nontransducing oncogenic retroviruses.

The first mechanism, **promoter insertion,** results in production of a chimeric RNA in which sequences transcribed from the proviral LTR are linked to cellular proto-oncogene sequences. If transcription originates from the left-end LTR, some viral coding sequences may be included. However, transcription from the right-end LTR seems to be more common, and in these cases the proviral left-end LTR has usually been deleted. Such deletion is probably important, because transcriptional read-through from the left-end promoter reduces transcription from the right-end LTR. Proviral integration often occurs within the cellular proto-oncogene, truncating cellular coding sequences and eliminating noncoding domains that may include negative regulatory sequences. Some chimeric transcripts formed by promoter insertion are analogous to the intermediates required for oncogene capture by the transducing retroviruses (compare Fig. 18.12 and 18.8). Indeed, it has been possible to isolate newly generated, oncogene-transducing retroviruses from tumors arising as a result of promoter insertion.

In the second type of insertional activation, **enhancer insertion,** viral and cellular transcripts are not fused. Instead, activation of the cellular gene is mediated by the strong viral enhancers, which increase transcription from the cellular promoter (Fig. 18.12). Because enhancer activity is independent of orientation and can be exerted over long distances, the provirus need not be oriented in the same direction as the proto-oncogene and may even lie downstream of it.

Viral Proteins That Alter Cellular Signaling Pathways

Some viruses alter the growth and proliferation of infected cells by the action of viral signal transduction proteins that are not obviously related in sequence to cellular proteins. Some of these viral proteins operate by mechanisms well established in studies of cellular components of signaling cascades, but others appear to modulate such cascades in virus-specific ways.

Constitutively active viral "receptors." The genomes of several gammaherpesviruses encode membrane proteins that initiate signal transduction (Table 18.8). The best-understood example of this mechanism is provided by Epstein-Barr virus latent membrane protein 1 (LMP-1), one of several viral gene products implicated in immortalization of human B lymphocytes (Table 16.4). This viral protein can also inhibit differentiation of epithelial cells in culture, and induce typical transformed phenotypes in established lines of rodent fibroblasts. LMP-1, which is an integral protein of the plasma membrane, functions as a constitutively active receptor. In the absence of any ligand, LMP-1 oligomerizes to form patches in the cellular membrane and activates the cellular transcriptional regulator Nf-κb. The viral protein binds to the same intracellular proteins as the active, ligand-bound form of members of the tumor necrosis factor receptor family (Fig. 18.13). When localized to the plasma membrane, the C-terminal segment of LMP-1 to which these proteins bind is sufficient for both immortalization of B cells and activation of cellular transcriptional regulators. It is therefore believed that LMP-1 activates the kinase cascade that normally releases Nf-κb from association with cytoplasmic inhibitors (Fig. 18.13).

When LMP-1 is made in the absence of other Epstein-Barr virus proteins, its constitutive signaling leads to many of the alterations in both cellular properties and cellular gene expression typically observed when primary B cells are infected with, and immortalized by, the virus. These changes include increased production of certain cell adhesion molecules and consequently increased cell adhesion and clumping. In infected cells, the EBNA-2 protein, which is a transcriptional activator, ensures efficient production of LMP-1. The EBNA-2 protein also stimulates transcription of several cellular genes that encode proteins that might be expected to influence cell growth, such as the Fgr tyrosine kinase (Table 18.6). A second viral membrane protein, LMP-2A, increases the stability of LMP-1 and therefore enhances its signaling.

Table 18.8 Constitutively active membrane receptors of gammaherpesviruses

Virus	Membrane protein	Mechanism and function
Epstein-Barr virus	LMP-1	Binds to cellular proteins that participate in signaling by tumor necrosis family members to activate Nf-κb and Jnk
Herpesvirus saimiri	Saimiri transformation proteins (STPs)	Bind to the same cellular proteins as LMP-1 to activate Nf-κb; required for induction of lymphomas (but not for replication) in marmosets and for transformation of rat cells in culture
Human herpesvirus 8	K1	Interacts with cellular signal transduction proteins, e.g., Vav and Syk kinase, via a motif present in cellular immune receptors to activate the transcriptional regulator for Nfat; when overproduced, can transform rat fibroblasts

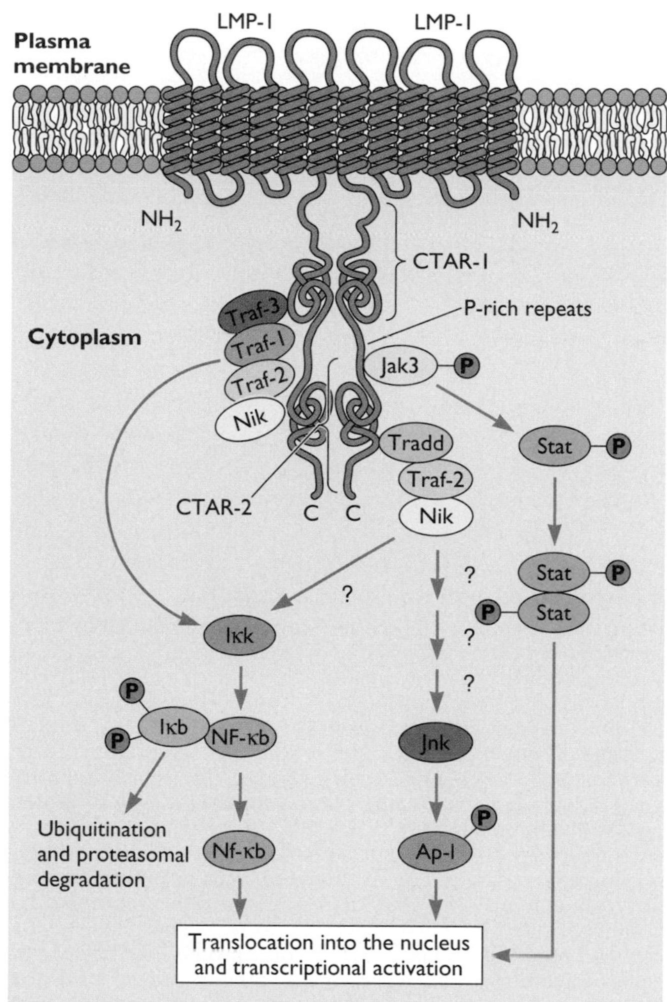

Figure 18.13 Constitutive signaling by Epstein-Barr virus latent membrane protein 1 (LMP-1). LMP-1, which possesses six membrane-spanning segments but no large extracellular domain, oligomerizes in the absence of ligand, a property represented by the LMP-1 dimer depicted. The long cytoplasmic C-terminal domain of the viral protein contains three segments implicated in activation of signaling, designated C-terminal activation regions (CTAR) 1 and 2, and proline (P)-rich repeats. As shown, multiple members of the tumor necrosis factor receptor-associated protein family (Trafs) bind to CTAR-1. Binding of Traf-2 leads to activation of Nf-κb, through the signaling relay that includes Nf-κb-inducing kinase (Nik) and Iκb kinase (Iκk), and induces release of Nf-κb from association with the cytoplasmic inhibitors. The same pathway is activated in uninfected cells by binding of tumor necrosis factor to its receptor. The CTAR-2 domain of LMP-1 is responsible for activation of Ap-1 via the Jun N-terminal kinase (Jnk) pathway. The first reaction appears to be indirect association of this region of LMP-1 with Traf-2 via a second cellular protein (Tradd), a reaction that may lead to activation of both Nf-κb and Ap-1, as shown. The latter effect is not essential for the ability of LMP-1 to alter the proliferation of B cells, but it may play an augmenting role. Binding of Janus kinase 3 (Jak3) to LMP-1 at a sequence within the P-rich repeat region induces tyrosine autophosphorylation of Jak3 and consequently activation of Stat transcriptional regulators. Whether activation of this pathway contributes to the transforming activity of LMP-1 is not known.

Viral adapter proteins. Members of both the *Polyomaviridae* and the *Herpesviridae* encode proteins that permanently activate cellular signal transduction pathways as a result of binding to Src family tyrosine kinases (Table 18.9). This mechanism was first encountered in studies of the mouse polyomavirus middle T protein (mT), a viral early gene product with no counterpart in the genome of the related polyomavirus simian virus 40 (Fig. 18.14A). This protein can transform established rodent cell lines (Table 18.5) and induce endotheliomas (Box 18.1) when overexpressed in transgenic animals.

mT antigen is located in the plasma membrane of transformed or infected cells, where it associates with Src or the related tyrosine kinases, Yes and Fyn. No increase in Src concentration is observed in mT-transformed cells, but the catalytic activity of Src bound to mT is increased by an order of magnitude. This change in tyrosine kinase activity is the result of inhibition of the autoregulatory mechanism, in which binding of the SH2 domain of Src to phos-

Table 18.9 Viral proteins that bind to Src family tyrosine kinases

Viral protein	Cellular tyrosine kinase(s) bound	Consequences of interaction
Epstein-Barr virus LMP-2	Fyn Lym	Inhibition of signal transduction via the cellular kinase; may prevent B-cell activation and the transition from latent to productive infection
Herpesvirus saimiri 484 tyrosine kinase-interacting protein (Tip)	Lck	Activation of the cellular kinase and signal transduction; transformation of human and simian T lymphocytes to interleukin-2-independent growth; Tip required for viral transformation of primary T cells in culture and oncogenic disease in vivo
Mouse polyomavirus mT	Src Fyn Yes	Activation of mT-bound kinase; tyrosine phosphorylation of mT and adaptor proteins bound to it; transformation of established cell lines

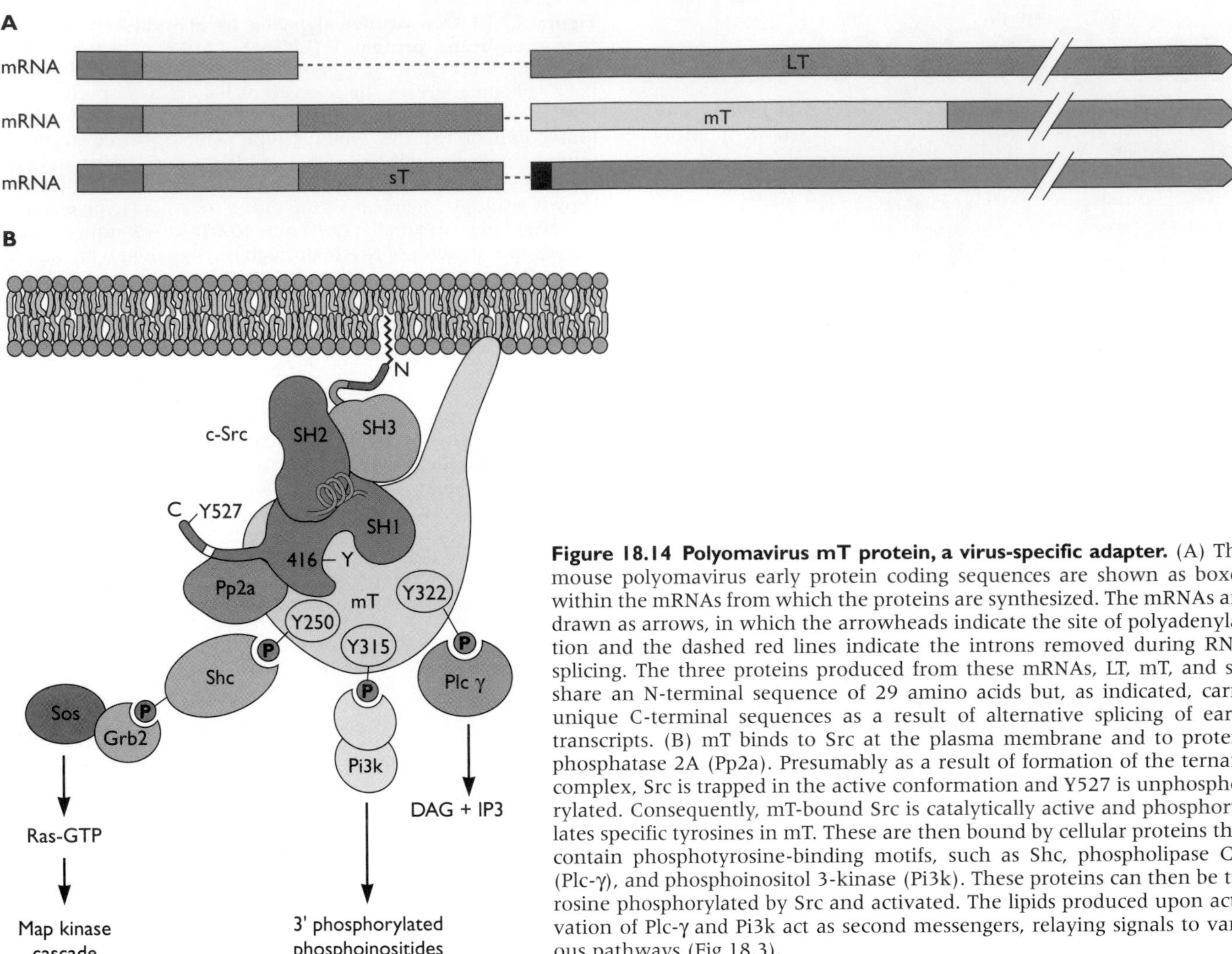

Figure 18.14 Polyomavirus mT protein, a virus-specific adapter. (A) The mouse polyomavirus early protein coding sequences are shown as boxes within the mRNAs from which the proteins are synthesized. The mRNAs are drawn as arrows, in which the arrowheads indicate the site of polyadenylation and the dashed red lines indicate the introns removed during RNA splicing. The three proteins produced from these mRNAs, LT, mT, and sT, share an N-terminal sequence of 29 amino acids but, as indicated, carry unique C-terminal sequences as a result of alternative splicing of early transcripts. (B) mT binds to Src at the plasma membrane and to protein phosphatase 2A (Pp2a). Presumably as a result of formation of the ternary complex, Src is trapped in the active conformation and Y527 is unphosphorylated. Consequently, mT-bound Src is catalytically active and phosphorylates specific tyrosines in mT. These are then bound by cellular proteins that contain phosphotyrosine-binding motifs, such as Shc, phospholipase C-γ (Plc-γ), and phosphoinositol 3-kinase (Pi3k). These proteins can then be tyrosine phosphorylated by Src and activated. The lipids produced upon activation of Plc-γ and Pi3k act as second messengers, relaying signals to various pathways (Fig 18.3).

phorylated Y527 inhibits kinase activity (Fig. 18.9). mT antigen appears to be able to sequester the Y527-containing segment of Src and induce its dephosphorylation (see below) to stabilize the active conformation of the kinase. However, it is unlikely that mT can disrupt preformed interaction of these Src domains, for only about 10% of the Src molecules, and a similar fraction of mT molecules, are bound to each other in transformed or infected cells. Such nonstoichiometric association suggests that mT can bind to Src molecules only during their transient activation by normal cellular mechanisms. While it remains a mystery why so little of the mT protein is associated with cellular tyrosine kinases, the permanent alteration of cell proliferation induced via such a small fraction of these cellular enzymes is remarkable.

The mT protein also binds to cellular protein phosphatase 2A via an N-terminal sequence that is also present in sT (Fig. 18.14A). Binding of mT to this cellular enzyme appears to be necessary for association of the viral protein with Src. This requirement would ensure that the phosphatase would be brought into close physical association with Src, and accounts for the fact that Y527 is dephosphorylated in mT-bound Src molecules. As we have seen in the context of the v-Src protein, the absence of phosphotyrosine at this position results in activation of the kinase.

It was initially surprising that mT-transformed cells do not contain elevated levels of phosphotyrosine, despite activation of Src family kinases. It is now clear that mT itself is a critical substrate of the cellular enzyme: phosphorylation of specific mT tyrosine residues by activated Src allows a number of cellular proteins that contain phosphotyrosine-binding domains to bind to mT (Fig. 18.14B). These proteins include the adapter Shc and phosphoinositol 3-kinase, an enzyme responsible for production of lipid second messengers (Fig. 18.3). In all cases, substitutions that disrupt binding of the cellular protein to mT impair the transforming activity of the viral protein. When bound to mT, such signaling proteins are tyrosine-phosphorylated by the activated Src kinase to trigger signal transduction, as illustrated for the activation of Ras and the Map kinase pathway in Fig. 18.14B. Consequently, mT both bypasses the normal mechanism by which the kinase activity of Src is regulated and also serves as a virus-specific adapter, bringing together cellular signal transduction proteins when they would not normally be associated.

The X protein of hepatitis B virus, which may contribute to the development of hepatocellular carcinoma, also activates Src and some other members of this family of tyrosine kinases. Production of the X protein results in prolonged enzyme activity and activation of the Map kinase pathway (Fig. 18.3). Whether the viral X protein, like polyomaviral mT, binds directly to Src to trigger this signal transduction cascade remains to be seen.

Alteration of the Activities of Cellular Signal Transduction Molecules

Alteration of cellular plasma membrane receptors. Many signal transduction cascades are initiated by binding of external growth factors to the extracellular portions of cell surface receptor tyrosine kinases. Such ligand binding induces oligomerization of receptor molecules and autophosphorylation of specific tyrosine residues in their intracellular domains (Fig. 18.3). Ligand-bound receptors are internalized rapidly (within 10 to 15 min) by endocytosis. Following acidification of the endosomes, the ligand is released and all but a small fraction of the receptor molecules are usually degraded. As a result, the initial signal is short-lived. The E5 protein of bovine papillomaviruses interferes with the normal mechanisms that control the function of this class of receptors.

The bovine papillomavirus type 1 E5 protein, a hydrophobic protein of only 44 amino acids, is the major transforming protein of this virus (Table 18.5). Transformation of fibroblasts by the E5 protein is in part the result of its binding to platelet-derived growth factor receptor. Because the E5 protein is a dimer, it induces ligand-independent dimerization of the receptor and hence activation of its tyrosine kinase, and the downstream signaling relay (Fig. 18.3). Synthesis of the E5 protein also increases the activity of phosphoinositol-3 kinase (Fig. 18.3), independently of effects on signaling from the platelet-derived growth factor receptor. These mechanisms are likely to be important in the oncogenicity of the virus in its natural hosts, in which it induces fibropapillomas (Box 18.1).

Other papillomaviruses, including many human serotypes, cause epithelial tumors by transforming cells that lack the platelet-derived growth factor receptor. Synthesis of the human papillomavirus type 16 E5 protein in cells containing the epidermal growth factor receptor stimulates signaling via downstream pathways, such as the Map kinase cascade (Fig. 18.3). Like its bovine papillomavirus counterpart, this viral protein can also modulate a cellular pathway that produces lipid second messengers.

Integration of proviruses of nontransducing retroviruses can also activate cell surface receptors, because the cellular gene products are altered. In certain chicken lines, Rous-associated virus-1 induces erythroblastosis instead of lymphoma (Box 18.1). These tumors contain intact, nondefective proviruses integrated in the cellular *erb*B gene, which encodes the cell surface receptor for epidermal growth factor. The proviral integrations are clustered in a region that encodes the extracellular portion of this receptor, and read-through transcription produces chimeric RNAs (Fig. 18.12). The proteins translated from these RNAs are truncated growth factor receptors that lack the ligand-binding domain. Consequently, they produce a constitutive mitogenic signal. The v-*erb*B gene captured by transducing retroviruses encodes a protein with a similar truncation.

Inhibition of cellular protein phosphatase 2A. In preceding sections, we have illustrated transformation by viral gene products as a result of permanent or prolonged activation of signal transduction pathways that control cell proliferation. In normal cells, such signaling is a transient process, for the molecular components are reset to the ground state once they have transmitted the signal. Inhibition of the reactions that reverse signaling therefore can also contribute to transformation, a mechanism exemplified by the sT protein of polyomaviruses such as simian virus 40.

sT protein is not necessary for transformation of many cell types, but can stimulate transformation by simian virus 40 LT. It is required for transformation of resting cells and stimulates them to proliferate. In both infected and transformed cells, the sT protein binds to protein phosphatase 2A, a widespread, abundant serine/threonine protein phosphatase. This enzyme is a heterotrimer composed of a two-subunit core enzyme bound to a regulatory subunit. Small T antigen binds to the core enzyme at sites that include those contacted by the regulatory subunit, and it partially inhibits enzyme activity. One important consequence of this interaction is failure to inactivate Map kinases, a process normally accomplished by the dephosphorylation of threonine or tyrosine residues (Fig. 18.15). Consequently, sT increases the activity of sequence-specific transcriptional activators that are substrates of Map kinases, such as activator protein 1 (Fig. 18.15). The viral protein also prevents dephosphorylation and inactivation of the cyclic AMP-responsive element-binding protein. The increased activities of these transcriptional stimulators lead to synthesis of a G_1-phase and an S-phase cyclin, cyclins D1 and A, respectively (Fig. 18.15), and account for the ability of the viral protein to circumvent the need for growth factors or other mitogens during transformation by simian virus 40.

Transformation via Cell Cycle Control Pathways

One end point of many cellular signal transduction pathways is the transcription of genes encoding cell cycle regulatory proteins. Consequently, permanent activation of such pathways by viral proteins, by any of the mechanisms described in the previous section, can result in an increased rate of cell growth and division or in proliferation of cells that would normally be in the resting state. Other viral proteins intervene directly in the intricate circuits by which cell cycle progression is mediated and regulated.

Abrogation of Restriction Point Control Exerted by the Rb Protein

The Restriction Point in Mammalian Cells

In mammalian cells, passage through G_1 into S and reentry into the cell cycle from G_0 depend on extracellular signals that regulate growth, termed **mitogens.** Late

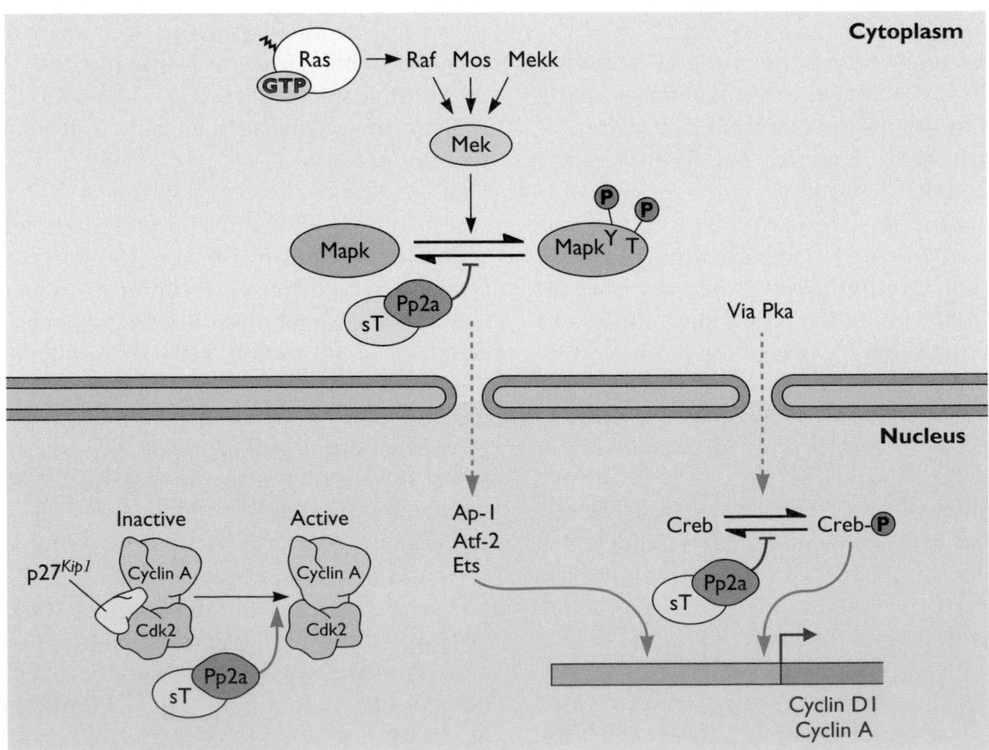

Figure 18.15 Inhibition of protein phosphatase 2A by simian virus 40 sT. Inhibition of the activity of protein phosphatase 2A (Pp2a) by sT results in activation of cellular transcriptional regulators, both via the Map kinase pathway [e.g., activator protein 1 (Ap-1) and activating transcription factor 2 (Atf-2)] and by inhibition of dephosphorylation of activated cyclic AMP response element-binding protein (Creb) within the nucleus. Production of sT within cells stimulates cyclin D1 and cyclin A transcription. Binding of sT to protein phosphatase 2A also induces a large increase in the activity of cyclin A-dependent Cdk2, concomitant with inhibition of dephosphorylation of the cyclin-Cdk inhibitor p27^{Kip1} (Fig. 18.5) and degradation of this protein.

in G_1, cells responding to such external cues become committed to enter S and to divide and complete the cell cycle; during this period, they are refractory to mitogens. Cells that have entered this state are said to have passed the G_1 **restriction point** (Fig. 18.5). Normal cells respond to mitogenic signals by mobilization of the G_1 Cdk complexes that contain D-type cyclins. Expression of genes that encode one or more of these cyclins is induced by such signals via Ras and Map kinases. The activation of Cdks by assembly with a cyclin D protein also depends on mitogenic stimulation. When such stimulation is continuous, Cdk activity appears at mid-G_1 and increases to a maximum near the G_1-to-S phase transition (Fig. 18.5). Such activity must be maintained until the restriction point has been passed, but then becomes dispensable. This property implies that the kinase activity of the cyclin D-dependent Cdks is necessary for exit from G_1. The best-characterized substrate of these kinases is the Rb protein. Many lines of evidence indicate that cyclin D-dependent Cdks initiate the transition through the restriction point solely by phosphorylation of Rb. For example, inhibition of cyclin D synthesis or function prevents entry into S phase in Rb-containing cells, but this cyclin is not required in Rb-negative cells. As discussed in Chapter 9, the Rb protein controls the activity of members of a family of sequence-specific transcriptional regulators known collectively as E2f.

Hypophosphorylated Rb present at the beginning of G_1 binds to specific members of the E2f family. Because Rb is a transcriptional repressor, these complexes inhibit transcription of E2f-responsive genes (Fig. 18.16A). The Rb protein is phosphorylated at numerous sites by G_1 cyclin-Cdk complexes. Such phosphorylated Rb can no longer bind to E2f, which therefore becomes available to activate transcription from E2f-responsive promoters (Fig. 18.16A). Such promoters include those of the genes encoding the kinase Cdk2, the cyclins that activate this kinase, and E2f proteins themselves. The initial release of E2fs from association with Rb therefore triggers a positive feedback loop that augments both Rb phosphorylation and release of E2fs. The result is a rapid increase in the concentrations of E2fs and cyclin E-Cdk2 complexes. In this way, cell cycle progression becomes independent of the mitogens necessary for the production of cyclin D-Cdk complexes. The regulatory circuits described so far suggest that cells that contain no Rb protein would proceed into S phase in the absence of mitogenic signals, but, in fact, such cells cannot pass the restriction point in the absence of external signals. Furthermore, cyclin E is required for entry into S phase in Rb-negative cells, indicating that reactions other than Rb phosphorylation must contribute to passage through G_1 and the restriction point.

The E2f proteins that accumulate upon Rb phosphorylation also stimulate transcription of genes encoding proteins needed for DNA synthesis (Chapter 9), allowing genome replication to take place in S phase. The cyclin A-Cdk2 produced in response to E2f phosphorylates and inhibits the ubiquitin ligase that marks cyclin B—which is required for entry into mitosis—for proteasomal degradation throughout much of the cell cycle (Fig. 18.16A). Consequently, cyclin B accumulates as S phase progresses. The modification of the Rb protein therefore ensures not only passage through the restriction point and entry into S phase, but also the coordination of these processes with later events in the cell cycle.

Although E2f proteins are the best-characterized targets of Rb, the latter protein can also bind to numerous other proteins that mediate or stimulate transcription, as well as to regulatory proteins such as the Abl tyrosine kinase (Table 18.6). These interactions can lead to activation or repression of transcription and, at least in some cases, have been implicated in inhibition of cell cycle progression by Rb. The Rb protein has also recently been shown to regulate DNA synthesis directly, at least in part by binding to a protein component of the complex that ensures that the genome is replicated only once per cell cycle.

Inhibition of Rb Function by Viral Proteins

The products of oncogenes of several DNA viruses bypass the sophisticated circuits that impose restriction point control, and hence the dependence of passage into S and M phases on environmental cues. The adenoviral E1A proteins, simian virus 40 LT, and the E7 proteins of oncogenic human papillomavirus (types 16 and 18) induce DNA synthesis and cell proliferation. Such mitogenic activity requires regions of the viral proteins that are necessary for their binding to hypophosphorylated Rb. All three viral proteins bind to the two noncontiguous regions by which Rb associates with E2f family members (regions A and B in Fig. 18.16B). These viral proteins sequester the inhibitory form of Rb and can disrupt Rb-E2f complexes. As a result, they induce transcription of E2f-dependent genes and inappropriate entry of cells into S phase (Fig. 18.16A).

Disruption of E2f-Rb complexes by the viral proteins appears to be an active process, rather than simply the result of passive competition for binding to Rb. Induction of cell cycle progression by simian virus 40 LT requires not only the Rb-binding site but also the N-terminal J domain located nearby

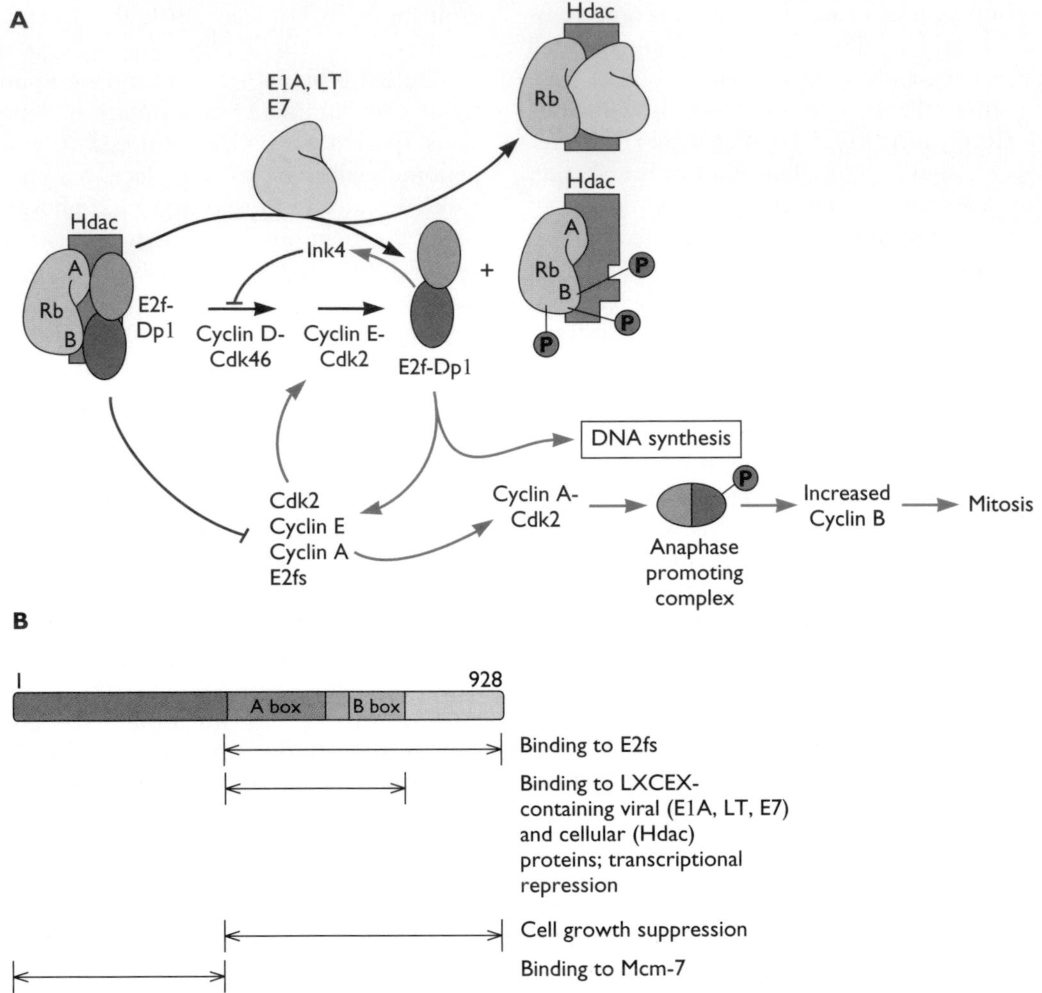

Figure 18.16 Regulation of E2f activity by the Rb protein. (A) Either phosphorylation of Rb by cyclin D- and cyclin E-dependent Cdks or binding by the adenoviral E1A, papillomaviral ET, or polyomaviral LT proteins disrupts binding of Rb to E2fs, which comprise heterodimers of an E2f protein and a Dp protein, such as Dp-1. In uninfected cells, Rb is phosphorylated at many sites by both cyclin D-Cdk4/6 and cyclin E-Cdk2. The latter cyclin, which appears in mid to late G_1 (Fig. 18.5), is required for entry into S phase and, when expressed ectopically, can drive quiescent cells to reenter the cell cycle. Its modification of Rb depends on the prior action of cyclin D-Cdk4/6. As discussed in the text, it appears that the viral proteins actively disrupt Rb-E2f complexes, rather than simply competing for binding to regions of Rb that contact E2fs. At least in some cases, the viral proteins also alter Rb metabolism. The human papillomavirus type 16 and 18 E7 proteins induce proteasome-mediated degradation of Rb in transformed epithelial cells, and can bind directly to one subunit of this multicatalytic complex. The E2f-Rb complex represses transcription when bound to E2f recognition sites in promoters of E2f-responsive genes. Rb is the transcriptional repressor, and this function requires the binding of histone deacetylases (Hdac). Binding to Hdac by the Rb family members p107 and p130 is also necessary for their repression of E2f-dependent transcription. Free E2f–Dp-1 heterodimers activate transcription from such promoters, including those of the genes encoding cyclins E and A, Cdk2, and E2f proteins themselves, to establish a positive autoregulatory loop. The positive feedback loop for activation of cyclin E-dependent kinases and E2fs late in G_1 is subject to several checks and balances imposed by inhibitory proteins (Fig. 18.5B). These inhibitory proteins must therefore be inactivated or destroyed to allow progression into S phase. The synthesis of at least one member of the Ink4 family of cyclin-Cdk inhibitors is also induced in response to free E2f. It is therefore believed that the accumulation of this inhibitor establishes a feedback loop that blocks the activity of the cyclin D-Cdks and hence the ability of cells to respond to mitogens, a characteristic property of cells that have passed the G_1 restriction point. (B) Functional domains of the human Rb protein are shown to scale. The A- and B-box regions form the so-called pocket domain, which is necessary for binding of Rb to both E2fs and the viral proteins described in the text. This

(continued on next page)

Figure 18.16 *(continued)* segment is also sufficient to repress transcription when fused to a heterologous DNA-binding domain, and it is required for binding to Hdac. Like the adenoviral, papillomaviral, and polyomaviral Rb-binding proteins, Hdac contains the motif LXCXE within the region that binds to Rb. Additional C-terminal sequences of Rb are required for transcriptional repression and binding to E2fs, as well as inhibition of cell growth. The N-terminal segment of the protein, which is also important for suppression of cell proliferation, binds to human Mcm-7, a component of a chromatin-bound complex required for DNA replication and control of initiation of DNA synthesis.

(Fig. 18.17). Because the J domain functions as a molecular chaperone (see Chapter 9), it has been suggested that LT actively dismantles the Rb-E2f complex. The CR1 sequence of the adenoviral E1A proteins fulfills a similar function (Fig. 18.18). Exactly how these viral proteins disrupt the association of Rb with E2f remains to be determined, but conformational alteration of Rb seems likely to be important.

Unless bound to Rb (or to the related proteins described in the next section), members of the E2f family such as E2f-1 and E2f-4 are unstable proteins that are degraded by the proteasome following ubiquitination. Such instability is dictated by a C-terminal sequence that lies close to the binding site for Rb, and appears to be shielded in Rb-E2f complexes. Recent experiments indicate that the adenoviral E1A proteins not only induce accumulation of E2f but also stabilize these proteins.

Inhibition of Negative Regulation by Rb-Related Proteins

The Rb protein is the founding member of a small family of related gene products. This family also contains the proteins p107 and p130, which were discovered by virtue of their binding to adenoviral E1A proteins (Fig. 18.18); they also bind to both simian virus 40 LT and human papillomavirus type 16 and 18 E7 proteins. The strong similarity of p107 and p130 to Rb is most pronounced in the A and B se-

Figure 18.17 Model for active dismantling of the Rb-E2f complex by simian virus 40 LT. The LT protein binds to the Rb A- and B-box domains via the sequence that contains the LXCXE motif (amino acids 101 to 118), designated R. The structure of a complex containing the Rb A- and B-box domain (blue) bound to the amino acids 7 to 117 of LT (pink), determined by X-ray crystallography, is shown as a surface representation. An extended segment of LT, which comprises two α-helices and a loop containing the LXCXE sequence, makes contact with an extensive surface of the Rb protein. The adjacent, N-terminal J domain (amino acids 1 to 82) of LT is not necessary for binding to Rb, but is required for induction of cell cycle progression. It has been proposed that the J domain recruits the cellular chaperone Hsc70 to the complex. The chaperone then acts to release E2f–Dp-1 heterodimers from their association with Rb, by a mechanism that is not yet understood, but that is believed to depend on ATP-dependent conformational change. Structural model adapted from M.-Y. Kim et al., *EMBO J.* **20**:295–304, 2001, with permission.

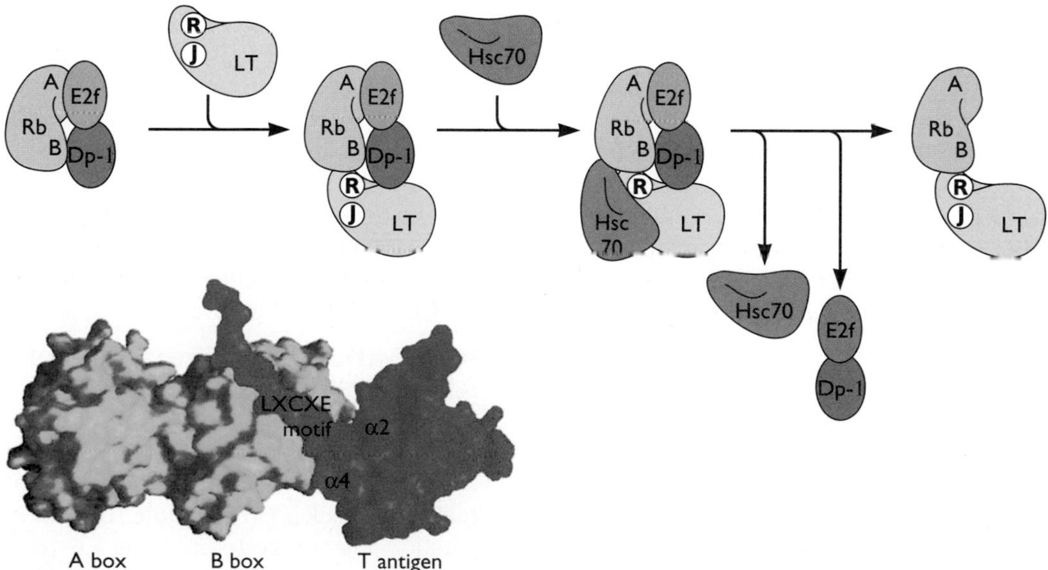

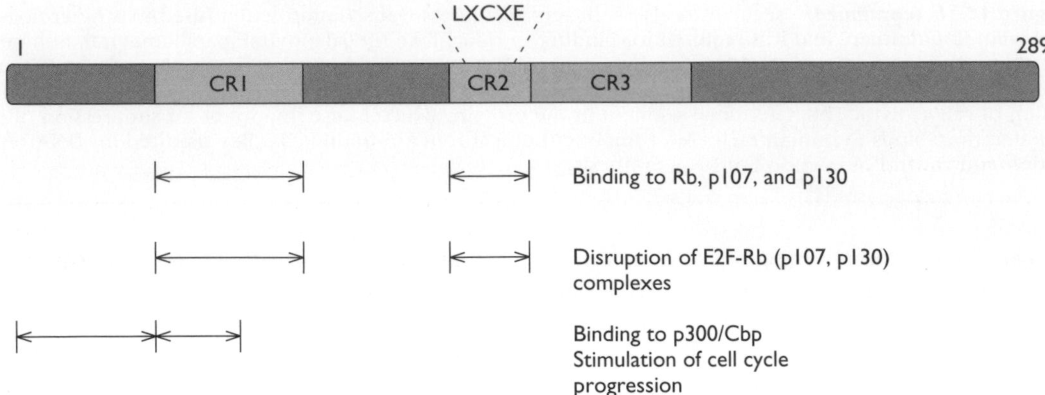

Figure 18.18 Organization of the larger adenoviral E1A protein. Regions of the protein are shown to scale. Those designated CR1 to CR3 are conserved in the E1A proteins of human adenoviruses. The CR3 region, most of which is absent from the smaller E1A protein because of alternative splicing, is not necessary for transformation. The locations of the Rb-binding motif and of the regions required for binding to the other cellular proteins discussed in the text are indicated.

quences needed for binding of Rb to both E2f and the viral transforming proteins (Fig. 18.16B). Indeed, the residues by which Rb contacts the common Rb-binding motif of the viral proteins (LXCXE) are invariant among the other family members. Binding of the viral proteins to p107 and p130 can make important contributions to transforming activities. For example, the LXCXE sequence of simian virus 40 LT is required for transformation of fibroblasts derived from *Rb*-null mice. Furthermore, the J domain of LT is also necessary and induces hypophosphorylation of p107 and p130, concomitant with their increased degradation.

The Rb, p107, and p130 proteins bind preferentially to different members of the E2f family during different phases of the cell cycle. Hypophosphorylated Rb binds primarily to E2f-1, -2, or -3 during the G_1 phase. In contrast, p107 is largely associated with E2f-4 during the G_1, S, and G_2 phases. Some fraction of such complexes also contain cyclin E-Cdk2 or cyclin A-Cdk2. These properties, and the targeting of p107 by transforming proteins of the smaller DNA viruses, indicate that inhibition (or more subtle regulation) of the activity of E2f-4 must be important for orderly progression through the cell cycle. Binding of p130 to E2f-4 and E2f-5 appears to be critical for maintaining cells in the quiescent state. Such complexes predominate in mammalian cells in G_0. Their disruption by adenoviral, papillomaviral, or polyomaviral transforming proteins is believed to allow such cells to reenter the cycle, in part because these E2f family members stimulate transcription of genes encoding both the E2f proteins and the cyclin-dependent kinase (Cdk2) needed for entry into S phase.

Production of Virus-Specific Cyclins

Human herpesvirus 8 and its close relative herpesvirus saimiri encode functional cyclins. The cyclin gene of human herpesvirus 8, designated v-*cyclin*, has 31% identity and 58% similarity to the human gene that encodes cyclin D2, and its product binds predominantly to Cdk6. Like its cellular counterpart, v-*cyclin* activates this protein kinase, which then phosphorylates the Rb protein. The viral cyclin also alters the substrate specificity of the kinase: the v-*cyclin*-Cdk6 complex phosphorylates proteins normally recognized by cyclin bound-Cdk2, but not by cyclin D-Cdk6. The viral cyclin and the active enzyme it forms with cellular Cdk6 are resistant to even high concentrations of the Cip/Kip and Ink4 proteins that inhibit cellular cyclin-Cdks (Fig. 18.5). Normally, synthesis of these inhibitory proteins blocks cell cycle progression (see "Activation and Functions of the Cellular p53 Protein" below for an example). However, neither Cip/Kip nor Ink4 family members bind well to the viral cyclin, because it differs from cellular cyclins in both the sequence and structure of the site to which the inhibitors bind. Synthesis of the viral cyclin can therefore overcome the G_1 arrest imposed when either type of inhibitory protein is made in human cells, and can induce cell cycle progression in quiescent cells. The specific advantages conferred by production of the viral cyclins during the infectious cycle have not been identified. However, it would be surprising if they do not contribute to the oncogenicity of these herpesviruses in their natural hosts.

Inactivation of Cyclin-Dependent Kinase Inhibitors

The production of viral cyclins in infected cells appears to be a unique property of certain herpesviruses, but other DNA viruses encode proteins that inactivate specific inhibitors of Cdks. For example, the E7 protein of human papillomavirus type 16 binds via a C-terminal sequence to the p21^{Cip1} protein and inactivates it. This member of the cellular Cip/Kip family inhibits G_1 cyclin-Cdk complexes

(Fig. 18.5). One event that results in accumulation of p21^{Cip1} is the increase in intranuclear concentrations of p53 triggered by unscheduled inactivation of the Rb protein (see next section). The ability of the papillomaviral E7 protein to inactivate both Rb and p21^{Cip1} would therefore appear to ensure entry into S phase of cells in which the viral protein is made. Indeed, both these functions of the E7 protein are necessary to induce S phase in differentiated human epithelial cells.

Inhibition of p53 Functions

As described in Chapter 15, the genomes of many viruses encode proteins that protect infected cells against an antiviral defense of last resort, programmed cell death (apoptosis). Among many other signals, this program is activated by damage to the cellular genome or unscheduled DNA synthesis. Consequently, viral transforming proteins that induce cells to enter S phase when they would not normally do so also promote the apoptotic response. This potentially fatal side effect of unscheduled passage through the cell cycle is counteracted by specific viral proteins to allow viral DNA replication or, under appropriate circumstances, survival of transformed cells.

Activation and Functions of the Cellular p53 Protein

Transformation by several DNA viruses requires inactivation of a second cellular tumor suppressor gene product, the p53 protein. This protein, first identified by virtue of its binding to simian virus 40 LT, is a critical component of regulatory circuits that determine the response of cells to damage to their genomes, as well as to low concentrations of nucleic acid precursors or hypoxia. The importance of this protein in the appropriate response to such damage or stress is emphasized by the fact that *p53* is the most frequently mutated gene in human tumors.

The intracellular concentration of p53 is normally very low, because the protein is targeted for nuclear export and proteasomal degradation by binding of the Mdm-2 protein to an N-terminal sequence (Fig. 18.19 and 18.20). However, DNA damage, such as double-strand breaks produced by γ-irradiation or the accumulation of DNA repair intermediates following ultraviolet irradiation, leads to the stabilization of p53 and a substantial increase in its concentration (Fig. 18.20). The rate of translation of the protein may also increase. Various proteins that appear to be important for stabilization of p53 have been identified, including the

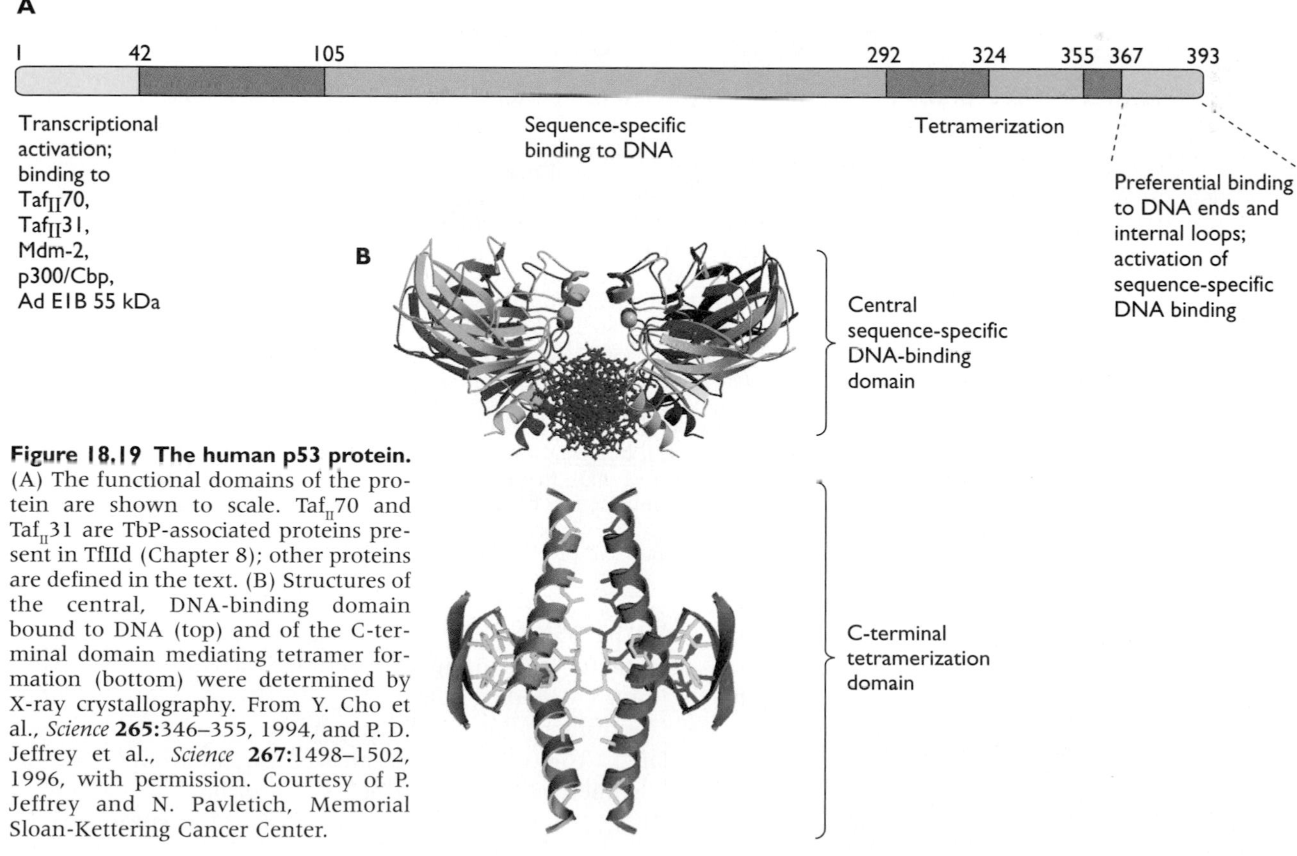

Figure 18.19 The human p53 protein. (A) The functional domains of the protein are shown to scale. $Taf_{II}70$ and $Taf_{II}31$ are TbP-associated proteins present in TfIId (Chapter 8); other proteins are defined in the text. (B) Structures of the central, DNA-binding domain bound to DNA (top) and of the C-terminal domain mediating tetramer formation (bottom) were determined by X-ray crystallography. From Y. Cho et al., *Science* **265:**346–355, 1994, and P. D. Jeffrey et al., *Science* **267:**1498–1502, 1996, with permission. Courtesy of P. Jeffrey and N. Pavletich, Memorial Sloan-Kettering Cancer Center.

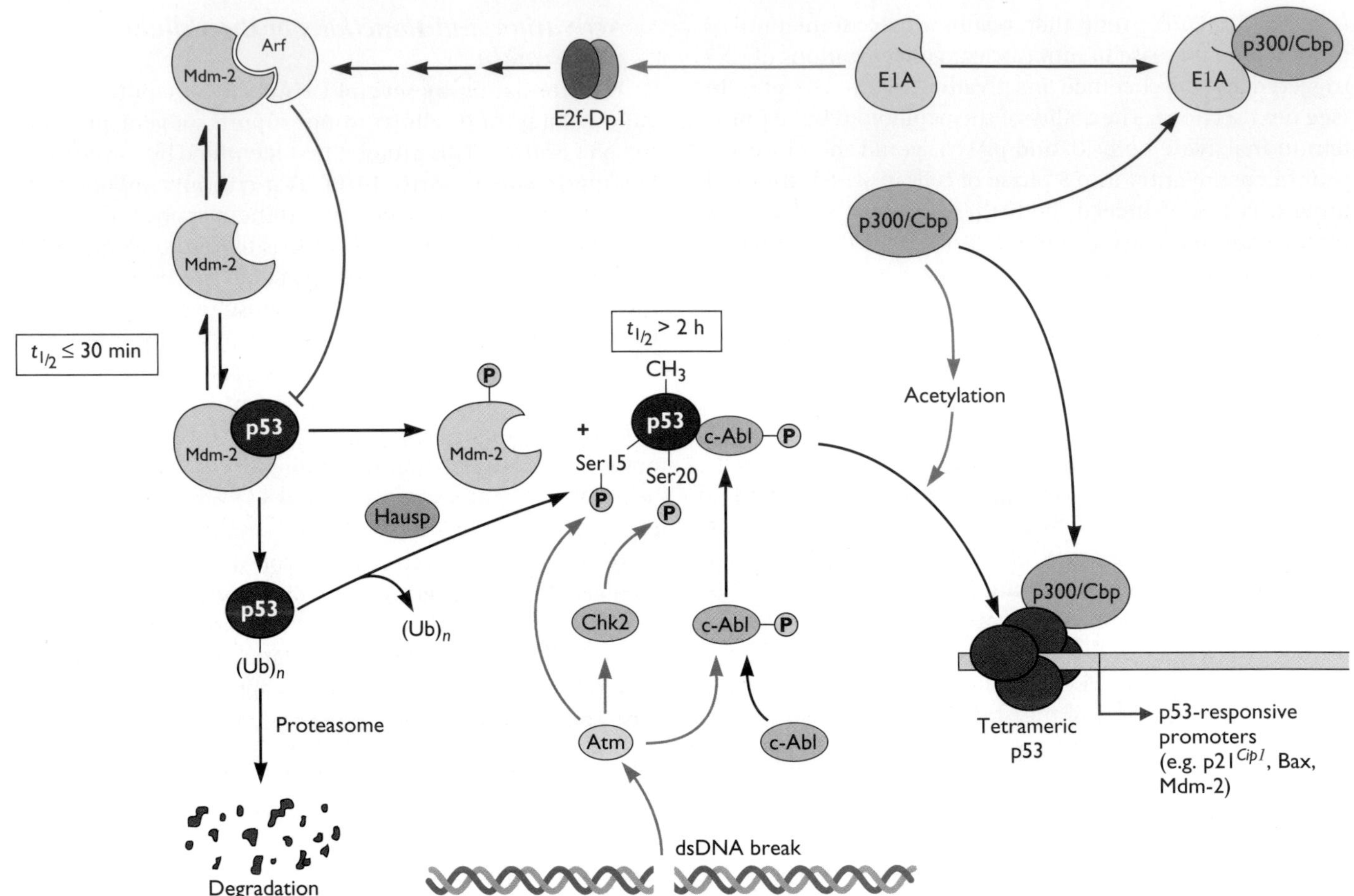

Figure 18.20 Regulation of the stability and activity of the p53 protein. Under normal conditions (left), cells contain only low concentrations of p53. This protein is unstable, turning over with a half-life of minutes, because it is targeted for proteasomal degradation by the Mdm-2 protein. Mdm-2 is a p53-specific E3 ubiquitin ligase that catalyzes polyubiquitination of p53, the signal that allows recognition by the proteasome. It binds to the N terminus of p53, an interaction that induces export of p53 from the nucleus and blocks the transcriptional activation domain of p53. When tethered to p53, Mdm-2 can actively repress transcription. The Mdm-2 protein therefore maintains inactive p53 at low concentrations. The availability and activity of Mdm-2 are also regulated, for example, by Arf proteins encoded by the *ink4a/arf* tumor suppressor gene, and by stimulation of Mdm-2 transcription by the p53 protein itself. Signaling pathways initiated in response to damage to the genome or other forms of stress lead to stabilization of p53. Such posttranscriptional regulation is believed to allow a very rapid response to conditions that could be lethal to the cell. As illustrated with pathways operating in response to DNA damage (double-strand [ds] breaks) caused by ionizing radiation or inappropriate production of active E2f family members when adenoviral E1A proteins are made, p53 is stabilized in multiple ways. These mechanisms include phosphorylation of p53 at specific serines by Atm (see text) and checkpoint kinase 2 (Chk2), binding to the c-Abl tyrosine kinase, sequestration of the Mdm-2 protein by Arf, and deubiquitination of p53 (in the presence of Mdm-2) by the herpesvirus-associated ubiquitin-specific protease (Hausp). Multiple mechanisms, some of which are listed, also stimulate the sequence-specific DNA-binding activity of p53 or its association with the transcriptional coactivators p300/Cbp, and hence transcription from p53-responsive promoters.

product of a human gene called *Atm* (ataxia telangiectasia mutated). The genetic disease caused by mutation in *Atm* is associated with a broad spectrum of defects, including hypersensitivity to X-ray and ionizing radiation. Atm recognizes potentially genotoxic DNA damage. Cells lacking the Atm protein do not accumulate the p53 protein, or arrest at the G_1/S boundary, in response to DNA damage.

The p53 protein is a sequence-specific transcriptional regulator, containing an N-terminal activation and a central DNA-binding domain (Fig. 18.19). Its ability to stimulate transcription of p53-responsive genes is tightly regulated. The p53 protein must form a tetramer via its oligomerization domain before it can bind specifically to DNA. The DNA-binding activity of p53 is stimulated by binding of the C-terminal domain to double-stranded DNA ends or sites of excision repair damage, as well as by various modifications within this short C-terminal sequence (Fig. 18.20). The central DNA-binding domain itself is frequently altered as a result of the changes that inactivate p53 in human tumors. The biochemical functions of the N-terminal transcriptional activation domain can also be regulated. For example, binding of the Mdm-2 protein to this segment inhibits p53-dependent transcription. The many mechanisms by which the accumulation or activity of p53 can be regulated (Fig. 18.20) provide the means to integrate the multiple signals that are monitored to ensure that this potent protein alters cell physiology only under extreme conditions.

In response to damage to the genome, or other inducing conditions, p53 can promote one of two responses, leading to either G_1/S arrest or apoptosis (Fig. 18.20 and 18.21). One important component of the former pathway is the p53-dependent stimulation of transcription of the gene that encodes the G_1 cyclin-dependent kinase inhibitor $p21^{Cip1}$. The p53 protein also stimulates transcription of a number of genes encoding proteins that participate in apoptosis, such as Bax and Fas (Fig. 18.21). However, it also induces this extreme cellular response without stimu-

Figure 18.21 Inactivation of the p53 protein by adenoviral, papillomaviral, and polyomaviral proteins. The synthesis of transforming proteins encoded within the genomes of these viruses in infected or transformed cells can activate p53. For example, the adenoviral E1A proteins induce increased concentrations of p53, and this effect depends on the regions of the viral protein required for induction of cellular DNA synthesis and cell proliferation. However, each of these viruses encodes proteins that interfere with the normal function of this critical cellular regulator. The E6 proteins of human papillomavirus types 16 and 18 bind to p53 via the cellular E6-associated protein (E6-Ap). The latter protein is a ubiquitin protein ligase that ubiquitinates p53 in the presence of the viral E6 protein, targeting p53 for degradation by the proteasome. Binding of simian virus 40 LT to p53, an interaction that is facilitated by sT, sequesters the cellular protein in inactive complexes. The adenoviral E1B 55-kDa and E4 ORF6 proteins bind to p53 at the N-terminal activation domain and a C-terminal region near the tetramerization domain (Fig. 18.20), respectively. In experimental systems, the former interaction converts p53 from an activator to a repressor of transcription. In transformed rodent cells, it also induces relocalization of p53 from the nucleus to a perinuclear, cytoplasmic body, as well as translocation of the product of a second cellular tumor suppressor protein, Wt1, to the same site. Binding of the E4 ORF6 protein increases the rate of degradation of p53. In infected cells, such increased degradation of p53 also requires the E1B 55-kDa protein.

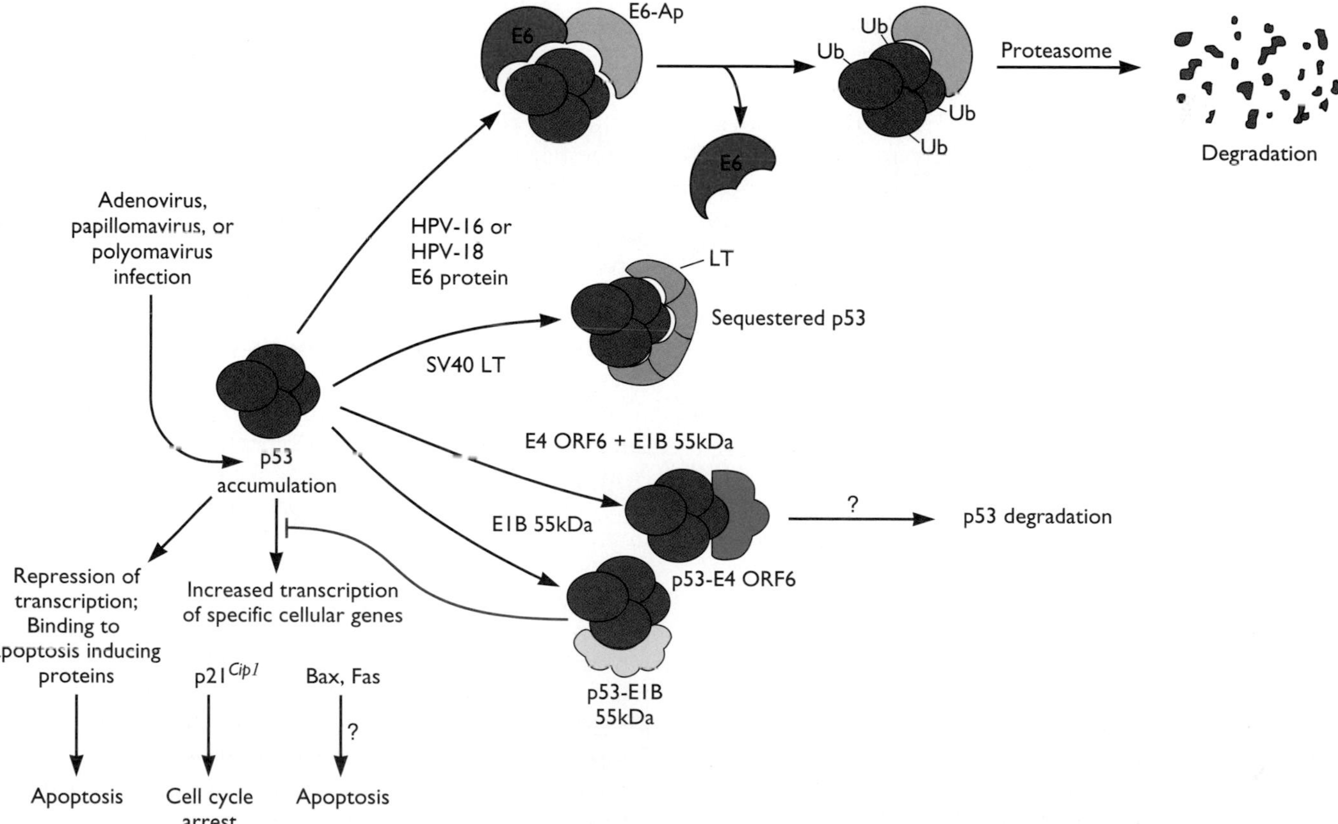

lating transcription. Such mechanisms remain incompletely understood, but they may include repression of transcription of genes that encode antiapoptotic proteins, and direct interaction with proteins that induce apoptosis. Whether p53 promotes cell cycle arrest or apoptosis is determined by numerous parameters, including the cell type, the nature of extracellular stimuli, and the concentration of the p53 protein itself. However, the apoptotic response prevails in many cell types under many circumstances: following DNA damage, in the absence of a hormone or growth factor necessary for cell survival, and following expression of viral oncogenes that induce entry into S phase.

Inactivation of p53 by Binding to Viral Proteins

The genomes of many viruses encode proteins that have been reported to interact with p53. However, the mechanisms by which the functions of this critical cellular regulator can be circumvented are best understood for the small DNA tumor viruses. In contrast to the common mechanism of inactivation of Rb by the different viral proteins discussed above, adenoviral, papillomaviral, and polyomaviral proteins block p53 function by different mechanisms (Fig. 18.21). The human papillomavirus type 16 or 18 E6 proteins bind to both p53 and a cellular ubiquitin protein ligase, the E6-associated protein, and thereby target p53 for proteasome-mediated destruction. Consequently, p53 is cleared from the infected or transformed cell. In contrast, simian virus 40 LT actually stabilizes the p53 protein upon binding to it, but sequesters this cellular regulator in inactive complexes. The adenoviral E1B 55-kDa protein binds specifically to the N-terminal activation domain of p53 and can convert the cellular protein from an activator to a repressor of transcription. In conjunction with the adenoviral E4 Orf6 protein, the E1B protein also induces increased turnover of the p53. Like the E1B protein, the E4 protein cannot alter cell growth or proliferation alone, but it can transform rodent cells in cooperation with viral E1A gene products. Furthermore, when made in adenovirus-transformed cells that do not induce tumors in nude mice, this protein converts them to tumorigenic cells.

Despite the mechanistic differences summarized in Fig. 18.21, cells infected by adenoviruses, papillomaviruses, or polyomaviruses lack functional p53 and are therefore refractory to stimuli that would normally trigger p53-mediated cell cycle arrest or apoptosis.

Alteration of p53 Activity via the p300/Cbp Proteins

The nuclear phosphoprotein p300 was first identified by virtue of its binding to adenoviral E1A proteins. This protein and the closely related transcriptional coactivator Creb-binding protein (Cbp) are required for stimulation of transcription by various sequence-specific DNA-binding proteins. It is believed that the coactivators improve access of components of the transcriptional machinery to DNA in chromatin templates by acetylating nucleosomal histones. Because the E1A proteins bind to the regions of p300/Cbp that contact the histone acetylase, they disrupt complexes containing these cellular proteins. Binding of the viral proteins to p300 correlates with induction of cell cycle progression and transformation. The region of simian virus 40 LT that contacts p53, which is required for transformation of many cell types, also interacts with p300. These properties suggest that inhibition of the histone-modifying functions of p300/Cbp by the viral proteins is important for the induction of cell cycle progression and transformation. The p53 protein has recently been identified as one important target of this mechanism.

As mentioned above, the p53 protein blocks cell cycle progression by stimulating transcription of genes that encode cyclin-Cdk-inhibitory proteins, such as $p21^{Cip1}$ (Fig. 18.21). Such stimulation of transcription is mediated by p300/Cbp, which binds to the activation domain of p53 via sequences that can also interact with the E1A proteins. Indeed, binding of E1A proteins to the coactivator inhibits p53-dependent transcription of *Cip1*. Consequently, this interaction suppresses the induction of G_1 arrest by p53 following DNA damage. It also blocks the apoptotic response by the mechanism that depends on transcriptional activation by p53. The E1A proteins therefore not only induce entry of cells into S phase but, by binding to p300/Cbp, also protect against deleterious consequences of the response to unscheduled DNA synthesis.

Other Mechanisms of Transformation and Oncogenesis

Nontransducing, Complex Oncogenic Retroviruses: Tumorigenesis with Very Long Latency

The prototype for this group of oncogenic retroviruses is human T-lymphotropic virus type 1, which is associated with adult T-cell lymphocytic leukemia (ATLL). This disease was first described in Japan in 1977, and has since been found in other parts of the world, including the Caribbean and areas of South America and Africa. The virus, which was isolated in 1980, is now classified as a deltaretrovirus, a group that also includes other retroviruses with complex genomes, such as bovine leukemia virus and simian T-cell leukemia virus.

Human T-lymphotropic virus is transmitted via the same routes as human immunodeficiency virus: during sexual intercourse, by intravenous drug use and blood transfusions, and from mother to child. Infection is usually asymptomatic, but can progress to ATLL in about 5% of infected individuals over a period of 30 to 50 years (see Chapter 16). There is no effective treatment for the disease, which is usually fatal within a year of diagnosis. The mechanism(s) by which the virus induces malignancies is still uncertain, but some of the features of ATLL are consistent with a role for a viral regulatory protein. A provirus is found at the same site in all leukemic cells from a given case of ATLL, but there are no preferred chromosomal locations for these integrations. Activation or inactivation of a specific cellular gene is not, therefore, a likely mechanism of transformation. As the genome of human T-lymphotropic virus type 1 does not contain **any** cell-derived nucleic acid, some viral sequences must be responsible for this activity. Surprisingly, however, proviral genes are not expressed in ATLL cells. But viral proteins are made if these cells are placed in culture. The last two features suggest an unusual mechanism of oncogenesis, in which the expression of viral genes is necessary for the initiation of transformation but not for its maintenance.

One of the best-studied regulatory proteins of human T-lymphotropic virus type 1 is the transcriptional activator Tax. A role for the *tax* gene in oncogenesis is suggested by the observation that transgenic mice carrying *tax* under the control of the viral LTR synthesize this viral protein in muscle tissue and develop multiple soft tissue sarcomas. One idea about how this virus might contribute to oncogenesis is that Tax alters the expression of cellular genes that encode proteins that regulate T-cell physiology. Infection by human T-lymphotropic virus type 1 does induce proliferation of T cells and increased synthesis of a number of cellular proteins, among them cytokines, and the product of the *fos* proto-oncogene. Furthermore, the Tax protein activates cellular Nf-κb, which is required for transcription of several of these genes. Another possibility is that expression of the viral regulatory protein induces genetic instability in the host cells. There is some evidence supporting both ideas. However, because the virus-induced oncogenic event occurs a long time before ATLL appears, it may be difficult to distinguish among these (and other) possibilities.

Oncogenesis by Hepatitis Viruses

Hepatitis B Virus

This virus is a member of the *Hepadnaviridae*. The major site of reproduction for all hepadnaviruses is the liver. Infections by these retroid viruses can be acute (3- to 12-month) or lifelong. In humans, the frequency of persistent infection ranges from 0.1 to 25% of the population in different parts of the world. Such long-term carriers are at high risk for developing hepatocellular carcinoma (Box 18.1), and as many as 1 million people die of this disease each year. However, study of hepatitis B virus oncogenesis in humans is difficult, because liver cancer arises only after a very long period (decades) of chronic infection. Woodchucks infected with woodchuck hepatitis B virus are used as an animal model for hepadnaviral oncogenesis. Normally, these animals experience only acute infections by the virus. However, if they are treated with an immunosuppressing drug (e.g., cyclosporine) before inoculation, the infections become persistent and almost all infected animals develop liver cancer by 2 to 4 years of age.

Sustained low-level liver damage of infected hepatocytes is characteristic of persistent infection by hepatitis B viruses. Almost all such damage can be attributed to attack by the host's immune system. The rate of hepatocyte proliferation must increase in such cases to compensate for cell loss. It is generally accepted that such an increased rate of proliferation over long periods is a major contributor to the development of liver cancer. In addition, the inflammation and phagocytosis that are integral to the immune response can result in high local concentrations of superoxides and free radicals. It is therefore possible that DNA damage and the resulting mutagenesis also play a role in hepadnavirus-induced hepatocellular carcinoma. Consequently, there is considerable incentive for developing antiviral therapies to treat persistent hepatitis B virus infection.

The almost universal presence of integrated fragments of hepadnaviral DNA suggests that this feature plays a role in the development of tumors. Indeed, in woodchucks activation of a member of the *myc* oncogene family, associated with the nearby integration of woodchuck hepatitis virus DNA, is observed in 90% of hepatocellular carcinomas. In contrast, insertional activation of a common oncogene has not been observed in hepatitis B virus-induced hepatocellular carcinomas in humans. Rather, there is accumulating evidence that viral proteins encoded by integrated viral DNA sequences contribute to carcinogenesis in humans. One such viral gene product is the X protein, which is synthesized from integrated viral DNA sequences in many hepatocellular carcinomas. As discussed in Chapter 8, the hepatitis B virus X protein stimulates transcription from many cellular genes (including proto oncogenes), both by altering the DNA binding of cellular transcriptional regulators and by activation of signaling via Nf-κb and other pathways. Activation of transcription is observed in transgenic mice that express the viral X gene in the liver, and the mice develop hepatocellular carcinoma when high concentrations of the X protein are made in liver cells. These properties, and the alterations in expression of cellular proto-oncogenes and tumor suppressor genes detected using microarray methods in hepatitis B virus-infected liver, suggest that activation of cellular gene expression by the viral X protein is important in the etiology of human hepatocellular carcinoma. The viral X protein can also increase susceptibility to chemical carcinogens, and thereby act as a tumor promoter, when made in transgenic mice. Under some conditions, it can inhibit apoptosis induced by external signals. It can also bind to the cellular p53 protein in hepatocellular carcinoma cells, although it remains unclear whether this activity blocks p53-dependent apoptosis.

The long time required for development of human liver cancer suggests that several low-probability reactions must take place. The viral X protein, and a second viral protein commonly made in hepatocellular carcinoma cells, could contribute to all such likely reactions (Fig. 18.22). Whether they in fact do so remains to be determined, as does the relative importance of the effects of viral protein and other contributing factors, such as the immune damage described above.

Hepatitis C Virus

As noted earlier, this virus is a member of the *Flaviviridae* and possesses a (+) strand RNA genome. Its discovery in 1989 established the etiology of what had been known previously as non-A, non-B hepatitis, a disease contracted by a small fraction of transfusion recipients who developed liver cancer years later. Routine screening of the blood supply has since reduced this mode of infection, but the virus is still transmitted by intravenous drug abusers and through the use of contaminated needles. Fortunately, the virus is not easily transmitted and seems to require direct blood-to-blood contact to be

Figure 18.22 Model for the role of hepatitis B virus proteins in development of hepatocellular carcinoma. In this model, the viral X (see text) or truncated middle surface proteins (MHBs) are proposed to facilitate every reaction in the likely sequence leading to metastatic hepatocellular carcinoma. The truncated MHBs proteins, which are unique to liver cells carrying integrated viral DNA sequence, can also alter transcription of cellular proto-oncogenes. When integration results in deletion of the sequences that encode C-terminal portions of the middle surface antigen (MHBs), stable, truncated proteins are made and accumulate in the endoplasmic reticulum (ER) or cytoplasm. These aberrant forms of MHBs can activate signaling via NF-κb and transcription of such genes as c-*fos*, c-*myc*, and c-*jun*, probably in part as a result of ER retention and "ER overload." Direct inhibition of DNA excision repair mechanisms by the X protein would increase susceptibility to chemical carcinogens and therefore the accumulation of mutations in cellular genes encoding proteins that control cell proliferation. Immune reactions to hepatitis B virus infection that produce DNA-damaging agents would also facilitate accumulation of mutations. Cells carrying appropriate mutations would then proliferate to undergo clonal expansion, processes that would be facilitated by activation of pathways that stimulate cell proliferation and inhibition of apoptosis by the X and truncated MHB proteins. The former protein could also promote the development of metastatic cancer cells as a result of modulation of interactions of cells in which it is made with the extracellular matrix. Adapted from C. Rabe et al., *Dig. Dis.* **19:**279–287, 2001, with permission.

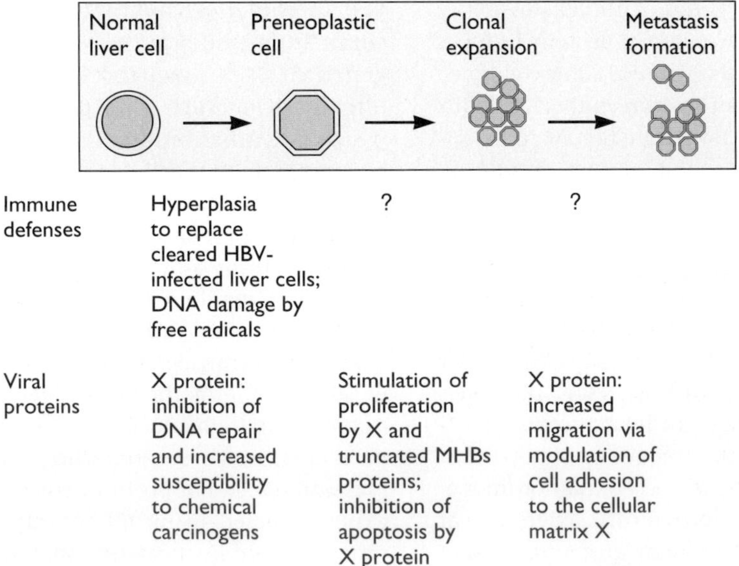

passed from one individual to another. Still, an estimated 170 million people are infected worldwide with one of six major genotypic variants (clades) that are specific for different geographical locations. Approximately 75 to 85% of infected individuals develop a persistent infection; among these, 1 to 5% will develop hepatocellular carcinoma. Although a small percentage, this still amounts to 1.3 million to 7.2 million cases of liver cancer worldwide.

Like hepatitis B virus, hepatitis C virus is hepatotropic. Chronic infection of hepatocytes leads to their destruction by the immune system and formation of fibrous scars that obstruct the passage of blood (a life-threatening condition known as cirrhosis). As not all patients with cirrhosis also develop cancer, host-specific factors must play a role. However, the possible roles of viral proteins and the host's immune system in the oncogenesis of hepatitis C virus have yet to be investigated.

"Hit-and-Run" Transformation by Herpes Simplex Viruses?

Since the first report over 30 years ago that herpes simplex viruses can transform rodent cells in culture, considerable effort has been devoted to identification of their transforming genes. The approaches to this question that proved so successful with smaller DNA viruses yielded a number of unexpected results. First, the regions required for transformation by herpes simplex virus types 1 and 2 do not coincide, even though the organization of their genomes is virtually identical. Second, these regions do not contain genes encoding proteins that would be expected to alter cellular signal transduction or cell cycle control mechanisms. For example, the two transforming regions of herpes simplex virus type 2 contain the genes specifying the two subunits of the viral ribonucleotide reductase. The viral A subunit does carry a unique N-terminal domain with serine/threonine kinase activity, which can induce activation of the cellular Ras protein and may facilitate transformation. However, unscheduled phosphorylation of cellular signal transduction proteins by the viral kinase cannot account for herpes simplex virus transformation. Most surprisingly, the minimal DNA sequences that exhibit transforming activity are very small (about 500 to 800 bp) and do not contain an intact coding sequence. Furthermore, no set of viral DNA sequences is consistently integrated into the cellular genome. Indeed, viral DNA sequences are not necessarily retained in transformed cells, indicating that they are not needed for permanent expression of transformed phenotypes.

These unusual properties of herpes simplex virus transformation have led to several models quite different from those describing transformation by the DNA viruses discussed in previous sections. Such models include mutation or rearrangement of cellular DNA and alteration of transcription of cellular genes by one of the minimal viral transforming segments, which acts as a transcriptional control region in transient assays. Such mechanisms resemble those by which nontransducing retroviruses can transform cells. More radical, with no precedent in other viral systems, is the hit-and-run hypothesis. This model proposes that once viral gene products have permanently altered the expression of cellular genetic information (e.g., by induction of mutations), viral genetic information is no longer required. This idea is attractive, because infection by herpes simplex viruses can lead to many types of chromosomal abnormality, including chromosomal breaks. However, neither this model nor the other possible mechanisms of transformation listed above have been placed on a firm molecular foundation, in part because there is no evidence that these viruses cause cancer in humans or other animals.

Perspectives

As noted at the beginning of this chapter, viruses do not cause cancer as an essential part of their infectious cycles. Nevertheless, the discovery that viruses can do so, initially made almost a century ago, was the harbinger of the spectacular progress toward understanding the molecular basis of oncogenesis and transformation made within the past 3 decades. Because tumor cells grow and divide when normal cells do not, elucidation of the mechanisms of transformation has been inevitably accompanied by the tracing of the intricate circuits that regulate cell proliferation in response to both external and internal signals. The remarkable discovery that the transforming gene of the acutely transforming retrovirus Rous sarcoma virus was a transduced cellular gene paved the way for identification of many cellular proto-oncogenes and the elucidation of the signal transduction pathways in which the proteins encoded by them function. Indeed, in several cases, we can now describe in atomic detail the mechanisms by which mutations introduced into these genes during or following capture into retroviral genomes lead to constitutive activation of signaling. These viral genes and their cellular counterparts that have acquired specific mutations in tumors are dominant oncogenes. In contrast, studies of a hereditary juvenile cancer in humans, retinoblastoma, had indicated that neoplastic disease can also develop following the loss of function of specific genes, which were therefore named tumor suppressor genes. Our current appreciation of the critical roles played by the products of such tumor suppressor genes in the control of cell cycle progression stems directly from studies of transforming proteins of adenoviruses, papillomaviruses, and polyomaviruses.

The initial cataloging of viral transforming genes and the properties of the proteins they encode suggested a be-

Table 18.10 The diversity of viral transformation mechanisms

Virus	Viral genetic information in transformed cells			Transformation mechanisms		
	Physical state	Lytic functions	Transforming functions	Cell cycle progression	Apoptosis	Immune modulation or stimulation
Adenovirus						
Human subgroup C and A serotypes	Integrated into host genome by random recombination	Lost during integration	E1A and E1B gene products; some E4 proteins	Deregulated; inactivation of pRb family proteins by E1A proteins	Blocked by inactivation of p53 (E1B 55-kDa and E4 Orf 6 proteins) and by the E1B 19-kDa protein, a viral Bcl-2 homolog	Inhibition of transcription of MHC class I genes by subgroup A E1A proteins
Flavivirus						
Hepatitis C virus	Not integrated	Noncytolytic, persistent infection of cells that are rapidly dividing	Not known	Not known	Not known	Immune systems kill infected cells; survivors are selected over years of replication; tumor cells escape elimination
Hepadnavirus						
Hepatitis B virus	Episomal; viral sequences are integrated in some hepatocellular carcinomas	None	X protein; large or truncated middle B surface proteins	Defects in DNA repair; altered transcription of proto-oncogenes (e.g., c-*fos*, c-*myc*) and transcriptional regulators (e.g., Nf-κb); direct interaction of X with proteins in growth signaling pathways	Both pro- and antiapoptotic effects are seen, depending on conditions; some may be mediated by X protein binding to p53	Continuous tissue damage due to immune cell attack drives abnormal liver proliferation
Herpesvirus						
Epstein-Barr virus	Episomal; replicates in concert with host cell genome	Repressed	LMP-1 protein; EBNA-2 and LMP-2 increase production and stability, respectively, of LMP-1	Deregulated by constitutive signaling from LMP-1	Blocked; LMP-1 induces synthesis of Bcl-2 family proteins and inhibits signaling from death domain proteins	Cytotoxic T cells recognizing EBNA-1 are not produced, because a sequence in this viral protein inhibits the proteasome; EBNA-2 blocks signal transduction cascade induced by IFN
Papillomavirus						
Human types 16 and 18	Integrated into host genome by random recombination	Lost during integration	E6 and E7 proteins	Deregulated by inactivation of Rb family proteins by E7 protein	Blocked by E6 protein-induced degradation of p53	None known

Table 18.10 *(continued)*

Virus	Viral genetic information in transformed cells			Transformation mechanisms		
	Physical state	Lytic functions	Transforming functions	Cell cycle progression	Apoptosis	Immune modulation or stimulation
Polyomavirus						
Simian virus 40	Integrated into host genome by random recombination	Lost during integration	LT; sT in some circumstances	Deregulated by inactivation of Rb family proteins by LT and stimulation of Mapk cause cascade signaling by sT	Blocked by inactivation of p53 by LT	None known
Retroviruses						
Transducing (e.g., Rous sarcoma virus)	Integrated	None	Virally transduced, cell-derived oncogenes	Loss of tissue, developmental, and cell cycle regulation by inappropriate expression of oncogenes	None	None known
Nontransducing (e.g., Rous-associated viruses 1 and 2)	Integrated	None	Insertional mutagenesis, alteration of cellular proto-oncogenes (or their expression) in vicinity of proviral integration site	Loss of tissue, developmental, and cell cycle regulation by inappropriate expression of wild-type or truncated proto-oncogenes	None	Viral superantigen (murine mammary tumor virus, Sag protein) stimulates T cells and induces proliferation of B cells
Long latency (e.g., human T-lymphotropic virus type 1)	Integrated	None	Viral transactivor proteins, Tax	Stimulates T-cell proliferation by induction of Nf-κb and IL-2	Not known	Abnormal T-cell proliferation is a critical early event, but other currently unknown changes are required

wildering variety of mechanisms of viral transformation. With the perspective provided by our present understanding of the circuits that control cell proliferation, we can now see that the great majority of these mechanisms fall into one of two general classes. Viral transformation can be the result of either constitutive activation of cytoplasmic signal transduction cascades or disruption of nuclear pathways that negatively regulate cell cycle progression. In both cases, viral proteins or transcriptional control signals override the finely tuned mechanisms that normally ensure that cells duplicate their DNA and divide only when external and internal conditions are propitious.

Such an integrated view of the mechanisms by which viruses belonging to very different families can transform cells is intellectually satisfying. Nevertheless, not all aspects of the complex interactions that can occur between viruses and their host cells are yet understood (Table 18.10). We do not fully appreciate why, in some cases, a single viral oncogene is sufficient for transformation of cells in culture whereas, in others, several viral functions are required. Nor do we understand why viral proteins that apparently alter the same cellular pathway can differ in their transforming potentials. For example, the constitutively active tyrosine kinase encoded in the v-*src* gene of Rous sarcoma virus is sufficient for both transformation of primary cells in culture and induction of sarcomas in animals. In contrast, activation of the c-Src tyrosine kinase by polyomaviral mT allows transformation of established cell lines, but the viral LT protein is also required for transformation of primary cells. As these examples illustrate, some of the more subtle aspects of the mechanisms of viral transformation have yet to be explained. However, the greatest challenge is posed by the mechanisms of tumorigenesis.

As indicated at the beginning of this chapter, transformation of cells in culture is not necessarily accompanied by acquisition of the ability to form tumors in animals. This dissociation is evident in the etiology of some human cancers associated with viral infections. For example, infection by Epstein-Barr virus, which immortalizes human B cells in culture by mechanisms that we can describe in some detail, is but one of several factors implicated in the development of Burkitt's lymphoma. Similarly, the increased rate of proliferation of liver cells following host immune responses to persistent hepatitis B virus infection makes a major contribution to the development of hepatocellular carcinoma. Furthermore, in many cases, cells that are fully transformed by viruses in culture do not induce tumors in animals, or do so only weakly, unless the host's immune defenses are compromised. A deeper appreciation of the parameters that determine a host's response to transformed cells will clearly be necessary if we are to understand the complex process of tumorigenesis. Oncogenic viruses and cells transformed by them will undoubtedly prove as valuable in this formidable endeavor as they have in elucidation of the molecular basis of control of cell proliferation, and the derangement of these processes during transformation.

References

Books

Cooper, G. M. 1995. *Oncogenes*, 2nd ed. Jones and Bartlett, Boston, Mass.

McCance, D. J. (ed.). 1998. *Human Tumor Viruses*. ASM Press, Washington, D.C.

Reddy, E. P., A. M. Skalka, and T. Curran. 1995. *The Oncogene Handbook*. Elsevier, New York, N.Y.

Weinberg, R., and M. Wigler (ed.). 1989. *Oncogenes and the Molecular Origins of Cancer*. Cold Spring Harbor Laboratory Press, Cold Spring Harbor, N.Y.

Chapters in Books

Mason, W. S., A. A. Evans, and W. T. London. 1998. Hepatitis B virus replication, liver disease, and hepatocellular carcinoma, p. 253–281. *In* D. J. McCance (ed.), *Human Tumor Viruses*. ASM Press, Washington, D.C.

Nevins, J. R., and P. K. Vogt. 1996. Cell transformation by viruses, p. 301–343. *In* B. N. Fields, D. M. Knipe, and P. M. Howley (ed.), *Fields Virology*, 3rd ed. Lippincott-Raven Publishers, Philadelphia, Pa.

Review Articles

Damania, B., J. K. Choi, and J. U. Jung. 2000. Signaling activities of gammaherpesvirus membrane proteins. *J. Virol.* **74:**1593–1601.

Dilworth, S. M. 1995. Polyomavirus middle T antigen: meddler or mimic? *Trends Microbiol.* **3:**31–35.

Ekholm, S. V., and S. I. Reed. 2000. Regulation of G(1) cyclin-dependent kinases in the mammalian cell cycle. *Curr. Opin. Cell Biol.* **12:**676–684.

Frame, M. C., V. J. Fincham, N. O. Carragher, and J. A. Wyke. 2002. v-Src's hold over actin and cell adhesions. *Nat. Rev. Mol. Cell. Biol.* **3:**233–245.

Ganem, D. 1997. KSHV and Kaposi's sarcoma: the end of the beginning? *Cell* **91:**157–160.

Goodman, R. H., and S. Smolik. 2000. CBP/p300 in cell growth, transformation, and development. *Genes Dev.* **14:**1553–1577.

Harbour, J. W., and D. C. Dean. 2000. The Rb/E2F pathway: expanding roles and emerging paradigms. *Genes Dev.* **14:**2393–2409.

Harper, J. W., and S. J. Elledge. 1996. Cdk inhibitors in development and cancer. *Curr. Opin. Genet. Dev.* **6:**56–64.

Hunter, T. 1995. Protein kinases and phosphatases: the yin and yang of protein phosphorylation and signaling. *Cell* **80:**225–236.

Laman, H., D. J. Mann, and N. C. Jones. 2000. Viral-encoded cyclins. *Curr. Opin. Genet. Dev.* **10:**70–74.

Marshall, C. J. 1995. Specificity of receptor tyrosine kinase signaling. *Cell* **80:**179–185.

Messerschmitt, A. S., N. Dunant, and K. Ballmer-Hofer. 1997. DNA tumor virus and Src family tyrosine kinases, an intimate relationship. *Virology* **227:**271–280.

Pawson, T., and J. D. Scott. 1997. Signaling through scaffold, anchoring and adaptor proteins. *Science* **278:**2075–2079.

Sherr, C. J. 1998. Tumor surveillance via the ARF-p53 pathway. *Genes Dev.* **12:**2984–2991.

Sionov, R. V., and Y. Haupt. 1999. The cellular response to p53: the decision between life and death. *Oncogene* **18:**6145–6157.

Sullivan, C. S., and J. M. Pipas. 2002. T antigens of simian virus 40: molecular chaperones for viral replication and tumorigenesis. *Microbiol. Mol. Biol. Rev.* **66:**179–202.

Thomas, M., D. Pim, and L. Banks. 1999. The role of the E6-p53 interaction in the molecular pathogenesis of HPV. *Oncogene* **18:**7690–7700.

Toker, A., and L. C. Cantley. 1997. Signaling through the lipid products of phosphoinositide-3-OH kinase. *Nature* **387:**673–676.

Papers of Special Interest

Transformation and Tumorigenesis by Retroviruses

Collett, M. S., and R. L. Erikson. 1978. Protein kinase activity associated with the avian sarcoma virus src gene product. *Proc. Natl. Acad. Sci. USA* **75:**2021–2024.

Dhar, R., R. W. Ellis, T. Y. Shih, S. Oroszlan, B. Shapiro, J. Maizel, D. Lowy, and E. Scolnick. 1982. Nucleotide sequence of the p21 transforming protein of Harvey murine sarcoma virus. *Science* **217:**945–946.

Ellermann, V., and O. Bang. 1908. Experimentelle Leukämie bei Huhnern. *Zentralbl. Bakteriol.* **46:**595–609.

Fasano, O., E. Taparowsky, J. Fiddes, M. Wigler, and M. Goldfarb. 1983. Sequence and structure of the coding region of the human H-ras-1 gene from T24 bladder carcinoma cells. *J. Mol. Appl. Genet.* **2:**173–180.

Hayward, W. S., B. G. Neel, and S. M. Astrin. 1981. Activation of a cellular onc gene by promoter insertion in ALV-induced lymphoid leukosis. *Nature* **290:**475–480.

Hunter, T., and B. Sefton. 1980. Transforming gene product of Rous sarcoma virus phosphorylates tyrosine. *Proc. Natl. Acad. Sci. USA* **77:**1311–1315.

Levinson, A. D., H. Opperman, L. Levintow, H. E. Varmus, and J. M. Bishop. 1978. Evidence that the transforming gene of avian sarcoma virus encodes a protein kinase associated with a phosphoprotein. *Cell* **15:**561–572.

Nusse, R., and H. E. Varmus. 1982. Many tumors induced by the mouse mammary tumor virus contain a provirus integrated in the same region of the host genome. *Cell* **31:**99–109.

Rous, P. 1910. A transmissible avian neoplasm: sarcoma of the common fowl. *J. Exp. Med.* **12:**696–705.

Sicheri, F., I. Moarefi, and J. Kuriyan. 1997. Crystal structure of the Src family tyrosine kinase Hck. *Nature* **385:**602–609.

Stehelin, D., H. E. Varmus, J. M. Bishop, and P. K. Vogt. 1976. DNA related to the transforming gene(s) of avian sarcoma viruses is present in normal avian DNA. *Nature* **260:**170–173.

Xu, W., S. C. Harrison, and M. J. Eck. 1997. Three-dimensional structure of the tyrosine kinase c-Src. *Nature* **385:**595–602.

Transformation by Polyomaviruses

Bolen, J. R., C. J. Thiele, M. A. Israel, W. Yonemoto, L. A. Lipsich, and J. S. Brugge. 1984. Enhancement of cellular src gene product-associated tyrosyl kinase activity following polyomavirus infection and transformation. *Cell* **38:**367–377.

DeCaprio, J. A., J. W. Ludlow, J. Figge, J. Y. Shew, C. M. Huang, W. H. Lee, E. Marsilio, E. Paucha, and D. M. Livingston. 1988. SV40 large tumor antigen forms a specific complex with the product of the retinoblastoma susceptibility gene. *Cell* **54:**275–283.

Lane, D. P., and L. V. Crawford. 1979. T antigen is bound to a host protein in SV40-transformed cells. *Nature* **278:**261–263.

Linzer, D. I., and A. J. Levine. 1979. Characterization of a 54K dalton cellular SV40 tumor antigen present in SV40-transformed cells and uninfected embryonal carcinoma cells. *Cell* **17:**43–52.

Sontag, E., S. Fedorov, C. Kamibayashi, D. Robbins, M. Cobb, and M. Mumby. 1993. The interaction of SV40 small tumor antigen with protein phosphatase 2A stimulates the map kinase pathway and induces cell proliferation. *Cell* **75:**887–897.

Symonds, H., L. Krall, L. Remington, M. Saenz-Robles, S. Lowe, T. Jacks, and T. VanDyke. 1994. p53-dependent apoptosis suppresses tumor growth and progression in vivo. *Cell* **78:**703–711.

Talmage, D. A., R. Freund, A. T. Young, J. Daki, C. J. Dawe, and T. L. Benjamin. 1989. Phosphorylation of middle T by pp60$^{c\text{-}src}$: a switch for binding of phosphatidylinositol 3-kinase and optimal tumorigenesis. *Cell* **54:**55–65.

Zalvide, J., H. Stubdal, and J. DeCaprio. 1998. The J domain of simian virus 40 large T antigen is required to functionally inactivate Rb family proteins. *Mol. Cell. Biol.* **18:**1408–1415.

Transformation by Papillomaviruses

Dyson, N., P. M. Howley, and E. Harlow. 1988. The human papillomavirus E7 oncoprotein is able to bind to the retinoblastoma gene product. *Science* **243:**934–937.

Lai, C. C., C. Henningson, and D. DiMaio. 1998. Bovine papillomavirus E5 protein induces oligomerization and trans-phosphorylation of the platelet-derived growth factor beta receptor. *Proc. Natl. Acad. Sci. USA* **95:**15241–15246.

Scheffner, M., B. A. Werness, J. M. Huibregtse, A. J. Levine, and P. M. Howley. 1990. The E6 oncoprotein encoded by human papillomavirus types 16 and 18 promotes the degradation of p53. *Cell* **63:**1129–1136.

Veldman, T., I. Horikawa, J. C. Barrett, and R. Schlegel. 2001. Transcriptional activation of the telomerase hTERT gene by human papillomavirus type 16 E6 oncoprotein. *J. Virol.* **75:**4467–4472.

Transformation by Adenoviruses

Egan, C., S. T. Bayley, and P. E. Branton. 1989. Binding of the Rb1 protein to E1A products is required for adenovirus transformation. *Oncogene* **4:**383–388.

Martin, M. E. D., and A. J. Berk. 1998. Adenovirus E1B 55K represses p53 activation in vitro. *J. Virol.* **72:**3146–3154.

Moore, M., N. Horikoshi, and T. Shenk. 1996. Oncogenic potential of the adenovirus Orf6 protein. *Proc. Natl. Acad. Sci. USA* **93:**11295–11301.

Rao, L., M. Debbas, D. Sabbatini, D. Hockenberry, S. Korsmeyer, and E. White. 1992. The adenovirus E1A proteins induce apoptosis which is inhibited by the E1B 19K and Bcl-2 proteins. *Proc. Natl. Acad. Sci. USA* **89:**7742–7746.

Somasundaram, K., and W. S. El-Deiry. 1997. Inhibition of p53-mediated transactivation and cell-cycle arrest by E1A through its p300/CBP-interacting region. *Oncogene* **14:**1047–1057.

Van der Eb, A. J., C. Mulder, F. L. Graham, and A. Houweling. 1977. Transformation with specific fragments of adenovirus DNAs. I. isolation of specific fragments with transforming activity of adenovirus 2 and 5. *Gene* **2:**133–146.

Whyte, P., N. M. Williamson, and E. Harlow. 1989. Cellular targets for transformation by the adenovirus E1A proteins. *Cell* **56:**67–75.

Transformation by Herpesviruses

Bias, C., B. Santomasso, O. Coso, L. Arvanitakis, E. Raaka, J. S. Gutkind, A. S. Asch, E. Cesarman, M. C. Gerhengorn, and E. A. Mesri. 1998. G-protein-coupled receptor of Kaposi's sarcoma-associated herpesvirus is a viral oncogene and angiogenesis activator. *Nature* **391:**86–89.

Ciufo, D. M., J. S. Cannon, L. J. Poole, F. Y. Wu, P. Murray, R. F. Ambinder, and G. S. Hayward. 2001. Spindle cell conversion by Kaposi's sarcoma-associated herpesvirus: formation of colonies and plaques with mixed lytic and latent gene expression in infected primary dermal microvascular endothelial cell cultures. *J. Virol.* **75:**5614–5626.

Gires, O., U. Zimber-Strobl, R. Gonnella, M. Ueffing, G. Marschall, R. Zeidler, D. Pick, and W. Hammerschmidt. 1997. Latent membrane protein 1 of Epstein-Barr virus mimics a constitutively active receptor molecule. *EMBO J.* **16:**8131–8140.

Kulwichi, W., R. H. Edwards, E. M. Davenport, J. F. Baskara, V. Godfrey, and N. Raab-Traub. 1998. Expression of the Epstein-Barr virus latent membrane protein 1 induces B cell lymphomas in transgenic mice. *Proc. Natl. Acad. Sci. USA* **95:**11963–11968.

Mann, D. J., E. S. Child, C. Swanton, H. Laman, and N. Jones. 1999. Modulation of p27(Kip1) levels by the cyclin encoded by Kaposi's sarcoma-associated herpesvirus. *EMBO J.* **18:**654–663.

Swanton, C., D. J. Mann, B. Fleckenstein, F. Neipel, G. Peters, and W. Jones. 1997. Herpesviral cyclin/cdk6 complexes evade inhibition by CDK inhibitor proteins. *Nature* **390:**184–187.

Sylla, B. D., S. C. Hung, D. M. Davidson, E. Hatzivassiliou, N. L. Malinin, D. Wallach, T. D. Gilmore, E. Kieff, and G. Mosialos. 1998. Epstein-Barr virus-transformation protein latent infection membrane protein 1 activates transcription factor Nf-κB inducing kinase and the IκB kinases Ikka and Ikkb. *Proc. Natl. Acad. Sci. USA* **95:**10106–10111.

Control and Evolution

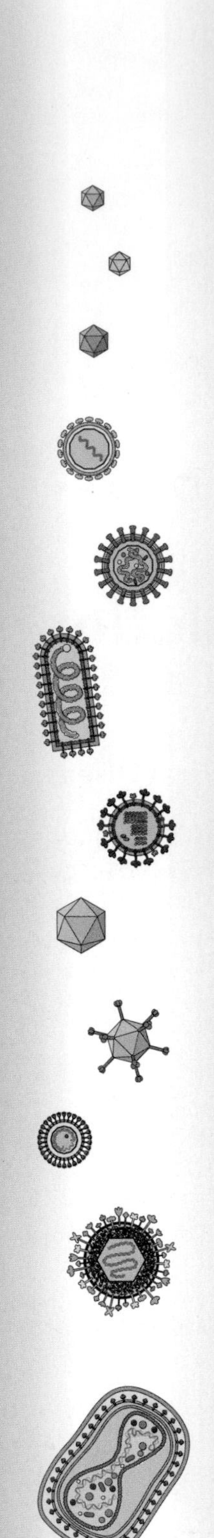

19

Prevention and Control of Viral Diseases

You can't go back and you can't stand still,
If the thunder don't get you, then the
lightning will.

JERRY GARCIA
The Wheel (words by Robert Hunter)

Introduction

The two main efforts of antiviral defense are prevention of viral infections by vaccines and treatment of viral diseases with drugs. When applied properly, both approaches can be remarkably effective. However, the proviso "when applied properly" warns of the complexity of host-virus interactions and the problems of controlling disease. Although we will discuss vaccines and antiviral drugs as distinct arms of antiviral defense, two common threads connect them. First, any agent or method that blocks viral replication, or reduces viral pathogenesis, imposes selection for viral mutants capable of bypassing the blocking agent. Second, as viruses are obligate intracellular parasites, any intervention affecting a host enzyme or process carries inherent risks. Indeed, drug resistance, antiviral drug toxicity, and immunopathology are major problems in any virus control program. While this discussion focuses on methods to control infections caused by common human viruses, the same principles apply to nonhuman viral infections.

Vaccines: a Proven Defense against Viral Infections

Smallpox: a Historical Perspective

Smallpox has been called the most destructive disease in history; smallpox virus has killed, crippled, or disfigured nearly one-tenth of all humankind. In the 20th century alone, more than 300 million people succumbed to this disease. Nevertheless, it was the first to be eliminated from the natural environment by directed human intervention. It is fascinating to follow the convoluted path of discovery that made it possible to orchestrate the demise of such a deadly agent.

Several thousand years ago, the Chinese knew and wrote about diseases that occur only once in a lifetime. The practical aspects of this observation were recognized by Chinese and Indian physicians of the 11th century who injected pus from smallpox lesions into healthy individuals in hope of inducing mild disease that would provide lifelong protection (a process later called **variolation**). Despite such early insight, little progress in controlling the disease was made for centuries. The dangerous practice of variolation was introduced into Europe in the beginning of the 18th century. Unfortunately, many people contracted smallpox from variolation, and the practice not only never became popular, but actually was prevented by law in many countries.

The horror of smallpox was vanishing from our collective consciousness until the specter of bioterrorism appeared in the late 20th century. Consequently, there is renewed interest in the virus and the vaccine that was so effective. The eradication story begins with Edward Jenner (1749–1823), a country doctor and naturalist, who was well known at the time for a seminal paper titled "Observations on the Natural History of the Cuckoo." At first glance, Jenner seems an unlikely candidate to conceive of, and establish, the means by which natural infection by smallpox was eventually eradicated. However, he was a careful and thoughtful observer of his patients. Jenner's insight was that milkmaids appeared to be spared the facial disfigurations of smallpox, because they had cowpox infections on their hands. He put this idea to the test on May 14, 1796, when he injected pus from a cowpox lesion on the finger of milkmaid Sarah Nelmes under the skin of James Phipps, a healthy young boy. Phipps developed a fever and a lesion typical of cowpox at the site of infection. Jenner then deliberately infected Phipps with smallpox 2 weeks later, and the boy survived this potentially lethal challenge. Needless to say, such an experiment would not be possible today.

Despite this amazing result, the Royal Society rejected Jenner's paper describing these experiments, and so he ultimately resorted to publishing his work privately. While it is Jenner's name that has been remembered, his friend William Woodville, a prominent physician, was responsible for the first large-scale clinical trial that confirmed Jenner's observations. Thanks to Woodville's careful clinical research, the idea of harnessing the body's natural defenses as a public health measure was placed on firm ground.

As is the case for many early discoveries, the scientific world was not prepared to exploit Jenner's new findings.

BOX 19.1

The current U.S. smallpox vaccine, 2002

The smallpox vaccine in the 2002 U.S. stockpile to protect civilian and military personnel against deliberate dissemination of smallpox virus is live vaccinia virus that was grown on the skin of calves.

This vaccine stockpile is more than 30 years old, although its potency is asserted to be high. This vaccine, called Dryvax, is the only currently licensed smallpox vaccine and was manufactured by Wyeth. It is a lyophilized preparation of the strain "New York City calf lymph," which was derived from a seed stock of the New York City Board of Health strain that was passed from 22 to 28 times on young calves. The lyophilized calf lymph is reconstituted with 50% glycerol, 0.25% phenol, and 0.005% brilliant green. Distribution of this vaccine was discontinued by Wyeth in 1983.

The vaccinia virus vaccine has been used in millions of vaccinations. The evidence of a productive immune response (called a "take") is a common local reaction of pustular lesions that occur 6 to 10 days after vaccination.

The vaccine causes rare, but serious, adverse reactions including dermatologic reactions, central nervous system disorders, and general disseminated disease. During the program in the United States in which every child was vaccinated, approximately seven to nine deaths per year were attributed to vaccination, with the highest risk for infants. Inadvertent administration of the vaccine to immunodeficient individuals, or to people with certain skin diseases, resulted in a significantly higher number of adverse reactions.

In 2002, the U.S. government announced plans to immunize military personnel and frontline civilian health care workers. Subsequently, the vaccine will be made available to the general public on a voluntary basis.

In today's world, with an AIDS pandemic and the common use of immunosuppressive drugs for transplant patients, the risk of inadvertent infection by the live vaccine virus and subsequent adverse reactions is considerably higher than it was in the 1950s. New vaccines that are safe and efficacious must be prepared to replace the aging stockpile.

Henderson, D. A. 1999. Smallpox: clinical and epidemiological features. *Emerg. Infect. Dis.* **5**:537–539.

Rosenthal, S., M. Merchlinsky, C. Kleppinger, and K. Goldenthal. 2001. Developing new smallpox vaccines. *Emerg. Infect. Dis.* **7**:920–926.

It took 100 more years before the next practical vaccine for a viral disease appeared. Louis Pasteur prepared a rabies vaccine from dried, infected rabbit spinal cord and introduced the term **vaccination** (from *vacca*, Latin for "cow") in honor of Jenner's pioneering work. Even with Pasteur's success, other antiviral vaccines were slow to follow, largely because viruses were difficult to identify, propagate, and study. Indeed, the next vaccines (against yellow fever and influenza viruses) did not appear until the mid-1930s.

Large-Scale Vaccination Programs Can Be Dramatically Effective

As illustrated by the eradication of smallpox, a vaccine is a remarkably effective antiviral defense (Box 19.1). Indeed, vaccines might lead to the eradication of diseases caused by poliovirus, measles virus, mumps virus, and rubella virus. The World Health Organization has made eradication of poliomyelitis and measles the highest priority. Significant progress with poliomyelitis has already been made (Fig. 19.1 and Box 19.2). As an example of the magnitude of the effort, the World Health Organization once immunized 127 million Indian children in more than 650,000 villages in a single day. Measles, which is one of the world's five major child killers, and subacute sclerosing panencephalitis (SSPE), a rare but lethal brain infection, also caused by measles virus, are rapidly disappearing in the United States because of effective vaccines (Fig. 19.1). Unlike the polio vaccine, which can be given orally, the current measles vaccine is injected. This requirement imposes a number of logistical and practical problems that will make worldwide eradication of this virus more difficult.

Eradicating a Viral Disease

The concept of eradicating a viral disease deserves careful contemplation. Viruses have survived countless bouts of selection during evolution, so the hubris of declaring a viral disease "eradicated" is obvious. Global eradication has been announced for smallpox, but some diseases caused by foot-and-mouth disease virus or African swine fever virus are also said to be eradicated from some, but not all, countries. What makes global or national eradication conceivable? A number of parameters are important, as summarized in Table 19.1, but two are absolutely essential for global eradication: the viral infectious cycle must take place in a single host, and infection (or vaccination) must induce lifelong immunity. By definition, a vaccine that renders the host population immune to subsequent infection by a virus **that can grow only in that host** effectively eliminates the virus.

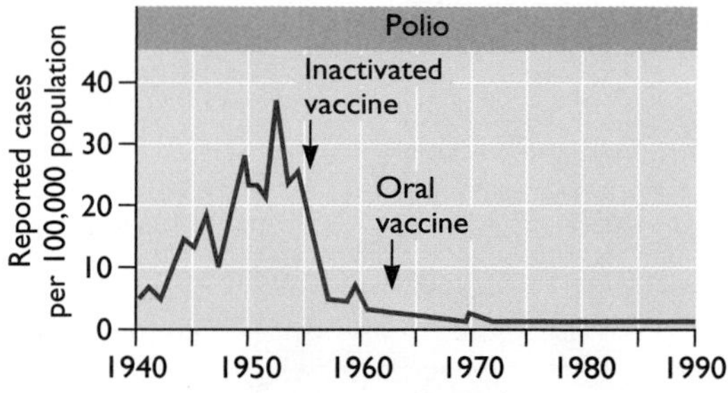

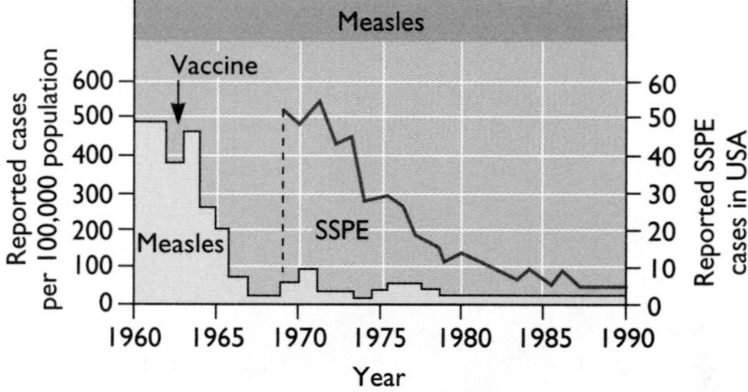

Figure 19.1 Profiles of successful vaccination campaigns. The consequences of poliovirus (top) and measles virus (bottom) infection have been reduced or eliminated in the United States after massive vaccination programs. SSPE is a rare but fatal brain disease that can occur 10 to 15 years after primary measles virus infection. Blocking primary measles virus infection by vaccination also eliminates SSPE. Adapted from C. A. Janeway, Jr., P. Travers, M. Walport, and M. Shlomchik, *Immunobiology: the Immune System in Health and Disease*. (Current Biology Limited, Garland Publishing Inc., New York, N.Y., 2001), with permission.

In contrast, a virus with alternative host species in which to propagate cannot be eliminated by vaccination of a single host population: other means of blocking viral spread are required.

National eradication programs typically are established for economically important livestock diseases. For example, foot-and-mouth disease has been declared eradicated in the United States and Canada but still occurs in parts of Europe and South America. Such efforts can be successful only when supplemented with broad governmental enforcement and border security, as animals in the virus-free country are constantly exposed to sources from outside the country. Obviously, other principles of control must be implemented. For example, surveillance and containment strategies must be mobilized quickly and aggressively to identify and stop the spread from niche outbreaks. A common practice is to slaughter every host animal in farms at increasing distances surrounding an outbreak site (the so-called "ring-slaughter" program

BOX 19.2

The poliomyelitis eradication effort: should vaccine eradication be next?

The worldwide effort to eradicate poliomyelitis is likely to reach its goal by 2005. The World Health Organization, in cooperation with the Centers for Disease Control and Prevention, has indicated that vaccination will cease thereafter, followed by the destruction of all poliovirus stocks. However, several aspects of these crucial final steps require reconsideration. Because the World Health Organization is relying on oral poliovirus vaccine, and because the Sabin oral strains readily revert to virulent forms, potentially pathogenic virions are still being released into aquifers. Immunocompromised individuals who have received live oral vaccine may also shed virions for years in the absence of clinical symptoms. We do not know how long such viruses, some of which may cause vaccine-associated poliomyelitis, will persist in the postvaccine era and pose a threat to unvaccinated individuals.

There have already been several sobering reminders of the genetic instability of the Sabin oral vaccine strains and its potential effect on the eradication campaign. In 2000–01, 6 years after the Americas had been certified as poliomyelitis-free, an outbreak occurred in the Dominican Republic and Haiti. There were a total of 21 confirmed cases, all but one of which occurred in unvaccinated or incompletely vaccinated children. The viruses responsible for the outbreak were derived from a single dose of Sabin poliovirus type 1 that had probably been administered in 1998–99. The neurovirulence and transmissibility of these viruses were indistinguishable from those of wild-type poliovirus type 1. All isolates from the outbreak were recombinants, with the 5′ untranslated region and the capsid coding sequence from Sabin type 1 poliovirus and the remainder of the viral genome from other enteroviruses. Furthermore, all the mutations associated with attenuation of the neurovirulence of Sabin type 1 poliovirus had either reverted or been removed by recombination. Evidence of circulating vaccine-derived poliovirus has also been found elsewhere. In Egypt, type 2 vaccine-derived poliovirus circulated from 1983 to 1993 and was associated with 32 reported cases. In these outbreaks, poor population immunity and the absence of circulating wild-type poliovirus were critical risk factors.

Before vaccination can be stopped, it will be necessary to destroy most existing viral stocks and restrict access to the remainder to prevent both accidental and deliberate release into the environment. In the case of smallpox, viral stocks were located at only a few institutions before eradication, and so control of the inventory seemed possible. There is no central record of poliovirus stocks, which are distributed among hundreds or possibly thousands of sites. Without an accurate inventory, it is unlikely that all viral stocks can be found and destroyed.

The recent outbreaks of poliomyelitis caused by circulating vaccine-derived virions, together with the other considerations discussed above, emphasize the need to conduct extensive global surveillance for poliovirus well after eradication has been achieved. In addition, emergency stockpiles of poliovirus vaccine must be maintained to control possible outbreaks caused by circulating vaccine-

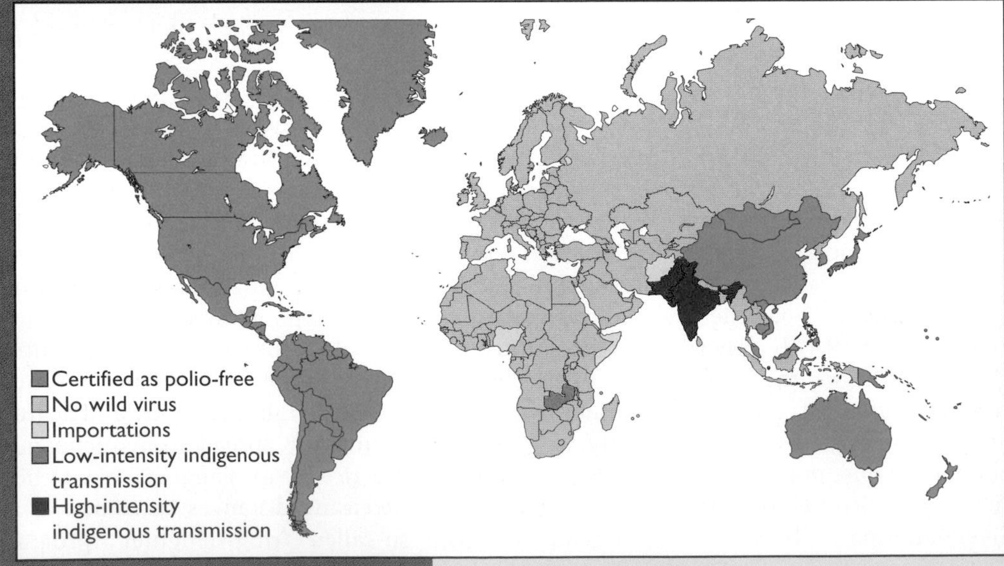

Globally reported incidence of poliomyelitis in 2001 and 2002. The Americas, Western Pacific, and Europe have been declared poliomyelitis-free by the World Health Organization. The number of cases has declined from an estimated 350,000 in 1988 to 470 in 2001. At the same time, the number of countries in which poliovirus is endemic has decreased from >125 to 10.

BOX 19.2 (continued)

derived poliovirus, a breach in poliovirus containment, or deliberate release.

Dove, A. W., and V. R. Racaniello. 1997. The polio eradication effort: should vaccine eradication be next? *Science* **277:**779–780.

Hull, H. F., and R. B. Aylward. 1997. Ending polio immunization. *Science* **277:**780.

Kew, O., V. Morris-Glasgow, M. Landaverde, C. Burns, J. Shaw, Z. Garib, J. Andre, E. Blackman, C. J. Freeman, J. Jorba, R. Sutter, G. Tambini, L. Venczel, C. Pedreira, F. Laender, H. Shimizu, T. Yoneyama, T. Miyamura, H. van Der Avoort, M. S. Oberste, D. Kilpatrick, S. Cochi, M. Pallansch, and C. de Quadros. 2002. Outbreak of poliomyelitis in Hispaniola associated with circulating type 1 vaccine-derived poliovirus. *Science* **296:**356–359.

Table 19.1 Features of smallpox that enabled its eradication

Virology and disease aspects
No secondary hosts: this is a human-only virus
Long incubation period
Infectious only after incubation period
Low communicability
No persistent infection
Subclinical infections are not a source of spread
Easily diagnosed
Immunology
Infection confers long-term immunity
One stable serotype
Effective vaccine available
Vaccine is stable and cheap
Social and political aspects
Severe disease with high morbidity and mortality
Considerable savings to developed, nonendemic countries
Eradication from developed countries demonstrated its feasibility
Few cultural or social barriers to case-tracing and control

[Box 19.3]). Because acute infections spread rapidly from the outbreak site by many routes, and do so before identifiable symptoms are visible, the ring-slaughter containment often is breached unknowingly before it is initiated. To deal with this fact, preemptive slaughter of **all** animals on "at-risk" farms may be required. Obviously, the faster an outbreak is identified, the more likely the success of containment actions. Unfortunately, on-farm diagnostic tools that provide guaranteed identification of pathogens before symptoms are visible are simply not available. A false-positive identification of an outbreak will have serious consequences.

The strategies devised originally for national veterinary virus control take on new significance when the possibility that human and animal viruses may be used for nefarious purposes is considered. We cannot be sure that deadly, frightening, or economically devastating viruses are not hidden away for use as weapons, or to inflict terror on an unsuspecting and unprotected public (Box 19.4).

Preparation of Vaccines

The Fundamental Challenge

Given the remarkable success of the smallpox and polio vaccines, it might seem reasonable to prepare vaccines against all viral diseases. Unfortunately, despite considerable progress in research, we cannot predict with confidence the efficacy or side effects of vaccines. We lack sufficiently detailed knowledge of the important mechanisms of immune protection against most viral infections. Even basic questions such as "Is a neutralizing antibody response important, or is a cytotoxic T-lymphocyte (CTL) response essential?" cannot be answered with certainty for

BOX 19.3

Stopping epidemics in agricultural animals by culling and slaughter

Vaccination of agriculturally important animals such as cattle and swine may not be cost-effective or may run afoul of government rules that block the shipping and sale of animals with antibodies to certain viruses. The recent foot-and-mouth disease epidemic in the United Kingdom provides a dramatic example of how viral disease is controlled when vaccination is not possible. The solution that stopped the epidemic was mass slaughter of **all animals** surrounding the affected areas and chemical decontamination of farms, followed by repopulation of herds. It is estimated that 6,131,440 animals were destroyed in less than a year before the spread of foot-and-mouth disease virus was contained.

Animal slaughter is often the only alternative available to officials dealing with potential epidemic spread. For example, in recent years, millions of chickens in Hong Kong were killed to stop an influenza virus epidemic with potential to spread to humans. In 2002, millions of chickens in California were slaughtered to stem the spread of Newcastle disease virus in major poultry factories.

Keeling, M. J., M. E. J. Woolhouse, R. M. May, G. Davies, and B. T. Grenfell. 2003. Modeling vaccination strategies against foot-and-mouth disease. *Nature* **421:**136–142.

Woolhouse, M., and A. Donaldson. 2001. Managing foot-and-mouth. The science of controlling disease outbreaks. *Nature* **410:**515–516.

BOX 19.4

Now that wild smallpox virus has been eliminated, should laboratory stocks of smallpox virus be destroyed?

The important issues are the following:

- Should we destroy biodiversity and gene pools that are not well understood?
- Is wild smallpox virus simply a bad collection of genes that should be destroyed?
- Are stocks of smallpox virus necessary for development of vaccines and antivirals?
- How do we ensure that all stocks have been destroyed?

Fenner, F. 1996. History of smallpox, p. 25–37. *In* H. Koprowski and M. B. A. Oldstone (ed.), *Microbe Hunters, Then and Now*. Medi-Ed Press, Bloomington, Ill.

Henderson, D. A. 1999. The looming threat of bioterrorism. *Science* **283:**1279–1282.

our most common viral infections. In fact, only when a vaccine is effective, or more often, when it fails, can we learn what actually does or does not constitute a protective response. The regulatory network of gene products required to reproduce an effective antiviral response is understood only in outline. The fundamental challenge now is to capitalize on the discoveries in molecular virology to make more effective and safe vaccines. Despite many problems, those involved in vaccine research and development remain optimistic, because of our remarkable history of success and our increasingly detailed knowledge of the molecular nature of viruses and viral disease.

Figure 19.2 Comparison of the predicted immune responses to live and inactivated viruses used in vaccine protocols. (Top) Immune responses plotted against time after injection of a killed virus vaccine (red curve). Three doses of inactivated virions were administered at the indicated positions. (Bottom) Results after injection of a live attenuated virus vaccine. A single dose was administered at the start of the experiment. The filled histogram (lavender-colored area) under the curve displays the titer of infectious attenuated virus. Redrawn from C. A. Mims et al., *Mims' Pathogenesis of Infectious Disease*, 4th ed. (Academic Press, Inc., Orlando, Fla., 1995), with permission.

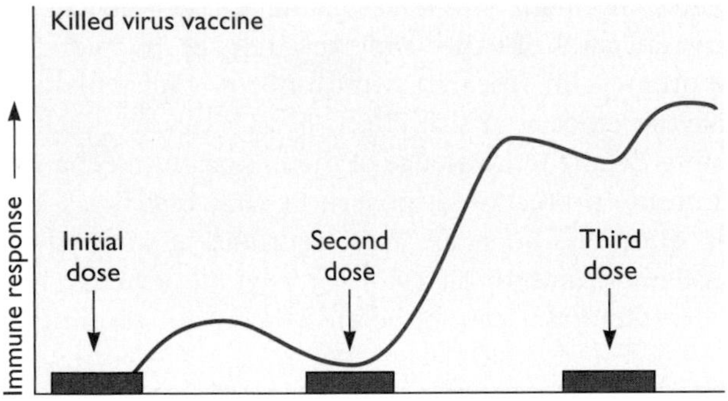

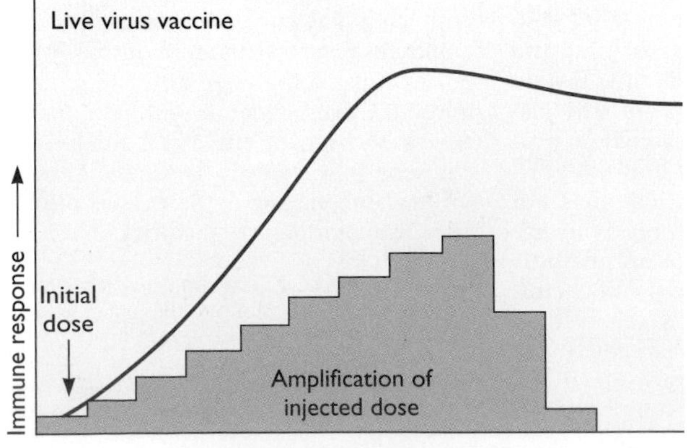

Vaccine Basics

Many successful commercial vaccines simply comprise attenuated or inactivated virions (Fig. 19.2); their preparation is based on principles that would be understood by Pasteur (Table 19.2 and Boxes 19.1 and 19.5). Vaccines can be classified broadly in three general groups: **attenuated viruses, inactivated (killed) virions,** and **purified viral components (subunit vaccines).**

Vaccines Stimulate Immune Memory

Vaccines work primarily because the immune system can recall the identity of a specific virus years after the initial encounter, a phenomenon called **immune memory** (Box 19.6). While the molecular aspects of the establishment and maintenance of memory continue to be active topics of research and debate, the resounding practical success of immunization in stimulating long-lived immune memory stands as one of humankind's greatest medical achievements.

In general, we define immune memory as the presence of specifically dedicated T and B lymphocytes that remain after an infection has waned and maintain a heightened ability to respond to, and proliferate after, subsequent infection (Fig. 19.3). Antiviral vaccines establish immunity and memory without the pathogenic events typical of the initial encounter with a virulent virus. Ideally, an effective vaccine is one that induces and maintains significant concentrations of specific antibodies (products of B cells) in serum and at points of viral entry. At the same time, T cells responsible for specific cellular immunity must be maintained in a precursor state, ready to make their lethal products (e.g., granzymes and perforins) when challenged by subsequent infection.

Table 19.2 Viral vaccines licensed in the United States[a]

Disease or virus	Type of vaccine	Indications for use	Schedule
Adenovirus	Live, attenuated, oral	Military recruits	One dose
Hepatitis A	Inactivated whole virus	Travelers, other high-risk groups	0, 1, and 6 mo
Hepatitis B	Yeast-produced recombinant surface protein	Universal in children, exposure to blood, sexual promiscuity	0, 1, 6, and 12 mo
Influenza	Inactivated viral subunits	Elderly and other high-risk groups	Two-dose primary series, then one seasonal dose
Japanese encephalitis	Inactivated whole virus	Travelers or inhabitants of high-risk areas in Asia	0, 7, and 30 days
Measles	Live attenuated	Universal vaccination of infants	12 mo; 2nd dose, 6–12 yr
Mumps	Live attenuated	Universal vaccination of infants	Same as measles, given as MMR[b]
Polio (inactivated)	Inactivated whole viruses of types 1, 2, and 3	Changing: commonly used for immunosuppressed where live vaccine cannot be used	2, 4, and 12 to 18 mo; then 4 to 6 yr
Polio (live)	Live, attenuated, oral mixture of types 1, 2, and 3	Universal vaccination; no longer used in United States	2, 4, and 6–18 mo
Rabies	Inactivated whole virus	Exposure to rabies, actual or prospective	0, 3, 7, 14, and 28 days after exposure
Rubella	Live attenuated	Universal vaccination of infants	Same as measles, given as MMR
Smallpox	Live vaccinia virus	Certain laboratory workers	One dose
Varicella	Live attenuated	Universal vaccination of infants	12–18 mo
Yellow fever	Live attenuated	Travel in areas where infection is common	One dose every 10 yr

[a]Adapted from D. D. Richman, R. J. Whitley, and F. G. Hayden (ed.), *Clinical Virology*, 2nd ed. (ASM Press, Washington, D.C., 2002), with permission.
[b]MMR, measles-mumps-rubella.

BOX 19.5

The response to infection and vaccination

After a natural infection, viral proteins that can be recognized by the immune system are made in the infected cell. The production of progeny virions and their subsequent spread to other cells amplifies the response. However, vaccination may not reproduce all aspects of this response. Several important variables include the following.

- Infection by live attenuated viruses may provoke an immune response that is qualitatively different from that stimulated by infection with virulent virus.
- In the case of killed vaccines or subunit vaccines, no new viral proteins are made after injection. The immune system therefore **must** recognize only the input material.
- Inactivated virus particles and virion subunits often induce an antibody response, but they rarely induce an effective CTL response.
- Inactivated virus particles or virion subunits may not persist in the body long enough to establish an effective immune memory response.

Ada, G. L. 1994. Vaccines and the immune response, p. 1503–1507. *In* R. G. Webster and A. Granoff (ed.), *Encyclopedia of Virology*. Academic Press, San Diego, Calif.

BOX 19.6

A natural "experiment" demonstrating immune memory

A striking example of immune memory is provided by a natural "experiment" in the 18th century on the Faroe Islands. In 1781 a devastating measles outbreak drastically reduced the islands' population. For the next 65 years, the islands remained measles free and the surviving population flourished. In 1846 measles struck again, infecting over 75% of the population with similar devastating results. An astute Danish physician noted that none of the aged people who survived the 1781 epidemic were infected. However, their age-matched peers who had not been infected earlier were ravaged by measles.

This natural experiment illuminates two important points: immune memory lasts a long time, and memory is maintained during this time without reexposure to another round of infection.

Ahmed, R., and D. Gray. 1996. Immunological memory and protective immunity: understanding their relation. *Science* **272**:54–60.

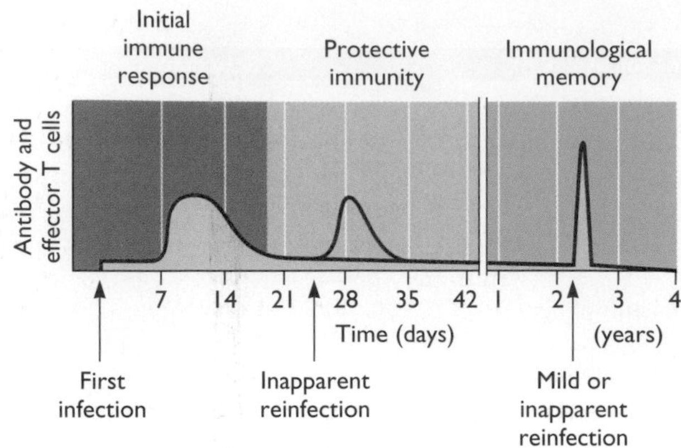

Figure 19.3 Antibody and effector T cells are the basis of protective immunity. The relative concentration of antibody and T cells is shown as a function of time after first (primary) infection. Antibody levels and activated T cells decline after the primary viral infection is cleared (purple). A second, inapparent infection is noted (tan) and recognized by the rapid increase in specific antibodies and T cells. This infection is "inapparent" because the patient shows no symptoms of the infection. The rapid immune response clears this infection with minimal effects. Reinfections at later times (years later), even if mild or inapparent, are marked by rapid and robust immune response because of "memory." Adapted from C. A. Janeway, Jr., P. Travers, M. Walport, and M. Shlomchik, *Immunobiology: the Immune System in Health and Disease* (Current Biology Limited, Garland Publishing Inc., New York, N.Y., 2001), with permission.

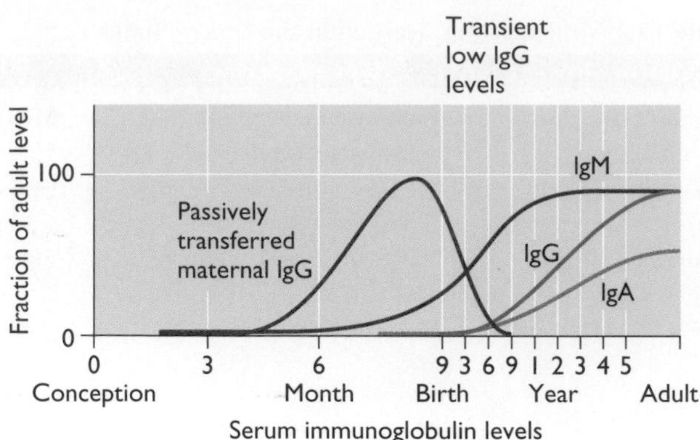

Figure 19.4 Passive transfer of antibody from mother to infant. The fraction of the adult concentration of various antibody classes is plotted as a function of time, from conception to adult. Newborn babies have high levels of circulating IgG antibodies derived from the mother during gestation (passively transferred maternal IgG), enabling the baby to benefit from the broad immune experience of the mother. This passive protection falls to low levels at about 6 months of age as the baby's own immune response takes over. Total antibody concentrations are low from about 6 months to 1 year after birth, which may lead to susceptibility to disease. Premature infants are particularly at risk for infections because the level of maternal IgG is lower and their immune system is less well developed. The time course of production of various isoforms of antibody (IgG, IgM, and IgA) synthesized by the baby is indicated. Adapted from C. A. Janeway, Jr., P. Travers, M. Walport, and M. Shlomchik, *Immunobiology: the Immune System in Health and Disease* (Current Biology Limited, Garland Publishing Inc., New York, N.Y., 2001), with permission.

Immunization Can Be Active or Passive

Active immunization with modified virions or purified viral proteins induces an immunologically mediated resistance to disease. In contrast, **passive immunization** introduces the **products** of the immune response (e.g., antibodies or stimulated immune cells) obtained from an appropriate donor(s) directly into the patient. Passive immunization is a preemptive response, usually given when a virus epidemic is suspected. In 1997, consumption of contaminated fruit led to a widespread outbreak of hepatitis A infections in the United States. Pooled human antibodies (also called immune globulin) were administered in an attempt to block the spread of infection and reduce disease. Immune globulin contains the collective experience of many individual infections and provides instant protection against some infections. The standard procedure for smallpox vaccination with live vaccinia virus requires that so-called "vaccine immune globulin" be available should disseminated vaccinia occur. When stimulated immune cells (e.g., T cells) are used, the process is called **adoptive transfer**. Passive immunization is expected to produce short-term effects, depending on the biological half-lives of the antibodies or immune cells. Mothers passively immunize their babies through colostrum (antibody-rich first milk), or by transfer of maternal antibody to the fetus via the placenta, providing a protective umbrella against a number of pathogens (Fig. 19.4). This protective effect can be detrimental if active immunization of infants is attempted too early, as maternal antibody may block the vaccine from stimulating immunity in the infant.

Vaccines Must Be Safe, Efficacious, and Practical

The major prerequisites for an effective vaccine are presented in Table 19.3. The overriding requirement for any vaccine is safety: the vaccine cannot cause undue harm. For example, it is imperative that infectious particles and viral nucleic acids be undetectable in vaccines containing inactivated virions or viral proteins. If a live attenuated vaccine is used, virulent revertants must be exceedingly rare, or undetectable. Contamination of vaccines with adventitious agents, such as other microbes introduced during production, must be avoided. These safety ideals are easy to state, but absolute safety is impossible to guarantee. Human error during vaccine production and testing, as well as limits of process biochemistry and cell biology,

Table 19.3 Requirements of an effective vaccine

Requirement	Comments
Safety	The vaccine must not cause disease. Side effects must be minimal.
Induction of protective immune response	Vaccinated individual must be protected from illness due to pathogen. Proper innate, cellular, and humoral responses must be evoked by vaccine.
Practical issues	Cost per dose must not be prohibitive. The vaccine should be biologically stable (no genetic reversion to virulence; able to survive use and storage in different surroundings). Vaccine should be easy to administer (oral delivery preferred to needles). The public must see more benefit than risk.

are simple facts of life. Rare side effects or immunopathology often can be identified only after millions of people have been vaccinated. Furthermore, live vaccine agents have the potential to spread to nonvaccinated individuals in a population, with potentially serious consequences. For example, the smallpox vaccine is not given intentionally to immunosuppressed individuals, because they cannot contain the infection and can die. These individuals obviously are at risk in a major vaccination program with the current smallpox vaccine (Box 19.1). Untoward side effects or other safety issues have been the death knell of many otherwise effective vaccines. Safety is paramount because vaccines are given to millions of people, many of whom may never be infected by the virus targeted by the vaccine.

The next requirement is that the vaccine must induce protective immunity **in the population as a whole.** Not every individual in the population need be immunized to stop viral spread, but a sufficient number must become immune to impede virus transmission. Infection stops when the viral load drops below the threshold required to sustain the infection in the population. This effect has been called **herd immunity**. The actual threshold is agent and population specific, but it generally corresponds to 80 to 95% of the population having vaccine-induced immunity. Achieving such high levels of immunity by vaccination is a daunting task. Moreover, if the virus remains in other populations, or in alternative hosts, reinfection is always possible. Closed populations (e.g., military training camps) can be immunized easily, but extended populations spread over large areas of the country present serious logistic and social problems. In addition, public complacency, or reluctance to be immunized, is dangerous to any vaccine program (Box 19.7). Unfortunately, epidemics can occur if the immunity of a population falls below a critical level because an available vaccine is not used.

The protection provided by a vaccine must also be long-term, lasting many years. Some vaccines cannot provide lifelong immunity after a single administration. In such cases, inoculations given after the initial vaccination (booster shots) to stimulate waning immunity are effective, but may be impractical to administer in large populations. Protective immunity also requires that the proper immune response be mounted at the site of potential viral infection in the host. Primary infection by some viruses, such as poliovirus, can be blocked only when a robust **antibody response** is evoked by vaccination. On the other hand, a potent **cellular immune response** is required for protection against herpesviral disease. To maximize effectiveness, a vaccine must be tailored to fit its viral target.

Outbred populations always have varied responses to vaccination. Some individuals exhibit a robust response, while others may not respond as well (a "poor take"). While many factors are responsible for this variability, the age and health of the recipient are major contributors. For example, the influenza virus vaccine available each year is far more effective in young adults than in the elderly. Weak immune responses to vaccination pose several problems. Obviously, protection against subsequent infection may be inadequate, but another concern is that upon such subsequent infection, replication will occur in the presence of the weak immune effectors. Mutants that can escape the host's immune response can then be selected, and may spread in the immunized population. Indeed, vaccine "escape" mutants (often called genetic drift) are well documented.

Important practical requirements for an effective vaccine include stability, ease of administration, and low cost. If a vaccine can be stored at room temperature, rather than refrigerated or frozen, it can be used where cold storage facilities are limited. If a vaccine can be administered orally, rather than by injection, more people will accept the vaccination. The World Health Organization estimates that vaccine costs must be less than $1 per dose if much of the world is to afford it. However, the research and development costs for a modern vaccine is in the range of $50 million to $350 million. Another, often prohibitive expense is covering the liability of the vaccine producer if there should be problems. Liability expenses can be astronomical in a litigious society and have forced many companies to forgo vaccine development completely. Unfortunately, there is an inherent conflict between providing a good return on investment to vaccine developers and supplying vaccines to people and government agencies with a limited ability to bear the cost.

BOX 19.7

National vaccine programs depend on public acceptance of their value

The measles-mumps-rubella (MMR) vaccine has proved to be effective in reducing the incidence of these highly contagious and serious diseases. The economic benefit in the United States from use of the MMR vaccine has been estimated to exceed $5 billion per year. However, in the past few years, the press has reported studies that link the MMR vaccine to autism. Infants can receive many immunizations early in life, and the symptoms of autism often first appear at the time of these immunizations. The MMR vaccine was singled out as a potential link to autism in some studies. Of prime concern was that thimerosal, a mercury compound with potential neurological effects, was used as a preservative in vaccines. As a result of the publicity and parental concern, public confidence has been shaken in most developed countries, and, in some, people are refusing to vaccinate their children. For example, the immunization rate has fallen significantly in the United Kingdom and, consequently, the incidence of measles infection is on the rise.

In the United States, the Centers for Disease Control and the National Institutes of Health recognized the need for an independent group to examine the hypothesized MMR-autism link and address other vaccine safety issues as well. The committee's assessment of the issue is described in the report titled *Immunization Safety Review: Measles-Mumps-Rubella Vaccine and Autism.*

The committee concluded in its report that "The evidence favors rejection of a causal relationship at the population level between MMR vaccine and autistic spectrum disorders (ASD)," and that "a consistent body of epidemiological evidence shows no association at a population level between MMR and ASD." Other leading medical groups, the American Academy of Pediatrics, the World Health Organization, and British health authorities have come to similar conclusions for largely the same reasons. In 1999, the Food and Drug Administration ordered the elimination of thimerosal from children's vaccines, despite evidence that the mercury exposure was too low to be responsible for neurological defects.

Unfortunately, public confidence in government proclamations on any subject is not always high. Moreover, members of the lay public are not well trained to analyze data collected from complex studies that seek to find correlations, or to determine cause and effect. It makes no difference if the public perceives that cancer is caused by high-tension lines or that autism is caused by vaccination: proving (or disproving) cause and effect to a disbelieving public is an exceedingly difficult task.

Information on the Immunization Safety Review Committee can be found at http://www.iom.edu/ImSafety. The full report can be purchased or read online at http://www.nap.edu/catalog/10101.html. Copies of *Immunization Safety Review: Measles-Mumps-Rubella Vaccine and Autism* are available for sale from the National Academy Press; call (800) 624-6242 or (202) 334-3313 (in the Washington, D.C., metropolitan area), or visit the National Academy Press home page at http://www.nap.edu.

Three Types of Vaccine

Live Attenuated Virus Vaccines

Virulent viruses can be made less virulent (attenuated) by growing them in cells other than those of the normal host, or by propagating them at nonphysiological temperatures (Fig. 19.5 and Box 19.8). Mutants able to propagate better than the wild-type virus under these selective conditions arise during viral replication. When such mutants are isolated, purified, and subsequently tested for pathogenicity in appropriate models, some may be less pathogenic than their parent. These **attenuated mutants** are good vaccine candidates, because they no longer grow well in the natural host, but replicate enough to stimulate an immune response. Temperature-sensitive and cold-adapted mutants are often less pathogenic than the parental virus, because of reduced capacity for replication and spread in the warm-blooded host. Cold-adapted influenza viruses with mutations in almost every gene are licensed. In the case of viruses with segmented genomes (e.g., arena-, orthomyxo-, bunya-, and reoviruses), attenuated, reassortant viruses may be obtained after mixed infections with pathogenic and nonpathogenic viruses.

Live oral poliovirus vaccine comprises three attenuated strains selected for their reduced neurovirulence. Types 1 and 3 vaccine strains were isolated by the passage of virulent viruses in different cells and tissues until mutants with reduced neurovirulence in laboratory animals were ob-

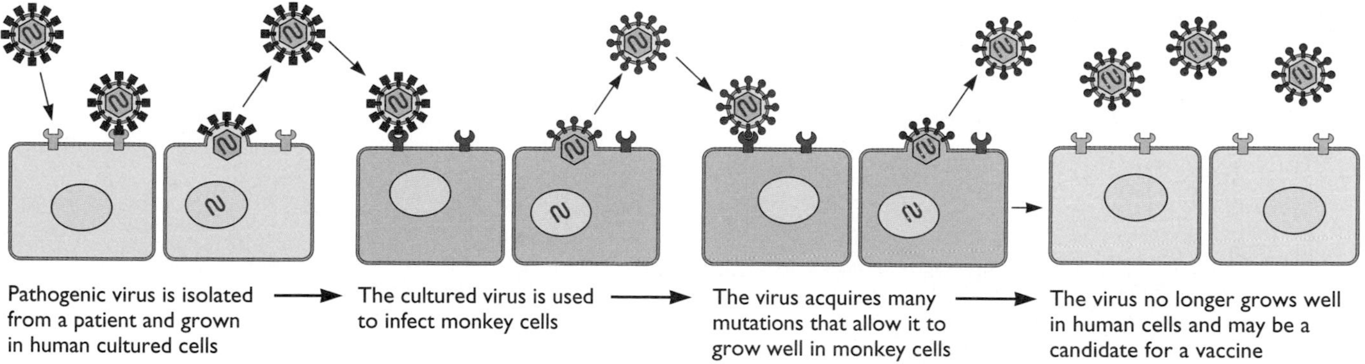
Pathogenic virus is isolated from a patient and grown in human cultured cells
The cultured virus is used to infect monkey cells
The virus acquires many mutations that allow it to grow well in monkey cells
The virus no longer grows well in human cells and may be a candidate for a vaccine

Figure 19.5 Viruses specific for humans may become attenuated by passage in nonhuman cell lines. The four panels show the process of producing an attenuated human virus by repeated transfers in culture cells. The first panel depicts isolation of the virus with cultured human cells (yellow). The second panel shows passage of the new virus in monkey cells (lavender). During the first few passages in nonhuman cells, virus yields may be low. Viruses that grow better can be selected by repeated passage, as shown in the third panel. These viruses usually have several mutations, facilitating growth in nonhuman cells. The last panel shows one outcome in which the monkey cell-adapted virus now no longer grows well in human cells. This virus may also be attenuated (have reduced ability to cause disease) after human infection. Such a virus may be a candidate for a live vaccine if it will induce immunity, but not disease. Adapted from C. A. Janeway, Jr., P. Travers, M. Walport, and M. Shlomchik, *Immunobiology: the Immune System in Health and Disease* (Current Biology Limited, Garland Publishing Inc., New York, N.Y., 2001), with permission.

tained (see Fig. 19.6 for a description of how type 3 was derived). The type 2 component was derived from a naturally occurring attenuated isolate. The mutations responsible for the attenuation phenotypes of all three serotypes are shown in Fig. 19.6. Curiously, each of the three serotypes contains a different mutation in the 5′ noncoding region that may affect messenger RNA (mRNA) translation. Mutations in capsid proteins are believed to influence viral assembly.

The live measles virus vaccine currently in use was derived from the original wild virus isolated in 1954 and called the Edmonston strain. Attenuated viruses were isolated following serial passages through human kidney, human amnion, sheep kidney, and chick embryo cells (Fig. 19.7). Even though this approach was empirical, the virions that were isolated replicated poorly at body temperature and caused markedly less disease in primates. The vaccine strain harbors a number of mutations, including several that affect the viral attachment protein.

Successful vaccines comprising live attenuated viruses are effective for at least three reasons: limited viral replication occurs, progeny virions are often contained at the site of replication, and the infection induces mild or inapparent disease. Live attenuated enteric viruses are administered by injection (e.g., measles-mumps-rubella vaccine) or by mouth (e.g., poliovirus, rotavirus, and adenovirus vaccines). The highly effective Sabin poliovirus vaccine is administered as drops to be swallowed, and enteric adenovirus vaccines are administered as virus-impregnated tablets. One virtue of the oral delivery method for enteric viruses is that it mimics the natural route of infection and, therefore, will induce the natural immune response. Another is that it bypasses the traditional need for hypodermic needles required for intramuscular or intradermal

BOX 19.8

Our best vaccines are based on old technology

It seems ironic in this age of modern biology that the mutations in many of our common vaccine strains were made, not by site-directed mutagenesis of genes known to be required for viral virulence, but rather by selection of mutants that could replicate in various cell types. The vaccines were produced with little bias as to how to reduce virulence. In current jargon, the vaccines were not the product of "reverse genetics."

Despite the old technology, the vaccines are relatively safe and remarkably effective. Consequently, their analysis has led to the identification of key attenuating mutations, as well as parameters affecting the protective immune response. The current vaccines, therefore, not only provide protection, but they also are the foundation for future vaccines.

A Derivation of Sabin type 3 attenuated poliovirus

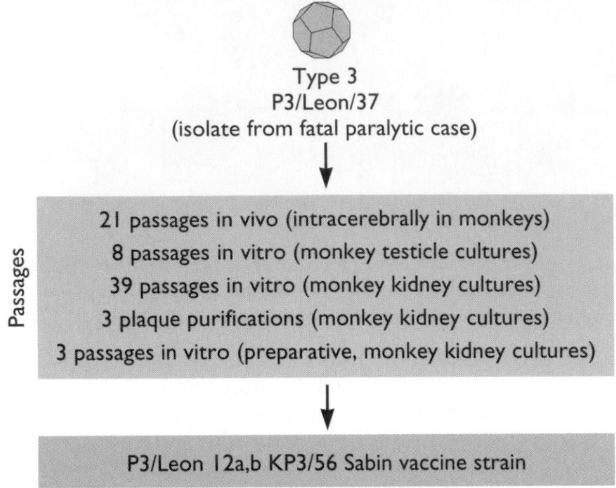

B Determinants of attenuation in the Sabin vaccine strains

Virus	Mutation (location/nucleotide position)
P1/Sabin	5'-UTR (480) VP1 (1106) VP1 (1134) VP3 (3225) VP4 (4065)
P2/Sabin	5'-UTR (481) VP1 (1143)
P3/Sabin	5'-UTR (472) VP3 (3091)

C Reversion of P3/Sabin

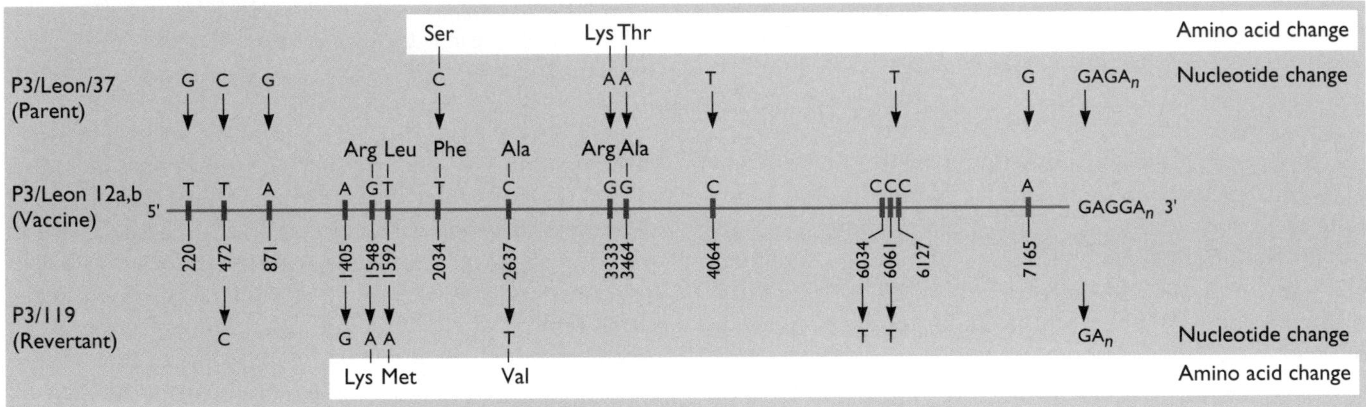

Figure 19.6 Live attenuated Sabin oral poliovirus vaccine and measles vaccines. (A) All three serotypes of poliovirus may cause poliomyelitis. Therefore, the Sabin vaccine is administered as a mixture of three different strains that are representatives of poliovirus serotypes 1, 2, and 3. Shown is the derivation of the type 3 vaccine strain, called P3/Sabin (the letter P means poliovirus). The parent of P3/Sabin is P3/Leon, a virus isolated from the spinal cord of an 11-year-old boy named Leon, who died of paralytic poliomyelitis in 1937 in Los Angeles. P3/Leon virus was passaged serially in monkeys, monkey testicle cell cultures, and monkey kidney cell cultures. At various intervals, viruses were cloned by limiting dilution, and the virulence of the virus was determined in monkeys. An attenuated strain was selected to be the final P3/Sabin strain included in the vaccine. (B) Determinants of attenuation in all three strains of the Sabin vaccine. The mutations responsible for the reduced neurovirulence of each serotype of the live poliovirus vaccine are shown. Mutations in the 5′ untranslated region (5′-UTR) are given as nucleotide numbers. Mutations in the capsid proteins VP1, VP3, and VP4 are given as a four-digit number in which the first digit is the capsid protein and the next three digits identify the amino acid within the capsid protein. (C) Reversion of P3/Sabin vaccine strain. Differences in the nucleotide sequences of the virulent P3/Leon strain and the attenuated P3/Sabin vaccine strain are shown above the cartoon of the parental viral RNA. A virus (P3/119) isolated from a case of vaccine-associated poliomyelitis is shown below. Comparison of the sequences demonstrates that P3/119 is a true revertant of P3/Sabin.

delivery. Live attenuated respiratory viruses can also be delivered by nasal spray, simulating the natural route of infection and immune response.

Live attenuated virus vaccines have inherent problems. For example, despite reduced spread of attenuated viruses in the vaccinee, some viral shedding occurs and these virions are available to infect unvaccinated individuals (Box 19.2). Given the high rate of mutation associated with RNA virus replication, a reversion to virulence should not surprise us. Such reversion is one of the main obstacles to

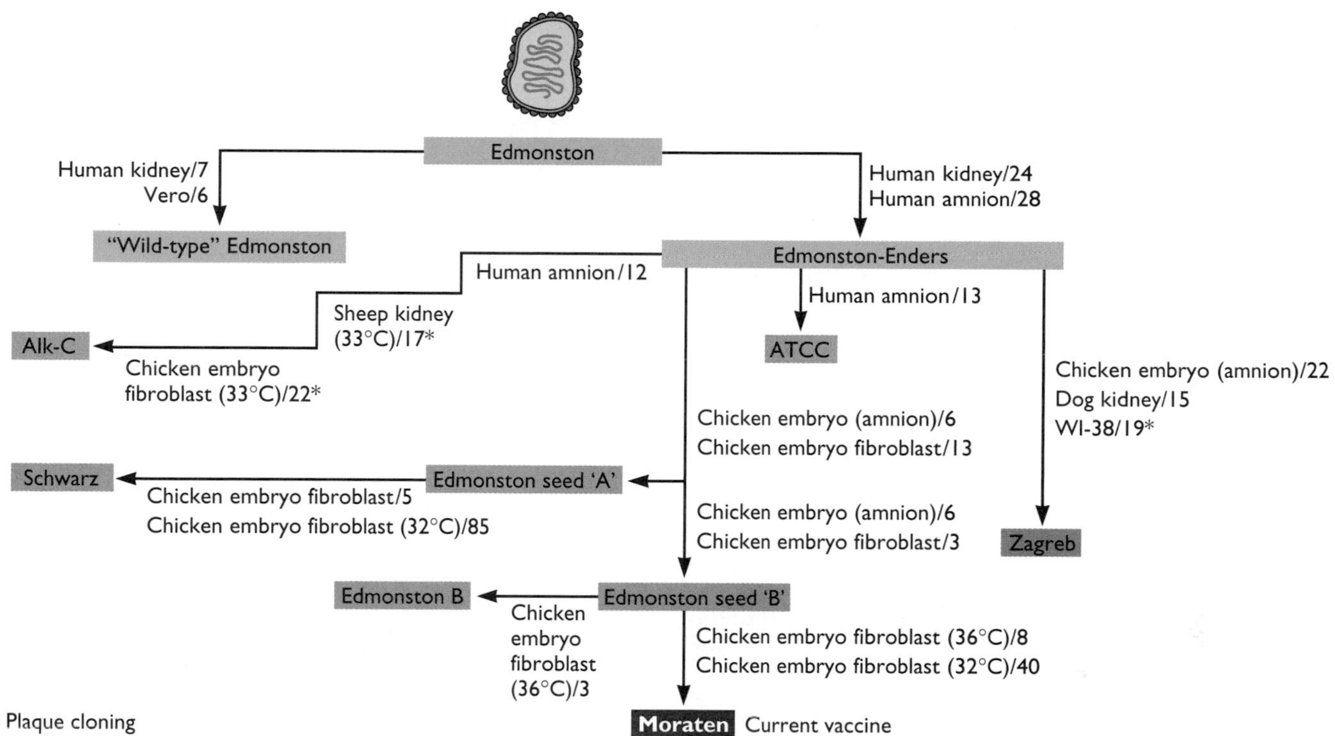

Figure 19.7 Live attenuated measles virus vaccine. Passage histories of live attenuated measles virus vaccines derived from John Enders' original isolate of Edmonston virus (top, blue). The current vaccine is called the Moraten strain (bottom, red). The cells used in passaging the virus to select attenuated mutants were as follows: human kidney, human amnion, Vero (African green monkey fibroblasts), chick embryo intra-amniotic cavity passage, chick embryo fibroblast, dog kidney, WI-38 (human diploid fibroblasts), and sheep kidney. The temperature during growth is given in parentheses, and the number of passages of virus in the particular cell or cell line follows the slash. An asterisk indicates that a single plaque was picked for further propagation. Adapted from D. D. Richman, R. J. Whitley, and F. G. Hayden (ed.), *Clinical Virology,* 2nd ed. (ASM Press, Washington, D.C., 2002), with permission.

developing effective live attenuated vaccines and is formally equivalent to the emergence of drug-resistant mutants (see "Resistance to Antiviral Drugs" below). While virulent revertants are a serious problem, considerable insight into viral pathogenesis and the immune response can be obtained by determining the changes responsible for increased virulence.

Because our knowledge about virulence is so limited, it is difficult to predict how a live attenuated virus will behave in individuals and in the population. As a result of the attenuating mutations, unexpected diversions from the natural infection and expected host response may occur. The attenuated virus may be eliminated from the vaccinated individual before it can induce a protective response, it may infect new niches in the host with unpredictable effects, or it may initiate atypical infections (e.g., slow or chronic infections) that can trigger immunopathological responses, such as **Guillain-Barré syndrome.** This syndrome may follow viral illness or vaccination and is characterized by rapidly progressing, symmetric weakness of the extremities. It is the most frequent cause of acute generalized paralysis. Vaccine side effects, whether real or not, often have a detrimental effect on public acceptance of national vaccine programs (Box 19.7).

Ensuring purity and sterility of the product is a problem inherent in the production of biological reagents on a large scale. If the cultured cells used to propagate attenuated viruses are infected with unknown or unexpected viruses, the vaccine may well contain these adventitious agents. For example, in the 1950s, early lots of poliovirus vaccine were grown in monkey cells that harbored the then unknown polyomavirus simian virus 40. It is estimated that from 10 million to 30 million persons received one or more doses of live simian virus 40 in contaminated vaccines. Many even developed antibody to simian virus 40 virion proteins. Some concern exists that rare tumors may be linked to this inadvertent infection, but proving cause and effect has been difficult (see Chapter 18).

Alternatives to the classical empirical approach to attenuation based on modern virology and recombinant DNA technology can now be applied. Viral genomes can be cloned, sequenced, and synthesized. Their genetic information can be analyzed using standard laboratory organisms, cultured cells, and animals. Viral genes required for pathogenesis in model systems can be identified in systematic fashion. In many cases, attenuation can be deliberately achieved by genetic manipulation. For example, deletion mutations with exceedingly low probabilities of reversion can be created (Fig. 19.8). In another approach that relies on genome segment reassortment of influenza viruses and reoviruses, genes contributing to virulence are replaced with those from related, but nonpathogenic viruses. No matter which technology is applied to achieve attenuation, the genetic engineer and the classical virologist must satisfy the same fundamental requirements: construction of an infectious agent with low pathogenic potential that is, nevertheless, capable of inducing a long-lived, protective immune response.

Figure 19.8 Construction of attenuated viruses by using recombinant DNA technology. Once the genome of a pathogenic virus is cloned in a suitable system, deletions, insertions, and point mutations can be introduced by standard recombinant DNA techniques. If the cloned genome is infectious or if mutations in plasmids can be transferred to infectious virus, it is possible to mutate viral genes systematically to find those required for producing disease. The virulence gene can then be isolated and mutated, and attenuated viruses can be constructed. Such viruses can be tested for their properties as effective live attenuated vaccines. The mutations in such attenuated viruses may be point mutations (e.g., temperature-sensitive mutations) or deletions. Multiple point mutations or deletions are preferred to reduce or eliminate the probability of reversion to virulence. Adapted from C. A. Janeway, Jr., P. Travers, M. Walport, and M. Shlomchik, *Immunobiology: the Immune System in Health and Disease* (Current Biology Limited, Garland Publishing Inc., New York, N.Y., 2001), with permission.

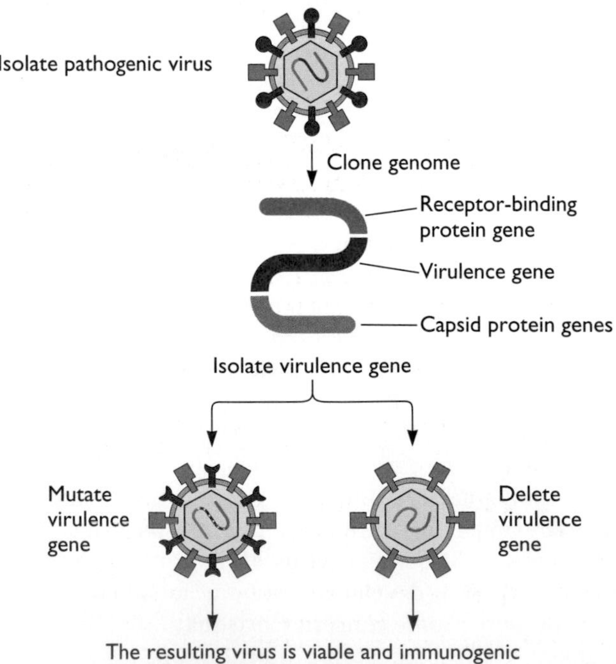

Inactivated or "Killed" Virus Vaccines

The inactivated poliovirus, influenza virus, hepatitis A virus, and rabies virus vaccines are examples of effective killed virus vaccines administered to humans (Table 19.2). In addition, inactivated virus vaccines are widely used in veterinary medicine. To construct an inactivated vaccine, virions of the virulent virus are isolated and inactivated by chemical or physical procedures. These treatments eliminate the infectivity of the virus, but do not compromise the antigenicity (i.e., the ability to induce the desired immune response). Common techniques include treatment with formalin or β-propriolactone, or extraction of enveloped virus particles with nonionic detergents.

Theoretically, killed virus vaccines are very safe, but accidents can and do happen. In the 1950s, a manufacturer of Salk polio vaccine did not inactivate it completely. As a result, more than 100 children developed disease after vaccination. Incomplete inactivation and contamination of vaccine stocks with potentially infectious viral nucleic acids have been singled out as major problems with this type of vaccine.

Vaccination is currently the most important measure for reducing influenza virus-induced morbidity and mortality. In the United States alone, influenza virus infections cause as many as 50,000 deaths every year and cost at least $12 billion in health care. Every year, millions of citizens seeking to avoid infection receive their "flu shot," which contains several inactivated influenza viruses that have been predicted to reach the United States that year. The magnitude of this undertaking is noteworthy. For example, between 75 million and 100 million doses of inactivated vaccine must be manufactured each year. Typically, these vaccines are formalin-inactivated, whole virion or detergent- or chemically disrupted virus (often called a "split" virus preparation). The viruses that are mass-produced in embryonated chicken eggs can be natural isolates or reassortant viruses constructed to express the appropriate hemagglutinin or neuraminidase gene from the virulent virus.

Currently, a typical influenza vaccine dose is standardized to comprise 15 μg of each viral hemagglutinin protein, but contains other viral structural proteins as well. Recent vaccines have contained antigens from two influenza A virus subtypes and antigens from one influenza B virus. The efficacy of inactivated influenza virus vaccines varies considerably. These vaccines are reportedly about 60 to 90% effective in healthy children and adults under the age of 65 after exposure to the homologous virus (the same virus that is in the vaccine). Inactivated vaccines are all highly immunogenic in young adults, but less so in the elderly, immunosuppressed individuals, and people with chronic illnesses. Protection against illness correlates with the concentration of serum antibodies that react with viral hemagglutinin (HA) and neuraminidase (NA) proteins produced after vaccination. Immunization may also stimulate limited mucosal antibody synthesis and CTL activities, but these responses vary widely.

The envelope proteins of influenza viruses change by antigenic drift and shift as the virus replicates in various animal hosts around the world (see Chapter 16). Consequently, protection one year **does not guarantee protection the following year.** To deal with this ever-changing agent, vaccine manufacturers must reformulate influenza virus vaccine every year so that it contains antigens from the predicted next generation of viruses. A committee of the Food and Drug Administration chooses the particular strains, in conjunction with World Health Organization-designated laboratories that monitor influenza infections. Timing is critical, as the final decision for the virus composition in the vaccine must be made within the first few months of each year to allow sufficient lead time for production of the vaccine. Any delay or error in the process, from prediction to manufacture, has far-reaching consequences, given the millions of people who are vaccinated and expect safe protection. Even if the vaccine contains the appropriate viral antigens, and is made promptly and safely, inactivated influenza virus vaccines have the potential to cause side effects in some individuals who are allergic to the eggs in which the vaccine strains are grown. As an example of other problems, the 1997 H5N1 avian virus that infected humans in Hong Kong was extraordinarily cytopathic to chicken embryos, making it difficult to propagate for vaccine purposes. Reassortants had to be constructed by placing the new H5N1 segments in viruses with less cytopathic viruses. The risk, of course, is that such reassortants will not provide the proper immune protection against the original strain.

Adjuvants

Inactivated virions often do not induce the same immune response as live attenuated preparations unless mixed with a substance that stimulates early processes in immune recognition, particularly the inflammatory response. Such immunostimulatory substances are called **adjuvants**. Adjuvants work in at least three different ways: by presentation of antigen as particles, by localization of antigen to the site of inoculation, and by direct stimulation of the innate immune response. The immune system can be stimulated directly when adjuvants mimic or induce cellular damage, stress, and release of heat shock proteins (sometimes called "danger" signals)

Adjuvants vary in composition from complex mixtures of killed mycobacteria and mineral oil (complete Freund's adjuvant), to lipid vesicles, or to mixtures of aluminum salts (Table 19.4). The discovery of adjuvants has been largely empirical, and it is not always clear how they potentiate the immune response. Some adjuvants, like alum

Table 19.4 Vaccine delivery systems and adjuvants[a]

Type of system or adjuvant	Characteristics
Aluminum salt	Aluminum hydroxide or phosphate. Forms precipitates with soluble antigen, making the complexes more immunogenic; antigen "depot" at site of injection; complement activation.
Emulsions	Freund's complete adjuvant: antigen suspended in water-mineral oil emulsion with killed *Mycobacterium tuberculosis* bacteria or muramyl di- or tripeptide to stimulate strong T-cell responses. Freund's incomplete adjuvant: antigen suspended in water-in-mineral oil emulsion.
Microspheres	Antigen encapsulated in polymers of lactic and glycolic acids. They are biodegradable and cause slow release of antigen.
ISCOMs	Immune-stimulating complexes composed of glycosides in an adjuvant called QuilA (a purified saponin from the plant *Quillaja saponaria*), cholesterol, phospholipids, and antigens. Form spheres of 30–40 nm in diameter that incorporate antigen.
Nucleic acid vaccines	Genes encoding antigens expressed from strong promoters are introduced directly to muscle or skin using physical methods or liposomes leading to intracellular protein production and presentation of antigen to the immune system.
Engineered viruses	Genes encoding foreign antigens are introduced into a viral genome (the vector) such that the new protein is made following infection. Common viral vectors are vaccinia virus, adenovirus, and baculovirus. Many other viruses can also be modified to express foreign genes.

[a]Adapted from C. A. Mims et al., *Mims' Pathogenesis of Infectious Disease*, 5th ed. (Academic Press, Inc., Orlando, Fla., 2001), with permission.

(microparticulate aluminum hydroxide gel), are widely used for human vaccines such as hepatitis A and B vaccines. Others, such as complete Freund's adjuvant, are used only in research. Freund's adjuvant is extremely potent, but is not licensed for use in humans, because it causes tissue damage and other undesirable effects. Two of the active components in Freund's adjuvant have been identified as muramyl dipeptide and lipid A, both potent activators of the inflammatory response. Less toxic derivatives, along with saponins and so-called block copolymers (linear polymers of clustered hydrophobic and hydrophilic monomers), are promising adjuvants. When purified, some viral structural proteins (e.g., hepatitis B virus and human papillomavirus capsid proteins) spontaneously form empty capsids that act as natural adjuvants. New particle-forming adjuvants including liposomes and lipid micelles 30 to 40 nm in diameter with viral proteins exposed on the surface (immune-stimulating complexes [ISCOMs]) currently are being evaluated in human trials (Table 19.4 and Fig. 19.9).

Some new adjuvants enable exogenous proteins to enter the major histocompatibility complex (MHC) class I pathway normally reserved for proteins synthesized de novo (Fig. 19.9). In addition, adjuvants of the saponin group, ISCOMs, and muramyl tripeptide-phosphatidylethanolamine also induce macrophages and dendritic cells to synthesize interleukin (IL)-12, which stimulates cellular immune responses. As a result, when mixed with purified viral proteins, these adjuvants have the potential to stimulate a Th1 response. Other proteins, including inactivated bacterial toxins, are being tested for their ability to induce mucosal immune responses.

Vaccine researchers speak of "tailoring" a vaccine by using different combinations of adjuvant and vaccine to induce a protective immune response. The principle is promising, but much work remains to be done to demonstrate the efficacy and safety of this approach.

Subunit Vaccines

It is possible to make a vaccine consisting of only a subset of viral proteins, as demonstrated by successful hepatitis B and influenza vaccines. Vaccines formulated with purified components of viruses, rather than the intact virion, are called **subunit vaccines**. We can deduce what viral proteins to include in a vaccine by determining what is recognized by the antibodies and CTLs found in individuals recovering from disease. Although the most obvious proteins would be those present on the virion surfaces, in fact, any viral protein could be a target.

Recombinant DNA methods. Recombinant DNA methods now allow cloning of appropriate viral genes into

Figure 19.9 ISCOMs as peptide delivery vehicles. ISCOMs are lipid micelles that can fuse with the plasma membranes of cells. The figure shows how an ISCOM can be used to introduce defined peptides into cells to activate cytotoxic $CD8^+$ T-cell responses. (A) When ISCOMs are formed with peptides in the solution, the resulting micelles have peptides trapped inside. (B) These trapped peptides can be delivered to the cytoplasm of antigen-presenting cells by fusion of the ISCOM with the cell membrane. (C) Some internalized peptide can be transported to the endoplasmic reticulum lumen by the Tap transporter system, where it can bind newly synthesized MHC class I proteins. (D) The peptide-MHC class I complex is then transported to the cell surface, where the peptide-MHC class I complex can be recognized by $CD8^+$ T cells. Adapted from C. A. Janeway, Jr., P. Travers, M. Walport, and M. Shlomchik, *Immunobiology: the Immune System in Health and Disease* (Current Biology Limited, Garland Publishing Inc., New York, N. Y., 2001), with permission.

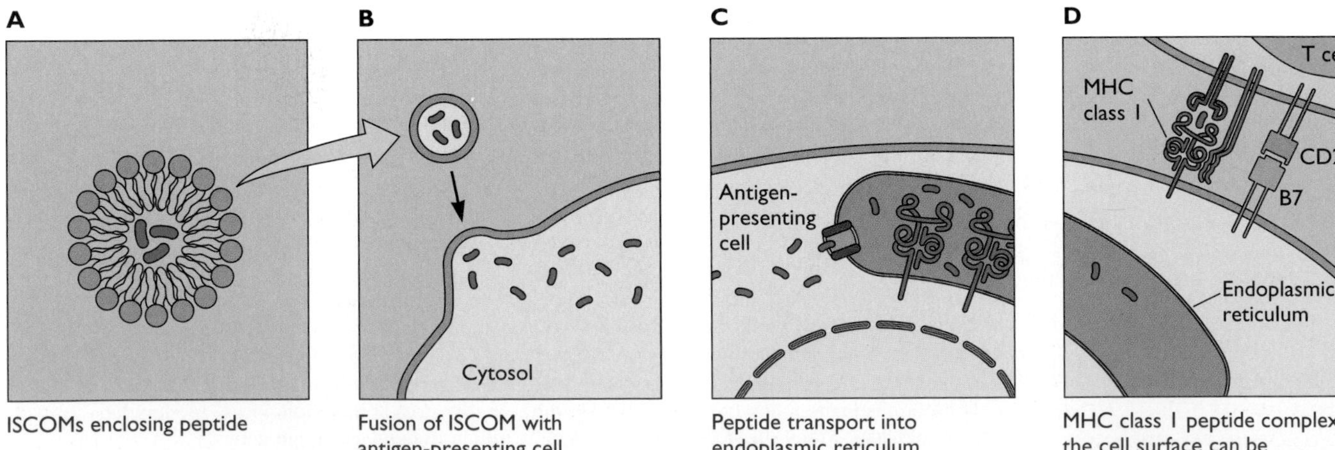

nonpathogenic viruses, bacteria, yeasts, insect cells, or plant cells (the last potentially yielding edible vaccines) to produce the immunogenic protein. As only a portion of the viral genome is required for such production, there can be no contamination of the resulting vaccine with the original virus, solving a major safety problem inherent in inactivated whole-virus vaccines. Viral proteins can be made inexpensively in quantity by engineered organisms under conditions that simplify purification and quality control. For example, problems with egg allergies after vaccination can be eliminated completely when influenza virus HA and NA proteins are synthesized in *Escherichia coli* or yeasts.

Unfortunately, even with adjuvants, many candidate subunit vaccines fail because they do not induce an immune response sufficient to protect against infection. Protection against infection (often called "challenge") is the "gold standard" of any vaccine. To achieve this standard, many variables must be considered: the nature, dose, and virulence of the challenging virus, as well as the route of immunization and the age and health of the host, come into play. The immune repertoire evoked by a live virus infection may be only partially represented in a response to a subunit vaccine. In particular, purified protein antigens rarely simulate the appearance of mucosal antibodies, particularly immunoglobulin A (IgA).

Hepatitis B subunit vaccine, produced by recombinant DNA technology, is an encouraging commercial success. This vaccine contains a single viral structural protein (hepatitis B virus surface antigen) that assembles spontaneously into virus-like particles, whether made in yeast, *E. coli*, or mammalian cell lines. Formation of particles appears to be important, as purified monomeric capsid protein does not induce an effective, protective immune response. In mice, as little as 0.025 μg of hepatitis B virus surface antigen particles can elicit antibody production without the presence of an adjuvant. Typically, in the human vaccine, 10 to 20 μg of empty capsids is administered in three doses over a 6-month period, and more than 95% of recipients develop antibody against the surface antigen. The capsid proteins of other viruses often self-assemble into virus-like structures (see Chapter 13), a phenomenon that may be useful in the development of effective vaccines.

Synthetic peptides. Synthetic peptides about 20 amino acids in length can induce specific antibody responses when chemically coupled to certain protein carriers that can be taken up, degraded, and presented by MHC class II proteins. In principle, synthetic peptides should be the basis for an extremely safe, well-defined vaccine. To date, however, peptide vaccines have had little commercial success, mainly because synthetic peptides are expensive to make in sufficient quantity, and the antibody response they elicit is often weak and short-lived. A weak immune response can be far worse than no response at all, because viral escape mutants may be selected. Given the simplicity of the antipeptide response (usually a single epitope is represented by the peptide), selection of escape mutants is highly probable.

Perhaps the most extensive work on peptide vaccines has been directed against foot-and-mouth disease virus. Economically, it is the most important disease of cattle and other farm animals throughout the world. Several countries have mounted a considerable effort in research and field work to evaluate foot-and-mouth disease virus peptide vaccines, but results have not been encouraging. The antipeptide response is not always predictable, partial protection is common, and viral mutants with amino acid substitutions at the major antigenic site represented in the peptide vaccines can be isolated.

New Vaccine Technology

Viral Vectors

Genomes of nonpathogenic viruses can be constructed to make selected viral proteins that can immunize a host against the pathogenic virus (see Chapter 2). This approach merges subunit and live attenuated virus vaccine technologies. In theory, such a system provides all the "benefits" of a viral infection with respect to the immune response with none of the pathogenesis associated with the virulent virus. The vector infection initiates a local inflammatory response such that the immune defenses are mobilized at this site. However, practical experience has demonstrated that viral vectors may produce pathogenic side effects, particularly if injected directly into organs or the bloodstream. The immune response is not always predictable, particularly if children, the elderly, and immunocompromised individuals are to be treated.

Vaccinia virus, the foundation of the smallpox eradication program, continues to serve as a useful vaccine research tool and as a vaccine vector. The virus was injected into millions of people around the world and has proved to be a relatively safe infectious agent (see Box 19.1). Nevertheless, side effects do occur, and infections of immunocompromised individuals or people with skin diseases can be life-threatening. Highly attenuated vaccinia virus mutants have been isolated and may be the basis for the next generation of vaccinia virus-based vaccines. Construction of recombinant viruses is straightforward. A wide variety of systems are available for the construction of vaccinia virus recombinants that cannot replicate in mammalian cells, but allow the efficient synthesis of

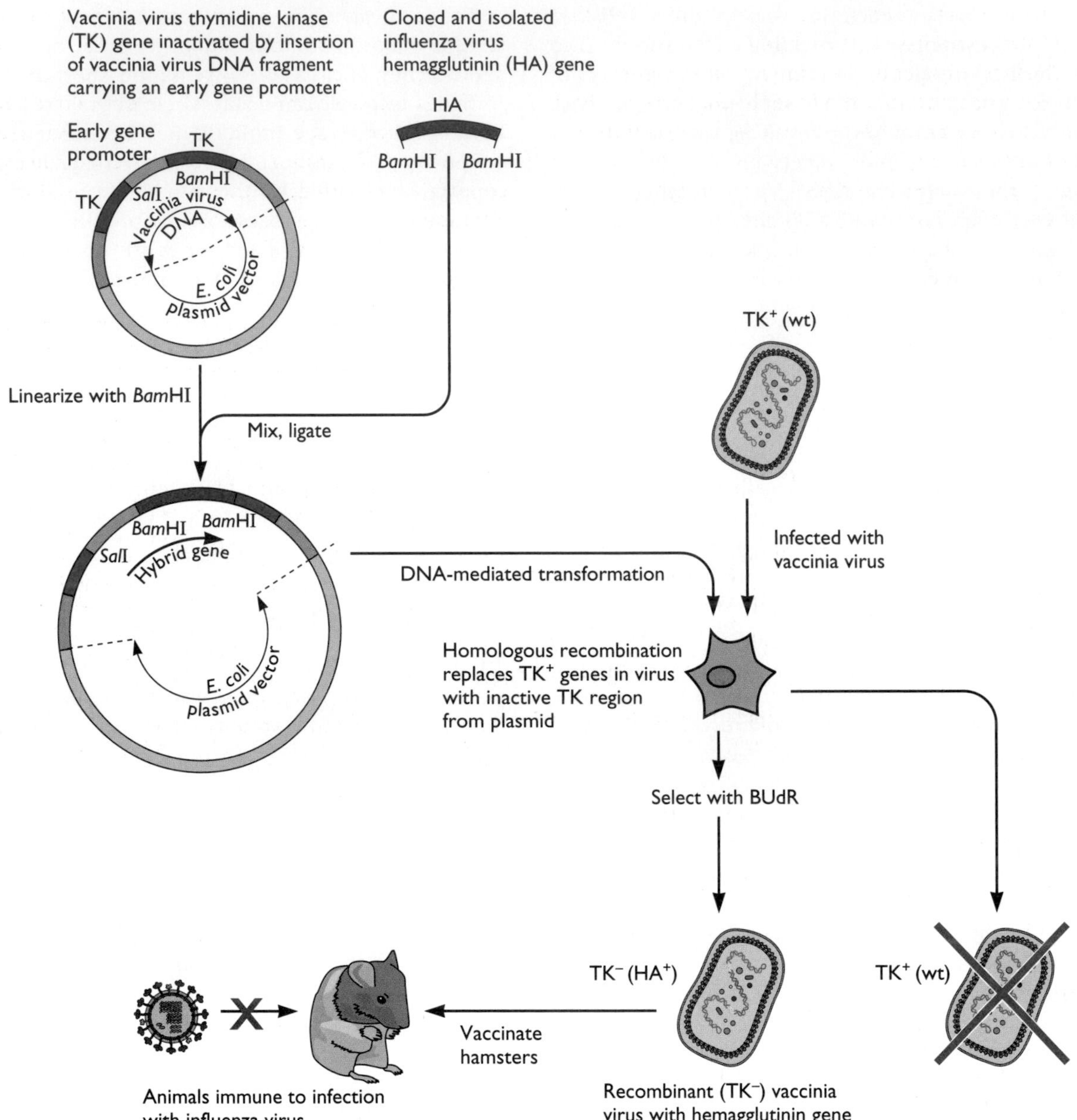

Figure 19.10 Construction of an infectious vaccinia virus capable of expressing the influenza virus HA gene. An *E. coli* plasmid vector containing a segment of vaccinia virus DNA encoding the TK gene (dark blue), as well as some contiguous viral sequences (yellow), is constructed. Next, a small segment of vaccinia virus DNA (a *Sal*I-*Bam*HI restriction fragment [light blue]) containing an early gene promoter is inserted into the TK gene, disrupting the gene. A *Bam*HI fragment of DNA encoding the influenza virus HA gene (red) is inserted into the *Bam*HI site immediately following the vaccinia virus early promoter. The HA gene can now be transcribed from the vaccinia virus early promoter by vaccinia virus RNA polymerase. This hybrid gene is transferred to a vaccinia virus genome by homologous recombination as follows. Plasmid DNA is introduced into a permissive cell line that has been infected by vaccinia virus with an intact TK gene. A few percent of the progeny virus will have replaced the functional viral TK gene with the inactive hybrid TK gene containing the influenza virus HA coding sequences. These TK$^+$ viruses can be selected from the mixture of progeny by treatment with bromodeoxyuridine (BUdR). BUdR, which is incorporated into DNA only when phosphorylated, is lethal to wild-type nonrecombinant viruses with an intact TK gene. The BUdR-resistant viruses are expected to express the influenza virus HA gene after infection. Animals infected with these recombinant viruses respond by making both antibodies and CTLs specific for the influenza HA protein. Such animals are protected from challenge by virulent influenza virus. Adapted from N. J. Dimmock and S. B. Primrose, *Introduction to Modern Virology*, 4th ed. (Blackwell Science, Cambridge, Mass., 1994), with permission.

cloned gene products that retain their immunogenicity. Attenuated vaccinia virus vectors can accommodate more than 25 kb of new genetic information. A simplified outline of how to construct a recombinant vaccinia virus that expresses the influenza virus HA gene is presented in Fig. 19.10. Vaccinia virus recombinants can also be used to dissect the immune response to a given protein from a pathogenic virus. This application is illustrated in Fig. 19.11. Other poxviruses, including raccoonpox, canarypox, and fowlpox viruses, offer additional possibilities, because they are able to infect, but not replicate in, humans.

The successful use of an oral rabies vaccine for wild animals in Europe and the United States demonstrates that recombinant vaccinia virus vaccines have considerable potential. Vaccinia virus genomes encoding the major envelope protein of rabies virus yield virions that are formulated

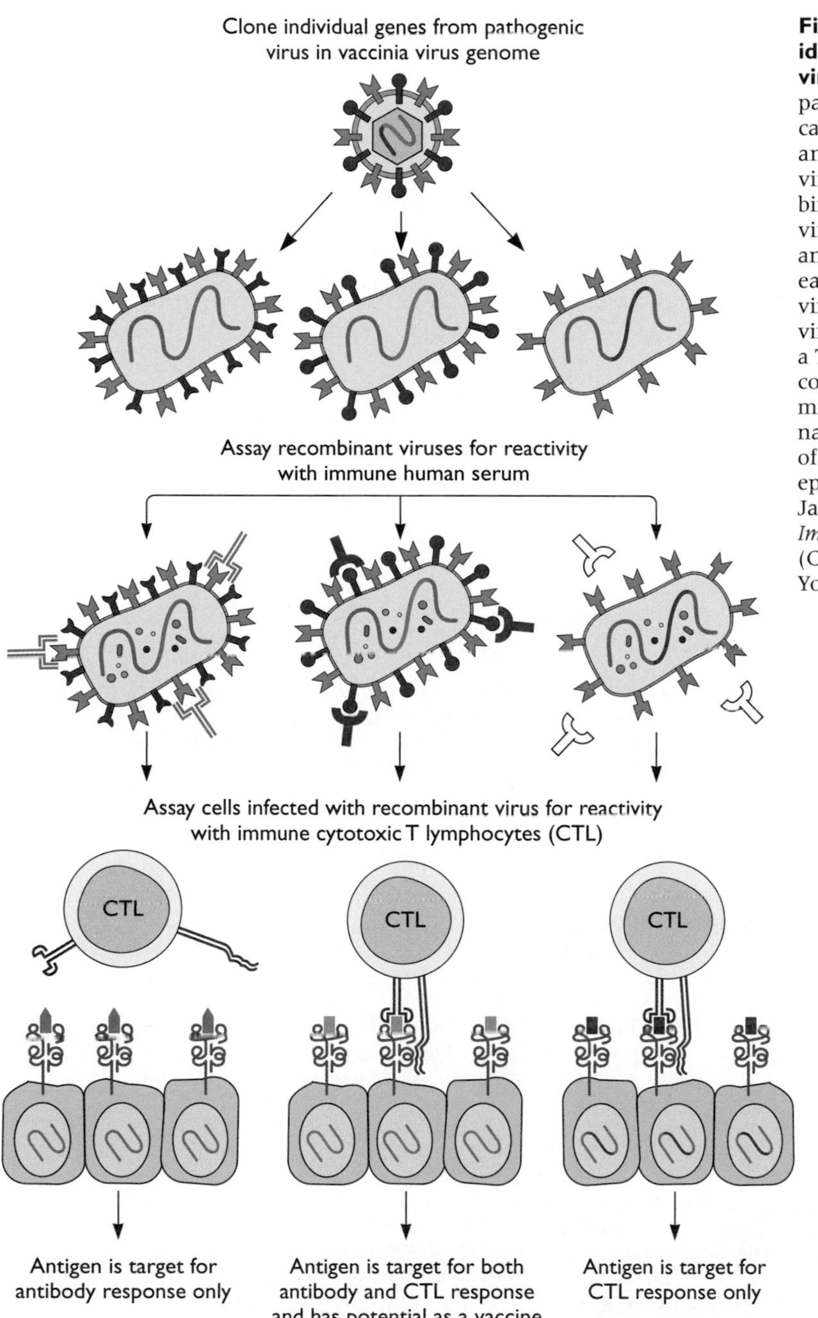
Clone individual genes from pathogenic virus in vaccinia virus genome
Assay recombinant viruses for reactivity with immune human serum
Assay cells infected with recombinant virus for reactivity with immune cytotoxic T lymphocytes (CTL)
CTL
CTL
CTL
Antigen is target for antibody response only
Antigen is target for both antibody and CTL response and has potential as a vaccine
Antigen is target for CTL response only

Figure 19.11 Use of recombinant vaccinia viruses to identify and analyze T- and B-cell epitopes from other viral pathogens. (Top) Individual genes from a viral pathogen suspected of encoding proteins that play critical roles in stimulating host immune defense are cloned and expressed from several recombinant vaccinia viruses by methods described in Fig. 19.10. The recombinant viruses are used to infect cells, and the resulting viruses or infected cells are tested with antiviral human antibodies or T cells (middle and bottom panels). As each recombinant virus expresses a single gene from the viral pathogen, it is possible to determine if a particular viral protein contains a B-cell epitope (binds antibodies), a T-cell epitope (recognized by CTLs), or both. This is accomplished by reacting antibodies and T cells from immune individuals with cells infected with each recombinant virus. Subsequent site-directed mutational analysis of the viral genes enables precise localization of these epitopes on the viral protein. Adapted from C. A. Janeway, Jr., P. Travers, M. Walport, and M. Shlomchik, *Immunobiology: the Immune System in Health and Disease* (Current Biology Limited, Garland Publishing Inc., New York, N.Y., 2001), with permission.

in edible pellets to be spread in the wild. The pellets are designed to attract the particular animal to be immunized (e.g., foxes or raccoons). Ideally, the animal eats the pellet, is infected by the recombinant virus, and responds with a protective immune response to subsequent rabies virus infection. Given the rare but serious side effects possible when vaccinia virus infects humans, inadvertent human infection by these wild-life vaccines must be avoided.

The rhabdovirus vesicular stomatitis virus is a particularly promising vaccine vector (see Fig. 2.17). Vesicular stomatitis virus causes an acute infection of short duration producing vesicular lesions in tongue, teats, and hooves of cattle, pigs, and horses. The virus also can infect humans, causing mild symptoms. Molecular virologists have studied this virus for many years and have developed techniques to construct attenuated mutants and recombinants that direct synthesis of almost any protein. Infection induces strong cellular and humoral responses in mice and rhesus monkeys to any foreign protein that has been made after viral infection. The virus is attractive for vaccine technology, because high titers can be obtained easily and it can be delivered by a mucosal route without injection. Initial experiments are promising: when rhesus monkeys were vaccinated with recombinant vesicular stomatitis viruses expressing the human immunodeficiency virus (HIV) *env* and *gag* genes, and then challenged with a strain of human virus that causes acquired immunodeficiency syndrome (AIDS) in monkeys, all animals remained healthy for over a year, with low or undetectable viral loads. In contrast, all nonvaccinated, challenged monkeys progressed to AIDS in less than 5 months.

Two general problems arise with live virus vector vaccines. First, the host usually is immunized against the viral vector, as well as the vaccine antigen. As a result, subsequent uses of the particular viral vector may result in a weak response, no response, or an immunopathological response. Second, introducing live viral vectors, even though they are attenuated, into the population at large may have long-term effects. For example, immunocompromised individuals may be infected, with adverse consequences.

DNA Vaccines

In 1992, scientists discovered a variation of the subunit vaccine approach: the vaccine did not comprise a protein, but rather highly purified DNA. The **DNA vaccine** consisted simply of a DNA plasmid encoding a viral gene that can be expressed inside cells of the animal to be immunized (Fig. 19.12). In the simplest case, the plasmid encodes only the immunogenic viral protein under the control of a strong eukaryotic promoter (the human cytomegalovirus immediate-early promoter is commonly used), and a selectable marker to facilitate maintenance of the plasmid in

Figure 19.12 Functional components of a DNA vaccine expression vector. The system is based on a double-stranded plasmid, typically from *E. coli*, carrying its own origin of replication and an antibiotic resistance gene as a selectable marker. A critical component is a strong eukaryotic promoter to initiate transcription of the cloned gene. Expression may be enhanced by including an intron sequence (to facilitate transport of mRNA from the nucleus). A multicloning site offers restriction enzyme sites to facilitate cloning of the antigen-encoding gene downstream of the strong promoter and intron. A short untranslated termination sequence provides a polyadenylation signal as well as stability to the antigen-encoding mRNA. CpG-rich sequences can be added and are often found naturally in the bacterial plasmid DNA. Such sequences function as an adjuvant to stimulate the immune response. Adapted from M. Oyaski and H. Ertl, *Sci. Med.* **7**:30–39, 2000, with permission.

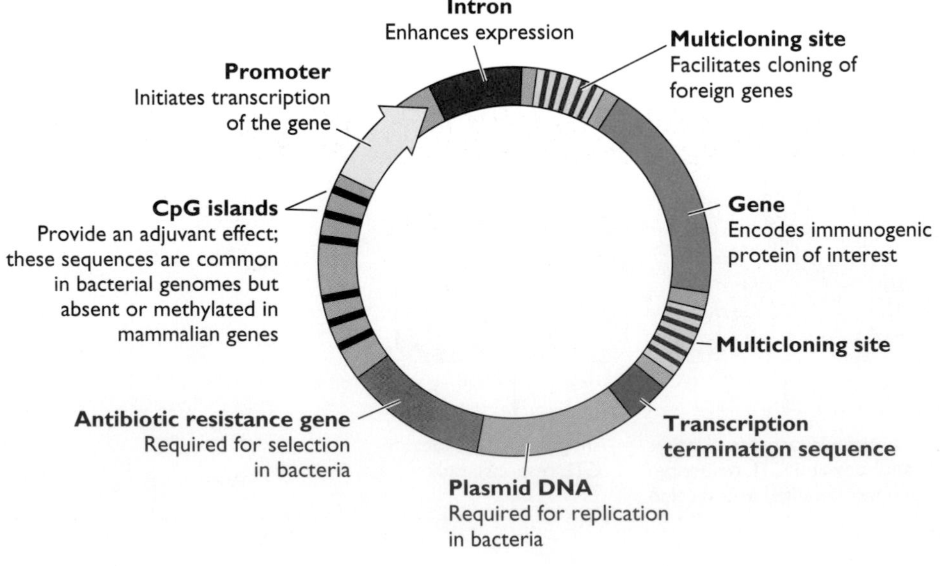

bacteria. The plasmid DNA, produced in bacteria, can be prepared free of contaminating protein and has no capacity to replicate in the vaccinated host. Remarkably, unlike the requirements for standard protein subunit vaccines, no adjuvants or special formulations are necessary to stimulate an immune response. The vaccine can be delivered by injection of an aqueous solution containing a few micrograms of the plasmid DNA into muscle or skin tissue. Another effective delivery method uses a "gene gun" that literally shoots DNA-coated microspheres through the skin to introduce the plasmid into dermal tissue.

Because the viral protein is made de novo inside a cell, a fraction of it is presented on the surfaces of producing cells by MHC class I molecules, and the cells are recognized by T cells. Both the humoral response and CTLs can be stimulated by DNA vaccination, and most important, the vaccine protects against challenge in some animal models (Table 19.5).

The striking property of DNA vaccination is that a relatively low dose of immunizing protein seems sufficient to induce long-lasting immune responses: the quantity of encoded protein produced is extremely small, perhaps in the nanogram range, as determined by sensitive reporter proteins such as luciferase. The type of immune response is dictated by the method of inoculation. A Th1 response predominates after injection of an aqueous DNA solution into muscle. In this method, DNA rapidly leaves the site of injection and travels to the spleen, where the immune response occurs. Animals vaccinated in this way usually synthesize high concentrations of gamma interferon (IFN-γ), IL-2, and IL-12, but only low levels of IL-4 or IL-10. In contrast, after DNA immunization by gene gun, the DNA either enters cells directly or is taken up by skin cells and keratinocytes to produce the viral protein. Langerhans cells in the skin then acquire the protein (but are not exposed to the DNA vaccine plasmid) and move to the draining lymph node. The response induced often is more typical of a Th2 response, with synthesis of IL-4 and antibodies, not IFN-γ and CTLs.

DNA vaccines are effective without addition of adjuvants. Remarkably, plasmid DNA itself has intrinsic adjuvant activity in mammalian cells. DNA motifs containing unmethylated CpG dinucleotides flanked by two 5′ purine nucleotides and two 3′ pyrimidines induce synthesis of immunomodulating cytokines, such as IL-6, IFN-γ, IL-12, and tumor necrosis factor alpha. We now know that the DNA of many microbes and invertebrates has adjuvant effects, whereas the DNA of many vertebrates is nonstimulatory. The strong adjuvant effects of complete Freund's adjuvant may be due at least in part to the mycobacterial DNA present in the emulsion. Recently, the mammalian protein that triggers the response to bacterial DNA has been identified as part of the innate immune system, the Toll-like receptor Tlr9 (see Chapter 15). When DNA with unmethylated CpG sequences binds Tlr9 on dendritic cells and monocytes, a burst of IL-12 and IFN-γ ensues and activates Th1 cells. These findings explain why vaccination techniques that bypass the Tlr9 receptor, such as gene guns that place DNA directly into target cells, often activate the Th2 antibody response rather than the Th1 CTL response.

Research and development in this area have only begun, and so we cannot be sure if DNA vaccine technology has a commercial future. From a basic research point of view, much remains to be investigated. Considerable data demonstrate that DNA vaccines can protect animals from appropriate virus challenge (Table 19.5), but we have little information about efficacy in humans. As a first step, DNA vaccines are being tested for safety in humans. Some possible dangers might be integration of plasmid DNA leading to insertional mutagenesis, induction of autoimmune responses such as anti-DNA antibodies, and induction of immune tolerance.

Table 19.5 Representative results of DNA vaccine trials[a]

Virus	Proteins	Induction of antibody	Induction of CTL response	Protection against challenge
Bovine herpesvirus	gD	+	ND	+ (cattle)
Hepatitis B virus	Surface and core antigens	+ (chimpanzees); ND (humans)	+ (chimpanzees)	+ (chimpanzees)
Hepatitis C virus	Nucleocapsid	+	+	+ (mice)
Herpes simplex virus type 1	gD, gB	+	+	+ (mice)
HIV type 1	Env, Gag, Rev	+	+	+ (rhesus macaques)
Influenza virus	HA, M1, Np	+	+	+ (chickens, mice)
Lymphocytic choriomeningitis virus	NP	+	+	+ (mice)
Rabies virus	Glycoprotein, NP	+	+	+ (cynomolgus monkeys)
Respiratory syncytial virus	Glycoprotein	+	+	+ (mice)

[a]Data from A. Reyes-Sandoval and H. C. Ertl, *Curr. Mol. Med.* **1**:217–243, 2001, with permission. ND, not detected.

An exciting possibility is a **genomic vaccine:** a library of all the pathogen's genes prepared in multiple DNA plasmid vectors. The entire plasmid mixture would be injected into an animal. Such a vaccine should be safe (no live organisms are used) and may be efficacious, because every gene product of the pathogen can be presented to the immune system. Moreover, from a basic research perspective, this approach has much to offer. Because each plasmid encodes a unique mRNA, it should be straightforward to determine the gene, or genes, necessary and sufficient to induce a protective immune response. A technique called **gene shuffling**, which can be applied to produce diverse coding sequences, may have utility in DNA vaccine technology (Box 19.9).

Vaccine Delivery Technology

Improvement of administration or delivery is an important goal of vaccine developers. At the present time, vaccines can be administered by a variety of methods, including the traditional hypodermic needle injection and the "air gun" injection of liquid vaccines under high pressure through the skin. Other formulation methods under consideration include new emulsions, artificial particles, and direct injection of fine powders through the skin. Oral delivery of vaccines can be effective in stimulating appearance of IgA antibodies at mucosal surfaces of the intestine and in inducing a more systemic response. However, oral delivery is not always possible, because the protective surfaces and enzymes of the oral cavity and alimentary tract block or destroy many vaccines. Specially constructed edible plants that make viral proteins represent an attractive, new approach for oral delivery. Transgenic plants can be engineered, or plant viruses with genomes encoding immunogenic proteins can be used to infect food plants. Early experiments are promising: when such a plant is eaten, antibodies to the viral structural protein can be demonstrated in the animal's serum. Another oral vaccine for-

BOX 19.9

DNA shuffling: directed molecular evolution

DNA shuffling enables scientists to assemble and analyze new combinations of DNA fragments not found in nature. The utility of this approach for vaccine development is illustrated by the following example. Many different clades of HIV exist, and neutralizing antibodies often recognize the Env protein from one clade, but not others. Some scientists suggest that an important component of a vaccine would be a protein that induces an antibody response capable of recognizing the Env proteins from all known viral clades.

DNA shuffling may provide technology to achieve this goal. DNA encoding Env proteins from various clades would be mixed, digested with restriction enzymes, denatured, reannealed, and amplified by the polymerase chain reaction. Some of the single-stranded DNA fragments from the *env* gene of one clade will anneal to similar (but not identical) DNA sequences derived from another *env* gene. The 3′ ends of these hybrids are extended in the polymerase chain reaction, and single-stranded regions are copied by DNA polymerase. A mixture of chimeric double-stranded DNA fragments is produced representing not only all combinations of *env* genes, but also many not found naturally. Expression libraries created by insertion of the new shuffled DNA segments into suitable plasmids would then be screened for production of new proteins that bind antibodies specific for Env proteins, or that induce antibodies capable of neutralizing a broad spectrum of virus isolates.

Optimists assert that DNA shuffling can be used to make effective, safe, multivalent DNA vaccines, and also can be used to boost the potency of known vaccines.

Stemmer, W. P. C. 1994. Rapid evolution of a protein in vitro by DNA shuffling. *Nature* **370:**389–391.

Whalen, R. G., R. Kaiwar, N. W. Soong, and J. Punnonen. 2001. DNA shuffling and vaccines. *Curr. Opin. Mol. Ther.* **3:**31–36.

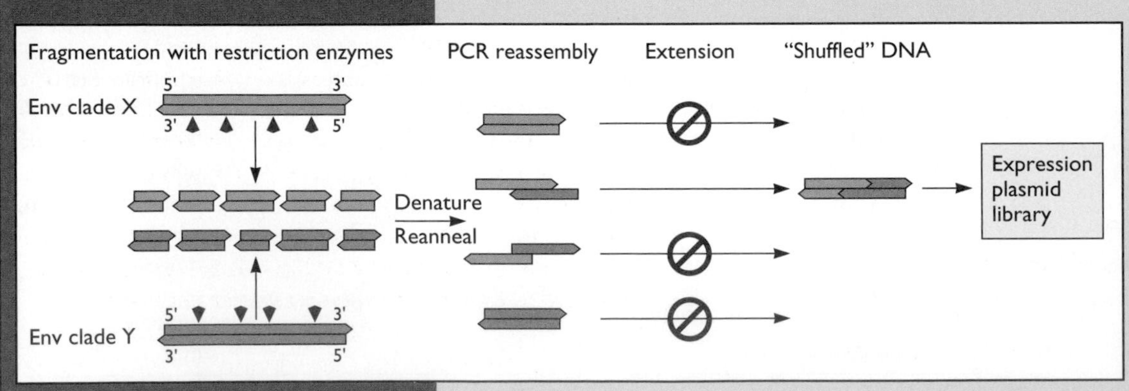

Shuffling of HIV Env coding DNA from various clades.

mulation technology is based on principles discovered by research with enteric viruses that survive passage through the stomach, despite its low pH and digestive enzymes. Biopolymers that mimic the protective action of enteric virus proteins can be formulated with vaccine components to resist the low pH of the stomach, but dissolve in the more hospitable regions of the upper intestinal tract. Such coating technology has been used for oral delivery of adenovirus vaccines and the vaccinia virus-based wildlife rabies vaccine discussed previously.

Immunotherapy

Patients already infected with viruses that cause persistent infections, or that reactivate from latency in the face of an immune response, present special problems. Common viruses that fall into this category are listed in Table 19.6. **Immunotherapy** provides the already-infected host with antiviral cytokines, other immunoregulatory agents, antibodies, or lymphocytes **over and above those provided by the normal immune response** in order to reduce viral pathogenesis. Immunotherapy can be administered by introduction of the purified compounds or of a gene encoding the immunotherapeutic molecule (a gene therapy approach). A live attenuated virus or a DNA vaccine can be modified to synthesize cytokines that stimulate the appropriate immune response. It is possible to isolate lymphocytes from patients, infect these cells with a defective virus vector (e.g., a retrovirus) carrying an activated, immunoregulatory molecule, and then introduce the transduced cells back into the patient. If these cells survive and express the immunotherapeutic gene, the patient's immune response may be bolstered. If stem cells are transduced, a long-term effect may be achieved, as these cells will replicate and propagate the transgene.

Table 19.6 Human viruses that cause persistent infections

Virus	Disease or consequence
Cytomegalovirus	Retinitis, pneumonia, encephalitis
Enteroviruses	Chronic recurrent meningoencephalitis
Epstein-Barr virus	Lymphoproliferative disorder
Hepatitis B virus	Chronic hepatitis, hepatocellular carcinoma
Hepatitis C virus	Chronic hepatitis, hepatocellular carcinoma
Herpes simplex virus types 1 and 2	Recurrent mucocutaneous infections, encephalitis
Human immunodeficiency virus	AIDS
Human T-lymphotropic virus	Leukemia
Measles virus	SSPE
Papillomavirus	Cervical carcinoma, warts, papillomatosis
Parvovirus B19	Agranulocytosis, granulocytic aplasia
Varicella-zoster virus	Recurrent infections (shingles or zoster)

We know that immunotherapy can be effective. Historically, it was pioneered by using interferons to treat a variety of diseases, including those caused by persistent viruses. For example, IFN-α-2b is one drug approved in the United States for treatment of chronic hepatitis caused by the hepadnavirus hepatitis B virus and the flavivirus hepatitis C virus. Its effect on chronic hepatitis B virus infection is remarkable: as many as 50% of the patients treated experience an apparent "cure." This result illustrates the potential of immunotherapy. However, treatment of hepatitis C virus infection with IFN-α-2b has been much less successful, for reasons that are not clear. Limitations of IFN therapy (and probably cytokine therapy in general) are that the biological activity of IFN is not sustained for a prolonged time, side effects are significant, patients over the age of 50 tend to be less responsive, and treatment is expensive.

Immunomodulating agents, including IFN, cytokines that stimulate the Th1 response (e.g., IL-2), and certain chemokines, are being studied individually and in combinations for their ability to reduce virus load and complications of persistent infections caused by papillomavirus and HIV. Cytokines that stimulate natural killer cells (e.g., IL-12 and IFN-γ) may have promise as well. An alternative to using cytokines or immunoregulatory molecules to stimulate the immune system is the introduction of more antibodies or activated lymphocytes into an infected individual. A fundamental problem with any immunotherapy approach is that immune stimulation can lead to immunopathology or provide new avenues for virus spread. The consequences of long-term synthesis of immune modulators, which are not known, must be investigated before this technology can be put into widespread practice.

Antiviral Drugs: Small Molecules That Block Viral Replication

Paradox: So Much Knowledge, So Few Antivirals

Vaccines have been very successful in preventing some viral diseases, however, they provide modest or no therapeutic effect in individuals who are already infected. Indeed, for some viral diseases like dengue fever, an immune response to one serotype can have adverse effects, if the individual is subsequently infected by a second serotype. Unfortunately, despite considerable effort, vaccines effective against major viral diseases caused by HIV and hepatitis C virus have been impossible to produce. Consequently, the second arm of antiviral defense is use of antiviral drugs that are capable of preventing an infection, or stopping it once started. However, despite almost 50 years of research, our armamentarium of antiviral drugs remains surprisingly small. The current arsenal holds only about 30 drugs, and most of these are directed against HIV and herpesviruses (see Table 19.9).

Because we know so much about some viruses, this dearth of antiviral drugs is perhaps unexpected. There are many reasons, but a primary one is that antiviral drugs must be safe. This simple goal is often unattainable, as viruses are parasites of cellular mechanisms, and compounds interfering with viral growth often have adverse effects on the host. Another reason is that antiviral compounds must be extremely potent, virtually 100% efficient in blocking viral growth. Even modest replication in the presence of an inhibitor provides the opportunity for resistant mutants to prosper. Achieving the potency to block viral replication **completely** is remarkably difficult. Additionally, many medically important viruses cannot be grown readily in the laboratory (e.g., hepatitis B virus, hepatitis C virus, and papillomaviruses); some have no available animal model of human disease (e.g., measles virus and hepatitis C virus). The testing and development of safe, potent, and efficacious compounds against infections by these viruses pose a significant challenge. Finally, as noted in Chapter 16, many acute infections are of short duration, and by the time the individual feels ill, viral replication is completed and infected cells are being cleared. Antiviral drugs must be given early in the infection, or prophylactically to populations at risk. It takes time not only to identify the specific virus (no broad-spectrum antiviral agents are currently available), but also to obtain and dispense the prescription antiviral drug. Under these circumstances, it is not surprising that the drug cannot be given, or is not administered in time to be useful (Box 19.10). The lack of rapid diagnostic reagents alone has hampered the development and marketing of antiviral drugs for many acute viral diseases, despite the existence of effective therapies.

Historical Perspective

The first antiviral drugs were derived from sulfonamide antibiotics in the early 1950s with the synthesis of thio-semicarbazones active against smallpox virus. In the 1960s and 1970s, spurred on by success in the treatment of bacterial infections with antibiotics, drug companies launched massive screening programs to find chemicals with antiviral activities. Despite much effort, there was relatively little success. One notable exception was amantadine (Symmetrel), approved in the late 1960s for treatment of influenza A

BOX 19.10

We can put a person on the moon, but we cannot cure the common cold

While the title phrase of this box is timeworn and trite, it contains elements of truth that underlie some basic problems of finding treatments to common, acute infections. The common cold is a syndrome caused by many different viruses, including rhinoviruses (about 50% of all common colds), adenoviruses, coronaviruses, and others. Several fundamental problems confound the quest for a cure for the common cold.

- Vaccines and most antiviral agents are virus specific, so curing the common cold is not a simple proposition.
- The common cold is not a life-threatening disease, but rather a mild, inconvenient illness. Antiviral drugs typically are not sold over the counter. Will people see their physicians to obtain a prescription every time they experience a drippy nose, congestion, or cough?
- By the time the symptoms of a common cold are evident, viral replication has reached its peak. Therefore, by the time an antiviral is given, it may be too late for significant effect.
- Drugs and vaccines may have to be given to millions of healthy people to be effective. Short-term and long-term safety of any treatment must be ensured with extreme confidence.

In 2001, a new antiviral compound called pleconaril was tested in humans. This compound is effective against rhinoviruses, but not coronaviruses or adenoviruses. It binds to the hydrophobic pocket under the canyon in the receptor binding site of many rhinoviruses (see Fig. 5.12). Pleconaril is thought to act by stabilizing the particle so that uncoating and genome release cannot occur. In human trials, symptoms were reduced and the time course of disease was shortened by a day or two. The compound has yet to be approved for widespread use.

An effective treatment of the common cold simply may be one that targets mechanisms of symptom production, not the viral proteins that promote infection and replication. As the symptoms are the result of immunopathology, targeting immune cells or soluble mediators may be possible. We have incomplete knowledge about the mediators of immune defense against cold viruses, but it may be that there will be overlap among the host responses to different viruses giving rise to the common symptoms.

Hayden, F. G., T. Coats, K. Kim, H. Hassman, M. M. Blatter, B. Zhang, and S. Liu. 2002. Oral pleconaril treatment of picornavirus-associated viral respiratory illness in adults: efficacy and tolerability in phase II clinical trials. *Antivir. Ther.* **7**:53–65.

virus infections. These antiviral discovery programs were called **blind screening,** because no effort was made to focus discovery on a particular virus or a virus-specific mechanism (now called mechanism-based screening). Instead, random chemicals and natural product mixtures were tested for their ability to block replication of a variety of viruses in cell culture systems. "Hits"—compounds or mixtures that blocked viral replication in cultured cells—were then purified, and fractions were tested in various cell and animal models for safety and efficacy. Promising molecules, called "leads," were modified systematically by medicinal chemists to reduce toxicity, increase solubility and bioavailability, or improve biological half-life. As a consequence, hundreds if not thousands of molecules were made and screened before a specific antiviral compound was tested in humans. Moreover, the mechanism by which these compounds inhibited the virus was often unknown. For example, in the case of amantadine, the mechanism of action was not deduced until the early 1990s.

Discovering Antiviral Compounds

With the advent of modern molecular virology and recombinant DNA technology, the random, blind-screening procedures described above have been all but discarded. Instead, viral genes essential for growth can be cloned and expressed in genetically tractable organisms, and their products can be purified and analyzed in molecular and atomic detail. The life cycles of many viruses are known, revealing many targets for intervention (Table 19.7). Because of recombinant DNA technology, inhibitors of these processes can be found, even for viruses that cannot be propagated in cultured cells.

Table 19.7 Some mechanisms targeted in antiviral drug discovery[a]

Replication target	Selected compounds	Virus
Attachment	Peptide analogs of attachment protein	HIV
Penetration and uncoating	Dextran sulfate, heparin	HIV, herpes simplex virus
	Amantadine, rimantadine	Influenza A virus
	Tromantadine	Herpes simplex virus
	Arildone, disoxaril, pleconaril	Picornaviruses
mRNA synthesis	Interferon	Hepatitis A, B, and C viruses; papillomavirus
	Antisense oligonucleotides	Papillomavirus, human cytomegalovirus
Protein synthesis		
Initiation	Interferon	Hepatitis A, B, and C viruses; papillomavirus
IRES elements	Ribozymes; antisense oligonucleotides	Flavi- and picornaviruses
DNA replication		
Polymerase	Nucleoside and nonnucleoside analogs	Herpesviruses, HIV, hepatitis B virus
	Phosphonoformate	Herpesviruses
Helicase/primase	Thiozole ureas	Herpes simplex virus
Processing/packaging	Benzimidazoles	Herpesviruses
Nucleoside biosynthesis		
Inosine monophosphate dehydrogenase	Ribavirin	Respiratory syncytial virus, Lassa fever virus
	Levovirin, viramidine	Hepatitis C virus
Nucleoside scavenging		
Thymidine kinase	Nucleoside analog	Herpes simplex virus, varicella-zoster virus
Ribonucleotide reductase	Inhibitors of protein-protein interaction of large and small subunits	Herpes simplex virus
Glycoprotein processing		
Assembly	No lead compounds	Enveloped viruses
Protease	Peptidomimetics	HIV
Virion integrity	Nonoxynol-9	HIV, herpes simplex virus
Lipid raft disruption	β-Cyclodextrins	HIV, herpes simplex virus

[a]Data from E. De Clercq, *Nat. Rev. Drug Discov.* **1**:13–25, 2002, and S.-L. Tan, A. Pause, Y. Shi, and N. Sonenberg, *Nat. Rev. Drug Discov.* **1**:867–881, 2002.

The New Lexicon of Antiviral Discovery

Mechanism-based screens, cell-based assays, high-throughput screens, rational drug design, bioinformatics, and combinatorial chemistry now describe the pathway of modern antiviral discovery (Fig. 19.13). This process does not find drugs (compounds approved and licensed for use in humans), but rather it detects hits and highlights strategies for developing them as leads. This procedure is used because the new methods cannot tell if a compound will get to the right place in the body at the appropriate concentration (**bioavailability**), or if it will persist in the body long enough to be effective (**pharmacokinetics**). They also fail to address questions of toxicity and specificity.

Viral Genetics and Antiviral Drug Discovery

With modern methods, viral genomes can be sequenced rapidly, and many viruses can be manipulated to test the proof of principle for an antiviral compound (Table 19.8). For example, if the hypothesis is that inhibition of a particular viral enzyme or process blocks virus replication and disease, then construction of a mutant virus in which the gene of interest is inactivated or deleted provides a rigorous test. If the mutant is unable to grow at all, or cannot cause disease in model systems, the investigator has increased confidence that an inhibitor will have antiviral activity.

Screening for Antiviral Compounds

Mechanism-Based Screens

As the name "mechanism-based screen" implies, the specific virus-encoded enzyme or molecular interaction to be inhibited is known (or predicted), and compounds that affect this particular mechanism are sought. Enzymes, transcriptional activators, cell surface receptors, and ion channels are popular targets. Often this screening is done with purified protein in formats that facilitate automated assay of many samples. Two examples of mechanism-based screens designed to identify inhibitors of a viral protease required for assembly are given in Fig. 19.14.

Cell-Based Screens

In cell-based assays, essential elements of the specific mechanism to be inhibited (e.g., a viral enzyme plus a readily assayable substrate) are engineered into an appropriate cell. Two examples of cell-based screens are illus-

Figure 19.13 Path of drug discovery. The flow of information and action followed by modern drug discovery programs that ultimately yield compounds that can be tested clinically for efficacy. The process begins with recognition of a disease or medical need that can be attacked with molecular biology research. Cloning, sequencing, protein purification, and gene manipulation are carried out such that defined assays with purified components can be established to identify the gene products responsible for the disease or medical need. Once identified, the molecular mechanisms can then be studied in detail by using structure-function experiments. Collections of small molecules from various sources can be tested to determine if they inhibit or activate the desired mechanism using various screening strategies. This is an iterative process because once a prospective compound is found (a hit), it can be modified and retested in the simple assays. After formulation of a hit to make it more soluble or tractable for further analysis, animal studies must be completed to determine various pharmacological parameters including toxicology and metabolism. Only then is a lead compound submitted for clinical testing. NMR, nuclear magnetic resonance.

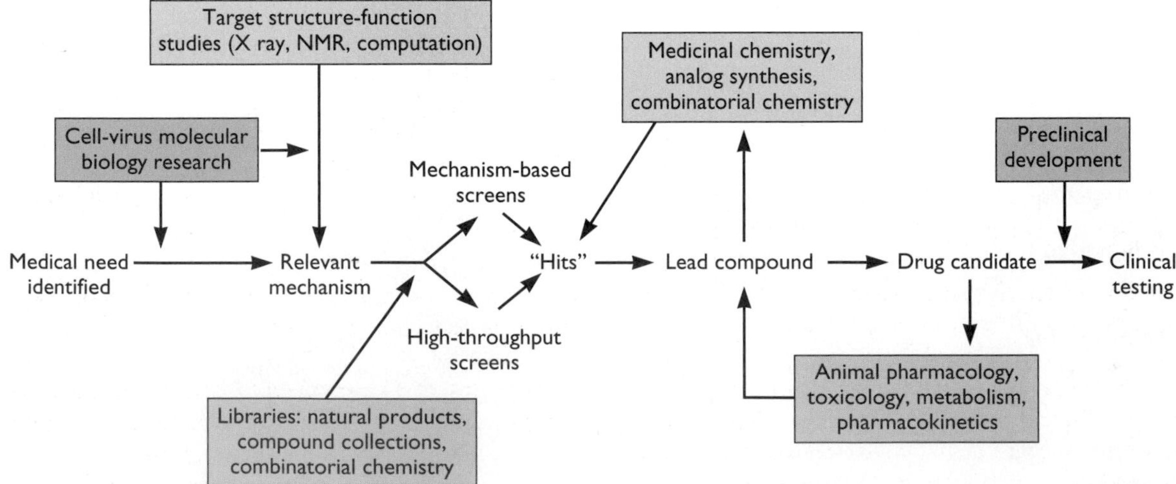

Table 19.8 Critical points for drug hunters seeking commercially viable antiviral compounds

Virology issues

Proof of principle must be obtained as soon as possible during discovery process.
- What is the molecular mechanism?
- Does the drug work in cell and animal models as predicted?

Compound should block viral spread early to limit cytopathology and host cytokine/inflammatory response.
- Drug must be potent and block replication completely; must not make an infection "persistent" by slow replication and spread.

Resistance to the antiviral drug must be manageable.
- Mutants resistant to any antiviral compound are guaranteed.
- Resistance mutations arise when any virus is permitted to replicate; antiviral compounds select for these spontaneous mutants.
- Viruses with resistance to other drugs must not be resistant to your drug.
- The virulence of resistant mutants must be understood.
- Noncompliance by patient may select for drug resistance or may encourage the persistent infection; multiple dosing, stringent dietary requirements, unpleasant taste or side effects affect compliance.

Toxicities should be anticipated, particularly for compounds given to control persistent infections.
- Long-term human toxicology studies (years) are difficult, expensive, and often not possible.
- Short-term toxicology studies may not predict side effects arising when drugs are taken over an extended time (years).

Business issues

Compound should be safe with no side effects.

Compound should be inexpensive to manufacture.

Compound should be easy to formulate and deliver; a pill to be swallowed is much preferred over injection.

Compound must satisfy an unmet medical need, i.e., must be better than any competitive drug or, better yet, have no competition.

Ultimately, a profit should be possible, so the market should be large enough to enable a profit to be made.

trated in Fig. 19.15. Such cell-based assays can provide information, not only about inhibition of the target reaction, but also about cytotoxicity and specificity, if suitable controls are included. New approaches now include technology to measure more than one event at a time by using several different reporter molecules, or optical methods to track movement of a labeled molecule inside the cell.

An interesting variation of the cell-based assay is the "**minireplicon.**" Certain viruses cannot be propagated in cultured cells (e.g., hepatitis C virus or human papillomavirus), or are dangerous enough to require high biological containment (e.g., the hemorrhagic fever viruses). These viruses present a variety of difficult problems for antiviral drug discovery. The minireplicon system comprises a set of plasmids which, when introduced into cells, sponsor the synthesis of viral replication proteins that promote replication of an engineered viral genome segment marked with a reporter gene (the minireplicon). Replication can be monitored by assaying for reporter gene expression. Inhibitors that block replication can be discovered, analyzed, and developed for therapeutic use.

High-Throughput Screens

High-throughput screens are mechanism-based or cell-based screens that allow very large numbers of compounds to be tested in an automated fashion. It is not unusual for pharmaceutical companies to process more than 10,000 compounds per assay per day, a rate inconceivable for early antiviral drug hunters. Compounds to be screened are usually arrayed in multiwell, interchangeable plastic dishes with a few microliters or less of compound solution per well. Robots then apply these compounds to other plastic dishes containing the cell-free or cell-based screen, and after incubation, the signal created by the reporter gene or other output is read and recorded. Data are stored in computers for retrieval and analysis. Not only numerical data, but also images of cells or reactions can be captured, stored, and analyzed.

Sources of Compounds Used in Screening

Many pharmaceutical and chemical companies maintain large libraries of chemical compounds. Usually, a sample of every compound synthesized by the company for

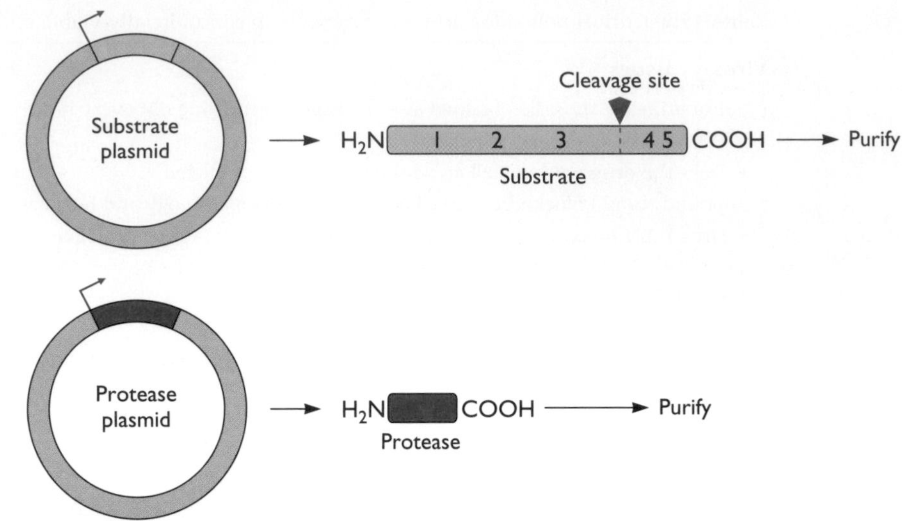

Assay:
1) Mix a constant amount of substrate with a concentration of protease.
2) Controls: protease and substrate alone.
3) Incubate for set periods under appropriate conditions with and without inhibitor.
4) Sample; separate proteins by electrophoretic mobility.

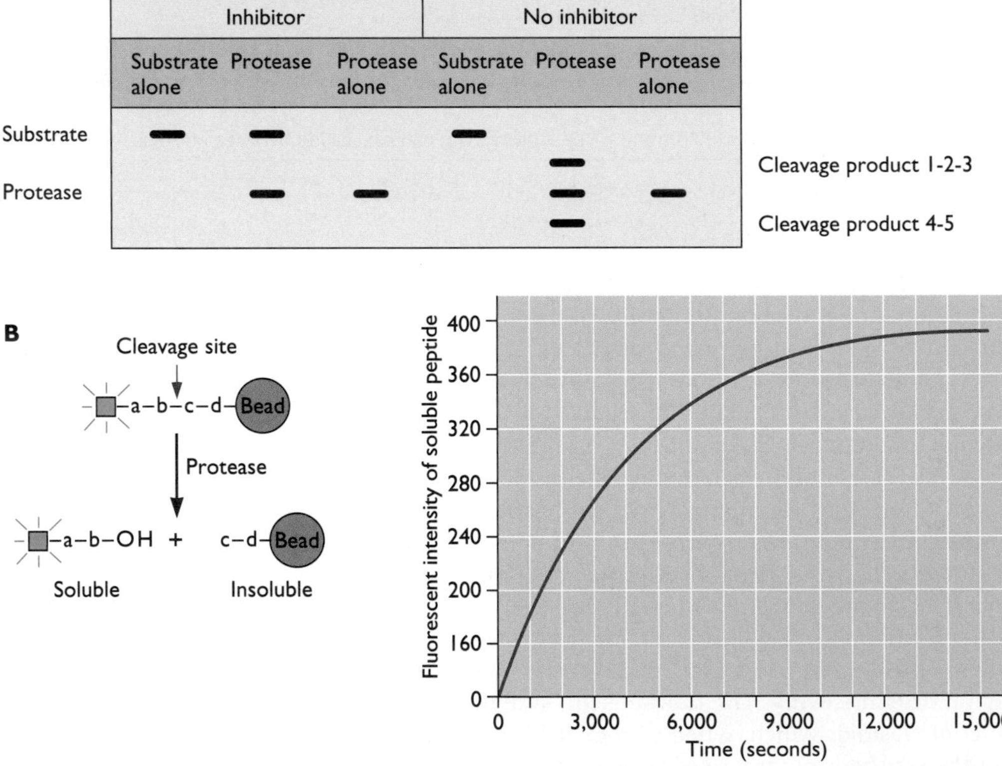

Figure 19.14 Screens for inhibitors of viral proteases. (A) A mechanism-based assay for activity of a viral protease. The plasmids used to produce the protease substrate and the protease are indicated by circles at the top of the figure. The red arrows indicate the promoter and direction of transcription. The resulting proteins are indicated on the right. The amino terminus (H_2N) and carboxy terminus (COOH) of each protein are indicated. The substrate protein has a defined cleavage site for the protease (red arrowhead). The protease is purified as an active enzyme from *E. coli* extracts. The assay is performed as indicated. A specific inhibitor of the protease activity is added to one set of reactions. A diagram of a typical electropherogram is shown at the bottom of the figure; the three samples on the left were from reactions containing the inhibitor, and the three samples on the right were not exposed to the inhibitor. The positions expected for the uncleaved substrate and the protease are shown on the left side of the image, and the specific cleavage products are noted on the right side. Experiments of this type can be used to determine kinetic parameters of enzyme activity and inhibition. (B) Fluorogenic, mechanism-based screen for a viral protease. The substrate is a short peptide encoding the protease cleavage site. A fluorogenic molecule is covalently joined to the N terminus of the peptide, and the entire complex is attached via its C terminus to a polystyrene bead. When the peptide/bead suspension is exposed to active protease, the peptide is cleaved such that the fluorogenic N terminus is released into the soluble fraction, which can be quickly and cleanly separated from the insoluble beads containing the nonfluorogenic product as well as the fluorogenic unreacted substrate. Protease activity is assayed by the appearance of soluble fluorescent peptide as a function of time.

any project is archived, and its history is stored in a database. Chemical libraries of half a million or more distinct compounds are not unusual for a large company. In addition, similar libraries of natural products collected from all over the world, including "broths" from microbial fermentations, extracts of plants and marine animals, and perfusions of soils containing diverse mixtures of unknown compounds, can be searched for components that may have antiviral activities. Another type of chemical library is now available from application of **combinatorial**

A

Active tetracycline efflux protein; insertion of protease site has no effect

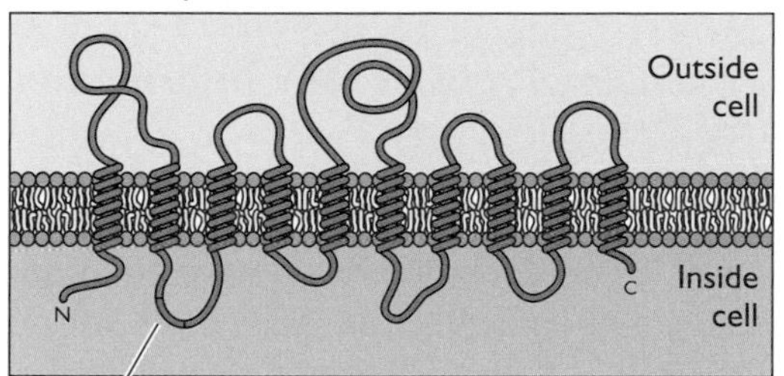
Outside cell
Inside cell
N
C
Tetracycline-resistant bacteria
Engineered HIV protease site

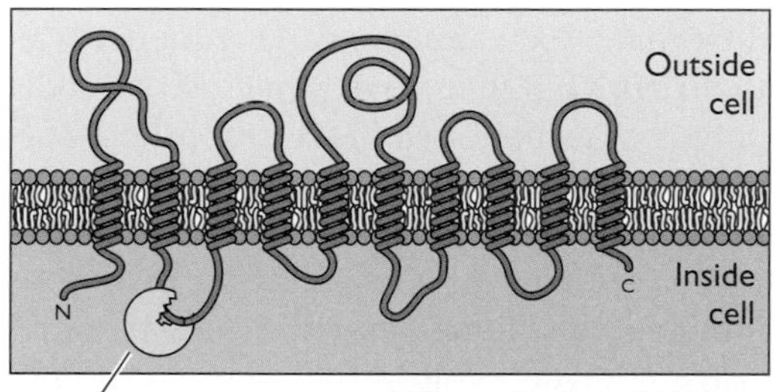
Outside cell
Inside cell
N
C
HIV protease
Coproduction of HIV protease leads to inactivation of the tetracycline efflux protein
Inactive tetracycline efflux protein

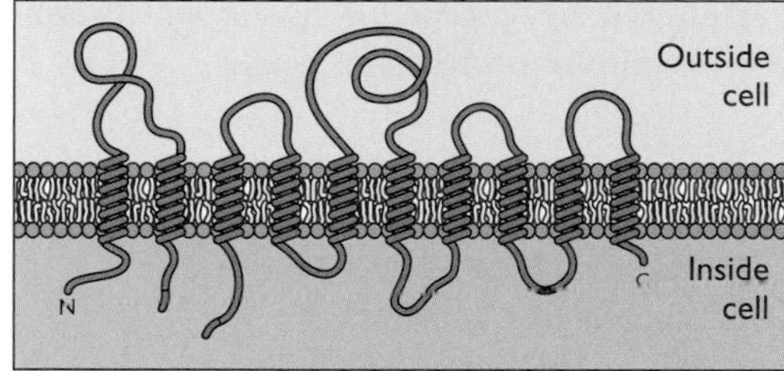
Outside cell
Inside cell
N
Tetracycline-sensitive bacteria
Active tetracycline efflux protein

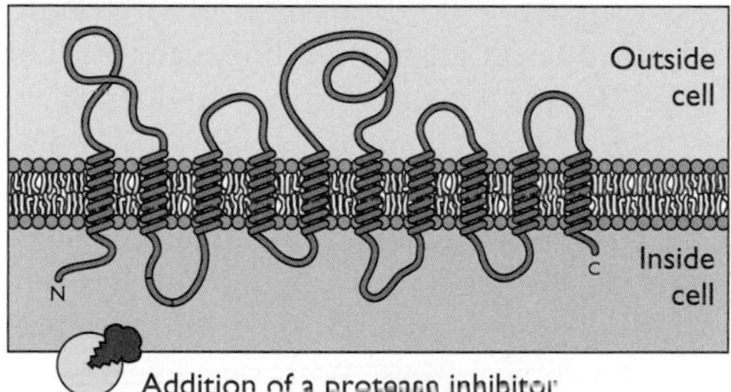
Outside cell
Inside cell
N
C
Tetracycline-resistant bacteria
Addition of a protease inhibitor blocks cleavage, leaving an active tetracycline efflux protein

B

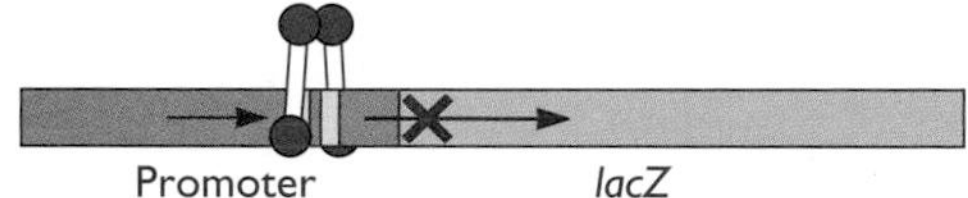
Transcription regulatory protein binding sites
Promoter
lacZ
A transcription regulatory protein, by virtue of its DNA binding ability, can block transcription in E. coli, leading to loss of reporter gene activity.

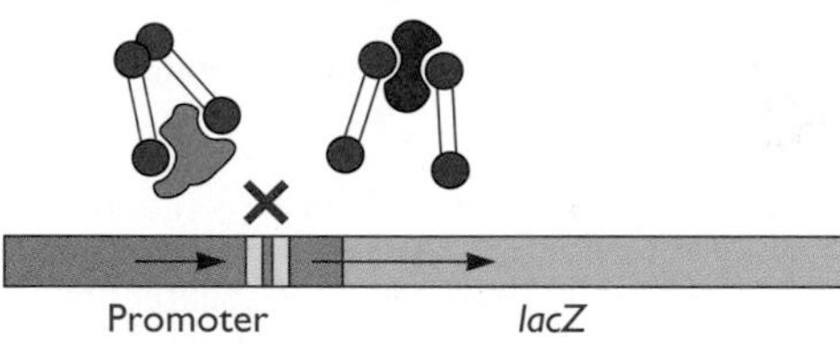
Promoter
lacZ
A small molecule that directly blocks DNA binding of the protein or blocks a protein-protein interaction required for the cooperative binding can be identified by the restoration of reporter gene activity.
Promoter
lacZ

Figure 19.15 Cell-based screens for a viral protease and a transcription regulatory protein. (A) Cell-based assay for HIV protease, using tetracycline resistance of genetically engineered bacteria. Tetracycline resistance in *E. coli* is mediated by the tetracycline efflux protein, which pumps the antibiotic out of the cell. The idea is to insert an HIV protease cleavage site into the tetracycline efflux protein such that the hybrid protein remains active and still confers tetracycline resistance (first panel). When the viral protease gene is made in the same cell, the hybrid tetracycline efflux protein is cleaved and no longer functions (second panel). As a result, the cell becomes tetracycline sensitive (third panel). When inhibitors are added to cells that synthesize the protease, cleavage may be blocked, leaving an active tetracycline efflux protein (fourth panel). Inhibitor activity is scored readily by the appearance of tetracycline-resistant bacteria. Inhibitors must be able to cross the cell membrane and not be toxic to the cell. To facilitate uptake of small molecules that might inhibit the protease, a variety of *E. coli* strains that have reduced permeability barriers are available. It is important to include many controls and secondary assays to identify false-positive and false-negative results. This assay is described in more detail in R. H. Grafstrom et al., *Adv. Exp. Med. Biol.* **312:**25–40, 1992. (B) Cell-based assay for inhibitors of a viral transcription regulatory protein. As in panel A, this assay requires construction of a specific expression cassette to be inserted into the test organism. This is accomplished by having the cassette as part of a plasmid or integrated into the test organism's genome. The system illustrated uses a *lacZ* reporter gene with a viral transcription protein binding site between the *lacZ* promoter and the *lacZ* open reading frame. When this specific viral protein is coproduced in the cell harboring this hybrid *lacZ* gene, the protein binds to its cognate site and represses transcription. The cell does not make β-galactosidase, and this phenotype can be scored by a variety of colony color assays on indicator media. A small molecule that blocks binding of the viral transcription protein to the DNA or protein-protein interaction can be identified through restoration of *lacZ* transcription and the appearance of active β-galactosidase. It is prudent to include a variety of controls and secondary assays to identify false-positive and false-negative results. Courtesy of R. Menzel, Optigenix, Inc.

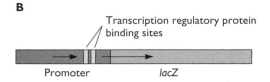

chemistry, a technology that provides drug hunters access to unprecedented numbers of small, synthetic molecules for screening (Fig. 19.16). Before implementation of this technology, a medicinal chemist could reliably make only about 50 compounds a year. Combinatorial chemistry can provide all possible combinations of a basic set of modular components, often on uniquely tagged microbeads or

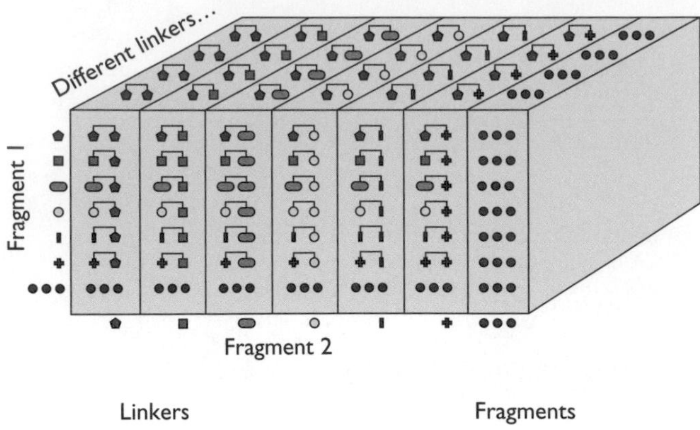

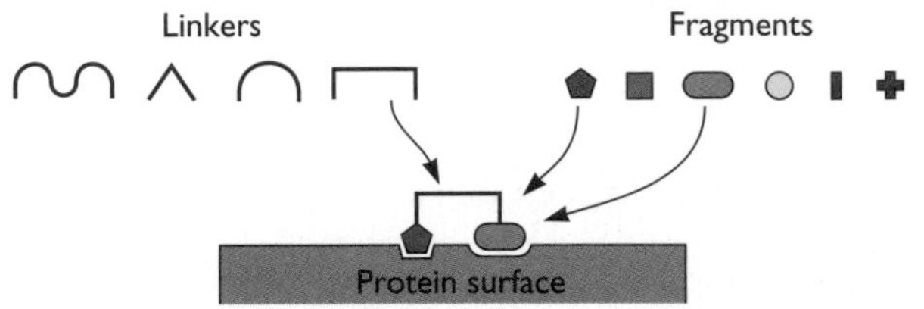

Figure 19.16 Combinatorial chemistry and the building block approach to chemical libraries. Small organic molecules predicted to bind to different pockets on the surfaces of proteins can be grouped into subsets of distinctive chemical structures, and these molecules can be synthesized. With automated procedures, these chemical entities (fragments) can be joined together by various chemical linkers to produce a large, but defined library of small compounds. For example, if assembled pairwise with 10 linkers, a collection of 10,000 small molecules yields a library of 1 billion new combinations. These defined chemical libraries allow a detailed exploration of the binding surfaces of complex proteins and are predicted to provide new antiviral compounds. To take advantage of these chemical libraries, technology is being developed to enable medicinal chemists to reconstruct the specific combination of linkers and fragments that gives rise to any useful compound found in the library. Adapted from P. J. Hajduk et al., *Science* **278:**497–499, 1997, with permission.

other chemical supports, such that active compounds in the mixtures can be traced, purified, and identified with relative ease. Making thousands of compounds in days is now routine with this technology.

Designer Antivirals and Computer-Based Searching

Structure-Based Drug Design

Structure-based design depends on having the atomic structure of the target molecule, usually provided by X-ray crystallography. Computer programs, known and predicted mechanisms of enzyme action, fundamental chemistry, and personal insight all aid an investigator in the design of ligands that will bind at a critical site and inhibit protein function. Currently, the atomic structures of tens of thousands of macromolecules, including important viral proteins, have been determined. An example of structure-based design is detailed below (see "Structure-Based Design of Influenza Virus Neuraminidase Inhibitors"). HIV protease inhibitors may be the most successful antiviral agents that have been designed by structure-based analyses (Fig. 19.17).

Genome Sequencing Provides New Information for Antiviral Drug Discovery

We now can anticipate having the entire sequence not only of any viral genome but also of the corresponding host genome. Such sequences will provide a wealth of information to define new targets for antiviral drug discovery. High-density arrays of DNA fragments on a solid surface can be used to assess the expression of thousands of genes in a single experiment (microarray analyses [see Chapter 2]). More than 10,000 unique gene sequences can be spotted onto a DNA chip the size of a microscope slide. Such global analysis of complicated processes in host-parasite interactions is unprecedented. This technology enables scientists to identify the changes in mRNA concentration that vary in response to viral infection in various tissues. It is likely that new human antiviral gene products and new defensive pathways will be identified in the not too distant future. In addition, microarray technology is assisting in the identification of toxic or adverse responses to lead antiviral compounds long before they are tested in animals. Similar approaches are being developed to analyze the entire repertoire of proteins in a cell.

The Difference between "R" and "D"

Antiviral Drugs Are Expensive To Discover, Develop, and Bring to the Market

It is not unusual for the cost of bringing an antiviral drug to market to exceed $100 million to $500 million. Even with modern methods, it is common for thousands of leads to yield but one promising candidate for further development (Fig. 19.18). Research and lead identification, the "R" of "R&D" (research and development), represent only the beginning of the process of producing a drug. The "D" of "R&D" is development, comprising all the steps necessary to take an antiviral lead compound through safety testing, scale-up of synthesis, formulation, pharmacokinetic studies, and clinical trials. With rare exceptions, it takes 5 to 10 years after the initial lead is found to get a drug to the market, even with the latest technology. Decisions made by drug companies often are influenced strongly by these remarkable costs (Box 19.11).

Antiviral Drugs Must Be Safe

As in vaccine development, safety is the overriding concern of any company developing an antiviral drug. Toxicity to test cells and animals is the first indication that

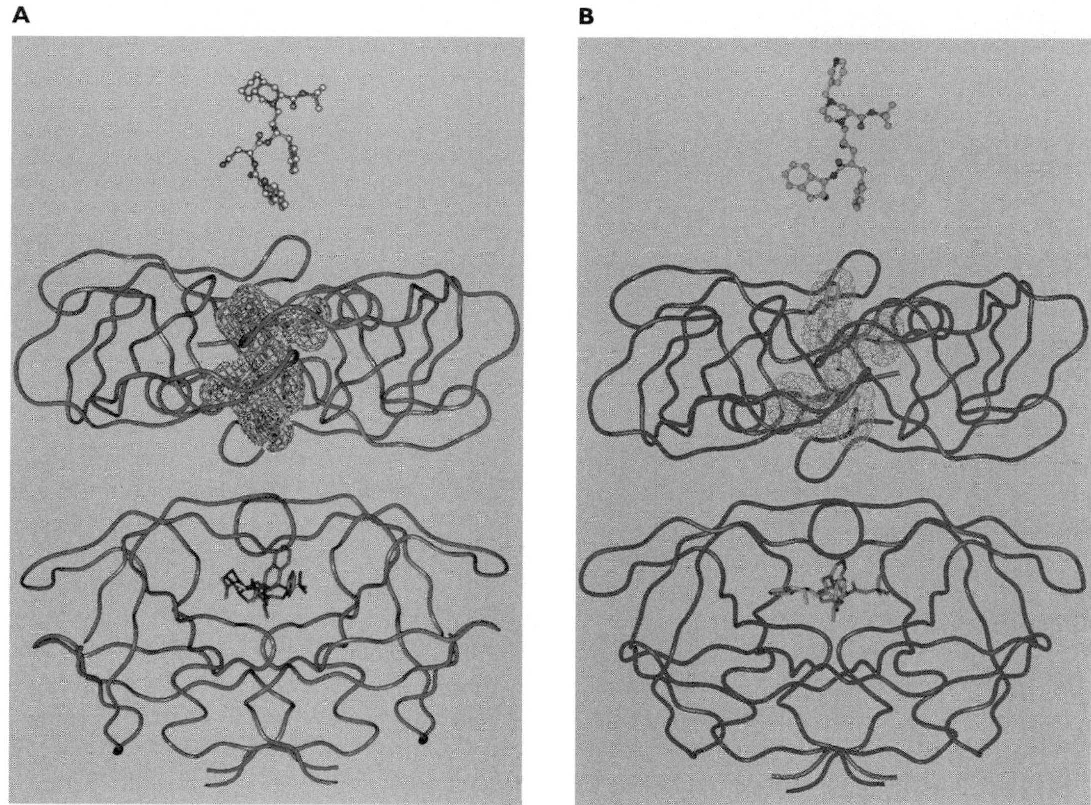

Figure 19.17 Structure of the HIV type 1 protease with the inhibitors indinavir and saquinavir. Structures of the viral protease dimer with bound AIDS drugs saquinavir (Roche) and indinavir (Merck), as well as the structures of these drugs by themselves, are shown. (A) Tracing of the main chain of the protease in two views rotated by 90° (middle and bottom), with the ball-and-stick model of saquinavir (top), saquinavir covered by a net representing the atomic surface (middle), and stick representation of saquinavir alone (bottom). (B) Similar images of indinavir. The images were prepared by J. Vondrasek and are reproduced from the protease database (http://srdata.nist.gov/hivdb/) (J. Vondrasek et al., *Nat. Struct. Biol.* **4**:8, 1997), with permission.

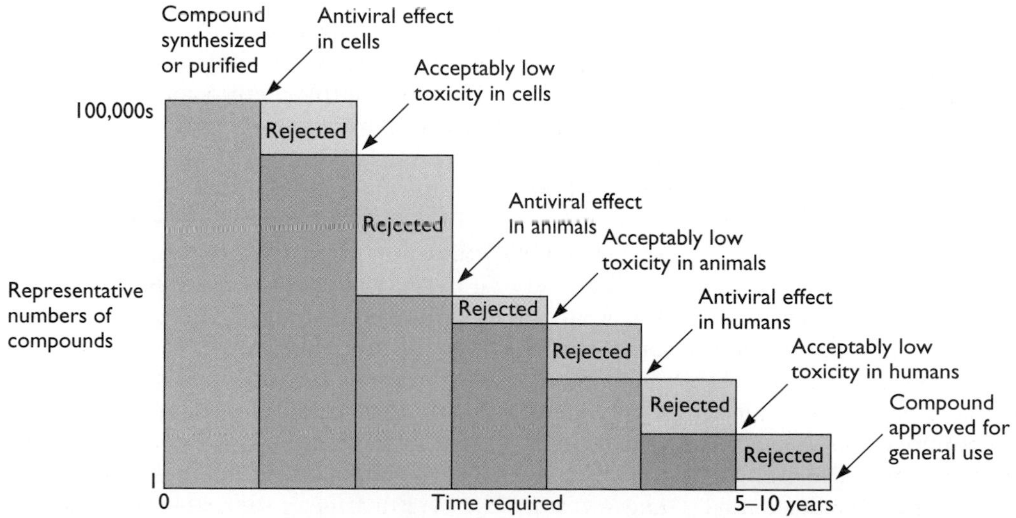

Figure 19.18 Up the down staircase of drug discovery. Many compounds must be tested before a commercially viable antiviral drug will become available. The attrition rate is very high, as hundreds of thousands of chemicals are tested in multiple steps taking several years before one compound emerges as a drug. A few of the significant hurdles in the process are illustrated.

BOX 19.11

New drugs, new mechanisms—no interest?

Recently, two pharmaceutical companies independently discovered a new class of drug that inhibits herpes simplex virus replication. These compounds are targeted to the DNA helicase-primase that is absolutely required for viral replication. They represent the first new anti-herpes simplex virus drugs since acyclovir was developed in the 1970s. The helicase-primase inhibitors are more potent than acyclovir and its derivatives in animal models, and have remarkable potential.

However, neither company has plans to develop the inhibitors or to test them in people. The reason is that acyclovir is a safe, effective drug, and the expense of taking a new drug through clinical trials is enormous. Marketing strategy asserts that it is not cost-effective to compete with a proven drug. The reality is that companies must make choices about where to put their resources.

Crumpacker, C. S., and P. A. Schaffer. 2002. New anti-HSV therapeutics target the helicase-primase complex. *Nat. Med.* **8:**327–328.

a compound may not be safe. More promising leads are discarded because of toxicity than because of any other problem. In the simplest case, an antiviral compound must be more toxic to the virus than to the cell (or to the host). This parameter is quantified as the **cytotoxic index** (for cells) or the **therapeutic index** (for hosts). These indices are defined as the dose that inhibits viral replication, divided by the dose that is toxic to cells or host. The smaller the index, the better; indices of 1/1,000 or smaller are preferred. Traditional approaches to determine toxicity and efficacy can be difficult and time-consuming for many reasons, including the fact that initial lead compounds often are not very soluble in aqueous solutions, or may have only modest antiviral activities. New technologies, such as DNA microarray analyses, may be valuable in identifying mechanisms and sources of toxicity.

No human trial can be initiated without detailed toxicology studies in diverse animal species. Compounds that may be used in long-term treatment of persistent infections must be evaluated for chronic toxicity, allergic effects, mutagenicity, teratogenicity, and carcinogenicity. Many drugs will be given to children, to pregnant women, and to individuals with reduced immune responses, all situations that present special problems. Safety overrides efficacy in most cases. On the other hand, when there are no other effective treatments, as in the early days of the AIDS epidemic, even drugs that caused some undesirable side effects can be licensed for human use (e.g., zidovudine).

Drug Formulation and Delivery

The science of formulation and delivery is an essential part of any antiviral discovery program. After introduction into a person, a drug must reach the proper place, at an effective concentration, and for an appropriate time to affect virus replication. A compound that cannot enter the bloodstream after ingestion is not likely to be effective. Simple entry into the bloodstream may not be sufficient, as many compounds bind to albumin or other proteins in the blood and are rendered ineffective. Other compounds may be metabolized as they pass through the liver, and therefore are inactivated and cleared rapidly from the body. Such problems are not easily predicted and often are discovered empirically. Insoluble compounds, or chemicals unable to enter the bloodstream after ingestion (poor bioavailability), can be modified by the addition of new side chains that may improve absorption from the intestine (see Fig. 19.22 for such a modification of acyclovir). In addition, new delivery vehicles like liposomes, minipumps, or slow-release capsules may improve bioavailability. Other desirable features include stability and cost-effective synthesis in quantity. Literally tons of precursor materials are needed to manufacture commercial quantities of an antiviral drug. Every time a lead compound falls short of a desirable feature, the research director must decide whether to rework the failed compound or to stop the project.

Examples of Some Approved Antiviral Drugs

The antiviral drugs approved for use are surprisingly few (about 30 compounds) and belong to a limited number of chemical classes (Fig. 19.19). While some of these approved drugs are safe and effective, many are marginally efficacious and have side effects that limit their use. Clearly there is room for improvement. Some of the better known antiviral drugs are discussed below.

Acyclovir

Acyclovir is an example of a specific, nontoxic drug that is highly efficacious against herpes simplex virus (genital and oral herpes) and, to some extent, varicella-zoster virus (chickenpox and shingles). It was discovered in 1974, but it was not until the mid-1980s that its full potential as an antiherpes drug was realized. It can be formulated to be administered by injection, as a topical cream, or as an ingestible capsule. Acyclovir is a nucleoside analog similar to guanosine, but contains an acyclic sugar group (Fig. 19.20). It is a prodrug—a precursor of the active antiviral compound (Fig. 19.21). Activation of

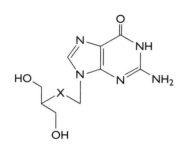

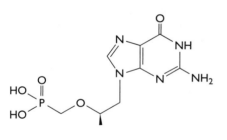

PVAS PVS

Viral adsorption inhibitors

TAK779 AMD3100

Viral-cell fusion inhibitors

Acyclic guanosine analog
(X = O, ganciclovir)
(X = CH_2, penciclovir)

Viral DNA polymerase inhibitor

NRTI
(X = CHN_3, AZT)

NNRTI (UC781)

Reverse transcriptase inhibitors

Tenofovir

Acyclic nucleoside phosphonate

Peptidomimetic Nonpeptidomimetic

Protease inhibitors

Oseltamivir

Neuraminidase inhibitor

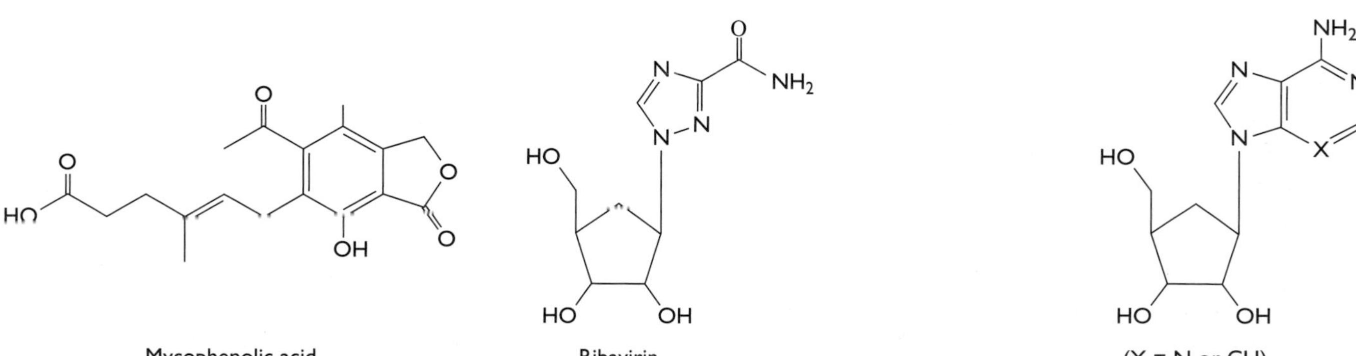

Mycophenolic acid Ribavirin (X = N or CH)

IMP dehydrogenase inhibitors **SAH hydrolase inhibitor**

Figure 19.19 The prototypic compounds (pharmacophores) of important classes of antiviral agents used today. AZT, azidothymidine; IMP, inosine monophosphate; NNRTI, nonnucleoside reverse transcriptase inhibitor; NRTI, nucleoside reverse transcriptase inhibitor; PVAS, polyvinylalcohol sulfate; PVS, polyvinylsulfonate; SAH, *S*-adenosylhomocysteine. TAK779, AMD3100, and UC781 are company designations for the compounds. Adapted from E. De Clercq, *Nat. Rev. Drug Discov.* **1**:13–25, 2002, with permission.

Figure 19.20 Many well-known antiviral compounds are nucleoside and nucleotide analogs. The four natural deoxynucleosides are highlighted in the yellow box. The chemical distinctions between the natural deoxynucleosides and antiviral drug analogs are highlighted by the red color. Arrows connect related drugs. Adapted from E. De Clercq, *Nat. Rev. Drug Discov.* **1:**13–25, 2002, with permission.

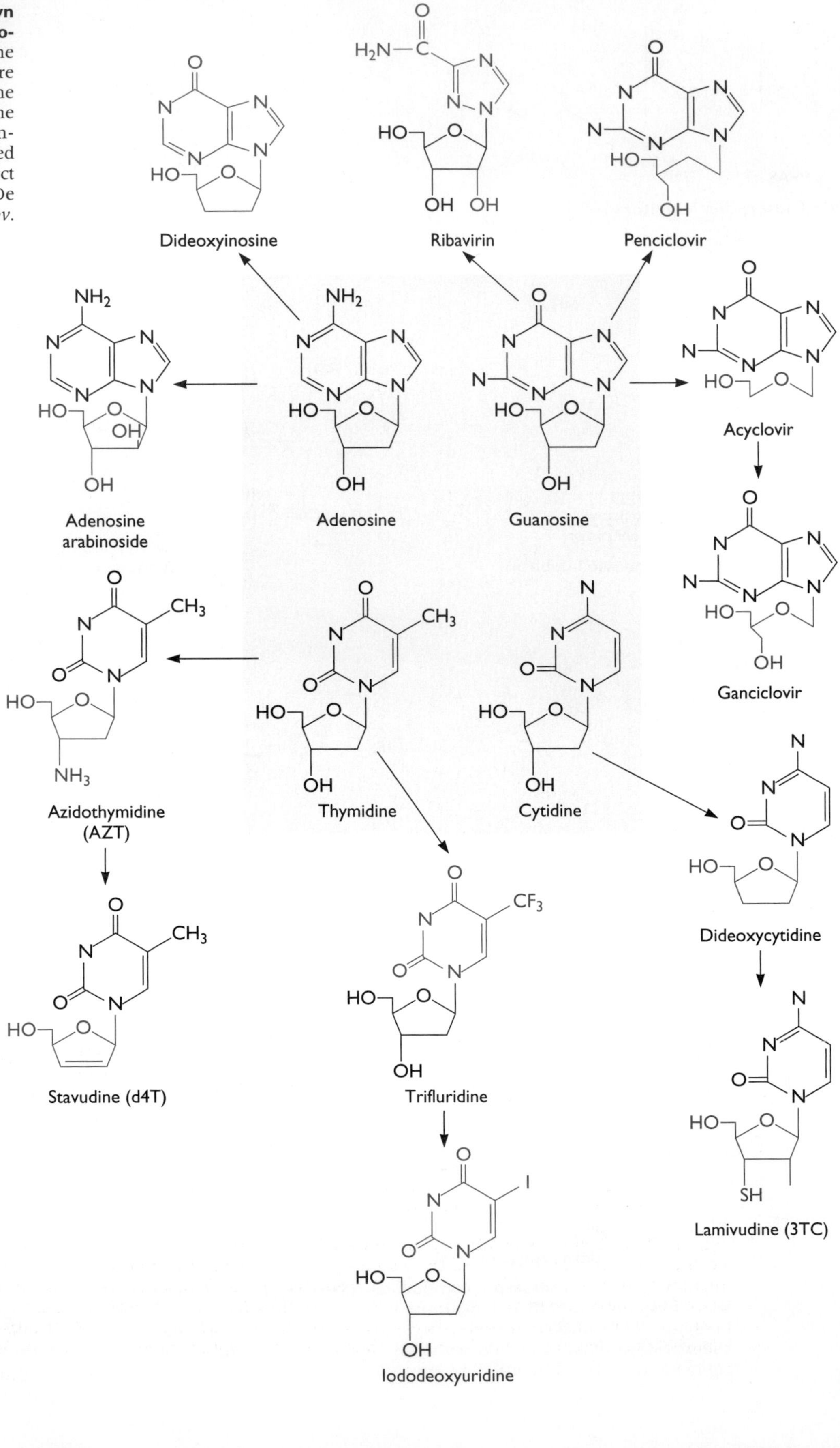

Dideoxyinosine

Ribavirin

Penciclovir

Adenosine arabinoside

Adenosine

Guanosine

Acyclovir

Ganciclovir

Azidothymidine (AZT)

Thymidine

Cytidine

Dideoxycytidine

Stavudine (d4T)

Trifluridine

Lamivudine (3TC)

Iododeoxyuridine

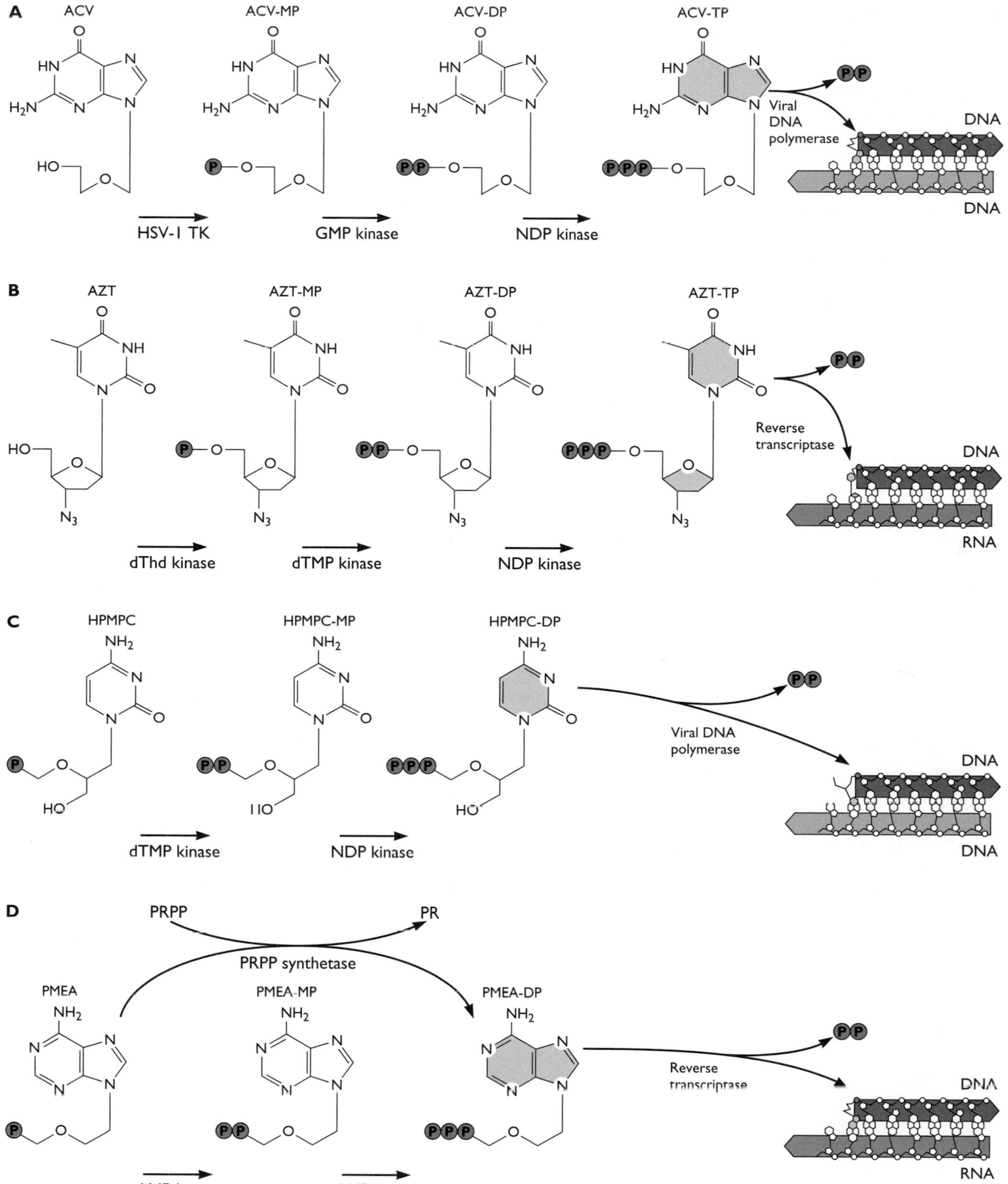

Figure 19.21 Chain termination by antiviral nucleoside analogs. (A) Acyclovir (ACV) is a prodrug that must be phosphorylated in the infected cell before it can inhibit viral DNA replication. The viral thymidine kinase (herpes simplex virus type 1 [HSV-1] TK), but not the cellular kinase, adds one phosphate (orange circle labeled P) to the 5′-hydroxyl group of ACV. The monophosphate compound is a substrate for cellular GMP kinases that convert the monophosphate compound to a diphosphate. Cellular NDP kinases then convert the diphosphate compound to acyclovir triphosphate. The triphosphate compound is recognized by the viral DNA

(continued on next page)

Figure 19.21 *(continued)* polymerase and incorporated into viral DNA. As acyclovir has no 3′-hydroxyl group, the growing DNA chain is terminated, and viral replication ceases. (B) Azidothymidine (AZT) targets the HIV reverse transcriptase. Azidothymidine must be phosphorylated by cellular kinases in three steps to the triphosphate compound, which is incorporated into the viral DNA to block reverse transcription. (C) Cidofovir [*S*-1-(3-hydroxy-2-phosphonylmethoxypropyl)cytosine (HPMPC)] is an acyclic nucleoside analog. In contrast to ACV and AZT, cidofovir requires only two phosphorylations by cellular kinases (typically dTMP kinase followed by NDP kinase) to be converted to the active triphosphate chain terminator. (D) Adefovir [9-(2-phosphonylmethoxyethyl)adenine (PMEA)] is an acyclic nucleoside analog and also requires only two phosphorylations by cellular AMP kinases to be converted to the active chain terminator. Through the action of phosphoribosyl pyrophosphate synthetase (PRPP), which forms the triphosphate from the monophosphate in one step, both cidofovir (panel C) and adefovir bypass the nucleoside-kinase reaction that limits the activity of dideoxynucleoside analogs such as AZT. DP, diphosphate; dThd, (2′-deoxy)-thymidine; MP, monophosphate; NDP, nucleoside 5′-diphosphate; PR, 5-phosphoribose; TP, triphosphate. Adapted from E. De Clercq, *Nat. Rev. Drug Discov.* **1**:13–25, 2002, with permission.

the drug requires three kinases to be present in the cell to convert acyclovir to a triphosphate derivative, the actual antiviral drug. The first kinase, which converts acyclovir to the monophosphate, is not found in an uninfected cell. When acyclovir enters a normal cell, it therefore has no effect on host DNA replication because it cannot be phosphorylated and incorporated into DNA. Herpes simplex virus and varicella-zoster virus genomes, however, encode a kinase that can phosphorylate the prodrug to produce acyclovir monophosphate. Cellular enzymes complete synthesis of acyclovir triphosphate, which is then incorporated into DNA by the viral DNA polymerase. As acyclovir lacks the 3′-OH group of the sugar ring, the growing DNA chain terminates, and further DNA replication is blocked. While this drug is highly specific for cells infected by herpes simplex virus and varicella-zoster virus, the specificity comes primarily from the virus-encoded kinase. This enzyme, thymidine kinase (TK), normally phosphorylates thymidine to thymidine monophosphate, but also phosphorylates a wide range of other substrates including acyclovir. Indeed, if herpes simplex virus TK is synthesized in an uninfected cell and acyclovir is added, the cell will die because its DNA replication will also be blocked by the chain-terminating, phosphorylated acyclovir base analog. This phenomenon is the basis of several strategies for selective cell killing during gene therapy and manipulation of embryonic stem cells.

Because herpesviral genomes encode the enzymes required for their DNA replication, this process has been a fruitful target for the discovery of antiherpes drugs (Table 19.9). Many variations of chain-terminating nucleoside analogs have been developed. Chemical modifications of these compounds can lead to dramatic increases in efficacy or bioavailability (Fig. 19.22). As a result, the latest variations of acyclovir—valacyclovir (Valtrex) and famciclovir (Famvir)—are much more effective and more widely used than their parent compound.

Ganciclovir (Cytovene)

Ganciclovir is another derivative of acyclovir, but this compound was developed to treat human cytomegalovirus infections. However, in contrast to acyclovir, initial formulations of ganciclovir were given intravenously and were quite toxic. Consequently, ganciclovir was used only for life-threatening human cytomegalovirus infections, including pneumonia in AIDS patients and transplant recipients, retinitis, and colitis. Recently, ganciclovir has been approved in an oral formulation that is effective for prophylaxis and long-term use for human cytomegalovirus infections in immunocompromised patients. The oral formulation appears to be much less toxic than the original form delivered intravenously.

Foscarnet (Foscavir, Trisodium Phosphonoformate)

Foscarnet is the only nonnucleoside DNA replication inhibitor of herpesviruses. It is a noncompetitive inhibitor of the pyrophosphate-binding site in herpesvirus DNA polymerase. It also inhibits hepatitis B virus polymerase and the reverse transcriptase of HIV. Foscarnet must be given intravenously. As it accumulates in bone and causes kidney toxicity, this drug is recommended only for life-threatening infections for which other antiviral drugs are no longer effective.

Ribavirin (Virazole)

Ribavirin, a nucleoside analog (Fig. 19.18), was synthesized in 1972 and was purported to have broad-spectrum activity against many DNA and RNA viruses. However, it is relatively toxic, and its development and indications for use have been controversial. In fact, this drug is not licensed in many countries. Despite its long history, its mechanism of antiviral action is not clear. Ribavirin monophosphate is a competitive inhibitor of cellular inosine monophosphate dehydrogenase, and such inhibition leads to reduced GTP (guanosine 5′-triphosphate) pools in

Table 19.9 The antiviral repertoire[a,b]

Approach	Viruses	Compounds approved[c]
Virus adsorption inhibitors	HIV, HSV, CMV, RSV, and other enveloped viruses	None
Virus-cell fusion inhibitors	HIV, RSV, and other paramyxoviruses	None
Virus-uncoating inhibitors	Influenza A virus	Amantadine, rimantadine
Viral DNA polymerase inhibitors	Herpesviruses (HSV-1, -2, VZV, CMV, EBV, HHV-6, -7, -8)	Acyclovir, valacyclovir, ganciclovir, valganciclovir, penciclovir, famciclovir, brivudin, foscarnet
Reverse transcriptase inhibitors	HIV	NRTIs; zidovudine, didanosine, zalcitabine, stavudine, lamivudine, abacavir NNRTIs: nevirapine, delavirdine, efavirenz
Acyclic nucleoside phosphonates	DNA viruses (polyoma-, papilloma-, herpes-, adeno-, and poxviruses), HIV, HBV	CMV: cidofovir HIV: tenofovir
Inhibitors of viral RNA synthesis	HIV, HCV	None
Viral protease inhibitors	HIV, herpesviruses, rhinoviruses, HCV	HIV: saquinavir, ritonavir, indinavir, nelfinavir, amprenavir, lopinavir
Viral neuraminidase inhibitors	Influenza A and B virus	Zanamivir, oseltamivir
Inosine monophosphate dehydrogenase inhibitors	HCV, RSV	Ribavirin
S-adenosylhomocysteine hydrolase inhibitors	(–) strand RNA hemorrhagic fever viruses, filoviruses	None

[a]Data from E. De Clercq, *Nat. Rev. Drug Discov.* **1**:13–25, 2002, with permission.

[b]Abbreviations: CMV, cytomegalovirus; EBV, Epstein-Barr virus; HBV, hepatitis B virus; HCV, hepatitis C virus; HHV, human herpesvirus; HSV, herpes simplex virus; NNRTI, nonnucleoside reverse transcriptase inhibitor; NRTI, nucleoside reverse transcriptase inhibitor; RSV, respiratory syncytial virus; VZV, varicella-zoster virus.

[c]Brivudin is approved in some countries, e.g., Germany. Lamivudine is also approved for the treatment of HBV. Ribavirin is used in combination with IFN-α for HCV.

Figure 19.22 Valacyclovir (Valtrex), an L-valyl ester derivative of acyclovir with improved oral bioavailability. Acyclovir is not taken up efficiently after oral ingestion. However, a derivative of acyclovir, valacyclovir, has as much as a fivefold-higher oral bioavailability than acyclovir (serum levels relative to the dose of drug given orally). The addition of a new side group to acyclovir allows increased passage of drug from the digestive tract to the circulation. These acyclovir derivatives are proprodrugs, which are converted to acyclovir inside the cell by cellular enzymes that cleave off the side chain. Acyclovir is then activated by the viral TK to inhibit viral DNA replication. Adapted from D. R. Harper, *Molecular Virology* (Bios Scientific Publishers, Ltd., Oxford, United Kingdom, 1994), with permission.

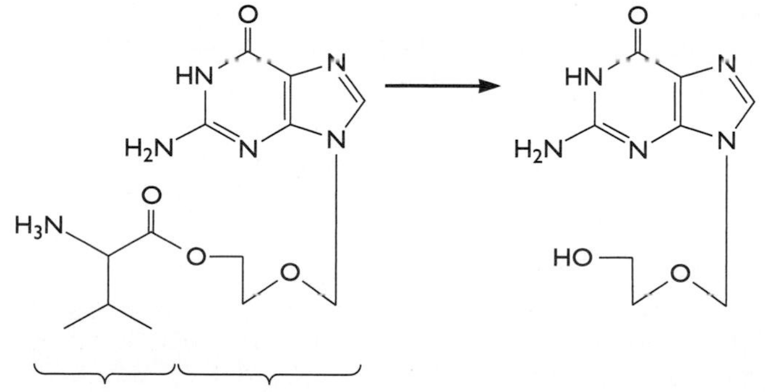

the cell. It also inhibits initiation and elongation by viral RNA-dependent polymerases and also interferes with capping of mRNA. Ribavirin is an RNA virus mutagen: once incorporated into a template, it pairs with C or U with equal efficiency. In some studies, its antiviral activity correlates directly with its mutagenic activity (see Box 20.2). Even with its unknown mechanism and toxicity, ribavirin is used as an aerosol for treatment of infants suffering from respiratory syncytial virus infection, as well as for treatment of Lassa fever virus and hantavirus infections. The combination of ribavirin and IFN has been modestly successful in combating hepatitis B and C viral infections. Viramidine and levovirin are analogs of ribavirin that are in clinical development for treatment of hepatitis C virus infections.

Lamivudine

Lamivudine is an orally delivered nucleoside analog requiring activation by intracellular kinases to act as a chain terminator. It has been effective in blocking the reverse transcriptases of hepatitis B virus and HIV. The drug is highly bioavailable and has low toxicity, as well as a long half-life, which makes it suitable for treatment of chronic hepatitis B virus infections.

Amantadine (Symmetrel)

The three-ringed symmetric amine known as amantadine was developed by DuPont chemists more than 30 years ago. It was the first highly specific, potent antiviral drug effective against any virus. At low concentrations, amantadine specifically inhibits influenza A virus. The target of the drug was shown only recently to be the M2 protein, a tetrameric, transmembrane ion channel that transports protons. The specificity of amantadine for influenza A virus and not influenza B virus can now be understood, as influenza B viral genomes do not encode an M2 protein. Influenza A virus mutants resistant to amantadine arise after therapy (Fig. 19.23). These mutants all have amino acid changes in the M2 transmembrane sequences predicted to form the ion channel, but it is unknown if the drug binds directly in the channel, or if a mutation that alters the channel affects binding of the drug to another site. In any case, amantadine blocks the channel, so that protons cannot enter the virion, effectively preventing the uncoating of influenza A virus.

Amantadine must be administered in the first 24 to 48 h of infection, and be given at high doses for at least 10 days. The drug is most effective when given to susceptible patients in anticipation of influenza virus infection. However, at high concentrations side effects are common, particularly those affecting the central nervous system. On follow-up of these side effects, amantadine was found to be useful for relieving symptoms of Parkinson's disease in some patients. Today, more amantadine is sold for central nervous system disease than for antiviral treatment. Rimantadine, a methylated derivative, cannot cross the blood-brain barrier and therefore has fewer central nervous system side effects.

Zanamivir and Oseltamivir

These antiviral drugs are inhibitors of the neuraminidase enzyme synthesized by influenza A and B viruses (Box 19.12). Zanamivir is delivered via inhalation, while oseltamivir can be given orally. When used within 48 h of onset of symptoms, the drugs reduce the median time to alleviation of symptoms by about 1 day compared to placebo. When used within 30 h of disease onset, the drugs reduce the course of symptoms by about 3 days.

Figure 19.23 Interaction of amantadine with the transmembrane domain of the influenza A virus M2 ion channel. At low concentrations (5 μM), amantadine is thought to exert its antiviral effect by blocking the virus-encoded M2 ion channel activity in infected cells. This inference is supported by genetic data and models of M2 protein, but the points in the virus life cycle critical for inhibition are still controversial. M2 protein is a tetramer with an aqueous pore in the middle of the four subunits. (A) Helical wheel representation of the transmembrane domain of M2 protein. The sequence Leu26-Leu43 of A/chicken/Germany/27(H7N7) is given in single-letter code. Polar amino acids (Ser, Thr, and Asn) and charged amino acids (His and Glu) are found on the right side of the helical wheel, while nonpolar amino acids cluster on the left. The four locations of single amino acid changes in different amantadine-resistant mutants are boxed, and the substitutions most commonly observed are shown in parentheses. (B) Diagram of the putative interaction of amantadine with two diagonally located α-helices of the M2 tetramer in a lipid bilayer. The polar faces of the helices form the interior of the proposed ion channel. Adapted from A. J. Hay, *Semin. Virol.* **3:**21–30, 1992, with permission. For new data, see D. A. Steinhauer and J. J. Skehel, *Annu. Rev. Genet.* **36:**305–332, 2002.

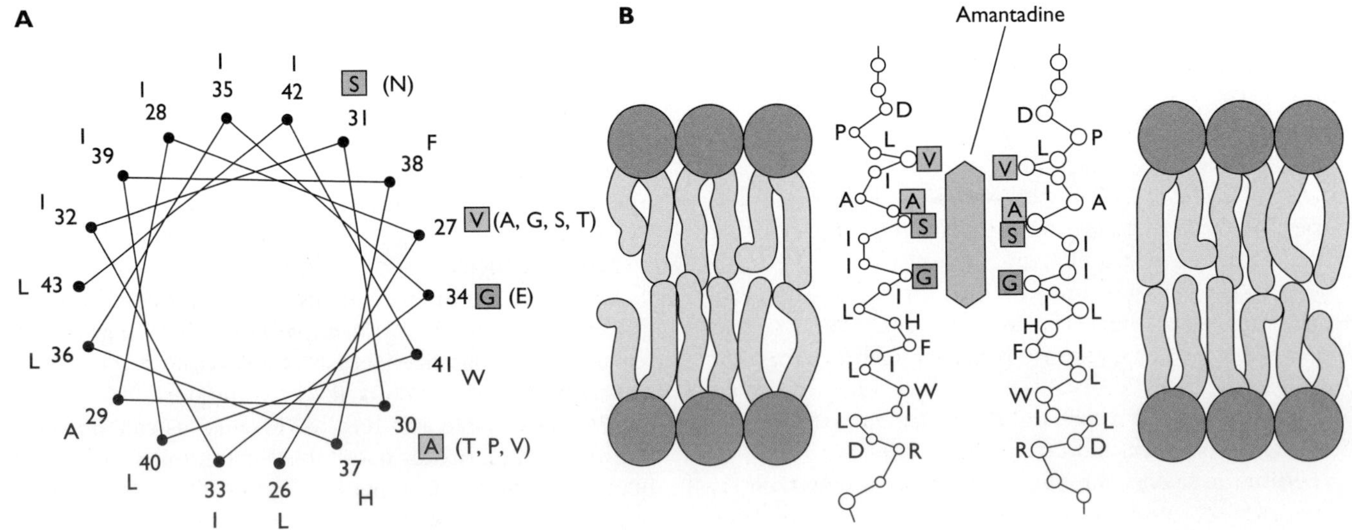

BOX 19.12

Designer drugs: inhibitors of influenza virus neuraminidase

Neuraminidase (NA) is a tetramer of identical subunits, each of which consists of six four-stranded antiparallel sheets arranged like the blades of a propeller (Fig. 19.26). The enzyme active site is a deep cavity lined by identical amino acids in all strains of influenza A and B viruses that have been characterized. Because of such invariance, compounds designed to fit in this cavity would be expected to inhibit the NA activity of all A and B strains of influenza virus, a highly desirable feature in an influenza antiviral drug. Moreover, as NA inhibitors are predicted to block spread, they may be effective in reducing symptoms after the infection has begun.

Sialic acid fits into the active site cleft in such a fashion that there is an empty pocket near the hydroxyl at the 4′ position on the sugar ring of sialic acid. On the basis of computer-assisted analysis, investigators predicted that replacement of this hydroxyl group with either an amino or guanidinyl group would fill the empty pocket and therefore would increase the binding affinity contacting one or more neighboring glutamic acid residues.

These predictions were verified by direct structural analysis of the new compounds bound to NA. Significantly, these compounds were inhibitors of both influenza A and B viruses in cultured cells. They did not inhibit other nonviral NAs, an important requirement for safety and lack of potential side effects.

Currently, two antiviral drugs, zanamivir and oseltamivir, that inhibit influenza virus A and B neuraminidase are licensed for use.

Varghese, J. N., V. C. Epa, and P. M. Colman. 1995. Three dimensional structure of the complex of 4-guanidino-Neu5Ac2en and influenza virus neuraminidase. *Protein Sci.* **4:**1081–1087.

von Itzstein, M., W. Y. Wu, G. B. Kok, M. S. Pegg, J. C. Dyason, B. Jin, T. Van Phan, M. L. Smythe, H. F. White, S. W. Oliver, P. M. Coleman, J. N. Varghese, D. M. Ryan, J. M. Woods, R. C. Bethell, V. J. Hotham, J. M. Cameron, and C. R. Penn. 1993. Rational design of potent sialidase-based inhibitors of influenza virus replication. *Nature* **363:**418–423.

The Search for New Antiviral Targets

Entry and Uncoating Inhibitors

The first step of viral replication has long been an attractive target, as virus-receptor interactions offer the promise of high specificity. Early enthusiasm for entry inhibitors came from experiments with monoclonal antibodies that inhibited virion attachment or entry in cultured cells. Specific receptor-virus interactions can be identified if one can prove that an antibody inhibits virion entry by direct means rather than by steric hindrance, or induction of conformational change. Passive immunization with these antibodies often can protect animals from challenge, suggesting that entry inhibitors may have possibilities as antiviral drugs. As it is usually impractical to develop antibodies as antiviral agents, these molecules have been most valuable in identifying proteins involved in entry and their mechanisms.

The antibody-antigen binding site of entry-specific antibodies provides a starting point for screening chemical libraries, or for design of small molecules that block viral entry. It is important that such compounds interfere with virus-host interactions, but not with the normal function of the cellular receptor. Obviously, an important assumption in this strategy is that no other routes of entry exist. However, alternative receptors are available for many viruses (e.g., herpesviruses and HIV), and blocking a single virion-receptor interaction may not be effective.

Membrane fusion, the ubiquitous process by which enveloped virions enter cells, is also a target for chemotherapeutic intervention. To identify inhibitors of influenza virus fusion, a computer program was first used to predict which small molecules might bind into a pocket of the HA molecule and possibly prevent low-pH-induced conformational change. From the list of molecules identified in this way, several benzoquinone- and hydroquinone-containing compounds were tested and found to inhibit HA-mediated membrane fusion at low pH. One of these inhibits influenza virus replication, and has been shown to block the low-pH-induced conformational changes required for fusion. Although it is not known how binding of the drug prevents fusion, it has been suggested that it acts as a "molecular glue" to prevent movement of the fusion peptide. Inhibitors of HIV entry are discussed in more detail below ("Examples of Anti-HIV Drugs").

Proteases

Viral proteases often cleave protein precursors to form functional units or to release structural components during ordered particle assembly (maturational proteases). As discussed in Chapter 13, all herpesviruses encode a serine protease that is absolutely required for formation of nucleocapsids. Many features of these enzymes and their substrates are conserved among the members of the family *Herpesviridae*. The structure of the human cytomegalovirus protease is shown in Fig. 19.24. Enthusiasm for this enzyme as a target is based on the unusual serine protease fold and the mechanism of catalysis.

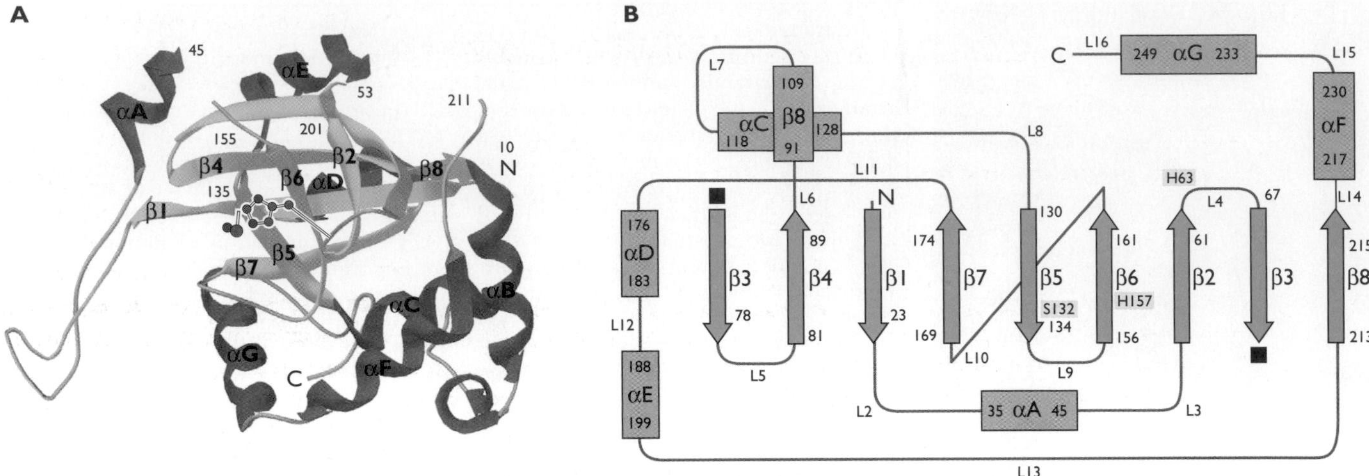

Figure 19.24 Human cytomegalovirus protease. (A) The β-strands are shown as arrowed ribbons in green, α-helices are orange, and the connecting loops are indicated in purple. The active site residues (Ser132, His63, and His157) are also indicated. The secondary-structure elements are labeled starting from the N terminus, the β-strands numerically (β1 to β8), and the α-helices alphabetically (αA to αG). The open end of the β-barrel is toward the top corner. (B) Topological drawing of the backbone fold of human cytomegalovirus protease. Strand β3 is broken at the crossover position, and the breakpoints are represented by black squares. The secondary-structure elements are labeled. The loops are numbered from the N terminus (L1 to L6). The locations of Ser132, His63, and His157 are also shown. Adapted from L. Tong et al., *Nature* **383:**272–275, 1996, with permission.

Hepatitis C virus infects more people than HIV in the United States, and infects millions more individuals around the world. As a result, an intensive antiviral drug discovery effort is in progress. The essential viral protease called NS3 is a major target (Fig. 19.25). Hepatitis C virus cannot be propagated in cultured cells, and so the effect of inhibiting NS3 protease on virus replication is tested by using minireplicons in cell-based assays. The crystal structure of NS3 protease has been solved by several companies, as have the structures of NS3 helicase and the NS5B polymerase; major efforts in structure-based design, as well as mechanism-based screening, have resulted in several NS3 and NS5B inhibitors currently in clinical trials.

Virus-Specific Nucleic Acid Synthesis and Processing

Viral replication enzymes are primary targets for antiviral drugs. From genetic analyses, we know that DNA polymerase accessory proteins, such as those that promote processivity or bind to viral origins, are essential for viral

Figure 19.25 The hepatitis C virus polyprotein is cleaved by several proteases. Hepatitis C virus is a human flavivirus with a (+) strand RNA genome. The viral proteins are encoded in one large open reading frame that is translated into a polyprotein. The polyprotein is processed by cellular and viral proteases to release the viral proteins. The numbers above the polyprotein indicate the amino acid marking the start of each viral protein. The last number, 3010, indicates the amino acid number of the C terminus of the polyprotein. The proteases and their cleavage sites are indicated by arrows. Signal peptidase is the host enzyme in the endoplasmic reticulum that cleaves signal peptides from membrane proteins. The red dotted lines represent cleavage sites that are inefficiently processed or not well studied. The viral metalloprotease that cleaves at position 1027 is viral protease 1. This 23-kilodalton autoprotease requires NS2 and the amino-terminal domain of NS3. The viral serine-type protease is the NS3 protein complexed with NS4A that cleaves at positions 1658, 1712, 1972, and 2420. The hepatitis C virus NS3 protease is essential for the assembly of virus particles.

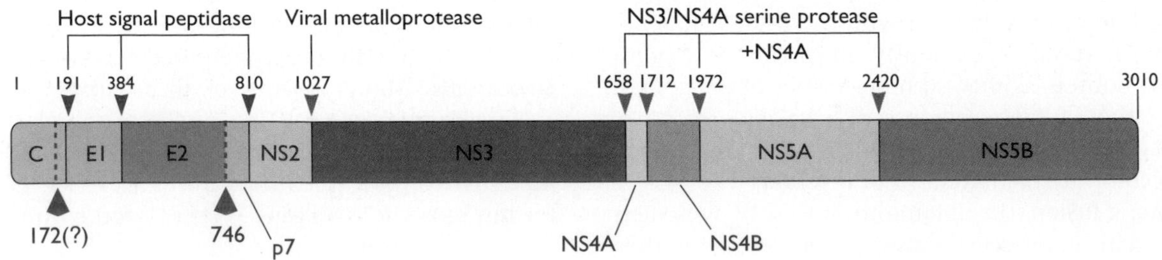

replication and represent attractive new targets for antiviral drugs. The RNA-dependent RNA polymerases of any RNA virus appear to be unique to the virus world—there are no known cellular counterparts in animal cells that can copy long RNA templates in animal cells. These polymerases and their varied virus-specific activities, including synthesis of primers, cap snatching, and recognition of RNA secondary structure in viral genomes, mark them as targets for drug discovery. The unique helicases encoded by many RNA viruses are already in many companies' high-throughput screens, but no lead compounds have yet progressed to the market.

Newly replicated herpesviral DNA is found in large head-to-tail polymers (concatemers) that are cleaved into monomeric units for packaging into virions. These processes are essential for viral replication, and are carried out by specific viral enzymes. They represent new targets for drug discovery with some promise. For example, 5-bromo-5,6-dichloro-1-β-D-ribofuranosyl benzimidazole binds to the human cytomegalovirus UL89 gene product, which is a component of the "terminase" complex responsible for cleaving and packaging replicated concatemeric DNA. Members of the UL89 gene family are highly conserved among all herpesviruses, and have homology with a known ATP-dependent endonuclease encoded in the genome of bacteriophage T4. This class of compounds may therefore be the basis for discovery of broad-based inhibitors of herpesvirus replication. Another class of compounds, the 4-oxo-dihydroquinolines, also have potential as broad-spectrum herpesvirus inhibitors. These nonnucleoside compounds interact with a common polymerase determinant that is not critical for binding of dNTPs or antiviral nucleoside analogs.

Regulatory Proteins

Control of the viral replication cycle often requires action of unusual regulatory proteins. Many of these proteins are known to be essential for viral growth and are prime targets for antiviral screens. Fomivirsen is the first licensed compound directed to inhibit the function of a viral regulatory protein. The drug is a phosphorothioate, antisense oligonucleotide and is approved to treat retinitis, a disease caused by human cytomegalovirus. The mechanism of action requires binding of the 21-nucleotide-long antisense molecule to the cytomegalovirus immediate-early 2 mRNA.

Structure-Based Design of Influenza Virus Neuraminidase Inhibitors

Influenza virus NA protein (Fig. 19.26) cleaves terminal sialic acid residues from glycoproteins, glycolipids, and oligosaccharides. It plays an important role in the spread of infection from cell to cell, because in cleaving sialic acid

Figure 19.26 Structure of influenza A virus NA. (A) A ribbon diagram of influenza A NA with α-sialic acid bound in the active site of the enzyme. The molecule is viewed down the fourfold axis of the tetramer. The molecule is an N2 subtype from A/Tokyo/3/67. The C terminus is on the outside surface near a subunit interface. The six β-sheets of the propeller fold are indicated in colors. (B) The packing of the N2 NA tetramer viewed from above and down the symmetry axis. Adapted from J. N. Varghese, p. 459–486, *in* P. Verrapandian (ed.), *Structure-Based Drug Design* (Marcel Dekker, New York, N.Y., 1997), with permission. Courtesy of J. Varghese.

A

B

residues, the enzyme releases virions bound to the surfaces of infected cells and facilitates viral diffusion through respiratory tract mucus. Moreover, the enzyme can activate transforming growth factor β by removing sialic acid from the inactive protein. Because the activated growth factor can induce apoptosis, NA may influence the host response to viral infection. Neuraminidase is a particularly intriguing target for drug hunters (Box 19.12).

Antiviral Gene Therapy and Transdominant Inhibitors

In 1988 a trio of scientists implanted a mutant herpes simplex virus gene in a cell, and that cell line subsequently proved to be resistant to further infection. Later, when the same mutant gene was inserted into the genome of mice, the mice became resistant to infection by herpes simplex virus. The gene encoded VP16, a viral tegument protein with powerful transcriptional activation properties. The mutation in the VP16 gene conferred a dominant-negative phenotype: the altered protein interfered with the function of the normal VP16 protein. Dominant-negative mutations are not new in genetics, but the idea that such an altered protein could be used to protect an animal against infection was novel. Inhibition of replication by a viral gene product conferring a dominant-negative phenotype can be obtained when the wild-type protein functions as a multimer. Mixed multimers containing a normal and a mutant protein may be inactive, depending on the nature of the mutation.

Many natural and engineered gene products that should block intracellular viral growth when present in an infected cell can be imagined, and some of them may yield effective antiviral therapies (Table 19.10). For example, RNA-binding proteins, DNA-binding proteins, ribozymes, and antisense oligonucleotides offer interesting possibilities. Some clever ideas are based on hybrid proteins composed of a target domain and a killer domain. The former brings the latter to the virus-infected cell. For example, nucleases fused to viral capsid proteins block viral propagation if the viral genome is degraded during virion assembly. Toxins have been fused to antibodies that bind to viral proteins present on the surfaces of infected cells. When the hybrid protein is taken up by endocytosis, it kills the infected cell.

The effectiveness of dominant-negative mutations and killer or suicide molecules has been validated in cell culture systems; indeed, the current strategies (now classified under the rubric of gene therapy) are too numerous to mention. While this approach often captures the fancy of molecular virologists, its practical utility in the treatment of disease remains uncertain. Such compounds or approaches provide interesting leads, but they face the same hurdles as more conventional antiviral lead compounds (Fig. 19.18). The basic requirements of safety and efficacy cannot be predicted with confidence until human trials are conducted.

Table 19.10 Molecules and gene products capable of inhibiting human immunodeficiency virus type 1 replication[a]

Molecule or product	Putative mechanism
Antisense RNA	Blocks translation of target HIV gene
Ribozymes/RNAi	Cleave HIV RNA
Dominant-negative mutants of *tat, rev, gag*	Inhibit normal function
RNA decoys (TAR, RRE)	Compete for essential viral regulatory proteins
Intracellular antibodies	Bind to HIV protein
Capsid-nuclease hybrid proteins	Destroy viral genome during packaging
Inducible toxins or killer proteins	HIV LTR fused to toxin; activated on HIV infection and kills infected cells (e.g., herpes simplex virus TK, diphtheria toxin, interferon); HIV infection-inducible host shutoff functions (e.g., herpes simplex virus *vhs* gene; destruction of many mRNAs in HIV-infected cells)

[a]Abbreviations: LTR, long terminal repeat; RNAi, RNA interference; RRE, Rev-responsive element; TAR, Tat activation region.

Resistance to Antiviral Drugs

Because viral replication is so efficient and is accompanied by moderate to high mutation frequencies, resistance to any antiviral drug must be anticipated. Drug resistance is of special concern during the extended therapy required for chronic infections. Indeed, viral mutants resistant to every antiviral drug manufactured to date have been detected in the clinic, a disconcerting fact because our arsenal of antiviral drugs is small (Table 19.9). Drug-resistant mutants are dangerous for patients, who cannot be treated for the resistant infection, and are potentially hazardous for the population who might be exposed to them. Nevertheless, drug-resistant mutants have value in the laboratory. A genetic analysis of resistance can provide powerful insight into the antiviral mechanism and may identify new strategies to reduce or circumvent the problem.

Consider acyclovir-resistant mutants of herpes simplex virus, which arise spontaneously during viral replication and are selected after exposure to the drug. These mutants are unable to phosphorylate the prodrug, or are unable to incorporate the phosphorylated drug into DNA. The majority of mutations that confer resistance are in the viral TK gene and inactivate kinase function. We can deduce that TK is essential for the action of the drug. As TK is not required for herpes simplex virus to grow in cultured cells, such mutants can be identified with ease. A subset of mutations leading to acyclovir resistance are found in the viral DNA polymerase gene. The altered polymerases possess reduced ability to incorporate phosphorylated drug into DNA. Similar patterns of resistance have been reported when other nucleoside analog inhibitors are used against varicella-zoster virus.

The appearance of TK mutants after acyclovir therapy is of concern in the clinic. Although TK-defective mutants have a markedly reduced ability to invade and replicate in the central nervous system in animal models, they can lead to devastating infections in AIDS patients. Such mutants are still able to cause disseminated disease, and are often resistant to other nucleoside analogs that share the same antiviral mechanism. Fortunately, foscarnet can be used, in some instances, to treat these infections. Acyclovir-resistant DNA polymerase mutants present a special problem because they can also be resistant to foscarnet, leaving essentially no recourse.

An unusual property of amantadine is the concentration dependence of its antiviral activity. The drug has broad antiviral effects at high concentrations, but at low concentrations, the compound is specific for influenza virus A, with no effect on B or C strains or other viruses. Analysis of resistant mutants provided insight into the apparent complex mechanism of action of amantadine. At concentrations of 100 μM or higher, the compound acts as a weak base and raises the pH of endosomes so that pH-dependent membrane fusion is blocked. Any virus with a pH-dependent fusion mechanism could be affected by high concentrations of amantadine. Resistant mutants of influenza A virus selected under these conditions in cultured cells harbor amino acid substitutions in HA that destabilize the protein and enable fusion at higher pH. Influenza virus mutants selected at concentrations of 5 μM or lower carried mutations in the M2 gene. Specifically, these mutations affected amino acids in the membrane-spanning region of the M2 ion channel protein (Fig. 19.23).

HIV and AIDS

The recognition that HIV was associated with AIDS in the early 1980s provided the impetus for unprecedented focus on antiviral drug discovery. As noted in Chapter 17, the first effective drugs acted on the viral reverse transcriptase. The strategy was based on the well-known ability of nucleoside analogs to block replication of herpesviruses and tumor cells. Researchers quickly deciphered many distinctive steps of the viral life cycle, identifying numerous additional targets (Fig. 19.27). As a result, not only new, ef-

Figure 19.27 Important steps in the replication of HIV. Five steps in the life cycle of HIV are highlighted as potential targets for antiviral compounds. Known or hypothetical inhibitory compounds are listed under each step. Adapted from D. D. Richman et al., *Clinical Virology* (Churchill Livingstone, New York, N.Y., 1997), with permission.

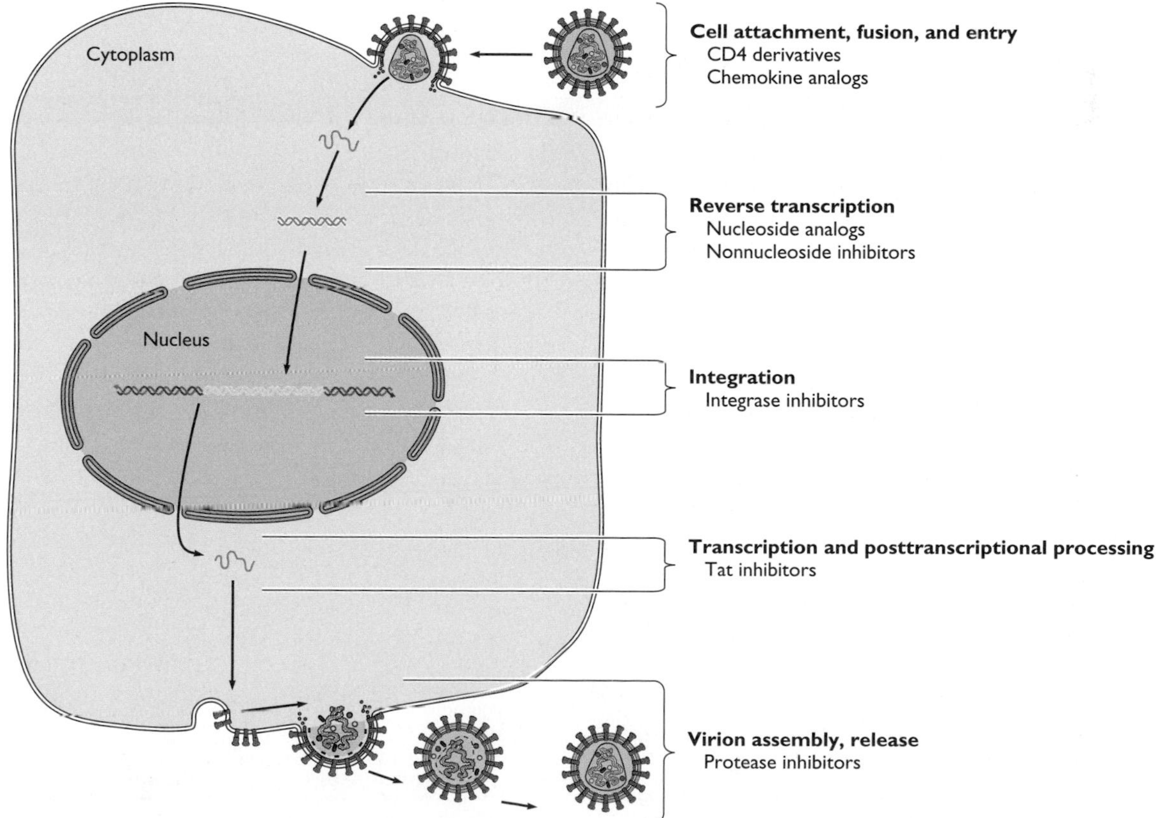

fective nucleoside inhibitors but also nonnucleoside inhibitors of reverse transcriptase and potent protease inhibitors became available in the clinic (Table 19.11). In 1997, a protocol known as highly active antiretroviral therapy was developed in which a combination of these inhibitors is used aggressively to treat patients. Scientists are also identifying more compounds directed to specific steps in the virus life cycle, including inhibitors of virus-cell interaction (attachment, fusion, and entry), integration of viral DNA, and nucleocapsid assembly.

Examples of Anti-HIV Drugs

Nucleoside Analogs

Azidodeoxythymidine (AZT), or zidovudine, was the first drug to be licensed. This drug was initially discovered during screens for antitumor cell compounds rather than for antiviral agents. Unlike acyclovir, which is phosphorylated by a viral kinase, AZT is phosphorylated to the active form by cellular enzymes (Fig. 19.21). Like acyclovir, phosphorylated AZT acts as a chain terminator if incorporated into DNA, because the 3′-OH group of adenosine is replaced by an azido (N_3) group. Phosphorylated AZT is not a good substrate for most cellular polymerases. However, it is incorporated into the DNA copy of the viral genome by the viral reverse transcriptase. The relative selectivity of this drug depends on the fact that reverse transcription takes place in the cytoplasm, where the drug appears first and in the highest concentration. Because AZT monophosphate competes with thymidine monophosphate for the formation of nucleoside triphosphate, its presence causes depletion of the intracellular pool of ribosylthymine 5′-triphosphate. Therefore, AZT is much less selective than acyclovir and has substantial side effects. AZT can be given orally and is absorbed quickly, but is also degraded rapidly by liver glucuronidation enzymes so that its effective half-life in a patient is only about 1 h. Patients must take the drug two or three times daily to maintain an effective antiviral concentration. If the illness is severe, AZT can be administered intravenously. As we will discuss below, a short half-life is problematic because if an effective concentration of any antiviral drug is not maintained, resistant mutants will be selected. At best, AZT provides modest and transient effects in infected adults with a concomitant increase in $CD4^+$ T-cell counts. However, it is effective in prophylactic treatment for accidental needle sticks, and for treatment of infected pregnant women, in whom treatment can reduce considerably the probability of delivering an HIV-infected baby.

AZT toxicity is of great concern when the drug must be given for long periods. AZT treatment can damage bone marrow, resulting in a reduction in the number of neutrophils. Treatment of this problem requires multiple transfusions of red blood cells. Muscle wasting, nausea,

Table 19.11 Some approved antiviral agents against human immunodeficiency virus

Target or mechanism	Generic name	Trade name	Manufacturer	Date approved
Nucleoside reverse transcriptase inhibitors	Zidovudine (AZT, ZDV)	Retrovir	Glaxo Wellcome	1987
	Didanosine (ddI)	Videx	Bristol-Myers Squibb	1991
	Zalcitabine (ddC)	Hivid	Roche	1992
	Stavudine (d4T)	Zerit	Bristol-Myers Squibb	1994
	Lamivudine (3TC)	Epivir	Glaxo Wellcome	1995
	AZT/3TC	Combivir	Glaxo Wellcome	1997
	Abacavir (ABC)	Ziagen	Glaxo Wellcome	1998
	AZT/3TC/ABC	Trizivir	Glaxo Wellcome	2000
	Tenofovir (TDF)	Viread	Gilead	2001
Nonnucleoside reverse transcriptase inhibitors	Nevirapine	Viramune	Roxane	1996
	Delavirdine	Rescriptor	Agouron	1997
	Efavirenz	Sustiva	Dupont	1998
Protease inhibitors	Saquinavir (hard gel)	Invirase	Roche	1995
	Saquinavir (soft gel)	Fortovase	Roche	1997
	Ritonavir	Norvir	Abbott	1996
	Indinavir	Crixivan	Merck	1996
	Nelfinavir	Viracept	Agouron	1997
	Amprenavir	Agenerase	Glaxo Wellcome	1999
	Lopinavir/ritonavir	Kaletra	Abbott	2000
Summary: three enzyme targets	**16 unique compounds**	**19 approved drugs**	**Nine companies**	**15 yr**

and severe headaches are but a few of AZT's other side effects that must be endured. Despite these problems, the drug has been used extensively, because, until recently, there simply was no alternative.

Considerable effort has been devoted to discovering alternatives to AZT. Particular emphasis has been placed on finding reverse transcriptase inhibitors that are effective against AZT-resistant mutants. Several nucleoside analogs that have therapeutic value are now available (Fig. 19.20).

Nonnucleoside Inhibitors of Reverse Transcriptase

Nonnucleoside inhibitors of viral reverse transcriptase do not bind at the nucleotide-binding site of the enzyme (Fig. 19.28). Examples of these compounds are nevirapine and the tetrahydroimidazobenzodiazepinone (TIBO) class of compounds. Initially, they offered the hope of complementing nucleoside analog inhibitors, but the rapid emergence of resistant mutants dampened enthusiasm. Although they are effective inhibitors of reverse transcriptase, a substitution in any of seven residues that line their binding site on the enzyme confers resistance (Fig. 19.28). Because resistant mutants are selected rapidly, nonnucleoside inhibitors cannot be used by themselves for the treatment of AIDS (monotherapy). However, as with AZT, nevirapine alone has value for treatment of pregnant women before delivery to prevent infection of the newborn. This use is now the preferred treatment in underdeveloped countries. These drugs have also proved valuable in combination therapy (see below).

Protease Inhibitors

HIV protease, which is encoded in the *pol* gene, is essential for production of mature infectious viral particles. The enzyme cleaves itself from the Gag-Pol precursor polyprotein and then cleaves at seven additional sites in Gag-Pol to yield six proteins (MA, CA, p2, NC, p1, and p6) and three enzymes (protease itself, reverse transcriptase, and integrase). Active protease has been produced in high yields in many recombinant organisms and has even been synthesized chemically. It was the first HIV enzyme to be crystallized and studied at the atomic level (Fig. 19.17). The active enzyme is a small (only 99 amino acids) dimeric aspartyl protease, similar to renin and pepsin.

The seven cleavage sites in HIV Gag-Pol are similar but not identical. In the early stages of research to find protease inhibitors, it was essential to understand how the enzyme bound to and cleaved these sites. In pursuing this

Figure 19.28 Structure of HIV type 1 reverse transcriptase highlighting the polymerase active site and the nonnucleoside reverse transcriptase inhibitor binding site. (A) Structure of the reverse transcriptase p66-p51 heterodimer complexed with a double-stranded DNA template-primer, showing the relative locations of the polymerase active site and the site for binding nonnucleoside reverse transcriptase inhibitors (NNRTIs). Data from A. Jacobo-Molina et al., *Proc. Natl. Acad. Sci. USA* **90:**6320–6324, 1993. (B) Structure of 8-CL TIBO (a prototype NNRTI) bound to reverse transcriptase. Amino acid side chains corresponding to sites of drug resistance mutations are shown in orange; residues with cyan side chains are sites at which NNRTI resistance mutations have not been detected. Data from J. Ding et al., *Nat. Struct. Biol.* **2:**407–415, 1995. Courtesy of K. Das and E. Arnold, Center for Advanced Biotechnology and Medicine, Rutgers University.

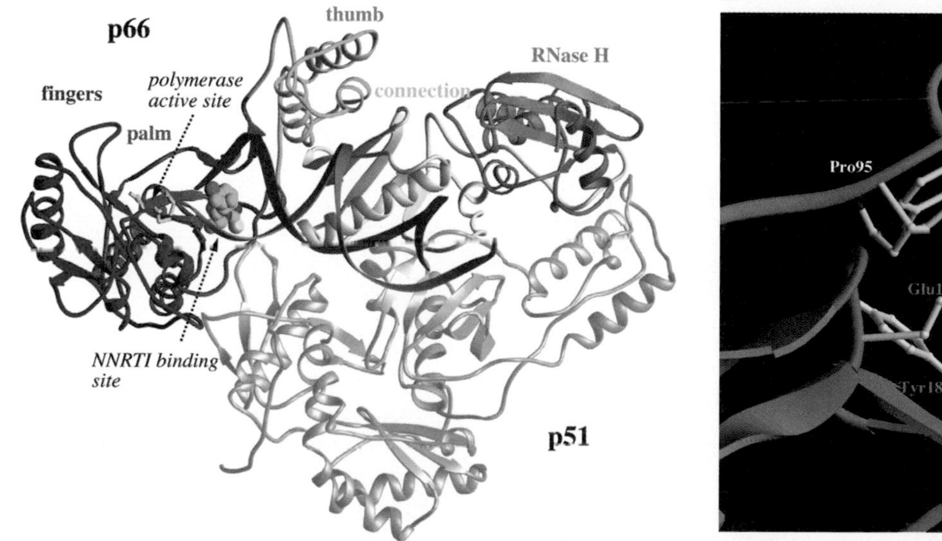

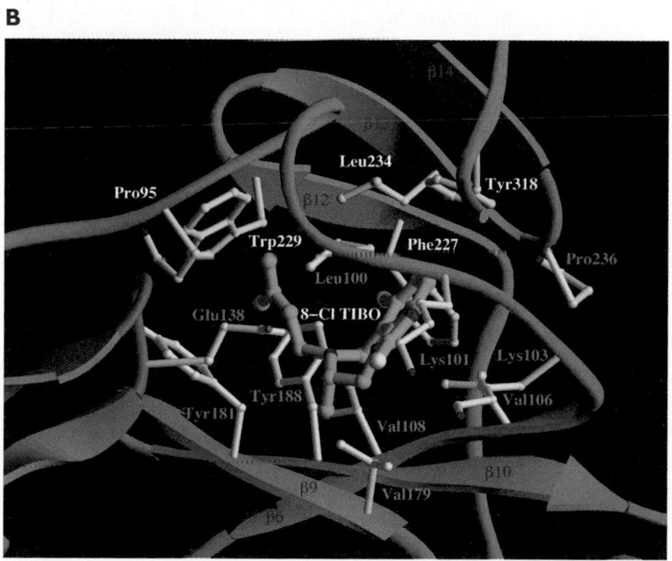

goal, an important discovery was made: the enzyme can recognize and cleave small peptide substrates in solution. It was subsequently determined that the active site is large enough to accommodate seven amino acids. The parameters of peptide binding and protease activity were determined by screening synthetic peptides containing variations of the seven cleavage sites for their ability to be recognized and cleaved by the enzyme. The first inhibitor leads were peptide mimics (peptidomimetics) modeled after inhibitors of other aspartyl proteases, such as human renin, an enzyme implicated in hypertension (Fig. 19.29). Subsequent screens for mechanism-based inhibitors and structure-based inhibitors designed de novo have yielded several powerful peptidomimetic inhibitors of the viral protease (Box 19.13).

Other Viral Targets

While reverse transcriptase and protease are proven targets for antiviral therapy, at least 12 other virus-specific proteins offer promising targets for drug hunters (see also Table 17.14). Integrase protein is particularly attractive, because biochemical and structural data are available. In the absence of integrase, viral DNA cannot be inserted into the host genome. As a result, the viral genome is unable to be expressed efficiently and propagate in dividing cells. Several companies have lead compounds under analysis, and rational design programs based on structures of integrase complexed with inhibitors are in progress. One overriding problem is to design inhibitors that are selective for HIV integrase but not other enzymes of DNA recombination (e.g., Rag recombinase; see Chapter 17).

Inhibitors of fusion and entry have considerable promise. The main neutralizing domain, identified by antibodies that block viral attachment, is the third variable domain (the so-called V3 loop) of the SU protein (Fig. 5.11). A variety of natural and synthetic molecules interfere with V3 loop activity. These compounds include specific antibodies and polysulfated or polyanionic compounds such as dextran sulfate, curdlan sulfate, and suramin. These compounds were identified early in the search for antiviral agents, but were found to be ineffective because of intolerable side effects such as anticoagulant activity. Although considerable effort was expended in the development of inhibitors of the SU-CD4 receptor interaction, including a "soluble CD4" that theoretically would act as a competitive inhibitor of infection, no effec-

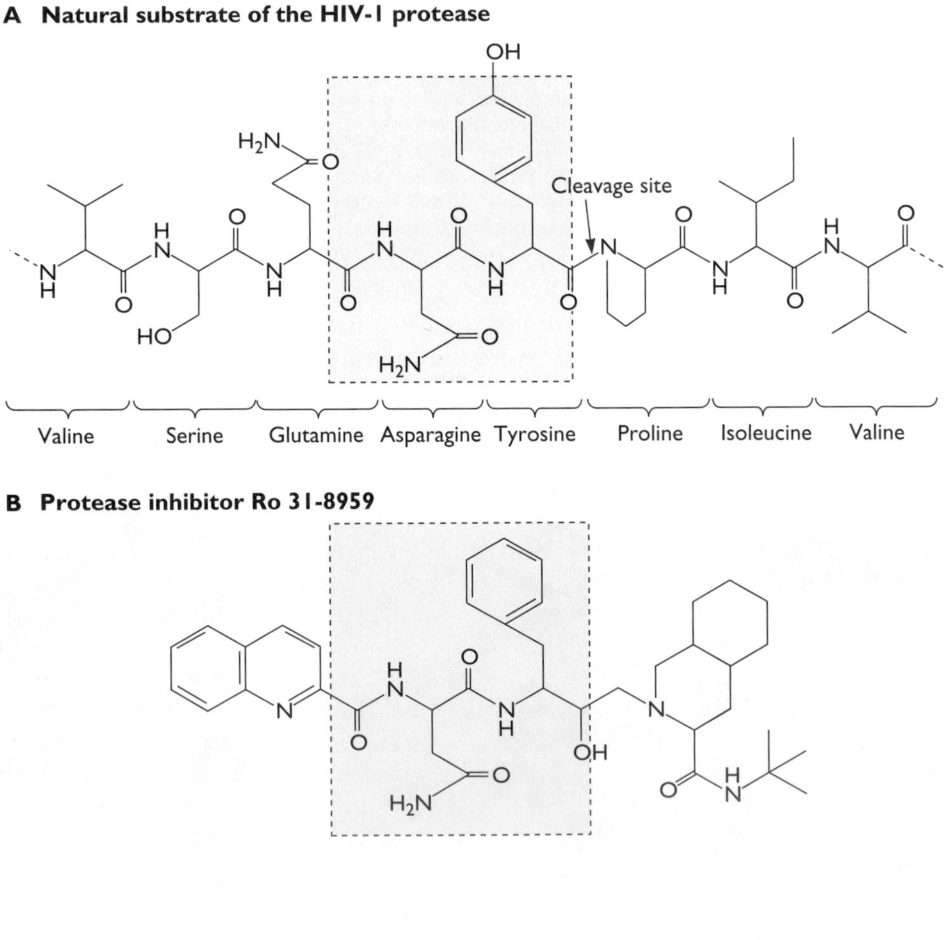

Figure 19.29 Comparison of one natural cleavage site for HIV protease with a peptidomimetic protease inhibitor. (A) The chemical structure of eight amino acids comprising one of the cleavage sites in the Gag-Pol polyprotein. The cleavage site between the tyrosine and proline is indicated by an arrow. (B) The chemical structure of an inhibitory Roche peptide mimic (Ro 31-8959). The dotted box indicates the region of similarity. Adapted from D. R. Harper, *Molecular Virology* (Bios Scientific Publishers Ltd., Oxford, United Kingdom, 1994), with permission.

BOX 19.13

Highly specific, designed inhibitors may have unexpected activities

The discovery and development of structure-based inhibitors of HIV protease have been pronounced a triumph of rational drug design. Structural biology and molecular virology came together to provide the protease inhibitors that anchor today's highly active antiretroviral therapy. However, patients receiving protease inhibitors often respond in unexpected ways, and the extent of immunological recovery with treatment is being debated. One idea was that protease inhibitors affected the immune response, in addition to blocking viral replication. Recently, it was discovered that the protease inhibitor ritonavir also inhibits the chymotrypsin-like activity of the proteasome. As a result, the protease inhibitor blocks the formation and subsequent presentation of peptides to CTLs by MHC class I proteins.

The challenge now is to determine if this secondary activity helps or hinders AIDS therapy. As discussed in Chapter 17, CTLs not only kill virus-infected cells, but also are responsible for significant immunopathology in persistent infections. Perhaps ritonavir blocks this immunopathology. On the other hand, reduction in immunosurveillance by CTLs potentiates persistent infections. In this case, the secondary activity of ritonavir may presage long-term problems. We now know that ritonavir and saquinavir inhibit proteasome activity, while indinavir and nelfinavir do not.

Therefore, it will be of some interest to follow, over time, the antiviral effects and lymphocyte functions in patients under treatment with these different protease inhibitors. As noted by the investigators who found this surprising activity, the modulation of antigen presentation by protease inhibitors can be exploited for treatment of autoimmune disease, chronic immunopathologies, and diseases of transplantation.

Andre, P., M. Groettrup, P. Klenerman, R. DeGiuli, B. L. Booth, Jr., V. Cerundolo, M. Bonneville, F. Jotereau, R. M. Zinkernagel, and V. Lotteau. 1998. An inhibitor of human immunodeficiency virus-1 protease modulates proteasome activity, antigen presentation, and T-cell responses. *Proc. Natl. Acad. Sci. USA* **95:**13120–13124.

tive antiviral agents have been found with this strategy. This lack of success can be attributed, in part, to the high concentration of SU on the virion, as well as to the alternative mechanisms for spread in an infected individual.

As often happens, the early research with failed Env inhibitors provided much insight into how virions enter cells, and has focused attention on other targets in the process. For example, it was curious that mutants resistant to neutralizing antibodies have clustered mutations in the V3 loop, yet virus-cell fusion is not affected. The implication was that CD4-V3 interactions were not involved in entry. As described in Chapter 5, entry is a multistep process requiring that the target cells synthesize not only CD4, but also any one of several chemokine receptors, such as CCr5 present on macrophages or CXCr4 found on T cells. Each step offers a potential target for an antiviral drug. Chemokine receptors are attractive targets because individuals homozygous for mutations in one such receptor (CCr5) are resistant to infection and suffer no apparent ill effects. Recent studies have shown that infected individuals carrying a mutation in the CCr5 chemokine receptor experience a delay of 2 to 4 years in the progression to AIDS. Development of a safe chemokine receptor antagonist that delays the onset of AIDS for a long period, if not indefinitely, may be feasible. We now know that the V3 loop of SU interacts with the chemokine receptor, exposing previously buried SU sequences required for membrane fusion. These transiently exposed surfaces that appear after chemokine receptor binding are excellent targets for antiviral agents. A 36-amino-acid synthetic peptide, termed T20 or pentafuside, derived from the second heptad repeat of HIV SU, binds to the exposed grooves on the surface of a transient triple-stranded coiled-coil and perturbs the transition of SU into the conformation active for fusion. At this time, T20 and similar compounds appear to be remarkably effective in reducing HIV loads in infected humans.

The Combined Problems of Treating a Persistent Infection and Emergence of Drug Resistance

The Heart of the Problem

The slow pattern of infection exemplified by HIV is characterized by an asymptomatic period that lasts for years to decades after primary infection (see Fig. 16.1 and Chapter 17). This asymptomatic phase was initially thought to reflect a quiescent period of no viral replication and was unfortunately called "latency." We now understand that there is extensive viral replication throughout this period and that such relentless replication is the heart of the problem facing effective antiviral therapy.

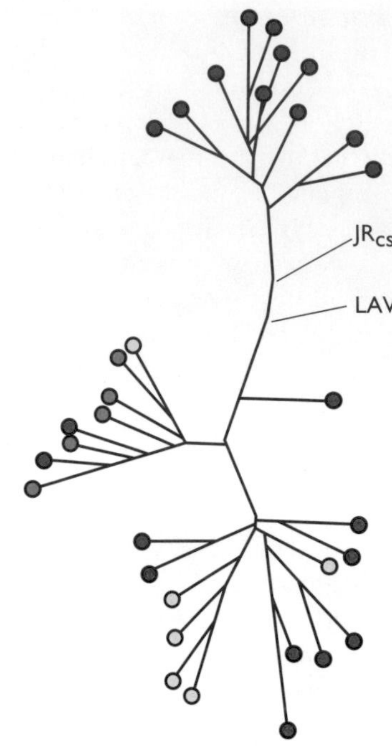

Figure 19.30 HIV type 1 evolution within a single patient. An unrooted phylogenetic tree showing relationships among 33 HIV type 1 variants coexisting within a single infected patient. Two prototype viruses, LAV and JR_{csf}, are shown in the center of the tree. Colors denote tissue sources for virus isolation: blue, lymph nodes; red, peripheral blood; yellow, spleen; green, lung; purple, spinal cord; brown, dorsal root ganglion. Adapted From M. Ait-Khaled et al., *AIDS* **9:**675–683, 1995, with permission.

Figure 19.31 Unselected mutations specifying resistance to HIV type 1 (HIV-1) reverse transcriptase inhibitors. In this study, investigators reported the nucleotide sequences of the *pol* gene from 60 HIV-1 genomes obtained directly from primary lymphocytes from infected individuals who were treated or not treated with inhibitors of reverse transcriptase (RT). Representative data are illustrated in this figure. The single-letter amino acid code is used. The reverse transcriptase inhibitors are indicated at the top left. NAN is a nonnucleoside inhibitor, and ddG is 2′,3′-dideoxyguanosine. The consensus amino acid sequence in the middle is that of HIV-1 group A reverse transcriptase. Amino acid replacements known to confer resistance to reverse transcriptase inhibitors are listed above the consensus sequence. A slash separating two amino acids denotes their association with resistance to the same inhibitor. The solid triangle corresponding to position 188 for NAN indicates that resistance has been associated with amino acids C, H, and L; the open triangle at position 190 indicates that resistance has been associated with amino acids A, E, and S. The green box below the consensus sequence indicates the amino acid changes found in individual viruses obtained from patients not exposed to reverse transcriptase inhibitors (untreated). Note the presence of amino acid changes that would result in resistance. Examples of viruses selected with antiviral drugs are indicated in the four groups in the blue boxes. Mutations related to drug resistance found in the absence of selection are the expected consequence of the statistical distribution of mutations along the *pol* gene. The presence of such important amino acid replacements in virus populations reminds us that viral quasispecies are reservoirs of diversity available for selection. Adapted from I. Najera et al., *J. Virol.* **69:**23–31, 1995, with permission.

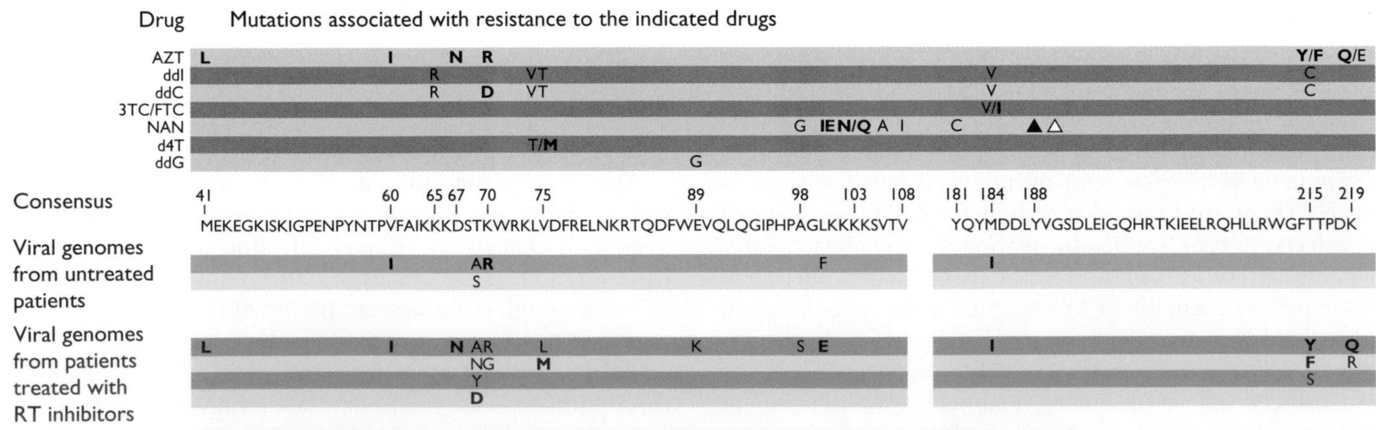

The Critical Parameters

About 0.1% of an infected individual's CD4+ T cells are replicating HIV at any time during the asymptomatic period, and at least 10% of the total CD4+ population contains viral nucleic acid. The steady-state level of HIV RNA detected in blood depends primarily on the rate of virion production in CD4+ T cells. Ultimately, in the untreated individual, viral RNA concentrations increase, CD4+ T-cell numbers decrease, and death is inevitable. The late stage of infection is characterized by high levels of replication (often more than 10^9 particles produced per day), a high turnover of viral particles (50% turnover in less than a day), and massive loss of CD4+ T cells (Fig. 17.12). The quantity of HIV RNA and the number of CD4+ T cells are therefore the most useful monitors of clinical status and effectiveness of antiviral therapy. More than 500 CD4+ lymphocytes per ml is considered a normal value; fewer than 200 is the definition of AIDS. Viral load is estimated from the number of copies of viral RNA per milliliter of serum or plasma. The lowest level detectable with current methodologies is about 50 copies, and values greater than 10^6 have been recorded.

The First Drug Resistant Mutants

Viral mutants that grew with impunity in the presence of AZT appeared almost immediately after the drug was approved for general use. The genomes of the mutants harbor single base pair changes at one of at least four sites in the reverse transcriptase gene. Reverse transcriptase enzymes bearing these substitutions no longer bind phosphorylated AZT, but they retain enzymatic activity. Mutants resistant to other nucleoside analogs, as well as to protease inhibitors, also arose with disheartening frequency. Resistance to protease inhibitors is found only after acquisition of several distinct amino acid substitutions, but mutants resistant to these drugs have been isolated from patients. These drug-resistant mutants were transmitted to new hosts and threatened to undermine the entire antiviral effort.

Because HIV replicates extensively during the so-called quiescent, asymptomatic period, mutations accumulate. On average every new HIV genome can be expected to carry at least one new mutation, and a single patient will harbor many new viral genomes in various tissues (Fig. 19.30). If an antiviral drug is given long after the primary infection, selection of drug-resistant mutants is inevitable (Fig. 19.31). Even genomes with detrimental mutations are maintained in the population, if their defects can be complemented. Drug resistance does not have to be absolute because any increase in fitness, no matter how subtle, can result in large changes in the virus population. The reality is that we do not know the principal factors controlling viral load in an infected patient. Future drug and vaccine trials providing quantitative data on immune responses will go far to solve this problem.

Antiviral Therapy Has the Potential To Promote or Prevent the Emergence of Resistant Viruses

Mutations appear only when the viral genome is replicated. Accordingly, if replication is blocked, no drug-resistant mutants can arise. If an individual harboring a small number of viral genomes with no relevant preexisting mutations is given sufficient drug to block all viral replication, the infection will be held in check (Fig. 19.32). In contrast, if the same antiviral drug is given after the viral population has expanded, or if the drug is given in quantities insufficient to block replication entirely, genomes that harbor mutations will survive and will continue to replicate and evolve. When the viral genome numbers are small, the infection may still be cleared by the host's immune system before resistant mutants take over. If resistance to an antiviral drug requires multiple mutations (e.g., resistance to

Figure 19.32 Virus load depends on the dose of antiviral drug. If virus replication is allowed in the presence of an antiviral drug, mutants resistant to that drug will be selected. This assertion is illustrated by plotting median virus load in relative units on the *y* axis as a function of time after exposure to a drug on the *x* axis. An antiviral drug is administered to a patient at a given time (drug given), and the concentration of virus in the blood or tissue sample is determined at various intervals thereafter. In the top curve (low dose), the concentration of antiviral drug is insufficient to block virus replication, and virus load is reduced transiently, if at all. Viruses that replicate may be enriched for resistant mutants. In the middle curve (intermediate dose), the concentration of antiviral drug appears to be successful in lowering viral load initially, indicating that some replication was blocked. In this example, virus replication was not blocked completely, and resistant viruses overwhelmed the patient. In the bottom curve (optimal dose), the concentration of the antiviral drug is such that all virus replication at the time of administration is blocked. As a consequence, no drug-resistant mutants can arise, and virus load drops dramatically. Redrawn from J. H. Condra and E. A. Emini, *Sci. Med.* **4**:14–23, 1997, with permission.

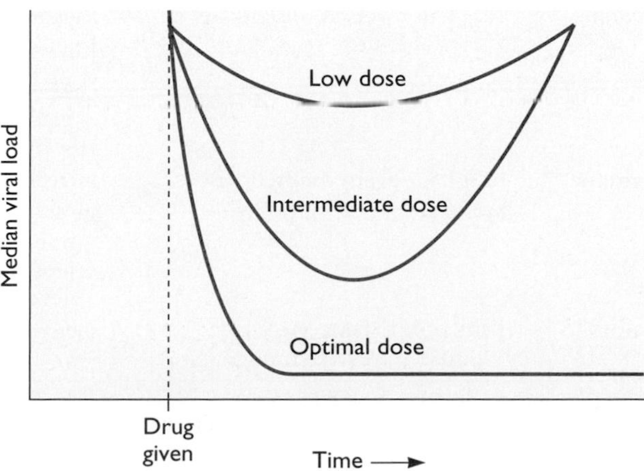

protease inhibitors), the chance that all mutations preexist in a single genome is lower than if only a single mutation is required. But if replication is allowed in the presence of the inhibitor, resistant mutants will accumulate.

Cross-Resistance to Similar Inhibitors

HIV mutants resistant to different nucleoside analogs often carry different amino acid substitutions in the reverse transcriptase. Furthermore, a mutation conferring resistance to one inhibitor may suppress resistance to another (Table 19.12). Consequently, combinations of nucleoside analogs were clinically tested with the expectation that double-resistance mutants would be rare, perhaps nonviable, or at least severely crippled. While initially promising, many combinations failed miserably, with mutants resistant to both drugs arising after less than a year of therapy. Furthermore, alterations in reverse transcriptase that conferred resistance to different nucleoside analogs appeared simultaneously. The barrier to resistance to many pairwise combinations of nucleoside and nonnucleoside inhibitors was higher than that for any single drug, but not high enough.

Experience with protease substrate analog inhibitors has been similar; resistance to two inhibitors emerges almost as fast as resistance to either one alone. As current protease inhibitors are all peptide mimics that bind to the substrate pocket of the enzyme (see Fig. 19.17), a change in substrate pocket residues can affect the binding of more than one inhibitor.

Treatment of a patient with one antiviral drug at a time clearly is of limited clinical value because of the possibility of selecting resistant mutants. Furthermore, the use of a single drug must be contemplated carefully because this same drug may be used for combination therapy in the future.

Combination Therapy

The Use of Two or More Antiviral Drugs To Combat the Resistance Problem

The use of two or more treatments simultaneously is well known in tuberculosis therapy and cancer treatment. Combining two mechanistically different treatments often leads to more effective killing of the bacterium or tumor, thereby circumventing the appearance of cells resistant to one treatment or the other. This principle also applies to antiviral therapy. In theory, if resistance to one drug occurs once in every 10^3 genomes, and resistance to a second occurs once in every 10^4, then the likelihood that a genome carrying both mutations will arise is the product of the two probabilities, or once in every 10^7.

Problems and Promise

Combination therapy does not always clear an infection. One obvious reason for resurgence of virion production is the appearance of drug-resistant mutants, if therapy is given late in the progression of disease, or if the combination is not sufficiently potent. Combination therapy is very demanding for physician and patient, and is not as easy to implement as it may first appear. Dozens of pills must be swallowed every day without fail. If other infections are being treated, as they almost always are in AIDS patients, then more than 50 pills a day may be required. Other problems arise because simple storage and keeping track of different medications is a daunting task for an ill patient. To compound the problems, every drug has side effects, and some are severe. For example, the gastrointestinal problems that accompany many protease inhibitors are particularly stressful for patients. Some side effects, such as the wasting of limbs and face with fat accumulation in the gut (lipodystrophy), may appear only after months of continuous use of

Table 19.12 Representative resistance mutation interactions[a]

Compound	Resistance substitution(s)	Resistance interaction
Zidovudine	T215F in reverse transcriptase[b]	Didanosine resistance mutation (L74V) restores zidovudine susceptibility.
		Lamivudine resistance mutation (M184V) restores zidovudine susceptibility.
		Nevirapine and loviride (α-APA) resistance mutation (Y181C) restores zidovudine susceptibility.
		Foscarnet resistance mutation (W88G) restores zidovudine susceptibility.
Delavirdine	P236L in reverse transcriptase	Increased susceptibility to nevirapine, R82913 (TIBO), and L-697,661 (pyridinone)
VX-478	M46I, I47V, I50V in protease	In the presence of saquinavir, VX-478 susceptibility is restored while saquinavir resistance develops. Distinct mutations include G48V, I50V, and I84L. In the presence of indinavir, VX-478 resistance persists while indinavir resistance develops. Distinct mutations include V32I and A71V.
Foscarnet	E89K, L92I, S56A, Q161L, H208Y in reverse transcriptase	Increased susceptibility to zidovudine, nevirapine, and R82150 (TIBO)

[a]Adapted from D. D. Richman, R. J. Whitley, and F. G. Hayden (ed.), *Clinical Virology,* 2nd ed. (ASM Press, Washington, D.C., 2002), with permission.
[b]T215F indicates that the threonine at position 215 is changed to a phenylalanine.

current antiprotease drugs. Because of these problems, some patients simply do not take their medication. The most insidious failure lies in wait when the patient begins to feel better and stops taking the pills. Viral replication resumes when the inhibitors are removed. Replication means mutation, and in such cases combination therapy may be ineffective if ever reinstated. Finally, combination therapy is very expensive, currently costing thousands of dollars per year for some regimens. In addition to the costs of medication, tests for viral load and CD4 counts must be performed regularly to monitor therapy. Such tests also are costly. Combination therapy is clearly not accessible to everyone.

Even in the face of these seemingly formidable issues, there is considerable optimism that combination therapy, particularly for long-term control of infection, will be effective. Many believe that the clinical success of combination therapy is truly remarkable and represents one of the high points in the battle against AIDS.

Strategic Treatment Interruption

The underlying premise of strategic treatment interruption is that combination drug treatment will stop replication and, as a consequence, the immune system will recover. The hypothesis is that when drug treatment is stopped, replication will begin again, but the relatively healthy immune system is primed to fight HIV. The hope is that cycles of drug therapy, bolstered by drug-free periods, will enable the patient's own immune defenses to clear the infection. There are many reasons why such a hypothesis is attractive. A primary one is that, given the toxicities of various potent drug combinations, drug holidays can be very attractive to patients. Initial studies in monkeys were encouraging when drug therapy was given very early in the acute infection. At these early times, the population of viral genomes in infected individuals is probably not as diverse, because it has not been subjected to many rounds of selection by the immune system. Consequently, replication could be held in check by the recovering immune system. However, studies with infected humans have not been as promising. Most AIDS patients have been infected for years, and the genetic diversity of viral genomes in their bodies is enormous. Indeed, the emergence of drug-resistant mutants, the small number of antiviral drugs available to treat resistant infections, and the appearance of CTL and antibody escape mutants present serious problems for widespread use of strategic treatment interruption.

Challenges and Lessons Learned from the Search for HIV Inhibitors

Several issues arise in the quest for effective anti-HIV therapy. Of utmost importance is that drugs must be potent if they are to reduce plasma levels of HIV RNA to fewer than 50 genome copies per ml (the current limit of detection). Such potency must be achieved to avoid selection of resistant mutants, and can be accomplished only with combinations of drugs. A potent drug must also reach the tissue compartment where viral genomes are replicating and be maintained at sufficient concentration and for a long enough period to block replication. Potency must be attained in the face of inevitable toxicity and unpredictable patient adherence, as well as individual differences in drug metabolism. A potent drug is of no use if the patient will not take it because the pills are too large, too numerous, or cause side effects.

With aggressive use of potent antiviral drugs, HIV replication can be suppressed, but the infection cannot be cured. Even when viral RNA has been undetectable for years during drug therapy, as soon as the drug is removed, replication begins again. We simply have no way to eliminate every last viral genome from the body of an infected individual. As years go by during this insidious, slow infection, the diversity of viral genomes increases as a result of mutation, recombination, and selection by drugs and an active immune system. While we understand in principle how this diversity arises, we have little insight about how to eliminate it.

At the moment, drug therapy is our only proven weapon. Therefore, it is prudent to avoid therapies in which resistance can be conferred by a single mutation in the viral genome. As a corollary, therapies that require multiple mutations for drug resistance to emerge should be used. Combinations of drugs will be most effective if the patient has not been treated previously with any drug.

The hope of any therapy is that as the cells of the immune system ultimately die and are replaced, the viral genome concentration will be reduced to a point at which the infection cannot be sustained. Even if a cure is wishful thinking, combination therapies, where available, have converted HIV infection to a chronic treatable disease rather than a death sentence.

The Quest for an AIDS Vaccine

In 1984, several years after HIV was identified, the U.S. government predicted that an AIDS vaccine would be available within 3 years. Although research has continued for more than 18 years with unprecedented intensity, we still do not know how to make a vaccine that will protect against AIDS. Such ill-founded optimism has a precedent in the history of vaccination. Poliovirus was isolated in 1908, and 3 years later Simon Flexner of the Rockefeller Institute announced that a vaccine would be prepared in 6 months. Fifty years of research on basic poliovirus biology was necessary to provide the knowledge of pathogenesis and immunity that allowed the development of effective poliomyelitis vaccines.

Preliminary experiments on HIV employing standard vaccine approaches (Table 19.13) have provided useful information but no clear indications of how to induce an effective immune defense against HIV infection of humans. Some of the problems to be surmounted include the antigenic drift of the virus, its infection of the immune system itself, and the fact that it establishes a persistent infection, giving rise to billions of particles per day for years. An effective vaccine therefore must protect against not only transmission of the genetically diverse population of virions that arise after every round of replication, but also the diverse quasispecies that now exist worldwide (see Box 20.8). Furthermore, such a vaccine must protect against virus-infected cells that can transmit the infection from person to person.

These problems are daunting, but there are others as well. The infection is maintained in the face of a vigorous immune response. For some scientists, this is the crux of the challenge: how do we induce an immune response **superior** to the normal response that apparently is not good enough to clear the infection? How do we eliminate immune escape mutants? Furthermore, testing of potential vaccines is a special problem, as animal models for viral replication and pathogenesis are not available. Current model systems based on simian immunodeficiency virus (HIV infection of primates) are expensive and do not reproduce all features of the human disease. There are complicated social, ethical, and political issues that must be addressed, if vaccines are to be tested in humans. For example, can vaccine trials be conducted so that appropriate placebo controls can be evaluated? Vaccine developers face costly litigation if unanticipated reactions occur. On a larger scale, significant political issues arise when decisions are made by Western leaders about vaccines that are to be used in developing countries. Resolution of these problems will not be easy.

Table 19.13 Possible human immunodeficiency virus vaccine approaches

Inactivated virus
Attenuated virus
Subunit vaccine
Envelope glycoproteins
Gag proteins
Viral vector producing selected immunogenic proteins or derivatives
Viral particles (Gag-Pol cores)
Synthetic peptides from immunogenic proteins
DNA vaccine producing immunogenic proteins
Combination: DNA vaccine to prime, and viral vector vaccine to boost

A number of approaches—inactivated virions, subunit vaccines based on single viral proteins, and passive immunization—have already been tested with no clear success. In particular, subunit vaccines, although capable of inducing strong antibody responses, are markedly inefficient in eliciting a CTL response. Therapeutic vaccines administered to already infected individuals may have value, if administered in combination with a drug regimen that protects the immune system. Live attenuated HIV vaccines, modeled after the successful live poliovirus vaccine, present difficult scientific and ethical problems: the risks associated with injecting thousands of healthy uninfected volunteers with a living (albeit attenuated) virus, are currently considered unacceptable. More promising, however, are vaccines based on recombinant poxvirus, vesicular stomatitis virus, and alphavirus genomes that encode various HIV proteins. Such recombinant vaccines have proved to be potent activators of CTLs as well as of humoral responses. A promising area of research is combination of a DNA vaccine to provide strong cellular and humoral responses (DNA priming) and a live attenuated virus vaccine (e.g., vector boosting).

Perspectives

Viral infection is a complex process making intervention a challenging enterprise. The built-in survival mechanisms of virus and host provide both opportunities and problems in our quest to control viral disease. The goal for vaccine research is to understand the molecular mechanisms of antiviral immune defense, so that vaccines can be tailored to particular infections. For example, we must find ways to deliver viral antigen into both the MHC class I and class II pathways in order to induce a well-tolerated, long-lasting immunity without significant vaccine-associated risk to the patient.

We possess a limited arsenal of antiviral drugs and an urgent need for more. Some of the most deadly viral diseases are caused by RNA viruses, yet few drugs are effective against any of these agents. Similarly, only one or two antiviral drugs are effective against acute smallpox virus infection. A formidable problem is that we are unable to diagnose acute infections accurately and in a timely fashion. Persistent viral infections present special problems for any antiviral intervention, as these infections may require the combined application of vaccines, immunotherapy, and antiviral drugs. It should be possible to reduce viral load by antiviral drugs and then clear the remaining infection with bolstered immune responses. However, mutants that are resistant to immune defenses, as well as to antiviral drugs, will be selected. This certainty is a specter that continues to haunt antiviral research and public health. As the search for effective therapy to treat HIV infection demonstrates, we still have much to learn.

A common perception is that effective prevention and treatment of viral diseases is possible only by means of vaccines and drugs. However, we must not forget that common sense, effective public health measures, proper nutrition, and simple personal hygiene remain fundamental contributors to the prevention of viral infections. Indeed, in some underdeveloped countries, these nontechnical solutions often are the **only** defense against viral diseases. The principle underlying the high and low technology approaches is remarkably similar: the sources and avenues of viral spread must be eliminated. Clean drinking water, adequate sewage disposal, insect control, good medical practice, and an uncontaminated blood supply can be powerful medicine. Simple personal activities such as washing hands, seeking protection from insects and rodents, and using condoms can be remarkably effective, low-cost antiviral measures. An often unappreciated antiviral defense is proper nutrition, because the malnourished have reduced innate and acquired immune defenses. Unfortunately, even these fundamental practices and policies cannot be put in place without basic infrastructure and social policy that provide public awareness of infectious disease, understanding of potential hazards and risks, appreciation of the importance of early detection, and reporting of the incidence of infections. A goal for any society is early education of the lay public to provide a basic understanding of biology and the scientific method.

References

Books

Blair, E., G. Darby, G. Gough, E. Littler, D. Rowlands, and M. Tisdale. 1998. *Antiviral Therapy*. Bios Scientific Publishers, Oxford, United Kingdom.

Janeway, C. A., Jr., P. Travers, M. Walport, and M. Shlomchik. 2001. *Immunobiology: the Immune System in Health and Disease*. Current Biology Limited, Garland Publishing Inc., New York, N.Y.

Mims, C., A. Nash, and J. Stephan. 2001. *Mims' Pathogenesis of Infectious Disease*, 5th ed. Academic Press, Inc., Orlando, Fla.

Richman, D. D., R. J. Whitley, and F. G. Hayden (ed.). 2002. *Clinical Virology*, 2nd ed. ASM Press, Washington, D.C.

Veerapandian, P. (ed.). 1997. *Structure-Based Drug Design: Diseases, Targets, Techniques and Developments*, vol. 1. Marcel Dekker, Inc., New York, N.Y.

Historical Papers

Jenner, E. 1788. Observations on the natural history of the cuckoo. *Philos. Trans. R. Soc.* **78:**219–237.

Jenner, E. 1798. *An Inquiry into the Causes and Effects of the Variolae Vaccinae, a Disease Discovered in Some of the Western Counties of England, especially Gloucestershire, and Known by the Name of Cow Pox*. Sampson Low, London, England.

Metchnikoff, E. 1905. *Immunity in the Infectious Diseases*, 1st ed. Macmillan Press, New York, N.Y.

Woodville, W. 1796. *The History of the Inoculation of the Smallpox in Great Britain; Comprehending a Review of all the Publications on the Subject: with an Experimental Inquiry into the Relative Advantages of Every Measure Which Has Been Deemed Necessary in the Process of Inoculation*. James Phillips, London, England.

Reviews

Antivirals and Drug Discovery

Air, G. M., A. Ghate, and S. Stray. 1999. Influenza neuraminidase as target for antivirals. *Adv. Virus Res.* **54:**375–402.

Coffin, J. 1995. Human immunodeficiency virus population dynamics in vivo: implications for genetic variation, pathogenesis and therapy. *Science* **267:**483–489.

Condra, J., and E. Emini. 1997. Preventing human immunodeficiency virus-1 drug resistance. *Sci. Med.* **4:**14–23.

Craigie, R. 2001. Human immunodeficiency virus integrase, a brief overview from chemistry to therapeutics. *J. Biol. Chem.* **276:**23213–23216.

De Clercq, E. 2002. Strategies in the design of antiviral drugs. *Nat. Rev. Drug Discov.* **1:**13–25.

Drews, J. 2000. Drug discovery: a historical perspective. *Science* **287:**1960–1964.

Elion, G. B. 1986. History, mechanisms of action, spectrum and selectivity of nucleoside analogs, p. 118–137. *In* J. Mills and L. Corey (ed.), *Antiviral Chemotherapy: New Directions for Clinical Application and Research*. Elsevier Science Publishing Co., New York, N.Y.

Evans, J., K. Lock, B. Levine, J. Champness, M. Sanderson, W. Summers, P. McLeish, and A. Buchan. 1998. Herpesviral thymidine kinases: laxity and resistance by design. *J. Gen. Virol.* **79:**2083–2092.

Fields, B. 1994. AIDS: time to turn to basic science. *Nature* **369:**95–96.

Hajduk, P., R. Meadows, and S. Fesik. 1997. Discovering high-affinity ligands for proteins. *Science* **278:**497–499.

Ikuta, K., S. Suzuki, H. Horikoshi, T. Mukai, and R. Luftig. 2000. Positive and negative aspects of the human immunodeficiency virus protease: development of inhibitors versus its role in AIDS pathogenesis. *Microbiol. Mol. Biol. Rev.* **64:**725–745.

Jackson, H., N. Roberts, Z. Wang, and R. Belshe. 2000. Management of influenza: use of new antivirals and resistance in perspective. *Clin. Drug Investig.* **20:**447–454.

Nolan, G. 1997. Harnessing viral devices as pharmaceuticals: fighting human immunodeficiency virus-1's fire with fire. *Cell* **90:**821–824.

Persaud, D., Y. Zhou, J. M. Siliciano, and R. F. Siliciano. 2003. Latency in human immunodeficiency virus type 1 infection: no easy answers. *J. Virol.* **77:**1659–1665.

Richman, D. 2001. Human immunodeficiency virus chemotherapy. *Nature* **410:**995–1001.

Ridky, T., and J. Leis. 1995. Development of drug resistance to human immunodeficiency virus-1 protease inhibitors. *J. Biol. Chem.* **270:**29621–29623.

Smith, K. 2001. To cure chronic human immunodeficiency virus infection, a new therapeutic strategy is needed. *Curr. Opin. Immunol.* **13:**617–624.

Tan, S.-L., A. Pause, Y. Shi, and N. Sonenberg. 2002. Hepatitis C therapeutics: current status and emerging strategies. *Nat. Rev. Drug Discov.* **1:**867–881.

Wade, R. C. 1997. "Flu" and structure-based drug design. *Structure* **5:**1139–1145.

Zivin, J. 2000. Understanding clinical trials. *Sci. Am.* 69-75.

Vaccines

Ada, G. L. 1994. Vaccines and immune response, p. 1503–1507. *In* R. G. Webster and A. Granoff (ed.), *Encyclopedia of Virology*. Academic Press, Inc., San Diego, Calif.

Ahmed, R., and D. Gray. 1996. Immunological memory and protective immunity: understanding their relation. *Science* **272:**54–60.

Baltimore, D. 1988. Intracellular immunization. *Nature* **335:**395–396.

Bendelac, A., and R. Medshitov. 2002. Adjuvants of immunity: harnessing innate immunity to promote adaptive immunity. *J. Exp. Med.* **195:**F19–F23.

DesJardin, J., and D. Snydman. 1998. Antiviral immunotherapy. A review of current status. *Biodrugs* **9:**487–507.

Domi, A., and B. Moss. 2002. Cloning the vaccinia virus genome as a bacterial artificial chromosome in Escherichia coli and recovery of infectious virus in mammalian cells. *Proc. Natl. Acad. Sci. USA* **99:**12415–12420.

Doms, R., and J. Moore. 2000. Human immunodeficiency virus-1 membrane fusion: targets of opportunity. *J. Cell Biol.* **151:**F9–14.

Ellis, R. 1996. The new generation of recombinant viral subunit vaccines. *Curr. Opin. Biotechnol.* **7:**646–652.

Hassett, D., and J. Whitton. 1996. DNA immunization. *Trends Microbiol.* **4:**307–311.

Moss, B. 1996. Genetically engineered poxviruses for recombinant gene expression, vaccination and safety. *Proc. Natl. Acad. Sci. USA* **93:**11341–11348.

Nabel, G. 2001. Challenges and opportunities for development of an AIDS vaccine. *Nature* **410:**1002–1007.

Oyaski, M., and H. Ertl. 2000. DNA vaccines. *Sci. Med.* **7:**30–39.

Restifo, N. 2000. Building better vaccines: how apoptotic cell death can induce inflammation and activate innate and adaptive immunity. *Curr. Opin. Immunol.* **12:**597–603.

Reyes-Sandoval, A., and H. C. Ertl. 2001. DNA vaccines. *Curr. Mol. Med.* **1:**217–243.

Salk, J., and D. Salk. 1977. Control of influenza and poliomyelitis with killed virus vaccines. *Science* **195:**834–847.

Schijns, V. 2000. Immunological concepts of vaccine adjuvant activity. *Curr. Opin. Immunol.* **12:**456–463.

Stewart, P., and G. Nemerow. 1997. Recent structural solutions for antibody neutralization of viruses. *Trends Microbiol.* **5:**229–233.

Walker, B., and B. T. Korber. 2001. Immune control of human immunodeficiency virus: the obstacles of HLA and viral diversity. *Nat. Immunol.* **2:**473–475.

Selected Papers

Antiviral Drug Discovery

Bonhoeffer, S., R. May, G. Shaw, and M. Nowak. 1997. Virus dynamics and drug therapy. *Proc. Natl. Acad. Sci. USA* **94:**6971–6976.

Coen, D., and P. Schaffer. 1980. Two distinct loci confer resistance to acycloguanosine in herpes simplex virus type 1. *Proc. Natl. Acad. Sci. USA* **77:**2265–2269.

Crotty, S., D. Maag, J. Arnold, W. Zhong, J. Lau, Z. Hong, R. Andino, and C. Cameron. 2000. The broad-spectrum antiviral ribonucleoside ribavirin is an RNA virus mutagen. *Nat. Med.* **6:**1375–1379.

Crute, J., C. Grygon, K. Hargrave, B. Simoneau, A. Faucher, G. Bolger, P. Kibler, M. Liuzzi, and M. Cordingley. 2002. Herpes simplex virus helicase-primase inhibitors are active in animal models of human disease. *Nat. Med.* **8:**386–391.

DeLucca, G., S. Erickson-Viitanen, and P. Lam. 1997. Cyclic human immunodeficiency virus protease inhibitors capable of displacing the active site structural water molecule. *Drug Discov. Today* **2:**6–18.

Kim, J., K. Morgenstern, C. Lin, T. Fox, M. Dwyer, J. Landro, S. Chambers, W. Markland, C. Lepre, E. O'Malley, S. Harbeson, C. Rice, M. Murcko, P. Caron, and J. Thompson. 1996. Crystal structure of the hepatitis C virus NS3 protease domain complexed with a synthetic NS4A cofactor peptide. *Cell* **87:**343–355.

Kleymann, G., R. Fischer, U. Betz, M. Hendrix, W. Bender, U. Schneider, G. Handke, P. Eckenberg, G. Hewlett, V. Pevzner, J. Baumeister, O. Weber, K. Henninger, J. Keldenich, A. Jensen, J. Kolb, U. Bach, A. Popp, J. Mäben, I. Frappa, D. Haebich, O. Lockhoff, and H. Rübsamen-Waigmann. 2002. New helicase-primase inhibitors as drug candidates for the treatment of herpes simplex disease. *Nat. Med.* **8:**392–398.

Perelson, A., P. Essunger, Y. Cao, M. Vesanen, A. Hurley, K. Saksela, M. Markowitz, and D. Ho. 1997. Decay characteristics of human immunodeficiency virus-1 infected compartments during combination therapy. *Nature* **387:**188–191.

Thomsen, D. R., N. L. Oien, T. A. Hopkins, M. L. Knechtel, R. J. Brideau, M. W. Wathen, and F. L. Homa. 2003. Amino acid changes within conserved region III of the herpes simplex virus and human cytomegalovirus DNA polymerases confer resistance to 4-oxo-dihydroquinolines, a novel class of herpesvirus antiviral agents. *J. Virol.* **77:**1868–1876.

Underwood, M., R. Harvey, S. Stanat, M. Hemphill, T. Miller, J. Drach, L. Townsend, and K. Biron. 1998. Inhibition of human cytomegalovirus DNA maturation by a benzimidazole ribonucleoside is mediated through the UL89 gene product. *J. Virol.* **72:**717–725.

van Zeijl, M., J. Fairhurst, T. Jones, S. Vernon, J. Morin, J. LaRocque, B. Feld, B. O'Hara, J. Bloom, and S. Johann. 2000. Novel class of thiourea compounds that inhibit herpes simplex virus type 1 DNA cleavage and encapsidation: resistance maps to the UL6 gene. *J. Virol.* **74:**9054–9061.

Varghese, J. N., V. C. Epa, and P. M. Colman. 1995. Three dimensional structure of the complex of 4-guanidino-Neu5Ac2en and influenza virus neuraminidase. *Protein Sci.* **4:**1081–1087.

von Itzstein, M., W. Wu, G. Kok, M. S. Pegg, J. Dyason, B. Jin, T. Van Phan, M. Smythe, H. White, S. Oliver, P. Colman, J. Varghese, D. Ryan, J. Woods, R. Bethell, V. Hotham, J. Cameron, and C. Penn. 1993. Rational design of potent sialidase-based inhibitors of influenza virus replication. *Nature* **363:**418–423.

Vaccine Discovery

Bolker, B. M., and B. T. Grenfell. 1996. Impact of vaccination on the spatial correlation and persistence of measles dynamics. *Proc. Natl. Acad. Sci. USA* **93:**12648–12653.

Borrow, P., H. Lewicki, X. Wei, M. S. Horwitz, N. Peffer, H. Meyers, J. A. Nelson, J. E. Gairin, B. H. Hahn, M. B. Oldstone, and G. M. Shaw. 1997. Antiviral pressure exerted by human immunodeficiency virus-1-specific cytotoxic T lymphocytes (CTLs) during primary infection demonstrated by rapid selection of CTL escape virus. *Nat. Med.* **3:**205–211.

Brochier, B., M. P. Kieny, F. Costy, P. Coppens, B. Baudilin, J. P. Lecocq, B. Languet, G. Chappuis, P. Desmettres, K. Afiademanyo, R. Lilbois, and P. P. Pastoret. 1991. Large scale eradication of rabies using recombinant vaccinia-rabies vaccine. *Nature* **354:**520–522.

Chen, D., R. Endres, C. Erickson, K. Weis, M. McGregor, Y. Kawaoka, and L. Payne. 2000. Epidermal immunization by a needle-free powder delivery technology: immunogenicity of influenza vaccine and protection in mice. *Nat. Med.* **6:**1187–1190.

Chu, R. S., O. S. Targoni, A. M. Krieg, P. V. Lehmann, and C. V. Harding. 1997. CpG oligodeoxynucleotides act as adjuvants that switch on T helper 1 (Th1) immunity. *J. Exp. Med.* **186:**1623–1631.

Eo, S., S. Lee, S. Chun, and B. Rouse. 2001. Modulation of immunity against herpes simplex virus infection via mucosal genetic transfer of plasmid DNA encoding chemokines. *J. Virol.* **75:**569–578.

Gans, H. A., A. M. Arvin, J. Galinus, L. Logan, R. Detlovitz, and Y. Maldonado. 1998. Deficiency of the humoral immune response to measles vaccine in infants immunized at age 6 months. *JAMA* **280:**527–532.

Hassett, D. E., J. Zhang, M. Slifka, and J. Whitton. 2000. Immune responses following neonatal DNA vaccination are long-lived, abundant, and qualitatively similar to those induced by conventional immunization. *J. Virol.* **74:**2620–2627.

Kuklin, N., M. Daheshia, K. Karem, E. Manickan, and B. Rouse. 1997. Induction of mucosal immunity against herpes simplex virus by plasmid DNA immunization. *J. Virol.* **71:**3138–3145.

Modelska, A., B. Dietzschold, and V. Yusibov. 1998. Immunization against rabies with plant-derived antigen. *Proc. Natl. Acad. Sci. USA* **95:**2481–2485.

Mortara, L., F. Letourneur, H. Gras-Masse, A. Venet, J.-G. Guillet, and I. Bourgault-Villada. 1998. Selection of virus variants and emergence of virus escape mutants after immunization with an epitope vaccine. *J. Virol.* **72:**1403–1410.

O'Neal, C. M., S. E. Crawford, M. K. Estes, and M. E. Conner. 1997. Rotavirus virus-like particles administered mucosally induce protective immunity. *J. Virol.* **71:**8707–8717.

Rodriguez, F., J. Zhang, and J. L. Whitton. 1997. DNA immunization: ubiquitination of a viral protein enhances cytotoxic T-lymphocyte induction and antiviral protection but abrogates antibody induction. *J. Virol.* **71:**8497–8503.

Rose, N., P. Marx, A. Luckay, D. Nixon, W. Moretto, S. Donahoe, D. Montefiori, A. Roberts, L. Buonocore, and J. Rose. 2001. An effective AIDS vaccine based on live attenuated vesicular stomatitis virus recombinants. *Cell* **106:**539–549.

Taboga, O., C. Tami, E. Carrillo, J. I. Nunez, A. Rodriguez, J. C. Saiz, E. Blanco, M. L. Valero, X. Roig, J. A. Camarero, D. Andreu, M. G. Mateu, E. Giralt, E. Domingo, F. Sobrino, and E. Palma. 1997. A large-scale evaluation of peptide vaccines against foot-and-mouth disease: lack of solid protection in cattle and isolation of escape mutants. *J. Virol.* **71:**2606–2614.

Xu, L., A. Sanchez, Z. Yang, S. R. Zaki, E. G. Nabel, S. T. Nichol, and G. Nabel. 1998. Immunization for Ebola virus infection. *Nat. Med.* **4:**37–42.

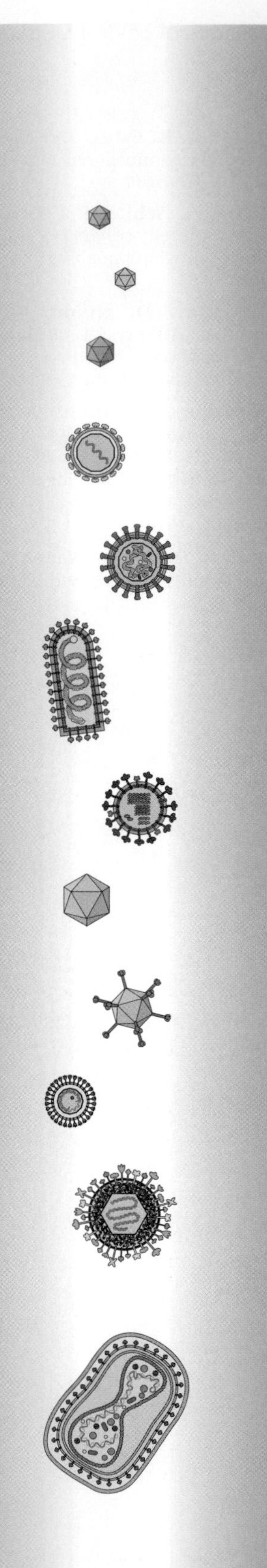

20

Evolution and Emergence

> *Anything produced by evolution is bound to be a bit of a mess.*
> SYDNEY BRENNER
>
> *Around here, it takes all the running you can do just to stay in the same place.*
> LEWIS CARROLL
> *The Red Queen to Alice in Alice in Wonderland*

Evolution. Derived from the Latin word *evolvere*—to unroll, to roll forth, to revolve. In classical usage, its noun *evolutio* acquired the special meaning of the unrolling of a scroll in order to read it, or "opening of the records" as we would say today.

Virus Evolution

The word "evolution" conjures up images of fossils, dusty rocks, and ancestral phylogenetic trees covering eons. Modern virology shatters this staid image and reminds us that evolution not only is contemporary (we are made aware of it every day), but also carries profound implications for the future. Indeed, as host populations grow and adapt, virus populations that can infect them are selected. It also works the other way: viral infections can be significant selective forces in the evolution of host populations. If a host population cannot adapt to a lethal viral infection, it may be exterminated.

The public is made aware of virus evolution by frequent, often sensationalized reports in the press of new viral diseases, superbugs, killer viruses, and the acquired immunodeficiency syndrome (AIDS) pandemic, not to mention our regular bouts with influenza and common cold viruses. Scientists are challenged regularly to devise new antiviral drugs, vaccines, diagnostics, and treatments for these emerging diseases, or to counter the possible threat of bioterrorism. In truth, we are only beginning to develop an understanding of the selective forces that drive virus evolution, the emergence of new and old infections, and the corresponding implications for our own survival.

How Do Viral Populations Evolve?

The harsh realities presented by a large, often dispersed host population in ever-changing environments appear to be overwhelming barriers to the survival of inanimate, submicroscopic, obligate intracellular pathogens. Obviously, such thinking must be flawed, as innumerable viruses abound on the planet. In this chapter, we discuss the primary reason for their remarkable success: despite a minimal set of genes, viral populations display spectacular diversity. It is such diversity, manifested in the large collection of genomic permutations that are present in any viral population at any given time, that provides constant opportunities for survival. The sources of this diversity are **mutation, recombination and reassortment,** and **selection.** The constant change of a viral population in the face of selective pressures is the definition of **virus evolution**.

Positive and negative selection for particular preexisting mutants can occur at any step in a viral life cycle. The selective forces acting on viral populations are imposed not only by the environment, but also by the limitations of the information encoded in the viral genome. The requirement to spread within the cells of an infected host and among individuals in the population exposes virions to a variety of host antiviral defenses. The population density, the social behavior, and the health of potential hosts clearly represent but a few of the powerful selective forces determining the survival of viral populations.

We define virus evolution in terms of a population, **not** an individual viral particle. Viral populations comprise diverse arrays of mutants that are produced in prodigious quantities. In most infections, thousands of progeny are produced after a single cycle of replication in one cell, and, because of error-prone copying, almost every new genome can differ from every other. Accordingly, it is misleading, and even confusing, to think of an individual particle as representing an average virion for that population. Virologists are population biologists whether they know it or not. As Stephen Jay Gould put it, the median is **not** the message when it comes to evolution. The diversity of the population provides surprising avenues for survival, yet every individual is a potential winner. Occasionally, even the most rare genotype in a current population may be the most common after a single selective event.

The trajectory of evolution has long been a subject of debate. Scientists and philosophers have considered many questions such as the following. Is there a predictable direction for evolution, and if so, what is the path? Does change occur in a series of small steps leading to higher and higher degrees of fitness? Does selection impose a landscape of distinct peaks and valleys representing various fitness niches in which single mutations can take a fit organism back into a valley to start over, perhaps on a new course? Are there really evolutionary dead ends? Virology provides a productive area for research and insight into these questions. A primary lesson is that we must avoid judging outcomes as "good or bad." While making such judgments is a uniquely human activity, anthropomorphic assessments and the comfort of analogies must be avoided, particularly when virus evolution is considered (see also Box 1.1). Furthermore, from the first principle that there is no goal but survival, we can deduce that evolution does not move a viral genome from "simple" to "complex," or along some trajectory aimed at "perfection." Rather, change is effected by elimination of the ill adapted of the moment, not on the prospect of building something better for the future.

Virus-Infected Cells Produce Large Numbers of Progeny

A single infectious particle generally gives rise to enormous numbers of progeny. For example, a single cell infected with poliovirus yields about 10,000 virus particles, and, in theory, three or four cycles of replication at this rate could produce sufficient virions to infect every cell in a human. Such overreplication does not happen for a variety of reasons, including a vigorous host defense and tropism for certain tissues. Nevertheless, high rates of replication over short periods and the resulting accumulation of large quantities of infectious particles are hallmarks of many common acute and persistent viral infections. This feature is illustrated in Table 20.1 for human immunodeficiency virus and hepatitis B virus. Total virion production in a single infected person can range from more than 10^9 particles per day for human immunodeficiency virus, and up to a staggering 10^{11} particles per day for hepatitis B virus. Moreover, these high rates of particle production can continue for years. In the case of human immunodeficiency virus, the time from release of a viral particle to infection and lysis of another target cell is estimated to be 2.6 days during the later stages of infection. What is more striking is the 50 to 90% turnover rate of this virus in plasma (replication minus elimination by host defenses). Such measurements reveal the prodigious, relentless production of new virus particles in the face of a vigorous host defense. The interface of host defense and virus replication is fertile ground for selection and evolution.

Table 20.1 In vivo dynamics of human immunodeficiency virus and hepatitis B virus

Characteristic	Hepatitis B virus	Human immunodeficiency virus
Virus in plasma		
Half-life	24 h	6 h
Daily turnover	50%	90%
Total production in blood	$>10^{11}$	$>10^{9}$
Virus in infected cell		
Half-life	10–100 days	2 days
Daily turnover	1–7%	30%

Large Numbers of Mutants Arise When Viral Genomes Replicate

Two simple principles are that evolution is possible only when mutants arise in a population, and that mutations are introduced during copying of any nucleic acid molecule. When viral genomes replicate, mutations invariably accumulate in their progeny (Box 20.1).

RNA viruses. Most viral RNA genomes are replicated with considerably less fidelity than those comprising DNA (see Chapters 6 and 7). The average error frequencies reported for copying of RNA genomes are about one misincorporation in 10^4 or 10^5 nucleotides polymerized, more than 1 million times greater than the rate for a host genome. Given a typical RNA viral genome of 10 kb, a mutation frequency of 1 in 10^5 nucleotides polymerized translates to about 1 mutation per 10 genomes; a 1 in 10^4 error rate corresponds to an average of 1 mutation in every replicated genome.

DNA viruses. The error rate of viral DNA replication is estimated to be about 300-fold less than that for RNA virus genomes described above. RNA polymerases lack error-correcting mechanisms, while most DNA polymerases can excise and replace misincorporated nucleotides. Some experimental data indicate that the replication of small single-stranded DNA virus genomes (e.g., *Parvoviridae* and *Circoviridae*) may be more error-prone than replication of the double-stranded DNA genomes of larger viruses. However, until experiments are conducted using identical cells and growth conditions, the validity of this suggestion cannot be determined.

RNA Viruses and the Quasispecies Concept

A 1978 paper described a detailed analysis of an RNA bacteriophage population (Qβ). The authors made a startling conclusion:

> A Qβ phage population is in a dynamic equilibrium with viral mutants arising at a high rate on the one hand, and being strongly selected against on the other. The genome of Qβ cannot be described as a defined unique structure, but rather as a weighted average of a large number of different individual sequences.
>
> **Domingo, E., D. Sabo, T. Taniguchi, and C. Weissmann.** 1978. Nucleotide sequence heterogeneity of an RNA phage population. *Cell* **13:**735–744.

This conclusion has been validated for many virus populations. Indeed, we now understand that virus populations exist as dynamic distributions of nonidentical but related replicons, often called **quasispecies**, a concept

BOX 20.1

Error rates are difficult to quantify

Estimates of mutation rates must be viewed with caution. Absolute error rates (measured as the number of misincorporations per nucleotide polymerized) for any nucleic acid polymerase are difficult, if not impossible, to determine. A variety of technical issues must be considered, including sampling problems and potential artifacts of experimental design. Estimates of error rate can vary substantially, depending on the experimental method by which they are assessed. For example, polymerase chain reaction (PCR) technology is commonly used to sample viral genomes, but the polymerase used in the amplification may itself introduce copying errors that must be factored into the analysis.

Another popular method makes use of reporter genes (e.g., the *lacZ* gene, which encodes β-galactosidase). Reporter genes encode proteins with activities that can be monitored easily. These genes can be inserted into a viral genome so that errors in the reporter gene can be scored by inspection or analysis of virus plaques. The error rate for the viral genome is then extrapolated from that determined for the reporter gene. While this method is relatively simple, it can yield misleading data, because errors of incorporation are not uniformly distributed as each genome is copied and are often dependent upon the particular sequence analyzed. For example, the reverse transcriptase from human immunodeficiency virus is inaccurate when measured in vitro, with an average error rate per nucleotide incorporated of 1 in 1,700. However, certain positions in this genome can be hot spots for mutation at which the error rate can be as high as 1 per 70 polymerized nucleotides.

Drake, J. 1992. Mutation rates. *Bioessays* **14:**137–140.

Drake, J. 1991. A constant rate of spontaneous mutation in DNA-based microbes. *Proc. Natl. Acad. Sci. USA* **88:**7160–7164.

Drake, J. 1993. Rates of spontaneous mutation among RNA viruses. *Proc. Natl. Acad. Sci. USA* **90:**4171–4175.

Hwang, Y. T., and C. B. C. Hwang. 2003. Exonuclease-deficient polymerase mutant of herpes simplex virus type 1 induces altered spectra of mutations. *J. Virol.* **77:**2946–2955.

developed by Manfred Eigen. The classical definition of a species (an interbreeding population of individuals) has little meaning when considering viruses. For example, most viral infections are initiated not by a single virion, but rather by a population of particles. The large number of progeny produced after such infections are complex products of intense selective forces that operate **inside an infected host.** The relatively few virions that successfully infect another host have been subjected to an entirely new set of **external** selective forces. A steady-state, equilibrium population of a given viral quasispecies, therefore, must comprise vast numbers of particles. Such an equilibrium cannot be attained in the small populations typically found after isolated infections in nature or in the laboratory. Consequently, when small populations of virus particles replicate, extreme fluctuations in genotype and phenotype are possible (Fig. 20.1).

For a given RNA virus population, the genome sequences cluster around a consensus or average sequence, but virtually every genome can be different from every other. A rare genome with a particular mutation may survive a selection event, and this mutation

Figure 20.1 Viral quasispecies, population size, bottlenecks, and fitness. When an RNA virus replicates, mutations accumulate in the progeny genomes. Genomes are indicated by the horizontal lines. Mutations are indicated by different symbols. A population of genomes, each member containing a characteristic set of mutations, is shown in the center. The consensus sequence for this population is shown as a single line at the bottom. Note that there are **no** mutations in the consensus sequence, despite their presence in every genome in the population (every genome is different). A population of genomes that emerges after repeated passage through bottlenecks is depicted on the right. The consensus sequence for this population is shown as a single line at the bottom. Note that in this example, three mutations are found in every member of the population, and these appear in the consensus sequence. If the large population is propagated without passage through bottlenecks (situation on the left), repeated passage enriches for mutant genomes that improve replication and increased fitness of the population. If the population continues to propagate through serial bottlenecks, accumulation of mutations approaches the error threshold for the genome and fitness plummets. Adapted from E. Domingo, C. Escarmis, L. Menendez-Arias, and J. Holland, p. 144, *in* E. Domingo, R. Webster, and J. Holland (ed.), *Origin and Evolution of Viruses* (Academic Press, Inc., San Diego, Calif., 1999), with permission.

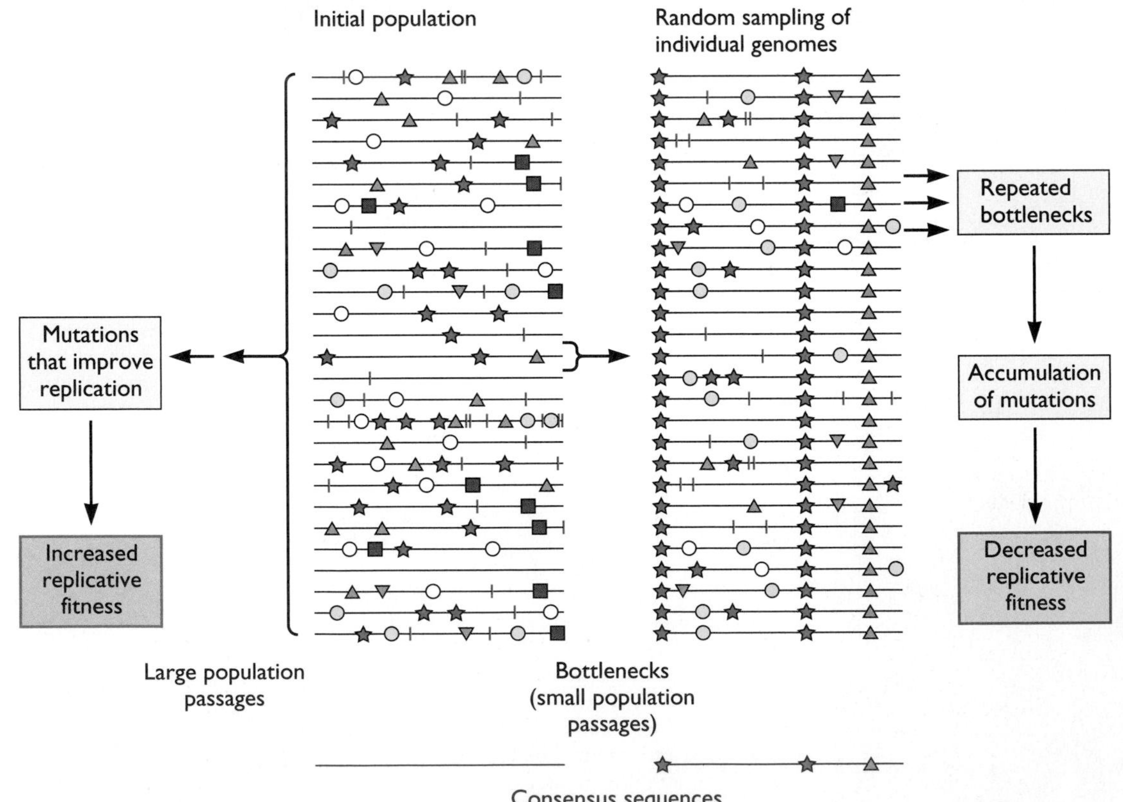

will be found in all progeny genomes. However, the linked but unselected mutations in that genome get a free ride. Consequently, the product of selection after replication is a new, diverse population that shares only the selected mutation. The ramifications of this phenomenon in virus evolution and pathogenesis are only now being appreciated.

Increased mutation rates are selected during virus evolution. Previously, we made the assertion that creation of diversity is the primary reason for virus survival. This conclusion follows in part from the identifications of certain mutations located in genes encoding viral polymerases that reduce the frequency of incorporation errors during growth in cultured cells. Such mutants are called **antimutators**. Certain spontaneous mutants of human immunodeficiency virus that are resistant to the reverse transcriptase inhibitor lamivudine exhibit a 3.2-fold reduction in error frequency. The seemingly modest increase in fidelity was associated with a significant growth disadvantage in infected individuals. As the wild-type viruses have maintained high mutation rates, we can infer that lower rates are neither advantageous nor selected in nature.

Sequence Conservation in Changing Genomes

Not all is in flux during viral genome replication. Despite high mutation rates, the so-called *cis*-acting sequences of RNA viruses change very little during propagation. These sequences are required for genome replication, messenger RNA (mRNA) synthesis, and genome packaging, and often are the binding sites for one or more viral or cell proteins. Any genome with mutations in such sequences, or in the gene that encodes the corresponding binding protein, is likely to be less fit, or may not replicate at all. Changes must occur in both partners for restoration of function. The tight, functional coupling of binding pro-

BOX 20.2

Pushing human immunodeficiency virus over the "error threshold": the Achilles' heel of the viral quasispecies?

At the end stage of AIDS, individuals infected with human immunodeficiency virus produce more than 10^{10} particles per day, and, on average, each genome contains one mutation. Diverse viral populations accumulate in the various tissues of every infected individual. Interestingly, most of the virions appear to be nonviable, suggesting that the population exists at its error threshold and cannot tolerate many additional mutations. What would happen if one could intentionally push replicating genomes over the error threshold with a mutagenic nucleoside analog? Is lethal mutagenesis of human immunodeficiency virus a practical approach to treating RNA virus infections?

In 1990, scientists first reported that it is difficult to increase the mutation rate for two RNA viruses using mutagens. Nine years later, another group selected antiviral nucleoside analogs that lacked toxicity for human cells but were incorporated into human immunodeficiency virus DNA by the viral reverse transcriptase. These are not chain-terminating compounds, but rather they promote base misincorporations when copied. One compound, 5-hydroxydeoxycytidine, was promising in that it was not toxic to human cells but was efficiently incorporated into replicating viral genomes.

The experiments comprised serial passage of virions in human cells in the presence of the nucleoside analog. In seven of nine experiments, there was a precipitous loss of viral replication after 9 to 24 serial passages in the presence of the compound. Sequence analysis showed that the progeny genomes accumulated G-to-A substitutions. Loss of viral replication was not observed in 28 control infections in which virions were serially passaged without the analog.

Similar results have been obtained after infection of cultured cells infected with a variety of RNA viruses. If this approach has merit for use in antiviral therapy, the safety of using mutagenic nucleoside analogs in uninfected as well as infected individuals must be examined closely.

Crotty, S., C. E. Cameron, and R. Andino. 2001. RNA virus error catastrophe: direct molecular test by using ribavirin. *Proc. Natl. Acad. Sci. USA* **98:**6895–6900.

Eigen, M. 2002. Error catastrophe and antiviral strategy. *Proc. Natl. Acad. Sci. USA* **99:** 13374–13376.

Holland, J. J., E. Domingo, J. C. de la Torre, and D. A. Steinhauer. 1990. Mutation frequencies at defined single codon sites in vesicular stomatitis virus and poliovirus can be increased only slightly by chemical mutagenesis. *J. Virol.* **64:**3960–3962.

Loeb, L. A., J. M. Essigmann, F. Kazazi, J. Zhang, K. D. Rose, and J. I. Mullins. 1999. Lethal mutagenesis of HIV with mutagenic nucleoside analogs. *Proc. Natl. Acad. Sci. USA* **96:**1492–1497.

Pareiente, N., S. Sierra, P. R. Lowenstein, and E. Domingo. 2001. Efficient virus extinction by combinations of a mutagen and antiviral inhibitors. *J. Virol.* **75:**9723–9730.

BOX 20.3

Influenza virus: the classical paradigm of antigenic drift and shift

Influenza, which has been recognized as a human disease for centuries, is notable because it causes a typical acute infection and elicits a strong immune response but continues to occur regularly. Influenza can occur in many nations almost simultaneously (pandemics) with serious consequences. After more than 65 years of research, we now have some reasonable ideas about why this virus causes epidemics annually and pandemics at frequent intervals. Influenza virus not only infects humans, but also infects several other animals in a complex cycle of dissemination.

Epidemiologists can classify a given influenza A virus by its antigenic composition, usually by serologic testing of the two envelope proteins hemagglutinin (HA or H) and neuraminidase (NA or N).

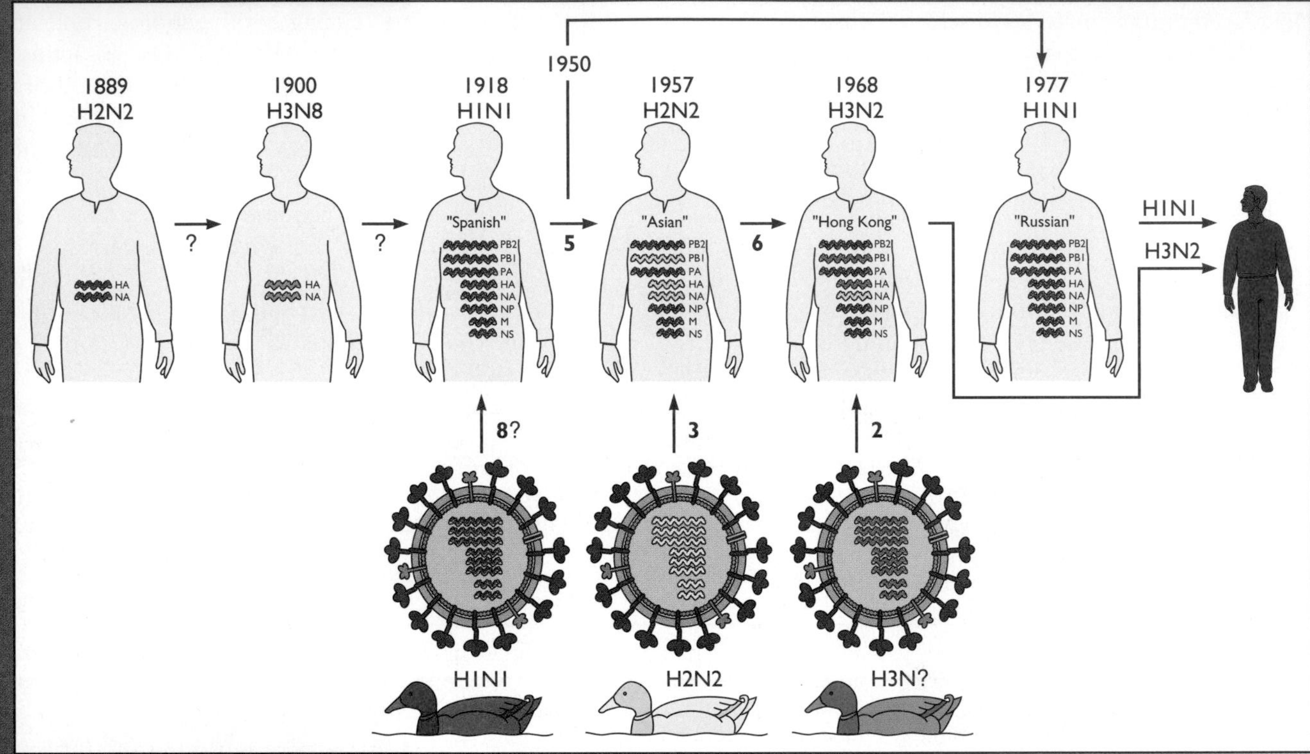

Postulated evolution of human influenza A viruses from 1889 to 1977. The figure depicts the appearance and transmission of distinct serotypes of influenza A virus in humans. The bottom part shows the nature of the avian influenza viruses that reassort with human viruses. The color of the genome segments represents a particular viral genotype. Segments of the predominant influenza virus genome and its gene products are indicated in each human silhouette for each year. The number next to the arrow indicates how many segments of the viral genome are known to have been transmitted. The earliest serology data are from 1889 and suggest that H2N2 was the predominant class in humans. In 1900 the predominant serotype was H3N8. No data exist for the other influenza virus genes present at these times, and these segments are not illustrated. Phylogenetic evidence is consistent with the appearance by 1918 of an influenza virus with eight distinct segments. This virus is thought to be of avian origin and is characterized by the H1N1 serotype (red). It was also found in pigs and was carried from North America to Europe by U.S. soldiers. It gained notoriety as the cause of the catastrophic Spanish influenza pandemic of 1918. In 1957 the Asian pandemic was caused by a virus that acquired three genes (PB1, H2, and N2) from avian viruses infecting wild ducks (yellow) and retained five other genes from the circulating human strain with H2N2 (red). As the Asian strains appeared, the H1N1 strains disappeared from the human population. In 1968, the Hong Kong pandemic viral genome acquired two new genes (PB1 and H3) from the wild duck reservoir (blue) and kept six genes that were circulating in human viruses (red and yellow). The pandemic virus had a characteristic H3N2 serotype. After the appearance of this virus, the Asian H2N2 strains could no longer be detected in humans. In 1977 the Russian H1N1 strain that had circulated in humans in the 1950s reappeared and infected young adults and children. One theory was that this virus escaped from a laboratory. It has continued to cocirculate with H3N2 influenza viruses in the human population. In 1997 and 1999, the avian viruses H5N1 and H9N2 were transmitted directly to humans. Adapted from R. G. Webster and Y. Kawaoka, *Semin. Virol.* **5**:103–111, 1994, with permission.

BOX 20.3 *(continued)*

A common simplification in nomenclature is to refer to combinations of H and N as H*x*N*y* (currently *x* ranges from 1 to 15, and *y* ranges from 1 to 9), e.g., H1N1 or H5N2. At least 15 subtypes of viral hemagglutinin are known in viruses that infect birds. Three of these subtypes are present in viruses that can infect humans (H1N1, H2N2, and H3N2), and at least two can infect pigs, horses, and aquatic mammals. For the past 25 years, H3N2 and H1N1 viruses have cocirculated in humans. In 1997 and 1999, the avian viruses H5N1 and H9N2, respectively, were transmitted directly to humans. The contribution of these serotypes to human disease is under close scrutiny.

tein and target sequence may be a marked constraint for evolution of these sequences. In some instances, these sequences are stable enough to represent lineage markers for molecular phylogeny. We have little information about how such interdependent pairs were established.

The Error Threshold

The capacity to sustain prodigious numbers of mutations is a powerful advantage. Yet, there must be a point at which selection and survival balance genetic fidelity and mutation rate. Intuitively, mutation rates higher than one error in 1,000 nucleotides incorporated must begin to challenge the very existence of the viral genome, as precious genetic information can be irreversibly lost. This limit is called the **error threshold**, a mathematical parameter that measures the complexity of the information that must be maintained to ensure survival of the population. RNA viruses tend to evolve close to their error threshold, while DNA viruses have evolved to exist far below it. We can infer these remarkable properties from experiments with mutagens. If one treats a cell culture infected with an RNA virus, such as vesicular stomatitis virus or poliovirus, with a base analog such as 5-azacytidine, virus titers drop dramatically and the error frequency per surviving genome increases only two- to threefold at best. In contrast, a similar experiment performed with a DNA virus, such as herpes simplex virus or simian virus 40, reveals an increase in single-site mutations of several orders of magnitude among survivors. One conclusion is that RNA viruses must exist at the precarious edge of their error threshold (Box 20.2).

The error threshold concept is more complicated than it might first appear. Many parameters contribute, including a complex property called **fitness**—the replicative adaptability of an organism to its environment. In the laboratory setting, fitness may be determined by comparison of growth rates or virus yields, but fitness is difficult to measure under more natural conditions, such as infection of complex organisms that live in large, interacting populations. Another essential component, equally difficult to measure, is the stability or predictability of the environment as it affects propagation of a virus. Host population dynamics and seasonal variation are but two examples of the many complicated environmental variables that exist. Finally, given the diversity in any viral population, determining the fitness of one population versus that of another depends on the mathematics of population genetics, a subject beyond the scope of this text.

Genetic Shift and Drift

In Chapter 16, we discussed the process of antigenic variation and its contribution to modulating the immune response. Selection of mutants resistant to elimination by antibodies or cytotoxic T lymphocytes is inevitable when sufficient virus replication occurs in an immunocompetent individual. The terms **genetic drift** and **genetic shift** describe distinct mechanisms of diversity production. Diversity arising from copying errors and immune selection (drift) is contrasted with diversity arising after recombination or reassortment of genomes and genome segments (shift). Drift is possible every time a genome replicates. Shift can only occur under certain circumstances and is relatively rare. For example, there are only six established instances of genetic shift for the influenza virus hemagglutinin gene since 1889. The combination of rapid drift and slow shift contributes to yet more diversity in populations (Boxes 20.3 and 20.4). In retrovirus infections, where multiple proviral genomes may be integrated in a single cell, genetic shift may be a frequent event, with ramifications only now being appreciated (Box 16.3).

Genetic Bottlenecks

Another critical parameter affecting virus evolution is the **genetic bottleneck**—extreme selective pressure on small populations that result in loss of diversity, accumulation of nonselected mutations, or both (Fig. 20.1). A simple experiment illustrating this idea is easily done in the laboratory. A single RNA virus plaque formed on a monolayer of cultured cells is picked and expanded. Next, a single plaque is picked from the expanded stock, and the process is repeated over and over. The bottleneck is the consequence of restricting further viral replication to the progeny found in a single plaque that contains a few thousand virions derived from a single infected cell. The perhaps surprising result is that after about 20 or 30 cycles of single-plaque amplification, many virus populations are barely able to grow. They are markedly less fit than the original population (Table 20.2). The environment is con-

BOX 20.4

Antigenic shift, not drift, was the driving force for pandemics of human influenza

Influenza viruses causing pandemics can be typed into H and N subtypes (Box 20.3). Six times since 1889, an influenza virus H subtype that had not been seen for years entered the human population. Three influenza virus H subtypes display a cyclic appearance, with the sequential introduction of H2 in 1889, H3 in 1900, H1 in 1918, H2 again in 1957, H3 again in 1968, and H1 again in 1977. With each H subtype introduction, the world experienced an influenza pandemic characterized by a new combination of H and N.

These dramatic shifts in H and N serotypes result from the exchange of genome segments by mammalian and avian influenza viruses. In general, influenza A viruses that grow well in birds are not efficient at infecting humans, and vice versa. This assertion is currently under scrutiny, as evidence of direct infection of humans by an avian influenza virus (H5N1) was documented in 1997. This H5N1 combination of antigens had never been observed previously in human infections, despite its occurrence in virulent avian viruses that have caused epidemics in domestic birds. Because no humans have immunity to the H5N1 viruses, the appearance of those viruses in humans was of major concern. It now appears that a pandemic was averted only because these viruses are incapable of efficient spread in human populations.

Virologists have demonstrated that certain combinations of H and N are better selected in avian hosts than in humans. An important observation was that both avian and human viruses replicate well in certain species such as pigs, no matter what the H/N composition. Indeed, the lining of the throats of pigs contains receptors for both human and avian influenza viruses, providing an environment in which both can flourish. As a result, the pig is a good nonselective host for mixed infection of avian and human viruses, in which reassortment of H and N segments can occur, creating new viruses that can reinfect the human population.

At first glance, one might think that this combination of human, bird, and pig infections must be extremely rare.

Genetic reassortment (shift) between avian and human influenza A viruses in swine. Studies of Italian pigs provide evidence for reassortment between avian and human influenza viruses. The figure shows how the avian H1N1 viruses in European pigs reassorted with H3N2 human viruses. The color of the segments of the influenza genome indicates the origin: blue segments are from avian viruses found in swine in 1979, and red segments are from human viruses found in swine in 1968. The host of origin of the influenza virus genes was determined by partial sequencing and phylogenetic analysis. These studies support the hypothesis that pigs can serve as an intermediate host in the emergence of new pandemic influenza viruses. Adapted from R. G. Webster and Y. Kawaoka, *Semin. Virol.* **5:**103–111, 1994, with permission. For more information, see J. S. Peiris et al., *J. Virol.* **75:**9679–9686, 2001.

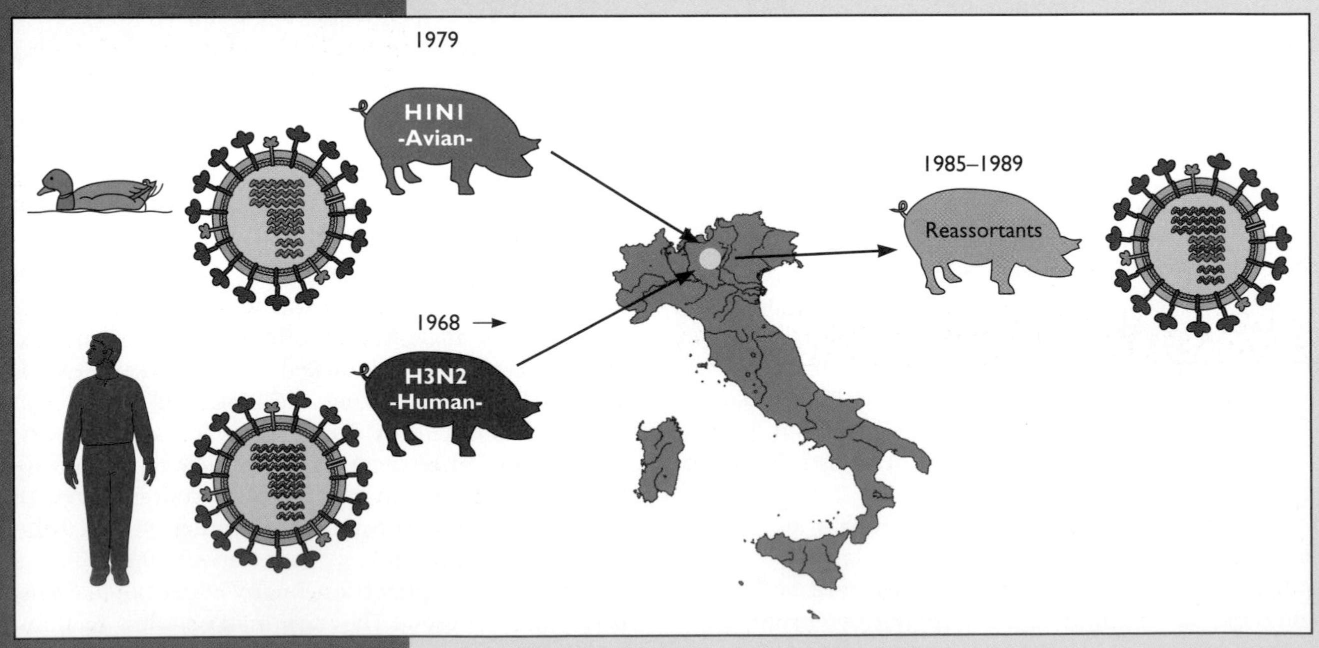

BOX 20.4 *(continued)*

However, the dense human populations in Southeast Asia that come in daily contact with domesticated pigs, ducks, and fowl remind us that these interactions are likely to be frequent. Indeed, epidemiologists can show that the 1957 and 1968 pandemic influenza A virus strains originated in the People's Republic of China and that the human H and N serotypes are circulating in wildfowl populations.

Southeast Asia should not be the only focus of attention, as intense production of domestic swine and turkeys in Europe and the United States, coupled with the major migratory paths of wild ducks and geese, is likely to make these regions centers of interspecies transfer as well.

Ito, T., J. N. S. S. Couceiro, S. Kelm, L. G. Baum, S. Krauss, M. R. Castrucci, I. Donatelli, H. Kida, J. C. Paulson, R. G. Webster, and Y. Kawaoka. 1998. Molecular basis for the generation in pigs of influenza A viruses with pandemic potential. *J. Virol.* **72:**7367–7373.

stant, and the only apparent selection is that imposed by the ability of the population of viruses from a single plaque to replicate. Why does fitness plummet?

The answer lies in a phenomenon dubbed **Muller's ratchet**: small, asexual populations decline in fitness over time if the mutation rate is high. The genomes of replicating RNA viruses accumulate many mutations; indeed, they survive close to their error threshold. By restricting population growth to serial single founders (the bottleneck) under otherwise nonselective conditions, mutations that exceed the threshold accumulate during replication and, consequently, fitness decreases.

The ratchet metaphor should be clear: a ratchet on a gear allows the gear to move forward, but not backward. After each round of replication, mutations accumulate but are not removed. Each round of error-prone replication works like a ratchet, "clicking" relentlessly as mutations accumulate at every replication cycle. Each mutation has the potential to erode the fitness of subsequent populations. Simple studies such as the serial plaque transfer experiment indicate that Muller's ratchet can be avoided if a more diverse viral population is replicated by serial passage (i.e., opening the bottleneck by increasing the number of clonal pools). One such study showed that pools of virus from 30 individual plaques were required in serial transfer to maintain the culture's original fitness. This observation can be explained as follows: more diversity in the replicating population facilitates construction of a mutation-free genome by recombination or reassortment, removing or compensating for mutations that affect growth adversely. Even if such a recombinant is rare, it has a powerful selective advantage in this experimental paradigm. Indeed, its progeny ultimately will predominate in the population. The message is simple but powerful: diversity of a viral population is important for the survival of individual members; remove diversity, and the population suffers.

Table 20.2 Fitness decline compared with initial virus clone after passage through a bottleneck[a]

Virus	No. of bottleneck passages	Avg % decrease in fitness
ϕ6 (bacteriophage)	40	22
Vesicular stomatitis virus	20	18
Foot-and-mouth disease virus	30	60
Human immunodeficiency virus	15	94
MS2 (bacteriophage)	20	17

[a]Data from A. Moya, S. Elena, A. Bracho, R. Miralles, and E. Barrio, *Proc. Natl. Acad. Sci. USA* **97:**6967–6973, 2000.

While the particular bottleneck of single-plaque passage is obviously artificial, infection by a small virus population and subsequent amplification are often found in nature. Examples include the small droplets of suspended virus particles during aerosol transmission, the activation of a latent virus from a limited population of cells, or the small volume of inoculum introduced in infection by insect bites. An important question is, how do viruses that spread in nature by these routes escape Muller's ratchet? They do so by exchanging genetic information.

Genetic Information Exchange

Genetic information is exchanged by recombination or by reassortment of genome segments (Fig. 20.1; see also Chapters 6, 7, and 9). In one step, recombination creates new combinations of many mutations that may be essential for survival under selective pressures. As discussed above, this process allows the construction of viable genomes from debilitated ones and avoids Muller's ratchet. Recombination occurs when the polymerase changes templates (copy choice) during replication or when nucleic acid segments are broken and rejoined. The former mechanism is common among RNA viruses, whereas the latter is more typical of double-stranded DNA virus recombination. Another mechanism for exchange of genetic information is reassortment among genomic segments when cells are coinfected with segmented RNA viruses. It is an important source of variation, as exemplified by orthomyxoviruses and reoviruses (Boxes 20.3 and 20.4; see also Box 16.3).

Insertion of nonviral nucleic acid into a viral genome (sequence-independent recombination) is well documented. Indeed, virus particles with genomes harboring cellular genetic information (transducing viruses) are not only a mainstay of viral genetics in the laboratory, but are also essential contributors to viral evolution. Incorporation of cellular sequences can lead to defective genomes or to more pathogenic viruses. Examples of such recombination include the appearance of a cytopathic virus in an otherwise nonpathogenic infection by the pestivirus bovine viral diarrhea virus (see Chapter 6) or the sudden appearance of oncogenic retroviruses in nononcogenic retroviruses (see "Host Range Can Be Expanded by Mutation or Recombination" below). The acquisition of activated oncogenes from the cellular genome is the hallmark of acutely transforming retroviruses such as Rous sarcoma virus (see Chapter 18). Poxvirus and gammaherpesvirus genomes carry virulence genes with sequence homology to host immune defense genes. These genes are usually found near the ends of the genome. One explanation for their location in the genome is that the process of DNA packaging (gammaherpesviruses) or initiation of DNA replication (poxviruses) stimulates viral-host recombination when viral DNA is cleaved.

Information can be exchanged in a variety of unexpected ways during viral infections. For example, a host can be infected or coinfected by many different viruses during its lifetime (Box 20.5). In fact, serial and concurrent infections are commonplace. Both have a profound effect on virus evolution. In the simplest case, propagation of a virus quasispecies in an infected individual allows coinfection of individual cells, phenotypic mixing, and genetic complementation. As a result, recessive mutations are not immediately eliminated, despite the haploid nature of most viral genomes. Of course, such coinfection also provides an opportunity for physical exchange of genetic information.

Viral infections occur as host defenses are modulated or bypassed. Those infections that actively suppress the immune response have a marked effect upon concurrent or subsequent infections of the same host by very different viruses. Indeed, human immunodeficiency virus and measles virus infections lead to serious disease and death of their hosts by facilitating subsequent infections by diverse pathogens (see Chapter 16). Animals infected with poxviruses are more susceptible to infection by a variety of pathogens, because infected cells secrete soluble inhibitors of interferons and other cytokines required for innate defense. Conversely, activation of host defenses by one viral infection can impair subsequent infection by a different virus. The existence of such multifactorial interchange among diverse viruses reinforces the idea that selection acts not only on one population of viruses, but also on interacting populations of different viruses.

Two General Pathways for Virus Evolution

As viruses are absolutely dependent on their hosts for replication, viral evolution tends to take one of two general pathways. In one pathway, viral populations coevolve with their hosts so that both share a common fate; as the host prospers, so does the viral population. However, given no other host, a bottleneck now exists: the entire viral population can be eliminated with antiviral measures or by loss of the host. In the other pathway, viral populations occupy broader niches and infect multiple host

BOX 20.5

Virus evolution by host switching and recombination

Viruses can be transmitted to completely new host species that have not experienced any prior infection. Usually host defenses stop the new infection before any replication and adaptation can take place. On rare occasions, a novel population of viruses arises in the new host. Past interspecies infections can occasionally be detected by sequence analyses. The data provide a glimpse of the rather amazing and unpredictable paths of virus evolution.

Circoviruses infect vertebrates and have small, circular, single-stranded genomes (Chapter 3). Nanoviruses have the same genome structure, but infect plants. The genes encoding the Rep protein of these viruses share significant sequence similarity in the 5′ coding sequences, and also exhibit homology in the 3′ sequences with an RNA binding protein encoded by caliciviruses.

The scientists who analyzed the DNA sequences suggested that two remarkable events occurred during the evolution of circo- and nanoviruses. At some time, a nanovirus was transferred from a plant to a vertebrate, perhaps when a vertebrate was exposed to sap from an infected plant. The virus adapted to vertebrates, and the circovirus family was established. As all known caliciviruses infect vertebrates, the recombination between circovirus and calicivirus sequences would have occurred after the host switch of nanoviruses to vertebrates.

Gibbs, M. J., and G. F. Weiller. 1999. Evidence that a plant virus switched hosts to infect a vertebrate and then recombined with a vertebrate-infecting virus. *Proc. Natl. Acad. Sci. USA* **96:**8022–8027.

Table 20.3 Three general theories of the origin of viruses

Regressive evolution: Viruses degenerated from previously independent life forms, lost many functions that other organisms possess, and have retained only what is needed for their parasitic lifestyle.
Cellular origins: Viruses are subcellular functional assemblies of macromolecules that have gained the capacity to move from cell to cell.
Independent entities: Viruses evolved, on a course parallel to that of cellular organisms, from the self-replicating molecules believed to have existed in the primitive prebiotic RNA world.

species. When one host species is compromised, the viral population can replicate in another. In general, as discussed below, the first pathway is typical of DNA viruses, whereas the second is common for RNA viruses. Both pathways provide excellent strategies for virus survival, and both suggest important avenues for interruption of the host-virus relationship.

The Origin of Viruses

We presented a brief introduction to the origin of viruses in Chapter 1. We can be sure of one basic fact: viruses cannot exist without living cells. As all cells on the planet today derive from preexisting cells, it follows that there was but one primordial cell. By this logic, all viruses in existence today are linked directly or indirectly to that primordial cell. The origin of the first viruses is a matter of conjecture and debate centering around three nonexclusive ideas (Table 20.3). The **regressive theory** suggests that viruses are derived from intracellular parasites that have lost all but the most essential genes encoding products required for replication and maintenance. The **cellular origin theory** proposes that viruses arose from cellular components that gained the ability to replicate autonomously within the host cell. The third theory postulates that viruses coevolved with cells from the origin of life itself. As there is no fossil record and there are few viral stocks more than 80 years old, we are not in a good position to test the three hypotheses for the primordial origins of viruses. Nevertheless, we can offer some insight into viral evolution by examining contemporary viruses. A particularly intriguing hypothesis is that DNA viruses of green algae may be the oldest eukaryotic viruses, because sequence analysis places their genomes near the root of most eukaryotic sequences (Box 20.6).

DNA Virus Relationships Deduced by Nucleic Acid Sequence Analysis

Nucleic acid sequence analyses have identified many relationships among different viral genomes, providing considerable insight into the origin of viruses. The her-

BOX 20.6

Chlorella viruses: clues to viral and eukaryotic origins?

Paramecium bursaria chlorella virus (PBCV-1) is one of the oldest eukaryotic viruses. This remarkable virus is a large, icosahedral, plaque-forming, double-stranded DNA virus that replicates in certain unicellular, eukaryotic chlorella-like green algae (family *Phycodnaviridae*, genus *Chlorovirus*). In nature, the chlorella host is a hereditary endosymbiont of the ciliated protozoan *P. bursaria*. The alga host can be grown in the laboratory in liquid and on solid media. Chloroviruses have been found in freshwater sources throughout the world, and many genetically distinct isolates usually can be found within the same sample. The titer of the viruses within a single water source can reach as high as 40,000 plaque-forming units per ml.

The 330,744-bp PBCV-1 genome is predicted to harbor ~375 protein-encoding genes and 10 transfer RNA genes. The predicted products of ~50% of these genes resemble proteins of known function. Besides their large genome, the chloroviruses have other unusual features, including multiple DNA methyltransferases and DNA site-specific endonucleases, the entire machinery to glycosylate viral glycoproteins, at least two types of introns (a self-splicing intron in a transcription regulatory gene and a spliceosome-processed intron in the viral DNA polymerase gene), a potassium ion channel, and the smallest known topoisomerase. Phylogenetic analyses based on DNA sequences and protein motifs indicate that these viral sequences lie near the root of most eukaryotic sequences. The implication is that the earliest eukaryotes were exchanging information with ancient members of the *Phycodnaviridae*.

Plugge, B., S. Gazzarrini, M. Nelson, R. Cerana, J. L. Van Etten, C. Derst, D. DiFrancesco, A. Moroni, and G. Thiel. 2000. A potassium channel protein encoded by chlorella virus PBCV-1. *Science* **287:**1641–1644.

Van Etten, J. L., and R. H. Meints. 1999. Giant viruses infecting algae. *Annu. Rev. Microbiol.* **53:**447–494.

Van Etten, J. L., M. V. Graves, D. G. Muller, W. Boland, and N. Delaroque. 2002. Phycodnaviridae—large DNA algal viruses. *Arch. Virol.* **147:**1479–1516.

Villarreal, L. P., and V. R. DeFilippis. 2000. A hypothesis for DNA viruses as the origin of eukaryotic replication proteins. *J. Virol.* **74:** 7079–7084.

pesviruses have been studied extensively by gene sequencing, and enough data have been accumulated for a detailed molecular phylogenetic analysis. The three main subfamilies of the family *Herpesviridae* (*Alphaherpesvirinae, Betaherpesvirinae,* and *Gammaherpesvirinae*) are easily distinguished by genome sequence analysis even though the original separation of these families was based on general, often arbitrary, biological properties (Fig. 20.2). By assuming a constant molecular clock (the relatively constant incorporation of errors over time by DNA replication systems) and by comparing the sequence of a gene shared by most herpesviruses (e.g., uracil-DNA glycosylase) with cellular homologs of this gene, researchers have estimated the time scale of herpesviral genome evolution. By correlating divergence of the shared gene sequence with time of host divergence based on more classical studies of fossils, these scientists proposed that for most herpesviruses, points of sequence divergence coincide with well-established points of host divergence. This property can be seen in the overlapping, branching pattern of the host and virus evolutionary trees. The conclusion is that an early herpesvirus infected an ancient host progenitor, and subsequent viruses developed by cospeciation with their hosts. Consistent with this conclusion, the genomes of all *Alpha-, Beta-,* and *Gammaherpesvirinae* that have been sequenced contain a core block of genes, often organized in similar clusters in the genome (Table 20.4).

Figure 20.2 Herpesvirus family relationships deduced by amino acid sequence comparison of the common enzyme uracil DNA glycosylase. The amino acid sequences were deduced from the DNA sequences of cloned genes from a variety of organisms. The figure is a neighbor-joining distance tree (from alignment with gap positions removed). It illustrates clusters of related sequences and the degree of their relatedness. The length of lines joining sequences illustrates the calculated divergence from a putative precursor. The three main subfamilies of the family *Herpesviridae* (*Alphaherpesvirinae, Betaherpesvirinae,* and *Gammaherpesvirinae*) are easily distinguished by this method even though the original separation of these families was based on general biological properties. Abbreviations: BHV-1, bovine herpesvirus 1; EBV, Epstein-Barr virus; EHV-1 and -2, equine herpesviruses 1 and 2; HCMV, human cytomegalovirus; HHV-6 and -8, human herpesviruses 6 and 8; HSV-1 and -2, herpes simplex virus types 1 and 2; HVS, herpesvirus saimiri; PRV, pseudorabies virus; VZV, varicella-zoster virus. Data from D. J. McGeoch et al., *J. Mol. Biol.* **247**:443–458, 1995, with permission.

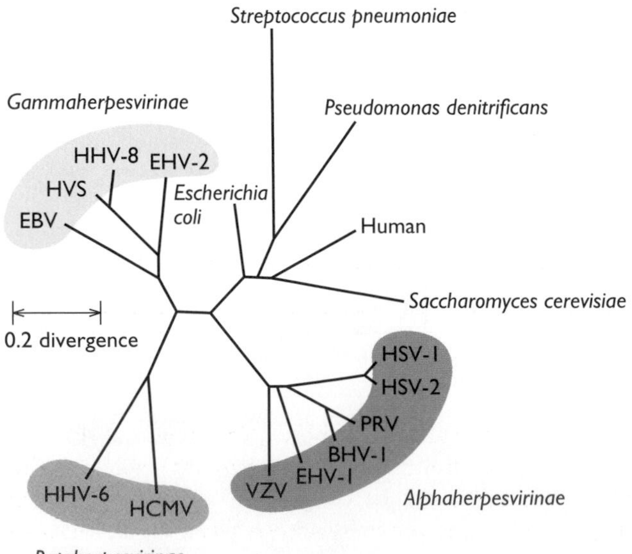

Our current best estimate is that the three major groups of herpesviruses arose approximately 180 million to 220 million years ago. These three subfamilies of viruses must therefore have been in existence before mammals spread over the planet 60 million to 80 million years ago. Perhaps surprisingly, fish, oyster, and amphibian herpesviruses have little or no sequence homology with the *Alpha-, Beta-,* and *Gammaherpesvirinae*. They are related only tenuously to the mammalian and avian herpesviruses by common virion architecture and must represent a very early branch of this ancient family.

Coevolution with a host is a characteristic not only of herpesviruses, but also of small DNA viruses (parvoviruses, polyomaviruses, and papillomaviruses). For these viruses, the evidence for coevolution comes not from comparison of host and viral genes, but rather from finding close association of a given viral DNA sequence with a particular host group. The linkage of host to virus was particularly striking when human papillomavirus types 16 and 18 were compared: the distribution of distinct viral genomes is congruent with the racial and geographic distribution of the human population. Another example of the same phenomenon is provided by JC virus, a ubiquitous human polyomavirus associated with a rare, fatal brain infection of oligodendrocytes. This virus exists as five or more genotypes identified in the United States, Africa, and parts of Europe and Asia. Recent polymerase chain reaction (PCR) analyses of these subtypes indicate that JC virus not only coevolved with humans but did so within specific human subgroups. Probably the most striking finding was that the JC virus of a particular group provides a convenient marker for human migrations from Asia to the Americas in both prehistoric and modern times.

How can virus evolution be linked to specific human populations in a manner akin to vertical transmission of a host gene? We can begin to appreciate this perhaps counterintuitive phenomenon from the unusual biology of human papillomaviruses. Infection of the basal keratinocytes of adult skin leads to viral replication, but only as the cells differentiate. Virus particles are assembled only

Table 20.4 Forty-three genes common to the alpha-, beta-, and gamma-herpesviruses; possible remnants of a common ancestor?[a]

Functional group	HSV-1 gene	Characteristics or function[b]
Control and modulation	UL13	Serine-threonine protein kinase; tegument protein
	UL54	Posttranscriptional regulator of gene expression
DNA replication machinery	UL30	Catalytic subunit of DNA polymerase; complexed with UL42 protein
	UL42	Processivity subunit of DNA polymerase; complexed with UL30 protein
	UL9	Binds to origins of DNA synthesis; helicase; α- and β_2-herpesviruses only
	UL5	Component of DNA helicase-primase complex; helicase
	UL8	Component of DNA helicase-primase complex
	UL52	Component of DNA helicase-primase complex; primase
	UL29	Single-stranded DNA-binding protein
Peripheral enzymes	UL23	Thymidine kinase; α- and γ-herpesviruses only
	UL39	Ribonucleotide reductase; large subunit; probably nonfunctional in β-herpesviruses
	UL40	Ribonucleotide reductase; small subunit (α- and γ-herpesviruses only)
	UL50	dUTPase; probably nonfunctional in β-herpesviruses
	UL2	Uracil-DNA glycosylase
Processing and packaging of DNA	UL12	Deoxyribonuclease; role in DNA maturation
	UL15	Putative ATPase subunit of terminase
	UL28	Putative subunit of terminase
	UL6	Minor capsid protein
	UL17	Tegument protein
	UL25	
	UL32	
	UL33	
Capsid assembly and structure	UL19	Major capsid protein; component of hexons and pentons
	UL18	Component of intercapsomeric triplex
	UL38	Component of intercapsomeric triplex
	UL35	Located on tips of hexons
	UL26	Protease and minor capsid scaffold protein
	UL26.5	Major capsid scaffold protein
Egress of capsids from nucleus	UL31	Nuclear matrix protein; interacts with UL34 protein
	UL34	Inner nuclear membrane protein; interacts with UL31 protein
Tegument	UL11	Myristoylated protein
	UL7	
	UL14	
	UL16	
	UL36	
	UL37	
	UL51	
Surface and membrane	UL27	Glycoprotein B
	UL1	Glycoprotein L; complexed with glycoprotein H
	UL22	Glycoprotein H; complexed with glycoprotein L
	UL10	Glycoprotein M; complexed with glycoprotein N
	UL49A	Glycoprotein N; complexed with glycoprotein M; not glycosylated in some herpesviruses
Unknown	UL24	

[a]Adapted from A. J. Davison, *Vet. Microbiol.* **2272:**1–20, 2002, with permission.
[b]A blank line indicates that a function has not been established.

as cells undergo terminal differentiation near the skin surface. Mothers infect newborns with high efficiency, because of close contact or reactivation of persistent virus during pregnancy or birth. The infection therefore appears to spread vertically, in preference to the more standard horizontal spread between hosts. This mode of transmission is the predominant mechanism for papillomavirus and stands in contrast to that of most acutely infecting viruses, which are spread by aerosols, contaminated water, or food.

RNA Virus Relationships Deduced by Genome Sequence Analysis

The relationships among RNA viruses can also be deduced from sequence analyses, but the high rates at which mutations accumulate impose some difficulties. Moreover, genomes of RNA viruses are often small and carry few if any nonessential genes that might be useful for comparative studies. And, unlike large DNA viruses, viral RNA genomes contain few, if any, genes that might be used to correlate virus and host evolution, i.e., genes in common with a host. Nevertheless, when nucleotide sequences of many (+) and (–) strand RNA viral genomes are compared, blocks of genes that encode proteins with similar functions can be defined. Common coding strategies can also be deduced. These groups are often called "superfamilies" because the similarities suggest a common ancestry (Fig. 20.3 and 20.4). Alternatively, similarities may result from convergent evolution with no implications of shared lineages.

If one examines the sequences of many (–) strand RNA genomes, the first obvious common feature is the limited number of encoded proteins (as few as 4 and not more than 13). These proteins can be arranged in one of three functional classes: core proteins that interact with the RNA genome, glycoproteins that are in the envelope and are required for attachment and penetration of virus particles, and a polymerase required for replication and mRNA synthesis (Fig. 20.3).

Figure 20.3 Genome organization among (–) strand RNA viruses. Gene maps of the *Rhabdoviridae, Paramyxoviridae, Bunyaviridae,* and *Arenaviridae* are aligned to maximize the similarity of gene products. The individual gene segments of the *Orthomyxoviridae* are arranged according to functional similarity to the two other groups of segmented viruses. Virus abbreviations: VSV, vesicular stomatitis virus; IHNV, infectious hematopoietic necrosis virus; RSV, respiratory syncytial virus; SV5, simian virus 5; SSH, snowshoe hare virus; UUK, Uukuniemi virus; LCM, lymphocytic choriomeningitis virus. Le is a nontranslated leader sequence. Gene product abbreviations: N, nucleoprotein; P, phosphoprotein; M (M1, M2), matrix proteins; G (G1, G2), membrane glycoproteins; F, fusion glycoprotein; HN, hemagglutinin/neuraminidase glycoprotein; L, replicase; NA, neuraminidase glycoprotein; HA, hemagglutinin glycoprotein; NS (NV, SH, NSs, and NSm), nonstructural proteins; PB1, PB2, and PA, components of the influenza virus replicase. Within a given genome, the genes are approximately to scale. Blue-outlined genes are those in which multiple proteins are synthesized from different open reading frames. Red-outlined genes are expressed by the ambisense strategy. Figure derived from J. H. Strauss and E. G. Strauss, *Microbiol. Rev.* **58:**491–562, 1994, with permission.

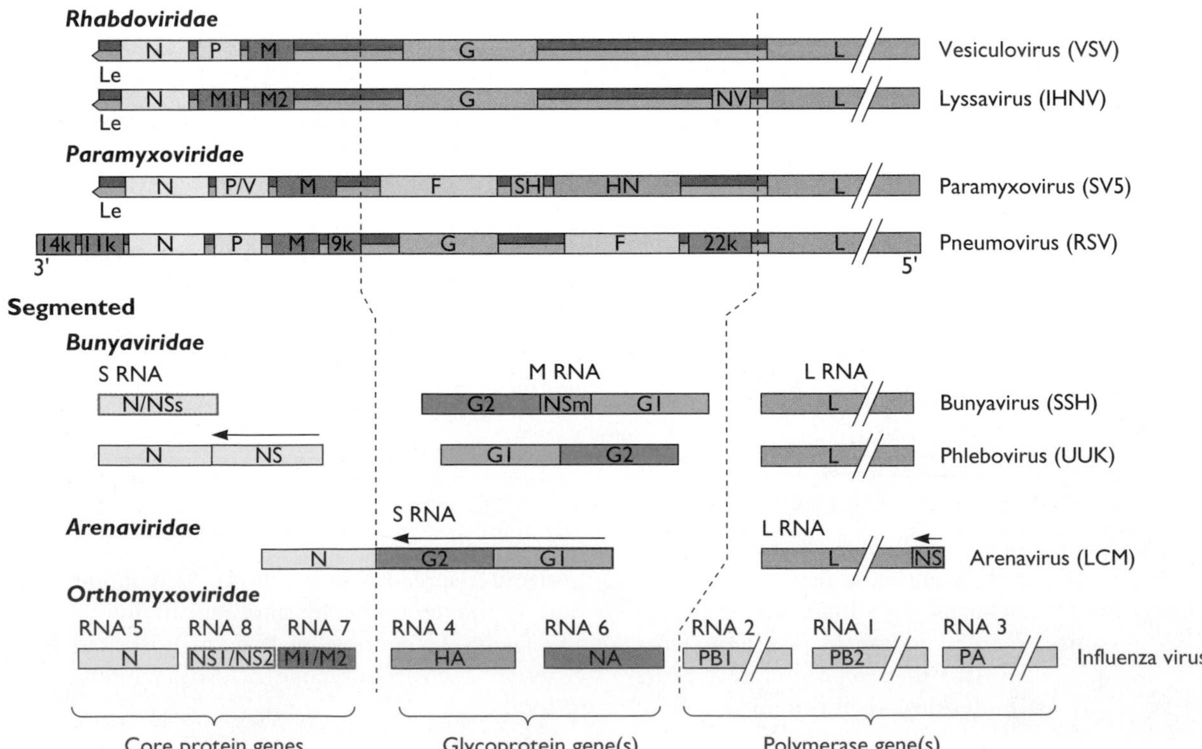

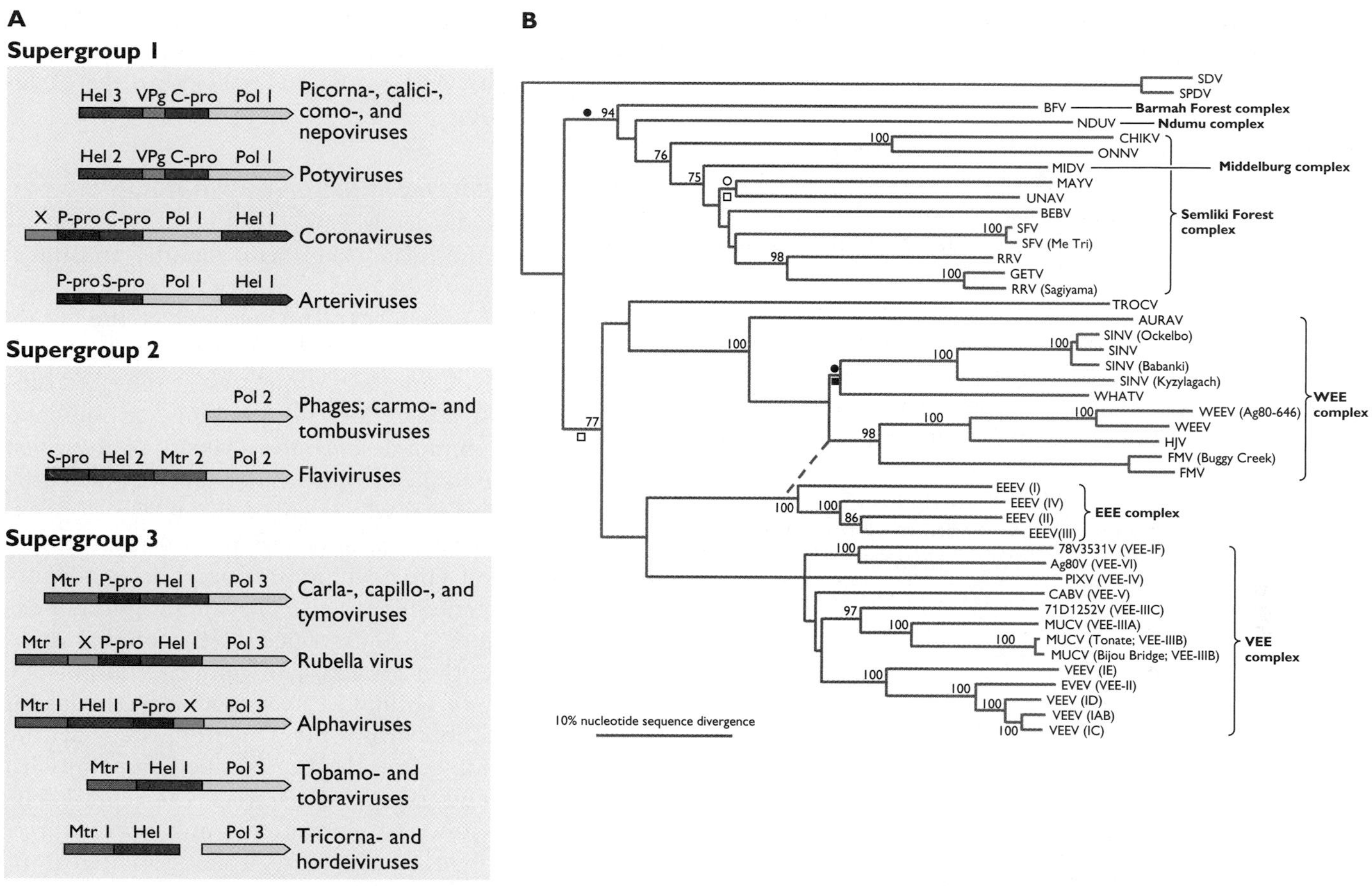

Figure 20.4 RNA virus genomes and evolution. (A) Organization of (+) strand RNA genomes. The genomes of (+) strand RNA viruses comprise several genes for replicative functions that have been mixed and matched in selected combinations. These functions include a helicase (purple), a genome-linked protein (orange), a chymotrypsinlike protease (red), a polymerase (yellow), a papainlike protease (brown), a methyltransferase (dark blue), and a region of unknown function, X (green). Some of these gene functions are encoded in more than one sequence (such as Pol 1, Pol 2, and Pol 3) as defined by amino acid sequence similarity. In this figure, the genes are not shown to scale and the structural proteins have been omitted for clarity. Derived from J. H. Strauss and E. G. Strauss, *Microbiol. Rev.* **58:**491–562, 1994, with permission. (B) Dendrogram depicting a proposed phylogenetic tree of modern alphaviruses based on nucleic acid sequence comparisons of the E1 gene. The modern lineages are shown on the right. The western equine encephalitis (WEE) virus subgroup arose by a recombination event between a member of the eastern equine encephalitis (EEE) virus lineage and the Sindbis virus lineage (dashed line). The bar indicates 10% nucleotide divergence. The open circle adjacent to a branch indicates a hypothetical Old to New World introduction, and the closed circle indicates New to Old World introduction, assuming a New World origin. The open square indicates Old to New World introduction, and the closed square indicates New to Old World introduction, assuming an Old World origin of the non-fish alphavirus clade. Abbreviations: SDV, sleeping disease virus (of trout); SPDV, salmon pancreas disease virus; BFV, Barmah Forest virus; NDUV, Ndumu virus; CHIKV, chikungunya virus; ONNV, O'nyong-nyong virus; MIDV, Middelburg virus; MAYV, Mayaro virus; UNAV, Una virus; BEBV, Bebaru virus; SFV, Semliki Forest virus; RRV, Ross River virus; GETV, Getah virus; TROCV, Trocara virus; AURAV, Aura virus; SINV, Sindbis virus; WHATV, Whataroa virus; WEEV, western equine encephalitis virus; HJV, Highlands virus; FMV, Fort Morgan virus; EEEV, eastern equine encephalitis virus; VEE and VEEV, Venezuelan equine encephalitis virus; PIXV, Pixuna virus; CABV, Cabassou virus; EVEV, Everglades virus. Redrawn from A. M. Powers, A. C. Brault, Y. Shirako, E. G. Strauss, W. Kang, J. H. Strauss, and S. C. Weaver, *J. Virol.* **75:**10118–10131, 2001, with permission.

The (+) strand RNA viruses (excluding the retroviruses) are the largest and most diverse subdivision of viruses: the genome sequences of more than 100 viruses representing at least 30 distinct groups are available for analysis. The number of proteins encoded by (+) strand RNA viruses ranges from 3 to more than 12, and, like the (–) RNA virus proteins, they can be divided into three groups by function: those required for RNA replication, encapsidation, and accessory functions. This comparison has resulted in the identification of three virus supergroups (Fig. 20.4A).

A unifying feature is that the RNA polymerase gene appears to be the most highly conserved, implying that it arose once in the evolution of these viruses. As each of the supergroups contains members that infect a broad variety of animals and plants, the primordial RNA polymerase gene might have been provided by an ancestor present before the separation of kingdoms. Alternatively, the ancestral (+) strand virus could have radiated horizontally among plants, animals, and bacteria.

It is instructive to consider the postulated evolution of supergroup 3, which contains the alphaviruses. Earlier work had indicated that these viruses arose less than 5,000 years ago, but certain assumptions were made about the rate of RNA genome divergence and the constancy of virus evolution in diverse hosts and environments. More recent analyses comparing protein and DNA sequences indicate that the assumption of uniform nucleotide substitution rates was invalid; nucleotide changes are far from uniform. With new data from E1 protein and gene sequences, a phylogenetic tree can be constructed, but no time estimates for the alphavirus progenitor are possible (Fig. 20.4B). The data do provide some interesting hypotheses about radiation of the species. The Old World viruses (e.g., the Semliki Forest complex) and the New World viruses (e.g., Venezuelan, eastern, and western equine encephalitis viruses) are clearly defined. The mosquito-borne alphaviruses could have arisen either in the Old or the New World, but at least two transoceanic introductions are required to account for their current distribution. Another conclusion is that recombination established the western equine encephalitis viruses as a new lineage. These viruses arose after a recombination event joined the E1 and E2 genes from a member of the eastern equine encephalitis virus lineage with the remaining genes from a member of the Sindbis virus lineage.

Japanese encephalitis virus, a flavivirus, is the most important cause of epidemic encephalitis in the world. The virus was first isolated in 1935 and has subsequently been found across most of Asia. The origins of the virus are uncertain, but two ideas are entertained. Phylogenetic comparisons with other flaviviruses are consistent with the hypothesis that it evolved as recently as a few centuries ago. Sequence analysis of all known Japanese encephalitis virus isolates indicates that the virus originated in the Indonesia-Malaysia region. Evolutionary biologists have long recognized that this area of the world is a unique environment with remarkable examples of divergent evolution of plants and animals. The number of insect species is particularly diverse, and it is likely that the diversity of arthropod-borne viruses also will be extensive. The remarkable variation among Murray Valley encephalitis virus isolates in New Guinea compared with Australia and the emergence of Nipah and Hendra viruses indicate that tropical Southeast Asia has a virus ecology worthy of detailed investigation.

Predictive Powers of Sequence Analyses

The phylogenetic dendrograms relating nucleic acid sequences depict the relationships as if founder and intermediary sequences were on a trajectory to the present sequences. A fair question is, can we predict the future trajectory of the dendrogram? What will comprise future branches of a given lineage? At the present time, we cannot answer these straightforward questions for one obvious reason: we cannot describe the diversity of any given virus population in the ecosystem, nor can we predict the selective pressures that will be imposed.

Sequence comparisons alone are not sufficient for an understanding of virus evolution. Comparison of function in diverse genomes can also provide insight. For example, in Chapter 3 we described seven genome replication strategies that are likely to comprise all possible solutions. We also outlined 12 expression strategies for protein production from these genomes. That the provenance of all viruses can be described by a finite list of replication and expression strategies is remarkable. Understanding how protein function and gene expression strategies evolved represents a new research frontier for which we have few data to guide us. Some initial studies are provocative. For example, RNA virus replication complexes described for different families may have fundamental similarities (Box 20.7). Localization of genomes to membrane sites or to assembling capsids leads to precise temporal and spatial organization of viral compartments important for gene expression, replication, and particle assembly. Are these overtly similar mechanisms products of convergent evolution and coincidence, or do they imply a common evolutionary origin for this abundant group of viral genomes?

The Protovirus Theory

The origin of retroviruses may center on the unusual enzyme reverse transcriptase, an idea embodied in the protovirus theory elaborated by Howard Temin, who shared the Nobel Prize for the discovery of reverse transcriptase. This theory proposed that a cellular reverse transcriptase-like enzyme copied segments of cellular RNA into DNAs that were then inserted into the genome to form retroelements. These DNA segments in turn acquired more complex sequences, including integrase and RNase H domains, regulatory sequences, and structural genes en-

BOX 20.7

Parallels in replication of (+) strand and double-stranded RNA genomes

The mRNA templates of viruses with double-stranded RNA (dsRNA) and (+) strand RNA genomes (including retroviruses) are sequestered in a multisubunit protein core that directs synthesis of the RNA or DNA intermediate from which more viral mRNA is made. Similarities in how the mRNA template and core proteins are assembled suggest that all three virus groups may share a common evolutionary history despite a complete lack of genome sequence homology. It is possible that this ancient replicative strategy provides RNA genomes with increased template specificity and retention of negative-strand products in the core or vesicle for template use. In addition, by sequestering dsRNA in vesicles or capsids, host defenses such as RNA interference, dsRNA-activated protein kinase, and RNase L are avoided.

Schwartz, M., J. Chen, J. Janda, M. Sullivan, J. den Boon, and P. Ahlquist. 2002. A positive-strand RNA virus replication complex parallels form and function of retrovirus capsids. *Mol. Cell* **9:**505–514.

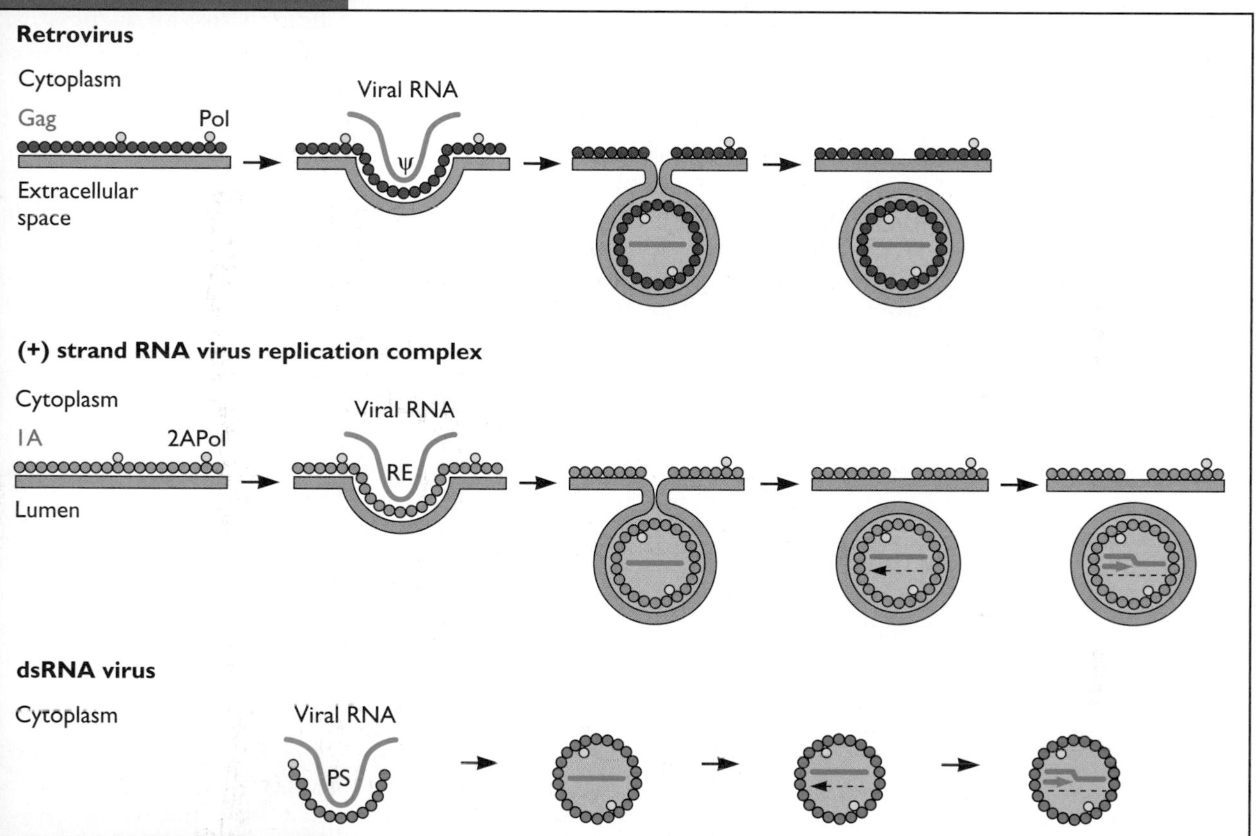

Similarities in replication and budding reactions. General parallels in assembly of mRNA- and polymerase-containing replication complexes for retroviruses, (+) strand RNA viruses, and dsRNA viruses. In the case of retroviruses, specific sequences on the RNA genome bind to Gag proteins that define the budding site. Gag proteins encapsidate viral RNA and reverse transcriptase with plasma membrane. Similarly, (+) strand RNA genomes are replicated on intracellular membrane vesicles that form in response to a viral protein that binds to membranes. Polymerase complexes and viral RNA templates are recruited to these vesicles. Replication of dsRNA genomes occurs in compartments formed by assembling capsid proteins that sequester single-stranded genome templates via specific protein-RNA interactions. Blue circles are Gag proteins (retrovirus), 1A protein for (+) strand RNA virus, and inner capsid protein for dsRNA virus. The polymerase protein (Pol or 2APol) interacts with Gag or 1A, respectively. The polymerase is part of the assembling capsid of dsRNA viruses. The RNA genome is indicated in green, and the binding sites for interaction with Gag, 1A, or capsid protein are ψ, RE, or PS, as indicated for each virus. The final reaction for retroviruses is the release of an enveloped particle with the genome and polymerase; for some (+) strand RNA viruses, the product is not an enveloped virion but rather an involuted vesicle or the surface of a membrane vesicle where mRNA synthesis, (–) strand genome template synthesis, and (+) strand genome synthesis occurs; for dsRNA viruses, the product is a capsid compartment within which mRNA synthesis and complementary strand genome replication occur.

coding capsid and envelope proteins (Fig. 20.5). This theory predicts that remnants of this process might still exist in the genomes of mammals and other species. Indeed, many of the predicted intermediates are found in abundance, including pseudogenes, retroposons, retrotransposons, and a variety of endogenous retroviruses (see Box 7.2). In humans, such endogenous retroviruses are surprisingly abundant, comprising several percent of the human genome, and may represent footprints of ancient germ cell infections.

Contemporary Virus Evolution

Even with sophisticated genome sequence analysis and powerful computers, it remains difficult, if not impossible, to answer the question of the origin of viruses because we lack physical evidence from ancient times. However, as discussed below, it may be possible to define modern precursors of new viruses by studying emerging viruses. In the final analysis, it is probably safe to assume that every living thing is infected with viruses. Studying the processes that result in the emergence of new contemporary viruses seems likely to prove more helpful in divining the present and future evolution of viruses than explaining evolution in their distant past.

If we concede that viruses are not likely to arise de novo, then modern and future viruses must arise from progenitors that already exist. Hence, even human immunodeficiency virus, which appeared within the past few decades, has many relatives in nature and probably descended from one, or a combination, of them. Therefore, the sources of new viruses are strictly limited to two possibilities: a mutant virus already existing in an infected host can be selected, or a virus can enter a naive population from an entirely different infected species. These two simple possibilities notwithstanding, the interactions of host and virus required to establish a stable relationship are remarkably complex.

Evolution Is Both Constrained and Driven by the Fundamental Properties of Viruses

We can recognize a herpesvirus, an adenovirus, a retrovirus, or an influenza virus genome by sequence analysis, no matter how much replication, mutation, and selection have occurred. This fact is emphasized by sequence analyses of human immunodeficiency virus strains isolated from patients around the world. As much as 10% of the total viral genome can vary from isolate to isolate, yet despite this, viral isolates fall into consistent subgroups called **clades.** Each clade differs from the others in amino acid sequence by at least 20% for Env proteins and 15% for Gag proteins. Differences within a clade can be as much as 8 to 10%, emphasizing the rather arbitrary delineation. As discussed in Chapter 18, these viruses have replicated in widely dispersed geographic locations and have very different histories, yet each sequence is clearly recognizable as the human immunodeficiency virus genome.

The important message is that viral populations frequently maintain quite stable master or consensus sequences, despite opportunities for extreme variation (see also Fig. 20.1). Diversity exists (indeed is necessary for virus survival), but the consensus sequence remains, despite many years of selection and growth. How is stability

Figure 20.5 The action of a primordial reverse transcriptase may drive evolution of retroelements. A speculative scheme shows the evolution of various retroelements through the action of a primordial reverse transcriptase enzyme. Dashed arrows depict the emergence of genetic elements by reverse transcription; the complementary DNA copies are shown integrated into the host genome (light blue). Solid arrows show the acquisition of new elements by recombination. cDNA, double-stranded DNA copied by reverse transcriptase; ORF, open reading frame; *gag*, capsid protein gene; *pol*, polymerase gene coding for reverse transcriptase, RNase H, integrase, and other enzymes; *env*, envelope protein gene; LTR, long terminal repeat; P, promoter; AAAAA, poly(A) tail; ψ, packaging signal; *, polypurine tract. Direct repeats of cellular DNA are indicated in gray. Adapted from R. J. Loewer et al., *Proc. Natl. Acad. Sci. USA* **93:**5177–5184, 1996, with permission.

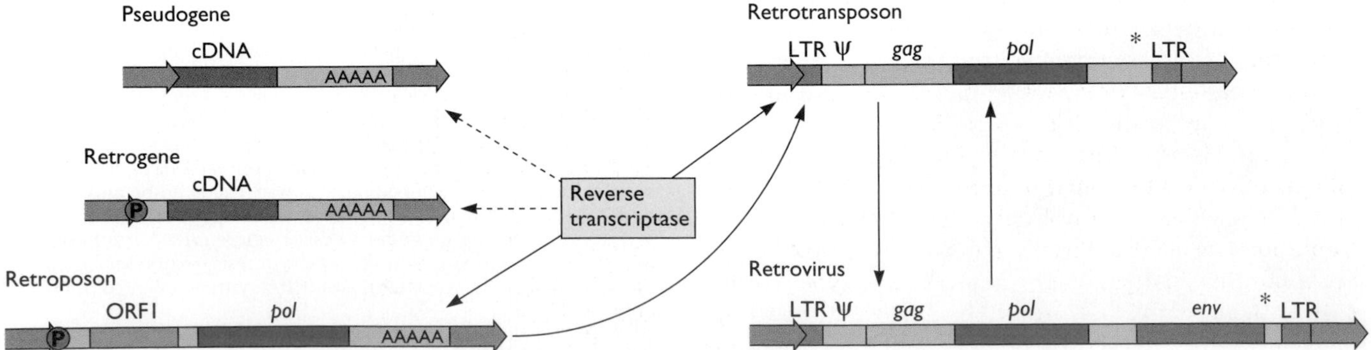

maintained in the face of mutation, recombination, and selection? One answer is that despite amazing diversity, all viruses share fundamental characteristics that define and constrain them. Those that can function within the constraints survive. Comprehending the evolution of viruses requires an understanding of these shared properties.

Fundamental Properties of Viruses Constrain Their Evolution

The very characteristics that enable us to define and classify viruses are the primary barriers to major genetic change. Extreme alterations in the viral consensus genome obviously do not survive selection. What are the constraints that define and confine virus evolution? We know that all viruses contain their nucleic acid genomes within particles assembled following limited symmetry principles. Therefore, the nature of the particle may constrain the quantity of nucleic acid sequence that can be packaged. For example, icosahedral capsids have a defined internal space that fixes the size of packaged nucleic acid. Once an icosahedral capsid is encoded, genome size is essentially fixed. Another constraint is that all genomes are based on nucleic acids: solutions to replication or the decoding of viral information are limited (see Chapter 3). Every step in viral replication requires direct and indirect interactions with host cell machinery. Any change in a viral component without a compensating change in the interacting host component may have a direct effect on the virus. Similarly, inappropriate synthesis, concentration, or location of a viral component is likely to be detrimental. Finally, all viral genomes encode products capable of modulating a broad spectrum of host defenses, including physical barriers to viral access, and the vertebrate immune system. A mutant that is too efficient in bypassing host defenses will kill its host and suffer the same fate as one that does not replicate efficiently enough: it will be eliminated.

Selection for Survival Inside a Host May Impose an Unexpected Constraint on Evolution

A thought-provoking finding from studies of human immunodeficiency virus infections indicates that additional constraints on virus evolution must be considered. Virions that initiate infections typically are macrophage tropic and engage the CCr5 chemokine receptor. At the end stage of disease, the infected individual is producing billions of virions that survive in the face of host defenses and antiviral therapy. Invariably, these virions are T cell tropic and engage the CXCr4 chemokine receptor. The diversity in this final population is a result of evolution **inside** a single individual. Amazingly, when virions from end-stage disease infect a new host, the first replicating viral genomes that can be detected are macrophage tropic, and engage the CCr5 receptor. The progeny genomes have passed through a bottleneck, and only a small subset of the diversity is passed on. The processes that select these variants from the previous T-cell-tropic population are not well understood. However, one conclusion is clear: the virions that ultimately devastate the immune system after years of replication and selection within a host are not the most fit for infection of new hosts.

Evolution of New Viruses

Even in the seemingly constrained context of a given virus, the number of all possible viable mutants is astronomical, if not inconceivable (Box 20.8). The number of possible mutants is so large, in fact, that all the possibilities can never be tested in nature. Sequence comparisons of several viral RNA genomes have demonstrated that well over half of all nucleotides can accommodate mutations. This property means that, for a 10-kb viral RNA genome, more than $4^{5,000}$ sequence permutations define all possible mutants. If one considers deletions, recombination, and reassortment, the numbers become even larger. Considering that there are roughly 4^{135} atoms in the visible universe, this is a large number indeed. Even with the high rates of replication and mutation characteristic of viruses, we can be sure that only a minuscule fraction of

BOX 20.8

The world's supply of human immunodeficiency virus genomes provides remarkable opportunity for selection

Tens of millions of humans are infected by human immunodeficiency virus. Before the end stage of disease, each infected individual produces billions of viral genomes per day. As a result, more than 10^{16} genomes are produced each day on the planet. Almost every genome has a mutation, and every infected human harbors viral genomes with multiple changes resulting from recombination and selection. Practically speaking, these large numbers provide an amazing pool of diversity. For example, mutants resistant to **every combination** of anti-reverse transcriptase and protease drugs in use, or in the pipeline, arise thousands of times each day, just by chance. See http://www.microbiology.wustl.edu/dept/fac/huang/ccas/mut/mut.html.

all possible viral genomes have arisen since life began: even the most efficiently replicating virus will fail to spawn all the possible permutations for a run through the gantlet of selection in nature. The conclusion is inescapable: virus evolution is relentless, new mutants will always arise, and the possibilities literally are unimaginable. Those who seek to predict the trajectory of virus evolution face significant challenges, as mutation rates are probabilistic and viral quasispecies are indeterminate. Nevertheless, we can be confident that all future viruses will arise from those now extant: they will be mutants, recombinants, and reassortants.

Unusual Infectious Agents

In Chapter 3, we discussed the functions specified by small and large viral genomes. A more fundamental question of size is what is the minimal genome size necessary to sustain an infectious agent? Taken to the extreme, the question becomes, could an infectious agent exist without any genome at all? The so-called subviral agents called viroids, satellites, and prions provide answers to these questions. The adjective subviral was coined, in part, because these agents did not fit into the standard taxonomy schemes for viruses. These widespread infectious oddities are thought provoking (Table 20.5). Perhaps more important, infection by some is deadly.

Viroids

Potato spindle tuber viroid was discovered in 1971 and was the prototype for the smallest known nucleic acid-based agents of infectious disease, the **viroids**. Viroids are unencapsidated, small, circular, single-stranded RNA molecules that replicate autonomously when mechanically introduced into host plants. Infection by some viroids causes economically important diseases of crop plants, while others appear to be benign, despite widespread presence in the plant world. Two examples of economically important viroids are coconut cadang-cadang viroid (the agent of a lethal infection of coconut palms) and apple scar skin viroid (the agent of an infection resulting in visually unappealing apples). Viroids have been classified in the families *Avsunviroidae* (group A viroids) and *Pospiviroidae* (group B viroids). The pathogenesis of some viroids has been correlated with activation of Pkr or p68 kinase, presumably by the viroid double-stranded RNA.

The circular single-stranded RNA molecules of all viroids range from 246 to 399 nucleotides in length and are extensively base paired, so that the RNA appears as a 50-nm rod in the electron microscope. There is no evidence that viroids encode proteins or mRNA. Unlike viruses that are parasites of host translation machinery, viroids are parasites of host transcription proteins: they depend on RNA polymerase II for replication. Several well-studied group A viroid RNA molecules are functional ribozymes, and this activity is essential for replication. Group B viroid RNA molecules have no detectable ribozyme activity, and instead require a host ribonuclease activity for replication. In general, replication occurs in the nucleus, and most viroids can be found in the nucleoli of infected cells.

Our current understanding is that the disease-causing viroids were transferred by chance from wild plants used for breeding modern crops. The widespread prevalence of viroids can be traced to use of genetically identical plants (monoculture), worldwide distribution of breeding lines, and mechanical transmission by contaminated farm machinery. As a consequence, these unusual pathogens now occupy niches around the planet never before available to them. As might be expected for an RNA replicon, viroid replication introduces mutations into the tiny genome at high frequency. In addition, the viroid genome encodes no gene products and infection requires no host receptors, allowing evolution to occur with amazing speed.

Satellites

Satellites are small, single-stranded RNA molecules that lack genes required for their replication, but do replicate in the presence of another virus (the **helper virus**). Unlike viroids, satellite genomes encode one or two proteins. Typical satellite genomes are 500 to 2,000 nucleotides of single-stranded RNA. We recognize two classes of satellites: viruses and nucleic acids. **Satellite viruses** are distinct particles that were discovered in preparations of their

Table 20.5 Viroids and satellites

Property	Viroids		Satellites
	Group A	Group B	
Requires coinfection with helper virus	No	No	Yes
Encodes protein	No	No	Yes
Replication	By host RNA polymerase II and: Viroid ribozyme	By host RNA polymerase II and: Host RNase	By helper virus replication proteins

helper viruses. These defective particles contain nucleic acid genomes that encode a structural protein that encapsidates the satellite genome. Satellite viruses are **not** defective viruses derived from the helper virus: their genomes have no homology to the helper. **Satellite nucleic acids** (sometimes called virusoids) are packaged by a capsid protein encoded in the helper virus genome.

Most satellites are associated with plant viruses, but a well-known example of a human satellite virus is hepatitis delta satellite virus with its helper, hepatitis B virus. In plants, satellites and satellite viruses cause disease symptoms not seen with the helper virus alone. However, despite being associated with human hepatitis, the contribution of hepatitis delta satellite virus to human disease remains controversial. Hepatitis delta satellite virus appears to be a hybrid between a viroid and a satellite. The genome is 1.7 kb of circular single-stranded RNA that folds upon itself in a tight rodlike structure (70% base paired). The RNA molecule is also a ribozyme required for replication. These properties resemble those of viroid genomes. On the other hand, the genome encodes a protein (delta antigen) that encapsidates the genome and therefore resembles satellite nucleic acids. Two functionally distinct forms of the delta protein are made as a result of RNA editing (Chapter 10). The hepatitis D satellite virion comprises the satellite nucleocapsid packaged within an envelope containing the surface antigen of the helper, hepatitis B virus.

The ecosystem abounds with RNA replicons that share structural and functional characteristics of satellites and viroids. Similarities with introns, transposons, and RNAs found in the signal recognition particle have been noted, but the origin of satellites and viroids remains an enigma. Moreover, we have only a modest understanding of the mechanisms by which these minimal pathogens cause disease. The striking speed of their divergence in the face of selection and their ease of manipulation make them useful models to study the evolution of infection and pathogenesis.

Prions and Transmissible Spongiform Encephalopathies

Can infectious agents exist without genomes? The question challenges our definitions of what constitutes an infectious agent, but the answer seems to be yes. The question arose with the discovery and characterization of infectious agents mediating one group of rare, slow diseases, now called **transmissible spongiform encephalopathies (TSEs)**. These diseases are rare, but always fatal, neurodegenerative disorders afflicting humans and other mammals (Table 20.6). Each year, thousands of individuals worldwide are diagnosed with spongiform encephalopathies, and about 1% of these cases arise by infection. There is increasing concern that this low rate of infection may increase, because some TSEs spread when animals ingest infected tissues. At the end of 2002, 120 individuals had succumbed to variant Creutzfeldt-Jakob disease, an affliction ascribed to consumption of meat from animals with bovine spongiform encephalopathy. Some investigators contend that viruses or virus-like particles cause TSEs, but most now argue that infectious proteins called **prions** cause them.

Table 20.6 Transmissible spongiform encephalopathies[a]

TSE diseases of animals
Bovine spongiform encephalopathy (BSE) ("mad cow disease")
Chronic wasting disease (CWD) (deer, elk)
Exotic ungulate encephalopathy (EUE) (nyala and greater kudu)
Feline spongiform encephalopathy (FSE) (domestic and great cats)
Scrapie in sheep and goats
Transmissible mink encephalopathy (TME)
TSE diseases of humans
Creutzfeldt-Jakob disease (CJD)
Fatal familial insomnia (FFI)
Gerstmann-Sträussler syndrome (GSS)
Kuru
Variant CJD (vCJD)

[a]From S. B. Prusiner (ed.), *Semin. Virol.* **7**:157–223, 1996.

Human TSEs

Initially, the human spongiform encephalopathies were not thought to be transmissible, but studies of the human disease **kuru,** a fatal encephalopathy found in the Fore people of New Guinea by Carleton Gajdusek and colleagues, proved otherwise. Kuru spread among women and children as a result of ritual cannibalism of the brains of deceased relatives. When cannibalism stopped, so did kuru. Insightful comparison of the pathology of scrapie-infected (see below) and kuru-infected brains by William Hadlow led to experiments by others that demonstrated the disease can be transmitted from humans to chimpanzees and other primates. Since 1957, the spongiform encephalopathies of animals and humans have been considered to have common features. In humans, the diseases fall into three classes, **infectious, familial,** and **sporadic,** distinguished by how the disease is initially acquired (Box 20.9).

Hallmarks of TSE Pathogenesis

For most TSEs, the presence of an infectious agent can be detected definitively only by injection of organ homogenates into susceptible recipient species. Clinical signs

BOX 20.9

Characteristics of the human spongiform encephalopathies

An **infectious** (or **transmissible**) **spongiform encephalopathy** is exemplified by kuru and iatrogenic spread of disease to healthy individuals by transplantation of infected corneas, the use of purified hormones, or transfusion with blood from patients with **Creutzfeldt-Jakob disease (CJD).** Recently, the epidemic spread of bovine spongiform encephalopathy (BSE, or mad cow disease) among cattle in Britain can be ascribed to the practice of feeding processed animal by-products to cattle as a protein supplement. Similarly, the new human disease, variant CJD, arose after consumption of beef from diseased animals.

Sporadic CJD is a disease affecting fewer than a million individuals worldwide, usually late in life (from age 50 to 70). About 65% of all spongiform encephalopathies are sporadic CJD. As the name indicates, the disease appears with no warning or epidemiological indications. Kuru may have originally been established in the small population of Fore people in New Guinea by eating the brain of an individual with sporadic CJD.

Familial spongiform encephalopathy is associated with an autosomal dominant mutation in the *prnp* gene. Familial CJD, for example, in contrast to sporadic CJD, is an inherited disease.

The important principle is that diseases of **all three classes** usually can be transmitted experimentally or naturally to primates by inoculation or ingestion of diseased tissue.

of infection commonly include cerebellar ataxia (defective motion or gait) and dementia, with death occurring after months or years. The infectious agent first accumulates in the lymphoreticular and secretory organs and then spreads to the nervous system. In model systems, spread of the disease from site of inoculation to other organs and the brain appears to require dendritic and B cells. The disease agent then appears to invade the peripheral nervous system and spreads from there to the spinal cord and brain. Once in the central nervous system, the characteristic pathology includes severe astrocytosis, vacuolization (hence the term "spongiform"), and loss of neurons. Occasionally, dense fibrils or aggregates (sometimes called plaques) can be detected in brain tissue at autopsy. There is little indication of inflammatory, antibody, or cellular immune responses. The time course, degree, and site of cytopathology within the central nervous system are dependent upon the particular TSE agent and the genetic makeup of the host.

Identification of the First Agent Causing TSE

One of the best-studied TSE diseases is **scrapie,** so called because infected sheep tend to scrape their bodies on fences so much that they rub themselves raw. A second characteristic symptom, tremors caused by skin rubbing over the flanks, led to the French name for the disease, *tremblant du mouton.* Soon after, motor disturbances appear as a wavering gait, staring eyes, and paralysis of the hindquarters. There is no fever, but infected sheep lose weight and die, usually within 4 to 6 weeks of the first appearance of symptoms. Scrapie has been recognized as a disease of European sheep for more than 200 years. It is endemic in some countries, e.g., the United Kingdom, where it affects 0.5 to 1% of the sheep population per year.

Sheep farmers discovered that animals from diseased herds could pass the affliction to a scrapie-free herd, implicating an infectious agent. In 1939, infectivity from extracts of scrapie-affected sheep brains was shown to pass through filters with pores small enough to retain all but viruses. In the 1970s, ultracentrifugation studies indicated that the agent was heterodisperse in size and density. Even to this day, purification of the infectious agent to homogeneity has not been achieved.

Physical Nature of the Scrapie Agent

A major point of contention is the physical nature of the infectious agent. Studies as early as 1966 showed that scrapie infectivity was considerably more resistant than that of most viruses to ultraviolet (UV) and ionizing radiation. For example, the scrapie agent is 200-fold more resistant to UV irradiation than polyomavirus and 40-fold more resistant than a mouse retrovirus. Other TSE agents exhibit similar UV resistance. The infectivity of scrapie agent is also more resistant to chemicals, such as 3.7% formaldehyde, and autoclaving routinely used to inactivate viruses. It is possible to reduce infectivity by 90 to 95% after several hours of such treatments. It has been argued that the inability to achieve complete inactivation implies novel properties of the agent. However, the resistant fraction provides no information about the physical properties of the fraction that was sensitive. On the basis of relative resistance to UV irradiation, some investigators have argued that TSE agents are viruses well shielded from irradiation. Others have claimed that TSE agents have little or no nucleic acid at all. Suffice it to say that TSE agents are not typical infectious agents.

Prions and the prnp *Gene*

The unconventional physical attributes and slow infection pattern have prompted many to argue that TSE agents are not viruses at all. For example, in 1967 the mathematician J. S. Griffith made three suggestions as to how scrapie may be mediated by a host protein, not by a nucleic acid-carrying virus. His thoughts were among the first of the "protein-only" hypotheses to explain TSE.

A significant breakthrough occurred in 1981 when characteristic protein complexes were visualized in infected brains. These complexes could be concentrated by centrifugation and remained infectious. Stanley B. Prusiner and colleagues developed an improved bioassay, as well as a fractionation procedure that allowed the isolation of a protein with unusual properties from scrapie-infected tissue. This protein is insoluble and relatively resistant to proteases. Sequence analysis led to the cloning of a gene called *prnp*, which is highly conserved in the genomes of many animals, including humans. Expression of the *prnp* gene is now known to be essential for the pathogenesis of common TSEs.

The *prnp* gene encodes a 35-kDa membrane-associated glycoprotein (PrP) found in many neurons. Nevertheless, mice lacking the *prnp* gene develop normally and have few obvious defects that can be directly attributed to lack of the gene. At least 18 specific mutations in the human *prnp* gene are associated with familial TSE diseases. Furthermore, specific *prnp* mutations appear to be associated with susceptibility to different strains of TSE (see below).

Prusiner named the scrapie infectious agent a **prion** and proposed that an altered form of the PrP protein causes the fatal encephalopathy characteristic of scrapie disease. Prusiner's protein-only hypothesis holds that the essential pathogenic component is the host-encoded PrP protein with an altered conformation, called PrPsc (PrP-scrapie; also called PrPres for protease-resistant form). Furthermore, in the simplest case, PrPsc is proposed to have the property of converting normal PrP protein into more copies of the pathogenic form. An important finding in this regard is that mice lacking both copies of their *prnp* gene are resistant to infection. In recognition of his work on this problem, Prusiner was awarded the Nobel Prize in physiology or medicine in 1997.

The conformational conversion of PrP to PrPsc can be described in two models: the **refolding model** or the **seeding model.** In the former, spontaneous conversion to PrPsc is virtually undetectable. However, interaction of PrP with **preformed** PrPsc facilitates refolding of PrP to PrPsc. PrPsc can be introduced from exogenous sources or by rare mutations that lower the activation energy of conversion. In the seeding model, PrP and PrPsc are in equilibrium, with the formation of PrP highly favored. PrPsc can be stabilized only when it binds to an aggregate (or seed) of PrPsc. Aggregates are rare, but once formed, PrPsc binds avidly and the aggregate grows rapidly by fragmentation and addition of new PrPsc.

Strains of Scrapie Prion

As a result of many serial infections of mice and hamsters with infected sheep brain homogenates, investigators have derived distinct strains of scrapie prion. Strains are distinguished by length of incubation time before the appearance of symptoms, brain pathologies, relative abundance of various glycoforms of PrP protein, and electrophoretic profiles of protease-resistant PrPsc. Some strains also have a different host range. For example, mouse-adapted scrapie prions cannot propagate in hamsters, but hamster-adapted scrapie prions can propagate in mice. A single amino acid substitution in the hamster protein enables it to be converted efficiently by mouse PrPsc into hamster PrPsc. Therefore, the barrier to interspecies transmission is in the sequence of the PrP protein. Bovine spongiform encephalopathy prions have an unusually broad host range, infecting a number of meat-eating animals, including domestic and wild cats, and humans. A striking finding is that different scrapie strains can be propagated in the same inbred line of mice, yet maintain their original phenotypes. Stable inheritance suggests to some that a nucleic acid must be an essential component and has been used as support for a viral etiology of TSE disease. The protein-only hypothesis explains the existence of strain variation by postulating that each strain represents a unique conformation of PrPsc. Each of these distinctive pathogenic conformations is then postulated to convert the normal PrP protein into a conformational image of itself.

While most investigators find the protein-only hypothesis compelling, a few maintain that the PrP protein is not the only participant in the TSE story and that an unrecognized virus causes the disease. In their view, the PrP protein would be a cofactor, a receptor, or a protein that determines susceptibility to infection. The experiment that would certainly resolve the basic controversy would be to produce the putative pathogenic conformation of PrP in vitro and demonstrate that it causes a specific TSE.

The existence of TSE as a human genetic disease and the presence of prions in the food supply are causes for concern, for we have only a modest understanding of either the molecular nature of the infectious agent or the processes of pathogenesis. Efforts are currently focused on basic research aimed at protection of the food supply and finding targets for therapeutic intervention. Remarkable progress is being made. For example, sensitive diagnostic tests for prions are now available and are being used to screen blood and tissues. An unexpected breakthrough came when researchers discovered that the antimalarial drug quinacrine blocked accumulation of infectious prions in cultured cells. The mechanism is unknown, but this lead and others are being pursued aggressively, and limited human trials of quinacrine in patients with advanced Creutzfeldt-Jakob disease are in progress. Another promising discovery is that in mice, monoclonal antibodies specific for PrP inhibit scrapie prion replication and delay development of prion disease. The hope is that similar immunotherapy strategies will benefit humans exposed to prions.

Emerging Viruses

We define an **emerging virus** as the causative agent of a new or hitherto unrecognized infection in a population. Occasionally, emerging infections are manifestations of expanded host range with an increase in disease that was not previously obvious. While the term became part of the popular press in the 1990s (usually with dire implications), emerging viruses are not new to virologists: they have long been recognized as an important manifestation of virus evolution. Some examples are given in Table 20.7. Each represents an example of contemporary virus evolution that, by definition, is the result of host-virus interactions on a large scale.

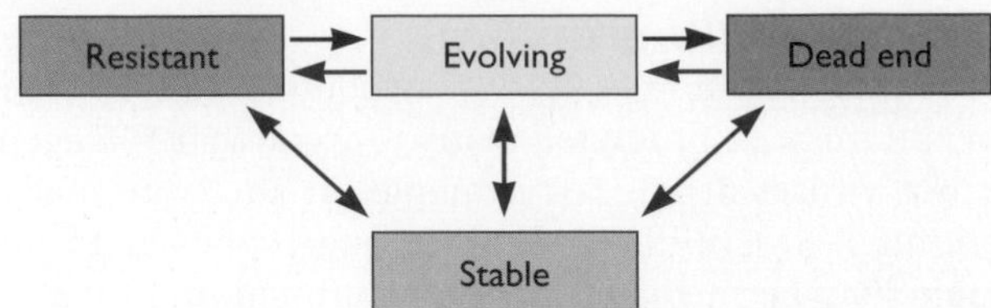

Figure 20.6 The general interactions of hosts and viruses. Four hypothetical host-virus interactions are indicated in the boxes. These are artificial constructs that serve to demonstrate the dynamics of viral interactions with the host population. The emergence of new viruses is one possible outcome. The **stable interaction** maintains the virus in the ecosystem. The **evolving interaction** describes the passage of a virus from "experienced" populations to naive populations in the same host species. The **dead-end interaction** represents one-way passage of a virus to different species. The host usually dies, or if it survives, the virus is not transmitted efficiently to the new host species. The **resistant host interaction** represents situations in which the host completely blocks infection. The arrows indicate possible flow of a virus from one situation to another or the possible transformation of one interaction into another. Mutation and selection, the environment, age of hosts, social behavior, floods, and wars all have a major influence on the establishment and maintenance of these interactions.

The Spectrum of Host-Virus Interactions

The spectrum of possible interactions among hosts and viruses can be complex, and the possible outcomes can contribute to the emergence of viruses (Fig. 20.6). To illustrate the principle, we define four hypothetical interactions: **stable, evolving, dead-end,** and **resistant.** These

Table 20.7 Examples of "emerging viruses"[a]

Virus	Family	Emergence factors
Canine parvovirus	*Parvoviridae*	Spontaneous mutation in feline panleukopenia virus (a parvovirus) resulted in new tropism and pathogenesis
Dengue virus	*Flaviviridae*	Urban population density; open water storage favors mosquito breeding (e.g., millions of used tires)
Ebola virus	*Filoviridae*	Human contact with unknown natural host in Africa; importation of monkeys in Europe and United States
Hantaan virus	*Bunyaviridae*	Agriculture techniques favor human contact with rodents
Human immunodeficiency virus	*Retroviridae*	Transfusions and blood products, sexual transmission, needle transfer during drug abuse
Human T-lymphotropic virus	*Retroviridae*	Transfusions and blood products, contaminated needles, social factors
Influenza virus	*Orthomyxoviridae*	Integrated pig/duck agriculture, mobile population
Junin virus	*Arenaviridae*	Agriculture techniques favor human contact with rodents
Machupo virus	*Arenaviridae*	Agriculture techniques favor human contact with rodents
Marburg virus	*Filoviridae*	Unknown; importation of monkeys in Europe
Norwalk/gastroenteritis viruses	*Caliciviridae*	New methods for detection, infectious diarrhea
Rift Valley virus	*Bunyaviridae*	Dams, irrigation
Sin Nombre virus	*Bunyaviridae*	Natural increase of deer mice and subsequent human-rodent contact
West Nile virus	*Flaviviridae*	Unknown route of introduction into United States

[a]Adapted from S. S. Morse (ed.), *Emerging Viruses* (Oxford University Press, Oxford, United Kingdom, 1993), with permission.

definitions are arbitrary snapshots of the extremes of dynamic host-virus interactions designed to emphasize essential concepts. In Fig. 20.6, arrows suggest the hypothetical flow of virions from one situation to another and stress the continuity of viral interactions in nature. In addition, it is important to understand that these definitions apply to large populations and **not** to a single virus-host interaction.

Stable Interactions

A stable host-virus interaction is one in which both members survive and multiply. As a result, the virus is maintained in the ecosystem. Such interactions are essential for the continued existence of the virus, and may influence host survival as well. This state is optimal for a host-parasite relationship, but the interaction need be neither benign nor permanent in an outbred population. Infected individuals can become ill, recover, develop immunity, or even die, yet in the long run, both populations survive. While often described as an equilibrium, the term is somewhat misleading. The interactions are dynamic and fragile, and certainly are rarely reversible. Viral populations may evolve to be more or less virulent if such a change enables them to be maintained in the population, while hosts may evolve to attenuate the more debilitating effects of the viruses that infect them.

Humans are thought to be the sole natural host for a variety of viruses, including measles virus, herpes simplex virus, and smallpox virus. Similarly, simian virus 40, monkeypox virus, and simian immunodeficiency virus infect only certain species of monkey. In contrast, some viruses, including influenza virus and togaviruses, are capable of propagating in a variety of species. In these instances, the host that maintains the virus may not be apparent or obvious. For example, influenza A virus infects humans, wild birds, and pigs, but birds may be its natural host. This conclusion is discussed in "Human Disease Resulting from Genetic Reassortment of Viruses in Animals" below.

The trajectory of evolution is unpredictable: what is successful today may be suicidal at another time. For example, once a virus population becomes completely dependent on one, and only one, host, it has entered a potential bottleneck that may constrain its further evolution. When the host becomes extinct, so does the virus. For example, if humans disappear from the planet, many virus populations, including poliovirus, measles virus, and several herpesviruses, will cease to exist. Eradication of natural smallpox virus was possible because humans are the only hosts and worldwide immunization was achieved.

As the arrows in Fig. 20.6 indicate, there is no reason to believe a priori that once a stable relationship is established, it is permanent. What we observe today is not necessarily the optimal state for any virus-host relationship.

The Evolving Host-Virus Interaction

The notion of an evolving host-virus interaction emphasizes that stability is unlikely to be established instantly. We highlight this interaction to illuminate the dynamic consequences of a virus spreading among populations of the same or perhaps closely related species. The hallmarks of this interaction are instability and unpredictability. These properties are to be expected, as selective forces are applied to both host and virus. For example, some host subpopulations may experience high infection rates, while others are unaffected. Outcome of infection may range from relatively benign symptoms to death. Such an interaction is typified by the introduction of smallpox and measles to natives of the Americas by Old World colonists and slave traders. Europeans previously had experienced the same horror when these diseases spread to Europe from Asia. Interestingly, in the case of human immunodeficiency virus, relatively resistant individuals who harbor mutations in chemokine or chemokine receptor genes have been identified. One possible selective force could have been applied by past infection with some other highly virulent virus for which the chemokine receptor was important. As the recent appearance of human immunodeficiency virus could not have been anticipated in evolution, the selective advantage that has maintained these mutations remains to be discovered. Other opportunities to enter the evolving host-virus interaction may arise if the virus in a stable interaction acquires a new property that increases its virulence or spread (as exemplified by expanded host range mutations of feline panleukopenia virus, see below) or if the host population suffers a far-reaching catastrophe that reduces resistance (e.g., famine or mass population changes during wars).

The Dead-End Interaction

A commonly encountered host-virus interaction is described rather vividly as a **dead-end interaction.** It has much in common with the evolving host-virus interaction, as both represent departures from a stable relationship, often with lethal consequences. A dead-end interaction is a frequent outcome of cross-species infection (but not of intraspecies infections, as noted in reference to the evolving host-virus interaction). Importantly, the dead-end interaction has captured the public's interest and provided inspiration for the more sensational implications of the term "emerging virus."

In a dead-end interaction, the host is often killed so quickly that there is little or no subsequent transmission of the virus to others. Classical examples of dead-end interactions are presented by the many viruses carried by arthropods like ticks and mosquitoes that cycle in the wild between insects and a vertebrate host in a stable relationship. Occasionally the infected insect bites a new species (e.g., humans) and transmits the virus (Fig. 20.7). As a consequence, the infection results in severe pathogenic effects in the infected human, but is not transmitted further; i.e., it reaches a dead end. As the human is not part of the natural, stable host-virus relationship, dead-end infections have little if any effect on the evolution of the virus and its natural host. To a first approximation, dead-end hosts are transparent participants. However, such logic may be too simplistic; infection by a less virulent mutant or infection of a more resistant individual may be the first step in establishing a new host-virus interaction.

Figure 20.7 A complex host-virus relationship. This arbovirus infection illustrates how multiple host species can maintain and transmit a virus in the ecosystem. In this example, the virus population is maintained in two different hosts (wild birds and domestic chickens) and is spread among individuals by a mosquito vector. The virus replicates in both species of bird and in the mosquito. Disease is likely to be nonexistent or mild in these species, as the hosts have adapted to the infection. A third host (in this example, horses or humans) occasionally may be infected when bitten by an infected mosquito that fed on an infected bird. Horses and humans are dead-end hosts and contribute little to the spread of the natural infection, but they may suffer from serious, life threatening disease. Occasionally another species of biting insect (e.g., other mosquito species) can feed on an infected individual (bird, horse, or human) and then transmit the infection to another species not targeted by the original mosquito vector.

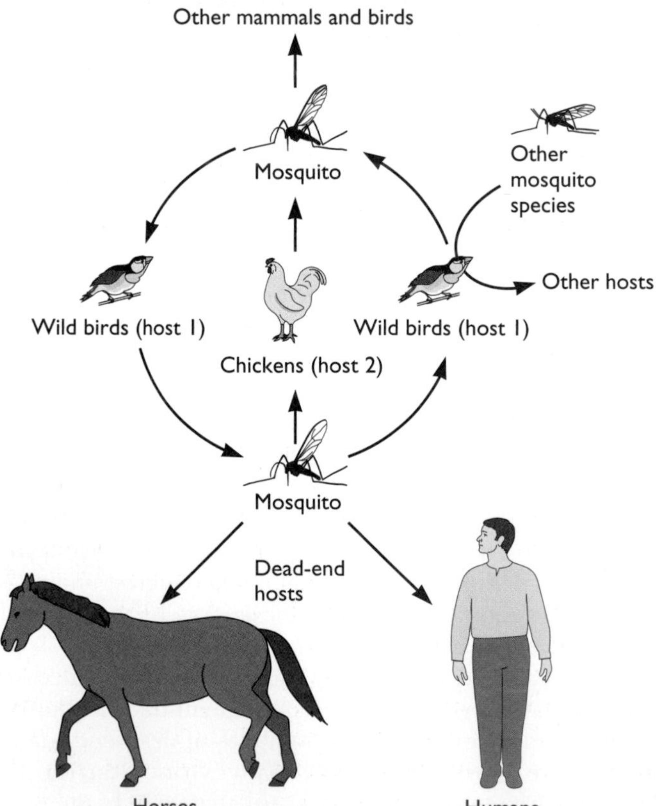

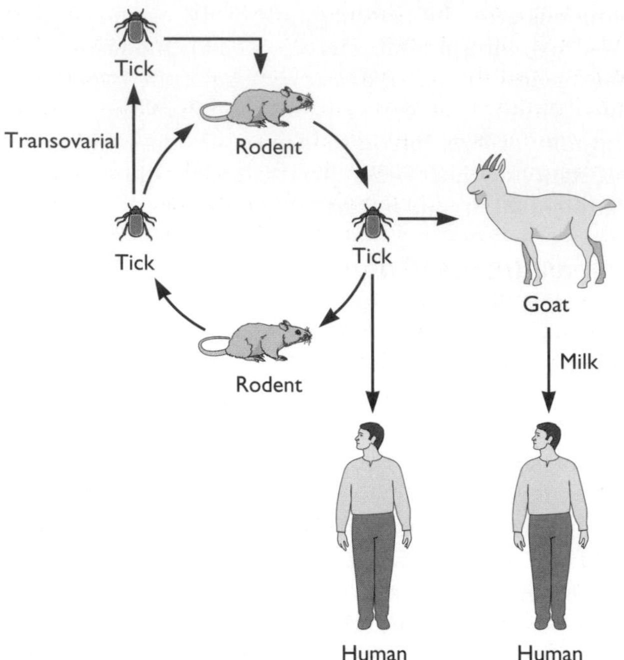

Figure 20.8 Infectious cycle of the central European tick-borne flavivirus. European tick-borne flavivirus infection is maintained and spread by multiple host infections. Congenital transmission in the tick maintains the virus in the tick population as they feed upon rodents. The newborn infected ticks have adapted to the infection and thrive. The virus is transmitted from the tick to a variety of animals, including cows, goats, and humans. Humans can also be infected by drinking milk from an infected goat, sheep, or cow.

Human diseases like yellow fever and dengue fever provide excellent examples of the dead-end interaction in nature. The viruses causing these diseases are endemic in the tropics and maintain a stable relationship with their natural host and insect vectors. When humans begin to develop a tropical area by clearing forests and building roads, dams, canals, and towns, they are at risk for being bitten by mosquitoes and other insects. The natural host-virus interaction that existed before the intrusion changes, and humans may experience new viral infections spread by insect vectors. A similar situation exists for some flaviviruses transmitted by ticks, which in turn are carried by rodents. Humans often experience European tick-borne encephalitis as dead-end hosts (Fig. 20.8).

Rodent vectors can play critical roles in the introduction of new viruses into populations in areas where these animals abound. Most hemorrhagic disease viruses, including

Lassa, Junin, and Sin Nombre viruses, are endemic in rodents, their natural hosts. The viruses establish a persistent infection, and the rodents show few if any ill effects. Substantial numbers of virus particles are excreted in urine, saliva, and feces to maintain the virus in the rodent population. However, infection by such rodent viruses can cause local, often lethal, outbreaks in humans as dead-end hosts. Humans become infected only because they happen to come in contact with rodent excretions containing infectious virus particles.

The lethality of the Marburg and Ebola filoviruses in humans is typical of a dead-end host infection. These are **the** emerging viruses in the eyes of the layperson, thanks to the popular press. Filoviruses are (–) single-stranded RNA viruses causing severe hemorrhagic manifestations in infected humans (Appendix B). Disease onset is sudden, with 25 to 90% fatality rates reported. Virus spreads through the blood and replicates in many organs, causing focal necrosis of the liver, kidneys, lymphatic organs, ovaries, and testes. Capillary leakage, shock, and acute respiratory disorders are observed in fatal cases. Patients usually die of intractable shock without evidence of an effective immune response. Even when recovery is under way, survivors do not have detectable neutralizing antibodies. The infection clearly overwhelms a particular individual, but apparently does not spread widely; human infections tend to cluster in local areas. Such viruses can be transmitted to other humans only by close personal contact with infected blood and tissue. Although the identities of the natural hosts for many filoviruses remain elusive, these viruses must have established a stable interaction with an animal host and infect humans only inadvertently. Some investigators have argued that the host is responsible for the apparent "geographic containment" of Ebola virus in central Africa.

Many animal models of disease might be considered examples of dead-end interactions. For example, herpes simplex virus is a human virus, but when it is introduced into mice, rabbits, or guinea pigs in the laboratory, these animals become infected and show pathogenic effects that mimic some aspects of the human disease. However, in their natural environment, these animals contribute nothing to the transmission or survival of the virus, and therefore are experimental models of the dead-end interaction.

The Resistant Host

It is instructive to include the resistant host when discussing viral evolution. All living things are exposed continuously to viruses of all types, yet the vast majority of these interactions are uneventful. This may be because the host cells are not susceptible, not permissive, or both. A more likely possibility is that the primary physical, intrinsic, and innate defenses are so strong that potential invaders are diverted or destroyed upon contact. In other cases, organisms may become infected and produce some virions, but the infection is cleared rapidly without activation of the host's acquired immune system (see Chapter 16). This outcome is in contrast to an inapparent infection, in which an immune response is mounted but the individual exhibits no signs of disease.

We include the resistant host in our analysis because the transition between inability and ability of the host to support viral replication need not be insurmountable. Indeed, **xenotransplantation** (the use of animal organs in humans) offers thought-provoking scenarios. Not only does transplantation bypass physical and innate defenses by surgery, but drugs also suppress the immune response. As a result, the virus particles or genomes in xenografts have direct access to the once-resistant host in the absence of crucial antiviral defenses. As many of these viruses can infect human cells or have close, human-adapted relatives, the xenotransplantation patient represents a source of new viral diversity. The outcome of such an experiment in viral evolution is not predictable, and deserves careful contemplation.

The Need for Experimental Analysis of Host-Virus Interactions

The definitions of different host-virus interactions listed above are useful for discussion, but we can learn only by doing real experiments and testing hypotheses. Unfortunately, such experiments are complicated by a variety of issues, including safety, cost, concerns for animal rights, and uncertainty of the relevance of laboratory models to natural disease and virus spread. Natural infections occur in outbred populations, a feature that is difficult to model in the laboratory. An example of how molecular techniques can be used to study the evolution of retroviral pathogenesis in an animal is shown in Box 20.10. Many important questions about the evolution of host-virus interactions can be posed. What is the relationship between viral virulence (ability to cause disease) and the rate of transmission in the population? How important is interspecies transmission in the establishment of new viruses in populations? Are quasispecies maintained during selection in a population? How important to survival is the ability of a virus to cause disease in its host?

We understand in principle that most virus populations survive in nature only because of **serial infections** (a chain of transmission) among individuals. Quantitative measures of these interactions, determination of the molecular mechanisms responsible for them, and de-

BOX 20.10

Studies of evolution of pathogenesis using molecular clones

In 1998, a collaboration of several research teams provided a detailed analysis of the molecular and population changes required for pathogenesis of simian immunodeficiency virus in macaques. Such infection is used as a model for lentivirus pathogenesis, transmission, tissue tropism, antiviral therapy, and virus evolution. After rounds of passage in cells, full-length DNA copies of viral RNA were prepared. Virus stocks derived from these clones were markedly attenuated. They failed to induce pathogenesis in macaques within a year. Scientists then tracked the natural history of infection of two such cloned attenuated strains as viruses derived from them gained virulence during repeated passages in macaques.

Detailed sequence comparisons of the original attenuated molecular clones and the newly evolved virulent strains yielded two important findings. First, different mutants predominated in different individuals infected with the same attenuated clone. Second, multiple and dispersed mutations were required in each new pathogenic virus clone, including elongation of TM and Nef coding sequences; changes in RNA splice donor and acceptor sites, the TATA sequence, and Sp1 binding sites; multiple changes in the V2 region of SU, including a consensus neutralization epitope; and five new N-linked glycosylation sites in SU.

The conclusion is that for these strains of simian immunodeficiency virus, evolution of virulence is a multistep, multigenic process understandable only in hindsight. The selective forces responsible for this process are not known.

Edmonson, P., M. Murphey-Corb, L. N. Martin, C. Delahunty, J. Heeney, H. Kornfeld, P. R. Donahue, G. H. Learn, L. Hood, and J. I. Mullins. 1998. Evolution of a simian immunodeficiency virus pathogen. *J. Virol.* **72:**405–414.

velopment of model systems that can predict them are sorely needed. Mathematical models are being developed to describe patterns of disease transmission in complex populations. Such models should help to determine the critical population size necessary to support the continuous transmission of viruses with differing incubation periods, and the dynamics of persistent viral infections. The value of these models remains to be seen, simply because it is difficult to perform a controlled experiment in nature.

Accidental Natural Infections

Our understanding of the dynamics of a viral infection in a large outbred human population is rudimentary, at best. What we do know is based predominantly on a limited number of accidental experiments. Two classic examples are provided by hepatitis B virus and poliovirus infections. During World War II, large doses of infectious hepatitis B virus were accidentally introduced into approximately 45,000 soldiers when they were injected with a contaminated yellow fever vaccine. Surprisingly, only 900 soldiers (2%) came down with clinical hepatitis, and fewer than 36 developed severe disease. Similarly, in 1955, 120,000 grade school-aged children were vaccinated with an improperly inactivated poliovirus vaccine. About half had preexisting antibodies to poliovirus thanks to inapparent infections by wild virus. Of the remainder, about 10 to 25% were infected by the vaccine, as determined by the appearance of antibodies. More than 60 cases of paralytic poliomyelitis were documented among these infected children, and the remainder escaped disease. These two experiments tell us that even when a large number of individuals are infected with a virulent virus, the outcome is unpredictable; we have only a rudimentary understanding of why this is so (Box 20.11).

Population Density and Health Are Important Factors

Two predominant parameters influencing spread of infection are the population density and the health of individuals in that population. High population density (not necessarily overall size) is an important parameter. Close personal contact, either by direct methods (e.g., aerosols or sexual contact) or by indirect action (e.g., water or sewage), is also required. Variables such as duration of immunity and the quantity of virions produced and shed from each individual have marked effects on spread of infection. As discussed in Chapter 16, at least half a million people in a more or less confined urban setting are required to ensure a large enough annual supply of susceptible hosts to maintain measles virus in a human population. When this large population of interacting hosts is not available, measles virus dies out. One can reduce the effective population by splitting the half million hosts into small groups physically separated from social discourse (exemplified by quarantine or the use of sanitariums to

BOX 20.11

A classic experiment in virus evolution: deliberate release of rabbitpox virus in Australia

In 1859, 24 European rabbits were introduced into Australia for hunting purposes and, lacking natural predators, the friendly rabbits reproduced to plague proportions. In 1907, the longest unbroken fence in the world (1,139 miles long) was built to protect portions of the country from invading rabbits. Such heroic actions were to no avail. As a last resort, the rabbitpox virus, myxoma virus, was released in Australia in the 1950s in an attempt to rid the continent of rabbits. The natural hosts of myxoma virus are the cottontail rabbit, the brush rabbit of California, and the tropical forest rabbit of Central and South America. The infection is spread by mosquitoes, and infected rabbits develop superficial warts on their ears. However, European rabbits are a different species and are killed rapidly by myxoma virus. In fact, the infection is 90 to 99% fatal!

In the first year, the infection was amazingly efficient in killing rabbits, with a 99.8% mortality rate. However, by the second year the mortality dropped dramatically to 25%. In subsequent years, the rate of killing was lower than the reproductive rate of the rabbits, and hopes for 100% eradication were dashed. Careful epidemiological analysis of this artificial epidemic provided important information about the evolution of viruses and hosts.

The infection spread rapidly during spring and summer, when mosquitoes are abundant, but slowly in winter, as expected. Given the large numbers of rabbits and virus particles, and the almost 100% lethal nature of the infection, advantageous mutations present in both the virus and host populations were quickly selected. Within 3 years, less virulent viruses appeared, as did rabbits that survived the infection. The host-virus interaction observed was that predicted for an evolving host coming to an equilibrium with the pathogen. A balance is struck: some infected rabbits die, but they die more slowly, and many rabbits survive.

What was learned? To paraphrase the Rolling Stones: you always get what you select, but you often don't get what you want. Probably the most obvious lesson was that the original idea to eliminate rabbits with a lethal viral infection was flawed. Powerful selective forces that could not be controlled or anticipated were at work. In this experiment, the viral genomes acquired mutations resulting in an attenuated infection: fewer rabbits were killed, infected rabbits were able to survive over the winter, and in the spring, mosquitoes spread the infection. Moreover, the rabbits evolved to become more resistant to, or tolerant of, the infection.

Surprisingly, more experiments in the biological control of rabbits are under way in Australia. One line of experimentation uses a lethal rabbit calicivirus, while another employs a genetically engineered myxoma virus designed to sterilize, but not kill, rabbits. The latter viruses encode a rabbit zona pellucida protein, and infected rabbits synthesize antibodies against their own eggs (so-called immunocontraception, also discussed in Box 14.9).

isolate infected patients) (Fig. 20.9B). Another way is by immunizing the group such that the large majority are immune and cannot propagate the virus. If these or similar actions are not taken, our ever-expanding and increasingly interactive human population will ensure the maintenance and continued evolution of measles virus.

The health of any potential host population is also important. For example, babies and the elderly are commonly more susceptible to a given virus than the general population. As a consequence, prevention of infection in these groups tends to reduce the overall infection rate. Similarly, individuals in poor health due to malnutrition are more susceptible to disease than those who are well fed.

Obviously, the health and genetic variation in any given outbred host population contribute to the unpredictable nature of viral infection. Perhaps not so obvious are the other variables that modulate infection. For example, what factors account for distinct seasonal variations? Respiratory infections caused by adenoviruses and rhinoviruses often occur in the spring, but those respiratory

A

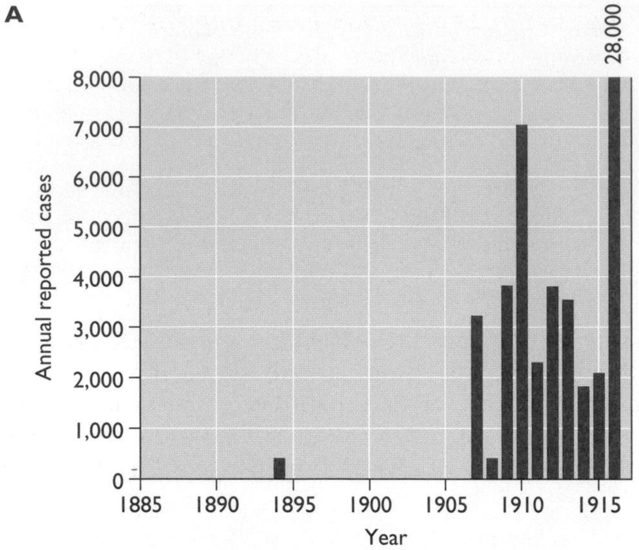

B

Figure 20.9 Poliovirus in the early 20th century. (A) The emergence of paralytic poliomyelitis in the United States, 1885 to 1915. From N. Nathanson, *ASM News* **63:**83–88, 1997, with permission. (B) Board of Health quarantine notice, San Francisco, Calif., circa 1910.

infections caused by coronaviruses and respiratory syncytial viruses tend to occur in the winter. Most arbovirus infections are experienced in the summer. While it is clear that insect-mediated infections are not likely to occur in climates where temperatures drop below freezing for months at a time, the seasonal variations for other viral diseases are not easy to explain (Chapter 14).

Changing Host Populations and Environment May Expand Viral Niches

A Disease of Modern Sanitation

Host populations change with time, and each change can have unpredictable effects on virus evolution. Analysis of poliomyelitis, a diseased caused by poliovirus infection, provides an instructive example. The disease is ancient, postulated by some to be present over 4,000 years ago (Fig. 1.1B). For centuries, the host-virus relationship was stable, and infection was endemic in the human population. Poliomyelitis epidemics were unheard of (or at least, not written about), but we imagine that occasional bouts of disease were obvious in scattered areas. This state of affairs changed radically in the first half of the 20th century, when large annual outbreaks of poliomyelitis appeared in Europe, North America, and Australia. Retrospective analysis reveals that the viral genome did not change or evolve substantially, yet poliovirus became a significant disease-causing agent (Fig. 20.9A). How can the emergence of epidemic poliomyelitis be explained?

The answer is that humans changed their lifestyle on an unprecedented scale. Poliomyelitis is caused by an enteric virus spread by oral-fecal contact. As a consequence, endemic disease was characteristic of life in rural communities, which generally had poor sanitation and small populations. Because virions circulated freely, most children were infected at an early age and developed antibodies to at least one of the serotypes. Maternal antibodies, which protected newborns, were also prevalent, as most mothers had experienced a poliovirus infection at least once. An important consideration is that most infected children do not develop paralysis, the most visible symptom of poliomyelitis. Paralysis is a more frequent result when older individuals are infected. Even the most virulent poliovirus strains cause 100 to 200 subclinical infections (inapparent infections) for every case of poliomyelitis. These inapparent infections in children provided a form of natural vaccination. As childhood disease and congenital malformations were not uncommon in rural populations, the few individuals who developed poliomyelitis were not seen as out of the ordinary. No one noticed endemic poliovirus.

However, during the 19th and 20th centuries, industrialization and urbanization changed the pattern of poliovirus transmission. Improved sanitation broke the normal pattern of virus transmission and effectively stopped natural vaccination. In addition, dense populations and increased travel provided new opportunities for rapid transmission of infection. As a result, children tended to encounter the virus for the first time at a later age, without the protection of maternal antibodies, and were at far greater risk for developing paralytic disease. Consequently, epidemic poliovirus infections emerged time and time again in communities across Europe, North America, and Australia. Until vaccines became available, quarantine was the only public health defense (Fig. 20.9B).

Widespread use of inexpensive, effective poliovirus vaccines has since controlled the epidemic. As the virus has no host other than humans, it should be possible to eradicate it by vaccinating sufficient people so that spread of the virus is impossible. Accordingly, the World Health Organization is targeting the eradication of poliovirus by

2005 with a massive worldwide vaccination program. For this eradication program to succeed, lessons learned from history must not be forgotten, and we must not drop our guard. Today, many parents do not have their children vaccinated because of fear of side effects. Complacency and the misguided idea that viral diseases are no longer a problem are also prevalent. Poverty, social problems, and economic conditions often conspire to prevent vaccination of children in inner cities and in poorer countries. There is hope that poliomyelitis will soon be a disease of the past, as a consequence of the worldwide poliovirus eradication program. However, given that it may be impossible to eliminate all sources of the virus (including computer files of the genome sequence), vaccination may be part of public health programs indefinitely.

Diseases of Exploration and Colonization

Explosive epidemic spread may occur when a virus enters a naive population (**the evolving host-virus interaction**) (Box 20.12). The resulting infections can be frightening, often devastating, and appear to "come out of the blue." Geographic isolation of human populations is a case in point. Consider the classic lethal epidemics of measles and smallpox caused by two diverse but well-known viral scourges of human populations throughout history. These viral infections require a high population density to be maintained. They are also human viruses with no other hosts. History records that smallpox reached Europe from the Far East in 710 A.D. and attained epidemic proportions in Europe in the 18th century as the population grew and became concentrated. The effects on

BOX 20.12

West Nile virus outbreak: watching a virus population establish a new geographic niche

The West Nile virus, an Old World flavivirus, discovered in 1937 in the West Nile district of Uganda, had never been isolated in the Western Hemisphere before 1999. In August 1999, six people were admitted to Flushing Hospital in Queens, N.Y., with similar symptoms of high fever, altered mental status, and headache. They were subsequently discovered to be infected with West Nile virus.

The virus has now spread from the Atlantic to the Pacific, as well as to Canadian provinces and territories. In the summer of 2002, it reached epidemic status, causing brain infections in hundreds of cases.

The New York isolate of West Nile virus is nearly identical to a virus isolated in 1998 from a domestic goose in Israel during an outbreak of the disease. The close relationship between these two isolates suggests that the virus was brought to New York City from Israel in the summer of 1999. How it crossed the Atlantic will probably never be known for sure, but it might have been via an infected bird, mosquito, human, horse, or other vertebrate host.

The infection spreads via many species of mosquitoes and is now circulating in wild birds. The virus or virus-specific antibodies have been found in more than 157 species of birds, but crows and jays appear to be particularly sensitive. Many zoos are reporting deaths of their exotic birds from West Nile virus infections. The virus has been found to infect 37 kinds of mosquitoes and 18 other vertebrates.

Humans and other animals acquire infections by mosquito bites after the insect has fed on infected birds. Human infections may be spread by transfusion from an infected donor, a possibility with far-reaching implications for our blood supply. Horses develop a lethal encephalitis, and hundreds of cases have been reported. A vaccine for horses is available.

About 20% of infected humans experience flu-like symptoms, but only 1 in 150 of these individuals develop meningitis, encephalitis, or poliomyelitis-like symptoms.

These events mark the first introduction in recent history of an Old World flavivirus into the New World. A fascinating, and yet unanswered, question is why the infection was established in New York City. The summer of 1999 was particularly hot and dry. Similar conditions spawn outbreaks of West Nile virus encephalitis in Africa, the Middle East, and the Mediterranean basin of Europe. Such conditions may promote mosquito breeding in polluted, standing water. In the absence of effective human vaccines or methods to stop natural spread in wild birds and animals, West Nile virus is likely to move throughout the Western Hemisphere. The consequences for public health cannot be predicted.

Brinton, M. A. 2002. The molecular biology of West Nile virus: a new invader of the Western Hemisphere. *Annu. Rev. Microbiol.* **56:**371–402.

Despommier. D. 2001. *West Nile Story.* Apple Trees Productions, LLC, New York, N.Y.

Petersen, L. R., and J. T. Roehrig. 2001. West Nile virus: a reemerging global pathogen. *Emerg. Infect. Dis.* **7:**611–614.

society are hard to imagine today, but as an example, at least five reigning monarchs died of smallpox.

Smallpox virus continued its spread around the world when European colonists and slave traders moved to the Americas and Australia. This viral infection changed the balance of human populations in the New World. Some say that viral infections were responsible for the elimination of a myriad of native languages and the nearly exclusive use of Spanish and Portuguese in South America. This suggestion may not be hyperbole. The first recorded outbreak of smallpox in the Americas occurred among African slaves on the island of Hispaniola in 1518, and the virus rapidly spread through the Caribbean islands. Within 2 years, this toehold of smallpox in the New World enabled the conquest of the Aztecs by European colonists. When Hernán Cortez first visited the Yucatan Peninsula in 1518 and began his conquest, his soldiers infected no one with smallpox virus. However, in 1520, smallpox reached the mainland from Cuba. Within 2 years, 3.5 million Aztecs were dead, more than could be accounted for by the bullets and swords of the small band of conquistadors. Smallpox spread like wildfire in the native population (which unfortunately was highly interactive and of sufficient density for efficient virus transmission), reaching as far as the Incas in Peru before Pizarro made his initial invasion in 1533. As is true in most smallpox epidemics, some Aztecs and Incas survived, but those who did were then devastated by measles virus, probably brought in by Cortez's and Pizarro's men. Conquest occurred by a one-two virological punch rather than by military prowess. Slave traders (who were most likely immune to infection) were populating Brazil with their infected human cargo at approximately the same time, with the same horrible result. The devastation of indigenous peoples by these viruses was documented in the colonization of North America and continued into the 20th century as contami-

Figure 20.10 Phylogenetic trees for 41 influenza A virus NP genes rooted to influenza B virus NP (B/Lee/40). The nucleotide tree is shown on the left. The horizontal distance is proportional to the minimum number of nucleotide differences to join nodes and NP sequences. Vertical lines are for spacing branches and labels and have no other meaning. Blue animal symbols denote the five host-specific lineages. Pink animal symbols denote viruses in the lineage that have been transmitted to other hosts. The amino acid tree is shown on the right. Adapted from R. G. Webster and Y. Kawaoka, *Semin. Virol.* **5:**103–111, 1994, with permission.

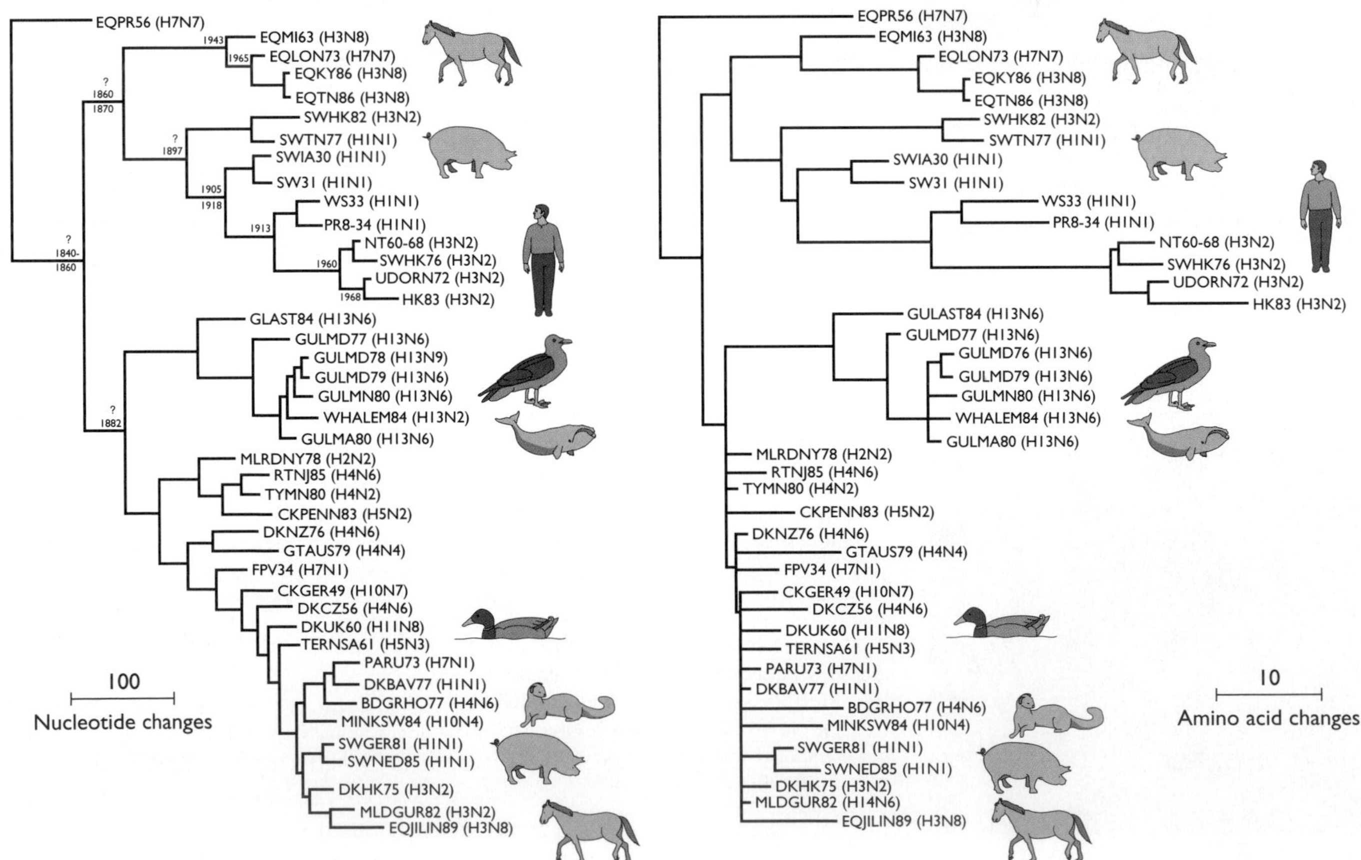

nated explorers infected isolated groups of Alaskan Inuit and native populations in New Guinea, Africa, South America, and Australia.

Human Disease Resulting from Genetic Reassortment of Viruses in Animals

Old diseases can be caused by new viruses. Influenza, a disease with symptoms that have remained unchanged for centuries, is caused by a virus with a constantly changing genome (Fig. 20.10). Influenza is the paradigm for the situation in which continued evolution of the virus in several host species is essential for its maintenance. The catastrophic pandemic in 1918 claimed over 25 million lives and affected more than 200 million people worldwide. It was estimated that 28% of the population of the United States was infected. Massive worldwide troop movements as a consequence of World War I played an important role in the pandemic. The unprecedented virulence of the 1918 influenza virus strain is still unexplained and remains of concern. Mortality rates were over 2.5%, compared with the normal mortality of less than 0.1%. Analysis of tissue from victims that succumbed during that epidemic has yet to provide an explanation of the remarkable virulence (Box 20.13). Despite the focus on the 1918 strain, current influenza virus serotypes are remarkably virulent. In 2002, epidemiologists in the United States estimated that influenza virus kills an average of 36,000 people per year, and the number may surge to 50,000 in severe seasons. Of these deaths, more than 90% of the victims are 65 and older. In the 1990s, three influenza A serotypes predominated, with the H3N2 serotype being particularly virulent.

The life cycle of influenza virus, while comparatively well understood at the molecular level, is remarkable for its complexity in nature. The virus can evolve rapidly by **antigenic shift** (genetic reassortment between a human and a nonhuman virus in a nonhuman host) and antigenic drift (accumulation of mutations that facilitate evasion of the hosts' antibody response). New influenza viruses constantly emerge from migratory aquatic bird populations to infect humans, swine, horses, domestic poultry, and aquatic mammals. In birds, influenza virus replicates in the gastrointestinal tract and is excreted in large quantities, a most efficient virus distribution system. The widespread dispersal of virus in water, the facile changing of hosts, and the ease of genetic reassortment form an engine for creation of new pathogenic strains. One hypothesis is that pigs are an important intermediate host between birds and humans, providing the mixing vessel for genetic exchange between avian and human viruses. Passage through pigs may also provide the selective pressures that enrich for virus populations with expanded host range. Recent evidence suggests that quail also may be intermediate hosts in which adaptation of duck influenza viruses results in viruses that can initiate transspecies infection.

Humans are not the only hosts that suffer from infections caused by this ever-changing virus. Outbreaks of swine influenza are economically devastating to pork producers. Despite large-scale immunization programs, virulent strains of swine influenza virus continue to emerge. When pigs are infected experimentally with an avirulent influenza virus mutant, a virulent strain can arise within a few days and cause swine disease. Poultry producers have similar experiences with avian influenza. A single mutation (changing a threonine to lysine in the hemagglutinin gene) in a relatively attenuated avian strain caused the 1983 influenza virus epidemic that devastated commercial chicken production in Pennsylvania. The direct transfer of a virulent avian virus to humans, first documented in 1997, reminds us how quickly host-virus interactions can change.

Even in this myriad of hosts and genetic exchanges, there is some remarkable stability that is crucial to the survival of influenza virus. One surprising finding is that the avian influenza virus genome has not changed much in more than 60 years, in contrast to the genomes of human and other nonavian viruses. Avian viral genomes exhibit

BOX 20.13

The hunt for the deadly 1918 influenza virus

Scientists want to know if the 1918 influenza pandemic was caused by a virus population with unprecedented virulence, or if other factors were responsible for this deadly event. Rather heroic efforts have been made to obtain answers. In one such study, scientists obtained archival samples of preserved tissue from several victims of the 1918 pandemic. Remnants of viral genomes were amplified by PCR and sequenced. Detailed sequence and functional analysis of recovered HA, NA, and NS1 genes provided essentially no information that could explain the high virulence of the 1918 influenza virus.

Reid, A., J. Taubenberger, and T. Fanning. 2001. The 1918 Spanish influenza: integrating history and biology. *Microbes Infect.* **3:**81–87.

Taubenberger, J., A. Reid, A. Kraft, K. Bijwaard, and T. Fanning. 1997. Initial genetic characterization of the 1918 "Spanish" influenza virus. *Science* **275:**1793–1796.

mutation and reassortment rates as high as those of human and swine viruses, but only those with neutral mutations are selected and maintained in the bird population. While virulent mutants do arise occasionally, in general, birds infected with avian influenza viruses experience no overt pathogenesis. Experts note that influenza virus appears to be in evolutionary stasis in birds. The avian host apparently provides the stable reservoir for influenza virus gene sequences that emerge as recombinants capable of transspecies infection.

Human Disease Resulting from Changing Climate and Animal Populations

A **zoonosis** is a disease or infection that is naturally transmitted between vertebrate animals and humans. Unanticipated viral zoonoses often are classic cases of emerging viral infections.

In 1993, a small but alarming epidemic of a highly lethal infectious disease occurred in the Four Corners area of New Mexico. Individuals who were in excellent health developed flu-like symptoms that were followed quickly by a variety of pulmonary disorders, including massive accumulation of fluid in the lungs, and death (Appendix B). Rapid action by local health officials and a prompt response by the Centers for Disease Control and Prevention were instrumental in discovering that these patients had low-level, cross-reacting antibodies to hantaviruses. These members of the family *Bunyaviridae* were previously associated with renal diseases in Europe and Asia and were well known to scientists, who associated them with viral hemorrhagic fever during the Korean War. Hantaviruses commonly infect rodents and are endemic in these populations around the world. Using powerful PCR technology, scientists from the Centers for Disease Control and Prevention found that the patients were infected with a new hantavirus. Field biologists discovered that this virus was found in a common New Mexican rodent called the deer mouse (*Peromyscus maniculatus*). Hantavirus pulmonary syndrome has invariably been associated with the presence of this virus in New Mexico as well as a few other isolated incidents around North America. The virus, which was given the name Sin Nombre virus (no-name virus), is an example of an emerging virus, endemic in rodents, that causes severe problems when it crosses the species barrier and infects humans.

The reason why humans became infected with Sin Nombre virus is being debated, but one popular idea is that a dramatic increase in the deer mouse population was an important factor. In 1992 and 1993, higher than normal rainfall resulted in a bumper crop of piñon nuts, a favorite deer mouse and local human food. Mouse populations rose in response, and contacts with humans inevitably increased as well. Hantavirus infection is asymptomatic in mice, but virions are excreted in large amounts in mouse urine and droppings, where they are quite stable. Human contact with contaminated blankets or dust from floors or food storage areas provided ample opportunities for infection. Hantavirus syndrome is rare because humans are not the natural host, and apparently are not efficient vehicles for virus spread. However, the hantavirus zoonotic infection serves as a warning that potential human diseases lurk in the wild if we inadvertently intrude upon another host-parasite relationship.

Caliciviruses represent an abundant, but underappreciated class of zoonotic viruses. These (+) strand RNA viruses are among the most common causes of virus-induced vomiting and diarrhea in humans (Appendix B). The Centers for Disease Control and Prevention estimates that 23 million cases of acute gastroenteritis in the United States are due to *Norovirus,* a genus in the *Caliciviridae* family. Remarkably, 50% of all food-borne outbreaks of gastroenteritis are caused by norovirus infection. Other common calicivirus-promoted symptoms include skin blistering, pneumonia, abortion, organ inflammation, and coagulation or hemorrhage. The epidemiology of calicivirus infections has been poorly understood, mainly because most caliciviruses cannot be propagated in the laboratory and diagnostic tools were rudimentary. Until recently, a common way to identify these viruses was by their unusual structure visible only by electron microscopy—hardly a facile diagnostic tool. Today, diagnosis of calicivirus infection is accomplished by accurate and rapid PCR technology. Humans often become infected via water and food sources, and we now understand that the ocean and its marine animals are a major reservoir of these interesting viruses. Four genera are recognized in the family *Caliciviridae: Norovirus* (formerly known as Norwalk agent or Norwalk-like virus), *Sapovirus* (formerly known as Sapporo-like virus), *Lagovirus,* and *Vesivirus.* Only noroviruses and sapoviruses cause human disease, and all have resisted attempts at cultivation. Only the marine caliciviruses can be propagated in vitro and are known to infect terrestrial hosts, including humans. The host range of these marine viruses is astounding. One virus serotype can infect five genera of seals, cattle, three genera of whales, donkeys, foxes, opaleye fish, horses, domestic swine, primates, and humans. The degree of exposure of land-based hosts to marine caliciviruses may be substantial, especially in confined areas such as shallow bays where aquatic mammals breed and calf, aquatic theme parks, and cruise ships. Infected whales excrete more than 10^{13} calicivirus particles daily, and the viruses remain viable more than 2 weeks in cold seawater. Despite our ignorance of this agent, human contact with contaminated water and sewage guarantees more outbreaks of calicivirus disease.

Host Range Can Be Expanded by Mutation or Recombination

Given the large numbers of viruses and ample opportunities for infection, we should expect to witness new host-virus interactions from time to time. Because the new host population will have had no experience with the virus, the infection will have unpredictable consequences (e.g., the dead-end or the evolving host-virus scenario). This invasion may be heralded by only an immune response and no obvious disease. Infection could also result in mild disease, the symptoms of which may or may not be noteworthy; occasionally, astute health care workers may notice unusual symptoms in local population centers. However, sometimes the invasion can be dramatic, as in the global epidemics caused by canine parvovirus and human immunodeficiency virus.

Cat-to-Dog Host Range Change by Two Mutations

In 1978 canine parvovirus, a member of the family *Parvoviridae,* was first identified simultaneously in several countries around the world as the cause of new enteric and myocardial disease in dogs. Canine parvovirus apparently evolved from the feline panleukopenia virus after two or three mutations made the latter pathogenic and highly transmissible in the dog population. Feline panleukopenia virus infects cats, mink, and raccoons, but not dogs. When the new canine virus emerged, it was able to infect feline cells in cultured cells, but did not replicate in cats. Because canine parvovirus appeared less than 30 years ago, it has been possible to analyze dog and cat tissue collected in Europe in the early 1970s to search for the progenitor canine parvovirus. Some fascinating studies have tracked and identified the virus mutants causing the epidemic. The ancestor of canine parvovirus began infecting dogs in Europe during the early 1970s, as deduced by the appearance of antibodies in dogs in Greece, The Netherlands, and Belgium. In 1978 evidence of virus infection was obvious in Japan, Australia, New Zealand, and the United States, suggesting that canine parvovirus had spread around the world in less than 6 months. The stability of the new virus, its efficient fecal-oral transmission, and the universal susceptibility and behavior of the world's dog population were important factors in the emergence of this new virus.

We now understand at the molecular level the genetic changes that created a new pathogen from a well-known virus. Only two amino acid substitutions in the VP2 capsid protein were necessary to change the tropism from cats to dogs. These substitutions are both necessary and sufficient to alter host range, because changing only the feline virus VP2 sequence to the canine virus sequence enabled the feline virus to replicate in cultured canine cells and in dogs. These critical amino acids are located on a raised region of the capsid that surrounds the threefold axis of icosahedral symmetry in the $T = 1$ capsid (Appendix A, Fig. 10). These differences in the capsid structure control the tropism for cells by affecting virion binding to host transferrin receptor, the protein used to establish infection. Feline panleukopenia virions bind only to the feline transferrin receptor. In contrast, canine parvovirus virions bind to both the feline and canine transferrin receptors.

Although the canine parvovirus DNA genome is replicated by host cell DNA polymerases and accumulates mutations less rapidly than RNA viruses, nevertheless, the viral genome continues to evolve rapidly. Indeed, the original antigenic variant, called type 2, was only a transient player in virus evolution. It was completely replaced within 2 years by new canine parvovirus strains called type 2a and 2b, which contain additional amino acid substitutions in the VP2 protein.

Host range switches by well-adapted viruses are rare events, but when a switch occurs, the outcome can be severe, as new variants can spread widely through the nonimmune and nonadapted population (the evolving host-virus interaction). The emergence of the canine parvovirus group provides an extraordinary opportunity to study virus-host adaptation and host-range shifts in the field as well as in the laboratory.

Host Range Change from Monkey to Human by Rapid, Multiple Changes

The origin of human immunodeficiency virus was an enigma until a remarkable study was published in 1998 (see also Chapter 17). A human serum sample taken in 1959 was found to be positive for human immunodeficiency virus type 1. This observation is noteworthy because the patient from whom the sample was taken had been infected over 20 years before the virus was identified. Careful PCR amplification of portions of the viral genome (called ZR59) yielded sufficient DNA for sequencing and analysis. Analysis of the ZR59 sequence revealed significant homology with all the major clades of this virus. Most significantly, it was clear that ZR59 was only a few nucleotides away from the putative progenitor of all human immunodeficiency virus strains (Chapter 17). Given the rapid rate of nucleotide change during virus replication, the founder virus is postulated to have emerged in the human population only 10 to 15 years before. Based on sequence homology between chimpanzee and human lentiviruses, it seems likely that the founder virus had its origin in chimpanzees. This precursor of contemporary human immunodeficiency virus acquired new mutations, giving it tropism for humans. A chance encounter of this mutant virus and a human established human immunodeficiency virus on the planet. The evolution of this virus led to escape from an impending bottleneck. The progenitor of human immunodeficiency virus was on the path to extinction because the population of wild chimpanzees had dropped to about 150,000 animals living in isolated troops. The new human host exceeds 6 billion individuals, and this niche is rapidly filling: more than 1 in 160 humans are now infected.

Shortly after the recognition of human immunodeficiency virus type 1 in the 1980s, a second related subtype, called type 2, was identified in Portugal and subsequently has been found in many countries (Chapter 17). Type 2 is not as widely distributed as type 1, being more commonly found in Africa. Interestingly, the membrane proteins of the type 2 human virus have homology to those of sooty mangabey and simian immunodeficiency viruses. In fact, humans infected with human immunodeficiency virus type 2 have antibodies that cross-react with the simian virus proteins. Moreover, the simian viruses can cause an AIDS-like disease in rhesus macaques. These observations have led some scientists to speculate that the human type 2 virus was derived from sooty mangabey or simian immunodeficiency virus mutants that crossed species into humans.

Pathogenicity Change by Recombination

Retroviruses provide instructive examples of the acquisition of entirely new genes by recombination with the host genome. Such recombination can yield viruses with unexpected, and frequently lethal, pathogenic potential. In general, a retrovirus can acquire a cellular oncogene and become an acutely oncogenic virus (see Chapter 18). As an example, feline leukemia virus is a well-known, highly conserved retrovirus infecting cats throughout the world. The virus causes lymphosarcomas and a range of degenerative diseases, including anemia and thymic atrophy. However, the sarcomas are caused not by feline leukemia virus but rather by new viruses with genomes that result from recombination of host and viral DNA sequences. These recombinant viruses are acutely oncogenic and are defective because their envelope genes have been replaced by cellular oncogenes. As a consequence of unregulated expression of the oncogene from the integrated defective virus, tumors arise rapidly, often with lethal consequences. Several host genes, including c-*abl*, c-*fes*, c-*fgr*, c-*fms*, c-*kit*, and c-*sis*, have been incorporated into the genomes of different feline sarcoma viruses.

Some Emergent Viruses Are Truly Novel

Classically, viruses were identified by the diseases they caused and, more recently, by growing them in culture to characterize their components. It is not difficult to imagine the existence of viruses that have never been described or grown in cultured cells. Many of these unculturable viruses cause serious, widespread disease. For example, rotaviruses infect every child under the age of 4 years in the United States, causing severe diarrhea. However, these viruses were not recognized until the 1970s simply because the technology for finding them did not exist and the viruses could not be propagated in the laboratory. Associating a disease or syndrome with a particular infectious agent used to be impossible if the agent could not be cultured. Fortunately, with powerful recombinant DNA, hybridization, and nucleic acid amplification technology, we are now able to detect and characterize such hitherto unknown viruses with comparative ease.

Discovery of New Hepatitis Viruses in the Blood Supply by Recombinant DNA Technology

One of the first examples of the contemporary quantum leap in diagnostic capacity occurred in the recognition of hepatitis C virus, a member of the family *Flaviviridae*. With the development of specific diagnostic tests for hepatitis A and B viruses in the 1970s, it became clear that most cases of hepatitis that occur after blood transfusion are caused by other agents. Even after more than 10 years of research, the identity of these so-called non-A, non-B (NANB) agents remained elusive. In the late 1980s, Chiron Corporation, a California biotechnology company, isolated a DNA copy of a fragment of the hepatitis C virus genome from a chimpanzee with NANB hepatitis. This remarkable feat was accomplished by examining DNA from about a million DNA clones from the chimp for production of proteins that were recognized by serum from a patient with chronic NANB hepatitis. The sequence analysis of these DNA clones identified a (+) strand RNA genome of about 10,000 nucleotides with high homology to known flaviviral genomes. The availability of the hepatitis C virus genome sequence made possible the development of diagnostic reagents that effectively eliminated the virus from the U.S. blood supply, reducing the incidence of transfusion-derived NANB hepatitis significantly. Viral proteins produced from the DNA clones were also instrumental in establishing mechanism-based screens for antiviral drugs (Chapter 19). These studies provide an excellent example of how a previously unknown human pathogen was identified and then analyzed by cloning technology. Interestingly, hepatitis C virus was not the only new virus hidden in the human blood supply; the same technology permitted the discovery of several viruses, including TT virus, a ubiquitous human circovirus of no known consequence.

Discovery through the PCR

A powerful technique for finding previously unknown nucleic acid sequences is PCR, which allows the amplification of minute quantities of any nucleic acid to concentrations that can be analyzed by standard recombinant DNA technologies. One useful PCR-based approach, **representational difference analysis,** compares DNA sequences present in diseased and nondiseased tissues. This technique was used to identify a new herpesvirus called human herpesvirus 8 present in Kaposi's sarcomas, a malignancy common in AIDS patients.

A Paradigm Shift in Diagnostic Virology

Recombinant DNA and PCR technology are not only expanding our abilities to define and understand the etiologies of complex diseases, but are also causing a revolution in diagnostic virology. New nucleic acid sequences can be associated with diseases and characterized in the absence of standard virological techniques. However, because of its sensitivity and the potential for contamination by adventitious viruses, care must be taken with any PCR-based method. Without proper controls, these techniques have the potential to associate a particular virus incorrectly with a disease, confounding the etiology. An excellent example was the spurious association of multiple sclerosis with infection by human T-lymphotropic virus type 1. Similarly, a variety of studies have been published in which PCR identified human herpesviruses in the brains of patients who had died of Alzheimer's disease, implying a causal link between the disease and the virus. These conclusions are called into question by the finding of the same viruses in the tissues of similarly aged patients who died of other causes. These examples illustrate the fact that the new technologies do not circumvent the time-honored Koch's postulates, which require that the infectious agent be isolated from a diseased patient, grown, and used to cause the same disease.

Clearly, the ability to detect new viruses by any technique has the potential to be a double-edged sword. While early detection and precise identification of viruses is important in public health and must be pursued with vigor, problems can arise if proper experimental controls are not applied or if no treatment or follow-up is available for the putative disease-causing agent. We do not have a robust armamentarium of antiviral drugs against even the most common pathogenic viruses. Therefore, developing treatments for previously unknown viruses discovered by these powerful techniques will be challenging.

Perceptions and Possibilities

Infectious Agents and Public Perceptions

While emerging virus infections are well known to virologists, in recent years they have become the subject of widespread public interest and concern. One reason why these infections have had such an impact on the public consciousness is that less than 30 years ago, many people were eager to close the book on infectious diseases. The public perception was that wonder drugs and vaccines had microbes under control. Obviously, in a few short years, this optimistic view has changed dramatically. Regular announcements of new and destructive viruses and bacteria appear with increasing frequency (Table 20.7). The discovery of human immunodeficiency virus and its effects at every level of society have attracted worldwide attention, while exotic viruses like Ebola virus capture front-page headlines. Movies and books bring viruses to the public consciousness more effectively than ever before. Consequently, it is important to separate fact from fiction and hyperbole. The widespread interest in emerging viruses reminds us that a little knowledge can be a dangerous thing: scientists, as well as the public, can be misled about host-virus interactions simply by semantics and jargon.

Virus Names Can Be Misleading

Unfortunately, it is common to name a virus by the host from which it was isolated. By calling the virus that causes AIDS **human** immunodeficiency virus, we give short shrift to its nonhuman origins. Canine parvovirus indeed causes disease in puppies, but it is clearly a feline virus that recently changed hosts. Similarly, canine distemper virus is not confined to dogs but can cause disease in lions, seals, and dolphins. Well-known viruses can cause new diseases when they change hosts. Much is implied, and more is ignored, about the host-virus interaction when the virus is given a host-specific name.

The Importance of Pathogenic and Nonpathogenic Viruses

It is not unusual to think that disease-causing viruses are important while nonpathogens are uninteresting and irrelevant. As we have seen, a virus that is stable in one host may be devastating when it enters a different species. Another misconception arises from human-centered thinking. How often do we think that viruses causing human diseases are more important than those that infect mammals, birds, fish, or other hosts? As noted above, naming viruses as "human" is not only inaccurate but also gives them inflated importance. Similarly, by thinking only of human needs we are blind to the multiple levels of interaction that constitute host-virus relationships. Indeed, viral particles have access to a broad smorgasbord of hosts, and humans often represent but one stop in the evolution of a virus.

Human Economic Interests Have Significant Effects

In a material society, it is routine to direct attention to viruses that have economic consequences as opposed to those that do not. Moreover, companies and governments generally concentrate resources on viruses that affect their economies, simply because they will be rewarded monetarily for their efforts. Consequently, research is focused on antiviral drugs and therapies that offer greater economic rewards as opposed to those aimed at smaller markets and needs. Such a bottom-line approach is short-sighted and does not address world health problems. Furthermore, because of the mobility of society and our expanding populations, economic impulses must be balanced more prudently by a larger worldview.

What Next?

Can We Predict the Next Viral Pandemic?

The world is currently in the midst of the AIDS pandemic, one of the most horrible viral pandemics in history. At the current rate of infection, more than 50 million people around the world will die within a year or two. This global devastation continues despite effective antiviral drug therapies. It is difficult to imagine an infection of greater consequence. However, it is instructive to consider the possibility that other viruses exist with similar potential to wreak havoc upon humans.

Upon reflection, the sobering reality is that some of the most serious threats come not from the popularized, highly lethal filoviruses (e.g., Ebola virus), the hemorrhagic disease viruses (e.g., Lassa virus), or an as yet undiscovered virus lurking in the wild. Rather, the most dangerous viruses are likely to be the well-adapted, evolving viruses already in the human population. Influenza virus is one that fits this description perfectly. Its yearly visits show no signs of diminishing, genes promoting pandemic spread and virulence are already circulating in the virus population, and the world is ever more prone to its dissemination. Indeed, a pandemic of influenza on the scale of the 1918–1919 outbreak is thought by many to be the most likely next emerging disease to affect humans on an unprecedented scale (Table 20.8). We were reminded of this possibility in 1997 with reports from Hong Kong that an avian strain of influenza virus (H5N1) apparently had spread directly to humans. The HA5 allele of viral hemagglutinin had never been isolated from an infected person.

Table 20.8 The 1918–1919 influenza pandemic: one of history's most deadly events

Event	Deaths (millions)
Influenza pandemic (1918–1919)	20–40[a]
Black Death (1348–1350)	20–25[a]
AIDS pandemic (through 2002)	20[b]
World War II (1937–1945)	15.9[a]
World War I (1914–1918, military deaths)	9.2[a]

[a]Data from the *New York Times*, 21 August 1998.
[b]Data from the World Health Organization.

Consequently no individual on the planet has an immune response directed toward this protein. Obviously, once such a virus infects the naive human population, pandemic spread is possible. Scientists from around the world are now following the trajectory of the H5N1 combination as it reassorts among the influenza virus population and interacts with the many hosts worldwide.

In this context, it is important to reflect upon the implications of the eradication of human viruses by vaccination, as exemplified by the successful campaign against naturally occurring smallpox virus. The human population is rapidly losing immunity to this virus, as would be predicted given that the natural virus has been eliminated. Ultimately, no one on the planet will be immune to smallpox infection, a condition that could be exploited in biowarfare. While this possibility is horrible to contemplate, we must recognize the danger and respond to it. Indeed, in recent years, vaccination programs for smallpox have been reactivated in some countries (see Chapter 19).

In early 2003, a new respiratory illness, severe acute respiratory syndrome (SARS), spread rapidly around the world. The agent responsible for SARS was rapidly identified as a coronavirus. It is unclear if this virus has crossed species barriers and is entering humans for the first time. In any case, this experience provides another reminder of our vulnerability to infectious agents.

Many Emerging Viral Infections Illuminate Immediate Problems and Issues

Obviously, it would be a mistake to concentrate on influenza virus or smallpox virus, to the exclusion of others, because many viral infectious diseases pose immediate and urgent problems for the world. The AIDS pandemic illustrates the ease with which a new virus can enter the human population. The secondary infections that accompany this viral infection have exposed the fragility of the world's health care system and the infrastructure of developing countries. The known outbreaks of new, exotic viruses, or well-known viruses that have invaded new geographic niches (Boxes 20.12 and 20.14), as well as the possible cross-species interactions promoted by new technologies, challenge the standard methods of diagnosis, epidemiology, treatment, and control.

Humans Are Constantly Providing New Ways To Meet Viruses

To understand and control emerging virus infections, humans must recognize the scale of the problem. The examples of poliovirus, measles virus, and smallpox virus remind us that viruses can suddenly cause illness and death on a catastrophic scale following a change in human behavior. Current human technological advances and changing social behaviors continue to influence the spread of viruses. Some

Table 20.9 Large-scale changes in virus ecology introduced by humans

Air travel
Dams and water impoundments
Irrigation
Rerouting of wildlife migration patterns
Wildlife parks
Hot tubs
Air conditioning
Blood transfusion
Xenotransplants
Long-distance transport of millions of head of livestock and birds
Moral and societal changes with regard to sex and drug abuse
Deforestation of massive proportions
Millions of used tires
Uncontrolled urbanization
Day care centers

human-introduced environmental changes that are significant to virology are listed in Table 20.9. Most did not exist 50 years ago, and each one brings humans and viruses into new situations. Probably the most critical fact is that the human population is as large as it ever has been, and there are few signs that our growth is slowing down. As a consequence, humans are interacting among themselves and with the environment on a scale unprecedented in history.

Many changes listed in Table 20.9 have major effects upon the transmission of viruses by vectors, such as insects and rodents. When humans interfere in a natural host-virus interaction, opportunities arise for cross-species infection. Population movements, the transport of livestock and birds, irrigation systems, and deforestation provide not only new contacts with mosquito and tick vectors, but also mechanisms for transport of infected hosts to new geographic areas (Box 20.14).

BOX 20.14

Yellow fever virus: humans change the pattern and pay the price

In the jungles, yellow fever virus is maintained by a monkey-mosquito cycle. Neither the monkey nor the mosquito is the worse for wear. Various mosquito species serve as vectors, including *Aedes* species in Africa and *Haemagogus* species in the Americas. If humans blunder into mosquito-infected areas, they stand a chance of being bitten by the infected insects and contracting the disease. Yellow fever came to the New World with the colonists and the slave trade, where it wreaked havoc among the indigenous human populations.

Humans were not the only species to suffer from the invasion. New World monkeys also died when infected with yellow fever virus, indicating that they were as unequipped to handle this new infection as were the indigenous humans.

Yellow fever virus and its mosquito vector spread by ship to the burgeoning populations of the U.S. South and East Coast with ease. Cities hit hard were New Orleans; Charleston, S.C.; Philadelphia; New York; and Boston. In 1800 Thomas Jefferson lamented, "Yellow Fever will discourage the growth of great cities in our nation." In a striking example of an emerging disease, 15% of the population of Philadelphia died of yellow fever in 1793. Neighboring New Jersey and Maryland attempted to bar panicked Philadelphians from entering their states.

The contribution of mosquitoes in the transmission of yellow fever was first glimpsed by Carlos Finlay in Cuba in 1880 and was established firmly by Walter Reed in 1900 when the disease was a problem during construction of the Panama Canal in the Central American jungle. A vaccine was developed in 1937 to contain the disease. Even with the vaccine, more than 10,000 people die of yellow fever every year in South America.

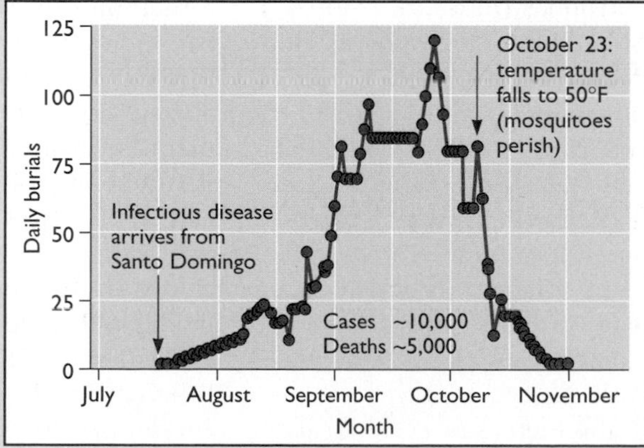

Yellow fever deaths in Philadelphia in the summer and early fall of 1793. From N. Nathanson, *ASM News* **63:**83–88, 1997, with permission.

Humans can provide novel habitats for viruses, as demonstrated by used tires, an unexpected vehicle for movement of viruses and their hosts. Several species of tropical mosquitoes (e.g., *Aedes* species) prefer to breed in small pockets of water that accumulate in tree trunks and flowers in the tropics. The used tire has provided a perfect mimic of this breeding ground and, as a consequence, the millions of used tires (almost all carrying a little puddle of water inside) accumulating around the world provide a mobile habitat for mosquitoes and their viruses. In the United States, such used tires are shipped all around the country for recycling, transmitting mosquito larvae to new environments literally overnight. As a result, the mosquito host for dengue and yellow fever viruses may be given the chance to establish a new range in New Jersey, thanks to the shipment of used tires for recycling from the South. It remains to be seen how the new insect vector will impinge on human health in the Northeast, but mosquito-borne viral diseases such as dengue fever clearly are of major concern.

Puerto Rico had five dengue fever epidemics in the first 75 years of the 20th century but had six epidemics in the last decade, with major economic costs. Simultaneously, Puerto Rico, Brazil, Nicaragua, and Cuba experienced their first dengue fever epidemics in 50 years. Brazil reported 180,000 cases in 1996. It is suspected that this viral infection is emerging because the mosquito host has new avenues for spread, and many governments have poor, or no, vector control programs.

Another contemporary example of humans moving viruses to new hosts comes from transport of livestock. African swine fever virus, a member of the family *Iridoviridae*, causes the most serious viral disease threatening the swine industries of developing and industrialized countries. Unlike any other DNA virus, it is spread by soft ticks of the genus *Ornithodoros*. African swine fever virus was spread from Africa to Portugal in 1957, to Spain in 1960, and to the Caribbean and South America in the 1960s and 1970s by long-distance transport of livestock and their resident infected arthropods (ticks).

The construction of dams and irrigation systems can change host-virus interactions on a large scale. The creation of vast areas of standing water has introduced new sources of viruses, hosts, and vectors. The 1987 outbreak of Rift Valley fever in Mauritania along the Senegal River was associated with the new Diama Dam, which created conditions ideal for mosquito propagation. Not only do dams and water impoundments affect insects, but they also alter population patterns of waterfowl and other animals and the viruses they carry. New hosts and new viruses are brought together as a consequence.

In industrialized countries, the increasing need for day care centers has led to new opportunities for viral transmission and other types of day care-associated illness. In the United States, more than 12 million children are in day care centers for several hours a day, and the vast majority are under 3 years of age. As most new parents can testify, respiratory and enteric infections are common, and infections spread easily to other children, day care workers, and the family at home.

What Can We Do about Emerging Virus Infections?

We have discussed how the modernization of society has led to spread of infection and selection of new virus variants. We cannot turn back the clock, but we can use common sense and learn from our experiences. The requirements for control of new viral infections range from the obvious to the idealistic. Implementation of these ideas will take money and integration of the public and private sectors at a level almost unprecedented today. The most obvious requirement is worldwide implementation of old-fashioned public health measures, including vector management (mosquito, tick, and rodent control), water and sewage management, and, perhaps most important, better nutrition so that natural human defenses are up to the task. Recognizing when human activities are likely to disrupt endemic disease patterns would be a major step forward. Infectious diseases are not contained by the boundaries between countries, nor the agencies that manage human and animal care and those that deal with public health. Therefore, an interesting idea is the promotion of more communication among these agencies to reflect more accurately the interaction between hosts and parasites in the evolution of disease. Because of the potential for rapid spread of viruses via air travel and urban development, a system of global surveillance and early warning is required for primary care physicians and health care workers. With modern sequencing technology, it should be possible to monitor all viral pathogens circulating in humans. When a new (or old) viral disease is suspected, the agent must be identified and characterized and the information must be shared widely. Active or passive vaccines should be made available, although the infrastructure for accomplishing anything more than local action is almost nonexistent and solutions are expensive to implement. Methods have to be developed for rapid diagnosis in the field, and primary health care workers must have access to databases. Certainly, baseline data on endemic disease and vector prevalence need to be compiled and included in this early warning database. Finally, responsible public and professional education is essential at all levels, as is more scientific research on host-parasite interactions. Inherent in this last proposal is the need for the scientific community to develop better mechanisms for sharing and developing the fruits of scientific research and making their work known, appreciated, and applied by the general public.

Viruses as Agents of War and Terror

Infectious agents have a documented capacity to cause harm, and can cause local epidemics as well as worldwide pandemics. Fearsome and deadly viruses abound, ranging from universal scourges such as smallpox virus, rabies virus, and influenza virus, to the less widely distributed, but no less deadly, hemorrhagic fever viruses. Any viral infection that can kill, maim, or debilitate humans, their crops, or their domesticated animals has the potential to be used as a biological weapon.

It is possible to make long lists of viruses that are readily available in nature, cause aerosol-mediated laboratory infections, and have case-fatality rates of about 30%. These viruses all are potential biowarfare agents (Table 20.10) Viruses such as smallpox virus, measles virus, and poliovirus that have been, or will be, declared eradicated may be prime candidates as bioweapons. Once immunization and natural infection cease, the entire population can become susceptible. Symptoms of infection would appear within days and require immediate attention by relatively skilled caregivers if fatalities were to be minimized. In addition, spread of some of these agents in a population is so rapid that the medical system would soon be overtaxed. A biological attack could also be designed to affect animals or plants. Such viruses could ruin food sources, creating havoc and disruption of populations as effectively as human viruses. Obviously, a biological attack need not cause mass destruction to be an instrument of terror, as was demonstrated by the far-reaching effects of the introduction of bacteria causing anthrax into the United States mail system. Such use of biological agents inspires fear and loathing in civilized societies.

Although their use has since been renounced, biological weapons, including viruses, were critical parts of the cold war armamentarium in the United States, the former Soviet Union, and other nations, and may well be retained in some countries. Documents from that time estimate that hundreds of thousands of deaths could be expected if virulent strains of smallpox virus or various hemorrhagic fever viruses were deployed by missiles, bombs, or other aerosol-generating methods. Anyone who has worked in a virology laboratory knows that almost any virus can be produced with ordinary equipment in simple facilities with relatively small risk. Although preparing this material for aerosol distribution does require specialized skills, it is sobering to contemplate that a determined, suicidal terrorist could easily initiate an epidemic (or the threat of an epidemic) after allowing him- or herself to become infected. The unintentional 1947 outbreak of smallpox in New York City originated from a single businessman who had acquired the disease in his travels. He died after infecting 12 others; to stop the epidemic, **over 6 million people** were vaccinated within a month.

Bioterrorism is a serious problem with no clear solutions. Some are concerned that publication of biological research data could aid terrorists, and measures of information control must be considered. Such concerns cloud another important issue. While the results of biological research may be used to harm, the same results may also have the potential to improve the human condition. Those responsible for national security call this an issue of "dual use" information. It is not easy to draw a red line separating the beneficial from the sinister when it comes to fun-

Table 20.10 Selected viruses with potential as biological weapons of terror or warfare

Family	Virus or disease	Target
Arenaviridae	Lassa fever	Humans
	South American hemorrhagic fever viruses	Humans
Bunyaviridae	Rift Valley fever virus	Humans, animals
	Crimean Congo hemorrhagic fever virus	Humans
	Hantavirus	Humans
Filoviridae	Marburg and Ebola hemorrhagic fevers	Humans
Flaviviridae	Yellow fever virus	Humans
	Tick-borne hemorrhagic fever virus	Humans
Orthomyxoviridae	Influenza virus	Humans
Paramyxoviridae	Measles virus	Humans
	Hendra and Nipah viruses	Animals, humans
Picornaviridae	Foot-and-mouth disease virus	Animals
	Poliovirus	Humans
Poxviridae	Smallpox virus, monkey poxvirus	Humans
Togaviridae	Venezuelan equine encephalitis virus	Humans

damental biological information. A helpful anticancer drug might be viewed as a nefarious "biological circuit disrupter"; unraveling the basis for an attenuated viral vaccine could be another person's entré to circumventing the vaccine. Pessimists and fearmongers can find malfeasance under every leaf. Valuable biological research is done not only in academia, but also in the pharmaceutical, biotechnology, and food industries. Biological research is an international enterprise and so is publication: no government can control the flow of scientific information. There are thousands of journals, and the Internet is always available for use by anyone with access to a computer.

As it is virtually impossible to prevent access to viral agents and information about biological processes, defensive strategies must be planned and implemented. Public health officials, first responders, medical service agencies, animal health officials, military, and police need to be part of an integrated effort. Important questions must be addressed. How will outbreaks of normal infections be distinguished from those of bioweapons? How will an unfamiliar viral infection be recognized and treated in the absence of diagnostic reagents and experience? What countermeasures should be employed? How will vaccines and antiviral drugs be chosen and administered? Given that there are few antiviral agents effective against the RNA viruses and no vaccines licensed for use against the hemorrhagic disease viruses, this last question appears to be of immediate concern. There is clearly much work ahead for all.

The bottom line is so simple that it is often ignored: the **best** way to defeat those who will use scientific information for nefarious purposes is to keep one step ahead. The only way to fight bioterrorism is with research. The battle has many fronts; suppression of information attacks the front that represents our greatest strength. As grim as it may be to contemplate, investing effort in addressing this enormous challenge is essential. We can be consoled by the knowledge that it will be a valuable effort, even if the world never experiences such a biological attack. Infectious diseases are still the second leading cause of death worldwide, and advances in modern transportation mean that any epidemic may be just one airplane seat away from any major city in the world. What we learn in attempting to counter a potential attack can help us to deal with the very real natural threats encountered every day in both developed and developing countries.

Perspectives

The relationships of viruses and their hosts are in constant flux. The perspective of an evolutionary biologist is required in order to understand why this is so. At present, the interplay of environment and genes, as well as the interactions of virus and host populations, can barely be articulated, much less studied in the laboratory. For viral infections, rapid production of huge numbers of progeny, the tolerance of amazing population fluctuations, and the capacity to produce enormous genetic diversity provide the adaptive palette that ensures survival. As Joshua Lederberg once said, "In the battle by attrition, humans have a real problem competing with microorganisms. Here we are, here are the bugs, they're looking for food, we're their meal in one sense or another, how do we compete?" One obvious reason for our success in the competition is that our intrinsic and immune defense systems are capable of recognizing and destroying invading viruses. A less obvious reason is that the current viral hosts on the planet represent progeny of survivors of past interactions with viruses. This experience is recorded in the vast gene pool of survivors. Indeed, heterozygosity at the many major histocompatibility complex alleles is known to confer a strong selective advantage on human populations in the battle against microbes. Viral infections have far-reaching effects ranging from shaping of the host immune system in survivors to eliminating entire populations. Clearly, survival of host and virus is poised on a delicate balance: once the balance is disrupted, the host, the virus, or both will be changed or even eliminated. Given the ever-changing viral populations and the drastic modifications of the ecosystem that have accompanied the current human population explosion, we are hard pressed to predict the future. Perhaps the most sobering fact for humans is that despite our intellect and ability to adapt, virus particles continue to infect and even kill us. In some senses, virus populations are, and probably always will be, at the top of the evolutionary ladder. The challenge is to keep them at bay and to strive to be at least one step ahead.

References

Books

Biddle, W. 1995. *A Field Guide to Germs.* Henry Holt and Company, New York, N.Y.

Bjornsson, J., R. I. Carp, A. Love, and H. M. Wisniewski (ed.). 1994. *Annals of the New York Academy of Sciences,* vol. 724. *Slow Infections of the Central Nervous System.* New York Academy of Sciences, New York, N.Y.

Diamond, J. 1997. *Guns, Germs and Steel: the Fates of Human Societies.* W. W. Norton and Company, New York, N.Y.

Domingo, E., R. W. Webster, and J. Holland (ed.). 1999. *Origin and Evolution of Viruses.* Academic Press, Inc., San Diego, Calif.

Ewald, P. W. 1994. *Evolution of Infectious Disease.* Oxford University Press, Oxford, United Kingdom.

Gould, S. J. 1996. *Full House.* Three Rivers Press, Pittsburgh, Pa.

Harvey, P. H., A. J. Leigh Brown, J. Maynard Smith, and S. Nee (ed.). 1996. *New Uses for New Phylogenies.* Oxford University Press, Oxford, United Kingdom.

Koprowski, H., and M. B. A. Oldstone (ed.). 1996. *Microbe Hunters Then and Now.* Medi-Ed Press, Bloomington, Ill.

Morse, S. S. (ed.). 1993. *Emerging Viruses*. Oxford University Press, Oxford, United Kingdom.

Scheld, W. M., D. Armstrong, and J. M. Hughes (ed.). 1998. *Emerging Infections 1*. ASM Press, Washington, D.C.

Scheld, W. M., W. A. Craig, and J. M. Hughes (ed.). 1999. *Emerging Infections 2*. ASM Press, Washington, D.C.

Reviews

Virus Evolution

Best, S. P., R. LeTissier, and J. P. Stoye. 1997. Endogenous retroviruses and the evolution of resistance to retroviral infection. *Trends Microbiol.* **5:**313–318.

Domingo, E., and J. J. Holland. 1997. RNA virus mutations and fitness for survival. *Annu. Rev. Microbiol.* **51:**151–178.

Dower, S. 2000. Cytokines, virokines and the evolution of immunity. *Nat. Immunol.* **1:**367–368.

Ebert, D. 1998. Experimental evolution of parasites. *Science* **282:**1432–1435.

Gould, E. A. 2002. Evolution of the Japanese encephalitis serocomplex viruses. *Curr. Top. Microbiol. Immunol.* **267:**391–404.

Loewer, R., J. Loewer, and R. Kurth. 1996. The viruses in all of us: characteristics and biological significance of human endogenous retrovirus sequences. *Proc. Natl. Acad. Sci. USA* **93:**5177–5184.

Malim, M. H., and M. Emerman. 2001. HIV-1 sequence variation: drift, shift, and attenuation. *Cell* **104:**469–472.

Palumbi, S. 2001. Humans as the world's greatest evolutionary force. *Science* **293:**1786–1790.

Shadan, F. F., and L. P. Villarreal. 1996. The evolution of small DNA viruses of eukaryotes: past and present considerations. *Virus Genes* **11:**239–257.

Smith, D. B., J. McAllister, C. Casino, and P. Simmonds. 1997. Virus "quasispecies": making a mountain out of a molehill? *J. Gen. Virol.* **78:**1511–1519.

Solomon, T., H. Ni, D. W. C. Beasley, M. Ekkelenkamp, M. J. Cardosa, and A. D. T. Barrett. 2003. Origin and evolution of Japanese encephalitis virus in southeast Asia. *J. Virol.* **77:**3091–3098.

Sturtevant, A. H. 1937. Essays on evolution. I. On the effects of selection on mutation rate. *Q. Rev. Biol.* **12:**464–467.

Van Regenmortel, M. H. V., D. H. L. Bishop, C. M. Fauquet, M. A. Mayo, J. Manilkoff, and C. H. Calisher. 1997. Guidelines to the demarcation of virus species. *Arch. Virol.* **142:**1505–1518.

Villarreal, L. P. 1997. On viruses, sex, and motherhood. *J. Virol.* **71:**859–865.

Worobey, M., and E. C. Holmes. 1999. Evolutionary aspects of recombination in RNA viruses. *J. Gen. Virol.* **80:**2535–2543.

Emerging Viruses

Brinton, M. A. 2002. The molecular biology of West Nile virus: a new invader of the Western Hemisphere. *Annu. Rev. Microbiol.* **56:**371–402.

Murphy, F. A., and N. Nathanson. 1994. New and emerging virus diseases. *Semin. Virol.* **5:**85–87.

Nathanson, N. 1997. The emergence of infectious diseases: societal causes and consequences. *ASM News* **63:**83–88.

Reid, A. H., J. K. Taubenberger, and T. G. Fanning. 2001. The 1918 Spanish influenza: integrating history and biology. *Microbes Infect.* **3:**81–87.

Steinhauer, D. A., and J. J. Skehel. 2002. Genetics of influenza viruses. *Annu. Rev. Genet.* **36:**305–332.

Strauss, J. H., and E. G. Strauss. 1997. Recombination in alphaviruses. *Semin. Virol.* **8:**85–94.

Weiss, R. A. 1998. Transgenic pigs and virus adaptation. *Nature* **391:**327–328.

Subviral Agents

Chesebro, B., and B. Caughey. 2002. Transmissible spongiform encephalopathies (prion protein diseases), p. 1241–1262. *In* D. D. Richman, R. J. Whitley, and F. G. Hayden (ed.), *Clinical Virology,* 2nd ed. ASM Press, Washington, D.C.

Cohen, F. E., and S. B. Prusiner. 1998. Pathological conformations of prion proteins. *Annu. Rev. Biochem.* **67:**793–819.

Diener, T. O. 1996. Origin and evolution of viroids and viroid-like satellite RNAs. *Virus Genes* **11:**119–131.

Farquhar, C. F., R. A. Somerville, and M. E. Bruce. 1998. Straining the prion hypothesis. *Nature* **391:**345–346.

Prusiner, S. B. 1998. Prions. *Proc. Natl. Acad. Sci. USA* **95:**13363–13383.

Weissmann, C., M. Enari, P.-C. Klohn, D. Rossi, and E. Flechsig. 2002. Transmission of prions. *Proc. Natl. Acad. Sci. USA* **99:**16378–16383.

Selected Papers

Virus Evolution

Agostini, H. T., R. Yanagihara, V. Davis, C. F. Ryschkewitsch, and G. L. Stoner. 1997. Asian genotypes of JC virus in Native Americans and in a Pacific island population: markers of viral evolution and human migration. *Proc. Natl. Acad. Sci. USA* **94:**14542–14546.

Baric, R. S., B. Yount, L. Hensley, S. A. Peel, and W. Chen. 1997. Episodic evolution mediates interspecies transfer of a murine coronavirus. *J. Virol.* **71:**1946–1955.

Bonhoeffer, S., and M. A. Nowak. 1994. Intra-host versus inter-host selection: viral strategies of immune function impairment. *Proc. Natl. Acad. Sci. USA* **91:**8062–8066.

Chao, L. 1990. Fitness of RNA virus decreased by Muller's ratchet. *Nature* **348:**454–455.

Chao, L. 1997. Evolution of sex and the molecular clock in RNA viruses. *Gene* **205:**301–308.

Clarke, D. K., E. A. Duarte, A. Moya, S. F. Elena, E. Domingo, and J. J. Holland. 1993. Genetic bottlenecks and population passages cause profound fitness differences in RNA viruses. *J. Virol.* **67:**222–228.

Clarke, D. K., E. A. Duarte, S. Elena, A. Moya, E. Domingo, and J. J. Holland. 1994. The Red Queen reigns in the kingdom of RNA viruses. *Proc. Natl. Acad. Sci. USA* **91:**4821–4824.

Domingo, E., D. Sabo, T. Taniguchi, and C. Weissmann. 1978. Nucleotide sequence heterogeneity of an RNA phage population. *Cell* **13:**735–744.

Drake, J. W., and E. F. Allen. 1968. Antimutagenic DNA polymerases of bacteriophage T4. *Cold Spring Harbor Symp. Quant. Biol.* **33:**339–344.

Eigen, M. 1971. Self organization of matter and the evolution of biological macromolecules. *Naturwissenschaften* **58:**465–523.

Eigen, M. 1996. On the nature of viral quasispecies. *Trends Microbiol.* **4:**212–214.

Guidotti, L. G., P. Borrow, M. V. Hobbs, B. Matzke, I. Gresser, M. B. A. Oldstone, and F. V. Chisari. 1996. Viral cross talk: intracellular inactivation of the hepatitis B virus during an unrelated viral infection of the liver. *Proc. Natl. Acad. Sci. USA* **93:**4589–4594.

Hall, J. D., D. M. Coen, B. L. Fisher, M. Weisslitz, S. Randall, R. E. Almy, P. T. Gelep, and P. A. Schaffer. 1984. Generation of genetic diversity in herpes simplex virus: an antimutator phenotype maps to the DNA polymerase locus. *Virology* **132:**26–37.

Herniou, E., J. Martin, and M. Tristem. 1998. Retroviral diversity and distribution in vertebrates. *J. Virol.* **72:**5955–5966.

Mayr, E. 1997. The objects of selection. *Proc. Natl. Acad. Sci. USA* **94:**2091–2094.

McGeoch, D. J., S. Cook, A. Dolan, F. E. Jamieson, and E. A. R. Telford. 1995. Molecular phylogeny and evolutionary timescale for the family of mammalian herpesviruses. *J. Mol. Biol.* **247:**443–458.

Muller, H. J. 1964. The relation of recombination to mutational advance. *Mutat. Res.* **1:**2–9.

Nichol, S. T. K., J. E. Rowe, and W. M. Fitch. 1993. Punctuated equilibrium and positive Darwinian evolution in vesicular stomatitis virus. *Proc. Natl. Acad. Sci. USA* **90:**10424–10428.

Oude Essink, B. B., N. K. T. Back, and B. Berkhout. 1997. Increased polymerase fidelity of the 3TC-resistant variants of HIV-1 reverse transcriptase. *Nucleic Acids Res.* **25:**3212–3217.

Powers, A. M., A. C. Brault, Y. Shirako, E. G. Strauss, W. Kang, J. H. Strauss, and S. C. Weaver. 2001. Evolutionary relationships and systematics of the alpha viruses. *J. Virol.* **75:**10118–10131.

Preston, B. D., B. J. Poiesz, and L. A. Loeb. 1988. Fidelity of HIV-1 reverse transcriptase. *Science* **242:**1168–1171.

Saiz, J.-C., and E. Domingo. 1996. Virulence as a positive trait in viral persistence. *J. Virol.* **70:**6410–6413.

Schrag, S. J., P. A. Rota, and W. Bellini. 1999. Spontaneous mutation rate of measles virus: direct estimation based on mutations conferring monoclonal antibody resistance. *J. Virol.* **73:**51–54.

Simmonds, P., and D. B. Smith. 1999. Structural constraints on RNA virus evolution. *J. Virol.* **73:**5787–5794

Spiegelman, S., N. R. Pace, D. R. Mills, R. Levisohn, T. S. Eikhom, M. M. Taylor, R. L. Peterson, and D. H. L. Bishop. 1968. The mechanism of RNA replication. *Cold Spring Harbor Symp. Quant. Biol.* **33:**101–124.

Tufariello, J., M. S. Cho, and M. S. Horwitz. 1994. Adenovirus E3 14.7-kilodalton protein, an antagonist of tumor necrosis factor cytolysis, increases the virulence of vaccinia virus in severe combined immunodeficient mice. *Proc. Natl. Acad. Sci. USA* **91:**10987–10991.

Wagenaar, T. R., V. T. K. Chow, C. Buranathai, P. Thawatsupha, and C. Grose. 2003. The out of Africa model of varicella-zoster virus evolution: single nucleotide polymorphisms and private alleles distinguish Asian clades from European/North American clades. *Vaccine* **21:**1072–1081.

Webster, R. G., W. G. Laver, G. M. Air, and G. C. Schild. 1982. Molecular mechanisms of variation in influenza viruses. *Nature* **296:**115–121.

Zhu, T., B. T. Korber, A. J. Nahmias, E. Hooper, P. M. Sharp, and D. D. Ho. 1998. An African HIV-1 sequence from 1959 and implications for the origin of the epidemic. *Nature* **391:**594–597.

Emerging Viruses and TSEs

Anderson, N. G., J. L. Gerin, and M. L. Anderson. 2003. Global screening for human viral pathogens. *Emerg. Infect. Dis.* **9:**768–773.

Chua, K. B., W. J. Bellini, P. A. Rota, B. H. Harcout, A. Tamin, S. K. Lam, T. G. Ksiazek, P. E. Rollin, S. R. Zaki, W.-J. Shieh, C. S. Goldsmith, D. J. Gubler, J. T. Roehrig, B. Eaton, A. R. Gould, J. Olson, H. Field, P. Daniels, A. E. Ling, C. J. Peters, L. J. Anderson, and B. W. J. Mahy. 2000. Nipah virus: a recently emergent deadly paramyxovirus. *Science* **288:**1432–1435.

Henige, D. 1986. When did smallpox reach the New World (and why does it matter)?, p. 11–26. *In* P. E. Lovejoy (ed.), *Africans in Bondage.* University of Wisconsin Press, Madison.

Jin, L., D. W. G. Brown, M. E. B. Ramsay, P. A. Rota, and W. J. Bellini. 1997. The diversity of measles virus in the United Kingdom, 1992–1995. *J. Gen. Virol.* **78:**1287–1294.

Leitner, T., S. Kumar, and J. Albert. 1997. Tempo and mode of nucleotide substitutions in gag and env gene fragments in human immunodeficiency virus type 1 populations with a known transmission history. *J. Virol.* **71:**4761–4770.

Morimoto, K., M. Patel, S. Corisdeo, D. C. Hooper, Z. F. Fu, C. E. Rupprecht, H. Koprowski, and B. Dietzschold. 1996. Characterization of a unique variant of bat rabies virus responsible for newly emerging human cases in North America. *Proc. Natl. Acad. Sci. USA* **93:**5653–5658.

Murray, K., P. Selleck, P. Hooper, A. Hyatt, A. Gould, L. Gleeson, H. Westbury, L. Hiley, L. Selvey, B. Rodwell, and P. Ketterer. 1995. A morbillivirus that caused fatal disease in horses and humans. *Science* **268:**94–97.

Parrish, C. R. 1997. How canine parvovirus suddenly shifted host range. *ASM News* **63:**307–311.

Reid, A. H., T. G. Fanning, J. V. Hultin, and J. K. Taubenberger. 1999. Origin and evaluation of the 1918 "Spanish" influenza virus hemagglutinin gene. *Proc. Natl. Acad. Sci. USA* **96:**1651–1656.

Sharp, G. B., Y. Kawaoka, D. J. Jones, W. J. Bean, S. P. Pryor, V. Hinshaw, and R. G. Webster. 1997. Co-infection of wild ducks by influenza A viruses: distribution patterns and biological significance. *J. Virol.* **71:**6128–6135.

Smith, A. W., D. E. Skilling, N. Cherry, J. H. Mead, and D. O. Matson. 1998. Calicivirus emergence from ocean reservoirs: zoonotic and interspecies movements. *Emerg. Infect. Dis.* **4:**13–20.

Subbarao, K., A. Klimov, J. Katz, H. Regnery, W. Lim, H. Hall, M. Perdue, D. Swayne, C. Bender, J. Huang, M. Hemphill, T. Rowe, M. Shaw, X. Xu, K. Fukuda, and N. Cox. 1998. Characterization of an avian influenza A (H5N1) virus isolated from a child with a fatal respiratory illness. *Science* **279:**393–396.

Taubenberger, J. K., A. H. Reid, A. E. Krafft, K. E. Bijwaard, and T. G. Fanning. 1997. Initial genetic characterization of the 1918 "Spanish" influenza virus. *Science* **275:**1793–1795.

White, A. R., P. Enever, M. Tayebi, R. Mushens, J. Linehan, S. Brandner, D. Anstee, J. Collinge, and S. Hawke. 2003. Monoclonal antibodies inhibit prion replication and delay the development of prion disease. *Nature* **422:**80–83.

Woolhouse, M. E. J. 2002. Population biology of emerging and re-emerging pathogens. *Trends Microbiol.* **10:**S3–S7.

APPENDIX A
Structure, Genome Organization, and Infectious Cycles

Adenoviruses

Family: *Adenoviridae*

Genus	Type species
Mastadenovirus	Human adenovirus C
Aviadenovirus	Fowl adenovirus A

Human serotypes are widespread in the population. Infection by these viruses is often asymptomatic, but can result in respiratory disease in children (members of subgenera B and C), conjunctivitis (members of subgroup B and D) and gastroenteritis (subgroup F serotypes 40 and 41). Human adenoviruses 40 and 41 are the second leading cause (after rotaviruses) of infantile viral diarrhea. These viruses share capsid morphology and linear double-stranded DNA genomes, but the members of the two genera differ in size, organization, and coding sequences. The *Mastadenovirinae* comprise some 50 human adenoviruses and adenoviruses of other mammals, including mice, sheep, and dogs, and some are oncogenic in rodents. Study of human adenovirus transformation has provided fundamental information about mechanisms that control progression of cells through the cell cycle and oncogenesis. Characteristic features of replication of these viruses include precise temporal control of viral gene expression and an unusual mechanism of initiation of viral DNA synthesis (protein priming). Mastadenoviral genomes also include genes transcribed by cellular RNA polymerase III.

Figure 1 Structure and genome organization of human adenovirus type 5. (A) The virion. (Left) Electron micrograph of a negatively stained human adenovirus type 5 particle showing the triangular faces of the icosahedral capsid and the fibers. Bar = 50 nm. Courtesy of M. Bisher, Princeton University. (Right) Diagram of the virion illustrating the major structural units of the capsid, hexons and pentons, additional proteins (e.g., VI and IX) that stabilize the capsid, and the viral core proteins. The latter are associated with the linear double-stranded DNA genome of nearly 36 kb, which carries covalently linked terminal protein at its 5′ ends. **(B) Genome organization.** The origins of replication (Ori) at each end of the genome, the terminal protein (TP) covalently attached to each end of the genome, and the eight RNA polymerase II (green arrows) and three RNA polymerase III (tan arrows) transcription units are shown. Arrows indicate the direction of transcription. The three small major late (ML) exons designated l-1, l-2, and l-3 are spliced to form the 202-nucleotide tripartite leader common to all messenger RNA (mRNA) species processed from major late primary transcripts. These mRNAs, which form the five families (L1, L2, L3, L4, and L5) defined by their common 3′ polyadenylation sites, encode all but one (polypeptide IX) of the virion structural proteins, as shown for several of the capsid proteins.

Figure 2 Single-cell reproductive cycle of human adenovirus type 2. The virus attaches to a permissive human cell via interaction between the fiber and the coxsackie-adenovirus receptor on the cell surface. The virus enters the cell via endocytosis **(1** and **2)**, a step that depends on the interaction of a second virion protein, penton base, with a cellular integrin protein. Partial disassembly takes place prior to entry of particles into the cytoplasm **(3)**. Further uncoating takes place and the viral genome associated with core protein VII is imported into the nucleus **(4)**. The host cell RNA polymerase II transcription system transcribes the immediate-early E1A gene **(5)**. Following alternative splicing and export of E1A mRNAs to the cytoplasm **(6)**, E1A proteins are synthesized by the cellular translation machinery **(7)**. These proteins, which are extensively modified by phosphorylation, are imported into the nucleus **(8)**, where they regulate transcription of both cellular and viral genes and the proliferation state of the host cell. The larger E1A protein stimulates transcription of the viral early genes by cellular RNA polymerase II **(9a)**. Transcription of the VA genes by host cell RNA polymerase III also begins during the early phase of infection **(9b)**. The early pre-mRNA species are processed, exported to the cytoplasm **(10)**, and translated **(11)**. These early proteins include the viral replication proteins, which are imported into the nucleus **(12)** and cooperate with a limited number of cellular proteins in viral DNA synthesis **(13)**. Replicated viral DNA molecules can serve as templates for further rounds of replication **(14)** or for transcription of late genes **(15)**. Some late promoters are activated simply by viral DNA replication, but maximally efficient transcription of the major late transcription unit requires the late IVa2 protein and a second, unidentified, infected-cell-specific protein. Processed late mRNA species are selectively exported from the nucleus as a result of the action of early E1B 55kDa and E4 Orf6 proteins **(16)**. Their efficient translation in the cytoplasm **(17)** requires the major VA RNA, VA RNA-I, which counteracts a cellular defense mechanism, and the late L4 100kDa protein. The latter protein also serves as a chaperone for assembly of trimeric hexons as they and the other structural proteins are imported into the nucleus **(18)**. Within the nucleus, capsids are assembled from these proteins and progeny viral genomes to form noninfectious immature virions **(19)**. Assembly requires a packaging signal located near the left end of the genome, as well as the IVa2 and L4 33kDa proteins. Immature virions contain the precursors of the mature forms of several proteins. Mature infectious virions are formed **(20)** when these precursor proteins are cleaved by the viral protease, encoded within the L3 region, and assembled into the virion core. Progeny virions are released **(21)**, usually upon destruction of the host cell via mechanisms that are not well understood.

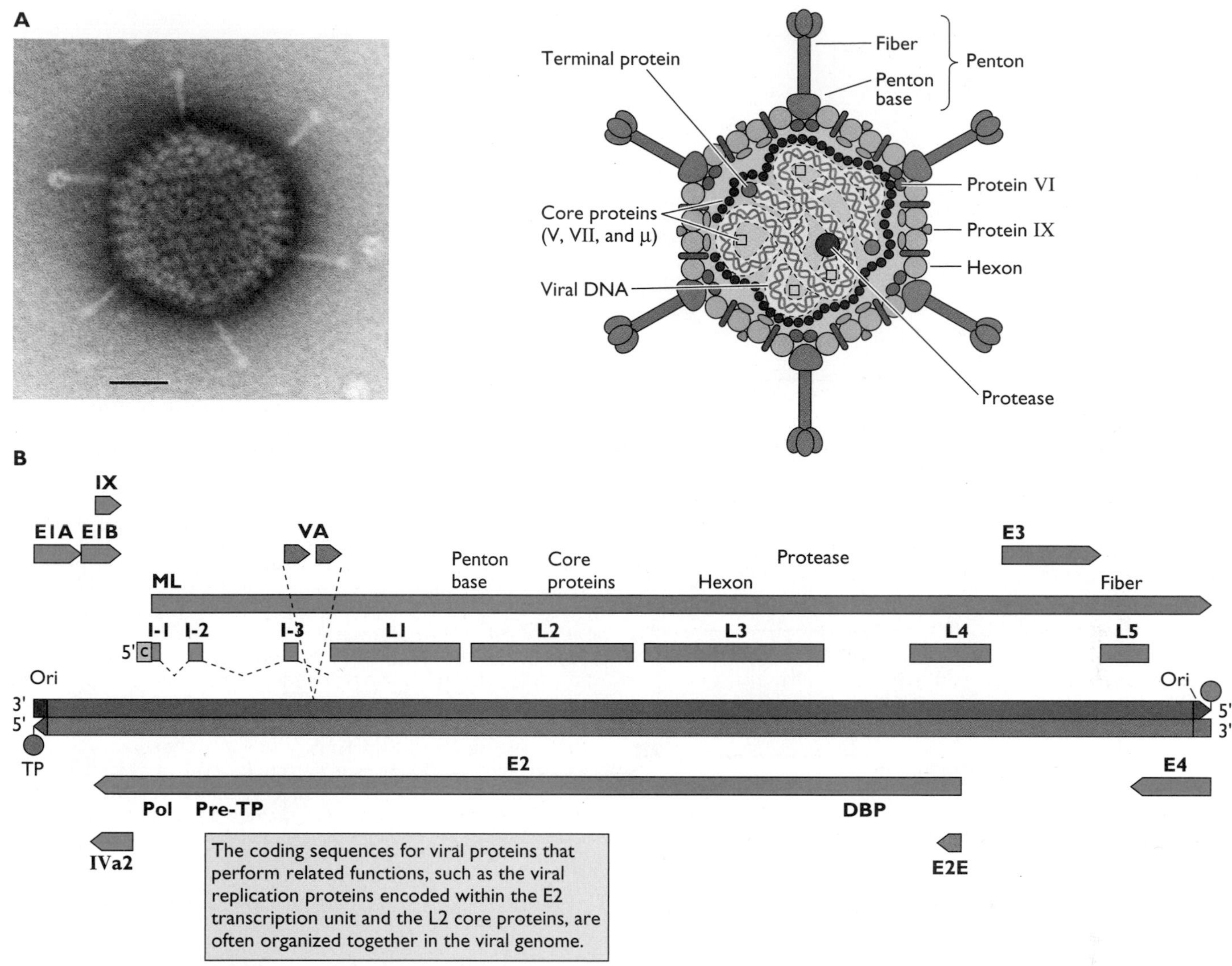

Figure 1 Structure and genome organization of human adenovirus type 5

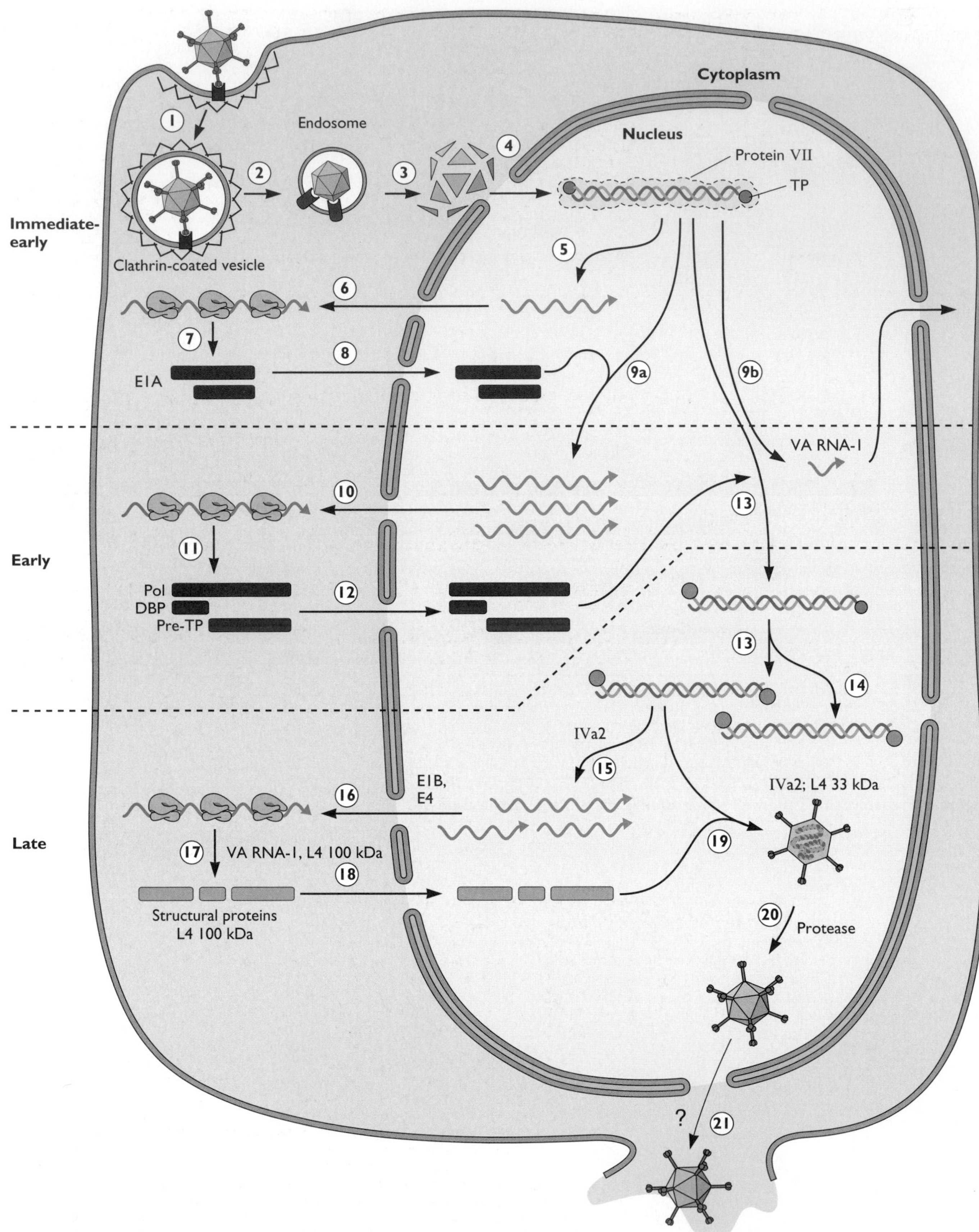

Figure 2 Single-cell reproductive cycle of human adenovirus type 2

Hepadnaviruses

Family: *Hepadnaviridae*

Genus	Type species
Orthohepadnavirus	Human hepatitis B virus
Avihepadnavirus	Duck hepatitis B virus

The hepadnaviruses all show very narrow host specificity and marked tropism for liver tissue. Some hepadnaviruses can replicate following inoculation of primary hepatocytes with virus-containing serum, but most hepadnaviruses cannot be propagated in cultured cells. However, replication can be initiated by transfection of culture cells with cloned viral DNA. Hepadnaviruses replicate via an RNA intermediate and, like the retroviruses, encode a reverse transcriptase. Both families are included in the group called **retroid viruses.** Natural infections may be acute or persistent, depending on host age, inoculum dose, and other (undefined) factors. Approximately 5% of the world's population has been infected with human hepatitis B virus; 250 million to 300 million are persistently infected. Persistent infection with the ortho- but not the avihepadnaviruses confers an increased risk for hepatocellular carcinoma.

Figure 3 Structure and genome organization of orthohepadnaviruses. (A) The virion. (Left) Electron micrograph of negatively stained woodchuck hepatitis virus, a mammalian hepadnavirus related to human hepatitis B virus. Courtesy of W. Mason and T. Gales, Fox Chase Cancer Center. (Right) Diagram of hepatitis B virus and hepatitis B virus-related particles that accumulate in the serum of infected individuals. **(B) Genome organization.** The viral reverse transcriptase (blue ball) is bound to the 5′ end of the (–) DNA strand, which includes a short terminal redundancy signified by the small overlap in the diagram. This strand contains a nick, whereas the (+) strand contains a large gap. Nucleotides are numbered from the single *Eco*RI site indicated. The (–) strand contains 3,227 nucleotides of unique sequence. Synthesis of subgenomic viral RNA transcripts is initiated at the several sites indicated, but all are polyadenylated at the polyadenylation site (*t*). The pregenome transcript is longer than genome length and begins at DR1 and ends at *t*. The positions of the open reading frames are indicated in the outer circle. A lot of information is packed into the hepadnaviral viral genome as a result of overlapping reading frames and multiple translational initiation sites. Every nucleotide in the hepadnaviral DNA is part of at least one open encoding frame, and every *cis*-acting sequence is also included within one or more open reading frames. Although this arrangement represents remarkable efficiency for the virus, it is a nightmare for the molecular biologist trying to manipulate the viral genome.

Figure 4 Single-cell reproductive cycle of hepatitis B virus. The virion attaches to a susceptible hepatocyte **(1)** through recognition of a cell surface receptor(s) that has yet to be identified. The mechanism of virus uptake **(2)** is also unknown, and repair of the gapped (+) DNA strand is accomplished **(3)** by as yet unidentified enzymes. The DNA is translocated to the nucleus **(4)**, where it is found in a covalently closed circular form called CCC DNA. The (–) strand of such CCC DNA is the template for transcription by cellular RNA polymerase II **(5)** of a longer-than-genome-length RNA called the pregenome and shorter, subgenomic transcripts (Fig. 3B), all of which serve as mRNAs. Viral mRNAs are transported from the nucleus **(6)**. Subgenomic viral mRNAs are translated by ribosomes bound to the endoplasmic reticulum (ER) **(7)**, and the proteins destined to become surface antigens (HBsAg) in the viral envelope enter the secretory pathway. The pregenome RNA is translated at low efficiency **(8)** to produce a 90-kDa polymerase protein, P, which possesses reverse transcriptase activity. This protein then binds to a specific site at the 3′ end of its own transcript, where viral DNA synthesis is eventually initiated. The pregenome RNA also serves as mRNA for the capsid protein **(9)**. Concurrently with capsid formation, the RNA-P protein complex is packaged **(10)** and reverse transcription begins with synthesis of (–) strand DNA **(11)**. At early times after infection, the DNA is recirculated to the nucleus **(12)**, where the process is repeated, resulting in the eventual accumulation of 10 to 30 molecules of CCC DNA and a concomitant increase in viral mRNA concentrations. At later times, and perhaps as a consequence of the accumulation of sufficient viral proteins, mature nucleocapsids are formed **(13)**. These structures acquire envelopes by budding into the ER, where viral maturation is completed **(14)**. Progeny enveloped virions are released from the cell by exocytosis **(15)**.

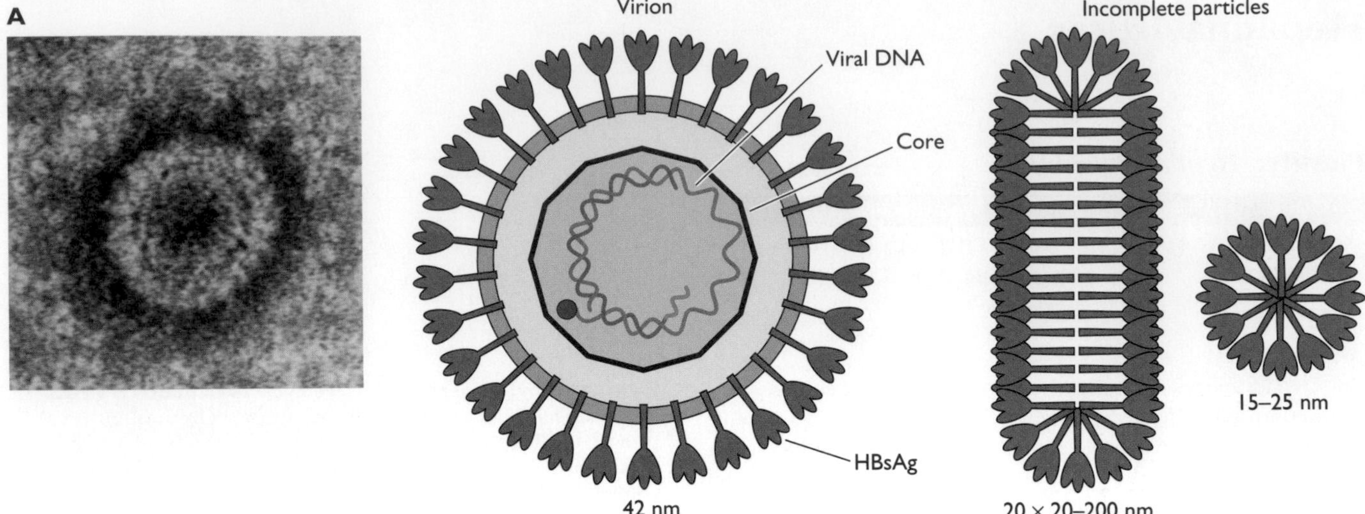

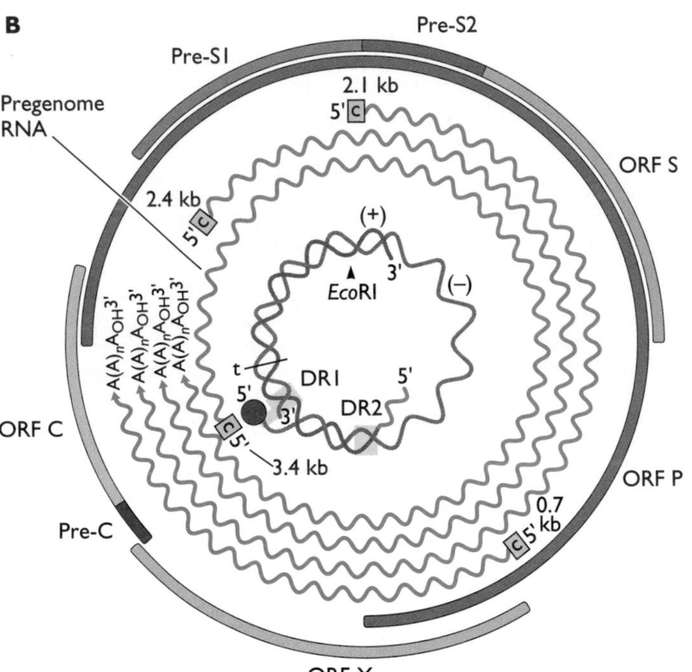

Figure 3 Structure and genome organization of orthohepadnaviruses

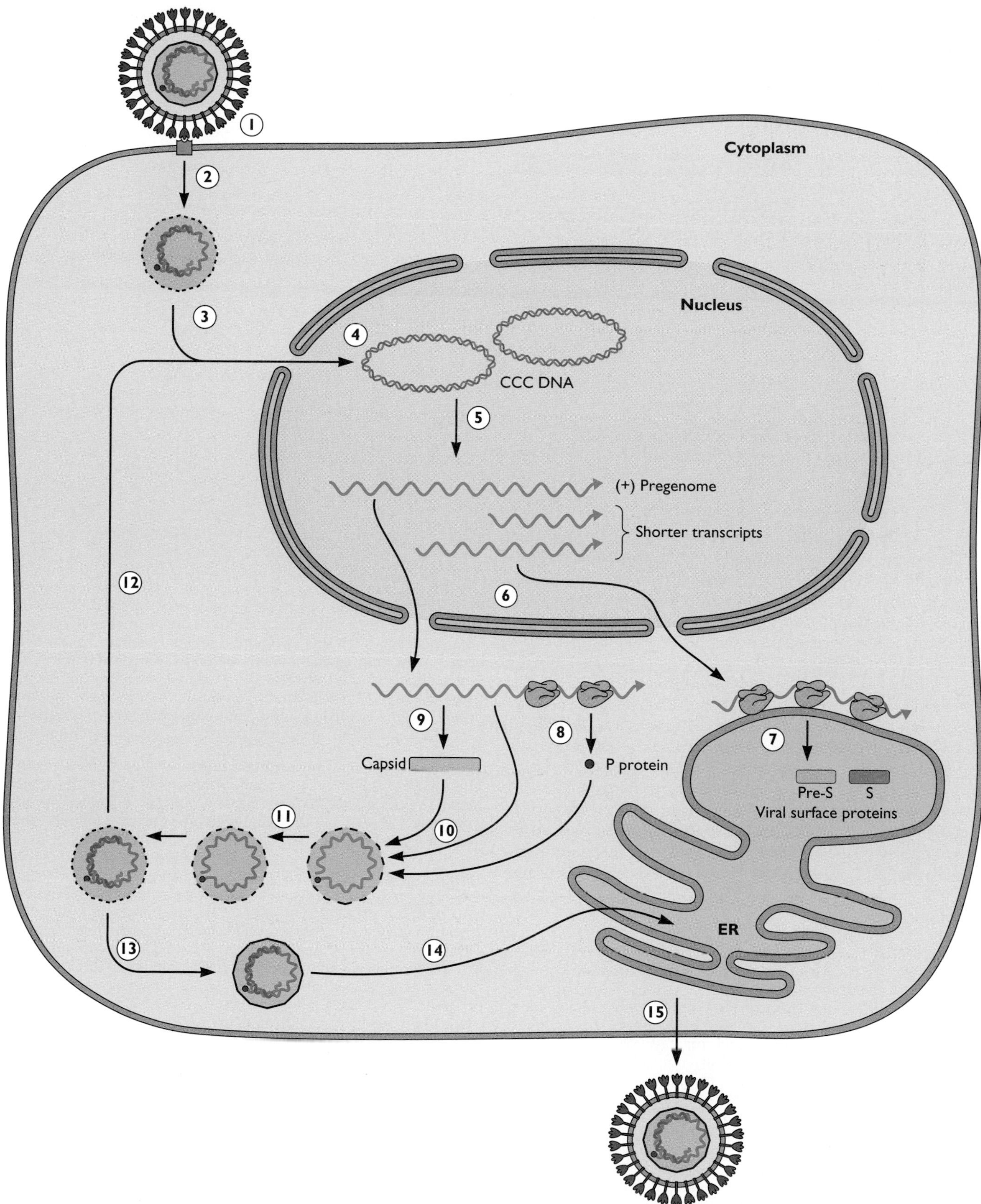

Figure 4 Single-cell reproductive cycle of hepatitis B virus

Herpesviruses

Family: *Herpesviridae*

Genus	Type species
Subfamily *Alphaherpesvirinae*	
Simplexvirus	Human herpes simplex virus type 1
Varicellovirus	Varicella-zoster virus
Subfamily *Betaherpesvirinae*	
Cytomegalovirus	Human cytomegalovirus
Muramegalovirus	Murine cytomegalovirus 1
Roseolovirus	Human herpesvirus 6
Subfamily *Gammaherpesvirinae*	
Lymphocryptovirus	Epstein-Barr virus
Rhadinovirus	Herpesvirus saimiri

Undefined subfamily: channel catfish virus, oyster herpesvirus

The family *Herpesviridae* comprises over 120 viruses that infect a wide range of vertebrates and at least one invertebrate (the oyster). While some herpesviruses have broad host ranges, most are restricted to infection of a single species and spread in the population by direct contact or aerosols. The hallmark of herpesvirus infections is the establishment of a lifelong, latent infection that can reactivate to cause one or more rounds of disease. Many herpesvirus infections are not apparent, but if the host's immune defenses are compromised, infections can be devastating. Humans are the natural hosts of eight different herpesviruses: herpes simplex virus types 1 and 2, varicella-zoster virus, Epstein-Barr virus, human cytomegalovirus, and human herpesvirus 6, 7, and 8 (also known as Kaposi's sarcoma-associated herpesvirus). Agricultural pathogens include the alphaherpesviruses pseudorabies virus, bovine herpes virus type 1, equine herpesvirus type 1, and Marek's disease virus types 1 and 2, and the gamma herpesviruses bovine herpesvirus 4 and equine herpesvirus 2.

Figure 5 Structure and genome organization of alphaherpesviruses. (A) The virion. (Left) Electron micrograph of a negatively stained pseudorabies virus. The virion is approximately 200 nm in diameter. Courtesy of T. Mettenleiter, Federal Research Center for Virus Diseases of Animals, Insel Riems, Germany. (Right) Diagram of the virion showing the lipid envelope studded with at least 10 viral glycoproteins, the tegument layer comprising at least 15 viral proteins, and the icosahedral nucleocapsid containing the linear double-stranded genome. **(B) Genome organization of herpes simplex virus type 1.** The genome is organized into a 126-kb long region and a 26-kb short region of double-stranded DNA. The unique sequences of the long and short regions are called UL (108 kb) and US (13 kb), respectively. The long region is bracketed by an inverted repeat sequence of about 9 kb. The terminal repeat is called TRL, and the internal repeat is called IRL. The short region is similarly bracketed by a different inverted repeat sequence (TRS and IRS) of about 7 kb. The herpes simplex virus type 1 genome can "isomerize" or recombine via the inverted repeat sequences such that any population within an infected cell, or in virions, consists of four equimolar isomers in which the UL and US sequences are inverted with respect to each other. The viral genome contains three origins of replication, one in the UL segment and two in a repeated sequence of the US region. There are at least 84 open reading frames in this genome. The approximate locations of genes encoding proteins with different functions during the infectious cycle and of the latency-associated transcription unit are indicated. Genes encoding common functions have the same color shading to illustrate the finding that gene functions tend to be dispersed, rather than clustered, in herpesviral genomes.

Figure 6 Single-cell reproductive cycle of herpes simplex virus type 1. (A) Productive infection. Virions bind to the extracellular matrix (e.g., heparan sulfate or chondroitin sulfate proteoglycans) via gB and gC **(1)**. Another viral membrane protein (gD) interacts with a second cellular receptor **(2)**. This contact initiates fusion of viral and plasma membranes mediated by viral membrane glycoproteins (gD, gB, gH, and gL) **(3)**. As a result of membrane fusion, some tegument proteins and the nucleocapsid are released into the cytoplasm **(4)**. Viral nucleocapsids attach to microtubules and are transported to the nucleus **(5a)**. Certain tegument proteins, like VP16, activate transcription of the viral genome and must also be transported to the nucleus **(5b)**. Other tegument proteins, such as Vhs, remain in the cytoplasm **(6)**. Viral nucleocapsid docks at the nuclear pore **(7)**, releasing viral DNA into the nucleus. VP16 interacts with host transcription proteins to stimulate transcription of immediate-early genes by host cell RNA polymerase II **(8)**. Immediate-early mRNAs are spliced and transported to the cytoplasm **(9)** where they are translated. The immediate-early proteins (α proteins) are transported to the nucleus where they activate transcription of early genes and regulate transcription of immediate-early genes **(10)**. Early gene transcripts, which are rarely spliced, are transported to the cytoplasm **(11)** where they are translated. The early proteins (β proteins) function primarily in DNA replication and production of substrates for DNA synthesis. Some β proteins are transported to the nucleus **(12)** and some function in the cytoplasm. Viral DNA synthesis is initiated from viral origins of replication **(13)**. DNA replication and recombination produces long, concatemeric DNA, the template for late gene

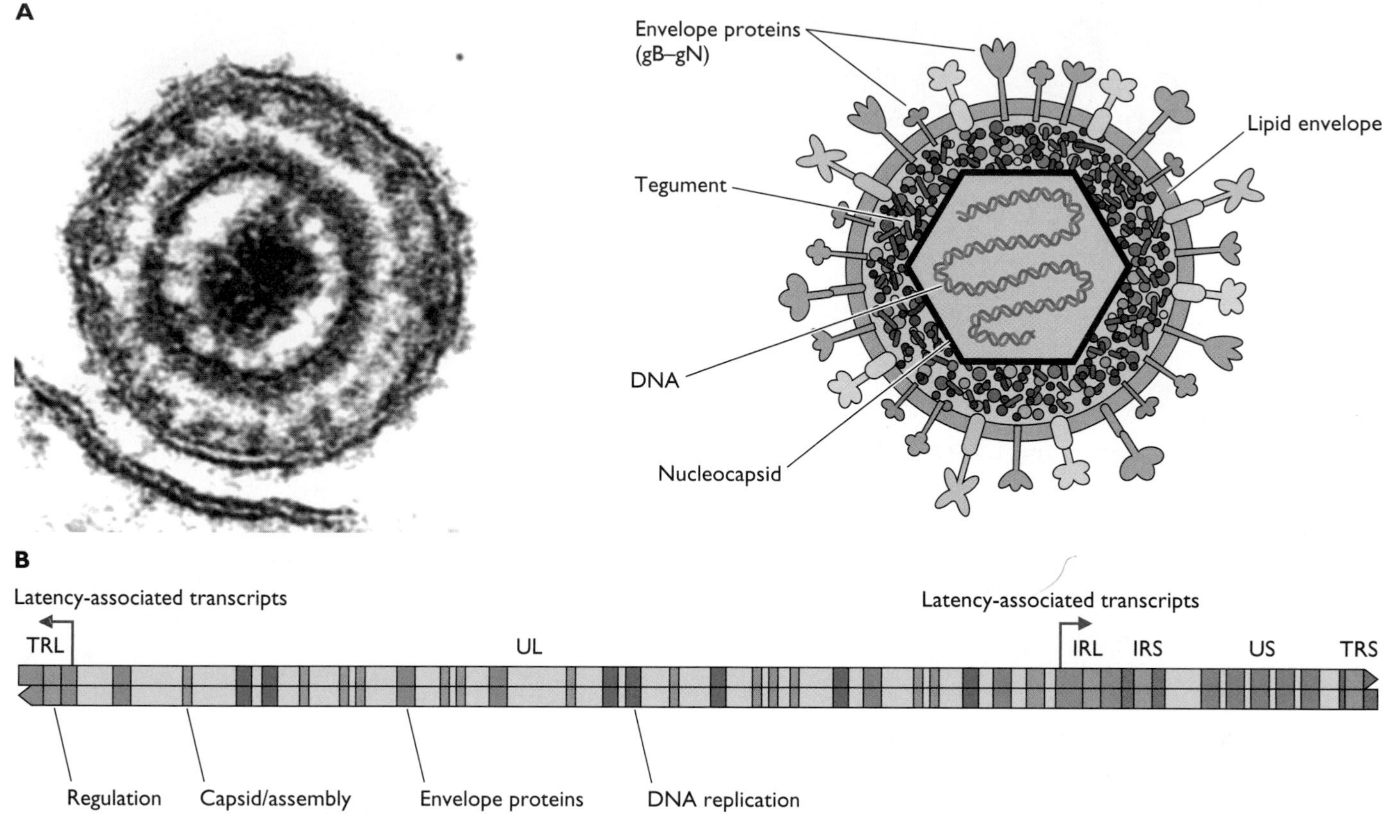

Figure 5 Structure and genome organization of alphaherpesviruses

expression **(14)**. Late mRNAs are rarely spliced, and are transported to the cytoplasm **(15)** where they are translated. Late proteins (γ proteins) primarily are virion structural proteins and additional proteins needed for virus assembly and particle egress. Some γ proteins are made on, and inserted into, membranes of the rough endoplasmic reticulum **(16a)**. Many of these membrane proteins are modified by glycosylation. Some precursor viral membrane proteins are thought to be localized both to the outer and inner nuclear membranes, as well as membranes of the endoplasmic reticulum **(16b)**. The precursor glycoproteins are also transported to the Golgi apparatus for further modification and processing **(16c)**. Mature glycoproteins are transported to the plasma membrane of the infected cell **(16d)**. Some γ proteins are transported to the nucleus for assembly of the nucleocapsid and DNA packaging, while some remain in the cytoplasm **(17)**. Newly replicated viral DNA is packaged into nucleocapsids **(18)**. DNA-containing nucleocapsids, together with some tegument proteins, bud from the inner nuclear membrane into the perinuclear lumen, acquiring an envelope thought to contain precursors to viral membrane proteins **(19)**. The UL34 membrane protein interacts with the UL31 protein to form the site of primary budding at the inner nuclear membrane. Immature enveloped virions fuse with the outer nuclear membrane from within **(20)**, leaving behind UL34 and UL31 proteins, while releasing the nucleocapsid into the cytoplasm. This structure is transported to a late *trans*-Golgi compartment or endosome that contains mature viral membrane proteins **(21)**. Tegument proteins added in the nucleus remain with the nucleocapsid, and others are added in the cytoplasm. As nucleocapsids bud into the late Golgi-endosome compartment, they acquire an envelope containing mature viral envelope proteins and the complete tegument layer (secondary envelopment; **22**). The enveloped virus particle then buds into a vesicle **(23)** that is transported to the plasma membrane for release by exocytosis **(24)**. When neurons are infected, transport vesicles with mature virions inside are not transported in the anterograde direction to axon termini. Rather the nucleocapsid and envelope proteins (both with associated tegument proteins) are transported separately, and secondary envelopment presumably occurs at an internal membrane compartment at the site of egress from axons.

Figure 7 Herpes simplex virus latent infection. Latent infection occurs primarily in neurons found in sensory and autonomic ganglia. In the simplest model, the initiation of latent infection occurs as in **steps 1 to 7** of the productive infection, except that the viral DNA circularizes in the nucleus. It is unclear if tegument proteins are transported into the nucleus **(8)**. How the normal transcriptional cascade is blocked remains a matter of research and debate. However, when the latent state is established, transcription is severely restricted such that a single pre-mRNA is produced from the latency associated transcript (LAT) promoter **(9)**. Low-level or sporadic transcription of immediate-early and early genes can occur, but is not sufficient to initiate a productive infection. The LAT RNA is spliced, and a stable intron in the form of a lariat, called the 2-kb LAT, is produced in the nucleus **(10)**. The spliced LAT mRNA is transported to the cytoplasm where several small open reading frames (e.g., OrfO and OrfP) may be translated into proteins **(11)**. The function of LAT RNAs and the production of LAT proteins are controversial. After months to years, changes in neuronal physiology induced by trauma, hormonal changes, and other stressful conditions render neurons permissive. The genome is then transcribed at higher levels and replicated, and progeny virions are produced as shown in Fig. 6.

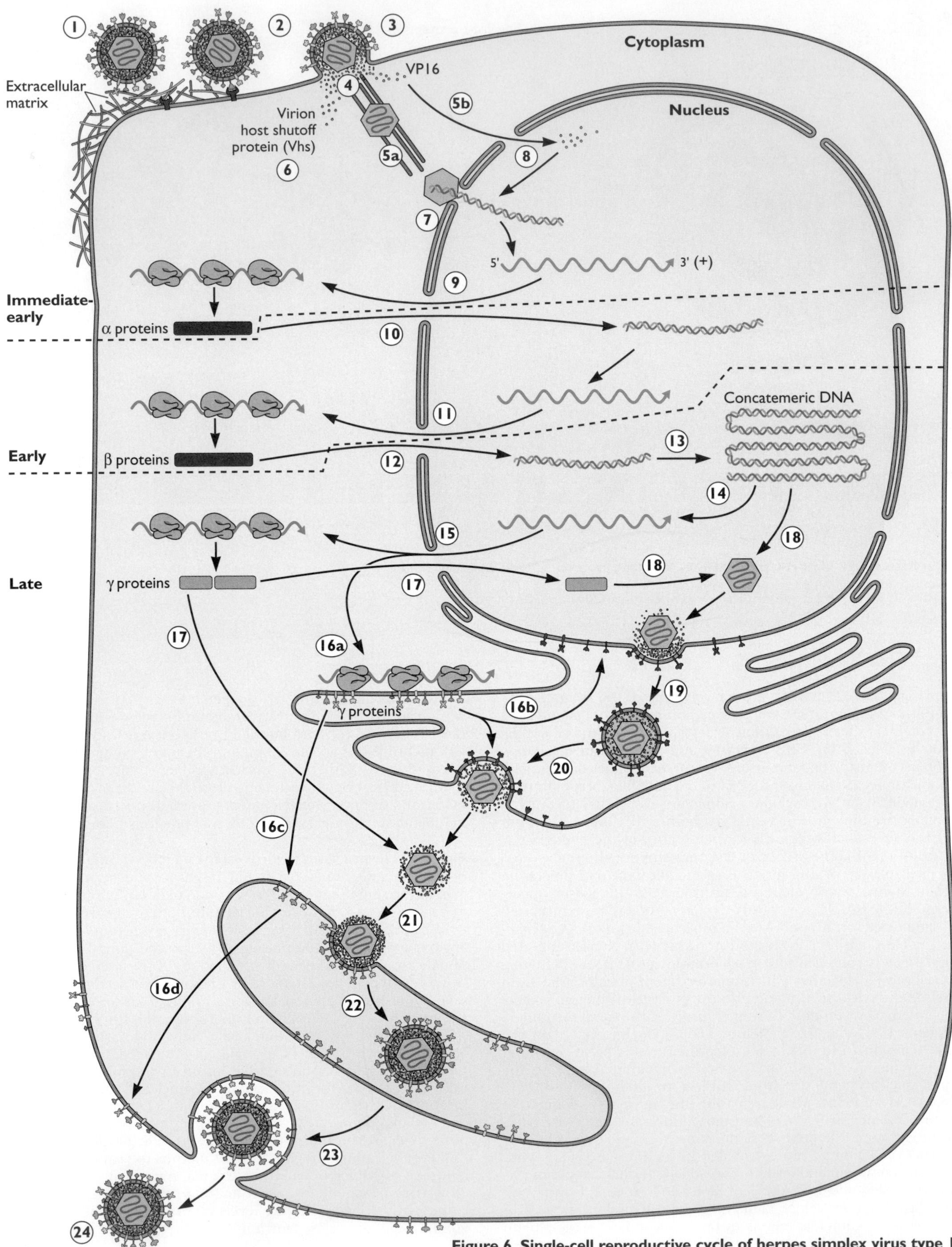

Figure 6 Single-cell reproductive cycle of herpes simplex virus type 1

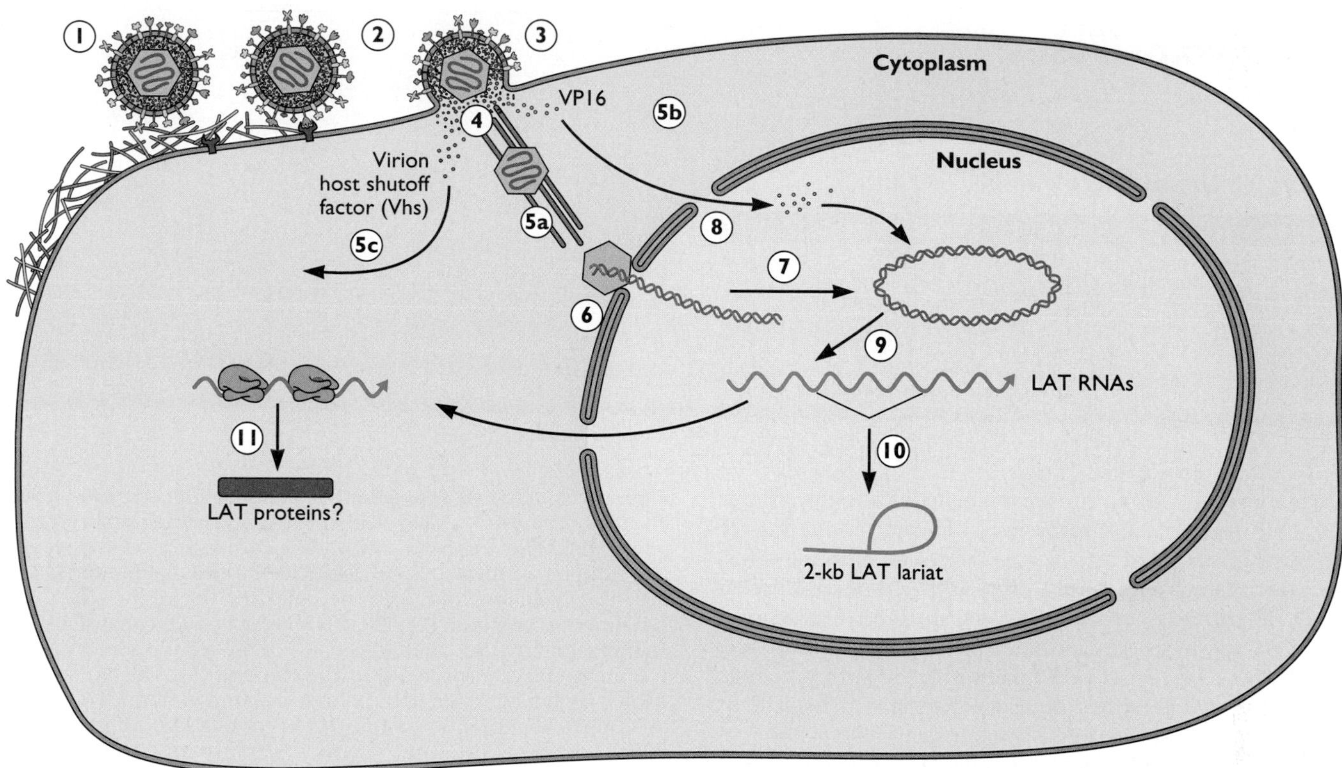

Figure 7 Herpes simplex virus latent infection

Orthomyxoviruses

Family: *Orthomyxoviridae*

Genus	Type species
Influenza A virus	A/PR/8/34(H1N1)
Influenza B virus	B/Lee/40
Influenza C virus	C/California/78
Thogotovirus	Thogoto virus
Isavirus	Infectious salmon anemia virus

Influenza viruses are the causative agents of a highly contagious, often serious acute respiratory illness. The importance of this disease has been an impetus for research on the virus. Influenza viruses are unusual among RNA viruses because all viral RNA synthesis occurs in the cell nucleus. Initiation of viral mRNA synthesis with a capped primer derived from host cell mRNA was first described in cells infected with influenza viruses. The viral genomes undergo extensive reassortment and variation, and are expressed via a remarkable panoply of unusual strategies, including RNA splicing, overlapping reading frames, and leaky scanning.

Figure 8 Structure and genomic organization of the orthomyxovirus influenza A virus. (A) The virion. (Left) Electron micrograph of negatively stained influenza A virus particles. Courtesy of P. Palese, Mt. Sinai School of Medicine. (Right) Diagram indicating the location of the nine virion proteins, the viral envelope, and the eight genomic RNA segments. **(B) Genome organization.** The (–) strand RNA genome comprises eight segments, each of which encodes at least one viral protein as shown. The two smallest genomic RNA segments, 7 and 8, each code for two proteins, because some fraction of the (+) strand mRNA copies are spliced by host cell enzymes. The NS1 (nonstructural) protein is so named because it is not incorporated into virus particles. An accessory protein with proapoptotic activity, PB1-F2, can be produced from the PB1 RNA. Blue boxes containing the letter C denote the capped host mRNA fragments that serve as primers during mRNA synthesis.

Figure 9 Single-cell reproductive cycle of influenza A virus. The virion binds to a sialic acid-containing cellular surface protein or lipid and enters the cell via receptor-mediated endocytosis **(1)**. Upon acidification of the vesicle, the viral membrane fuses with the membrane of the vesicle, releasing the viral nucleocapsids into the cytoplasm **(2)**. The viral nucleocapsids containing (–) strand genomic RNA, multiple copies of the NP protein, and the P proteins are transported into the nucleus **(3)**. The (–) strand RNA is copied by virion RNA polymerase into viral mRNA, using the capped 5′ ends of host pre-mRNAs (or mRNAs) as primers to initiate synthesis **(4)**. The mRNAs are transported to the cytoplasm **(5)**, following splicing in the case of the mRNAs encoding NEP and M2 **(6)**. The mRNAs specifying the viral membrane proteins (HA, NA, M2) are translated by ribosomes bound to the endoplasmic reticulum (ER) **(7)**. These proteins enter the host cell's secretory pathway, where HA and NA are glycosylated. All other mRNAs are translated by ribosomes in the cytoplasm **(8** and **9)**. The PA, PB1, PB2, and NP proteins are imported into the nucleus **(10a)**, where they participate in the synthesis of full-length (+) strand RNAs **(11)** and then (–) strand genomic RNAs **(12)**, both of which are synthesized in the form of nucleocapsids. Some of the newly synthesized (–) strand RNAs enter the pathway for mRNA synthesis **(13)**. The M1 protein and the NS1 protein are transported into the nucleus **(10b)**. Binding of the M1 protein to newly synthesized (–) strand RNAs shuts down viral mRNA synthesis and, in conjunction with the NEP protein, induces export of progeny nucleocapsids to the cytoplasm **(14)**. The HA, NA, and M2 proteins are transported to the cell surface **(15)** and become incorporated into the plasma membrane **(16)**. The virion nucleocapsids associated with the M1 protein **(17)**, and the NEP protein **(18)**, are transported to the cell surface and associate with regions of the plasma membrane that contain the HA and NA proteins. Assembly of virions is completed at this location by budding from the plasma membrane **(19)**.

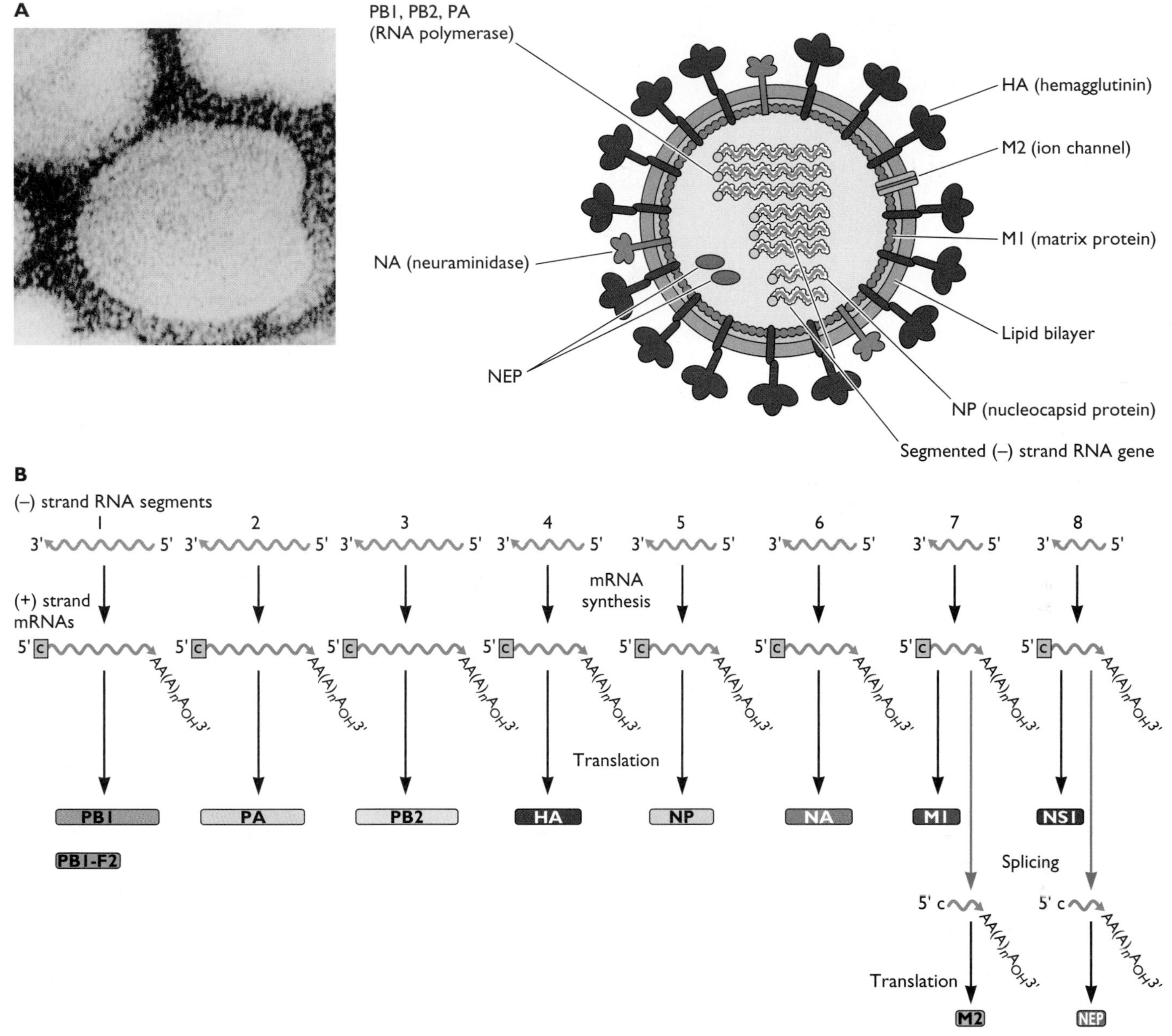

Figure 8 Structure and genomic organization of the orthomyxovirus influenza A virus

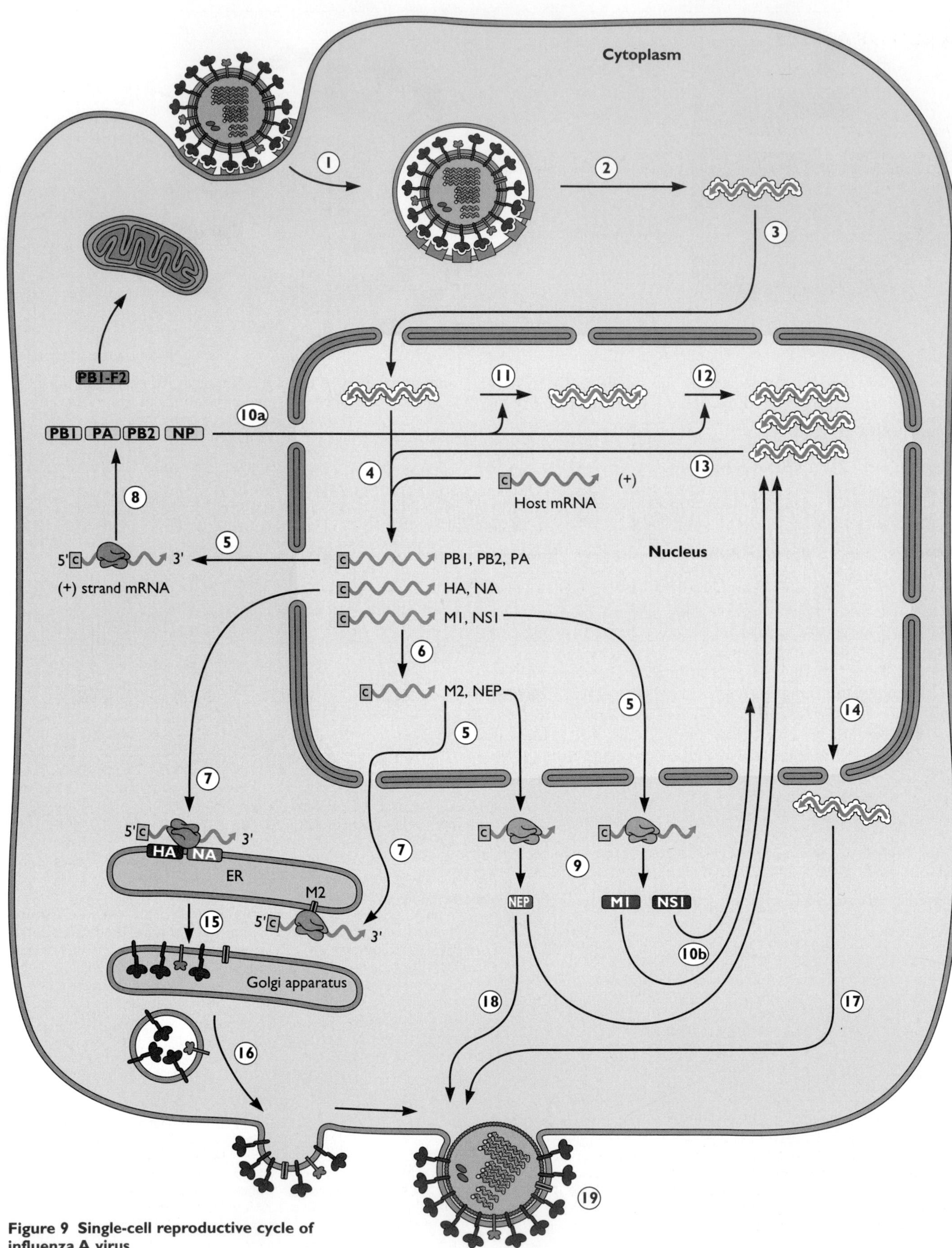

Figure 9 Single-cell reproductive cycle of influenza A virus

Parvoviruses

Family: *Parvoviridae*

Genus	Type species
Subfamily *Densovirinae*	
(3 genera; insect viruses)	
Subfamily *Parvovirinae*	
Parvovirus	Minute virus of mice
Erythrovirus	Human B19 virus
Dependovirus	Human adeno-associated viruses

Members of this family are among the smallest of the DNA animal viruses. They are of special interest because of their unique genomic DNA structure and mechanism of replication. Most parvoviruses, such as the well-studied minute virus of mice, can propagate autonomously, although they require the host cell to go through S phase for replication. Replication of dependoviruses requires a helper, adeno- or herpesvirus, to induce S phase and for expression of components that promote dependovirus replication. These viruses can establish a latent infection during which their DNA is integrated into the host cell genome in an inactive state, to be activated upon subsequent infection with a helper. The genome of the dependovirus featured here, adeno-associated virus (prototype AAV2), is integrated in a specific locus in the long arm of human chromosome 19. Because of its ability to persist, and its lack of pathogenicity, human adeno-associated virus has been of significant interest as a potential vector for gene therapy. Very little is known about the life cycle of adeno-associated virus in an infected individual; the steps described in this appendix have been elucidated primarily from studies of infection in cell culture.

Figure 10 Structure and genome organization of adeno-associated virus type 2 (AAV2). (A) The virion. (Left) Electron micrograph of recombinant AAV empty and DNA-filled (dense) particles. Courtesy of M. Linden, Mount Sinai School of Medicine, New York, N.Y. (Right) The three-dimensional structure of canine parvovirus from cryo-electron microscopy rendered to 21-Å resolution. The icosahedral two-, three- and five-fold axes are indicated. Prepared by Mavis Agbandje-McKenna (University of Florida). From N. Muzyczka and K. I. Berns, *in* D. M. Knipe et al. (ed.), *Fields Virology*, 4th ed. (Lippincott Williams & Wilkins, Philadelphia, Pa., 2001), with permission. The capsid is a naked icosahedron with $T = 1$ symmetry, ~25 nm in diameter and comprising 60 protein subunits, primarily (~90%) VP2. Virions contain equivalent numbers of (+) or (–) single-stranded DNA in separate particles. **(B) Genomic organization.** The AAV2 DNA genome of 4,679 nucleotides, corresponding to 100 map units, includes terminal repeats (TR) of 145 nucleotides, the first 125 of which form a palindromic sequence. The TR is required in *cis* for genome replication, transcription, and encapsidation, and plays a role in integration into the host DNA during establishment of a latent infection. Two open reading frames, *rep* and *cap*, encode replication and capsid proteins, respectively. Promoters for transcription are located at map positions 5, 19, and 40, giving rise to three transcripts; in each case, unspliced forms are translated. Nucleotide positions of splice sites and initiation and termination codons are indicated. The most abundant mRNA is the 2.3-kb species that specifies VP2 and VP3. Colored bars indicate coding sequences, and the protein product of each mRNA is listed at the left. Adapted from R. M. Linden and K. Berns, p. 68–84, *in* S. Faisst and J. Rommelaere (ed.), *Parvoviruses: From Molecular Biology to Pathology and Therapeutic Uses*, vol. 4, *Contributions to Microbiology* (S. Karger, Basel, Switzerland, 2000), with permission.

Figure 11 Latent infection. Heparan sulfate proteoglycans are the primary cell surface receptors for AAV2. However, the processes of adsorption **(1)**, uncoating **(2)**, and entry of the DNA into the nucleus **(3)** are poorly understood for all *Parvoviridae*. In the absence of helper virus, the viral DNA is integrated into human host cell DNA, establishing a latent infection. A small amount of replication of the single-stranded genome must occur prior to integration **(4)** to produce a double-stranded DNA template for transcription from the p5 promoter **(5)**. Translation of this transcript produces Rep proteins **(6)** that are required to both facilitate integration **(7)** and suppress further transcription from all three promoters. The junction with cellular DNA is usually within the terminal repeat (TR) or near a Rep-binding site within the p5 promoter. Neither the actual site in the viral DNA nor the cellular sequence at the junction is unique. However, 70 to 100% of the integration events occur within a region of several hundred nucleotides in chromosome 19q13.3-qter. This region-specific integration is facilitated by viral Rep proteins which bind to a specific sequence present in both the chromosome 19 target and the viral genome, bringing them together via Rep-Rep contacts **(8)**. In most latently infected cells the viral DNA is integrated as a tandem (head-to-tail) array of several genome equivalents. This unexpected arrangement indicates that the mechanism of integration is likely to be distinct from that of viral DNA replication (see Fig. 12, step 16). The integrated viral genome(s) can remain dormant through many cell cycles. However, upon superinfection with a helper virus, the AAV genome is rescued and a productive infection is initiated. The molecular details of the rescue process have not yet been elucidated.

Figure 12 Single-cell reproductive cycle. Upon coinfection with the helper virus, or rescue by superinfection with a helper virus **(9)**, adeno-associated virus undergoes a productive infection. With an adenovirus helper, this response is dependent on the expression of early genes E1A, E1B, E4, and E2A **(10)**, which induce S phase and the concomitant production of cellular DNA replication proteins needed for viral DNA replication **(11)**. The adenovirus EIA transcriptional activator also induces transcription from the p5 promoter **(12)**, leading to the production of large amounts of Rep78/68 mRNA and proteins **(13)**. These proteins then function as powerful transcriptional activators rather than repressors as in latency, and they induce transcription from both the p5 and p19 promoters. Viral DNA is replicated by a single-strand displacement mechanism that is initiated by recognition of the terminal resolution site (*trs*) by the Rep78/68 proteins, which remain linked covalently to the DNA through subsequent steps of DNA synthesis **(14)**. A very large number of replicating forms (ca. 10^6 double-stranded genomes/cell) can be produced within a short time **(15)**. Some dimeric head-to-head and tail-to tail forms are also observed **(16)**. The capsid proteins produced in the cytoplasm **(17)** self-associate in the nucleus during virion assembly **(18)**. As with the adenovirus helper, progeny virions are released **(19)** usually upon destruction of the cell.

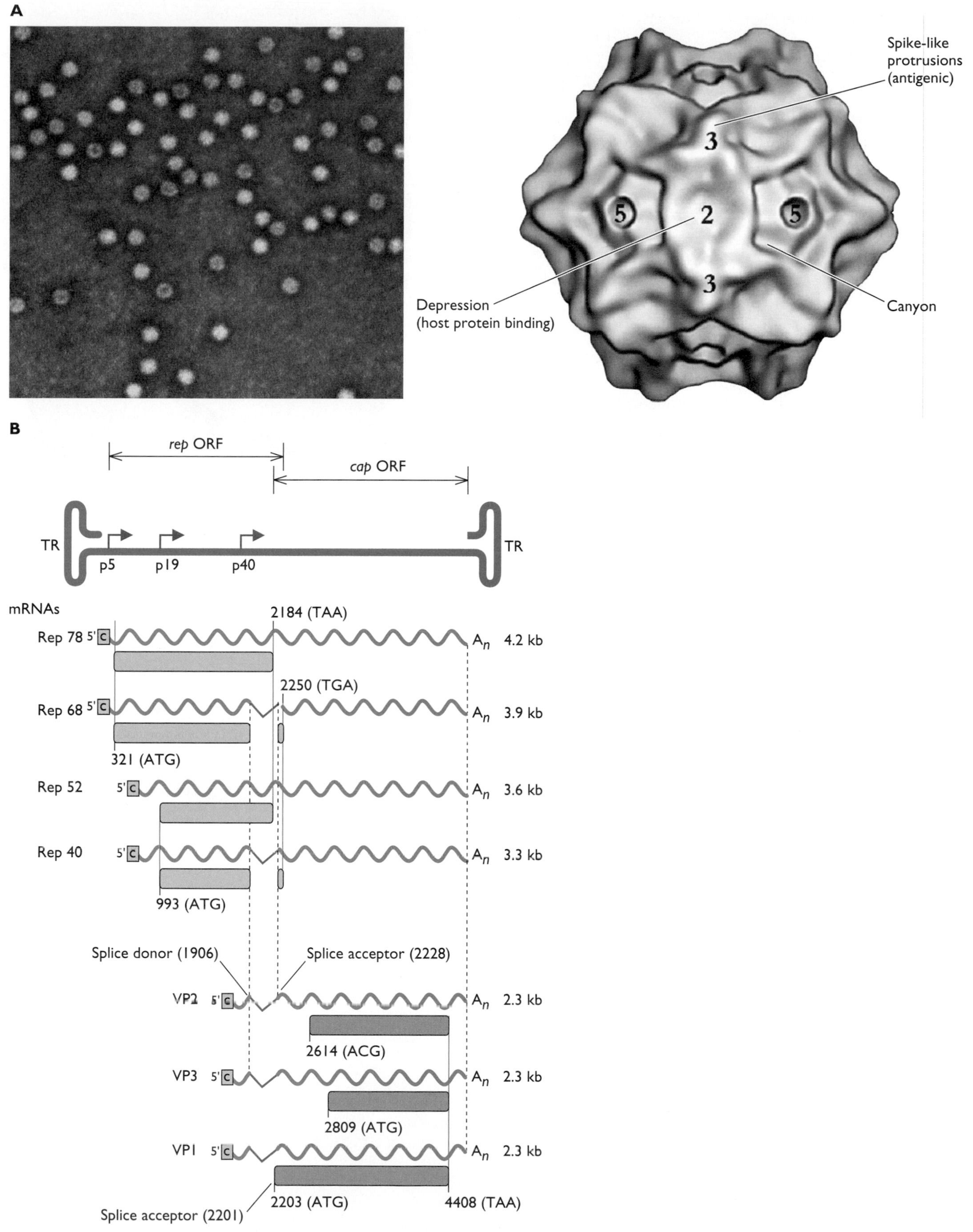

Figure 10 Structure and genome organization of adeno-associated virus type 2

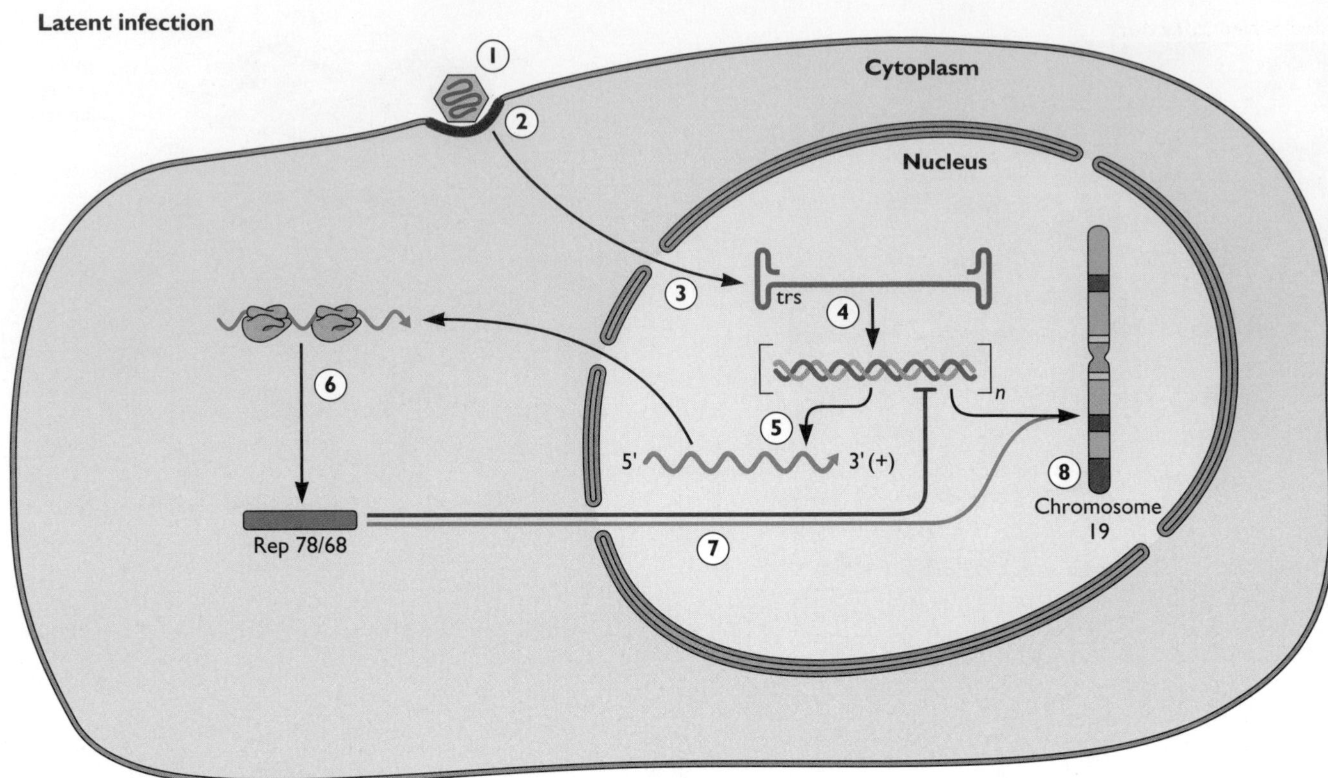

Figure 11 Latent infection

Productive infection

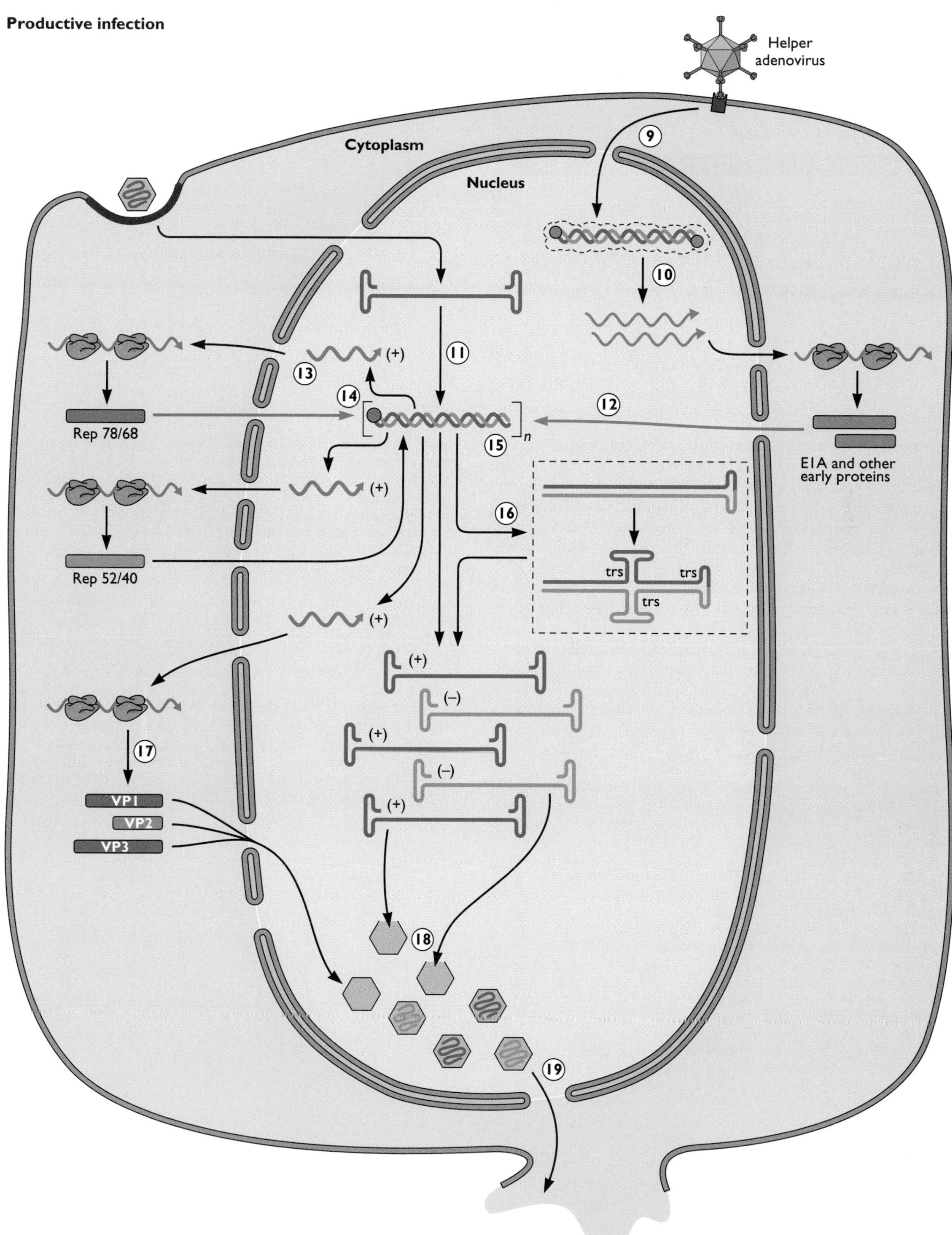

Figure 12 Single-cell reproductive cycle

Picornaviruses

Family: *Picornaviridae*

Genus	Type species
Enterovirus	Poliovirus
Rhinovirus	Human rhinovirus A
Cardiovirus	Encephalomyocarditis virus
Aphthovirus	Foot-and-mouth disease virus
Hepatovirus	Hepatitis A virus
Parechovirus	Human parechovirus
Erbovirus	Equine rhinitis B virus
Kobuvirus	Aichi virus
Teschovirus	Porcine teschovirus 1

The family *Picornaviridae* includes many important human and animal pathogens, such as poliovirus, hepatitis A virus, foot-and-mouth disease virus, and rhinovirus. Because they cause serious disease, poliovirus and foot-and-mouth disease viruses have been the best-studied picornaviruses. Study of these two viruses has had important roles in the development of virology: the first animal virus discovered, in 1898, was foot-and-mouth disease virus. The plaque assay was developed using poliovirus, and the first RNA-dependent RNA polymerase discovered was poliovirus $3D^{pol}$. Polyprotein synthesis was discovered in poliovirus-infected cells, as was translation by internal ribosome entry. The first infectious DNA clone of an animal RNA virus was that of the poliovirus genome, and the first three-dimensional structures of animal viruses determined by X-ray crystallography were those of poliovirus and rhinovirus.

Figure 13 Structure and genomic organization. (A) The virion. (Left) Electron micrograph of negatively stained poliovirus. Courtesy of N. Cheng and D. M. Belnap, National Institutes of Health. (Right) Diagram of the virion, showing the names and locations of component proteins and genomic RNA. The capsid consists of 60 structural units (each made up of a single copy of VP1, VP2, VP3, and VP4, colored blue, yellow, red, and green, respectively) arranged in 12 pentamers. One of the icosahedral faces has been removed to illustrate the locations of VP4 and the viral RNA. **(B) Genome organization.** Polioviral RNA is shown with the VPg protein covalently attached to the 5′ end. The genome is of (+) polarity and encodes a polyprotein precursor. The polyprotein is cleaved during translation by two virus-encoded proteases, $2A^{pro}$ and $3C^{pro}$, to produce structural and nonstructural proteins. The P1 protein is cleaved into the virion capsid proteins, while the P2 and P3 proteins are cleaved to form the proteases and the proteins that participate in viral RNA synthesis.

Figure 14 Single-cell reproductive cycle. The virion binds to a cellular receptor **(1)**; the mechanism and site of uncoating of the RNA genome are unknown **(2)**. The VPg protein, depicted as a small orange circle at the 5′ end of the virion RNA, is removed, and the resulting RNA associates with ribosomes **(3)**. Translation is initiated at an internal site 741 nucleotides from the 5′ end of the viral mRNA, and a polyprotein precursor is synthesized **(4)**. The polyprotein is cleaved during and after its synthesis to yield the individual viral proteins **(5)**. Only the initial cleavages are shown here. The proteins that participate in viral RNA synthesis are transported to membrane vesicles **(6)**. RNA synthesis occurs on the surfaces of these infected-cell-specific membrane vesicles. The (+) strand RNA is transported to these membrane vesicles **(7)**, where it is copied into (–) strands carrying VPg at their 5′ ends **(8)**. These (–) strands serve as templates for the synthesis of (+) strand genomic RNAs **(9)**. Some of the newly synthesized (+) strand RNA molecules enter the translational system after the removal of VPg **(10)**. Structural proteins formed by partial cleavage of the P1 precursor **(11)** associate with (+) strand RNA molecules that retain VPg to form progeny virions **(12)**, which are released from the cell upon lysis **(13)**.

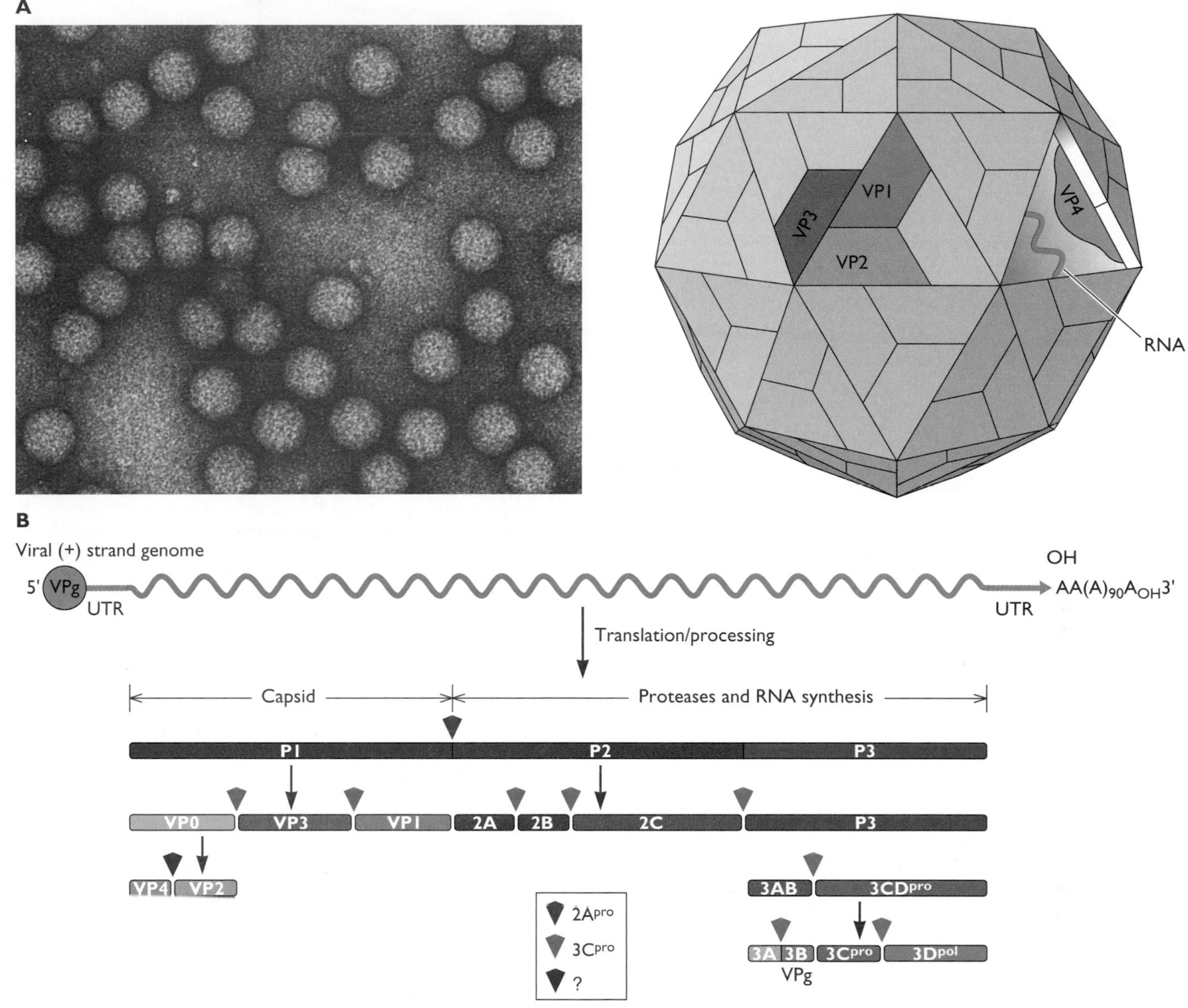

Figure 13 Structure and genomic organization

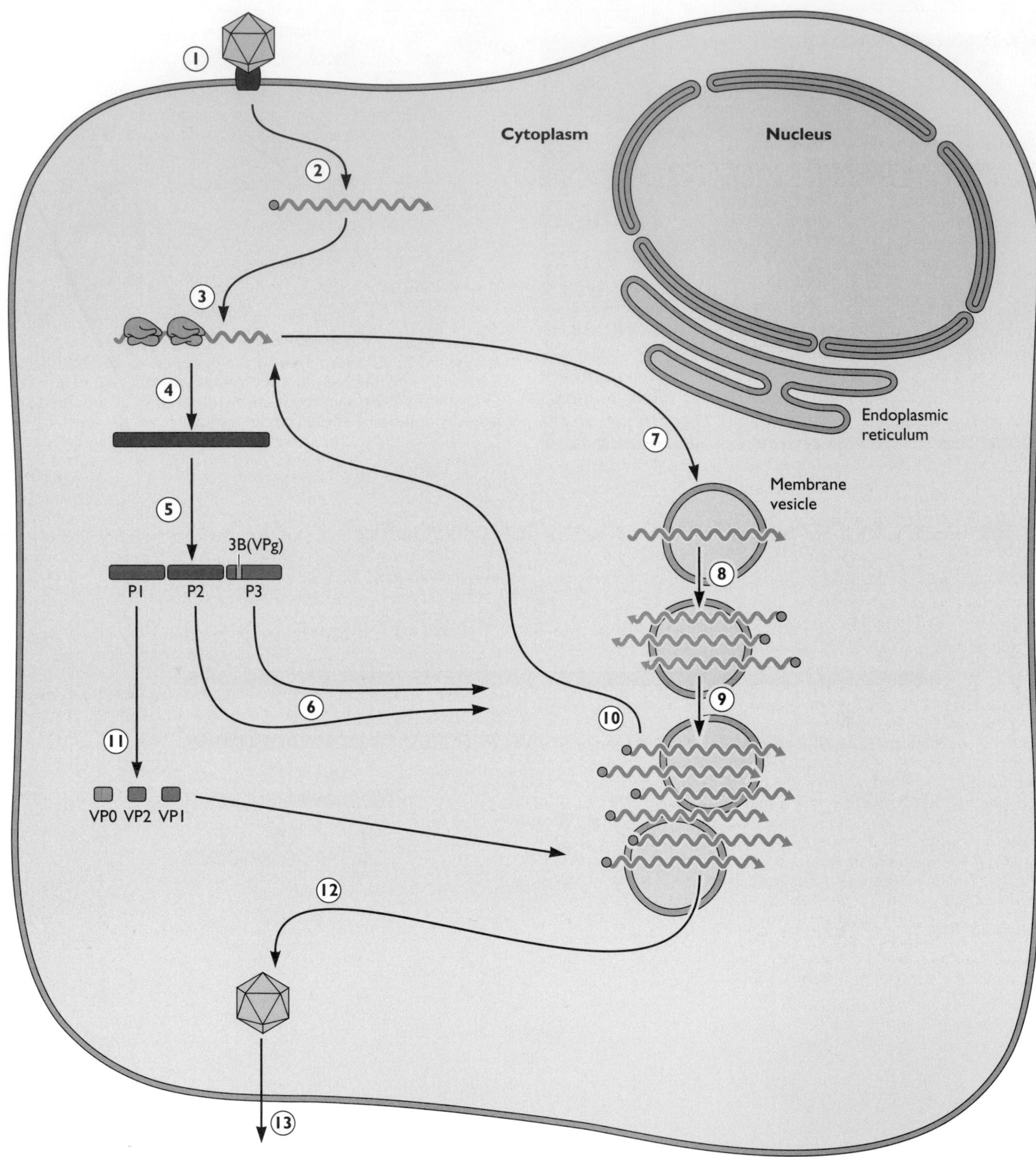

Figure 14 Single-cell reproductive cycle

Polyomaviruses

Family: *Polyomaviridae*

Genus	Type species
Polyomavirus	Simian virus 40

The family *Polyomaviridae* includes mouse polyomaviruses and two human viruses, JC and BK viruses, which were isolated from a patient with progressive multifocal leukencephalopathy and an immunosuppressed recipient of a kidney transplant, respectively. Under some conditions, mouse polyomavirus infection of the natural host results in formation of a wide variety of tumors (hence the name). A characteristic property of the members of this family is an ability to transform cultured cells or to induce tumors in animals. Investigation of such transforming activity has provided much information about mechanisms of oncogenesis, including the discovery of the cellular tumor suppressor protein p53. These viruses, particularly simian virus 40, have also been important in elucidation of cellular mechanisms of transcription and its regulation. For example, the simian virus 40 enhancer was the first member of this class of regulatory sequences to be identified.

Figure 15 Structure and genome organization. (A) The virion. (Left) Electron micrograph of negatively stained simian virus 40 virions. From F. A. Andered et al., *Virology* **32:**511–523, 1967, with permission. (Right) Diagram of the virion, showing the names and locations of virion proteins and the organization of the 5,243-bp circular, double-stranded DNA genome into approximately 25 nucleosomes by the cellular histones H2A, H2B, H3, and H4 (the core histones). One molecule of either VP2 or VP3, which possess a common C-terminal sequence, is associated with each VP1 pentamer. **(B) Genome organization.** Locations of the origin of viral DNA synthesis (Ori) and of the early and late mRNA sequences encoding the large and small T antigens (LT and sT) and the virion structural proteins VP1, VP2, and VP3 are indicated. The late mRNA species generally contain additional open reading frames in their 5′-terminal exons, such as that encoding leader protein 1 (LP1).

Figure 16 Single-cell reproductive cycle of simian virus 40. The virion attaches to permissive monkey cells upon binding of VP1 to a major histocompatibility complex (MHC) class I molecule on the surface of the cell. The virion is then endocytosed in caveolae **(1** and **2)**, is transported to the endoplasmic reticulum, and enters that organelle **(3)**. It is then transported to the nucleus and uncoated by unknown mechanisms **(4)**. The viral genome packaged by cellular nucleosomes is found within the nucleus **(5)**. The early transcription unit is transcribed by host cell RNA polymerase II **(6)**. After alternative splicing and export to the cytoplasm **(7)**, the early mRNAs are translated by cytoplasmic ribosomes to produce the early proteins LT and sT **(8)**. The former is imported into the nucleus **(9)**, where it binds to the simian virus 40 origin of replication to initiate DNA synthesis **(10)**. Apart from LT, all components needed for viral DNA replication are provided by the host cell. As they are synthesized, daughter viral DNA molecules associate with cellular nucleosomes to form the viral nucleoproteins often called minichromosomes. LT also stimulates transcription of the late gene from replicated viral DNA templates **(11)**. Processed late mRNAs are exported to the cytoplasm **(12)**, and translated to produce the virion structural proteins VP1, VP2, and VP3 **(13)**. These structural proteins are imported into the nucleus **(14)** and assemble around viral minichromosomes to form progeny virions **(15)**. Virions are released by an unknown mechanism **(16)**.

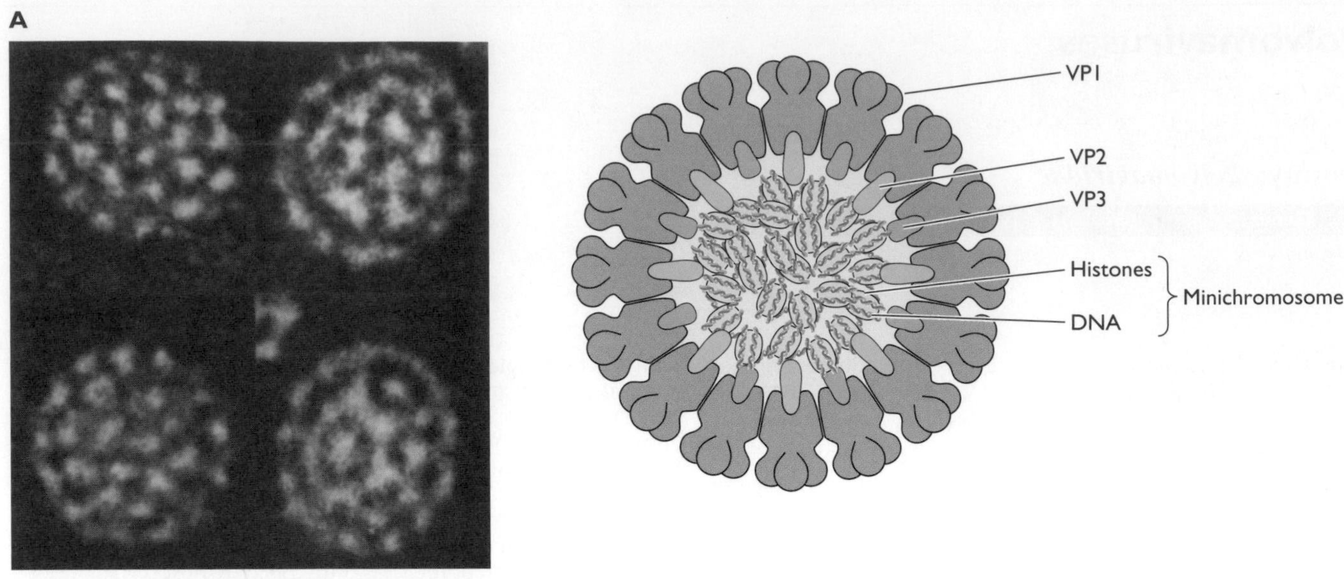

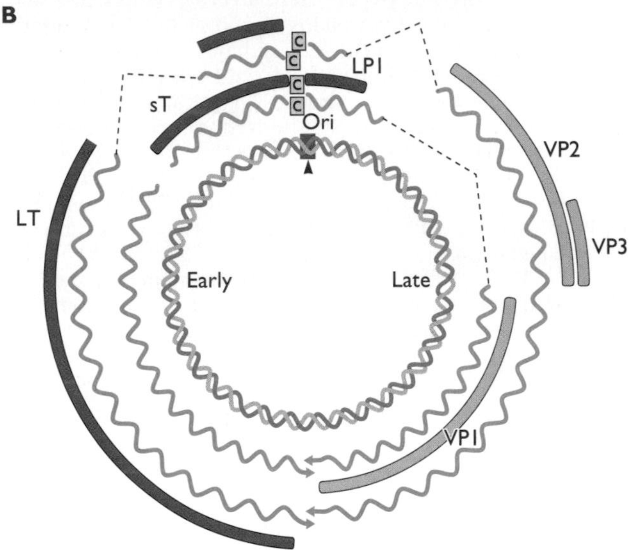

Figure 15 Structure and genome organization

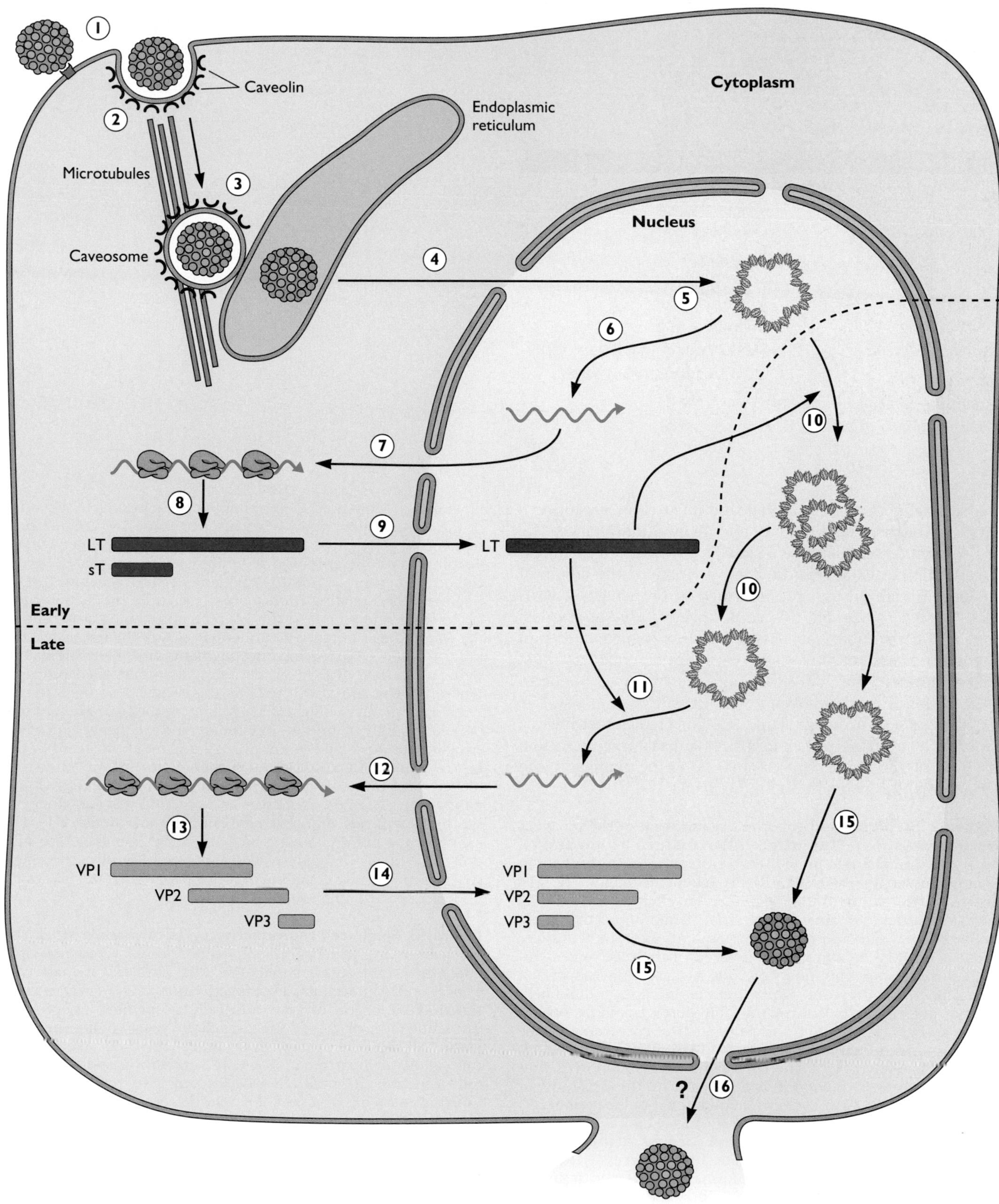

Figure 16 Single-cell reproductive cycle of simian virus 40

Poxviruses

Family: *Poxviridae*

Genus	Type species
Subfamily *Chordopoxvirinae*	
Orthopoxvirus	Vaccinia virus
Parapoxvirus	Orf virus
Avipoxvirus	Fowlpox virus
Capripoxvirus	Sheeppox virus
Leporipoxvirus	Myxoma virus
Suipoxvirus	Swinepox virus
Molluscipoxvirus	Molluscum contagiosum virus
Yabapoxvirus	Yaba monkey tumor virus
Subfamily *Entomopoxvirinae*	
Three genera of viruses that infect insects	

Poxviruses infect most vertebrate and invertebrates, causing a variety of diseases of veterinary and medical importance. The best-known poxviral disease is smallpox, a devastating human disease that has been eradicated by vaccination. The origins of modern vaccinia virus, the virus used in the smallpox virus vaccine, are obscure, but this virus is widely used as a model poxvirus in the laboratory. Myxoma virus, which causes an important disease of domestic rabbits, was described in 1896. Rabbit fibroma virus, which was first described by Shope in 1932, is the first virus proven to cause tissue hyperplasia. The genomes of poxviruses are large DNA molecules that include genes for all proteins needed for DNA synthesis and production of viral mRNAs. These viruses replicate in the cytoplasm and are minimally dependent on the host cell.

Figure 17 Structure and genome organization of the poxvirus vaccinia virus. (A) The intracellular mature virion (IMV). (Left) Electron micrograph of IMV in cross section. Courtesy of R. Condit, University of Florida. (Right) Schematic diagram of IMV structure. The nature of the membrane envelope(s) surrounding the IMV particle remains controversial. Some argue that there are two closely opposed membranes and others argue that there is one. The IMV membrane(s) observed in the electron microscope form the outer boundary of a 300 Å-thick surface layer that surrounds the inner core, which frequently appears dumbbell-shaped. The outer row of reproducibly observed knobs on the core wall has been termed the palisade. The core contains the double-stranded viral DNA genome and virion enzymes, including DNA-dependent RNA and DNA polymerases, and RNA-processing enzymes. A second infectious particle, the extracellular enveloped virion (EEV) contains an additional outer membrane. Poxvirions contain no obvious helical or icosahedral nucleocapsid. **(B) Structural features and organization of the genome.** The size of the genome and inverted repeat elements are strain-specific. We illustrate information obtained from the Copenhagen strain of vaccinia virus. The 191-kb genome, depicted at the top, is a double-stranded DNA molecule with ends that are covalently connected by hairpin loops of 101 nucleotides. The sequences that form the hairpins are AT-rich and lie within the 12-kb inverted terminal repetition (ITR). As indicated in the expansion of the left end of the genome, the ITR contains multiple repeated sequences. Determination of the complete sequence of the Copenhagen strain has identified some 185 unique protein-coding sequences and provided a detailed genetic map. Features of the organization of viral genes are illustrated by the coding sequences of *Hin*dIII fragment D, shown below. The D fragment open reading frames are depicted by the arrows indicating their orientation in the genome and named according to the *Hin*dIII fragment in which the first ATG of the frame lies, the order of their ATGs within the fragment, and their orientation (left [L] or right [R]). The identities of the encoded proteins, where known, are also listed. As this segment of the vaccinia virus genetic map illustrates, coding sequences are densely packed and may be present in either strand. Sequences encoding structural proteins and essential enzymes are all clustered in the central ~120 kb, whereas genes encoding proteins that affect virulence, host range, or immunomodulation are found predominantly near the ends. Proteins that fulfill different functions are synthesized during different phases of the infectious cycle. For example, the mRNA capping enzyme, an RNA polymerase subunit, and uracil DNA glycosylase are products of early genes, whereas the genes encoding structural proteins are expressed only during the late phase of infection.

Figure 18 Single-cell reproductive cycle of vaccinia virus. The mechanisms by which either infectious form of vaccinia virus attaches to and enters susceptible host cells **(1)** are not well understood. Extracellular enveloped virion (EEV) entry is illustrated. After fusion of viral and cellular membranes, primary uncoating takes place and the viral core is released into the cytoplasm. All subsequent steps in the infectious cycle take place in this cellular compartment. The core contains, in addition to the viral genome, the viral DNA-dependent RNA polymerase and initiation proteins necessary for specific recognition of the promoters of viral early genes, as well as several RNA-processing enzymes that modify viral transcripts. Upon its release into the host cell's cytoplasm, the core synthesizes viral early mRNAs **(2)**, which exhibit the features typical of cellular mRNAs and are translated by the cellular protein-synthesizing machinery **(3)**. About half the viral genes are expressed during the early phase of infection. Some early proteins have sequence similarity to cellular growth factors, which can induce proliferation of neighboring host cells following their secretion, or proteins that counteract host immune defense mechanisms **(4)**. Synthesis of early proteins also induces a second uncoating reaction in which a nucleoprotein complex containing the genome is released from the core **(5)**. Such core disassembly leads to inhibition of viral early gene expression. Other early proteins mediate replication of the viral DNA genome **(6)**. Newly synthesized viral DNA molecules can serve as templates for additional cycles of genome replication **(7)** and are the templates for transcription of viral intermediate genes **(8)**. Transcription of intermediate genes requires viral initiation proteins, which confer specificity for intermediate promoters on the viral RNA polymerase, the products of early genes, and a cellular protein (Vitf2), which relocates from the infected cell nucleus to the cytoplasm. Proteins made upon translation of intermediate mRNAs **(9)** include those necessary for transcription of late genes **(10)**. The latter genes encode the proteins from which virions are built and the virion enzymes and other essential proteins, such as the early initiation proteins, that must be incorporated into virus particles during assembly. As these late proteins are synthesized by the cellular translation machinery **(11)**, assembly of progeny virus particles begins, probably in association with internal membranes of the infected cell **(12)**. The initial assembly reactions result in formation of the immature virion **(13)**, a spherical particle that is believed to possess a double membrane acquired upon wrapping of the membranes of an early compartment of the cellular secretory pathway about the assembling particle. This virus particle then matures into the brick-shaped intracellular mature virion (IMV) **(14)**, which is released only upon cell lysis **(15)**. However, this particle can acquire a second, double membrane from a *trans*-Golgi or early endosomal compartment to form the intracellular enveloped virion (IEV) **(16)**. The latter particles move to the cell surface on microtubules where fusion with the plasma membrane forms cell-associated virions **(17)**. Those particles induce actin polymerization for direct transfer to surrounding cells **(18)** or dissociate from the membrane as the EEV.

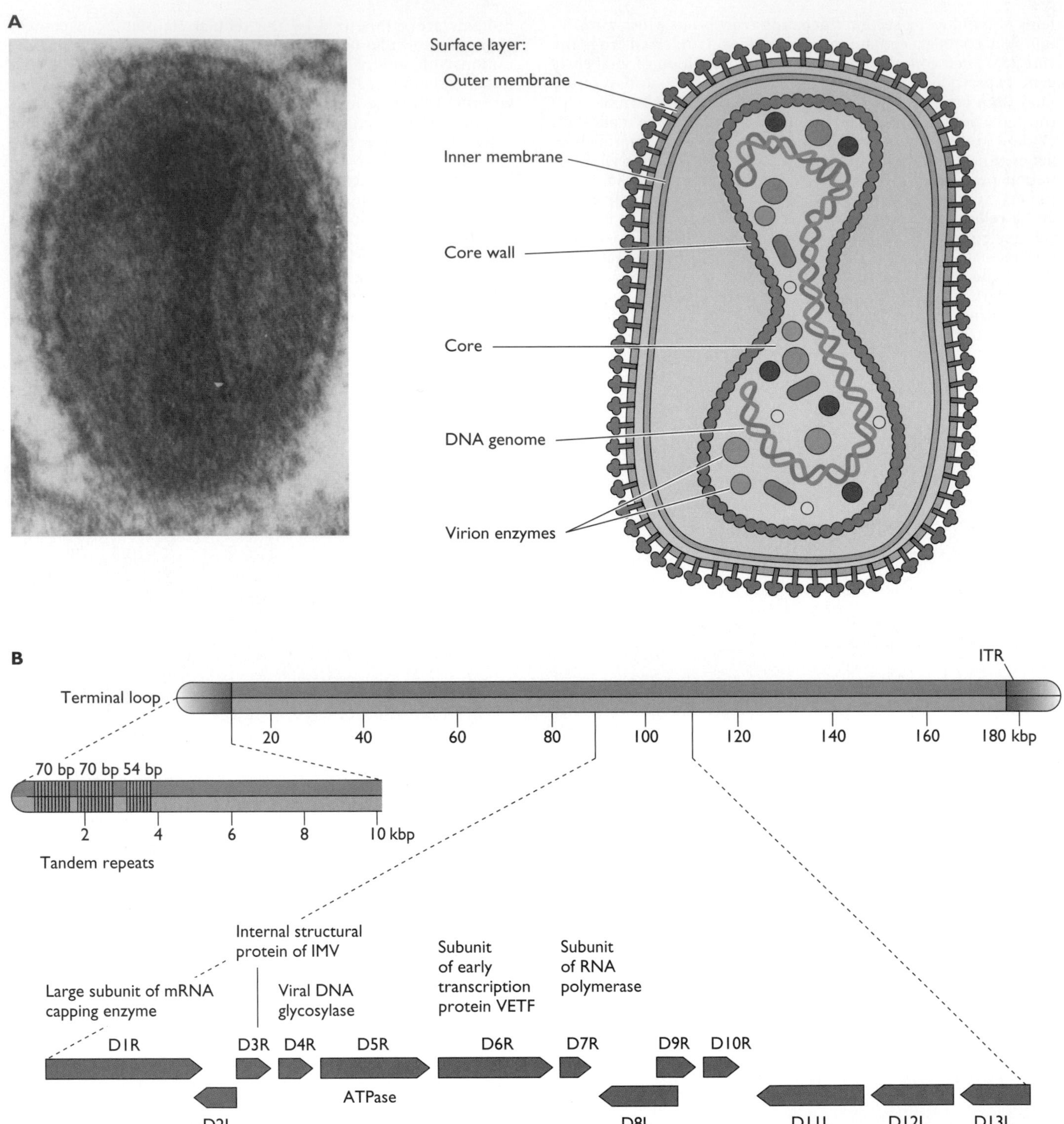

Figure 17 Structure and genome organization of the poxvirus vaccinia virus

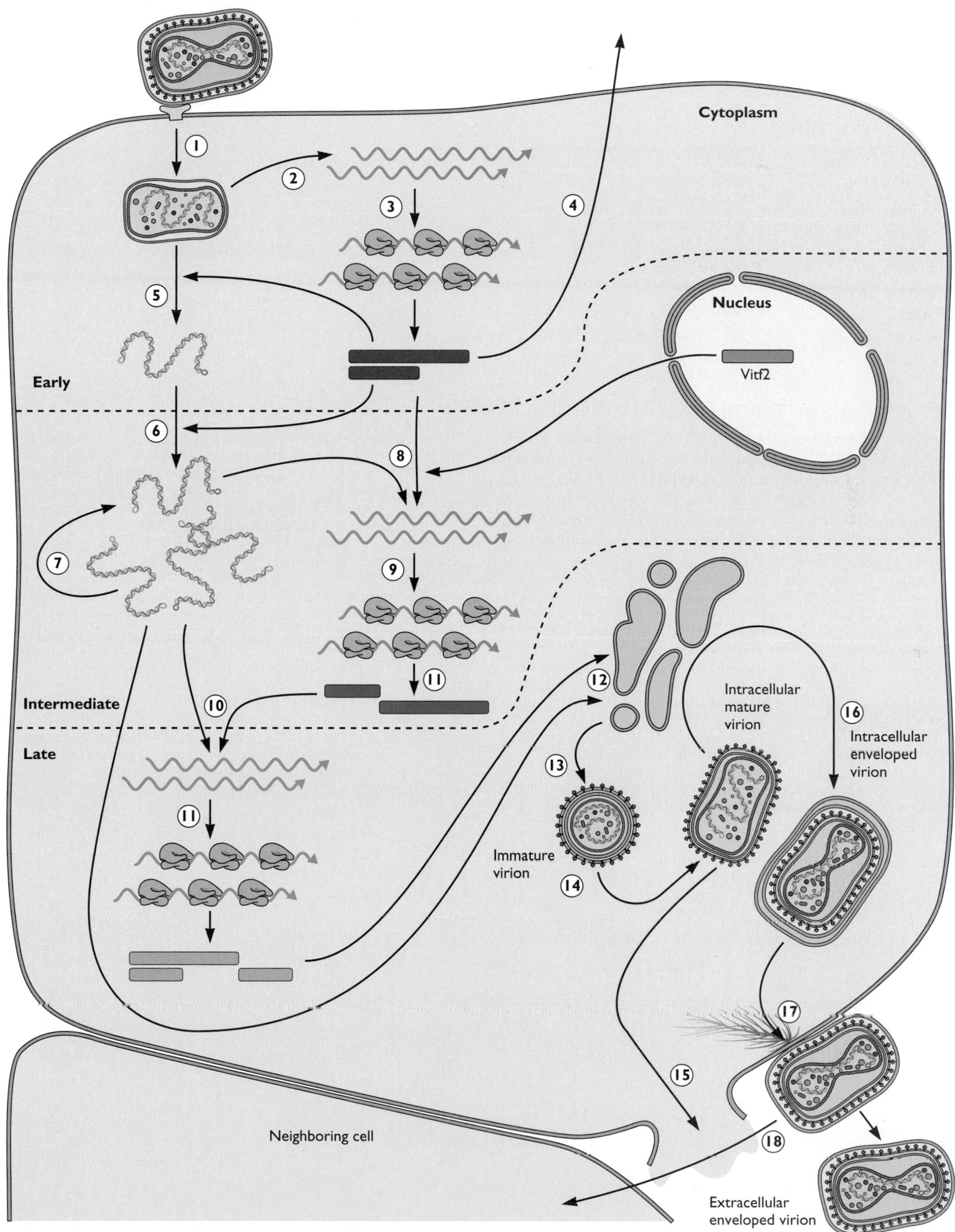

Figure 18 Single-cell reproductive cycle of vaccinia virus

Reoviruses

Family: *Reoviridae*

Genus	Type species
Orthoreovirus	Mammalian orthoreovirus
Orbivirus	Bluetongue virus
Rotavirus	Rotavirus A
Coltivirus	Colorado tick fever virus
Aquareovirus	Aquareovirus A
Cypovirus	Cypovirus 1
Fijivirus	Fijivirus 1
Phytoreovirus	Rice dwarf virus
Oryzavirus	Rice ragged stunt virus

Reoviridae is one of six families of viruses with double-stranded RNA genomes. Included in the *Reoviridae* are the human pathogens rotaviruses and Colorado tick fever virus. Reoviruses are the best studied of all the double-stranded RNA viruses. Some of the first in vitro work on RNA synthesis was done using reoviruses, and the 5′-terminal cap structure of mRNA was discovered on reovirus mRNAs.

Figure 19 Structure and genomic organization of an orthoreovirus. (A) The virion. (Left) Electron micrograph of a negatively stained reovirus. Courtesy of S. McNulty, Queen's University, Belfast, United Kingdom. (Right) Diagram of the virion (left), infectious subviral particle (ISVP, middle), and core (right) indicating the location of six virion proteins. **(B) Genome organization.** The double-stranded genome comprises 10 segments, named according to size: large (L), medium (M), and small (S). Each RNA segment encodes at least one viral protein. The S1 RNA encodes two proteins: σ1s protein is translated from a second initiation codon in a different reading frame from σ1. Two proteins are also produced from the M3 RNA. Protein μNSC is produced by translation at a second initiation codon in the same reading frame as μNS.

Figure 20 Single-cell reproductive cycle of orthoreovirus. The virion **(1)** or an ISVP **(4)** derivative **(3)** binds to a cellular receptor **(2, 5)** and enters the cell via receptor-mediated endocytosis **(6)**. In endosomes and lysosomes, the virion undergoes acid-dependent proteolytic cleavage **(7)** and can then penetrate the vacuolar membrane **(8, 9)**. Under some conditions (e.g., in the lumen of the intestine) the ISVP may be produced by extracellular proteolysis **(3)**. The entry of ISVPs **(5, 10)** may not depend on endocytosis. Synthesis of 10 capped viral mRNAs begins within the core particle **(11)**, which is derived from the ISVP. These mRNAs are translated and associate with newly synthesized viral proteins **(12)** to form RNase-sensitive subviral particles in which reassortment may occur **(13)**. Each of the 10 mRNAs is a template for (–) strand RNA synthesis, leading to the production of an RNase-resistant subviral particle that contains 10 double-stranded RNAs **(14)**. Viral mRNAs produced within subviral particles **(15)** are used for the synthesis of viral proteins and the assembly of additional virus particles. In the final steps of capsid assembly, preformed complexes of outer capsid proteins are added to subviral particles **(16)**. Mature virus particles are released from the cell by lysis **(17)**.

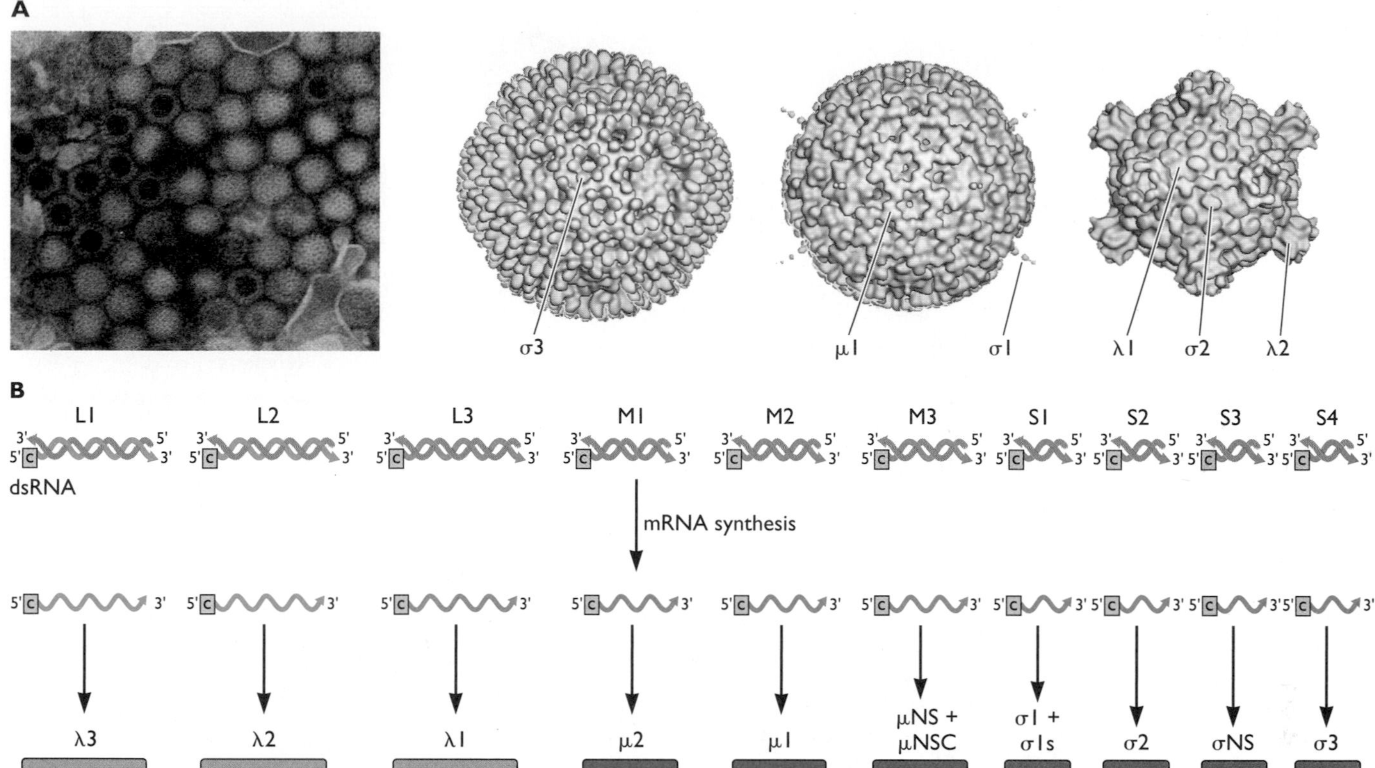

Figure 19 Structure and genomic organization of an orthoreovirus

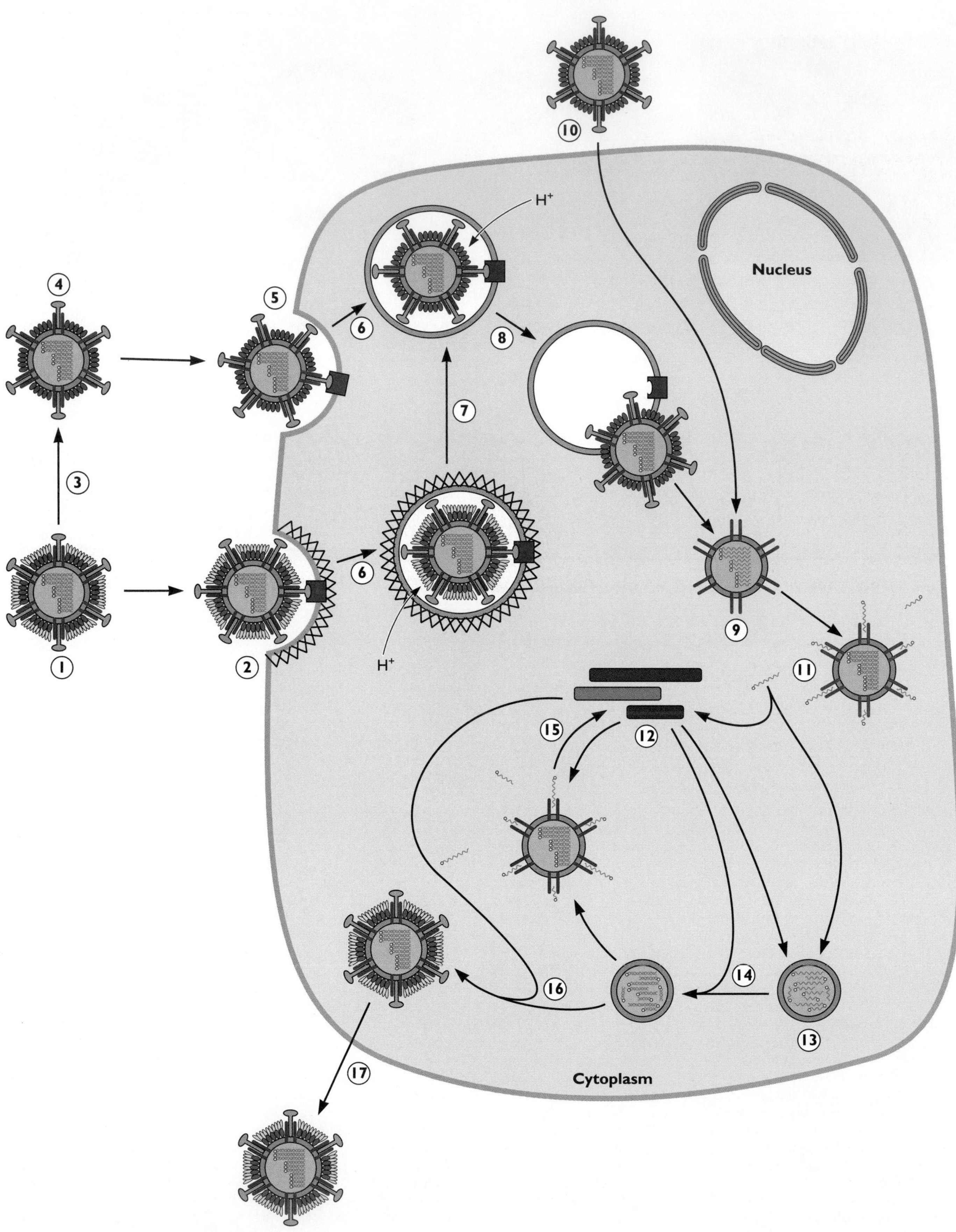

Figure 20 Single-cell reproductive cycle of orthoreovirus

Retroviruses

Family: *Retroviridae*

Genus	Type species
Subfamily *Orthoretrovirinae*	
Alpharetrovirus	Avian leukosis virus
Betaretrovirus	Mouse mammary tumor virus
Gammaretrovirus	Murine leukemia virus
Deltaretrovirus	Bovine leukemia virus
Epsilonretrovirus	Walleye dermal sarcoma virus
Lentivirus	Human immunodeficiency virus type 1
Subfamily *Spumavirinae*	
Spumavirus	Chimpanzee foamy virus

Retrovirus particles contain the enzyme reverse transcriptase, which mediates synthesis of a double-stranded DNA copy of the viral RNA genome. Although once thought to be unique to this family, similar enzymes are now known to be encoded in other viral genomes (i.e., hepadnaviruses and caulimoviruses), and the term **retroid viruses** has been coined to include these families. Retrovirus particles contain a second enzyme, integrase, that mediates the insertion of the viral DNA into essentially random sites in host DNA. The retroviruses can be propagated as integrated elements (called proviruses) that are transmitted in the germ line or as exogenous infectious agents. They infect a wide range of animal hosts and can cause cancer by multiple mechanisms. Some retroviruses, i.e., alpha-, beta-, and gammaretroviruses, have **simple** genomes that encode only the three genes common to all retroviruses—*gag, pol,* and *env*. All of the others have more **complex** genomes, which include auxiliary or accessory genes that encode nonstructural proteins that affect viral gene expression and/or pathogenesis.

Figure 21 Structure and genomic organization. (A) The virion. (Left) Electron micrograph of a negatively stained alpharetrovirus, Rous sarcoma virus. Courtesy of R. Katz and T. Galcs, Fox Chase Cancer Center. The method of preparation for electron microscopy (staining of thin sections) does not allow visualization of the envelope protein projections. (Right) Diagram of the alpharetrovirus, avian leukosis virus (ALV), indicating the names and locations of the component proteins, genomic RNA, and envelope. **(B) Genomic organization.** (Left) A retrovirus with a simple genome, here denoted "simple retrovirus." (Top) Genetic map of the avian leukosis virus provirus. Colored boxes delineate open reading frames, which, as indicated, are overlapping. (Bottom) The map of genomic RNA shows the genes common to all replication-competent retroviruses, *gag, pol,* and *env*. The ends of the RNA include short terminal repeats. Near the termini of the genomic RNA, sections containing *cis*-acting sequences that are required in replication are also shown (in orange): U5 (unique to the 5′ end) and U3 (unique to the 3′ end). In the integrated proviral DNA, the U5 and U3 sections are duplicated and comprise long terminal repeats (LTRs) at the borders between cellular and viral DNA. Initiation of DNA transcription and polyadenylation of pre-mRNA occur within the LTRs to produce progeny viral genomes and mRNAs. The viral mRNAs are translated to produce the indicated polyprotein precursors, which are eventually processed proteolytically to form the mature viral proteins. (Right) A retrovirus with a more complex genome, here denoted "complex retrovirus." (Top) Genetic map of the lentivirus human immunodeficiency virus type 1 (HIV-1) provirus. Genes are encoded in all three reading frames, as indicated by the overlaps. (Bottom) Human immunodeficiency virus type 1 mRNAs, which fall into one of three classes. The first type is an unspliced transcript of 9.1 kb, identical in function to that of the simple retrovirus. The second type consists of five singly spliced mRNAs (average length, 4.3 kb) that result from splicing from a 5′ splice site upstream of *gag* to any one of a number of 3′ splice sites near the center of the genome. Among these mRNAs is one specifying the Env polyprotein precursor, as illustrated for the simple retrovirus. The others specify the human immunodeficiency virus type 1 accessory proteins Vif, Vpr, and Vpu. Both unspliced RNA and singly spliced mRNAs require the Rev protein for transport out of the nucleus. This transport is mediated by the binding of Rev to the Rev-responsive element (RRE) in the *env* gene. The third type of mRNAs, multiply spliced molecules, average 1.8 kb in length; all lack the region of *env* that includes the RRE. This group includes a complex class of 16 mRNAs derived by exhaustive splicing from 5′ and 3′ splice sites throughout the genome. Such multiply spliced products are the first mRNAs to accumulate after infection; they specify the regulatory proteins Tat and Rev and the accessory protein Nef. As Rev protein accumulates, more and more of the 9.1- and 4.3-kb mRNAs are exported from the nucleus, allowing the production of virion proteins and formation of progeny virus.

Figure 22 Single-cell reproductive cycle of a simple retrovirus. The virus attaches by binding of SU and TM to specific receptors on the surface of the cell **(1)**. The identities of receptors (which are normal components of the cell surface) are known for several retroviruses. The viral core is deposited into the cytoplasm **(2)** following viral protein-assisted fusion of the virion and cell envelopes. The viral RNA genome is reverse transcribed by the virion RT **(3)** within a subviral particle. The product is a linear double-stranded viral DNA with ends that are shown juxtaposed in preparation for integration. The viral preintegration complex, which includes viral DNA and IN, enters the nucleus **(4)**, where it gains access to the host chromatin. Integrative recombination **(5)**, catalyzed by IN, results in site-specific insertion of the viral DNA ends, which can take place at numerous (essentially random) sites in the host genome. Integrated viral DNA is called the provirus. Transcription of proviral DNA by host cell RNA polymerase II **(6)** produces full-length RNA transcripts, which are used for multiple purposes. Some full-length RNA molecules are

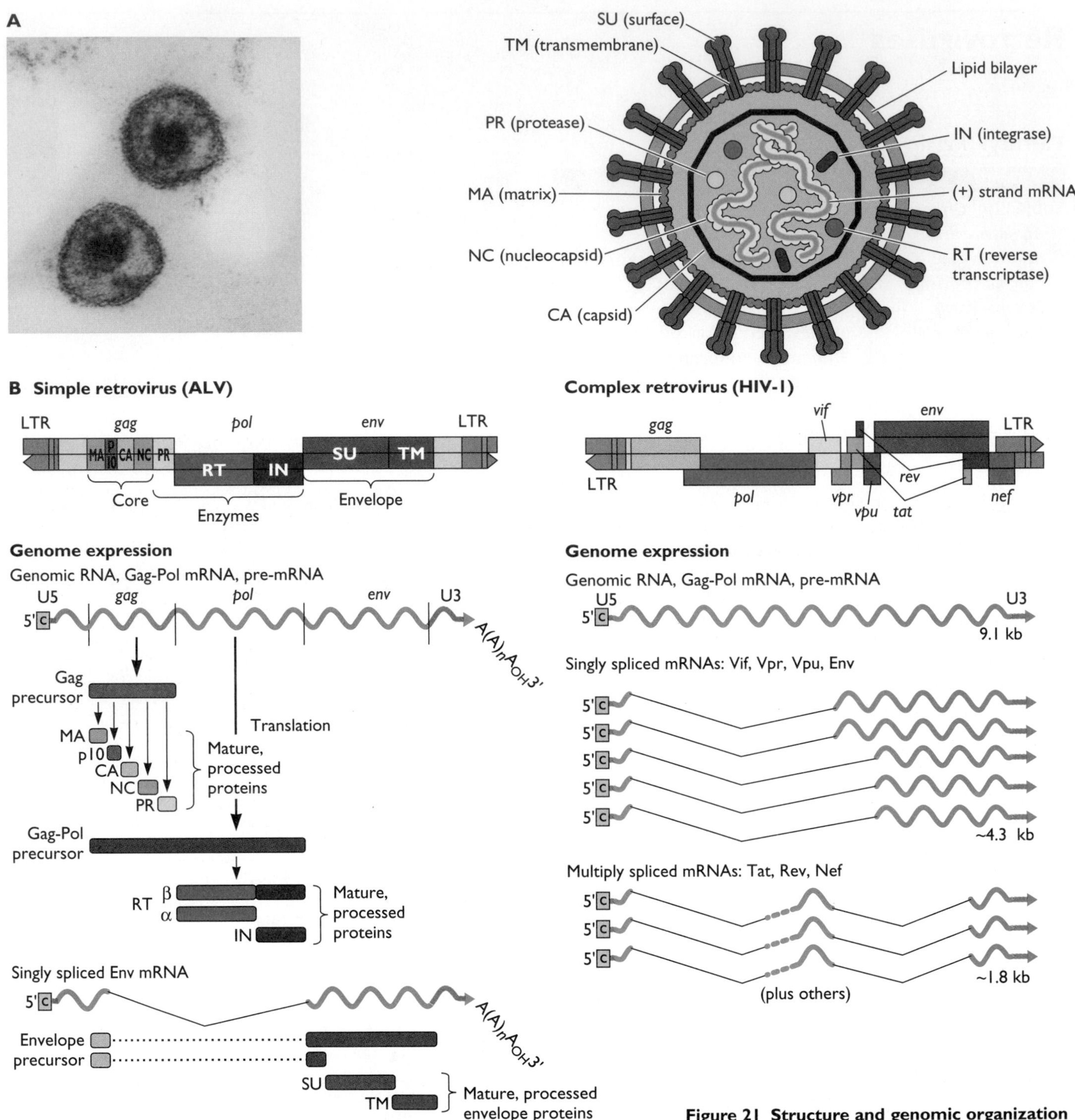

Figure 21 Structure and genomic organization

exported from the nucleus and serve as mRNAs **(7)**, which are translated by cytoplasmic ribosomes to form the viral Gag and Gag-Pol polyprotein precursors **(8)**. Some full-length RNA molecules become encapsidated as progeny viral genomes following transport into the cytoplasm **(9)**. Other full-length RNA molecules are spliced within the nucleus **(10)** to form mRNA for the Env polyprotein precursor proteins. Env mRNA is translated by ribosomes bound to the endoplasmic reticulum (ER) **(11)**. The Env proteins are transported through the Golgi apparatus **(12)**, where they are glycosylated and cleaved to form the mature SU-TM complex. These mature envelope proteins are then delivered to the surface of the infected cell **(13)**. Virion components (viral RNA, Gag and Gag-Pol precursors, and SU-TM) assemble at budding sites **(14)** with the help of *cis*-acting signals encoded in each. Type C retroviruses (e.g., alpharetroviruses and lentiviruses) assemble at the inner face of the plasma membrane, as illustrated. Other types (A, B, and D) assemble on internal cellular membranes. The nascent virions bud from the surface of the cell **(15)**. Maturation (and infectivity) requires the action of the virus-encoded protease (PR), which is itself a component of the core precursor polyprotein. During or shortly after budding, PR cleaves at specific sites within the Gag and Gag-Pol precursors **(16)** to produce the mature virion proteins. This process causes a characteristic condensation of the virion cores.

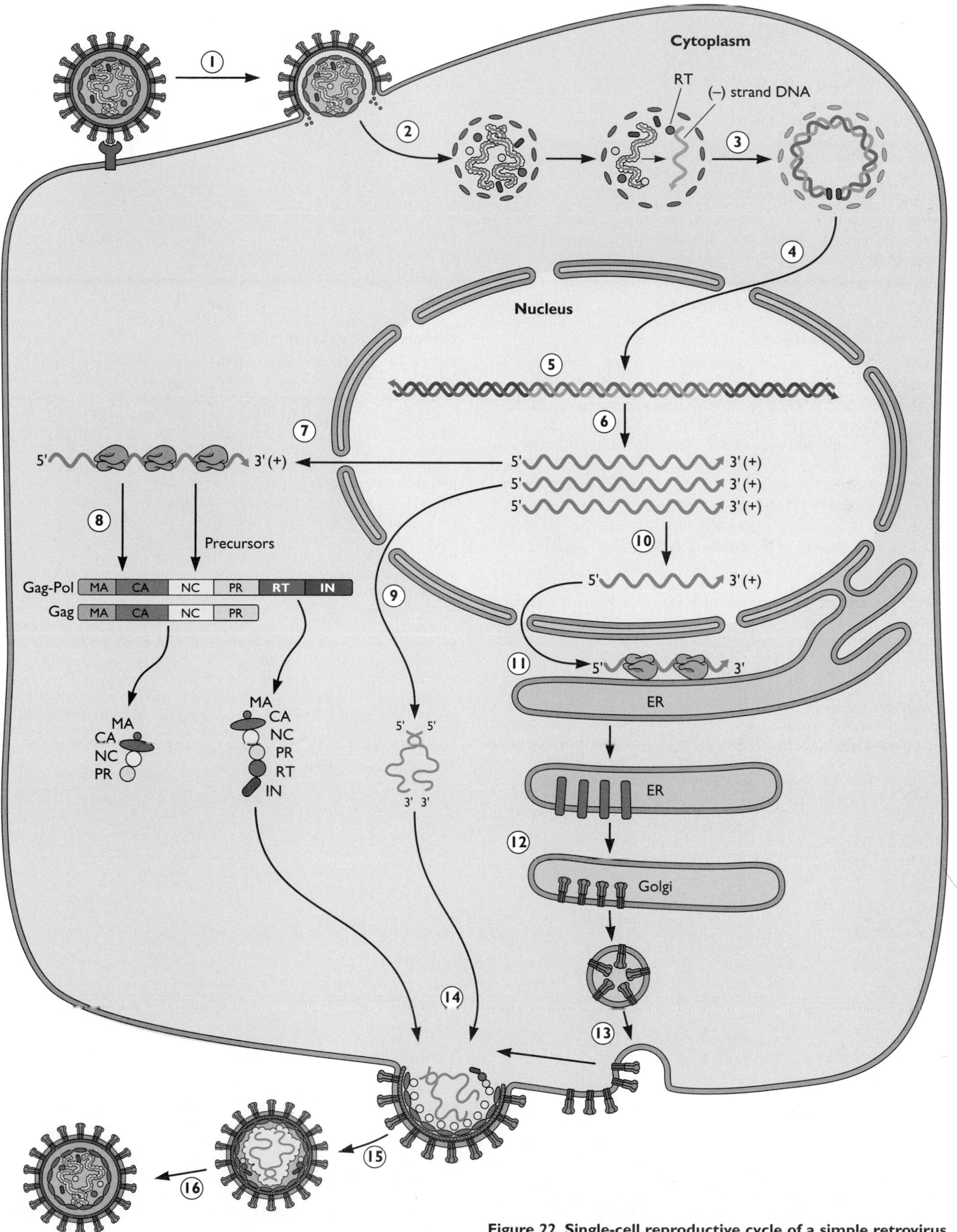

Figure 22 Single-cell reproductive cycle of a simple retrovirus

Rhabdoviruses

Family: *Rhabdoviridae*

Genus	Type species
Vesiculovirus	Vesicular stomatitis virus Indiana
Lyssavirus	Rabies virus
Ephemerovirus	Bovine ephemeral fever virus
Novirhabdovirus	Infectious hematopoietic necrosis virus
Cytorhabdovirus	Lettuce necrotic yellows virus
Nucleorhabdovirus	Potato yellow dwarf virus

Among the 175 known rhabdoviruses are the causative agents of rabies, one of the oldest recognized infectious diseases, and economically important diseases of fish. The host range of these viruses is very broad: they infect many vertebrates, invertebrates, and plants. The genome of vesicular stomatitis virus has been a model for the replication and expression of viral genomes that consist of a single molecule of (–) strand RNA. The first RNA-dependent RNA polymerase discovered in a virus particle was that of vesicular stomatitis virus. The mapping of vesicular stomatitis virus gene order with ultraviolet light remains a classic study.

Figure 23 Structure and genomic organization of vesicular stomatitis virus. (A) The virion. (Left) Electron micrograph of negatively stained vesicular stomatitis virus. Courtesy of J. Rose, Yale University School of Medicine. (Right) Diagram of the virion, indicating the locations of the five virion proteins, the nonsegmented (–) strand genomic RNA, and the viral envelope. **(B) Genome organization.** Starting at the 3′ end, the (–) strand RNA genome encodes a small leader (l) RNA and the N, P, M, G, and L proteins. The (–) strand RNA is the template for synthesis of leader RNA and five monocistronic mRNAs (capped and polyadenylated) encoding the five viral proteins.

Figure 24 Single-cell reproductive cycle. The virion binds to a cellular receptor and enters the cell via receptor-mediated endocytosis **(1)**. The viral membrane fuses with the membrane of the endosome, releasing the helical viral nucleocapsid **(2)**. This structure comprises (–) strand RNA coated with nucleocapsid protein molecules and a small number of L and P protein molecules, which catalyze viral RNA synthesis. The (–) strand RNA is copied into five subgenomic mRNAs by the L and P proteins **(3)**. The N, P, M, and L mRNAs are translated by free cytoplasmic ribosomes **(4)**, while G mRNA is translated by ribosomes bound to the endoplasmic reticulum **(5)**. Newly synthesized N, P, and L proteins participate in viral RNA replication. This process begins with synthesis of a full-length (+) strand copy of genomic RNA, which is also in the form of a ribonucleoprotein containing the N, L, and P proteins **(6)**. This RNA in turn serves as a template for the synthesis of progeny (–) strand RNA in the form of nucleocapsids **(7)**. Some of these newly synthesized (–) strand RNA molecules enter the pathway for viral mRNA synthesis **(8)**. Upon translation of G mRNA, the G protein enters the secretory pathway **(9)**, in which it becomes glycosylated and travels to the plasma membrane **(10)**. Progeny nucleocapsids and the M protein are transported to the plasma membrane **(11** and **12)**, where association with regions containing the G protein initiates assembly and budding of progeny virions **(13)**.

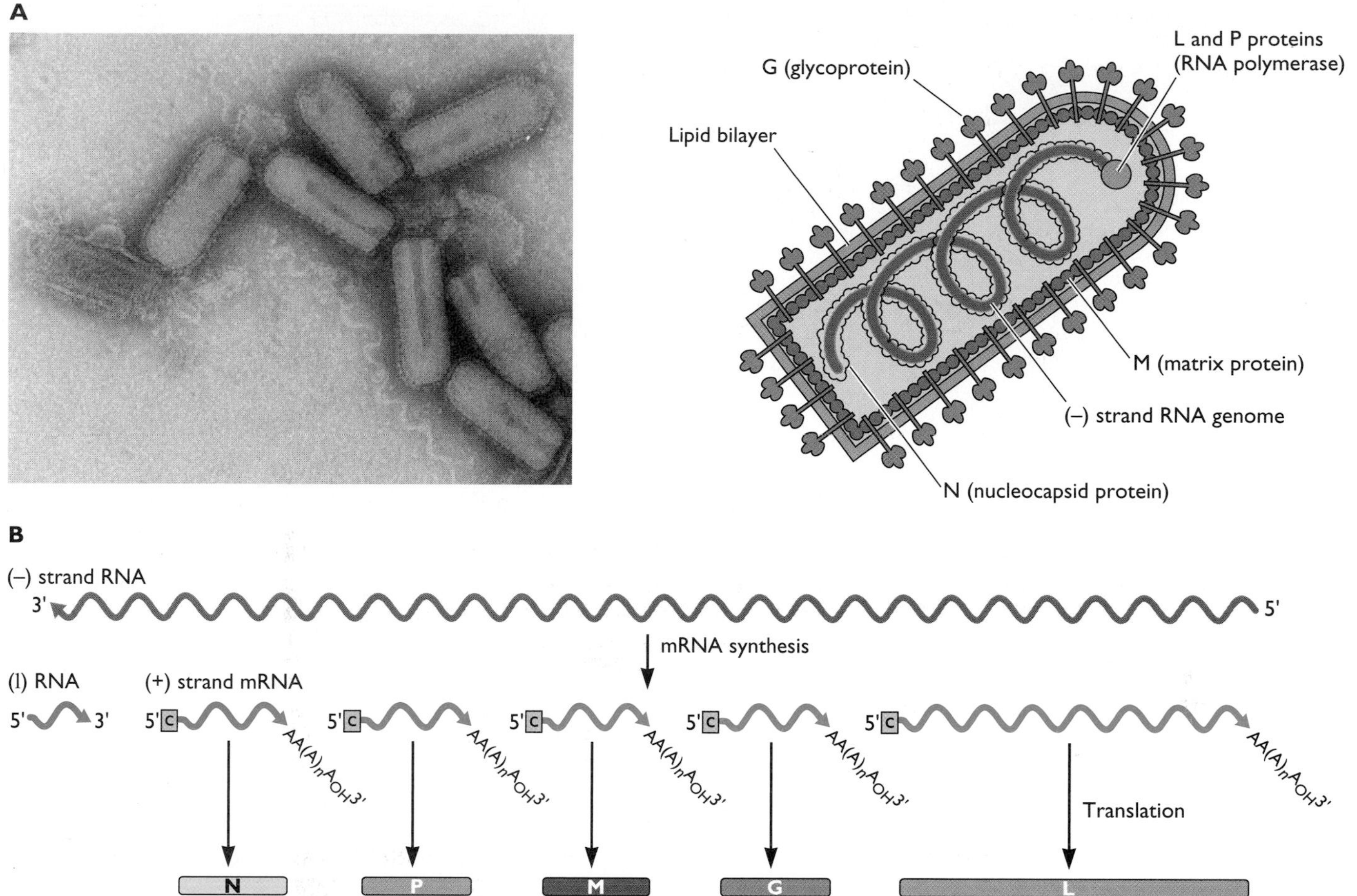

Figure 23 Structure and genomic organization of vesicular stomatitis virus

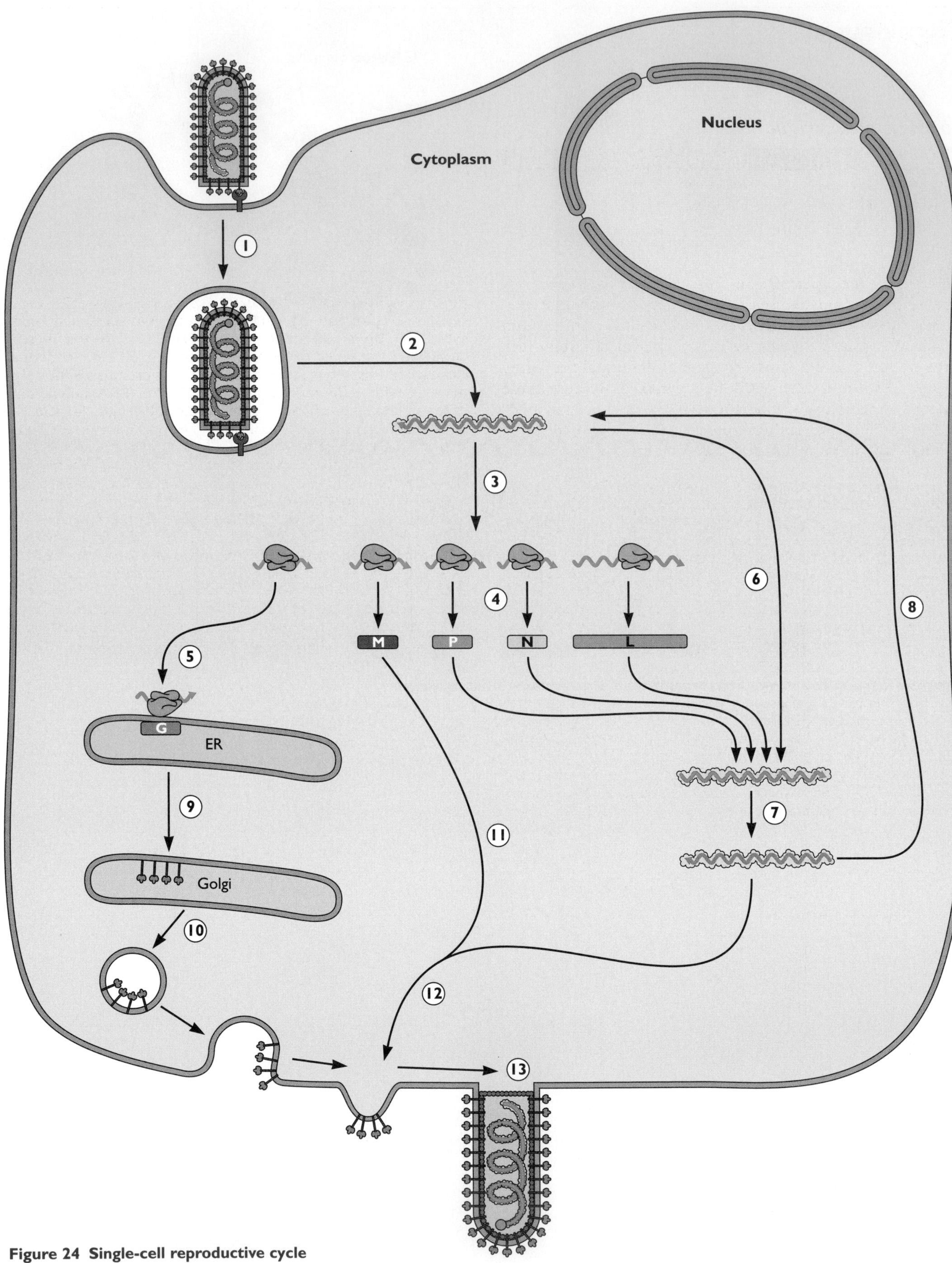

Figure 24 Single-cell reproductive cycle

Togaviruses

Family: *Togaviridae*

Genus	Type species
Alphavirus	Sindbis virus
Rubivirus	Rubella virus

Members of the *Togaviridae* are responsible for two very different kinds of human disease. The alphaviruses are all transmitted by arthropods and cause encephalitis, arthritis, and rashes in humans. Rubella virus is the agent of a mild rash disease that can also cause congenital abnormalities when acquired by the mother early in pregnancy. Because these virions have a lipid envelope, they have been important models for studying the synthesis, posttranslational modification, and localization of membrane glycoproteins.

Figure 25 Structure and genomic organization. (A) The virion. (Left) Cryoelectron micrograph of the alphavirus Ross River virus. Courtesy of N. Olson, Purdue University. (Right) Three-dimensional image reconstruction of Sindbis virus (courtesy of B. V. V. Prasad, Baylor College of Medicine), showing the intact virus (yellow, with glycoprotein spikes visible) and the nucleocapsid only (blue). Also shown is a cross section of the particle (green), illustrating the relationship among the spike glycoproteins (S), the lipid membrane (M), the capsid (C), and the viral RNA genome (RNA). **(B) Genome organization.** The first two-thirds of togavirus genomic RNA, which is of (+) polarity and carries a 5′ cap, is translated to produce the polyproteins P123 and P1234. The latter is the precursor of the RNA polymerase. The P1234 polyprotein is produced by translational suppression of a stop codon located at the end of the nsP3 coding region. The proteins encoded in the terminal one-third of the genome are produced from a subgenomic mRNA that is copied from a full-length (–) strand RNA intermediate. The subgenomic mRNA encodes the structural proteins.

Figure 26 Single-cell reproductive cycle. The virion binds to a cellular receptor and enters the cell via receptor-mediated endocytosis **(1)**. Upon acidification of the vesicle, viral RNA is uncoated **(2)** and translated to form the polyprotein P1234. Sequential cleavage of this polyprotein at different sites produces RNA polymerases with different specificities. **(3)**. The RNA polymerase produced initially copies (+) strands into full-length (–) and (+) strands **(4)** and catalyzes synthesis of the subgenomic mRNA **(5)**. This mRNA is translated by free cytoplasmic ribosomes to produce the capsid protein **(6)**; proteolytic cleavage to liberate the capsid protein exposes a hydrophobic sequence of PE2 that induces the ribosomes to associate with the endoplasmic reticulum (ER) **(7)**. As a result, the PE2, 6K, E1 polyprotein enters the secretory pathway. The membrane topology of these proteins is shown in detail, with cleavage sites for cellular proteases indicated by red arrows. The glycoproteins are transported to the cell surface **(8** and **9)**. The capsid protein and (+) strand genomic RNA assemble to form capsids **(10)** that migrate to the plasma membrane and associate with viral glycoproteins **(11)**. The capsid acquires an envelope by budding at this site **(12)**, and mature virions are released **(13)**.

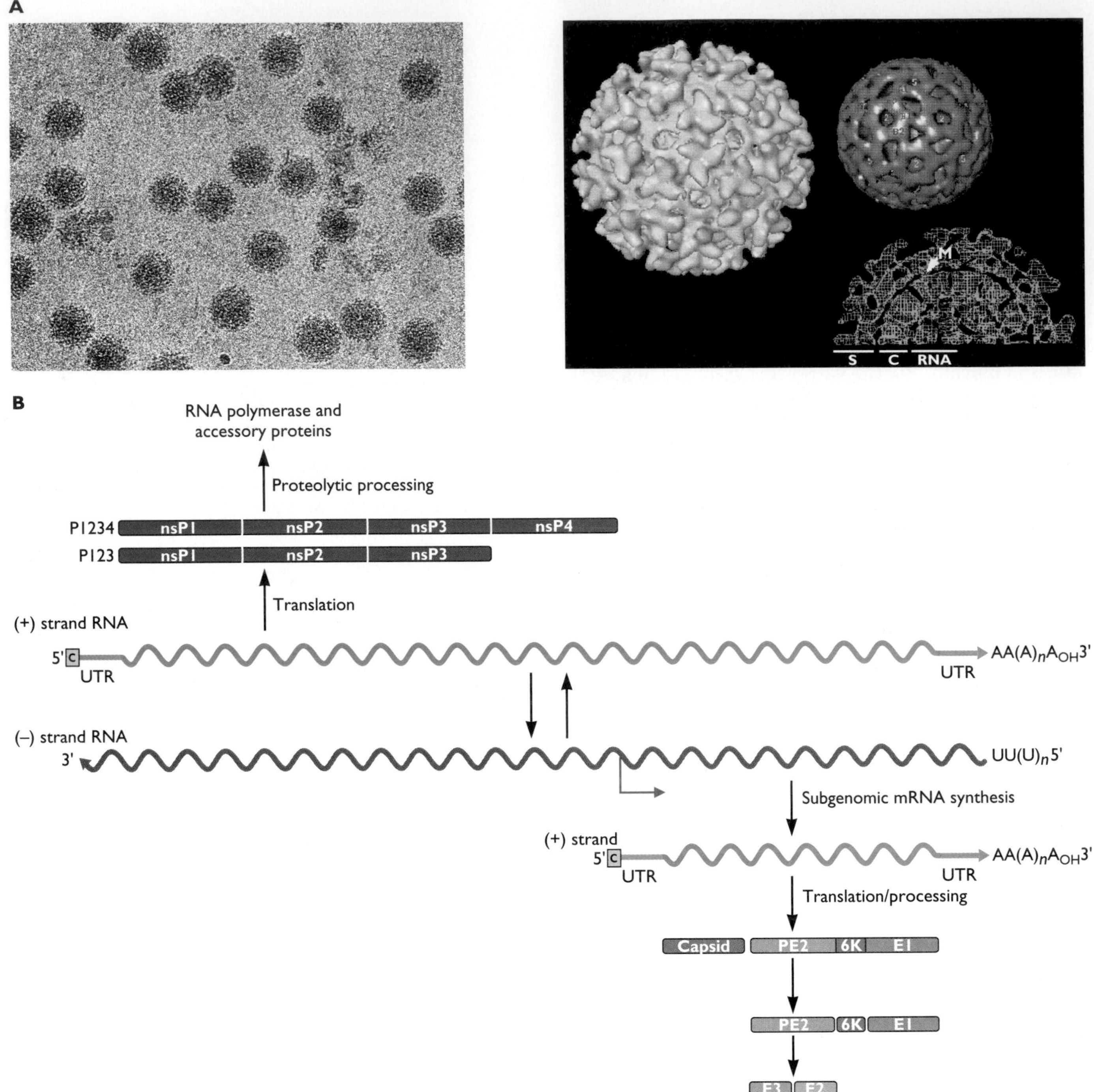

Figure 25 Structure and genomic organization

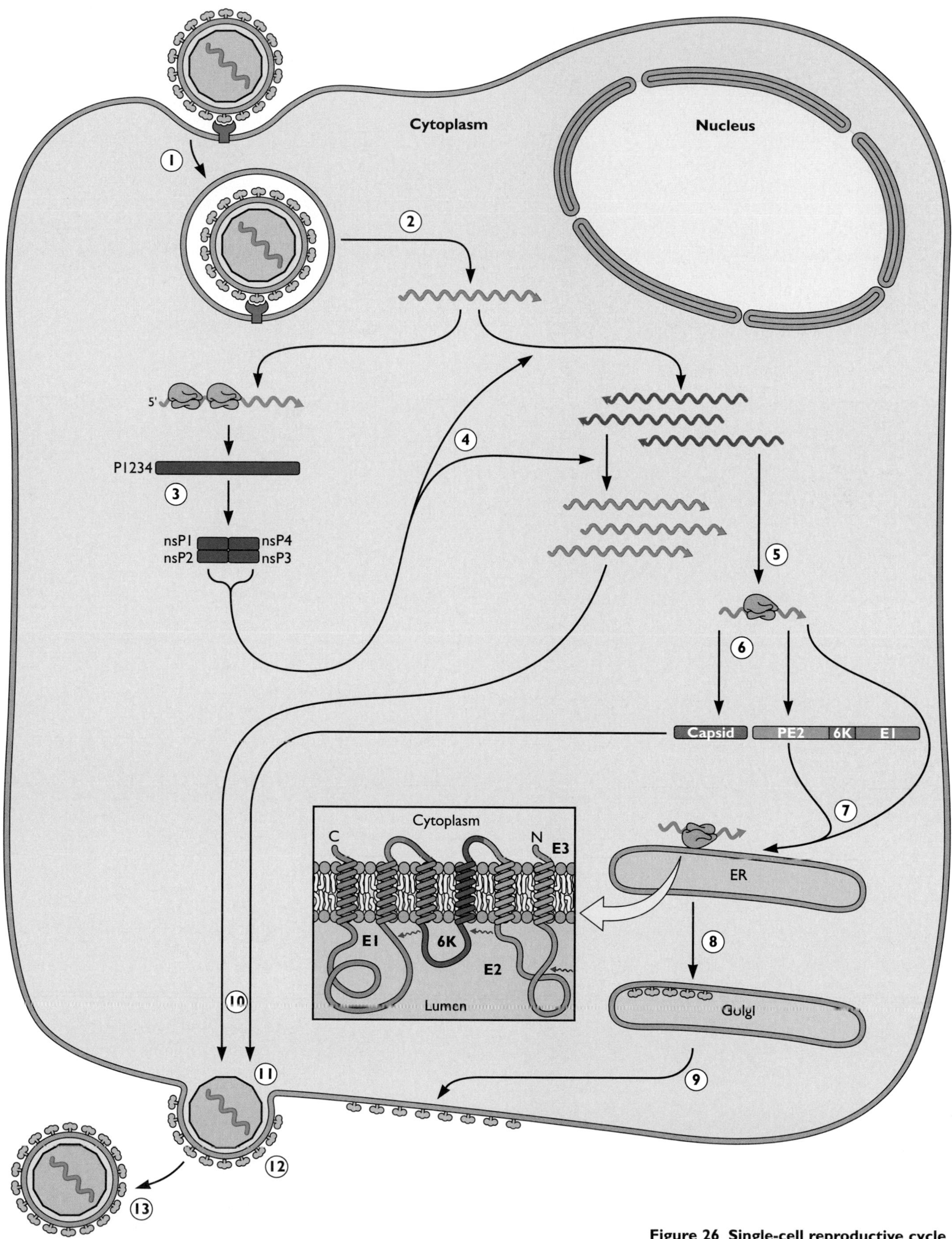

Figure 26 Single-cell reproductive cycle

APPENDIX B
Diseases, Epidemiology, and Disease Mechanisms of Selected Animal Viruses Discussed in This Book

This appendix presents key facts about the pathogenesis of selected animal viruses that cause human disease. Information about each virus or virus group is presented in three sections. In the first section, the viruses and associated diseases are listed. The second section, "Epidemiology," is outlined in four key areas: transmission, distribution of virus, those at risk or risk factors, and vaccines or antiviral drugs. The third section, "Disease mechanisms," provides simple images that will enable the reader to visualize infection and the resulting pathogenesis. Each of the three sections can be made into a slide for lectures or teaching, providing a "snapshot" of the pathogenesis of a specific virus.

Adenoviruses

Virus	Disease
47 adenovirus serotypes that infect humans, classified into six subgroups	Respiratory diseases • Febrile upper tract infection • Pharyngoconjunctival fever • Acute disease • Pertussis-like disease • Pneumonia Other diseases • Acute hemorrhagic cystitis • Epidemic keratoconjunctivitis • Gastroenteritis

Epidemiology

Transmission
- Respiratory droplets, fecal matter, fomites
- Close contact
- Poorly sanitized swimming pools

Distribution of virus
- Ubiquitous
- No seasonal incidence

At risk or risk factors
- Children aged <14 years
- Day care centers, military camps, swimming clubs

Vaccines or antiviral drugs
- Live, attenuated vaccine, serotypes 4 and 7 for the military

Disease mechanisms

Transmitted by **aerosol**, **close contact**, **fecal-oral route**, or **fingers** and **ophthalmologic instruments** (eye infections)

Virus infects mucoepithelial cells of respiratory and gastrointestinal tract, conjunctiva, cornea

Virus persists in lymphoid tissue (tonsils, adenoids, Peyer's patches)

Antibody is essential for recovery from infection

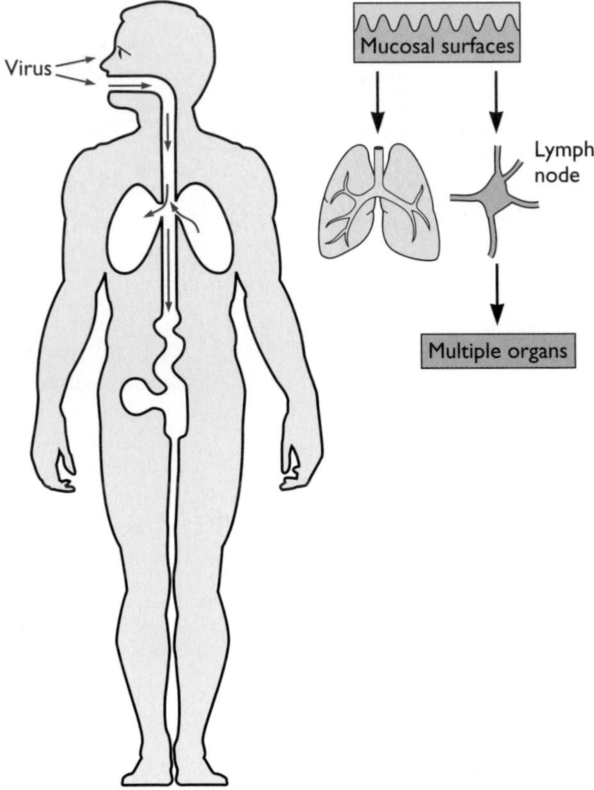

Arenaviruses

Virus	Disease
Lymphocytic choriomeningitis virus	Febrile, flu-like myalgia; meningitis
Lassa virus	Lassa fever: severe systemic illness, increased vascular permeability, shock
Junin virus	Argentine hemorrhagic fever: similar to Lassa fever but more extensive bleeding
Machupo virus	Bolivian hemorrhagic fever

Epidemiology

Transmission
- Contact with infected rodents or their secretions or body fluids

At risk or risk factors
- Lymphocytic choriomeningitis virus: contact with pet hamsters, areas with rodent infestation
- Other arenaviruses: habitat of rodents

Distribution of virus
- Lymphocytic choriomeningitis virus: hamsters and house mice in Europe, Americas, Australia, possibly Asia
- Other arenaviruses: Africa, South America, United States
- No seasonal incidence

Vaccines or antiviral drugs
- No vaccines
- Antiviral drug ribavirin

Disease mechanisms

Transmitted from persistently infected rodents (zoonoses)

Persistent infection of rodents caused by neonatal infection and induction of immune tolerance

Viruses infect macrophages and release mediators of cell and vascular damage

Tissue destruction caused by T-cell immunopathology

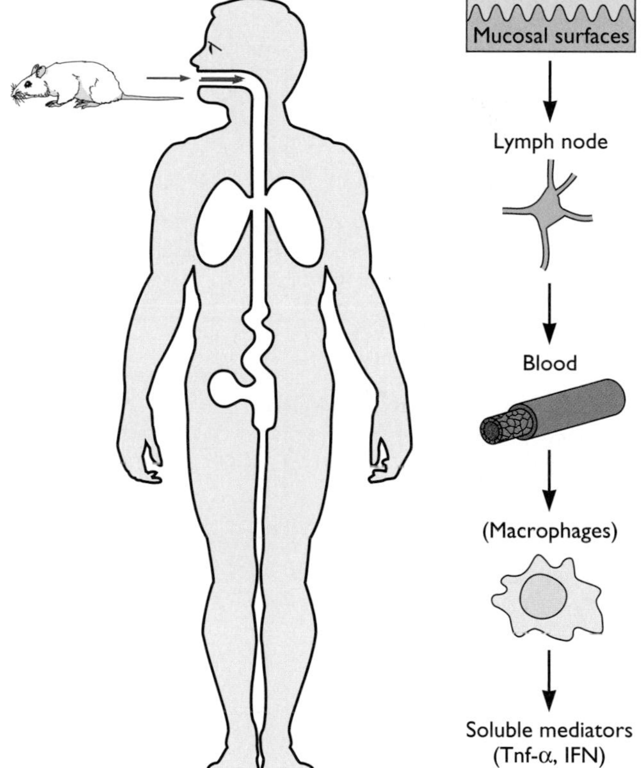

Bunyaviruses

Virus	Vector	Disease
Bunyavirus (150 members) • Bunyamwera virus • California encephalitis virus • La Crosse virus • Oropouche virus	Mosquito	Febrile illness, encephalitis, febrile rash
Phlebovirus (36 members) • Rift Valley fever virus • Sandfly fever virus	Fly	Sandfly fever, hemorrhagic fever, encephalitis, conjunctivitis, myositis
Nairovirus (6 members) • Crimean-Congo hemorrhagic fever virus	Tick	Hemorrhagic fever
Uukuvirus (7 members) • Uukuniemi virus	Tick	
Hantavirus • Hantaan virus	None	Hemorrhagic fever with renal syndrome, adult respiratory distress syndrome
• Sin Nombre virus	None	Hantavirus pulmonary syndrome, shock, pulmonary edema

Epidemiology

Transmission
- Arthropod bite
- Rodent excreta

Distribution of virus
- Depends on distribution of vector or rodents
- Disease more common in summer

At risk or risk factors
- People in area of vector, e.g., campers, forest rangers, woodspeople

Vaccines or antiviral drugs
- None

Disease mechanisms

Transmitted by arthropod bite or rodent excreta

Primary viremia, then secondary viremia leads to virus spread to target tissues, including central nervous system, various organs, vascular endothelium

Antibody essential for controlling viremia

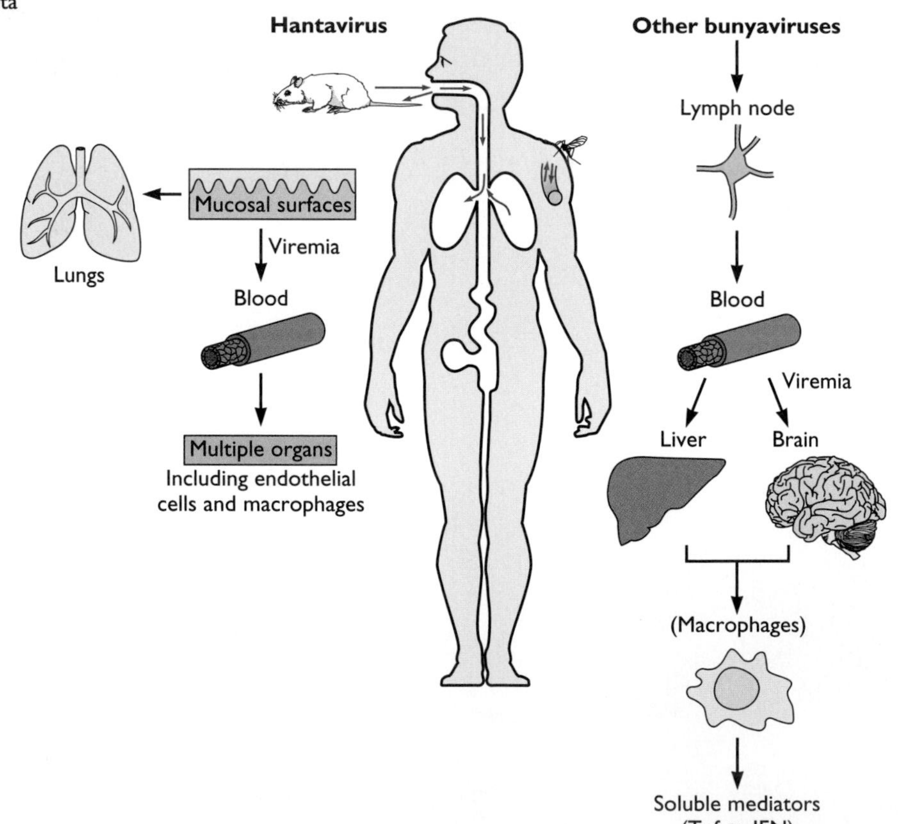

Caliciviruses

Virus	Disease
Norovirus • Desert Shield virus • Lordsdale virus • Mexico virus • Norwalk virus • Hawaii virus • Snow Mountain virus • Southampton virus *Sapovirus* • Sapporo virus	Gastroenteritis

Epidemiology

Transmission
- Fecal-oral route from contaminated water and food

Distribution of virus
- Ubiquitous
- No seasonal incidence

At risk or risk factors
- Children in day care centers
- Schools, resorts, hospitals, nursing homes, restaurants, cruise ships (due to infected food handlers)

Vaccines or antiviral drugs
- None

Disease mechanisms

Viruses are resistant to detergents, drying, acid

Transmitted by fecal-oral route (contaminated water and food)

Viruses infect intestinal brush border, preventing proper absorption of water and nutrients

Viruses cause diarrhea, vomiting, abdominal cramps, nausea, headache, malaise, fever

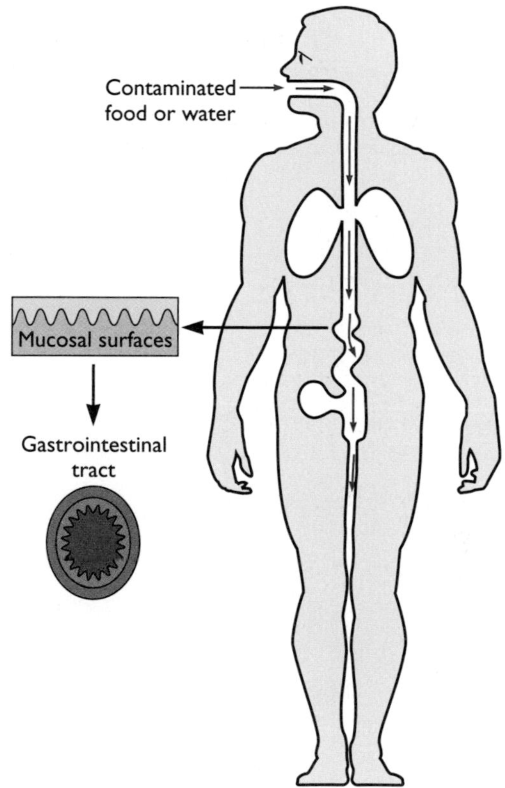

Filoviruses

Virus	Disease
Marburg virus	Hemorrhagic fever
Ebola virus	Hemorrhagic fever

Epidemiology

Transmission
- Contact with infected monkeys, or their tissues, secretions, or body fluids
- Contact with infected humans
- Accidental injection, contaminated syringes

Distribution of virus
- Endemic in monkeys in Africa
- No seasonal incidence

At risk or risk factors
- Monkey handlers
- Health care workers attending sick

Vaccines or antiviral drugs
- None

Disease mechanisms

Acquired from monkeys or infected humans

Virus replication causes necrosis in liver, spleen, lymph nodes, and lungs

Hemorrhage causes edema and shock

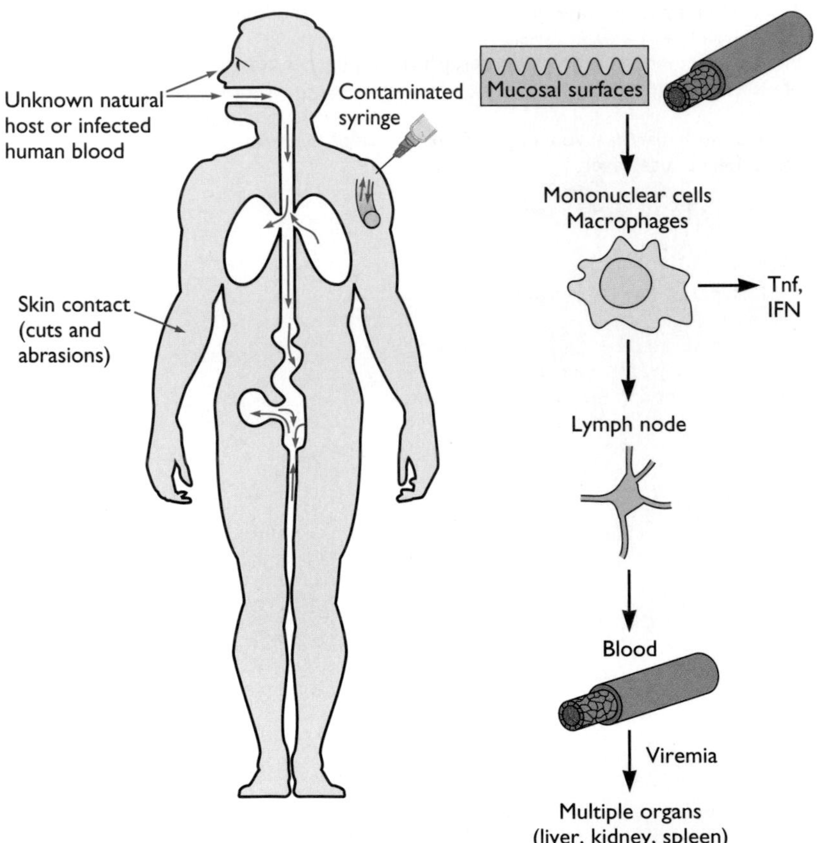

Flaviviruses

Virus	Vector	Disease
Dengue virus	*Aedes* mosquitoes	Mild systemic; breakbone fever, dengue hemorrhagic fever, dengue shock syndrome
Yellow fever virus	*Aedes* mosquitoes	Hepatitis, hemorrhagic fever
Japanese encephalitis virus	*Culex* mosquitoes	Encephalitis
West Nile virus	*Culex* mosquitoes	Fever, encephalitis, hepatitis
St. Louis encephalitis virus	*Culex* mosquitoes	Encephalitis
Russian spring-summer encephalitis virus	*Ixodes, Dermacentor* ticks	Encephalitis
Powassan virus	*Ixodes* ticks	Encephalitis
Hepatitis C virus	None	Hepatitis (see "Hepatitis viruses," next page)

Epidemiology

Transmission
- Mosquito or tick vectors

Distribution of virus
- Determined by habitat of vector
  - *Aedes* mosquito: urban areas
  - *Culex* mosquito: forest, urban areas
- More common in summer

At risk or risk factors
- People in niche of vector

Vaccines or antiviral drugs
- Live, attenuated vaccines for yellow fever and Japanese encephalitis
- No antivirals

Disease mechanisms

Viruses are cytolytic

Viruses cause viremia, systemic infection

Nonneutralizing antibodies can facilitate infection of monocytes/macrophages via Fc receptors

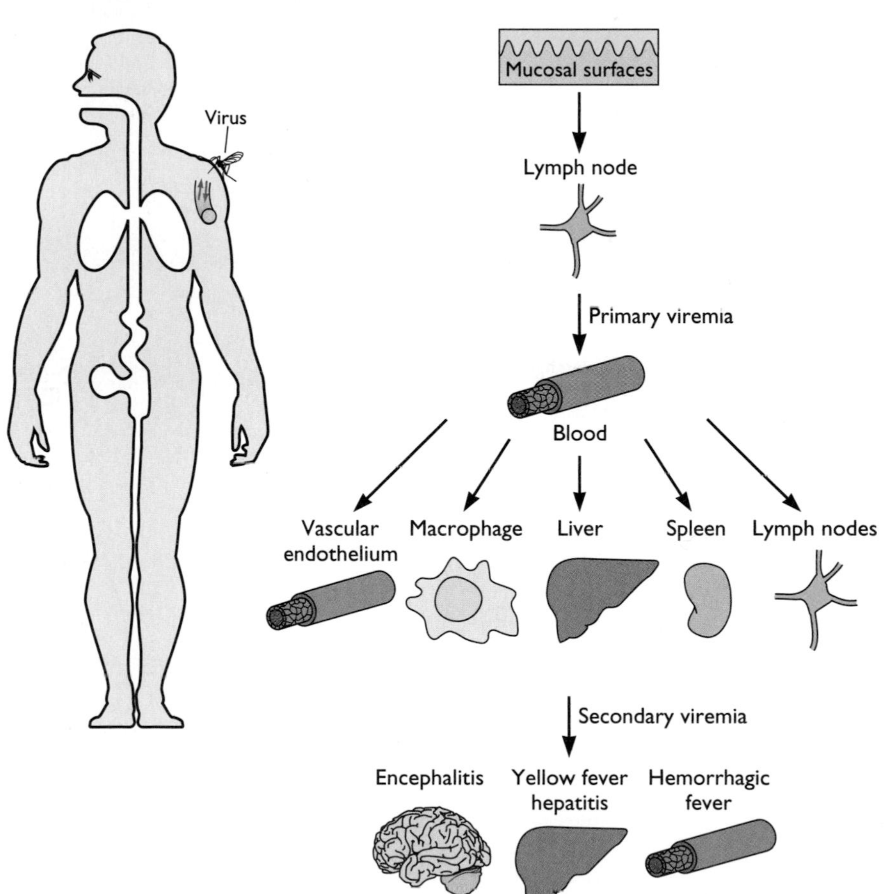

Hepatitis viruses

Virus	Transmission	Incubation period	Mortality rate	Persistent infections	Other diseases
Hepatitis A virus (picornavirus)	Fecal-oral	15–50 days	<0.5%	No	None
Hepatitis B virus (hepadnavirus)	Parenteral, sexual	45–160 days	1–2%	Yes	Primary hepatocellular carcinoma, cirrhosis
Hepatitis C virus (flavivirus)	Parenteral, sexual	14–180 days	~4%	Yes	Primary hepatocellular carcinoma, cirrhosis
Hepatitis D virus (viroid-like; contribution to hepatitis is controversial)	Parenteral, sexual	15–64 days			
Hepatitis E virus (calicivirus)	Fecal-oral	15–50 days	1–2%; pregnant women, 20%	No	None

Epidemiology

Transmission
- Hepatitis A and E viruses
  - Fecal-oral route (contaminated food, water)
  - Transmitted by food handlers, day care workers, children
- Hepatitis B, C, and D viruses
  - Blood, semen, vaginal secretions
  - Transfusions, needle injury, drug paraphernalia, sex, breast-feeding
- Hepatitis B virus
  - Saliva, mother's milk

Distribution of virus
- Ubiquitous
- No seasonal incidence

At risk or risk factors
- Hepatitis A and E viruses
  - Children (mild disease)
  - Adults (abrupt-onset hepatitis)
  - Pregnant women (high mortality with hepatitis E virus)
- Hepatitis B, C, and D viruses
  - Children (mild, chronic infection)
  - Adults (insidious hepatitis)
  - Adults with chronic hepatitis
- Hepatitis B or C virus (primary hepatocellular carcinoma)

Vaccines or antiviral drugs
- Hepatitis A virus: inactivated vaccine
- Hepatitis B virus: subunit vaccine
- Hepatitis C virus: ribavirin + IFN-α

Disease mechanisms

Viruses cause viremia, replicate in liver; may enter liver in peripheral blood cells

Generally not cytolytic; tissue damage caused by cell-mediated immune response

Viruses cause acute infections

Hepatitis B and C virus chronic infections can lead to hepatocellular carcinoma

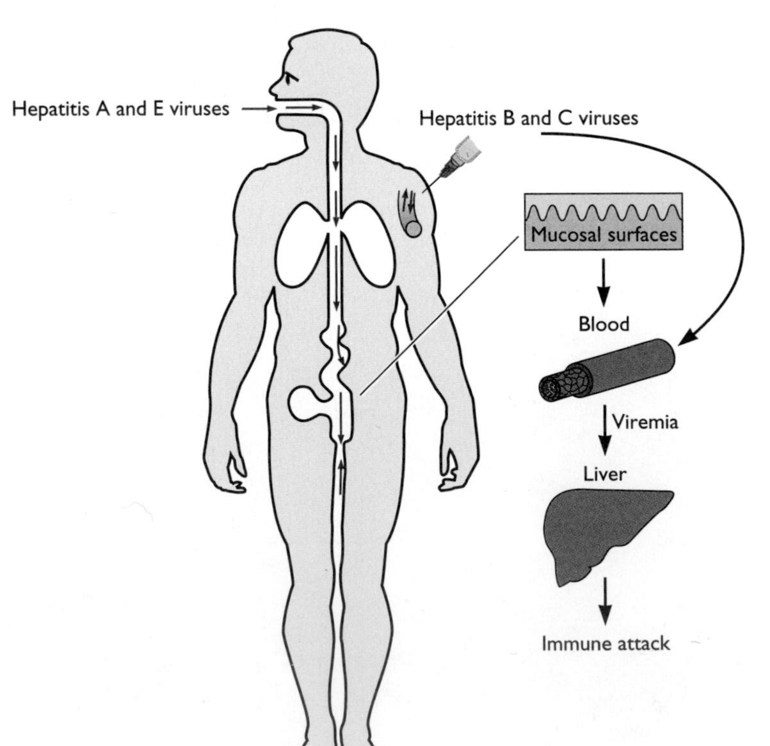

Herpesviruses

Herpes simplex virus

Virus	Disease
Alphaherpesviruses	
• **Herpes simplex virus types 1 and 2**	Mucosal lesions, encephalitis
• Varicella-zoster virus	Chickenpox, shingles
Betaherpesviruses	
• Cytomegalovirus	Congenital defects; opportunistic pathogen in immunocompromised patients
• Human herpesvirus 6	Roseola
• Human herpesvirus 7	Orphan virus
Gammaherpesviruses	
• Epstein-Barr virus	Infectious mononucleosis; associated with a variety of lymphomas
• Human herpesvirus 8 (Kaposi's sarcoma-related virus)	Kaposi's sarcoma, rare B-cell lymphoma

Epidemiology

Transmission
- Saliva, vaginal secretions, lesion fluid
- Into eyes and breaks in skin
- Herpes simplex virus type 1 mainly oral, herpes simplex virus type 2 mainly sexual

Distribution of virus
- Ubiquitous
- No seasonal incidence

At risk or risk factors
- Children (type 1) and sexually active people (type 2)
- Physicians, nurses, dentists, and those in contact with oral and genital secretions (herpetic whitlow, infection of finger)
- Immunocompromised and neonates (disseminated, life-threatening disease)

Vaccines or antiviral drugs
- No vaccine
- Antiviral drugs: acyclovir, penciclovir, valacyclovir, famciclovir, adenosine arabinoside, iododeoxyuridine, trifluridine

Disease mechanisms

Transmitted by **direct contact**: type 1, above the waist; type 2, below the waist

Virus spreads cell to cell, not neutralized by antibody

Cell-mediated immunopathology contributes to symptoms

Cell-mediated immunity is required for resolution of infection

Virus establishes latency in neurons

Virus is reactivated from latency by stress or immune suppression

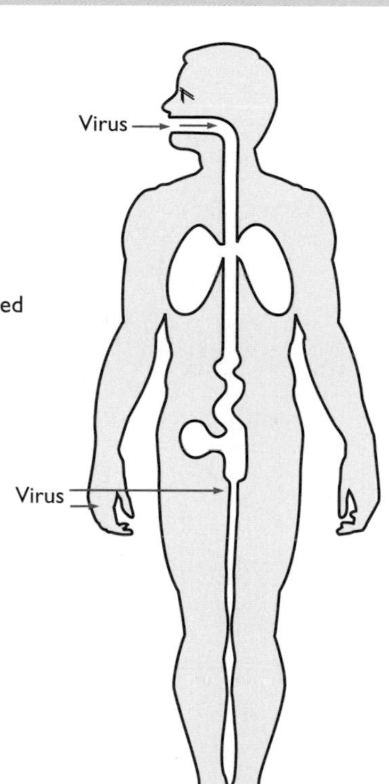

Normal maintenance cycle

Pathways leading to serious disease

Primary infection

Disseminated infection

Mucosal surfaces

Viremia

Organ systems

Latent infection

Reactivation

Brain and spinal cord

Sensory and autonomic ganglia

PNS

CNS

Deadly CNS infections

Herpesviruses

Cytomegalovirus

Virus	Disease
Alphaherpesviruses	
• Herpes simplex virus types 1 and 2	Mucosal lesions, encephalitis
• Varicella-zoster virus	Chickenpox, shingles
Betaherpesviruses	
• Cytomegalovirus	Congenital defects; opportunistic pathogen in immunocompromised patients
• Human herpesvirus 6	Roseola
• Human herpesvirus 7	Orphan virus
Gammaherpesviruses	
• Epstein-Barr virus	Infectious mononucleosis; associated with a variety of lymphomas
• Human herpesvirus 8 (Kaposi's sarcoma-related virus)	Kaposi's sarcoma, rare B-cell lymphoma

Epidemiology

Transmission
- Blood, tissue, and body secretions (urine, saliva, semen, cervical secretions, breast milk, tears)

Distribution of virus
- Ubiquitous
- No seasonal incidence

At risk or risk factors
- Babies whose mothers become infected during pregnancy (congenital defects)
- Sexual activity
- Blood and transplant recipients
- Burn victims
- Immunocompromised (recurrent disease)

Vaccines or antiviral drugs
- No vaccines
- Antiviral drugs: acyclovir, ganciclovir, valganciclovir, foscarnet, cidofovir, formivirsen

Disease mechanisms

Transmitted by blood, tissue, and body secretions

Infects epithelial and other cells

Mainly causes subclinical infections

Cell-mediated immunity required for resolution of infection

Latent infection in T cells, macrophages, other cells

Suppression of cell-mediated immunity leads to recurrence and severe disease

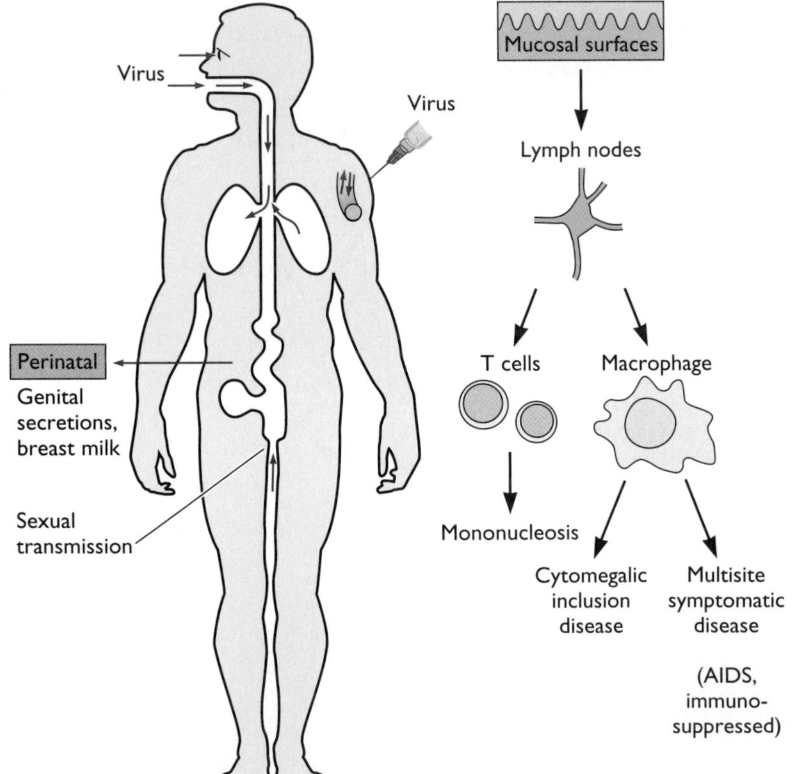

Herpesviruses

Epstein-Barr virus

Virus	Disease
Alphaherpesviruses	
• Herpes simplex virus types 1 and 2	Mucosal lesions, encephalitis
• Varicella-zoster virus	Chickenpox, shingles
Betaherpesviruses	
• Cytomegalovirus	Congenital defects; opportunistic pathogen in immunocompromised patients
• Human herpesvirus 6	Roseola
• Human herpesvirus 7	Orphan virus
Gammaherpesviruses	
• Epstein-Barr virus	Infectious mononucleosis; associated with a variety of lymphomas
• Human herpesvirus 8 (Kaposi's sarcoma-related virus)	Kaposi's sarcoma, rare B-cell lymphoma

Epidemiology

Transmission
- Saliva, close oral contact, or shared items (cup, toothbrush)

At risk or risk factors
- Children (asymptomatic or mild symptoms)
- Teenagers (infectious mononucleosis)
- Immunocompromised (fatal neoplastic disease)
- Malaria (Burkitt's lymphoma)

Distribution of virus
- Ubiquitous
- No seasonal incidence

Vaccines or antiviral drugs
- None

Disease mechanisms

Transmitted in saliva

Infects oral epithelial cells, B cells

Immortalizes B cells

T cells are required to control infection

T cells contribute to symptoms of **infectious mononucleosis**

Associated with lymphoma in immunosuppressed patients, Burkitt's lymphoma, and nasopharyngeal carcinoma

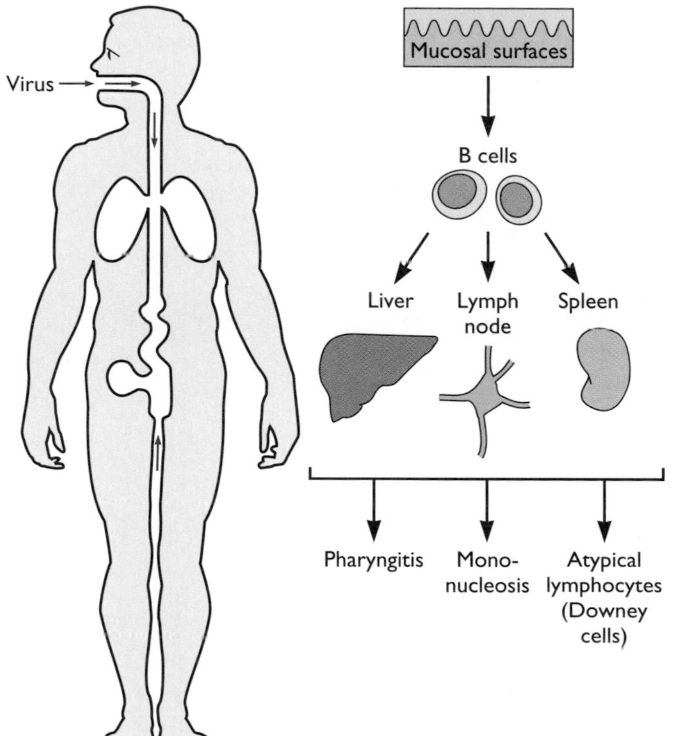

Herpesviruses

Varicella-zoster virus

Virus	Disease
Alphaherpesviruses	
• Herpes simplex virus types 1 and 2	Mucosal lesions, encephalitis
• **Varicella-zoster virus**	Chickenpox, shingles
Betaherpesviruses	
• Cytomegalovirus	Congenital defects; opportunistic pathogen in immunocompromised patients
• Human herpesvirus 6	Roseola
• Human herpesvirus 7	Orphan virus
Gammaherpesviruses	
• Epstein-Barr virus	Infectious mononucleosis; associated with a variety of lymphomas
• Human herpesvirus 8 (Kaposi's sarcoma-related virus)	Kaposi's sarcoma, rare B-cell lymphoma

Epidemiology

Transmission
- Virus is transmitted by respiratory droplets or contact

At risk or risk factors
- Children (ages 5–9 years) (mild disease)
- Teenagers and adults (more severe disease, possibly pneumonia)
- Immunocompromised or neonates (fatal pneumonia, encephalitis, disseminated varicella)
- Elderly, immunocompromised (recurrent zoster)

Distribution of virus
- Ubiquitous
- No seasonal incidence

Vaccines or antiviral drugs
- Live vaccine (Oka strain) available
- Antiviral drugs: acyclovir, foscarnet

Disease mechanisms

Transmitted by **respiratory** route

Infects epithelial cells and fibroblasts, spread by viremia to skin, causes lesions of **chicken pox**

Cell-mediated immunopathology contributes to symptoms

Cell-mediated immunity is required for resolution of infection

Latent infection in neurons

Reactivation by immune suppression

Reactivation leads to **zoster** or **shingles**, formation of lesions over entire dermatome

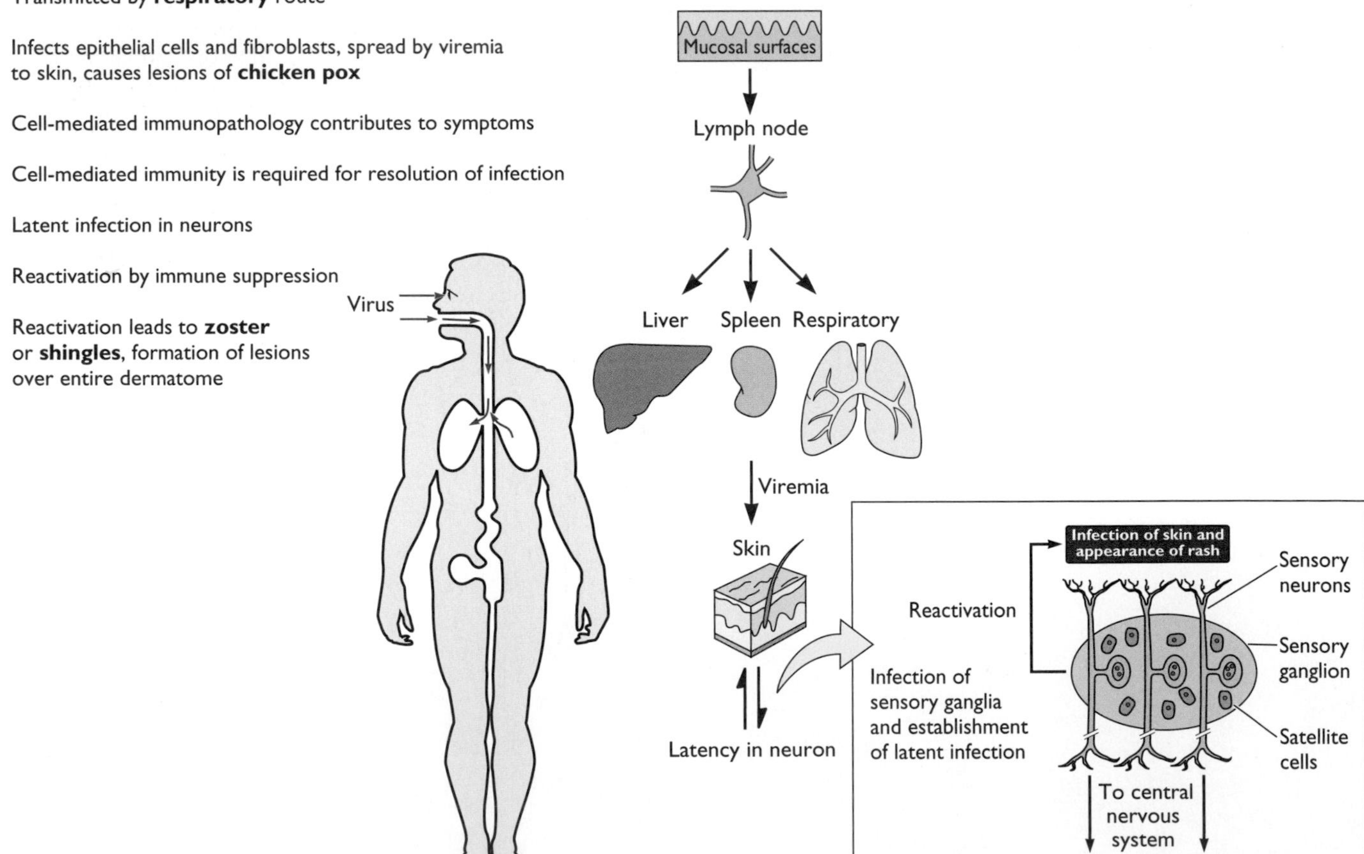

Orthomyxoviruses

Virus	Disease
Influenza A, B, and C viruses	Influenza • Acute febrile respiratory tract infection • Rapid onset of fever, malaise, sore throat, cough • Children may also have abdominal pain, vomiting, otitis media, myositis, croup Complications • Primary viral pneumonia • Myositis and cardiac involvement • Guillain-Barré syndrome • Encephalopathy • Encephalitis • Reye's syndrome

Epidemiology

Transmission
- Inhalation of small aerosol droplets
- Widely spread by schoolchildren

At risk or risk factors
- Adults (typical "flu" syndrome)
- Children (asymptomatic to severe infections)
- Elderly, immunocompromised, and those with cardiac or respiratory problems (high risk)

Distribution of virus
- Ubiquitous; local epidemics, global pandemics
- More common in winter

Vaccines or antiviral drugs
- Killed vaccine against annual strains of influenza A and B viruses
- Live, attenuated influenza A and B vaccine (nasal spray)
- Antiviral drugs: amantadine, rimantadine, zanamivir, oseltamivir

Disease mechanisms

Infects upper and lower respiratory tract

Pronounced systemic symptoms caused by cytokine response to infection

Antibodies against hemagglutinin and neuraminidase (HA and NA) are important for protection against infection

Recovery depends upon interferon and cell-mediated immune response

Susceptibility to bacterial superinfection due to loss of natural epithelial barriers

HA and NA of influenza A virus undergo major and minor antigenic changes, leading to new susceptible hosts

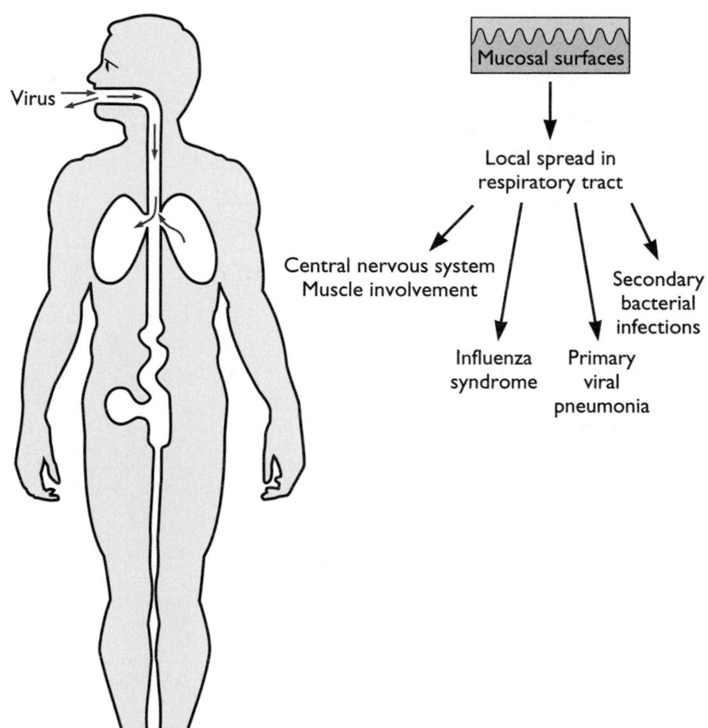

Papillomaviruses

Virus	Disease
Papillomavirus • 90 genotypes	Skin warts: plantar, common, flat warts, epidermodysplasia verruciformis Benign head and neck tumors: laryngeal, oral, and conjunctival papillomas Anogenital warts: condyloma acuminatum, cervical intra-epithelial neoplasia, cancer

Epidemiology

Transmission
- Direct contact, sexual contact
- During birth, from infected birth canal

At risk or risk factors
- Sexually active

Distribution of virus
- Ubiquitous
- No seasonal incidence

Vaccines or antiviral drugs
- A vaccine against type 11 virus is being considered for licensure

Disease mechanisms

Transmitted by **close contact**

Infect epithelial cells of skin, mucous membranes

Replication depends on stage of epithelial cell differentiation

Cause benign outgrowth of cells into warts

Some types are associated with dysplasia that may become cancerous

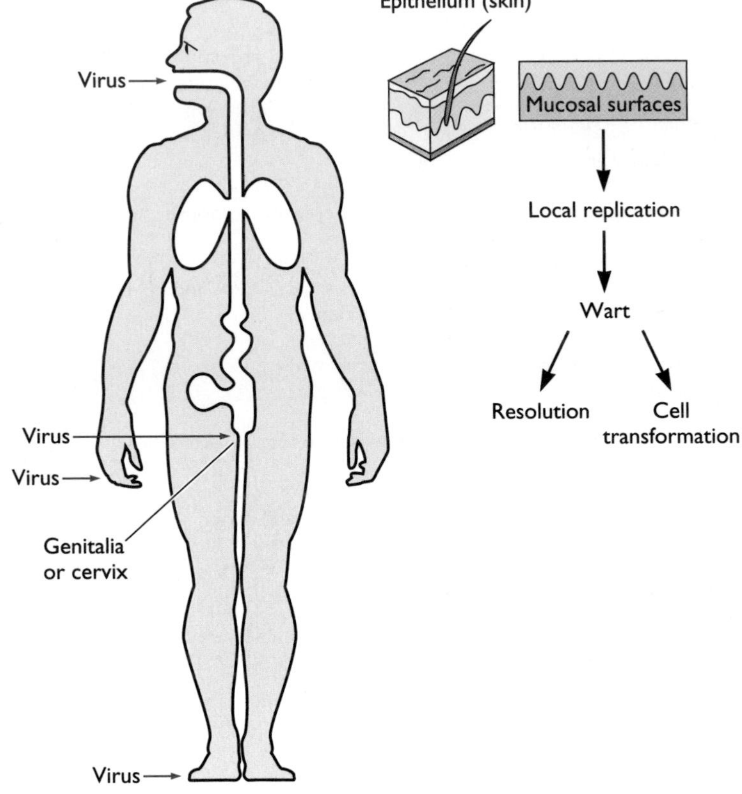

Paramyxoviruses

Measles virus

Virus	Disease
Morbilliviruses	
• **Measles virus**	Measles Complications: otitis media, croup, bronchopneumonia, encephalitis Subacute sclerosing panencephalitis
Paramyxoviruses	
• Parainfluenza virus types 1–4	Cold-like symptoms, bronchitis, croup
• Mumps virus	Mumps
Pneumoviruses	
• Respiratory syncytial virus	Bronchiolitis, pneumonia, febrile rhinitis, pharyngitis, common cold

Epidemiology

Transmission
- Inhalation of large-droplet aerosols
- Highly contagious

At risk or risk factors
- Adults and children
- Immunocompromised persons (more serious outcomes)

Distribution of virus
- Ubiquitous
- Endemic from autumn to spring

Vaccines or antiviral drugs
- Live, attenuated vaccine
- No antiviral drugs

Disease mechanisms

Transmitted by respiratory secretions

Infects epithelial cells of respiratory tract, spreads in lymphocytes and by viremia

Replicates in conjunctivae, respiratory tract, urinary tract, lymphatic system, blood vessels, central nervous system

T-cell response to virus-infected capillary endothelial cells causes rash

Cell-mediated immunity is required to control infection

Complications are due to immunopathogenesis (postinfectious measles encephalitis) or viral mutants (subacute sclerosing panencephalitis)

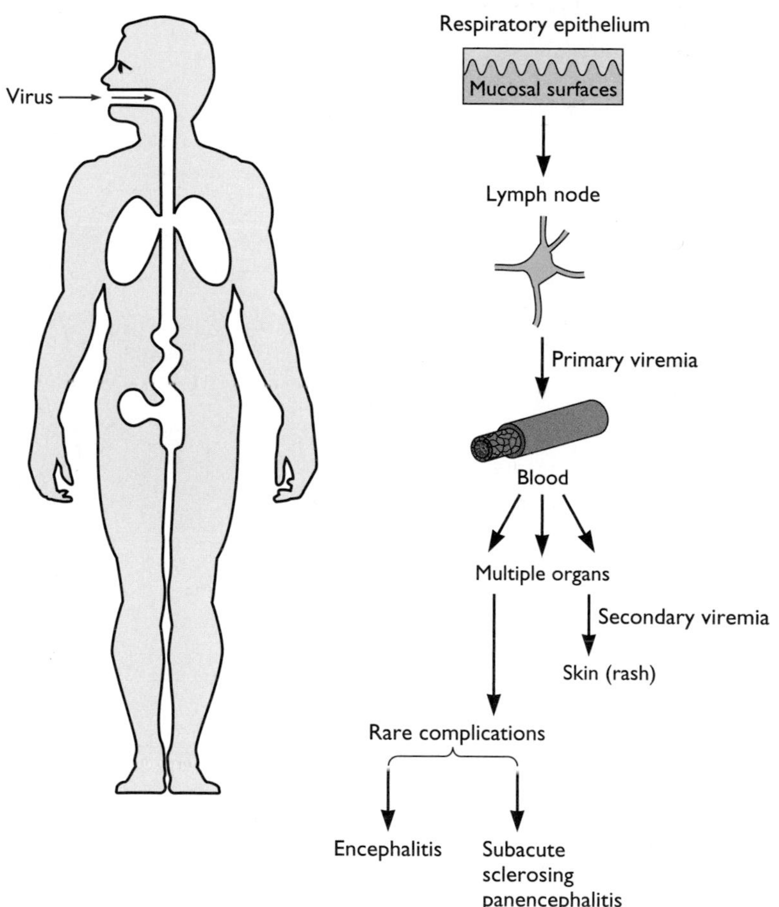

Paramyxoviruses

Mumps virus

Virus	Disease
Morbilliviruses	
• Measles virus	Measles Complications: otitis media, croup, bronchopneumonia, encephalitis Subacute sclerosing panencephalitis
Paramyxoviruses	
• Parainfluenza virus types 1–4	Cold-like symptoms, bronchitis, croup
• Mumps virus	Mumps
Pneumoviruses	
• Respiratory syncytial virus	Bronchiolitis, pneumonia, febrile rhinitis, pharyngitis, common cold

Epidemiology

Transmission
- Inhalation of large-droplet aerosols
- Highly contagious

At risk or risk factors
- Adults and children
- Immunocompromised persons (more serious outcomes)

Distribution of virus
- Ubiquitous
- Endemic in late winter, early spring

Vaccines or antiviral drugs
- Live, attenuated vaccine
- No antiviral drugs

Disease mechanisms

Transmitted by respiratory secretions

Infects epithelial cells of respiratory tract, spreads by viremia

Replicates in salivary glands, testes, respiratory tract, central nervous system

Cell-mediated immunity is required to control infection

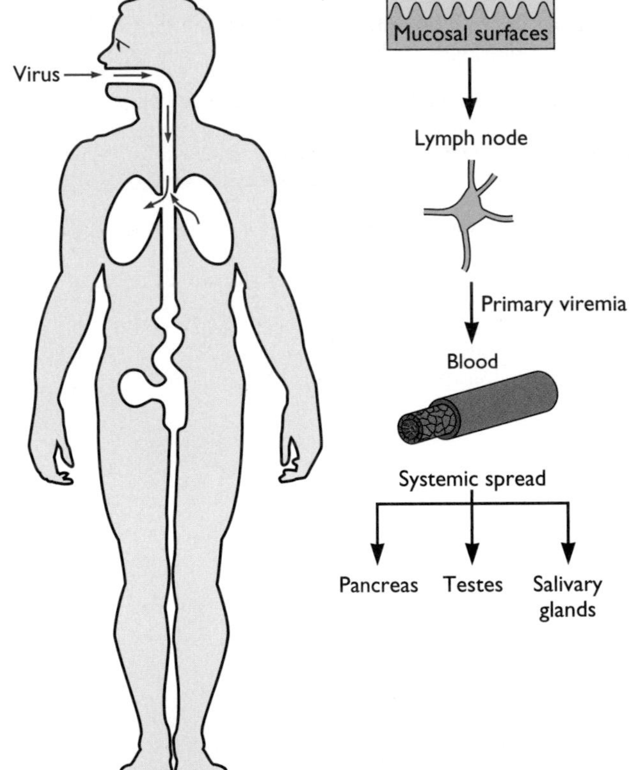

Paramyxoviruses

Respiratory syncytial virus

Virus	Disease
Morbilliviruses	
• Measles virus	Measles Complications: otitis media, croup, bronchopneumonia, encephalitis Subacute sclerosing panencephalitis
Paramyxoviruses	
• Parainfluenza virus types 1–4	Cold-like symptoms, bronchitis, croup
• Mumps virus	Mumps
Pneumoviruses	
• **Respiratory syncytial virus**	Bronchiolitis, pneumonia, febrile rhinitis, pharyngitis, common cold

Epidemiology

Transmission
- Inhalation of large-droplet aerosols

At risk or risk factors
- Infants (bronchiolitis, pneumonia)
- Children (mild disease to pneumonia)
- Adults (mild symptoms)

Distribution of virus
- Ubiquitous
- Incidence seasonal

Vaccines or antiviral drugs
- No vaccine
- Antiviral drugs: ribavirin for infants

Disease mechanisms

Infects the respiratory tract, does not spread systemically

May cause bronchitis, febrile rhinitis, phraryngitis, common cold, or pneumonia

Bronchiolitis probably caused by the host immune response

In newborns, the infection may be fatal because narrow airways are blocked by virus-induced pathology

Infants are not protected from infection by maternal antibody

Reinfection may occur after a natural infection

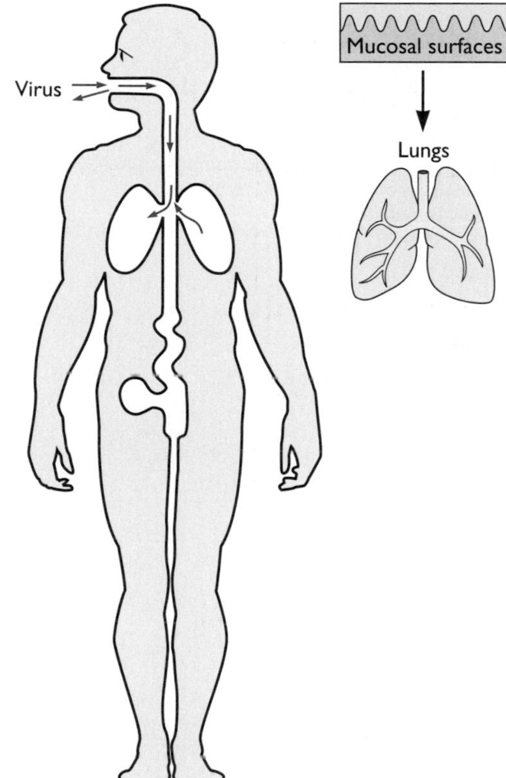

Parvoviruses

Virus	Disease
B19 parvovirus	Erythema infectiosum (fifth disease) Aplastic crisis in patients with chronic hemolytic anemia Acute polyarthritis Abortion
Adeno-associated virus	Commonly infects humans, not associated with illness

Epidemiology

Transmission
- Respiratory and oral droplets

At risk or risk factors
- Children in elementary school (fifth disease)
- Parents of infected children
- Pregnant women (fetal infection and disease)
- Patients with chronic anemia (aplastic crisis)

Distribution of virus
- Ubiquitous
- Fifth disease most common in late winter and spring

Vaccines or antiviral drugs
- None

Disease mechanisms

Transmitted by respiratory and oral secretions

In utero infection

Virus infects mitotically active erythroid precursor cells in bone marrow

Biphasic disease
- Flu-like phase, viral shedding during viremia
- Later phase: erythematous maculopapular rash, arthralgia, and arthritis caused by circulating virus-antibody immune complexes

Aplastic crisis in patients with chronic hemolytic anemia is caused by depletion of erythroid precursors and destabilization of erythrocytes

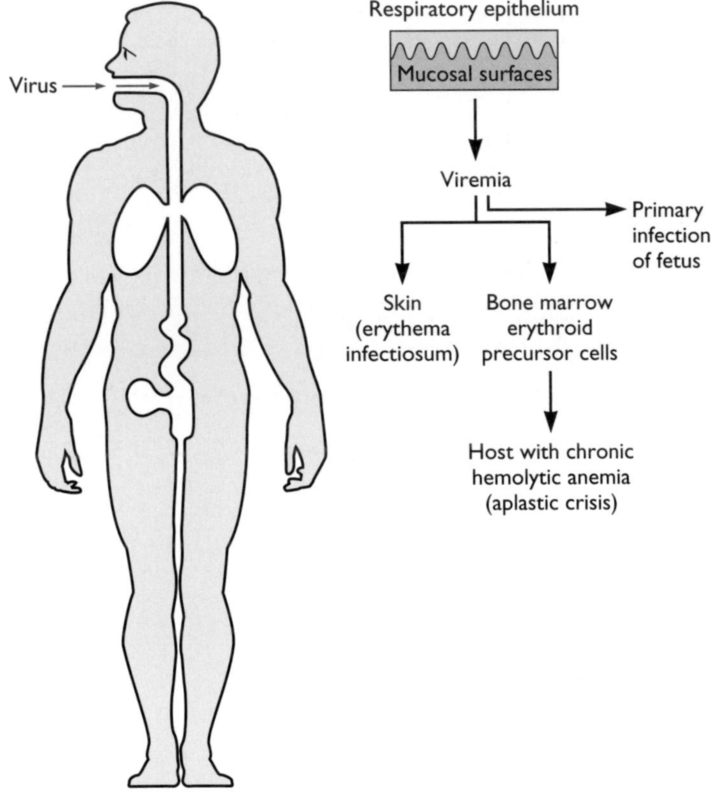

Picornaviruses

Virus	Paralytic disease	Encephalitis, meningitis	Carditis	Neonatal disease	Pleurodynia	Herpangina	Hand-foot-and-mouth disease	Rash disease	Acute hemorrhagic conjunctivitis
Poliovirus 1–3	+	+							
Coxsackie A viruses 1–24	+	+	+			+	+	+	+
Coxsackie B viruses 1–6	+	+	+	+	+			+	
Echoviruses 1–33	+	+	+	+				+	
Enterovirus 70	+								+
Enterovirus 71	+	+					+		
Parechoviruses 1–3	+	+							
Rhinoviruses 1–100									

Virus	Respiratory tract infections	Undifferentiated fever	Diarrhea, gastrointestinal disease	Diabetes, pancreatitis	Orchitis	Disease in immunodeficient patients	Congenital anomalies
Poliovirus 1–3	+	+				+	
Coxsackie A viruses 1–24	+	+				+	+
Coxsackie B viruses 1–6		+		+	+		+
Echoviruses 1–33	+	+	+			+	
Enterovirus 70							
Enterovirus 71							
Parechoviruses 1–3	+	+	+				
Rhinoviruses 1–100	+						

Epidemiology

Transmission
- Enteroviruses: fecal-oral
- Rhinoviruses: inhalation of droplets, contact with contaminated hands

At risk or risk factors
- Poliovirus
    - Young children (asymptomatic or mild disease)
    - Older children, adults (asymptomatic to paralytic disease)
- Coxsackievirus and enterovirus (newborns and neonates at highest risk for serious disease)
- Rhinovirus (all ages)

Distribution of virus
- Ubiquitous; poliovirus is nearly eradicated
- Enteroviruses: disease more common in summer
- Rhinovirus: disease more common in early autumn, late spring

Vaccines or antiviral drugs
- Poliovirus: live oral or inactivated polio vaccines
- No vaccines for other enteroviruses or rhinoviruses
- No antiviral drugs

Disease mechanisms

Virus	Disease
Enteroviruses	Enter oropharyngeal or intestinal mucosa
	Secretory immunoglobulin A can prevent infection
	Spread by viremia to target tissues
	Serum antibody blocks spread
	Virus shed in feces
	High asymptomatic infection rate

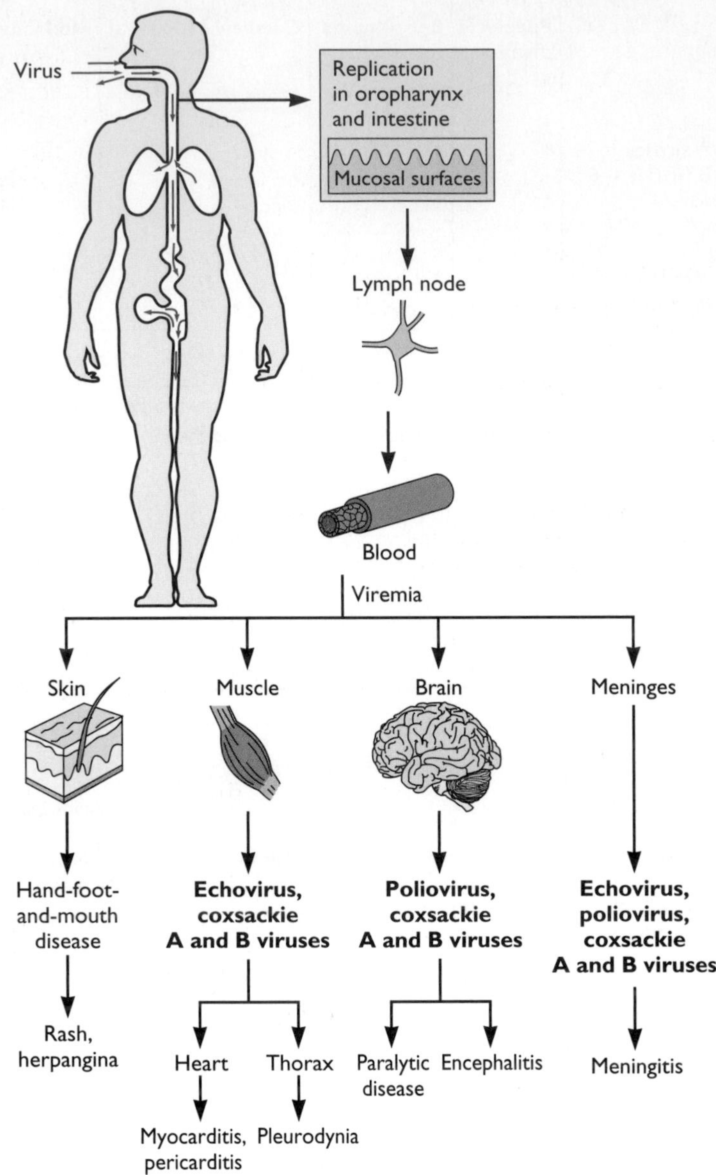

Virus	Disease
Rhinoviruses	Enter upper respiratory tract, where infection is usually limited
	Major factor in asthma exacerbations

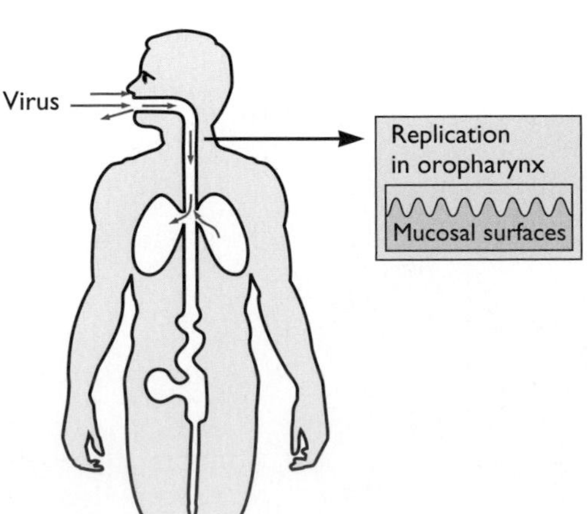

Polyomaviruses

Virus	Disease
Polyomavirus	
• BK virus	Renal disease in immunosuppressed patients
• JC virus	Progressive multifocal leukoencephalopathy in immunosuppressed patients

Epidemiology

Transmission
• Inhalation of infectious aerosols

Distribution of virus
• Ubiquitous
• No seasonal incidence

At risk
• Immunocompromised persons

Vaccines or antiviral drugs
• None

Disease mechanisms

Likely acquired through the respiratory route, spread by viremia to kidneys early in life

Infections are usually asymptomatic

Virus establishes persistent and latent infection in organs such as the kidneys and lungs

In immunocompromised people, JC virus is activated, spreads to the brain, and causes progressive multifocal leukoencephalopathy; oligodendrocytes are killed, causing demyelination

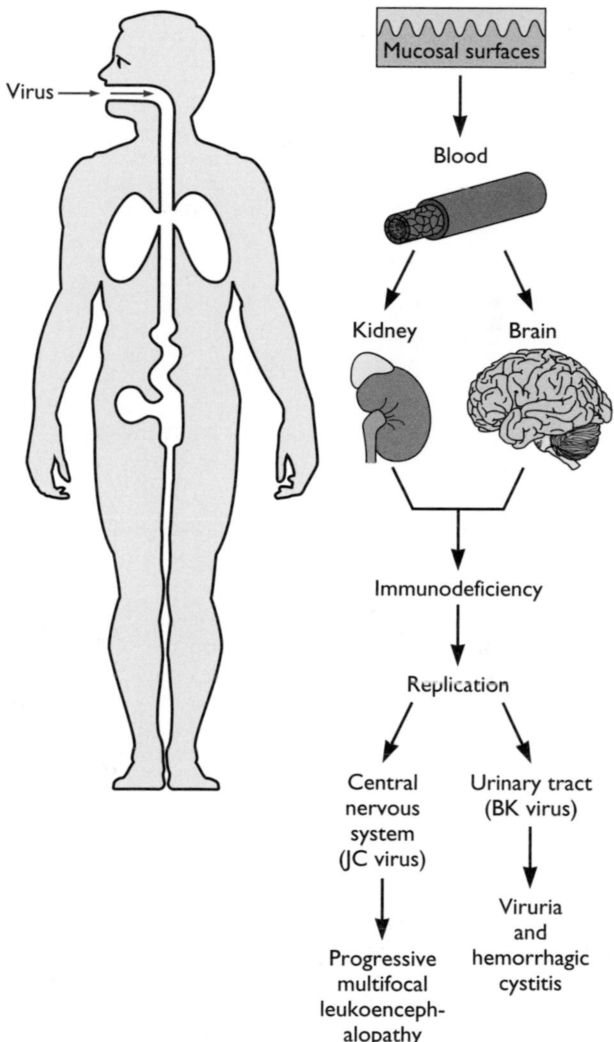

Poxviruses

Virus	Disease
Variola virus	Smallpox
Vaccinia virus (smallpox vaccine)	Encephalitis and vaccinia necrosum (complications of vaccination)
Orf virus	Localized lesion
Cowpox virus	Localized lesion
Pseudocowpox virus	Milker's nodule
Monkeypox virus	Generalized disease
Bovine papular stomatitis virus	Localized lesion
Tanapox virus	Localized lesion
Yaba monkey tumor virus	Localized lesion
Molluscum contagiosum virus	Disseminated skin lesions

Epidemiology

Transmission
- Smallpox: respiratory droplets, contact with virus on fomites
- Other poxviruses: direct contact or fomites

At risk or risk factors
- Molluscum contagiosum: sexual contact, wrestling
- Zoonoses: animal handlers (contact with lesion)

Distribution of virus
- Ubiquitous
- No seasonal incidence
- Natural smallpox has been eradicated

Vaccines or antiviral drugs
- Live vaccine against smallpox (vaccinia virus)
- No antiviral drugs

Disease mechanisms

Smallpox transmitted by respiratory tract secretions, spreads through lymphatics and blood

Molluscum contagiosum and zoonoses transmitted by contact

Sequential infection of multiple organs

Cell-mediated and humoral immunity important to resolve infection

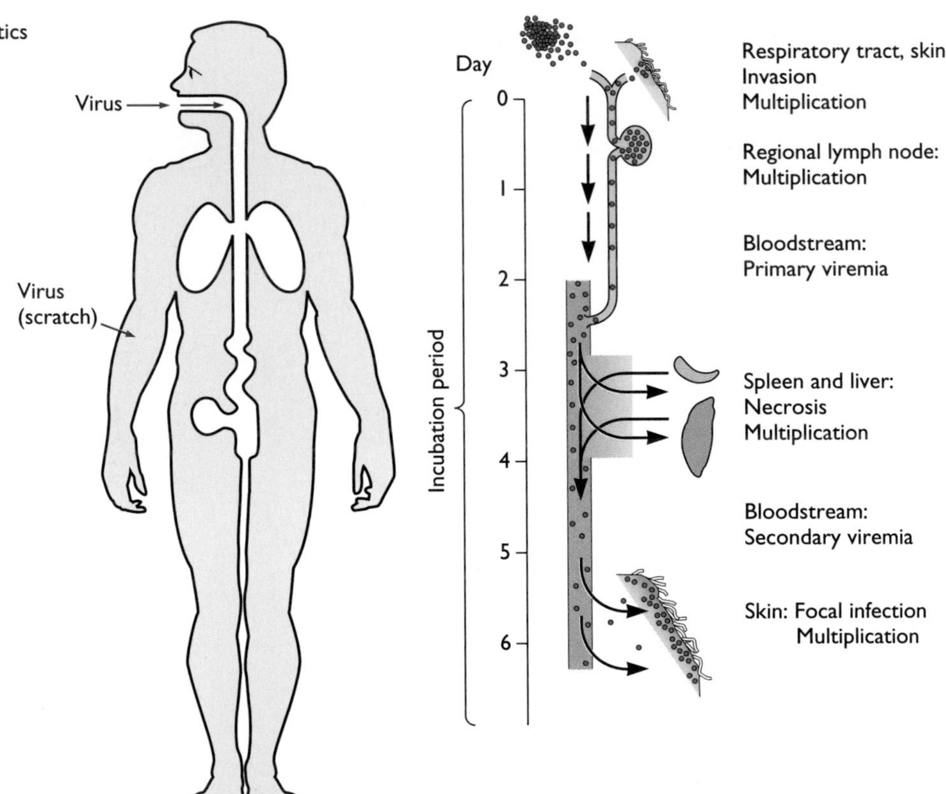

Reoviruses

Rotavirus

Virus	Disease
Orthoreovirus	Mild upper respiratory tract disease, gastroenteritis, biliary atresia
Orbivirus/coltivirus	Colorado tick fever: febrile disease, headache, myalgia (zoonosis)
Rotavirus	Gastroenteritis

Epidemiology

Transmission
- Fecal-oral route

At risk
- Rotavirus type A
  - Infants <24 months of age (gastroenteritis, dehydration)
  - Older children (mild diarrhea)
  - Undernourished persons in underdeveloped countries (diarrhea, dehydration, death)
- Rotavirus type B
  - Infants, older children, adults in China (severe gastroenteritis)

Distribution of virus
- Ubiquitous (type A)
- Less common in summer

Vaccines or antiviral drugs
- None

Disease mechanisms

Transmitted by fecal-oral route

nsP4 is a viral enterotoxin that causes diarrhea

Disease is serious in infants <24 months old, asymptomatic in adults

Large quantities of virions released in diarrhea

Immunity to infection depends on IgA in gut lumen

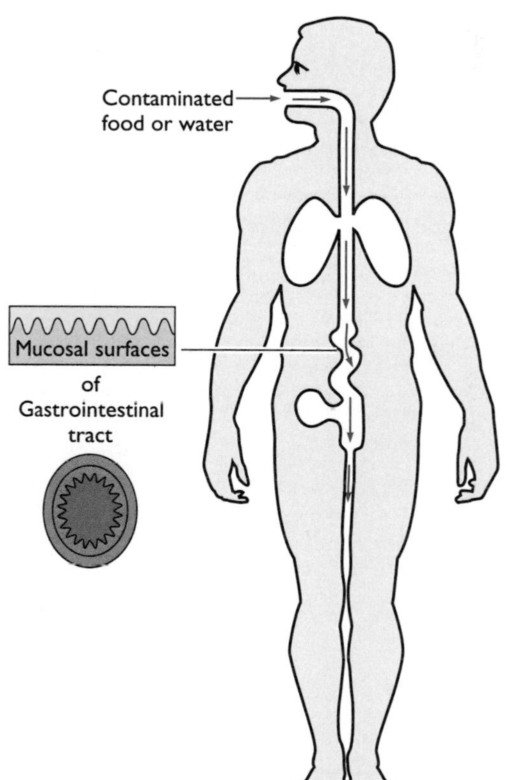

Retroviruses

Human T-lymphotropic virus type 1

Virus	Disease
Deltaretrovirus	
• **Human T-lymphotropic virus type 1**	Adult T-cell leukemia Tropical spastic paraparesis
• Human T-lymphotropic virus type 2	Hairy-cell leukemia
• Human T-lymphotropic virus type 5	Malignant cutaneous lymphoma
Lentivirus	
• Human immunodeficiency virus types 1 and 2	Acquired immune deficiency syndrome

Epidemiology

Transmission
- Virus in blood
  Transfusions, needle sharing among drug users; infected lymphocytes must be present
- Virus in semen
  Anal and vaginal intercourse
- Perinatal transmission
  Transplacental passage of infected maternal lymphocytes, infected lymphocytes in breast milk

Distribution of virus
- Ubiquitous
- No seasonal incidence

At risk
- Intravenous drug users
- Homosexuals and heterosexuals with many partners
- Prostitutes
- Newborns of virus-positive mothers

Vaccines or antiviral drugs
- No vaccines
- Antiviral drugs: azidothymidine plus IFN

Disease mechanisms

Infects T lymphocytes

Remains latent or replicates slowly, induces clonal outgrowth of T-cell clones

Long latency period (30 years) before onset of leukemia

Infection leads to immunosuppression

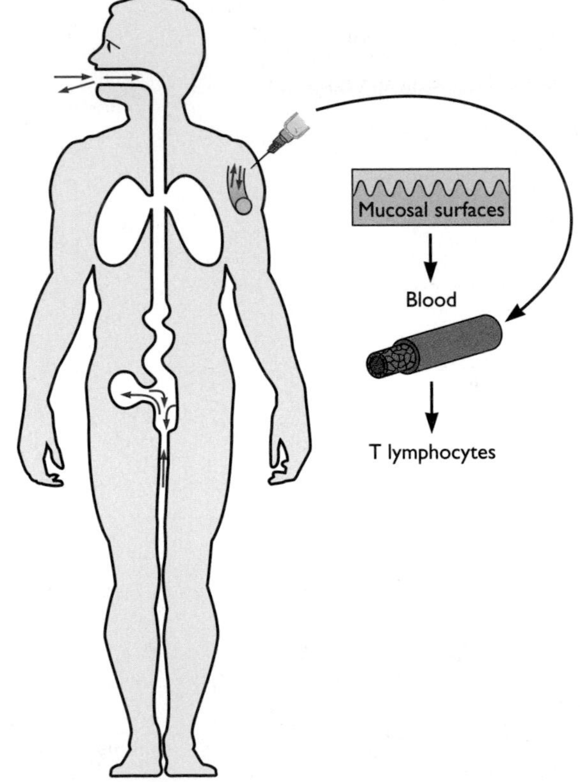

Retroviruses

Human immunodeficiency virus types I and 2

Virus	Disease
Deltaretrovirus	
• Human T-lymphotropic virus type I	Adult T-cell leukemia Tropical spastic paraparesis
• Human T-lymphotropic virus type 2	Hairy-cell leukemia
• Human T-lymphotropic virus type 5	Malignant cutaneous lymphoma
Lentivirus	
• **Human immunodeficiency virus types I and 2**	Acquired immune deficiency syndrome

Epidemiology

Transmission
- Virus in blood
  - Transfusions, needle sharing among drug users, needle sticks in health care workers, tattoo needles
- Virus in semen and vaginal secretions
  - Anal and vaginal intercourse
- Perinatal transmission
  - Intrauterine and peripartum transmission; breast milk

Distribution of virus
- Ubiquitous
- No seasonal incidence

At risk
- Intravenous drug users
- Homosexuals and heterosexuals with many partners
- Prostitutes
- Newborns of virus-positive mothers

Vaccines or antiviral drugs
- No vaccines
- Antiviral drugs
  - Nucleoside analog reverse transcriptase inhibitors (e.g., azidothymidine, dideoxycytidine)
  - Nonnucleoside reverse transcriptase inhibitors (e.g., nevirapine, delavirdine)
  - Protease inhibitors (e.g., saquinavir, ritonavir)

Disease mechanisms

Infects mainly $CD4^+$ T cells and macrophages

Lyses $CD4^+$ T cells, persistently infects macrophages

Infection alters T-cell and macrophage function; immunosuppression leads to secondary infection and death

Infects long-lived cells, establishing reservoir for persistent infection

Infected monocytes spread to brain, causing dementia

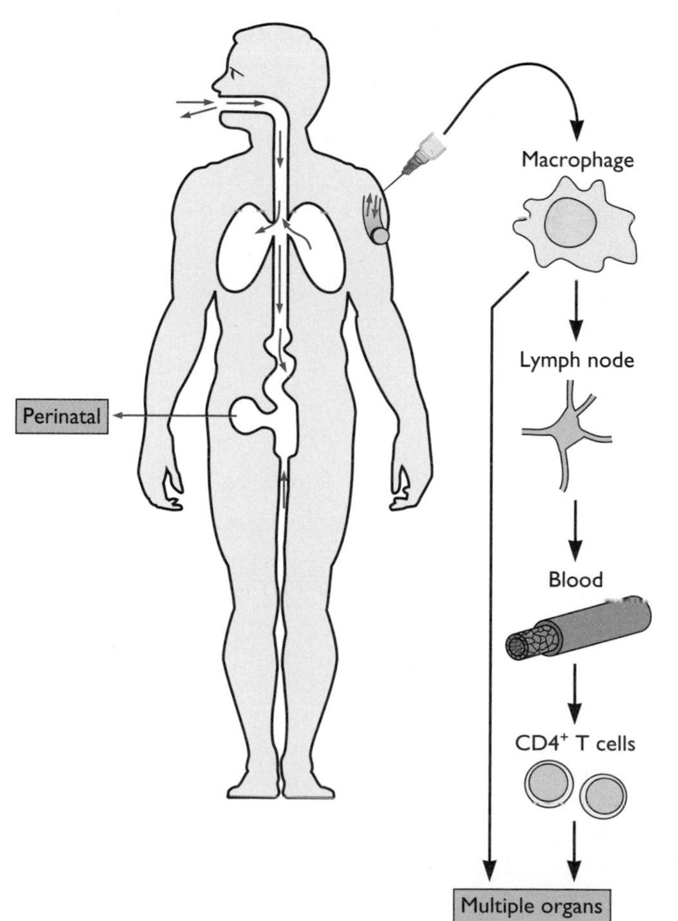

Rhabdoviruses

Rabies virus

Virus	Disease
Lyssavirus	
• **Rabies virus**	Rabies
• Related viruses of rodents and bats	Rarely cause rabies-like encephalitis
Vesiculovirus	
• Vesicular stomatitis virus	Flu-like illness

Epidemiology

Transmission
- Reservoir: wild animals
- Vectors: wild animals, unvaccinated dogs, cats
- Bite of rabid animal (virus in saliva) or aerosols in caves harboring rabid bats

Distribution of virus
- Ubiquitous, except certain islands
- No seasonal incidence

At risk or risk factors
- Animal handlers, veterinarians
- Those in countries with no pet vaccinations or quarantine

Vaccines or antiviral drugs
- Vaccines for pets and wild animals
- Inactivated virus vaccine for at-risk personnel, postexposure prophylaxis
- No antiviral drugs

Disease mechanisms

Transmitted by saliva through bite of rabid animal or by aerosols in caves populated by infected bats

Replicates in muscle at bite site

Incubation period of weeks to months, depending on inoculum and distance of bite from central nervous system

Infects peripheral nerves and travels to brain

Replication in brain causes hydrophobia, seizures, hallucinations, paralysis, coma, and death

Spreads to salivary glands from where it is transmitted

Postexposure immunization can prevent disease due to long incubation period

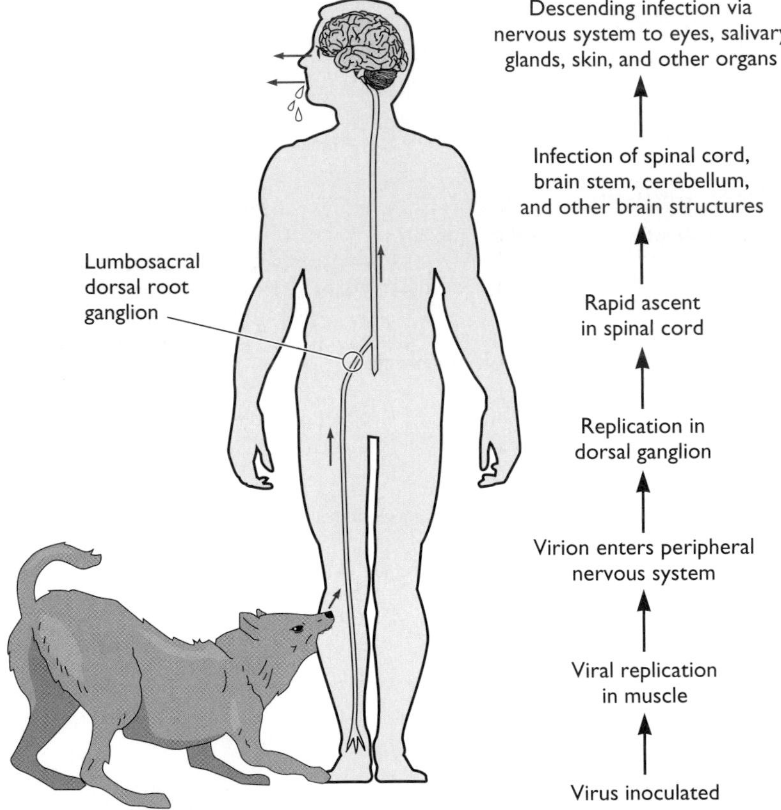

Togaviruses

Virus	Vector	Disease
Alphaviruses		
• Sindbis virus	*Aedes* mosquitoes	Subclinical
• Semliki Forest virus	*Aedes* mosquitoes	Subclinical
• Venezuelan equine encephalitis virus	*Aedes, Culex* mosquitoes	Mild systemic; severe encephalitis
• Eastern equine encephalitis virus	*Aedes, Culiseta* mosquitoes	Mild systemic; encephalitis
• Western equine encephalitis virus	*Culex, Culiseta* mosquitoes	Mild systemic; encephalitis
• Chikungunya virus	*Aedes* mosquitoes	Fever, arthralgia, arthritis
Rubella virus	None	Rubella

Epidemiology

Transmission
- Mosquito vectors
- Rubella virus: respiratory route

At risk
- Arthropod-borne viruses
  People in niche of vector
- Rubella virus
  Neonates <20 weeks old (congenital defects)
  Children (mild rash)
  Adults (more severe disease, arthritis, arthralgia)

Distribution of virus
- Arthropod-borne viruses: determined by habitat of vector
  Aedes mosquito: urban areas
  Culex mosquito: forest, urban areas
  More common in summer
- Rubella virus: ubiquitous

Vaccines or antiviral drugs
- Live, attenuated vaccine for rubella virus
- No antiviral drugs

Disease mechanisms

Viruses are cytolytic (except rubella virus)

Cause viremia, systemic infection

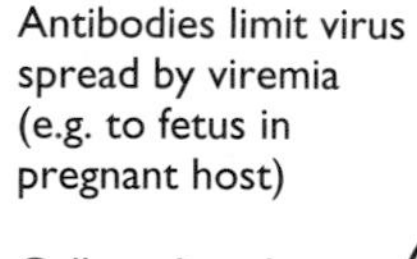

Antibodies limit virus spread by viremia (e.g. to fetus in pregnant host)

Cell-mediated immunity important to resolve infection

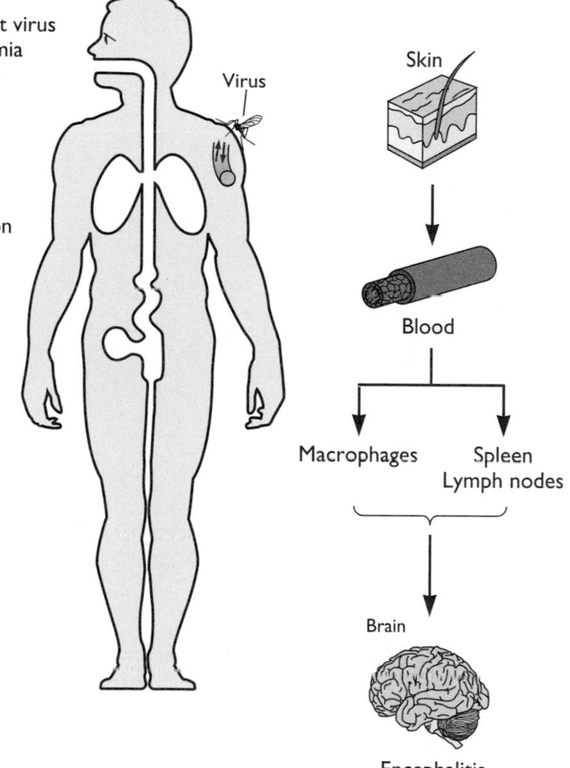

Mosquito borne

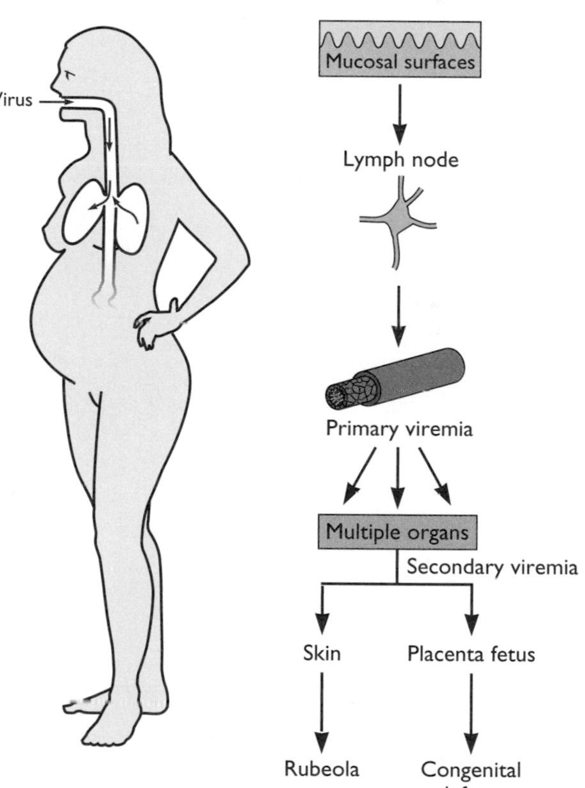

Rubella

Glossary

Abortive infection An incomplete infectious cycle; virions infect a susceptible cell or host but do not complete replication, usually because an essential viral or cellular gene is not expressed. *(Chapter 16)*

Absolute efficiency of plating The plaque titer divided by the number of virus particles in the sample. *(Chapter 2)*

Accessibility An attribute of tropism that describes the physical accessibility of cells to virions. *(Chapter 14)*

Acquired immune response *See* Adaptive response

Active immunization The process of inducing an immune response by exposure to a vaccine. *(Chapter 19)*

Active viremia The presence of newly synthesized virions in the blood. *(Chapter 14)*

Acute infections Common infections in which virions are produced rapidly, and the infection is resolved and cleared quickly by the immune system; survivors usually are immune to subsequent infection. *(Chapter 16)*

Acylated Containing saturated or unsaturated fatty acids added posttranslationally; said of a protein. *(Chapter 12)*

Adaptive response The action of antigen-specific white blood cells and antibodies in response to infection and the development of immune memory. Also called acquired immune response. *(Chapter 15)*

ADCC *See* Antibody-dependent cell-mediated cytotoxicity

Adjuvants Substances administered with antigen that stimulate early processes in immune recognition, particularly the inflammatory response. *(Chapter 19)*

Adoptive transfer The transfer of cells, usually lymphocytes, from an immunized donor to a nonimmune recipient. *(Chapters 15 and 19)*

Alternative pathway Complement pathway initiated by direct recognition of a microbial invader by C1q or by C3b proteins; the alternative pathway is not triggered by antibody, in contrast to the classical pathway of complement activation. *(Chapter 15)*

Alternative splicing Splicing of different combinations of exons in a pre-mRNA, generally leading to synthesis of mRNAs with different protein coding sequences. *(Chapter 10)*

Ambisense Term to describe a viral genome in which mRNAs are produced both from (–) strand genomic RNA and from complementary (+) strands. *(Chapters 1 and 6)*

Anchorage independent Describes the ability of cells to grow in the absence of a surface on which to adhere; often detected by the ability to form colonies in semisolid media. *(Chapter 18)*

Aneuploid Abnormal in chromosome morphology and number. *(Chapter 2)*

Antibody-dependent cell-mediated cytotoxicity The killing of antibody-coated target cells by cells carrying Fc receptors that bind to the constant region of bound antibody; natural killer (NK) cells carrying specific Fc receptors on their surfaces often mediate this reaction. Abbreviated ADCC. *(Chapter 15)*

Antigenic drift The appearance of virions with a slightly altered surface protein (antigen) structure caused by accumulation of point mutations following passage and immune selection in the natural host. *(Chapter 16)*

Antigenic shift A major change in the surface of a virion when completely new surface proteins are acquired during infection; this process occurs when viruses with segmented genomes exchange segments, or when nonsegmented viral genomes recombine after coinfection. *(Chapters 16 and 20)*

Antigenic variation The display on virions or infected cells of new protein sequences that are not recognized by antibodies or T cells that responded to previous infections. *(Chapter 16)*

Antiviral state A property of cells that bind and respond to interferon; interferon-stimulated cells produce many gene products that block the propagation of diverse viruses that may infect subsequently. *(Chapter 11)*

Apical surface The surface of a cell exposed to the environment. *(Chapters 5 and 12)*

Apoptosis A sequence of tightly regulated reactions that ensue in response to external or internal stimuli that signal DNA damage or other forms of stress; characterized by chromosome degradation, nuclear degeneration, and cell lysis; a natural process in development and the immune system, but also an intrinsic defense of cells to viral infection. Also called programmed cell death. *(Chapter 15)*

Attenuated Describes an outcome of infection in which normally severe symptoms or pathology is mild or inconsequential; having reduced virulence. *(Chapters 1 and 14)*

Attenuated mutants Viral mutants that give rise to an attenuated infection compared with the parental strain; virulence is reduced by mutation of the viral genome. *(Chapter 19)*

Autocrine growth stimulation Stimulation of cell growth by growth factors produced and sensed by the same cell. *(Chapter 18)*

Autophagy A process whereby cells are induced to degrade the bulk of their cellular contents by formation of specialized membrane compartments related to lysosomes. *(Chapter 15)*

Auxiliary proteins Nonstructural lentiviral proteins that perform regulatory (e.g., HIV Tat and Rev) or accessory (e.g., human immunodeficiency virus Nef, Vif, Vpr, and Vpu) functions; the latter may not be required for propagation in tissue culture. *(Chapter 17)*

Avirulent Causing no, or mild, disease. *(Chapter 14)*

Axonal transport Transport within neurons of vesicles, macromolecules, and viral components by molecular motors that move along microtubules in axons. *(Chapter 12)*

Bacteriophages Viruses that infect bacteria; derived from the Greek word for eating, "phage." *(Chapter 1)*

Basal lamina A thin layer of extracellular matrix bound tightly to the basolateral surface of cells; the basal lamina is linked to the basolateral membrane by integrins. *(Chapter 5)*

Base-pairing-dependent recombination Exchange of nucleotide sequences between molecules that have a high percentage of nucleotide identity. *(Chapter 6)*

Base-pairing-independent recombination Exchange of nucleotide sequences between molecules that are very different. *(Chapter 6)*

Basolateral surface The nonspecialized surface of a cell that contacts an internal basal lamina or adjacent or underlying cells in the tissue. *(Chapters 5 and 12)*

β-barrel jelly roll A wedge-shaped protein domain formed by an eight-stranded β-sheet. This domain is present in the capsid proteins of several RNA and DNA viruses. *(Chapter 4)*

Breakage-induced template exchange A mechanism of recombination in which breakage of the RNA template that is being copied promotes exchange of one template for another. *(Chapter 6)*

Bulge loops Short, single-stranded regions within a base-paired stem region of RNA secondary structure. *(Chapter 6)*

C3 convertase A central protease activity of the complement pathways; the C3 convertase of the classical pathway is formed from membrane-bound C4b protein and the C2b protease; the C3 convertase of the alternative pathway comprises membrane-bound C3b protein and the Bb protease. *(Chapter 15)*

Cap snatching Cleavage of cellular RNA polymerase II transcripts by a viral endonuclease to produce capped primers for viral mRNA synthesis. *(Chapter 6)*

Capping The addition of m^7G via a 5′–5′ phosphodiester bond to the 5′ ends of cellular and viral transcripts made in eukaryotic cells. *(Chapter 10)*

Capsid The outer shell of viral proteins that surrounds the genome in a virus particle. *(Chapters 1 and 3)*

Caspases A family of closely related cysteine proteases that cleave polypeptides after aspartate residues. They play important roles in apoptosis. *(Chapter 15)*

Caveolae Flask-shaped invaginations in the plasma membrane of many cell types that contain the protein caveolin and are rich in lipid rafts; caveolae internalize membrane components, extracellular ligands, bacterial toxins, and some animal viruses. *(Chapter 5)*

$CD4^+$ T cells T lymphocytes that carry the coreceptor protein CD4 on their surfaces. *(Chapter 15)*

$CD8^+$ T cells T lymphocytes that carry the coreceptor CD8 on their surfaces. *(Chapter 15)*

Cell cycle The orderly and reproducible sequence in which cells increase in size, duplicate the genome, segregate duplicated chromosomes, and divide. *(Chapter 18)*

Cell-mediated response The combined actions of specific helper and effector T lymphocytes; the lymphocyte-mediated response. Also called cell-mediated immunity. *(Chapter 15)*

Centrosome The major microtubule organizing center in mammalian cells, located near the nucleus; microtubules are organized with their (–) ends at the centrosome. *(Chapter 5)*

Chaperones Proteins that facilitate the folding of other polypeptide chains, the assembly of multimeric proteins, or the formation of macromolecular assemblies. *(Chapter 13)*

Chemokines Small proteins that attract and stimulate cells of the immune defense system; produced by many cells in response to infection. *(Chapter 5)*

Chronic infections Persistent infections that are eventually cleared. *(Chapter 16)*

Clades Subtypes of human immunodeficiency virus that are prevalent in different geographic areas. *(Chapter 17)*

Classical pathway A pathway of complement activation initiated by the binding of C1q to antibody-antigen complexes on the surface of an invader or an infected cell. *(Chapter 15)*

Clinical latency A state of persistent viral infection in which no clinical symptoms are manifested. *(Chapter 17)*

Coactivators Proteins that stimulate transcription by RNA polymerase II without binding to a specific DNA sequence; generally interact with sequence-specific transcriptional activators. *(Chapter 8)*

Codon Three contiguous bases in an mRNA template that specify the amino acids incorporated into protein. *(Chapter 11)*

Complement system A set of plasma proteins that act in a concerted fashion to destroy extracellular pathogens and infected cells; originally defined as a heat-labile activity that lysed bacteria in the presence of antibody (it "complemented" antibody action); the activated complement pathway also stimulates phagocytosis, chemotaxis, and inflammation. *(Chapter 15)*

c-Oncogenes The normal cellular genes that are the origin of v-oncogenes; their products can be oncogenic when altered or synthesized inappropriately. Also called proto-oncogenes. *(Chapter 18)*

Contact inhibition Cessation of cell division when cells make physical contact, as occurs at high density in a culture dish. *(Chapter 18)*

Continuous cell lines Cell cultures of a single cell type that can be propagated indefinitely in culture. *(Chapter 2)*

Copy choice A mechanism of recombination in which an RNA polymerase first copies the 3′ end of one parental strand and then exchanges one template for another at the corresponding position on a second parental strand. *(Chapters 6 and 7)*

Core promoters The minimal set of DNA sequences required for accurate initiation of transcription by RNA polymerase II. *(Chapter 8)*

Coreceptor A cell surface molecule that, in addition to the receptor, is required for entry of virus particles into cells. *(Chapter 5)*

CTL escape mutants Mutations in the viral coding sequences for immunodominant peptides that render infected cells invisible to the T-cell response. *(Chapter 16)*

Cutaneous immune system The unique collection of skin-specific lymphocytes and dendritic cells (Langerhans cells) that are important in the initial response and resolution of skin infections. Also called skin-associated lymphoid tissue, or SALT. *(Chapter 15)*

Cytokines Soluble proteins produced by cells in response to various stimuli, including virus infection; they affect the behavior of other cells both locally and at a distance, by binding to specific cytokine receptors. *(Chapter 15)*

Cytopathic effects The morphological changes induced in cells by viral infection. *(Chapter 2)*

Cytopathic viruses Viruses that kill a cell rapidly while producing a burst of new infectious particles. *(Chapter 16)*

Cytoskeleton The intracellular structural network composed of actin filaments, microtubules, and intermediate filaments. *(Chapter 5)*

Defective-interfering RNAs Subgenomic RNAs that replicate more rapidly than full-length RNA, and therefore compete for the components of the RNA synthesis machinery and interfere with the replication of full-length RNAs. *(Chapter 6)*

Delayed-type hypersensitivity Cell-mediated immunity caused by $CD4^+$ T cells that recognize antigens in the skin; the reaction typically occurs hours to days after antigen is injected, hence its name; it is partially responsible for characteristic local reactions to virus infection, such as a rash. *(Chapter 15)*

Deletion mutation Removal of nucleic acid sequences. *(Chapter 2)*

Dendritic cells Migratory, phagocytic cells; immature dendritic cells, which are found in the periphery of the body, around body cavities, and under mucosal surfaces, can synthesize copious quantities of cytokines, including interferon, and take up soluble proteins avidly. These proteins are retained in endosomes for hours or days until the dendritic cell matures; mature dendritic cells migrate to local lymph nodes, where their antigen cargo is scrutinized by lymphocytes of the adaptive immune system. *(Chapter 15)*

Dermis The layer of skin beneath the epidermis that supports the basement membrane or vascular network; it is composed of a dense connective tissue that provides support and elasticity to the skin. *(Chapter 14)*

Diapedesis The process by which viruses cross the vascular endothelium, while being carried within monocytes or lymphocytes. *(Chapter 14)*

Difference imaging The subtraction of density maps of related proteins or assemblies to reveal structural features of the more complex structure. *(Chapter 4)*

Diploid cell strains Homogeneous populations of a single type that can divide up to 100 times before dying. *(Chapter 2)*

Direct immunostaining A method for visualizing viral proteins in infected cells or tissues in which an antibody that recognizes a viral antigen is coupled directly to an indicator, such as a fluorescent dye or an enzyme. *(Chapter 2)*

Disseminated Describes an infection that spreads beyond the primary site of infection; often includes a viremia and infection of major organs such as liver, lungs, and kidneys. *(Chapter 14)*

DNA microarrays Hundreds or thousands of unique DNA sequences fixed on a glass slide in aligned rows (a DNA microarray chip); by using differentially labeled hybridization probes made from uninfected and virus-infected cell mRNAs, it is possible to determine whether production of specific mRNAs is induced, repressed, or unchanged after virus infection. *(Chapter 2)*

DNA shuffling A technique that enables scientists to assemble and analyze new combinations of DNA fragments not found in nature. *(Chapter 19)*

DNA synthesis phase The phase of the cell cycle in which the cell's genome is duplicated. Also called S phase. *(Chapter 18)*

DNA vaccine A bacterial plasmid that encodes an antigenic protein; the protein is synthesized upon introduction of the plasmid DNA into cells, and an immune response is directed to this protein; a variation of the subunit vaccine. *(Chapter 19)*

Domains Parts of a protein than can fold independently of other segments into stable structures with specific functions, e.g., a DNA-binding domain. *(Chapter 4)*

Draining lymph nodes Small nodules found in many peripheral sites at which lymphatic vessels converge and the adaptive immune responses are initiated; lymph nodes have direct connections to the circulatory system. *(Chapter 15)*

Eclipse period The phase of viral infection during which the viral nucleic acid is uncoated from its protective shell and no infectious virus can be detected inside cells. *(Chapter 2)*

Elongation The stepwise incorporation of ribonucleoside or deoxyribonucleoside monophosphates (NMPs or dNMPs) into the 3′-OH end of a growing RNA or DNA chain. *(Chapters 6, 8, and 9)*

Emerging virus A viral population responsible for a marked increase in disease incidence, usually as a result of changed societal, environmental, or population factors. *(Chapter 20)*

Endemic Typical of a particular geographic area; persisting in a population for a long period without reintroduction of the causative virus from outside sources. *(Chapter 1)*

Endogenous pathway of antigen presentation The cellular process whereby viral proteins are degraded inside the infected cell and the resulting peptides are loaded onto major histocompatibility complex (MHC) class I molecules that move to the cell surface. *(Chapter 15)*

Endogenous proviruses Proviruses that enter the germ line at some point in the history of an organism and are thereafter inherited in normal Mendelian fashion by every cell in that organism and by its progeny. *(Chapter 7)*

Endogenous reactions Reverse transcription reactions that take place in purified virions (sometimes after permeabilization with detergents), using the viral RNA genome as a template. *(Chapter 7)*

Endosomes Vesicles that transport molecules from the plasma membrane to the cell interior. *(Chapter 5)*

Enhancer A DNA sequence containing multiple elements that can stimulate RNA polymerase II transcription over long distances, independently of orientation or location relative to the site of transcriptional initiation. *(Chapter 8)*

Enhancing antibodies Antibodies that can facilitate viral infection by allowing virions to which they bind to enter susceptible cells. *(Chapter 17)*

Enterotropic Describes a virus that replicates in tissues of the gastrointestinal tract. *(Chapter 14)*

Envelope The host cell-derived lipid bilayer carrying viral glycoproteins that forms the outer layer of many virus particles. *(Chapters 1, 4, 12, and 13)*

Epidemic A pattern of disease characterized by a rapid and sudden appearance of cases spreading over a wide area. *(Chapter 1)*

Epidermis The external surface of the skin, composed of a keratinized, stratified squamous epithelium. *(Chapter 14)*

Episome An exogenous genetic element not necessary for cell survival; often applied to viral genomes that can be maintained in cells by autonomous replication. *(Chapters 1 and 9)*

Epitope A short contiguous sequence or unique conformation of a macromolecule that can be recognized by the immune system; also called an antigenic determinant. A T-cell epitope is a short peptide recognized by a particular T-cell receptor; a B-cell epitope is recognized by the antigen-binding domain of antibody and is part of an intact protein. *(Chapter 2)*

ER lumen The interior of the membrane-bound compartment of the endoplasmic reticulum (ER); a distinctive chemical environment topologically equivalent to the exterior of the cell. *(Chapter 12)*

Error threshold The point at which accumulated mutations reduce fitness (replication competence); more precisely, a mathematical parameter that measures the complexity of the information to be maintained in the genome of an individual virus and the positive selective features of the resultant population. *(Chapter 20)*

Evolution The change of a population over time in response to selective processes. *(Chapter 20)*

Exogenous antigen presentation The cellular process whereby viral proteins are taken up from the outside of the cell and digested and the resulting peptides are loaded onto MHC class II molecules that move to the cell surface. *(Chapter 15)*

Exons Blocks of noncontiguous coding sequences (generally short) present in many cellular and viral pre-mRNAs. *(Chapter 10)*

Fitness The replicative adaptability of an organism to its environment. *(Chapter 20)*

Foci Clusters of cells that are derived from a single progenitor and share properties, such as unregulated growth, that cause them to pile up on one another. *(Chapters 18)*

Fomites Inanimate objects that may be contaminated with microorganisms and become vehicles for transmission. *(Chapter 14)*

Fusion peptide A short hydrophobic amino acid sequence (20 to 30 amino acids) that is believed to insert into target membranes to initiate fusion. *(Chapter 5)*

Fusion pore An opening between two lipid bilayers formed by the action of fusion proteins that allows exchange of material across membranes. *(Chapter 5)*

G proteins Membrane-associated guanine nucleotide-binding proteins with GTPase activity. *(Chapter 18)*

G_0 *See* Resting state

Gap phases The cell cycle phases before (G_1) or after (G_2) the phase in which DNA synthesis (S) takes place. *(Chapter 18)*

Genetic bottleneck Extreme selective pressures on small populations that result in loss of diversity, accumulation of nonselected mutations, or both. *(Chapter 20)*

Germ line transmission Transfer of viral genomes from an organism to its offspring as part of the host genome. *(Chapter 14)*

Glycoforms The total set of forms of a protein that differ in the number, location, and nature of oligosaccharide chains. *(Chapter 12)*

Glycoproteins Proteins carrying covalently linked sugar chains (oligosaccharides). *(Chapter 4)*

Hairpin loops The single-stranded regions of RNA that join stem regions of RNA secondary structure. *(Chapter 6)*

Half-life The time required for 50% of the cytoplasmic poof of an mRNA to be degraded, when synthesis of the mRNA is inhibited. *(Chapter 10)*

Helical symmetry The symmetry of regularly wound structures defined by the relationship $P = \mu \times \rho$, where P = pitch of the helix, μ = the number of structural units per turn, and ρ = the axial rise per unit. *(Chapter 4)*

Helper virus A virus that provides viral proteins needed for the replication of a coinfecting defective virus. *(Chapter 6)*

Hemagglutination Linking of multiple red blood cells by virus particles, resulting in a lattice. *(Chapter 2)*

Hematogenous spread Spread of virus particles through the bloodstream. *(Chapter 14)*

Hepatitis Inflammation and necrosis of the liver. *(Chapter 14)*

Herd immunity The immune status of a population, rather than an individual. *(Chapter 19)*

Heteromeric proteins Proteins comprising multiple different subunits. *(Chapter 4)*

Homomeric proteins Proteins comprising multiple copies of a single subunit. *(Chapter 4)*

Horizontal transmission Transfer of viral infections by means other than between parent and offspring. *(Chapter 14)*

Host range A listing of species (hosts) that are susceptible to and permissive for infection. *(Chapter 5)*

Humoral response The actions of antibodies; immunity can be transferred to nonimmunized recipients by purified antibody or by serum from immunized donors. Also called humoral immunity. *(Chapter 15)*

Iatrogenic Transmitted from a health care worker to a patient; said of an infectious disease. *(Chapter 14)*

Icosahedral symmetry The symmetry of the icosahedron, the solid with 20 faces and 12 vertices related by axes of two-, three- and fivefold rotational symmetry. *(Chapter 4)*

Immortality The capacity of cells to grow and divide indefinitely. *(Chapter 18)*

Immune response The highly coordinated interaction of cytokines and effector white blood cells; the sum total of host defense mechanisms. *(Chapter 15)*

Immunoblotting A method in which proteins are fractionated by electrophoresis in a polyacrylamide gel, transferred to a thin, synthetic membrane that has a strong affinity for proteins, and then detected by immunostaining. Also called Western blot analysis. *(Chapter 2)*

Immunodominant Describes the peptides and epitopes that are recognized most efficiently by cytotoxic T lymphocytes and antibodies. *(Chapter 16)*

Immunological synapse The focal collection of coreceptors around T-cell receptors and their respective binding partners in the target cell; this large structure is characterized by a rearrangement of the cytoskeleton and focused release of effector molecules at the site of contact with the target cell. *(Chapter 15)*

Immunotherapy A treatment that provides an infected host with exogenous antiviral cytokines, other immunoregulatory agents, antibodies, or lymphocytes in order to reduce viral pathogenesis. *(Chapter 19)*

Inapparent infections Asymptomatic acute infections that are recognized by the presence of virus-specific antibodies in individuals with no reported history of disease. *(Chapter 16)*

Inclusion bodies Intracellular granules that contain virions or unassembled viral components in the nucleus and/or cytoplasm. *(Chapters 2 and 6)*

Incubation period The period before the initial appearance of characteristic symptoms of a disease. *(Chapter 16)*

Indirect immunostaining A method for visualizing viral proteins in infected cells or tissues; an antibody that recognizes a viral antigen is bound by a second antibody that is coupled to an indicator, such as a fluorescent dye or an enzyme. *(Chapter 2)*

Indirectly anchored proteins Proteins that are indirectly bound to the plasma membrane by interacting with either integral membrane proteins or the charged sugars of membrane glycolipids. *(Chapter 5)*

Infectious mononucleosis A common disease caused by Epstein-Barr virus; infection of B cells in an individual with a normal immune system induces infected cells to divide, resulting in substantial immune and cytokine responses. *(Chapter 16)*

Inflammation A general term for the complex response that gives rise to local accumulation of white blood cells and fluid; initiated by local infection or damage; many different forms of this response, characterized by the degrees of tissue damage, capillary leakage, and cellular infiltration, occur after infection with pathogens. *(Chapter 15)*

Initiation of transcription All reactions necessary to complete synthesis of the first phosphodiester bond in an RNA transcript of a DNA template. *(Chapter 8)*

Initiator sequences Short DNA sequences that can direct accurate initiation of RNA polymerase II transcription in the absence of any other promoter sequences. *(Chapter 8)*

Innate response A coordinated, immediate response to infection, in part mediated by local sentinel cells (dendritic cells and macrophages) that respond to infected cells and release cytokines; by a complex collection of serum proteins termed complement, which when activated destroys infected cells and virus particles; and by cytolytic lymphocytes called NK cells, which recognize and destroy infected cells. *(Chapter 15)*

Insertion mutation Addition of nucleic acid sequences. *(Chapter 2)*

Insertional activation Mechanism of oncogenesis by nontransducing retroviruses; integration of a proviral promoter or enhancer in the vicinity of a c-oncogene results in inappropriate transcription of that gene. *(Chapter 18)*

Insertional mutagenesis Mutation in a genome caused by the integration of viral DNA or the DNA of a transposable element. *(Chapter 1)*

Instability elements *cis*-acting AU-rich sequences present in unspliced or singly spliced human immunodeficiency virus mRNA that impair gene expression by impeding RNA transport to the cytoplasm or decreasing RNA stability. *(Chapter 17)*

Integral membrane proteins Proteins embedded in a lipid bilayer, with external and internal domains connected by one or more membrane-spanning domains. *(Chapters 4 and 5)*

Interfering antibodies Antibodies that can bind to virions or infected cells and block interaction with neutralizing antibodies. *(Chapter 17)*

Interferons Specific cytokines that, when bound to their receptors, stimulate an antiviral state in the cell; interferons α and β are produced mainly by leukocytes and fibroblasts, whereas interferon γ is produced by T cells and NK cells. *(Chapter 15)*

Interior loops Unpaired sequences between stem regions of RNA secondary structure. *(Chapter 6)*

Intrinsic cellular defenses The conserved cellular programs that respond to various stresses, such as starvation, irradiation, and infection; intrinsic defenses include apoptosis, autophagy, and RNA interference. *(Chapter 15)*

Introns Noncoding sequences that separate coding sequences (exons) in many cellular and viral pre-mRNAs. *(Chapter 10)*

Koch's postulates Criteria developed by the German physician Robert Koch in the late 1800s to determine if a given agent is the cause of a specific disease. *(Chapter 1)*

Kupffer cells Mononuclear phagocytic cells in the liver. *(Chapter 14)*

Latency-associated transcripts RNAs produced during a latent infection by herpes simplex virus. Abbreviated LATs. *(Chapters 8 and 16)*

Latent infections Persistent infections that last the life of the host; few or no virions can be detected, despite continuous presence of the viral genome. *(Chapters 8 and 16)*

Latent period The phase of viral infection during which no extracellular virus can be detected. *(Chapter 2)*

Leaky scanning A mechanism for producing multiple proteins from a single mRNA, by initiation of translation at different in-frame initiation codons. *(Chapter 11)*

Lipid raft A microdomain of the plasma membrane that is enriched in cholesterol and saturated fatty acids, with a distinct protein composition; lipid rafts participate in such processes as cell movement, protein sorting, and signal transduction. *(Chapters 5 and 12)*

Long-latency virus Retrovirus that causes cancer in a host many years after infection; the viral genome does not encode cellular oncogenes, nor does it cause cancer by perturbing the expression of cellular oncogenes. *(Chapter 18)*

Lymphocytes A class of white blood cells that initiate immune responses and carry antigen-specific receptors on their surfaces; B cells, T cells, and NK cells are the primary classes of lymphocytes. *(Chapter 15)*

Lysogenic Describes a bacterium that carries the genetic information of a quiescent bacteriophage, which can be induced to reproduce and subsequently lyse the bacterium. *(Chapter 1)*

Lysogens Lysogenic bacteria. *(Chapter 1)*

Lysogeny The phenomenon by which the lysogenic state is established and maintained in bacteria. *(Chapter 1)*

Lysosomes Cellular vesicles containing enzymes that degrade carbohydrates, proteins, nucleic acids, and lipids. *(Chapter 5)*

M cell A specialized microfold or membranous epithelial cell of mucosal surfaces, found mainly in intestines, that is important in uptake of proteins and pathogens; these cells deliver ingested antigens to lymphoid cells in close proximity to their basal surfaces. *(Chapter 15)*

M phase *See* Mitosis

Macules Flat, colored skin lesions caused by virus replication in the dermis. *(Chapter 14)*

Mannan-binding pathway A mechanism that triggers complement action via the interaction of a mannan-binding lectin similar to C1q with mannose-containing carbohydrates on bacteria, virus particles, or virus-infected cells. *(Chapter 15)*

Marker rescue Replacement of all local nucleic acid, including a mutation, with wild-type nucleic acid. *(Chapter 2)*

Marker transfer Introduction of a mutation by replacement of a segment of viral nucleic acid with one containing the mutation. *(Chapter 2)*

Membrane-spanning domains Segments of integral membrane proteins spanning the lipid bilayer, typically α-helical. *(Chapters 4 and 5)*

Memory cells A subset of lymphocytes maintained after each encounter with a foreign antigen; these cells survive for years in the body and are ready to respond immediately to any subsequent encounter by rapid proliferation and efficient production of their defensive products. *(Chapter 15)*

Metastable structures Structures that are stable because they possess low (but not minimal) free energy; they are separated from the minimal free energy state by an energetic or kinetic barrier. *(Chapter 4)*

Metastases Secondary tumors, often at distant sites, that arise from the cells of a malignant tumor. *(Chapter 18)*

Missense mutation A change in a single nucleotide or codon that results in the production of a protein with a single amino acid substitution. *(Chapter 2)*

Mitosis The phase of the cell cycle in which newly duplicated chromosomes are distributed to two new daughter cells as a result of cell division. Abbreviated M phase. *(Chapter 18)*

Molecular chaperones Proteins that assist the folding and oligomerization of nascent proteins. *(Chapter 12)*

Monocistronic mRNA mRNA that encodes one polypeptide. *(Chapter 11)*

Monoclonal antibodies Antibodies of a single specificity made by a clone of antibody-producing cells. *(Chapter 2)*

Monoclonal antibody-resistant mutants Viral mutants selected to propagate in the presence of neutralizing monoclonal antibodies; these mutants often carry mutations in viral genes that encode virion proteins. *(Chapter 15)*

Monolayer A single layer of cultured cells growing in a cell culture dish. *(Chapter 2)*

Morphological units Surface structures of virus particles defined by electron microscopy. Also called capsomeres. *(Chapter 4)*

Motifs Combinations of a small number of protein secondary structure units in specific geometric arrangements. *(Chapter 4)*

Mucosal immune system The lymphoid tissues below the mucosa of the gastrointestinal and respiratory tracts. Also called mucosa-associated lymphoid tissue, or MALT. *(Chapter 15)*

Muller's ratchet The dictum that small, asexual populations decline in fitness over time if the mutation rate is high. *(Chapter 20)*

Multibranched loops Unpaired loops that are connected to multiple stem regions in RNA secondary structure. *(Chapter 6)*

Multiplicity of infection The number of virions added per cell. *(Chapter 2)*

Mutagen An agent that causes base changes in nucleic acids. *(Chapter 2)*

Myelomonocytes Monocytes, macrophages, dendritic cells, and a variety of granulocytes derived from myeloid progenitor cells in the bone marrow. *(Chapter 15)*

Naked Lacking an envelope derived from a host cell membrane; said of a virus particle. *(Chapter 1)*

Natural killer cells An abundant lymphocyte population that comprises large, granular lymphocytes; distinguished from others by the absence of B- and, in most cases, T-cell antigen receptors; NK cells are a component of the innate defense system. *(Chapter 15)*

Negative [(–)] strand The strand of DNA or RNA that is complementary in sequence to the (+) strand. *(Chapter 1)*

Neuroinvasive Able to spread to the central nervous system (brain and spinal cord) after infection of a peripheral site. *(Chapter 14)*

Neurotropic Describes a virus that replicates in neural tissue. *(Chapter 14)*

Neurovirulent Able to cause disease in the nervous system. *(Chapter 14)*

Neutralize Block the infectivity of virus particles; a term applied to appropriate antibodies. *(Chapter 2)*

NK cells *See* Natural killer cells

Noncytopathic viruses Viruses that infect cells and actively produce infectious particles without causing immediate host cell death. *(Chapter 16)*

Nonsense mutations Substitution mutations that produce a translation termination codon. *(Chapter 2)*

Nontransducing oncogenic retroviruses Retroviruses that do not encode cell-derived oncogene sequences, but can cause cancer (at low efficiency) when their DNA becomes integrated in the vicinity of a cellular oncogene, thereby perturbing its expression. *(Chapter 18)*

Northern blot hybridization A method in which RNA preparations are fractionated by gel electrophoresis, denatured within the gel, and transferred to a membrane to which they bind strongly; the bound RNAs are then detected by hybridization. *(Chapter 2)*

Nosocomial Describes transmission of an infectious disease in a hospital or health care facility. *(Chapter 14)*

Nuclear localization signals Amino acid sequences that are necessary and sufficient for protein import into the nucleus. *(Chapter 5)*

Nucleocapsid Nucleic acid-protein assembly packaged within the virion; used when this complex is a discrete substructure of a complex particle. *(Chapters 1 and 4)*

Obligate parasites Entities or organisms that are dependent on another living organism for reproduction. *(Chapter 1)*

Oligonucleotides Short DNA or RNA chains, typically 2 to 100 bases in length. *(Chapter 2)*

Oligosaccharides Short linear or branched chains of sugar residues (monosaccharides). *(Chapter 4)*

Oncogenes Genes encoding proteins that cause cellular transformation or tumorigenesis. *(Chapter 18)*

Oncogenesis The development of cancer. *(Chapter 18)*

One-hit kinetics The number of plaques or lesions is directly proportional to the first power of the concentration of the inoculum, i.e., if the concentration is doubled, the number of plaques or lesions is doubled; this property indicates that one infectious particle is sufficient to initiate infection. *(Chapter 2)*

Origin for plasmid maintenance The origin of Epstein-Barr virus DNA active in latently infected but not productively infected cells. *(Chapter 9)*

Origin of replication Site at which replication of a DNA genome, or a segment of a genome, begins. *(Chapter 9)*

Packaging Incorporation of a viral nucleic acid genome into a virus particle during assembly. *(Chapter 13)*

Packaging signals Nucleic acid sequences or structural features directing incorporation of a viral genome into a virus particle. *(Chapter 13)*

Pandemic Worldwide epidemic. *(Chapter 1)*

Pantropic Describes a virus that replicates in many tissues and cell types. *(Chapter 14)*

Papules Slightly raised skin lesions caused by virus replication in the dermis. *(Chapter 14)*

Particle-to-PFU ratio The inverse value of the absolute efficiency of plating (see above). *(Chapter 2)*

Passive immunization Direct administration of the products of the immune response (e.g., antibodies or stimulated immune cells) obtained from an appropriate donor(s) to a patient. *(Chapter 19)*

Passive viremia Introduction of virus particles into the blood without viral replication at the site of entry. *(Chapter 14)*

Pathogen Disease-causing virus or other microorganism. *(Chapter 1)*

Pattern recognition receptors Unique protein receptors of the innate immune system that bind common molecular structures on the surfaces of pathogens; some reside on the cell surfaces of sentinel cells, such as immature dendritic cells and macrophages, and others are soluble initiator proteins of the complement system. *(Chapter 15)*

PCR *See* Polymerase chain reaction

Permissive Describes a cell that can support viral replication, because it possesses the necessary intracellular components. *(Chapter 5)*

Permissivity Requirement of a virus for differentially expressed cellular gene products to complete the infection. *(Chapter 14)*

Persistent infections Infections in which infected cells or virions are not cleared efficiently by the adaptive immune response, and virus particles or viral gene products continue to be produced for long periods. *(Chapter 16)*

PFU/ml *See* Plaque-forming units per milliliter

Plaque A circular zone of infected cells that can be distinguished from the surrounding monolayer. *(Chapter 2)*

Plaque purified A virus stock prepared from a single plaque; when one infectious virus particle initiates a plaque, the viral progeny within the plaque are clones. *(Chapter 2)*

Plaque-forming units per milliliter A measure of infectivity. Abbreviated PFU/ml. *(Chapter 2)*

Polar Describes the differential distribution of proteins and lipids in the plasma membranes that creates apical and basolateral domains of certain cell types. *(Chapter 5)*

Polarized cells Differentiated cells with surfaces divided into functionally specialized regions. *(Chapter 12)*

Polyadenylation The addition of ~200 adenylate (A) residues to the 3′ ends of cellular and viral transcripts made in eukaryotic cells. *(Chapter 10)*

Polycistronic mRNA mRNA that encodes several polypeptides. *(Chapter 11)*

Polyclonal antibodies The antibody repertoire against the many epitopes of an antigen produced in an animal. *(Chapter 2)*

Polymerase chain reaction Amplification of viral (or of any) DNA sequences by using thermostable DNA polymerases and specific oligonucleotides. Abbreviated PCR. *(Chapter 2)*

Polymorphic Describes a gene with many allelic forms, as found in outbred populations. *(Chapter 15)*

Portal Site of entry of a viral genome into a preassembled protein shell. *(Chapter 4)*

Positive [(+)] strand The strand of DNA or RNA with the sequence corresponding to that of the mRNA. Also called the "sense" strand. *(Chapter 1)*

Pregenomic RNA The hepadnaviral mRNA that is reverse transcribed to produce the DNA genome. *(Chapter 7)*

Preinitiation complexes Promoter-bound complexes of an RNA polymerase and initiator proteins competent to initiate transcription. *(Chapter 8)*

Primary cell cultures Cell cultures prepared from animal tissues; these cultures include several cell types and have a limited life span, usually no more than 5 to 20 cell divisions. *(Chapter 2)*

Primary cells Cells that have been freshly derived from an organ or tissue. *(Chapter 1)*

Primary structure The amino acid sequence of a protein, listed from the N to the C terminus. *(Chapter 4)*

Primary viremia Progeny virions released into the blood after initial replication at the site of entry. *(Chapter 14)*

Primer Free 3′-OH group required for initiation of synthesis of DNA from DNA or RNA templates and for initiation of synthesis of some viral RNA genomes. *(Chapters 6 and 9)*

Prions Infectious agents comprising an abnormal isoform of a normal cellular protein and no nucleic acid; implicated as the causative agents of transmissible spongiform encephalopathies. *(Chapters 1 and 20)*

Procapsids Closed, protein-only structures into which viral genomes are inserted; precursors to capsids or nucleocapsids. *(Chapter 13)*

Processivity Ability of an enzyme to copy a nucleic acid template over long distances from a single site of initiation. *(Chapters 8 and 9)*

Professional antigen-presenting cells Cells that are specially equipped to initiate immediate immune defense by secreting interferon and to convey information of the attack to the adaptive immune system; the main classes are dendritic cells and B cells. *(Chapter 15)*

Programmed cell death *See* Apoptosis

Promoter Set of DNA sequences necessary for initiation of transcription by a DNA-dependent RNA polymerase. *(Chapter 8)*

Proofreading Correction of mistakes made during chain elongation by exonuclease activities of DNA-dependent DNA polymerases. *(Chapters 6 and 9)*

Prophage The genome of the quiescent bacteriophage in a lysogenic bacterium. *(Chapter 1)*

Proteasome Complex containing multiple proteases that is responsible for degradation of proteins tagged with polyubiquitin. *(Chapter 8)*

Proteoglycans Proteins linked to glycosaminoglycans, which are unbranched polysaccharides made of repeating disaccharides. *(Chapter 5)*

Proto-oncogenes *See* c-Oncogenes

Proviral DNA Retroviral DNA that is integrated into its host cell genome and is the template for formation of retroviral mRNAs and genomic RNA. Also termed a provirus. *(Chapter 7)*

Provirion Noninfectious precursor to a mature virion. *(Chapter 13)*

Pseudodiploid Having two RNA genomes per virion but giving rise to only one DNA copy, as is the case for retroviruses. *(Chapter 7)*

Pseudoreversion Phenotypic reversion caused by second-site mutation. Also called suppression. *(Chapter 2)*

Pustules Skin lesions derived from a vesicle in which secondary infiltration of leukocytes occurs. *(Chapter 14)*

Quality control Mechanisms that detect unfolded or misfolded proteins in the ER and promote folding of such proteins or target them for ER exit and degradation. *(Chapter 12)*

Quasiequivalence Arrangement of structural units in a virus particle that allows similar interactions among them. *(Chapter 4)*

Quasispecies Dynamic distribution of nonidentical but related viral replicons; a term often constrained to RNA virus populations. *(Chapters 6 and 20)*

Quaternary structure The number, arrangement, and interactions among the subunits (individual folded polypeptide chains) in a multimeric protein. *(Chapter 4)*

Reactivation Switch from a latent to a productive infection; usually applied to herpesviruses. *(Chapters 8 and 16)*

Reassortants Viral genomes that have exchanged segments after coinfection of cells with viruses with segmented genomes. *(Chapter 2)*

Reassortment The exchange of entire RNA molecules between genetically related viruses with segmented genomes. *(Chapters 2 and 6)*

Receptor The cellular molecule to which a virion attaches to initiate replication. *(Chapter 5)*

Receptor-mediated endocytosis The uptake of molecules into the cell from the extracellular fluid; in this process, the molecule binds a cell surface receptor, and the complex is taken into the cell by invagination of the membrane and formation of a vesicle. *(Chapter 5)*

Relative efficiency of plating A ratio of viral titers obtained on two different cell types; this number may be more or less than 1, depending on how well the virus grows in the different host cells. *(Chapter 2)*

Replication forks The sites of synthesis of nascent DNA chains that move away from an origin as replication proceeds. *(Chapter 9)*

Replication intermediates Incompletely replicated DNA molecules containing newly synthesized DNA. *(Chapter 9)*

Replication licensing Mechanisms that ensure replication of cellular DNA is initiated at each origin once, and only once, per cell cycle. *(Chapter 9)*

Replicons Units of replication in large genomes, defined by discrete origin and termini. *(Chapter 9)*

Reservoir A host population in which a viral infection is maintained in the environment, and from which it is spread to other hosts. *(Chapter 14)*

Resolution The minimal size of an object that can be distinguished by microscopy or other methods of structural analysis. *(Chapter 4)*

Resting state A state in which the cell has ceased to grow and divide and has withdrawn from the cell cycle. Also called G_0. *(Chapter 18)*

Restriction point A point in the G_1 phase of the mammalian cell cycle beyond which cells will not respond to extracellular growth-regulating stimuli, but are committed to progress into DNA synthesis (S) phase. *(Chapter 18)*

Retroid viruses Viruses that replicate their genomes via reverse transcription. *(Chapter 7)*

Revert Change to the parental, or wild-type, genotype or phenotype. *(Chapter 2)*

Rev-responsive element A structural element in *env* RNA that is recognized by the human immunodeficiency virus Rev protein, which mediates its export from the nucleus. *(Chapter 17)*

Ribosome shunting A mechanism of translation initiation in which ribosomes physically bypass, or shunt over, parts of the 5′ untranslated region to reach the initiation codon. *(Chapter 11)*

Ribozyme An RNA molecule with catalytic activity. *(Chapter 10)*

RNA editing The introduction into an RNA molecule of nucleotides that are not specified by a cellular or viral gene. *(Chapter 10)*

RNA interference A sequence-specific mechanism of RNA degradation found in plants and animals; it may have evolved to protect against foreign nucleic acid. Abbreviated RNAi. *(Chapter 10)*

RNA processing The series of co-, or post-, transcriptional, covalent modifications that produce mature mRNAs from primary transcripts. *(Chapter 10)*

RNA pseudoknot RNA secondary structure formed when a single-stranded loop region base pairs with a complementary sequence outside the loop. *(Chapter 6)*

RNA-dependent RNA polymerase The protein assembly required to carry out RNA synthesis from an RNA template. *(Chapter 6)*

RNAi *See* RNA interference

Rough ER ER membranes to which polyribosomes synthesizing proteins destined to enter the secretory pathway are bound. *(Chapter 12)*

RRE *See* Rev-responsive element

Rule of six The requirement that the (–) strand RNA genome of paramyxoviruses is copied efficiently only when its length in nucleotides is a multiple of 6. *(Chapter 6)*

S phase *See* DNA synthesis phase

Satellite nucleic acids Small nucleic acids that require a helper virus for replication and are packaged by a capsid protein encoded in the helper virus genome. Also called virusoids. *(Chapter 20)*

Satellite virus A satellite with a genome that encodes one or two proteins. *(Chapters 1 and 20)*

Satellites Subviral agents that lack genes that encode proteins required for replication or propagation; satellite replication depends on coinfection of the host cell with a helper virus that can supply the missing proteins. *(Chapters 1 and 20)*

Scaffolding proteins Viral proteins required for assembly of an icosahedral protein shell, but absent from mature virions. *(Chapter 13)*

Second messengers Small molecules, such as cyclic nucleotides and lipids, that are synthesized by some membrane-bound proteins in a signal transduction cascade; they act as diffusible components in a signal relay. *(Chapter 18)*

Secondary structure Local, regular structure stabilized by hydrogen bonding, e.g., α-helix and β-sheet. *(Chapter 4)*

Secondary viremia Delayed appearance of a high concentration of infectious virus in the blood as a consequence of disseminated infections. *(Chapters 14 and 16)*

Secretory pathway The series of membrane-demarcated compartments (e.g., ER and Golgi), tubules, and vesicles through which secreted and membrane proteins travel to the cell surface. *(Chapter 12)*

Semiconservative replication Production of two daughter DNA molecules, each containing one strand of the parental template and a newly synthesized complementary strand. *(Chapter 9)*

Sentinel cells Dendritic cells and macrophages; these migratory cells are found in the periphery of the body and can take up proteins and cell debris for presentation of peptides derived from them on MHC molecules; they also respond to recognition of a pathogen by synthesizing cytokines such as interferons. *(Chapter 15)*

Serology The study of the properties and reaction of sera; the use of antibodies in sera to study properties of antigens. *(Chapter 2)*

Serotypes Virus types as defined with neutralizing antibodies. *(Chapter 2)*

Shedding Release of virions from an infected host. *(Chapter 14)*

Signal peptidase The ER protease that removes signal peptides from many proteins as they are translocated into the ER. *(Chapter 12)*

Signal peptide A short sequence (generally hydrophobic) that directs nascent proteins to the ER. The signal may be removed, or retained as a transmembrane domain. *(Chapter 12)*

Signal recognition particle A complex containing a small RNA molecule and several proteins that binds to signal peptides of nascent proteins destined for the secretory pathway to halt translation and allow binding to the ER membrane. Abbreviated SRP. *(Chapter 12)*

Signal transduction cascade A chain of sequential physical interactions among, and biochemical modifications of, membrane-bound and cytoplasmic proteins. *(Chapter 18)*

Single-exon mRNAs mRNAs produced without splicing because their precursors lack introns and splice sites. *(Chapter 10)*

Sinusoids Small blood vessels characterized by a discontinuous basal lamina, with no significant barrier between the blood plasma and the membranes of surrounding cells. *(Chapter 14)*

Slow infections Extreme variants of the persistent pattern of infection; long incubation period (years) from the time of initial infection until the appearance of recognizable symptoms. *(Chapter 16)*

Slow viruses Viruses characterized by long incubation periods, typical for the genus *Lentivirus* in the family *Retroviridae*. *(Chapter 17)*

Snares Soluble Nsf attachment protein receptors; proteins present on membranes of compartments of the secretory pathway (target, or t-Snares) and transport vesicles (vesicle, or v-Snares); important determinants of the specificity of vesicular transports. *(Chapter 12)*

Southern blot hybridization A method in which DNA preparations are digested with a restriction endonuclease, fractionated by gel electrophoresis, denatured within the gel, and transferred to a membrane to which they bind strongly; the bound DNAs are then detected by hybridization. *(Chapter 2)*

Sphingolipid-cholesterol rafts *See* Lipid raft

Spliceosome The large complex that assembles on an intron-containing pre-mRNA prior to splicing. In mammalian cells, it comprises the small nuclear ribonucleoproteins containing U1, U2, U4, U5, and U6 small nuclear RNAs and ~150 proteins. *(Chapter 10)*

Splicing The precise ligation of blocks of noncontiguous coding sequences (exons) in cellular or viral pre-mRNAs with excision of the intervening noncoding sequences (introns). *(Chapter 10)*

SRP *See* Signal recognition particle

SSPE *See* Subacute sclerosing panencephalitis

Stem regions Regions of RNA secondary structure where complementary RNA sequences base pair. *(Chapter 6)*

Stop transfer signal A hydrophobic sequence that halts translocation of a nascent protein across the ER membrane; serves as a transmembrane domain. *(Chapter 12)*

Structural plasticity The resistance of the structure and function of proteins or virus particles to amino acid substitutions, such as those resulting from antigenic variation. *(Chapter 16)*

Structural units The units from which capsids or nucleocapsids of virus particles are built. Also called protomers or asymmetric units. *(Chapter 4)*

Subacute sclerosing panencephalitis A rare and often lethal brain disease caused by measles virus; a result of a slow infection that occurs when intracellular host proteins in cells of the central nervous system interfere with acute infection by inhibiting viral gene expression. Abbreviated SSPE. *(Chapter 16)*

Substitution mutation Replacement of one or more nucleotides in a nucleic acid. *(Chapter 2)*

Subunit vaccines Vaccines formulated with purified components of virus particles, rather than intact virions. *(Chapter 19)*

Subunits Single folded polypeptide chains. *(Chapter 4)*

Suppression Phenotypic reversion caused by second-site mutation. Also called pseudoreversion. *(Chapter 2)*

Susceptible Describes a cell that produces the receptor(s) required for virus entry. *(Chapter 5)*

Susceptibility The presence of cell receptors for virus entry. *(Chapter 14)*

Suspension cultures Cells propagated in suspension, in which a spinning magnet continuously stirs the cells. *(Chapter 2)*

Syncytia Fused cells with multiple nuclei. *(Chapter 2)*

Synergism Cooperative activity of two proteins that is greater than the product of their individual activities. *(Chapter 8)*

Systemic Describes an infection that results in spread to many organs of the body. *(Chapter 14)*

Systemic inflammatory response syndrome The large-scale production and systemic release of inflammatory cytokines and stress mediators that may overwhelm and kill an infected host; sometimes referred to as a "cytokine storm"; similar syndromes include toxic shock promoted by certain bacterial pathogens. *(Chapter 15)*

Tegument The layer interposed between the nucleocapsid and the envelope of herpesvirus particles. *(Chapter 4)*

Telomeres Specialized DNA structures comprising simple repeated sequences that are present at the ends of linear chromosomal DNA and copied by an RNA-templated mechanism. *(Chapter 9)*

Termini Sites at which DNA replication stops. *(Chapter 9)*

Tertiary structure The folded structure of a protein chain. *(Chapter 4)*

Tight junctions The area of contact between adjacent cells, circumscribing the cells at the apical edges of their lateral membranes. *(Chapter 5)*

Toll-like receptors Type I transmembrane proteins found on sentinel cells; these proteins are pattern recognition receptors and, when bound to particular microbial components, activate a signal transduction pathway to increase cytokine production and NF-κB activity. *(Chapter 15)*

Topology The geometrical arrangement of, and connections among, secondary structure units and motifs of a protein. *(Chapter 4)*

Topology diagram Two-dimensional representation of protein secondary structure units, and their orientations and connections, in a motif, domain, or protein. *(Chapter 4)*

Transcriptional control region Local and distant DNA sequences necessary for initiation of transcription and regulation of transcription. *(Chapter 8)*

Transcytosis A mechanism of transport in which material in the intestinal lumen is endocytosed by M cells, transported to the basolateral surface, and released to the underlying tissues. *(Chapters 5 and 14)*

Transducing oncogenic retroviruses Retroviruses that include oncogenic, cell-derived sequences in their genomes and carry these sequences to each newly infected cell; such viruses are, therefore, highly oncogenic. *(Chapter 18)*

Transduction The transfer of genes from one cell to another via viral vectors. *(Chapter 1)*

Transfection Introduction of viral nucleic acid into cells by transformation, resulting in the infection of cells. *(Chapter 2)*

Transformed Having changed growth properties and morphology as a consequence of infection with certain oncogenic viruses, introduction of oncogenes, or exposure to chemical carcinogens. *(Chapter 18)*

Transforming infection A class of persistent infection in which cells infected by certain DNA viruses or retroviruses may exhibit altered growth properties and proliferate more rapidly than uninfected cells. *(Chapter 16)*

Transmissible spongiform encephalopathies Fatal neurodegenerative disorders that are transmitted by prions. Abbreviated TSEs. *(Chapter 20)*

Transport vesicles Membrane-bound structures with external protein coats that bud from compartments of the secretory pathway and carry cargo in anterograde or retrograde directions. *(Chapter 12)*

Triangulation Division of the triangular face of a large icosahedral structure into smaller triangles. *(Chapter 4)*

Triangulation number The number of structural units per face of a capsid or nucleocapsid with icosahedral symmetry. Abbreviated *T*. *(Chapter 4)*

Tropism The predilection of a virus to invade, and replicate in, a particular cell type or tissue. *(Chapter 5)*

TSEs *See* Transmissible spongiform encephalopathies

Tumor A mass of cells originating from abnormal growth. *(Chapter 18)*

Tumor suppressor genes Cellular genes encoding proteins that negatively regulate cell proliferation; mutational inactivation of both copies of the genes is associated with tumor development. *(Chapter 18)*

Two-hit kinetics The number of plaques or lesions is directly proportional to the ½ power of the concentration of the inoculum, i.e., the number of plaques or lesions doubles when 4 times the concentration of virus particles is inoculated; this property indicates that two different types of virus particle must infect a cell to ensure replication. *(Chapter 2)*

Type-specific antigens Epitopes, defined by neutralizing antibodies, that define viral serotypes (e.g., poliovirus types 1, 2, and 3). *(Chapter 2)*

Uncoating The release of viral nucleic acid from its protective protein coat or lipid envelope; in some cases, the liberated nucleic acid remains bound to viral proteins. *(Chapter 5)*

Vaccination Inoculation of healthy individuals with attenuated or related microorganisms, or their antigenic products, in order to elicit an immune response that will protect against later infection by the corresponding pathogen. *(Chapters 1 and 19)*

Variolation Inoculation of healthy individuals with material from a smallpox pustule, or in modern times from an attenuated cowpox (vaccinia) virus preparation, through a scratch on the skin (called scarification). *(Chapters 1 and 19)*

Vertical transmission Transfer of viral genomes between parent and offspring. *(Chapter 14)*

Vesicles Focal necroses that occur when an infection spreads from the capillaries to the superficial layers of the skin and replicates in the epidermis; a vesicle usually contains inflammatory fluids. *(Chapter 14)*

Viral pathogenesis The processes by which viral infections cause disease. *(Chapters 1, 2, and 14)*

Viremia The presence of infectious virions in the blood. *(Chapter 14)*

Virion An infectious virus particle. *(Chapters 1 and 4)*

Viroceptors Viral proteins that modulate cytokine signaling or cytokine production by mimicking host cytokine receptors. *(Chapters 14 and 15)*

Viroids Unencapsidated, small, circular, single-stranded RNAs that replicate autonomously when inoculated into plant cells; some are pathogenic and of economic importance. *(Chapters 1 and 20)*

Virokines Secreted viral proteins that mimic cytokines, growth factors, or similar extracellular immune regulators. *(Chapters 14 and 15)*

Virulence The relative capacity of a viral infection to cause disease. *(Chapter 14)*

Virulent Describes an infection that causes significant damage, in contrast to "attenuated." *(Chapter 14)*

Viruria Presence of virus particles in the urine. *(Chapter 14)*

Virus evolution The change of a viral population over time in response to selection processes. *(Chapter 20)*

v-Oncogenes Oncogenes that are encoded in viral genomes. *(Chapter 18)*

Western blot analysis *See* Immunoblotting

Wild type The original (often laboratory-adapted) virus from which mutants are selected, and which is used as the basis for comparison. *(Chapter 2)*

Zoonoses Diseases that are naturally transmitted between humans and other vertebrates. *(Chapter 14)*

Index

D

F

G

H

J

K

L

N

S

U

W

X

Y

Z